How to Work an Active Example

Step 1: When you come to an example, locate the point in the left column at which the first blue-shaded background appears. Use this shield to cover all of the blue-shaded boxes in the left column.

Step 2: Read the problem statement. Write any answers or calculations needed in the blank space where the pencil icon is located. Note that the “Think Before You Write” instructions are different for each Active Example.

Step 3: Move the shield down to reveal the first blue-shaded box.

Step 4: Compare your answer to the one you can now read in the book. Be sure you understand the example up to that point before going on.

Step 5: Repeat the procedure until you finish the example.

How to Work an Active Example

Step 1: When you come to an example, locate the point in the left column at which the first blue-shaded background appears. Use this shield to cover all of the blue-shaded boxes in the left column.

Step 2: Read the problem statement. Write any answers or calculations needed in the blank space where the pencil icon is located. Note that the “Think Before You Write” instructions are different for each Active Example.

Step 3: Move the shield down to reveal the first blue-shaded box.

Step 4: Compare your answer to the one you can now read in the book. Be sure you understand the example up to that point before going on.

Step 5: Repeat the procedure until you finish the example.

Current U.S. usage (1A, 2A, 3B … 8A) / IUPAC notation (1–18)

1A 1	2A 2	3B 3	4B 4	5B 5	6B 6	7B 7	8B 8	8B 9	8B 10	1B 11	2B 12	3A 13	4A 14	5A 15	6A 16	7A 17	8A 18
1 **H** 1.008																1 **H** 1.008	2 **He** 4.003
3 **Li** 6.94	4 **Be** 9.012											5 **B** 10.81	6 **C** 12.01	7 **N** 14.01	8 **O** 16.00	9 **F** 19.00	10 **Ne** 20.18
11 **Na** 22.99	12 **Mg** 24.31											13 **Al** 26.98	14 **Si** 28.09	15 **P** 30.97	16 **S** 32.06	17 **Cl** 35.45	18 **Ar** 39.95
19 **K** 39.10	20 **Ca** 40.08	21 **Sc** 44.96	22 **Ti** 47.87	23 **V** 50.94	24 **Cr** 52.00	25 **Mn** 54.94	26 **Fe** 55.85	27 **Co** 58.93	28 **Ni** 58.69	29 **Cu** 63.55	30 **Zn** 65.38	31 **Ga** 69.72	32 **Ge** 72.63	33 **As** 74.92	34 **Se** 78.96	35 **Br** 79.90	36 **Kr** 83.80
37 **Rb** 85.47	38 **Sr** 87.62	39 **Y** 88.91	40 **Zr** 91.22	41 **Nb** 92.91	42 **Mo** 95.96	43 **Tc** (98)	44 **Ru** 101.1	45 **Rh** 102.9	46 **Pd** 106.4	47 **Ag** 107.9	48 **Cd** 112.4	49 **In** 114.8	50 **Sn** 118.7	51 **Sb** 121.8	52 **Te** 127.6	53 **I** 126.9	54 **Xe** 131.3
55 **Cs** 132.9	56 **Ba** 137.3	57 **La** 138.9	72 **Hf** 178.5	73 **Ta** 180.9	74 **W** 183.8	75 **Re** 186.2	76 **Os** 190.2	77 **Ir** 192.2	78 **Pt** 195.1	79 **Au** 197.0	80 **Hg** 200.6	81 **Tl** 204.4	82 **Pb** 207.2	83 **Bi** 209.0	84 **Po** (209)	85 **At** (210)	86 **Rn** (222)

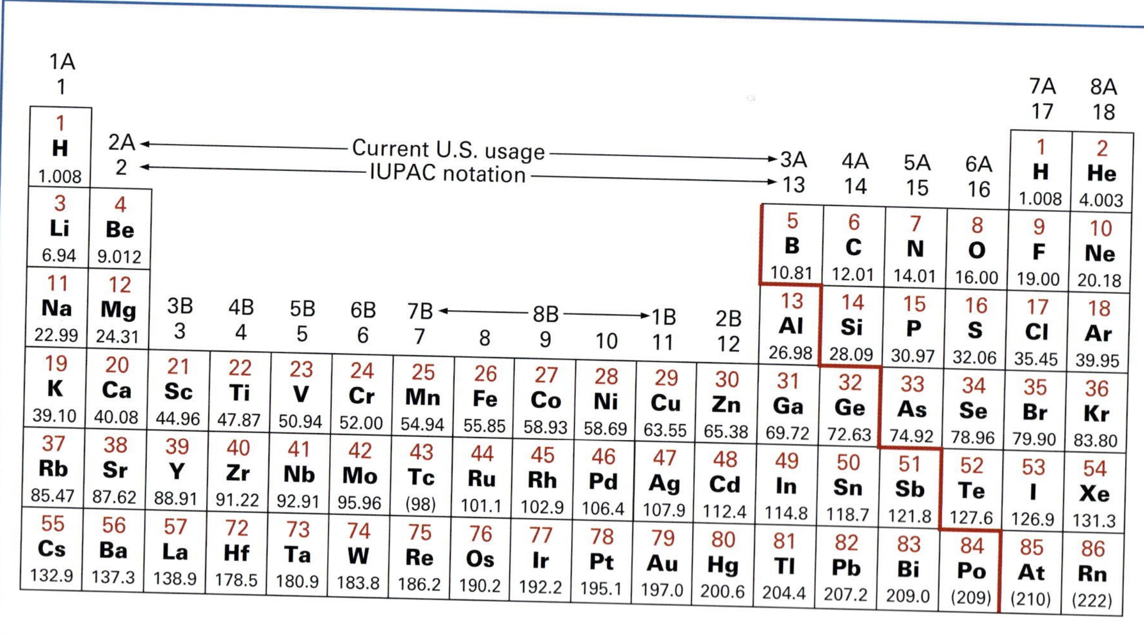

Current U.S. usage (1A, 2A, 3B … 8A) / IUPAC notation (1–18)

1A 1	2A 2	3B 3	4B 4	5B 5	6B 6	7B 7	8B 8	8B 9	8B 10	1B 11	2B 12	3A 13	4A 14	5A 15	6A 16	7A 17	8A 18
1 **H** 1.008																1 **H** 1.008	2 **He** 4.003
3 **Li** 6.94	4 **Be** 9.012											5 **B** 10.81	6 **C** 12.01	7 **N** 14.01	8 **O** 16.00	9 **F** 19.00	10 **Ne** 20.18
11 **Na** 22.99	12 **Mg** 24.31											13 **Al** 26.98	14 **Si** 28.09	15 **P** 30.97	16 **S** 32.06	17 **Cl** 35.45	18 **Ar** 39.95
19 **K** 39.10	20 **Ca** 40.08	21 **Sc** 44.96	22 **Ti** 47.87	23 **V** 50.94	24 **Cr** 52.00	25 **Mn** 54.94	26 **Fe** 55.85	27 **Co** 58.93	28 **Ni** 58.69	29 **Cu** 63.55	30 **Zn** 65.38	31 **Ga** 69.72	32 **Ge** 72.63	33 **As** 74.92	34 **Se** 78.96	35 **Br** 79.90	36 **Kr** 83.80
37 **Rb** 85.47	38 **Sr** 87.62	39 **Y** 88.91	40 **Zr** 91.22	41 **Nb** 92.91	42 **Mo** 95.96	43 **Tc** (98)	44 **Ru** 101.1	45 **Rh** 102.9	46 **Pd** 106.4	47 **Ag** 107.9	48 **Cd** 112.4	49 **In** 114.8	50 **Sn** 118.7	51 **Sb** 121.8	52 **Te** 127.6	53 **I** 126.9	54 **Xe** 131.3
55 **Cs** 132.9	56 **Ba** 137.3	57 **La** 138.9	72 **Hf** 178.5	73 **Ta** 180.9	74 **W** 183.8	75 **Re** 186.2	76 **Os** 190.2	77 **Ir** 192.2	78 **Pt** 195.1	79 **Au** 197.0	80 **Hg** 200.6	81 **Tl** 204.4	82 **Pb** 207.2	83 **Bi** 209.0	84 **Po** (209)	85 **At** (210)	86 **Rn** (222)

Periodic Table of the Elements

1A 1	2A 2	3B 3	4B 4	5B 5	6B 6	7B 7	8B 8	9	10	1B 11	2B 12	3A 13	4A 14	5A 15	6A 16	7A 17	8A 18
1 **H** 1.008																	2 **He** 4.003
3 **Li** 6.94	4 **Be** 9.012											5 **B** 10.81	6 **C** 12.01	7 **N** 14.01	8 **O** 16.00	9 **F** 19.00	10 **Ne** 20.18
11 **Na** 22.99	12 **Mg** 24.31											13 **Al** 26.98	14 **Si** 28.09	15 **P** 30.97	16 **S** 32.06	17 **Cl** 35.45	18 **Ar** 39.95
19 **K** 39.10	20 **Ca** 40.08	21 **Sc** 44.96	22 **Ti** 47.87	23 **V** 50.94	24 **Cr** 52.00	25 **Mn** 54.94	26 **Fe** 55.85	27 **Co** 58.93	28 **Ni** 58.69	29 **Cu** 63.55	30 **Zn** 65.38	31 **Ga** 69.72	32 **Ge** 72.63	33 **As** 74.92	34 **Se** 78.97	35 **Br** 79.90	36 **Kr** 83.80
37 **Rb** 85.47	38 **Sr** 87.62	39 **Y** 88.91	40 **Zr** 91.22	41 **Nb** 92.91	42 **Mo** 95.95	43 **Tc** (98)	44 **Ru** 101.1	45 **Rh** 102.9	46 **Pd** 106.4	47 **Ag** 107.9	48 **Cd** 112.4	49 **In** 114.8	50 **Sn** 118.7	51 **Sb** 121.8	52 **Te** 127.6	53 **I** 126.9	54 **Xe** 131.3
55 **Cs** 132.9	56 **Ba** 137.3	57 * **La** 138.9	72 **Hf** 178.5	73 **Ta** 180.9	74 **W** 183.8	75 **Re** 186.2	76 **Os** 190.2	77 **Ir** 192.2	78 **Pt** 195.1	79 **Au** 197.0	80 **Hg** 200.6	81 **Tl** 204.4	82 **Pb** 207.2	83 **Bi** 209.0	84 **Po** (209)	85 **At** (210)	86 **Rn** (222)
87 **Fr** (223)	88 **Ra** (226)	89 † **Ac** (227)	104 **Rf** (265)	105 **Db** (268)	106 **Sg** (271)	107 **Bh** (270)	108 **Hs** (277)	109 **Mt** (276)	110 **Ds** (281)	111 **Rg** (280)	112 **Cn** (285)	113 **Nh** (286)	114 **Fl** (289)	115 **Mc** (288)	116 **Lv** (292)	117 **Ts** (294)	118 **Og** (294)

*Lanthanoids	58 **Ce** 140.1	59 **Pr** 140.9	60 **Nd** 144.2	61 **Pm** (145)	62 **Sm** 150.4	63 **Eu** 152.0	64 **Gd** 157.3	65 **Tb** 158.9	66 **Dy** 162.5	67 **Ho** 164.9	68 **Er** 167.3	69 **Tm** 168.9	70 **Yb** 173.1	71 **Lu** 175.0
†Actinoids	90 **Th** 232.0	91 **Pa** 231.0	92 **U** 238.0	93 **Np** (237)	94 **Pu** (244)	95 **Am** (243)	96 **Cm** (247)	97 **Bk** (247)	98 **Cf** (251)	99 **Es** (252)	100 **Fm** (257)	101 **Md** (258)	102 **No** (259)	103 **Lr** (262)

All atomic masses have been rounded to four significant figures. (The natural composition of lithium isotopes limits its atomic mass to three significant figures.) Atomic masses in parentheses are the mass number of the longest-lived isotope. The current U.S. common usage group number is listed above the IUPAC format for group number. *Sources:* IUPAC Standard Atomic Weights and Conventional Atomic Weights 2009 and updates after 2009.

TABLE OF ELEMENTS

ELEMENT	SYMBOL	ATOMIC NUMBER	ATOMIC MASS
Actinium	Ac	89	(227)
Aluminum	Al	13	26.98
Americium	Am	95	(243)
Antimony	Sb	51	121.8
Argon	Ar	18	39.95
Arsenic	As	33	74.92
Astatine	At	85	(210)
Barium	Ba	56	137.3
Berkelium	Bk	97	(247)
Beryllium	Be	4	9.012
Bismuth	Bi	83	209.0
Bohrium	Bh	107	(270)
Boron	B	5	10.81
Bromine	Br	35	79.90
Cadmium	Cd	48	112.4
Cesium	Cs	55	132.9
Calcium	Ca	20	40.08
Californium	Cf	98	(251)
Carbon	C	6	12.01
Cerium	Ce	58	140.1
Chlorine	Cl	17	35.45
Chromium	Cr	24	52.00
Cobalt	Co	27	58.93
Copernicium	Cn	112	(285)
Copper	Cu	29	63.55
Curium	Cm	96	(247)
Darmstadtium	Ds	110	(281)
Dubnium	Db	105	(268)
Dysprosium	Dy	66	162.5
Einsteinium	Es	99	(252)
Erbium	Er	68	167.3
Europium	Eu	63	152.0
Fermium	Fm	100	(257)
Flerovium	Fl	114	(289)
Fluorine	F	9	19.00
Francium	Fr	87	(223)
Gadolinium	Gd	64	157.3
Gallium	Ga	31	69.72
Germanium	Ge	32	72.63
Gold	Au	79	197.0
Hafnium	Hf	72	178.5
Hassium	Hs	108	(277)
Helium	He	2	4.003
Holmium	Ho	67	164.9
Hydrogen	H	1	1.008
Indium	In	49	114.8
Iodine	I	53	126.9
Iridium	Ir	77	192.2
Iron	Fe	26	55.85
Krypton	Kr	36	83.80
Lanthanum	La	57	138.9
Lawrencium	Lr	103	(262)
Lead	Pb	82	207.2
Lithium	Li	3	6.94
Livermorium	Lv	116	(292)
Lutetium	Lu	71	175.0
Magnesium	Mg	12	24.31
Manganese	Mn	25	54.94
Meitnerium	Mt	109	(276)
Mendelevium	Md	101	(258)
Mercury	Hg	80	200.6
Molybdenum	Mo	42	95.96
Moscovium	Mc	115	(288)
Neodymium	Nd	60	144.2
Neon	Ne	10	20.18
Neptunium	Np	93	(237)
Nickel	Ni	28	58.69
Nihonium	Nh	113	(286)
Niobium	Nb	41	92.91
Nitrogen	N	7	14.01
Nobelium	No	102	(259)
Oganesson	Og	118	(294)
Osmium	Os	76	190.2
Oxygen	O	8	16.00
Palladium	Pd	46	106.4
Phosphorus	P	15	30.97
Platinum	Pt	78	195.1
Plutonium	Pu	94	(244)
Polonium	Po	84	(209)
Potassium	K	19	39.10
Praseodymium	Pr	59	140.9
Promethium	Pm	61	(145)
Protactinium	Pa	91	231.0
Radium	Ra	88	(226)
Radon	Rn	86	(222)
Rhenium	Re	75	186.2
Rhodium	Rh	45	102.9
Roentgenium	Rg	111	(280)
Rubidium	Rb	37	85.47
Ruthenium	Ru	44	101.1
Rutherfordium	Rf	104	(265)
Samarium	Sm	62	150.4
Scandium	Sc	21	44.96
Seaborgium	Sg	106	(271)
Selenium	Se	34	78.96
Silicon	Si	14	28.09
Silver	Ag	47	107.9
Sodium	Na	11	22.99
Strontium	Sr	38	87.62
Sulfur	S	16	32.06
Tantalum	Ta	73	180.9
Technetium	Tc	43	(98)
Tellurium	Te	52	127.6
Tennessine	Ts	117	(294)
Terbium	Tb	65	158.9
Thallium	Tl	81	204.4
Thorium	Th	90	232.0
Thulium	Tm	69	168.9
Tin	Sn	50	118.7
Titanium	Ti	22	47.87
Tungsten	W	74	183.8
Uranium	U	92	238.0
Vanadium	V	23	50.94
Xenon	Xe	54	131.3
Ytterbium	Yb	70	173.1
Yttrium	Y	39	88.91
Zinc	Zn	30	65.38
Zirconium	Zr	40	91.22

Introductory Chemistry

SEVENTH EDITION

An Active Learning Approach

Introductory Chemistry

SEVENTH EDITION

An Active Learning Approach

Mark S. Cracolice
University of Montana

Edward I. Peters

Australia • Brazil • Mexico • Singapore • United Kingdom • United States

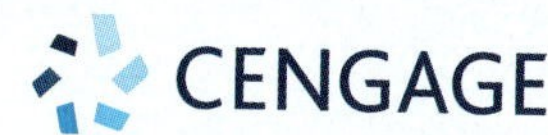

***Introductory Chemistry: An Active Learning Approach*, Seventh Edition**

Mark S. Cracolice, Edward I. Peters

SVP, Higher Education & Skills Product: Erin Joyner

VP, Product Management: Thais Alencar

Product Manager: Katherine Caudill-Rios

Product Assistant: Nellie Mitchell

Director, Learning Design: Rebecca von Gillern

Senior Manager, Learning Design: Melissa Parkin

Learning Designer: Peter McGahey

Marketing Director: Janet del Mundo

Marketing Manager: Timothy Cali

Content Creation Manager: Andrea Wagner

Senior Content Manager: Meaghan Tomaso

Digital Delivery Lead: Beth McCracken

Art Designer: Lizz Anderson

Text Designer: Lizz Anderson

Cover Designer: Lizz Anderson

Cover image(s):

Background Image: Antenna/fStop/Getty Images

Molecule illustrations: Cengage/Dragonfly Media Group

Interior Images:

Stubby well used pencil - Lucidio Studio Inc/ Getty images

Confused student facing chalkboard in a classroom by himself - PNC/Stockbyte/ Getty Images

Periodic table of elements and salt - Tetra Images/Getty Images

Library of Congress Control Number: 2019917479

Student Edition
ISBN: 978-0-357-36366-9

Loose-leaf Edition
ISBN: 978-0-357-36391-1

Cengage
200 Pier 4 Boulevard
Boston, MA 02210
USA

Cengage is a leading provider of customized learning solutions with employees residing in nearly 40 different countries and sales in more than 125 countries around the world. Find your local representative at **www.cengage.com.**

Cengage products are represented in Canada by Nelson Education, Ltd.

To learn more about Cengage platforms and services, register or access your online learning solution, or purchase materials for your course, visit **www.cengage.com.**

Printed in the United States of America
Print Number: 01 Print Year: 2020

Dedication

This book is dedicated to the memory of my late mother, Marjorie Sharp, the Worthy Advisor.

Contents Overview

Contents

Mindaugas Dulinskas/Shutterstock.com

NASA

Bloomberg/Getty Images

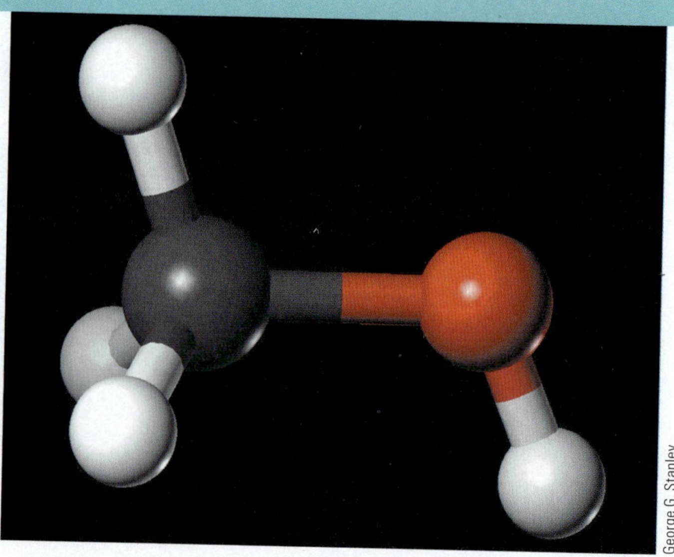

George G. Stanley

Candyfloss Film/Shutterstock.com

Charles D. Winters

Charles D. Winters

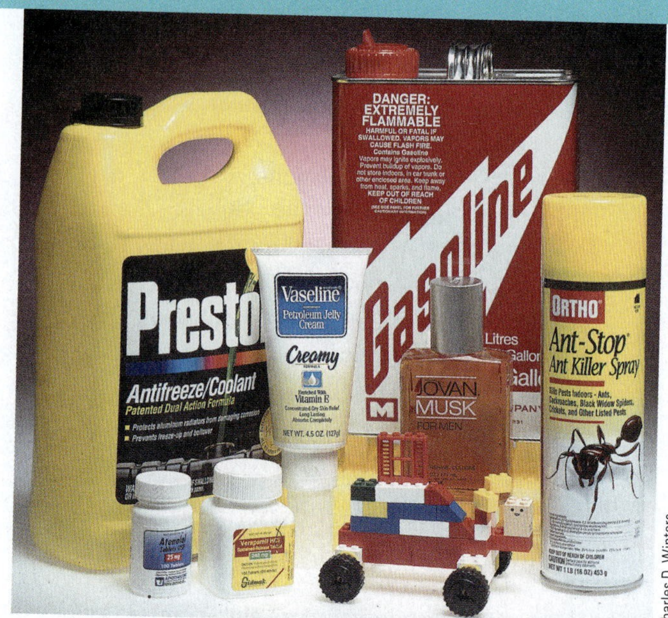

Charles D. Winters

21 Organic Chemistry 817

22 Biochemistry 875

Preface

Audience

The seventh edition of *Introductory Chemistry: An Active Learning Approach* is written for a college-level introductory or preparatory chemistry course for students who next will take a college general chemistry course. It is also appropriate for the first-term general portion of a two-term general, organic, and biological chemistry (GOB) course. The textbook is written with the assumption that this is a student's first chemistry course, or if there has been a prior chemistry course, it has not adequately prepared the student for general or GOB chemistry.

Overarching Goals

Introductory Chemistry was written with the following broad-based goals. Upon completing the course while using this textbook, our aim is that students will be able to:

1. Read, write, and talk about chemistry, using a basic chemical vocabulary.
2. Write routine chemical formulas and equations.
3. Set up and solve chemistry problems.
4. Think about fundamental chemistry on an atomic or molecular level and visualize what happens in a chemical change.

To reach these goals, *Introductory Chemistry* helps students deal with three common problems: developing good learning skills, overcoming a weak background in mathematics, and overcoming difficulties in reading scientific material. The first problem is addressed beginning in Sections 1.4–1.6, which together make up an "Introduction to Active Learning." These sections describe the pedagogical features of the textbook and how to use them effectively to learn chemistry as efficiently as possible.

Introductory Chemistry deals with a weak quantitative problem-solving background beginning in Chapter 3. Algebra, including the use of conversion factors, is presented as a problem-solving method that can be used for nearly all of the quantitative problems in the textbook. The thought processes introduced in Chapter 3 are used throughout the text, constantly reinforcing the student's ability to solve quantitative problems.

We address difficulties in reading scientific material via many of the features of the textbook. Clearly stated learning goals lead to carefully written narratives, which are then often summarized in a numbered list. Key words are printed in bold, summarized at the end of each chapter, and collected into a glossary. Chapter summaries are used to help students review as they complete each chapter. Active learning techniques are used throughout to keep students engaged in learning while they are reading.

Active Learning

The *An Active Learning Approach* portion of the title of the textbook refers to what general cognitive science and applied chemistry education research indicate is the best curricular approach to facilitate construction of *procedural* knowledge.

When we use the term *procedural* knowledge, we are referring to knowledge of how to do something, such as solve quantitative chemistry problems, as opposed to *declarative* knowledge, which is knowledge of facts. Both types of knowledge are important in an introductory chemistry course; students must learn facts such as the symbol for the element hydrogen is H, *and* they should learn how to calculate the amount of water that will be produced when a given mass of hydrogen is reacted with excess oxygen. However, declarative knowledge is relatively straightforward to teach; it is mostly a matter of organization and making connections. Procedural knowledge is relatively difficult to teach. It requires a curriculum centered on active learning.

Evidence in support of our claim about active learning is strong. The work of Scott Freeman of the University of Washington and associates provides an excellent example.* They used a statistical approach called a *meta-analysis* that combines results from many individual studies. This technique provides stronger evidence than any given individual study. They compared active-learning-centered classrooms with those that primarily relied on expository teaching, finding that the active learning classrooms produced both better exam performance *and* lower failure rates. Specifically, student performance on exams was about one-half standard deviation higher in active learning classrooms and failure rates in expository courses were 1.5 times the rate in active learning courses.

Active learning means that the student spends as much of his or her time as possible invested into studying actively, working to construct knowledge. Most textbooks engage students in active learning only while answering end-of-chapter questions. Our book engages students in active learning while answering end-of-chapter questions *and* studying the body of the chapter. We next examine how we accomplish this goal.

Active Examples

The examples in our textbook are written in a question-and-answer format in which the student actively learns chemistry *while* studying an assignment, rather than studying now with the intent to learn later. A typical example leads students through a series of steps where they "listen" to the authors tutor them as they work the solution, step-by-step. As students solve the example problem, they actively write for themselves each step in the solution, covering the authors' answer with the shield provided in the book. This example format turns the common passive "read the authors' solution" approach to an active "you solve the problem while we tutor you" methodology.

To serve as an example of and explanation about this methodology, let's break down Active Example 10.4, the first mass-to-mass stoichiometry example in the textbook. The problem statement comes first. The examples are numbered and titled for easy reference.

Active Example 10.4 Mass–Mass Stoichiometry II

What is the mass in grams of CO_2 that will be produced by burning 66.0 g C_7H_{16} by the same reaction as in Active Example 10.3, $C_7H_{16}(\ell) + 11\ O_2(g) \rightarrow 7\ CO_2(g) + 8\ H_2O(\ell)$?

The next portion of the Active Example is titled Think Before You Write. This feature has two purposes. One is to teach students to engage the portion of the

*Freeman, S., Eddy, S. L., McDonough, M., Smith, M. K., Okoroafor, N., Jordt, H., & Wenderoth, M. P. (2014). Active learning increases student performance in science, engineering, and mathematics. *Proceedings of the National Academy of Sciences of the United States of America, 111*(23), 8410–8415.

brain used for higher-order thinking, avoiding reacting impulsively.[*] The other purpose is to help students learn how to extract the relevant information from a problem statement, focusing on the deep structure rather than on the surface features.[†] Here, we begin by discussing how to analyze the problem statement from the deep structure perspective. We then discuss the problem-solving approach from the general perspective, divorced from the context of this specific problem. At the end of the box, students are reminded to actively work the example for themselves, covering our answers until they have produced their own.

Think Before You Write You are given the mass of one species in a chemical change, and you are asked to determine the mass of another species. Thus, you will switch from the macroscopic mass quantity to the particulate number of moles, use the mole ratio in the chemical equation to determine the amount in moles of the wanted quantity, and then switch back to the macroscopic level and determine the mass of that amount in grams. This is illustrated in Figure 10.4.

Answers *Cover the left column with your cut-out shield. Reveal each answer only after you have written your own answer in the right column.*

When appropriate, quantitative Active Examples are solved using a four-step problem-solving approach: analyze, identify, construct, and check. In the *analyze* step, students identify the given quantity and the unit of the wanted quantity. Space is provided for students to write under the pencil icon.

Students literally write their responses, making a commitment to reveal their present state of understanding and recording it.

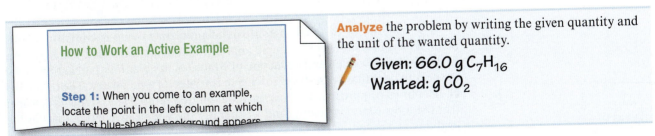

They then reveal the authors' answer, comparing their answer to that of an expert. If the answers match, their correct thinking is reinforced. If the answers don't match, students get immediate feedback at the specific point at which they don't correctly understand the problem-solving process. Earlier in the textbook, we gave overarching guidance to students to go back to the narrative before the Active Example when this occurs and figure out what is wrong.

---

*Wright, S. B., Matlen, B. J., Baym, C. L., Ferrer, E., & Bunge, S. (2007). Neural correlates of fluid reasoning in children and adults. *Frontiers in Human Neuroscience, 1*(8), doi: 10.3389/neuro.09.008.2007.

†Chi, M. T. H., & VanLehn, K. A., (2012). Seeing deep structure from the interactions of surface features. *Educational Psychologist, 47*(3), 177–188.

Given: 66.0 g C_7H_{16}
Wanted: g CO_2

Analyze the problem by writing the given quantity and the unit of the wanted quantity.

Given: 66.0 g C_7H_{16}
Wanted: g CO_2

The second step in our four-step approach is the *identify* step. Here we introduce the unit path approach, where students first write the units for each step in the solution setup. We then instruct them to identify the equivalency that connects each pair of units. After that, the equivalencies are changed to conversion factors. Equivalency and conversion factors are introduced in Section 3.3 and used continuously from that point forward.

$$\text{g } C_7H_{16} \rightarrow \text{mol } C_7H_{16} \rightarrow \text{mol } CO_2 \rightarrow \text{g } CO_2$$

$$1 \text{ mol } C_7H_{16} = 100.20 \text{ g } C_7H_{16}$$

$$7 \text{ mol } CO_2 = 1 \text{ mol } C_7H_{16}$$

$$44.01 \text{ g } CO_2 = 1 \text{ mol } CO_2$$

$$\frac{1 \text{ mol } C_7H_{16}}{100.20 \text{ g } C_7H_{16}} \quad \frac{7 \text{ mol } CO_2}{1 \text{ mol } C_7H_{16}} \quad \frac{44.01 \text{ g } CO_2}{1 \text{ mol } CO_2}$$

The next step is to **identify** the equivalencies. With multiple-step problems such as this, it can be helpful to write the units for each step before you write the equivalencies:

$$\text{g } C_7H_{16} \rightarrow \text{mol } C_7H_{16} \rightarrow \text{mol } CO_2 \rightarrow \text{g } CO_2$$

Each arrow in this **unit path** requires an equivalency. Change the equivalencies to conversion factors.

$$\text{g } C_7H_{16} \rightarrow \text{mol } C_7H_{16} \rightarrow \text{mol } CO_2 \rightarrow \text{g } CO_2$$

$$1 \text{ mol } C_7H_{16} = 100.20 \text{ g } C_7H_{16}$$

$$7 \text{ mol } CO_2 = 1 \text{ mol } C_7H_{16}$$

$$44.01 \text{ g } CO_2 = 1 \text{ mol } CO_2$$

$$\frac{1 \text{ mol } C_7H_{16}}{100.20 \text{ g } C_7H_{16}} \quad \frac{7 \text{ mol } CO_2}{1 \text{ mol } C_7H_{16}} \quad \frac{44.01 \text{ g } CO_2}{1 \text{ mol } CO_2}$$

The majority of the challenging part of problem solving is complete at this point. Through our Active Example approach, students learn that identification of the given and wanted and deduction of equivalencies that link pairs of units are the keys to quantitative problem solving in introductory chemistry. The third step of the four-step approach is to *construct* the solution setup. Here, students confirm that the units cancel correctly, and we literally show the cancellation lines in the textbook and encourage students to do the same, and then they calculate the value (quantity × unit) of the answer.

$$66.0 \cancel{\text{g } C_7H_{16}} \times \frac{1 \cancel{\text{mol } C_7H_{16}}}{100.20 \cancel{\text{g } C_7H_{16}}} \times \frac{7 \cancel{\text{mol } CO_2}}{1 \cancel{\text{mol } C_7H_{16}}} \times \frac{44.01 \text{ g } CO_2}{1 \cancel{\text{mol } CO_2}} = 203 \text{ g } CO_2$$

Construct the solution setup and determine the answer.

$$66.0 \cancel{\text{g } C_7H_{16}} \times \frac{1 \cancel{\text{mol } C_7H_{16}}}{100.20 \cancel{\text{g } C_7H_{16}}} \times \frac{7 \cancel{\text{mol } CO_2}}{1 \cancel{\text{mol } C_7H_{16}}} \times \frac{44.01 \text{ g } CO_2}{1 \cancel{\text{mol } CO_2}} = 203 \text{ g } CO_2$$

The fourth step in the four-step approach has two parts. One aim is to have students do mental arithmetic to the point where the answer obtained from a calculator is verified as reasonable, and a second aim is to teach students to reflect on how they have improved their problem-solving skills. Here, we guide students to be flexible in their choices in doing the calculation *check*.

$(60 \times 7) \times 50 \div 100 = 420 \times (50 \div 100) = 420 \times 0.5 = 210$, OK. $(70 \times 7) \times 40 \div 100 = 490 \times (40 \div 100) \approx 500 \times 0.4 = 200$, OK.	The **check** of a setup with a large number of values becomes a bit challenging. You have to round the numbers, aiming to round in opposite directions. For example, this setup could be estimated as 60 (round down) $\times$ 7 $\times$ 50 (round up) $\div$ 100 or 70 (round up) $\times$ 7 $\times$ 40 (round down) $\div$ 100. Remember that the goal is to be sure you in are in the ballpark, not to calculate the *exact* answer in your head. $(60 \times 7) \times 50 \div 100 = 420 \times (50 \div 100) = 420 \times 0.5 = 210$, OK.

The second part of the fourth step is to encourage students to think about the purpose of the Active Example and to contemplate if they have successfully achieved that purpose. This step is designed to invoke metacognition* so that students become explicitly aware of and make conscious the thought processes that they just learned.

You improved your skill at solving mass–mass stoichiometry problems.	What did you learn by solving this Active Example? I am beginning to understand the mass given $\rightarrow$ amount given $\rightarrow$ amount wanted $\rightarrow$ mass wanted problem solving strategy.

Finally, each Active Example is followed by a practice exercise that is based on the deep structure of the example that comes immediately before it. This allows the student who correctly solved the example to receive reinforcement and the student who did not solve the example correctly an opportunity to solve a parallel problem correctly before moving to the next topic. Solutions to the practice exercises are at the end of each chapter.

Practice Exercise 10.4

What mass of fluorine is formed when 3.0 grams of bromine trifluoride decomposes into its elements?

---

*Flavell, J. H. (1979). Metacognition and cognitive monitoring: A new area of cognitive-developmental inquiry. *American Psychologist, 34*(10), 906–911.

Rickey, D., & Stacy, A. M. (2000). The role of metacognition in learning chemistry. *Journal of Chemical Education, 77*(7), 915–920.

Target Checks

When a multistep Active Example is not warranted, we use another active learning feature termed *Target Checks.* These are just-in-time, fundamental questions, primarily utilized with nonquantitative topics, that help students monitor their progress as they work instead of waiting for the end-of-chapter questions so that they can identify and diagnose incomplete understandings or misunderstandings *as they study.* As an example, Target Check 14.1 is from the "Avogadro's Law" section:

Target Check 14.1

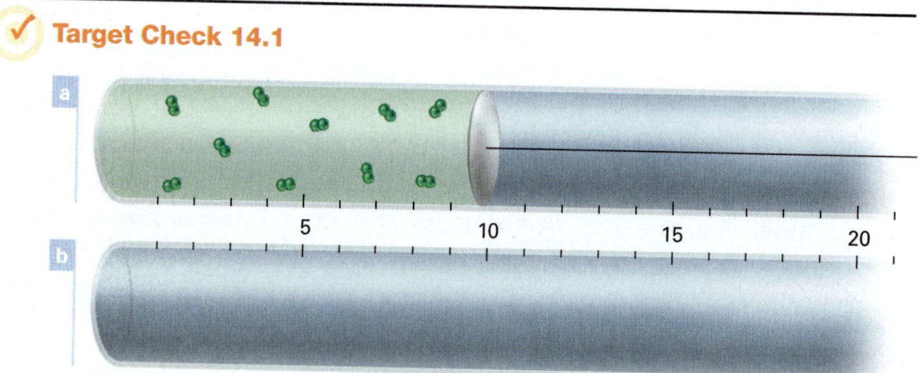

A horizontal cylinder (a) is closed at one end by a piston that moves freely left or right, depending on the pressure exerted by the enclosed gas. The gas consists of 10 two-atom molecules. A reaction occurs in which five of the molecules separate into one-atom particles. In cylinder (b), sketch the position to which the piston would move as a result of the reaction. Pressure and temperature remain constant throughout the process. (Hint: How many total particles would be present after the reaction? Include them in your sketch.)

Order of Coverage: A Flexible Format

Topics in a preparatory course or the general portion of a general–organic–biological chemistry course may be presented in several logical sequences, one of which is the order in which they appear in this textbook. However, it is common for individual instructors to prefer a different organization. *Introductory Chemistry* has been written to accommodate these different preferences by carefully writing each topic so that regardless of when it is assigned, it never assumes knowledge of any concept that an instructor might reasonably choose to assign later in the course. If some prior information is needed at a given point, it may be woven into the text as a ***Preview*** to the extent necessary to ensure continuity for students who have not seen it before, while affording a brief ***Review*** for those who have. (See the following P/Review.) At other times, margin notes are used to supply the needed information. Occasionally, digressions in small print are inserted for the same purpose. There is also an *Option* feature that actually identifies the alternatives for some topics. In essence, we have made a conscious effort to be sure that all students have all the background they need for any topic whenever they reach it.

◀ **P/Review** Information and section references are provided in the narrative or as a note in the margin showing students where to find relevant information before or after a given section.

Introductory Chemistry also offers choices in how some topics are presented. The most noticeable example of this is the coverage of gases, which is spread over two chapters. Chapter 4 introduces the topic through the P-V-T combined gas laws. This allows application of the problem-solving principles from Chapter 3 immediately after they are taught. Then the topic is picked up again in Chapter 14, which introduces the Ideal Gas Law. An instructor is free to move Chapter 4 to immediately precede Chapter 14, should a single "chapter" on gases be preferred.

We have a two-chapter treatment of chemical reactivity with a qualitative emphasis, preceding the quantitative chapter on stoichiometry. Chapter 8 provides an introduction to chemical reactivity, with an emphasis on writing and balancing chemical equations and recognizing reaction types based on the nature of the equation. After students have become confident with the fundamentals, we then increase the level of sophistication of our presentation on chemical change by introducing solutions of ionic compounds and net ionic equations. Chapter 9 on chemical change in solution may be postponed to any point after Chapter 8. Chapter 8 alone provides a sufficient background in chemical equation writing and balancing to allow students to successfully understand stoichiometry, the topic of Chapter 10. You may wish to combine Chapter 9 with Chapter 16 on solutions.

Chapter 14 features sections that offer *alternative* ways to solve gas stoichiometry problems at given temperatures and pressures. You can choose the section that you want to assign. Section 14.9 is based on what we call the *molar volume method*, where molar volume is used as a conversion factor to change between amount of substance in moles and volume. Section 14.10 is based on what we term the *ideal gas equation method*, where $PV = nRT$ and algebra is the method to make the amount–volume conversion.

On a smaller scale, there are minor concepts that are commonly taught in different ways. These may be identified specifically in the book, or mentioned only briefly, but always with the same advice to the student: *Learn the method that is presented in lecture. If your instructor's method is different from anything in the book, learn it the way your instructor teaches it.* Our aim is to have the book support the classroom presentation, whatever it may be.

Readability

We aim to help students overcome difficulties in reading scientific material by discussing chemistry in simple, direct, and user-friendly language. Maintaining the book's readability continues to be a primary focus in this edition. The book features relatively short sections and chapters to facilitate learning and to provide flexibility in ordering topics.

Features

Active Examples Active Examples were described in detail previously. An Active Example is an active learning feature that is formatted in two columns. The left column (the authors' answers) is to be covered by students while they write their own answers in the space provided in the right column. As students actively work through and complete the solution in the right column, they can reveal the solution to each step in the left column, thereby receiving *immediate feedback* about their understanding of the concept as it is being formed. Each example is titled so that students can better identify the concept or problem-solving skill they are learning. This will be useful when reviewing for exams.

Practice Exercises Each Active Example is followed immediately by a parallel Practice Exercise designed to firm up the potentially fragile new knowledge that was just constructed during the process of completing the companion Active Example. The Practice Exercises cover the same concept as the Active Examples, but they are typically slightly more challenging, leading students toward improved conceptual understanding and problem-solving skills. Solutions to the Practice Exercises are provided at the end of each chapter.

Target Checks Target Checks were described in detail previously. Target Check questions are an active learning feature that enable students to test their understanding immediately after studying a topic. Target Checks are most prominent in the qualitative chapters where the material often does not align well with Active Examples.

Reference Pages Cut-out cards may be used as shields to cover step-by-step answers while solving Active Examples. One side of each card has a partial periodic table that gives students ready access to all the information that table provides. The reverse side of each card contains instructions on how to use it in solving examples.

We also include a larger, complete version of the Periodic Table and an alphabetical listing of the elements in another cut-out card. In addition, the information on the inside covers of the book comprises a summary of nomenclature rules, selected numbers and constants, definitions, and equations, and a mini-index of important text topics, all keyed to the appropriate section number in the text.

Section 1 Each chapter except for Chapter 1 begins with an introductory section designed to *engage* students in thinking about an issue related to the major topic of the chapter. Our goal here is to pique students' curiosity and generate interest. We sometimes discuss topics that are being actively researched at the moment. We seek to convey some of the excitement that comes from using the scientific method to seek the creation of new knowledge about the natural world. We do not include end-of-chapter questions for these sections in order to keep the focus on engagement of student interest.

Goals Learning objectives, identified simply as Goals, appear at the beginning of the section in which each topic is introduced. They focus attention on what students are expected to learn or the skill they are expected to develop while studying the section.

Emphasis on Mental Arithmetic To address the issue of insufficient mathematical preparation, we have an emphasis on estimating and verifying the reasonableness of calculation results. All Active Examples that include a calculation include an arithmetic check step. At a minimum, we aim to instill students with the philosophy that all results displayed on a calculator must be mentally challenged. Ideally, we hope they will embrace these estimation steps and improve their skill at doing mental arithmetic through practice. You may instruct students to omit these calculation verification steps, should your educational philosophy be such that you do not wish to require them in your course.

Thinking About Your Thinking Boxes This feature helps students think about more than just the content of the chemical concepts; it gives them a broader view of the thinking skills used in chemistry. By focusing on how chemists think, students cannot only learn the context in which material is presented but also improve their competence with the more general skill. These broad thinking skills can then be applied to new contexts in their future chemistry courses, in other academic disciplines, and throughout their lives.

P/Review The flexible format of this book is designed so that any common sequence of topics will be supported. A cross reference called *P/Review* refers to a topic already studied or one that is yet to be studied. Our aim is to provide a textbook that will work for your curriculum, as opposed to a book that dictates the curriculum design. We therefore assume that the chapters will not necessarily be assigned in numerical order. The P/Reviews help to allow flexibility in chapter order.

***a summary of...* and *how to...* Boxes** Clear in-chapter summaries and listings of steps that explain how to carry out a procedure appear throughout the text.

These boxes allow students to reflect on what they've just studied and give them supplementary structure for learning in their first college chemistry course.

Everyday Chemistry All chapters have an Everyday Chemistry section that moves chemistry out of the textbook and classroom and into the daily experience of students. This feature gives students a concrete application of a principle within each chapter.

Everyday Chemistry Quick Quizzes Each Everyday Chemistry essay is followed by two questions about the essay. Assignment of these questions is optional. Answers are provided in the Instructor's Manual.

Art and Photography We have maintained the large number of photographs in the book, illustrating the chemistry that is also described in words. We have also retained and revised high-quality art pieces, with an emphasis on simple color schemes, plentiful macro-to-micro art, and instructional descriptions.

Chapter Summaries Each chapter includes a summary immediately following the last narrative section. It presents a list of the chapter goals, and each goal is matched to a summary of the key concepts associated with the goal, with key terms in bold. These summaries can be used as a preview to help students organize their learning before new material is introduced in the lecture portion of the course, and they serve as a review source during the term, as well as a comprehensive review source for the final exam.

The chapter summaries, when combined with worked examples and some end-of-chapter questions, would constitute a study guide for the textbook. Our aim is for the book to effectively serve as a combined study guide and textbook integrated into a single package.

Glossary An important feature for a preparatory chemistry course is a glossary. With each end-of-chapter summary of Key Terms, we remind students to use their glossary regularly. The glossary provides definitions of many of the terms used in the textbook, and it is a convenient reference source to use to review vocabulary from past chapters.

Frequently Asked Questions This end-of-chapter feature has two main purposes: (1) to identify particularly important ideas and offer suggestions on how they can be mastered and (2) to alert students to some common mistakes so they can avoid making them.

Concept-Linking Exercises An isolated concept in chemistry often lacks meaning to students until they understand how that concept is related to other concepts. Concept-Linking Exercises ask students to write a brief description of the relationships among a small group of terms of phrases. If they can express those relationships correctly in their own words, they understand the concepts. Explicitly writing these connections also helps with long-term retention of the concepts.

Small-Group Discussion Questions A growing number of courses feature some sort of group work formally integrated within the curriculum. We believe that the end-of-chapter questions typically used as homework are best for individual study, so each chapter has a set of questions for that were designed with group work in mind. These questions are typically more conceptual, more challenging, and, potentially, more lengthy than the average end-of-chapter questions. We have not provided solutions to these questions in the hope of removing the temptation for students to give up too quickly and look at the solution as a method of learning how to answer the questions.

Questions, Exercises, and Problems Each chapter includes an abundant supply of questions, exercises, and except for Chapter 1, the problems arranged in three categories. There are questions grouped according to sections in the chapter, General Questions from any section in the chapter and, finally, More Challenging Problems. Complete solutions (not just answers) for all blue-numbered questions appear at the end of the chapter. Solutions for to the black-numbered questions are in the Instructor's Manual.

End-of-Chapter Illustrations Well over 100 photographs and line drawings appear in the end-of-chapter Questions, Exercises, and Problems primarily to better illustrate the macroscopic aspects of chemistry. Students will be able to see physical and chemical changes and common forms of industrial manufacturing processes, as well as better visualize the scenarios described in the questions.

Appendices Appendix I includes a general review of arithmetic, scientific notation, algebra, and logarithms as they are used in this book. Appendix II gives a more complete treatment of the SI system than in Chapter 3.

New to This Edition

Revised Approach to Biochemistry (Chapter 22) We believe that previous editions of this chapter had a mismatch between the level of coverage that is appropriate for an *Introductory Chemistry* course and the presentation in Chapter 22. Thus, we rewrote the chapter with the intention of keeping an emphasis on only the major concepts. All of the minor details have been removed. To accomplish this, the chapter narrative was nearly completely rewritten. In addition to the revised level of coverage, we believe that the narrative is now more appropriately sequential and therefore more pedagogically suitable for a student who has not yet taken a general chemistry course.

Revision to Accommodate the Revised International System of Units (SI) In November 2018, the Member States of the Bureau International des Poids et Mesures unanimously voted to adopt a revised SI, changing the definitions of three units central to introductory chemistry, the mole, the kilogram, and the kelvin, and one unit usually not included in an introductory chemistry course, the ampere, effective on May 20, 2019. All of the SI revisions are integrated throughout the textbook and Appendix II on the SI System of Units.

Section 1 We have added a new first section to each chapter that emphasizes big picture topics that have a connection to a topic in each chapter. These sections were written with the philosophy that the first step in each topical cycle of learning should be to intellectually and emotionally engage the student. Thus, we give a chemist's perspective on big picture questions on topics such as the origins of the elements, the universal nature of chemical change, the existence of water on other planets, and the origins of life. No Target Check or end-of-chapter questions are associated with these sections to clearly convey to students that the sections are meant to help inspire them to wonder about the nature of the universe, with no pressure to be responsible for related textbook questions.

Improved Photography Program We sought to improve the quality of as many of the hundreds of photographs in the book as possible. We were actually surprised to come to the realization that many photos from the previous edition were quite good! Nonetheless, we looked at numerous alternatives for many photos, changing to images that more clearly illustrated a concept or reflected a more modern perspective when possible.

Video Solutions to Active Examples (eBook Only) Students at the University of Montana have been featured in dozens of videos that mirror what an in-person tutor would do if a student asked for help with understanding an Active Example. With the assistance of the editorial and production teams, textbook author Mark

Cracolice wrote scripts and directed students as they performed in these short audiovisual performances. We believe that many students will find these videos more engaging than the print programmed examples used in the book, but the print examples are still available for students who prefer learning via this format. The content of the video solutions and the Active Examples are equivalent, providing no disadvantage to students who prefer one format to the other.

Looseleaf Edition

Loose-Leaf Edition for Introductory Chemistry: An Active Learning Approach, 7e.

ISBN: 9780357363911

A loose-leaf (unbound, three-hole-punched) version of *Introductory Chemistry: An Active Learning Approach 7e,* which can be inserted in a binder, is also available.

OWLv2

The OWL online learning system offers additional practice exercises. OWLv2 also contains a complete range of practice exercises to supplement the end-of-chapter problems found in the book. In addition, the chemical input tools have been improved to allow students to create more accurate chemical symbols, formulas, and equations. OWLv2 offers a range of study and planning tools that can be adjusted as a student progresses through the course topics.

Students can use this ISBN at www.CENGAGEbrain.com to purchase instant access to OWLv2, the most trusted online learning solution for chemistry. Featuring chemist-developed content, OWLv2 is the only system designed to elevate thinking through Mastery Learning, allowing students to work at their own pace until they understand each concept and skill. Each time a student tries a problem, OWLv2 changes the chemicals, values, and sometimes even the wordings of the question to ensure students are learning the concepts and not cheating the system. With detailed, instant feedback and interactive learning resources, students get the help they need when they need it. Now with improved student and instructor tools and greater functionality, OWLv2 is more robust than ever. Visit www.CENGAGE.com/owlv2 to learn more.

MindTap eBook

MindTap™ is an interactive online learning management system. The MindTap™ edition of this book has clickable answers for every Active Example problem, as well as clickable key terms and figure callouts. Students are able to create personalized Learning Paths with MindTap™ Reader that are flexible and easy to follow.

Instructor Companion Site

The instructor supplements and supporting materials are available to qualified adopters on the Instructor Companion Site. Go to login.cengage.com, find this textbook, and choose Instructor Companion Site to see samples of these materials, request a desk copy, and locate your sales representative.

- **PowerPoint® lecture slides** written for this text that instructors can customize by importing their own slides or other materials.
- **Image libraries** that contain digital files for figures, photographs, and numbered tables from the text.
- The **Instructor's Manual** provides for each chapter authors' comments, answers to Everyday Chemistry Quick Quiz questions, and solutions to black-numbered end of chapter questions.

- **Test bank questions** including dozens of multiple choice questions per chapter.
- **Chemistry Multimedia Library** of lecture-ready animations, simulations, and movies.

Cengage Testing, powered by Cognero® for Cracolice/Peters' *Introductory Chemistry: An Active Learning Approach*

Cengage Learning Testing Powered by Cognero is a flexible, online system that allows you to author, edit, and manage test bank content from multiple Cengage Learning solutions, create multiple test versions in an instant, and deliver tests from your LMS, your classroom, or wherever you want.

Acknowledgments

At Cengage, we appreciate Maureen McLaughlin, former Senior Product Manager for Chemistry, for supporting the production of a new edition. Liz Woods, who worked her way up through the ranks at Cengage, serving in various roles in previous editions, was the Learning Designer who solicited reviews, worked with us to develop a revision plan, and created the initial schedule. We are thankful for her valuable contributions. Peter McGahey took over as Learning Designer early in the project, and we are grateful for his efforts, particularly in helping to make the print and online versions integrate so well. We also thank Meaghan Tomaso, Senior Content Manager, for all of her work in coordination of all of the people who collaborate to produce a modern textbook.

At Lumina Datamatics, we are appreciative of the work of Arul Joseph Raj, who was instrumental in transforming a series of word processing files, art renderings, and photographs into the beautiful book you are reading.

Our accuracy reviewer was Dr. David Shinn, Associate Professor in the Department of Math and Science at the United States Merchant Marine Academy. David had the challenging task of reviewing every word, every number, every photograph, and every illustration in the textbook while under considerable time pressure. We appreciate his attention to detail. David's suggestions led to a number of improvements to the initial draft of the textbook.

The reviewers of the seventh edition helped to shape our thinking, and for that, we are most appreciative. They include:

Chester Dabalos, University of Hawaii at Manoa

Michael Hauser, St. Louis Community College–Meramec

Ling Huang, Sacramento City College

Tara Hurt, East Mississippi Community College

E. Kay Sutton, Campbellsville University

We are also grateful to the faculty and student users of the first through sixth editions of *Introductory Chemistry*. Their comments and suggestions over the past 20 years have led to significant improvements in this book.

We thank the reviewers of the previous editions:

Melvin T. Arnold, Adams State College

Joe Asire, Cuesta College

Caroline Ayers, East Carolina University

Bob Blake, Texas Tech University

Juliette A. Bryson, Las Positas College

Sharmaine Cady, East Stroudsburg State College

K. Kenneth Caswell, University of South Florida

Bill Cleaver, University of Vermont
Pam Coffin, University of Michigan–Flint
Claire Cohen-Schmidt, The University of Toledo
Mapi Cuevas, Santa Fe Community College
Jan Dekker, Reedley College
Michelle Driessen, University of Minnesota
Jerry A. Driscoll, University of Utah
Jeffrey Evans, University of Southern Mississippi
Coretta Fernandes, Lansing Community College
Donna G. Friedman, St. Louis Community College at Florissant Valley
Galen C. George, Santa Rosa Junior College
Carol J. Grimes, Golden West College
Alton Hassel, Baylor University
Randall W. Hicks, Michigan State University
Ling Huang, Sacramento City College
William Hunter, Illinois State University
Jeffrey A. Hurlburt, Metropolitan State College
C. Fredrick Jury, Collin County Community College
Jane V. Z. Krevor, California State University, San Francisco
Rebecca Krystyniak, St. Cloud State University
Joseph Ledbetter, Contra Costa College
Jerome Maas, Oakton Community College
Kenneth Miller, Milwaukee Area Technical College
James C. Morris, The University of Vermont
Felix N. Ngassa, Grand Valley State University
Bobette D. Nourse, Chattanooga State Technical Community College
Brian J. Pankuch, Union County College
Erin W. Richter, University of Northern Iowa
Jan Simek, California Polytechnic State University, San Luis Obispo
John W. Singer, Alpena Community College
David A. Stanislawski, Chattanooga State Technical Community College
Linda Stevens, Grand Valley State University
David Tanis, Grand Valley State University
Amy Waldman, El Camino College
Andrew Wells, Chabot College
Linda Wilson, Middle Tennessee State University
David L. Zellmer, California State University, Fresno

We continue to be very much interested in your opinions, comments, critiques, and suggestions about any feature or content in the book. Please feel free to e-mail us directly or through Cengage.

Mark S. Cracolice

Department of Chemistry and Biochemistry
University of Montana
Missoula, MT 59812
mark.cracolice@umontana.edu

1 Introduction to Chemistry and Introduction to Active Learning

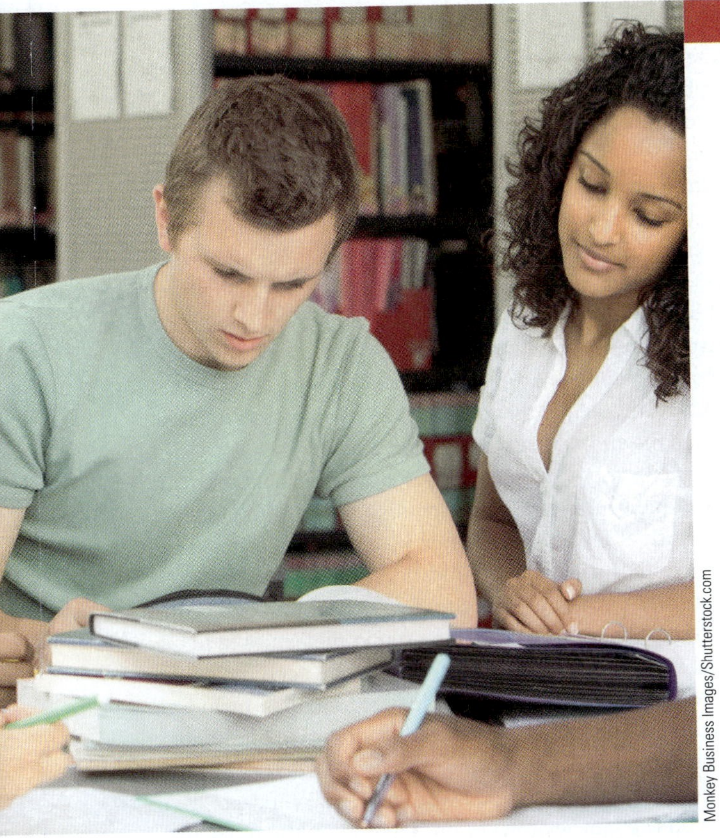

Monkey Business Images/Shutterstock.com

◀ How many students in a typical Introductory Chemistry course are chemistry majors? Usually it is only a small fraction. How many students in a typical Introductory Chemistry course need chemistry for their major? All of them—that is why the students gathered around this table in their school library are studying chemistry together. In fact, all educated members of the society need to know the fundamentals of chemistry to understand the natural world. In this chapter, we introduce you to the science and study of chemistry and all of the learning tools available to you, including this textbook.

CHAPTER CONTENTS

Welcome to your first college chemistry course! Chemistry is the gateway to careers in scientific research and human and animal health. You may be wondering why you, as a biology, premedicine, pharmacy, nursing, or engineering major—or as someone with any major other than chemistry—are required to take this course. The answer is that all matter is made up of molecules, and chemistry is the science that studies how molecules behave. If you need to understand matter, you need to know chemistry.

What lies before you is a fascinating new perspective on nature. You will learn to see the universe through the eyes of a chemist, as a place where you can think of all things large or small as being made up of extremely tiny molecules. Let's start by taking a brief tour of some of the amazing variety of molecules in our world.

First consider the simple hydrogen molecules in **Figure 1.1(a)**. This shows you what you would see if you could take a molecular-level look at a cross section from a cylinder filled with pure hydrogen. The molecules are moving incredibly fast—more than 4000 miles

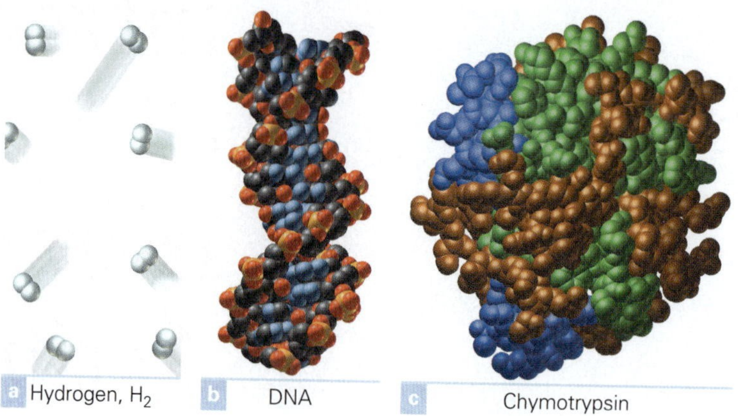

Figure 1.1 A sampling from the amazing variety of molecules. (a) A molecular-level view of a tiny sample of pure hydrogen. Each hydrogen molecule is made up of two hydrogen atoms. Hydrogen is a gas (unless pressurized and cooled to a very low temperature), so the molecules are independent of one another and travel at very high speeds. (b) A molecule of deoxyribonucleic acid, more commonly known as DNA. Notice how the molecule twists around a central axis. Also observe the repeating units of the pattern within the molecule. (c) The protein chymotrypsin, which is one of approximately 100,000 different types of protein molecules in the human body. The function of this molecule is to speed up chemical reactions.

per hour when the gas is at room temperature! The individual molecule is two hydrogen atoms attached by the interaction between minute, oppositely charged particles within the molecule. Even though the hydrogen molecule is simple, it is the high-energy fuel that powers the sun and other stars. It is the ultimate source of most of the energy on earth. Hydrogen is found everywhere in the universe. It is part of many molecules in your body. Hydrogen is also the favorite molecule of theoretical chemists, who take advantage of its simplicity and use it to investigate the nature of molecules at the most fundamental level.

Now look at the DNA molecule (**Figure 1.1[b]**). DNA is nature's way of storing instructions for the molecular makeup of living beings. At first glance, it seems complex, but on closer inspection you can see a simple pattern that repeats to make up a larger molecule. This illustrates one of the mechanisms by which nature works—a simple pattern repeats many times to make up a larger structure. DNA is an abbreviation for **d**eoxyribo**n**ucleic **a**cid, a compound name that identifies the simpler patterns within a molecule.

Even this relatively large molecule is very tiny in comparison with objects that can be directly observed. Five million DNA molecules can fit side-by-side across your smallest fingernail. (By the way, if you are a health or life sciences major, we think you'll agree that understanding the DNA molecule is a critical part of your education.)

Speaking of fingernails, they are made of the protein keratin. The human body contains about 100,000 different kinds of protein molecules. Some protein molecules in living organisms act to speed up chemical reactions. **Figure 1.1(c)** shows one such molecule, known as chymotrypsin. Proteins have many other essential biological functions, including being the primary components of skin, hair, and muscles, as well as serving as hormones.

Before you can truly understand the function of complex molecules such as DNA or proteins, you will have to understand and link together many fundamental concepts. This book and course are your first steps on the journey toward understanding the molecular nature of matter.

Now that you've had a look into the future of your chemistry studies, let's step briefly back to the past and consider the time when the science now called chemistry began.

1.1 Introduction to Chemistry: Lavoisier and the Beginning of Experimental Chemistry

ACTIVE MUSEUM/Alamy Stock Photo

Figure 1.2 Antoine Lavoisier and his wife, Marie. They were married in 1771 when he was 28 and she was only 14. Marie was Antoine's laboratory assistant and secretary.

Antoine Lavoisier (1743–1794) is often referred to as the father of modern chemistry (**Figure 1.2**). His book *Traité Élémentaire de Chime*, published in 1789, marks the beginning of chemistry as we know it today, in the same way Darwin's *Origin of Species* forever changed the science of biology.

Lavoisier's experiments and theories revolutionized thinking that had been accepted since the time of the early Greeks. Throughout history, a simple observation defied explanation: When you burn a wooden log, all that remains is a small amount of ash. What happens to the rest of the log? Johann Becher (1635–1682) and Georg Stahl (1660–1734) proposed an answer to the question. They accounted for the "missing" weight of the log by saying that *phlogiston* was given off during burning. In essence, wood was made up of two things: phlogiston, which was lost in burning, and ash, which remained after. In general, Becher and Stahl proposed that *all* matter that had the ability to burn was able to do so because it contained phlogiston.

Lavoisier doubted the phlogiston theory. He knew that matter loses weight when it burns. He also knew that when a candle burns inside a sealed jar, the flame eventually goes out. The larger the jar, the longer it takes for the flame to disappear. How does the phlogiston theory account for these observable facts? If phlogiston is given off in burning, the air must absorb the phlogiston. Apparently, a given amount of air can absorb only so much phlogiston. When that point is reached, the flame is extinguished. The more air that is available, the longer the flame burns.

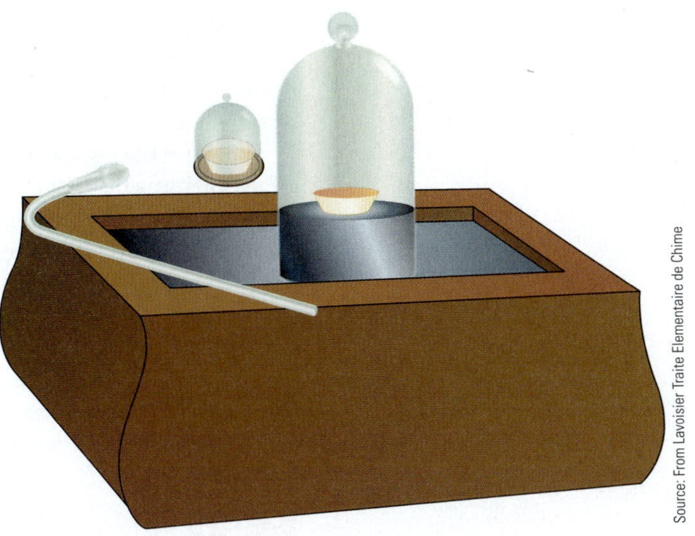

Source: From Lavoisier Traite Elementaire de Chime

Figure 1.3 Lavoisier's phosphorus-burning experiment, as illustrated in his book *Traité Élémentaire de Chime*. A sample of solid phosphorus was placed in the dish inside the bell jar and ignited. The ash that remained after burning weighed more than the original sample. The quantity of oxygen gas in the bell jar decreased. How could phosphorus lose phlogiston but weigh more?

So far, so good—no contradictions. Still, Lavoisier doubted. He tested the phlogiston theory with a new experiment. Instead of a piece of wood or a candle, he burned some phosphorus. Moreover, he burned it in a bell jar filled with oxygen (**Figure 1.3**). When the phosphorus burned, its ash appeared as smoke. The smoke was a finely divided powder, which Lavoisier collected and weighed. Curiously, the ash weighed *more* than the original phosphorus. What's more, the liquid level in the bell jar increased in height, indicating that there was less oxygen in the jar after burning than before.

What happened to the phlogiston? What was the source of the additional weight? Why did the volume of oxygen in the jar decrease when it was supposed to be absorbing phlogiston? Is it possible that the phosphorus absorbed something from the oxygen, instead of the oxygen absorbing something (phlogiston) from the phosphorus? Whatever the explanation, something was very wrong with the theory of phlogiston.

Lavoisier needed new answers and new ideas. He sought them in the chemist's workshop: the laboratory. He devised a new experiment in which he burned liquid mercury in air. This formed a solid red substance (**Figure 1.4**). The result resembled that of the phosphorus experiment. The red powder formed weighed more than the original mercury. Lavoisier then heated the red powder by itself. It decomposed, reforming the original mercury and a gas. The gas turned out to be oxygen, which had been discovered and identified just a few years earlier.

These experiments—burning phosphorus and mercury, both in the presence of oxygen and both resulting in an increase in weight—disproved the phlogiston theory. A new hypothesis took its place: When a substance burns, it combines with oxygen. This hypothesis has been confirmed many times. It is now accepted as the correct explanation of the process known as burning.

But wait a moment. What about the ash left after a log burns? It does weigh less than the log. What happened to the lost weight? We'll leave that to you to think about for a while. You probably have a good idea about it already, but (also

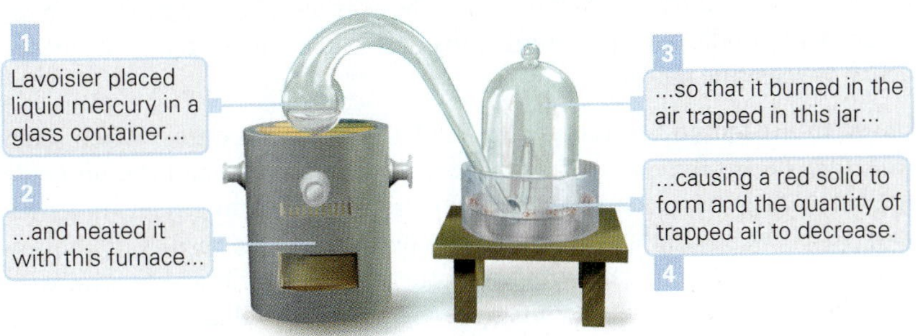

1
Lavoisier placed liquid mercury in a glass container...
2
...and heated it with this furnace...
3
...so that it burned in the air trapped in this jar...
...causing a red solid to form and the quantity of trapped air to decrease.
4

Figure 1.4 Lavoisier's apparatus for investigating the reaction of mercury and oxygen, as illustrated in his book *Traité Élémentaire de Chime*.

probably) you aren't really sure. If you were Lavoisier and you wondered about the same thing, what would you have done? Another experiment, perhaps? We won't ask you to perform an experiment to find out what happens to the lost weight. We'll tell you—but not now. The answer is explained in Chapter 9.

Before leaving Lavoisier, let's briefly visit a spin-off of his phosphorus experiment. Lavoisier was the first chemist to measure the weights of chemicals in a reaction. The concept of measuring weight may seem obvious to you today, but it was revolutionary in the 1700s. We have already noted that the phosphorus gained weight. The weight gained by the phosphorus was "exactly" the same as the weight lost by the oxygen. "Exactly" is in quotation marks because the weighing was only as exact as Lavoisier's scales and balances were able to measure. As you will see in Chapter 3, no measurement can be said to be "exact." In Chapter 2, you will see the modern-day conclusion of Lavoisier's weight observations. It is commonly known as the Law of Conservation of Mass. It says that mass is neither gained nor lost in a chemical change.

1.2 Introduction to Chemistry: Science and the Scientific Method

Key terms are indicated with **boldface print** throughout the textbook.

We have selected a few of Antoine Lavoisier's early experiments to illustrate what has become known as the **scientific method** (Figure 1.5). Examining the history of physical and biological sciences reveals features that occur repeatedly. They show how science works, develops, and progresses. They include the following:

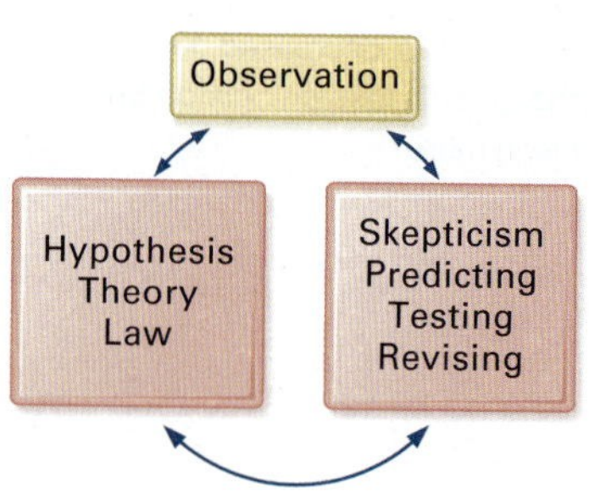

Figure 1.5 The scientific method.

1. *Observing.* A wooden log loses weight when it burns.
2. *Proposing a hypothesis.* A **hypothesis** is a tentative *explanation* for observations. The initial hypothesis posed by scientists before Lavoisier was that wood—and everything else that burns—contains phlogiston. When something burns, it loses phlogiston.
3. *Divorcing yourself from bias and personal beliefs.* You must also minimize the role of bias in evaluating the work of others. Lavoisier was skeptical of the phlogiston hypothesis because metals gained weight when strongly heated. If this process was similar to burning wood, why was the phlogiston not lost?
4. *Predicting an outcome that should result if the hypothesis is true.* When phosphorus burns, it should lose weight.
5. *Testing the prediction by an experiment.* Lavoisier burned phosphorus. It gained weight instead of losing it. The hypothesis is refuted. A new hypothesis is required.
6. *Revising or changing the hypothesis.* Lavoisier proposed that burning combines the substance burned and oxygen. (How did Lavoisier know about oxygen?)
7. *Testing the revised or new hypothesis and predicting a new experimental outcome.* The new hypothesis was supported when Lavoisier burned mercury and it gained weight.
8. *Upgrading the hypothesis to a theory by more experiments.* Lavoisier and others performed many more experiments. (How did others get into the process?) All the experiments supported the explanation that burning involves combining with oxygen in the air. When a hypothesis is tested and confirmed by many experiments under varying conditions, without contradiction, it becomes a **theory** or **scientific model**.

The scientific method is not a rigid set of rules or procedures. When scientists get ideas, they most often try to determine if anyone else has had the same idea or perhaps has done some research on it. They do this by reading relevant articles in the many scientific journals in which researchers report the results of their work. Modern scientists communicate with each other through technical literature. Scientific periodicals are also a major source of new ideas, as well as talks and presentations at scientific professional meetings.

Figure 1.6 Chemical Abstracts Service, a division of the American Chemical Society, is located in Columbus, Ohio. They maintain a database of chemical substances. You can search about 7,900 common chemicals at http://commonchemistry.org/. Your college or university library may have subscriptions to more powerful database searching tools.

Communication is not usually included in the scientific method but it should be. Lavoisier knew about oxygen because he read the published reports of Joseph Priestley and Carl Wilhelm Scheele, who discovered oxygen independently in the early 1770s. In turn, other scientists learned of Lavoisier's work and confirmed it with their own experiments. Today, communication is responsible for the explosive growth in scientific knowledge (**Figure 1.6**). It is estimated that the total volume of published scientific literature in the world doubles every 8 to 10 years.

Another term used to describe patterns in nature in a general way is *law.* In science, a **law** is a summary of a pattern of regularity detected in nature. Probably the best known is the law of gravity: Objects are attracted to one another. If you release a rock above the surface of the earth, it will fall to the earth. No rock has ever "fallen" upward.

A scientific law does not explain anything, as a hypothesis, theory, or scientific model might. A law simply expresses a pattern. Although laws cannot be proved, they form the foundation of scientific knowledge. The only justification for such confidence is that in order for a law to be so classified, it must have no known exceptions.

1.3 Introduction to Chemistry: The Science of Chemistry Today

Chemists study matter and its changes from one substance to another by probing the smallest basic particles of matter to understand how these changes occur. Chemists also investigate energy transferred in chemical change—heat, electrical, mechanical, and other forms of energy.

Chemistry has a unique, central position among the sciences (**Figure 1.7**). It is so central that much research in chemistry today overlaps physics, biology, geology, and other sciences. You will frequently find both chemists and physicists, or chemists and biologists, working on the same research problems. Scientists often refer to themselves with compound words or phrases that include the suffix or word *chemist:* biochemist, geochemist, physical chemist, medicinal chemist, and so on.

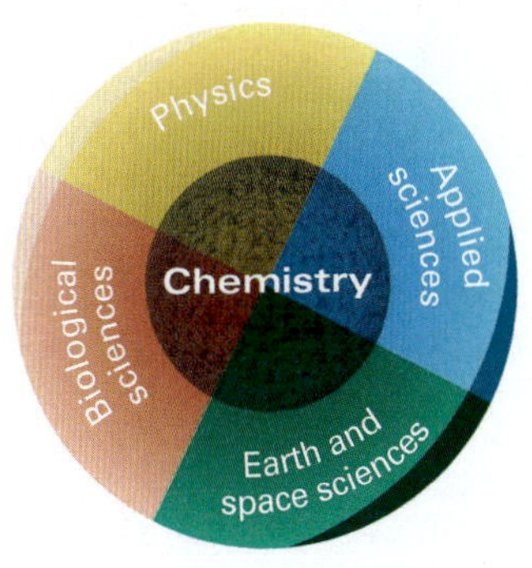

Figure 1.7 Chemistry is the central science. Imagine all sciences as a sphere. This cross section of the science sphere shows chemistry at the core. If you view the other sciences as surface-to-center samples, each contains a chemistry core.

Chemistry has traditionally been classified into five subdivisions: analytical, biological, organic, inorganic, and physical. Analytical chemistry is the study of what (qualitative analysis) and how much (quantitative analysis) are in a sample of matter. Biological chemistry—biochemistry—is concerned with living systems and is by far the most active area of chemical research today. Organic chemistry is the study of the properties and reactions of compounds that contain carbon. Inorganic chemistry is the study of all substances that are not organic. Physical chemistry examines the physics of chemical change.

Figure 1.8 Chemists at work.

Figure 1.9 Polypropylene plant. Plastics are the substances produced in the greatest quantity by the chemical industry. This plastic manufacturing facility is located in Tobolsk, Russia (a historic capital of Siberia).

You will find chemists—the people who practice chemistry—in many fields (**Figure 1.8**). Probably the chemists most familiar to you are those who teach and do chemical research in colleges and universities. Many industries employ chemists for research, product development, quality control, production supervision, sales, and other tasks. The petroleum industry is the largest single employer of chemists, but chemists are also highly visible in medicine, government, chemical manufacturing, the food industry, and mining.

Chemical manufacturers produce many things that we buy and take for granted today. They convert raw materials available in nature, such as oil, coal, and natural gas, into products such as plastics, fertilizers, and pharmaceutical drugs. The most commonly produced products are plastics, such as plastic bags, bottles, and packaging (**Figure 1.9**). Another familiar and important category of manufactured goods from the chemical industry is health products, such as pharmaceuticals and nutritional supplements. Millions of people are employed worldwide by the chemical industry. The German-based company BASF is the largest chemical company in the world, with well over 70 billion dollars in annual sales, and employing more than 100,000 people.

1.4 Introduction to Active Learning: Learning How to Learn Chemistry

Here is your first chemistry "test" question:

Which of the following is your primary goal in this introductory chemistry course?

A. To learn all the chemistry that I can in the coming term

B. To spend as little time as possible studying chemistry

C. To get a good grade in chemistry

D. All of the above

If you answered A, you have the ideal motive for studying chemistry—and any other course for which you have the same goal. Nevertheless, this is not the best answer.

If you answered B, we have a simple suggestion: Drop the course. Mission accomplished.

If you answered C, you have acknowledged the greatest short-term motivator of many college students.

Fortunately, most students have a more meaningful purpose for taking a course.

If you answered D, you have chosen the best answer.

Let's examine answers A, B, and C in reverse order.

C: There is nothing wrong in striving for a good grade in any course, just as long as it is not your major objective. A student who has developed a high level of skill in cramming for and taking tests can get a good grade even though he or she has not learned much. That helps the grade point average, but it can lead to trouble in the next course of a sequence, not to mention the trouble it can cause when you graduate and aren't prepared for your career. It is better to regard a good grade as a reward earned for good work.

B: There is nothing wrong with spending "as little time as possible studying chemistry," as long as you *learn* the needed amount of chemistry in the time spent. Soon we'll show why the amount of time required to *learn* (not just study) chemistry depends on *when* you study *and* learn. They should occur simultaneously. Reducing the time required to complete any task satisfactorily is a worthy objective. It even has a name: *efficiency.*

A: There is nothing wrong with learning all the chemistry you can learn in the coming term, as long as it doesn't interfere with the rest of your schoolwork and the rest of your life. The more time you spend studying chemistry, the more you will learn. College is the last period in the lives of most people in which the majority of their time can be devoted to intellectual development and the acquisition of knowledge, and you should take advantage of the opportunity. But maintain balance. Mix some of answer B in your endeavor to learn. Again, the key is efficiency.

To summarize, the best goal for this chemistry course—and for all courses—is to learn as much as you can possibly learn in the smallest *reasonable* amount of time.

The rest of this section identifies choices that you need to make to ensure that you will reach your goal.

Choice 1: Commit to Sufficient Time Outside of Class

A rule of thumb for college coursework is that an average student in an average course should spend two hours outside of class for every hour in class. Are you ready to *choose* to make this commitment? You may have to spend more time outside of class if your math skills are weak, if you have not recently had a good high school chemistry course, if English is not your native language, or if you have been out of school for some time. To keep your out-of-class time to an efficient minimum, you must study regularly, doing each assignment before the next class meeting. Chemistry builds on itself. If you don't complete today's assignment before the next class meeting, you will not be ready to learn the new material. Many successful students schedule regular study time, just as they would schedule a class. *Failure to commit sufficient time outside of class is the biggest problem when it comes to learning chemistry.*

Choice 2: Commit to Quality Time When Studying

Efficient learning means learning at the time you are studying. It does not mean just reading your notes or the book and deciding to come back and learn the material later. It takes longer to *learn now* than it does to passively read the textbook, but the payoff comes with all the time you save by not having to learn later. This is so important that we have special *Learn It Now!* reminders throughout the textbook. Are you ready to *choose* to commit to making your study time high quality? If so, you should also commit to studying without distractions—without sounds, sights, people, or thoughts that take your attention away from learning. Turn your cell phone off for at least a half hour at a time while studying. Every minute your mind wanders while you study must be added to your total study time. Your time is limited, and that wasted minute is lost forever.

Choice 3: Commit to Utilizing All Learning Resources

College chemistry courses typically have a multitude of learning resources, which may include lecture, this textbook and its accompanying online learning tools, laboratory exercises, discussion sections, help centers, tutors, instructor office hours, Internet resources, and your school library. Are you ready to *choose* to commit to taking advantage of all of the learning tools provided in your course? Let's consider some of these tools in more detail.

Lecture Although it is obviously the wrong way to learn, some students choose to skip lectures occasionally. Don't be one of those students. *Attend every lecture* (**Figure 1.10**). If you miss just one lecture per month in a semester course, you will probably miss 10% of the material. That is a reduction of one letter grade worth of content in a typical course.

You need to learn the role of lecture in your course. If your instructor expects you to listen to his or her discussion and watch presentation slides and/or material written on the board or an overhead projector, you will need to take notes. We recommend that your note-taking procedure follow these general steps: (1) Preview the material by skimming the textbook. Usually, this only needs to be done every few lectures as a new chapter is about to be introduced. Look in particular for new words and the major concepts so that you are not caught unprepared when they are introduced in lecture. (2) Concentrate during lecture and take notes. Don't fool yourself; concentrating over an extended period of time is hard work. Focus on what is being shown and said, and work to transcribe as much material as accurately and quickly as you can. Use a notebook that is exclusively for chemistry lecture. (3) Organize your notes as soon as possible after lecture. Organization is the key. During a classic lecture, you often are mostly working to transcribe the material. True learning occurs when you work to make sense of the material and try to analyze the relationships among the concepts that were discussed. (4) Study the textbook, work the assigned problems, and look for connections between the lecture and the textbook. You will often find that seeing the material presented in a slightly different way is the key to helping you make sense of a concept. Combining your organized lecture notes with the textbook presentation of the same topic is a powerful learning technique.

Textbook This book is a central learning resource in your chemistry course. We will help you to become familiar with its structure in the next section.

Laboratory If your course includes a laboratory, learn what each experiment is designed to teach. Relate the experiment to the lecture and textbook coverage of the

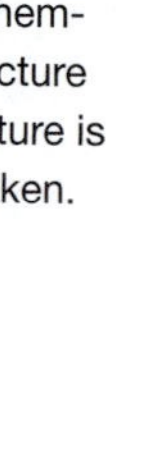

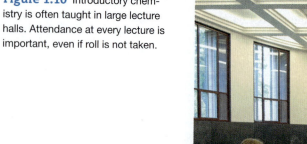

Figure 1.10 Introductory chemistry is often taught in large lecture halls. Attendance at every lecture is important, even if roll is not taken.

iStock.com/thelinke

same topic. Seeing something in the laboratory and getting a hands-on experience is often just what you need to fully understand what you read in the textbook and see and hear in the lecture.

Instructor Office Hours Many chemistry instructors are available for help outside of class. If your instructor is not, you likely have a teaching assistant with office hours or a tutoring center that you can visit instead. No matter the quality of digital or print instructional resources available to you, human help is occasionally needed to accomplish your learning goals. We recommend that you develop a list of questions and/or sample problems that you cannot solve before you attend office hours.

Internet The Internet provides you with an abundance of information related to introductory chemistry. When a topic presented in class or this textbook is unclear, clarification may be available by doing a search for the topic to see if an alternative perspective helps you learn. A well-designed website can often have the information you need to solidify your understanding of a concept. However, you should use the Internet with a healthy dose of skepticism. Most websites lack the sequencing, structure, and integration of topics that your instructor, your course curriculum, and this textbook provide. Also be sure that you choose reputable websites to ensure that you are not led astray by incorrect or incomplete information.

Library or Learning Center Many college libraries and learning centers have Internet resources, computer programs, workbooks, and other learning aids that are helpful for practice with using chemical formulas, balancing equations, solving problems, and other routine skills. Find out what is available for your course and use it as needed. Some instructors will also put supplementary materials on reserve. Take advantage of these, if provided.

Choice 4: Commit to Improvement

By definition, you are changed as a result of learning. You need to be willing to open your mind to new, more powerful ways of thinking about the natural world and the process of personal intellectual development. The purpose of your college education is to make you a better person. Are you willing to *choose* to commit to improving the way you understand nature, becoming a better learner, and developing your intellect? Let's look at some ways to do this within the framework of this chemistry course.

Think Like a Chemist The perspective of the chemist is unique, as is the perspective of the philosopher, the mathematician, the geographer, or the linguist. Each course you take in college will expose you to a different way of thinking about the world. In this chemistry course, you should work to understand the distinctive viewpoint of a chemist. In particular, focus on the relationships among the macroscopic, directly observable natural world; the abstract, particulate makeup of those macroscopic materials; and the symbols that chemists use to represent both the macroscopic and particulate world, as illustrated in **Figure 1.11**.

Think Conceptually A trap that some students fall into while solving quantitative chemistry problems is to mindlessly crunch numbers without thinking about the underlying concept. Almost certainly, there will be a few routine types of quantitative problem setups that you should master without the need to reinvent the procedure each time you solve such a problem. But many other problems will be more complex. With these more complex problems, it is critical to understand the underlying concept. If you can imagine the particulate-level process described in the problem statement, do so. Remember that it is not the answer that is important when you tackle difficult problems but rather the process that should be your focus.

Embrace Multiple Ways of Knowing This chemistry course will expose you to many ways of obtaining new knowledge. You will likely need to learn (in order of increasing

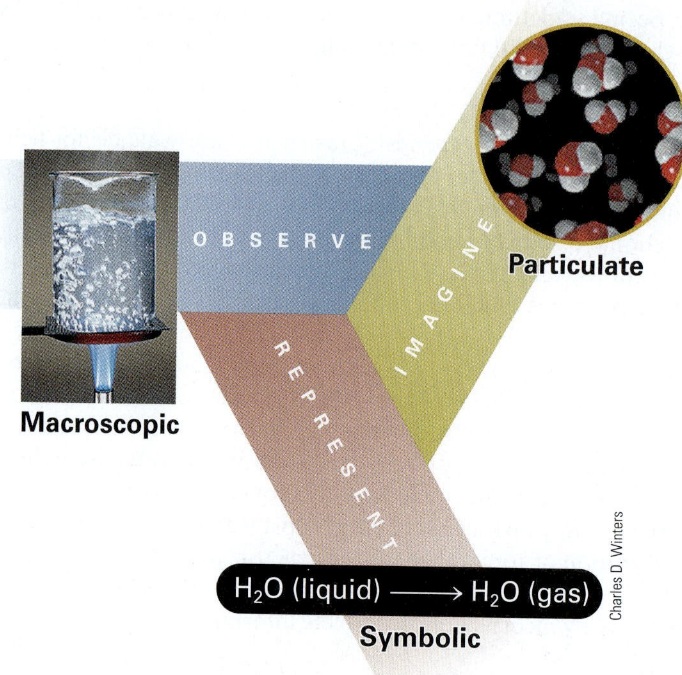

Figure 1.11 How to think like a chemist. You are familiar with the macroscopic view of matter, as seen in this container filled with boiling water. A key characteristic of thinking like a chemist is imagining how the water would appear if you could see it at the particulate level. The particulate circle shows how a chemist views water. To express this viewpoint in writing, chemists use symbols. The symbols in the formula H_2O describe the particulate-level composition of each water molecule.

complexity) facts, rules, concepts, and problem solving. Facts are things that you need to memorize, such as the fact that the symbol for hydrogen is H. Rules are connections between things, and they are often expressed as mathematical relationships. For example, the volume of a pure solid substance is directly proportional to its mass, which can be expressed in symbols as $V \propto m$. Rules are also often expressed in the form of if/then statements. *If* an element forms a monatomic anion, *then* the name of the anion is the name of the element, changed to end in *-ide.* Concepts are mental models of the natural world. We will present relatively simple conceptual models in this introductory course, and as you learn more about chemistry in future courses, you will find that you will need to revise and increase the complexity of your conceptual models. Problem solving is a skill that you learn through coaching and practice. Good problem solvers are highly regarded in all aspects of professional life. We will help guide you in developing your problem-solving skills in this textbook, but you will also need to put in a good deal of practice time to become a skilled problem solver. You will likely have your favorite type of learning, and that will probably shape your decision about your major and, ultimately, your career path, but recognize that each mode of learning has its importance in your education. Embrace the opportunity to become a more skilled learner in each type of knowing.

Think About Your Thinking It is important not only to learn chemistry content while in this course but also to work to develop the thinking skills that are used by chemists. An example of a thinking skill is proportional reasoning, which is the ability to recognize and apply relationships between two variables that are directly proportional to one another. If you learn to see these types of relationships beyond their immediate application, you will be able to utilize these skills in solving problems in many other contexts. We will discuss this further in the next section.

Utilize Feedback in a Positive Manner All courses will provide you with feedback on your performance in some way. Typically, courses have exams and/or quizzes that assess your learning. This textbook has many end-of-chapter questions, exercises, and problems that are accompanied by answers at the end of each chapter. You can choose to use such feedback as merely a descriptor of your learning history, such as "I earned an 80 on the gases chapter test," or "I got that problem wrong," or you can use the feedback in a positive manner by thinking, "What did I do wrong, and how

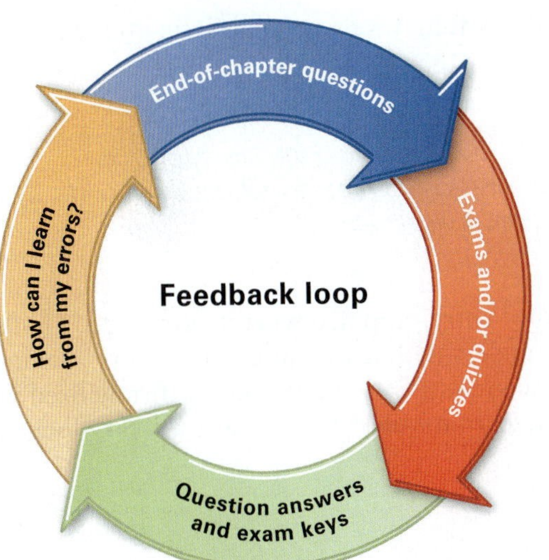

Figure 1.12 The feedback loop. Learning from your mistakes is an essential part of the knowledge-building process.

can I improve?" A critical element of the process of learning is to learn from your mistakes (**Figure 1.12**). When you receive a corrected exam or quiz, look at your errors and make a commitment to change your thinking so that you don't repeat the same error. When you solve an end-of-chapter problem incorrectly, assess what you did wrong and restudy the appropriate material so that you can replace the misconception with a more accurate understanding of the concept or procedure.

1.5 Introduction to Active Learning: Your Textbook

The most important tool in most college courses is the textbook. It is worth taking a few minutes to examine this book and learn about its unique learning aids. In this section, we'll show you the book's features that are designed specifically to help you learn chemistry as efficiently as possible.

Section-by-Section Goals

Goal
1. Read, write, and discuss chemistry using a fundamental and scientifically accurate and precise chemical vocabulary.
2. Write a fundamental set of inorganic chemical formulas and write names of substances when formulas are given.
3. Write, balance, and interpret chemical equations.
4. Set up and solve elementary chemical problems.
5. "Think" chemistry in some of the relatively simple theoretical areas and visualize what happens at the particulate level.
6. Improve your scientific thinking skills, particularly in proportional reasoning and mental modeling.

The goals listed here are not only for a section but also for this entire book and the course in which you will use it. They tell you what you will be able to do when you complete the course.

As you approach most sections in this text, you will find one or more goals. They tell you what you should be able to do after you study the section and complete the end-of-chapter questions. If you focus your attention on learning what is in the goals, you will learn more in less time.

Few chemistry textbooks include section-by-section goals, although they sometimes appear in study guides that accompany those books. When you move on to the next chemistry course required for your major, it becomes your responsibility to write the goals yourself—to figure out what understanding or ability you are expected to gain in your study. Literally writing your own goals is an excellent way to prepare for an exam.

Learn It Now! *In the previous section, "Choice 2: Commit to Quality Time When Studying," we discussed the importance of learning efficiently. We noted that we would provide you with* Learn It Now! *reminders throughout the textbook, printed in red.*

When you come to a *Learn It Now!* entry, stop. Do what it says to do. Think about it. Make a conscious effort to understand, learn, and, if necessary, memorize what is being presented. When you are satisfied that this idea is firmly fixed in your mind, then continue on. In short, learn it—*now*! Tomorrow it will take longer. Tomorrow is too late.

Active Examples

As you study this book, you will acquire certain "chemical skills." These include writing chemical names and formulas, writing and interpreting chemical equations, and solving chemical problems—the things listed previously as Goals 2, 3, and 4. You will develop these skills by studying and actively working the examples in the text.

Active Example 1.1 How to Work an Active Example

What do you do when you come to an Active Example in this textbook?

Think Before You Write All active examples begin with a brief discussion designed to help you think about the nature of the problem statement. This helps you to think carefully about approaching the problem and to avoid acting impulsively. It also allows you to activate and engage the location in your brain where scientific thinking is processed.

Answers *Cover the left column with your cut-out shield. Reveal each answer only after you have written your own answer in the right column.*

Our answer is provided in the blue-shaded box immediately to the left of where you write your answer. *Always keep this box covered until you have written your answer.* Use the cut-out shield provided in the book for this purpose. You will maximize the utility of this book by writing your answers first and then comparing your answer with ours. *When our answer needs additional explanation, the discussion appears in this style of print in a separate paragraph.*	The remaining frames lead you step-by-step through the thought process needed to answer the question or solve the problem. These mimic what a personal tutor would be doing if you were working one-on-one. To take full advantage of the Active Examples, use your cut-out shield to cover the left column and literally write your answers in the space provided. This process—writing your own responses *before you look at ours*—is a powerful and efficient learning technique.

Practice Exercise 1.1

All Active Examples end with an exercise similar to the problem that you just solved for yourself with our guidance. These Practice Exercises allow you to see if you can answer a similar question or solve a similar problem without our stepwise assistance. They will help you reinforce what you just learned and extend and refine your knowledge.

Most Active Examples in this book take you through a series of questions and answers. Space is provided for you to actually write in a formula or equation or to solve a problem yourself. A small pencil icon appears within each space in which you should be writing answers. The next Active Example in this book appears in Section 2.3. At that point you will find detailed instructions for working this kind of example problem.

If you are to learn from Active Examples, *you* need to work through each one as you come to it. Never weaken the learning process that occurs when *you* work through an example by simply passively reading *our* answer.

Also, you will often see that what you learn in an Active Example is used immediately in the next section. Keep in mind that knowledge of chemistry is acquired through a building process. You will not be able to understand that next section without understanding the concept from the current Active Example. *Learn it now!*

Target Checks

Chemical principles, models, and theories are introduced with words and illustrations. Ideally, you will learn and understand these ideas as you study the text and figures. Use a Target Check question to find out if you have caught on to the main ideas immediately after they appear in the text. A Target Check is identified by the heading

 **Target Check**

Like Active Examples, Target Checks should be completed as you reach them. Answers to Target Checks can be found at the end of each chapter. If you answer a Target Check incorrectly, go back and restudy the targeted material before moving on to the next section or example.

The best thing that you can do to maximize your learning from the textbook is to take written notes, work the Active Examples, and complete the Target Checks *while you study*. This means writing in the book and not simply highlighting the textbook. A heavily highlighted textbook is nothing more than a brightly colored list of things you plan to learn later. It is just the opposite of the *Learn It Now!* philosophy. The very act of reading something, thinking about what it means, summarizing it, and answering a Target Check question in your own words actively generates learning.

P/Review

Often in the study of chemistry you see some term or concept that was introduced earlier in the course. To understand the idea in its new context, you may wish to review it as it was presented earlier. At other times, a topic is introduced briefly to meet a present need, even though it may not be necessary to understand it fully. That comes later; the present introduction is a preview.

This book has an optional order of topics, so the same item may be a preview with one instructor and a review with another. We therefore identify this kind of cross-reference as a P/Review. Each P/Review is carefully worded so that as a review, it gives you the information you must recall immediately but not so much that it will be confusing if it's read as a preview. A P/Review usually appears in the margin closest to where this icon appears ▶, and it always includes a specific chapter or section number that you may refer to, if you wish.

◀ **P/Review** This is what a P/Review looks like in the margin.

In-Chapter *a summary of...* and *how to...* Features

Throughout this book, you will find summaries and step-by-step procedures that are headed as follows:

Format of *a summary of...* and *how to...*

Each a summary of... or how to... explanation or procedure is highlighted like this paragraph. These give you, in relatively few words, the main ideas and methods you should learn from a more general discussion nearby in the text. They should help you clinch your understanding of the topic. Occasionally, summaries are in the form of a table or illustration; some even combine the two. These forms are particularly helpful in reviewing for a test. Not only do they review the topic briefly but they also create a mental image that may be recalled during an exam.

Thinking About Your Thinking

You will find passages throughout the text that look like the following:

Thinking About Your Thinking

Name of Skill

One goal of this textbook is to help you learn the thinking skills that chemists and other scientists commonly use. In this feature, we discuss thinking skills themselves, somewhat removed from the content of the surrounding text, so that you can clearly think about the skill itself, learn it, and apply it in any context. This feature will help you to learn to think about chemistry—and many other subjects—far beyond the days you spend in this course.

When you come to a Thinking About Your Thinking discussion, take a few moments to read it and reflect on the thinking skill it discusses. Ask yourself, "Could I apply this skill in any other context?" Perhaps you've used the skill before in a math or physics course. Maybe you've used it before in this course. The greater the number of contexts in which you can imagine applying a skill, the more generalizable your thinking skills will become. In this way, you grow intellectually.

Key Terms

Immediately after the last section of each chapter, you will find the first end-of-chapter feature, which is the key terms list. Key terms and concepts appear in the Glossary. Use your Glossary regularly.

Frequently Asked Questions

Immediately after the Key Terms, you will find the second end-of-chapter feature, which provides answers to questions that students often ask. These are written in a question-and-answer format, with a question that is similar to those frequently asked of instructors during office hours, followed by an answer. The questions typically cover both "how to study" topics and common mistakes and how to avoid them. You will find it helpful to read these questions and answers after you have studied each chapter. It will take only a couple of minutes, and you will likely find the answer to a question that you may have, or you may learn to avoid a potential pitfall before it becomes an issue.

Concept-Linking Exercises

After completing your study of a chapter, but before you begin to work on the Questions, Exercises, and Problems (described shortly), you will need to have a firm understanding of the key terms and concepts from the chapter and the relationships among them. To help you learn the relationships among the concepts, most chapters include Concept-Linking Exercises. These exercises consist of groups of concepts that you link together with a brief description of how they are related. Answers to the Concept-Linking Exercises can be found after the answers to the Target Checks at the end of each chapter.

Small-Group Discussion Questions

After the Concept-Linking Exercises, you will find a group of questions that are written so that they are best solved in collaboration with a small group of classmates.

They should be attempted after you have completed your independent work studying the chapter. Whereas the end-of-chapter questions (discussed in the next subsection) are designed for individual work, the Small-Group Discussion Questions are generally more complex and are aimed at helping you to "think chemistry" at a deeper level than you might in the absence of support and assistance from a group of peers in the same course. We do not provide answers to the Small-Group Discussion Questions. The process of verifying your answers to the questions without "the" answer being readily provided will help you to better understand how scientists actually work.

End-of-Chapter Questions, Exercises, and Problems

At the end of all the chapters, you will find Questions, Exercises, and Problems that are grouped by the section in the chapter to which they apply. Some questions are relatively straightforward, similar to the Active Examples and Target Checks. Others are more demanding. You may have to analyze a situation, apply a chemical principle, and then explain or predict some event or calculate some result. General Questions that may be drawn from any section in the chapter follow the section-specific questions. After these are the More Challenging Questions, designed to stretch you beyond the goals listed for the chapter.

The Questions, Exercises, and Problems generally are in matched pairs in which the consecutive odd–even numbered combinations involve similar reasoning, and in the case of exercises, similar calculations. All of the odd-numbered questions out of the matched-pair groups are answered at the end of the chapter. Solutions to exercises and problems include calculation setups. Most General Questions and More Challenging Questions, odd- and even-numbered, are also answered at the end of the chapter. Numbers of answered questions are printed in blue; numbers of unanswered questions are printed in **black**.

As you solve problems in the textbook, remember that your *main* objective is to understand the principle on which the problem is based, not to get a correct answer. Even when your answer is correct, stop and think about it for a moment. Don't leave the problem until you feel confident that you will recognize any new problem that is worded differently but that requires reasoning based on the same principle. Then be confident that you can solve such a problem.

Even more important is what you do when you do *not* get the correct answer to a problem. You may be tempted to return to the Active Examples and Target Checks, find one that matches your problem, and then solve the assigned problem step by step as in the example. *You should resist this temptation.* If you get stuck on a problem, it means that you did not truly *learn* from the earlier examples. Leave the problem. Turn back to the corresponding section. Study it again, by itself, until you understand it thoroughly. Then return to the assigned problem with a fresh start and work on it to the end without further reference to the section. Finally, work the remainder of the problems in the group until you are confident that you can solve this problem type.

Appendices

The Appendix of this book has three parts:

1. *Appendix I: Chemical Calculations.* Here you will find suggestions on how to use a calculator specifically to solve chemistry problems. There is also a general review of the arithmetic and algebraic operations used in this book. You will find these quite helpful if your math skills need dusting off before you can use them.
2. *Appendix II: The SI System of Units.* This explains the units in which quantities are measured and expressed in current science textbooks and other scientific publications.

3. *Glossary.* Like other fields of study, chemistry has its own special language, in which common words have very specialized and specific meanings. The Glossary lists these words in alphabetical order so you can find them easily and learn to use them. All of the boldfaced words in the text are in the Glossary. Use the Glossary regularly; it's a real time saver.

Figure 1.13 "With self-discipline most anything is possible." —Theodore Roosevelt (1858–1919), the 26th president of the United States.

1.6 A Choice

Discipline is the bridge between goals and accomplishments.
—Author unknown

You have a choice to make. You can choose to continue learning as you did before this moment, or you can choose to improve your learning skills (**Figure 1.13**). Even if those skills are already good, they can be improved. This chapter gives you some specific suggestions on how to do this. It also helps you to upgrade your study habits, beginning here and continuing throughout the book.

If you ever begin to feel that chemistry is a difficult subject, read this chapter again. Then ask yourself, and give an honest answer: "Do I have trouble because the subject is difficult, or is it because I did not choose to improve my learning skills?" Your honest answer will tell you what to do next.

At all stages of our lives we make choices. We then live with the consequences of those choices. Choose wisely—and enjoy learning chemistry.

Key Terms

Most of the key terms and concepts and many others appear in the Glossary. Use your Glossary regularly.

Hypothesis p. 4
Law p. 5
Scientific method p. 4
Scientific model p. 4
Theory p. 4

Questions, Exercises, and Problems

1. Develop a study plan for your chemistry course. Your instructor has probably already explicitly or implicitly informed the class about what she or he considers to be the most important out-of-class activity needed for success in the course. What activity is most important in your course? Write down all out-of-class activities that will be part of your study plan. Things to consider include previewing chapters to identify key terms; spending time in class, time in lab, and/or time in discussion section (if your course includes these); organizing lecture notes; studying the textbook; doing end-of-chapter questions, exercises, and problems; studying and completing online assignments; consulting academic websites on the Internet; reviewing and learning from your mistakes on returned quizzes and exams; reviewing for exams (including the final exam, if any); writing lab reports (if applicable); and accessing human, in-person help via activities such as attending instructor office hours, review sessions, tutoring, and working with other students in the course. Estimate the quantity of time needed per week for each activity you include in your study plan. Adjust your study plan as the academic term progresses and as you learn more about the keys to success in your course.
2. Develop a weekly calendar for the quarter or semester. Search the Internet for "study schedule template" or something similar to find a form that you can print, copy, or download. As you fill in the schedule, be honest and realistic. Include time for personal activities such as meals, time with family and friends, exercise, TV, games, Internet surfing, clubs and organizations, religious activities, routine household chores, and so on. Enter your typical work hours if you are employed. Enter all class and lab hours. Most importantly, add time for out-of-class study. Use the rule of thumb that you should spend two hours outside of class for each hour in class. Is your schedule realistic? If not, consider how you might change it to give yourself the maximum opportunity to succeed in college while still living a balanced lifestyle. As the academic term progresses, revisit your calendar and adjust it as you gain experience in learning the time demands of a typical week.

2 Matter and Energy

◀ Understanding the chemistry of living organisms, such as these dolphins, is the goal of many 21st century chemists. All living organisms—bacteria, plants, and animals—share a set of tiny particles that change from one substance to another by remarkably similar chemical processes.

CHAPTER CONTENTS

Chemistry is the study of substances and the energy associated with their change. In this chapter, we explain some fundamental features of matter and energy and introduce the vocabulary scientists use to communicate about their transformations.

2.1 What Makes Up the Universe?

Figure 2.1 illustrates what the universe is made of. It shows that we understand only 5% of the universe! If we add together all of the forms of understood energy, such as light energy and heat energy, and all of the forms of understood matter, such as plants and animals, planets, stars, and galaxies, we can account for only 5% of the universe.

Learn It Now! *Textbook authors often use illustrations and photographs to help teach concepts. Due to the constraints of page layout, these figures are often displaced a slight distance on the page relative to where the text discusses the figure. Always study a figure at the exact point it is discussed in the text. Don't separate your study of written words from the accompanying illustrations and photographs. Integrate the words and images in your mind to form a cohesive whole.*

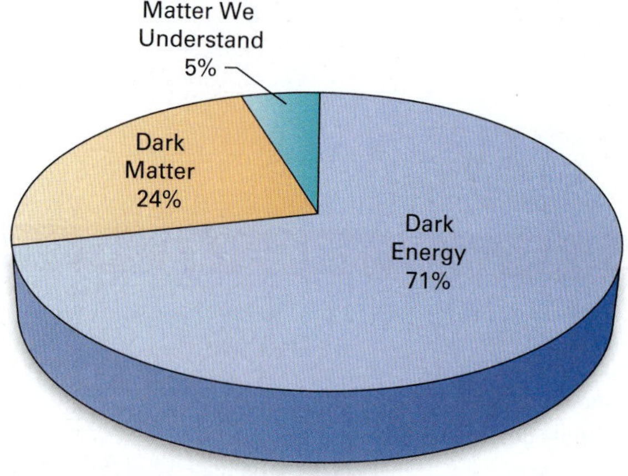

Figure 2.1 Composition of the universe.

How do we know that we don't know what most of the universe is made of? Most of the gravity we measure in the universe *has no known origin.* One way through which this is determined is by studying the rotation of spiral galaxies—those that are largely flat and rotating, with a high concentration of stars in the center. When accounting for all of the visible matter in these galaxies, it can be calculated that they should fly apart. But they don't. Therefore, something that we can't see—something that does not interact with light—is holding them together. This "something" has been named dark matter. The American scientist Vera Rubin made the measurements and performed the calculations that convinced the scientific community of the mismatch between gravity and visible mass (*weight* in everyday language) (**Figure 2.2**).

Similarly, we observe that the universe is expanding at an accelerating rate. When we account for the known energy in the universe, it is not enough to account for this acceleration. There must be "something" in the universe that we do not yet understand, currently labeled as dark energy.

As you embark upon your study of chemistry in this academic term, be aware that although it will seem at times as if the human race has mastered a scientific understanding of everything in the natural universe and the amount of chemistry you are being challenged to learn may occasionally seem overwhelming, the truth is that human race actually has much to learn. We do know that we can categorize the universe as matter, which are objects that have mass, and energy, the ability to do work or transfer heat. Thus, we will begin your study of the science of chemistry—using the scientific method to investigate matter and its transformation—by introducing some features of matter and energy that we believe *are* currently well understood.

Archives and Special Collections, Vassar College Library

Figure 2.2 Vera Rubin (1928–2016) was the first person to publish data that provided evidence of the existence of dark matter. Her findings revolutionized the field of cosmology. She was awarded the 1993 National Medal of Science for the Physical Sciences.

2.2 Representations of Matter: Models and Symbols

Goal 1 Identify and explain the differences between interpreting and describing matter at the macroscopic, microscopic, and particulate levels.

2 Define the term *model* as it is used in chemistry to represent pieces of matter too small to see.

Learn It Now! *The numbered Goals tell you what you should be able to do after you study a section. Always focus your study on the goals. When you complete this section and the corresponding end-of-chapter questions, you should know the differences between the macroscopic, microscopic, and particulate levels when interpreting and describing matter, and you should be able to define the term* model *as it is used by scientists.*

Everything that has mass is **matter**. The warmth of the sun and the light it gives off are not forms of matter, but the substances the sun is made of are matter. All physical objects are matter.

We can describe some forms of matter by observation with the naked eye. These **macroscopic** samples of matter include a huge range of sizes, varying from mountains, rocky cliffs, huge boulders, and all sizes of rocks and stone to gravel and tiny grains of sand. Geologists often study matter at this level. The macroscopic size

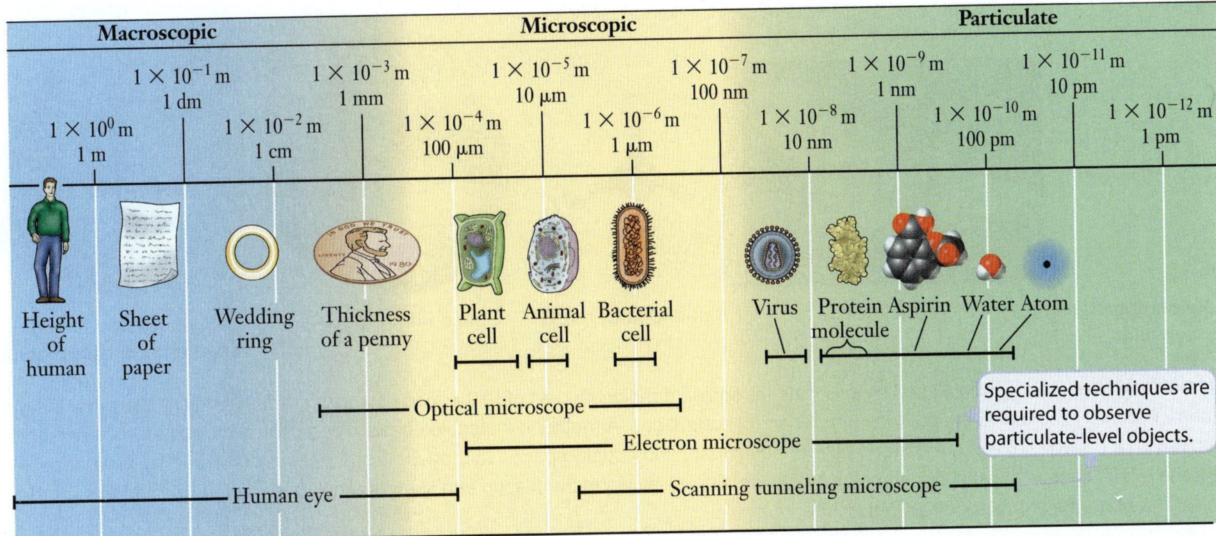

Figure 2.3 Macroscopic, microscopic, and particulate forms of matter. The particulate-level drawings are *models* of types of matter too small to see with the human eye or an optical microscope. Electron microscopes "shine" a beam of electrons through a sample in much the same way as optical microscopes shine light through the specimen. Scanning-tunneling microscopes depend on electrical properties of matter to create computer-generated images.

range illustrated in **Figure 2.3** shows additional examples of macroscopic samples of matter that are visible with the human eye.

It is common for middle school and high school biology students to use microscopes to observe types of matter that are too small to be seen with the unaided human eye ▶. You've probably used a microscope to observe tiny animals or plants, or perhaps the tiny crystals on the surface of a polished rock. These are examples of **microscopic** samples of matter. The middle section of Figure 2.3 illustrates some microscopic samples of matter.

The prefix *macro-* means large, and the prefix *micro-* means small. The suffix *-scopic* refers to viewing or observing.

Chemists often think about matter that is too small to be seen even with the most powerful optical microscope. A chemist considers the behavior and transformations of the tiny particles that make up matter. Thinking this way is thinking at the **particulate** level. One of the most valuable skills you will learn in this course is to think about the particulate nature of matter. The particulate range illustrated in Figure 2.3 is not visible, even with a microscope. Chemists use specialized instruments to collect information that is used to help imagine what the particles would look like if they were visible.

Most of the time, chemists and biologists work with macroscopic samples in the laboratory, but they imagine what happens at the particulate level while they do so. By understanding and directing the behavior of particles, the chemist and biologist control the macroscopic behavior of matter. This is, in fact, a distinguishing characteristic of chemistry, biochemistry, and molecular biology. A chemist, biochemist, or molecular biologist imagines the nature of the behavior of the tiny particles that make up matter, and then she or he applies this knowledge to carry out changes from one type of matter to another.

Because matter at the particulate level is too small to see, chemists use models to represent the particles. A **model** is a representation of something. Chemists use models of atoms and molecules—tiny particulate-level entities—that are based on experimental data. We cannot see atoms and molecules directly, so we use data from experiments to infer what they would look like if they were much, much larger. We then construct physical models that match the data.

The two most common molecular models used by chemists are ball-and-stick models and space-filling models, which are illustrated in **Figure 2.4**. The **ball-and-stick model** shows atoms as balls and linking electrons as sticks connecting the atoms.

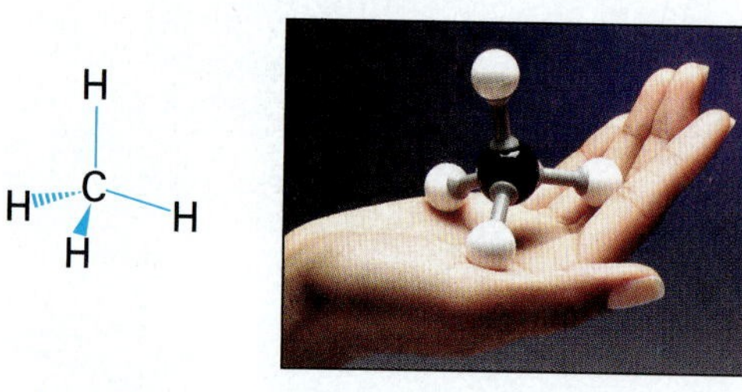

Simple perspective drawing

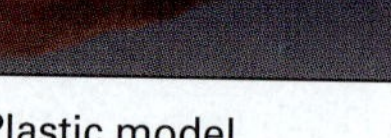

Plastic model

Charles D. Winters

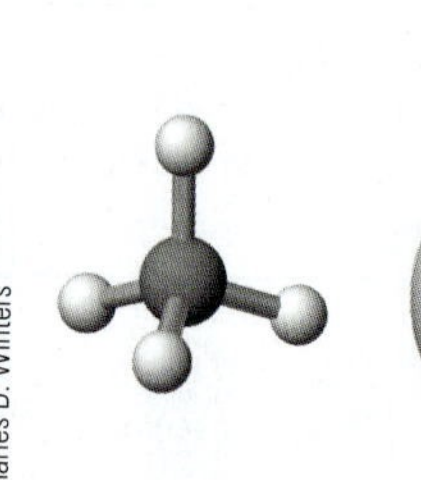

Ball-and-stick model

Space-filling model

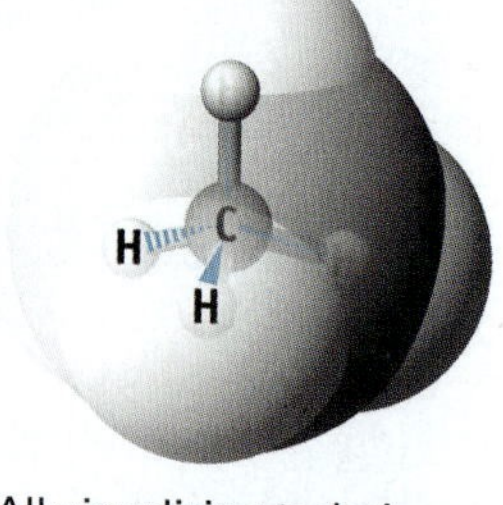

All visualizing techniques represent the same molecule

Figure 2.4 Models and symbols used by chemists to represent molecules. All of these models and symbols represent the same thing: a methane molecule. Methane is the primary component of natural gas. A methane molecule consists of a central carbon atom, represented by the symbol C or a black sphere, surrounded by four equidistant and equally spaced hydrogen atoms, each represented by the symbol H or a white sphere. The lines in the simple perspective drawing and the sticks in the plastic and sketched ball-and-stick models represent the tiny particles that bond the atoms together to form the molecule.

A. Barrington Brown/Science Source

Figure 2.5 James D. Watson (b. 1928) (*left*) and Francis H. C. Crick (1916–2004) (*right*) posing with a model of the DNA molecule. Physical models (such as this one) can be valuable tools in chemistry ▼. Today's chemists, however, usually use computer animations of molecules to model matter at the particulate level.

The **space-filling model** shows the outer boundaries of the particle in three-dimensional space.

The renowned scientists James Watson and Francis Crick (**Figure 2.5**) used physical models to deduce the structure of the DNA molecule, the molecular storehouse of genetic information. They used numerical information from experiments, such as the distances between atoms, to construct models of the pieces that make up the DNA molecule. Working with these physical representations of the small pieces, they were able to reason about how the pieces would logically fit together, forming the double-helix structure of DNA. They completed the scientific research cycle by using their model to predict experimental outcomes, and the actual experimental outcomes were indeed consistent with the model.

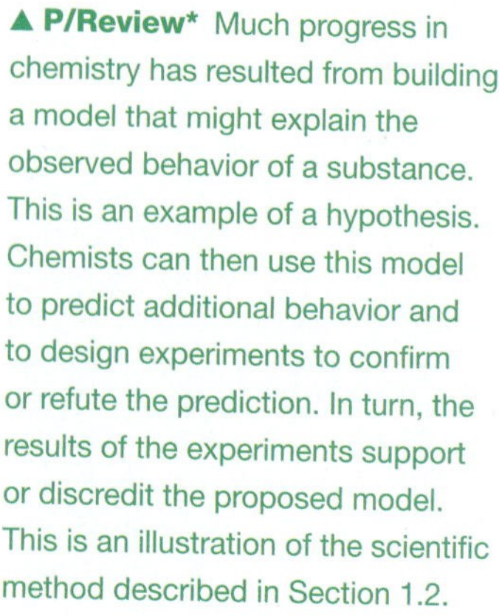

▲ **P/Review*** Much progress in chemistry has resulted from building a model that might explain the observed behavior of a substance. This is an example of a hypothesis. Chemists can then use this model to predict additional behavior and to design experiments to confirm or refute the prediction. In turn, the results of the experiments support or discredit the proposed model. This is an illustration of the scientific method described in Section 1.2.

Thinking About Your Thinking

Mental Models

A *model* is a depiction of something. You are probably familiar with models of the Earth. Globe models are common in geology and earth science classrooms and laboratories. We use these models because the Earth is so large. A globe allows us to mentally picture something that otherwise is too big to understand well. Models of molecules are used in chemistry for exactly the opposite reason: Molecules are too small to observe with the naked eye. Chemists use concrete models to understand the behavior of molecules. Figure 2.4 helps you form a mental image of a methane molecule. Many other molecular models will be shown throughout this book.

As you study chemistry, you will develop the skill of visualizing the particulate-level behavior of molecules. Imagine the models of the molecules in three dimensions. Try to think about how they move. When we describe chemical processes in words and two-dimensional illustrations, translate these words and figures into three-dimensional "mental movies" in which the molecules are in motion.

We will remind you periodically throughout the book about forming these mental models. You will see the words "Thinking About Your Thinking: *Mental Models*" whenever you should think about this process. Doing so will greatly improve your skill in thinking as a chemist.

---

*P/Reviews are references to items covered in another chapter in the textbook. For a description of P/Reviews, see Section 1.5.

Models are often represented by a simple **chemical symbol** that can be easily written or typed. The symbol H is used to represent a hydrogen atom, and the symbol O represents an oxygen atom. You are probably familiar with how these symbols are combined to form the symbolic representation of a water molecule. Its chemical formula is H_2O. This formula tells us that a water molecule is composed of two hydrogen atoms and one oxygen atom.

A chemist can think about matter at many levels, often switching among different representations in one discussion. When talking about a drop of water, for example, as illustrated in **Figure 2.6**, a chemist may initially think about the macroscopic characteristics of the drop, such as its shape, lack of color, and size. She may then form a mental model of the water molecules in the drop and imagine the invisible forces that make the molecules stick together, as shown in the thought cloud in Figure 2.6. To express her mental model in two dimensions on the board in Figure 2.6, she will then use symbols to represent her three-dimensional model in the form of letters, dashes, and dots. One of our major goals in this book is to help you think as a chemist by using models and symbolic representations of the particulate nature of matter.

Figure 2.6 Imagining macroscopic matter as particulate-level mental models and using symbols to represent those particles. Chemists frequently make mental transformations between visible macroscopic matter: in this case, the drops of water, and the particulate-level model of the molecules that make up the matter, as shown in the instructor's thought cloud. Written symbols, as shown on the board, serve as simpler representations of the particulate-level models.

Learn It Now! *Always complete a Target Check as soon as you come upon one. Then check the answer at the end of the chapter. If your answer is correct, then move on to the next section. If your answer is not correct, then figure out why—immediately. Reread the text until you understand how the correct answer was derived. If restudying the text does not help you understand the answer, then write a note about what you still cannot figure out. As soon as possible, ask a classmate or see your instructor (or teaching assistant) to get in-person help.*

Target Check 2.1

Consider the photograph and illustration of table salt. Do they include a model? Do they include a depiction of matter at the macroscopic, microscopic, and/or particulate levels? Explain your answers.

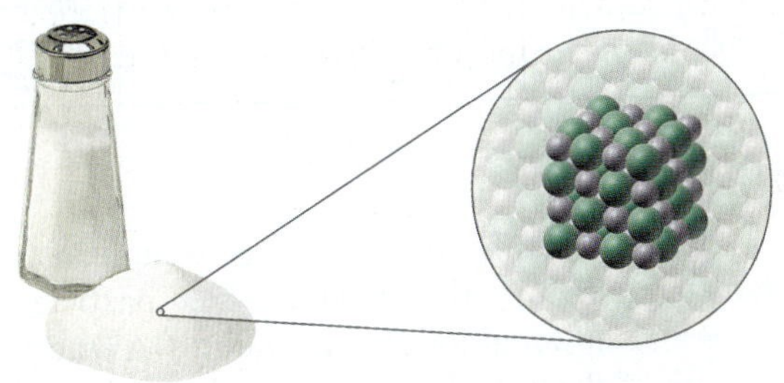

2.3 States of Matter

Goal 3 Identify and explain the differences among gases, liquids, and solids in terms of (a) the macroscopic-level properties shape and volume; (b) particle movement; and (c) particle spacing.

The air you breathe, the water you drink, and the foods you eat are examples of the **states of matter** called *gases*, *liquids*, and *solids* ▶. Water is a common substance that is familiar to us in all three states, as **Figure 2.7** depicts. We can explain the differences among gases, liquids, and solids in terms of the **kinetic molecular theory**. According to this theory—a hypothesis that explains a broad class of related phenomena—all matter consists of extremely tiny particles that are in constant motion. *Kinetic* refers to motion; *molecular* comes from **molecule**, the smallest individual particle in one kind of matter. Oxygen, water, and sugar are examples of three common molecular substances.

Another state of matter is plasma, which consists of positive ions and free electrons in a gaslike state. Examples of plasmas include the substances inside fluorescent lights and neon signs.

Molecules are attracted to one another. The strength of these attractive forces varies from substance to substance. For example, the tungsten atoms in a lightbulb filament are strongly attracted to one another. In contrast, the helium atoms in a helium balloon are very weakly attracted to one another.

	Gas	Liquid	Solid
Water as an example:	 Gaseous water (steam)	 Liquid water	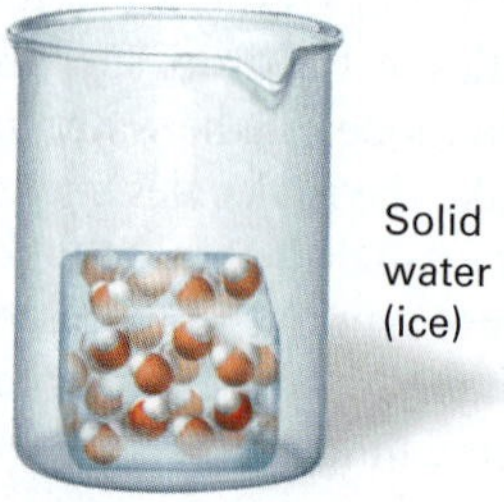 Solid water (ice)
Shape	Variable—same as a closed container	Variable—same as the bottom of the container	Constant—rigid, fixed
Volume	Variable—same as a closed container	Constant	Constant
Particle Movement	Completely independent (random); each particle may go anyplace in a closed container	Independent beneath the surface, limited to the volume of the liquid and the shape of the bottom of the container	Vibration in fixed position
Particle Spacing	Very far apart—particle interactions are negligible	Close—particle interactions are important	Close—particle interactions are important

Figure 2.7 Three states of matter illustrated by water.

Opposing this attractive force is molecular motion. According to the kinetic molecular theory, the speed at which particles move is faster at higher temperatures and slower at lower temperatures. As the particles in a sample move faster, the motion of the particles overcomes the tendency for particles to stick together. When particles are moving fast enough to overcome the attractive forces, the sample exists as a **gas**. In the gaseous state, the particles move so far apart from one another that we can ignore the interactions between them.

When a gas is in a closed container, the particles move in straight lines until they interact with something else, either another particle or the walls of the container. This causes the shape of a gas to be the same shape as the container in which it is held. The volume of a gas is also the same as the volume of the container in which it is held.

If we imagine a gas at a high temperature in a closed container, the particle–particle attractions have almost no effect. If the temperature decreases, the particles move more slowly. Therefore, the attractions have more of an effect, and the particles clump together to form a **liquid** drop. The drops fall to the bottom of the container, taking on the shape of the bottom of the container. At the particulate level, the molecules in a liquid touch one another, but their movement is sufficiently rapid to allow them to move freely among themselves. Since the particles are touching one another, the volume of the liquid is constant.

If we continue to imagine further reducing the temperature of our liquid substance, particle movement becomes more and more sluggish. Eventually, the particles no longer move among one another. Their movement is reduced to vibrating, or shaking, in fixed positions relative to one another. This is how we model the **solid** state at the particulate level. Like a liquid, a solid has a constant volume. But unlike a liquid, a solid has its own unique shape that remains the same wherever the sample may be placed.

Learn It Now! *Figure 2.7 summarizes the states of matter. Study it carefully to be sure that you can satisfy Goal 3.*

The particles in the solid state can arrange into an orderly pattern. Imagine how a brick wall is typically laid, with the pattern repeating in every other row. When the molecules of a substance are arranged in an orderly pattern, the substance is called a **crystalline solid**. Common crystalline solids include table salt and snowflakes. Chemists use the term *crystalline* to refer to the orderly arrangement at the particulate level, as illustrated in **Figure 2.8**. This orderly arrangement tends to be "patchy," with orderly portions interspersed with breaks.

An **amorphous** solid has no long-range order. The beginning of the word, *a–*, means *a lack of*, and the middle of the word, *–morph–*, means *shape*. An amorphous solid has a lack of a pattern in its particle arrangement (Figure 2.8). Common examples of amorphous solids include rubber and many types of plastic and glass.

Most of the examples in this book are written in an active-learning, self-teaching style in which a series of questions and answers guide you to understanding a problem. To reach that understanding, you should answer each question before looking at the answer. This requires a shield to cover that answer while you consider the question. Cut-out shields for this purpose are provided in the book.

Find them now and cut one out. On one side you will find instructions on how to use the shield, copied from this section. On the other side is a periodic table that you can use for reference.

The examples provided in this textbook are designed so that you can actively work them, writing your response to each question. This does not mean that the examples are optional or are something to do after you read the text. On the contrary, they are an integral part of the textbook. To maximize your learning, work each Active Example when you come to it.

We understand the temptation to just read the textbook in the same way you might read a novel. It's certainly much easier than doing the work of writing answers to our questions, comparing your answers to ours, thinking about your thinking, and working to improve. But research on human learning clearly shows that to only *read* a textbook results in almost no meaningful long-term learning. You have to actively work at learning. This textbook provides that opportunity. A saying often attributed to Confucius is, "I hear and I forget, I see and I remember, I do and I understand." Our textbook follows the "I do and I understand" philosophy.

This is the point in this course where you need to ask yourself, "Do I want to learn chemistry as a result of my study? Do I want to learn the thinking skills chemists use?" If the answer is yes, put your wants into action. Make the commitment now to work each example when it is presented. There is no better way to learn the content and process of chemistry than by actively answering questions and solving examples ▶. This is why we've titled the textbook *Introductory Chemistry: An Active Learning Approach.*

Active learners learn permanently and grow intellectually. Passive learners remember temporarily and then forget, without mental growth.

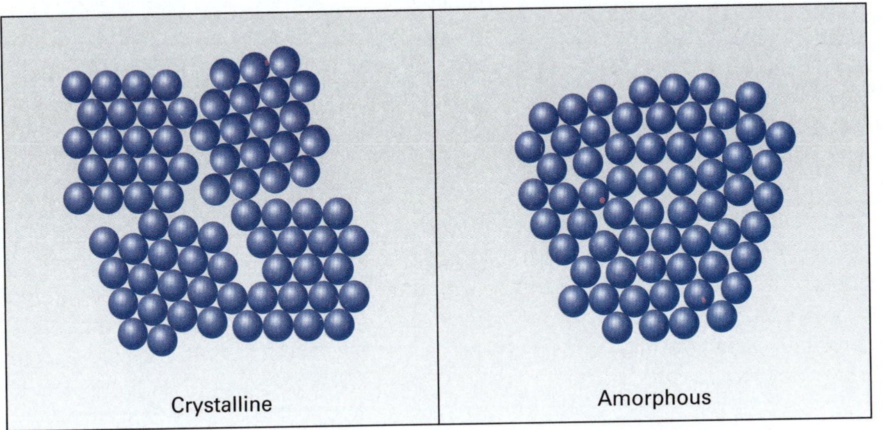

Figure 2.8 Crystalline and amorphous solids. At the macroscopic level, you cannot necessarily distinguish between a crystalline and an amorphous solid by visual observation alone. At the particulate level, the difference is apparent. A crystalline solid has an orderly particle arrangement, whereas an amorphous solid has no long-range consistency in structure.

how to... Work an Active Example

Step 1: When you come to an example, locate the point in the left column at which the first blue-shaded background appears. Use the shield to cover all of the blue-shaded boxes in the left column.

Step 2: Read the problem statement. Write any answers or calculations needed in the blank space where the pencil icon is located. Note that the "Think Before You Write" instructions are different for each Active Example.

Step 3: Move the shield down to reveal the first blue-shaded box.

Step 4: Compare your answer to the one you can now read in the book. Be sure you understand the example up to that point before going on.

Step 5: Repeat the procedure until you finish the example.

Active Example 2.1 States of Matter

In the top box, draw a particulate-level illustration of a substance in the solid state. Model the particles as spheres, which can be drawn as simple circles. Assume that the box represents a tiny, open container that holds the particles. Use "cartoonist's lines" to illustrate the motion of the particles. In the center box, show the same particles after they have completely melted to become a liquid. In the bottom box, draw a particulate-level illustration of the same substance after it is further heated and becomes a gas. Use words to describe your drawings.

Think Before You Write Place your shield over the left column. Imagine what the appearance of a solid would be if you could see molecules. Think about how this appearance changes as time passes.

Answers *Cover the left column with your cut-out shield. Reveal each answer only after you have written your own answer in the right column.*

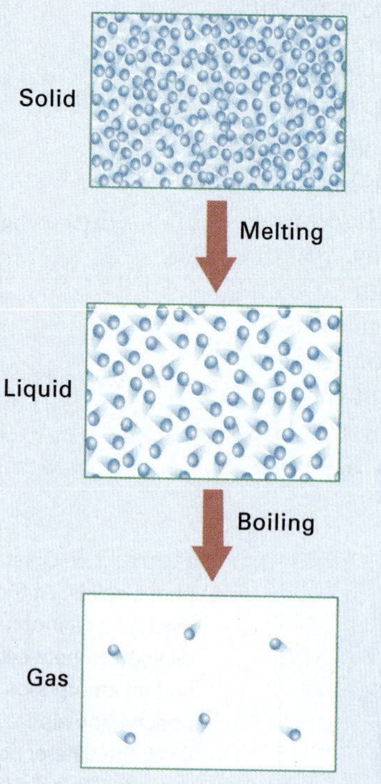

Draw in the boxes provided. When you are finished, move the shield to reveal our solution, and compare your answer with ours.

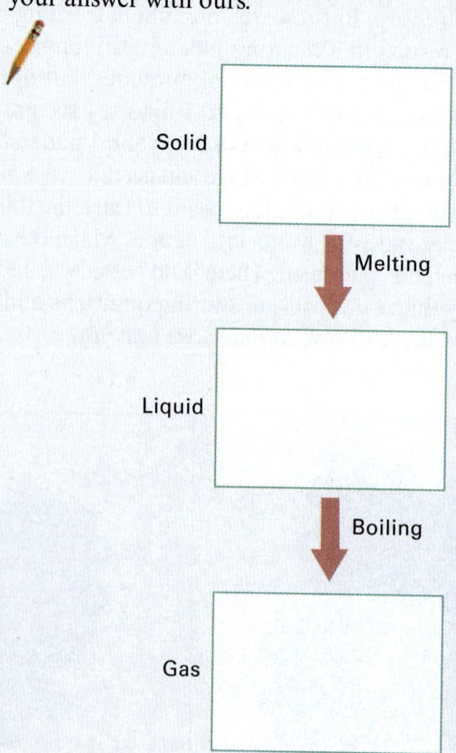

At low temperatures, the molecular movement is relatively sluggish. The molecules do not move past one another, remaining surrounded by the same set of neighbors. They vibrate, but they stay in the same positions relative to one another. The space between the particles approaches its minimum.	Use words to explain the nature of particle movement in the solid state and the relative amount of space between the particles. 
The molecules begin to overcome the attractive forces of neighboring particles. Although the particles are moving more slowly than in the gaseous state, they are moving fast enough to move past one another while remaining close enough to touch their neighbors. The molecules rotate and slip around one another.	Imagine that the solid melts, forming a liquid. How do the particles of a liquid move? 
In the gaseous state, the molecules are far apart and moving very rapidly in straight lines until they interact with one another or the walls of the container. They then move off in a straight line in another direction.	Additional heat energy is transferred to the sample, changing it to the gas phase. How do the molecules behave? 
You deepened your understanding of the particulate-level nature of the phases of matter. You also improved your ability to express in writing the key characteristics of your mental model of the states of matter. *All substances have similar particulate-level behavior with respect to how they change between the three states of matter and behave in each state. As you continue in this course, practice using your mental model of the states of matter each time the opportunity arises.*	As you complete each exercise or end-of-chapter problem in this textbook, you should ask yourself, "What new knowledge or skill did I obtain or improve by doing this exercise or problem?" 

Practice Exercise 2.1

Describe the relative strengths of the interparticle attractive forces for water in the solid, liquid, and gas phases. What is the relative nature of particle motion in the three phases of matter? Explain how these opposing forces contribute to the characteristics of each state of matter.

Learn It Now! *We cannot overemphasize how important it is that you answer each part of an Active Example question yourself before you look at the answer. This is the most important of all the* Learn It Now! *statements.*

2.4 Physical and Chemical Properties and Changes

Goal 4 Distinguish between physical and chemical properties at both the particulate level and the macroscopic level.

5 Distinguish between physical and chemical changes at both the particulate level and the macroscopic level.

If you were asked to describe a substance, you might list its color, feel, and smell. Charcoal is black; sulfur is yellow. Glass is hard; bread dough is soft. The smell of a rose is pleasant; the odor of ammonia is disagreeable.

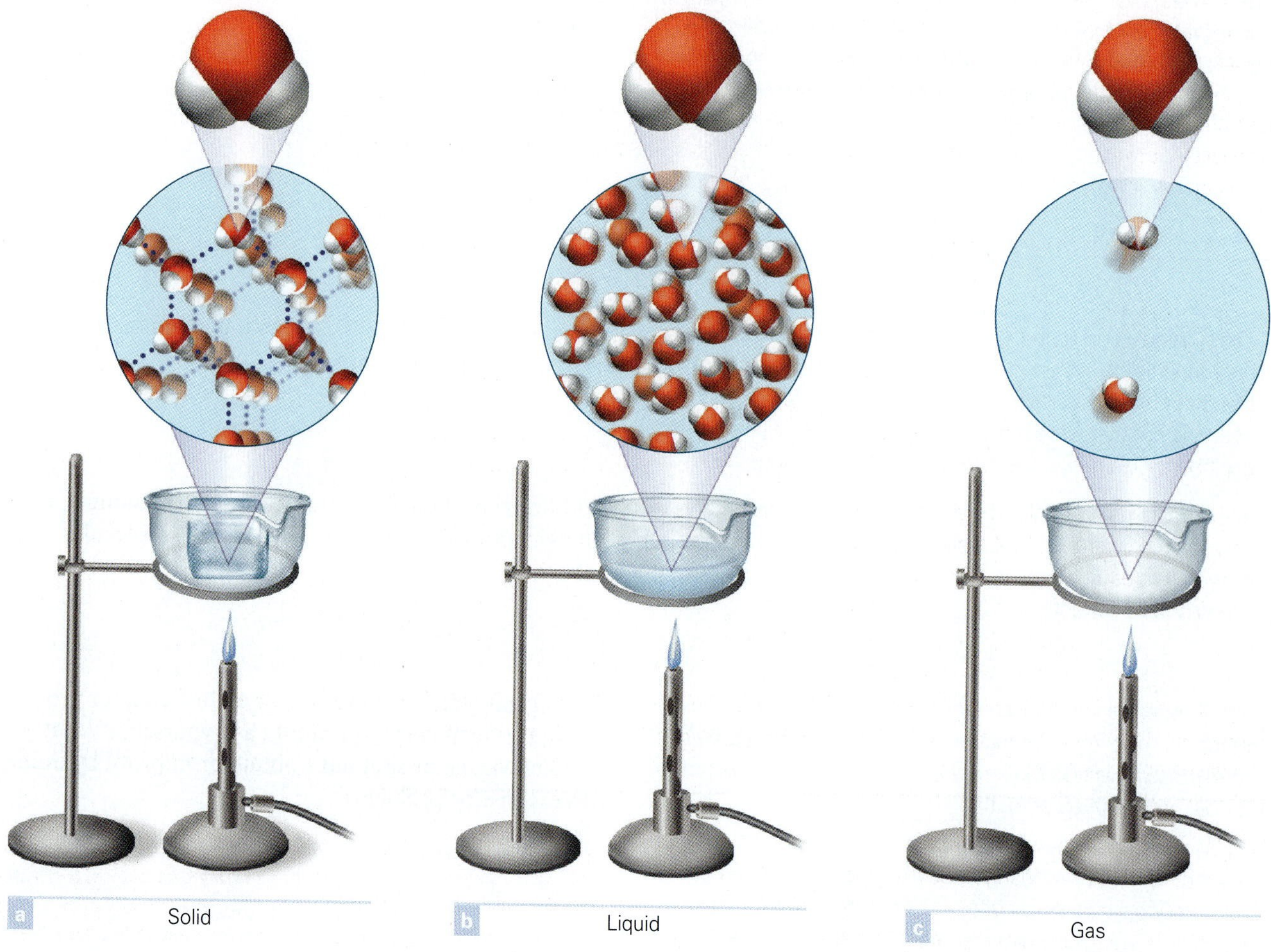

Figure 2.9 Physical changes. Changing ice to liquid water to steam is a physical change. At the macroscopic level, the form of water changes, but in all three states the substance is water. At the particulate level, the smallest particle of matter—the molecule—is unchanged. Water molecules are still water molecules whether in the solid, liquid, or gaseous state.

To be more thorough, you could continue in the laboratory. For example, you could determine the temperature at which a substance boils (the boiling point) or melts (the melting point). You could see if it is attracted to a magnet.

Color, feel, smell, boiling point, melting point, and magnetism are all **physical properties**: characteristics that can be observed and measured without altering the identity of the substance.

If you were asked to change the physical form of a sample of matter without changing its chemical identity, you might melt an ice cube. You could then refreeze the liquid water. The substance is still water after it melts and after it refreezes. Another choice that you might make is to dissolve sugar in water. The sugar may seem to disappear, but if you taste the sugar water solution, you'll know the sugar is there. You could then recover the dissolved sugar by evaporating the water. Change of state and change of appearance are examples of **physical change**: a change in the form of a substance without changing its chemical identity (Figure 2.9). Changes in size or shape are other examples of physical changes.

A **chemical change** occurs when the chemical identity of a substance is destroyed and a new substance forms. For example, at the macroscopic level, when liquid water is subjected to an input of electrical energy from batteries, it decomposes the liquid from the mixture to pass through and prevents passage of the gravel, sand, and dirt in the sediment.

Coffee drinkers are familiar with one of the most common applications of filtration: separating coffee grounds from a coffee solution with filter paper.

Filtration is based on the physical properties of the components of a mixture. The particle sizes of the components of the mixture must be significantly larger or smaller than the pore size of the filtration medium. The smaller particles pass through the filter, and the larger particles are left behind.

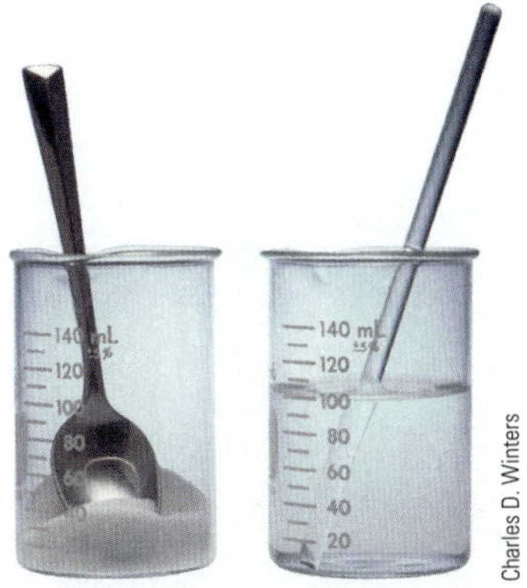

Table salt (*left*) and a solution of table salt dissolved in water (*right*).

Target Check 2.5

Table salt from the beaker on the left in the photograph is added to water, forming the solution on the right of the photograph. If you want to separate the mixture, would the distillation apparatus in Figure 2.18 or the filtration apparatus in Figure 2.19 be the better choice? Explain.

2.7 Elements and Compounds

Goal 9 Distinguish between elements and compounds.

10 Distinguish between elemental symbols and the formulas of chemical compounds.

11 Distinguish between atoms and molecules.

12 State the Law of Definite (or Constant) Composition, and explain its implication for how compounds and mixtures differ in composition.

Silver represents one of the two kinds of pure substances. Like all pure substances, it has its own unique set of physical and chemical properties that are unlike the properties of any other substance. Among its chemical properties is that silver cannot be decomposed or separated into other stable pure substances (**Figure 2.20**). This identifies silver as an **element**.

Water represents the second kind of pure substance. Unlike silver, it *can be* decomposed into other pure substances. Go back and look at Figure 2.10, which illustrates the decomposition of water. Any pure substance that can be decomposed by a chemical change into two or more other pure substances is a **compound**. **Figure 2.21** diagrams how a compound is separated into other pure substances by a chemical change (e.g., water is separated by electrolysis into hydrogen and oxygen). Figure 2.21 also contrasts a compound with a mixture, which may be separated (see Section 2.6) into the components of the mixture by a physical change (e.g., pure water can be distilled from a natural water solution).

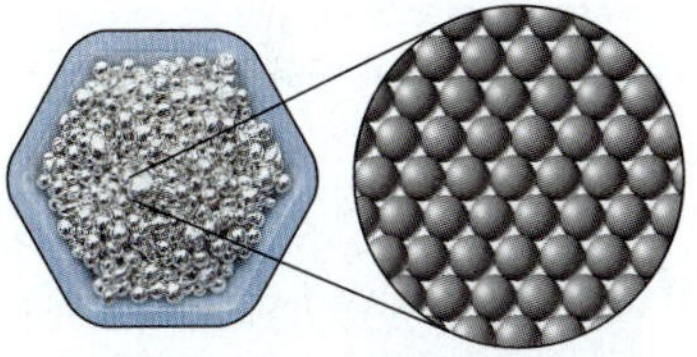

Figure 2.20 The element silver. Silver cannot be decomposed into other stable pure substances because all atoms are the same element. Atoms of one element cannot be changed to atoms of another element under normal conditions. The particulate-level view suggests that the silver atoms are arranged in a regularly repeating pattern.

Learn It Now! *Be sure to catch the distinction between separating a compound into other pure substances by chemical means and separating a mixture into its components by physical means.*

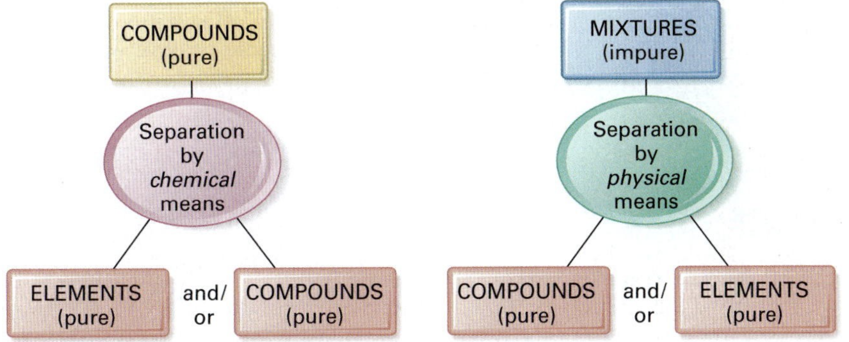

Figure 2.21 Separation of a compound versus separation of a mixture. A compound—a pure substance—may be separated into other pure substances by chemical means, using a chemical change. A mixture—an impure substance—may be separated into pure substances by physical means, using a physical change or by use of a physical property.

Thinking About Your Thinking

Mental Models

Imagine the different forms an element can take at the particulate level. It can be a collection of single atoms. It can be made up of molecules of two atoms, four atoms, eight atoms, or more. But in any case, all of the atoms in the smallest particle of an element are the same. Now compare your mental model of an element to your model of a compound. At least one of the atoms in a molecule or unit of a compound is different from the other atoms.

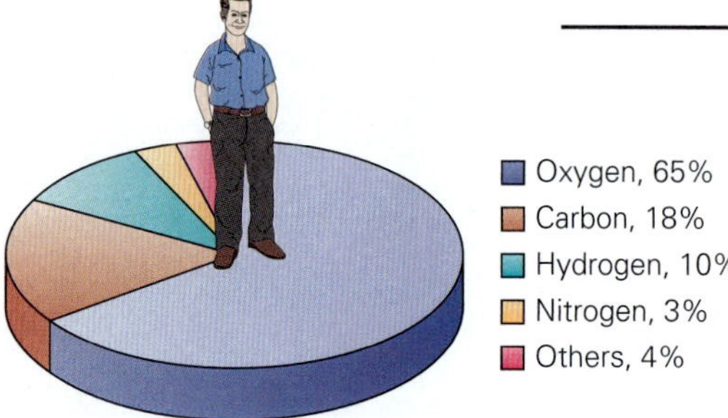

Figure 2.22 Major elements of the human body by mass percent. Of the mass of your body, 96% comes from atoms of just four elements: oxygen, carbon, hydrogen, and nitrogen.

To date, scientists have discovered 98 elements on Earth. Copper, silver, gold, and sulfur are among the few well-known solid elements that sometimes occur uncombined in nature. At common temperatures and pressures, 11 elements occur as gases, 2 (mercury and bromine) occur as liquids, and the remainder are solids. Most of the human body is made up of compounds made from just four elements (**Figure 2.22**).

To distinguish between an element and a compound, at present, you may use the number of words in the name of a substance to predict its categorization. The name of an element is always a single word, such as *oxygen* or *iron.* The names of many compounds have two words, such as *sodium chloride* (the main ingredient in table salt) and *calcium carbonate* (limestone). A few

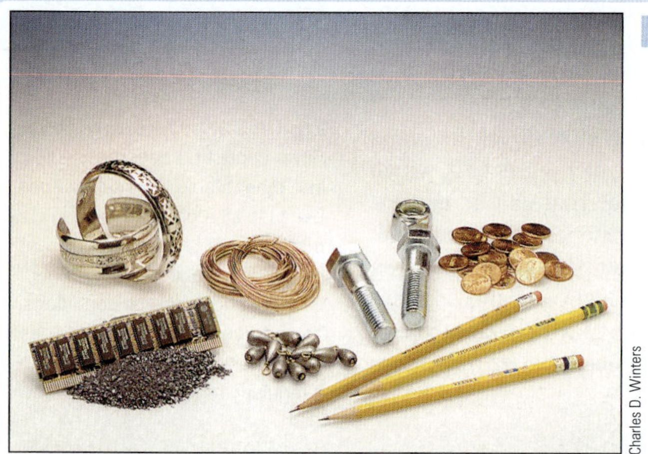

a Familiar objects that have a significant component that is a pure or a nearly pure element (*approximately clockwise from the right rear*): U.S. one cent coins, copper plating on a zinc coin; pencils, the core is a mixture of carbon (in its graphite form) and clay; nuts and bolts, a steel core that is largely iron with a zinc coating; fishing sinkers, lead or tungsten; copper wire, copper; silicon, used in the semiconductors in computer hardware; bracelets, mostly silver.

Charles D. Winters

b Familiar substances that have a significant component that is a pure or a nearly pure compound (*approximately clockwise from the center rear*): photographic fixer, a solution of sodium thiosulfate; Rolaids brand antacid, largely calcium carbonate and magnesium hydroxide; baking soda, sodium hydrogen carbonate; milk of magnesia tablets, magnesium hydroxide; blackboard chalk sticks, calcium sulfate dihydrate; water; solid ant killer, boric acid; lye, sodium hydroxide.

Charles D. Winters

Figure 2.23 Common elements and compounds.

familiar compounds have one-word names, such as *water* and *ammonia*. **Figure 2.23** shows some well-known elements and compounds.

Chemists represent each element in print or writing with an **elemental symbol**. The first letter of the name of the element, written as a capital, is often its symbol. If more than one element begins with the same letter, a second letter, written in lowercase, is added. Thus, H is the symbol for hydrogen and He is the symbol for helium; O is the symbol for oxygen and Os is the symbol for osmium; C is the symbol for carbon and Cl is the symbol for chlorine. The symbols of most elements correspond to their English names, but the symbols of some elements are derived from their Latin names, such as Na for sodium (from *natrium*) and Fe for iron (from *ferrum*).

On one side of the cut-out reference page is a Table of Elements that lists the symbols of all the elements and their names, in addition to other information about them. The symbols of the elements also appear on the **periodic table of the elements** ▶.

◀ **P/Review** The periodic table is a remarkable source of information that you will use throughout your study of chemistry. Your first use of it will be as an aid to learning the names and symbols of the elements in Section 5.8.

In the next few paragraphs, we describe the particulate character of several pure substances. These are illustrated in **Figure 2.24**, which also includes, whenever possible, photographs of macroscopic samples of the substances. We suggest that you refer to Figure 2.24 as you read. It will help you visualize the particulate nature of the substances we name.

Thinking About Your Thinking

Mental Models

Figure 2.24 includes particulate-level illustrations to help you form mental models of atoms, molecules, and crystalline solids.

The symbolic representation of a molecule or formula unit of a pure substance is its **chemical formula**. A formula is a combination of the symbols of the elements in the substance with subscript numbers to show the number of atoms of each element in a molecule or formula unit.

The formula of most elements is the same as the symbol of the element. This indicates that the stable, particulate-level composition of the element is single atoms. Helium (He), sodium (Na), and barium (Ba) are examples. Other elements that exist in nature have stable, distinct, and independent molecules that consist of two or more atoms chemically bonded to one another.* Hydrogen, oxygen, and chlorine are three such elements. Their *elemental symbols* are H, O, and Cl, respectively, but their *chemical formulas* are H_2, O_2, and Cl_2. The subscript "2" indicates in each case that a molecule of the element has two atoms. Figure 2.24 shows all of these elements.

The formulas of multielement molecules also use elemental symbols and subscript numbers to show the number of atoms of each element in a molecule. Hydrogen and chlorine, for example, form a compound whose molecules consist of one atom of each element, a ratio of 1:1. Its chemical formula is therefore HCl. A molecule of water contains two atoms of hydrogen and one atom of oxygen, a ratio of two hydrogens to one oxygen. Its formula is therefore H_2O. Notice that if there is only one atom of an element in a molecule, the subscript for that element is omitted. Figure 2.24 shows both of these compounds.

Another type of pure substance exists as an orderly, repeating pattern of two or more elements rather than as independent molecules. The chemical formula

---

*A molecule is the smallest individual particle in a pure substance that retains the identity of that substance. Therefore, the term *molecule* is not restricted to molecules of two or more atoms. It also refers to elemental particles that are stable as individual atoms. In this sense, helium atoms are monatomic molecules. *Mono-* is a prefix that means *one*.

Figure 2.24 Particulate and macroscopic views of elements and compounds discussed in Section 2.7. Some elements, such as helium, sodium, and barium, occur in nature as one-atom molecules. Other elements, such as hydrogen, oxygen, and chlorine, occur in nature as multiatom molecules. Hydrogen chloride, water, sodium chloride, and barium chloride are compounds—substances that consist of two or more elements. The term *crystalline* refers to solids with particles that are arranged in a regularly repeating pattern.

Substance	Symbol or Formula	Natural Form at Room Temperature	Particulate Illustration (not to scale)	Macroscopic Photograph
Helium	He	Atom/gas		A colorless gas
Sodium	Na	Atom/ crystalline solid		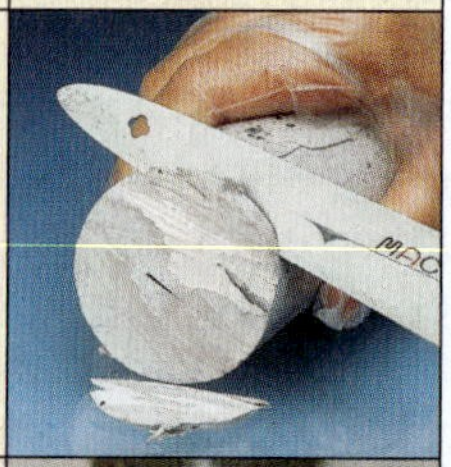 Charles D. Winters
Barium	Ba	Atom/ crystalline solid		 Charles D. Winters
Hydrogen	H_2	Two-atom molecule/gas		A colorless gas
Oxygen	O_2	Two-atom molecule/gas		A colorless gas
Chlorine	Cl_2	Two-atom molecule/gas		 Charles Steele
Hydrogen chloride	HCl	Two-atom molecule/gas		A colorless gas
Water	H_2O	Three-atom molecule/ liquid		 Charles D. Winters
Sodium chloride	NaCl	Crystalline solid		 George Semple
Barium chloride	$BaCl_2$	Crystalline solid		 Charles D. Winters

expresses the simplest ratio of particles in the solid. Atoms of sodium and chlorine combine in a 1:1 ratio to form sodium chloride, commonly known as additive-free table salt. This 1:1 ratio is expressed in the formula $NaCl$. Barium atoms and chlorine atoms combine in a 1:2 ratio to form barium chloride. Its formula is $BaCl_2$.

The precise ratio of atoms of different elements in a compound is responsible for the **Law of Definite Composition**, also called the **Law of Constant Composition**: Any compound is always made up of elements in the same proportion by mass (weight). This mass proportion is a direct consequence of the atom proportion in a molecule of the compound. For example, a water molecule is made up of two atoms of hydrogen and one atom of oxygen. It is not possible to have a fraction of an atom. Water molecules have exactly a 2-to-1 hydrogen-to-oxygen atom ratio, never a 2.1-to-1 or 2-to-1.05 atom ratio, and so forth. Since the atom ratio is fixed, and since atoms have a characteristic average weight, the weight ratio of elements in a compound must also be fixed.

A compound has a definite composition. Contrast this fact with the composition of a mixture, which depends on the relative quantities of the components that make up a mixture. For example, a water–alcohol mixture can be 10% water and 90% alcohol, 20% water and 80% alcohol, and an infinite number of other proportions. Thus, a characteristic that distinguishes compounds from mixtures is that a compound has definite composition, whereas a mixture has variable composition.

The properties of compounds are *always different and completely independent* from the properties of the elements of which they are formed. Sodium is a shiny metal that reacts vigorously when exposed to air or water (see Figure 2.24); chlorine is a yellow-green gas that was the first poison gas used in World War I (also shown in Figure 2.24). Yet the compound formed from these two elements, sodium chloride (commonly known as table salt; also shown in Figure 2.24), is safe to eat and even essential in the diets of many animals, including humans.

Figure 2.25 summarizes the classification system for matter.

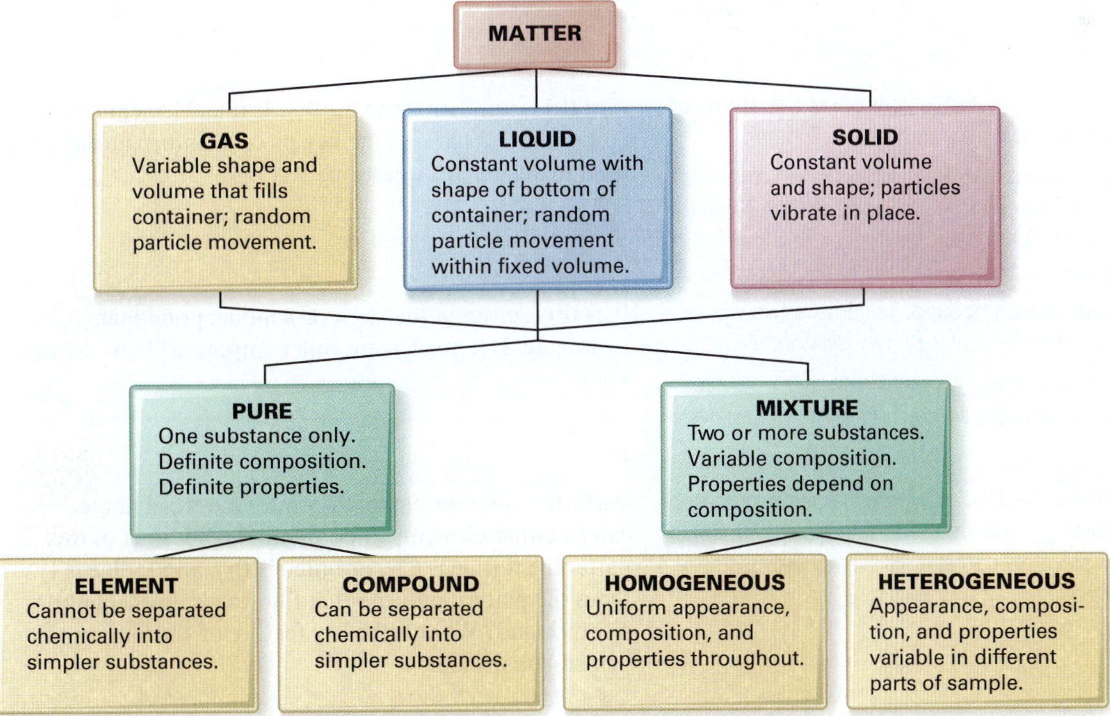

Figure 2.25 Summary of the classification system for matter.

Learn It Now! *Figure 2.25 is designed to help you organize your knowledge of topics related to matter into a cohesive whole. The states of matter—gases, liquids, and solids—were introduced in Section 2.3. Pure substances and mixtures and homogeneous and heterogeneous matter were introduced in Section 2.5. Elements and compounds are introduced here in Section 2.7. You now need to consider the relationships among all of these concepts. If you can name a few examples that fit into each box in Figure 2.25, you are well on your way to organizing your knowledge into a connected system, and the development of these systems of concepts is an essential component of learning.*

Active Example 2.2 Elements and Compounds

Which of the following are compounds, and which are elements? Explain the reasoning you used to arrive at each answer. (a) Na_2S, (b) Br_2, (c) potassium hydroxide, (d) fluorine,

(e) 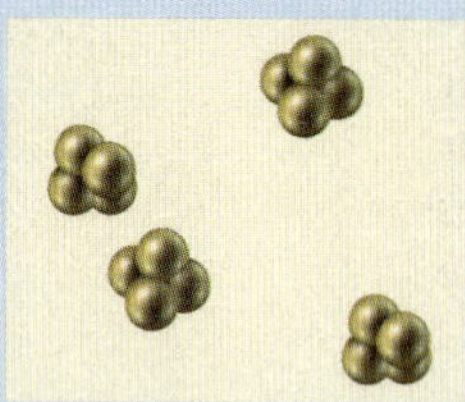

(f) 

Think Before You Write Place your shield over the left column. Think about the answers to the following questions before you work on each part of this Active Example: What is an element? What is a compound? What are the key characteristics that allow one to distinguish between an element and a compound? When you are confident in your answers to these questions, proceed as usual with the Active Example.

Answers *Cover the left column with your cut-out shield. Reveal each answer only after you have written your own answer in the right column.*

Na_2S is a compound. Its chemical formula indicates that it is made of two different elements, Na and S. Thus, it can be decomposed to its elements, which makes it a compound.	The substance in part (a) is described in the form of a formula: Na_2S. Is it an element or a compound? How did you decide? 
Br_2 is an element. The Br_2 molecule consists of two atoms *of the same element* chemically bonded to one another. *Br_2 exists in nature as a two-atom molecule, as its formula reflects. If we were to decompose the molecule, the separate atoms would rebond to each other. Br_2 is therefore not a compound because it cannot be changed to other substances.*	Part (b) is another formula: Br_2. Is it an element or a compound? What was the key piece of information you used to answer the question? 
Potassium hydroxide is a compound. Its name is two words, indicating that it is made up of at least two different elements. *Potassium hydroxide, KOH, is actually made up of three elements: potassium, oxygen, and hydrogen.*	Part (c) is given in the form of a name: potassium hydroxide. Is it an element or a compound? How do you know? 
Fluorine is an element. Cl_2, Br_2, and I_2 are elements, and their names have the form ___–ine, so these data indicate that fluo*rine* is likely a two-atom elemental molecule as well. *Fluorine is an element with the formula F_2.*	Part (d) is another name: fluorine. One-word names can be either elements or compounds. But look at this pattern: chlorine is Cl_2, bromine is Br_2, and iodine is I_2. Does the pattern indicate that fluorine is an element or a compound? What is the key feature of the pattern that helped you decide? 

The molecules illustrated are those of an element. All of the atoms in the molecule are of the same element. *One form of the element phosphorus exists as P_4 molecules.*	Part (e) asks you to think from a different perspective. It is an illustration of molecules of a substance. Are the molecules of an element or a compound? What was the basis of your decision? 
The molecules illustrated are of a compound. Two different atom types combine to form the molecule, so the molecule could be decomposed to its elements. *Chemists often use the convention of using black to represent carbon atoms and white to represent hydrogen atoms, and thus the molecules illustrated are CH_4, methane, the primary component of natural gas.*	Part (f) is an illustration that shows a black central atom bound to four white atoms. Are the molecules of an element or a compound? Explain. 
You improved your knowledge of how to distinguish between an element and a compound, based on formula, name, and molecular illustration.	Reflect. What new knowledge or skill did you obtain or improve by doing this exercise? 

Practice Exercise 2.2

Is a compound a pure substance or a mixture? On what differences between pure substances and mixtures is your answer based?

2.8 The Electrical Character of Matter

Goal 13 Match electrostatic forces of attraction and repulsion with combinations of positive and negative charges.

If you release an object held above the floor, it falls to the floor. This is the result of gravity, a familiar invisible attractive force between the object and the Earth. Gravity is one of four fundamental forces that govern the operation of the universe. Forces are sensed as pushes or pulls on an object. Another less familiar force is the electromagnetic force ▶. Electricity and magnetism are each a part of the electromagnetic force. A physical property of matter is **electric charge**, which we will now consider more closely.

The four fundamental forces in the universe are gravity, the electromagnetic force, the strong force, and the weak force. The strong and weak forces operate within atoms.

Figure 2.26 illustrates an experiment that demonstrates the nature of electric charge. If a glass rod is rubbed with a silk cloth, the rod gains a positive charge.

When a silk cloth is rubbed on a glass rod, electrons are transferred from the rod to the cloth, leaving the rod positively charged.

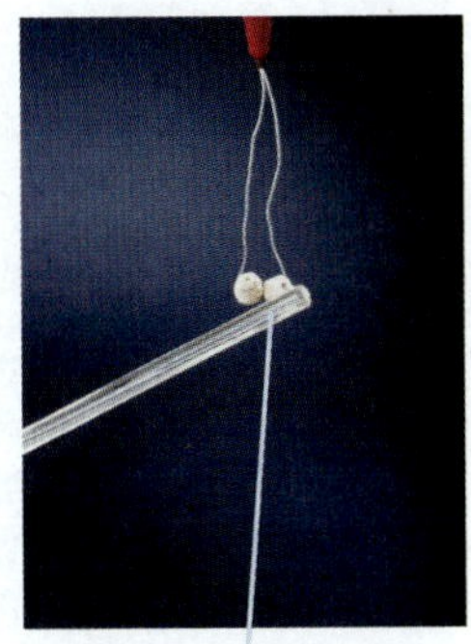

When the positively charged rod touches two pith balls, it removes some electrons from the pith balls, leaving them with positive charges.

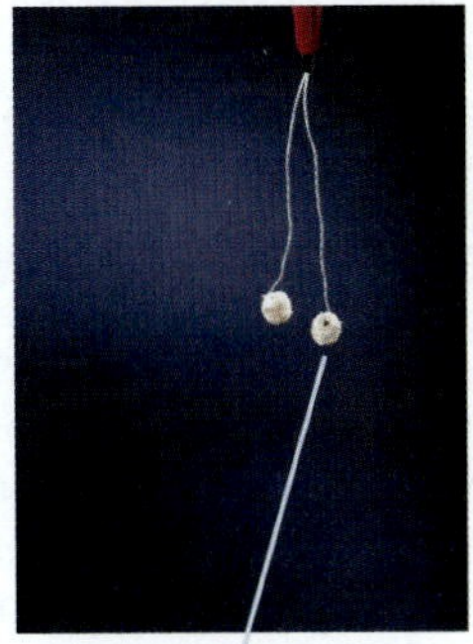

Each pith ball has a positive charge. This shows evidence that like charges repel each other.

When fur is rubbed on a rubber rod, electrons are transferred from the fur to the rod, leaving the rod negatively charged.

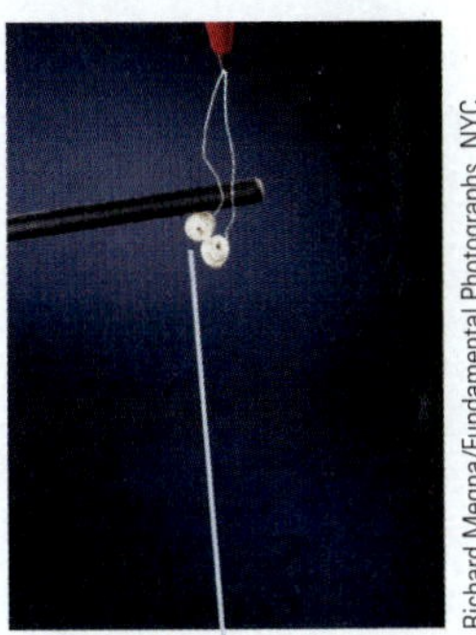

The rod has a negative charge and the pith balls have positive charge. This shows evidence that opposite charges attract each other.

Figure 2.26 Electrostatic attraction and repulsion. Pith is a spongy tissue extracted from plants. It is lightweight, and it transfers electrical charge very readily.

If a pith ball (a small spongy ball made of plant fiber) is touched with a positively charged rod, the pith ball itself becomes positively charged. When two pith balls that are positively charged are suspended close to one another, they repel each other.

A hard rubber rod that is rubbed with fur acquires a negative charge. The ebonite rod shown in Figure 2.26 is made from a very hard rubber used to make bowling balls. The positively charged pith balls are attracted to the negatively charged rubber rod.

These charges are like those you develop if you scrape your feet across a rug on a dry day. You can discharge yourself by touching another person, and each of you receives a mild shock in the process. In each of these situations—you rubbing your feet on a rug and a pith ball being touched with a positively charged glass rod—the object acquires an electrical charge that does not move over a distance. It is a static, or unmoving, charge. The electrical force is known as **static electricity**. The force is also called an **electrostatic force**.

These experiments and numerous other similar experiments illustrate the nature of electric charge:

a summary of... **The Electrical Character of Matter**

1. Electric charge is a physical property of matter.
2. There are only two types of electric charge: positive and negative.
3. Two objects having the *same charge*—either both positive or both negative—exert a repulsive force on one another. Like charges repel.
4. Two objects having *unlike charges*—one positive and one negative—exert an attractive force on one another. Unlike charges attract.

▶ **P/Review** The *quantity* of energy transferred in a chemical change is considered in Sections 10.8 through 10.10.

Electrical forces show that matter has electrical properties. These forces are responsible for the energy transferred in chemical changes, which will be discussed further in the next section ◀.

 **Target Check 2.6**

Identify the net electrical force—attraction, repulsion, or none—between the following pairs:

a) Two positively charged table tennis balls
b) A negatively charged piece of dust and a positively charged dust particle
c) A positively charged sodium ion and a positively charged potassium ion (ions are charged particles similar to atoms).

2.9 Characteristics of a Chemical Change

Chemical Equations

Goal 14 Distinguish between reactants and products in a chemical equation.

In Section 2.4, we said that a chemical change occurs when a substance is destroyed and a new substance forms. Chemists describe such a change in a convenient, compact form by writing a **chemical equation**. The formulas of the original substances, called **reactants**, are written to the left of an arrow that points to the formulas of the new substances formed, called **products**. The chemical equation for the reaction of the element carbon with the element oxygen to form the compound carbon dioxide is

$$C + O_2 \rightarrow CO_2$$

Notice how the chemical equation is a symbolic representation of the chemical change: The reactants carbon (C) and oxygen (O_2) no longer exist after the chemical change occurs, as symbolized by the arrow, and the product carbon dioxide (CO_2) forms.

The decomposition of water, as illustrated in Figure 2.10, is a chemical change that we can use as another example to illustrate how a chemical equation is written. The original compound, water (H_2O), the reactant, decomposes into the elements hydrogen (H_2) and oxygen (O_2), the products. The equation is

$$2\ H_2O \rightarrow 2\ H_2 + O_2$$

As noted in Figure 2.10 caption, atoms are neither created nor destroyed in a chemical change. To represent this in a chemical equation, coefficients may be used to achieve atom balance in the **equation**. The "2" before the formula of water and the "2" before the formula of hydrogen are the coefficients needed to achieve atom balance, 4 total hydrogen atoms before and after the change and 2 total oxygen atoms before and after the change. We will discuss coefficients and the balancing of chemical equations in more detail in Chapter 8.

Energy in Chemical Change

Goal 15 Distinguish between exothermic and endothermic changes.
16 Distinguish between kinetic energy and potential energy.

If you ignite a butane lighter and move your finger close to the flame, you will quickly sense the heat energy being transferred to your finger. **Energy** is defined as the ability to do work or transfer heat. In this case, energy in the form of heat is transferred from the chemical system—the burning reaction—to your finger (and the surrounding atmosphere). A chemical change that transfers energy to its surroundings is called an **exothermic reaction**.

Sometimes energy terms are included in chemical equations. Burning reactions combine the fuel with oxygen (O_2) from the air, producing invisible carbon dioxide (CO_2) and water (H_2O) gases. In the case of an ignited butane (C_4H_{10}) lighter, we have

$$2\ C_4H_{10} + 13\ O_2 \rightarrow 8\ CO_2 + 10\ H_2O + \text{energy}$$

Although the coefficients in this reaction appear complicated, you will learn in Chapter 8 that it is a relatively straightforward process to balance this equation!

In an exothermic reaction, the energy term appears as a product; heat energy is transferred from the system.

The reason that you sense the heat transferred from a burning reaction is that the molecules in your finger move faster due to the energy gained. **Kinetic energy** is the energy due to the motion of an object. Thus, what your brain interprets as *hot!* is the increased motion of the molecules near and within the nerves in your finger.

Although most chemical reactions result in the transfer of energy to the surroundings, some chemical changes result in the transfer of energy from the surroundings to the chemical system. An **endothermic reaction** is a chemical change that removes energy from the surroundings. The photosynthesis reaction by which plants convert light energy to chemical energy is perhaps the most important endothermic reaction that occurs on Earth:

$$6\ CO_2 + 6\ H_2O + \text{energy} \rightarrow C_6H_{12}O_6 + 6\ O_2$$

The products of photosynthesis are a sugar found in plants, $C_6H_{12}O_6$, and oxygen. Notice how it is essentially the opposite of a burning reaction.

The energy in plant sugar is stored as a form of **potential energy**: the energy due to the arrangement of the charged particles in a system.* Each atom in a molecule is made up of miniscule particles even smaller than the atom. Some of these particles are charged, with both positive charge and negative charge. We will formally introduce these particles in Chapter 5. When like-charged particles are moved

You rely on the potential energy in a sugar molecule to sustain your life when you metabolize it and use the energy transferred from the reaction of the sugar molecule to power bodily functions.

*Another form of potential energy is gravitational potential energy: the energy of an object due to its position in a gravitational field. This form of energy is primarily affected by the very large mass of Earth. For chemical change, the tiny particles exert a trivial gravitational force on one another, and thus the only important form of potential energy for chemical change is the interaction among charged particles.

closer to one another within a molecule or when unlike-charged particles are moved apart, the potential energy of the molecule is increased.

Chemical systems tend to change in a way that reduces their total energy. For example, a fuel will burn on its own, transferring energy to the surroundings, after the burning reaction is initiated with a spark. The chemical energy within molecules comes largely from the arrangement of minuscule charged particles within the molecules. Reduction of the energy in a chemical system to the smallest amount possible is one of the driving forces that causes chemical changes to occur. You will see this mentioned again from time to time in this textbook.

✓ Target Check 2.7

a) Is the process of boiling water exothermic or endothermic with respect to the water?

b) A charged object is moved closer to another object that has the same charge. The energy of the system changes. Is it a change in kinetic energy or potential energy? Is the energy change an increase or a decrease?

Learn It Now! *There are three pairs of terms in Section 2.9 that represent concepts that you need to understand: reactant and product; exothermic and endothermic; and kinetic and potential energy. Be sure that you understand each individual term as well as how it is related to the term it is paired with.*

2.10 Conservation Laws and Chemical Change

The Law of Conservation of Mass and Energy

In 1905, Albert Einstein (**Figure 2.27**) published a paper in a scientific journal (see the discussion about communication in science in Section 1.2) that proposed that a fundamental principle of nature is the sameness between energy and mass. This **mass-energy equivalence** may be expressed as the equation

$$E = mc^2$$

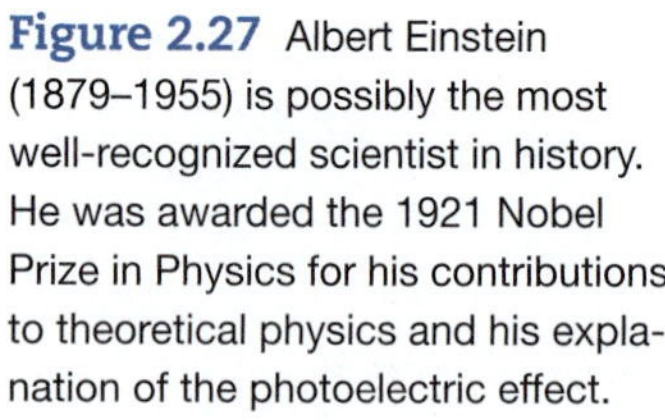

Chronicle/Alamy Stock Photo

Figure 2.27 Albert Einstein (1879–1955) is possibly the most well-recognized scientist in history. He was awarded the 1921 Nobel Prize in Physics for his contributions to theoretical physics and his explanation of the photoelectric effect.

where E is energy, m is mass,* and c is the speed of light. The relationship shown in the equation tells us that mass and energy are the same thing that can be expressed as the same quantity. A pattern in nature is that the total quantity of mass and energy in the universe is fixed and does not change. This is the **Law of Conservation of Mass and Energy**.

The Law of Conservation of Mass

Goal 17 State the meaning of and draw conclusions based on the Law of Conservation of Mass.

As discussed in Chapter 1, early chemists who studied burning wood hypothesized that because the ash remaining was so much lighter than

---

**Mass* is quantity of matter. It is closely related to the more familiar term *weight*. Section 3.6 explains the difference between mass and weight.

the object burned, something called *phlogiston* was lost in the reaction. Lavoisier proposed an alternate hypothesis: Burning is a chemical reaction between the burning object and oxygen in the air. His experiments demonstrated that oxygen in the air, which could not be seen, was a reactant and that carbon dioxide and water vapor, also invisible, were the products. Lavoisier showed that

$$\text{Mass of (wood + oxygen)} = \text{Mass of (ash + carbon dioxide + water vapor)}$$

More generally, we can state that, for any chemical change,

$$\text{Total mass of reactants} = \text{Total mass of products}$$

This equation is the **Law of Conservation of Mass**: The total mass of the reactants in a chemical change is equal to the total mass of the products ▶.

◀ **P/Review** Lavoisier's experiments, the disproving of the phlogiston theory, and the proposal of the Law of Conservation of Mass are described more fully in Section 1.1.

Given the equivalence between mass and energy, the Law of Conservation of Mass and Energy and the Law of Conservation of Mass are statements of the same principle.

The Law of Conservation of Energy

Goal 18 State the meaning of and draw conclusions based on the Law of Conservation of Energy.

Energy changes take place all around us—and within us—all the time. Driving an automobile starts with the chemical energy of a battery and fuel. This changes into kinetic, potential, sound, and heat energy as the car moves up and down hills and into light energy when the brake lights go on. The cell phone alarm clock that works silently next to your bed converts electrical energy to chemical energy in the battery and then to light energy and sound energy when the night is past. Even as you sleep, your body is processing the food you ate into heat energy that maintains your body temperature. Figure 2.28 shows other energy conversions.

Careful study of energy conversions shows that the energy lost or used in one form is always exactly equal to the energy gained in another form. This leads to another conservation law, the **Law of Conservation of Energy**: The quantity of energy within an isolated system does not change. Even when a chemical change occurs within that system, the energy of the system is conserved. It is neither created nor destroyed.

Again, given the equivalence between mass and energy, the Law of Conservation of Mass and Energy, the Law of Conservation of Mass, and the Law of Conservation of Energy are statements of the same principle.

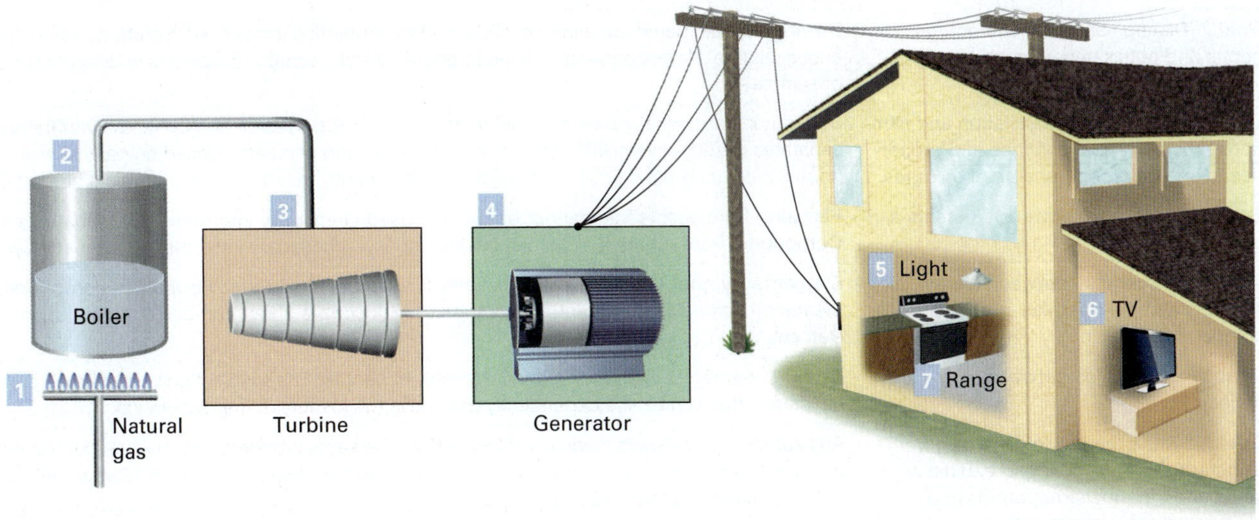

Figure 2.28 Energy conversions. Common events in which energy changes from one form to another.
1. Chemical energy of fuel changes to heat energy.
2. Heat energy changes to higher kinetic and potential energy of steam compared to liquid water.
3. Kinetic energy changes to rotating mechanical energy in a turbine.
4. Mechanical energy is transmitted to a generator where it changes to electrical energy.
5. Electrical energy changes to heat and light energy.
6. Electrical energy is changed to light, sound, and heat energy.
7. Electrical energy changes to heat energy.

Learn It Now! *The term conserve is used in the sciences in a very specific, unique way. Be sure that you understand the concept of conservation as it is applied in the sciences.*

Target Check 2.8

In everyday language, the term *conserve* usually refers to protecting something. (It is important to conserve natural resources.) What does the term *conserve* mean in scientific language?

Mindaugas Dulinskas/Shutterstock.com

CHAPTER 2 IN REVIEW: MATTER AND ENERGY

Goal 1 Identify and explain the differences between interpreting and describing matter at the macroscopic, microscopic, and particulate levels.	**Matter**, anything that has mass, can be studied at three levels: macroscopic (seen with the human eye), microscopic (seen with a light microscope), and particulate (cannot be directly seen).
Goal 2 Define the term *model* as it is used in chemistry to represent pieces of matter too small to see.	Chemists use symbols and **models** to represent particulate matter (Figure 2.4): chemical formula, Lewis diagram, ball-and-stick model, space-filling model.
Goal 3 Identify and explain the differences among gases, liquids, and solids in terms of (a) the macroscopic-level properties shape and volume; (b) particle movement; and (c) particle spacing.	**Kinetic molecular theory:** Particles of matter are always moving. Kinetic means motion. **Gas:** Variable shape and volume, independent particle movement, particles very far apart. **Liquid:** Variable shape, constant volume, independent particle movement beneath the surface, particles close. **Solid:** Constant shape and volume, particles vibrate in fixed positions, particles close.
Goal 4 Distinguish between physical and chemical properties at both the particulate level and the macroscopic level.	Matter is described by its properties, which may be physical or chemical. Changes in matter may also be physical or chemical. **Physical property:** Can be observed and measured without a chemical change. **Chemical property:** The chemical changes possible for a substance.
Goal 5 Distinguish between physical and chemical changes at both the particulate level and the macroscopic level.	**Physical change:** Alteration in physical form without change in identity. **Chemical change:** Identity of original substance changed to something new.
Goal 6 Distinguish between a pure substance and a mixture at both the particulate level and the macroscopic level.	**Pure substance:** One chemical with distinct set of physical and chemical properties; cannot be separated by physical changes. **Mixture:** Two or more pure substances; properties vary, depending on relative amounts of pure substances; components can be separated by physical changes.
Goal 7 Distinguish between homogeneous and heterogeneous matter.	**Homogeneous:** Same appearance, composition, properties throughout. **Solution:** A homogenous mixture. **Heterogeneous:** Different phases visible, variable properties in different parts of sample.
Goal 8 Describe how distillation and filtration rely on physical changes and properties to separate components of mixtures.	Separations are usually based on different physical properties of components. **Distillation** separates based on volatilities of components. **Filtration** separates based on particle size. A porous medium is used to separate mixture components based on size (a physical property).
Goal 9 Distinguish between elements and compounds.	**Element:** Pure substance, cannot be decomposed chemically into other pure substances. **Compound:** Pure substance that can be decomposed chemically into other pure substances.
Goal 10 Distinguish between elemental symbols and the formulas of chemical compounds.	**Elemental symbol:** Capital letter, sometimes followed by small letter. **Formula of compound:** Elemental symbols of elements in compound. Subscripts show number of atoms of each element.
Goal 11 Distinguish between atoms and molecules.	**Atom:** The smallest unit particle of an element. **Molecule:** The smallest unit particle of a pure substance that can exist independently and retain the identity of that substance.
Goal 12 State the Law of Definite (or Constant) Composition, and explain its implication for how compounds and mixtures differ in composition.	Any compound is always made up of elements in the same proportion by mass. A compound has a definite composition. The composition of a mixture depends on the relative quantities of the components that make up the mixture.

Goal 13 Match electrostatic forces of attraction and repulsion with combinations of positive and negative charges.	Objects with the same charge repel. Objects with opposite charges attract.
Goal 14 Distinguish between reactants and products in a chemical equation.	Equation: Reactants → Products
Goal 15 Distinguish between exothermic and endothermic changes.	**Exothermic:** Transfers energy to surroundings. **Endothermic:** Removes energy from surroundings.
Goal 16 Distinguish between kinetic energy and potential energy.	**Potential energy:** Energy due to the arrangement of the charged particles in a system. Tendency toward reduction of energy to the smallest amount possible is a driving force for chemical reactions. **Kinetic energy:** Energy of motion.
Goal 17 State the meaning of and draw conclusions based on the Law of Conservation of Mass.	**The Law of Conservation of Mass and Energy:** Total mass + Energy in the universe is constant. **The Law of Conservation of Mass:** Mass is conserved in chemical change; mass is neither created nor destroyed.
Goal 18 State the meaning of and draw conclusions based on the Law of Conservation of Energy.	**The Law of Conservation of Energy:** Quantity of energy within an isolated system does not change; energy is neither created nor destroyed.

Key Terms

Most of the key terms and concepts and many others appear in the Glossary. Use your Glossary regularly.

Amorphous p. 23
Ball-and-stick model p. 19
Chemical change p. 26
Chemical equation p. 42
Chemical formula p. 37
Chemical properties p. 27
Chemical reaction p. 27
Chemical symbol p. 21
Compound p. 35
Crystalline solid p. 23
Distillation p. 33
Distilled water p. 33
Electric charge p. 41
Electrostatic force p. 42
Element p. 35
Elemental symbol p. 37
Endothermic reaction p. 43
Energy p. 43
Equation p. 43
Exothermic reaction p. 43
Filtration p. 34
Gas p. 22
Heterogeneous mixture p. 32
Homogeneous mixture p. 32
Kinetic energy p. 43
Kinetic molecular theory p. 21
Law of Conservation of Energy p. 45
Law of Conservation of Mass p. 45
Law of Conservation of Mass and Energy p. 44
Law of Constant Composition p. 39
Law of Definite Composition p. 39
Liquid p. 22
Macroscopic p. 18
Mass-energy equivalence p. 44
Matter p. 18
Microscopic p. 19
Mixture p. 31
Model p. 19
Molecule p. 21
Particulate p. 19
Periodic table of the elements p. 37
Phases p. 32
Physical change p. 26
Physical properties p. 26
Potential energy p. 43
Products p. 42
Pure substance p. 30
Reactants p. 42
Solid p. 22
Solution p. 32
Space-filling model p. 20
States of matter p. 21
Static electricity p. 42

Frequently Asked Questions

How can I determine if a change is chemical or physical?
If the description of the change is at the macroscopic level, consider whether or not the identity of the substance is destroyed. The change is chemical if the original substance is destroyed and a new substance is formed. The change is physical if the substance is the same but in a different form. If the description of the change is at the particulate level, the change is chemical if the molecules are changed, and the change is physical if the molecules remain the same.

I don't understand why a compound isn't a mixture.
Think of an example when you think about the terms *pure substance* and *mixture* and *element* and *compound.* Let's use water to help clarify these terms. Water is a pure substance because it is one kind of matter, simply water. Water is also a compound because its molecules are made up of two different types of atoms. Even though a water molecule is composed of two hydrogen atoms attached to an oxygen atom, once the atoms become attached to one another, the molecule behaves as the smallest individual particle of a substance. When water is subjected to an electric current, the input of energy will cause the water molecules to decompose, forming hydrogen molecules and oxygen molecules. Neither the hydrogen molecules nor the oxygen molecules will undergo decomposition, so each substance is an element. A container holding both hydrogen gas and oxygen gas holds a mixture: a combination of two or more pure substances that can have variable composition. A glass filled with water is not holding a mixture because it contains a single pure substance: water.

Concept-Linking Exercises

Write a brief description of the relationships among each of the following groups of terms or phrases. Answers to the Concept-Linking Exercises are given at the end of the chapter.

Example: Natural sciences, physical sciences, biological sciences, chemistry, physics, botany, zoology.

Solution: The natural sciences can be divided into two general categories: physical sciences (the study of matter and energy) and biological sciences (the study of living organisms). Botany and zoology are biological sciences. Physics and chemistry are physical sciences, although chemistry overlaps the biological sciences in the fields of biochemistry, biological chemistry, and chemical biology.

1. Matter, state of matter, kinetic molecular theory, gas, liquid, solid
2. Homogeneous, heterogeneous, pure substance, mixture
3. Element, compound, atom, molecule
4. Physical property, physical change, chemical property, chemical change
5. Conservation of mass, conservation of energy, conservation of mass and energy
6. Energy, kinetic energy, potential energy, endothermic change, exothermic change

Small-Group Discussion Questions

Small-Group Discussion Questions are for group work, either in class or under the guidance of a leader during a discussion section.

1. A model often is simpler than the natural phenomenon that it represents. How is this an advantage for thinking about matter? How is it a disadvantage?
2. Viscosity is defined as the resistance of a substance to flow. Explain why each state of matter either does or does not have the property of viscosity. How do you suppose viscosity occurs at the particulate level?
3. Describe as many chemical and physical changes and properties as possible that are given in, or that you can deduce from, the following: At the end of a day of hiking in the mountains, you set up camp. You gather dry sticks and logs to build a fire, and you light the fire with a match. After roasting marshmallows over the fire, you douse it with cold water from a nearby stream, causing a cloud of steam to form as the fire goes out.
4. If you were to go on a hunt around campus for pure substances, what would you find? List at least as many pure substances as there are members of your group, and for each substance listed, explain why it is pure.
5. Consider the distillation apparatus in Figure 2.18. Explain how you can use the change illustrated as evidence to deduce the composition of the gas that forms in the distillation flask.
6. If you were to go on a hunt around campus for elements, what would you find? List at least as many elements as there are members of your group, and for each element listed, explain why it is an element. How many elements on your list are pure substances? Explain.
7. Compare and contrast the electrical character of matter with the magnetic character of bar magnets. How are they same? How are they different?
8. Can you have a chemical change without an accompanying physical change? Explain, defining both *chemical change* and *physical change*. Give an example and a counterexample, if possible, to support your explanation.
9. If matter is indeed conserved, how can you explain the fact that a glass of water left alone on a kitchen counter will eventually empty? The outside of a glass of cold water typically becomes wet. What substance wets the glass, and where does it come from?

Questions, Exercises, and Problems

Solutions for blue-numbered questions are at the end of the chapter. Questions other than those in the General Questions and More Challenging Problems sections are paired in consecutive odd–even number combinations; the odd-numbered questions are blue, and the even-numbered questions are black in these pairs.

Section 2.2: Representations of Matter: Models and Symbols

1. Identify the following samples of matter as macroscopic, microscopic, or particulate: (a) a human skin cell; (b) a sugar molecule; (c) a blade of grass; (d) a helium atom; (e) a single-celled plant too small to see with the unaided eye.
2. Classify each of the following as macroscopic, microscopic, or particulate: (a) a cell membrane; (b) a silver atom; (c) iron filings.
3. Suggest a reason for studying matter at the particulate level, given that it is too small to see.
4. How does a chemist think about particles that are so small that they are impossible to see with the naked eye or with even the most powerful optical microscope?

Section 2.3: States of Matter

5. Using spheres to represent individual atoms, sketch particulate illustrations of a substance as it is heated from the solid to the liquid and to the gaseous state.
6. Describe a piece of ice at the particulate level. Then describe what happens to the ice as it is heated until it melts and eventually boils.

7. The word *pour* is commonly used in reference to liquids but not to solids or gases. Can you pour a solid or a gas? Why or why not? If either answer is yes, can you give an example?

8. The slogan "When it rains, it pours" has been associated with a brand of table salt for decades. How can salt, a solid, be poured? What unique feature—unique at one time, but not today—do you suppose was being emphasized by the slogan? In other words, under what circumstances would one brand of salt "pour" whereas another brand would not, and why?

9. Which of the three states of matter is most easily compressed? Suggest a reason for this.

10. Compare the volumes occupied by the same sample of matter when in the solid, liquid, and gaseous states.

Section 2.4: Physical and Chemical Properties and Changes

11. Classify each of the following properties as chemical or physical: (a) hardness of a diamond; (b) combustibility of gasoline; (c) corrosive character of an acid; (d) elasticity of a rubber band; (e) taste of chocolate.

12. Classify the *italicized* property as chemical or physical: (a) a shiny piece of iron metal *gets rusty* when left outside; (b) the purple crystalline solid potassium permanganate *forms a purple solution* when dissolved in water; (c) the shiny metal mercury *is a liquid at room temperature*.

13. Which among the following are physical changes: (a) blowing glass; (b) fermenting grapes; (c) forming a snowflake; (d) evaporating dry ice; (e) decomposing a substance by heating it?

14. Classify each of the following changes as chemical or physical: (a) grilling a steak; (b) souring of milk; (c) removing nail polish.

15. Is the change illustrated here a physical change or a chemical change? Explain your answer.

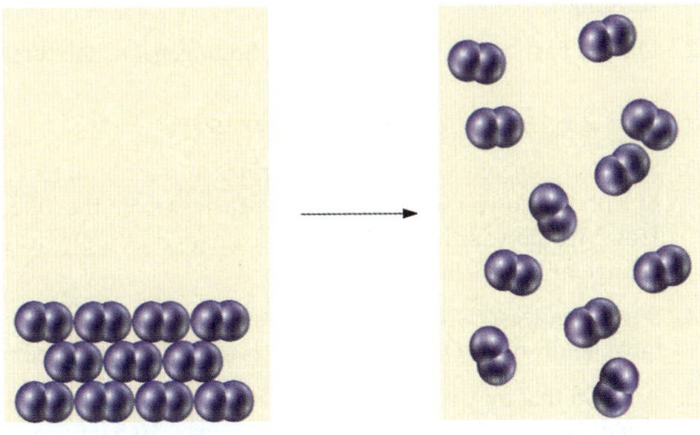

16. Is the change in the illustration here a physical change or a chemical change? Explain your answer.

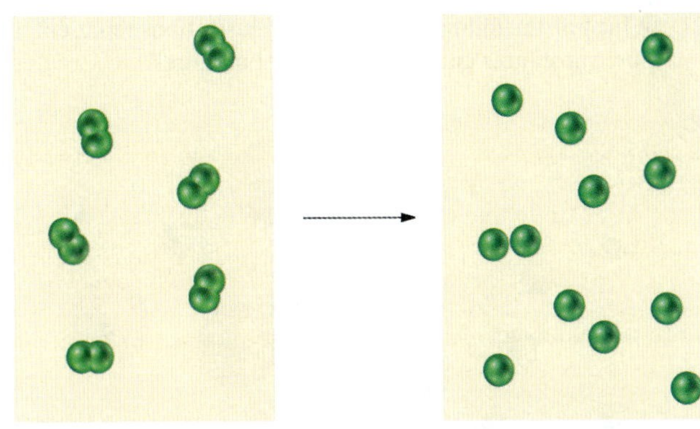

Section 2.5: Pure Substances and Mixtures

17. Diamonds and graphite are two forms of carbon. Carbon is an element. Chunks of graphite are sprinkled among the diamonds on a jeweler's display tray. Is the material on the tray a pure substance or a mixture? Is the display homogeneous or heterogeneous? Justify both answers.

18. Aspirin is a pure substance. If you had the choice of buying a widely advertised brand of aspirin whose effectiveness is well known or the generic product of a new manufacturer at half the price, which would you buy? Explain.

19. The substance in the glass here is from a kitchen tap. Is it a pure substance or a mixture? What if it came from a bottle of distilled water?

Charles D. Winters

20. Are the contents of the bottle in the picture here a pure substance or a mixture?

Charles D. Winters

21. Which of the following particulate illustrations represent pure substances and which represent mixtures?

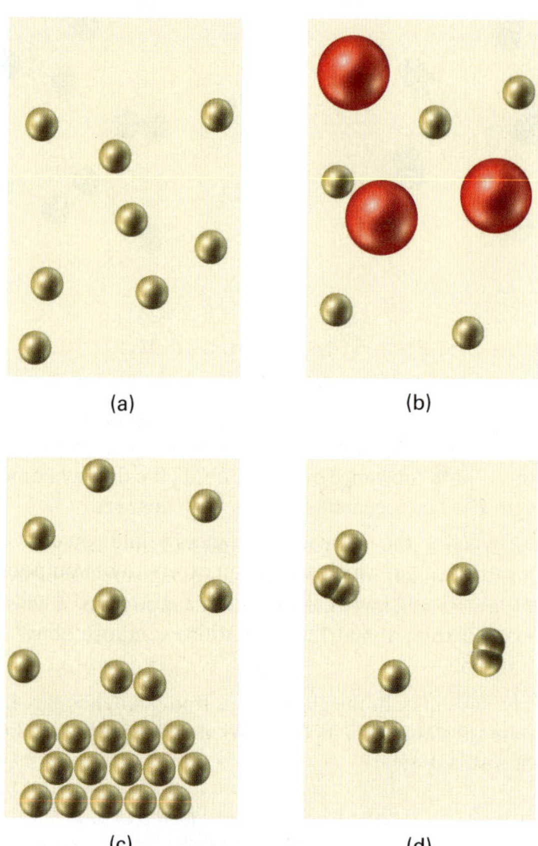

(a) (b)

(c) (d)

22. Which of the following particulate illustrations represent pure substances and which represent mixtures?

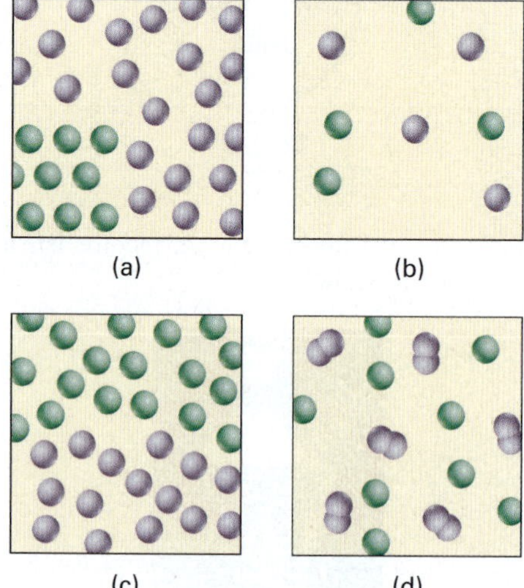

(a) (b)

(c) (d)

23. Which of the following are pure substances and which are mixtures: (a) table salt; (b) tap water; (c) clean, dry air; (d) steam?

24. Which of the substances here are pure and which are mixtures? Which could be either? Explain your answers.

(a)

(b)

(c)

Charles D. Winters

25. Apart from food, list five things in your home that are homogeneous.

26. Can the terms homogeneous and heterogeneous be applied to pure substances as well as to mixtures? Explain.

27. Which items in the following list are heterogeneous: (a) sterling silver; (b) freshly opened root beer; (c) popcorn; (d) scrambled eggs; (e) motor oil?

28. Classify each of the following mixtures as either homogeneous or heterogeneous: (a) apple juice; (b) concrete; (c) gin.

29. Some ice cubes are homogeneous and some are heterogeneous. Into which group do ice cubes from your home refrigerator fall? If homogeneous ice cubes are floating on water in a glass, are the contents of the glass homogeneous or heterogeneous? Justify both answers.

30. The freshly polished brass cylinder in the picture here is a mixture of copper and zinc. Is the cylinder a homogeneous or heterogeneous substance?

Charles D. Winters

31. Draw a particulate-level sketch of a heterogeneous pure substance.

32. Draw a particulate-level sketch of a homogeneous mixture.

Section 2.6: Separation of Mixtures

33. Suppose someone emptied ball bearings into a container of salt. Could you separate the ball bearings from the salt? How? Would your method involve no change, be a physical change or be a chemical change?

34. Suggest at least two ways to separate ball bearings from table tennis balls. On what property is each method based?

35. A liquid that may be either pure or a mixture is placed in a distillation apparatus (see Figure 2.18). The liquid is allowed to boil, and some condenses in the receiving flask. The remaining liquid is then removed and frozen, and the freezing point is found to be lower than the freezing point of the original liquid. Is the original liquid pure or a mixture? Explain.

36. You receive a mixture of table salt and sand and have to separate the mixture into pure substances. Explain how you would carry out this task. Is your method based on physical or chemical properties? Explain.

Section 2.7: Elements and Compounds

37. Classify the following as compounds or elements: (a) silver bromide (used in photography); (b) calcium carbonate (limestone); (c) sodium hydroxide (lye); (d) uranium; (e) tin; (f) titanium.

38. Classify each of the following pure substances as either an element or a compound: (a) silicon dioxide; (b) tungsten; (c) silver.

39. Which of the following are elements and which are compounds: (a) NaOH; (b) $BaCl_2$; (c) He; (d) Ag; (e) Fe_2O_3?

40. Classify each of the following pure substances as either an element or a compound: (a) C; (b) C_2H_5OH; (c) Cl_2.

41. Classify each substance in the illustrations here as an element or a compound.

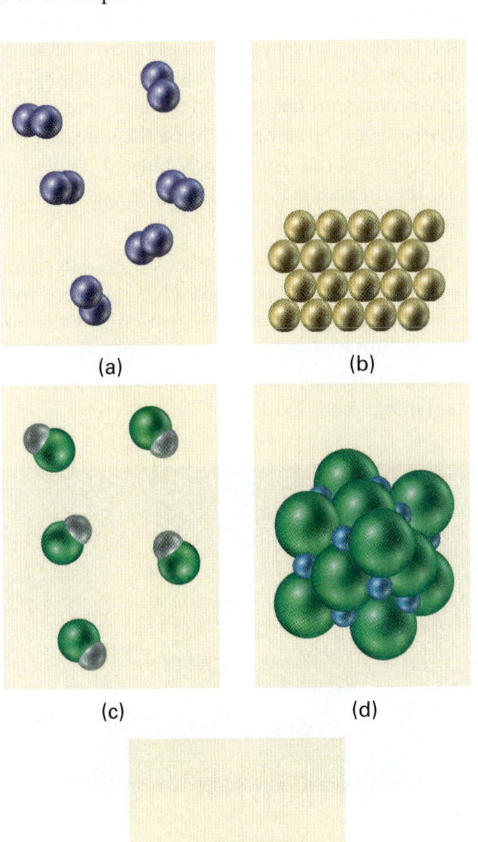

42. Does each of the particulate-level models here depict an element or a compound?

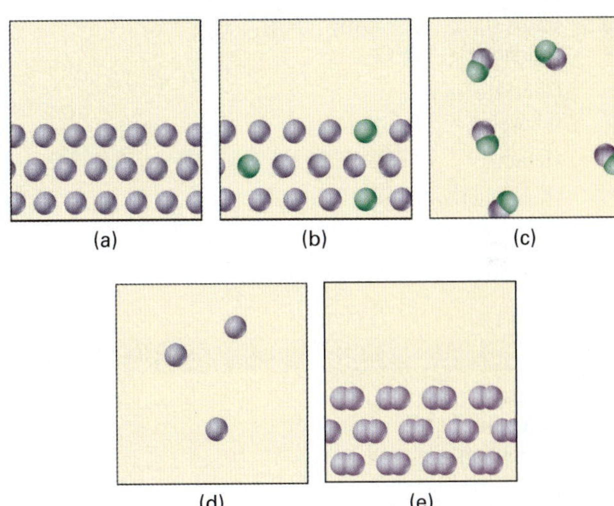

43. (a) Which of the following substances would you expect to be elements and which would you expect to be compounds: (1) calcium carbonate; (2) arsenic; (3) uranium; (4) potassium chloride; (5) chloromethane? (b) On what general rule do you base your answers to part (a)? Can you name any exceptions to this general rule?

44. (a) Which of the following substances would you expect to be elements and which would you expect to be compounds: (1) aluminum sulfate; (2) osmium; (3) radon; (4) lithium carbonate; (5) dimethylhydrazine? (b) On what general rule do you base your answers to part (a)? Can you name any exceptions to this general rule for compounds?

45. Metal A dissolves in nitric acid solution. You can recover the original metal if you place Metal B in the solution. Metal A becomes heavier after prolonged exposure to air. The procedure is faster if the metal is heated. From the evidence given, can you tell if Metal A definitely is or could be an element or a compound? If you cannot, what other information do you need to make that classification?

46. A white, crystalline material that looks like table salt gives off a gas when heated under certain conditions. There is no change in the appearance of the solid that remains, but it does not taste the same as it did originally. Was the beginning material an element or a compound? Explain your answer.

Questions 47 and 48: Samples of matter may be classified in several ways, including gas, liquid, or solid (G, L, S); pure substance or mixture (P, M); homogeneous or heterogeneous (Hom, Het); and, for pure substances, element or compound (E, C). For each substance in the left column of the tables shown, place in the other columns the symbol from the top of the column that best describes the substance in its most common state at room temperature and pressure. Assume that the material is clean and uncontaminated. (The first box is filled in as an example.)

47.

	G, L, S	P, M	Hom, Het	E, C
Factory smokestack emissions	All, but mostly G			
Concrete (in a sidewalk)				
Helium				
Hummingbird feeder solution				
Table salt				

48.

	G, L, S	P, M	Hom, Het	E, C
Limestone (calcium carbonate)				
Lead				
Freshly squeezed orange juice				
Oxygen				
Butter in the refrigerator				

Section 2.8: The Electrical Character of Matter

49. What is the main difference between electrostatic forces and gravitational forces? Which is more similar to the magnetic force? Can two or all three of these forces be exerted between two objects at the same time?

50. Identify the net electrostatic force (attraction, repulsion, or none) between the following pairs of substances: (a) a small, negatively charged piece of paper and a small, positively charged piece of paper; (b) two positively charged lint balls; (c) a positively charged sodium ion and a negatively charged oxide ion.

Section 2.9: Characteristics of a Chemical Change

51. Identify the reactants and products in the following equation: $AgNO_3 + NaCl \rightarrow AgCl + NaNO_3$.

52. In the following equation for a chemical reaction, the notation (s), (ℓ), or (g) indicates whether the substance is in the solid, liquid, or gaseous state: $2\ H_2S(g) + 3\ O_2(g) \rightarrow 2\ H_2O(g) + 2\ SO_2(g) +$ energy. Identify each of the following as a product or reactant: (a) $SO_2(g)$; (b) $H_2S(g)$; (c) $O_2(g)$; (d) $H_2O(g)$. When the reaction takes place, is energy released or absorbed? Is the reaction endothermic or exothermic?

53. In the equation: $Ni + Cu(NO_3)_2 \rightarrow Ni(NO_3)_2 + Cu$, which of the reactants is/are elements, and which of the products is/are compounds?

54. Write the formulas of the elements that are products and the formulas of the compounds that are reactants in the following equation: $2\ Na + 2\ H_2O \rightarrow 2\ NaOH + H_2$.

55. Which of the following processes is/are exothermic: (a) water freezing; (b) water vapor in the air changing to liquid water droplets on a windowpane; (c) molten iron solidifying; (d) chocolate candy melting?

56. Classify each of the following changes as endothermic or exothermic with respect to the *italicized* object: (a) cooling a *beer*; (b) burning *leaves*; (c) cooking a *hamburger*.

57. As a child plays on a swing, at what point in her movement is her kinetic energy the greatest? At what point is potential energy at its maximum?

58. A bicycle accelerates from 5 miles per hour to 15 miles per hour. Does its energy increase or decrease? Is the change in potential energy or kinetic energy?

Section 2.10: Conservation Laws and Chemical Change

59. After solid limestone is heated, the rock that remains weighs less than the original limestone. What do you conclude has happened?

60. Before electronic flashes were commonly used in photography, a darkened area was lit by a device known as a flashbulb. This one-use device was essentially a glass bulb filled with oxygen that encased a metal wire. An electrical discharge from the camera ignited the wire, causing a brief flash of light as the wire quickly burned. How would you expect the mass of a flashbulb before use to compare with its mass after use? Explain.

61. The photograph here shows a beaker of water and a sugar cube, the combined mass of which is balanced by the weights on the right pan. The sugar cube is then placed in the water, and it dissolves completely. Do weights need to be added to or taken away from the right pan to keep the system in balance? Explain.

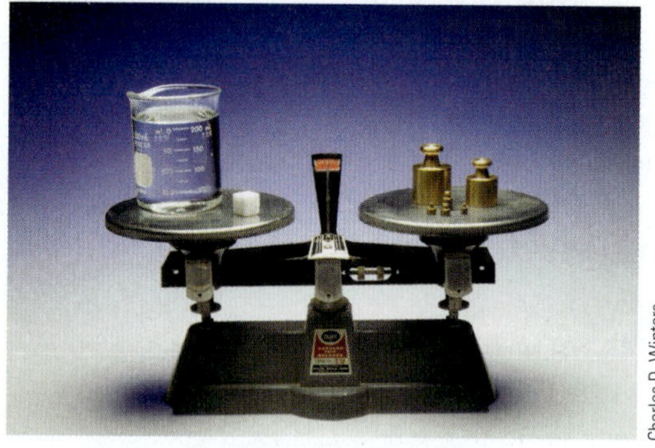

Charles D. Winters

62. Plants manufacture their own food. What is the source of energy for this process? Explain how energy is conserved as a plant makes its food.

63. Identify several energy conversions that occur regularly in your home. State whether each is useful, wasteful, or sometimes useful and sometimes wasteful.

64. List the energy conversions that occur in the process from the time water is about to enter a hydroelectric dam to the burning of an electric lightbulb in your home.

General Questions

65. Distinguish precisely and in scientific terms the differences among items in the following groups.
a) Macroscopic matter, microscopic matter, particulate matter
b) Physical change, physical property, chemical change, chemical property
c) Gases, liquids, solids
d) Element, compound
e) Atom, molecule
f) Pure substance, mixture
g) Homogeneous matter, heterogeneous matter
h) Reactant, product
i) Exothermic change, endothermic change
j) Potential energy, kinetic energy

66. Determine whether each of the following statements is true or false:
a) The fact that paper burns is a physical property.
b) Particles of matter are moving in gases and liquids, but not in solids.
c) A heterogeneous substance has a uniform appearance throughout.
d) Compounds are impure substances.
e) If one sample of sulfur dioxide is 50% sulfur and 50% oxygen, then all samples of sulfur dioxide are 50% sulfur and 50% oxygen.
f) A solution is a homogeneous mixture.
g) Two positively charged objects attract each other, but two negatively charged objects repel each other.
h) Mass is conserved in an endothermic chemical change but not in an exothermic chemical change.
i) Potential energy can be related to positions in an electrical field.
j) Chemical energy can be converted to kinetic energy.
k) Potential energy is more powerful than kinetic energy.
l) A chemical change always destroys something and always creates something.

67. A natural-food store advertises that no chemicals are present in any food sold in the store. If the ad is true, what do you expect to find in the store?

68. Name some things you have used today that are not the result of human-made chemical change.

69. Name some pure substances you have used today.

70. How many homogeneous substances can you reach without moving from where you are sitting right now?

71. Which of the following can be pure substances: mercury, milk, water, a tree, ink, iced tea, ice, carbon?

72. Can you have a mixture of two elements as well as a compound of the same two elements?

73. Can you have more than one compound made of the same two elements? If yes, try to give an example.

74. Rainwater comes from the oceans. Is rainwater more pure, less pure, or of the same purity as ocean water? Explain.

75. A large box contains a white powder of uniform appearance. One sample is taken from the top of the box and another is taken from the bottom. Analysis reveals that the percentage of oxygen in the sample from the top is 48.2%, whereas in the sample from the bottom it is 45.3%. Answer each question here independently and give a reason that supports your answer.
a) Is the powder an element or a compound?
b) Are the contents of the box homogeneous or heterogeneous?
c) Can you be certain that the contents of the box are either a pure substance or a mixture?

76. If energy cannot be created or destroyed, as the Law of Conservation of Energy states, why are we so concerned about wasting our energy resources?

77. Consider the sample of matter in the illustration here.

Answer each question independently and explain your answers.
a) Is the sample homogeneous or heterogeneous?
b) Is the sample a pure substance or a mixture?
c) Are the particles elements or compounds?
d) Are the particles atoms or molecules?
e) Is the sample a gas, a liquid, or a solid?

78. A particulate-level illustration of the reaction AB + CD → AD + CB is shown here.

a) Identify the reactants and products in this reaction.
b) Is the change shown chemical or physical?
c) Is the mass of the product particles less than, equal to, or greater than the mass of the reactant particles?
d) If the reaction takes place in a container that allows no energy to enter or to leave, how does the total energy in the container after the reaction compare with the total energy in the container before the reaction?

More Challenging Problems

79. A clear, colorless liquid is distilled in an apparatus similar to that shown in Figure 2.18. The temperature remains constant throughout the distillation process. The liquid leaving the condenser is also clear and colorless. Both liquids are odorless, and they have the same freezing point. Is the starting liquid a pure substance or a mixture? What single bit of evidence in the preceding description is the most convincing reason for your answer?

80. The density of a liquid is determined in the laboratory. The liquid is left in an open container overnight. The next morning the density is measured again and found to be greater than it was the day before. Is the liquid a pure substance or a mixture? Explain your answer.

81. There is always an increase in potential energy when an object is raised higher above the surface of the Earth, that is, when the distance between the Earth and the object increases. Increasing the distance between two electrically charged objects, however, may raise or lower potential energy. How can this be?

82. In the gravitational field of the Earth, an object always falls until some physical object prevents it from falling farther. Two electrically charged objects, each of which is made up of unequal numbers of both positive and negative charges, will reach a certain separation distance and stay there without physical support. Can you suggest an explanation for this?

83. Particles in the illustration here undergo a chemical change.

Which among the remaining boxes, (a) through (d), can represent the products of the chemical change? If a box cannot represent the products of the chemical change, explain why.

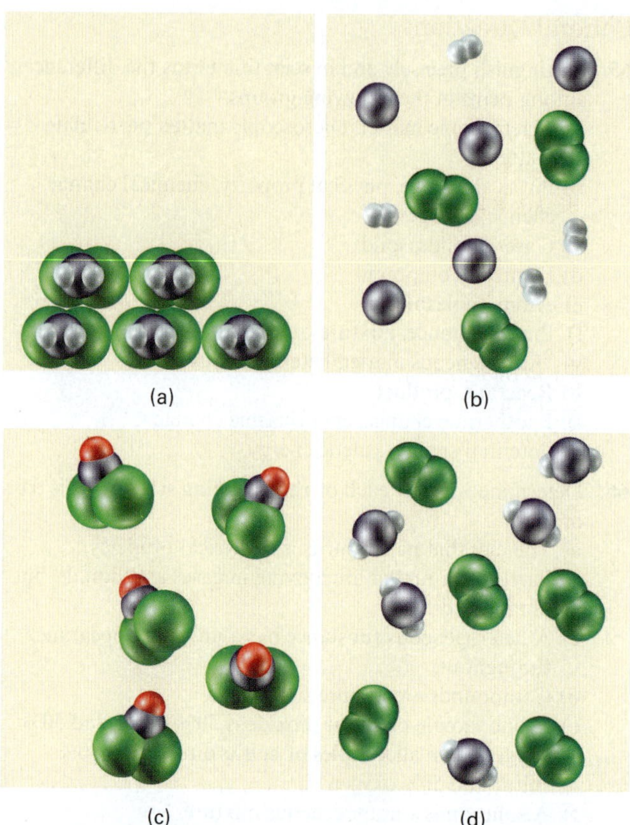

84. Draw a particulate illustration of five particles in the gas phase in a box. Show the particles at a lower temperature, in the liquid phase, in a new box. Now show the particles at an even lower temperature, in the solid phase, in a new box. Write a description of your illustrations in terms of the kinetic molecular theory.

Answers to Practice Exercises

1. The attractive forces between particles for a sample of water—or any other pure substance—are the same no matter the state of matter. Particle motion increases in the order solid < liquid < gas. In the solid phase, the tendency for particles to stick together indicates that the particle motion is insufficient to overcome the attractive forces. In the liquid state, the particle motion is sufficient to overcome the attractive forces to the extent that the particles move among themselves. In the gas phase, the particle motion is great enough to completely overcome the attractive forces, and the particles move independently of one another.

2. A compound is a pure substance because it has definite physical and chemical properties.

Answers to Target Checks

1. The art depicts a model of table salt at the particulate level. The photograph shows salt at the macroscopic level.
2. (a) and (d) are chemical changes; (b) and (c) are physical changes.
3. Beaker B holds a pure substance because its specific gravity, a physical property, is constant. Beakers A and C hold mixtures because their specific gravities are variable.
4. (b), (c), and (d) are heterogeneous; (a) is homogeneous.
5. The distillation apparatus is the better choice. The filtration apparatus will not work to separate the components of the salt water solution because the solution is homogeneous. The filtration apparatus separates solid from liquid in a heterogeneous mixture.
6. a and c, repulsion; b, attraction.
7. (a) Boiling water is endothermic with respect to the water. Energy must be transferred to the water in order to boil. (b) The change is an increase in potential energy.
8. In a scientific context, the term *conserve* means that the quantity of something remains constant before and after a change.

Answers to Concept-Linking Exercises

You may have found more relationships or relationships other than the ones given in these answers.

1. Matter is whatever has mass. The kinetic molecular theory explains that matter consists of particles in constant motion. The strength of the attractive forces among particles versus the amount of motion determines the state of matter: A gas has the greatest amount of motion, a solid has the least, and the amount of motion in a liquid is intermediate.
2. A homogeneous substance has a uniform appearance and composition throughout. It may be pure, consisting of only one substance, or it may be a mixture of two or more substances. A heterogeneous substance has different phases.
3. An element is a pure substance that cannot be changed into a simpler pure substance. A compound can be changed into simpler pure substances. An atom is the smallest particle of an element. Molecules are the smallest individual particles in a pure substance.
4. A chemical change occurs when one substance disappears and another substance appears. The chemical properties of a substance are the chemical changes that are possible for the substance. A physical change is a change in the form of a substance without a change in its identity. Physical properties can be measured or detected with the five physical senses.
5. The Law of Conservation of Mass and Energy states that the total of all mass and energy is conserved in all changes. The individual Laws of Conservation of Mass and Energy hold that both mass and energy are conserved independently.
6. Energy is the ability to do work or transfer heat. Kinetic energy is associated with molecules or other objects in motion. Potential energy is related to the position of charged particles in a system. A change is exothermic if it transfers energy to the surroundings and endothermic if energy is transferred from the surroundings.

Answers to Blue-Numbered Questions, Exercises, and Problems

1. Macroscopic: (c). Microscopic: (a), (e). Particulate: (b), (d).
3. An advantage is that understanding the behavior of particles allows us to predict the macroscopic behavior of samples of matter made from those particles. Chemists can then design particles to exhibit desired macroscopic characteristics, as seen in drug design and synthesis, for example.
5. Your illustration should resemble the particulate view in Figure 2.7.
7. A dense gas that is concentrated at the bottom of a container can be poured because its particles can move relative to each other. Chunks of solids, such as sugar crystals, can be poured.
9. Gases are most easily compressed because of the large spaces between molecules.
11. Chemical: (b), (c). Physical: (a), (d), (e).
13. (a), (c), (d).
15. Physical—the particles simply change state.
17. The material is a pure substance, one kind of matter. However, the display is heterogeneous, consisting of two visibly different forms or phases of carbon.
19. Tap water is a mixture of water and dissolved minerals and gases. Distilled water is pure water.
21. Pure substances: (a), (c). Mixtures: (b), (d).
23. Pure substances: (a), (d). Mixtures: (b), (c).
25. Examples include glass products, plastic products, aluminum foil, cleaning and grooming solutions, and the air.
27. (b), (c), (d).
29. Ice cubes from a home refrigerator are usually heterogeneous, containing trapped air. Homogeneous cubes in liquid water are heterogeneous, having visible solid and liquid phases.
31. Your sketch should show one type of particle (a pure substance) but in more than one state of matter or molecular form (heterogeneous).
33. Pick out the ball bearings (no change); use the magnetic property of steel to pick up the ball bearings with a magnet (no change); dissolve the salt in water and filter or pick out the ball bearings (physical change).
35. The original liquid must be a mixture because the freezing point changed when some of the liquid was removed. The freezing point of a pure substance is the same no matter how much of the substance you have.
37. Compounds: (a), (b), (c). Elements: (d), (e), (f).
39. Elements: (c), (d). Compounds: (a), (b), (e).
41. Elements: (a), (b), (e). Compounds: (c), (d).

43. (a) Elements: 2, 3; compounds: 1, 4, 5. (b) In general, if there are two or more words in the name, the substance is a compound. However, many compounds are known by one-word common names, and one-word names for many compounds that contain carbon are assembled from prefixes, suffixes, and special names for recurring groups. Chloromethane is such a compound. The name of an element is always a single word.

45. There is no evidence that A is a compound because it has not been broken down into two or more other pure substances by a chemical or physical change. However, only two methods have been tried. A is most likely an element, but the evidence is not conclusive. More tests need to be conducted.

47.

	G, L, S	P, M	Hom, Het	E, C
Factory smokestack emissions	All, but mostly G	M	Het	
Concrete (in a sidewalk)	S	M	Het	
Helium	G	P	Hom	E
Hummingbird feeder solution	L	M	Hom	
Table salt	S	P	Hom	C

49. Gravitational forces are attractive only; electrostatic forces can be attractive or repulsive. Magnetic forces can be attractive or repulsive also. All three can act simultaneously.

51. Reactants: $AgNO_3$, NaCl. Products: AgCl, $NaNO_3$.

53. The reactant Ni is an element; the product $Ni(NO_3)_2$ is a compound.

55. (a), (b), (c).

57. Kinetic energy is greatest when the swing moves through its lowest point. Potential energy is at a maximum when the swing is at its highest point.

59. A gaseous substance has been driven off by the heating process.

61. The pans will balance without changing the weights. The Law of Conservation of Mass states that mass is neither created nor destroyed in a change.

63. Examples include electrical energy being converted to mechanical energy (washing machine), light energy (lightbulb), or heat energy (oven). These changes are useful because they are advantageous to you, but they are wasteful because they are not 100% efficient and thus are an imperfect use of energy.

66. True: e, f, i, j, l. False: a, b, c, d, g, h, k.

67. Nothing.

71. Mercury, water, ice, carbon.

72. Yes; nitrogen and oxygen in air are a mixture of two elements. Compounds of nitrogen and oxygen, such as nitrogen dioxide, also exist.

73. Yes; nitrogen oxides, for example, occur as at least six different compounds. You also may have thought of carbon monoxide and carbon dioxide.

74. Rainwater is more pure. Ocean water is a solution of salt and other substances. Ocean water is distilled by evaporation and condensed into rain.

75. (a) The powder is neither an element nor a compound, both of which have a fixed composition. (b) The contents of the box are heterogeneous because, although the powder has a uniform appearance, it lacks constant composition. (c) The contents must be a mixture of varying composition.

76. The sources of usable energy now available are limited. If we change them into forms that we cannot use, we risk having an energy shortage in the future.

77. (a) Neither: The distinction between homogeneous and heterogeneous is not a particulate property. (b) The sample is a mixture because it consists of two different particle types. (c) The particles are compounds because they consist of more than one type of atom. (d) The particles are molecules because they are made up of more than one atom. (e) The sample is a gas because the particles are completely independent of one another.

78. (a) Reactants: AB, CD. Products: AD, CB. (b) A chemical change is shown. (c) The masses of product particles and the reactant particles are equal. (d) The energy in the container is the same before and after the reaction.

79. Pure substance. A mixture would have changed boiling temperature during distillation because of a change in composition of the mixture.

80. The substance is a mixture. If it was a pure substance, its density would not change.

81. If the objects have opposite charge, there is an attraction between them. An increase in separation will be an increase in potential energy. If they have the same charge (repulsion), the greater distance will be lower in potential energy.

82. Because each object contains particles with both positive and negative charges, there are both attractive and repulsive forces between the objects. If the net force is one of attraction, the particles move toward each other; if the net force is one of repulsion, the particles separate. When the two forces are balanced, the net force is zero and the particles remain separated at a constant distance.

83. (a) This is a physical change. (b) The number of particles is not conserved. (c) New particles appear from nowhere. (d) This could be the product of a chemical change.

3 Measurement and Chemical Calculations

CHAPTER CONTENTS

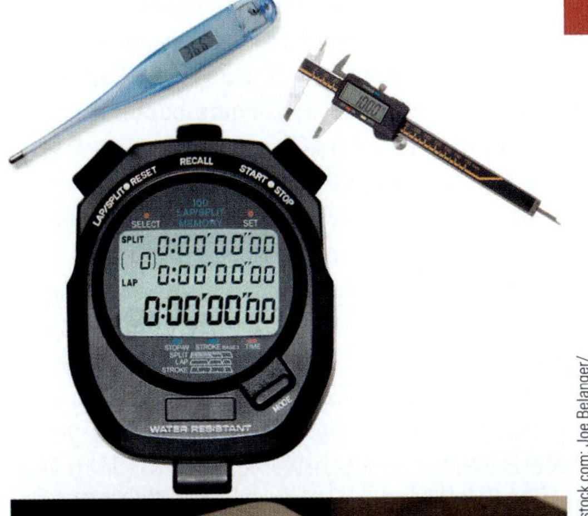

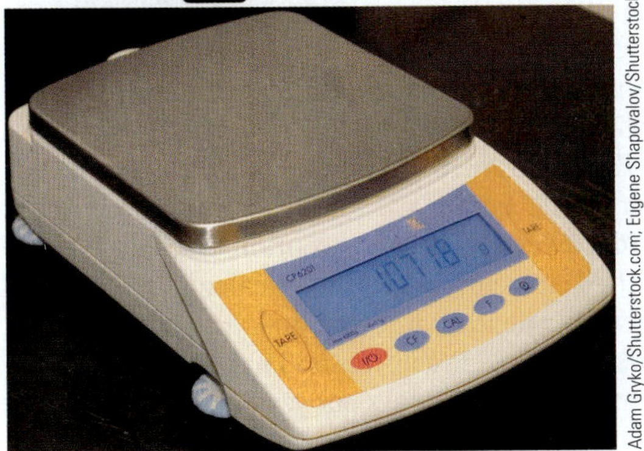

Adam Gryko/Shutterstock.com; Eugene Shapovalov/Shutterstock.com; Joe Belanger/Shutterstock.com; Ron Kloberdanz/Shutterstock.com

◀ In Chapter 1, you learned that French scientist Antoine Lavoisier was the first person to measure the weights of substances before and after chemical changes. Lavoisier is sometimes called the father of modern chemistry because of his demonstration of the usefulness of measurement in the science of chemistry. It has now become routine to accept the central role of accurate and precise measurement in scientific investigations. The mechanical balance that Lavoisier used to revolutionize the nature of scientific investigation has been improved upon by the electronic balance shown in the photograph. Also shown here are other modern instruments used to measure length, time, and temperature.

The science of chemistry is both qualitative and quantitative. In its qualitative role, it explains *how* and *why* chemical and physical changes occur. Quantitatively, it considers the amount of a substance measured, used, or produced. Determining the amount involves both measuring and performing calculations, which are the subjects of this chapter. Performing calculations requires you to develop your problem-solving skills; therefore, providing the opportunity for you to do so is also a major focus of this chapter.

3.1 How Is Time Measured?

People have long had a need to measure time. How old am I? How long can I expect to live? When will spring come so that I can safely plant this year's crops? Will my stored food last for the remainder of winter? Historically, it has been particularly important to have knowledge of when the seasons will reoccur. This type of measurement of time is done with an organizational system that we call a calendar. Today, we largely take for granted the nature of the calendar, but construction of a calendar is not as simple of a process as it may first appear.

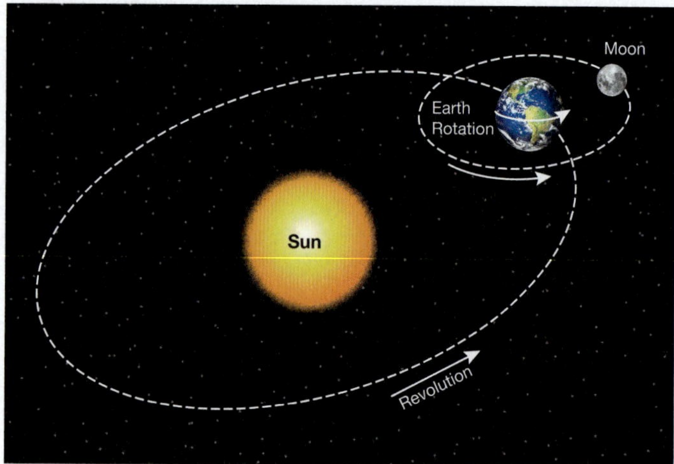

Figure 3.1 The day and the year. A day is the time period during which Earth makes one rotation around its axis. A year is the time period during which Earth makes one revolution around the sun. One year and 365.2422 days are the same quantity of time.

The primary challenge in calendar design is reconciliation of the lengths of the day and the year. **Figure 3.1** illustrates the two time spans. A day is the time needed for Earth to make one complete rotation on its axis. A year is the time needed for Earth to make one complete revolution around the sun. The ratio of these two quantities of time is 365.2422 days/year. How do you design a calendar to account for fractional number of days in a year?

The current names of the months trace back to the Roman calendar, which may have originated as far back as the 8th century BCE. It was organized into 38 eight-day weeks, totaling 304 days that spanned over 10 months of 30 or 31 days each, beginning with March (the month of Mars, the Roman god of war) and ending with December (tenth month: deca—from the Latin *decas*, ten). The remaining 61 days were not part of the calendar. Of course, with the Roman calendar and its 61 omitted days, it was impossible to accurately keep track of the quantity of time that had passed over periods of more than a few years.

In the 1st century BCE, the Roman dictator Julius Caesar ordered the reform of the calendar so that it would be consistent with the revolution of Earth about the sun. Astronomers set out to meet the demands of their leader, coming up with what is now called the Julian calendar. Their idea was to have two different types of years: standard years of 365 days that occurred for three consecutive years and a leap year every fourth year with 366 days. This resulted in an average year of 365.25 days, which is very close to the actual 365.2422 days.

The 0.0078 day (11 minutes) difference between the natural year and the Julian year, although a small error, eventually caused the seasons to creep out of alignment over time. In 1582, Pope Gregory XIII proclaimed that a new calendar be initiated. It followed the fundamental pattern of the Julian calendar except that leap years exactly divisible by 100 become standard years with the exception of century years exactly divisible by 400. Thus, 3 of the leap years in the Julian calendar were eliminated every 400 years. In the Gregorian calendar, in every 400-year period, there are 303 ($300 + 3$) standard years and 97 ($100 - 3$) leap years. That is $(303 \times 365) + (97 \times 366) = 146{,}097$ days per 400 years, or 365.2425 days per year on average. That is an error of only 0.00008%!

The Gregorian calendar continues to be in use today. The rotation of Earth does not occur in exactly the same absolute time period each year, so continued attempts to revise the calendar would be largely futile. Instead, leap seconds are added to the clock on occasion, every couple of years or so.

Later in this chapter, you will see that the history of many measurement standards follows a pattern similar to the history of the measurement of time. Measurement initially begins as based on what is thought to be a simple standard, such as the quantity of time it takes Earth to revolve around the sun, and then the standard becomes sufficiently inaccurate that society needs to change it. Measurement is important to the nonscientist and scientist alike, as you will see as you study this chapter.

3.2 Scientific Notation

Goal 1 Write in scientific notation a number given in ordinary decimal form; write in ordinary decimal form a number given in scientific notation.

2 Use a calculator to add, subtract, multiply, and divide numbers expressed in scientific notation.

Chemistry calculations and measurements sometimes involve very large or very small numbers. For example, the mass (weight) of a helium atom is

0.0000000000000000000000665 gram (a gram is 1/454 pound). In 1 liter of helium at 0°C and 1 atmosphere of pressure (that's about a quart of the gas at 32°F and the atmospheric pressure found at sea level on a sunny day), there are 26,880,000,000,000,000,000,000 helium atoms. These are two very good reasons to use **scientific notation** for very large and very small numbers (**Figures 3.2** and **3.3**). It is also a good reason to devote a section to reviewing calculation methods using these numbers.

Space Science Institute/JPL/NASA

Figure 3.2 Scientists work with large numbers. This picture was taken by the Casini spacecraft on May 7, 2004, when it was 2.82×10^7 kilometers (17.6 million miles) from Saturn.

A number may be written in **scientific notation** as follows:

$$\text{a.bcd} \times 10^{e}$$

Number equal to or greater than 1 and less than 10 — Power of 10

where a.bcd is the **coefficient** and 10^e is an **exponential**. The coefficient a.bcd may have as many digits as necessary after the decimal point, or it may have no digits after the decimal point. For example, 1, 3.4, 8.87, and 4.232990 all are acceptable as coefficients, but 0.23, 12.5, and 200 usually are not. Occasionally, the coefficient may be written in a nonstandard format, outside of the standard range of being equal to or greater than 1 and less than 10. The **exponent**, e, is a whole number (integer); it may be positive or negative.

When an exponent is positive, it indicates that the coefficient is to be multiplied by 1 multiplied by 10 e times, as in the following examples:

$$1.234 \times 10^0 = 1.234 \times 1 = 1.234$$
$$1.234 \times 10^1 = 1.234 \times 1 \times 10 = 12.34$$
$$1.234 \times 10^2 = 1.234 \times 1 \times 10 \times 10 = 123.4$$
$$1.234 \times 10^3 = 1.234 \times 1 \times 10 \times 10 \times 10 = 1234$$

When an exponent is negative, it indicates that the coefficient is to be multiplied by 1 multiplied by 1/10 e times, as in the following examples:

$$5.678 \times 10^{-1} = 5.678 \times 1 \times \frac{1}{10} = 0.5678$$
$$5.678 \times 10^{-2} = 5.678 \times 1 \times \frac{1}{10} \times \frac{1}{10} = 0.05678$$
$$5.678 \times 10^{-3} = 5.678 \times 1 \times \frac{1}{10} \times \frac{1}{10} \times \frac{1}{10} = 0.005678$$

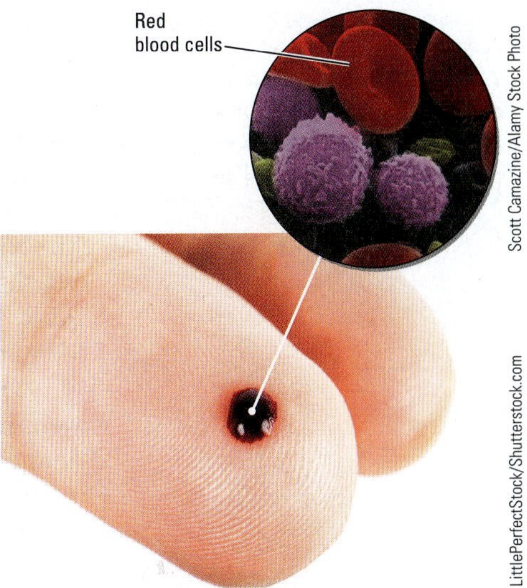

Scott Camazine/Alamy Stock Photo

LittlePerfectStock/Shutterstock.com

Figure 3.3 Scientists work with small numbers. At the microscopic level, blood is a mixture of numerous components. The diameter of a typical human red blood cell is 7×10^{-6} meter (3×10^{-4} inch).

Do you see the pattern in the relationship between the exponent and the movement of the decimal place in the coefficient? If not, go back and look at our examples again. This leads to what we call the larger/smaller approach to changing decimal numbers to scientific notation.

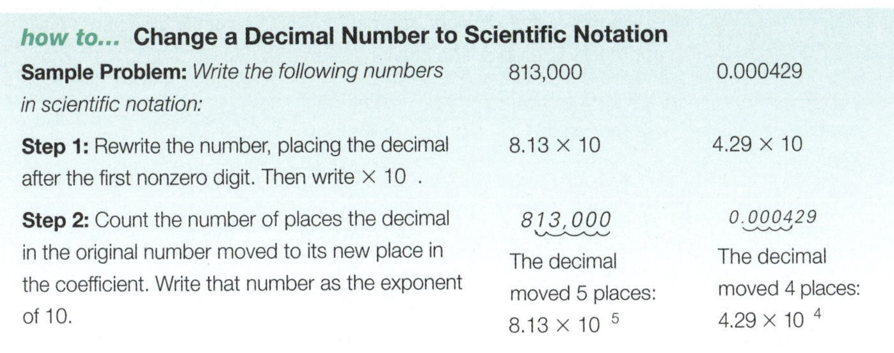

how to... **Change a Decimal Number to Scientific Notation**

Sample Problem: *Write the following numbers in scientific notation:*	813,000	0.000429
Step 1: Rewrite the number, placing the decimal after the first nonzero digit. Then write × 10 .	8.13 × 10	4.29 × 10
Step 2: Count the number of places the decimal in the original number moved to its new place in the coefficient. Write that number as the exponent of 10.	813,000 The decimal moved 5 places: 8.13×10^5	0.000429 The decimal moved 4 places: 4.29×10^4

Step 3: Compare the original number with the coefficient in Step 1.

a) If the coefficient in Step 1 is *smaller* than the original number, the exponent is *larger* than 0; it has a positive value. It is not necessary to write the + sign. — $8.13 \times 10^{+5} = 8.13 \times 10^{5}$

b) If the coefficient in Step 1 is *larger* than the original number, the exponent is *smaller* than 0; it has a negative value. Insert a minus sign in front of the exponent. — 4.29×10^{-4}

In a previous course, you may have learned that the exponent is positive if the decimal moves left and negative if it moves right. This rule is easy to learn but just as easy to reverse in one's memory. We have found that students are more successful when thinking with the larger/smaller approach.

The larger/smaller approach works no matter which way the decimal moves ◀. You can also use it for relocating the decimal of a number already in scientific notation or for changing a number in scientific notation to ordinary decimal form.

Most calculators will convert between scientific notation and ordinary decimal form. To change a number in scientific notation to decimal form, simply enter the number in scientific notation and hit the equals key. For example, to change 2.18×10^3 to decimal form, enter 2.18, then press the EE or EXP key, type 3, and then press the equals key: 2180 shows on the display. To change a number in decimal form to scientific notation, enter the number and then press the scientific notation key, typically labeled SCI (it is often at the second or third function level). For example, to convert 12,985 to scientific notation, enter 12985 and press SCI. The display shows 1.2985 ^{04}, indicating 1.2985×10^4. If our instructions do not work on your calculator, you should use an online web search engine or consult your instruction book to learn how to do this now.

Thinking About Your Thinking

Equilibrium

The larger/smaller approach has the mathematical form a × b = constant. This type of reasoning is called equilibrium, and it is used frequently in science and in other aspects of everyday life. This is one of the reasons you should use it instead of the positive/negative, left/right approach.

If a goes up, b must come down proportionally; if b goes up, a must come down by a reciprocal quantity. The product of the two quantities must always be the same; it is constant. When applying the equilibrium thinking skill to scientific notation, the constant is the value of the number itself. The product of the coefficient and the exponential must equal the number, no matter whether it is in ordinary decimal form or in scientific notation. This leads to the larger/smaller approach. Let's look at a previous example:

a	×	b	=	constant	
coefficient	×	exponential	=	constant	
0.000318	×	10^0	=	0.000318	($10^0 = 1$)
|		|		|	
larger		*smaller*		*unchanged*	This is *equilibrium.*
↓		↓		↓	
3.18	×	10^{-4}	=	0.000318	

As with all Thinking About Your Thinking skills, we will revisit equilibrium later in this book.

Read the next five paragraphs if your instructor has assigned this chapter before Chapter 2. If you have already covered Chapter 2, skip the material in this shaded box and work Active Example 3.1 now.

Most of the examples in this book are written in an active-learning, self-teaching style in which a series of questions and answers guide you to understanding a problem. To reach that understanding, you should answer each question before looking at the answer. This requires a shield to cover that answer while you consider the question. Cut-out shields for this purpose are provided in the book.

Find them now and cut one out. On one side you will find instructions on how to use the shield, copied from this section. On the other side is a periodic table that you can use for reference.

The examples provided in this textbook are designed so that you can actively work them, writing your response to each question. This does not mean that the examples are optional or are something to do after you read the text. On the contrary, they are an integral part of the textbook. To maximize your learning, work each Active Example when you come to it.

We understand the temptation to just read the textbook in the same way you might read a novel. It's certainly much easier than doing the work of writing answers to our questions, comparing your answers to ours, thinking about your thinking, and working to improve. But research on human learning clearly shows that to only *read* a textbook results in almost no real long-term learning. You have to actively work at learning. This textbook provides that opportunity. A saying often attributed to Confucius is, "I hear and I forget, I see and I remember, I do and I understand." Our textbook follows the "I do and I understand" philosophy.

This is the point in this course where you need to ask yourself, "Do I want to learn chemistry as a result of my study? Do I want to learn the thinking skills chemists use?" If the answer is yes, put your wants into action. Make the commitment now to work each example when it is presented. There is no better way to learn the content and process of chemistry than by actively answering questions and solving examples ▶. This is why we've titled the textbook *Introductory Chemistry: An Active Learning Approach.*

Active learners learn permanently and grow intellectually. Passive learners remember temporarily and then forget, without mental growth.

how to... Work an Active Example

The procedure for solving an Active Example is as follows:

Step 1: When you come to an example, locate the point in the left column at which the first blue-shaded background appears. Use the shield to cover all of the blue-shaded boxes in the left column.

Step 2: Read the problem statement. Write any answers or calculations needed in the blank space where the pencil icon is located. Note that the "Think Before You Write" instructions are different for each Active Example.

Step 3: Move the shield down to reveal the first blue-shaded box.

Step 4: Compare your answer to the one you can now read in the book. Be sure you understand the example up to that point before going on.

Step 5: Repeat the procedure until you finish the example.

Learn It Now! *We cannot overemphasize how important it is that you answer the question yourself before you look at the answer in the book. This is the most important of all the* Learn It Now! *statements.*

Active Example 3.1 Conversion from Decimal Form to Scientific Notation

Write each of the following in scientific notation: (a) 3,672,199; (b) 0.000098; (c) 0.00461; (d) 198.75.

Think Before You Write Place your shield over the left column. Use larger/smaller reasoning to make the conversions.

Answers *Cover the left column with your cut-out shield. Reveal each answer only after you have written your own answer in the right column.*

(a) 3,672,199 = 3.672199×10^6
(b) 0.000098 = 9.8×10^{-5}
(c) 0.00461 = 4.61×10^{-3}
(d) 198.75 = 1.9875×10^2

Write your answers to all parts of the question in the space provided under the pencil icon below. When you are finished, move the shield to reveal our solution, and compare your answer with ours.

(a) Stepwise thinking: (1) 3,672,199 becomes 3.672199 × 10 .
(2) The decimal moved six places, giving 3.672199 × 10 ^{6}.
(3) The coefficient is smaller than the original number. The exponential must then be larger than 1. The sign is + and therefore omitted, yielding the final answer 3.672199×10^6.

(b) Stepwise thinking: (1) 0.000098 becomes 9.8 × 10 .
(2) The decimal moved five places, giving 9.8 × 10 ^{5}.
(3) The coefficient is larger than the original number. The exponential must then be smaller than 1. The sign is −, yielding the final answer 9.8×10^{-5}.

(c) (1) 4.61 × 10 . (2) 4.61 ×10 ^{3}. (3) 4.61 is larger than 0.00461, 10^{-3} is smaller than 1.

(d) (1) 1.9875 × 10 . (2) 1.9875 × 10 ^{2}. (3) 1.9875 is smaller than 198.75, 10^2 is larger than 1.

There are no more light gray answer spaces, so this is the end of the example. Check your answers against ours. If any answer is different, find out why before proceeding.

Practice Exercise 3.1

Write each of the following in scientific notation: (a) 3,887.5; (b) 409,809,089 (c) 0.000022; (d) 0.0000005.

If you still have some doubt about whether you understand how to convert from ordinary decimal form to scientific notation, go back and study the textbook again until you are confident that you have learned this procedure. If (or when) you are satisfied with your learning, continue on with your studies.

Now that you've completed Active Example 3.1, you can see how this textbook promotes active learning. Congratulations! You've just taken the most important step—the first one—toward becoming an active learner.

To change a number written in scientific notation to ordinary decimal form, simply perform the indicated multiplication. The size of the exponent tells you how many places to move the decimal point. A positive exponent indicates a large number, so the coefficient is made larger—the decimal is moved to the right. A negative exponent says the number is small, so the coefficient is made smaller—the decimal is moved to the left. Thus the positive exponent in 7.89×10^5 says the ordinary decimal number is larger than the coefficient, so the decimal is moved five places to the right: 789,000. The negative exponent in 5.37×10^{-4} indicates that the ordinary decimal number is smaller than the coefficient, so the decimal moves four places to the left: 0.000537.

Active Example 3.2 Conversion from Scientific Notation to Decimal Form

Write each of the following numbers in ordinary decimal form: (a) 3.49×10^{-11}, (b) 3.75×10^{-1}, (c) 5.16×10^4, (d) 43.71×10^{-4}.

Think Before You Write Move the decimal point the number of places and in the direction indicated by the exponent.

Answers *Cover the left column with your cut-out shield. Reveal each answer only after you have written your own answer in the right column.*

(a) $3.49 \times 10^{-11} = 0.0000000000349$
(b) $3.75 \times 10^{-1} = 0.375$
(c) $5.16 \times 10^{4} = 51,600$
(d) $43.71 \times 10^{-4} = 0.004371$

Complete all four parts.

Practice Exercise 3.2

Write each of the following numbers in ordinary decimal form: (a) 0.011×10^{-3}; (b) 14.3×10^{-2}; (c) 0.00477×10^{5}; (d) 5.00858585×10^{6}.

3.3 Conversion Factors

Goal 3 Convert an equivalency to two conversion factors.

A Brief Algebra Refresher

Think back to when you first learned division. You probably started with a simple problem like this:

Divide 8 diamonds into groups of 2. How many groups are there?

Initially, you literally circled groups of 2 diamonds and then counted them, finding that there are 4. This approach taught you that 8 is made up of 4 groups of 2. You then translated that into an equation: $8 = 4 \times 2$. You then learned that the process you were doing is called division, and you changed the "how many groups of 2 are in 8?" question into an equation: $8 \div 2 = ?$

Since 8 is the same as 4×2, you eventually learned to solve the problem by factoring:

$$8 \div 2 = \frac{8}{2} = \frac{4 \times 2}{2} = 4 \times \frac{2}{2} = 4 \times 1 = 4$$

You then learned that since $2 \div 2 = 1$, you could solve the same problem more efficiently by canceling the factors that appear on the top and the bottom of a fraction (for multiplication only; not addition or subtraction):

$$\frac{8}{2} = \frac{4 \times \cancel{2}}{\cancel{2}} = 4$$

As you moved along in your mathematical comprehension, you learned that variables—quantities that can have any value—may be canceled in the same way as numbers. As examples,

$$\frac{6 \times \cancel{b}}{\cancel{b}} = 6 \quad \text{and} \quad \frac{a \times \cancel{b}}{3 \times \cancel{b} \times c} = \frac{a}{3 \times c} \quad \text{and} \quad \frac{6 \times a \times b}{3 \times b \times c} = \frac{\cancel{3} \times 2 \times a \times \cancel{b}}{\cancel{3} \times \cancel{b} \times c} = \frac{2 \times a}{c}$$

In summary, when you have one or a series of factors multiplied by one another, and these are divided by one or a series of factors multiplied by one another, the factors common to the numerator (top of the fraction) and denominator (bottom of the fraction) may be canceled because each pair of common divided factors is equal to 1.

Let's now extend this idea. If someone asks how long your chemistry class is, you don't answer "50," instead you say "50 minutes." When we express quantities, we must use both a value and a unit:

$$\text{Quantity} = \text{Value} \times \text{Unit}$$

If you conceptualize a **quantity** as the product of a **value** (50) and a **unit** (minutes), a powerful problem-solving tool is at your disposal. ***When you solve quantitative chemistry problems, think of a quantity as the product of a value and a unit.***

Let's see how this works. How long is your chemistry class in seconds? You know that it is 50 minutes, but you want to express this amount of time in seconds.

You know that there are 60 seconds in a minute. Before you took this course, you may have said that your class is scheduled for $50 \times 60 = 3000$ seconds. But look at what happens if we treat this conversion like an algebra problem in which each quantity is expressed as the product of a value and a unit:

$$50 \times \cancel{\text{minutes}} \times \frac{60 \times \text{seconds}}{1 \times \cancel{\text{minute}}} = 50 \times 60 \times \text{seconds} = 3000 \times \text{seconds}$$

Minutes, as units, cancel, just as numbers and variables cancel.

In algebra, you learned that you could write $3 \times b$ as just 3b. Similarly, we can remove the value $\times$ unit multiplication symbols from our minutes-to-seconds setup:

$$50\ \cancel{\text{minutes}} \times \frac{60\ \text{seconds}}{1\ \cancel{\text{minute}}} = 3000\ \text{seconds}$$

Notice how the units cancel in exactly the same way when we omit the value $\times$ unit multiplication symbols.

But wait a moment. It was easy to convert minutes to seconds simply by multiplying by 60. Why go to all of this trouble of constructing an algebraic setup with quantities treated as the products of values and units? The eventual answer will be that we will use this strategy to solve chemistry problems, where units are much less familiar to you than are common time units. We'll also be working on problems that require more than one step, and canceling the units will be a productive way to make sure that each step is making sense. Given that, let's take a quick look at how this strategy can help you become a better problem solver.

Active Example 3.3 Using "Quantity = Value × Unit" to Solve a Problem

The third oldest horse racing track of its kind in the United States is the 8-furlong Pleasanton Fairgrounds Racetrack in California. What is the length of the track in rods? 0.025 furlong is equal to 1 rod.

Think Before You Write You're likely unfamiliar with the length units furlongs and rods, which makes it challenging to quickly do the calculation in your head. Is the solution 8×0.025, $8 \div 0.025$, 0.025×8, or $0.025 \div 8$? How can you be sure?

Answers *Cover the left column with your cut-out shield. Reveal each answer only after you have written your own answer in the right column.*

$8\ \cancel{\text{furlongs}} \times \frac{1\ \text{rod}}{0.025\ \cancel{\text{furlong}}} = 320\ \text{rods}$ *The units furlongs must appear in the numerator and the denominator in order to cancel. The units of the answer must be rods, so they must be in the numerator.*	Use the cancellation of units to guide you. You need to convert from furlongs to rods, and thus furlongs need to be canceled, appearing in both the numerator and denominator, leaving rods as the surviving unit. Set up the solution so that the units work out appropriately.

Practice Exercise 3.3

An acre used to be defined as the area 1 furlong long and 4 rods wide. Area is calculated by multiplying length times width. What is the area of an acre expressed in square rods (rods × rods)?

Equivalency and Conversion Factors

The relationship "0.025 furlong is equal to 1 rod" from Active Example 3.3 is an example of an **equivalency**, two quantities that are equivalent in value. In this case, 0.025 furlong and 1 rod represent the same physical length, expressed in different units. Any equivalency can be expressed in the form of an equation or as a fraction:

$$0.025\ \text{furlong} = 1\ \text{rod} \quad \text{and} \quad \frac{0.025\ \text{furlong}}{1\ \text{rod}} \quad \text{and} \quad \frac{1\ \text{rod}}{0.025\ \text{furlong}}$$

When an equivalency is expressed as a fraction, we call it a **conversion factor**. Two conversion factors result from each equivalency. In Active Example 3.3, you used the conversion factor $\frac{1 \text{ rod}}{0.025 \text{ furlong}}$ to convert a length expressed in furlongs to a length expressed in rods.

In Active Example 3.3, the equivalency from which you derived the conversion factor needed to solve the problem was given to you. More commonly, you will encounter problems where the equivalency is not explicitly stated. It may be something you already know or a new equivalency you will need to learn. We will help you learn many new equivalencies in the course in which this textbook is being used. However, before we begin to introduce new equivalencies, let's look at an example of a problem that you can already solve that does not explicitly state the needed equivalency: How many days are there in three weeks?

What unstated equivalency is needed to solve this problem? You have probably already figured this out: 7 days = 1 week. This yields the two conversion factors:

$$\frac{7 \text{ days}}{1 \text{ week}} \quad \text{and} \quad \frac{1 \text{ week}}{7 \text{ days}}$$

The quantity given in the problem statement is 3 weeks, which is in the numerator. Therefore, we want to use the conversion factor with weeks in the denominator so that the units will cancel:

$$3 \cancel{\text{weeks}} \times \frac{7 \text{ days}}{1 \cancel{\text{week}}} = 21 \text{ days}$$

We recommend that you literally draw cancellation lines through units, just as we are doing in our examples.

What would have happened if you had selected the wrong conversion factor, the reciprocal of 7 days/week? Let's see:

$$3 \text{ weeks} \times \frac{1 \text{ week}}{7 \text{ days}} = \frac{0.42857\ldots \text{week}^2}{\text{day}}$$

It wouldn't take long to recognize that this answer is wrong—even if you saw the numerical answer on your calculator! First of all, you know that the number of days in 3 weeks can't be 0.42. . . . The number of days has to be greater than the number of weeks. Second, what is a "week^2/day"? This unit makes no sense. In the incorrect setup, the weeks don't cancel to leave only the wanted days, as they should. Any time your calculation setup yields nonsensical units, you can be sure that the answer is wrong in both *value* and *units.* If you get an answer with nonsensical units, you *know* you have made a mistake. *Always include units in your calculation setups.*

Thinking About Your Thinking

Proportional Reasoning

The mathematical requirement for an equivalency and the resulting conversion factors is that the two quantities are directly proportional to each other. What you pay for raw potatoes at the grocery store in units of money, for example, is directly proportional to the amount you buy in units of weight. Two pounds cost twice as much as 1 pound. If potatoes are priced at 25 cents per pound, 3 pounds—three times as many pounds—cost 75 cents, three times as many cents (**Figure 3.4**).

iStock.com/Timotale

Figure 3.4 Direct proportionalities. You use direct proportionalities almost every day. When in the grocery store, you use the price per pound of raw potatoes as a direct proportionality to determine the cost of a measured weight. The cost in cents and the weight in pounds of the potatoes are directly proportional: cost $\propto$ weight.

Learn It Now! *Two variables, x and y, are directly proportional if there is a nonzero constant k that relates them in the form y = kx. This relationship can also be expressed as y $\propto$ x. The symbol $\propto$ is the "is proportional to" sign.*

In general, if a variable, y, is directly proportional to another variable, x, we express this mathematically as

$$y \propto x$$

The symbol $\propto$ represents "is proportional to." This means that any change in either x or y will be accompanied by a corresponding change in the other: double x, and y is also doubled; reduce y by 1/3, and x will also be reduced by 1/3.

A proportionality can be converted into an equality by inserting a proportionality constant, m:

$$y \propto x \xrightarrow{\text{the proportionality changes to an equality}} y = m \times x$$

In the "days in 3 weeks" example, y represents days, x represents weeks, and m is the proportionality constant. Proportionality constants sometimes, but not always, express a meaningful physical relationship. This one does. It can be found by solving y = m × x for the constant, m, and inserting the units of the variables:

$$m = \frac{y}{x} = \frac{y \text{ days}}{x \text{ weeks}} = \frac{7 \text{ days}}{1 \text{ week}}$$

In this form, the proportionality constant is a conversion factor.

Active Example 3.4 Equivalency and Conversion Factors

Write the equivalency and conversion factors for each of the following: (a) 1 day is 24 hours, (b) 1 dime is 0.1 dollar, (c) grapes are \$2.99 per pound, (d) my car averages 17 miles per gallon.

Think Before You Write You are being asked to write three things for each statement: (1) the equivalency, (2) one of the two conversion factors, and (3) the reciprocal of the first conversion factor you wrote.

Answers *Cover the left column with your cut-out shield. Reveal each answer only after you have written your own answer in the right column.*

1 day = 24 hours $\qquad \dfrac{1 \text{ day}}{24 \text{ hours}} \qquad \dfrac{24 \text{ hours}}{1 \text{ day}}$

Your equivalency states that 1 day and 24 hours represent the same amount of time, expressed in different units.

Let's take this one part at a time. Remember that a quantity is a value times a unit, but you don't have to show the multiplication sign. Write all three items for part (a).

1 dime = 0.1 dollar $\qquad \dfrac{1 \text{ dime}}{0.1 \text{ dollar}} \qquad \dfrac{0.1 \text{ dollar}}{1 \text{ dime}}$

If you multiply both sides of the equivalency by 10, or if you multiply the numerator and denominator of the fractions by 10, you get the more familiar 10 dimes = 1 dollar. Either form is expressing the same relationship.

In part (b), one of the values is fractional, but that does not make a difference in how you write the equivalency and conversion factors.

\$2.99 = 1 pound $\qquad \dfrac{\$2.99}{1 \text{ pound}} \qquad \dfrac{1 \text{ pound}}{\$2.99}$

The dollar sign in U.S. currency traditionally comes before the value, but value × unit = unit × value, so you can use either \$2.99 or 2.99 \$ in a conversion factor that has dollars at the unit.

The relationship in part (c) is only valid for the period of time at the grocer during which it applied, but the same procedure applies in writing the three items.

17 miles = 1 gallon $\qquad \dfrac{17 \text{ miles}}{1 \text{ gallon}} \qquad \dfrac{1 \text{ gallon}}{17 \text{ miles}}$

As long as the conditions of the equivalence are maintained (either implied or explicitly stated), the conversion factors remain valid. For this car under these conditions, if it has traveled 17 miles, it has used 1 gallon of gas, and if it has used 1 gallon of gas, it has traveled 17 miles.

The relationship in part (d) is an average for a specific automobile over a specific time period, but again, you can still write an equivalency and conversion factors.

Practice Exercise 3.4

Write the equivalency and conversion factors for each of the following: (a) each tablet of vitamin C is 500 milligrams of calcium ascorbate, (b) there is a total volume of 288 fluid ounces in the case of soda, (c) 10 nickels are the same amount of money as 2 quarters, (d) the recipe calls for 2¾ cups of flour to make 24 cupcakes.

3.4 A Strategy for Solving Quantitative Chemistry Problems

Goal 4 Learn and apply the algorithm for using conversion factors to solve quantitative problems.

A major goal of the course in which this book is being used is to help you as you work to improve your skills at solving quantitative chemistry problems. To achieve this goal, two overarching efforts must interact with one another. One effort is the responsibility of the teaching team: your course instructor or instructors and the authors of this textbook. We have researched what is known about chemistry teaching and learning, and we have designed a curriculum that will maximize the potential for you to learn. We will also coach you as you work through the process of learning chemistry.

The other required effort is yours. You must be willing to keep trying even when the concepts you are working to learn are challenging. You must be willing to go back and review when feedback from the course indicates that you have not yet fully understood a concept, and, most importantly, you must commit the time needed to learn chemistry. As the old joke goes, "How do you get to Carnegie Hall?....Practice, practice, practice"—there is no better alternative than to practice solving problems while working to improve your problem-solving skills.

This book and the course in which it is being used are set up to provide you with a big-picture, good-quality curriculum and continual coaching. This section in particular is a mini version of the course and textbook. We will first provide you with a procedure to use to solve quantitative chemistry problems, and then we will have you practice while being assisted with our coaching. Let's dive in!

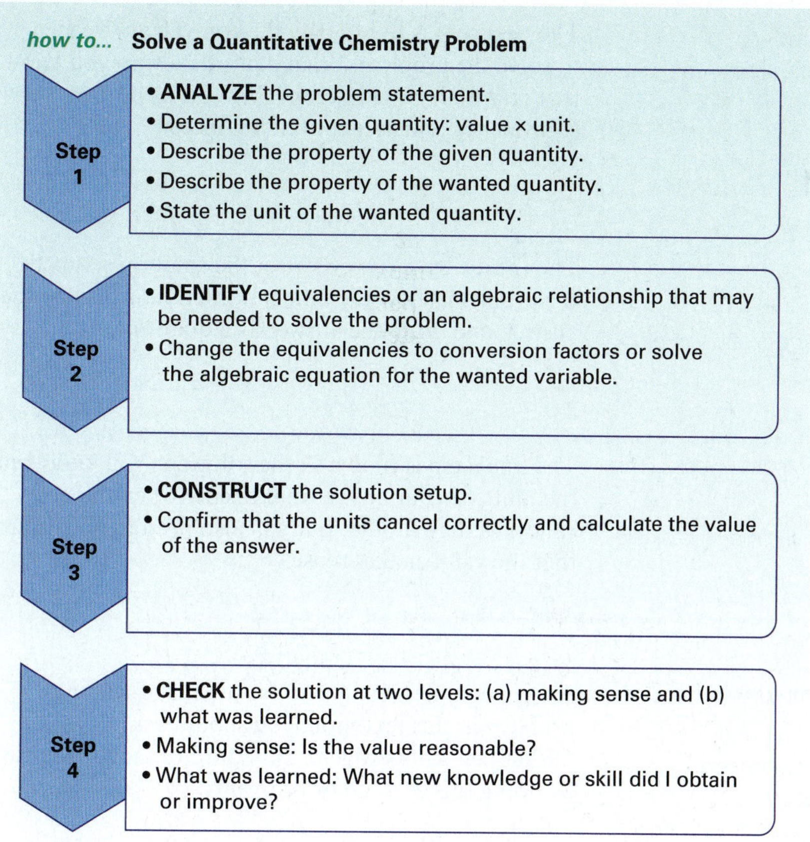

how to... Solve a Quantitative Chemistry Problem
Step 1
• ANALYZE the problem statement.
• Determine the given quantity: value × unit.
• Describe the property of the given quantity.
• Describe the property of the wanted quantity.
• State the unit of the wanted quantity.
Step 2
• IDENTIFY equivalencies or an algebraic relationship that may be needed to solve the problem.
• Change the equivalencies to conversion factors or solve the algebraic equation for the wanted variable.
Step 3
• CONSTRUCT the solution setup.
• Confirm that the units cancel correctly and calculate the value of the answer.
Step 4
• CHECK the solution at two levels: (a) making sense and (b) what was learned.
• Making sense: Is the value reasonable?
• What was learned: What new knowledge or skill did I obtain or improve?

Don't try to memorize this *how to*.... You will learn it by applying it while practicing problem solving. Reading through it once is all that is necessary for now. Let's start with a simple problem just to illustrate the methodology.

Active Example 3.5 An Introduction to Solving Quantitative Problems

How many weeks are in 35 days?

Think Before You Write We understand that you can already solve this problem, but work through it with the purpose of learning to apply the quantitative problem-solving procedure. We will guide you through a number of small steps.

Answers *Cover the left column with your cut-out shield. Reveal each answer only after you have written your own answer in the right column.*

Answers	
35 days *Always give both the value and the unit when you state a quantity.*	The first step is to **analyze** the problem statement. Recall: *quantity* = *value* × *unit.* The given quantity is typically something that has been measured. Write the given quantity.
Time units	Continue your analysis by describing the property of the given quantity. When we say "property," we mean the general attribute of the quantity. For example, 12 inches and 1 foot have the property of being U.S. length units.
Time units, weeks	State the property of the wanted quantity, and state the units in which it is to be expressed. 
7 days = 1 week $\frac{1 \text{ week}}{7 \text{ days}}$ *We wrote the conversion factor with days in the denominator because we want to cancel the unit days in the given quantity.*	The next step is to **identify** the equivalency needed to solve the problem. What equivalency do you know that relates days with weeks? Write it as both as an equivalency and as a conversion factor. 
$35 \text{ days} \times \frac{1 \text{ week}}{7 \text{ days}} = 5 \text{ weeks}$ *The unit cancellation verifies the correctness of your setup.*	The third step is to **construct** the solution setup. Set up the problem, show the unit cancellation, calculate the answer, and write the answer as a quantity. 
35 days is a larger value with a smaller unit than 5 weeks. The answer quantity makes sense. *We will discuss this method for checking in more detail immediately following this Active Example.*	The final step is to **check** the solution. You know that the units of your answer make sense when the units cancel as they should, but you also need to make sure that the value makes sense. 
With this exercise, you began to work to improve your skill at solving quantitative problems. *Improvement of a skill requires deliberate practice: repeated practice with that skill with a goal of achieving performance slightly beyond your current level of proficiency.*	As you complete each exercise or end-of-chapter problem in this textbook, you should ask yourself, "What new knowledge or skill did I obtain or improve by doing this exercise or problem?" 

Practice Exercise 3.5

How many seconds are in 12 minutes?

The important final step in solving any problem is to be sure the answer makes sense. Let's use Active Example 3.5 to establish a method of checking conversions between units. The two equal quantities, 5 weeks and 35 days, are made up of a smaller value with a proportionally larger unit and a larger value with a smaller unit:

$$5 \text{ weeks} = 35 \text{ days}$$

smaller value × larger unit = larger value × smaller unit

This is the same larger/smaller reasoning we used when moving a decimal point in working with scientific notation in the last section.

Thinking About Your Thinking

Equilibrium

The equilibrium thinking skill has the form a × b = constant. No matter whether we call it 5 weeks, 35 days, 840 hours, 50,400 minutes, or 3,024,000 seconds, we are describing the exact same amount of time—a constant quantity. Therefore, if we use a relatively large unit, such as the week, it must be associated with a relatively small value of those larger units. If we express the same amount of time in a smaller unit (days in this case), we must have a larger value of those smaller units to describe the same number of "ticks of the clock." A larger *value* of time units compensates for a smaller *unit* in which time is expressed and vice versa.

Active Example 3.6 Multiple-Step Conversions

Calculate the number of weeks in 672 hours.

Think Before You Write When you read the problem statement, you should think about the relationship between weeks and hours, and (for most people) you will recognize that you don't know an equivalency. When this occurs, think about the equivalencies that you do know that can relate the units in multiple steps.

Answers *Cover the left column with your cut-out shield. Reveal each answer only after you have written your own answer in the right column.*

Given: 672 hours, time units

Wanted: Time units, weeks

Analyze the problem. Write the given quantity and its property, and then write the wanted property and its unit.

24 hours = 1 day $\quad \dfrac{1 \text{ day}}{24 \text{ hours}}$

7 days = 1 week $\quad \dfrac{1 \text{ week}}{7 \text{ days}}$

If you had to take a couple of wrong turns before coming up with these equivalencies and conversion factors, don't be concerned. It's a normal part of the problem-solving process. The key is to keep working and keep trying.

Identify the equivalencies needed to solve the problem. If you don't immediately see the equivalencies you need, brainstorm. You can work from hours to larger units and/or from weeks to smaller units. Write as many equivalencies as you need until you find a combination that works. Then change them into the conversion factors that you will use.

$$672 \cancel{\text{hours}} \times \frac{1 \cancel{\text{day}}}{24 \cancel{\text{hours}}} \times \frac{1 \text{ week}}{7 \cancel{\text{days}}} = 4 \text{ weeks}$$

Construct the solution. Set up the problem, show the unit cancellations, calculate the answer, and write the answer as a quantity.

Larger value × Smaller unit (672 hours) = Smaller value × Larger unit (4 weeks). The value of the answer is reasonable.

You improved your skill at solving quantitative problems with more than one conversion factor.

What you learn from each Active Learning Exercise won't always match what we targeted, but keep in mind that the point of reflection is to be conscious of your learning and intellectual growth as you progress in this course.

Check. Use larger/smaller reasoning to decide if the value of the answer is reasonable. Reflect on what you learned by working this exercise.

Practice Exercise 3.6

Determine the number of seconds in 15 hours.

Some conversion factors are established by definition, such as 1 week ≡ 7 days, or 1 minute ≡ 60 seconds. In these examples, **the symbol** ≡ may be read as "is defined as" or "is identical to": "1 week is defined as 7 days" or "1 minute is identical to 60 seconds." The relationship never changes. The units of measurement are for the same thing—time, in this case.

Other conversion relationships are temporary. They may have one value in one problem and another value in another problem. Speed, measured in kilometers per hour, is an example (**Figure 3.5**). If we drive a car at an average of 80 kilometers per hour, we can calculate the distance traveled in 2, 3, or any number of hours. As long as the speed remains the same, the number of kilometers driven is proportional to the number of hours, and 1 hour of driving is equivalent to 80 kilometers. Thus, the equivalence between kilometers and hours is 80 kilometers = 1 hour *at 80 kilometers per hour.* But if you slow to an average of 50 km/hr, then 50 kilometers = 1 hour.

Better Stock/Shutterstock.com

Figure 3.5 Temporary relationships can be expressed as direct proportionalities. The average speed of a car is a direct proportionality between distance and time.

Active Example 3.7 Temporary Relationships as Conversion Factors

What distance, expressed in miles, will an automobile travel in 3 hours at an average speed of 62 miles per hour?

Think Before You Write The average speed in miles per hour is an equivalency that states the relationship between the quantities 62 miles and 1 hour.

Answers *Cover the left column with your cut-out shield. Reveal each answer only after you have written your own answer in the right column.*

Given: 3 hours, time units **Wanted:** distance (length) units, miles	**Analyze** the problem statement. Write the given quantity and its property, and then write the wanted property and its unit.
$\frac{62 \text{ miles}}{1 \text{ hour}}$	**Identify** the equivalency. In this example, the equivalency is given in the problem statement. Write the conversion factor.
$3 \text{ hours} \times \frac{62 \text{ miles}}{1 \text{ hour}} = 186 \text{ miles}$	**Construct** the solution. Set up the problem, show the unit cancellations, calculate the answer, and write the answer as a quantity.
We expect a car to travel a greater number of miles than the number of hours that it has been traveling, so the answer makes sense. You learned how temporary relationships can be used as conversion factors in problem solving.	**Check.** Does the answer make sense? Reflect on what you learned by working this exercise.

Practice Exercise 3.7

Japanese bullet trains travel 345 miles between Tokyo and Osaka at an average speed of 150 miles per hour. If you leave Tokyo at 9:00 AM, at what time will you arrive in Osaka?

The decimal-based monetary system in the United States yields calculations that are just like calculations with metric units, which are introduced in Section 3.6.

Active Example 3.8 Decimal System of Units I

If you went into a bank and obtained 325 dollars in dimes, how many dimes would you receive?

Think Before You Write The equivalency needed to solve this problem is not given in the problem statement. Instead, it is something that you already know.

Answers *Cover the left column with your cut-out shield. Reveal each answer only after you have written your own answer in the right column.*

Given: 325 dollars, U.S. money units **Wanted:** U.S. money units, dimes	**Analyze** the problem statement. Write the given quantity and its property, and then write the wanted property and its unit.
10 dimes = 1 dollar $\frac{10 \text{ dimes}}{1 \text{ dollar}}$	**Identify** and write the equivalency needed to convert between the given and wanted units, and then write the conversion factor.

$325 \text{ dollars} \times \frac{10 \text{ dimes}}{\text{dollar}} = 3250 \text{ dimes}$

Multiplication by 10 moves the decimal place one position to the right to make the value larger. 325.0 becomes 3250.

Construct the solution. Set up the problem, show the unit cancellations, calculate the answer, and write the answer as a quantity. If you can calculate the value of the answer without a calculator, do so.

Smaller value × Larger unit (325 dollars) = Larger value × Smaller unit (3250 dimes). The value of the answer is reasonable.

You learned that all you had to do to find the answer was move the decimal point, and the setup shows you which way to move the decimal point.

With a decimal system of units, such as the U.S. money system, conversion among units is relatively simple because of the ease of moving a decimal point.

Check the solution at two levels. Use larger/smaller reasoning to decide if the value of the answer is reasonable. Reflect on what you learned by working this exercise.

Practice Exercise 3.8

How many pennies would you receive if you cashed in 135 dimes?

Active Example 3.9 Decimal System of Units II

A little girl broke open her piggy bank. Inside she found 2608 pennies. How many dollars did she have?

Think Before You Write This is the last problem in this section that emphasizes learning about problem solving. Even though you probably know the answer, work through all of the steps in our problem-solving procedure. There is a reason; you will see why in the next section.

Answers *Cover the left column with your cut-out shield. Reveal each answer only after you have written your own answer in the right column.*

Given: 2608 pennies, U.S. money units
Wanted: U.S. money units, dollars

100 pennies = 1 dollar

$\frac{1 \text{ dollar}}{100 \text{ pennies}}$

Analyze the problem statement. Write the given quantity and its property, and then write the wanted property and its unit.

Again notice that the equivalency is not given in the problem statement. **Identify** and write the equivalency needed to convert between the given and wanted units, and then write the conversion factor.

$2608 \text{ pennies} \times \frac{1 \text{ dollar}}{100 \text{ pennies}} = 26.08 \text{ dollars}$

Larger value × Smaller unit (2608 pennies) = Smaller value × Larger unit (26.08 dollars). OK.

A decimal-based system makes answer values simple to calculate.

Construct the solution. Set up the problem, show the unit cancellations, calculate the answer, and write the answer as a quantity. Calculate the value of the answer without using a calculator.

Check. Use larger/smaller reasoning to decide if the value of the answer is reasonable. Reflect on what you learned by working this exercise.

Practice Exercise 3.9

How many five-dollar bills should you receive in exchange for 800 nickels?

3.5 Introduction to Measurement

Goal 5 Explain why the metric system of measurement is used in the sciences.

Andrey_Popov/Shutterstock.com

Figure 3.6 The measurement of time is typically the first measured quantity you encounter in a day. Measured quantities then continue to be an important part of your daily routine.

Take a moment to think about how measurement is a recurrent part of your life. When your alarm went off this morning, it was measuring time (**Figure 3.6**). You may have then adjusted your thermostat to a comfortable awakening temperature, which is another measurement. The orange juice that you may have had with your breakfast may have come in a one-quart container, a measurement of volume. If you used that orange juice to wash down a multivitamin, the pill had measured amounts of each vitamin by weight. There are many more examples. You might consider spending the rest of the day making mental notes of how central measurement is to your daily activities.

Although measurement is already familiar to you, what may be partially new in this introductory chemistry course, if you live in the United States, are the units in which scientific measurements are expressed. Scientific measurements—and most everyday measurements in the rest of the world—are made in the **metric system**. Scientists prefer the metric system because it is internationally standardized. All scientists everywhere in the world agree on the definitions of metric units. Additionally, scientists adopted the metric system for their standard of measurement because it is decimal based. Units differ by powers of 10. There are 10 millimeters in a centimeter, 10 centimeters in a decimeter, and 10 decimeters in a meter. Compare that to the awkward 12 inches per foot, 3 feet per yard, 1760 yards per mile, and so on, of the U.S. Customary System.

Modern scientists often use **SI units**, which are a subset of units in the metric system. SI is an abbreviation for the French name for the International System of Units ▶. The SI system describes seven **base units** that are derived from values of constants. This chapter describes three of these—units of mass, length, and temperature. Other quantities are made up of combinations of base units; these are called **derived units**. Two of these, units for volume and density, appear in this chapter.

A summary of the SI system appears in Appendix II.

3.6 Metric Units

Goal 6 State and write with appropriate metric prefixes the relationship between any metric unit and its corresponding kilounit, centiunit, and milliunit.

7 Using **Table 3.1**, state and write with appropriate metric prefixes the relationship between any metric unit and other larger and smaller metric units.

8 Distinguish between mass and weight.

9 Identify the metric units of mass, length, and volume.

Table 3.1 Metric Prefixes*

LARGE UNITS			SMALL UNITS		
Metric Prefix	Metric Symbol	Multiple	Metric Prefix	Metric Symbol	Multiple
tera-	T	10^{12}	Unit (gram, meter, liter)		$1 = 10^0$
giga-	G	10^9	deci-	d	$0.1 = 10^{-1}$
mega-	M	$1,000,000 = 10^6$	**centi-**	**c**	$\mathbf{0.01 = 10^{-2}}$
kilo-	**k**	$\mathbf{1,000 = 10^3}$	**milli-**	**m**	$\mathbf{0.001 = 10^{-3}}$
hecto-	h	$100 = 10^2$	micro-	μ	$0.000001 = 10^{-6}$
deca-	da	$10 = 10^1$	nano-	n	10^{-9}
Unit (gram, meter, liter)		$1 = 10^0$	pico-	p	10^{-12}

*The most important prefixes are printed in **boldface**. ▶

Your instructor will probably tell you which prefixes from this table to memorize. If not, memorize the three prefixes in **boldface**. You need to recall and use the memorized prefixes, and you need to be able to use the other prefixes, given their value.

Metric Prefixes

In the metric system, units that are larger than the basic unit are larger by multiples of 10—that is, 10 times larger, 100 times larger, 1000 times larger, and so on. Similarly, smaller units are 1/10 as large, 1/100 as large, and so forth. This is what makes the metric system so easy to work with. To convert from one unit to another, all you have to do is move the decimal point.

Larger and smaller metric units are identified by metric prefixes. The prefix for the unit 1000 times larger than the basic unit is *kilo-,* and its symbol is k. Let's apply this using the symbol $ for U.S. dollars. When the *kilo-* symbol, k, is combined with the symbol for dollars, $, you have the symbol for *kilo*dollar, k$. One kilodollar is a thousand dollars, so if we applied metric prefixes to the money system, a thousand-dollar bill would be called a kilodollar. Similarly, *centi-,* symbol c, is the prefix for the unit that is 1/100 as large as the basic unit. Thus, 1/100 of a dollar is 1 centidollar, c$. One centidollar is one cent, a penny.

It is *essential* that you use upper- and lowercase metric prefixes and symbols appropriately. For example, the symbol for the metric prefix *mega-*, 1,000,000 times larger than the basic unit, is M (uppercase), and the symbol for *milli-*, 1000 times smaller than the basic unit, is m (lowercase).

Table 3.1 lists many metric prefixes and their symbols. Entries for the kilo-, centi-, and milli- units are shown in boldface. Memorize these; you should be able to apply them to any metric unit. We will have fewer occasions to use the prefixes and symbols for other units, but by referring to the table you should be able to work with them, too.

Mass and Weight

Consider a tool carried to the International Space Station (ISS) by scientists. Suppose that tool weighs 6 ounces on the earth. Released in the near-weightless environment of the ISS, it will remain there, floating, until moved by someone to some other location. Yet in both locations it would be the same tool, having a constant quantity of matter.

Mass is a measure of quantity of matter. **Weight** is a measure of the force of gravitational attraction. Weight is proportional to mass, but the ratio between them depends on where in the universe you happen to be. This proportionality is essentially constant over the surface of Earth. Therefore, when you weigh something—that is, measure the force of gravity on the object—you can express this weight in terms of mass. In effect, weighing an object is one way of measuring its mass. In the laboratory, mass is measured on a balance (Figure 3.7).

The SI unit of mass is the **kilogram (kg)**. We will first state its technical definition, and then we'll "translate," so keep reading after the next sentence! The kilogram is defined by taking the fixed numerical value of the Planck constant (h) to be $6.62607015 \times 10^{-34}$ when expressed in the unit $J \cdot s$, which is equal to $kg \cdot m^2 \cdot s^{-1}$, where the meter and the second are defined in terms of c and Δv_{Cs}. This means

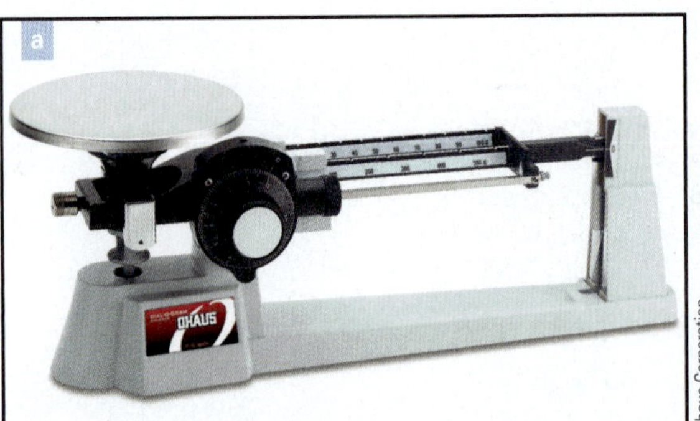

Ohaus Corporation

Ohaus Corporation

Ohaus Corporation

Figure 3.7 Three examples of laboratory balances. (a) A triple-beam balance measures mass with an error of ±0.01 g. It is usually used when high accuracy is not required, such as in nonchemistry applications. (b) This top-loading balance measures mass with an error of ±0.001 g. It has sufficient accuracy for most introductory chemistry applications. (c) An analytical balance measures mass with an error of ±0.0001 g. It is usually used in more advanced courses and in scientific laboratories.

that a kilogram is defined in terms of three fundamental constants: (1) the Planck constant (h), which describes the amount of energy in the energy packet that atoms use to transfer energy; (2) the speed of light (c); and (3) the natural microwave radiation of an atom of the element cesium (Δv_{Cs}). You will learn more about these constants in more advanced courses.

A kilogram weighs about 2.2 pounds. The basic unit for mass is the **gram (g)**. There are 1000 grams in a kilogram.

Figure 3.8 Length measurements: inches, centimeters, and millimeters. This illustration is very close to full scale. The numbers and the long lines on the top scale are inches, and the numbers and the long lines on the bottom scale are centimeters. One inch is defined as 2.54 centimeters, which is the same as 25.4 millimeters (shorter unnumbered lines).

Length

The SI unit of length is the **meter***; its abbreviation is **m**. The meter is defined in terms of two constants: (1) the same natural microwave radiation of an atom of the element cesium used in the definition of the kilogram and (2) defining the speed of light as 299,792,458 meters per second.

The meter is 39.37 inches long—about 3 inches longer than a yard. The common longer length unit, the kilometer (km) (1000 meters), is about 0.6 mile. Both the centimeter and the millimeter are used for small distances. A centimeter (cm) is about the width of a fingernail; a millimeter (mm) is roughly the thickness of a dime. **Figure 3.8** compares small metric and U.S. length units. A very tiny unit of length is illustrated in **Figure 3.9**.

Volume

The SI volume unit is the cubic meter, m^3. This is a derived unit because it consists of three base units, all meters, multiplied by each other. A cubic meter is too large a volume—larger than a cube whose sides are 3 feet long—to use for typical chemistry laboratory measurements. A more practical unit is the **cubic centimeter (cm^3)**. It is the volume of a cube with an edge of 1 cm (**Figure 3.10**). A teaspoon holds about 5 cm^3.

Liquids and gases are not easy to weigh, so we usually measure them in terms of the volumes they occupy. The basic unit for expressing their volumes is the **liter (L)**, which is defined as exactly 1000 cubic centimeters. Thus there are 1000 cm^3/L. This

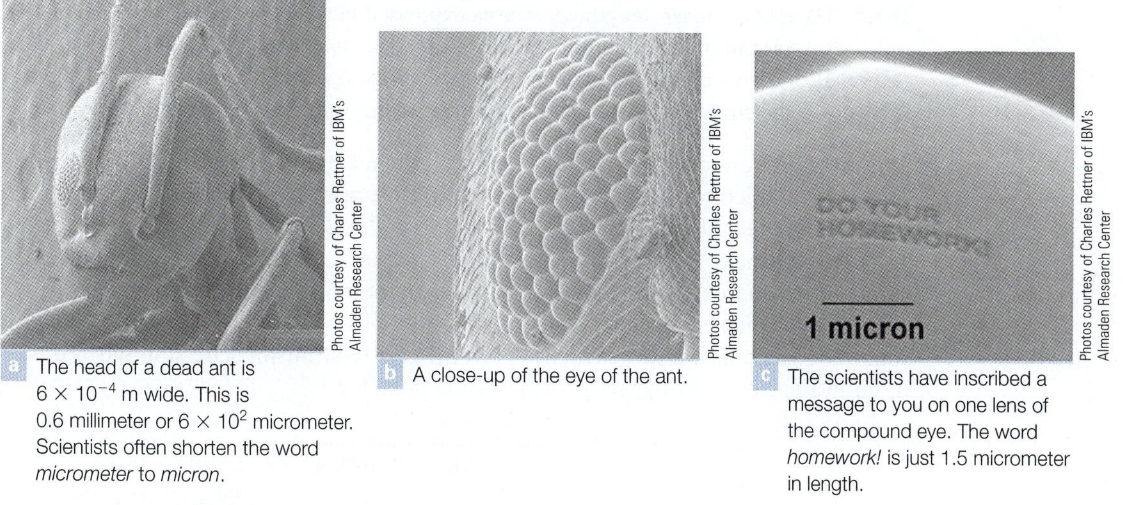

a The head of a dead ant is 6×10^{-4} m wide. This is 0.6 millimeter or 6×10^{2} micrometer. Scientists often shorten the word *micrometer* to *micron*.

b A close-up of the eye of the ant.

c The scientists have inscribed a message to you on one lens of the compound eye. The word *homework!* is just 1.5 micrometer in length.

Figure 3.9 The micron. Scientists at the IBM Almaden Research Center in California used a scanning electron microscope to produce these images.

*Outside the United States, the length unit is spelled *metre*, and the liter, the volume unit that we will discuss shortly, is spelled *litre*. These spellings and their corresponding pronunciations come from France, where the metric system originated. In this book, we use the U.S. spellings, which match their pronunciations in the English language.

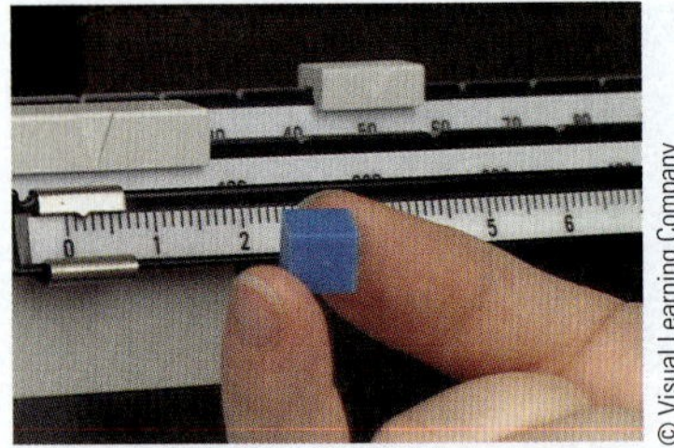

Figure 3.10 One cubic centimeter. One milliliter and one cubic centimeter are the same volume, 1 mL ≡ 1 cm^3. The milliliter and the cubic centimeter are the most common volume units you will use in introductory chemistry.

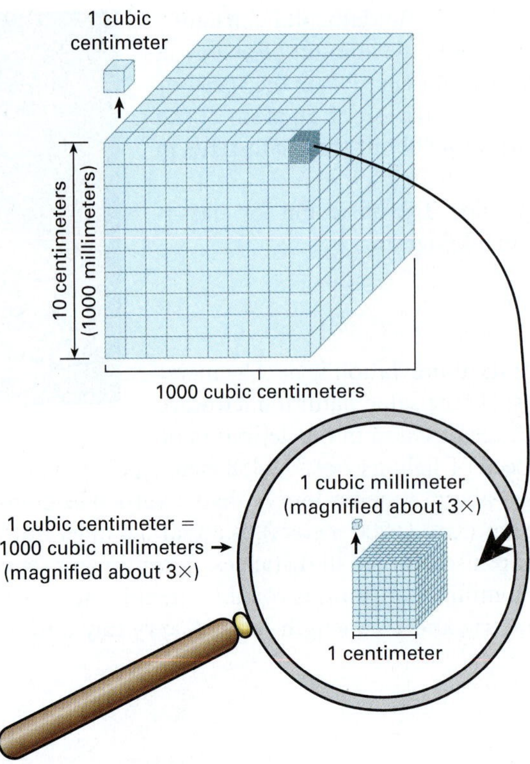

Figure 3.11 The relationship among 1000 cubic centimeters (a liter), the cubic centimeter (a milliliter), and the cubic millimeter. 1 L = 1000 cm^3 = 1000 mL, and 1000 mm^3 = 1 cm^3.

volume is equal to 1.06 U.S. quarts. Smaller volumes are expressed using the **milliliter (mL)**. Notice that there are 1000 mL in 1 liter (there are always 1000 milliunits in a unit), and 1 liter is 1000 cm^3. This makes 1 mL and 1 cm^3 exactly the same volume (**Figure 3.11**):

$$1 \text{ mL} = 0.001 \text{ L} = 1 \text{ cm}^3$$

Learn It Now! *This simple relationship is often missed. There is 1 mL in 1 cm³, not 1000.*

Unit Conversions within the Metric System

Goal 10 Given a mass, length, or volume expressed in basic metric units, kilounits, centiunits, or milliunits, express that quantity in other three units.

11 Given a mass, length, or volume expressed in any metric unit and Table 3.1 or the equivalent, express that quantity in any other metric unit.

Conversions from one metric unit to another are accomplished by recalling or looking up the needed equivalency, changing it to the appropriate conversion factor, and performing the multiplication. Goal 10 says you should memorize the values needed to be able to make these conversions among the unit, kilounit, centiunit, and milliunit. In this context, unit (u) may be gram (g), meter (m), or liter (L). Goal 11 says that for any other metric prefix, given the value of the prefix, you should be able to make conversions to another metric prefix. These relationships are summarized here as equivalencies and their resulting conversion factors:

Equivalency	Conversion factor	Conversion factor
1000 units (u) = 1 kilounit (ku)	$\frac{1000\text{ u}}{1\text{ ku}}$	$\frac{1\text{ ku}}{1000\text{ u}}$
100 centiunits (cu) = 1 unit (u)	$\frac{100\text{ cu}}{1\text{ u}}$	$\frac{1\text{ u}}{100\text{ cu}}$
1000 milliunits (mu) = 1 unit (u)	$\frac{1000\text{ mu}}{1\text{ u}}$	$\frac{1\text{ u}}{1000\text{ mu}}$

Your instructor may add other units to those you are required to know from memory. If you can look at Table 3.1, you should be able to convert from any unit in the table to any other unit.

Thinking About Your Thinking

Proportional Reasoning

Conversions between metric units are an example of proportional reasoning. Let's examine the 1000 units = 1 kilounit equivalency more closely. In any measured quantity, the number of units is directly proportional to the number of kilounits: (# of units) $\propto$ (# of kilounits) (**Figure 3.12**).

Changing to the form of an equality, (# of units) = m × (# of kilounits). The graph shows that the quantities have a straight-line relationship that passes through the origin and that the slope, m, is 1000 units/kilounit, so (# of units) = 1000 units/kilounit × (# of kilounits).

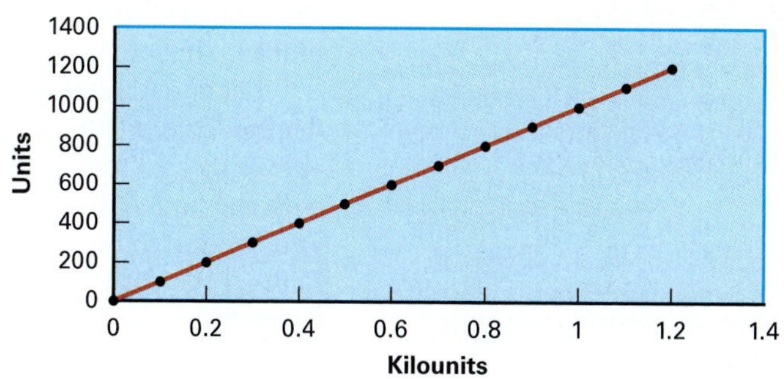

Figure 3.12 A plot of units versus kilounits.

Active Example 3.10 Unit Conversions within the Metric System: Mass

How many grams are in 2608 centigrams?

Think Before You Write You have memorized the relationships among kilounits, centiunits, milliunits, and basic units. This problem requires you to use the appropriate equivalency from among these.

Answers *Cover the left column with your cut-out shield. Reveal each answer only after you have written your own answer in the right column.*

Given: 2608 centigrams (cg), metric mass unit
Wanted: metric mass unit, grams (g)

100 cg = 1 g

$$\frac{1\text{ g}}{100\text{ cg}}$$

Analyze the problem. Write the given quantity and its property, and then write the wanted property and its unit.

Identify and write the equivalency needed to convert between the given and wanted units, and then write the conversion factor.

$$2608\ \cancel{\text{cg}} \times \frac{1\text{ g}}{100\ \cancel{\text{cg}}} = 26.08\text{ g}$$

Larger value × Smaller unit (2608 cg) = Smaller value × Larger unit (26.08 g). OK.

You have improved your skill at performing unit conversions in the metric system.

Construct the solution. Set up the problem, show the unit cancellations, calculate the answer, and write the answer as a quantity. Calculate the value of the answer without using a calculator.

Check. Use larger/smaller reasoning to decide if the value of the answer is reasonable. Reflect on what you learned by working this exercise.

Practice Exercise 3.10

How many grams are in 0.711 kilogram?

A critical difference between novice problem solvers and expert problem solvers is that the experts are constantly checking and verifying the reasonableness of their work. Novices tend to answer as quickly as possible without conducting checks. Continually develop your level of expertise by learning to check your results!

When did you first *know* the answer to the preceding example? Was it as soon as you read the question? Look back to Active Example 3.9. You probably knew the answer to that question just as quickly. In essence, that problem was, "How many dollars are in 2608 pennies?" Aside from the units, Active Examples 3.9 and 3.10 are exactly the same problem. In both examples, all you had to do was move the decimal point.

The most common error made in metric–metric conversions is moving the decimal the wrong way. The best protection against that mistake is to avoid shortcuts and set up the problem, including all units. Always check your result with the larger/smaller rule: If your quantities have a large value and small units and a small value and large units, and if you've moved the decimal the correct number of places, which can be verified on your calculator, the answer should be correct ◀.

Active Example 3.11 Unit Conversions within the Metric System: Length

How many millimeters are in 3.04 centimeters?

Think Before You Write If you are sufficiently familiar with the metric system, you can solve this problem in one step. However, at this point we recommend that you convert from the given unit to the basic unit and then from the basic unit to the wanted unit.

Answers *Cover the left column with your cut-out shield. Reveal each answer only after you have written your own answer in the right column.*

Given: centimeters (cm), metric length unit
Wanted: metric length unit, millimeters (mm)
Given to basic: 100 cm = 1 m
Basic to Wanted: 1000 mm = 1 m

$$\frac{1\text{ m}}{100\text{ cm}}$$

$$\frac{1000\text{ mm}}{1\text{ m}}$$

Analyze the problem. Write the given quantity and its property, and then write the wanted property and its unit.

Identify and write the equivalency needed to convert between the given and basic unit, and then write the equivalency needed to convert between the base unit and the wanted unit. Write both conversion factors.

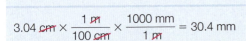

$$3.04\text{ }\cancel{\text{cm}} \times \frac{1\text{ }\cancel{\text{m}}}{100\text{ }\cancel{\text{cm}}} \times \frac{1000\text{ mm}}{1\text{ }\cancel{\text{m}}} = 30.4\text{ mm}$$

Smaller value × Larger Unit (3.04 cm) = Larger value × Smaller unit (30.4 mm). OK.

You have improved your skill at performing unit conversions in the metric system.

Construct the solution. Set up the problem, show the unit cancellations, calculate the answer, and write the answer as a quantity. Calculate the value of the answer without using a calculator.

Check. Use larger/smaller reasoning to decide if the value of the answer is reasonable. Reflect on what you learned by working this exercise.

Practice Exercise 3.11

Convert 5.25×10^5 milligrams to kilograms.

Active Example 3.12 Unit Conversions within the Metric System: Volume

A fruit drink is sold in bottles that contain 1892 milliliters. Express the volume in cubic centimeters and in liters. (See Figure 3.13 for equipment that is used to measure liquid volumes in a chemistry laboratory.)

Think Before You Write Go back and review the volume subsection, if necessary, if you need a reminder about how cubic centimeters are related to other metric volume units.

Answers *Cover the left column with your cut-out shield. Reveal each answer only after you have written your own answer in the right column.*

1892 mL = 1892 cm^3

1 mL and 1 cm^3 are the same volume.

Analyze the problem statement. There are two problems here. One is to convert the given volume to cubic centimeters, and the other is to convert the given volume to liters. Complete the milliliters-to-cubic-centimeters conversion first.

Given: milliliters (mL), metric volume unit

Wanted: metric volume unit, liters

1000 mL = 1 L

$$\frac{1\text{ L}}{1000\text{ mL}}$$

$$1892\text{ mL} \times \frac{1\text{ L}}{1000\text{ mL}} = 1.892\text{ L}$$

Larger value × Smaller Unit (1892 mL) = Smaller value × Larger unit (1.892 L). OK.

You have improved your skill at performing unit conversions in the metric system.

Now work on the milliliters-to-liters conversion. Write the given quantity and its property, and then write the wanted property and its unit.

Identify and write the equivalency needed to convert between the given and wanted units, and then write the conversion factor.

Construct the solution. Set up the problem, show the unit cancellations, calculate the answer, and write the answer as a quantity. Calculate the value of the answer without using a calculator.

Check. Use larger/smaller reasoning to decide if the value of the answer is reasonable. Reflect on what you learned by working this exercise.

Practice Exercise 3.12

Convert 7.05 × 10^3 cubic centimeters to centiliters.

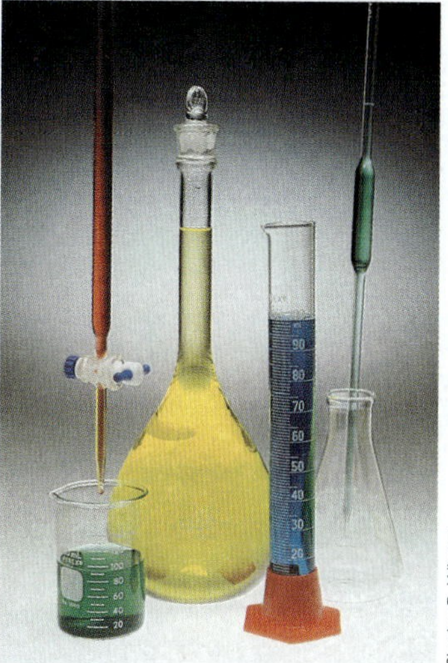

Charles D. Winters

Figure 3.13 Volumetric glassware. The beaker is only for estimating volumes. The tall graduated cylinder is used to measure volume more accurately. The flask with the tall neck (volumetric flask) and the pipet are used to obtain samples of fixed but precisely measured volumes. The buret is used to dispense variable volumes with high precision.

The uncertain digit is also called the doubtful or estimated digit.

arek_malang/Shutterstock.com

Figure 3.14 Grocery store produce scale. If a scientist were to record the weight of an object measured by this grocery store scale, she would write the value so that it expresses both the weight of the object and the associated uncertainty in the measurement.

3.7 Significant Figures

Goal 12 Given a description of a measuring instrument and an associated measurement, express the measured quantity with the uncertain digit in the correct location in the value.

It is important in scientific work to make accurate measurements and to record them correctly. The recorded measurement should indicate the size of its uncertainty. One way to do this is to attach a ± value to the recorded number. For example, if a grocery store scale indicates weight correctly to within one-tenth of a pound, and a shopper reads the scale at 2.4 pounds, this weight would properly be expressed as 2.4 ± 0.1 pound (**Figure 3.14**). The last digit, 4, is the **uncertain digit**.

Another way to indicate uncertainty is to use **significant figures**. The number of significant figures in a quantity is the number of digits that are known accurately plus the uncertain digit. It follows that if a quantity is recorded correctly in terms of significant figures, *the uncertain digit is the last digit written*. In 2.4 pounds, 4 is the uncertain digit. **Uncertainty in measurement** is illustrated in **Figure 3.15**. Study the figure carefully now, before proceeding to the next paragraph.

Learn It Now! *Textbook authors sometimes use illustrations to introduce major concepts. If the text that accompanies the figure is long, it may be written into the figure. This keeps the text and the picture on the same page when the book is printed. When this is done, the figure becomes the text. Figure 3.15 is such an illustration. In fact, it is the only text that addresses Goal 12.*

a

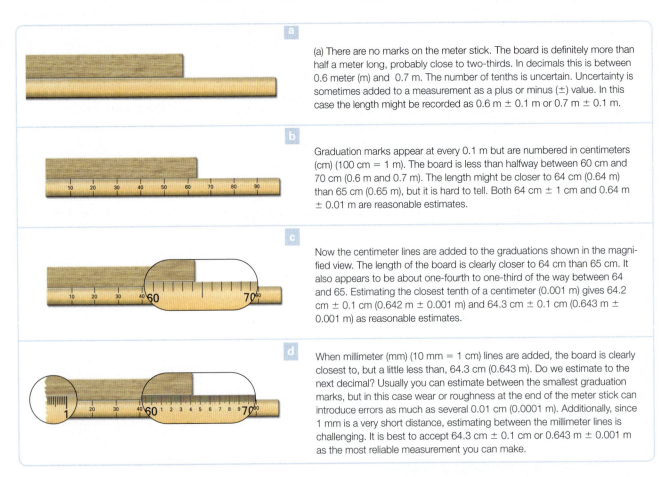

(a) There are no marks on the meter stick. The board is definitely more than half a meter long, probably close to two-thirds. In decimals this is between 0.6 meter (m) and 0.7 m. The number of tenths is uncertain. Uncertainty is sometimes added to a measurement as a plus or minus (±) value. In this case the length might be recorded as 0.6 m ± 0.1 m or 0.7 m ± 0.1 m.

b

Graduation marks appear at every 0.1 m but are numbered in centimeters (cm) (100 cm = 1 m). The board is less than halfway between 60 cm and 70 cm (0.6 m and 0.7 m). The length might be closer to 64 cm (0.64 m) than 65 cm (0.65 m), but it is hard to tell. Both 64 cm ± 1 cm and 0.64 m ± 0.01 m are reasonable estimates.

c

Now the centimeter lines are added to the graduations shown in the magnified view. The length of the board is clearly closer to 64 cm than 65 cm. It also appears to be about one-fourth to one-third of the way between 64 and 65. Estimating the closest tenth of a centimeter (0.001 m) gives 64.2 cm ± 0.1 cm (0.642 m ± 0.001 m) and 64.3 cm ± 0.1 cm (0.643 m ± 0.001 m) as reasonable estimates.

d

When millimeter (mm) (10 mm = 1 cm) lines are added, the board is clearly closest to, but a little less than, 64.3 cm (0.643 m). Do we estimate to the next decimal? Usually you can estimate between the smallest graduation marks, but in this case wear or roughness at the end of the meter stick can introduce errors as much as several 0.01 cm (0.0001 m). Additionally, since 1 mm is a very short distance, estimating between the millimeter lines is challenging. It is best to accept 64.3 cm ± 0.1 cm or 0.643 m ± 0.001 m as the most reliable measurement you can make.

Figure 3.15 Uncertainty in measurement. The length of a board is measured (estimated) by comparing it with meter sticks that have different graduation marks.

Active Example 3.13 The Location of the Uncertain Digit in a Measured Quantity

A graduated cylinder is used in the laboratory to measure the volume of liquids. Graduated cylinders are calibrated to read at the bottom of the curved surface. What is the volume of the blue liquid in the cylinder?

Think Before You Write When a quantity is recorded in scientific writing, its value expresses both the measurement and its uncertainty.

Answers *Cover the left column with your cut-out shield. Reveal each answer only after you have written your own answer in the right column.*

The volume is definitely between 34 mL and 35 mL.	You are to report the volume with all digits known with certainty plus one uncertain digit. We'll start with the digits known with certainty. The volume of the liquid is definitely between what two whole-number quantities? 
34.5 mL *You may have answered 34.4 mL, 34.5 mL, or 34.6 mL; all answers are acceptable because the uncertain digit—the last digit in the value—is estimated.*	Now determine the uncertain digit. Estimate the volume to the nearest one-tenth of a milliliter. Express the quantity with the appropriate number of significant figures.

Practice Exercise 3.13

What is the volume of liquid in the buret in the illustration? (Note that the volumes on a buret, expressed in milliliters, are sequenced in the opposite direction of a graduated cylinder.)

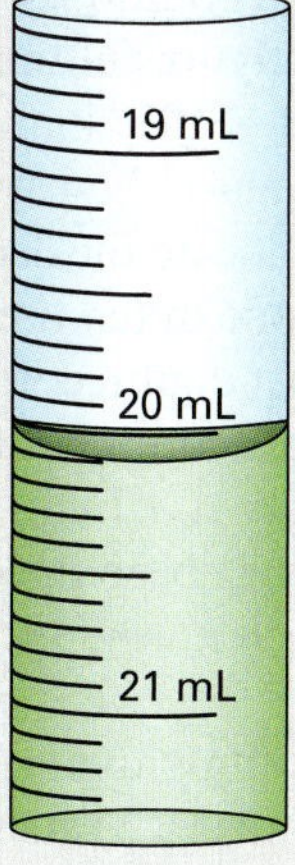

Charles Steele

Figure 3.16 Exact numbers. There is no uncertainty in exact numbers, so significant figures do not apply. There are 12 items in a dozen.

Significant figures do not apply to **exact numbers**. An exact number has no uncertainty; it is infinitely precise. Counting numbers are exact. A bicycle has exactly two wheels. Numbers fixed by definition are exact. There are exactly 60 minutes in an hour and exactly 12 eggs in 1 dozen eggs (**Figure 3.16**).

Counting Significant Figures

Goal 13 State the number of significant figures in a given quantity.

Reconsider Figure 3.15. It also illustrates how to count the number of significant figures in a measured quantity. In part (a), where the board is 0.6 m long, the first nonzero digit is 6, so counting starts there. The last digit shown, the same 6, is uncertain, so counting stops there. Therefore 0.6 m has one significant figure. In part (b), the length of the board is 0.64 m or 64 cm. The first nonzero digit is 6, so counting starts there. The last digit shown, 4, is uncertain, so counting stops there. Both 0.64 and 64 have two significant figures. In parts (c) and (d), the measuring instrument allows us to express the board's length as 0.643 m or 64.3 cm. The first nonzero digit is 6, so counting starts there. The last digit shown, 3, is uncertain, so counting stops there. Both 0.643 and 64.3 have three significant figures.

Based on this pattern, we can state the rule for counting the number of significant figures in any quantity:

***how to...* Count Significant Figures**

Begin with the first nonzero digit and end with the uncertain digit—the last digit shown.

It should be no surprise that both 0.643 m and 64.3 cm in Figure 3.15(d) have three significant figures. Both quantities came from the same measurement. Therefore, they should have the same uncertainty, the same uncertain digit. Only because the units are different are the decimal points in different places. Therefore, the important conclusion is: *The measurement process, not the unit in which a result is expressed, determines the number of significant figures in a quantity.*

When authors emphasize a point so strongly, there must be a reason. Many errors are made when students relate significant figures to the location of the decimal point. Do not allow yourself to be included among such students!

If 0.643 m is written in kilometers (km) (1000 m = 1 km), it is 0.000643 km. This is also a three-significant-figure number. The first three zeros after the decimal point are not significant, but they are required to locate the decimal point. Counting still begins with the first nonzero digit, 6. *Do not begin counting at the decimal point.*

If 0.643 m is written in nanometers (nm) (1 m = 1×10^9 nm), it is 643,000,000 nm. This is still a three-significant-figure number, but the uncertainty is 1,000,000 nm. This time we have six zeros that are not significant, but they are required to locate the decimal point. How do we end the recorded value with the uncertain digit, 3? Write it in scientific notation: 6.43×10^8 nm. The coefficient shows clearly the number of significant figures in the quantity. Scientific notation works with very small numbers, too: 6.43×10^{-4} km.

The last two values show that in very large and very small numbers, zeros whose only purpose is to locate the decimal point are not significant.

Sometimes—one in ten times, on average—the uncertain digit is a zero. If so, it still must be the last digit recorded. Suppose, for example, that the length of a board is 75 centimeters plus or minus 0.1 cm. To record this as 75 cm is incorrect; it implies that the uncertainty is ±1 cm. The correct way to write this number is 75.0 cm. If the measurement has been uncertain to 0.01 cm, it would be recorded as 75.00 cm. *The uncertain digit is always the last digit written, even if it is a zero to the right of the decimal point.*

When the uncertain digit is a zero, it is best to have it to the right of the decimal point. If 75.0 cm was to be written in the next smaller decimal unit, it would be 750 mm. The reader of this measurement is faced with the question, "Is the zero significant, or is it a placeholder for the decimal point?" With 75.0 cm or 7.50×10^2 mm, there is no question: The zero is significant.

a summary of... **The Number of Significant Figures in a Measurement**

1. Significant figures are applied to measurements and quantities calculated from measurements. They do not apply to exact numbers.
2. The number of significant figures in a quantity is the number of digits that are known accurately plus the one that is uncertain—the uncertain digit.
3. The measurement process, not the unit in which the result is expressed, determines the number of significant figures in a quantity.
4. The uncertain digit is the last digit written. If the uncertain digit is a zero to the right of the decimal point, that zero must be written.
5. Scientific notation must be used for very large numbers to show if final zeros are significant.

Some instructors prefer a different method to indicate the significance of numbers ending in zero. If yours is among them, we recommend that you follow your instructor's method rather than ours.

Active Example 3.14 The Number of Significant Figures in a Measurement

How many significant figures are in each of the following quantities?
(a) 45,261 ft, (b) 0.109 in., (c) 0.00025 kg, (d) 2.3600 × 10^{-3} cm

Think Before You Write Review the summary of the number of significant figures in a measurement, if necessary. Be sure that you understand the role of zeroes in counting significant figures.

Answers *Cover the left column with your cut-out shield. Reveal each answer only after you have written your own answer in the right column.*

(a) 45,261 ft: 5 significant figures, (b) 0.109 in.: 3 significant figures, (c) 0.00025 kg: 2 significant figures (begin counting significant figures with the first nonzero digit, 2), (d) 2.3600 × 10^{-3} cm, 5 significant figures (the zeroes to the right of the decimal point are there because the uncertainty of the measuring process was ±0.0001 × 10^{-3} cm)

Answer all four parts for yourself before you look at our answer.

Practice Exercise 3.14

State the number of significant figures in each of the following quantities: (a) 52.0 cm^3; (b) 0.00020 kL; (c) 8.50 × 10^4 mm; (d) 300.0 cg.

3.8 Significant Figures in Calculations

Chemists often calculate amounts of reactants or products in a chemical change based on how much product is desired or how much reactant is used initially. For example, how much starting material is needed to produce 5000 tablets of a 425-milligram dose of medication? These calculations involve a measured quantity, and thus the uncertainty of the measurement needs to continue to be reflected in the result of the calculation. In this section, we investigate how to express calculated quantities with the correct number of significant figures.

Rounding Off

Goal 14 Round off given values to a specified number of significant figures.

Sometimes when you add, subtract, multiply, or divide experimentally measured quantities, the result contains digits that are not significant. When this happens, you must **round off** the result. Rules for rounding are:

how to... **Round Off a Calculated Result**

Step 1: If the first digit to be dropped is less than 5, leave the digit before it unchanged.

Example: If we round 5.9936 to three significant figures, we obtain 5.99.

Step 2: If the first digit to be dropped is 5 or greater, increase the digit before it by 1.

Example: If we round 34.581 to three significant figures, we obtain 34.6.

For any individual round off by any method, every rule has a 50% chance of being "more correct." Even when "wrong," the rounded off result is acceptable because only the uncertain digit is affected.

Other rules for rounding vary if the first digit to be dropped is exactly 5. We recommend that you follow your instructor's advice if it differs from ours.

Active Example 3.15 Rounding Off a Calculated Value

Round off each of the following quantities to three significant figures:
(a) 1.42752 cm^3, (b) 45,853 cm, (c) 643.349 cm^2, (d) 0.03944498 m

Think Before You Write Review the *how to...* box that describes how to round off a calculated number, if necessary. Keep in mind that the key is to know what to do when the first digit to be dropped is less than 5 versus 5 or greater.

Answers *Cover the left column with your cut-out shield. Reveal each answer only after you have written your own answer in the right column.*

(a) 1.43 cm^3, (b) 4.59×10^4 cm, (c) 643 cm^2, (d) 0.0394 m or 3.94×10^{-2} m	Complete the problem.

Practice Exercise 3.15

Round off each of the following quantities to two significant figures: (a) 25.55 mL; (b) 0.00254 m; (c) 1.491×10^5 mg; (d) 199 Gg.

Figure 3.17 Weighing a substance added to a beaker on a balance.

Addition and Subtraction

Goal 15 Add or subtract given measured quantities and express the result in the proper number of significant figures.

The **significant figure rule for addition and subtraction** is:

***how to...* Round Off a Sum or Difference**

Round off the answer to the first column that has an uncertain digit. (Active Example 3.16 shows how to apply this rule.)

Active Example 3.16 The Significant Figure Rule for Addition and Subtraction

A student weighs a beaker and four different substances using different balances (Figure 3.17). The individual masses and their sum are as follows:

Beaker	319.5 g
Substance A	20.460 g
Substance B	0.0639 g
Substance C	45.642 g
Substance D	4.173 g
Total	389.8389 g

Express the sum to the proper number of significant figures.

Think Before You Write Review the procedure that describes how to round off a sum or difference, if necessary. The key is to express the sum with no more digits than the value with an uncertain digit in the highest-value column.

Answers *Cover the left column with your cut-out shield. Reveal each answer only after you have written your own answer in the right column.*

Tenths. The uncertain digit in 319.5 g is in the tenths column. In all other numbers the uncertain digit is in the hundredths column or smaller.	The sum is to be rounded to the first column that has an uncertain digit. What column is this: hundreds, tens, ones, tenths, hundredths, thousandths, or ten thousandths? Explain. 
389.8 g	According to the addition and subtraction rule, the answer must now be rounded off to the nearest number of tenths. What answer should be reported? 

Practice Exercise 3.16

Express the following sum to the proper number of significant figures: 2.3×10^3 mL + 4.22×10^4 mL + 9.04×10^3 mL + 8.71×10^5 mL.

We can use this example to justify the rule for addition and subtraction. A sum or difference digit must be uncertain if any number entering into that sum or difference is uncertain or unknown. In the left addition that follows, all uncertain digits are shown in blue, and all digits to the right of a colored digit are simply unknown:

Left addition	Right addition
319.5	319.5 |
20.460	20.4 | 60
0.0639	0.0 | 639
45.642	45.6 | 42
4.173	4.1 | 73
389.8389 = 389.8	389.8 | 389 = 389.8

In the left addition, the 8 in the tenths column is the first uncertain digit.

The addition at the right shows a mechanical way to locate the first uncertain digit in a sum. Draw a vertical line after the last column in which every space is occupied. The uncertain digit in the sum will be just left of that line.

The same rule, procedure, and rationalization hold for subtraction.

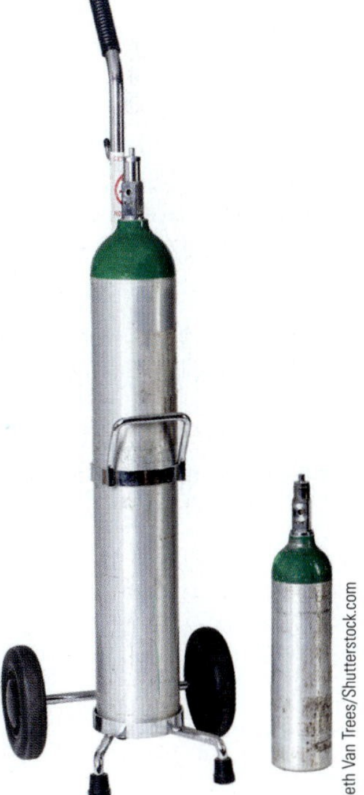

Figure 3.18 Compressed oxygen for medical use is stored in aluminum cylinders with green tops. If you know the mass of a specified volume of oxygen, you can calculate the mass of any other volume of oxygen at the same pressure and temperature.

Multiplication and Division

Goal 16 Multiply or divide given measured quantities and express the result in the proper number of significant figures.

The **significant figure rule for multiplication and division** is:

how to... **Round Off a Product or Quotient**

Round off the answer to the same number of significant figures as the smallest number of significant figures in any measured quantity. (Active Example 3.17 shows an application of this rule.)

Active Example 3.17 The Significant Figure Rule for Multiplication and Division I

If the mass of 1.000 L of a gas is 1.436 g, what is the mass of 0.0573 L of the gas (Figure 3.18)?

Think Before You Write This Active Example provides an opportunity to check your progress in developing your problem-solving skills. Do you recognize the equivalency given in the problem statement?

Answers *Cover the left column with your cut-out shield. Reveal each answer only after you have written your own answer in the right column.*

Given: 0.0573 L, metric volume unit

Wanted: mass, no unit is specified

When no unit is specified, you can use the most convenient unit or the unit that is used conventionally. In this case, the mass given in the problem statement is in grams, so answering in grams will be convenient.

Analyze the problem statement: Write the given quantity, its property, the property of the wanted quantity, and its unit.

1.000 L = 1.436 g

$$\frac{1.436\text{ g}}{1.000\text{ L}}$$

Notice that we retained the volume value intact from the problem statement as 1.000 L. The measurement was made to four significant figures, and that information would be lost if we wrote 1 L.

Identify the equivalency and write the conversion factor.

$$0.0573\cancel{\text{L}} \times \frac{1.436\text{ g}}{1.000\cancel{\text{L}}} = 0.0823\text{ g}$$

Construct the setup and write the answer. Show unit cancellation. Make sure that the answer is expressed with the correct number of significant figures.

We are multiplying 0.05… by about $1\frac{1}{2}$, so we expect something close to 0.075 ($5 \times 1\frac{1}{2} = 7.5$). The answer makes sense. The smallest number of significant figures in any factor is 3, from 0.0573 L. Three significant figures are expressed in the answer, which is correct.

You improved your skill at applying the significant figure rule for multiplication and division.

Check the value of the answer. Does it make sense? Are the significant figures correct? What did you learn from this Active Example?

Practice Exercise 3.17

At a certain temperature, 0.878 g of a pure liquid substance occupies 1.00 mL. What is the volume of 33 grams of liquid?

Active Example 3.17 and its product

$$0.0573 \times 1.436 = 0.0822828 = 0.0823 \text{ in three significant figures}$$

may be used to justify the rule for multiplication and division. Suppose these measurements are the *true* values, and they are correctly expressed in terms of significant figures—that is, the final digit is the uncertain digit in each value. Now consider what the product would be if the actual measurements were both 1 higher in the uncertain digit. In that case, the multiplication would be

$$0.0574 \times 1.437 = 0.0824838 = 0.0825 \text{ in three significant figures}$$

If the actual measurements were both 1 lower in the uncertain digit, the multiplication would be

$$0.0572 \times 1.435 = 0.0820820 = 0.0821 \text{ in three significant figures}$$

Compare the three results in three significant figures: 0.0823, 0.0825, and 0.0821. With the third nonzero digit uncertain, these three three-significant-figure products are equal. They might be expressed as 0.0823 ± 0.0002.

Alternately, each product number reached with an uncertain multiplier must itself be uncertain. Blue numbers indicate the uncertain digits in the detailed multiplication:

1.436	4 significant digits
$\times\ 0.0573$	3 significant digits
4308	
10052	
7180	
$0.0822828 = 0.0823$	3 significant digits (multiplication/division rule)

Active Example 3.18 The Significant Figure Rule for Multiplication and Division II

How many hours are in 6.924 days? Express the answer in the proper number of significant figures.

Think Before You Write What equivalency is needed to solve the problem?

Answers *Cover the left column with your cut-out shield. Reveal each answer only after you have written your own answer in the right column.*

Given: 6.924 days, time units
Wanted: time units, hours

24 hours = 1 day

$$\frac{24\ \text{h}}{1\ \text{day}}$$

Analyze the problem by writing the given quantity and its property and the wanted property and unit. **Identify** the equivalency needed to solve the problem, and write the conversion factor.

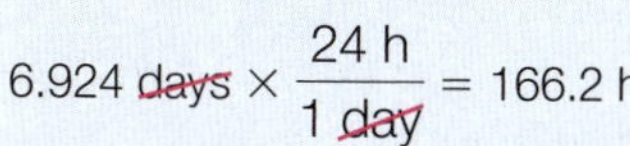

$$6.924\ \text{days} \times \frac{24\ \text{h}}{1\ \text{day}} = 166.2\ \text{h}$$

Smaller value × Larger unit = Larger value × Smaller unit, OK. There are 4 significant figures in the measured quantity, and significant figures do not apply to the definition 24 hours = 1 day, so there should be 4 significant figures in the answer, OK.

You improved your skill at applying the significant figure rule for multiplication and division.

Construct the setup, and **check** your answer. What did you learn by doing this example?

Practice Exercise 3.18

How many days are in 55 weeks?

Addition/Subtraction and Multiplication/Division Combined

When a calculation contains both addition/subtraction and multiplication/division, you must apply each individual rule for significant figures separately. If two numbers are to be added and their sum is to be divided by another number, such as in $\frac{(34.49 + 7.3)}{13.80}$, first perform the addition to the correct number of significant figures. Then perform the division, applying the multiplication/division significant-figure rule.

For the addition, we obtain

$34.4 \mid 9$	← uncertain two digits past the decimal point
$+\ 7.3 \mid$	← uncertain one digit past the decimal point
$41.7 \mid 9 = 41.8$	← uncertain one digit past the decimal point (addition and subtraction rule)

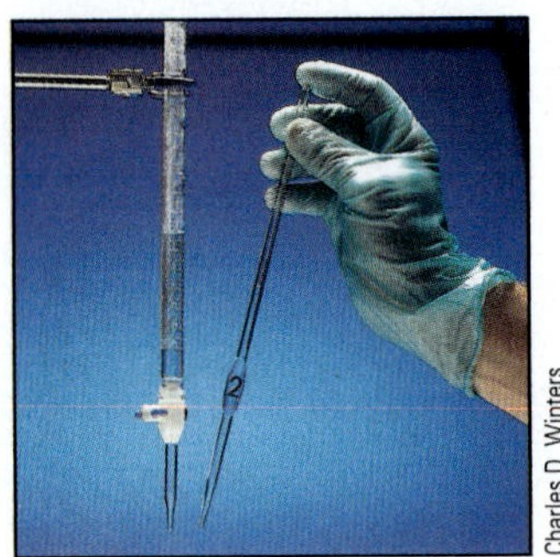

Charles D. Winters

Figure 3.19 A buret and a pipet. The buret on the left is used to deliver variable volumes to a precision of ±0.05 mL. The pipet on the right is used to dispense a fixed volume of liquid with a precision of ±0.03 mL. These instruments are common in introductory chemistry laboratories, but most research labs use automated titrators and automatic pipets.

Now apply the multiplication/division rule:

Three significant figures in the numerator ↓

$$\frac{41.8}{13.80} = 3.03 \leftarrow$$ Three significant figures (the smallest number of significant figures in any factor)

↑ Four significant figures in the denominator

Active Example 3.19 Addition/Subtraction and Multiplication/Division Combined

The density of a substance is given by: density $\equiv \frac{\text{mass}}{\text{volume}}$.

A pipet (Figure 3.19) is used to deliver a volume of 5.00 mL of a liquid to a beaker that has a mass of 36.3 g. The mass of the liquid plus the beaker is 42.54 g. What is the density of the liquid?

Think Before You Write The problem is solved by first calculating the mass of the liquid, which is the difference between the mass of the liquid plus beaker and the mass of the empty beaker. This difference is then divided by the volume of the liquid.

Answers *Cover the left column with your cut-out shield. Reveal each answer only after you have written your own answer in the right column.*

1.2 g/mL

If you got the correct answer, good. Before showing the setup that produces that answer, though, let's look at a few wrong answers and the common errors that produce them.

1.248 g/mL comes from ignoring significant figures altogether and writing the number displayed on your calculator.

1.25 g/mL comes from not recognizing that the numerator has only two significant figures after applying the addition/subtraction rule:

$$\begin{array}{r|l} 42.5 & 4\text{ g} \\ -36.3 & \text{ g} \\ \hline 6.2 & 4\text{ g} = 6.2\text{ g} \end{array}$$

Using 6.24 g in the numerator and 5.00 mL in the denominator of the density fraction yields 6.24 g (3 SF) ÷ 5.00 mL (3 SF) = 1.248 g/mL, which rounds to 1.25 g/mL (3 SF).

1.24 g/mL comes from rounding off the numerator correctly to two significant figures but not recognizing that those two significant figures limit the final answer to two significant figures: 6.2 g ÷ 5.00 mL = 1.24 g/mL.

The correct answer comes from recognizing that the mass of liquid is expressed in two significant figures because of the addition/subtraction rule, and then the density quotient is expressed in two significant figures because of the multiplication/division rule.

Perform the addition (or subtraction, in this case) first and then do the division. Take it all the way to a calculated answer, expressed in the proper number of significant figures.

You improved your skill at applying the significant-figure rule for a problem with both addition/subtraction and multiplication/division.

What did you learn from working this Active Example?

Practice Exercise 3.19

A solid metallic object is dropped into a graduated cylinder that contains 25.00 mL of water. The volume increases to 37.25 mL. The mass of the object is 96 grams. Use the definition density ≡ mass ÷ volume to calculate the density of the metal.

Everyday Chemistry 3.1

SHOULD THE UNITED STATES CONVERT TO METRIC UNITS? AN EDITORIAL

In the 1970s, the four major English-speaking nations of the world—the United States, Great Britain, Canada, and Australia—took action to replace what was then called the British system of units with the metric system. Three of these nations have made the change successfully. Anyone living in the United States knows which country did not. In fact, the name of the U.S. measurement system has been changed to the United States Customary System (USCS) because the British no longer use it.

Why does the United States cling to USCS units? Simply because of resistance to change. We've grown up with USCS units. For most Americans, there is no advantage to metrics. One kilogram of potatoes is no more convenient a quantity than 2.2 pounds of potatoes, so why bother to change? Americans can get used to metric quantities, however. We now buy soft drinks in 1- or 2-liter bottles (**Figure 3.20**). Is 2 liters less convenient than 2.11 quarts, the equivalent USCS volume? The number is not important once you get used to it, until you get to calculating.

If you must work with measured quantities—calculate as we must in the sciences, or buy, sell, and build as they do in commerce and industry—metrics are so much easier and so much more logical that the choice is obvious. If, in solving the problems in this book, you had to work with quantities expressed in USCS units, your chemistry course would be much more difficult, not because of the chemistry, but because of the USCS units!

Figure 3.20 Americans are accustomed to buying soft drinks in containers with metric volumes.

The fact is that most people have little day-to-day need to calculate beyond simple arithmetic. And the general public in the United States can vote out of office any politician who proposes a mandatory change to metrics.

Changing to metric units is not nearly as difficult as the opponents of change would have us think. The Canadians, British, and Australians did it. Consider all the immigrants to the United States who have changed from their native metric system to the USCS system—a much more difficult change than from USCS to metric. If they can do that, Americans are surely capable of making the easier change in the other direction. So has hope for conversion to metrics been lost in the United States? Not really. Economic motives for the conversion could be stronger than public opposition to change. If by adopting metrics, and paying the one-time costs that go along with it, a company can (a) make more money, (b) save money, or (c) avoid losing money, you can be quite sure that metrics will be adopted. Here are a few examples.

One major manufacturing firm delivered a large number of appliances to a Middle Eastern country, only to have them rejected because the connecting cord was 6 feet instead of 2 meters long.

European countries have told the United States that they may, as a matter of policy, refuse U.S. imports that fail to meet metric standards. In regard to mechanical equipment, their position is, "You may build it with USCS dimensions, but the tools we use to fix it are metric (**Figure 3.21**). If we can't fix it, we won't buy it." It's hard to argue against that.

All equipment purchased under U.S. Defense Department contracts must meet metric specifications. The conversion to metrics is at or close to 100% in the U.S. automobile and drug industries.

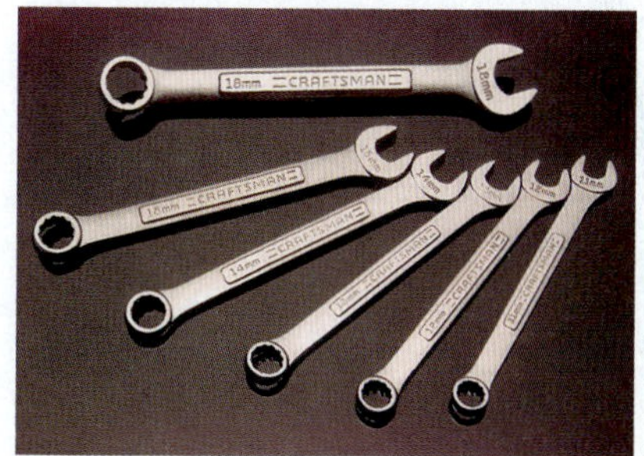

Figure 3.21 When industrial parts are manufactured in metric sizes, the tools needed to fix them are available everywhere in the world.

One major U.S. manufacturer has reduced the number of screw sizes it holds in inventory from 70 to 15 by using only metric products. Another large exporter reports "saving tens of millions of dollars by avoiding double inventory costs and operating all our 32 domestic and foreign plants on one system."

In the ten years after one U.S. firm introduced metrics, its number of employees doubled to meet its new export demand. In a two-year slack domestic market, those exports were the only thing that kept the company afloat.

Another American company adopted metrics in order to hold a few Canadian customers. The change opened so many new markets that the company now ships 28% of its output abroad.

Many U.S. companies that have converted to metrics report that the cost of conversion was much lower than expected—less than half in some cases—and that those costs have been recovered quickly and often unexpectedly.

Will the United States ever become a fully metric nation? As these examples attest, it is already a metric nation where it really counts. Industry and government could no longer wait around for a reluctant populace to do what had to be done. So we are, in effect, a nation of two systems of units. Maybe that's the way it's supposed to be. But metrics are so much easier, so logical, and have so many advantages.... Oh, well!

Quick Quiz

1. How many fluid ounces are in a quart? How many feet are in a mile? How many milliliters are in a liter? How many millimeters are in a meter? Which conversions are easiest to learn and remember if you were to start with no knowledge of measurement systems?
2. What is *your* answer to the question that titles this Everyday Chemistry essay: Should the United States Convert to Metric Units? List at least two pros and cons for each side of the argument.

In performing the arithmetic in Active Example 3.19, it is not necessary to round off the subtraction from 6.24 to 6.2. You may never see 6.2 in your calculator display or written on a piece of paper. Both $6.24 \div 5.00 = 1.248$ and $6.2 \div 5.00 = 1.24$ round off to 1.2 with two significant figures. It is necessary to recognize that the numerator in the final division has only two significant figures, and therefore, the answer should be rounded to two significant figures.

Significant Figures and This Book

Be wary of additions and subtractions. They sometimes increase or reduce the number of significant figures in an answer to greater or fewer than the number in any measured quantity.

Calculators report answers in all the digits they are able to display, usually eight or more. Such answers are unrealistic; never use them. Calculations in this book have generally been made using all digits given, and only the final answer has been rounded off. If, in a problem with several steps, you round off at each step, your answers may differ slightly from those in the book. The difference will be in the uncertain digit; both answers are acceptable.

3.9 Metric–USCS Conversions

Goal 17 Given a metric–USCS conversion factor and a quantity expressed in any unit in Table 3.2, express that quantity in the corresponding unit in the other system.

The length and mass equivalencies given in Table 3.2 define the USCS units. Since volume is a derived unit, it cannot be a definition. The volume equivalency given in Table 3.2 is exact, however.

Table 3.2 gives the defining or exact equivalencies between measurement units in the United States Customary System (USCS) and metric system. All countries of the world use the metric system, except for the United States, which is only partly metric. The USCS, formerly the British system of units, is the system used in the United States. **Table 3.3** gives equivalencies between USCS measurement units.

Table 3.2 Metric–USCS Equivalencies

Length	1 in. ≡ 2.54 cm (definition)
Mass	1 lb ≡ 453.59237 g (definition)
Volume	1 gal ≡ 3.785411784 L (exactly)

Table 3.3 USCS–USCS Equivalencies

Length
1 ft ≡ 12 in.
1 yd ≡ 3 ft
1 mi ≡ 5280 ft
Mass
1 lb ≡ 16 oz
Volume
1 qt ≡ 32 fl oz
1 gal ≡ 4 qt

Thinking About Your Thinking

Proportional Reasoning

The relationships between the number of metric units and the number of USCS units are direct proportionalities. You will find it useful to memorize only one conversion in each of three categories: length, mass (weight), and volume, as given in Table 3.2. You can then use familiar metric–metric and/or USCS–USCS equivalencies to change units within each system of measurement. Although it can add a few steps to a problem, this approach minimizes the amount of memorization necessary.

In Table 3.2, notice two things about the length equivalency 1 in. ≡ 2.54 cm: (1) The symbol for the USCS unit is in., including the period ▶. The period distinguishes the symbol for inch from the common word *in.* (2) The ≡ symbol indicates a definition and may be read "one inch is defined as 2.54 centimeters." Thus exactly 1 in. is equal to exactly 2.54 cm, and the numbers are infinitely precise and therefore have an infinite number of significant figures in the equivalency. There are other definitions and exact conversions in Tables 3.2 and 3.3, also indicated by the ≡ symbol. Since the mass and volume equivalencies in Table 3.2 have a lot of digits, we will follow the practice of rounding the conversion factor to one significant digit more than the measured quantity that limits the significant digits in the calculation.

USCS units are defined in terms of metric units. Note the definitions of the inch and the pound in Table 3.2.

Active Example 3.20 Metric–USCS Conversions: Length

What is the length in meters of an object that is 38 inches long?

Think Before You Write Imagine that you read this problem without knowing it was in a metric–USCS conversion section of a textbook. How would you classify it? In other words, how would you describe the generalized characteristics of the problem?

Answers *Cover the left column with your cut-out shield. Reveal each answer only after you have written your own answer in the right column.*

Given: 38 in., USCS length unit **Wanted:** metric length unit, m	**Analyze** the problem statement: List the given quantity and property and the wanted quantity and property.
1 in. ≡ 2.54 cm; 100 cm = 1 m $\frac{2.54 \text{ cm}}{1 \text{ in.}}; \frac{1 \text{ m}}{100 \text{ cm}}$ *The defining equivalency for metric–USCS length takes you from the given USCS length unit in. to metric length, but in cm. You need to use the memorized metric relationship 100 centiunits = 1 unit to convert from cm to the wanted unit m.*	**Identify** the equivalencies needed. Table 3.2 has the metric–USCS equivalency you need. You will need an additional equivalency. Write the conversion factors. 
$38 \cancel{\text{in.}} \times \frac{2.54 \cancel{\text{cm}}}{1 \cancel{\text{in.}}} \times \frac{1 \text{ m}}{100 \cancel{\text{cm}}} = 0.97 \text{ m}$ *The number of inches is the only measured quantity, so there are two significant figures in the final answer.*	**Construct** the solution setup. Carefully consider significant figures when you express the value of the answer.

If you know that a meter and a yard (36 in.) are about the same length, the answer is reasonable in magnitude.

You improved your skill in solving metric–USCS conversion problems.

Check. Does the answer make sense? What skill did you improve?

Practice Exercise 3.20

Express 2.50 kilometers as miles.

Active Example 3.21 Metric–USCS Conversions: Mass

The mass of a 250-mL glass beaker is listed as 108,255 mg. Express this in ounces for a coworker who needs to calculate a shipping cost based on weight.

Think Before You Write A feature of this problem statement is that it contains a quantity that is irrelevant in the solution of the problem. Do you recognize it?

Answers *Cover the left column with your cut-out shield. Reveal each answer only after you have written your own answer in the right column.*

Given: 108,255 mg, metric mass
Wanted: USCS weight, oz

Analyze the problem by writing the given quantity, its property, the property of the wanted quantity, and the wanted unit.

The metric–USCS mass equivalency is 1 lb ≡ 453.59237 g. The given mass is mg, and g is in the metric–USCS equivalency, so we need 1000 mg = 1 g. The wanted weight is oz, and lb is in the metric–USCS equivalency, so we need 1 lb = 16 oz (from Table 3.3).

$$\frac{1\ \text{lb}}{453.59237\ \text{g}};\ \frac{1\ \text{g}}{1000\ \text{mg}};\ \frac{16\ \text{oz}}{1\ \text{lb}}$$

Identify the equivalencies needed for the solution. Recall that you can find metric–USCS equivalencies in Table 3.2. That equivalency will guide you to the others you need. Write the conversion factors.

$$108{,}255\ \cancel{\text{mg}} \times \frac{1\ \cancel{\text{g}}}{1000\ \cancel{\text{mg}}} \times \frac{1\ \cancel{\text{lb}}}{453.59237\ \cancel{\text{g}}} \times \frac{16\ \text{oz}}{1\ \cancel{\text{lb}}}$$
$$= 3.81858\ \text{oz}$$

Note how this solution setup progresses from the given metric mass unit to the metric mass unit in the metric–USCS equivalency to the USCS unit in that equivalency, and then to the wanted USCS unit. The key element is to identify the metric–USCS equivalency, and then the other conversions are simply metric–metric or USCS–USCS.

Construct the solution setup and answer. Look carefully at significant figures.

It is not straightforward to mentally verify the value of the answer, but 4 oz is a reasonable beaker weight.

You improved your skill in solving metric–USCS conversion problems.

Check. Is the value of the answer reasonable? What did you learn by doing this exercise?

Practice Exercise 3.21

A liquid sample has a weight of 408 ounces. What is the mass of the sample in kilograms?

Active Example 3.22 Metric–USCS Conversions: Volume

A soft drink manufacturer is converting from USCS units to metric units. If the quart bottle initially is to remain at the same volume, what number of milliliters should be printed on the label?

Think Before You Write When you read this problem, did you think, "a USCS volume unit is given, and a metric volume unit is wanted"?

Answers *Cover the left column with your cut-out shield. Reveal each answer only after you have written your own answer in the right column.*

Given: 1.00 qt, USCS volume
Wanted: metric volume, mL

We made the assumption that the dispensing equipment can measure the volume delivered to ±0.01 qt.

Analyze the problem by writing the given quantity, its property, the property of the wanted quantity, and the unit of the wanted quantity.

1 gal = 3.785 L; 4 qt = 1 gal; 1000 mL = 1 L

$\frac{1 \text{ gal}}{4 \text{ qt}}; \frac{3.785 \text{ L}}{1 \text{ gal}}; \frac{1000 \text{ mL}}{1 \text{ L}}$

We used 3.785 L/gal instead of the full 3.785411784 L/gal given in Table 3.2 because the given quantity has three significant figures, and thus the answer is limited to three significant figures. Using a conversion factor with at least one more significant figure than the given quantity ensures that the uncertain digit in the final answer will not be affected by the number of significant figures in the conversion factor. A conversion factor should never limit the number of significant figures in a calculation.

Identify the equivalencies needed to solve the problem. You first need to identify the USCS–metric volume conversion factor. You may also possibly need to make USCS–USCS and/or metric–metric conversions. After you write the equivalencies, change them to conversion factors.

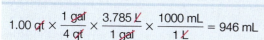

$1.00 \text{ qt} \times \frac{1 \text{ gal}}{4 \text{ qt}} \times \frac{3.785 \text{ L}}{1 \text{ gal}} \times \frac{1000 \text{ mL}}{1 \text{ L}} = 946 \text{ mL}$

Construct the setup and determine the answer.

A quart and a liter (1000 mL) are about the same volume, so the answer makes sense.

You improved your skill in solving metric–USCS conversion problems.

Check to be sure the answer makes sense. What did you learn from this Active Example?

Practice Exercise 3.22

Express 45 milliliters as fluid ounces.

3.10 Temperature

Goal 18 Given a temperature in either Celsius or Fahrenheit degrees, convert it to the other scale.

19 Given a temperature in Celsius degrees or kelvins, convert it to the other scale.

The familiar temperature scale in the United States is the **Fahrenheit scale**. All scientists, including those in the United States, use the **Celsius scale** (Figure 3.22). Both scales are based on the temperature at which water freezes and boils at standard atmospheric pressure (see Section 4.4). In the Fahrenheit scale, the freezing point of water is 32°F, and boiling occurs at 212°F, a range of 180 degrees. The Celsius scale divides the range into 100 degrees, from 0°C to 100°C. Figure 3.23 compares these scales.

Note the two reference points on the thermometers in Figure 3.23: the boiling point and the freezing point of water. These are the same no matter what temperature

Science History Images/Alamy Stock Photo

Figure 3.22 Anders Celsius (1701–1744), a Swedish scientist, first proposed setting the freezing point of water at 100°C and the boiling point at 0°C. The scale was reversed after his death.

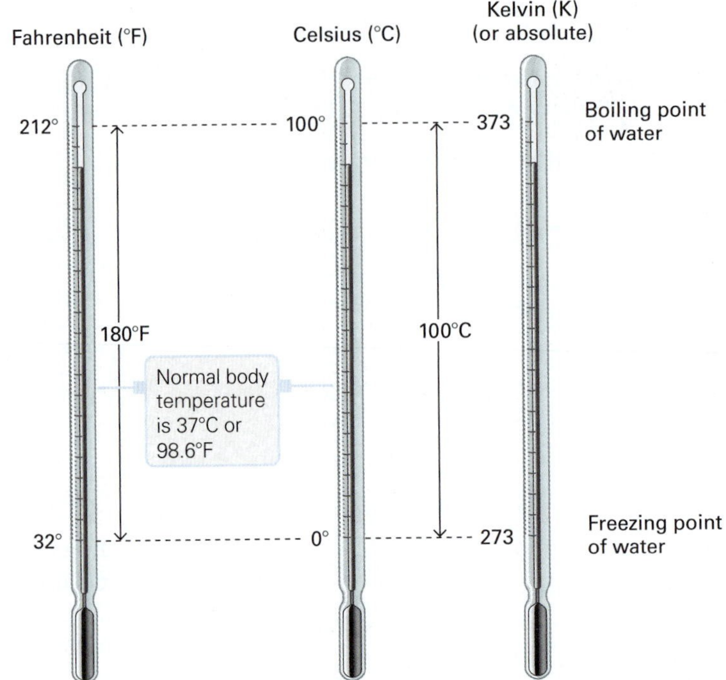

Figure 3.23 A comparison of Fahrenheit, Celsius, and Kelvin temperature scales. The reference or starting point for the Kelvin scale is absolute zero (0 K = −273°C), the lowest temperature theoretically obtainable. Note that the abbreviation K for the Kelvin unit is used without the degree sign (°). Also note that 1 Celsius degree = 1 Kelvin unit = 180/100 or 9/5 or 1.8 Fahrenheit degree.

scale is used to measure them. The Fahrenheit scale divides the boiling-to-freezing temperature difference into 180 degrees (212 − 32 = 180), and the Celsius scale divides the same temperature difference into 100 degrees (100 − 0 = 100). This means that 180 Fahrenheit degrees and 100 Celsius degrees are equivalent: 180 Fahrenheit degrees = 100 Celsius degrees. Dividing both sides of the equivalency by 100, we have 1.8 Fahrenheit degrees = 1 Celsius degree.

The other critical feature illustrated by Figure 3.23 is that the freezing point of water is 0° on the Celsius scale and 32° on the Fahrenheit scale. In fact, neither of these represents the absence of temperature in the same sense that 0 meter indicates no length and 0 gram indicates no mass. Neither 0°C nor 32°F is the true zero point of temperature. Thus, the equivalency 1.8 Fahrenheit degrees = 1 Celsius degree is not a direct proportionality. We also must adjust for the fact that 0°C = 32°F. Thus, to change a temperature in degrees Fahrenheit to degrees Celsius, we must first subtract 32 to adjust for the difference in freezing points and then divide by 1.8 (180 ÷ 100) to adjust for the difference in degree size:

$$T_{°C} = \frac{T_{°F} - 32}{1.8}$$

Thinking About *Your Thinking*

Proportional Reasoning

Rearranged, the Fahrenheit–Celsius relationship equation is $T_{°F} = 1.8\ T_{°C} + 32$, which corresponds to the slope-intercept form of the equation of a straight line, $y = mx + b$. A y-versus-x plot of that equation does not pass through the origin. When one variable is zero, the other is not. Thus y and x increase and decrease together, but not in the form of a direct proportionality, $y = mx$. Celsius–Fahrenheit conversions therefore cannot be done by multiplying by a conversion factor. They require the use of an algebraic equation.

SI units include a third temperature scale known as the **Kelvin temperature scale** or **absolute temperature scale**. The temperature unit on the Kelvin scale is the same size as a Celsius degree. The origin of the Kelvin scale is discussed in Chapter 4. It is based on zero at the lowest temperature possible, which is 273° below zero on the Celsius scale. The two scales are therefore related by the equation

$$T_K = T_{°C} + 273$$

K in this equation represents kelvins, the actual temperature unit. The degree symbol, °, is not used for Kelvin temperatures.

Converting a temperature from one scale to another is done by algebra. In this book we will always solve algebra problems by first solving the equation *algebraically* for the wanted quantity. When the unknown is the only term on one side of the equation, given values are substituted on the other side, and the result is calculated. The advantage of this procedure will become clear as you gain experience in solving algebra-based chemistry problems.

Active Example 3.23 Temperature Conversions I

What is the Celsius temperature on a comfortable 72°F day? What is the Kelvin temperature?

Think Before You Write When two quantities have a relationship in the form $y = mx + 0$, you can use a conversion factor. Since $T_{°F}$ and $T_{°C}$ are related by $T_{°F} = 1.8\,T_{°C} + 32$, or in general, $y = mx + b$, where $b \neq 0$, use of a conversion factor does not apply. You must use an algebraic equation to solve the problem.

Answers *Cover the left column with your cut-out shield. Reveal each answer only after you have written your own answer in the right column.*

Given: 72°F, USCS temperature
Wanted: metric temperature in °C and K

Analyze the problem statement. What is the given quantity and property? What is the property of the wanted quantities and what units are wanted?

$$T_{°C} = \frac{T_{°F} - 32}{1.8};\ T_K = T_{°C} + 273$$

Both equations are already written in the form where the wanted variable is by itself on one side of the equals sign.

Identify the algebraic relationships needed to solve the problem. If necessary, solve for the wanted variable.

$$T_{°C} = \frac{T_{°F} - 32}{1.8} = \frac{72 - 32}{1.8} = 22°C$$

$$T_K = T_{°C} + 273 = 22 + 273 = 295\ K$$

*Scientists usually measure temperature directly in °C with a Celsius thermometer (**Figure 3.24**). However, if you are ever asked to convert between $T_{°F}$ and T_K, you must determine $T_{°C}$ as an intermediate step, as you did in this example.*

Construct the solution setups.

You improved your skill at performing conversions among temperature scales.

Don't worry about a **check** of the value of the answers for now; we'll discuss them in more detail immediately following the Active Example. What did you learn from solving this problem?

Practice Exercise 3.23

Convert 288 K to its Fahrenheit scale equivalent.

Charles D. Winters

Figure 3.24 An infrared thermometer. When conventional temperature sensors cannot be used, a noncontact device such as this infrared thermometer is used instead. A lens focuses on the infrared light emitted from a heated object onto a detector, which is calibrated to convert the wavelength of light to an electronic signal that can be displayed as temperature.

A larger/smaller check of the answer is not easy to do for Active Example 3.23, or for most problems solved algebraically. It is true, however, that there is a convenient room temperature equivalence point that is relatively easy to remember: 68°F = 20°C. Beyond that, checking the answer for reasonableness involves some mental arithmetic. The numerator is 72 − 32, which is 40. The denominator is about 2. Dividing 40 by 2 gives 20, which is close to the calculated answer 22. You will learn more about estimating answers in Appendix I, Part D and in future chapters.

Active Example 3.24 Temperature Conversions II

From each temperature given in the following table, calculate the equivalent temperatures in the other scales.

Think Before You Write *This summary problem will test if you can convert from °F to °C and from °C to °F and in both directions in the °C–K relationship.*

Answers *Cover the left column with your cut-out shield. Reveal each answer only after you have written your own answer in the right column.*

°F	°C	K
−13	−25	248
105	41	314
196	91	364

Complete the table below.

°F	°C	K
	−25	
105		
		364

You improved your skill at performing conversions among temperature scales.

Be sure you are able to convert any Fahrenheit temperature to its Celsius equivalent and vice versa before moving on. What did you learn from this Active Example?

Practice Exercise 3.24

At what temperature is the Fahrenheit temperature the same as the Celsius temperature?

3.11 Proportionality and Density

Goal 20 Write a mathematical expression indicating that one quantity is directly proportional to another quantity.

21 Use a proportionality constant to convert a proportionality to an equality.

22 Given the values of two quantities that are directly proportional to each other, calculate the value of the proportionality constant, including its units.

23 Write the defining equation for a proportionality constant and identify units in which it might be expressed.

24 Given two of the following for a sample of a pure substance, calculate the third: mass, volume, and density.

Consider the following experiment. The mass of a clean, dry container for measuring liquid volume was measured and found to be 26.42 g. Then 5.2 mL of cooking oil was placed into the container. The mass of the oil and the container was 31.06 g.

The mass of the 5.2 mL of oil in the container was found by subtracting the mass of the empty container from the mass of the container plus the oil:

	Mass of 5.2 mL of oil + container	31.06 g
−	Mass of empty container	− 26.42 g
	Mass of 5.2 mL of oil	4.64 g

This procedure was repeated several times. The resulting data are shown in the table and summarized in the graph in **Figure 3.25**.

We can use the results of this experiment to find the relationship between the mass of the cooking oil and its volume. Whenever a graph of two related measurements is a straight line that passes through the origin, the measured quantities are **directly proportional** to each other. More formally, a direct proportionality exists between two quantities when they increase or decrease at the same rate. "At the same rate" means that if one quantity doubles, the other quantity doubles; if one triples, the other triples; if one is reduced by 20%, the other is reduced by 20%, and so on.

We use direct proportionalities every day. The relationships between defined units of mass, length, and time are proportionalities: examples include grams per kilogram, inches per foot, and seconds per hour. Speed is a proportionality, whether measured in kilometers per hour, feet per second, or any other distance per time unit. The prices we pay for things are based on proportionalities: dollars per pound or per kilogram, cents per liter or quart, cents per bunch (of carrots), and dollars per six-pack are all money units per mass, volume, or counting unit of some substance.

Thinking About *Your Thinking*

Proportional Reasoning

The discussion in this part of the text further explains the relationship between direct proportionalities and equivalencies. You will find yourself doing this type of thinking continually throughout this course, in many other science courses, and, as discussed in the body of the text, in everyday situations. Be sure to *Learn It Now!*

a

Volume (mL)	Mass (g)	Mass ÷ Volume (g/mL)
5.2	4.64	0.89
7.7	6.91	0.90
10.4	9.28	0.892
13.8	12.35	0.895
17.6	15.72	0.893

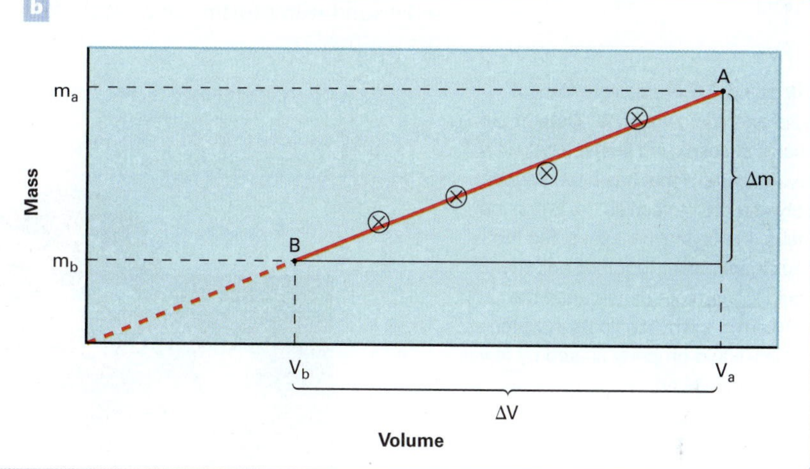

Figure 3.25 Mass–volume data for a sample of cooking oil. (a) The third column is the result of dividing the measured mass by the measured volume. All quotients are rounded to the number of significant figures justified by the data. (b) The slope of the line between any two points, A and B, on the line, $\Delta m/\Delta V$, is the density of the oil. Notice that the slope of the line of the plot in part (b) is the same as the mass/volume ratio in the third data column in part (a).

From the preceding discussion we can list several conclusions:

1. We can describe direct proportionalities between measured quantities with equivalencies.
2. Direct proportionalities between measured quantities yield two conversion factors between the quantities.
3. Given either quantity in a direct proportionality and the conversion factor between the quantities, we can calculate the other quantity.

A direct proportionality between two variables, such as mass (m) and volume (V), is indicated by $m \propto V$, where the symbol $\propto$ means "is proportional to." A proportionality can be changed into an equation by inserting a multiplier called a **proportionality constant**. If we let D be the proportionality constant,

$$m \propto V \xrightarrow{\text{the proportionality changes to an equality}} m = D \times V$$

Solving for the proportionality constant, D, yields the **defining equation** for a physical property of a pure substance called its **density**:

$$m = D \times V \xrightarrow{\text{divide both sides by V}} \frac{m}{V} = \frac{D \times V}{V} \xrightarrow{\text{cancel V's and reverse order}} D \equiv \frac{m}{V}$$

In words, the density of a substance is its mass per unit volume:

$$\text{Density} \equiv \frac{\text{mass}}{\text{volume}}$$

The symbol $\equiv$ identifies a definition.

Some proportionality constants have physical significance. Density, a physical property, is among them. The density of cooking oil in the experiment described at the beginning of this section is equal to the slope of the line in Figure 3.25.

We can think of density as a measure of the relative "heaviness" of a substance, in the sense that a block of iron is heavier than a block of aluminum of the same size. **Table 3.4** lists the densities of some common materials.

Table 3.4 Densities of Some Common Substances (g/cm^3 at 20°C and 1 atm)

Substance	Density
Helium	0.00017
Air	0.0012
Lumber	
Pine	0.5
Maple	0.6
Oak	0.8
Water	1.0
Glass	2.5
Aluminum	2.7
Iron	7.8
Copper	9.0
Silver	10.5
Lead	11.4
Mercury	13.6
Gold	19.3

You are familiar with the fact that ice—solid water—floats on liquid water. Why is this? Ice is *less dense* than liquid water. What does this mean on the particulate level? Considering the definition of density, a given volume of ice must have less mass than the same volume of liquid water. In other words, if all water molecules have the same mass and volume, whether liquid or solid, the molecules in liquid water must pack more closely together than molecules in ice. **Figure 3.26** shows how solid water forms ice crystals with spaces between the molecules, whereas liquid

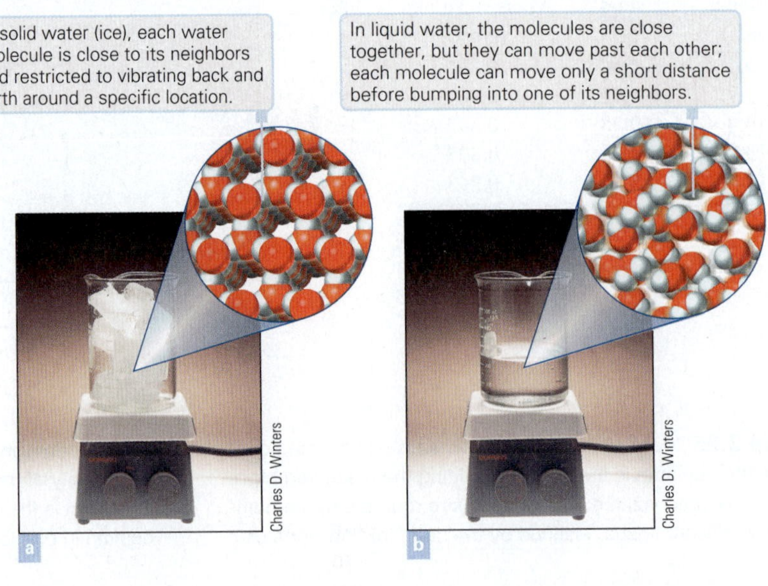

In solid water (ice), each water molecule is close to its neighbors and restricted to vibrating back and forth around a specific location.
In liquid water, the molecules are close together, but they can move past each other; each molecule can move only a short distance before bumping into one of its neighbors.
Charles D. Winters
Charles D. Winters
a
b

Figure 3.26 Particulate view of solid and liquid water. (a) Water molecules in solid form (ice) are held in a crystal pattern that has open spaces between the molecules. (b) When ice melts, the crystal collapses, the molecules are closer together, and the liquid is therefore denser than the solid. This is why ice floats in water, a solid–liquid property shared by few other substances.

Figure 3.27 Densities of solids and liquids. (a) Water is unusual in that its solid phase, ice, will float on its liquid phase. (b) Solid ethanol sinks to the bottom of the liquid. The solid form of almost all substances is denser than the liquid phase.

molecules have fewer and smaller open spaces. So, at the particulate level, the density of a given pure substance in a given state of matter is a measure of how tightly packed the molecules are in that state. Water is an unusual substance, and the fact that its solid phase is less dense than its liquid phase is just one of its many unusual properties. For almost all other substances, the solid phase is denser than the liquid phase (**Figure 3.27**).

The definition of density establishes its units. Mass is commonly measured in grams (g); volume is often measured in cubic centimeters (cm^3). Therefore, according to the equation $D \equiv m/V$ and the mass-per-unit-volume definition, the units of density can be grams per cubic centimeter, g/cm^3. Liquid densities are often given in grams per milliliter, g/mL. (Recall that 1 milliliter is exactly 1 cubic centimeter by definition.) There are, of course, other units in which density can be expressed, but they all must reflect the definition in terms of mass ÷ volume. Examples are grams/liter, usually used for gases because their densities are so low, and the USCS pounds/cubic foot.

To find the density of a substance, you must know both the mass and volume of a sample of that substance. Dividing the mass by volume yields density.

Active Example 3.25 Determination of Density from Mass and Volume

The mass of a 12.0-cm^3 piece of magnesium is 20.9 grams. Find the density of magnesium.

Think Before You Write When you read the problem statement, think about the properties of the measured quantities. In this case, 12.0-cm^3 describes a metric volume. What property is described by 20.9 grams? What is the definition of density?

Answers *Cover the left column with your cut-out shield. Reveal each answer only after you have written your own answer in the right column.*

Given: 12.0 cm^3, metric volume; 20.9 g, metric mass **Wanted:** density, mass ÷ volume, assume that units g ÷ cm^3 are OK	**Analyze** the problem by writing the given quantities, their properties, the property of the wanted quantity, and the wanted units.
$D \equiv \frac{m}{V}$	**Identify** the definition that is needed to solve the problem.
$D \equiv \frac{m}{V} = \frac{20.9\ g}{12.0\ cm^3} = 1.74\ g/cm^3$	**Construct** the solution setup.

Units OK, 20 ÷ 10 = 2, which is close to 1.74, OK.

You improved your skill at solving density problems.

Check the value of the answer to be sure it is reasonable, and state what you learned from this Active Example.

Practice Exercise 3.25

The mass of a 50.00-milliliter sample of methanol is 39.59 grams. Determine the density of methanol.

Active Example 3.26 Determination of Volume from Mass and Density

The density of a certain machine oil is 0.910 g/mL. Find the volume occupied by 196 g of that oil.

Think Before You Write When a problem statement includes a density, think of that density as the equivalency it represents. In this case, think "0.910 g = 1 mL."

Answers *Cover the left column with your cut-out shield. Reveal each answer only after you have written your own answer in the right column.*

Given: g, metric mass
Wanted: volume, assume mL

We assume that milliliters are the appropriate unit in which to express volume because it is the volume unit within the given density.

Analyze the problem statement by writing the given quantity, its property, the property of the wanted quantity, and its unit.

0.910 g = 1 mL

$$\frac{1 \text{ mL}}{0.910 \text{ g}}$$

Identify the equivalency given in the problem statement, and turn it into the appropriate conversion factor.

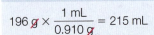

$$196 \text{ g} \times \frac{1 \text{ mL}}{0.910 \text{ g}} = 215 \text{ mL}$$

Construct the setup, and determine the volume.

The value of the answer makes sense: 200 divided by a value a little less than 1 should be a little more than 200.

You improved your skill at solving density problems.

Check the solution. Does it make sense? What did you learn from this example?

Practice Exercise 3.26

What is the mass of 975 mL of carbon disulfide, which has a density of 1.26 g/mL?

The setup in Active Example 3.26 provides a good opportunity to illustrate what happens when you invert the conversion factor. If you wrote $\frac{0.910 \text{ g}}{1 \text{ mL}}$, your setup would be

$$196 \text{ g} \times \frac{0.910 \text{ g}}{1 \text{ mL}} = 178 \text{ g}^2/\text{mL}$$

The difference in the values of the answers, 215 versus 178, is not large enough to necessarily make you realize that the answer is wrong. However, the units are indisputably wrong. The wanted volume is certainly not expressed in g^2/mL. This again illustrates the importance of including units in all setups.

3.12 Thoughtful and Reflective Practice

The only way to learn how to solve problems is to solve them yourself. If you have followed the question-and-answer approach to the Active Examples in this chapter and have covered the answers to the steps in the examples until you have figured out the answer for yourself, *you* have solved the problems and not merely looked at how *we* solved them. Your knowledge and problem-solving ability is growing. But you still need more practice.

The end-of-chapter problems with answers give you lots of opportunities to solve problems with immediate feedback on the correctness of your methods. Be sure to keep in mind your reason for solving problems. It is not to get the answer we got, but to *learn how to solve the problem.*

Solving a problem with one finger at its solution in the answer section so you can check your progress is neither solving the problem nor *learning how to solve it.* When you tackle a problem, solve it completely. If you get stuck at any point, put the problem aside and check the part of the chapter that covers that point. Learn there what you need to learn. *Do not check the answer section.* Return to your solution of the problem and complete it. Then compare your solution to the one at the end of the chapter. If they do not agree, find out why. Be sure you understand the problem before going to the next one.

We close this chapter with a few more examples that you can practice with while we guide you in following the four-part strategy for solving quantitative problems.

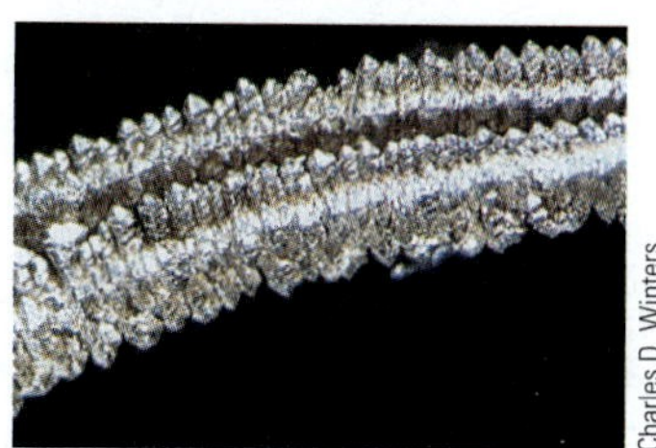

Figure 3.28 Solid silver metal. Silver occurs as an uncombined element in nature.

Active Example 3.27 Solving Quantitative Problems I

Calculate the mass of 7.04 cm^3 of silver (Figure 3.28).

Think Before You Write At first glance, it might seem that you don't have enough information to solve this problem. When this situation occurs, start by assessing what you know from the problem statement. This will often help you discover what else you need to know to solve the problem.

Answers *Cover the left column with your cut-out shield. Reveal each answer only after you have written your own answer in the right column.*

Given: 7.04 cm^3, metric volume **Wanted:** mass, assume g *We assume that grams are the appropriate unit for mass in this problem because the gram is the basic metric unit for mass.*	**Analyze** the problem by writing the given quantity, its property, the property of the wanted quantity, and the units of the wanted quantity.
Density. *Density is mass per unit volume.*	The next step in thinking about this problem is to ask yourself if you know or can find a relationship that connects what you know to what you want. You know volume; you want mass. What property of a substance provides a connection between mass and volume?
10.5 g = 1 cm^3 $\frac{10.5\ \text{g}}{1\ \text{cm}^3}$	Chemistry problems often do not include all of the information necessary to find their solutions. It is typically assumed that you should either know or are able to look up information such as the density of a pure substance, as in this case, or a USCS–metric conversion, as in Section 3.9. In this book, selected densities are given in Table 3.4. When you have what you need, **identify** the equivalency needed to solve the problem, and change it to the conversion factor.

$$7.04\ \cancel{cm^3} \times \frac{10.5\ g}{1\ \cancel{cm^3}} = 73.9\ g$$

Construct the solution setup.

$7 \times 10 = 70$, so the value of the answer is reasonable.

You improved your skill at solving quantitative problems.

Check. Is the value of the answer reasonable? What did you learn from this Active Example?

Practice Exercise 3.27

Determine the mass in kilograms of a lead brick that is 15 cm long, 8 cm wide, and 3 cm tall.

Active Example 3.28 Solving Quantitative Problems II

A container label states that it has a volume of 1.06 quarts. Express this volume in milliliters.

Think Before You Write When you encounter a problem that does not immediately bring a solution setup to mind, a key issue is to realize that you have the ability to solve the problem. Your instructor will not assign "impossible" problems. Simply work through the quantitative problem solving strategy and be persistent.

Answers *Cover the left column with your cut-out shield. Reveal each answer only after you have written your own answer in the right column.*

Given: 1.06 qt, USCS volume
Wanted: metric volume, mL

Analyze the problem statement by writing the given quantity, its property, the property of the wanted quantity, and the units of the wanted quantity.

1 gal = 3.785 L

Since the given quantity has three significant figures, we rounded off the equivalency to four significant figures to insure that the measured quantity limits the number of significant figures in the answer, rather than the conversion factor.

Your analysis revealed an important piece of information: You need to do a metric–USCS volume conversion. Look at Table 3.2, if necessary, to identify the appropriate equivalency.

4 qt = 1 gal

1000 mL = 1 L

Now it's a matter of doing the needed USCS–USCS and metric–metric conversions. **Identify** those equivalencies.

$$1.06\ \cancel{qt} \times \frac{1\ \cancel{gal}}{4\ \cancel{qt}} \times \frac{3.785\ \cancel{L}}{1\ \cancel{gal}} \times \frac{1000\ mL}{1\ \cancel{L}} = 1.00 \times 10^3\ mL$$

Construct the setup by changing the equivalencies to conversion factors, inserting them directly into the setup. Determine the answer.

A quart and a liter (1.00×10^3 = 1000 mL) are about the same volume. The answer makes sense.

You improved your skill at solving quantitative problems.

Check. Does the answer make sense? What did you learn by working this Active Example?

Practice Exercise 3-28

What volume, in cubic feet, is a 591 mL bottle?

You can use equivalencies and conversion factors to solve everyday problems as well as those associated with chemistry. The final example is one you may be able to identify with personally.

Active Example 3.29 Solving Quantitative Problems III

Suppose that you have just landed a part-time job that pays \$9.25 per hour. You will work five shifts each week, and the shifts are 4 hours long. You plan to save all of your earnings to pay cash for a 55-inch 1080p Curved Smart OLED TV that costs \$1082.49, tax included. You are paid weekly. How many weeks must you work in order to save enough money to buy the television? You might also be interested in knowing how much cash you will have left for video games or other goodies.

Think Before You Write This example requires a multistep solution, and we ask you to solve it completely and without help.

Answers *Cover the left column with your cut-out shield. Reveal each answer only after you have written your own answer in the right column.*

Given: 1082.49 \$, money units
Wanted: time units, weeks

9.25 \$ = 1 h; 1 shift = 4 h; 5 shifts = 1 week

$$1082.49\ \cancel{\$} \times \frac{1\ \cancel{h}}{9.25\ \cancel{\$}} \times \frac{1\ \cancel{shift}}{4\ \cancel{h}} \times \frac{1\ week}{5\ \cancel{shifts}} = 5.85\ldots\ weeks = 6\ weeks$$

All of the numbers in the calculation are exact numbers, but it will take your sixth paycheck to get the \$1082.49 you need.

Develop your plan, set up the problem, and calculate the number of weeks.

$$6\ \cancel{weeks} \times \frac{5\ \cancel{shifts}}{1\ \cancel{week}} \times \frac{4\ \cancel{h}}{1\ \cancel{shift}} \times \frac{9.25\ \$}{1\ \cancel{h}} \times = \$1110.00$$

\$1110.00 earned − \$1082.49 = \$27.51 cash remaining

Again, all values are exact, to the penny.

What will be your total pay at the end of the sixth week? And how much excess cash will there be?

Practice Exercise 3.29

A college chemistry laboratory experiment requires 15 grams of sodium nitrate for each student. If an introductory chemistry course runs 24 laboratory sections each semester, each with 20 students, how many 2.5-pound bottles of sodium nitrate should be ordered for the two-semester academic year?

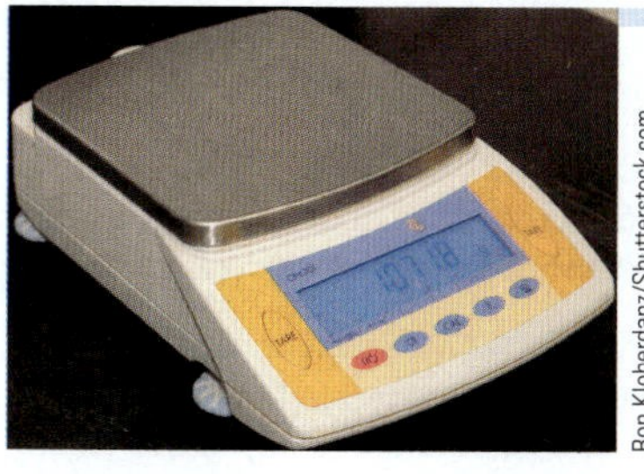

Ron Kloberdanz/Shutterstock.com

CHAPTER 3 IN REVIEW: MEASUREMENT AND CHEMICAL CALCULATIONS

Goal 1 Write in scientific notation a number given in ordinary decimal form; write in ordinary decimal form a number given in scientific notation.

Any decimal number can be written in **scientific notation**. Scientific notation expresses a number as a coefficient C (between 0 and 9.99...) multiplied by 10 raised to the e power, in general, $C \times 10^e$. When e is larger than 0, 10^e is larger than 1; when e is smaller than 0, 10^e is smaller than 1. (Remember that $10^0 = 1$.)

Goal 2 Use a calculator to add, subtract, multiply, and divide numbers expressed in scientific notation.	To add, subtract, multiply, or divide numbers in exponential notation, following the instructions that are appropriate for your calculator.
Goal 3 Convert an equivalency to two conversion factors.	An **equivalency** is two quantities that are equivalent in value. When an equivalency is expressed as a fraction, it is called a **conversion factor**. Two reciprocal conversion factors result from each equivalency.
Goal 4 Learn and apply the algorithm for using conversion factors to solve quantitative problems.	(1) **Analyze** the problem statement by writing the given quantities, their properties, the property of the wanted quantity, and its unit. (2) **Identify** the equivalencies or the algebraic relationship needed to solve the problem. (3) **Construct** the solution setup. (4) **Check** the solution at two levels: (a) making sense and (b) what was learned.
Goal 5 Explain why the metric system of measurement is used in the sciences.	The metric system is internationally standardized and decimal based.
Goal 6 State and write with appropriate metric prefixes the relationship between any metric unit and its corresponding kilounit, centiunit, and milliunit.	The important metric prefixes for this course are *kilo-* (1000), *centi-* (0.01) and *milli-* (0.001).
Goal 7 Using Table 3.1, state and write with appropriate metric prefixes the relationship between any metric unit and other larger and smaller metric units.	Given the value of a metric prefix other than the three of which you have memorized their values, you should be able to write an equivalency between a quantity with a unit with that prefix and the basic unprefixed unit.
Goal 8 Distinguish between mass and weight.	The **mass** of an object *does not change* in different gravitational fields; the **weight** of that object *does change*.
Goal 9 Identify the metric units of mass, length, and volume.	The SI metric unit of mass is the **kilogram (kg)**. The SI metric unit of length is the **meter (m)**. The SI unit of volume is the cubic meter (m^3).
Goal 10 Given a mass, length, or volume expressed in basic metric units, kilounits, centiunits, or milliunits, express that quantity in the other three units.	You should memorize the values needed to be able to make conversions among the unit, kilounit, centiunit, and milliunit.
Goal 11 Given a mass, length, or volume expressed in any metric units and Table 3.1 or the equivalent, express that quantity in any other metric unit.	For any metric prefix other than *kilo-, centi-,* or *milli-,* given the value of the prefix, you should be able to make conversions to another metric prefix.
Goal 12 Given a description of a measuring instrument and an associated measurement, express the measured quantity with the uncertain digit in the correct location in the value.	The number of significant figures in a quantity is the number of digits that are known accurately plus the uncertain digit; the uncertain digit is the last digit written.
Goal 13 State the number of significant figures in a given quantity.	To count significant figures, begin with the first nonzero digit and end with the uncertain digit—the last digit shown.
Goal 14 Round off given values to a specified number of significant figures.	To round off a number to the proper number of significant figures, leave the uncertain digit unchanged if the digit to its right is less than 5. Increase the uncertain digit by one if the digit to its right is 5 or more.
Goal 15 Add or subtract given measured quantities and express the result in the proper number of significant figures.	In addition and subtraction, round off a sum or difference to the first column that has an uncertain digit.
Goal 16 Multiply or divide given measured quantities and express the result in the proper number of significant figures.	In multiplication and division, round off a product or quotient to the same number of significant figures as the smallest number of significant figures in any factor.
Goal 17 Given a metric–USCS conversion factor and a quantity expressed in any unit in Table 3.2, express that quantity in the corresponding unit in the other system.	A metric–USCS equivalency from Table 3.2 is converted to the appropriate conversion factor and used in a setup to calculate a quantity in the other system of measurement.
Goal 18 Given a temperature in either Celsius or Fahrenheit degrees, convert it to the other scale.	The Celsius and Fahrenheit temperature scales are related by the equation $T_{°C} = \frac{T_{°F} - 32}{1.8}$.

Goal 19 Given a temperature in Celsius degrees or kelvins, convert it to the other scale.	The Celsius and Kelvin temperature scales are related by the equation $T_K = T_{°C} + 273$.
Goal 20 Write a mathematical expression indicating that one quantity is directly proportional to another quantity.	If y is directly proportional to x, it is expressed mathematically as $y \propto x$.
Goal 21 Use a proportionality constant to convert a proportionality to an equality.	To convert the proportionality $y \propto x$ into an equality, insert a proportionality constant, k: $y = kx$.
Goal 22 Given the values of two quantities that are directly proportional to each other, calculate the value of the proportionality constant, including its units.	If $y = kx$, then $k = y \div x$.
Goal 23 Write the defining equation for a proportionality constant and identify units in which it might be expressed.	Units are set by definition and the defining equation. For example, for density, the units are mass units over volume units. Examples: kg/m^3, g/cm^3, g/mL, g/L.
Goal 24 Given two of the following for a sample of a pure substance, calculate the third: mass, volume, and density.	The defining equation for density is: **Density** $\equiv \frac{\textbf{mass}}{\textbf{volume}}$. Because the defining equation for density is a conversion factor, density problems are solved with that conversion factor or its reciprocal.

Key Terms

Most of the key terms and concepts and many others appear in the Glossary. Use your Glossary regularly.

≡ (symbol) p. 70
analyze (a problem statement) p. 67
base units p. 73
Celsius scale p. 93
check (a solution) p. 67
coefficient p. 59
construct (a solution) p. 67
conversion factor p. 64
cubic centimeter (cm^3) p. 75
defining equation p. 98
density p. 98
derived units p. 73
directly proportional p. 97
equivalency p. 64
exact numbers p. 82
exponent p. 59
exponential p. 59
Fahrenheit scale p. 93
gram (g) p. 75
identify (equivalencies or an equation) p. 67
Kelvin temperature scale or absolute temperature scale p. 95
kilogram (kg) p. 74
liter (L) p. 76
mass p. 74
meter (m) p. 75
metric system p. 73
milliliter (mL) p. 76
proportionality constant p. 98
quantity p. 63
round off p. 83
scientific notation p. 59
SI units p. 73
significant figure rule for addition and subtraction p. 84
significant figure rule for multiplication and division p. 85
significant figures p. 80
uncertain digit p. 80
uncertainty (in measurement) p. 80
unit p. 63
value p. 63
weight p. 74

Frequently Asked Questions

How can I stop making mistakes while changing ordinary decimal numbers to scientific notation?
Scientific notation is not usually a problem, except for careless errors when relocating the decimal in the coefficient and adjusting the exponent. These errors will not occur if you make sure the exponent and the coefficient move in opposite directions, one larger and one smaller. It sometimes helps to think about an ordinary decimal number as being written in scientific notation in which the exponential is 10^0. Thus 0.0024 becomes 0.0024×10^0, and the larger/smaller changes in the coefficient and exponent are clear when changing to 2.4×10^{-3}.

What is the best overarching approach to solving quantitative chemistry problems?
Include units in every problem you solve. This is the best advice that we can give to someone learning how to solve quantitative chemistry problems. Always treat a quantity as the product of a value and a unit: Quantity = Value × Unit. Challenge every problem answer in both value and units.

Even though I get the setup right, I sometimes calculate an answer incorrectly. How can I be sure that I'm doing the calculation correctly?
Many errors would never been seen by a test grader if the test taker had simply checked the reasonableness of an answer. Part D in Appendix I offers some suggestions on how to estimate the numerical result in a problem. Please read this section and put it into practice with every problem you solve.

I like doing metric-to-metric conversions in my head, but I sometimes move the decimal point the wrong way. How can I avoid this?
A common error in metric-to-metric conversions is moving the decimal point the wrong way. To avoid this, fully write out the setup. Then challenge your answer. Use the larger/smaller rule.

For a given amount of anything, the number of larger units is smaller and the number of smaller units is larger.

I'm losing significant figures points on every exam! How can I get better at expressing the answer quantity with the correct number of significant figures?
Significant figures can indeed be troublesome, but they need not be if you learn to follow a few basic rules. There are four common errors to watch for:

1. Starting to count significant figures at the decimal point of a very small number instead of at the first nonzero digit.
2. Using the significant figure rule for multiplication/division when rounding off an addition or subtraction result.
3. Failing to show an uncertain tail-end zero on the right-hand side of the decimal.
4. Failing to use scientific notation when writing larger numbers, thereby causing the last digit shown to be other than the uncertain digit.

How do I stop confusing the addition/subtraction and multiplication/division significant figures rules?
Most of the arithmetic operations you will perform are multiplications and divisions. Students often learn the rule for those operations well but then erroneously apply that rule to the occasional addition or subtraction problem that comes along. Products and quotients have the same number of significant figures as the smallest number in any factor. Sums can have more significant figures than the largest number in any number added. Example: $68 + 61 = 129$. Differences can have fewer significant figures than the smallest number in either number in the subtraction. Example: $68 - 61 = 7$.

Concept-Linking Exercises

Write a brief description of the relationships among each of the following groups of terms or phrases. Answers to the Concept-Linking Exercises are given at the end of the chapter.

1. Equivalency, conversion factor, solving quantitative problems, analyze, identify, construct, check
2. Metric system, SI units, derived unit, base unit
3. Mass, weight, kilogram, gram, pound
4. Uncertainty in measurement, uncertain digit, significant figures, exact numbers
5. Direct proportionality, inverse proportionality, proportionality constant

Small-Group Discussion Questions

Small-Group Discussion Questions are for group work, either in class or under the guidance of a leader during a discussion section.

1. How do you change an ordinary decimal number to scientific notation? Change 3,876,989 to scientific notation, and use this example while explaining the general procedure. How do you change a number in scientific notation to ordinary decimal form? Change 3.99×10^{-5} to ordinary decimal form, and use this example while explaining the general procedure.
2. Write at least two equivalencies in each of the following categories: (a) USCS–USCS units, (b) metric–metric units, (c) USCS–metric units, (d) neither USCS nor metric units, and (e) temporary relationships. Convert each equivalency into two conversion factors.
3. The life expectancy for a person born in the United States in 2000 is 74.3 years for males and 79.7 years for females. How many more times will the heart of an average female beat during her lifespan than an average male? To answer this question, explicitly (a) analyze the problem statement, (b) identify equivalencies that are needed to solve the problem, (c) construct the solution setup, and (d) check the solution.
4. How have you used measurement in your life in the past week? List at least as many measurements as the number of people in your group. What instrument did you use for each measurement?
5. A cube measures 1 m × 2 m × 3 m. What is the volume of the cube in liters? If the cube is made of a pure substance that has a density of 2.5 g/mL, what is its mass in kilograms?
6. How many significant figures are usually implied when a person expresses her or his (a) weight, (b) height, (c) age, (d) pulse rate, and (e) number of fingers? Explain each answer.
7. When a three-significant-figure quantity is multiplied by a four-significant-figure quantity, how many significant figures are justified in the product? When a three-significant-figure quantity is subtracted from a four-significant-figure quantity, how many significant figures are justified in the difference? Explain.
8. The original definition of the meter was one ten-millionth of the length of the meridian through Paris from pole to the equator. Based on this definition, how many miles is it from the North Pole to the equator? How many milliliters are in a cup? A metric ton is 1000 kilograms. How many USCS tons (2000 pounds) are equal to a metric ton?
9. Which is larger, a Celsius degree or a Fahrenheit degree? What is the temperature change in Fahrenheit when the temperature increases by 14°C? At what point do the Celsius and Fahrenheit temperatures have the same value? At what point do the Kelvin and Celsius temperatures have the same value?
10. The circumference of a circle is directly proportional to its diameter. Write this as a mathematical statement. Change the proportionality to an equation by inserting the appropriate proportionality constant. Write the defining equation for the proportionality constant. What are the units of the proportionality constant?

11. Determine the volume in cubic inches of a gold bar that weighs 1.0 troy ounce. 12 troy ounces equal 16 ounces. The density of gold is 19.3 g/cm^3.

12. A school supplies pencils to its students. The pencils are packaged in boxes of 1 gross, where 1 gross = 12 dozen. The historical average for the school's pencil needs has been 8.7 pencils per student. If the projected enrollment of the school is 932 students, how many boxes should be ordered for the next school year?

Questions, Exercises, and Problems

Interactive versions of these problems may be assigned by your instructor. Solutions for blue-numbered questions are at the end of the chapter. Questions other than those in the General Questions and More Challenging Problems sections are paired in consecutive odd-even number combinations; solutions for the odd-numbered questions are also at the end of the chapter.

Section 3.2: Scientific Notation

1. Write the following numbers in scientific notation: (a) 0.000322, (b) 6,030,000,000, (c) 0.00000000000619.

2. Write each of the following numbers in scientific notation: (a) 70,300, (b) 0.0231, (c) 0.000154, (d) 5040.

3. Write the following numbers in ordinary decimal form: (a) 5.12×10^6, (b) 8.40×10^{-7}, (c) 1.92×10^{21}.

4. Write the ordinary form of the following numbers: (a) 2.32×10^{-2}, (b) 9.27×10^4, (c) 2.54×10^3, (d) 8.96×10^{-4}.

5. Complete the following operations:
   a) $(7.87 \times 10^4)(9.26 \times 10^{-8}) =$
   b) $(5.67 \times 10^{-6})(9.05 \times 10^{-7}) =$
   c) $(309)(9.64 \times 10^6) =$
   d) $(4.07 \times 10^3)(8.04 \times 10^{-8})(1.23 \times 10^{-2}) =$

6. Complete the following operations:
   a) $(5.08 \times 10^{-5})(1.83 \times 10^{-7}) =$
   b) $\dfrac{9.42 \times 10^{-4}}{5.98 \times 10^{-4}} =$
   c) $\dfrac{(2.33 \times 10^6)(9.61 \times 10^4)}{(1.83 \times 10^{-7})(8.76 \times 10^4)} =$

7. Complete the following operations:
   a) $\dfrac{6.18 \times 10^4}{817} =$
   b) $\dfrac{4.91 \times 10^6}{5.22 \times 10^5} =$
   c) $\dfrac{4.60 \times 10^7}{1.42 \times 10^3} =$
   d) $\dfrac{9.32 \times 10^4}{6.24 \times 10^7} =$

8. Complete the following operations:
   a) $(2.34 \times 10^6)(4.23 \times 10^5) =$
   b) $\dfrac{8.60 \times 10^{-4}}{1.72 \times 10^{-4} - 9.25 \times 10^{-7}} =$
   c) $\dfrac{(7.54 \times 10^{-5})(1.72 \times 10^{-4})}{8.60 \times 10^{-4}} =$

9. Complete the following operations:
   a) $\dfrac{9.84 \times 10^3}{(6.12 \times 10^3)(4.27 \times 10^7)} =$
   b) $\dfrac{(4.36 \times 10^8)(1.82 \times 10^3)}{0.0856(4.7 \times 10^6)} =$

10. Complete the following operations:
    a) $9.25 \times 10^{-7} + 8.60 \times 10^{-4} =$
    b) $\dfrac{8.60 \times 10^{-4} + 4.23 \times 10^5}{1.72 \times 10^{-4}} =$
    c) $\dfrac{7.54 \times 10^{-5}}{(9.25 \times 10^{-7})(8.60 \times 10^{-4})} =$

11. Complete the following operations:
    a) $6.38 \times 10^7 + 4.01 \times 10^8 =$
    b) $1.29 \times 10^{-6} - 9.94 \times 10^{-7} =$

12. Complete the following operations:
    a) $8.63 \times 10^5 + 1.80 \times 10^{-4} =$
    b) $\dfrac{1.80 \times 10^{-4} + 2.90 \times 10^{-7}}{9.53 \times 10^4} =$
    c) $\dfrac{1.06 \times 10^{-5}}{(8.63 \times 10^5)(1.80 \times 10^{-4})} =$

Section 3.3: Conversion Factors

13. What is the mathematical criterion for two quantities to be equivalent?

14. Sixty seconds and one minute are equivalent quantities. Explain why this is true.

15. Write the equivalency for each conversion factor:
    a) $\dfrac{\text{3 feet}}{\text{1 yard}}$
    b) $\dfrac{\text{1 meter}}{\text{100 centimeters}}$
    c) $\dfrac{\text{1 milliliter}}{\text{20 drops}}$

16. Write the equivalency for each conversion factor:
    a) $\dfrac{\text{16 fluid ounces}}{\text{1 pint}}$
    b) $\dfrac{\text{1 liter}}{\text{1000 cubic centimeters}}$
    c) $\dfrac{\text{28 miles}}{\text{1 gallon}}$

17. Write the equivalency and both conversion factors for each relationship:
    a) Ten dimes is one dollar.
    b) A football field is 100 yards long.
    c) 16 ounces is 1 pound.
18. Write the equivalency and both conversion factors for each relationship:
    a) A pack of gum is 9 pieces.
    b) Two quarters are equal in value to five dimes.
    c) One mile is 5280 feet.
19. Write the equivalency and both conversion factors that relate each pair of units:
    a) Minutes and seconds
    b) Inches and feet
    c) Dollars and cents
20. Write the equivalency and both conversion factors that relate each pair of units:
    a) Hours and days
    b) Gallons and quarts
    c) Quarters and dollars

Section 3.4: A Strategy for Solving Quantitative Chemistry Problems

Use the questions in this section to practice your problem-solving skills in an everyday context. Show all setups and unit cancellations. Our answers are rounded off according to the rules given in Section 3.8. Your unrounded answers are acceptable only if you complete these questions before studying Section 3.8.

21. How long will it take to travel the 406 miles between Los Angeles and San Francisco at an average speed of 48 miles per hour?
22. A student who is driving home for the holidays averages 70.5 miles per hour. How many miles will the student travel if the trip lasts 8.73 hours?
23. How many minutes does it take a car traveling 88 kilometers per hour to cover 4.3 kilometers?
24. How many days are in 89 weeks?
25. What will be the cost in dollars for nails for a fence 62 feet long if you need 9 nails per foot of fence, there are 36 nails in a pound, and they sell for 69 cents per pound?
26. A student working for Stop and Shop is packing eggs into cartons that contain a dozen eggs. How many eggs will the student need in order to pack 72 cartons?
27. An American tourist in Mexico was startled to see $259 on a menu as the price for a meal. However, that dollar sign refers to Mexican pesos, which on that day had an average rate of 13 pesos per American dollar. How much did the tourist pay for the meal in American funds?
28. How many nickels should you receive in exchange for 89 quarters?
29. How many weeks are in a decade?
30. How many seconds are in the month of January?

Section 3.5: Introduction to Measurement

31. List at least two measurements that would be routinely made in a health care clinic.
32. List at least two measurements that would be made by a team of chemists studying the world's oceans.

Section 3.6: Metric Units

33. A woman stands on a scale in an elevator in a tall building. The elevator starts going up, rises rapidly at constant speed for half a minute, and then slows to a stop. Compare the woman's weight as recorded by the scale and her mass while the elevator is standing still during the starting period, during the constant rate period, and during the slowing period.
34. A person can pick up a large rock that is submerged in water near the shore of a lake but may not be able to pick up the same rock from the beach. Compare the mass and the weight of the rock when in the lake and when on the beach.
35. What is the metric unit of length?
36. What is the metric unit of mass?
37. *Kilobuck* is a slang expression for a sum of money. How many dollars are in a kilobuck? How about a megabuck (see Table 3.1)?
38. What is the difference between the terms *kilounit* and *kilogram*?
39. One milliliter is equal to how many liters?
40. How many centimeters are in a meter?
41. Which unit, megagrams or grams, would be more suitable for expressing the mass of an automobile? Why?
42. What is the name of the unit whose symbol is nm? Is it a long distance or a short distance? How long or how short?

Questions 43–50: Make each conversion indicated. Use scientific notation to avoid long integers or decimal fractions. Write your answers without looking at a conversion table.

43. (a) 5.74 cg to g, (b) 1.41 kg to g, (c) 4.54×10^8 cg to mg
44. (a) 15.3 kg to g and mg, (b) 80.5 g to kg and mg, (c) 58.5 mg to kg and g
45. (a) 21.7 m to cm, (b) 517 m to km, (c) 0.666 km to cm
46. (a) 90.4 mm to m and cm, (b) 11.9 m to mm and cm, (c) 53.6 cm to mm and m
47. (a) 494 cm^3 to mL, (b) 1.91 L to mL, (c) 874 cm^3 to L
48. (a) 90.8 mL to L and cm^3, (b) 16.9 L to mL and cm^3, (c) 65.4 cm^3 to mL and L

Questions 49 and 50: Refer to Table 3.1 for less common prefixes in these metric conversions.

49. (a) 7.11 hg to g, (b) 5.27×10^{-7} m to pm, (c) 3.63×10^6 g to dag
50. (a) 0.194 Gg to g, (b) 5.66 nm to m, (c) 0.00481 Mm to cm

Section 3.7: Significant Figures

51. State the volume of liquid in each graduated cylinder in the figure here and explain how you decided upon the appropriate number of significant figures. The markings are calibrated to be read at the bottom of the curved surface, and the numbers represent milliliters.

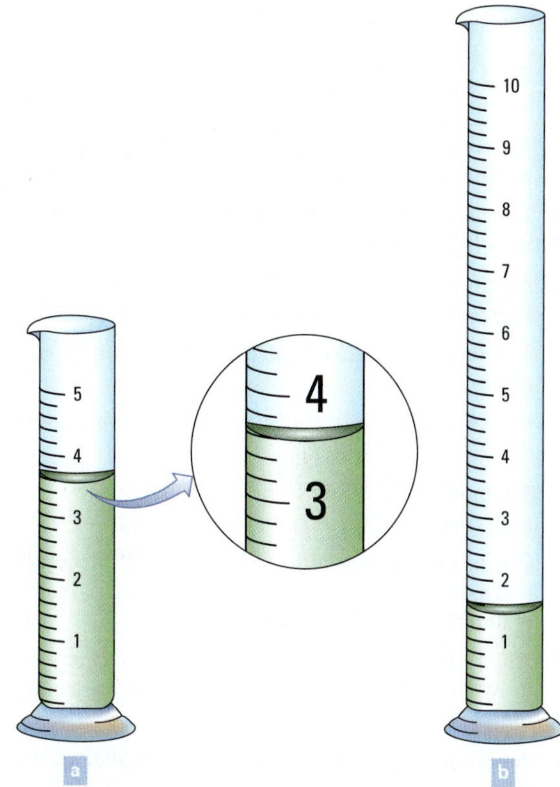

52. How long is the object measured with the rulers shown? The ruler is calibrated in inches. Explain why you selected the number of digits in your length values.

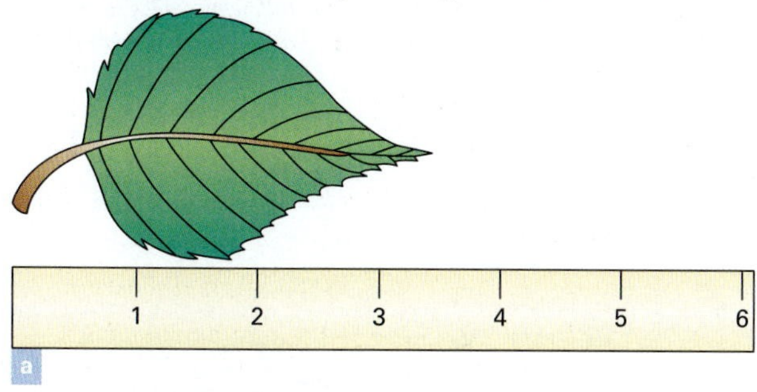

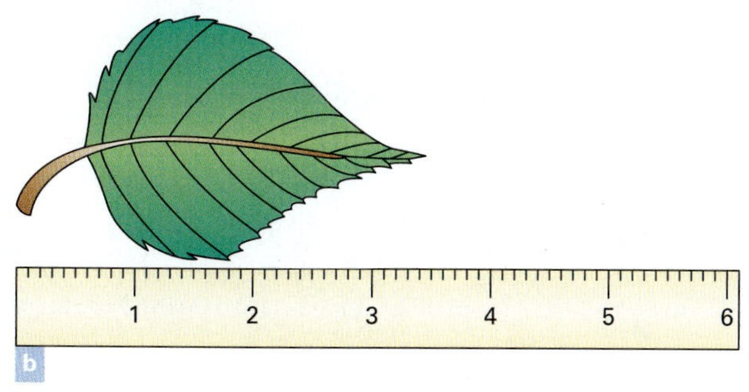

53. The same volume of liquid is in each measuring instrument depicted here (volume delivered by the buret on the right). Explain why the quantities are expressed with different values.

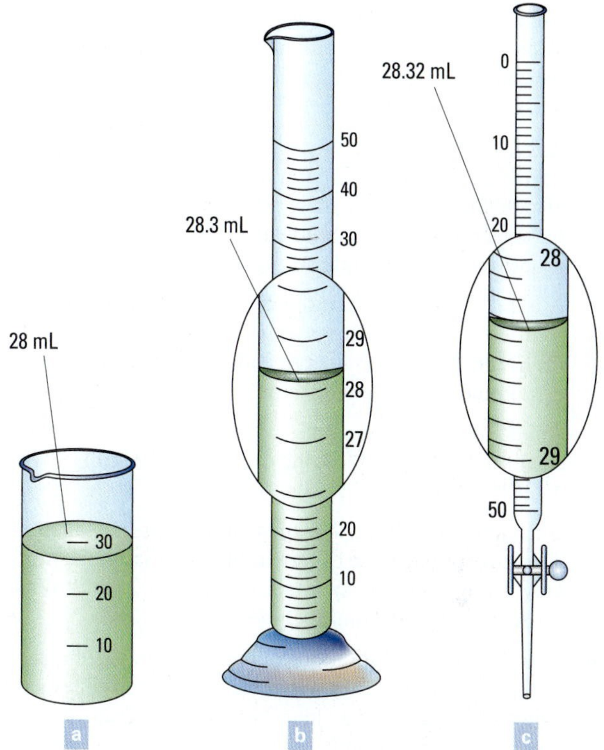

54. Why is the length of the line in the illustration here reported as 2.55 in. with one ruler and as 2.5 in. with another?

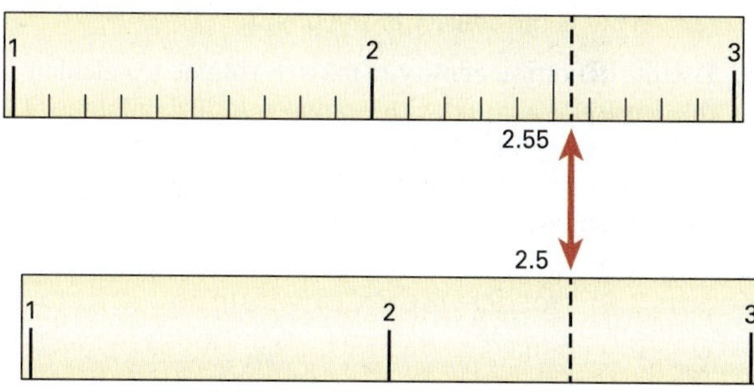

Questions 55 and 56: To how many significant figures is each quantity expressed?

55. (a) 75.9 g sugar, (b) 89.583 mL weed killer, (c) 0.366 in. diameter glass fiber, (d) 48,000 cm wire, (e) 0.80 ft spaghetti, (f) 0.625 kg silver, (g) 9.6941×10^6 cm thread, (h) 8.010×10^{-3} L acid

56. (a) 4.5609 g salt, (b) 0.10 in. diameter wire, (c) 12.3×10^{-3} kg fat, (d) 5310 cm^3 copper, (e) 0.0231 ft licorice, (f) 6.1240×10^6 L salt brine, (g) 328 mL ginger ale, (h) 1200.0 mg dye

Section 3.8: Significant Figures in Calculations

Questions 57 and 58: Round off each quantity to three significant figures.

57. (a) 6.398×10^{-3} km rope, (b) 0.0178 g silver nitrate, (c) 79,000 m cable, (d) 42,150 tons fertilizer, (e) $649.85

58. (a) 52.20 mL helium, (b) 17.963 g nitrogen, (c) 78.45 mg MSG, (d) 23,642,000 mm wavelength, (e) 0.0041962 kg lead

59. A moving-van crew picks up the following items: a couch that weighs 147 pounds, a chair that weighs 67.7 pounds, a piano at 3.6×10^2 pounds, and several boxes having a total weight of 135.43 pounds. Calculate and express in the correct number of significant figures the total weight of the load.

60. A solution is prepared by dissolving 2.86 grams of sodium chloride, 3.9 grams of ammonium sulfate, and 0.896 grams of potassium iodide in 246 grams of water. Calculate the total mass of the solution and express the sum in the proper number of significant figures.

61. A buret contains 22.93 milliliters of sodium hydroxide solution. A few minutes later, the volume is down to 19.4 milliliters because of a small leak. How many milliliters of solution have drained from the buret?

62. An empty beaker has a mass of 94.33 grams. After some chemical has been added, the mass is 101.209 grams. What is the mass of the chemical in the beaker?

63. The mole is the SI unit for the amount of a substance. The mass of one mole of pure table sugar is 342.3 grams. How many grams of sugar are in exactly 1/2 mole? What is the mass of 0.764 mole?

64. Exactly 1 liter of a solution contains 31.4 grams of a certain dissolved substance. What mass in grams is in exactly 2 liters? How about 7.37 liters? Express the results in the proper number of significant figures.

65. An empty beaker with a mass of 42.3 grams is filled with a liquid, and the resulting mass of the liquid and the beaker is 62.87 grams. The volume of this liquid is 19 milliliters. What is the density of the liquid?

66. Use the definition density ≡ mass ÷ volume to calculate the density of a liquid with a volume of 50.6 milliliters if that liquid is placed in an empty beaker with a mass of 32.344 grams and the mass of the liquid plus the beaker is 84.64 grams.

Section 3.9: Metric–USCS Conversions

Questions 67–84: You may consult Table 3.2 while answering these questions.

67. 0.0715 gal = _____ cm^3 2.27×10^4 mL = _____ gal

68. 19.3 L = _____ gal 0.461 qt = _____ L

69. A popular breakfast cereal comes in a box containing 515 grams. How many pounds (lb) of cereal is this?

70. A copy of your chemistry textbook is found to have a mass of 2.60×10^3 grams. What is the mass of this copy of your chemistry textbook in ounces?

71. The payload of a small pickup truck is 1450 pounds (assume three significant figures). What is this in kilograms?

72. The Hope diamond is the world's largest blue diamond. It weighs 45.52 carats. If 1 carat is defined as 200 milligrams, calculate the mass of the diamond in grams and in ounces.

73. There is 115 milligrams of calcium in a 100-gram serving of whole milk. How many grams of calcium is this? How many pounds?

74. The largest recorded difference of weight between spouses is 922 pounds. The husband weighed 1020 pounds, and his wife, 98 pounds. Express this difference in kilograms.

75. An Austrian boxer reads 69.1 kilograms when he steps on a balance (scale) in his gymnasium. Should he be classified as a welterweight (136 to 147 pounds) or a middleweight (148 to 160 pounds)?

76. A woman gives birth to a 7.5-pound baby. How would a hospital using metric units record this baby's mass?

77. The height of Angel Falls in Venezuela is 979 meters. How high is this in (a) yards; (b) feet?

78. A penny is found to have a length of 1.97 centimeters. What is the length of this penny in inches?

79. The Willis Tower in Chicago is 1451 feet tall. How high is this in meters?

80. What is the length of the Mississippi River in kilometers if it is 2.3×10^3 miles long?

81. The summit of Mount Everest is 29,029 ft above sea level. Express this height in kilometers.

82. One of the smallest brilliant-cut diamonds ever crafted had a diameter of about 1/50 inch (0.02 inch). How many millimeters is this?

83. An office building is heated by oil-fired burners that draw fuel from a 619-gallon storage tank. Calculate the tank volume in liters.

84. A gas can is found to have a volume of 9.10 liters. What is the volume of this gas can in gallons?

Section 3.10: Temperature

Questions 85 and 86: Fill in the spaces in the following tables so that each temperature is expressed in all three scales. Round off answers to the nearest degree.

85.

Celsius	Fahrenheit	Kelvin
69		
	−29	
		111
	36	
		358
−141		

86.

Celsius	Fahrenheit	Kelvin
	40	
		590
−13		
		229
440		
	−314	

87. "Normal" body temperature is 98.6°F. What is this temperature in Celsius degrees?

88. In the winter, a heated home in the Northeast might be maintained at a temperature of 74°F. What is this temperature on the Celsius and Kelvin scales?

89. Energy conservationists suggest that air conditioners should be set so that they do not turn on until the temperature tops 78°F. What is the Celsius equivalent of this temperature?

90. The melting point of an unknown solid is determined to be 49.0°C. What is this temperature on the Fahrenheit and Kelvin scales?

91. The world's highest shade temperature was recorded in Libya at 58.0°C. What is its Fahrenheit equivalent?

92. The boiling point of a liquid is calculated to be 454 K. What is this temperature on the Celsius and Fahrenheit scales?

Section 3.11: Proportionality and Density

93. Consider the graph. (a) Determine the value of m and b in the equation y = mx + b. (b) Are x and y directly proportional? Explain. (c) What is the value of y when x = −3.00?

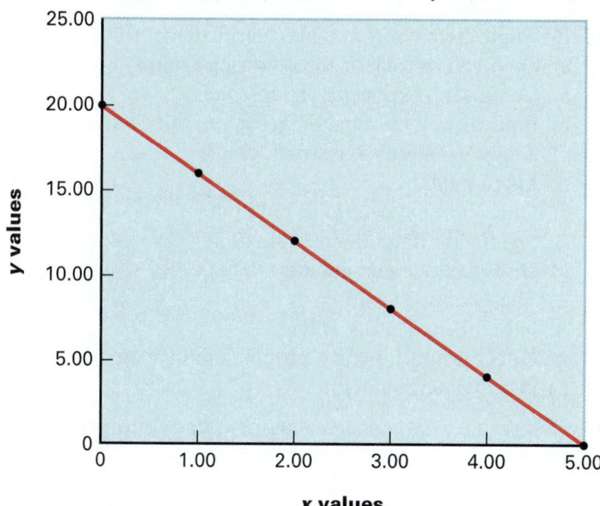

94. Consider the graph. (a) Determine the value of m and b in the equation y = mx + b. (b) Are x and y directly proportional? Explain. (c) What is the value of x when y = 25.00?

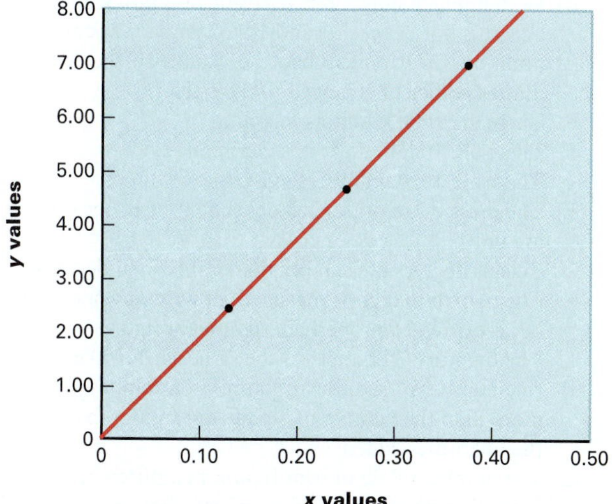

95. The amount of heat (q) absorbed when a pure substance melts is proportional to the mass of the sample (m). Express this proportionality in mathematical form. Change it into an equation, using the symbol ΔH_{fus} for the proportionality constant. This constant is the heat of fusion of a pure substance. If heat is measured in calories, what are the units of heat of fusion? Write a word definition of heat of fusion.

96. The distance, d, traveled by an automobile moving at an average speed of s is directly proportional to the time, t, spent traveling. The proportionality constant is the average speed. (a) Express this proportionality in mathematical form. (b) If it takes the automobile 3.88 hours to travel a distance of 239 miles, what are the value and units of the proportionality constant? (c) Assuming the same average speed as in part (b), how long will it take the automobile to travel a distance of 659 miles?

97. It takes 7.39 kilocalories to melt 92 grams of ice. Calculate the heat of fusion of water. (*Hint:* See Question 95. Careful on the units; the answer will be in calories per gram.)

98. If the pressure of a sample of gas is held constant, its volume, V, is directly proportional to the absolute temperature of the gas, T. (a) Write an equation for the proportionality between V and T, in which b is the proportionality constant. (b) For 48.0 g of oxygen gas at a pressure of 0.373 atmosphere, V is observed to be 93.4 L when T is 283 K. What are the value and units of the proportionality constant, b? (c) What volume will this gas sample occupy at a temperature of 375 K?

99. If the temperature and amount of a gas are held constant, the pressure (P) it exerts is inversely proportional to volume (V). This means that pressure is directly proportional to the inverse of volume, or 1/V. Write this as a proportionality and then as an equation with k′ as the proportionality constant. What are the units of k′ if pressure is in atmospheres and volume is in liters?

100. The mass, m, of a piece of metal is directly proportional to its volume, V, where the proportionality constant is the density, D, of the metal. (a) Write an equation that represents this direct proportion, in which D is the proportionality constant. (b) The density of iron metal is 7.88 g/cm^3. What is the mass of a piece of iron that has a volume of 23.5 cm^3? (c) What is the volume of a piece of iron metal that has a mass of 172 g?

101. Give a particulate-level explanation of why ice (solid water) is less dense than liquid water.

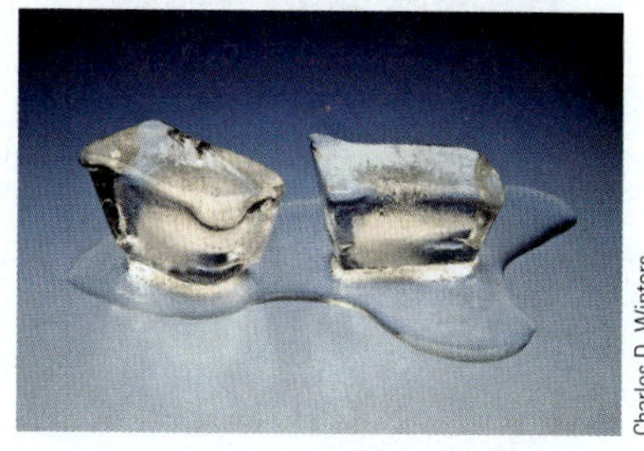
Charles D. Winters

102. Rank the substances in the photograph from least dense to most dense. Explain your reasoning.

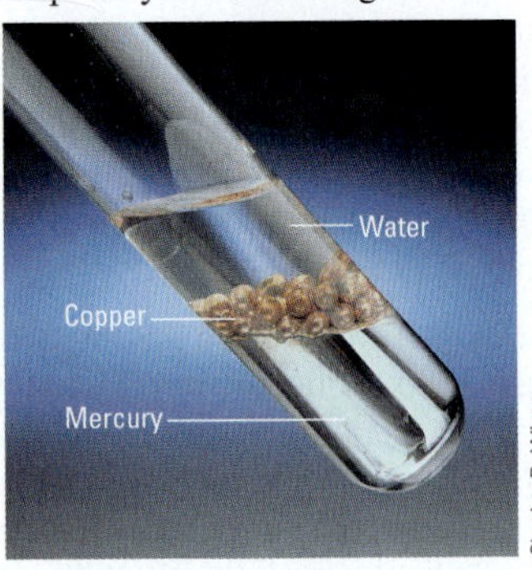

Charles D. Winters

103. Calculate the density of benzene, a liquid used in chemistry laboratories, if 166 g of benzene fills a graduated cylinder to the 188-mL mark.

104. A general chemistry student found a chunk of metal in the basement of a friend's house. To figure out what it was, she used the ideas just developed in class about density. She measured the mass of the metal to be 175.2 grams. Then she dropped the metal into a measuring cup and found that it displaced 15.3 milliliters of water. Calculate the density of the metal. What element is this metal most likely to be?

105. Densities of gases are usually measured in grams per liter (g/L). Calculate the density of air if the mass of 15.7 L is 18.6 g.

106. A rectangular block of iron 4.60 cm × 10.3 cm × 13.2 cm has a mass of 4.92 kg. Find its density in g/cm^3.

107. Ether, a well-known anesthetic, has a density of 0.736 g/cm^3. What is the volume of 471 g of ether?

108. Calculate the mass of 17.0 mL of aluminum, which has a density of 2.72 g/mL.

109. Determine the mass of 2.0 L rubbing alcohol, which has a density of 0.786 g/mL.

110. Calculate the volume occupied by 15.4 g of nickel, which has a density of 8.91 g/mL.

111. The mass of the liquid in the graduated cylinder shown here is 7.304 g. What is the density of the liquid?

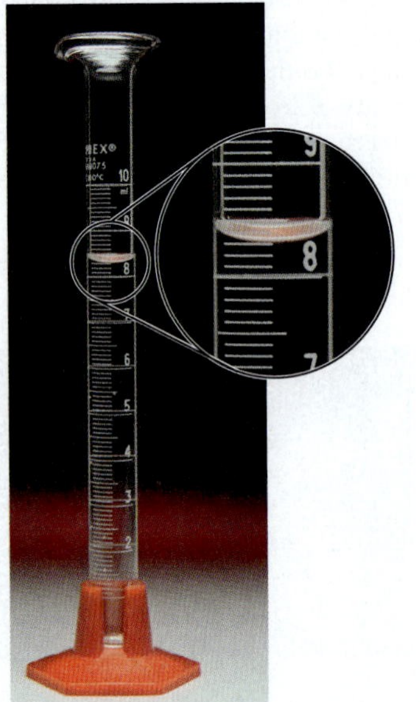

Charles D. Winters

112. What is the density of the metal shown here?

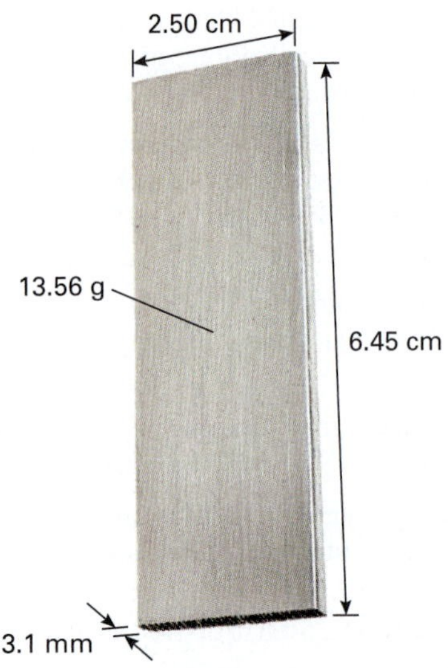

General Questions

113. Distinguish precisely and in scientific terms the differences among items in each of the following groups.

a) Coefficient, exponent, exponential
b) Equivalency, conversion factor, quantity, value, unit
c) Analyze, identify, construct, check
d) Mass, weight
e) Unit, kilounit, centiunit, milliunit
f) Significant figures, uncertain digit
g) Uncertainty, exact number
h) The symbols = and ≡
i) Fahrenheit, Celsius, kelvin
j) Direct proportionality, proportionality constant
k) Density, mass, volume

114. Determine whether each statement that follows is true or false:

a) The SI system includes metric units.
b) If two quantities are expressed in an equivalency, they are directly proportional to each other.
c) The scientific notation form of a number smaller than 1 has a positive exponent.
d) In changing a number in scientific notation whose coefficient is not between 1 and 10 to standard scientific notation, the exponent becomes smaller if the decimal in the coefficient is moved to the right.
e) There are 1000 kilounits in a unit.
f) There are 10 milliunits in a centiunit.
g) There are 1000 milliliters in a cubic centimeter.
h) The mass of an object is independent of its location in the universe.
i) Celsius degrees are smaller than Fahrenheit degrees.
j) The uncertain digit is the last digit written when a number is expressed properly in significant figures.
k) The quantity 76.2 g means the same as 76.200 g.
l) The number of significant figures in a sum may be more than the number of significant figures in any of the quantities added.
m) The number of significant figures in a difference may be fewer than the number of significant figures in any of the quantities subtracted.

n) The number of significant figures in a product may be more than the number of significant figures in any of the quantities multiplied.

o) The process of analysis of a problem statement includes describing the properties of the given and wanted quantities.

p) If the quantity in the answer to a problem is familiar, it is not necessary to check to make sure the answer is reasonable.

q) Conversion factors can be used to change from one unit to another only when the quantities are directly proportional.

r) When you are learning chemistry, you should check the solution to each problem you solve at two levels: (1) Is the value reasonable? (2) What new knowledge or skill did I obtain or improve?

s) There is no advantage to using units in a problem that is solved by algebra.

t) A Fahrenheit temperature can be changed to a Celsius temperature by multiplying by a conversion factor.

115. How tall are you in (a) meters; (b) decimeters; (c) centimeters; (d) millimeters? Which of the four metric units do you think would be most useful in expressing people's heights without resorting to decimal fractions?

116. What do you weigh in (a) milligrams; (b) grams; (c) kilograms? Which of these units do you think is best for expressing a person's weight? Why?

117. Standard printer and copier paper is the United States is 8½ in. by 11 in. What are these dimensions in centimeters?

118. The density of aluminum is 2.7 g/cm^3. An ecology-minded student has gathered 126 empty aluminum cans for recycling. If there are 21 cans per pound, how many cubic centimeters and grams of aluminum does the student have?

More Challenging Problems

119. What is the average density of a single marble shown here? The total mass of the three marbles is 96.5 grams. The photograph on the left shows the graduated cylinder filled with 61 milliliters of water before the marbles were added.

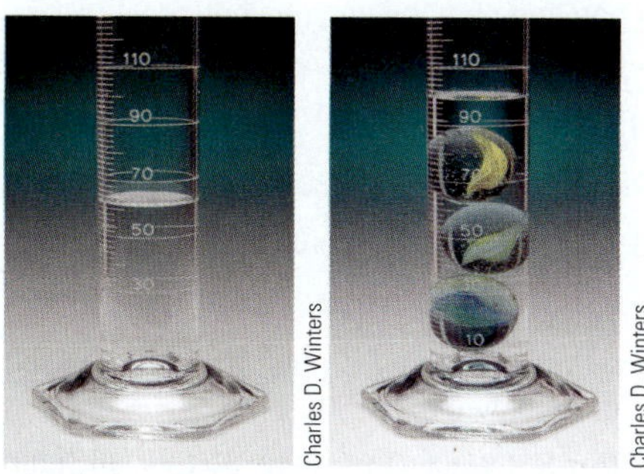

120. A woman has just given birth to a bouncing 6 pounds, 7 ounces baby boy. How should she describe the weight of her child to her sister, who lives in Sweden—in metric units, of course?

121. A student's driver's license lists her height as 5 feet, 5 inches. What is her height in meters?

122. How many grams of milk are in a 12.0-fluid-ounce glass? The density of milk is 64.4 lb/ft^3. There are 7.48 gal/ft^3; and, by definition, there are 4 qt/gal and 32 fl oz/qt.

123. The fuel tank in an automobile has a capacity of 11.8 gallons. If the density of gasoline is 42.0 lb/ft^3, what is the mass of fuel in kilograms when the tank is full?

124. A welcome rainfall caused the temperature to drop by 33°F after a sweltering day in Chicago. What is this temperature drop in degrees Celsius?

125. At high noon on the lunar equator the temperature may reach 243°F. At night the temperature may sink to −261°F. Express the temperature difference in degrees Celsius.

126. A recipe calls for a quarter cup of butter. Calculate its mass in grams if its density is 0.86 g/cm^3 (1 cup = 0.25 qt).

127. Calculate the mass in pounds of 1 gallon of water, given that the density of water is 1.0 g/mL.

128. In Active Example 3.29 you calculated that you would have to work six weeks to earn enough money to buy a \$1082.49 television. You would be working five shifts of four hours each at \$9.25/hr. But, alas, when you received your first paycheck, you found that exactly 23% of your earnings had been withheld for social security, federal and state income taxes, and workers' compensation insurance. Taking these into account, how many weeks will it take to earn the \$1082.49?

Answers to Practice Exercises

1. (a) 3.8875×10^3; (b) 4.09809089×10^8; (c) 2.2×10^{-5}; (d) 5×10^{-7}

2. (a) 0.000011; (b) 0.143; (c) 477: (d) 5,008,585.85

3. $\text{Area} = \text{length} \times \text{width} = 1 \text{ furlong} \times \dfrac{1 \text{ rod}}{0.025 \text{ furlong}} \times 4 \text{ rods} = 160 \text{ rods} \times \text{rods} = 160 \text{ rods}^2 = 160 \text{ square rods}$

4. (a) $1 \text{ tablet} = 500 \text{ milligrams}, \dfrac{1 \text{ tablet}}{500 \text{ milligrams}}, \dfrac{500 \text{ milligrams}}{1 \text{ tablet}}$

(b) $288 \text{ fluid ounces} = 1 \text{ case}, \dfrac{288 \text{ fluid ounces}}{1 \text{ case}}, \dfrac{1 \text{ case}}{288 \text{ fluid ounces}}$

(c) $10 \text{ nickels} = 2 \text{ quarters}, \dfrac{10 \text{ nickels}}{2 \text{ quarters}}, \dfrac{2 \text{ quarters}}{10 \text{ nickels}}$

(d) $2\frac{3}{4} \text{ cups} = 24 \text{ cupcakes}, \dfrac{2\frac{3}{4} \text{ cups}}{24 \text{ cupcakes}}, \dfrac{24 \text{ cupcakes}}{2\frac{3}{4} \text{ cups}}$

5. $12 \text{ minutes} \times \dfrac{60 \text{ seconds}}{\text{minute}} = 720 \text{ seconds}$

6. $15 \text{ hours} \times \dfrac{60 \text{ minutes}}{\text{hour}} \times \dfrac{60 \text{ seconds}}{\text{minute}} = 54{,}000 \text{ seconds}$

7. $345 \text{ miles} \times \dfrac{1 \text{ hour}}{150 \text{ miles}} = 2.3 \text{ hours}; 0.3 \text{ hour} \times \dfrac{60 \text{ minutes}}{\text{hour}} = 18 \text{ minutes}; 11{:}18 \text{ AM}$

8. $135 \text{ dimes} \times \frac{10 \text{ pennies}}{\text{dime}} = 1350 \text{ pennies}$

9. $800 \text{ nickels} \times \frac{1 \text{ five-dollar bill}}{100 \text{ nickels}} = 8 \text{ five-dollar bills}$

10. $0.711 \text{ kg} \times \frac{1000 \text{ g}}{\text{kg}} = 711 \text{ g}$

11. $5.25 \times 10^5 \text{ mg} \times \frac{1 \text{ g}}{1000 \text{ mg}} \times \frac{1 \text{ kg}}{1000 \text{ g}} = 0.525 \text{ kg}$

12. $7.05 \times 10^3 \text{ cm}^3 \times \frac{1 \text{ mL}}{1 \text{ cm}^3} \times \frac{1 \text{ L}}{1000 \text{ mL}} \times \frac{100 \text{ cL}}{\text{L}} = 705 \text{ cL}$

13. 20.0 mL

14. (a) 3; (b) 2; (c) 3; (d) 4

15. (a) 26 mL; (b) 0.0025 m; (c) 1.5×10^5 mg; (d) 2.0×10^2 Gg

16. $2.3 \times 10^3 \text{ mL} + 4.22 \times 10^4 \text{ mL} + 9.04 \times 10^3 \text{ mL} + 8.71 \times 10^5 \text{ mL} = 2.3 \times 10^3 \text{ mL} + 42.2 \times 10^3 \text{ mL} + 9.04 \times 10^3 \text{ mL} + 871 \times 10^3 \text{ mL} = 925 \times 10^3 \text{ mL} = 9.25 \times 10^5 \text{ mL}$

17. $33 \text{ g} \times \frac{1.00 \text{ mL}}{0.878 \text{ g}} = 38 \text{ mL}$

18. $55 \text{ weeks} \times \frac{7 \text{ days}}{1 \text{ week}} = 3.9 \times 10^2 \text{ days}$

19. $\frac{96 \text{ g}}{(37.25 - 25.00) \text{ mL}} = 7.8 \text{ g/mL}$

20. $2.50 \text{ km} \times \frac{1000 \text{ m}}{\text{km}} \times \frac{100 \text{ cm}}{\text{m}} \times \frac{1 \text{ in.}}{2.54 \text{ cm}} \times \frac{1 \text{ ft}}{12 \text{ in.}} \times \frac{1 \text{ mi}}{5280 \text{ ft}} = 1.55 \text{ mi}$

21. $408 \text{ oz} \times \frac{1 \text{ lb}}{16 \text{ oz}} \times \frac{453.59237 \text{ g}}{1 \text{ lb}} \times \frac{1 \text{ kg}}{1000 \text{ g}} = 11.6 \text{ kg}$

22. $45 \text{ mL} \times \frac{1 \text{ L}}{1000 \text{ mL}} \times \frac{1 \text{ gal}}{3.785 \text{ L}} \times \frac{4 \text{ qt}}{1 \text{ gal}} \times \frac{32 \text{ fl oz}}{1 \text{ qt}} = 1.5 \text{ fl oz}$

23. $T_{°C} = T_K - 273 = 288 - 273 = 15°C$; $T_{°F} = 1.8\, T_{°C} + 32 = (1.8 \times 15) + 32 = 59°F$

24. $T_{°F} - 32 = 1.8\, T_{°C}$; $x - 32 = 1.8x$; $0.8x = -32$; $x = -40°F = -40°C$

25. $D \equiv \frac{m}{V} = \frac{39.59 \text{ g}}{50.00 \text{ mL}} = 0.7918 \text{ g/mL}$

26. $975 \text{ mL} \times \frac{1.26 \text{ g}}{\text{mL}} = 1.23 \times 10^3 \text{ g}$

27. $15 \text{ cm} \times 8 \text{ cm} \times 3 \text{ cm} \times \frac{11.4 \text{ g}}{\text{cm}^3} \times \frac{1 \text{ kg}}{1000 \text{ g}} = 4 \text{ kg}$

28. $591 \text{ mL} \times \frac{1 \text{ cm}^3}{1 \text{ mL}} \times \frac{1 \text{ in.}^3}{(2.54)^3 \text{ cm}^3} \times \frac{1 \text{ ft}^3}{(12)^3 \text{ in.}^3} = 0.0209 \text{ ft}^3$

29. $1 \text{ AY} \times \frac{2 \text{ sem}}{\text{AY}} \times \frac{24 \text{ sections}}{\text{semester}} \times \frac{20 \text{ students}}{\text{section}} \times \frac{15 \text{ g}}{\text{semester}} \times \frac{1 \text{ lb}}{454 \text{ g}} \times \frac{1 \text{ bottle}}{2.5 \text{ lb}} = 13 \text{ bottles}$

Answers to Concept-Linking Exercises

You may have found more relationships or relationships other than the ones given in these answers.

1. An equivalency is an expression of two quantities that are equivalent, or essentially equal. For example, 12 inches = 1 foot, 1000 milligrams = 1 gram, and 1 cubic centimeter = 1 milliliter. An equivalency can be expressed in the form of two conversion factors; for example, the equivalency 12 inches = 1 foot yields $\frac{12 \text{ inches}}{1 \text{ foot}}$ and $\frac{1 \text{ foot}}{12 \text{ inches}}$. The use of equivalencies and conversion factors is central to solving quantitative problems in chemistry. A four-step method for solving quantitative problems is: (1) analyze the problem statement, (2) identify the equivalencies needed to solve the problem, (3) construct the solution setup, and (4) check the solution.

2. Scientific measurements are made using the metric system of measurement. SI units are included in the metric system. SI is an abbreviation for the French name for the international system of units. The SI system defines seven base units. Examples are kilograms (for mass) and meters (for length). Other quantities are made up of combinations of base units; these are called derived units. Examples are volume and density.

3. The kilogram is the unit of mass in the SI system, and the gram is 1/1000 of a kilogram; the smaller unit is more commonly used in the laboratory. Weight is a measure of gravitational attraction that is proportional to mass. A pound is a weight unit.

4. There is some degree of uncertainty in every physical measurement. In scientific work, measurements are expressed in all digits known accurately plus one digit that is uncertain, which is known as the uncertain digit. Collectively, these digits are significant figures. Significant figures are not applied to exact numbers, which have no uncertainty.

5. Two quantities are directly proportional if they increase or decrease at the same rate; the ratio of one to the other is constant. Two related variables are inversely proportional to each other if one increases and the other decreases in such a way that their product is a constant. A proportionality is indicated by the operator symbol $\propto$: $A \propto B$ is the symbolic expression of "A is proportional to B." A proportionality may be converted to an equation by inserting a proportionality constant, often symbolized as k: $A = kB$.

Answers to Blue-Numbered Questions, Exercises, and Problems

1. (a) 3.22×10^{-4}; (b) 6.03×10^9; (c) 6.19×10^{-12}

3. (a) 5,120,000; (b) 0.000000840; (c) 1,920,000,000,000,000,000,000

5. (a) 7.29×10^{-3}; (b) 5.13×10^{-12}; (c) 2.98×10^9; (d) 4.02×10^{-6}

7. (a) 75.6; (b) 9.41; (c) 3.24×10^4; (d) 1.49×10^{-3}

9. (a) 3.77×10^{-8}; (b) 2.0×10^6

11. (a) 4.65×10^8; (b) 3.0×10^{-7}

13. Two quantities, x and y, must be related in the form $y = mx + b$, where $b \neq 0$.

15. (a) 3 feet = 1 yard; (b) 1 meter = 100 centimeters; (c) 1 milliliter = 20 drops

17. (a) 10 dimes = 1 dollar, $\frac{10 \text{ dimes}}{1 \text{ dollar}}, \frac{1 \text{ dollar}}{10 \text{ dimes}}$
(b) 1 field = 100 yards, $\frac{1 \text{ field}}{100 \text{ yards}}, \frac{100 \text{ yards}}{1 \text{ field}}$

(c) 16 ounces = 1 pound, $\frac{16 \text{ ounces}}{1 \text{ pound}}, \frac{1 \text{ pound}}{16 \text{ ounces}}$

19. (a) 1 minute = 60 seconds, $\frac{1 \text{ minute}}{60 \text{ seconds}}, \frac{60 \text{ seconds}}{1 \text{ minute}}$
(b) 12 inches = 1 foot, $\frac{12 \text{ inches}}{1 \text{ foot}}, \frac{1 \text{ foot}}{12 \text{ inches}}$
(c) 1 dollar = 100 cents, $\frac{1 \text{ dollar}}{100 \text{ cents}}, \frac{100 \text{ cents}}{1 \text{ dollar}}$

21. $406 \text{ mi} \times \frac{1 \text{ h}}{48 \text{ mi}} = 8.5 \text{ h}$

23. $4.3 \text{ km} \times \frac{1 \text{ h}}{88 \text{ km}} \times \frac{60 \text{ min}}{\text{h}} = 2.9 \text{ min}$

25. $62 \text{ ft} \times \frac{9 \text{ nails}}{\text{foot}} \times \frac{1 \text{ lb}}{36 \text{ nails}} \times \frac{69 \text{ cents}}{\text{lb}} \times \frac{1 \text{ dollar}}{100 \text{ cents}} = 11 \text{ dollars}$

27. $259 \text{ pesos} \times \frac{1 \text{ dollar}}{13 \text{ pesos}} = 2.0 \times 10^1 \text{ dollars}$

29. $1 \text{ decade} \times \frac{10 \text{ years}}{\text{decade}} \times \frac{52 \text{ weeks}}{\text{year}} = 5.2 \times 10^2 \text{ weeks}$

31. A patient's temperature, blood pressure, weight, height, and so on; the volume (liquid) or mass (solid) of a medication dosage, and so on.

33. Her weight is the same when the elevator is standing still as when it moves at a constant rate. It decreases when the elevator slows; it increases when the elevator accelerates. Her mass is constant no matter what the elevator is doing.

35. The meter.

37. A kilobuck is \$1000. A megabuck is \$1,000,000.

39. 1 mL = 0.001 L.

41. Megagrams, because a gram is a very small unit when compared to the mass of an automobile.

43. (a) 0.0574 g; (b) 1.41×10^3 g; (c) 4.54×10^9 mg

45. (a) 2.17×10^3 cm; (b) 0.517 km; (c) 6.66×10^4 cm

47. (a) 494 mL; (b) 1.91×10^3 mL; (c) 0.874 L

49. (a) 711 g; (b) 5.27×10^5 pm; (c) 3.63×10^5 dag

51. (a) 3.5 mL; each mark represents 0.2 mL, and the volume is estimated to be about halfway between the 3.4 mL mark and the 3.6 mL mark. (b) 1.5 mL; the volume is estimated to be about 1/2 of the distance between the 1.4 mL and the 1.6 mL marks.

53. The number of digits in a measured value depends on the measuring process. In the beaker, the tens place is known with certainty, and the ones place is estimated. In the graduated cylinder, the ones place is known with certainty, and the tenths place is estimated. In the buret, the tenths place is known with certainty, and the hundredths place is estimated.

55. (a) 3; (b) 5; (c) 3; (d) uncertain—2 to 5; (e) 2; (f) 3; (g) 5; (h) 4

57. (a) 6.40×10^{-3} km; (b) 0.0178 g; (c) 7.90×10^4 m; (d) 4.22×10^4 tons; (e) 6.50×10^2 dollars

59. $147 \text{ lb} + 67.7 \text{ lb} + 3.6 \times 10^2 \text{ lb} + 135.43 \text{ lb} = 7.1 \times 10^2 \text{ lb}$

61. 22.93 mL − 19.4 mL = 3.5 mL

63. $\frac{1}{2} \text{ mol} \times \frac{342.3 \text{ g}}{\text{mol}} = 171.2 \text{ g}; 0.764 \text{ mol} \times \frac{342.3 \text{ g}}{\text{mol}} = 262 \text{ g}$

65. $\frac{(62.87 \text{ g} - 42.3 \text{ g})}{19 \text{ mL}} = 1.1 \text{ g/mL}$

67. $0.0715 \text{ gal} \times \frac{3.785 \text{ L}}{\text{gal}} \times \frac{1000 \text{ cm}^3}{\text{L}} = 271 \text{ cm}^3;$

$2.27 \times 10^4 \text{ mL} \times \frac{1 \text{L}}{1000 \text{ mL}} \times \frac{1 \text{ gal}}{3.785 \text{ L}} = 6.00 \text{ gal}$

69. $515 \text{ g} \times \frac{1 \text{ lb}}{453.59 \text{ g}} = 1.14 \text{ lb}$

71. $1.45 \times 10^3 \text{ lb} \times \frac{453.59 \text{ g}}{\text{lb}} \times \frac{1 \text{ kg}}{1000 \text{ g}} = 658 \text{ kg}$

73. $115 \text{ mg} \times \frac{1 \text{ g}}{1000 \text{ mg}} = 0.115 \text{ g}; 0.115 \text{ g} \times \frac{1 \text{ lb}}{453.59 \text{ g}} = 2.54 \times 10^{-4} \text{ lb}$

75. $69.1 \text{ kg} \times \frac{1000 \text{ g}}{\text{kg}} \times \frac{1 \text{ lb}}{453.59 \text{ g}} = 152 \text{ lb}$, a middleweight

77. $979 \text{ m} \times \frac{100 \text{ cm}}{\text{m}} \times \frac{1 \text{ in.}}{2.54 \text{ cm}} \times \frac{1 \text{ ft}}{12 \text{ in.}} (= 3.21 \times 10^3 \text{ ft}) \times \frac{1 \text{ yd}}{3 \text{ ft}} = 1.07 \times 10^3 \text{ yd}$

79. $1451 \text{ ft} \times \frac{12 \text{ in.}}{\text{ft}} \times \frac{2.54 \text{ cm}}{\text{in.}} \times \frac{1 \text{ m}}{100 \text{ cm}} = 442.3 \text{ m}$

81. $29{,}029 \text{ ft} \times \frac{12 \text{ in.}}{\text{ft}} \times \frac{2.54 \text{ cm}}{\text{in.}} \times \frac{1 \text{ m}}{100 \text{ cm}} \times \frac{1 \text{ km}}{1000 \text{ cm}} = 8.8480 \text{ km}$

83. $619 \text{ gal} \times \frac{3.785 \text{ L}}{\text{gal}} = 2.34 \times 10^3 \text{ L}$

85.

Celsius	Fahrenheit	Kelvin
69	156	342
−34	−29	239
−162	−260	111
2	36	275
85	185	358
−141	−222	132

87. 37.0°C

89. 26°C

91. 136°F

93. (a) $m = \frac{\Delta y}{\Delta x} = \frac{0 - 20.00}{5.00 - 0} = -4.00; (y - y_1) = m(x - x_1);$
$(y - 20.00) = -4.00(x - 0); y = -4.00x + 20.00$
(b) No, a direct proportionality must have the form $y = mx + 0$. (c) $y = -4.00(-3.00) + 20.00 = 32.00$

95. $q \propto m; q = \Delta H_{fus} \times m$; cal/g; heat energy lost or gained per gram while changing state from liquid to solid or vice versa

97. $\Delta H_{fus} = \frac{q}{m} = \frac{7.39 \text{ kcal}}{92 \text{ g}} \times \frac{1000 \text{ cal}}{\text{kcal}} = 8.0 \times 10^1 \text{cal/g}$

99. $P \propto \frac{1}{V}; P = k' \times \frac{1}{V};$ atm · L

101. See Figure 3.26.

103. $D \equiv \frac{m}{V} = \frac{166\ g}{188\ mL} = 0.883\ g/mL$

105. $D \equiv \frac{m}{V} = \frac{18.6\ g}{15.7\ L} = 1.18\ g/L$

107. $471\ g \times \frac{1\ cm^3}{0.736\ g} = 6.40 \times 10^2\ cm^3$

109. $2.0\ L \times \frac{0.786\ g}{mL} \times \frac{1000\ mL}{L} = 1.6 \times 10^3\ g$

111. $D \equiv \frac{m}{V} = \frac{7.304\ g}{8.30\ mL} = 0.880\ g/mL$

114. True: a, b, d, f, h, j, l, m, o, q, r. False: c, e, g, i, k, n, p, s, t.

115. A 6-foot-tall person is 1.83 m, 18.3 dm, 183 cm, and 1.83×10^3 mm. The centimeter is thus generally accepted as the preferred unit to express human height without using decimal fractions.

116. Sample calculation for a 150-lb person: $150\ lb \times \frac{453.59\ g}{lb} \times \frac{1\ kg}{1000\ g} = 68.0\ kg = 68{,}000\ g = 68{,}000{,}000\ mg$. Kilograms are best because the number of g and mg are inconveniently large.

117. $8.5\ in. \times \frac{2.54\ cm}{1\ in.} = 22\ cm$; $11\ in. \times \frac{2.54\ cm}{1\ in.} = 28\ cm$

118. $126\ cans \times \frac{1\ lb}{21\ cans} \times \frac{454\ g}{lb} (= 2.7 \times 10^3\ g) \times \frac{1\ cm^3}{2.7\ g}$

$= 1.0 \times 10^3\ cm^3$

119. $D \equiv \frac{m}{V} = \frac{96.5\ g}{99\ mL - 61\ mL} = 2.5\ g/mL$

120. $7\ oz \times \frac{1\ lb}{16\ oz} = 0.4\ lb$; $6.4\ lb \times \frac{454\ g}{lb} = 2.9 \times 10^3\ g = 2.9\ kg$

122. $12.0\ fl\ oz \times \frac{1\ qt}{32\ fl\ oz} \times \frac{1\ gal}{4\ qt} \times \frac{1\ ft^3}{7.48\ gal} \times \frac{64.4\ lb}{ft^3} \times \frac{454\ g}{lb} = 366\ g$

124. $33°F \times \frac{100\ Celsius\ degrees}{180\ Fahrenheit\ degrees} = 18°C$

126. $0.25\ cup \times \frac{0.25\ qt}{cup} \times \frac{1\ gal}{4\ qt} \times \frac{3.785\ L}{gal} \times \frac{1000\ mL}{L}$

$\times \frac{1\ cm^3}{1\ mL} \times \frac{0.86\ g}{cm^3} = 51\ g$

128. $\frac{\$9.25\ earned}{hr} \times \frac{\$(100 - 23)\ take\ home}{\$100\ earned}$

$= \$7.12/hour\ take\text{-}home\ pay$

$1082.49\ \$ \times \frac{1\ hr}{7.12\ \$} \times \frac{1\ shift}{4\ hr} \times \frac{1\ week}{5\ shifts} = 7.60\ weeks = 8\ weeks$

4 Introduction to Gases

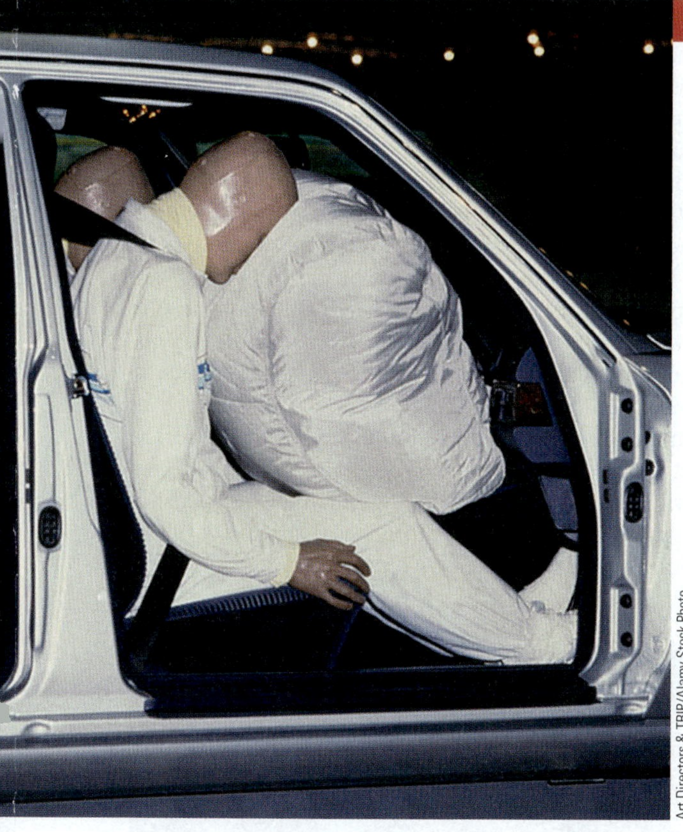
Art Directors & TRIP/Alamy Stock Photo

◀ Automobile airbags are an everyday application of the chemical principles you will study in this course. In Chapter 2, you learned that the particles in a gas are independent of one another, with a relatively large volume of empty space between them. The volume occupied by a substance in the gas phase is much greater than the volume occupied by the same particles in the solid or liquid state, generally by a factor of 1000 times or more. When an automobile airbag system sensor detects rapid deceleration, as in a collision, a small volume of a solid compound in the steering wheel or dashboard undergoes a reaction to form a relatively large volume of a gas, inflating the airbag.

CHAPTER CONTENTS

Most gases are invisible, and therefore you probably don't think about them very often. Because they are invisible, you might guess that they are difficult to model at the particulate level. In fact, the opposite is true. The particulate-level characteristics of substances in the gas phase make the development of a model of a gas more straightforward than models of the other phases of matter. Chemists were able to develop a model for the gas phase relatively early in the history of chemistry; therefore, we will follow a historical thread in this chapter.

This chapter provides an opportunity for you to apply the chemical calculation skills you learned as a result of studying Chapter 3. Scientific notation, conversion factors, metric units, significant figures, temperature, proportionality, and density are needed to understand the concepts and to work the problems in this introduction to gases. You may find that you occasionally need to review Chapter 3 as you study this chapter. If so, don't be concerned. All successful science students review and refine their understanding of prior material—even content from prior coursework—as they learn new ideas. In fact, we selected the topics of this chapter in part to give you a chance to apply your calculating skills immediately after you learned them.

4.1 Are the Gas Giants... Gas Giants?

Figure 4.1 is an approximately-to-scale drawing of the relative sizes of the planets of the solar system. Observe the patterns. One that quickly jumps out is that the four planets closest to the sun are relatively small in comparison with the outer four planets. Also note in the table below the illustration that hydrogen is the major component of the atmospheres of the large outer planets, whereas carbon dioxide or nitrogen is the principal component of the atmospheres of the small inner planets. The small inner planets are primarily solid; the large outer planets are mostly fluid. What is the cause of this pattern?

The term *gas giant* is a nonformal, loosely defined term sometimes used to refer to the outer planets Jupiter, Saturn, Uranus, and Neptune. But it is a bit of a misnomer. The gas giants do have the distinction of being gigantic relative to the inner planets, but they are not giant balls of gas, as we will define the word *gas* in this chapter. We used the word *fluid* in Figure 4.1 to be more precise. At the high temperature and pressure characteristic of these planets, much of the matter of which they are made is in a state rarely found on Earth, with properties in between those of liquids and gases. The particles of the fluid can freely rotate around one another as in a liquid, but they are much closer together than in a conventional gas. The matter that makes up the planet is solid at its core, fluid when getting closer to the "surface," and gas at the largest distance from the center.

The most abundant element in the universe is hydrogen. It is logical, then, that the most abundant element in a planet's atmosphere is also hydrogen. So, the atmospheres of the gas giants make sense. But why is hydrogen not predominant in the inner planets? Gases of different elements at the same temperature have molecules that move at different average speeds. All of these molecules move very fast in comparison to things like

Mercury	Venus	Earth	Mars	Jupiter	Saturn	Uranus	Neptune
Size (Earth = 1.0)							
0.4	0.9	1.0	0.5	11.2	9.1	4.0	3.9
Major Component of Atmosphere							
No atmosphere	Carbon dioxide	Nitrogen	Carbon dioxide	Hydrogen	Hydrogen	Hydrogen	Hydrogen
Primary Planet State of Matter							
Solid	Solid	Solid	Solid	Fluid	Fluid	Fluid	Fluid

Figure 4.1 The planets of the solar system. The inner four planets are of similar size, as are the outer four planets, but the outer planets are much larger than the inner planets. The major component of the atmospheres of the large outer planets is hydrogen. The atmospheres of the small inner planets primarily consist of a gas other than hydrogen. The small inner planets are mostly solid, and the large outer planets are fluid.

cars and people on bikes and so on. The less massive the molecule, the faster the molecules move. A particle moving quickly has the potential to reach what is called escape velocity, which is the speed at which the particle will escape from the gravitational attraction of a planet. Hydrogen molecules move sufficiently fast to escape from the atmospheres of the lower-mass inner planets but not from the higher-gravity, high-mass, giant outer planets. Molecules also move faster at higher temperatures. Thus, Mercury, being both the smallest planet and the one closest to the sun, has no atmosphere.

The gas giants are not exclusively made of gas. They are a mix of all three common phases of matter found on Earth, plus the fluid state, which is rare on Earth. But if we were to use the term *fluid giant*, it would probably be confusing to those who have no knowledge of science, and gas giant at least conveys that the outer planets are not primarily solid.

In Chapter 2, you studied the three common states of matter on Earth—gases, liquids, and solids. You learned that they could be distinguished at the macroscopic level by their shape and volume characteristics and at the particulate level by particle movement and spacing characteristics. In this chapter, you've already seen that pressure and temperature influence state of matter. We will now continue your study of the gaseous state of matter by pondering some macroscopic characteristics of gases and then their particulate-level causes.

4.2 Characteristics of Gases

Goal 1 Describe five macroscopic characteristics unique to the gas phase of matter.

In Chapter 2, we stated that the air you breathe is an example of the gaseous state of matter. We also reminded you that the shape and volume of a sample of a gas are both variable, in contrast to the liquid and solid states of matter. We will now examine five additional characteristics of gases that are unique to this phase of matter.

1. *Gases may be compressed.* **Figure 4.2(a)** shows a fixed amount of a gas in a cylinder with a leak-proof piston that can be moved to change the volume occupied by the enclosed air, which is shaded in pink so that its relative concentration is visible. If you push the piston down by applying more force by adding additional weights to the top of the piston, the volume of the air is reduced, as illustrated in moving from Figure 4.2(a) to **Figure 4.2(b)**. This demonstrates that a fixed amount of a gas may be compressed to a smaller volume by applying pressure.
2. *Gases may be expanded.* Moving from Figure 4.2(b) to **Figure 4.2(c)** illustrates what happens when weights are removed from the top of the piston. This reduction in the force applied to the gas results in the gas responding instantaneously, pushing the piston upward, increasing the volume of gas. If the weights are removed and the piston is pulled upward, as illustrated in **Figure 4.2(d)**, the gas again expands, filling the additional volume. We may conclude that a fixed amount of a gas may be expanded to a larger volume by reducing pressure.
3. *Gases have low densities.* Recall from Chapter 3 that density is the ratio of mass to volume for a substance. Therefore, a substance with a low density has a relatively small mass when comparing equal volumes. **Figure 4.3** shows the relative densities of gaseous air (in a balloon), liquid water, and a solid nail made of iron. The density of air is 0.0012 g/cm^3. The density of water is 830 times greater than the density of air, and iron is 6600 times denser than air, when all are at room temperature. In comparison with liquids and solids, gases have low densities.
4. *Gases may be mixed in a fixed volume.* You may add the same or a different gas to a rigid container of fixed volume that already is occupied by a gas, as illustrated in **Figure 4.4**.
5. *Gases exert constant pressure on the walls of their container uniformly in all directions.* **Figure 4.5** compares the measurement of pressure at various points in a container of a gas versus a container of liquid. In the gas container, you

Figure 4.2 Compression and expansion properties of gases. ▼ A gas within a cylinder with a movable piston demonstrates that a gas may be compressed and expanded, always filling its container.

▲ **P/Review** The distinction between macroscopic and particulate views of matter is introduced in Section 2.2 and first illustrated in Figure 2.6.

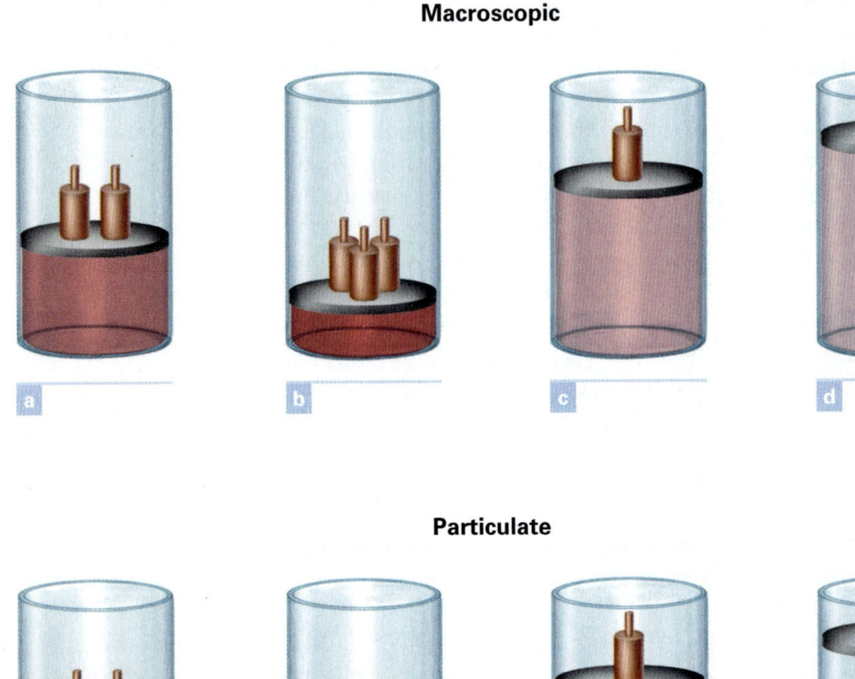

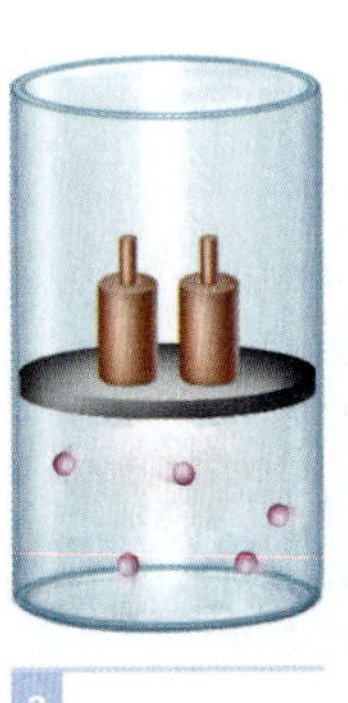

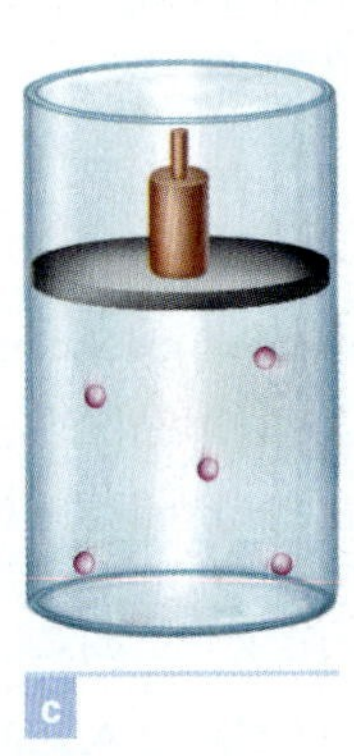

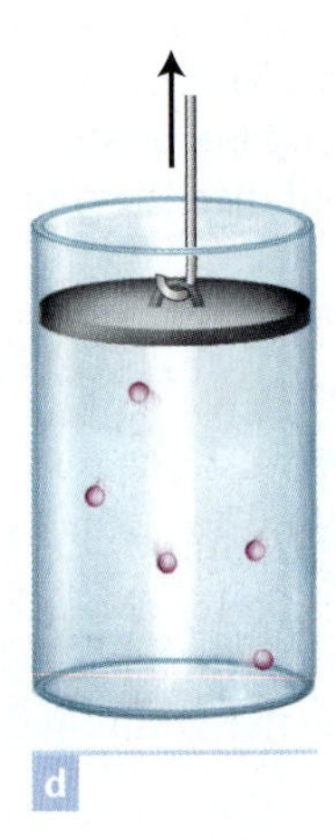

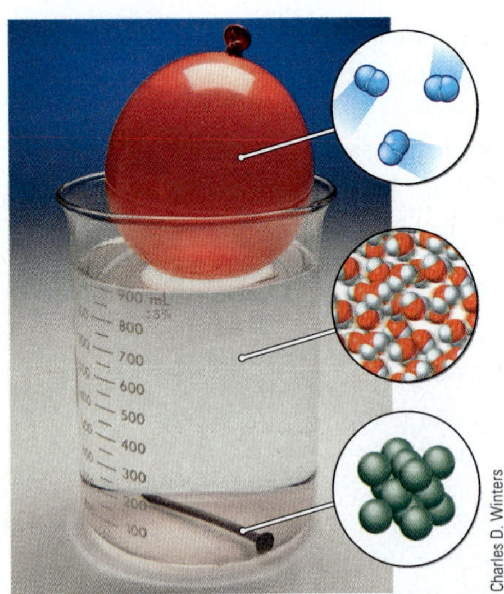

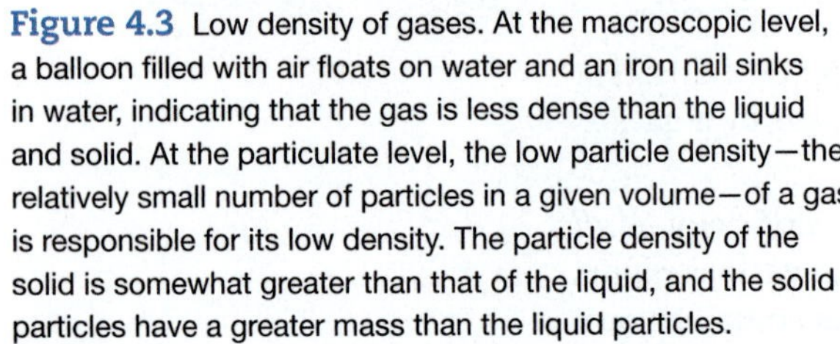

Figure 4.3 Low density of gases. At the macroscopic level, a balloon filled with air floats on water and an iron nail sinks in water, indicating that the gas is less dense than the liquid and solid. At the particulate level, the low particle density—the relatively small number of particles in a given volume—of a gas is responsible for its low density. The particle density of the solid is somewhat greater than that of the liquid, and the solid particles have a greater mass than the liquid particles.

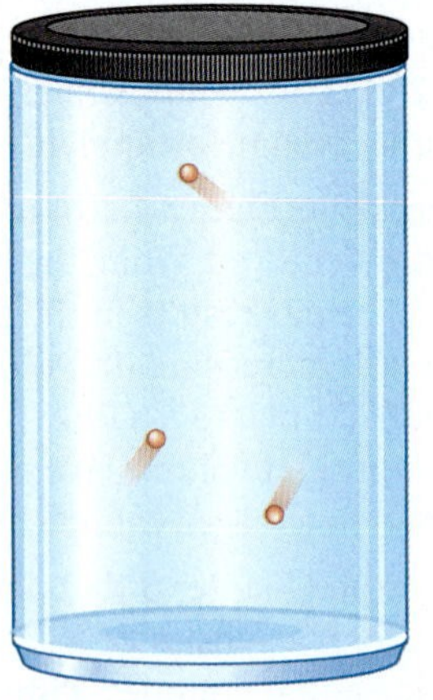

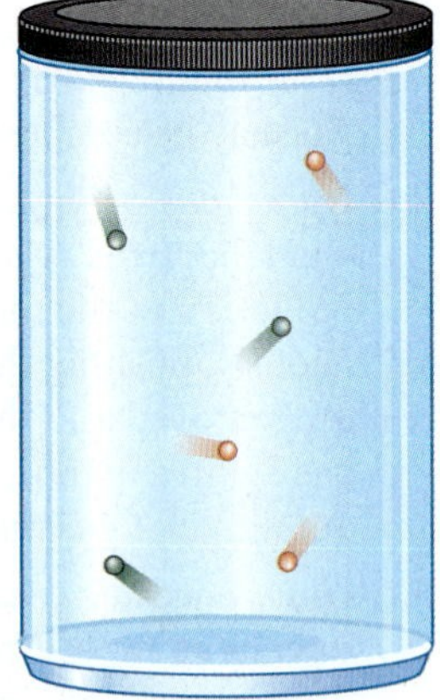

Figure 4.4 Mixing gases. At the macroscopic level, it is possible to add more gas to a container already holding a gas. At the particulate level, the volume of matter in a container holding a gas is very tiny compared with the volume of the container. If another gas or more of the same gas is added to the container, it occupies only a small fraction of the large volume of space that is available.

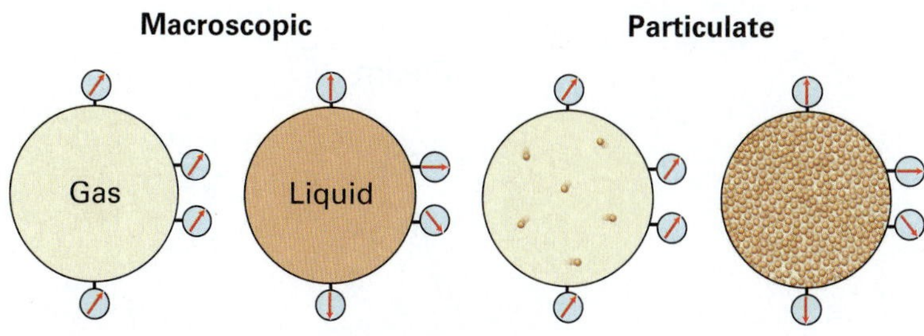

Figure 4.5 Pressure in gases and liquids. Each container has four pressure gauges, one on the top, one on the bottom, and two on the side. Note how the four gauges on the gas container all read the same pressure, but the liquid gauges show that pressure increases with increasing depth. Gas pressures are exerted uniformly in all directions; liquid pressures depend on the depth of the liquid.

can see that the pressure is equal at all locations. In contrast, in the liquid container, the pressure increases with increasing depth. For small samples of enclosed gases, such measurements indicate that gas pressure is independent of external factors such as gravitational forces.

4.3 A Particulate-Level Explanation of the Characteristics of Gases

Goal 2 Use the postulates of the kinetic molecular theory to explain the reasons for the macroscopic characteristics unique to the gas phase of matter.

Why does the gas phase of matter have the characteristics described in Section 4.2? Why are these *not* characteristics of liquids or solids? Scientists who pondered these questions hundreds of years ago recognized that the **kinetic molecular theory** (first introduced in Section 2.3) could be developed in greater detail to serve as a particulate-level explanation of the macroscopic characteristics of gases.

Let's start with the *molecular* part of kinetic molecular theory first:

1. *A gas consists of molecules and empty space.* The physical substance that makes up a gas is molecules: the smallest pieces of a pure substance that retain the identity of the substance. In between the molecules is nothing.
2. *The volume occupied by the molecules in a gas is negligible when compared with the volume of the space they occupy.* One gram of liquid water at the boiling point occupies 1.04 cm^3. When changed to steam at the same temperature, the same number of molecules fills 1700 cm^3, an expansion of over 1600 times (Figure 4.6)! The total volume of the molecules in 1 gram of water is approximately the same regardless of its state of matter. If we assume that water molecules are touching one another in the liquid state, we can conclude that they must be very widely separated in the gaseous state. The 1.04-cm^3 volume of water in the liquid state is 0.06% of the 1700 cm^3 total volume the molecules occupy as a gas. Negligible means that a quantity is small enough to ignore. In almost all situations, we can ignore the 0.06% of the volume of a gas that is composed of molecules and consider the volume of the container to be available to each molecule in a sample of a gas.
3. *The attractive forces among molecules of a gas are negligible.* In Chapter 2, we said that molecules are attracted to one another. Consider the pressure in an automobile or bicycle tire. The attractive forces among molecules of a gas cause pressure to be lower than it would be in the absence of these attractive forces. When we account for the attractive forces among the gas molecules within a tire, the pressure is about 0.1% less than the pressure would be if there were no attractive forces among the molecules. This qualifies as negligible. Thus, for a gas we can ignore the effect of the attractive forces among the molecules.

Now let's consider the *kinetic* part of kinetic molecular theory.

4. *The average kinetic energy of gas molecules is proportional to the temperature, expressed in kelvins.* The average kinetic energy of gas molecules is proportional

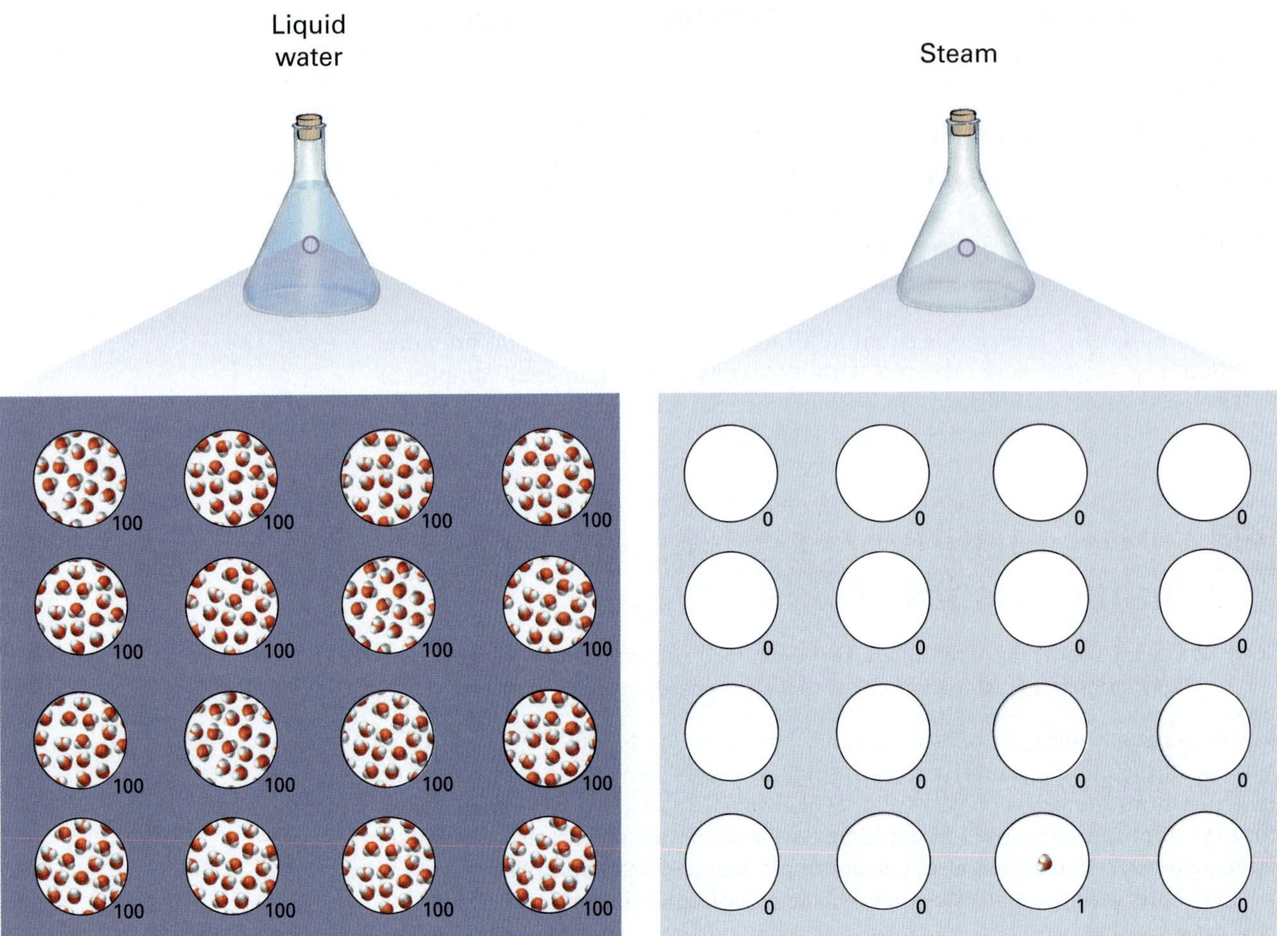

Figure 4.6 A gas is mostly empty space. Suppose you were to select a sample of liquid water that contained 100 molecules. If you then selected 15 more samples—for a total of 16—of the same volume, each sample would contain 100 molecules. If you were to then select 16 separate samples of steam, each sample having the same volume as each sample of liquid water, how many water molecules would be in each sample? On average, 15 of the sample volumes of steam would be empty—no molecules—and the 16th sample volume would contain only 1 water molecule. This shows that almost all of the volume of a gas is empty space, just like the sample volumes illustrated in this figure.

to both their mass and their speed. Heavier particles move slower, and lighter particles move more quickly, but all molecules are continually moving extremely fast. The average molecule in the air in the library or home in which you are now studying is moving at over 1000 miles per hour! At higher temperatures, the average speed of the molecules increases; at lower temperatures, the average speed decreases. The molecules move in straight paths until they interact with another molecule or a container wall, and then they are deflected and move in a different straight path, independent of the path of any other molecule.

5. *Molecules interact with one another and with the container walls without loss of total kinetic energy*. Kinetic energy may be transferred from one molecule to another as they interact, but the total energy of the molecules is the same before and after the interaction. Because of the relationship between temperature and average molecular speed, temperature would drop if energy were lost in molecular interactions. Any enclosed gas would eventually become a liquid because of this loss of energy. But these things do not happen, so we conclude that energy is not lost in molecular interactions, either with the molecules that make up the walls of the container or among gas molecules.

Thinking About Your Thinking

Mental Models

Section 4.2 discussed the familiar characteristics of gases. Section 4.3 explains the reason for those characteristics as a chemist does, by using descriptions of particulate-level behavior. Imagine the behavior of gas particles in the form of a mental movie in your mind. Make sure you can use your mental movie to explain each of the macroscopic characteristics of a gas.

✓ Target Check 4.1

For each of the macroscopic characteristics unique to the gas phase of matter described in Section 4.2—(a) compressibility, (b) expandability, (c) low density, (d) may be mixed in a fixed volume, and (e) uniform, constant pressure on container walls—describe how a postulate of the kinetic molecular theory explains the reason for the characteristic.

4.4 Gas Pressure

Goal 3 Define *pressure* and interpret statements in which the term *pressure* is used.
4 Explain the cause of the pressure of a gas.

Throughout most of early human history, the prevailing intuitive belief was that the atmosphere did not have weight. However, Italian scientist Evangelista Torricelli (1608–1647) (**Figure 4.7**) doubted this belief. Torricelli was a student of Galileo, as well as a friend and a literary assistant to Galileo in the waning months of his life. After Galileo's passing, Torricelli took on a problem that Galileo had once worked on: Why could water not be pumped from deep mines? In an effort to measure the weight of the atmosphere, Torricelli invented the mercury **barometer**. He filled a long glass tube with liquid mercury, placed his finger over the end, inverted the tube, and placed it in a dish partially filled with mercury. The height of the column of mercury was about 30 inches (76 centimeters), although it varied slightly from day to day. **Figure 4.8** is an illustration of Torricelli's barometer, and the figure caption explains the operational principles that govern the barometer.

Torricelli proposed that the height of the mercury in the column was due to the weight of the atmosphere pushing the liquid up the tube. Torricelli's hypothesis was confirmed in part by the experiments of the French scientist Blaise Pascal (1623–1662). Pascal found that when a barometer was taken up a mountain, where the atmosphere weighs less than at lower altitudes, the height of the mercury column decreased. Thus the barometer is weighing the atmosphere.

The force exerted by the weight of the atmosphere on the area of the surface of the mercury outside the tube in a barometer is its pressure. By definition, **pressure** is force per unit area:

$$\text{Pressure} \equiv \frac{\text{Force}}{\text{Area}} \quad \text{or, in symbols,} \quad \text{P} \equiv \frac{\text{F}}{\text{A}}$$

The cause of gas pressure follows from the definition. Pressure is the effect of the *force* of the interactions of the huge number of rapidly moving molecules of a gas as they collide with the surface *area* of an object in contact with that gas.

Three factors affect the pressure exerted by a gas sample. Consider the particulate-level illustration of a gas in a cylinder with a movable leak-proof piston in Figure 4.2. Imagine what happens as the volume of the container increases as in Figure 4.2(b) through (d) while the temperature is held constant. Any given molecule

Figure 4.7 Evangelista Torricelli (1608–1647), the inventor of the mercury barometer. Torricelli is known for the statement, "We live submerged at the bottom of an ocean of air."

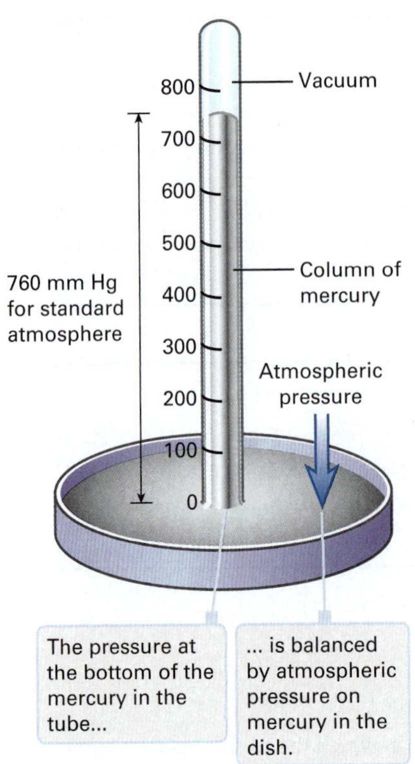

Figure 4.8 The mercury barometer. The total pressure at any point in a liquid system is the sum of the pressures of each gas or liquid phase above that point. All points on the liquid surface outside the tube are at the same level as the mercury at the bottom inside the tube.

The total pressures at any two points at the same level in a liquid system are always equal. The only thing exerting downward pressure on the surface outside the tube is the atmosphere; the downward arrow represents atmospheric pressure. The only thing exerting downward pressure at the bottom of the mercury inside the tube is the mercury above that point. Both points are at the same level; therefore the pressures at these points are equal: $P_{atmosphere} = P_{mercury}$.

in the container must travel over a longer path, on average, before it collides with the container wall. This means that there will be less cumulative force exerted on the wall because of less-frequent collisions, and the pressure will decrease. Thus, pressure and volume are inversely proportional. As the volume increases, pressure decreases, and vice versa.

Another factor that affects the pressure of a gas is the amount of gas. Consider the particulate-level illustrations of different amounts of gas in the same container in Figure 4.4. In the container with more particles, there will be more frequent collisions, and therefore more cumulative force exerted on the wall, and pressure will be higher. Pressure and amount of gas are directly proportional. As the number of gas molecules increases, pressure increases. If the number of gas molecules is reduced, pressure decreases.

Temperature is the other factor that affects the pressure of a gas. Recall that the temperature of a gas is proportional to the average kinetic energy of the molecules in the gas sample. Thus, as temperature goes up, the average kinetic energy of the molecules increases. Now think about the kinetic energy of any given gas molecule. Kinetic energy is energy of motion, and therefore when a molecule experiences an increase in kinetic energy, it moves faster. When the molecules in a gas sample are moving faster, the average molecule will take less time to travel along any given path before colliding with a container wall. The cumulative force exerted on the wall will be higher because of more frequent collisions, and the pressure will increase. Furthermore, the faster moving molecules will impact the wall with greater force, adding to the pressure-increasing effect. Pressure and temperature are directly proportional. As the temperature increases, pressure increases, and vice versa.

The properties of a gas are summarized in **Table 4.1**.

✓ Target Check 4.2

a) If the volume of a sealed container holding a gas is increased and the temperature is increased, will the pressure of the gas sample increase or decrease? Explain.

b) If more gas molecules are injected into a container of fixed volume that already holds a gas and the temperature is increased, will the pressure of the gas sample increase or decrease? Explain.

Table 4.1 Properties of a Gas

Property	Description	Relationship to Gas Pressure
Pressure	Force ÷ Area	
Volume	The quantity of three-dimensional space occupied by a gas sample	$\text{Pressure} \propto \frac{1}{\text{Volume}}$
Amount	The number of gas molecules in a sample	Pressure ∝ Amount
Temperature	The average kinetic energy of the gas molecules in a sample	Pressure ∝ Temperature

Pressure Units

Goal 5 Express the relationship among the following gas pressure units: atmospheres, torr, millimeters of mercury, inches of mercury, pascals, kilopascals, bars, or pounds per square inch.

A mercury barometer effectively serves as a balance that weighs the atmosphere. The weight of Earth's atmosphere—atmospheric pressure—varies depending upon elevation. At the top of Mount Everest, 29,029 ft (8848 m) above sea level, the atmospheric pressure is about one-third of the pressure that is typically found at sea level; at the surface of the Dead Sea, 1388 ft (423 m) below sea level, the atmospheric pressure is about 5% greater than at sea level. Atmospheric pressure also varies as the weather changes (see the Everyday Chemistry box following this section). Because of this variance in atmospheric pressure, the scientific community decided that it needed to create a standard for atmospheric pressure. They agreed to choose the average atmospheric pressure at sea level at the latitude of Paris, France, to be the standard. When the height of the mercury column in a barometer is measured in millimeters, this average is 760 mm. Thus, one standard **atmosphere (atm) (pressure unit)** is 760 millimeters of mercury.

$$1 \text{ atm} = 760 \text{ mm Hg}$$

Hg is the elemental symbol for mercury.

These pressure units, atm and mm Hg, are both based on reading a barometer. Another unit, the **torr**, named to honor Torricelli's work, is defined as 1/760 atmosphere, and thus it is equal to a **millimeter of mercury**:

$$1 \text{ atm} = 760 \text{ mm Hg} = 760 \text{ torr}$$

The SI unit of pressure is the **pascal (Pa)**, which is defined as 1 kg/m · s^2. Although the unit is not based on a barometer, it is preferred by SI because it is derived from base SI units. The unit is named to honor the work of Blaise Pascal. It is a very small amount of pressure, so the kilopascal, kPa, is typically used instead. It is the common unit of pressure in most countries. The atmosphere is defined in terms of pascals: 1 atm = 101,325 Pa.

Most chemists have agreed to express pressures using the **bar** when communicating with one another in scientific journals and at scientific meetings: 1 bar = 100 kPa = 100,000 Pa.

In the United States, barometric pressure at weather websites and on news reports is typically given in **inches of mercury**:

$$760 \cancel{\text{mm Hg}} \times \frac{1 \cancel{\text{cm Hg}}}{10 \cancel{\text{mm Hg}}} \times \frac{1 \text{ in. Hg}}{2.54 \cancel{\text{cm Hg}}} = 29.92 \text{ in. Hg}$$

In North America, tire pressure is usually expressed in **pounds per square inch (psi)**. Note that weight in pounds is a force, and inches squared is an area, so this unit is a direct application of the definition of pressure. One standard atmosphere is equal to 14.70 psi.

The pressure units introduced in this section, their definitions, and their relationship to the standard atmosphere pressure unit are given in Table 4.2. Your instructor may require you to memorize the relationships among the complete set of pressure units or some subset of units, or your instructor may not require you to memorize any pressure unit relationships. Be sure that you learn the expectations of your instructor with regard to pressure units.

Pressure Measurement

Goal 6 Describe the operational principle of an open-end manometer.

7 Given the pressure difference in an open-end manometer and the atmospheric pressure, determine the pressure of the gas in the manometer.

8 Given the gauge pressure of a gas and the atmospheric pressure, determine the absolute pressure of the gas.

Table 4.2 Units for Pressure Quantities

Unit	Standard Abbreviation	Definition	Relationship to the Standard Atmosphere
Atmosphere	atm	101,325 Pa	
Millimeters of mercury	mm Hg	133.322387415 Pa (exactly)	760 mm Hg
Torr	torr	1/760 atm	760 torr
Pascal	Pa	$1\ kg/m \cdot s^2$	101,325 Pa
Kilopascal	kPa	1000 Pa	101.325 kPa
Bar	bar	100,000 Pa	1.01325 bar
Inches of mercury	in. Hg	The pressure exerted by a 1-inch circular column of mercury 1 inch in height at 32°F and $9.80865\ m/s^2$ gravitational acceleration	29.92 in. Hg
Pounds per square inch	psi	A force of 1 pound applied to an area of 1 square inch	14.70 psi

A wide variety of instruments that measure pressure have been developed. A simple, nonmechanical instrument for measuring pressure is the open-end **manometer**, illustrated in **Figure 4.9**. An enclosed gas pushes on one side of the column of mercury and the atmosphere pushes on the other side. The difference in pressure is literally measured in millimeters of mercury.

Outside the laboratory, mechanical or electronic gauges are used to measure gas pressure. The operation of most mechanical gauges is based on a pressure-sensing element that varies in shape as pressure varies (**Figure 4.10**). Electronic gauges operate based on the mechanical strain of a material and the resulting variation in electrical resistivity of that material.

You probably own a tire pressure gauge, and you use it occasionally to check the pressure in your car and bicycle tires (**Figure 4.11**). Tire pressure gauges show the pressure above atmospheric pressure, rather than the absolute pressure measured by manometer. Even a flat tire contains air that exerts pressure. If it did not, the whole tire would collapse, not just the bottom. The pressure of gas remaining in a flat tire is equal to atmospheric pressure. If a tire pressure gauge shows 32 psi, that is the **gauge pressure** of the gas (air) in the tire, that is, the pressure *above* atmospheric pressure. The **absolute pressure** is about 47 psi—the 32 psi shown by the gauge plus about 15 psi (1 atm = 14.70 psi) from the atmosphere.

Pressure Unit Conversions

Goal 9 Given a pressure in atmospheres, torr, millimeters (or centimeters) of mercury, inches of mercury, pascals, kilopascals, bars, or pounds per square inch, express pressure in each of the other units.

The quantitative problem-solving methods you learned in Chapter 3 are used to convert from one pressure unit to another. The relationships given in Table 4.2 are the equivalencies that you use.

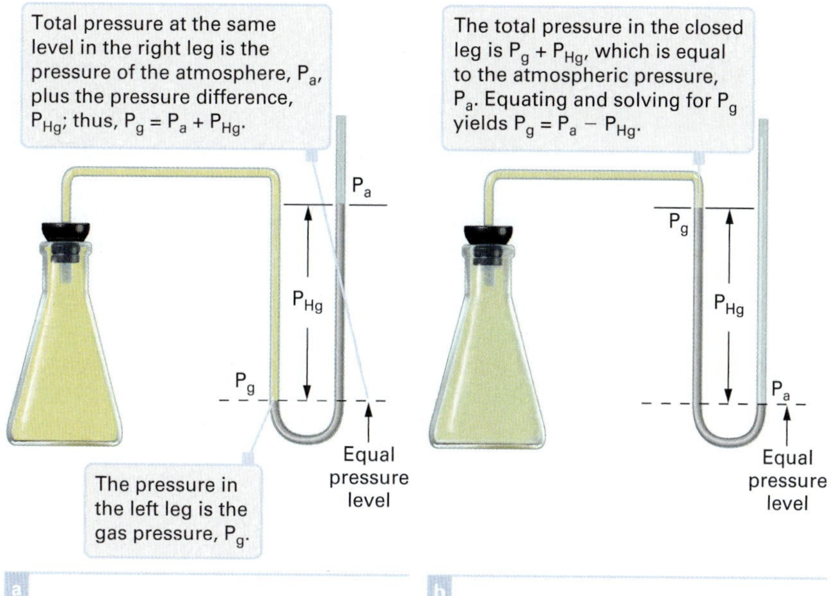
Total pressure at the same level in the right leg is the pressure of the atmosphere, P_a, plus the pressure difference, P_{Hg}; thus, $P_g = P_a + P_{Hg}$.
P_a
P_{Hg}
P_g
Equal pressure level
The pressure in the left leg is the gas pressure, P_g.
a
The total pressure in the closed leg is $P_g + P_{Hg}$, which is equal to the atmospheric pressure, P_a. Equating and solving for P_g yields $P_g = P_a - P_{Hg}$.
P_g
P_{Hg}
P_a
Equal pressure level
b

Figure 4.9 Open-end manometers. Open-end manometers are governed by the same principles as mercury barometers (Figure 4.8). The pressure of the gas, P_g, is exerted on the mercury surface in the closed left leg of the manometer. Atmospheric pressure is exerted on the mercury surface in the open right leg. With a ruler, the difference between these two pressures, P_{Hg}, may be measured directly in millimeters of mercury. Gas pressure is determined by equating the total pressures at the lower liquid mercury level, indicated with a dashed line.

In (a), the enclosed gas is exerting more pressure than the atmosphere, so the quantities are added to result in a higher pressure. In (b), the enclosed gas exerts less pressure than the atmosphere, and a lower pressure is obtained by subtracting the pressure difference from the atmospheric pressure.

You can determine the pressure of a gas, as measured by a manometer, by adding the pressure difference to, or subtracting the pressure difference from, atmospheric pressure: $P_g = P_a \pm P_{Hg}$.

Change
Rain
Fair
Stormy
Very dry

Ungor/Shutterstock.com

Figure 4.10 A mechanical gauge used to measure atmospheric pressure. In general, mechanical gauges work via an air-filled tube that changes shape with changing pressure. Levers and/or gears move in response to the tube, and a pointing needle and scale or an electronic display is calibrated to show the corresponding atmospheric pressure.

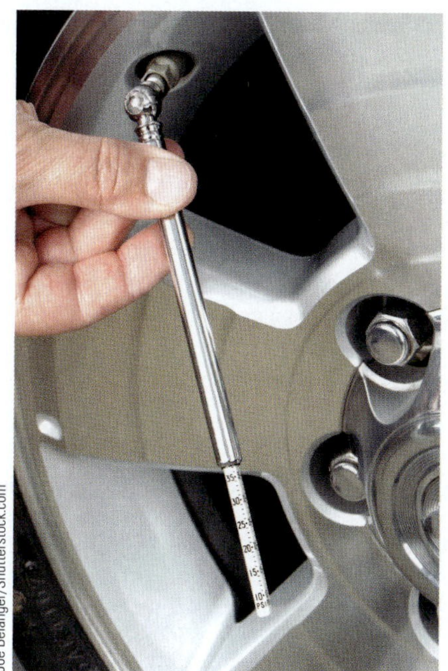

Joe Belanger/Shutterstock.com

Figure 4.11 Tire pressure gauge. Most tire pressure gauges have a piston inside the casing that is attached to a spring. A plastic rod rides on top of the piston. When the gauge is placed on a tire valve stem, the air in the tire pushes on the piston, which pushes the calibrated rod out. When the gauge is removed, the spring pushes the piston back, but the rod remains in the same position to allow you to read the tire pressure.

Active Example 4.1 Pressure Unit Conversions

The pressure inside a steam boiler is 1427 psi. Express this pressure in atmospheres.

Think Before You Write If it has been some time since you studied Section 3.4, A Strategy for Solving Quantitative Chemistry Problems, you might want to go back and take a brief look at the procedure for using conversion factors to solve quantitative problems.

Answers *Cover the left column with your cut-out shield. Reveal each answer only after you have written your own answer in the right column.*

Given: 1427 psi, pressure unit **Wanted:** pressure unit, atm	**Analyze** the problem statement: Write the given quantity and its property, and write the property of the wanted quantity and its unit. With pressure units, you don't need to state whether they are metric or USCS. 
1 atm = 14.70 psi $\frac{1 \text{ atm}}{14.70 \text{ psi}}$	**Identify** the equivalency needed to solve the problem. If you are required to know the relationship between psi and atmospheres, write it from memory; if you are allowed to use a reference source for the relationship, get it from Table 4.2. Change the equivalency to the needed conversion factor. 
$1427 \text{ psi} \times \frac{1 \text{ atm}}{14.70 \text{ psi}} = 97.07 \text{ atm}$	**Construct** the solution setup. Cancel units, and calculate the answer. 
An atmosphere is a larger unit than a psi, so there should be a smaller number of atmospheres in a given pressure than the number of psi. The number is reasonable (1400 ÷ 14 = 100), and the unit is what was wanted.	**Check** the answer. Is the value reasonable? Is the unit correct? 
You improved your skill at solving pressure unit conversion problems.	What did you learn by solving this active example?

Practice Exercise 4.1

Convert 741 torr to bar and inches of mercury.

Active Example 4.2 Gauge Pressure and Pressure Unit Conversions

You measure the pressure in an automobile tire as 29 psi on a day when the weather report states that the barometric pressure is 30.09 in. Hg. How would you report the absolute pressure in the tire to a colleague in Canada in kPa?

Think Before You Write Notice that the problem statement does not directly tell you that 29 psi is gauge pressure. You have to infer that from the context.

Answers *Cover the left column with your cut-out shield. Reveal each answer only after you have written your own answer in the right column.*

Given: 29 psi, pressure unit; 30.09 in. Hg, pressure unit
Wanted: pressure unit, kPa

Analyze the problem by writing the given quantity and its property and the wanted property and its unit.

14.70 psi = 1 atm & 1 atm = 101.325 kPa
29.92 in. Hg = 1 atm & 1 atm = 101.325 kPa

$$\frac{101.325\,\text{kPa}}{14.70\,\text{psi}};\ \frac{101.325\,\text{kPa}}{29.92\,\text{in. Hg}}$$

Each pressure conversion can be solved in two steps, one for each equivalency, but you can see by the way we wrote the equivalencies that if x = y and y = z, then x = z. That allows us to condense the two conversion factors into one.

Identify the equivalencies needed to solve the problem, and change them to the appropriate conversion factors. We will change both given pressures to kPa.

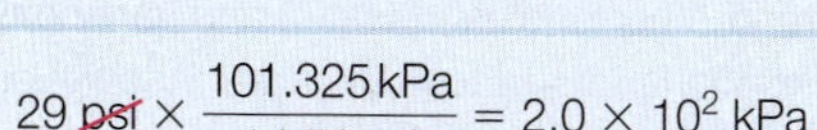

$$29\,\cancel{\text{psi}} \times \frac{101.325\,\text{kPa}}{14.70\,\cancel{\text{psi}}} = 2.0 \times 10^2\,\text{kPa}$$

$$30.09\,\cancel{\text{in. Hg}} \times \frac{101.325\,\text{kPa}}{29.92\,\cancel{\text{in. Hg}}} = 101.9\,\text{kPa}$$

$P_{absolute} = P_{gauge} + P_{atmosphere} =$
$2.0 \times 10^2\,\text{kPa} + 101.9\,\text{kPa} = 3.0 \times 10^2\,\text{kPa}$

Construct the solution, cancel units, and calculate the answer.

$$\frac{30 \times 100}{15} = 200,\ \text{OK};\ \frac{30 \times 100}{30} = 100,\ \text{OK}$$

200 + 100 = 300, OK

Check the solution. Are the values reasonable?

You improved your understanding of gauge pressure, and you improved your skill at solving pressure unit conversion problems.

What did you learn by solving this Active Example?

Practice Exercise 4.2

The absolute pressure in a hybrid bicycle tire is 515 kPa when the atmospheric pressure is 29.54 in. Hg. What will be the reading on a tire gauge calibrated in pounds per square inch when it is used to measure the pressure in this tire?

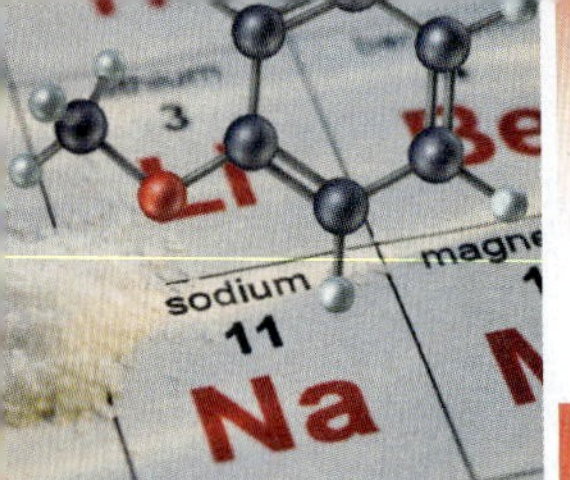

Everyday Chemistry 4.1

THE WEATHER MACHINE

You have probably heard a television weather reporter talking about areas of high pressure and low pressure. These uneven areas of pressure in the lower levels of Earth's atmosphere are one cause of weather. In fact, the atmosphere is sometimes referred to as "the weather machine" (**Figure 4.12**).

There are two key elements of air movement in the atmosphere: air masses and fronts. An air mass is a relatively large sphere of air that has about the same temperature and water content throughout for a given altitude. A typical cold air mass in the winter may stretch from Montana to Minnesota and from the Canadian border to the Mexican border. In contrast, a front is a small section of the lower atmosphere, usually 1/10 to 1/100 of the size of an air mass. Fronts are where the action is for changes in the weather. The temperature and humidity can vary greatly within a front, and the relatively stable air masses on either side of a front are often very different from one another.

When people refer to an area of high pressure, they are talking about an air mass that has a relatively high pressure when compared with surrounding air (see **Figure 4.13**). This air, initially cold, descends toward the surface of the earth, compressing and warming. This inhibits cloud formation. High-pressure areas are therefore generally associated with clear weather. Conversely, air in a low-pressure mass rises and cools, which often leads to clouds, rain, or snow.

SPL/Science Source

Figure 4.12 Earth's atmosphere as seen from space.

Weather is a very complex phenomenon, affected by numerous variables, such as the sun, Earth's tilt, the oceans and other bodies of water, and variations in the land surface. Ultimately, however, each complex variable can be described in terms of simpler concepts such as temperature, pressure, and gas density. Weather experts are known as meteorologists, and their college preparation to understand the science of the atmosphere includes courses in chemistry, physics, geology, and mathematics, as well as advanced courses in meteorology itself. So the next time you hear a weather forecaster discussing high- and low-pressure systems, remember that the functioning of the weather machine is ultimately governed by the basic principles of the gaseous state of matter.

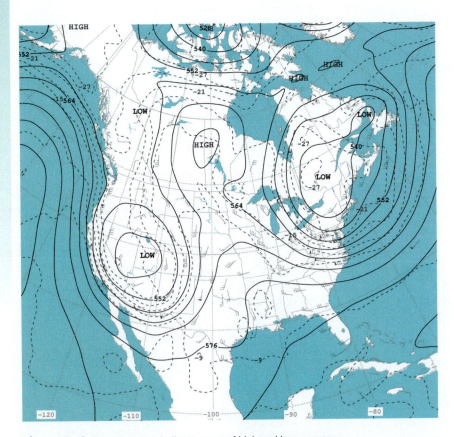

Figure 4.13 Weather maps indicate areas of high and low pressure.

Quick Quiz

1. Which are larger: air masses or air fronts? Are air masses or air fronts shown on a weather map?
2. Would an air mass in an area of high pressure have a relatively high density or a low density? Would the air mass be expected to sink or rise? Explain. Does this match what actually happens to a high-pressure air mass?

4.5 Charles's Law: Volume and Temperature

Temperature

Goal 10 When plotting the relationship between gas volume and temperature, explain how the straight line can be extrapolated to determine the temperature at which the gas has zero volume, and explain the significance of this temperature.

In Section 3.10, we promised that we would explain the origin of the Kelvin temperature scale (or absolute temperature scale) in this chapter. This explanation is rooted in the experiment described in **Figure 4.14**. Study the figure and its caption before moving on to the next paragraph.

The results of the experiment illustrated in Figure 4.14 are observed again and again if the experiment is repeated. Combined with other experiments using many

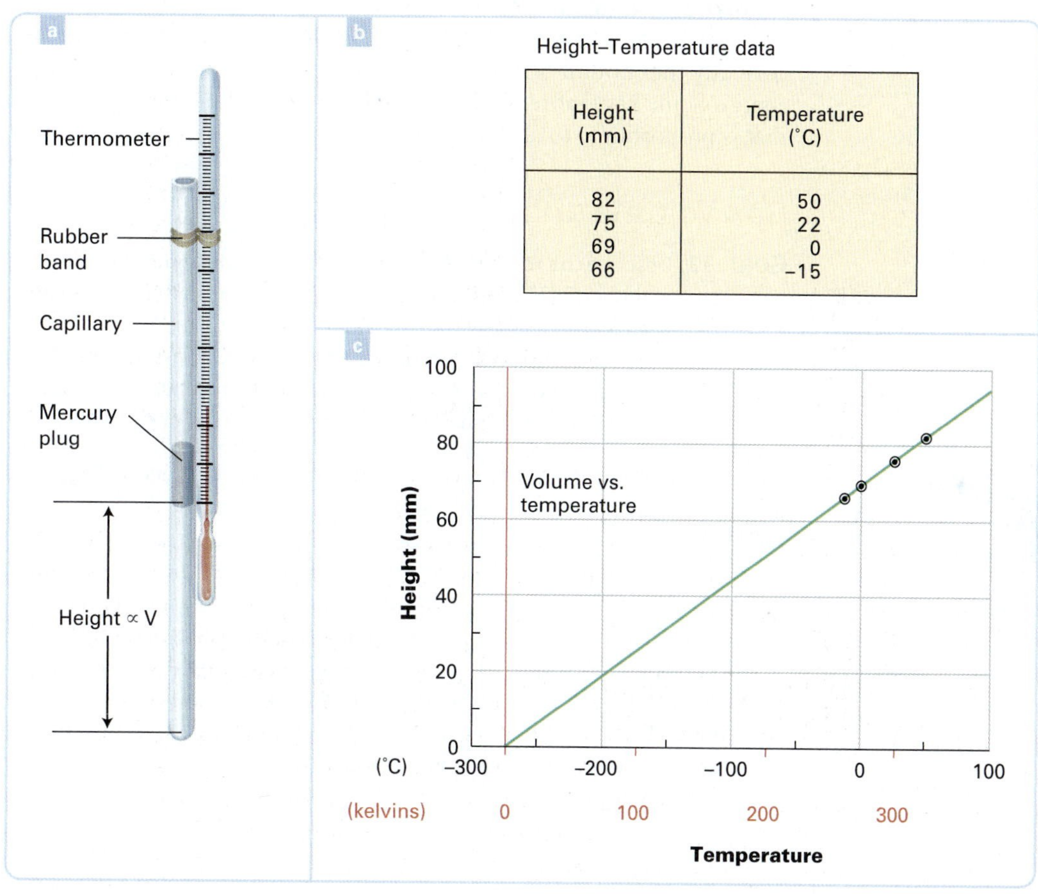

Height (mm)	Temperature (°C)
82	50
75	22
69	0
66	−15

Figure 4.14 Volume and temperature of a gas. A student performed an experiment to discover the relationship between the volume and temperature of a fixed amount of gas at constant pressure. A small plug of liquid mercury was placed in a glass capillary tube sealed at the lower end, as illustrated in (a). The amount (number of molecules) of gas (air) trapped between the sealed end and the mercury plug is unchanged when the mercury plug moves within the glass tube. Since the circumference of the tube is constant, the height of the gas column beneath the mercury plug is proportional to the volume of the gas. (a) also shows that the tube is attached to a thermometer with a rubber band. To conduct the experiment, the student submerged the tube and thermometer apparatus into liquid baths at different temperatures.

The student measured the height of the gas column at each temperature. Her data are shown in (b).

(c) is a graph of the data. On the graph, she extended the straight line that fits her data to the point where it intersects the x-axis. This process is called extrapolation. To extrapolate is to infer by extending beyond the range of your data. By extrapolating to a height of zero, which corresponds to a volume of zero, the trend line reaches the x-axis at −273°C. Below the °C points on the x-axis, she added another temperature scale where −273°C corresponds to zero. This is the Kelvin scale: $T_K = T_{°C} + 273$. Zero kelvin and −273°C are the temperature at which the volume of the gas would become zero.

kinds and quantities of gases, all yielding the same temperature that corresponds with zero gas volume, we have overwhelming evidence that there is an **absolute zero** of temperature—a lowest temperature. This zero is a true zero for temperature, similar to how zero meter or feet indicates a true absence of length. The temperatures 0°C and 0°F are not the lowest temperature possible; they are arbitrary zero points because lower temperatures are possible. In contrast, zero on the Kelvin scale indicates a true absence of temperature.

In order to understand *why* there is an absolute zero of temperature, we need to know just what temperature measures. Kinetic molecular theory postulates that the temperature of a substance is a measure of the average kinetic energy of the molecules in the sample. Kinetic energy is the energy of motion as a particle goes from one place to another. It is expressed mathematically as $\frac{1}{2}mv^2$, where m is the mass of the molecule and v is its velocity or speed.

Since the mass of the particles in a pure gas sample are nearly constant, the particle speed must be higher at high temperatures and lower at low temperatures. If the speed reaches zero—if the molecule stops moving through space—the absolute temperature becomes zero.

Notice the word *average* in the phrase "average kinetic energy." It suggests correctly that not all of the molecules in a sample of matter have the same kinetic energy. Some have more, some have less, and as a whole they have an average energy that is proportional to absolute temperature.

Gas Volume and Temperature

Goal 11 Describe the relationship between the volume and temperature of a fixed amount of gas at constant pressure, and express that relationship as a proportionality, an equality, and a graph.

12 Given the initial volume (or temperature) and the initial and final temperatures (or volumes) of a fixed amount of gas at constant pressure, calculate the final volume (or temperature).

Consider the demonstration shown in the photographs in **Figure 4.15**. About a dozen rubber balloons were blown up and tied shut. A substance that stays in the liquid state at low temperatures was poured into the beaker. The balloons were then dipped into the liquid in the beaker. As soon as the gaseous contents of each balloon experienced a temperature decrease, its volume decreased, as shown in the photograph in Figure 4.15(a). Once their volume decreased, all of the balloons fit into the beaker, as shown in the photograph in Figure 4.15(b). The photograph in Figure 4.15(c) shows the volume of the gas in the balloons increasing after they were placed back on the table at room temperature. This shows that the volume of a gas and its temperature vary in the same direction, even when demonstrated with familiar everyday objects outside of the laboratory.

Figure 4.15 Gas volume and temperature. Rubber balloons were blown up and then submerged into a very cold liquid (a). Their volume decreased substantially; all of the balloons fit into the beaker (b). When the balloons were removed from the cold bath, they began to inflate back to their original volume (c). As the temperature of a fixed amount of gas decreases at constant pressure, its volume decreases. As the temperature increases, the volume increases.

a

b

c

French scientist Jacques Charles (1746–1823) (**Figure 4.16**) is the first person known to perform experiments investigating the relationship between the volume and temperature of gases within balloons. Charles's curiosity about the volume–temperature relationship was rooted in his interest in taking flight in a hot air balloon, and his work led to the first successful manned hydrogen balloon flight. Accordingly, the volume–temperature relationship is named to honor Charles. In science, a law is a statement of a pattern found in nature. **Charles's Law** states that the volume of a fixed amount of gas at constant pressure is directly proportional to absolute temperature. This pattern is shown on the graph in Figure 4.14, and it is explained at the particulate level in **Figure 4.17**.

We can abbreviate the proportional relationship between the volume and Kelvin temperature of a gas with mathematical symbols:

$$V \propto T$$

where V is the volume of the container holding a fixed amount of gas at constant pressure and T is the temperature, expressed in kelvins. We can change the proportionality to an equality by inserting a proportionality constant:

$$V \propto T \xrightarrow{\text{the proportionality changes to an equality}} V = \text{constant} \times T$$

Figure 4.16 Jacques Charles (1746–1823). The world's first hydrogen-filled balloon was designed and launched by Charles in collaboration with Anne-Jean Robert and his brother, Nicholas-Louis Robert. The gaseous hydrogen was generated by pouring a sulfuric acid solution onto iron.

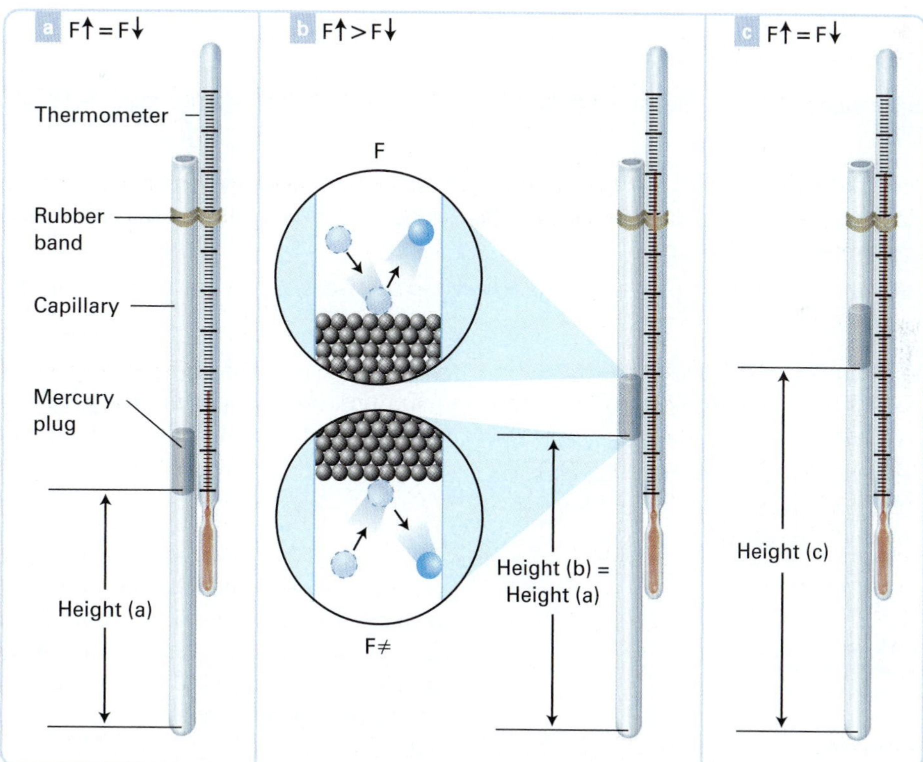

Figure 4.17 A particulate-level view of Charles's Law. (a) When the temperature of the gas trapped between the sealed end of the capillary and the mercury plug is the same as the temperature of the outside air, the average kinetic energy of the outside air molecules and the trapped gas molecules is the same. The force exerted upward on the mercury plug by the trapped gas molecules (F↑) is equal to the force exerted downward on the mercury plug by the outside air molecules (F↓). (This discussion assumes that the downward force exerted by the mercury is negligible.)

(b) As the trapped gas molecules are heated, their average kinetic energy increases, so they collide with more force against the mercury plug. More importantly, at higher temperatures, the gas molecules move faster. This increases the frequency with which molecules collide with the bottom of the mercury plug, which also adds to F↑. Since F↑ is now greater than F↓, the plug moves upward.

(c) As the plug rises and the gas molecules distribute themselves over the larger volume available to them, the frequency of molecules colliding with the bottom of the plug decreases, reducing the force, decreasing the pressure of the trapped gas. When the pressure drops to the point such that it again equals the outside pressure, the plug stops moving. Thus the volume of gas is higher at the higher temperature.

Dividing both sides of the V = constant × T equation by T gives

$$V = \text{constant} \times T \xrightarrow{\text{divide both sides by T}} \frac{V}{T} = \frac{\text{constant} \times \cancel{T}}{\cancel{T}} \xrightarrow{\text{cancel the common factor}} \frac{V}{T} = \text{constant}$$

Writing the equation in this form helps to illustrate the nature of a direct proportionality. The ratio of V to T must stay the same; the ratio must stay equal to the constant, a fixed, unchanging quantity. Thus, if V increases by 50%, T must also increase by 50% to keep the ratio the same. Similarly, if V decreases by $\frac{1}{6}$, T must also decrease by $\frac{1}{6}$.

Therefore, for one pair of volume and temperature quantities,

$$\frac{V_1}{T_1} = \text{constant}$$

and for a different set of volume and temperature conditions for the same fixed amount of gas at constant pressure,

$$\frac{V_2}{T_2} = \text{constant}$$

This leads to the fact that

$$\frac{V_1}{T_1} = \text{constant} = \frac{V_2}{T_2} \quad \text{or, more simply,} \quad \frac{V_1}{T_1} = \frac{V_2}{T_2}$$

Active Example 4.3 Charles's Law

A gas with an initial volume of 1.67 L, measured at 32°C, is heated to 55°C at constant pressure. What is the new volume of the gas?

Think Before You Write Keep in mind that when working gas law problems involving temperature, the proportional relationships apply to temperatures expressed in kelvins and not to Celsius temperatures. As part of a gas law problem setup, you must convert Celsius temperatures to kelvins.

Answers *Cover the left column with your cut-out shield. Reveal each answer only after you have written your own answer in the right column.*

Given: 1.67 L, initial volume (V_1); 32°C, initial Celsius temperature (T_1); 55°C, final Celsius temperature (T_2)
Wanted: final volume, assume L (V_2)

We assumed that the final volume should be expressed in liters because the problem statement did not specify a unit and the initial volume was given in liters.

Analyze the problem statement by writing the given quantities, their properties, the property of the wanted quantity, and its unit. Label each quantity as V_1, V_2, T_1, and so on.

$$\frac{V_1}{T_1} = \frac{V_2}{T_2} \xrightarrow{\text{multiply both sides by } T_2}$$

$$\frac{V_1}{T_1} \times T_2 = \frac{V_2}{\cancel{T_2}} \times \cancel{T_2} \xrightarrow{\text{cancel the common factor}}$$

$$\frac{V_1}{T_1} \times T_2 = V_2 \xrightarrow{\text{reverse the order}} V_2 = \frac{V_1}{T_1} \times T_2$$

We reversed the order in the last step of solving the equation for the wanted variable so that we can extend the equation to the right when we enter the given quantities.

Identify the algebraic relationship needed to solve the problem. In the analysis, you simplified the problem to the fact that you are given an initial V_1 and T_1 and a final T_2, and you need to determine the final V_2. After stating the algebraic relationship, solve it for V_2, the wanted variable.

$$V_2 = \frac{V_1}{T_1} \times T_2 = \frac{1.67\,\text{L}}{(32 + 273)\,\cancel{\text{K}}} \times (55 + 273)\,\cancel{\text{K}} = 1.80\,\text{L}$$

Construct the solution setup by extending your equation with the given quantities, and then cancel units and calculate the answer. Remember to change temperature in °C to temperature in K.

The fraction $\frac{(55 + 273)\,K}{(32 + 273)\,K}$ is greater than one, so the final volume should be larger than the initial volume. The value of the answer appears to be reasonable.	**Check** the value of your answer by looking at the temperatures in the numerator and denominator of your setup. Is the fraction greater than one or less than one? Should your final volume be larger than the initial volume or smaller? 
You improved your skill at solving gas volume–temperature problems.	What did you learn by working through this Active Example? 

Practice Exercise 4.3

A container with flexible walls holds a gas that is cooled from 65°C to 22°C with no change in pressure. If the container volume was 4.5 L initially, what is its volume after being cooled?

There are two ways to solve gas law problems: by algebra, as you just did in Active Example 4.3, and by "reasoning." Reasoning is based on the proportionality between the variables. In the case of Charles's Law, the variables are volume and temperature, which are directly proportional to one another. This means that they change in the same direction; if one goes up, the other goes up (and vice versa). **Figure 4.18** shows how volume increases with increasing temperature.

Reconsider the equation you arrived at after solving the Charles's Law equation for V_2 in Active Example 4.3:

$$V_2 = \frac{V_1}{T_1} \times T_2$$

We can rearrange this equation so that both temperatures are in a fraction:

$$V_2 = \frac{V_1}{T_1} \times T_2 \xrightarrow{\text{rearrange}} V_2 = V_1 \times \frac{T_2}{T_1}$$

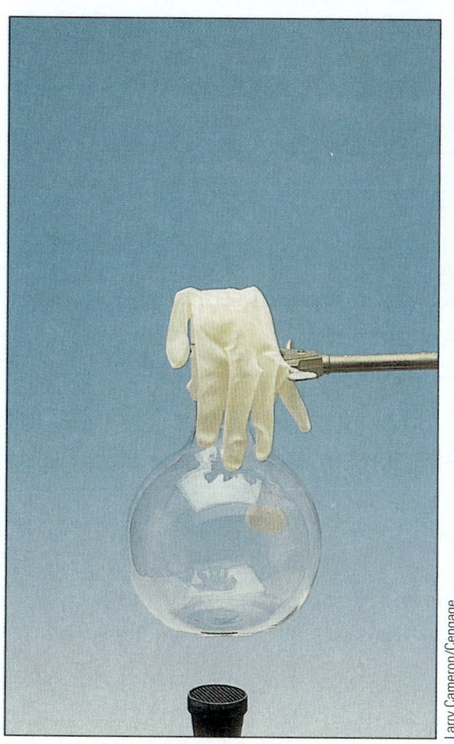

Larry Cameron/Cengage

Larry Cameron/Cengage

Figure 4.18 Volume and temperature of a gas. For a fixed amount of a gas at constant pressure, the volume and the Kelvin temperature of the gas are directly proportional to each other. As the temperature increases, the volume increases proportionally.

When set up this way, you can clearly see that the new volume, V_2, is the initial volume, V_1, multiplied by a ratio of temperatures—a temperature fraction.

If T_2 is greater than T_1, the temperature fraction has a value greater than 1. Multiplication of the initial volume by a value greater than 1 will result in a final volume larger than the initial volume. If T_2 is less than T_1, the temperature fraction has a value less than 1. Multiplication of the initial volume by a value less than 1 will result in a final volume smaller than the initial volume.

Now let's connect the mathematical reasoning with what we know about a gas. If temperature goes up, the volume goes up. So if you are given an initial volume of a gas and the temperature goes up, the final volume must be greater than the initial volume. To achieve that numerically, you have to multiply the initial volume by a value greater than 1—the temperature fraction needs to be greater than 1.

If you are given an initial volume and the temperature goes down, the final volume must be less than the initial volume. You need to multiply the initial volume by a temperature fraction that is less than 1. Let's revisit Active Example 4.3 to see how this reasoning is applied. You have 1.67 L of gas at 32°C, and the temperature changes to 55°C. Temperature goes up, so volume must go up (think about the balloons in Figure 4.15 and the rubber glove in Figure 4.18). Therefore, the initial volume must be multiplied by a temperature fraction greater than 1:

$$V_2 = V_1 \times \frac{T_2}{T_1} = 1.67 \text{ L} \times \frac{(55 + 273)\text{K}}{(32 + 273)\text{K}} = 1.80 \text{ L}$$

The temperature fraction reasoning approach leads to exactly the same answer as did the algebraic approach. It has to, of course, because the calculation setup is equivalent.

We suggest that you solve the gas law problems in the remainder of this chapter using algebra and then check your setup with reasoning. ◀ We will guide you through this process in the remaining Active Examples.

Some instructors prefer that their students solve gas law problems by reasoning rather than by algebra. If yours is among them, we recommend that you follow your instructor's lead rather than ours.

Your Thinking

Thinking About

Proportional Reasoning

The reasoning approach discussed here is the essence of thinking about your thinking on the subject of proportional reasoning. When all other variables are held constant, cooling a gas reduces the volume, so the temperature fraction (the proportionality) by which the initial volume is multiplied must be less than 1. Heating a gas increases the volume, which means that the ratio of temperatures multiplier must be greater than 1.

✓ Target Check 4.3

If the final temperature of a gas is less than the initial temperature (pressure and amount constant), how will the value of the final volume compare with the initial volume? Will the ratio of temperatures by which the initial volume is multiplied be greater than, equal to, or less than 1?

4.6 Boyle's Law: Volume and Pressure

Goal 13 Describe the relationship between the volume and pressure of a fixed amount of a gas at constant temperature, and express that relationship as a proportionality, an equality, and a graph.

14 Given the initial volume (or pressure) and initial and final pressures (or volumes) of a fixed amount of gas at constant temperature, calculate the final volume (or pressure).

Have you ever used a hand-pumped bicycle tire pump (**Figure 4.19**)? If so, you've felt the resistance that occurs as you push down on the handle, particularly as the

The fraction $\frac{(55 + 273)\text{K}}{(32 + 273)\text{K}}$ is greater than one, so the final volume should be larger than the initial volume. The value of the answer appears to be reasonable.

Check the value of your answer by looking at the temperatures in the numerator and denominator of your setup. Is the fraction greater than one or less than one? Should your final volume be larger than the initial volume or smaller?

You improved your skill at solving gas volume–temperature problems.

What did you learn by working through this Active Example?

Practice Exercise 4.3

A container with flexible walls holds a gas that is cooled from 65°C to 22°C with no change in pressure. If the container volume was 4.5 L initially, what is its volume after being cooled?

There are two ways to solve gas law problems: by algebra, as you just did in Active Example 4.3, and by "reasoning." Reasoning is based on the proportionality between the variables. In the case of Charles's Law, the variables are volume and temperature, which are directly proportional to one another. This means that they change in the same direction; if one goes up, the other goes up (and vice versa). **Figure 4.18** shows how volume increases with increasing temperature.

Reconsider the equation you arrived at after solving the Charles's Law equation for V_2 in Active Example 4.3:

$$V_2 = \frac{V_1}{T_1} \times T_2$$

We can rearrange this equation so that both temperatures are in a fraction:

$$V_2 = \frac{V_1}{T_1} \times T_2 \xrightarrow{\text{rearrange}} V_2 = V_1 \times \frac{T_2}{T_1}$$

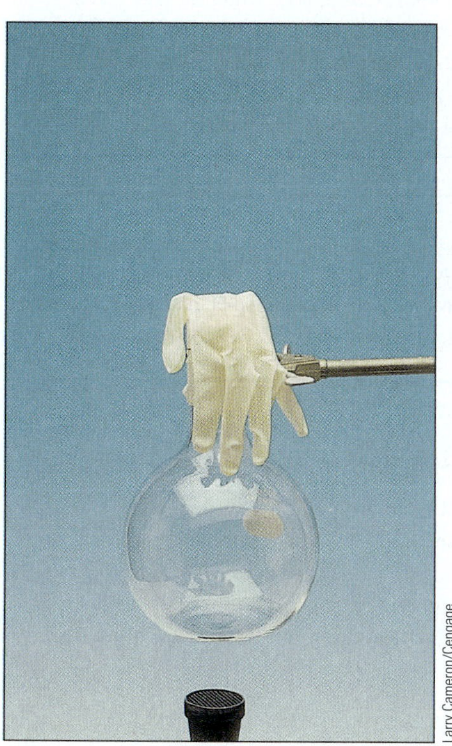

Figure 4.18 Volume and temperature of a gas. For a fixed amount of a gas at constant pressure, the volume and the Kelvin temperature of the gas are directly proportional to each other. As the temperature increases, the volume increases proportionally.

When set up this way, you can clearly see that the new volume, V_2, is the initial volume, V_1, multiplied by a ratio of temperatures—a temperature fraction.

If T_2 is greater than T_1, the temperature fraction has a value greater than 1. Multiplication of the initial volume by a value greater than 1 will result in a final volume larger than the initial volume. If T_2 is less than T_1, the temperature fraction has a value less than 1. Multiplication of the initial volume by a value less than 1 will result in a final volume smaller than the initial volume.

Now let's connect the mathematical reasoning with what we know about a gas. If temperature goes up, the volume goes up. So if you are given an initial volume of a gas and the temperature goes up, the final volume must be greater than the initial volume. To achieve that numerically, you have to multiply the initial volume by a value greater than 1—the temperature fraction needs to be greater than 1.

If you are given an initial volume and the temperature goes down, the final volume must be less than the initial volume. You need to multiply the initial volume by a temperature fraction that is less than 1. Let's revisit Active Example 4.3 to see how this reasoning is applied. You have 1.67 L of gas at 32°C, and the temperature changes to 55°C. Temperature goes up, so volume must go up (think about the balloons in Figure 4.15 and the rubber glove in Figure 4.18). Therefore, the initial volume must be multiplied by a temperature fraction greater than 1:

$$V_2 = V_1 \times \frac{T_2}{T_1} = 1.67 \text{ L} \times \frac{(55 + 273)\text{K}}{(32 + 273)\text{K}} = 1.80 \text{ L}$$

The temperature fraction reasoning approach leads to exactly the same answer as did the algebraic approach. It has to, of course, because the calculation setup is equivalent.

We suggest that you solve the gas law problems in the remainder of this chapter using algebra and then check your setup with reasoning. ◀ We will guide you through this process in the remaining Active Examples.

Some instructors prefer that their students solve gas law problems by reasoning rather than by algebra. If yours is among them, we recommend that you follow your instructor's lead rather than ours.

Your Thinking

Thinking About

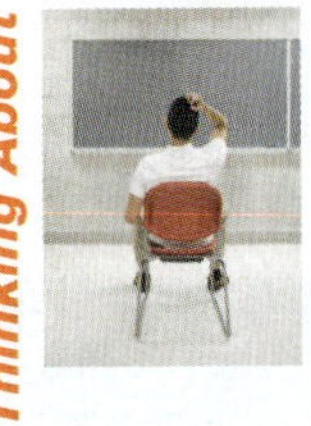

Proportional Reasoning

The reasoning approach discussed here is the essence of thinking about your thinking on the subject of proportional reasoning. When all other variables are held constant, cooling a gas reduces the volume, so the temperature fraction (the proportionality) by which the initial volume is multiplied must be less than 1. Heating a gas increases the volume, which means that the ratio of temperatures multiplier must be greater than 1.

✓ Target Check 4.3

If the final temperature of a gas is less than the initial temperature (pressure and amount constant), how will the value of the final volume compare with the initial volume? Will the ratio of temperatures by which the initial volume is multiplied be greater than, equal to, or less than 1?

4.6 Boyle's Law: Volume and Pressure

Goal 13 Describe the relationship between the volume and pressure of a fixed amount of a gas at constant temperature, and express that relationship as a proportionality, an equality, and a graph.

14 Given the initial volume (or pressure) and initial and final pressures (or volumes) of a fixed amount of gas at constant temperature, calculate the final volume (or pressure).

Have you ever used a hand-pumped bicycle tire pump (**Figure 4.19**)? If so, you've felt the resistance that occurs as you push down on the handle, particularly as the

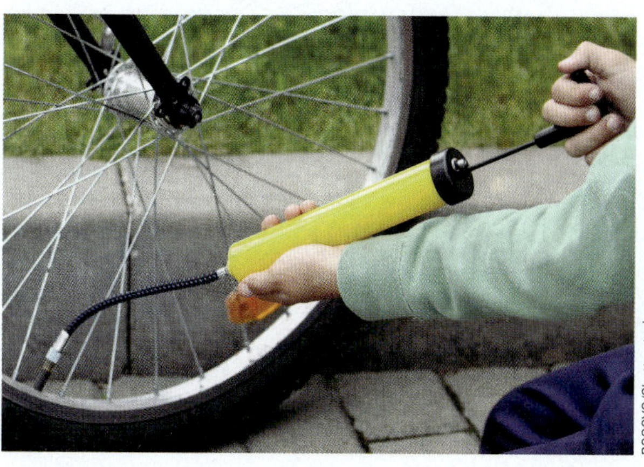

BOOCYS/Shutterstock.com

Figure 4.19 Bicycle tire pump. When the handle is pulled up, a valve located near the point where the hose connects to the cylinder closes, and air is drawn into the cylinder. When the handle is pushed down, the valve opens, and the volume of the pump-hose-tire system decreases, causing an increase in the pressure of the enclosed gas (which is air).

pressure in the tire nears the recommended quantity. That resistance that you feel is a force pushing back on the surface of the plunging mechanism—it is air pressure, the ratio of force to area. You can feel the pressure increasing. When you push the handle down, you are decreasing the volume of air in the connected pump-hose-tire system. Thus, when you use a hand-operated tire pump, you are experiencing the fact that as the volume of a gas decreases, the pressure increases.

The first person known to investigate the quantitative relationship between the pressure and the volume of a fixed amount of gas at constant temperature was 17th-century Irish scientist Robert Boyle (**Figure 4.20**). A modern laboratory experiment to investigate this relationship utilizes a mercury-filled manometer such as that shown in **Figure 4.21(a)**. The data collected by manipulating this instrument are given in the first two columns of the table in **Figure 4.21(b)**. These data are plotted in the graph of pressure versus volume in **Figure 4.21(c)**. The shape of the curve in this graph suggests an inverse proportionality between the variables. Expressed mathematically,

$$\text{Volume} \propto \frac{1}{\text{Pressure}} \quad \text{or, in symbols,} \quad V \propto \frac{1}{P}$$

Portrait of the Hon. Robert Boyle, c.1689 (oil on canvas)/Lely, Peter (1618-80) (school of)/CHATSWORTH/© The Devonshire Collections, Chatsworth/Bridgeman Images

Figure 4.20 Robert Boyle (1627–1691). Boyle's book *The Sceptical Chymist* argued for an experimental approach to scientific investigation. Although this seems to be obviously necessary today, it was a radical position in 1661.

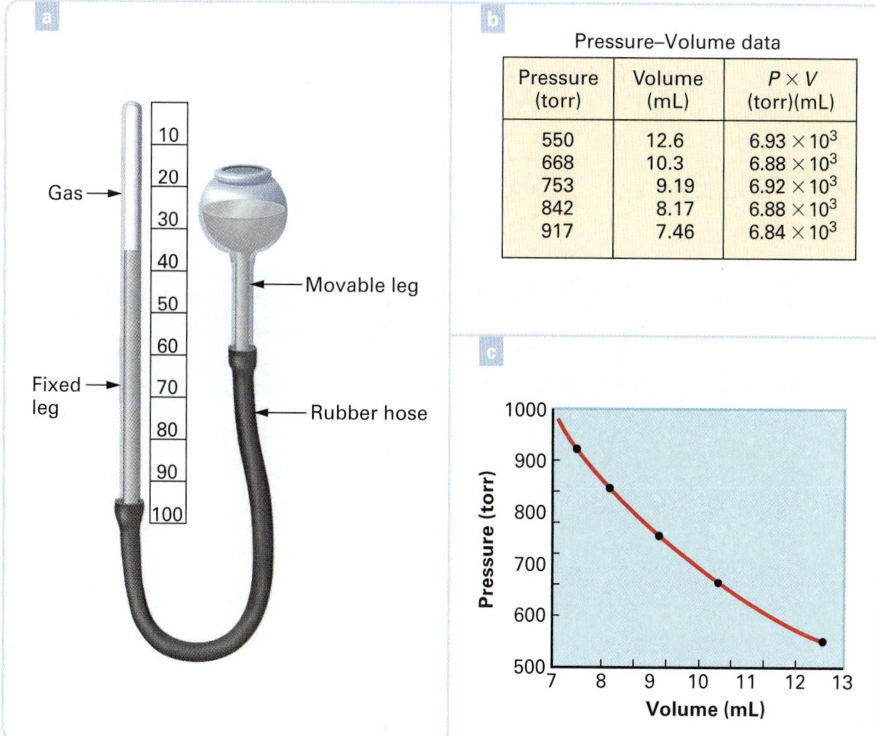

Pressure–Volume data

Pressure (torr)	Volume (mL)	$P \times V$ (torr)(mL)
550	12.6	6.93×10^3
668	10.3	6.88×10^3
753	9.19	6.92×10^3
842	8.17	6.88×10^3
917	7.46	6.84×10^3

Figure 4.21 Boyle's Law. A student performed an experiment to find the relationship between the pressure and the volume of a fixed amount of gas at constant temperature. The student raised or lowered the movable leg of the apparatus (a) and found the volume of the trapped gas at different pressures. The first two columns of the table (b) give the student's data. A graph (c) indicates that pressure and volume are inversely proportional to each other. This is confirmed by the constant product of pressure × volume (within experimental error, ±0.7% from the average value), shown in the third column of the table.

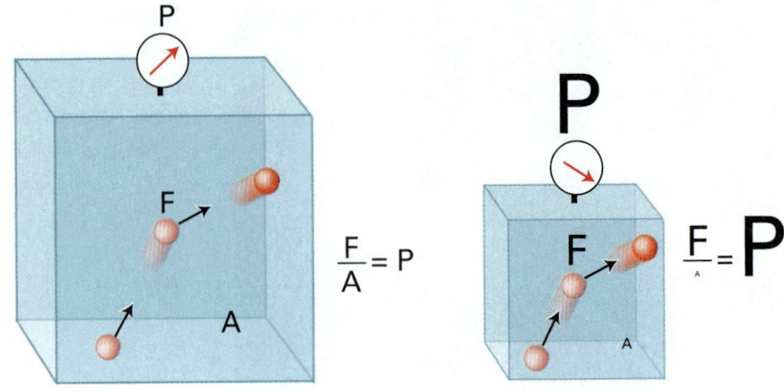

Figure 4.22 Explanation of Boyle's Law. At constant temperature, the particles in a fixed amount of gas have a constant average kinetic energy. Recall that the defining equation for pressure is $P \equiv F \div A$, where P is pressure, F is force, and A is area. When the volume of a container is reduced, the particles strike a smaller area with greater frequency. Let's assume that the length, width, and height of the cube shown on the left in the illustration are each reduced by half. The new area is $(1/2)^2 = 1/4$ of the initial area and the new volume is $(1/2)^3 = 1/8$ of the initial volume. The force is two times greater because the particles strike the container walls with double the frequency. The combination of the increased force and reduced area (symbolized by the change in the height of the letters in the illustration on the right) leads to increased pressure.

Introducing a proportionality constant gives

$$V \propto \frac{1}{P} \xrightarrow{\text{the proportionality changes to an equality}} V = \text{constant} \times \frac{1}{P}$$

Multiplying both sides of the equation by P yields

$$V = \text{constant} \times \frac{1}{P} \xrightarrow{\text{multiply both sides by P}}$$

$$P \times V = \cancel{P} \times \text{constant} \times \frac{1}{\cancel{P}} \xrightarrow{\text{cancel the common factor}} P \times V = \text{constant}$$

Within experimental error, the product of pressure and volume is indeed a constant, as shown in the third column of the table in Figure 4.21(b).

Boyle's Law, which this experiment illustrates, states that for a fixed amount of gas at constant temperature, pressure is inversely proportional to volume. **Figure 4.22** explains Boyle's Law at the particulate level, and **Figure 4.23** shows an application of Boyle's Law. Since the product of P and V is constant, when one factor increases, the other must decrease. This is what is meant by an inverse proportionality. Notice the difference between an inverse proportionality and the direct proportionality of Charles's Law, in which both variables—volume and temperature—increase or decrease together.

Also note the shape of the curve in Figure 4.21(c). It is characteristic of an inverse proportionality. Contrast the plot of an inverse proportionality with that of a direct proportionality (Figure 4.14[c]).

For a fixed amount of gas at constant temperature at two different pressure–volume conditions, we have

$$P_1 \times V_1 = \text{constant} \quad \text{and} \quad P_2 \times V_2 = \text{constant}, \quad \text{therefore} \quad P_1 \times V_1 = P_2 \times V_2$$

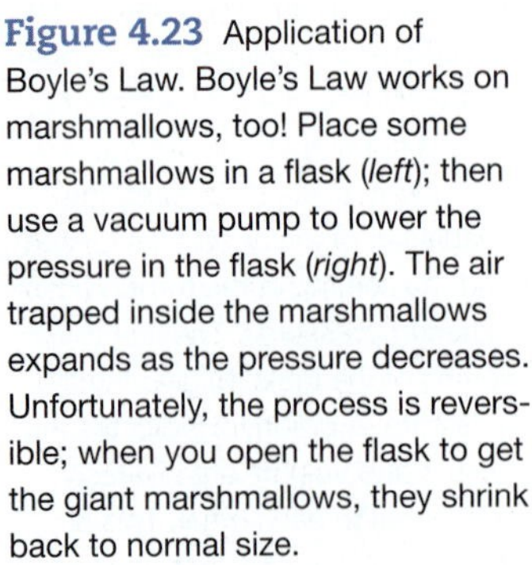

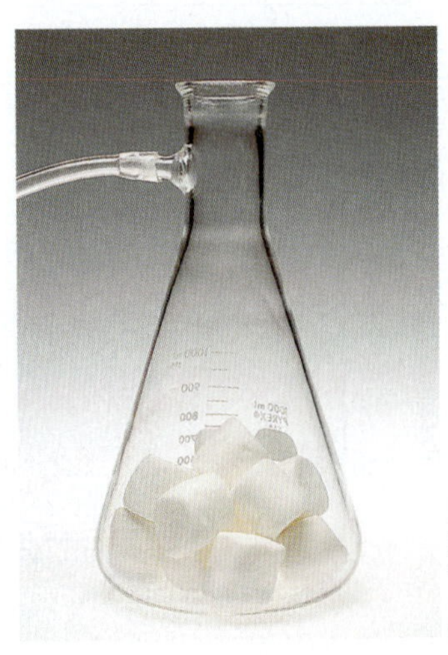

Figure 4.23 Application of Boyle's Law. Boyle's Law works on marshmallows, too! Place some marshmallows in a flask (*left*); then use a vacuum pump to lower the pressure in the flask (*right*). The air trapped inside the marshmallows expands as the pressure decreases. Unfortunately, the process is reversible; when you open the flask to get the giant marshmallows, they shrink back to normal size.

where subscripts 1 and 2 refer to the first and second measurements of pressure and volume at constant temperature.

Thinking About Your Thinking

Proportional Reasoning

An inverse proportionality has the algebraic form $y \propto \frac{1}{x}$. A common example of an inverse proportionality is the relationship between the time needed to travel between two points and the speed at which you drive. Mathematically, the proportionality is $\text{time} \propto \frac{1}{\text{speed}}$. If you increase your speed, you decrease the value of $\frac{1}{\text{speed}}$ and therefore decrease the time needed for the trip. If your speed doubles, the time is cut in half. To change an inverse proportionality to an equality, a proportionality constant is introduced: $y = k\frac{1}{x}$. The product of the two variables is therefore a constant: $xy = k$. Since the product must always equal a fixed number, when one value goes up, the other must go down proportionally.

✓ Target Check 4.4

Assume constant temperature and amount of gas. (a) If the final pressure is less than the initial pressure, how will the final volume compare with the initial volume? (b) If the final volume is greater than the initial volume, how will the final pressure compare with the initial pressure?

Active Example 4.4 Boyle's Law

A gas sample occupies 5.18 liters at 776 torr. Find the volume of the gas if the pressure is changed to 827 torr. Temperature and amount remain constant.

Think Before You Write Ignore the fact that we are in the Boyle's Law section of the textbook. How can you recognize that the solution of this problem requires application of Boyle's Law? Notice that volume and pressure are the only two variables in the problem, and Boyle's Law states the relationship between volume and pressure: they are inversely proportional.

Answers *Cover the left column with your cut-out shield. Reveal each answer only after you have written your own answer in the right column.*

Given: 5.18 L, initial volume (V_1); 776 torr, initial pressure (P_1); 827 torr, final pressure (P_2)
Wanted: final volume, assume L (V_2)

Analyze the problem statement. What are the given quantities, the wanted unit, and the properties of all of the quantities? Label each quantity as V_1, V_2, T_1, and so on.

$$P_1 \times V_1 = P_2 \times V_2 \xrightarrow{\text{divide both sides by } P_2}$$

$$\frac{P_1}{P_2} \times V_1 = \frac{\cancel{P_2}}{\cancel{P_2}} \times V_2 \xrightarrow{\text{cancel the common factor}}$$

$$\frac{P_1}{P_2} \times V_1 = V_2 \xrightarrow{\text{rearrange}} V_2 = V_1 \times \frac{P_1}{P_2}$$

Note that we directly set up pressure fractions when we divided by P_2. This makes the check more straightforward, and it makes cancellation of the common factor easier.

Identify the algebraic relationship between the variables, and solve for the wanted variable.

$$V_2 = V_1 \times \frac{P_1}{P_2} = 5.18\text{ L} \times \frac{776\,\cancel{\text{torr}}}{827\,\cancel{\text{torr}}} = 4.86\text{ L}$$

Construct the solution setup, cancel units, and calculate the value of the answer.

Because volume varies inversely with pressure, the volume should decrease. In an inverse proportion, the variables go in opposite ways: one goes up and the other goes down. Therefore, the pressure fraction ratio should be less than one, with the lower pressure on top, so that the final volume is less than the initial volume. The pressure fraction checks.	**Check** your setup. First, the pressure of the gas went up. In what direction, up or down, should volume go? Is your pressure fraction set up to increase or decrease volume? 
Yes. The fraction is somewhat smaller than one, and 4.86 is somewhat smaller than 5.18.	**Check** the value of the answer. Is the result, in general, reasonable? 
You improved your skill at solving gas volume–pressure problems.	What did you learn by working this Active Example? 

Practice Exercise 4.4

A fixed amount of gas occupies 557 mL and exerts a pressure of 2.4 bar. What will be the new volume if the pressure is changed to 3.0 atm while the temperature is kept constant?

4.7 The Combined Gas Law: Volume, Temperature, and Pressure

Goal 15 For a fixed amount of a confined gas, given the initial volume, pressure, and temperature and the final values of any two variables, calculate the final value of the third variable.

16 State the values associated with standard temperature and pressure (STP) for gases.

As you ascend a mountain while you drive up a steep mountain road, you can feel the temperature of the air decrease as you gain altitude. Your ears may also pop as they adjust to the decreasing pressure. This prompts the question: How is the volume of a gas affected by *both* a change in temperature *and* a change in pressure?

In Section 4.5, you built an understanding of Charles's Law: The volume and the Kelvin temperature of a fixed amount of a gas at constant pressure are directly proportional, $V \propto T$. In Section 4.6, you worked on acquiring knowledge about Boyle's Law: The volume and the pressure of a fixed amount of a gas at constant temperature are inversely proportional, $V \propto 1/P$. When a quantity is proportional to two other quantities, it is proportional to the product of those quantities:

$$V \propto T \quad \text{and} \quad V \propto \frac{1}{P} \quad \text{therefore} \quad V \propto T \times \frac{1}{P}$$

The proportionality becomes an equality upon the introduction of a proportionality constant:

$$V \propto T \times \frac{1}{P} \xrightarrow{\text{the proportionality changes to an equality}} V = \text{constant} \times T \times \frac{1}{P}$$

We can rearrange the equation to isolate the constant on one side of the equals sign:

$$V = \text{constant} \times T \times \frac{1}{P} \xrightarrow{\text{multiply both sides by } \frac{P}{T}} \frac{P}{T} \times V = \text{constant} \times \cancel{T} \times \frac{1}{\cancel{P}} \times \frac{\cancel{P}}{\cancel{T}}$$

$$\xrightarrow{\text{cancel the common factor}} \frac{P}{T} \times V = \text{constant} \xrightarrow{\text{rearrange}} \frac{PV}{T} = \text{constant}$$

Again, using subscripts 1 and 2 for initial and final values of all variables, we obtain

$$\frac{P_1V_1}{T_1} = \frac{P_2V_2}{T_2}$$

This relationship is called the **Combined Gas Law**. It is an algebraic combination of Charles's Law and Boyle's Law.

Given five of the six variables in the combined gas law equation, the remaining variable can be calculated using algebra. For example, if you know the volume occupied by a gas at one temperature and pressure (V_1, T_1, and P_1), you can find the volume that the gas will occupy (V_2) at another temperature and pressure (T_2 and P_2).

Active Example 4.5 The Combined Gas Law I

A cylinder in an automobile engine has a volume of 352 cm³. This engine takes in air at 21°C and 0.945 atm pressure. The compression stroke compresses and heats this gas until the pressure is 4.95 atm and the temperature is 95°C. What is the final volume in the cylinder?

Think Before You Write You are given the volume, temperature, and pressure of a fixed amount of a gas and the new temperature and pressure. You are asked to find the new volume. Thus, the solution to this problem requires application of the Combined Gas Law.

Answers *Cover the left column with your cut-out shield. Reveal each answer only after you have written your own answer in the right column.*

Analyze the problem statement. What are the given quantities, the wanted unit, and the properties of all of the quantities? Label each quantity as V_1, V_2, T_1, and so on.

Given: 352 cm^3, initial volume (V_1); 21°C, initial temperature (T_1); 0.945 atm, initial pressure (P_1); 4.95 atm, final pressure (P_2); 95°C, final temperature (T_2).
Wanted: final volume (V_2), assume cm^3

Identify the algebraic relationship needed to solve the problem. Solve it for the wanted variable.

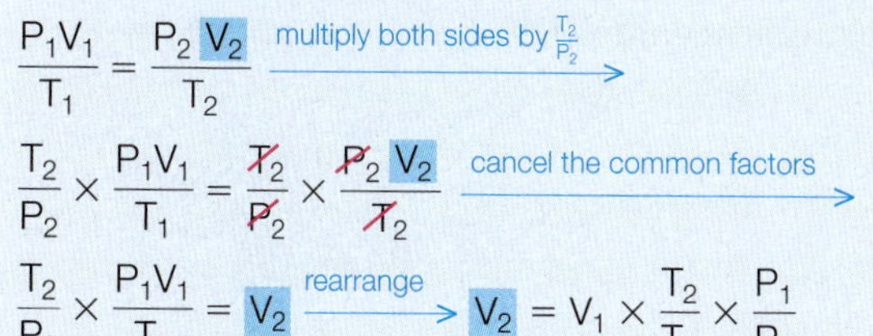

$$\frac{P_1V_1}{T_1} = \frac{P_2V_2}{T_2} \xrightarrow{\text{multiply both sides by } \frac{T_2}{P_2}}$$

$$\frac{T_2}{P_2} \times \frac{P_1V_1}{T_1} = \frac{\cancel{T_2}}{\cancel{P_2}} \times \frac{\cancel{P_2}V_2}{\cancel{T_2}} \xrightarrow{\text{cancel the common factors}}$$

$$\frac{T_2}{P_2} \times \frac{P_1V_1}{T_1} = V_2 \xrightarrow{\text{rearrange}} V_2 = V_1 \times \frac{T_2}{T_1} \times \frac{P_1}{P_2}$$

When we rearranged the equation, we set apart a temperature fraction and a pressure fraction to allow use of the reasoning approach in the check step.

Construct the solution setup. Cancel units and calculate the value of the answer.

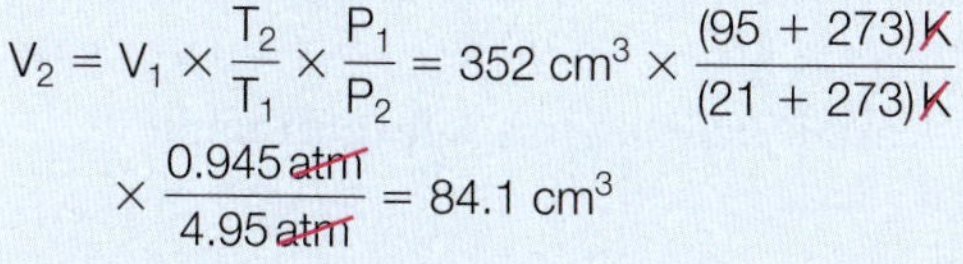

$$V_2 = V_1 \times \frac{T_2}{T_1} \times \frac{P_1}{P_2} = 352\ \text{cm}^3 \times \frac{(95 + 273)\cancel{\text{K}}}{(21 + 273)\cancel{\text{K}}} \times \frac{0.945\ \cancel{\text{atm}}}{4.95\ \cancel{\text{atm}}} = 84.1\ \text{cm}^3$$

Check the temperature fraction. Does the problem statement indicate that the temperature is increasing or decreasing? How will that affect volume? Does the temperature fraction in your setup correspond accordingly?

Temperature is increasing (from 21°C to 95°C), so volume should increase because of this. The temperature fraction is greater than 1, as it should be.

Check the pressure fraction. Is it correct? Explain.

Pressure is increasing (from 0.945 atm to 4.95 atm), so volume should decrease because of this. The pressure fraction is less than 1. It is correct.

You improved your skill at solving gas volume–pressure–temperature problems.	What did you learn by doing this Active Example?

Practice Exercise 4.5

A confined gas in a cylinder with a moveable piston occupies 122 mL at 45°F and 55.1 in. Hg. What volume will the gas occupy at 255°F and 29.9 in. Hg?

The volume of a fixed amount of gas depends on its temperature and pressure. Therefore, it is not meaningful to state the quantity of gas in volume units without also specifying the temperature and pressure. The International Union of Pure and Applied Chemistry (IUPAC), an international organization of chemists (that, among other things, seeks to bring consistency in how chemical quantities and names are expressed), recommends 0°C and 1 bar as **standard temperature and pressure (STP)**. Having this standard helps in comparison of gas volumes by establishing a set of temperature and pressure conditions at which volumes should be reported. Thus, it is sometimes necessary to change gas volumes at the conditions at which they were measured to STP.

Active Example 4.6 The Combined Gas Law II

The volume of a sample of nitrogen gas is measured as 3.62 liters when it is at 0.843 bar and 16°C. What is the STP volume of the nitrogen gas sample?

Think Before You Write STP is a shorthand way of giving you the value of two variables. This is therefore a problem of the type "given the value of five variables in a six-variable equation, find the value of the sixth."

Answers *Cover the left column with your cut-out shield. Reveal each answer only after you have written your own answer in the right column.*

Given: 3.62 L, initial volume (V_1); 0.843 bar, initial pressure (P_1); 16°C, initial temperature (T_1); 0°C, final temperature (T_2); 1 bar, final pressure (P_2) **Wanted:** final volume (V_2), assume L $V_2 = V_1 \times \frac{T_2}{T_1} \times \frac{P_1}{P_2} = 3.62\ \text{L} \times \frac{(0 + 273)\ \cancel{\text{K}}}{(16 + 273)\ \cancel{\text{K}}}$ $\times \frac{0.843\ \cancel{\text{bar}}}{1\ \cancel{\text{bar}}} = 2.88\ \text{L}$ Temperature decreases, decreasing volume, so temperature fraction should be less than 1, OK. Pressure increases, decreasing volume, so pressure fraction should be less than 1, OK.	Solve the problem completely.
You improved your skill at solving gas volume–temperature–pressure problems that include STP conditions.	What did you learn by solving this Active Example?

Practice Exercise 4.6

The STP volume of a gas is 78 mL. What volume will the gas occupy if the temperature is changed to 72°F and the pressure is 29.7 inches of mercury?

Now that you have learned the Combined Gas Law, we can show you how to use it to derive the other gas laws. Start with the Combined Gas Law:

$$\frac{P_1V_1}{T_1} = \frac{P_2V_2}{T_2}$$

If pressure is constant, $P_1 = P_2$, so we can divide both sides by P:

$$\frac{\cancel{P_1}V_1}{T_1} = \frac{\cancel{P_2}V_2}{T_2} \xrightarrow{\text{divide both sides by P}} \frac{V_1}{T_1} = \frac{V_2}{T_2} \text{ (constant P)}$$

The relationship that results is Charles's Law.

If temperature is constant, $T_1 = T_2$, and multiplying both sides by T yields:

$$\frac{P_1V_1}{\cancel{T}_1} = \frac{P_2V_2}{\cancel{T}_2} \xrightarrow{\text{multiply both sides by T}} P_1V_1 = P_2V_2 \text{ (constant T)}$$

Boyle's Law results.

Another gas law can be derived for situations where V is constant, that is, where $V_1 = V_2$:

$$\frac{P_1\cancel{V}_1}{T_1} = \frac{P_2\cancel{V}_2}{T_2} \xrightarrow{\text{divide both sides by V}} \frac{P_1}{T_1} = \frac{P_2}{T_2} \text{ (constant V)}$$

This equation shows that pressure and Kelvin temperature are directly proportional, $P \propto T$. Thus, we can state that the pressure exerted by a fixed amount of gas at constant volume is directly proportional to Kelvin temperature. This relationship is called **Amontons's Law** or Gay-Lussac's Law.

You now have a broad picture of the gas laws that describe pressure, volume, and temperature relationships for a fixed amount of gas. The Combined Gas Law is the most generalized equation, and the other gas laws can be derived from it by holding the value of one of the variables constant and canceling it algebraically. The gas laws for a fixed amount of gas are summarized in **Table 4.3**.

Learn It Now! *Remembering one equation and understanding the algebraic cancellation of variables is easier and will be remembered longer than trying to remember multiple equations. Be sure that you can derive Charles's Law, Boyle's Law, and Amontons's Law from the Combined Gas Law before you proceed further with your studies.*

Table 4.3 Gas Laws for a Fixed Amount of Gas

Name	Relationship	Variables	Constants	Relationship to the Combined Gas Law
Combined Gas Law	$\frac{P_1V_1}{T_1} = \frac{P_2V_2}{T_2}$	P, V, T	Amount	
Charles's Law	$\frac{V_1}{T_1} = \frac{V_2}{T_2}$	V, T	Amount, P	$\frac{\cancel{P}_1V_1}{T_1} = \frac{\cancel{P}_2V_2}{T_2}$
Boyle's Law	$P_1V_1 = P_2V_2$	V, P	Amount, T	$\frac{P_1V_1}{\cancel{T}_1} = \frac{P_2V_2}{\cancel{T}_2}$
Amontons's Law	$\frac{P_1}{T_1} = \frac{P_2}{T_2}$	P, T	Amount, V	$\frac{P_1\cancel{V}_1}{T_1} = \frac{P_2\cancel{V}_2}{T_2}$

Art Directors & TRIP/Alamy Stock Photo

CHAPTER 4 IN REVIEW: INTRODUCTION TO GASES

Goal 1 Describe five macroscopic characteristics unique to the gas phase of matter.

Five macroscopic characteristics unique to the gas phase of matter are as follows:

1. Gases may be compressed.
2. Gases may be expanded.
3. Gases have low densities.
4. Gases may be mixed in a fixed volume.
5. Gases exert constant pressure on the walls of their container uniformly in all directions.

Goal 2 Use the postulates of the kinetic molecular theory to explain the reasons for the macroscopic characteristics unique to the gas phase of matter.	The reasons for the macroscopic characteristics unique to the gas phase of matter are as follows: **1.** A gas consists of molecules and empty space. **2.** The volume occupied by the molecules in a gas is negligible when compared with the volume of the space they occupy. **3.** The attractive forces among molecules of a gas are negligible. **4.** The average kinetic energy of gas molecules is proportional to the temperature, expressed in kelvins. **5.** Molecules interact with one another and with the container walls without loss of total kinetic energy.
Goal 3 Define *pressure* and interpret statements in which the term *pressure* is used.	**Pressure** is defined as force per unit area.
Goal 4 Explain the cause of the pressure of a gas.	Pressure is the effect of the force of the interactions of the huge number of rapidly moving molecules of a gas as they collide with the surface area of an object in contact with that gas.
Goal 5 Express the relationship among the following gas pressure units: atmospheres, torr, millimeters of mercury, inches of mercury, pascals, kilopascals, bars, or pounds per square inch.	Common pressure units and their relationships to one another are: 1 atm = 760 torr = 760 mm Hg = 29.92 in. Hg = 1.013×10^5 Pa = 101.3 kPa = 1.013 bar = 14.69 psi.
Goal 6 Describe the operational principle of an open-end manometer.	In an **open-end manometer**, total pressures are equal at the lower mercury level. The pressure of the gas is exerted on the mercury surface in the closed leg; the pressure of the atmosphere is exerted on the mercury surface in the open leg.
Goal 7 Given the pressure difference in an open-end manometer and the atmospheric pressure, determine the pressure of the gas in the manometer.	$P_{gas} = P_{atmosphere} \pm P_{mercury}$
Goal 8 Given the gauge pressure of a gas and the atmospheric pressure, determine the absolute pressure of the gas.	$P_{absolute} = P_{gauge} + P_{atmosphere}$
Goal 9 Given a pressure in atmospheres, torr, millimeters (or centimeters) of mercury, inches of mercury, pascals, kilopascals, bars, or pounds per square inch, express pressure in each of the other units.	The quantitative problem-solving methods you learned in Chapter 3 are used to convert from one pressure unit to another.
Goal 10 When plotting the relationship between gas volume and temperature, explain how the straight line can be extrapolated to determine the temperature at which the gas has zero volume, and explain the significance of this temperature.	Gas volume and absolute temperature are directly proportional for a fixed amount of gas at constant pressure. When the relationship is extrapolated to zero volume, the trend line reaches the temperature axis at 0 K, –273°C. This is **absolute zero**, the lowest temperature.
Goal 11 Describe the relationship between the volume and temperature of a fixed amount of a gas at constant pressure, and express that relationship as a proportionality, an equality, and a graph.	**Charles's Law** states that the volume of a fixed quantity of gas at constant pressure is directly proportional to absolute temperature, $V \propto T$ and $V = kT$. The plot of volume versus temperature is a straight line that passes through the origin.
Goal 12 Given the initial volume (or temperature) and the initial and final temperatures (or volumes) of a fixed amount of gas at constant pressure, calculate the final volume (or temperature).	For a fixed amount of gas at constant pressure, $\frac{V_1}{T_1} = \frac{V_2}{T_2}$.
Goal 13 Describe the relationship between the volume and pressure of a fixed amount of a gas at constant temperature, and express that relationship as a proportionality, an equality, and a graph.	**Boyle's Law** states that for a fixed amount of gas at constant temperature, pressure is inversely proportional to volume, $P \propto 1/V$ and $P = k(1/V)$ or $PV = k$. The plot of pressure versus the inverse of volume is a straight line that passes through the origin.
Goal 14 Given the initial volume (or pressure) and initial and final pressures (or volumes) of a fixed amount of gas at constant temperature, calculate the final volume (or pressure).	For a fixed amount of gas at constant temperature, $P_1V_1 = P_2V_2$.

Goal 15 For a fixed amount of a confined gas, given the initial volume, pressure, and temperature and the final values of any two variables, calculate the final value of the third variable.	The **Combined Gas Law** states that for a fixed amount of gas, $\frac{P_1V_1}{T_1} = \frac{P_2V_2}{T_2}$. Given the initial (or final) values of all three variables and the final (or initial) values of two, the unknown value is calculated with the Combined Gas Law equation.
Goal 16 State the values associated with standard temperature and pressure (STP) for gases.	Standard temperature and pressure (**STP**) is defined as **273 K (0°C)** and **1 bar** pressure.

Key Terms

Most of the key terms and concepts and many others appear in the Glossary. Use your Glossary regularly.

absolute pressure p. 126
absolute zero p. 132
Amontons's Law p. 143
atmosphere (atm) (pressure unit) p. 125
bar p. 125
barometer p. 123
Boyle's Law p. 138
Charles's Law p. 133
Combined Gas Law p. 141
gauge pressure p. 126
inches of mercury (in. Hg) p. 125
kinetic molecular theory p. 121
manometer p. 126
millimeters of mercury (mm Hg) p. 125
pascal (Pa) p. 125
pounds per square inch (psi) p. 125
pressure p. 123
standard temperature and pressure, STP p. 142
torr p. 125

Frequently Asked Questions

I'm confused about direct and inverse proportionalities; can you explain these in a different way?
Mathematically, a direct proportionality exists between two variables x and y if $y = k \times x$, where $k \neq 0$. When y increases, x must also increase. For example, if you were to exchange the quarters in your change jar for dollar bills on an equal value basis, the number of quarters you cash in is directly proportional to the number of dollar bills you receive: quarters $\propto$ dollar bills and quarters $= 4 \times$ dollar bills. In this chapter, Charles's Law and Amontons's Law are direct proportionalities: $V \propto T$ (fixed P) and $P \propto T$ (fixed V).

An inverse proportionality has the form $x \times y = k$, where $k \neq 0$, which can also be expressed as $y = k \div x$. When y increases, x must decrease. For example, the time needed to count the number of quarters in your change jar is inversely proportional to the number of people doing the counting: time $\times$ number of people = constant. More time is needed when fewer people are counting; less time is required when the number of people counting increases. In this chapter, Boyle's Law is an inverse proportionality: $V \propto 1/P$ (fixed T).

Why don't the gas laws work with Celsius temperatures?
Take a look at Figure 4.14 again. A direct proportionality must include the (0,0) point on a graph. In the case of quarters $\propto$ dollar bills proportionality in the previous answer, you get 0 dollar bills for 0 quarters. The graph in Figure 4.14 shows that the volume of a gas goes to zero when the temperature goes to zero on the Kelvin scale, not the Celsius scale. The Kelvin scale assigns zero to what is in nature a true absence of temperature, based on what occurs at the particulate level. In contrast, other scales, such as the Celsius and Fahrenheit scales, use a zero that is convenient to humans at the macroscopic level.

When I solve $P_1V_1/T_1 = P_2V_2/T_2$, I sometimes mess up the algebra. How can I avoid this?
Use the reasoning approach first discussed after Active Example 4.3. If any of the fractions in your algebraic setup don't make sense, then the algebra must be wrong. Continually perform mental checks on your work as you solve problems.

When I solve $P_1V_1/T_1 = P_2V_2/T_2$, I sometimes assign the 1's and 2's incorrectly. How can I avoid this?
Use a table like this to change the problem statement into given and wanted variables:

	Volume	Temperature	Pressure
Initial Value (1)			
Final Value (2)			

Concept-Linking Exercises

Write a brief description of the relationships among each of the following groups of terms or phrases. Answers to the Concept-Linking Exercises are given at the end of the chapter.

1. Gaseous state of matter, compressibility, density, mixability
2. Kinetic molecular theory, particulate behavior
3. Pressure, barometer, Earth's atmosphere, gauge pressure
4. Temperature, average kinetic energy, absolute zero, kelvin (unit)
5. Charles's Law, Boyle's Law, Combined Gas Law

Small-Group Discussion Questions

Small-Group Discussion Questions are for group work, either in class or under the guidance of a leader during a discussion section.

1. Early scientists did not believe that gases were matter. What apparent properties of gases led to this belief? How do properties of gases contrast with properties of liquids and solids?
2. Use kinetic molecular theory to formulate a particulate-level explanation for each of the following:
   a) All gases behave approximately the same at common temperatures and pressures.
   b) Gases may be compressed.
   c) A tiny quantity of gas will completely fill a large container.
   d) The density of gases is relatively low.
   e) Gases can be mixed in a fixed volume.
   f) A confined gas exerts the same pressure on all container walls.
3. Compare Torricelli's mercury barometer with a barometer filled with a liquid exactly half as dense as mercury. Describe how the barometers would differ and how they would be the same.
4. Show how a volume-versus-temperature graph of a fixed amount of a gas at constant pressure can be used to discover the absolute zero of temperature.
5. Use the data in Figure 4.21 to construct a plot of volume-versus-1/pressure. Explain the nature of the plot in terms of Boyle's Law.
6. Describe how a change in each variable in the Combined Gas Law can be explained by kinetic molecular theory. For example, if you decrease the pressure of a confined gas at constant temperature, what happens at the particulate level to cause the volume to change?

Questions, Exercises, and Problems

Interactive versions of these problems may be assigned by your instructor. Solutions for blue-numbered questions are at the end of the chapter. Questions other than those in the General Questions and More Challenging Problems sections are paired in consecutive odd-even number combinations; solutions for the odd-numbered questions are also at the end of the chapter.

Section 4.3: A Particulate-Level Explanation of the Characteristics of Gases

1. What properties of gases are the result of the kinetic character of a gas? Explain.
2. What is the meaning of *kinetic* as it is used in describing gases and molecular theory? What is kinetic energy?
3. State how and explain why the pressure exerted by a gas is different from the pressure exerted by a liquid or solid.
4. What causes pressure in a gas?
5. What are the desirable properties of air in an air mattress used by a camper? Show which part of kinetic molecular theory is related to each property.
6. List some properties of air that make it suitable for use in automobile tires. Explain how each property relates to kinetic molecular theory.

Questions 7 through 14: Explain how the physical phenomenon described is related to one or more features of kinetic molecular theory.

7. Pressure is exerted on the top of a tank holding a gas, as well as on its sides and bottom.
8. Gases with a distinctive odor can be detected some distance from their source.
9. Balloons expand in all directions when blown up, not just at the bottom as when filled with water.
10. Steam bubbles rise to the top in boiling water.
11. Even though an automobile tire is "filled" with air, more air can always be added without increasing the volume of the tire significantly.
12. The density of liquid oxygen is about 1.4 g/cm^3. Vaporized at 0°C and 760 torr, this same 1.4 g occupies 980 cm^3, an expansion of 700 times the volume of the liquid.
13. Gas bubbles always rise through a liquid and become larger as they move upward.
14. Any container, regardless of size, will be completely filled by 1 gram of hydrogen.

Section 4.4: Gas Pressure

15. What does pressure measure? What does temperature measure?
16. Four properties of gases may be measured. Name them.
17. Explain how a manometer works.
18. Explain how a barometer works.

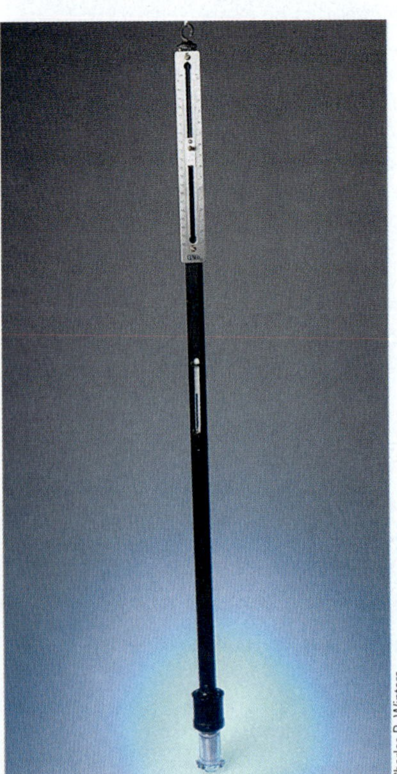

Charles D. Winters

A mercury-free barometer is often used in introductory chemistry laboratories.

Questions 19 and 20: Complete the table by converting the given pressure to each of the other pressures.

19.

atm					
psi					
in. Hg					
cm Hg					
mm Hg	785				
torr		124			
Pa			1.18×10^5		
kPa				91.4	
bar					0.977

20.

atm	1.84			
psi		13.9		
in. Hg			28.7	
cm Hg				74.8
mm Hg				
torr				
Pa				
kPa				
bar				

21. Find the pressure of the gas in the mercury manometer shown if the atmospheric pressure is 747 torr.

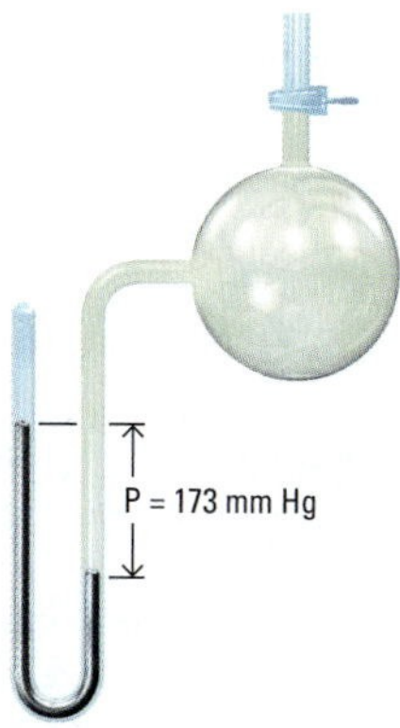

22. A mercury manometer is used to measure pressure in the apparatus illustrated here. Calculate the pressure exerted by the gas if atmospheric pressure is 752 torr.

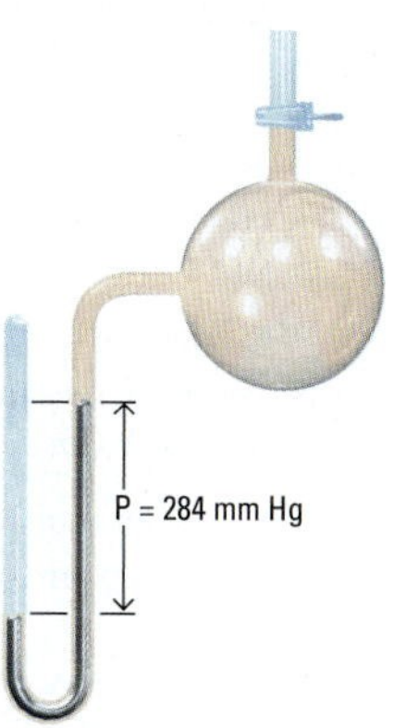

Section 4.5: Charles's Law: Volume and Temperature

23. A student records a temperature of −18 K in an experiment. What is the nature of things at that temperature? What would you guess the student meant to record? What absolute temperature corresponds to the temperature the student meant to record?

24. What does the term *absolute zero* mean? What physical condition is presumed to exist at absolute zero?

25. If the temperature in the room is 31°C, what is the equivalent Kelvin temperature?

26. Substances like sulfur dioxide, which is a gas at room temperature and pressure, can often be liquefied or solidified only at very low temperatures. At a pressure of 1 atm, SO_2 does not condense to a liquid until −10.1°C and does not freeze until −72.7°C. What are the equivalent Kelvin temperatures?

27. Hydrogen remains a gas at very low temperatures. It does not condense to a liquid until the temperature is −253°C, and it freezes shortly thereafter, at −259°C. What are the equivalent Kelvin temperatures?

28. Many common liquids have boiling points that are less than 110°C, whereas most metals are solids at room temperature and have much higher boiling points. The boiling point of bromoethane, C_2H_5Br, is 38°C. What is the equivalent Kelvin temperature? The boiling point of aluminum is 2740 K. What is the equivalent temperature on the Celsius scale?

29. Hydrogen cyanide is the deadly gas used in some execution chambers. It melts at 259 K and changes to a gas at 299 K. What are the Celsius temperatures at which hydrogen cyanide changes state?

30. Many common liquids have boiling points that are less than 110°C, whereas most metals are solids at room temperature and have much higher boiling points. The boiling point of propanol, C_3H_7OH, is 370 K. What is the equivalent Celsius temperature? The boiling point of nickel is 3003 K. What is the equivalent temperature on the Celsius scale?

31. A variable-volume container holds 24.3 L of gas at 55°C. If pressure remains constant, what will the volume be if the temperature falls to 17°C?

32. A sample of argon gas at a pressure of 715 mm Hg and a temperature of 26°C occupies a volume of 8.97 L. If the gas is heated at constant pressure to a temperature of 71°C, what will be the volume of the gas sample?

33. A spring-loaded closure maintains constant pressure on a gas system that holds a fixed amount of gas, but a bellows allows the volume to adjust for temperature changes. From a starting point of 1.26 L at 19°C, what Celsius temperature will cause the volume to change to 1.34 L?

34. A sample of hydrogen gas at a pressure of 0.520 atm and a temperature of 151°C occupies a volume of 857 mL. If the gas is heated at constant pressure until its volume is 1.02×10^3 mL, what will be the Celsius temperature of the gas sample?

Section 4.6: Boyle's Law: Volume and Pressure

35. If you squeeze the bulb of a dropping pipet (eye dropper) when the tip is below the surface of a liquid, bubbles appear. When you release the bulb, liquid flows into the pipet. Explain why in terms of Boyle's Law.

36. Squeezing a balloon is one way to burst it. Why?

37. The pressure of 648 mL of a gas is changed from 772 torr to 695 torr. What is the volume at the new pressure?

38. A sample of carbon dioxide gas at a pressure of 0.556 atm and a temperature of 21.2°C occupies a volume of 12.9 L. If the gas is allowed to expand at constant temperature to a volume of 22.3 L, what will be the pressure of the gas sample (in atmospheres)?

39. A cylindrical gas chamber has a piston at one end that can be used to compress or expand the gas. If the gas is initially at 1.22 atm when the volume is 7.26 L, what will the pressure be if the volume is adjusted to 3.60 L?

40. A sample of krypton gas at a pressure of 905 torr and a temperature of 28.4°C occupies a volume of 631 mL. If the gas is allowed to expand at constant temperature until its pressure is 606 torr, what will be the volume of the gas sample (in liters)?

Section 4.7: The Combined Gas Law: Volume, Temperature, and Pressure

41. The gas in a 0.717 L cylinder of a diesel engine exerts a pressure of 744 torr at 27°C. The piston suddenly compresses the gas to 48.6 atm, and the temperature rises to 547°C. What is the final volume of the gas?

42. A sample of krypton gas occupies a volume of 6.68 L at 64°C and 361 torr. If the volume of the gas sample is increased to 8.99 L while its temperature is increased to 114°C, what will be the resulting gas pressure (in torr)?

43. A collapsible balloon for carrying meteorological testing instruments aloft is partly filled with 626 L of helium, measured at 25°C and 756 torr. Assuming the volume of the balloon is free to expand or contract according to changes in pressure and temperature, what will be its volume at an altitude where the temperature is 258°C and the pressure is 0.641 atm?

44. A sample of neon gas occupies a volume of 7.68 L at 59°C and 0.634 atm. If it is desired to increase the volume of the gas sample to 9.92 L while decreasing its pressure to 0.436 atm, what will be the temperature of the gas sample at the new volume and pressure? Answer in degrees Celsius.

45. Why have the arbitrary conditions of STP been established? Are they realistic?

46. What is the meaning of STP?

47. If 1 cubic foot—28.3 L—of air at common room conditions of 23°C and 0.985 bar is adjusted to STP, what does the volume become?

48. A sample of helium gas has a volume of 7.06 L at 45°C and 2.12 bar. What would be the volume of this gas sample at STP?

49. An experiment is designed to yield 44.5 mL oxygen, measured at STP. If the actual temperature is 28°C and the actual pressure is 0.894 bar, what volume of oxygen will result?

50. A sample of krypton gas has a volume of 9.92 L at STP. What would be the volume of this gas sample at 40°C and 1.31 bar?

51. What pressure (in bar) will be exerted by a tank of natural gas used for home heating if its volume is 19.6 L at STP and it is compressed to 6.85 L at 24°C?

52. A gas occupies 2.33 L at STP. What pressure (bar) will this gas exert if it is expanded to 6.19 L and warmed to 17°C?

53. A container with a volume of 56.2 L holds helium at STP. The gas is compressed to 23.7 L and 2.09 bar. To what must the temperature change (°C) to satisfy the new volume and pressure?

54. At STP, a sample of neon fills a 4.47-L container. The gas is transferred to a 6.05 L container, and its temperature is adjusted until the pressure is 0.736 bar. What is the new temperature?

General Questions

55. Distinguish precisely and in scientific terms the differences among items in each of the following groups:
    a) Particulate-level explanation of characteristics of gases, kinetic molecular theory
    b) Pascal, mm Hg, torr, atmosphere, psi, bar
    c) Barometer, manometer
    d) Absolute pressure, gauge pressure
    e) Boyle's Law, Charles's Law, Amontons's Law, Combined Gas Law
    f) Celsius and Kelvin temperature scales
    g) Temperature and pressure, standard temperature and pressure, STP, $T_{°C}$, $T_{°F}$, T_K

56. Determine whether each of the following statements is true or false:
    a) The total kinetic energy of two molecules in the gas phase is the same before and after they collide with each other.
    b) Gas molecules are strongly attracted to each other.
    c) Gauge pressure is always greater than absolute pressure, except in a vacuum.
    d) For a fixed amount of gas at constant temperature, if volume increases, pressure decreases.
    e) For a fixed amount of gas at constant pressure, if temperature increases, volume decreases.
    f) For a fixed amount of gas at constant volume, if temperature increases, pressure increases.
    g) At a given temperature, the number of degrees Celsius is larger than the number of kelvins.
    h) Both temperature and pressure ratios are larger than 1 when calculating the final gas volume as conditions change from STP to 15°C and 0.834 bar.

More Challenging Problems

Questions 57 and 58: Explain how the given physical phenomenon is related to one or more of the features of kinetic molecular theory.

57. Very small dust particles, seen in a beam of light passing through a darkened room, appear to be moving about erratically.

58. Properties of gases become less "ideal"—the substance adopts behavior patterns not typical of gases—when subjected to very high pressures so that the individual molecules are close to each other.

59. The volume of the air chamber of a bicycle pump is 0.26 L. The volume of a bicycle tire, including the hose between the pump and the tire, is 1.80 L. If both the tire and the air in the pump chamber begin at 743 torr, what will be the pressure in the tire after a single stroke to the pump?

60. A 1.91 L gas chamber contains air at 959 torr. It is connected through a closed valve to another chamber with a volume of 2.45 L. The larger chamber is evacuated to negligible pressure. What pressure will be reached in both chambers if the valve between them is opened and the air occupies the total volume?

61. A hydrogen cylinder holds gas at 3.67 atm in a laboratory where the temperature is 25°C. To what quantity will the pressure change when the cylinder is placed in a storeroom where the temperature drops to 7°C?

62. Air in a steel cylinder is heated from 19°C to 42°C. If the initial pressure was 4.26 atm, what is the final pressure?

63. A gas storage tank is designed to hold a fixed volume and amount of gas at 1.74 atm and 27°C. To prevent excessive pressure due to overheating, the tank is fitted with a relief valve that opens at 2.00 atm. To what temperature (°C) must the gas rise in order to open the valve?

64. A gas in a steel cylinder shows a gauge pressure of 355 psi while sitting on a loading dock in the winter when the temperature is −18°C. What pressure (in psi) will the gauge show when the tank is brought inside and its contents warm up to 23°C? The pressure inside the laboratory is 14.7 psi.

65. If 1.62 m^3 of air at 12°C and 738 torr is compressed into a 0.140 m^3 tank, and the temperature is raised to 28°C while the external pressure is 14.7 psi, what pressure (in psi) will show on the gauge?

66. The pressure gauge reads 125 psi on a 0.140 m^3 compressed air tank when the gas is at 33°C. To what volume will the contents of the tank expand if they are released to an atmospheric pressure of 751 torr and the temperature is 13°C?

67. The compression ratio in an automobile engine is the ratio of the gas pressure at the end of the compression stroke to the pressure at the beginning. Assume that compression occurs at constant temperature. The total volume of a cylinder in an automobile is 350 cm^3, and the displacement (the reduction in volume during the compression stroke) is 309 cm^3. What is the compression ratio in that engine?

Answers to Target Checks

1. (a) Compressibility is due to the fact that there is a huge volume of empty space relative to the volume occupied by the molecules. When a gas is compressed, the relative volume of empty space decreases somewhat, but it is still relatively large. (b) Expandability is due to the fact that the molecules of a gas are in constant, rapid motion. It is also due to the fact that a gas is primarily composed of empty space. When a container holding an enclosed gas is expanded, the molecules continue to fill the volume within the container, and the proportion of empty space increases somewhat. (c) Density is defined as the ratio of mass to volume. Gases, with large empty spaces between the molecules, have a low amount of matter and therefore a relatively low mass within a given volume. In contrast, the molecules in a liquid or solid are touching, and therefore they have a relatively large quantity of matter in a given volume. (d) Gases can be mixed in a fixed volume because of the relatively large quantity of empty space between the molecules. (e) The uniform, constant pressure on the walls of a container in an enclosed gas results from particle motion. When an individual gas molecule interacts with the molecules that make up a container wall, it exerts a force at the point of interaction. When this is added to billions upon billions of similar interactions occurring continuously, the total effect is the steady force that is responsible for gas pressure.

2. (a) There is not enough information given to predict how pressure will be affected. Increasing the volume of the container will lead to a pressure decrease, and increasing the temperature of the gas will lead to a pressure increase. (b) The pressure will increase. The addition of gas particles will increase the pressure, as will the increase in temperature.

3. If $T_2 < T_1$, $V_2 < V_1$, because $V \propto T$. The ratio of temperatures must be < 1.

4. Remember that Boyle's Law is an inverse proportion: $V \propto 1/P$. (a) Because $P_2 < P_1$, $V_2 > V_1$. (b) If $V_2 > V_1$, then $P_2 < P_1$.

Answers to Practice Exercises

1. $741 \text{ torr} \times \frac{1.013 \text{ bar}}{760 \text{ torr}} = 0.988 \text{ bar};$

$$741 \text{ torr} \times \frac{29.92 \text{ in. Hg}}{760 \text{ torr}} = 29.2 \text{ in. Hg}$$

2. $29.52 \text{ in. Hg} \times \frac{101.325 \text{ kPa}}{29.92 \text{ in. Hg}} = 99.97 \text{ kPa};$

$$515 \text{ kPa} - 99.97 \text{ kPa} = 415 \text{ kPa};$$

$$415 \text{ kPa} \times \frac{14.70 \text{ psi}}{101.325 \text{ kPa}} = 60.2 \text{ psi}$$

3. $V_2 = V_1 \times \frac{T_2}{T_1} = 4.5 \text{ L} \times \frac{(22 + 273) \text{ K}}{(65 + 273) \text{ K}} = 3.9 \text{ L}$

4. $V_2 = V_1 \times \frac{P_1}{P_2} = 557 \text{ mL} \times \frac{2.4 \text{ bar}}{3.0 \text{ atm}} \times \frac{1 \text{ atm}}{1.013 \text{ bar}}$

$$= 4.4 \times 10^2 \text{ mL}$$

5. $T_{°C} = \frac{T_{°F} - 32}{1.8} = \frac{45 - 32}{1.8} = 7.2°C$; $7.2 + 273 = 280\ K$
$= 2.80 \times 10^2\ K$

$T_{°C} = \frac{T_{°F} - 32}{1.8} = \frac{255 - 32}{1.8} = 124°C$; $124 + 273 = 397\ K$

$V_2 = V_1 \times \frac{P_1}{P_2} \times \frac{T_2}{T_1} = 122\ mL \times \frac{55.1\ in.\ Hg}{29.9\ in.\ Hg} \times \frac{397\ K}{280\ K}$
$= 319\ mL$

6. $T_{°C} = \frac{T_{°F} - 32}{1.8} = \frac{72 - 32}{1.8} = 22°C$; $22 + 273 = 295\ K$

$V_2 = V_1 \times \frac{P_1}{P_2} \times \frac{T_2}{T_1} = 78\ mL \times \frac{1\ bar}{29.7\ in.\ Hg} \times \frac{295\ K}{273\ K}$
$\times \frac{29.92\ in.\ Hg}{1.013\ bar} = 84\ mL$

Answers to Concept-Linking Exercises

You may have found more relationships or relationships other than the ones given in these answers.

1. Substances in the gaseous state of matter can be compressed easily, they have low density, and they can be mixed in all proportions in a fixed volume. All of these properties, to some extent, result from the relatively large open spaces between gas molecules.
2. The kinetic molecular theory describes all matter as made up of tiny particles that are in constant motion. The particle behavior of a gas has particles completely independent of each other, filling the container, and moving in straight lines until colliding or striking the container walls with no loss of energy.
3. Pressure is the force exerted on a unit area of a surface. Earth is surrounded by gases (the atmosphere) that are drawn to it by gravity, creating atmospheric pressure. A barometer measures pressure directly. Gauge pressure, measured by a mechanical gauge that records zero at atmospheric pressure, measures only the pressure above atmospheric pressure.
4. Temperature is a measure of the average kinetic energy of particles in a sample. The kelvin is the SI unit for expressing temperature. The Kelvin temperature is absolute zero, 0 K, when no energy can be removed from a system.
5. For a fixed amount of gas: (a) Charles's Law states that volume is directly proportional to Kelvin temperature, $V \propto T$. (b) Boyle's Law states that volume is inversely proportional to pressure, $V \propto 1/P$. (c) These two proportionalities are joined in the Combined Gas Law, which is expressed in the equation $\frac{P_1V_1}{T_1} = \frac{P_2V_2}{T_2}$.

Solutions to Blue-Numbered Questions, Exercises, and Problems

1. All gas properties relate in some way to the kinetic character. Specifically, particle motion explains why gases fill their containers. Also, pressure results from the large numbers of particle collisions with the container walls.
3. Gas molecules move independently of one another in all directions. Therefore, they exert pressure on the container walls, including the top, uniformly in all directions. Liquid and solid pressures are the result of gravity, so they exert a downward pressure on the bottom of the container. Liquid particles can move amongst themselves, so they also exert horizontal pressures against the walls of their container. Solid particles lack this freedom of horizontal movement, so they exert no horizontal pressure on the walls of a container.
5. Because gas molecules are widely spaced, air is compressible, making for a soft and comfortable surface to lie on. The uniform pressure of air in the mattress keeps the person on the mattress off the ground. The low density of a gas allows the entire volume inside the mattress to be filled with a small mass of air, so the mattress can be filled by mouth rather than by a pump.
7. Gas particles collide with the walls of the container, exerting pressure uniformly in all directions, including the top of the tank and its sides.
9. Gas particles collide with the walls of the container, exerting pressure uniformly in all directions.
11. Gas particles are very widely spaced, which leaves room for adding more.
13. Gas particles are always in motion, "pushing" back the surrounding water. As the bubble rises, there is less liquid pushing on the bubble, so the gas volume increases because the gas particles are pushing against less force. Gas particles are widely spaced; thus gases have lower densities than liquids, so the bubble rises.
15. Pressure measures force per unit area. Temperature measures the average kinetic energy of the particles in a sample.
17. Figure 4.9 explains manometer operation.
19.

atm	1.03	0.163	1.16	0.902	0.964
psi	15.2	2.40	17.1	13.3	14.2
in. Hg	30.9	4.88	34.9	27.0	28.9
cm Hg	78.5	12.4	88.5	68.6	73.3
mm Hg	785	124	885	686	733
torr	785	124	885	686	733
Pa	1.05×10^5	1.65×10^4	1.18×10^5	9.14×10^4	9.77×10^4
kPa	105	16.5	118	91.4	97.7
bar	1.05	0.165	1.18	0.914	0.977

21. $747\ torr + 173\ torr = 920\ torr = 9.20 \times 10^2\ torr$

23. It is impossible to have a negative Kelvin temperature. 0 K, absolute zero, is the lowest temperature that can be approached. The student probably meant to record −18°C. $-18 + 273 = 255$ K.

25. $273 + 31 = 304$ K

27. $273 - 253 = 20\text{ K} = 2.0 \times 10^1\text{ K}$; $273 - 259 = 14$ K

29. $259 - 273 = -14°\text{C}$; $299 - 273 = 26°\text{C}$

31. $24.3\text{ L} \times \frac{(17 + 273)\text{ K}}{(55 + 273)\text{ K}} = 21.5\text{ L}$

33. $(19 + 273)\text{ K} \times \frac{1.34\text{ L}}{1.26\text{ L}} = 311\text{ K}$; $311 - 273 = 38°\text{C}$

35. Squeezing the bulb reduces gas volume, increasing pressure and forcing some air bubbles out of the pipet. Releasing the bulb increases volume and decreases pressure. Liquid pressure is then greater than gas pressure, so liquid enters the pipet, reducing gas volume until liquid and gas pressures are equal.

37. $648\text{ mL} \times \frac{772\text{ torr}}{695\text{ torr}} = 720\text{ mL} = 7.20 \times 10^2\text{ mL}$

39. $1.22\text{ atm} \times \frac{7.26\text{ L}}{3.60\text{ L}} = 2.46\text{ atm}$

41. $0.717\text{ L} \times \frac{744\text{ torr}}{48.6\text{ atm}} \times \frac{(547 + 273)\text{ K}}{(27 + 273)\text{ K}} \times \frac{1\text{ atm}}{760\text{ torr}} = 0.0395\text{ L}$

43. $626\text{ L} \times \frac{(-58 + 273)\text{K}}{(25 + 273)\text{K}} \times \frac{756\text{ torr}}{0.641\text{ atm}} \times \frac{1\text{ atm}}{760\text{ torr}} = 701\text{ L}$

45. STP conditions have been established so that all data are reported at the same conditions. One bar pressure is easily achieved in a laboratory, but 0°C is not a convenient temperature.

47. $28.3\text{ L} \times \frac{273\text{K}}{(23 + 273)\text{K}} \times \frac{0.985\text{ bar}}{1\text{ bar}} = 25.7\text{ L}$

49. $44.5\text{ mL} \times \frac{(28 + 273)\text{K}}{273\text{K}} \times \frac{1\text{ bar}}{0.894\text{ bar}} = 54.9\text{ mL}$

51. $1\text{ bar} \times \frac{19.6\text{ L}}{6.85\text{ L}} \times \frac{(24 + 273)\text{ K}}{273\text{ K}} = 3.11\text{ bar}$

53. $273\text{ K} \times \frac{27.3\text{L}}{56.2\text{L}} \times \frac{2.09\text{ bar}}{1\text{ bar}} = 241\text{ K}$; $241 - 273 = -32°\text{C}$

56. True: a, d, f, h. False: b, c, e, g.

57. Dust particles are being pushed around by moving gas particles, which are too small to see.

59. $743\text{ torr} \times \frac{(1.80 + 0.26)\text{ L}}{1.80\text{ L}} = 850\text{ torr} = 8.50 \times 10^2\text{ torr}$

61. $3.67\text{ atm} \times \frac{(7 + 273)\text{ K}}{(25 + 273)\text{ K}} = 3.45\text{ atm}$

63. $(27 + 273)\text{ K} \times \frac{2.00\text{ atm}}{1.74\text{ atm}} = 345\text{ K}$; $345 - 273 = 72°\text{C}$

65. $738\text{ torr} \times \frac{1.62\text{ m}^3}{0.140\text{ m}^3} \times \frac{(28 + 273)\text{K}}{(12 + 273)\text{K}} = 9.02 \times 10^3\text{ torr}$;

$9.02 \times 10^3\text{ torr} \times \frac{14.7\text{ psi}}{760\text{ torr}} = 174\text{ psi absolute}$;

$174 - 14.7 = 159\text{ psi gauge}$

67. Volume at start: 350 cm^3. Volume at end: $350 - 309 = 41$ cm^3. Volume is inversely proportional to pressure.

$\text{Compression ratio} = \frac{350\text{ cm}^3}{41\text{ cm}^3} = 8.5$

5 Atomic Theory: The Nuclear Model of the Atom

ELEMENTS
Hydrogen 1
Azote 5
Carbon 54
Oxygen 7
Phosphorus 9
Sulphur 13
Magnesia 20
Lime 24
Soda 28
Potash 42
Strontian 46
Barytes 68
Iron 50
Zinc 56
Copper 56
Lead 90
Silver 190
Gold 190
Platina 190
Mercury 167

Science Source/Science Source

◀ John Dalton's 1808 book *A New System of Chemical Philosophy* set the stage for the modern science of chemistry by reviving the ancient Greek concept that matter was made of elementary particles called atoms. Dalton added to the Greek idea by proposing that atoms of any one element are different from atoms of other elements. This illustration from Dalton's book shows his proposed system of "arbitrary marks or signs," as he wrote, to represent the elements. In this chapter, you will learn about Dalton's atomic theory and the modern system of "arbitrary marks or signs" used by today's scientists to represent the elements.

CHAPTER CONTENTS

In this chapter you will begin studying the atom, the smallest particle of any element. You'll learn that the atom is made up of three smaller particles. You will also see that different combinations of these smaller particles account for atoms of different elements. In addition, we introduce the arrangement of elements into groups that have similar properties.

5.1 Have the Elements Always Existed?

Recall that you learned in Section 2.7 that an element is a substance that cannot be decomposed into other stable pure substances because all atoms are the same element. It is currently believed that there are 94 different naturally occurring elements on Earth. Have all 94 of the elements existed since the beginning of the universe? If not, how were they formed?

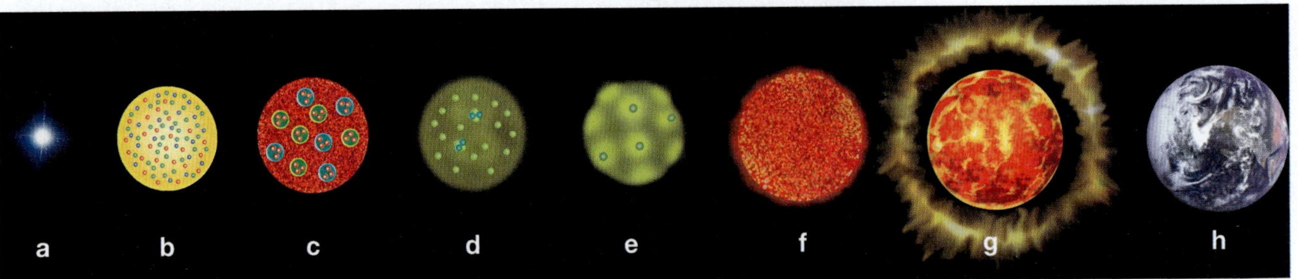

Figure 5.1 The formation of the elements. (a) The origin of the universe, t = 0. (b) The most fundamental particles form, t = 0.000001 s. (c) Protons and neutrons form, t = 0.001 s. (d) Helium nuclei form, t = 3 min. (e) Neutral atoms form, t = 400,000 yr. (f) Stars form, t = 1 billion yr. (g) Heavier elements are synthesized and dispersed, t > 1 billion yr. (h) The solar system is formed, t = 7 to 8 billion yr.

Figure 5.1 illustrates how the elements were formed. Look at each part of the figure as we explain the series of illustrations. (a) If we extrapolate backward in time current measurements of how the universe is changing, we conclude that 14 billion years ago, all of the matter and energy of the universe were contained in a volume just slightly greater than zero. Before this time, we have no knowledge of the nature of the universe. The universe began to expand rapidly in a process termed the *Big Bang*. (b) At 0.000001 second into the expansion, the most fundamental of the elementary particles formed. (c) At 0.001 second, some of the original fundamental particles condensed into particles smaller than atoms called *protons* and *neutrons*, which we will describe in Section 5.4. (d) At three minutes, the cores of helium atoms formed. We will consider the characteristics of this atomic core in Section 5.4 as well. (e) After 400,000 years of additional cooling and expansion, the first atoms formed! Most of these atoms were hydrogen and helium. In today's universe, it is estimated that 74% by mass of the known matter is hydrogen and 24% is helium. A small amount of lithium, beryllium, and boron likely also formed in this era. No other elements were directly formed as a result of the Big Bang. (f) After a billion years, the universe cooled enough to allow matter to be attracted together by the gravitational force and condense into galaxies. Each galaxy contained billions of stars. The stars that were more than 10 times the mass of our sun had such high temperatures and pressures in their cores that the hydrogen and helium they were originally made from fused together to make elements with atoms heavier than their parent atoms. These elements included carbon, nitrogen, oxygen, and others. (g) Large stars eventually exhaust their natural lifespan, and they die via an explosion, scattering their atoms throughout the galaxy. Some of those heavier atoms became part of other stars. After cycles of subsequent star formation and explosion, all of the additional heavier elements were formed. The elements continued to be scattered throughout the universe. (h) After 7 or 8 billion years of the production of heavier elements in the stars of our galaxy and others, our solar system was formed. As the planets accumulated matter and cooled, one of the eight planets was Earth. Conditions on Earth, such as its distance from the sun, were such that life evolved. The molecules of the living organisms on Earth contain atoms of many of the elements.

Thus, all of the lighter atoms in the molecules in your body, primarily atoms of the element hydrogen (helium atoms do not combine with other atoms to form molecules), were formed in the Big Bang. Each of your heavier atoms, primarily atoms of the elements carbon, nitrogen, and oxygen, was generated in stars. Think about this for a moment: The hydrogen atoms in your body are about 14 billion years old! Your carbon atoms are at least a billion years younger! The American scientist Carl Sagan wrote in the conclusion of his 1980 book entitled *Cosmos*, "For we are the local embodiment of a Cosmos grown to self-awareness. We have begun to contemplate our origins: starstuff pondering the stars; organized assemblages of ten billion billion billion atoms considering the evolution of atoms; tracing the long journey by which, here at least, consciousness arose."

5.2 Dalton's Atomic Theory

Goal 1 Identify the main features of Dalton's atomic theory.
2 State the meaning of, or draw conclusions based on, the Law of Multiple Proportions.

Have you ever wondered if there is a limit to the size of a piece of matter? If you were to cut a sheet of paper in half, and then in half again, and if you continued cutting each new half in half, would you get to a smallest piece of paper that still had the properties of paper, or (with the appropriate technology) could you keep dividing the paper into increasingly smaller pieces forever?

The Ancient Greek philosopher Leucippus (first half of 5th century BCE) and his student Democritus (c. 460–c. 370 BCE) were the first people in history to write about the possibility of a smallest piece of matter. They proposed that matter is composed of tiny indivisible, indestructible particles that they named *atoms,* from the Greek word *atomos,* meaning indivisible. Their logic was that the properties of substances must come from something that permanently stores those properties, even after a substance undergoes a chemical or physical change.

World-renowned Ancient Greek philosopher Aristotle (384–322 BCE) rejected the atom concept. He believed that matter was continuous: There was no smallest particle of a substance. Any piece of matter could be continually divided into smaller pieces. Given that Aristotle was considered to be among the greatest philosophers that ever lived, his opinions were of great influence, and the atom concept received no further serious contemplation until the 17th century.

In the early 1800s, the English scientist John Dalton (**Figure 5.2**) began to think about the atom concept. His focus was on three forms of evidence that came from experiments conducted in the late 1700s and early 1800s:

1. *The Law of Conservation of Mass.* The total mass of the reactants in a chemical change is equal to the total mass of the products (Chapter 2, Section 2.10).
2. *The Law of Definite Composition.* The percentage by mass of the elements in a compound is always the same (Chapter 2, Section 2.7).
3. *The* ***Law of Multiple Proportions****.* When two elements combine to form more than one compound, the different masses of one element that combine with the same mass of the other element are in a simple ratio of whole numbers (**Figure 5.3**). This is like threading one, two, or three identical nuts onto the same bolt. The mass of the bolt is constant. The mass of two nuts is twice the mass of one; the mass of three nuts is three times the mass of one. There are no fractional nuts, only whole nuts. The masses of nuts are in a simple ratio of whole numbers, 1:2:3.

E. F. Smith Memorial Collection in the History of Chemistry, Department of Special Collections, Van Pelt Dietrich Library, University of Pennsylvania.

Figure 5.2 John Dalton (1766–1844) devised the atomic theory while trying to formulate a hypothesis to explain the Law of Multiple Proportions and other related patterns seen in chemical investigations.

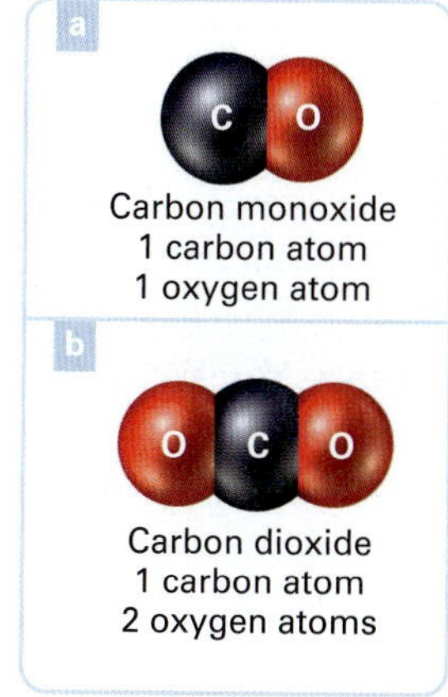

Figure 5.3 Explanation of the Law of Multiple Proportions. Carbon and oxygen combine to form more than one compound. Assume, according to the atomic theory, that carbon and oxygen atoms combine in a 1:1 ratio to form one compound and in a 1-carbon-to-2-oxygen-atom ratio to form another compound. If this assumption at the particulate level is true, the mass of the two oxygen atoms in the 1:2 compound is twice the mass of the one oxygen atom in the 1:1 compound. The oxygen mass ratio is 2:1 in the two compounds with the fixed mass of one carbon atom. This analysis at the particulate level predicts a similar result at the macroscopic level, where masses of combining elements can be measured. Going to the laboratory, we find that (a) 1.0 gram of carbon combines with 1.3 grams of oxygen to form carbon monoxide, CO, and (b) 1.0 gram of carbon combines with 2.6 grams of oxygen to form carbon dioxide, CO_2. The macroscopic mass ratio of oxygen that combines with 1.0 gram of carbon in the two compounds is 2.6/1.3, which reduces to 2/1, exactly as predicted by the atomic theory at the particulate level.

An element is made up of atoms. All atoms of a given element are identical.

Atoms cannot be destroyed or created.

All atoms of any given element have the same properties, including mass. Atoms of different elements have different masses.

Atoms of different elements may combine in a ratio of small, whole numbers to form compounds.

Figure 5.4 Atoms according to Dalton's atomic theory. Note that individual atoms cannot be struck with a hammer or placed on the pan of a balance! These caricaturizations are drawn only to help you remember the principles of Dalton's atomic theory.

Dalton realized that the three laws could be explained if matter was made of atoms. **Dalton's atomic theory** accounts for chemical reactions in this way: Before the reaction, the reacting substances contain a certain number of atoms of different elements. As the reaction proceeds, the atoms are rearranged to form the products. The atoms are neither created nor destroyed but are simply arranged differently. The starting arrangement is destroyed (reacting substances are destroyed in a chemical change), and a new arrangement is formed (new substances form). The main features of Dalton's atomic theory are as follows (**Figure 5.4**):

a summary of... **Dalton's Atomic Theory**

1. Each element is made up of tiny, individual particles called *atoms*.
2. Atoms are indivisible; they cannot be created or destroyed.
3. All atoms of any one element are identical in every respect.
4. Atoms of one element are different from atoms of any other element.
5. Atoms of one element combine with atoms of other elements to form chemical compounds.

As with many new ideas, Dalton's theory was not immediately accepted. However, over time, an overwhelming quantity of evidence has accumulated to erase any doubt about the existence of atoms. Relatively recently, in 2013, scientists were able to capture the first images of atomic theory in action. **Figure 5.5** shows a molecule with 26 carbon atoms and 14 hydrogen atoms undergoing a chemical change to become a different molecule composed of the same atoms.

Active Example 5.1 Dalton's Atomic Theory

A scientist finds that sulfur and oxygen form two different compounds. In each experiment, 1.0 g of sulfur is allowed to react with oxygen. Under one set of conditions, 0.5 g of oxygen reacts, and under another set of conditions, 1.0 g of oxygen reacts. Do these data support Dalton's atomic theory? If so, explain how. If not, explain why not.

Think Before You Write Don't get caught up in the details in the problem statement. Look for the underlying structure. Instead of thinking about 1.0 g of *sulfur,* think about 1.0 g of *one element.* Similarly, oxygen is another element; the fact that its identity is oxygen is not relevant to this problem.

Answers *Cover the left column with your cut-out shield. Reveal each answer only after you have written your own answer in the right column.*

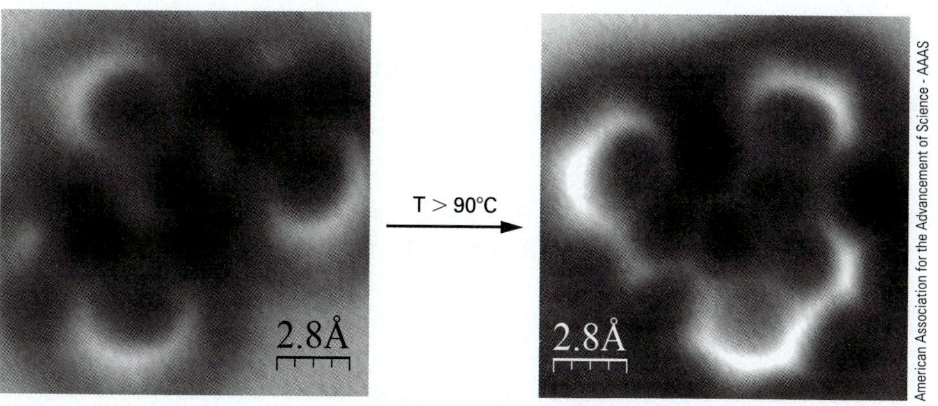

Figure 5.5 Images of atoms changing from one molecule to another. These are the first high-resolution images of a molecule changing to a different molecule made of the same set of atoms. Dalton proposed that atoms are neither created nor destroyed in a chemical change; both molecules have 26 carbon atoms and 14 hydrogen atoms. You can see how the arrangement of those 40 atoms changes as the temperature rises above 90°C. The reference line on each photograph is 2.8 Å long, where Å is 10^{-10} meter. Each molecule is therefore about 10×10^{-10} meter across, or 1×10^{-9} meter, which is one-billionth of a meter. You may be wondering why the edges appear somewhat blurred; this is because the molecules shown are interacting with neighboring molecules not shown in the image.

The mass of sulfur in each experiment is the same.	Compare the masses of sulfur in each of the two experiments. What relationship exists?
$\frac{0.5\text{ g}}{1.0\text{ g}} = \frac{1}{2}$ or $\frac{1.0\text{ g}}{0.5\text{ g}} = \frac{2}{1}$ The masses of oxygen are in a 1:2 (or 2:1) ratio. *The comparison can be thought of as either 1:2 or 2:1. The sequence of how the data are presented in the problem statement is not important.*	Compare the masses of oxygen in each of the two experiments. What quantitative relationship exists?
The different masses of one element (oxygen) that combine with the same mass of the other element (sulfur) are in a simple ratio of whole numbers, 2:1 (or 1:2).	The scenario given in the problem statement is that two elements combine to form more than one compound. What collective numerical pattern exists for the masses of the elements oxygen and sulfur?
The Law of Multiple Proportions.	What name is used for the pattern that you just described?
If matter is made of atoms, then when two elements combine to form more than one compound, a fixed number of atoms of one element will combine in a simple ratio of whole numbers with atoms of the other element.	Describe the relationship between the Law of Multiple Proportions and Dalton's atomic theory at the particulate level. Start with "*If* matter is made of atoms, *then*...."

A fixed number of atoms of one element is measured as a fixed mass of that element. A simple ratio of whole numbers of atoms of the other element is measured as a simple, whole number mass ratio.	If atoms combine as you just stated, what will be observed with measurements at the macroscopic level? 
Yes, what Dalton's atomic theory predicts is matched by experimental evidence.	Do these data support Dalton's atomic theory? Explain the relationship between the theory and the experimental data. 
The ratio could have any value. It would not be restricted to a simple ratio of whole numbers. It is the fact that matter is made of atoms that restricts the ratio to whole numbers because two atoms have two times the mass as one atom, three atoms have three times the mass as one atom, and so on. If matter was not made of atoms, any amount of one element could combine with a fixed mass of another element, for example, 1.43-to-1, 0.22-to-1, and so on.	What would be observed experimentally when two elements combine to form more than one compound if matter was continuous—not made of whole atom units—for the ratio of one element that combines with a fixed mass of another element? Explain. 
You improved your knowledge of experimental evidence of the existence of atoms.	What did you learn by solving this Active Example? 

Practice Exercise 5.1

A 5.00-g sample of liquid water is decomposed by subjecting it to an electrical current. All of the water disappears, and 0.56 g of hydrogen gas and 4.44 g of oxygen gas form. Do these data support Dalton's atomic theory? If so, explain how. If not, explain why not.

Oesper Collection in the History of Chemistry/University of Cincinnati

Figure 5.6 A photograph of Michael Faraday (1791–1867) was one of three photos that Albert Einstein kept in his study. The other two portraits were of Isaac Newton and James Clerk Maxwell.

5.3 The Electron

Goal 3 Describe the electron by charge and approximate mass, expressed as a comparison with the mass of a hydrogen atom, and write the symbol for the electron.

Dalton hypothesized that an atom of an element was an indestructible, permanent particle. The invention of the battery in 1800 paved the way for a series of investigations that led to doubts about the indestructibility of the atom. In 1834, English scientist Michael Faraday (**Figure 5.6**) discovered scientific laws that led to the hypothesis that electricity was carried by particles. In other words, Faraday's laws could be explained if a smallest unit of electricity existed, just like the atom represents the smallest unit of an elemental substance.

By 1875, English scientist William Crookes (**Figure 5.7**) had invented a high-quality glass tube with metal electrodes in each end that could be attached to a battery. He was able to pump almost all of the air out of the tube, creating a near-vacuum within. When the electrodes were attached to a high-voltage battery, electric current moved through the near-vacuum from the negative pole to the positive. It was hypothesized that this could be evidence that electrical current was a stream of negatively charged particles.

North Wind/North Wind Picture Archives

Figure 5.7 William Crookes's (1832–1919) discovery of what is now called the Crookes tube was a critical early step that led to the discovery of both the electron and x-rays.

In 1897, English scientist J. J. Thomson was able to construct an improved version of Crookes's vacuum tube. He reasoned that if the flow moving through the tube was composed of negatively charged particles, then it would be deflected by an electric field. Alternatively, if it was a form of light, it would not be deflected by an electric field. His experiment showed that the stream was deflected away from a negatively charged electric plate, providing convincing evidence that the substance flowing through the tube was indeed negatively charged particles.

Thomson then measured the deflection of the stream in a magnetic field. The amount of deflection depends on the mass of the particle *and* the magnitude of its charge. Thus, the angle of deflection was proportional to the mass-to-charge ratio; but neither the mass nor the charge itself could be determined from the results of this experiment. The result indicated that the particles were either of very low mass or very high charge.

American scientist Robert Millikan recognized that if either the mass or the charge of the particles was measured, then both would be known since the value of their ratio was known. In 1909, he designed an apparatus where an electrical charge could be placed on tiny oil droplets, which were then suspended in an electric field as they descended through air because of the force of gravity. By measuring the electrical force needed to prevent the drop from falling, Millikan was able to calculate the charge on a drop. The charge was always a whole-number multiple of a certain value, and thus that fundamental value was the charge of the particle, which is now called an **electron** (so named to combine the terms *electric* as *electr-* and *ion*, or "to go," as *-on*).

The combination of the values of Thomson's mass-to-charge ratio and Millikan's charge yields the mass of the electron. It is very tiny: 1/1837 of the mass of a hydrogen atom. That's just 0.05% of the mass of the atom. A hydrogen atom has the smallest mass of atoms of any element, so the electron is much, much lighter than the smallest atom. The electron is called a **subatomic particle** because its size is below (the prefix *sub-* indicates below or under) that of an atom. The charge on the electron is the smallest charge of any particle of matter and thus is used as the basis for comparison with other charges ▶. It has been assigned a value of 1−, with the minus sign indicating it is a negative charge. The symbol for an electron is $\mathbf{e^-}$. J. J. Thomson is typically credited as the person who discovered the electron.

◀ **P/Review** Electric charge was introduced in Section 2.8. There are only two types of electric charge, positive and negative. In this section, you are learning that an electron is a tiny particle of matter that possesses the smallest quantity of negative charge.

Target Check 5.1

A hydrogen atom is electrically neutral. The works of Faraday, Crookes, Thomson, Millikan, and others indicated that 1/1837 of the mass of the hydrogen atom is due to the electron, a particle with a charge of 1−. What does this imply about the charge of the other 1836/1837 of the mass of the hydrogen atom? Explain your reasoning.

5.4 The Nuclear Atom and Subatomic Particles

Goal 4 Describe and/or interpret the Rutherford scattering experiments and the nuclear model of the atom.

5 Identify the three major subatomic particles by symbol, charge, location within the nuclear atom, and approximate atomic mass, expressed relative to the mass of a hydrogen atom.

The discovery of the electron introduced more questions than it answered. Dalton's hypothesis that the atom was indestructible was clearly wrong. So what, then, is the

Bettmann/Getty Images

Figure 5.8 J. J. Thomson (1856–1940) (*left*) discovered the electron, and Ernest Rutherford (1871–1937) (*right*) developed the nuclear model of the atom and discovered the proton.

structure of an atom? How is the negative charge of the electron balanced within an atom? How are electrons removed from atoms when converted to an electrical current?

Thomson proposed a hypothesis: An atom is like a spherical serving of Jell-O with a small number of grains of sand equally distributed within it. The Jell-O has a positive charge equal to the negative charge contributed by the electrons, which are modeled as the grains of sand. Thomson's hypothesis is called the **plum pudding model**, named after the traditional English Christmas dessert of fruitcake, in which "plums" are the pieces of fruit (the grains of sand in our model) and "pudding" is the cake (the Jell-O).

In the period 1909–1911, New Zealand-born scientist Ernest Rutherford (**Figure 5.8**), in collaboration with Hans Geiger, a German scientist, and Ernest Marsden, then a student at the University of Manchester in England, designed a series of experiments usually referred to as the **Rutherford scattering experiments**. Their experiments were designed to test Thomson's plum pudding model hypothesis.

Research opportunities for undergraduate students may be available at your college or university. Marsden continued his studies and then went on to become a professor of physics in New Zealand, where he was among the most prominent of the scientists of his time.

Figure 5.9 illustrates the experimental apparatus used in the Rutherford laboratory. A narrow beam of positively charged alpha particles (helium atoms stripped of their negatively charged electrons) from a radioactive source was directed at a very thin gold foil. When alpha particles strike a fluorescent screen, they cause the screen to glow, and thus the gold foil was surrounded by a fluorescent screen to detect the location of the alpha particles after they interacted with the foil. If Thomson's plum pudding model was correct, then the alpha particles would act like a tiny bullet passing through Jell-O. The predicted result was that all of the alpha particles should hit the screen on the opposite side of the source of the alpha particle beam.

The experimental results were that most of the articles passed straight through the foil. However, some particles were deflected through moderate angles. Furthermore, a very small number of particles were deflected at large angles: 0.001% of the total were reflected at angles of greater than 90° and returned back toward the alpha particle source. When other metal foils were substituted for the gold foil, similar results were recorded.

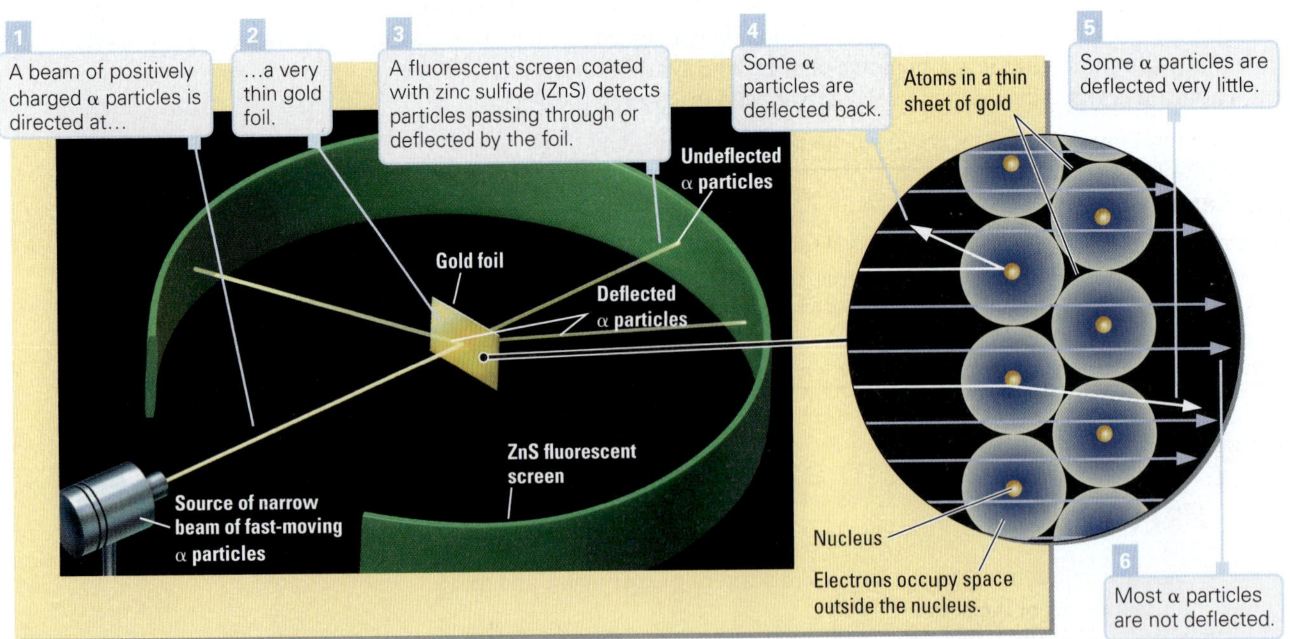
1
A beam of positively charged α particles is directed at…
2
…a very thin gold foil.
3
A fluorescent screen coated with zinc sulfide (ZnS) detects particles passing through or deflected by the foil.
4
Some α particles are deflected back.
Atoms in a thin sheet of gold
5
Some α particles are deflected very little.
Undeflected α particles
Gold foil
Deflected α particles
ZnS fluorescent screen
Source of narrow beam of fast-moving α particles
Nucleus
Electrons occupy space outside the nucleus.
6
Most α particles are not deflected.

Figure 5.9 Rutherford scattering experiments. A narrow beam of positively charged alpha particles (helium atoms stripped of their negatively charged electrons) from a radioactive source was directed at a very thin gold foil. 0.001% of the alpha particles were reflected back at angles of more than 90°. These data did not match what was expected to result from the experimental hypothesis. A greater percentage of the particles were deflected through moderate angles. Most of the particles passed straight through the foil, striking the screen at a 180° angle to the source of the beam.

The research team was shocked by the results! What was repelling the alpha particle "bullets"? The plum pudding model was clearly wrong, but what new hypothesis could replace it?

Rutherford expressed his surprise at the results by saying, "It was almost as incredible as if you fired a 15-inch shell at a piece of tissue paper and it came back and hit you."

Rutherford worked on a new model of the atom for months, and eventually he developed a new hypothesis that fits all of the data known to date. It is illustrated on the right side of Figure 5.9. An atom is modeled as consisting mostly of open space. At the center is an extremely tiny and extremely dense **nucleus** that contains all of the atom's positive charge and nearly all of its mass. Rutherford selected the term *nucleus* because it is the Latin word for the kernel of a nut. Calculations based on the results of the experiments indicated that the diameter of the open-space portion of the atom is about 10,000 times greater than the diameter of the nucleus. The electrons are thinly distributed throughout the open space.

Let's now consider how this **nuclear model of the atom** fits the data collected in the scattering experiments. Most of the positively charged alpha particles pass through the open space undeflected, not coming near any positively charged nuclei. This accounts for the fact that most alpha particles were not deflected. The small fraction of alpha particles that pass fairly close to a nucleus are repelled by positive–positive electrostatic forces and thereby deflected a little. The very few particles that are on a collision course with the nuclei were repelled backward. The experimental data from the Rutherford scattering experiments provide compelling evidence in support of the nuclear model of the atom. A great deal of modern evidence continues to support the validity of this model.

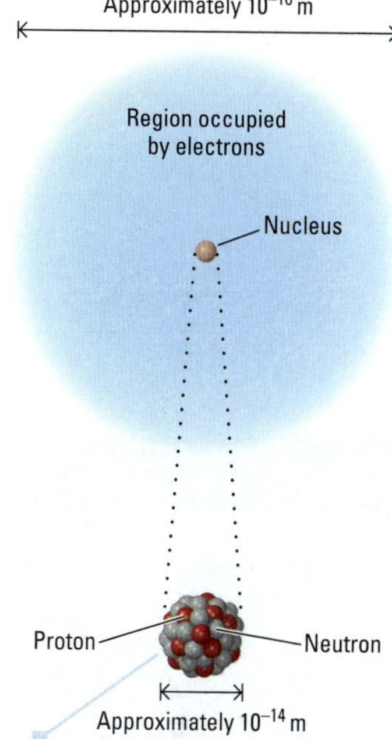
Approximately 10^{-10} m
Region occupied by electrons
Nucleus
Proton
Neutron
Approximately 10^{-14} m
This nucleus is shown 1 cm in diameter. The diameter of the region occupied by its electrons would be 100 m.

Figure 5.10 Relative sizes of an atom and its nucleus. The diameter of an atom is approximately 10,000 times the diameter of its nucleus.

a summary of... The Nuclear Model of the Atom

1. Every atom contains an extremely small, extremely dense nucleus.
2. All of the positive charge and nearly all of the mass of an atom are concentrated in the nucleus.
3. The tiny nucleus is surrounded by a much larger volume of nearly empty space that makes up the majority of the volume of an atom.
4. The space outside the nucleus is very thinly populated by electrons, the total charge of which exactly balances the positive charge of the nucleus.

The emptiness of the atom may be difficult to visualize. Look at the nucleus illustrated in **Figure 5.10**. If the nucleus of an atom were that size, the diameter of an atom would be the length of a football field. Between this nucleus and its nearest neighboring nucleus would be almost nothing—only a small number of electrons of negligible volume and mass. If it were possible to eliminate all of this nearby empty space and pack nothing but nuclei into a sphere the size of a period on this page, that sphere could, for some elements, weigh as much as 1 million tons!

Thinking About Your Thinking

Mental Models

The preceding paragraph and the Figure 5.10 caption about the emptiness of the atom were written to help you form a mental model of the atom. Take a moment to mentally picture the vast amount of empty space that makes up most of the volume of an atom. Keep in mind that a nucleus 1 cm in diameter would be at the center of an atom with the diameter of a football field.

After evidence supporting the nuclear model of the atom was disseminated to the scientific community, scientists quickly started investigating the nature of the nucleus. What was the composition of this portion of the atom that accounted for all of the positive charge and almost all of the mass of the atom but only a tiny fraction of its volume? Is the nucleus itself a subatomic particle or is it composed of other subatomic particles? Furthermore, if electrons are a fundamental subatomic particle common to atoms of all elements, what is in the nucleus that accounts for the different elements?

Rutherford himself set out to investigate the structure of the nucleus, continuing his collaboration with Marsden. They knew that a gold atom was relatively large and that its nucleus would have a relatively large positive charge. The gold atom nucleus would strongly repel alpha particles. Therefore they initiated investigations of the interaction of alpha particles with smaller atoms with nuclei of lesser charge and lesser repulsion, starting with hydrogen. Their results indicated that when a hydrogen atom collided with an alpha particle, a hydrogen nucleus was produced.

Marsden left the University of Manchester to serve in World War I, so Rutherford continued the research independently as his war-related obligations subsided. In 1919, Rutherford discovered that when alpha particles interacted with nitrogen, hydrogen nuclei were produced. This, combined with other evidence, led Rutherford to conclude that the hydrogen nucleus was a fundamental subatomic particle. It has a charge equal in magnitude to that of the electron but of opposite sign, 1+. Rutherford named this particle the **proton**, based on the Greek *protos,* meaning *first.* Since a hydrogen atom is composed of a proton and an electron, the mass of the proton is essentially the same as the mass of the hydrogen atom because the electron contributes a negligible mass.

Just after his discovery of the proton, Rutherford realized that the masses of nuclei were too large to be composed of only protons or a combination of protons and electrons. In 1920, he hypothesized that there must be yet another subatomic particle that has no electrical charge. This as-yet-undetected particle was named the **neutron** based on the combination of *neutr*(al) and (prot)*on.* In 1930, yet another interaction with alpha particles, this time with the element beryllium, yielded what was either radiation or subatomic particles as a product. In 1932, James Chadwick (**Figure 5.11**), an English scientist working in Rutherford's laboratory, correctly interpreted experiments conducted by others between 1930 and 1932, plus his own experimental results, and proved the existence of neutrons. The neutron has almost the same mass as a proton (and a hydrogen atom).

Pantheon/Superstock

Figure 5.11 James Chadwick (1891–1974) discovered the neutron in 1932. He was awarded the 1935 Nobel Prize in Physics for his discovery.

We now know that the proton and the neutron are made up of even smaller elementary particles (the first particles formed at 0.000001 second into the expansion of the universe, as described in Section 5.1) and that the electron is an elementary particle. However, to understand chemical principles, the proton, the neutron, and the electron are all we need to consider. All atoms are composed of this set of subatomic particles, which is summarized in **Table 5.1**.

The number of protons in an atom establishes its identity as an element. For example, *all* hydrogen atoms have one proton, *all* helium atoms have two protons, and *all* gold atoms have 79 protons. An atom has no net electrical charge. This means that the number of electrons must be equal to the number of protons. The number of neutrons in an atom varies (we will discuss this topic in more detail in the next section).

Table 5.1 Subatomic Particles

Subatomic Particle	Symbol	Charge	Mass in Comparison with Mass of Hydrogen Atom	Location	Discovered
Electron	e^-	1–	1/1837th	Outside nucleus	1897 by Thomson
Proton	p or p^+	1+	Same	Inside nucleus	1919 by Rutherford
Neutron	n or n^0	0	Same	Inside nucleus	1932 by Chadwick

When protons and neutrons were discovered, experimental evidence clearly indicated that these relatively massive subatomic particles make up the nucleus of the atom. This leads to the following question: How do electrons behave in the vast open space that they occupy? The leading hypothesis was that electrons traveled in circular orbits around the nucleus, much as planets move in orbits around the sun. We will discuss experimental evidence testing this hypothesis in more detail in Chapter 11.

Target Check 5.2

Identify the true statements and rewrite the false statements to make them true.

a) Atoms are like small, hard spheres.
b) An atom is electrically neutral.
c) An atom consists mostly of empty space.
d) The subatomic particles of an atom are electrons, protons, and neutrons.
e) The mass of an electron is less than the mass of a proton.
f) The mass of a proton is about the same as the mass of a hydrogen atom.
g) Electrons, protons, and neutrons are electrically charged.

5.5 Isotopes

Goal 6 Explain what isotopes are and how they differ from one another.

7 For an isotope whose chemical symbol is known, given one of the following, state the other two: (a) nuclear symbol, (b) number of protons and neutrons, (c) atomic number and mass number.

8 Identify the features of Dalton's atomic theory that are no longer considered valid and explain why.

From 1912 to 1913, while working on research investigating the nature of the atomic nucleus, J. J. Thomson improved a newly invented apparatus that stripped gaseous-state molecules of an electron, subjected them to a magnetic field, and then recorded their deflection because of the influence of that field. The larger the mass of the molecule, the smaller the deflection caused by a magnetic field when charge is equal. (Imagine a moving bowling ball and a moving Ping-Pong ball encountering a perpendicular gust of wind. The heavier bowling ball will be deflected the least.)

When Thomson allowed *pure* neon gas to stream through the apparatus, there were *two* locations to which atoms were deflected. These data showed that there are two types of neon atoms, one lighter and one heavier. The location of neon atoms that were deflected the most, the lighter atoms, had about 10 times the number of impacts as the other location. This indicated that the least massive type of neon atom was 10 times more abundant in nature as the neon atoms of larger mass (**Figure 5.12**). (We now know that there is a third neon atom of mass in between the two discovered by Thomson but less than 1% of the atoms in a natural neon sample have this mass.) The experimental results were always the same with samples of neon purified from various sources, indicating that the variation in mass of neon atoms is a natural characteristic.

Every atom of a particular element has the same number of protons. This number is called the **atomic number** of the element. It is represented by the symbol **Z**. Atoms are electrically neutral, so the number of electrons must be the same as the number of protons. It follows that the total contribution to the mass of an atom from protons and electrons is the same for every atom of the element. That leaves neutrons. We conclude that the mass differences between atoms of an element must

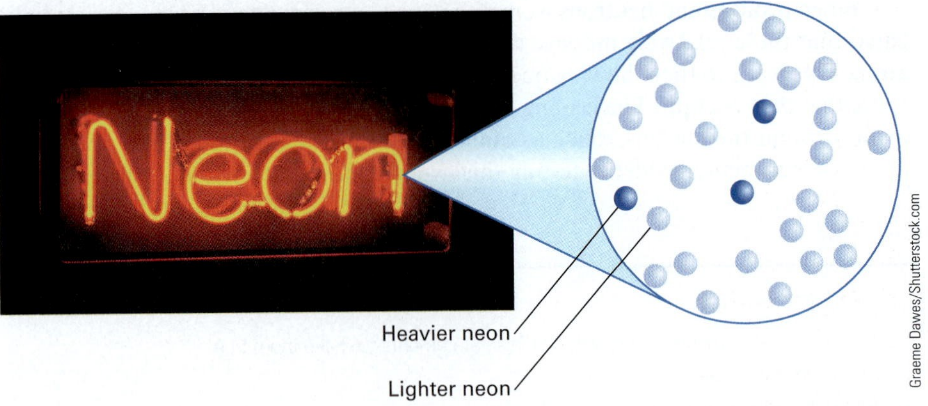

Figure 5.12 Variation in mass of neon atoms in a pure sample. Naturally occurring pure neon is about 90% atoms of one mass and 10% atoms of another, heavier mass. Less than 1% of neon atoms have an intermediate mass.

be caused by different numbers of neutrons. Atoms of the same element that have different masses—different numbers of neutrons—are called **isotopes.***

An isotope is identified by its **mass number (A)**, the total number of protons and neutrons in the nucleus:

$$\begin{array}{ccccc} \text{Mass number} & = & \text{number of protons} & + & \text{number of neutrons} \\ A & = & Z & + & N \end{array}$$

The name of an isotope is its elemental name followed by its mass number. Thus, an oxygen atom that has 8 protons and 8 neutrons has a mass number of 8 + 8, or 16, and its name is "oxygen sixteen." It is written "oxygen-16."

An isotope is represented by a **nuclear symbol** that has the form

$$^{\text{number of protons + number of neutrons}}_{\text{number of protons}}\text{Sy} \quad \text{or} \quad ^{\text{mass number}}_{\text{atomic number}}\text{Sy} \quad \text{or} \quad ^{A}_{Z}\text{Sy}$$

Sy is the chemical symbol of the element. The symbol and the mass number are actually all we need to identify an isotope, so the atomic number, Z, is sometimes omitted. The symbol for oxygen-16 is $^{16}_{8}O$ or ^{16}O.

Two natural isotopes of carbon are $^{12}_{6}O$ and $^{13}_{6}C$, carbon-12 and carbon-13. From the name and symbol of the isotopes and from the relationship A = Z + N, you can find the number of neutrons in each nucleus. In carbon-12, if you subtract the atomic number (protons) from the mass number (protons + neutrons), you get the number of neutrons:

$$\begin{array}{rlr} \text{Mass number} = & \text{protons} + \text{neutrons} = & 12 \\ -\text{ atomic number} = & -\text{protons} \qquad\qquad = & -6 \\ \hline & \text{neutrons} = & 6 \end{array}$$

In carbon-13 there are 7 neutrons: 13 − 6 = 7.

▶ **P/Review** The symbols of the elements (Section 2.7) are shown in the alphabetical list of elements on the Reference Page. They also appear on the periodic table on your shield cards.

You can find the mass number and nuclear symbol of an isotope from the number of protons and neutrons. A nucleus with 12 protons and 14 neutrons has the atomic number 12, the same as the number of protons. From the relationship A = Z + N, the mass number is 12 + 14 = 26. The symbol of the element may be found by searching for 12 in the periodic table on your shield ◀. The number at

*The term *isotope* refers to an atom of an element that has more than one variation in number of neutrons, and, technically, the term *nuclide* describes atoms with specific numbers of protons and neutrons. However, in practice, many chemists use the term isotope to refer to both isotopes and nuclides.

the top of each box is the atomic number. The elemental symbol corresponding to $Z = 12$ is Mg for magnesium. The isotope is therefore magnesium-26, and its nuclear symbol is $^{26}_{12}Mg$.

ATOMIC-THEORY-BEFORE-CALCULATIONS OPTION If your instructor has chosen to schedule this chapter before Chapters 2 and 3, this may be your first encounter with Active Examples in this book. If so, turn to Chapter 2 and read the discussion that precedes Active Example 2.1. It explains how our self-teaching Active Examples are designed to make you an active learner of chemistry.

Learn It Now! *We cannot overemphasize the value of answering each part of an Active Example for yourself before you look at our answer.*

Active Example 5.2 Isotopes

Fill in each of the blanks in the following table. Use the table of the elements on the Reference Page and the relationship $A = Z + N$ for needed information. The number at the top of each box in the periodic table is the atomic number of the element whose symbol is in the middle of the box.

Name of Element	Elemental Symbol	Atomic Number, Z	Number of Protons	Number of Neutrons, N	Mass Number, A	Nuclear Symbol	Name of Isotope
Barium						$^{138}_{56}Ba$	
Oxygen				10			
		82			206		
							zinc-66

Think Before You Write We'll help you work through the first line of the table. This will help you learn the thought process needed to satisfy Goal 7.

Answers *Cover the left column with your cut-out shield. Reveal each answer only after you have written your own answer in the right column.*

$^{\text{mass number}}_{\text{atomic number}}Sy = {}^{A}_{Z}Sy$ Mass number, A = 138 Atomic number, Z = 56 Elemental symbol = Ba	The nuclear symbol in the first line, $^{138}_{56}Ba$, gives you three pieces of information: the superscript mass number, A, the subscript atomic number, Z, and the elemental symbol on the right. Fill in those three boxes for just the Barium line.
Isotope name = barium-138 Number of protons = 56	The name of the isotope is the elemental name, followed by a hyphen and then the mass number. Also, the atomic number is the number of protons. Fill in those two boxes.
Number of neutrons = Mass number − Number of protons = 138 − 56 = 82	One box remains, number of neutrons. The mass number is the total number of protons plus neutrons. You know both the mass number and the number of protons. Complete the final box in the Barium line.
	Fill in the remaining blanks; complete the table.

Name of Element	Elemental Symbol	Atomic Number, Z	Number of Protons	Number of Neutrons, N	Mass Number, A	Nuclear Symbol	Name of Isotope
Barium	Ba	56	56	82	138	$^{138}_{56}Ba$	barium-138
Oxygen	O	8	8	10	18	$^{18}_{8}O$	oxygen-18
Lead	Pb	82	82	124	206	$^{206}_{82}Pb$	lead-206
Zinc	Zn	30	30	36	66	$^{66}_{30}Zn$	zinc-66

The atomic number, Z, and the number of protons are the same in each case, by definition. Also by definition, the mass number, A, is equal to the sum of the number of protons and the number of neutrons. For Z = 82, you must search for 82 in the atomic number column in the table of elements to identify the element as lead.

You improved your skill at solving isotope problems.

What did you learn by solving this Active Example?

Practice Exercise 5.2

What is the number of each type of subatomic particle in a potassium-41 atom? Write its nuclear symbol in both commonly used forms. State the mass number and atomic number of the isotope.

✓ Target Check 5.3

Identify the true statements and rewrite the false statements to make them true.

a) The atomic number of an element is the number of protons in its nucleus.

b) All atoms of a specific element have the same number of protons.

c) The difference among isotopes of an element is a difference in the number of neutrons in the nucleus.

d) The mass number of an atom is always equal to or larger than the atomic number.

5.6 Atomic Mass

Goal 9 Define and use the atomic mass unit (u).

10 Given the relative abundances of the natural isotopes of an element and the atomic mass of each isotope, calculate the atomic mass of the element.

The mass of a single atom is very small—much too small to be measured on a balance. Nevertheless, early chemists did find ways to isolate samples of elements that contained the same number of atoms. These samples were weighed and compared. The ratio of the masses of equal numbers of atoms of different elements is the same as the ratio of the masses of individual atoms ◀. From this, a scale of relative atomic masses was developed. Chemists didn't know about isotopes at that time, so they applied the idea of what was then called *atomic weight* to all of the natural isotopes of an element.

▶ **P/Review** Equal volumes of gases at the same temperature and pressure have the same number of molecules (Section 14.3).

The masses of atoms are typically not expressed in grams or kilograms. Instead, a very small mass unit has been defined that is more convenient for expressing the masses of atoms: The **atomic mass unit (u)** is defined as exactly $\frac{1}{12}$ the mass of a carbon-12 atom.

$$1\ \text{u} \equiv \frac{1}{12}\ \text{the mass of one carbon-12 atom}$$

Multiplying both sides by 12, it follows that the mass of one carbon-12 atom is 12 u. Since the mass of an electron is nearly zero and a carbon-12 atom has a total of 6 protons plus 6 neutrons, and the mass of a proton is almost the same as the mass of a neutron, both protons and neutrons have atomic masses very close to 1 u. We can also think of the u as simply another mass or weight unit, like the gram, kilogram, or pound ▶. To three significant figures, the relationship between the gram and the atomic mass unit is

$$1 \text{ g} = 6.02 \times 10^{23} \text{ u}$$

◀ **P/Review** The relationship between the grams and the atomic mass unit is derived from the SI unit for amount of substance, called the *mole*. Section 7.4 describes the mole in detail.

$$\frac{12 \text{ g } ^{12}\text{C}}{\text{mol } ^{12}\text{C}} \times \frac{1 \text{ mol } ^{12}\text{C}}{6.02 \times 10^{23}\ ^{12}\text{C atoms}} \times \frac{\frac{1}{12}\ ^{12}\text{C atom}}{1 \text{ u}} = \frac{1}{6.02 \times 10^{23}} \text{ g/u} = 1.66 \times 10^{-24} \text{ g/u}$$

Samples of most pure elements consist of two or more isotopes—atoms that have different masses because they have varying numbers of neutrons. Thus, the mass of the atoms of an element is averaged based on the masses of the isotopes and their percentage abundance in nature. The **atomic mass** of an element is the average mass of all atoms of an element as they are distributed in nature. Keep in mind that atomic mass is always measured relative to the mass of an atom of carbon-12, which is defined as 12 u.

To find the (average) atomic mass of an element, you must know the atomic mass of each isotope and the fraction of each isotope in a sample (**Figure 5.13**). **Table 5.2** gives the percentage abundance of the natural isotopes of some common elements. The following Active Example shows you how to calculate the atomic mass of an element from these data.

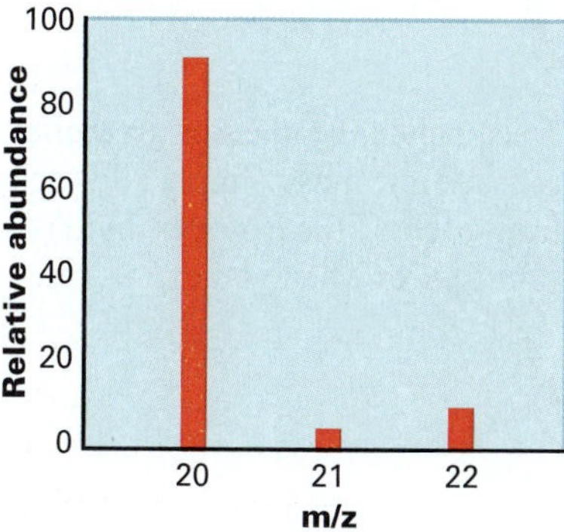

Figure 5.13 Mass spectrum of neon (1+ ions only). The relative abundance plotted on the y-axis is the percentage of each isotope found in a natural sample of the pure element. The mass-to-charge ratio, m/z, is plotted on the x-axis. Since this plot shows 1+ ions only, m/z is the same as the mass of the isotope, expressed in atomic mass units. Neon contains three isotopes, of which neon-20 is by far the most abundant (90.5%). The mass of that isotope is 19.9924356 u.

Table 5.2 Percentage Abundance of Some Natural Isotopes

Symbol	Mass (u)	Percentage	Symbol	Mass (u)	Percentage
$^{1}_{1}\text{H}$	1.0078250321	99.9885	$^{18}_{8}\text{O}$	17.9991604	0.205
$^{2}_{1}\text{H}$	2.0141017780	0.0115	$^{19}_{9}\text{F}$	18.99840320	100
$^{3}_{2}\text{He}$	3.0160293097	0.000137	$^{32}_{16}\text{S}$	31.97207069	94.93
$^{4}_{2}\text{He}$	4.0026032497	99.999863	$^{33}_{16}\text{S}$	32.97145850	0.76
$^{9}_{4}\text{Be}$	9.0121821	100	$^{34}_{16}\text{S}$	33.96786683	4.29
$^{12}_{6}\text{C}$	12 (exactly)	98.93	$^{36}_{16}\text{S}$	35.96708088	0.02
$^{13}_{6}\text{C}$	13.0033548378	1.07	$^{35}_{17}\text{Cl}$	34.96885271	75.78
$^{14}_{7}\text{N}$	14.0030740052	99.632	$^{37}_{17}\text{Cl}$	36.96590260	24.22
$^{15}_{7}\text{N}$	15.0001088984	0.368	$^{39}_{19}\text{K}$	38.9637069	93.2581
$^{16}_{8}\text{O}$	15.9949146221	99.757	$^{40}_{19}\text{K}$	39.96399867	0.0117
$^{17}_{8}\text{O}$	16.99913150	0.038	$^{41}_{19}\text{K}$	40.96182597	6.7302

ATOMIC-THEORY-BEFORE-CALCULATIONS OPTION The rules of significant figures are given in Section 3.7. If you have not yet studied Chapter 3, follow the directions of your instructor regarding significant figures.

Active Example 5.3 Atomic Mass I

The natural distribution of the isotopes of boron is 19.9% $^{10}_{5}B$ at a mass of 10.0129370 u and 80.1% $^{11}_{5}B$ at 11.0093055 u. Calculate the atomic mass of boron.

Think Before You Write The atomic mass of an element is the sum of the products of the fractional abundance times the mass of each of the naturally occurring isotopes.

Answers *Cover the left column with your cut-out shield. Reveal each answer only after you have written your own answer in the right column.*

$0.801 \times 11.0093055\ u = 8.82\ u$ *You multiplied a three-significant-figure fractional abundance by a nine-significant-figure mass, so, by following the multiplication and division rule for significant figures, three significant figures are justified in the solution.*	Part of the "average" boron atom consists of 19.9% of an atom with mass 10.01294 u. Therefore, it contributes $(19.9 \div 100 =)\ 0.199 \times 10.0129370\ u = 1.99\ u$ to the mass of the average boron atom. Perform a similar calculation for the other isotope. 
$1.99\ u + 8.82\ u = 10.81\ u$ *Both masses are known to the hundredths place, so the sum of their masses is expressed to the hundredths place according to the addition and subtraction rule for significant figures. Note that we gained a significant figure in the addition process because we ended with two digits to the left of the decimal point in the sum. The currently accepted value of the atomic mass of boron is 10.81 u, which matches your calculated value to four significant figures.*	To calculate the mass of an average atom of boron, which is its atomic mass, simply add together the contributions from each of the isotopes. Don't forget to consider significant figures when you express the sum. 
You improved your understanding of the atomic mass concept and you improved your skill at calculating the atomic mass of an element.	What did you learn by solving this Active Example? 

Practice Exercise 5.3

Naturally occurring lithium is composed of two isotopes: 7.59% at 6.0151224 u and 92.41% at 7.0160041 u. Use these data to determine the atomic mass of lithium.

One solution of Active Example 5.3 on a calculator gives 10.81102817 u as the answer. Notice that your calculator does not tell you the column to which the result should be rounded off. You must determine that yourself by applying the rules of significant figures to each step in the problem.

You are now ready to complete an atomic mass calculation on your own.

Active Example 5.4 Atomic Mass II

Calculate the atomic mass of potassium (symbol K), using data from Table 5.2.

Think Before You Write This time there are three isotopes, but the procedure remains the same: Calculate the fractional abundance times the mass for each isotope, and then sum the products. Be very deliberate with significant figures.

Answers *Cover the left column with your cut-out shield. Reveal each answer only after you have written your own answer in the right column.*

0.932581 × 38.9637069 = 36.3368 u 0.000117 × 39.96399867 = 0.00468 u 0.067302 × 40.96182597 = 2.7568 u 39.0983 u *The accepted value of the atomic mass of potassium is 39.0983 u.*	Complete the problem.
You improved your understanding of the atomic mass concept, and you improved your skill at calculating the atomic mass of an element.	What did you learn by solving this Active Example?

Practice Exercise 5.4

Use the data in Table 5.2 to determine the atomic mass of sulfur, symbol S.

5.7 The Periodic Table

Goal 11 Distinguish between groups and periods in the periodic table and identify them by number (and letter).

12 Given the atomic number of an element, use a periodic table to find the symbol and atomic mass of that element, and identify the period and group in which it is found.

13 Given an elemental symbol or information from which it can be identified, classify the element as either a main group or transition element and as a metal, nonmetal, or metalloid.

During the time of early research on the atom, before subatomic particles were identified, some chemists searched for an order among elements. In 1869, two men found an order, independently of each other. Dmitri Mendeleev (**Figure 5.14**) and Lothar Meyer (**Figure 5.15**) observed that when elements are arranged according to their atomic masses, certain properties repeat at regular intervals.

Mendeleev and Meyer arranged the elements in tables so that elements with similar properties were in the same column or row. These were the first periodic tables of the elements. The arrangements were not perfect. For all elements to fall into the proper groups, it was necessary to switch a few of them in a way that interrupted the orderly increase in atomic masses. Of the two reasons for this, one was anticipated at that time: There were errors in atomic weights as they were known in 1869. The second reason was more important: About 50 years later, it was found that the correct ordering property is the atomic number (Z) rather than the atomic mass.

Mendeleev realized that he had to put blank spaces in his table. He reasoned that the blank spaces belonged to elements that were yet to be discovered. By averaging the properties of elements above and below or on each side of the blanks, he predicted the properties of the unknown elements. Germanium is one of the elements about which he made these predictions. **Table 5.3** summarizes the predicted properties and their currently accepted values.

Oesper Collection in the History of Chemistry/University of Cincinnati.

Figure 5.14 Element 101, symbol Md, is named Mendelevium in honor of Dmitri Mendeleev (1834–1907) for his development of the periodic table of the elements.

E. F. Smith Memorial Collection in the History of Chemistry, Department of Special Collections, Van Pelt Dietrich Library, University of Pennsylvania.

Figure 5.15 J. Lothar Meyer's (1830–1895) periodic table predated Mendeleev's by five years, but it had fewer elements, and it did not make predictions about corrected atomic masses and new elements, as Mendeleev's did, so Mendeleev is generally given sole credit for the development of the first periodic table.

Thinking About Your Thinking

Classification

There are many ways to classify a collection of objects. Different criteria satisfy different purposes. Mendeleev arranged his periodic table based on two criteria: The atomic masses of the elements increased across a row and the chemical properties of the elements in a column were similar. Mendeleev is more famous than Meyer as the founder of the periodic table primarily because he subsequently used his classification scheme to make predictions about unknown elements. His predictions were later found to be true.

When you formulate a classification scheme, its power is that it allows you to fill in gaps in your knowledge, just as Mendeleev filled in gaps in his periodic table. If you can deduce a classification pattern, you can interpolate and extrapolate to make predictions about unknown things or events in the future. (Interpolation is predicting something within the range of your data. Extrapolation is predicting something beyond the range of your data.)

When you practice understanding classification schemes made up by others, such as the periodic table in chemistry, it helps you develop your skill in formulating your own classifications in all aspects of your life.

Periodic means at regular intervals. When Mendeleev predicted the properties of germanium and other elements not yet discovered, he demonstrated that the properties of the elements, when ordered by atomic mass (we now know the sequence is by atomic number), followed a regular pattern. Although the amazing accuracy of Mendeleev's predictions showed that the periodic table made sense, nobody understood why; that came later. The reason for the shape of the table is explained in Chapter 11.

Figure 5.16 is a modern periodic table. It also appears on your shield. You will find yourself referring to the periodic table throughout your study of chemistry. This is why a periodic table is printed on the shields provided for working the Active Examples.

The number at the top of each box in our periodic table is the atomic number of the element. The chemical symbol is in the middle, and the atomic mass, rounded to four significant figures,* is at the bottom (**Figure 5.17**) ◀. The boxes are arranged in horizontal rows called **periods**. Periods are numbered from top to bottom, but the numbers are not usually printed. Periods vary in length: The first period has 2 elements; the second and third have 8 elements each; and the fourth and fifth each have 18. Period 6 has 32 elements, including atomic numbers 58 to 71, which are printed separately at the bottom to keep the table from becoming too wide. Period 7 also has 32 elements.

▶ **P/Review** Some atomic masses of elements in the periodic table are in parentheses. These elements are radioactive, with unstable nuclei, and there is no atomic mass in the sense that we have defined it. Instead, parentheses enclose the mass number of a single isotope, the one that is the most stable. Radioactivity is discussed in Chapter 20.

*The magnitude of the uncertainty in the natural abundance of the lithium isotopes limits its atomic mass to three significant figures.

Table 5.3 Predicted and Observed Properties of Germanium

Property	Predicted by Mendeleev	Currently Accepted Values
Atomic weight	72 g/mol[†]	72.64 g/mol[†]
Density of metal	5.5 g/cm^3	5.32 g/cm^3
Color of metal	Dark gray	Gray-white
Formula of oxide	GeO_2	GeO_2
Density of oxide	4.7 g/cm^3	4.25 g/cm^3
Formula of chloride	$GeCl_4$	$GeCl_4$
Density of chloride	1.9 g/cm^3	1.90 g/cm^3
Boiling point of chloride	Below 100°C	87°C
Formula of ethyl compound	$Ge(C_2H_5)_4$	$Ge(C_2H_5)_4$
Boiling point of ethyl compound	160°C	163°C
Density of ethyl compound	0.96 g/cm^3	1.00 g/cm^3

[†]The mole is the SI unit for amount of substance. The mole is introduced in Chapter 7.

Current U.S. Usage
IUPAC notation

1A 1	2A 2	3B 3	4B 4	5B 5	6B 6	7B 7	8B 8	8B 9	8B 10	1B 11	2B 12	3A 13	4A 14	5A 15	6A 16	7A 17	8A 18
1 **H** 1.008																	2 **He** 4.003
3 **Li** 6.94	4 **Be** 9.012											5 **B** 10.81	6 **C** 12.01	7 **N** 14.01	8 **O** 16.00	9 **F** 19.00	10 **Ne** 20.18
11 **Na** 22.99	12 **Mg** 24.31											13 **Al** 26.98	14 **Si** 28.09	15 **P** 30.97	16 **S** 32.06	17 **Cl** 35.45	18 **Ar** 39.95
19 **K** 39.10	20 **Ca** 40.08	21 **Sc** 44.96	22 **Ti** 47.87	23 **V** 50.94	24 **Cr** 52.00	25 **Mn** 54.94	26 **Fe** 55.85	27 **Co** 58.93	28 **Ni** 58.69	29 **Cu** 63.55	30 **Zn** 65.38	31 **Ga** 69.72	32 **Ge** 72.63	33 **As** 74.92	34 **Se** 78.97	35 **Br** 79.90	36 **Kr** 83.80
37 **Rb** 85.47	38 **Sr** 87.62	39 **Y** 88.91	40 **Zr** 91.22	41 **Nb** 92.91	42 **Mo** 95.95	43 **Tc** (98)	44 **Ru** 101.1	45 **Rh** 102.9	46 **Pd** 106.4	47 **Ag** 107.9	48 **Cd** 112.4	49 **In** 114.8	50 **Sn** 118.7	51 **Sb** 121.8	52 **Te** 127.6	53 **I** 126.9	54 **Xe** 131.3
55 **Cs** 132.9	56 **Ba** 137.3	57 ***La** 138.9	72 **Hf** 178.5	73 **Ta** 180.9	74 **W** 183.8	75 **Re** 186.2	76 **Os** 190.2	77 **Ir** 192.2	78 **Pt** 195.1	79 **Au** 197.0	80 **Hg** 200.6	81 **Tl** 204.4	82 **Pb** 207.2	83 **Bi** 209.0	84 **Po** (209)	85 **At** (210)	86 **Rn** (222)
87 **Fr** (223)	88 **Ra** (226)	89 **†Ac** (227)	104 **Rf** (265)	105 **Db** (268)	106 **Sg** (271)	107 **Bh** (270)	108 **Hs** (277)	109 **Mt** (276)	110 **Ds** (281)	111 **Rg** (280)	112 **Cn** (285)	113 **Nh** (286)	114 **Fl** (289)	115 **Mc** (288)	116 **Lv** (292)	117 **Ts** (294)	118 **Og** (294)

*Lanthanoids	58 **Ce** 140.1	59 **Pr** 140.9	60 **Nd** 144.2	61 **Pm** (145)	62 **Sm** 150.4	63 **Eu** 152.0	64 **Gd** 157.3	65 **Tb** 158.9	66 **Dy** 162.5	67 **Ho** 164.9	68 **Er** 167.3	69 **Tm** 168.9	70 **Yb** 173.1	71 **Lu** 175.0
†Actinoids	90 **Th** 232.0	91 **Pa** 231.0	92 **U** 238.0	93 **Np** (237)	94 **Pu** (244)	95 **Am** (243)	96 **Cm** (247)	97 **Bk** (247)	98 **Cf** (251)	99 **Es** (252)	100 **Fm** (257)	101 **Md** (258)	102 **No** (259)	103 **Lr** (262)

Primary source: IUPAC Standard Atomic Weights and Conventional Atomic Weights 2009.

All atomic masses have been rounded to four significant figures. (The natural composition of lithium isotopes limits its atomic mass to three significant figures.) Atomic masses in parentheses are the mass number of the longest-lived isotope. The current U.S. common usage group number is listed above the IUPAC format for group number.

Figure 5.16 Periodic table of the elements.

Elements with similar properties are placed in a vertical column called a **group** or **chemical family**. Groups are identified by two rows of numbers across the top of the table. The top row shows the group numbers commonly used in the United States. European chemists use the same numbers but a different arrangement of As and Bs. The International Union of Pure and Applied Chemistry (IUPAC) has approved a compromise that simply numbers the columns in order from left to right. This is the second row of numbers at the top of Figure 5.16.

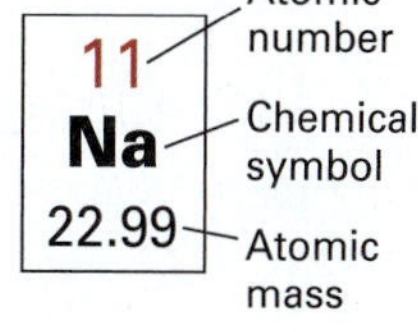

Figure 5.17 Sample box from the periodic table, representing sodium.

We will use both sets of numbers, leaving it to your instructor to recommend which you should use. When we have occasion to refer to a group number, we will have the U.S. number first, followed by the IUPAC number after a slash. Thus, the column headed by carbon, Z = 6, is Group 4A/14.

Two other regions in the periodic table separate the elements into special classifications. Elements from one of the A groups (1, 2, and 13 to 18) are called **main group elements**. Main group elements are also known as **representative elements**. Similarly, elements in the B groups (3 to 12) are known as **transition elements**, or **transition metals**.

The stair-step line that begins between atomic numbers 4 and 5 in Period 2 and ends between 84 and 85 in Period 6 separates the **metals** on the left from the **nonmetals** on the right. Six elements that border the stair-step line have some properties of both metals and nonmetals: B, Si, Ge, As, Sb, and Te. These elements are classified as **metalloids**. Chemical reasons for these classifications appear in Chapter 11.

Active Example 5.5 The Periodic Table

List the atomic number, chemical symbol, and atomic mass of the third-period element in Group 6A/16. Classify the element as either a main group or transition element and either a metal, nonmetal, or metalloid.

Think Before You Write The location of an element in the periodic table is given by its period and group numbers. Thus, all information requested is found in the periodic table.

Answers *Cover the left column with your cut-out shield. Reveal each answer only after you have written your own answer in the right column.*

Z = 16; symbol, S; atomic mass, 32.06 u; main group element; nonmetal *In Group 6A/16, the third column from the right side of the table, you find Z = 16 in Period 3. The element is in a U.S. A group, so it is a main group element, and because it is to the right of the stair-step line, it is a nonmetal.* *The element is sulfur.*	List the requested information.
You improved your understanding of the periodic table and the information that can be found on it.	What did you learn by solving this Active Example?

Practice Exercise 5.5

Complete the following table:

Symbol	Atomic Number	Atomic Mass	Period	Group	Main Group or Transition Element	Metal, Nonmetal, or Metalloid
N						
	12					
		52.00 u				

Target Check 5.4

a) How many Group 3A/13 elements are metals?

b) How many Period 4 elements are transition metals?

5.8 Elemental Symbols and the Periodic Table

Goal 14 Given the name or the symbol of an element in Figure 5.20, write the other.

The periodic table contains a large amount of information, far more than is indicated by the three items in each box. Its usefulness will become apparent gradually as your study progresses. In this section, we will use the periodic table to learn the names and symbols of 35 elements. Learning symbols and names is much more efficient if you learn the location of the elements in the periodic table at the same time. Here's how to do it.

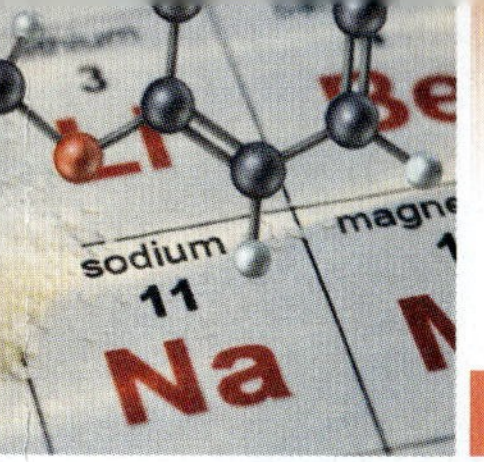

Everyday Chemistry 5.1

INTERNATIONAL RELATIONS AND THE PERIODIC TABLE

When a new element is discovered, it has been customary to allow the person who made the discovery to name the element. When an artificial element is newly synthesized, a similar custom has been followed. The person who leads the team of scientists working on the synthesis proposes the name of the new element. That is, this has been the custom until recently.

Because of the expense of the equipment and the highly specialized expertise of research personnel, at present only three laboratories in the world work on synthesizing new elements: the Lawrence Berkeley National Laboratory in the United States (**Figure 5.18**), the Society for Heavy-Ion Research in Germany, and the Joint Institute for Nuclear Research in Russia. All three laboratories are making similar progress, although they usually use slightly different methods to approach the same problem. You might be able to guess how this leads to conflict. More than once, two different labs have claimed to have synthesized a new element at about the same time. Who gets to choose the name when each lab claims that it was the first? Problems with international relations are not limited to politicians; chemists struggle with issues of national pride as well!

A source of great controversy was the naming of element 106. U.S. chemists endorsed the name seaborgium, in honor of the U.S. chemist Glenn Seaborg (**Figure 5.19**), who over his career led teams of scientists that synthesized ten new elements. No person has ever equaled this achievement, so the Americans were confident that their proposal for the name of element 106 would easily gain acceptance from the worldwide scientific community. To their dismay, however, the International Union of Pure and Applied Chemistry (IUPAC) endorsed the name rutherfordium in honor of Ernest Rutherford (see Section 5.4) for element 106. Moreover, the U.S. chemists were shocked by the IUPAC proposal that element 104 be named dubnium in honor of achievements at the research laboratory in Dubna, Russia. There were serious doubts as to the validity of the Russian chemists' data.

As with many political debates, the controversy was finally settled with a compromise. Element 106 was named seaborgium. Element 104 is now called rutherfordium, and element 105 is dubnium. All parties got "their" name assigned to an element; there was just some shuffling over which element received which name. You can see that the common conception that science is divorced from emotion and governed only by cold logic is a myth. Scientific disciplines are subject to the frailties and strengths of the human character in the same way as any other human endeavor.

Lawrence Berkeley National Laboratory

Figure 5.19 Glenn Seaborg (1912–1999) was awarded the Nobel Prize in Chemistry in 1951 for his contributions leading to the discovery of many elements. In this photograph, he points to element 106, named seaborgium in honor of his work.

Quick Quiz

1. What countries conduct research on the synthesis of new elements?
2. Why is the element with atomic number 106 named seaborgium?

Lawrence Berkeley National Laboratory

Figure 5.18 The Lawrence Berkeley National Laboratory is located in the hills just east of San Francisco Bay, next to the University of California, Berkeley campus.

Common Elements

Atomic Number	Symbol	Element	Atomic Number	Symbol	Element	Atomic Number	Symbol	Element
1	H	Hydrogen	13	Al	Aluminum	28	Ni	Nickel
2	He	Helium	14	Si	Silicon	29	Cu	Copper
3	Li	Lithium	15	P	Phosphorus	30	Zn	Zinc
4	Be	Beryllium	16	S	Sulfur	35	Br	Bromine
5	B	Boron	17	Cl	Chlorine	36	Kr	Krypton
6	C	Carbon	18	Ar	Argon	47	Ag	Silver
7	N	Nitrogen	19	K	Potassium	50	Sn	Tin
8	O	Oxygen	20	Ca	Calcium	53	I	Iodine
9	F	Fluorine	24	Cr	Chromium	56	Ba	Barium
10	Ne	Neon	25	Mn	Manganese	80	Hg	Mercury
11	Na	Sodium	26	Fe	Iron	82	Pb	Lead
12	Mg	Magnesium	27	Co	Cobalt			

a

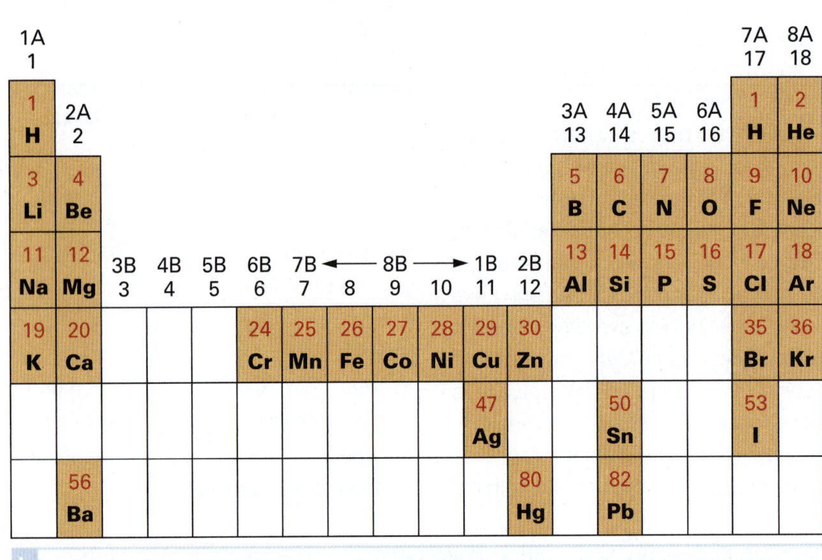

Figure 5.20 (a) Table of common elements, with symbols and atomic numbers. (b) Partial periodic table showing the symbols and locations of the common elements. The table in Part (a) and the list below identify the elements that you should be able to recognize or write, referring only to a complete periodic table. Associating the names and symbols with the table makes learning them much easier. The elemental names are as follows:

aluminum	fluorine	nickel
argon	helium	nitrogen
barium	hydrogen	oxygen
beryllium	iodine	phosphorus
boron	iron	potassium
bromine	krypton	silicon
calcium	lead	silver
carbon	lithium	sodium
chlorine	magnesium	sulfur
chromium	manganese	tin
cobalt	mercury	zinc
copper	neon	

Part (a) of **Figure 5.20**, the table of common elements, gives the name, symbol, and atomic number of the elements whose names and symbols you should learn.* Part (b) is a partial periodic table showing the atomic numbers and symbols of the same elements. Their names are listed in alphabetical order in the caption.

Study Part (a) briefly. Try to learn the symbol that goes with each element, but don't spend more than a few minutes doing this. Then cover Part (a) and look at the periodic table in Part (b). Run through the symbols mentally and see how many elements you can name. If you can't name one, glance through the alphabetical list in the caption and see if it jogs your memory. If you still can't get it, note the atomic number and check Part (a) for the elemental name. Do this a few times until you become fairly quick in naming most of the elements from their symbols.

*Your instructor may require you to learn different names and symbols, or perhaps more or fewer. If so, follow your instructor's directions. If they include elements not among our 35, we recommend that you add or subtract them to both parts of Figure 5.20 and also to the caption.

Next, reverse the process with Part (a) still covered. Look at the alphabetical list in the caption. For each name, mentally "write"—in other words, think—the symbol. Glance up to the periodic table and find the element. Again, use Part (a) as a temporary help only if necessary. Repeat the procedure several times, taking the elements in random order. Move in both directions, from name to symbol in the periodic table and from symbol to name.

When you feel reasonably sure of yourself, try the following Active Example. Do not refer to Figure 5.20. Instead, use only the periodic table that is on your shield.

Thinking About Your Thinking

Memory

The greater the number of mental pathways you can form to access memorized information, the more likely you are to recall it. We suggest that you memorize name–symbol pairs *and* the location of the element in the periodic table at the same time so that you have an additional mental pathway associated with each elemental symbol.

Active Example 5.6 Names and Symbols of Elements

For each elemental symbol listed, write the name; for each name, write the symbol: N, F, I, carbon, aluminum, copper. Use no reference other than the periodic table on your shield.

Think Before You Write If you have memorized the 35 name–symbol pairs and quickly can find the symbol of any given element in the periodic table, you are ready to complete this Active Example.

Answers *Cover the left column with your cut-out shield. Reveal each answer only after you have written your own answer in the right column.*

N, nitrogen; F, fluorine; I, iodine; carbon, C; aluminum, Al; copper, Cu.	Write the names and symbols.
You improved your skill at writing the name–symbol pairs of 35 common elements.	What did you learn by solving this Active Example?

Practice Exercise 5.6

For each elemental symbol listed, write the name; for each name, write the symbol: Pb, Hg, Mn, sodium, phosphorus, tin.

Look closely at your symbols for aluminum and copper in Active Example 5.6. If you wrote AL or CU, the symbol is wrong. The letters are right, but the symbol is not. Whenever a chemical symbol has two letters, the first letter is always capitalized, but *the second letter is always written in lowercase*, or as a small letter. You can enjoy a long and happy life with a pile of Co in your house, but CO is a potentially serious problem in homes. Co is the metal cobalt, which is sometimes used in steel and pottery, among other things. CO is the deadly gas carbon monoxide, which is present in automobile exhaust and tobacco smoke. Carbon monoxide poisoning is the number one cause of accidental poisoning deaths in the world.

Science Source/Science Source

CHAPTER 5 IN REVIEW: ATOMIC THEORY: THE NUCLEAR MODEL OF THE ATOM

Goal 1 Identify the main features of Dalton's atomic theory.

The main features of **Dalton's atomic theory** are as follows:

1. Each element is made up of tiny, individual particles called *atoms*.
2. Atoms are indivisible; they cannot be created or destroyed.
3. All atoms of any one element are identical in every respect.
4. Atoms of one element are different from atoms of any other element.
5. Atoms of one element combine with atoms of other elements to form chemical compounds.

Goal 2 State the meaning of, or draw conclusions based on, the Law of Multiple Proportions.

The **Law of Multiple Proportions** states that when two elements combine to form more than one compound, the different weights of one element that combine with the same weight of the other element are in a simple ratio of whole numbers.

Goal 3 Describe the electron by charge and approximate mass, expressed as a comparison with the mass of a hydrogen atom, and write the symbol for the electron.

The **electron,** symbol e^-, has a mass that is 1/1837 of the mass of a hydrogen atom. It has been assigned a charge of 1−.

Goal 4 Describe and/or interpret the Rutherford scattering experiments and the nuclear model of the atom.

The **Rutherford scattering experiments** were designed so that positively charged alpha particles were directed at thin metal foils. Most particles passed through the foils, but some were deflected at large angles. The main features of the **nuclear model of the atom** are as follows:

1. Every atom contains an extremely small, extremely dense nucleus.
2. All of the positive charge and nearly all of the mass of an atom are concentrated in the nucleus.
3. The tiny nucleus is surrounded by a much larger volume of nearly empty space that makes up the majority of the volume of an atom.
4. The space outside the nucleus is very thinly populated by electrons, the total charge of which exactly balances the positive charge of the nucleus.

Goal 5 Identify the three major subatomic particles by symbol, charge, location within the nuclear atom, and approximate atomic mass, expressed relative to the mass of a hydrogen atom.

Subatomic Particle	Symbol	Charge	Location	Mass Relative to H Atom
Electron	e^-	1−	Outside nucleus	1/1837th
Proton	p or p^+	1+	Inside nucleus	Same
Neutron	n or n^0	0	Inside nucleus	Same

Goal 6 Explain what isotopes are and how they differ from one another.

Atoms of the same element that have different masses are called **isotopes.** The mass differences between atoms of an element are caused by different numbers of neutrons.

Goal 7 For an isotope whose chemical symbol is known, given one of the following, state the other two: (a) nuclear symbol, (b) number of protons and neutrons, (c) atomic number and mass number.

An isotope can be represented by a **nuclear symbol** that has the form ${}^{A}_{Z}\mathrm{Sy}$, where A is the **mass number** of the element, Z is the **atomic number** of the element, and Sy is the chemical symbol of the element. The mass number is the sum of the number of protons plus the number of neutrons. The atomic number is the number of protons.

Goal 8 Identify the features of Dalton's atomic theory that are no longer considered valid and explain why.

Two features of Dalton's atomic theory are no longer considered valid are as follows:

1. *Atoms are indivisible.* The existence of ions, charged particles that form when atoms gain or lose electrons, introduced in Chapter 6, invalidates this postulate.
2. *Atoms of an element are identical.* The existence of isotopes invalidates this postulate.

Goal 9 Define and use the atomic mass unit (u).	1 **atomic mass unit (u)** $\equiv \frac{1}{12}$ the mass of one carbon-12 atom
Goal 10 Given the relative abundances of the natural isotopes of an element and the atomic mass of each isotope, calculate the atomic mass of the element.	The average mass of atoms of an element as found in nature is called the **atomic mass.** The atomic mass may be calculated from the masses of the natural isotopes of that element and the percentage abundance of each isotope.
Goal 11 Distinguish between groups and periods in the periodic table and identify them by number (and letter).	The periodic table arranges the elements into seven **periods** (horizontal rows) and 18 **groups** (vertical columns) in order of atomic numbers and periodic recurrence of physical and chemical properties.
Goal 12 Given the atomic number of an element, use a periodic table to find the symbol and atomic mass of that element, and identify the period and group in which it is found.	Each box in the periodic table gives the atomic number (Z), the elemental symbol, and the atomic mass of an element.
Goal 13 Given an elemental symbol or information from which it can be identified, classify the element as either a main group or transition element and as a metal, nonmetal, or metalloid.	Elements in the A groups (1, 2, and 13 to 18) of the periodic table are called **main group elements.** Elements in the B groups (3 to 12) are called **transition elements.** The stair-step line that begins between atomic numbers 4 and 5 in Period 2 and ends between 84 and 85 in Period 6 separates the **metals** on the left from the **nonmetals** on the right.
Goal 14 Given the name or the symbol of an element in Figure 5.20, write the other.	The names and symbols of 35 common elements are to be learned, using the periodic table as a memory aid: H, He, Li, Be, B, C, N, O, F, Ne, Na, Mg, Al, Si, P, S, Cl, Ar, K, Ca, Cr, Mn, Fe, Co, Ni, Cu, Zn, Br, Kr, Ag, Sn, I, Ba, Hg, Pb.

Key Terms

Most of the key terms and concepts and many others appear in the Glossary. Use your Glossary regularly.

atomic mass p. 167
atomic mass unit (u) p. 166
atomic number (Z) p. 163
chemical family p. 171
Dalton's atomic theory p. 156
electron p. 159
group p. 171
isotopes p. 164
Law of Multiple Proportions p. 155
main group elements p. 171
mass number (A) p. 164
metal p. 171
metalloid p. 171
neutron p. 162
nonmetal p. 171
nuclear model of the atom p. 161
nuclear symbol p. 164
nucleus p. 161
periods p. 170
plum pudding model p. 160
proton p. 162
representative elements p. 171
Rutherford scattering experiments p. 160
subatomic particle p. 159
transition elements p. 171
transition metals p. 171

Frequently Asked Questions

What is the difference between the mass of an isotope and atomic mass?
Students often have no trouble with questions about isotopes when studying Section 5.5. In particular, questions such as those posed in Active Example 5.2 are generally easy to answer at first. However, after studying Section 5.6, students may become confused about the difference between the mass of a specific isotope of an element and the average atomic mass of all of the naturally occurring isotopes of an element. For example, 12.01 u is the atomic mass of carbon, but it is neither the mass number of carbon-12 nor its mass. Additionally, you cannot use the atomic mass of an element to determine the number of neutrons in an isotope of that element. The number of neutrons is the difference between the mass number (A) of a specified isotope and the atomic number (Z) for the element.

What is the purpose of locating elements in the periodic table while learning name–symbol pairs?
There are a couple of reasons why this is beneficial. One is that it is a memory aid. Even though it seems simpler to memorize names and symbols without referring to a periodic table, research on memory shows that the opposite is true. When you associate symbols with their positions in the periodic table, it helps you remember their names. Another reason for doing this is that the value of the periodic table cannot be overstated. You will be using it soon in Chapter 6 to learn the system of naming chemical compounds; after that, it will help you in many other ways. For example, you will need atomic masses to solve some of the problems in this course. The periodic table is a readily available source of these values. Other applications of the periodic table will appear as the course progresses. The faster you

are at locating an element's position within the table, the more efficient you will be at using the periodic table in these upcoming applications.

What common errors should I look to avoid when learning name–symbol pairs?
The most common error while learning to write symbols and formulas is writing both letters in a two-letter elemental symbol as capitals. The first letter is always a capital letter. If a second letter is present, it is *always* written in lowercase. The language of chemistry is very precise, and correctly written symbols are part of that language. It is also important to learn the correct spelling of elemental names as you come to them. *Flourine* instead of *fluorine* is the most common misspelling of an elemental name.

Concept-Linking Exercises

Write a brief description of the relationships among each of the following groups of terms or phrases. Answers to the Concept-Linking Exercises are given at the end of the chapter.

1. Dalton's atomic theory, nuclear model of the atom, Rutherford scattering experiments
2. Electron, proton, neutron, subatomic particles
3. Atomic mass unit, carbon-12, gram
4. Isotopes, neutron, proton, mass number, atomic number, atomic mass
5. Periodic table, groups, periods
6. Main group elements, transition elements, transition metals, metals, nonmetals, metalloids
7. Atomic mass of an element, atomic mass of an isotope, average atomic mass

Small-Group Discussion Questions

Small-Group Discussion Questions are for group work, either in class or under the guidance of a leader during a discussion section.

1. Define the Law of Definite Composition, the Law of Conservation of Mass, and the Law of Multiple Proportions. Explain how these laws lead to Dalton's atomic theory.
2. Fill in the blanks in the following table. Do not use any references.

	Symbol	Charge Relative to an Electron at 1−	Mass Relative to a Hydrogen Atom	Is It a Nuclear Particle? (Y/N)
Electron				
Proton				
Neutron				

3. If the nucleus of a single atom was scaled up to a diameter of 1 inch, what would be the diameter of an atom? Answer in USCS units.
4. When expressed to the full, unrounded correct number of significant figures, the atomic mass of zinc is 65.38 u and the atomic mass of nickel is 58.6934 u. Why does the number of significant figures vary among elements? Nickel has five naturally occurring isotopes, nickel-58, -60, -61, -62, and -64. Which is most abundant? How do you know? How many electrons, protons, and neutrons are in each nickel isotope?
5. A balloon is filled with nitrogen gas and placed in a refrigeration unit set at 2°C. The atmospheric pressure is 1 bar. The volume of the balloon is decreased until it is 1 L. The balloon and its contents are weighted, and the gas is found to have a mass of 1.2 g. The experiment is repeated with oxygen, fluorine, and chlorine gases. What is the mass of the gas in each trial? Equal volumes of gases at the same pressure and temperature have equal numbers of molecules.
6. How many elements fall into each of the following categories? Use the periodic table on your shield when answering this question. Period 4, main group elements, metals, U.S. Group 8B, transition elements, Period 7, nonmetals, IUPAC Group 16, representative elements.
7. Construct a set of flash cards with the symbols of each of the 35 elements to be memorized on one side and their names on the other side. (Make more or fewer cards if your instructor requires you to memorize more or fewer elemental symbols.) Split into pairs and practice giving names for symbols and symbols for names until each person can repeat the entire set rapidly and accurately in both directions.

Questions, Exercises, and Problems

Interactive versions of these problems may be assigned by your instructor. Solutions for blue-numbered questions are at the end of the chapter. Questions other than those in the General Questions and More Challenging Problems sections are paired in consecutive odd–even number combinations; solutions for the odd numbered questions are also at the end of the chapter.

Section 5.2: Dalton's Atomic Theory

1. According to Dalton's atomic theory, can more than one compound be made from atoms of the same two elements?

2. List the major points in Dalton's atomic theory.

3. Show that Dalton's atomic theory explains the Law of Definite Composition.

4. How does Dalton's atomic theory account for the Law of Conservation of Mass?

5. The chemical name for limestone, a compound of calcium, carbon, and oxygen, is calcium carbonate. When heated, limestone decomposes into solid calcium oxide and gaseous carbon dioxide. From the names of the products, tell where you might find the atoms of each element after the reaction. How does Dalton's atomic theory explain this?

6. The brilliance with which magnesium burns makes it ideal for use in marine flares and fireworks. Compare the mass of magnesium that burns with the mass of magnesium in the magnesium oxide ash that forms. Explain this in terms of atomic theory.

Moura Pereira/Shutterstock.com

The white light in fireworks can result from burning magnesium.

7. Sulfur and fluorine form at least two compounds—SF_4 and SF_6. Explain how these compounds can be used as an example of the Law of Multiple Proportions.

8. When 10.0 g of chlorine reacts with mercury under varying conditions, the reaction consumes either 28.3 g or 56.6 g of mercury. No other combinations occur. Explain these observations in terms of the Law of Multiple Proportions.

Section 5.3: The Electron

9. Advances in technology and science often progress hand-in-hand. What advance in technology was necessary to set the stage for the discovery of the electron?

10. When J. J. Thomson measured the deflection of a stream of electrons in a magnetic field, he was able to demonstrate that electrons were particles, but he was unable to determine the mass of an electron. Explain.

11. The mass of an electron is 1/1837 of the mass of a hydrogen atom. Express this fraction as a percentage. What percentage of the mass of a hydrogen atom is due to the remaining subatomic particle(s)?

12. What is meant by stating that the charge of an electron is 1−? What is the symbol of the electron?

Section 5.4: The Nuclear Atom and Subatomic Particles

13. How can we account for the fact that, in the Rutherford scattering experiments, some of the alpha particles were deflected from their paths through the gold foil, and some were even deflected back at various angles?

14. How can we account for the fact that most of the alpha particles in the Rutherford scattering experiments passed directly through a solid sheet of gold?

15. What do we call the central part of an atom?

16. What major conclusions were drawn from the Rutherford scattering experiments?

17. Describe the activity of electrons according to the planetary model of the atom that appeared after the Rutherford scattering experiments.

18. The Rutherford experiments were performed and conclusions were reached before protons and neutrons were discovered. When they were found, why was it believed that they were in the nucleus of the atom?

19. Compare the three major parts of an atom in charge and mass.

20. Which of the following applies to the electron: (a) Charge = 1 − ; (b) Mass = 1 u; (c) Charge = 0; (d) Mass = 0 u; (e) Charge = 1+?

Section 5.5: Isotopes

21. Can two different elements have the same atomic number? Explain.

22. Compare the number of protons and electrons in an atom, the number of protons and neutrons, and the number of electrons and neutrons.

23. Explain why isotopes of different elements can have the same mass number, but isotopes of the same element cannot.

24. How many protons, neutrons, and electrons are there in a neutral atom of the isotope represented by ${}^{8}_{5}B$?

Questions 25 and 26: From the information given in the following tables, fill in as many blanks as you can without looking at any reference. If there are unfilled spaces, continue by referring to your periodic table. As a last resort, check the table of elements on the Reference Page. All atoms are neutral.

25.

Name of Element	Nuclear Symbol	Atomic Number	Mass Number	Protons	Neutrons	Electrons
					24	21
	${}^{76}_{32}Ge$					
			122		72	
			37			17
		11			12	

26.

Name of Element	Nuclear Symbol	Atomic Number	Mass Number	Protons	Neutrons	Electrons
	$^{138}_{57}La$					
Phosphorus			28			
				57	82	
		29				36
	$^{63}_{29}Cu$					
Aluminum			25			

Section 5.6: Atomic Mass

Although this set of questions is based on material in Section 5.6, some parts of some questions assume that you have also studied Section 5.7 and can use the periodic table as a source of atomic masses.

27. What advantage does the atomic mass unit have over grams when speaking of the mass of an atom or a subatomic particle?

28. What is an atomic mass unit?

29. The mass of an average atom of a certain element is 6.66 times as great as the mass of an atom of carbon-12. Using either the periodic table or the table of elements, identify the element.

30. The average mass of boron atoms is 10.81 u. How would you explain what this means to a friend who had never taken chemistry?

31. The atomic masses of the natural isotopes of neon are 19.99244 u, 20.99395 u, and 21.99138 u. The average of these three masses is 20.99259 u. The atomic mass of neon is listed as 20.1797 u on the periodic table. Which isotope do you expect is the most abundant in nature? Explain.

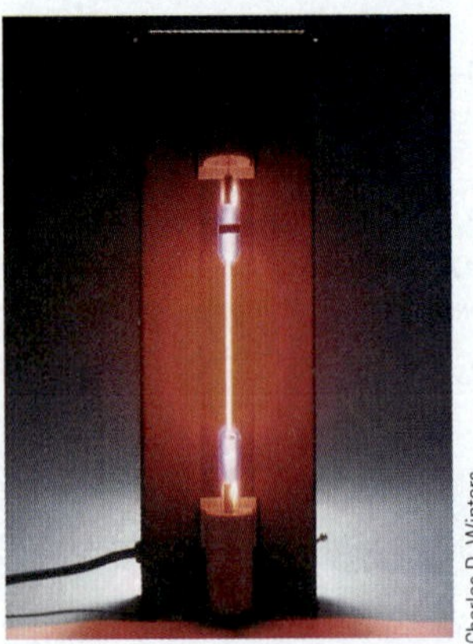

Charles D. Winters

When neon gas is within an activated gas discharge tube, reddish-orange light is emitted.

32. A certain element consists of two stable isotopes. The first has an atomic mass of 137.9068 u and a percentage natural abundance of 0.09%. The second has an atomic mass of 138.9061 u and a percentage natural abundance of 99.91%. What is the atomic mass of the element?

33. The mass of 60.4% of the atoms of an element is 68.9257 u. There is only one other natural isotope of that element, and its atomic mass is 70.9249 u. Calculate the average atomic mass of the element. Using the periodic table and/or the table of the elements, write its symbol and name.

34. Isotopic data for boron allow the calculation of its atomic mass to the number of significant figures justified by the measurement process. One analysis showed that 19.78% of boron atoms have an atomic mass of 10.0129 u, and the remaining atoms have an atomic mass of 11.00931 u. Find the average mass in as many significant figures as those data will allow.

Questions 35 through 40: Percentage abundances and atomic masses (u) of the natural isotopes of an element are given. (a) Calculate the atomic mass of each element from these data. (b) Using other information that is available to you, identify the element.

	Percentage Abundance	Atomic Mass (u)
35.	51.82	106.9041
	48.18	108.9047
36.	69.09	62.9298
	30.91	64.9278
37.	57.25	120.9038
	42.75	122.9041
38.	37.07	184.9530
	62.93	186.9560
39.	0.193	135.907
	0.250	137.9057
	88.48	139.9053
	11.07	141.9090
40.	67.88	57.9353
	26.23	59.9332
	1.19	60.9310
	3.66	61.9283
	1.08	63.9280

Section 5.7: The Periodic Table

41. How many elements are in Period 5 of the periodic table? Write the atomic numbers of the elements in Group 3B/3.

42. Write the symbol of the element in each given group and period: (a) Group 1A/1, Period 6; (b) Group 6A/16, Period 3; (c) Group 7B/7, Period 4; (d) Group 1A/1, Period 2.

43. Locate in the periodic table each element whose atomic number is given, and identify first the number of the period it is in and then the number of the group: (a) 20; (b) 14; (c) 43.

44. List the symbols of the elements of each of the following: (a) transition metals in the fourth period; (b) metals in the third period; (c) nonmetals in Group 6A/16 or 7A/17 with $Z < 40$; (d) main group metals in the sixth period.

45. Using only a periodic table for reference, list the atomic masses of the elements whose atomic numbers are 29, 55, and 82.

46. Write the atomic number of the element in each given group and period: (a) Group 4A/14, Period 3; (b) Group 4A/14, Period 2; (c) Group 1B/11, Period 4; (d) Group 2A/2, Period 3.

47. Write the atomic masses of helium and aluminum.

48. Give the atomic mass of the element in each given group and period: (a) Group 8A/18, Period 3; (b) Group 3A/13, Period 4; (c) Group 4B/4, Period 4; (d) Group 2A/2, Period 3.

Section 5.8: Elemental Symbols and the Periodic Table

49. The names, atomic numbers, or symbols of some of the elements in Figure 5.20 are given in **Table 5.4**. Fill in the open spaces, referring only to a periodic table for any information that you need.

Table 5.4 Table of Elements

Name of Element	Atomic Number	Symbol of Element
		Mg
	8	
Phosphorus		
		Ca
Zinc		
		Li
Nitrogen		
	16	
	53	
Barium		
		K
	10	
Helium		
		Br
		Ni
Tin		
	14	

50. The names, atomic numbers, or symbols of some of the elements in Figure 5.20 are given in **Table 5.5**. Fill in the open spaces, referring only to a periodic table for any information you need.

Table 5.5 Table of Elements

Name of Element	Atomic Number	Symbol of Element
Sodium		
		Pb
Aluminum		
	26	
		F
Boron		
	18	
Silver		
	6	
Copper		
		Be
Krypton		
Chlorine		
	1	
		Mn
	24	
Cobalt		
	80	

General Questions

51. Distinguish precisely and in scientific terms the differences among items in each of the following groups:

a) Atom, subatomic particle
b) Electron, proton, neutron
c) Nuclear model of the atom, nucleus
d) Atomic number, mass number
e) Chemical symbol of an element, nuclear symbol
f) Atom, isotope
g) Atomic mass, atomic mass unit
h) Atomic mass of an element, atomic mass of an isotope
i) Period, group, or family (in the periodic table)
j) Main group element, transition element

52. Determine whether each statement that follows is true or false.

a) Dalton proposed that atoms of different elements always combine on a one-to-one basis.
b) According to Dalton, all oxygen atoms have the same diameter.
c) The mass of an electron is about the same as the mass of a proton.
d) There are subatomic particles in addition to the electron, proton, and neutron.
e) The mass of an atom is uniformly distributed throughout the atom.
f) Most of the particles fired into the gold foil in the Rutherford experiment were not deflected.
g) The masses of the proton and electron are equal but opposite in sign.
h) Isotopes of an element have different electrical charges.
i) The atomic number of an element is the number of particles in the nucleus of an atom of that element.
j) An oxygen-16 atom has the same number of protons as an oxygen-17 atom.
k) The nuclei of nitrogen atoms have a different number of protons from the nuclei of any other element.
l) Neutral atoms of sulfur have a different number of electrons from neutral atoms of any other element.
m) Isotopes of different elements that exhibit the same mass number exhibit similar chemical behavior.
n) The mass number of a carbon-12 atom is exactly 12 g.
o) Periods are arranged vertically in the periodic table.
p) The atomic mass of the second element in the farthest right column of the periodic table is 10 u.
q) Nb is the symbol of the element for which $Z = 41$.
r) Elements in the same column of the periodic table have similar properties.
s) The element for which $Z = 38$ is in both Group 2A/2 and the fifth period.

53. The first experiment to suggest that an atom consisted of smaller particles showed that one particle had a negative charge. From that fact, what could be said about the charge of other particles that might be present?

More Challenging Problems

54. Sodium oxide and sodium peroxide are two compounds made up of the elements sodium and oxygen. Sixty-two grams of sodium oxide contains 46 g of sodium and 16 g of oxygen; 78 g of sodium peroxide has 46 g of sodium and 32 g of oxygen. Show how these figures confirm the Law of Multiple Proportions.

55. Two compounds of mercury and chlorine are mercury(I) chloride and mercury(II) chloride. The amount of mercury(I) chloride that contains 71 g of chlorine has 402 g of mercury; the amount of mercury(II) chloride that has 71 g of chlorine has 201 g of mercury. Show how the Law of Multiple Proportions is illustrated by these quantities.

56. The *CRC Handbook*, a large reference book of chemical and physical data, lists two isotopes of rubidium (Z = 37). The atomic mass of 72.15% of rubidium atoms is 84.9118 u. Through a typographical oversight, the atomic mass of the second isotope is not printed. Calculate that atomic mass.

57. The element lanthanum has two stable isotopes, lanthanum-138 with an atomic mass of 137.9071 u and lanthanum-139 with an atomic mass of 138.9063 u. From the atomic mass of La, 138.9 u, what conclusion can you make about the relative percentage abundance of the isotopes?

58. The atomic mass of lithium on a periodic table is 6.94 u. Lithium has two natural isotopes with atomic masses of 6.10512 u and 7.01600 u. Calculate the percentage distribution between the two isotopes.

Charles D. Winters

Pure lithium is composed of two isotopes.

59. When Thomson identified the electron, he found that the ratio of its charge to its mass (the e/m ratio) was the same regardless of the element from which the electron came. This showed that the electron is a unique particle that is found in atoms of all elements. Positively charged particles found at about the same time did not all have the same e/m ratio. (Later it was found that even different atoms of the same element contain positive particles that have different e/m ratios.) What does that suggest about the mass, particle charge, and minimum number of positive particles from different elements?

60. Why were scientists inclined to think of an atom as a miniature solar system in the nuclear model of the atom? What are the similarities and differences between electrons in orbit around a nucleus and planets in orbit around the sun?

61. Isotopes were unknown until nearly a century after Dalton proposed the atomic theory. When they were discovered, it was through experiments more closely associated with physics than with chemistry. What does this suggest about the chemical properties of isotopes?

62. The element carbon occurs in two crystal forms, diamond and graphite. The density of the diamond form is 3.51 g/cm^3, and of graphite, 2.25 g/cm^3. The volume of a carbon atom is $1.9 \times 10^{-24} \text{ cm}^3$. As stated in Section 5.6, $1 \text{ g} = 6.02 \times 10^{23} \text{ u}$.
    a) Calculate the average density of a carbon atom.
    b) Suggest a reason for the density of the atom being so much larger than the density of either form of carbon.
    c) The radius of a carbon atom is roughly 1×10^5 times larger than the radius of the nucleus. What is the volume of that nucleus? (*Hint:* Volume is proportional to the cube of the radius.)
    d) Calculate the average density of the nucleus.
    e) The radius of a period on this page is about 0.02 cm. The volume of a sphere that size is $4 \times 10^{-5} \text{ cm}^3$. Calculate the mass of that sphere if it were completely filled with carbon nuclei. Express the mass in tons.

Answers to Target Checks

1. The remainder of the hydrogen atom must have a charge of 1+ to balance the 1− charge of the electron, (1+) + (1−) = 0.
2. a) Atoms are mostly made up of empty space, so they are not hard. They are small, and their shape is spherical. (g) Electrons and protons are electrically charged, but neutrons have no charge (or zero charge). True: b, c, d, e, f.
3. All statements are true.
4. (a) Four Group 3A/13 elements are metals (Al, Ga, In, Tl). (b) Ten Period-4 elements are transition metals (Sc through Zn).

Answers to Practice Exercises

1. The total mass of the reactant in this chemical change, 5.00 g, is equal to the total mass of the products, 0.56 g + 4.44 g = 5.00 g. If matter is made of atoms, the Law of Conservation of Mass must occur since a chemical change is a rearrangement of the same set of atoms. The set of atoms have the same mass before and after the change. However, if matter is not made of atoms and any elemental ratio can occur to form a compound, mass would still be conserved in chemical change because matter is not destroyed in a chemical change. Thus, the Law of Conservation of Mass must occur if matter is made of atoms, but it alone is not sufficient to prove Dalton's atomic theory.
2. 19 protons, 19 electrons, 22 neutrons; ${}^{41}_{19}\text{K}$ and ${}^{41}\text{K}$; A = 41, Z = 19.

3. (0.0759 × 6.0151223 u) + (0.9241 × 7.0160041 u) = 6.940 u

4. (0.9493 × 31.97207070 u) + (0.0076 × 32.97145843 u) + (0.0429 × 33.96786665 u) + (0.0002 × 35.96705062 u) = 32.07 u

5.

Symbol	Atomic Number	Atomic Mass	Period	Group	Main Group or Transition Element	Metal, Nonmetal, or Metalloid
N	7	14.01 u	2	5A/15	Main group	Nonmetal
Mg	12	24.31 u	3	2A/2	Main group	Metal
Cr	24	52.00 u	4	5B/5	Transition	Metal

6. Pb, lead; Hg, mercury; Mn, manganese; sodium, Na; phosphorus, P; tin, Sn

Answers to Concept-Linking Exercises

You may have found more relationships or relationships other than the ones given in these answers.

1. Dalton's atomic theory proposed that matter is composed of indivisible atoms. The nuclear model of the atom resulted from Rutherford's scattering experiments. The model pictures the atom as a dense nucleus surrounded by electrons moving in orbit in the otherwise empty space outside the nucleus.

2. An atom contains many particles (subatomic particles), the most important of which are electrons, protons, and neutrons.

3. One atom of carbon-12 has been assigned a mass of 12 atomic mass units, u. Thus 1 u is, by definition, exactly 1/12 of the mass of a carbon-12 atom. Both the atomic mass unit and the gram are mass units. 6.02×10^{23} u = 1 g.

4. All atoms of an element have the same number of protons; this is the atomic number of the element. Atoms of the same element may have different numbers of neutrons and therefore different atomic masses. Atoms of an element with different numbers of neutrons are isotopes. The mass number of an atom is the sum of the number of protons plus the number of neutrons.

5. Horizontal rows in the periodic table are periods, and vertical columns are groups.

6. Main group elements are those in the A groups (U.S.) of the periodic table, and transition elements are those in the B groups. In the periodic table, metals are to the left of the stair-step line beginning between atomic numbers 4 and 5 in Period 2 and ending between atomic numbers 84 and 85 in Period 6, and the nonmetals are to the right of this line. The metalloids, with properties of both metals and nonmetals, include elements of atomic number 5, 14, 32, 33, 51, and 52. Transition metals are the elements in the B groups, an area that includes all the transition elements.

7. The term *atomic mass* actually refers to average atomic mass, but the word *average* is omitted. The atomic mass of an element is the average mass of all of its natural isotopes.

The atomic mass of an individual isotope of an element is the mass of an atom of that particular isotope. All atoms of a particular isotope have the same mass.

Solutions to Blue-Numbered Questions, Exercises, and Problems

1. Yes, see Figure 5.3.

3. The Law of Definite Composition says that any compound is always made up of elements in the same proportion by mass. Dalton's atomic theory explains this by stating that atoms of different elements combine to form compounds.

5. The calcium atoms and some of the oxygen atoms are in the calcium oxide; the carbon atoms and some of the oxygen atoms are in the carbon dioxide.

7. The Law of Multiple Proportions in this case states that the same mass of sulfur combines with masses of fluorine in the ratio of simple whole numbers, 4 to 6.

9. The invention of the Crookes tube, a high-quality vacuum tube.

11. $\frac{1}{1837} \times 100\% = 0.05444\%$; $100.00000 - 0.05444 = 99.94556\%$

13. Alpha particles and atomic nuclei are positively charged. As an alpha particle approached a nucleus, the repulsion between the positive charges deflected the alpha particle from its path.

15. The nucleus.

17. The electrons were thought to travel in circular orbits around the nucleus.

19. See Table 5.1

21. No, the atomic number is the number of protons, and all atoms of an element have the same number of protons.

23. Mass number is the sum of protons plus neutrons. Isotopes of the same element have different numbers of neutrons but the same number of protons. The sums must be different. Atoms of different elements *must* have different numbers of protons, and they may have different numbers of neutrons. An atom with one less proton than another may have one more neutron than the other, so their mass numbers would be the same. Example: carbon-14 (6 protons, 8 neutrons) and nitrogen-14 (7 protons, 7 neutrons).

25.

Name of Element	Nuclear Symbol	Atomic Number	Mass Number	Protons	Neutrons	Electrons
Scandium	$^{45}_{21}Sc$	21	45	21	24	21
Germanium	$^{76}_{32}Ge$	32	76	32	44	32
Tin	$^{122}_{50}Sn$	50	122	50	72	50
Chlorine	$^{37}_{17}Cl$	17	37	17	20	17
Sodium	$^{23}_{11}Na$	11	23	11	12	11

27. The mass of a proton and neutron is close to 1 u, and 6.02×10^{23} u = 1 g. It is more convenient to use the u because mass values expressed in u are not tiny fractions that must be expressed in scientific notation.

29. 6.66 × 12.0 u = 79.9 u, Br

31. The atomic mass of neon is less than the average of the three atomic masses, so the isotope with the lowest mass must be present in the greatest abundance.

33. $0.604 \times 68.9257\ u + (1.000 - 0.604) \times 70.9249\ u = 69.7\ u$, Ga, gallium

35. $0.5182 \times 106.9041\ u + 0.4818 \times 108.9047\ u = 107.9\ u$, Ag, silver

37. $0.5725 \times 120.9038\ u + 0.4275 \times 122.9041\ u = 121.8\ u$, Sb, antimony

39. $0.00193 \times 135.907\ u + 0.00250 \times 137.9057\ u + 0.8848 \times 139.9053\ u + 0.1107 \times 141.9090\ u = 140.1\ u$, Ce, cerium

41. 18; 21, 39, 57, 89

43. (a) Period 4, Group 2A/2; (b) Period 3, Group 4A/14; (c) Period 5, Group 7B/7

45. Z = 29, 63.55 u; Z = 55, 132.9 u; Z = 82, 207.2 u

47. He, 4.003 u; Al, 26.98 u

49.

Name of Element	Atomic Number	Symbol of Element
Magnesium	12	Mg
Oxygen	8	O
Phosphorus	15	P
Calcium	20	Ca
Zinc	30	Zn
Lithium	3	Li
Nitrogen	7	N
Sulfur	16	S
Iodine	53	I
Barium	56	Ba
Potassium	19	K
Neon	10	Ne
Helium	2	He
Bromine	35	Br
Nickel	28	Ni
Tin	50	Sn
Silicon	14	Si

52. True: b,* d, f, j, k, l, q, r, s. False: a, c, e, g, h, i, m, n, o, p.

53. What was left had to have a positive charge to account for the neutrality of the complete atom.

54. Sixteen grams of oxygen combines with 46 grams of sodium in sodium oxide, and 32 grams of oxygen combines with 46 grams of sodium in sodium peroxide. The ratio 16/32 reduces to 1/2, a ratio of small, whole numbers.

56. $0.7215 \times 84.9118\ u + (1 - 0.7215) \times x\ u = 85.4678\ u$; $x = 86.91\ u$

58. $y \times 6.10512\ u + (1 - y) \times 7.01600\ u = 6.941\ u$; $y = 0.08234$; 8.234% at 6.10512 u; $1 - 0.08234 = 0.91766$; 91.77% at 7.01600 u

59. Different e/m ratios for positively charged particles from different elements indicate that, unlike the electron, all positively charged particles are not alike. The charge, the mass, or both must vary from element to element. This suggests the presence of at least two particles in varying number ratios. One or both must have a positive charge; others could be electrically neutral.

60. The nuclear model of the atom is similar to the solar system in that electrons orbit the nucleus as planets orbit the sun. Both models are similar in terms of the size of the nucleus/sun being large compared with the electrons/planets and the vast amount of empty space in the atom/solar system.

61. Chemical properties of isotopes of an element are identical.

62. (a) $\dfrac{12.01\ u}{1\ C\ atom} \times \dfrac{1\ g}{6.02 \times 10^{23}\ u} \times \dfrac{1\ C\ atom}{1.9 \times 10^{-24}\ cm^3} = 1.1 \times 10^{1}\ g/cm^3$

(b) In packing carbon atoms into a crystal there are spaces between the atoms. There are no voids in a single atom. (In fact, voids in diamond account for 66% of the total volume, and in graphite, 78%.)

(c) $\dfrac{1.9 \times 10^{-24}\ cm^3}{(1 \times 10^5)^3} = 2 \times 10^{-39}\ cm^3$

(d) $\dfrac{12.01\ u}{1\ C\ nucleus} \times \dfrac{1\ g}{6.02 \times 10^{23}\ u} \times \dfrac{1\ C\ nucleus}{2 \times 10^{-39}\ cm^3} = 1 \times 10^{16}\ g/cm^3$

(e) $4 \times 10^{-5}\ cm^3 \times \dfrac{1 \times 10^{16}\ g}{1\ cm^3} \times \dfrac{1\ lb}{454\ g} \times \dfrac{1\ ton}{2000\ lb} = 4 \times 10^{5}\ tons$

---

*Dalton apparently did not make any specific comment about the diameter of an atom, but he did propose that all atoms of an element are identical in every respect. This would include diameters.

6 Chemical Nomenclature

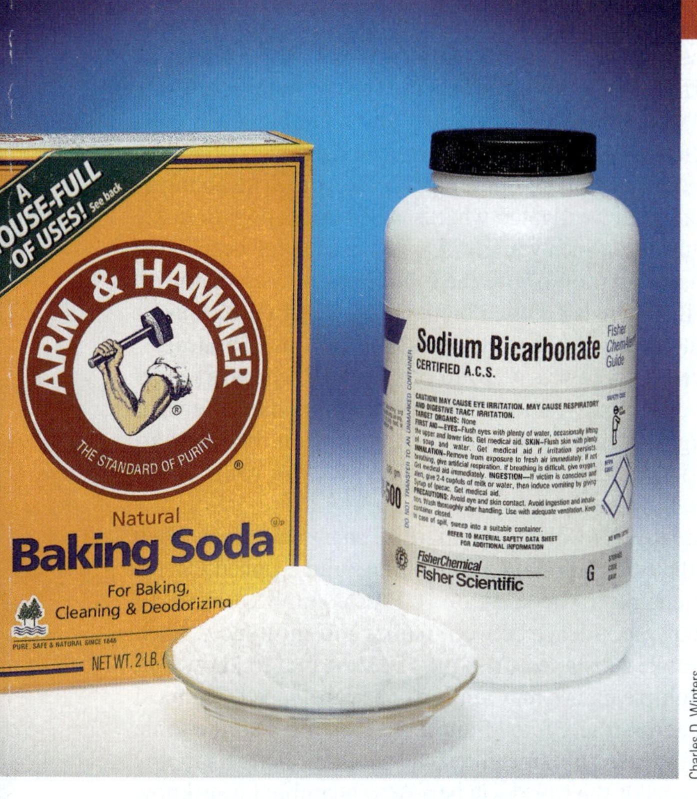

Charles D. Winters

◀ The same substance may be known by several names, but it can have only one chemical formula. Baking soda is the common name of $NaHCO_3$ because of its widespread use in baking. Its old chemical name, bicarbonate of soda, was replaced by a similar name, sodium bicarbonate. Its official name today is sodium hydrogen carbonate. The current name and formula are part of a system of chemical names and formulas that you will learn in this chapter.

CHAPTER CONTENTS

Imagine that you go to a party, are introduced to hundreds of people you have never seen before, and then spend the rest of the evening trying to remember their names. This awkward situation could be avoided altogether if you had access to a *system* that enabled you to memorize the names of the people instead of using brute-force drill-and-practice over and over again for each individual. You'd still have to memorize the system, but that's much easier than memorizing hundreds of names.

In this chapter we will introduce nearly 1,500 chemical names for elements and compounds. You will memorize a few, but you will know the names of the others simply by learning and applying the system. After you have studied the chapter, if you are given only the name of an element or compound, you will be able to supply the chemical formula, and vice versa. Furthermore, you will be able to do this with hundreds of compounds that aren't even directly covered in the chapter. But you will be able to do these remarkable things only if you *learn the system*.

The term **nomenclature** comes from a Latin word meaning "calling by name." Thus, chemical nomenclature is the system of naming chemicals. In this chapter we present an introduction to the language of chemistry as used by most chemists today.

6.1 Is It Soda or Pop or Coke?

What general term do you use for sweetened, flavored, carbonated beverages? If you live in the United States, your answer to this question likely depends on where you grew up or where you now live. If you live in Los Angeles, you probably call it "soda." In Atlanta, you most likely call it "coke," generically meaning cola, root beer, lemon-lime soda, and so on, collectively. "Pop" is used in Minneapolis.

Figure 6.1 shows the use of the term most popularly used for a sweetened, carbonated beverage by county in the United States. You can see that there is a regionality. If you order a "coke" in a restaurant in the South, the server will likely ask, "What kind?" If you order a coke in a California restaurant, the server will either not question your order and bring a Coca-Cola or ask if Pepsi is acceptable.

Imagine that you are a scientist doing research on carbonated beverages. You wish to communicate your findings to people who are interested, so you prepare a manuscript to be submitted to a research journal. The journal restricts the type of manuscript you are writing to no more than 2,200 words. How do you clearly and unambiguously convey that your work was with "sweetened (both natural and artificial sweeteners), flavored (both natural and artificial flavorings) carbonated beverages"? Do you write an introductory sentence that explains that you investigated what is called soda, pop, or coke, depending on the region of the country in which you live? Do you use the full descriptive phrase each time? If you write an explanation or use the full phrase, you use many of your limited words unnecessarily. But there is no unambiguous single term used universally. And how do you communicate clearly to scientists outside of the United States?

Your problem would be solved if there was a group of scientists who worked to standardize terminology in your discipline. If they decided something like "the term *pop* shall be used refer to beverages intended for human consumption that contain carbonated water, flavoring, and a sweetener," then everyone reading your report would understand the precise meaning of the term.

There is such a group of chemists who work to standardize chemical names. They belong to an organization called the International Union of Pure and Applied Chemistry, abbreviated as IUPAC. The chemists who serve on IUPAC committees work to ensure that there is a one-to-one correspondence between names and substances. Each name uniquely corresponds to a single substance. (You may be surprised to know that this correspondence is one way. Multiple names may apply to one substance!) Chemical names are not only standardized across the United States, but they are also the same worldwide. In this chapter you will learn to name chemicals so that scientists in any region, in any industry, and in any part of the world will be able to unambiguously know the exact composition of the substance that you name.

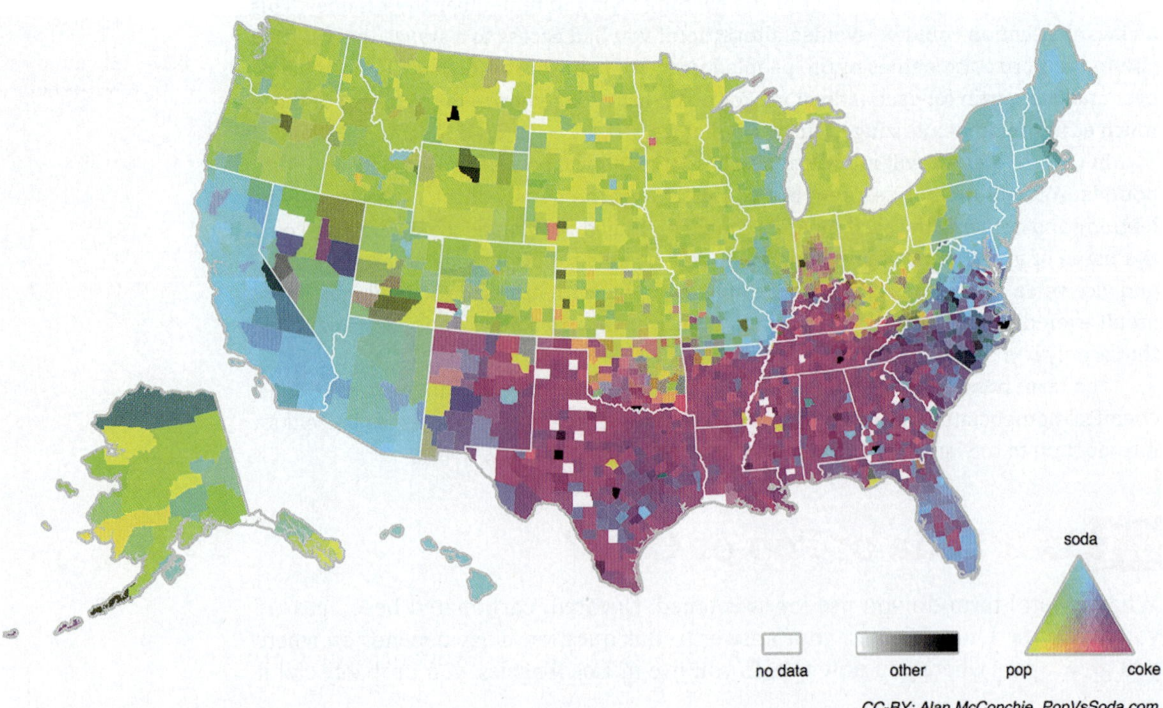

Figure 6.1 Most popular term for a sweetened, carbonated beverage by county in the United States.

6.2 Review of Selected Concepts Related to Nomenclature

Goal 1 Given a representation or a written description of the particulate-level composition of an element or compound, write the chemical formula of that substance.

In Chapter 2, you learned that there are two kinds of pure substances: elements, composed of single atoms or two or more atoms of the element joined together, and compounds, which are made up of atoms of different elements. Chemical formulas are used to represent the particles of an element or compound in written form. The formula of a substance includes the symbols of the elements in that substance. You memorized the symbols of 35 common elements while studying Chapter 5.

In a chemical formula, a subscript number after a symbol shows the number of atoms of the element in the formula unit of the substance. If there is only one atom of an element in the formula, the subscript is omitted.

Reconsider Figure 2.10, which is duplicated here as **Figure 6.2**. The illustration shows how liquid water will decompose into gaseous oxygen and hydrogen when exposed to an electrical current. In the particulate-level illustrations, the red spheres represent oxygen atoms and the white spheres represent hydrogen atoms. You can see that the water molecules consist of two hydrogen atoms attached to one oxygen atom. The chemical formula of the compound water is therefore H_2O, comprised of the elemental symbol for hydrogen, H, followed by a subscript 2, and then the elemental symbol for oxygen, O, with no subscript, which indicates one atom. The chemical formula describes the composition of the molecule in terms of the number of each type of atom that makes up the particle.

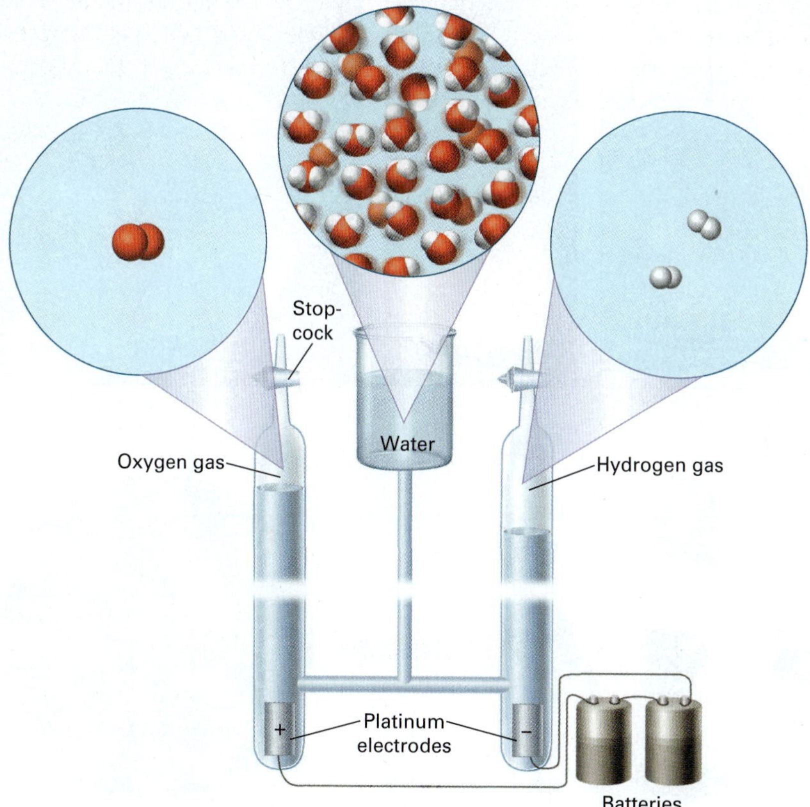

Figure 6.2 The decomposition of liquid water to its elements. The chemical formulas of the substances in this illustration must match their particulate-level composition. Because water molecules are composed of two hydrogen atoms and one oxygen atom, its formula is H_2O. Both oxygen and hydrogen consist of two-atom molecules, so their formulas are O_2 and H_2. (Only the top and bottom of the apparatus are shown to conserve vertical space. The middle of the tubes have been omitted.)

The element oxygen on the left side of Figure 6.2 is shown to consist of oxygen molecules, which are made up of two attached oxygen atoms. Its formula is therefore O_2, indicating that particles of oxygen occur as two attached oxygen atoms. Elements may be composed of single-atom molecules, two-atom molecules, or more complex multi-atom molecules, but if the substance is an element, all atoms are of the same element. Similarly, the illustration of the white hydrogen molecules on the right shows that hydrogen occurs in nature, under normal conditions, as a two-atom hydrogen molecule. The chemical formula of hydrogen is therefore H_2.

Now consider **Figure 6.3**. First observe the photographs of the reaction at the macroscopic level at the top of the figure. On the left is sodium, which is a silvery-white solid at room conditions. Moving to the right, the yellow-green gas in the flask is chlorine. The next photograph shows the reaction of sodium and chlorine. A close-up of the product of the reaction, solid sodium chloride, is shown in the photograph on the far right.

Below the macroscopic-level photographs, you see particulate-level illustrations of each substance. The illustration of sodium shows that it is composed of sodium atoms arranged in a definite, repeating pattern—a crystalline solid. The chemical formula for sodium is therefore simply Na, indicating that the element consists of sodium (Na) atoms. In the illustration of chlorine, you can see that it is composed of two-atom molecules, so its formula is Cl_2. The illustration of sodium chloride shows that it does not consist of individual molecules; rather, it is a crystal made up of particles in a definite ratio—in this case, 1-to-1. The formula of sodium chloride is therefore NaCl, showing that for each particle coming from a sodium atom, there is one particle that comes from chlorine.

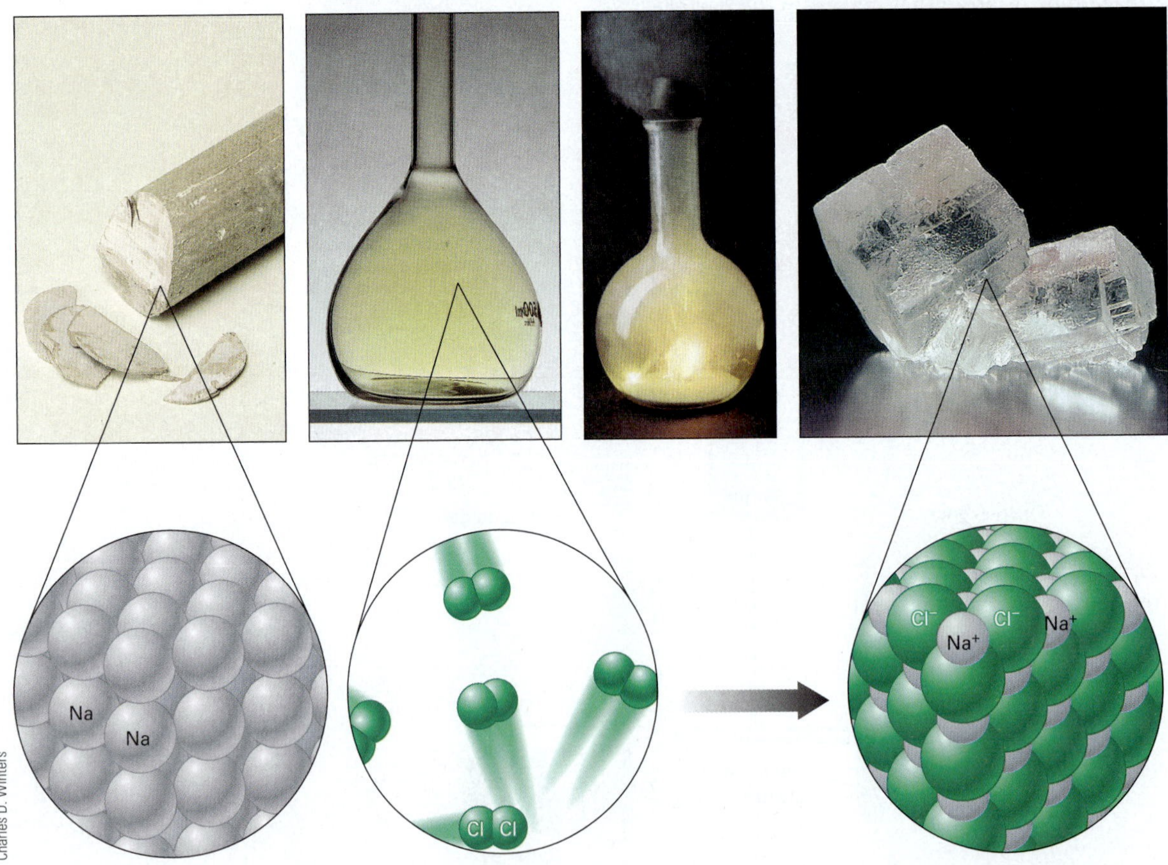

Figure 6.3 Sodium and chlorine react to form sodium chloride. The chemical formula of sodium is Na, indicating that the particulate-level composition of sodium is individual sodium atoms. Chlorine is made up of two-atom molecules, so its formula is Cl_2. Sodium chloride is a collection of particles that come from sodium atoms and particles that come from chlorine atoms in a 1-to-1 ratio, as reflected by the formula Na_1Cl_1, or simply NaCl.

Active Example 6.1 Writing Chemical Formulas

Write the chemical formula of each of the following. In Parts a–d, the black spheres represent carbon atoms; the red, oxygen atoms; and the white, hydrogen atoms.

e) The compound made up of molecules with two nitrogen atoms and five oxygen atoms.

f) The compound made up of a crystal with one particle coming from a magnesium atom for every two particles coming from fluorine atoms.

Think Before You Write Keep in mind that a chemical formula is simply a shorthand description of the particulate-level composition of an element or compound.

Answers *Cover the left column with your cut-out shield. Reveal each answer only after you have written your own answer in the right column.*

a) CO; b) CO_2; c) CH_4; d) CH_4O	Begin with Parts a–d. Based on the particulate-level illustrations, write the formula of each. Oxygen is written last in these formulas.
e) N_2O_5; f) MgF_2	Finish the example by writing the formulas for Parts e and f.
You strengthened your understanding of the concept that a chemical formula is a shorthand symbolic representation of the particulate-level composition of a substance.	What did you learn by solving this Active Example?

Practice Exercise 6.1

Write the chemical formula of each of the following. In Parts a–d, the black spheres represent carbon atoms; the red, oxygen atoms; and the white, hydrogen atoms.

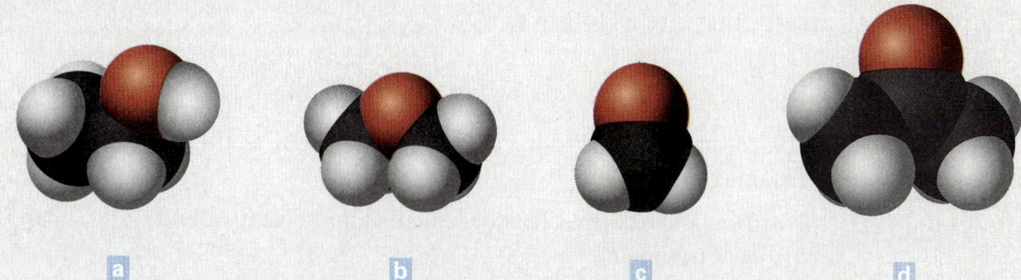

e) The compound made up of molecules with one nitrogen atom and three hydrogen atoms.

f) The compound made up of a crystal with two particles coming from aluminum atoms for every three particles coming from oxygen atoms.

Developing an understanding of basic chemical nomenclature requires that you master a series of facts, rules, and procedures. Thus far in this section we have reviewed some of the key prerequisite topics. You may also need to review additional topics as necessary. This checklist will guide you to the relevant sections that you may need to review:

- ❑ You know the definitions of *pure substance* (Section 2.5), *element, compound, atom, molecule, elemental symbol, chemical formula*, and the *Law of Definite Composition* (Section 2.7).
- ❑ You know that an atom is electrically neutral because the number of protons is equal to the number of electrons (Section 5.4).
- ❑ You can use the periodic table to help with the identification of elemental symbols, main group and transition elements, and metals, metalloids, and nonmetals (Section 5.7).
- ❑ You have memorized 35 name–symbol pairs (Section 5.8).
- ❑ You understand what a chemical formula represents (Section 6.2).

Learn It Now! *If you are not confident in your current understanding of any of the concepts in the review checklist, go back and take a few minutes—now!—to reexamine the topic. Review of previously covered concepts is something that you should make a habit of in all of your courses in which the material builds on itself.*

6.3 Formulas of Elements

Goal 2 Given a name or formula of an element in Figure 5.20, write the other.

In Chapter 5 you learned the symbols of 35 elements and their location in a partial periodic table that contained only those symbols. That periodic table was Figure 5.20. In this section you will learn the chemical formulas of those elements, reflecting their natural particulate-level composition under normal conditions.

The vast majority of the elements exist as single-atom particles. Helium is an example of an element that exists as single atoms in the gas phase at common temperatures (**Figure 6.4[a]**), and lithium is an example of an atomic element normally found in the solid phase (**Figure 6.4[b]**). The formula of these one-atom elements is simply the elemental symbol, reflecting the fact that the basic unit is one uncombined atom. Thus, the chemical formula of helium is He, and the chemical formula of lithium is Li.

Seven elements exist as two-atom, or **diatomic**, molecules, at common temperatures. *Di-* is a prefix that means two. ◀ Thus, a diatomic molecule consists of two atoms that are chemically bonded to one another, forming the smallest particle of that substance found in nature, the molecule. An example is hydrogen (**Figure 6.4[c]**). Its chemical formula is H_2.

Some elements have additional natural molecular forms that consist of more than one or two atoms. We will not consider these in this introductory course.

You must be able to recognize the seven elements that occur as diatomic molecules and write their formulas correctly. Listed in a way that will help you remember them, they are as follows:

Elements	Formulas
Hydrogen, the element that makes up about 90% of the atoms in the universe	H_2
Nitrogen and oxygen, the two elements that make up about 98% of the atmosphere	N_2, O_2
Fluorine, chlorine, bromine, and iodine, four consecutive elements in Group 7A/17	F_2, Cl_2, Br_2, I_2

a

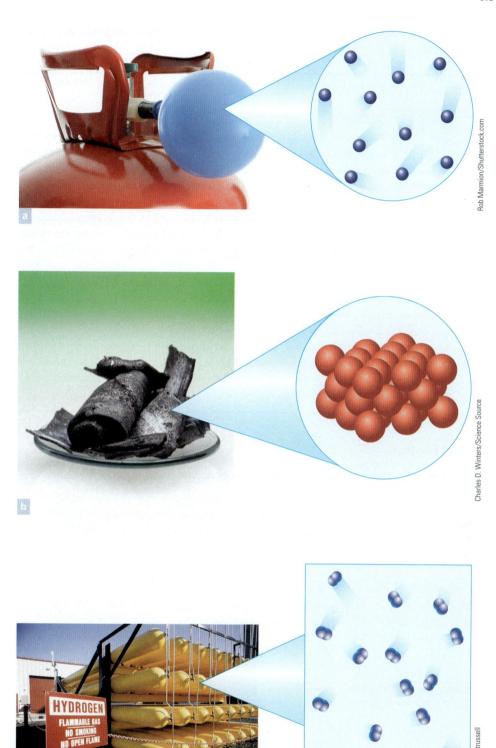

Rob Marmion/Shutterstock.com

b

Charles D. Winters/Science Source

c

iStock.com/bentrussell

Figure 6.4 (a) Helium at the macroscopic and particulate levels. The smallest unit of helium at the particulate level is a one-atom molecule, and thus its formula is He. (b) Lithium at the macroscopic and particulate levels. The smallest unit of lithium at the particulate level is a one-atom molecule, and thus its formula is Li. (c) Hydrogen at the macroscopic and particulate levels. The smallest unit of hydrogen at the particulate level is a two-atom molecule, and thus its formula is H_2.

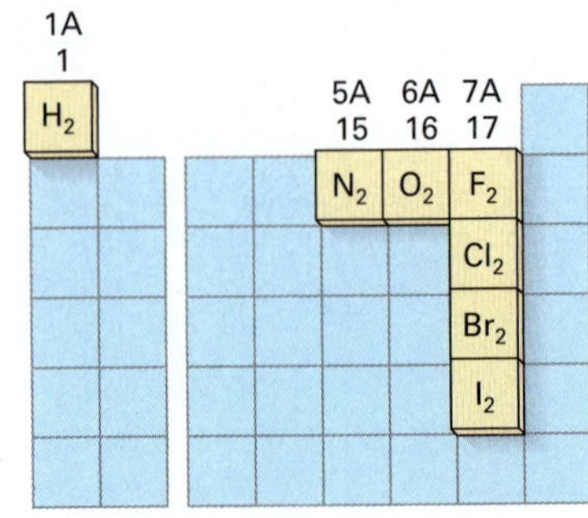

Figure 6.5 The seven elements that exist as stable diatomic molecules and their positions in the periodic table.

The formulas of these elements are shown in their positions in the periodic table in Figure 6.5.

Thinking About Your Thinking

Memory

A mnemonic (pronounced neh-MAH-nick) is something that helps you remember. Perhaps the first mnemonic device you learned was the alphabet song. By matching the letters of the alphabet with the musical notes in the song, it became easier for you to recall the letters in their proper order. Even though you must remember more information, it helps to associate one thing with another. There are many other mnemonics, such as mental images, pictures, catchy sayings, and jingles. Advertisers often use mnemonics to help you remember the names of their products. You may be able to think of a few advertising jingles right now; notice how the jingle helps you remember the company's slogan.

A mnemonic device to remember the seven diatomic elements is the saying: **H**orses **N**eed **O**ats **F**or **Cl**ear **Br**own **I**'s. The first letter(s) of each word is the elemental symbol of one of the seven elements, following the pattern of their sequence in the periodic table. Remembering this phrase is often easier than learning the elemental symbols alone.

As you gain experience in writing the formulas of the elements, your reliance on this mnemonic will gradually decrease, something that occurs through repeated practice. When you automatically think "O_2" or "Br_2" without the mnemonic, you know that you are improving in your command of chemical facts and information.

Active Example 6.2 Formulas of Elements

Write the chemical formulas of the following elements: potassium, fluorine, hydrogen, nitrogen, calcium.

Think Before You Write First, note the difference between being asked to write an elemental symbol and a chemical formula. An elemental symbol is the one- or two-letter abbreviation used in place of the full name of the element. A chemical formula uses elemental symbols to describe the particulate-level composition of the substance as it occurs in nature. The formulas of most elements are monatomic (the prefix *mono-* means one). If you have memorized the seven elements that occur as diatomic molecules, you are ready to complete this example.

Answers *Cover the left column with your cut-out shield. Reveal each answer only after you have written your own answer in the right column.*

Potassium, K; fluorine, F_2; hydrogen, H_2; nitrogen, N_2; calcium, Ca	Complete the Active Example.
You improved your skill at writing the chemical formulas of the elements.	What did you learn by solving this Active Example?

Practice Exercise 6.2

Write the chemical formulas of chlorine, argon, bromine, and krypton.

6.4 Compounds Made from Two Nonmetals

Goal 3 Given the name or formula of a binary molecular compound, write the other.

4 Given the name or the formula of water, write the other; given the name or the formula of ammonia, write the other.

The stair-step line in the periodic table that begins between atomic numbers 4 and 5 and ends between 84 and 85 separates elements on the left that are metals, such as iron, copper, and lead, from elements on the right that are nonmetals, such as hydrogen, oxygen, and nitrogen (**Figure 6.6**). Several elements bordering on this stair-step line are called *metalloids*. Many compounds are formed by two nonmetal elements or by one metalloid and one nonmetal element. These are called **binary molecular compounds**. A binary molecular compound contains *two different elements*, both nonmetals (or one can be a metalloid). There may be more than one atom of each element.

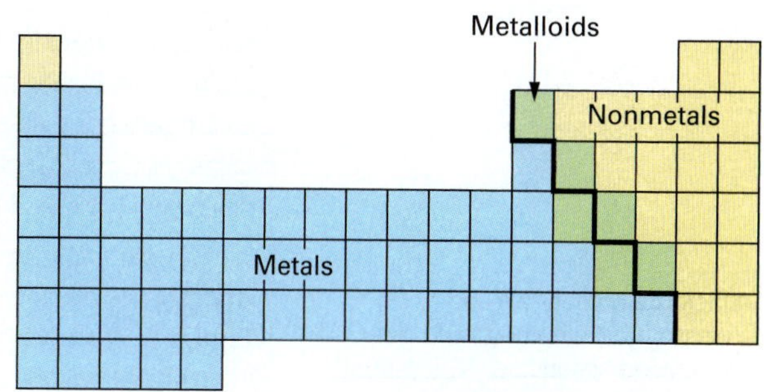

Figure 6.6 Metals and nonmetals. Green identifies elements that are metalloids, which have properties that are intermediate between those of metals and nonmetals.

how to... Write the Name of a Binary Molecular Compound

1. The name of a binary molecular compound has two words.
2. The first word is the name of the element appearing first in the chemical formula, including a prefix to indicate the number of atoms of that element in the molecule.
3. The second word is the name of the element appearing second in the chemical formula, changed to end in *-ide*, and also including a prefix to indicate the number of atoms in the molecule.

Table 6.1 gives the first 10 number prefixes. ▶ The letter *o* at the end of the prefix *mono-* and the letter *a* in the prefixes from 4 to 10 are omitted if the resulting word sounds better. This usually occurs when the next letter is a vowel. For example, a compound with a formula ending in O_5 is a *pentoxide* rather than a *pentaoxide*.

The Frequently Asked Questions section at the end of this chapter gives some suggestions to help you memorize the prefixes in Table 6.1.

The same two elements often form more than one binary compound. Their names are distinguished by the prefixes in Table 6.1. Silicon and chlorine form silicon tetrachloride, $SiCl_4$, and disilicon hexachloride, Si_2Cl_6. The prefix *tetra-* identifies four chlorine atoms in a molecule of $SiCl_4$. In Si_2Cl_6, *di-* indicates two silicon atoms, and *hexa-* shows six chlorine atoms in the molecule. Technically, $SiCl_4$ should be monosilicon tetrachloride, but the prefix *mono-* for "one" is usually omitted in the first word. If an elemental name has no prefix in a binary molecular compound, you may assume that there is only one atom of that element in the molecule.

Active Example 6.3 Names and Formulas of Binary Molecular Compounds

For each name, write the formula; for each formula, write the name: nitrogen monoxide, dinitrogen monoxide, dinitrogen pentoxide, NO_2, N_2O_3, N_2O_4.

Think Before You Write Be sure that you have memorized the ten number prefixes used in chemical names.

Answers *Cover the left column with your cut-out shield. Reveal each answer only after you have written your own answer in the right column.*

Each compound consists of two elements, nitrogen and oxygen. Both nitrogen and oxygen are nonmetals because they are to the right of the stair-step line in the periodic table. The compounds are binary (two element) molecular (all nonmetal atoms) compounds.	Look at the six compounds. How many elements are in each compound? Are those elements metals, nonmetals, or metalloids? How do you know? What is the name of this type of compound?

Table 6.1 Number Prefixes Used in Chemical Names

Number	Prefix
1	*mono-*
2	*di-*
3	*tri-*
4	*tetra-*
5	*penta-*
6	*hexa-*
7	*hepta-*
8	*octa-*
9	*nona-*
10	*deca-*

Nitrogen monoxide, NO (when there is no prefix, there is one atom of that element); dinitrogen monoxide, N_2O; dinitrogen pentoxide, N_2O_5; NO_2, nitrogen dioxide; N_2O_3, dinitrogen trioxide; N_2O_4, dinitrogen tetroxide *Nitrogen dioxide could also be correctly identified as mononitrogen dioxide.* ◀ **Figure 6.7** *shows nitrogen monoxide reacting with the oxygen in the air, yielding nitrogen dioxide.*	Complete the Active Example. 
You improved your skill at writing the names and formulas of binary molecular compounds.	What did you learn by solving this Active Example?

Chemists sometimes use nonsystematic names that have been used historically. N_2O is often called *nitrous oxide* or *laughing gas*, and NO is also known as nitric oxide.

Practice Exercise 6.3

For each name, write the formula; for each formula, write the name: dinitrogen pentasulfide, tetrasilicon hexahydride, iodine heptafluoride, I_4O_9, Br_3O_8, PF_5.

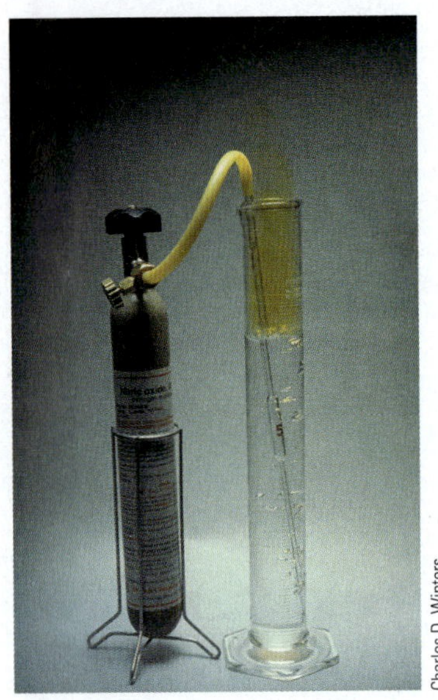
Charles D. Winters

Figure 6.7 Nitrogen monoxide and nitrogen dioxide. The gas cylinder contains colorless nitrogen monoxide, which can be seen as bubbles in the liquid. When the nitrogen monoxide in the bubbles contacts the oxygen in the air, a reaction occurs, producing red-brown nitrogen dioxide.

Two compounds are so common that they are always called by their traditional names rather than their chemical names. H_2O is always called *water* rather than dihydrogen oxide, and NH_3 is always called *ammonia* rather than nitrogen trihydride. These traditional names and formulas are important; memorize them.

6.5 Names and Formulas of Monatomic Ions: Group 1A/1 and 2A/2 Metals and the Nonmetals

Goal 5 Given the name or formula of an ion in Figure 6.8, write the other.

A neutral atom contains the same number of protons (positive charges) and electrons (negative charges). That makes the atom electrically neutral; there is no net charge. But an atom can gain or lose one or more electrons. When it does, the balance between positive and negative charges is upset. Thus, the particle acquires a net electrical charge that is equal to the number of electrons gained or lost. If electrons are lost, the charge on the particle is positive; if electrons are gained, the charge is negative.

The charged particle formed when an atom gains or loses electrons is an **ion**. If the ion has a positive charge, it is a **cation** (pronounced CAT-ion, not ca-SHUN). If the ion has a negative charge, it is an **anion** (AN-ion). An ion that is formed from a single (*mono-*) atom is a **monatomic ion**.

The rules for naming monatomic ions depend on where the element is located in the periodic table. The rules for naming ions formed from main group elements are different from the rules for most transition elements. We will first consider how to name ions formed from main group elements. ◀

▶ **P/Review** In Section 5.7, the elements in A groups (1, 2, and 13 to 18) of the periodic table were identified as *main group elements*, and the elements in B groups (3 to 12) were called *transition elements*.

The formation of a monatomic ion from an atom can be illustrated with the nuclear model of the atom. For example, a sodium atom forms a sodium ion by losing one electron.

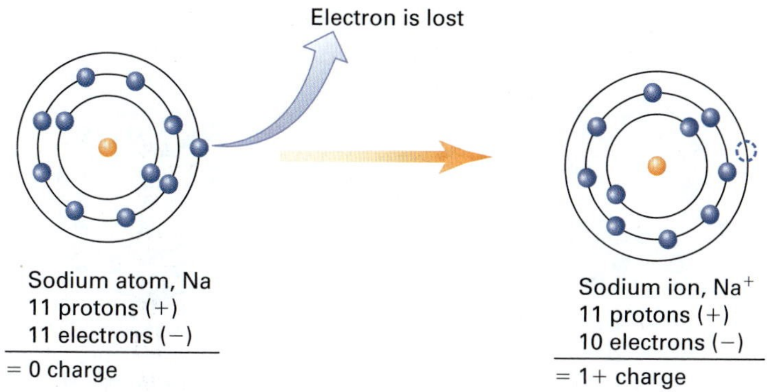

If we write this in the form of an equation, it is $Na \rightarrow Na^+ + e^-$. The illustration and equation lead to the procedure for writing the names and formulas of monatomic cations.

how to... Write the Name and Formula of a Monatomic Cation

1. The name of a monatomic cation is the name of the element followed by the word *ion*.
2. The formula of a monatomic cation is the elemental symbol followed by its electrical charge, written in superscript, number first, + sign second. If the charge is 1+, the number is omitted.

The periodic table is a guide to determining the electrical charge on many monatomic ions. *All* of the elements in Group 1A/1 form monatomic ions by losing one electron. This loss of a single electron in a chemical change is what gives the elements of Group 1A/1 their similar chemical properties. This concept also applies to other groups in the periodic table. Group 2A/2 elements form monatomic ions by losing two electrons per atom. ▶

Examples of Period 3 elements are as follows:

Group	Atom Symbol	Atom Name	Ion Symbol	Ion Name
1A/1	Na	Sodium	Na^+	Sodium ion
2A/2	Mg	Magnesium	Mg^{2+}	Magnesium ion

◀ **P/Review** The relationship between charges on monatomic ions and position in the periodic table is explained in Section 11.5. These charges arise when an atom loses or gains one, two, or three electrons to become isoelectronic with a noble-gas atom, that is, when the number and arrangement of electrons are the same as that found in a noble-gas (Group 8A/18) atom.

Note the differences between the atom symbols and names and the ion symbols and names, highlighted in blue. The formulas of the ions are the elemental symbol followed by the ionic charge, written in superscript. The names of the ions are the names of the atoms, followed by the word *ion*. Also note that the quantity of positive charge on a monatomic cation of a Group 1A/1 or Group 2A/2 element is the same as the group number (U.S. usage) or the last digit of the group number (IUPAC usage) of the element in the periodic table.

Let's now consider negatively charged monatomic ions. Again, we will illustrate this with the nuclear model of the atom. For example, a chlorine atom forms a chloride ion by gaining one electron.

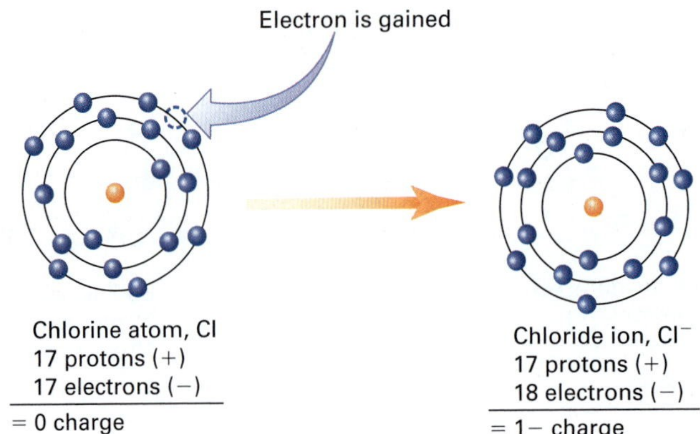

If we write this in the form of an equation, it is $Cl + e^- \rightarrow Cl^-$. The illustration and equation lead to the procedure for writing the names and formulas of monatomic anions.

how to... Write the Name and Formula of a Monatomic Anion

1. The name of a monatomic anion is the name of the element, changed to end in *-ide*, followed by the word *ion*.
2. The formula of a monatomic anion is the elemental symbol followed by its electrical charge, written in superscript, number first, − sign second. If the charge is 1− the number is omitted.

Again, the periodic table is a guide to determining the electrical charge on monatomic ions. Atoms of Group 7A/17 elements form ions by gaining one electron, resulting in a 1− charge, the nonmetal Group 6A/16 elements gain two electrons per atom, and Group 5A/15 nonmetals gain three electrons per atom.

Examples of Period 3 elements are as follows:

Group	Atom Symbol	Atom Name	Ion Symbol	Ion Name
7A/17	Cl	Chlorine	Cl^-	Chloride ion
6A/16	S	Sulfur	S^{2-}	Sulfide ion
5A/15	P	Phosphorus	P^{3-}	Phosphide ion

The formulas of the ions are the elemental symbol followed by the ionic charge, written in superscript, shown in blue. The names of the ions are the names of the atoms changed to end in *-ide*, followed by the word *ion*, also shown in blue.

Figure 6.8 shows the formulas of the monatomic main group ions in the positions of their parent atoms on the periodic table. Again notice the left-to-right 1+, 2+ pattern and the right-to-left 1−, 2−, 3− pattern. Also notice that the positively charged cations are all to the left of the red stairstep that separates the metals on the left from the nonmetals on the right. (The exception is hydrogen, which is a nonmetal.) The negatively charged anions are all on the nonmetal side of the line. This is one of the features that distinguish metals from nonmetals.

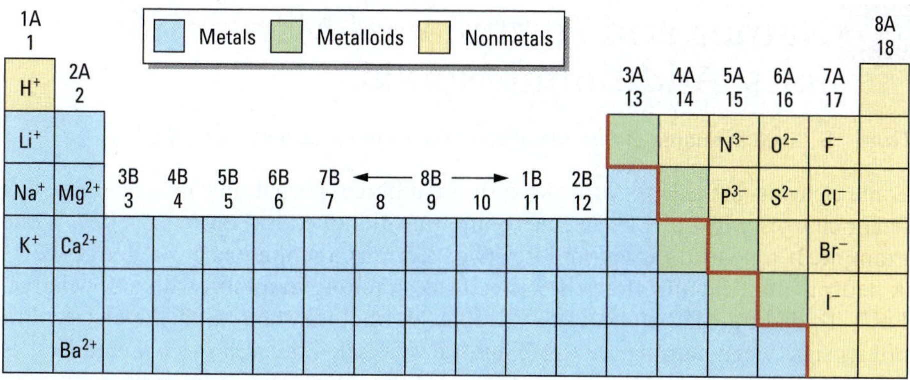

Figure 6.8 Partial periodic table with charges of ions of main group elements. The periodic table is a guide to the charge of many ions related to main group elements. The ions in Group 1A/1 have 1+ charges, and the ions in Group 2A/2 have 2+ charges. Reading from right to left, the ions in Group 7A/17 have 1− charges, those in Group 6A/16 have 2− charges, and those in 5A/15 have 3– charges. The position of the element in a group in the periodic table is a guide to the charge on the ions; you do not need to memorize these charges. You are not required to memorize the symbols of the elements of unfilled boxes in Groups 1A/1, 2A/2, and 6A/16, but when given the elemental symbol or atomic number of these elements, you can write the chemical formulas of their ions based on their positions in the periodic table.

Active Example 6.4 Monatomic Ions of Main Group Elements

Look only at a clean periodic table as you write (a) the names of Br^- and Ba^{2+} and (b) the formulas of the potassium and fluoride ions.

Think Before You Write If you understand how to write the name and formula of Group 1A/1 through Group 2A/2 cations and Group 5A/15 through Group 7A/17 anions, you are ready to complete the Active Example.

Answers *Cover the left column with your cut-out shield. Reveal each answer only after you have written your own answer in the right column.*

Br^- and fluoride ion are monatomic anions because they are formed from a single (*mono-*) atom and they have a negative charge (fluorine is a nonmetal, so it forms negatively charged ions). Ba^{2+} and potassium ion are monatomic (formed from a single atom) cations (positively charged; potassium is a metal).	What type of species (e.g., diatomic molecule) is Br^- and fluoride ion? What type of species is Ba^{2+} and potassium ion? Explain each answer. 
(a) Bromide ion and barium ion (b) K^+ and F^- *Notice that the formula of the fluoride ion is F^-, not F_2^-. Fluorine occurs as a diatomic molecule, F_2, when it is uncombined, but it is never a diatomic ion.*	Write the names and formulas. 
You improved your skill at writing the names and formulas of monatomic ions of main group elements.	What did you learn by solving this Active Example? 

Practice Exercise 6.4

For each name, write the formula; for each formula, write the name: iodide ion, calcium ion, phosphide ion, Li^+, S^{2-}, Al^{3+}.

Charles D. Winters

Figure 6.9 Many colored substances contain transition-element ions. The contents of all these paint pigments include ions of elements in the B groups of the periodic table.

6.6 Names and Formulas of Monatomic Ions: Additional Metals

Goal 6 Given the name or formula of an ion in Figure 6.10, write the other.

Some elements are able to form two or more different monatomic ions that have different charges. Most of these elements are transition metals (**Figure 6.9**). Iron is one example. If a neutral iron atom loses two electrons, the ion has a 2+ charge, Fe^{2+}. A neutral atom can also lose three electrons, resulting in an ion with a 3+ charge, Fe^{3+}. To distinguish between the two ions, we include the size of the charge, but not its sign, when naming the ion. Thus, Fe^{2+} is called the *iron two ion*, and Fe^{3+} is called the *iron three ion*. In writing, the ion charge appears in Roman numerals and is enclosed in parentheses: Fe^{2+} is the iron(II) ion, and $Fe^{3+}$is the iron(III) ion. Note that there is no space between the last letter of the elemental name and the opening parenthesis.

Active Example 6.5 Monatomic Ions of Elements with Variable Charges

Refer only to a periodic table, if necessary, and write (a) the names of Cr^{2+} and Cr^{3+} and (b) the formulas of copper(I) ion and copper(II) ion.

Think Before You Write If you understand how to use Roman numerals in naming transition element ions, you are ready to complete this Active Example.

Answers *Cover the left column with your cut-out shield. Reveal each answer only after you have written your own answer in the right column*

When the ions of an element exhibit more than one common charge, the size of the charge is included in the name in the form of Roman numerals in parentheses. When the ion of an element has just one common charge, the size of the charge is not included in the name.	How is the name of an ion of an element that commonly exists with different charges different from an ion of an element that has only one common charge? 
(a) Chromium(II) ion and chromium(III) ion (b) Cu^{+} and Cu^{2+}	Write the names and formulas. 
You improved your skill at writing the names and formulas of monatomic ions of elements with variable charges.	What did you learn by solving this Active Example?

Practice Exercise 6.5

For each name, write the formula; for each formula, write the name: Mn^{2+}, Mn^{3+}, nickel(II) ion, nickel(III) ion.

Notice that when we named ions formed by Groups 1A, 2A (1–2), and 5A–7A (15–17) nonmetal elements, the charge was *not* included in the name of the ion. Na^{+} was named "sodium ion." When we named Fe^{2+} and Fe^{3+}, the charge was included in the name to distinguish between the iron(II) ion and the iron(III) ion. This leads to an important point:

The charge is included in the name of an ion only when the ions of an element exhibit more than one common charge.

Charge is not indicated in the name of the sodium ion because all sodium ions have the same charge, 1+.

Metals | Metalloids | Nonmetals

1A 1	2A 2	3B 3	4B 4	5B 5	6B 6	7B 7	← 8B 8	9	→ 10	1B 11	2B 12	3A 13	4A 14	5A 15	6A 16	7A 17	8A 18
H^+																	
Li^+														N^{3-}	O^{2-}	F^-	
Na^+	Mg^{2+}											Al^{3+}		P^{3-}	S^{2-}	Cl^-	
K^+	Ca^{2+}				Cr^{2+} Cr^{3+}	Mn^{2+} Mn^{3+}	Fe^{2+} Fe^{3+}	Co^{2+} Co^{3+}	Ni^{2+} Ni^{3+}	Cu^+ Cu^{2+}	Zn^{2+}					Br^-	
										Ag^+			Sn^{2+} Sn^{4+}			I^-	
	Ba^{2+}										Hg_2^{2+} Hg^{2+}		Pb^{2+} Pb^{4+}				

Figure 6.10 Partial periodic table with charges of common monatomic ions. This figure includes the ions from Figure 6.8 in Section 6.5, plus it includes additional metal cations in Groups 6B through 4A (6 through 14). Some of these additional cations have variable charge, while others have only one common charge. Any of the elements from the set of 35 common elements with no ion shown here do not form ions.

Figure 6.10 shows all of the common ions from the elements that form ions among the set of 35 common elements first introduced in Section 5.8 (some elements do not form ions, such as Be, B, and C). In Section 6.5, you learned that Group 1A/1 ions have a charge of 1+, Group 2A/2 ions have a charge of 2+, Group 7A/17 ions have a charge of 1−, Group 6A/16 nonmetal ions have a charge of 2−, and Group 5A/15 nonmetal ions have a charge of 3−.

Let's now turn our attention to the ions of metal elements in Groups 6B through 4A (6 through 14). Figure 6.10 shows that most of these elements form ions with more than one common charge. The magnitude of the charges must therefore be included in the name of the ions. You do not need to memorize these charges; they are given in the name. For example, the name chromium(II) ion clearly indicates that the charge on the ion is 2+, Cr^{2+}.

Figure 6.10 shows that there are three common cations not in Groups 1A/1 or 2A/2 with only one charge. This means that their names do *not* include the magnitude of the charge. For example, Ag^+ is named "silver ion." If you are given the name of one of these three ions—silver ion, zinc ion, and aluminum ion—how do you know the magnitude of the charge? *You must memorize the charges of these three ions.* There are a couple of things that make this easier than it may look at first glance. First, notice that the three ions are on a diagonal. Their positions in the periodic table will serve as a memory aid. Second, Ag^+ is in Group 1B/11, with the number 1 beginning the U.S. group number and 1 ending the IUPAC group number, reminding you that silver has a 1+ charge. Zinc ion, Zn^{2+}, with its 2+ charge, is in Group 2B (U.S) and 12 (IUPAC). Similarly, aluminum ion, Al^{3+} which has a 3+ charge, is in Group 3A (U.S.) and 13 (IUPAC).

Martyn F. Chillmaid/Science Source

Figure 6.11 Water solutions that contain nickel(II) ion are green.

Active Example 6.6 Monatomic Ions of Groups 6B Through 4A (6 Through 14) Metals

(a) Write the names of Ag^+ and Ni^{2+} (Figure 6.11). (b) Write the formulas of zinc ion and manganese(II) ion. Refer only to a periodic table.

Think Before You Write Be sure that you have memorized the charges of the three metal ions in Groups 1B, 2B, and 3A (11–13) that have only one common charge.

Answers *Cover the left column with your cut-out shield. Reveal each answer only after you have written your own answer in the right column.*

Ag^+ and zinc ion have only one charge under common conditions. The names of these ions do not include the size of the charge. Ni^{2+} and manganese(II) are like most transition metal ions in that they are one of multiple ions of the same element, and the charge must be included in their names.	What is different about Ag^+ and zinc ion in comparison with Ni^+ and manganese(II) ion?
(a) Silver ion and nickel(II) ion (b) Zn^{2+} and Mn^{2+}	Complete the Active Example.
You improved your skill at writing the names and formulas of monatomic ions of Groups 6B through 4A (6 through 14) metals.	What did you learn by solving this Active Example?

Practice Exercise 6.6

For each name, write the formula; for each formula, write the name: Co^{3+}, Zn^{2+}, aluminum ion, chromium(III) ion.

The Figure 6.10 entry for mercury deserves special comment. Mercury forms a monatomic ion, Hg^{2+}, the mercury(II) ion. This monatomic ion follows the pattern we have seen before. However, mercury also forms a *diatomic ion.* In $Hg_2{}^{2+}$ the 2+ charge is shared by two mercury atoms, just as if each atom were contributing 1+ to the total charge of the ion. Accordingly, its name is mercury(I) ion. (Hg^+ does not exist under normal circumstances.) The mercury(I) ion is the only case of a diatomic elemental ion that we will consider in this introductory course.

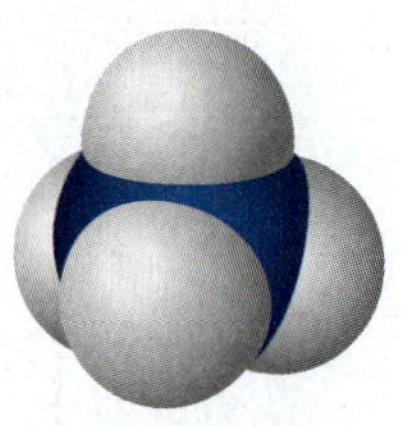

Figure 6.12 The ammonium ion, $NH_4{}^+$ consists of four hydrogen atoms (white) attached to one nitrogen atom (blue). The particle has 11 protons, 7 in the nucleus of the nitrogen atom and 1 each in the nuclei of the hydrogen atoms, but it possesses just 10 electrons, giving it a 1+ charge.

6.7 Formulas of Ionic Compounds

Goal 7 Given the formula (or name) of the ammonium ion or hydroxide ion, write the corresponding name (or formula).

8 Given the name of any ionic compound made up of identifiable ions, or other ions whose formulas are given, write the formula of that compound.

Not all ions are monatomic. In fact, many common compounds include **polyatomic ions** (*poly-* means many). In future sections we will look at a systematic method for naming polyatomic ions. For now, we will introduce just two common polyatomic ions.

The ammonium ion, $NH_4{}^+$, is a positively charged polyatomic ion (**Figure 6.12**). The nitrogen atom is chemically bonded to the four hydrogen atoms, and this collection of five atoms has one fewer electron than the total number of protons. The ammonium ion is an important part of the nitrogen cycle, the process by which atmospheric nitrogen becomes nitrogen compounds in plants.

Figure 6.13 The hydroxide ion, OH^-, is composed of an oxygen atom (red) attached to a hydrogen atom (white). The particle has 9 protons—8 in the nucleus of the oxygen atom and 1 in the nucleus of the hydrogen atom—and 10 electrons, resulting in an overall net 1− charge.

The hydroxide ion, OH^-, is a negatively charged diatomic ion (**Figure 6.13**). The oxygen atom and the hydrogen atom are chemically bonded to one another, and the particle formed by the combination of two atoms has one more electron than the total number of protons, giving it a negative charge. Hydroxide ions are found in all water solutions and in pure water, so all living organisms possess huge quantities of hydroxide ions, and they are also present in the water solutions found on almost all parts of the surface of the Earth.

Now that you've learned the formulas of three dozen monatomic ions and two polyatomic ions, you are ready to learn how to write the formulas of **ionic compounds** (**Figure 6.14**). Ions are charged particles, and collections of particles that have only like charges are never found in nature. For example, you will not find nor can you create a water solution of just ammonium ions. The solution must be electrically neutral. An equal number of 1− chloride ions, for example, must exactly balance the

charge of the 1+ ammonium ions. If we were to evaporate the water from the solution of ammonium ions and chloride ions, a white solid would remain, consisting of equal numbers of ammonium ions and chloride ions. Not coincidentally, that compound is called ammonium chloride. Furthermore, it is the interaction between the positively and negatively charged ions that causes an ionic compound to be stable.

The formula of an ionic compound expresses the ratio of positive to negative ions in the substance. Since there are no discrete molecules, the formula is the lowest whole-number ratio of ions in the compound. This will necessarily be the smallest electrically neutral collection of ions, which is called a **formula unit**. The total positive charge in a formula unit is equal to the total negative charge. Understanding this is the key to writing formulas of ionic compounds.

Figure 6.14 Ionic compounds are hard and brittle solids at room temperature. This photograph shows the ionic compounds (clockwise from the rear center) sodium chloride, calcium fluoride, iron(III) oxide, and copper(II) bromide.

how to... Write the Formula of an Ionic Compound

1. Write the formula of the cation and the formula of the anion.
2. Mentally balance the charges. Decide what the fewest number of ions is that will make the compound electrically neutral.
3. Write the formula, cation first, anion second, each with subscripts as needed for charge balance.
   a) If only one ion is needed, omit the subscript.
   b) If a polyatomic ion is needed more than once, enclose the formula of the ion in parentheses and place the subscript after the closing parenthesis.

Active Example 6.7 Formulas of Ionic Compounds I

Write the formulas of calcium chloride and calcium hydroxide. (A method for identification of the calcium ion and other Group 2A/2 cations is shown in Figure 6.15.)

Think Before You Write Review the *how to...* box that describes how to write the formula of an ionic compound if you get stuck on any step in this Active Example.

Answers *Cover the left column with your cut-out shield. Reveal each answer only after you have written your own answer in the right column.*

Ca^{2+} Cl^- OH^-	Step 1 is to write the formulas of the individual ions. Write the formulas of the three ions in the two compounds. 
1 Ca^{2+} ion + 2 Cl^- ions; 1 Ca^{2+} ion + 2 OH^- ions *$(2+) + 2 \times (1-) = 0$ for both compounds.*	Step 2 is to balance the charges. Decide how many Ca^{2+} ions must combine with how many Cl^- ions to produce a total charge of zero. Then do the same for Ca^{2+} and OH^- ions. 
Calcium chloride, $CaCl_2$ Calcium hydroxide, $Ca(OH)_2$	Step 3 is to write the formulas of the compounds. Be careful about how and where you use parentheses.

Figure 6.15 Cations can be identified by the characteristic color of their flames. Calcium chloride, strontium chloride, and barium chloride were dissolved in methanol (a colorless flammable liquid), and the solutions were ignited. Calcium ion gives a yellow-orange flame, strontium ion gives a red flame, and the barium ion flame is yellow-green. (*Caution:* This demonstration should be performed only in a fume hood.)

You improved your skill at writing the formulas of ionic compounds.	What did you learn by solving this Active Example? 

Practice Exercise 6.7

Write the formulas of chromium(III) fluoride and sodium sulfide.

Active Example 6.7 illustrates the following two important points:

1. *Parentheses are used in a chemical formula to enclose a polyatomic ion that is used more than once.* That is the case with calcium hydroxide. Two OH^- ions are needed to combine with the one Ca^{2+} ion: $Ca^{2+} + OH^- + OH^-$. There are two tempting but incorrect ways to write the formulas of calcium hydroxide. First, writing CaO_2H_2 gives a correct atom count in the formula, 1 Ca, 2 O, and 2 H, but the identity of the hydroxide ion, OH^-, has been lost. Second, writing $CaOH_2$ gives an atom count of 1 Ca, 1 O, and 2 H, which is not correct.
2. *Parentheses are used only with* polyatomic *ions.* Notice that calcium chloride is $CaCl_2$, not $Ca(Cl)_2$. Even though its symbol has two letters, Cl^- is a monatomic ion.

Active Example 6.8 Formulas of Ionic Compounds II

Write the formulas of ammonium bromide and ammonium sulfide.

Think Before You Write Keep in mind that parentheses are used in a chemical formula to enclose a polyatomic ion that is used more than once and parentheses are used only with polyatomic ions.

Answers *Cover the left column with your cut-out shield. Reveal each answer only after you have written your own answer in the right column.*

NH_4^+ Br^- S^{2-}	Complete Step 1: Write the formulas of the three ions. 
Ammonium bromide: 1 NH_4^+ ion + 1 Br^- ion *(1+) + (1−) = 0* Ammonium sulfide: 2 NH_4^+ ions + 1 S^{2-} ion *2 × (1+) + (2−) = 0*	Now work on Step 2: What will be the numbers of cations and anions in each formula? 
Ammonium bromide, NH_4Br Ammonium sulfide, $(NH_4)_2S$ *With ammonium sulfide, the subscript 4 is part of the ammonium ion formula, so it goes inside the parentheses. The two ammonium ions are indicated by the subscript 2 after the closing parenthesis.*	Complete the exercise by writing the formulas, Step 3. Be thoughtful about how you use parentheses. 
You improved your skill at writing the formulas of ionic compounds.	What did you learn by solving this Active Example?

Practice Exercise 6.8

Write the formulas of ammonium oxide and zinc hydroxide.

When one ion has a charge of 3 and the other has a charge of 2, the lowest common multiple is 6, so to obtain balance you must have two ions with a charge of 3 and three ions with a charge of 2. The next Active Example illustrates this.

Active Example 6.9 Formulas of Ionic Compounds III

Write the formulas of aluminum oxide (Figures 6.16 and 6.17) and barium nitride.

Think Before You Write The smallest whole number ratio of ions that forms a neutral compound is used in writing formulas of ionic compounds.

Answers *Cover the left column with your cut-out shield. Reveal each answer only after you have written your own answer in the right column.*

Al_2O_3 *The ions are Al^{3+} and O^{2-}. Two Al^{3+} ions and three O^{2-} ions give a charge balance of $2 \times (3+) + 3 \times (2-) = 0$ with the least number of each ion.* *If you wrote Al_4O_6, you have the charge balanced, but you did not do it with the fewest number of ions possible. Both 4 and 6 are divisible by 2 to give Al_2O_3.*	Start with aluminum oxide. Work through all three steps that lead to the formula of the compound. 
Ba_3N_2 *Ba^{2+} and N^{3-} are the ions. Three Ba^{2+} + two N^{3-} give a charge balance of $3 \times (2+) + 2 \times (3-) = 0$.*	Now write the formula of barium nitride. 
You improved your skill at writing the formulas of ionic compounds.	What did you learn by solving this Active Example? 

Practice Exercise 6.9

Write the formulas of potassium nitride and cobalt(III) oxide.

Charles D. Winters

Figure 6.16 Corundum (*left*), ruby (*top right*), and sapphire (*bottom right*) are mostly aluminum oxide. Small quantities of transition metal ions give the minerals their colors.

John Luke/Photolibrary/Getty Images

Figure 6.17 Even though airplanes are frequently exposed to rain and snow, they never rust. A thin layer of aluminum oxide forms on the surface of the aluminum metal used for the exterior skin of the aircraft. The aluminum oxide serves as a chemically inert, protective, rust-resistant shield.

6.8 Names of Ionic Compounds

Goal 9 Given the formula of an ionic compound made up of identifiable ions, write the name of the compound.

If you recognize the names of each the two ions in the formula of an ionic compound, you have the name of the ionic compound. Note that *prefixes are not included* in the names of ionic compounds.

how to... **Write the Name of an Ionic Compound**

1. Write the name of the cation.
2. Write the name of the anion.

In writing or speaking the name of an ionic compound containing a metal that commonly is *capable of* having more than one ionic charge, *the compound name includes the charge of that metal.* For example, what is the name of $FeCl_3$? Iron chloride is not an adequate answer. It fails to distinguish between the two possible charges on the iron ion. Is $FeCl_3$ iron(II) chloride or iron(III) chloride? To decide, you must reason from the known charge on the chloride ion, 1−, and the fact that the total charge on the compound is zero. The formula has three chloride ions, so the total negative charge is 3−. This must be balanced by three positive charges from the iron ion, so it must be an iron(III) ion. The compound is iron(III) chloride. If the formula had been $FeCl_2$, you would have reached the name iron(II) chloride by recognizing that the 2− of the two chloride ions is balanced by the 2+ of a single iron(II) ion.

Active Example 6.10 Names of Ionic Compounds

Write the name of each of the ionic compounds listed in the box below.

Think Before You Write Review the *how to...* box about how to write the name of an ionic compound, if necessary.

Answers *Cover the left column with your cut-out shield. Reveal each answer only after you have written your own answer in the right column.*

LiBr is composed of a metal and a nonmetal, and therefore, it is an ionic compound. It is named as an ionic compound: name of cation followed by name of anion, lithium bromide. Prefixes are not used in naming ionic compounds.

NO is composed of two nonmetal atoms, and therefore, it is a molecular compound; more specifically, it is a binary molecular compound. It is named as a binary molecular compound: prefix–name of first element followed by prefix–name of second element changed to end in -ide, mononitrogen monoxide. Following the practice of omitting *mono-* in the first word, we have nitrogen monoxide.

Before you start writing the names of the compounds, take a moment to recognize how to identify an ionic compound as ionic. The first formula listed below is LiBr. Contrast it with another two-atom compound, NO. What critical difference is there between the two compounds that you must recognize in order to name them correctly? Name each compound.

$LiBr$, lithium bromide

MgI_2, magnesium iodide

Ag_3N, silver nitride

$MnCl_2$, manganese(II) chloride

Hg_2Br_2, mercury(I) bromide

In $MnCl_2$, two 1− charges from two Cl^- ions require 2+ from the manganese ion, so it is the manganese(II) ion.

In Hg_2Br_2, the two 1− charges from two Br^- ions are balanced by the 2+ charge from the diatomic mercury(I) ion.

Write each name.

$LiBr$ ______________________

MgI_2 ______________________

Ag_3N ______________________

$MnCl_2$ ______________________

Hg_2Br_2 ______________________

You improved your skill at writing the names of ionic compounds.

What did you learn by solving this Active Example?

Practice Exercise 6.10

Write the name of each of the following: CuI, Cr_2O_3, $Fe(OH)_2$, $SnBr_4$, $BaCl_2$

Everyday Chemistry 6.1

COMMON NAMES OF CHEMICALS

Although chemists agree that the system of nomenclature presented in this chapter gives the official names of compounds, many practicing scientists still use *un*official, or *trivial*, names when talking about many substances. Just as there are variations in the English language that depend on whether you are in Australia, Great Britain, Canada, or the United States, or even in a region within the United States, there are variations in the language of chemistry that depend on an individual's specialization within chemistry, or on whether the person is a biologist, another type of scientist, or a nonscientist.

Alcohol is an interesting example of a term that has many meanings. In everyday language, alcohol refers to what is officially called *ethanol*, CH_3CH_2OH. However, many chemists use the similar term *ethyl alcohol* for this compound. Additionally, a chemist would *never* refer to this substance simply as alcohol when discussing chemistry. In chemistry, the term *alcohol* refers to an entire class of compounds that have a certain carbon–oxygen–hydrogen arrangement within a molecule. Prohibition-era bootleggers were careful to distinguish between the very poisonous but legal *wood alcohol* (*methanol* or *methyl alcohol* to a chemist) (**Figure 6.18**) and *grain alcohol*, which is ethanol. Wood alcohol can be produced from wood smoke, and grain alcohol can be made by fermenting grain, and thus they are aptly named. Yet another common alcohol, *rubbing alcohol*, is used to cool the skin rapidly. A chemist would probably call this substance *isopropyl alcohol*, although its official name is *isopropanol*.

Sugar is similar to *alcohol* in that it refers to a specific substance in everyday language but to an entire class of compounds in scientific language. Everyday usage of the term *sugar* refers to the substance *sucrose*, $C_{12}H_{22}O_{11}$, which has the official name *α-D-glucopyranosyl-(1 → 2)-β-D-fructofuranoside*. We doubt that you'll be surprised to learn that most chemists avoid the official name! Ordinary table sugar is also called *cane sugar* (when it is processed from sugar cane) or *beet sugar* (when it comes from sugar beets). Whatever its origin, pure table sugar is pure sucrose. To a chemist, the term *sugar* refers to a class of compounds that have the general formula $C_nH_{2n}O_n$, where n varies from three to nine, or $C_{12}H_{22}O_{11}$. There are literally dozens of common sugars, such as lactose (milk sugar) (**Figure 6.19**), glucose (blood sugar), fructose (fruit sugar), and invert sugar (a mixture of glucose and fructose) (**Figure 6.20**).

People joke about alternative names for *water*, which in itself appears to be a trivial name, but because it is so entrenched in our language, it is also the official chemical name for H_2O. Since it is a binary molecular compound, it could be called *dihydrogen monoxide*. Using the letter abbreviations for the prefixes and stem words, dihydrogen monoxide can be abbreviated DHMO. You may have seen some of the many "warnings" about DHMO circulated by email or at websites on the Internet. These are just poking fun at those who are unaware of this way of (improperly) naming water.

Many of these trivial names have been used for a century or more, and many will continue to be used well into the future in spite of attempts to standardize the system for naming chemical compounds. **Table 6.2** has many more common names of chemicals.

Figure 6.18 Methanol, also known as methyl alcohol or wood alcohol, is used as a chemical building block in the manufacture of wood paneling, paints, adhesives, and fuels. It is poisonous when consumed by humans.

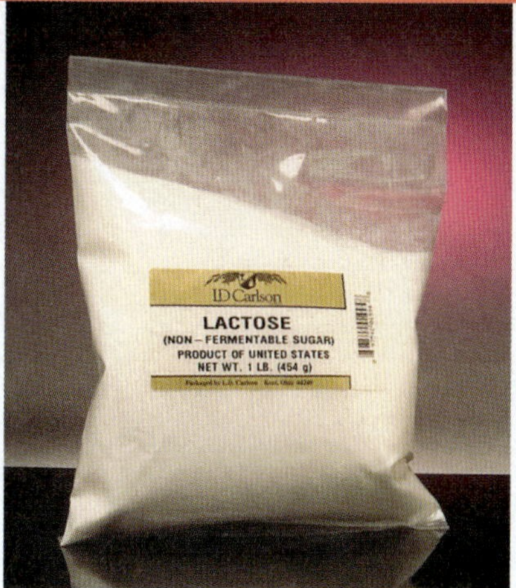

Figure 6.19 Lactose is the predominant sugar in milk.

Figure 6.20 Invert sugar is produced by the "inversion" (chemical breakdown) of table sugar, resulting in a mixture of glucose and fructose.

Table 6.2 Common Names of Chemicals		
Common Name	**Chemical Name**	**Formula**
Alumina	Aluminum oxide	Al_2O_3
Baking soda	Sodium hydrogen carbonate	$NaHCO_3$
Bleach (liquid)	Hydrogen peroxide or sodium hypochlorite	H_2O_2 $NaClO$
Bleach (solid)	Sodium perborate	$NaBO_2 \cdot H_2O_2 \cdot 3\ H_2O$
Bluestone	Copper(II) sulfate pentahydrate	$CuSO_4 \cdot 5\ H_2O$
Borax	Sodium tetraborate decahydrate	$Na_2B_4O_7 \cdot 10\ H_2O$
Brimstone	Sulfur	S
Carbon tetrachloride	Tetrachloromethane	CCl_4
Chile saltpeter	Sodium nitrate	$NaNO_3$
Chloroform	Trichloromethane	$CHCl_3$
Cream of tartar	Potassium hydrogen tartrate	$KHC_4H_4O_6$
Diamond	Carbon	C
Dolomite	Calcium magnesium carbonate	$CaCO_3 \cdot MgCO_3$
Epsom salt	Magnesium sulfate heptahydrate	$MgSO_4 \cdot 7\ H_2O$
Freon (refrigerant)	Dichlorodifluoromethane	CCl_2F_2
Galena	Lead(II) sulfide	PbS
Grain alcohol	Ethyl alcohol; ethanol	C_2H_5OH
Graphite	Carbon	C
Gypsum (**Figure 6.21**)	Calcium sulfate dihydrate	$CaSO_4 \cdot 2\ H_2O$
Hypo	Sodium thiosulfate	$Na_2S_2O_3$
Laughing gas	Dinitrogen monoxide	N_2O
Lime	Calcium oxide	CaO
Limestone	Calcium carbonate	$CaCO_3$
Lye	Sodium hydroxide	$NaOH$
Marble	Calcium carbonate	$CaCO_3$
MEK	Methyl ethyl ketone	$CH_3COC_2H_5$
Milk of magnesia	Magnesium hydroxide	$Mg(OH)_2$
Muriatic acid	Hydrochloric acid	HCl
Oil of vitriol	Sulfuric acid (conc.)	H_2SO_4
Plaster of Paris	Calcium sulfate 1/2-hydrate	$CaSO_4 \cdot 1/2\ H_2O$
Potash	Potassium carbonate	K_2CO_3
Pyrite (fool's gold)	Iron(II) disulfide	FeS_2 ◀
Quartz	Silicon dioxide	SiO_2
Quicksilver	Mercury	Hg
Rubbing alcohol	Isopropyl alcohol; isopropanol	$(CH_3)_2CHOH$
Sal ammoniac	Ammonium chloride	NH_4Cl
Salt	Sodium chloride	$NaCl$
Salt substitute	Potassium chloride	KCl
Saltpeter	Potassium nitrate	KNO_3
Slaked lime	Calcium hydroxide	$Ca(OH)_2$

Charles D. Winters

Figure 6.21 Gypsum is the common name for $CaSO_4 \cdot 2\ H_2O$, which has the chemical name calcium sulfate dihydrate.

The disulfide ion, S_2^{2-}, is chemically similar to the peroxide ion, O_2^{2-}.

Table 6.2 Common Names of Chemicals (*continued*)

Common Name	Chemical Name	Formula
Sugar	Sucrose	$C_{12}H_{22}O_{11}$
TSP (trisodium phosphate)	Sodium phosphate	Na_3PO_4
Washing soda	Sodium carbonate decahydrate	$Na_2CO_3 \cdot 10\ H_2O$
Wood alcohol	Methyl alcohol; methanol	CH_3OH

Quick Quiz

1. List two examples of names that usually refer to a specific substance in everyday language and to a class of substances in scientific language. For each, give two examples of everyday names of variants within the class of substances.
2. Some of the hazards of DHMO include the following: (a) can cause an electrical system to short circuit, (b) is the major component of acid rain, and (c) can be deadly if inhaled. Should the U.S. Food and Drug Administration ban DHMO? Explain.

6.9 The Nomenclature of Oxoacids

Goal 10 Given the name (or formula) of an acid in Table 6.3, write its formula (or name).

You were probably somewhat familiar with a few acids from your everyday experiences, even before you began this chemistry course. When you were young, you may have experienced heartburn, and your parent may have told you that you have too much acid in your digestive system, so you were given an antacid tablet (*anti* plus *acid*). You may have learned that the source of the sharp or tangy taste in cola is the presence of an acid. Phosphoric acid is added to colas to give them that aspect of their taste.

In this section, you will learn the names and formulas of 27 acids. This will be accomplished by memorizing just 5 acid names and formulas, and then you will learn a systematic method for deriving the names of the other 22 acids. Furthermore, you will learn the systematic method for deriving the names and formulas of ions that are related to the 27 acids, and then you will see how those ions can occur in ionic compounds.

Table 6.6 shows all of the acids and ions you will learn to name in this section and the next. Page or scroll forward to take a quick look at it. The key to learning the names and formulas in Table 6.6 is to follow our three-step process: (1) memorize just five acid names and formulas, (2) understand how chemical family relationships are related to names and formulas in groups in the periodic table, and (3) learn the system for changing names as the number of oxygen atoms in the acid molecule varies. If you follow this learning procedure in sequence, one step at a time, you will soon be able to write the names and formulas of literally hundreds of compounds!

Acids

Early humans discovered that a number of solutions exhibited similar properties. For example, vinegar, lemon juice, and aged milk all have a sour taste. Even though these substances are derived from distinctly different sources (vinegar can be made from many different fruits), they have a common property. Thus, it is logical to classify them into a common category. In this case, the modern term *acid* is based on a Latin word that literally means "sour."

As chemists began to experiment with acids, they discovered many more properties shared by solutions classified within this category of substances. For example, when an acid solution is poured on limestone, a common mineral found in rocks across the world, the liquid in contact with the rock will bubble because a gas is being produced. When chemists tested the gas, they discovered that it was carbon dioxide, which is the one of the gases found in the mixture of gases released when humans exhale.

Chemists wondered what *caused* the properties of acids. In 1884, Swedish scientist Svante Arrhenius published the first recorded particulate-level model that explained acid properties. He proposed that acids were molecules that have a proton that can be stripped away from the acid molecule in the presence of water. Let's consider an example of Arrhenius's hypothesis. Hydrogen chloride is a gas at room conditions. Each hydrogen chloride molecule consists of a hydrogen atom bound to a chlorine atom. When hydrogen chloride gas is bubbled through a water solution, water molecules remove the proton from the hydrogen atom in the hydrogen chloride molecules.

A hydrogen atom consists of a proton and an electron. We can think of it as H^+, a hydrogen nucleus (1 proton) with no electron, plus an electron: H atom = proton + electron = $H^+ + e^-$.

One product of the reaction of hydrogen chloride and water is a water molecule with the proton it removed from the acid molecule: $H_2O + H^+ = H_3O^+$. The H_3O^+ ion is called the **hydronium ion**. The other product is a chlorine atom with the electron that was left behind: $Cl + e^- = Cl^-$, a chloride ion.

Summarizing the reaction of hydrogen chloride and water using symbols and drawings, we have the following:

$$HCl + H_2O \rightarrow Cl^- + H_3O^+$$

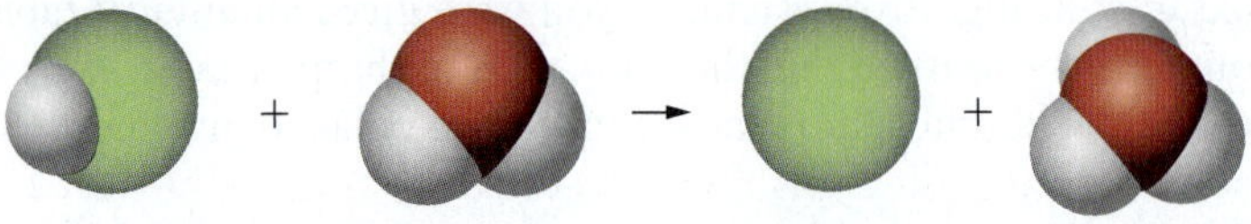

Now consider another example. Dihydrogen sulfide is the gas responsible for the odor of rotten eggs. When it is dissolved in water, water molecules remove a proton from the dihydrogen sulfide molecules.

$$H_2S + H_2O \rightarrow HS^- + H_3O^+$$

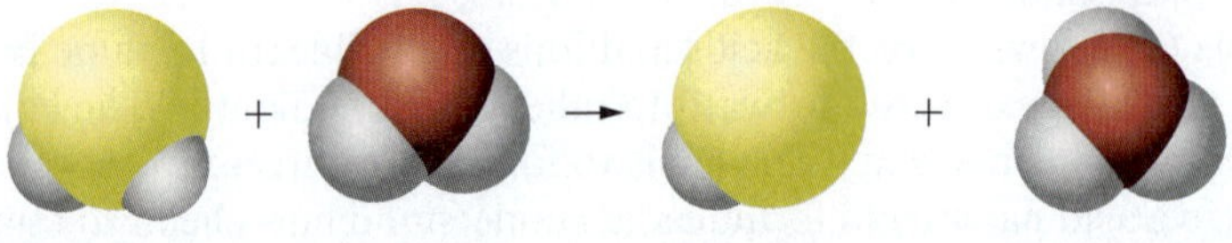

Look at the product in common in the two reactions. It is H_3O^+, the hydronium ion. If we were to look at more examples of substances that exhibit the macroscopic properties of acids, in each case we would find that the hydronium ion is a product of the reaction of the acid molecule with water. The hydronium ion is formed by the reaction of a proton in the acid molecule with a water molecule. Thus, we can conclude that an **acid** is a substance that has a proton—an H^+ ion—that can be removed by a water molecule when in a water solution.

Many substances are not acids, but their molecules include hydrogen atoms. Methane, CH_4, the primary component of the mixture that is known as natural gas, is an example. When methane is bubbled through water, some will dissolve, but the methane molecules do not release a proton to the water molecules. To distinguish between hydrogen-atom-containing molecules that are *not* acidic and those that *are* acidic, chemists use the convention of writing the hydrogen atoms first in the formulas of acids and later in the formulas of substances that are not acids. Thus, HCl and H_2S are acids, CH_4 is not an acid, and $HC_2H_3O_2$ (acetic acid, the acid in vinegar) has one acidic proton that can be removed by water molecules and three nonacidic protons that cannot.

Our purpose in this section is to familiarize you with acids to the point where you will be able to write names and formulas of acids and their corresponding anions. We have now provided sufficient information to accomplish that goal. You will learn more about acids as you continue your study of chemistry.

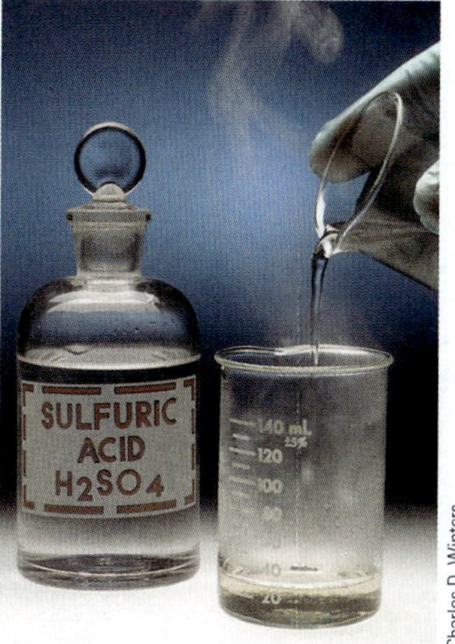

Charles D. Winters

Figure 6.22 Sulfuric acid, H_2SO_4. Sulfuric acid is one of the five *-ic* acids whose name and formula must be memorized. A great deal of heat is evolved when sulfuric acid is added to water. Acids are usually purchased in concentrated form and then diluted for laboratory use. An acid is always diluted by adding it to water to prevent splattering. Never add water to a concentrated acid!

Five Acid Names and Formulas to Be Memorized

Our goal in this nomenclature chapter is to maximize the number of names and formulas you can write with the minimum possible amount of memorization. In learning the nomenclature of acids and their corresponding anions, that minimum quantity of memorization involves learning five acid names and formulas. You will soon see that two of these five acids stand alone in terms of leading to other acid names, but the other three will be the basis of knowing the names and formulas of dozens of other acids and ions without the need for additional memorization.

Memorize the five acid names and formulas in **Table 6.3**. Learn it *now*, before moving on in this section. You need to learn these "cold"—that is, without using any memory aids or tricks. For example, when you see the words "carbonic acid," you need to be able to immediately think "H_2CO_3"; when you see "H_2SO_4," you need to think "sulfuric acid" (**Figure 6.22**); and when you see "HNO_3," you need to have "nitric acid" (**Figure 6.23**) at your immediate disposal in your working memory.

Learn It Now! *Memorize the five acid names and formulas in Table 6.3 now. Don't procrastinate. Before you continue studying this section, be sure that you can instantly recall the formula when given the name of one of the acids and vice versa. This small quantity of memorization now will serve as the foundation for the ability to write hundreds of names and formulas with no additional memorization in the near future.*

Look at the patterns in the formulas of the five acids in Table 6.3. Hydrogen is the first elemental symbol in each formula, conforming to the convention of writing H first in acid formulas. Each formula has three different elements: a variable number of hydrogen atoms, a single nonmetal atom, and a variable number of oxygen atoms. An acid with at least one hydrogen atom in its molecule that can be removed by water molecules, plus oxygen and at least one other type of atom, is called an **oxoacid** (some chemists use *oxyacid* as a synonym).

Now look at the patterns in the names of the five memorized acids. Note that the beginning part of the name identifies the single non-oxygen nonmetal atom (e.g., carbon—for carbon, nitr—for nitr-ogen) and the end of the name is always *-ic*. We will take advantage of these patterns in the system for developing other names and formulas from the five memorized acids.

Table 6.3 Five Acid Names and Formulas

Periodic Table Group Number of Central Nonmetal Atom	Acid Name	Acid Formula
4A/14	Carbonic acid	H_2CO_3
5A/15	Nitric acid	HNO_3
5A/15	Phosphoric acid	H_3PO_4
6A/16	Sulfuric acid	H_2SO_4
7A/17	Chloric acid	$HClO_3$

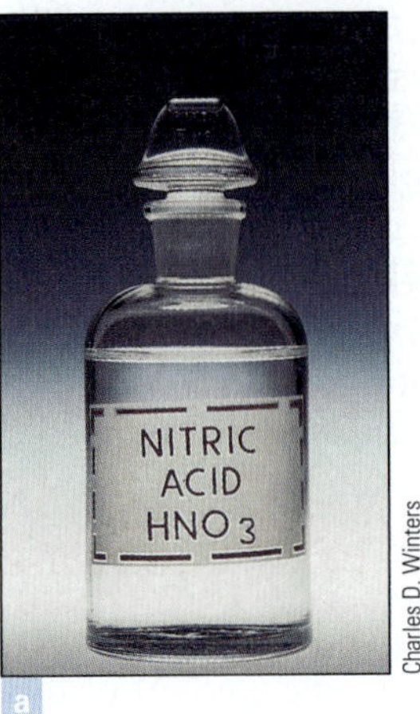

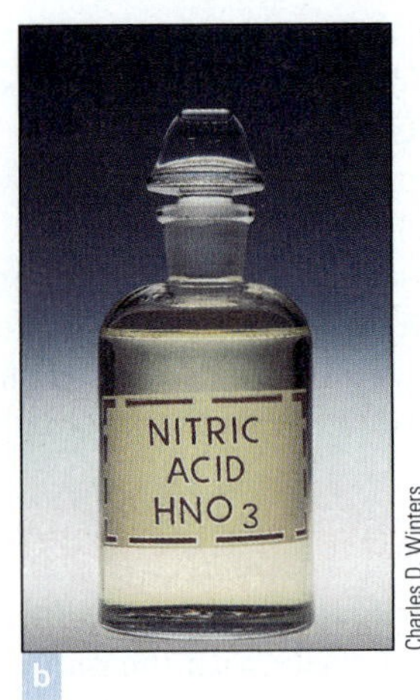

Figure 6.23 Nitric acid, HNO_3. (a) Nitric acid solution, like most acid solutions, is colorless. (b) When exposed to sunlight, the solution turns yellow because nitric acid decomposes, forming nitrogen dioxide as one product, which colors the solution.

Family Relationships in Acid Names and Formulas

In Section 5.7, we introduced the concept of **chemical families**, which are groups of elements with similar properties analogous to human family relationships. Members of a chemical family are readily identifiable because they are positioned in a vertical sequence in a group in the periodic table. The members of the **halogen family** are found in Group 7A/17: fluorine, chlorine, bromine, iodine, and astatine (Z = 85). The similar properties of these elements lead to similar formulas for the oxoacids of Group 7A/17. For example, since the (memorized) formula of chloric acid is $HClO_3$, the formula of bromic acid has the same number of hydrogen atoms and oxygen atoms: $HBrO_3$. Their names are also based on the same rule: the beginning part of the name identifies the single non-oxygen nonmetal atom. If $HClO_3$ is *chlor*-ic acid, $HBrO_3$ is *brom*-ic acid.

Two other chemical family sequences are part of our nomenclature system. In Group 6A/16, sulfur, selenium (Z = 34), and tellurium (Z = 52) form a family. Sulfuric acid, H_2SO_4, is the model for naming acids of selenium and tellurium. In Group 5A/15, phosphorus and arsenic (Z = 33) have similar chemical properties, and thus phosphoric acid, H_3PO_4, is the model for naming acids of arsenic. Active Example 6.11 provides you with an opportunity to test your understanding of family relationships in acid names and formulas.

Active Example 6.11 Family Relationships in Acid Names and Formulas

For each given name, write the formula; for each formula, write the name.

Think Before You Write Be sure you (a) have memorized the five acid names and formulas and (b) understand how family relationships in the periodic table are related to nomenclature.

Answers *Cover the left column with your cut-out shield. Reveal each answer only after you have written your own answer in the right column.*

Telluric acid, H_2TeO_4 HIO_3, iodic acid Arsenic acid, H_3AsO_4 H_2SeO_4, selenic acid	Write each formula or name. 
For the given telluric acid, you use the periodic table to locate Te in Group 6A/16, and then you use the memorized model for that group, sulfuric acid, H_2SO_4, as the basis for the formula, substituting Te for S in the formula.	Telluric acid (tellurium, Z = 52) ________ HIO_3 ________

HIO_3, with I in its formula, follows the model set by the memorized $HClO_3$, chloric acid, for Group 7A/17 elements. Chlor-ic acid changes to iod-ic acid.

For arsenic acid, you find As in Group 5A/15, so it is based on the model set by the memorized phosphoric acid, H_3PO_4, yielding H_3AsO_4 as the formula.

With H_2SeO_4, you find Se in Group 6A/16, so you follow the pattern of the memorized H_2SO_4, sulfur-ic acid, yielding selen-ic acid.

Arsenic acid (arsenic, Z = 33) ____________________

H_2SeO_4 (Se, selenium) ____________________

You improved your understanding of family relationships in acid names and formulas.

What did you learn by solving this Active Example?

Practice Exercise 6.11

a) Given that the name of H_2SO_3 is sulfurous acid, what is the name of H_2TeO_3 (Te, tellurium)?

b) Given that the formula of perchloric acid is $HClO_4$, what is the formula of perbromic acid?

The Prefix–Suffix System for Acid Nomenclature

We now consider the nomenclature of acids, as the number of oxygens varies in relation to the memorized *-ic* acid. Five chlorine-based acids will serve as our introductory example. **Table 6.4** illustrates a system of nomenclature that is based on the *-ic* acid, the name and formula of which you have memorized. From there on, it is a system of beginnings (prefixes) and endings (suffixes) built on the number of oxygens in the *-ic* acid.

The prefix–suffix system for acid nomenclature is summarized in both Table 6.4 and the following box:

a summary of... **The Prefix–Suffix System for Acid Nomenclature**

1. The system starts with the memorized number of oxygens in any of the five *-ic* acids. Example: $HClO_3$ is chlor**ic** acid.

From that starting point...

2. If the number of oxygens is **one more** than the number in the *-ic* acid, the prefix *per-* is added to the acid name. Example: $HClO_4$ is **per**chloric acid.
3. If the number of oxygens is **one fewer** than the number in the *-ic* acid, the suffix *-ic* is replaced with *-ous*. Example: $HClO_2$ is chlor**ous** acid.
4. If the number of oxygens is **two fewer** than the number in the *-ic* acid, the prefix *hypo-* is added and the suffix *-ic* is replaced with *-ous*. Example: HClO is **hypo**chlor**ous** acid.
5. If the acid has **no oxygen,** the prefix *hydro-* is added to the acid name. Example: HCl is **hydro**chloric acid.

Table 6.4 The Prefix–Suffix System for Acid Nomenclature

Number of Oxygen Atoms Compared with an *-ic* Acid	Example	Acid Prefix	Acid Suffix
Same (the memorized *-ic* acid)	$HClO_3$, chloric acid	none	*-ic*
One more	$HClO_4$, **per**chloric acid	*per-*	*-ic*
One fewer	$HClO_2$, chlor**ous** acid	none	*-ous*
Two fewer	HClO, **hypo**chlor**ous** acid	*hypo-*	*-ous*
No oxygen	HCl, **hydro**chloric acid	*hydro-*	*-ic*

Active Example 6.12 The Prefix–Suffix System for Acid Nomenclature

(a) What is the name of HNO_2? (b) What is the formula of hydrosulfuric acid?

Think Before You Write Be sure you have the prefix–suffix system for acid nomenclature in your mind before you proceed.

Answers *Cover the left column with your cut-out shield. Reveal each answer only after you have written your own answer in the right column.*

Nitrous acid *The memorized -ic acid is nitric acid, HNO_3. In comparison, HNO_2 has one fewer oxygen. If the number of oxygens is one fewer than the number in the -ic acid, the suffix -ic is replaced with -ous.*	Start by deducing the name of HNO_2. Recall the memorized *-ic* acid that has the formula type HNO_x. Compare the number of oxygens in the memorized *-ic* acid to the number in HNO_2. Then change the name of the *-ic* acid accordingly. 
H_2S *The memorized -ic acid with sulfur is sulfuric acid, H_2SO_4. When the hydro- prefix is added and the -ic suffix is unchanged, the acid has no oxygen.*	To write the formula of hydrosulfuric acid, recall the name and formula of the *-ic* acid with sulfur. From there, consider the prefix and suffix of the given name, and then modify the number of oxygens as prescribed by the system. 
You improved your understanding of the prefix–suffix system for acid nomenclature.	What did you learn by solving this Active Example? 

Practice Exercise 6.12

(a) What is the name of H_2SO_3? (b) You know that the formula of bromic acid is $HBrO_3$, so what should be the formula of hypobromous acid?

Family Relationships and the Prefix–Suffix System for Acid Nomenclature

The prefix–suffix system for acid nomenclature extends to family relationships between the memorized *-ic* acids and other acids in their families. For example, to derive the formula of perbromic acid, you first recognize that bromine, Br, is in Group 7A/17, and the memorized *-ic* acid in the halogen family is chloric acid, $HClO_3$. Adding a *per-* prefix indicates adding one more oxygen than the number in the *-ic* acid, so perchloric acid is $HClO_4$. The formula of perbromic acid is therefore the parallel $HBrO_4$.

Active Example 6.13 Family Relationships and the Prefix–Suffix System for Acid Nomenclature

(a) What is the name of HI(aq)? (b) What is the formula of selenous acid (selenium, Z = 34)?

Think Before You Write To answer this question, you must have mastery of the following ideas: (a) the five memorized *-ic* acids, (b) how family relationships lead to parallel names and formulas, and (c) the prefix–suffix system. Review any idea that you are unsure of, and then proceed.

Answers *Cover the left column with your cut-out shield. Reveal each answer only after you have written your own answer in the right column.*

Hydroiodic acid *The -iod- middle part of the name of the acid indicates iodine, which is in Group 7A/17. You've memorized that chloric acid, $HClO_3$, is the representative compound for Group 7A/17 acids. The* hydro- *prefix indicates no oxygen, so HCl is hydrochloric acid and -chlor- changes to -iod-.*	First, work on writing the name of HI. Think about the memorized *-ic* acid that serves as the basis for naming acids in Group 7A/17, compare the number of oxygens in the given formula with the number in the *-ic* acid, and then write the name. 
H_2SeO_3 *Selenium is in Group 6A/16, so the beginning point is to recall that the formula of sulfuric acid is H_2SO_4. The* -ous *suffix indicates one fewer oxygen, and H_2SO_3 is sulfurous acid. Changing S to Se completes the formula.*	Now write the formula of selenous acid. Find selenium (Z = 34) on the periodic table, and take it from there. 
You improved your understanding of family relationships and the prefix–suffix system for acid nomenclature.	What did you learn by solving this Active Example? 

Practice Exercise 6.13

(a) What is the name of $HBrO_2$? (b) What is the formula of tellurous acid (tellurium, Z = 52)?

6.10 The Nomenclature of Oxoanions

Goal 11 Given the name (or formula) of an anion in Table 6.5, write its formula (or name).

When an oxoacid is in a water solution, protons are removed from the acid molecules by water molecules. Recall that a hydrogen atom is made up of a proton and an electron. When a proton is removed from a molecule, the species that remains keeps the electron that used to be part of the hydrogen atom. (In this book, **species** is used as a generic term for any chemical particle, such as atom, ion, or molecule.) It therefore acquires a negative charge because it has one more electron than the total number of protons in the formerly neutral molecule. The species formed is an ion, a charged particle. More specifically, it is an **oxoanion**, an oxygen-containing (*oxo-*) negatively charged (*-anion*) species.

Our objective in this section is to write the formulas and names of oxoanions. Let's return to the familiar chloric acid, $HClO_3$, to see how this is done. Recall that we can think of a hydrogen atom as H^+, a hydrogen nucleus (1 proton) with no electron, plus an electron: H atom = $H^+ + e^-$. When a chloric acid molecule reacts with a water molecule, the proton is removed:

$$(H^+)\text{—}(e^-)\text{—}ClO_3 + H_2O \rightarrow (H^+)\text{—}H_2O + (e^-)\text{—}ClO_3$$

Condensing the formulas, we have

$$HClO_3 + H_2O \rightarrow H_3O^+ + ClO_3^-$$

On the product side, there is the water molecule in possession of the proton it removed, H_3O^+, a water molecule, H_2O, plus a proton, H^+, and the chloric acid molecule minus the proton, but in possession of the electron that used to be part of the hydrogen atom, ClO_3^-.

Table 6.5 The Prefix–Suffix System for Acid and Anion Nomenclature

Number of Oxygen Atoms Compared with an *-ic* Acid	Acid Example	Acid Prefix	Acid Suffix	Anion Suffix	Anion Example
Same (the memorized *-ic* acid)	$HClO_3$ chloric acid	none	*-ic*	*-ate*	ClO_3^- chlorate ion
One more	$HClO_4$ perchloric acid	*per-*	*-ic*	*-ate*	ClO_4^- perchlorate ion
One fewer	$HClO_2$ chlorous acid	none	*-ous*	*-ite*	ClO_2^- chlorite ion
Two fewer	$HClO$ hypochlorous acid	*hypo-*	*-ous*	*-ite*	ClO^- hypochlorite ion
No oxygen	HCl hydrochloric acid	*hydro-*	*-ic*	*-ide* (no *hypo-* prefix)	Cl^- chloride ion

The formula of an oxoanion is therefore the formula of the corresponding oxoacid minus one H^+ for each proton that is removed. Let's look at some examples:

Oxoacid	Oxoanion Formed from Removal of All Hydrogen Ions
$HClO_3$	ClO_3^-
H_2SO_4	SO_4^{2-}
H_3PO_4	PO_4^{3-}

In each case, the formula of the oxoanion is the formula of the oxoacid minus the number of H^+ ions that are in the acid molecule. Always recall that water molecules remove *only the proton* from acid molecules. For each proton removed, an electron is left behind in the oxoanion. Thus, for *each* H removed from the acid formula, a 1− charge is added to the anion formula because of the remaining electron.

The name of an oxoanion is derived by modifying the ending—the suffix—of the parent oxoacid. If the acid ends in *-ic*, the anion ends in *-ate*. $HClO_3$ is chlor**ic** acid, and therefore, ClO_3^- is chlor**ate** ion. If the acid ends in *-ous*, the anion ends in *-ite*. $HClO_2$ is chlor**ous** acid, and therefore, ClO_2^- is chlor**ite** ion.

Thinking About Your Thinking

Memory

Four suffixes are used in the prefix–suffix system for acid and anion nomenclature: *-ic* (acid) and *-ate* (ion) for the "normal" number of oxygens, and *-ous* (acid) and *-ite* (ion) for one fewer oxygen than in the *-ic* or *-ate* species. A mnemonic to remember these in order is: **Ick**! I **ate** a hide**ous** **bite**!

The *-ic* → *-ate* and *-ous* → *-ite* acid → anion suffix system works the same when the acids and ions have *per-* or *hypo-* prefixes. $HClO_4$ is perchlor**ic** acid, and therefore, ClO_4^- is perchlor**ate** ion. $HClO$ is hypochlor**ous** acid, and therefore, ClO^- is hypochlor**ite** ion.

The prefix–suffix system for acid and anion nomenclature is summarized in both **Table 6.5** and the following summary:

a summary of... The Prefix–Suffix System for Acid and Anion Nomenclature

1. The system is based on memorizing the number of oxygens in the five *-ic* acids. The formula of the anion is the formula of the acid minus the H^+ ion(s). Example: $HClO_3$ is chloric acid. The name of the anion is the name of the acid with the *-ic* suffix changed to *-ate*. Example: ClO_3^- is chlorate ion.
2. If the number of oxygens is **one more** than the number in the *-ic* acid, the prefix *per-* is added to the acid name. Example: $HClO_4$ is perchloric acid. The name of the anion is the name of the acid with the *-ic* suffix changed to *-ate*. Example: ClO_4^- is perchlorate ion (**Figure 6.24**).
3. If the number of oxygens is **one fewer** than the number in the *-ic* acid, the suffix *-ic* is replaced with *-ous*. Example: $HClO_2$ is chlorous acid. The name of the anion is the name of the acid with the *-ous* suffix changed to *-ite*. Example: ClO_2^- is chlorite ion.
4. If the number of oxygens is **two fewer** than the number in the *-ic* acid, the prefix *hypo-* is added and the suffix *-ic* is replaced with *-ous*. Example: HClO is hypochlorous acid. The name of the anion is the name of the acid with the *-ous* suffix changed to *-ite*. Example: ClO^- is hypochlorite ion (**Figure 6.25**).
5. If the acid has **no oxygen**, the prefix *hydro-* is added to the acid name. Example: HCl is hydrochloric acid. HCl is not an oxoacid, so oxoanion naming rules do not apply. Removal of an H^+ from HCl yields Cl^-, a monatomic anion. Monatomic anions are named by changing the name of the element to end in *-ide*. Example: Cl^- is the chloride ion.

NASA

Figure 6.24 Perchlorate ion was one of the components in the solid fuel that propels the booster rockets of the Space Shuttle.

Charles D. Winters

Figure 6.25 Hypochlorite ion is the ingredient in bleach responsible for its whitening power.

Active Example 6.14 The Prefix–Suffix System for Acid and Anion Nomenclature

(a) Write the names of NO_2^- and SO_3^{2-}. (b) Write the formulas of perbromate ion and hypoiodite ion.

Think Before You Write Always start with the memorized *-ic* acid related to the species for which you are writing the name or formula. From there, apply family relationships and/or the prefix–suffix system, as necessary.

Answers *Cover the left column with your cut-out shield. Reveal each answer only after you have written your own answer in the right column.*

Nitrite ion and sulfite ion

With NO_2^-, the starting point is the memorized -ic acid, nitric acid, HNO_3. The ion has one fewer oxygen, so the corresponding acid is HNO_2, nitrous acid. An -ous acid has an -ite ion.

The logic is similar for SO_3^{2-}. The -ic acid with sulfur is H_2SO_4, sulfuric acid. There is one fewer oxygen, so the acid H_2SO_3, sulfurous acid, loses two H^+ to yield SO_3^{2-}, and -ous changes to -ite.

Begin with deriving and writing the names of NO_2^- and SO_3^{2-}.

BrO_4^- and IO^-

The memorized -ic acid in Group 7A/17 is chloric acid, $HClO_3$.

The per- and -ate prefix–suffix on perbromate ion indicate one more oxygen, and substituting Br for Cl, we have $HBrO_4$, perbromic acid. Removing the H^+ gives the ion formula.

Now write the formulas of perbromate ion and hypoiodite ion.

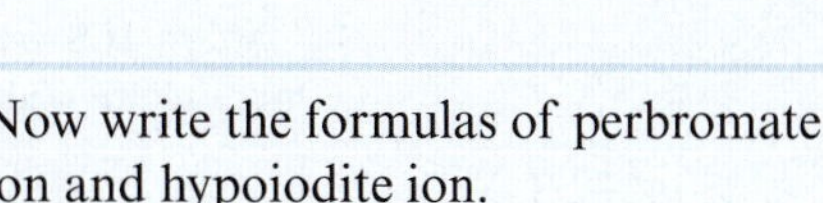

In hypoiodite ion, the hypo- *and* -ite *prefix–suffix indicate two fewer oxygens, and substituting I for Cl, we have HIO, hypoiodous acid. Removing the H^+ gives the ion formula.*	
You improved your understanding of the prefix–suffix system for acid and anion nomenclature.	What did you learn by solving this Active Example?

Practice Exercise 6.14

(a) What is the name of IO_2^-? (b) What is the formula of the periodate ion?

Charles D. Winters

Figure 6.26 Nitric acid, one of the five memorized *-ic* acids, can be prepared from the reaction of sulfuric acid and sodium nitrate.

Table 6.6 summarizes the prefix–suffix system for acid and anion nomenclature. It is in the form of Groups 4A/14 through 7A/17 of the periodic table. Each entry in the table shows the formula of an acid to the left of an arrow. To the right is the formula of the anion that results from the total ionization—the removal of all protons that can be removed by water molecules—of the acid. The table includes all the acids and anions whose names and formulas can be figured out by the nomenclature system we have described. The key acids and anions—the ones that are the basis of this part of the nomenclature system—are highlighted in yellow. If you have memorized the key acids and understand the system, you should be able to figure out any name or formula in this table, given the formula or name.

Table 6.6 Acids and Anions Derived from Their Total Ionization

4A 14	5A 15	6A 16	7A 17
$H_2CO_3 \rightarrow CO_3^{2-}$ carbonic acid → carbonate ion	$HNO_3 \rightarrow NO_3^-$ nitric acid (Figure 6.26) → nitrate ion		$HOF \rightarrow OF^-$ hypofluorous acid → hypofluorite ion
	$HNO_2 \rightarrow NO_2^-$ nitrous acid → nitrite ion		$HF \rightarrow F^-$ hydrofluoric acid → fluoride ion
	$H_3PO_4 \rightarrow PO_4^{3-}$ Phosphoric acid → phosphate ion	$H_2SO_4 \rightarrow SO_4^{2-}$ Sulfuric acid → sulfate ion	$HClO_4 \rightarrow ClO_4^-$ perchloric acid → perchlorate ion
		$H_2SO_3 \rightarrow SO_3^{2-}$ sulfurous acid → sulfite ion	$HClO_3 \rightarrow ClO_3^-$ chloric acid → chlorate ion
		$H_2S \rightarrow S^{2-}$ hydrosulfuric acid → sulfide ion	$HClO_2 \rightarrow ClO_2^-$ chlorous acid → chlorite ion
			$HClO \rightarrow ClO^-$ hypochlorous acid → hypochlorite ion
			$HCl \rightarrow Cl^-$ hydrochloric acid → chloride ion

Table 6.6 Acids and Anions Derived from Their Total Ionization (*continued*)

4A 14	5A 15	6A 16	7A 17
	$H_3AsO_4 \rightarrow AsO_4^{3-}$ arsenic acid → arsenate ion	$H_2SeO_4 \rightarrow SeO_4^{2-}$ selenic acid → selenate ion	$HBrO_4 \rightarrow BrO_4^-$ perbromic acid → perbromate ion
		$H_2SeO_3 \rightarrow SeO_3^{2-}$ selenous acid → selenite ion	$HBrO_3 \rightarrow BrO_3^-$ bromic acid → bromate ion
		$H_2Se \rightarrow Se^{2-}$ hydroselenic acid → selenide ion	$HBrO_2 \rightarrow BrO_2^-$ bromous acid → bromite ion
			$HBrO \rightarrow BrO^-$ hypobromous acid → hypobromite ion
			$HBr \rightarrow Br^-$ hydrobromic acid → bromide ion
		$H_2TeO_4 \rightarrow TeO_4^{2-}$ telluric acid → tellurate ion	$HIO_4 \rightarrow IO_4^-$ periodic acid → periodate ion
		$H_2TeO_3 \rightarrow TeO_3^{2-}$ tellurous acid → tellurite ion	$HIO_3 \rightarrow IO_3^-$ iodic acid → iodate ion
		$H_2Te \rightarrow Te^{2-}$ hydrotelluric acid → telluride ion	$HIO_2 \rightarrow IO_2^-$ iodous acid → iodite ion
			$HIO \rightarrow IO^-$ hypoiodous acid → hypoiodite ion
			$HI \rightarrow I^-$ hydroiodic acid → iodide ion

The following Active Example gives you the opportunity to practice what you have learned about oxoanion nomenclature. If you have memorized what is necessary and know how to apply the rules, you will be able to write the required names and formulas with reference to nothing other than a periodic table.

Active Example 6.15 Nomenclature of Ionic Compounds with Oxoanions

For each of the following names, write the formula; for each formula, write the name.

Think Before You Write This Active Example provides you with an excellent opportunity to test your nomenclature knowledge. If you find it straightforward, congratulations! You have done well in learning nomenclature. If you struggle with some of the names and formulas, go back and review the pertinent section(s). A brief review is often needed to cement newly learned concepts.

Use no reference other than a "clean" periodic table—one that has no information other than what is found on your shield.

Answers *Cover the left column with your cut-out shield. Reveal each answer only after you have written your own answer in the right column.*

$Zn(ClO)_2$, zinc hypochlorite	$Zn(ClO)_2$ ________
$Sn(NO_3)_2$, tin(II) nitrate	$Sn(NO_3)_2$ ________
$Hg(ClO_4)_2$, mercury(II) perchlorate	$Hg(ClO_4)_2$ ________
Ammonium iodite, NH_4IO_2	Ammonium iodite ________
Silver sulfite, Ag_2SO_3	Silver sulfite ________
Chromium(III) carbonate, $Cr_2(CO_3)_3$	Chromium(III) carbonate ________
You improved your skill at writing the names and formulas of ionic compounds with oxoanions.	What did you learn by solving this Active Example?

Practice Exercise 6.15

For each of the following names, write the formula; for each formula, write the name: (a) $Mg(BrO_4)_2$; (b) $Fe(NO_2)_2$; (c) $Cu(IO)_2$; (d) aluminum chlorite; (e) lead(IV) phosphate; (f) potassium selenate.

6.11 The Nomenclature of Acid Anions

Goal 12 Given the name (or formula) of an ion formed by the step-by-step ionization of a polyprotic acid from a Group 4A/14, 5A/15, or 6A/16 element, write its formula (or name).

Polyprotic acids do not give up their hydrogen ions all at once, but rather one at a time. The intermediate anions produced are stable chemical species that are the negative ions in many ionic compounds. The hydrogen-bearing **acid anion**, as it is called, releases a hydrogen ion when dissolved in water, just like any other acid.

Baking soda, commonly found in kitchen cabinets, contains the acid anion HCO_3^-. It can be regarded as the intermediate step in the ionization of carbonic acid:

$$\underset{\text{carbonic acid}}{H_2CO_3} \xrightarrow{-H^+} \underset{\text{hydrogen carbonate ion}}{HCO_3^-} \xrightarrow{-H^+} \underset{\text{carbonate ion}}{CO_3^{2-}}$$

HCO_3^- is the hydrogen carbonate ion—a logical name, since the ion is literally a hydrogen ion bonded to a carbonate ion.

A carbonic acid–hydrogen carbonate ion system and a dihydrogen phosphate ion–hydrogen phosphate ion system are two of the three major chemical systems that regulate the acidity of your blood.

Phosphoric acid, H_3PO_4, has three steps in its ionization process:

$$\underset{\text{phosphoric acid}}{H_3PO_4} \xrightarrow{-H^+} \underset{\text{dihydrogen phosphate ion}}{H_2PO_4^-} \xrightarrow{-H^+} \underset{\text{(mono)hydrogen phosphate ion}}{HPO_4^{2-}} \xrightarrow{-H^+} \underset{\text{phosphate ion}}{PO_4^{3-}}$$

$H_2PO_4^-$ is the dihydrogen phosphate ion, signifying two hydrogen ions attached to a phosphate ion. HPO_4^- is the monohydrogen phosphate ion, or simply the hydrogen phosphate ion. It is essential that the prefix *di-* be used in naming the $H_2PO_4^-$ ion to distinguish it from HPO_4^{2-}, but the prefix *mono-* is usually omitted in naming HPO_4^{2-}.

If you recognize the logic of this part of the nomenclature system, you will be able to extend it to intermediate ions from the step-by-step ionization of hydrosulfuric, sulfuric, and sulfurous acids. All of these are shown in Table 6.7.

Table 6.7 Names and Formulas of Anions Derived from the Step-by-Step Ionization of Acids

Acid	Ion	Names of Ions	
		Preferred	Other
H_2CO_3	HCO_3^-	Hydrogen carbonate	Bicarbonate; acid carbonate
H_2S	HS^-	Hydrogen sulfide	Bisulfide; acid sulfide
H_2SO_4	HSO_4^-	Hydrogen sulfate	Bisulfate; acid sulfate
H_2SO_3	HSO_3^-	Hydrogen sulfite	Bisulfite; acid sulfite
H_3PO_4	$H_2PO_4^-$	Dihydrogen phosphate	Monobasic phosphate
$H_2PO_4^-$	HPO_4^{2-}	Hydrogen phosphate	Dibasic phosphate

Active Example 6.16 Nomenclature of Acid Anions

What is the name of $Fe(HCO_3)_3$? What is the formula of ammonium hydrogen phosphate?

Think Before You Write When going from name to formula, you will be less likely to make a mistake if you first write the formula of the cation, then write the formula of the anion, and then combine them to produce the formula of the ionic compound.

Answers *Cover the left column with your cut-out shield. Reveal each answer only after you have written your own answer in the right column.*

Iron(III) hydrogen carbonate; $(NH_4)_2HPO_4$

For $Fe(HCO_3)_3$, you know that carbonic acid is H_2CO_3, so carbonate ion is CO_3^{2-}. Adding a hydrogen ion to the carbonate ion yields HCO_3^-. Three 1− ions require a 3+ charge, so the iron ion species must be iron(III) ion.

The formula of the ammonium ion is NH_4^+ and the formula of the phosphate ion is PO_4^{3-} (phosphoric acid is H_3PO_4), and adding an H^+ yields HPO_4^{2-} as the formula of the hydrogen phosphate ion. Two 1+ ammonium ions are needed to balance the 2− hydrogen phosphate ion.

Complete the Active Example.

You improved your skill at writing the names and formulas of acid anions.

What did you learn by solving this Active Example?

Practice Exercise 6.16

What is the name of $NaHSO_4$? What is the formula of sodium hydrogen sulfide?

6.12 The Nomenclature of Hydrates

Goal 13 Given the formula of a hydrate, state the number of water molecules associated with each formula unit of the anhydrous compound.

14 Given the name (or formula) of a hydrate, write its formula (or name). (This goal is limited to hydrates of ionic compounds for which a name and formula can be written based on the rules of nomenclature presented in this book.)

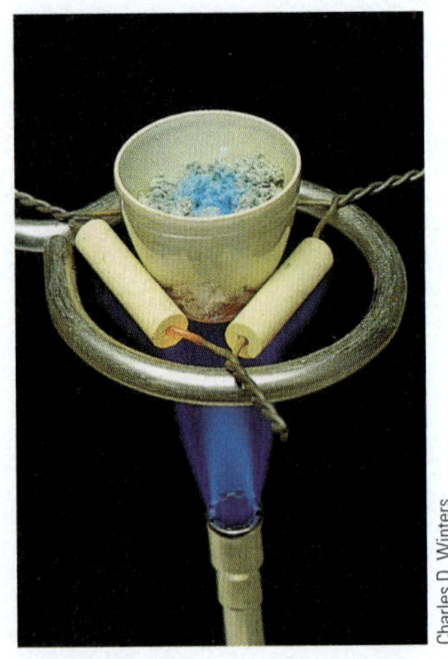

Figure 6.27 A sample of hydrated and anhydrous copper(II) sulfate pentahydrate. The blue compound in the center of the crucible is $CuSO_4 \cdot 5\,H_2O$; the near-white compound near the edges is $CuSO_4$.

Some compounds, when crystallized from water solutions, form solids that include water molecules as part of the crystal structure. Such water is referred to as **water of crystallization** or **water of hydration**. The compound is said to be **hydrated** and is called a **hydrate**. Hydration water can usually be driven from a compound by heating, leaving the **anhydrous compound**.

Copper(II) sulfate is an example of a hydrate. The anhydrous compound, $CuSO_4$, is a nearly white powder. Each formula unit of $CuSO_4$ combines with five water molecules in the hydrate, which is a blue crystal (**Figure 6.27**). Its formula is $CuSO_4 \cdot 5\,H_2O$. The number of water molecules that crystallize with each formula unit of the anhydrous compound is shown after the anhydrous formula, separated by a dot. This number is 5 for the hydrate of copper(II) sulfate. The equation for the dehydration of this compound is

$$CuSO_4 \cdot 5\,H_2O \rightarrow CuSO_4 + 5\,H_2O$$

Just as in binary molecular compounds, prefixes are used to indicate the number of water molecules in a formula unit of a hydrate (see Table 6.1). By this system, $CuSO_4 \cdot 5\,H_2O$ is called *copper(II) sulfate pentahydrate*, since *penta-* is the prefix for 5.

Active Example 6.17 **Nomenclature of Hydrates**

a) How many water molecules are associated with each formula unit of anhydrous sodium carbonate in $Na_2CO_3 \cdot 10\,H_2O$? Name the hydrate.

b) Write the formula of nickel(II) chloride hexahydrate.

Think Before You Write Before beginning this exercise, mentally recite the number prefixes used in binary molecular compounds (and now, hydrates) from 1 to 10. If you can't do it, go back to Table 6.1 and review.

Answers *Cover the left column with your cut-out shield. Reveal each answer only after you have written your own answer in the right column.*

a) Ten; sodium carbonate decahydrate b) $NiCl_2 \cdot 6\,H_2O$	Complete the Active Example.
You improved your skill at writing the names and formulas of hydrates.	What did you learn by solving this Active Example?

Practice Exercise 6.17

What is the name of $Al(BrO_3)_3 \cdot 9\,H_2O$? What is the formula of cobalt(II) perchlorate pentahydrate?

6.13 Summary of the Nomenclature System

Throughout this chapter we have emphasized memorizing a relatively small set of names and formulas and some prefixes and suffixes. They are the basis for the system of chemical nomenclature. Once you have memorized them, you apply the system to the different names and formulas you encounter. **Table 6.8** summarizes all the ideas that have been presented. It should help you review the nomenclature system.

Table 6.8 Summary of Nomenclature System

Substance	Name	Formula
Element	Name of element	Symbol of element; exceptions: H_2, N_2, O_2, F_2, Cl_2, Br_2, I_2
Compounds made up of two nonmetals	First element in formula followed by second, changed to end in *-ide,* each element preceded by prefix to show the number of atoms in the molecule	Symbol of first element in name followed by symbol of second element, with subscripts to show number of atoms in molecule
Monatomic cation	Name of element followed by ion; if element forms more than one monatomic cation, elemental name is followed by ion charge in Roman numerals and in parentheses	Symbol of element followed by superscript to indicate charge
Monatomic anion	Name of element changed to end in *-ide*	Symbol of element followed by superscript to indicate charge
Other polyatomic ions	Ammonium ion Hydroxide ion	NH_4^+ OH^-
Ionic compound	Name of cation followed by name of anion	Formula of cation followed by formula of anion, each taken as many times as necessary to yield a net charge of zero (polyatomic ion formulas enclosed in parentheses if taken more than once)
Acid	Most common: middle element changed to end in *-ic* One more oxygen than *-ic* acid: add prefix *per-* to name of *-ic* acid One fewer oxygen than *-ic* acid: change ending of *-ic* acid to *-ous* Two fewer oxygens than *-ic* acid: add prefix *hypo-* to name of *-ous* acid No oxygen: Prefix *hydro-* followed by name of second element changed to end in *-ic*	H followed by symbol of nonmetal followed by O (if necessary), each with appropriate subscript *Memorize the following:* Carbonic acid, H_2CO_3 Nitric acid, HNO_3 Phosphoric acid, H_3PO_4 Sulfuric acid, H_2SO_4 Chloric acid, $HClO_3$
Polyatomic anion from total ionization of oxyacid	Replace *-ic* in acid name with *-ate,* or replace *-ous* in acid name with *-ite,* followed by ion	Acid formula without hydrogen plus superscript showing negative charge equal to number of hydrogens removed from acid formula
Polyatomic anion from step-by-step ionization of oxyacid	Hydrogen followed by name of ion from total ionization of acid (dihydrogen in the case of $H_2PO_4^-$)	Acid formula minus one (or two for H_3PO_4) hydrogen(s), plus superscript showing negative charge equal to number of hydrogen removed from acid formula
Hydrate	Name of anhydrous compound followed by (number prefix) hydrate, where (number prefix) indicates the number of water molecules associated with one formula unit of anhydrous compound	Formula of anhydrous compound followed by "· n H_2O" where n is number of water molecules associated with one formula unit of anhydrous compound

Charles D. Winters

CHAPTER 6 IN REVIEW: CHEMICAL NOMENCLATURE

Goal 1 Given a representation or a written description of the particulate-level composition of an element or compound, write the chemical formula of that substance.

A **chemical formula** describes the composition of the molecule in terms of the number of each type of atom that makes up the particle. The number of each type of atom is written in a formula as a subscript; if that number is 1, it is omitted.

Goal 2 Given a name or formula of an element in Figure 5.20, write the other.

The names and formulas of the 35 elements in Figure 5.20 should already be in memory. The **seven diatomic elements** H_2, N_2, O_2, F_2, Cl_2, Br_2, and I_2 must be learned. Remember: **H**orses **N**eed **O**ats **F**or **Cl**ear **Br**own **I**'s.

Goal 3 Given the name or formula of a binary molecular compound, write the other.

Two nonmetals or a nonmetal and a metalloid form chemical bonds with each other to form **binary molecular compounds**. The name of a binary molecular compound is the name of the first element followed by the name of the second element, modified with an *-ide* suffix. Prefixes are used to indicate the number of atoms of each element in the molecule. Memorize the **number prefixes** used in chemical names: *mono-* = 1, *di-* = 2, *tri-* = 3, *tetra-* = 4, *penta-* = 5, *hexa-* = 6, *hepta-* = 7, *octa-* = 8, *nona-* = 9, *deca-* = 10.

Goal 4 Given the name or the formula of water, write the other; given the name or the formula of ammonia, write the other.

Two common binary molecular compounds with nonsystematic names are **water, H_2O,** and **ammonia, NH_3.**

Goal 5 Given the name or formula of an ion in Figure 6.8, write the other.

Ions are charged particles. A **cation** has a positive charge; an **anion** has a negative charge. The name of a monatomic cation is the name of the element, followed by the word *ion*. The name of a monatomic anion is the name of the element, changed to end in *-ide*, followed by the word *ion*. The formula of a monatomic ion is the symbol of the element followed by its electrical charge, written in superscript. The charge of ions formed from many main-group elements corresponds to the group number: 1A/1, 1+; 2A/2, 2+; 5A/15, 3−; 6A/16, 2−; 7A/17, 1−.

Goal 6 Given the name or formula of an ion in Figure 6.10, write the other.

Some elements commonly form more than one ion. For these ions, its oxidation state is added to the elemental name. The oxidation state is written in parentheses immediately after the name. For example, the formula of the iron(II) ion is Fe^{2+}. There are three common ions not in Groups 1A/1 or 2A/2 with only one charge: silver ion, Ag^+; zinc ion, Zn^{2+}; and aluminum ion, Al^{3+}. These three formulas must be memorized.

Goal 7 Give the formula (or name) of the ammonium ion or hydroxide ion, write the corresponding name (or formula).

The ammonium ion is $\mathbf{NH_4^+}$; the hydroxide ion is $\mathbf{OH^-}$.

Goal 8 Given the name of any ionic compound made up of identifiable ions, or other ions whose formulas are given, write the formula of that compound.

To write the **formula of an ionic compound:**

1. Write the formula of the cation and the formula of the anion.
2. Mentally balance the charges. Decide what the fewest number of ions is that will make the compound electrically neutral.
3. Write the formula, cation first, anion second, each with subscripts as needed for charge balance. (a) If only one ion is needed, omit the subscript. (b) If a polyatomic ion is needed more than once, enclose the formula of the ion in parentheses and place the subscript after the closing parenthesis.

Goal 9 Given the formula of an ionic compound made up of identifiable ions, write the name of the compound.

To write the **name of an ionic compound:**

1. Write the name of the cation.
2. Write the name of the anion.

Goal 10 Given the name (or formula) of an acid in Table 6.3, write its formula (or name).

An **acid** ionizes in water to give H_3O^+ and an anion. A general equation used to describe acid ionization is $HX + H_2O \rightarrow H_3O^+ + X^-$. Five ***-ic* acids** must be memorized: carbonic acid, H_2CO_3; nitric acid, HNO_3; phosphoric acid, H_3PO_4; sulfuric acid, H_2SO_4; chloric acid, $HClO_3$.

Goal 11 Given the name (or formula) of an anion in Table 6.5, write its formula (or name).

A system is used to name oxoacids and oxoanions with different numbers of oxygens than the related *-ic* acids and *-ate* anions.

Number of Oxygen Atoms Compared with *-ic* Acid	Acid Prefix and/or Suffix	Anion Prefix and/or Suffix
One more	*per- -ic*	*per- -ate*
Same	*-ic*	*-ate*
One fewer	*-ous*	*-ite*
Two fewer	*hypo- -ous*	*hypo- -ite*
No oxygen	*hydro- -ic*	*-ide*

Remember: **Ic**k! I **ate** a hide**ous** bite!

Goal 12 Given the name (or formula) of an ion formed by the step-by-step ionization of a polyprotic acid from a Group 4A/14, 5A/15, or 6A/16 element, write its formula (or name).

Acid anions are named in the same way as oxoanions, with the term *hydrogen* or *dihydrogen* added to indicate the number of hydrogen ions bonded to the oxoanion.

Goal 13 Given the formula of a hydrate, state the number of water molecules associated with each formula unit of the anhydrous compound.

Hydrates are ionic compounds that exist with a definite number of water molecules in their crystal structure. **Waters of hydration** are indicated by the "•" symbol before the number of water molecules.

Goal 14 Given the name (or formula) of a hydrate, write its formula (or name). (This goal is limited to hydrates of ionic compounds for which a name and formula can be written based on the rules of nomenclature presented in this book.)

Prefixes are used to indicate the number of water molecules in a formula unit of a hydrate. These are the same as the prefixes used in naming binary molecular compounds.

Key Terms

Most of the key terms and concepts and many others appear in the Glossary. Use your Glossary regularly.

acid p. 208
acid anion p. 218
anhydrous compound p. 220
anion p. 194
binary molecular compounds p. 193
cation p. 194
chemical families p. 210
diatomic p. 190
formula unit p. 201
halogen family p. 210
hydrate p. 220
hydrated p. 220
hydronium ion p. 208
ion p. 194
ionic compounds p. 200
monatomic ion p. 194
nomenclature p. 185
oxoacid p. 209
oxoanion p. 213
polyatomic ions p. 200
species p. 213
water of crystallization p. 220
water of hydration p. 220

Frequently Asked Questions

I'm overwhelmed with the large number of names and formulas in this chapter. What is the best way to deal with learning the names and formulas of hundreds of substances?
As noted in the introduction to this chapter, the most important thing you can do to learn nomenclature is to *learn the system.* The system is based on some rules, prefixes, and suffixes that you must learn or memorize. You can then apply the rules in writing the names and formulas of hundreds of chemical substances. This is by far the easiest and quickest way to learn how to write chemical names and formulas.

I'm still confused about elements. Do the diatomic elements have to be diatomic when part of a compound?
Remember that the *elements* hydrogen, nitrogen, oxygen, fluorine, chlorine, bromine, and iodine exist as diatomic molecules at common temperatures. Note the limitation; it refers to the uncombined *elements*, not to compounds—either molecules or ions—in which the elements may be present. Hydrogen is H_2, chlorine is Cl_2, but hydrogen chloride is HCl.

I'm having trouble remembering the number prefixes used in binary molecular compounds and hydrates. Is there a way to relate them to something I already know?
Here are a few memory aids that may help you learn the number prefixes in Table 6.1: A *mono*poly is when *one* company controls an economic product or service. A *two*-wheel cycle is a *bi*cycle, but a chemist might call it a *di*cycle. No problem with *three* wheels: it's a *tri*cycle. No help on *tetra-* for four, unless you happen to remember that a *four*-sided solid is called a *tetra*hedron. The *Penta*gon is the *five*-sided building in Washington

that serves as headquarters for U.S. military operations. *Six* and *hex-* are the only number/prefix combination that has the letter *x*. If you change the *s* to an *h* in September, the *Hept*ember, *Oct*ober, *Nov*ember, and *Dec*ember list the beginnings of what were once the *seven*th, *eigh*th, *nin(e)*th, and *ten*th months of the year.

How are the number prefixes used in writing the names of ionic compounds?
Number prefixes are *almost never* used in naming ionic compounds. The *di*hydrogen phosphate ion is the only exception in this chapter. Number prefixes were used in the past, but today exam graders will find use for their red pens if you write about aluminum trichloride.

When are parentheses used and not used in writing formulas of ionic compounds?
Parentheses enclose *polyatomic* ions used more than once but never a monatomic ion. Examples are $BaCl_2$, not $Ba(Cl)_2$; $Ba(OH)_2$, not $BaOH_2$; *but* NaOH, not Na(OH).

When is a charge included in the formula of an ion? When is a charge included in the formula of an ionic compound?
A charge, written as a superscript, is included in the formula of *every* ion. Without a charge, it is not an ion. Do not include ionic charge in the formula of an ionic compound, however: Na_2S, not $Na^+{}_2S^{2-}$.

I wrote "nitrogen dioxide" as the name of NO_2^- on a quiz, and it was marked wrong. I thought that the mono- *prefix could be left off, and di- is the prefix for two. Why was it marked wrong?*
The rules for naming binary molecular compounds are sometimes confused with those for naming polyatomic oxyanions. The critical difference in the formulas of the two species is that ions have charges and molecules do not. NO_2 is uncharged, and therefore, it is a binary molecular compound. Its name is nitrogen dioxide. NO_2^- has a charge; therefore, it is an oxyanion. Its name is nitrite ion.

I wrote "hydrogen nitrate" as the name of HNO_3 on a quiz, and it was marked wrong. H^+ is the hydrogen ion, and NO_3^- is the nitrate ion. Why was it marked wrong?
An oxoacid should be named as an acid, not as an ionic compound. Ionic compounds have a *metal* ion and a nonmetal ion or a group of nonmetal elements bound together that collectively form an ion. HNO_3 is nitric acid, not hydrogen nitrate, because H^+ is an ion of a *nonmetal* element.

You said to memorize the charges on silver ion and zinc ion (and aluminum ion). Why can't we treat them like other transition element ions and include their charge in their names?
The charge of a cation is used when naming an ionic compound *if* the element forms more than one monatomic ion, as with iron(II) ion, Fe^{2+}, and iron(III) ion, Fe^{3+}. Do *not* use the charge if the element forms only one common cation, as with silver ion, Ag^+, and zinc ion, Zn^{2+}. Since these ions have only one common charge, there is no ambiguity when writing their names without their charges.

What is the best method for studying this chapter?
Learning the nomenclature system correctly is the first of two steps. The second is applying the system correctly. To develop this skill, you must practice, practice, and then practice some more until you can write names and formulas almost automatically. The end-of-chapter questions give ample opportunity for practice. Take full advantage of them. In particular, perfect your skill in writing formulas of ionic compounds by completing Questions 47 and 54. Your ultimate self-test lies in answering the General Questions, in which different kinds of substances are mixed. You must first identify the kind of substance the compound is, then select the proper rule to apply, and then apply the rule correctly. If you make a mistake, use Table 6.8 to trace the error in your logic. Remember that making mistakes while practicing is normal. Mistakes can even be helpful, as long as you use them as a learning tool.

Concept-Linking Exercises

Write a brief description of the relationships among each of the following groups of terms or phrases. Answers to the Concept-Linking Exercises are given at the end of the chapter.

1. Atom, molecule, formula unit
2. Ion, cation, anion, monatomic ion, polyatomic ion, acid anion
3. Acid, hydrogen ion, hydronium ion, hydrated hydrogen ion
4. Water of hydration, hydrate, anhydrous compound

Small-Group Discussion Question

Small-Group Discussion Questions are for group work, either in class or under the guidance of a leader during a discussion section.

1. Construct a set of flash cards with the formulas of a variety of elements, binary molecular compounds, ions, acids, ionic compounds, and hydrates on one side and their names on the other side. Split into pairs and practice giving names for formulas and formulas for names until each person can repeat the entire set rapidly and accurately in both directions.

Questions, Exercises, and Problems

Interactive versions of these problems may be assigned by your instructor. Solutions for blue-numbered questions are at the end of the chapter. Questions other than those in the General Questions and More Challenging Problems sections are paired in consecutive odd–even number combinations; solutions for the odd numbered questions are also at the end of the chapter.

General Instructions: Most of the questions in this chapter ask that you write the name of any species if the formula is given or the formula if the name is given. You should try to follow these instructions without reference to anything except a clean periodic table, one that has nothing written on it. Names and/or atomic numbers are given in questions involving elements not shown in Figure 5.20.

Section 6.2: Review of Selected Concepts Related to Nomenclature

1. Write the chemical formula of each of the following. The blue spheres represent nitrogen atoms and the red spheres represent oxygen atoms. Oxygen is written last in the formulas that include oxygen.

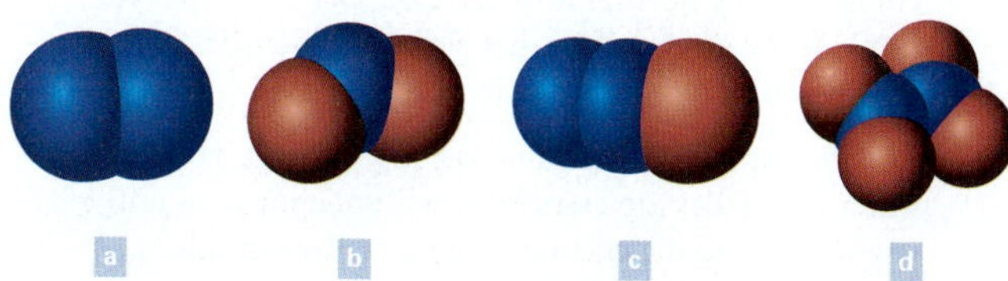

2. Write the chemical formula of each of the following. The yellow spheres represent fluorine atoms and the green spheres represent chlorine atoms. Fluorine is written last in these formulas.

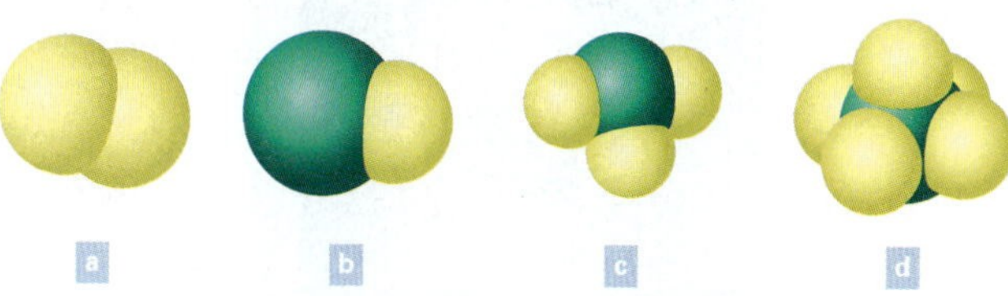

3. Write the chemical formula of each of the following:
   a) The compound made up of a crystal with one particle coming from a nickel atom for every two particles coming from chlorine atoms.
   b) The compound made up of a crystal with two particles coming from silver atoms for every one particle coming from an oxygen atom.
   c) The compound made up of molecules with 6 carbon atoms, 12 hydrogen atoms, and 6 oxygen atoms.
   d) The compound made up of molecules with two hydrogen atoms, one sulfur atom, and four oxygen atoms.
4. Write the chemical formula of each of the following:
   a) The compound made up of a crystal with two particles coming from chromium atoms for every three particles coming from oxygen atoms.
   b) The compound made up of a crystal with one particle coming from a barium atom for every two particles coming from chlorine atoms.
   c) The compound made up of molecules with 12 carbon atoms, 22 hydrogen atoms, and 11 oxygen atoms.
   d) The compound made up of molecules with three hydrogen atoms, one phosphorus atom, and four oxygen atoms.

Section 6.3: Formulas of Elements

5. The elements of Group 8A/18 are stable as monatomic atoms. Write their formulas.
6. Write the formulas of all elements that exist as diatomic molecules under normal conditions.

Questions 7 through 12: Given names, write formulas; given formulas, write names.

7. Fluorine, boron, nickel, sulfur
8. Germanium (Z = 32), fluorine, argon, gallium (Z = 31)
9. Cr, Cl_2, Be, Fe
10. O_2, Ca, Ba, Ag
11. Carbon, silicon, tin, lead

Carbon (*right*), silicon (*bottom*), tin (*left*), and lead (*top*).

12. Chlorine, magnesium, nitrogen, vanadium (Z = 23)

Section 6.4: Compounds Made from Two Nonmetals

13. Dichlorine monoxide, tribromine octoxide, HBr(g), P_2O_3
14. CCl_4, CBr_4, NO, dinitrogen monoxide, sulfur dioxide, sulfur hexafluoride

Section 6.5: Names and Formulas of Monatomic Ions: Groups 1A/1 and 2A/2 Metals and the Nonmetals

15. Explain how monatomic anions are formed from atoms, in terms of protons and electrons.
16. Write an equation that shows the formation of a rubidium ion from a neutral rubidium atom (Z = 37).
17. What are the formulas of lithium ion and the nitride ion?
18. What are the names of O^{2-} and I^-?

Section 6.6: Names and Formulas of Monatomic Ions: Additional Metals

19. What are the formulas of cobalt(III) ion and aluminum ion; what are the names of Fe^{3+} and Hg_2^{2+}?
20. What are the formulas of tin(IV) ion and manganese(III) ion; what are the names of Pb^{2+} and Co^{2+}?

Section 6.7: Formulas of Ionic Compounds

21. What are the formulas of nickel(III) oxide, cobalt(II) chloride, silver sulfide, and mercury(II) iodide?

22. What are the formulas of lithium nitride, chromium(II) chloride, aluminum phosphide, and tin(IV) bromide?

Section 6.8: Names of Ionic Compounds

23. What are the names of AgBr, SnF_2, Fe_2O_3, and $CuCl_2$?

24. What are the names of MnF_3, NiS, PbO_2, and Co_2S_3?

Section 6.9: The Nomenclature of Oxoacids

25. How do you recognize the formula of an acid?

26. What element is present in all acids discussed in this chapter?

27. Write the names and formulas of the five *-ic* acids that must be memorized before learning the acid nomenclature system.

28. Which of the five memorized *-ic* acids serve as models for family relationships in the nomenclature of acids, and which stand alone? For those that serve as models, what non-hydrogen, non-oxygen element or elements follow(s) the model of each model acid?

29. Fill in the blanks in the following table. Selenium is Z = 34, Se, and tellurium is Z = 52, Te.

Acid Name	Acid Formula	Acid Name	Acid Formula
Sulfuric acid		Selenous acid	
	H_2CO_3		H_2TeO_3
Chloric acid		Hypoiodous acid	
	HF		HBrO
Bromic acid		Telluric acid	
	H_2SO_3		$HBrO_4$
Arsenic acid		Hydrobromic acid	
	HIO_4		

30. Fill in the blanks in the following table. Selenium is Z = 34, Se, and tellurium is Z = 52, Te.

Acid Name	Acid Formula	Acid Name	Acid Formula
	HCl		$HClO_4$
Nitric acid		Hydroiodic acid	
	H_3PO_4		$HClO_2$
Hydrosulfuric acid		Selenic acid	
	HNO_2		$HBrO_2$
Hydroselenic acid			

Section 6.10: The Nomenclature of Oxoanions

31. What are the formulas of carbonate ion and selenite ion; what is the name of BrO_3^-?

32. What is the formula of arsenate ion; what are the names of TeO_3^{2-} and IO_3^-?

33. What is the formula of sodium tellurate; what are the names of $Mg(BrO)_2$ and $FePO_4$?

34. What are the formulas of zinc nitrate and calcium selenate; what is the name of NH_4BrO_2?

Section 6.11: The Nomenclature of Acid Anions

35. Explain how an anion can behave like an acid. Is it possible for a cation to be an acid?

36. Write equations for the first and second steps in the step-by-step ionization of selenic acid (selenium, Z = 34).

Questions 37 through 40: Given names, write formulas; given formulas, write names.

37. Hydrogen sulfite ion, hydrogen carbonate ion

38. Dihydrogen phosphate ion, hydrogen sulfate ion, hydrogen tellurite ion (tellurium, Z = 52)

39. $HSeO_3^-$, HTe^-

40. HCO_3^-, $HTeO_4^-$, HSO_3^-

Section 6.12: The Nomenclature of Hydrates

41. Among the following, identify all hydrates and anhydrous compounds: $NiSO_4 \cdot 6\ H_2O$, KCl, $Na_3PO_4 \cdot 12\ H_2O$.

42. It is often possible to change a hydrate into an anhydrous compound by heating it to drive off the water (dehydration). Write an equation that shows the dehydration of potassium fluoride dihydrate.

43. Epsom salt has the formula $MgSO_4 \cdot 7\ H_2O$. How many water molecules are associated with one formula unit of $MgSO_4$? Write the chemical name of Epsom salt.

Charles D. Winters/Science Source

Epsom salt.

44. How many water molecules are associated with one formula unit of calcium chloride in $CaCl_2 \cdot 2\ H_2O$? Write the name of this compound.

45. Write the formulas of ammonium phosphate trihydrate and potassium sulfide pentahydrate.

46. The compound barium perchlorate forms a hydrate with three water molecules per formula unit. What is the name of hydrate? What is its formula?

General Questions

47. In each box, write the chemical formula of the compound formed by the cation at the head of the column and the anion at the left of the row. Refer only to a clean periodic table when completing this exercise.

Ions	Potassium	Calcium	Chromium(III)	Zinc	Silver	Iron(III)	Aluminum	Mercury(I)
Nitrate								
Sulfate								
Hypochlorite								
Nitride								omit
Hydrogen sulfide								
Bromite								
Hydrogen phosphate								
Chloride								
Hydrogen carbonate								

48. Calcium hydroxide, ammonium bromide, potassium sulfate
49. Barium nitrite, sodium hydrogen carbonate, aluminum nitrate, calcium chromate (chromate ion, CrO_4^{2-})
50. Magnesium oxide, aluminum phosphate, sodium sulfate, calcium sulfide
51. Aluminum bromite, potassium dichromate (dichromate ion, $Cr_2O_7^{2-}$), barium dihydrogen phosphate, rubidium selenate (rubidium, Z = 37; selenium, Z = 34)
52. Barium sulfite, chromium(III) oxide, potassium periodate, calcium hydrogen phosphate
53. Cobalt(II) hydroxide, zinc chloride, lead(II) nitrate, copper(I) nitrate
54. In each box, write the chemical formula of the compound formed by the cation at the head of the column and the anion at the left of the row, and then write the name of the compound. Refer only to a clean periodic table when completing this exercise.

Ions	Na^+	Hg^{2+}	NH_4^+	Pb^{2+}	Mg^{2+}	Fe^{3+}	Cu^{2+}
OH^-							
BrO^-							
CO_3^{2-}							
ClO_3^-							
HSO_4^-							
Br^-							
PO_4^{3-}							
IO_4^-							
S^{2-}							

Write the name of each compound in Questions 55–60.

55. Li_3PO_4, $MgCO_3$, $Ba(NO_3)_2$
56. PbS, AgOH, $CoBr_3$
57. KF, NaOH, CaI_2, $Al_2(CO_3)_3$
58. $Mg(OH)_2$, Na_2CO_3, $AgNO_2$, $Co_3(PO_4)_2$
59. $CuSO_4$, $Cr(OH)_3$, $Hg2I_2$
60. $Ba(BrO_3)_2$, $K_2C_2O_4$ ($C_2O_4^{2-}$, oxalate ion), $Al(HTeO_4)_3$ (Te, tellurium, Z = 52), K_2SeO_3 (Se, selenium, Z = 34)

61. In each box, when given a formula, write the name; given a name, write the formula. Refer to nothing but the periodic table printed on your shield.

Formula	Name	Name	Formula
OH^-		Sodium bromide	
PbO		Ammonium phosphate	
SO_4^{2-}		Bromine	
Cl_2		Hydrogen phosphate ion	
Cl_2O		Nitrite ion	
$Cr(ClO_4)_3$		Iron(III) phosphate	
$H_2PO_4^-$		Nickel(II) sulfide	
Al_2S_3		Nitrogen	
$CuSO_4$		Copper(II) bromide	
HF(aq)		Oxygen difluoride	
C		Potassium ion	

62. In each box, when given a formula, write the name; given a name, write the formula. Refer to nothing but the periodic table printed on your shield.

Formula	Name	Name	Formula
SeO_4^{2-} (Se, selenium, Z = 34)		Gallium sulfate (Gallium, Z = 31)	
HCO_3^-		Perchloric acid	
Ne		Lithium	
N_2O_5		Cobalt(II) chloride hexahydrate	
HNO_2		Barium dihydrogen phosphate	
CI_4		Hydrosulfuric acid	

BaH_2	Magnesium nitride
$CaTeO_3$ (Te, tellurium, Z = 52)	Selenic acid (Selenium, Z = 34)
HBrO	Calcium sulfite
$Fe(NO_3)_2$	Sodium hydride
$MgSO_4 . 7\,H_2O$	Mercury(I) chloride

Questions 63–92: Given a formula, write the name; given a name, write the formula. Refer to nothing but the periodic table printed on your shield.

63. Perchlorate ion, barium carbonate, NH_4I, PCl_3
64. Calcium fluoride, sodium bromide, aluminum iodide, HS^-
65. $MgBr_2$, aluminum nitrate, oxygen difluoride, Na_3PO_4
66. $CaCO_3$, mercury(I) ion, cobalt(II) chloride, SiO_2
67. $LiNO_2$, SO_3, Na_2O, ammonium sulfate
68. Dinitrogen tetroxide, N^{3-}, $Ca(ClO_3)_2$, sulfur

Charles D. Winters

Measuring the mass of a sulfur sample.

69. Phosphorus pentachloride, HPO_4^{2-}, CuO, ammonia
70. Tin(II) fluoride, $FeCO_3$, hypochlorous acid, chromium(II) bromide
71. $KHCO_3$, HNO_2, $Zn(HSO_4)_2$, copper(I) fluoride
72. Co_2O_3, Na_2SO_3, mercury(II) iodide, aluminum hydroxide
73. Magnesium nitride, lithium bromite, $NaHSO_3$, calcium dihydrogen phosphate
74. NH_4IO_3, H_2SeO_4 (Se is selenium, Z = 34), $Ni(HCO_3)_2$, CuS
75. Chromium(III) iodate, potassium hydrogen phosphate, Hg_2Cl_2, HIO_4
76. Cobalt(II) sulfate, lead(II) nitrate, selenium dioxide (selenium, Z = 34), magnesium nitrite
77. $FeBr_2$, Ag_2O, tetraphosphorus heptasulfide, $MnCl_2$
78. $NaClO_2$, SnO, K_2TeO_4 (Te is tellurium, Z = 52), $ZnCO_3$
79. Chromium(II) chloride, $Ni(ClO_3)_2$, cobalt(III) phosphate, calcium periodate
80. Cobalt(III) sulfate, iron(III) iodide, $Cu_3(PO_4)_2$, $Mn(OH)_2$
81. Calcium sulfite, CuCl, $AgNO_3$, Al_2Se_3 (Se is selenium, Z = 34)
82. $MgHPO_4$, potassium perchlorate, bromous acid, $NiCO_3$
83. Iron(II) oxide, hydrosulfuric acid, strontium iodate (strontium, Z = 38), sodium hypochlorite
84. Rb_2SO_4 (rubidium, Z = 37), P_2O_5, zinc phosphide, cesium nitrate (cesium, Z = 55)
85. S_2F_{10}, ICl, lead(II) dihydrogen phosphate, gallium fluoride (gallium, Z = 31)
86. N_2O_3, $Mn(OH)_3$, Magnesium sulfate, mercury(II) bromite

Answers to Practice Exercises

1. a) C_2H_6O; b) C_2H_6O; c) CH_2O; d) C_3H_6O; e) NH_3; f) Al_2O_3
2. Chlorine, Cl_2; argon, Ar; bromine, Br_2; krypton, Kr
3. N_2S_5, Si_4H_{10}, IF_7, tetraiodine nonoxide, tribromine octoxide, phosphorus pentafluoride
4. I^-, Ca^{2+}, P^{3-}, lithium ion, sulfide ion, aluminum ion
5. (a) Manganese(II) ion and manganese(III) ion; (b) Ni^{2+} and Ni^{3+}
6. Cobalt(III) ion, zinc ion, Al^{3+}, Cr^{3+}
7. CrF_3, Na_2S
8. $(NH_4)_2O$; $Zn(OH)_2$
9. K_3N; Co_2O_3
10. Copper(I) iodide, chromium(III) oxide, iron(II) hydroxide, tin(IV) bromide, barium chloride
11. (a) Tellurous acid, (b) $HBrO_4$
12. (a) Sulfurous acid; (b) HBrO
13. (a) Bromous acid; (b) H_2TeO_3
14. (a) Iodite ion; (b) IO_4^-
15. (a) Magnesium perbromate; (b) iron(II) nitrite; (c) copper(II) hypoiodite; (d) $Al(ClO_2)_3$; (e) $Pb_3(PO_4)_4$; (f) K_2SeO_4
16. Sodium hydrogen sulfate; NaHS
17. Aluminum bromate nonahydrate, $Co(ClO_4)_2 \cdot 5\,H_2O$

Answers to Concept-Linking Exercises

You may have found more relationships or relationships other than the ones given in these answers.

1. An atom is the smallest unit particle of an element. Two or more atoms can combine chemically to form molecules. A chemical formula represents a formula unit, which may be an actual particle, as with molecular compounds, or an electrically neutral combination of charged particles, as with ionic compounds.
2. An ion is an atom or chemically bound group of atoms that has an electric charge because of an excess or deficiency of electrons compared with protons. A cation has a positive charge, and an anion has a negative charge. A monatomic ion has only one atom, while a polyatomic ion is made up of two or more atoms. An acid anion is a polyatomic anion that includes an ionizable hydrogen atom.
3. An acid is a hydrogen-bearing molecular compound that reacts with water to produce a hydrated hydrogen ion and an anion. The hydrated hydrogen ion, $H^+ \cdot (H_2O)_x$, is sometimes represented by the hydronium ion, H_3O^+, or simply by the hydrogen ion, H^+.

4. Some solid crystal structures include water molecules. These crystals are called *hydrates*, and the water molecules in the crystal are water of hydration. A compound whose crystals contain no water molecules is an anhydrous compound.

Answers to Blue-Numbered Questions, Exercises, and Problems

1. a) N_2; b) NO_2; c) N_2O; d) N_2O_4
3. a) $NiCl_2$; b) Ag_2O; c) $C_6H_{12}O_6$; d) H_2SO_4
5. He, Ne, Ar, Kr, Xe, Rn
7. F_2, B, Ni, S (or S_8)
9. Chromium, chlorine, beryllium, iron
11. C, Si, Sn, Pb
13. Cl_2O, Br_3O_8, hydrogen bromide, diphosphorus trioxide
15. When an atom gains electrons, the particle that results is a monatomic anion. The ion has a negative charge because it has more electrons than protons.
17. Li^+ and N^{3-}
19. Co^{3+} and Al^{3+}; iron(III) ion and mercury(I) ion
21. Ni_2O_3, $CoCl_2$, Ag_2S, HgI_2
23. Silver bromide, tin(II) fluoride, iron(III) oxide, and copper(II) chloride
25. The formula of an acid usually begins with H, and the other elements are nonmetals.
27. Carbonic acid, H_2CO_3; nitric acid, HNO_3; phosphoric acid, H_3PO_4; sulfuric acid, H_2SO_4; chloric acid, $HClO_3$

29.

Acid Name	Acid Formula	Acid Name	Acid Formula
Sulfuric acid	H_2SO_4	Selenous acid	H_2SeO_3
Carbonic acid	H_2CO_3	Tellurous acid	H_2TeO_3
Chloric acid	$HClO_3$	Hypoiodous acid	HIO
Hydrofluoric acid	HF	Hypobromous acid	HBrO
Bromic acid	$HBrO_3$	Telluric acid	H_2TeO_4
Sulfurous acid	H_2SO_3	Perbromic acid	$HBrO_4$
Arsenic acid	H_3AsO_4	Hydrobromic acid	HBr
Periodic acid	HIO_4		

31. CO_3^{2-} and SeO_3^{2-}; bromate ion
33. Na_2TeO_4; magnesium hypobromite and iron(III) phosphate
35. An anion or cation that contains an ionizable hydrogen, such as HSO_4^- and NH_4^+, can release the hydrogen ion to a water molecule and thus behave as an acid.
37. HSO_3^-, HCO_3^-
39. Hydrogen selenite ion, hydrogen telluride ion
41. Hydrates: $NiSO_4 \cdot 6\,H_2O$, $Na_3PO_4 \cdot 12\,H_2O$; anhydrous compound: KCl
43. 7; magnesium sulfate heptahydrate
45. $(NH_4)_3PO_4 \cdot 3\,H_2O$, $K_2S \cdot 5\,H_2O$
47.

Ions	Potassium	Calcium	Chromium(III)	Zinc	Silver	Iron(III)	Aluminum	Mercury(I)
Nitrate	KNO_3	$Ca(NO_3)_2$	$Cr(NO_3)_3$	$Zn(NO_3)_2$	$AgNO_3$	$Fe(NO_3)_3$	$Al(NO_3)_3$	$Hg_2(NO_3)_2$
Sulfate	K_2SO_4	$CaSO_4$	$Cr_2(SO_4)_3$	$ZnSO_4$	Ag_2SO_4	$Fe_2(SO_4)_3$	$Al_2(SO_4)_3$	Hg_2SO_4
Hypochlorite	KClO	$Ca(ClO)_2$	$Cr(ClO)_3$	$Zn(ClO)_2$	AgClO	$Fe(ClO)_3$	$Al(ClO)_3$	$Hg_2(ClO)_2$
Nitride	K_3N	Ca_3N_2	CrN	Zn_3N_2	Ag_3N	FeN	AlN	omit
Hydrogen sulfide	KHS	$Ca(HS)_2$	$Cr(HS)_3$	$Zn(HS)_2$	AgHS	$Fe(HS)_3$	$Al(HS)_3$	$Hg_2(HS)_2$
Bromite	$KBrO_2$	$Ca(BrO_2)_2$	$Cr(BrO_2)_3$	$Zn(BrO_2)_2$	$AgBrO_2$	$Fe(BrO_2)_3$	$Al(BrO_2)_3$	$Hg_2(BrO_2)_2$
Hydrogen phosphate	K_2HPO_4	$CaHPO_4$	$Cr_2(HPO_4)_3$	$ZnHPO_4$	Ag_2HPO_4	$Fe_2(HPO_4)_3$	$Al_2(HPO_4)_3$	Hg_2HPO_4
Chloride	KCl	$CaCl_2$	$CrCl_3$	$ZnCl_2$	AgCl	$FeCl_3$	$AlCl_3$	Hg_2Cl_2
Hydrogen carbonate	$KHCO_3$	$Ca(HCO_3)_2$	$Cr(HCO_3)_3$	$Zn(HCO_3)_2$	$AgHCO_3$	$Fe(HCO_3)_3$	$Al(HCO_3)_3$	$Hg_2(HCO_3)_2$

48. $Ca(OH)_2$, NH_4Br, K_2SO_4

50. MgO, $AlPO_4$, Na_2SO_4, CaS

52. $BaSO_3$, Cr_2O_3, KIO_4, $CaHPO_4$

54.

Na^+	Hg^{2+}	NH_4^+
$NaOH$, sodium hydroxide $NaBrO$, sodium hypobromite Na_2CO_3, sodium carbonate $NaClO_3$, sodium chlorate $NaHSO_4$, sodium hydrogen sulfate $NaBr$, sodium bromide Na_3PO_4, sodium phosphate $NaIO_4$, sodium periodate Na_2S, sodium sulfide	$Hg(OH)_2$, mercury(II) hydroxide $Hg(BrO)_2$, mercury(II) hypobromite $HgCO_3$, mercury(II) carbonate $Hg(ClO_3)_2$, mercury(II) chlorate $Hg(HSO_4)_2$, mercury(II) hydrogen sulfate $HgBr_2$, mercury(II) bromide $Hg_3(PO_4)_2$, mercury(II) phosphate $Hg(IO_4)_2$, mercury(II) periodate HgS, mercury(II) sulfide	NH_4OH, ammonium hydroxide NH_4BrO, ammonium hypobromite $(NH_4)_2CO_3$, ammonium carbonate NH_4ClO_3, ammonium chlorate NH_4HSO_4, ammonium hydrogen sulfate NH_4Br, ammonium bromide $(NH_4)_3PO_4$, ammonium phosphate NH_4IO_4, ammonium periodate $(NH_4)_2S$, ammonium sulfide
Pb^{2+}	Mg^{2+}	
$Pb(OH)_2$, lead(II) hydroxide $Pb(BrO)_2$, lead(II) hypobromite $PbCO_3$, lead(II) carbonate $Pb(ClO_3)_2$, lead(II) chlorate $Pb(HSO_4)_2$, lead(II) hydrogen sulfate $PbBr_2$, lead(II) bromide $Pb_3(PO_4)_2$, lead(II) phosphate $Pb(IO_4)_2$, lead(II) periodate PbS, lead(II) sulfide	$Mg(OH)_2$, magnesium hydroxide $Mg(BrO)_2$, magnesium hypobromite $MgCO_3$, magnesium carbonate $Mg(ClO_3)_2$, magnesium chlorate $Mg(HSO_4)_2$, magnesium hydrogen sulfate $MgBr_2$, magnesium bromide $Mg_3(PO_4)_2$, magnesium phosphate $Mg(IO_4)_2$, magnesium periodate MgS, magnesium sulfide	
Fe^{3+}	Cu^{2+}	
$Fe(OH)_3$, iron(III) hydroxide $Fe(BrO)_3$, iron(III) hypobromite $Fe_2(CO_3)_3$, iron(III) carbonate $Fe(ClO_3)_3$, iron(III) chlorate $Fe(HSO_4)_3$, iron(III) hydrogen sulfate $FeBr_3$, iron(III) bromide $FePO_4$, iron(III) phosphate $Fe(IO_4)_3$, iron(III) periodate Fe_2S_3, iron(III) sulfide	$Cu(OH)_2$, copper(II) hydroxide $Cu(BrO)_2$, copper(II) hypobromite $CuCO_3$, copper(II) carbonate $Cu(ClO_3)_2$, copper(II) chlorate $Cu(HSO_4)_2$, copper(II) hydrogen sulfate $CuBr_2$, copper(II) bromide $Cu_3(PO_4)_2$, copper(II) phosphate $Cu(IO_4)_2$, copper(II) periodate CuS, copper(II) sulfide	

55. Lithium phosphate, magnesium carbonate, barium nitrate

57. Potassium fluoride, sodium hydroxide, calcium iodide, aluminum carbonate

59. Copper(II) sulfate, chromium(III) hydroxide, mercury(I) iodide

61.

Formula	Name	Name	Formula
OH^-	Hydroxide ion	Sodium bromide	$NaBr$
PbO	Lead(II) oxide	Ammonium phosphate	$(NH_4)_3PO_4$
SO_4^{2-}	Sulfate ion	Bromine	Br_2
Cl_2	Chlorine	Hydrogen phosphate ion	HPO_4^{2-}
Cl_2O	Dichlorine monoxide	Nitrite ion	NO_2^-
$Cr(ClO_4)_3$	Chromium(III) perchlorate	Iron(III) phosphate	$FePO_4$
$H_2PO_4^-$	Dihydrogen phosphate ion	Nickel(II) sulfide	NiS
Al_2S_3	Aluminum sulfide	Nitrogen	N_2
$CuSO_4$	Copper(II) sulfate	Copper(II) bromide	$CuBr_2$
$HF(aq)$	Hydrofluoric acid	Oxygen difluoride	OF_2
C	Carbon	Potassium ion	K^+

63. ClO_4^-, $BaCO_3$, ammonium iodide, phosphorus trichloride

65. Magnesium bromide, $Al(NO_3)_3$, OF_2, sodium phosphate

67. Lithium nitrite, sulfur trioxide, sodium oxide, $(NH_4)_2SO_4$

69. PCl_5, hydrogen phosphate ion, copper(II) oxide, NH_3

71. Potassium hydrogen carbonate, nitrous acid, zinc hydrogen sulfate, CuF

73. Mg_3N_2, $LiBrO_2$, sodium hydrogen sulfite, $Ca(H_2PO_4)_2$

75. $Cr(IO_3)_3$, K_2HPO_4, mercury(I) chloride, periodic acid

77. Iron(II) bromide, silver oxide, P_4S_7, manganese(II) chloride

79. $CrCl_2$, nickel(II) chlorate, $CoPO_4$, $Ca(IO_4)_2$

81. $CaSO_3$, copper(I) chloride, silver nitrate, aluminum selenide

83. FeO, $H_2S(aq)$ [the state designation (aq) distinguishes hydrosulfuric acid from dihydrogen sulfide, $H_2S(g)$], $Sr(IO_3)_2$, NaClO

85. Disulfur decafluoride, iodine monochloride, $Pb(H_2PO_4)_2$, GaF_3

7 Chemical Formula Relationships

Charles D. Winters

◀ How much time would you have to invest to count the grains in this sample of copper? It would take a while, wouldn't it? But how about counting all the *atoms* in this sample of copper? Even if all the people who ever lived assisted you, it would be impossible. Yet you can effectively count atoms if you weigh the sample. A major goal of this chapter is for you to learn how to count atoms or molecules by weighing samples.

CHAPTER CONTENTS

In Chapter 2, we introduced a critical component of a chemist's way of thinking: Chemists study the behavior of the tiny molecules that make up matter, and then they apply what they learn to carry out changes from one type of macroscopic matter to another. In this chapter, we introduce quantitative methods to connect the particulate and macroscopic views of matter. You may be returning to quantitative problem solving as you start this chapter. Recall the four-step approach that is outlined in Section 3.4:

how to... Solve a Quantitative Chemistry Problem

Step 1
- **ANALYZE** the problem statement.
- Determine the given quantity: value × unit.
- Describe the property of the given quantity.
- Describe the property of the wanted quantity.
- State the unit of the wanted quantity.

Step 2
- **IDENTIFY** equivalencies or an algebraic relationship that may be needed to solve the problem.
- Change the equivalencies to conversion factors or solve the algebraic equation for the wanted variable.

Step 3
- **CONSTRUCT** the solution setup.
- Confirm that the units cancel correctly and calculate the value of the answer.

Step 4
- **CHECK** the solution at two levels: (a) making sense and (b) what was learned.
- Making sense: Is the value reasonable?
- What was learned: What new knowledge or skill did I obtain or improve?

Note the emphasis on units in Steps 1 and 3. They are an essential part of solving all quantitative chemistry problems.

7.1 How Do You Weigh Something Too Small to Weigh?

Reexamine the photograph on the first page of Chapter 5 that is of a page from John Dalton's book *A New System of Chemical Philosophy.* Note that in addition to Dalton's "arbitrary marks or signs" that served as the original elemental symbols, there is a number next to each elemental name. These numbers were Dalton's experimentally determined relative weights of atoms of the elements. Recall that a postulate of his atomic theory was that all atoms of each element were identical and atoms of one element were different from atoms of other elements. One of these differences was that atoms of each element had their unique characteristic weights.

How could Dalton, in the early 1800s, weigh something that is too small to be perceived by the human senses? Of course, he couldn't. At least not in the conventional sense. What Dalton *could* do was develop measurements of *relative* weights.

Dalton relied on the measurements of others as the primary source of his data. As an example, in 1783, French scientist Antoine Lavoisier (Section 1.1) read a report to the Royal Academy of Sciences about his measurements of the reaction of hydrogen and oxygen gases to form liquid water. He had developed an apparatus that allowed premeasured amounts of the gases to burn very slowly in the ratio that produced complete combustion, and the liquid water product was captured so that it could be collected and weighed (**Figure 7.1**). In one trial, 5.0 weight units of water was formed from 0.6 units of hydrogen and 4.4 units of oxygen. (Note that these data support the validity of the Law of Conservation of Mass.)

To develop a list of relative atomic weights from data like these, Dalton made two assumptions. First, he assigned a weight of 1 to hydrogen. He did so because hydrogen always weighed less than other elements in experiments such as Lavoisier's and because hydrogen had the lowest density of any gas known at the time. This would make his system express weights in the form of a multiple of the weight of what he hypothesized was the lightest element. Second, Dalton hypothesized that compounds were formed from the simplest possible combination of atoms of the elements. For a two-element compound such as water, the simplest combination was one atom of hydrogen and one atom of oxygen.

Thus, the relative weight of oxygen based on Lavoisier's data was the ratio of the weight of oxygen to the weight of hydrogen: 4.4 units ÷ 0.6 unit = 7. The other

From Robert Routledge, "A Popular History of Science." London, 1881.

Figure 7.1 Lavoisier's apparatus for measurement of the amount of hydrogen and oxygen gases combusted and the amount of liquid water formed.

weights in Dalton's book were derived from similar experiments with substances such as ammonia, establishing the weight of nitrogen, and nitrogen monoxide, corroborating the weights of nitrogen and oxygen.

Soon after Dalton's book was published, experimental evidence that we will examine in Chapter 14 indicated that water consisted of a 2:1 ratio of hydrogen atoms to oxygen atoms. If 0.6 unit of hydrogen is the weight of two atoms, then one atom weighs 0.3 unit, and the relative weight of oxygen is 4.4 units $\div$ 0.3 unit $=$ 15. Higher quality measurements soon thereafter indicated that the relative weight of oxygen was 16, matching today's measurements.

Although Dalton did not know about the correct atom ratios in compounds in the early 1800s, his system of relative atomic weights is still used today. We assign a weight—technically a mass—to a particle and then express the mass of other particles in the form of a multiple of the mass of the standard. This chapter introduces the modern system of relative atomic masses and a number of interrelated concepts.

7.2 The Number of Atoms in a Formula

Goal 1 Given the formula of a chemical compound (or a name from which the formula may be written), state the number of atoms of each element in the formula unit.

In writing the formula of a substance, subscript numbers are used to indicate the number of atoms or groups of atoms of each element in the formula unit of the substance. There is one exception: If that number is 1, it is omitted. The formula of sodium nitrate is $NaNO_3$. The formula unit contains one sodium atom, one nitrogen atom, and three oxygen atoms.

The formula of calcium nitrate is $Ca(NO_3)_2$ (**Figure 7.2**). How many atoms of each element are in this formula? Applying the information from the preceding paragraph, you know that there is one calcium atom. There is one nitrogen atom in *each* of the two nitrate ions, so there are $2 \times 1 = 2$ nitrogen atoms. (This is like asking, "How many seats are on two tricycles?") There are three oxygen atoms in *each* of the two nitrate ions, so there are $2 \times 3 = 6$ oxygen atoms. (How many wheels are on two tricycles?)

In some examples in this chapter, you will be asked to solve quantitative problems that have only formula names in the question. Part of the problem solution is to write the chemical formula as you learned to do in Chapter 6. This will give you an opportunity to continue to improve your formula-writing skills. In every case, the correct formula is given in the first step of the example solution so that you may verify it before proceeding with the remainder of the problem. (You may skip the formula-writing step if your course hasn't yet covered Chapter 6 and you are not yet responsible for writing formulas.)

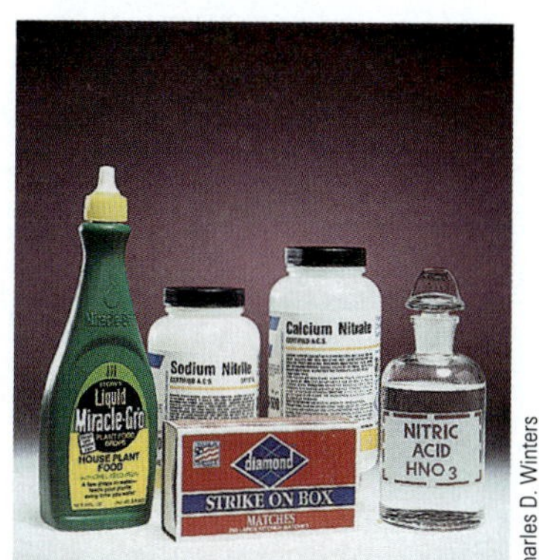

Figure 7.2 Sodium nitrate and calcium nitrate and products made from them. Sodium nitrate is also called "Chile saltpeter"—it occurs as a natural mineral in Chile and saltpeter means "salt of the rock." It is used in manufacturing sulfuric acid as well as other products. Calcium nitrate is used in explosives and as a corrosion inhibitor in diesel fuels. Both compounds are used in fertilizers and matches and in the manufacture of nitric acid.

Active Example 7.1 Number of Atoms in a Formula

How many atoms of each element are in a formula unit of (a) magnesium chloride and (b) barium iodate?

Think Before You Write If you understand what parentheses symbolize in a chemical formula, you are ready to complete this Active Example.

Answers *Cover the left column with your cut-out shield. Reveal each answer only after you have written your own answer in the right column.*

$MgCl_2$ and $Ba(IO_3)_2$ *$MgCl_2$ comes from the Mg^{2+} ion giving a 2+ charge and two 1− Cl^- ions giving a total 2− charge. The formula of barium (Ba^{2+}) iodate (IO_3^-) is developed in the same way.*	Before you can answer these questions, you need the formulas of magnesium chloride and barium iodate. Use only a periodic table as a guide.

$MgCl_2$: 1 magnesium atom, 2 chlorine atoms $Ba(IO_3)_2$: 1 barium atom, 2 iodine atoms, 6 oxygen atoms *We find the two iodine atoms and six oxygen atoms in $Ba(IO_3)_2$ the same way we found the two nitrogen atoms and six oxygen atoms in $Ca(NO_3)_2$: 2 × 1 for iodine, and 2 × 3 for oxygen.*	How many atoms of each element are in each formula? 
You improved your skill at determining the number of atoms in a formula.	What did you learn by solving this Active Example? 

Practice Exercise 7.1

How many atoms of each element are in a formula unit of (a) ammonium sulfate and (b) aluminum carbonate?

7.3 Molecular Mass and Formula Mass

Goal 2 Distinguish among atomic mass, molecular mass, and formula mass.
3 Calculate the formula (molecular) mass of any compound whose formula is given (or known).

In Section 5.6, you learned that the atomic mass of an element is the average mass of its atoms, usually expressed in atomic mass units. Similarly, for compounds, the average mass of a molecule or a formula unit is called its **molecular mass** (for molecules) or its **formula mass** (for any substance for which a chemical formula can be written). ◀

▶ **P/Review** The term *molecular mass* applies to compounds that have covalent bonds, and the term *formula mass* applies to all compounds, but it is most commonly used with compounds with at least one ionic bond (Sections 12.3 and 12.4).

The formula mass of a compound is equal to the sum of all the atomic masses in its chemical formula. For example, consider carbon dioxide, CO_2. There are one carbon atom and two oxygen atoms in the molecule. The molecular mass is the sum of the atomic masses of these three atoms (**Figure 7.3**):

Element	Atoms in Formula		Atomic Mass		Mass in Formula
Carbon	1	×	12.01 u	=	12.01 u
Oxygen	2	×	16.00 u	=	32.00 u
			Total molecular mass:		44.01 u

16.00 u + 12.01 u + 16.00 u = 44.01 u

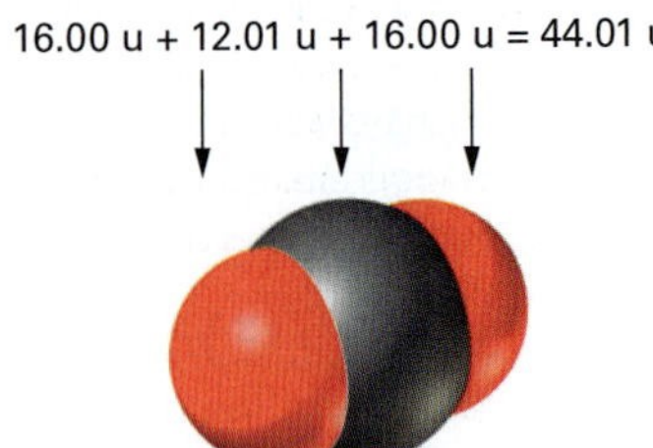

Figure 7.3 Molecular (or formula) mass. Molecular mass is the sum of the atomic masses of each atom in the molecule. The molecular mass of a carbon dioxide molecule is the sum of the masses of one atom of carbon and two atoms of oxygen.

Active Example 7.2 Molecular Mass and Formula Mass

Calculate the formula mass of ammonium nitrate (Figure 7.4).

Think Before You Write This is likely to be the first time in a while that you have needed to apply the addition/subtraction significant figure rule. Review it (Section 3.8) if you can't recall how it is applied.

Answers *Cover the left column with your cut-out shield. Reveal each answer only after you have written your own answer in the right column.*

NH_4NO_3 *In writing the formula of an ionic compound, the cation always comes first, followed by the anion. Both ions in this compound are polyatomic.* *Both the ammonium ion, NH_4^+, and the nitrate ion, NO_3^-, contain nitrogen. They are not combined to $N_2H_4O_3$ in writing the formula, so that each ion keeps its identity in the compound formula.*	First, you need the formula of ammonium nitrate. 

There are two N atoms, four H atoms, and three O atoms in NH_4NO_3:

$$
\begin{array}{r@{\;}l}
2 \times 14.01\ \ \text{u} = & 28.02\ \ \ \ \text{u} \\
4 \times \ 1.008\ \text{u} = & \ \ 4.03\,2\ \text{u} \\
3 \times 16.00\ \ \text{u} = & 48.00\ \ \ \ \text{u} \\
\hline
& 80.05\,2\ \text{u} = 80.05\ \text{u}
\end{array}
$$

If you had any difficulty getting the correct answer on your calculator, take a few minutes to learn the technique. Learn it now!

Now count up the atoms of each element in NH_4NO_3 and calculate its formula mass. Use a vertical setup for your addition to help in applying the addition/subtraction rule for significant figures. ▶

You improved your skill at calculating molecular or formula mass.

What did you learn by solving this Active Example?

Practice Exercise 7.2

Calculate the formula mass of sulfuric acid.

When considering significant figures in formula and molecular masses, always think in terms of the addition/subtraction rule. For example, for P_4, think 30.97 u + 30.97 u + 30.97 u + 30.97 u = 123.88 u.

7.4 Stoichiometric Amount

Goal 4 Define the term *mole*. Identify the number of objects that correspond to 1 mole.

5 Given the number of moles (or units) in any sample, calculate the number of units (or moles) in the sample.

A single molecule is almost unimaginably small. This makes molecular mass inconvenient to use in the laboratory, where we work with macroscopic quantities of matter, not individual molecules. There are huge numbers of molecules in one milliliter of water or one gram of lithium, for example. This makes it fitting to use a grouping unit that corresponds to a number of molecules that typically would be in a macroscopic sample. The unit used in the sciences to represent a certain large collection of particles is called a **mole**. In calculation setups, the mole is abbreviated as **mol**.

A mole contains exactly $6.02214076 \times 10^{23}$ particles. This number is established by definition. We will often round it to three significant figures in calculation setups, writing the equivalency as 1 mol (of a substance) = $\mathbf{6.02 \times 10^{23}}$ (type of particles) (of a substance). ▶ As examples, 1 mol H_2O = 6.02×10^{23} molecules H_2O and 1 mol Li = 6.02×10^{23} atoms Li.

Why did the scientific community choose the number $6.02214076 \times 10^{23}$ as the definition of the number of particles in a mole? It corresponds to the measured number of atoms in 12 grams of carbon-12. Twelve grams is a convenient macroscopically sized sample. If you were able to measure out 12 grams of a pure carbon-12 sample, your sample would be 1 mole of carbon atoms.

$6.02214076 \times 10^{23}$/mol is called **Avogadro's number**, or the **Avogadro constant,** N_A, in honor of Italian scientist Amedeo Avogadro (Figure 7.5), whose experiments with gases eventually led to the mole definition. The amount in moles of a substance may be called the **stoichiometric amount**. The term *stoichiometry* was coined from the Greek *stoikheion,* meaning "elements," and the English -metry, meaning "to measure." Hence, the stoichiometric amount is a measure of the amount, typically the number of moles, of an element, and by extension, combinations of elements in the form of compounds.

$6.02214076 \times 10^{23}$/mol is a huge number. To get some appreciation of the size of the Avogadro constant, if you were to try to count 6×10^{23} objects and proceeded at the rate of 100 objects per minute without interruption, 24 hours per day,

AP images/David Longstreath

Figure 7.4 The Alfred P. Murrah Federal Building in Oklahoma City soon after the April 19, 1995 bombing. Seven thousand pounds of ammonium nitrate was one component of the explosive used in the tragic bombing. It also is used in making many common products, such as anesthetics, matches, fireworks, and fertilizer.

Our three-significant-figure roundoff will be adequate for most of the calculations in this book. However, a definition should never limit the number of significant figures in a calculation. Only measured quantities should limit the number of significant figures to which a calculation is expressed.

E. F. Smith Collection/Van Pelt Library/University of Pennsylvania.

Figure 7.5 Amedeo Avogadro's (1776–1856) hypothesis that equal volumes of gases at the same temperature and pressure contain the same number of molecules provided the conceptual foundation on which the definition of the mole and the Avogadro constant were built.

365 days per year, the task would require 11,500,000,000,000,000 years. This is about 2 million times as long as Earth has been in existence!

It is convenient to think of the mole as the number 6.02×10^{23} just as we think of a dozen as the number 12. In this sense, as one dozen eggs is 12 eggs, one mole of eggs is 6.02×10^{23} eggs. If we wanted to find the number of eggs in three dozen eggs, we would set up the problem this way:

$$3\ \cancel{\text{doz eggs}} \times \frac{12\ \text{eggs}}{1\ \cancel{\text{doz eggs}}} = 36\ \text{eggs}$$

Similarly, the number of carbon atoms in three moles of carbon atoms is:

$$3\ \cancel{\text{mol C atoms}} \times \frac{6.02 \times 10^{23}\ \text{C atoms}}{1\ \cancel{\text{mol C atoms}}} = 1.81 \times 10^{24}\ \text{C atoms}$$

Thinking About Your Thinking

Proportional Reasoning

The number of eggs in a dozen and the number of particles in a mole are both examples of direct proportionalities used in proportional reasoning: (number of eggs) = 12 × (dozens of eggs) and (number of particles) = 6.02×10^{23} × (moles of particles).

Active Example 7.3 The Avogadro Constant

What is the stoichiometric amount of water, expressed in moles, in a sample of 1.49×10^{21} water molecules?

Think Before You Write How would you calculate the number of dozens of eggs in 48 eggs? This problem is solved in exactly the same way.

Answers *Cover the left column with your cut-out shield. Reveal each answer only after you have written your own answer in the right column.*

Given: 1.49×10^{21} molecules water **Wanted:** stoichiometric amount, mol water	**Analyze** the problem statement by writing the given quantity and the property and unit of the wanted quantity.
1 mol water = 6.02×10^{23} molecules water $\frac{1\ \text{mol water}}{6.02 \times 10^{23}\ \text{molecules water}}$	**Identify** the equivalency that relates the given and wanted units. Change it to the needed conversion factor.
$1.49 \times 10^{21}\ \cancel{\text{molecules water}} \times \frac{1\ \text{mol water}}{6.02 \times 10^{23}\ \cancel{\text{molecules water}}} = 0.00248\ \text{mol water}$ *You may be interested to know that 0.00248 mole of water is about one drop.*	**Construct** the solution setup. Cancel units and calculate the value of the answer. 
$\frac{1.5 \times 10^{21}}{6.0 \times 10^{23}} = 0.25 \times 10^{-2} = 0.0025$ The value of the answer is reasonable, and the units canceled to give the wanted unit. You improved your skill at converting between number of particles and number of particles grouped in moles.	**Check** the solution. Is the value reasonable? Are the units correct? What did you learn by solving this Active Example? 

Practice Exercise 7.3

A sample consists of 0.25 mole of ammonia. How many ammonia molecules is this?

7.5 Molar Mass

Goal 6 Define *molar mass* or interpret statements in which the term *molar mass* is used.
7 Calculate the molar mass of any substance whose chemical formula can be written or is given.

Calculation of the atomic, molecular, and formula masses of atoms, molecules, and compounds has a limited usefulness because we cannot isolate and weigh an individual particle. To make measurements of mass useful, we must express chemical quantities that can be measured at the macroscopic level. The bridge between the particulate and the macroscopic levels is **molar mass**, mass divided by stoichiometric amount. Mass is typically expressed in grams, and stoichiometric amount is expressed in moles. The units of molar mass follow: grams per mole (g/mol).

The definitions of atomic mass, the mole, and molar mass are all directly or indirectly related to carbon-12. Consider two definitions:

1. 1 atomic mass unit $\equiv \frac{1}{12}$ the mass of one carbon-12 atom (Section 5.6).
2. 1 mole $\equiv$ exactly $6.02214076 \times 10^{23}$ particles = the number of atoms in 12 g of carbon-12 (Section 7.4).

This leads to two important facts:

1. The mass of one atom of carbon-12—the atomic mass of carbon-12—is exactly 12 atomic mass units.
2. The mass of 1 mole of carbon-12 atoms is 12 g; its molar mass is 12 grams per mole.

Notice that the atomic mass and the molar mass of carbon-12 are *numerically* equal. They differ only in units; atomic mass is measured in atomic mass units, and molar mass is measured in grams per mole. The same relationships exist between atomic and molar masses of elements, between molecular masses and molar masses of molecular substances, and between formula masses and molar masses of ionic compounds. In other words,

The molar mass of any substance in grams per mole is numerically equal to the atomic, molecular, or formula mass of that substance in atomic mass units.

Once you've determined the atomic, molecular, or formula mass of a substance, change the units from u to g/mol and you have its molar mass.

Active Example 7.4 Molar Mass

Calculate the molar mass of (a) fluorine and (b) calcium fluoride (Figure 7.6).

Think Before You Write If you are confident that you learned how to calculate formula mass in Section 7.3, you are ready to proceed with this Active Example. If you have any doubt about your ability to calculate formula mass, review Section 7.3, and then return to this Active Example.

Answers *Cover the left column with your cut-out shield. Reveal each answer only after you have written your own answer in the right column.*

F_2 and CaF_2	Write the formulas of fluorine and calcium fluoride.

Figure 7.6 Calcium fluoride occurs in nature as the mineral known as fluorite or fluorspar. It is used in making steel.

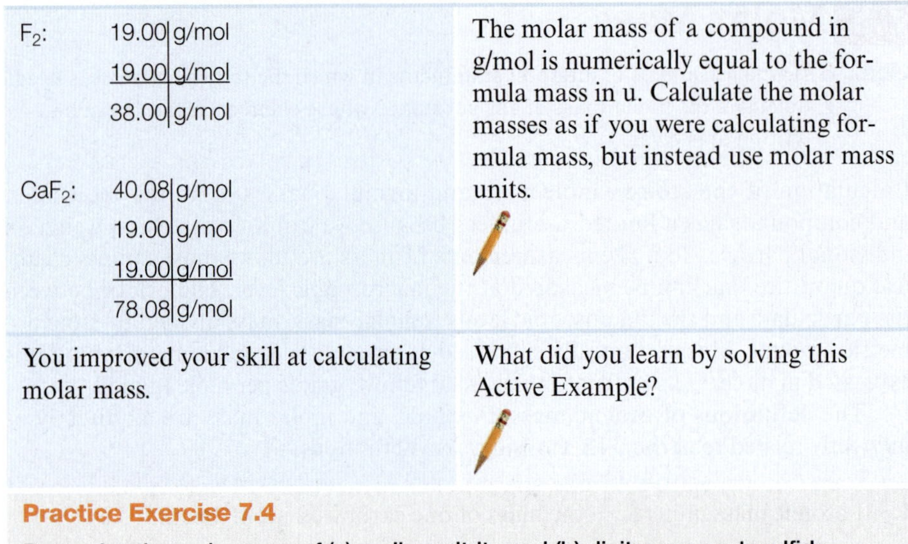

F_2: 19.00 g/mol 19.00 g/mol 38.00 g/mol CaF_2: 40.08 g/mol 19.00 g/mol 19.00 g/mol 78.08 g/mol	The molar mass of a compound in g/mol is numerically equal to the formula mass in u. Calculate the molar masses as if you were calculating formula mass, but instead use molar mass units.
You improved your skill at calculating molar mass.	What did you learn by solving this Active Example?

Practice Exercise 7.4

Determine the molar mass of (a) sodium nitrite and (b) dinitrogen pentasulfide.

There is one mole of each substance in **Figure 7.7**. This means that each sample has the same number of atoms, molecules, or formula units. The mass of each sample in grams is numerically the same as the molar mass of each substance.

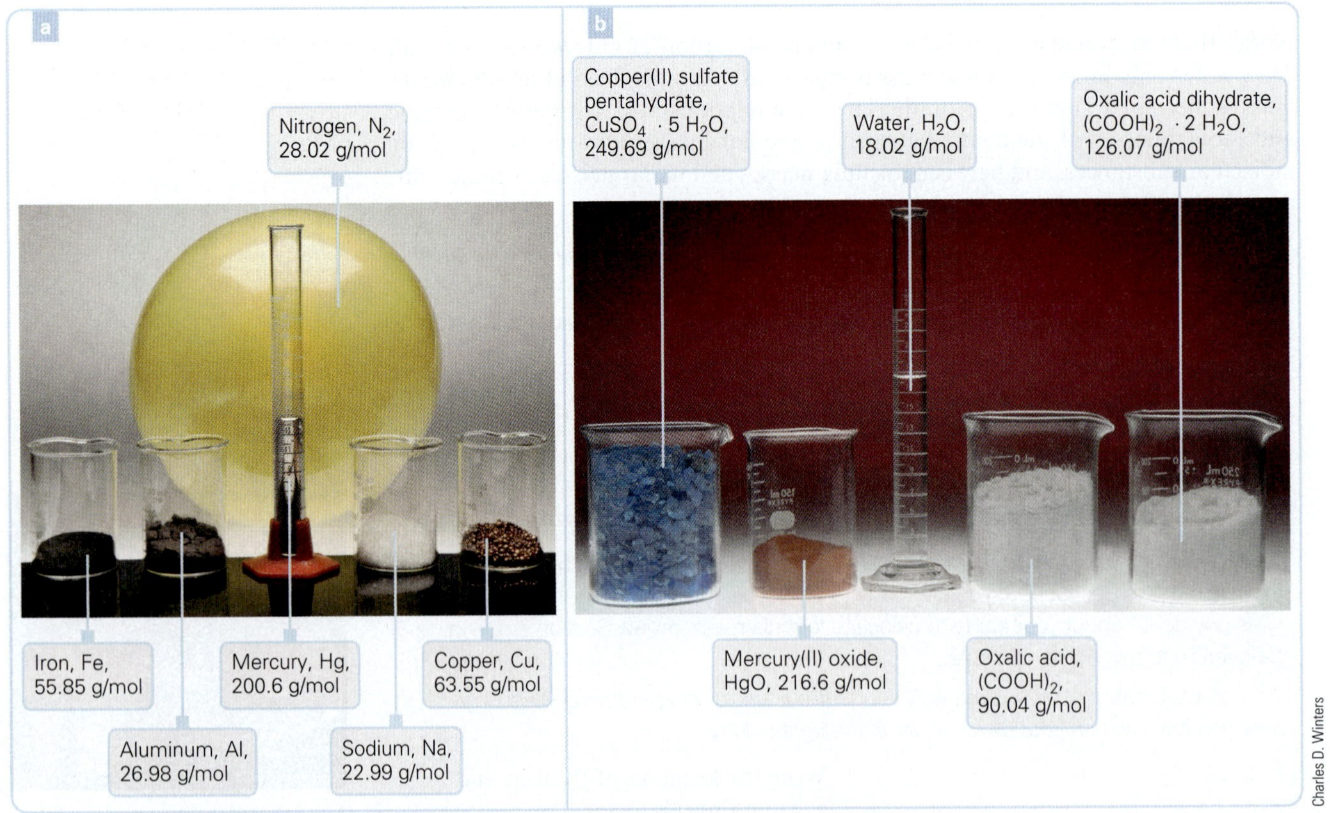

Figure 7.7 (a) One mole of elements and (b) one mole of molecular and ionic compounds.

7.6 Conversion Among Mass, Amount in Moles, and Number of Units

Goal 8 Given any one of the following for a substance with a given (or known) formula, calculate the other two: (a) mass; (b) amount in moles; (c) number of formula units, molecules, or atoms.

Molar mass is a conversion factor that allows you to convert from grams to moles or from moles to grams. This one-step conversion is probably used more often than any other conversion in an introductory chemistry course because mass is measured on the macroscopic level, typically in grams. However, chemical reactions take place on the particulate level, and moles are a grouping unit that express the number of particles.

Active Example 7.5 Conversion Between Amount and Mass

You are carrying out a laboratory reaction that requires 0.0350 mole of barium chloride. How many grams of the compound do you measure out?

Think Before You Write Molar mass is a conversion factor that allows you to convert between mass in grams and amount in number of moles.

Answers *Cover the left column with your cut-out shield. Reveal each answer only after you have written your own answer in the right column.*

Given: 0.0350 mole of barium chloride **Wanted:** mass in grams	**Analyze** the problem statement. What is the given quantity? What is the property and unit of the wanted quantity?
Molar mass	**Identify** the type of equivalency needed to convert between quantity in moles and mass in grams. Don't determine its value yet; just state the general term used to describe the needed conversion factor.
$BaCl_2$	Before you can use molar mass as a conversion factor, you must calculate its value. This first requires the formula of barium chloride.
$BaCl_2$: 137.3 g/mol 35.45 g/mol 35.45 g/mol 208.20 g/mol = 208.2 g/mol	You are now ready to calculate the molar mass of barium chloride.
$0.0350 \text{ mol BaCl}_2 \times \frac{208.2 \text{ g BaCl}_2}{1 \text{ mol BaCl}_2} = 7.29 \text{ g BaCl}_2$	**Construct** the solution setup, cancel units, and calculate the value of the answer.
$(3.5 \times 10^{-2}) \times (2 \times 10^{2}) = 7 \times 10^{0} = 7$ The value is reasonable. You improved your skill at making conversions between amount and mass.	**Check** the solution to be sure it makes sense, and state what you learned from working this Active Example.

Practice Exercise 7.5

What amount, expressed in moles, of phosphorus trichloride molecules is in 765 mg of the pure substance?

Thinking About Your Thinking

Proportional Reasoning

The idea of counting by weighing was introduced in the photograph and caption on the first page of this chapter. Many grocery items are sold by weight instead of number because of the inconvenience of counting small objects. Cookies, candies, nuts, and fruits, for example, could be sold by the piece, but instead they are usually sold by the pound in the United States.

Let's say that you decide to eat two dozen cherries a day for the next week. When you go to the grocery store, you will not find a price per cherry or per dozen cherries, but rather a price per pound of cherries. Even though you are thinking about the number of cherries, grouped in dozens, your grocer thinks about the number of pounds of cherries. You need to understand both methods for expressing the amount of cherries.

Similarly, in chemistry, you have to learn to think about both the amount of particles, grouped in moles, and the mass of those particles. Considering moles of particles is useful when you are thinking about models of the particles and their reactivity. On the other hand, when you want a certain number of particles for a chemistry experiment, you will often weigh a specified amount of the macroscopic solid. So thinking about mass is important, too. You have to understand both methods for expressing the amount of a substance.

The link between the amount in moles of particles and their mass is molar mass, expressed in grams per mole. The mass in grams of a pure substance is directly proportional to the stoichiometric amount in moles of that substance. The examples that follow will provide you with practice in thinking as a chemist by using proportional reasoning to convert between measured macroscopic quantities and particulate quantities.

In Active Example 7.3, you converted number of molecules into moles using the Avogadro constant. In Active Example 7.5, you converted moles into grams using molar mass. Notice that the mole is present in both of these changes. In fact, the mole is the connecting link between the macroscopic world, in which we often measure quantities in grams, and the particulate world, in which we count the number of atoms, molecules, or formula units, which we will refer to in general as number of units. Using N_A for the Avogadro constant, 6.02×10^{23}/mol, and MM for molar mass in g/mol, we have

$$\text{units} \xrightarrow{N_A} \text{mol} \xrightarrow{MM} \text{g} \quad or \quad \text{g} \xrightarrow{MM} \text{mol} \xrightarrow{N_A} \text{units}$$

Changing from units to mass or vice versa is a two-step conversion: Change the given quantity to moles, and then change moles to the wanted quantity. Figure 7.8 summarizes the process of conversion among mass, amount in moles, and number of unit particles.

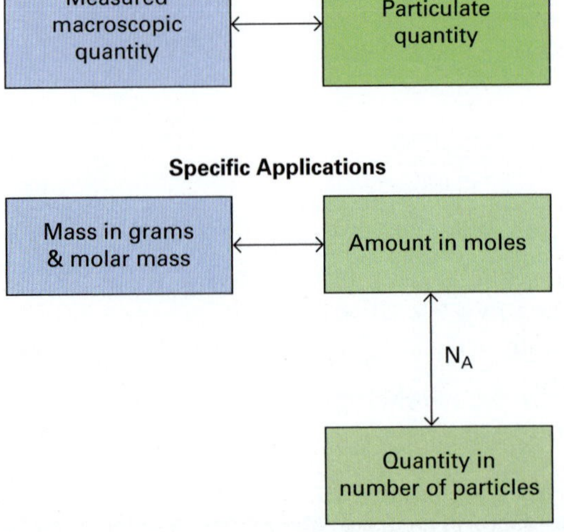

Figure 7.8 Conversion among grams, moles, and number of particles. Molar mass, MM, is a conversion factor that links mass in grams, a measurable macroscopic quantity, with amount in moles, the number of particles in the sample. The Avogadro constant, N_A, links the number of particles grouped in moles to the absolute number of particles. The overarching pattern, illustrated by the two boxes at the top of the figure, is that a measured macroscopic quantity in a blue box (e.g., mass in grams, pounds) can be converted to a particulate quantity in a green box.

Active Example 7.6 Conversion Between Mass and Number of Particles

How many molecules are in 454 g of water? (454 g equals 1 lb)

Think Before You Write Look at Figure 7.8. You are given mass in grams in this problem, so your solution path will move from the blue mass in grams and molar mass box to the green box on the right, amount in moles, and then to the green box below it, quantity in number of particles, which are molecules in this case. In the generalized case (top boxes), you are converting a macroscopic measured quantity, mass in grams, to a particulate quantity, number of molecules.

Answers *Cover the left column with your cut-out shield. Reveal each answer only after you have written your own answer in the right column.*

Given: 454 g water
Wanted: number of molecules

Analyze the problem statement. Write the given quantity and the unit of the wanted quantity.

H_2O:

$$\begin{array}{r|l} 1.00 & 8 \text{ g/mol} \\ 1.00 & 8 \text{ g/mol} \\ \underline{16.00} & \underline{\ \ \text{g/mol}} \\ 18.01 & 6 \text{ g/mol} = 18.02 \text{ g/mol} \end{array}$$

$$\frac{6.02 \times 10^{23} \text{ molecules } H_2O}{1 \text{ mol } H_2O}$$

Identify the equivalencies needed to solve the problem and change them to (or write them directly as) conversion factors. Recall from the discussion earlier and Figure 7.8 that you need to make two conversions.

$$454 \text{ g } H_2O \times \frac{1 \text{ mol } H_2O}{18.02 \text{ g } H_2O} \times \frac{6.02 \times 10^{23} \text{ molecules } H_2O}{1 \text{ mol } H_2O}$$

$$= 1.52 \times 10^{25} \text{ molecules } H_2O$$

Construct the solution setup, cancel units, and determine the answer.

$$\frac{(5 \times 10^2)(6 \times 10^{23})}{2 \times 10^1} = 15 \times 10^{24} = 1.5 \times 10^{25}$$

The value of the answer is reasonable.

You improved your skill at solving problems involving conversions between measurable macroscopic quantities and number of units.

Check the solution. Is the value of the answer reasonable? What did you learn by solving this Active Example?

Practice Exercise 7.6

Determine the mass in grams of 3.2×10^{24} molecules of sulfur tetrafluoride.

7.7 Mass Relationships Among Elements in a Compound: Percentage Composition by Mass

Goal 9 Calculate the percentage composition by mass of any compound whose formula is given (or known).

10 Given the mass of a sample of any compound with a given (or known) formula, calculate the mass of any element in the sample, or, given the mass of any element in the sample, calculate the mass of the sample or the mass of any other element in the sample.

The term *cent* refers to 100. For example, there are 100 cents in a dollar and 100 years in a century. Percent, therefore, means "per 100." Thus, **percent** is the amount of one part of a mixture per 100 total parts in the mixture. If the part whose percentage we wish to identify is A, then

$$\% \text{A} \equiv \frac{\text{parts of A in the mixture}}{\text{100 total parts in the mixture}} \tag{7.1}$$

Equation 7.1 is a defining equation for percentage. To calculate percentage, we use a more convenient form that is derived from Equation 7.1:

$$\% \text{A} = \frac{\text{parts of A}}{\text{total parts}} \times 100\%$$

The ratio (parts of A/total parts) is the fraction of the sample that is A. Multiplying that fraction by 100 gives the percentage of A. To illustrate, in Active Example 7.4 you calculated the molar mass of calcium fluoride. The calculation setup was

$$40.08 \text{ g/mol Ca} + 2(19.00 \text{ g/mol F}) = 78.08 \text{ g/mol } CaF_2$$

Thus, for one mole of calcium fluoride, the total mass is 78.08 g, with 40.08 g due to the calcium ions. The mass fraction of a mole that is calcium is consequently 40.08 g/78.08 g. The mass percentage of calcium is therefore

$$\% \text{ Ca by mass} = \frac{\text{parts of Ca}}{\text{total parts}} \times 100\% = \frac{40.08 \text{ g Ca}}{78.08 \text{ g } CaF_2} \times 100\% = 51.33\% \text{ Ca}$$

Active Example 7.7 Percentage Composition by Mass I

Calculate the mass percentage of fluorine in calcium fluoride.

Think Before You Write This is an application of $\% \text{A} = \frac{\text{parts of A}}{\text{total parts}} \times 100\%$

Answers *Cover the left column with your cut-out shield. Reveal each answer only after you have written your own answer in the right column.*

Given: 2(19.00 g F); 78.08 g CaF_2
Wanted: % F by mass

$$\% \text{ F by mass} = \frac{\text{g F}}{\text{g } CaF_2} \times 100\% = \frac{2(19.00 \text{ g F})}{78.08 \text{ g } CaF_2} \times 100\% = 48.67\% \text{ F}$$

$$\frac{2 \times 20 \times 100}{80} = \frac{40 \times 100}{80} = \frac{1}{2} \times 100 = 50$$

The value of the answer is reasonable.

Notice that you are finding the percentage of the element fluorine in the compound, which is % F, not % F_2.

Set up and solve.

You improved your skill at calculating percentage composition by mass.

What did you learn by solving this Active Example?

Practice Exercise 7.7

Determine the mass percentage of chlorine in aluminum chlorate.

The **percentage composition by mass** of a compound is the percentage by mass of each element in the compound. The percentage composition by mass of calcium fluoride is 51.33% calcium and 48.67% fluorine. As you have seen with CaF_2, percentage composition by mass can be calculated from the same numbers that are used to find the molar mass of a compound.

If you calculate the percentage composition by mass of a compound correctly, the sum of all percents must be 100%. This fact can be used to check your work. With calcium fluoride, 51.33% + 48.67% = 100.00%. When you apply this check, don't be concerned if you are high or low by ±1 or even ±2 in the uncertain digit of the sum of the percents. This can result from legitimate roundoffs along the way.

Active Example 7.8 Percentage Composition by Mass II

Calculate the percentage composition by mass of aluminum sulfate (Figure 7.9).

Think Before You Write When you are asked to determine the percentage composition by mass of a compound, it means that you need to calculate the mass percentage of each of the elements in the compound.

Answers *Cover the left column with your cut-out shield. Reveal each answer only after you have written your own answer in the right column.*

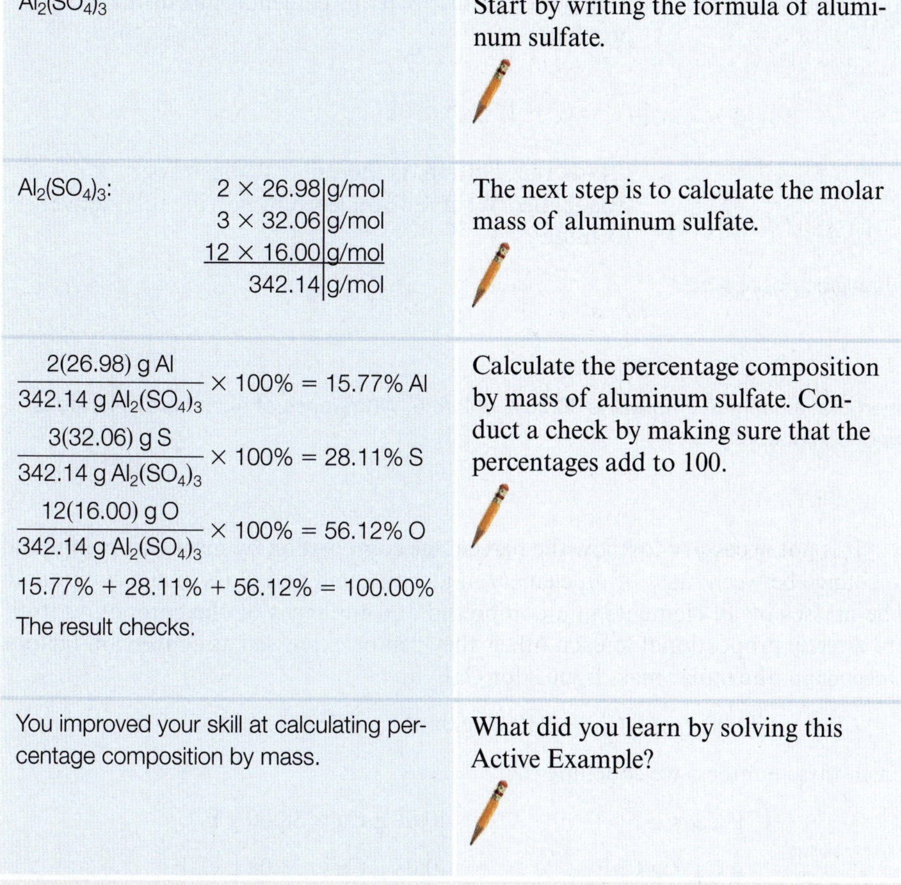

$Al_2(SO_4)_3$	Start by writing the formula of aluminum sulfate.
$Al_2(SO_4)_3$: 2×26.98 g/mol; 3×32.06 g/mol; 12×16.00 g/mol; 342.14 g/mol	The next step is to calculate the molar mass of aluminum sulfate.
$\frac{2(26.98)\text{ g Al}}{342.14\text{ g }Al_2(SO_4)_3} \times 100\% = 15.77\%\text{ Al}$ $\frac{3(32.06)\text{ g S}}{342.14\text{ g }Al_2(SO_4)_3} \times 100\% = 28.11\%\text{ S}$ $\frac{12(16.00)\text{ g O}}{342.14\text{ g }Al_2(SO_4)_3} \times 100\% = 56.12\%\text{ O}$ $15.77\% + 28.11\% + 56.12\% = 100.00\%$ The result checks.	Calculate the percentage composition by mass of aluminum sulfate. Conduct a check by making sure that the percentages add to 100.
You improved your skill at calculating percentage composition by mass.	What did you learn by solving this Active Example?

Figure 7.9 Aluminum sulfate is used in manufacturing paper. Unfortunately, moisture in the air reacts with the compound, forming an acid that causes paper to yellow and eventually disintegrate.

Practice Exercise 7.8

Determine the percentage composition by mass of sodium sulfide nonahydrate.

If you have already calculated the percentage composition of a compound, you can find the amount of any element in a known amount of the compound. One way to do this is to use percentage as a conversion factor, grams of the element per 100 g of the compound. For example, because aluminum sulfate is 15.77% aluminum, the mass of aluminum in 88.9 g $Al_2(SO_4)_3$ is

$$88.9\ \cancel{\text{g } Al_2(SO_4)_3} \times \frac{15.77\text{ g Al}}{100\ \cancel{\text{g } Al_2(SO_4)_3}} = 14.0\text{ g Al}$$

Are there three significant figures in the denominator, 100 g $Al_2(SO_4)_3$? The 100 is a defined quantity, like 12 inches is equal to 1 foot. The total percentage of anything is defined to be exactly 100. The denominator, therefore, has an infinite number of significant figures, more than any other measured quantity in the calculation. The significant figures are limited by the measured quantity with the smallest number of significant digits in the multiplication/division setup, which in this case is the three-significant-figure measured quantity 88.9 g $Al_2(SO_4)_3$.

Active Example 7.9 Percentage Composition by Mass III

What is the mass in grams of fluorine in 216 g of calcium fluoride?

Think Before You Write The key concept is to use percentage as a conversion factor, grams of the element per 100 g of the compound.

Answers *Cover the left column with your cut-out shield. Reveal each answer only after you have written your own answer in the right column.*

Given: 216 g CaF_2

Wanted: g F

$48.67 \text{ g F} = 100 \text{ g } CaF_2$

$$216 \text{ g } \cancel{CaF_2} \times \frac{48.67 \text{ g F}}{100 \text{ g } \cancel{CaF_2}} = 105 \text{ g F}$$

In Active Example 7.7 you found that calcium fluoride is 48.67% fluorine. Use this percentage to solve the problem.

$$\frac{200 \times 50}{100} = 2 \times 50 = 100$$

The value of the answer is reasonable.

You improved your skill at using percentage composition by mass as a conversion factor.

Check the solution. Is the value of the answer reasonable? What did you learn by solving this Active Example?

Practice Exercise 7.9

In Practice Exercise 7.7, you determined that aluminum chlorate is 38.35% chlorine. What mass of aluminum chlorate is needed as a source of 50.0 milligrams of chlorine?

It is not necessary to know the percentage composition by mass of a compound to change between mass of an element in a compound and mass of the compound. The masses of all elements in a compound and the mass of the compound itself are directly proportional to each other; they can be expressed as conversion factors. Once again, the molar mass figures for CaF_2 are:

$$40.08 \text{ g/mol Ca} + 2(19.00 \text{ g/mol F}) = 78.08 \text{ g/mol } CaF_2$$

From these numbers we conclude that

$\text{g Ca} \propto \text{g F}$	$40.08 \text{ g Ca} \propto 38.00 \text{ g F}$
$\text{g Ca} \propto \text{g } CaF_2$	$40.08 \text{ g Ca} \propto 78.08 \text{ g } CaF_2$
$\text{g F} \propto \text{g } CaF_2$	$38.00 \text{ g F} \propto 78.08 \text{ g } CaF_2$

Any of these proportionalities, or the equivalencies that result from them, may be changed into a conversion factor from the mass of one species to the mass of the other. For instance, to find the mass of CaF_2 that contains 3.55 g Ca from these numbers, we calculate

$$3.55 \text{ g } \cancel{Ca} \times \frac{78.08 \text{ g } CaF_2}{40.08 \text{ g } \cancel{Ca}} = 6.92 \text{ g } CaF_2$$

Active Example 7.10 Percentage Composition by Mass IV

Calculate the mass in grams of fluorine in a sample of calcium fluoride that contains 2.01 g of calcium.

Think Before You Write The molar mass figures for calcium fluoride are given in the foregoing text. All masses are proportional to one another.

Answers *Cover the left column with your cut-out shield. Reveal each answer only after you have written your own answer in the right column.*

Given: 2.01 g Ca **Wanted:** g F	**Analyze** the problem statement by writing the given quantity and wanted unit.
38.00 g F = 40.08 g Ca $\frac{38.00 \text{ g F}}{40.08 \text{ g Ca}}$	**Identify** the equivalency needed to solve the problem. Write it as a conversion factor.
$2.01 \text{ g Ca} \times \frac{38.00 \text{ g F}}{40.08 \text{ g Ca}} = 1.91 \text{ g F}$ The fraction is about equal to 1, so the value of the answer should be close to 2. It makes sense.	**Construct** the solution setup, cancel units, and calculate the answer. **Check** the value of the answer to be sure it makes sense.
You improved your skill at using composition by mass as a conversion factor.	Reflect on what you learned by solving this Active Example.

Practice Exercise 7.10

A sample of sodium hydrogen carbonate contains 447 milligrams of carbon. What is the mass of oxygen in the sample?

7.8 Empirical Formula of a Compound

Empirical Formulas and Molecular Formulas

Goal 11 Distinguish between an empirical formula and a molecular formula.

Where do chemical formulas come from? They come from the same source as any fundamental chemical information—from experiments, usually performed in the laboratory. Among other things, chemical analysis can give us the percentage composition by mass of a compound. Such data can be used to give us the empirical formula of the compound. *Empirical* is a term that means "experimentally determined."

The percentage composition by mass of ethylene is 85.6% carbon and 14.4% hydrogen. Its chemical formula is C_2H_4. The percentage composition by mass of propylene, formula C_3H_6, is also 85.6% carbon and 14.4% hydrogen. These are, in fact, two of a whole series of compounds having the general formula C_nH_{2n}, where n is an integer. In ethylene, n = 2, and in propylene, n = 3. All compounds with the general formula C_nH_{2n} have the same percentage composition by mass.

C_2H_4 and C_3H_6, ethylene and propylene, are the building blocks from which the plastics polyethylene (plastic bags, plastic bottles, etc.) and polypropylene (laboratory plastic ware, bottle tops, etc.) are made.

C_2H_4 and C_3H_6 are typical molecular formulas of real chemical substances. If, in the general formula, we let n = 1, the result is CH_2. This is the empirical formula for all compounds having the general formula C_nH_{2n}. The **empirical formula** shows the simplest ratio of atoms of the elements in the compound. All subscripts are reduced to their lowest terms; they have no common divisor.

Empirical formulas may or may not be molecular formulas of real chemical compounds. For example, there happens to be no stable compound with the formula CH_2—and there is a good reason to believe that no such compound can exist. On the other hand, the molecular formula of dinitrogen tetroxide is N_2O_4. The subscripts have a common divisor, 2. Dividing by 2 gives the empirical formula, NO_2. This is also the molecular formula of a real substance, nitrogen dioxide. In other words, NO_2 is both the empirical formula and the molecular formula of nitrogen dioxide, as well as the empirical formula of dinitrogen tetroxide.

Nitrogen dioxide, NO_2, is responsible for one kind of chemical smog that produces a brown haze in the atmosphere.

Active Example 7.11 Empirical and Molecular Formulas

Write EF after each formula in the list here that is an empirical formula. Write the empirical formula after each compound whose formula is not already an empirical formula.

Think Before You Write An empirical formula expresses the lowest whole-number ratio of atoms of the elements in a compound.

Answers *Cover the left column with your cut-out shield. Reveal each answer only after you have written your own answer in the right column.*

Answers		Fill in the blanks.	
C_4H_{10}: C_2H_5	C_2H_6O: EF	C_4H_{10}: ________	C_2H_6O: ________
Hg_2Cl_2: HgCl	C_6H_6: CH	Hg_2Cl_2: ________	C_6H_6: ________

Practice Exercise 7.11

Write the molecular and empirical formula for each compound named: (a) tetranitrogen tetrasulfide, (b) disilicon hexabromide, (c) tetrasilicon decahydride, (d) tetraphosphorus hexoxide.

Determination of an Empirical Formula

Goal 12 Given data from which the mass of each element in a sample of a compound can be determined, find the empirical formula of the compound.

To find the empirical formula of a compound, you must find the whole-number ratio of atoms of the elements in a sample of the compound. When the numbers in the ratio are reduced to their lowest terms—that is, when they have no common divisor—they are the subscripts in the empirical formula. The following procedure shows how this is done.

how to... Determine an Empirical Formula

Step 1: Determine the percentage composition by mass or the mass of each element in a sample of the compound.

Step 2: Convert the masses into amount in moles of atoms of the different elements.

Step 3: Determine the amount ratio of moles of atoms.

Step 4: Express the amounts of atoms as the smallest possible ratio of integers.

Step 5: Write the empirical formula, using the number for each atom in the integer ratio as the subscript in the formula.

It is usually helpful in an empirical formula problem to organize the calculations in a table with the following headings:

Element	Mass (g)	Amount (mol)	Amount Ratio	Formula Ratio

Elena Elisseeva/Shutterstock.com

Figure 7.10 Ethylene is the building block from which the widely used plastic polyethylene is made. Liquids are often packaged or stored in polyethylene bottles.

We will use ethylene (**Figure 7.10**) to show how to find the empirical formula of a compound from its percentage composition by mass. As noted, the compound is 85.6% carbon and 14.4% hydrogen. In Step 1 of the procedure, if we think of percent as the number of grams of one element per 100 g of the compound, then a 100-g sample must contain 85.6 g of carbon and 14.4 g of hydrogen. From this, we see that *percentage composition by mass quantities represent the masses of each* *element in a 100-g sample of the compound*. These quantities complete Step 1 of the procedure. They are entered into the first two columns of the table:

Element	Mass (g)	Amount (mol)	Amount Ratio	Formula Ratio
C	85.6			
H	14.4			

We are now ready to find the number of moles of atoms of each element—Step 2 in the procedure. This is a one-step conversion between grams and moles, using molar mass as the conversion factor.

Element	Mass (g)	Amount (mol)	Amount Ratio	Formula Ratio
C	85.6	$85.6\,\cancel{g\,C} \times \frac{1\text{ mol C}}{12.01\,\cancel{g\,C}} = 7.13\text{ mol C}$		
H	14.4	$14.4\,\cancel{g\,H} \times \frac{1\text{ mol H}}{1.008\,\cancel{g\,H}} = 14.3\text{ mol H}$		

The ratio of these moles of atoms must now be determined—Step 3 in the procedure. This is most easily done by dividing each number of moles by the smallest number of moles. In this problem, the smallest number of moles is 7.13. Thus:

Element	Mass (g)	Amount (mol)	Amount Ratio	Formula Ratio
C	85.6	7.13	$\frac{7.13\,\cancel{mol}}{7.13\,\cancel{mol}} = 1.00$	
H	14.4	14.3	$\frac{14.3\,\cancel{mol}}{7.13\,\cancel{mol}} = 2.01$	

The ratio of *atoms* of the elements in a compound is the same as the ratio of *moles* of atoms in the compound. To see this in a more familiar setting, the ratio of seats to wheels in a bicycle is

$$\frac{1\text{ seat}}{2\text{ wheels}}\text{ or, numerically, }\frac{1}{2}$$

On four dozen bicycles, there are four dozen seats and eight dozen wheels. This yields a ratio that can be reduced as follows:

$$\frac{4\text{ doz seats}}{8\text{ doz wheels}} = \frac{4 \times \cancel{12}\text{ seats}}{8 \times \cancel{12}\text{ wheels}} = \frac{4\text{ seats}}{8\text{ wheels}} = \frac{1\text{ seats}}{2\text{ wheels}}\text{ or, numerically, }\frac{1}{2}$$

Dariush M/Shutterstock.com

Figure 7.11 The ratio of seats to wheels on a bicycle is the same as the ratio of dozens of seats to dozens of wheels. In both cases, it is 1:2.

When reduced to lowest terms, the ratio of seats to wheels is the same as the ratio of dozens of seats to dozens of wheels (**Figure 7.11**). Similarly, the numbers in the Amount Ratio column are in the same ratio as the subscripts in the empirical formula.

When used in a formula, the numbers must be integers. The numbers represent atoms, and a fraction of an atom does not exist. Accordingly, small roundoffs may be necessary to compensate for experimental errors. In this problem, 1.00/2.01 becomes 1/2. Step 4 is now finished. Step 5 is to write the empirical formula. The formula ratio tells us that there is 1 carbon atom for every 2 hydrogen atoms, so the empirical formula is CH_2. The entire procedure is now complete and summarized in the following table:

Element	Mass (g)	Amount (mol)	Amount Ratio	Formula Ratio
C	85.6	7.13	1.00	1
H	14.4	14.3	2.01	2

Active Example 7.12 Determination of an Empirical Formula I

A sample of a pure compound is made up of 1.61 g of phosphorus and 2.98 g of fluorine. Find the empirical formula of the compound.

Think Before You Write The goal of an empirical formula problem is to determine the ratio of atoms of the elements in the compound. The ratio of moles of atoms is the same as the ratio of atoms, so we need an amount in moles ratio expressed as lowest-possible whole numbers.

Answers *Cover the left column with your cut-out shield. Reveal each answer only after you have written your own answer in the right column.*

$$1.61\ \cancel{g\,P} \times \frac{1\ \text{mol P}}{30.97\ \cancel{g\,P}} = 0.0520\ \text{mol P}$$

$$2.98\ \cancel{g\,F} \times \frac{1\ \text{mol F}}{19.00\ \cancel{g\,F}} = 0.157\ \text{mol F}$$

Element	Mass (g)	Amount (mol)	Amount Ratio	Formula Ratio
P	1.61	0.0520		
F	2.98	0.157		

The information given in the problem statement is already placed in the following table. Calculate the number of moles of each element in the space above the table and add the results to the table.

Element	Mass (g)	Amount (mol)	Amount Ratio	Formula Ratio
P	1.61			
F	2.98			

$$\frac{0.0520\ \cancel{\text{mol}}}{0.0520\ \cancel{\text{mol}}} = 1.00 \qquad \frac{0.157\ \cancel{\text{mol}}}{0.0520\ \cancel{\text{mol}}} = 3.02$$

Element	Mass (g)	Amount (mol)	Amount Ratio	Formula Ratio
P	1.61	0.0520	1.00	
F	2.98	0.157	3.02	

Recalling that the amount ratio quantities are obtained by dividing each number of moles by the smallest number of moles, calculate those numbers and place them in the preceding table.

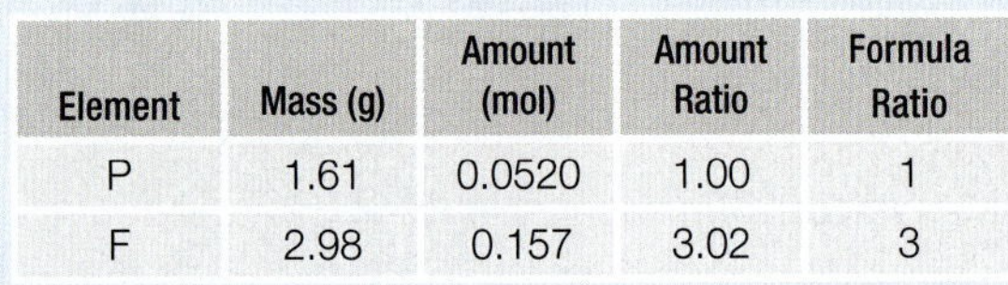

Element	Mass (g)	Amount (mol)	Amount Ratio	Formula Ratio
P	1.61	0.0520	1.00	1
F	2.98	0.157	3.02	3

Now round off the amount ratio quantities to get the whole-number formula ratio. Write the whole numbers in the preceding table.

PF_3

Use the formula ratio numbers as subscripts and write the empirical formula.

You improved your skill at determining the empirical formula of a compound.

What did you learn by solving this Active Example?

Practice Exercise 7.12

What is the empirical formula of a pure substance if an analysis indicates that a sample contains 0.292 g of hydrogen, 3.50 g of oxygen, and 2.04 g of nitrogen?

If either quotient in the Amount Ratio column is not close to a whole number, the Formula Ratio may be found by multiplying both the quotients by a small integer. Typical roundoffs are of the form 0.5/1, or ½/1, which is multiplied by 2/2 to become 1/2; 0.33/1, or ⅓/1, which is multiplied by 3/3 to become 1/3; or 0.25/1, or ¼/1,which is multiplied by 4/4 to become 1/4. However, don't expect amount ratios to always be exactly 0.50, 0.33, or 0.25 to 1. Slight variances in the uncertain digit are common. You will be guided in learning how to do this in the next Active Example.

Active Example 7.13 Determination of an Empirical Formula II

The mass of a piece of iron is 1.62 g. If the iron is exposed to oxygen under conditions in which oxygen combines with all of the iron to form a pure oxide of iron, the final mass increases to 2.31 g (Figure 7.12). Find the empirical formula of the compound.

Think Before You Write You need to know the masses of the elements in the compound, and those are not given directly in the problem statement. This time you must determine one mass from the data.

Answers *Cover the left column with your cut-out shield. Reveal each answer only after you have written your own answer in the right column.*

Mass of iron oxide = mass of iron + mass of oxygen

$$\text{Mass of oxygen} = \text{mass of iron oxide} - \text{mass of iron}$$
$$= 2.31\text{ g} - 1.62\text{ g}$$
$$= 0.69\text{ g}$$

The number of grams of iron in the final compound is the same as the number of grams at the start. The rest is oxygen. How many grams of oxygen combined with 1.62 g of iron if the iron oxide produced has a mass of 2.31 g?

$$1.62\ \cancel{\text{g Fe}} \times \frac{1\text{ mol Fe}}{55.85\ \cancel{\text{g Fe}}} = 0.0290\text{ mol Fe}$$

$$0.69\ \cancel{\text{g O}} \times \frac{1\text{ mol O}}{16.00\ \cancel{\text{g O}}} = 0.043\text{ mol O}$$

Element	Mass (g)	Amount (mol)	Amount Ratio	Formula Ratio
Fe	1.62	0.0290		
O	0.69	0.043		

The table is started here, with the symbols and masses of elements already entered. Step 2 is to calculate the amount in moles of atoms of each element. Do so, and then put the results in the table.

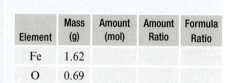

Element	Mass (g)	Amount (mol)	Amount Ratio	Formula Ratio
Fe	1.62			
O	0.69			

$$\frac{0.0290\ \cancel{\text{mol}}}{0.0290\ \cancel{\text{mol}}} = 1.00 \qquad \frac{0.043\ \cancel{\text{mol}}}{0.0290\ \cancel{\text{mol}}} = 1.5$$

Element	Mass (g)	Amount (mol)	Amount Ratio	Formula Ratio
Fe	1.62	0.0290	1.00	
O	0.69	0.043	1.5	

Now calculate the amount ratios in the space here and place the results in the foregoing table.

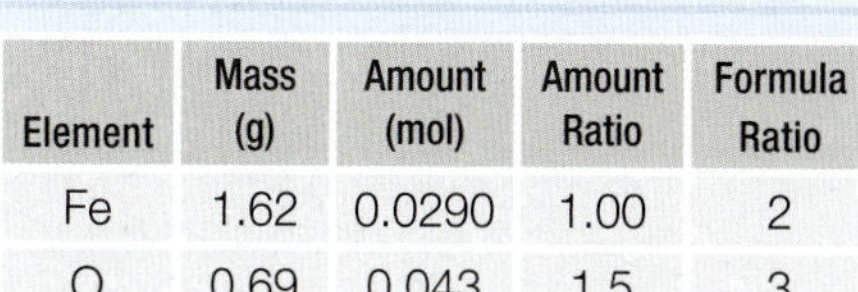

Element	Mass (g)	Amount (mol)	Amount Ratio	Formula Ratio
Fe	1.62	0.0290	1.00	2
O	0.69	0.043	1.5	3

Fe_2O_3

Multiplying the mole ratio numbers by 2 yields 1.00 × 2 = 2 and 1.5 × 2 = 3, both whole numbers without a common divisor.

This time the numbers in the Amount Ratio column are not both integers or very close to integers. But they can be changed to integers and kept in the same ratio by multiplying both of them by the same small integer. Find the smallest whole number that will yield integers when used as a multiplier for 1.00 and 1.5. Use it to obtain the formula-ratio numbers. Complete the table and write the empirical formula of the compound.

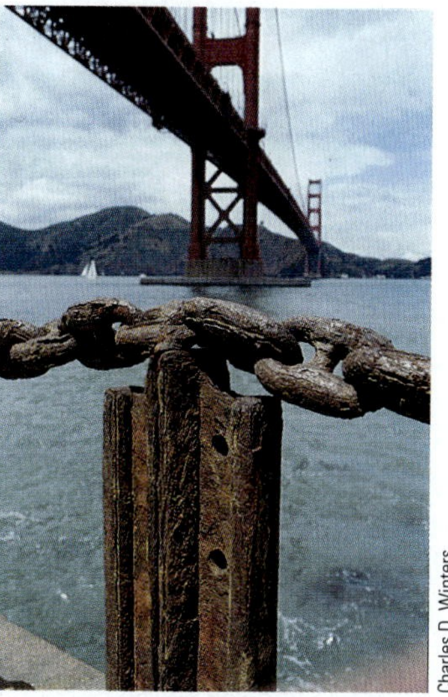

Charles D. Winters

Figure 7.12 Iron metal rusts when exposed to the moisture and oxygen in the air. Various compounds are collectively called rust; the variant that forms depends on the type of exposure to water and oxygen.

You improved your skill at determining the empirical formula of a compound.	What did you learn by solving this Active Example?

Practice Exercise 7.13

A compound is determined to be 0.135 gram of iron and 0.115 gram of sulfur. Determine the empirical formula of the compound.

The procedure for determining an empirical formula is the same for compounds containing more than two elements.

Active Example 7.14 Determination of an Empirical Formula III

A compound is found to contain 20.0% carbon, 2.2% hydrogen, and 77.8% chlorine. Determine the empirical formula of the compound.

Think Before You Write When percentage composition by mass is given, assume 100 g of the compound. This makes the percentages the same as the masses in a 100-g sample.

Answers *Cover the left column with your cut-out shield. Reveal each answer only after you have written your own answer in the right column.*

$C_3H_4Cl_4$

Set up the table and complete the problem.

Element	Mass (g)	Amount (mol)	Amount Ratio	Formula Ratio
C	20.0	1.67	1.00	3
H	2.2	2.2	1.3	4
Cl	77.8	2.19	1.31	4

Mass column to Amount column:

$$20.0\ \cancel{g\,C} \times \frac{1\ \text{mol C}}{12.01\ \cancel{g\,C}} = 1.67\ \text{mol C}$$

$$2.2\ \cancel{g\,H} \times \frac{1\ \text{mol H}}{1.008\ \cancel{g\,H}} = 2.2\ \text{mol H}$$

$$77.8\ \cancel{g\,Cl} \times \frac{1\ \text{mol Cl}}{35.45\ \cancel{g\,Cl}} = 2.19\ \text{mol Cl}$$

Amount column to Amount Ratio column:

$$\frac{1.67\ \text{mol}}{1.67\ \text{mol}} = 1.00 \quad \frac{2.2\ \text{mol}}{1.67\ \text{mol}} = 1.3 \quad \frac{2.19\ \text{mol}}{1.67\ \text{mol}} = 1.31$$

Multiplying the Amount Ratio figures by 3 yields integers from the Formula Ratio column:

$1.00 \times 3 = 3$; $1.3 \times 3 = 3.9$ or 4; $1.31 \times 3 = 3.93$ or 4

You improved your skill at determining the empirical formula of a compound.	What did you learn by solving this Active Example?

Practice Exercise 7.14

Nicotine is 74.1% carbon, 8.64% hydrogen, and 17.3% nitrogen. What is its empirical formula?

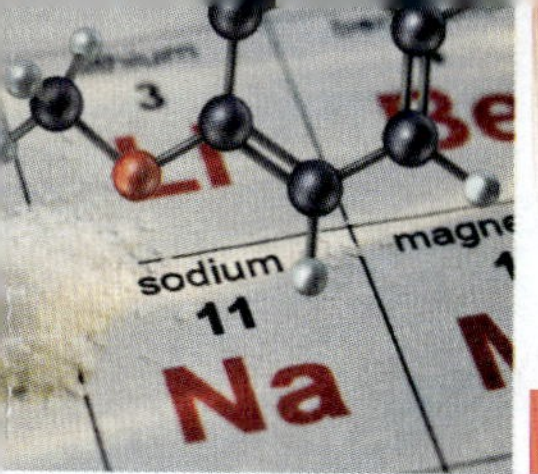

Everyday Chemistry 7.1

HOW TO READ A FOOD LABEL

Hundreds of thousands of food products are available in the United States, courtesy of a multibillion-dollar-a-year industry. The Nutrition Labeling and Education Act of 1990 (NLEA) requires food manufacturers to list amounts of total fat, saturated fat, cholesterol, sodium, total carbohydrates, dietary fiber, sugars, protein, vitamin A, vitamin C, calcium, and iron in each serving of their products. Total Calories and Calories from fat must also be listed, as well as the ingredients. Food labels include the amount per serving of each nutrient (except vitamins and minerals) and the amount of each nutrient as a percentage of a daily value based on a 2000-Calorie diet.

The U.S. Food and Drug Administration (FDA) and the American Heart Association recommend a daily diet in which no more than 30% of the Calories come from fat. Fat is listed on a label in grams per serving and as a percentage of daily value. Each gram of fat accounts for 9 Calories. These data can be used to calculate the percentage of fat from any food serving. For example, according to the label on a carton of whole milk, one serving accounts for a total of 160 Calories and has 9 g of fat. Thus,

$$\frac{9\ \cancel{\text{g fat}}}{\text{serving}} \times \frac{9\ \text{Cal}}{\cancel{\text{g fat}}} = \frac{81\ \text{Cal}}{\text{serving}}\ \text{from fat}$$

$$\%\ \text{Calories from fat} = \frac{81\ \text{fat Cal/serving}}{160\ \text{total Cal/serving}} \times 100\% = 51\%$$

If you were consciously watching the fat content of your diet, would you buy the whole milk? That 51% of Calories from fat seems excessive, doesn't it? But put it into perspective. If that serving of milk is poured over a serving of corn flakes (100 Calories per serving), the 81 fat Calories are now distributed among 260 total Calories. The percentage of Calories from fat drops to (81 ÷ 260) × 100% = 31%, which is essentially at the recommended 30%. Add some fruit to the breakfast and you add more total Calories without increasing fat Calories, and the percentage drops some more. Watch out, though, if you eat a couple of pieces of buttered toast with your meal.

Let's say that you want buttered toast, so you decide to reduce the fat from the milk by buying "2% fat" milk. You figure 2% is *much* lower than 51%. Unfortunately, it is not that simple. The fat content of 2% milk is lower, but not by the amount those numbers suggest. The 2% is the percentage by *mass* of the milk that is fat. If you check the label you will see that a serving of 2% milk has 5 g of fat (5 g × 9 Cal/g = 45 Cal from fat) and 140 Calories. The label also shows that this means 32% of the Calories are from fat (you can check the calculation). That's the number to be compared with 51% for whole milk. Go another step to "1% fat" milk; it has 15% of its Calories from fat. Skim milk has close to 0% Calories from fat.

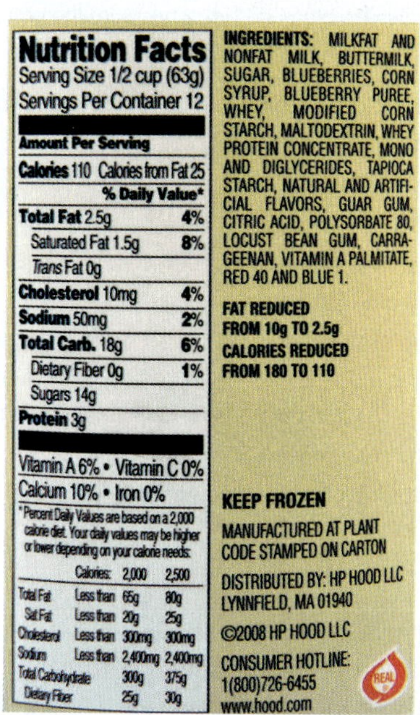

Charles D. Winters

Figure 7.13 Label from a carton of "light" ice cream.

Figure 7.13 shows a label from a carton of "light" ice cream that complies with the NLEA. It clearly states the percentage of fat by mass and the number of Calories from fat.

Isn't there an easier way to arrange a healthful diet than dealing with all these percentages? Yes, there is. The FDA diet suggests a daily fat allowance of 65 g. This is essentially the same as the recommended 30% of Calories from fat: 65 g × 9 Calories/g = 585 Calories, and (585 ÷ 2000) × 100% = 29%. Now you can think of your glass of whole milk simply as 9 of the 65 g, whether you drink the milk or put it on your corn flakes. Add the grams of fat, regardless of percentages, from everything else you eat. When you have had 65 g of fat, eat only nonfat foods for the rest of the day and your diet will be acceptable—at least from a fat standpoint.

Perhaps the best advice is still the old saying, "Eat to live; don't live to eat." And read the labels.

Quick Quiz

1. List at least two nutrients required to appear on a food label that you should seek to limit and two that you need to be sure you get enough of.

2. Compare the following data about a 1-cup serving of 2% milk and nonfat milk:

	2% Milk	Nonfat Milk
Calories	120	80
Total fat	5 g	None
Saturated fat	3 g	None
Protein	9 g	9 g
Calcium	300 mg	300 mg

Which is healthier? Which do you drink? Why do you choose your particular milk preference?

7.9 Determination of a Molecular Formula

Goal 13 Given the molar mass and empirical formula of a compound, or information from which they can be found, determine the molecular formula of the compound.

At the beginning of Section 7.8, you learned that there is a series of compounds having the general formula C_nH_{2n}, where n is an integer. This can also be written $(CH_2)_n$. CH_2 is the empirical formula of all compounds in the series. The molar mass of the empirical formula unit is 12.01 g/mol C + 2(1.008 g/mol H) = 14.03 g/mol CH_2. If the actual compound contains two empirical formula units, that is, if n = 2, the molar mass of the real compound is 2 × 14 g/mol = 28 g/mol. If n = 3, the molar mass of the compound is 3 × 14 g/mol = 42 g/mol, and so forth.

The purpose of Equation 7.2 is to find the number of empirical formula units in a molecule. This must be a counting number. The value of the molar mass of a compound comes from a measurement, and it includes its associated uncertainty, so don't necessarily expect the result of the division to be precisely 1.000..., 2.000..., and so on. Some variance in the uncertain digit may occur.

Now reverse the process. Suppose an experiment determines that the molar mass of a compound with the empirical formula CH_2 is 70.2 g/mol. To find n, find the number of 14.03 g/mol empirical formula units in one 70.2 g/mol molecular formula unit (that is, find how many 14s are in 70). That number is five: 70 ÷ 14 = 5. The real compound is $(CH_2)_5$, or C_5H_{10}. In general,

$$n = \text{empirical formula units in 1 molecule} = \frac{\text{molar mass of compound}}{\text{molar mass of empirical formula}} \quad (7.2)$$

Note that n *must* be an integer. If any problem yields an n that is not an integer, or very close to an integer, either the empirical formula or the molar mass is incorrect.

Spencer Jones/Getty Images

Figure 7.14 Fructose is a sugar found in honey, fruit, and some vegetables. It is much sweeter than table sugar.

how to... **Find the Molecular Formula of a Compound**

Sample Problem: *A compound with the empirical formula C_2H_5 has a molar mass of 58.12 g/mol. Find the molecular formula of the compound.*

Step 1: Determine the empirical formula of the compound.	*The empirical formula is given in this example.*
Step 2: Calculate the molar mass of the empirical formula unit.	2(12.01 g/mol C) + 5(1.008 g/mol H) = 29.06 g/mol
Step 3: Determine the molar mass of the compound (which will be given in this book).	Given as 58.12 g/mol
Step 4: Divide the molar mass of the compound by the molar mass of the empirical formula unit to get n, the number of empirical formula units per molecule.	$\frac{58.12\text{ g/mol}}{29.06\text{ g/mol}} = 2$
Step 5: Write the molecular formula.	$(C_2H_5)_2 = C_4H_{10}$

Active Example 7.15 Determination of a Molecular Formula

Fructose is the sugar found in honey and fruits (Figure 7.14). It is commonly known as fruit sugar. Its percentage composition is 40.0% carbon, 6.71% hydrogen, and the remainder is oxygen. The molar mass of fructose is 180.16 g/mol. Find the empirical and molecular formulas of the compound.

Think Before You Write Review the *how to...* box, if necessary, before working on this Active Example. The key steps in solving this problem are finding the empirical formula and then determining the number of empirical formula units in the molecule.

Answers *Cover the left column with your cut-out shield. Reveal each answer only after you have written your own answer in the right column.*

CH_2O

Element	Mass (g)	Amount (mol)	Amount Ratio	Formula Ratio
C	40.0	3.33	1.00	1
H	6.71	6.66	2.00	2
O	53.3	3.33	1.00	1

Mass column to Amount column:

$$40.0\ \cancel{g\ C} \times \frac{1\ mol\ C}{12.01\ \cancel{g\ C}} = 3.33\ mol\ C$$

$$6.71\ \cancel{g\ H} \times \frac{1\ mol\ H}{1.008\ \cancel{g\ H}} = 6.66\ mol\ H$$

$$53.3\ \cancel{g\ O} \times \frac{1\ mol\ O}{16.00\ \cancel{g\ O}} = 3.33\ mol\ O$$

Amount column to Amount Ratio column:

$$\frac{3.33\ mol}{3.33\ mol} = 1.00 \quad \frac{6.66\ mol}{3.33\ mol} = 2.00 \quad \frac{3.33\ mol}{3.33\ mol} = 1.00$$

The Amount Ratio figures are whole numbers, so they are not changed in the Formula Ratio column.

Start by finding the empirical formula.

CH_2O: 12.01 g/mol
1.008 g/mol
1.008 g/mol
16.00 g/mol
30.026 g/mol = 30.03 g/mol

To use Equation 7.2, you must have the mass of the empirical formula unit. Calculate the molar mass of CH_2O.

$n = 180.16\ g/mol \div 30.03\ g/mol = 5.999 = 6$

$(CH_2O)_6 = C_6H_{12}O_6$

Determine the number of empirical formula units in the molecule and write the molecular formula.

You improved your skill at determining the empirical formula and the molecular formula of a compound.

What did you learn by solving this Active Example?

Practice Exercise 7.15

A compound has a molar mass of 292 g/mol. Its percentage composition by mass is 49.3% carbon, 2.1% hydrogen, and 48.6% chlorine. What is its molecular formula?

Charles D. Winters

CHAPTER 7 IN REVIEW: CHEMICAL FORMULA RELATIONSHIPS

Goal 1 Given the formula of a chemical compound (or a name from which the formula may be written), state the number of atoms of each element in the formula unit.

A **chemical formula** expresses the number of atoms of each element present in the formula unit of a substance. The number of each atom is given by a subscript following the symbol of that atom or a group of atoms. If the number is one, it is omitted in the formula.

Goal 2 Distinguish among atomic mass, molecular mass, and formula mass.	**Atomic mass** is the average mass of all atoms of an element as they occur in nature. It is measured relative to the assignment of a mass of 12 u to an atom of carbon-12. **Molecular (or formula) mass** is the average mass of molecules (or formula units) compared with the mass of an atom of carbon-12, which is 12 atomic mass units.
Goal 3 Calculate the formula (molecular) mass of any compound whose formula is given (or known).	The formula mass of a compound is equal to the sum of all of the atomic masses in the formula unit: **Formula mass = Σ atomic masses in the formula unit.**
Goal 4 Define the term *mole.* Identify the number of objects that corresponds to 1 mole.	One **mole** of anything contains exactly $6.02214076 \times 10^{23}$ particles. This value is the **Avogadro constant, N_A, 6.02×10^{23},** to three significant figures.
Goal 5 Given the number of moles (or units) in any sample, calculate the number of units (or moles) in the sample.	Use conversion factors to **convert between moles and number of units:** $\text{\# of moles} \times \frac{6.02 \times 10^{23}\text{ units}}{\text{mol}} = \text{\# of units}$ $\text{\# of units} \times \frac{1\text{ mol}}{6.02 \times 10^{23}\text{ units}} = \text{\# of moles}$
Goal 6 Define *molar mass* or interpret statements in which the term *molar mass* is used.	**Molar mass** is mass divided by stoichiometric amount of substance. Mass is typically expressed in grams and amount is expressed in moles, yielding the most common units of molar mass: grams per mole (g/mol).
Goal 7 Calculate the molar mass of any substance whose chemical formula can be written or is given.	The molar mass of any substance in grams per mole is numerically equal to the atomic, molecular, or formula mass of that substance in atomic mass units.
Goal 8 Given any one of the following for a substance with a given (or known) formula, calculate the other two: (a) mass; (b) amount in moles; (c) number of formula units, molecules, or atoms.	The mole is the connecting link between the macroscopic level of observation and measurement of matter, in which we measure quantities in grams, and the particulate level, in which we count the number of units, usually grouped in moles. Using N_A for the Avogadro constant and MM for molar mass, $\text{units} \xrightarrow{N_A} \text{mol} \xrightarrow{MM} \text{g}$ *or* $\text{g} \xrightarrow{MM} \text{mol} \xrightarrow{N_A} \text{units}$ Changing from units to mass or vice versa is a two-step process, requiring two conversion factors.
Goal 9 Calculate the percentage composition by mass of any compound whose formula is given (or known).	The **percentage composition by mass** of a compound is the percentage by mass of each element in the compound. **Percent** is the amount of one part of a mixture per 100 total parts in the mixture. To calculate the percentage of each element, $\%\text{ Element} = \frac{\text{Total molar mass of element in compound}}{\text{Molar mass of compound}} \times 100\%$ If you calculate the percentage composition of a compound correctly, the sum of all percentages must be 100%.
Goal 10 Given the mass of a sample of any compound with a given (or known) formula, calculate the mass of any element in the sample, or, given the mass of any element in the sample, calculate the mass of the sample or the mass of any other element in the sample.	To find the amount of any element in a known amount of compound, use percentage as a conversion factor, based on the equivalency: grams of the element per 100 grams of the compound.
Goal 11 Distinguish between an empirical formula and a molecular formula.	An **empirical formula** shows the simplest whole-number ratio of atoms of the elements in a compound. Empirical formulas are calculated from percentage composition data. They may also be calculated from the mass of each element in a sample of a compound. Empirical formulas may or may not be the actual molecular formulas of compounds. The molar mass of the compound is needed to determine molecular formulas from empirical formulas.
Goal 12 Given data from which the mass of each element in a sample of a compound can be determined, find the empirical formula of the compound.	To **determine an empirical formula:** **1.** Determine the percentage composition by mass or the mass of each element in a sample of the compound. **2.** Convert the masses into amount in moles of atoms of the different elements. **3.** Determine the amount ratio of moles of atoms. **4.** Express the amounts of atoms as the smallest possible ratio of integers. **5.** Write the empirical formula, using the number for each atom in the integer ratio as the subscript in the formula.

Goal 13 Given the molar mass and empirical formula of a compound, or information from which they can be found, determine the molecular formula of the compound.

To **determine the molecular formula of a compound:**

1. Determine the empirical formula of the compound.
2. Calculate the molar mass of the empirical formula unit.
3. Determine the molar mass of the compound (which will be given in this book).
4. Divide the molar mass of the compound by the molar mass of the empirical formula unit to get n, the number of empirical formula units per molecule.
5. Write the molecular formula.

Key Terms

Most of the key terms and concepts and many others appear in the Glossary. Use your Glossary regularly.

6.02×10^{23} p. 237
Avogadro constant, N_A p. 237
Avogadro's number p. 237
empirical formula p. 247
formula mass p. 236
molar mass p. 239
mole, mol p. 237
molecular mass p. 236
percent p. 243
percentage composition by mass p. 244
stoichiometric amount p. 237

Frequently Asked Questions

Is the mole a number?
Yes. The Avogadro constant is defined as 6.02×10^{23}/mole. You may have learned in previous courses that the mole is *not* a number, but that changed in 2019. You will see in future chapters that you will frequently use the mole in chemical calculations, but you will rarely use the Avogadro constant.

Is the molar mass of hydrogen 1.008 g/mol or (2 × 1.008) g/mol? What about the other diatomic elements?
Always calculate molar masses that correspond to the chemical formula. For example, the molar mass of H_2 is 2×1.008 g/mol = 2.016 g/mol; the molar mass of F_2 is 2×19.00 g/mol = 38.00 g/mol; and the molar mass of hydrogen in HF is 1.008 g/mol.

Sometimes I get confused about deciding whether to multiply by molar mass or divide by it. How can I get this straight?
Use conversion factors in solving problems and label each entry completely. Specifically, include the chemical formula of each substance in the calculation setup. Then cancel units in your setups to be sure they are correct.

What multiplier should I use when the mole ratio in an empirical formula problem is 0.12-to-1 or something similar?
When you calculate a mole ratio in an empirical formula problem that cannot easily be converted into whole numbers, you should carefully check your work in the previous part of the problem. Unless your instructor tells you otherwise, mole ratios in this introductory chemistry course will be 0.5-to-1, 0.33-to-1, and 0.25-to-1.

Concept-Linking Exercises

Write a brief description of the relationships among each of the following groups of terms or phrases. Answers to the Concept-Linking Exercises are given at the end of the chapter.

1. Atomic mass, molecular mass, formula mass, molar mass
2. Mole, 6.02×10^{23}, carbon-12, Avogadro constant

Small-Group Discussion Questions

Small-Group Discussion Questions are for group work, either in class or under the guidance of a leader during a discussion section.

1. How many sodium ions and how many chloride ions are in a 0.01-g grain of table salt? How is this ion ratio expressed in the formula of the compound? In general, what is the meaning of a formula of an ionic compound?
2. Compare and contrast the terms *atomic mass*, *molecular mass*, *molar mass*, and *formula mass*. Write the formulas of five examples that fit into each category. Are there examples that fit two categories simultaneously? Can a single species be described by three or all four types of mass? If yes, give examples.
3. The mole is defined as exactly $6.02214076 \times 10^{23}$ particles. This corresponds with the number of atoms in 12 grams of carbon-12. What is the relationship between the definition of the mole and the number of atoms in 12 grams of carbon-12?
4. Carbon-12 serves as the basis of a number of interrelated definitions central to chemistry. Atomic mass, the mole, and molar mass are all directly or indirectly related to carbon-12. Write a definition of each. Explain how these definitions lead to the facts: (a) the atomic mass of carbon-12 is exactly 12 u and (b) the molar mass of carbon-12 is exactly 12 g/mol.

5. What experimental procedure would you use to determine the number of atoms in a pure gold coin? Explain, and give an example of the predicted outcome of your experiment for any selected coin weight.
6. What is the definition of the *percentage composition by mass* of a compound? How does this compare with the percentage by number of atoms of each element in a compound? Explain in detail.
7. Explain the relationship between the percentage composition by mass of a compound and its empirical formula. If you know the percentage composition of a compound, can you determine its empirical formula? If you know the empirical formula of a compound, can you determine its percentage composition by mass? If the answer to either or both questions is yes, illustrate with an example.
8. What experimental information is needed to determine the molecular formula of a compound?

Questions, Exercises, and Problems

Interactive versions of these problems may be assigned by your instructor. Solutions for blue-numbered questions are at the end of the chapter. Questions other than those in the General Questions and More Challenging Problems sections are paired in consecutive odd–even number combinations; solutions for the odd numbered questions are also at the end of the chapter.

Many questions in this chapter are written with the assumption that you have studied Chapter 6 and can write the required formulas from their chemical names. If this is not the case, we have placed a list of all chemical formulas needed to answer this chapter's questions at the end of the Questions, Exercises, and Problems.

If you have studied Chapter 6 and you get stuck while answering a question because you cannot write a formula, we urge you to review the appropriate section of Chapter 6 before continuing. Avoid the temptation to "just peek" at the list of formulas. Developing skill in chemical nomenclature is part of your learning process in this course, and in many cases, you truly learn the material from Chapter 6 as you apply it here in this chapter and throughout your study of chemistry.

Section 7.2: The Number of Atoms in a Formula

1. How many atoms of each element are in a formula unit of aluminum nitrate?
2. How many atoms of each element are in a formula unit of ammonium phosphate?

Section 7.3: Molecular Mass and Formula Mass

3. Why is it proper to speak of the molecular mass of water but not of the molecular mass of sodium nitrate?
4. It may be said that because atomic, molecular, and formula masses are all based on carbon-12, they are conceptually alike. What then are their differences?
5. Which of the three terms *atomic mass*, *molecular mass*, or *formula mass* is most appropriate for each of the following: ammonia, calcium oxide, barium, chlorine, sodium carbonate?
6. In what units are atomic, molecular, and formula mass expressed? Define those units.
7. Find the formula mass of each of the following substances:
   a) Lithium chloride
   b) Aluminum carbonate
   c) Ammonium sulfate
   d) Butane, C_4H_{10} (molecular mass)
   e) Silver nitrate
   f) Manganese(IV) oxide
   g) Zinc phosphate
8. Determine the formula or molecular mass of each substance in the following list:
   a) Nitrogen trifluoride
   b) Barium chloride
   c) Lead(II) phosphate
9. What is the molecular mass of each of the following compounds?
   a) Sulfur trioxide, an atmospheric pollutant that reacts with water to form acid rain.

   b) Hydrogen fluoride, which will react with SbF_5 to form the strongest known superacid, $HSbF_6$.

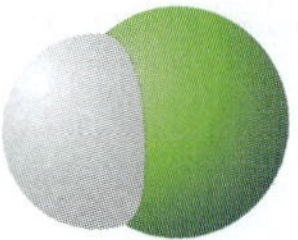

10. Calculate the mass of each of the following molecules.
   a) Nitrogen dioxide, an atmospheric pollutant that causes the brown coloring of smog.

   b) Dichlorine monoxide, used industrially in the process of manufacturing bleaching agents.

Section 7.4: Stoichiometric Amount

11. What do quantities representing 1 mole of iron atoms and 1 mole of ammonia molecules have in common?

12. Explain what the term *mole* means. Why is it used in chemistry?

13. Is the mole a number? Explain.

14. Give the name and value of the number that defines the mole.

15. Determine how many atoms, molecules, or formula units are in each of the following:
a) 7.75 moles of methane, CH_4
b) 0.0888 mole of carbon monoxide
c) 57.8 moles of iron
d) 0.81 mole of magnesium chloride

16. a) How many molecules of boron trifluoride are present in 1.25 moles of this compound?
b) How many moles of boron trifluoride are present in 7.04×10^{22} molecules of this compound?
c) How many moles of nitrogen dioxide are present in 7.89×10^{22} molecules of this compound?
d) How many molecules of nitrogen dioxide are present in 1.60 moles of this compound?

17. Calculate the amount in moles in each of the following:
a) 2.45×10^{23} acetylene molecules, C_2H_2
b) 6.96×10^{24} sodium atoms

18. a) How many atoms of hydrogen are present in 2.69 moles of water?
b) How many moles of oxygen are present in 1.39×10^{22} molecules of water?

Section 7.5: Molar Mass

19. In what way are the molar mass of atoms and atomic mass the same?

20. How does molar mass differ from molecular mass?

21. Find the molar mass of each of the following substances:
a) C_3H_8
b) C_6Cl_5OH
c) Nickel(II) phosphate
d) Zinc nitrate

22. Calculate the molar mass of each of the following:
a) Chromium(II) iodide
b) Silicon dioxide
c) Carbon tetrafluoride

Section 7.6: Conversion Among Mass, Amount in Moles, and Number of Units

Questions 23 to 26: Find the amount in moles for each mass of substance given.

23. a) 6.79 g oxygen
b) 9.05 g magnesium nitrate
c) 0.770 g aluminum oxide
d) 659 g C_2H_5OH
e) 0.394 g ammonium carbonate
f) 34.0 g lithium sulfide

24. a) 53.8 g beryllium
b) 781 g $C_3H_4Cl_4$
c) 0.756 g calcium hydroxide
d) 9.94 g cobalt(III) bromide
e) 8.80 g ammonium dichromate (dichromate ion, $Cr_2O_7^{2-}$)
f) 28.3 g magnesium perchlorate

25. a) 0.797 g potassium iodate
b) 68.6 g beryllium chloride
c) 302 g nickel(II) nitrate

26. a) 91.9 g sodium hypochlorite
b) 881 g aluminum acetate (acetate ion, $C_2H_3O_2^-$)
c) 0.586 g mercury(I) chloride

Questions 27 to 30: Calculate the mass of each substance from the amount in moles given.

27. a) 0.769 mol lithium chloride
b) 57.1 mol acetic acid, $HC_2H_3O_2$
c) 0.68 mol lithium
d) 0.532 mol iron(III) sulfate
e) 8.26 mol sodium acetate (acetate ion, $C_2H_3O_2^-$)

28. a) 0.542 mol sodium hydrogen carbonate
b) 0.0789 mol silver nitrate
c) 9.61 mol sodium hydrogen phosphate
d) 0.903 mol calcium bromate
e) 1.14 mol ammonium sulfite

29. a) 0.379 mol lithium sulfate
b) 4.82 mol potassium oxalate (oxalate ion, $C_2O_4^{2-}$)
c) 0.132 mol lead(II) nitrate

30. a) 0.819 mol manganese(IV) oxide
b) 8.48 mol aluminum chlorate
c) 0.926 mol chromium(II) chloride

Questions 31 to 34: Calculate the number of atoms, molecules, or formula units that are in each given mass.

31. a) 29.6 g lithium nitrate
b) 0.151 g lithium sulfide
c) 457 g iron(III) sulfate

32. a) 85.5 g beryllium nitrate
b) 9.42 g manganese
c) 0.0948 g C_3H_7OH

33. a) 0.0023 g iodine molecules
b) 114 g $C_2H_4(OH)_2$
c) 9.81 g chromium(III) sulfate

34. a) 7.70 g iodine atoms
b) 0.447 g C_9H_{20}
c) 72.6 g manganese(II) carbonate

Questions 35 and 36: Calculate the mass of each of the following.

35. a) 4.30×10^{21} molecules of $C_{19}H_{37}COOH$
b) 8.67×10^{24} atoms of fluorine
c) 7.23×10^{23} formula units of nickel(II) chloride

36. a) 2.58×10^{23} formula units of iron(II) oxide
b) 8.67×10^{24} molecules of fluorine
c) 7.36×10^{23} atoms of gold (Z = 79)

37. On a certain day a financial website quoted the price of gold at $1,258 per troy ounce (1 troy ounce = 31.1 g). What is the price of a single atom of gold (Z = 79)?

38. How many carbon atoms has a gentleman given his bride-to-be if the engagement ring has a 0.500 carat diamond? There are 200 mg in a carat. (The price of diamonds doesn't seem so high when figured at dollars per atom.)

Elnur/Shutterstock.com

How many carbon atoms are in this 0.500 carat diamond?

39. A person who sweetens coffee with two teaspoons of sugar, $C_{12}H_{22}O_{11}$, uses about 0.65 g. How many sugar molecules is this?

40. The mass of 1 gallon of gasoline is about 2.7 kg. Assuming the gasoline is entirely octane, C_8H_{18}, calculate the number of molecules in the gallon.

The stable form of certain elements is a two-atom molecule. Fluorine and nitrogen are two of those elements. As you answer Questions 41 and 42, keep in mind that the chemical formula of a molecule identifies precisely what is in the individual molecule.

41. a) What is the mass of 4.12×10^{24} N atoms?
b) What is the mass of 4.12×10^{24} N_2 molecules?
c) How many atoms are in 4.12 g N?
d) How many molecules are in 4.12 g N_2?
e) How many atoms are in 4.12 g N_2?

42. a) How many molecules are in 3.61 g F_2?
b) How many atoms are in 3.61 g F_2?
c) How many atoms are in 3.61 g F?
d) What is the mass of 3.61×10^{23} F atoms?
e) What is the mass of 3.61×10^{23} F_2 molecules?

Section 7.7: Mass Relationships Among Elements in a Compound: Percentage Composition by Mass

Questions 43 and 44: Calculate the percentage composition by mass of each compound.

43. a) Ammonium nitrate
b) Aluminum sulfate
c) Ammonium carbonate
d) Calcium oxide
e) Manganese(IV) sulfide

44. a) Magnesium nitrate
b) Sodium phosphate
c) Copper(II) chloride
d) Chromium(III) sulfate
e) Silver carbonate

45. Lithium fluoride is used as a flux when welding or soldering aluminum. How many grams of lithium are in 1.00 lb (454 g) of lithium fluoride?

46. Ammonium bromide is a raw material in the manufacture of photographic film. What mass of bromine is found in 7.50 g of the compound?

47. Potassium sulfate is found in some fertilizers as a source of potassium. How many grams of potassium can be obtained from 57.4 g of the compound?

48. Magnesium oxide is used in making bricks to line very high temperature furnaces. If a brick contains 1.82 kg of the oxide, what is the mass of magnesium in the brick?

49. Zinc cyanide (cyanide ion, CN^-), is a compound used in zinc electroplating. How many grams of the compound must be dissolved in a test bath in a laboratory to introduce 146 g of zinc into the solution?

50. An experiment requires that enough $C_5H_{12}O$ be used to yield 19.7 g of oxygen. How much $C_5H_{12}O$ must be weighed out?

51. Molybdenum (Z = 42) is an element used in making steel alloys. It primarily comes from an ore called *molybdenite*, MoS_2. What mass of pure molybdenite must be treated to obtain 201 kg Mo?

bonchan/Shutterstock.com

Molybdenite is the primary commercial source of molybdenum.

52. How many grams of nitrogen monoxide must be weighed out to get a sample of nitrogen monoxide that contains 14.7 g of oxygen?

53. How many grams of the insecticide calcium chlorate must be measured if a sample is to contain 4.17 g chlorine?

54. If a sample of carbon dioxide contains 16.4 g of oxygen, how many grams of carbon dioxide does it contain?

Section 7.8: Empirical Formula of a Compound

55. Explain why C_6H_{10} must be a molecular formula, whereas C_7H_{10} could be a molecular formula, an empirical formula, or both.

56. From the following list, identify each formula that could be an empirical formula. Write the empirical formulas of any compounds that are not already empirical formulas: C_2H_6O; Na_2O_2; $C_2H_4O_2$; N_2O_5.

57. A certain compound is 52.2% carbon, 13.0% hydrogen, and 34.8% oxygen. Find the empirical formula of the compound.

58. A compound is found to contain 15.94% boron and 84.06% fluorine by mass. What is the empirical formula for this compound?

59. A researcher exposes 11.89 g of iron to a stream of oxygen until it reacts to produce 16.99 g of a pure oxide of iron. What is the empirical formula of the product?

60. A compound is found to contain 39.12% carbon, 8.772% hydrogen, and 52.11% oxygen by mass. What is the empirical formula for this compound?

61. A compound is 17.2% C, 1.44% H, and 81.4% F. Find its empirical formula.

62. A compound is found to contain 21.96% sulfur and 78.04% fluorine by mass. What is the empirical formula for this compound?

Section 7.9: Determination of a Molecular Formula

63. An antifreeze and coolant widely used in automobile engines is 38.7% carbon, 9.7% hydrogen, and 51.6% oxygen. Its molar mass is 62.0 g/mol. What is the molecular formula of the compound?

Ethylene glycol is a widely used automotive antifreeze and coolant.

64. A compound is found to contain 31.42% sulfur, 31.35% oxygen, and 37.23% fluorine by mass. What is the empirical formula for this compound? The molar mass for this compound is 102.1 g/mol. What is the molecular formula for this compound?

65. A compound is 73.1% chlorine, 24.8% carbon, and the balance is hydrogen. If the molar mass of the compound is 97 g/mol, find the molecular formula.

66. A compound is found to contain 25.24% sulfur and 74.76% fluorine by mass. What is the empirical formula for this compound? The molar mass for this compound is 254.1 g/mol. What is the molecular formula for this compound?

General Questions

67. Distinguish precisely and in scientific terms the differences among items in each of the following groups:
a) Atomic mass, molecular mass, formula mass, molar mass
b) Molecular formula, empirical formula

68. Classify each of the following statements as true or false:
a) The term *molecular mass* applies mostly to ionic compounds.
b) Molar mass is measured in atomic mass units.
c) Grams are larger than atomic mass units; therefore, molar mass is numerically larger than atomic mass.
d) The molar mass of hydrogen is read directly from the periodic table, whether it is monatomic hydrogen, H, or hydrogen gas, H_2.
e) An empirical formula is always a molecular formula, although a molecular formula may or may not be an empirical formula.

69. Would you need a truck to transport 10^{25} atoms of copper? Explain.

70. The stable form of elemental phosphorus is a tetratomic molecule. Calculate the number of molecules and atoms in 85.0 g P_4.

71. Is it reasonable to set a dinner table with 1 mole of salt, NaCl, in a salt shaker with a capacity of 2 oz? How about 1 mole of sugar, $C_{12}H_{22}O_{11}$, in a sugar bowl with a capacity of 10 oz?

More Challenging Problems

72. The quantitative significance of "take a deep breath" varies, of course, with the individual. When one person did so, he found that he inhaled 2.95×10^{22} molecules of the mixture of mostly nitrogen and oxygen, which we call air. Assuming this mixture has an average molar mass of 29 g/mol, what is his apparent lung capacity in grams of air?

73. Assuming gasoline to be pure octane, C_8H_{18} (actually, it is a mixture of many substances), an automobile getting 25.0 miles per gallon would consume 5.62×10^{23} molecules per mile. Calculate the mass of this amount of fuel.

74. A researcher took 27.37 g of a certain compound containing only carbon and hydrogen and burned it completely in pure oxygen. All the carbon was changed to 85.9 g of CO_2, and all the hydrogen was changed to 35.5 g of H_2O. What is the empirical formula of the original compound? (*Hint:* Find the mass in grams of carbon and hydrogen in the original compound.)

75. $Co_aS_bO_c \cdot X\,H_2O$ is the general formula of a certain hydrate. When 43.0 g of the compound is heated to drive off the water, 26.1 g of anhydrous compound is left. Further analysis shows that the percentage composition by mass of the anhydrate is 42.4% Co, 23.0% S, and 34.6% O. Find the empirical formula of (a) the anhydrous compound and (b) the hydrate. (*Hint:* Treat the anhydrous compound and water just as you have treated elements in calculating X in the formula of the hydrate.)

Formulas

These formulas are provided in case you are studying this chapter before you study Chapter 6. You should not use this list unless your instructor has not yet assigned Chapter 6 or otherwise indicated that you may use the list.

1. $Al(NO_3)_3$
2. $(NH_4)_3PO_4$
7. $LiCl$, $Al_2(CO_3)_3$, $(NH_4)_2SO_4$, $AgNO_3$, MnO_2, $Zn_3(PO_4)_2$
8. NF_3, $BaCl_2$, $Pb_3(PO_4)_2$
9. SO_3, HF
10. NO_2, Cl_2O
15. CO, Fe
16. BF_3, NO_2
17. Na
18. H_2O
21. $Ni_3(PO_4)_2$, $Zn(NO_3)_2$
22. CrI_2, SiO_2, CF_4
23. O_2, $Mg(NO_3)_2$, Al_2O_3, $(NH_4)_2CO_3$, Li_2S
24. Be, $Ca(OH)_2$, $CoBr_3$, $(NH_4)_2Cr_2O_7$, $Mg(ClO_4)_2$
25. KIO_3, $BeCl_2$, $Ni(NO_3)_2$
26. NaClO, $Al(C_2H_3O_2)_3$, Hg_2Cl_2
27. LiCl, Li, $Fe_2(SO_4)_3$, $NaC_2H_3O_2$
28. $NaHCO_3$, $AgNO_3$, Na_2HPO_4, $Ca(BrO_3)_2$, $(NH_4)_2SO_3$
29. Li_2SO_4, $K_2C_2O_4$, $Pb(NO_3)_2$
30. MnO_2, $Al(ClO_3)_3$, $CrCl_2$
31. $LiNO_3$, Li_2S, $Fe_2(SO_4)_3$
32. $Be(NO_3)_2$, Mn
33. I_2, $Cr_2(SO_4)_3$
34. I, $MnCO_3$
35. F, $NiCl_2$
36. FeO, F_2, Au
37. Au
38. C
43. NH_4NO_3, $Al_2(SO_4)_3$, $(NH_4)_2CO_3$, CaO, MnS_2
44. $Mg(NO_3)_2$, Na_3PO_4, $CuCl_2$, $Cr_2(SO_4)_3$, Ag_2CO_3
45. LiF
46. NH_4Br
47. K_2SO_4
48. MgO
50. O_2
52. NO
53. $Ca(ClO_3)_2$
54. CO_2
69. Cu

Answers to Practice Exercises

1. (a) $(NH_4)_2SO_4$: 2 nitrogen atoms, 8 hydrogen atoms, 1 sulfur atom, 4 oxygen atoms;
   (b) $Al_2(CO_3)_3$: 2 aluminum atoms, 3 carbon atoms, 9 oxygen atoms
2. H_2SO_4: 2(1.008 u) + 32.06 u + 4(16.00 u) = 98.08 u
3. $$0.25 \text{ mol NH}_3 \times \frac{6.02 \times 10^{23} \text{ NH}_3 \text{ molecules}}{\text{mol NH}_3}$$
   $$= 1.5 \times 10^{23} \text{ NH}_3 \text{ molecules}$$
4. (a) $NaNO_2$: 22.99 g/mol Na + 14.01 g/mol N + 2(16.00 g/mol O) = 69.00 g/mol $NaNO_2$
   (b) N_2S_5: 2(14.01 g/mol N) + 5(32.06 g/mol S) = 188.32 g/mol N_2S_5
5. $$765 \text{ mg PCl}_3 \times \frac{1 \text{ g PCl}_3}{1000 \text{ mg PCl}_3} \times \frac{1 \text{ mol PCl}_3}{137.32 \text{ g PCl}_3}$$
   $$= 5.57 \times 10^{-3} \text{ mol PCl}_3$$
6. $$3.2 \times 10^{24} \text{ molecules SF}_4 \times \frac{1 \text{ mol SF}_4}{6.02 \times 10^{23} \text{ molecules SF}_4}$$
   $$\times \frac{108.06 \text{ g SF}_4}{1 \text{ mol SF}_4} = 5.7 \times 10^2 \text{ g SF}_4$$
7. $Al(ClO_3)_3$: 26.98 g/mol Al + 3(35.45 g/mol Cl) + 9(16.00 g/mol O) = 277.33 g/mol $Al(ClO_3)_3$
   $$\% \text{ Cl} = \frac{\text{g Cl}}{\text{g Al(ClO}_3)_3} \times 100\% = \frac{3(35.45) \text{ g Cl}}{277.33 \text{ g Al(ClO}_3)_3}$$
   $$\times 100\% = 38.35\% \text{ Cl}$$
8. $Na_2S \cdot 9\ H_2O$: 2 (22.99 g/mol Na) + 32.06 g/mol S + 18(1.008 g/mol H) + 9(16.00 g/mol O)
   = 240.18 g/mol $Na_2S \cdot 9\ H_2O$
   $$\frac{2(22.99) \text{ g Na}}{240.18 \text{ g Na}_2\text{S} \cdot 9 \text{ H}_2\text{O}} \times 100\% = 19.14\% \text{ Na}$$
   $$\frac{32.06 \text{ g S}}{240.18 \text{ g Na}_2\text{S} \cdot 9 \text{H}_2\text{O}} \times 100\% = 13.35\% \text{ S}$$
   $$\frac{18(1.008) \text{ gH}}{240.18 \text{ g Na}_2\text{S} \cdot 9 \text{H}_2\text{O}} \times 100\% = 7.554\% \text{ H}$$
   $$\frac{9(16.00) \text{ g O}}{240.18 \text{ g Na}_2\text{S} \cdot 9 \text{ H}_2\text{O}} \times 100\% = 59.96\% \text{ O}$$
   19.14% + 13.35% + 7.554% + 59.96% = 100.00%
9. $$50.0 \text{ mg Cl} \times \frac{100 \text{ mg Al(ClO}_3)_3}{38.35 \text{ mg Cl}} = 1.30 \times 10^2 \text{ mg Al(ClO}_3)_3$$
10. $NaHCO_3$: 22.99 g/mol Na + 1.008 g/mol H + 12.01 g/mol C + 3(16.00 g/mol O)
    $$447 \text{ mg C} \times \frac{3(16.00 \text{ g O})}{12.01 \text{ g C}} = 1.79 \times 10^3 \text{ mg O}$$
11. (a) N_4S_4, NS; (b) Si_2Br_6, $SiBr_3$; (c) Si_4H_{10}, Si_2H_5; (d) P_4O_6, P_2O_3

12.

	Element	Mass (g)	Amount (mol)	Amount Ratio	Formula Ratio
$H_4O_2N_2$	H	0.292	0.290	1.99	4
	O	3.50	0.219	1.50	3
	N	2.04	0.146	1.00	2

13.

	Element	Mass (g)	Amount (mol)	Amount Ratio	Formula Ratio
Fe_2S_3	Fe	0.135	0.00242	1.00	2
	S	0.115	0.00359	1.48	3

14.

	Element	Mass (g)	Amount (mol)	Amount Ratio	Formula Ratio
C_6H_7N	C	74.1	6.17	5.02	5
	H	8.64	8.57	6.97	7
	N	17.3	1.23	1.00	1

15.

	Element	Mass (g)	Amount (mol)	Amount Ratio	Formula Ratio
EF =	C	49.3	4.10	2.99	6
$C_6H_3Cl_2$	H	2.1	2.1	1.5	3
	Cl	48.6	1.37	1.00	2

$\frac{292 \text{ g/mol}}{145.98 \text{ g/mol}} = 2$; molecular formula is $C_{12}H_6Cl_4$

Answers to Concept-Linking Exercises

You may have found more relationships or relationships other than the ones given in these answers.

1. The terms *atomic mass*, *molecular mass*, and *formula mass* refer to masses measured on the particulate level. Atomic mass is the mass of an atom, which is usually expressed in atomic mass units (u). Molecular and formula mass refer to the mass of a molecule and a formula unit, respectively. They are also typically expressed in atomic mass units. Molar mass, the mass in grams of 1 mole of a substance, is a unit that bridges the particulate and macroscopic worlds. Molar mass is numerically equal to atomic, molecular, or formula mass, but it is expressed in grams per mole.
2. One mole is, to three significant figures, is 6.02×10^{23} particles. This definition is based on the number of atoms in 12 grams of carbon-12. The Avogadro constant is $N_A = 6.02 \times 10^{23}$/mol.

Solutions to Blue-Numbered Questions, Exercises, and Problems

1. $Al(NO_3)_3$: 1 aluminum atom, 3 nitrogen atoms, 9 oxygen atoms

3. *Molecular mass* is a term properly applied only to substances that exist as molecules. Sodium nitrate is an ionic compound.

5. Atomic mass, Ba; molecular mass, NH_3 and Cl_2 because these are molecular substances; formula mass, all substances, but particularly CaO and Na_2CO_3 because neither of the other two terms technically fit these ionic compounds.

7. a) 6.94 u Li + 35.45 u Cl = 42.39 u LiCl
   b) 2(26.98) u Al + 3(12.01 u C) + 9(16.00 u O) = 233.99 u $Al_2(CO_3)_3$
   c) 2(14.01 u N) + 8(1.008 u H) + 32.06 u S + 4(16.00 u O) = 132.14 u $(NH_4)_2SO_4$
   d) 4(12.01 u C) + 10(1.008 u H) = 58.12 u C_4H_{10}
   e) 107.9 u Ag + 14.01 u N + 3(16.00 u O) = 169.9 u $AgNO_3$
   f) 54.94 u Mn + 2(16.00 u O) = 86.94 u MnO_2
   g) 3(65.38 u Zn) + 2(30.97 u P) + 8(16.00 u O) = 386.08 u $Zn_3(PO_4)_2$

9. a) 32.06 u S + 3(16.00 u O) = 80.06 u SO_3
   b) 1.008 u H + 19.00 u F = 20.01 u HF

11. The quantity of particles of each is the same—that is, the number of iron atoms and ammonia molecules is the same.

13. Yes. The mole is defined as, to three significant figures, 6.02×10^{23} particles.

15. a) $7.75 \text{ mol } CH_4 \times \frac{6.02 \times 10^{23} \text{ molecules } CH_4}{\text{mol } CH_4} =$ 4.67×10^{24} molecules CH_4
    b) $0.0888 \text{ mol CO} \times \frac{6.02 \times 10^{23} \text{ molecules CO}}{\text{mol CO}} =$ 5.35×10^{22} molecules CO
    c) $57.8 \text{ mol Fe} \times \frac{6.02 \times 10^{23} \text{ atoms Fe}}{\text{mol Fe}} =$ 3.48×10^{25} atoms Fe
    d) $0.81 \text{ mol } MgCl_2 \times \frac{6.02 \times 10^{23} \text{ fu } MgCl_2}{\text{mol } MgCl_2} =$ 4.88×10^{23} formula units $MgCl_2$

17. a) $2.45 \times 10^{23} \text{ molecules } C_2H_2 \times \frac{1 \text{ mol } C_2H_2}{6.02 \times 10^{23} \text{ molecules } C_2H_2} = 0.407 \text{ mol } C_2H_2$
    b) $6.96 \times 10^{24} \text{ atoms Na} \times \frac{1 \text{ mol Na}}{6.02 \times 10^{23} \text{ atoms Na}} =$ 11.6 mol Na

19. Numerically.

21. a) 3(12.01 g/mol C) + 8(1.008 g/mol H) = 44.09 g/mol C_3H_8
    b) 6(12.01 g/mol C) + 5(35.45 g/mol Cl) + 16.00 g/mol O + 1.008 g/mol H = 266.32 g/mol C_6Cl_5OH
    c) 3(58.69 g/mol Ni) + 2(30.97 g/mol P) + 8(16.00 g/mol O) + 366.01 g/mol $Ni_3(PO_4)_2$
    d) 65.38 g/mol Zn + 2(14.01 g/mol N) + 6(16.00 g/mol O) = 189.40 g/mol $Zn(NO_3)_2$

23. a) $6.79 \text{ g } O_2 \times \frac{1 \text{ mol } O_2}{32.00 \text{ g } O_2} = 0.212 \text{ mol } O_2$

b) $9.05 \text{ g } Mg(NO_3)_2 \times \frac{1 \text{ mol } Mg(NO_3)_2}{148.33 \text{ g } Mg(NO_3)_2} =$
$0.0610 \text{ mol } Mg(NO_3)_2$

c) $0.770 \text{ g } Al_2O_3 \times \frac{1 \text{ mol } Al_2O_3}{101.96 \text{ g } Al_2O_3} = 0.00755 \text{ mol } Al_2O_3$

d) $659 \text{ g } C_2H_5OH \times \frac{1 \text{ mol } C_2H_5OH}{46.07 \text{ g } C_2H_5OH} =$
$14.3 \text{ mol } C_2H_5OH$

e) $0.394 \text{ g } (NH_4)_2CO_3 \times \frac{1 \text{ mol } (NH_4)_2CO_3}{96.09 \text{ g } (NH_4)_2CO_3} =$
$0.00410 \text{ mol } (NH_4)_2CO_3$

f) $34.0 \text{ g } Li_2S \times \frac{1 \text{ mol } Li_2S}{45.94 \text{ g } Li_2S} = 0.740 \text{ mol } Li_2S$

25. a) $0.797 \text{ g } KIO_3 \times \frac{1 \text{ mol } KIO_3}{214.0 \text{ g } KIO_3} = 0.00372 \text{ mol } KIO_3$

b) $68.6 \text{ g } BeCl_2 \times \frac{1 \text{ mol } BeCl_2}{79.91 \text{ g } BeCl_2} = 0.858 \text{ mol } BeCl_2$

c) $302 \text{ g } Ni(NO_3)_2 \times \frac{1 \text{ mol } Ni(NO_3)_2}{182.71 \text{ g } Ni(NO_3)_2} =$
$1.65 \text{ mol } Ni(NO_3)_2$

27. a) $0.769 \text{ mol LiCl} \times \frac{42.39 \text{ g LiCl}}{\text{mol LiCl}} = 32.6 \text{ g LiCl}$

b) $57.1 \text{ mol } HC_2H_3O_2 \times \frac{60.05 \text{ g } HC_2H_3O_2}{\text{mol } HC_2H_3O_2} =$
$3.43 \times 10^3 \text{g } HC_2H_3O_2$

c) $0.68 \text{ mol Li} \times \frac{6.94 \text{ g Li}}{\text{mol Li}} = 4.7 \text{ g Li}$

d) $0.532 \text{ mol } Fe_2(SO_4)_3 \times \frac{399.91 \text{ g } Fe_2(SO_4)_3}{\text{mol } Fe_2(SO_4)_3} =$
$213 \text{ g } Fe_2(SO_4)_3$

e) $8.26 \text{ mol } NaC_2H_3O_2 \times \frac{82.03 \text{ g } NaC_2H_3O_2}{\text{mol } NaC_2H_3O_2} =$
$678 \text{ g } NaC_2H_3O_2$

29. a) $0.379 \text{ mol } Li_2SO_4 \times \frac{109.94 \text{ g } Li_2SO_4}{\text{mol } Li_2SO_4} = 41.7 \text{ g } Li_2SO_4$

b) $4.82 \text{ mol } K_2C_2O_4 \times \frac{166.22 \text{ g } K_2C_2O_4}{\text{mol } K_2C_2O_4} = 801 \text{ g } K_2C_2O_4$

c) $0.132 \text{ mol } Pb(NO_3)_2 \times \frac{331.2 \text{ g } Pb(NO_3)_2}{\text{mol } Pb(NO_3)_2} =$
$43.7 \text{ g } Pb(NO_3)_2$

31. a) $29.6 \text{ g } LiNO_3 \times \frac{1 \text{ mol } LiNO_3}{68.95 \text{ g } LiNO_3} \times \frac{6.02 \times 10^{23} \text{ fu } LiNO_3}{\text{mol } LiNO_3} =$
2.58×10^{23} formula units $LiNO_3$

b) $0.151 \text{ g } Li_2S \times \frac{1 \text{ mol } Li_2S}{45.94 \text{ g } Li_2S} \times \frac{6.02 \times 10^{23} \text{ fu } Li_2S}{\text{mol } Li_2S} =$
1.98×10^{21} formula units Li_2S

c) $457 \text{ g } Fe_2(SO_4)_3 \times \frac{1 \text{ mol } Fe_2(SO_4)_3}{399.88 \text{ g } Fe_2(SO_4)_3} \times$
$\frac{6.02 \times 10^{23} \text{ fu } Fe_2(SO_4)_3}{\text{mol } Fe_2(SO_4)_3} = 6.88 \times 10^{23}$ formula units
$Fe_2(SO_4)_3$

33. a) $0.0023 \text{ g } I_2 \times \frac{1 \text{ mol } I_2}{253.8 \text{ g } I_2} \times \frac{6.02 \times 10^{23} \text{ molecules } I_2}{\text{mol } I_2} =$
5.5×10^{18} molecules I_2

b) $114 \text{ g } C_2H_4(OH)_2 \times \frac{1 \text{ mol } C_2H_4(OH)_2}{62.07 \text{ g } C_2H_4(OH)_2} \times$
$\frac{6.02 \times 10^{23} \text{ molecules } C_2H_4(OH)_2}{\text{mol } C_2H_4(OH)_2} = 1.11 \times 10^{24}$
molecules $C_2H_4(OH)_2$

c) $9.81 \text{ g } Cr_2(SO_4)_3 \times \frac{1 \text{ mol } Cr_2(SO_4)_3}{392.18 \text{ g } Cr_2(SO_4)_3} \times$
$\frac{6.02 \times 10^{23} \text{ fu } Cr_2(SO_4)_3}{\text{mol } Cr_2(SO_4)_3} = 1.51 \times 10^{22}$ formula units
$Cr_2(SO_4)_3$

35. a) 4.30×10^{21} molecules $C_{19}H_{37}COOH \times$
$\frac{1 \text{ mol } C_{19}H_{37}COOH}{6.02 \times 10^{23} \text{ molecules } C_{19}H_{37}COOH} \times$
$\frac{310.50 \text{ g } C_{19}H_{37}COOH}{\text{mol } C_{19}H_{37}COOH} = 2.22 \text{ g } C_{19}H_{37}COOH$

b) $8.67 \times 10^{24} \text{ atoms F} \times \frac{1 \text{ mol F}}{6.02 \times 10^{23} \text{ atoms F}} \times$
$\frac{19.00 \text{ g F}}{\text{mol F}} = 274 \text{ g F}$

c) $7.23 \times 10^{23} \text{ fu } NiCl_2 \times \frac{1 \text{ mol } NiCl_2}{6.02 \times 10^{23} \text{ fu } NiCl_2} \times$
$\frac{129.59 \text{ g } NiCl_2}{\text{mol } NiCl_2} = 156 \text{ g } NiCl_2$

37. $1 \text{ atom Au} \times \frac{1 \text{ mol Au}}{6.02 \times 10^{23} \text{ atoms Au}} \times \frac{197.0 \text{ g Au}}{\text{mol Au}} \times$
$\frac{1 \text{ troy oz}}{31.1 \text{ g}} \times \frac{\$1{,}258}{\text{troy oz}} = \$1.32 \times 10^{-20}$

39. $0.65 \text{ g } C_{12}H_{22}O_{11} \times \frac{1 \text{ mol } C_{12}H_{22}O_{11}}{342.20 \text{ g } C_{12}H_{22}O_{11}} \times$
$\frac{6.02 \times 10^{23} \text{ molecules } C_{12}H_{22}O_{11}}{\text{mol } C_{12}H_{22}O_{11}} =$
1.1×10^{21} $C_{12}H_{22}O_{11}$ molecules

41. a) 4.12×10^{24} N atoms $\times \frac{1 \text{ mol N atom}}{6.02 \times 10^{23} \text{ N atoms}} \times \frac{14.01 \text{ g N}}{\text{mol N atom}} = 95.9 \text{ g N}$

b) 4.12×10^{24} N_2 molecules $\times \frac{1 \text{ mol } N_2 \text{ molecules}}{6.02 \times 10^{23} \text{ } N_2 \text{ molecules}} \times \frac{28.02 \text{ g } N_2}{\text{mol } N_2 \text{ molecules}} = 192 \text{ g } N_2$

c) $4.12 \text{ g N} \times \frac{1 \text{ mol N}}{14.01 \text{ g N}} \times \frac{6.02 \times 10^{23} \text{ atoms N}}{\text{mol N}} =$ 1.77×10^{23} atoms N

d) $4.12 \text{ g } N_2 \times \frac{1 \text{ mol } N_2}{28.02 \text{ g } N_2} \times \frac{6.02 \times 10^{23} \text{ molecules } N_2}{\text{mol } N_2} =$ 8.85×10^{22} N_2 molecules

e) $4.12 \text{ g } N_2 \times \frac{1 \text{ mol } N_2}{28.02 \text{ g } N_2} \times \frac{6.02 \times 10^{23} \text{ molecules } N_2}{\text{mol } N_2} \times \frac{2 \text{ atoms N}}{\text{molecule } N_2} = 1.77 \times 10^{23} \text{ atoms N}$

43. a) NH_4NO_3: $\frac{2(14.01 \text{ g N})}{80.05 \text{ g } NH_4NO_3} \times 100\% = 35.00\% \text{ N}$

$\frac{4(1.008 \text{ g H})}{80.05 \text{ g } NH_4NO_3} \times 100\% = 5.037\% \text{ H}$

$\frac{3(16.00 \text{ g O})}{80.05 \text{ g } NH_4NO_3} \times 100\% = 59.96\% \text{ O}$

$35.00\% + 5.037\% + 59.96\% = 100.00\%$

b) $Al_2(SO_4)_3$: $\frac{2(26.98 \text{ g Al})}{342.14 \text{ g } Al_2(SO_4)_3} \times 100\% = 15.77\% \text{ Al}$

$\frac{3(32.06 \text{ g S})}{342.14 \text{ g } Al_2(SO_4)_3} \times 100\% = 28.11\% \text{S}$

$\frac{12(16.00 \text{ g O})}{342.14 \text{ g } Al_2(SO_4)_3} \times 100\% = 56.12\% \text{ O}$

$15.77\% + 28.11\% + 56.12\% = 100.00\%$

c) $(NH_4)_2CO_3$: $\frac{2(14.01 \text{ g N})}{96.09 \text{ g } (NH_4)_2CO_3} \times 100\% = 29.16\% \text{ N}$

$\frac{8(1.008 \text{ g H})}{96.09 \text{ g } (NH_4)_2CO_3} \times 100\% = 8.392\% \text{ H}$

$\frac{12.01 \text{ g C}}{96.09 \text{ g } (NH_4)_2CO_3} \times 100\% = 12.50\% \text{ C}$

$\frac{3(16.00 \text{ g O})}{96.09 \text{ g } (NH_4)_2CO_3} \times 100\% = 49.95\% \text{ O}$

$29.16\% + 8.392\% + 12.50\% + 49.95\% = 100.00\%$

d) CaO: $\frac{40.08 \text{ g Ca}}{56.08 \text{ g CaO}} \times 100\% = 71.47\% \text{ Ca}$

$\frac{16.00 \text{ g O}}{56.08 \text{ g CaO}} \times 100\% = 28.53\% \text{ O}$

$71.47\% + 28.53\% \text{ O} = 100.00\%$

e) MnS_2: $\frac{54.94 \text{ g Mn}}{119.06 \text{ g } MnS_2} \times 100\% = 46.14\% \text{ Mn}$

$\frac{2(32.06 \text{ g S})}{119.06 \text{ g } MnS_2} \times 100\% = 53.86\% \text{ S}$

$46.14\% + 53.86\% = 100.00\%$

45. $454 \text{ g LiF} \times \frac{6.941 \text{ g Li}}{25.94 \text{ g LiF}} = 121 \text{ g Li}$

47. $57.4 \text{ g } K_2SO_4 \times \frac{2(39.10 \text{ g K})}{174.26 \text{ g } K_2SO_4} = 25.8 \text{ g K}$

49. $146 \text{ g Zn} \times \frac{117.42 \text{ g Zn(CN)}_2}{65.38 \text{ g Zn}} = 262 \text{ g Zn(CN)}_2$

51. $201 \text{ kg Mo} \times \frac{160.08 \text{ kg } MoS_2}{95.96 \text{ kg Mo}} = 335 \text{ kg } MoS_2$

53. $4.17 \text{ g Cl} \times \frac{206.98 \text{ g Ca(ClO}_3)_2}{2(35.45 \text{ g Cl})} = 12.2 \text{ g Ca(ClO}_3)_2$

55. C_6H_{10} must be a molecular formula because both 6 and 10 are divisible by 2. Its empirical formula is C_3H_5. There is no common divisor for 7 and 10, so C_7H_{10} can be an empirical formula. An empirical formula can also be a molecular formula.

	Element	Mass (g)	Amount (mol)	Amount Ratio	Formula Ratio	Empirical Formula	Molecular Formula
57.	C	52.2	4.35	2.00	2		
	H	13.0	12.9	5.92	6	C_2H_6O	
	O	34.8	2.18	1.00	1		
59.	Fe	11.89	0.213	1.00	2	Fe_2O_3	
	O	5.10	0.319	1.50	3		
61.	C	17.2	1.43	1.00	1		
	H	1.44	1.43	1.00	1	CHF_3	
	F	81.4	4.28	2.99	3		
63.	C	38.7	3.22	1.00	1		$\frac{62.0}{31.0} = 2$
	H	9.7	9.6	3.0	3	CH_3O	
	O	51.6	3.23	1.00	1		$C_2H_6O_2$
65.	Cl	73.1	2.06	1.00	1		$\frac{97}{49} = 2$
	C	24.8	2.06	1.00	1	ClCH	
	H	2.1	2.1	1.0	1		$Cl_2C_2H_2$

68. All statements are false.

69. Hardly—the mass of 10^{25} atoms of copper is 2 pounds:

$$10^{25}\text{ atoms Cu} \times \frac{1\text{ mol Cu}}{6.02 \times 10^{23}\text{ atoms Cu}} \times \frac{63.55\text{ g Cu}}{\text{mol Cu}} \times \frac{1\text{ lb Cu}}{454\text{ g Cu}} = 2\text{ lb Cu}$$

70. $85.0\text{ g }P_4 \times \frac{1\text{ mol }P_4}{123.88\text{ g }P_4} \times \frac{6.02 \times 10^{23}\text{ molecules }P_4}{\text{mol }P_4} =$

$4.13 \times 10^{23}\ P_4$ molecules

$$85.0\text{ g }P_4 \times \frac{1\text{ mol }P_4}{123.88\text{ g }P_4} \times \frac{6.02 \times 10^{23}\text{ molecules }P_4}{\text{mol }P_4} \times \frac{4\text{ atoms P}}{\text{molecule }P_4} = 1.65 \times 10^{24}\text{ P atoms}$$

71. It is a close call: 2 oz is about the capacity of a typical salt shaker; 12 oz is too much for a 10-oz sugar bowl, but only by a little.

$$1\text{ mol NaCl} \times \frac{58.44\text{ g NaCl}}{\text{mol NaCl}} \times \frac{1\text{ lb NaCl}}{454\text{ g NaCl}} \times \frac{16\text{ oz NaCl}}{\text{lb NaCl}} = 2\text{ oz NaCl}$$

$$1\text{ mol }C_{12}H_{22}O_{11} \times \frac{342.30\text{ g }C_{12}H_{22}O_{11}}{\text{mol }C_{12}H_{22}O_{11}} \times \frac{1\text{ lb }C_{12}H_{22}O_{11}}{454\text{ g }C_{12}H_{22}O_{11}} \times \frac{16\text{ oz }C_{12}H_{22}O_{11}}{\text{lb }C_{12}H_{22}O_{11}} = 12\text{ oz }C_{12}H_{22}O_{11}$$

72. $$2.95 \times 10^{22}\text{ molecules air} \times \frac{1\text{ mol air}}{6.02 \times 10^{23}\text{ molecules air}} \times \frac{29\text{ g air}}{\text{mol air}} = 1.42\text{ g air}$$

73. $$5.62 \times 10^{23}\text{ molecules }C_8H_{18} \times \frac{1\text{ mol }C_8H_{18}}{6.02 \times 10^{23}\text{ molecules }C_8H_{18}} \times \frac{114.22\text{ g }C_8H_{18}}{\text{mol }C_8H_{18}} = 107\text{ g }C_8H_{18}$$

74. $$85.9\text{ g }CO_2 \times \frac{12.01\text{ g C}}{44.01\text{ g }CO_2} = 23.4\text{ g C}$$

$$35.5\text{ g }H_2O \times \frac{2(1.008\text{ g H})}{18.02\text{ g }H_2O} = 3.97\text{ g H}$$

Element	Mass (g)	Amount (mol)	Amount Ratio	Formula Ratio	Empirical Formula
C	23.4	1.95	1.00	1	CH_2
H	3.97	3.94	2.02	2	

75.

	Element	Mass (g)	Amount (mol)	Amount Ratio	Formula Ratio	Empirical Formula
a)	Co	42.4	0.719	1.00	1	
	S	23.0	0.717	1.00	1	$CoSO_3$
	O	34.6	2.16	3.01	3	
b)	$CoSO_3$	26.1	0.188	1.00	1	$CoSO_3 \cdot 5\,H_2O$
	H_2O	16.9	0.938	4.99	5	

43.0 g hydrate − 26.1 g anhydrate = 16.9 g H_2O

8 Chemical Reactions

◄ A huge quantity of energy was necessary to provide enough power for the space shuttle to escape Earth's gravitational pull. The reactants in this exothermic reaction are liquid hydrogen and liquid oxygen; the product, water. Rather than using words, a chemist might write a chemical equation for this combination reaction: $2\ H_2(\ell) + O_2(\ell) \rightarrow 2\ H_2O(g)$. In this chapter, you will learn about many kinds of chemical reactions and how to write equations for them.

NASA

CHAPTER CONTENTS

We will now begin to consider chemical reactions in detail. The overarching objective in this chapter is to learn how to write chemical equations, that is, symbolic representations of those reactions. Recall from Section 2.9 that an equation shows the formulas of the **reactants**—the starting substances that will be destroyed in the chemical change—written on the left side of an arrow, →. The formulas of the **products** of the reaction—the new substances formed in the chemical change—are written on the right side of the arrow.

In this chapter, you will learn to identify four different patterns seen in chemical equations. These four patterns are based on how atoms are rearranged in the chemical equation. They are not a complete description of the chemical change. However, when you combine knowledge of these rearrangements with an understanding of the particulate-level driving forces for reactions (which are introduced in Chapter 9), you will have learned the most common reactions that occur in living organisms, industry, the laboratory, and nature. Writing a chemical equation is easier if you can classify the chemical equation as a certain type. You will be able to look at a particular combination of reactants, recognize what kind of reaction is possible (if any), and predict what products will be formed. The equation follows.

With these facts in mind, we suggest that you set your sights on the following goals as you begin this chapter:

1. Understand how a chemical equation serves as a particulate-level, symbolic representation of a macroscopic-level process.
2. Learn the mechanics of writing a chemical equation.
3. Learn how to identify four different patterns in chemical equations.
4. Learn how to predict the products of each kind of reaction and write the formulas of those products.
5. Given potential reactants, write the equations for the probable reaction.

8.1 Do Chemical Reactions Occur Outside of Earth?

How do we know whether or not chemical reactions occur outside of Earth? In Section 5.1, we discussed how all of the heavier elements are synthesized in stars and are then scattered throughout the universe when the stars explode and die. So, elements do exist in space. But do they undergo reactions?

To know if species react in space, we must have a method for observing their behavior. An instrument that enables observation of tiny particles at huge distances is the Hubble Space Telescope. A space telescope is an instrument located outside of the Earth's atmosphere that detects and measures light, including light beyond the range of that which can be detected by human senses. The advantage of placing the telescope in space is that the Earth's atmosphere severely reduces the quality of light that can be observed.

The successor to the Hubble Telescope will be the James Webb Space Telescope (JWST) (**Figure 8.1**), which is scheduled to be launched in 2021. The JWST will make the same type of measurements as the Hubble, but its capabilities will be much greater. Stefanie Milam, PhD, is a chemist who serves as the James Webb Space Telescope Deputy Project Scientist for Planetary Science at the Goddard Space Flight Center of the National Aeronautics and Space Administration, Greenbelt, Maryland (**Figure 8.2**). She explains use of the JWST as a tool for studying the "fingerprints" of molecules. These fingerprints give information about the amount of any specific species in space, its temperature, and its phase of matter. When multiple fingerprints are combined, the chemistry of the matter in space can be deduced. One thing that the JWST will allow us to study in new ways is the composition of comets in the outer solar system.

NASA

Figure 8.1 The James Webb Space Telescope, scheduled to be launched in 2021.

NASA

Figure 8.2 Stefanie Milam, PhD, is a chemist who works at NASA studying the chemistry that occurs in outer space.

What have we learned from space telescopes about extraterrestrial chemical reactions thus far? As an example, one molecule whose fingerprint can be measured is HCN, hydrogen cyanide. HCN has been detected throughout the universe: on Earth, in the atmosphere of Saturn's moon Titan, in comets, in the atmosphere of planets in other solar systems, in stars, and in the space between star systems in galaxies. It forms in interstellar clouds by multiple pathways, one of which is the reaction of CH_2 and N, yielding HCN and H. So, yes, chemical reactions occur in places other than on Earth! All laws of science that we have discovered on Earth appear to be universal. In this chapter, you will begin to learn about common types of chemical reactions, primarily those that have only been detected on Earth to date, but again, we have no reason to believe that the chemistry on distant Earth-like planets is different from the chemistry on today's Earth.

8.2 Evidence of a Chemical Change

Goal 1 Describe five types of evidence detectable by human senses that usually indicate a chemical change.

In Chapter 2, we stated that a chemical change occurs when the chemical identity of the reacting substances is destroyed and new substances form. ▶ Particles of matter are literally changed. The number and type of *atoms* that make up molecules are the same before and after a chemical change, but the number and type of *molecules* change. If we could observe matter at the particulate level, we would have a simple method for detecting chemical change. Of course, we cannot directly see what happens at the particulate level, so we must rely on indirect evidence of particulate-level rearrangements.

◀ **P/Review** Physical and chemical changes were introduced in Section 2.4. A physical change alters the physical form of matter without changing its chemical identity (e.g., a piece of paper is shredded). A chemical change occurs when the chemical identity of a substance is destroyed and a new substance forms (e.g., a piece of paper is burned).

We also stated in Chapter 2 that one or more of our five physical senses can usually detect chemical change. This is true because particulate-level changes are often accompanied by macroscopic changes that we can sense by sight, hearing, touch, smell, or taste.* A **change of color** is almost always evidence that a chemical reaction has occurred. Figure 8.3 shows a violet-colored dye reacting with hydroxide ion to form a colorless product. The original dye molecules react with the hydroxide ions to form new molecules that do not exhibit color in a water solution.

Another visible form of evidence of a chemical change is the **formation of a solid product** when clear solutions are combined (Figure 8.4). Substances in each of the reacting solutions combine to form the solid product. **Formation of a gas**, visible as bubbles forming in a liquid, is also evidence of a chemical reaction (Figure 8.5).

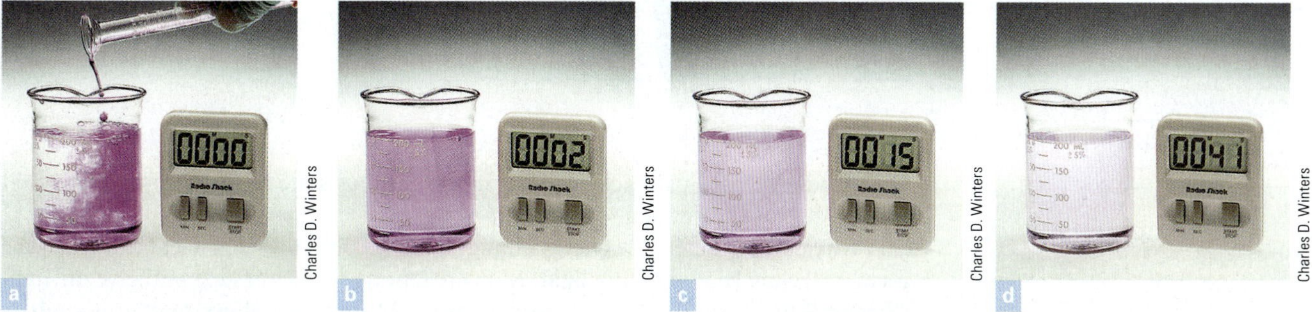

Figure 8.3 Color change as evidence of a chemical reaction. When a solution containing a violet-colored dye is added to a solution containing hydroxide ion, the intensity of the color decreases with time until it disappears. The product of the reaction is colorless in water solution.

*Tasting is not an acceptable method for detecting chemical change in the laboratory. Many early chemists sacrificed their health and even their lives by using this sense. The other senses must also be used with caution and proper technique.

Charles D. Winters

Figure 8.4 Formation of a solid product as evidence of a chemical reaction. When two clear, colorless solutions are combined, one containing barium ion and the other containing sulfate ion, solid barium sulfate is formed. The white solid will eventually settle to the bottom of the test tube.

Charles D. Winters

Figure 8.5 Formation of a gas as evidence of a chemical reaction. Alka-Seltzer antacid tablets contain citric acid and sodium hydrogen carbonate. When added to water, these two reactants combine to form carbon dioxide gas as one of the products.

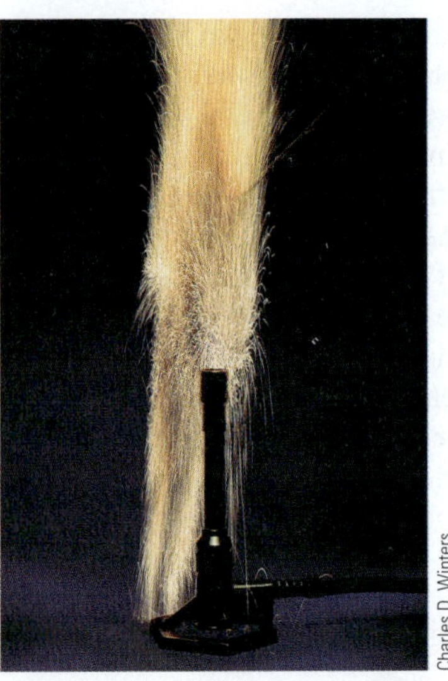
Charles D. Winters

Figure 8.6 Evolution of heat and light as evidence of a chemical reaction. This photograph shows aluminum reacting with iron(III) oxide to produce molten iron and aluminum oxide. The quantity of heat evolved from this chemical change is large enough to cause the iron produced to be in the liquid (molten) state.

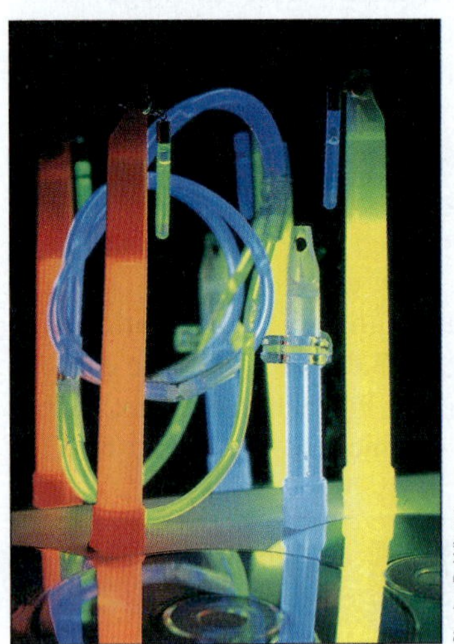
Charles D. Winters

Figure 8.7 Emission of light as evidence of a chemical reaction. The light emitted from a light stick is a form of energy released in a multistep chemical change that involves several reactants. One of the reactants is stored in a fragile glass vial inside of the larger plastic tube. When you bend the plastic tube, you break the glass vial, mixing the reactants.

One product of such a reaction exists in the gaseous state at room conditions, so it bubbles out of the liquid as soon as it forms.

We can often detect the energy changes that accompany a chemical change in the form of feeling **heat** or seeing **light**. Many reactions both emit heat and give off light (**Figure 8.6**). Burning is a common example. More unusual are reactions such as those that occur in light sticks, which give off light without getting hot (**Figure 8.7**). Some reactions absorb heat, such as the reaction of hydrogen and iodine gases to form hydrogen iodide gas.

One must always be cautious when using sensory evidence of a chemical change. Many chemical reactions are not visible, and their heat energy changes are too small to feel. Also, sensory evidence can indicate a physical change, rather than one that is chemical. For example, we feel the coldness when ice is placed on our skin, but there is no chemical change when ice—solid water—changes to the liquid state. Similarly, when water boils, we see bubbles of steam rising through the liquid, but the liquid-to-vapor change is also physical. Water molecules are unchanged, whether in the solid (ice), liquid, or gaseous (steam) state.

Another example of sensory evidence indicating a physical change is the coldness you feel when an instant cold pack is placed on swollen tissue (**Figure 8.8**). Most cold packs contain a small, sealed container of solid ammonium nitrate that breaks when the larger packet is squeezed, allowing the ionic solid to dissolve in water via an endothermic process. This is an example of an endothermic physical change rather than a chemical change because the ammonium ions and the nitrate ions are the same, regardless of whether they are in the solid state or the aqueous state.

Table 8.1 summarizes the five types of evidence that indicate the possibility of a chemical change.

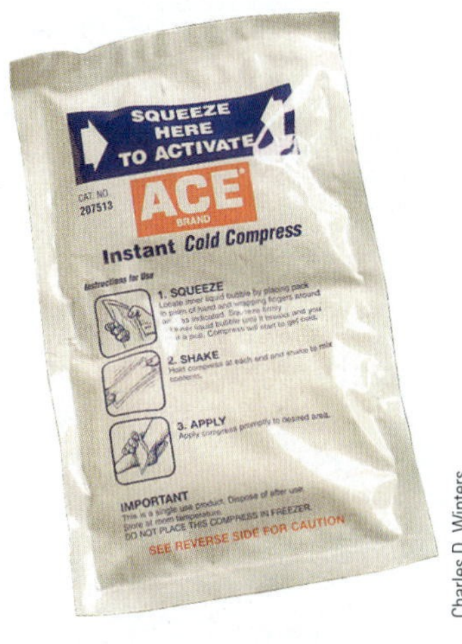

Charles D. Winters

Figure 8.8 Heat transfer as evidence of a physical change. An endothermic physical change causes the transfer of heat from the surroundings, making the compress cold. The cold compress can then be placed on skin to help minimize swelling.

Target Check 8.1

Are any of the following a chemical change? Give evidence for each answer.

a) A firecracker explodes.

b) A puddle of water evaporates.

c) The process shown here:

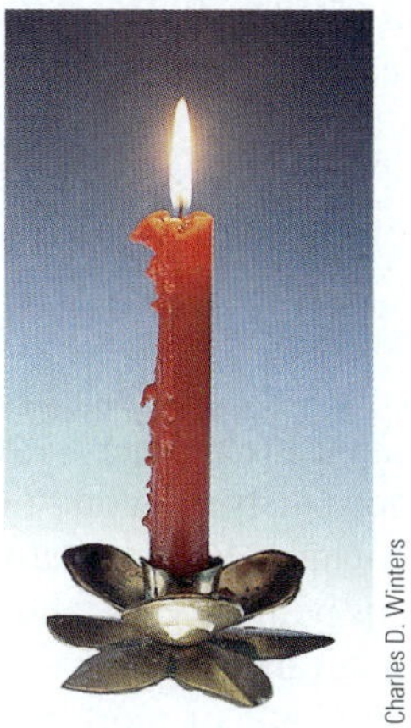

Charles D. Winters

d) The process shown here:

YAKOBCHUK VASYL/Shutterstock.com

Table 8.1	Evidence of Chemical Change
1.	Color change
2.	Formation of a solid
3.	Formation of a gas
4.	Absorption or release of heat energy
5.	Emission of light energy

8.3 Evolution of a Chemical Equation

Goal 2 Distinguish between an unbalanced and a balanced chemical equation, and explain why a chemical equation needs to be balanced.

If a piece of sodium is dropped into water, a vigorous reaction occurs (**Figure 8.9**). A full, qualitative description of the chemical change is "solid sodium plus liquid water yields hydrogen gas plus sodium hydroxide solution plus heat." ▶ That sentence may be abbreviated by using symbols to write a **chemical equation**:

◀ **P/Review** We examine the energy factor in a chemical reaction in Sections 11.11 and 11.12.

$$Na(s) + H_2O(\ell) \rightarrow H_2(g) + NaOH(aq) + heat \quad (8.1)$$

The (s) after the symbol of sodium indicates it is a solid. Similarly, the (ℓ) after H_2O and the (g) after H_2 show that they are a liquid and a gas, respectively. When a substance is dissolved in water, the mixture is an **aqueous solution** and is identified by (aq). (*Aqueous* comes from the Latin word for water, *aqua*.) These **state symbols** are sometimes omitted when writing equations, but they are included in most of the

Figure 8.9 Sodium reacting with water. (a) A small piece of sodium is dropped into a test tube of water containing phenolphthalein, a substance that turns pink in a solution of a metallic hydroxide. (b) Sodium forms a "ball" that dashes erratically over the water surface, releasing hydrogen as it reacts. Pink color near the sodium indicates that a sodium hydroxide (NaOH) solution is formed in the region. (c) Dissolved NaOH is now distributed uniformly through the solution, which is hot because of the heat released in the reaction. *Warning:* Do not try this experiment yourself; it is dangerous, potentially splattering hot alkali into your eyes and onto your skin and clothing.

Table 8.2 State Symbols and Their Meanings

Symbol	Meaning
(s)	Solid
(ℓ)	Liquid
(g)	Gas
(aq)	Aqueous (dissolved in water)

equations in this book. They are discussed in more detail in Chapter 9. We suggest that whether you do use or do not use them at this time is according to the directions of your instructor. State symbols are summarized in **Table 8.2**.

Nearly all chemical reactions involve some transfer of energy, usually in the form of heat. Generally, we omit energy terms from equations unless there is a specific reason for including them.

The Law of Conservation of Mass (Section 2.10) is that the total mass of the products of a reaction is the same as the total mass of the reactants. Atomic theory explains this by saying that atoms involved in a chemical change are neither created nor destroyed but are simply rearranged. Equation 8.1 does not satisfy this condition. There are two hydrogen atoms in H_2O in the left side of the equation but three atoms of hydrogen on the right—two in H_2 and one in NaOH. At this point, the equation is only a *qualitative* description of the reaction. The equation is not **balanced**.

An equation is balanced by placing a **coefficient** in front of one or more of the formulas, indicating that it is used more than once. Once balanced, an equation is both a *qualitative* and a *quantitative* description of the reaction.

Hydrogen is lacking on the left side of Equation 8.1, so let's try two water molecules:

$$Na(s) + 2\,H_2O(\ell) \rightarrow H_2(g) + NaOH(aq) \qquad \textbf{(8.2)}$$

At first glance, this hasn't helped; indeed, it seems to have made matters worse. The hydrogen is still out of balance (four on the left, three on the right), and furthermore, oxygen is now unbalanced (two on the left, one on the right). We are short one oxygen atom and one hydrogen atom on the right side. But look closely: Oxygen and hydrogen are part of the same compound on the right, and there is one atom of each in that compound. If we take two NaOH units

$$Na(s) + 2\,H_2O(\ell) \rightarrow H_2(g) + 2\,NaOH(aq) \qquad \textbf{(8.3)}$$

there are four hydrogens and two oxygens on both sides of the equation. These elements are now in balance. Unfortunately, the sodium is now *un*balanced. This condition is corrected by adding a coefficient of 2 to the sodium

$$2\,Na(s) + 2\,H_2O(\ell) \rightarrow H_2(g) + 2\,NaOH(aq) \qquad \textbf{(8.4)}$$

The equation is now balanced. Note that in the absence of a numerical coefficient, as with H_2, the coefficient is assumed to be 1.

Balancing an equation involves some important do's and don'ts that are apparent in this example:

DO: Balance the equation entirely by using coefficients placed before the different chemical formulas.

DON'T: Change a correct chemical formula in order to make an element balance.

DON'T: Add some real or imaginary chemical species to either side of the equation just to make an element balance.

A moment's thought shows why the two "don'ts" are improper. The first step in writing an equation is to write formulas of reactants and products. These formulas describe the substances in the reaction, and these do not change. Changing a formula or adding another formula would change the qualitative description of the reaction.

In other words, writing and balancing a chemical equation require two main steps.

how to... Write and Balance a Chemical Equation

Sample Problem: *An aluminum strip reacts with the oxygen in air to form solid aluminum oxide. Write and balance a chemical equation for the reaction.*

Step 1: Write a qualitative description of the reaction. In this step you write the formulas of the given reactants to the left of an arrow and the formula of the given or predicted products to the right.

$Al(s) + O_2(g) \rightarrow Al_2O_3(s)$

Step 1: Step 2: Quantify the description by balancing the equation. Do this by adding coefficients. Do not change the qualitative description of the reaction by adding, removing, or altering any formula.

$4\ Al(s) + 3\ O_2(g) \rightarrow 2\ Al_2O_3(s)$

✓ Target Check 8.2

Are the following true or false? Explain your reasoning.

a) The equation $C_2H_4O + 3\ O_2 \rightarrow 2\ CO_2 + 2\ H_2O$ is balanced.

b) The equation $H_2 + O_2 \rightarrow H_2O$ may be balanced by changing it to $H_2 + O_2 \rightarrow H_2O_2$.

8.4 Balancing Chemical Equations

Goal 3 Given an unbalanced chemical equation, balance it by inspection.

The balancing procedure in the preceding section is sometimes called *balancing by inspection*. It is a trial-and-error method that succeeds in nearly all the reactions you are likely to encounter in an introductory or general chemistry course. Most equations can be balanced without following a set series of steps. However, if you prefer a formal written procedure, the following works well, even with equations that are quite complicated.

how to... Balance a Chemical Equation: A Formal Approach

Step 1: Place a "1" in front of the formula with the largest number of atoms. If two formulas have the same number of atoms, select the one with the greater number of elements. We will call this formula the **starting formula** in the discussion and Active Examples that follow.

Step 2: Insert coefficients that balance the elements *that appear in compounds.* Use fractional coefficients, if necessary. Do not balance element-only formulas, such as Na or O_2, at this time. We call these *uncombined elements.* Choosing elements in the following order is usually easiest:

a) Elements in the starting formula that are in only one other compound
b) All other elements from the starting formula
c) All other elements *in compounds*

Step 3: Place coefficients in front of formulas of uncombined elements that balance those elements. Use fractional coefficients, if necessary.

Step 4: Clear fractions, if any, by multiplying all coefficients by the lowest common denominator. Remove any "1" coefficients that remain.

Step 5: Check to be sure that the final equation is balanced.

In following this procedure, don't change a coefficient once written, except to clear fractions. *Never* change or add a formula to balance an equation.

We now apply this procedure to the sodium-plus-water reaction in Section 8.3. Each step is listed, accompanied by comments or explanations.

Step 1: Place the coefficient "1" in front of the starting formula.

NaOH is the starting formula in this example because it has three atoms and three elements. — $Na + H_2O \rightarrow H_2 + 1\ NaOH$

Step 2: Balance elements in compounds.

Start by balancing the elements in the starting formula that are in only one other compound formula. Oxygen appears only in H_2O. The 1 O in NaOH is balanced by 1 O in H_2O. — $Na + 1\ H_2O \rightarrow H_2 + 1\ NaOH$

Step 3: Balance uncombined elements.

The 1 Na in NaOH on the right requires 1 Na on the left. — $1\ Na + 1\ H_2O \rightarrow H_2 + 1\ NaOH$

Only 1 H atom in H_2O comes from NaOH, so the other must come from H_2. To get 1 atom of H from H_2, we need 1/2 of an H_2 unit. — $1\ Na + 1\ H_2O \rightarrow 1/2\ H_2 + 1\ NaOH$

Step 4: Clear fractions; remove 1s.

Multiply by 2 to clear the 1/2. — $2 \times (1\ Na + 1\ H_2O \rightarrow 1/2\ H_2 + 1\ NaOH)$

$2\ Na + 2\ H_2O \rightarrow H_2 + 2\ NaOH$

Step 5: Check your work. — 2 Na, 4 H, 2 O on each side. ✓

Notice that the procedure in this section is not the same as the thought process by which we balanced the same equation in Section 8.3. There is no one "correct" way to balance an equation, but there are techniques that can shorten the process. Look for them in the Active Examples later in this chapter.

Is $Na(s) + H_2O(\ell) \rightarrow 1/2\ H_2(g) + 1\ NaOH(aq)$ a legitimate equation? Yes and no. If you think of the equation as "1 Na atom reacts with 1 H_2O molecule to produce 1/2 an H_2 molecule and 1 NaOH unit," the equation is not legitimate. There is no such thing as "1/2 an H_2 molecule," any more than a chicken can lay half an egg. But if you think about "1 mole of Na atoms reacts with 1 mole of H_2O molecules to produce 1/2 mole of hydrogen molecules and 1 mole of NaOH units," the fractional coefficient is reasonable. There can be a half mole of hydrogen molecules just as a chicken can lay a half dozen eggs.

It is sometimes preferable to use fractional coefficients, but few appear in this book. We stay with the standard practice of writing equations with whole-number coefficients. These coefficients should be written in the lowest terms possible; they should not have a common divisor. For example, although $4\ Na(s) + 4\ H_2O(\ell) \rightarrow 2\ H_2(g) + 4\ NaOH(aq)$ is a legitimate equation, it can and should be reduced to Equation 8.4 by dividing all coefficients by 2.

Notice that you can treat chemical equations exactly the same way you treat algebraic equations. Chemical formulas replace x, y, or other variables. You can multiply or divide an equation by some number by multiplying or dividing each term by that number. The order in which you write reactants or products may change; in an equation, $2\ H_2O + 2\ Na$ is the same as $2\ Na + 2\ H_2O$.

Active Example 8.1 Balancing Chemical Equations I

Consider the reaction of methane, CH_4, and oxygen, forming carbon dioxide and water, at a temperature at which all reactants and products are in the gas phase (Figure 8.10). Write a balanced equation and show that the number of atoms of each element remains the same before and after the reaction.

Think Before You Write Recall the two big-picture steps in writing a chemical equation: (1) Write a qualitative description of the reaction with formulas only first and then (2) add coefficients to balance the equation. We will guide you through the more formal approach in this Active Example.

Answers *Cover the left column with your cut-out shield. Reveal each answer only after you have written your own answer in the right column.*

1 $CH_4(g)$ + __ $O_2(g)$ → __ $CO_2(g)$ + __ $H_2O(g)$ *CH_4 has five atoms, more than any other molecule.*	Step 1 in the formal approach is to place a "1" in front of the formula with the greatest number of atoms. Begin there. __ $CH_4(g)$ + __ $O_2(g)$ → __ $CO_2(g)$ + __ $H_2O(g)$ + ⟶ +
1 $CH_4(g)$ + __ $O_2(g)$ → **1** $CO_2(g)$ + **2** $H_2O(g)$	The next step is to balance elements in the starting formula that appear in compounds. Carbon is in CO_2 and hydrogen is in H_2O, so balance both elements by adding the appropriate coefficients to the product compounds. 1 $CH_4(g)$ + __ $O_2(g)$ → __ $CO_2(g)$ + __ $H_2O(g)$

Figure 8.10 Natural gas burning. Methane, CH_4, is the principal component of natural gas. When natural gas burns, methane reacts with the oxygen in the air, forming the gaseous products carbon dioxide and water vapor.

1 $CH_4(g)$ + **2** $O_2(g) \rightarrow$ 1 $CO_2(g)$ + 2 $H_2O(g)$

Two oxygens in CO_2 plus one from each of two water molecules makes a total of four oxygen atoms.

Now balance the remaining species, which is an uncombined element.

$$1\ CH_4(g) + \_\_\ O_2(g) \rightarrow 1\ CO_2(g) + 2\ H_2O(g)$$

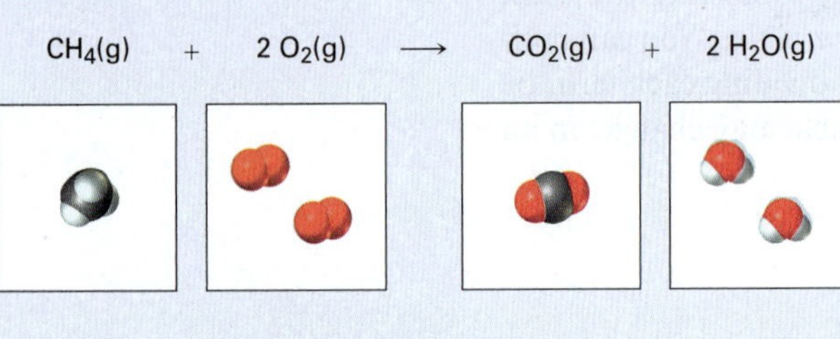

The final balancing step is to remove the "1" coefficients that remain. We've done that for you in the following equation. Draw the number of molecules of each species indicated by the coefficients in the balanced equation in the boxes below the equation. Use our molecular drawings from the first step as a guide.

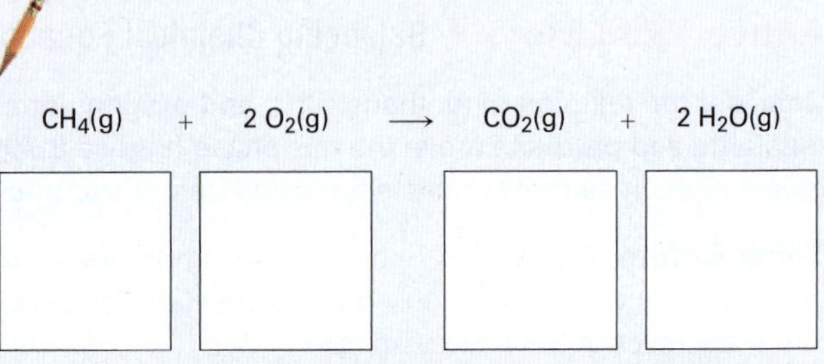

Yes, the atoms balance, 1 C, 4 H, and 4 O, on each side.

In a chemical change, the atoms are the same before and after the change, in both number and identity. Their new arrangement in molecules is the essence of the change.

Does the number of atoms balance? To answer this question, count the number of each type of atom in your boxes.

You improved your skill at balancing chemical equations.

What did you learn by solving this Active Example?

Practice Exercise 8.1

Sulfur dioxide, water, and calcium sulfide react to yield sulfur and calcium hydroxide. Write a balanced chemical equation for the reaction without state symbols.

Active Example 8.2 Balancing Chemical Equations II

Balance the equation for the reaction of solid phosphorus pentachloride with liquid water, yielding aqueous solutions of phosphoric acid and hydrochloric acid.

Think Before You Write If you found out that you didn't fully understand the procedure for balancing a chemical equation while working Active Example 8.1, review the *how to...* box before doing this example. When you are ready to continue, be sure to work this Active Example yourself, writing your answer at each step before you look at ours.

Answers *Cover the left column with your cut-out shield. Reveal each answer only after you have written your own answer in the right column.*

$\_\_\ PCl_5(s) + \_\_\ H_2O(\ell) \rightarrow \_\_\ H_3PO_4(aq) + \_\_\ HCl(aq)$

If you are still struggling with chemical nomenclature, go back to Chapter 6 and review the summary section, and, if necessary, the section of the chapter that has a more detailed explanation of how to write formulas.

Begin with the qualitative description of the reaction, with just the formulas of the reactants written to the left of an arrow and the formulas of the products to the right.

Answers	Your work
$\_\,PCl_5(s) + \_\,H_2O(\ell) \rightarrow$ **1** $H_3PO_4(aq) + \_\,HCl(aq)$	Begin the balancing process by identifying the starting formula—the formula with the greatest number of atoms and/or elements. Give it a coefficient of 1.  $\_\,PCl_5(s) + \_\,H_2O(\ell) \rightarrow \_\,H_3PO_4(aq) + \_\,HCl(aq)$
1 $PCl_5(s)$ + **4** $H_2O(\ell) \rightarrow 1\,H_3PO_4(aq) + \_\,HCl(aq)$ *Both phosphorus and oxygen are in one other formula. One P in H_3PO_4 requires 1 PCl_5, and 4 O in H_3PO_4 is satisfied by 4 H_2O. Hydrogen is in two other compounds, so we save that until later.*	Now balance the elements in the starting formula that are in only one other formula. $\_\,PCl_5(s) + \_\,H_2O(\ell) \rightarrow 1\,H_3PO_4(aq) + \_\,HCl(aq)$
$1\,PCl_5(s) + 4\,H_2O(\ell) \rightarrow 1\,H_3PO_4(aq)$ + **5** $HCl(aq)$ *Chlorine is the only element appearing in compounds on both sides of the equation. Five Cl in PCl_5 are balanced by 5 HCl.*	Are there any elements in only one compound on each side of the equation? Balance them. $1\,PCl_5(s) + 4\,H_2O(\ell) \rightarrow 1\,H_3PO_4(aq) + \_\,HCl(aq)$
$PCl_5(s) + 4\,H_2O(\ell) \rightarrow H_3PO_4(aq) + 5\,HCl(aq)$ There are 1 P, 5 Cl, 8 H, and 4 O on each side.	Hydrogen remains, and hydrogen is already balanced at 8 H on each side. When no uncombined elements are present, the last element should already be balanced when you reach it. If not, look for an error on some earlier element. There are no fractional coefficients, so remove the 1s and make a final check. Write the final balanced equation and show the atom balance.
You improved your skill at balancing chemical equations.	What did you learn by solving this Active Example?

Practice Exercise 8.2

Sulfur dioxide, oxygen, and calcium oxide react to form calcium sulfate. Write a balanced chemical equation for the reaction. State symbols are not required.

Active Example 8.3 Balancing Chemical Equations III

Balance the equation that represents the reaction of liquid propionic acid, $C_2H_5COOH(\ell)$, with gaseous oxygen to form gaseous carbon dioxide and liquid water.

Think Before You Write You are ready to balance an equation in just two steps: writing the formulas and then adding the coefficients.

Answers *Cover the left column with your cut-out shield. Reveal each answer only after you have written your own answer in the right column.*

Answers	Your work
$\_\,C_2H_5COOH(\ell) + \_\,O_2(g) \rightarrow \_\,CO_2(g) + \_\,H_2O(\ell)$	The formula of propionic acid, C_2H_5COOH, is commonly written in this way to suggest the arrangement of the atoms in the molecule. Write the complete unbalanced equation.

$2\,C_2H_5COOH(\ell) + 7\,O_2(g) \rightarrow 6\,CO_2(g) + 6\,H_2O(\ell)$ 1 before starting formula: **1** $C_2H_5COOH(\ell)$ + __ $O_2(g) \rightarrow$ __ $CO_2(g)$ + __ $H_2O(\ell)$ Balance C: 1 $C_2H_5COOH(\ell)$ + __ $O_2(g) \rightarrow$ **3** $CO_2(g)$ + __ $H_2O(\ell)$ Balance H: 1 $C_2H_5COOH(\ell)$ + __ $O_2(g) \rightarrow 3\,CO_2(g)$ + **3** $H_2O(\ell)$ Balance O: 1 $C_2H_5COOH(\ell)$ + **7/2** $O_2(g) \rightarrow 3\,CO_2(g) + 3\,H_2O(\ell)$ Clear fractions, remove 1s: $2\,C_2H_5COOH(\ell) + 7\,O_2(g) \rightarrow 6\,CO_2(g) + 6\,H_2O(\ell)$ Final check: There are 6 C, 12 H, and 18 O on each side. ✓ *In the formula C_2H_5COOH, the symbols of all three elements appear twice. Students usually count carbon and hydrogen correctly, but they often overlook the oxygen in the original compound when selecting the coefficient for O_2.*	Take it to completion. __ $C_2H_5COOH(\ell)$ + __ $O_2(g) \rightarrow$ __ $CO_2(g)$ + __ $H_2O(\ell)$
You improved your skill at balancing chemical equations.	What did you learn by solving this Active Example?

Practice Exercise 8.3

$Fe_3O_4(s)$ reacts with solid aluminum to yield solid iron and solid aluminum oxide. Write the balanced chemical equation for the reaction with state symbols.

The reason for balancing uncombined elements last is that you can insert *any* coefficient that is needed without *un*balancing any element that has already been balanced.

8.5 Interpreting Chemical Equations

Goal 4 Given a balanced chemical equation or information from which it can be written, describe its meaning on the particulate, molar, and macroscopic levels.

In the previous two sections, you learned that a chemical equation contains a great deal of information. First, it gives a qualitative description of a reaction. The equation $2\,H_2(g) + O_2(g) \rightarrow 2\,H_2O(g)$ says that hydrogen reacts with oxygen to form water and that both reactants and the product are in the gas phase. Second, the equation gives quantitative information about the reaction. You can interpret this quantitative information on a number of different levels.

We stated earlier that standard practice is to write equations with the lowest whole-number coefficients possible. This is because equations are often interpreted on the *particulate level*. Thus, $2\,H_2(g) + O_2(g) \rightarrow 2\,H_2O(g)$ means "two molecules of hydrogen react with one oxygen molecule to form two water molecules." The symbols in the equation represent the particles, shown in **Figure 8.11** as space-filling models.

We can scale up our particulate interpretation to any desired multiple of the coefficients in the balanced equation. We can think of the equation in terms of dozens. For example, "two *dozen* molecules of hydrogen react with one *dozen* oxygen molecules to form two *dozen* water molecules" (**Figure 8.12**).

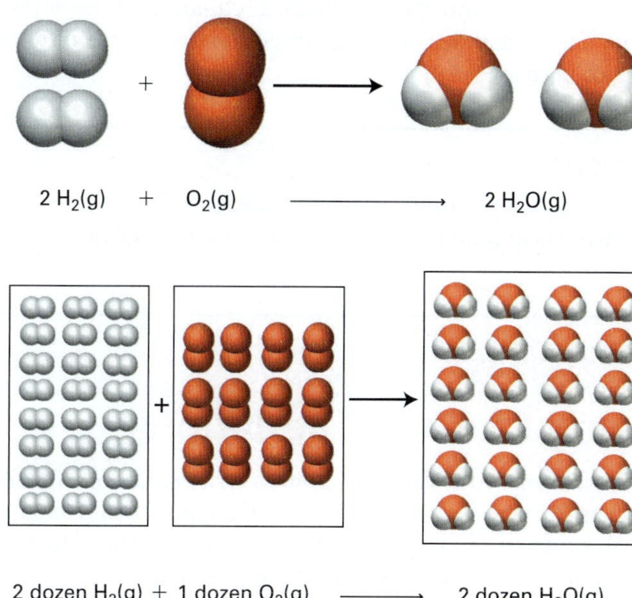

Figure 8.11 The particulate-level representation of the reaction of hydrogen and oxygen, forming water.

Figure 8.12 The grouping-unit representation of the reaction of hydrogen and oxygen, forming water. The 2 molecules of hydrogen–1 molecule of oxygen–2 molecules of water *ratio* is retained, no matter the grouping unit.

We can also choose the "chemistry dozen," the mole, as our grouping unit: "Two *moles* of hydrogen molecules react with one *mole* of oxygen molecules to form two *moles* of water molecules." We simply change the grouping unit from 12 to 6.02×10^{23}.

The *molar* interpretation of a chemical equation involves reading the coefficients as the amounts in moles of the reactants and products. This is still a particulate-level explanation, but we are grouping the particles into counting units that make it easier to translate into a macroscopic-level interpretation. On the molar level, fractional coefficients are acceptable. $H_2(g) + 1/2\ O_2(g) \rightarrow H_2O(g)$ can be read as "one mole of hydrogen molecules reacts with one-half mole of oxygen molecules to form one mole of water molecules." **Figure 8.13** shows a macroscopic demonstration of energy released in this reaction.

Hulton Archive/Getty Images

Figure 8.13 The end of the *Hindenburg* dirigible in May 1937. The explosion demonstrates the tremendous amount of energy released when hydrogen reacts with oxygen to form water. This disaster ended the use of hydrogen in lighter-than-air aircraft. Today, the unreactive Group 8A/18 gas, helium, is used in blimp airships—from which aerial views of major sporting events are displayed on television.

Thinking About Your Thinking

Mental Models

The illustrations that accompany the preceding representations of the $2\ H_2(g) + O_2(g) \rightarrow 2\ H_2O(g)$ reaction are designed to help you form mental models of chemical reactivity. If you can imagine what the particles look like as they react—and if you can see in your mind how the various combinations of atoms occur—you are making good progress toward thinking like a chemist. In the following sections, we will introduce four types of chemical reactions. We provide a particulate-level illustration of each reaction type. Use those drawings to form mental models of each reaction type.

Literally counting particles is impossible. Instead, we measure the mass or some other measurable *macroscopic* property of the sample. But the reaction still occurs as a combination of particles in the ratio given in the chemical equation. The balanced equation does *not* directly give information about masses. We *cannot* say, "two grams of hydrogen reacts with one gram of oxygen to form two grams of water." We do know, however, that molar mass provides a link between the number of moles of particles and their masses. ▶ We will consider this concept in more detail in Chapter 10.

◀ **P/Review** Molar mass is the mass in grams of one mole of a substance (Section 7.5).

You need to be able to interpret a chemical equation on both the particulate and molar levels. The remainder of this chapter and other, future chapters provide additional practice in this.

Target Check 8.3

Consider the chemical equation for the decomposition of hydrogen peroxide solution: $2\ H_2O_2(aq) \rightarrow O_2(g) + 2\ H_2O(\ell)$. Write a word description of this equation on the particulate and molar levels.

8.6 Writing Chemical Equations

In the next four sections, you will learn how to write equations for four kinds of chemical reactions for which you may be given the names or formulas of the reactants only. It will be up to you to predict the formulas of the products. Your ability to classify a reaction as a certain type will be a big help in predicting what products will form. Thus, we modify our two main steps for writing a chemical equation from Section 8.3 by adding a new first step. The overall procedure for writing a chemical equation thus becomes:

how to... **Write and Balance a Chemical Equation**

Step 1: Classify the reaction type.

Step 2: Write a qualitative description of the reaction. In this step, you write the formulas of the given reactants to the left of an arrow and the formulas of the given or predicted products to the right.

Step 3: Quantify the description by balancing the equation. Do this by adding coefficients. Do not change the qualitative description of the reaction by adding, removing, or altering any formula.

We will no longer show the detailed, formal approach to balancing equations unless the reaction is complex.

8.7 Combination Reactions

Goal 5 Write the equation for the reaction in which a single product compound is formed by the combination of two or more substances.

A reaction in which two or more substances combine to form a single product is a **combination reaction** or **synthesis reaction**. The reactants are often elements but sometimes are compounds or are both elements and compounds. The general equation for a combination reaction is as follows:

$$A + X \longrightarrow AX$$

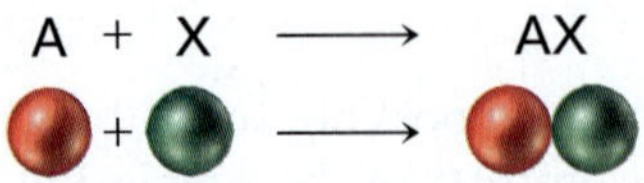

The reaction between sodium and chlorine to form sodium chloride is a combination reaction: $2\ Na(s) + Cl_2(g) \rightarrow 2\ NaCl(s)$.

Active Example 8.4 Combination Reactions I

Write the equation for the formation of solid sodium peroxide, Na_2O_2 (Figure 8.14), by direct combination of solid sodium and gaseous oxygen.

Think Before You Write You can identify this combination reaction by the fact that a single product is formed.

Answers *Cover the left column with your cut-out shield. Reveal each answer only after you have written your own answer in the right column.*

_ Na(s) + _ O_2(g) → _ Na_2O_2(s) *Did you remember to show oxygen as a diatomic molecule? If so, great! If not, review formulas of elements in Section 6.3.*	Write the formulas of the elements on the left side of an arrow and the formula of the product on the right. In other words, write the qualitative description of the reaction.
2 Na(s) + O_2(g) → Na_2O_2(s)	Balance the equation.
You improved your skill at balancing chemical equations in general, and specifically, balancing combination reaction equations.	What did you learn by solving this Active Example? 

Practice Exercise 8.4

Write the equation for the combination of solid aluminum and gaseous oxygen to form solid aluminum oxide.

Turtle Rock Scientific/Science Source

Figure 8.14 Sodium peroxide is used to bleach animal and vegetable fibers, feathers, bones, and ivory. The textile industry uses it in dyeing. It is used to maintain breathable air in confined areas that develop a high level of carbon dioxide, such as submarines and diving bells, by the reaction: 2 Na_2O_2 + 2 CO_2 → 2 Na_2CO_3 + O_2.

Elena Elisseeva/Shutterstock.com

Figure 8.15 Backyard barbecue. A simple combination reaction occurs in this widely used device.

Active Example 8.5 Combination Reactions II

Carbon dioxide gas is formed when charcoal (carbon) burns in air, as in a backyard barbecue (Figure 8.15). Write the equation for the reaction.

Think Before You Write A word description of a reaction sometimes assumes that you already know something about it. This Active Example assumes you know that when something burns in air, it is reacting chemically with the oxygen in the air. In other words, gaseous oxygen is an unidentified reactant, so it should appear on the left side of the equation along with carbon.

Answers *Cover the left column with your cut-out shield. Reveal each answer only after you have written your own answer in the right column.*

C(s) + O_2(g) → CO_2(g)	Write the qualitative, or unbalanced, equation.
C(s) + O_2(g) → CO_2(g) *Sometimes balancing an equation is easy, as when all coefficients are 1!*	Now balance the equation.
You improved your skill at balancing chemical equations in general, and specifically, balancing combination reaction equations.	What did you learn by solving this Active Example?

Practice Exercise 8.5

Write the equation for the formation of solid iron(III) hydroxide from its elements.

a summary of... Combination Reactions

Reactants:	Any combination of elements and/or compounds
Reaction type:	Combination
Equation type:	A + X ⟶ AX
Products:	One compound

Learn It Now! *The four points in this summary are components in the thought process you should use to write an equation. Step 1 in the equation-writing procedure is classifying the reaction. After you examine the reactants, you will be able to decide the reaction type and what kind of equation it is. This also enables you to predict the products. This summary and those that follow it are gathered as a chapter summary in Section 8.11.*

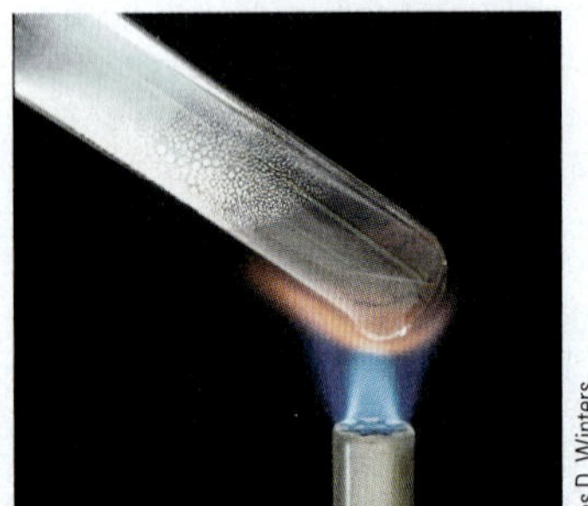
Charles D. Winters

Figure 8.16 Decomposition of mercury(II) oxide. On heating, the bright orange-red powder undergoes a decomposition reaction to form silver liquid mercury and colorless oxygen gas.

8.8 Decomposition Reactions

Goal 6 Given a single compound that is decomposed into two or more substances, either compounds or elements, write the equation for the reaction.

A **decomposition reaction** is the opposite of a combination reaction, in that a single reactant compound breaks down into two or more product substances. The products may be any combination of elements and compounds. The general decomposition equation is as follows:

$$AX \longrightarrow A + X$$

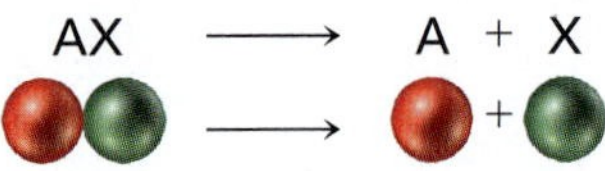

A typical decomposition reaction occurs when mercury(II) oxide is heated: $2\,HgO(s) \rightarrow 2\,Hg(\ell) + O_2(g)$ (**Figure 8.16**).

The decomposition of mercury(II) oxide was a key reaction used by Lavoisier in disproving the phlogiston theory described in Chapter 1.

Most decomposition reactions require an energy input of some sort. The decomposition of water (Active Example 8.6) is accomplished electrolytically—that is, with electrical energy, as shown in Figure 8.17. The reaction in the forthcoming Active Example 8.7 is achieved by heating.

Active Example 8.6 Decomposition Reactions I

Water decomposes into its elements when electrical energy is added (Figure 8.17). Write the equation.

Think Before You Write You can identify this decomposition reaction by the fact that a single reactant is changed into two (or more) products.

Answers *Cover the left column with your cut-out shield. Reveal each answer only after you have written your own answer in the right column.*

$\_\_\,H_2O(\ell) \rightarrow \_\_\,H_2(g) + \_\_\,O_2(g)$	Start with the formulas of the reactants and products. 
$2\,H_2O(\ell) \rightarrow 2\,H_2(g) + O_2(g)$ *This reaction is literally the reverse of the reaction discussed in Section 8.5.*	Now balance the equation. 

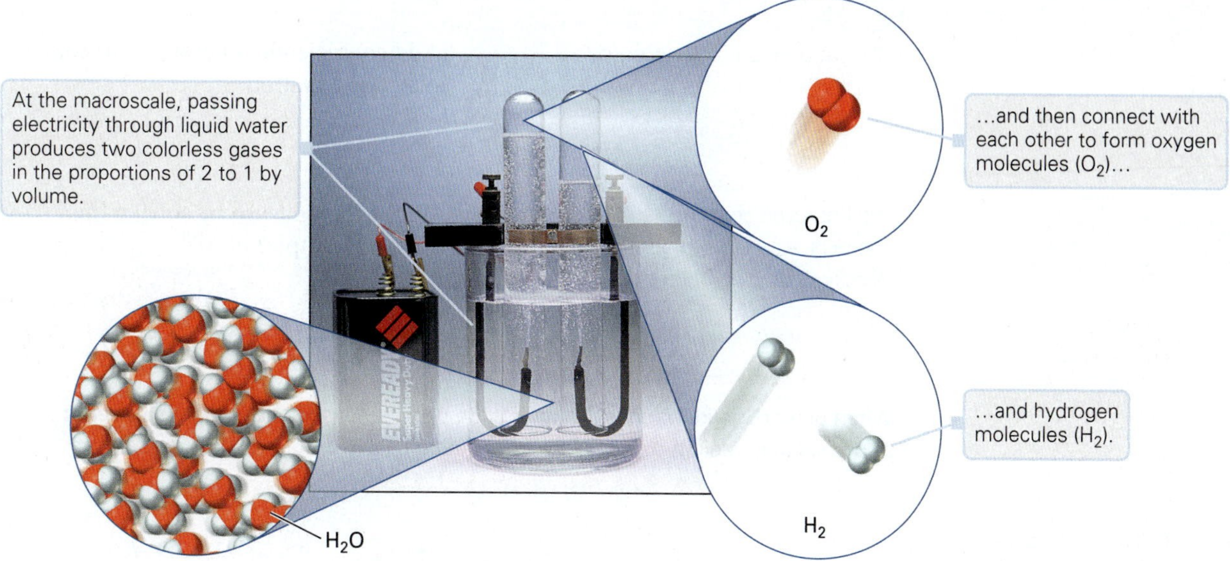

At the particulate level, hydrogen atoms and oxygen atoms originally connected in water molecules (H_2O) separate…
At the macroscale, passing electricity through liquid water produces two colorless gases in the proportions of 2 to 1 by volume.
…and then connect with each other to form oxygen molecules (O_2)…
O_2
…and hydrogen molecules (H_2).
H_2
H_2O
EVEREADY
Charles D. Winters

Figure 8.17 Decomposition of water. Electrical energy is used to decompose liquid water to its elements, hydrogen and oxygen.

You improved your skill at balancing chemical equations in general, and specifically, balancing decomposition reaction equations.	What did you learn by solving this Active Example?

Practice Exercise 8.6

Write the equation that symbolizes the decomposition of mercury(II) oxide into its elements. State symbols are not required.

Many chemical changes can be made to go in either direction, as the reversibility of the "hydrogen plus oxygen to form water" reaction suggests. This type of reaction is called a **reversible reaction**. A double arrow is often used to indicate reversibility: $\rightleftharpoons$. Thus,

$$2\ H_2O(\ell) \rightleftharpoons 2\ H_2(g) + O_2(g) \quad \textit{and} \quad 2\ H_2(g) + O_2(g) \rightleftharpoons 2\ H_2O(\ell)$$

are equivalent reversible equations. ▶

◀ **P/Review** Reversible changes, both physical and chemical, are quite common. Changes of state, such as freezing or boiling, are reversible physical changes. So is the process by which a substance dissolves in water. When a reversible change occurs in both directions at the same time and at equal rates, equilibrium is established. Physical equilibria are described in Sections 15.5 and 16.4, and all of Chapter 18 is devoted to chemical equilibrium.

Active Example 8.7 Decomposition Reactions II

Lime, CaO(s), and carbon dioxide gas are the products of thermal decomposition of limestone, solid calcium carbonate (Figure 8.18). (*Thermal* refers to heat. The reaction occurs at high temperature.) ▶ Write the equation.

Think Before You Write Sometimes the written description of a chemical change lists the products first. Don't let that change the way you write the reaction, reactants → products. Again, note that there is a single reactant in chemical change described in the problem statement, indicating that it is a decomposition reaction.

Answers *Cover the left column with your cut-out shield. Reveal each answer only after you have written your own answer in the right column.*

Calcium oxide is an industrial chemical used in making construction materials such as bricks, mortar, plaster, and stucco. It is also used in manufacturing sodium carbonate, one of the most widely used of all chemicals, and in manufacturing steel, magnesium, and aluminum.

__ $CaCO_3(s) \rightarrow$ __ $CaO(s) +$ __ $CO_2(g)$	Start with the unbalanced equation.
$CaCO_3(s) \rightarrow CaO(s) + CO_2(g)$	Add coefficients as necessary to complete the equation.
You improved your skill at balancing chemical equations in general, and specifically, balancing decomposition reaction equations.	What did you learn by solving this Active Example?

Practice Exercise 8.7

When heated, solid magnesium hydroxide will decompose, forming water vapor and solid magnesium oxide. Write the equation.

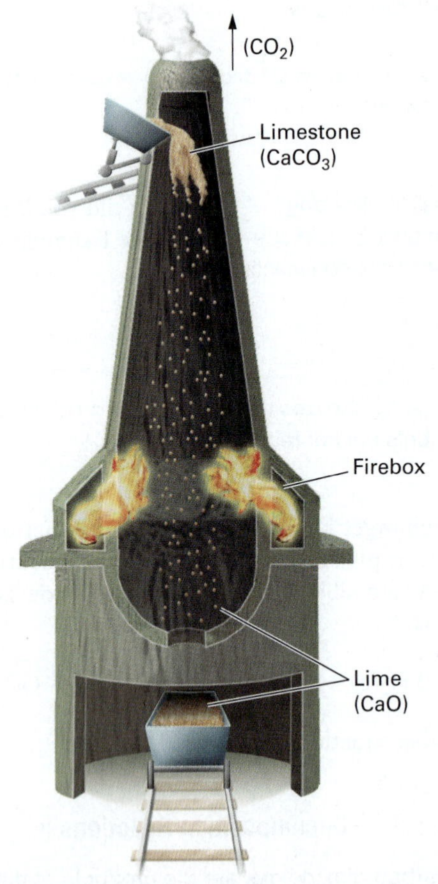

Figure 8.18 Thermal decomposition of limestone (calcium carbonate). Calcium oxide (lime) is prepared by decomposing calcium carbonate in a large kiln at 825°C to 1000°C. Calcium oxide is among the most widely used chemicals in the world, with annual consumption measured in the hundreds of millions of tons. Some of the CaO output is used in the steel industry, and much of the remainder is used to make "slaked lime," $Ca(OH)_2$, by reaction with water.

Learn It Now! *Compare the summary of... Combination Reactions with the summary of... Decomposition reactions.*

a summary of... **Decomposition Reactions**

Reactants: One compound

Reaction type: Decomposition

Equation type: $AX \longrightarrow A + X$

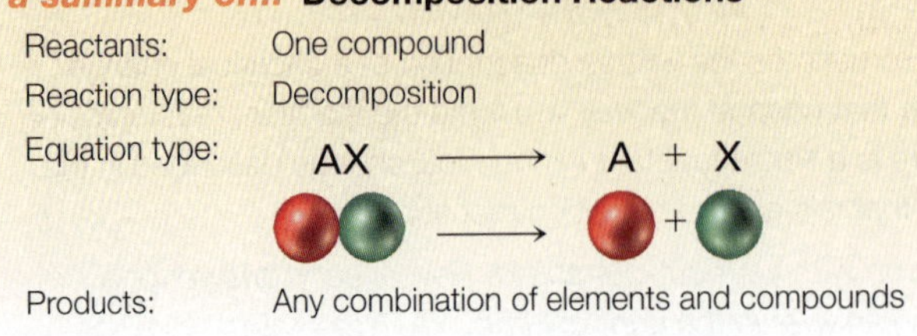

Products: Any combination of elements and compounds

Everyday Chemistry 8.1

FEMTOCHEMISTRY

At this point in your study of chemistry, you have recognized that chemical reactions are a continual part of your everyday life. Human life itself depends on literally billions of reactions per second. Until recently, though, chemists have had an incomplete idea of how a chemical reaction proceeds. We know a great deal about reactants and products in chemical change, but the nature of the pathway between the two has been a difficult problem to study. This is because chemical change occurs so quickly. Many reactions are complete in as little as 10×10^{-15} second.

In the late 1980s, groundbreaking research into the fundamentals of chemical reactivity began. The instrumentation used for this research was, in essence, a high-speed camera designed to take snapshots of a reaction with a "shutter speed" fast enough to capture a chemical change as it occurs. This is accomplished with lasers that emit light flashes on the femtosecond timescale, where *femto-* is the metric prefix for 10^{-15}. The prefix *femto-* was derived from the Scandinavian word for *fifteen.* The first reaction studied with this new technique was a decomposition reaction: $ICN \rightarrow I + CN$. This reaction is complete in 200 femtoseconds.

A century earlier, in the late 1880s, an unanswered question was: Does a horse always have at least one hoof on the ground while trotting? This question was answered with a relatively new technology for the time—photography. A camera was designed to take a series of photographs at a fixed time interval, every 0.052 second. As you can see in **Figure 8.19**, the photographic evidence shows that all four hooves are off the ground at once.

Analogously, femtosecond lasers can be used to "photograph" molecules as they undergo chemical change. However, instead of recording the image every 0.052 second, the image needs to be documented every 0.000000000000001 second! At this miniscule timescale, we can follow the real-time progress of a chemical reaction. **Figure 8.20** starts by showing a complex molecule formed by the association of an iodine molecule with a ring molecule, benzene, C_6H_6. In the second frame, the complex molecule is subjected to a flash from a femtosecond laser. An electron jumps from the benzene molecule to the iodine molecule. Even if the electron leaps back to the benzene molecule, the iodine molecule is decomposed into individual iodine atoms in less than 1500×10^{-15} second.

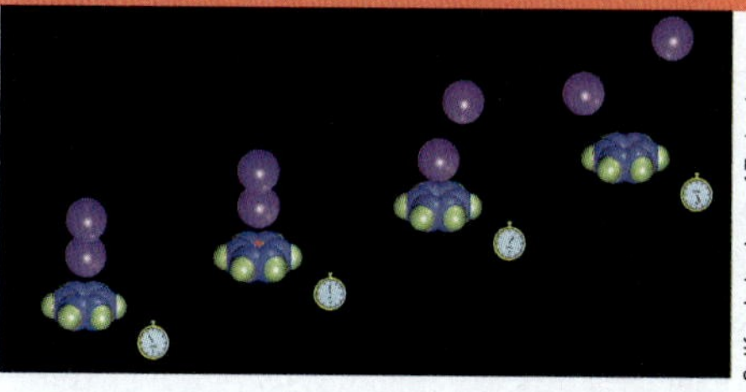

California Institute of Technology

Figure 8.20 Femtosecond technology reveals the mechanism of how an iodine molecule is split into iodine atoms via the momentary transfer of an electron from the benzene molecule.

Ahmed Zewail was the scientist who pioneered the field now known as *femtochemistry* (**Figure 8.21**). He was awarded the 1999 Nobel Prize in Chemistry for his work.

Femtochemistry research holds great promise. Chemists see enormous potential in using it to gain greater control over chemical reactions. Biologists are doing femtochemistry research on photosynthesis. There is a possibility that artificial photosynthesis may be possible, in which light energy is converted to energy contained in chemical bonds. Engineers and materials scientists hope to use femtochemistry technology in creating faster and higher capacity electronics. Medical researchers are investigating the use of drugs activated by light to selectively destroy cancerous tumors; femtochemistry techniques have the potential to help researchers study the reactions that occur. All of these areas of research show immense promise in using femtochemistry to significantly improve our lives in the near future!

Quick Quiz

1. **How many femtoseconds are in a millisecond?**
2. **What type of energy transformation is the target of investigation by research on artificial photosynthesis?**

Eadweard Muybridge/Getty Images

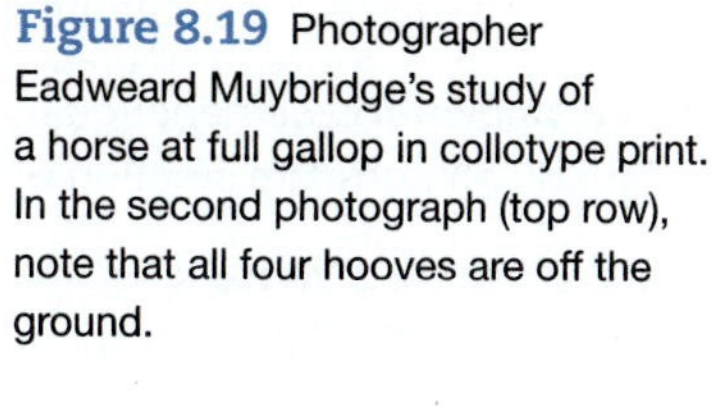

Figure 8.19 Photographer Eadweard Muybridge's study of a horse at full gallop in collotype print. In the second photograph (top row), note that all four hooves are off the ground.

AP Images/HO

Figure 8.21 Ahmed Zewail (1946–2016) was a professor at Caltech who led a team of scientists who studied the behavior of matter at the timescale at which chemical reactions occur.

8.9 Single-Replacement Reactions

Goal 7 Given the reactants of a single-replacement reaction, write the equation for the reaction.

▶ **P/Review** *Oxidation* and *reduction* transfer are terms that refer to the transfer of electrons in a chemical reaction. Oxidation, or burning reactions, and many of the combination and decomposition reactions of the previous sections are also oxidation–reduction reactions. "Redox" reactions, as oxidation–reduction reactions are called, are discussed in detail in Chapter 19.

Many elements are capable of replacing ions of other elements in aqueous solution. This is one kind of **oxidation–reduction reaction**, or **redox** reaction. ◀ The equation for such a reaction looks as if one element is replacing another in a compound. It is a **single-replacement equation**. This type of reaction is sometimes called a **single-replacement reaction**. The general equation is as follows:

$$A + BX \longrightarrow AX + B$$

The element sodium is able to replace the element hydrogen in the compound water in a single-replacement reaction: $2\,Na(s) + 2\,HOH(\ell) \rightarrow 2\,NaOH(aq) + H_2(g)$ (**Figure 8.22**).

Reactants in a single-replacement equation are always an element and a compound. If the element is a metal, it replaces the metal or hydrogen in the compound. If the element is a nonmetal, it replaces the nonmetal in the compound.

$$2\,Na(s) + 2\,H_2O(\ell) \longrightarrow 2\,NaOH(aq) + H_2(g)$$

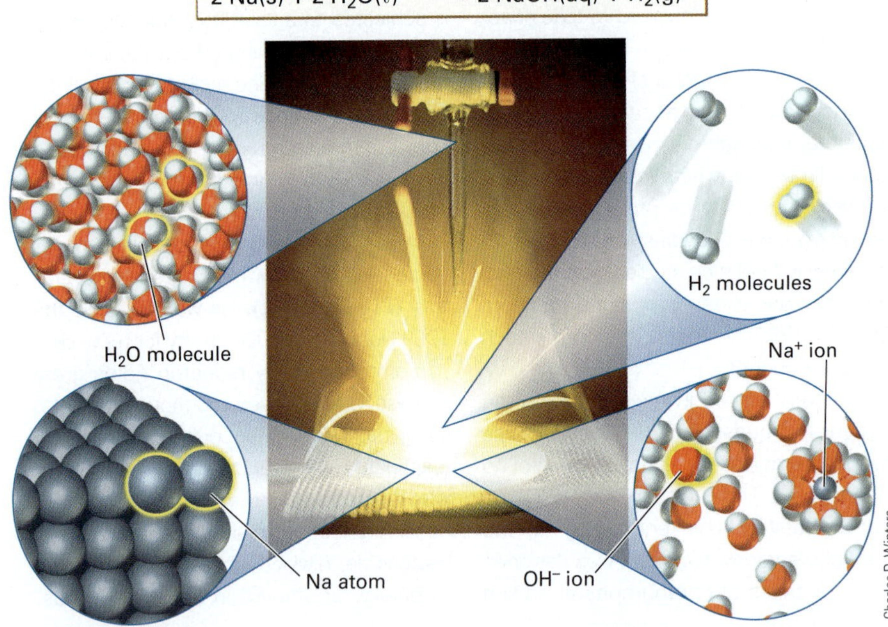

Figure 8.22 Reaction of sodium with water. A single-replacement reaction occurs when sodium, an element, reacts with water, a compound. The metal sodium appears to replace hydrogen in water: $2\,Na + 2\,HOH\ (A + BX) \rightarrow 2\,NaOH + H_2\ (AX + B)$. The products of the reaction are a different compound, sodium hydroxide, and a different element, hydrogen. Sodium hydroxide is a soluble ionic compound, so it exists as sodium ions and hydroxide ions in water solution.

Figure 8.23 The reaction of calcium metal and hydrochloric acid solution.

Active Example 8.8 Single-Replacement Reactions I

Write the single-replacement equation for the reaction between solid elemental calcium and an aqueous solution of hydrochloric acid (Figure 8.23).

Think Before You Write First, recognize that you have a single-replacement reaction, an element (Ca) plus a compound with positive and negative ions $(H^+)(Cl^-)$. Second, you will have to decide whether the element will be transformed into a positive ion or a negative ion. That decision will guide you to replacing the correct ion.

Answers *Cover the left column with your cut-out shield. Reveal each answer only after you have written your own answer in the right column.*

$\_\ Ca(s) + \_\ HCl(aq) \rightarrow$	Begin by writing the formulas of just the reactants to the left of the arrow. 

__ $Ca(s)$ + __ $HCl(aq) \rightarrow$ __ $CaCl_2(aq)$ + __ $H_2(g)$ A + BX → AX + B *The elemental metal reactant, A, becomes the positive ion in the product aqueous solution, AX. The positive ion in the reactant aqueous solution, BX, becomes an uncombined element as a product, B. B is usually a metal, but if the reactant solution is an acid, B is hydrogen. At this time, you have no reason to know that AX ($CaCl_2$, in this case) will be in aqueous solution. However, you do need to recognize that the 2+ calcium ion, Ca^{2+}, requires two 1− chloride ions, Cl^-, for the correct formula for calcium chloride, $CaCl_2$. You also have to recognize that uncombined hydrogen exists in nature as a diatomic molecule, H_2.* *We will discuss how you can decide upon the state symbol of ionic compounds in Section 9.7. For now, we will assume that ionic reactants and products are in aqueous solution unless stated otherwise.*	Now decide which element in the compound, hydrogen or chlorine, will be replaced by the calcium. Reread the paragraph before this Active Example if you need help. Then write the formulas of the products on the right side of the equation that you started earlier.
$Ca(s) + 2\,HCl(aq) \rightarrow CaCl_2(aq) + H_2(g)$	Balance the equation.
You improved your skill at balancing chemical equations in general, and specifically, balancing single-replacement reaction equations.	What did you learn by solving this Active Example?

Practice Exercise 8.8

Write the equation that represents the reaction that occurs when chlorine gas is bubbled through an aqueous solution of sodium bromide. Do not use state symbols.

Active Example 8.9 Single-Replacement Reactions II

A solid strip of copper reacts with a solution of silver nitrate (Figure 8.24). Write the equation for the reaction.

Think Before You Write In this case an elemental metal, copper, is replacing a metal ion, silver ion in silver nitrate, in an aqueous solution. The key to developing the equation is to notice that the reactants match the pattern A + BX →.

Answers *Cover the left column with your cut-out shield. Reveal each answer only after you have written your own answer in the right column.*

__ $Cu(s)$ + __ $AgNO_3(aq) \rightarrow$	Write the formulas of the reactants. 
__ $Cu(s)$ + __ $AgNO_3(aq) \rightarrow$ __ $Cu(NO_3)_2(aq)$ + __ $Ag(s)$	Complete the unbalanced equation. The copper ion that forms is a copper(II) ion.

Charles D. Winters

Charles D. Winters

Figure 8.24 The reaction of copper and a solution of silver nitrate. (a) A copper strip is placed in a clear, colorless solution of silver nitrate. (b) After about an hour at room temperature, the products of the reaction become visible. The copper strip is covered with solid silver metal. The solution becomes blue because of the presence of hydrated copper(II) ions.

Just because an equation for a reaction can be written does not mean the reaction will occur. For example, if a piece of solid silver and a solution of copper(II) nitrate are given as reactants, you would produce the equation $2\ Ag(s) + Cu(NO_3)_2(aq) \rightarrow 2\ AgNO_3(aq) + Cu(s)$, the reverse of the equation in Active Example 8.9. This reaction does not occur spontaneously, although it can be forced with some help from a source of electrical current. To find which reaction works, you must try the reactions in the laboratory. In Chapter 9, we show how to use the results of many experiments to predict which reactions will occur and which will not.

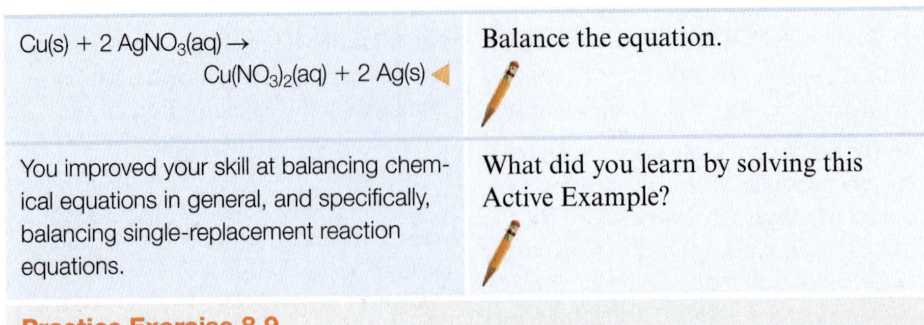

$Cu(s) + 2\ AgNO_3(aq) \rightarrow Cu(NO_3)_2(aq) + 2\ Ag(s)$	Balance the equation.
You improved your skill at balancing chemical equations in general, and specifically, balancing single-replacement reaction equations.	What did you learn by solving this Active Example?

Practice Exercise 8.9

Lead reacts with a solution of copper(II) nitrate. Lead(II) ions form. Write the equation.

Active Example 8.9 gives us an opportunity to show you a balancing technique that can save you some time. Whenever a polyatomic ion is unchanged in a chemical reaction, the entire ion can be balanced as a unit. The nitrate ion is unchanged in the following equation:

$$Cu(s) + 2\ AgNO_3(aq) \rightarrow Cu(NO_3)_2(aq) + 2\ Ag(s)$$

Start with $Cu(NO_3)_2$. It has three ions: one copper(II) ion and two nitrate ions. Cu is already balanced. $AgNO_3$ has one nitrate ion, so it takes two $AgNO_3$ to balance the nitrate ions in one $Cu(NO_3)_2$. This, in turn, requires two Ag on the right. When you learn this technique, you will find it quicker and easier than balancing each element in a polyatomic ion separately. But remember the condition: *All of the ions must be unchanged.* This technique will not work, for instance, if there is an NO_3^- compound on one side and an NO_3^- plus an NO or some other nitrogen species on the other side.

Learn It Now! *Compare* a summary of... *Combination Reactions with* a summary of... *Decomposition Reactions with* a summary of... *Single-Replacement Reactions.*

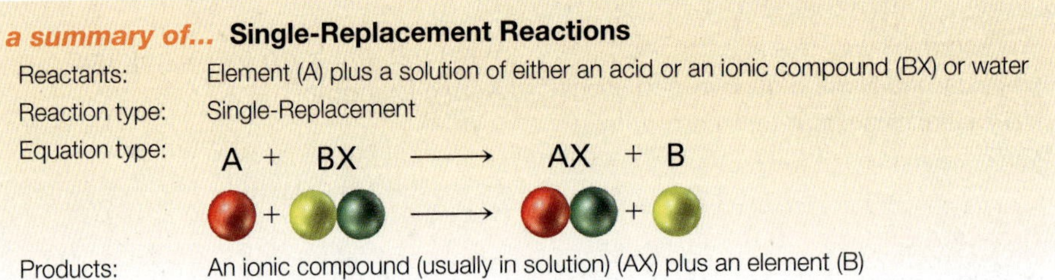

***a summary of...* Single-Replacement Reactions**

Reactants:	Element (A) plus a solution of either an acid or an ionic compound (BX) or water
Reaction type:	Single-Replacement
Equation type:	A + BX ⟶ AX + B
Products:	An ionic compound (usually in solution) (AX) plus an element (B)

8.10 Double-Replacement Reactions

Goal 8 Given the reactants in a double-replacement precipitation or neutralization reaction, write the equation.

When solutions of two compounds are mixed, a positive ion from one compound may combine with the negative ion from the other compound to form a solid compound that settles to the bottom. The solid is a **precipitate**; the reaction is a **precipitation reaction**. In the equation for a precipitation reaction, ions of the two reactants appear to change partners. The equation, and sometimes the reaction itself, is a **double-replacement equation** or a **double-replacement reaction**. The general equation, with bridges to show the rearrangement of ions, is as follows:

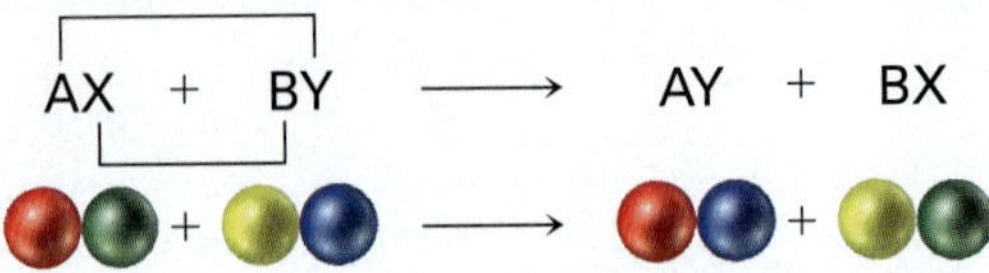

A typical precipitation reaction occurs between solutions of calcium chloride and sodium fluoride: $CaCl_2(aq) + 2\ NaF(aq) \rightarrow CaF_2(s) + 2\ NaCl(aq)$.

Active Example 8.10 Double-Replacement Reactions I

Silver chloride precipitates when solutions of sodium chloride and silver nitrate are combined (Figure 8.25). Write the equation for the reaction.

Think Before You Write The problem statement identifies silver chloride as one of the products. It forms when the silver ion from silver nitrate combines with the chloride ion from sodium chloride. The only formula that you need to deduce is the second product.

Answers *Cover the left column with your cut-out shield. Reveal each answer only after you have written your own answer in the right column.*

__ $NaCl(aq)$ + __ $AgNO_3(aq) \rightarrow$	Start by writing the formulas of the reactants. 
__ $NaCl(aq)$ + __ $AgNO_3(aq) \rightarrow$ __ $NaNO_3(aq)$ + __ $AgCl(s)$ *The Na^+ ion from NaCl is paired with the NO_3^- ion from $AgNO_3$ for the other product, $NaNO_3$, which remains in solution. The nitrate ion is unchanged in the reaction, so it may be balanced as a unit, as explained after Active Example 8.9. Note that the solid that forms, AgCl(s), has the state symbol (s) to indicate that it is the precipitate.*	Remember that positive ions combine with negative ions to form products. Complete the unbalanced equation. 
$NaCl(aq) + AgNO_3(aq) \rightarrow NaNO_3(aq) + AgCl(s)$ *The skeleton equation is balanced, all coefficients being 1. In checking the balance, the nitrate ion again may be thought of as a unit. There is one nitrate ion on each side of the equation.*	Now balance the equation. 
You improved your skill at balancing chemical equations in general, and specifically, balancing double-replacement reaction equations.	What did you learn by solving this Active Example? 

Figure 8.25 The precipitation of silver chloride.

Practice Exercise 8.10

Copper(II) hydroxide precipitates when solutions of copper(II) nitrate and sodium hydroxide are combined. Write the equation.

An **acid** is a compound that has a hydrogen ion, H^+, that can be removed by a water molecule. A substance that contains hydroxide ions, OH^-, is a **base**. When an acid is added to an equal amount of base, each hydrogen ion reacts with a hydroxide ion to form a molecule of water. ▶ The acid and the base **neutralize** each other in a **neutralization reaction**.* An ionic compound is also formed; it usually remains in solution. The general equation, using M to represent a metal ion, is as follows:

$$\underset{\text{acid}}{HX(aq)} + \underset{\text{base}}{MOH(s)} \rightarrow \underset{\text{water}}{H_2O(\ell)} + \underset{\text{ionic compound}}{MX(aq)}$$

◀ **P/Review** Acids are discussed in more detail in Section 6.9 and in Chapter 17.

*There are acids that do not contain hydrogen ions and bases that do not contain hydroxide ions. Reactions between them are also called *neutralization reactions*. The H^+-plus-OH^- neutralization is the most common and the only one we will consider here.

Neutralization reactions are described by double-replacement equations, although it might not seem that way at first glance. The water molecule forms when the hydrogen ion from the acid combines with the hydroxide ion from the base. The double-replacement character of the equation becomes clear if you write the formula of water as HOH rather than H_2O. For the neutralization of hydrochloric acid by solid sodium hydroxide, the equation is as follows:

$$HCl(aq) + NaOH(s) \rightarrow HOH(\ell) + NaCl(aq)$$

With the formula of water in conventional form, we have:

$$HCl(aq) + NaOH(s) \rightarrow H_2O(\ell) + NaCl(aq)$$

We suggest that when you are balancing a neutralization reaction equation, begin by writing the formula of water as HOH. This will help you see the double-replacement pattern of the reaction. When you've completely balanced the equation, change the formula of water to H_2O.

Most hydroxides do not dissolve in water, but they do appear to "dissolve" in acids. What really happens is that they react with the acid in a double-replacement neutralization reaction. The next Active Example shows one such reaction.

Active Example 8.11 Double-Replacement Reactions II

Write the equation for the reaction between sulfuric acid solution and solid aluminum hydroxide.

Think Before You Write The important item to identify from the problem statement is that both reactants consist of positive ions and negative ions. After you have recognized this, you know to switch the positive and negative partners on the product side of the equation.

Answers *Cover the left column with your cut-out shield. Reveal each answer only after you have written your own answer in the right column.*

Answers	Questions
$\_\_ H_2SO_4(aq) + \_\_ Al(OH)_3(s) \rightarrow$	What are the formulas of the reactants?
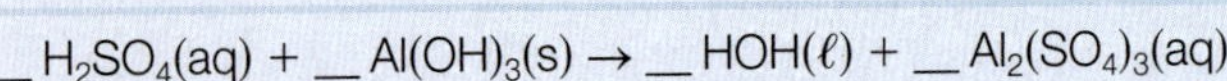 $\_\_ H_2SO_4(aq) + \_\_ Al(OH)_3(s) \rightarrow \_\_ HOH(\ell) + \_\_ Al_2(SO_4)_3(aq)$	Complete the unbalanced equation by writing the formulas of the products. 
$3\ H_2SO_4(aq) + \_\_ Al(OH)_3(s) \rightarrow \_\_ HOH(\ell) + \_\_ Al_2(SO_4)_3(aq)$ *Three sulfate ions in $Al_2(SO_4)_3$ require 3 H_2SO_4.*	We will have you balance the equation in a series of steps. First, balance the sulfate ions.
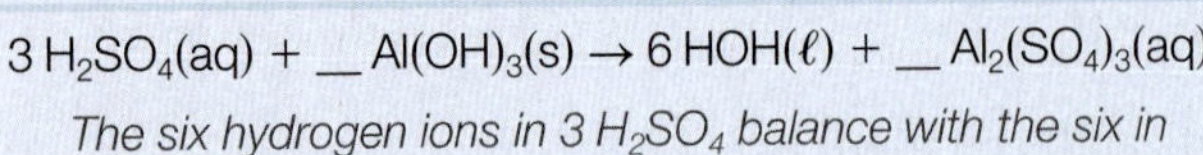 $3\ H_2SO_4(aq) + \_\_ Al(OH)_3(s) \rightarrow 6\ HOH(\ell) + \_\_ Al_2(SO_4)_3(aq)$ *The six hydrogen ions in 3 H_2SO_4 balance with the six in 6 HOH.*	Balance the hydrogen ions. Temporarily treat water as "hydrogen hydroxide" for balancing purposes.
$3\ H_2SO_4(aq) + 2\ Al(OH)_3(s) \rightarrow 6\ HOH(\ell) + \_\_ Al_2(SO_4)_3(aq)$ *Six hydroxide ions in 6 HOH require 2 $Al(OH)_3$ to achieve six on the left.*	Now balance the hydroxide ions while continuing to treat water as "hydrogen hydroxide."
$3\ H_2SO_4(aq) + 2\ Al(OH)_3(s) \rightarrow 6\ H_2O(\ell) + Al_2(SO_4)_3(aq)$ 6 H, 3 SO_4, 2 Al, and 6 OH on each side; or 12 H, 3 S, 18 O, and 2 Al on each side.	Aluminum ions are the only thing left unchecked. Balance the Al, restore the formula of water to its correct form, and make a final atom and/or ion balance check.

You improved your skill at balancing chemical equations in general, and specifically, balancing double-replacement reaction equations.

What did you learn by solving this Active Example?

Practice Exercise 8.11

Barium hydroxide solution is neutralized with sulfuric acid. A precipitate forms. Write the equation.

a summary of... **Double-Replacement Reactions**

Reactants: Solutions of two compounds, each with positive and negative ions (AX + BY)

Reaction type: Double-Replacement

Equation type:

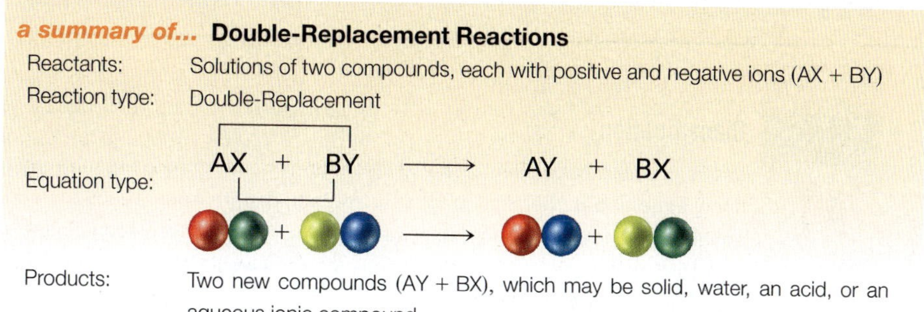

Products: Two new compounds (AY + BX), which may be solid, water, an acid, or an aqueous ionic compound

Learn It Now! *Compare* a summary of... *Combination Reactions with* a summary of... *Decomposition Reactions with* a summary of... *Single-Replacement Reactions with* a summary of... *Double-Replacement Reactions.*

8.11 Summary of Reactions and Equations

All of the information from the equation-writing summaries at the ends of Sections 8.7 through 8.10 has been assembled into **Table 8.3**. The reactants (first column) are shown for each reaction type (second column). Each reaction type has a certain "equation type" (third column) that yields predictable products (fourth column).

Reading the column heads from left to right follows the "thinking order" by which reactants and products are identified. It will help you organize your approach to writing equations. Given the reactants of a specific chemical change, you can fit them into one of the reactant boxes in the table and thereby determine the reaction type. Once you know what kind of reaction it is, you know the type of equation that describes it. You can then write the formulas of the products on the right. Balance the equation, and you are finished.

Table 8.3 Summary of Types of Reactions and Equations

Reactants	Reaction Type	Equation Type	Products
Any combination of elements and compounds that form one product	Combination	A + X ⟶ AX	One compound
One compound	Decomposition	AX ⟶ A + X	Any combination of elements and compounds
Element + ionic compound or acid or water	Single-Replacement	A + BX ⟶ AX + B	Element + ionic compound
Solutions of two compounds, each with positive and negative ions	Double-Replacement	AX + BY ⟶ AY + BX	Two new compounds, which may be a solid, water, an acid, or an aqueous ionic compound

We indicated earlier that just because an equation can be written, it does not necessarily mean that the reaction will happen. By using the results of experiments performed over many years, we can make reliable predictions. For example, we can predict with confidence that zinc will replace copper in a copper sulfate solution but that silver will not or that pouring calcium nitrate solution into sodium fluoride solution will yield a precipitate but that pouring it into sodium bromide solution will not. We have deliberately refrained from making these predictions in this chapter; learning to write equations is enough for now.

Thinking About Your Thinking

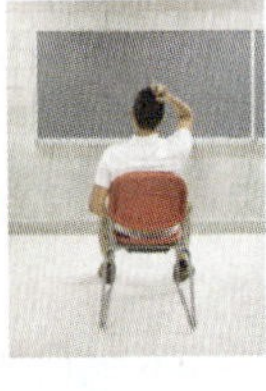

Classification

In this chapter, we presented a classification scheme to help organize your knowledge about chemical reactions. You could learn about reactions and equations without an organizing system, but our classification scheme helps you learn more easily and more efficiently than you would without a method of organizing your newly obtained knowledge.

There are many ways to classify chemical changes, and our scheme is just one of them. This is the essence of the classification thinking skill: As what you know changes, you need to practice reorganizing your knowledge in different ways. Later in your study of chemistry, you will find that there are three broad classifications of chemical change. One is metathesis reactions, the formation of a solid or molecular substance. Another category is proton-transfer reactions, in which a hydrogen ion is transferred. Finally, there are electron-transfer reactions, which are the redox reactions introduced in Section 8.9.

Is our four-category classification scheme better than the three-category scheme? No, it's just better for now, given your present understanding of chemistry. Later, you will have a great deal more knowledge about many different types of chemical change, and you will probably find that the three-category scheme is useful when you are considering broad, general differences in reactivity. The four-category scheme you are learning now will be useful in predicting products of reactions. Both schemes will be helpful in organizing your knowledge, and your choice of mode of classification will depend on the situation. You will be forming a hierarchical system, one in which different needs lead to different classifications. In doing so, your knowledge will be organized in such a way that the whole is greater than the sum of the parts because of the added value of the classification systems.

Always be looking for classification systems to organize your knowledge, and don't hesitate to change your systems when new information yields a better method of classification.

NASA

CHAPTER 8 IN REVIEW: CHEMICAL REACTIONS

The **overarching goals** for this chapter are the following:

1. Understand how a chemical equation serves as a particulate-level, symbolic representation of a macroscopic-level process.
2. Learn the mechanics of writing a chemical equation.
3. Learn how to identify four different patterns in chemical equations.
4. Learn how to predict the products of each kind of reaction and write the formulas of those products.
5. Given potential reactants, write the equations for the probable reaction.

Goal 1 Describe five types of evidence detectable by human senses that usually indicate a chemical change.

Five types of evidence indicate the possibility of a chemical change:

1. A color change
2. The formation of a solid
3. The formation of a gas
4. The absorption or release of heat energy
5. The emission of light energy

Goal 2 Distinguish between an unbalanced and a balanced chemical equation, and explain why a chemical equation needs to be balanced.	A **chemical equation** is a shorthand description of a chemical reaction. A **balanced** chemical equation reflects the Law of Conservation of Mass. The subscripts in a chemical equation may never be changed simply to balance the equation. Changing a subscript in the formula of a substance changes the chemical identity of that substance (Law of Definite Composition).
Goal 3 Given an unbalanced chemical equation, balance it by inspection.	A **formal approach to balancing a chemical equation** is: 1. Place a "1" in front of the formula with the largest number of atoms. If two formulas have the same number of atoms, select the one with the greater number of elements. 2. Insert coefficients that balance the elements *that appear in compounds.* Use fractional coefficients, if necessary. Choosing elements in the following order is usually easiest: (a) elements in the starting formula that are in only one other compound, (b) all other elements from the starting formula, and (c) all other elements *in compounds.* 3. Place coefficients in front of formulas of uncombined elements that balance those elements. Use fractional coefficients, if necessary. 4. Clear fractions, if any, by multiplying all coefficients by the lowest common denominator. Remove any "1" coefficients that remain. 5. Check to be sure that the final equation is balanced.
Goal 4 Given a balanced chemical equation or information from which it can be written, describe its meaning on the particulate, molar, and macroscopic levels.	The **coefficients** in a balanced chemical equation have two common interpretations: 1. **Particulate-level interpretation:** The coefficients represent the number of atoms, molecules, or formula units of each species. 2. **Molar interpretation:** The coefficients represent the amounts in moles of each species. To interpret coefficients at the macroscopic level, you need to combine the molar interpretation with the molar mass of each species.
Goal 5 Write the equation for the reaction in which a single product compound is formed by the combination of two or more substances.	A reaction in which two or more substances combine to form a single product is a **combination reaction** or synthesis reaction. Reactants: Any combination of elements and/or compounds Product: One compound Equation type: $A + X \rightarrow AX$
Goal 6 Given a single compound that is decomposed into two or more substances, either compounds or elements, write the equation for the reaction.	A **decomposition reaction** occurs when a single compound breaks down into simpler substances. Reactant: One compound Products: Any combination of elements and compounds Equation type: $AX \rightarrow A + X$
Goal 7 Given the reactants of a single-replacement reaction, write the equation for the reaction.	In a **single-replacement reaction,** it looks as if one element is replacing another in a compound. If the reactant element is a metal, it replaces the metal or hydrogen in the compound. If the element is a nonmetal, it replaces the nonmetal in a compound. Reactants: Element (A) plus a solution of either an acid or an ionic compound (BX) or water Products: An ionic compound (usually in solution) (AX) plus an element (B) Equation type: $A + BX \rightarrow AX + B$
Goal 8 Given the reactants in a double-replacement precipitation or neutralization reaction, write the equation.	In a **double-replacement reaction,** it looks as if the ions of the two reactants change partners. When one (or more) product of a double-replacement reaction is a solid that is formed from reactants in solution, the reaction is **a precipitation reaction**. When one reactant is an **acid** and the other is a **base**, the double-replacement reaction yields water as one product, and the reaction is a **neutralization reaction**. Reactants: Solutions of two compounds, each with positive and negative ions $(AX + BY)$ Products: Two new compounds $(AY + BX)$, which may be a solid, water, an acid, or an aqueous ionic compound Equation type: $AX + BY \rightarrow AY + BX$

Key Terms

Most of the key terms and concepts and many others appear in the Glossary. Use your Glossary regularly.

double-replacement equation p. 288
double-replacement reaction p. 288
formation of a gas p. 269
formation of a solid product p. 269
heat p. 270
light p. 270
neutralization reaction p. 289
neutralize p. 289
oxidation–reduction reaction p. 286
precipitate p. 288
precipitation reaction p. 288
products p. 267
reactants p. 267
redox p. 286
reversible reaction p. 283
single-replacement equation p. 286
single-replacement reaction p. 286
starting formula p. 274
state symbols p. 271
synthesis reaction p. 280

Frequently Asked Questions

What are the mistakes to avoid while balancing equations?
Be careful about the "don'ts" in Section 8.3. It is very tempting to balance an equation quickly by changing a correct formula or adding one that doesn't belong. Another device some creative students invent is slipping what should be a coefficient into the middle of a correct formula, such as changing $NaNO_3$ to $Na2NO_3$ to balance the nitrates in $Ca(NO_3)_2$ on the other side. That doesn't work either. There is no such thing as $Na2NO_3$.

What other mistakes should I avoid while balancing equations?
Don't forget the "hidden" (unnamed) reactants and products. "Bubbles form…" indicates a gaseous product, such as H_2, O_2, or CO_2. In the word description of a reaction, ionic products are often unnamed if their formulas can be derived from the reactants.

What process should I use to ensure that I'm approaching equation balancing correctly?
Remember to write the qualitative description of the reaction first and then balance the equation—*using coefficients*—without changing the qualitative description.

How can I polish my fundamental equation-balancing skills before I work through the Questions, Exercises, and Problem?
An equation-classification exercise and an equation-balancing exercise follow these study hints. We suggest you use them to practice your skills.

What is the best way to learn how to predict products when given only the reactants in a chemical change?
Look for the "big picture." Knowing how things fit together helps in learning the details. Look at the *reactants.* They should tell you the *reaction type.* Each reaction type has a certain *equation type*, and that gives you the *products*. In order from left to right, these are the column heads in the classification exercise table.

I am getting pretty good at writing and balancing equations, so what should I do to polish and refine my skills before the chapter exam?
After completing the equation-classification and equation-balancing exercises and the section-specific questions, you might wish to practice further by referring to the unclassified reactions beginning with Question 31. Based on the reactants described in each question, see if you can determine mentally what kind of reaction it is and what its products are. Run through the whole list this way without writing equations until you feel sure of your ability. Then write and balance equations until you are completely confident that you can write any equation without hesitation.

Equation-Classification Exercise

Step 1 of our general equation-writing procedure is to "classify the reaction type." In this exercise, we provide the chemical formulas of reactants in a reaction. Without writing the formulas of the products or balancing the equation, classify the following as one of the four reaction types given in this chapter by placing the appropriate abbreviation after each. The first two are completed as examples. Answers to 3–12 appear at the end of the chapter.

Reaction Type	Equation Type	Abbreviation
Combination	A + X ⟶ AX	Comb
Decomposition	AX ⟶ A + X	Decomp
Single-Replacement	A + BX ⟶ AX + B	SR
Double-Replacement	AX + BY ⟶ AY + BX	DR

1. $Mg + N_2 \rightarrow$ **Comb**
2. $Zn(OH)_2 + H_2SO_4 \rightarrow$ **DR**
3. $Zn + I_2 \rightarrow$
4. $PbO_2 \rightarrow$
5. $Al + CuSO_4 \rightarrow$
6. $FeCl_2 + Na_3PO_4 \rightarrow$
7. $Al + HCl \rightarrow$
8. $HgO \rightarrow$
9. $Fe + O_2 \rightarrow$
10. $Fe_2(SO_4)_3 + Ba(OH)_2 \rightarrow$
11. $HNO_3 + CsOH \rightarrow$ (Cs, Z = 55)
12. $Cu + AgNO_3$

Equation-Balancing Exercise

Balance the following equations, for which correct chemical formulas are already written. We have provided space for you to insert coefficients directly into the unbalanced equations. Balanced equations are given in the answer section at the end of the chapter.

1. $Na + \quad O_2 \rightarrow \quad Na_2O$
2. $H_2 + \quad Cl_2 \rightarrow \quad HCl$
3. $P + \quad O_2 \rightarrow \quad P_2O_3$
4. $KClO_4 \rightarrow \quad KCl + \quad O_2$
5. $Sb_2S_3 + \quad HCl \rightarrow \quad SbCl_3 + \quad H_2S$
6. $NH_3 + \quad H_2SO_4 \rightarrow \quad (NH_4)_2SO_4$
7. $CuO + \quad HCl \rightarrow \quad CuCl_2 + \quad H_2O$
8. $Zn + \quad Pb(NO_3)_2 \rightarrow \quad Zn(NO_3)_2 + \quad Pb$
9. $AgNO_3 + \quad H_2S \rightarrow \quad Ag_2S + \quad HNO_3$
10. $Cu + \quad S \rightarrow \quad Cu_2S$
11. $Al + \quad H_3PO_4 \rightarrow \quad H_2 + \quad AlPO_4$
12. $NaNO_3 \rightarrow \quad NaNO_2 + \quad O_2$
13. $Mg(ClO_3)_2 \rightarrow \quad MgCl_2 + \quad O_2$
14. $H_2O_2 \rightarrow \quad H_2O + \quad O_2$
15. $BaO_2 \rightarrow \quad BaO + \quad O_2$
16. $H_2CO_3 \rightarrow \quad H_2O + \quad CO_2$
17. $Pb(NO_3)_2 + \quad KCl \rightarrow \quad PbCl_2 + \quad KNO_3$
18. $Al + \quad Cl_2 \rightarrow \quad AlCl_3$
19. $C_6H_{14} + \quad O_2 \rightarrow \quad CO_2 + \quad H_2O$
20. $NH_4NO_3 \rightarrow \quad N_2O + \quad H_2O$
21. $H_2 + \quad N_2 \rightarrow \quad NH_3$
22. $Fe + \quad Cl_2 \rightarrow \quad FeCl_3$
23. $H_2S + \quad O_2 \rightarrow \quad H_2O + \quad SO_2$
24. $MgCO_3 + \quad HCl \rightarrow \quad MgCl_2 + \quad CO_2 + \quad H_2O$
25. $P + \quad I_2 \rightarrow \quad PI_3$

Small-Group Discussion Questions

Small-Group Discussion Questions are for group work, either in class or under the guidance of a leader during a discussion section.

1. What is the meaning of the arrow in a balanced chemical equation? Is the meaning different when considering the particulate level versus the macroscopic level? Write a definition of the arrow in a chemical equation that would be appropriate for a chemical dictionary.
2. Chemical equations are usually interpreted at two different levels. Using the reaction of dihydrogen sulfide and oxygen to form sulfur dioxide and water as an example, explain each interpretation of the balanced equation.
3. Write and balance the equation for the decomposition of hydrogen peroxide, H_2O_2, to water and oxygen. Is the number of molecules the same before and after the chemical change? Does your answer to the previous question depend on the initial number of reactant molecules? Is the number of atoms the same before and after the chemical change? Justify your answers in terms of the Law of Conservation of Mass.
4. Explain *why* the subscripts in a chemical formula cannot be changed to balance a chemical equation.
5. Write the phrase "chemical equation" on the board. Each student should jot down at least one word or phrase that comes to mind that is related to chemical equations. After the group has come up with a list of related words, separate them into three categories. Each word or phrase should fit only one category. Now give each category a name. Explain the relationship between each category and the phrase "chemical equation." Summarize the overall organization scheme in a few sentences.
6. Write and balance the equation for the reaction of hydrogen and oxygen to form water. If 2 moles of hydrogen react with 1 mole of oxygen, how many moles of water are formed? Explain how you know the answer to this question. How many moles of water will form when 2 moles of hydrogen react with 2 moles of oxygen? Explain. Investigate the number of moles of water that will form from a variety of combinations of numbers of moles of hydrogen and oxygen, such as 14 moles of hydrogen and 20 moles of oxygen, or 21 moles of hydrogen and 8 moles of oxygen. In each case, explain the role of the chemical equation in determining the number of moles of water produced.

Questions, Exercises, and Problems

Interactive versions of these problems may be assigned by your instructor. Solutions for blue-numbered questions are at the end of the chapter. Questions other than those in the General Questions and More Challenging Problems sections are paired in consecutive odd–even number combinations; solutions for the odd numbered questions are also at the end of the chapter.

Many questions in this chapter are written with the assumption that you have studied Chapter 6 and can write the required formulas from their chemical names. If this is not the case, we have placed a list of all chemical formulas needed to answer this chapter questions at the end of the Questions, Exercises, and Problems.

Section 8.5: Interpreting Chemical Equations

1. Consider the following particulate-level representation of a chemical equation:

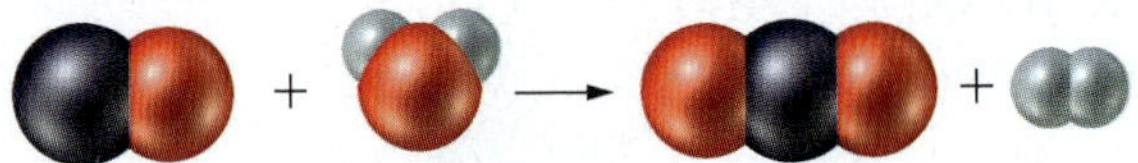

The white spheres represent hydrogen atoms, the black sphere represents a carbon atom, and the red spheres represent oxygen atoms. (a) Write a balanced chemical equation representing this reaction. (b) Write a word description of the reaction on the particulate and molar levels.

2. Give a description, using words and chemical formulas, of the following chemical equation on both the particulate and molar levels: $C_2H_4 + H_2 \rightarrow C_2H_6$.

3. The left box of the following diagram shows the hypothetical elements A (green atoms) and B (purple diatomic molecules) before they react. Write a balanced chemical equation, using the lowest possible whole-number coefficients, for the reaction that occurs to form the product in the right box.

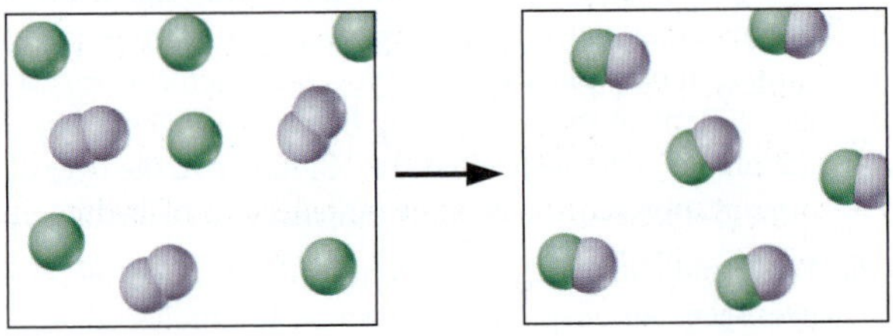

4. Draw a box and then sketch five space-filling models of diatomic molecules within it, similar to those in Question 3. Draw an arrow and a product box to the right of your arrow. Add the remaining space-filling models necessary to depict the balanced reaction $X_2 + Y_2 \rightarrow XY_3$ (the reaction equation is not yet balanced), in which your five molecules represent the species X_2.

5. A particulate-level sketch of a reaction shows four spheres representing atoms of element A and four spheres depicting atoms of element B before reacting and four models of the molecule AB after the reaction.

Explain why $4\ A + 4\ B \rightarrow 4\ AB$ is not the correct equation for the reaction.

6. Consider the reaction of the elements antimony and chlorine, forming antimony trichloride. Write and balance the chemical equation for this reaction, and then state which of the following boxes best represent the reactants and product of the reaction.

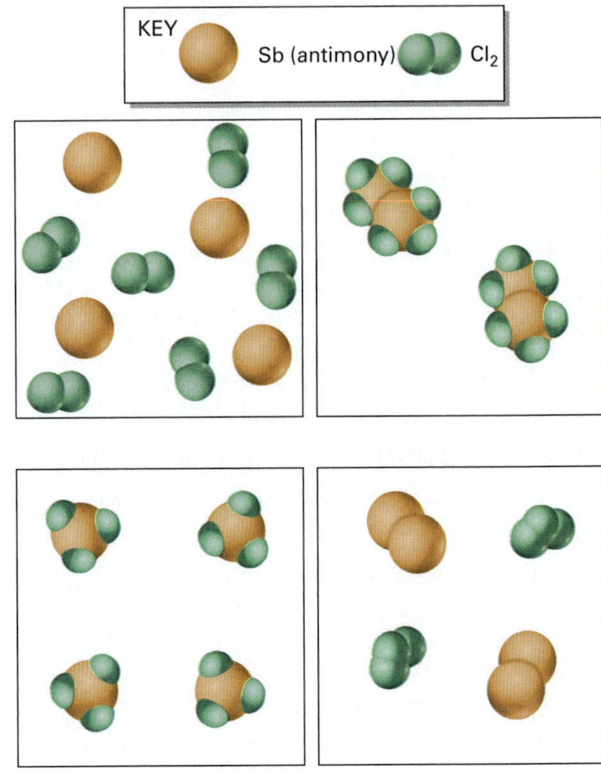

7. Write a balanced chemical equation to represent the reaction illustrated here.

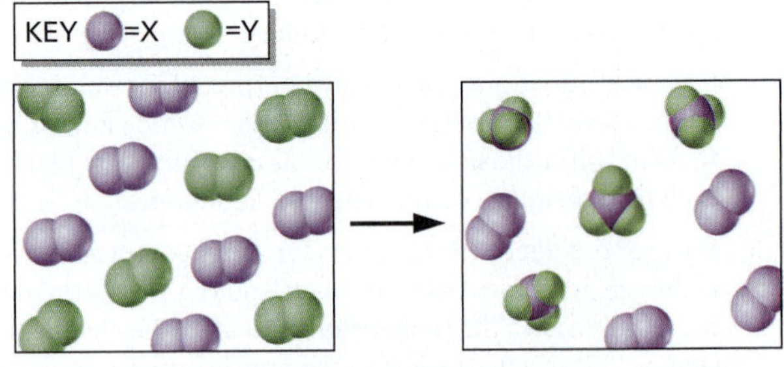

8. Write and balance the equation for the reaction of butane, C_4H_{10}, with oxygen to form carbon dioxide and water. Use your equation to help answer the following questions: (a) Write, in words, a description of the reaction on the particulate level. (b) If you were to build physical ball-and-stick models of the reactants and products, what minimum number of balls representing atoms of each element

do you need if you show both reactants and products at the same time? (c) What if the models of the reactants from Part (b) were built and then rearranged to form products? How many balls would you need? (d) Use words to interpret the equation on the molar level. (e) Use the molar-level interpretation of the equation from Part (d) and molar masses rounded to the nearest gram to show that mass is indeed conserved in this reaction.

Questions 9 to 30: Write the equation for each reaction described. Follow your instructor's advice about whether state symbols should be included.

Section 8.7: Combination Reactions

9. Lithium combines with oxygen to form lithium oxide.
10. When nitrogen combines with hydrogen, ammonia is formed.
11. Boron combines with oxygen to form diboron trioxide.
12. When nitrogen monoxide combines with oxygen, nitrogen dioxide is formed.
13. Aluminum combines with bromine to make aluminum bromide.

Aluminum reacts with bromine.

14. When phosphorus (P_4) combines with chlorine, phosphorus trichloride is formed.

Phosphorus reacts with chlorine.

Section 8.8: Decomposition Reactions

15. Pure hydrogen iodide decomposes spontaneously to its elements.
16. When sodium nitrate decomposes, sodium nitrite and oxygen are formed.
17. Barium peroxide, BaO_2, breaks down into barium oxide and oxygen.
18. Pure bromine trifluoride decomposes to its elements.

Section 8.9: Single-Replacement Reactions

19. Calcium reacts with hydrobromic acid.
20. Magnesium is placed into sulfuric acid.
21. Chlorine gas is bubbled through an aqueous solution of potassium iodide.
22. Copper metal combines with aqueous silver nitrate.

Copper reacts with silver nitrate solution.

Section 8.10: Double-Replacement Reactions

It is not necessary for you to identify the precipitates formed in Questions 23 to 26.

23. Calcium chloride and potassium fluoride solutions react to form a precipitate.
24. A precipitate forms when aqueous solutions of cobalt(III) iodide and lead(II) nitrate are combined.
25. Milk of magnesia is the precipitate that results when sodium hydroxide and magnesium bromide solutions are combined.
26. A precipitate forms when aqueous solutions of chromium(III) nitrate and sodium hydroxide are combined.
27. Sulfuric acid reacts with barium hydroxide solution.
28. A reaction occurs when aqueous solutions of perchloric acid and sodium hydroxide are combined.
29. Sodium hydroxide is added to phosphoric acid.
30. A reaction occurs when aqueous solutions of hydrosulfuric acid and sodium hydroxide are combined.

Unclassified Reactions

Questions 31 to 66: Write the equation for the reaction described or for the most likely reaction between given reactants.

31. Lead(II) nitrate solution reacts with a solution of sodium iodide.
32. A precipitate forms when aqueous solutions of calcium nitrate and potassium carbonate are combined.
33. The fuel butane, C_4H_{10}, burns in oxygen, forming carbon dioxide and water.
34. Acetaldehyde, CH_3CHO, a raw material used in manufacturing vinegar, perfumes, dyes, plastics, and other organic materials, is oxidized, reacting with oxygen, forming carbon dioxide and water.

35. Sulfurous acid decomposes spontaneously to sulfur dioxide and water.

36. Carbonated beverages contain carbonic acid, an unstable compound that decomposes to carbon dioxide and water.

37. Calcium reacts with water. (*Hint:* Think of water as an acid with the formula HOH.)

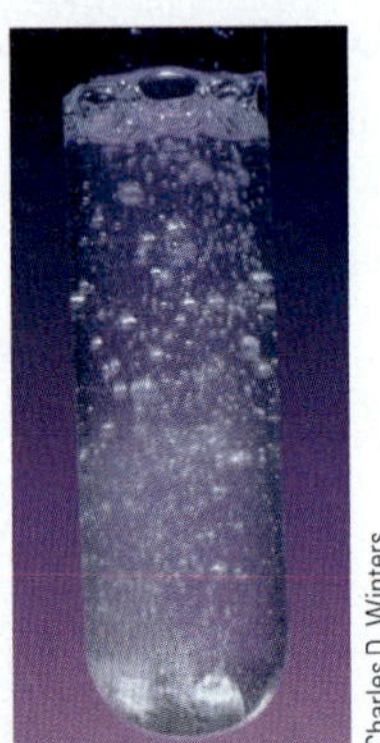

Calcium reacts with water.

38. Bubbles form when metallic barium is placed in water. (*Hint:* Think of water as an acid with the formula HOH.)

39. Zinc metal is placed in a silver chlorate solution.

40. Chlorine gas combines with solid potassium.

Chlorine and potassium react.

41. Ammonium sulfide is added to a solution of copper(II) nitrate.

42. Silver nitrate is added to a solution of sodium sulfide.

43. Phosphorus tribromide is produced when phosphorus reacts with bromine.

44. When iron reacts with oxygen, iron(III) oxide is formed.

45. When calcium hydroxide (sometimes called slaked lime) is heated, it forms calcium oxide (lime) and water vapor.

46. Hydrogen peroxide, H_2O_2, the familiar bleaching compound, decomposes slowly into water and oxygen.

47. Glycerine, $C_3H_8O_3$—used in making soap, cosmetics, and explosives—reacts with oxygen to form carbon dioxide and water.

48. Aqueous solutions of hydrobromic acid and potassium hydroxide are combined.

49. Powdered antimony (Z = 51) ignites when sprinkled into chlorine gas, producing antimony(III) chloride.

50. Fluorine reacts spontaneously with nearly all elements. Oxygen difluoride is produced when fluorine reacts with oxygen.

51. A solution of potassium hydroxide reacts with a solution of zinc chloride.

52. A precipitate forms when aqueous solutions of silver nitrate and magnesium sulfate are combined.

53. Solid ammonium dichromate, $(NH_4)_2Cr_2O_7$, decomposes to solid chromium(III) oxide, nitrogen, and steam.

Solid ammonium dichromate decomposes.

54. Magnesium nitride is formed from its elements.

55. A solution of lithium sulfite is mixed with a solution of sodium phosphate.

56. Aqueous solutions of sulfuric acid and zinc hydroxide are combined.

57. A solution of chromium(III) nitrate is one of the products of the reaction between metallic chromium and aqueous tin(II) nitrate.

58. Lithium is added to a solution of manganese(II) chloride.

59. Sulfuric acid is produced when sulfur trioxide reacts with water.

60. Solid zinc and aqueous copper(II) sulfate are combined.

The remaining reactions are not readily placed into one of the four classifications used in this chapter. Nevertheless, enough information is given for you to write the equations.

61. A solid oxide of iron, Fe_3O_4, and hydrogen are the products of the reaction between iron and steam.

62. Metallic zinc reacts with steam at high temperatures, producing zinc oxide and hydrogen.

63. Aluminum carbide, Al_4C_3, reacts with water to form aluminum hydroxide and methane, CH_4.

64. When solid barium oxide is placed into water, a solution of barium hydroxide is produced.

65. Magnesium nitride and hydrogen are the products of the reaction between magnesium and ammonia.

66. Solid iron(III) oxide reacts with gaseous carbon monoxide to produce iron and carbon dioxide.

General Questions

67. Distinguish precisely and in scientific terms the differences among items in each of the following pairs or groups:
   a) Reactant, product
   b) (g), (ℓ), (s), (aq)
   c) Combination reaction, decomposition reaction
   d) Single replacement, double replacement
   e) Acid, base
   f) Precipitation, neutralization

68. Classify each of the following statements as true or false:
   a) In a chemical reaction, reacting substances are destroyed and new substances are formed.
   b) A chemical equation expresses the quantity relationships between reactants and products in terms of amount in moles.
   c) Elements combine to form compounds in combination reactions.
   d) Compounds decompose into elements in decomposition reactions.
   e) A nonmetal cannot replace a nonmetal in a single-replacement reaction.

69. Each reactant or pair of reactants listed is *potentially* able to participate in one type of reaction described in this chapter. In each case, name the type of reaction and complete the equation. In Parts (a) and (j), the 2+ ion is formed from the reaction of the corresponding metal. In Part (c), carbon dioxide and water are the products of the reaction (you do not have to name this reaction type).
   a) $Pb + Cu(NO_3)_2 \rightarrow$
   b) $Mg(OH)_2 + HBr \rightarrow$
   c) $C_5H_{10}O + O_2 \rightarrow$
   d) $Na_2CO_3 + CaSO_4 \rightarrow$
   e) $LiBr \rightarrow$
   f) $NH_4Cl + AgNO_3 \rightarrow$
   g) $Ca + Cl_2 \rightarrow$
   h) $F_2 + NaI \rightarrow$
   i) $Zn(NO_3)_2 + Ba(OH)_2 \rightarrow$
   j) $Cu + NiCl_2 \rightarrow$

70. Hydrogen, nitrogen, oxygen, fluorine, chlorine, bromine, and iodine exist as diatomic molecules at the temperatures and pressures at which most reactions occur. Under these normal conditions, when may the formulas of these elements be written without a subscript 2 in a chemical equation?

71. Acid rain is rainfall that contains sulfuric acid originating from organic fuels that contain sulfur. The process occurs in three major steps. The sulfur burns first, forming sulfur dioxide. In sunlight, the sulfur dioxide reacts with oxygen in the air to produce sulfur trioxide. When rainwater falls through the sulfur trioxide, the reaction produces sulfuric acid. Write the equation for each step in the process and state what type of reaction it is.

72. One of the harmful effects of acid rain is its reaction with structures made of limestone, which include marble structures, ancient ruins, and many famous statues. Write the equation you would expect for the reaction between acid rain, which contains sulfuric acid, and limestone, solid calcium carbonate. Write a second equation with the same reactants, showing that the expected but unstable carbonic acid decomposes to carbon dioxide and water.

Ted Spiegel/Getty Images

The effect of acid rain on a marble statue.

73. The tarnish that appears on silver is silver sulfide, which is formed when silver is exposed to sulfur-bearing compounds in the presence of oxygen in the air. When hydrogen sulfide is the sulfur-bearing reacting compound, water is the second product. Write the equation for the reaction.

74. Sulfur combines directly with three of the halogens but not with iodine. The products, however, do not have similar chemical formulas. When sulfur reacts with fluorine, sulfur hexafluoride is the most common product; with chlorine, the product is usually sulfur dichloride; and reaction with bromine usually yields disulfur dibromide. Write the equation for each reaction.

75. One source of the pure tungsten (Z = 74) filament used in conventional light bulbs is tungsten(VI) oxide. It is heated with hydrogen at high temperatures. The hydrogen reacts with the oxygen in the oxide, forming steam. Write the equation for the reaction, and classify it as one of the reaction types discussed in this chapter.

Questions 76 to 79: Write the equation for each reaction described.

76. Write the equation for the neutralization reaction in which barium nitrate is the ionic compound formed.

77. Lithium sulfate is one of the products of a neutralization reaction.

78. Only the one hydrogen is removed in the reaction between sulfamic acid, HNH_2SO_3, and potassium hydroxide. (Figure out the formula of the anion from the acid and use it to write the formula of the ionic compound formed.)

79. The concentration of sodium hydroxide solution can be found from reacting it with oxalic acid, $H_2C_2O_4$. (The formula of oxalic acid should lead you to the formula of the oxalate ion and then to the formula of the ionic compound formed in the reaction.)

Formulas

These formulas are provided in case you are studying this chapter before you study Chapter 6. You should not use this list unless your instructor has not yet assigned Chapter 6 or otherwise indicated that you may use the list.

9. Li, O_2, Li_2O

10. N_2, H_2, NH_3

11. B, O_2, B_2O_3

12. NO, O_2, NO_2

13. Al, Br_2, $AlBr_3$

14. Cl_2, PCl_3

15. HI

16. $NaNO_3$, $NaNO_2$, O_2

17. BaO, O_2

18. BrF_3

19. Ca, HBr

20. Mg, H_2SO_4

21. Cl_2, KI

22. Cu, $AgNO_3$

23. $CaCl_2$, KF

24. CoI_3, $Pb(NO_3)_2$

25. NaOH, $MgBr_2$

26. $Cr(NO_3)_3$, NaOH

27. H_2SO_4, $Ba(OH)_2$

28. $HClO_4$, NaOH

29. NaOH, H_3PO_4

30. H_2S, NaOH

31. $Pb(NO_3)_2$, NaI

32. $Ca(NO_3)_2$, K_2CO_3

33. O_2, CO_2, H_2O

34. O_2, CO_2, H_2O

35. H_2SO_3, SO_2, H_2O

36. H_2CO_3, CO_2, H_2O

37. Ca, HOH

38. Ba, HOH

39. Zn, $AgClO_3$

40. Cl_2, K

41. $(NH_4)_2S$, $Cu(NO_3)_2$

42. $AgNO_3$, Na_2S

43. PBr_3, P, Br_2

44. Fe, O_2, Fe_2O_3

45. $Ca(OH)_2$, CaO, H_2O

46. H_2O, O_2

47. O_2, CO_2, H_2O

48. HBr, KOH

49. Sb, Cl_2, $SbCl_3$

50. F_2, OF_2, O_2

51. KOH, $ZnCl_2$

52. $AgNO_3$, $MgSO_4$

53. Cr_2O_3, N_2, H_2O

54. Mg_3N_2

55. Li_2SO_3, Na_3PO_4

56. H_2SO_4, $Zn(OH)_2$

57. $Cr(NO_3)_3$, Cr, $Sn(NO_3)_2$

58. Li, $MnCl_2$

59. H_2SO_4, SO_3, H_2O

60. Zn, $CuSO_4$

61. H_2, Fe, H_2O

62. Zn, H_2O, ZnO, H_2

63. H_2O, $Al(OH)_3$

64. BaO, H_2O, $Ba(OH)_2$

65. Mg_3N_2, H_2, Mg, NH_3

66. Fe_2O_3, CO, Fe, CO_2

71. S, SO_2, SO_3, H_2O, H_2SO_4, O_2

72. H_2SO_4, $CaCO_3$, H_2CO_3, CO_2, H_2O

73. Ag_2S, Ag, O_2, H_2S, H_2O

74. S, F_2, SF_6, Cl_2, SCl_2, Br_2, S_2Br_2

75. W, WO_3, H_2, H_2O

76. $Ba(NO_3)_2$

77. Li_2SO_4

78. KOH

79. NaOH

Answers to Target Checks

1. (a) Yes; evidence includes sound, light, smell, and heat emitted. (b) No; physical transformation of liquid water to water vapor. (c) Yes; evidence includes light and heat emitted. (d) No; not a transformation of matter.
2. (a) False. The equation has seven oxygen atoms on the left and six on the right. (b) False. Never change a chemical formula to balance an equation. H_2O_2 is a real compound, but it has nothing to do with this reaction.
3. (a) Two hydrogen peroxide molecules decompose to form one oxygen molecule and two water molecules. (b) Two moles of hydrogen peroxide molecules decompose to form one mole of oxygen molecules and two moles of water molecules.

Answers to Practice Exercises

1. $SO_2 + 2\,H_2O + 2\,CaS \rightarrow 3\,S + 2\,Ca(OH)_2$
2. $2\,SO_2 + O_2 + 2\,CaO \rightarrow 2\,CaSO_4$
3. $3\,Fe_3O_4(s) + 8\,Al(s) \rightarrow 9\,Fe(s) + 4\,Al_2O_3(s)$
4. $4\,Al(s) + 3\,O_2(g) \rightarrow 2\,Al_2O_3(s)$
5. $2\,Fe(s) + 3\,O_2(g) + 3\,H_2(g) \rightarrow 2\,Fe(OH)_3(s)$
6. $2\,HgO \rightarrow 2\,Hg + O_2$

7. $Mg(OH)_2(s) \rightarrow H_2O(g) + MgO(s)$
8. $Cl_2 + 2\ NaBr \rightarrow Br_2 + 2\ NaCl$
9. $Pb(s) + Cu(NO_3)_2(aq) \rightarrow Cu(s) + Pb(NO_3)_2(aq)$
10. $Cu(NO_3)_2(aq) + 2\ NaOH(aq) \rightarrow Cu(OH)_2(s) + 2\ NaNO_3(aq)$
11. $Ba(OH)_2(aq) + H_2SO_4(aq) \rightarrow BaSO_4(s) + 2\ HOH(\ell)$

Answers to Equation-Classification Exercise

3. Comb
4. Decomp
5. SR
6. DR
7. SR
8. Decomp
9. Comb
10. DR
11. DR
12. SR

Answers to Equation-Balancing Exercise

1. $4\ Na + O_2 \rightarrow 2\ Na_2O$
2. $H_2 + Cl_2 \rightarrow 2\ HCl$
3. $4\ P + 3\ O_2 \rightarrow 2\ P_2O_3$
4. $KClO_4 \rightarrow KCl + 2\ O_2$
5. $Sb_2S_3 + 6\ HCl \rightarrow 2\ SbCl_3 + 3\ H_2S$
6. $2\ NH_3 + H_2SO_4 \rightarrow (NH_4)_2SO_4$
7. $CuO + 2\ HCl \rightarrow CuCl_2 + H_2O$
8. $Zn + Pb(NO_3)_2 \rightarrow Zn(NO_3)_2 + Pb$
9. $2\ AgNO_3 + H_2S \rightarrow Ag_2S + 2\ HNO_3$
10. $2\ Cu + S \rightarrow Cu_2S$
11. $2\ Al + 2\ H_3PO_4 \rightarrow 3\ H_2 + 2\ AlPO_4$
12. $2\ NaNO_3 \rightarrow 2\ NaNO_2 + O_2$
13. $Mg(ClO_3)_2 \rightarrow MgCl_2 + 3\ O_2$
14. $2\ H_2O_2 \rightarrow 2\ H_2O + O_2$
15. $2\ BaO_2 \rightarrow 2\ BaO + O_2$
16. $H_2CO_3 \rightarrow H_2O + CO_2$
17. $Pb(NO_3)_2 + 2\ KCl \rightarrow PbCl_2 + 2\ KNO_3$
18. $2\ Al + 3\ Cl_2 \rightarrow 2\ AlCl_3$
19. $2\ C_6H_{14} + 19\ O_2 \rightarrow 12\ CO_2 + 14\ H_2O$
20. $NH_4NO_3 \rightarrow N_2O + 2\ H_2O$
21. $3\ H_2 + N_2 \rightarrow 2\ NH_3$
22. $2\ Fe + 3\ Cl_2 \rightarrow 2\ FeCl_3$
23. $2\ H_2S + 3\ O_2 \rightarrow 2\ H_2O + 2\ SO_2$
24. $MgCO_3 + 2\ HCl \rightarrow MgCl_2 + CO_2 + H_2O$
25. $2\ P + 3\ I_2 \rightarrow 2\ PI_3$

Answers to Blue-Numbered Questions, Exercises, and Problems

1. (a) $CO + H_2O \rightarrow CO_2 + H_2$. (b) Particulate: One molecule of carbon monoxide reacts with one molecule of water to form one molecule of carbon dioxide and one molecule of hydrogen. Molar: One mole of carbon monoxide reacts with one mole of water to form one mole of carbon dioxide and one mole of hydrogen.
3. $2\ A + B_2 \rightarrow 2\ AB$
5. Equations use the lowest whole-number coefficients possible to give the ratio of species in the reaction. $A + B \rightarrow AB$ is the appropriate equation for this reaction.
7. $X_2 + 3\ Y_2 \rightarrow 2\ XY_3$
9. $4\ Li(s) + O_2(g) \rightarrow 2\ Li_2O(s)$
11. $4\ B(s) + 3\ O_2(g) \rightarrow 2\ B_2O_3(s)$
13. $Al(s) + Br_2(\ell) \rightarrow AlBr_3(s)$
15. $2\ HI(g) \rightarrow H_2(g) + I_2(s)$
17. $2\ BaO_2(s) \rightarrow 2\ BaO(s) + O_2(g)$
19. $Ca(s) + 2\ HBr(aq) \rightarrow CaBr_2(aq) + H_2(g)$
21. $Cl_2(g) + 2\ KI(aq) \rightarrow 2\ KCl(aq) + I_2(s)$
23. $CaCl_2(aq) + 2\ KF(aq) \rightarrow CaF_2(s) + 2\ KCl(aq)$
25. $2\ NaOH(aq) + MgBr_2(aq) \rightarrow 2\ NaBr(aq) + Mg(OH)_2(s)$
27. $H_2SO_4(aq) + Ba(OH)_2(aq) \rightarrow 2\ H_2O(\ell) + BaSO_4(s)$
29. $3\ NaOH(s) + H_3PO_4(aq) \rightarrow Na_3PO_4(aq) + 3\ H_2O(\ell)$
31. $Pb(NO_3)_2(aq) + 2\ NaI(aq) \rightarrow PbI_2(s) + 2\ NaNO_3(aq)$
33. $2\ C_4H_{10}(\ell) + 13\ O_2(g) \rightarrow 8\ CO_2(g) + 10\ H_2O(g)$
35. $H_2SO_3(aq) \rightarrow SO_2(aq) + H_2O(\ell)$
37. $Ca(s) + 2\ HOH(\ell) \rightarrow Ca(OH)_2(aq) + H_2(g)$
39. $Zn(s) + 2\ AgClO_3(aq) \rightarrow Zn(ClO_3)_2(aq) + 2\ Ag(s)$
41. $(NH_4)_2S(s) + Cu(NO_3)_2(aq) \rightarrow 2\ NH_4NO_3(aq) + CuS(s)$
43. $2\ P(s) + 3\ Br_2(\ell) \rightarrow 2\ PBr_3(\ell)$ *or* $P_4(s) + 6\ Br_2(\ell) \rightarrow 4\ PBr_3(\ell)$
45. $Ca(OH)_2(s) \rightarrow CaO(s) + H_2O(g)$
47. $2\ C_3H_8O_3(\ell) + 7\ O_2(g) \rightarrow 6\ CO_2(g) + 8\ H_2O(g)$
49. $2\ Sb(s) + 3\ Cl_2(g) \rightarrow 2\ SbCl_3(s)$
51. $2\ KOH(aq) + ZnCl_2(aq) \rightarrow 2\ KCl(aq) + Zn(OH)_2(s)$
53. $(NH_4)_2Cr_2O_7(s) \rightarrow Cr_2O_3(s) + N_2(g) + 4\ H_2O(g)$
55. $3\ Li_2SO_3(aq) + 2\ Na_3PO_4(aq) \rightarrow 2\ Li_3PO_4(s) + 3\ Na_2SO_3(aq)$
57. $2\ Cr(s) + 3\ Sn(NO_3)_2(aq) \rightarrow 2\ Cr(NO_3)_3(aq) + 3\ Sn(s)$
59. $SO_3(g) + H_2O(\ell) \rightarrow H_2SO_4(aq)$
61. $3\ Fe(s) + 4\ H_2O(g) \rightarrow Fe_3O_4(s) + 4\ H_2(g)$
63. $Al_4C_3(s) + 12\ H_2O(\ell) \rightarrow 4\ Al(OH)_3(s) + 3\ CH_4(g)$
65. $3\ Mg(s) + 2\ NH_3(g) \rightarrow Mg_3N_2(s) + 3\ H_2(g)$
68. True: a, b, c, d. False: e. Note: Statement d is true, but it is not "the whole truth." The products of a decomposition reaction may be an element and a compound or two or more compounds.

69. a) Single replacement, $Pb + Cu(NO_3)_2 \rightarrow Pb(NO_3)_2 + Cu$
b) Double replacement, $Mg(OH)_2 + 2\,HBr \rightarrow MgBr_2 + 2\,H_2O$
c) $C_5H_{10}O + 7\,O_2 \rightarrow 5\,CO_2 + 5\,H_2O$
d) Double replacement, $Na_2CO_3 + CaSO_4 \rightarrow Na_2SO_4 + CaCO_3$
e) Decomposition, $2\,LiBr \rightarrow 2\,Li + Br_2$
f) Double replacement, $NH_4Cl + AgNO_3 \rightarrow NH_4NO_3 + AgCl$
g) Combination, $Ca + Cl_2 \rightarrow CaCl_2$
h) Single replacement, $F_2 + 2\,NaI \rightarrow 2\,NaF + I_2$
i) Double replacement, $Zn(NO_3)_2 + Ba(OH)_2 \rightarrow Zn(OH)_2 + Ba(NO_3)_2$
j) Single replacement, $Cu + NiCl_2 \rightarrow CuCl_2 + Ni$

70. Under the circumstances discussed in this book, you may never write the formula of a diatomic molecule without a subscript 2 in its formula.

71. (1) $S + O_2 \rightarrow SO_2$, Combination; (2) $2\,SO_2 + O_2 \rightarrow 2\,SO_3$, Combination; (3) $SO_3 + H_2O \rightarrow H_2SO_4$, Combination

72. $H_2SO_4 + CaCO_3 \rightarrow CaSO_4 + H_2CO_3$; $H_2SO_4 + CaCO_3 \rightarrow CaSO_4 + H_2O + CO_2$

73. $4\,Ag + 2\,H_2S + O_2 \rightarrow 2\,Ag_2S + 2\,H_2O$

74. $S + 3\,F_2 \rightarrow SF_6$ *or* $S_8 + 24\,F_2 \rightarrow 8\,SF_6$

$S + Cl_2 \rightarrow SCl_2$ *or* $S_8 + 8\,Cl_2 \rightarrow 8\,SCl_2$

$2\,S + Br_2 \rightarrow S_2Br_2$ *or* $S_8 + 4\,Br_2 \rightarrow 4\,S_2Br_2$

75. $WO_3 + 3\,H_2 \rightarrow W + 3\,H_2O$, single-replacement redox

76. $Ba(OH)_2(aq) + 2\,HNO_3(aq) \rightarrow Ba(NO_3)_2(aq) + 2\,H_2O(\ell)$

78. $HNH_2SO_3(aq) + KOH(aq) \rightarrow H_2O(\ell) + KNH_2SO_3(aq)$

9 Chemical Change

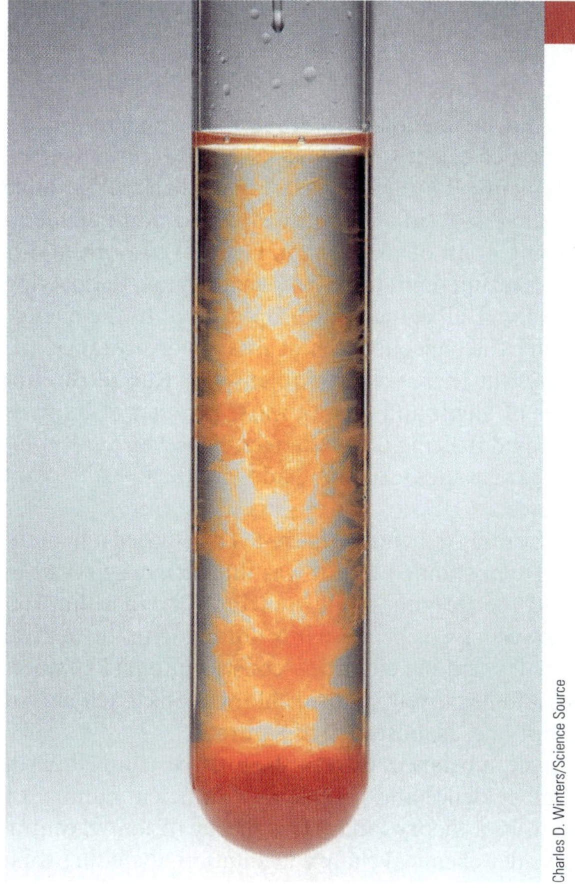

Charles D. Winters/Science Source

◀ This photograph shows the mixing of solutions of iron(III) nitrate and sodium hydroxide. If you were to write the equation for the reaction, as you learned how to do in Chapter 8, you would get this:

$$Fe(NO_3)_3(aq) + 3\ NaOH(aq) \rightarrow Fe(OH)_3(s) + 3\ NaNO_3(aq)$$

Is this an accurate and real equation? Real, yes—and the simplest equation you can use to calculate quantity relationships among the species in the reaction. Accurate—well, not exactly. In this chapter, you will learn why this equation is not altogether correct and how to write an equation that tells exactly what species take part in a chemical change and nothing more.

CHAPTER CONTENTS

In Chapter 8, you learned how to write chemical equations for reactions, some of which occur in water solutions. These equations, however, are not entirely accurate in describing the ions in the solutions and the chemical changes that occur. Net ionic equations identify precisely the species in the solution that experience a change and the species that are produced. In this chapter, you will refine and improve your equation-writing skill so that you can more accurately describe aqueous solutions and chemical change.

Our focus in this chapter will be primarily on chemical change that occurs with substances dissolved in water, or aqueous solutions. Chemists define a **solution** as a homogeneous mixture. ▶ An aqueous solution, therefore, is a homogeneous mixture of substances dissolved in water.

◀ **P/Review** Solutions are discussed in more detail in Chapter 16.

9.1 Why Is Salt Solution Different from Sugar Solution?

You add a teaspoon of sugar to a glass of water. The solid completely dissolves, and the sugar solution that forms looks indistinguishable from the water before the sugar was added. Where did the now-invisible sugar go? If you taste the solution, you immediately know that the sugar is in it. The sweet taste of the solution is the same as when you put solid sugar in your mouth. To explain this phenomenon at the particulate level, it is logical to hypothesize that macroscopically visible solid sugar crystals must separate into particles too small to be seen when in solution.

If we perform the same activity with salt, the result is similar. The salt solution looks the same as water, but the taste of the solution provides evidence that the salt is in the solution. Again, a logical hypothesis is that when dissolved, the macroscopically visible solid salt crystals form particles that human eyes can't see due to their tininess.

Up until 1878, it was logical to believe that sugar particles and salt particles in solution were similar. After all, at the macroscopic level, their appearance and behavior is similar. But then the French scientist François-Marie Raoult published a description of puzzling measurements. Raoult discovered that when solid soluble substances—those that dissolve—are dissolved in water, the freezing point temperature is always lower than the freezing point of plain water. He went to work to discover the factors that affected this freezing point lowering phenomenon, discovering that the concentration of the dissolved substance was proportional to the change in freezing temperature. The greater the concentration of dissolved substance, the greater the freezing point change. But there was a problem that Raoult did not understand. Equal concentrations of sugar and urea (a waste product of humans that is dissolved in urine) yielded equal freezing point changes. However, a salt solution equal in concentration to a sugar (or urea) solution had twice the freezing point lowering! What caused this difference?

In 1884, the Swedish scientist Svante Arrhenius (**Figure 9.1**) proposed a hypothesis to explain Raoult's findings. He postulated that it was the *number of particles* dissolved in solution that affected the freezing point lowering. The fact that salt caused twice the freezing point lowering was because it is not made up of neutral molecules, as are sugar and urea. Salt—sodium chloride—consists of equal numbers of positively charged sodium ions and negatively charged chloride ions. Each unit of sodium chloride is two particles, not one, as illustrated in **Figure 9.2**.

Arrhenius's hypothesis that ionic substances exist in the form of charged particles has been supported by enough evidence and it explains a sufficient number of observations that it is considered to be a theory today. This theory of ions is central in this chapter, where you will consider chemical change in solution, including solutions of ionic compounds.

Pictorial Press/Pictorial Press Ltd/Alamy Stock Photo

Figure 9.1 The Swedish scientist Svante Arrhenius (1859–1927) was awarded the 1903 Nobel Prize for Chemistry for proposing the hypothesis that ionic compounds exist in the form of charged particles.

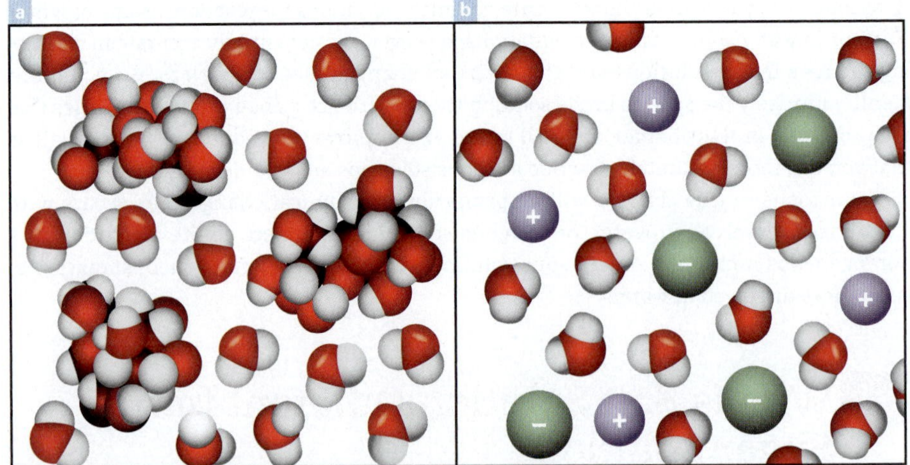

Figure 9.2 A particulate-level view of molecular versus ionic compounds in solution. (a) Sugar exists in the form of neutral molecules. When dissolved in water, the sugar molecules spread out among the water molecules. (b) Sodium chloride exists in the form of positively charged sodium ions and negatively charged chloride ions. When dissolved in water, the ions spread out among the water molecules.

9.2 Electrolytes and Solution Conductivity

Goal 1 Distinguish between strong electrolytes, weak electrolytes, and nonelectrolytes.

Consider the apparatus shown in **Figure 9.3**. It is made of two metal strips called **electrodes**, an electrical cord attached to a battery, and a light bulb. When the two electrodes touch, the light bulb glows. If the electrodes are not in contact, electrons cannot flow through the apparatus, and the bulb does not light.

Suppose we place the apparatus in a solution of a molecular compound dissolved in water, as in **Figure 9.4(a)**, which shows drinking alcohol (ethanol) dissolved in water. Nothing happens. But if the liquid is a solution of an ionic compound, the bulb glows brightly, as in **Figure 9.4(b)**, which shows copper(II) chloride dissolved in water. ▶

◀ **P/Review** Formulas of ionic compounds were introduced in Section 6.7. Chemical compounds are electrically neutral. A net zero charge is achieved by combining cations and anions in numbers such that positive and negative charges are balanced.

What is it about the solution of ions that allows it to conduct an electric current? Why doesn't the solution of a molecular compound conduct electricity? When the two metal strips in the conductivity apparatus touch, electrons have a path to flow through the filament in the light bulb and thus make it glow. When the strips are separated, the electrons cannot flow. From this, we can reason that when the apparatus is placed in a solution containing ions, there must be a way for electrons to move from one metal strip to the other. The solution of a molecular substance does not have this property.

A solid ionic substance is made up of positively and negatively charged ions that are held in fixed positions. ▶ These ions become free to move when they are dissolved in water. It is the movement of ions that makes up an electric current in a solution—a process called **electrolysis**. In fact, the ability of a solution to conduct electricity is regarded as positive evidence that it contains ions (**Figure 9.5**).

◀ **P/Review** Figure 2.7 (Section 2.3) shows that solid particles are held in fixed position relative to one another, whereas liquid particles are free to move. Ions in an ionic solid are arranged in crystals, as shown in Figures 12.6 and 12.7 (Section 12.3). When the solid is dissolved in water, the ions are released from the crystal lattice, and they are free to move among the water molecules (as described in Section 16.4).

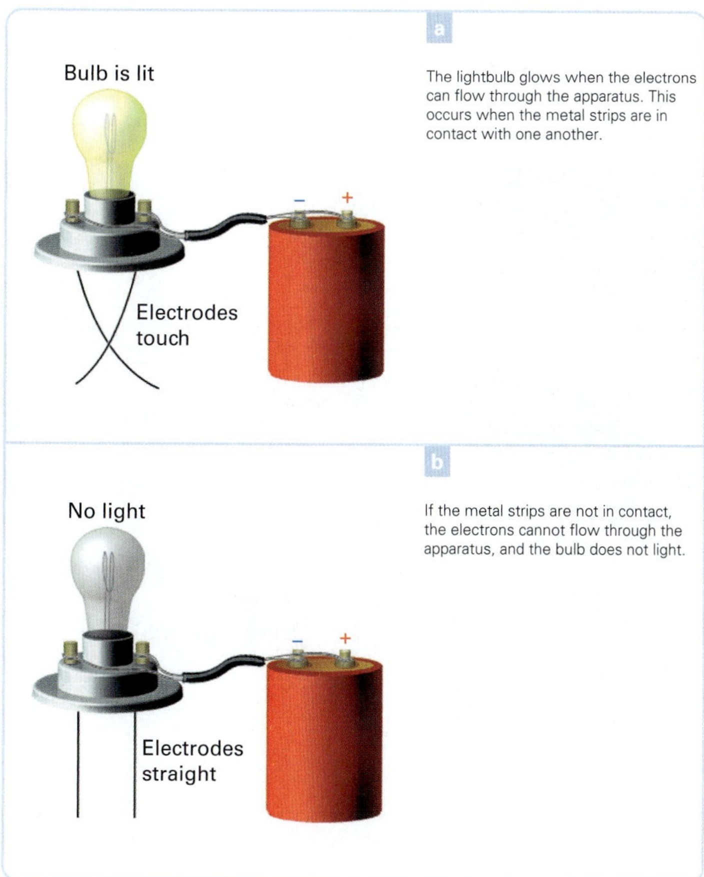

a
Bulb is lit
The lightbulb glows when the electrons can flow through the apparatus. This occurs when the metal strips are in contact with one another.
–
+
Electrodes touch
b
No light
If the metal strips are not in contact, the electrons cannot flow through the apparatus, and the bulb does not light.
–
+
Electrodes straight

Figure 9.3 A light bulb conductivity apparatus.

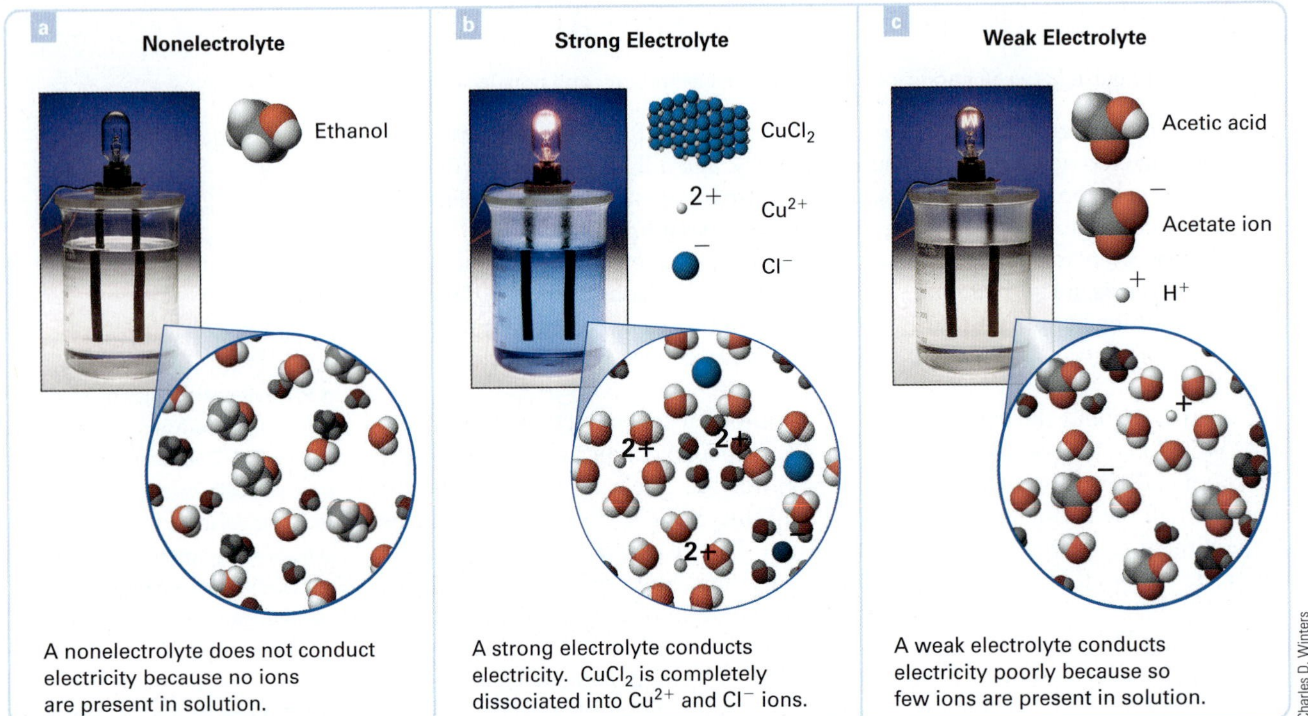
a
Nonelectrolyte
Ethanol
A nonelectrolyte does not conduct electricity because no ions are present in solution.
b
Strong Electrolyte
$CuCl_2$
Cu^{2+}
Cl^-
A strong electrolyte conducts electricity. $CuCl_2$ is completely dissociated into Cu^{2+} and Cl^- ions.
c
Weak Electrolyte
Acetic acid
Acetate ion
H^+
A weak electrolyte conducts electricity poorly because so few ions are present in solution.
Charles D. Winters

Figure 9.4 Nonelectrolytes, strong electrolytes, and weak electrolytes. The liquid in the beaker is used to "close" the electrical circuit. (a) If the liquid is pure water or a solution of molecular compounds, the bulb does not light. Solutes whose solutions do not conduct electricity are nonelectrolytes. (b) A soluble ionic salt is a strong electrolyte because its solution makes the bulb glow brightly. (c) Some solutes are called weak electrolytes because their solutions conduct poorly, and the bulb glows dimly.

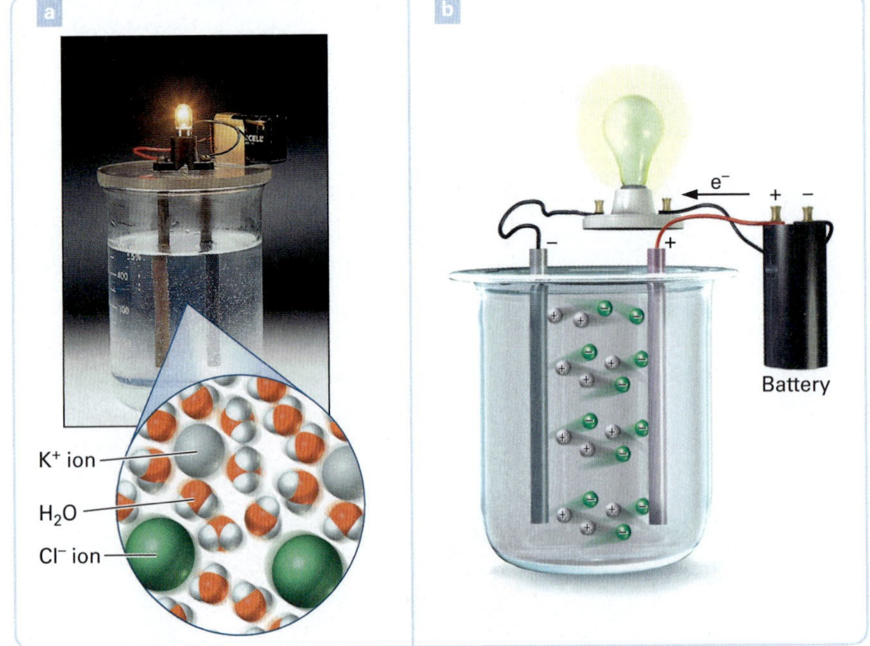
a
K^+ ion
H_2O
Cl^- ion
b
e^-
Battery
Charles D. Winters

Figure 9.5 Conductivity in an ionic solution. (a) The illuminated bulb shows that this solution, made by dissolving the solid ionic compound potassium chloride in water, is indeed a conductor. If a solution conducts electricity, it is tangible evidence that mobile ions are present. At the particulate level, we model the solution showing potassium ions and chloride ions free to move among the water molecules. (b) A simplified model of the electrical circuit. Positively charged ions are attracted to the negatively charged metal strip. Similarly, negatively charged ions move to the positively charged metal strip. Note how electrons flow through the system.

Thinking About Your Thinking

Mental Models

Forming a mental model of the particulate-level process that occurs in a solution is a goal for you to work toward in Section 9.2. You need to incorporate Figures 9.3–9.5 into this mental model. The key part of forming your model is to make Figure 9.5 into a dynamic, moving, three-dimensional mental image. Ions must be present in solution in order for the solution to conduct electricity. The current is carried by ions. The word *ion* comes from a Greek word meaning "to go." These charged particles move when an external electrical force is applied.

Solutions of molecular compounds do not conduct electricity because no ions are present. The ethanol and water molecules are electrically neutral. Since they are neutral, they do not move toward either metal strip. Even if they did move, there would be no current because the molecules have no charge.

A substance whose solution is a good conductor is called a **strong electrolyte**. Copper(II) chloride is an example. Ethanol, whose solution is a nonconductor, is an example of a **nonelectrolyte**. A substance can also be a **weak electrolyte**. Their solutions conduct electricity, but poorly, permitting only a dim glow of the light bulb, as in **Figure 9.4(c)**, which shows acetic acid dissolved in water. The poor conductivity of weak electrolytes is the result of low concentrations of ions in their solutions. The term *electrolyte* is also applied generally to the solution through which a current passes. The acid solution in an automobile battery is an electrolyte in this sense.

✓ Target Check 9.1

Three compounds, P, G, and N, are dissolved in water, and the solutions are tested for electrical conductivity. Solution P is a poor conductor, G is a good conductor, and N does not conduct. Classify P, G, and N as electrolytes (strong, weak, or non-), and state the significance of the conductivities of their solutions.

9.3 Solutions of Ionic Compounds

Goal 2 Given the formula of an ionic compound (or its name), write the formulas of the ions present when it is dissolved in water.

When an ionic compound dissolves in water, its solution consists of water molecules and ions. The ions are identified simply by separating the compound into its ions. When sodium chloride dissolves, the solution consists of water molecules, sodium ions, and chloride ions (**Figure 9.6**):

$$NaCl(s) \xrightarrow{H_2O} Na^+(aq) + Cl^-(aq)$$

The H_2O above the arrow indicates that the reaction occurs in the presence of water. We will be concerned primarily with reactions that occur in water in this chapter, so from this point onward we simply assume the presence of water molecules when we discuss solutions.

If the dissolved compound is copper(II) chloride, the solution contains copper(II) ions and chloride ions (**Figure 9.7**):

$$CuCl_2(s) \rightarrow Cu^{2+}(aq) + 2\,Cl^-(aq)$$

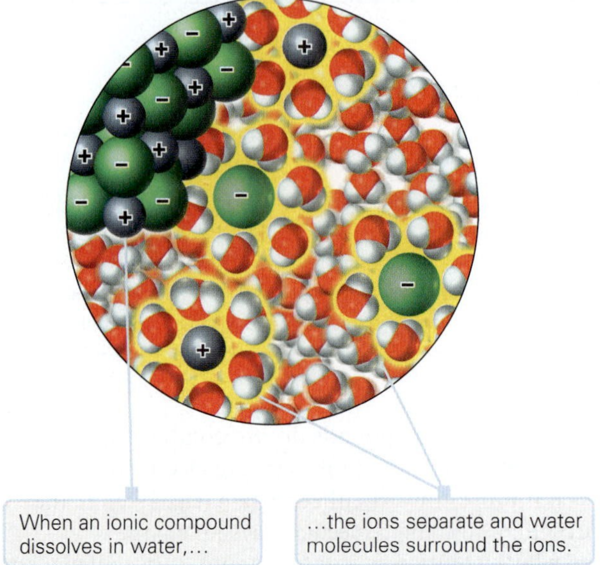

When an ionic compound dissolves in water,...
...the ions separate and water molecules surround the ions.

Figure 9.6 Sodium chloride dissolves in water. The solution that results consists of water molecules, sodium ions, and chloride ions.

Notice that no matter where a chloride ion comes from, its formula is always Cl^-, never Cl_2^- or Cl_2^{2-}. The subscript after an ion in a formula, such as the 2 in $CuCl_2$, tells us how many ions are present in the formula unit. This subscript is not part of the ion formula.

Thinking About Your Thinking

Mental Models

We first discussed forming a mental model of the particulate-level process that occurs when a solid ionic solute is dissolved in water in Section 9.2. That model is applied here in Section 9.3. This section asks you to use symbols—chemical formulas of ions—to describe the species in solution when an ionic compound dissolves. These symbols are a written representation of the composition of the solution at the particulate level. Be sure that your mental model matches the symbols you use to describe the particles in solution. Figures 9.6 and 9.7 will help you form your mental models.

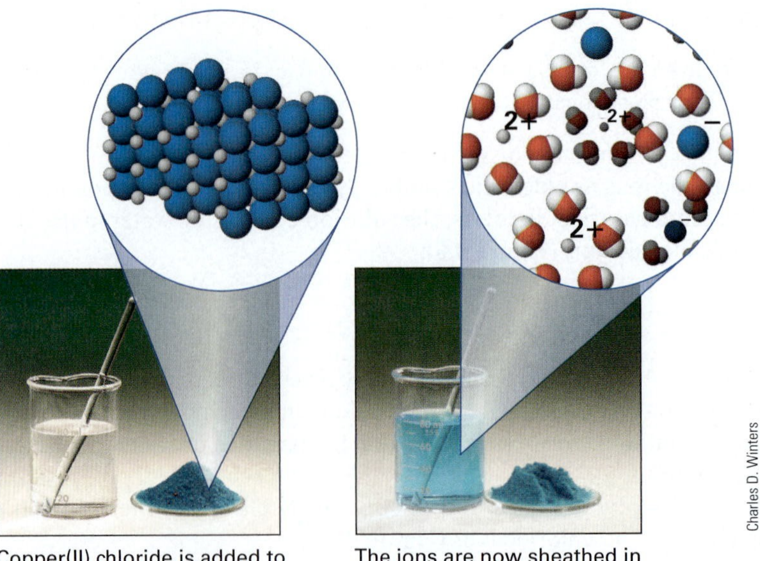

2+
2+
2+
Charles D. Winters
Copper(II) chloride is added to water. Interactions between water and the Cu^{2+} and Cl^- ions allow the solid to dissolve.
The ions are now sheathed in water molecules.

Figure 9.7 Copper(II) chloride is dissolved in water. The solution is made up of water molecules, copper(II) ions, and chloride ions. The copper(II) ions and chloride ions are in a 1:2 ratio.

In this chapter, you should include state symbols for all species in equations because it helps in writing net ionic equations. Remember also to include the charge every time you write the formula of an ion.

Active Example 9.1 Solution Inventory I

For the following ionic compounds, write the formulas of all species in solution by writing their dissolving equations: sodium hydroxide, potassium sulfate, ammonium carbonate.

Think Before You Write Polyatomic and monatomic ions are handled in exactly the same way in writing the formulas of ions in solution.

Answers *Cover the left column with your cut-out shield. Reveal each answer only after you have written your own answer in the right column.*

$NaOH(s) \rightarrow Na^{+}(aq) + OH^{-}(aq)$ $K_2SO_4(s) \rightarrow 2\ K^{+}(aq) + SO_4^{2-}(aq)$ $(NH_4)_2CO_3(s) \rightarrow 2\ NH_4^{+}(aq) + CO_3^{2-}(aq)$ *In $(NH_4)_2CO_3$, notice that the subscript outside the parentheses tells us how many ammonium ions are present, and the subscript 4 inside the parentheses is part of the polyatomic ion formula.*	Remember to include the charge of each ion and the (aq) state symbol. $NaOH(s) \rightarrow$ $K_2SO_4(s) \rightarrow$ $(NH_4)_2CO_3(s) \rightarrow$
You improved your understanding of the contents of a solution of an ionic compound.	What did you learn by working this Active Example?

Practice Exercise 9.1

Write the dissolving equation for each compound: barium chlorate, cobalt(II) iodide, magnesium nitrite.

Active Example 9.2 Solution Inventory II

Write the formulas of the ions present in solutions of the following ionic compounds *without* writing the dissolving equations: magnesium sulfate, calcium nitrate, aluminum bromide, iron(III) sulfate.

Think Before You Write This question is similar to Active Example 9.1. It asks simply for the formulas of the dissolved ions without writing an equation. This is how you will write these ion formulas later.

Answers *Cover the left column with your cut-out shield. Reveal each answer only after you have written your own answer in the right column.*

$MgSO_4$: $Mg^{2+}(aq) + SO_4^{2-}(aq)$ $AlBr_3$: $Al^{3+}(aq) + 3\ Br^{-}(aq)$ $Ca(NO_3)_2$: $Ca^{2+}(aq) + 2\ NO_3^{-}(aq)$ $Fe_2(SO_4)_3$: $2\ Fe^{3+}(aq) + 3\ SO_4^{2-}(aq)$	Be sure to show the relative number of each kind of ion released by a formula unit—the coefficient if you were writing the dissolving equation. Also remember state symbols. $MgSO_4$: $AlBr_3$: $Ca(NO_3)_2$: $Fe_2(SO_4)_3$:
You improved your understanding of the contents of a solution of an ionic compound.	What did you learn by solving this Active Example?

Practice Exercise 9.2

Write the formulas of the ions present and the ratio in which they will occur in solutions of the following ionic compounds without writing the dissolving equations: nickel(II) chloride, lithium carbonate, lead(II) nitrite, ammonium phosphate.

9.4 Strong and Weak Acids

Goal 3 Explain why the solution of an acid may be a good conductor or a poor conductor of electricity.

4 Given the formula or the name of a soluble acid, write the major and minor species present when it is dissolved in water.

▶ **P/Review** Lone hydrogen ions do not exist by themselves in solution. H^+ is always bound to a variable number of water molecules in the form $H^+ \cdot (H_2O)_x$. The hydrated hydrogen ion is usually represented by the hydronium ion, H_3O^+, which is one hydrogen ion combined with one water molecule. Lewis diagrams (Section 13.2) can be used to show the ionization of an acid and the bonds (Section 12.4) broken and formed in the process:

$$\mathrm{H{:}\ddot{O}{:}\,(H) + H{:}\ddot{X}{:} \longrightarrow [H{:}\ddot{O}{:}H\,(H)]^+ + [{:}\ddot{X}{:}]^-}$$

water acid hydronium ion anion

See Section 6.9 and Chapter 17 for additional information about hydronium ions and their formation.

▶ **P/Review** A reversible reaction is one in which the products, as an equation is written, change back into the reactants. Reversibility is indicated by a double arrow, one pointing in each direction, $\rightleftharpoons$. Reversible changes are introduced in Section 8.8. Other reversible changes involve liquid–vapor equilibria (Section 15.5) and the formation of solutions (Section 16.4); others are found throughout Chapter 18 on chemical equilibrium.

An **acid** is a substance that has a proton that can be removed by a water molecule when in a water solution. ◀ Its formula is HX, H_2X, or H_3X, where X is any negatively charged ion produced when the acid ionizes:

$$HX + H_2O \rightarrow H_3O^+ + X^-$$

X may be a monatomic ion, as when HCl ionizes and leaves Cl^-:

$$HCl + H_2O \rightarrow H_3O^+ + Cl^-$$

X may contain oxygen, as does the NO_3^- from HNO_3:

$$HNO_3 + H_2O \rightarrow H_3O^+ + NO_3^-$$

X may be an ion that contains hydrogen. Acids that have two or more ionizable hydrogens release them in steps. For sulfuric acid, this is:

$$H_2SO_4 + H_2O \rightarrow H_3O^+ + HSO_4^-$$

$$HSO_4^- + H_2O \rightarrow H_3O^+ + SO_4^{2-}$$

HSO_4^- is an example of an ion that behaves as an acid.

Carbon-containing acids usually contain hydrogen that is not ionizable. Acetic acid, $HC_2H_3O_2$, for example, ionizes to H^+ and the acetate ion, $C_2H_3O_2^-$:

$$HC_2H_3O_2 + H_2O \rightleftharpoons H_3O^+ + C_2H_3O_2^- \blacktriangleleft$$

Do not be concerned if the anion (negatively charged ion) is not familiar. It will behave just like the anion from any other acid.

Although protons do not exist as independent species in water solutions, the chemical reactivity of the hydronium ion is due to the proton. Accordingly, we can simplify reaction equations involving acids by subtracting a water molecule from each side of the equation. The general reaction of an acid is as follows:

$$HX + H_2O \rightarrow H_3O^+ + X^- \quad \textit{or} \quad HX + H_2O \rightarrow H^+\text{–}H_2O + X^-$$

This becomes

$$HX \rightarrow H^+ + X^-$$

This simplification allows us to focus on the reactivity of the proton—the H^+ ion—as it interacts with another molecule. We will use this abbreviated form for acids when we are considering a reaction of the acid and not just the behavior of the acid itself.

Acids are classified as strong or weak, depending on the extent of the reaction of the acid molecules with water molecules. In a dilute solution of a **strong acid**, almost all of the molecules of the original compound react; very few remain as unionized molecules. We say that the **major species** present are ions and the **minor species** are unionized molecules. Consequently, strong acids are excellent conductors of electricity (**Figure 9.8**[*left*]).

A **weak acid**, on the other hand, is only slightly ionized in solution. The major species in a weak acid are unionized molecules. Some ions are present, however; they are the minor species in solution. Because of the low concentration of ions, weak acids are poor conductors of electricity (Figure 9.8[*right*]).

The differences between strong acids and weak acids are illustrated by comparing the acids formed by two very similar compounds, HCl and HF. **Table 9.1** makes this comparison. Notice the four facts listed at the end of the table.

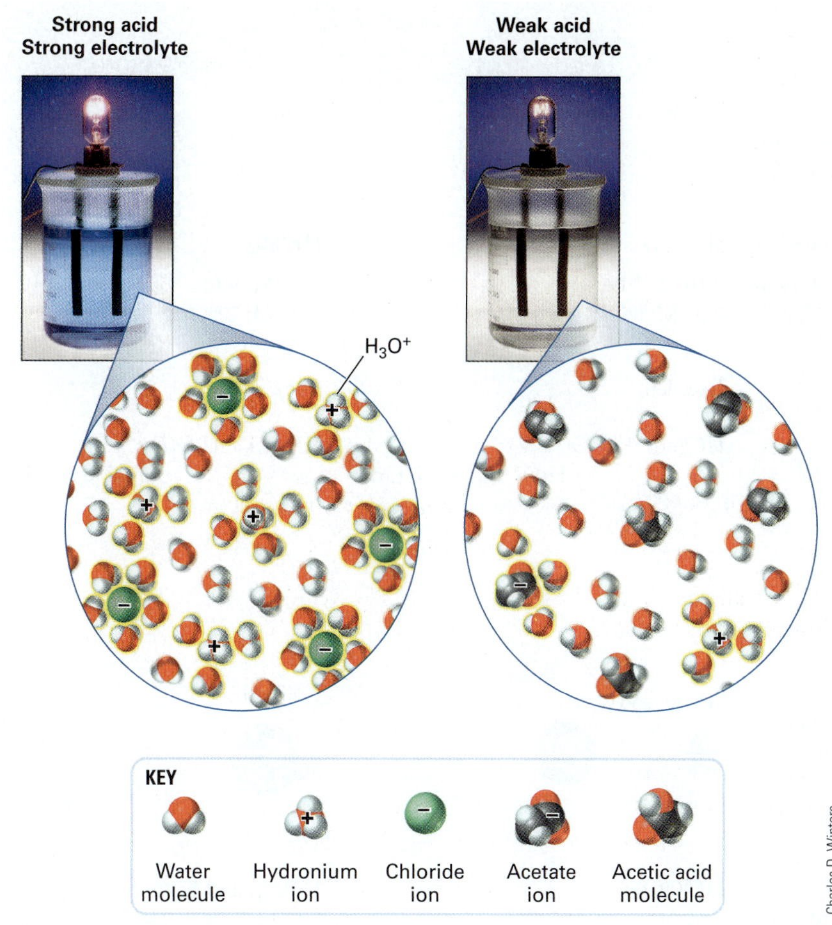

Figure 9.8 Strong and weak acids at the macroscopic and particulate levels. Strong acids are strong electrolytes because the major species in solution are ions. For example, a hydrochloric acid solution primarily consists of water molecules, hydronium ions, and chloride ions. Weak acids are weak electrolytes because the major species in solution are neutral molecules. Acetic acid is an example of a weak acid. Its solution primarily consists of water molecules and acetic acid molecules. Some acid molecules ionize, however, providing for the weak conductivity of the solution.

There are **seven common strong acids**. You must memorize their names and formulas. It helps to group them into three classifications:

Seven Common Strong Acids

Two Well-Known Acids	Three Group 7A/17 Acids	Two Chlorine Oxoacids
Nitric acid, $HNO_3(aq)$	Hydrochloric acid, $HCl(aq)$	Chloric acid, $HClO_3(aq)$
Sulfuric acid, $H_2SO_4(aq)$	Hydrobromic acid, $HBr(aq)$	Perchloric acid, $HClO_4(aq)$
	Hydroiodic acid, $HI(aq)$	

To decide whether an acid is strong or weak, ask yourself, "Is it one of the seven strong acids?" If it is listed here, it is strong. If it is not one of these, it is weak.

Learn It Now! *The primary focus of the remainder of this chapter is learning to write net ionic equations. To accomplish this, you need to have memorized the seven common strong acids. Don't put off this task. Learn It Now!*

The dividing line between strong and weak acids is arbitrary, and some acids are marginal in their classifications. Sulfuric acid is definitely strong in its first ionization step but marginal in the second. At present, we will avoid questions involving the classification of acids on the borderline between strong and weak.

Note that the terms *strong* and *weak* do *not* refer to the corrosiveness of the acid or its reactivity. Hydrofluoric acid is weak, but it cannot be stored in glass bottles because it will react with the glass. The terms *do* refer to the extent to which the acid ionizes in water solution.

Table 9.1 Strong and Weak Acids

	Hydrochloric Acid	Hydrofluoric Acid
Strength of Acid More than half the molecules of a strong acid are ionized. More than half the molecules in a weak acid are in molecular form.	Strong	Weak
Formula as it is written in a conventional equation	HCl(aq)	HF(aq)
Percentage ionization in 0.1- and 0.01-M solutions M refers to concentration. (In a 0.1-M solution, 1 liter of solution contains 0.1 mole of acid.)	0.10 M: 79% 0.01 M: 99+%	0.10 M: 9.6% 0.01 M: 22%
Electrical conductivity	Excellent	Poor
Ionization equation The relative lengths of the arrows are a visual indication of the extent of ionization. If the longer arrow points to the right, it means more than half of the acid molecules are separated into ions. If the longer arrow points left, more than half of the acid molecules are present as unionized molecules.	$HCl(aq) + H_2O(\ell) \rightleftarrows H_3O^+(aq) + Cl^-(aq)$	$HF(aq) + H_2O(\ell) \leftrightarrows H_3O^+(aq) + F^-(aq)$
Major species These are the species present in greater abundance in the reaction; they can be found on the side of the equation pointed to by the longer arrow. Formulas of the major species are written in the net ionic equations you are about to write.	$H^+(aq)$ and $Cl^-(aq)$	HF(aq)
Minor species These are the species present in lesser abundance in the reaction; they can be found on the side of the equation pointed to by the shorter arrow. Even though these species are present in a reaction vessel, they do not take part in a reaction in the form shown. Therefore, formulas of the minor species do not appear in the net ionic equation.	HCl(aq)	$H^+(aq)$ and $F^-(aq)$

- HCl(aq), which is 79% ionized in a 0.10-M solution (0.10 mole of HCl in 1 liter of solution), is a strong acid, whereas 0.10-M HF, at 9.6% ionized, is a weak acid.
- The double arrows in the ionization equations show that the ionization process is reversible. The longer arrow pointing to the right indicates that it is much more likely for HCl molecules to break into ions than for the ions to combine and form dissolved HCl molecules.
- The major species in the strong acid are the ions, but in the weak acid the major species is the acid molecule. The formulas of the major species appear in the net ionic equation.
- The minor species are present in the reaction vessel but do not participate in the reaction as shown. Minor species are not included in a net ionic equation.

Active Example 9.3 Solution Inventory III

Write the formulas of the major species in solutions of the following acids: nitrous acid, propanoic acid ($HC_3H_5O_2$), chloric acid, hydroiodic acid.

Think Before You Write If you have the seven common strong acids memorized, and if you understand the difference in how to write the major species for strong versus weak acids, you are ready to complete the Active Example.

Answers *Cover the left column with your cut-out shield. Reveal each answer only after you have written your own answer in the right column.*

Nitrous acid: $HNO_2(aq)$ Propanoic acid, $HC_3H_5O_2$: $HC_3H_5O_2(aq)$ Chloric acid: $H^+(aq) + ClO_3^-(aq)$ Hydroiodic acid: $H^+(aq) + I^-(aq)$ *Chloric acid and hydroiodic acid are in the group of seven common strong acids that you have memorized. Nitrous acid and propanoic acid are not in the group of seven strong acids.*	Write each given species as the major species that occur in solution. Use $H^+(aq)$ instead of $H_3O^+(aq)$ as the ion formed by the ionization of an acid. Include state symbols.  Nitrous acid: Propanoic acid, $HC_3H_5O_2$: Chloric acid: Hydroiodic acid:
You improved your understanding of the contents of a solution of an ionic compound.	What did you learn by working this Active Example?

Practice Exercise 9.3

Write the formulas of the major species in solutions of the following acids: hydrochloric acid, hydrosulfuric acid, perchloric acid, hypochlorous acid.

We can now summarize the major species in solutions as they have been described in this section and the last. ▶

The major species in solution are the species that appear in net ionic equations, which we examine in the remainder of the chapter.

a summary of... **Identifying the Major Species in Solution**

1. **Ions** are the major species in the solutions of two kinds of substances:
   - All soluble ionic compounds
   - The seven strong acids
2. **Neutral molecules** are the major species in solutions of everything else, primarily:
   - Compounds with all nonmetal atoms (except strong acids)
   - Weak acids
   - Water

9.5 Net Ionic Equations: What They Are and How to Write Them

Goal 5 Distinguish among conventional, total ionic, and net ionic equations.

In Section 8.10, we showed how to write the double-replacement equation for a precipitation reaction between solutions of two ionic compounds. Such a reaction occurs when a solution of lead(II) nitrate is added to a solution of sodium chloride:

$$Pb(NO_3)_2(aq) + 2\,NaCl(aq) \rightarrow PbCl_2(s) + 2\,NaNO_3(aq) \qquad \textbf{(9.1)}$$

In this chapter we call this kind of equation a **conventional equation**.

A conventional equation serves many useful purposes, including its essential role in understanding quantitative relationships in chemical change. ▶ However, it falls short in describing precisely the reaction that occurs in water solution. Usually, it does not describe the form of the reactants or products correctly. Rarely does it accurately describe the chemical changes that occur.

◀ **P/Review** The quantitative relationships between substances involved in a chemical reaction are explored in Chapter 10.

The shortcomings of a conventional equation are illustrated by this fact: The solutions that react in Equation 9.1 contain no substances with the formulas $Pb(NO_3)_2$ or NaCl. Actually present are the major species, Pb^{2+} and NO_3^- in one

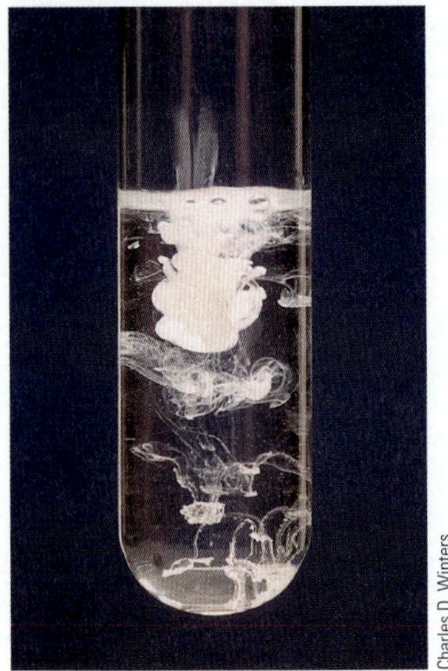
Charles D. Winters

Figure 9.9 Precipitation of lead(II) chloride. When clear, colorless solutions of lead(II) nitrate and sodium chloride are mixed, white lead(II) chloride can be seen precipitating from the solution. Nitrate ions and sodium ions remain in solution.

solution and Na^+ and Cl^- in the other. The conventional equation doesn't tell you that. Nothing with the formula $NaNO_3$ is formed in the reaction. The Na^+ and NO_3^- ions are still there in solution after the reaction. The conventional equation kept that a "secret," too. The only substance in Equation 9.1 that is *really there* is solid lead(II) chloride, $PbCl_2(s)$. If you perform the reaction, you can see the precipitate (**Figure 9.9**).

To write an equation that describes the reaction in Equation 9.1 more accurately, we replace the formulas of the dissolved substances with the major species in solution. This produces the **total ionic equation**:

$$Pb^{2+}(aq) + 2\ NO_3^-(aq) + 2\ Na^+(aq) + 2\ Cl^-(aq) \rightarrow PbCl_2(s) + 2\ Na^+(aq) + 2\ NO_3^-(aq) \quad \textbf{(9.2)}$$

Notice that in order to keep the total ionic equation balanced, we must include the coefficients from the conventional equation. One formula unit of $Pb(NO_3)_2$ gives one Pb^{2+} ion and two NO_3^- ions, two formula units of NaCl give two Na^+ ions and two Cl^- ions, and two formula units of $NaNO_3$ give two Na^+ ions and two NO_3^- ions.

A total ionic equation tells more than just what species take part in a chemical change. It includes **spectator ions**, or simply **spectators**. A spectator is an ion that is present at the scene of a reaction but experiences no chemical change. It appears on both sides of the total ionic equation. $Na^+(aq)$ and $NO_3^-(aq)$ are spectators in Equation 9.2. To change a total ionic equation into a **net ionic equation**, you remove the spectators:

$$Pb^{2+}(aq) + 2\ Cl^-(aq) \rightarrow PbCl_2(s) \quad \textbf{(9.3)}$$

A net ionic equation indicates exactly what chemical change takes place, and nothing else.

In Equations 9.1–9.3 you have the three steps to follow in writing a net ionic equation. They are:

how to... Write a Net Ionic Equation

Step 1: Write the conventional equation, including state symbols—(g), (ℓ), (s), and (aq). Balance the equation.

$$Pb(NO_3)_2(aq) + 2\ NaCl(aq) \rightarrow PbCl_2(s) + 2\ NaNO_3(aq)$$

Step 2: Write the total ionic equation by replacing each aqueous (aq) substance that is a strong acid or a soluble ionic compound with its major species. *Do not separate a weak acid into ions*, even though its state is aqueous (aq). Also, never change solids (s), liquids (ℓ), or gases (g) into ions. Be sure the equation is balanced in both atoms and charge. (Charge balance is discussed in more detail in the next section.)

$$Pb^{2+}(aq) + 2\ NO_3^-(aq) + 2\ Na^+(aq) + 2\ Cl^-(aq) \rightarrow PbCl_2(s) + 2\ Na^+(aq) + 2\ NO_3^-(aq)$$

Step 3: Write the net ionic equation by removing the spectators from the total ionic equation. Reduce coefficients to lowest terms, if necessary. Be sure the equation is balanced in both atoms and charge.

$$Pb^{2+}(aq) + 2\ Cl^-(aq) \rightarrow PbCl_2(s)$$

We will have you practice the net ionic equation-writing procedure in the next two Active Examples.

Active Example 9.4 Writing a Net Ionic Equation I

When nickel(II) nitrate and sodium carbonate solutions are combined, solid nickel(II) carbonate precipitates, leaving a solution of sodium nitrate. Write the conventional equation, total ionic equation, and net ionic equation for this reaction.

Think Before You Write The key to writing net ionic equations is to not try to rush the process. Follow each of the three steps in the procedure. When you write the total ionic equation, replace each aqueous substance that is a strong acid or soluble ionic compound with its major species.

Answers *Cover the left column with your cut-out shield. Reveal each answer only after you have written your own answer in the right column.*

$Ni(NO_3)_2(aq) + Na_2CO_3(aq) \rightarrow NiCO_3(s) + 2\ NaNO_3(aq)$

Step 1 in the procedure is to write the conventional equation, including state symbols. The problem statement gives you all the necessary information to complete this step.

$Ni^{2+}(aq) + 2\ NO_3^{-}(aq) + 2\ Na^{+}(aq) + CO_3^{2-}(aq) \rightarrow NiCO_3(s) + 2\ Na^{+}(aq) + 2\ NO_3^{-}(aq)$

Note how the two nitrate and two sodium ions on the left come from the compound formula unit and the two on the right result from the coefficients in the balanced equation.

The next step is to write the total ionic equation. The species that occur as ions in solution are written *as ions* rather than as ionic compounds, which is how they appear in the conventional equation. In other words, you replace each aqueous substance except weak acids with its major species. There are no weak acids in this Active Example, so separate each aqueous substance into its ions. Don't forget to account for the number of ions in the formula unit of each compound as well as the coefficients in the balanced conventional equation.

$Ni^{2+}(aq) + CO_3^{2-}(aq) \rightarrow NiCO_3(s)$

The net ionic equation tells us that nickel(II) ions from one solution combine with the carbonate ions from the other solution to form solid nickel(II) carbonate. The sodium ions and nitrate ions that were in the separate solutions remain as ions in the combined solution.

The final step in the procedure is to write the net ionic equation. This is accomplished by identifying and eliminating any spectator ions. Cross out the ions that appear on both sides of the equation, and the net ionic equation will be what remains.

Charge balance: The two plus charges from $Ni^{2+}(aq)$ cancel the two minus from $CO_3^{2-}(aq)$, leaving a net zero charge on the left to balance the zero charge on the right.

Atom/ion balance: There are one nickel and one carbonate on each side. The equation remains balanced.

Check to be sure your final equation is correct. Are the charges balanced? Are the atoms and ions balanced?

You improved your skill at writing net ionic equations.

What did you learn by working this Active Example?

Practice Exercise 9.4

Aluminum nitrate and sodium hydroxide solutions are combined. Solid aluminum hydroxide forms, and a solution of sodium nitrate remains. Write the conventional equation, total ionic equation, and net ionic equation for this reaction.

Active Example 9.5 Writing a Net Ionic Equation II

When a strip of magnesium metal is placed in an iron(III) chloride solution, a magnesium chloride solution results and solid iron also forms. Develop the net ionic equation for this reaction.

Think Before You Write Again, the key to writing net ionic equations is to work stepwise, faithfully following each step in the procedure. The most important part of the procedure is correctly identifying and replacing each aqueous substance that is a strong acid or soluble ionic compound with its major species. Review Sections 9.3 and 9.4, if necessary.

Answers *Cover the left column with your cut-out shield. Reveal each answer only after you have written your own answer in the right column.*

$3\ Mg(s) + 2\ FeCl_3(aq) \rightarrow 3\ MgCl_2(aq) + 2\ Fe(s)$ *The three Cl on the left and two Cl on the right require a total of six Cl on each side to balance the equation. The coefficients on Mg and Fe follow.*	Start with the conventional equation. Always use state symbols when your goal is to write the net ionic equation. 
$3\ Mg(s) + 2\ Fe^{3+}(aq) + 6\ Cl^-(aq) \rightarrow 3\ Mg^{2+}(aq) + 6\ Cl^-(aq) + 2\ Fe(s)$ *The six $Cl^-(aq)$ on each side come from multiplying the coefficient in the balanced equation by the number of ions in the formula unit of the compound.*	Now go for the total ionic equation. Neither of the aqueous species is a weak acid, so they separate into their ions. 
$3\ Mg(s) + 2\ Fe^{3+}(aq) \rightarrow 3\ Mg^{2+}(aq) + 2\ Fe(s)$ The six plus charge on the left balances the six plus charge on the right. Three Mg and two Fe on each side give atom balance. *In this case, the net ionic equation tells you that magnesium metal is changed to magnesium ions in solution while the opposite process occurs for iron. The iron ions that were in solution were changed to iron atoms, which collectively form solid iron. The chloride ions in the solution are unchanged while the metals react, and thus, they do not appear in the net ionic equation.*	Remove the spectator ions to arrive at the net ionic equation. Check for charge and atom balance. 
You improved your skill at writing net ionic equations.	What did you learn by working this Active Example? 

Practice Exercise 9.5

A piece of solid zinc is dropped into hydrochloric acid solution. Gaseous hydrogen bubbles out, and a solution of zinc chloride remains. Write the conventional equation, total ionic equation, and net ionic equation for this reaction.

9.6 Single-Replacement Oxidation–Reduction (Redox) Reactions

Goal 6 Given two substances that may engage in a single-replacement redox reaction and an activity series by which the reaction may be predicted, write the conventional, total ionic, and net ionic equations for the reaction that will occur, if any. ◀

▶ **P/Review** A single-replacement redox reaction is one for which the conventional equation has the form A + BX → AX + B. An uncombined element, A, appears to replace another element, B, in a compound, BX. See Section 8.9 and Table 8.3.

An iron nail is dropped into a solution of hydrochloric acid. Hydrogen gas bubbles out (**Figure 9.10**). When the reaction ends, the test tube contains a solution of iron(II) chloride. The question is, "What happened?" The answer lies in the net ionic equation.

The conventional equation (*Step 1* in the procedure from Section 9.5) is a single-replacement equation:

$$Fe(s) + 2\ HCl(aq) \rightarrow H_2(g) + FeCl_2(aq) \quad \textbf{(9.4)}$$

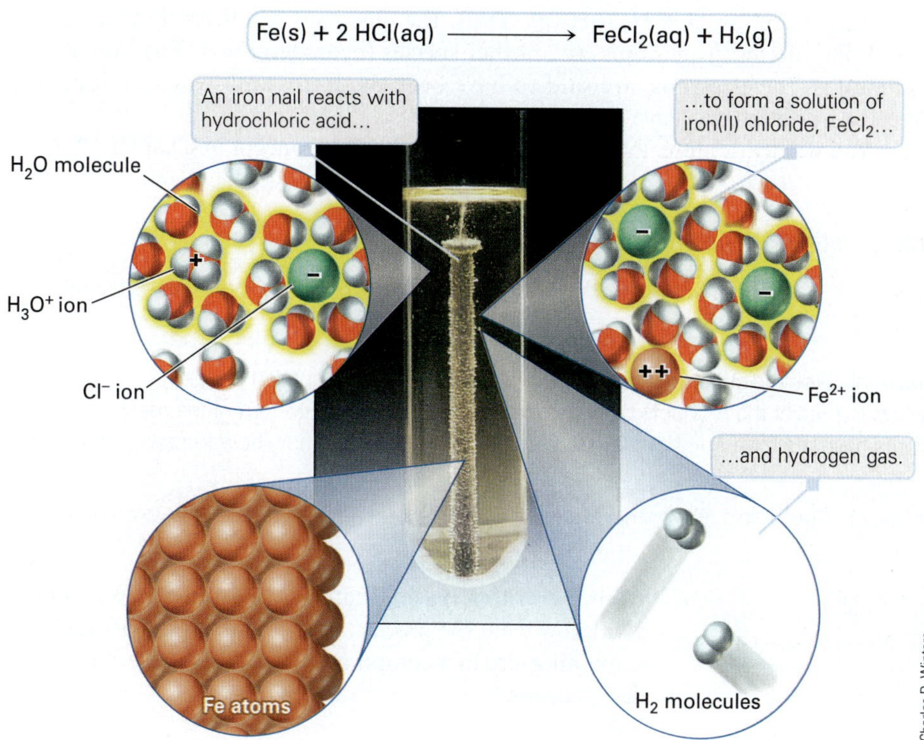
Fe(s) + 2 HCl(aq) ⟶ FeCl2(aq) + H2(g)
An iron nail reacts with hydrochloric acid...
...to form a solution of iron(II) chloride, FeCl2...
H2O molecule
H3O+ ion
Cl− ion
Fe2+ ion
...and hydrogen gas.
Fe atoms
H2 molecules
Charles D. Winters

Figure 9.10 The reaction between iron and hydrochloric acid. This is an example of an oxidation–reduction, or redox, reaction. Redox reactions are electron-transfer reactions.

HCl is a strong acid, and $FeCl_2$ is a soluble ionic compound. Both have ions as the major species in their solutions. In the total ionic equation, the formulas of these ions are written, replacing the formulas of the compounds (*Step 2* in the procedure for writing net ionic equations):

$$Fe(s) + 2\,H^+(aq) + 2\,Cl^-(aq) \rightarrow H_2(g) + Fe^{2+}(aq) + 2\,Cl^-(aq) \qquad \textbf{(9.5)}$$

The total ionic equation remains balanced in both atoms and charge. The net charge is zero on both sides.

The final step (*Step 3*) in writing a net ionic equation is to rid the total ionic equation of spectators. Chloride ions are the only spectators. Taking them away gives:

$$Fe(s) + 2\,H^+(aq) \rightarrow H_2(g) + Fe^{2+}(aq) \qquad \textbf{(9.6)}$$

This is the net ionic equation. This is what happened—and no more.

Notice that all equations are balanced. Balancing atoms is discussed in Chapter 8; balancing charge is something new. Neither protons nor electrons are created or destroyed in a chemical change, so the total charge among the products must equal the total charge among the reactants. In Equation 9.5, two plus charges in $2\,H^+(aq)$ added to two negative charges in $2\,Cl^-(aq)$ give a net zero charge on the left. Two plus charges in $Fe^{2+}(aq)$ and two negative charges in $2\,Cl^-(aq)$ on the right also total zero. The equation is balanced in charge.

Notice also that the net charge does not have to be zero on each side for the equation to be balanced. The charges must be *equal*. In Equation 9.6 the net charge is 2+ on each side.

Equation 9.6 illustrates why this is an oxidation–reduction, or redox, reaction. Redox reactions are electron-transfer reactions: Electrons are transferred from one species to another. In Equation 9.6, a neutral iron atom becomes an iron(II) ion with a 2+ charge. It does this by losing two electrons: $Fe(s) \rightarrow Fe^{2+}(aq) + 2\,e^-$. Where do the electrons go? One goes to each of two hydrogen ions so they become neutral hydrogen atoms, and the atoms combine to form a diatomic molecule:

$2\,H^+(aq) + 2\,e^- \rightarrow H + H \rightarrow H_2(g)$. Thus, the electrons are literally transferred from one species (iron atoms) to another species (hydrogen ions). The iron atoms, which lost the electrons, are said to have been oxidized, and hydrogen ions, the receivers of electrons, have been reduced.

You will learn more about the oxidation–reduction concept in Chapter 19.

Active Example 9.6 Net Ionic Equations: Redox Reactions I

A reaction occurs when a piece of zinc is dipped into a solution of copper(II) nitrate (Figure 9.11). Write the conventional, total ionic, and net ionic equations.

Think Before You Write Notice how this problem statement differs from the other net ionic equation problem statements you've seen thus far: You are being asked to predict the products of a reaction. Recognition of the pattern of the reactant formulas is key. In this case, the reactants have the form A + BX. Review Section 8.9 if you don't know how to predict the products.

Answers *Cover the left column with your cut-out shield. Reveal each answer only after you have written your own answer in the right column.*

$Zn(s) + Cu(NO_3)_2(aq) \rightarrow Zn(NO_3)_2(aq) + Cu(s)$ *In this single-replacement equation, zinc appears to replace copper in $Cu(NO_3)_2$.*	Begin by writing the conventional equation (*Step 1*). Do you know what the products are? Something must replace something else in a compound. Don't forget the state designations.
$Zn(s) + Cu^{2+}(aq) + 2\,NO_3^-(aq) \rightarrow Zn^{2+}(aq) + 2\,NO_3^-(aq) + Cu(s)$ *The equation is balanced. There is one zinc on each side, an atom on one side and an ion on the other. The same is true for copper. Finally, there are two nitrate ions on each side. The net charge is zero on each side of the equation.*	Now write the total ionic equation (*Step 2*), replacing formulas of the appropriate aqueous compounds in solution with their ions. Only dissolved ionic compounds and strong acids are divided into ions. Be sure your equation is balanced in both atoms and charge.
$Zn(s) + Cu^{2+}(aq) \rightarrow Zn^{2+}(aq) + Cu(s)$ *Removing two nitrate ions from each side yields the net ionic equation.*	Examine the total ionic equation. Are there any spectators? If so, remove them and write the net ionic equation.
Zinc and copper are balanced as before. This time the net charge is 2+ on each side of the equation. The equation is balanced in atoms and charge.	Recheck atom and charge balance to complete the process.
You improved your skill at writing net ionic equations in general and writing net ionic equations for redox reactions in particular.	What did you learn by working this Active Example?

Practice Exercise 9.6

Chlorine gas is bubbled through a sodium bromide solution, forming liquid bromine as one of the products. Write the conventional, total ionic, and net ionic equations.

If you were to place a strip of copper into a solution of zinc nitrate and go through the identical thought process, the equations would be exactly the reverse of those in Active Example 9.6. Does the reaction occur in both directions? If not, in which way does it occur? How can you tell?

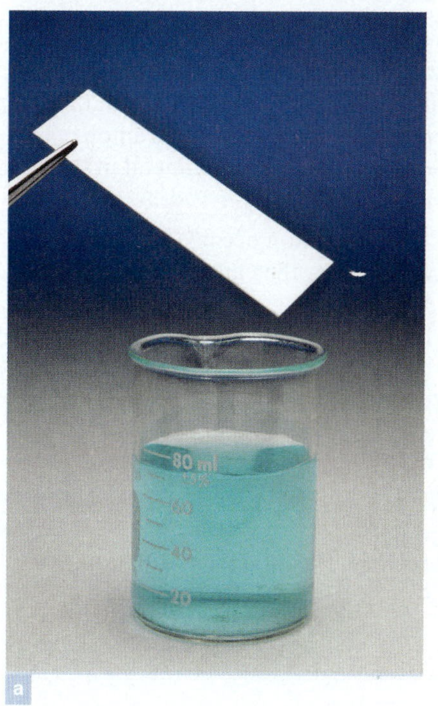

Charles D. Winters

Figure 9.11 The reaction between zinc and a solution of copper(II) nitrate. (a) The zinc strip is shiny white before the reaction. (b) After the zinc is dipped into the solution for about two seconds, the strip is covered with tiny, finely divided particles of copper that appear almost black when wet with the solution.

Technically, both reactions occur, or can be made to occur. Only the reaction in Active Example 9.6 takes place without an outside source of energy to drive the reaction. The best way to find out which of two reversible reactions occurs is to try them and see. These experiments have been done, and the results are summarized in the **activity series** in Table 9.2. Under normal conditions, any element in the table will replace the dissolved ions of any element beneath it. Zinc is above copper in the table, so zinc will replace $Cu^{2+}(aq)$ ions in the solution. Copper, being below zinc, will not replace $Zn^{2+}(aq)$ in solution.

Use Table 9.2 to predict whether or not a single-replacement redox reaction will take place. If asked to write the equation for a reaction that does not occur, write NR for "no reaction" on the product side: $Cu(s) + Zn^{2+}(aq) \rightarrow NR$.

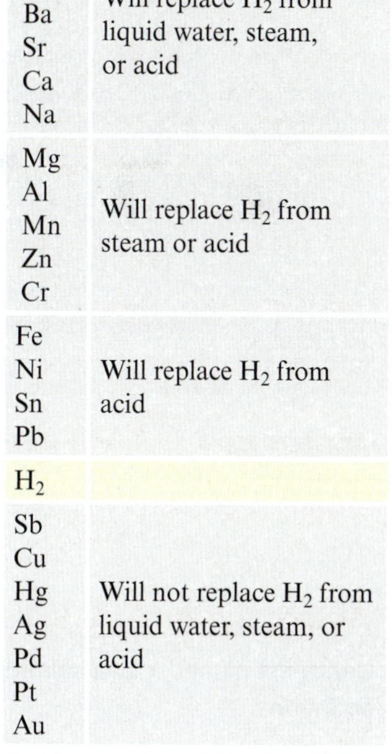

Table 9.2 Activity Series

Li K Ba Sr Ca Na	Will replace H_2 from liquid water, steam, or acid
Mg Al Mn Zn Cr	Will replace H_2 from steam or acid
Fe Ni Sn Pb	Will replace H_2 from acid
H_2	
Sb Cu Hg Ag Pd Pt Au	Will not replace H_2 from liquid water, steam, or acid

Active Example 9.7 Net Ionic Equations: Redox Reactions II

Write the conventional, total ionic, and net ionic equations for the reaction that occurs, if any, between calcium and hydrobromic acid solution.

Think Before You Write The formula of hydrobromic acid is not found in the activity series. Remember from Section 9.4 that hydrobromic acid is a strong acid, and thus, there are no HBr molecules in solution. Instead, when prompted, look in the activity series for the elemental form of the species that actually exists in solution.

Answers *Cover the left column with your cut-out shield. Reveal each answer only after you have written your own answer in the right column.*

$H^+(aq) + Br^-(aq)$, so $H^+(aq)$ is the potential reactant, and H_2 is the corresponding elemental formula.	Write the abbreviated form of the formulas of the species that are in a hydrobromic acid solution. What dissolved ion will potentially be a reactant in this reaction? What corresponding elemental formula will be in the activity series?

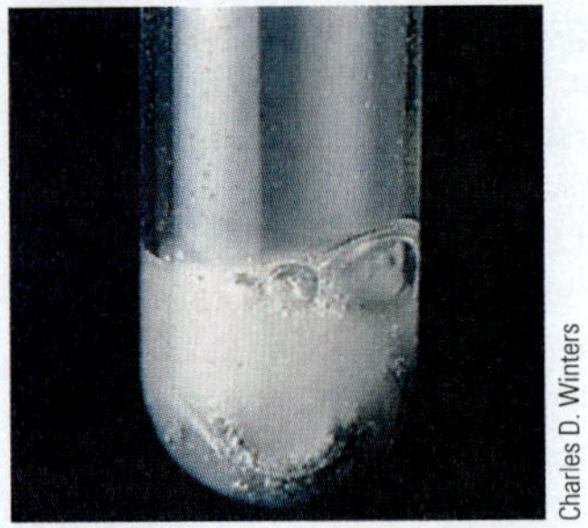

Charles D. Winters

Figure 9.12 Calcium reacts with hydrobromic acid.

Answer	Step
Yes. Calcium is above hydrogen in the activity series, so calcium ions will replace hydrogen ions in solution. *Figure 9.12 is a photograph of the reaction taking place.*	In a single-replacement reaction, if the element is a metal (Ca is a metal), it replaces the metal or hydrogen in the compound. If the element is a nonmetal, it replaces the nonmetal in the compound. Will a reaction occur? Check the activity series, then answer and explain.
$Ca(s) + 2\ HBr(aq) \rightarrow CaBr_2(aq) + H_2(g)$	Write the conventional equation, including state symbols.
$Ca(s) + 2\ H^+(aq) + 2\ Br^-(aq) \rightarrow Ca^{2+}(aq) + 2\ Br^-(aq) + H_2(g)$ *Each HBr(aq) on the left yields one $H^+(aq)$ and one $Br^-(aq)$. The conventional equation has two HBr(aq), so there will be 2 $H^+(aq)$ and 2 $Br^-(aq)$ in the total ionic equation.*	Now write the total ionic equation.
$Ca(s) + 2\ H^+(aq) \rightarrow Ca^{2+}(aq) + H_2(g)$ *Bromide ion, $Br^-(aq)$, is the only spectator. The equation is balanced with 1 Ca and 2 H on each side and a 2+ charge on each side.*	Eliminate the spectators and write the net ionic equation. Check atom and charge balance.
You improved your skill at writing net ionic equations in general and writing net ionic equations for redox reactions in particular.	What did you learn by working this Active Example?

Practice Exercise 9.7

Write the conventional, total ionic, and net ionic equations for the reaction that occurs, if any, when solid barium is added to liquid water.

Active Example 9.8 Net Ionic Equations: Redox Reactions III

Copper is placed into a solution of silver nitrate, forming copper(II) ions as one of the products. Write the three equations.

Think Before You Write This will be your third net ionic equation for a single-replacement redox reaction, so you are ready to complete the process with minimal guidance.

Answers *Cover the left column with your cut-out shield. Reveal each answer only after you have written your own answer in the right column.*

Answer	Step
Cu is above Ag in the activity series, so copper atoms will change to copper(II) ions and silver ions will change to silver atoms: $Cu(s) + 2\ AgNO_3(aq) \rightarrow Cu(NO_3)_2(aq) + 2\ Ag(s)$. *The reaction is shown in Figure 9.13.*	Check the activity series. Will a reaction occur? If so, write the conventional equation.

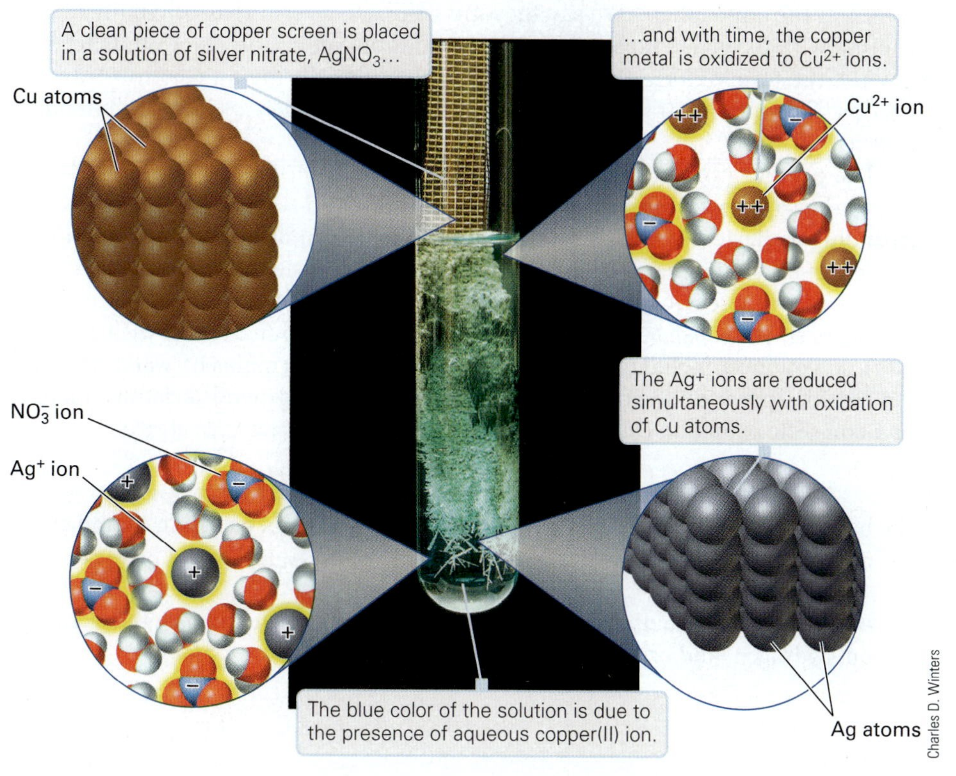
A clean piece of copper screen is placed in a solution of silver nitrate, AgNO3...
...and with time, the copper metal is oxidized to Cu2+ ions.
Cu atoms
Cu2+ ion
NO3- ion
Ag+ ion
The Ag+ ions are reduced simultaneously with oxidation of Cu atoms.
The blue color of the solution is due to the presence of aqueous copper(II) ion.
Ag atoms
Charles D. Winters

Figure 9.13 The reaction of copper atoms and silver ions at the macroscopic level and the particulate level.

$Cu(s) + 2\,Ag^+(aq) + 2\,NO_3^-(aq) \rightarrow Cu^{2+}(aq) + 2\,NO_3^-(aq) + 2\,Ag(s)$ $Cu(s) + 2\,Ag^+(aq) \rightarrow Cu^{2+}(aq) + 2\,Ag(s)$	Write both the total ionic and the net ionic equation.
There are one Cu and two Ag atoms on each side and a 2+ charge on each side. The net ionic equation is balanced. The reactants in Figure 9.13 are Cu atoms and Ag^+ ions. The products are Cu^{2+} ions and Ag atoms. The net ionic equation is an accurate symbolic representation of the particulate-level process.	Check your net ionic equation for atom and charge balance, and then look at the particulate-level illustrations in Figure 9.13. Does your net ionic equation accurately describe the reaction that occurs?
You improved your skill at writing net ionic equations in general and writing net ionic equations for redox reactions in particular.	What did you learn by working this Active Example?

Practice Exercise 9.8

Iron metal is added to a solution of copper(II) sulfate, forming a solution of iron(II) sulfate as one of the products. Write the three equations.

9.7 Oxidation–Reduction Reactions of Some Common Organic Compounds

Goal 7 Write the equation for the complete oxidation or burning of any compound containing only carbon and hydrogen or only carbon, hydrogen, and oxygen.

Another type of electron-transfer reaction occurs when a substance burns. A large number of compounds, including fuels, petroleum products, alcohols, some acids, and sugars, consist of only two or three elements: carbon and hydrogen or carbon, hydrogen, and oxygen. When such compounds are burned in air, they react with oxygen in the atmosphere. We say they are oxidized, a term that originally meant reacting with oxygen. The reaction is an oxidation–reduction reaction. Chemists now use the term *oxidized* to mean loss of electrons during an electron-transfer reaction. Electrons are transferred from the carbon–hydrogen–oxygen compound to oxygen.

The products of a complete burning or oxidation of these compounds are always the same: gaseous carbon dioxide, $CO_2(g)$, and water, $H_2O(g)$ or $H_2O(\ell)$, depending on the temperature at which the product is examined. The distinction is not important in this chapter, so we will use $H_2O(\ell)$ consistently.

Both carbon dioxide gas, $CO_2(g)$, and steam, $H_2O(g)$, are invisible. The white "smoke" commonly seen rising from chimneys and smokestacks is really tiny drops of condensed liquid water, $H_2O(\ell)$. Black smoke comes from carbon that is not completely burned.

In writing these equations, you will be given only the identity of the compound that *burns* or is *oxidized*. These words tell you the compound reacts with oxygen, $O_2(g)$, so you must include it as a second reactant. The formulas of water and carbon dioxide appear on the right side of the equation. The general skeleton equation for a complete oxidation (burning) reaction, without coefficients, is always:

$$C_xH_yO_z + O_2(g) \rightarrow CO_2(g) + H_2O(\ell) \text{ [or } H_2O(g)]$$

The burning of methane, the principal component of natural gas, is an example:

$$CH_4(g) + 2\,O_2(g) \rightarrow CO_2(g) + 2\,H_2O(\ell)$$

As a rule, these equations are most easily balanced if you balance the elements carbon, hydrogen, and oxygen in that order.

Active Example 9.9 Complete Burning (Oxidation) Reactions of C/H and C/H/O Compounds

Write the equation for the complete burning of ethane, $C_2H_6(g)$.

Think Before You Write The phrase "complete burning" is your cue that the substance reacts with oxygen, forming carbon dioxide and water as products. Remember, balance elements in the order C first, H second, and O last.

Answers *Cover the left column with your cut-out shield. Reveal each answer only after you have written your own answer in the right column.*

$2\,C_2H_6(g) + 7\,O_2(g) \rightarrow 4\,CO_2(g) + 6\,H_2O(\ell)$ *Balancing C:* $\_\_\,C_2H_6(g) + \_\_\,O_2(g) \rightarrow 2\,CO_2(g) + \_\_\,H_2O(\ell)$ *Balancing H:* $\_\_\,C_2H_6(g) + \_\_\,O_2(g) \rightarrow 2\,CO_2(g) + 3\,H_2O(\ell)$ *Balancing O:* $\_\_\,C_2H_6(g) + 7/2\,O_2(g) \rightarrow 2\,CO_2(g) + 3\,H_2O(\ell)$ *Clearing the fraction by multiplying all coefficients by 2 produces the final equation (written earlier).*	Complete the active example.
You improved your skill at balancing chemical equations in general, and specifically, balancing complete burning reactions of C/H and C/H/O compounds.	What did you learn by solving this Active Example?

Practice Exercise 9.9

Write the chemical equation that represents the reaction that occurs when liquid ethanol, $CH_3CH_2OH(\ell)$, is completely burned.

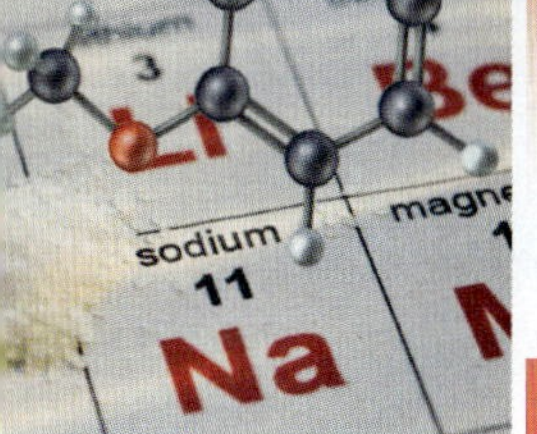

Everyday Chemistry 9.1

AN EVERY-MOMENT TYPE OF CHEMICAL REACTION

There is one type of chemical reaction that affects your life each and every day, no matter who you are, no matter where you live, from the day you are born to the day you die. This reaction type is the one found in the preceding section, the complete oxidation, or burning, of organic compounds.

If you live in an urban area, your home is probably heated with natural gas, which is mostly methane. The reaction that occurs in your furnace is:

$$CH_4(g) + 2\ O_2(g) \rightarrow CO_2(g) + 2\ H_2O(g) + \text{heat}$$

If you live in a rural area where there are no natural gas pipelines, you probably use liquid propane as a source of energy (**Figure 9.14**):

$$C_3H_8(\ell) + 5\ O_2(g) \rightarrow 3\ CO_2(g) + 4\ H_2O(g) + \text{heat}$$

Even in the summer when you don't use your furnace, you probably use one of these reactions to heat the water in your hot water heater. Both reactions are the oxidation of an organic compound.

You may ask, "What about homes that use electric heat?" Well, it's likely that the power used to generate the electricity was generated either by burning coal, a form of carbon, or by burning natural gas. However, if you live in the western United States, your electricity may be generated by dams, in which case no burning reaction is necessary. Nuclear power is also used to generate electricity without burning organic compounds.

Figure 9.14 Rural homes that do not have access to natural gas pipelines can use liquid propane as a source of fuel for their furnaces and hot water heaters. The propane is stored in large tanks that are refilled periodically.

Every time you drive a car (**Figure 9.15**), gasoline, diesel fuel, or, less commonly, other forms of carbon–hydrogen or carbon–hydrogen–oxygen compounds are oxidized. A major component in gasoline is octane, C_8H_{18}:

$$2\ C_8H_{18}(\ell) + 25\ O_2(g) \rightarrow 16\ CO_2(g) + 18\ H_2O(g) + \text{heat}$$

But what if you are far away from civilization, where there are no home heaters, hot water heaters, or automobiles? Didn't we say that oxidization is an every-moment chemical reaction? Although there are many steps in what is overall a complex process, the energy you use to sustain your life is released through oxidation reactions. The overall simple form of the reaction by which humans convert blood sugar into energy is

$$C_6H_{12}O_6(aq) + 6\ O_2(g) \rightarrow 6\ CO_2(g) + 6\ H_2O(\ell) + \text{energy}$$

We breathe in—inhale—oxygen from the atmosphere, the source of one of the two reactants. This oxygen then reacts with the blood sugar (**Figure 9.16**), and we breathe out—exhale—carbon dioxide and some of the water formed, which are products of the reaction.

It may be difficult to imagine that burning gasoline and metabolizing food occur via the same reaction type, but it is true. One major difference is that you can "burn" food in your body at 37°C only with the help of specialized molecules known as *enzymes*. We'll discuss these molecules, which are needed for low-temperature burning, in more detail in Chapter 22, which discusses the chemistry of biological molecules, an area of study known as biochemistry. In the meantime, remember that burning reactions occur in you and around you throughout every moment of your life!

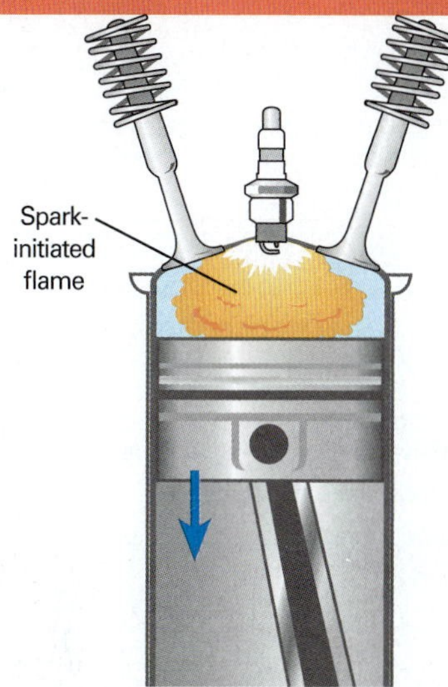

Figure 9.15 Within an automobile engine, gasoline and air are mixed, and the mixture is burnt, causing the production of high-temperature gaseous products that push the cylinder down and power the engine.

Quick Quiz

1. **What are the products of the complete burning reaction of organic compounds?**
2. **How can a human body burn sugar at body temperature, given that sugar outside of the body does not burn at 37°C?**

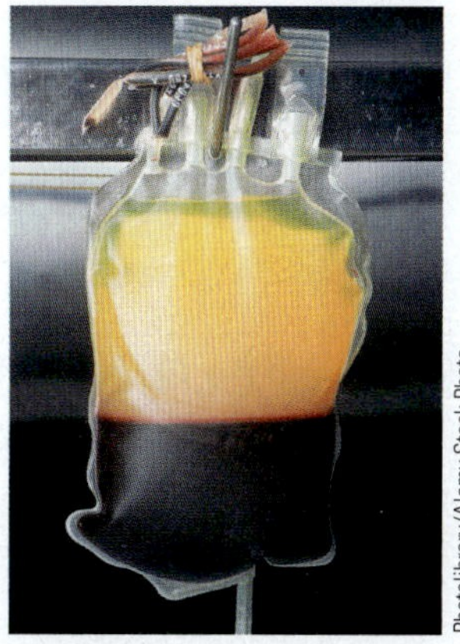

Figure 9.16 Glucose—blood sugar—is one of the substances dissolved in the blood plasma, a straw-colored fluid that comprises 55% of the total blood volume. Glucose is the primary source of energy for the cells of the body.

9.8 Double-Replacement Precipitation Reactions

Goal 8 Predict whether a precipitate will form when known solutions are combined; if a precipitate forms, write the net ionic equation. (Reference to a solubility table or a solubility guidelines list may or may not be allowed, depending upon the preference of your instructor.)

9 Given the product of a precipitation reaction, write the net ionic equation.

An **ion-combination reaction** occurs when the cation (positively charged ion) from one reactant combines with the anion (negatively charged ion) from another to form a particular kind of product compound. The conventional equation is a double-replacement type in which the ions appear to "change partners": $AX + BY \rightarrow AY + BX$. In this section, the product is an insoluble ionic compound that settles to the bottom of mixed solutions. A solid formed this way is called a **precipitate**; the reaction is a **precipitation reaction**. ◀

▶ **P/Review** Double-replacement equations were introduced to describe precipitation reactions in Section 8.10. See Table 8.3.

Active Example 9.10 Net Ionic Equations: Double-Replacement Precipitation Reactions I

Mixing sodium chloride and silver nitrate solutions produces a white precipitate of silver chloride. Develop the net ionic equation for the reaction.

Think Before You Write The same three steps you used on single-replacement redox reactions are applied to precipitation reactions: (1) conventional, (2) total ionic, and (3) net ionic. In this case, the products can be predicted by "changing partners" of the reacting ions.

Answers *Cover the left column with your cut-out shield. Reveal each answer only after you have written your own answer in the right column.*

$NaCl(aq) + AgNO_3(aq) \rightarrow$	Start by writing the formulas of the reactants and their state symbols as they will appear in the conventional equation.
$NaCl(aq) + AgNO_3(aq) \rightarrow NaNO_3(aq) + AgCl(s)$	Complete the conventional equation by switching the cation–anion pairs. The problem statement gives you the state of one of the products; the other is aqueous. 
$Na^+(aq) + Cl^-(aq) + Ag^+(aq) + NO_3^-(aq) \rightarrow Na^+(aq) + NO_3^-(aq) + AgCl(s)$ *Solids are unchanged in moving from the conventional to the total ionic equation.*	Write the total ionic equation.
$Cl^-(aq) + Ag^+(aq) \rightarrow AgCl(s)$ *$Na^+(aq)$ and $NO_3^-(aq)$ are spectators in the total ionic equation.* *The reaction is illustrated in* **Figure 9.17**.	Write the net ionic equation.
There is 1 Cl and 1 Ag on each side; the atoms balance. The charge on each side is zero and therefore balanced. The product of the reaction is a solid that resulted from the mixing of solutions; such a solid is called a *precipitate*.	Check the net ionic equation for atom and charge balance. Explain why the reaction is called a *precipitation* reaction. 

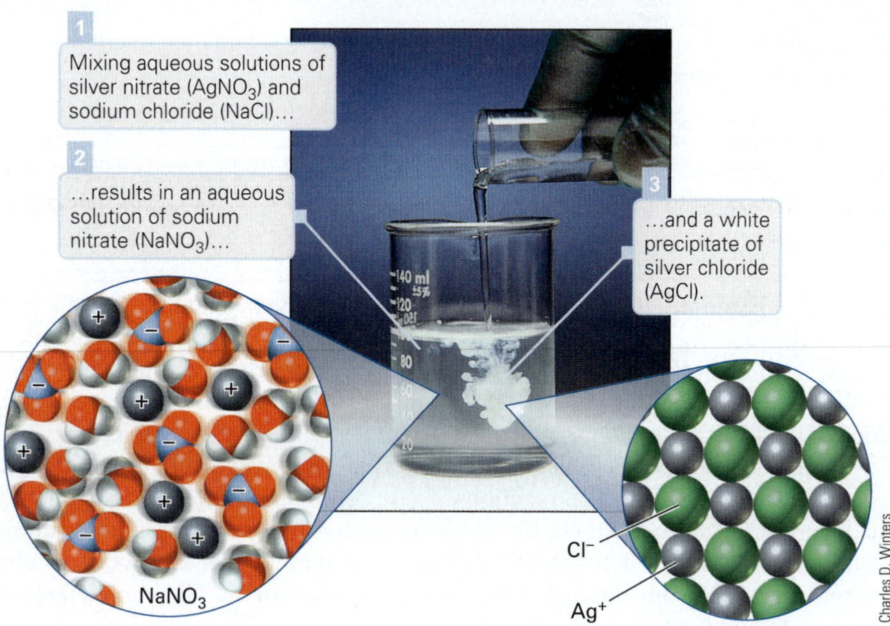

1
Mixing aqueous solutions of silver nitrate ($AgNO_3$) and sodium chloride (NaCl)...
2
...results in an aqueous solution of sodium nitrate ($NaNO_3$)...
3
...and a white precipitate of silver chloride (AgCl).
$NaNO_3$
Cl^-
Ag^+
Charles D. Winters

Figure 9.17 Sodium chloride and silver nitrate solutions are mixed, producing a white precipitate of silver chloride. Sodium ions and nitrate ions remain in solution.

You improved your skill at writing net ionic equations in general and writing net ionic equations for double-replacement precipitation reactions in particular.	What did you learn by working this Active Example?

Practice Exercise 9.10

Write the net ionic equation for the reaction between solutions of lead(II) nitrate and potassium iodide, yielding a precipitate of lead(II) iodide.

Let's try another Active Example that has two interesting features.

Active Example 9.11 Net Ionic Equations: Double-Replacement Precipitation Reactions II

When solutions of silver chlorate and aluminum chloride are combined, silver chloride precipitates. Write the three equations.

Think Before You Write Reread *Step 3* of the *how to...* Write a Net Ionic Equation procedure in Section 9.5 before working on this Active Example.

Answers *Cover the left column with your cut-out shield. Reveal each answer only after you have written your own answer in the right column.*

$3\ AgClO_3(aq) + AlCl_3(aq) \rightarrow 3\ AgCl(s) + Al(ClO_3)_3(aq)$	Start with the conventional equation. 
$3\ Ag^+(aq) + 3\ ClO_3^-(aq) + Al^{3+}(aq) + 3\ Cl^-(aq) \rightarrow$ $3\ AgCl(s) + Al^{3+}(aq) + 3\ ClO_3^-(aq)$	Write the total ionic equation. 
$3\ Ag^+(aq) + 3\ Cl^-(aq) \rightarrow 3\ AgCl(s)$ $Ag^+(aq) + Cl^-(aq) \rightarrow AgCl(s)$ *After elimination of the spectator ions, the net ionic equation has 3 for the coefficient of all species. Step 3 of the procedure says to reduce coefficients to lowest terms.* *The second promised interesting feature is described in the paragraph following this Active Example.*	Write the net ionic equation. 

You improved your skill at writing net ionic equations in general and writing net ionic equations for double-replacement precipitation reactions in particular.

What did you learn by working this Active Example?

Practice Exercise 9.11

Write the net ionic equation for the reaction that occurs, if any, when solutions of nickel(II) bromide and ammonium sulfite are mixed, yielding a precipitate of nickel(II) sulfite.

Active Examples 9.10 and 9.11 produced the same net ionic equation, even though the reacting solutions were completely different. Actually, only the spectators were different, but they are not part of the chemical change. Eliminating them shows that both reactions are exactly the same. It will be this way whenever a solution containing $Ag^{+}(aq)$ ion is added to a solution containing $Cl^{-}(aq)$ ion.

If we knew in advance which combinations of ions yield insoluble compounds, we could predict precipitation reactions. These compounds have been identified in the laboratory. **Table 9.3** shows the results of such experiments for a large number of ionic compounds. Their solubilities have been summarized in a set of solubility guidelines that your instructor may ask you to memorize. These guidelines are shown in **Figure 9.18**.

Table 9.3 Solubilities of Ionic Compounds*

Ions	Acetate	Bromide	Carbonate	Chlorate	Chloride	Fluoride	Hydrogen Carbonate	Hydroxide	Iodide	Nitrate	Nitrite	Phosphate	Sulfate	Sulfide	Sulfite
Aluminum	slightly soluble	soluble	does not exist	soluble	soluble	slightly soluble		nearly insoluble	soluble (hydrate)	soluble		nearly insoluble	soluble	decomposes	
Ammonium	soluble	soluble	soluble	soluble	soluble	soluble	soluble	soluble	soluble	soluble	soluble	soluble	soluble	soluble	soluble
Barium	soluble	soluble	nearly insoluble	soluble	soluble	slightly soluble		soluble	soluble	soluble	soluble	nearly insoluble	nearly insoluble	decomposes	nearly insoluble
Calcium	soluble	soluble	nearly insoluble	soluble	soluble	slightly soluble	exists only in solution	slightly soluble	soluble	soluble	soluble	nearly insoluble	slightly soluble	decomposes	nearly insoluble
Cobalt(II)	soluble	soluble	nearly insoluble	soluble	soluble	nearly insoluble		nearly insoluble	soluble	soluble	slightly soluble	nearly insoluble	soluble	nearly insoluble	nearly insoluble
Copper(II)	soluble	soluble	nearly insoluble	soluble	soluble	soluble		nearly insoluble	decomposes	soluble		nearly insoluble	soluble	nearly insoluble	
Iron(II)	soluble	soluble	nearly insoluble	soluble	soluble	soluble		nearly insoluble	soluble	soluble		nearly insoluble	soluble	nearly insoluble	nearly insoluble
Iron(III)		soluble	does not exist		soluble	slightly soluble		nearly insoluble	does not exist	soluble		nearly insoluble	soluble (hydrate)	decomposes	
Lead(II)	soluble	slightly soluble	nearly insoluble	soluble	slightly soluble	slightly soluble		nearly insoluble	slightly soluble	soluble	soluble	nearly insoluble	nearly insoluble	nearly insoluble	nearly insoluble
Lithium	soluble	soluble	soluble	soluble	soluble	soluble	soluble	soluble	soluble	soluble	soluble	slightly soluble	soluble	soluble	soluble
Magnesium	soluble	soluble	nearly insoluble	soluble	soluble	slightly soluble		nearly insoluble	soluble	soluble	soluble	nearly insoluble	soluble	decomposes	soluble
Nickel(II)	soluble	soluble	nearly insoluble	soluble	soluble	soluble		nearly insoluble	soluble	soluble		nearly insoluble	soluble	nearly insoluble	nearly insoluble
Potassium	soluble	soluble	soluble	soluble	soluble	soluble	soluble	soluble	soluble	soluble	soluble	soluble	soluble	soluble	soluble
Silver	slightly soluble	nearly insoluble	nearly insoluble	soluble	nearly insoluble	soluble		does not exist	nearly insoluble	soluble	slightly soluble	nearly insoluble	slightly soluble	nearly insoluble	nearly insoluble
Sodium	soluble	soluble	soluble	soluble	soluble	soluble	soluble	soluble	soluble	soluble	soluble	soluble	soluble	soluble	soluble
Zinc	soluble	soluble	nearly insoluble	soluble	soluble	soluble (hydrate)		nearly insoluble	soluble	soluble		nearly insoluble	soluble	nearly insoluble	nearly insoluble

*Compounds having solubilities of 0.1 mole or more in 1 L of water at 25°C are listed as soluble. If the solubility is less than 0.1 mole per liter of water, the compound is listed as slightly soluble or nearly insoluble, as appropriate. Both slightly soluble and nearly insoluble compounds are solids in net ionic equations. No ionic compound is entirely insoluble in water. A blank space indicates a lack of data. In writing equations for reactions that occur in water solution, slightly soluble and nearly insoluble substances (indicated with a pink background in this table) have the state symbol of a solid (s). Soluble substances (those with a green background) are designated by (aq) in an equation.

SILVER COMPOUNDS

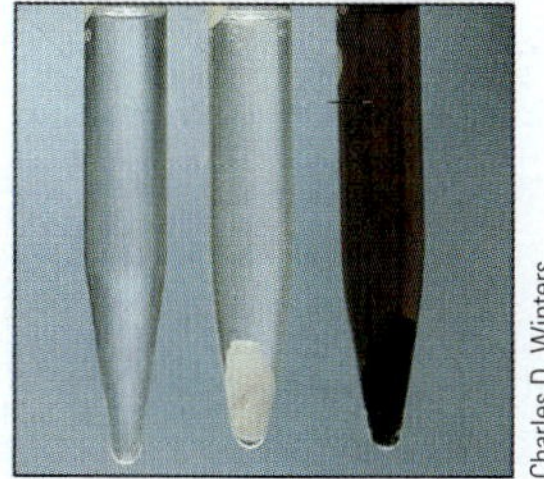

$AgNO_3$ $AgCl$ $AgOH$

Nitrates are generally soluble, as are chlorides (except AgCl). Hydroxides are generally not soluble.

SULFIDES

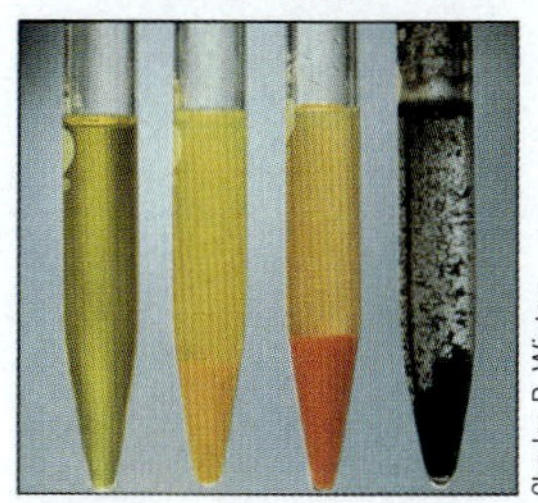

$(NH_4)_2S$ CdS Sb_2S_3 PbS

Sulfides are generally not soluble (exceptions include compounds with NH_4^+ and Na^+).

HYDROXIDES

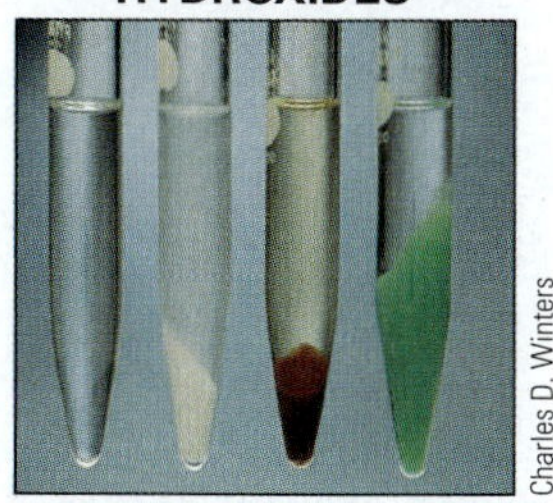

$NaOH$ $Ca(OH)_2$ $Fe(OH)_3$ $Ni(OH)_2$

Hydroxides are generally not soluble except when the cation is a Group 1A/1 metal.

Solubility Guidelines for Ionic Compounds*

Soluble Compounds	
Compounds with	**Exceptions**
ClO_3^-, HCO_3^-, NO_3^-	
$C_2H_3O_2^-$	$Al(C_2H_3O_2)_3$, $AgC_2H_3O_2$
Br^-, Cl^-, I^-	$PbBr_2$, $PbCl_2$, PbI_2, $AgBr$, $AgCl$, AgI
F^-	AlF_3, BaF_2, CaF_2, CoF_2, FeF_3, PbF_2, MgF_2
NO_2^-	$Co(NO_2)_2$, $AgNO_2$
SO_4^{2-}	$BaSO_4$, $CaSO_4$, $PbSO_4$, Ag_2SO_4

Insoluble Compounds	
Compounds with	**Exceptions†**
CO_3^{2-}, PO_4^{3-}, S^{2-}	
OH^-	$Ba(OH)_2$
SO_3^{2-}	$MgSO_3$

*Organized based on common anions.
†Compounds with the cations NH_4^+, K^+, and Na^+ are soluble. Compounds with the cation Li^+ are soluble except for Li_3PO_4.

Figure 9.18 Solubility guidelines for ionic compounds. If a compound contains at least one of the ions in the Soluble Compounds list, apart from the exceptions listed, it is likely to be at least moderately soluble in water. Compounds with at least one ion in the Insoluble Compounds list are poorly soluble in water (again, with the exception of the few compounds in the Exceptions list).

In the remaining Active Examples in this section and for the end-of-chapter questions, use the table suggested by your instructor—either Table 9.3 or Figure 9.18 (or memorize the solubility guidelines in Figure 9.18, if required)—to predict precipitation reactions.

Active Example 9.12 Net Ionic Equations: Double-Replacement Precipitation Reactions III

Solutions of lead(II) nitrate and sodium bromide are combined. Write the net ionic equation for any precipitation reaction that may occur.

Think Before You Write Your instructor will probably tell you if you will have access to Table 9.3, Figure 9.18, or the equivalent, or if you need to memorize the solubility guidelines. Complete this Active Example while following the policy of your instructor. Continue to follow this policy when you work on the questions, exercises, and problems at the end of the chapter.

Answers *Cover the left column with your cut-out shield. Reveal each answer only after you have written your own answer in the right column.*

$Pb(NO_3)_2(aq) + 2\ NaBr(aq) \rightarrow PbBr_2(\ ) + 2\ NaNO_3(\ )$	Start with the conventional equation. Leave blank the state symbols of the products for now. 
$Pb(NO_3)_2(aq) + 2\ NaBr(aq) \rightarrow PbBr_2(s) + 2\ NaNO_3(aq)$ *In Table 9.3, the intersection of the lead(II)-ion row and the bromide-ion column shows that lead(II) bromide is slightly soluble; lead(II) bromide forms a solid precipitate, $PbBr_2(s)$. Only 0.8 g of lead(II) bromide dissolves in 100 mL of water at 20°C when excess solid is present; this is an example of what we mean by slightly soluble. You will see solid lead(II) bromide form if you perform this reaction.* *The intersection of the sodium-ion row and the nitrate-ion column shows that sodium nitrate is soluble in water; sodium nitrate remains in aqueous solution, $NaNO_3(aq)$.* *Both conclusions can also be reached from the solubility guidelines: Lead(II) bromide is one of the exceptions to the "compounds with bromide ion are soluble" guideline, and all sodium and nitrate compounds are soluble.*	You must determine whether the product compounds are soluble or insoluble. If soluble, the state symbol should be (aq), if insoluble, (s). Fill in the state symbols to complete your conventional equation. 
$Pb^{2+}(aq) + 2\ NO_3^-(aq) + 2\ Na^+(aq) + 2\ Br^-(aq) \rightarrow PbBr_2(s) + 2\ Na^+(aq) + 2\ NO_3^-(aq)$ $Pb^{2+}(aq) + 2\ Br^-(aq) \rightarrow PbBr_2(s)$	Complete the Active Example by writing the total ionic and net ionic equations. 
You improved your skill at writing net ionic equations in general and writing net ionic equations for double-replacement precipitation reactions in particular.	What did you learn by working this Active Example? 

Practice Exercise 9.12

Write the net ionic equation for any precipitation reaction that may occur when solutions of aluminum bromide and sodium fluoride are combined.

9.9 Double-Replacement Molecule-Formation Reactions

Goal 10 Given reactants for a double-replacement reaction that yield a molecular product, write the conventional, total ionic, and net ionic equations.

The reaction of an acid often leads to an ion combination that yields a **molecular product** instead of a precipitate. Except for the difference in the product, the equations are written in exactly the same way. Just as you had to recognize an insoluble product and not break it up when writing the total ionic equation, you must now recognize a molecular product and not break it into ions. Water and weak acids are the two kinds of molecular products you will find.

▶ **P/Review** The neutralization reactions in Section 8.10 are between acids and hydroxide bases. The products are water and an ionic compound. The general equation is $HX + MOH \rightarrow HOH + MX$.

A **neutralization reaction** is the most common type of molecular-product reaction, in which an acid is reacted with its chemical opposite, a base, to yield water, which is neutral in the sense that it is neither an acid nor a base. ◀

Everyday Chemistry 9.2

GREEN CHEMISTRY

Green chemistry is a phrase you'll probably hear with increasing frequency in the near future. It refers to carrying out chemical changes in a way that avoids environmentally hazardous substances (**Figure 9.19**). This can be accomplished by finding new ways of reacting chemicals, conducting reactions in environmentally friendly solvents, and incorporating more reactant atoms in the final product, eliminating waste.

A central issue in making chemistry "greener" is the efficiency of incorporating the reactant atoms into the product compound. Any atoms that are used in the reaction but that are not part of the product are wasted, and this waste must subsequently be disposed of. If chemists can design reactions that do not have wasted atoms, there are no disposal issues and no potential for environmental impact. This type of efficiency can be measured in terms of *percentage atom utilization:*

$$\% \text{ atom utilization} = \frac{\text{molar mass of target product}}{\text{molar mass of all products}} \times 100\%$$

Figure 9.19 Paint pigments. Many paint pigments, such as chrome yellow, lead red, and mercadmium orange, used to contain toxic heavy metals such as chromium, lead, and cadmium, as their names imply. Green chemistry involves developing new methods to produce the same colors without toxic ingredients.

One well-known case of improving atom economy in the pharmaceutical drug industry occurred in the 1990s when a new method of synthesizing ibuprofen (sold in generic form and under the brand names Advil, Medipren, Motrin, and Nuprin) was developed (**Figure 9.20**). The tried-and-true method was originally developed in the 1960s; it involved six steps and generated a significant quantity of waste. It has a percentage atom utilization of only about 40%! The newer green synthesis requires only three steps and has 99% atom utilization, partly because one by-product can be recovered.

Another example of green chemistry is the development of hydrogen fuel cells. A typical fuel cell is designed to allow the exothermic reaction of hydrogen and oxygen without explosive combustion, producing water as its only product with 100% atom utilization. Some major automobile manufacturers are currently manufacturing passenger cars powered with hydrogen fuel cells (**Figure 9.21**).

Even though fuel cells are beginning to appear in cars, the primary problem with this technology is producing the hydrogen fuel itself. It is expensive; its production usually involves a pollution-generating process; and the hydrogen requires a large storage volume compared with gasoline. Research is ongoing to find solutions to these problems.

Green chemists continue to work on methods of making product compounds that use all of the reactant atoms. The conventional and net ionic equations for these reactions are usually classified as combination reactions, which were first introduced in Section 8.7. Research on safer chemicals, better reaction conditions, and more efficient reaction pathways continues to make chemical manufacturing more healthful for humans and more friendly to the environment.

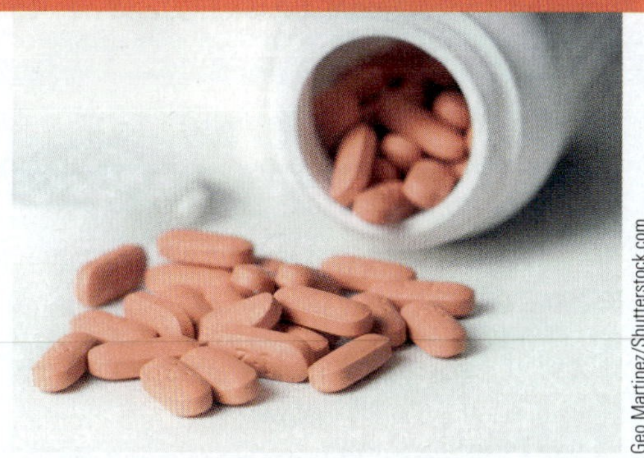

Figure 9.20 Ibuprofen is a pain reliever and fever reducer. Its official chemical name is *(RS)*-2-(4-(2-methylpropyl)phenyl)propanoic acid. It used to be called **i**so-**bu**tyl-**pro**panoic-**phen**olic acid, from which the name ibuprofen was derived.

Quick Quiz

3. What is *green chemistry*?
4. What condition is required to achieve 100% atom utilization? Explain.

Figure 9.21 The Toyota Fuel Cell Vehicle, the Mirai (Japanese for *future*). The technology is currently available for large-scale production of automobiles powered by hydrogen fuel cells. The infrastructure for hydrogen fueling is still very limited, however.

Active Example 9.13 Net Ionic Equations: Double-Replacement Molecule-Formation Reactions I

Write the conventional, total ionic, and net ionic equations for the reaction between solutions of hydrochloric acid and sodium hydroxide.

Think Before You Write In this type of double-replacement reaction, you must recognize a molecular product, either water or a weak acid (you can recognize a weak acid by it not being among the memorized seven common strong acids), and not break it into ions in the total ionic equation.

Answers *Cover the left column with your cut-out shield. Reveal each answer only after you have written your own answer in the right column.*

$HCl(aq) + NaOH(aq) \rightarrow HOH(\ell) + NaCl(aq)$ *You need to recognize that water is a product, no matter whether you write it as HOH or H_2O, and that it occurs as a liquid at ordinary temperatures.*	Write the conventional equation. Watch your state designations.
$H^+(aq) + Cl^-(aq) + Na^+(aq) + OH^-(aq) \rightarrow HOH(\ell) + Na^+(aq) + Cl^-(aq)$	Now proceed just as you did for precipitation reactions with the total ionic equation.
$H^+(aq) + OH^-(aq) \rightarrow H_2O(\ell)$ 2 H and 1 O on each side; 0 charge on each side; the equation is balanced. *A particulate-level illustration of this reaction is presented in* **Figure 9.22**.	Write the net ionic equation and check for atom and charge balance. Change water to its correct formula if you initially wrote it as HOH. 

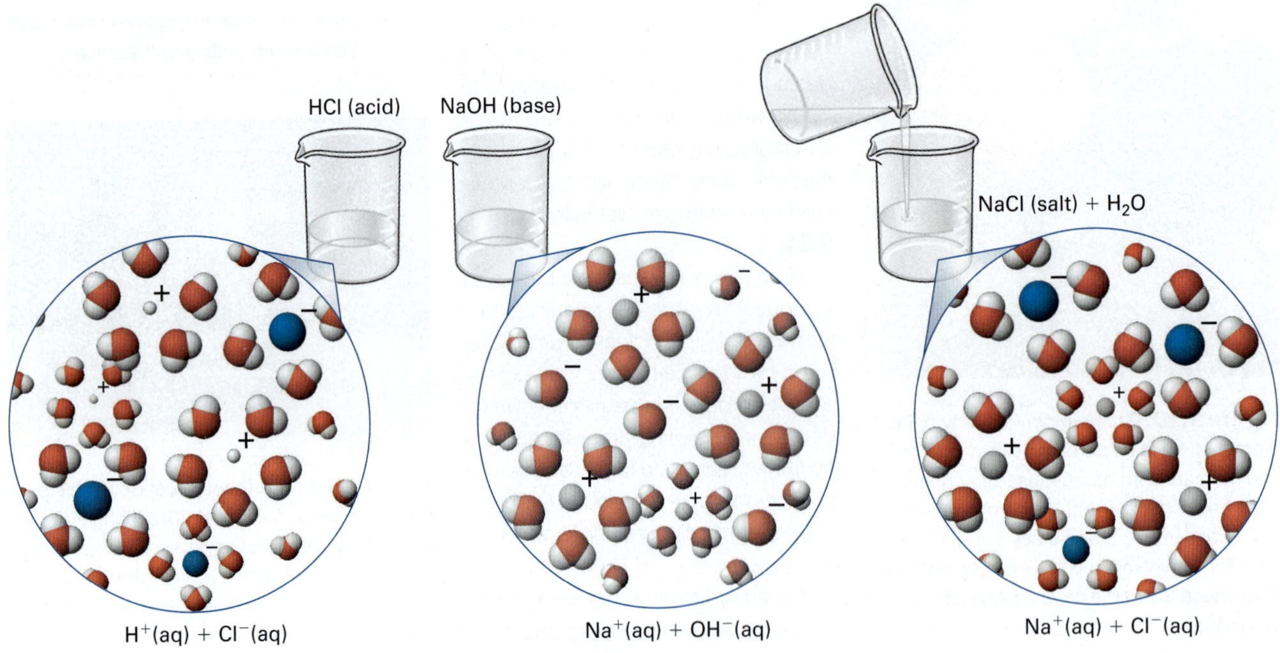

Figure 9.22 The reaction between solutions of hydrochloric acid and sodium hydroxide. Hydrogen ions from one solution combine with hydroxide ions from the other to form water molecules. Sodium ions and chloride ions are spectators.

You improved your skill at writing net ionic equations in general and writing net ionic equations for double-replacement molecule-formation reactions in particular.	What did you learn by working this Active Example?

Practice Exercise 9.13

Solutions of hydrobromic acid and barium hydroxide are combined. Write the conventional, total ionic, and net ionic equations.

When the reactants are a strong acid and an ionic compound that includes the anion of a weak acid, that weak acid is formed as the molecular product. You must recognize it as a weak acid (not one of the seven strong acids) and leave it in molecular form in the total ionic and net ionic equations.

Active Example 9.14 Net Ionic Equations: Double-Replacement Molecule-Formation Reactions II

Develop the net ionic equation for the reaction between hydrochloric acid solution and a solution of sodium acetate. The formula of the acetate ion is $C_2H_3O_2^-(aq)$.

Think Before You Write A weak acid primarily exists in its molecular form, so it remains molecular in the total and net ionic equations, no matter on which side of the equation it appears.

Answers *Cover the left column with your cut-out shield. Reveal each answer only after you have written your own answer in the right column.*

$HCl(aq) + NaC_2H_3O_2(aq) \rightarrow HC_2H_3O_2(aq) + NaCl(aq)$	Write the conventional equation.
$H^+(aq) + Cl^-(aq) + Na^+(aq) + C_2H_3O_2^-(aq) \rightarrow HC_2H_3O_2(aq) + Na^+(aq) + Cl^-(aq)$ *The weak acid $HC_2H_3O_2(aq)$ remains in its molecular form in the total ionic equation.*	Write the total ionic equation.
$H^+(aq) + C_2H_3O_2^-(aq) \rightarrow HC_2H_3O_2(aq)$ Atom balance: 4 H, 2 C, and 2 O on each side. Charge balance: 0 charge on each side.	Write the net ionic equation and verify that it is balanced.
You improved your skill at writing net ionic equations in general and writing net ionic equations for double-replacement molecule-formation reactions in particular.	What did you learn by working this Active Example?

Practice Exercise 9.14

Solutions of hydroiodic acid and sodium fluoride are combined. Develop the net ionic equation.

Compare the reactions in Active Examples 9.13 and 9.14: $H^+(aq) + OH^-(aq) \rightarrow H_2O(\ell)$ and $H^+(aq) + C_2H_3O_2^-(aq) \rightarrow HC_2H_3O_2(aq)$. The only difference between them is that one has the hydroxide ion as a reactant and the other has the acetate ion as a reactant. In the first case, the molecular product is water, formed when the hydrogen ion bonds to the hydroxide ion. In the second case, the molecular product is acetic acid, a weak acid formed when the hydrogen ion bonds to the acetate ion.

The acid in a neutralization reaction may be a weak acid. You must continue to recall that the major species in a weak acid solution is the weak acid molecule; it is not broken into ions, regardless of whether it is a reactant or a product.

Active Example 9.15 Net Ionic Equations: Double-Replacement Molecule-Formation Reactions III

Write the three equations leading to the net ionic equation for the reaction between acetic acid, $HC_2H_3O_2(aq)$, and a solution of sodium hydroxide.

Think Before You Write Remember that a weak acid (an acid that is not among the seven memorized strong acids) primarily exists in its molecular form, so it remains molecular no matter whether the reaction is conventional, total ionic, or a net ionic equation.

Answers *Cover the left column with your cut-out shield. Reveal each answer only after you have written your own answer in the right column.*

$HC_2H_3O_2(aq) + NaOH(aq) \rightarrow HOH(\ell) + NaC_2H_3O_2(aq)$	The conventional equation, please ...
$HC_2H_3O_2(aq) + Na^+(aq) + OH^-(aq) \rightarrow HOH(\ell) + Na^+(aq) + C_2H_3O_2^-(aq)$ *The weak acid molecule remains in its molecular form in the total ionic equation.*	Develop the total ionic equation.
$HC_2H_3O_2(aq) + OH^-(aq) \rightarrow H_2O(\ell) + C_2H_3O_2^-(aq)$ Atom balance: 5 H, 2 C, 3 O on each side. Charge balance: 1− on each side.	Write the net ionic equation and check it for balance.
You improved your skill at writing net ionic equations in general and writing net ionic equations for double-replacement molecule-formation reactions in particular.	What did you learn by working this Active Example?

Practice Exercise 9.15

Write the conventional, total ionic, and net ionic equations that represent the reaction that occurs when solutions of chlorous acid and potassium hydroxide are combined.

There are two features that allow you to identify a molecular-product reaction: (1) one reactant is an acid, usually strong, and (2) one product is water or a weak acid. One of the most common mistakes in writing net ionic equations is failing to recognize a weak acid as a molecular product. If one reactant in a double-replacement equation is a strong acid, you can be sure there will be a molecular product. If it isn't water, look for a weak acid.

9.10 Double-Replacement Reactions That Form Unstable Products

Goal 11 Given reactants that form H_2CO_3, H_2SO_3, or "NH_4OH" by ion combination, write the net ionic equation for the reaction.

Three ion combinations yield molecular products that are not the products you would expect.

One expected product is carbonic acid. If hydrogen ions from one reactant react with carbonate ions from another, carbonic acid, H_2CO_3, may be predicted to form:

$$2\,H^+(aq) + CO_3^{2-}(aq) \rightarrow H_2CO_3(aq)$$

But carbonic acid is unstable and decomposes to carbon dioxide gas and water. The correct net ionic equation is, therefore,

$$2\,H^+(aq) + CO_3^{2-}(aq) \rightarrow CO_2(g) + H_2O(\ell)$$

Sulfurous acid, H_2SO_3, decomposes in the same way to sulfur dioxide and water, but the sulfur dioxide gas remains in solution:

$$2\,H^+(aq) + SO_3^{2-}(aq) \rightarrow \text{“}H_2SO_3\text{”} \rightarrow SO_2(aq) + H_2O(\ell)$$

The third ion combination that initially yields an incorrect product occurs when ammonium and hydroxide ions react:

$$NH_4^+(aq) + OH^-(aq) \rightarrow \text{“}NH_4OH\text{”}$$

In spite of the existence of printed labels, laboratory bottles with NH_4OH etched on them, and wide use of the name ammonium hydroxide, no substance having the formula “NH_4OH” exists in significant quantities at ordinary temperatures. The actual product is a solution of ammonia molecules, $NH_3(aq)$. The proper net ionic equation is, therefore,

$$NH_4^+(aq) + OH^-(aq) \rightleftharpoons NH_3(aq) + H_2O(\ell)$$

The reaction is reversible and yields a solution in which $NH_3(aq)$ is the major species and $NH_4^+(aq)$ and $OH^-(aq)$ are minor species (**Figure 9.23**).

There is no system by which these three unstable product reactions can be recognized. You simply must be alert to them and catch them when they appear. Once again, the predicted but unstable formulas are H_2CO_3, H_2SO_3, and NH_4OH.

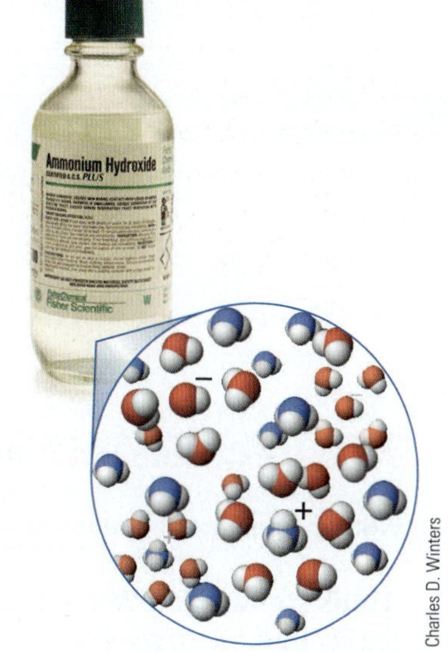

Figure 9.23 Aqueous ammonia at the macroscopic and particulate levels. In spite of the fact that bottles are labeled with the name *ammonium hydroxide* (NH_4OH), the major species in solution are ammonia molecules and water molecules [$NH_4OH \rightarrow NH_3(aq) + H_2O(\ell)$]. Ammonium ions and hydroxide ions are minor species.

Active Example 9.16 Net Ionic Equations: Double-Replacement Reactions That Form Unstable Products

Write the conventional, total ionic, and net ionic equations for the reaction between solutions of sodium carbonate and hydrochloric acid (Figure 9.24).

Think Before You Write If you know the predicted but unstable double-replacement reaction formulas and the actual products formed, you are ready to complete this Active Example.

Answers *Cover the left column with your cut-out shield. Reveal each answer only after you have written your own answer in the right column.*

$Na_2CO_3(aq) + 2\ HCl(aq) \rightarrow$ $2\ NaCl(aq) + CO_2(g) + H_2O(\ell)$ *The standard double-replacement equation would predict NaCl and H_2CO_3 as the products of this reaction, but H_2CO_3 is one of the unstable ion combinations. We therefore break it down into its decomposition products, CO_2 and H_2O.*	Write the conventional equation.
$2\ Na^+(aq) + CO_3^{2-}(aq) + 2\ H^+(aq) + 2\ Cl^-(aq) \rightarrow$ $2\ Na^+(aq) + 2\ Cl^-(aq) + CO_2(g) + H_2O(\ell)$	Write the total ionic equation.
$CO_3^{2-}(aq) + 2\ H^+(aq) \rightarrow CO_2(g) + H_2O(\ell)$ 1 C, 3 O, and 2 H on each side; 0 charge on each side.	Write the net ionic equation and verify the balance.
You improved your skill at writing net ionic equations in general and writing net ionic equations for double-replacement reactions that form unstable products in particular.	What did you learn by working this Active Example?

Practice Exercise 9.16

Write the conventional, total ionic, and net ionic equations for the reaction between solutions of ammonium nitrate and sodium hydroxide.

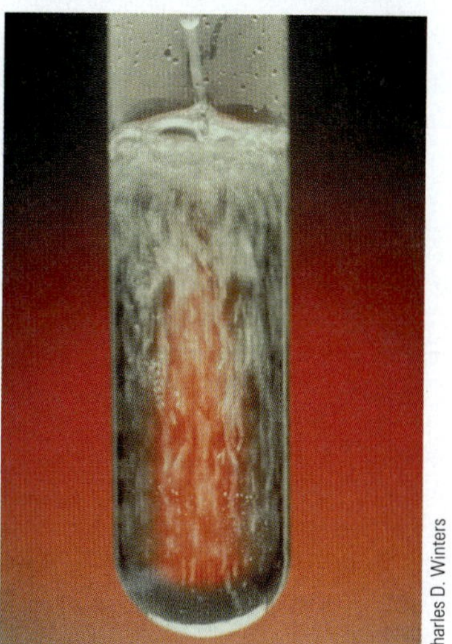

Figure 9.24 The reaction between hydrochloric acid and a solution of sodium carbonate.

9.11 Double-Replacement Reactions with Undissolved Reactants

In every ion-combination reaction considered so far, it has been assumed that both reactants are in solution. This is not always the case. Sometimes the description of the reaction will indicate that a reactant is a solid, liquid, or gas, even though it may be soluble in water. In such a case, write the correct state symbol after the formula in the conventional equation and carry the formula through all three equations unchanged.

A common example of this kind of reaction occurs with a compound that is insoluble in water but reacts with acids. The net ionic equation shows why.

Active Example 9.17 Net Ionic Equations: Double-Replacement Reactions with Undissolved Reactants

Write the net ionic equation to describe the reaction when solid aluminum hydroxide "dissolves" in hydrochloric acid.

Think Before You Write This is a neutralization reaction between a strong acid and a solid hydroxide—that is, a hydroxide ion containing ionic compound that is insoluble in water. Don't let the presence of a solid reactant throw you off. Just follow the procedure for writing net ionic equations.

Answers *Cover the left column with your cut-out shield. Reveal each answer only after you have written your own answer in the right column.*

$3\ HCl(aq) + Al(OH)_3(s) \rightarrow 3\ H_2O(\ell) + AlCl_3(aq)$	Write the conventional equation. 
$3\ H^+(aq) + 3\ Cl^-(aq) + Al(OH)_3(s) \rightarrow 3\ H_2O(\ell) + Al^{3+}(aq) + 3\ Cl^-(aq)$	Write the total ionic equation. Remember that you replace only aqueous substances that exist in solution as ions with the symbols of their ions. 
$3\ H^+(aq) + Al(OH)_3(s) \rightarrow 3\ H_2O(\ell) + Al^{3+}(aq)$ 6 H, 1 Al, 3 O on each side; 3+ charge on each side.	Finish with the net ionic equation and balance check. 
You improved your skill at writing net ionic equations in general and writing net ionic equations for double-replacement reactions with undissolved reactants in particular.	What did you learn by working this Active Example?

Practice Exercise 9.17

A nitric acid solution is poured onto solid nickel(II) carbonate. Write the conventional, total ionic, and net ionic equations for the reaction that occurs.

9.12 Other Double-Replacement Reactions

Active Example 9.18 Net Ionic Equations: Other Double-Replacement Reactions

Write the net ionic equation for any reaction that would occur when solutions of sodium chloride and potassium nitrate are combined.

Think Before You Write As with writing all net ionic equations, follow the procedure in a stepwise fashion.

Answers *Cover the left column with your cut-out shield. Reveal each answer only after you have written your own answer in the right column.*

$Na^+(aq) + Cl^-(aq) + K^+(aq) + NO_3^-(aq) \rightarrow NR$ *The total ionic equation is* *$Na^+(aq) + Cl^-(aq) + K^+(aq) + NO_3^-(aq) \rightarrow$ $Na^+(aq) + Cl^-(aq) + K^+(aq) + NO_3^-(aq)$* *You don't necessarily have a reaction every time you mix solutions of ionic compounds! There is no net ionic equation for this combination because there is no reaction.*	Complete the Active Example.
You improved your skill at writing net ionic equations in general and understanding what happens when solutions are mixed and in which no reaction occurs in particular.	What did you learn by working this Active Example?

Practice Exercise 9.18

Write the net ionic equation that represents what happens when solutions of sodium nitrate and ammonium sulfate are mixed.

9.13 Summary of Net Ionic Equations

Table 9.4 summarizes this chapter. The blue area is essentially the same as the last three rows of Table 8.3 in Section 8.11 (the table that summarized writing conventional equations).

Table 9.4 Summary of Net Ionic Equations

Reactants (Conventional)	Reaction Type	Equation Type (Conventional)	Products (Conventional)	Reactants (Net Ionic)	Products (Net Ionic)
Element + ionic compound *or* Element + strong acid	Oxidation–reduction	Single-replacement	Element + ionic compound	Element + ion	Element + ion
Two ionic compounds *or* ionic compound + strong acid *or* ionic compound + hydroxide base	Precipitation	Double-replacement	Two ionic compounds	Two ions	Ionic precipitate
Strong acid + hydroxide base	Molecule formation (H_2O), neutralization	Double-replacement	Ionic compound + H_2O	$H^+ + OH^-$	H_2O
Weak acid + hydroxide base	Molecule formation (H_2O), neutralization	Double-replacement	Ionic compound + H_2O	Weak acid + OH^-	H_2O + anion from weak acid
Strong acid + cation and anion of weak acid	Molecule formation (weak acid)	Double-replacement	Ionic compound + weak acid	H^+ + anion of weak acid	Weak acid
Strong acid + carbonate *or* hydrogen carbonate	Unstable product + decomposition	Double-replacement + decomposition	Ionic compound + $H_2O + CO_2$ Ionic compound + $H_2O + CO_2$	$H^+ + CO_3^{2-}$ $H^+ + HCO_3^-$	$H_2O + CO_2$ $H_2O + CO_2$
Strong acid + sulfite *or* hydrogen sulfite	Unstable product + decomposition	Double-replacement + decomposition	Ionic compound + $H_2O + SO_2$ Ionic compound + $H_2O + SO_2$	$H^+ + SO_3^{2-}$ $H^+ + HSO_3^-$	$H_2O + SO_2$ $H_2O + SO_2$
Ammonium ion compound + hydroxide base	"NH_4OH" + decomposition	Double-replacement + decomposition	Ionic compound + $NH_3 + H_2O$	$NH_4^+ + OH^-$	$H_2O + NH_3$

Thinking About Your Thinking

Classification

The classification thinking skill calls on your ability to change the way you see relationships among associated concepts when there is a better way of organizing them. This chapter presents a new way of classifying chemical equations that includes the ideas from Chapter 8 and also moves beyond them. Table 9.4 is designed to help you to reorganize your mental arrangement of the relationships among types of equations. Notice, in particular, how the equation types from Chapter 8 fit into the new scheme.

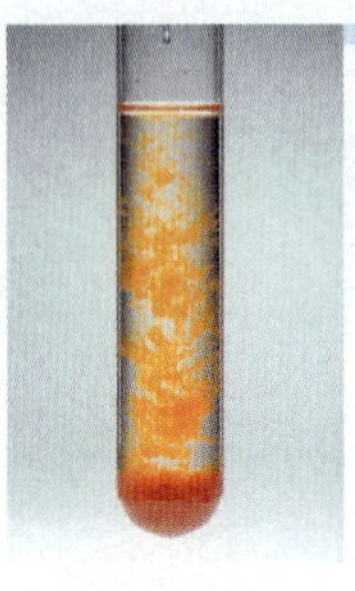

Charles D. Winters/Science Source

CHAPTER 9 IN REVIEW: CHEMICAL CHANGE

Goal 1 Distinguish between strong electrolytes, weak electrolytes, and nonelectrolytes.

A **solution** is a homogeneous mixture. A solute may be classified as a strong **electrolyte**, a weak electrolyte, or a nonelectrolyte according to the ability of its water solution to conduct electricity strongly, weakly, or not at all. If a solution conducts electricity, ions must be present as solute particles.

Goal 2 Given the formula of an ionic compound (or its name), write the formulas of the ions present when it is dissolved in water.

When an ionic compound dissolves in water, its solution consists of water molecules and ions surrounded by water molecules. The species in solution are aqueous ions, occurring in the same ratio as in the formula of the ionic compound.

Goal 3 Explain why the solution of an acid may be a good conductor or a poor conductor of electricity.

An **acid** is a substance that has a proton that can be removed by a water molecule when in a water solution. Acids are classified as strong or weak, depending on the extent to which the original compound ionizes when dissolved in water. When a **strong acid** dissolves, it dissociates into ions. The **major species** present in the solution are ions, and the **minor species** present are unionized molecules. Strong acids are strong electrolytes because the major species in solution are ions, and the presence of ions in a solution makes the solution a good conductor. When a **weak acid** dissolves, it does not dissociate into ions to a large extent. The solution inventory is mainly unionized molecules, and the solution is a poor conductor.

Goal 4 Given the formula or the name of a soluble acid, write the major and minor species present when it is dissolved in water.

There are seven common strong acids. Their names and formulas must be memorized: nitric acid, HNO_3; sulfuric acid, H_2SO_4; hydrochloric acid, HCl; hydrobromic acid, HBr; hydroiodic acid, HI; chloric acid, $HClO_3$; perchloric acid, $HClO_4$. If an acid is not one of the seven strong acids, it is a weak acid. Ions are the major species in the solutions of two kinds of substances: (1) all soluble ionic compounds and (2) the seven strong acids. Neutral molecules are the major species in solutions of everything else, primarily (1) compounds with all nonmetal atoms (except strong acids), (2) weak bases, and (3) water.

Goal 5 Distinguish among conventional, total ionic, and net ionic equations.

A **conventional equation** shows the formulas of the reactants written on the left side of an arrow and the formulas of the products on the right side of the arrow. Strong acids and ionic compounds designated (aq) in a conventional equation are rewritten in a **total ionic equation** with the formulas of the major species in solution. A total ionic equation is made into a **net ionic equation** by removing the spectators, those species that are on both sides of the total ionic equation. The procedure for writing a net ionic equation is as follows:

1. Write the conventional equation, including state symbols—(g), (ℓ), (s), and (aq). Balance the equation.
2. Write the total ionic equation by replacing each aqueous (aq) substance that is a strong acid or a soluble ionic compound with its major species. *Do not separate a weak acid into ions,* even though its state is aqueous (aq). Also, never change solids (s), liquids (ℓ), or gases (g) into ions. Be sure the equation is balanced in both atoms and charge.
3. Write the net ionic equation by removing the spectators from the total ionic equation. Reduce coefficients to lowest terms, if necessary. Be sure the equation is balanced in both atoms and charge.

Goal 6 Given two substances that may engage in a single-replacement redox reaction and an activity series by which the reaction may be predicted, write the conventional, total ionic, and net ionic equations for the reaction that will occur, if any.

A **single-replacement redox reaction** has the form A + BX → AX + B. The compounds BX and AX are usually aqueous and the elements A and B are usually solid metals or $H_2(g)$. Prediction of a single-replacement redox reaction is made by referring to an activity series. A more active element will replace the dissolved ions of any less active element.

Goal 7 Write the equation for the complete oxidation or burning of any compound containing only carbon and hydrogen or only carbon, hydrogen, and oxygen.

The general equation for a **complete oxidation (burning)** of a compound that consists of only carbon and hydrogen or only carbon, hydrogen, and oxygen is $C_xH_yO_z + O_2(g) \rightarrow CO_2(g) + H_2O(\ell)$. As a rule, these equations are most easily balanced if you balance the elements carbon, hydrogen, and oxygen in that order.

Goal 8 Predict whether a precipitate will form when known solutions are combined; if a precipitate forms, write the net ionic equation. (Reference to a solubility table or a solubility guidelines list may or may not be allowed, depending upon the preference of your instructor.)

An **ion-combination reaction** occurs when the cation (positively charged ion) from one reactant combines with the anion (negatively charged ion) from another to form a product compound. The conventional equation is a **double-replacement** type in which the ions appear to change partners: AX + BY → AY + BX. When a product in this type of reaction is an insoluble ionic compound, the solid is called a **precipitate** and the reaction is a **precipitation reaction.**

Goal 9 Given the product of a precipitation reaction, write the net ionic equation.

Your instructor will have you memorize **solubility guidelines** or allow you to have access to the guidelines or a solubility table. Use the method suggested by your instructor to predict precipitation reactions.

Goal 10 Given reactants for a double-replacement reaction that yield a molecular product, write the conventional, total ionic, and net ionic equation.

The reaction of an acid often leads to an ion combination that yields a **molecular product** instead of a precipitate. Except for the difference in the product, the equations are written in exactly the same way. Just as you had to recognize an insoluble product and not break it up into total ionic equations, you must recognize a molecular product and not break it into ions. Water and weak acids are the two kinds of molecular products you will find. **Neutralization** reactions are the most common molecular-product reactions, HX + MOH → HOH + MX. There are two points by which you can identify a molecular-product reaction: (1) one reactant is an acid, usually strong, and (2) one product is water or a weak acid.

Goal 11 Given reactants that form H_2CO_3, H_2SO_3, or "NH_4OH" by ion combination, write the net ionic equation for the reaction.

Three ion combinations yield molecular products that are not the products you would expect:

$$2\,H^+(aq) + CO_3^{2-}(aq) \rightarrow H_2CO_3(aq) \rightarrow CO_2(g) + H_2O(\ell)$$

$$2\,H^+(aq) + SO_3^{2-}(aq) \rightarrow H_2SO_3(aq) \rightarrow SO_2(aq) + H_2O(\ell)$$

$$NH_4^+(aq) + OH^-(aq) \rightarrow \text{“}NH_4OH\text{”} \rightarrow NH_3(aq) + H_2O(\ell)$$

When ion combinations yield **unstable** substances, the right side of the net ionic equation has the formulas of the stable decomposition products.

Key Terms

Most of the key terms and concepts and many others appear in the Glossary. Use your Glossary regularly.

acid p. 310
activity series p. 319
conventional equation p. 313
electrodes p. 305
electrolysis p. 305
ion-combination reaction p. 324
ions p. 313
major species p. 310
minor species p. 310
molecular product p. 328
net ionic equation p. 314
neutral molecules p. 313
neutralization reaction p. 328
nonelectrolyte p. 307
precipitate p. 324
precipitation reaction p. 324
seven common strong acids p. 311
solution p. 303
spectator ions p. 314
spectators p. 314
strong acid p. 310
strong electrolyte p. 307
total ionic equation p. 314
weak acid p. 310
weak electrolyte p. 307

Frequently Asked Questions

What should be the focal point for my review of this chapter?
Table 9.4 summarizes this chapter. It should be a focal point of your study of net ionic equations. The upper-left corner of the table is taken from Table 8.3; it describes conventional equations. Be sure to see the connection between these equations and the expanded Table 9.4.

What is the most critical step in the procedure for writing net ionic equations that I should focus on while doing the end-of-chapter questions, exercises, and problems?
If a conventional equation, *including states*, can be written, it can be converted into a net ionic equation by the three steps discussed in Section 9.5. In order to write these three steps, you must understand how to change a conventional equation into a total ionic equation. This is the critical step in the procedure. The electrolyte-classification exercise that follows will help you master this concept.

How do I decide if there is no reaction after being given potential reactants?
Check the activity series, check the solubility table or guidelines, or look for molecular products (water or a weak acid) to be sure there is a reaction.

What are the pitfalls to avoid in this chapter?
Several pitfalls await you in this chapter. Be careful of these:

1. Not watching out for double reactions. A double-replacement reaction can yield *two* nonionic products.
2. Failing to recognize unstable products (H_2CO_3, H_2SO_3, or "NH_4OH").
3. Incorrect or missing states of reactants and products. Many incorrect formulas are written because students do not recognize the states of some species.
4. Failing to recognize weak acids as molecular products that *do not* separate into ions. This wins the title of "most common error in writing net ionic equations."
5. "Inventing" diatomic ions because there are two atoms of an element in a compound. H_2^+ is the most common error.
6. Insufficient practice. Writing net ionic equations is a learning-by-doing skill. Making mistakes and learning from those mistakes is the usual route to success. Students who complete *both* steps *before* the exam are happier students.

Concept-Linking Exercises

Write a brief description of the relationships among each of the following groups of terms or phrases. Answers to the Concept-Linking Exercises are given at the end of the chapter.

1. Electrolyte, nonelectrolyte, strong electrolyte, weak electrolyte
2. Conventional equation, total ionic equation, net ionic equation, spectator ion
3. Redox reaction, ion-combination reaction, precipitation reaction, molecule-formation reaction, neutralization reaction

Electrolyte-Classification Exercise

A skill you need for writing net ionic equations successfully is the ability to identify the major species in an aqueous solution. For each solute formula (in aqueous solution) given in the Formula column, (a) classify it as a strong electrolyte, a weak electrolyte, or a nonelectrolyte; (b) identify the major species in solution; and (c) identify the minor species in solution (if any). The first two rows are completed as examples. Answers appear at the end of the chapter.

Formula	Electrolyte Classification	Major Species	Minor Species
1. HI	Strong electrolyte	$H^+(aq)$, $I^-(aq)$	HI(aq)
2. Na_2S	Strong electrolyte	$Na^+(aq)$, $S^{2-}(aq)$	none
3. $C_{12}H_{22}O_{11}$			
4. HNO_2			
5. HF			
6. LiF			
7. $HClO_4$			
8. $HCHO_2$			
9. NH_4NO_3			
10. $HC_2H_3O_2$			
11. HCl			
12. $C_6H_{12}O_6$			

Small-Group Discussion Questions

Small-Group Discussion Questions are for group work, either in class or under the guidance of a leader during a discussion section.

1. Draw particulate-level sketches of a strong electrolyte, a weak electrolyte, and a nonelectrolyte dissolved in water. Explain how each sketch fits the corresponding definition. Describe the conductivity of each solution. Write the formulas of at least three substances that fit each category.
2. Draw a particulate-level sketch of a tiny fraction of a grain of table salt dissolving in water. Start with separate salt and water, then show the salt dissolving, and finally, illustrate the salt water solution.
3. Draw particulate-level sketches of solutions of a strong acid and a weak acid. Based on your sketches, how are the terms *strong* and *weak* used to differentiate between these two acid classifications? How can you use symbols to illustrate the difference between chemical equations illustrating the dissociation of strong and weak acids?
4. A silver-white strip of zinc is placed in a blue solution of copper(II) nitrate. Describe the macroscopic changes that occur as the products of the reaction form. Zinc ion is colorless in aqueous solution, and finely divided copper appears black when wet. Draw a particulate-level sketch of this reaction. Explain how your macroscopic description is related to your particulate-level sketch.
5. Gasoline is a mixture of many different compounds. It is made by boiling and cooling crude oil to separate the components of crude oil into fractions, each of which represents a small range of boiling points. Small quantities of additives are mixed into the final product, which contains more than 150 different compounds. Major components include butane (C_4H_{10}), pentane (C_5H_{12}), hexane (C_6H_{14}), heptane (C_7H_{16}), toluene ($C_6H_5CH_3$), and xylene ($C_6H_4C_2H_6$). Write the equation for the complete oxidation of each of these components. Then write a procedure for writing and balancing hydrocarbon oxidation equations that could be used by a beginning chemistry student.
6. Cobalt(II) chloride and sodium carbonate solutions are combined, forming a precipitate. Explain how you can determine the identity of the precipitate. Draw particulate-level sketches of the separated solutions, and then illustrate the chemical change that occurs when the solutions are combined.
7. How are double-replacement molecule-formation reactions and double-replacement precipitation reactions similar? How are they different? Draw a particulate-level sketch of the formation of water from the combination solutions of two soluble ionic compounds. Compare this sketch with your sketch of the formation of a precipitate.
8. Copy Table 9.4 onto the board in your classroom. For each of the eight reaction types, give an example of a conventional, total ionic, and net ionic equation. Explain how each of your example reactions satisfies the description in each of the six columns in the table.

Questions, Exercises, and Problems

Interactive versions of these problems may be assigned by your instructor. Solutions for blue-numbered questions are at the end of the chapter. Questions other than those in the General Questions and More Challenging Problems sections are paired in consecutive odd-even number combinations; solutions for the odd numbered questions are also at the end of the chapter.

Many questions in this chapter are written with the assumption that you have studied Chapter 6 and can write the required formulas from their chemical names. If this is not the case, we have placed a list of all chemical formulas needed to answer this chapter's questions at the end of the Questions, Exercises, and Problems.

If you have studied Chapter 6 and you get stuck while answering a question because you cannot write a formula, we urge you to review the appropriate section of Chapter 6 before continuing. Avoid the temptation to "just peek" at the list of formulas. Developing skill in chemical nomenclature is part of your learning process in this course, and in many cases, you truly learn the material from Chapter 6 as you apply it here in this chapter and throughout your study of chemistry.

Section 9.2: Electrolytes and Solution Conductivity

1. How does a weak electrolyte differ from a nonelectrolyte?
2. Solid A is a strong electrolyte, and solid B is a nonelectrolyte. State how these two compounds differ.
3. How can it be that all soluble ionic compounds are electrolytes but soluble molecular compounds may or may not be electrolytes?
4. Compare the passage of electricity through a wire and through a solution. What conclusion may you draw if a liquid is able to carry an electrical current?

Sections 9.3 and 9.4: Solutions of Ionic Compounds; Strong and Weak Acids

Questions 5 through 12: Write the major species in the water solution of each substance given. All ionic compounds given are soluble.

5. $(NH_4)_2SO_4$, $MnCl_2$
6. $Mg(NO_3)_2$, $FeCl_3$
7. $NiSO_4$, K_3PO_4
8. HCN, $NaNO_3$, $HClO_4$
9. HNO_3, HBr
10. HCl, $HCHO_2$, KI
11. $H_2C_4H_4O_4$, HF
12. $NaClO_4$, HI, $HC_2H_3O_2$

Section 9.5: Net Ionic Equations: What They Are and How to Write Them

Questions 13 through 18: For each reaction described, write the net ionic equation.

13. A zinc chloride solution is mixed with a sodium phosphate solution, forming a precipitate of solid zinc phosphate and a sodium chloride solution.

14. When aqueous solutions of nickel(II) chloride and ammonium phosphate are combined, solid nickel(II) phosphate and a solution of ammonium chloride are formed.

15. Solid iron metal is dropped into a solution of hydrochloric acid. Hydrogen gas bubbles out, leaving a solution of iron(III) chloride.

16. When solid nickel metal is put into an aqueous solution of copper(II) sulfate, solid copper metal and a solution of nickel(II) sulfate result.

17. A solution of potassium nitrate and a solid precipitate of silver phosphate form when aqueous solutions of silver nitrate and potassium phosphate are mixed.

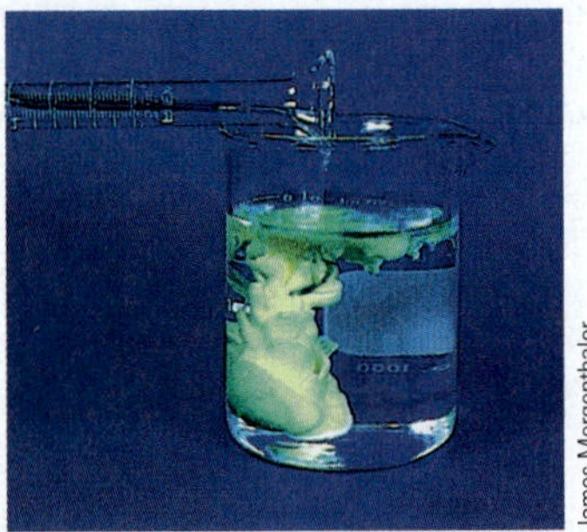
James Morgenthaler

Aqueous solutions of silver nitrate and potassium phosphate are combined.

18. When aqueous solutions of sodium nitrite and hydrobromic acid are mixed, an aqueous solution of sodium bromide and nitrous acid results.

Section 9.6: Single-Replacement Oxidation–Reduction (Redox) Reactions

Questions 19 through 24: For each pair of reactants, write the net ionic equation for any single-replacement redox reaction that may be predicted by Table 9.2 (Section 9.6). If no redox reaction occurs, write NR.

19. $Cu(s) + Li_2SO_4(aq)$

20. $Sn(s) + Zn(NO_3)_2(aq)$

21. $Ba(s) + HCl(aq)$

22. $Mg(s) + CuSO_4(aq)$

23. $Ni(s) + CaCl_2(aq)$

24. $Zn(s) + HNO_3(aq)$

Section 9.7: Oxidation–Reduction Reactions of Some Common Organic Compounds

Questions 25 through 28: Write the equation for each reaction described.

25. Propane, C_3H_8, a component of "bottled gas," is burned as a fuel in heating homes.

26. Butane, C_4H_{10}, is burned as lighter fluid in disposable lighters.

Charles D. Winters

Butane is the fuel in disposable lighters.

27. Ethanol, C_2H_5OH, the alcohol in alcoholic beverages, is oxidized.

28. Acetylene, C_2H_2, is burned in welding torches.

Section 9.8: Double-Replacement Precipitation Reactions

Questions 29 through 36: For each pair of reactants given, write the net ionic equation for any precipitation reaction that may be predicted by Table 9.3 or Figure 9.18 (Section 9.8). If no precipitation reaction occurs, write NR.

29. $Pb(NO_3)_2(aq) + KI(aq)$

30. $Co(NO_3)_2(aq) + K_3PO_4(aq)$

31. $KClO_3(aq) + Mg(NO_2)_2(aq)$

32. $Ba(NO_3)_2(aq) + Fe_2(SO_4)_3(aq)$

33. $AgNO_3(aq) + LiBr(aq)$

34. $NiBr_2(aq) + KOH(aq)$

35. $ZnCl_2(aq) + Na_2SO_3(aq)$

36. $NaNO_3(aq) + Ba(OH)_2(aq)$

37. Write the net ionic equations for the precipitation of each of the following insoluble ionic compounds from aqueous solutions: $PbCO_3$; $Ca(OH)_2$.

38. Write the net ionic equations for the precipitation of each of the following insoluble ionic compounds from aqueous solutions: lead(II) sulfide, copper(II) carbonate.

Section 9.9: Double-Replacement Molecule-Formation Reactions

Questions 39 through 44: For each pair of reactants given, write the net ionic equation for the molecule-formation reaction that will occur.

39. $NaNO_2(aq) + HI(aq)$

40. $KNO_2(aq) + HCl(aq)$

41. $KC_3H_5O_3(aq) + HClO_4(aq)$

42. $NaOH(aq) + HI(aq)$

43. $RbOH(aq) + HF(aq)$

44. $HC_2H_3O_2(aq) + Ba(OH)_2(aq)$

Section 9.10: Double-Replacement Reactions That Form Unstable Products

Questions 45 through 48: For each pair of reactants given, write the net ionic equation for the reaction that will occur.

45. $MgCO_3(aq) + HCl(aq)$
46. $HCl(aq) + CaSO_3(s)$
47. $Na_2SO_3(aq) + HClO_3(aq)$
48. $(NH_4)_2SO_4(aq) + NaOH(aq)$

Section 9.13: Summary of Net Ionic Equations

The remaining questions include all types of reactions discussed in this chapter. Use the activity series and solubility guidelines to predict whether redox or precipitation reactions will take place. If a reaction will take place, write the net ionic equation; if not, write NR.

49. Barium chloride and sodium sulfite solutions are combined in an oxygen-free atmosphere.
50. Aqueous solutions of nickel(II) chloride and potassium hydroxide are added together.

Solid nickel(II) hydroxide precipitates from solution.

51. Aqueous solutions of iron(III) sulfate and cobalt(II) nitrate are combined.
52. Copper(II) sulfate and sodium hydroxide solutions are combined.
53. Bubbles appear as hydrochloric acid is poured onto solid magnesium carbonate.
54. A piece of solid nickel metal is put into an aqueous solution of lead(II) nitrate.
55. Nitric acid appears to "dissolve" solid lead(II) hydroxide.
56. Solutions of hydrobromic acid and potassium carbonate are combined.
57. Liquid methanol, CH_3OH, is burned.
58. Aqueous solutions of sodium hydroxide and hydrochloric acid are combined.
59. Benzoic acid, $HC_7H_5O_2(s)$, is neutralized by sodium hydroxide solution.
60. Aqueous solutions of sodium hydroxide and potassium nitrate are combined.
61. Aluminum is placed into hydrochloric acid.

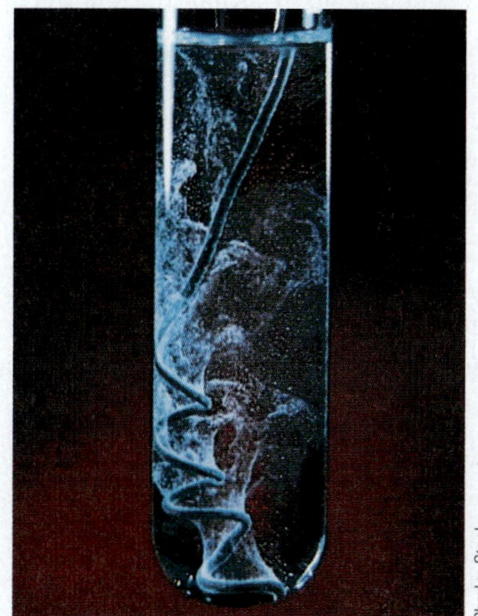

Aluminum in a hydrochloric acid solution.

62. A piece of solid copper metal is put into an aqueous solution of lead(II) nitrate.
63. Hydrochloric acid is poured into a solution of sodium hydrogen sulfite.
64. Aqueous solutions of sodium cyanide (cyanide ion, CN^-) and hydrobromic acid are mixed.
65. Solutions of magnesium sulfate and ammonium bromide are combined.
66. Aqueous solutions of sodium hydroxide and ammonium iodide are combined.
67. Magnesium ribbon is placed in hydrochloric acid.
68. Aqueous solutions of barium hydroxide and calcium iodide are combined.
69. Solid nickel(II) hydroxide is apparently readily "dissolved" by hydrobromic acid.
70. A piece of solid lead metal is put into an aqueous solution of nitric acid.
71. Sodium fluoride solution is poured into nitric acid.

72. Nitric acid and solid manganese(II) hydroxide are combined.

73. Silver wire is dropped into hydrochloric acid.

74. Aqueous solutions of silver nitrate and nickel(II) bromide are combined.

75. When solid lithium is added to water, hydrogen is released.

76. Nitric acid and solid calcium sulfite are combined.

77. Aluminum shavings are dropped into a solution of copper(II) nitrate.

78. Aqueous solutions of hydrofluoric acid and potassium hydroxide are combined.

General Questions

79. Distinguish precisely, and in scientific terms, the differences among items in each of the following pairs or groups:
   a) Strong electrolyte, weak electrolyte
   b) Electrolyte, nonelectrolyte
   c) Strong acid, weak acid
   d) Conventional, total ionic, and net ionic equations
   e) Burn, oxidize
   f) Ion-combination, precipitation, and molecule-formation reactions
   g) Molecule-formation and neutralization reactions
   h) Acid, base, salt

80. Classify each of the following statements as true or false:
   a) The solution of a weak electrolyte is a poor conductor of electricity.
   b) Ions must be present if a solution conducts electricity.
   c) Ions are the major species in a solution of a soluble ionic compound.
   d) There are no ions present in the solution of a weak acid.
   e) Only seven "important" acids are weak.
   f) Hydrofluoric acid, which is used to etch glass, is a strong acid.
   g) Spectators are included in a net ionic equation.
   h) A net ionic equation for a reaction between an element and an ion is the equation for a single-replacement redox reaction.
   i) A compound that is insoluble forms a precipitate when its ions are combined.
   j) Precipitation and molecule-formation reactions are both ion-combination reactions having double-replacement conventional equations.
   k) Neutralization is a special case of a molecule-formation reaction.
   l) One product of a molecule-formation reaction is a strong acid.
   m) Ammonium hydroxide is a possible product of a molecule-formation reaction.
   n) Carbon is changed to carbon dioxide when a carbon-containing compound burns completely.

Formulas

These formulas are provided in case you are studying this chapter before you study Chapter 6. You should not use this list unless your instructor has not yet assigned Chapter 6 or otherwise indicated that you may use the list.

13. $ZnCl_2$, Na_3PO_4, $Zn_3(PO_4)_2$, $NaCl$
14. $NiCl_2$, $(NH_4)_3PO_4$, $Ni_3(PO_4)_2$, NH_4Cl
15. Fe, HCl, H_2, $FeCl_3$
16. Ni, $CuSO_4$, Cu, $NiSO_4$
17. $H_2C_2O_4$, $NaCl$, $Na_2C_2O_4$, HCl
18. $NaNO_2$, HBr, $NaBr$, HNO_2
38. PbS, $CuCO_3$
49. $BaCl_2$, Na_2SO_3
50. $NiCl_2$, KOH
51. $Fe_2(SO_4)_3$, $Co(NO_3)_2$
52. $CuSO_4$, $NaOH$
53. HCl, $MgCO_3$
54. Ni, $Pb(NO_3)_2$
55. HNO_3, $Pb(OH)_2$
56. HBr, K_2CO_3
58. $NaOH$, HCl
59. $NaOH$
60. $NaOH$, KNO_3
61. Al, HCl
62. Cu, $Pb(NO_3)_2$
63. HCl, $NaHSO_3$
64. $NaCN$, HBr
65. $MgSO_4$, NH_4Br
66. $NaOH$, NH_4I
67. Mg, HCl
68. $Ba(OH)_2$, CaI_2
69. $Ni(OH)_2$, HBr
70. Pb, HNO_3
71. NaF, HNO_3
72. HNO_3, $Mn(OH)_2$
73. Ag, HCl
74. $AgNO_3$, $NiBr_2$
75. Li, H_2O, H_2
76. HNO_3, $CaSO_3$
77. Al, $Cu(NO_3)_2$
78. HF, KOH

Answer to Target Check

1. P is a weak electrolyte because its solution is a poor conductor. This indicates the presence of ions but in relatively low concentration. G is a strong electrolyte because its solution is a good conductor. The solution contains ions in relatively high concentration. N is a nonelectrolyte because it does not conduct. Its solution contains no ions.

Answers to Practice Exercises

1. $Ba(ClO_3)_2(s) \rightarrow Ba^{2+}(aq) + 2\,ClO_3^-(aq)$; $CoI_2(s) \rightarrow Co^{2+}(aq) + 2\,I^-(aq)$; $Mg(NO_2)_2(s) \rightarrow Mg^{2+}(aq) + 2\,NO_2^-(aq)$
2. $NiCl_2$: $Ni^{2+}(aq) + 2\,Cl^-(aq)$; Li_2CO_3: $2\,Li^+(aq) + CO_3^{2-}(aq)$; $Pb(NO_2)_2$: $Pb^{2+}(aq) + 2\,NO_2^-(aq)$; $(NH_4)_3PO_4$: $3\,NH_4^+(aq) + PO_4^{3-}(aq)$
3. HCl: $H^+(aq) + Cl^-(aq)$; H_2S: $H_2S(aq)$; $HClO_4$: $H^+(aq) + ClO_4^-(aq)$; HClO: HClO(aq)
4. $Al(NO_3)_3(aq) + 3\,NaOH(aq) \rightarrow Al(OH)_3(s) + 3\,NaNO_3(aq)$
   $Al^{3+}(aq) + 3\,NO_3^-(aq) + 3\,Na^+(aq) + 3\,OH^-(aq) \rightarrow Al(OH)_3(s) + 3\,Na^+(aq) + 3\,NO_3^-(aq)$
   $Al^{3+}(aq) + 3\,OH^-(aq) \rightarrow Al(OH)_3(s)$
5. $Zn(s) + 2\,HCl(aq) \rightarrow H_2(g) + ZnCl_2(aq)$
   $Zn(s) + 2\,H^+(aq) + 2\,Cl^-(aq) \rightarrow H_2(g) + Zn^{2+}(aq) + 2\,Cl^-(aq)$
   $Zn(s) + 2\,H^+(aq) \rightarrow H_2(g) + Zn^{2+}(aq)$
6. $Cl_2(g) + 2\,NaBr(aq) \rightarrow Br_2(\ell) + 2\,NaCl(aq)$
   $Cl_2(g) + 2\,Na^+(aq) + 2\,Br^-(aq) \rightarrow Br_2(\ell) + 2\,Na^+(aq) + 2\,Cl^-(aq)$
   $Cl_2(g) + 2\,Br^-(aq) \rightarrow Br_2(\ell) + 2\,Cl^-(aq)$
7. $Ba(s) + 2\,HOH(\ell) \rightarrow Ba(OH)_2(aq) + H_2(g)$
   $Ba(s) + 2\,H_2O(\ell) \rightarrow Ba^{2+}(aq) + 2\,OH^-(aq) + H_2(g)$
8. $Fe(s) + CuSO_4(aq) \rightarrow Cu(s) + FeSO_4(aq)$
   $Fe(s) + Cu^{2+}(aq) + SO_4^{2-}(aq) \rightarrow Cu(s) + Fe^{2+}(aq) + SO_4^{2-}(aq)$
   $Fe(s) + Cu^{2+}(aq) \rightarrow Cu(s) + Fe^{2+}(aq)$
9. $CH_3CH_2OH(\ell) + 3\,O_2(g) \rightarrow 2\,CO_2(g) + 3\,H_2O(\ell)$
10. $Pb(NO_3)_2(aq) + 2\,KI(aq) \rightarrow PbI_2(s) + 2\,KNO_3(aq)$
    $Pb^{2+}(aq) + 2\,NO_3^-(aq) + 2\,K^+(aq) + 2\,I^-(aq) \rightarrow PbI_2(s) + 2\,K^+(aq) + 2\,NO_3^-(aq)$
    $Pb^{2+}(aq) + 2\,I^-(aq) \rightarrow PbI_2(s)$
11. $NiBr_2(aq) + (NH_4)_2SO_3(aq) \rightarrow NiSO_3(s) + 2\,NH_4Br(aq)$
    $Ni^{2+}(aq) + 2\,Br^-(aq) + 2\,NH_4^+(aq) + SO_3^{2-}(aq) \rightarrow NiSO_3(s) + 2\,NH_4^+(aq) + 2\,Br^-(aq)$
    $Ni^{2+}(aq) + SO_3^{2-}(aq) \rightarrow NiSO_3(s)$
12. $AlBr_3(aq) + 3\,NaF(aq) \rightarrow AlF_3(s) + 3\,NaBr(aq)$
    $Al^{3+}(aq) + 3\,Br^-(aq) + 3\,Na^+(aq) + 3\,F^-(aq) \rightarrow AlF_3(s) + 3\,Na^+(aq) + 3\,Br^-(aq)$
    $Al^{3+}(aq) + 3\,F^-(aq) \rightarrow AlF_3(s)$
13. $2\,HBr(aq) + Ba(OH)_2(aq) \rightarrow 2\,HOH(\ell) + BaBr_2(aq)$
    $2\,H^+(aq) + 2\,Br^-(aq) + Ba^{2+}(aq) + 2\,OH^-(aq) \rightarrow 2\,HOH(\ell) + Ba^{2+}(aq) + 2\,Br^-(aq)$
    $2\,H^+(aq) + 2\,OH^-(aq) \rightarrow 2\,H_2O(\ell)$
    $H^+(aq) + OH^-(aq) \rightarrow H_2O(\ell)$
14. $HI(aq) + NaF(aq) \rightarrow HF(aq) + NaI(aq)$
    $H^+(aq) + I^-(aq) + Na^+(aq) + F^-(aq) \rightarrow HF(aq) + Na^+(aq) + I^-(aq)$
    $H^+(aq) + F^-(aq) \rightarrow HF(aq)$
15. $HClO(aq) + KOH(aq) \rightarrow HOH(\ell) + KClO(aq)$
    $HClO(aq) + K^+(aq) + OH^-(aq) \rightarrow HOH(\ell) + K^+(aq) + ClO^-(aq)$
    $HClO(aq) + OH^-(aq) \rightarrow H_2O(\ell) + ClO^-(aq)$
16. $NH_4NO_3(aq) + NaOH(aq) \rightarrow NH_3(aq) + H_2O(\ell) + NaNO_3(aq)$
    $NH_4^+(aq) + NO_3^-(aq) + Na^+(aq) + OH^-(aq) \rightarrow NH_3(aq) + H_2O(\ell) + Na^+(aq) + NO_3^-(aq)$
    $NH_4^+(aq) + OH^-(aq) \rightarrow NH_3(aq) + H_2O(\ell)$
17. $2\,HNO_3(aq) + NiCO_3(s) \rightarrow H_2O(\ell) + CO_2(g) + Ni(NO_3)_2(aq)$
    $2\,H^+(aq) + 2\,NO_3^-(aq) + NiCO_3(s) \rightarrow H_2O(\ell) + CO_2(g) + Ni^{2+}(aq) + 2\,NO_3^-(aq)$
    $2\,H^+(aq) + NiCO_3(s) \rightarrow H_2O(\ell) + CO_2(g) + Ni^{2+}(aq)$
18. $2\,NaNO_3(aq) + (NH_4)_2SO_4(aq) \rightarrow Na_2SO_4(aq) + 2\,NH_4NO_3(aq)$
    $2\,Na^+(aq) + 2\,NO_3^-(aq) + 2\,NH_4^+(aq) + SO_4^{2-}(aq) \rightarrow 2\,Na^+(aq) + SO_4^{2-}(aq) + 2\,NH_4^+(aq) + 2\,NO_3^-(aq)$
    $2\,Na^+(aq) + 2\,NO_3^-(aq) + 2\,NH_4^+(aq) + SO_4^{2-}(aq) \rightarrow NR$

Answers to Concept-Linking Exercises

You may have found more relationships or relationships other than the ones given in these answers.

1. An electrolyte is a substance that, when dissolved, yields a solution that conducts electricity. A nonelectrolyte yields a solution that does not conduct electricity. The terms *electrolyte* and *nonelectrolyte* also refer to solutions of dissolved substances. A strong electrolyte is a substance that forms a solution that is a good conductor of electricity. A weak electrolyte conducts electricity but does so poorly.
2. A conventional equation shows the complete formulas of the compounds and elements in a chemical reaction. Substances that exist as ions in water solutions are not shown in their ionic form, but rather as the hypothetical unit particle represented by the formula. A total ionic equation replaces the formulas of the dissolved substances with the major species in solution, and it includes spectator ions—those that experience no chemical change. The net ionic equation is written by removing the spectators from the total ionic equation, leaving only the species that undergo a chemical change.
3. A redox reaction is an electron-transfer reaction. Electrons are transferred from one reacting species to another. An uncombined element appears to replace another element in a compound to yield a single-replacement equation. In ion-combination reactions, the cation from one reactant compound combines with the anion from another. These reactions have double-replacement equations. A precipitation reaction is an ion-combination reaction that yields an insoluble ionic compound. A molecule-formation reaction is an ion-combination reaction that yields a molecular product, usually water or a weak acid. If the product of an ion-combination reaction is water, formed by the reaction of an acid and a hydroxide base, the reaction is a neutralization reaction.

Answers to Electrolyte-Classification Exercise

Formula	Electrolyte Classification	Major Species	Minor Species*
1. HI	Strong electrolyte	$H^+(aq)$, $I^-(aq)$	HI(aq)
2. Na_2S	Strong electrolyte	$Na^+(aq)$, $S^{2-}(aq)$	None
3. $C_{12}H_{22}O_{11}$	Nonelectrolyte	$C_{12}H_{22}O_{11}(aq)$	None
4. HNO_2	Weak electrolyte	$HNO_2(aq)$	$H^+(aq)$, $NO_2^-(aq)$
5. HF	Weak electrolyte	HF(aq)	$H^+(aq)$, $F^-(aq)$
6. LiF	Strong electrolyte	$Li^+(aq)$, $F^-(aq)$	None
7. $HClO_4$	Strong electrolyte	$H^+(aq)$, $ClO_4^-(aq)$	$HClO_4(aq)$
8. $HCHO_2$	Weak electrolyte	$HCHO_2(aq)$	$H^+(aq)$, $CHO_2^-(aq)$
9. NH_4NO_3	Strong electrolyte	$NH_4^+(aq)$, $NO_3^-(aq)$	None
10. $HC_2H_3O_2$	Weak electrolyte	$HC_2H_3O_2(aq)$	$H^+(aq)$, $C_2H_3O_2^-(aq)$
11. HCl	Strong electrolyte	$H^+(aq)$, $Cl^-(aq)$	HCl(aq)
12. $C_6H_{12}O_6$	Nonelectrolyte	$C_6H_{12}O_6(aq)$	None

*No ionic compound is completely insoluble in water. There are, in fact, very small concentrations of calcium and carbonate ions in solution in the presence of solid calcium carbonate and very small concentrations of aluminum and phosphate ions in solution when that solid is placed in water. Additionally, there are very small concentrations of hydrogen and hydroxide ions present in water or any water solution.

Answers to Blue-Numbered Questions, Exercises, and Problems

1. Solutions of weak electrolytes conduct electricity poorly because only a small quantity of ions forms when the electrolyte dissolves. Solutions of nonelectrolytes do not conduct electricity because essentially no ions exist in the solution.

3. The movement of ions makes up an electric current in solution. Soluble molecular compounds are generally neutral and are usually nonelectrolytes. Some molecular compounds can react with water, forming ions as a product, and thus can act as electrolytes.

5. $NH_4^+(aq)$, $SO_4^{2-}(aq)$; $Mn^{2+}(aq)$, $Cl^-(aq)$

7. $Ni^{2+}(aq)$, $SO_4^{2-}(aq)$; $K^+(aq)$, $PO_4^{3-}(aq)$

9. $H^+(aq)$, $NO_3^-(aq)$; $H^+(aq)$, $Br^-(aq)$

11. $H_2C_4H_4O_4(aq)$, HF(aq)

13. $3\ Zn^{2+}(aq) + 2\ PO_4^{3-}(aq) \rightarrow Zn_3(PO_4)_2(s)$

15. $2\ Fe(s) + 6\ H^+(aq) \rightarrow 3\ H_2(g) + 2\ Fe^{3+}(aq)$

17. $Na_2C_2O_4(s) + 2\ H^+(aq) \rightarrow H_2C_2O_4(aq) + 2\ Na^+(aq)$

19. $Cu(s) + Li_2SO_4(aq) \rightarrow NR$

21. $Ba(s) + 2\ H^+(aq) \rightarrow H_2(g) + Ba^{2+}(aq)$

23. $Ni(s) + CaCl_2(aq) \rightarrow NR$

25. $C_3H_8(\ell) + 5\ O_2(g) \rightarrow 3\ CO_2(g) + 4\ H_2O(\ell)$

27. $C_2H_5OH(\ell) + 3\ O_2(g) \rightarrow 2\ CO_2(g) + 3\ H_2O(\ell)$

29. $Pb^{2+}(aq) + 2\ I^-(aq) \rightarrow PbI_2(s)$

31. $KClO_3(aq) + Mg(NO_2)_2(aq) \rightarrow NR$

33. $Ag^+(aq) + Br^-(aq) \rightarrow AgBr(s)$

35. $Zn^{2+}(aq) + SO_3^{2-}(aq) \rightarrow ZnSO_3(s)$

37. $Pb^{2+}(aq) + CO_3^{2-}(aq) \rightarrow PbCO_3(s)$; $Ca^{2+}(aq) + 2\ OH^-(aq) \rightarrow Ca(OH)_2(s)$

39. $H^+(aq) + NO_2^-(aq) \rightarrow HNO_2(aq)$

41. $H^+(aq) + C_3H_5O_3^-(aq) \rightarrow HC_3H_5O_3(aq)$

43. $OH^-(aq) + HF(aq) \rightarrow H_2O(\ell) + F^-(aq)$

45. $2\ H^+(aq) + CO_3^{2-}(aq) \rightarrow H_2O(\ell) + CO_2(aq)$

47. $2\ H^+(aq) + SO_3^{2-}(aq) \rightarrow H_2O(\ell) + SO_2(aq)$

49. $Ba^{2+}(aq) + SO_3^{2-}(aq) \rightarrow BaSO_3(s)$

51. $Fe_2(SO_4)_3(aq) + Co(NO_3)_2(aq) \rightarrow NR$

53. $2\ H^+(aq) + MgCO_3(s) \rightarrow Mg^{2+}(aq) + H_2O(\ell) + CO_2(g)$

55. $2\ H^+(aq) + Pb(OH)_2(s) \rightarrow Pb^{2+}(aq) + 2\ H_2O(\ell)$

57. $2\ CH_3OH(\ell) + 3\ O_2(g) \rightarrow 2\ CO_2(g) + 4\ H_2O(\ell)$

59. $HC_7H_5O_2(s) + OH^-(aq) \rightarrow C_7H_5O_2^-(aq) + H_2O(\ell)$

61. $2\ Al(s) + 6\ H^+(aq) \rightarrow 2\ Al^{3+}(aq) + 3\ H_2(g)$

63. $H^+(aq) + HSO_3^-(aq) \rightarrow H_2O(\ell) + SO_2(aq)$

65. $MgSO_4(aq) + NH_4Br(aq) \rightarrow NR$

67. $Mg(s) + 2\ H^+(aq) \rightarrow Mg^{2+}(aq) + H_2(g)$

69. $Ni(OH)_2(s) + 2\ H^+(aq) \rightarrow Ni^{2+}(aq) + 2\ H_2O(\ell)$

71. $F^-(aq) + H^+(aq) \rightarrow HF(aq)$

73. $Ag(s) + HCl(aq) \rightarrow NR$

75. $2\ Li(s) + 2\ H_2O(\ell) \rightarrow 2\ Li^+(aq) + 2\ OH^-(aq) + H_2(g)$

77. $2\ Al(s) + 3\ Cu^{2+}(aq) \rightarrow 2\ Al^{3+}(aq) + 3\ Cu(s)$

80. True: a, b, c, h, i, j, k, n. False: d, e, f, g, l, m.

Answers to Practice Exercises

1. $Ba(ClO_3)_2(s) \rightarrow Ba^{2+}(aq) + 2\ ClO_3^-(aq)$; $CoI_2(s) \rightarrow Co^{2+}(aq) + 2\ I^-(aq)$; $Mg(NO_2)_2(s) \rightarrow Mg^{2+}(aq) + 2\ NO_2^-(aq)$

2. $NiCl_2$: $Ni^{2+}(aq) + 2\ Cl^-(aq)$; Li_2CO_3: $2\ Li^+(aq) + CO_3^{2-}(aq)$; $Pb(NO_2)_2$: $Pb^{2+}(aq) + 2\ NO_2^-(aq)$; $(NH_4)_3PO_4$: $3\ NH_4^+(aq) + PO_4^{3-}(aq)$

3. HCl: $H^+(aq) + Cl^-(aq)$; H_2S: $H_2S(aq)$; $HClO_4$: $H^+(aq) + ClO_4^-(aq)$; HClO: HClO(aq)

4. $Al(NO_3)_3(aq) + 3\ NaOH(aq) \rightarrow Al(OH)_3(s) + 3\ NaNO_3(aq)$
   $Al^{3+}(aq) + 3\ NO_3^-(aq) + 3\ Na^+(aq) + 3\ OH^-(aq) \rightarrow Al(OH)_3(s) + 3\ Na^+(aq) + 3\ NO_3^-(aq)$
   $Al^{3+}(aq) + 3\ OH^-(aq) \rightarrow Al(OH)_3(s)$

5. $Zn(s) + 2\ HCl(aq) \rightarrow H_2(g) + ZnCl_2(aq)$
   $Zn(s) + 2\ H^+(aq) + 2\ Cl^-(aq) \rightarrow H_2(g) + Zn^{2+}(aq) + 2\ Cl^-(aq)$
   $Zn(s) + 2\ H^+(aq) \rightarrow H_2(g) + Zn^{2+}(aq)$

6. $Cl_2(g) + 2\ NaBr(aq) \rightarrow Br_2(\ell) + 2\ NaCl(aq)$
   $Cl_2(g) + 2\ Na^+(aq) + 2\ Br^-(aq) \rightarrow Br_2(\ell) + 2\ Na^+(aq) + 2\ Cl^-(aq)$
   $Cl_2(g) + 2\ Br^-(aq) \rightarrow Br_2(\ell) + 2\ Cl^-(aq)$

7. $Ba(s) + 2\ HOH(\ell) \rightarrow Ba(OH)_2(aq) + H_2(g)$
   $Ba(s) + 2\ H_2O(\ell) \rightarrow Ba^{2+}(aq) + 2\ OH^-(aq) + H_2(g)$

8. $Fe(s) + CuSO_4(aq) \rightarrow Cu(s) + FeSO_4(aq)$
   $Fe(s) + Cu^{2+}(aq) + SO_4^{2-}(aq) \rightarrow Cu(s) + Fe^{2+}(aq) + SO_4^{2-}(aq)$
   $Fe(s) + Cu^{2+}(aq) \rightarrow Cu(s) + Fe^{2+}(aq)$

9. $CH_3CH_2OH(\ell) + 3\ O_2(g) \rightarrow 2\ CO_2(g) + 3\ H_2O(\ell)$

10. $Pb(NO_3)_2(aq) + 2\ KI(aq) \rightarrow PbI_2(s) + 2\ KNO_3(aq)$
    $Pb^{2+}(aq) + 2\ NO_3^-(aq) + 2\ K^+(aq) + 2\ I^-(aq) \rightarrow PbI_2(s) + 2\ K^+(aq) + 2\ NO_3^-(aq)$
    $Pb^{2+}(aq) + 2\ I^-(aq) \rightarrow PbI_2(s)$

11. $NiBr_2(aq) + (NH_4)_2SO_3(aq) \rightarrow NiSO_3(s) + 2\ NH_4Br(aq)$
    $Ni^{2+}(aq) + 2\ Br^-(aq) + 2\ NH_4^+(aq) + SO_3^{2-}(aq) \rightarrow NiSO_3(s) + 2\ NH_4^+(aq) + 2\ Br^-(aq)$
    $Ni^{2+}(aq) + SO_3^{2-}(aq) \rightarrow NiSO_3(s)$

12. $AlBr_3(aq) + 3\ NaF(aq) \rightarrow AlF_3(s) + 3\ NaBr(aq)$
    $Al^{3+}(aq) + 3\ Br^-(aq) + 3\ Na^+(aq) + 3\ F^-(aq) \rightarrow AlF_3(s) + 3\ Na^+(aq) + 3\ Br^-(aq)$
    $Al^{3+}(aq) + 3\ F^-(aq) \rightarrow AlF_3(s)$

13. $2\ HBr(aq) + Ba(OH)_2(aq) \rightarrow 2\ HOH(\ell) + BaBr_2(aq)$
    $2\ H^+(aq) + 2\ Br^-(aq) + Ba^{2+}(aq) + 2\ OH^-(aq) \rightarrow 2\ HOH(\ell) + Ba^{2+}(aq) + 2\ Br^-(aq)$
    $2\ H^+(aq) + 2\ OH^-(aq) \rightarrow 2\ H_2O(\ell)$
    $H^+(aq) + OH^-(aq) \rightarrow H_2O(\ell)$

14. $HI(aq) + NaF(aq) \rightarrow HF(aq) + NaI(aq)$
    $H^+(aq) + I^-(aq) + Na^+(aq) + F^-(aq) \rightarrow HF(aq) + Na^+(aq) + I^-(aq)$
    $H^+(aq) + F^-(aq) \rightarrow HF(aq)$

15. $HClO(aq) + KOH(aq) \rightarrow HOH(\ell) + KClO(aq)$
    $HClO(aq) + K^+(aq) + OH^-(aq) \rightarrow HOH(\ell) + K^+(aq) + ClO^-(aq)$
    $HClO(aq) + OH^-(aq) \rightarrow H_2O(\ell) + ClO^-(aq)$

16. $NH_4NO_3(aq) + NaOH(aq) \rightarrow NH_3(aq) + H_2O(\ell) + NaNO_3(aq)$
    $NH_4^+(aq) + NO_3^-(aq) + Na^+(aq) + OH^-(aq) \rightarrow NH_3(aq) + H_2O(\ell) + Na^+(aq) + NO_3^-(aq)$
    $NH_4^+(aq) + OH^-(aq) \rightarrow NH_3(aq) + H_2O(\ell)$

17. $2\ HNO_3(aq) + NiCO_3(s) \rightarrow H_2O(\ell) + CO_2(g) + Ni(NO_3)_2(aq)$
    $2\ H^+(aq) + 2\ NO_3^-(aq) + NiCO_3(s) \rightarrow H_2O(\ell) + CO_2(g) + Ni^{2+}(aq) + 2\ NO_3^-(aq)$
    $2\ H^+(aq) + NiCO_3(s) \rightarrow H_2O(\ell) + CO_2(g) + Ni^{2+}(aq)$

18. $2\ NaNO_3(aq) + (NH_4)_2SO_4(aq) \rightarrow Na_2SO_4(aq) + 2\ NH_4NO_3(aq)$
    $2\ Na^+(aq) + 2\ NO_3^-(aq) + 2\ NH_4^+(aq) + SO_4^{2-}(aq) \rightarrow 2\ Na^+(aq) + SO_4^{2-}(aq) + 2\ NH_4^+(aq) + 2\ NO_3^-(aq)$
    $2\ Na^+(aq) + 2\ NO_3^-(aq) + 2\ NH_4^+(aq) + SO_4^{2-}(aq) \rightarrow NR$

Answers to Concept-Linking Exercises

You may have found more relationships or relationships other than the ones given in these answers.

1. An electrolyte is a substance that, when dissolved, yields a solution that conducts electricity. A nonelectrolyte yields a solution that does not conduct electricity. The terms *electrolyte* and *nonelectrolyte* also refer to solutions of dissolved substances. A strong electrolyte is a substance that forms a solution that is a good conductor of electricity. A weak electrolyte conducts electricity but does so poorly.

2. A conventional equation shows the complete formulas of the compounds and elements in a chemical reaction. Substances that exist as ions in water solutions are not shown in their ionic form, but rather as the hypothetical unit particle represented by the formula. A total ionic equation replaces the formulas of the dissolved substances with the major species in solution, and it includes spectator ions—those that experience no chemical change. The net ionic equation is written by removing the spectators from the total ionic equation, leaving only the species that undergo a chemical change.

3. A redox reaction is an electron-transfer reaction. Electrons are transferred from one reacting species to another. An uncombined element appears to replace another element in a compound to yield a single-replacement equation. In ion-combination reactions, the cation from one reactant compound combines with the anion from another. These reactions have double-replacement equations. A precipitation reaction is an ion-combination reaction that yields an insoluble ionic compound. A molecule-formation reaction is an ion-combination reaction that yields a molecular product, usually water or a weak acid. If the product of an ion-combination reaction is water, formed by the reaction of an acid and a hydroxide base, the reaction is a neutralization reaction.

Answers to Electrolyte-Classification Exercise

Formula	Electrolyte Classification	Major Species	Minor Species*
1. HI	Strong electrolyte	$H^+(aq)$, $I^-(aq)$	HI(aq)
2. Na_2S	Strong electrolyte	$Na^+(aq)$, $S^{2-}(aq)$	None
3. $C_{12}H_{22}O_{11}$	Nonelectrolyte	$C_{12}H_{22}O_{11}(aq)$	None
4. HNO_2	Weak electrolyte	$HNO_2(aq)$	$H^+(aq)$, $NO_2^-(aq)$
5. HF	Weak electrolyte	HF(aq)	$H^+(aq)$, $F^-(aq)$
6. LiF	Strong electrolyte	$Li^+(aq)$, $F^-(aq)$	None
7. $HClO_4$	Strong electrolyte	$H^+(aq)$, $ClO_4^-(aq)$	$HClO_4(aq)$
8. $HCHO_2$	Weak electrolyte	$HCHO_2(aq)$	$H^+(aq)$, $CHO_2^-(aq)$
9. NH_4NO_3	Strong electrolyte	$NH_4^+(aq)$, $NO_3^-(aq)$	None
10. $HC_2H_3O_2$	Weak electrolyte	$HC_2H_3O_2(aq)$	$H^+(aq)$, $C_2H_3O_2^-(aq)$
11. HCl	Strong electrolyte	$H^+(aq)$, $Cl^-(aq)$	HCl(aq)
12. $C_6H_{12}O_6$	Nonelectrolyte	$C_6H_{12}O_6(aq)$	None

*No ionic compound is completely insoluble in water. There are, in fact, very small concentrations of calcium and carbonate ions in solution in the presence of solid calcium carbonate and very small concentrations of aluminum and phosphate ions in solution when that solid is placed in water. Additionally, there are very small concentrations of hydrogen and hydroxide ions present in water or any water solution.

Answers to Blue-Numbered Questions, Exercises, and Problems

1. Solutions of weak electrolytes conduct electricity poorly because only a small quantity of ions forms when the electrolyte dissolves. Solutions of nonelectrolytes do not conduct electricity because essentially no ions exist in the solution.

3. The movement of ions makes up an electric current in solution. Soluble molecular compounds are generally neutral and are usually nonelectrolytes. Some molecular compounds can react with water, forming ions as a product, and thus can act as electrolytes.

5. $NH_4^+(aq)$, $SO_4^{2-}(aq)$; $Mn^{2+}(aq)$, $Cl^-(aq)$

7. $Ni^{2+}(aq)$, $SO_4^{2-}(aq)$; $K^+(aq)$, $PO_4^{3-}(aq)$

9. $H^+(aq)$, $NO_3^-(aq)$; $H^+(aq)$, $Br^-(aq)$

11. $H_2C_4H_4O_4(aq)$, HF(aq)

13. $3\,Zn^{2+}(aq) + 2\,PO_4^{3-}(aq) \rightarrow Zn_3(PO_4)_2(s)$

15. $2\,Fe(s) + 6\,H^+(aq) \rightarrow 3\,H_2(g) + 2\,Fe^{3+}(aq)$

17. $Na_2C_2O_4(s) + 2\,H^+(aq) \rightarrow H_2C_2O_4(aq) + 2\,Na^+(aq)$

19. $Cu(s) + Li_2SO_4(aq) \rightarrow NR$

21. $Ba(s) + 2\,H^+(aq) \rightarrow H_2(g) + Ba^{2+}(aq)$

23. $Ni(s) + CaCl_2(aq) \rightarrow NR$

25. $C_3H_8(\ell) + 5\,O_2(g) \rightarrow 3\,CO_2(g) + 4\,H_2O(\ell)$

27. $C_2H_5OH(\ell) + 3\,O_2(g) \rightarrow 2\,CO_2(g) + 3\,H_2O(\ell)$

29. $Pb^{2+}(aq) + 2\,I^-(aq) \rightarrow PbI_2(s)$

31. $KClO_3(aq) + Mg(NO_2)_2(aq) \rightarrow NR$

33. $Ag^+(aq) + Br^-(aq) \rightarrow AgBr(s)$

35. $Zn^{2+}(aq)+ SO_3^{2-}(aq) \rightarrow ZnSO_3(s)$

37. $Pb^{2+}(aq) + CO_3^{2-}(aq) \rightarrow PbCO_3(s)$; $Ca^{2+}(aq) + 2\,OH^-(aq) \rightarrow Ca(OH)_2(s)$

39. $H^+(aq) + NO_2^-(aq) \rightarrow HNO_2(aq)$

41. $H^+(aq) + C_3H_5O_3^-(aq) \rightarrow HC_3H_5O_3(aq)$

43. $OH^-(aq) + HF(aq) \rightarrow H_2O(\ell) + F^-(aq)$

45. $2\,H^+(aq) + CO_3^{2-}(aq) \rightarrow H_2O(\ell) + CO_2(aq)$

47. $2\,H^+(aq) + SO_3^{2-}(aq) \rightarrow H_2O(\ell) + SO_2(aq)$

49. $Ba^{2+}(aq) + SO_3^{2-}(aq) \rightarrow BaSO_3(s)$

51. $Fe_2(SO_4)_3(aq) + Co(NO_3)_2(aq) \rightarrow NR$

53. $2\,H^+(aq) + MgCO_3(s) \rightarrow Mg^{2+}(aq) + H_2O(\ell) + CO_2(g)$

55. $2\,H^+(aq) + Pb(OH)_2(s) \rightarrow Pb^{2+}(aq) + 2\,H_2O(\ell)$

57. $2\,CH_3OH(\ell) + 3\,O_2(g) \rightarrow 2\,CO_2(g) + 4\,H_2O(\ell)$

59. $HC_7H_5O_2(s) + OH^-(aq) \rightarrow C_7H_5O_2^-(aq) + H_2O(\ell)$

61. $2\,Al(s) + 6\,H^+(aq) \rightarrow 2\,Al^{3+}(aq) + 3\,H_2(g)$

63. $H^+(aq) + HSO_3^-(aq) \rightarrow H_2O(\ell) + SO_2(aq)$

65. $MgSO_4(aq) + NH_4Br(aq) \rightarrow NR$

67. $Mg(s) + 2\,H^+(aq) \rightarrow Mg^{2+}(aq) + H_2(g)$

69. $Ni(OH)_2(s) + 2\,H^+(aq) \rightarrow Ni^{2+}(aq) + 2\,H_2O(\ell)$

71. $F^-(aq) + H^+(aq) \rightarrow HF(aq)$

73. $Ag(s) + HCl(aq) \rightarrow NR$

75. $2\,Li(s) + 2\,H_2O(\ell) \rightarrow 2\,Li^+(aq) + 2\,OH^-(aq) + H_2(g)$

77. $2\,Al(s) + 3\,Cu^{2+}(aq) \rightarrow 2\,Al^{3+}(aq) + 3\,Cu(s)$

80. True: a, b, c, h, i, j, k, n. False: d, e, f, g, l, m.

10 Quantity Relationships in Chemical Reactions

Bloomberg/Getty Images

◀ Chlorine is the most produced inorganic chemical in the world. It is usually manufactured in large industrial plants by subjecting a salt water solution to an electrical current, yielding chlorine, hydrogen, and a solution of sodium hydroxide: $2\ NaCl(s) + 2\ H_2O(\ell) \rightarrow Cl_2(g) + H_2(g) + 2\ NaOH(aq)$. The type of question that this chapter asks—and answers—is, "What mass of sodium chloride is needed for the manufacture of a specified quantity of chlorine?"

CHAPTER CONTENTS

In this chapter, you will develop an understanding of the **stoichiometry** concept. A stoichiometry problem asks, "How much, or how many?" How many tons of sodium chloride must be electrolyzed to produce 10 tons of sodium hydroxide? How many kiloliters of chlorine at a certain temperature and pressure will be produced at the same time? How much energy is needed to do the job? These are only a few examples of the types of quantity relationships in chemical reactions that you will learn how to setup and calculate.

10.1 Okay, Houston, We've Had a Problem Here

The 1995 Academy Award winning film *Apollo 13* starring Tom Hanks and directed by Ron Howard made famous the plight of the April 1970 three-man mission to the moon. The Apollo missions were conducted by the National Aeronautics and Space Administration (NASA) with the goal of landing people on the moon, allowing them to make measurements and collect samples, and returning them safely to Earth. At almost 56 hours after launch on the 80-hour Apollo 13 trip to the moon, the astronauts heard a loud, abnormal sound and then, among numerous other malfunctions, the volume reading of one of the two oxygen tanks in the service module was zero and the reading on the second tank was rapidly decreasing.

Command module pilot Jack Swigert informed the Mission Operations Control Room about the malfunction with the renowned statement "Okay, Houston, we've had a problem here."*

The damage to the spacecraft was sufficiently severe to make returning to Earth as quickly as possible the only choice, and the challenge for the NASA team was to protect the lives of the men for the remaining 87 hours needed to fly around the moon and use its gravity to redirect the spacecraft and return to Earth. Among the many problems that needed to be solved on the fly, one of the simpler ones was the issue of the loss of the two oxygen tanks. There was sufficient oxygen in the portion of the spacecraft designed for the lunar descent—the lunar module—to sustain the men on the return trip. They would simply have to move from the command module to the lunar module. However, a related issue was a major problem. The carbon dioxide exhaled by the astronauts needed to be removed from the air in the lunar module. This was done by the reaction of carbon dioxide with solid lithium hydroxide: $CO_2 + 2\,LiOH \rightarrow Li_2CO_3 + H_2O$.

The lunar module remained attached to the remainder of the spacecraft. But it was designed for relatively short trips from the orbiting command module to the moon and back. It had enough lithium hydroxide to sustain two men for 30 hours. (How did NASA scientists know the mass of lithium hydroxide needed to destroy the amount of carbon dioxide that was anticipated to be exhaled?) There was a sufficient supply of lithium hydroxide canisters for the return trip in the command module, but those canisters were square and the environmental control system in the lunar module was designed to hold round canisters. The challenge was to figure out how to adapt the system using only the materials available on the spacecraft, namely, tape, cardboard, and plastic bags.

Immediately after the malfunction occurred and the decision to move the men to the lunar module was made, Mission Control quickly assembled a team of engineers, gave them a set of materials that matched those on the spacecraft, and challenged them to solve the life-or-death problem before the carbon dioxide levels in the lunar module built up to the point at which the astronauts would be affected. The engineering team quickly figured out how to use the square canisters in the round-based environmental control system, Mission Control relayed instructions on how to construct the adaptation, and the astronauts successfully modified the system. **Figure 10.1** shows Deke Slayton, the Flight Crew Operations Director, explaining how the modified system will work to other members of the team on Earth. **Figure 10.2** is a photograph taken on the Apollo 13 spacecraft, showing a square lithium hydroxide canister in the back, with astronauts Fred Haise (*left,* hands only) and Jack Swigert (*right*) making modifications. Because of this modification and many other clever adjustments, the crew was safely returned to Earth.

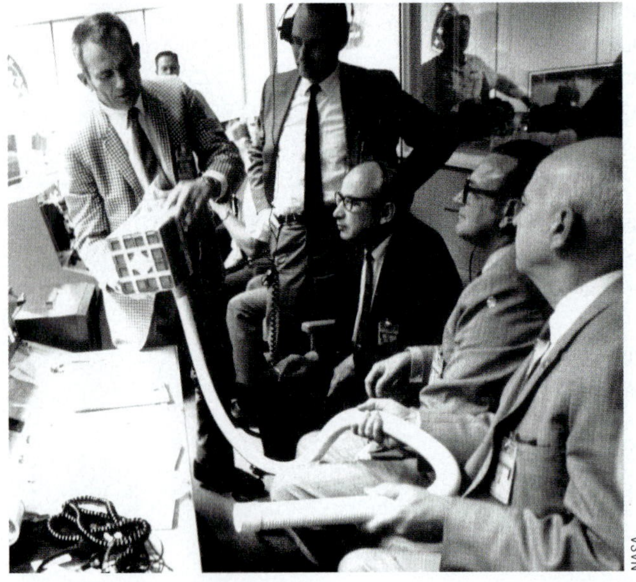

Figure 10.1 NASA Director of Flight Crew Operations Deke Slayton (1924–1993) shows the lithium hydroxide canister modification system to other members of the ground crew on April 15, 1970.

Let's now return to the question: How did NASA scientists know the mass of lithium hydroxide needed to destroy the amount of carbon dioxide that was anticipated to be exhaled? You will learn the fundamental principle on which the answer to the question is based here in this chapter. That principle, combined with additional knowledge of properties of gases that will come in Chapter 14, will allow you to do the same calculation as did the Apollo missions NASA teams!

*The line recited by actor Tom Hanks, who portrays Mission Commander Jim Lovell in the film *Apollo 13,* is "Houston, we have a problem."

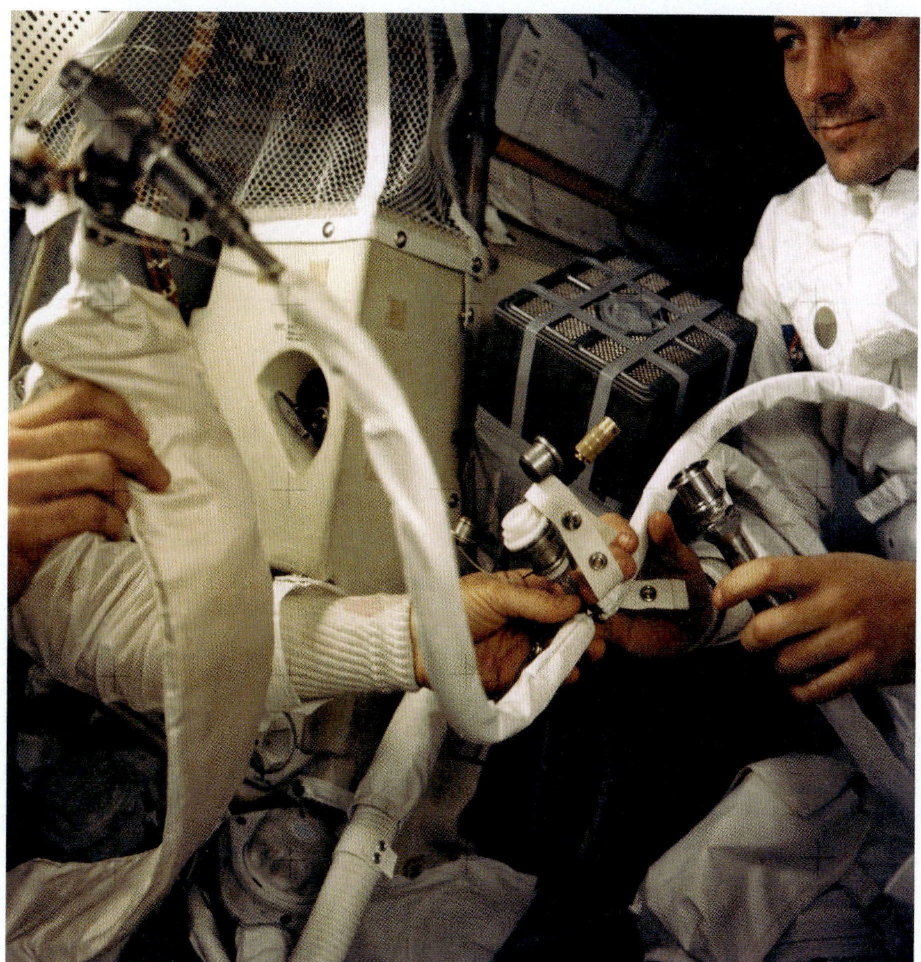

Figure 10.2 Apollo 13 astronauts Fred Haise (b. 1933) (*left*) and Jack Swigert (1931–1982) (*right*) assembling the modified environmental control system in the lunar module. A cubical lithium hydroxide canister is in the back.

10.2 Conversion Factors from a Chemical Equation

Goal 1 Given a chemical equation, or a reaction for which the equation can be written, and the amount in moles of one species in the reaction, calculate the amount in moles of any other species.

In Section 8.5, we introduced the two levels on which the coefficients of a chemical equation may be interpreted: molecule ratios and mole ratios. Let's briefly review these two levels. Consider the following equation:

$$4\ NH_3(g) \quad + \quad 5\ O_2(g) \ \rightarrow \ 4\ NO(g) \ + \ 6\ H_2O(g) \qquad \textbf{(10.1)}$$

To think about molecule ratios, this equation is read, "Four NH_3 molecules react with 5 O_2 molecules to produce 4 NO molecules and 6 H_2O molecules."

The coefficients in a chemical equation give us the conversion factors to convert from the number of particles of one substance to the number of particles of another substance in a reaction. For example, Equation 10.1 shows that five molecules of O_2 are needed to react with every four molecules of NH_3. In other words, the reaction uses five molecules of O_2 per four molecules of NH_3, which can be expressed as the conversion factor $\frac{5 \text{ molecules } O_2}{4 \text{ molecules } NH_3}$.

Thinking About Your Thinking

Proportional Reasoning

Each conversion factor from a chemical equation is a direct proportionality. Therefore, in the reaction $4\ NH_3(g) + 5\ O_2(g) \rightarrow 4\ NO(g) + 6\ H_2O(g)$, $5 \text{ mol } O_2 \propto 4 \text{ mol } NH_3$, $5 \text{ mol } O_2 \propto 4 \text{ mol NO}$, $5 \text{ mol } O_2 \propto 6 \text{ mol } H_2O$, and so on.

As an example, consider the question: How many O_2 molecules are required to react with 308 NH_3 molecules? The answer to this question comes from applying the conversion factor:

$$308 \cancel{\text{molecules } NH_3} \times \frac{5 \text{ molecules } O_2}{4 \cancel{\text{molecules } NH_3}} = 385 \text{ molecules } O_2$$

Equation 10.1 can also be interpreted as, "Four *moles* of NH_3 molecules react with 5 *moles* of O_2 molecules to produce 4 *moles* of NO molecules and 6 *moles* of H_2O molecules.". If you know the stoichiometric amount in moles of any species in a reaction, reactant or product, there is a one-step conversion to the amount of any other species, as illustrated in **Figure 10.3**. If 3.20 moles of NH_3 react according to Equation 10.1, how many moles of H_2O will be produced? Of the several mole ratio relationships that are available in the equation, the conversion factor needed is $\frac{6 \text{ mol } H_2O}{4 \text{ mol } NH_3}$:

$$3.20 \cancel{\text{mol } NH_3} \times \frac{6 \text{ mol } H_2O}{4 \cancel{\text{mol } NH_3}} = 4.80 \text{ mol } H_2O$$

Note that the 6 and 4 are exact numbers, and therefore, they do not affect the significant figures in the final answer.

This method may be applied to any equation.

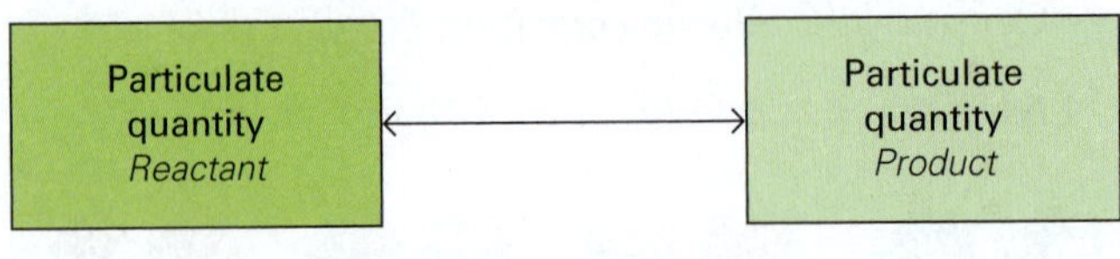

Specific Applications

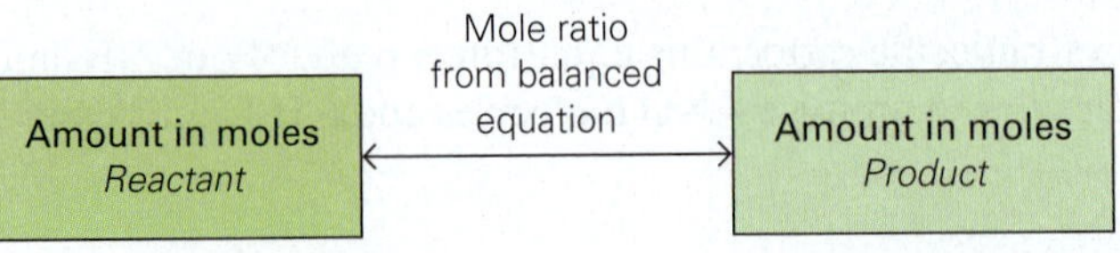

Figure 10.3 Conversion between amount in moles of one species and moles of another species in a chemical change. The mole ratio between the two species, as given in the balanced chemical equation, is the conversion factor that links the amount in moles of one species to moles of another. The overarching pattern, illustrated by the two boxes at the top of the figure, is that one particulate quantity in a green box is linked to another.

Active Example 10.1 Mole Ratios I

How many moles of oxygen are required to burn 10.11 moles of ethane, $C_2H_6(g)$?

Think Before You Write When you read this problem statement, the first thing you should note is that a chemical reaction is occurring. You are given information about one species in the chemical change, and you are being asked for information about another species. This means that you need to know the mole ratio between the two species.

Answers *Cover the left column with your cut-out shield. Reveal each answer only after you have written your own answer in the right column.*

$2\ C_2H_6(g) + 7\ O_2(g) \rightarrow 4\ CO_2(g) + 6\ H_2O(\ell)$	To find the mole ratio, you need a chemical equation for this burning reaction. You wrote equations for burning reactions in Section 9.7. Recall the unnamed second reactant and the two products that are always formed. Write and balance the equation.
Given: 10.11 mol C_2H_6 **Wanted:** mol O_2	**Analyze** the problem statement by writing the given quantity and the unit of the wanted quantity. Include the formula of each species as part of the units.
$7\text{ mol } O_2 = 2\text{ mol } C_2H_6$ $\dfrac{7\text{ mol } O_2}{2\text{ mol } C_2H_6}$	**Identify** the equivalency needed to solve the problem. Change it to the conversion factor that you will use in the setup.
$10.11\ \cancel{\text{mol } C_2H_6} \times \dfrac{7\text{ mol } O_2}{2\ \cancel{\text{mol } C_2H_6}} = 35.39\text{ mol } O_2$	**Construct** the solution setup, cancel units, and calculate the value of the answer.
$10 \times 3.5 = 35$. The value of the answer is OK.	**Check** the solution. Does the value of the answer make sense?
You improved your skill at converting between the stoichiometric amount in moles of one species in a chemical change to the amount in moles of another species in that change.	What did you learn by solving this Active Example?

Practice Exercise 10.1

Solutions of zinc bromide and sodium hydroxide are mixed. How many moles of sodium hydroxide are required to form 0.033 mole of zinc hydroxide?

Active Example 10.2 Mole Ratios II

Ammonia is formed directly from its elements. How many moles of hydrogen are needed to produce 4.0 moles of ammonia?

Think Before You Write As with Active Example 10.1, the key to knowing what to do after you read this problem statement is to recognize that a chemical change is described. You are given the stoichiometric amount in moles of one species in that chemical change, and you are being asked to calculate the amount in moles of another species in that change.

Answers *Cover the left column with your cut-out shield. Reveal each answer only after you have written your own answer in the right column.*

Given: 4.0 mol NH_3 **Wanted:** mol H_2 $N_2 + 3\,H_2 \rightarrow 2\,NH_3$	**Analyze** the problem statement by writing the given quantity and the unit of the wanted quantity. Write and balance the equation that will give you the needed mole ratio between the two species.
$3\text{ mol }H_2 = 2\text{ mol }NH_3$ $\frac{3\text{ mol }H_2}{2\text{ mol }NH_3}$	**Identify** the equivalency needed to solve the problem. Change it to the conversion factor that you will use in the setup.
$4.0\text{ }\cancel{\text{mol }NH_3} \times \frac{3\text{ mol }H_2}{2\text{ }\cancel{\text{mol }NH_3}} = 6.0\text{ mol }H_2$	**Construct** the solution setup, cancel units, and calculate the value of the answer.
$4 \times 1.5 = 6$. The value of the answer is correct.	**Check** the solution. Does the value of the answer make sense?
You improved your skill at converting the stoichiometric amount in moles of one species in a chemical change to the amount in moles of another species in that change.	What did you learn by solving this Active Example?

Practice Exercise 10.2

How many moles of water are produced when 2.90×10^{-5} mole of liquid diethyl ether, $CH_3CH_2OCH_2CH_3$, is completely oxidized?

10.3 Mass–Mass Stoichiometry

Goal 2 Given a chemical equation, or a reaction for which the equation can be written, and the mass or amount in moles of one species in the reaction, find the mass or amount in moles of any other species.

You are now ready to solve the problem that underlies the manufacture of chemicals and the design of many laboratory experiments: How much product will you get from a certain amount of raw material, or how much raw material will you need to obtain a specific amount of product? In solving this type of problem, you will tie together several skills:

You will write chemical formulas (Chapter 6) (if you are required to know nomenclature at this point in your study of chemistry).

You will calculate molar masses from chemical formulas (Section 7.5).

You will use molar masses to change mass to amount in moles and amount to mass (Section 7.6).

You will write and balance chemical equations [Chapters 8 and (optional) 9].

You will use a balanced chemical equation to change from moles of one species to moles of another (Section 10.2).

And you will do all these things in one problem setup!

The preceding list should impress upon you how much the solving of stoichiometry problems depends on other skills. If you have any doubt about these skills, you will find it helpful to review the sections or chapters listed.

To solve any stoichiometry problem, you must have the balanced equation for the reaction and the conversion factors between amount in moles and quantities of given and wanted substances. Molar mass is the conversion factor in this section; others will appear later.

The solution of a stoichiometry problem usually falls into a three-step "mass-to-mass" path. The **mass-to-mass stoichiometry path** is:

$$\text{Mass of Given} \longrightarrow \text{Moles of Given} \longrightarrow \text{Moles of Wanted} \longrightarrow \text{Mass of Wanted}$$

$$\text{Mass Given} \times \underset{\text{Step 1}}{\frac{\text{mol Given}}{\text{mass Given}}} \times \underset{\text{Step 2}}{\frac{\text{mol Wanted}}{\text{mol Given}}} \times \underset{\text{Step 3}}{\frac{\text{mass Wanted}}{\text{mol Wanted}}} = \text{Mass Wanted}$$

◀ **P/Review** The mass-to-mass stoichiometry path will be expanded to include gases in Chapter 14 and again in Chapter 16 to include solutions. The method by which those chapters' problems are solved is the same as the method for solving the problems in this section. *Learn it Now!* and you will be ready to use it later.

Thinking About Your Thinking

Proportional Reasoning

Each step in the mass-to-mass stoichiometry path involves a conversion based on a direct proportionality.

In words, these three steps are as follows:

how to... **Solve a Stoichiometry Problem: The Mass-to-Mass Stoichiometry Path**

Step 1: Change the mass of the given species to amount in moles (Section 7.6).

Step 2: Change the amount in moles of the given species to amount in moles of the wanted species (Section 10.2).

Step 3: Change the amount in moles of the wanted species to mass (Section 7.6).

Occasionally, you may be given the mass of one substance and asked to find the amount in moles of a second substance. In this case, *Step 1* and *Step 2* of the mass-to-mass path complete the problem. You may also be given the amount in moles of one substance and asked to find the mass of another. *Step 2* and *Step 3* solve this problem.

Figure 10.4 ties together the grams–moles–number-of-particles conversions from Figure 7.8, the mole–mole conversions from Figure 10.3, and the mass-to-mass stoichiometry path. The four boxes in the middle row illustrate the mass-to-mass stoichiometry path. If you tie in what you learned in Chapter 7, you see that you can calculate the number of particles of any species if you can calculate the amount in moles of that species. The top of Figure 10.4 illustrates the general stoichiometry path. The boxes in the far left and far right columns symbolize macroscopic quantities, which are what we can measure in the laboratory. The middle two boxes symbolize particulate quantities, which are the quantities that govern the nature of the chemical change.

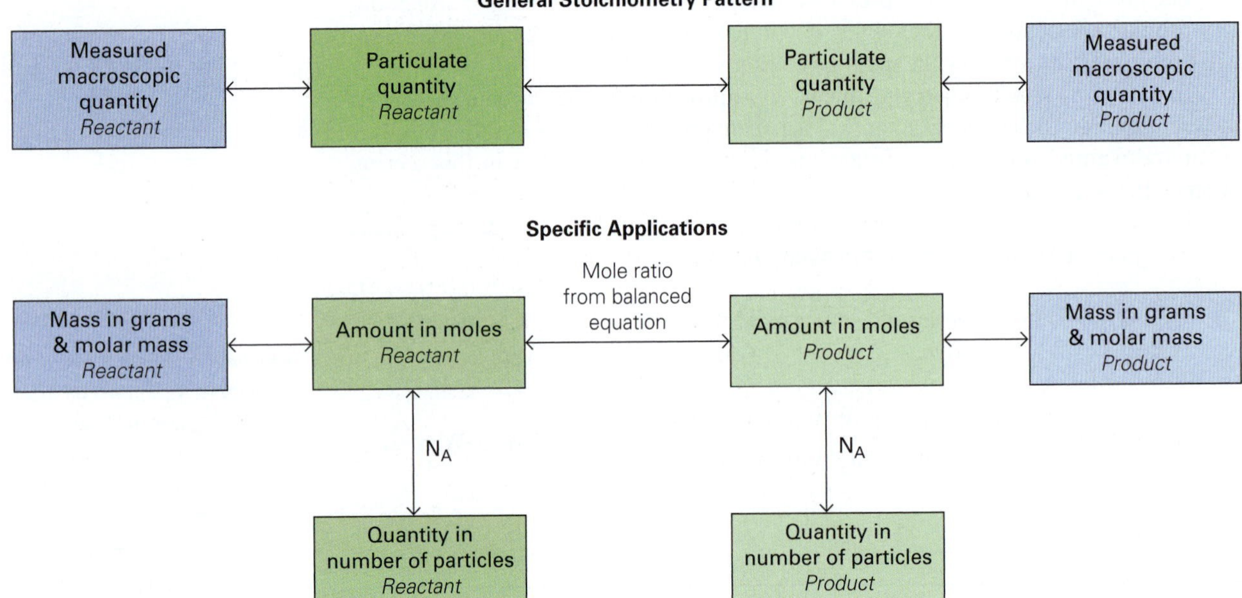

Figure 10.4 Conversion among grams, moles, and number of particles for two species in a chemical change. This figure combines Figure 7.8 and Figure 10.3. Molar mass is a conversion factor that links mass in grams, a measurable macroscopic quantity, with amount in moles, the number of particles in the sample. The Avogadro constant, N_A, links the number of particles grouped in moles to the absolute number of particles. The mole ratio between the two species, as given in the balanced chemical equation, is the conversion factor that links amount in moles of one species to amount of another. The overarching pattern, illustrated by the four boxes at the top of the figure, is that a measured macroscopic quantity in the far left or far right box (e.g., mass in grams, pounds) can be converted to a particulate quantity in a middle box, and one particulate quantity can be converted to any other by using the mole ratio from the balanced chemical equation.

Lynne Furrer/Shutterstock.com

Figure 10.5 Heptane is present in gasoline, so the reaction in Active Examples 10.3 and 10.4 occurs in automobile engines. A fuel made of 90% iso-octane and 10% heptane has an octane rating of 90.

Active Example 10.3 Mass–Mass Stoichiometry I

What mass in grams of oxygen is required to burn 3.50 moles of liquid heptane, $C_7H_{16}(\ell)$ (Figure 10.5)?

Think Before You Write First, recognize that the problem statement describes a burning reaction—one type of a chemical change. Thus, you know that you will need to do a mole-to-mole conversion between two species in the change. In this case, the given quantity is amount in *moles* of heptane. In other words, *Step 1* of the mass-to-mass stoichiometry path is completed. This problem, therefore, requires *Step 2*, the conversion of amount in moles of heptane to moles of oxygen, and *Step 3*, the conversion of moles of oxygen to grams of oxygen.

Answers *Cover the left column with your cut-out shield. Reveal each answer only after you have written your own answer in the right column.*

Given: 3.50 mol C_7H_{16} **Wanted:** g O_2 $C_7H_{16}(\ell) + 11\ O_2(g) \rightarrow 7\ CO_2(g) + 8\ H_2O(\ell)$	**Analyze** the problem statement by writing the given quantity and the unit of the wanted quantity. Also write a balanced equation for the reaction. 

$11\ mol\ O_2 = 1\ mol\ C_7H_{16}$ $32.00\ g\ O_2 = 1\ mol\ O_2$ $\frac{11\ mol\ O_2}{1\ mol\ C_7H_{16}}$ $\frac{32.00\ g\ O_2}{1\ mol\ O_2}$	**Identify** the equivalencies needed to solve the problem. There are two of them, one for the mole–mole conversion and one for the mole–gram conversion. Change each to their corresponding conversion factors.
$3.50\ \cancel{mol\ C_7H_{16}} \times \frac{11\ \cancel{mol\ O_2}}{1\ \cancel{mol\ C_7H_{16}}} \times \frac{32.00\ g\ O_2}{1\ \cancel{mol\ O_2}}$ $= 1.23 \times 10^3\ g\ O_2$	**Construct** the solution setup, cancel units, and calculate the value of the answer.
$4 \times 10 \times 30 = 40 \times 30 = 1200 = 1.2 \times 10^3$. The value is reasonable.	**Check** the value of the solution. (Hint: $4 \times 10 \times 30$.)
You improved your skill at solving mass–mass stoichiometry problems.	What did you learn by solving this Active Example?

Practice Exercise 10.3

How many moles of copper are formed when a 2.688-g ribbon of zinc is placed in excess* copper(II) nitrate solution?

Now you will perform the complete three-step mass-to-mass stoichiometry path.

Active Example 10.4 Mass–Mass Stoichiometry II

What is the mass in grams of CO_2 that will be produced by burning 66.0 g C_7H_{16} by the same reaction as in Active Example 10.3, $C_7H_{16}(\ell) + 11\ O_2(g) \rightarrow 7\ CO_2(g) + 8\ H_2O(\ell)$?

Think Before You Write You are given the mass of one species in a chemical change, and you are asked to determine the mass of another species. Thus, you will switch from the macroscopic mass quantity to the particulate number of moles, use the mole ratio in the chemical equation to determine the amount in moles of the wanted quantity, and then switch back to the macroscopic level and determine the mass of that amount in grams. This is illustrated in Figure 10.4.

Answers *Cover the left column with your cut-out shield. Reveal each answer only after you have written your own answer in the right column.*

Given: 66.0 g C_7H_{16} **Wanted:** g CO_2	**Analyze** the problem by writing the given quantity and the unit of the wanted quantity.

*The word *excess* as used here means "more than enough," just as there are "more than enough" oxygen molecules to react with a compound burning in air. The atmosphere effectively provides an inexhaustible supply of oxygen molecules. In this case, there is "more than enough" copper(II) nitrate in the solution to react with all the zinc in a 2.688-g sample.

$g\ C_7H_{16} \rightarrow mol\ C_7H_{16} \rightarrow mol\ CO_2 \rightarrow g\ CO_2$

$1\ mol\ C_7H_{16} = 100.20\ g\ C_7H_{16}$

$7\ mol\ CO_2 = 1\ mol\ C_7H_{16}$

$44.01\ g\ CO_2 = 1\ mol\ CO_2$

$$\frac{1\ mol\ C_7H_{16}}{100.20\ g\ C_7H_{16}} \quad \frac{7\ mol\ CO_2}{1\ mol\ C_7H_{16}} \quad \frac{44.01\ g\ CO_2}{1\ mol\ CO_2}$$

The next step is to **identify** the equivalencies. With multiple-step problems such as this, it can be helpful to write the units for each step before you write the equivalencies:

$$g\ C_7H_{16} \rightarrow mol\ C_7H_{16} \rightarrow mol\ CO_2 \rightarrow g\ CO_2$$

Each arrow in this **unit path** requires an equivalency. Change the equivalencies to conversion factors.

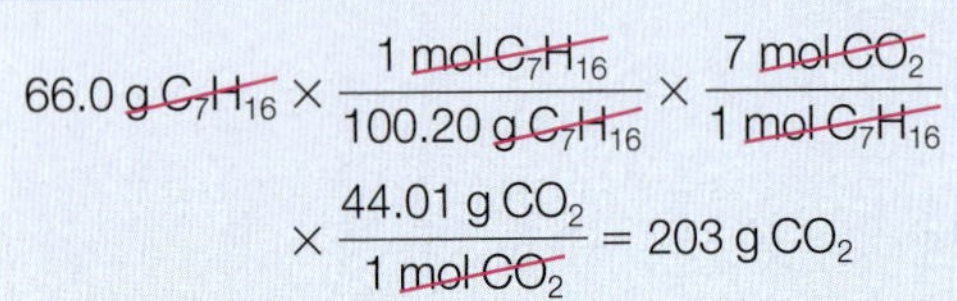

$$66.0\ g\ C_7H_{16} \times \frac{1\ mol\ C_7H_{16}}{100.20\ g\ C_7H_{16}} \times \frac{7\ mol\ CO_2}{1\ mol\ C_7H_{16}} \times \frac{44.01\ g\ CO_2}{1\ mol\ CO_2} = 203\ g\ CO_2$$

Construct the solution setup and determine the answer.

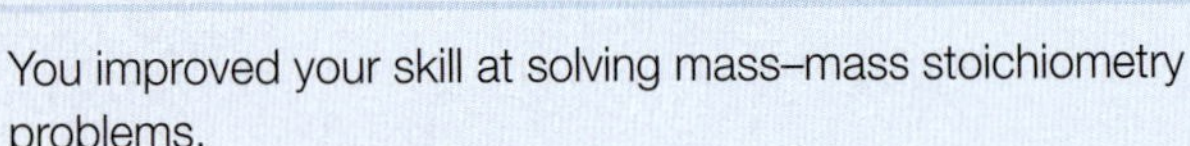

$(60 \times 7) \times 50 \div 100 = 420 \times (50 \div 100) = 420 \times 0.5 = 210$, OK.

$(70 \times 7) \times 40 \div 100 = 490 \times (40 \div 100) \approx 500 \times 0.4 = 200$, OK.

The **check** of a setup with a large number of values becomes a bit challenging. You have to round the numbers, aiming to round in opposite directions. For example, this setup could be estimated as 60 (round down) $\times$ 7 $\times$ 50 (round up) $\div$ 100 or 70 (round up) $\times$ 7 $\times$ 40 (round down) $\div$ 100. Remember that the goal is to be sure you in are in the ballpark, not to calculate the *exact* answer in your head.

You improved your skill at solving mass–mass stoichiometry problems.

What did you learn by solving this Active Example?

Practice Exercise 10.4

What mass of fluorine is formed when 3.0 grams of bromine trifluoride decomposes into its elements?

Sometimes it is more convenient to measure mass in larger or smaller units. Most analytical and academic chemists, for example, work in milligrams and millimoles of substance, whereas an industrial chemical engineer is more likely to think in kilograms and kilomoles. The mass-to-mole and mole-to-mass conversions in *Step 1* and *Step 3* are performed in exactly the same way. The conversion factor 44.01 g CO_2/mol CO_2 is equal to 44.01 kg CO_2/kmol CO_2:

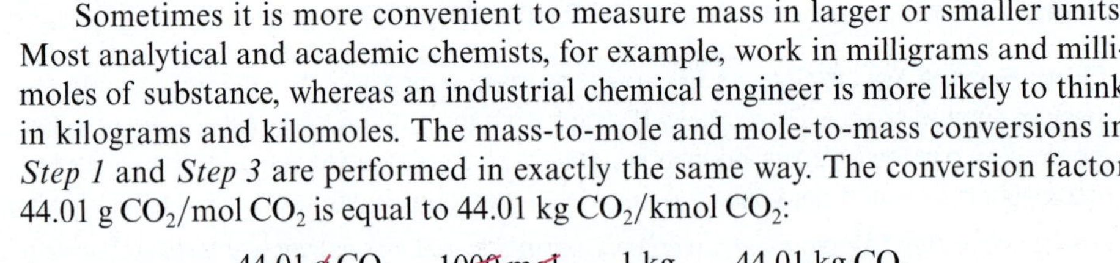

$$\frac{44.01\ g\ CO_2}{1\ mol\ CO_2} \times \frac{1000\ mol}{1\ kmol} \times \frac{1\ kg}{1000\ g} = \frac{44.01\ kg\ CO_2}{kmol\ CO_2}$$

The factors of 1000 cancel in changing from units to kilounits. The same is true if milliunits are used; 44.01 mg CO_2/mmol CO_2 is equal to 44.01 g CO_2/mol CO_2.

Chaykovsky Igor/Shutterstock.com

Figure 10.6 Nickel(II) chloride solution has a distinctive green color due to the nickel(II) ion. It can be used as the starting material for the process of electrically depositing a layer of nickel on a metal object.

Active Example 10.5 Mass–Mass Stoichiometry III

What mass in milligrams of nickel(II) chloride (Figure 10.6) is in a solution if 503 mg of silver chloride is precipitated in the reaction of silver nitrate and nickel(II) chloride solutions?

Think Before You Write Did you recognize that this is a three-step stoichiometry problem? If so, you are well on your way to understanding the stoichiometry concept. If not, go back and revisit Figure 10.4. Think about how the pattern in Figure 10.4 applies to the solution setup for this problem as you work the Active Example.

Answers *Cover the left column with your cut-out shield. Reveal each answer only after you have written your own answer in the right column.*

Given: 503 mg AgCl
Wanted: mg $NiCl_2$

Analyze the problem statement by writing the given and wanted.

mg AgCl → mmol AgCl → mmol $NiCl_2$ → mg $NiCl_2$

1 mmol AgCl = 143.4 mg AgCl

$$2\ AgNO_3(aq) + NiCl_2(aq) \rightarrow 2\ AgCl(s) + Ni(NO_3)_2(aq)$$

1 mmol $NiCl_2$ = 2 mmol AgCl

129.59 mg $NiCl_2$ = 1 mmol $NiCl_2$

$$\frac{1\ \text{mmol AgCl}}{143.4\ \text{mg AgCl}} \quad \frac{1\ \text{mmol NiCl}_2}{2\ \text{mmol AgCl}} \quad \frac{129.59\ \text{mg NiCl}_2}{1\ \text{mmol NiCl}_2}$$

To **identify** the needed equivalencies, write the unit path that you will follow in your setup. Remember to stay in milliunits throughout. Then identify the appropriate equivalency for each arrow in the unit path. Finally, change your equivalencies to conversion factors.

$$503\ \cancel{\text{mg AgCl}} \times \frac{1\ \cancel{\text{mmol AgCl}}}{143.4\ \cancel{\text{mg AgCl}}} \times \frac{1\ \cancel{\text{mmol NiCl}_2}}{2\ \cancel{\text{mmol AgCl}}} \times \frac{129.59\ \text{mg NiCl}_2}{1\ \cancel{\text{mmol NiCl}_2}} = 227\ \text{mg NiCl}_2$$

Construct the setup, cancel units, and calculate the answer.

503 × 130 ÷ (143 × 2) = 503 × (130 ÷ 143) ÷ 2 ≈ 500 × (1) ÷ 2 = 250, OK.

If you recognize that 129.59 ÷ 143.4 is about 1, the **check** is relatively straightforward.

You improved your skill at solving mass–mass stoichiometry problems.

What did you learn by solving this Active Example?

Practice Exercise 10.5

What mass of nitrogen dioxide (in kilograms) can be formed from the reaction of 50.0 kg of nitrogen monoxide and excess oxygen?

Learn It Now! *You have reached a critical point in your study of chemistry. Students usually have no difficulty changing grams to moles, moles of one substance to moles of another, and moles to grams when those operations are presented as separate problems. When the operations are combined in a single problem, however, trouble sometimes follows. It doesn't help when (as in the following sections) many other ideas are added. In other words, you will not understand the following concepts unless you have already mastered this one.*

Just studying the foregoing Active Examples will not give you the skill that you need. That comes only with practice. You should solve some end-of-chapter problems now—before you begin the next section. We strongly recommend that, at the very least, you solve Problems 1, 13, 17, 21, 23, and 27 at this time. If you understand and can solve these problems, you are ready to proceed to the next section.

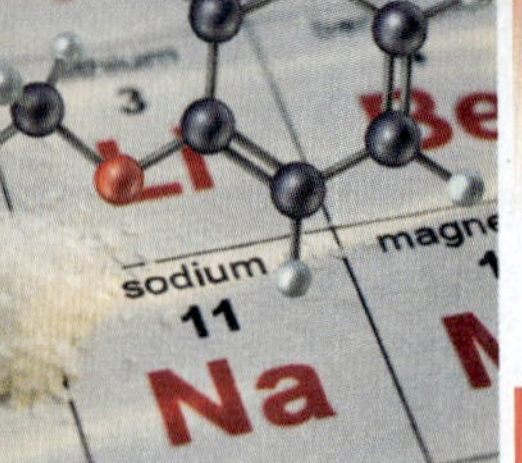

Everyday Chemistry 10.1

THE STOICHIOMETRY OF CO_2 EMISSIONS IN AUTOMOBILE EXHAUST

A common concern of environmentalists, politicians, and most citizens of developed nations is the extent to which automobile exhaust contributes carbon dioxide gas to the atmosphere. There is evidence that excessive CO_2 emissions can contribute to what is known as the greenhouse effect,* a process by which some atmospheric gases effectively store heat energy and increase the Earth's temperature relative to what it would be in the absence of these gases.

The greenhouse effect is necessary to support life on Earth as we know it. Some atmospheric gases keep Earth's temperature much warmer than it would be if the atmosphere was devoid of greenhouse gases (**Figure 10.7**). The question presently being debated is, "Are we over-warming Earth because of human-made industrial and automotive emissions of greenhouse gases, the most abundant of which is carbon dioxide?" This is a difficult question to answer, and many scientists are presently studying the effects of the byproducts of modern civilization on Earth's average temperature.

Basic stoichiometry, such as you are now learning, is one part of understanding how automotive emissions contribute to the CO_2 quantity in the atmosphere. We'll show you some very approximate calculations

NASA

Figure 10.7 Greenhouse gases such as water vapor, carbon dioxide, and methane keep Earth's temperature higher than it otherwise would be in the absence of these gases in the atmosphere.

*The term *global warming* refers to the gradual rise in the near-surface temperature of the Earth that has been occurring over the past century. Scientists are presently debating whether this is due to natural influences or an increasing greenhouse effect.

so that you can appreciate the amount of CO_2 emitted by cars. Let's make some assumptions about an average car: It gets 25 miles per gallon of gasoline (which we'll consider to be pure octane, C_8H_{18}, with a density of 0.7 g/mL), and it is driven at an average of 50 miles per week. How much carbon dioxide does this car emit in a year? Let's see.

First, we need the reaction equation: $2\ C_8H_{18}(\ell) + 25\ O_2(g) \rightarrow 16\ CO_2(g) + 18\ H_2O(\ell)$. Now we'll determine the mass of CO_2 emitted by this single car in a year:

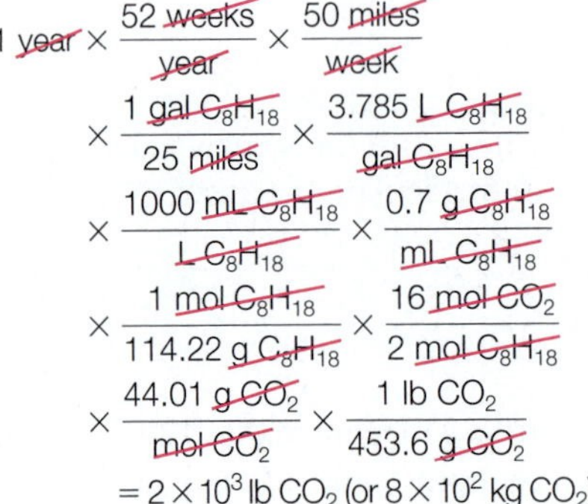

$$1\ \text{year} \times \frac{52\ \text{weeks}}{\text{year}} \times \frac{50\ \text{miles}}{\text{week}} \times \frac{1\ \text{gal } C_8H_{18}}{25\ \text{miles}} \times \frac{3.785\ \text{L } C_8H_{18}}{\text{gal } C_8H_{18}} \times \frac{1000\ \text{mL } C_8H_{18}}{\text{L } C_8H_{18}} \times \frac{0.7\ \text{g } C_8H_{18}}{\text{mL } C_8H_{18}} \times \frac{1\ \text{mol } C_8H_{18}}{114.22\ \text{g } C_8H_{18}} \times \frac{16\ \text{mol } CO_2}{2\ \text{mol } C_8H_{18}} \times \frac{44.01\ \text{g } CO_2}{\text{mol } CO_2} \times \frac{1\ \text{lb } CO_2}{453.6\ \text{g } CO_2} = 2 \times 10^3\ \text{lb } CO_2\ (\text{or } 8 \times 10^2\ \text{kg } CO_2)$$

That's 2000 pounds—1 ton—of carbon dioxide emitted into the atmosphere each year by an "average" car (**Figure 10.8**). Now, how many cars are there in your city, your state, or in the entire world? How many exceed our 50-miles-per-week estimate? What about all of the trucks, buses, airplanes, ships, and other forms of mass transportation? There's no doubt that industrialization is putting a great deal more CO_2 into the atmosphere today than before the internal combustion engine was invented.

On the other hand, plants capture carbon dioxide from the atmosphere and convert it into their biomass and oxygen through photosynthesis (**Figure 10.9**). As long as we grow more vegetation and forests, any additional CO_2 produced can, in turn, be consumed. Scientists have calculated that a little more than an acre of trees will consume all of the CO_2 emissions one automobile driver can produce in a lifetime. Use this figure with caution, however, because the world's population is over 7 billion. Even if only one in a hundred people in the world drive a car, 70 million acres of trees—30 times the size of Yellowstone National Park—are needed to consume the CO_2. That's a huge number of forested acres that would need to be planted! You are also partly responsible for carbon dioxide emission from power plants and factories that produce the energy you use and the products you buy.

barteverett/Shutterstock.com

Figure 10.8 Each car emits about a ton of carbon dioxide into the atmosphere each year.

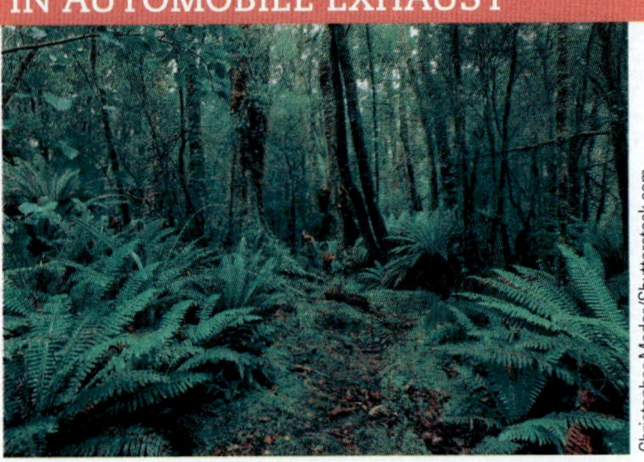

Christopher Meder/Shutterstock.com

Figure 10.9 Plants remove carbon dioxide from the atmosphere by reacting it with water to produce sugar and oxygen.

Efforts to understand the effects of CO_2 emissions on the atmosphere will no doubt continue throughout your lifetime. It is a complex problem, but our understanding of it continues to progress. The next time you read an article on climate change or the greenhouse effect, keep in mind that basic stoichiometric concepts are at the heart of all the lengthy calculations in the various sophisticated mathematical models used in scientific studies.

Quick Quiz

1. **What is the greenhouse effect?**
2. **What role does basic stoichiometry have in the study of the effect of carbon dioxide emissions from automobiles and how these emissions affect the atmosphere?**

10.4 Percentage Yield

Goal 3 Given two of the following, or information from which two of the following may be determined, calculate the third: ideal yield, actual yield, percentage yield.

In Active Example 10.4, you found that burning 66.0 grams of C_7H_{16} will produce 203 grams of CO_2. This is the **ideal yield**, the amount of product ideally formed from the *complete* conversion of the given amount of reactant to product. Ideal yield is always a calculated quantity, calculated by the principles of stoichiometry. In actual practice, factors such as impure reactants, incomplete reactions, and side reactions cause the **actual yield** to be different from the ideal yield. The actual yield is a measured quantity, determined by experiment or experience.

If you know the actual yield and the ideal yield, you can find the **percentage yield**. This is the actual yield expressed as a percentage of ideal yield (**Figure 10.10**). Percentage yield, as with all percentages, is the part quantity—actual yield in this specific case—over the whole quantity—ideal yield—times 100:

$$\% \text{ yield} \equiv \frac{\text{actual yield}}{\text{ideal yield}} \times 100\% \blacktriangleright$$

◀ **P/Review** This is a specific form of the general Equation 7.2 in Section 7.7:

$$\% \text{ of A} = \frac{\text{parts of A}}{\text{total parts}} \times 100\%$$

When a part quantity and a whole quantity are both given, percentage is calculated by substitution into this equation.

If in Active Example 10.4 only 181 grams of CO_2 had been produced, instead of the calculated 203 grams, the percentage yield would be

$$\% \text{ yield} \equiv \frac{\text{actual yield}}{\text{ideal yield}} \times 100\% = \frac{181 \not{g}}{203 \not{g}} \times 100\% = 89.2\%$$

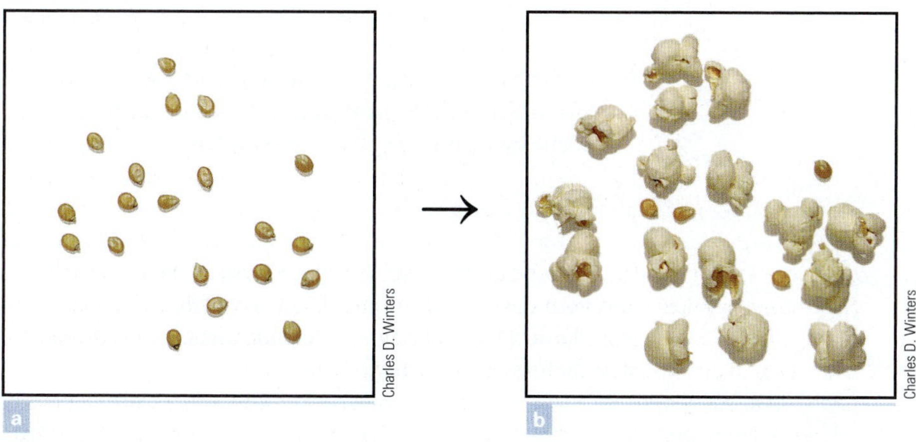

Figure 10.10 We began with 20 popcorn kernels and found that only 16 of them popped. The percentage yield of popcorn from our "reaction" is (16 ÷ 20) × 100 = 80%. (Since we can count the kernels, this is *exactly* 80.000...%.)

Active Example 10.6 Percentage Yield I

A solution containing 3.18 g of barium chloride is added to a second solution containing excess sodium sulfate. (If you did not complete Practice Exercise 10.3, see its footnote for an explanation of the term *excess* as it is used in this context.) Barium sulfate precipitates (Figures 10.11 and 10.12). (a) Calculate the ideal yield of barium sulfate. (b) If the actual yield is 3.37 g, calculate the percentage yield.

Think Before You Write Calculating ideal yield is a typical stoichiometry problem. Part (b) is solved by substitution into the equation $\% \text{ yield} \equiv \frac{\text{actual yield}}{\text{ideal yield}} \times 100\%$.

Answers *Cover the left column with your cut-out shield. Reveal each answer only after you have written your own answer in the right column.*

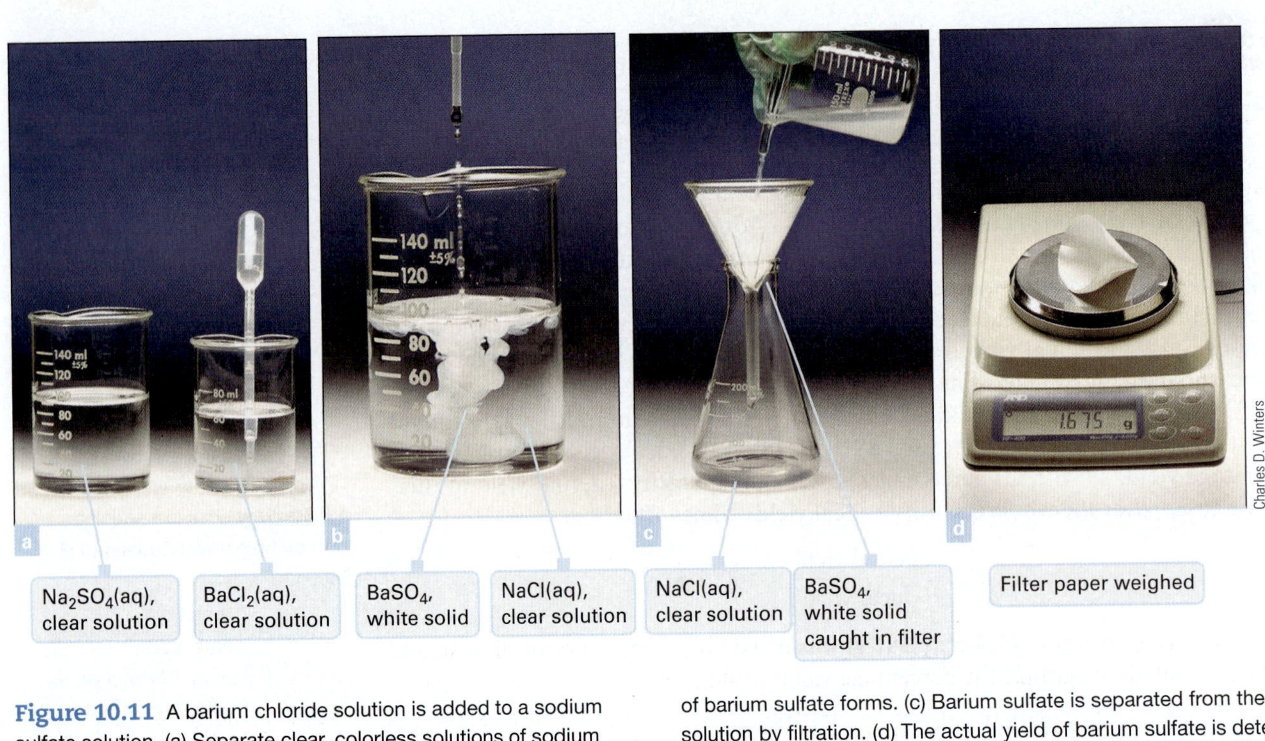

Figure 10.11 A barium chloride solution is added to a sodium sulfate solution. (a) Separate clear, colorless solutions of sodium sulfate and barium chloride before mixing. (b) The barium chloride solution is added to the sodium sulfate solution. A white precipitate of barium sulfate forms. (c) Barium sulfate is separated from the solution by filtration. (d) The actual yield of barium sulfate is determined by drying the product and subtracting the mass of the filter paper from the combined mass of the filter paper and product.

Given: 3.18 g $BaCl_2$ **Wanted:** g $BaSO_4$	Let's start with the ideal yield, which is a typical mass-to-mass stoichiometry problem. **Analyze** the problem statement by listing the given and wanted.
g $BaCl_2$ → mol $BaCl_2$ → mol $BaSO_4$ → g $BaSO_4$ 1 mol $BaCl_2$ = 208.2 g $BaCl_2$ $BaCl_2$(aq) + Na_2SO_4(aq) → $BaSO_4$(s) + 2 NaCl(aq) 1 mol $BaSO_4$ = 1 mol $BaCl_2$ 233.4 g $BaSO_4$ = 1 mol $BaSO_4$	**Identify** the equivalencies by first writing the unit path and then deriving the equivalency for each unit conversion. Hold off on changing the equivalencies to conversion factors until the next step.
$3.18\ \cancel{\text{g BaCl}_2} \times \frac{1\ \cancel{\text{mol BaCl}_2}}{208.2\ \cancel{\text{g BaCl}_2}} \times \frac{1\ \cancel{\text{mol BaSO}_4}}{1\ \cancel{\text{mol BaCl}_2}} \times \frac{233.4\ \text{g BaSO}_4}{1\ \cancel{\text{mol BaSO}_4}} = 3.56\ \text{g BaSO}_4$	**Construct** the solution setup by changing the equivalences to conversion factors and inserting the conversion factors directly into the setup. Cancel units and calculate the value of the answer.
3.18 × (A little more than 1) = A little more than 3.18, OK.	**Check** the value of the answer. Note that (233.4 ÷ 208.2) is a little more than 1.

$$\% \text{ yield} \equiv \frac{\text{actual yield}}{\text{ideal yield}} \times 100\%$$

$$= \frac{3.37\ \cancel{g}}{3.56\ \cancel{g}} \times 100\% = 94.7\%$$

The actual yield from the problem statement is given as 3.37 g, and your stoichiometry calculation found that the ideal yield is 3.56 g.

Determine the percentage yield.

You improved your skill at solving percentage yield stoichiometry problems.

What did you learn by solving this Active Example?

Practice Exercise 10.6

What is the percentage yield if 2.801 g of calcium phosphate is recovered from the reaction of a solution containing 3.215 g of sodium phosphate with excess calcium nitrate solution?

If you know the percentage yield, you can use it as a conversion factor. ▶ Percentage yield is mass actual yield per 100 units of mass ideal yield. For example, assume that a manufacturer of magnesium hydroxide knows from experience that the percentage yield is 81.3% from the production process. This can be used as either of two conversion factors, $\frac{81.3 \text{ g actual}}{100 \text{ g ideal}}$ or $\frac{100 \text{ g ideal}}{81.3 \text{ g actual}}$. How many grams of product are expected if the ideal yield is 697 g? Using percentage yield as a conversion between g actual and g ideal, we obtain:

$$697 \cancel{\text{g Mg(OH)}_2 \text{ ideal}} \times \frac{81.3 \text{ g Mg(OH)}_2 \text{ actual}}{100 \cancel{\text{g Mg(OH)}_2 \text{ ideal}}} = 567 \text{ g Mg(OH)}_2 \text{ actual}$$

Note that the g ideal units cancel, leaving g actual as the remaining unit.

Charles D. Winters

Figure 10.12 Much of the world's production of barium sulfate is used as a component of a fluid used when drilling oil wells. When mixed with water to form a suspension, barium sulfate is used as a contrasting agent to improve the visibility of x-ray or computed tomography (CT) scans.

◀ **P/Review** Percentage is used as a conversion factor in the solution of Active Example 7.9 in Section 7.7. This is a specific example of a conversion factor obtained from a defining equation, as introduced in Section 3.11.

Thinking About Your Thinking

Proportional Reasoning

Thinking about percentage as a direct proportionality between the number of units in the percentage and 100 total units is a powerful thinking tool that you can use in a variety of problem-solving situations. In this section, we translate 81.3% yield into the conversion factors $\frac{81.3 \text{ g actual}}{100 \text{ g ideal}}$ and $\frac{100 \text{ g ideal}}{81.3 \text{ g actual}}$.

A number of other conversion factors can result from the same percentage yield, such as $\frac{81.3 \text{ mg actual}}{100 \text{ mg ideal}}$, $\frac{81.3 \text{ kg actual}}{100 \text{ kg ideal}}$, or even $\frac{81.3 \text{ lb actual}}{100 \text{ lb ideal}}$. Your choice of units in the conversion factor depends on the other units given in the problem statement.

In Chapter 7, we used percentage composition by mass in the same way. For example, sodium chloride is 39.34% sodium. This can be interpreted as $\frac{39.34 \text{ g Na}}{100 \text{ g NaCl}}$, $\frac{100 \text{ g NaCl}}{39.34 \text{ g Na}}$, and through many other equivalent conversion factors with different mass units.

Remember that when you see a percentage in other contexts, you can always choose to write it as a conversion factor in terms of units in the percentage per 100 total units.

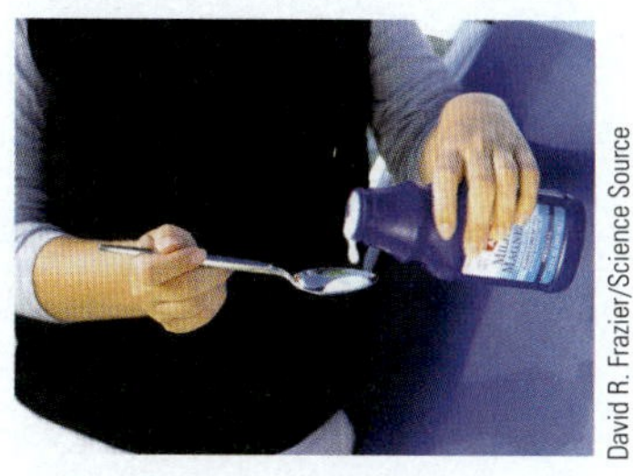

David R. Frazier/Science Source

Figure 10.13 A water mixture of magnesium hydroxide is commonly known as milk of magnesia.

A more likely problem for this maker of magnesium hydroxide is finding out how much raw material is required to obtain a certain amount of product.

Active Example 10.7 Percentage Yield II

A manufacturer wants to prepare 8.00×10^2 kg of $Mg(OH)_2$ by the reaction $MgO(s) + H_2O(\ell) \rightarrow Mg(OH)_2(s)$ (Figure 10.13). Previous production experience shows that the process has an 81.3% yield calculated from the initial MgO. How much MgO should the manufacturer use?

Think Before You Write Think of this as a two-part problem. In the first part, you are given the actual yield, 8.00×10^2 kg of $Mg(OH)_2$ (act). However, the stoichiometry setup for finding mass of MgO must be based on ideal yield. Thus, you need to determine the ideal yield. When you are given percentage yield, think of it as an equivalency that can be changed into a conversion factor. In this case, 81.3 kg actual = 100 kg ideal.

Answers *Cover the left column with your cut-out shield. Reveal each answer only after you have written your own answer in the right column.*

Given: 8.00×10^2 kg $Mg(OH)_2$ actual
Wanted: kg $Mg(OH)_2$ ideal

$$\frac{100 \text{ kg Mg(OH)}_2 \text{ ideal}}{81.3 \text{ kg Mg(OH)}_2 \text{ actual}}$$

Again, the first part of the problem is a conversion from kg $Mg(OH)_2$ (act) to kg $Mg(OH)_2$ (ideal). **Analyze** the problem and change the already-**identified** equivalency to the conversion factor.

$$8.00 \times 10^2 \text{ kg Mg(OH)}_2 \text{ actual} \times \frac{100 \text{ kg Mg(OH)}_2 \text{ ideal}}{81.3 \text{ kg Mg(OH)}_2 \text{ actual}} = 984 \text{ kg Mg(OH)}_2 \text{ ideal}$$

$800 \times (\sim 1.25) = 800 + 200 = 1000$, OK.

Construct the solution setup for the first part of the problem and mentally **check** your calculated answer.

Given: 984 kg $Mg(OH)_2$
Wanted: kg MgO

kg $Mg(OH)_2$ → kmol $Mg(OH)_2$ → kmol MgO → kg MgO

1 kmol $Mg(OH)_2$ = 58.33 kg $Mg(OH)_2$

1 kmol MgO = 1 kmol $Mg(OH)_2$

40.31 kg MgO = 1 kmol MgO

From this point forward, you are solving a standard three-step mass-to-mass stoichiometry problem. Your goal is to convert 984 kg $Mg(OH)_2$ to the number of kilograms of magnesium oxide needed for its production. **Analyze** the problem statement by listing the given and wanted and then **identify** the equivalencies by first writing the unit path and then deriving the equivalency for each unit conversion. Don't change the equivalencies to conversion factors until the next step.

$$984 \text{ kg Mg(OH)}_2 \times \frac{1 \text{ kmol Mg(OH)}_2}{58.33 \text{ kg Mg(OH)}_2} \times \frac{1 \text{ kmol MgO}}{1 \text{ kmol Mg(OH)}_2} \times \frac{40.31 \text{ kg MgO}}{1 \text{ kmol MgO}} = 6.80 \times 10^2 \text{ kg MgO}$$

$1000 \times (4/6) = 1000 \times (2/3) = 667$, OK.

Construct the solution setup, changing the equivalencies directly to conversion factors in the setup, and **check** the value of the answer to be sure it makes sense.

You improved your skill at solving percentage yield stoichiometry problems.	What did you learn by solving this Active Example?

Practice Exercise 10.7

An aqueous solution of sodium sulfate and liquid water are the products of the reaction of solid sulfur, gaseous oxygen, and an aqueous solution of sodium hydroxide. If the process has a 91.9% yield based on the mass of sodium hydroxide that reacts, determine the number of milligrams of sodium hydroxide needed to produce 5.00×10^2 milligrams of sodium sulfate (Figure 10.14).

The problem in Active Example 10.7 can be solved with a single calculation setup by adding the actual → ideal conversion to the stoichiometry path. The 984 kg of $Mg(OH)_2$ that starts the preceding setup is replaced by the calculation that produced the number—the setup that changed actual yield to ideal yield. The two unit paths

kg $Mg(OH)_2$ actual → kg $Mg(OH)_2$ ideal *and*

kg $Mg(OH)_2$ ideal → kmol $Mg(OH)_2$ → kmol MgO → kg MgO

are combined to give

kg $Mg(OH)_2$ actual → kg $Mg(OH)_2$ ideal → kmol $Mg(OH)_2$ → kmol MgO → kg MgO

The calculation setup becomes:

$$800 \text{ kg Mg(OH)}_2 \text{ actual} \times \frac{100 \text{ kg Mg(OH)}_2 \text{ ideal}}{81.3 \text{ kg Mg(OH)}_2 \text{ actual}} \times \frac{1 \text{ kmol Mg(OH)}_2}{58.33 \text{ kg Mg(OH)}_2} \times \frac{1 \text{ kmol MgO}}{1 \text{ kmol Mg(OH)}_2} \times \frac{40.31 \text{ kg MgO}}{1 \text{ kmol MgO}} = 6.80 \times 10^2 \text{ kg MgO}$$

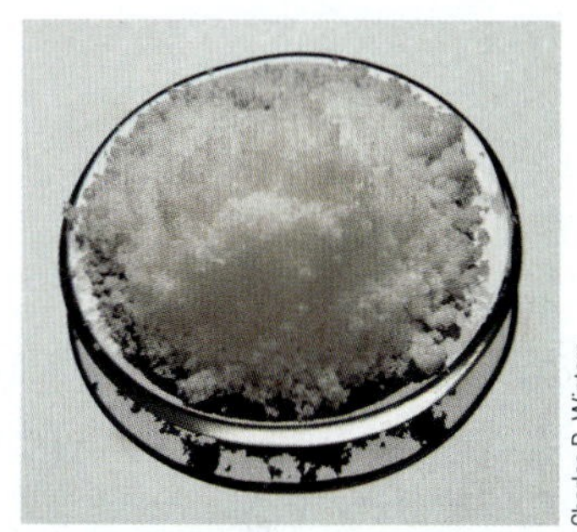

Figure 10.14 Sodium sulfate is used in dyeing textiles and in manufacturing glass and paper pulp.

Charles D. Winters

After conversion from actual yield to ideal yield, there is no need to continue to use the term *ideal* in the basic stoichiometry path.

You may solve percentage yield problems such as this as two separate problems or as a single extended setup. Either way, note that percentage *yield* refers to the *product,* not to a reactant. *Yield always refers to a product.* Accordingly, your percentage yield conversion should always be between actual and ideal product quantities, not reactant quantities. Sometimes the conversion is at the beginning of the setup, as in Active Example 10.7, and sometimes it is at the end.

Active Example 10.8 Percentage Yield III

Solid sodium nitrate decomposes to solid sodium nitrite and oxygen gas upon heating. One set of reaction conditions gives a 95% yield. What mass of sodium nitrite will be produced by the decomposition of 1.50 grams of sodium nitrate?

Think Before You Write You are given the mass of one species in a chemical change and asked for the mass of another species, so this is a stoichiometry problem. The percentage yield tells you that you get 95 grams of product for every 100 grams that you would get under ideal conditions.

Answers *Cover the left column with your cut-out shield. Reveal each answer only after you have written your own answer in the right column.*

Given: 1.50 g $NaNO_3$
Wanted: mass $NaNO_2$ (assume g)

g $NaNO_3$ → mol $NaNO_3$ → mol $NaNO_2$ → g $NaNO_2$ ideal → g $NaNO_2$ actual

1 mol $NaNO_3$ = 85.00 g $NaNO_3$

$2\ NaNO_3(s) \rightarrow 2\ NaNO_2(s) + O_2(g)$

2 mol $NaNO_2$ = 2 mol $NaNO_3$

69.00 g $NaNO_2$ = 1 mol $NaNO_2$

95 g $NaNO_2$ actual = 100 g $NaNO_2$ ideal

Analyze the problem statement by listing the given and wanted and then **identify** the equivalencies by first writing the unit path and then deriving the equivalency for each unit conversion. Don't change the equivalencies to conversion factors until the next step.

$$1.50\ \text{g NaNO}_3 \times \frac{1\ \text{mol NaNO}_3}{85.00\ \text{g NaNO}_3} \times \frac{2\ \text{mol NaNO}_2}{2\ \text{mol NaNO}_3} \times \frac{69.00\ \text{g NaNO}_2}{\text{mol NaNO}_2} \times \frac{95\ \text{g NaNO}_2\ \text{actual}}{100\ \text{g NaNO}_2\ \text{ideal}} = 1.2\ \text{g NaNO}_2\ \text{actual}$$

69.00 ÷ 85.00 is a little less than 1, and 95 ÷ 100 is a little less than 1 as well. Thus, the value of answer should be a little less than 1.5, OK.

Construct the solution setup, changing the equivalencies directly to conversion factors in the setup, and **check** the value of the answer to be sure it makes sense.

You improved your skill at solving percentage yield stoichiometry problems.

What did you learn by solving this Active Example?

Practice Exercise 10.8

Under the conditions utilized by one manufacturer, the percentage yield of gaseous sulfur dioxide is 92.9% when it is made by burning liquid carbon disulfide. Carbon dioxide is a second "waste" product of the reaction. How many kilograms of sulfur dioxide will result from burning 5.00 × 10^2 kg of carbon disulfide?

10.5 Limiting Reactants: The Problem

Goal 4 Identify, describe, and explain limiting reactants and excess reactants.

How many pairs of gloves can you assemble from 20 left gloves and 30 right gloves? Picture the solution to this question in your mind. How many unmatched gloves will be left over? Which hand will these unmatched gloves fit? The answers are that 20 pairs of gloves can be assembled and that 10 right gloves will remain.

The same reasoning can be applied to a chemistry question. Carbon reacts with oxygen to form carbon dioxide: $C(g) + O_2(g) \rightarrow CO_2(g)$. Suppose you put three atoms of carbon and two molecules of oxygen in a reaction vessel and cause them to react until either carbon or oxygen is totally used up. How many carbon dioxide molecules will result? How many particles of which element will remain unreacted? The answers to these questions are drawn in **Figure 10.15**.

The reactants combine in a one-to-one ratio. If you start with three atoms of carbon and two molecules of oxygen, the reaction will stop when the two oxygen molecules are used up. Oxygen, the reactant that is completely used up by the reaction, is called the **limiting reactant**. One atom of carbon, the **excess reactant**, will remain unreacted.

It is very unusual for reactants in a chemical change to be present in the exact quantities that will react completely with each other. This condition is approached, however, in the process of titration, which is considered in Sections 16.13 and 16.14.

Since we cannot count individual particles, we again must turn to our macroscopic-particulate link—the mole. The amount of product is limited by the amount in *moles* of the limiting reactant, and it must be calculated from that number of moles. If we start with 3 moles of carbon and 2 moles of oxygen, we have the same ratio of particles as when we start with three carbon atoms and two oxygen molecules. If 2 moles of the limiting reactant—oxygen—react, 2 moles of carbon

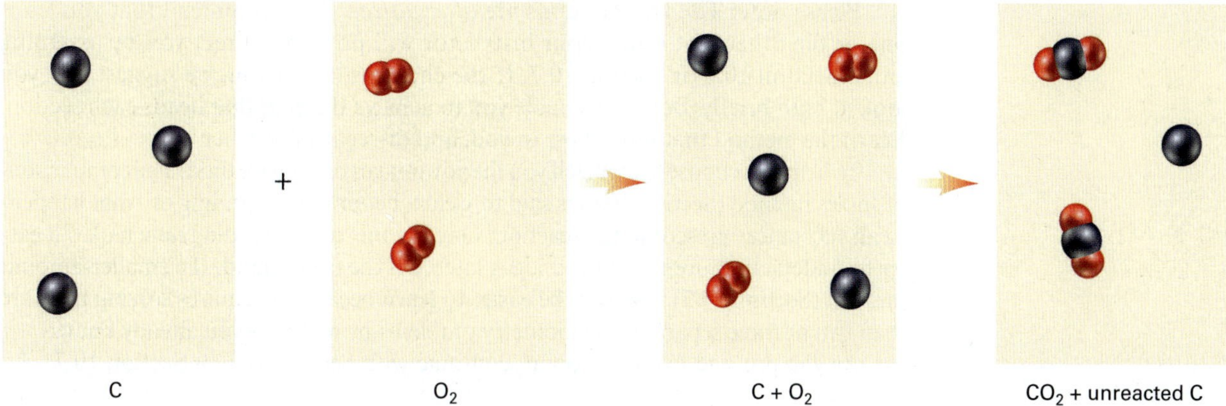

Figure 10.15 The reaction of carbon and oxygen gases at the particulate level.

dioxide are produced. According to the equation, each mole of carbon dioxide produced requires 1 mole of carbon to react. The entire analysis on the molar level may be summarized as follows:

	C	+	O_2	→	CO_2
Amount at start (mol)	3		2		0
Amount used (−) or produced (+) (mol)	−2		−2		+2
Amount after the reaction (mol)	1		0		2

Target Check 10.1

Consider the hypothetical reaction 2 A + B → A_2B. The illustration in **Figure 10.16** shows a mixture of particles before the reaction, in which atoms of A are in gold and atoms of B are in black. Draw a particulate representation after a complete reaction. Explain your reasoning.

Limiting Reactants: A Choice

In practice, you must approach analyzing and solving limiting-reactant problems with measurable units, usually mass units. There are two ways to solve these problems. In Section 10.6, we describe the comparison-of-moles method, which is a continuation of the analysis we have been doing. Section 10.7 presents the smaller-amount method, which solves the problem entirely in terms of mass units.

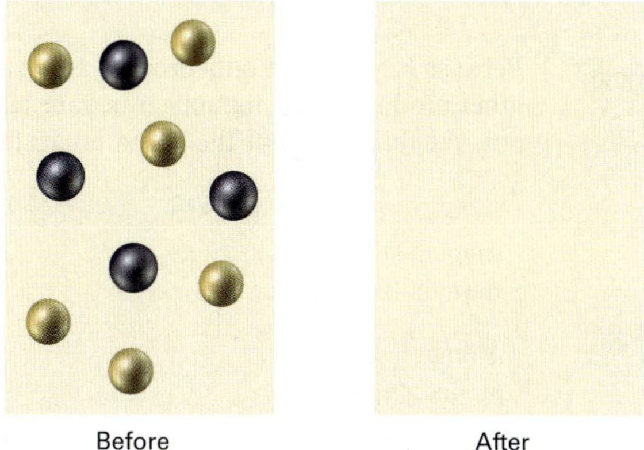

Figure 10.16 The mixture before the hypothetical reaction of A (gold) and B (black).

Please note: The two methods are *alternatives*. We recommend that you learn one or the other, not both. Your instructor will probably direct you by assigning either Section 10.6 or Section 10.7. If the choice is left to you, we suggest that you look at both briefly. Both will enable you to achieve the goal that heads each section. Learn the method that looks best to you, and disregard the other.

To help you choose, we will tell you the advantages of each method. The comparison-of-moles method (Section 10.6) is said to yield a better understanding of what happens to all substances present in the reaction. Gaining this understanding now makes it easier to understand more advanced ideas later. On the other hand, the smaller-amount method (Section 10.7) is said to be easier to learn because it requires solving no more than two or three separate stoichiometry problems by methods you already know.

As you proceed from this point, continue with Section 10.6 *or* Section 10.7.

Figure 10.17 Potassium nitrate is the major component of tree stump remover. It is also used in making matches and explosives, including fireworks, gunpowder, and blasting powder.

10.6 Limiting Reactants: Comparison-of-Moles Method

Goal 5 Given a chemical equation, or information from which it may be determined, and initial quantities of two or more reactants, (a) identify the limiting reactant; (b) calculate the ideal yield of a specified product, assuming complete use of the limiting reactant; and (c) calculate the quantity of the reactant initially in excess that remains unreacted.

Limiting-reactant problems are more complicated when the quantities are expressed in measurable mass units. The "3 moles of carbon react with 2 moles of oxygen" question becomes 36.03 grams of carbon (3 moles) combines with 64.00 grams of oxygen (2 moles) until one is totally used up in the reaction $C(g) + O_2(g) \rightarrow CO_2(g)$. What mass in grams of carbon dioxide, CO_2, results? Also, what mass in grams of which element remains unreacted?

The following two Active Examples show you what to do, first, when the reactants do not react in a 1:1 mole ratio, and then when the given quantities are in mass units.

Active Example 10.9 Limiting Reactants: Comparison-of-Moles Method I

What amount in moles of the tree stump remover potassium nitrate (Figure 10.17) can be made from 7.94 moles of potassium chloride and 9.96 moles of nitric acid solution by the reaction $3\ KCl(s) + 4\ HNO_3(aq) \rightarrow 3\ KNO_3(s) + Cl_2(g) + NOCl(g) + 2\ H_2O(\ell)$? Also, what amount in moles of which reactant will be unused?

Think Before You Write You are given the quantities of both reacting species. Before now, one has always been in excess; this time, you cannot directly tell when one is in excess. Thus, this will be a limiting reactant stoichiometry problem. Setting up the table that we describe here effectively plans this kind of problem.

Answers *Cover the left column with your cut-out shield. Reveal each answer only after you have written your own answer in the right column.*

	$3\ KCl$	+	$4\ HNO_3$	→	$3\ KNO_3$
Amount at start (mol)	7.94		9.96		0
Amount used (−) or produced (+) (mol)					
Amount after the reaction (mol)					

Because KNO_3 is the only product asked about, the other products need not appear in your tabulation. Fill in only the three boxes of the "Amount at start (mol)" line.

	$3\ KCl$	+	$4\ HNO_3$	→	$3\ KNO_3$
Amount at start (mol)					
Amount used (−) or produced (+) (mol)					
Amount after the reaction (mol)					

The limiting reactant is not immediately recognizable. One way to identify it is to select either reactant and ask, "If Reactant A is the limiting reactant, how many moles of Reactant B are needed to react with all of Reactant A?" If the number of moles of Reactant B needed is more than the number available, Reactant B is the limiting reactant. If the number of moles of Reactant B needed is smaller than the number present, Reactant B is the excess reactant and Reactant A is the limiting reactant. The conclusion is the same regardless of which reactant you select as A and which you select as B. Let's see what happens if we guess that KCl is the limiting reactant.

How many moles of HNO_3 are required to react with all of the KCl? Completely set up and solve this one-step conversion.

$$7.94 \text{ mol KCl} \times \frac{4 \text{ mol HNO}_3}{3 \text{ mol KCl}} = 10.6 \text{ mol HNO}_3$$

If KCl is the limiting reactant, 10.6 mol of HNO_3 is required to react with all of the KCl. Only 9.96 mol of HNO_3 is present. This is not enough HNO_3, so HNO_3 is the limiting reactant.

You now know that HNO_3 is the limiting reactant, so determine the number of moles of KCl that will be used to react with all of the 9.96 moles of HNO_3 that are available.

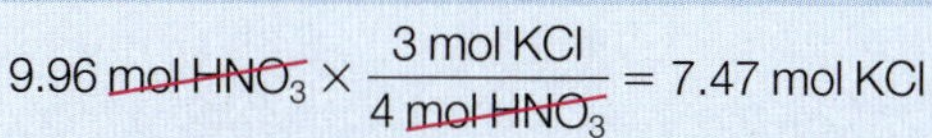

$$9.96 \text{ mol HNO}_3 \times \frac{3 \text{ mol KCl}}{4 \text{ mol HNO}_3} = 7.47 \text{ mol KCl}$$

This calculation tells you that 7.47 mol of KCl is required to react with all of the HNO_3. There is 7.94 mol of KCl present. This is more than enough KCl, so it verifies that HNO_3 is the limiting reactant.

You now have enough information to fill in the second line of the table. The 3 mol KCl-to-3 mol KNO_3 ratio allows you to fill in the amount in moles of KNO_3 produced without having to do a calculation. Use + and − signs to indicate whether the number of moles is produced or used. Complete the second line.

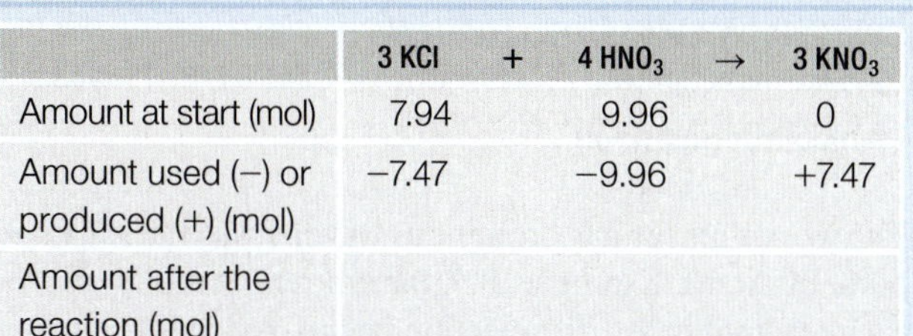

	3 KCl	+	4 HNO_3	→	3 KNO_3
Amount at start (mol)	7.94		9.96		0
Amount used (−) or produced (+) (mol)	−7.47		−9.96		+7.47
Amount after the reaction (mol)					

The amount in moles of each species at the end is simply the sum of the amount at the start and the amount used or produced. Complete the problem by filling in the third line of the table.

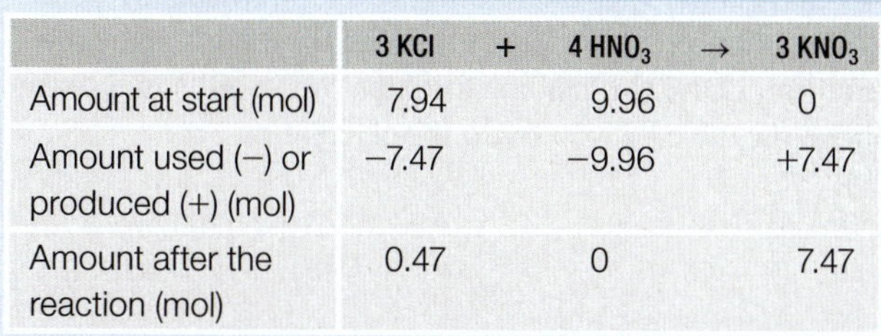

	3 KCl	+	4 HNO_3	→	3 KNO_3
Amount at start (mol)	7.94		9.96		0
Amount used (−) or produced (+) (mol)	−7.47		−9.96		+7.47
Amount after the reaction (mol)	0.47		0		7.47

7.47 moles of potassium nitrate is produced, and 0.47 mole of potassium chloride remains unreacted.

What did you learn by solving this Active Example?

You improved your skill at solving limiting reactant stoichiometry problems.

Practice Exercise 10.9

What amount in moles of lead(II) iodide will precipitate when 1.10 moles of sodium iodide and 0.761 moles of lead(II) nitrate are dissolved in the same beaker of water? What amount in moles of which reactant will remain unreacted?

Reactant and product quantities are not usually expressed in moles, but rather in grams. This adds one or more steps before and after the sequence in Active Example 10.9.

Active Example 10.10 Limiting Reactants: Comparison-of-Moles Method II

Calculate the mass of antimony(III) iodide, SbI_3, that can be produced by the reaction of 129 g of antimony, Sb (Z = 51), and 381 g of iodine (Figure 10.18). Also, find the mass in grams of the element that will be left.

Think Before You Write You are given the mass of both reactants in this chemical change, and you are asked to find the mass of the product. Therefore, this is a limiting reactant stoichiometry problem.

Answers *Cover the left column with your cut-out shield. Reveal each answer only after you have written your own answer in the right column.*

$$129\ \cancel{g\ Sb} \times \frac{1\ mol\ Sb}{121.8\ \cancel{g\ Sb}} = 1.06\ mol\ Sb$$

$$381\ \cancel{g\ I_2} \times \frac{1\ mol\ I_2}{253.8\ \cancel{g\ I_2}} = 1.50\ mol\ I_2$$

	2 Sb	+	$3\ I_2$	→	$2\ SbI_3$
Mass at start (g)	129		381		0
Molar mass (g/mol)	121.8		253.8		502.5
Amount at start (mol)	1.06		1.50		0

This time the table begins with the starting *masses* of all species, rather than moles. You must convert the masses to moles. It is convenient to add a line to the table for molar mass, too. The first two lines are completed for you. Do the two mass-to-moles conversions in the space below, and then fill in the third line of the table.

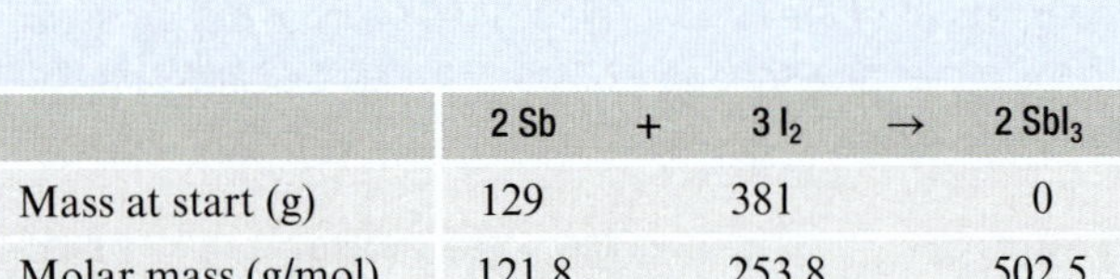

	2 Sb	+	$3\ I_2$	→	$2\ SbI_3$
Mass at start (g)	129		381		0
Molar mass (g/mol)	121.8		253.8		502.5
Amount at start (mol)					

$2\ Sb + 3\ I_2 \rightarrow 2\ SbI_3$

$$1.06\ \cancel{mol\ Sb} \times \frac{3\ mol\ I_2}{2\ \cancel{mol\ Sb}} = 1.59\ mol\ I_2$$

1.59 mol I_2 > 1.50 mol I_2 available, so I_2 is the limiting reactant.

$$1.50\ \cancel{mol\ I_2} \times \frac{2\ mol\ Sb}{3\ \cancel{mol\ I_2}} = 1.00\ mol\ Sb\ used$$

	2 Sb	+	$3\ I_2$	→	$2\ SbI_3$
Mass at start (g)	129		381		0
Molar mass (g/mol)	121.8		253.8		502.5
Amount at start (mol)	1.06		1.50		0
Amount used (−) or produced (+) (mol)	−1.00		−1.50		+1.00
Amount at end (mol)	0.06		0		1.00
Mass at end (g)					

The work thus far has brought us to what was the starting point of Active Example 10.9, the amount in moles of each reactant before the reaction begins. Extend the table through the next *two* lines to find the moles of product and excess reactant after the limiting reactant is used up. To do this, you'll need to write and balance the equation for the reaction. Leave the last line for later. We've left some space above the table for you to do the calculation.

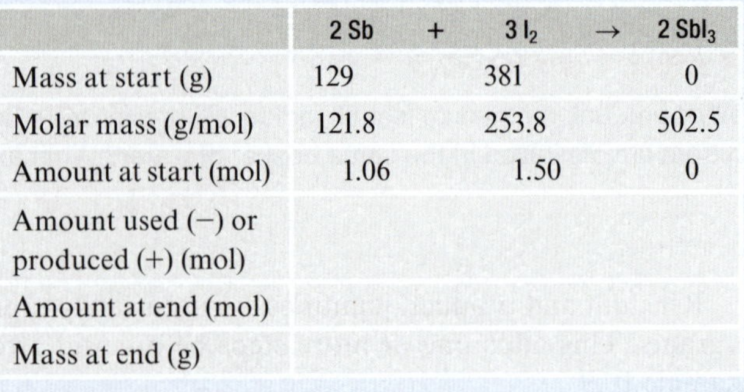

	2 Sb	+	$3\ I_2$	→	$2\ SbI_3$
Mass at start (g)	129		381		0
Molar mass (g/mol)	121.8		253.8		502.5
Amount at start (mol)	1.06		1.50		0
Amount used (−) or produced (+) (mol)					
Amount at end (mol)					
Mass at end (g)					

$$0.06 \text{ mol Sb} \times \frac{121.8 \text{ g Sb}}{\text{mol Sb}} = 7 \text{ g Sb}$$

Amount at end (mol)	0.06	0	1.00
Mass at end (g)	7	0	503

503 g of antimony(III) iodide is produced, and 7 g of antimony remains unreacted.

The final step is to change amount in moles to mass in grams by means of molar mass. Set up the calculation in the space below (you should be able to fill in two of the three columns without a setup), and then fill in the blanks in the table to complete the problem.

You improved your skill at solving limiting reactant stoichiometry problems.

What did you learn by solving this Active Example?

Practice Exercise 10.10

A solution containing 43.5 g of calcium nitrate is added to a solution containing 39.5 g of sodium fluoride. What mass (grams) of calcium fluoride precipitates? What mass (grams) of which reactant is in excess?

We are now ready to summarize the overall procedure for solving a limiting-reactant problem.

how to... **Solve a Limiting-Reactant Problem**

1. Convert the mass in grams of each reactant to amount in moles.
2. Identify the limiting reactant.
3. Calculate the amount in moles of each species that reacts or is produced.
4. Calculate the amount in moles of each species that remains after the reaction.
5. Change the amount in moles of each species to mass in grams.

Figure 10.18 Iodine is a bluish-black lustrous solid at room conditions, as seen on the right. It readily sublimates—changes directly from solid to gas—to a blue-violet gas, as seen on the left.

10.7 Limiting Reactants: Smaller-Amount Method

Goal 5 Given a chemical equation, or information from which it may be determined, and initial quantities of two or more reactants, (a) identify the limiting reactant, (b) calculate the ideal yield of a specified product, assuming complete use of the limiting reactant, and (c) calculate the quantity of the reactant initially in excess that remains unreacted.

Limiting-reactant problems are more complicated when the quantities are expressed in measurable mass units. The "3 moles of carbon react with 2 moles of oxygen" question becomes 36.03 grams of carbon (3 moles) combines with 64.00 grams of oxygen (2 moles) until one is totally used up in the reaction $C(g) + O_2(g) \rightarrow CO_2(g)$. What mass in grams of carbon dioxide, CO_2, results? Also, what mass in grams of which element remains unreacted?

A type of reasoning frequently used in science involves "what if?" thinking. A scientist will *assume* that a particular condition or set of conditions is true and then calculate the answer based on this assumption. If the answer proves to be impossible—if it contradicts known facts—that answer is discarded and another is considered. In limiting-reactant problems, we need to identify the limiting reactant. We can answer this question by assuming that each reactant is limiting.

First, what if carbon is the limiting reactant? What mass in grams of CO_2 will be produced? In other words, if all the carbon is used up in the reaction, what mass of CO_2 will be made? This is a fundamental three-step stoichiometry problem:

$$36.03 \text{ g C} \times \frac{1 \text{ mol C}}{12.01 \text{ g C}} \times \frac{1 \text{ mol } CO_2}{1 \text{ mol C}} \times \frac{44.01 \text{ g } CO_2}{1 \text{ mol } CO_2} = 132.0 \text{ g } CO_2$$

Now, what if oxygen is the limiting reactant? What mass in grams of CO_2 will be produced by 64.00 g of O_2?

$$64.00 \text{ g } O_2 \times \frac{1 \text{ mol } O_2}{32.00 \text{ g } O_2} \times \frac{1 \text{ mol } CO_2}{1 \text{ mol } O_2} \times \frac{44.01 \text{ g } CO_2}{1 \text{ mol } CO_2} = 88.02 \text{ g } CO_2$$

We interpret these two results by saying that we have enough carbon to produce 132.0 g of CO_2. However, it is impossible to form more than 88.02 g CO_2 from 64.00 g O_2. Therefore, oxygen is the limiting reactant, and the reaction stops when the limiting reactant is used up. *The limiting reactant is always the one that yields the smaller amount of product.*

To find the mass of excess reactant that remains, calculate how much of that reactant will be used by the entire amount of limiting reactant:

$$64.00 \text{ g } O_2 \times \frac{1 \text{ mol } O_2}{32.00 \text{ g } O_2} \times \frac{1 \text{ mol C}}{1 \text{ mol } O_2} \times \frac{12.01 \text{ g C}}{1 \text{ mol C}} = 24.02 \text{ g C}$$

We started with 36.03 g C. Reaction with the limiting reactant used up 24.02 g C. The mass that remains is the starting mass minus the mass used:

$$36.03 \text{ g C (initial)} - 24.02 \text{ g C (used)} = 12.01 \text{ g C left}$$

The smaller-amount method can be summarized as follows:

Notice that this procedure refers to "mass" of reactant. In this section, we are expressing amount in terms of grams. It can be expressed as amount in moles just as well. In fact, that is exactly how amount is described in Section 10.4, in which the limiting-reactant concept was introduced.

how to... Solve a Limiting-Reactant Problem

1. Calculate the mass of product that can be formed by the initial mass of each reactant.
   a) The reactant that yields the smaller mass of product is the limiting reactant.
   b) The smaller mass of product is the mass that will be formed when all of the limiting reactant is used up.
2. Calculate the mass of excess reactant that is used by the total mass of limiting reactant.
3. Subtract from the mass of excess reactant present initially the mass that is used by all of the limiting reactant. The difference is the amount of excess reactant that is left.

Active Example 10.11 Limiting Reactants: Smaller-Amount Method

Calculate the mass of antimony(III) iodide, SbI_3, that can be produced by the reaction of 129 g of antimony, Sb (Z = 51), and 381 g of iodine (Figure 10.18). Also find the mass in grams of the element that will be left.

Think Before You Write You are given the mass of both reactants in this chemical change, and you are asked to find the mass of the product. Therefore, this is a limiting reactant stoichiometry problem.

Answers *Cover the left column with your cut-out shield. Reveal each answer only after you have written your own answer in the right column.*

Given: 129 g Sb **Wanted:** g SbI_3 g Sb → mol Sb → mol SbI_3 → g SbI_3 1 mol Sb = 121.8 g Sb $2 \text{ Sb} + 3 \text{ I}_2 \rightarrow 2 \text{ SbI}_3$ 2 mol SbI_3 = 2 mol Sb 502.5 g SbI_3 = 1 mol SbI_3	What if antimony is the limiting reactant? What mass in grams of SbI_3 can be produced with 129 g Sb? **Analyze** this subproblem by listing the given and wanted and then **identify** the equivalencies by first writing the unit path and then deriving the equivalency for each unit conversion.

$$129\ \text{g Sb} \times \frac{1\ \text{mol Sb}}{121.8\ \text{g Sb}} \times \frac{2\ \text{mol SbI}_3}{2\ \text{mol Sb}} \times \frac{502.5\ \text{g SbI}_3}{1\ \text{mol SbI}_3} = 532\ \text{g SbI}_3$$

(129 ÷ 121.8) × (2 ÷ 2) × 502.5 = (A little more than 1) × 1 × ~500 = A little more than 500, OK.

Construct the solution setup, changing the equivalencies directly to conversion factors in the setup, and **check** the value of the answer to be sure it makes sense.

Given: 381 g I_2
Wanted: g SbI_3

g I_2 → mol I_2 → mol SbI_3 → g SbI_3

1 mol I_2 = 253.8 g I_2

$2\ Sb + 3\ I_2 \rightarrow 2\ SbI_3$

2 mol SbI_3 = 3 mol I_2

502.5 g SbI_3 = 1 mol SbI_3

What if the iodine is the limiting reactant? What mass in grams of SbI_3 can be produced with 381 g I_2? **Analyze** this subproblem by listing the given and wanted and then **identify** the equivalencies by first writing the unit path and then deriving the equivalency for each unit conversion.

$$381\ \text{g I}_2 \times \frac{1\ \text{mol I}_2}{253.8\ \text{g I}_2} \times \frac{2\ \text{mol SbI}_3}{3\ \text{mol I}_2} \times \frac{502.5\ \text{g SbI}_3}{1\ \text{mol SbI}_3} = 503\ \text{g SbI}_3$$

381 × (2 ÷ 3) × (502.5 ÷ 253.8) ≈ 360 × (2 ÷ 3) × (500 ÷ 250) = 360 × (2 ÷ 3) × 2 = (360 ÷ 3) × 2 × 2 = 120 × 4 = 480, OK.

We rounded 381 to 360 because 360 is easy to use in estimates because it is evenly divisible by 2, 3, 4, 6, 8, and so on.

Construct the solution setup, changing the equivalencies directly to conversion factors in the setup, and **check** the value of the answer to be sure it makes sense.

The limiting reactant is iodine, I_2.

129 g Sb is enough antimony to produce 532 g SbI_3, but there is only enough I_2 to produce 503 g SbI_3. Therefore, iodine is the limiting reactant, and 503 g SbI_3 is produced.

You are now able to identify the limiting reactant. What is it? Explain how you know. What mass of SbI_3 is produced?

Given: 381 g I_2
Wanted: g Sb

g I_2 → mol I_2 → mol Sb → g Sb

1 mol I_2 = 253.8 g I_2

$2\ Sb + 3\ I_2 \rightarrow 2\ SbI_3$

2 mol Sb = 3 mol I_2

121.8 g Sb = 1 mol Sb

Antimony will react until there's no more iodine available. Some antimony will be left. How much?

You know how much antimony you started with. If you knew how much you used, you could subtract what was used from what you had at the beginning to find out how much is left. Well, how much antimony *is* used to react with 381 g I_2? This is another stoichiometry question.

Figure 10.18 Iodine is a bluish-black lustrous solid at room conditions, as seen on the right. It readily sublimates—changes directly from solid to gas—to a blue-violet gas, as seen on the left.

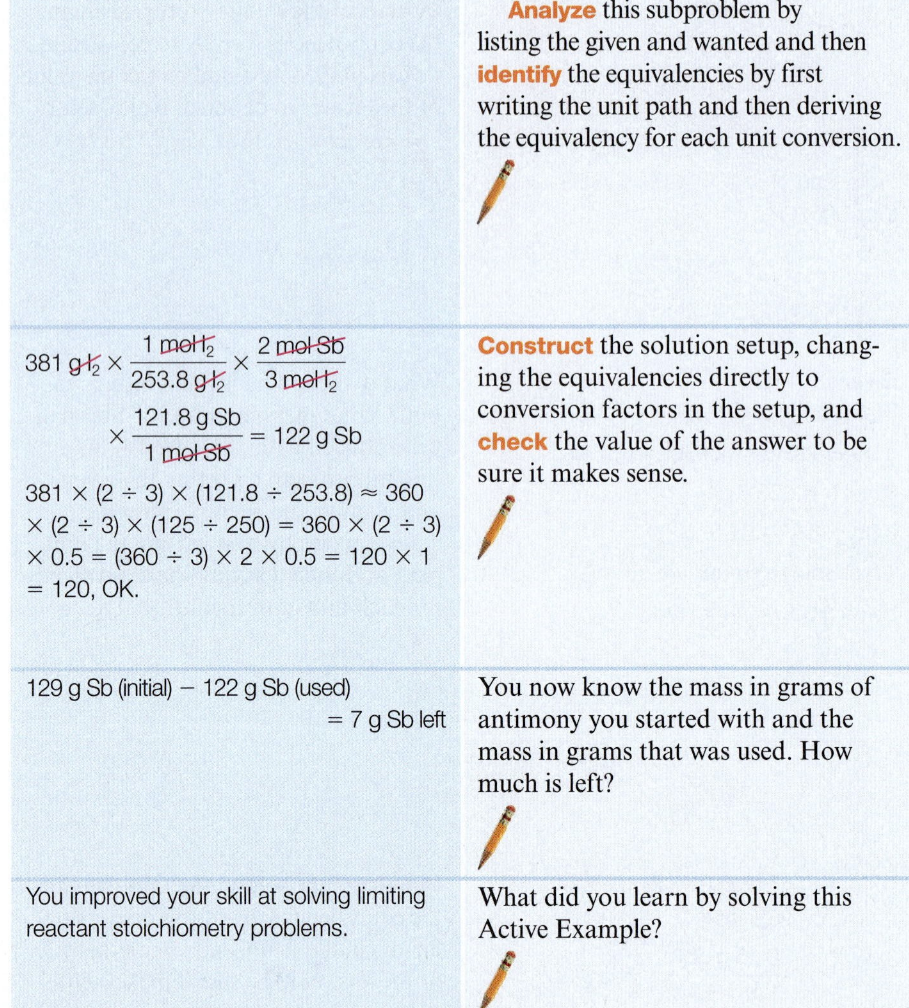

	Analyze this subproblem by listing the given and wanted and then **identify** the equivalencies by first writing the unit path and then deriving the equivalency for each unit conversion.
$381\ g\ I_2 \times \frac{1\ mol\ I_2}{253.8\ g\ I_2} \times \frac{2\ mol\ Sb}{3\ mol\ I_2} \times \frac{121.8\ g\ Sb}{1\ mol\ Sb} = 122\ g\ Sb$ 381 × (2 ÷ 3) × (121.8 ÷ 253.8) ≈ 360 × (2 ÷ 3) × (125 ÷ 250) = 360 × (2 ÷ 3) × 0.5 = (360 ÷ 3) × 2 × 0.5 = 120 × 1 = 120, OK.	**Construct** the solution setup, changing the equivalencies directly to conversion factors in the setup, and **check** the value of the answer to be sure it makes sense.
129 g Sb (initial) − 122 g Sb (used) = 7 g Sb left	You now know the mass in grams of antimony you started with and the mass in grams that was used. How much is left?
You improved your skill at solving limiting reactant stoichiometry problems.	What did you learn by solving this Active Example?

Practice Exercise 10.11

A solution containing 43.5 g of calcium nitrate is added to a solution containing 39.5 g of sodium fluoride. What mass (grams) of calcium fluoride precipitates? What mass (grams) of which reactant is in excess?

Nutrition Facts	
4 servings per container	
Serving size 1 1/2 cup (208g)	
Amount per serving	
Calories	**240**
	% Daily Value*
Total Fat 4g	**5%**
Saturated Fat 1.5g	**8%**
Trans Fat 0g	
Cholesterol 5mg	**2%**
Sodium 430mg	**19%**
Total Carbohydrate 46g	**17%**
Dietary Fiber 7g	**25%**
Total Sugars 4g	
Includes 2g Added Sugars	**4%**
Protein 11g	
Vitamin D 2mcg	10%
Calcium 260mg	20%
Iron 6mg	35%
Potassium 240mg	6%

* The % Daily Value (DV) tells you how much a nutrient in a serving of food contributes to a daily diet. 2,000 calories a day is used for general nutrition advice.

Figure 10.19 A food Calorie is 1000 thermochemical calories. A U.S. federal law, the Nutrition Labeling and Education Act, requires that most food packages display nutritional information such as the caloric value of a food serving.

10.8 Energy

Goal 6 Given an energy quantity in one of the following units, plus variations created by adding metric prefixes, calculate the other two: joules, calories, and food Calories.

In Section 8.3, we pointed out that nearly all chemical changes involve an energy transfer, usually in the form of heat. We now wish to consider the amount of energy that is transferred in a reaction. To do that, we must introduce the common units in which energy is measured and expressed.

The SI unit for energy is the **joule (J)** (rhymes with *pool*), which is defined in terms of three base units as $1\ kg \cdot m^2/s^2$ (s is an abbreviation for the time unit second). An older energy unit is the **calorie (cal)**, originally defined as the amount of energy required to raise the temperature of 1 gram of water by 1°C. The calorie has now been redefined in terms of joules:

$$1\ \text{calorie} \equiv 4.184\ \text{joules} \qquad 4.184\ \text{J/cal}$$

Both the calorie and the joule are small amounts of energy, so the kilojoule (kJ) and kilocalorie (kcal) are often used instead. ◀ It follows that

$$1\ \text{kcal} = 4.184\ \text{kJ} \qquad 4.184\ \text{kJ/kcal}$$

▶ **P/Review** *Kilo-* is the metric prefix for the unit 1000 times larger than the base unit (Section 3.6). Thus, 1000 calories = 1 kilocalorie and 1000 joules = 1 kilojoule.

Active Example 10.12 Energy Unit Conversions

Calculate the number of kilocalories, calories, and joules that are equivalent to 42.5 kilojoules.

Think Before You Write As long as you know that a calorie is 4.184 joules, you are ready to solve this problem.

Answers *Cover the left column with your cut-out shield. Reveal each answer only after you have written your own answer in the right column.*

Given: 42.5 kJ
Wanted: kcal

1 kcal = 4.184 kJ

$$\frac{1\text{ kcal}}{4.184\text{ kJ}}$$

Consider the kJ → kcal conversion first. **Analyze** this subproblem by listing the given and wanted and then **identify** the equivalency. Change the equivalency to the needed conversion factor.

$$42.5\text{ kJ} \times \frac{1\text{ kcal}}{4.184\text{ kJ}} = 10.2\text{ kcal}$$

$$(4 \times 10^1) \times \frac{1}{4} = 1 \times 10^1 = 10\text{, OK}$$

Construct the solution setup, and **check** the value of the answer to be sure it makes sense.

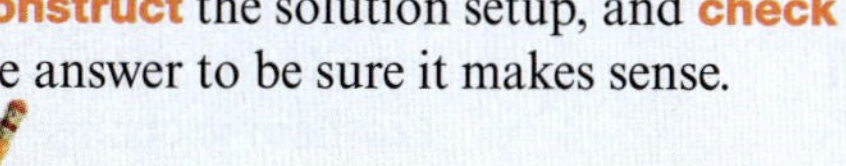

42.5 kJ = 42,500 J = 4.25×10^4 J

10.2 kcal = 10,200 cal = 1.02×10^4 cal

The larger/smaller rule (Section 3.2) states that any quantity may be expressed as a large number of small units or a small number of large units. In this example, this translates into 1000 units in 1 kilounit. Therefore, the number of smaller units (cal and J) is larger than the number of larger units (kcal and kJ). The decimal must move right three places.

Alternatively, conversion factors may be used:

$$42.5\text{ kJ} \times \frac{1000\text{ J}}{\text{kJ}} = 4.25 \times 10^4\text{ J}$$

$$10.2\text{ kcal} \times \frac{1000\text{ cal}}{\text{kcal}} = 1.02 \times 10^4\text{ cal}$$

Now change 42.5 kJ to J and 10.2 kcal to cal. Try to do it by just moving the decimal point.

You improved your skill at converting between energy units.

What did you learn by solving this Active Example?

Practice Exercise 10.12

A 1-oz serving of almonds has 1.6×10^2 Calories. Express this in joules and kilojoules.

The **Calorie** used when talking about food energy is actually the kilocalorie. It is written with a capital C to distinguish it from the thermochemical calorie, which is written with a lowercase c. Caloric food requirements vary considerably among individuals, and the caloric value of food is easy to track with nutrition labeling (**Figure 10.19**). A typical diet provides about 2000 Calories (kcal) per day.

10.9 Thermochemical Equations

Goal 7 Given a chemical equation, or information from which it may be written, and the heat (enthalpy) of reaction, write the thermochemical equation either with $\Delta_r H$ (a) to the right of the conventional equation or (b) as a reactant or product.

The quantity of heat transferred in a chemical reaction can be measured in the laboratory. We sometimes call this change the **heat of reaction**; more formally, it is the

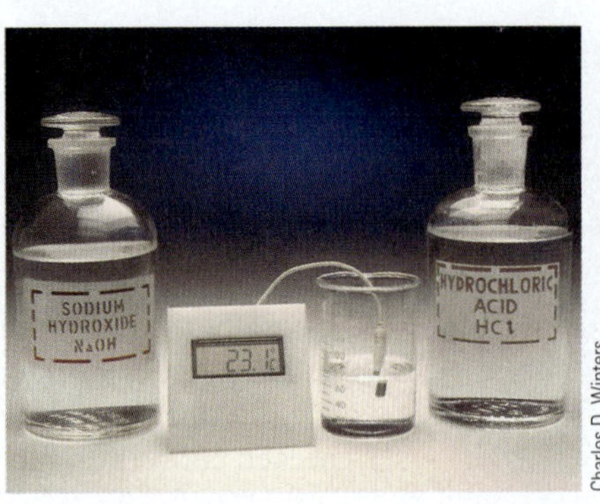

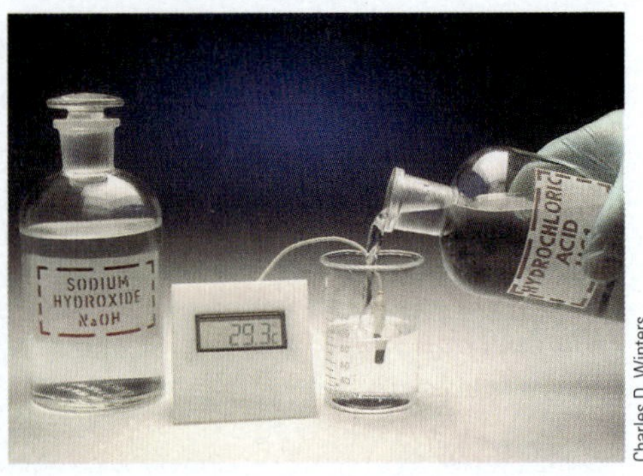

Figure 10.20 An exothermic reaction. (a) Room temperature sodium hydroxide solution at 23.1°C is poured into a beaker. (b) Room temperature hydrochloric acid solution is added to the beaker. The reaction between solutions of hydrochloric acid and sodium hydroxide is exothermic, releasing heat and causing the temperature of the solution to rise to 29.3°C.

enthalpy of reaction, $\Delta_r H$.* H is the symbol for enthalpy (or heat), Δ, the Greek uppercase delta, indicates change (final value minus initial value), and the subscript r stands for reaction. When a system transfers heat to the surroundings—an exothermic change (**Figure 10.20**)—the enthalpy of the system goes down, and $\Delta_r H$ has a negative value. When heat is transferred to a system because of a reaction—an endothermic change—enthalpy increases, and $\Delta_r H$ is positive.

A chemical equation that includes a change in energy is a **thermochemical equation**. There are two kinds of thermochemical equations. One simply writes the $\Delta_r H$ of the reaction to the right of the conventional equation. For example, if you burn 2 moles of ethane, C_2H_6, 2855 kJ of heat is transferred. This is an exothermic reaction, so $\Delta_r H$ is negative: $\Delta_r H = -2855$ kJ. The thermochemical equation is:

$$2\,C_2H_6(g) + 7\,O_2(g) \rightarrow 4\,CO_2(g) + 6\,H_2O(g) \quad \Delta_r H = -2855\text{ kJ}$$

The second form of a thermochemical equation includes energy as if it were a reactant or product. In an exothermic reaction, heat is "produced," so it appears as a positive quantity on the product side of the equation:

$$2\,C_2H_6(g) + 7\,O_2(g) \rightarrow 4\,CO_2(g) + 6\,H_2O(g) + 2855\text{ kJ}$$

In an endothermic reaction, energy must be transferred to the reactants to make the reaction happen. Heat is a "reactant." The decomposition of ammonia into its elements is an example:

$$2\,NH_3(g) + 92\text{ kJ} \rightarrow N_2(g) + 3\,H_2(g)$$
$$2\,NH_3(g) \rightarrow N_2(g) + 3\,H_2(g) \quad \Delta_r H = +92\text{ kJ}$$

When you write thermochemical equations, you *must* use the state symbols (g), (ℓ), (s), and (aq). The equation is meaningless without them because the size of the enthalpy change depends on the states of the reactants and products. If the equation for the combustion of 2 moles of ethane is written with water in the liquid state,

$$2\,C_2H_6(g) + 7\,O_2(g) \rightarrow 4\,CO_2(g) + 6\,H_2O(\ell) \quad \Delta_r H = -3119\text{ kJ}$$

the value of $\Delta_r H$ is −3119 kJ. Unless stated otherwise, we will assume $H_2O(\ell)$ to be the product of burning reactions because data measured for thermochemical processes are usually recorded in reference sources for the process as it takes place at 25°C, and water is a liquid at this temperature (25°C = 77°F).

Active Example 10.13 Thermochemical Equations

The thermal decomposition of limestone, solid calcium carbonate, $CaCO_3(s)$, to lime, solid $CaO(s)$, and gaseous carbon dioxide, $CO_2(g)$, is an endothermic reaction requiring the transfer of 178 kJ per mole of calcium carbonate decomposed. Write the thermochemical equation in two forms.

*These terms may lead you to think that heat and enthalpy are the same. They are not, but the difference between them is beyond the scope of an introductory text. Under certain common conditions, however, heat transferred into or out of a system is equal to the enthalpy change. We will limit ourselves to such reactions.

Think Before You Write Thermochemical equations can be written with the $\Delta_r H$ term as part of the equation itself or as a separate term. If you know both ways of writing these equations, you are ready for this Active Example.

Answers *Cover the left column with your cut-out shield. Reveal each answer only after you have written your own answer in the right column.*

$CaCO_3(s) \rightarrow CaO(s) + CO_2(g)$ $\Delta_r H = +178\ kJ$ $CaCO_3(s) + 178\ kJ \rightarrow CaO(s) + CO_2(g)$	Write the two equations.
You improved your skill at writing thermochemical equations.	What did you learn by solving this Active Example?

Practice Exercise 10.13

When gaseous hydrogen and oxygen react to form 1 mole of liquid water, 286 kJ of energy is released (Figure 10.21). Write the thermochemical equation in two forms.

Space Exploration Technologies Corp

Figure 10.21 An enormous amount of energy was expended by the reaction of hydrogen and oxygen in a SpaceX Falcon Heavy launch. Would you classify the fuel's reaction as exothermic or endothermic?

10.10 Thermochemical Stoichiometry

Goal 8 Given a thermochemical equation, or information from which it may be written, calculate the amount of energy transferred for a given amount of reactant or product; alternately, calculate the mass of reactant required to transfer a given amount of energy.

If you burn twice as much fuel, it is logical to expect that twice as much energy should be transferred. The proportional relationships between amount in moles of different substances in a chemical equation, expressed by their coefficients, extend to energy terms, as illustrated in Figure 10.22. Note that the general stoichiometry pattern remains the same.

General Stoichiometry Pattern

Measured macroscopic quantity *Reactant* ↔ Particulate quantity *Reactant* ↔ Particulate quantity *Product* ↔ Measured macroscopic quantity *Product*

Specific Applications

Energy & $\Delta_r H$ *Reactant* ↔ Amount in moles *Reactant*

Mass in grams & molar mass *Reactant* ↔ Amount in moles *Reactant*

Amount in moles *Reactant* ↔ (Mole ratio from balanced equation) ↔ Amount in moles *Product*

Amount in moles *Product* ↔ Energy & $\Delta_r H$ *Product*

Amount in moles *Product* ↔ Mass in grams & molar mass *Product*

Amount in moles *Reactant* ↔ (N_A) ↔ Quantity in number of particles *Reactant*

Amount in moles *Product* ↔ (N_A) ↔ Quantity in number of particles *Product*

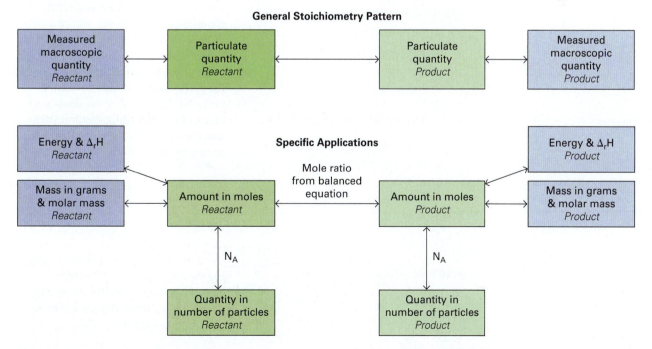

Figure 10.22 Conversion among mass in grams and molar mass or energy and $\Delta_r H$, amount in moles, and quantity in number of particles for two species in a chemical change. This figure adds a new measured macroscopic quantity, energy and $\Delta_r H$, to Figure 10.4. The link between energy and the amount in moles is the balanced thermochemical equation, which includes $\Delta_r H$. The overarching pattern, illustrated by the four boxes at the top of the figure, remains the same.

The equation for burning ethane from Section 10.9 indicates that for every 2 moles burned, 3119 kJ of energy is transferred: $\frac{3119\ \text{kJ}}{2\ \text{mol}\ C_2H_6}$ or $\frac{2\ \text{mol}\ C_2H_6}{3119\ \text{kJ}}$. Similar conversion factors may be written between energy transferred in kilojoules and amount in moles of any other substance in the equation. These factors are used in solving **thermochemical stoichiometry** problems.

Thinking About Your Thinking

Proportional Reasoning

The direct proportionalities between the amount in moles of the reactants and products in a chemical change and the quantity of energy transferred allow these relationships to be used as conversion factors in problem setups.

Figure 10.23 Natural gas is about 5 to 9% ethane, C_2H_6. Therefore, this reaction occurs in many urban homes every time a gas furnace, a gas range, or a gas water heater is used.

Active Example 10.14 Thermochemical Stoichiometry I

What quantity of energy in kilojoules is transferred when 73.0 grams of $C_2H_6(g)$ burns according to $2\ C_2H_6(g) + 7\ O_2(g) \rightarrow 4\ CO_2(g) + 6\ H_2O(\ell) + 3119\ \text{kJ}$ (Figure 10.23)?

Think Before You Write You are given the mass of a species undergoing a chemical change, and you are asked for the quantity of energy transferred. The balanced thermochemical equation expresses the proportionality between amount in moles of each species and quantity of energy in kJ. Thus, this is a thermochemical stoichiometry problem.

Answers *Cover the left column with your cut-out shield. Reveal each answer only after you have written your own answer in the right column.*

Given: 73.0 g C_2H_6
Wanted: kJ

g C_2H_6 → mol C_2H_6 → kJ

1 mol C_2H_6 = 30.07 g C_2H_6

3119 kJ = 2 mol C_2H_6

This is a two-step problem setup. You first convert mass in grams to amount in moles. Once you reach moles of any substance, you can go directly to quantity of energy via the thermochemical equation.

Analyze the problem by listing the given quantity and the wanted unit and then **identify** the equivalencies by first writing the unit path and then deriving the equivalency for each unit conversion.

$$73.0\ \text{g}\ C_2H_6 \times \frac{1\ \text{mol}\ C_2H_6}{30.07\ \text{g}\ C_2H_6} \times \frac{3119\ \text{kJ}}{2\ \text{mol}\ C_2H_6} = 3.79 \times 10^3\ \text{kJ}$$

$(73.0 \div 2) \times (3119 \div 30.07) \approx (74 \div 2) \times (3000 \div 30) = 37 \times 100 = 3.7 \times 10^3$, OK.

Construct the solution setup, changing the equivalencies directly to conversion factors in the setup, and **check** the value of the answer to be sure it makes sense.

You improved your skill at solving thermochemical stoichiometry problems.

What did you learn by solving this Active Example?

Practice Exercise 10.14

When gaseous hydrogen and oxygen react to form 1 mole of liquid water, 286 kJ of energy is released. What quantity of energy is released when 15.0 g of hydrogen burns in excess oxygen?

Active Example 10.15 Thermochemical Stoichiometry II

What mass of liquid *n*-octane, $C_8H_{18}(\ell)$, a component of gasoline, must be burned to transfer 9.05×10^4 kJ of heat? $\Delta_rH = -5471$ kJ/mol C_8H_{18} burned.

Think Before You Write The problem statement gives you the relationship between quantity of energy and the amount in moles of the substance that you are being asked about. Thus, the information that you would need to extract from the thermochemical equation has already been provided for this thermochemical stoichiometry problem.

Answers *Cover the left column with your cut-out shield. Reveal each answer only after you have written your own answer in the right column.*

Given: 9.05×10^4 kJ

Wanted: mass C_8H_{18} (assume g)

kJ → mol C_8H_{18} → g C_8H_{18}

1 mol C_8H_{18} = 5471 kJ

114.22 g C_8H_{18} = 1 mol C_8H_{18}

Analyze the problem by listing the given quantity and the wanted unit and then **identify** the equivalencies by first writing the unit path and then deriving the equivalency for each unit conversion.

$$9.05 \times 10^4\ \cancel{kJ} \times \frac{1\ \cancel{mol\ C_8H_{18}}}{5471\ \cancel{kJ}} \times \frac{114.22\ g\ C_8H_{18}}{\cancel{mol\ C_8H_{18}}} = 1.89 \times 10^3\ g\ C_8H_{18}$$

$[9.05 \times 10^4 \times 114.22 \div 5471] \approx [10 \times 10^4 \times 100 \div 5000]$

$= [10 \times 10^4 \times 10^2 \div (5 \times 10^3)] = [10^7 \div 5 \div 10^3]$

$= (1 \div 5) \times 10^4 = 0.2 \times 10^4 = 2 \times 10^3$, OK.

Construct the solution setup, changing the equivalencies directly to conversion factors in the setup, and **check** the value of the answer to be sure it makes sense.

You improved your skill at solving thermochemical stoichiometry problems.

What did you learn by solving this Active Example?

Practice Exercise 10.15

What mass of methane, CH_4, must be burned to transfer 5.00×10^5 kJ of heat to the surroundings? The heat of combustion—the energy released when the compound completely burns in oxygen—of methane is 889 kJ/mol.

In these examples, we have been able to disregard the sign of Δ_rH because of the way the questions were worded. The question, "How much heat…?" is answered simply with a quantity of kilojoules. The wording of the question tells whether the heat is transferred into or out of the chemical system. However, if a question is of the form, "What is the value of Δ_rH?" the algebraic sign is an essential part of the answer.

Bloomberg/Getty Images

CHAPTER 10 IN REVIEW: QUANTITY RELATIONSHIPS IN CHEMICAL REACTIONS

Goal 1 Given a chemical equation, or a reaction for which the equation can be written, and the amount in moles of one species in the reaction, calculate the amount in moles of any other species.

The **coefficients** in a chemical equation express the **mole relationships** between the different substances in the reaction. The coefficients may be used in a conversion factor from moles of one substance to moles of another.

Goal 2 Given a chemical equation, or a reaction for which the equation can be written, and the mass or amount in moles of one species in the reaction, find the mass or amount in moles of any other species.

The **mass-to-mass stoichiometry path** is the following:

1. Change the mass of the given species to amount in moles.
2. Change the amount in moles of the given species to moles of the wanted species.
3. Change the amount in moles of the wanted species to mass.

This three-step method is at the heart of almost all stoichiometry problems.

Goal 3 Given two of the following, or information from which two of the following may be determined, calculate the third: ideal yield, actual yield, percentage yield.

The efficiency of a reaction is stated in **percentage yield**, in which the **actual yield** is expressed as a percent of the **ideal yield** calculated by stoichiometry:

$$\%\ \text{yield} \equiv \frac{\text{actual yield}}{\text{ideal yield}} \times 100\%$$

You need to be able to solve three types of percentage yield problems:

1. Given: Actual and theoretical yields — Wanted: Percentage yield

   Solve by: $\%\ \text{yield} \equiv \frac{\text{actual yield}}{\text{ideal yield}} \times 100\%$

2. Given: Reactant quantity and percentage yield — Wanted: Product quantity

   Solve by: Use percentage yield as a conversion factor

3. Given: Product quantity and percentage yield — Wanted: Reactant quantity

   Solve by: Use percentage yield as a conversion factor

Goal 4 Identify, describe, and explain limiting reactants and excess reactants.

A **limiting reactant** is the reactant totally consumed in a reaction, thereby determining the maximum yield possible. An **excess reactant** is a reactant that has a quantity that is in excess of the amount needed to completely react with the limiting reactant.

Goal 5 Given a chemical equation, or information from which it may be determined, and initial quantities of two or more reactants, (a) identify the limiting reactant; (b) calculate the ideal yield of a specified product, assuming complete use of the limiting reactant; and (c) calculate the quantity of the reactant initially in excess that remains unreacted.

The **comparison-of-moles method** for solving limiting reactant problems is the following:

1. Convert the mass in grams of each reactant to amount in moles.
2. Identify the limiting reactant.
3. Calculate the amount in moles of each species that reacts or is produced.
4. Calculate the amount in moles of each species that remains after the reaction.
5. Change the amount in moles of each species to mass in grams.

The **smaller-amount method** for solving limiting reactant problems is the following:

1. Calculate the mass of product that can be formed by the initial mass of each reactant. (a) The reactant that yields the smaller mass of product is the limiting reactant. (b) The smaller mass of product is the mass that will be formed when all of the limiting reactant is used up.
2. Calculate the mass of excess reactant that is used by the total mass of limiting reactant.
3. Subtract from the mass of excess reactant present initially the mass that is used by all of the limiting reactant. The difference is the amount of excess reactant that is left.

Goal 6 Given an energy quantity in one of the following units, plus variations created by adding metric prefixes, calculate the other two: joules, calories, and food Calories.	The SI unit of energy is the **joule**, a force of one newton applied for a distance of one meter. Another energy unit used by chemists is the **calorie**, which is equal to 4.184 joules. The joule (J) and the calorie (cal) are small energy units, so units 1000 times larger, the **kilojoule, kJ**, and the **kilocalorie, kcal**, are often used. The food energy **Calorie** is the thermochemical kilocalorie.
Goal 7 Given a chemical equation, or information from which it may be written, and the heat (enthalpy) of reaction, write the thermochemical equation either with Δ_rH (a) to the right of the conventional equation or (b) as a reactant or product.	The amount of heat transferred in a chemical reaction is called the **enthalpy of reaction**, symbolized by $\mathbf{\Delta_rH}$. The Δ_rH of a reaction may be included in the chemical equation as a reactant or a product, or it may be written next to the equation. The equation is then called a **thermochemical equation**. For an endothermic reaction, Δ_rH is positive; heat is a reactant in the thermochemical equation. For an exothermic reaction, Δ_rH is negative; heat is a product in the thermochemical equation.
Goal 8 Given a thermochemical equation, or information from which it may be written, calculate the amount of energy transferred for a given amount of reactant or product; alternately, calculate the mass of reactant required to transfer a given amount of energy.	Using the Δ_rH of a thermochemical equation and the coefficient of any substance in that equation, you can convert in either direction from amount in moles of that substance to amount of heat transferred. **Summary of stoichiometry pattern:**

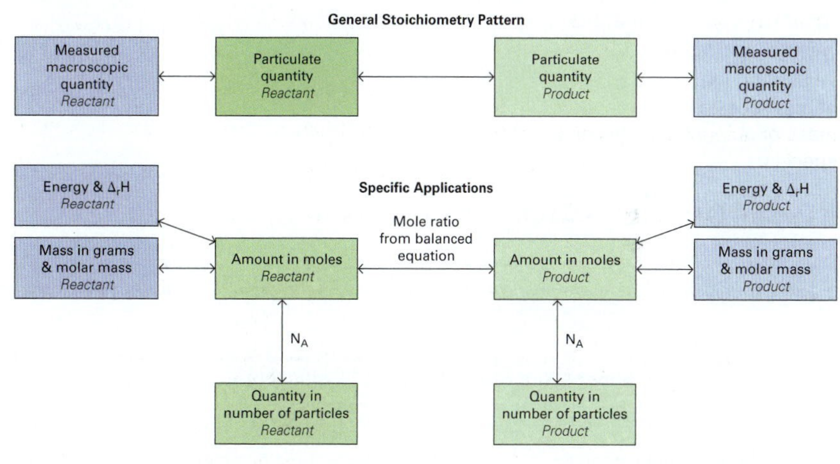

Key Terms

Most of the key terms and concepts and many others appear in the Glossary. Use your Glossary regularly.

actual yield p. 357
calorie (cal) p. 370
Calorie (food, kcal) p. 371
enthalpy of reaction, Δ_rH p. 372
excess reactant p. 362
heat of reaction, Δ_rH p. 371
ideal yield p. 357
joule (J) p. 370
limiting reactant p. 362
mass-to-mass stoichiometry path p. 351
percentage yield p. 357
stoichiometry p. 345
thermochemical equation p. 372
thermochemical stoichiometry p. 374
unit path p. 354

Frequently Asked Questions

What is the most important big-picture issue to focus on while studying this chapter?
Understanding the stoichiometry concept is probably the most important overarching idea in the course in which this book is being used, and thus, you should work toward fully comprehending the general stoichiometry pattern diagramed at the top of Figures 10.4 and 10.22. In terms of skills to develop, the ability to solve stoichiometry problems is probably the most important problem-solving skill you can cultivate in a beginning chemistry course. Work to understand the steps in the quantity-to-quantity path, and not just to be able to "do" the steps from memory or by juggling units. You have not finished a problem when you reach an answer, whether it is right or wrong. You must understand each problem. Be sure that you do so before going on to the next problem.

What should I focus on learning to make sure that I will be able to apply the stoichiometry concept in future courses?
Take a moment to think through the logic of the calculation sequence in stoichiometry. Look particularly at the mole-to-mole conversion in the middle. If you understand the process, apart from a specific problem, you will be able to recognize and solve other kinds of stoichiometry problems in chapters to come and in future courses.

How can I figure out where the percentage yield conversion factor should go in a solution setup?
There are three kinds of percentage yield problems in this book. You should understand the similarities and differences among them. They are summarized in the following table:

Given Quantity or Calculated from the Given Quantity	Wanted	Solve by
Actual and ideal yields	Percentage yield	$\% \text{ yield} = \dfrac{\text{actual yield}}{\text{ideal yield}} \times 100\%$
Reactant quantity and percentage yield	Product quantity	Equivalencies
Product quantity and percentage yield	Reactant quantity	Equivalencies

When the product quantity is wanted, the percentage yield is applied to the product at the end of the calculation setup. When the reactant quantity is wanted, the percentage conversion is applied to the product at the beginning of the setup. It will help you keep things straight if you always apply the percentage conversion to the *product* and then distinguish between actual and ideal product in your setup.

What strategy should be used with thermochemical stoichiometry problems?
Thermochemical stoichiometry problems have one less step than other stoichiometry problems because they involve only one substance. There is no mole-to-mole conversion, but rather a mole-to-energy change between the single substance and the $\Delta_r H$ of the reaction. Watch the sign of $\Delta_r H$ if the wording of the problem is such that it must be taken into account.

Problem-Classification Exercises

Four examples from the chapter are repeated here. Each example represents one kind of stoichiometry problem. If you can classify a problem as one of these types, you will find it easier to select the correct procedure for solving it. The problem types are summarized in Table 10.1.

Table 10.1 Summary of Stoichiometry Classifications

Classification	Given	Wanted
Mass Stoichiometry	Mass of reactant or product	Mass of reactant or product
Percentage yield	Actual and ideal yields	Percentage yield
	Reactant quantity and percentage yield	Product quantity
	Product quantity and percentage yield	Reactant quantity
Limiting reactant	Masses of two reactants	Mass of product and mass of unreacted reactant
Thermochemical stoichiometry	Mass of reactant or product	Quantity of heat energy
	Quantity of heat energy	Mass of reactant or product

Mass Stoichiometry

Active Example 10.5: What mass in milligrams of nickel(II) chloride are in a solution if 503 mg of silver chloride is precipitated in the reaction of silver nitrate and nickel(II) chloride solutions?
General format: Given one mass, find another mass.

Percentage Yield

Active Example 10.8: Solid sodium nitrate decomposes to solid sodium nitrite and oxygen gas upon heating. One set of reaction conditions gives a 95% yield. What mass of sodium nitrite will be produced by the decomposition of 1.50 g of sodium nitrate?
General format: (1) Given actual and ideal yields, find percentage yield or (2) given percentage yield and either the reactant quantity or the product quantity, find the unknown quantity.

Limiting Reactant

Active Examples 10.10 and 10.11: Calculate the mass of antimony(III) iodide, SbI_3, that can be produced by the reaction of 129 g of antimony, Sb ($Z = 51$), and 381 g of iodine. Also, find the mass in grams of the element that will be left.
General format: Given masses of two reactants, find the mass of product and the unreacted mass of the excess reactant.

Thermochemical Stoichiometry

Active Example 10.14: What quantity of energy in kilojoules of is transferred when 73.0 grams of $C_2H_6(g)$ burns according to
$2\ C_2H_6(g) + 7\ O_2(g) \rightarrow 4\ CO_2(g) + 6\ H_2O(\ell) + 3119\text{ kJ}$?
General format: For a reaction with a known $\Delta_r H$, given mass, find heat, or, given heat, find mass.

Eight problem-classification examples follow. Test your classification skill by deciding which kind of problem each one represents. It is not necessary to set up and solve the problem now; this exercise is primarily concerned with problem classification.

1. What mass of magnesium hydroxide will precipitate if 2.09 g of potassium hydroxide is added to a magnesium nitrate solution?
2. What mass in grams of octane, a component of gasoline, would you have to burn in your car to liberate 9.48×10^5 kJ of energy? $\Delta_r H = -1.09 \times 10^4$ kJ for the following reaction: $2\ C_8H_{18}(\ell) + 25\ O_2(g) \rightarrow 16\ CO_2(g) + 18\ H_2O(\ell)$.
3. A mixture of tetraphosphorus trisulfide and powdered glass is in the white tip of strike-anywhere matches. The compound is made by the direct combination of the elements: $8\ P_4 + 3\ S_8 \rightarrow 8\ P_4S_3$. If 133 g of phosphorus is mixed with the full contents of a 4-oz (126 g) bottle of sulfur, what mass in grams of the compound can be formed? How much of which element will be left over?
4. What mass in grams of sodium hydroxide is needed to neutralize completely 32.6 g of phosphoric acid?
5. The Haber process for making ammonia from nitrogen in the air is given by the equation $N_2 + 3\ H_2 \rightarrow 2\ NH_3$. Calculate the mass in kilograms of hydrogen that must be supplied to make 5.00×10^2 kg ammonia in a system that has an 88.8% yield.
6. Lime, the common name for calcium oxide, CaO, is made by heating limestone, $CaCO_3$, in slowly rotating kilns, which are a type of oven. The reaction is $CaCO_3(s) + 178\ kJ \rightarrow CaO(s) + CO_2(g)$. What quantity of energy in kilojoules is required to decompose 5.80 kg of limestone?
7. A solution containing 1.63 g barium chloride is added to a solution containing 2.40 g sodium chromate (chromate ion, CrO_4^{2-}). Find the mass in grams of barium chromate that can precipitate.
8. Ethylacetate, $CH_3COOC_2H_5$, is manufactured by the reaction between acetic acid, CH_3COOH, and ethanol, C_2H_5OH, by the equation $CH_3COOH + C_2H_5OH \rightarrow CH_3COOC_2H_5 + H_2O$. What mass in kilograms of acetic acid must be used to get 62.5 kg of ethylacetate if the percentage yield is 69.1%?

Small-Group Discussion Questions

Small-Group Discussion Questions are for group work, either in class or under the guidance of a leader during a discussion section.

1. Natural gas is usually composed of about 80% methane, CH_4, 5% ethane, C_2H_6, and the balance is other gases. Assume that we have a natural gas sample that is 90% methane and 10% ethane in this problem. (a) Write a balanced chemical equation for the reaction of each gas. (b) What total mass of oxygen is required to burn 1 kg of natural gas? (c) What total mass of products will be formed when 1 kg of natural gas is burned?
2. Titanium tetrachloride, $TiCl_4$, is a colorless liquid at room conditions. It is used in the plastic, electronics, metals, ceramics, and leather industries, and it is sometimes sold in drums that contain 300 kg of titanium tetrachloride. When exposed to water, the liquid reacts to form solid titanium dioxide, TiO_2, and hydrogen chloride gas. What volume of water is needed to completely react with a drum of titanium tetrachloride? How many pounds of titanium dioxide will be formed?
3. A mass of 0.500 g of a metal element reacts with oxygen to yield 0.909 g of the metal oxide MO_2. What is the metal?
4. An Alka-Seltzer tablet contains 325 mg aspirin, 1916 mg sodium hydrogen carbonate, and citric acid. When dissolved in water, the sodium hydrogen carbonate reacts with citric acid, $H_3C_6H_5O_7$, to form sodium citrate, water, and carbon dioxide. The aspirin does not react, and it serves as a pain reliever; the sodium citrate serves as an antacid. Assuming that the tablets are manufactured with the correct stoichiometric quantity of citric acid, determine the mass of citric acid per tablet.
5. Chemists consider changes in matter at three levels: the macroscopic, the particulate, and the symbolic, which are symbols that are used to represent matter and its changes (you may wish to review Figure 1.11). Consider the reaction of charcoal, which is mostly carbon, and oxygen from the air. Carbon dioxide is the product of this reaction. (a) Sketch and describe this reaction at the macroscopic level. Explain why it might lead you to believe that the Law of Conservation of Mass is not correct. (b) Sketch and describe this reaction at the particulate level. (c) Use symbols to describe the reaction, and explain how they are describing the reaction at the macroscopic level and the particulate level.
6. Limestone is a rock that is largely composed of calcium carbonate. When limestone is heated in a lime kiln (essentially a large oven), it decomposes to lime, which is calcium oxide, and carbon dioxide. A 150.8-kg sample of limestone was heated, yielding 56.1 kg of carbon dioxide. Determine the percent limestone in the original sample.
7. Gold occurs naturally in low amounts in hard rock deposits in some locations in the world. This gold can be extracted from the rock by reacting the gold with sodium cyanide solution (cyanide ion, CN^-), oxygen, and water, yielding $NaAu(CN)_2$ and sodium hydroxide in solution. The complex is then reacted with zinc to yield gold metal. One metric ton (1000 kg) of ore that contains 0.02% gold is to be reacted to remove the precious metal. What mass of sodium cyanide is needed for the process?
8. Malonic acid is 34.62% carbon, 3.88% hydrogen, and 61.50% oxygen by mass. Write the balanced equation for its complete combustion, and then determine the masses of all additional reactants and products consumed and produced when 1 pound of malonic acid is burned in air.

9. A solution with 24.99 mg of sodium carbonate is combined with a solution containing 16.70 mg of silver nitrate. Determine the mass of each species in the flask (other than water) after the two solutions are combined.

10. One process for producing sodium sulfate is known as the Hargreaves method. Sodium chloride, sulfur dioxide, water, and oxygen react to form hydrogen chloride and sodium sulfate. Water is inexpensive, and oxygen is obtained from the air. The two reactants that control the cost of the manufacturing process are sodium chloride and sulfur dioxide. How many pounds of sodium chloride are needed for each pound of sulfur dioxide consumed in the reaction if excess water and oxygen are available?

Questions, Exercises, and Problems

Interactive versions of these problems may be assigned by your instructor. Solutions for blue-numbered questions are at the end of the chapter. Questions other than those in the General Questions and More Challenging Problems sections are paired in consecutive odd-even number combinations; solutions for the odd numbered questions are also at the end of the chapter.

Many questions in this chapter are written with the assumption that you have studied Chapter 6 and can write the required formulas from their chemical names. If this is not the case, we have placed a list of all chemical formulas needed to answer this chapter questions at the end of the Questions, Exercises, and Problems.

If you have studied Chapter 6 and you get stuck while answering a question because you cannot write a formula, we urge you to review the appropriate section of Chapter 6 before continuing. Avoid the temptation to "just peek" at the list of formulas. Developing skill in chemical nomenclature is part of your learning process in this course, and in many cases, you truly learn the material from Chapter 6 as you apply it here in this chapter and throughout your study of chemistry.

Section 10.2: Conversion Factors from a Chemical Equation

1. The first step in the Ostwald process for manufacturing nitric acid is the reaction between ammonia and oxygen described by the equation $4\ NH_3 + 5\ O_2 \rightarrow 4\ NO + 6\ H_2O$. Use this equation to answer all parts of this question.
   a) What amount in moles of ammonia will react with 95.3 moles of oxygen?
   b) What amount in moles of nitrogen monoxide will result from the reaction of 2.89 moles of ammonia?
   c) If 3.35 moles of water is produced, what amount in moles of nitrogen monoxide will also be produced?

2. When dihydrogen sulfide reacts with oxygen, water and sulfur dioxide are produced. The balanced equation for this reaction is $2\ H_2S(g) + 3\ O_2(g) \rightarrow 2\ H_2O(\ell) + 2\ SO_2(g)$. For all parts of this question, consider what will happen if 4 moles of dihydrogen sulfide react.
   a) What amount in moles of oxygen is consumed?
   b) What amount in moles of water is produced?
   c) What amount in moles of sulfur dioxide is produced?

3. Magnesium hydroxide is formed from the reaction of magnesium oxide and water. What amount in moles of magnesium oxide is needed to form 0.884 mole of magnesium hydroxide, when the oxide is added to excess water?

4. In our bodies, sugar is broken down by reacting with oxygen to produce water and carbon dioxide. What amount in moles of carbon dioxide will be formed upon the complete reaction of 0.424 moles of glucose sugar ($C_6H_{12}O_6$) with excess oxygen gas?

5. When sulfur dioxide reacts with oxygen, sulfur trioxide forms. What amount in moles of sulfur dioxide is needed to produce 3.99 moles of sulfur trioxide if the reaction is carried out in excess oxygen?

6. Aqueous solutions of potassium hydrogen sulfate and potassium hydroxide react to form aqueous potassium sulfate and liquid water. What amount in moles of potassium hydroxide is necessary to form 0.636 mole of potassium sulfate?

Section 10.3: Mass–Mass Stoichiometry

7. The first step in the Ostwald process for manufacturing nitric acid is the reaction between ammonia and oxygen described by the equation $4\ NH_3 + 5\ O_2 \rightarrow 4\ NO + 6\ H_2O$. Use this equation to answer all parts of this question.
   a) What amount in moles of ammonia can be oxidized by 268 g of oxygen?
   b) If the reaction consumes 31.7 moles of ammonia, what mass in grams of water will be produced?
   c) What mass in grams of ammonia is required to produce 404 g of nitrogen monoxide?
   d) If 6.41 g of water results from the reaction, what will be the yield of nitrogen monoxide (in grams)?

8. Butane, C_4H_{10}, is a common fuel used for heating homes in areas not served by natural gas. The equation for its combustion is $2\ C_4H_{10} + 13\ O_2 \rightarrow 8\ CO_2 + 10\ H_2O$. All parts of this question are related to this reaction.
   a) What mass in grams of butane can be burned by 1.42 moles of oxygen?
   b) If 9.43 g of oxygen is used in burning butane, what amount in moles of water result?
   c) Calculate the mass in grams of carbon dioxide that will be produced by burning 78.4 g of butane.
   d) What mass in grams of oxygen is used in a reaction that produces 43.8 g of water?

9. The explosion of nitroglycerin is described by the equation $4\ C_3H_5(NO_3)_3 \rightarrow 12\ CO_2 + 10\ H_2O + 6\ N_2 + O_2$. What mass in grams of carbon dioxide is produced by the explosion of 21.0 g of nitroglycerin?

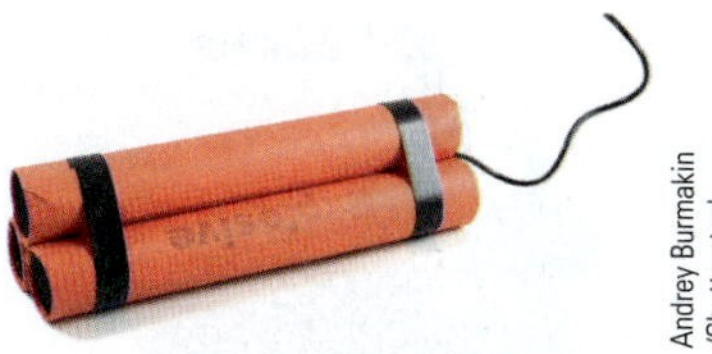

Andrey Burmakin /Shutterstock.com

The interior of a stick of dynamite is composed of liquid nitroglycerin contained within an absorbent material, such as sawdust.

10. According to the reaction $2\ AgNO_3 + Cu \rightarrow Cu(NO_3)_2 + 2\ Ag$, what mass in grams of copper(II) nitrate will be formed upon the complete reaction of 26.8 g of copper with excess silver nitrate?

11. Soaps are produced by the reaction of sodium hydroxide with naturally occurring fats. The equation for one such reaction is $C_3H_5(C_{17}H_{35}COO)_3 + 3\ NaOH \rightarrow C_3H_5(OH)_3 + 3\ C_{17}H_{35}COONa$. The last compound is the soap. Calculate the mass in grams of sodium hydroxide required to produce 323 g of soap by this method.

12. According to the reaction $CH_4 + CCl_4 \rightarrow 2\ CH_2Cl_2$, what mass in grams of carbon tetrachloride is required for the complete reaction of 24.7 g of methane, CH_4?

13. One way to make sodium thiosulfate, which is used as an antidote to cyanide poisoning, is described by the equation $Na_2CO_3 + 2\ Na_2S + 4\ SO_2 \rightarrow 3\ Na_2S_2O_3 + CO_2$. What mass in grams of sodium carbonate is required to produce 681 g of sodium thiosulfate?

14. One of the methods for manufacturing sodium sulfate, used in the manufacturing of detergents, involves the reaction $4\ NaCl + 2\ SO_2 + 2\ H_2O + O_2 \rightarrow 2\ Na_2SO_4 + 4\ HCl$. Calculate the mass in kilograms of sodium chloride required to produce 5.00 kg of sodium sulfate.

15. The hard water scum that forms a ring around the bathtub is an insoluble soap, $Ca(C_{18}H_{35}O_2)_2$. It is formed when a soluble soap, $NaC_{18}H_{35}O_2$, reacts with the calcium ion that is responsible for the hardness in water: $2\ NaC_{18}H_{35}O_2 + Ca^{2+} \rightarrow Ca(C_{18}H_{35}O_2)_2 + 2\ Na^+$. What mass in milligrams of scum can form from 616 mg of $NaC_{18}H_{35}O_2$?

Flirt/SuperStock

Hard water contains calcium ion, which reacts with the stearate ion in soap to form bathtub scum.

16. Trinitrotoluene is the chemical name for the explosive commonly known as TNT. Its formula is $C_7H_5N_3O_6$. TNT is manufactured by the reaction of toluene, C_7H_8, with nitric acid: $C_7H_8 + 3\ HNO_3 \rightarrow C_7H_5N_3O_6 + 3\ H_2O$.

a) What mass in kilograms of nitric acid is needed to react completely with 1.90 kg of toluene?

b) What mass in kilograms of TNT can be produced in the reaction?

17. Pig iron from a blast furnace contains several impurities, one of which is phosphorus. Additional iron ore, Fe_2O_3, is included with pig iron in making steel. The oxygen in the ore oxidizes the phosphorus by the reaction $12\ P + 10\ Fe_2O_3 \rightarrow 3\ P_4O_{10} + 20\ Fe$. If a sample of the remains from the furnace contains 802 mg of tetraphosphorus decoxide, what mass in grams of Fe_2O_3 was used in making it?

18. According to the reaction $N_2 + O_2 \rightarrow 2\ NO$, what mass in grams of nitrogen monoxide will be formed upon the complete reaction of 25.4 g of oxygen gas with excess nitrogen gas?

The Solvay process is a multistep industrial method for the manufacture of sodium carbonate, Na_2CO_3, which is also known as washing soda. Although most of the industrialized world utilizes the Solvay process for production of sodium carbonate, it is not manufactured in the United States because it can be obtained at lower cost from a large natural deposit in Wyoming. Questions 19 through 22 are based on reactions in that process.

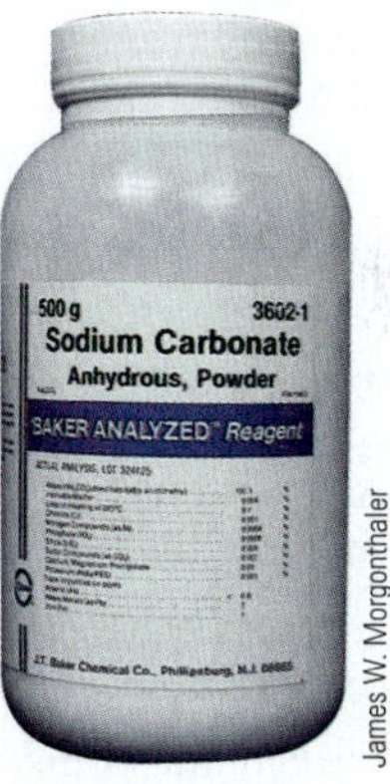

James W. Morgonthaler

Sodium carbonate is used for many purposes, including glass manufacturing.

19. What mass in grams of NaCl is needed to react completely with 83.0 g of ammonium hydrogen carbonate in $NaCl + NH_4HCO_3 \rightarrow NaHCO_3 + NH_4Cl$?

20. What mass in grams of ammonium hydrogen carbonate will be formed by the reaction of 81.2 g of ammonia in the reaction $NH_3 + H_2O + CO_2 \rightarrow NH_4HCO_3$?

21. By-product ammonia is recovered from the Solvay process by the reaction $Ca(OH)_2 + 2\ NH_4Cl \rightarrow CaCl_2 + 2\ H_2O + 2\ NH_3$. What mass in grams of calcium chloride can be produced along with 62.0 g of ammonia?

22. What mass of $NaHCO_3$ must decompose to produce 448 g of Na_2CO_3 in $2\ NaHCO_3 \rightarrow Na_2CO_3 + H_2O + CO_2$?

23. What mass in grams of sodium hydroxide is needed to neutralize completely 32.6 g of phosphoric acid?

24. Solid ammonium chloride decomposes to form ammonia and hydrogen chloride gases. What mass in grams of ammonia will be formed upon the complete reaction of 28.3 g of ammonium chloride?

Solid ammonium chloride decomposes upon heating, forming gaseous ammonia and hydrogen chloride. As the two gases cool, they react to re-form solid ammonium chloride, which appears as smoke.

25. What mass in grams of magnesium hydroxide will precipitate if 2.09 g of potassium hydroxide is added to a magnesium nitrate solution?

26. Under specialized conditions, an aqueous solution of hydrobromic acid can decompose to hydrogen gas and liquid bromine. What mass in grams of hydrobromic acid is needed to form 31.5 g of bromine?

27. An experimenter recovers 0.521 g of sodium sulfate from the neutralization of sodium hydroxide by sulfuric acid. What mass in grams of sulfuric acid reacted?

28. Solid sulfur reacts with carbon monoxide gas to form gaseous sulfur dioxide and solid carbon. What mass in grams of carbon monoxide is required for the complete reaction of 25.5 g of sulfur?

29. The reaction of a dry cell battery may be represented as follows: Zinc reacts with ammonium chloride to form zinc chloride, ammonia, and hydrogen. Calculate the mass in grams of zinc consumed during the release of 7.05 g of ammonia in such a cell.

30. Solid calcium carbonate decomposes to solid calcium oxide and carbon dioxide gas. What mass in grams of calcium carbonate is needed to form 22.8 g of carbon dioxide?

Section 10.4: Percentage Yield

31. The function of sodium thiosulfate in photographic developing is to remove excess silver bromide by the reaction $2\ Na_2S_2O_3 + AgBr \rightarrow Na_3Ag(S_2O_3)_2 + NaBr$. What is the percentage yield if the reaction of 8.18 g of sodium thiosulfate produces 2.61 g of sodium bromide?

32. Butane gas is used as a fuel for camp stoves, transferring heat by the reaction $2\ C_4H_{10} + 13\ O_2 \rightarrow 8\ CO_2 + 10\ H_2O$. When 4.14 g of butane reacts with excess oxygen gas, the reaction yields 11.0 g of carbon dioxide. What is the percentage yield in this reaction?

Small, portable stoves used for camping often utilize butane as a fuel.

33. Calcium cyanamide is a common fertilizer. When mixed with water in the soil, it reacts to produce calcium carbonate and ammonia: $CaCN_2 + 3\ H_2O \rightarrow CaCO_3 + 2\ NH_3$. What mass in grams of ammonia can be obtained from 7.25 g of calcium cyanamide in a laboratory experiment in which the percentage yield is 92.8%?

34. Hydrochloric acid is known as an enemy of stainless steel, the primary component of which is iron. The reaction is $2\ HCl + Fe \rightarrow FeCl_2 + H_2$. What will be the actual yield of iron(II) chloride when 3.17 g of iron reacts with excess hydrochloric acid? Assume that the percentage yield of iron(II) chloride is 76.6%.

35. The Haber process for making ammonia from nitrogen in the air is given by the equation $N_2 + 3\ H_2 \rightarrow 2\ NH_3$. Calculate the mass in kilograms of hydrogen that must be supplied to make 5.00×10^2 kg of ammonia in a system that has an 88.8% yield.

36. Hydrogen iodide is a colorless, nonflammable, corrosive gas with a penetrating and suffocating odor. It can be prepared by direct reaction of its elements: $H_2 + I_2 \rightarrow 2\ HI$. It is desired to produce 331 g of hydrogen iodide in a process in which the percentage yield is 71.1%. What mass in grams of hydrogen gas will need to be reacted?

37. Calculate the percentage yield in the photosynthesis reaction by which carbon dioxide is converted to sugar if 7.03 g of carbon dioxide yields 3.92 g of $C_6H_{12}O_6$. The equation is $6\ CO_2 + 6\ H_2O \rightarrow C_6H_{12}O_6 + 6\ O_2$.

38. Consider the reaction between ethene and hydrogen in the presence of nickel: $CH_2CH_2 + H_2 \xrightarrow{Ni} CH_3CH_3$. When 5.20 g of ethene reacts with excess hydrogen gas, the reaction yields 4.75 g of ethane. What is the percentage yield for this reaction?

39. Ethylacetate, $CH_3COOC_2H_5$, is manufactured by the reaction between acetic acid, CH_3COOH, and ethanol, C_2H_5OH: $CH_3COOH + C_2H_5OH \rightarrow CH_3COOC_2H_5 + H_2O$. What mass in kilograms of acetic acid must be used to get 62.5 kg of ethylacetate if the percentage yield is 69.1%?

40. Nitrogen monoxide is produced by combustion in an automobile engine. It can then react with oxygen: $2\ NO + O_2 \rightarrow 2\ NO_2$. If 5.74 g of oxygen gas reacts with

excess nitrogen monoxide in a system in which the percentage yield of nitrogen dioxide is 80.2%, what is the actual yield of nitrogen dioxide?

Section 10.5: Limiting Reactants: The Problem

41. Ammonia can be formed from a combination reaction of its elements. A small fraction of an unreacted mixture of elements is illustrated in the following diagram, in which white spheres represent hydrogen atoms and blue spheres represent nitrogen atoms. The temperature is such that all species are gases.

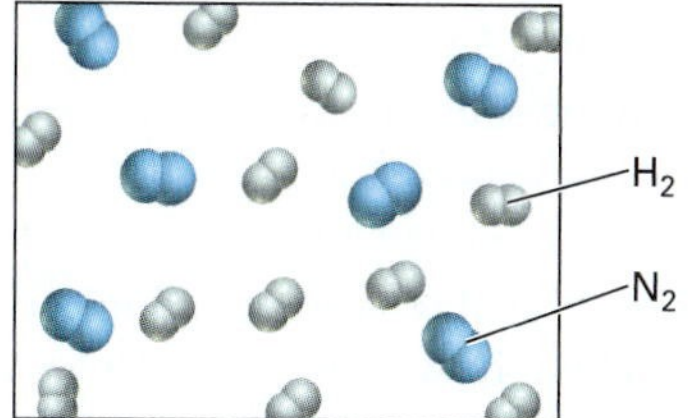

a) Write and balance the equation for the reaction.
b) Which of the following correctly represents the product mixture?

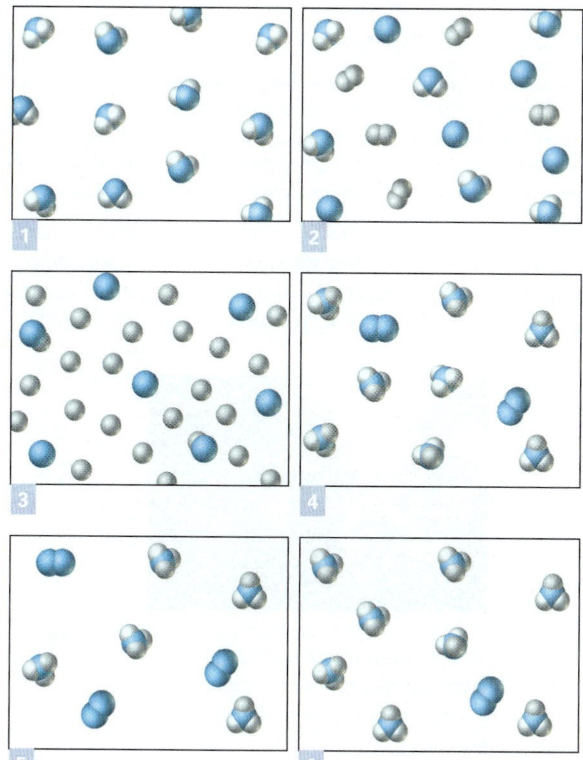

c) Which species is the limiting reactant? Explain.

42. Carbon monoxide reacts with oxygen to form carbon dioxide at a temperature at which all species are in the gas phase. A tiny sample of the prereaction mixture is shown in the following diagram, in which purple spheres represent oxygen atoms and green spheres represent carbon atoms. Draw the product mixture. Give a written description of your reasoning.

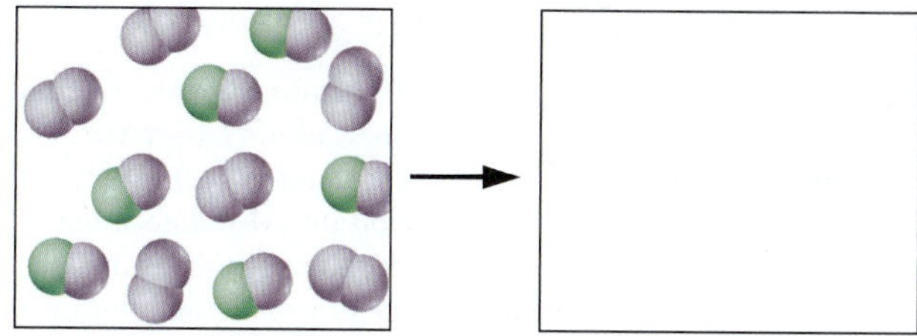

43. An experiment is conducted in which varying amounts of solid iron are added to a fixed volume of liquid bromine. The product of the reaction is a single compound, which can be separated from the product mixture and weighed. The graph shows the relationship between the mass of iron in each trial versus the mass of the product compound. Explain why the graph has a positive slope for low masses of iron and a zero slope when the mass of iron added becomes larger.

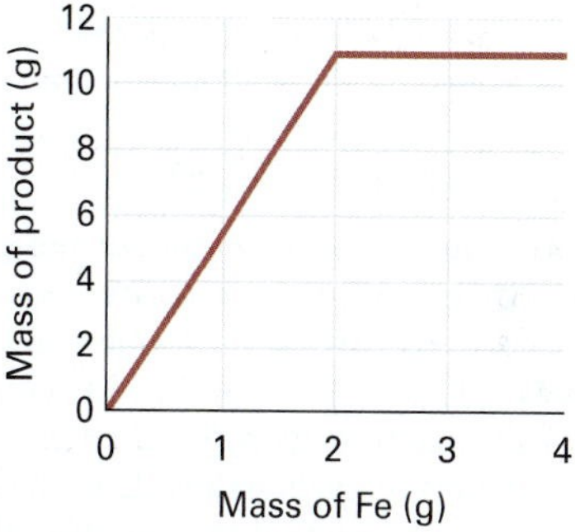

44. The flasks below illustrate three trials of a reaction between varying amounts of zinc and a constant volume of hydrochloric acid solution of constant concentration. The table gives the initial amount of zinc added and the observations at the conclusion of each reaction.

Charles D. Winters

Mass of Zinc Added (g)	Observations at Conclusion of Reaction
6.10	Balloon inflated completely Some unreacted zinc remains
2.37	Balloon inflated completely No zinc remains
0.41	Balloon not inflated completely No zinc remains

The reaction produces hydrogen gas and an aqueous solution of zinc chloride. Explain the observed results.

Sections 10.6 and 10.7: Limiting Reactants

45. A solution containing 1.63 g of barium chloride is added to a solution containing 2.40 g of sodium chromate (chromate ion, CrO_4^{2-}). Find the mass in grams of barium chromate that can precipitate. Also determine which reactant was in excess, as well as the mass in grams in excess of the amount required by the limiting reactant.

46. Carbon dioxide can be removed from the air by reaction with potassium hydroxide: $CO_2 + 2\ KOH \rightarrow K_2CO_3 + H_2O$. If 15.6 g of carbon dioxide reacts with 43.0 g of potassium hydroxide, what mass of potassium carbonate will be formed? What amount of excess reactant remains after the reaction is complete?

47. The equation for one method of preparing iodine is $2\ NaIO_3 + 5\ NaHSO_3 \rightarrow I_2 + 3\ NaHSO_4 + 2\ Na_2SO_4 + H_2O$. If 6.00 kg of sodium iodate is reacted with 7.33 kg of sodium hydrogen sulfite, what mass in kilograms of iodine can be produced? Which reactant will be left over? What mass in kilograms will be left?

48. Carbon monoxide is a waste product of the gasoline combustion process. Auto manufacturers use catalytic converters to convert carbon monoxide into carbon dioxide: $2\ CO + O_2 \rightarrow 2\ CO_2$. When 12.6 g of carbon monoxide is allowed to react with 5.22 g of oxygen gas, what mass in grams of carbon dioxide is formed? What amount of the excess reactant remains after the reaction is complete?

49. A mixture of tetraphosphorus trisulfide and powdered glass is in the white tip of strike-anywhere matches. The P_4S_3 is made by the direct combination of the elements: $8\ P_4 + 3\ S_8 \rightarrow 8\ P_4S_3$. If 133 g of phosphorus is mixed with the full contents of a 126-g (4-oz) bottle of sulfur, what mass in grams of the compound can be formed? How much of which element will be left over?

Charles Steele

The tips of strike-anywhere matches contain tetraphosphorus trisulfide, which reacts with the oxygen in the air in an exothermic change to initiate the burning reaction.

50. Sodium carbonate can neutralize nitric acid by the reaction $2\ HNO_3 + Na_2CO_3 \rightarrow 2\ NaNO_3 + H_2O + CO_2$. Is 135 g of sodium carbonate enough to neutralize a solution that contains 188 g of nitric acid? What mass in grams of carbon dioxide will be released in the reaction?

Section 10.8: Energy

51. a) How many kilojoules are equal to 0.731 kcal?
    b) What quantity of calories is the same as 651 J?
    c) Determine the quantity of kilocalories that is equivalent to 6.22×10^3 J.

52. Complete the following:
    a) 504 J = ______ cal = ______ kJ
    b) 192 cal = ______ J = ______ kJ
    c) 0.423 kJ = ______ J = ______ cal

53. When 15 g of carbon is burned, 493 kJ of energy is released. Calculate the equivalent number of calories and kilocalories this represents.

54. A list of the Calorie content of foods indicates that a piece of lemon meringue pie contains 333 Calories. Express this quantity in kilojoules and in joules.

55. Each day, 5.8×10^2 kcal of heat is transferred out of the body by evaporation. How many kilojoules are transferred in a year?

56. Burning sucrose transfers 3.94×10^3 cal/g to the atmosphere. Calculate the amount of energy in kilojoules transferred when 56.7 g of sucrose is burned.

Section 10.9: Thermochemical Equations

Questions 57 through 62: Thermochemical equations may be written in two ways, either with an energy term as a part of the equation or with $\Delta_r H$ set apart from the regular equation. In the questions that follow, write both forms of the equations for the reactions described. Recall that state designations are required for all substances in a thermochemical equation.

57. Energy is transferred from sunlight in the photosynthesis reaction in which carbon dioxide and water vapor combine to produce sugar, $C_6H_{12}O_6$, and release oxygen. The amount of energy is 2.82×10^3 kJ per mole of sugar formed.

Elenamiv/Shutterstock.com

The photosynthesis reaction is endothermic, converting energy from sunlight to energy stored in chemical bonds.

58. When gaseous nitrogen reacts with gaseous oxygen to form nitrogen dioxide gas, 66.4 kJ of energy is transferred for each mole of nitrogen that reacts.

59. The electrolysis of water is an endothermic reaction, transferring 286 kJ for each mole of liquid water decomposed to its elements.

60. When ammonia and oxygen react at a temperature at which both compounds are in the gaseous state, 226 kJ of

energy is transferred for each mole of ammonia that reacts. Gaseous nitrogen monoxide and steam are the products of the reaction.

61. The reaction in an oxyacetylene torch is highly exothermic, releasing 1.31×10^3 kJ of heat to the environment for every mole of acetylene, $C_2H_2(g)$, burned. The end products are gaseous carbon dioxide and liquid water.

62. Carbon dioxide gas reacts with hydrogen gas to form carbon monoxide gas and steam, and 41.2 kJ of energy is transferred for each mole of carbon dioxide that reacts.

Section 10.10: Thermochemical Stoichiometry

63. Lime, a common name for calcium oxide, is made by heating limestone, which is calcium carbonate, in slowly rotating kilns. Kilns are a type of oven. The reaction is $CaCO_3(s) + 178\ kJ \rightarrow CaO(s) + CO_2(g)$. What quantity of energy in kilojoules is required to decompose 5.80 kg of limestone?

64. What mass in grams of hydrogen has to react to transfer 71.9 kJ of energy from the reaction $2\ H_2(g) + O_2(g) \rightarrow 2\ H_2O(g) + 484\ kJ$?

65. The lime produced in Question 63 is frequently converted to calcium hydroxide, sometimes called slaked lime, by an exothermic reaction with water: $CaO(s) + H_2O(\ell) \rightarrow Ca(OH)_2(s) + 65.3\ kJ$. What mass in grams of lime was processed in a reaction that transferred 291 kJ of energy?

66. $\Delta_rH = -75.8$ kJ for the reaction $S(s) + 2\ CO(g) \rightarrow SO_2(g) + 2\ C(s)$. When 10.1 g of sulfur reacts with excess carbon monoxide, what quantity of energy, expressed in kilojoules, is transferred?

67. What mass in grams of octane, a component of gasoline, would you have to burn in your car to transfer 9.48×10^5 kJ of energy? $\Delta_rH = -1.09 \times 10^4$ kJ for the following reaction: $2\ C_8H_{18}(\ell) + 25\ O_2(g) \rightarrow 16\ CO_2(g) + 18\ H_2O(\ell)$?

68. Calculate the quantity of energy (kJ) transferred when 10.4 g of carbon dioxide reacts with excess hydrogen if $\Delta_rH = 41.2$ kJ for the reaction $CO_2(g) + H_2(g) \rightarrow CO(g) + H_2O(g)$.

General Questions

69. Distinguish precisely and in scientific terms the differences among items in each of the following pairs or groups.
    a) Ideal, actual, and percentage yield
    b) Limiting reactant, excess reactant
    c) Heat of reaction, enthalpy of reaction
    d) Chemical equation, thermochemical equation
    e) Stoichiometry, thermochemical stoichiometry
    f) Joule, calorie

70. Classify each of the following statements as true or false:
    a) Coefficients in a chemical equation express the molar proportions among both reactants and products.
    b) A stoichiometry problem can be solved with an unbalanced equation.
    c) In solving a stoichiometry problem, the change from quantity of given substance to quantity of wanted substance is based on masses.
    d) Percentage yield is actual yield expressed as a percentage of ideal yield.
    e) The quantity of product of any reaction can be calculated only through the amount in moles of the limiting reactant.
    f) Δ_rH is positive for an endothermic reaction and negative for an exothermic reaction.

71. One of the few ways of "fixing" nitrogen, that is, making a nitrogen compound from the elemental nitrogen in the atmosphere, is by the reaction $Na_2CO_3 + 4\ C + N_2 \rightarrow 2\ NaCN + 3\ CO$. Calculate the mass in grams of Na_2CO_3 required to react with 35 g N_2.

72. What mass in grams of calcium phosphate will precipitate if excess calcium nitrate is added to a solution containing 3.98 g of sodium phosphate?

73. Emergency oxygen masks contain potassium superoxide, KO_2, pellets. When exhaled CO_2 passes through the KO_2, the following reaction occurs: $4\ KO_2(s) + 2\ CO_2(g) \rightarrow 2\ K_2CO_3(s) + 3\ O_2(g)$. The oxygen produced can then be inhaled, so no air from outside the mask is needed. If the mask contains 125 g of KO_2, what mass in grams of oxygen can be produced?

MSA AUER GmbH

Solid potassium superoxide is used in emergency oxygen masks. It reacts with exhaled carbon dioxide, yielding breathable oxygen.

74. Baking cakes and pastries involves the production of CO_2 to make the batter rise. For example, citric acid, $H_3C_6H_5O_7$, in lemon or orange juice can react with baking soda, $NaHCO_3$, to produce carbon dioxide gas: $H_3C_6H_5O_7(aq) + 3\ NaHCO_3(aq) \rightarrow Na_3C_6H_5O_7(aq) + 3\ CO_2(g) + 3\ H_2O(\ell)$.
    a) If 6.00 g of $H_3C_6H_5O_7$ reacts with 20.0 g of $NaHCO_3$, what mass in grams of carbon dioxide will be produced?
    b) What mass (in grams) of which reactant will remain unreacted?
    c) Can you name $Na_3C_6H_5O_7$? Remember, it comes from citr*ic* acid.

75. A laboratory test of 12.8 g of aluminum ore yields 1.68 g of aluminum. If the aluminum compound in the ore is Al_2O_3 and it is converted to the pure metal by the reaction $2\ Al_2O_3(s) + 3\ C(s) \rightarrow 4\ Al(s) + 3\ CO_2(g)$, what is the percentage of Al_2O_3 in the ore?

76. What quantity of energy in kilojoules is required to decompose 1.42 g of $KClO_3$ according to the equation: $2\ KClO_3(s) \rightarrow 2\ KCl(s) + 3\ O_2(g)\ \Delta_rH = 89.5$ kJ?

More Challenging Problems

77. A phosphate rock quarry yields rock that is 79.4% calcium phosphate, the raw material used in preparing $Ca(H_2PO_4)_2$, a fertilizer known as *superphosphate.* The rock reacts with sulfuric acid according to the equation $Ca_3(PO_4)_2 + 2\ H_2SO_4 \rightarrow Ca(H_2PO_4)_2 + 2\ CaSO_4$. What is the smallest possible mass of rock (in kilograms) that must be processed to yield 0.500 ton of fertilizer?

78. Carborundum, SiC, is widely used as an abrasive in industrial grinding wheels. It is prepared by the reaction of sand, SiO_2, with the carbon in coke: $SiO_2 + 3\ C \rightarrow SiC + 2\ CO$. What mass in kilograms of carborundum can be prepared from 727 kg of coke that is 88.9% carbon?

79. A sludge containing silver chloride is a waste product from making mirrors. The silver may be recovered by dissolving the silver chloride in a sodium cyanide (NaCN) solution, and then reducing the silver with zinc. The overall equation is $2\ AgCl + 4\ NaCN + Zn \rightarrow 2\ NaCl + Na_2Zn(CN)_4 + 2\ Ag$. What minimum mass in kilograms of sodium cyanide is needed to dissolve all the silver chloride from 40.1 kg of sludge that is 23.1% by mass silver chloride?

80. The chemical equation that describes what happens in an automobile storage battery as it generates electrical energy is $PbO_2 + Pb + 2\ H_2SO_4 \rightarrow 2\ PbSO_4 + 2\ H_2O$.
    a) What fraction of the lead in $PbSO_4$ comes from PbO_2, and what fraction comes from elemental lead?
    b) If the process uses 29.7 g $PbSO_4$, what mass in grams of PbO_2 must have been consumed?

81. Hydrogen and chlorine are produced simultaneously in commercial quantities by the electrolysis of salt water: $2\ NaCl + 2\ H_2O \rightarrow 2\ NaOH + H_2 + Cl_2$. Yield from the process is 61% by mass. What mass in kilograms of solution that is 9.6% by mass NaCl must be processed to obtain 105 kg of chlorine?

82. A dry mixture of hydrogen chloride and air is passed over a heated catalyst in the Deacon process for manufacturing chlorine. Oxidation occurs by the following reaction: $4\ HCl + O_2 \rightarrow 2\ Cl_2 + 2\ H_2O$. If the conversion is 63% complete, what weight in tons of chlorine can be recovered from 1.4 tons of HCl? (*Hint:* Whatever you can do with moles, kilomoles, and millimoles, you can also do with ton-moles.)

83. Fluorides retard tooth decay by forming a hard, acid-resisting calcium fluoride layer in the reaction $SnF_2 + Ca(OH)_2 \rightarrow CaF_2 + Sn(OH)_2$. If at the time of a treatment there are 239 mg $Ca(OH)_2$ on the teeth and the dentist uses a mixture that contains 305 mg SnF_2, has enough of the mixture been used to convert all of the $Ca(OH)_2$? If no, what minimum additional amount should have been used? If yes, by what mass in milligrams was the amount in excess?

The fluoride ion in toothpaste replaces some of the surface hydroxide ions in teeth, forming calcium fluoride, which is more resistant to tooth decay than the natural calcium hydroxide in tooth enamel.

84. In the recovery of silver from silver chloride waste (see Question 79), a certain quantity of waste material is estimated to contain 184 g of silver chloride, AgCl. The treatment tanks are charged with 45 g of zinc and 145 g of sodium cyanide, NaCN. Is there enough of the two reactants to recover all of the silver from the AgCl? If no, what mass in grams of silver chloride will remain? If yes, what additional mass in grams of silver chloride could have been treated by the available Zn and NaCN?

85. In 1866, a young chemistry student conceived the electrolytic method of obtaining aluminum from its oxide. This method is still used. $\Delta_rH = 1.97 \times 10^3$ kJ for the reaction $2\ Al_2O_3(s) + 3\ C(s) \rightarrow 4\ Al(s) + 3\ CO_2(g)$. The large amount of electrical energy required limits the process to areas of relatively inexpensive power. How many kilowatt-hours of energy are needed to produce 1 pound (454 g) of aluminum by this process? 1 kw-h = 3.60×10^3 kJ.

Most residents of the United States use items made of aluminum on a daily basis.

86. Nitroglycerine is the explosive ingredient in industrial dynamite. Much of its destructive force comes from the sudden creation of large volumes of gaseous products. A great deal of energy is transferred, too. $\Delta_rH = -6.17 \times 10^3$ kJ for the equation $4\ C_3H_5(NO_3)_3(\ell) \rightarrow 12\ CO_2(g) + 10\ H_2O(g) + 6\ N_2(g) + O_2(g)$. Calculate the weight in pounds of nitroglycerine that must be used in a blasting operation that requires 5.88×10^4 kJ of energy.

87. A student was given a 1.6240-g sample of a mixture of sodium nitrate and sodium chloride and was asked to find the percentage of each compound in the mixture. She dissolved the sample and added a solution that contained an excess of silver nitrate. The silver ion precipitated all of the chloride ion in the mixture as silver chloride. It was filtered, dried, and weighed. Its mass was 2.056 g. What was the percentage of each compound in the mixture?

88. A researcher dissolved 1.382 g of impure copper in nitric acid to produce a solution of $Cu(NO_3)_2$. The solution went through a series of steps in which $Cu(NO_3)_2$ was changed to $Cu(OH)_2$, then to CuO, and then to a solution of $CuCl_2$. This was treated with an excess of a soluble phosphate, precipitating all the copper in the original sample as pure $Cu_3(PO_4)_2$. The precipitate was dried and weighed. Its mass was 2.637 g. Find the percentage by mass of copper in the original sample.

89. What mass in grams of magnesium nitrate must be used to precipitate as magnesium hydroxide all of the hydroxide ion in 50.0 mL of 17.0% by mass NaOH, the density of which is 1.19 g/mL? The precipitation reaction is $2\ NaOH + Mg(NO_3)_2 \rightarrow Mg(OH)_2 + 2\ NaNO_3$.

90. If a solution of silver nitrate is added to a second solution containing a chloride, bromide, or iodide, the silver ion from the first solution will precipitate the halide as silver chloride, silver bromide, or silver iodide. If excess $AgNO_3(aq)$ is added to a mixture of halides, it will precipitate them. A solution contains 0.230 g of NaCl and 0.771 g of NaBr. What is the smallest quantity of $AgNO_3$ that is required to precipitate both halides completely?

Formulas

These formulas are provided in case you are studying this chapter before you study Chapter 6. You should not use this list unless your instructor has not yet assigned Chapter 6 or otherwise indicated that you may use the list.

3. $Mg(OH)_2$, MgO, H_2O
4. O_2, H_2O, CO_2
5. SO_2, O_2, SO_3
6. $KHSO_4$, KOH, K_2SO_4, H_2O
23. NaOH, H_3PO_4
24. NH_4Cl, NH_3, HCl
25. $Mg(OH)_2$, KOH, $Mg(NO_3)_2$
26. HBr, H_2, Br_2
27. Na_2SO_4, NaOH, H_2SO_4
28. S, CO, SO_2, C
29. Zn, NH_4Cl, $ZnCl_2$, NH_3, H_2
30. $CaCO_3$, CaO, CO_2
41. NH_3, N_2, H_2
42. CO, O_2, CO_2
45. $BaCl_2$, Na_2CrO_4, $BaCrO_4$
57. CO_2, H_2O, O_2
58. N_2, O_2, NO_2
59. H_2O, H_2, O_2
60. NH_3, O_2, NO, H_2O
61. CO_2, H_2O
62. CO_2, H_2, CO, H_2O
72. $Ca_3(PO_4)_2$, $Ca(NO_3)_2$, Na_3PO_4
87. $NaNO_3$, NaCl, $AgNO_3$, AgCl

Answer to Target Check

1. According to the equation, two particles of A react for every particle of B. There are six particles of A and four particles of B available to react. When three particles of B react, all six particles of A will also react, forming three particles of A_2B. There will be one particle of B left.

Answers to Practice Exercises

1. $ZnBr_2(aq) + 2\ NaOH(aq) \rightarrow Zn(OH)_2(s) + 2\ NaBr(aq)$

$$0.033\ \text{mol}\ Zn(OH)_2 \times \frac{2\ \text{mol NaOH}}{1\ \text{mol}\ Zn(OH)_2} = 0.066\ \text{mol NaOH}$$

2. $CH_3CH_2OCH_2CH_3(\ell) + 6\ O_2(g) \rightarrow 4\ CO_2(g) + 5\ H_2O(\ell)$

$$2.90 \times 10^{-5}\ \text{mol}\ CH_3CH_2OCH_2CH_3 \times \frac{5\ \text{mol}\ H_2O}{1\ \text{mol}\ CH_3CH_2OCH_2CH_3} = 1.45 \times 10^{-4}\ \text{mol}\ H_2O$$

3. $Zn(s) + Cu(NO_3)_2(aq) \rightarrow Cu(s) + Zn(NO_3)_2(aq)$

$$2.688\ \text{g Zn} \times \frac{1\ \text{mol Zn}}{65.38\ \text{g Zn}} \times \frac{1\ \text{mol Cu}}{1\ \text{mol Zn}} = 0.04111\ \text{mol Cu}$$

4. $2\ BrF_3 \rightarrow Br_2 + 3\ F_2$

$$3.0\ \text{g}\ BrF_3 \times \frac{1\ \text{mol}\ BrF_3}{136.90\ \text{g}\ BrF_3} \times \frac{3\ \text{mol}\ F_2}{2\ \text{mol}\ BrF_3} \times \frac{38.00\ \text{g}\ F_2}{1\ \text{mol}\ F_2} = 1.2\ \text{g}\ F_2$$

5. $2\ NO + O_2 \rightarrow 2\ NO_2$

$$50.0\ \text{kg NO} \times \frac{1\ \text{kmol NO}}{30.01\ \text{kg NO}} \times \frac{2\ \text{kmol}\ NO_2}{2\ \text{kmol NO}} \times \frac{46.01\ \text{kg}\ NO_2}{1\ \text{kmol}\ NO_2} = 76.7\ \text{kg}\ NO_2$$

6. $2\ Na_3PO_4(aq) + 3\ Ca(NO_3)_2(aq) \rightarrow 6\ NaNO_3(aq) + Ca_3(PO_4)_2(s)$

$$3.215\ g\ Na_3PO_4 \times \frac{1\ mol\ Na_3PO_4}{163.94\ g\ Na_3PO_4} \times \frac{1\ mol\ Ca_3(PO_4)_2}{2\ mol\ Na_3PO_4} \times \frac{310.18\ g\ Ca_3(PO_4)_2}{1\ mol\ Ca_3(PO_4)_2} = 3.041\ g\ Ca_3(PO_4)_2\ \text{ideal};\ \frac{2.801\ g}{3.041\ g} \times 100\% = 92.11\%$$

7. $S(s) + 3/2\ O_2(g) + 2\ NaOH(aq) \rightarrow Na_2SO_4(aq) + H_2O(\ell)$

$$5.00 \times 10^2\ mg\ Na_2SO_4\ \text{actual} \times \frac{100\ mg\ Na_2SO_4\ \text{ideal}}{91.9\ mg\ Na_2SO_4\ \text{actual}} \times \frac{1\ mmol\ Na_2SO_4}{142.05\ mg\ Na_2SO_4} \times \frac{2\ mmol\ NaOH}{1\ mmol\ Na_2SO_4} \times \frac{40.00\ mg\ NaOH}{1\ mmol\ NaOH} = 306\ mg\ NaOH$$

8. $CS_2(\ell) + 3\ O_2(g) \rightarrow 2\ SO_2(g) + CO_2(g)$

$$5.00 \times 10^2\ kg\ CS_2 \times \frac{1\ kmol\ CS_2}{76.13\ kg\ CS_2} \times \frac{2\ kmol\ SO_2}{1\ kmol\ CS_2} \times \frac{64.06\ kg\ SO_2\ \text{ideal}}{1\ kmol\ SO_2} \times \frac{92.9\ kg\ SO_2\ \text{actual}}{100\ kg\ SO_2\ \text{ideal}} = 782\ kg\ SO_2\ \text{actual}$$

9.

	$2\ NaI(aq)$	+	$Pb(NO_3)_2(aq)$	→	$PbI_2(s)$	+	$2\ NaNO_3(aq)$
Amount at start (mol)	1.10		0.761		0		
Amount used (−) or produced (+) (mol)	−1.10		−0.550		+0.550		
Amount after the reaction (mol)	0		0.211		0.550		

10.

	$Ca(NO_3)_2(aq)$	+	$2\ NaF(aq)$	→	$CaF_2(s)$	+	$2\ NaNO_3(aq)$
Mass at start (g)	43.5		39.5		0		
Molar mass (g/mol)	164.10		41.99		78.08		
Amount at start (mol)	0.265		0.941		0		
Amount used (−) or produced (+) (mol)	−0.265		−0.530		+0.265		
Amount at end (mol)	0		0.411		0.265		
Mass at end (g)	0		17.3		20.7		

11. $Ca(NO_3)_2(aq) + 2\ NaF(aq) \rightarrow CaF_2(s) + 2\ NaNO_3(aq)$

$$43.5\ g\ Ca(NO_3)_2 \times \frac{1\ mol\ Ca(NO_3)_2}{164.10\ g\ Ca(NO_3)_2} \times \frac{1\ mol\ CaF_2}{1\ mol\ Ca(NO_3)_2} \times \frac{78.08\ g\ CaF_2}{1\ mol\ CaF_2} = 20.7\ g\ CaF_2$$

$$39.5\ g\ NaF \times \frac{1\ mol\ NaF}{41.99\ g\ NaF} \times \frac{1\ mol\ CaF_2}{1\ mol\ NaF} \times \frac{78.08\ g\ CaF_2}{1\ mol\ CaF_2} = 36.7\ g\ CaF_2$$

20.7 g CaF_2 precipitates

$$43.5\ g\ Ca(NO_3)_2 \times \frac{1\ mol\ Ca(NO_3)_2}{164.10\ g\ Ca(NO_3)_2} \times \frac{2\ mol\ NaF}{1\ mol\ Ca(NO_3)_2} \times \frac{41.99\ g\ NaF}{1\ mol\ NaF} = 22.3\ g\ NaF$$

39.5 g NaF − 22.3 g NaF = 17.2 g NaF in excess

12. $1.6 \times 10^2\ kcal \times \frac{4.184\ kJ}{1\ kcal} = 6.7 \times 10^2\ kJ = 6.7 \times 10^5\ J$

13. $H_2(g) + 1/2\ O_2(g) \rightarrow H_2O(\ell)\ \Delta_rH = -286\ kJ$

$H_2(g) + 1/2\ O_2(g) \rightarrow H_2O(\ell) + 286\ kJ$

14. $H_2(g) + 1/2\ O_2(g) \rightarrow H_2O(\ell) + 286\ kJ$

$$15.0\ g\ H_2 \times \frac{1\ mol\ H_2}{2.016\ g\ H_2} \times \frac{286\ kJ}{1\ mol\ H_2} = 2.13 \times 10^3\ kJ$$

15. $5.00 \times 10^5\ kJ \times \frac{1\ mol\ CH_4}{889\ kJ} \times \frac{16.04\ g\ CH_4}{1\ mol\ CH_4} = 9.02 \times 10^4\ g\ CH_4$

Answers to Problem-Classification Exercises

1. Mass stoichiometry
2. Thermochemical stoichiometry
3. Limiting reactant
4. Mass stoichiometry
5. Percentage yield
6. Thermochemical stoichiometry
7. Limiting reactant
8. Percentage yield

Answers to Blue-Numbered Questions, Exercises, and Problems

1. a) $95.3\ \text{mol}\ O_2 \times \frac{4\ \text{mol}\ NH_3}{5\ \text{mol}\ O_2} = 76.2\ \text{mol}\ NH_3$

b) $2.89\ \text{mol}\ NH_3 \times \frac{4\ \text{mol NO}}{4\ \text{mol}\ NH_3} = 2.89\ \text{mol NO}$

c) $3.35\ \text{mol}\ H_2O \times \frac{4\ \text{mol NO}}{6\ \text{mol}\ H_2O} = 2.23\ \text{mol NO}$

3. $MgO + H_2O \rightarrow Mg(OH)_2$

$0.884\ \text{mol}\ Mg(OH)_2 \times \frac{1\ \text{mol MgO}}{1\ \text{mol}\ Mg(OH)_2} = 0.884\ \text{mol MgO}$

5. $2\ SO_2 + O_2 \rightarrow 2\ SO_3$

$3.99\ \text{mol}\ SO_3 \times \frac{2\ \text{mol}\ SO_2}{2\ \text{mol}\ SO_3} = 3.99\ \text{mol}\ SO_2$

7. a) $268\ \text{g}\ O_2 \times \frac{1\ \text{mol}\ O_2}{32.00\ \text{g}\ O_2} \times \frac{4\ \text{mol}\ NH_3}{5\ \text{mol}\ O_2} = 6.70\ \text{mol}\ NH_3$

b) $31.7\ \text{mol}\ NH_3 \times \frac{6\ \text{mol}\ H_2O}{4\ \text{mol}\ NH_3} \times \frac{18.02\ \text{g}\ H_2O}{1\ \text{mol}\ H_2O} = 857\ \text{g}\ H_2O$

c) $404\ \text{g NO} \times \frac{1\ \text{mol NO}}{30.01\ \text{g NO}} \times \frac{4\ \text{mol}\ NH_3}{4\ \text{mol NO}} \times \frac{17.03\ \text{g}\ NH_3}{1\ \text{mol}\ NH_3} =$

$229\ \text{g}\ NH_3$

d) $6.41\ \text{g}\ H_2O \times \frac{1\ \text{mol}\ H_2O}{18.02\ \text{g}\ H_2O} \times \frac{4\ \text{mol NO}}{6\ \text{mol}\ H_2O} \times \frac{30.01\ \text{g NO}}{1\ \text{mol NO}}$

$= 7.12\ \text{g NO}$

9. $21.0\ \text{g}\ C_3H_5(NO_3)_3 \times \frac{1\ \text{mol}\ C_3H_5(NO_3)_3}{227.10\ \text{g}\ C_3H_5(NO_3)_3} \times$

$\frac{12\ \text{mol}\ CO_2}{4\ \text{mol}\ C_3H_5(NO_3)_3} \times \frac{44.01\ \text{g}\ CO_2}{1\ \text{mol}\ CO_2} = 12.2\ \text{g}\ CO_2$

11. $323\ \text{g}\ C_{17}H_{35}COONa \times \frac{1\ \text{mol}\ C_{17}H_{35}\ COONa}{306.45\ \text{g}\ C_{17}H_{35}\ COONa} \times$

$\frac{3\ \text{mol NaOH}}{3\ \text{mol}\ C_{17}H_{35}COONa} \times \frac{40.00\ \text{g NaOH}}{1\ \text{mol NaOH}} =$

$42.2\ \text{g NaOH}$

13. $681\ \text{g}\ Na_2S_2O_3 \times \frac{1\ \text{mol}\ Na_2S_2O_3}{158.10\ \text{g}\ Na_2S_2O_3} \times \frac{1\ \text{mol}\ Na_2CO_3}{3\ \text{mol}\ Na_2S_2O_3} \times$

$\frac{105.99\ \text{g}\ Na_2CO_3}{1\ \text{mol}\ Na_2CO_3} = 152\ \text{g}\ Na_2CO_3$

15. $616\ \text{mg}\ NaC_{18}H_{35}O_2 \times \frac{1\ \text{mmol}\ NaC_{18}H_{35}O_2}{306.45\ \text{mg}\ NaC_{18}H_{35}O_2}$

$\times \frac{1\ \text{mmol}\ Ca(C_{18}H_{35}O_2)_2}{2\ \text{mmol}\ NaC_{18}H_{35}O_2} \times \frac{607.00\ \text{mg}\ Ca(C_{18}H_{35}O_2)_2}{1\ \text{mmol}\ Ca(C_{18}H_{35}O_2)_2}$

$= 6.10 \times 10^2\ \text{mg}\ Ca(C_{18}H_{35}O_2)_2$

17. $802\ \text{mg}\ P_4O_{10} \times \frac{1\ \text{mmol}\ P_4O_{10}}{283.88\ \text{mg}\ P_4O_{10}} \times \frac{10\ \text{mmol}\ Fe_2O_3}{3\ \text{mmol}\ P_4O_{10}} \times$

$\frac{159.70\ \text{mg}\ Fe_2O_3}{1\ \text{mmol}\ Fe_2O_3} = 1.50 \times 10^3\ \text{mg}\ Fe_2O_3 = 1.50\ \text{g}\ Fe_2O_3$

19. $83.0\ \text{g}\ NH_4HCO_3 \times \frac{1\ \text{mol}\ NH_4HCO_3}{79.06\ \text{g}\ NH_4HCO_3} \times$

$\frac{1\ \text{mol NaCl}}{1\ \text{mol}\ NH_4HCO_3} \times \frac{58.44\ \text{g NaCl}}{1\ \text{mol NaCl}} = 61.4\ \text{g NaCl}$

21. $62.0\ \text{g}\ NH_3 \times \frac{1\ \text{mol}\ NH_3}{17.03\ \text{g}\ NH_3} \times \frac{1\ \text{mol}\ CaCl_2}{2\ \text{mol}\ NH_3} \times$

$\frac{110.98\ \text{g}\ CaCl_2}{1\ \text{mol}\ CaCl_2} = 202\ \text{g}\ CaCl_2$

23. $H_3PO_4 + 3\ NaOH \rightarrow 3\ H_2O + Na_3PO_4$

$32.6\ \text{g}\ H_3PO_4 \times \frac{1\ \text{mol}\ H_3PO_4}{97.99\ \text{g}\ H_3PO_4} \times \frac{3\ \text{mol NaOH}}{1\ \text{mol}\ H_3PO_4} \times$

$\frac{40.00\ \text{g NaOH}}{1\ \text{mol NaOH}} = 39.9\ \text{g NaOH}$

25. $2\ KOH + Mg(NO_3)_2 \rightarrow 2\ KNO_3 + Mg(OH)_2$

$2.09\ \text{g KOH} \times \frac{1\ \text{mol KOH}}{56.11\ \text{g KOH}} \times \frac{1\ \text{mol}\ Mg(OH)_2}{2\ \text{mol KOH}} \times$

$\frac{58.33\ \text{g}\ Mg(OH)_2}{1\ \text{mol}\ Mg(OH)_2} = 1.09\ \text{g}\ Mg(OH)_2$

27. $2\ NaOH + H_2SO_4 \rightarrow Na_2SO_4 + 2\ H_2O$

$0.521\ \text{g}\ Na_2SO_4 \times \frac{1\ \text{mol}\ Na_2SO_4}{142.04\ \text{g}\ Na_2SO_4} \times \frac{1\ \text{mol}\ H_2SO_4}{1\ \text{mol}\ Na_2SO_4} \times$

$\frac{98.08\ \text{g}\ H_2SO_4}{1\ \text{mol}\ H_2SO_4} = 0.360\ \text{g}\ H_2SO_4$

29. $Zn + 2\ NH_4Cl \rightarrow ZnCl_2 + 2\ NH_3 + H_2$

$7.05\ \text{g}\ NH_3 \times \frac{1\ \text{mol}\ NH_3}{17.03\ \text{g}\ NH_3} \times \frac{1\ \text{mol Zn}}{2\ \text{mol}\ NH_3} \times \frac{65.38\ \text{g Zn}}{1\ \text{mol Zn}} =$

$13.5\ \text{g Zn}$

31. $8.18\ \text{g}\ Na_2S_2O_3 \times \frac{1\ \text{mol}\ Na_2S_2O_3}{158.10\ \text{g}\ Na_2S_2O_3} \times \frac{1\ \text{mol NaBr}}{2\ \text{mol}\ Na_2S_2O_3} \times$

$\frac{102.89\ \text{g NaBr}}{1\ \text{mol NaBr}} = 2.66\ \text{g NaBr (ideal)}$

$\frac{2.61\ \text{g NaBr (act)}}{2.66\ \text{g NaBr (ideal)}} \times 100\% = 98.1\%\ \text{yield}$

33. $7.25\ \text{g}\ CaCN_2 \times \frac{1\ \text{mol}\ CaCN_2}{80.11\ \text{g}\ CaCN_2} \times \frac{2\ \text{mol}\ NH_3}{1\ \text{mol}\ CaCN_2} \times$

$\frac{17.03\ \text{g}\ NH_3}{1\ \text{mol}\ NH_3} \times \frac{92.8\ \text{g}\ NH_3\ \text{(act)}}{100\ \text{g}\ NH_3\ \text{(ideal)}} = 2.86\ \text{g}\ NH_3\ \text{(act)}$

35. $5.00 \times 10^2\ \text{kg}\ NH_3\ \text{(act)} \times \frac{100\ \text{kg}\ NH_3\ \text{(ideal)}}{88.8\ \text{kg}\ NH_3\ \text{(act)}} \times$

$\frac{1\ \text{kmol}\ NH_3}{17.03\ \text{kg}\ NH_3} \times \frac{3\ \text{kmol}\ H_2}{2\ \text{kmol}\ NH_3} \times \frac{2.016\ \text{kg}\ H_2}{1\ \text{kmol}\ H_2}$

$= 1.00 \times 10^2\ \text{kg}\ H_2$

37. $7.03 \text{ g } CO_2 \times \frac{1 \text{ mol } CO_2}{44.01 \text{ g } CO_2} \times \frac{1 \text{ mol } C_6H_{12}O_6}{6 \text{ mol } CO_2} \times \frac{180.16 \text{ g } C_6H_{12}O_6}{1 \text{ mol } C_6H_{12}O_6} = 4.80 \text{ g } C_6H_{12}O_6 \text{ (ideal)}$

$\frac{3.92 \text{ g } C_6H_{12}O_6 \text{ (act)}}{4.80 \text{ g } C_6H_{12}O_6 \text{ (ideal)}} \times 100\% = 81.7\% \text{ yield}$

39. $62.5 \text{ kg } CH_3COOC_2H_5 \text{ (act)} \times \frac{100 \text{ kg } CH_3COOC_2H_5 \text{ (ideal)}}{69.1 \text{ kg } CH_3COOC_2H_5 \text{ (act)}} \times \frac{1 \text{ kmol } CH_3COOC_2H_5}{88.10 \text{ kg } CH_3COOC_2H_5} \times \frac{1 \text{ kmol } CH_3COOH}{1 \text{ kmol } CH_3COOC_2H_5} \times$

$\frac{60.05 \text{ kg } CH_3COOH}{1 \text{ kmol } CH_3COOCH} = 61.7 \text{ kg } CH_3COOH$

41. a) $N_2(g) + 3\,H_2(g) \rightarrow 2\,NH_3(g)$
b) 4
c) Hydrogen is the limiting reactant. The 12 H_2 molecules in the reaction mixture are shown to yield 8 NH_3 molecules. The 6 N_2 molecules in the reaction mixture can yield 12 NH_3 molecules. The excess nitrogen, 2 N_2 molecules, is shown as part of the product mixture.

43. The positive slope occurs in trials in which iron is the limiting reactant. Thus, the mass of product increases as the mass of the limiting reactant increases. When the mass of iron is greater than 2 g, the mass of product is constant, and the slope of the line is zero. This indicates that bromine is the limiting reactant in these trials.

Questions 45–49: Both the comparison-of-moles and smaller-amount methods are shown. Differences in answers between the methods are caused by roundoffs in calculations.

45.

Comparison-of-moles method	$BaCl_2$	+	Na_2CrO_4	→	$BaCrO_4$	+	2 NaCl
Mass at start (g)	1.63		2.40		0		
Molar mass (g/mol)	208.2		161.98		253.3		
Amount at start (mol)	0.00783		0.0148		0		
Amount used (–) or produced (+) (mol)	−0.00783		−0.00783		+0.00783		
Amount at end (mol)	0		0.0070		0.00783		
Mass at end (g)	0		1.13		1.98		

Smaller-amount method:

$1.63 \text{ g } BaCl_2 \times \frac{1 \text{ mol } BaCl_2}{208.2 \text{ g } BaCl_2} \times \frac{1 \text{ mol } BaCrO_4}{1 \text{ mol } BaCl_2} \times \frac{253.3 \text{ g } BaCrO_4}{1 \text{ mol } BaCrO_4} = 1.98 \text{ g } BaCrO_4$

$2.40 \text{ g } Na_2CrO_4 \times \frac{1 \text{ mol } Na_2CrO_4}{161.98 \text{ g } Na_2CrO_4} \times \frac{1 \text{ mol } BaCrO_4}{1 \text{ mol } Na_2CrO_4} \times \frac{253.3 \text{ g } BaCrO_4}{1 \text{ mol } BaCrO_4} = 3.75 \text{ g } BaCrO_4$

$BaCl_2$ is the limiting reactant. The yield is the smaller amount, 1.98 g $BaCrO_4$.

$1.63 \text{ g } BaCl_2 \times \frac{1 \text{ mol } BaCl_2}{208.2 \text{ g } BaCl_2} \times \frac{1 \text{ mol } Na_2CrO_4}{1 \text{ mol } BaCl_2} \times \frac{161.98 \text{ g } Na_2CrO_4}{1 \text{ mol } Na_2CrO_4} = 1.27 \text{ g } Na_2CrO_4$

2.40 g Na_2CrO_4 (initial) − 1.27 g Na_2CrO_4 (used) = 1.13 g Na_2CrO_4 left

47.

Comparison-of-moles method	2 $NaIO_3$	+	5 $NaHSO_3$	→	I_2	+	others
Mass at start (kg)	6.00		7.33		0		
Molar mass (kg/kmol)	197.9		104.06		253.3		
Amount at start (kmol)	0.0303		0.0704		0		
Amount used (–) or produced (+) (kmol)	−0.0282		−0.0704		+0.0141		
Amount at end (kmol)	0.0021		0		0.0141		
Mass at end (kg)	0.42		0		3.58		

Smaller-amount method:

$6.00 \text{ kg } NaIO_3 \times \frac{1 \text{ kmol } NaIO_3}{197.9 \text{ kg } NaIO_3} \times \frac{1 \text{ kmol } I_2}{2 \text{ kmol } NaIO_3} \times \frac{253.8 \text{ kg } I_2}{1 \text{ kmol } I_2} = 3.85 \text{ kg } I_2$

$$7.33 \text{ kg NaHSO}_3 \times \frac{1 \text{ kmol NaHSO}_3}{104.06 \text{ kg NaHSO}_3} \times \frac{1 \text{ kmol I}_2}{5 \text{ kmol NaHSO}_3} \times \frac{253.8 \text{ kg I}_2}{1 \text{ kmol I}_2} = 3.58 \text{ kg I}_2$$

$NaHSO_3$ is the limiting reactant. The yield is the smaller amount, 3.58 kg I_2.

$$7.33 \text{ kg NaHSO}_3 \times \frac{1 \text{ kmol NaHSO}_3}{104.06 \text{ kg NaHSO}_3} \times \frac{2 \text{ kmol NaIO}_3}{5 \text{ kmol NaHSO}_3} \times \frac{197.9 \text{ kg NaIO}_3}{1 \text{ kmol NaIO}_3} = 5.58 \text{ kg NaIO}_3$$

6.00 kg $NaIO_3$ (initial) − 5.58 kg $NaIO_3$ (used) = 0.42 kg $NaIO_3$ left

49.

Comparison-of-moles method	$8\ P_4$	+	$3\ S_8$	→	$8\ P_4S_3$
Mass at start (g)	133		126		0
Molar mass (g/mol)	123.88		256.48		220.06
Amount at start (mol)	1.07		0.491		0
Amount used (−) or produced (+) (mol)	−1.07		−0.401		+1.07
Amount at end (mol)	0		0.090		1.07
Mass at end (g)	0		23.1		235

Smaller-amount method:

$$133 \text{ g P}_4 \times \frac{1 \text{ mol P}_4}{123.88 \text{ g P}_4} \times \frac{8 \text{ mol P}_4\text{S}_3}{8 \text{ mol P}_4} \times \frac{220.06 \text{ g P}_4\text{S}_3}{1 \text{ mol P}_4\text{S}_3} = 236 \text{ g P}_4\text{S}_3$$

$$126 \text{ g S}_8 \times \frac{1 \text{ mol S}_8}{256.48 \text{ g S}_8} \times \frac{8 \text{ mol P}_4\text{S}_3}{3 \text{ mol S}_8} \times \frac{220.06 \text{ g P}_4\text{S}_3}{1 \text{ mol P}_4\text{S}_3} = 288 \text{ g P}_4\text{S}_3$$

P_4 is the limiting reactant. The yield is the smaller amount, 236 g P_4S_3.

$$133 \text{ g P}_4 \times \frac{1 \text{ mol P}_4}{123.88 \text{ g P}_4} \times \frac{3 \text{ mol S}_8}{8 \text{ mol P}_4} \times \frac{256.48 \text{ g S}_8}{1 \text{ mol S}_8} = 103 \text{ g S}_8$$

126 g S_8 (initial) − 103 g S_8 (used) = 23 g S_8 left

51. a) $0.731 \text{ kcal} \times \frac{4.184 \text{ kJ}}{1 \text{ kcal}} = 3.06 \text{ kJ}$

b) $651 \text{ J} \times \frac{1 \text{ cal}}{4.184 \text{ J}} = 156 \text{ cal}$

c) $6.22 \times 10^3 \text{ J} \times \frac{1 \text{ cal}}{4.184 \text{ J}} \times \frac{1 \text{ kcal}}{1000 \text{ cal}} = 1.49 \text{ kcal}$

53. $493 \text{ kJ} \times \frac{1 \text{ kcal}}{4.184 \text{ kJ}} = 118 \text{ kcal}$

$118 \text{ kcal} \times \frac{1000 \text{ cal}}{1 \text{ kcal}} = 1.18 \times 10^5 \text{ cal}$

55. $\frac{5.8 \times 10^2 \text{ kcal}}{\text{day}} \times \frac{4.184 \text{ kJ}}{1 \text{ kcal}} \times \frac{365 \text{ days}}{\text{year}} = 8.9 \times 10^5 \text{ kJ/year}$

57. $2.82 \times 10^3 \text{ kJ} + 6\ CO_2(g) + 6\ H_2O(\ell) \rightarrow C_6H_{12}O_6(s) + 6\ O_2(g)$

$6\ CO_2(g) + 6\ H_2O(\ell) \rightarrow C_6H_{12}O_6(s) + 6\ O_2(g)$
$\Delta_rH = +2.82 \times 10^3 \text{ kJ}$

59. $286 \text{ kJ} + H_2O(\ell) \rightarrow H_2(g) + 1/2\ O_2(g)$

$H_2O(\ell) \rightarrow H_2(g) + 1/2\ O_2(g)\ \Delta H = +286 \text{ kJ}$

61. $C_2H_2(g) + 5/2\ O_2(g) \rightarrow 2\ CO_2(g) + H_2O(\ell) + 1.31 \times 10^3 \text{ kJ}$

$C_2H_2(g) + 5/2\ O_2(g) \rightarrow 2\ CO_2(g) + H_2O(\ell)$
$\Delta_rH = -1.31 \times 10^3 \text{ kJ}$

63. $5.80 \text{ kg CaCO}_3 \times \frac{1000 \text{ g CaCO}_3}{\text{kg CaCO}_3} \times \frac{1 \text{ mol CaCO}_3}{100.09 \text{ g CaCO}_3} \times \frac{178 \text{ kJ}}{1 \text{ mol CaCO}_3} = 1.03 \times 10^4 \text{ kJ}$

65. $291 \text{ kJ} \times \frac{1 \text{ mol CaO}}{65.3 \text{ kJ}} \times \frac{56.08 \text{ g CaO}}{1 \text{ mol CaO}} = 2.50 \times 10^2 \text{ g CaO}$

67. $9.48 \times 10^5 \text{ kJ} \times \frac{2 \text{ mol C}_8\text{H}_{18}}{1.09 \times 10^4 \text{ kJ}} \times \frac{114.22 \text{ g C}_8\text{H}_{18}}{1 \text{ mol C}_8\text{H}_{18}} = 1.99 \times 10^4 \text{ g C}_8\text{H}_{18}$

70. True: a, d, e, f. False: b, c.

71. $35 \text{ g N}_2 \times \frac{1 \text{ mol N}_2}{28.02 \text{ g N}_2} \times \frac{1 \text{ mol Na}_2\text{CO}_3}{1 \text{ mol N}_2} \times \frac{105.99 \text{ g Na}_2\text{CO}_3}{1 \text{ mol Na}_2\text{CO}_3} = 1.3 \times 10^2 \text{ g Na}_2\text{CO}_3$

72. $3\ Ca(NO_3)_2 + 2\ Na_3PO_4 \rightarrow Ca_3(PO_4)_2 + 6\ NaNO_3$

$3.98 \text{ g Na}_3\text{PO}_4 \times \frac{1 \text{ mol Na}_3\text{PO}_4}{163.94 \text{ g Na}_3\text{PO}_4} \times \frac{1 \text{ mol Ca}_3(\text{PO}_4)_2}{2 \text{ mol Na}_3\text{PO}_4} \times \frac{310.18 \text{ g Ca}_3(\text{PO}_4)_2}{1 \text{ mol Ca}_3(\text{PO}_4)_2} = 3.77 \text{ g Ca}_3(\text{PO}_4)_2$

73. $125 \text{ g KO}_2 \times \frac{1 \text{ mol KO}_2}{71.10 \text{ g KO}_2} \times \frac{3 \text{ mol O}_2}{4 \text{ mol KO}_2} \times \frac{32.00 \text{ g O}_2}{1 \text{ mol O}_2} = 42.2 \text{ g O}_2$

74. a) $6.00 \text{ g H}_3\text{C}_6\text{H}_5\text{O}_7 \times \frac{1 \text{ mol H}_3\text{C}_6\text{H}_5\text{O}_7}{192.12 \text{ g H}_3\text{C}_6\text{H}_5\text{O}_7} \times \frac{3 \text{ mol NaHCO}_3}{1 \text{ mol H}_3\text{C}_6\text{H}_5\text{O}_7} \times \frac{84.01 \text{ g NaHCO}_3}{1 \text{ mol NaHCO}_3} = 7.87 \text{ g NaHCO}_3$

20.0 g $NaHCO_3$ is available; $H_3C_6H_5O_7$ is limiting

$$6.00 \text{ g } H_3C_6H_5O_7 \times \frac{1 \text{ mol } H_3C_6H_5O_7}{192.12 \text{ g } H_3C_6H_5O_7} \times \frac{3 \text{ mol } CO_2}{1 \text{ mol } H_3C_6H_5O_7} \times \frac{44.01 \text{ g } CO_2}{1 \text{ mol } CO_2} = 4.12 \text{ g } CO_2$$

b) 20.0 g $NaHCO_3$ − 7.87 g $NaHCO_3$ = 12.1 g $NaHCO_3$ unreacted

c) When changing the name of an *-ic* acid to an ion, *-ic* changes to *-ate*. Thus, citr*ic* acid, $H_3C_6H_5O_7$, becomes citr*ate* ion, $C_6H_5O_7^{3-}$. $Na_3C_6H_5O_7$ is therefore sodium citrate.

75. $$1.68 \text{ g Al} \times \frac{1 \text{ mol Al}}{26.98 \text{ g Al}} \times \frac{2 \text{ mol } Al_2O_3}{4 \text{ mol Al}} \times \frac{101.96 \text{ g } Al_2O_3}{1 \text{ mol } Al_2O_3} = 3.17 \text{ g } Al_2O_3$$

$$\frac{3.17 \text{ g } Al_2O_3}{12.8 \text{ g ore}} \times 100\% = 24.8\% \ Al_2O_3 \text{ in the ore}$$

76. $$1.42 \text{ g } KClO_3 \times \frac{1 \text{ mol } KClO_3}{122.55 \text{ g } KClO_3} \times \frac{89.5 \text{ kJ}}{2 \text{ mol } KClO_3} = 0.519 \text{ kJ}$$

77. $$0.500 \text{ ton } Ca(H_2PO_4)_2 \times \frac{2000 \text{ lb}}{1 \text{ ton}} \times \frac{453.6 \text{ g}}{1 \text{ lb}} \times \frac{1 \text{ kg}}{1000 \text{ g}} \times \frac{1 \text{ kmol } Ca(H_2PO_4)_2}{234.05 \text{ kg } Ca(H_2PO_4)_2} \times \frac{1 \text{ kmol } Ca_3(PO_4)_2}{1 \text{ kmol } Ca(H_2PO_4)_2} \times \frac{310.18 \text{ kg } Ca_3(PO_4)_2}{1 \text{ kmol } Ca_3(PO_4)_2} \times \frac{100 \text{ kg rock}}{79.4 \text{ kg } Ca_3(PO_4)_2} = 759 \text{ kg rock}$$

79. $$40.1 \text{ kg sludge} \times \frac{23.1 \text{ kg AgCl}}{100 \text{ kg sludge}} \times \frac{1 \text{ kmol AgCl}}{143.4 \text{ kg AgCl}} \times \frac{4 \text{ kmol NaCN}}{2 \text{ kmol AgCl}} \times \frac{49.01 \text{ kg NaCN}}{1 \text{ kmol NaCN}} = 6.33 \text{ kg NaCN}$$

81. $$105 \text{ kg } Cl_2 \text{ (act)} \times \frac{100 \text{ kg } Cl_2\text{(ideal)}}{61 \text{ kg } Cl_2\text{(act)}} \times \frac{1 \text{ kmol } Cl_2}{70.90 \text{ kg } Cl_2} \times \frac{2 \text{ kmol NaCl}}{1 \text{ kmol } Cl_2} \times \frac{58.44 \text{ kg NaCl}}{1 \text{ kmol NaCl}} \times \frac{100 \text{ kg solution}}{9.6 \text{ kg NaCl}} = 3.0 \times 10^3 \text{ kg solution}$$

83. $$239 \text{ mg } Ca(OH)_2 \times \frac{1 \text{ mmol } Ca(OH)_2}{74.10 \text{ mg } Ca(OH)_2} \times \frac{1 \text{ mmol } SnF_2}{1 \text{ mmol } Ca(OH)_2} \times \frac{156.7 \text{ mg } SnF_2}{1 \text{ mmol } SnF_2} = 505 \text{ mg } SnF_2 \text{ is}$$

needed to treat 239 mg $Ca(OH)_2$;

505 mg SnF_2 needed − 305 mg used = 2.00×10^2 mg SnF_2 additional should be used

85. $$454 \text{ g Al} \times \frac{1 \text{ mol Al}}{26.98 \text{ g Al}} \times \frac{1.97 \times 10^3 \text{ kJ}}{4 \text{ mol Al}} \times \frac{1 \text{ kw-h}}{3.60 \times 10^3 \text{ kJ}} = 2.30 \text{ kw-h}$$

87. $NaCl + AgNO_3 \rightarrow NaNO_3 + AgCl$

$$2.056 \text{ g AgCl} \times \frac{1 \text{ mol AgCl}}{143.4 \text{ g AgCl}} \times \frac{1 \text{ mol NaCl}}{1 \text{ mol AaCl}} \times \frac{58.44 \text{ g NaCl}}{1 \text{ mol NaCl}} = 0.8379 \text{ g NaCl}$$

1.6240 g mixture − 0.8379 g NaCl = 0.7861 g $NaNO_3$

$$\frac{0.8379 \text{ g NaCl}}{1.6240 \text{ g mixture}} \times 100\% = 51.59\% \text{ NaCl}$$

$$\frac{0.7861 \text{ g } NaNO_3}{1.6240 \text{ g mixture}} \times 100\% = 48.41\% \ NaNO_3$$

88. From the formula, there is 3 mol Cu in 1 mol $Cu_3(PO_4)_2$

$$2.637 \text{ g } Cu_3(PO_4)_2 \times \frac{1 \text{ mol } Cu_3(PO_4)_2}{380.59 \text{ g } Cu_3(PO_4)_3} \times \frac{3 \text{ mol Cu}}{1 \text{ mol } Cu_3(PO_4)_2} \times \frac{63.55 \text{ g Cu}}{1 \text{ mol Cu}} = 1.321 \text{ g Cu}$$

$$\frac{1.321 \text{ g Cu}}{1.382 \text{ g sample}} \times 100\% = 95.59\% \text{ Cu}$$

89. $$50.0 \text{ mL} \times \frac{1.19 \text{ g soln}}{1 \text{ mL}} \times \frac{17.0 \text{ g NaOH}}{100 \text{ g soln}} \times \frac{1 \text{ mol NaOH}}{40.00 \text{ g NaOH}} \times \frac{1 \text{ mol } Mg(NO_3)_2}{2 \text{ mol NaOH}} \times \frac{148.33 \text{ g } Mg(NO_3)_2}{1 \text{ mol } Mg(NO_3)_2} = 18.8 \text{ g } Mg(NO_3)_2$$

90. $AgNO_3 + NaCl \rightarrow AgCl + NaNO_3$

$AgNO_3 + NaBr \rightarrow AgBr + NaNO_3$

$$0.230 \text{ g NaCl} \times \frac{1 \text{ mol NaCl}}{58.44 \text{ g NaCl}} \times \frac{1 \text{ mol } AgNO_3}{1 \text{ mol NaCl}} \times \frac{169.9 \text{ g } AgNO_3}{1 \text{ mol } AgNO_3} = 0.669 \text{ g } AgNO_3$$

$$0.771 \text{ g NaBr} \times \frac{1 \text{ mol NaBr}}{102.89 \text{ g NaBr}} \times \frac{1 \text{ mol } AgNO_3}{1 \text{ mol NaBr}} \times \frac{169.9 \text{ g } AgNO_3}{1 \text{ mol } AgNO_3} = 1.27 \text{ g } AgNO_3$$

0.669 g $AgNO_3$ + 1.27 g $AgNO_3$ = 1.94 g $AgNO_3$

11 Atomic Theory: The Quantum Model of the Atom

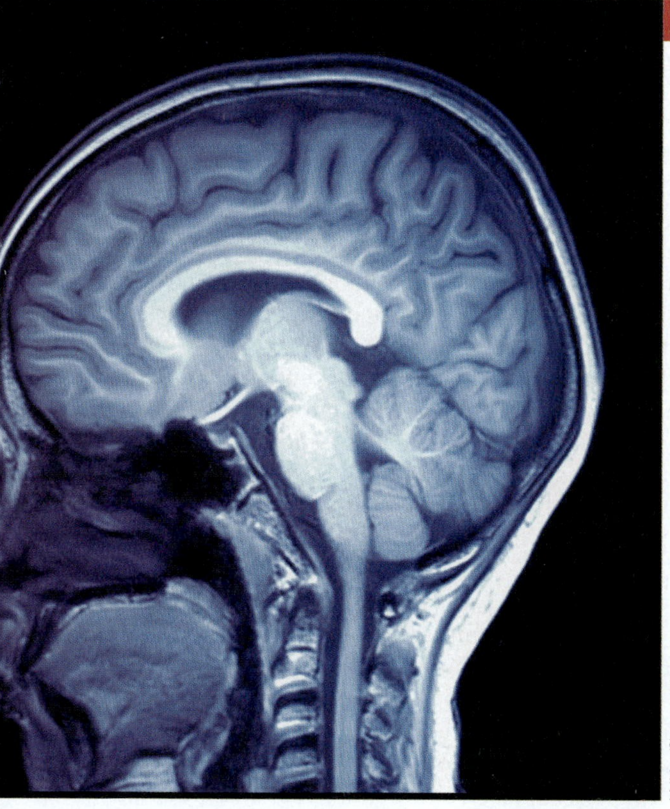

◀ Magnetic resonance imaging (MRI) scans, such as this computer-enhanced scan of a human head, are possible because of scientists' understanding of the quantum mechanical model of the atom. To make such an image, the MRI patient is placed in the opening of a long, narrow tube that is surrounded by a magnet. The magnetic field then interacts with the hydrogen atom protons in the part of the body being scanned. The protons first absorb energy from the magnet, and then they release the energy in such a way that the characteristics of their signals can be used to form an image. This image shows a normal, healthy brain.

MriMan/Shutterstock.com

CHAPTER CONTENTS

The four decades from 1890 to 1930 were a period of rapid progress in learning about the atom. Chapter 5 discusses the first half of this period; in this chapter, we cover the second half. In 1911, the data from Rutherford's alpha-particle scattering experiments were challenging to interpret. In the Rutherford model of the atom, all the positive charge was crammed into the dense, tiny nucleus. Like charges repel, so the nucleus of the atom should not have been stable, yet it was. The relationships of classical physics that worked so well in explaining large-scale systems did not work on atom-sized systems. Thus, scientists had to develop a new approach to understanding the atom and subatomic particles. The breakthrough that was needed was the development of the field of study now known as **quantum mechanics.**

11.1 What Causes the Northern Lights?

Source: Johnny Henriksen/ Spaceweather.com/NASA

Figure 11.1 The northern lights, as photographed by Johnny Henriksen on October 8, 2015 in Harstad, Norway.

Figure 11.1 is a photograph of the northern lights, also called the *aurora borealis*. It is a light-show-in-the-sky phenomenon that occurs on some nights in the northern latitudes (e.g., Alaska, Canada, Sweden, Finland), primarily on cloud-free nights with no moonlight in September through March. The term *aurora borealis* is derived from the Latin, meaning "northern dawn." What is the cause of the northern lights? To answer this question, we must first consider the magnetism phenomenon.

Figure 11.2 shows the pattern that occurs when tiny pieces of iron called *iron filings* are sprinkled on a piece of paper laid over a bar magnet. The filings are not attracted to the magnetic force uniformly; instead, they are attracted to the force along lines. Notice how the lines converge at each end of the bar magnet. These are called *magnetic poles*.

Figure 11.3 shows how charged particles emitted by the sun interact with the Earth's magnetic field. The Earth essentially acts like a giant magnet. The charged particles are attracted in the same pattern to the Earth's magnetic field as iron filings are attracted to a bar magnet, except that the solar particles originate from one direction, and then move toward and past the Earth, so the pattern is elongated on the side opposite of the sun. Some of the particles emitted by the sun are directed to the north and south poles of the Earth by the Earth's magnetic field.

The intensity of particle emission by the sun varies, making the northern lights an intermittent phenomenon. But when the particle intensity is sufficient and the Earth's atmosphere allows a person on the surface to see high into the night sky, the northern lights may occur. How do the solar particles cause the lights? The particles collide with the nitrogen and oxygen in the atmosphere, transferring their energy to those molecules and atoms (some of the gas in the upper atmosphere exists in the form of atoms), causing their electrons to gain energy and move farther away from the nucleus. The electrons then move back to their normal distance from the nucleus. That previously gained energy must be transferred somewhere. That energy is transferred in the form of light, causing the northern lights! In this chapter, we will investigate the fundamentals of the nature of the interaction between movement of electrons away and toward the nucleus and the emission of energy in the form of light. This, in turn, will allow us to develop a more complete model of the atom that accounts for the behavior of electrons.

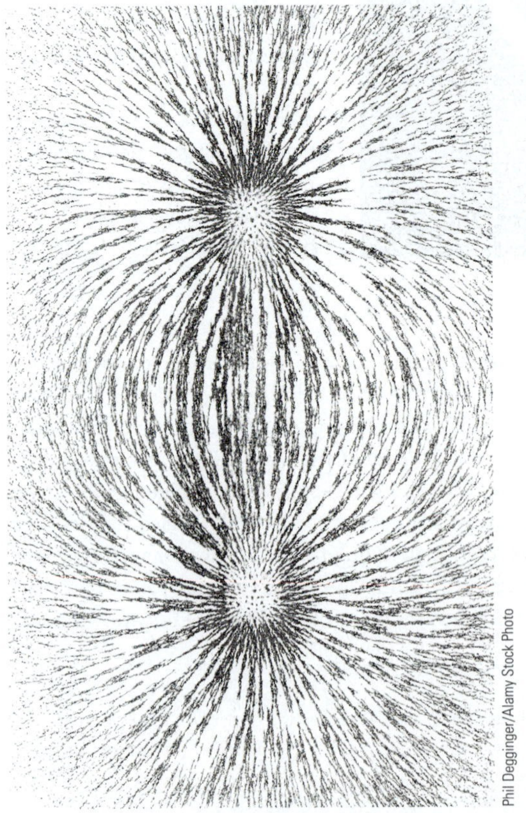

Figure 11.2 Electromagnetism is one of the four fundamental forces in the universe. The magnetic properties of materials (*–magnetism*) results from the arrangement and behavior of their electrons (*electro–*). Magnetic materials produce a magnetic field that attracts certain substances, such as iron. When solid iron is finely divided and sprinkled onto a piece of paper placed over a bar magnet, the lines of force of the magnetic field become visible. There are poles on magnets where the field lines converge.

Phil Degginger/Alamy Stock Photo

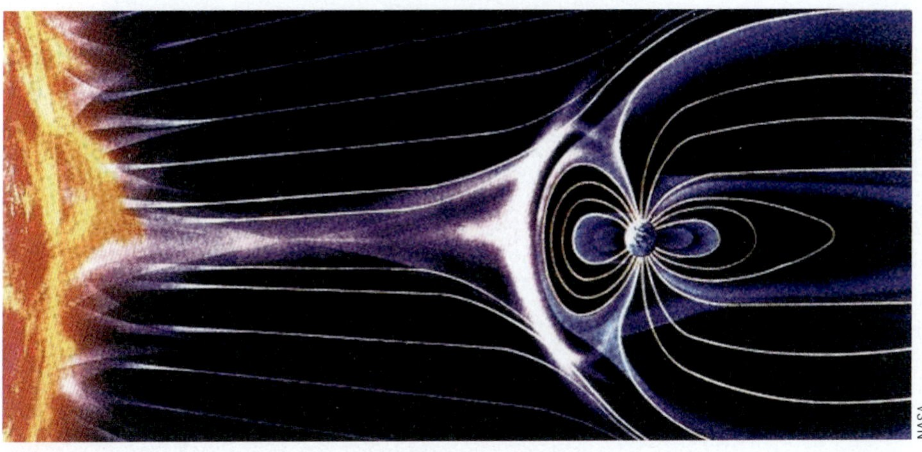

NASA

Figure 11.3 Charged particles are emitted by the sun and travel in all directions in the solar system. Some travel toward the Earth and interact with its magnetic field. Some solar particles are directed by the magnetic field toward the Earth's magnetic poles. In turn, the particles interact with the nitrogen and oxygen atoms and molecules in the atmosphere, transferring energy to their electrons, moving them away from the nucleus. As the electrons move back to their stable positions, energy is transferred out of the atoms and molecules in the form of light.

11.2 Electromagnetic Radiation

Goal 1 Define and describe electromagnetic radiation.
2 Distinguish between continuous and line spectra.

When you first open your eyes in the morning after a night's rest, the light that strikes your eyes stimulates nerve cells, which send signals to your brain that you interpret as the scene you see. Simultaneously, additional energy in the same form as the visible light—radio and television waves—also strikes your eyes, but that energy does not stimulate your optical nerve cells. Visible light, radio and television waves, and microwaves are some of the variations of the same type of energy, called **electromagnetic radiation** (Figure 11.4). Our eyes have evolved to detect just a small fraction of the entire **electromagnetic spectrum** (Figure 11.5), in a manner similar to how your radio is engineered to detect only radio waves without responding to television waves, microwaves, and visible light. Other parts of the electromagnetic spectrum include gamma rays, x-rays, ultraviolet radiation, and infrared radiation.

Anton Ryabtsov/Shutterstock.com

Figure 11.4 Bee's eyes. The eyes of the honey bee have evolved to detect electromagnetic radiation in the ultraviolet range, at wavelengths that are not visible to humans. However, bees cannot see red.

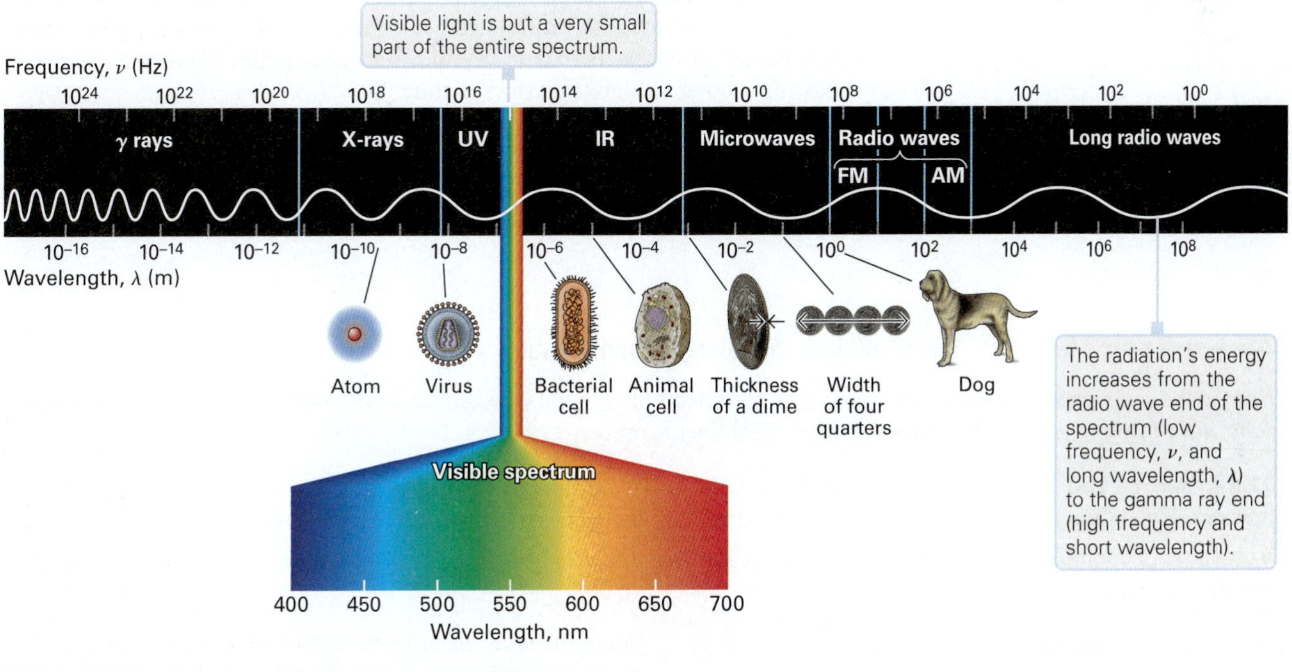

Figure 11.5 The electromagnetic spectrum. Visible light is only a small portion of the electromagnetic spectrum, covering a wavelength range of about 400 to 700 nm. There seems to be no upper or lower limit to the theoretically possible lengths of electromagnetic waves, although experimentally measured waves vary between about 10^{11} m and 10^{-15} m.

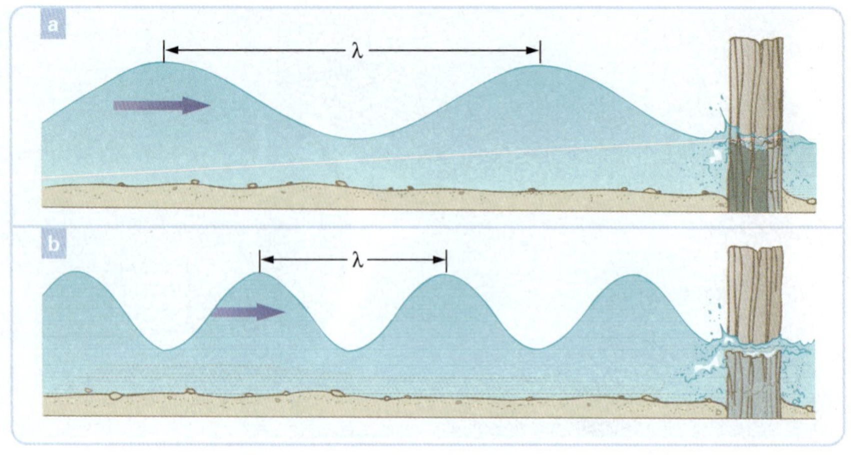

Figure 11.6 Wave properties. Water waves are similar to electromagnetic waves in that they can be described by properties such as velocity, wavelength, and frequency. Velocity, v, is the linear speed of a point on a wave. In this illustration, the two waves are traveling at the same velocity. Wavelength, λ, is the distance between corresponding points on a wave. The wavelength in (a) is longer than that in (b). Frequency, ν, is the number of complete waves passing a point per second. In this illustration, the wave crests in (b) will hit the post more often per unit of time than those in (a), indicating a greater frequency.

Electromagnetic radiation is a form of energy that consists of both electric and magnetic fields. Although transmission of energy in the form of electromagnetic radiation seems instantaneous in our everyday experiences, it actually travels at a finite speed, 3.00×10^8 meters per second (186,000 miles per hour) to three significant figures. This is called the **speed of light** and usually appears in equations with the symbol **c.** This speed is so great that we cannot sense the slight delay between the time that light reflects off an object and when our brains interpret it.

Light takes a little more than 8 minutes to travel from the sun to the Earth and just over 1 second to travel from the moon to the Earth. The timespan for light to travel short distances, such as from this page to your eye, is very, very tiny and beyond the limit of human perception.

Electromagnetic radiation has wavelike properties. Waves can be mathematically described by a **wave equation**. Every wave can be described by its properties, which include velocity (v or, specifically for electromagnetic radiation, c), wavelength (λ), and frequency (ν) (**Figure 11.6**). λ and ν are the Greek lowercase letters lambda (pronounced *lam-duh*) and nu (pronounced *new*), respectively. Wavelength is the distance between identical parts of a wave, and frequency is the number of complete waves passing a point in a given period of time.

The velocity of an electromagnetic wave is related to its wavelength and frequency by the equation $c = \lambda \times \nu$. Since the speed of light is a fixed quantity, the wavelength and frequency of electromagnetic waves are inversely proportional. As the wavelength increases, the frequency decreases, and vice versa.

Active Example 11.1 Speed of Light, Wavelength, and Frequency

What is the mathematical relationship among (a) the speed of light, (b) wavelength, and (c) frequency of an electromagnetic wave? Define each of the three terms and give their symbols.

Think Before You Write Review this section, as necessary, if you cannot answer the questions. Figure 11.6 is helpful in connecting something in your everyday experience to the more abstract electromagnetic wave concept.

Answers *Cover the left column with your cut-out shield. Reveal each answer only after you have written your own answer in the right column.*

Speed of light = wavelength × frequency	State the relationship among the three properties as a mathematical expression, using their full, written names. 

Speed of light = c = 3.00×10^8 m/s	What is the symbol used to represent the speed of light? What is the three-significant-figure value of the speed of light in meters per second?
Wavelength = λ = the distance between identical parts of a wave	Write the symbol used for the wavelength of electromagnetic radiation and explain what wavelength is.
Frequency = ν = the number of complete waves that pass a point in a given period of time, typically one second	What symbol is used to represent the frequency of an electromagnetic wave? What is frequency?
You improved your understanding of electromagnetic waves and wave properties.	What did you learn by solving this Active Example?

Practice Exercise 11.1

Explain why a decrease in frequency of electromagnetic radiation corresponds to an increase in wavelength.

Light from standard light bulbs, known as *white light*, produces a **continuous spectrum** when passed through a prism (**Figure 11.7**). However, when a gaseous pure element is placed in a glass container and subjected to an electrical discharge, it glows, or emits light, and when that light is passed through a prism, the light forms a **line spectrum**, as in **Figure 11.8**. These separate lines of color are known as **discrete** lines to indicate that they are individually distinct. Each element has a unique line spectrum (**Figure 11.9**).

Why does the light from an element produce a discrete line spectrum? We won't be able to answer that question fully until Section 11.3, but an important piece of the answer was provided in 1900 by Max Planck, a German scientist (**Figure 11.10**).

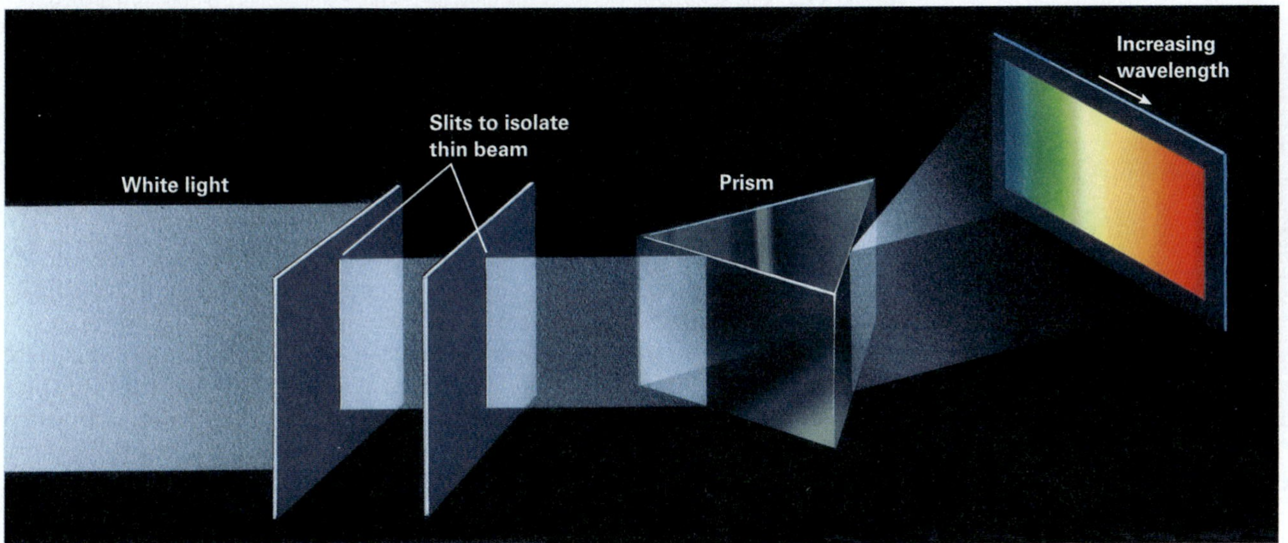

Figure 11.7 Dispersion of white light by a prism. White light is passed through slits and then through a prism. It is separated into a continuous spectrum of all wavelengths of visible light. This corresponds to the visible spectrum in Figure 11.5.

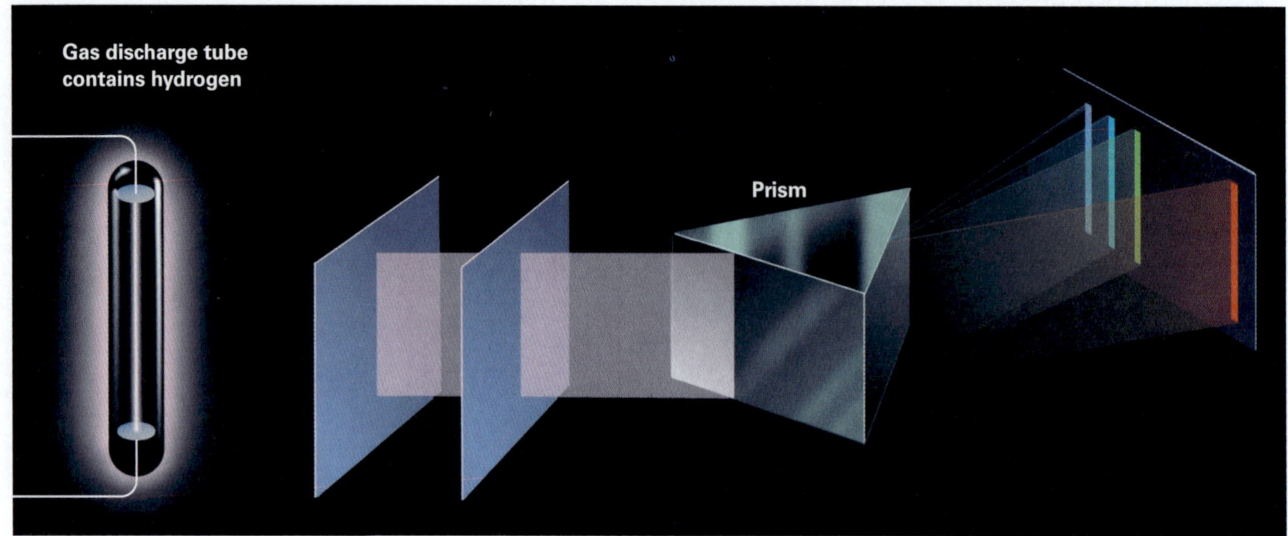

Figure 11.8 Dispersion of light from a gas discharge tube filled with hydrogen. This is like a neon light, except that neon gives red light. The magenta light from hydrogen is passed through slits and then through a prism. It is separated into a line spectrum made up of four wavelengths of visible light.

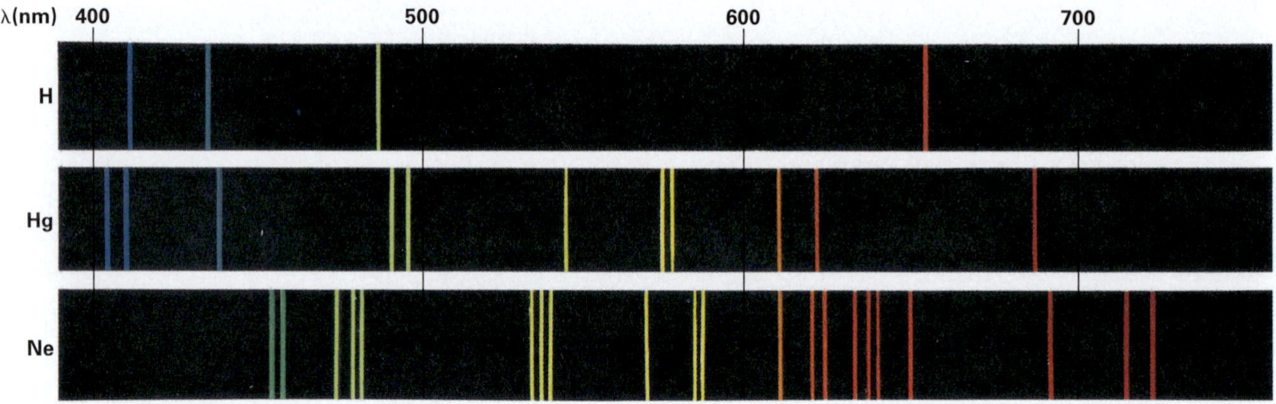

Figure 11.9 Line spectra of hydrogen, mercury, and neon. Each element produces a unique spectrum that can be used to identify the element. The hydrogen spectrum corresponds with that shown in Figure 11.8. The neon spectrum is a combination of the colors we see as the red light of a neon sign.

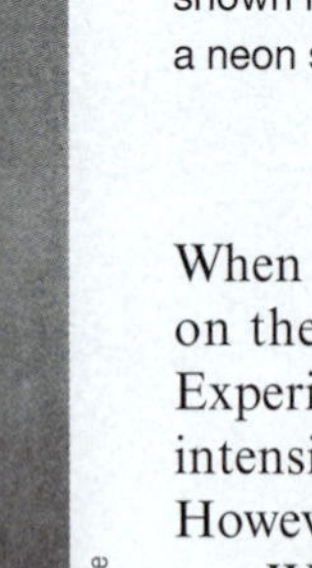

AIP/Science Source

Figure 11.10 Max Planck (1858–1947). Planck was awarded the 1918 Nobel Prize in Physics for his discovery of the quantization of energy.

When a substance is heated sufficiently, it emits light. You see this when you turn on the burner of an electric stove. The burner emits red light as it gets hotter. Experimental scientists in Planck's time had measured the relationship between the intensity of the emitted light and its frequency and had characterized the pattern. However, no one was able to explain *why* this pattern existed.

While working on trying to understand the pattern, Planck reasoned that since an antenna can be a source of radio waves, objects such as red-hot metals can be a source of light waves. Furthermore, since electrons moving in certain patterns within the atoms of the substance that made up an antenna were believed to be the source of radio waves, he assumed that the source of the emitted light was also electrons moving in a particular way. He hypothesized that the frequency at which the electrons oscillated, or varied in position relative to a fixed point, was proportional to their energy: The greater the energy of the electron (E), the greater the frequency of its oscillation (ν). Whereas it seemed to Planck that this hypothesis was very logical, using E ∝ ν did not fit the experimental data. In the sciences, when a model does not fit measured observations, the model is *wrong*, no matter how sensible it may seem.

Planck eventually noticed something that was never seen before in the sciences. If he inserted a positive integer (1, 2, 3, …) into his E ∝ ν equation, he was able to derive an equation that fits the experimental data perfectly! The correct energy–frequency relationship for the oscillating electrons was $E \propto n \times \nu$, where $n = 1, 2, 3, \ldots$. With this concept, Planck essentially originated the science of quantum mechanics, although at the time he didn't understand why the whole-number multiplier was needed in the equation that matched the experimental data. At this point in history, the community of scientists had measured observations but no explanatory model.

Albert Einstein interpreted Planck's equation by proposing that light waves are not *only* waves. He believed that $E \propto n \times \nu$ was evidence that light is emitted in the form of particles. The positive integer in Planck's equation, n, corresponds to the number of particles of light emitted of any specific wave frequency. The energy released by electrons occurs in the form of massless "packets" of electromagnetic radiation known as *photons*. A **photon** is a particle of light.

You may be wondering why we originally said that light is a wave, and now we are saying that light is a particle. If so, you are in good company, because many knowledgeable scientists of the early 20th century were also intrigued by what is often referred to as the **wave-particle duality**. Today, we recognize that not only do electromagnetic waves have particle-like properties, but particles also have wavelike properties! As you will see later in this chapter, the nature of the universe as we observe it at the macroscopic level differs from what we know of how photons of light and tiny particles behave. Light has properties that have no analogy at the macroscopic level; thus, we have to combine two different ideas to describe its behavior. When light moves through space, it acts like a wave, but when it is first emitted or it is absorbed, it acts like a particle.

Active Example 11.2 Electromagnetic Radiation and Continuous and Line Spectra

What is the significance of the wave-particle duality in terms of the mental model you should be developing for electromagnetic radiation?

Think Before You Write Consider the facts that electromagnetic radiation corresponds to the wave equation and the relationship $E \propto n \times \nu$.

Answers *Cover the left column with your cut-out shield. Reveal each answer only after you have written your own answer in the right column.*

The velocity of an electromagnetic wave, the speed of light, is the product of the wave's wavelength and frequency. *The fact that electromagnetic radiation follows the wave equation relationship is evidence that it behaves as a wave.*	Explain what it means to say that electromagnetic radiation has wavelike properties.
Since $E \propto n \times \nu$, there must be a reason for n to need to be in the equation. The integer n is the number of particles of light of a given frequency.	Explain what it means to say that electromagnetic radiation has particle-like properties. 
Electromagnetic waves have both wavelike properties and particle-like properties.	What is the *wave-particle duality*? 
Electromagnetic radiation acts like a wave when it moves through space. It acts like a particle when it is emitted or absorbed.	Under what circumstances does electromagnetic radiation best fit a wave model? When does it fit the photon model?

You improved your understanding of the nature of electromagnetic radiation.	What did you learn by solving this Active Example?

Practice Exercise 11.2

How does a continuous spectrum differ from a line spectrum? What is the source of each?

11.3 The Bohr Model of the Hydrogen Atom

Goal 3 Describe the Bohr model of the hydrogen atom.

4 Explain the meaning of quantized energy levels in an atom and show how these levels relate to the discrete lines in the spectrum of that atom.

5 Distinguish between ground state and excited state.

In 1913, Niels Bohr, a Danish scientist (**Figure 11.11**), suggested that an atom consists of an extremely dense nucleus that contains all of the atom's positive charge and nearly all of its mass. Negatively charged electrons of very small mass travel in orbits around the nucleus ◀ The orbits are huge compared to the nucleus, which means that most of the atom is empty space. Bohr's model of the atom was based on the results of many scientific investigations that were known at the time, including Rutherford's nuclear atom, Planck's relationship postulating that a whole-number multiplier was needed when describing electron energies, and Einstein's particles of light.

▶ **P/Review** Two objects having opposite charges, one positive and one negative, attract each other. Two objects with the same charge repel each other. See Section 2.8.

Science History Images/Alamy Stock Photo

Figure 11.11 Niels Bohr (1885–1962). Bohr won the 1922 Nobel Prize in Physics for his work in investigating the structure of the atom.

Bohr reasoned that, for a hydrogen atom, both the energy of its electron and the radius of its orbit are **quantized**. An amount that is quantized is limited to specific values; it may never be between two of those values. By contrast, an amount is **continuous** if it can have any value; between any two values there is an infinite number of other acceptable values (**Figure 11.12**). A line spectrum is quantized, but the spectrum of white light is continuous. Bohr hypothesized that the electron in the hydrogen atom has **quantized energy levels**. This means that at any instant, the electron may have one of several possible energies, but at no time may it have an energy that is between any two possible energies.

Bohr calculated the values of the quantized energy levels by using an equation that contains an integer, n, that is, 1, 2, 3, ..., and so forth. The results for the integers 1 to 4 are shown in **Figure 11.13**.

As previously noted, the radius of the electron orbit is also quantized in the Bohr model of the hydrogen atom. This leads to one of the most interesting and

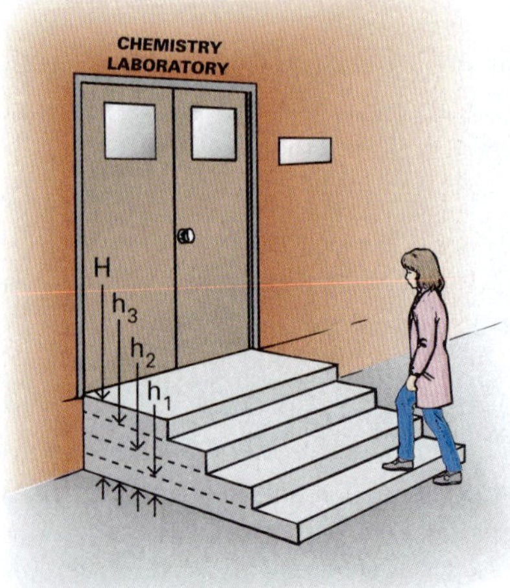

Figure 11.12 The quantum concept. A woman on a ramp can stop at any level above ground. Her elevation is a continuous quantity. A woman on the stairs can stop only on a step. Her elevation is quantized at h_1, h_2, h_3, or H.

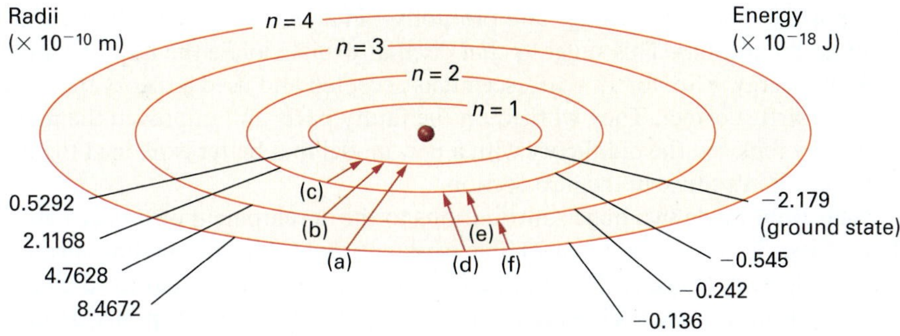

Figure 11.13 The Bohr model of the hydrogen atom. The electron is allowed to circle the nucleus only at certain radii and with certain energies, the first four of which are shown. An electron in the ground state level, $n = 1$, can absorb the exact amount of energy to raise it to any other level, such as $n = 2$, 3, or 4. An electron at such an excited state is unstable and drops back to the $n = 1$ level in one or more steps. Electromagnetic energy is radiated with each step dropped. Jumps *a*, *b*, and *c* are in the ultraviolet portion of the spectrum, *d* and *e* are in the visible range, and *f* is in the infrared region.

strangest results from Bohr's model: The electron can orbit the nucleus at certain specified distances, but *it is never found between them.* This means that the electron simply disappears from one orbit and reappears in another! The process by which an electron moves between orbits is called a **quantum jump** or a **quantum leap**.

The electron is normally found in its **ground state**, the condition when the electron is in $n = 1$ (for a hydrogen atom) or when all electrons in an atom occupy the lowest possible energy levels (for atoms with multiple electrons). If an atom absorbs energy, an electron can be raised to an **excited state**, the condition at which one or more electrons in an atom has an energy level above ground state.

An electron in an excited state is unstable. It falls back to the ground state, sometimes in one quantum jump, sometimes in two or more. In doing this, it releases a photon of light energy with a frequency proportional to the energy difference between the two levels (**Figure 11.14**). This energy release appears as a line in the spectrum of the element.

Using different values of n in his equation, Bohr was able to calculate the energies of all known lines in the spectrum of the hydrogen atom. He also predicted additional lines and their energies. When the lines were found, his predictions were proved to be correct. In fact, all of Bohr's calculations correspond with measured values to within one part per thousand. It certainly seemed that Bohr had discovered the structure of the atom.

There were problems, however. First, hydrogen is the *only* atom that fits the Bohr model. The model fails for any atom with more than one electron. Second, it is a fact that a charged body moving in a circle radiates energy. This means the

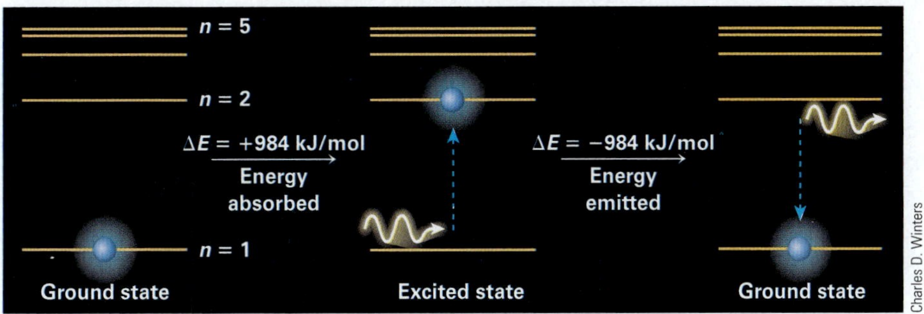

Figure 11.14 Ground and excited states. When an electron in the ground state absorbs energy, it is promoted to an excited state. When the electron returns to the ground state, energy can be released in the form of a photon of electromagnetic radiation that corresponds to one of the lines in that element's line spectrum. ΔE is the amount of energy absorbed or emitted.

electron itself should lose energy and promptly—in about 0.00000000001 second!—crash into the nucleus. This suggests that circular orbits violate the Law of Conservation of Energy. ◀ So, for 13 years, scientists accepted and used a theory they knew was only partly correct. They worked on the faulty parts and improved them, and, finally, they replaced the old theory with a new model that better explained the data, which we will introduce in the next section.

▶ **P/Review** The Law of Conservation of Energy expresses a pattern seen in nature. The quantity of energy within an isolated system does not change (Section 2.10).

Niels Bohr made two huge contributions to the development of modern atomic theory. First, he suggested a reasonable explanation for the atomic line spectra in terms of electron energies. Second, he introduced the idea of quantized electron energy levels in the atom. These levels appear in modern theory as **principal energy levels**; they are identified by the **principal quantum number, *n*.**

Active Example 11.3 The Quantum Concept

Which of the following are quantized? Explain each answer. (a) The speed of automobiles on a highway, (b) paper money in the United States, (c) the weight of canned soup of a grocery store shelf, (d) the volume of water coming from a faucet, (e) a person's height.

Think Before You Write An amount that is quantized is limited to specific values. It may never be between those values.

Answers *Cover the left column with your cut-out shield. Reveal each answer only after you have written your own answer in the right column.*

The speed of automobiles on a highway is *not* quantized. They may be traveling at any speed from slightly above 0 to the maximum achievable speed.	We'll take this one part at a time. In each case, apply the definition of quantized and explain your logic. Is the speed of automobiles on a highway quantized? 
Paper money in the United States *is quantized.* It is only printed in \$1, \$2, \$5, \$10, \$20, \$50, and \$100 denominations (larger denominations are in circulation but are no longer printed). It is limited to specific values.	Explain why paper money in the United States is or is not quantized. 
The weight of canned soup on a grocery store shelf *is quantized.* This conclusion is restricted to a few significant figures, however. When considering canned soup weights to fractions of a milligram, there are nearly an infinite, continuous number of possible weights.	Consider the weight of canned soup on a grocery store shelf. Is it quantized? Explain. 
The volume of water coming from a faucet is *not* quantized. The volume is not limited to specific values.	Explain whether or not the volume of water coming from a faucet is quantized. 
A person's height is *not* quantized. A continuous range of values occurs even though we usually record and report human heights in increments of centimeters or feet and inches.	Is the height of a human quantized? Explain. 
You improved your understanding of the quantum concept.	What did you learn by solving this Active Example?

Practice Exercise 11.3

Which of the following are quantized in the Bohr model of the hydrogen atom? Explain each answer: (a) the number of protons, (b) the energy of the electron, (c) the radius of the orbit of the electron.

11.4 The Quantum Mechanical Model of the Atom

In 1924, Louis de Broglie, a French scientist (**Figure 11.15**), hypothesized that matter in motion has properties that were formerly thought to be only associated with waves. He also calculated that these properties are especially significant in subatomic particles. Beginning in January 1926, Erwin Schrödinger, an Austrian scientist (**Figure 11.16**), applied the principles of wave mechanics to atoms and developed the **quantum mechanical model of the atom**. This model has been tested for more than 90 years. It explains more satisfactorily than any other theory all observations to date, and no exceptions have appeared. In fact, attempts to disprove the quantum mechanical model have served only to reinforce it. Today, it is the generally accepted model of the atom.

The quantum mechanical model is both mathematical and conceptual. It keeps the quantized energy levels that Bohr introduced. In fact, it uses four *quantum numbers* to describe electron energy. These refer to (1) the principal energy level; (2) the sublevel; (3) the orbital; and (4) the number of electrons in an orbital.* The model is summarized at the end of this section. You might find it helpful to keep a finger at that summary and refer to it as details of the model are developed.

Figure 11.15 Louis de Broglie (1892–1987) earned the 1929 Nobel Prize in Physics for his discovery of the wave characteristics of electrons.

Principal Energy Levels

Goal 6 Identify the principal energy levels in an atom and state the energy trend among them.

Following the Bohr model, principal energy levels are identified by the principal quantum number, n. The first principal energy level is $n = 1$, the second is $n = 2$, and so on. Mathematically, there is no end to the number of principal energy levels, but the seventh level is the highest occupied by ground state electrons in any element now known. The energy possessed by an electron depends on the principal energy level it occupies. In general, energies increase as the principal quantum numbers increase:

$$n = 1 < n = 2 < n = 3 < \ \dots \ < n = 7$$

Figure 11.16 Erwin Schrödinger (1887–1961) shared the 1933 Nobel Prize in Physics for his contributions to atomic theory.

Sublevels

Goal 7 For each principal energy level, state the number of sublevels, identify them, and state the energy trend among them.

For each principal energy level, there are one or more **sublevels**. They are the ***s*, *p*, *d*, and *f* sublevels**, using initial letters that come from terms formerly used in spectroscopy, the study of the interaction of matter and electromagnetic radiation.** A specific sublevel is identified by both the principal energy level and sublevel. Thus, the *p* sublevel in the third principal energy level is the 3*p* sublevel. An electron that is in the 3*p* sublevel may be referred to as a "3*p* electron."

The total number of sublevels within a given principal energy level is equal to n, the principal quantum number. For $n = 1$, there is one sublevel, designated 1*s*. At $n = 2$, there are two sublevels, 2*s* and 2*p*. When $n = 3$, there are three sublevels, 3*s*, 3*p*, and 3*d*; $n = 4$ has four sublevels, 4*s*, 4*p*, 4*d*, and 4*f*. Quantum theory describes sublevels beyond *f* when $n = 5$ or more, but these are not needed for the ground state electron configurations of elements known today.

---

*The formal names of these numbers are principal, azimuthal, magnetic, and electron spin. We use the name and number of the principal quantum number but not the other three. All, however, are described to the extent necessary to specify the distribution of electrons in an atom.

**The terms are sharp, principal, diffuse, and fundamental. They describe the appearance of the spectral lines in a line spectrum.

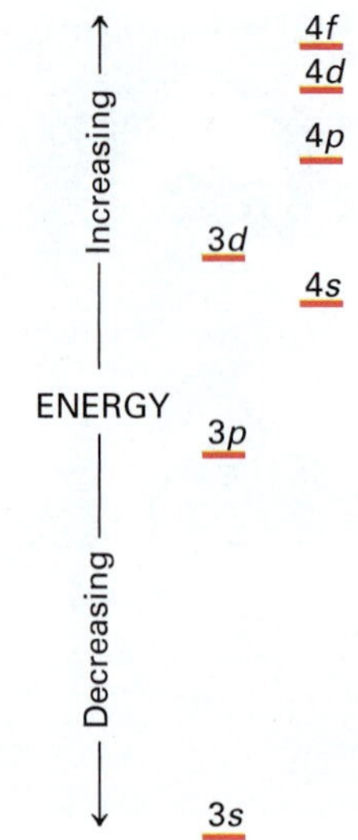

Figure 11.17 Energies of $n = 3$ and $n = 4$ sublevels.

For elements other than hydrogen, the energy of each principal energy level spreads over a range related to the sublevels. These energies increase in the order *s*, *p*, *d*, *f*. Thus,

at $n = 2$, the increasing order of energy is $2s < 2p$

at $n = 3$, the increasing order of energy is $3s < 3p < 3d$

at $n = 4$, the increasing order of energy is $4s < 4p < 4d < 4f$

Beginning with principal quantum numbers 3 and 4, the energy ranges overlap. This is shown in **Figure 11.17**. When "plotted" vertically, the highest $n = 3$ electrons (3*d*) are at higher energy than the lowest $n = 4$ electrons (4*s*). Note, however, that for the same sublevel, $n = 3$ electrons always have lower energy than $n = 4$ electrons: $3s < 4s$; $3p < 4p$; $3d < 4d$.

Electron Orbitals

Goal 8 Sketch the shapes of s and p orbitals.

9 State the number of orbitals in each sublevel.

According to the quantum mechanical model, it is not possible to know at the same time both the position of an electron in an atom and its velocity. This means it is not possible to describe the path an electron travels. There are no clearly defined orbits, as in the Bohr atom. However, we can describe mathematically a region in space around a nucleus in which there is a high probability of finding an electron. These regions are called **orbitals**. Notice the uncertainty of the *orbital*, stated in terms of "probability," compared to a Bohr *orbit* that states exactly where the electron is, where it was, and where it is going (**Figure 11.18**).

Each sublevel has a certain number of orbitals. There is only one orbital for every *s* sublevel. All *p* sublevels have three orbitals, all *d* sublevels have five, and all *f* sublevels have seven. This 1–3–5–7 sequence of odd numbers continues through higher sublevels.

Figure 11.19 shows the shapes of the *s*, *p*, and *d* orbitals. The seven *f* orbitals have even more complex shapes. The *x*, *y*, and *z* axes around which these shapes are drawn are from the mathematics of the quantum theory. We will be concerned with the shapes of only the *s* and *p* orbitals.

All *s* orbitals are spherical. As the principal quantum number increases, the size of the orbital increases. Thus, a 2*s* orbital is larger than a 1*s* orbital, 3*s* is larger than 2*s*, and so forth (**Figure 11.20**). Similar increases in size through constant shapes are present with *p*, *d*, and *f* orbitals at higher principal energy levels.

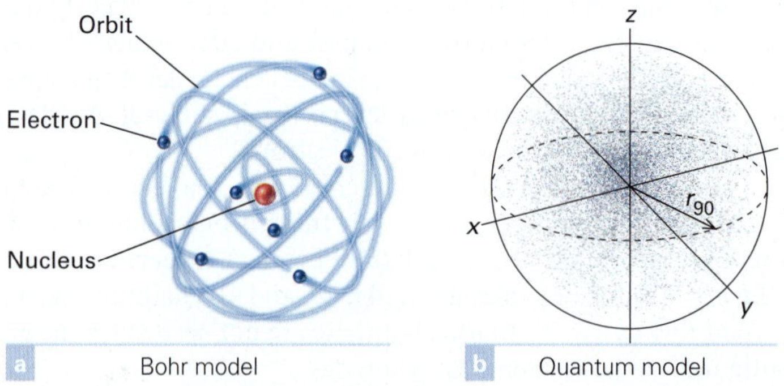

Figure 11.18 The Bohr model of the atom compared with the quantum model. (a) Bohr's hypothesis proposed that the electron was moving in circular orbits of fixed radii around the nucleus. (b) The quantum model says nothing about the precise location of the electron or the path in which it moves. Instead, each dot represents a possible location for the electron. The higher the density of dots, the higher the probability that an electron is in that region. The r_{90} dimension is a radius; 90% of the time the electron is inside the dashed line, and 10% of the time it is farther from the nucleus.

Figure 11.19 Shapes of electron orbitals according to the quantum mechanical model of the atom.

Thinking About Your Thinking

Probability

The quantum mechanical model is probabilistic in nature, as opposed to the Bohr model, which is deterministic. A probabilistic phenomenon is one for which we cannot predict the future with certainty, even if we know all there is to know about the system at the present moment. A deterministic phenomenon is one for which we can determine the future state of the system if we know all of the system's present conditions. For example, in the Bohr model, if we know the speed of an electron and its orbit path, we can predict exactly where the electron will be at some specified future time. This is a deterministic phenomenon. The quantum mechanical model is an example of a probabilistic phenomenon.

Even if we know everything that can be simultaneously measured about an electron at a given moment in time, we cannot predict things such as the future position or speed of the electron. We can only predict *probabilities* about the future. Thus, quantum mechanics gives us *orbitals*, regions in space where an electron is likely to be found at a certain probability level, rather than the deterministic orbits of the Bohr model.

Many natural phenomena are probabilistic in nature. This is one concept that distinguishes modern science from the thinking accepted at the beginning of the 20th century, when scientists believed that, given knowledge of all present conditions, the future conditions could always be predicted. Alas, nature is not that simple. Look for more examples of probabilistic relationships and phenomena in your future studies in science.

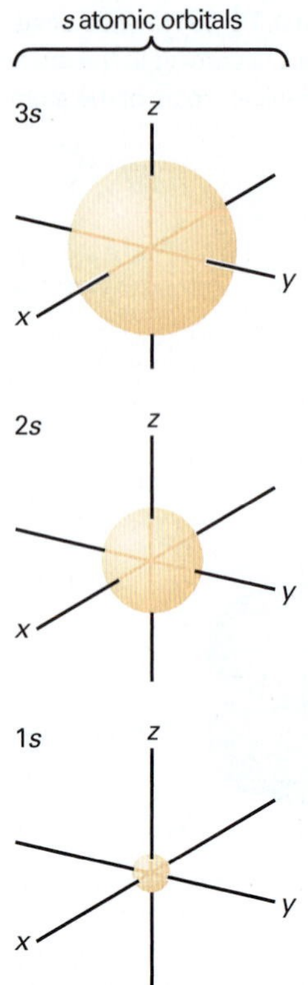

Figure 11.20 Relative sizes of the 1s, 2s, and 3s orbitals according to the quantum mechanical model of the atom.

Figure 11.21 Wolfgang Pauli (1900–1958) discovered the exclusion principle and was awarded the 1945 Nobel Prize in Physics for his work.

The Pauli Exclusion Principle

Goal 10 State the restrictions on the electron population of an orbital.

The last detail of the quantum mechanical model of the atom comes from the **Pauli exclusion principle**, named after the Austrian scientist Wolfgang Pauli (**Figure 11.21**). Its effect is to limit the population of any orbital to two electrons. At any instant, an orbital may be (1) unoccupied, (2) occupied by one electron, or (3) occupied by two electrons. No other occupancy is possible.

Summary of the Quantum Mechanical Model of the Atom

The quantum mechanical model of the atom is summarized in the *a summary of...* box that follows. Refer back to it as you study the next section, in which we will tie together the quantum model and the periodic table. You will find that this connection will help you better understand the quantum model. **Figure 11.22** presents an analogy that will also help you learn the quantum model.

a summary of... The Quantum Mechanical Model of the Atom

Principal Energy Levels

Principal energy levels are identified by the principal quantum number, *n*, in a series of counting numbers: $n = 1, 2, 3, \ldots, 7$. Generally, energy increases with increasing n: $n = 1 < n = 2 < n = 3, \ldots$.

Sublevels

Each principal energy level—each value of *n*—has *n* sublevels. These sublevels are identified by the principal quantum number followed by the letter *s*, *p*, *d*, or *f*. Sublevels that are not needed for the ground state electron configurations of elements known today appear in blue.

Energy Trend	*n*	Number of Sublevels	Identification of Sublevels
Increasing Energy ↓	1	1	1*s*
	2	2	2*s*, 2*p*
	3	3	3*s*, 3*p*, 3*d*
	4	4	4*s*, 4*p*, 4*d*, 4*f*
	5	5	5*s*, 5*p*, 5*d*, 5*f*, 5*g*
	6	6	6*s*, 6*p*, 6*d*, 6*f*, 6*g*, 6*h*
	7	7	7*s*, 7*p*, 7*d*, 7*f*, 7*g*, 7*h*, 7*i*

For any given value of *n*, energy increases through the sublevels in the order of *s*, *p*, *d*, *f*: $2s < 2p$; $3s < 3p < 3d$; $4s < 4p < 4d < 4f$; and so on.

The range of energies in consecutive principal energy levels may overlap. Example: $4s < 3d < 4p$. However, for any given sublevel, energy and orbital size increase with increasing *n*: $1s < 2s < 3s \ldots$, $2p < 3p < 4p \ldots$, and so on.

Orbitals and Orbital Occupancy

Each kind of sublevel contains a definite number of orbitals that begin with 1 and increase in order with odd numbers: *s*, 1; *p*, 3; *d*, 5; *f*, 7. An orbital may be unoccupied or occupied by 1 or 2 electrons, but never more than 2. Therefore, the maximum number of electrons in a sublevel is twice the number of orbitals in the sublevel.

Sublevel	Orbitals	Maximum Electrons per Sublevel
s	1	$1 \times 2 = 2$
p	3	$3 \times 2 = 6$
d	5	$5 \times 2 = 10$
f	7	$7 \times 2 = 14$

Thinking About Your Thinking

Formal Models

The quantum mechanical model of the atom is an example of a formal model, one in which the concept being modeled has abstract parts that have to be imagined. Figures 11.18 and 11.19 are the keys to building a model in your mind that you can use to think about the quantum mechanical model.

Figure 11.18 is designed to help you think about how electrons behave in atoms. The Bohr model is an analogy to the solar system. Electrons orbit the nucleus like planets orbit the sun. This model is incorrect. The quantum model, on the right of Figure 11.18, is the mental model you should form. Each blue dot represents the position of a single electron at one point in time. Are you wondering why the dots appear as a scattering with areas that are dense and areas that are diffuse? Imagine that you can take a photograph of the single electron in a hydrogen atom. You center your camera so that the nucleus is at the convergence of the *x*, *y*, and *z* axes in the figure. When you take a picture, you make an overhead transparency of the single blue dot. Now you overlap a few hundred such transparencies. This will result in the quantum model illustration you see. The dashed line that is labeled as the r_{90} distance is the radius that encloses 90% of the dots. In other words, 90% of the time, you will find the electron within this radius.

Now look at the *s* orbital in Figure 11.19. It represents the same thing as the quantum model illustration in Figure 11.18. The radius of the sphere in Figure 11.19 corresponds to the r_{90} radius of the circle in Figure 11.18. It illustrates in three dimensions the limit of a region in space, inside of which there is a 90% probability of finding an electron at any given instant. The *p* and *d* orbitals of Figure 11.19 also represent the regions in which the *p* and *d* electrons may be found 90% of the time. Your mental models of the *s* and *p* electron orbitals, plus your model of the behavior of an electron in an atom, combine to make the quantum mechanical model you should hold in your mind at this point in your chemistry studies.

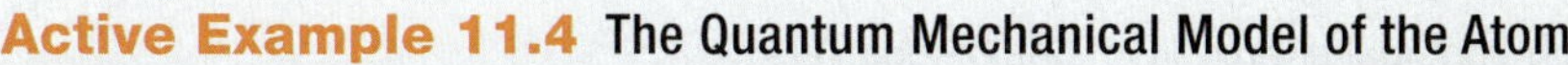

Active Example 11.4 The Quantum Mechanical Model of the Atom

Identify the true statements and explain, in general terms, why they are true. Rewrite the false statements to make them true and explain your logic in general terms.

a) There is one *s* orbital when $n = 1$, two *s* orbitals when $n = 2$, three *s* orbitals when $n = 3$, and so on.

b) All $n = 3$ orbitals are at lower energy than all $n = 4$ orbitals.

c) There is no *d* sublevel when $n = 2$.

d) There are five *d* orbitals at both the fourth and sixth principal energy levels.

Think Before You Write Reread the *a summary of...* the quantum mechanical model box one more time before working this Active Example. It usually takes at least a couple of read throughs and attempts to make mental models based on the figures before you can begin to conceptualize the quantum model. Given that it is so different from our everyday experiences in the macroscopic world, you will likely have to invest additional time relative to your standard study time investment to meet the chapter goals to this point.

Answers *Cover the left column with your cut-out shield. Reveal each answer only after you have written your own answer in the right column.*

No, there is one *s* orbital at each principal energy level—that is, for each value of *n*. The 1*s* orbital is at $n = 1$, the 2*s* orbital is at $n = 2$, the 3*s* orbital is at $n = 3$, and so on. See the *Identification of Sublevels* column in the *Sublevels* section of *a summary of...* box.	Is there one *s* orbital when $n = 1$, two *s* orbitals when $n = 2$, three *s* orbitals when $n = 3$, and so on? Explain your reasoning. 

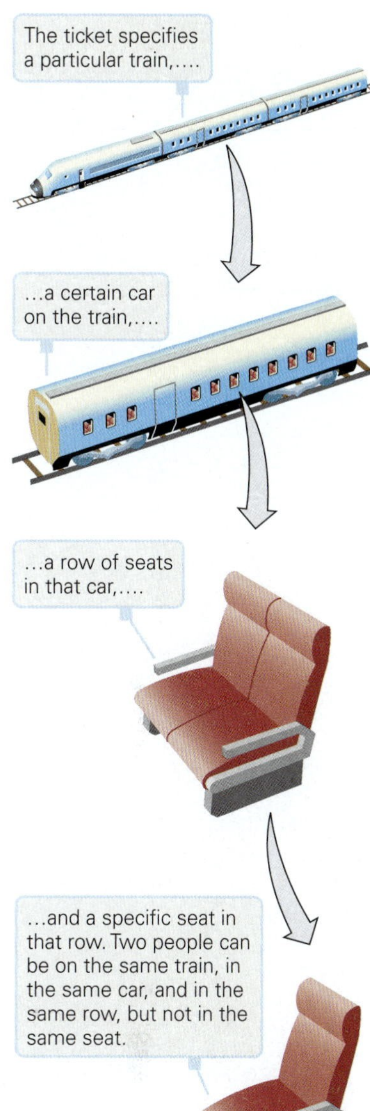

The ticket specifies a particular train,....
...a certain car on the train,....
...a row of seats in that car,....
...and a specific seat in that row. Two people can be on the same train, in the same car, and in the same row, but not in the same seat.

Figure 11.22 An analogy between the quantum mechanical model of the atom and a seat on a train. Principal energy levels are like the many trains that run on a track; specifying a particular train is similar to specifying a group of electrons occupying a principal energy level. Sublevels have a likeness to the cars on the train; when you specify a car, you now describe only the electrons in that sublevel. Electron orbitals match up with a row of seats; the two-seat row is like the two electrons that can occupy an orbital. A particular electron is completely described when you describe a specific seat; no other electron in an atom has the same four quantum characteristics.

No, beginning with $n = 3$ and $n = 4$, the energy ranges of the sublevels begin to overlap. Figure 11.17 shows that the energy of the 4*s* sublevel is less than that of the 3*d* sublevel.	Are all $n = 3$ orbitals at lower energy than all $n = 4$ orbitals? Explain your response. 
This statement is true. *d* orbitals do not begin to appear until $n = 3$. See the *Identification of Sublevels* column in the *Sublevels* section of *a summary of...* box.	Is there no *d* sublevel when $n = 2$? Explain. 
There are five *d* orbitals at each sublevel that has *d* orbitals. *d* orbitals exist at $n = 3$ and higher, so there are five *d* orbitals at $n = 4$ and $n = 6$.	Are there five *d* orbitals at both the fourth and sixth principal energy levels? Explain. 
You improved your understanding of the quantum mechanical model of the atom.	What did you learn by solving this Active Example? 

Practice Exercise 11.4

Identify the true statements and explain, in general terms, why they are true. Rewrite the false statements to make them true and explain your logic in general terms.

a) A 1*s* orbital is larger than a 2*s* orbital, which is larger than a 3*s* orbital, and so on.

b) *s* orbitals cannot have different orientations in space, and *p* orbitals can have different orientations in space.

c) Each principal energy level has *n* sublevels.

d) An orbital may be either unoccupied or occupied by an electron.

11.5 Electron Configuration

Electron Configurations and the Periodic Table

Goal 11 Use a periodic table to list electron sublevels in order of increasing energy.

Many of the chemical properties of an atom or ion depend on its **electron configuration**, the ground state distribution of electrons among the orbitals of the species. Two rules guide the assignments of electrons to orbitals:

1. At ground state the electrons fill the *lowest* energy orbitals available.
2. No orbital can have more than two electrons.

Although the first periodic tables were based on properties of the elements, quantum mechanics shows that its true basis lies in the arrangement of electrons as predicted by the quantum model. As the number of electrons in an atom increases, specific sublevels are filled that correspond to different regions of the periodic table. This is indicated by color in **Figure 11.27**. In Groups 1A/1 and 2A/2, the *s* sublevels are the highest *occupied* energy sublevels. ◀ Other sublevels have higher energy, but they have no electrons in them, and thus they are *not occupied.* The *p* orbitals are filled in order across Groups 3A/13 to 8A/18. The *d* electrons appear in the B groups (3 to 12). Finally, *f* electrons show up in the lanthanide and actinide series.

▶ **P/Review** We are showing both the traditional U.S. A–B numbering scheme for groups in the periodic table and the 1-to-18 system that has been approved by the International Union of Pure and Applied Chemistry (IUPAC) (Section 5.7). Use—and read—the one selected by your instructor.

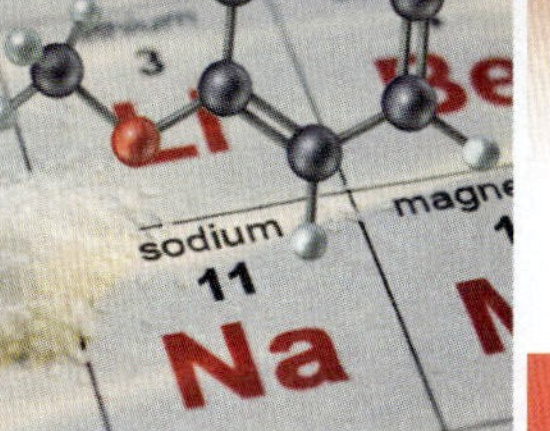

Everyday Chemistry 11.1

SIMPLY PURE DARN FOOLISHNESS?

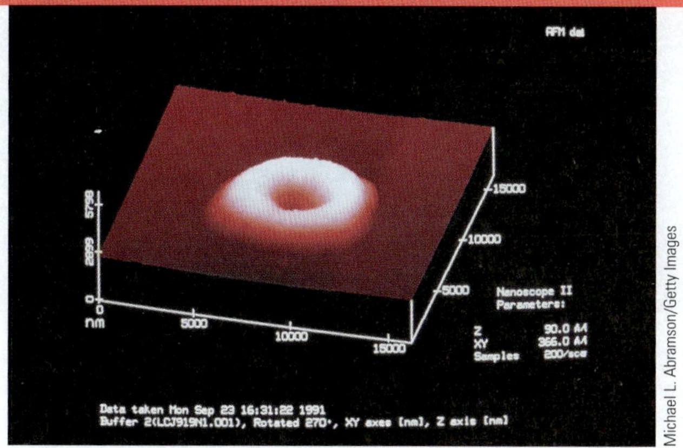

Michael L. Abramson/Getty Images

Figure 11.24 A single red blood cell as imaged by an atomic force microscope. The atomic force microscope (AFM), a modification of the scanning tunneling microscope, allows us to see groups of atoms. This is a single red blood cell. The AFM can also slice the cell to reveal individual protein molecules inside.

The rules of quantum mechanics have no direct correlation in physical systems much larger than atoms. It is easy to believe that this chapter has nothing to do with "real life" and is just something else to memorize today and forget later. Some students would argue that the *s*, *p*, *d*, and *f* classifications for electron orbitals stand for "**s**imply ***p***ure ***d***arn ***f***oolishness."

Nothing could be further from the truth.

Because it's so different, quantum mechanics has been tested by many scientists since Erwin Schrödinger proposed it in January 1926. It has passed all the tests. Indeed, Albert Einstein spent portions of the last 30 years of his professional life futilely searching for an alternative explanation of atomic structure.

The most rigorous test a scientific model can pass is to predict the results of an experiment and then have those results confirmed in the laboratory. The first major prediction of quantum mechanics that came true was the transistor, invented in 1947 at Bell Laboratories. The small size and low power consumption of the transistor make possible the complicated electrical circuits in the personal computer microprocessor (**Figure 11.23**). This textbook was typed on a desktop computer, a tool that has immensely changed writers' lives for the better since the 1980s.

Today, microprocessors are everywhere. Of course, they are in your computer and calculator, but microprocessors are also in many simple things we take for granted. Portable audio and video players, appliances with digital displays, automobiles, and smart phones all are controlled by microprocessors. One of the most common uses for microprocessors in the United States today is for controlling traffic lights.

Look back at Figure 11.19. See the *p* orbitals with the two lobes. You know that those orbitals are high-probability regions, but you may not have known that the lobes of a *p* orbital do not touch each other. Between those two lobes is a node, a region in which there is zero probability of finding an electron. Yet we know from experiment that a single 2*p* electron is in *both* lobes at the same time! Quantum mechanics holds that an electron can tunnel from one lobe to the other without going through the node that is between them. This is unbelievable on the macroscopic level but true at the level of subatomic particles.

In 1981, Gerd Binnig and Heinrich Rohrer, two researchers at IBM in Zurich, Switzerland, invented the scanning tunneling microscope (STM) (**Figure 11.24**). They shared the 1986 Nobel Prize in Physics (with Ernst Ruska, inventor of the electron microscope in 1933) for this invention, which is rapidly changing the way chemists think about atoms. The STM probes surfaces with a tungsten needle that is at most a few atoms wide. At very short distances, electrons tunnel from the needle across the gap between the needle and the surface under study. This generates a minute current that can be converted into images of individual atoms.

After mapping individual atoms, scientists using STMs learned they could apply a voltage pulse at the needle to pick up atoms one at a time and move them! Their first effort was an IBM logo (**Figure 11.25**). The dots are individual atoms of xenon that were moved approximately 4×10^{-7} millimeter per second. The process took about 22 hours. Carbon monoxide man was drawn by arranging 28 CO molecules in the stick figure shown (**Figure 11.26**).

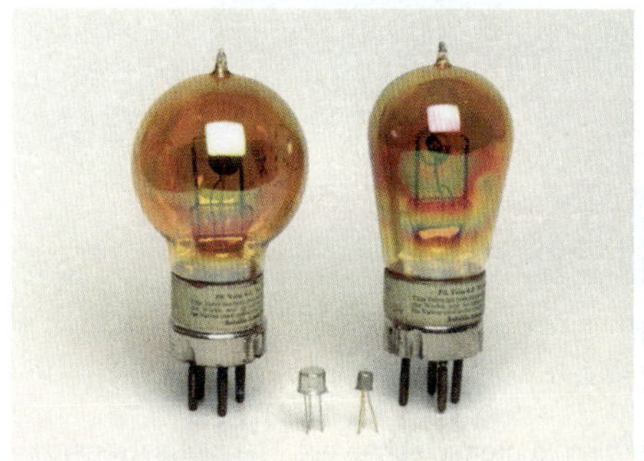

sciencephotos/Alamy Stock Photo

Joe White/Shutterstock.com

Figure 11.23 Vacuum tubes and transistors. (a) The transistor (*center*) replaced the vacuum tube (*left* and *right*) on approximately a 1-to-1 basis, making battery-powered portable radios possible in the 1950s. (b) A modern integrated circuit (*far right*) contains millions of transistors.

Figure 11.25 Arrangement of xenon atoms to spell IBM. The logo is composed of 35 xenon (Z = 54) atoms, *individually* moved using a scanning tunneling microscope. The space between the atoms is about 1.3×10^{-7} cm. As printed, the magnification is 4.7 million.

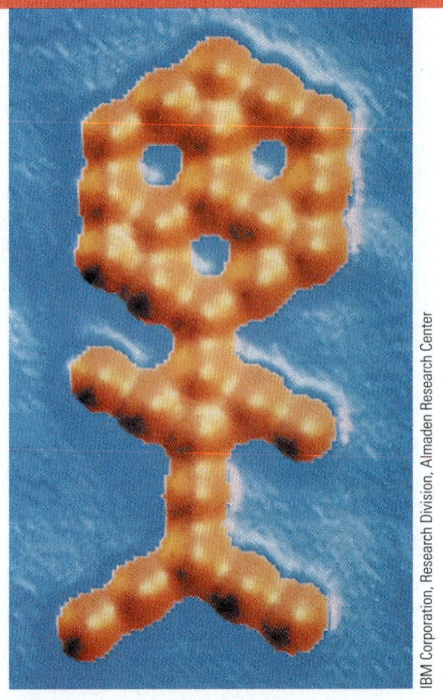

Figure 11.26 Carbon monoxide man. This image is made of 28 carbon monoxide molecules on a platinum surface. The carbon and oxygen atoms are stacked vertically. The carbon atoms are attached to the platinum surface. The stick figure is all of 5×10^{-7} cm tall.

The implications of moving individual atoms are stunning. For example, as scientists mapping out the full set of human chromosomes learn which human diseases are caused by molecular errors in a chromosome, tools such as the STM could theoretically be used to move individual atoms to fix the error and cure the disease.

Hang on to your hats! Your study of chemistry places you in the middle of the most exciting new developments in medicine, science, and technology. You ain't seen nothin' yet!

Quick Quiz

1. **What was the point of titling this essay as "Simply Pure Darn Foolishness?"**
2. **Does a scanning tunneling microscope allow users to see atoms? Explain.**

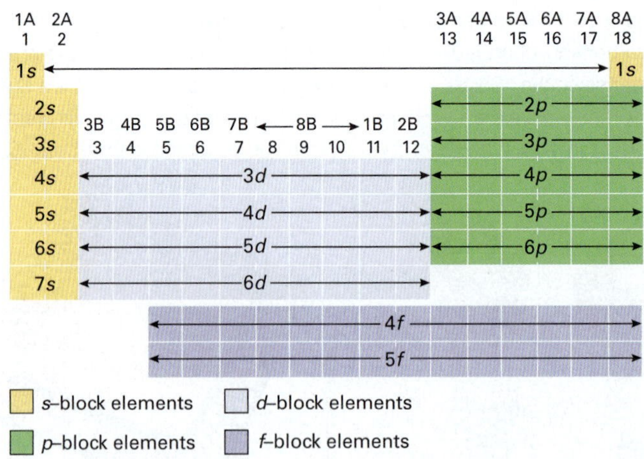

Figure 11.27 Arrangement of periodic table according to atomic sublevels. The highest-energy sublevels occupied at ground state are *s* sublevels in Groups 1A/1 and 2A/2. This region of the periodic table is the *s*-block. Similarly, *p* sublevels are the highest occupied sublevels in Groups 3A/13 to 8A/18, the *p*-block. The *d*-block includes the B Groups (3 to 12), whose highest occupied energy sublevels are *d* sublevels. Finally, the *f*-block is made up of the elements whose *f* sublevels hold the highest-energy electrons.

When you read the periodic table from left to right across the periods in Figure 11.27, you get the order of increasing sublevel energy. The first period gives only the 1*s* sublevel. The 1*s* orbital is the lowest-energy orbital in any atom. Period 2 takes you through 2*s* and 2*p*. Similarly, the third period covers 3*s* and 3*p*. Period 4 starts with 4*s*, follows with 3*d*, and ends up with 4*p*, and so forth. Ignoring minor variations in Periods 6 and 7, the complete list of sublevel energies is given in **Figure 11.28**, in order of increasing sublevel energy.

The periodic table, therefore, is a guide to the order of increasing sublevel energy. Think, for a moment, how remarkable this is. Mendeleev and Meyer developed their periodic tables from the physical and chemical properties of the elements. They knew nothing of electrons, protons, nuclei, wave functions, or quantized energy levels. Yet, when these things were found some 60 years later, the match between the first periodic tables and the quantum mechanical model of the atom was nearly perfect.

If you are ever required to list the sublevels in order of increasing energy without reference to a periodic table, the following diagram taken from the summary of the quantum model may be helpful. Beginning at the upper left, the diagonal lines pass through the sublevels in the sequence required.

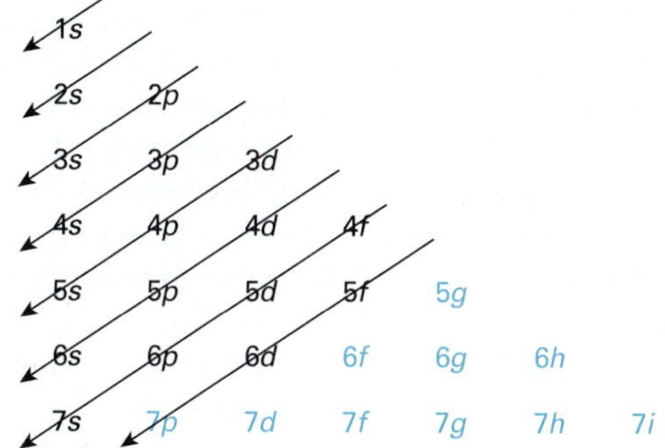

The sublevels shown in color are not needed for the elements known today, but the mathematics of the quantum mechanical model predict the order of increasing energy indefinitely.

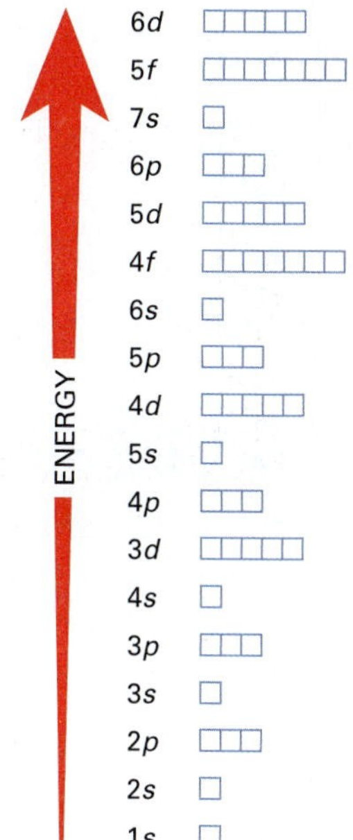

Figure 11.28 Sublevel energy diagram. All sublevels are positioned vertically, as represented by the column of sublevels in the middle and the red arrow to the left showing increasing energy. Each box represents an orbital. When orbitals are filled from the lowest energy level, each orbital will generally hold two electrons before any orbital higher in energy accepts an electron.

◀ **P/Review** Z is the atomic number of an element. It stands for the number of protons in the nucleus of each atom of the element. See Section 5.5.

Let's look at examples of ground state electron configurations of atoms.

The one electron of a hydrogen atom (H, Z = 1) occupies the lowest-energy orbital in any atom. ▶ Both Figures 11.27 and 11.28 show that this is the 1*s* orbital. The total number of electrons in any sublevel is shown by a superscript number. Therefore, the electron configuration of hydrogen is $1s^1$ (**Figure 11.29**). Helium (He, Z = 2) has two electrons, and both fit into the 1*s* orbital. The helium configuration is $1s^2$.

These and other electron configurations to be developed appear in the first four periods of the periodic table in **Figure 11.30**.

Lithium (Li, Z = 3) has three electrons. The first two fill the 1*s* orbital, as before. The third electron goes to the next orbital up the energy scale that has a vacancy. According to the order of increasing energy derived from reading the periodic table, as in Figure 11.27, this is the 2*s* orbital. The electron configuration for lithium is, therefore, $1s^22s^1$. Similarly, beryllium (Be, Z = 4) divides its four electrons between the two lowest-energy orbitals, filling both: $1s^22s^2$. These configurations are also in Figure 11.30.

The first four electrons of boron (B, Z = 5) fill 1*s* and 2*s* orbitals. The fifth electron goes to the next highest level, 2*p*, according to the periodic table. The configuration for boron is $1s^22s^22p^1$. Similarly, carbon (C, Z = 6) has a $1s^22s^22p^2$ configuration.* The next four elements increase the number of electrons in the three

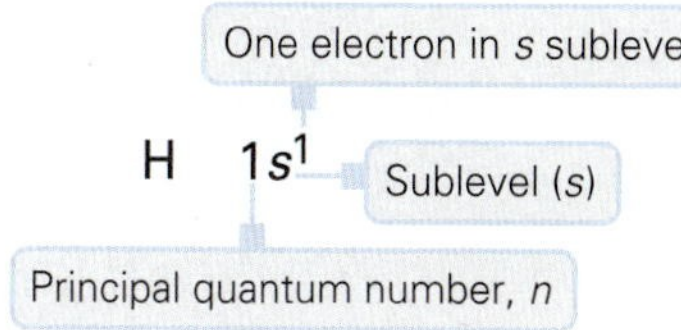

Figure 11.29 Electron configuration notation. An electron configuration specifies the number of electrons in each sublevel for any specified atom. The value of *n* that represents the principal energy level is written first, followed by the letter that indicates the sublevel. The number of electrons in each sublevel is written as a superscript.

*Although we will not emphasize the point, the two 2*p* electrons occupy different 2*p* orbitals. In general, all orbitals in a sublevel are half-filled before any orbital is completely filled.

1 **H** $1s^1$																	2 **He** $1s^2$
3 **Li** $1s^2$ $2s^1$	4 **Be** $1s^2$ $2s^2$											5 **B** $1s^2$ $2s^2 2p^1$	6 **C** $1s^2$ $2s^2 2p^2$	7 **N** $1s^2$ $2s^2 2p^3$	8 **O** $1s^2$ $2s^2 2p^4$	9 **F** $1s^2$ $2s^2 2p^5$	10 **Ne** $1s^2$ $2s^2 2p^6$
11 **Na** [Ne] $3s^1$	12 **Mg** [Ne] $3s^2$											13 **Al** [Ne] $3s^2 3p^1$	14 **Si** [Ne] $3s^2 3p^2$	15 **P** [Ne] $3s^2 3p^3$	16 **S** [Ne] $3s^2 3p^4$	17 **Cl** [Ne] $3s^2 3p^5$	18 **Ar** [Ne] $3s^2 3p^6$
19 **K** [Ar] $4s^1$	20 **Ca** [Ar] $4s^2$	21 **Sc** [Ar] $4s^2 3d^1$	22 **Ti** [Ar] $4s^2 3d^2$	23 **V** [Ar] $4s^2 3d^3$	24 **Cr** [Ar] $4s^1 3d^5$	25 **Mn** [Ar] $4s^2 3d^5$	26 **Fe** [Ar] $4s^2 3d^6$	27 **Co** [Ar] $4s^2 3d^7$	28 **Ni** [Ar] $4s^2 3d^8$	29 **Cu** [Ar] $4s^1 3d^{10}$	30 **Zn** [Ar] $4s^2 3d^{10}$	31 **Ga** [Ar] $4s^2$ $3d^{10} 4p^1$	32 **Ge** [Ar] $4s^2$ $3d^{10} 4p^2$	33 **As** [Ar] $4s^2$ $3d^{10} 4p^3$	34 **Se** [Ar] $4s^2$ $3d^{10} 4p^4$	35 **Br** [Ar] $4s^2$ $3d^{10} 4p^5$	36 **Kr** [Ar] $4s^2$ $3d^{10} 4p^6$

Figure 11.30 Ground state electron configurations of neutral atoms of the first 36 elements.

$2p$ orbitals until they are filled with six electrons for neon (Ne, Z = 10). All of these configurations appear in Figure 11.30.

The first 10 electrons of sodium (Na, Z = 11) are distributed in the same way as the ten electrons in neon. The 11th sodium electron is a $3s$ electron: $1s^2 2s^2 2p^6 3s^1$. The configurations for all elements with atomic numbers greater than 10 begin with the neon configuration, $1s^2 2s^2 2p^6$. This part of the configuration is often shortened to the **neon core**, represented by [Ne]. For sodium, this becomes [Ne]$3s^1$; for magnesium (Mg, Z = 12), [Ne]$3s^2$; for aluminum (Al, Z = 13), [Ne]$3s^2 3p^1$; and so on to argon (Ar, Z = 18), [Ne]$3s^2 3p^6$. The sequence follows the same pattern as in Period 2. The neon core is used for Period 3 in Figure 11.30.

Potassium (K, Z = 19) repeats at the $4s$ level the development of sodium at the $3s$ level. Its complete configuration is $1s^2 2s^2 2p^6 3s^2 3p^6 4s^1$. All configurations for atomic numbers greater than 18 distribute their first 18 electrons in the configuration of argon, $1s^2 2s^2 2p^6 3s^2 3p^6$. This may be shortened to the **argon core**, [Ar]. Accordingly, the configuration for potassium may be written [Ar]$4s^1$ and calcium (Ca, Z = 20) is [Ar]$4s^2$.

The periodic table tells us that five $3d$ orbitals are next available for electron occupancy. The next three elements fill in order, as predicted, to vanadium (V, Z = 23): [Ar]$4s^2 3d^3$.* Chromium (Cr, Z = 24) is the first element to break the orderly sequence in which the lowest-energy orbitals are filled. Its configuration is [Ar]$4s^1 3d^5$, rather than the expected [Ar]$4s^2 3d^4$. This is attributed to an extra stability found when all orbitals in a sublevel are half-filled or completely filled. Manganese (Mn, Z = 25) puts us back on the track, only to be derailed again at copper (Cu, Z = 29): [Ar]$4s^1 3d^{10}$. Zinc (Zn, Z = 30) has the expected configuration: [Ar]$4s^2 3d^{10}$. Examine the sequence for atomic numbers 21 to 30 in Figure 11.30 and note the two exceptions, chromium and copper.

By now the pattern should be clear. Atomic numbers 31 to 36 fill in sequence in the next orbitals available, which are the $4p$ orbitals. This is shown in Figure 11.30.

Our consideration of electron configuration ends with atomic number 36, krypton. If we were to continue, we would find the higher s and p orbitals fill just as they do in Periods 2 to 4. The $4d$, $4f$, $5d$, and $5f$ orbitals have several variations like those for chromium and copper, so their configurations must be looked up. But you should be able to reproduce the configurations for the first 36 elements—*not from memory or from Figure 11.30, but by referring to a periodic table.*

*Some chemists prefer to write this configuration [Ar]$3d^3 4s^2$, putting the $3d$ before the $4s$. This is equally acceptable. There is, perhaps, some advantage at this time in listing the sublevels in the order in which they fill, which can be "read" from the periodic table, as you will see shortly. This same idea continues as the f sublevels are filled but with frequent irregularities. We will not be concerned with f-block elements.

Writing Electron Configurations

Goal 12 Referring only to a periodic table, write the ground state electron configuration of an atom of any element up to atomic number 36.

To write the electron configuration of an atom with help from the periodic table, you must be able to:

1. list the sublevels in order of increasing energy (Goal 11);
2. state the maximum number of electrons that occupy each sublevel; and
3. establish the number of electrons in the highest occupied energy sublevel of the atom.

The number of electrons in the highest occupied energy sublevel of an atom is related to the position of the element in the periodic table. For atoms of all elements in Group 1A/1, that number is 1. This can be written as ns^1, where n is the highest occupied principal energy level. For hydrogen, n is 1; for lithium, n is 2; for sodium, 3; and so forth. In Group 2A/2, the highest occupied sublevel is ns^2. In Groups 3A/13 to 8A/18, the number of p electrons is found by counting from the left from 1 to 6. For the main group elements: ▶

◀ **P/Review** Main group elements are those in the A groups of the periodic table, or Groups 1, 2, and 13 to 18 by the IUPAC system. The B groups (3 to 12) are transition elements (Section 5.7).

Group (U.S.):	1A	2A		3A	4A	5A	6A	7A	8A
Group (IUPAC):	1	2		13	14	15	16	17	18
***s* electrons:**	1	2	***p* electrons:**	1	2	3	4	5	6
Electron configuration:	ns^1	ns^2		np^1	np^2	np^3	np^4	np^5	np^6

A similar count-from-the-left order appears among the transition elements, in which the d sublevels are filled. There are interruptions, however. Among the $3d$ electrons, the interruptions appear at chromium (Z = 24) and copper (Z = 29).

Group (U.S.):	3B	4B	5B	6B	7B	←	8B	→	1B	2B
Group (IUPAC):	3	4	5	6	7	8	9	10	11	12
***d* electrons:**	1	2	3	5	5	6	7	8	10	10
Electron configuration:	$3d^1$	$3d^2$	$3d^3$	$3d^5$	$3d^5$	$3d^6$	$3d^7$	$3d^8$	$3d^{10}$	$3d^{10}$

Fix these electron populations and their positions in the periodic table firmly in your thoughts now. Then cover both of these summaries and refer only to a full periodic table as you try the following Active Example.

Active Example 11.5 Writing Electron Configurations I

Write the electron configuration of the highest occupied energy sublevel for each of the following elements: beryllium, phosphorus, manganese.

Think Before You Write The key to solving this problem is to have the location of the *s*-, *p*-, *d*-, and *f*-block elements in the periodic table set in your mind, as illustrated in Figure 11.27.

Answers *Cover the left column with your cut-out shield. Reveal each answer only after you have written your own answer in the right column.*

Beryllium, Be, $2s^2$ Phosphorus, P, $3p^3$ Manganese, Mn, $3d^5$ *Counting from the left, beryllium is in the second box (Group 2A/2) among the 2s sublevel elements, so its configuration is $2s^2$. Phosphorus is in the third box (Group 5A/15) among the 3p sublevel elements, so its configuration is $3p^3$. Manganese is in the fifth box (Group 7B/7) among the 3d sublevel elements, so its configuration is $3d^5$.*	As an example, if the question asked for the electron configuration of the highest occupied energy sublevel for boron, you would write $2p^1$. Complete the example.
You improved your skill at writing electron configurations.	What did you learn by solving this Active Example?

Practice Exercise 11.5

Write the electron configuration of the highest occupied energy sublevel for each of the following elements: bromine, sodium, oxygen.

You are now ready to write electron configurations. The procedure follows.

how to... **Write Electron Configurations**

Sample Problem: *Write the complete electron configuration for chlorine (Cl, Z = 17).*

Step 1: Locate the element in the periodic table. From its position in the table, identify and write the electron configuration of its highest occupied energy sublevel. (Leave room for writing lower-energy sublevels to its left.)	From its position in the periodic table (Group 7A/17, Period 3), the electron configuration of the highest occupied energy sublevel of chlorine is $3p^5$.
Step 2: To the left of what has already been written, list all lower-energy sublevels in order of increasing energy.	The sublevels having lower energies than 3*p* can be "read" across the periods from left to right in the periodic table, as in Figure 11.27: 1*s* 2*s* 2*p* 3*s* $3p^5$. If the neon core were to be used, these would be represented by [Ne]3*s* $3p^5$.
Step 3: For each filled lower-energy sublevel, write as a superscript the number of electrons that fill that sublevel. (Two electrons fill an *s* sublevel, ns^2; six electrons fill a *p* sublevel, np^6; and, for elements with *d* sublevels, ten electrons fill a *d* sublevel, nd^{10}. Exceptions: For chromium and copper, the 4*s* sublevel has only one electron, $4s^1$.)	A filled *s* sublevel has two electrons, and a filled *p* sublevel has six. Filling in these numbers yields $1s^22s^22p^63s^23p^5$, or $[Ne]3s^23p^5$.
Step 4: Confirm that the total number of electrons is the same as the atomic number.	2 + 2 + 6 + 2 + 5 = 17 = Z

The last step checks the correctness of your final result. The atomic number is the number of protons in the nucleus of the atom, which, for a neutral atom, is equal to the number of electrons. Therefore, the sum of the superscripts in an electron configuration, which is the total number of electrons in the atom, must be the same as the atomic number. For example, the electron configuration of oxygen (Z = 8) is $1s^22s^22p^4$. The sum of the superscripts is 2 + 2 + 4 = 8, the same as the atomic number.

Active Example 11.6 Writing Electron Configurations II

Write the electron configuration (both without and with a Group 8A/18 core) for potassium (K, Z = 19).

Think Before You Write Review the procedure describing *how to...* Write Electron Configurations, as necessary, and be sure that you can mentally picture the *s*, *p*, and *d* blocks in the periodic table, as illustrated in Figures 11.27 and 11.30. Then proceed with completing the Active Example.

Answers *Cover the left column with your cut-out shield. Reveal each answer only after you have written your own answer in the right column.*

$4s^1$ *Group 1A/1 elements have one s electron, and K is in the fourth period.*	First, what is the electron configuration of the highest occupied energy sublevel? (When you write the answer, leave space for the lower-energy sublevels.)
$1s\ 2s\ 2p\ 3s\ 3p\ 4s^1$	Now list to the left of $4s^1$ all lower-energy sublevels in order of increasing energy. Leave space to fill in the superscripts.
$1s^22s^22p^63s^23p^64s^1$ $2 + 2 + 6 + 2 + 6 + 1 = 19$ ✓	Finally, add the superscripts that show how many electrons fill the lower-energy sublevels. Check the final result.
$[Ar]4s^1$	Rewrite the configuration with a core from the closest Group 8A/18 element that has a smaller atomic number.
You improved your skill at writing electron configurations.	What did you learn by solving this Active Example?

Practice Exercise 11.6

Write the complete electron configuration for sulfur.

Active Example 11.7 Writing Electron Configurations III

Develop the electron configuration for cobalt (Co, Z = 27).

Think Before You Write This is your first example with *d* electrons. The procedure is the same.

Answers *Cover the left column with your cut-out shield. Reveal each answer only after you have written your own answer in the right column.*

$1s^22s^22p^63s^23p^64s^23d^7$ and $[Ar]4s^23d^7$ *By steps,* ***Step 1:*** *$3d^7$* ***Step 2:*** *$1s\ 2s\ 2p\ 3s\ 3p\ 4s\ 3d^7$ and $[Ar]4s\ 3d^7$* ***Step 3:*** *$1s^22s^22p^63s^23p^64s^23d^7$ and $[Ar]4s^23d^7$*	Write both a complete configuration and one with a Group 8A/18 core.

Step 4: $2 + 2 + 6 + 2 + 6 + 2 + 7 = 27 = Z$ ✓
and $18 + 2 + 7 = 27 = Z$ ✓

We show two different ways to calculate Step 4. The first sum is the addition of the superscripts in the full electron configuration. The second sum is based on the configuration with a Group 8A/18 core. Argon, Ar, has $Z = 18$, so we started with 18 in the second sum. We then added the remaining superscripts to complete the check.

You improved your skill at writing electron configurations.

What did you learn by solving this Active Example?

Practice Exercise 11.7

Write the electron configuration for chromium.

11.6 Valence Electrons

Goal 13 Using *n* for the highest occupied energy level, write the configuration of the valence electrons of any main group element.

14 Write the Lewis (electron dot) symbol for an atom of any main group element.

▶ **P/Review** One way that valence electrons act in forming chemical compounds, and thereby determine the chemical properties of an element, is described briefly in the next section. The topic is discussed more fully in Chapter 12.

▶ **P/Review** The vertical groups in the periodic table make up families of elements that have similar chemical properties (Section 5.7).

▶ **P/Review** Elements in different areas of the periodic table are identified in Section 5.7. Among them are the main group elements that are found in the A groups of the table (1, 2, and 13 to 18) and the transition elements in the B groups (3 to 12).

Many of the similar chemical properties of elements in the same column of the periodic table are related to the total number of *s* and *p* electrons in the highest occupied energy level. These are called **valence electrons.** ◀ In sodium, $1s^2 2s^2 2p^6 3s^1$, the highest occupied energy level is three. There is a single *s* electron in that sublevel, and there are no *p* electrons. Thus, sodium has one valence electron. With phosphorus, $1s^2 2s^2 2p^6 3s^2 3p^3$, the highest occupied energy level is again three. There are two *s* electrons and three *p* electrons, a total of five valence electrons.

Using *n* for any principal quantum number, we note that ns^1 is the configuration of the highest occupied principal energy level for all Group 1A/1 elements. All members of this family have one valence electron. ◀ Similarly, all elements in Group 5A/15 have the general configuration ns^2np^3, and they have five valence electrons.

The highest occupied sublevels of all families of main group elements can be written in the form ns^xnp^y. ◀ These are shown in the third row of **Table 11.1**. In all cases, the number of valence electrons (fourth row) is the sum of the superscripts, $x + y$. Notice that for every group the number of valence electrons is the same as the group number in the U.S. system, or the same as the only or the last digit in the IUPAC system.

Table 11.1 Valence Electrons and Lewis Symbols of the Elements

Group (U.S.)	1A	2A	3A	4A	5A	6A	7A	8A
Group (IUPAC)	1	2	13	14	15	16	17	18
Highest-Energy Electron Configuration	ns^1	ns^2	ns^2np^1	ns^2np^2	ns^2np^3	ns^2np^4	ns^2np^5	ns^2np^6
Number of Valence Electrons	1	2	3	4	5	6	7	8
Lewis Symbol of Third-Period Element	Na·	Mg:	Al:	·Si:	·P:	·S:	:Cl:	:Ar:

Active Example 11.8 **Valence Electrons I**

a) Write the electron configuration for the highest occupied energy level for Group 6A/16 elements.

b) Identify the group whose electron configuration for the highest occupied energy level is ns^2np^2.

Think Before You Write Try to answer these questions without referring to anything; if you cannot, use only a full periodic table. Review Table 11.1, if necessary.

Answers *Cover the left column with your cut-out shield. Reveal each answer only after you have written your own answer in the right column.*

(a) ns^2np^4 (b) Group 4A/14 *(a) A Group 6A/16 element has six valence electrons, as indicated by the group number. The first two must be in the ns sublevel, and the remaining four must be in the np sublevel.* *(b) The total number of valence electrons is 2 + 2 = 4. The group is therefore 4A/14.*	Complete both parts.
You improved your understanding of the valence electrons concept.	What did you learn by solving this Active Example?

Practice Exercise 11.8

a) Write the electron configuration for the highest occupied energy level for Group 3A/13 elements.

b) Identify the group whose electron configuration for the highest occupied energy level is ns^2np^3.

Another way to show valence electrons uses **Lewis symbols**, which are also called **electron dot symbols**. The symbol of the element is surrounded by the number of dots that matches the number of valence electrons. Dot symbols for the main group elements in Period 3 are given in Table 11.1. Paired electrons, those that occupy the same orbital, are usually placed on the same side of the symbol, and single occupants of one orbital are by themselves. This is not a fixed rule; exceptions are common if other positions better serve a particular purpose.

Group 8A/18 atoms have a full set of eight valence electrons, two in the *s* orbital and six in the *p* orbitals. This is sometimes called an **octet of electrons**. Elements in Group 8A/18 are particularly unreactive; only a few compounds of these elements are known. The filled octet is responsible for this chemical property.

Active Example 11.9 **Valence Electrons II**

Write electron dot symbols for the elements whose atomic numbers are 38 and 52.

Think Before You Write The procedure to complete this exercise is as follows: (1) Locate in the periodic table the elements whose atomic numbers are 38 and 52. (2) Write their symbols. (3) Determine the number of valence electrons from the group each element is in. (4) Surround its symbol with the number of dots that matches the number of valence electrons.

Answers *Cover the left column with your cut-out shield. Reveal each answer only after you have written your own answer in the right column.*

Sr: ·Ṫe: (Te with six dots) *The periodic table gives Sr for the symbol of atomic number 38. The element is strontium, which is in Group 2A/2, indicating two valence electrons. Te (tellurium) is the symbol for Z = 52. It is in Group 6A/6, so there are six electron dots.*	Write both electron dot symbols.
You improved your understanding of the valence electrons concept, and you improved your skill at writing electron dot symbols.	What did you learn by solving this Active Example?

Practice Exercise 11.9

Write electron dot symbols for iodine and sodium.

11.7 Trends in the Periodic Table

Mendeleev and Meyer developed their periodic tables by trying to organize some recurring physical and chemical properties of the elements. Some of these properties are examined in this section.

Atomic Size

Goal 15 Predict how and explain why atomic size varies with position in the periodic table.

Figure 11.31 shows the sizes of atoms of main group elements. With a few exceptions, two trends can be identified. Moving across the table from left to right, the atoms become smaller. Moving down the table in any group, atoms increase in size. These observations are believed to be primarily the result of two influences:

1. *Highest occupied principal energy level.* As valence electrons occupy higher and higher principal energy levels, they are generally farther from the nucleus, and the atoms become larger. For example, the valence electron of a lithium atom is a 2*s* electron, whereas the valence electron of a sodium atom is a 3*s* electron. As the principal quantum number increases, the size of an orbital increases. Sodium atoms are, therefore, larger.
2. *Nuclear charge.* Within any period, the valence electrons are all in the same principal energy level. As the number of protons in an atom increases, the positive charge in the nucleus also increases. This pulls the valence electrons closer to the nucleus, so the atom becomes smaller. For example, the atomic number of sodium is 11 (11 protons in the nucleus), and the atomic number of magnesium is 12 (12 protons). The 12 protons in a magnesium atom attract the 3*s* valence electrons more strongly than the 11 protons of a sodium atom. The magnesium atom is, therefore, smaller.

a summary of... Atomic Size

Atomic size generally increases from right to left across any row of the periodic table and from top to bottom in any column. The smallest atoms are toward the upper-right corner of the table, and the largest are toward the bottom left corner (**Figure 11.32**).

1A/1	2A/2	3A/13	4A/14	5A/15	6A/16	7A/17	8A/18
H 37							He 31
Li 152	Be 111	B 80	C 77	N 74	O 73	F 72	Ne 71
Na 186	Mg 160	Al 143	Si 118	P 110	S 103	Cl 100	Ar 98
K 227	Ca 197	Ga 125	Ge 122	As 120	Se 119	Br 114	Kr 112
Rb 248	Sr 215	In 167	Sn 140	Sb 140	Te 142	I 133	Xe 131
Cs 265	Ba 222	Tl 170	Pb 146	Bi 150	Po 168	At (140)	Rn (141)

Main group metals
Metalloids
Transition metals
Nonmetals

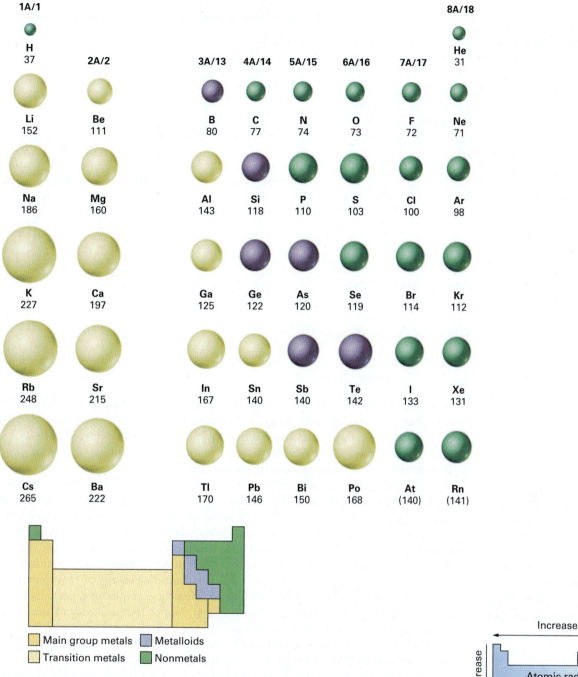

Figure 11.31 Sizes of atoms of main group elements, expressed in picometers (1 pm ≡ 10^{-12} m). Note the position of these elements in the periodic table silhouette; the transition metal sizes are excluded. Trends to observe: (1) Sizes of atoms increase with increasing atomic number in any group—that is, going down any column. (2) Sizes of atoms decrease with increasing atomic number within a given period—that is, going from left to right across any row.

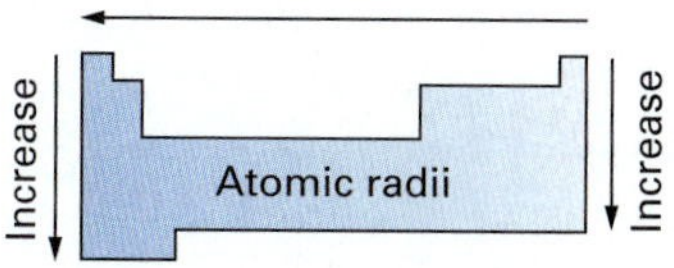

Figure 11.32 General trends in atomic radii with position in the periodic table.

Active Example 11.10 Periodic Trends: Atomic Size

Referring only to a periodic table, list atomic numbers 15, 16, and 33 in order of increasing atomic size.

Think Before You Write The preceding discussion of the two primary influences on atomic size and Figure 11.31 should guide you into selecting the smallest and the largest of the three atoms.

Answers *Cover the left column with your cut-out shield. Reveal each answer only after you have written your own answer in the right column.*

16 < 15 < 33 *The smallest atom is toward the upper right (Z = 16) and the largest is toward the bottom left (Z = 33). Specifically, Z = 16 (sulfur) atoms are smaller than Z = 15 (phosphorus) atoms because sulfur atoms have a higher nuclear charge to attract the highest-energy 3s and 3p electrons. The highest occupied energy level in a phosphorus atom is n = 3, but for a Z = 33 (arsenic) atom it is n = 4. Therefore, the phosphorus atom is smaller than the arsenic atom.*	List the atomic numbers from smallest atom to largest.
You improved your understanding of periodic trends in general, and specifically, you improved your understanding of periodic trends in atomic size.	What did you learn by solving this Active Example?

Practice Exercise 11.10

a) Explain why a potassium atom is larger than a sodium atom.

b) Explain why a chlorine atom is smaller than a sulfur atom.

Ionization Energy

Goal 16 Predict how and explain why first ionization energy varies with position in the periodic table.

A sodium atom (Z = 11) has 11 protons and 11 electrons. One of the electrons is a valence electron. Mentally separate the valence electron from the other 10. This is pictured in the larger block of **Figure 11.33**. The valence electron, with its 1− charge, is still part of the neutral atom. The rest of the atom has 11 protons and 10 electrons (11 plus charges and 10 minus charges), giving it a net charge of 1+. If we take away the valence electron, the particle that is left keeps that 1+ charge. This particle is a sodium ion, Na^+. An **ion** is an atom or group of atoms that has an electrical charge because of a difference in the number of protons and electrons. ◀

▶ **P/Review** A monatomic (one atom) ion is an atom that has gained or lost one, two, or three electrons. If the atom loses electrons, the ion has a positive charge and is a cation; if the atom gains electrons, it becomes a negatively charged anion (Section 6.5).

It takes work to remove an electron from a neutral atom. Energy must be transferred to overcome the attraction between the negatively charged electron and the positively charged ion that is left. The energy required to remove one electron from a neutral gaseous atom of an element is the **first ionization energy** of that element.

First ionization energy is one of the more striking examples of a periodic property, particularly when graphed (**Figure 11.34**). Notice the similarity of the shape of the graph between Li and Ne (Period 2 in the periodic table), between Na and Ar (Period 3), and between K and Kr (Period 4). Notice also that the peaks are elements in Group 8A/18, and the low points are from Group 1A/1.

Continue observing the trends in Figure 11.34. As atomic number increases within a period, the general trend in ionization energy is an increase. The general

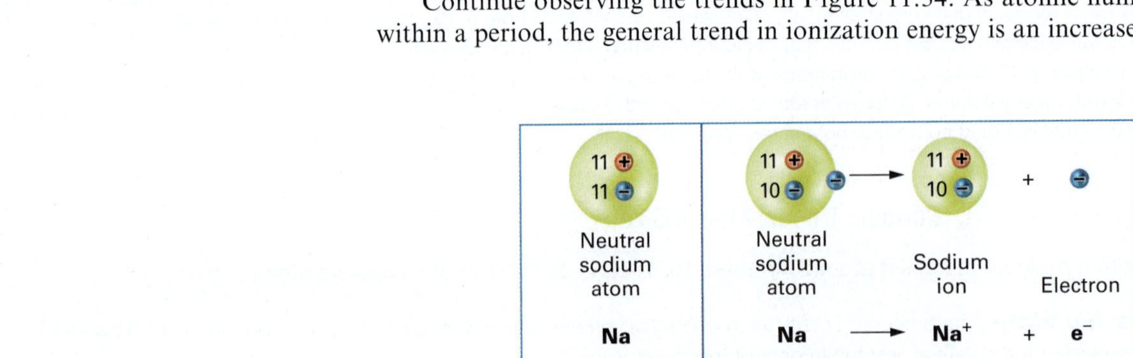

Figure 11.33 The formation of a sodium ion from a sodium atom.

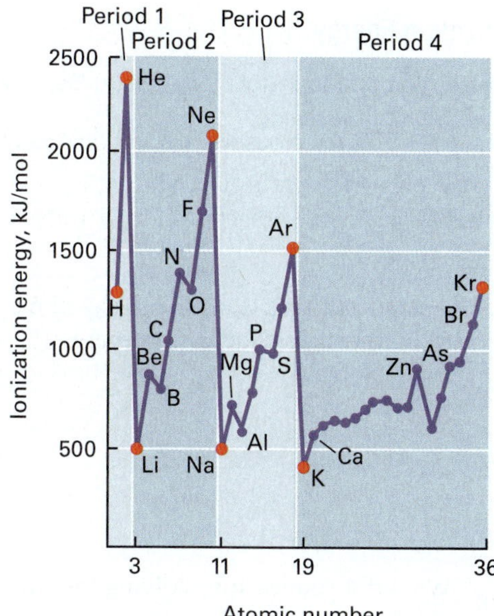

Figure 11.34 First ionization energy plotted as a function of atomic number, to show periodic properties of elements. Ionization energies of elements in the same period generally increase as atomic number increases. Ionization energies of elements in the same group generally decrease as atomic number increases.

trend among elements in the same group in the periodic table is a decrease. You can see this by looking at the elements in Group 1A/1 (Li, Na, and K), 2A/2 (Be, Mg, and Ca), and 8A/18 (He, Ne, Ar, and Kr). Ionization energies are lower as the atomic number increases within the group. If the graph is extended to the right, we find the same general shapes and trends in all periods.

The energy required to remove a second electron from the 1+ ion of an element is its **second ionization energy**, and removing a third electron requires the **third ionization energy**. In all cases, the ionization energy displays a big jump in size when the valence electrons are gone, and the next electron must be removed from a full octet of electrons. This adds to our belief that valence electrons are largely responsible for the chemical properties of an element. This also provides evidence to support the validity of the quantum model of the atom.

Periodic trends in first ionization energies are explained by the same influences that explain atomic size trends:

1. *Highest occupied principal energy level.* First ionization energy decreases down a group in the periodic table because the electron that is removed is farther from the nucleus. The attractive force between the electrons and the nucleus decreases with increasing distance.
2. *Nuclear charge.* The outermost electrons in atoms of main-group elements within a period are in the same principal energy level. First ionization energy increases across a period because the electron that is removed is attracted by an increasing quantity of positive charge. As the attractive force becomes stronger, the energy required to remove the electron will increase.

a summary of... First Ionization Energy

First ionization energy generally increases from left to right across any row of the periodic table and from bottom to top in any column. The atoms with the highest ionization energy are toward the upper-right corner, and those with the lowest ionization energy are toward the bottom-left corner (**Figure 11.35**).

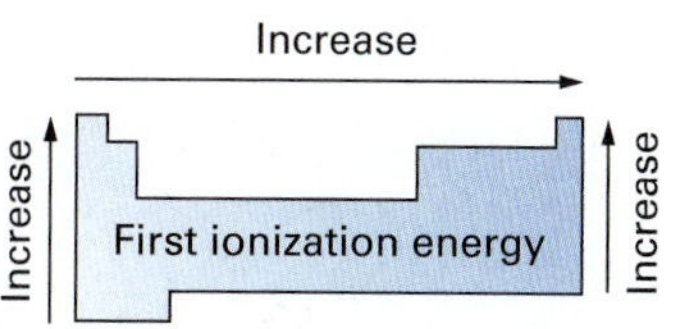

Figure 11.35 General trends in first ionization energy with position in the periodic table.

Active Example 11.11 Periodic Trends: First Ionization Energy

List each of the following groups of elements in order of increasing first ionization energy: (a) Be, N, F; (b) F, Cl, Br.

Think Before You Write You are ready to take this all the way. Review the preceding discussion if you are unsure about how to answer.

Answers *Cover the left column with your cut-out shield. Reveal each answer only after you have written your own answer in the right column.*

(a) Be < N < F (b) Br < Cl < F *In part (a), ionization energy increases across a period because of the increasing nuclear charge holding electrons that are in the same principal energy level.* *In part (b), ionization energy increases up a group because the highest occupied principal energy level is decreasing and the electrons are closer to the positive charge of the nucleus.*	Answer both parts, listing the symbols of the elements from lowest first ionization energy to highest.
You improved your understanding of periodic trends in general, and specifically, you improved your understanding of periodic trends in first ionization energy.	What did you learn by solving this Active Example?

Practice Exercise 11.11

a) Explain why the first ionization energy of potassium is less than that of sodium.

b) Explain why the first ionization energy of chlorine is greater than that of sulfur.

Chemical Families

Goal 17 Explain, from the standpoint of electron configuration, why certain groups of elements make up chemical families.

18 Identify in the periodic table the following chemical families: alkali metals, alkaline earths, halogens, noble gases.

Elements with similar chemical properties appear in the same group, or vertical column, in the periodic table. Several of these groups form **chemical families**. Family trends are most apparent among the main group elements. **Figure 11.36** shows family trends in chemical properties for two groups. In this text, we will consider four families: the alkali metals in Group 1A/1; the alkaline earths in Group 2A/2; the halogens in Group 7A/17; and the noble gases in Group 8A/18. As these families are discussed, the symbol X is used to refer to any member of the family.

Alkali Metals—Valence Electrons: ns^1 X· With the exception of hydrogen, Group 1A/1 elements are known as **alkali metals** (Figures 11.36, **11.37**, and **11.38**). The single valence electron may be removed with a relatively small quantity of energy, forming an ion with a 1+ charge. All ions with a 1+ charge, such as the Na^+ ion described earlier, tend to combine with other elements in the same way. This is why the chemical properties of the elements in a family are similar.

Figure 11.34 shows that ionization energies of the alkali metals decrease as the atomic number increases. The higher the energy level in which the negatively charged ns^1 electron is located, the farther it is from the positively charged nucleus, and therefore less energy is required for its removal. As a direct result of this, the **reactivity** of the element—that is, its tendency to react with other elements to form compounds—increases as you go down the column.

When an alkali metal atom loses its valence electron, the ion formed is **isoelectronic** with a noble gas atom. (The prefix *iso-* means "same.") For example, the electron configuration of sodium is $1s^22s^22p^63s^1$ or [Ne]$3s^1$. If the $3s^1$ electron is

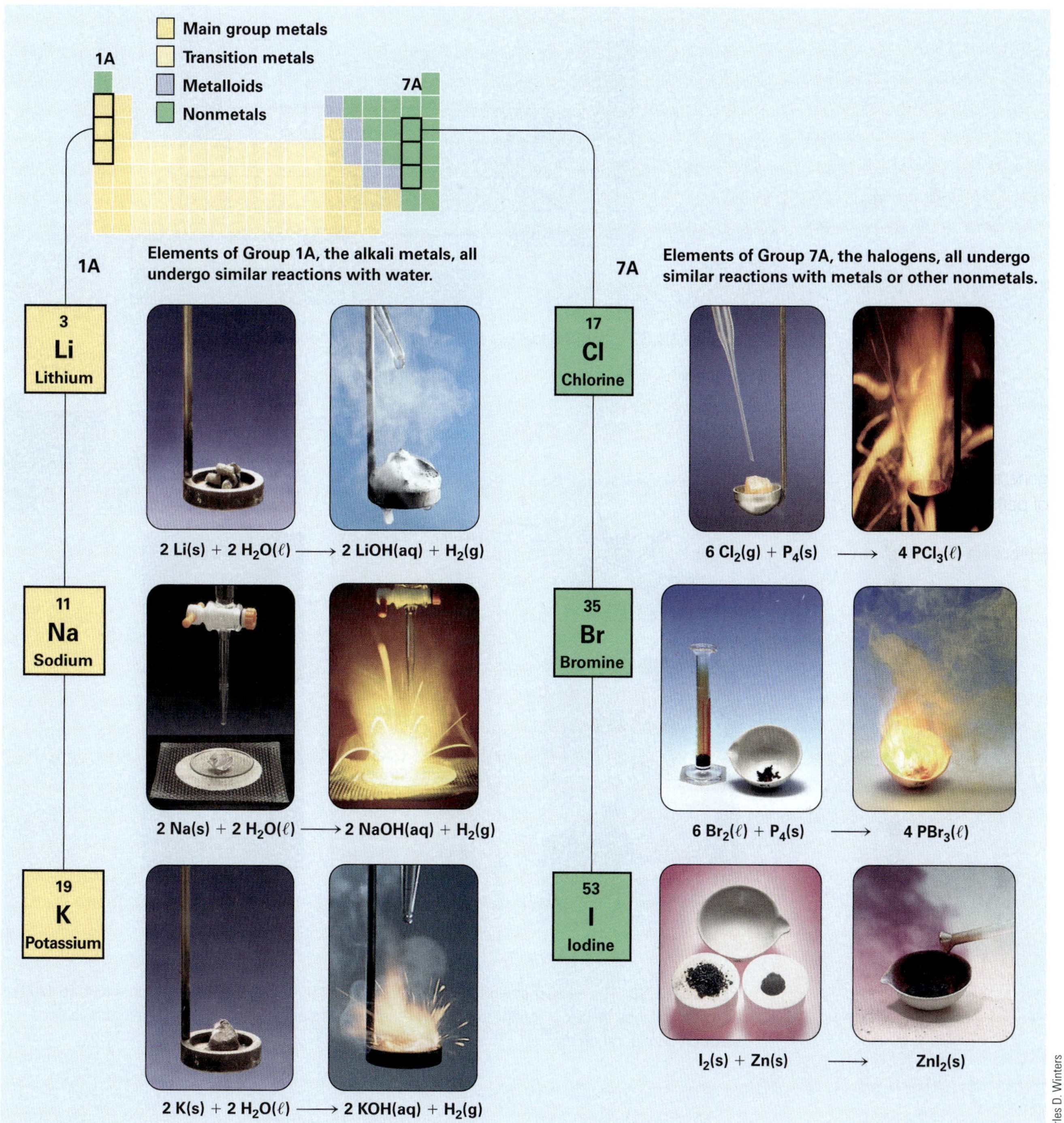

Figure 11.36 Chemical families. Elements with similar chemical properties form chemical families. These elements are in the same group in the periodic table. Lithium, sodium, and potassium are three members of Group 1A/1, and they undergo similar reactions with water. Atoms of each element in the group have one valence electron. Chlorine, bromine, and iodine, members of Group 7A/17, each have atoms with seven valence electrons. This leads to similar reactivity among all members of the chemical family.

removed, $1s^22s^22p^6$ is left. This is the configuration of the noble gas neon. In each case, the alkali metal ion reaches the same configuration as the noble gas just before it in the periodic table. Its highest energy octet is complete; all electron orbitals are filled. This is a highly stable electron distribution. The chemical properties of most

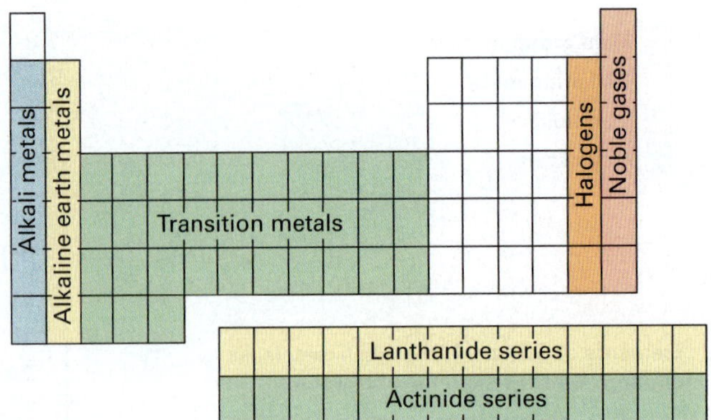

Figure 11.37 Chemical families and regions in the periodic table.

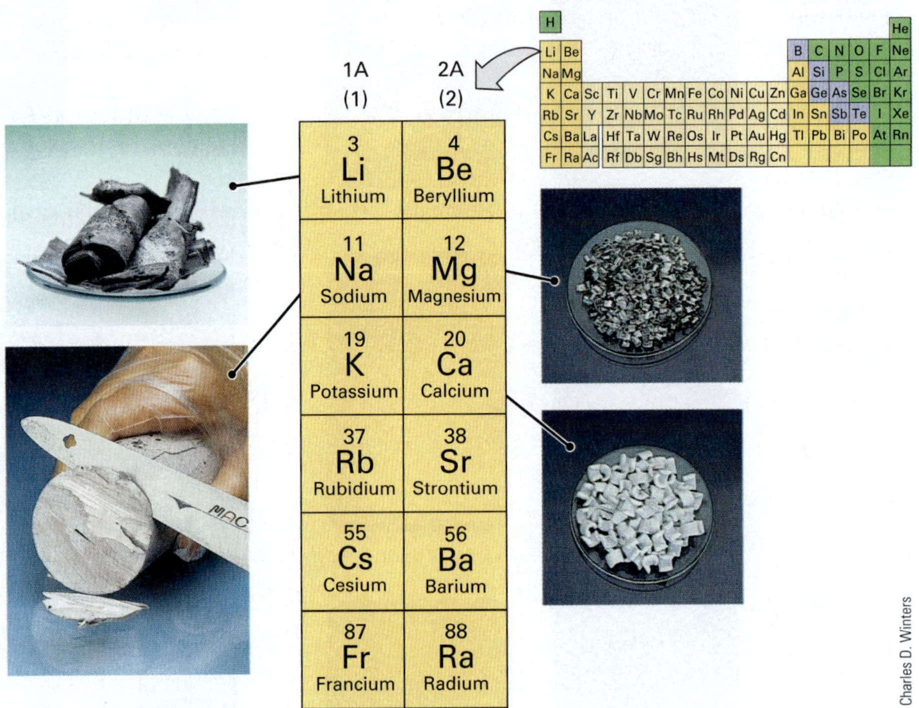

Figure 11.38 The alkali metal and alkaline earth metal families. Alkali metal atoms (Group 1A/1) have the valence electron configuration ns^1. Alkaline earth atoms (Group 2A/2) have the valence electron configuration ns^2.

main group elements can be explained in terms of their atoms becoming isoelectronic with a noble gas atom.

Alkali metals do not normally look like common, everyday metals. This is because they are so reactive that they combine with oxygen in the air to form an oxide coating, which hides the bright metallic luster that can be seen in a freshly cut sample. These elements do possess other common metallic properties, however. For example, they are good conductors of heat and electricity, and they are easy to form into wires and thin foils.

We can see distinct trends in the physical properties of alkali metals. Their densities increase as atomic number increases. Boiling and melting points generally decrease as you go down the periodic table. The single exception is cesium ($Z = 55$), which boils at a temperature slightly higher than the boiling point of rubidium ($Z = 37$).

Alkaline Earths—Valence Electrons: ns^2 X: Group 2A/2 elements are called **alkaline earths** or **alkaline earth metals** (see Figures 11.37 and 11.38). Both the first and second ionization energies are relatively low, so the two valence electrons are given up readily to form ions with a 2+ charge. Again, the ions have the configuration of a noble gas. If magnesium, [Ne]$3s^2$, loses two electrons, only the [Ne] core is left. Trends like those noted with the alkali metals are also seen with the alkaline earths. Reactivity again increases as you go down the column in the periodic table. Physical property trends are less evident among the alkaline earths.

Halogens—Valence Electrons: ns^2np^5 :Ẋ: The elements in Group 7A/17 make up the family known as the **halogens**, or "salt formers" (see Figures 11.36, 11.37, and 11.39). Halogens have seven valence electrons. The lowest-energy process for a halogen to reach a full octet of electrons is to gain one. This gives it the configuration of a noble gas and forms an ion with a 1− charge. The tendency to gain an electron is greater in small atoms, in which the added electron is closer to the nucleus. Consequently, reactivity is greatest for fluorine at the top of the group and least for iodine at the bottom.

Density, melting point, and boiling point all increase steadily with increasing atomic number among the halogens.

Noble Gases—Valence Electrons: ns^2np^6 (for He, Z = 2, ns^2) :Ẍ: The elements of Group 8A/18 are the **noble gases** (see Figures 11.37 and 11.40). In chemistry, the word *noble* means having "a reluctance to react." Only a few hundred compounds of the noble gases have been synthesized, and none occur naturally.

The inactivity of the noble gases is believed to be the result of their filled valence electron sublevels. The two electrons of helium fill the 1*s* orbital, the only valence orbital helium has. All other elements in the group have a full octet of electrons—completely filled *s* and *p* valence electron orbitals. This configuration apparently

Figure 11.39 The halogen family. Halogen atoms (Group 7A/17) have the valence electron configuration ns^2np^5, for a total of seven valence electrons.

Figure 11.40 An airship. The "lighter than air" feature of an airship comes from the low-density helium that fills it. Helium is safe for this purpose because it is unreactive, unlike even lower density hydrogen. See Figure 8.14.

▶ **P/Review** In Section 2.9, the expression "reduction of the energy in a chemical system to the smallest amount possible" was introduced to identify a natural tendency for chemical and physical changes to occur spontaneously if the result of that change is a lower total energy within the system. That is why objects fall in a gravitational field, and why oppositely charged particles attract each other and similarly charged particles repel each other in an electrical field.

represents a "reduction of energy to the smallest amount possible" arrangement of electrons that is very stable. ◀ The high ionization energies of a full octet have already been noted, so the noble gases resist forming positively charged ions. Nor do the atoms tend to gain electrons to form negatively charged ions.

The noble gases provide excellent examples of periodic trends in physical properties. Without exception, the densities, melting points, and boiling points increase as you move down the column in the periodic table.

Hydrogen—Valence Electron: $1s^1$ H· You have probably wondered why hydrogen appears twice in our periodic table, at the tops of Groups 1A/1 and 7A/17. Hydrogen is neither an alkali metal nor a halogen, although it shares some properties with both groups. Hydrogen combines with some elements in the same ratio as alkali metals, but the way the compounds are formed is different. Hydrogen atoms can also gain an electron to form an ion with a 1− charge, similar to a halogen. But other properties of hydrogen differ from those of the halogens. The way the periodic table is used makes it handy to have hydrogen in both positions, although it really stands alone as an element.

Metals and Nonmetals

Goal **19** Identify metals and nonmetals in the periodic table.

20 Predict how and explain why metallic character varies with position in the periodic table.

Both physically and chemically, the alkali metals, alkaline earths, and transition elements are metals. At the particulate level, an element is a **metal** if it can lose one or more electrons and become a positively charged ion. An element that lacks this quality is a **nonmetal**. The larger the atom, the more easily the outermost electron is removed. Therefore, the **metallic character** of elements in a group increases as you go down a column in the periodic table.

An atom becomes smaller as the nuclear charge increases across a period in the table. The larger number of protons holds the outermost electrons more strongly, making it more difficult for them to be lost. This makes the metallic character of elements decrease as you go from left to right across the period.

Table 11.2 compares the properties of metals and nonmetals. Chemically, the distinction between metals and nonmetals—elements that lose electrons in chemical reactions and those that do not—is not sharp. It can be drawn roughly as a stair-step line beginning between atomic numbers 4 and 5 in Period 2 and ending between 84 and 85 in Period 6 (**Figure 11.41**). Elements to the left of the line are metals; those to the right are nonmetals.

Table 11.2 Some Physical and Chemical Properties of Metals and Nonmetals

Metals	Nonmetals
Tend to lose electrons to form cations	Tend to gain electrons to form anions
1, 2, or 3 valence electrons	4 or more valence electrons
Low ionization energies	High ionization energies
Form compounds with nonmetals but not with other metals	From compounds with metals and with other nonmetals
High electrical conductivity	Poor electrical conductivity (carbon in the form of graphite is an exception)
High thermal conductivity	Poor thermal conductivity; good insulator
Malleable (can be hammered into sheets)	Brittle
Ductile (can be drawn into wires)	Nonductile

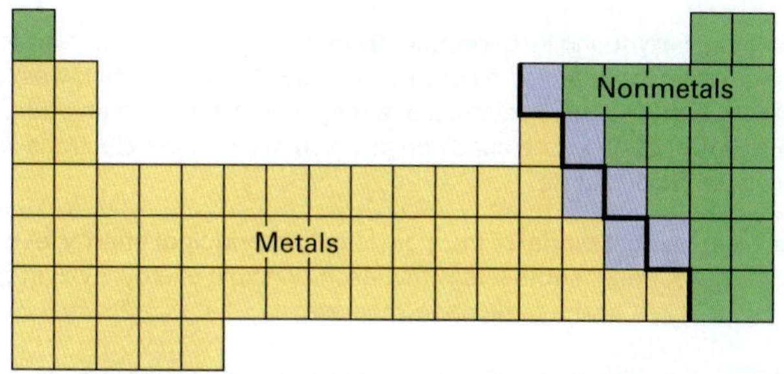

Figure 11.41 Metals and nonmetals. Purple identifies elements that are metalloids, which have properties that are intermediate between those of metals and nonmetals.

Most of the elements next to the stair-step line have some properties of both metals and nonmetals. They are often called **metalloids**. Included in the group are silicon and germanium, the semiconductors on which the electronics industry has been built. Indeed, silicon is so important to the industry that the area south of San Francisco where many major electronics manufacturers are located is known as *Silicon Valley*.

a summary of... Metallic Character

Metallic character generally increases from right to left across any row of the periodic table and from top to bottom in any column. The least metallic character is toward the upper-right corner of the table, and the most is toward the bottom-left corner (**Figure 11.42**).

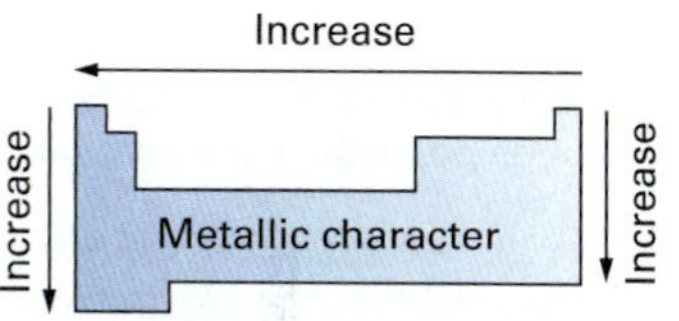

Figure 11.42 General trends in metallic character with position in the periodic table.

Target Check 11.1

a) List the following in order of increasing ionization energy: C, N, F.

b) Write the symbol of the Period 2 element in each of the following chemical families: halogens, alkali metals, noble gases, alkaline earths.

c) List the following in order of increasing atomic size: N, O, P.

d) Which Period 3 elements are metals? Which are nonmetals? Which are metalloids?

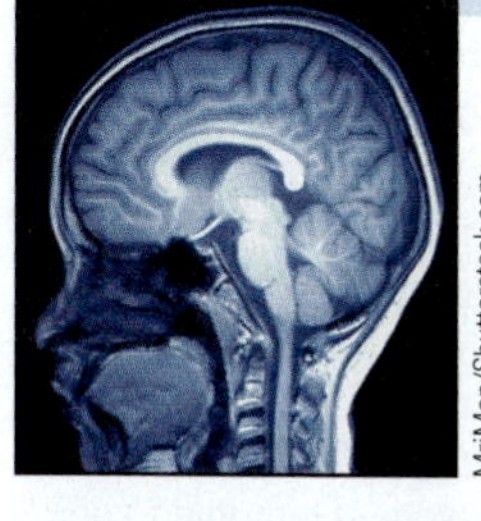

MriMan/Shutterstock.com

CHAPTER 11 IN REVIEW: ATOMIC THEORY: THE QUANTUM MODEL OF THE ATOM

Goal 1 Define and describe electromagnetic radiation.	**Electromagnetic radiation** is a form of energy that has wavelike properties that includes gamma rays, x-rays, ultraviolet radiation, visible light, infrared radiation, microwaves, and radio waves. It travels at the **speed of light, c:** $c = \lambda$ (wavelength) $\times$ ν (frequency).
Goal 2 Distinguish between continuous and line spectra.	A **continuous spectrum** is a spectrum that is distributed over a continuous range of wavelengths. A **line spectrum** has **discrete** lines.
Goal 3 Describe the Bohr model of the hydrogen atom.	The **Bohr model of the hydrogen atom** features a relatively small volume, extremely dense nucleus that contains all of the atom's positive charge and nearly all of its mass. The negatively charged electron of the hydrogen atom has a very small mass, and it travels in one of a series of circular orbits around the nucleus.
Goal 4 Explain the meaning of quantized energy levels in an atom and show how these levels relate to the discrete lines in the spectrum of that atom.	The Bohr model of the hydrogen atom restricts the electron to certain **quantized energy levels**. The electron can have certain definite energies, but never may it have an energy between the quantized values. The spectrum of the atoms of an element is the result of energy released as electrons in an excited state drop to a lower energy level via a **quantum jump** or **quantum leap**.

Goal 5 Distinguish between ground state and excited state.	The electron is normally found in its **ground state**, the condition when the electron is in $n = 1$ (for a hydrogen atom) or when all electrons in an atom occupy the lowest possible energy levels (for atoms with multiple electrons). If energy is transferred to an atom, an electron can be raised to an **excited state**, the condition at which one or more electrons in an atom has an energy level above ground state.
Goal 6 Identify the principal energy levels in an atom and state the energy trend among them.	The **quantum mechanical model** of the atom identifies **principal energy levels,** and for atoms with two or more electrons, **sublevels** within each principal energy level. In general, energies increase as the principal quantum numbers increase: $n = 1 < n = 2 < n = 3 \ldots < n = 7$.
Goal 7 For each principal energy level, state the number of sublevels, identify them, and state the energy trend among them.	The total number of sublevels within a given principal energy level is equal to *n*, the principal quantum number. For any given value of *n*, energy increases through the sublevels in the order $\mathbf{s < p < d < f}$.
Goal 8 Sketch the shapes of *s* and *p* orbitals.	Figure 11.19 shows the shapes of the *s* and *p* (and *d*) orbitals. An *s* orbital is spherical. A *p* orbital is dumbbell shaped.
Goal 9 State the number of orbitals in each sublevel.	There is one orbital for every *s* sublevel. All *p* sublevels have three orbitals, all *d* sublevels have five, and all *f* sublevels have seven. This **1–3–5–7 sequence of odd numbers** continues through higher sublevels.
Goal 10 State the restrictions on the electron population of an orbital.	An **orbital** may be unoccupied, occupied by one electron, or occupied by two electrons.
Goal 11 Use a periodic table to list electron sublevels in order of increasing energy.	Two rules guide the assignments of electrons to orbitals: **1.** At ground state the electrons fill the *lowest*-energy orbitals available. **2.** No orbital can have more than two electrons. The periodic table is a guide to the order of **increasing energy of sublevels:** $1s < 2s < 2p < 3s < 3p < 4s < 3d < 4p$ and so on.
Goal 12 Referring only to a periodic table, write the ground state electron configuration of an atom of any element up to atomic number 36.	The procedure for writing an **electron configuration** follows: **1.** Locate the element in the periodic table. From its position in the table, identify and write the electron configuration of its highest occupied energy sublevel. **2.** To the left of what has already been written, list all lower-energy sublevels in order of increasing energy. **3.** For each filled lower-energy sublevel, write as a superscript the number of electrons that fill that sublevel. **4.** Confirm that the total number of electrons is the same as the atomic number.
Goal 13 Using *n* for the highest occupied energy level, write the configuration of the valence electrons of any main group element.	Atoms in the same group of the periodic table (same column) have the same highest occupied sublevel electron configurations, or the same ns^xnp^y **valence electron** configurations. These range from ns^1 for Group 1A/1 to ns^2np^6 for Group 8A/18. The highest occupied principal energy level value (*n*) increases as you go down a group.
Goal 14 Write the Lewis (electron dot) symbol for an atom of any main group element.	Valence electrons are depicted by **Lewis symbols**, which are also called *electron dot symbols*. The symbol of the element is surrounded by the number of dots that matches the number of valence electrons.
Goal 15 Predict how and explain why atomic size varies with position in the periodic table.	The **sizes of atoms** in the periodic table increase as you move down a group, but decrease as you move left to right across a period. This is explained by the highest occupied energy level and nuclear charge.
Goal 16 Predict how and explain why first ionization energy varies with position in the periodic table.	**First ionization energy**, the energy required to remove an electron from a neutral atom, decreases as you move down a group (highest occupied energy level is farther from the nucleus) and increases as you move across a period (more positive charge in the nucleus).
Goal 17 Explain, from the standpoint of electron configuration, why certain groups of elements make up chemical families.	Groups of elements in the periodic table exhibit similar behavior and are called **chemical families.** The chemical properties of the elements in a family are similar because they have the same valence electron configuration.
Goal 18 Identify in the periodic table the following chemical families: alkali metals, alkaline earths, halogens, noble gases.	Alkali metals, ns^1 Group 1A/1; alkaline earths, ns^2, Group 2A/2; halogens, ns^2np^5, Group 7A/17; noble gases, ns^2np^6, Group 8A/18.

Goal 19 Identify metals and nonmetals in the periodic table.	The elements in the periodic table are classified as **metals or nonmetals,** based on their chemical behavior. Metals are to the left of the stair-step line, nonmetals are to the right, and six **metalloids** hug the line.
Goal 20 Predict how and explain why metallic character varies with position in the periodic table.	**Metallic character** increases from right to left across any row of the periodic table and from top to bottom in any group. A metal can lose one or more electrons and become a positively charged ion. The lower the energy required to remove an electron, the greater the metallic character of an element.

Key Terms

Most of the key terms and concepts and many others appear in the Glossary. Use your Glossary regularly.

alkali metals p. 422
alkaline earth metals p. 425
alkaline earths p. 425
argon core p. 412
chemical families p. 422
continuous p. 400
continuous spectrum p. 397
discrete p. 397
electromagnetic radiation p. 395
electromagnetic spectrum p. 395
electron configuration p. 408
electron dot symbols p. 417
excited state p. 401
first ionization energy p. 420
ground state p. 401
halogens p. 425
ion p. 420
isoelectronic p. 422
Lewis symbols p. 417
line spectrum p. 397
metal p. 426
metallic character p. 426
metalloids p. 427
neon core p. 412
noble gases p. 425
nonmetal p. 426
octet of electrons p. 417
orbitals p. 404
Pauli exclusion principle p. 406
photon p. 399
principal energy levels p. 402
principal quantum number, *n* p. 402
quantized p. 400
quantized energy levels p. 400
quantum jump p. 401
quantum leap p. 401
quantum mechanical model of the atom p. 403
quantum mechanics p. 393
reactivity p. 422
***s, p, d,* and *f* sublevels** p. 403
second ionization energy p. 421
speed of light p. 396
sublevels p. 403
third ionization energy p. 421
valence electrons p. 416
wave equation p. 396
wave-particle duality p. 399

Frequently Asked Questions

What is the best way to review the quantum mechanical model of the atom?
It takes many words to describe the quantum model of the atom, even at this introductory level. The words are not easy to remember unless they are organized into some kind of pattern. The summary near the end of Section 11.4 gives you this organization.

What vocabulary terms used to describe the quantum mechanical model of the atom are potential pitfalls to be avoided?
Be sure you know what *quantized* means. Also, understand the difference between a Bohr *orbit* (a fixed path the electron travels around the nucleus) and the quantum *orbital* (a mathematically defined region in space in which there is a high probability of finding the electron).

What is the best way to learn to write electron configurations?
In writing electron configurations, we recommend that you use the periodic table to list the sublevels in increasing energy rather than the slanting-line memory device. The periodic table is the greatest organizer of chemical information there is, and every time you use it you strengthen your ability to use it in all other ways.

What is the most important concept to take away from the section on periodic trends?
Understand well the two influences that determine atomic size. The same thinking explains other periodic trends.

Concept-Linking Exercises

Write a brief description of the relationships among each of the following groups of terms or phrases. Answers to the Concept-Linking Exercises are given at the end of the chapter.

1. Electromagnetic spectrum, continuous spectrum, line spectrum
2. Quantized energy levels, ground state, excited state, continuous values
3. Bohr model of the hydrogen atom, orbit, orbital
4. Quantum mechanical model of the atom, principal energy level, sublevel, electron orbital, Pauli exclusion principle
5. Valence electrons, Lewis symbols, octet of electrons
6. Ionization energy, second ionization energy, third ionization energy
7. Chemical families, alkali metals, alkaline earths, halogens, noble gases, hydrogen
8. Atomic size, highest occupied principal energy level, nuclear charge
9. Metal, nonmetal, metalloid

Small-Group Discussion Questions

Small-Group Discussion Questions are for group work, either in class or under the guidance of a leader during a discussion section.

1. Terrestrial radio stations identify themselves by the frequency at which they broadcast, such as 98.5 FM, which stands for a signal of 98.5 megahertz carried by an FM (frequency modulation) wave. (A hertz is a wave cycle per second.) Could radio stations identify themselves by wavelength? If yes, what wavelength is emitted by 98.5 FM? If no, why not?
2. Explain how the existence of line spectra for elements leads to the Bohr model of the atom. Is a quantum jump a large or small quantity of energy? How does this quantity of energy compare with the everyday usage of the term *quantum leap*? Why does an electron in a Bohr orbit farther from the nucleus have more energy than one that is closer to the nucleus?
3. Carefully define each and distinguish between *orbit* and *orbital*. How many orbitals are within each of the seven ground state principal energy levels?
4. Using nothing but a periodic table, write the ground state electron configuration for each of the first 36 elements on the periodic table.
5. Write the electron dot symbol for each of the first 18 elements in the periodic table. Explain the relationship between each electron configuration from Question 4 and the electron dot symbol.
6. In a plot of ionization energy versus atomic number, ionization energy tends to increase within a period in the periodic table. However, in Period 2, there are two instances where ionization energy decreases: Be to B and N to O. Write the electron configurations of these elements and use them to propose an explanation for these disruptions in the general trend.

Questions, Exercises, and Problems

Interactive versions of these problems may be assigned by your instructor. Solutions for blue-numbered questions are at the end of the chapter. Questions other than those in the General Questions and More Challenging Problems sections are paired in consecutive odd-even number combinations; solutions for the odd numbered questions are also at the end of the chapter.

Section 11.2: Electromagnetic Radiation

1. The visible spectrum is a small part of the whole electromagnetic spectrum. Name several other parts of the whole spectrum that are included in our everyday vocabulary.

Dgrilla/Shutterstock.com

A rainbow displays the visible spectrum.

2. What is meant by a discrete line spectrum? What kind of spectra do not have discrete lines? Have you ever seen a spectrum that does not have discrete lines? Have you ever seen one with discrete lines?
3. Identify measurable wave properties that are used in describing light.
4. What do visible light and radio waves have in common?

Section 11.3: The Bohr Model of the Hydrogen Atom

5. Which among the following are *not* quantized: (a) number of plastic bags in a box; (b) cars passing through a toll plaza in a day; (c) birds in an aviary; (d) flow of a river in m^3/h; (e) percentage of salt in a solution?
6. Which of the following are quantized: (a) canned soup from a grocery store; (b) weight of jelly beans; (c) elevation of a person on a ramp?
7. What kind of light would atoms emit if the electron energy were not quantized? Why?
8. In the Bohr model of the hydrogen atom, the electron occupies distinct energy states. In one transition between energy states, an electron moves from $n = 1$ to $n = 2$. Is energy absorbed or emitted in the process? Does the electron move closer to or farther from the nucleus?
9. What must be done to an atom, or what must happen to an atom, before it can emit light?
10. What experimental evidence leads us to believe that electron energy levels are quantized?
11. Which atom is more apt to emit light, one in the ground state or one in an excited state? Why?
12. What is meant when an atom is "in its ground state"?
13. Using a sketch of the Bohr model of an atom, explain the source of the observed lines in the spectrum of hydrogen.
14. Draw a sketch of an atom according to the Bohr model. Describe the atom with reference to the sketch.
15. Identify the major advances that came from the Bohr model of the atom.
16. Identify the shortcomings of the Bohr model of the atom.

Section 11.4: The Quantum Mechanical Model of the Atom

17. Compare the relative energies of the principal energy levels within the same atom.
18. What is the meaning of the principal energy levels of an atom?
19. How many sublevels are present in each principal energy level?
20. How many sublevels are there in an atom with $n = 4$?

21. How many orbitals are in each of the following: the *s* sublevel, the *p* sublevel, the *d* sublevel, and the *f* sublevel?

22. What is an orbital? Describe the shapes of *s* and *p* orbitals using words and sketches.

23. Each *p* sublevel contains six orbitals. Is this statement true or false? Comment on this statement.

24. How many orbitals are there in an atom with $n = 3$?

25. The principal energy level with $n = 6$ contains six sublevels, although not all six are occupied for any element now known. Is this statement true or false? Comment on this statement.

26. Although we may draw the 4*s* orbital with the shape of a ball, there is some probability of finding the electron outside the ball we draw. Is this statement true or false? Comment on this statement.

27. An orbital may hold one electron or two electrons, but no other number. Is this statement true or false? Comment on this statement.

28. What is the significance of the Pauli exclusion principle?

29. What is your opinion of the common picture showing one or more electrons whirling around the nucleus of an atom?

testing/Shutterstock.com

Companies sometimes use an illustration of the Bohr model of the atom to promote their products.

30. An electron in an orbital of a *p* sublevel follows a path similar to the shape of the numeral 8. Comment on this statement.

31. What is the largest number of electrons that can occupy each of the following: the 4*p* orbitals, the 3*d* orbitals, and the 5*f* orbitals?

32. How many 4*f* orbitals are there in an atom?

33. Energies of the principal energy levels of an atom sometimes overlap. Explain how this is possible.

34. What general statement may be made about the energies of the principal energy levels? About the energies of the sublevels?

35. Which of the following statements is true: The quantum mechanical model of the atom includes (a) all of the Bohr model; (b) part of the Bohr model; (c) no part of the Bohr model? Justify your answer.

36. Is the quantum mechanical model of the atom consistent with the Bohr model? Why or why not?

Section 11.5: Electron Configuration

37. What do you conclude about the symbol $2d^5$?

38. What is the meaning of the symbol $3p^4$?

39. What element has the electron configuration $1s^22s^22p^63s^23p^4$? In which period and group of the periodic table is it located?

40. In what group and period do you find elements with the electron configurations given?
    a) $1s^22s^22p^63s^24s^23d^8$
    b) $1s^22s^22p^63s^2$

41. What is the argon core? What is its symbol, and what do you use it for?

42. What is meant by [Ne] in [Ne]$3s^23p^1$?

Questions 43 through 46: Identify the elements whose electron configurations are given.

43. a) $1s^22s^22p^63s^23p^64s^23d^{10}4p^4$
    b) $1s^22s^22p^4$
    c) $1s^22s^22p^63s^2$

44. a) $1s^22s^22p^63s^23p^64s^23d^2$
    b) $1s^22s^22p^6$

45. a) [He]$2s^22p^3$
    b) [Ne]$3s^23p^1$
    c) [Ar]$4s^23d^7$

46. a) [He]$2s^22p^2$
    b) [Ne]$3s^23p^5$
    c) [Ar]$4s^23d^{10}4p^3$

Questions 47 through 50: Write complete ground state electron configurations of an atom of the elements shown. Do not use a noble gas core.

47. Magnesium and nickel

48. Bromine and manganese

49. Chromium and selenium (Z = 34)

50. Calcium and copper

51. If it can be done, rewrite the electron configurations in Questions 47 and 49 with a neon or argon core.

52. a) Write the complete ground state electron configuration for the silicon atom.

Charles D. Winters

Silicon is the primary component from which computer chips are made.

b) Use a noble gas core to write the electron configuration of phosphorus.

White phosphorus is one of several forms of elemental phosphorus. It is stored in water because it will ignite in air.

53. Use a noble gas core to write the electron configuration of vanadium (Z = 23).

54. a) Write the complete ground state electron configuration for the scandium (Z = 21) atom.
b) Use a noble gas core to write the electron configuration of iron.

Section 11.6: Valence Electrons

55. Why are valence electrons important?

56. What are valence electrons?

57. What are the valence electrons of aluminum? Write both ways that they may be represented.

58. Write the name of the element with the valence electron configuration given.
a) $2s^2$
b) $4s^24p^4$

59. Using n for the principal quantum number, write the electron configuration for the valence electrons of Group 6A/16 atoms.

60. In what group and period do you find elements with the valence electron configurations given?
a) $2s^22p^2$
b) $4s^24p^5$

Section 11.7: Trends in the Periodic Table

61. Identify an atom of a main-group element in Period 4 that is (a) larger than an atom of arsenic (Z = 33) and (b) smaller than an atom of arsenic.

62. Using only the periodic table, arrange the following elements in order of increasing atomic radius: aluminum, boron, indium (Z = 49), gallium (Z = 31).

63. Why does atomic size decrease as you go left to right across a row in the periodic table?

64. Using only the periodic table, arrange the following elements in order of increasing atomic radius: sulfur, magnesium, sodium, chlorine.

65. Even though an atom of germanium (Z = 32) has more than twice the nuclear charge of an atom of silicon (Z = 14), the germanium atom is larger. Explain.

Germanium had not yet been discovered when Mendeleev published his periodic table, and his accurate prediction of its properties illustrated the significance of the periodic table.

66. The text describes the formation of a sodium ion as the removal of one electron from a sodium atom. Predict the size of a sodium ion compared to the size of a sodium atom. Explain your prediction.

67. What is the general trend in ionization energies of elements in the same chemical family?

68. Using only the periodic table, arrange the following elements in order of increasing ionization energy: radon (Z = 86), helium, neon, xenon (Z = 54).

69. Explain the reason for the general trend of increasing ionization energy across a period.

70. Using only the periodic table, arrange the following elements in order of increasing ionization energy: phosphorus, aluminum, silicon, chlorine.

71. What is it about strontium (Z = 38) and barium that makes them members of the same chemical family?

72. Give the symbol for an element that is: (a) a halogen; (b) an alkali metal; (c) a noble gas; (d) an alkaline earth metal.

73. To what family does the electron configuration ns^2np^5 belong?

74. a) What is the name of the alkali metal that is in Period 6?
b) What is the name of the halogen that is in Period 6?
c) What is the name of the alkaline earth metal that is in Period 3?
d) What is the name of the noble gas that is in Period 3?

75. Expressed as ns^xnp^y, what electron configuration is isoelectronic with a noble gas?

76. Which of the following describes the element Ba? Choose all that apply: (a) is very reactive as a metal; (b) consists of diatomic molecules in elemental form; (c) forms an ion with a charge of 2+; (d) forms a cation that is isoelectronic with a noble gas; (e) reacts with alkali metals to form salts; (f) is one of the group of the least reactive elements.

77. Identify the chemical families in which (a) krypton (Z = 36) and (b) beryllium are found.

78. Which of the following describes the element bromine? Choose all that apply: (a) reacts vigorously as a metal; (b) belongs to a group in the periodic table consisting entirely of gases; (c) consists of diatomic molecules in elemental form; (d) is one of the group of least reactive elements; (e) reacts vigorously with alkali metals to form salts; (e) forms a cation with a charge of 1+.

79. Account for the chemical similarities among chlorine, bromine, and iodine in terms of their electron configurations.

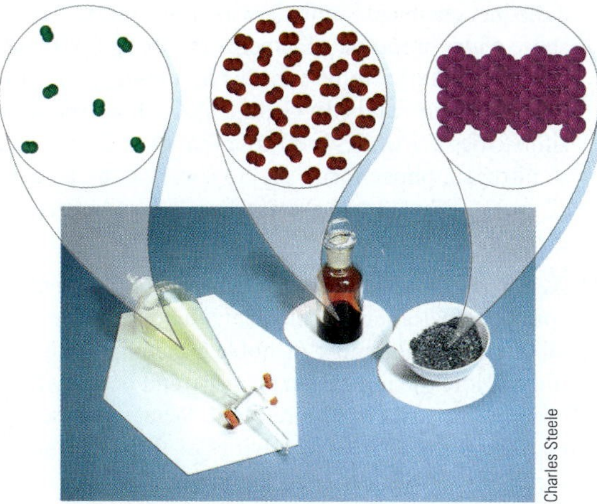

Group 7A/17 is the only group in the periodic table that has elements that occur in all three states of matter at common temperatures and pressures, as illustrated here by *(left to right)* gaseous chlorine, liquid bromine, and solid iodine.

80. Where in the periodic table do you find the transition elements? Are they metals or nonmetals?

81. What property of atoms of an element determines that the element is a metal rather than a nonmetal?

82. Arrange the following elements in order of increasing metallic character: sodium, aluminum, phosphorus, silicon.

83. What are *metalloids*? How do their properties compare with those of metals and nonmetals?

84. Arrange the following elements in order of increasing metallic character: nitrogen, antimony (Z = 51), arsenic (Z = 33), phosphorus.

Use the periodic table in **Figure 11.43** *to answer Questions 85 through 90. Answer with letters (D, E, X, Z, and so on) from the figure.*

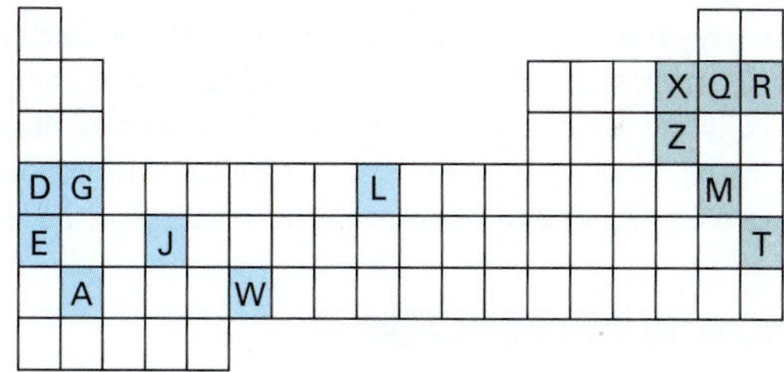

Figure 11.43 Periodic table for Questions 85 to 90.

85. Give the letters that are in the positions of (a) alkaline earth metals and (b) noble gases.

86. Give the letters that are in the positions of (a) halogens and (b) alkali metals.

87. List the letters that correspond to nonmetals.

88. Give the letters that correspond to transition elements.

89. List the elements Q, X, and Z in order of increasing atomic size (smallest atom first).

90. List the elements D, E, and G in order of decreasing atomic size (largest atom first).

General Questions

91. Distinguish precisely, and in scientific terms, the differences among items in each of the following pairs or groups.
   a) Continuous spectrum, discrete line spectrum
   b) Quantized, continuous
   c) Ground state, excited state
   d) Principal energy level, principal quantum number
   e) Bohr model of the atom, quantum mechanical model of the atom
   f) Principal energy level, sublevel, orbital
   g) *s*, *p*, *d*, *f* (sublevels)
   h) Orbit, orbitals
   i) First, second, third ionization energies
   j) Metal, nonmetal, metalloid, semimetal

92. Determine whether each statement that follows is true or false:
   a) Electron energies are quantized in excited states but not in the ground state.
   b) Line spectra of the elements are experimental evidence of the quantization of electron energies.
   c) Energy is released as an electron passes from ground state to an excited state.
   d) The energy of an electron may be between two quantized energy levels.
   e) The Bohr model explanation of line spectra is still thought to be correct.
   f) The quantum mechanical model of the atom describes orbits in which electrons travel around the nucleus.
   g) Orbitals are regions in which there is a high probability of finding an electron.
   h) All energy sublevels have the same number of orbitals.
   i) The 3*p* orbitals of an atom are larger than its 2*p* orbitals but smaller than its 4*p* orbitals.
   j) At a given sublevel, the maximum number of *d* electrons is 5.
   k) The halogens are found in Group 7A/17 of the periodic table.
   l) The dot structure of the alkaline earths is X•, where X is the symbol of any element in the family.
   m) Stable ions formed by alkaline earth metals are isoelectronic with noble gas atoms.
   n) Atomic numbers 23 and 45 both belong to transition elements.
   o) Atomic numbers 52, 35, and 18 are arranged in order of increasing atomic size.
   p) Atomic numbers 7, 16, and 35 are all nonmetals.

93. Figure 11.9 shows the spectra of hydrogen and neon. Hydrogen has four lines, one of which is sometimes difficult to see. Yet we say there are more lines in the hydrogen spectrum. Why do we not see them?

94. What is meant by the statement that something behaves like a wave?

More Challenging Problems

95. Write the electron configurations you would expect for barium and technetium (Z = 43).

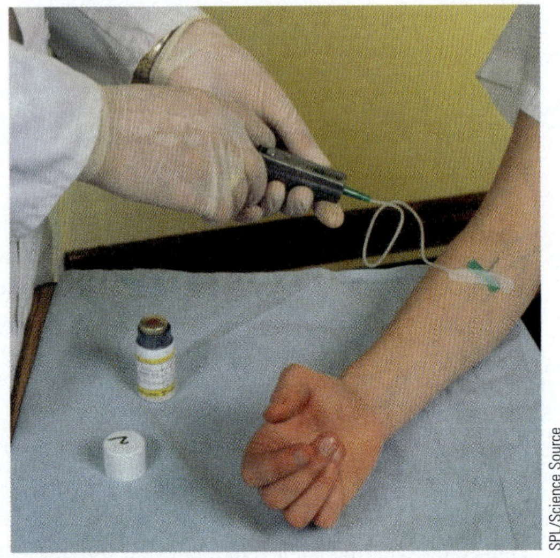

SPL/Science Source

A nuclear medicine technologist injects technetium into a patient with a shielded syringe. Technetium is the element with the lowest atomic number that has only radioactive isotopes.

96. Write the electron configurations you would expect for iodine and tungsten (Z = 74).

97. Why do you suppose the second ionization energy of an element is always greater than the first ionization energy?

98. Why is the definition of atomic number based on the number of protons in an atom rather than the number of electrons?

99. Compare the ionization energies of aluminum and chlorine. Why are these values as they are?

100. Suggest why the ionization energy of magnesium is greater than the ionization energies of sodium, aluminum, and calcium.

101. Do elements become more metallic or less metallic as you (a) go down a group in the periodic table, (b) move left to right across a period in the periodic table? Support your answers with examples.

102. One of the successes of the Bohr model of the atom was its explanation of the lines in atomic spectra. Does the quantum mechanical model also have a satisfactory explanation for these lines? Justify your answer.

103. Although the quantum model of the atom makes predictions for atoms of all elements, most of the quantitative confirmation of these predictions is limited to substances whose formulas are H, He^+, and H_2^+. From your knowledge of electron configurations and the limited information about chemical formulas given in this chapter, can you identify a single feature that all three substances have that makes them unique and makes them relatively easy to investigate?

104. What do you suppose are the electron configuration and the formula of the monatomic ion formed by scandium (Z = 21)?

105. Carbon does not form a stable monatomic ion. Suggest a reason for this.

106. Consider the block of elements in Periods 2 to 6 and Groups 5A/15 to 7A/17. In Group 7A/17 are the halogens, a distinct chemical family in which the different elements share many properties. Elements in Group 6A/16 have enough similarity to be considered a family; they are called the chalcogen family. In Group 5A/15, however, family similarities are weak, and those that exist belong largely to nitrogen, phosphorus, and to some extent, arsenic (Z = 33). Why do you suppose family similarities break down in Group 5A/15?

107. Xenon (Z = 54) was the first noble gas to be chemically combined with another element. Xenon is present in nearly all of the small number of noble gas compounds known today. Note the ionization energy trend begun with the other noble gases in Figure 11.34. What do these facts suggest about the relative reactivities of the noble gases and the character of noble gas compounds?

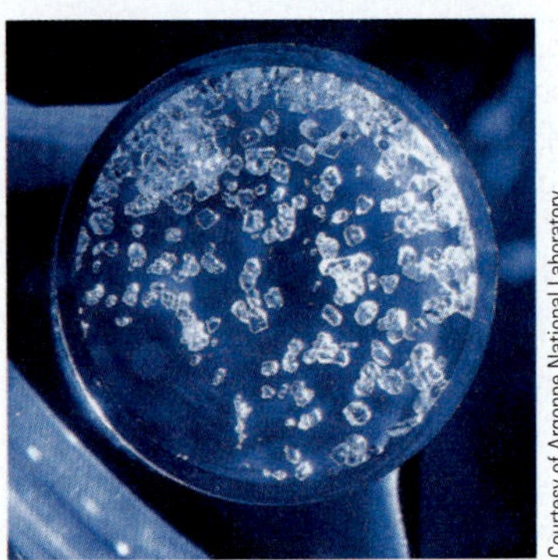

Courtesy of Argonne National Laboratory

Xenon tetrafluoride is one of the noble gas compounds that have been synthesized.

108. Iron forms two monatomic ions, Fe^{3+} and Fe^{2+}. From which sublevels do you expect electrons are lost in forming these ions? (*Hint:* It is possible for electrons other than those in the *s* and *p* sublevels to be involved in forming ions.)

109. Figure 11.34 indicates a general increase in the first ionization energy from left to right across a row of the periodic table. However, there are two sharp breaks in Periods 2 and 3. (a) Suggest why ionization energy should increase as atomic number increases within a period. (b) Can you correlate features of electron configurations with the locations of the breaks?

Answer to Target Check

1. (a) C < N < F. (b) halogen, F; alkali metal, Li; noble gas, Ne; alkaline earth, Be. (c) O < N < P. (d) metals, Na, Mg, Al; nonmetals, P, S, Cl, Ar; metalloid, Si.

Answers to Practice Exercises

1. The product of the wavelength and frequency of electromagnetic radiation is a fixed quantity, the speed of light: $\lambda \times \nu = c$. As frequency decreases, wavelength must increase so that their product remains equal to 3.00×10^8 m/s.

2. A continuous spectrum contains a rainbow of colors that result from energy at all wavelengths visible to humans. It has no lines or bands. A line spectrum has discrete lines of color that correspond to only some of the wavelengths in the visible spectrum. Continuous spectra result from passing white light through a prism. (An example of white light is the light generated by a standard light bulb.) Line spectra are the product of passing through a prism the light emitted by pure elemental substances in gas discharge tubes.

3. (a) The number of protons in a hydrogen atom is always one. It is limited to a single specific value and therefore quantized. (b) The energy of the electron in a hydrogen atom is restricted to certain energies, as evidenced by hydrogen's line spectrum. It is quantized. (c) The radius of the orbit of the electron in a hydrogen atom is restricted to certain radii, as proposed by the Bohr model. This is inferred from hydrogen's line spectrum.

4. (a) A 1*s* orbital is *smaller* than a 2*s* orbital, which is *smaller* than a 3*s* orbital, and so on. As the principal quantum number increases, the size of the orbital increases. See Figure 11.20. (b) The statement is true. *s* orbitals are spherical, so they cannot point in a direction in space. *p* orbitals are dumbbell shaped, which allows them to point in the *x, y,* and *z* directions in space. See Figure 11.19. (c) The statement is true. At $n = 1$, there is 1 sublevel, 1*s*. At $n = 2$, there are two sublevels, 2*s* and 2*p*. At $n = 3$ there are three sublevels, 3*s*, 3*p*, and 3*d*. The pattern continues for increasing values of *n*. See the *Identification of Sublevels* column in the *Sublevels* section of *a summary of…* The Quantum Mechanical Model of the Atom. (d) An orbital may be unoccupied, occupied by 1 electron, or occupied by 2 electrons. See the *Orbitals and Orbital Occupancy* section of the summary.

5. Bromine: $4p^4$; sodium, $3s^1$; oxygen: $2p^4$

6. $1s^22s^22p^63s^23p^4$

7. $1s^22s^22p^63s^23p^64s^13d^5$

8. (a) ns^2np^1; (b) 5A/15

9. $:\ddot{\underset{..}{I}}\cdot$ Na·

10. (a) A potassium atom is larger than a sodium atom because the highest occupied principal energy level in potassium is $n = 4$, and the highest occupied principal energy level in sodium is $n = 3$. (b) A chlorine atom is smaller than a sulfur atom because chlorine has 17 positively charged protons in its nucleus, but sulfur has only 16 protons.

11. (a) The first ionization energy of potassium is less than that of sodium atom because the highest occupied principal energy level in potassium is $n = 4$, and the highest occupied principal energy level in sodium is $n = 3$. The outermost electron is farther from the nucleus. (b) The first ionization energy of chlorine is greater than that of sulfur atom because chlorine has 17 positively charged protons in its nucleus, but sulfur has only 16 protons. The higher positive charge increases the energy needed to remove the outermost electron.

Answers to Concept-Linking Exercises

You may have found more relationships or relationships other than the ones given in these answers.

1. The electromagnetic spectrum refers to the whole range of electromagnetic waves, which includes visible light. In a continuous spectrum of light, the colors blend into each other; there are no separations. In a line spectrum, only separate lines of color appear.

2. If energy levels change gradually in continuous values of energy, theoretically, between any two values, there is an infinite number of other values. Quantized energy levels have only certain values of energy; "in between" values do not exist. In a one electron system, the ground state is when the electron is in the lowest quantized energy level, $n = 1$. If the electron is in any level above $n = 1$, it is in an excited state.

3. The Bohr model of the hydrogen atom pictures the electron in the atom moving in a precisely defined *orbit* around the nucleus. An *orbital* is a region of space around the nucleus in which there is a high probability of finding the electron. An orbital does not define the path of the electron.

4. The principal energy level (*n*), sublevel (*s, p, d, f*), and electron orbitals (1 *s* orbital, 3 *p* orbitals, 5 *d* orbitals, and 7 *f* orbitals) represent values that are described mathematically by the quantum mechanical model of the atom. The Pauli exclusion principle limits the occupancy of an orbital to two electrons.

5. Valence electrons are the electrons in the highest occupied principal energy level when the atom is at ground state. Lewis symbols show the valence electrons as dots placed around the symbol of the element. A filled set of valence electrons, in which the *s* orbital and all three *p* orbitals at the highest occupied principal energy level contain two electrons, is referred to as an octet of electrons.

6. The ionization energy of an element is the amount of energy required to remove one electron from an atom of that element. The second and third ionization energies are the energies required to remove the second and third electrons, respectively, from the atom.

7. Chemical families are groups of elements that have similar chemical properties. They are located in groups in the periodic table: alkali metals in Group 1A/1; alkaline earths in Group 2A/2; halogens in Group 7A/17; and noble gases in Group 8A/18. Though hydrogen forms compounds that correspond to compounds formed by alkali metals and halogens, other properties exclude hydrogen from either of those families.

8. Atomic size is affected by the highest occupied principal energy level (the higher the level, the larger the atom) and the nuclear charge (the higher the charge for a given principal energy level, the smaller the atom).

9. Physically, metals are distinguished from nonmetals by certain properties, such as luster and good electrical conductivity. The character of the elements changes from metal to nonmetal from left to right in any row of the periodic table. The stair-step line in the periodic table separates the metals (left) from the nonmetals (right). The properties of the elements close to the line are neither distinctly metallic nor nonmetallic. These elements are metalloids.

Answers to Blue-Numbered Questions, Exercises, and Problems

1. Gamma rays, x-rays, ultraviolet radiation, infrared radiation, microwaves, and radio waves are other parts of the electromagnetic spectrum. x-rays, microwaves, and radio waves are commonly part of everyday vocabulary.
3. Wavelength, frequency, velocity (speed of light, c)
5. d and e
7. All colors of light could be emitted because of the infinite number of energy differences possible. White light consists of all visible colors. Infrared and ultraviolet light would also be emitted.
9. An atom must absorb energy before it can release that energy in the form of light.
11. An atom with its electron(s) in the ground state cannot emit light because there is no lower energy level to which the electron may fall. Only an atom with electrons in excited states can emit light.
13. Individual lines result when an electron jumps from $n = 2$ to $n = 1$, $n = 3$ to $n = 2$, $n = 3$ to $n = 1$, and so on.
15. The Bohr model provided an explanation for atomic line spectra in terms of electron energies. It also introduced the idea of quantized electron energy levels in the atom.
17. In general, energies increase as the principal energy level increases: $n = 1 < n = 2 < n = 3 \ldots < n = 7$.
19. The total number of sublevels within a given principal energy level is equal to *n,* the principal quantum number.
21. *s:* 1; *p:* 3; *d:* 5; *f:* 7
23. False. Each *p* sublevel contains three orbitals, each of which can hold two electrons, for a maximum of six electrons in the *p* sublevel.
25. True. Each principal energy level has *n* sublevels.
27. False. An orbital may hold one or two electrons, or it may be empty.
29. See Figure 11.18.
31. 6; 10; 14
33. For elements other than hydrogen, the energy of each principal energy level spreads over a range related to the sublevels.
35. (b) Bohr's quantized electron energy levels appear in quantum theory as principal energy levels.
37. $n = 2$ has 2 sublevels, *s* and *p,* so there is no such thing as a 2*d* electron. It is an incorrect symbol.
39. Sulfur, which is in Period 3, Group 6A/16.
41. [Ar] substitutes for $1s^22s^22p^63s^23p^6$, and it is used to write shorthand electron configurations for elements with $19 < Z < 35$.
43. (a) selenium; (b) oxygen; (c) magnesium
45. (a) nitrogen; (b) aluminum; (c) cobalt
47. Mg: $1s^22s^22p^63s^2$; Ni: $1s^22s^22p^63s^23p^64s^23d^8$
49. Cr: $1s^22s^22p^63s^23p^64s^13d^5$; Se: $1s^22s^22p^63s^23p^64s^23d^{10}4p^4$
51. Mg: [Ne]$3s^2$; Ni: [Ar]$4s^23d^8$; Cr: [Ar]$4s^13d^5$; Se: [Ar]$4s^23d^{10}4p^4$
53. [Ar]$4s^23d^3$
55. Many of the similar chemical properties of elements in the same column of the periodic table are related to the valence electrons.
57. $3s^23p^1$ or $\dot{\text{A}}\text{l}$:
59. ns^2np^4
61. (a) Z = 19, 20, 31, 32. (b) Z = 34, 35, 36.
63. As you go from left to right across a row in the periodic table, the valence electrons are all in the same principal energy level. As the number of protons in an atom increases, the positive charge in the nucleus increases. This pulls the valence electrons closer to the nucleus, so the atom becomes smaller.
65. Germanium has $n = 4$ valence electrons; silicon has $n = 3$ valence electrons. When comparing atoms in the same group, the number of occupied energy levels is more influential than nuclear charge in determining size.
67. Ionization energies of elements in the same chemical family decrease as atomic number increases.
69. Within a period, the valence electrons are in the same principal energy level. As the number of protons increases across a period, the positive nuclear charge increases, exerting a greater pull on the electrons. Therefore, ionization energy must increase to pull an electron away.
71. They both have the valence electron configuration ns^2.
73. Halogens
75. ns^2np^6 (Helium is a noble gas with the electron configuration $1s^2$.)
77. (a) noble gases; (b) alkaline earths
79. Chlorine, bromine, and iodine have seven valence electrons: ns^2np^5.
81. Generally, an element is known as a *metal* if it can lose one or more electrons and become a positively charged ion.
83. Most of the elements next to the stair-step line in the periodic table have some properties of both metals and nonmetals. These elements are the metalloids.
85. (a) G, A; (b) R, T
87. X, Q, R, Z, M, T
89. Q < X < Z
92. True: b, e, g, i, k, m, n, p. False: a, c, d, f, h, j, l, o.
93. The other lines are outside the visible spectrum, in the infrared or ultraviolet regions.

94. Something that behaves like a wave has properties normally associated with waves, some of which appear in Figure 11.6.

95. Ba: [Xe]$6s^2$, Tc: [Kr]$5s^2 4d^5$

99. Aluminum atoms have a lower nuclear charge—fewer protons in the nucleus—than chlorine atoms. It is therefore easier to remove a 3*p* electron from an aluminum atom than from a chlorine atom, so the ionization energy of aluminum is lower.

102. The quantum and Bohr model explanations of atomic spectra are essentially the same.

103. All species have a single electron. Species with two or more electrons are far more complex.

104. Sc^{3+} is isoelectronic with an argon atom, $1s^2 2s^2 2p^6 3s^2 3p^6$.

105. To form a monatomic ion, carbon would have to lose four electrons. The fourth ionization energy of any atom is very high.

106. The smaller atoms in Group 5A/15 tend to complete their octets by gaining or sharing electrons, which is a characteristic of nonmetals. Larger atoms in the group tend to lose their highest-energy *s* electrons and form positively charged ions, a characteristic of metals.

107. Xenon has a low ionization energy and high reactivity, relative to the other noble gases. This is the same ionization energy trend seen in all groups in the periodic table.

108. Iron loses two electrons from the 4*s* orbital to form Fe^{2+} and a third from a 3*d* orbital to form Fe^{3+}. This is an example of *d* electrons contributing to the chemical properties of an element.

109. (a) Ionization energy increases across a period of the periodic table because of increasing nuclear charge and the same principal energy level of the outermost electrons. (b) The breaks in ionization energy trends across Periods 2 and 3 occur just after the *s* orbital is filled and just after the *p* orbitals are half-filled.

12 Chemical Bonding

CHAPTER CONTENTS

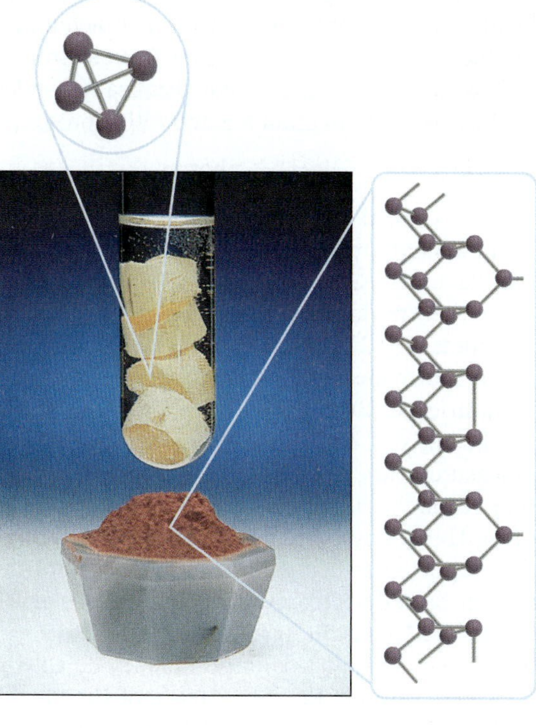

Charles D. Winters

◀ These two substances appear very different, but they are the same pure element—phosphorus. On top is white phosphorus, P_4, stored in water because it catches fire when exposed to air. Notice that its particulate-level structure is a collection of discrete four-atom molecules. When white phosphorus is heated, it changes to red phosphorus, as shown. This substance is stable in air. At the particulate level, red phosphorus is a network of phosphorus atoms with no individual molecules. How can two forms of the same pure element be so different? The answer lies in understanding the differing arrangements of the chemical bonds, the subject of this chapter.

In his atomic theory, John Dalton proposed that atoms of different elements combine to form compounds. He didn't hypothesize about how they combined, or why. We now believe we understand *how* most chemical compounds form. *Why* atoms combine was touched on in Section 2.9: "Reduction of the energy in a chemical system to the smallest amount possible is one of the driving forces that causes chemical changes to occur." This reduction refers to lowering the potential energy resulting from the attractions and repulsions among charged particles in the structure of the compound. The term **chemical bond** describes the forces that hold atoms together to form molecules or polyatomic ions, that hold atoms together in metals, or that hold oppositely charged ions together to form ionic compounds. Chemical bonds can break and reform in new combinations when atoms, molecules, or ions collide. The first contact in such a collision is between the outermost electrons of two particles. These outermost electrons are the valence electrons of the bonded atoms. Our study of chemical bonds, therefore, focuses on the role of valence electrons in forming a bond between two atoms. ▶

◀ **P/Review** The valence electrons of an atom (Section 11.6) are the electrons in the *s* and *p* sublevels of the highest occupied principal energy level when an atom is in the ground state. When these sublevels contain a total of eight electrons, two in the *s* sublevel and six in the *p*, we say the octet of electrons is filled. Two particles are isoelectronic if they have identical electron configurations (Sections 11.5 and 11.7).

12.1 How Did the Chemistry of the Universe Begin?

In Section 5.1, we explained how present-day measurements of the universe can be extrapolated back 14 billion years to deduce the nature of the universe at the point at which it began. At 0.001 second after the beginning, some of the original fundamental particles condensed into protons and neutrons. After a few minutes, the nuclei of helium atoms formed. After 400,000 years of additional cooling and expansion, the first stable atoms formed in significant quantities. Most of these atoms were hydrogen and helium. A part of the story that we did not discuss in Section 5.1 is that before the first permanent atoms formed at a large scale, tiny amounts of the first atoms and even the first chemical bond came into being! This event marks the beginning of the chemistry of the universe.

At 100,000 years after the Big Bang, small fractions of the universe cooled sufficiently to allow atoms and other species to form. One such species was formed when a helium nucleus, He^{2+}, combined with a free electron to form He^+. This helium 1+ ion was then able to combine with another free electron to form a neutral helium atom, He. The first chemically bonded species resulted when a helium atom reacted with a proton:

$$He + H^+ \rightarrow HeH^+ + \text{photon}$$

The HeH^+ species is called the *helium hydride ion*, which is a logical choice for a binary molecular ion, although its IUPAC name is hydridohelium(1+).

The reaction to form this first compound in the early universe cannot be understood by direct measurement, of course. We must take currently measurable phenomena and project them backward in time. Even though these extrapolations are logical on the basis of corroborating numerous types of present-day measurements, one thing in particular had frustrated Big Bang research scientists: a lack of detection of naturally occurring HeH^+ anywhere in the universe. Although HeH^+ was first reported as being made in the laboratory in 1925, it had never been detected other than what was artificially synthesized by humans on Earth...until April 2019.

NASA/Jim Ross

Figure 12.1 The Stratospheric Observatory for Infrared Astronomy (SOFIA). It is a collaborative project of the U. S. National Aeronautics and Space Administration (NASA) and the German Aerospace Center, Deutsches Zentrum für Luft- und Raumfahrt e.V. (DLR). SOFIA is a modified Boeing 747 jet.

With today's technology, it is impossible to detect the fingerprint of HeH^+ through the interference of the Earth's lower atmosphere. Thus, to look for HeH^+ in the distant reaches of the universe, scientists placed a telescope on a flying observatory, the Stratospheric Observatory for Infrared Astronomy (SOFIA) (**Figure 12.1**), which can fly above 99% of the Earth's atmosphere. They pointed the telescope toward NGC 7027 (**Figure 12.2**), a planetary nebula 3,000 light years from Earth. This nebula—a cloud of gases and dust in space—was selected because it presently exists with many of the characteristics of the early universe: hot, dense, and young.

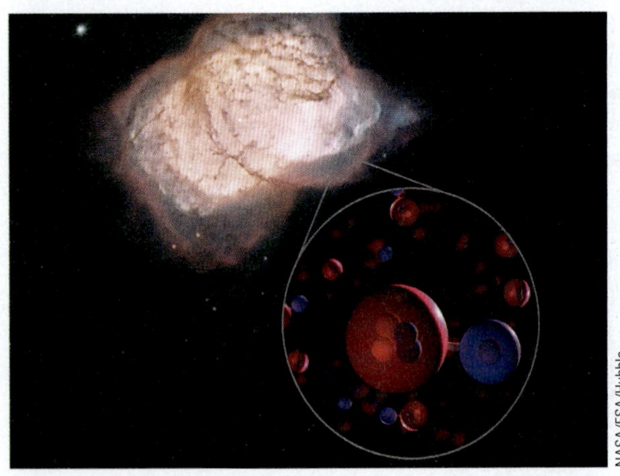

NASA/ESA/Hubble

Figure 12.2 A macroscopic image of NGC 7027, a planetary nebula 2×10^{16} miles (3×10^{16} km) from Earth, with an illustration of particulate-level models of HeH^+, which was first discovered in the natural universe in the nebula. In the particulate-level models, hydrogen atoms are represented by the smaller blue spheres and helium atoms are illustrated as the larger red spheres. HeH^+ was the first chemically bonded species in the history of the universe.

The data definitively indicated the presence of HeH^+ in the nebula. The HeH^+ dilemma with the Big Bang Theory has now been resolved. Nonetheless, research still continues to be done to work toward an increasingly better understanding of the origins of chemical change.

The first time the reaction $He + H^+ \rightarrow HeH^+ + \text{photon}$ occurred, the universe's first chemical bond was formed. This chapter is an introduction to the chemical bonding process, by which atoms or ions become attached in such a way that they form a particle that has a unique identity that is separate from the properties of the reacting species. Also in this chapter, you will see that the quantum mechanical model of the atom, as introduced in Chapter 11, is the fundamental model used to understand how chemical bonds form.

12.2 Monatomic Ions with Noble Gas Electron Configurations

Goal 1 Define and distinguish between cations and anions.

2 Identify the monatomic ions that are isoelectronic with a given noble gas atom and write the electron configuration of these ions.

Elements in the same chemical family usually form monatomic ions having the same charge (Section 6.5). They do this by gaining or losing valence electrons until they become isoelectronic with a noble gas atom and the octet of electrons is complete. ▶ If an atom loses one or more electrons, the ion has a positive charge and is called a **cation.** If the atom gains electrons, the ion is negatively charged and is called an **anion.**

◀ **P/Review** In Section 11.7, an ion is defined as an atom or group of atoms that has an electrical charge because of a difference in the number of protons and electrons. A discussion of the formation of both cations and anions appears in Section 6.5.

Table 12.1 shows how electrons are gained and lost when nitrogen, oxygen, fluorine, sodium, magnesium, and aluminum form monatomic ions that are isoelectronic with a neon atom. The pattern built around neon is duplicated for other noble gases. Thus, phosphorus, sulfur, chlorine, potassium, calcium, and scandium ($Z = 21$) are the elements that form ions that are isoelectronic with argon, and five of the elements on either side of krypton form ions that duplicate the electron configuration of krypton.

The monatomic hydride and lithium ions, H^- and Li^+, duplicate the electron configuration of helium with just two electrons: $1s^2$. **Figure 12.3** is a periodic table that summarizes the elements that form monatomic ions that are isoelectronic with noble gas atoms. Beryllium and boron do not appear in Figure 12.3 because they tend to form covalent bonds by sharing electrons rather than forming ions. We will look at covalent bonds later in this chapter.

Elements below aluminum on the periodic table—gallium (Ga), indium (In), and thallium (Tl)—are not included in Figure 12.3 because these elements are uncommon and have an increasing tendency to form 1+ ions as you move down the group.

You may wonder about the hydrogen ion, H^+. This ion does not normally exist by itself, but rather exists as a hydrated hydrogen ion, which means that the hydrogen ion is associated with one or more molecules of water. ▶ This can be represented as $H^+ \cdot (H_2O)_x$, where x is some number of water molecules. One form of the hydrated hydrogen ion is commonly called the **hydronium ion** and written H_3O^+ (**Figure 12.4**). The ion is not truly monatomic, and therefore, it is not properly included in this section.

◀ **P/Review** The hydrated hydrogen ion is present in water and water solutions. If a solution has a greater hydrogen ion concentration than hydroxide ion concentration, it is acidic. In acidic solutions, the hydrated hydrogen ion forms when a hydrogen-bearing compound reacts with water (Section 6.9).

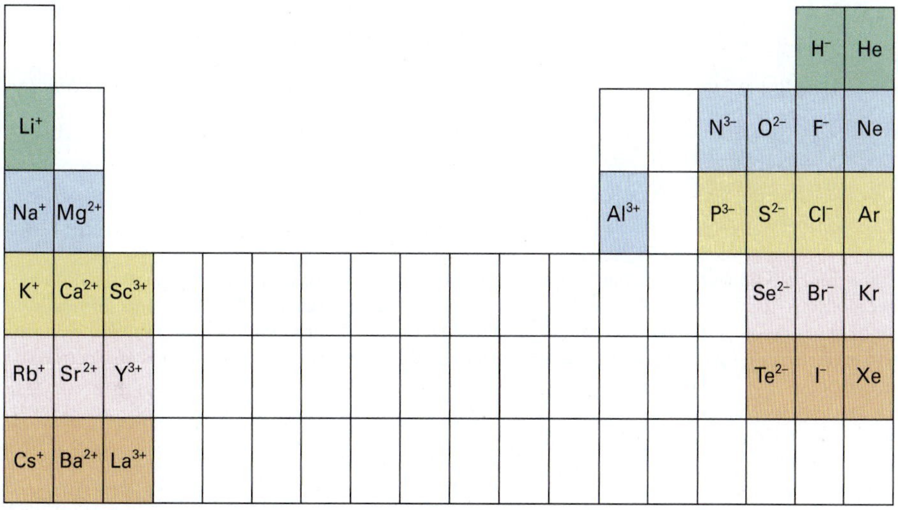

																H^-	He
Li^+														N^{3-}	O^{2-}	F^-	Ne
Na^+	Mg^{2+}											Al^{3+}		P^{3-}	S^{2-}	Cl^-	Ar
K^+	Ca^{2+}	Sc^{3+}													Se^{2-}	Br^-	Kr
Rb^+	Sr^{2+}	Y^{3+}													Te^{2-}	I^-	Xe
Cs^+	Ba^{2+}	La^{3+}															

Figure 12.3 Monatomic ions with noble gas electron configurations. Each color group includes one noble gas atom and the monatomic ions that are isoelectronic with that atom. Beryllium and boron are not included because these elements more commonly form covalent bonds (Section 12.4).

Figure 12.4 The hydronium ion. Hydrogen ions do not exist as independent particles under normal conditions. When in a water solution, one form in which hydrogen ions exist is in a combination with water molecules, forming hydronium ions, H_3O^+.

Table 12.1 Formation of Monatomic Ions That Are Isoelectronic with Neon Atoms*

Element	Atom		Electron(s)		Monatomic Ion	Atom/Ion Electron Count		
Nitrogen	7 p^+ 7 e^-	+	e^- + e^- + e^-	→	7 p^+ 10 e^-	*Start*	*Change*	*Final*
Z = 7						7	+3	10
Group 5A/15	·Ṅ:	+	·(e^-) + ·(e^-) + ·(e^-)	→	[:N̈:]$^{3-}$ N^{3-}			
	$1s^22s^22p^3$	+	3 e^-	→	$1s^22s^22p^6$			
Oxygen	8 p^+ 8 e^-	+	e^- + e^-	→	8 p^+ 10 e^-	*Start*	*Change*	*Final*
Z = 8						8	+2	10
Group 6A/16	·Ȯ:	+	·(e^-) + ·(e^-)	→	[:Ö:]$^{2-}$ O^{2-}			
	$1s^22s^22p^4$	+	2 e^-	→	$1s^22s^22p^6$			
Fluorine	9 p^+ 9 e^-	+	e^-	→	9 p^+ 10 e^-	*Start*	*Change*	*Final*
Z = 9						9	+1	10
Group 7A/17	:Ḟ:	+	·(e^-)	→	[:F̈:]$^-$ F^-			
	$1s^22s^22p^5$	+	e^-	→	$1s^22s^22p^6$			
Neon	10 p^+ 10 e^-				10 p^+ 10 e^-	*Start*	*Change*	*Final*
Z = 10						10		10
Group 8A/18	:N̈e:				Ne			
	$1s^22s^22p^6$				$1s^22s^22p^6$			
Sodium	11 p^+ 11 e^-	−	e^-	→	11 p^+ 10 e^-	*Start*	*Change*	*Final*
Z = 11						11	−1	10
Group 1A/1	Na·	−	·(e^-)	→	Na^+			
	$1s^22s^22p^63s^1$	−	e^-	→	$1s^22s^22p^6$			
Magnesium	12 p^+ 12 e^-	−	e^- + e^-	→	12 p^+ 10 e^-	*Start*	*Change*	*Final*
Z = 12						12	−2	10
Group 2A/2	Mg· (with one dot below)	−	·(e^-) + ·(e^-)	→	Mg^{2+}			
	$1s^22s^22p^63s^2$	−	2 e^-	→	$1s^22s^22p^6$			
Aluminum	13 p^+ 13 e^-	−	e^- + e^- + e^-	→	13 p^+ 10 e^-	*Start*	*Change*	*Final*
Z = 13						13	−3	10
Group 3A/13	·Al· (with one dot below)	−	·(e^-) + ·(e^-) + ·(e^-)	→	Al^{3+}			
	$1s^22s^22p^63s^23p^1$	−	3 e^-	→	$1s^22s^22p^6$			

*This table shows how monatomic anions (pink) and cations (yellow) that are isoelectronic with neon atoms (blue) are formed. A neon atom has ten electrons, including a full octet of valence electrons. Its electron configuration is $1s^22s^22p^6$. Nitrogen, oxygen, and fluorine atoms form anions by gaining enough electrons to reach the same configuration. Sodium, magnesium, and aluminum atoms form cations by losing valence electrons to reach the same configuration. Dots around each elemental symbol represent valence electrons. The notation "·(e^-)" represents an electron added to or subtracted from an electron-dot symbol.

Active Example 12.1 Electron Configurations of Monatomic Ions

Write the electron configurations for calcium and chloride ions. With what noble gases are these ions isoelectronic?

Think Before You Write To write the electron configuration for an ion, start with the electron configuration of the parent atom. Then add or subtract electrons, as appropriate.

Answers *Cover the left column with your cut-out shield. Reveal each answer only after you have written your own answer in the right column.*

Ca: $1s^22s^22p^63s^23p^64s^2$ **Calcium ion:** Ca^{2+} *The electron configuration is developed as explained in Section 11.5.* *Calcium is in Group 2A/2 of the periodic table, so it forms a 2+ ion.*	Write the electron configuration of a calcium atom, and also write the *formula* of a calcium ion. 
Ca^{2+}: $1s^22s^22p^63s^23p^6$	The formula of the calcium ion tells you that it has two fewer electrons than a calcium atom. Remove the two highest-energy electrons from the electron configuration of the calcium atom to produce the electron configuration of the calcium ion. 
The calcium ion has 18 electrons, which makes it isoelectronic with argon, Ar, Z = 18.	Count the number of electrons in the configuration of the calcium ion. With which neutral atom is the calcium ion isoelectronic? 
Cl: $1s^22s^22p^63s^23p^5$ **Chloride ion:** Cl^- **Cl^-:** $1s^22s^22p^63s^23p^6$ The chloride ion has 18 electrons, which makes it isoelectronic with argon, Ar, Z = 18.	Repeat the process that you just followed to derive the electron configuration of the chloride ion. With which neutral atom is the chloride ion isoelectronic? 
You improved your skill at writing electron configurations of monatomic ions, and you improved your understanding of the concept of isoelectronic species.	What did you learn by solving this Active Example? 

Practice Exercise 12.1

Write the electron configurations for the oxide and potassium ions. With what noble gas is each ion isoelectronic?

Target Check 12.1

Which ions among the following are isoelectronic with noble gas atoms?

Cu^{2+} S^{2-} Fe^{3+} Ag^+ Ba^{2+}

▶ **P/Review** This section explains why the formulas of ionic compounds are what they are. Writing formulas of ionic compounds is discussed in Section 6.7. The total positive charge in the formula unit of an ionic compound is equal to the total negative charge.

Figure 12.5 The reaction of sodium and chlorine. Sodium metal burns with a bright yellow flame as it reacts violently with chlorine gas to form sodium chloride.

12.3 Ionic Bonds

Goal 3 Use Lewis symbols to illustrate how an ionic bond can form between monatomic cations from Groups 1A, 2A, and 3A (1, 2, 13) and anions from Groups 5A, 6A, and 7A (15–17) of the periodic table.

We have been discussing the formation of monatomic ions as neutral atoms that gain or lose electrons. For most elements, this is not a common event, but an accomplished fact. The natural occurrence of many elements is in **ionic compounds**—compounds made up of ions—or solutions of ionic compounds. ◀ Almost nowhere in nature, for example, are sodium atoms or chlorine molecules to be found, but there are large natural deposits of sodium chloride (table salt) that are made up of sodium ions and chloride ions. The compound, along with other ionic compounds, may also be obtained by evaporating seawater, which contains the ions in solution.

Sodium and chlorine are both highly reactive elements. If they are brought together, after having been prepared from any natural source, they will react vigorously to form the compound sodium chloride (**Figure 12.5**). In that reaction, a sodium atom literally loses an electron to become a sodium ion, Na^+, and a chloride ion gains an electron to become a chloride ion, Cl^-, as shown in **Figure 12.6**. Lewis symbols can also be used to illustrate the electron transfer:

$$\text{Na}\cdot + \cdot\ddot{\underset{..}{\text{Cl}}}: \longrightarrow \text{Na}^+ + \left[:\ddot{\underset{..}{\text{Cl}}}:\right]^- \longrightarrow \text{NaCl crystal}$$

Ions are not always present in a 1:1 ratio, as in sodium chloride. Calcium atoms have two valence electrons to lose to form a calcium ion, Ca^{2+}, but a chlorine atom

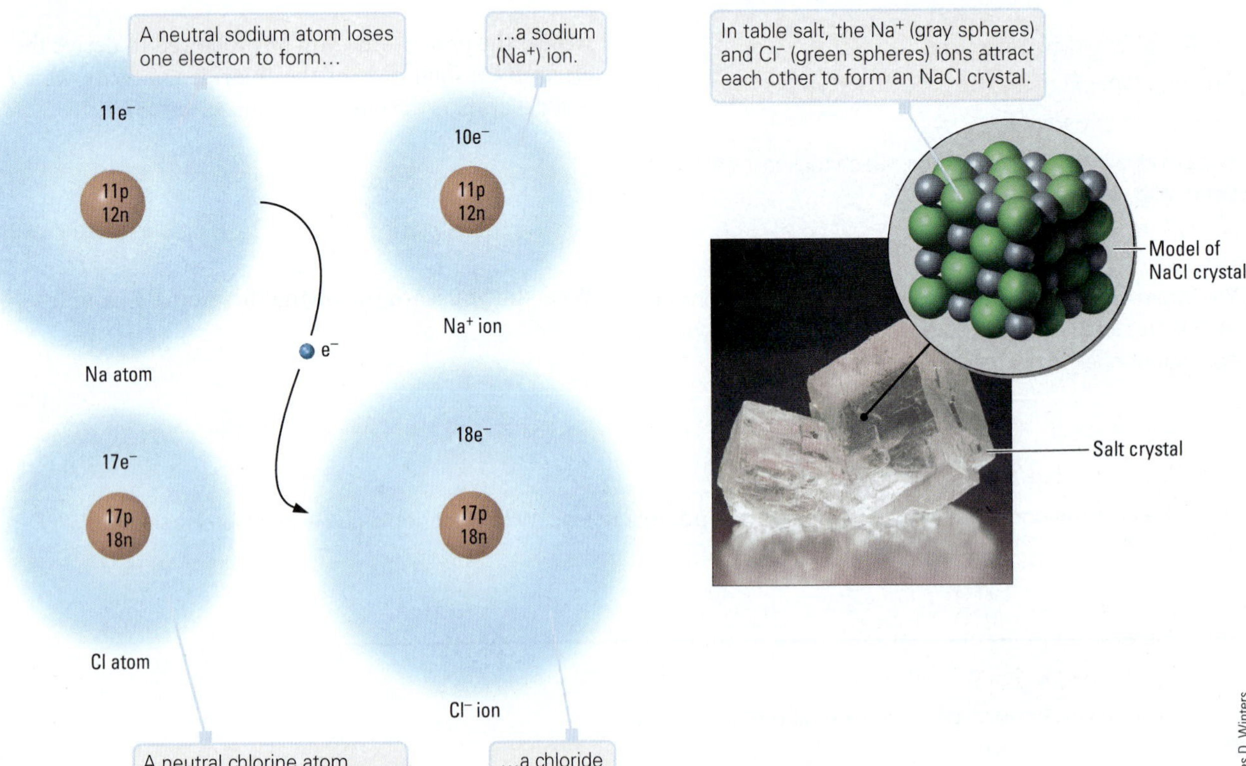

Figure 12.6 The formation of the ionic compound sodium chloride from sodium and chlorine atoms.

receives only one electron when forming a chloride ion, Cl^-. Therefore, it takes two chlorine atoms to receive the two electrons from a single calcium atom:

$$\mathrm{Ca\!:} + \begin{matrix}\cdot\ddot{\underset{..}{\mathrm{Cl}}}: \\ \cdot\ddot{\underset{..}{\mathrm{Cl}}}:\end{matrix} \longrightarrow \mathrm{Ca}^{2+} + 2\left[:\ddot{\underset{..}{\mathrm{Cl}}}:\right]^- \longrightarrow \mathrm{CaCl_2\ crystal}$$

The 1:2 ratio of calcium ions to chloride ions is reflected in the formula of the compound formed, calcium chloride, $CaCl_2$. The combinations of charges in ionic compounds are always in such numbers that the compound is electrically neutral.

Nearly all ionic compounds are solids at normal temperatures and pressures. The solid has a definite geometric structure called a **crystal lattice** (Figure 12.7). Ions in a crystal lattice are arranged so that the potential energy resulting from the attractions and repulsions between them is at a minimum (Section 2.9). The precise form of the lattice depends on the kinds of ions in the compound, their sizes, and the ratio in which they appear. The strong electrostatic forces that hold the ions in fixed position in the crystal are called **ionic bonds.** ▶

◀ **P/Review** Electrostatic forces are unmoving electrical forces (Section 2.8).

Ionic crystals are not limited to monatomic ions; polyatomic ions—ions consisting of two or more atoms—also form crystal structures. Atoms within a polyatomic ion are held together by covalent bonds (Section 12.4). Figure 12.8 shows calcium carbonate at the macroscopic level, and Figure 12.9 is a particulate-level model of a calcium carbonate crystal. The formula of a carbonate ion—one of which is circled in Figure 12.9—is CO_3^{2-}. The carbon atom is surrounded by three covalently bonded oxygen atoms. The carbonate ion is a distinct unit in the structure of the crystal, and it also behaves as a unit in many chemical changes. ▶

◀ **P/Review** The unit-like behavior of polyatomic ions is used when balancing chemical equations (Section 8.9).

The bonds in an ionic crystal are very strong, which is why nearly all ionic compounds are solids at room temperature. A high temperature is required to break the many ionic bonds, free the ions from one another, and melt the crystal to become a liquid. Solid ionic compounds are poor conductors of electricity because the ions are locked in place in the crystal. When the substance is melted or dissolved, the crystal is destroyed. The ions are then free to move and are able to carry electric current (Figure 12.10). ▶ Liquid ionic compounds and water solutions of ionic compounds are good conductors.

◀ **P/Review** Solutions that conduct electric current because of the movement of ions are called *electrolytes* (Section 9.2).

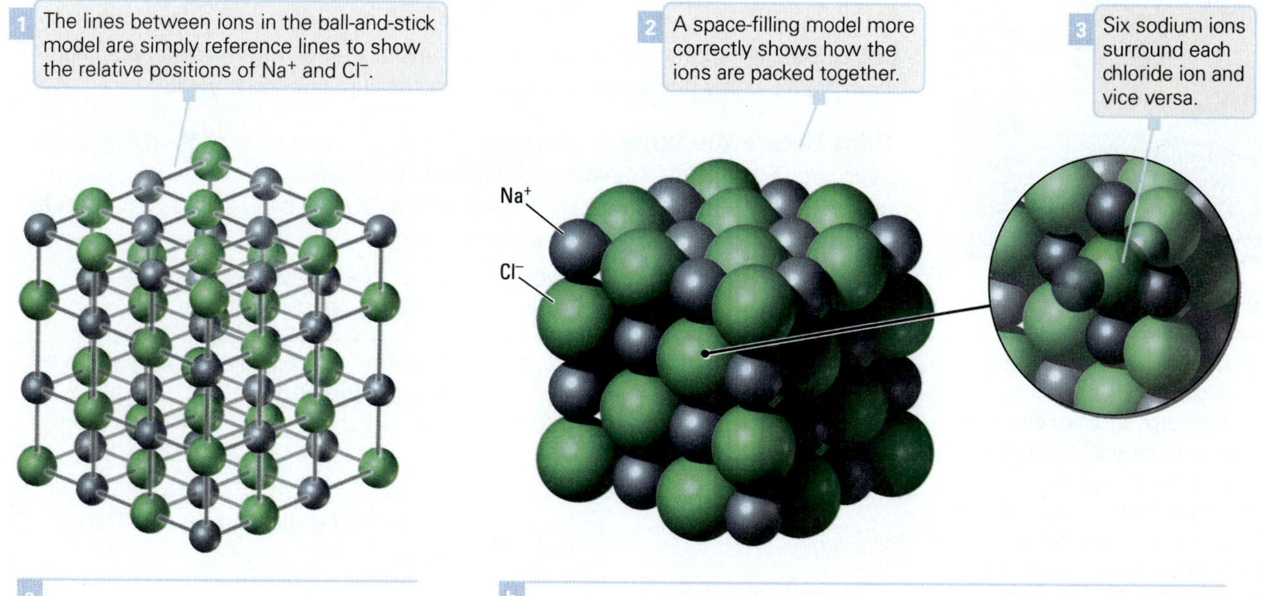

1 The lines between ions in the ball-and-stick model are simply reference lines to show the relative positions of Na⁺ and Cl⁻.
2 A space-filling model more correctly shows how the ions are packed together.
3 Six sodium ions surround each chloride ion and vice versa.
Na⁺
Cl⁻
a
b

Figure 12.7 The sodium chloride crystal lattice. (a) The ball-and-stick model shows the arrangement of the ions very clearly. However, it misrepresents the distances among the ions. (b) The space-filling model shows the way that the ions are packed together, but it does not allow you to easily see the pattern in which the ions are positioned.

Figure 12.8 The many forms of calcium carbonate. (*clockwise from top*) (a) An abalone shell is composed of thin, overlapping layers of calcium carbonate, interspersed with a protein (a class of macromolecules described in Chapter 22). (b) Limestone is a mixture of calcium carbonate and other compounds found in sedimentary rocks. (c) Aragonite is one of the crystalline solid forms of calcium carbonate, characterized by the arrangement of carbonate ions in two planes that point in opposite directions. (d) Blackboard chalk is often incorrectly believed to be made from the mineral chalk, which is a form of calcium carbonate, but in fact it is made from calcium sulfate. (e) Iceland spar is the clear form of the mineral calcite, which is calcium carbonate arranged in a different crystal structure from that of aragonite.

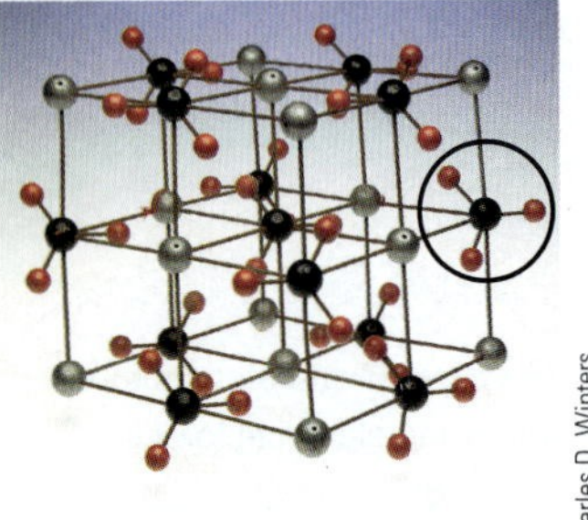

Figure 12.9 Model of a calcium carbonate crystal. The gray spheres represent calcium ions, Ca^{2+}. The black spheres with three red spheres attached (see circle) are carbonate ions, CO_3^{2-}. There are equal numbers of calcium and carbonate ions, yielding a compound that is electrically neutral.

Thinking About Your Thinking

Mental Models

Figures 12.6, 12.7, and 12.9 should help you to form a mental model of an ionic compound. A key characteristic of your mental model needs to be an image of the ion itself as an individual particle. There are no such things as "sodium chloride molecules." The ionic compound sodium chloride consists of equal numbers of sodium ions and chloride ions, but there are no discrete sodium chloride units. Each positively charged sodium ion is surrounded by negatively charged chloride ions, and each chloride ion is surrounded by sodium ions.

Recall that the process of dissolving sodium chloride in water is a physical change. (Physical and chemical changes are described in Section 2.4.) The sodium ions and chloride ions originally locked in the solid crystal are themselves unchanged as they become sodium ions and chloride ions surrounded by water molecules. If the water evaporates, you have the same sodium ions and chloride ions, but they are in a solid crystalline state once again. No chemical change has occurred; the sodium ions and chloride ions remain unchanged throughout the process. If you can imagine how this occurs at the particulate level, you have a good understanding of the structure of ionic compounds.

Figure 12.10 Electrical conductivity of a liquid ionic compound. A lightbulb conductivity apparatus (review Figure 9.5 if necessary) shows that when an ionic compound is heated until the solid changes to the liquid state, it conducts electric current. We interpret this observation at the particulate level (shown in Figure 9.5) by concluding that the ions must be free to move relative to one another when in the liquid state, allowing them to carry current between the electrodes.

Active Example 12.2 Ionic Bonds

Use Lewis diagrams to show the electron transfer that occurs when a potassium atom reacts with a fluorine atom to form a potassium fluoride crystal.

Think Before You Write In this Active Example, you will use symbols to represent a particulate-level process. Chemists commonly express their mental models with written symbols, and thus this example provides an opportunity for you to think as a chemist as well as improve your understanding of ionic bonding.

Answers *Cover the left column with your cut-out shield. Reveal each answer only after you have written your own answer in the right column.*

K· *Potassium is in Group 1A/1, so it has one valence electron.*	Begin by writing the Lewis diagram of a potassium atom. 
·F: (with dots above and below F) *Fluorine is in Group 7A/17, so it has seven valence electrons. It does not matter which of the four sides has one electron.*	Write the Lewis diagram of a fluorine atom. 

$K\cdot + \cdot\ddot{F}: \longrightarrow K^+ + [:\ddot{F}:]^-$ *Potassium has one excess electron and fluorine is one electron short of an octet, so the electron transfers in such a way to give both products an octet of electrons.*	Use an arrow to illustrate the electron transfer and draw the Lewis diagrams of the species that result. Don't forget to indicate the charges of charged particles. Imagine the particulate-level process in your mind as you write the symbols.
$K\cdot + \cdot\ddot{F}: \longrightarrow K^+ + [:\ddot{F}:]^-$ $\longrightarrow$ KF crystal	The final step is to think about the huge numbers of potassium ions and fluoride ions that form a potassium fluoride crystal. The crystal cannot be conveniently illustrated by a Lewis diagram, so finish the illustration by drawing another arrow and writing "KF crystal" to the right.
You improved your skill at using symbols to represent mental models of particulate-level processes, and you improved your understanding of ionic bonding.	What did you learn by solving this Active Example?

Practice Exercise 12.2

Use Lewis diagrams to show the electron transfer that occurs when a magnesium atom reacts with fluorine atoms to form a magnesium fluoride crystal.

12.4 Covalent Bonds

Goal 4 Describe, use, and explain each of the following with respect to forming a covalent bond: electron cloud, charge cloud, or charge density; valence electrons; half-filled electron orbital; filled electron orbital; electron sharing; orbital overlap; octet rule or rule of eight.

5 Use Lewis symbols to show how covalent bonds are formed between two nonmetal atoms.

6 Distinguish between bonding electron pairs and lone pairs.

Let's consider two familiar white crystals—salt and sugar. Both dissolve easily in water, but the salt water conducts electricity, whereas the sugar water does not (**Figure 12.11**). Salt melts at 801°C to give a colorless liquid, but sugar doesn't melt at all. Instead, at only 160°C, sugar begins to char, emitting the aroma of caramel. (Indeed, that's how caramel is made.)

We know that the properties of table salt and other ionic compounds are explained by ionic bonding. Other compounds, such as sugar, have properties so different from those of ionic compounds that a different type of bonding must be responsible for holding together atoms in these compounds.

Water (H_2O), hydrogen fluoride (HF), and methane (CH_4) are compounds of the "other kind." Hydrogen fluoride and methane are gases at room temperature and pressure, and water is a liquid. When hydrogen fluoride and methane are condensed to liquids, they are nonconductors, like water. These are **molecular compounds**, whose ultimate structural unit is an individual particle known as a **molecule**. The physical and chemical properties of a molecule are different from the properties of the atoms that make up the molecule.

Figure 12.11 Conductivity of solutions. Ions are held together by ionic bonds in salt, but another type of bonding must be responsible for the properties of sugar.

For electrical current to flow and light the bulb, the liquid in which the electrodes are immersed must contain ions, which carry electrical charge.

Pure water does not contain ions and thus does not light the bulb.

a Pure water

Marna G. Clarke

The solution of sucrose (table sugar) and pure water also lacks ions, and fails to light the bulb.

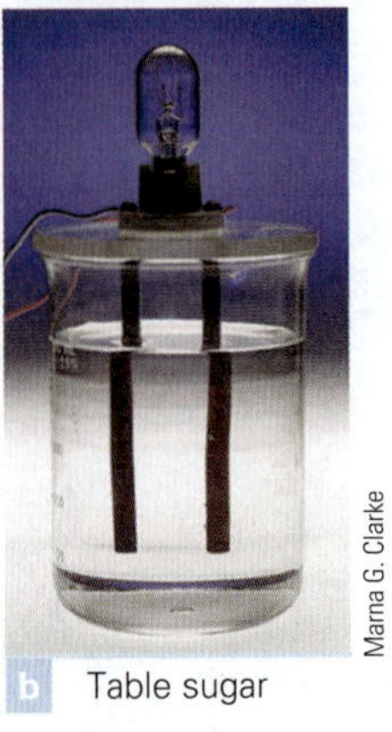

b Table sugar

Marna G. Clarke

The solution of sodium chloride (NaCl) and pure water does contain ions, and thus lights the bulb.

c Table salt

Charles D. Winters/Science Source

Oesper Collection in the History of Chemistry/ University of Cincinnati.

Figure 12.12 Gilbert N. Lewis (1875–1946). Electron dot structures are called Lewis diagrams in his honor.

In 1916, Gilbert N. Lewis, an American scientist (**Figure 12.12**), proposed that two atoms in a molecule are held together by a **covalent bond** in which they share one or more pairs of electrons. The idea is that when the bonding electrons can spend most of their time between two atomic nuclei, they attract *both* positively charged nuclei and "couple" the atoms to each other, much like a car and a trailer are held together by the trailer hitch between them. The result is a bond that is permanent until broken by a chemical change.

The simplest covalent bond appears in hydrogen, H_2. Using Lewis symbols, the formation of a molecule of H_2 can be represented as:

$$\mathrm{H\cdot + \cdot H \longrightarrow H:H} \quad \textit{or} \quad \mathrm{H{-}H}$$

The two dots or the straight line drawn between the two atoms represent the covalent bond that holds the atoms together. In quantum mechanical terms, we say that the **electron cloud** or **charge density** formed by the two electrons is concentrated in the region between the two nuclei. This is where there is the greatest probability of locating the bonding electrons. The atomic orbitals of the hydrogen atoms **overlap**, coupling them together to form the hydrogen molecule.

The formation of a bond between two hydrogen atoms is shown in **Figure 12.13**. The caption describes and explains in some detail how and why this bond forms. Study both the illustration and the caption carefully. In particular, notice the last sentence. When bonding electrons are *between two nuclei*, both nuclei are attracted to the electrons. The electrons link the nuclei together; the electrons are the "glue" that bonds the atoms to each other.

A similar approach shows the formation of the covalent bond between two fluorine atoms to form a molecule of F_2 and between one hydrogen atom and one fluorine atom to form an HF molecule:

$$:\ddot{\underset{..}{\mathrm{F}}}\cdot + \cdot\ddot{\underset{..}{\mathrm{F}}}: \longrightarrow :\ddot{\underset{..}{\mathrm{F}}}:\ddot{\underset{..}{\mathrm{F}}}: \quad \textit{or} \quad :\ddot{\underset{..}{\mathrm{F}}}{-}\ddot{\underset{..}{\mathrm{F}}}: \quad \textit{or} \quad \mathrm{F{-}F}$$

$$\mathrm{H}\cdot + \cdot\ddot{\underset{..}{\mathrm{F}}}: \longrightarrow \mathrm{H}:\ddot{\underset{..}{\mathrm{F}}}: \quad \textit{or} \quad \mathrm{H}{-}\ddot{\underset{..}{\mathrm{F}}}: \quad \textit{or} \quad \mathrm{H{-}F}$$

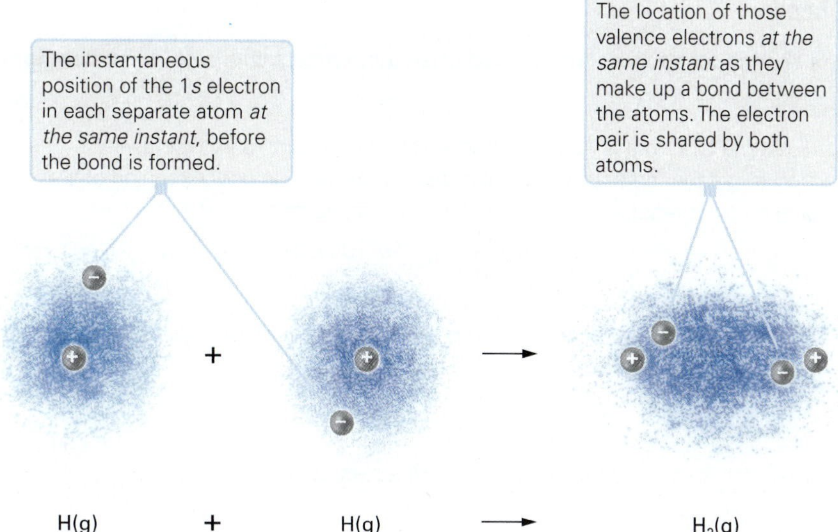

The instantaneous position of the 1s electron in each separate atom at the same instant, before the bond is formed.
The location of those valence electrons at the same instant as they make up a bond between the atoms. The electron pair is shared by both atoms.
H(g) + H(g) → H₂(g)

Figure 12.13 The formation of a hydrogen molecule from two hydrogen atoms. ▶ Each location in the electron cloud represents an instantaneous position of the electron in the atom. The charge clouds of the half-filled 1*s* orbitals of the two hydrogen atoms overlap in the combined atoms in such a way that the two electrons now fill the overlapping 1*s* orbitals of both atoms. The bonding electrons spend most of their time between the two nuclei, as suggested by the heavier density of electron-position dots in that area.

To understand why covalent bonds form, consider the attraction and repulsion forces both within each atom and between the two atoms when the atoms are separated and after the bond has been formed. Before bonding (left of the arrow), attractions and repulsions between the separated atoms are negligible. Within the atoms, there is an attraction between the electron and the nucleus. After bonding (right of the arrow), the atoms are close to each other, and the electrons are attracted to *both* nuclei. *It is this attraction of the bonding electrons to two nuclei that holds the nuclei together in a covalent bond.*

Illustrations and their captions are sometimes used to explain major concepts. That's what is being done here.

Fluorine has seven valence electrons. The 2*s* orbital and two of the 2*p* orbitals are filled, but the remaining 2*p* orbital has only one electron. The F_2 bond is formed by the overlap of the half-filled 2*p* orbitals of two fluorine atoms. In the HF molecule, the bond forms from the overlap of the half-filled 1*s* orbital of a hydrogen atom with the half-filled 2*p* orbital of a fluorine atom.

When used to show the bonding arrangement between atoms in a molecule, an electron dot symbol is commonly called a **Lewis diagram**, **Lewis formula**, or **Lewis structure**. Notice that the unshared electron pairs of fluorine are shown for two of the Lewis diagrams for F_2 and HF, but they are omitted in the rightmost F—F and H—F diagrams. Technically, they should always be shown, but they are sometimes omitted when not absolutely needed. Unshared electron pairs are often called **lone pairs**.

When two atoms share two bonding electrons, the electrons effectively "belong" to both atoms. They count as valence electrons for each bonded atom. Thus, each hydrogen atom in H_2 and the hydrogen atom in HF have two electrons, the same number as an atom of the noble gas helium. Each fluorine atom in F_2 as well as the fluorine atom in HF has eight valence electrons, matching neon and the other noble gas atoms.

These and many similar observations lead us to believe that the stability of a noble gas electron configuration is the result of a minimization of energy associated with that configuration. This generalization is known as the **octet rule** (*octa-* = eight; eight valence electrons) because each atom has "completed its octet." The tendency toward a complete octet of electrons in a bonded atom reflects the natural tendency of a system to move to the lowest-energy state possible. ▶

In Section 2.9, minimization of energy was identified as "one of the driving forces that causes chemical changes to occur." Many laboratory measurements show that the energy of a system is reduced as bonds form that reach the noble gas electron configuration.

Active Example 12.3 Covalent Bonds

Use Lewis diagrams to show the electron sharing that occurs when a hydrogen atom reacts with a chlorine atom to form a hydrogen chloride molecule.

Think Before You Write As with Active Example 12.2, in this Active Example, you will use symbols to represent a particulate-level process. The point is not simply to get the "right answer," but instead to develop an understanding of the covalent bonding process. Think about how the particles react as you use symbols to illustrate that reaction.

Answers *Cover the left column with your cut-out shield. Reveal each answer only after you have written your own answer in the right column.*

Answers	
$\text{H}\cdot$ *A hydrogen atom has one electron, so its Lewis diagram has a single dot.*	Begin with the Lewis diagram of the hydrogen atom.
$\cdot\underset{\cdot\cdot}{\ddot{\text{Cl}}}:$ *Chlorine is in Group 7A/17, so it has seven valence electrons.*	Draw the Lewis diagram of a chlorine atom.
$\text{H}\cdot + \cdot\underset{\cdot\cdot}{\ddot{\text{Cl}}}: \longrightarrow \text{H}:\underset{\cdot\cdot}{\ddot{\text{Cl}}}:$	Show the separated atoms to the left of a reaction arrow and show the covalently bonded hydrogen chloride molecule to the right.
The attraction of the negatively charged bonding electron pair to the two positively charged nuclei holds the nuclei together to form the molecule.	What holds the atoms together in the hydrogen chloride molecule?
You improved your skill at using symbols to represent mental models of particulate-level processes, and you improved your understanding of covalent bonding.	What did you learn by solving this Active Example?

Practice Exercise 12.3

Use Lewis diagrams to show the electron transfer that occurs when hydrogen atoms react with an oxygen atom to form a water molecule.

12.5 Polar and Nonpolar Covalent Bonds

Goal
7 Distinguish between polar and nonpolar covalent bonds.
8 Predict which end of a polar bond between identified atoms is positive and which end is negative.
9 Rank bonds in order of increasing or decreasing polarity based on periodic trends in electronegativity values or actual values, if given.

The two electrons joining the atoms in the H_2 molecule are shared equally by the two nuclei. A bond in which bonding electrons are shared equally is a **nonpolar covalent bond**. A more formal way of saying this is that the charge density is centered in the region between the bonded atoms. A bond between identical atoms, as in H_2 or F_2, is always nonpolar.

The fluorine atom in an HF molecule has a stronger attraction for the bonding electron pair than the hydrogen atom has. The bonding electrons, therefore, spend

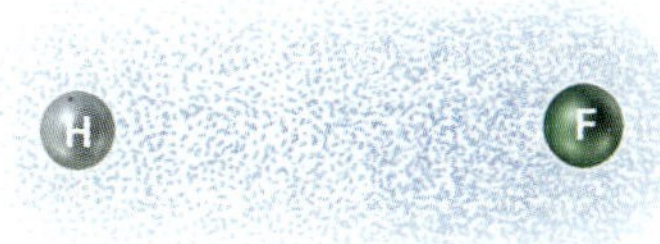

Figure 12.14 A polar bond. Fluorine in a molecule of HF has a higher electronegativity than hydrogen. The bonding electron pair is therefore shifted toward fluorine. The uneven distribution of charge yields a polar bond.

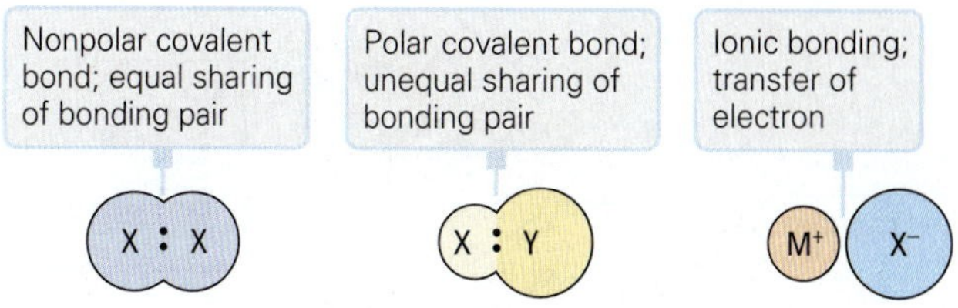

Nonpolar covalent bond; equal sharing of bonding pair

X : X

Polar covalent bond; unequal sharing of bonding pair

X : Y

Ionic bonding; transfer of electron

M^+ X^-

Figure 12.15 A comparison of nonpolar covalent, polar covalent, and ionic bonding.

more time nearer the fluorine atom nucleus than the hydrogen atom nucleus. A bond in which bonding electrons are shared, but shared unequally, is a **polar covalent bond**. The charge density is shifted toward the fluorine atom and away from the hydrogen atom. Because the negative charge density is closer to the fluorine atom, that atom acts as a negative pole, and the hydrogen atom acts as a positive pole (**Figure 12.14**). When the charge density shift is extreme, the bonding electrons are effectively transferred from one atom to another, and an ionic bond results (**Figure 12.15**).

Bond polarity in covalent bonds may be described in terms of the electronegativities of the bonded atoms. American chemist Linus Pauling (**Figure 12.16**) originally described the **electronegativity** of an element as the ability of an atom of that element in a molecule to attract bonding electron pairs to itself. High electronegativity identifies an element with a strong attraction for bonding electrons.

AP Images/ROD SEARCY

Figure 12.16 Linus Pauling (1901–1994). The concept of electronegativity was proposed by Pauling, who was awarded the 1954 Nobel Prize in Chemistry and the 1963 Nobel Peace Prize.

Thinking About Your Thinking

Mental Models

Figures 12.13–12.15 are designed to help you form a mental model of a covalent bond. Note, in particular, how the blue electron-position dots in Figures 12.13 and 12.14 are concentrated between the nuclei of the atoms that are bonding. Nuclei are made up of positively charged protons and neutral neutrons. The net effect is a strong positive charge for a nucleus. Electrons carry a negative charge, and oppositely charged particles attract one another. Therefore, both positively charged nuclei linked by a covalent bond are attracted to the negatively charged electrons that are most likely to be found between the nuclei. This is why a covalent bond is also called an *electron-sharing bond*.

The covalent bond in H_2, shown in Figure 12.13, has an even distribution of electron density. This illustration should be in your mind as the model for any nonpolar covalent bond. The distribution of electron density is uneven in the polar covalent bond in HF, shown in Figure 12.14. Your mental model should be similar, with the electron density shifted toward the more electronegative atom. The greater the electronegativity difference, the more extreme the shift.

Electronegativity values of the elements are shown in **Figure 12.17**. Notice that electronegativities tend to be greater at the top of any column for the main group elements. In a smaller atom, the bonding electrons are closer to the nucleus and therefore are attracted by it more strongly. Electronegativities also increase from left to right across any row of the periodic table. This matches the increase in nuclear charge among atoms whose bonding electrons are in the same principal energy level. ▶ In general, electronegativities are highest in the upper-right region of the periodic table and lowest in the lower-left region.

◀ **P/Review** These explanations of electronegativity trends are identical with those given for atomic size (Section 11.7).

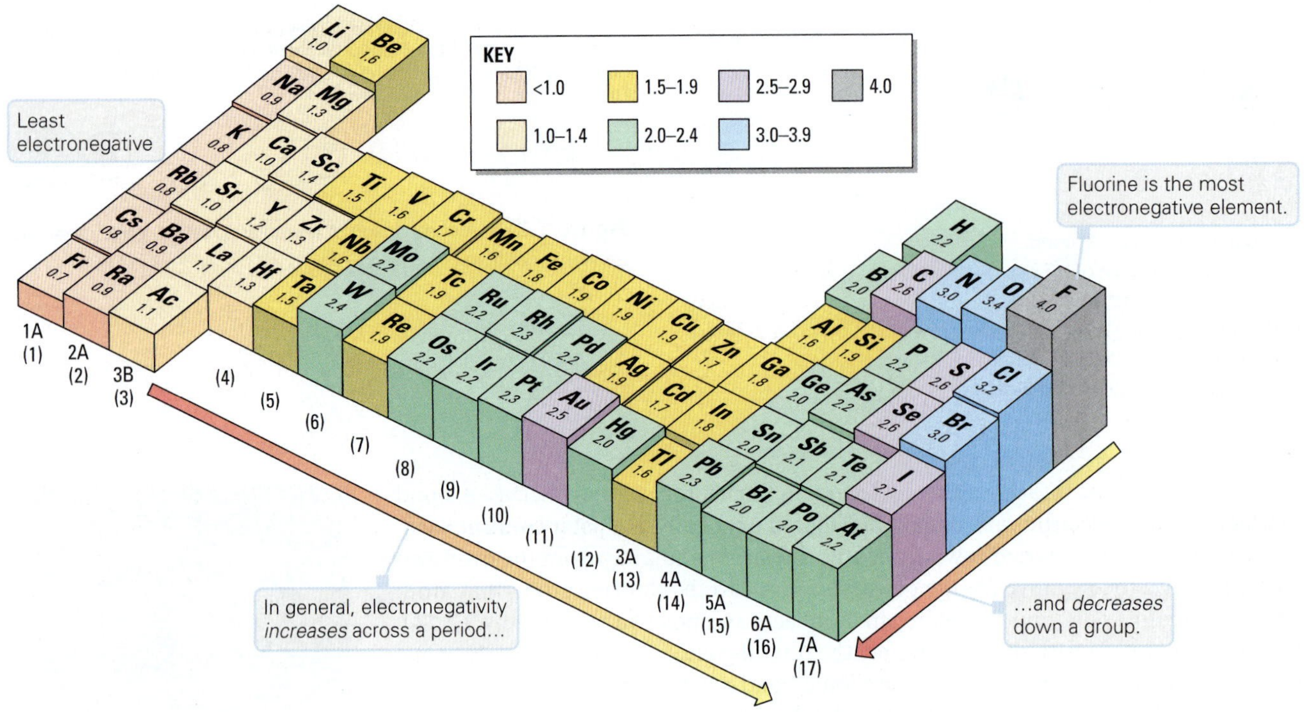

Figure 12.17 Periodic trends in electronegativity values. Electronegativity values generally increase from left to right across any period of the table, and they decrease from the top to the bottom of any group for the main group elements. The values given are on the Pauling scale, expressed in Pauling units.

▶ **P/Review** Bond polarities are used to predict the polarities of molecules (Section 13.6). Molecular polarity is largely responsible for the physical properties of many compounds (Section 15.4).

You can estimate the polarity of a bond by calculating the difference between the electronegativity values for the bonded elements: The greater the difference, the more polar the bond. ◀ In H_2 and F_2, in which two atoms of the same element are bonded, the electronegativity differences are zero and the bonds are nonpolar. In HF, the electronegativity difference is 1.8 (4.0 for fluorine minus 2.2 for hydrogen). A bond between carbon and chlorine, for example, with an electronegativity difference of $3.2 - 2.6 = 0.6$, is more polar than an H—H bond, but less polar than an H—F bond.

The bonding electrons are displaced toward the more electronegative element, which acts as the negative pole in a polar covalent bond. The less electronegative element is the positive pole. This is sometimes indicated by using an arrow rather than a simple dash, with the arrow pointing to the negative pole. In a bond between hydrogen and fluorine, this is $\mathrm{H} \mapsto \mathrm{F}$. Another representation is δ^- written in the region of the negative pole and δ^+ in the area with the positive pole. The character δ is a lowercase Greek delta. In this use, it represents a partial negative or partial positive charge—less than the full charge found on ions. Thus, for hydrogen fluoride $^{\delta+}\mathrm{H} \mapsto \mathrm{F}^{\delta-}$.

Active Example 12.4 Polarity of Covalent Bonds

Using data from Figure 12.17, arrange the following bonds in order of increasing polarity and place an arrowhead pointing toward the element that will act as the negative pole: H–O, H–S, H–P, H–C.

Think Before You Write To predict the magnitude of polarity, you need to calculate the electronegativity difference between the bonded elements. The greater the electronegativity difference, the more polar the bond. The most electronegative element is the negative pole.

Answers *Cover the left column with your cut-out shield. Reveal each answer only after you have written your own answer in the right column.*

Locate the elements in the table in Figure 12.17, and calculate and write down the differences in electronegativity.

H–O: 3.4 − 2.2 = 1.2
H–S: 2.6 − 2.2 = 0.4
H–P: 2.2 − 2.2 = 0.0
H–C: 2.6 − 2.2 = 0.4

Now arrange the bonds in order from the least polar to the most polar.

H–P < H–S ≈ H–C < H–O

The H–P bond is essentially nonpolar (electronegativity difference = 0.0 to the precision of the data in Figure 12.17). H–O is the most polar bond because it has the largest electronegativity difference (1.2). The other bonds, with equal electronegativity differences, have about the same polarity.

Now place an arrowhead on each bond that points toward the negative pole.

Since sulfur, carbon, and oxygen are all more electronegative than hydrogen, the electron density in these bonds is shifted away from hydrogen toward the other element. There is essentially no electronegativity difference in the H–P bond, so neither atom will act as a negative pole.

What did you learn by solving this Active Example?

You improved your understanding of the concept of covalent bond polarity.

Practice Exercise 12.4

Arrange the following bonds in order of decreasing polarity: C–C, C–H, C–N, C–O, C–S. State the element that is the negative pole in each. You may refer to Figure 12.17.

12.6 Multiple Bonds

Goal 10 Distinguish among single, double, and triple bonds, and identify these bonds in a Lewis diagram.

So far our consideration of covalent bonds has been limited to the sharing of one pair of electrons by two bonded atoms. Such a bond is called a **single bond**. In many molecules, we find two atoms bonded by two pairs of electrons; this is a **double bond**. When two atoms are bonded by three pairs of electrons, the bond is called a **triple bond**. ▶ All four electrons in a double bond and all six electrons in a triple bond are counted as valence electrons for each of the bonded atoms.

In special situations, evidence indicates that quadruple bonds, quintuple bonds, and sextuple bonds exist in compounds.

The most abundant substance containing a triple bond is nitrogen, N_2. Its Lewis diagram may be thought of as the combination of two nitrogen atoms, each with three unpaired electrons:

$$:\dot{\underset{\cdot}{N}}\cdot + \cdot\dot{\underset{\cdot}{N}}: \longrightarrow :N\vdots\vdots N: \quad \textit{or} \quad :N{\equiv}N:$$

Counting the bonding electrons for both atoms, each nitrogen atom is satisfied with a full octet of electrons.

Experimental evidence supports the idea of **multiple bonds**, a general term that includes double and triple bonds. A triple bond is stronger and the distance between bonded atoms is shorter than the same measurements for a double bond between the same atoms, and a double bond is stronger and shorter than a single bond. Bond strength is measured as the energy required to break a bond. The greater the energy needed to break a bond, the stronger the bond. The triple bond in N_2 is among the strongest bonds known. This is one of the reasons elemental nitrogen is so stable and unreactive in Earth's atmosphere. ▶

Nitrogen makes up 78% of Earth's atmosphere by volume.

Target Check 12.2

Which among the following have a double bond? Which have a triple bond? Which have a multiple bond?

(I) H₃C—C≡N: — $H{-}\underset{H}{\overset{H}{C}}{-}C{\equiv}N{:}$

(II) $H{-}\overset{H}{C}{=}\overset{H}{C}{-}C{\equiv}N{:}$

(III) $H{-}\underset{H}{\ddot{N}}{-}\overset{:O:}{\overset{\|}{C}}{-}\underset{H}{\ddot{N}}{-}H$

(IV) $H{-}\underset{H}{\overset{H}{C}}{-}\overset{:O:}{\overset{\|}{C}}{-}\underset{H}{\ddot{N}}{-}H$

12.7 Atoms That Are Bonded to Two or More Other Atoms

Using hydrogen and fluorine as examples, we have seen that two atoms that have a single unpaired valence electron are able to form a covalent bond by sharing those electrons. What if an atom has two unpaired valence electrons? Can it form two bonds with two different atoms? Yes, it can. In fact, that is how a water molecule forms. A hydrogen atom forms a bond with one of the two unpaired valence electrons in an oxygen atom, and a second hydrogen atom does the same with the second unpaired oxygen electron:

$$H\cdot + \cdot\ddot{\underset{..}{O}}\cdot + \cdot H \longrightarrow H:\ddot{\underset{..}{O}}:H \quad or \quad H{-}\ddot{\underset{..}{O}}{-}H$$

This is the same as a hydrogen atom forming a bond with another hydrogen atom or with a fluorine atom.

A nitrogen atom has five valence electrons, three of which are unpaired. It therefore forms bonds with three hydrogen atoms to produce a molecule of ammonia, NH_3. Carbon has four valence electrons, only two of which are unpaired. This would lead us to expect that a carbon atom can form bonds with only two hydrogen atoms. In fact, all four electrons form bonds. The compound produced is methane, CH_4, the principal component of the natural gas burned as fuel in many homes. The Lewis diagrams of ammonia and methane are:

All expectations should be checked experimentally. That carbon forms only two bonds is an example of a logical prediction that, when tested in the laboratory, is not confirmed. All life on Earth has as its basis the ability of the carbon atom to form four bonds.

$$H{-}\underset{H}{\ddot{N}}{-}H \qquad H{-}\underset{H}{\overset{H}{C}}{-}H$$

ammonia methane

Many polyatomic molecules contain multiple bonds. Ethylene, C_2H_4, the structural unit of the plastic polyethylene, has a double bond between two carbon atoms. There is a triple bond between carbon atoms in acetylene—C_2H_2—the fuel used in a welder's torch. The carbon atom in carbon dioxide, CO_2, is double bonded to two oxygen atoms. The Lewis diagrams for these compounds are:

$$H_2C{=}CH_2 \qquad H{-}C{\equiv}C{-}H \qquad \ddot{\underset{..}{O}}{=}C{=}\ddot{\underset{..}{O}}$$

ethylene acetylene carbon dioxide

Notice that if both lone pairs and bonding electron pairs are counted, the octet rule is satisfied for all atoms except hydrogen in all Lewis diagrams in this section. Hydrogen, as usual, duplicates the two electron count of the noble gas helium.

Target Check 12.3

a) Is it possible, under the octet rule, for a single atom to be bonded by double bonds to each of three other atoms? Explain your answer.

b) What is the maximum number of atoms that can be bonded to the same atom and have that central atom conform to the octet rule? Explain.

12.8 Exceptions to the Octet Rule

Odd-Electron Molecules

Not all substances conform to the octet rule. Two common oxides of nitrogen, NO and NO_2, have an odd number of electrons. Therefore, it is impossible to write Lewis diagrams for these compounds in which each atom is surrounded by eight valence electrons. Their Lewis diagrams are drawn with one single electron in a space where an electron pair would otherwise appear:

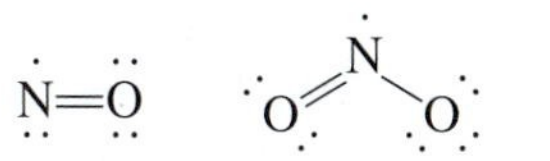

N=O O=N—O

Molecules with More Than Four Electron Pairs Around the Central Atom

All of the bonds we have encountered thus far have been formed by the valence electrons, the highest-energy *s*- and *p*-orbital electrons in the atom. Atoms of elements in the third period and higher can have more than four electron pairs surrounding them. This is accomplished by involving *d*-orbitals in the formation of covalent bonds. ▶ Phosphorus pentafluoride, PF_5, places five electron-pair bonds around the phosphorus atom; six pairs surround sulfur in SF_6:

◀ **P/Review** Electrons first appear in *d*-orbitals in the third principal energy level (Section 11.4).

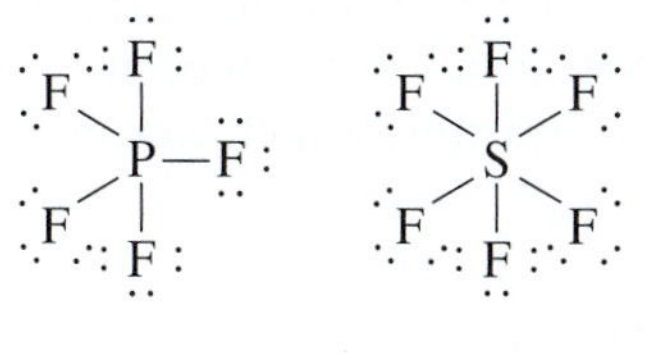

Molecules with Fewer Than Four Electron Pairs Around the Central Atom

Two other substances for which satisfactory octet rule diagrams can be drawn, but which are contradicted experimentally, are the fluorides of beryllium and boron. We might even expect BeF_2 and BF_3 to be ionic compounds, but laboratory evidence strongly supports covalent structures having the Lewis diagrams:

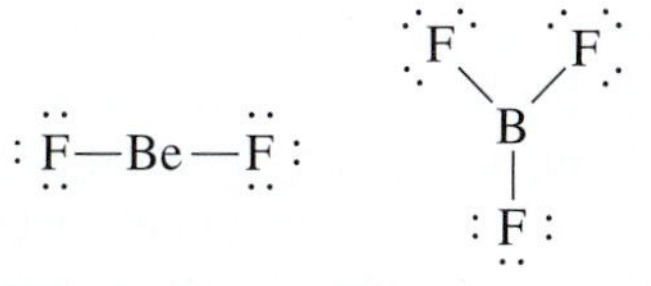

:F—Be—F:

In these compounds, the beryllium and boron atoms are surrounded by two and three pairs of electrons, respectively, rather than four.

Figure 12.18 Physical properties of oxygen. When pure gaseous oxygen is cooled to temperatures between −183°C and −218°C, it becomes a liquid (*left*). Liquid oxygen has a pale blue color (*center*). If the liquid is poured into the field of a strong magnet, some of it is trapped and held between the poles of the magnet (*right*). This is due to unpaired electrons in the oxygen molecule.

Oxygen

Certain molecules whose Lewis diagrams obey the octet rule do not have the properties that would be predicted based on those diagrams. On paper, oxygen, O_2, appears to have an ideal double bond:

$$\ddot{\underset{\cdot\cdot}{O}}{=}\ddot{\underset{\cdot\cdot}{O}}$$

But liquid oxygen is attracted by a magnetic field. (See **Figure 12.18**, which shows liquid oxygen held in place between the poles of a magnet.) This is a characteristic of molecules that have unpaired electrons, a fact that might suggest a Lewis diagram with each oxygen surrounded by seven electrons:

$$:\ddot{\underset{\cdot}{O}}{-}\ddot{\underset{\cdot}{O}}:$$

However, this conflicts with other evidence that the oxygen atoms are connected by something other than a single bond. In essence, it is impossible to write a single Lewis diagram that satisfactorily explains all the properties of molecular oxygen.

Oxygen and the fluorides of beryllium and boron are additional examples showing that all predictions must be confirmed experimentally. A specific step in the progress of chemistry begins and ends in the laboratory, but each ending is a new beginning, leading to new predictions and new experiments to confirm them. In this way our knowledge of chemistry increases.

12.9 Metallic Bonds

Goal 11 Describe how metallic bonding differs from ionic and covalent bonding.

12 Sketch a particulate-level illustration of the electron-sea model of metallic bonding.

Another familiar group of substances has properties quite different from those of either ionic compounds or molecular substances. These are the metals, the elements to the left of the stair-step line in the periodic table (Section 5.7).

Most metal atoms have one, two, or three valence electrons. These electrons are loosely held, which is why they are easily removed to form cations. When surrounded entirely by other metal atoms, there is no place for the valence electrons to go. They cannot transfer to half-filled *p*-electron orbitals in nonmetal atoms to form anions, and there is no way they can establish covalent bonds by sharing electron pairs with other atoms. A particulate-level model of metals consists of metal ions—the positively charged ions that result from the loss of the outermost *s* and *p* valence electrons—submersed in a freely moving "sea of electrons," an "ocean" of negative charge, as shown in **Figure 12.19**.

Tin (Z = 50) and lead (Z = 82) are notable exceptions to the common pattern of metal atoms having one, two, or three valence electrons. They have four valence electrons.

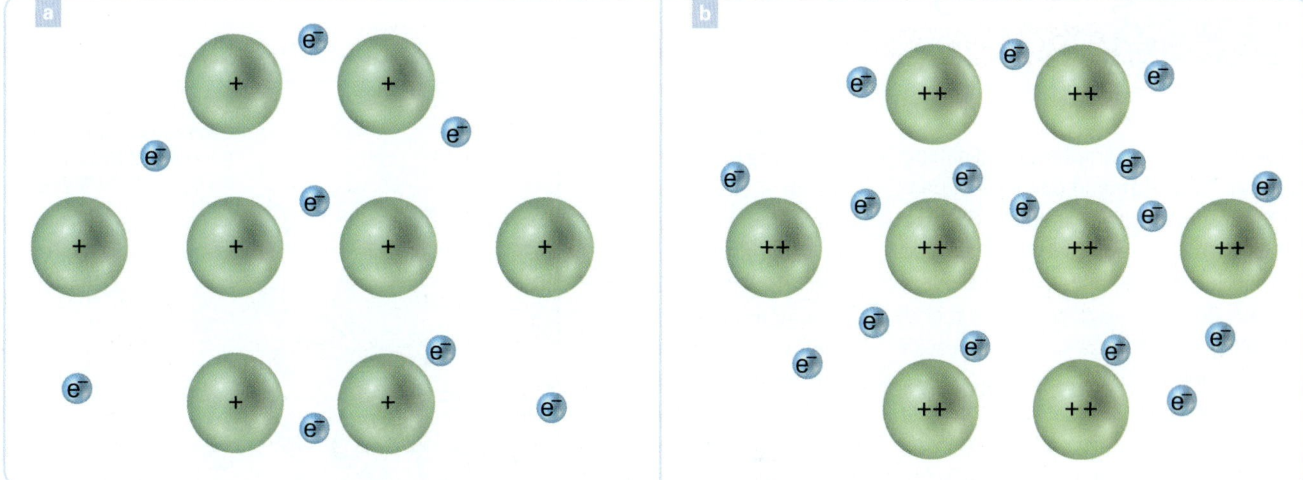

Figure 12.19 The electron-sea model of a metallic crystal. The monatomic ions formed by the metal remain fixed in a definite crystal pattern, but the highest-energy valence electrons are relatively free to move, which explains the high electrical conductivity of metals. (a) This might represent metallic sodium, or another Group 1A/1 element, with its highest-energy ns^1 electrons, and (b) might represent magnesium, or another Group 2A/2 element, with its highest-energy ns^2 electrons.

Thinking About *Your Thinking*

Mental Models

Figure 12.19 is the starting point from which you can form a mental model of the subatomic structure of metals. Notice how the highest-energy (outermost) valence electron(s) in a metal atom is (are) separated from the remainder of the atom. ▶ These electrons have much more freedom of movement than those that remain localized near a specific atom in the crystal. The positively charged metal ions are held in fixed positions in the crystal because of their attraction to the negatively charged valence electrons that move among the ions.

The mobility of the valence electrons is what gives metals many of their characteristic macroscopic properties. Mentally imagine how the valence electrons can carry charge and be responsible for electric current. Even though a metal is electrically neutral, charged particles—electrons—can carry current through a sample.

◀ **P/Review** The energy required to remove a valence electron from an atom is called its ionization energy. Figure 11.34 in Section 11.7 shows that the ionization energies of metals are lower than the ionization energies of nonmetals.

A **metallic bond** occurs because of the attractive forces between the positively charged metal ions and the negatively charged valence electrons that move among them. This is like the attractive forces caused by an electron pair spending most of its time between atoms in a covalent bond. In covalent bonds, we say the bonding electrons are **localized electrons**, restricted to occupying the space between two specific atoms. In contrast, electrons in a metallic bond are **delocalized electrons** because the electrons do not stay near any single atom or pair of atoms.

The nature of the metallic bond explains many properties of metals. An electrical current is made up of moving electrons. Because electrons are free to move within the metallic crystal, metals are good conductors of electricity. Metals bend instead of break because their atoms can rearrange themselves in the electron sea.

Metals can combine with other metals in many different ways. These combinations are not compounds; they lack a constant composition. They are mixtures rather than pure substances because their properties are not constant but instead depend on the relative composition of the mix. ▶ Steel is made up of iron and other elements. Brass, a mixture of copper and zinc, and bronze, a combination of copper and tin, are common alloys (Table 12.2). An **alloy** is a solid mixture of two or more elements that has macroscopic metallic properties.

◀ **P/Review** Recall from Section 2.5 that constant physical and chemical properties are what distinguish pure substances from those that are impure—mixtures. Pure substances are either elements or compounds.

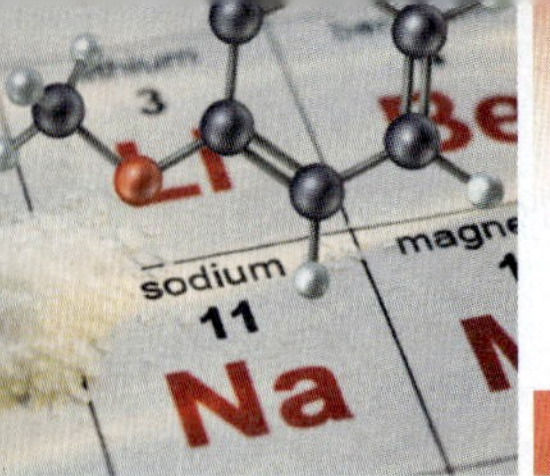

Everyday Chemistry 12.1

THE INFLUENCE OF BONDING ON MACROSCOPIC PROPERTIES

Look at the photographs of a diamond (**Figure 12.20[a]**) and one type of pencil "lead" that contains graphite (**Figure 12.20[b]**). (The original lead pencil was replaced by the graphite pencil. Modern pencils use a mixture of graphite and clay. The different hardness of pencils results from varying the mixture of graphite to clay.) Would you believe that both diamond and graphite are pure carbon? Yes, it's true—both substances are pure carbon. The particulate-level composition of each is nothing but carbon atoms. Why do they appear so very different? The answer lies in how the carbon atoms bond to one another.

Diamond has a particulate-level arrangement in which each carbon atom is covalently bonded to four other carbon atoms (**Figure 12.21**). Look at a single carbon atom "ball" in the figure and count the number of bonding "sticks." Count the number of bonds that are connected to another atom. Note how each carbon atom in diamond has four bonds. This arrangement, in which each atom is bonded to four atoms, is responsible for the physical properties of diamond. It is one of the hardest natural materials known.

Now look at the arrangement of carbon atoms in

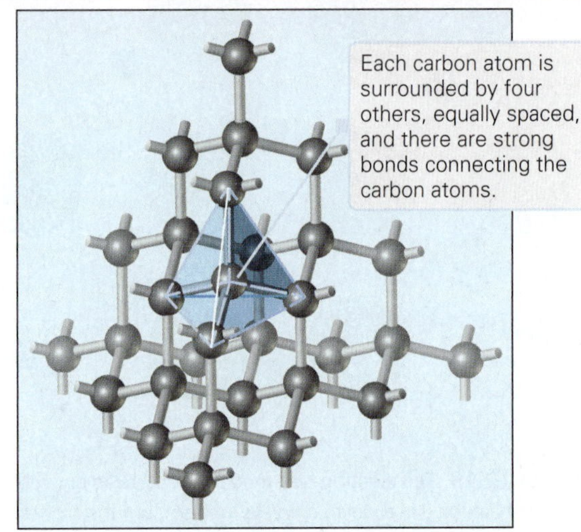

Each carbon atom is surrounded by four others, equally spaced, and there are strong bonds connecting the carbon atoms.

Figure 12.21 A particulate-level model of diamond. Each carbon atom is bonded to four other carbon atoms. This bonding arrangement leads to the characteristic physical properties of diamond at the macroscopic level.

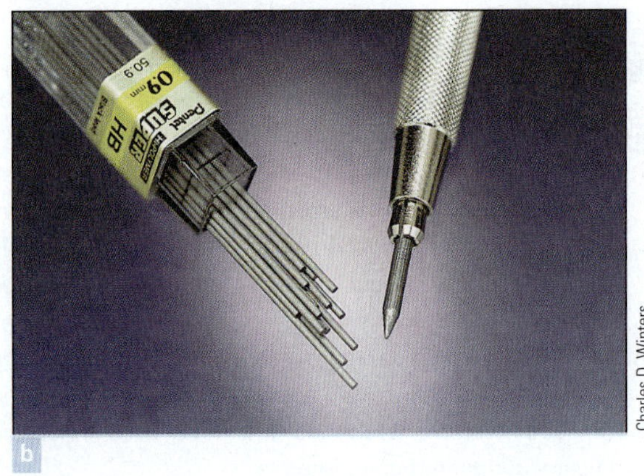

Figure 12.20 These two forms of carbon are physically very different. (a) A diamond. The physical properties of a diamond include a density of 3.5 g/cm^3, lack of color, and extreme hardness. (b) Graphite. The physical properties of graphite include a density of 2.3 g/cm^3, black in color, and relative softness.

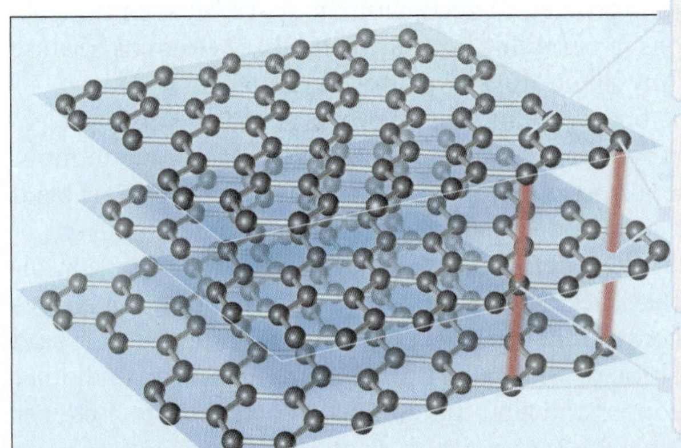

Each carbon atom is surrounded by three others in a flat sheet, and there are strong bonds connecting the carbon atoms within the same sheet.
The flat sheets stack to form the three-dimensional structure. Some carbon atoms in one sheet are aligned with the centers of rings in the next sheet.
Some carbon atoms in one sheet are aligned with carbon atoms in the next sheet.

Figure 12.22 A particulate-level model of graphite. Each carbon atom is bonded to three other carbon atoms in a flat sheet. Very weak chemical forces loosely hold the sheets together until an external force peels them apart. This bonding arrangement leads to the characteristic physical properties of graphite at the macroscopic level.

graphite as shown in **Figure 12.22**. How many bonds does each carbon atom have? You can see that each atom in graphite is bonded to three other atoms. Notice how all of the bonded atoms are in a single plane. Only very weak attractive forces exist between each "sheet" of bonded atoms in graphite. This bonding arrangement is what allows sheets of graphite to stick to paper when you write with a graphite pencil. The bonded atoms stay bonded together, but the force of pushing the pencil on paper is sufficient to slide these sheets apart.

A third form of pure carbon was discovered in the 1980s. It was called *buckminsterfullerene*, or, as this form of carbon is more commonly known, *buckyballs*. Sixty carbon atoms are arranged in a cagelike structure (**Figure 12.23[a]**), with a carbon atom at the equivalent of each corner of a soccer ball (**Figure 12.23[b]**). Each atom is bonded to three other atoms to form the C_{60} molecule shown as a model in Figure 12.23(a). This form of carbon is found in the soot made by burning carbon-containing substances in a low-oxygen environment.

The macroscopic properties of a substance depend on much more than just the number and type of atoms that make up the molecule. Bonding has a tremendous influence, too. If you ever have any doubt, compare the price of the two common forms of pure carbon—diamond and pencil "lead"—and you'll quickly understand the economic value of chemical bonds!

Quick Quiz

1. **What do diamond, graphite, and buckminsterfullerene have in common at the particulate level? What are their differences?**
2. **Diamond is hard, and graphite is soft. Give a particulate-level explanation for the macroscopic physical properties of the two substances.**

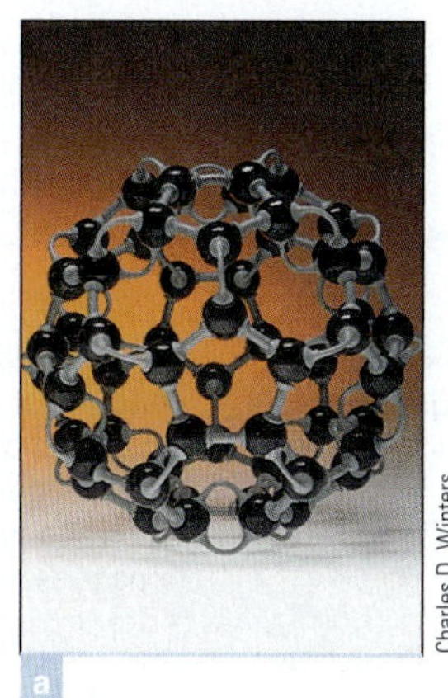

Figure 12.23 Buckminsterfullerene models. (a) A particulate-level model of buckminsterfullerene. Each carbon atom is bonded to three other carbon atoms in a cagelike sphere. The physical properties of buckyballs are very different from those of either diamond or graphite. (b) A soccer ball is another model for the particulate-level arrangement of carbon atoms in buckminsterfullerene. A carbon atom lies at the intersection of each set of three seams on the ball. The result is alternating five-membered (black) and six-membered (white) rings.

Table 12.2 Composition of Some Common Alloys

Alloy	Composition
18 K gold	75% gold, 12.5% silver, 12.5% copper
Brass	60–95% copper, 5–40% zinc
Bronze	90% copper, 10% tin
Carbon steel	99% iron, 0.2–1.5% carbon
Pewter	91% tin, 7.5% antimony, 1.5% copper
Stainless steel	About 70% iron, 18–20% chromium, 8–12% nickel; may have small quantities of other elements
Sterling silver	92.5% silver, 7.5% copper

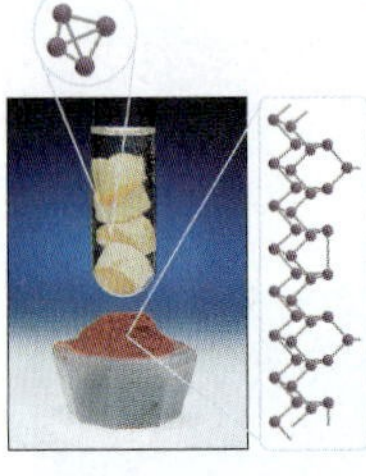

Charles D. Winters

CHAPTER 12 IN REVIEW: CHEMICAL BONDING

Goal 1 Define and distinguish between cations and anions.	A **cation** is a positively charged ion; an **anion** is a negatively charged ion.
Goal 2 Identify the monatomic ions that are isoelectronic with a given noble gas atom and write the electron configuration of these ions.	The formation of monatomic ions that are **isoelectronic** with neon atoms illustrates the pattern that is duplicated for other noble gases. A neon atom has 10 electrons, including a full octet of valence electrons. Its electron configuration is $1s^22s^22p^6$. Nitrogen, oxygen, and fluorine atoms form anions by gaining enough electrons to reach the same configuration. Sodium, magnesium, and aluminum atoms form cations by losing valence electrons to reach the same configuration.
Goal 3 Use Lewis symbols to illustrate how an ionic bond can form between monatomic cations from Groups 1A, 2A, and 3A (1, 2, 13) and anions from Groups 5A, 6A, and 7A (15–17) of the periodic table.	Ionic bonds between atoms form when atoms of a metal lose one, two, or three electrons to form a cation that is isoelectronic with a noble gas and atoms of a nonmetal gain one, two, or three electrons to form an anion that is also isoelectronic with a noble gas. An **ionic bond**, also called an *electron-transfer bond*, is formed because of the electrostatic attraction between oppositely charged ions.
Goal 4 Describe, use, and explain each of the following with respect to forming a covalent bond: electron cloud, charge cloud, or charge density; valence electrons; half-filled electron orbital; filled electron orbital; electron sharing; orbital overlap; octet rule or rule of eight.	Two atoms in a molecule are held together by a **covalent bond** when they share one or more pairs of electrons. The **electron cloud** or **charge cloud** formed by the two bonding electrons is concentrated in the region between the two nuclei. The bonding electrons count as **valence electrons** for each bonded atom. Covalent bonds form by the **overlap** of half-filled electron orbitals. The stability of a noble-gas electron configuration—the **octet rule** or **rule of eight**—is a result of the minimization of energy associated with that configuration.
Goal 5 Use Lewis symbols to show how covalent bonds are formed between two nonmetal atoms.	A **covalent bond** is formed between two nonmetal atoms, both of which have atoms that are one, two, or even three electrons short of a noble gas electron configuration. This covalent bonding process is characterized by valence electron pairs that are **shared.**
Goal 6 Distinguish between bonding electron pairs and lone pairs.	**Bonding electron pairs** are represented in a Lewis diagram as two dots or a straight line drawn between atoms. Both formats represent the covalent bond that holds the atoms together. Unshared pairs are also shown in Lewis diagrams. These are called **lone pairs.**
Goal 7 Distinguish between polar and nonpolar covalent bonds.	In a **nonpolar covalent bond,** the bonding electrons are shared equally by the bonded atoms. In a **polar covalent bond**, the nucleus of one atom attracts the shared electrons more strongly than the other.
Goal 8 Predict which end of a polar bond between identified atoms is positive and which end is negative.	The relative ability of atoms of an element to attract electron pairs in covalent bonds is expressed by the **electronegativity** of the element. The **polarity** of a bond is estimated by the difference in electronegativities of the bonded atoms. The atom with the higher electronegativity is the negative end of the bond. The atom with the lower electronegativity is the positive end of the bond.
Goal 9 Rank bonds in order of increasing or decreasing polarity based on periodic trends in electronegativity values or actual values, if given.	There is a periodic trend in the electronegativity of elements. In general, electronegativity increases across a period and decreases down a group. The greater the difference in electronegativity between two bonded elements, the more polar the bond.
Goal 10 Distinguish among single, double, and triple bonds, and identify these bonds in a Lewis diagram.	The sharing of one pair of electrons by two bonded atoms is called a **single bond.** When two atoms are bonded by two pairs of electrons, it is a **double bond.** When two atoms are bonded by three pairs of electrons, the bond is called a **triple bond.** All four electrons in a double bond and all six electrons in a triple bond are counted as valence electrons for the bonded atoms. **Multiple bonds** is a general term that includes double and triple bonds.
Goal 11 Describe how metallic bonding differs from ionic and covalent bonding.	**Metallic bonding** occurs because of attractive forces between negatively charged valence electrons moving among positively charged metal ions. In covalent bonds, the bonding electrons are localized between two specific atoms. Electrons in a metallic bond are **delocalized** because the bonding electrons do not stay near any single atom or pair of atoms. The nature of the metallic bond explains many of the properties of metals, such as electrical conductivity and the ability to bend and be stretched into thin wires.

Goal 12 Sketch a particulate-level illustration of the electron-sea model of metallic bonding.

The **electron-sea model** of a metallic crystal is characterized by monatomic ions in a crystal pattern with the highest-energy valence electrons free to move among the ions. The positively charged metal ions are held in fixed positions in the crystal because of their attraction to the negatively charged valence electrons that move among the ions.

Key Terms

Most of the key terms and concepts and many others appear in the Glossary. Use your Glossary regularly.

alloy p. 457
anion p. 441
cation p. 441
charge density p. 448
chemical bond p. 439
covalent bond p. 448
crystal lattice p. 445
delocalized electrons p. 457
double bond p. 453
electron cloud p. 448
electronegativity p. 451
hydronium ion p. 441
ionic bonds p. 445
ionic compounds p. 444
Lewis diagram p. 449
Lewis formula p. 449
Lewis structure p. 449
localized electrons p. 457
lone pairs p. 449
metallic bond p. 457
molecular compounds p. 447
molecules p. 447
multiple bonds p. 453
nonpolar covalent bond p. 450
octet rule p. 449
overlap p. 448
polar covalent bond p. 451
single bond p. 453
triple bond p. 453

Frequently Asked Questions

How can you tell if the bond between atoms of two elements is most likely to be ionic, covalent, or metallic?
Metal atoms have one, two, or three valence electrons. Monatomic cations form when metal atoms lose these electrons, usually reaching an outer electron configuration that is isoelectronic with a noble gas. Nonmetal atoms in Groups 5A/15 through 7A/17 have five, six, or seven valence electrons. These atoms form monatomic anions by gaining enough electrons to reach an electron configuration that matches that of a noble gas. These oppositely charged ions form ionic bonds between themselves, yielding the crystalline structure that is typical of an ionic compound. Conclusion: *Ionic bonds generally form between a metal and a nonmetal.*

Nonmetal atoms have too many valence electrons to lose them all and become cations, so an anion and a cation cannot both come from two nonmetal atoms. By sharing electrons, however, they form covalent bonds in which each atom reaches a noble gas structure. The compounds formed in this way are molecular. Conclusion: *The bond between two nonmetals is generally covalent.*

To form a metallic bond, valence electrons must be shared by all "metal ions" in a crystal. Atoms that have one, two, or three valence electrons, when surrounded by other similar atoms, are apt to donate these electrons to a "sea of electrons" in which the metal ions are located. With a few exceptions, only *metals* have one, two, or three valence electrons. Conclusion: *The bonding among metal atoms is metallic.*

Given that it has a single valence electron, how do hydrogen atoms form bonds?
You've noted that hydrogen is a nonmetal that has only one valence electron rather than four, five, six, or seven. Nevertheless, a hydrogen atom reaches the noble gas structure of helium with one additional electron. Hydrogen, therefore, forms either ionic or covalent bonds in the same way as other nonmetals.

When given pairs of bonded atoms, how can I predict which bond is the most or least polar?
You are not expected to memorize the electronegativities of the elements in Figure 12.17 (unless your instructor states otherwise). However, you should know the periodic trend in electronegativity values: They are higher at the top of any column (group) and at the right of any row (period) in the periodic table. From this, you can often predict which of two bonds is more polar. The greater the electronegativity difference, the greater the polarity of the bond will be.

Concept-Linking Exercises

Write a brief description of the relationships among each of the following groups of terms or phrases. Answers to the Concept-Linking Exercises are given at the end of the chapter.

1. Chemical bond, atoms, molecules, ionic compounds, valence electron
2. Ionic bond, electron transfer, crystal, cation, anion
3. Covalent bond, electron cloud, overlap, molecule
4. Polar bond, nonpolar bond, electronegativity, charge density, distribution of bonding electron charge
5. Metallic bond, electron-sea model, valence electrons
6. Ionic bond, covalent bond

Small-Group Discussion Questions

Small-Group Discussion Questions are for group work, either in class or under the guidance of a leader during a discussion section.

1. Why do Group 1A/1 elements tend to form ions with a 1+ charge? Provide similar explanations for the common ions formed from third-period elements from Groups 2A/2, 3A/13, 5A/15, 6A/16, and 7A/17. Why do Group 8A/18 elements not exist as ions?
2. Consider the particulate-level depiction of the arrangement of ions in sodium chloride illustrated in Figure 12.7. Why are the sodium ions smaller than the chloride ions? Why are there equal numbers of sodium ions and chloride ions? Draw a sketch of the arrangement of ions in magnesium sulfide. How is it similar and how is it different from the sodium chloride illustration?
3. Be as specific as possible and state which orbitals overlap in forming each of the following covalent bonds: H—F, H—H, F—F, H—Cl, Cl—Cl. Explain your reasoning in each case.
4. What is the definition of *electronegativity*? What is the periodic trend in electronegativity values? In Section 11.7, you studied the periodic trends in a number of properties of elements. How are these other trends related to periodic trends in electronegativity values? Can any other periodic trend explain the periodic trend in electronegativity or vice versa?
5. Nitrogen molecules contain a triple bond. What orbitals from each nitrogen atom overlap to form the three bonds? Explain.
6. The Lewis diagrams for ammonia, methane, ethylene, acetylene, and carbon dioxide are given in Section 12.7. For each atom in each molecule, show how the octet rule is satisfied. For the hydrogen atoms, explain how the hydrogen atom achieves the electron configuration of a helium atom.
7. Describe the evidence that indicates that oxygen molecules do not have a double bond. Show how the Lewis diagram for oxygen with a double bond between the two atoms satisfies the octet rule. Explain how oxygen illustrates the limitations of Lewis diagrams and the need for experimental investigation of the structure of molecules.
8. One well-known macroscopic property of metals is their ability to conduct electricity. Explain how the particulate-level electron-sea model of metals provides an explanation for this macroscopic property.
9. Draw, sketch, and/or illustrate each of the following bond types at the particulate level: ionic, nonpolar covalent, polar covalent, metallic. Write a brief statement that describes the critical similarities and differences in the bond types.

Questions, Exercises, and Problems

Interactive versions of these problems may be assigned by your instructor. Solutions for blue-numbered questions are at the end of the chapter. Questions other than those in the General Questions and More Challenging Problems sections are paired in consecutive odd-even number combinations; solutions for the odd-numbered questions are also at the end of the chapter.

Section 12.2: Monatomic Ions with Noble Gas Electron Configurations

1. Write the electron configuration for the ions of the third period elements that form monatomic ions that are isoelectronic with a noble gas atom. Also, identify the noble gas atoms having those configurations.
2. A monatomic ion with a charge of 1− has an electronic configuration of $1s^22s^22p^63s^23p^6$. (a) Is this ion a cation or an anion? (b) With what noble gas is it isoelectronic? (c) What is the symbol of the ion?
3. Identify by symbol two positively charged monatomic ions that are isoelectronic with argon.

Argon emits a violet color when in an activated gas discharge tube.

4. Considering only ions with charges of 1+, 2+, 1−, and 2−, or neutral atoms, give the symbols for four species that are isoelectronic with the chloride ion, Cl^-.
5. Write the symbols of two ions that are isoelectronic with the barium ion.
6. Considering only ions with charges of 1+, 2+, 1−, and 2−, or neutral atoms, give the symbols for four species that are isoelectronic with xenon, Xe.

Section 12.3: Ionic Bonds

7. Aluminum oxide is an ionic compound. Sketch the transfer of electrons from aluminum atoms to oxygen atoms that accounts for the chemical formula of the compound Al_2O_3.
8. Using Lewis symbols, show how atoms of sulfur and sodium form ionic bonds, leading to the correct formula of sodium sulfide, Na_2S.
9. When potassium and chlorine react and form an ionic compound, why is there only one chlorine atom for each potassium atom instead of two?
10. Fill in the blanks with the smallest integers possible. When gallium (Z = 31) reacts with sulfur to form an ionic compound, each metal atom loses _____ electron(s) and each nonmetal atom gains _____ electron(s). There must be _____ gallium atom(s) for every _____ sulfur atom(s) in the reaction.

Section 12.4: Covalent Bonds

11. What is the meaning of *orbital overlap* in the formation of a covalent bond?

12. What is an electron cloud?

13. Sketch the formation of two covalent bonds by an atom of sulfur in making a molecule of hydrogen sulfide, H_2S.

14. How many covalent bond(s) would the element germanium (Z = 32) be expected to form in order to obey the octet rule? Use the octet rule to predict the formula of the compound that would form between germanium and hydrogen, if the molecule contains only one germanium atom and only single bonds are formed.

15. Circle the lone electron pairs in the Lewis diagram for hydrogen chloride: $\text{H—}\ddot{\underset{..}{\text{Cl}}}\text{:}$

16. How many covalent bond(s) would the element tellurium (Z = 52) be expected to form in order to obey the octet rule? Use the octet rule to predict the formula of the compound that would form between tellurium and hydrogen, if the molecule contains only one tellurium atom and only single bonds are formed.

Susan E. Degginger/Alamy Stock Photo

Tellurium is among the rarest of the elements in Earth's crust.

17. Show how atoms achieve the stability of noble gas atoms in forming covalent bonds.

18. Does the energy of a system tend to increase, decrease, or remain unchanged as two atoms form a covalent bond?

Section 12.5: Polar and Nonpolar Covalent Bonds

19. Compare the electron cloud formed by the bonding electron pair in a polar bond with that in a nonpolar bond.

20. What is meant by saying that a bond is polar or nonpolar? What bonds are completely nonpolar?

21. Refer to Figure 12.17 and list the following bonds in order of decreasing polarity: S–O, N–Cl, C–C.

22. Consider the following bonds: Ge–Se, Br–Se, Br–Ge. Indicate the direction of polarity of each. Which bond is expected to be the most polar?

23. For each bond in Question 21, identify the positive pole, if any.

24. Consider the following bonds: Te–Se, O–Te, O–Se. Indicate the direction of polarity of each. Which bond is expected to be the most polar?

25. Identify the trends in electronegativities in the periodic table.

26. Arrange the following elements in order of increasing electronegativity: silicon, chlorine, sulfur, phosphorus.

Section 12.6: Multiple Bonds

27. Atoms with double bonds and triple bonds can conform to the octet rule. Could an atom with a quadruple (four) bond obey that rule? Why or why not?

28. What is a multiple bond? Distinguish among single, double, and triple bonds.

Section 12.7: Atoms That Are Bonded to Two or More Other Atoms

29. An atom, X, is bonded to another atom by a double bond. What is the largest number of *additional atoms*—don't count the one to which it is already bonded—to which X may be bonded and still conform to the octet rule? What is the minimum number? Justify your answers.

30. What is the maximum number of atoms to which a central atom in a molecule can bond and still conform to the octet rule? What is the minimum number?

31. A molecule contains a triple bond. Theoretically, what are the maximum and minimum numbers of atoms that can be in the molecule and still conform to the octet rule? Sketch Lewis diagrams that justify your answer.

32. Draw Lewis diagrams of *central* atoms of molecules showing all possible combinations of single, double, and triple bonds that can form around an atom that conforms to the octet rule. It is not necessary to show the atoms to which the central atom is bonded—just the bonds.

Section 12.8: Exceptions to the Octet Rule

33. Because nitrogen has five valence electrons, it is sometimes difficult to fit a nitrogen atom into a Lewis diagram that obeys the octet rule. Why is this so? Without actually drawing them, can you tell which of the following species do not have a Lewis diagram that satisfies the octet rule? N_2O, NO_2, NF_3, NO, N_2O_3, N_2O_4, NOCl, NO_2Cl.

34. Why is it impossible to draw a Lewis diagram that obeys the octet rule if the species has an odd number of electrons?

Section 12.9: Metallic Bonds

35. What are the meanings of the terms *localized* and *delocalized* when used to describe electrons in a compound?

36. Is a metallic bond more similar to an ionic bond or a covalent bond? Explain.

37. How would a particulate-level illustration differ if you were to draw a calcium crystal as compared with a potassium crystal?

38. Sketch a particulate-level illustration, similar to Figure 12.19, which shows the electron-sea model for a potassium crystal.

39. Are alloys pure substances or mixtures? Are they compounds? Explain your answers.

40. What is an alloy? Give an example other than bronze.

Todd Taulman/Shutterstock.com

Bronze is an alloy, usually of copper and tin. Some people in the United States bronze their baby's first pair of shoes.

General Questions

41. Distinguish precisely, and in scientific terms, the differences among items in each of the following pairs or groups.
a) Ionic compound, molecular compound
b) Ionic bond, covalent bond
c) Lone pair, bonding pair (of electrons)
d) Nonpolar bond, polar bond
e) Single, double, triple, multiple bonds

42. Classify each of the following statements as true or false:
a) A single bond between carbon and nitrogen is polar.
b) A bond between phosphorus and sulfur will be less polar than a bond between phosphorus and chlorine.
c) The electronegativity of calcium is less than the electronegativity of aluminum.
d) Strontium (Z = 38) ions, Sr^{2+}, are isoelectronic with bromide ions, Br^-.
e) The monatomic ion formed by selenium (Z = 34) is expected to be isoelectronic with a noble gas atom.
f) Most elements in Group 4A/14 do not normally form monatomic ions.
g) Multiple bonds can form only between atoms of the same element.
h) If an atom is triple-bonded to another atom, it may still form a bond with one additional atom.
i) An atom that conforms to the octet rule can bond to no more than three other atoms if one bond is a double bond.
j) Electrons are localized between atoms in a metal.
k) Alloys are pure substances.

43. Explain why ionic bonds are called *electron-transfer bonds* and covalent bonds are known as *electron-sharing bonds*.

44. Explain why the total energy of a system changes in the formation of (a) an ionic bond and (b) a covalent bond.

45. Compare the bond between potassium and chlorine in potassium chloride with the bond between two chlorine atoms in chlorine gas. Which bond is ionic, and which is covalent? Describe how each bond forms.

46. Considering bonds between the following pairs of elements, which are most apt to be ionic and which are most apt to be covalent: sodium and sulfur; fluorine and chlorine; oxygen and sulfur? Explain your choice in each case.

47. What is the electron configuration of the hydrogen ion, H^+? Explain your answer.

48. Identify the pairs among the following that are *not* isoelectronic: (a) Ne and Na^+; (b) S^{2-} and Cl^-; (c) Mg^{2+} and Ar; (d) K^+ and S^{2-}; (e) Ba^{2+} and Te^{2-} (Te, Z = 52).

49. Which orbitals of each atom overlap in forming a bond between bromine and oxygen?

50. Is there any such thing as a completely nonpolar bond? If yes, give an example.

51. If you did not have an electronegativity table, could you predict the relative electronegativities of elements whose positions are Ⓑ—Ⓐ (B lower left, A upper right) in the periodic table? What about elements whose positions are Ⓧ—Ⓨ (X upper left, Y lower right)? In both cases, explain why or why not.

More Challenging Problems

52. A monatomic ion with a 2− charge has the electron configuration $1s^22s^22p^63s^23p^64s^23d^{10}4p^6$. (a) What neutral noble gas atom has the same electron configuration? (b) What is the monatomic ion with a 2− charge that has this configuration? (c) Write the symbol of an ion with a 1+ charge that is isoelectronic with the species in (a) and (b).

53. If the monatomic ions in Question 52 (b) and (c) combine to form a compound, what is the formula of that compound?

54. "The bond between a metal atom and a nonmetal atom is most apt to be ionic, whereas the bond between two nonmetal atoms is most apt to be covalent." Explain why this statement is true.

55. The metallic bond is neither ionic nor covalent. Explain, according to the octet rule, why this is so.

56. How do the energy and stability of bonded atoms and noble gas electron configurations appear to be related in forming covalent and ionic bonds?

57. Which bond, F—Si or O—P, is more polar? You may look at a full periodic table in answering this question, but do not look at any source of electronegativity values.

58. Arrange the following bonds in order of increasing polarity: Na—O; Al—O; S—O; K—O; Ca—O. If the polarity of any two bonds cannot be positively placed relative to each other based on periodic trends, explain why.

59. There are two iodides of arsenic (Z = 33), AsI_3 and AsI_5. A Lewis diagram that conforms to the octet rule can be drawn for one of these but not for the other. Draw the diagram that is possible. From that diagram, see if you can figure out how the second molecule might be formed by covalent bonds, even though it violates the octet rule. Draw the Lewis diagram for the second molecule.

60. How is it possible for central atoms in molecules to be surrounded by five or six bonding electron pairs when there are only four valence electrons from the *s* and *p* orbitals of any atom?

61. Suggest why BF_3 behaves as a molecular compound, whereas AlF_3 appears to be ionic.

Answers to Target Checks

1. S^{2-} and Ba^{2+}. Copper(II), iron(III), and silver ions are all positive ions, which means that they lose electrons to form ions. If they lose electrons, they potentially can be isoelectronic only with the nearest noble gas atom with a lower atomic number. All have *d* sublevel electrons not present in the noble gas atom with a lower atomic number.
2. **II**, **III**, and **IV** have double bonds. **I** and **II** have triple bonds. All have multiple bonds.
3. (a) No, it is not possible, under the octet rule, for a single atom to be bonded by double bonds to each of three other atoms. Three double bonds would be 6 electron pairs, placing 12 electrons around the central atom. (b) Four atoms can be bonded by single bonds to the same central atom. At 2 electrons per bond, there would be 8 electrons around the atom, a full octet.

Answers to Practice Exercises

1. O^{2-}: $1s^2 2s^2 2p^6$, isoelectronic with neon. K^+: $1s^2 2s^2 2p^6 3s^2 3p^6$, isoelectronic with argon.
2. $Mg: + 2\,\cdot\ddot{\underset{\cdot\cdot}{F}}: \longrightarrow Mg^{2+} + 2\left[:\ddot{\underset{\cdot\cdot}{F}}:\right]^- \longrightarrow MgF_2$ crystal
3. $2\,H\cdot + \cdot\dot{\underset{\cdot\cdot}{O}}: \longrightarrow H:\ddot{\underset{\cdot\cdot}{O}}:H$
4. $C{-}O(\delta^-)\ (\Delta EN = 0.8) > C{-}N(\delta^-)\ (\Delta EN = 0.4) \approx (\delta^-)C{-}H\ (\Delta EN = 0.4) > C{-}S\ (\Delta EN \approx 0) \approx C{-}C\ (\Delta EN = 0)$

Answers to Concept-Linking Exercises

You may have found more relationships or relationships other than the ones given in these answers.

1. Chemical bonds are forces that hold together: (1) atoms in molecules; (2) atoms in polyatomic ions; (3) ions in ionic compounds; (4) atoms in metals. Valence electrons, the outermost electrons, are responsible for these bonding forces.
2. Ionic bonds are the forces that hold ions in fixed positions in a crystal—a solid with a definite geometric structure. The positively charged ions, cations, and negatively charged ions, anions, arrange themselves in the crystal in a manner that minimizes the potential energy of the crystal. An ionic bond can be thought of as an electron-transfer bond because the atoms that form cations do so by transferring electrons to other atoms to change them into anions.
3. A covalent bond is formed by the overlap of two atomic orbitals. This produces an electron cloud that is concentrated between the nuclei being bonded. Atoms in molecules are held together by covalent bonds.
4. Bonding electrons are shared equally in a nonpolar bond; they are shared unequally in a polar bond. Electronegativity is a measure of the ability of an atom to attract bonding electrons toward itself. The distribution of bonding electron charge will be unequal in a polar bond, with the charge density shifted toward the more electronegative atom.
5. On the particulate level, a metal can be thought of as positive ions surrounded by a sea of freely moving electrons. In this electron-sea model, the electrons that are free to move are the outermost electrons of the metal atoms—the valence electrons. In a metallic bond, a sea of electrons is shared among positively charged metal ions.
6. An ionic bond forms between oppositely charged ions when the forces of attraction and repulsion balance in the lowest-energy state, forming an ionic crystal. A covalent bond forms between positively charged nuclei when the nuclei share negatively charged electron pairs.

Answers to Blue-Numbered Questions, Exercises, and Problems

1. All monatomic ions of third-period elements that are isoelectronic with a noble gas atom have the electron configuration of Ne (cations) or Ar (anions): $1s^2 2s^2 2p^6$ or $1s^2 2s^2 2p^6 3s^2 3p^6$.

3. Any two of K^+, Ca^{2+}, Sc^{3+}.

5. Any two of Te^{2-}, I^-, Cs^+.

7. 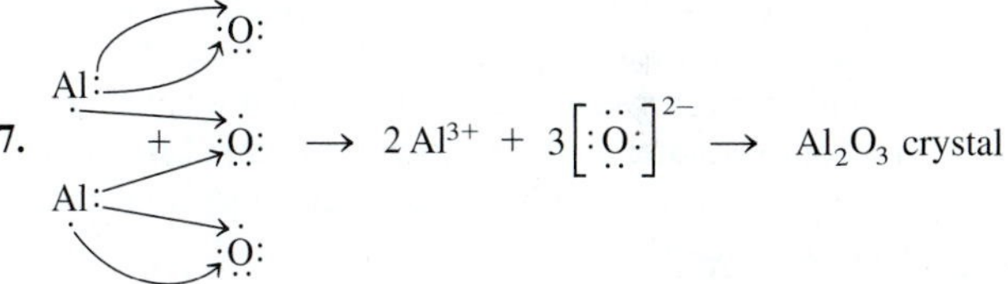

9. A potassium atom forms a K^+ ion by losing one electron. A chlorine atom can accept one electron to form a Cl^- ion, so it takes one potassium atom to donate the one electron to a single chlorine atom.

11. Orbital overlap refers to the covalent bond formed when the atomic orbitals of individual atoms extend over one another so that the two nuclei in the bond share electrons.

13. $2\,H\cdot + \cdot\ddot{\underset{\cdot\cdot}{S}}: \rightarrow H:\ddot{\underset{\cdot\cdot}{S}}:H$ *or* $H{-}\ddot{\underset{\cdot\cdot}{S}}{-}H$ *or* $H{-}S{-}H$

15. You should circle the three lone pairs (··) around the chlorine atom.

17. The energy of a system is reduced when bonds form that reach the noble gas electron configuration. In a covalent (or electron-sharing) bond, atoms share electrons to achieve the noble gas configuration. An example is shown in the answer to Question 13. The hydrogen atoms obtain the electron configuration of helium, and the sulfur atom obtains the electron configuration of argon.

19. In a nonpolar bond, the charge density of the electron cloud is centered in the region between the bonded atoms. In a polar bond, this charge density is shifted toward the more electronegative atom.

21. $S{-}O > N{-}Cl > C{-}C$

23. S in S–O

25. Electronegativity values increase from left to right across any row of the table, and they increase from the bottom to top of any column.

27. An atom bonded by a quadruple bond can conform to the octet rule if it has no unshared electron pairs.

29. X can be bonded to the maximum of two additional atoms by single bonds. X can be bonded to the minimum of no additional atoms if it has two lone pairs.

31. The molecule can theoretically have a maximum of an infinite number of atoms: A—B≡C—D.... The minimum number is two: :A≡B:

33. NO_2 and NO have an odd number of electrons and thus cannot satisfy the octet rule.

35. Localized electrons are those that stay near a single atom or pair of atoms; delocalized electrons do not.

37. The calcium ions in the crystal have a 2+ charge, and there are two electrons for each ion, whereas with potassium metal, the potassium ions have a 1+ charge and there is a 1:1 ion-to-electron ratio.

39. Alloys are mixtures. They are neither pure substances nor compounds. They lack constant composition and have variable physical properties.

42. True: a, b, c, d, e, f, h, i. False: g, j, k.

43. Ions are formed when neutral atoms lose or gain electrons. The electron(s) that is (are) lost by one atom is (are) transferred to another atom. The attraction between the ions is an ionic bond. Covalent bonds are formed when a pair of electrons is shared by the two bonded atoms. Effectively, the electrons belong to both atoms, spending some time near each nucleus.

45. The K—Cl bond is ionic, formed by "transferring" an electron from a potassium atom to a chlorine atom. The Cl—Cl bond is covalent, formed by two chlorine atoms sharing a pair of electrons.

47. The H^+ ion has no electrons, so it has no electron configuration.

48. (c) Mg^{2+} is isoelectronic with Ne, not Ar.

49. 4*p* from bromine and 2*p* from oxygen.

50. A bond between identical atoms is completely nonpolar. Their attractions for the bonding electrons are equal.

51. Electronegativities are highest at the upper-right corner of the periodic table and lowest at the lower-left corner. Therefore, the electronegativity of A is higher than the electronegativity of B. Because X is higher in the table than Y, the electronegativity of X should be larger than that of Y, but because Y is farther to the right, the electronegativity of X should be smaller than Y. Therefore, no prediction can be made for X and Y.

52. (a) Kr, krypton, Z = 36; (b) The 2− ion had two electrons added to the neutral atom. The neutral atom is therefore Z = 36 − 2 = 34. Z = 34 is Se, selenium. The ion is Se^{2-}, selenide ion. (c) A 1+ ion has one electron removed from the neutral atom, so the neutral atom is Z = 36 + 1 = 37. Z = 37 is Rb, rubidium. The ion is Rb^+, rubidium ion.

54. Nonmetal atoms generally have 4, 5, 6, 7, or 8 valence electrons. Fewer electrons have to be added to a nonmetal atom to achieve an octet than would have to be subtracted. For example, a sulfur atom can theoretically achieve an octet by gaining two electrons or losing six. There is a smaller energy change for nonmetal atoms to gain electrons to achieve an octet. When two nonmetal atoms combine, the lowest-energy way for both atoms to reach the octet is to share each other's electrons, forming a covalent bond. If the second atom is a metal, however, it has one, two, or possibly three electrons more than an octet. It reaches the octet by releasing its electrons to the nonmetal, becoming a positive ion itself, and making the nonmetal atom a negative ion. The two atoms form an ionic bond.

57. An F—Si bond is more polar than an O—P bond. F has a higher electronegativity than O, and Si has a lower electronegativity than P, based on their relative positions in the periodic table (high at the upper right, low at the lower left). Therefore, the *difference* in electronegativities is largest for F—Si, which makes it the more polar bond.

59. AsI_3 conforms to the octet rule. AsI_5 can be formed if each of the lone-pair electrons forms a bonding pair with one electron from an I atom. :I—As—I: (with :I: bonded below As) :I—As—I: (with two I atoms bonded above As and :I: bonded below As)

(The angles at which these bonds are drawn are related to the spatial arrangements of the bonded atoms. This is considered in Chapter 13. In this chapter, you may limit your interpretations of these diagrams to what atom is bonded to what atom.)

61. One explanation is that a boron atom is smaller than an aluminum atom. The valence electrons in boron are therefore closer to the nucleus than they are in aluminum. That makes it more difficult to remove the electrons to form an ion.

13 Structure and Shape

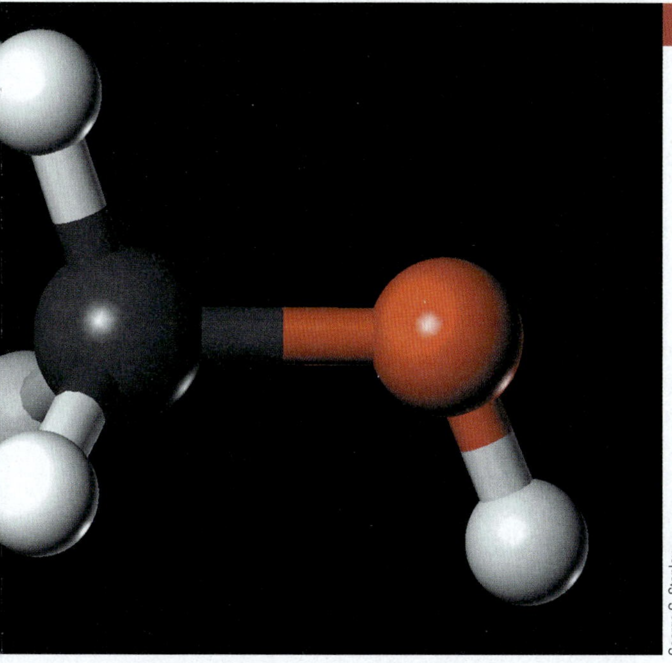

◀ This model represents the structure and shape of a molecule of methanol, CH_3OH. Hydrogen atoms are shown in white, the carbon atom is shown in black, and the oxygen atom is shown in red. The sticks connecting the balls represent single bonds. In this chapter, you will learn to predict and sketch the three-dimensional shapes of molecules, starting with nothing more than their formulas.

George G. Stanley

CHAPTER CONTENTS

You learned in Chapter 12 that atoms in molecular compounds and in polyatomic ions are held together by covalent bonds. Lewis diagrams show, in two dimensions, how the atoms are connected. However, Lewis diagrams do not show how the atoms are *arranged* in three dimensions—the actual shape of the molecule. In this chapter, you will learn how the distribution of electron pairs leads to the distribution of atoms, which in turn leads to the structure and shape of molecules.

13.1 How Is Genetic Information Stored in Molecules?

It has been known throughout human history that the offspring of humans and other animals resemble their parents and even share personality traits. Many scientists throughout history have pondered the cause of this pattern. The Russian scientist Nikolai Koltsov (1872–1940) was likely the first person to propose that the mechanism of inheritance was *molecular* in nature. Alas, Koltsov was assassinated by the Soviet government and was unable to investigate his hypothesis. In Koltsov's time, most scientists believed that a class of molecules known as *proteins*—large, complex molecules with a modular sequence—was the most likely type of molecule of inheritance.

In 1943, the Canadian-born scientist Oswald Avery, Jr. (1877–1955) discovered that the molecular hypothesis about inheritance was indeed correct, but the molecule was not a protein. His work with isolation of types of molecules in bacteria led to the conclusion that deoxyribonucleic acid (DNA) was the basis of molecular inheritance.

Soon thereafter, the race to discover *how* DNA carried genetic information was on. What was the structure of the DNA molecule? The odds were likely that if the structure of DNA could be discovered, the mechanism of inheritance would follow. Three teams of scientists were the primary contenders to discover the answer. The team that appeared to be the most likely to win the race was led by Linus Pauling (1901–1994) at Caltech. Pauling had an extensive record of success with scientific discovery and theory-building by the early 1950s. The next most-qualified team was led by Maurice Wilkins (1916–2004) and Rosalind Franklin (1920–1958) at King's College London. They were experts in the use of projecting x-ray radiation at molecules and analyzing the patterns that resulted from the interaction, which allowed deduction of the structure of the molecule. (Pauling was also an x-ray expert.) The underdogs in the race were Francis Crick (1916–2004) and James Watson (b. 1928) at the University of Cambridge. Crick also had expertise in use of x-rays to study molecular structure, but he had switched his specialization to biology from physics only a few years before working on the DNA puzzle. When Watson arrived at Cambridge, he was only 23 years old and had just earned his Ph.D. a year earlier.

Pauling's group went public with their proposed structure first. They proposed a triple helix, three strands of covalently bonded atoms that were intertwined in a twisted curving pattern. When Watson read an advance copy of Pauling's manuscript, he knew that Pauling's structure was wrong because he and Crick had arrived at almost the same structure earlier and deduced that the model did not match the data. This illustrates one of the most important tenants of the scientific method: No matter how famous you are or what you've accomplished, if your hypothesis does not fit the data, it is wrong.

The team led by Wilkins and Franklin specialized in producing the highest-quality photographic images at that point in history that resulted from the interaction of x-rays and DNA molecules. They learned how to obtain the best DNA samples, how to prepare the samples, how to get the best equipment, how to photograph the x-ray–DNA interaction, and how to do the math needed to transform the photographs into data about molecular structure.

In 1953, Watson visited King's College and Wilkins showed him what was likely the best x-ray photograph of DNA ever developed to date, labeled *photograph 51*, which was taken by Franklin (**Figure 13.1**). Crick was also able to obtain through a third party a progress report written by Franklin about her conclusions about the information gleaned thus far about DNA from x-ray photograph data. (You may be wondering about the ethics of Watson and Crick indirectly obtaining Franklin's work, and if so, you are not alone. This is an issue that continues to be discussed today.*)

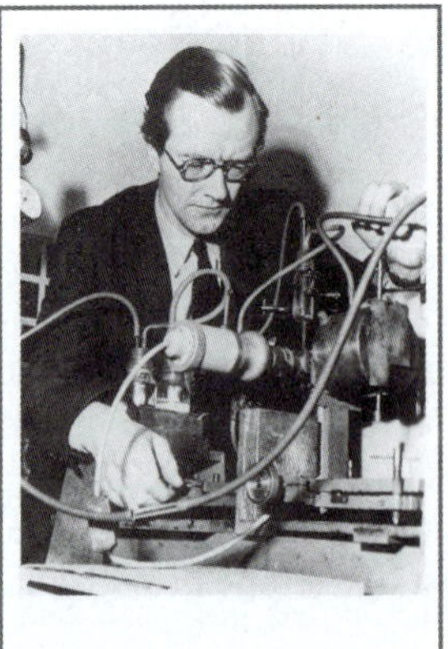

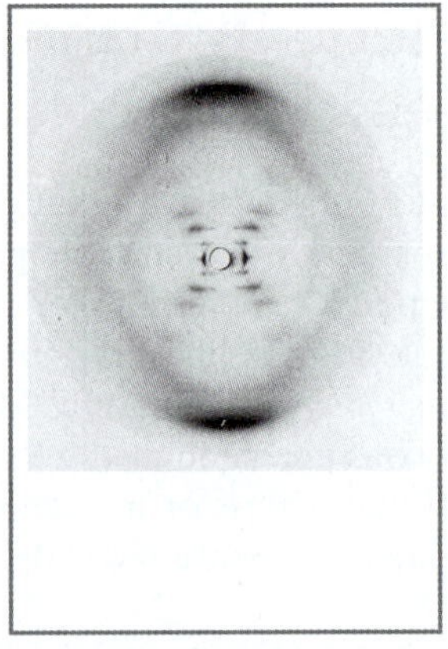

King's College London Archives

Figure 13.1 Maurice Wilkins (*left*), *photograph 51* (*center*), and Rosalind Franklin (*right*).

*In the scientific publication in which Watson and Crick proposed the structure of DNA and discussed how it was likely the mechanism of inheritance, they wrote, "We have also been stimulated by a knowledge of the general nature of the unpublished experimental results and ideas of Dr. M. H. F. Wilkins, Dr. R. E. Franklin and their co-workers at King's College, London."

With the data from the Franklin and Wilkins team, Watson and Crick continued to work on the structure of DNA, knowing that they were likely getting close. A critical element of the process was that they transformed the numeric data into physical models (**Figure 13.2**). Models were not commercially available at this point in history, of course, so they put in a work order to have the university's machine shop fabricate metal models that fit the structure of the parts of the DNA molecule to scale. They were so anxious to build models that they decided to not wait for the machine shop, and they built models out of cardboard. Using their knowledge of the chemistry of intermolecular forces (which you will learn about in Chapter 15) and by manipulating the to-scale physical models, they realized that their models were the correct structure of DNA. Given the nature of the two-stranded molecule, they also realized that they had discovered how genetic information could be copied and passed from parents to offspring.

The 1962 Nobel Prize in Physiology or Medicine was awarded to Crick, Watson (**Figure 13.3**), and Wilkins for their discovery of DNA and their follow-up work. Franklin had passed away in 1958, and Nobel Prizes are not awarded posthumously.

This chapter is an introduction to the nature of and the fundamental causes of molecular structure and shape. Later in this course, or perhaps in a future course, you will see that molecular structure and shape determines the macroscopic properties of substances and their chemical reactivity. Unlike it was in 1953 when Watson and Crick deduced the structure of DNA, commercial models are now available that allow you to construct physical models of molecules. Even given the high-quality illustrations and computer-generated images available to you, we recommend that you build physical models of the molecules presented in this chapter. As Watson and Crick demonstrated, there is no substitute for holding a three-dimensional model in your hands and rotating it in helping you form accurate mental models of structure and shape.

Wikimedia Commons

Figure 13.2 The DNA model built by Crick and Watson in 1953, photographed while on display in the National Science Museum of London.

UniversalImagesGroup/Getty Images

Central Press/Hulton Archive/Getty Images

Figure 13.3 Francis Crick (*left*) and James Watson (*right*) shared the 1962 Nobel Prize in Physiology or Medicine with Maurice Wilkins for their contributions to our knowledge about DNA and other similar molecules.

13.2 Drawing Lewis Diagrams

Goal 1 Draw the Lewis diagram for any molecule or polyatomic ion made up of main group elements.

Determination of the shape of a molecule begins with the Lewis diagram, and in case it has been some time since you studied Lewis diagrams, we will review them briefly. Important terms are printed in italics.

Lewis symbols for elements (also called *electron dot symbols*) were introduced in Section 11.6. The number of dots equal to the number of *valence electrons* is placed around the chemical symbol of the element. In Section 12.4, these symbols were used to show how two electrons are shared in forming a chemical bond between two atoms. The electrons can be shown as two dots, but usually a bonding pair appears as a dash between the symbols of the bonded atoms.

Two atoms bonded to each other by one electron pair are connected by a *single bond.* Sometimes, two atoms are bonded by more than one electron pair. This is shown by two or three dashes, which represent *double* or *triple bonds*, respectively.

Bonded atoms may have valence electrons that are not used for bonding; these are represented by dots. In reaching a stable *octet of electrons*, these nonbonding electrons occur in pairs called *lone pairs*.

Once you are confident in your understanding of these Chapter 12 concepts, you are ready to learn how to draw Lewis diagrams. You may use the procedure that follows to sketch the Lewis diagram for any species that obeys the octet rule. Each step is illustrated by drawing the Lewis diagram for carbon tetrachloride.

Learn It Now! *Matching a main group element with its group number in the periodic table is a quick way to count the valence electrons in atoms of that element.*

Step 1: *Calculate the total number of valence electrons in the molecule or ion.* Note that the number of valence electrons for an atom of a main group element is the same as the element's column number in the periodic table (or the final digit of the column number if you are using IUPAC group numbers). If the species is an ion, the number of valence electrons must be adjusted to account for the charge on the ion. For each positive charge, subtract one electron; for each negative charge, add one electron.

In CCl_4, there are four valence electrons from the carbon atom and seven from each chlorine atom:

$$4\ (\mathrm{C}) + 4 \times 7\ (\mathrm{Cl}) = 32$$

▶ **P/Review** Electronegativity values increase from left to right across any row of the periodic table and from the bottom to top of any column. See Section 12.5. The atoms in a chemical formula are usually written from left to right in order of increasing electronegativity. Thus, the least electronegative element is usually the first one written in a formula. There are exceptions, however, so be cautious in assuming that the first element listed is the least electronegative.

Step 2: *Determine the central atom(s) of the molecule or ion.* The central atom of the molecule or ion is usually the least electronegative atom of the species. ◀ The most common exception is the hydrogen atom. Hydrogen is never the central atom in a molecule.

Carbon is less electronegative than chlorine, so the central atom in CCl_4 is C. Carbon is always a central atom in a species that contains carbon.

Step 3: *Draw a tentative diagram for the molecule or ion,* joining atoms by single bonds and placing electron dots around each symbol except hydrogen, so the total number of electrons for each atom is eight. In some cases, only one arrangement of atoms is possible. In others, two or more diagrams are possible. Ultimately, chemical or physical evidence must be used to decide which diagram is correct. A few general rules will help you make diagrams that are most likely to be correct:

a) A hydrogen atom always forms one bond and has no lone pairs of electrons. Hydrogen is always a **terminal atom** in a Lewis diagram—an atom that is bonded to only one other atom.

b) A carbon atom is always surrounded by four electron pairs: four single bonds, a double bond and two single bonds, two double bonds, and so on. Carbon is always a **central atom**—an atom that is bonded to two or more other atoms—in a Lewis diagram that has three or more atoms.

c) When several carbon atoms appear in the same molecule, they are often bonded to one another.

d) Make your diagram as balanced as possible. In particular, a compound or ion having two or more oxygen atoms and one atom of another nonmetal usually has the oxygen atoms arranged around the central nonmetal atom. If hydrogen is also present, it is usually bonded to an oxygen atom, which is then bonded to the nonmetal with a single bond: X—O—H, where X is the nonmetal. ◀

The Active Examples that follow show you how to place hydrogen, oxygen, and other elements in Lewis diagrams.

CCl_4 is described by Steps 3b and 3d. The four chlorine atoms are placed around the carbon atom. Here is the tentative diagram:

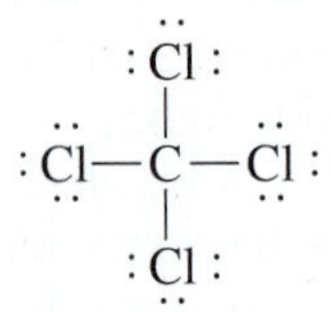

```
          ..
         :Cl:
   ..     |     ..
  :Cl  —  C  —  Cl:
   ..     |     ..
         :Cl:
          ..
```

Step 4: *Compare the number of valence electrons you have available from Step 1 to the number that you used in your tentative Lewis diagram. If they are not equal, replace two lone pairs with one bonding pair. Repeat if necessary.* When your tentative Lewis diagram has more electrons than are needed, erase two lone pairs of electrons, one from the central atom and one from a terminal atom, and replace them with one bonding electron pair between those two atoms. ◀

You will see how to do this in Active Example 13.3.

There are four bonds and 12 unshared pairs in the tentative diagram:

$$4 \times 2 \text{ (bonds)} + 12 \times 2 \text{ (lone pairs)} = 32$$

These 32 electrons are equal to the total number of electrons counted in Step 1. No further adjustment is necessary.

Step 5: *Check to be sure that each atom other than hydrogen has four electron pairs and that hydrogen has only one electron pair.* Remember that a bonding pair is counted for both bonded atoms. This final step will ensure that you have drawn a correct Lewis diagram.

Checking each atom in the molecule:

```
   |        ..        |         ..       ..
 — C —    : Cl —    : Cl :    — Cl :   : Cl :
   |        ..        ..        ..       |
```

All atoms have four electron pairs. The Lewis diagram is now complete.

The steps in drawing a Lewis diagram are as follows:

how to... Draw a Lewis Diagram

Step 1: Count the total number of valence electrons. Adjust for charge on ions.

Step 2: Place the least electronegative atom(s) in the center of the molecule.

Step 3: Draw a tentative diagram. Join atoms by single bonds. Add unshared pairs to complete the octet around all atoms except hydrogen.

Step 4: Calculate the number of valence electrons in your tentative diagram and compare it with the actual number of valence electrons. If the tentative diagram has too many electrons, remove a lone pair from the central atom and from a terminal atom, and replace them with an additional bonding pair between those atoms. If the tentative diagram still has too many electrons, repeat the process.

Step 5: Check the Lewis diagram. Hydrogen atoms must have only one bond, and all other atoms should have a total of four electron pairs.

Active Example 13.1 Drawing Lewis Diagrams I

Draw Lewis diagrams for ammonia, NH_3, and phosphorus tribromide, PBr_3.

Think Before You Write Faithfully follow the five-step procedure for Drawing a Lewis Diagram step-by-step as you learn this important process. Avoid shortcuts.

Answers *Cover the left column with your cut-out shield. Reveal each answer only after you have written your own answer in the right column.*

5 (N) + 3 × 1 (H) = 8

First consider the ammonia molecule. Start by counting the number of valence electrons.

```
   ..
H—N—H
  |
  H
```

A hydrogen atom is never central, so the nitrogen atom must be central. Nitrogen has four pairs, and each hydrogen has one pair.

Determine the central atom and draw the tentative diagram.

Eight electrons in the tentative diagram match the number available. The diagram is complete.	Compare the number of electrons in your tentative diagram with the number available. 
5 (P) + 3 × 7 (Br) = 26 valence electrons :B̤r̈—P̈—B̤r̈: | :B̤r̈: *Phosphorus is less electronegative than bromine, so P is the central atom. The number of electrons in the tentative diagram matches the number available.* *Notice the similarity between this diagram and the one for* NH_3. *The central elements are both in Group 5A/15 and have five valence electrons. The other element in each case is single-bonded to the central atom.*	Now complete all of the steps while drawing the Lewis diagram for phosphorus tribromide.
You improved your skill at drawing Lewis diagrams.	What did you learn by solving this Active Example? 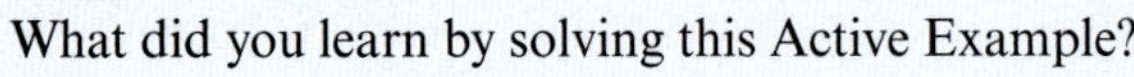

Practice Exercise 13.1

Draw Lewis diagrams for water and dihydrogen sulfide.

Chlorine monofluoride is a colorless gas at room conditions. When cooled below its boiling point of −100°C, it becomes a yellow liquid.

Active Example 13.2 Drawing Lewis Diagrams II

Write Lewis diagrams for the ClF molecule and the ClO^- **ion.**

Think Before You Write Faithfully follow each of the five steps in the procedure for Drawing a Lewis Diagram.

Answers *Cover the left column with your cut-out shield. Reveal each answer only after you have written your own answer in the right column.*

ClF: 7 (Cl) + 7 (F) = 14 ClO^-: 7 (Cl) + 6 (O) + 1 (charge) = 14 *Chlorine and fluorine, both in Group 7A/17, have seven valence electrons. Oxygen in Group 6A/16 has six. In* ClO^- *there is one additional electron to account for the 1− charge on the ion.*	This time, we'll work on the two diagrams simultaneously. *Step 1* is to count the valence electrons for each species. 
:C̤l̈—F̤̈: :C̤l̈—Ö̤:	*Step 2* is to determine the central atom. With two atoms, the only possible diagram has them bonded to each other. *Step 3* is to draw the tentative diagrams. Join the atoms by single bonds and complete the octet for each atom.

1 × 2 (bond) + 6 × 2 (lone pairs) = 14

The number of available valence electrons equals the number of electrons in the tentative diagrams. No modifications are necessary.

The final step, Step 5, is to check that each atom has an octet. The diagrams check. In both cases there is just the right number of electrons to complete the octet for both atoms. The diagrams are therefore complete. In the case of ClO^- the diagram is enclosed in brackets and a charge is shown because it is an ion:

$$\left[:\ddot{\underset{..}{Cl}}-\ddot{\underset{..}{O}}:\right]^-$$

Step 4 is to compare the actual number of valence electrons available to the number in the tentative diagrams. Each bond accounts for two of the total number of electrons. How many valence electrons are in each tentative diagram? Are modifications necessary?

You improved your skill at drawing Lewis diagrams.

What did you learn by solving this Active Example?

Practice Exercise 13.2

Write Lewis diagrams for hydrogen fluoride and hypobromite ion.

Did you notice how each step was the same for the two species in Active Example 13.2? This is because any two species that have (1) the same number of atoms and (2) the same number of valence electrons also have similar Lewis diagrams, whether they are molecules or polyatomic ions.

In all examples so far, covalent bonds have been formed when each atom contributes one electron to the bonding pair. This is not always the case. Many bonds are formed in which one atom contributes both electrons and the other atom offers only an empty orbital.* To illustrate, an ammonium ion is produced when a hydrogen ion is bonded to the unshared electron pair of the nitrogen atom in an ammonia molecule:

$$H^+ + :\underset{\underset{H}{|}}{\overset{\overset{H}{|}}{N}}-H \longrightarrow \left[H-\underset{\underset{H}{|}}{\overset{\overset{H}{|}}{N}}-H\right]^+$$

ammonia ammonium ion

The four bonds in an ammonium ion are identical. This shows that a bond formed from an electron pair and an empty orbital is the same as a bond formed by one electron from each atom.

Active Example 13.3 Drawing Lewis Diagrams III

Draw the Lewis diagram for the hydrogen carbonate ion, HCO_3^-.

Think Before You Write Continue to build your skill by following each step in the procedure. You will find that this stepwise approach pays off both now and, in particular, when you begin to work on more complex diagrams.

Answers *Cover the left column with your cut-out shield. Reveal each answer only after you have written your own answer in the right column.*

*This is called a *coordinate covalent bond.*

$1 (H) + 4 (C) + 3 \times 6 (O) + 1 (\text{charge}) = 24$

In *Step 1*, you count the total number of valence electrons. Don't forget to account for the charge.

Carbon is the central atom.

Hydrogen is always a terminal atom. Carbon is less electronegative than oxygen, so the central atom is C. Carbon will always be a central atom in a species that contains carbon.

Now determine the central atom (*Step 2*).

H—O—C—O:
:O:

The hydrogen atom can be attached to any of the oxygen atoms. Hydrogen has one electron pair, and all other atoms have four.

Step 3 is to draw the tentative diagram for HCO_3^-. Remember that oxygen is usually arranged around the central nonmetal atom and that hydrogen is usually bonded to oxygen.

There are 26 valence electrons.

The tentative diagram has four bonds and nine unshared pairs: 4×2 (bonds) $+ 9 \times 2$ (lone pairs) $= 26$.

Count the number of electrons in your tentative diagram.

H—O—C=O:
:O:

In *Step 1*, you determined that 24 valence electrons are available, so you have two more electrons in the tentative diagram than you need. Complete *Step 4* by removing one lone pair from the central atom (carbon) and one lone pair from either terminal oxygen atom. Then place another bonding pair between those atoms to form a double bond.

All atoms check.

Checking each atom in the ion:

H— —O— —C= =O: :O:

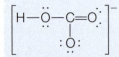

The hydrogen has one electron pair, and all other atoms have four.

Check to be sure that all atoms except hydrogen have four electron pairs.

[H—O—C=O: / :O:]⁻

Lewis diagrams of ions are enclosed in square brackets with the charge indicated as a superscript on the upper right. Complete the diagram.

You improved your skill at drawing Lewis diagrams.

What did you learn by solving this Active Example?

Practice Exercise 13.3

Draw the Lewis diagram for the nitrate ion.

1×2 (bond) + 6×2 (lone pairs) = 14 The number of available valence electrons equals the number of electrons in the tentative diagrams. No modifications are necessary. *The final step, Step 5, is to check that each atom has an octet. The diagrams check. In both cases there is just the right number of electrons to complete the octet for both atoms. The diagrams are therefore complete. In the case of ClO^- the diagram is enclosed in brackets and a charge is shown because it is an ion:* $[:\ddot{\underset{..}{Cl}}-\ddot{\underset{..}{O}}:]^-$	*Step 4* is to compare the actual number of valence electrons available to the number in the tentative diagrams. Each bond accounts for two of the total number of electrons. How many valence electrons are in each tentative diagram? Are modifications necessary?
You improved your skill at drawing Lewis diagrams.	What did you learn by solving this Active Example?

Practice Exercise 13.2

Write Lewis diagrams for hydrogen fluoride and hypobromite ion.

Did you notice how each step was the same for the two species in Active Example 13.2? This is because any two species that have (1) the same number of atoms and (2) the same number of valence electrons also have similar Lewis diagrams, whether they are molecules or polyatomic ions.

In all examples so far, covalent bonds have been formed when each atom contributes one electron to the bonding pair. This is not always the case. Many bonds are formed in which one atom contributes both electrons and the other atom offers only an empty orbital.* To illustrate, an ammonium ion is produced when a hydrogen ion is bonded to the unshared electron pair of the nitrogen atom in an ammonia molecule:

$$H^+ + :\underset{H}{\overset{H}{\underset{|}{\overset{|}{N}}}}-H \longrightarrow \left[H-\underset{H}{\overset{H}{\underset{|}{\overset{|}{N}}}}-H\right]^+$$

ammonia　　ammonium ion

The four bonds in an ammonium ion are identical. This shows that a bond formed from an electron pair and an empty orbital is the same as a bond formed by one electron from each atom.

Active Example 13.3 Drawing Lewis Diagrams III

Draw the Lewis diagram for the hydrogen carbonate ion, HCO_3^-.

Think Before You Write Continue to build your skill by following each step in the procedure. You will find that this stepwise approach pays off both now and, in particular, when you begin to work on more complex diagrams.

Answers *Cover the left column with your cut-out shield. Reveal each answer only after you have written your own answer in the right column.*

---

*This is called a *coordinate covalent bond.*

1 (H) + 4 (C) + 3 × 6 (O) + 1 (charge) = 24	In *Step 1*, you count the total number of valence electrons. Don't forget to account for the charge. 
Carbon is the central atom. *Hydrogen is always a terminal atom. Carbon is less electronegative than oxygen, so the central atom is C. Carbon will always be a central atom in a species that contains carbon.*	Now determine the central atom (*Step 2*). 
H—Ö—C̈—Ö: (with :Ö: bonded below C) *The hydrogen atom can be attached to any of the oxygen atoms. Hydrogen has one electron pair, and all other atoms have four.*	*Step 3* is to draw the tentative diagram for HCO_3^-. Remember that oxygen is usually arranged around the central nonmetal atom and that hydrogen is usually bonded to oxygen. 
There are 26 valence electrons. *The tentative diagram has four bonds and nine unshared pairs: 4 × 2 (bonds) + 9 × 2 (lone pairs) = 26.*	Count the number of electrons in your tentative diagram. 
H—Ö—C=Ö: (with :Ö: bonded below C)	In *Step 1*, you determined that 24 valence electrons are available, so you have two more electrons in the tentative diagram than you need. Complete *Step 4* by removing one lone pair from the central atom (carbon) and one lone pair from either terminal oxygen atom. Then place another bonding pair between those atoms to form a double bond. 
All atoms check. *Checking each atom in the ion:* H— —Ö— —C= =Ö: :Ö: *The hydrogen has one electron pair, and all other atoms have four.*	Check to be sure that all atoms except hydrogen have four electron pairs. 
[H—Ö—C=Ö: (with :Ö: bonded below C)]$^-$	Lewis diagrams of ions are enclosed in square brackets with the charge indicated as a superscript on the upper right. Complete the diagram. 
You improved your skill at drawing Lewis diagrams.	What did you learn by solving this Active Example? 

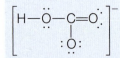

Practice Exercise 13.3

Draw the Lewis diagram for the nitrate ion.

The carbon–oxygen double bond in Active Example 13.3 could have been placed between the carbon atom and either of the terminal oxygen atoms. When changing only the positions of electrons produces two or more equivalent Lewis diagrams for a molecule or ion, the diagrams represent **resonance structures**. On paper, the bonds between the central carbon and the outlying oxygens look different. In the molecule itself, they are identical. Moreover, the bond strengths and lengths are between those found in true single and double bonds connecting the same two atoms. The actual molecule is an average of the resonance structures, and it is called a **resonance hybrid**.

It is customary to place a two-headed arrow between resonance structures:

```
[      ..     ..  ]-          [      ..     ..  ]-
[ H—O—C=O:        ]    <-->   [ H—O—C—O:        ]
[      ..  |  ..  ]           [      ..  ||  .. ]
[         :O:     ]           [          O      ]
[          ..     ]           [         ..  ..  ]
```

However, further discussion of resonance is beyond the scope of an introductory text. Therefore, when we encounter a resonance structure, we will simply show one of the alternative diagrams.

Active Example 13.4 Drawing Lewis Diagrams IV

Draw the Lewis diagram for the sulfite ion, SO_3^{2-}.

Think Before You Write Remember to add one valence electron for *each* negative charge on an anion.

Answers *Cover the left column with your cut-out shield. Reveal each answer only after you have written your own answer in the right column.*

6 (S) + 3 × 6 (O) + 2 (charge) = 26 Sulfur is less electronegative than oxygen, so it is the central atom. :Ö—S̈—Ö: (with :O: bonded below S, each O bearing three unshared pairs and S one unshared pair) *The ion has a 2– charge, so two electrons must be added to the valence electrons of the atoms themselves. This diagram has the three oxygen atoms distributed around sulfur as the central (least electronegative) atom, conforming to Steps 2 and 3d in the earlier detailed procedure.*	Do *Steps 1, 2,* and *3*: Get the total valence electron count, determine the central atom of the molecule, and draw a tentative diagram.
3 × 2 (bonds) + 10 × 2 (unshared pairs) = 26 The number of valence electrons available is equal to the number in the tentative diagram. No further modification is necessary.	Calculate the number of valence electrons in the tentative diagram, compare these with the number available, and modify the diagram if necessary.

$$\left[:\ddot{O}-\ddot{S}-\ddot{O}: \atop \quad\ \ |\ \ \atop \quad :\underset{..}{O}: \right]^{2-}$$

Every atom is surrounded by four electron pairs.

The "finishing touch" is to surround the diagram with brackets and indicate the charge. You might wish to compare the SO_3^{2-} diagram with the diagram for PBr_3 in Active Example 13.1. Both species have four atoms and 26 electrons, so the electron distribution in their diagrams is the same.

Finally, check the diagram to be sure that every atom has an octet and add the "finishing touch" to complete the diagram that you started in the space near the beginning of the Active Example.

You improved your skill at drawing Lewis diagrams.

What did you learn by solving this Active Example?

Practice Exercise 13.4

Draw the Lewis diagram for the sulfate ion.

Active Example 13.5 Drawing Lewis Diagrams V

Draw the Lewis diagram for sulfur dioxide, SO_2.

Think Before You Write Recall that when the number of valence electrons available is less than the number in your tentative diagram, you replace two lone pairs with one bonding pair, and you repeat if necessary.

Answers *Cover the left column with your cut-out shield. Reveal each answer only after you have written your own answer in the right column.*

Total valence electrons: 6 (S) + 2 × 6 (O) = 18

S is less electronegative than O, so it is the central atom.

$$:\ddot{\underset{..}{O}}-\ddot{\underset{..}{S}}-\ddot{\underset{..}{O}}:$$

The tentative diagram has 20 valence electrons.

Complete the procedure through the point at which you compare the number of valence electrons available with the number in the tentative diagram.

$$:\ddot{\underset{..}{O}}-\underset{..}{S}=\dot{\underset{.}{O}}\dot{} \quad \text{or} \quad \dot{}\dot{\underset{.}{O}}=\underset{..}{S}-\ddot{\underset{..}{O}}:$$

These Lewis diagrams are resonance structures. Either diagram is acceptable.

This time there are too many electrons in the tentative diagram. Modify the tentative diagram by removing two lone pairs and replacing them with one bonding pair.

You improved your skill at drawing Lewis diagrams.

What did you learn by solving this Active Example?

Practice Exercise 13.5

Draw the Lewis diagram for carbon dioxide.

The guidelines we are following are readily applied to simple organic molecules, which always contain carbon and hydrogen atoms, and may contain atoms of other elements, notably oxygen. ▶ If oxygen is present in an organic compound, it usually forms two bonds. If you remember that carbon forms four bonds and hydrogen forms one, and that two or more carbon atoms often bond to one another (Steps 3a–c at the beginning of this section), your tentative diagrams are likely to be correct.

◀ **P/Review** Organic chemistry is the chemistry of carbon compounds. Chapter 21 is an introduction to this branch of chemistry.

Active Example 13.6 Drawing Lewis Diagrams VI

Write the Lewis diagram for propane, C_3H_8.

Think Before You Write Recall *Step 3c* in the procedure for drawing a Lewis diagram: When several carbon atoms are in the same molecule, they are often bonded to one another.

Answers *Cover the left column with your cut-out shield. Reveal each answer only after you have written your own answer in the right column.*

Total valence electrons: 3×4 (C) + 8×1 (H) = 20

The carbon atoms will be central.

```
    H   H   H
    |   |   |
H — C — C — C — H
    |   |   |
    H   H   H
```

All 20 of the valence electrons are needed for the 10 single bonds in the diagram. There are no lone pairs.

Complete the example.

You improved your skill at drawing Lewis diagrams.

What did you learn by solving this Active Example?

Practice Exercise 13.6

What is the Lewis diagram of butane, C_4H_{10}?

Active Example 13.7 Drawing Lewis Diagrams VII

Draw the Lewis diagram for acetylene, C_2H_2.

Think Before You Write Continue to follow the procedure step-by-step, and you will produce correct Lewis diagrams.

Answers *Cover the left column with your cut-out shield. Reveal each answer only after you have written your own answer in the right column.*

Total valence electrons: 2×4 (C) + 2×1 (H) = 10

The hydrogen atoms must be terminal, so the carbon atoms are central:

```
    ..  ..
H — C — C — H
    ..  ..
```

The tentative diagram has 14 valence electrons.

Take it through the point at which you compare the number of valence electrons available with the number in the tentative diagram.

H—C≡C—H *Twice replacing two lone pairs with one shared bonding pair yields a triple bond.*	There are four too many electrons in the tentative diagram this time; however, the remedy is the same. Erase two lone pairs from the carbons and replace them with a bonding pair to reduce the electron count by two. Then do it again to reduce the electron count by a total of four.
You improved your skill at drawing Lewis diagrams.	What did you learn by solving this Active Example?

Practice Exercise 13.7

Draw the Lewis diagram for ethylene, C_2H_4 (Figure 13.4).

Darengphoto/Shutterstock.com

Figure 13.4 Ethylene is the most produced carbon–hydrogen compound in the world. Its major use is to make polyethylene, the world's most widely used plastic. Many items are made of polyethylene, such as numerous types of plastic containers, packaging materials, Saran™ wrap, plastic pipes, and the plastic shopping bags shown here.

Active Example 13.8 Drawing Lewis Diagrams VIII

Draw a Lewis diagram for C_2H_6O.

Think Before You Write Take this one all the way with no further guidance from us.

Answers *Cover the left column with your cut-out shield. Reveal each answer only after you have written your own answer in the right column.*

Valence electron count: $2 \times 4\text{ (C)} + 6 \times 1\text{ (H)} + 6\text{ (O)} = 20$ The carbon atoms are central. H H | | H—C—C—Ö—H (with two lone pairs on O) | | H H *The tentative diagram has the correct number of valence electrons, 20. You may have switched the –O–H group with any of the hydrogen atoms in this diagram, and if so, your Lewis diagram is also correct. By our insisting that the carbons be bonded to each other, the oxygen had to go between a carbon atom and a hydrogen atom.*	Bond the carbon atoms to each other.
You improved your skill at drawing Lewis diagrams.	What did you learn by solving this Active Example?

Practice Exercise 13.8

Draw two different Lewis diagrams for C_3H_8O where both have the carbon atoms bonded to one another in a chain.

The compound in Active Example 13.8 is ethanol. If we had not insisted that the carbon atoms be bonded to each other, the oxygen atom might have been placed between them:

```
     H       H
     |  ..   |
 H — C — O — C — H
     |  ..   |
     H       H
```

This is another well-known compound, dimethyl ether. Note that both compounds have the same molecular formula, C_2H_6O, but different structures. Compounds that have the same molecular formulas but different structures are called **isomers** of one another. Isomers are distinctly different substances; each isomer has its own unique set of properties. For example, ethanol boils at 78°C and is a liquid at room temperature, whereas dimethyl ether boils at −24° C and is a gas at room temperature.

13.3 Electron-Pair Repulsion: Electron-Pair Geometry

Goal 2 Describe the electron-pair geometry when a central atom is surrounded by two, three, or four regions of electron density.

The shape of a molecule plays a major role in determining the macroscopic properties of a substance. We examine this role in other chapters in this book. To understand and predict the shape–property relationship, you first need to know what is responsible for molecular shape. This is the focus of this section and the next. Discussion in these sections is limited to molecules having only single bonds. We then expand our consideration to molecules with multiple bonds in Section 13.5.

No single theory or model yet developed succeeds in explaining all the molecular shapes observed in the laboratory. A theory that explains one group of molecules cannot explain another group. Each model has its advantages and limitations. Chemists, therefore, use them all within the areas to which they apply, fully recognizing that there is still much to learn about how atoms are assembled in molecules.

In this text, we will explore one of the models used to explain **molecular geometry**, the more precise term used to describe the shape of a molecule. It is called the **valence shell electron-pair repulsion theory** or **VSEPR theory**. VSEPR theory applies primarily to substances in which a second-period atom is bonded to two, three, or four other atoms. You may wonder why we focus so much attention on so few elements. The answer is that the second period includes carbon, nitrogen, and oxygen. Carbon alone is present in about 90% of all known compounds, and a large percentage of those include oxygen or nitrogen or both. These elements warrant this kind of attention.

The basic idea of VSEPR is that the electron pairs we draw in Lewis diagrams repel each other in real molecules. ▶ Therefore, they distribute themselves in positions around the central atom that are as far away from one another as possible. These are the locations of lowest potential energy; they satisfy the "minimization of energy" tendency that, we have noted, is one driving force in nature. ▶ This arrangement of electron pairs is called **electron-pair geometry**. The electron pairs may be shared in a covalent bond, or they may be lone pairs; it makes no difference.

◀ **P/Review** The forces that electrically charged particles exert on each other are described in Section 2.8. A particle with a positive charge is attracted to a particle with a negative charge. Two particles with the same charge, both positive or both negative, repel each other.

◀ **P/Review** Minimization of energy is mentioned as a driving force for change in many places in this book, notably in Section 2.9.

Earlier, we drew Lewis diagrams in which carbon, nitrogen, or oxygen was the central atom. In all cases, the central atom was surrounded by four pairs of electrons. In Section 12.8, we showed how beryllium and boron—also Period 2 elements—do not conform to the octet rule. The beryllium atom in BeF_2 is flanked by only two electron pairs; in BF_3 the boron atom has three electron pairs around it. Our question then is this: How do two, three, or four electron pairs distribute themselves

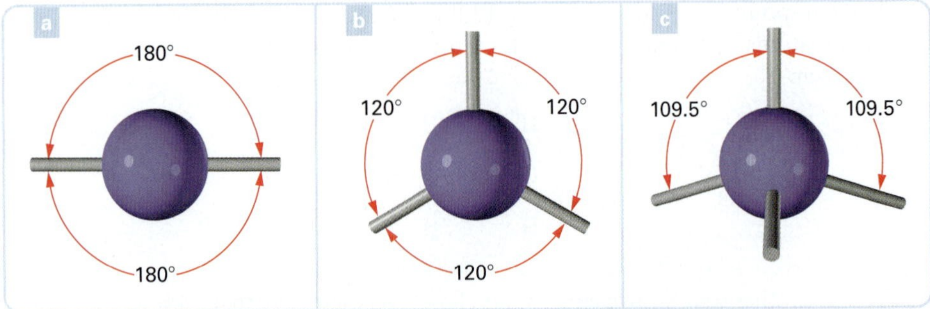

Figure 13.5 Electron-pair geometry. Ball-and-stick models show the arrangement of two, three, and four electron pairs (sticks) around a central atom (ball). (a) According to the electron-pair repulsion principle, two sticks are as far from each other as possible when they are diametrically opposite each other. The geometry is linear, and the angle formed is 180°. (b) Three sticks are as far from one another as possible when equally spaced on a circumference of the ball. The sticks and the center of the ball are in the same plane and the angles are 120°. The geometry is trigonal planar. (c) Four electron pairs are as far from one another as possible when arranged to form a tetrahedron. Each angle is 109.5°, which is called the *tetrahedral angle*.

around a central atom so they are as far apart as possible? This question is answered by identifying the **electron-pair angle**, the angle formed by any two electron pairs and the central atom.

When electron pairs are as far apart as possible, all electron-pair angles around the central atoms are equal. The electron-pair geometries that result from two, three, or four electrons pairs are shown in **Figure 13.5**. The bond angles are derived by geometry.

Can we find these angles in our familiar, large-scale world? Indeed, we can. If balloons of *similar size and shape* are tied together, they naturally arrange themselves in the same way (**Figure 13.6**). These arrangements are their minimum energy positions. The balloons are like identically sized and shaped electron orbitals. Atom-size (particulate) properties are reproduced naturally on a larger (macroscopic) scale.

Electron-pair geometries are summarized in **Table 13.1**.

Table 13.1 and the Figure 13.5 caption introduce the words **tetrahedron** and **tetrahedral**. A tetrahedron is the simplest regular solid. A regular solid is a solid figure with identical faces. A cube is a regular solid that has six identical squares as its faces. A tetrahedron has four identical equilateral triangles for its faces, as shown in **Figure 13.7**.

Figure 13.6 Electron–pair geometries for two to four electron pairs. If one ties together balloons of similar size and shape, they will naturally assume the geometries shown. Negatively charged electron pairs are attracted to the positively charged nucleus. This is analogous to the balloons being tied together at their tied-off openings. Negatively charged electron pairs are also repelled by one another and therefore move to positions as far apart from one another while still being attracted to the nucleus. This is analogous to the bodies of the balloons taking positions as far apart as possible while still being tethered at a central point.

Table 13.1 Electron-Pair Geometries		
Electron Pairs	Geometry	Electron-Pair Angles
2	Linear	180°
3	Trigonal planar	120°
4	Tetrahedral	109.5°

This geometric figure appears in all molecules in which carbon forms single bonds with four other atoms. Notice that a tetrahedron is a three-dimensional figure. It can only be drawn in perspective on the two-dimensional plane of a book page.

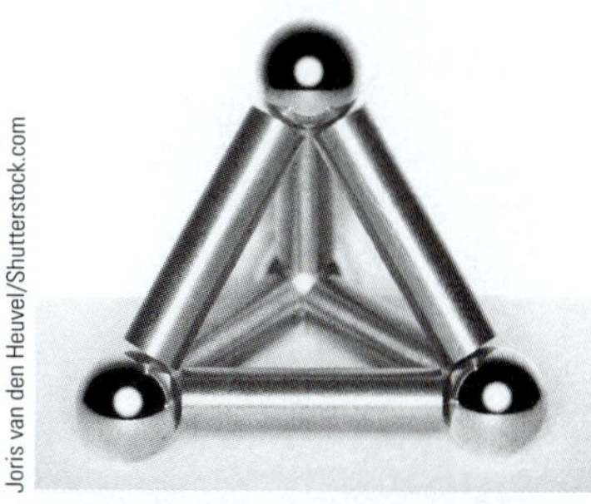

Joris van den Heuvel/Shutterstock.com

Charles D. Winters

Figure 13.7 Tetrahedral models. The metal figure is a tetrahedron. Its four faces are identical equilateral triangles. The model of methane, CH_4, has a tetrahedral structure. The carbon atom (black) is in the middle of the tetrahedron, and a hydrogen atom (white) is found at each of the four corners.

13.4 Molecular Geometry

Goal 3 Given or having derived the Lewis diagram of a molecule or polyatomic ion in which a central atom is surrounded by two, three, or four regions of electron density, predict and sketch the molecular geometry around that atom.

4 Draw a wedge-and-dash diagram of any molecule for which a Lewis diagram can be drawn.

Molecular geometry describes the shape of a molecule and the arrangement of atoms around a central atom. You might think of it as an "atom geometry," in the same sense that the arrangement of electron pairs is the electron-pair geometry. Thus, the **bond angle** is the angle between two bonds formed by the same central atom, as shown in **Figure 13.8**.

When all the electron pairs around a central atom are bonding pairs—that is, when there are no lone pairs—the bond angles are the same as the electron-pair angles. The molecular geometries are the same as the electron-pair geometries described previously. The same terms are also used to describe the shapes of the molecules. If a molecule contains one or two lone pairs, the bond angles are close to the electron-pair angles predicted by the VSEPR theory, but we need different terms to describe the molecular geometries of these molecules.

We now describe the molecular geometries for six combinations of regions of electron density and atoms that are connected to the central atom. A single bond, a double bond, a triple bond, and a lone pair are each considered to be one region of electron density. These molecular geometry descriptions are illustrated and summarized in **Table 13.2**. A discussion of each line in this table follows.

Line 1: Two Regions of Electron Density, Two Bonded Atoms Two regions of electron density, both bonding, yield the same electron-pair and molecular geometries: **linear**. A linear geometry has a 180° bond angle.

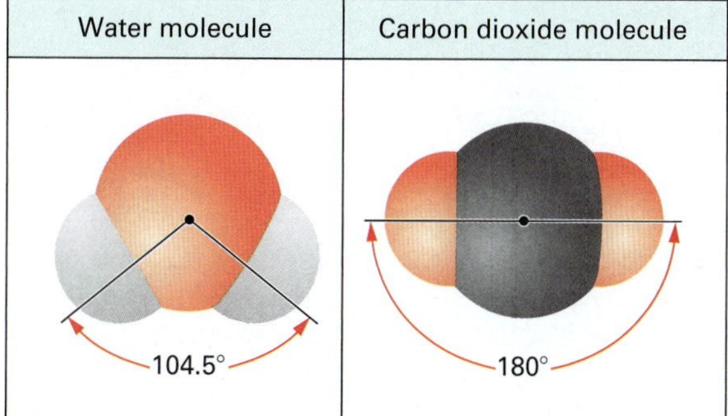

Figure 13.8 Bond angle. If an atom forms bonds with two other atoms, the angle between the bonds is the bond angle. In a water molecule (*left*), the bonds form at an angle of 104.5°. In a carbon dioxide molecule (*right*), the bonds lie in a straight line and the bond angle is 180°.

Table 13.2 Electron-Pair and Molecular Geometries

Line	Regions of Electron Density*	Bonded Regions	Electron-Pair Geometry	Ball-and-Stick Model	Electron-Pair and Bond Angle†	Molecular Geometry	Lewis Diagram	Ball-and-Stick Model	Space-Filling Model	Example
1	2	2	Linear		180°	Linear	A—B—A		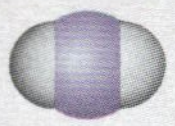	BeF_2
2	3	3	Trigonal planar	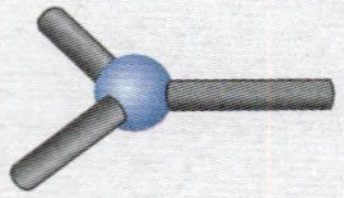	120°	Trigonal planar	A—B(—A)—A	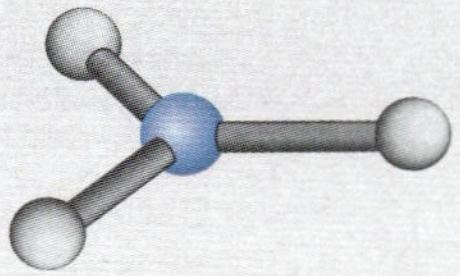		BF_3
3	3	2	Trigonal planar	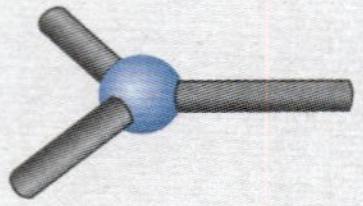	120°	Angular or bent	A═B̈—A	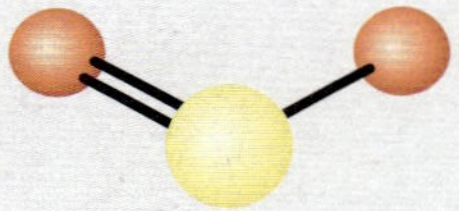		SO_2
4	4	4	Tetrahedral	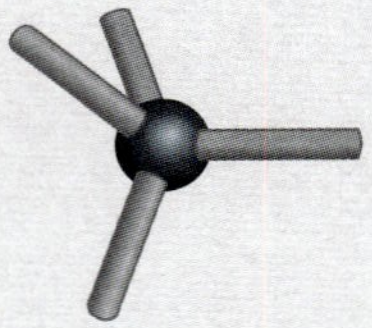	109.5°	Tetrahedral	A—B(—A)(—A)—A	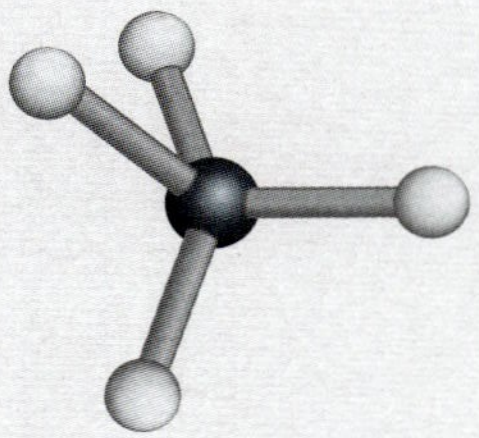	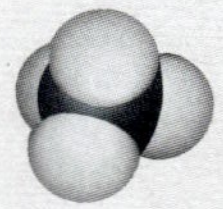	CH_4
5	4	3	Tetrahedral	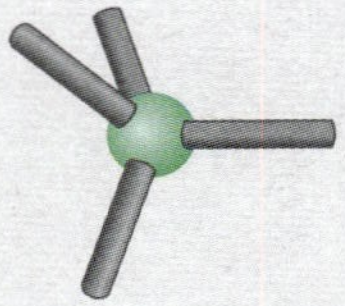	109.5°	Trigonal pyramidal	A—B̈(—A)—A	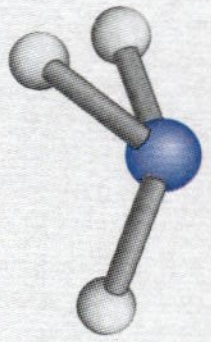		NH_3
6	4	2	Tetrahedral	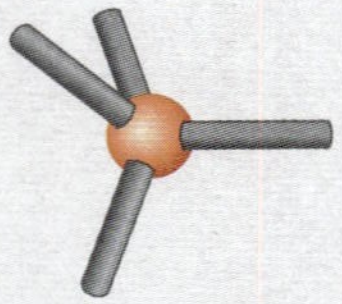	109.5°	Bent or angular	A—B̈—A (with two lone pairs) or A—B̈:(—A)	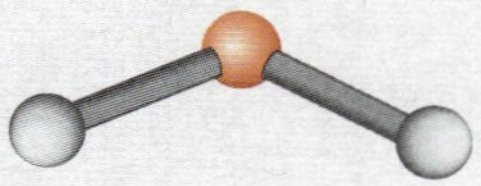	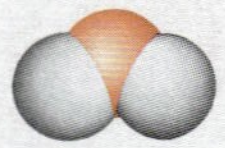	H_2O

*A region of electron density is a single bond, a double bond, a triple bond, or a lone pair. This is explained in more detail in Section 13.5.

†All bond angles are ideal, based on ideal electron-pair geometries.

Line 2: Three Regions of Electron Density, Three Bonded Atoms Three regions of electron density, all bonding, yield the same electron-pair and molecular geometries: **trigonal planar**. Each bond angle is 120°.

Line 3: Three Regions of Electron Density, Two Bonded Atoms The three regions of electron density retain their trigonal planar geometry, but only two are bonded to atoms. The resulting molecular geometry is called **angular** or **bent**, and the bond angle is 120°.*

Practicing chemists do not consistently use a single term to describe the shape of a molecule with three regions of electron density and two bonded atoms. Follow the advice of your instructor as to whether you should use *angular* or *bent* to describe this shape.

Line 4: Four Regions of Electron Density, Four Bonded Atoms Four regions of electron density, all bonding, yield the same electron-pair and molecular geometries: tetrahedral. The tetrahedral methane molecule, CH_4, looks like a tall pyramid with a triangular base (Figure 13.9[a]). Each bond angle is 109.5°—the tetrahedral angle.

Line 5: Four Regions of Electron Density, Three Bonded Atoms The four regions of electron density retain their tetrahedral geometry, which is modified because only three of the regions of electron density form bonds to other atoms. The resulting shape is like a "squashed-down" pyramid, called **trigonal pyramidal** (Figure 13.9[b]).

Line 6: Four Regions of Electron Density, Two Bonded Atoms Again, the tetrahedral electron-pair geometry is predicted. The molecular geometry is bent (or angular) (Figure 13.9[c]).

Chemists also interchange *bent* and *angular* for four regions of electron density/two bonded atoms. As always, follow your instructor's preference for the term to use in describing this shape.

With the help of the six preceding paragraphs and their summaries in Table 13.2, you are now ready to predict and name some electron-pair and molecular geometries around a central atom and sketch three-dimensional representations

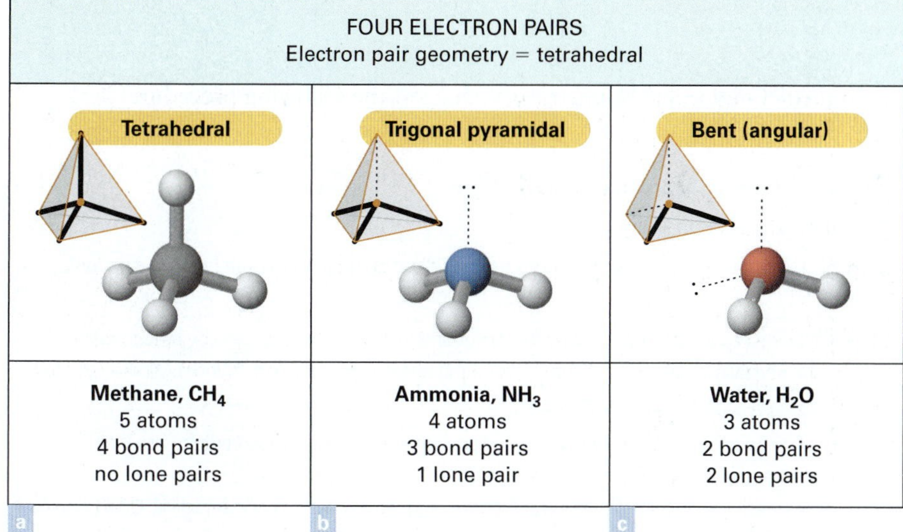

Figure 13.9 Molecular geometries based on four regions of electron density around the central atom. (a) In methane, CH_4, the four hydrogen atoms are at the corners of a tetrahedron, and the carbon atom is at its center. If the top hydrogen atom and the carbon atom are in the plane of the page, the front hydrogen atom in the base is in front of the page and the other two hydrogen atoms are behind the page. (b) Ammonia, NH_3, has the shape of a pyramid with a triangular base. The nitrogen atom is in the plane of the page, the front hydrogen atom is in front of the page, and the other two hydrogen atoms are behind the page. Like the carbon atom in methane, the nitrogen atom in ammonia is surrounded by four electron pairs. (c) Water, H_2O, has a bent (angular) shape. With only three atoms, the bond angle of the water molecule lies in two dimensions. The oxygen atom in water is surrounded by four electron pairs. These pairs are *not* in the same plane as the three atoms; one pair is above that plane and the other is beneath it.

*In this book, we will disregard minor changes in bond angles due to lone pairs.

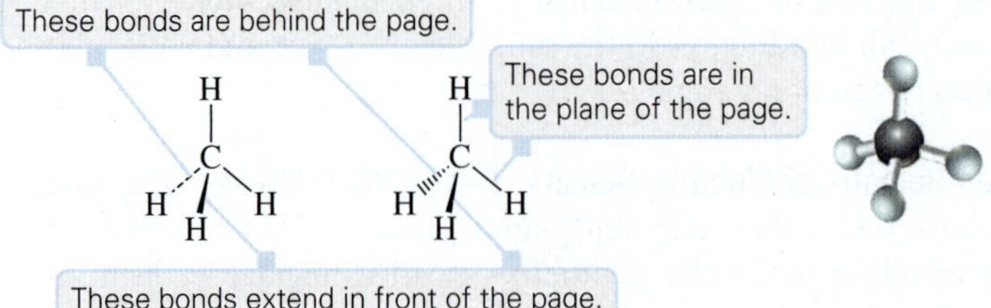

Figure 13.10 Conventions used for drawing wedge-and-dash diagrams. A dashed line indicates a bond that leads to an atom behind the plane of the page. A solid wedge-shaped line symbolizes a bond that leads to an atom in front of the page. Atoms connected by standard lines are in the plane of the page.

of molecules. Chemists often use **wedge-and-dash diagrams** to draw the three-dimensional structure of molecules.

We will follow these conventions, which are illustrated in **Figure 13.10**.

how to... Draw a Wedge-and-Dash Diagram

1. When two atoms are in the same plane as the page, connect them with a solid line of uniform width (–).
2. When an atom is behind the plane of the page, connect it to the central atom by a line that is dashed (IIIII). Increase the width of the dashed line as it moves away from the central atom.
3. When an atom is in front of the plane of the page, connect it to the central atom by a line that is wedge-shaped (◀). Increase the width of the wedge-shaped line as it moves away from the central atom.

To predict molecular geometries, we suggest the following procedure:

how to... Predict Molecular Geometries

Step 1: Draw the Lewis diagram.

Step 2: Count the regions of electron density around the central atom, both bonding and lone pairs.

Step 3: Determine electron-pair and molecular geometries. This is best done by reason rather than by memorization. Ask yourself, and picture the answer in your mind: "Where will the regions of electron density go to be as far apart as possible?" There are three answers:

a) Two regions of electron density: Electron-pair and molecular geometries are both linear. Bond angle is 180°.

b) Three regions of electron density: Electron-pair geometry is trigonal planar. Bond angles are 120°.

   i) All regions of electron density bonding: Molecular geometry is trigonal planar.

   ii) Two regions of electron density bonding, one lone pair: Molecular geometry is angular (bent).

c) Four regions of electron density: Electron-pair geometry is tetrahedral. Bond angles are tetrahedral (109.5°).

   i) All regions of electron density bonding: Molecular geometry is tetrahedral.

   ii) Three regions of electron density bonding, one lone pair: Molecular geometry is trigonal pyramidal.

   iii) Two regions of electron density bonding, two lone pairs: Molecular geometry is bent (angular).

Step 4: Sketch the wedge-and-dash diagram. This should match the mental picture you formed in Step 3.

Thinking About Your Thinking

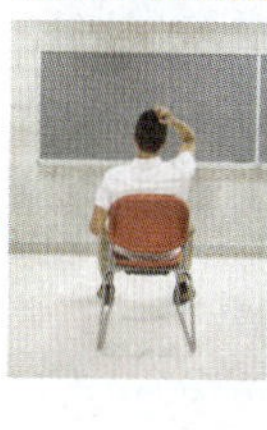

Mental Models

Learning to form three-dimensional mental models of molecules and polyatomic ions is a skill you should develop when studying this chapter. The Lewis diagram is a convenient two-dimensional paper-and-pencil representation of the distribution of electrons in a molecule, but it has limited information. A goal to work toward is to be able to look at a Lewis diagram and then see a model of that species in your mind. Our emphasis on three-dimensional wedge-and-dash diagrams in the Active Examples that follow and in the end-of-chapter questions will help you make the connection between Lewis diagrams and these mental models. If you have a molecular model kit, your models and the three-dimensional sketches you draw should match.

Active Example 13.9 Molecular Geometry I

Predict the electron-pair and molecular geometries and sketch the shape of a carbon tetrachloride molecule, CCl_4.

Think Before You Write Be sure that you are familiar with the information in Table 13.2 before proceeding with this Active Example. Also be sure that you understand the conventions illustrated in Figure 13.10.

Answers *Cover the left column with your cut-out shield. Reveal each answer only after you have written your own answer in the right column.*

The Lewis diagram is shown below. From this, you should establish the number of regions of electron density around the central atom and the number of atoms bonded to the central atom. Both geometries follow. Write their names.

```
      ..
     :Cl:
  ..  |  ..
 :Cl—C—Cl:
  ..  |  ..
     :Cl:
      ..
```

With four electron pairs around the carbon atom, all bonded to other atoms, both geometries are tetrahedral.

Use words to describe the shape of a molecule with a tetrahedral molecular geometry. What are the bond angles?

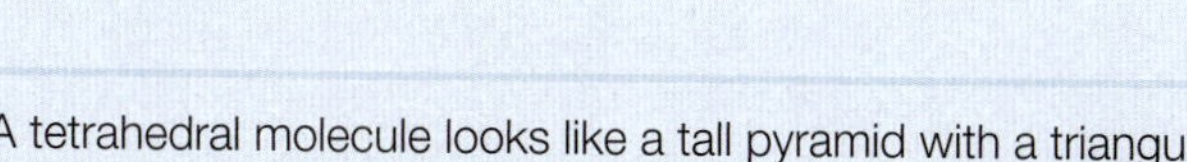

A tetrahedral molecule looks like a tall pyramid with a triangular base. Each bond angle is 109.5°—the tetrahedral angle.

Change your word description into a corresponding three-dimensional wedge-and-dash diagram.

What did you learn by solving this Active Example?

You improved your skill at predicting molecular geometries and sketching the three-dimensional shapes of molecules.

Practice Exercise 13.9

Draw the Lewis diagram of the chlorate ion. Describe its electron-pair and molecular geometry, and then sketch a three-dimensional wedge-and-dash diagram of the ion.

Active Example 13.10 Molecular Geometry II

Describe the shape of a molecule of boron trihydride, BH_3, and sketch the molecule.

Think Before You Write Notice the location of boron on the periodic table. It is in Group 3A/13. Remember that boron forms molecules that are exceptions to the octet rule. Boron atoms have only three valence electrons to contribute to covalent bonds.

Answers *Cover the left column with your cut-out shield. Reveal each answer only after you have written your own answer in the right column.*

```
H
 \
  B—H
 /
H
```

Three regions of electron density yield both an electron-pair geometry and a molecular geometry that are trigonal planar, with 120° bond angles.

First draw the Lewis diagram. From the structure, write your description of the shape by naming the electron-pair and molecular geometries and stating the bond angles.

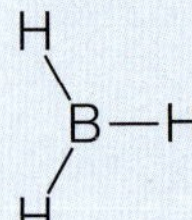

When all four atoms and all three bonds lie in the same plane, there is no need for wedges or dashes. It is simplest to put the entire molecule in the plane of the page.

Complete the Active Example with a wedge-and-dash diagram of the molecule. Remember that all atoms are in the same plane.

You improved your skill at predicting molecular geometries and sketching the three-dimensional shapes of molecules.

What did you learn by solving this Active Example?

Practice Exercise 13.10

Draw the Lewis diagram for carbon disulfide. State the electron-pair and molecular geometries of the molecule and sketch a wedge-and-dash diagram.

Active Example 13.11 Molecular Geometry III

At ordinary conditions, dichlorine monoxide is a pale orange-yellow gas. Its boiling point is 2°C.

Predict the electron-pair geometry and shape of a molecule of dichlorine monoxide, Cl_2O. Draw a wedge-and-dash representation of the molecule.

Think Before You Write In problems that ask you to predict and draw the shape of a molecule, you must start with the Lewis diagram. From there, the number of regions of electron density around the central atom gives you the electron-pair geometry, and then the number of bonded atoms gives you the molecular geometry. Keep in mind that the three-dimensional structure of a molecule is not necessarily reflected in its two-dimensional Lewis diagram.

Answers *Cover the left column with your cut-out shield. Reveal each answer only after you have written your own answer in the right column.*

Start with the electron-pair and molecular geometries. Draw the Lewis diagram and then name each. Even though oxygen is more electronegative than chlorine, this molecule is an exception to the rule and the oxygen atom is central.

```
 ..  ..
:Cl—O:
 ..  |
    :Cl:
     ..
```

The electron-pair geometry is tetrahedral, and the molecular geometry is bent.

Oxygen has four electron pairs around it, yielding an electron-pair geometry that is tetrahedral. Only two of the electron pairs are bonded to other atoms, so the molecule is bent. The structure is similar to that of water. Even if you drew the correct Lewis diagram as

```
 ..  ..  ..
:Cl—O—Cl:
 ..  ..  ..
```

you need to recognize that the four electron pairs around the central O lead to a tetrahedral electron-pair geometry and a bent molecular geometry. It is a good idea to draw the Lewis diagram of the molecule showing the bent geometry.

To draw the wedge-and-dash diagram, carefully consider the bent shape. It is preferable to have as many coplanar atoms as possible in your final sketch. Unshared pairs are not shown in a wedge-and-dash diagram.

```
Cl—O
    \
     Cl
```

A bent molecular geometry includes the tetrahedral angle, 109.5°. All three atoms can be drawn in the same plane.

What did you learn by solving this Active Example?

You improved your skill at predicting molecular geometries and sketching the three-dimensional shapes of molecules.

Practice Exercise 13.11

In the gas phase, tin(II) chloride is a covalently bonded compound. The number of valence electrons contributed by the tin atom is what you expect from its group in the periodic table. Draw the Lewis diagram for tin(II) chloride, state its electron-pair and molecular geometries, and construct a three-dimensional sketch of the molecule.

Active Example 13.12 Molecular Geometry IV

Draw the wedge-and-dash diagram for nitrogen trichloride, NCl_3. ▶ Name the electron-pair and molecular geometries.

Think Before You Write When a question asks for a wedge-and-dash diagram, you need to recognize that you first need to determine the molecular geometry, which, as a prerequisite, requires the electron-pair geometry, which, of course, requires the Lewis diagram.

Answers *Cover the left column with your cut-out shield. Reveal each answer only after you have written your own answer in the right column.*

The flour used in baking has a natural yellowish color. It is made white during the manufacturing process by bleaching. Nitrogen trichloride used to be one of the bleaching agents used, but it is now prohibited for a number of reasons, including the fact that it is a dangerous explosive.

$:\ddot{\text{Cl}}-\ddot{\text{N}}-\ddot{\text{Cl}}:$ with $:\ddot{\text{Cl}}:$ bonded below N With four electron pairs surrounding the central atom, the electron-pair geometry is tetrahedral. The three bonded atoms yield a trigonal pyramidal molecular geometry. N bonded to Cl, Cl, Cl (wedge-and-dash) *The wedge-and-dash diagram reflects the tetrahedral electron-pair geometry with one lone pair of electrons.*	Take it all the way. Draw the Lewis diagram and, from that, state the electron-pair and molecular geometries. Then draw the wedge-and-dash diagram of the molecule. 
You improved your skill at predicting molecular geometries and sketching the three-dimensional shapes of molecules.	What did you learn by solving this Active Example? 

Practice Exercise 13.12

State the electron-pair and molecular geometries of the BF_4^- ion. Draw its Lewis diagram and its wedge-and-dash diagram.

13.5 The Geometry of Multiple Bonds

Goal 5 For a molecule with more than one central atom and/or multiple bonds, draw the Lewis diagram and predict and sketch the molecular geometry around each central atom, and draw a wedge-and-dash diagram of the molecule.

Experimental evidence shows that the two or three electron pairs in a multiple bond behave as a single electron pair in establishing molecular geometry. This appears if we compare beryllium difluoride, carbon dioxide, and hydrogen cyanide, whose Lewis diagrams are:

$$:\ddot{\text{F}}-\text{Be}-\ddot{\text{F}}: \qquad \ddot{\text{O}}=\text{C}=\ddot{\text{O}} \qquad \text{H}-\text{C}\equiv\text{N}:$$

All three molecules are linear; their bond angles are 180°. The two electron pairs in BeF_2 are as far from each other as possible. According to the VSEPR principle, this is responsible for the 180° bond angle in that compound. Carbon is flanked by two double bonds in CO_2 and one single bond and one triple bond in HCN. Evidently, the second and third electron pairs in double and triple bonds don't affect the molecular geometry.

Further evidence supporting this conclusion comes from comparing the bond angles in boron trifluoride and formaldehyde:

$$:\ddot{\text{F}}-\text{B}(\ddot{\text{F}}:)_2 \qquad \text{H}-\text{C}(=\ddot{\text{O}})-\text{H}$$

Everyday Chemistry 13.1

CHIRALITY

The word *chiral* (pronounced KYE-rull) refers to an object that cannot be superimposed on its mirror image. Your hands are an example of chiral objects. Hold your left hand next to your right hand. Your right hand is a mirror image of your left and vice versa. If you try to stack your hands on one another, you see that they cannot be superimposed, or matched up to one another (**Figure 13.11**). In fact, the word *chiral* is derived from the Greek word *cheir*, which means *hand*. A chiral object is similar to a hand. Feet and ears are other body parts that are also chiral.

Achiral (not chiral) objects include anything with a mirror image that is superimposable. Your pencil, a blank piece of paper, your pants, and perhaps your chair are examples of achiral objects near you now.

Charles D. Winters

Figure 13.12 A seashell as an example of a chiral object. The spiral pattern in this seashell has a handedness. This is a right-handed shell.

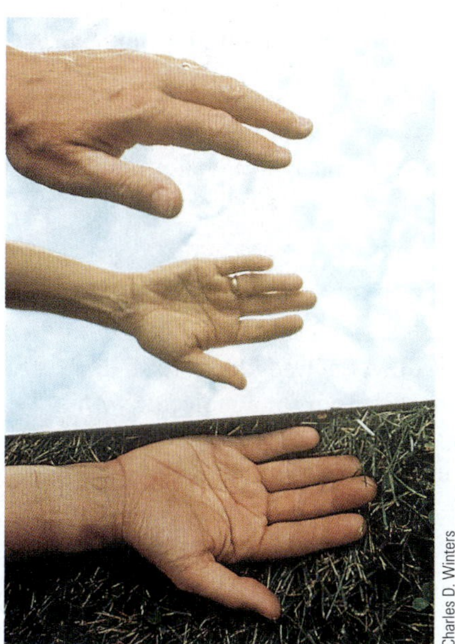

Charles D. Winters

Figure 13.11 Hands as an example of chiral objects. When you look at your left hand in a mirror, you see an image of your right hand. If you superimpose your palm-down left hand over your palm-down right hand, however, you see that they are not the same. Thus, hands are nonsuperimposable mirror images of one another, and therefore they are chiral objects.

Nature has many chiral objects. Seashells are chiral, and furthermore, they are almost always right-handed. **Figure 13.12** shows a right-handed shell. Snail shells are also chiral. Vine plants will exhibit chirality as they grow up a pole. A number of human-made objects are chiral. Many screws, shoes, propellers, and spiral notebook binders are chiral.

The chirality found in the macroscopic world also occurs naturally at the particulate level. Many molecules have chirality. Fascinatingly, as with the case of seashells, nature also has a preferred handedness for molecules. The molecular building blocks of the proteins in your body are known as *amino acids*. When synthesized in a laboratory, the product amino acid molecules are half left-handed and half right-handed. However, the amino acids in your body and in all living systems are exclusively left-handed.

Chiral drugs are an important issue in pharmacy and medicine. Ibuprofen, the active ingredient in the pain relievers Advil, Motrin, and Nuprin, is an equal mixture of left-handed and right-handed molecules. Yet, only the left-handed molecule is medicinally active. The body converts the right-handed molecules into left-handed molecules. In other prescription drugs, one handedness will be active and the other will be inert or even harmful. A significant research effort in chemistry is presently directed toward synthesizing and purifying single-handed drugs and other compounds.

You will study chirality in more detail if you take a course in organic chemistry. Knowledge of a chemical formula alone is insufficient to completely identify many molecules. If you take chirality into consideration, even the Lewis diagram by itself does not tell the whole story! The structure and shape must be specified to the type of handedness for chiral molecules.

Quick Quiz

1. **List some examples of everyday objects that are chiral.**
2. **The two mirror-image forms of chiral molecules have the same physical properties. Explain why this makes separation of a mixture of chiral molecules difficult.**

ilya akinshin/Shutterstock.com

Figure 13.13 Teflon, the nonstick coating applied to cookware, is made from tetrafluoroethylene. The relatively high strength of the carbon–fluorine bonds leads to the macroscopic properties of Teflon.

The shapes are both trigonal planar with 120° bond angles. This is the angle predicted for three electron pairs under the VSEPR theory.

We can conclude that the *number of regions of electron density* that surround a central atom determines the electron-pair geometry around that atom. A region of electron density can be a single, double, or triple bond, or a lone pair. No matter the number of pairs of bonding electrons between two atoms, each region of electron density is distributed as far away from other regions of electron density as possible, as predicted by VSEPR theory.

Active Example 13.13 Molecular Geometry: Multiple Bonds

Determine the molecular geometry of tetrafluoroethylene, C_2F_4 (Figure 13.13). You will need to describe the geometry around each carbon atom. Sketch a wedge-and-dash diagram of the molecule.

Think Before You Write The first step in determining the molecular geometry around an atom in a molecule is to sketch the Lewis diagram. You then determine the electron-pair geometry, and finally, the molecular geometry. The number of regions of electron density that surround a central atom determines the electron-pair geometry around that atom, no matter whether that region of electron density is a single bond, a double bond, a triple bond, or a lone pair.

Answers *Cover the left column with your cut-out shield. Reveal each answer only after you have written your own answer in the right column.*

:F̤:   :F̤:
  \\   /
   C=C
  /   \\
:F̤:   :F̤:
 	
The electron-pair and molecular geometries are trigonal planar around each carbon atom.
 	
For each carbon atom, there are three regions of electron density, making the electron-pair geometry trigonal planar.
	Start with the Lewis diagram and then determine and state the electron-pair and molecular geometries around *each* carbon atom.

F F \\ / C=C / \\ F F *With all atoms in the same plane, no wedges or dashes are needed. The Lewis diagram and the wedge-and-dash diagram are the same, with the exception of the lone pairs on the fluorine atoms.*	Now draw the wedge-and-dash diagram. All atoms in a molecule with a trigonal *planar* electron-pair geometry lie in the same plane. 
You improved your skill at predicting molecular geometries and sketching the three-dimensional shapes of molecules.	What did you learn by solving this Active Example? 

Practice Exercise 13.13

Determine the molecular geometry around each carbon atom in propylene, C_3H_6. Sketch a wedge-and-dash diagram of the molecule.

Active Example 13.14 Molecular Geometry: Molecules With More Than One Central Atom

Describe the molecular geometry around each carbon atom in hexafluoroethane, C_2F_6 (Figure 13.14). Draw a wedge-and-dash diagram that illustrates the geometry.

Think Before You Write When sketching wedge-and-dash diagrams of molecules with more than one central atom, maximize the number of atoms in the plane of the page. This makes the drawing process more straightforward.

Answers *Cover the left column with your cut-out shield. Reveal each answer only after you have written your own answer in the right column.*

2 × 4 (C) + 6 × 7 (F) = 50 valence electrons :F̤: :F̤: :F̤—C—C—F̤: :F̤: :F̤:	Determine the number of valence electrons and draw the Lewis diagram.
Tetrahedral around each carbon atom. *Each carbon atom is surrounded by four regions of electron density, four bonded.*	What is the electron-pair and molecular geometry around each carbon atom?
 *We placed both carbon atoms and a fluorine atom bonded to each carbon atom in the plane of the page to maximize the number of atoms in the plane of the page. That makes it necessary to draw one fluorine atom out of the page and one fluorine atom behind the page on each carbon atom. Variations on this diagram are acceptable as long as they correctly reflect the tetrahedral geometry around each carbon atom, as seen in this illustration:* 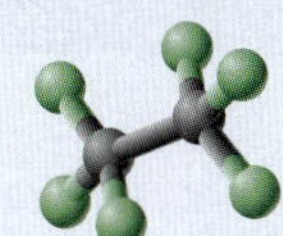	Finish the Active Example by constructing the wedge-and-dash diagram. You have two central atoms, each with a tetrahedral molecular geometry. The diagram is easiest to draw and interpret if you put both carbon atoms in the plane of the page. 
You improved your skill at predicting molecular geometries and sketching the three-dimensional shapes of molecules.	What did you learn by solving this Active Example?

Practice Exercise 13.14

Describe the molecular geometry around each carbon atom in acetonitrile, CH_3CN. Draw a wedge-and-dash diagram that illustrates the geometry.

Dmitry Kalinovsky/Shutterstock.com

Figure 13.14 Hexafluoroethane is a colorless, odorless gas used in the semiconductor manufacturing process.

13.6 Polarity of Molecules

Goal 6 Given or having determined the Lewis diagram of a molecule, predict whether the molecule is polar or nonpolar and indicate the positive and negative poles on its Lewis diagram.

▶ **P/Review** A polar covalent bond is defined in Section 12.5 as a bond in which bonding electrons are shared unequally. The charge density of the bonding electrons is shifted toward the more electronegative atom. For hydrogen fluoride, fluorine is the more electronegative atom:

We previously discussed the polarity of covalent bonds. ◀ Now that you have some idea about how atoms are arranged in molecules, you are ready to learn about the polarity of molecules themselves. A **polar molecule** is one in which there is an asymmetrical* distribution of charge, resulting in + and − poles. A simple example is the HF molecule. The fact that the bonding electrons are closer to the fluorine atom gives the fluorine end of the molecule a partial negative charge, whereas the hydrogen end acts as a positive pole. (See the illustration in the P/Review.)

In general, any diatomic molecule in which the two atoms differ from each other will be at least slightly polar. Examples are HCl and BrCl. In both of these molecules, the more electronegative chlorine atom acts as a negative pole. In an electric field, polar molecules tend to line up with the more electronegative atoms pointing toward the plate with the positive charge and the less electronegative (more electropositive) atoms pointing toward the plate with the negative charge (**Figure 13.15**).

When a molecule has more than two atoms, we must know something about the bond angles in order to decide whether the molecule is polar or nonpolar. Consider, for example, the two triatomic molecules CO_2 and H_2O. Despite the presence of two polar bonds, the linear CO_2 *molecule* is **nonpolar**, as illustrated in **Figure 13.16(a)**. We say that the cancellation of the polar bonds results in no overall (net) regions of positive and negative charge (a dipole), or no net dipole moment (*moment* refers to the combined effect of a quantity and a force). Since the oxygen atoms are symmetric around the carbon atom, the two polar C=O bonds cancel each other.

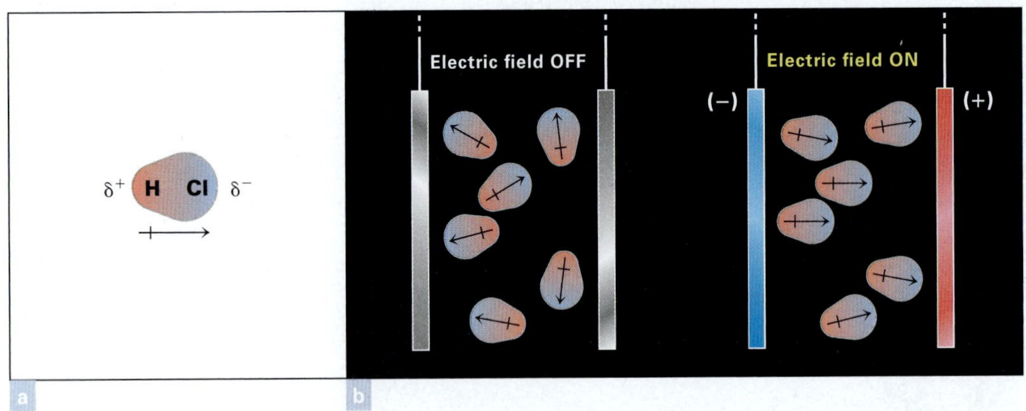

Figure 13.15 Orientation of polar molecules in an electric field. (a) HCl is an example of a polar molecule. The partially positive end is shaded in red, and the partially negative side is blue. The arrow is drawn with a plus sign at the positive end and an arrowhead at the negative end. The arrow points in the direction of greatest charge density for the bonding electrons. (b) Two plates are connected through a switch to a source of an electric field. With the switch open, the orientation of the molecules is random (*left*). When the switch is closed (*right*), the molecules line up with the positive end (*red*) toward the negative plate and the negative pole (*blue*) toward the positive plate.

**Symmetry* refers to balance. As we use the word, a molecule has an *asymmetrical distribution of charge* if charge distribution is unbalanced. There is a point where positive charge appears to be concentrated and a different point where negative charge appears to be concentrated.

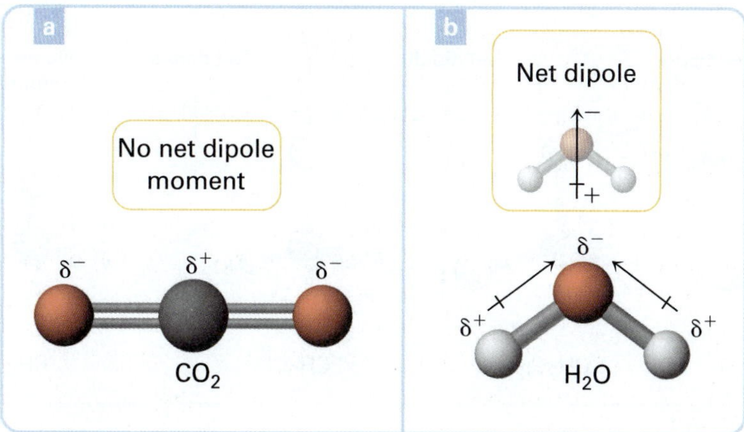

Figure 13.16 Polarity of triatomic molecules. Molecular polarity depends on both bond polarity and molecular geometry.

In contrast, the bent water molecule is polar; the two polar bonds do not cancel each other because the molecule is not symmetrical around a horizontal axis (**Figure 13.16[b]**). The bonding electrons spend more time near the more electronegative oxygen atom, which is the negative pole. The positive pole is midway between the two hydrogen atoms.

Another molecule that is nonpolar despite the presence of polar bonds is CCl_4. The four C—Cl bonds are themselves polar, with the bonding electrons displaced toward the chlorine atoms. But because the four chlorines are symmetrically distributed about the central carbon atom (**Figure 13.17**), the polar bonds cancel one another. If one of the chlorine atoms in CCl_4 is replaced by hydrogen, the symmetry of the molecule is destroyed. As an example, the chloroform molecule, $CHCl_3$, is polar. Figure 13.17 shows other similar tetrahedral molecules and an analysis of their polarities.

From these observations, we can state an easy way to decide whether a simple molecule is polar or nonpolar. If the central atom has no lone pairs and all atoms bonded to it are identical, the molecule is nonpolar. If these conditions are not met, the molecule is polar.

Thinking About Your Thinking

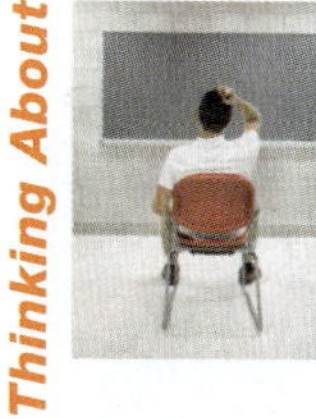

Mental Models

Forming mental models of molecules will help you understand the polarity concept. The challenge in this section is to combine the concept of polar and nonpolar bonds with that of polar and nonpolar molecules. Begin by imagining an asymmetric distribution of bonding electrons as illustrated in Section 12.5 and the P/Review at the beginning of this section. The bonding electrons are displaced toward the atom of the more electronegative element.

Now expand this mental model to a molecule with more than one polar bond. Imagine a central atom bonded to four terminal atoms that are the same, as in Figure 13.17. The electrons in the bonds are shifted toward the atom of the more electronegative element, but the unequal distribution of charge is evenly spread out in opposing directions within the molecule. All of the pushes or pulls cancel one another out, and the molecule itself is nonpolar. Now imagine that you destroy the symmetry of the model in your mind by changing one of the terminal atoms to something different, as illustrated in Figure 13.17. Can you mentally visualize the lack of symmetry in the polar bonds? Think about how the molecule has an area that is electron-rich and an opposite area that is relatively electron-poor. This is a polar molecule.

Practice looking at the polar bonds within molecules and their net effect on overall molecular polarity whenever you draw Lewis diagrams or three-dimensional wedge-and-dash diagrams.

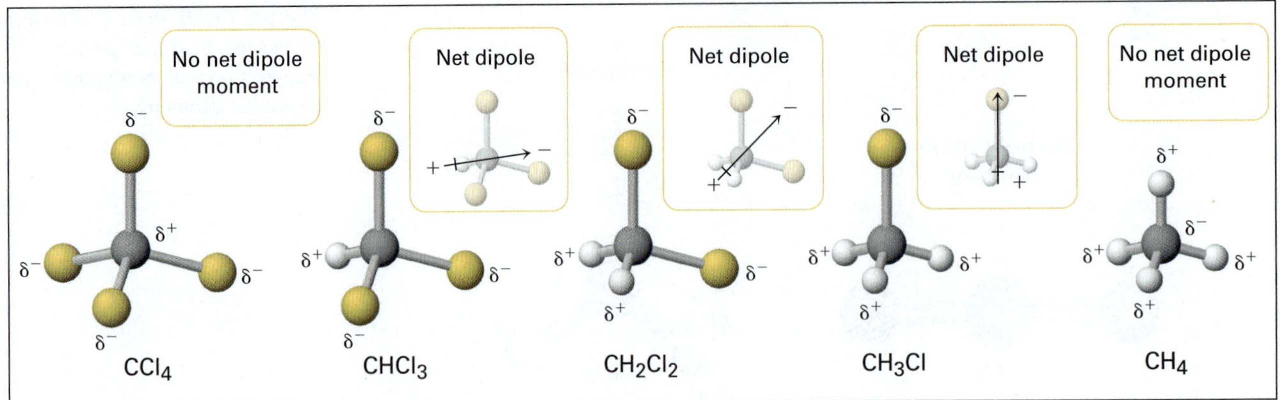

Figure 13.17 Polar and nonpolar molecules. The carbon–chlorine bond is polar, with chlorine as the end toward which bonding electrons are displaced. The carbon–hydrogen bond is also polar, with carbon at the more negative end. CCl_4 and CH_4 are nonpolar because the dipoles from the polar bonds cancel due to the molecular symmetry. $CHCl_3$, CH_2Cl_2, and CH_3Cl are polar because the dipoles from the polar bonds do not cancel.

Active Example 13.15 Polarity of Molecules

Is the BF_3 molecule polar? Is the NH_3 molecule polar?

Think Before You Write Always consider two questions when analyzing molecular polarity: Does the molecule have polar bonds? Are the polar bonds arranged symmetrically? To be able to answer these questions, you'll start by drawing the Lewis diagram.

Answers *Cover the left column with your cut-out shield. Reveal each answer only after you have written your own answer in the right column.*

Consider BF_3 first. Sketch the Lewis diagram. Place arrows on each of the bonding electron pairs pointing to the more electronegative element. Are the bonds polar? Are the bonds symmetrically arranged? Is the molecule polar?

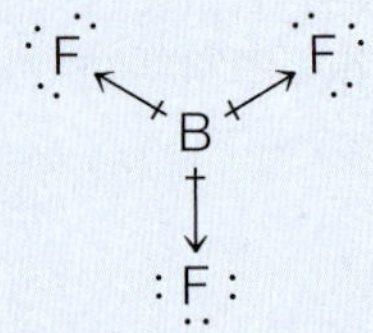

The B–F bonds are polar. The bonds are symmetrically arranged. Therefore, the BF_3 molecule is nonpolar.

Even though fluorine is more electronegative than boron, the three fluorine atoms are arranged symmetrically around the boron atom. The polar bonds cancel.

Furthermore, since the boron atom has no lone pairs and all atoms bonded to it are identical, the molecule must be nonpolar.

Draw the Lewis diagram for NH_3, with the arrows pointing to the more electronegative element. Also sketch the wedge-and-dash diagram of the trigonal pyramidal structure of the molecule with the same arrows included. Is NH_3 polar or nonpolar?

δ−
H⟷N⟷H
H
δ+

N
H H H

The bonding electrons in ammonia are displaced toward the more electronegative nitrogen atom. The bonds do not cancel in the asymmetrical pyramidal shape, so the molecule is polar. The wedge-and-dash diagram, which illustrates the molecular shape, better suggests the charge displacement toward the nitrogen atom.

You improved your skill at identifying polar molecules, and you improved your understanding of the molecular polarity concept.

What did you learn by solving this Active Example?

Practice Exercise 13.15

Is the difluoromethane molecule polar or nonpolar? Explain.

```
       H
       |
 :F̤—C—F̤:
       |
       H
```

13.7 The Structures of Some Organic Compounds (Optional)

Goal 7 Distinguish between organic compounds and inorganic compounds.
8 Distinguish between hydrocarbons and other organic compounds.
9 On the basis of structure and the geometry of the identifying group, distinguish among alcohols, ethers, and carboxylic acids.

The Bonding Capabilities of the Carbon Atom

Organic chemistry is the chemistry of carbon compounds. The vast majority of compounds that have been synthesized, characterized, and catalogued by chemists are carbon compounds. Carbonates, cyanides (compounds containing the cyano group, CN), oxides of carbon, and a few other carbon-containing compounds are exceptions that are often classified as inorganic.

The property of carbon that qualifies it to define a whole branch of chemistry is the bonding capability of the carbon atom. Carbon atoms have four valence electrons that enable them to form up to four covalent bonds with other atoms. The carbon atom is also the right size for these bonds to be of just the right strength—strong enough to form stable molecules but weak enough to undergo reactions. Most important is the fact that these carbon atoms may be bonded to other carbon atoms. As a result, many organic compounds have extremely long chains and/or rings of carbon atoms. No other element has the bonding characteristics of carbon.

The structure of molecules is a major concern of organic research chemists. They build on the previously discovered physical and chemical characteristics of a variety of structures. From this starting point, they "create," on paper or on a computer, Lewis diagrams and molecular models of molecules that should have certain desirable properties. Then, beginning with the structures of chemicals that are known and available, they figure out chemical reactions that change the existing structures to those that are wanted. ▶ This kind of research has led to many things we use daily, including synthetic fabrics, plastics, and medical products.

◀ **P/Review** The structure of different kinds of organic compounds is explored in greater detail in Chapter 21.

Hydrocarbons

Hydrocarbons are compounds made up of only hydrogen and carbon. One type of hydrocarbon is an **alkane**, in which each carbon atom forms four single bonds. Alkanes differ from one another by the number of carbon atoms that bond together to form a chain. Methane, CH_4, is the simplest alkane, with only one carbon atom. Ethane, C_2H_6, has two carbon atoms bonded to one another; propane, C_3H_8, has three carbon atoms in a chain; and butane, C_4H_{10}, has four carbon atoms bonded together. Compare the Lewis diagrams with the three-dimensional representations of the three compounds in **Figure 13.18**.

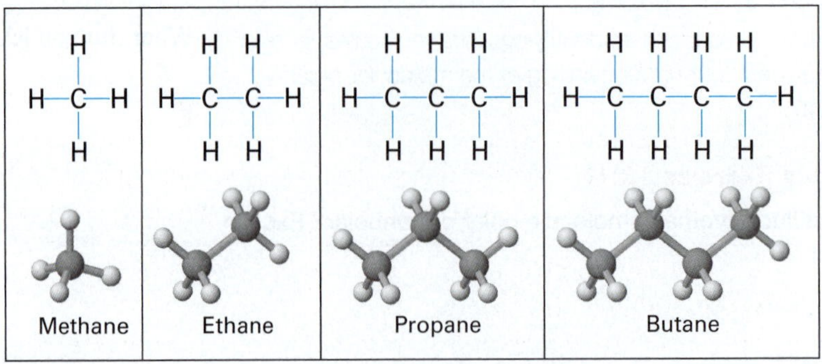

Figure 13.18 Alkanes. An alkane is a hydrocarbon in which each carbon atom forms four bonds.

The formulas for ethane and propane can be written in two different ways. Ethane is C_2H_6 or CH_3CH_3. Propane is C_3H_8 or $CH_3CH_2CH_3$. C_2H_6 and C_3H_8 are molecular formulas. CH_3CH_3 and $CH_3CH_2CH_3$ are called **condensed structural formulas**. They suggest the structure of the molecule. Ethane, CH_3CH_3, is made up of two $-CH_3$ groups bonded together. Divide the Lewis diagram between the carbon atoms and that's exactly what you have. The condensed structural formula for propane, $CH_3CH_2CH_3$, suggests correctly that the molecule is made up of two $-CH_3$ groups with a $-CH_2-$ group between them. Any number of $-CH_2-$ groups can be placed between two $-CH_3$ groups.

Butane is the simplest alkane that exists as two distinct isomers—that is, as molecules with the same molecular formula but different arrangements of atoms (**Figure 13.19[a]**). The two butanes are distinctly different chemical compounds, each having its own set of physical and chemical properties. Notice that the name *butane* is modified to distinguish between the isomers. Pentane, C_5H_{12}, exists as three isomers, as shown in **Figure 13.19(b)**. Again, each molecule has its own distinct set of properties. Even though isomers are made from the same set of atoms, their differing arrangement into molecules makes them different substances.

As the number of carbon atoms in an alkane increases, the number of isomers increases dramatically. There are 5 isomers possible for 6-carbon alkanes, 9 isomers possible for 7-carbon alkanes, and 75 isomers possible for 10-carbon alkanes. It is possible to draw more than 300,000 isomeric structures for $C_{20}H_{42}$ and more than 100 million for $C_{30}H_{62}$. As you may guess, not all of them have been prepared and identified! This does give us some idea, though, why there are many more known organic compounds than inorganic compounds.

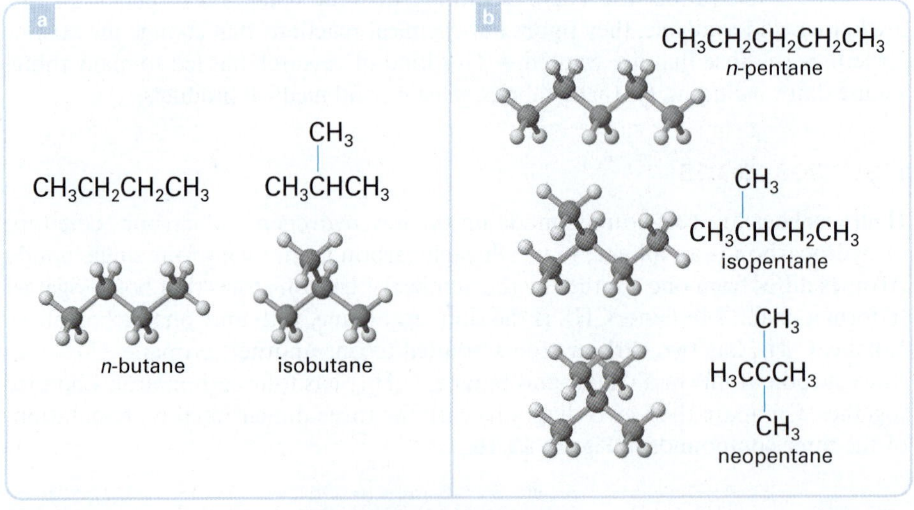

Figure 13.19 The isomers of butane and pentane. (a) Butane has two isomers, *n*-butane and isobutane. (b) Pentane has three isomers, *n*-pentane, isopentane, and neopentane.

Thinking About Your Thinking

Mental Models

As the number of atoms in a molecule increases, you need to practice to become proficient at forming mental models after drawing a two-dimensional Lewis diagram. You will spend a good deal of time developing this skill in a future course. At this point in your study of chemistry, we will help you visualize organic compounds by showing particulate-level illustrations of most molecules along with their Lewis diagrams.

Compare the Lewis diagrams in Figures 13.18 and 13.19 with the three-dimensional representations of the molecules. Notice that the molecules do not look like their two-dimensional Lewis diagrams. All bond angles are tetrahedral. Thus, what can be drawn as a straight line of carbon atoms in a Lewis diagram is actually a zigzag chain of carbon atoms in the molecule.

Alcohols and Ethers

In Active Example 13.8, you drew the Lewis diagram for ethanol, and you were shown the diagram for its isomer, dimethyl ether. These diagrams, plus the Lewis diagram for water, along with their three-dimensional drawings, are shown in **Figure 13.20**. All three molecules include an oxygen atom, with two lone pairs of electrons, that is single-bonded to two other atoms. This, as you learned in Section 13.4, yields a bent structure around the oxygen atom. In Section 13.6, you saw that a molecule with this structure is polar. This polarity is present in all three molecules, but to a diminishing extent from water to ethanol to dimethyl ether.

The —OH part of an alcohol is called a **hydroxyl group**. Essentially, an alcohol is an alkane in which a hydrogen atom is replaced by a hydroxyl group (**Figure 13.21**).

H_2O *or* HOH Water	C_2H_6O *or* C_2H_5OH Ethyl alcohol or ethanol	C_2H_6O *or* CH_3OCH_3 Dimethyl ether

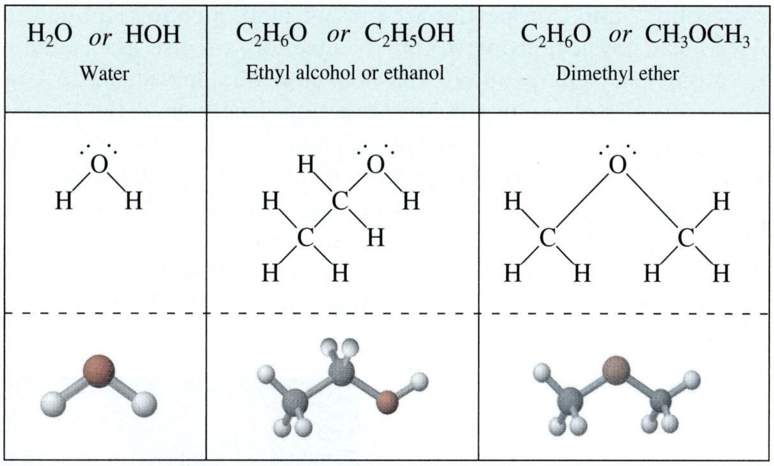

Figure 13.20 Water, ethanol, and dimethyl ether. All molecules have a bent structure around an oxygen atom with two unshared electron pairs and single bonds to two other atoms.

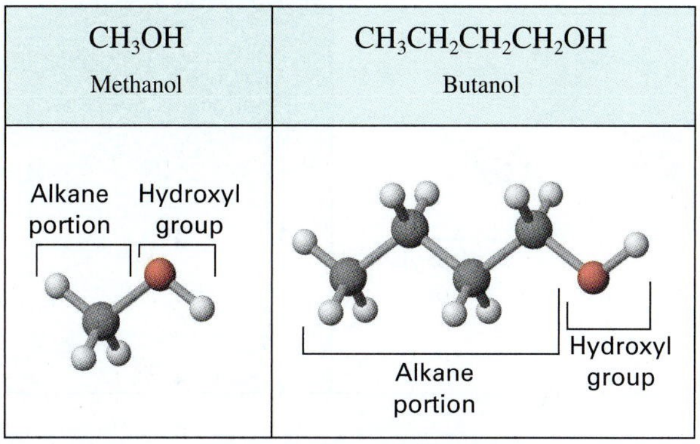

CH_3OH Methanol	$CH_3CH_2CH_2CH_2OH$ Butanol
Alkane portion; Hydroxyl group	Alkane portion; Hydroxyl group

Figure 13.21 Alcohols. An alcohol is an alkane in which a hydroxyl group replaces a hydrogen atom.

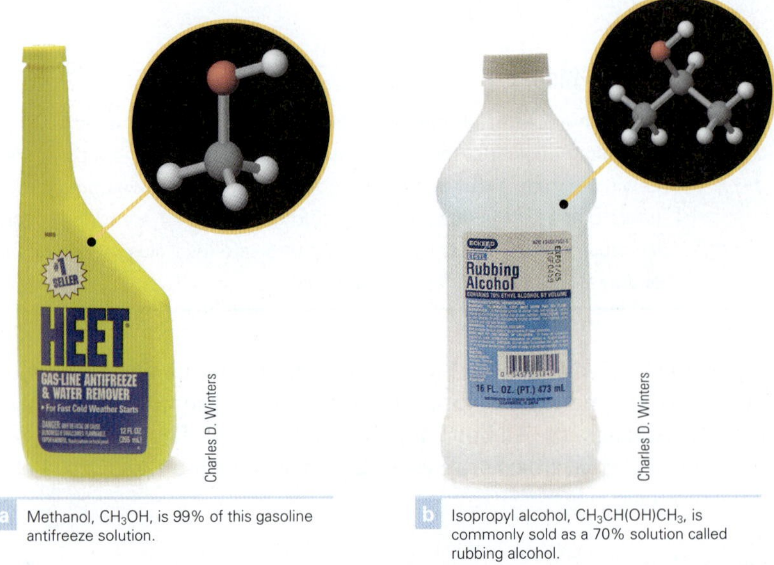

Figure 13.22 Commercial products that include methanol and isopropyl alcohol as the major ingredient.

a Methanol, CH_3OH, is 99% of this gasoline antifreeze solution.

b Isopropyl alcohol, $CH_3CH(OH)CH_3$, is commonly sold as a 70% solution called rubbing alcohol.

An alcohol can also be thought of as a derivative of water in which one of the H_2O hydrogens is replaced with a hydrocarbon group. The —OH group may be anyplace in the molecule, and there may be more than one. Another example of an alcohol is methanol, also called *methyl alcohol* or *wood alcohol*, which is used as a solvent, as a starting substance from which more complex molecules are produced, and as a gasoline antifreeze (Figure 13.22[a]). Similarly, 2-propanol, often called *isopropyl alcohol* or *rubbing alcohol*, is also used as a solvent and a reactant in the manufacture of other substances, in addition to its most well-known use as a disinfectant and to cool and soothe skin (Figure 13.22[b]). Ethylene glycol is an example of a diol, an alcohol with two hydroxyl groups. It is the major component in most automobile antifreeze solutions (Figure 13.23).

The chemical properties of an alcohol are the chemical properties of the hydroxyl group. Similar properties are present in all alcohols, though to different degrees. The physical properties of the alcohols are also associated with the hydroxyl group, particularly among the smaller alcohol molecules. In large molecules, the properties of the hydrocarbon section usually have the greatest influence on the physical properties of the overall molecule. For example, alcohols with shorter hydrocarbon portions, such as CH_3OH and CH_3CH_2OH, dissolve in water in all proportions, whereas alcohols with longer hydrocarbon portions, such as $CH_3CH_2CH_2CH_2CH_2CH_2CH_2CH_2OH$, are immiscible in water.

An **ether** is a compound that has two hydrocarbon groups bonded to an oxygen atom. Ethers can be thought of as water molecules in which both hydrogen atoms

Figure 13.23 Ethylene glycol, $HOCH_2CH_2OH$, is the major component in most automobile antifreeze solutions. The molecule has two hydroxyl groups, making it a dialcohol, or a diol.

$C_4H_{10}O$ *or* $(C_2H_5)_2O$ *or* $C_2H_5—O—C_2H_5$ Diethyl ether	C_7H_8O *or* $CH_3OC_6H_5$ *or* $CH_3—O—C_6H_5$ Methyl phenyl ether

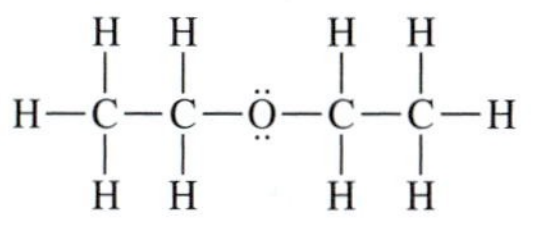

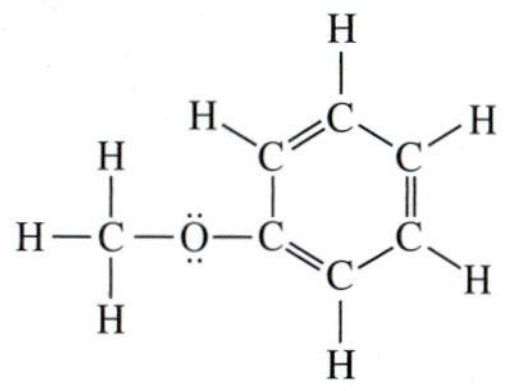

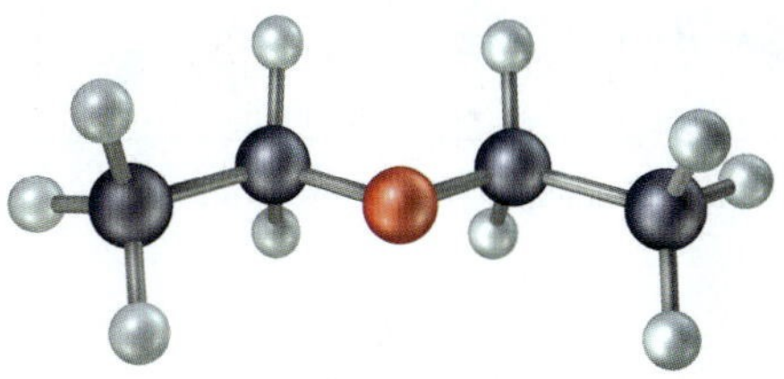

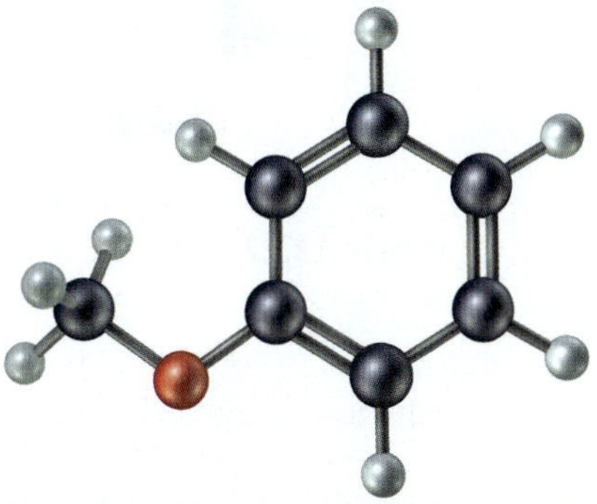

Figure 13.24 Ethers. An ether is a compound that has two hydrocarbon groups bonded to an oxygen atom. In everyday language, the term *ether* refers to diethyl ether, which was once commonly used as an anesthetic. Scientists must use more precise language, because *ether* actually refers to an entire class of molecules.

are replaced by hydrocarbon groups. Either or both of the $-CH_2CH_3$ groups in diethyl ether (**Figure 13.24**) may be replaced by longer carbon chains. An additional example of an ether is also shown in Figure 13.24.

Carboxylic Acids

The formulas of formic acid, H–COOH (**Figure 13.25**), acetic acid, CH_3–COOH, and propanoic acid, CH_3CH_2–COOH, suggest the structure of the first three members of a series of **carboxylic acids**, which contain the **carboxyl group**, –COOH.

Charles D. Winters

Figure 13.25 Many insects, including the red ant *Formica rufa*, from which the acid was named, use formic acid as a chemical weapon, a communication system, and for protection against parasites.

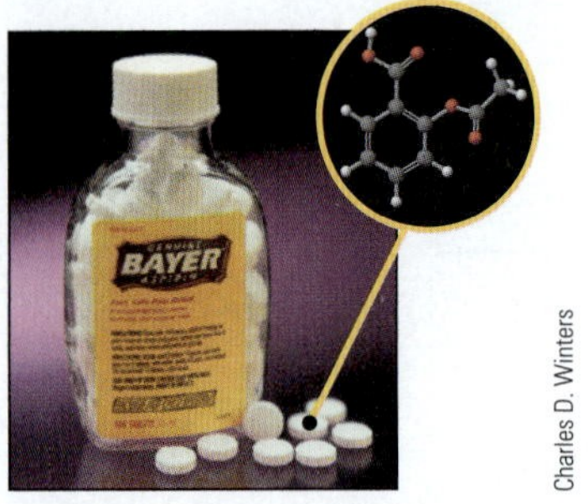

Figure 13.26 Aspirin is acetylsalicylic acid, a carboxylic acid.

$HCOOH$ *or* $HCHO_2$	CH_3COOH *or* $HC_2H_3O_2$	C_2H_5COOH *or* $HC_3H_5O_2$
Formic acid	Acetic acid	Propanoic acid

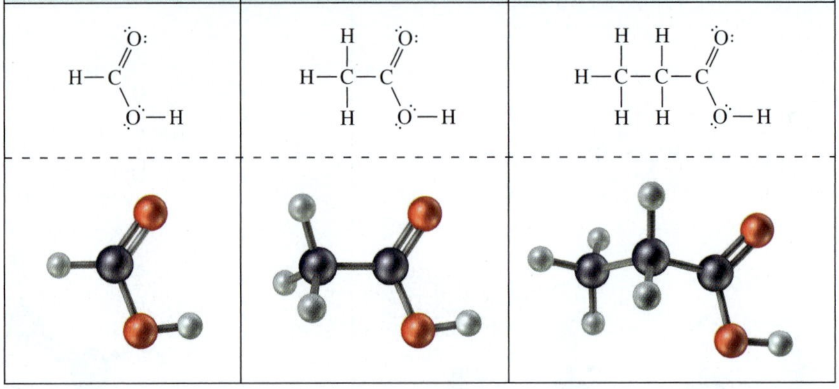

Figure 13.27 Carboxylic acids. A carboxylic acid is characterized by the presence of the carboxyl group, —COOH.

One of the most widely used medications in the world is aspirin, which has the trivial name acetylsalicylic acid because it is a carboxyl*ic acid* (**Figure 13.26**). Notice that the names of carboxylic acid always end in *–ic acid.* The Lewis diagrams and ball-and-stick models of the one-, two-, and three-carbon carboxylic acids are shown in **Figure 13.27**. As with hydrocarbons, alcohols, and ethers, the carbon chain may extend indefinitely.

Acetic acid is the best known of the carboxylic acids. Given the structure of acetic acid (Figure 13.27 [*center*]), you can see the difference between the ionizable hydrogen, which is bonded to an oxygen atom, and the other hydrogens, which are bonded to the carbon atom. The organic chemist's way of writing the ionization equation, along with Lewis diagrams, is:

$$CH_3COOH \longrightarrow H^+ + CH_3COO^-$$

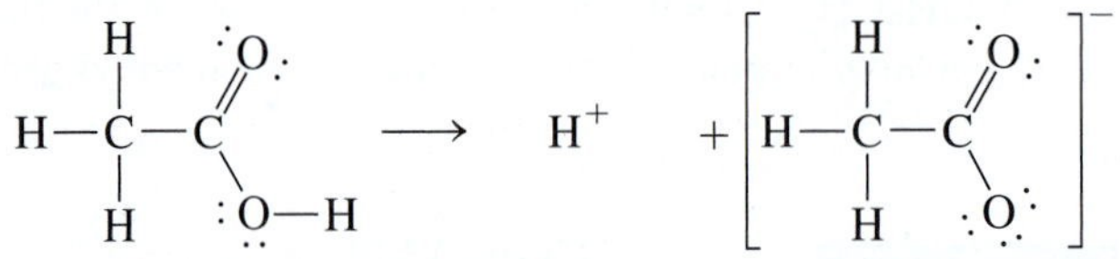

acetic acid ⟶ hydrogen ion + acetate ion

Carboxylic acids are weak acids, which means that they ionize only slightly in water. To the extent they ionize, however, it is always the carboxylic hydrogen that reacts when they react with bases in neutralization reactions. ◀

▶ **P/Review** Writing the equation for the reaction between an acid and a hydroxide base is discussed in Section 8.10. It is the most common form of a neutralization reaction.

The geometry of the carboxyl group is shown in Figure 13.27 and the preceding Lewis diagrams. The molecule is bent around the oxygen atom that is single-bonded to both the carbon atom and the ionizable hydrogen atom, forming a tetrahedral angle. In determining the geometry around the carbon atom, the double bond counts as one region of electron density. The geometry is trigonal planar with 120° bond angles. Bond angles around carbon atoms in the alkane part of the molecule are tetrahedral.

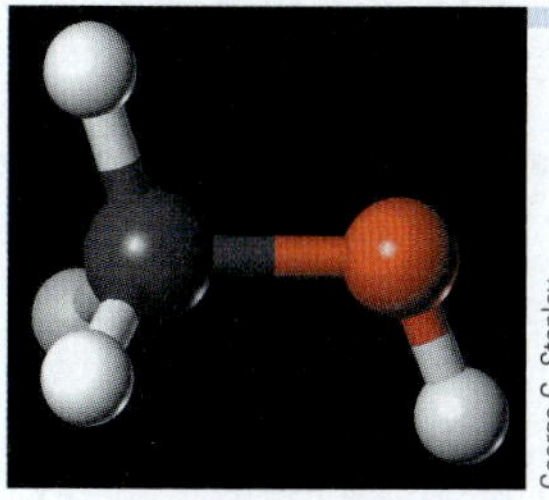
George G. Stanley

CHAPTER 13 IN REVIEW: STRUCTURE AND SHAPE

Goal 1 Draw the Lewis diagram for any molecule or polyatomic ion made up of main group elements.

The procedure for **drawing a Lewis diagram** is:

1. Count the total number of valence electrons. Adjust for charge on ions.
2. Place the least electronegative atom(s) in the center of the molecule.
3. Draw a tentative diagram. Join atoms by single bonds. Add unshared pairs to complete the octet around all atoms except hydrogen.
4. Calculate the number of valence electrons in your tentative diagram and compare it with the actual number of valence electrons. If the tentative diagram has too many electrons, remove a lone pair from the central atom and from a terminal atom, and replace them with an additional bonding pair between those atoms. If the tentative diagram still has too many electrons, repeat the process.
5. Check the Lewis diagram. Hydrogen atoms must have only one bond, and all other atoms should have a total of four electron pairs.

Goal 2 Describe the electron-pair geometry when a central atom is surrounded by two, three, or four regions of electron density.

Electron-pair geometry describes the arrangement of two, three, or four regions of electron density around a central atom.

Electron Pairs	Geometry	Electron-Pair Angles
2	Linear	180°
3	Trigonal planar	120°
4	Tetrahedral	109.5°

Goal 3 Given or having derived the Lewis diagram of a molecule or polyatomic ion in which a second-period central atom is surrounded by two, three, or four regions of electron density, predict and sketch the molecular geometry around that atom.

Molecular geometry describes the arrangement of two, three, or four atoms around a central atom to which they are all bonded.

Electron Pairs	Bonded Atoms	Molecular Geometry	Bond Angle
2	2	Linear	180°
3	3	Trigonal planar	120°
3	2	Angular	120°
4	4	Tetrahedral	109.5°
4	3	Trigonal pyramidal	109.5°
4	2	Bent	109.5°

Goal 4 Draw a wedge-and-dash diagram of any molecule for which a Lewis diagram can be drawn.

The procedure for **drawing a wedge-and-dash diagram** is:

1. When two atoms are in the same plane as the page, they are connected with a solid line of uniform width.
2. When an atom is behind the plane of the page, it is connected to the central atom by a line that is dashed. The width of the dashed line increases as it moves away from the central atom.
3. When an atom is in front of the plane of the page, it is connected to the central atom by a line that is wedge-shaped. The width of the wedge-shaped line increases as it moves away from the central atom.

Goal 5 For a molecule with more than one central atom and/or multiple bonds, draw the Lewis diagram and predict and sketch the molecular geometry around each central atom, and draw a wedge-and-dash diagram of the molecule.

It is the **number of regions of electron density** that surround a central atom that determines the electron-pair geometry around that atom. A region of electron density can be a single, double, or triple bond, or a lone pair. No matter the number of pairs of bonding electrons between two atoms, each region of electron density is distributed as far away from other regions of electron density as possible, as predicted by VSEPR theory.

Goal 6 Given or having determined the Lewis diagram of a molecule, predict whether the molecule is polar or nonpolar and indicate the positive and negative poles on its Lewis diagram.

Molecular polarity depends on both bond polarity and molecular geometry. A **polar molecule** is one in which there is an asymmetrical distribution of charge, resulting in positive and negative poles. A **nonpolar molecule** either has nonpolar bonds or has polar bonds that cancel, resulting in no overall regions of positive and negative charge. If the central atom of a molecule has no lone pairs and all atoms bonded to it are identical, the molecule is nonpolar. If these conditions are not met, the molecule is polar.

Goal 7 Distinguish between organic compounds and inorganic compounds.	The majority of all chemical compounds that have been characterized are classified as **organic compounds**—compounds based upon the carbon atom. Inorganic compounds are those without carbon atoms, plus the following carbon-containing compounds: carbonates, cyanides, and oxides of carbon.
Goal 8 Distinguish between hydrocarbons and other organic compounds.	**Hydrocarbons** are made of only carbon and hydrogen. The **alkanes** are a hydrocarbon family with all single bonds.
Goal 9 On the basis of structure and the geometry of the identifying group, distinguish among alcohols, ethers, and carboxylic acids.	If a hydrogen atom in a hydrocarbon, CH_4, for example, is replaced by a hydroxyl group, $-OH$, the resulting molecule is an **alcohol**, CH_3-OH. An alcohol may be thought of as a water molecule ($H-OH$) in which one hydrogen atom is replaced by a hydrocarbon group (CH_3-OH). An **ether** may be thought of as a water molecule ($H-O-H$) in which both hydrogen atoms are replaced by hydrocarbon groups (CH_3-O-CH_3). Alcohols and ethers with the same number of carbon atoms are isomers. If a hydrogen atom in a hydrocarbon is replaced by a carboxyl group, $-COOH$, the resulting molecule is a **carboxylic acid**. The most common carboxylic acid is acetic acid, written as $HC_2H_3O_2$ or CH_3COOH.

Key Terms

Most of the key terms and concepts and many others appear in the Glossary. Use your Glossary regularly.

angular p. 483
bent p. 483
bond angle p. 481
central atom p. 470
electron-pair angle p. 480
electron-pair geometry p. 479
isomers p. 479
linear p. 481
molecular geometry p. 479
nonpolar p. 492
polar molecule p. 492
resonance hybrid p. 475
resonance structures p. 475
terminal atom p. 470
tetrahedral p. 480
tetrahedron p. 480
trigonal planar p. 483
trigonal pyramidal p. 483
valence shell electron-pair repulsion theory or VSEPR theory p. 479
wedge-and-dash diagrams p. 484

Terms from Optional Section 13.7

alkane p. 495
carboxyl group p. 499
carboxylic acids p. 499
condensed structural formulas p. 496
ether p. 498
hydrocarbons p. 495
hydroxyl group p. 497
organic chemistry p. 495

Frequently Asked Questions

What are the key guidelines to follow to ensure that I'm writing Lewis diagrams correctly?
There are many molecules for which you can draw two or more Lewis diagrams that satisfy the octet rule. You can also draw incorrect diagrams that satisfy the octet rule. Your diagram is most likely to be correct if you remember that: (1) hydrogen always forms one bond and carbon is always surrounded by four electron pairs (usually four single bonds or a double bond and two single bonds); (2) two or more oxygen atoms are distributed around a central atom; (3) an oxygen atom is between a hydrogen atom and another nonmetal atom; and (4) your diagram should be as symmetric as possible.

What are most common errors in Lewis diagrams?
The most common errors in Lewis diagrams are bonding oxygen atoms to one another and surrounding a central atom with three or five electron pairs. There are some compounds in which oxygen atoms are bonded to one another, but not many. One of these appears in the Questions, Exercises, and Problems section that follows. The errors involving three or five electron pairs most often occur when double bonds are present. Always check your final diagram to be sure all atoms conform to the octet rule.

What exceptions to the octet rule are we responsible for knowing?
Two exceptions to the octet rule in which central atoms are surrounded by fewer than four pairs of electrons are compounds of beryllium and boron, with two and three electron pairs, respectively, found in stable compounds.

What can I do to improve my skill at drawing wedge-and-dash diagrams?
Wedge-and-dash diagrams are based on three basic structures: (1) a linear geometry with a 180° angle and all atoms in the same plane; (2) a trigonal planar geometry with 120° angles and all atoms in the same plane; and (3) a tetrahedral geometry with 109.5° angles between the atoms. The tetrahedral diagram is most easily drawn by starting with three coplanar atoms—the central atom and two of the terminal atoms at a 109.5° angle—and then drawing one atom coming forward out of the plane of the page with a wedge, and finally drawing one atom going behind the plane of the page with a dash. More complex molecules are drawn as combinations of the three basic structures. The most common error in wedge-and-dash diagrams is failing to account for lone pairs. Lone pairs are not shown in a wedge-and-dash diagram, yet they are there in the actual molecule, and they affect the molecular geometry.

Why don't we just use planar and pyramidal instead of trigonal planar and trigonal pyramidal?
Geometry places limits on the shapes of some molecules. If there are only two atoms, the geometry is linear; two points determine a line. If there are three atoms, they are either in a straight line (linear) or they are not (angular or bent). Four atoms take you into the *possibility* of a three-dimensional molecule. That is why you must distinguish between trigonal planar and trigonal pyramidal. The adjective *trigonal* is necessary because some elements in the third and later periods form square planar and square pyramidal structures.

How do you distinguish between polar and nonpolar molecules?
To distinguish between polar and nonpolar molecules, test the molecule for two conditions that are required for nonpolarity. First, all atoms bonded to the central atom must be the same element. Second, there can be no lone pairs. If the molecule passes both tests, it is nonpolar; if it fails either test, it is polar.

Lewis Diagram Recognition Exercises

Classify each of the following Lewis diagrams as either acceptable or unacceptable. For those that are unacceptable, explain why and draw an acceptable diagram.

1. BCl_3

```
 ..       ..
:Cl—B—Cl:
 ..  |    ..
    :Cl:
     ..
```

2. NO_3^-

```
 ┌ ..          ┐ −
 │:N—O=O       │
 │ ..  |       │
 │    :O:      │
 └     ..      ┘
```

3. $HClO_3$

```
        H
 ..     |    ..
:O—Cl—O:
 ..     |    ..
       :O:
        ..
```

4. C_3H_6

```
     H         H
     |   ..    |
H—C—C—C—H
     |         |
     H         H
```

5. HCN

```
        ..
H—C=N:
```

6. BeF_2

```
 ..        ..
:F—Be—F:
 ..        ..
```

7. C_2H_6O

```
     H
     |   ..
H—C—O:
     |   |
     H   |
         |
     H—C—H
         |
         H
```

8. CH_3Br

```
     H
     |   ..
H—C=Br
     |   ..
     H
```

9. CO

```
C≡O
```

10. H_2O_2

```
 ..        ..
:O=H=O—H
 .
```

Small-Group Discussion Questions

Small-Group Discussion Questions are for group work, either in class or under the guidance of a leader during a discussion section.

1. Molecules with multiple carbon atoms sometimes exist in a configuration where the carbon atoms are in a closed loop, bonded to one another. Cyclopropane, C_3H_6, cyclobutane, C_4H_8, cyclopentane, C_5H_{10}, and cyclohexane, C_6H_{12}, are examples of this type of molecule, called *cycloalkanes*. Draw the Lewis diagram of each.
2. Five "-*ic* acids" form the basis for the nomenclature system presented in Chapter 6. Draw the Lewis diagram of each acid. Why are these molecules called *oxoacids*? What bonding arrangement is responsible for the acidic hydrogen? How are nonacidic hydrogens different?
3. How many distinctly different Lewis diagrams can you draw for each of the following formulas: C_4H_{10}, C_5H_{12}, C_6H_{14}?
4. How many resonance structures can be drawn for each of the following: ozone, O_3, carbonate ion, benzene, C_6H_6, and cyanate ion, CNO^-?
5. We suggested that you might think of molecular geometry as an "atom geometry." Why, then, do unshared electron pairs on a central atom influence molecular geometry, given that they are not bonded to an atom?
6. In the end-of-chapter Questions, Exercises, and Problems, you are asked to sketch the wedge-and-dash diagram for Questions 19–22. Sketch the wedge-and-dash diagrams for the remaining questions in that section, Questions 23–30.
7. Rank the following in order of increasing polarity: methane (CH_4), carbon disulfide, boron trifluoride, nitrogen trichloride. Explain your reasoning.
8. Draw the Lewis diagram of a four-carbon molecule that is an example of each of the following: hydrocarbon, alcohol, ether, carboxylic acid. Then draw a second example of a four-carbon molecule from each class that is different from your first example.
9. What is the role of the quantum model of the atom in the determination of molecular geometry? Explain in as much detail as possible, discussing the links and relationships among quantum mechanics, bonding, structure, and shape.

Questions, Exercises, and Problems

Interactive versions of these problems may be assigned by your instructor. Solutions for blue-numbered questions are at the end of the chapter. Questions other than those in the General Questions and More Challenging Problems sections are paired in consecutive odd-even number combinations; solutions for the odd-numbered questions are also at the end of the chapter.

Section 13.2: Drawing Lewis Diagrams

Draw the Lewis diagrams for each of the following sets of molecules.

1. HI, H_2O, NCl_3
2. HF, OF_2, NF_3
3. CO_2, SF_2, BrO_3^-
4. NH_4^+, PH_3, NHF_2
5. BrO^-, $H_2PO_4^-$, ClO_4^-
6. C_3H_8, C_3H_6 (do not consider ring structures)
7. $CHFCl_2$, CHF_3, $CClI_3$
8. C_2H_6O (bond O to atoms of two different elements), C_2H_6O (bond O to atoms of the same element)

There are two or more acceptable diagrams for most species in Questions 9 through 18.

9. $C_2H_2Cl_4$, $C_2H_2Cl_2F_2$, $C_3H_4Br_3I$
10. $C_2H_4Br_2$, C_2H_4BrF, $C_3H_5FBr_2$
11. C_4H_8, C_2H_6O, $C_3H_8O_2$
12. C_3H_8, C_3H_6, C_3H_6O
13. C_6H_{14}, C_4H_6O, $C_2H_2F_2$
14. C_5H_{12}, C_4H_8O, C_4H_6
15. Butanoic acid, C_3H_7COOH
16. Acetic acid, CH_3COOH
17. NO_2^+, N_2O, NO^+
18. Hydroxide ion, OH^-; water, H_2O; methanol, CH_3OH

Section 13.3: Electron-Pair Repulsion: Electron-Pair Geometry

Section 13.4: Molecular Geometry

Section 13.5: The Geometry of Multiple Bonds

Questions 19 through 30: For each molecule or ion, or for the atom specified in a molecule or ion, write the Lewis diagram, then describe (a) the electron-pair geometry and (b) the molecular geometry predicted by the valence shell electron-pair repulsion theory. Also sketch the wedge-and-dash diagram of each molecule or ion in Questions 19–22.

19. BCl_3, PH_3, H_2S
20. BF_4^-, CCl_4
21. BrO^-, ClO_3^-, PO_4^{3-}
22. PCl_3, $SeOF_2$
23. Each carbon atom in C_3H_7OH
24. Each carbon atom in CH_3CH_2COOH
25. Nitrogen atom in $C_2H_5NH_2$
26. Each carbon atom and the oxygen atom in $CH_3CH_2OCH_2CH_3$
27. Each carbon atom in C_2H_2
28. Each carbon atom and the nitrogen atom in CH_3CONH_2
29. Carbon atom in HCN
30. Carbon atom in $OCCl_2$
31. The Lewis diagram of a certain compound has the element E as its central atom. The bonding and lone-pair electrons around E are shown. What is the molecular geometry around E?

```
  |
—E—
  ..
```

32. Draw a central atom with whatever combination of bonding and lone-pair electrons is necessary to yield an angular or bent structure around that atom.

Questions 33 through 38: For each space-filling or ball-and-stick model shown, identify the electron-pair and molecular geometry. There are no "hidden" atoms in any of the models.

33. 

34. 

35. 

36. 

37. 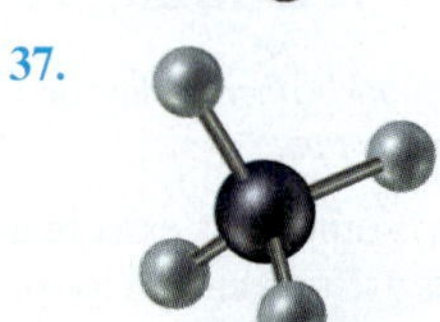

38. 

Questions 39 through 42: Estimate the value of the bond angles identified by letters (A, B, C, D) in each Lewis diagram. Explain your reasoning in each case.

39. Aspirin, acetylsalicylic acid:

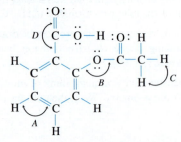

```
            :O:
             ‖
         D ( C—Ö—H  :O:  H
             |  ..   ‖   |
      H      C      Ö—C—C—H
        \  //  \   ..(B)  |
         C       C        H ) C
         |       ‖
         C       C
       //  \   /   \
     H  (A)  C       H
             |
             H
```

40. Lactic acid:

```
A ⌒ H   H  :O:
    |   |   ||    ..
H — C — C — C — O — H
    |   |       ..
    H  :O:   C
        |  B
        H
```

41. Cinnamaldehyde:

```
              H          B        :O: C
          A   |                    ||
              |       H            C—H
   H     C     C  C=C
    C        C        H
    |        ||
    C        C
 H      C        H
        |
        H
```

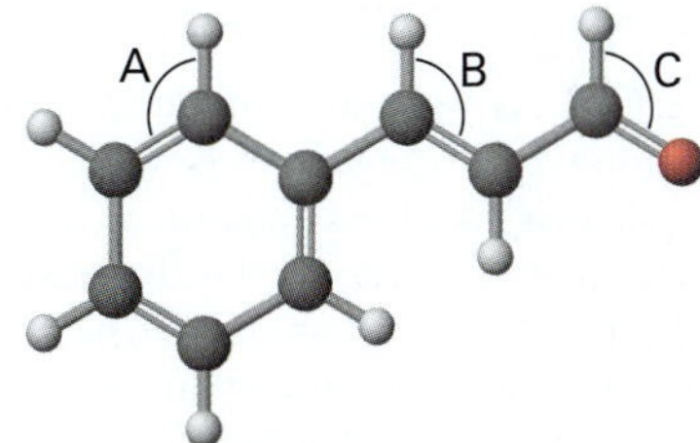

42. Amphetamine:

```
        H        H  H
        |        | B|
   H  A C        C——C——CH3
    C       C    |  |
    |       ||   H :N—H
    C       C        C
 H      C      H    H
        |
        H
```

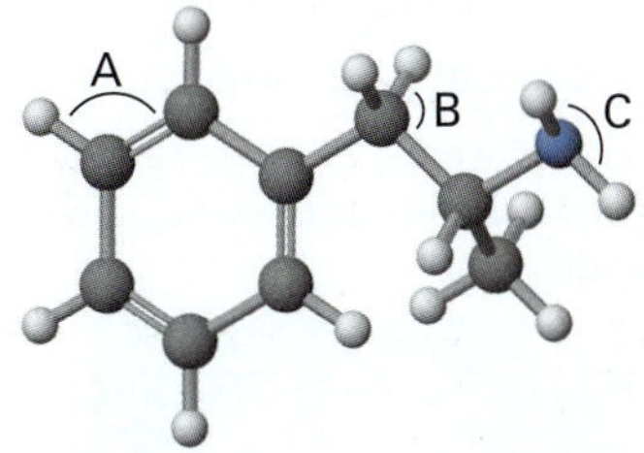

Section 13.6: Polarity of Molecules

Consider the following general Lewis diagrams for Questions 43 and 44.

a) A—B̈—A (with a lone pair above and below B) b) A—B—A

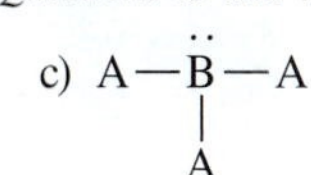

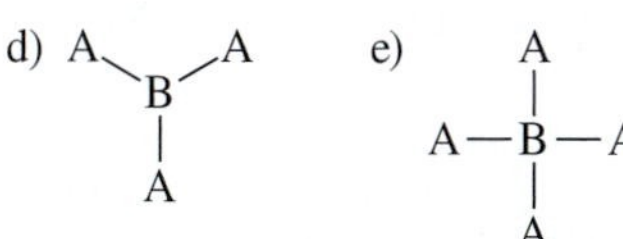

43. Which of the molecules are nonpolar with polar bonds?

44. Which of the molecules are polar?

45. Is the carbon tetrachloride molecule, CCl_4, which contains four polar bonds (C–Cl electronegativity difference 0.6) polar or nonpolar? Explain.

46. For each formula given, draw the Lewis diagram, determine the molecular geometry, and determine whether the molecule is polar or nonpolar.

a) BF_3 b) NH_3 c) GeH_4

47. Describe the shapes and compare the polarities of HF and HBr molecules. In each case, identify the end of the molecule that is more positive.

48. For each formula given, draw the Lewis diagram, determine the molecular geometry, and determine whether the molecule is polar or nonpolar.

a) BeF_2 b) H_2O c) BF_3

49. Sketch the Lewis diagram of the water molecule, paying particular attention to the bond angle and using arrows to indicate the polarity of each bond. Then sketch the methanol molecule, $HOCH_3$, again using arrows to show bond polarity. Predict the approximate shape of both molecules around the oxygen atom. Also predict relative polarities of the two molecules and explain your prediction.

50. For each formula given, draw the Lewis diagram, determine the molecular geometry, and determine whether the molecule is polar or nonpolar.

a) $SiCl_4$ b) SeH_2 c) CH_2O

Select the answers for Questions 51 through 54 from the group of Lewis diagrams that follows. Each question may have more than one answer.

a) [H—B—H with H above and below B]$^-$ b) :C̈l—S̈—C̈l: c) :C̈l—P̈—C̈l: with :C̈l: below P

d) [H—N—H with H above and below N]$^+$ e) :F̈—Be—F̈: f) :F̈: above B, with :F̈ and F̈: below B

51. Which species have tetrahedral shapes?

52. Which species are linear?

53. Which neutral molecules are polar?

54. Identify all species that have trigonal planar geometries and all whose shapes are trigonal pyramidal.

Section 13.7: The Structures of Some Organic Compounds (Optional)

55. What distinguishes an organic compound from an inorganic compound?

56. What features of the carbon atom account for its ability to form so many chemical compounds?

57. Identify the hydrocarbons among the following: CH_3OH, $CH_3(CH_2)_6CH_3$, C_6H_6, $CH_2(NH_2)_2$, C_8H_{18}.

58. *Hydrocarbon* and *carbohydrate* are similar terms. Both are made by combining words or parts of words. Can you guess what those words or parts of words are in each case? Can a carbohydrate be an example of a hydrocarbon, or vice versa?

Charles D. Winters

Carbohydrates, a source of dietary energy, are found in grain products such as breads, pastas, and rices.

59. Distinguish between the *structure* of an organic molecule, as given by its Lewis diagram, and the three-dimensional *shape* of the molecule.

60. How are molecular structure and molecular shape shown in Lewis diagrams?

61. Why can a zigzag chain of carbon atoms be drawn as a straight line in a Lewis diagram?

62. What is a condensed structural formula? Why is it useful in drawing Lewis diagrams of organic molecules?

63. How are alcohols and ethers similar to water in structure and shape? What distinguishes between alcohols and ethers?

64. Name and describe the group that is present in all alcohols. What does the presence of this group contribute to an alcohol molecule and its properties?

65. How do carboxylic acids differ from other acids, such as hydrochloric, sulfuric, and carbonic acids?

66. What are carboxylic acids? Identify the group that is present in carboxylic acids. Write a Lewis diagram for the group and describe the geometry around all central atoms within the group.

General Questions

67. Distinguish precisely, and in scientific terms, the differences among items in each of the following pairs or groups.
a) Central atom, terminal atom
b) Molecular geometry, electron-pair geometry
c) Angular geometry, trigonal planar geometry
d) Trigonal planar geometry, trigonal pyramidal geometry
e) Trigonal pyramidal geometry, tetrahedral geometry
f) Polar molecule, nonpolar molecule

68. Classify each of the following statements as true or false:
a) Molecular geometry around an atom may or may not be the same as electron-pair geometry around the atom.
b) Electron-pair geometry is the direct effect of molecular geometry.
c) If the geometry of a molecule is linear, the molecule must have at least one double bond.
d) A molecule with a double bond cannot have a trigonal pyramidal geometry around the double-bonded atom.
e) A CO_2 molecule is linear, but an SO_2 molecule is bent.
f) A molecule is polar if it contains polar bonds.
g) A molecule with a central atom that has one lone pair of electrons is always polar.
h) A molecule with a central atom that has two lone pairs and two bonded pairs of electrons is always polar.
i) Carbon atoms normally form four bonds.
j) Hydrogen atoms never form double bonds.

69. One kind of C_6H_{12} molecule has its carbon atoms in a ring. Draw the Lewis diagram.

70. Draw Lewis diagrams for these five acids of bromine: HBr, HBrO, $HBrO_2$, $HBrO_3$, $HBrO_4$.

71. Draw a wedge-and-dash representation of the molecule shown below, then criticize the statement, "Carbon atoms in the molecule lie in a straight line."

```
    H   H   H
    |   |   |
H — C — C — C — H
    |   |   |
    H   H   H
```

More Challenging Problems

72. Benzene, C_6H_6, is a planar compound in which the carbon atoms are arranged in a hexagon (six-sided regular polygon). A Lewis diagram of this molecule has a geometry that accounts for the observed *shape* of the molecule, including bond angles. Draw this diagram. (Experimental evidence shows that the diagram does not describe the *structure* of benzene accurately.)

73. Describe the shapes of C_2H_6 and C_2H_4. In doing so, explain why one molecule is planar and why the other molecule cannot be planar.

74. $NaCHO_2$ is the way an inorganic chemist is most likely to write the formula of sodium formate. Write the formula of the formate ion as the inorganic chemist is likely to write it. Now draw a Lewis diagram of the formate ion. Compare the diagram with the Lewis diagram of formic acid, $HCHO_2$. Is your formate ion diagram what you would expect from the ionization of formic acid?

75. Draw two different Lewis diagrams of C_4H_6.

76. $C_4H_{10}O$ is the formula of diethyl ether. The same group of atoms is attached on either side of the oxygen atom. Draw the Lewis diagram.

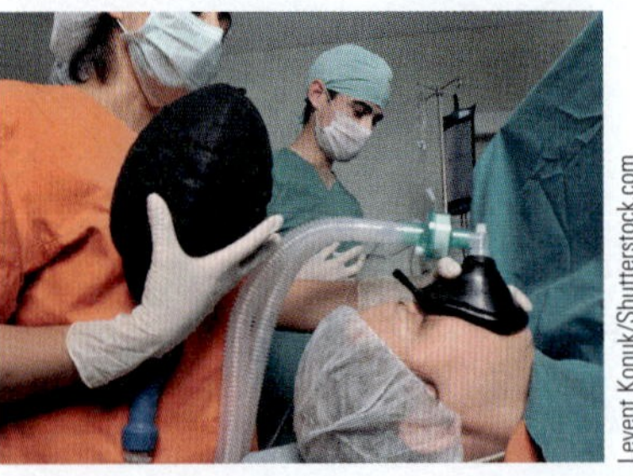

Levent Konuk/Shutterstock.com

Diethyl ether was among the first anesthetics used in surgery, but it has been largely replaced by safer anesthetics today.

77. Compare Lewis diagrams for CCl_4, SO_4^{2-}, ClO_4^-, and PO_4^{3-}. Identify two things that are alike about these diagrams and the way they are drawn. From these generalizations, can you predict the Lewis diagram of SeO_4^{2-} and CI_4?

78. What are the shapes of the following: (a) 1,2-dibromoethene, $C_2H_2Br_2$ (the bromine atoms are on different carbon atoms); (b) acetylene, C_2H_2; (c) *n*-butane, $CH_3CH_2CH_2CH_3$ (the carbon atoms are in a chain)?

Answers to Practice Exercises

1. H—O—H H—S—H

2. H—F [Br—O]$^{-}$

3. [O—N(=O)—O]$^{-}$

4. [O—S(O)(O)—O]$^{2-}$

5. O=C=O

6. CH_3—CH_2—CH_2—CH_3 (H—C—C—C—C—H, each C bonded to H atoms)

7. $H_2C{=}CH_2$

8. H—C—C—C—O—H ($CH_3CH_2CH_2OH$); H—C—C(O—H)—C—H ($CH_3CH(OH)CH_3$)

9. Electron-pair geometry: tetrahedral; molecular geometry: trigonal pyramidal [O—Cl(O)—O]$^{-}$

10. Electron-pair geometry: linear; molecular geometry: linear S=C=S

11. Electron-pair geometry: trigonal planar; molecular geometry: angular Cl—Sn—Cl

12. Electron-pair geometry: tetrahedral; molecular geometry: tetrahedral [F—B(F)(F)—F]$^{-}$

13. Each of the double-bonded carbon atoms are trigonal planar in both electron-pair and molecular geometry; the single-bonded carbon atom is tetrahedral in both electron-pair and molecular geometry.

14. Carbon with four single bonds: electron-pair and molecular geometry is tetrahedral; carbon with a single and triple bond: electron-pair and molecular geometry is linear. H—C(H)(H)—C≡N:

15. Polar. The bonding electrons are displaced toward the portion of the molecule where the fluorine atoms are located. (H, H, F, F on C; δ−)

Answers to Lewis Diagram Recognition Exercises

1. Acceptable. Boron is an exception to the octet rule and requires only three bonds.

2. Unacceptable. Nitrogen is less electronegative than oxygen, so nitrogen should be the central atom. [O—N(O)=O]$^{-}$

3. Unacceptable. In a compound with hydrogen, two or more oxygen atoms, and one atom of another nonmetal, hydrogen is bonded to an oxygen atom. O—Cl(O)—O—H

4. Unacceptable. The middle carbon atom is surrounded by only three electron pairs. Carbon atoms almost always form four bonds and thus usually have no lone pairs.

```
    H  H  H
    |  |  |
H—C=C—C—H
          |
          H
```

5. Unacceptable. Carbon does not have a complete octet. H—C≡N:

6. Acceptable. Beryllium is an exception to the octet rule and requires only two bonds.

7. Acceptable.

8. Unacceptable. A carbon atom must follow the octet rule.

```
     H
     |    ..
  H—C—Br:
     |    ..
     H
```

9. Unacceptable. Although carbon usually forms four bonds, in this case it cannot, given the 10 valence electrons. This structure accounts for only eight valence electrons. :C≡O:

10. Unacceptable. Hydrogen is never the central atom in a molecule, and it needs only one bond. H—Ö—Ö—H

Answers to Blue-Numbered Questions, Exercises, and Problems

1.

```
   ..        ..       ..  ..  ..
H—I:      H—O:      :Cl—N—Cl:
   ..        |        ..  |   ..
             H           :Cl:
                          ..
```

3.

```
..     ..      ..  ..       [  ..    ..   .. ]-
O=C=O        :F—S:        [ :O—Br—O:    ]
..     ..      ..  |        [  ..  |    ..  ]
                  :F:       [     :O:       ]
                   ..       [      ..       ]
```

5.

```
                     ..          -          ..      -
  ..  ..  -      [  :O:        ]       [   :O:    ]
[:Br—O:]         [   ..|  ..   ]       [ .. |  .. ]
  ..  ..         [ H—O—P—O—H   ]       [ :O—Cl—O: ]
                 [   ..|  ..   ]       [ .. |  .. ]
                 [    :O:      ]       [   :O:    ]
                      ..                    ..
```

7.

```
                                        ..
     H             H                   :Cl:
     |             |                    |
  ..     ..     ..     ..           ..      ..
:Cl—C—F:       :F—C—F:             :I—C—I:
  ..  |  ..     ..  |  ..           ..  |   ..
     :Cl:          :F:                 :I:
      ..            ..                  ..
```

Questions 9–17: When two or more acceptable diagrams are possible, only one representative diagram is shown.

9.

```
  ..   ..          ..   ..                  ..
 :Cl: :Cl:        :Cl: :Cl:        H   H   :Br:
  |    |           |    |          |   |    |
H—C————C—H       H—C————C—H      H—C———C————C—H
  |    |           |    |          |   |    |
 :Cl: :Cl:        :F:  :F:        :I: :Br: :Br:
  ..   ..          ..   ..         ..   ..   ..
```

11.

```
  H H  H  H             H  H                   H  H  H
  | |  |  |             |  |     ..         ..  |  |  |  ..
H—C=C—C—C—H          H—C—C—O—H          H—O—C—C—C—O—H
       |  |             |  |     ..         ..  |  |  |  ..
       H  H             H  H                   H  H  H
```

13.

```
  H  H  H  H  H  H         H  H  H  ..            H :F:
  |  |  |  |  |  |         |  |  | :O—H           |  |   ..
H—C—C—C—C—C—C—H         H—C=C—C=C—H           H—C=C—F:
  |  |  |  |  |  |                                      ..
  H  H  H  H  H  H
```

15.

```
  H  H  H :O:
  |  |  |  ||   ..
H—C—C—C—C—O—H
  |  |  |       ..
  H  H  H
```

17.

```
  ..      ..  +     ..      ..
[ O=N=O ]       N=N=O     [:N≡O:]+
  ..      ..        ..      ..
```

	Substance	Lewis Diagram	Electron-Pair Geometry	Molecular Geometry	Wedge-and-Dash Diagram
19.	BCl_3	:Cl—B—Cl: | :Cl:	Trigonal planar	Trigonal planar	Cl B Cl | Cl
	PH_3	H—P—H | H	Tetrahedral	Trigonal pyramidal	P H H H
	H_2S	H—S: | H	Tetrahedral	Bent	H | S \\ H
21.	BrO^-	[:Br—O:]$^-$	Linear	Linear	Br–O
	ClO_3^-	[:O—Cl—O: | :O:]$^-$	Tetrahedral	Trigonal pyramidal	Cl O O O
	PO_4^{3-}	[:O: | :O—P—O: | :O:]$^{3-}$	Tetrahedral	Tetrahedral	O | P O O O
23.	C in C_3H_7OH	H H H | | | H—C—C—C—H | | | H H :O—H	Tetrahedral	Tetrahedral	
25.	N in $C_2H_5NH_2$	H H | | H—C—C—H | | H :N—H | H	Tetrahedral	Trigonal pyramidal	
27.	C in C_2H_2	H—C≡C—H	Linear	Linear	
29.	C in HCN	H—C≡N:	Linear	Linear	

31. Trigonal pyramidal

33. Electron-pair and molecular: trigonal planar

35. Electron-pair: tetrahedral; molecular: trigonal pyramidal

37. Electron-pair and molecular: tetrahedral

39. A: 120°, three regions of electron density around C; B: 109°, four regions of electron density around O; C: 109°, four regions of electron density around C; D: 120°, three regions of electron density around C

41. A: 120°, three regions of electron density around C; B: 120°, three regions of electron density around C; C: 120°, three regions of electron density around C

43. (b), (d), and (e)

45. CCl_4 molecules are nonpolar because the polar bonds are symmetrically arranged, which makes the molecule itself nonpolar.

47. Both molecules are linear. HF is more polar than HBr. The H end of each is more positive.

49. Both molecules are polar because of the bent structure around the oxygen atom and the concentration of negative charge near the highly electronegative oxygen atom. The carbon–oxygen electronegativity difference is slightly less than the hydrogen–oxygen electronegativity difference, so CH_3OH is slightly less polar than H_2O.

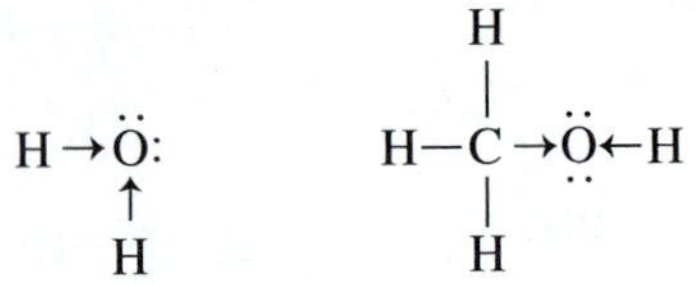

51. (a) and (d)

53. (b) and (c)

55. Organic compounds all contain carbon; inorganic compounds (with some exceptions) do not contain carbon.

57. $CH_3(CH_2)_6CH_3$, C_6H_6, and C_8H_{18}

59. The structure of a molecule simply describes in two dimensions the relative positions of how its atoms are bonded. The three-dimensional shape of a molecule results from its structure and is the arrangement of the atoms in real three-dimensional space.

61. A straight line of carbon atoms in a Lewis diagram simply indicates that the atoms are bonded to one another. Single-bonded carbon atoms form tetrahedral bond angles, which form a zigzag chain when in the same plane.

63. Alcohols, ethers, and water molecules all include an oxygen atom with two lone pairs of electrons that are single-bonded to two other atoms. In an alcohol, one of the bonds to the oxygen atom is to a hydrogen atom, and the other is to a carbon atom. In an ether, both bonds to the oxygen atom are to carbon atoms.

65. Carboxylic acids contain the carboxyl group, –COOH.

68. True: (a), (d), (e), (g), (h)*, (i), (j)

69.

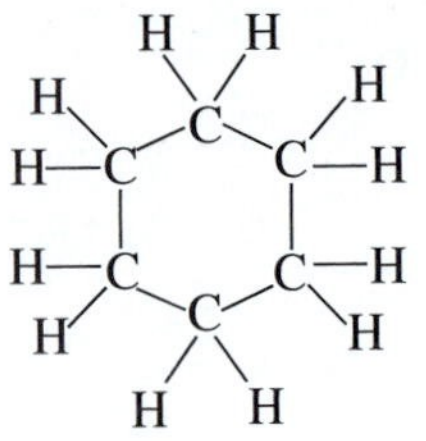

70.

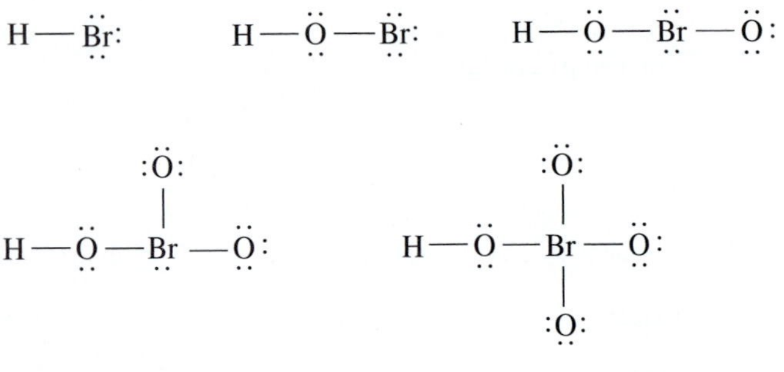

71.

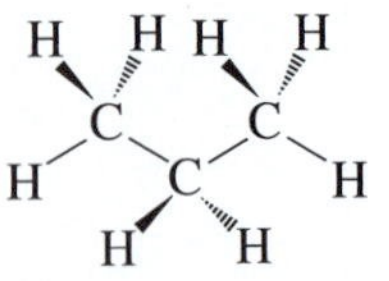

Bond angles between carbon atoms in an alkane are tetrahedral, so the atoms cannot lie in a straight line.

72.

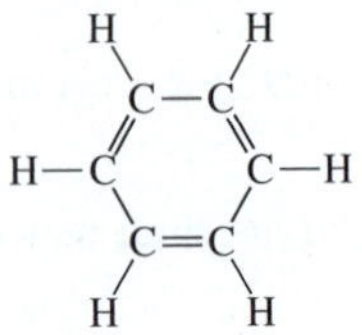

74.

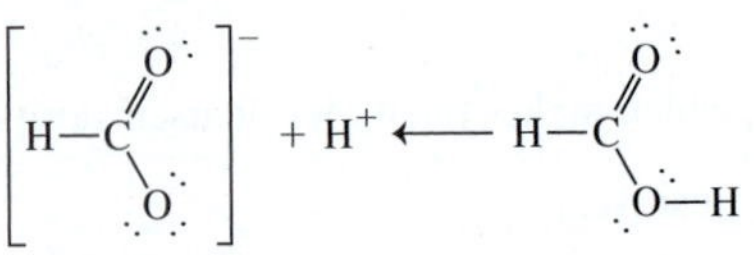

formate ion, CHO_2^- formic acid

*Statement h is true if central atoms are limited to four electron pairs, as they are in this book.

75. Any two from:

```
H         H  H           H   H      H
|         |  |           |   |      |
C=C=C—C—H          C=C—C=C
|            |           |        |   |
H            H           H        H   H

        H   H                 H           H
        |   |                 |           |
H—C≡C—C—C—H           H—C—C≡C—C—H
        |   |                 |           |
        H   H                 H           H
```

76.

```
   H  H     H  H
   |  |     |  |
H—C—C—O—C—C—H
   |  |     |  |
   H  H     H  H
```

77. All species in the question, including SeO_4^{2-} and CI_4, have 32 valence electrons and four atoms to be distributed around a central atom. Consequently, they all have the same tetrahedral shape.

```
     :Cl:             [    :O:    ]2−        [    :O:     ]−
      |               [     |     ]          [     |      ]
:Cl— C —Cl:           [:O — S — O:]          [:O — Cl — O:]
      |               [     |     ]          [     |      ]
     :Cl:             [    :O:    ]          [    :O:     ]

[    :O:    ]3−       [    :O:     ]2−            :I:
[     |     ]         [     |      ]              |
[:O — P — O:]         [:O — Se — O:]         :I — C — I:
[     |     ]         [     |      ]              |
[    :O:    ]         [    :O:     ]              :I:
```

78. (a) Trigonal planar with 120° angles around both carbon atoms. (b) Linear. (c) Zigzag carbon chain; all bond angles are tetrahedral.

14 The Ideal Gas Law and Its Applications

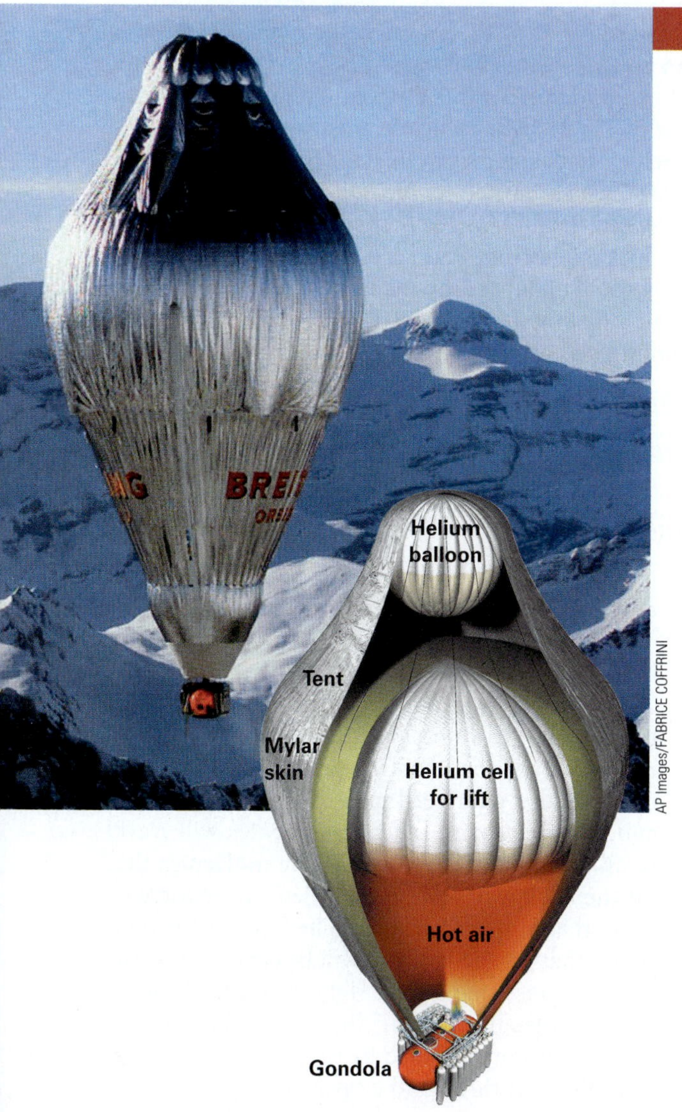

AP Images/FABRICE COFFRINI

◀ In March 1999, the Breitling Orbiter 3 became the first balloon to make a nonstop flight around the world. The very first manned balloon flight occurred in 1783. Why was there a 216-year time lag between the first flight and the first successful circumnavigation of the globe? The answer lies in understanding the engineering and technology necessary to design a balloon capable of flying for 20 continuous days while covering a distance of 28,431 miles. In this chapter, you will learn about the relationship between the temperature and the density of gases, which, in part, governs the flight of hot air balloons.

CHAPTER CONTENTS

In Chapter 4, you learned about several proportionalities among the volume, pressure, and temperature of a fixed amount of gas. In this chapter, the amount becomes a fourth variable. The result of combining the proportionalities is a single mathematical relationship, the Ideal Gas Law, which summarizes all the measurable properties of gases.

NASA

Figure 14.1 The launch of Apollo 17 on December 7, 1972. The height of this spacecraft is greater than the length of an American football field, and it weighed 6.5 million pounds when fully fueled at launch.

14.1 How Are Tiny Gas Molecules Capable of Launching a Rocket?

The only vehicle that has ever carried humans beyond low-Earth orbit (about 1200 miles or 2000 kilometers) is the Saturn V (V is the Roman number five), used starting with the Apollo 4 mission in 1967 and ending with the Skylab launch in 1973. The need for this rocket came from the challenge of launching a 100,000-pound spacecraft from the surface of the earth and propelling it to the moon. The Saturn V rocket used to launch Apollo 17, the last manned mission to the moon, is shown in **Figure 14.1**. One of the most viewed photographs in history was taken by a member of the Apollo 17 crew as the ship sped toward the moon at the point when it was 18,000 miles from Earth. **Figure 14.2** is what is known as *The Blue Marble* photograph. Many people have found inspiration from this photograph to develop a new respect for the fact that all people of Earth share a single planet.

What fuel is capable of propelling such a massive vehicle? The rocket was made up of three parts, or stages, each of which burned its fuel load and then detached from the remaining parts in sequence. The fuel for the first stage was called *Rocket Propellant-1*. It is a refined product made from kerosene, which is a mixture of carbon–hydrogen molecules. The most abundant component of Rocket Propellant-1 is called *2,9-dimethyldecane*, which has the formula $C_{12}H_{26}$. It was reacted with liquid oxygen, and thus one reaction that occurred in the first stage was:

$$2\ C_{12}H_{26}(\ell) + 37\ O_2(\ell) \rightarrow 24\ CO_2(g) + 26\ H_2O(g)$$

The fuel for the second and third stages was hydrogen. This was also reacted with liquid oxygen:

$$2\ H_2(\ell) + O_2(\ell) \rightarrow 2\ H_2O(g)$$

Note that in both reactions, liquid reactants react to form gaseous products.

How can these oxidation reactions propel a rocket with enough force to lift millions of pounds? The key concept is that the liquid reactants change to gaseous products. We can model the reaction chamber to which the gaseous products are released as a cylinder that is open at the bottom end. As they form, the gas molecules will travel in all directions. The force of the molecules that hit the back wall will be cancelled by the force of those that hit the front wall. The force of the molecules that hit the left wall will be cancelled by the force of those that hit the right wall. But the force of the molecules that hit the top of the reaction chamber is not cancelled by the force of the molecules that exit through the open bottom end. A net upward force results, propelling the rocket upward. Each individual gas molecule is very tiny, but the molecules move very fast at high temperatures and there are enormous numbers of them. Collectively, the gas molecules provide the force that lifts the rocket out of Earth's gravitational hold. In this chapter, you will learn how to determine the number of gaseous product molecules that can be formed by any given weight or mass of Rocket Propellant or hydrogen that reacts with a given weight or mass of oxygen. You will also improve your knowledge of other properties of gases.

NASA

Figure 14.2 *The Blue Marble*. This photograph was taken from Apollo 17 when it was 18,000 miles from Earth. In this perspective, the Antarctic is at the bottom, the African continent is in the middle, and the Arabian Peninsula is at the top. The photograph was taken on December 7, 1972 by either Harrison Schmitt or Ron Evans.

14.2 Gases Revisited

These properties of gases were introduced in Section 4.2:

- Gases may be compressed (**Figure 14.3**).
- Gases expand to fill their containers uniformly.
- Gases have low densities in comparison with densities of liquids and solids.
- Gases may be mixed in a fixed volume.
- Gases exert constant, uniform pressure in all directions on the walls of their containers.

These properties are accounted for in the kinetic molecular theory (Section 4.3). This set of postulates describes the particulate behavior of gases as follows:

- A gas consists of molecules and empty space.
- The volume occupied by the molecules in a gas is negligible when compared with the volume of the space the molecules occupy.
- The attractive forces among molecules of a gas are negligible.
- The average kinetic energy of gas molecules is proportional to the temperature, expressed in kelvins.
- Molecules interact with one another and with the container walls without loss of total kinetic energy.

Gas measurements (Sections 4.4 and 4.5) and the units in which they are usually expressed include:

- pressure (P), expressed in torr, atmospheres (atm), or bar
- volume (V), expressed in liters (L)
- temperature (T), expressed in degrees Celsius (°C) or kelvins (K)
- amount (n), expressed in moles (Section 7.4)

The relationship that shows how a change in pressure, volume, or temperature of a gas sample (n is constant) affects the other two quantities is given in the equation for the Combined Gas Law (Section 4.7):

$$\frac{P_1V_1}{T_1} = \frac{P_2V_2}{T_2}$$

Subscripts 1 and 2 refer to initial and final values of pressure, temperature, and volume, respectively.

Given five of the six variables in the combined gas law equation, the remaining variable can be calculated using algebra. For example, if you know the volume

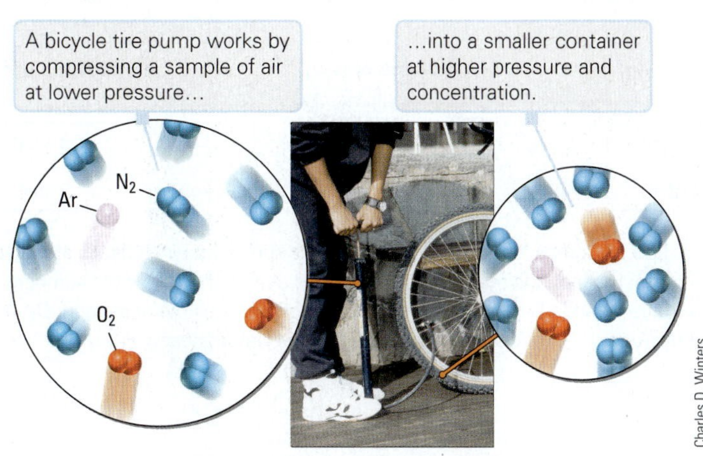

A bicycle tire pump works by compressing a sample of air at lower pressure...
...into a smaller container at higher pressure and concentration.
Ar
N2
O2
Charles D. Winters

Figure 14.3 The compression of gases. The huge quantity of empty space between gas particles allows gases to be compressed. Liquids and solids do not have this property.

occupied by a gas at one temperature and pressure (V_1, T_1, and P_1), you can find the volume that the gas will occupy (V_2) at another temperature and pressure (T_2 and P_2):

$$V_2 = V_1 \times \frac{P_1}{P_2} \times \frac{T_2}{T_1}$$

Gas volumes in some problems are given at 0°C (273 K) and 1 bar. These conditions are known as *standard temperature and pressure (STP)*.

14.3 Avogadro's Law

Goal 1 State how volume and amount of gas are related when pressure and temperature are constant and explain phenomena or make predictions based on that relationship.

Early in the 19th century, French scientist Joseph Gay-Lussac (Figure 14.4) observed that when gases react with one another, the reacting volumes are always in the ratio of small whole numbers *if the volumes are measured at the same temperature and pressure.* This observation is known as the **Law of Combining Volumes**. It extends to gaseous products, too. An example appears in Figure 14.5.

Shortly after Gay-Lussac's observations became known, Italian scientist Amedeo Avogadro (Figure 14.6) reasoned that the Law of Combining Volumes could be explained if equal volumes of all gases *at the same temperature and*

Kean Collection/Getty Images

Figure 14.4 Joseph Gay-Lussac (1778–1850). Gay-Lussac's investigations of the gaseous state of matter led to the Law of Combining Volumes.

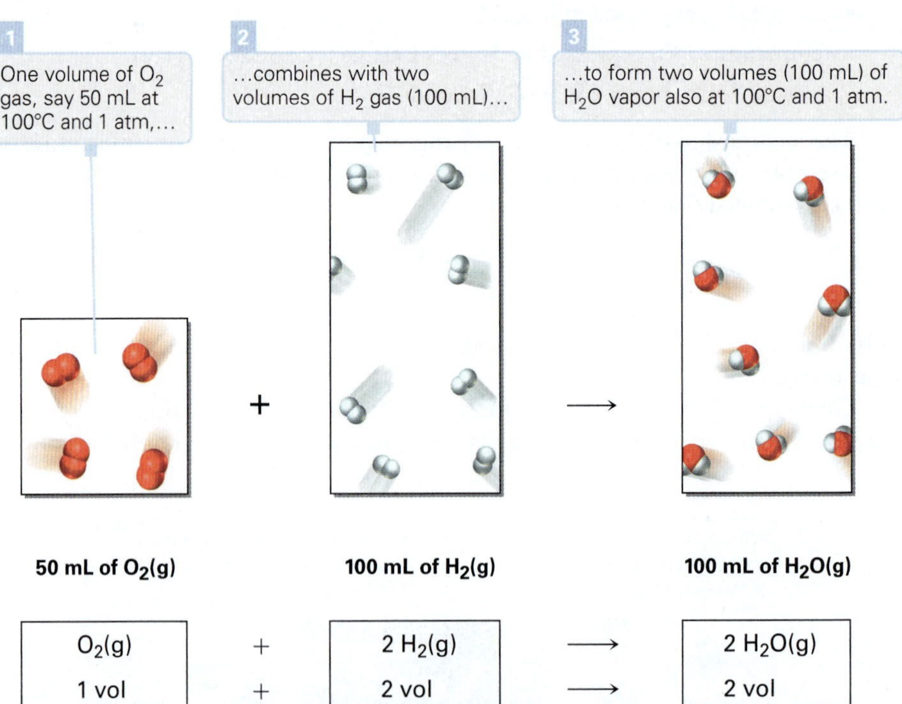
1
One volume of O_2 gas, say 50 mL at 100°C and 1 atm,...
2
...combines with two volumes of H_2 gas (100 mL)...
3
...to form two volumes (100 mL) of H_2O vapor also at 100°C and 1 atm.
+
→
50 mL of $O_2(g)$
100 mL of $H_2(g)$
100 mL of $H_2O(g)$
$O_2(g)$ + $2\ H_2(g)$ → $2\ H_2O(g)$
1 vol + 2 vol → 2 vol

Figure 14.5 Law of Combining Volumes. Experiments show that when gases at the same temperature and pressure react, the reacting volumes are in a ratio of small whole numbers. The small whole number ratio extends to products, too. In the reaction, 1 volume of oxygen, O_2, reacts with 2 volumes of hydrogen, H_2, to form 2 volumes of gaseous water (steam), H_2O.

"Volumi eguali di gas nelle stesse condizioni di temperatura e di pressione contengono lo stesso numero di molecole,,
1856
1956
AMEDEO AVOGADRO
POSTE ITALIANE L25

Stamp collection of Professor C. M. Lang, photo by Gary J. Shulfer, University of Wisconsin-Stevens Point.

Figure 14.6 Count Amedeo Avogadro (1776–1856), depicted on a 1956 Italian postage stamp. The quote says, "Equal volumes of gas in the same conditions of temperature and of pressure contain the same number of molecules."

14.2 Gases Revisited

These properties of gases were introduced in Section 4.2:

- Gases may be compressed (**Figure 14.3**).
- Gases expand to fill their containers uniformly.
- Gases have low densities in comparison with densities of liquids and solids.
- Gases may be mixed in a fixed volume.
- Gases exert constant, uniform pressure in all directions on the walls of their containers.

These properties are accounted for in the kinetic molecular theory (Section 4.3). This set of postulates describes the particulate behavior of gases as follows:

- A gas consists of molecules and empty space.
- The volume occupied by the molecules in a gas is negligible when compared with the volume of the space the molecules occupy.
- The attractive forces among molecules of a gas are negligible.
- The average kinetic energy of gas molecules is proportional to the temperature, expressed in kelvins.
- Molecules interact with one another and with the container walls without loss of total kinetic energy.

Gas measurements (Sections 4.4 and 4.5) and the units in which they are usually expressed include:

- pressure (P), expressed in torr, atmospheres (atm), or bar
- volume (V), expressed in liters (L)
- temperature (T), expressed in degrees Celsius (°C) or kelvins (K)
- amount (n), expressed in moles (Section 7.4)

The relationship that shows how a change in pressure, volume, or temperature of a gas sample (n is constant) affects the other two quantities is given in the equation for the Combined Gas Law (Section 4.7):

$$\frac{P_1V_1}{T_1} = \frac{P_2V_2}{T_2}$$

Subscripts 1 and 2 refer to initial and final values of pressure, temperature, and volume, respectively.

Given five of the six variables in the combined gas law equation, the remaining variable can be calculated using algebra. For example, if you know the volume

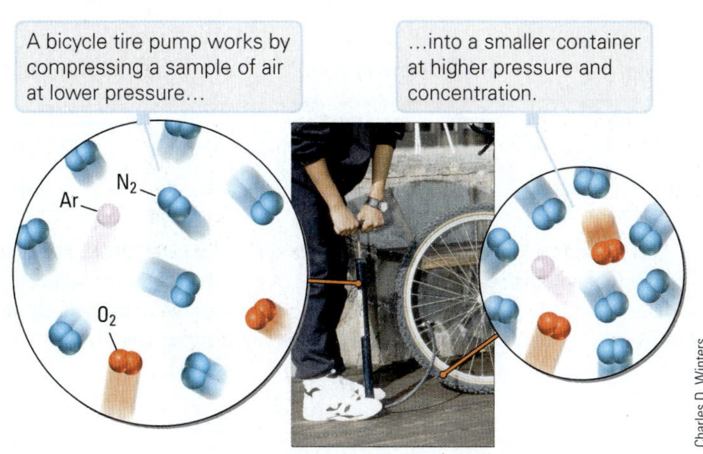
A bicycle tire pump works by compressing a sample of air at lower pressure...
...into a smaller container at higher pressure and concentration.
Ar
N2
O2
Charles D. Winters

Figure 14.3 The compression of gases. The huge quantity of empty space between gas particles allows gases to be compressed. Liquids and solids do not have this property.

occupied by a gas at one temperature and pressure (V_1, T_1, and P_1), you can find the volume that the gas will occupy (V_2) at another temperature and pressure (T_2 and P_2):

$$V_2 = V_1 \times \frac{P_1}{P_2} \times \frac{T_2}{T_1}$$

Gas volumes in some problems are given at 0°C (273 K) and 1 bar. These conditions are known as *standard temperature and pressure (STP)*.

14.3 Avogadro's Law

Goal 1 State how volume and amount of gas are related when pressure and temperature are constant and explain phenomena or make predictions based on that relationship.

Early in the 19th century, French scientist Joseph Gay-Lussac (**Figure 14.4**) observed that when gases react with one another, the reacting volumes are always in the ratio of small whole numbers *if the volumes are measured at the same temperature and pressure.* This observation is known as the **Law of Combining Volumes**. It extends to gaseous products, too. An example appears in **Figure 14.5**.

Shortly after Gay-Lussac's observations became known, Italian scientist Amedeo Avogadro (**Figure 14.6**) reasoned that the Law of Combining Volumes could be explained if equal volumes of all gases *at the same temperature and*

Kean Collection/Getty Images

Figure 14.4 Joseph Gay-Lussac (1778–1850). Gay-Lussac's investigations of the gaseous state of matter led to the Law of Combining Volumes.

"Volumi eguali di gas nelle stesse condizioni di temperatura e di pressione contengono lo stesso numero di molecole".
1856
1956
AMEDEO AVOGADRO
POSTE ITALIANE L25

Stamp collection of Professor C. M. Lang, photo by Gary J. Shulfer, University of Wisconsin-Stevens Point.

Figure 14.6 Count Amedeo Avogadro (1776–1856), depicted on a 1956 Italian postage stamp. The quote says, "Equal volumes of gas in the same conditions of temperature and of pressure contain the same number of molecules."

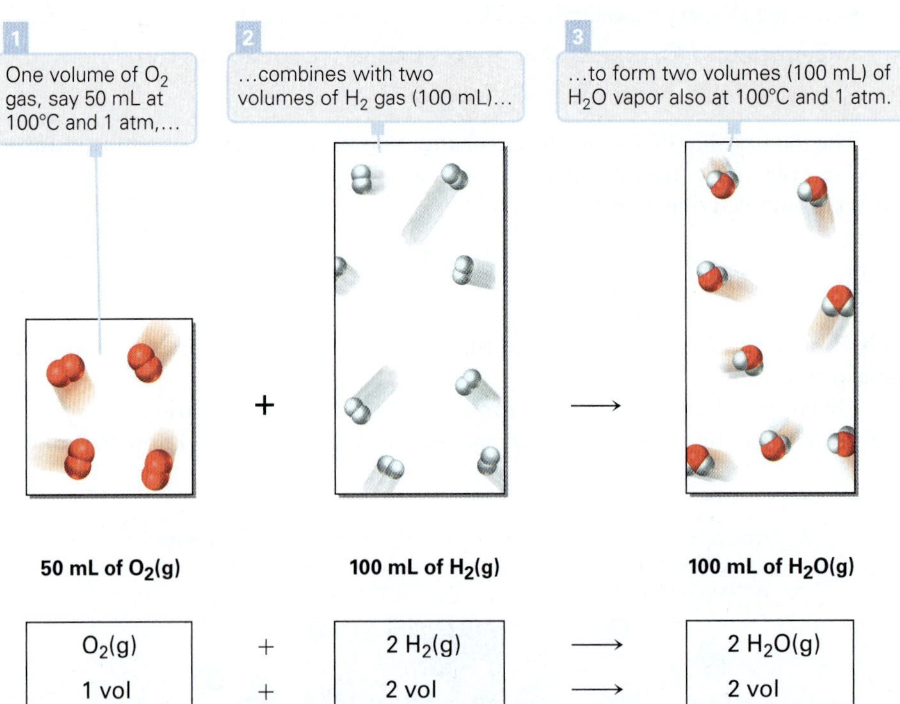
1
One volume of O_2 gas, say 50 mL at 100°C and 1 atm,...
2
...combines with two volumes of H_2 gas (100 mL)...
3
...to form two volumes (100 mL) of H_2O vapor also at 100°C and 1 atm.
+
→
50 mL of $O_2(g)$
100 mL of $H_2(g)$
100 mL of $H_2O(g)$
$O_2(g)$ + $2\ H_2(g)$ → $2\ H_2O(g)$
1 vol + 2 vol → 2 vol

Figure 14.5 Law of Combining Volumes. Experiments show that when gases at the same temperature and pressure react, the reacting volumes are in a ratio of small whole numbers. The small whole number ratio extends to products, too. In the reaction, 1 volume of oxygen, O_2, reacts with 2 volumes of hydrogen, H_2, to form 2 volumes of gaseous water (steam), H_2O.

pressure contain the same number of molecules (**Figure 14.7**). If the reacting molecules react in a 1:1 ratio, and the reacting volumes also have a 1:1 ratio, then the equal volumes of the different gases must have the same number of molecules. If both ratios are 1:2, then the larger volume must have twice as many molecules as the smaller volume of the other gas. It follows that $V \propto n$ at constant temperature and pressure. These statements are known as **Avogadro's Law** (**Figure 14.8**).

Avogadro's Law as illustrated in Figure 14.7 gives us a hint about why the formulas of hydrogen and chlorine are H_2 and Cl_2, not simply H and Cl. According to Avogadro's Law, one molecule of hydrogen divides to form two molecules of hydrogen chloride. It follows that the hydrogen molecule contains an even number of hydrogen atoms; if the starting number of hydrogen atoms divides into two equal parts, the starting number must be even. The same applies to chlorine. Other experiments prove that, for both elements, the even number is 2 and not 4, 6, or some higher even number. Similar results show that nitrogen, oxygen, fluorine, bromine, and iodine form diatomic (two-atom) molecules with the formulas N_2, O_2, F_2, Br_2, and I_2, respectively.

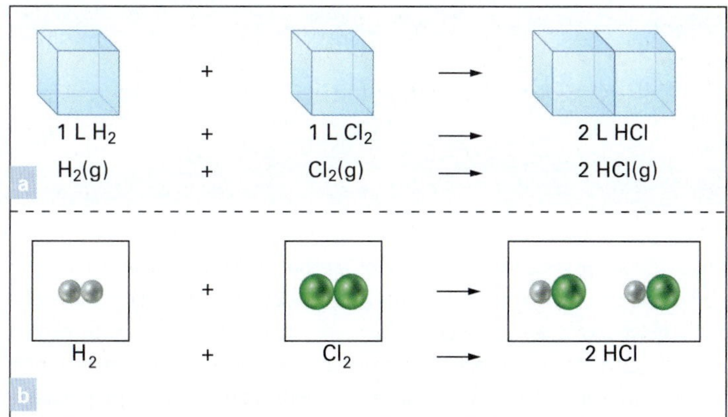

Figure 14.7 The logic of Avogadro's reasoning. (a) Experimentally, it is found that, for gases at the same temperature and pressure, 1 liter of hydrogen reacts with 1 liter of chlorine to form 2 liters of hydrogen chloride. The logic behind Avogadro's reasoning appears in the subsequent particulate-level illustration. In (b), the same numbers of molecules in equal-reacting volumes use all the molecules of both gases.

Figure 14.8 Avogadro's Law. When temperature and pressure conditions are constant, the volume of a container is directly proportional to the number of particles in the container.

Thinking About *Your Thinking*

Proportional Reasoning

Avogadro's Law is an example of a direct proportionality. This means that the volume of a gas at constant temperature and pressure is equal to a proportionality constant times the number of gas particles in the container. Also, the volume is zero when the number of particles is zero. If the number of particles is doubled, the volume is doubled; when the number of particles is tripled, the volume is tripled, and so on. Therefore, a conversion factor can be written that allows for a one-step conversion between container volume and number of particles for a gas at constant T and P.

Target Check 14.1

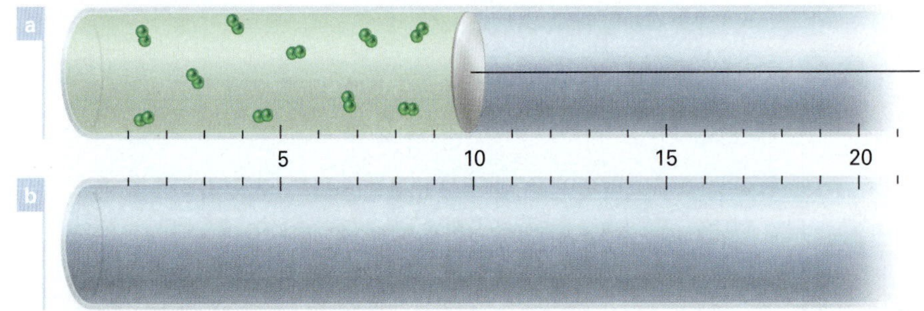

A horizontal cylinder (a) is closed at one end by a piston that moves freely left or right, depending on the pressure exerted by the enclosed gas. The gas consists of 10 two-atom molecules. A reaction occurs in which 5 of the molecules separate into one-atom particles. In cylinder (b), sketch the position to which the piston would move as a result of the reaction. Pressure and temperature remain constant throughout the process. (Hint: How many total particles would be present after the reaction? Include them in your sketch.)

14.4 The Ideal Gas Law

Goal 2 Explain how the ideal gas equation can be constructed by combining Charles's Law (Section 4.5), Boyle's Law (Section 4.6), and Avogadro's Law (Section 14.3), and explain how the ideal gas equation can be used to derive each of the three two-variable laws.

Having accumulated several different proportionalities between volume and the three other measurable properties of a gas, we now look for a single equation that ties them all together. We have seen from Charles's Law that $V \propto T$, from Boyle's Law that $V \propto \frac{1}{P}$, and from Avogadro's Law that $V \propto n$. If volume is proportional to three different quantities, it is logical to assume that it is proportional to their product, $n \times T \times \frac{1}{P}$ or

$$V \propto \frac{nT}{P}$$

This assumption is indeed verified experimentally.

Inserting a proportionality constant, R, yields an equality:

$$V \propto \frac{nT}{P} \xrightarrow{\text{the proportionality changes to an equality}} V = \mathbf{R} \times \frac{nT}{P}$$

Rearranging gives the **ideal gas equation** in its most common form:

$$V = R \times \frac{nT}{P} \xrightarrow{\text{multiply both sides by P}} \mathbf{P}V = \mathbf{R}n\mathrm{T} \xrightarrow{\text{multiplication is commutative}} PV = \mathbf{n}RT$$

The relationship symbolized in this equation is called the **Ideal Gas Law**. If you are a science or engineering major, you will probably encounter the ideal gas equation in at least half a dozen other courses. Knowing the relationships among the four variables is useful, and you should memorize the ideal gas equation.

An **ideal gas** is one whose variable quantities fit the ideal gas equation. This is most likely to happen when a gas is made up of identical particles that occupy minimal volume with minimal attractive forces between one another. The physical conditions that lead to a gas behaving ideally are high temperature and low pressure.

Now that you have seen the logic in developing the Ideal Gas Law, you will find that it is convenient to derive Boyle's, Charles's, or Avogadro's Law, as necessary, by starting from the ideal gas equation. For example, if you have a fixed quantity of gas at constant temperature, n and T in the ideal gas equation are constants, making the entire nRT product a constant, which we can symbolize with k:

$$PV = \mathbf{nRT} \xrightarrow{\text{nRT = a constant = k}} PV = \mathbf{k}$$

Rearranging to isolate V,

$$\mathbf{P}V = k \xrightarrow{\text{multiply both sides by 1/P}} V = k \times \frac{1}{P} \xrightarrow{\text{change the equality to a proportionality}} V \propto \frac{1}{P}$$

$V \propto \frac{1}{P}$ is Boyle's Law. **Figure 14.9** shows how the other laws are derived from the Ideal Gas Law through a similar approach.

The proportionality constant in the ideal gas equation is the **gas constant (R)**. Its value is the same for any gas or mixture of gases that behaves like an ideal gas.* It is significant to note that both the ideal gas equation and the value of R can be derived by applying the laws of physics to the ideal gas model. The fact that the same conclusion can be reached by theoretical calculations and by experiments makes us quite confident that the ideal gas model is correct.

The value of R may be found experimentally…

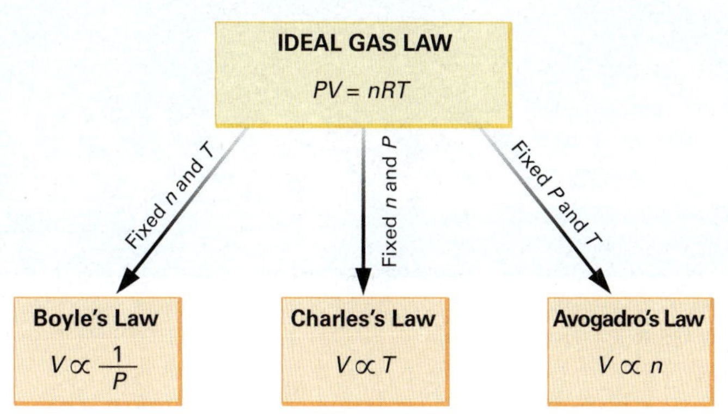

Figure 14.9 Derivation of Boyle's, Charles's, and Avogadro's Laws from the Ideal Gas Law.

*Most real gases behave like an ideal gas, particularly under low-pressure and high-temperature conditions. We will not consider the differences between real and ideal gases in this book.

Active Example 14.1 Determination of the Value of the Gas Constant I

A 0.592-mole sample of helium (Figure 14.10) was found to occupy 14.7 liters when measured at 12°C and 0.942 atm. From these experimental data, calculate the value of R.

Think Before You Write The problem statement gives the values of n, V, T, and P. You are asked to find R. Thus, you must recall the ideal gas equation and then solve it for the wanted quantity.

Answers *Cover the left column with your cut-out shield. Reveal each answer only after you have written your own answer in the right column.*

Given: $n = 0.592$ mol; $V = 14.7$ L; $T = (12 + 273) = 285$ K; $P = 0.942$ atm
Wanted: R

Analyze the problem statement by listing the given quantities and symbol of the wanted quantity. Change the given Celsius temperature to kelvins in this step.

$PV = nRT$

$$PV = \mathbf{nRT} \xrightarrow{\text{divide both sides by nT}} \frac{PV}{\mathbf{nT}} = R$$

$$R = \frac{PV}{nT} = P \times V \times \frac{1}{n} \times \frac{1}{T} = 0.942 \text{ atm} \times 14.7 \text{ L} \times \frac{1}{0.592 \text{ mol}} \times \frac{1}{285 \text{ K}} = 0.0821 \text{ L} \cdot \text{atm/mol} \cdot \text{K}$$

Identify the algebraic relationship needed to solve the problem, and solve it for the wanted variable. Construct the setup and solve.

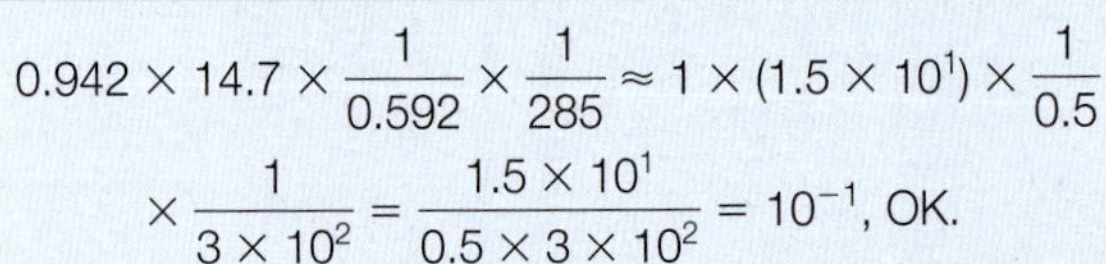

$$0.942 \times 14.7 \times \frac{1}{0.592} \times \frac{1}{285} \approx 1 \times (1.5 \times 10^1) \times \frac{1}{0.5} \times \frac{1}{3 \times 10^2} = \frac{1.5 \times 10^1}{0.5 \times 3 \times 10^2} = 10^{-1}, \text{ OK.}$$

With four values in the setup, the **check** is challenging. Just be sure that the value is in the ballpark.

You improved your skill at solving ideal gas equation problems.

What did you learn by solving this Active Example?

Practice Exercise 14.1

Use the following data to determine the value of the gas constant in L · bar/mol · K: A 3.50-L sample of a gas at 23°C and 0.93 bar contains 0.132 mole of molecules.

Figure 14.10 Helium-gas-filled sign. When a glass tube is filled with helium at low pressure and a high voltage is applied, current will flow and the ionizing gas will glow, producing a color characteristic of the element. Helium-filled gas discharge tubes emit a yellow glow.

Karen Perhus/Shutterstock.com

The value of the gas constant that you just derived, 0.0821 L · atm/mol · K, is an important quantity in the physical sciences. Your instructor may require you to memorize it.

Active Example 14.2 Determination of the Value of the Gas Constant II

At 23°C, a 656-mL sample of argon (Figure 14.11) exerted a pressure of 839 torr. It was determined that the sample contained 0.0298 mole of the gas. Use these data to find the value of R in L · torr/mol · K.

Think Before You Write As in Active Example 14.1, it is relatively straightforward to recognize that this problem requires the ideal gas equation. You are given T, V, P, and n, and you are asked to find R.

Answers *Cover the left column with your cut-out shield. Reveal each answer only after you have written your own answer in the right column.*

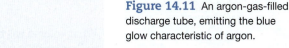

Given: T = (23 + 273) = 296 K;
V = 656 mL; P = 839 torr; n = 0.0298 mol
Wanted: R in L · torr/mol · K

$$PV = \mathbf{nRT} \xrightarrow{\text{divide both sides by nT}} \frac{PV}{\mathbf{nT}} = R$$

$$R = \frac{PV}{nT} = P \times V \times \frac{1}{n} \times \frac{1}{T} = 839 \text{ torr} \times 656 \text{ mL} \times \frac{1}{0.0298 \text{ mol}} \times \frac{1}{296 \text{ K}} \times \frac{1 \text{ L}}{1000 \text{ mL}} = 62.4 \text{ L} \cdot \text{torr/mol} \cdot \text{K}$$

We needed to include the 1 L/1000 mL conversion factor because the given volume was in mL, but the problem asked for the value of R in L · torr/mol · K, and the volume unit in that collection of units is L.

Complete the solution.

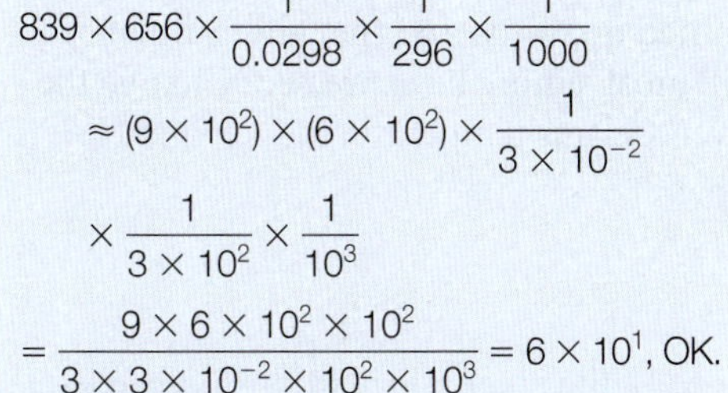

$$839 \times 656 \times \frac{1}{0.0298} \times \frac{1}{296} \times \frac{1}{1000} \approx (9 \times 10^2) \times (6 \times 10^2) \times \frac{1}{3 \times 10^{-2}} \times \frac{1}{3 \times 10^2} \times \frac{1}{10^3} = \frac{9 \times 6 \times 10^2 \times 10^2}{3 \times 3 \times 10^{-2} \times 10^2 \times 10^3} = 6 \times 10^1, \text{ OK.}$$

Check to be sure that the value of your answer is approximately correct.

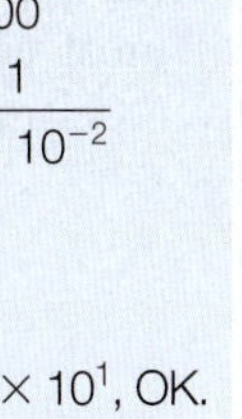

You improved your skill at solving ideal gas equation problems.

What did you learn by solving this Active Example?

Practice Exercise 14.2

A laboratory determines that a gas sample contains 0.659 mole of gas when a 21.0-L container is at 33°C and 23.6 inches of mercury pressure. Determine the value of R in L · in. Hg/mol · K.

Charles D. Winters/Science Source

Figure 14.11 An argon-gas-filled discharge tube, emitting the blue glow characteristic of argon.

You can choose to memorize both 0.0821 L · atm/mol · K and 62.4 torr · L/mol · K and use the appropriate value in ideal gas equation problems depending on whether the given pressure is in atm or torr (or mm Hg), or you can just memorize one value and use the 1 atm = 760 torr relationship as a conversion factor when necessary.

14.5 The Ideal Gas Equation: Determination of a Single Variable

Goal 3 Given values for all except one of the variables in the ideal gas equation, calculate the value of the unknown variable.

If you know values for all variables except one in the ideal gas equation, you can use algebra to find the value of the unknown variable. In fact, that's how you just calculated R. As with all problems to be solved algebraically, first solve the equation for the wanted quantity. Then substitute the quantities that are given, *including units*, calculate the answer, check that your answer is reasonable, and reflect on what you have learned by solving the problem.

Carlos E. Santa Maria/Shutterstock.com

Figure 14.12 The Federal Aviation Administration requires that the tires of large commercial jet aircraft be filled with nitrogen because it will not support combustion if a tire fire occurs.

Active Example 14.3 Determination of a Single Variable in the Ideal Gas Equation

What volume will be occupied by 0.393 mole of nitrogen (Figure 14.12) at 0.971 atm and 24°C?

Think Before You Write Now that you have derived the value and units of the gas constant R, you can treat it as a given quantity in ideal gas equation problems. This problem therefore has R, n, P, and T as givens.

Answers *Cover the left column with your cut-out shield. Reveal each answer only after you have written your own answer in the right column.*

Given: n = 0.393 mol; P = 0.971 atm; T = (24 + 273) = 297 K; R = 0.0821 L · atm/mol · K
Wanted: V (assume L)

We selected liters as the volume unit because the volume unit in the gas constant is liters.

Analyze the problem statement by listing the given quantities, the symbol of the wanted quantity, and the unit in which it will most simply be expressed. Change the given Celsius temperature to kelvins.

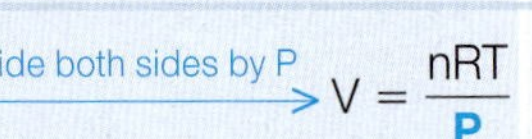

$$PV = nRT \xrightarrow{\text{divide both sides by P}} V = \frac{nRT}{P}$$

Identify the algebraic relationship needed to solve the problem, and solve it for the wanted variable.

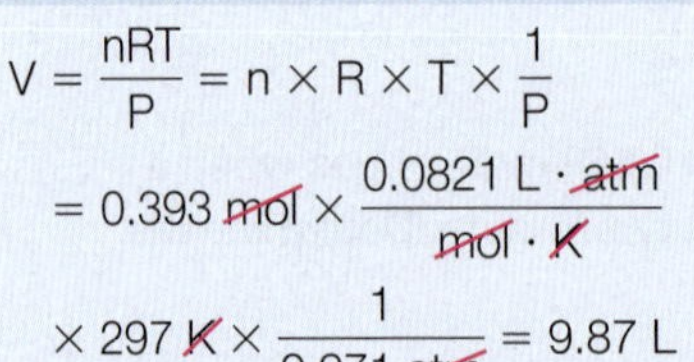

$$V = \frac{nRT}{P} = n \times R \times T \times \frac{1}{P}$$
$$= 0.393\ \cancel{\text{mol}} \times \frac{0.0821\ \text{L} \cdot \cancel{\text{atm}}}{\cancel{\text{mol}} \cdot \cancel{\text{K}}} \times 297\ \cancel{\text{K}} \times \frac{1}{0.971\ \cancel{\text{atm}}} = 9.87\ \text{L}$$

Construct the solution setup and calculate the value of the answer. Be sure that the units cancel appropriately.

Remember to always include units when solving problems. You will be performing many algebraic manipulations when solving problems involving the ideal gas equation. Your calculation will have many units. One term alone, R, includes four units. There is opportunity for routine errors in algebra. However, if you include units in your setups, you will catch those errors before you even pick up your calculator.

$$0.393 \times 0.0821 \times 297 \times \frac{1}{0.971}$$
$$\approx (4 \times 10^{-1}) \times (8 \times 10^{-2}) \times (3 \times 10^{2}) \times 1 = 96 \times 10^{-1} = 9.6, \text{OK.}$$

Check to be sure that the value of your answer is in the ballpark.

You improved your skill at solving ideal gas equation problems.	What did you learn by solving this Active Example?

Practice Exercise 14.3

What is the pressure of 0.22 mole of an ideal gas in a 1.0-gallon container at 19°C? Answer in mm Hg.

A useful variation of the ideal gas equation replaces n, the amount in moles, by the mass of a sample, m, divided by molar mass, MM. ▶ Thus:

$$n = \frac{\text{mass}}{\text{molar mass}} = \frac{m}{MM} = m \times \frac{1}{MM} = g \times \frac{1}{\frac{g}{mol}} = \cancel{g} \times \frac{mol}{\cancel{g}} = mol$$

$$PV = nRT = \frac{m}{MM}RT$$

Some instructors prefer that the conversion between mass and moles be done independently of the ideal gas equation rather than substituting m/MM. As usual, our recommendation is that you follow the directions of your instructor.

One useful application of PV = (m/MM)RT is in determining the molar mass of an unknown substance. If it is a liquid or a solid at room conditions, it may be heated beyond its boiling point to change it to a gas so that the gas laws will apply. The next Active Example illustrates the procedure.

Active Example 14.4 Determination of Molar Mass from the Ideal Gas Equation

A researcher vaporizes 1.67 g of a liquid at a temperature of 125°C. Its volume is 0.421 L at 749 torr. Calculate its molar mass.

Think Before You Write The problem statement gives you the quantities m, T, V, and P. MM is wanted. Thus, you need the PV = (m/MM)RT version of the ideal gas equation.

Answers *Cover the left column with your cut-out shield. Reveal each answer only after you have written your own answer in the right column.*

Given: m = 1.67 g; T = (125 + 273) = 398 K; V = 0.421 L; P = 749 torr
Wanted: MM (g/mol)

Analyze the problem statement by listing the given quantities and the symbol of the wanted quantity. Also list the units of the wanted quantity.

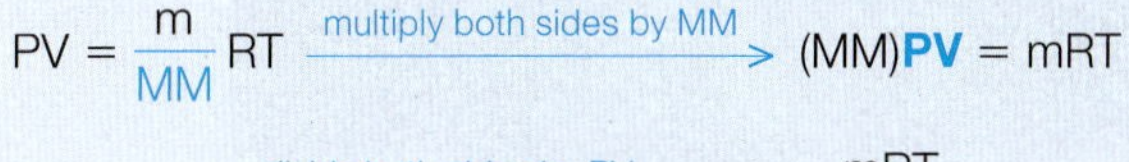

$$PV = \frac{m}{MM}RT \xrightarrow{\text{multiply both sides by MM}} (MM)PV = mRT$$

$$\xrightarrow{\text{divide both sides by PV}} MM = \frac{mRT}{PV}$$

Identify the algebraic relationship needed to solve the problem, and solve it for the wanted variable.

$$MM = \frac{mRT}{PV} = m \times R \times T \times \frac{1}{P} \times \frac{1}{V} = 1.67\ g \times \frac{62.4\ \cancel{L} \cdot \cancel{torr}}{mol \cdot \cancel{K}} \times 398\ \cancel{K} \times \frac{1}{749\ \cancel{torr}} \times \frac{1}{0.421\ \cancel{L}} = 132\ g/mol$$

We used the 62.4 L · torr/mol · K version of the gas constant because the given pressure was in torr. You may have used the 0.0821 L · atm/mol · K version and added a 760 torr/atm conversion factor to the setup. That approach is equally correct.

Construct the solution setup and calculate the value of the answer. Cancel units, and be sure that the remaining uncanceled units match what you expect.

$1.67 \times 62.4 \times 398 \times \frac{1}{749} \times \frac{1}{0.421} \approx 1.5 \times (7.5 \times 10^1)$ $\times (4 \times 10^2) \times \frac{1}{7.5 \times 10^2} \times \frac{1}{5 \times 10^{-1}} = \frac{1.5 \times 4 \times 10^3}{5 \times 10^2 \times 10^{-1}}$ $= 1.2 \times 10^2$, OK.	**Check** to be sure that the value of your answer is approximately correct.
You improved your skill at solving ideal gas equation problems.	What did you learn by solving this Active Example?

Practice Exercise 14.4

At a time at which the temperature is 22°C and the pressure is 0.988 atm, the air in a 0.50-liter container is found to have a mass of 0.59 g. Find the effective molar mass of air—the mass per mole of mixed molecules in the container.

14.6 Gas Density

Goal 4 Calculate the density of a known gas at any specified temperature and pressure.
5 Given the density of a pure gas at specified temperature and pressure, or information from which the density may be found, calculate the molar mass of that gas.

In this section, we consider the density of a gas. The most common unit for expressing the density of a liquid or solid is grams per cubic centimeter, g/cm^3. The number of grams per cubic centimeter of a gas is so small, however, that gas densities are usually given in grams per liter, g/L (**Figure 14.13**). The quantities represented by these units, mass (m) and volume (V), both appear in the equation PV = (m/MM)RT. The equation can, therefore, be solved for the density of a gas, m/V, in terms of pressure, temperature, molar mass, and the ideal gas constant:

$$PV = \frac{m}{MM}RT \xrightarrow{\text{multiply both sides by MM}} (MM)PV = mRT$$

$$\xrightarrow{\text{divide both sides by V}} (MM)P = \frac{m}{V}RT$$

$$\xrightarrow{\text{divide both sides by RT}} \frac{(MM)P}{RT} = \frac{m}{V} = D$$

Figure 14.13 The density of the carbon dioxide gas being expelled from this fire extinguisher is greater than the density of air. Therefore, the carbon dioxide flows downward, depriving the fire of the oxygen that it needs to continue burning. Gaseous carbon dioxide is invisible; the white "smoke" is water vapor condensing from the air.

Charles D. Winters

Thinking About Your Thinking

Proportional Reasoning

The density of a gas is related to four quantities:

$$D = \frac{(MM)P}{RT}$$

If density is measured at constant temperature and pressure, three of those quantities are constant, T, P, and R:

$$D = MM \times \frac{P}{RT} = MM \times \text{constant}$$

Therefore, at constant T and P, density is directly proportional to the molar mass of the gas. The greater the density is, the greater the molar mass of the gas must be. In the laboratory, measurement of gas density (actually, measurement of mass and volume) is a method of determining molar mass.

Active Example 14.5 Determination of Gas Density from the Ideal Gas Equation

Calculate the density of nitrogen at 0.632 atm and 44°C.

Think Before You Write At first glance, it may seem as if there is not enough information to perform this calculation. The PV = (m/MM)RT form of the ideal gas equation has five variables, and only two quantities are given. But you are being asked to calculate density, which is the ratio of mass to volume, so those two variables essentially become one. The problem statement also assumes that you know how to determine the molar mass of a substance from its name, giving you another value.

Answers *Cover the left column with your cut-out shield. Reveal each answer only after you have written your own answer in the right column.*

Given: P = 0.632 atm; T = (44 + 273) = 317 K; MM of N_2 = (2 × 14.01 g/mol) = 28.02 g/mol
Wanted: Density (assume g/L)

Analyze the problem statement by listing the given quantities, including the molar mass of nitrogen, the name of the wanted quantity, and the unit in which it will most simply be expressed. Change the given Celsius temperature to kelvins.

$$PV = nRT \xrightarrow{m/MM = n} PV = \frac{m}{MM}RT \xrightarrow{\text{multiply both sides by MM}}$$

$$(MM)PV = mRT \xrightarrow{\text{divide both sides by V}} (MM)P = \frac{m}{V}RT$$

$$\xrightarrow{\text{divide both sides by RT}} \frac{(MM)P}{RT} = \frac{m}{V}$$

Identify the algebraic relationship needed to solve the problem, substitute m/MM for n, and then rearrange the equation to isolate m/V on one side of the equals sign.

$$D = \frac{m}{V} = \frac{(MM)P}{RT} = MM \times P \times \frac{1}{R} \times \frac{1}{T}$$

$$= \frac{28.02\ \text{g}}{\cancel{\text{mol}}} \times 0.632\ \cancel{\text{atm}} \times \frac{\cancel{\text{mol}} \cdot \cancel{\text{K}}}{0.0821\ \text{L} \cdot \cancel{\text{atm}}} \times \frac{1}{317\ \cancel{\text{K}}}$$

$$= 0.680\ \text{g/L}$$

Construct the solution setup and calculate the value of the answer. Cancel units, and be sure that the remaining uncanceled units match what you expect.

$$28.02 \times 0.632 \times \frac{1}{0.0821} \times \frac{1}{317} \approx (3 \times 10^1) \times (6 \times 10^{-1})$$

$$\times \frac{1}{8 \times 10^{-2}} \times \frac{1}{3 \times 10^2} = \frac{3 \times 6 \times 10^1 \times 10^{-1}}{8 \times 3 \times 10^{-2} \times 10^2}$$

= 0.75, OK.

Check to be sure that the value of your answer is approximately correct.

You improved your skill at solving ideal gas equation problems.	What did you learn by solving this Active Example?

Practice Exercise 14.5

Determine the density of oxygen at 25°C and 741 torr.

Active Example 14.6 Determination of Molar Mass from Density and the Ideal Gas Equation

At 125°C and 749 torr, the density of an unknown gas is 3.97 g/L. Find the molar mass of that gas.

Think Before You Write Before you start to plan the solution to this problem, let's think about the $PV = \frac{m}{MM}RT$ form of the ideal gas equation and how we manipulated it to solve for density. We first multiplied both sides by MM to get rid of the fraction: $(MM)PV = mRT$. (In general, you should get rid of fractions as a first step in manipulating equations until you become very proficient at algebra. It helps in the prevention of errors.) We then divided both sides by V to place the m/V ratio on one side of the equals sign: $(MM)P = \frac{m}{V}RT$. We finished by dividing both sides by RT to isolate m/V, the definition of density, on one side: $\frac{(MM)P}{RT} = \frac{m}{V}$. The $\frac{m}{V}$ term is a *ratio*. We don't need to know *actual values* of both mass and volume of a gas sample; we just need to know their ratio. If we know the density of a sample, we know the m/V ratio (**Figure 14.14**). Thus, we can use the $PV = \frac{m}{MM}RT$ form of the ideal gas equation to find molar mass when density, temperature, and pressure are given.

Answers *Cover the left column with your cut-out shield. Reveal each answer only after you have written your own answer in the right column.*

Given: T = (125 + 273) = 398 K; P = 749 torr; m/V = 3.97 g/L; R = 62.4 L · torr/mol · K
Wanted: MM (g/mol)

Analyze the problem statement by listing the given quantities, the symbol of the wanted quantity, and the unit in which it will most simply be expressed. Change the given Celsius temperature to kelvins.

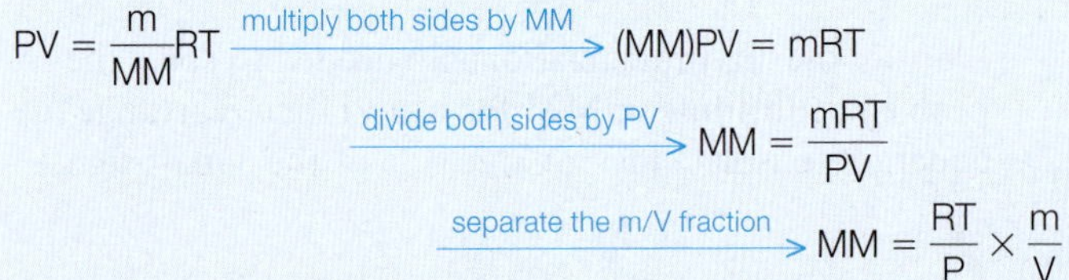

$$PV = \frac{m}{MM}RT \xrightarrow{\text{multiply both sides by MM}} (MM)PV = mRT \xrightarrow{\text{divide both sides by PV}} MM = \frac{mRT}{PV} \xrightarrow{\text{separate the m/V fraction}} MM = \frac{RT}{P} \times \frac{m}{V}$$

Identify the algebraic relationship needed to solve the problem, substitute m/MM for n, and then rearrange the equation to isolate molar mass on one side of the equals sign. Then take the equation one step further and separate out the m/V (density) fraction.

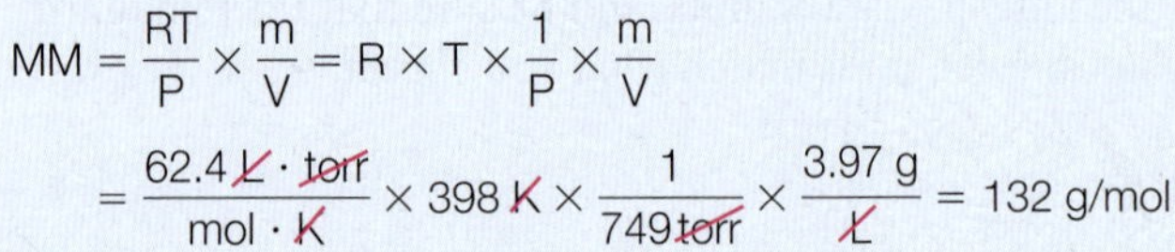

$$MM = \frac{RT}{P} \times \frac{m}{V} = R \times T \times \frac{1}{P} \times \frac{m}{V}$$
$$= \frac{62.4\,\text{L} \cdot \text{torr}}{\text{mol} \cdot \text{K}} \times 398\,\text{K} \times \frac{1}{749\,\text{torr}} \times \frac{3.97\,\text{g}}{\text{L}} = 132\ \text{g/mol}$$

Construct the solution setup and calculate the value of the answer. Cancel units, and be sure that the remaining uncanceled units match what you expect.

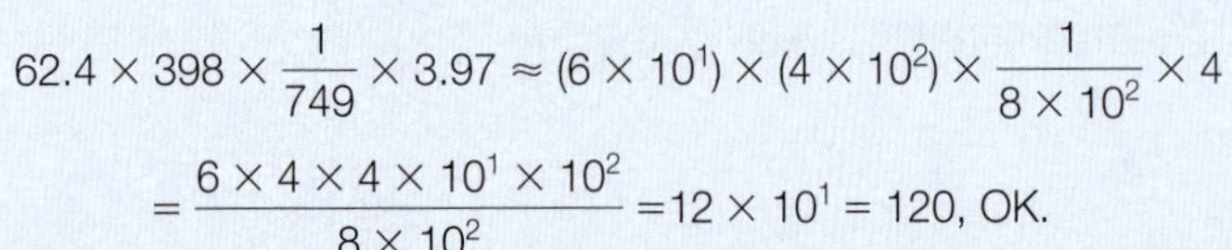

$$62.4 \times 398 \times \frac{1}{749} \times 3.97 \approx (6 \times 10^1) \times (4 \times 10^2) \times \frac{1}{8 \times 10^2} \times 4$$
$$= \frac{6 \times 4 \times 4 \times 10^1 \times 10^2}{8 \times 10^2} = 12 \times 10^1 = 120, \text{OK.}$$

Check to be sure that the value of your answer is reasonable.

Figure 14.14 Gas density and floating balloons. (a) A helium-filled balloon floats, but air-filled balloons lie on the surface of the table. The density of helium is lower than that of air at the same temperature and pressure. (b) Hot air balloons float because the higher temperature of the air enclosed in the balloon makes it less dense than the surrounding air.

You improved your skill at solving ideal gas equation problems.	What did you learn by solving this Active Example?

Practice Exercise 14.6

The density of a gas is 2.2 g/L at 29°C and 0.966 bar. What is its molar mass?

Active Example 14.6 may have looked familiar. It's the same problem as Active Example 14.4. The only difference is that in Active Example 14.6, *density* is given, and in Active Example 14.4 the numbers that make up the density are given: 1.67 g/0.421 L = 3.97 g/L. Notice Goal 5: "Calculate the density…*or information from which the density may be found*… ."

14.7 Molar Volume

Goal 6 Calculate the molar volume of any gas at any given temperature and pressure.

7 Given the molar volume of a gas at any specified temperature and pressure, or information from which the molar volume may be determined, and either the amount in moles or the volume of a sample of that gas, calculate the other quantity.

The **molar volume (MV)** of a gas is the volume occupied by one mole of gas molecules:

$$MV \equiv \frac{V}{n}$$

MV ≡ V/n is the defining equation for molar volume. Defining equations and their corresponding conversion factors are summarized in Section 3.11.

The ideal gas equation, PV = nRT, can be solved for the ratio of volume to amount in moles, V/n:

$$PV = nRT \xrightarrow{\text{divide both sides by P}} V = \frac{nRT}{P} \xrightarrow{\text{divide both sides by n}} \frac{V}{n} = \frac{RT}{P} = MV$$

There is no widespread agreement on *standard* temperature and pressure. IUPAC defines STP as 0°C and 1 bar, which leads to a molar volume of 22.7 L/mol for a gas at standard conditions. However, before 1982, IUPAC used 0°C and 1 atmosphere, which led to 22.4 L/mol as molar volume at standard conditions. As always, follow your instructor's guidelines as to the best choice for standard conditions and the resulting molar volume of a gas. Your instructor will likely also let you know if you need to memorize a STP value.

The relationship MV = RT/P shows that molar volume depends on temperature and pressure; T and P are both variables in the equation. All gases have the same molar volume at any given temperature and pressure. The gas molecules may all be the same—a pure substance—or they may be a mixture of two or more gases. The value of molar volume can be calculated for any temperature and pressure by substituting their values into MV = RT/P. For example, at the common reference conditions of standard temperature (0°C, or 273 K) and standard pressure (1 bar),

$$MV \equiv \frac{V}{n} = \frac{RT}{P} = \frac{0.0821\ \text{L} \cdot \cancel{\text{atm}}}{\text{mol} \cdot \cancel{\text{K}}} \times 273\ \cancel{\text{K}} \times \frac{1}{1\ \cancel{\text{bar}}} \times \frac{1.013\ \cancel{\text{bar}}}{\cancel{\text{atm}}} = 22.7\ \text{L/mol}$$

This value, **22.7 L/mol**, is used in problems in which gas volumes are expressed at standard temperature and pressure, STP. A 22.7-L box is shown in **Figure 14.15**.

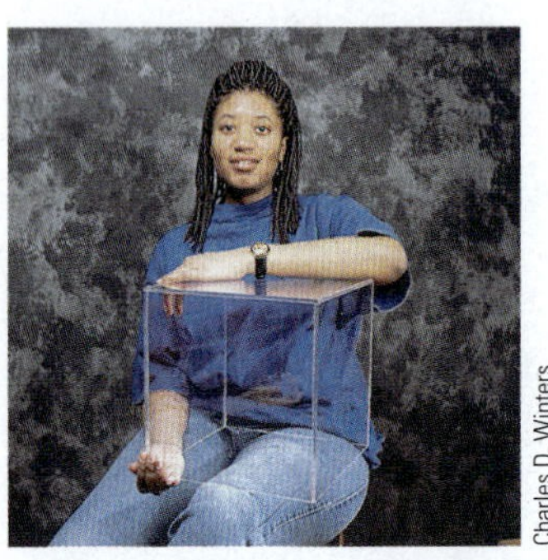

Charles D. Winters

Figure 14.15 Molar volume at standard temperature and pressure (STP). The volume of the box in front of the student is 22.7 liters, the volume occupied by 1 mole of gas at 1 bar and 0°C.

Learn It Now! *The STP molar volume of an ideal gas is 22.7 L/mol. This conversion factor provides you with a one-step conversion factor linking the macroscopic volume of a gas at 0°C and 1 bar and the particulate-level amount of particles, grouped in moles.*

Active Example 14.7 Determination of Molar Volume from the Ideal Gas Equation

Calculate the molar volume of SO_2 at 8°C and 0.568 atm.

Think Before You Write You are asked for molar volume, and you are given temperature and pressure. Therefore, this is a plug-in to MV ≡ V/n = RT/P.

Answers *Cover the left column with your cut-out shield. Reveal each answer only after you have written your own answer in the right column.*

Given: T = (8 + 273) = 281 K; P = 0.568 atm; R = 0.0821 L · atm/mol · K

Wanted: MV (assume L/mol)

Analyze the problem statement by listing the given quantities, the symbol of the wanted quantity, and the unit in which it will most simply be expressed. Change the given Celsius temperature to kelvins.

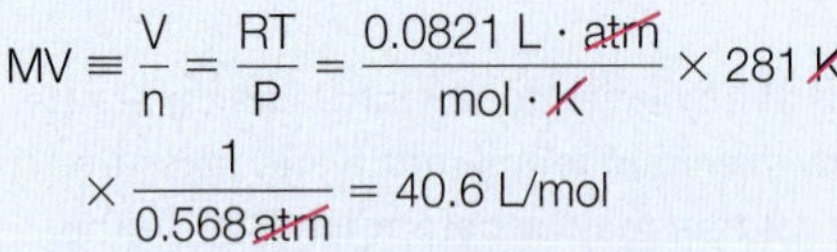

$$MV \equiv \frac{V}{n} = \frac{RT}{P} = \frac{0.0821\ \text{L} \cdot \cancel{\text{atm}}}{\text{mol} \cdot \cancel{\text{K}}} \times 281\ \cancel{\text{K}} \times \frac{1}{0.568\ \cancel{\text{atm}}} = 40.6\ \text{L/mol}$$

Notice that the identity of the gas, SO_2, has no bearing on the problem. At a given temperature and pressure, all gases have the same molar volume.

We have already **identified** the algebraic relationship needed to solve the problem, so go ahead and **construct** the solution setup and calculate the value of the answer. Cancel units, and be sure that the remaining uncanceled units match what you expect.

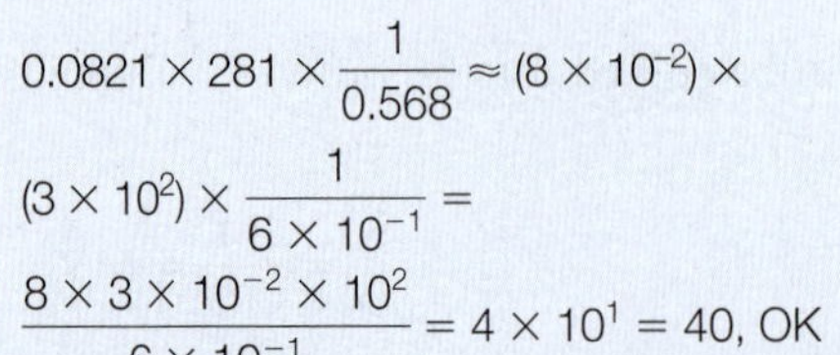

$$0.0821 \times 281 \times \frac{1}{0.568} \approx (8 \times 10^{-2}) \times (3 \times 10^{2}) \times \frac{1}{6 \times 10^{-1}} = \frac{8 \times 3 \times 10^{-2} \times 10^{2}}{6 \times 10^{-1}} = 4 \times 10^{1} = 40,\ \text{OK}$$

Check to be sure that the value of your answer is reasonable.

You improved your skill at solving molar volume problems.	What did you learn by solving this Active Example?

Practice Exercise 14.7

A carbon dioxide sample is stored at –15°C and 9.0 bar What is its molar volume at these conditions?

Note the similarity between molar volume, L/mol, and molar mass, g/mol. You have used molar mass to convert between mass of a substance and amount in moles. In the same way, you can use molar volume to convert between volume of a gas and amount in moles. ▶ There is one *very* important difference, however: The molar mass of a substance is always the same, but the molar volume of a gas is *not.* Molar volume is pressure- and temperature-dependent. Be sure you use the molar volume at the given temperature and pressure when using molar volume for this conversion.

Here again, you come upon the macroscopic and particulate character of matter. Mass and volume are macroscopic; they can be seen and measured. Molar volume and molar mass provide ways to measure macroscopic quantities of substances that correspond to a specified number of particles.

Active Example 14.8 Using Molar Volume to Determine Gas Volume

A reaction requires 4.21 moles of ethane, C_2H_6. What volume of the pure gas should be used (a) at standard temperature and pressure and (b) at 8°C and 0.568 atm (the temperature and pressure from Active Example 14.7), for which you determined that the molar volume is 40.6 L/mol?

Think Before You Write You are given amount in moles in this problem statement, and volume is wanted. Two different molar volumes are also given. You know that molar volume is a conversion factor that lets you convert between volume in liters and amount in moles.

Answers *Cover the left column with your cut-out shield. Reveal each answer only after you have written your own answer in the right column.*

Given: 4.21 mol **Wanted:** volume (assume L) 22.7 L/mol	**Analyze** the problem statement and **identify** the equivalency for part (a).
$4.21\ \cancel{mol} \times \frac{22.7\ L}{\cancel{mol}} = 95.6\ L$ $4.21 \times 22.7 \approx 4 \times (2.5 \times 10^1) = 10 \times 10^1 = 100$, OK.	**Construct** the solution setup, determine the answer, and **check** its value.
Given: 4.21 mol **Wanted:** volume (assume L) 40.6 L/mol $4.21\ \cancel{mol} \times \frac{40.6\ L}{\cancel{mol}} = 171\ L$ $4.21 \times 40.6 \approx 4 \times (4 \times 10^1) = 16 \times 10^1 = 160$, OK.	Complete part (b) in its entirety.
You improved your skill at solving molar volume problems.	What did you learn by solving this Active Example?

Practice Exercise 14.8

What is the STP volume of 5.0 g of nitrogen?

14.8 Gas Stoichiometry at Standard Temperature and Pressure

Goal 8 Given a chemical equation, or a reaction for which the equation can be written, and the mass or amount in moles of one species in the reaction, or the STP volume of a gaseous species, find the mass or amount in moles of another species, or the STP volume of another gaseous species.

In Section 10.3, you learned how to calculate the mass of one species in a chemical reaction with a known equation from a given mass of another species in three steps that we called a *mass-to-mass stoichiometry path:* (1) convert the mass of the given species to moles; (2) from the chemical equation, convert the moles of given species to moles of wanted species; (3) convert the moles of wanted species to grams. In this section, you will learn how to solve the same problem if the given and/or wanted species is the volume of a gas at STP. In Section 14.9 or 14.10, you will learn how to solve the problem if the given and/or wanted species is the volume of the gas at nonstandard conditions.

In Chapter 10, the quantities in stoichiometry problems are expressed in mass units, usually grams. The quantities can also be measured in volume units when the reactants or products are gases. But the volume of a gas depends on temperature and pressure. At 0°C and 1 bar pressure (STP), the molar volume of all gases is 22.7 L/mol.

The ability to convert between gas volume at STP and amount in moles means we can expand our stoichiometry pattern to include gas volume as a macroscopic quantity, as illustrated in **Figure 14.16**.

In this section, the "given temperature and pressure" are STP. The calculation setup for mass stoichiometry is modified here for gases at STP:

$$\cancel{\text{g Given}} \times \frac{\cancel{\text{mol Given}}}{\cancel{\text{g Given}}} \times \frac{\cancel{\text{mol Wanted}}}{\cancel{\text{mol Given}}} \times \frac{22.7 \text{ L Wanted}}{\cancel{\text{mol Wanted}}} = \text{L Wanted}$$

$$\cancel{\text{L Given}} \times \frac{\cancel{\text{mol Given}}}{22.7 \ \cancel{\text{L Given}}} \times \frac{\cancel{\text{mol Wanted}}}{\cancel{\text{mol Given}}} \times \frac{\text{g Wanted}}{\cancel{\text{mol Wanted}}} = \text{g Wanted}$$

To solve a stoichiometry problem that involves a gas whose volume is measured at STP, simply use 22.7 L/mol where you would have used molar mass if the amount had been given in grams.

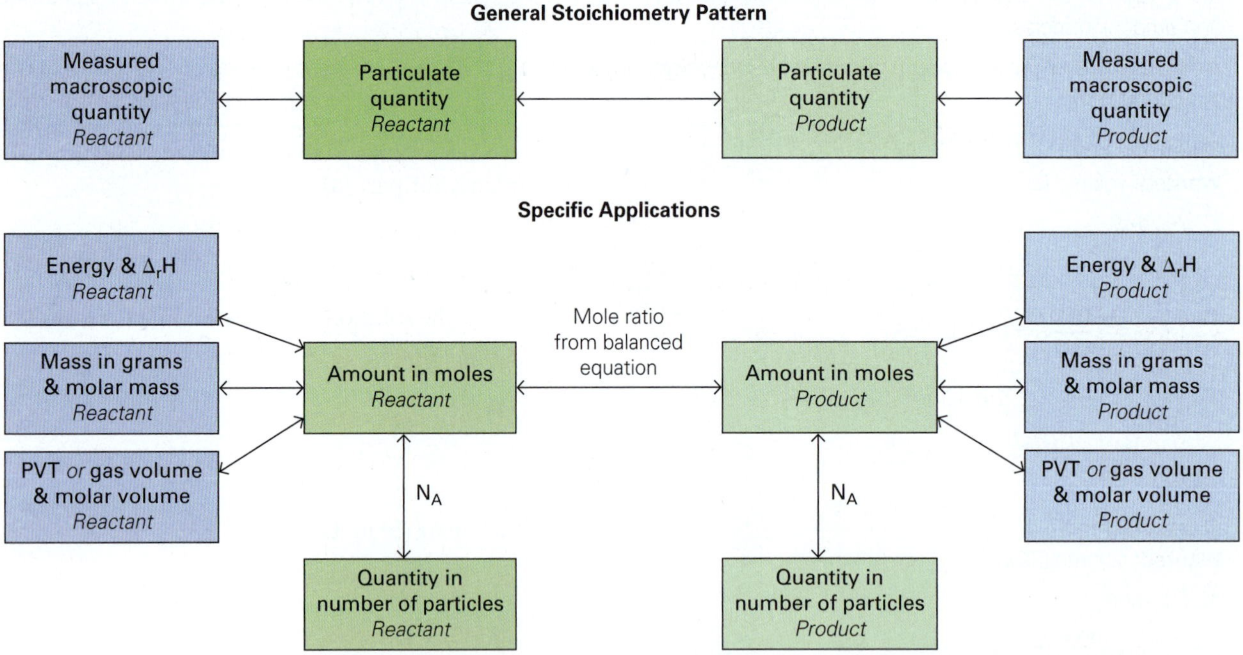

General Stoichiometry Pattern

Measured macroscopic quantity *Reactant* ↔ Particulate quantity *Reactant* ↔ Particulate quantity *Product* ↔ Measured macroscopic quantity *Product*

Specific Applications

Energy & Δ_rH *Reactant*; Mass in grams & molar mass *Reactant*; PVT *or* gas volume & molar volume *Reactant* ↔ Amount in moles *Reactant* ↔ (Mole ratio from balanced equation) ↔ Amount in moles *Product* ↔ Energy & Δ_rH *Product*; Mass in grams & molar mass *Product*; PVT *or* gas volume & molar volume *Product*

Amount in moles *Reactant* ↔ (N_A) ↔ Quantity in number of particles *Reactant*

Amount in moles *Product* ↔ (N_A) ↔ Quantity in number of particles *Product*

Figure 14.16 For two species in a chemical change (a reactant and a product), conversion between the following: mass in grams and molar mass; energy and Δ_rH; gas pressure, volume, and temperature (or gas volume and molar volume); and amount in moles and number of particles. In Chapter 10, we introduced the general stoichiometry pattern with mass, measured in grams. We now add gas volume at a known temperature and pressure as another measured macroscopic quantity. In general, a measured macroscopic quantity is changed to the number of particles in the sample, grouped in moles, and then moles of one species in the chemical change are converted to moles of another species using the mole ratio from the balanced equation. The resulting particulate quantity can be converted to a measured macroscopic quantity.

Active Example 14.9 Gas Stoichiometry at Standard Temperature and Pressure

What volume of hydrogen, measured at STP, can be released by 42.7 g of zinc as it reacts with hydrochloric acid? The equation is $Zn(s) + 2\ HCl(aq) \rightarrow H_2(g) + ZnCl_2(aq)$.

Think Before You Write You can recognize this as a stoichiometry problem because you are given the quantity of one species in a chemical change and you are asked to determine a quantity of another species. More specifically, you are being asked to determine the volume of a gas measured at STP, which should make you think that you need to use 22.7 L/mol.

Answers *Cover the left column with your cut-out shield. Reveal each answer only after you have written your own answer in the right column.*

Given: 42.7 g Zn **Wanted:** volume $H_2(g)$, assume L	**Analyze** the problem by writing the given quantity and the property and the assumed unit of the wanted quantity.
$g\ Zn \rightarrow mol\ Zn \rightarrow mol\ H_2 \rightarrow L\ H_2$ $1\ mol\ Zn = 65.38\ g\ Zn$ $1\ mol\ H_2 = 1\ mol\ Zn$ $22.7\ L\ H_2 = 1\ mol\ H_2$	**Identify** the equivalencies by first writing the unit path and then deriving the equivalency for each unit conversion. Hold off on changing the equivalencies to conversion factors until the next step.
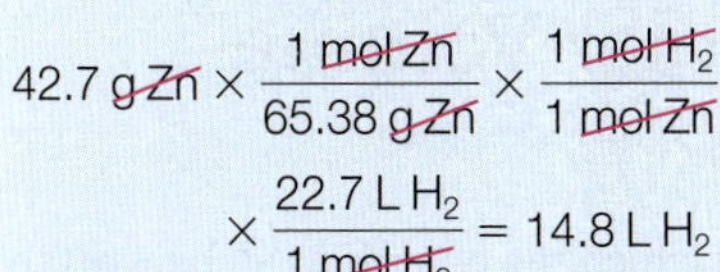 $42.7\ \cancel{g\ Zn} \times \frac{1\ \cancel{mol\ Zn}}{65.38\ \cancel{g\ Zn}} \times \frac{1\ \cancel{mol\ H_2}}{1\ \cancel{mol\ Zn}} \times \frac{22.7\ L\ H_2}{1\ \cancel{mol\ H_2}} = 14.8\ L\ H_2$	**Construct** the solution setup by changing the equivalences to conversion factors and inserting the conversion factors directly into the setup. Cancel units and calculate the value of the answer.
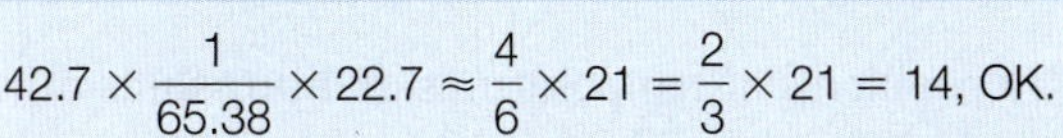 $42.7 \times \frac{1}{65.38} \times 22.7 \approx \frac{4}{6} \times 21 = \frac{2}{3} \times 21 = 14$, OK.	**Check** the value of the answer. Note that $42.7 \div 65.38 \approx 40 \div 60 = 2 \div 3$.
You improved your skill at solving gas stoichiometry problems.	What did you learn by solving this Active Example?

Practice Exercise 14.9

Hydrogen peroxide solution, $H_2O_2(aq)$, spontaneously decomposes into liquid water and gaseous oxygen. What STP volume of oxygen is produced from the decomposition of 5.00×10^2 g of a 3.0% H_2O_2 solution?

GAS STOICHIOMETRY OPTIONS The next two sections offer *alternative* ways to solve gas stoichiometry problems at given temperatures and pressures.

Option 1

In the **molar volume method** (Section 14.9), the ideal gas equation is solved for the molar volume at the given temperature and pressure. The calculated molar volume is then used to solve the problem by the three steps in the stoichiometry path, just as 22.7 L/mol is used to solve a problem at STP. All problems are solved in the same way.

Option 2

In the **ideal gas equation method** (Section 14.10), there are two procedures: (1) If the given quantity is a gas, the ideal gas equation is solved for n to change the given volume to amount in moles. The problem is completed by the second and third steps in the stoichiometry path. (2) If the wanted quantity is a gas, the amount in moles of wanted quantity is calculated by the first and second steps in the stoichiometry path. The ideal gas equation is then solved for V to convert the amount in moles of gas to volume in liters.

***IMPORTANT:* Your instructor will probably assign one section and tell you to skip the other.** If you must choose between the sections, we suggest that you compare the Active Examples and end-of-chapter answers for both methods—the same problems are used—and choose the one that looks best for you. Whether your instructor chooses or you do, we recommend that you learn one method and disregard the other.

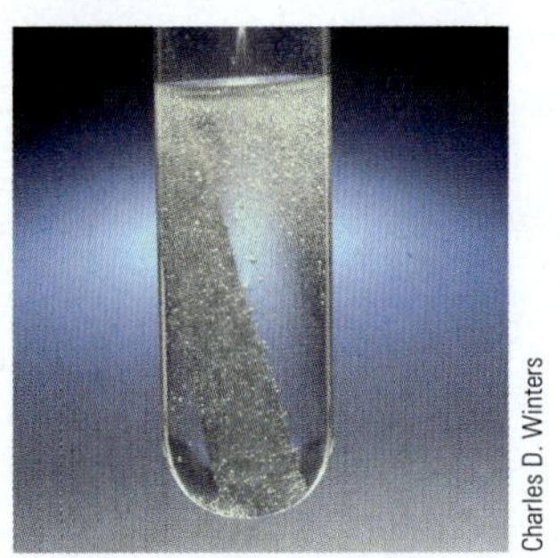

Charles D. Winters

Figure 14.17 A zinc strip reacts with hydrochloric acid solution, forming bubbles of hydrogen gas and a zinc chloride solution.

14.9 Gas Stoichiometry: Molar Volume Method (Option 1)

Goal 9 Given a chemical equation, or a reaction for which the equation can be written, and the mass or amount in moles of one species in the reaction, or the volume of any gaseous species at a given temperature and pressure, find the mass or amount in moles of any other species, or the volume of any other gaseous species at a given temperature and pressure.

We now consider gas stoichiometry problems at temperatures and pressures other than STP. At STP, you know that molar volume is 22.7 L/mol, and you have used it to solve those problems. In this section, you must first calculate the molar volume at the given temperature and pressure. Once you know the molar volume, you can solve all gas stoichiometry problems in exactly the same way.

Active Example 14.10 Gas Stoichiometry: Molar Volume Method I

What volume of hydrogen, measured at 739 torr and 21°C, can be released by 42.7 g of zinc as it reacts with hydrochloric acid (Figure 14.17)? The equation is $Zn(s) + 2\ HCl(aq) \rightarrow H_2(g) + ZnCl_2(aq)$.

Think Before You Write This is the same problem statement as in Active Example 14.9, except that the temperature and pressure are not the standard values. So the first step is to find the molar volume at the conditions of the problem, 739 torr and 21°C.

Answers *Cover the left column with your cut-out shield. Reveal each answer only after you have written your own answer in the right column.*

Given: P = 739 torr; T = (21 + 273) = 294 K **Wanted:** MV = V/n (L/mol) $PV = \mathbf{n}RT \xrightarrow{\text{divide both sides by nP}} \frac{\mathbf{V}}{\mathbf{n}} = \frac{RT}{P}$ $MV \equiv \frac{V}{n} = \frac{RT}{P} = R \times T \times \frac{1}{P} = \frac{62.4\ L \cdot \cancel{torr}}{mol \cdot \cancel{K}} \times 294\ \cancel{K}$ $\times \frac{1}{739\ \cancel{torr}} = 24.8\ L/mol$ $62.4 \times 294 \times \frac{1}{739} \approx (6 \times 10^1) \times (3 \times 10^2) \times \frac{1}{0.75 \times 10^3}$ $= \frac{6 \times 3 \times 10^1 \times 10^2}{\frac{3}{4} \times 10^3} = \frac{4 \times 6 \times 3 \times 10^1 \times 10^2}{3 \times 10^3}$ $= 24 \times 10^0 = 24$, OK.	Consider just the first step: finding the molar volume of hydrogen at 739 torr and 21°C. **Analyze** the problem, **identify** the algebraic equation, solve it for the wanted variable, and **construct** the complete solution, including the **check.** 
Given: 42.7 g Zn **Wanted:** volume H_2(g), assume L g Zn → mol Zn → mol H_2 → L H_2 1 mol Zn = 65.38 g Zn 1 mol H_2 = 1 mol Zn 24.8 L H_2 = 1 mol H_2 $42.7\ \cancel{g\ Zn} \times \frac{1\ \cancel{mol\ Zn}}{65.38\ \cancel{g\ Zn}} \times \frac{1\ \cancel{mol\ H_2}}{1\ \cancel{mol\ Zn}} \times \frac{24.8\ L\ H_2}{1\ \cancel{mol\ H_2}} = 16.2\ L\ H_2$ $42.7 \times \frac{1}{65.38} \times 24.8 \approx \frac{2}{3} \times 24 = 16$, OK.	Complete the problem exactly as you did in Active Example 14.9, but use your newly calculated molar volume. 
You improved your skill at solving gas stoichiometry problems.	What did you learn by solving this Active Example?

Practice Exercise 14.10

Hydrogen peroxide solution, H_2O_2(aq), spontaneously decomposes into liquid water and gaseous oxygen. What volume of oxygen, measured at 27°C and 733 torr, is produced from the decomposition of 5.00×10^2 g of a 3.0% by mass H_2O_2 solution?

how to... **Solve a Gas Stoichiometry Problem**

Step 1: Use the ideal gas equation to find the molar volume at the given temperature and pressure: V/n = RT/P.

Step 2: Use the molar volume to calculate the wanted quantity by all three steps of the stoichiometry path.

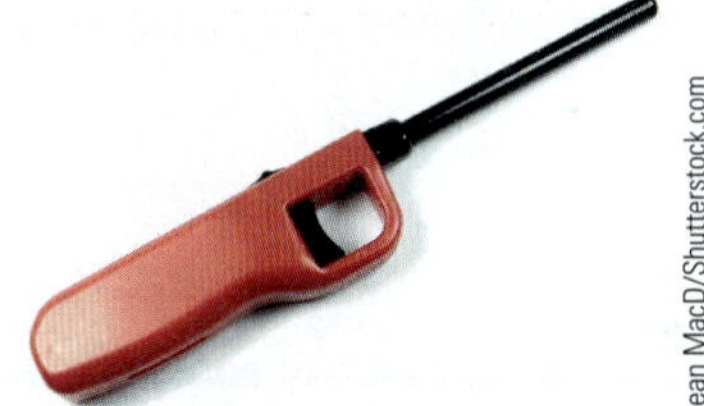

Sean MacD/Shutterstock.com

Figure 14.18 Most utility lighters use butane as a fuel.

Active Example 14.11 Gas Stoichiometry: Molar Volume Method II

What volume in liters of CO_2, measured at 744 torr and 131°C, will be produced by the complete burning of 16.2 g of gaseous butane, C_4H_{10} (Figure 14.18)? The equation is $2\ C_4H_{10}(g) + 13\ O_2(g) \rightarrow 8\ CO_2(g) + 10\ H_2O(\ell)$.

Think Before You Write The problem statement asks for a macroscopic measurable quantity of a substance that is produced by a chemical change. You are given the quantity of another substance in the change. Thus, this is a stoichiometry problem. One substance is a gas, so your macroscopic ↔ particulate conversion factor for the gas is molar volume.

Answers *Cover the left column with your cut-out shield. Reveal each answer only after you have written your own answer in the right column.*

Given: P = 744 torr; T = (131 + 273) = 404 K
Wanted: MV = V/n (L/mol)

$$PV = nRT \xrightarrow{\text{divide both sides by nP}} \frac{V}{n} = \frac{RT}{P}$$

$$MV \equiv \frac{V}{n} = \frac{RT}{P} = R \times T \times \frac{1}{P} = \frac{62.4\ L \cdot torr}{mol \cdot K} \times 404\ K \times \frac{1}{744\ torr} = 33.9\ L/mol$$

$$62.4 \times 404 \times \frac{1}{744} \approx (6 \times 10^1) \times (4 \times 10^2) \times \frac{1}{0.75 \times 10^3}$$

$$= \frac{6 \times 4 \times 10^1 \times 10^2}{\frac{3}{4} \times 10^3} = \frac{4 \times 6 \times 4 \times 10^1 \times 10^2}{3 \times 10^3}$$

$$= 32 \times 10^0 = 32,\ \text{OK.}$$

First, calculate the molar volume at the given temperature and pressure, and then use the molar volume as a conversion factor in the stoichiometry setup. To start, for just the molar volume calculation, **analyze** the problem, **identify** the algebraic equation, solve it for the wanted variable, and **construct** the complete solution, including the **check.**

Given: 16.2 g C_4H_{10}
Wanted: L $CO_2(g)$

g C_4H_{10} → mol C_4H_{10} → mol CO_2 → L CO_2

1 mol C_4H_{10} = 58.12 g C_4H_{10}

8 mol CO_2 = 2 mol C_4H_{10}

33.9 L CO_2 = 1 mol CO_2

$$16.2\ g\ C_4H_{10} \times \frac{1\ mol\ C_4H_{10}}{58.12\ g\ C_4H_{10}} \times \frac{8\ mol\ CO_2}{2\ mol\ C_4H_{10}} \times \frac{33.9\ L\ CO_2}{1\ mol\ CO_2} = 37.8\ L\ CO_2$$

$$16.2 \times \frac{1}{58.12} \times \frac{8}{2} \times 33.9 \approx \frac{15 \times 8}{60 \times 2} \times 34$$

$$= \frac{1}{4} \times 4 \times 34 = 34,\ \text{OK.}$$

Complete the solution by setting up and solving the stoichiometry calculation.

You improved your skill at solving gas stoichiometry problems.

What did you learn by solving this Active Example?

Practice Exercise 14.11

A 0.25-g piece of sodium is dropped into a large container filled with water. How many milliliters of hydrogen gas, measured at 765 torr and 18°C, will be produced?

14.10 Gas Stoichiometry: Ideal Gas Equation Method (Option 2)

Goal 9 Given a chemical equation, or a reaction for which the equation can be written, and the mass or number of moles of one species in the reaction, or the volume of any gaseous species at a given temperature and pressure, find the mass or number of moles of any other species, or the volume of any other gaseous species at a given temperature and pressure.

The stoichiometry path may be summarized as given quantity → mol given → mol wanted → wanted quantity. In a gas stoichiometry problem, the first or third step in the path is a conversion between amount in moles and volume in liters of gas at a given temperature and pressure. If you are given volume, you must convert to amount in moles; if you find amount in moles of wanted substance, you must convert to volume. These conversions are made with the ideal gas equation, PV = nRT. You have already made conversions like these. For example, in Active Example 14.3, you calculated the volume occupied by 0.393 mole of N_2 at 24°C and 0.971 atm. You used the ideal gas equation solved for V.

The ideal gas equation method for solving gas stoichiometry problems combines use of the ideal gas equation (Sections 14.4 and 14.5) and the stoichiometry path (Section 10.3). Let's see how in the following Active Example.

Figure 14.17 A zinc strip reacts with hydrochloric acid solution, forming bubbles of hydrogen gas and a zinc chloride solution.

Active Example 14.12 Gas Stoichiometry: Ideal Gas Equation Method I

What volume of hydrogen, measured at 739 torr and 21°C, can be released by 42.7 g of zinc as it reacts with hydrochloric acid (Figure 14.17)? The equation is $Zn(s) + 2\ HCl(aq) \rightarrow H_2(g) + ZnCl_2(aq)$.

Think Before You Write This is the same problem statement as in Active Example 14.9, except that the temperature and pressure are not the standard values. You are asked to determine the volume of a gas, and volume can be found with V = nRT/P. R is known, of course, and you are given P and T. Thus, n is needed.

Answers *Cover the left column with your cut-out shield. Reveal each answer only after you have written your own answer in the right column.*

Given: 42.7 g Zn
Wanted: mol H_2

g Zn → mol Zn → mol H_2

1 mol Zn = 65.38 g Zn

1 mol H_2 = 1 mol Zn

$$42.7\ \cancel{g\ Zn} \times \frac{1\ \cancel{mol\ Zn}}{65.38\ \cancel{g\ Zn}} \times \frac{1\ mol\ H_2}{1\ \cancel{mol\ Zn}} = 0.653\ mol\ H_2$$

$$42.7 \times \frac{1}{65.38} \approx \frac{40}{60} = \frac{2}{3} = 0.67\text{, OK.}$$

The first part of the solution is to change the given mass of zinc to amount in moles of zinc and then to amount in moles of hydrogen. **Analyze** the problem, **identify** the needed equivalencies, and **construct** the complete solution, including the **check.**

Given: n = 0.653 mol, P = 739 torr,
T = (21 + 273) = 294 K
Wanted: V (assume L)

$$PV = nRT \xrightarrow{\text{divide both sides by P}} \mathbf{V} = \frac{nRT}{P}$$

$$V = \frac{nRT}{P} = n \times R \times T \times \frac{1}{P} = 0.653\ \cancel{mol} \times \frac{62.4\ L \cdot \cancel{torr}}{\cancel{mol} \cdot \cancel{K}} \times 294\ \cancel{K} \times \frac{1}{739\ \cancel{torr}} = 16.2\ L$$

$$0.653 \times 62.4 \times 294 \times \frac{1}{739} \approx (7 \times 10^{-1}) \times (6 \times 10^{1}) \times (3 \times 10^{2}) \times \frac{1}{7 \times 10^{2}} = \frac{7 \times 6 \times 3 \times 10^{-1} \times 10^{1} \times 10^{2}}{7 \times 10^{2}} = 18\text{, OK.}$$

Now the problem has become, "What is the volume of 0.653 mol H_2, measured at 739 torr and 21°C?" This is just like Active Example 14.3. Write the given and wanted, solve the ideal gas equation for V, plug in the numbers, calculate the answer, and do the **check.**

You improved your skill at solving gas stoichiometry problems.	What did you learn by solving this Active Example? 

Practice Exercise 14.12

Hydrogen peroxide solution, $H_2O_2(aq)$, spontaneously decomposes into liquid water and gaseous oxygen. What volume of oxygen, measured at 27°C and 733 torr, is produced from the decomposition of 5.00×10^2 grams of a 3.0% by mass H_2O_2 solution?

In a gas stoichiometry problem, either the given quantity or the wanted quantity is a gas at specified temperature and pressure. The problem is usually solved in two steps, the order of which depends on whether the gas volume is the wanted quantity or the given quantity.

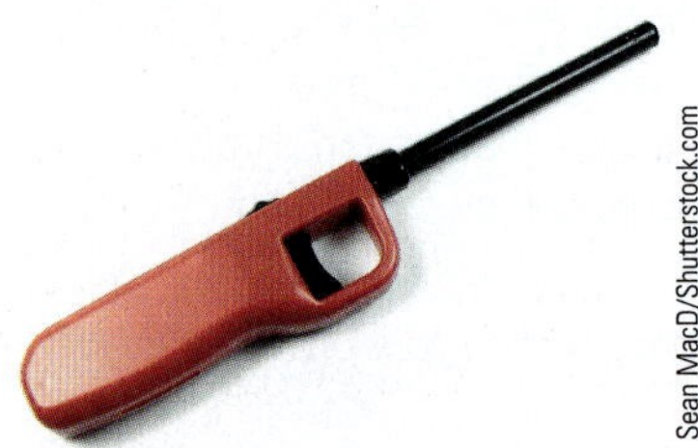

Sean MacD/Shutterstock.com

Figure 14.18 Most utility lighters use butane as a fuel.

how to... **Solve a Gas Stoichiometry Problem**

Volume Given	**Volume Wanted**
Step 1: Use the ideal gas equation to change given volume to amount in moles: n = PV/RT.	**Step 1:** Calculate amount in moles of the wanted substance using Steps 1 and 2 of the stoichiometry path.
Step 2: Use the result in Step 1 to calculate the wanted quantity using Steps 2 and 3 of the stoichiometry path.	**Step 2:** Use the ideal gas equation to change amount in moles calculated above to volume: V = nRT/P.

Active Example 14.13 Gas Stoichiometry: Ideal Gas Equation Method II

What volume in liters of CO_2, measured at 744 torr and 131°C, will be produced by the complete burning of 16.2 g of gaseous butane, C_4H_{10} (Figure 14.18)? The equation is $2\ C_4H_{10}(g) + 13\ O_2(g) \rightarrow 8\ CO_2(g) + 10\ H_2O(\ell)$.

Think Before You Write The problem statement asks for a macroscopic measurable quantity of a substance that is produced by a chemical change. You are given the quantity of another substance in the change. Thus, this is a stoichiometry problem. The given mass of C_4H_{10} can be converted to amount in moles of C_4H_{10}. The balanced equation gives the information needed to determine the amount in moles of CO_2. You then have P, n, R, and T, and you are asked to find V.

Answers *Cover the left column with your cut-out shield. Reveal each answer only after you have written your own answer in the right column.*

Given: 16.2 g C_4H_{10}
Wanted: mol CO_2

g $C_4H_{10} \rightarrow$ mol $C_4H_{10} \rightarrow$ mol CO_2

1 mol C_4H_{10} = 58.12 g C_4H_{10}

8 mol CO_2 = 2 mol C_4H_{10}

$$16.2\ \cancel{g\ C_4H_{10}} \times \frac{1\ \cancel{mol\ C_4H_{10}}}{58.12\ \cancel{g\ C_4H_{10}}} \times \frac{8\ \cancel{mol\ CO_2}}{2\ \cancel{mol\ C_4H_{10}}} = 1.11\ \cancel{mol\ CO_2}$$

$$16.2 \times \frac{1}{58.12} \times \frac{8}{2} \approx 16 \times \frac{1}{60} \times 4 = \frac{64}{60} \approx 1, \text{OK.}$$

For the amount in moles of CO_2 calculation, **analyze** the problem, **identify** the conversion factors, and **construct** the complete solution, including the **check.**

The stoichiometry path may be summarized as given quantity → mol given → mol wanted → wanted quantity. In a gas stoichiometry problem, the first or third step in the path is a conversion between amount in moles and volume in liters of gas at a given temperature and pressure. If you are given volume, you must convert to amount in moles; if you find amount in moles of wanted substance, you must convert to volume. These conversions are made with the ideal gas equation, PV = nRT. You have already made conversions like these. For example, in Active Example 14.3, you calculated the volume occupied by 0.393 mole of N_2 at 24°C and 0.971 atm. You used the ideal gas equation solved for V.

The ideal gas equation method for solving gas stoichiometry problems combines use of the ideal gas equation (Sections 14.4 and 14.5) and the stoichiometry path (Section 10.3). Let's see how in the following Active Example.

Charles D. Winters

Figure 14.17 A zinc strip reacts with hydrochloric acid solution, forming bubbles of hydrogen gas and a zinc chloride solution.

Active Example 14.12 Gas Stoichiometry: Ideal Gas Equation Method I

What volume of hydrogen, measured at 739 torr and 21°C, can be released by 42.7 g of zinc as it reacts with hydrochloric acid (Figure 14.17)? The equation is $Zn(s) + 2\,HCl(aq) \rightarrow H_2(g) + ZnCl_2(aq)$.

Think Before You Write This is the same problem statement as in Active Example 14.9, except that the temperature and pressure are not the standard values. You are asked to determine the volume of a gas, and volume can be found with V = nRT/P. R is known, of course, and you are given P and T. Thus, n is needed.

Answers *Cover the left column with your cut-out shield. Reveal each answer only after you have written your own answer in the right column.*

Given: 42.7 g Zn
Wanted: mol H_2

g Zn → mol Zn → mol H_2

1 mol Zn = 65.38 g Zn

1 mol H_2 = 1 mol Zn

$$42.7\ \cancel{\text{g Zn}} \times \frac{1\ \cancel{\text{mol Zn}}}{65.38\ \cancel{\text{g Zn}}} \times \frac{1\ \text{mol H}_2}{1\ \cancel{\text{mol Zn}}} = 0.653\ \text{mol H}_2$$

$$42.7 \times \frac{1}{65.38} \approx \frac{40}{60} = \frac{2}{3} = 0.67\text{, OK.}$$

The first part of the solution is to change the given mass of zinc to amount in moles of zinc and then to amount in moles of hydrogen. **Analyze** the problem, **identify** the needed equivalencies, and **construct** the complete solution, including the **check.**

Given: n = 0.653 mol, P = 739 torr,
T = (21 + 273) = 294 K
Wanted: V (assume L)

$$\text{P}\mathbf{V} = \text{nRT} \xrightarrow{\text{divide both sides by P}} \mathbf{V} = \frac{\text{nRT}}{\text{P}}$$

$$\text{V} = \frac{\text{nRT}}{\text{P}} = \text{n} \times \text{R} \times \text{T} \times \frac{1}{\text{P}} = 0.653\ \cancel{\text{mol}} \times \frac{62.4\ \text{L} \cdot \cancel{\text{torr}}}{\cancel{\text{mol}} \cdot \cancel{\text{K}}}$$

$$\times\ 294\ \cancel{\text{K}} \times \frac{1}{739\ \cancel{\text{torr}}} = 16.2\ \text{L}$$

$$0.653 \times 62.4 \times 294 \times \frac{1}{739} \approx (7 \times 10^{-1})$$

$$\times\ (6 \times 10^{1}) \times (3 \times 10^{2}) \times \frac{1}{7 \times 10^{2}}$$

$$= \frac{7 \times 6 \times 3 \times 10^{-1} \times 10^{1} \times 10^{2}}{7 \times 10^{2}} = 18\text{, OK.}$$

Now the problem has become, "What is the volume of 0.653 mol H_2, measured at 739 torr and 21°C?" This is just like Active Example 14.3. Write the given and wanted, solve the ideal gas equation for V, plug in the numbers, calculate the answer, and do the **check.**

You improved your skill at solving gas stoichiometry problems.

What did you learn by solving this Active Example?

Practice Exercise 14.12

Hydrogen peroxide solution, $H_2O_2(aq)$, spontaneously decomposes into liquid water and gaseous oxygen. What volume of oxygen, measured at 27°C and 733 torr, is produced from the decomposition of 5.00×10^2 grams of a 3.0% by mass H_2O_2 solution?

In a gas stoichiometry problem, either the given quantity or the wanted quantity is a gas at specified temperature and pressure. The problem is usually solved in two steps, the order of which depends on whether the gas volume is the wanted quantity or the given quantity.

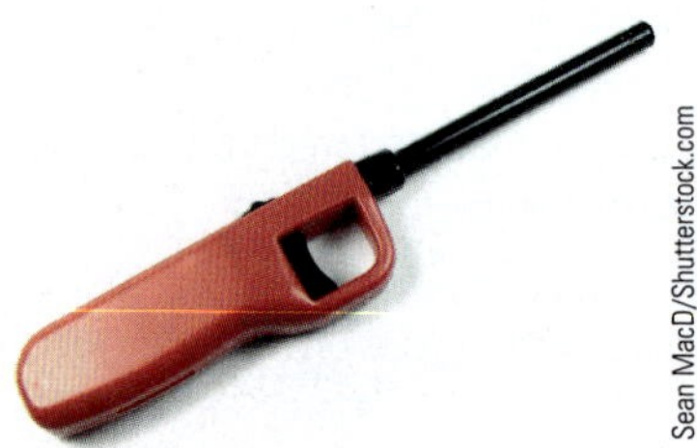

Sean MacD/Shutterstock.com

Figure 14.18 Most utility lighters use butane as a fuel.

how to... **Solve a Gas Stoichiometry Problem**

Volume Given	Volume Wanted
Step 1: Use the ideal gas equation to change given volume to amount in moles: n = PV/RT.	**Step 1:** Calculate amount in moles of the wanted substance using Steps 1 and 2 of the stoichiometry path.
Step 2: Use the result in Step 1 to calculate the wanted quantity using Steps 2 and 3 of the stoichiometry path.	**Step 2:** Use the ideal gas equation to change amount in moles calculated above to volume: V = nRT/P.

Active Example 14.13 Gas Stoichiometry: Ideal Gas Equation Method II

What volume in liters of CO_2, measured at 744 torr and 131°C, will be produced by the complete burning of 16.2 g of gaseous butane, C_4H_{10} (Figure 14.18)? The equation is $2\ C_4H_{10}(g) + 13\ O_2(g) \rightarrow 8\ CO_2(g) + 10\ H_2O(\ell)$.

Think Before You Write The problem statement asks for a macroscopic measurable quantity of a substance that is produced by a chemical change. You are given the quantity of another substance in the change. Thus, this is a stoichiometry problem. The given mass of C_4H_{10} can be converted to amount in moles of C_4H_{10}. The balanced equation gives the information needed to determine the amount in moles of CO_2. You then have P, n, R, and T, and you are asked to find V.

Answers *Cover the left column with your cut-out shield. Reveal each answer only after you have written your own answer in the right column.*

Given: 16.2 g C_4H_{10}
Wanted: mol CO_2

g $C_4H_{10} \rightarrow$ mol $C_4H_{10} \rightarrow$ mol CO_2

1 mol C_4H_{10} = 58.12 g C_4H_{10}

8 mol CO_2 = 2 mol C_4H_{10}

$$16.2\ \cancel{\text{g C}_4\text{H}_{10}} \times \frac{1\ \cancel{\text{mol C}_4\text{H}_{10}}}{58.12\ \cancel{\text{g C}_4\text{H}_{10}}} \times \frac{8\ \cancel{\text{mol CO}_2}}{2\ \cancel{\text{mol C}_4\text{H}_{10}}} = 1.11\ \cancel{\text{mol CO}_2}$$

$$16.2 \times \frac{1}{58.12} \times \frac{8}{2} \approx 16 \times \frac{1}{60} \times 4 = \frac{64}{60} \approx 1, \text{OK.}$$

For the amount in moles of CO_2 calculation, **analyze** the problem, **identify** the conversion factors, and **construct** the complete solution, including the **check**.

Given: n = 1.11 mol, P = 744 torr, T = (131 + 273) = 404 K
Wanted: V (L)

$$PV = nRT \xrightarrow{\text{divide both sides by P}} V = \frac{nRT}{P}$$

$$V = \frac{nRT}{P} = n \times R \times T \times \frac{1}{P} = 1.11\ \cancel{\text{mol}} \times \frac{62.4\ \text{L}\cdot\cancel{\text{torr}}}{\cancel{\text{mol}}\cdot\cancel{\text{K}}} \times 404\ \cancel{\text{K}} \times \frac{1}{744\ \cancel{\text{torr}}} = 37.6\ \text{L}$$

$$1.11 \times 62.4 \times 404 \times \frac{1}{744} \approx 1 \times (6 \times 10^1) \times (4 \times 10^2) \times \frac{1}{\frac{3}{4} \times 10^3} = \frac{6 \times 4 \times 10^1 \times 10^2}{\frac{3}{4} \times 10^3}$$

$$= \frac{4 \times 6 \times 4 \times 10^1 \times 10^2}{3 \times 10^3} = 32 \times 10^0 = 32\text{, OK.}$$

Step 2 is to use the ideal gas equation to change 1.11 mol CO_2 to volume. Complete the solution.

You improved your skill at solving gas stoichiometry problems.

What did you learn by solving this Active Example?

Practice Exercise 14.13

A 0.25-g piece of sodium is dropped into a large container filled with water. How many milliliters of hydrogen gas, measured at 765 torr and 18°C, will be produced?

14.11 Volume–Volume Gas Stoichiometry

Goal 10 Given a chemical equation, or a reaction for which the equation can be written, and the volume of any gaseous species at a given temperature and pressure, find the volume of any other gaseous species at a given temperature and pressure.

Avogadro's Law (Section 14.3) states that gas volume is directly proportional to the amount in moles of the gas at constant temperature and pressure, $V \propto n$. This means that the ratio of volumes of gases in a reaction is the same as the ratio of moles, *provided that the gas volumes are measured at the same temperature and pressure*. The ratio of moles comes from the coefficients in the chemical equation. It follows that the coefficients give us a ratio of gas volumes, too. This is illustrated by the "volume equation" and its corresponding "molar equation" in Figure 14.5 in Section 14.3. The volumes of the individual gases are in the same ratio as the numbers of moles in the equation.

Thinking About *Your Thinking*

Proportional Reasoning

When gas volumes are measured at the same temperature and pressure, the ratio of volumes of gases in a reaction is the same as the ratio of moles as given by the coefficients in a balanced chemical equation. The gas volumes are directly proportional to one another. Any conversion factor that relates mole ratios can also be written as a volume ratio for a gas-phase reaction when volumes are measured at the same T and P.

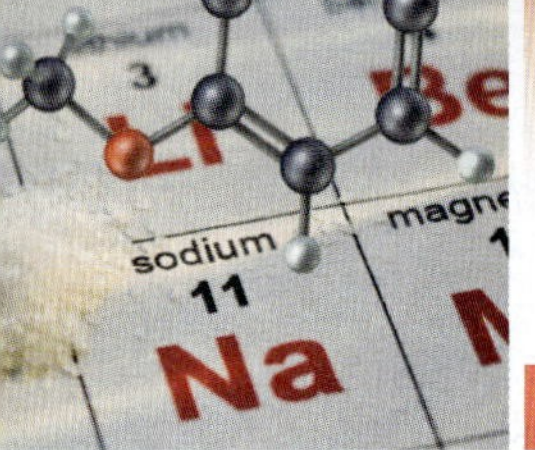

Everyday Chemistry 14.1

AUTOMOBILE AIR BAGS

Since 1998, all cars sold in the United States have been required to have driver and passenger air bags (**Figure 14.19**). Research by the U.S. Department of Transportation shows that air bags reduce the risk of death in a frontal collision by about 30%. An air bag plus wearing lap and shoulder belts reduces the likelihood of moderate injury in front-end crashes by about 60%.

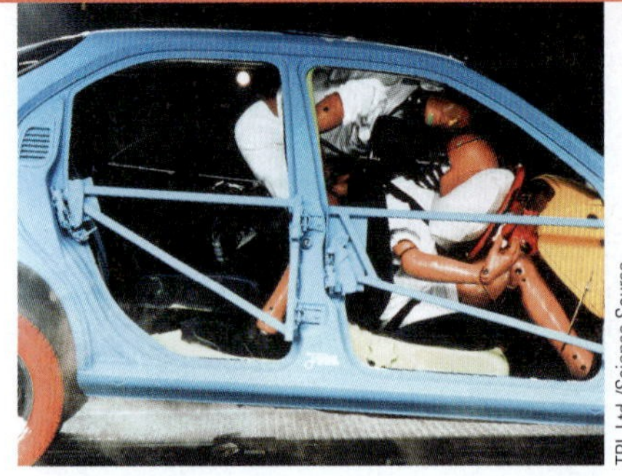

TRL Ltd./Science Source

Figure 14.19 The crash test dummy in the driver's seat is restrained by a seat belt and an inflated air bag. The dummy in the back seat, not wearing a seat belt, is being propelled through the cabin toward the front windshield.

An air bag system has three components (**Figure 14.20**). The bag itself is made of nylon fabric. Nylon was the first completely synthetic fiber chemists manufactured. The second component in the system is a sensor that sends a signal to inflate the bag when a front-end collision occurs with sufficient force. This is to prevent the bag from inflating because of a minor fender bender. The third component is the inflation system, which is an application of the principles of gas stoichiometry presented in this chapter.

The inflation system is often a solid that reacts to release nitrogen gas. Modern air bags usually use a mixture of a number of substances that typically includes ammonium nitrate, NH_4NO_3, and/or nitroguanidine, $(NH_2)_2CNNO_2$, as the solid from which nitrogen gas forms after the chemical change.

Before the 1990s, a solid mixture of sodium azide, NaN_3, potassium nitrate, KNO_3, and silicon dioxide, SiO_2, was in the core of the bag, surrounding an electronic igniter. When the sensor detected a collision, the igniter detonated the solids, forming gaseous nitrogen, inflating the bag. The primary source of nitrogen was the decomposition reaction of sodium azide: $2\ NaN_3 \rightarrow 2\ Na + 3\ N_2$.

The sodium that was generated was very reactive and could not be allowed to remain in the bag after detonation. Potassium nitrate was added to the solid mixture to react with the sodium: $10\ Na + 2\ KNO_3 \rightarrow K_2O + 5\ Na_2O + N_2$. Notice that this reaction produced even more nitrogen gas.

The final reaction involved the potassium oxide and sodium oxide produced in the second reaction. The silicon dioxide in the original solid mixture is a type of sand, and when this sand reacts with Group 1A/1 oxides, a type of glass is produced. This glass is nonflammable and unreactive, so it is a safe final product. After the air bag is deployed, it deflates through small holes that allow the nitrogen gas to leak out, leaving behind a deflated bag containing inert substances.

Research is continuing to find the most effective method to employ side air bags to protect passengers from nonfrontal collisions. The primary challenge is the short time lag between the collision and the need for the inflated air bag. In a front-end collision, the relatively long distance between the front bumper and the passengers allows sufficient time for the bag to fully inflate before the passengers strike the dashboard or windshield. With side collisions, the distance from the car body to the passengers is much smaller.

Quick Quiz

1. **What are the three components of an air bag system, and which component is an application of gas stoichiometry principles?**
2. **In what state of matter is the reactant in the inflation system in an air bag, and what is the typical product of the reaction and its state of matter after ignition? Why is each state of matter used in an airbag system?**

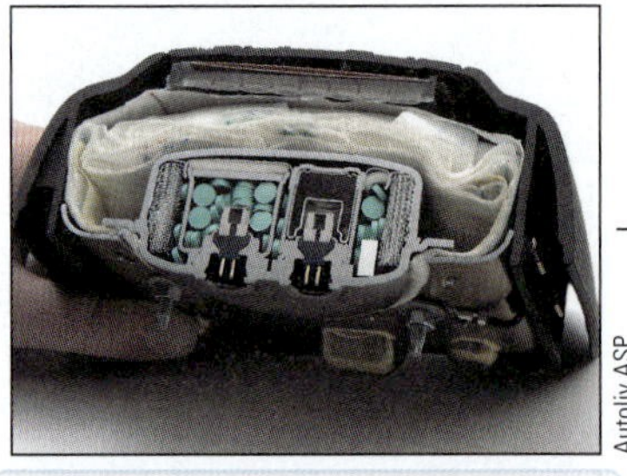

Autoliv ASP

When a car decelerates in a collision, an electrical contact is made in the sensor unit. The propellant (green solid) detonates, releasing nitrogen gas, and the folded nylon bag explodes out of the plastic housing.

1

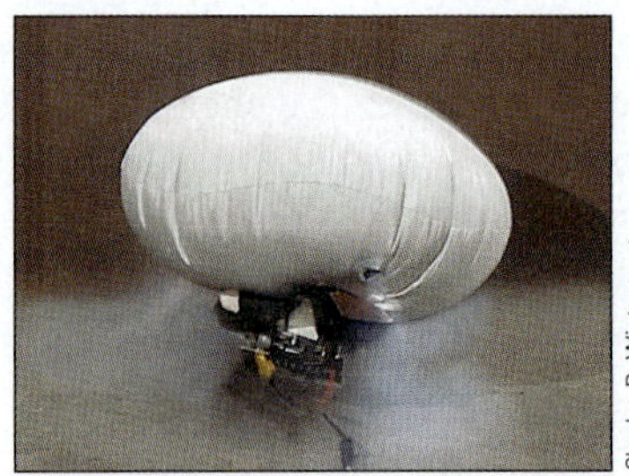

Charles D. Winters

Driver side air bags inflate with 35–70 L of N_2 gas, whereas passenger air bags hold about 60–160 L.

2

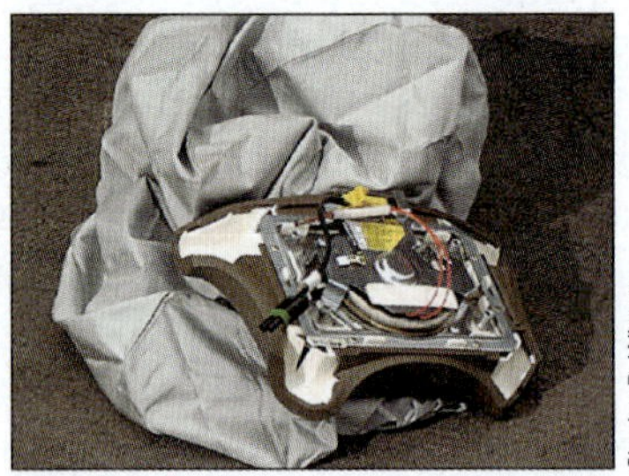

Charles D. Winters

The bag deflates within 0.2 s, the gas escaping through holes in the bottom of the bag.

3

Figure 14.20 Operation of the three components of an air bag system.

This volume ratio is useful for stoichiometry problems when both the given and wanted quantities are gases measured at the same temperature and pressure. For example, consider the reaction $3\ H_2(g) + N_2(g) \rightarrow 2\ NH_3(g)$. Let's calculate the volume of ammonia that will be produced by the reaction of 4 L of N_2 with excess hydrogen, with both gases measured at STP. The equation coefficients, interpreted for gas volumes, tell us that 2 L of ammonia are formed per 1 L of nitrogen used. The one-step unit path is $\text{L } N_2 \rightarrow \text{L } NH_3$:

$$4\ \cancel{\text{L } N_2} \times \frac{2\ \text{L } NH_3}{1\ \cancel{\text{L } N_2}} = 8\ \text{L } NH_3$$

The full stoichiometry setup for the problem confirms this result:

$$4\ \cancel{\text{L } N_2} \times \frac{1\ \cancel{\text{mol } N_2}}{\cancel{22.7}\ \cancel{\text{L } N_2}} \times \frac{2\ \cancel{\text{mol } NH_3}}{1\ \cancel{\text{mol } N_2}} \times \frac{\cancel{22.7}\ \text{L } NH_3}{1\ \cancel{\text{mol } NH_3}} = 8\ \text{L } NH_3$$

Notice that the 22.7s cancel. Molar volumes will always cancel if the given and wanted gases are measured at the same temperature and pressure.

Be sure to recognize the restriction on the volume-to-volume conversion with coefficients from the equation: *Both gas volumes must be measured at the same temperature and pressure.* They don't have to be at STP, but they must be the same to make their molar volumes identical.

Active Example 14.14 Volume–Volume Gas Stoichiometry I

A scientist completely burns 1.30 L of gaseous ethylene, C_2H_4. What volume of oxygen is required if both gas volumes are measured at STP? What if both volumes are measured at 22°C and 748 torr? The equation is $C_2H_4(g) + 3\ O_2(g) \rightarrow 2\ CO_2(g) + 2\ H_2O(g)$.

Think Before You Write You are given the quantity of one species in a chemical change, and you are asked to calculate the quantity of another species in that change, so this is a stoichiometry problem. This problem is different than the ones you solved previously in this chapter in that *both* the given and wanted species are gases.

Answers *Cover the left column with your cut-out shield. Reveal each answer only after you have written your own answer in the right column.*

Given: 1.30 L C_2H_4 **Wanted:** Volume O_2 (assume L) $3\ \text{L } O_2 = 1\ \text{L } C_2H_4$ $1.30\ \cancel{\text{L } C_2H_4} \times \frac{3\ \text{L } O_2}{1\ \cancel{\text{L } C_2H_4}} = 3.90\ \text{L } O_2$ $1.30 \times 3 = 3.90$, OK.	For the STP calculation, **analyze** the problem, **identify** the conversion factors, and **construct** the complete solution, including the **check.** Remember that $V \propto n$ when the volumes are measured at the same temperature and pressure.
The answer is 3.90 L O_2, calculated by the same setup. *The volume ratio may be used whenever both gases are measured at the* same temperature and pressure. *It does not have to be STP.*	Now solve the problem when both gas volumes are measured at 22°C and 748 torr.
You improved your skill at solving gas stoichiometry problems.	What did you learn by solving this Active Example? 

Practice Exercise 14.14

Consider the complete burning of pure gaseous methane, CH_4. What volume of carbon dioxide is produced when 355 mL of methane burns? Both gases are measured at 0.97 bar and 21°C.

More often than not, the given and wanted gas volumes are at different temperatures and pressures. The convenient L (given) → L (wanted) cannot be used—yet. First, you use the Combined Gas Law equation ($P_1V_1/T_1 = P_2V_2/T_2$) solved for V_2 to change the volume of the given gas from its initial temperature and pressure to what that volume would be at the temperature and pressure of the wanted gas. Then both gases are at the same temperature and pressure, and you can use the L (given) → L (wanted) proportionality from the balanced chemical equation.

Active Example 14.15 Volume–Volume Gas Stoichiometry II

It is found that 1.75 L O_2, measured at 24°C and 755 torr, is used in burning sulfur (Figure 14.21). At one point in the exhaust hood, the sulfur dioxide produced is at 165°C and 785 torr. Find the volume of SO_2 at those conditions. The equation is $S_8(s) + 8\ O_2(g) \rightarrow 8\ SO_2(g)$.

Think Before You Write You are given P, V, and T for one gas in a chemical change, and you are asked for V for another gas in that change at a different P and T. This is a gas stoichiometry problem. If P and T were the same for both gases, you could use a one-step conversion as in Active Example 14.14. However, in this problem, the P and T conditions are different.

Answers *Cover the left column with your cut-out shield. Reveal each answer only after you have written your own answer in the right column.*

Oxygen is the gas whose volume you know. You need to change the known oxygen volume to what it would be at the temperature and pressure at which the wanted gas volume is to be expressed. **Analyze** the problem statement by filling in the table below.

	Volume	Temperature	Pressure
Initial values (1)			
Final values (2)			

Answer:

	Volume	Temperature	Pressure
Initial values (1)	1.75 L	24 + 273 = 297 K	755 torr
Final values (2)	V_2	165 + 273 = 438 K	785 torr

Identify the algebraic relationship needed to solve this part of the problem, solve it for the wanted variable, **construct** the solution setup, and calculate and **check** the value of the answer.

Answer:

$$\frac{P_1V_1}{T_1} = \frac{P_2\mathbf{V_2}}{T_2} \xrightarrow{\text{multiply both sides by } T_2/P_2}$$

$$\frac{T_2}{P_2} \times \frac{P_1V_1}{T_1} = \frac{\cancel{T_2}}{\cancel{P_2}} \times \frac{\cancel{P_2}\mathbf{V_2}}{\cancel{T_2}} \xrightarrow{\text{cancel the common factors}}$$

$$\frac{T_2}{P_2} \times \frac{P_1V_1}{T_1} = \mathbf{V_2} \xrightarrow{\text{rearrange}}$$

$$\mathbf{V_2} = V_1 \times \frac{T_2}{T_1} \times \frac{P_1}{P_2}$$

$$V_2 = V_1 \times \frac{T_2}{T_1} \times \frac{P_1}{P_2} = 1.75\ \text{L} \times \frac{438\ \cancel{\text{K}}}{297\ \cancel{\text{K}}} \times \frac{755\ \cancel{\text{torr}}}{785\ \cancel{\text{torr}}} = 2.48\ \text{L}$$

$$1.75 \times \frac{438 \times 755}{297 \times 785} \approx 2 \times \frac{4 \times 10^2}{3 \times 10^2} \times \frac{8 \times 10^2}{8 \times 10^2} = \frac{2 \times 4}{3} = \frac{8}{3} = 2\frac{2}{3}, \text{ OK.}$$

Figure 14.21 Elemental sulfur is found in nature as a yellow solid. At the particulate level, solid sulfur occurs as a variety of S_n molecules, where n has selected values between 6 and 20, most commonly S_8.

Given: 2.48 L O_2
Wanted: volume SO_2 (assume L)

8 L SO_2 = 8 L O_2

$$2.48\ \text{L O}_2 \times \frac{8\ \text{L SO}_2}{8\ \text{L O}_2} = 2.48\ \text{L SO}_2$$

The problem now becomes, "What volume of SO_2 is produced by the reaction of 2.48 L O_2, with both gases at the same temperature and pressure?" This makes the volume-conversion factor the same as the mole ratio from the equation. Complete all of the remaining steps in the solution of the problem.

You improved your skill at solving gas stoichiometry problems.

What did you learn by solving this Active Example?

Practice Exercise 14.15

A scientist burns 125 mL of ethane gas, C_2H_6, which was measured at 18°C and 0.94 atm. What volume of carbon dioxide gas is produced if it is measured at 224°C and 0.90 atm?

You did not have to calculate the numerical value of the volume of oxygen at the conditions at which the SO_2 volume is to be measured. The calculation setup alone is sufficient because it is equal to the calculated volume. The conversion from given substance to wanted can be tacked on as a final step:

$$1.75\ \text{L O}_2 \times \frac{755\ \text{torr}}{785\ \text{torr}} \times \frac{438\ \text{K}}{297\ \text{K}} \times \frac{8\ \text{L SO}_2}{8\ \text{L O}_2} = 2.48\ \text{L SO}_2$$

CHAPTER 14 IN REVIEW: THE IDEAL GAS LAW AND ITS APPLICATIONS

Important ideas to review from Chapter 4:

Charles's Law states that at constant pressure, the volume of a fixed amount of a gas is directly proportional to the absolute temperature, $V \propto T$.

Boyle's Law states that at constant temperature, the volume of a fixed amount of a gas is inversely proportional to its pressure, $V \propto 1/P$.

Charles's and Boyle's Laws can be coupled as the **Combined Gas Law:** $\frac{P_1V_1}{T_1} = \frac{P_2V_2}{T_2}$.

Goal	Summary
Goal 1 State how volume and amount of gas are related when pressure and temperature are constant and explain phenomena or make predictions based on that relationship.	**Avogadro's Law** states that equal volumes of two gases at the same temperature and pressure contain the same number of molecules, $V \propto n$.
Goal 2 Explain how the ideal gas equation can be constructed by combining Charles's Law (Section 4.5), Boyle's Law (Section 4.6), and Avogadro's Law (Section 14.3), and explain how the ideal gas equation can be used to derive each of the three two-variable laws.	Since $V \propto T$, $V \propto 1/P$, and $V \propto n$, it follows that $V \propto T \times (1/P) \times n$. Inserting a proportionality constant R, $V = RT(1/P)n$, or, rearranging, $\mathbf{PV = nRT}$. When pressure and amount are constant, $V = kT$, which is Charles's Law. When temperature and amount are constant, $PV = k$, which is Boyle's Law. When pressure and temperature are constant, $V = kn$, which is Avogadro's Law.
Goal 3 Given values for all except one of the variables in the ideal gas equation, calculate the value of the unknown variable.	The **ideal gas equation** is $PV = nRT$. Two values of the **gas constant, R**, are 0.0821 L · atm/mol · K and 62.4 L · torr/mol · K. Given all the values in the ideal gas equation except one, the unknown value may be calculated. Substituting m/MM for its equivalent, n, in the ideal gas equation gives $PV = \frac{m}{MM}RT$.
Goal 4 Calculate the density of a known gas at any specified temperature and pressure.	Solving the $PV = (m/MM)RT$ form of the ideal gas equation for m/V, which is **density,** yields $D = \frac{m}{V} = \frac{(MM)P}{RT}$.
Goal 5 Given the density of a pure gas at specified temperature and pressure, or information from which the density may be found, calculate the molar mass of that gas.	The density of a gas at constant temperature and pressure is directly proportional to its molar mass, $\mathbf{D \propto MM}$. Either molar mass or density can be calculated from the other using the ideal gas equation.
Goal 6 Calculate the molar volume of any gas at any given temperature and pressure.	**Molar volume** is the volume occupied by one mole of a gas, **V/n**. The molar volume of any ideal gas at STP (0°C and 1 bar) is 22.7 L/mol. This quantity is useful in calculations involving amount in moles, mass, volume, density, and molar mass of a gas measured at STP.
Goal 7 Given the molar volume of a gas at any specified temperature and pressure, or information from which the molar volume may be determined, and either the amount in moles in or the volume of a sample of that gas, calculate the other quantity.	Molar volume is $MV = \frac{V}{n} = \frac{RT}{P}$. Once molar volume is determined, it can be used to convert between the macroscopic volume of a gas and the particulate-level number of particles, grouped in moles.
Goal 8 Given a chemical equation, or a reaction for which the equation can be written, and the mass or amount in moles of one species in the reaction, or the STP volume of a gaseous species, find the mass or amount in moles of another species, or the STP volume of another gaseous species.	22.7 liters = 1 mole is an equivalency that can be used to convert between the volume of a gas at STP and the amount in moles of that gas. 22.7 L/mol can be used *only* for ideal gases at STP. If your stoichiometry skills are rusty, review Section 10.3.
Goal 9 Given a chemical equation, or a reaction for which the equation can be written, and the mass or amount in moles of one species in the reaction, or the volume of any gaseous species at a given temperature and pressure, find the mass or amount in moles of any other species, or the volume of any other gaseous species at a given temperature and pressure.	A **gas stoichiometry** problem at non-STP conditions can be solved by finding the molar volume of the gas and then following the stoichiometry path (molar volume method presented in Section 14.9) or by applying the ideal gas equation and then following the stoichiometry path or by following the stoichiometry path and then applying the ideal gas equation (ideal gas equation method presented in Section 14.10).
Goal 10 Given a chemical equation, or a reaction for which the equation can be written, and the volume of any gaseous species at a given temperature and pressure, find the volume of any other gaseous species at a given temperature and pressure.	The ratio of volumes of gases in a reaction is the same as the ratio of moles, provided that the gas volumes are measured at the same temperature and pressure. Thus, the coefficients in a balanced chemical equation can be used to convert between volumes, as long as the volumes are at the same temperature and pressure.

Key Terms

Most of the key terms and concepts and many others appear in the Glossary. Use your Glossary regularly.

22.7 L/mol p. 528
Avogadro's Law p. 517
gas constant (R) p. 519
ideal gas p. 519
ideal gas equation p. 519
ideal gas equation method p. 532
Ideal Gas Law p. 519
Law of Combining Volumes p. 516
molar volume (MV) p. 527

Frequently Asked Questions

I understand the concepts in this chapter, but I'm struggling with the problem solving; can you help me figure out why?
Your ability to solve the gas problems in this chapter depends largely on your algebra skills. Most students find it easiest to determine what is wanted and then solve the ideal gas equation for that variable. If the wanted quantity is density or molar volume, solve the equation for the combination of variables that represents the desired property. Then substitute the known variables, *including units*, and calculate the answer. Units are important: If they don't come out right, you know there is an error in the algebra.

Should I solve gas stoichiometry problems in one step or two steps?
Gas stoichiometry problems are usually solved in two steps. However, using RT/P for molar volume makes it possible to solve the problem in a single setup. If you're comfortable with this procedure, use it. If two separate steps are easier for you, solve the problem that way. Either way, be careful when you must divide by RT/P. Division appears in the setup as multiplication by P/RT. Again, the consistent use of units will help you to avoid a mistake.

I can shorten my setups by omitting the units; is that a more efficient way of problem solving?
No! Just as you check the value of your answers with an estimate, you must also check the units of your answers with unit cancellation. If you have the impression that we regard the use of units in calculation setups as very important—well, you're right!

Concept-Linking Exercises

Write a brief description of the relationships among each of the following groups of terms or phrases. Answers to the Concept-Linking Exercises are given at the end of the chapter.

1. Law of Combining Volumes, Avogadro's Law, chemical equation
2. Ideal Gas Law, ideal gas equation, gas constant
3. Moles, atmospheres, torr, liters, kelvins
4. 22.7, 62.4, 0.0821

Small-Group Discussion Questions

Small-Group Discussion Questions are for group work, either in class or under the guidance of a leader during a discussion section.

1. Describe how kinetic molecular theory (Section 4.3) explains each of the five macroscopic characteristics of gases listed in Section 4.2 and summarized at the beginning of Section 14.2.
2. Define the Law of Combining Volumes and Avogadro's Law. Explain how the Law of Combining Volumes is explained by Avogadro's Law. In light of the Law of Conservation of Mass, how can 50 mL of oxygen combine with 100 mL of hydrogen to produce 100 mL of steam?
3. Derive the Ideal Gas Law from Boyle's Law, Charles's Law, and Avogadro's Law. Then show how the Ideal Gas Law yields each of the three two-variable gas laws when the other two variables are held constant.
4. What is the Fahrenheit temperature of 1.0-ft^3 sample of gas that has a pressure of 15 psi if the container holds 0.1 mole of the gas?
5. The appearance of the atmosphere of Jupiter has long fascinated astronomers and many hobbyists because of its Giant Red Spot and colorful cloud layer. The atmospheric pressure builds up from its outer boundary toward the surface. At 1 bar pressure, the temperature is $-108°C$, and the density of the atmospheric gas mixture is 0.16 kg/m^3. What is the average molar mass of the mixture? The atmosphere is mostly composed of hydrogen and helium. Use the average molar mass to find the percentage of each of the two major components of the Jovian atmosphere.
6. Compare and contrast molar volume and molar mass. How are they similar? How are they different?
7. Two premises of the ideal gas model are usually not precisely correct for real gases. One is the independence of the particles. In a real gas, the particles are attracted to one another, sticking together for brief periods of time. How do you suppose that affects gas pressure? Another characteristic of real gases is that they have particle volume. How does the container volume available to a real gas compare with the volume available to an ideal gas? What assumption about container volume do you make when you apply the ideal gas model?
8. Acetylene, the fuel for welder's torches, can be made by the reaction of calcium carbide and water at high temperature: $CaC_2(s) + 2\,H_2O(\ell) \rightarrow C_2H_2(g) + Ca(OH)_2(s)$. Determine the percent yield of acetylene for a laboratory-scale reaction of 5.30 g of calcium carbide with excess water, producing 1595 mL of acetylene measured at the temperature and pressure of the lab, 25°C and 733 mm Hg.
9. Chlorine and sodium chlorite react to form chlorine dioxide and sodium chloride: $Cl_2(g) + 2\,NaClO_2(s) \rightarrow 2\,ClO_2(g) + 2\,NaCl(s)$. If 0.73 L of chlorine at 22°C and 1.0×10^3 torr reacts with 7.8 g of sodium chlorite, what mass in grams of chlorine dioxide will be produced? What volume or mass of reactant will remain unreacted?

Questions, Exercises, and Problems

Interactive versions of these problems may be assigned by your instructor. Solutions for blue-numbered questions are at the end of the chapter. Questions other than those in the General Questions and More Challenging Problems sections are paired in consecutive odd-even number combinations; solutions for the odd-numbered questions are also at the end of the chapter.

Section 14.3: Avogadro's Law

1. Compare the volumes of 1×10^{23} hydrogen molecules, 1×10^{23} oxygen molecules, and 2×10^{23} nitrogen molecules, all at the same temperature and pressure.
2. Which of the following gas samples would have the largest volume, if all samples are at the same temperature and pressure? (a) 263 g of Xe; (b) 5×10^{23} molecules of H_2; (c) 4.00 moles of CO_2; (d) they would all have the same volume.

Section 14.5: The Ideal Gas Equation: Determination of a Single Variable

3. Find the pressure in torr produced by 0.0888 mol of carbon dioxide in a 5.00-L vessel at 36°C.
4. A 0.917-mol sample of hydrogen gas at a temperature of 25.0°C is found to occupy a volume of 21.7 L. What is the pressure of this gas sample in torr?
5. The pressure exerted by 6.04 mol of nitrogen monoxide at a temperature of 18°C is 17.2 atm. What is the volume of the gas in liters?
6. A 0.512-mol sample of argon gas is collected at a pressure of 872 torr and a temperature of 18°C. What is the volume (L) of the sample?
7. A 784-mL hydrogen lecture bottle is left with the valve slightly open. Assuming no air has mixed with the hydrogen, what amount in moles of hydrogen is left in the bottle after the pressure has become equal to an atmospheric pressure of 752 torr at a temperature of 22°C?

Gencho Petkov/Shutterstock.com

A lecture bottle is a relatively small container that holds a compressed gas.

8. A sample of xenon gas collected at a pressure of 1.18 atm and a temperature of 18°C is found to occupy a volume of 26.7 L. What amount in moles of xenon gas is in the sample?
9. At what temperature (°C) will 0.810 mol of chlorine in a 15.7-L vessel exert a pressure of 756 torr?
10. A 0.142-mol sample of argon gas has a volume of 834 mL at a pressure of 4.83 atm. What is the temperature of the argon gas sample on the Celsius scale?
11. What amount in moles of carbon monoxide must be placed into a 40.0-L tank to develop a pressure of 965 torr at 18°C?
12. A sample of neon gas collected at a pressure of 0.946 atm and a temperature of 276 K is found to occupy a volume of 712 mL. What amount in moles of neon gas is in the sample?
13. Find the volume of 0.621 mol of helium at −32°C and 0.771 atm.
14. A sample of neon gas collected at a pressure of 531 mm Hg and a temperature of 291 K has a mass of 10.2 g. What is the volume (L) of the sample?

Section 14.6: Gas Density

15. The STP density of an unknown gas is found to be 2.32 g/L. What is the molar mass of the gas?
16. What is the density of a sample of oxygen gas at a pressure of 1.40 atm and a temperature of 49°C?
17. The "effective" molar mass of air is 29 g/mol. Use this value to calculate the density of air at (a) STP and (b) 20°C and 751 torr.
18. A 2.94-g sample of an unknown gas is found to occupy a volume of 2.06 L at a pressure of 1.16 atm and a temperature of 46°C. What is the molar mass of the unknown gas?
19. If the density of an unknown gas at 41°C and 2.61 atm is 1.61 g/L, what is its molar mass?
20. A sample of an unknown gas is found to have a density of 2.00 g/L at a pressure of 0.939 atm and a temperature of 40.0°C. What is the molar mass of the unknown gas?
21. The gas in an 8.07-L cylinder at 13°C has a mass of 33.5 g and exerts a pressure of 3.25 atm. Find the molar mass of the gas.
22. A 0.201-mol sample of an unknown gas contained in a 5.00 L flask is found to have a density of 1.72 g/L. What is the molar mass of the unknown gas?
23. NO_2 and N_2O_4 both have the same empirical (simplest) formula. At a temperature and pressure at which both substances are gases, can you tell without calculating which gas is more dense? Explain.

Charles D. Winters

The gas in the blue balloon is less dense than air; the gas in the red balloon is denser than air.

24. A 0.284-mol sample of xenon (Z = 54) gas is contained in a 7.00-L flask at room temperature and pressure. What is the density of the gas, in grams/liter, under these conditions?

Section 14.7: Molar Volume

25. Compare the molar volumes of helium and neon at 30°C and 1.10 torr. What are their values in L/mol?

26. What is the molar volume of carbon dioxide gas at a pressure of 0.974 atm and a temperature of 43°C?

27. Find the molar volume of acetylene, C_2H_2, at 21°C and 0.908 atm.

28. The molar volume for oxygen gas at a pressure of 0.684 atm and a temperature of 31°C is 36.5 L/mol. What is the volume occupied by 1.31 moles of oxygen gas at the same temperature and pressure?

Section 14.8: Gas Stoichiometry at Standard Temperature and Pressure

29. One small-scale laboratory method for preparing oxygen is to heat potassium chlorate in the presence of a catalyst: $2\,KClO_3(s) \rightarrow 2\,KCl(s) + 3\,O_2(g)$. Find the STP volume of oxygen that can be produced by 5.74 grams of $KClO_3$.

30. What volume of carbon dioxide is produced at 0°C and 1 atm when 64.2 g of calcium carbonate reacts completely according to the reaction $CaCO_3(s) \rightarrow CaO(s) + CO_2(g)$?

31. The reaction used to produce chlorine in the laboratory is $2\,KMnO_4(aq) + 16\,HCl(aq) \rightarrow 2\,MnCl_2(aq) + 5\,Cl_2(aq) + 8\,H_2O(aq) + 2\,KCl(aq)$. Calculate the mass in grams of potassium permanganate, $KMnO_4$, that is needed to produce 9.81 L of chlorine, measured at STP.

32. What mass in grams of carbon (graphite) is required to react completely with 42.5 L of oxygen gas at 0°C and 1 atm according to the reaction $C(s) + O_2(g) \rightarrow CO_2(g)$?

Section 14.9: Gas Stoichiometry: Molar Volume Method (Option 1)

Section 14.10: Gas Stoichiometry: Ideal Gas Equation Method (Option 2)

Questions 33–40 may be solved by the molar volume method (Section 14.9) or by the ideal gas equation method (Section 14.10). In the answer section, the setups are given first for the molar volume method and then for the ideal gas equation method. Check your work according to the section you studied.

33. One source of sulfur dioxide used in making sulfuric acid comes from sulfide ores by the reaction $4\,FeS_2(s) + 11\,O_2(g) \rightarrow 2\,Fe_2O_3(s) + 8\,SO_2(g)$. What volume in liters of SO_2, measured at 983 torr and 214°C, is produced by the reaction of 598 g FeS_2?

34. What volume of chlorine gas at 42.0°C and 1.33 atm is required to react completely with 41.5 g of phosphorus according to the reaction $P_4(s) + 6\,Cl_2(g) \rightarrow 4\,PCl_3(\ell)$?

35. What mass in grams of water must decompose by electrolysis to produce 23.9 L H_2, measured at 28°C and 728 torr?

36. What volume of hydrogen gas at 29°C and 0.809 atm is produced when 31.8 g of iron reacts completely according to the reaction $Fe(s) + 2\,HCl(aq) \rightarrow FeCl_2(aq) + H_2(g)$?

Charles D. Winters

When iron and hydrochloric acid react, the hydrogen gas produced can be collected by displacement of water. See Figure 15.5.

37. The reaction chamber in a modified Haber process for making ammonia by the direct combination of its elements is operated at 575°C and 248 atm. What volume in liters of nitrogen, measured at these conditions, will react to produce 9.16×10^3 grams of ammonia?

38. What mass in grams of iron is required to react completely with 5.49 L of oxygen gas at 41°C and 1.41 atm according to the reaction $2\,Fe(s) + O_2(g) \rightarrow 2\,FeO(s)$?

39. When properly detonated, ammonium nitrate explodes violently, releasing hot gases: $NH_4NO_3(s) \rightarrow N_2O(g) + 2\,H_2O(g)$. If the total volume of gas produced, both dinitrogen oxide and steam, is 82.3 L at 447°C and 896 torr, what mass in grams of NH_4NO_3 exploded?

40. What mass in grams of sodium is needed to produce 25.3 L of hydrogen gas at 43°C and 0.886 atm according to the reaction $2\,Na(s) + 2\,H_2O(\ell) \rightarrow 2\,NaOH(aq) + H_2(g)$?

Section 14.11: Volume–Volume Gas Stoichiometry

41. Sulfur burns to SO_2 with a beautiful deep blue–purple flame, but with a foul, suffocating odor. (a) What volume in liters of O_2 is needed to form 35.2 L of SO_2, both gases being measured at 741 torr and 26°C? (b) What if only the SO_2 is at those conditions, but the O_2 is at 17°C and 847 torr?

Charles D. Winters/Science Source

Sulfur burns in air.

42. Consider the following gas-phase reaction: $4\ NH_3(g) + 5\ O_2(g) \rightarrow 4\ NO(g) + 6\ H_2O(g)$. If 6.98 L of $O_2(g)$ gas at 129°C and 0.836 atm is reacted, what volume of $H_2O(g)$ will be produced if it is collected at 74°C and 0.630 atm?

43. Gaseous chlorine dioxide, ClO_2, is used to bleach wood pulp and in water treatment. It is produced by the reaction of chlorine with sodium chlorite: $Cl_2 + 2\ NaClO_2 \rightarrow 2\ ClO_2 + 2\ NaCl$. What volume in liters of ClO_2, measured at 0.961 atm and 31°C, will be produced by 283 L Cl_2 at 2.92 atm and 21°C?

44. Consider the following gas-phase reaction: $N_2(g) + 2\ O_2(g) \rightarrow 2\ NO_2(g)$. If 8.44 L of $N_2(g)$ gas at 127°C and 0.638 atm is used, what volume of $NO_2(g)$ gas will be formed if it is collected at 187°C and 0.982 atm?

45. In the natural oxidation of hydrogen sulfide released by decaying organic matter, the following reaction occurs: $2\ H_2S + 3\ O_2 \rightarrow 2\ SO_2 + 2\ H_2O$. What volume in milliliters of hydrogen sulfide, measured at 19°C and 549 torr, will be used in a reaction that also uses 704 mL O_2 at 159 torr and 26°C?

46. Consider the following gas-phase reaction: $H_2(g) + Cl_2(g) \rightarrow 2\ HCl(g)$. If 8.36 L of $Cl_2(g)$ gas at 113°C and 0.693 atm is used, what volume of hydrogen chloride gas will be formed if it is collected at 64°C and 0.355 atm?

General Questions

47. Distinguish precisely, and in scientific terms, the differences among items in each of the following pairs or groups.
    a) Avogadro's Law, Law of Combining Volumes
    b) Gas density, molar volume, molar mass
    c) Ideal gas equation, combined gas equation

48. Determine whether each of the following statements is true or false:
    a) The molar volume of a gas at 1.06 atm and 212°C is less than 22.7 L/mol.
    b) The mass of 5.00 L of NH_3 is the same as the mass of 5.00 L of CO if both volumes are measured at the same temperature and pressure.
    c) At a given temperature and pressure, the densities of two gases are proportional to their molar masses.
    d) To change volume of a gas to amount in moles, multiply by RT/P.

49. Calculate the volume of 6.74 g $C_2H_4(g)$ at 41°C and 733 torr.

50. Find the mass of 57.9 L of krypton at 775 torr and 6°C.

51. What is the density of H_2S at 0.972 atm and 14°C?

52. At 17°C and 0.835 atm, 16.2 L of ammonia has a mass of 9.68 g. What is the molar volume of ammonia at those conditions? (This is easier than it may seem!)

53. A 7.60-g sample of pure liquid is vaporized at 183°C and 179 torr. At these conditions it occupies 3.87 L. What is the molar mass of the substance?

More Challenging Problems

54. The density of nitrogen at 0.913 atm and 18°C is 1.07 g/L. Explain how this shows that the formula of nitrogen is N_2 rather than just N.

The answer to Question 56 follows from the answer to Question 55. Answer Question 55 first.

55. At a given temperature and pressure, what mathematical relationship exists between the density and molar mass of a gas? Explain your answer.

56. Labels have become detached from cylinders of two gases, one of which is known to be propane, C_3H_8, and the other butane, C_4H_{10}. The densities of the two gases are compared at the same temperature and pressure. The density of gas A is 1.37 g/L, and the density of gas B is higher. Which gas is A and which is B? What is the density of B?

Charles D. Winters

Both propane and butane are used as fuels in small, portable stoves.

57. An organic chemist has produced a solid that she believes to be pure; she expects a molar mass of 346 g/mol. Using 4.08 g of the solid, she melts and then boils it in a 3.36-L vacuum chamber at 117 torr and 243°C. She is disappointed; the molar mass is close to what she expected, but not close enough. (a) What molar mass did she find? (b) Her finding suggested to her what her mistake might have been. Does it suggest anything to you? [*Hint:* Part (b) is beyond the scope of this chapter, but you might have an inspiration if you have studied Chapter 13.]

Answer to Target Check

1. In your sketch, the piston should be at the 15 mark on the scale. The 5 two-atom particles that separate form 10 one-atom particles. Five particles remain unseparated, for a total of 15 particles after the reaction. Since $V \propto n$ at constant pressure and temperature,

$$10 \text{ volume units} \times \frac{15 \text{ particles}}{10 \text{ particles}} = 15 \text{ volume units}$$

Solutions to Practice Exercises

1. $R = \frac{PV}{nT} = 0.93\text{ bar} \times 3.50\text{ L} \times \frac{1}{0.132\text{ mol}} \times \frac{1}{(23 + 273)\text{ K}} = 0.083\text{ L} \cdot \text{bar/mol} \cdot \text{K}$

2. $R = \frac{PV}{nT} = 23.6\text{ in. Hg} \times 21.0\text{ L} \times \frac{1}{0.659\text{ mol}} \times \frac{1}{(33 + 273)\text{ K}} = 2.46\text{ L} \cdot \text{in. Hg/mol} \cdot \text{K}$

3. $P = \frac{nRT}{V} = 0.22\text{ mol} \times \frac{62.4\text{ L} \cdot \text{mm Hg}}{\text{mol} \cdot \text{K}} \times (19 + 273)\text{ K} \times \frac{1}{1.0\text{ gal}} \times \frac{1\text{ gal}}{3.785\text{ L}} = 1.1 \times 10^3\text{ mm Hg}$

4. $MM = \frac{mRT}{PV} = m \times R \times T \times \frac{1}{P} \times \frac{1}{V} = 0.59\text{ g} \times \frac{0.0821\text{ L} \cdot \text{atm}}{\text{mol} \cdot \text{K}} \times 295\text{ K} \times \frac{1}{0.988\text{ atm}} \times \frac{1}{0.50\text{ L}} = 29\text{ g/mol}$

5. $D \equiv \frac{m}{V} = \frac{(MM)P}{RT} = \frac{32.00\text{ g}}{\text{mol}} \times 741\text{ torr} \times \frac{\text{mol} \cdot \text{K}}{62.4\text{ L} \cdot \text{torr}} \times \frac{1}{(25 + 273)\text{ K}} = 1.28\text{ g/L}$

6. $MM = R \times T \times \frac{1}{P} \times \frac{m}{V} = \frac{0.0821\text{ L} \cdot \text{atm}}{\text{mol} \cdot \text{K}} \times (29 + 273)\text{ K} \times \frac{1}{0.966\text{ bar}} \times \frac{2.2\text{ g}}{\text{L}} \times \frac{1.013\text{ bar}}{\text{atm}} = 57\text{ g/mol}$

7. $MV \equiv \frac{V}{n} = \frac{RT}{P} = \frac{0.0821\text{ L} \cdot \text{atm}}{\text{mol} \cdot \text{K}} \times (-15 + 273)\text{ K} \times \frac{1}{9.0\text{ bar}} \times \frac{1.013\text{ bar}}{\text{atm}} = 2.4\text{ L/mol}$

8. $5.0\text{ g } N_2 \times \frac{1\text{ mol } N_2}{28.02\text{ g } N_2} \times \frac{22.7\text{ L}}{\text{mol } N_2} = 4.1\text{ L}$

9. $2\,H_2O_2 \rightarrow 2\,H_2O + O_2$

$5.00 \times 10^2\text{ g solution} \times \frac{3.0\text{ g } H_2O_2}{100\text{ g solution}} \times \frac{1\text{ mol } H_2O_2}{34.02\text{ g } H_2O_2} \times \frac{1\text{ mol } O_2}{2\text{ mol } H_2O_2} \times \frac{22.7\text{ L } O_2}{\text{mol } O_2} = 5.0\text{ L } O_2$

10. $2\,H_2O_2 \rightarrow 2\,H_2O + O_2$

$MV \equiv \frac{V}{n} = \frac{RT}{P} = \frac{62.4\text{ L} \cdot \text{torr}}{\text{mol} \cdot \text{K}} \times (27 + 273)\text{ K} \times \frac{1}{733\text{ torr}} = 25.5\text{ L/mol}$

$5.00 \times 10^2\text{ g solution} \times \frac{3.0\text{ g } H_2O_2}{100\text{ g solution}} \times \frac{1\text{ mol } H_2O_2}{34.02\text{ g } H_2O_2} \times \frac{1\text{ mol } O_2}{2\text{ mol } H_2O_2} \times \frac{25.5\text{ L } O_2}{\text{mol } O_2} = 5.6\text{ L } O_2$

11. $2\text{ Na} + 2\,H_2O \rightarrow 2\text{ NaOH} + H_2$

$MV \equiv \frac{V}{n} = \frac{RT}{P} = \frac{62.4\text{ L} \cdot \text{torr}}{\text{mol} \cdot \text{K}} \times (18 + 273)\text{ K} \times \frac{1}{765\text{ torr}} = 23.7\text{ L/mol}$

$0.25\text{ g Na} \times \frac{1\text{ mol Na}}{22.99\text{ g Na}} \times \frac{1\text{ mol } H_2}{2\text{ mol Na}} \times \frac{23.7\text{ L } H_2}{\text{mol } H_2} \times \frac{1000\text{ mL } H_2}{\text{L } H_2} = 1.3 \times 10^2\text{ mL } H_2$

12. $2\,H_2O_2 \rightarrow 2\,H_2O + O_2$

$5.00 \times 10^2\text{ g solution} \times \frac{3.0\text{ g } H_2O_2}{100\text{ g solution}} \times \frac{1\text{ mol } H_2O_2}{34.02\text{ g } H_2O_2} \times \frac{1\text{ mol } O_2}{2\text{ mol } H_2O_2} = 0.22\text{ mol } O_2$

$V = \frac{nRT}{P} = 0.22\text{ mol } O_2 \times \frac{62.4\text{ L} \cdot \text{torr}}{\text{mol} \cdot \text{K}} \times (27 + 273)\text{ K} \times \frac{1}{733\text{ torr}} = 5.6\text{ L } O_2$

13. $2\text{ Na} + 2\,H_2O \rightarrow 2\text{ NaOH} + H_2$

$0.25\text{ g Na} \times \frac{1\text{ mol Na}}{22.99\text{ g Na}} \times \frac{1\text{ mol } H_2}{2\text{ mol Na}} = 0.0054\text{ mol } H_2$

$V = \frac{nRT}{P} = 0.0054\text{ mol} \times \frac{62.4\text{ L} \cdot \text{torr}}{\text{mol} \cdot \text{K}} \times (18 + 273)\text{ K} \times \frac{1}{765\text{ torr}} \times \frac{1000\text{ mL } H_2}{\text{L } H_2} = 1.3 \times 10^2\text{ mL } H_2$

14. $CH_4 + 2\,O_2 \rightarrow CO_2 + 2\,H_2O$; $355\text{ mL } CH_4 \times \frac{1\text{ mL } CO_2}{1\text{ mL } CH_4} = 355\text{ mL } CH_4$

15. $2\,C_2H_6 + 7\,O_2 \rightarrow 4\,CO_2 + 6\,H_2O$

$V_2 = V_1 \times \frac{P_1}{P_2} \times \frac{T_2}{T_1} = 125\text{ mL} \times \frac{0.94\text{ atm}}{0.90\text{ atm}} \times \frac{(224 + 273)\text{ K}}{(18 + 273)\text{ K}} = 223\text{ mL}$

$223\text{ mL } C_2H_6 \times \frac{4\text{ mL } CO_2}{2\text{ mL } C_2H_6} = 446\text{ mL } CO_2$

Answers to Concept-Linking Exercises

You may have found more relationships or relationships other than the ones given in these answers.

1. The Law of Combining Volumes states that when gases at the same temperature and pressure react, they do so in ratios of small whole numbers. Avogadro's Law states that the volume occupied by a gas at a fixed temperature and pressure is directly proportional to the amount of gas present, expressed in number of particles (moles). Therefore, the molar ratios that appear in chemical equations are equal to the ratio of volumes of gaseous reactants and products, provided that they are measured at the same temperature and pressure.

2. The Ideal Gas Law is based on the ideal gas model and the experimentally observed proportionalities among gas pressure, temperature, volume, and amount (number of particles). The constant that combines these proportionalities into the ideal gas equation is called the *gas constant*.

3. The units in which the variables in the ideal gas equation are expressed are atmospheres or torr (pressure), liters (volume), kelvins (temperature), and moles (amount).

4. The most common value of the gas constant is 0.0821 L · atm/mol · K. If pressure is expressed in torr, the value of the constant is 62.4 L · torr/mol · K. If the ideal gas equation is solved for liters per mole, the result is V/n = RT/P. Substituting standard temperature (273 K) and standard pressure (1 bar) into the right side of the equation gives 22.7 L/mol. This is the molar volume of an ideal gas at STP.

Solutions to Blue-Numbered Questions, Exercises, and Problems

1. $n = \frac{PV}{RT}$, so since temperature and pressure are the same for all three samples, n is directly proportional to V. The volume of 1×10^{23} hydrogen molecules is the same as the volume of 1×10^{23} oxygen molecules and half the volume of 2×10^{23} nitrogen molecules.

3. $P = \frac{nRT}{V} = 0.0888\text{ mol} \times \frac{62.4\text{ L} \cdot \text{torr}}{\text{mol} \cdot \text{K}} \times (36 + 273)\text{ K} \times \frac{1}{5.00\text{ L}} = 342\text{ torr}$

5. $V = \frac{nRT}{P} = 6.04\text{ mol} \times \frac{0.0821\text{ L} \cdot \text{atm}}{\text{mol} \cdot \text{K}} \times (18 + 273)\text{ K} \times \frac{1}{17.2\text{ atm}} = 8.39\text{ L}$

7. $n = \frac{PV}{RT} = 752\text{ torr} \times 784\text{ mL} \times \frac{\text{mol} \cdot \text{K}}{62.4\text{ L} \cdot \text{torr}} \times \frac{1}{(22 + 273)\text{ K}} \times \frac{1\text{ L}}{1000\text{ mL}} = 0.0320\text{ mol}$

9. $T = \frac{PV}{nR} = 756\text{ torr} \times 15.7\text{ L} \times \frac{1}{0.810\text{ mol}} \times \frac{\text{mol} \cdot \text{K}}{62.4\text{ L} \cdot \text{torr}} = 235\text{ K}; 235 - 273 = -38°\text{C}$

11. $n = \frac{PV}{RT} = 965\text{ torr} \times 40.0\text{ L} \times \frac{\text{mol} \cdot \text{K}}{62.4\text{ L} \cdot \text{torr}} \times \frac{1}{(18 + 273)\text{ K}} = 2.13\text{ mol}$

13. $V = \frac{nRT}{P} = 0.621\text{ mol} \times \frac{0.0821\text{ L} \cdot \text{atm}}{\text{mol} \cdot \text{K}} \times (-32 + 273)\text{ K} \times \frac{1}{0.771\text{ atm}} = 15.9\text{ L}$

15. $MM = \frac{mRT}{PV} = \frac{2.32\text{ g}}{\text{L}} \times \frac{0.0821\text{ L} \cdot \text{atm}}{\text{mol} \cdot \text{K}} \times 273\text{ K} \times \frac{1}{1\text{ bar}} \times \frac{1.013\text{ bar}}{\text{atm}} = 52.7\text{ g/mol}$

17. (a) $D = \frac{m}{V} = \frac{(MM)P}{RT} = \frac{29\text{ g}}{\text{mol}} \times 1\text{ bar} \times \frac{\text{mol} \cdot \text{K}}{0.0821\text{ L} \cdot \text{atm}} \times \frac{1}{273\text{ K}} \times \frac{1\text{ atm}}{1.013\text{ bar}} = 1.3\text{ g/L}$

(b) $D = \frac{m}{V} = \frac{(MM)P}{RT} = \frac{29\text{ g}}{\text{mol}} \times 751\text{ torr} \times \frac{\text{mol} \cdot \text{K}}{62.4\text{ L} \cdot \text{torr}} \times \frac{1}{(20 + 273)\text{K}} = 1.2\text{ g/L}$

19. $MM = \frac{mRT}{PV} = \frac{m}{V} \times \frac{RT}{P} = \frac{1.61\text{ g}}{\text{L}} \times \frac{0.0821\text{ L} \cdot \text{atm}}{\text{mol} \cdot \text{K}} \times (41 + 273)\text{ K} \times \frac{1}{2.61\text{ atm}} = 15.9\text{ g/mol}$

21. $MM = \frac{mRT}{PV} = 33.5\text{ g} \times \frac{0.0821\text{ L} \cdot \text{atm}}{\text{mol} \cdot \text{K}} \times (13 + 273)\text{ K} \times \frac{1}{3.25\text{ atm}} \times \frac{1}{8.07\text{ L}} = 30.0\text{ g/mol}$

23. $D = \frac{m}{V} = \frac{(MM)P}{RT}$, so when temperature and pressure are constant, density is directly proportional to molar mass. N_2O_4 is denser than NO_2.

25. $MV = \frac{V}{n} = \frac{RT}{P} = \frac{62.4\text{ L} \cdot \text{torr}}{\text{mol} \cdot \text{K}} \times (30 + 273)\text{ K} \times \frac{1}{1.10\text{ torr}} = 1.72 \times 10^4\text{ L/mol}$
Molar volume is independent of the identity of the gas.

27. $MV = \frac{V}{n} = \frac{RT}{P} = \frac{0.0821\text{ L} \cdot \text{atm}}{\text{mol} \cdot \text{K}} \times (21 + 273)\text{ K} \times \frac{1}{0.908\text{ atm}} = 26.6\text{ L/mol}$

29. $5.74\text{ g } KClO_3 \times \frac{1\text{ mol } KClO_3}{122.55\text{ g } KClO_3} \times \frac{3\text{ mol } O_2}{2\text{ mol } KClO_3} \times \frac{22.7\text{ L } O_2}{\text{mol } O_2} = 1.59\text{ L } O_2$

31. $9.81\text{ L } Cl_2 \times \frac{1\text{ mol } Cl_2}{22.7\text{ L } Cl_2} \times \frac{2\text{ mol } KMnO_4}{5\text{ mol } Cl_2} \times \frac{158.04\text{ g } KMnO_4}{\text{mol } KMnO_4} = 27.3\text{ g } KMnO_4$

Questions 33–40 may be solved by the molar volume method (Section 14.9) or by the ideal gas equation method (Section 14.10). Answer setups are given first for the molar volume method in 33, 35, 37, and 39. Then, the answers are given for the ideal gas equation method in 33, 35, 37, and 39, again. Check your work according to the section you studied.

Solutions for Molar Volume Method (Section 14.9):

33. $MV = \frac{V}{n} = \frac{RT}{P} = \frac{62.4\text{ L} \cdot \text{torr}}{\text{mol} \cdot \text{K}} \times (214 + 273)\text{ K} \times \frac{1}{983\text{ torr}} = 30.9\text{ L/mol}$

$598\text{ g } FeS_2 \times \frac{1\text{ mol } FeS_2}{119.97\text{ g } FeS_2} \times \frac{8\text{ mol } SO_2}{4\text{ mol } FeS_2} \times \frac{30.9\text{ L } SO_2}{\text{mol } SO_2} = 308\text{ L } SO_2$

35. $2\, H_2O \rightarrow 2\, H_2 + O_2$

$MV = \frac{V}{n} = \frac{RT}{P} = \frac{62.4\text{ L} \cdot \text{torr}}{\text{mol} \cdot \text{K}} \times (28 + 273)\text{ K} \times$

$\frac{1}{728 \text{ torr}} = 25.8 \text{ L/mol}$

$23.9 \text{ L } H_2 \times \frac{1 \text{ mol } H_2}{25.8 \text{ L } H_2} \times \frac{2 \text{ mol } H_2O}{2 \text{ mol } H_2} \times \frac{18.02 \text{ g } H_2O}{\text{mol } H_2O} =$

$16.7 \text{ g } H_2O$

37. $N_2 + 3\, H_2 \rightarrow 2\, NH_3$

$MV = \frac{V}{n} = \frac{RT}{P} = \frac{0.0821 \text{ L} \cdot \text{atm}}{\text{mol} \cdot \text{K}} \times (575 + 273) \text{ K} \times$

$\frac{1}{248 \text{ atm}} = 0.281 \text{ L/mol}$

$9.16 \times 10^3 \text{ g } NH_3 \times \frac{1 \text{ mol } NH_3}{17.03 \text{ g } NH_3} \times \frac{1 \text{ mol } N_2}{2 \text{ mol } NH_3} \times$

$\frac{0.281 \text{ L } N_2}{\text{mol } N_2} = 75.6 \text{ L } N_2$

39. $MV = \frac{V}{n} = \frac{RT}{P} = \frac{62.4 \text{ L} \cdot \text{torr}}{\text{mol} \cdot \text{K}} \times (447 + 273) \text{ K} \times$

$\frac{1}{896 \text{ torr}} = 50.1 \text{ L/mol}$

$82.3 \text{ L gases} \times \frac{1 \text{ mol gases}}{50.1 \text{ L}} \times \frac{1 \text{ mol } NH_4NO_3}{3 \text{ mol gases}} \times$

$\frac{80.05 \text{ g } NH_4NO_3}{\text{mol } NH_4NO_3} = 43.8 \text{ g } NH_4NO_3$

Solutions for Ideal Gas Equation Method (Section 14.10):

33. $598 \text{ g } FeS_2 \times \frac{1 \text{ mol } FeS_2}{119.97 \text{ g } FeS_2} \times \frac{8 \text{ mol } SO_2}{4 \text{ mol } FeS_2} = 9.97 \text{ mol } SO_2$

$V = \frac{nRT}{P} = 9.97 \text{ mol} \times \frac{62.4 \text{ L} \cdot \text{torr}}{\text{mol} \cdot \text{K}} \times (214 + 273) \text{ K} \times$

$\frac{1}{983 \text{ torr}} = 308 \text{ L } SO_2$

35. $2\, H_2O \rightarrow 2\, H_2 + O_2$

$n = \frac{PV}{RT} = 728 \text{ torr} \times 23.9 \text{ L} \times \frac{\text{mol} \cdot \text{K}}{62.4 \text{ L} \cdot \text{torr}} \times$

$\frac{1}{(28 + 273)\text{K}} = 0.926 \text{ mol } H_2$

$0.926 \text{ mol } H_2 \times \frac{2 \text{ mol } H_2O}{2 \text{ mol } H_2} \times \frac{18.02 \text{ g } H_2O}{\text{mol } H_2O} = 16.7 \text{ g } H_2O$

37. $N_2 + 3\, H_2 \rightarrow 2\, NH_3$

$9.16 \times 10^3 \text{ g } NH_3 \times \frac{1 \text{ mol } NH_3}{17.03 \text{ g } NH_3} \times \frac{1 \text{ mol } N_2}{2 \text{ mol } NH_3} = 269 \text{ mol } N_2$

$V = \frac{nRT}{P} = 269 \text{ mol} \times \frac{0.0821 \text{ L} \cdot \text{atm}}{\text{mol} \cdot \text{K}} \times (575 + 273) \text{ K} \times$

$\frac{1}{248 \text{ atm}} = 75.5 \text{ L } N_2$

39. $n = \frac{PV}{RT} = 896 \text{ torr} \times 82.3 \text{ L} \times \frac{\text{mol} \cdot \text{K}}{62.4 \text{ L} \cdot \text{torr}} \times$

$\frac{1}{(447 + 273)\text{K}} = 1.64 \text{ mol}$

$1.64 \text{ mol gases} \times \frac{1 \text{ mol } NH_4NO_3}{3 \text{ mol gases}} \times \frac{80.05 \text{ g } NH_4NO_3}{\text{mol } NH_4NO_3} =$

$43.8 \text{ g } NH_4NO_3$

41. $S + O_2 \rightarrow SO_2$

a) $35.2 \text{ L } SO_2 \times \frac{1 \text{ L } O_2}{1 \text{ L } SO_2} = 35.2 \text{ L } O_2$

b) $V_2 = V_1 \times \frac{P_1}{P_2} \times \frac{T_2}{T_1} = 35.2 \text{ L} \times \frac{741 \text{ torr}}{847 \text{ torr}} \times \frac{(17 + 273)\text{K}}{(26 + 273)\text{K}} =$

$29.9 \text{ L } SO_2$

$29.9 \text{ L } SO_2 \times \frac{1 \text{ L } O_2}{1 \text{ L } SO_2} = 29.9 \text{ L } O_2$

43. $V_2 = V_1 \times \frac{P_1}{P_2} \times \frac{T_2}{T_1} = 283 \text{ L} \times \frac{2.92 \text{ atm}}{0.961 \text{ atm}} \times$

$\frac{(31 + 273)\text{K}}{(21 + 273)\text{K}} = 889 \text{ L}$

$889 \text{ L } Cl_2 \times \frac{2 \text{ L } ClO_2}{1 \text{ L } Cl_2} = 1.78 \times 10^3 \text{ L } ClO_2$

45. $V_2 = V_1 \times \frac{P_1}{P_2} \times \frac{T_2}{T_1} = 704 \text{ mL} \times \frac{159 \text{ torr}}{549 \text{ torr}} \times$

$\frac{(19 + 273)\text{K}}{(26 + 273)\text{K}} = 199 \text{ mL}$

$199 \text{ mL } O_2 \times \frac{2 \text{ L } H_2S}{3 \text{ L } O_2} = 133 \text{ mL } H_2S$

48. True: (c). False: (a), (b), (d).

49. $V = \frac{mRT}{(MM)P} = 6.74 \text{ g} \times \frac{62.4 \text{ L} \cdot \text{torr}}{\text{mol} \cdot \text{K}} \times \frac{(41 + 273)\text{K}}{733 \text{ torr}} \times$

$\frac{1 \text{ mol}}{28.05 \text{ g}} = 6.42 \text{ L}$

50. $m = \frac{PV(MM)}{RT} = 775 \text{ torr} \times 57.9 \text{ L} \times \frac{83.80 \text{ g}}{\text{mol}} \times$

$\frac{\text{mol} \cdot \text{K}}{62.4 \text{ L} \cdot \text{atm}} \times \frac{1}{(6 + 273)\text{K}} = 216 \text{ g}$

51. $D = \frac{m}{V} = \frac{(MM)P}{RT} = \frac{34.08 \text{ g}}{\text{mol}} \times \frac{0.972 \text{ atm}}{287 \text{ K}} \times$

$\frac{\text{mol} \cdot \text{K}}{0.0821 \text{ L} \cdot \text{atm}} = 1.41 \text{ g/L}$

52. $MV = \frac{V}{n} = \frac{RT}{P} = \frac{0.0821 \text{ L} \cdot \text{atm}}{\text{mol} \cdot \text{K}} \times (17 + 273) \text{ K} \times$

$\frac{1}{0.835 \text{ atm}} = 28.5 \text{ L/mol}$

At a given temperature and pressure, molar volume is the same for all gases. The identity of the gas, the volume, and the mass are not needed to solve the problem.

53. $MM = \frac{mRT}{PV} = 7.60 \text{ g} \times \frac{62.4 \text{ L} \cdot \text{torr}}{\text{mol} \cdot \text{K}} \times (183 + 273) \text{ K} \times$

$\frac{1}{179 \text{ torr}} \times \frac{1}{3.87 \text{ L}} = 312 \text{ g/mol}$

54. $MM = \frac{mRT}{PV} = \frac{1.07 \text{ g}}{\text{L}} \times \frac{0.0821 \text{ L} \cdot \text{atm}}{\text{mol} \cdot \text{K}} \times (18 + 273) \text{ K} \times$

$\frac{1}{0.913 \text{ atm}} = 28.0 \text{ g/mol}$

The molar mass of nitrogen is twice its atomic mass. Therefore, there must be two atoms in the molecule, N_2.

55. In algebra, if $x \propto y$, then $x = ky$, where k is a proportionality constant. At a given temperature and pressure, solving $PV = (m/MM)RT$ for density, m/V, yields

$D = \frac{m}{V} = \frac{P(MM)}{RT} = \frac{P}{RT} \times MM = k \times MM$

$D = k \times MM$

In this equation, P/RT has the role of a proportionality constant. Hence, molar mass and density are directly proportional to each other at a given temperature and pressure.

56. Butane, C_4H_{10}, has a higher molar mass, 58.12 g/mol, than propane, C_3H_8, 44.09 g/mol. At a given temperature and pressure, density is proportional to molar mass. Therefore, B, the higher-density gas, must be butane. Its density is

$$1.37 \text{ g/L} \times \frac{58.12 \text{ g/mol}}{44.09 \text{ g/mol}} = 1.81 \text{ g/L}$$

57. (a) $\text{MM} = \frac{\text{mRT}}{\text{PV}} = 4.08 \text{ g} \times \frac{62.4 \text{ L} \cdot \text{torr}}{\text{mol} \cdot \text{K}} \times$

$$(243 + 273) \text{ K} \times \frac{1}{117 \text{ torr}} \times \frac{1}{3.36 \text{ L}} = 334 \text{ g/mol}$$

(b) The difference between the expected molar mass and the molar mass that was found experimentally is 346 − 334 = 12 g/mol. This is the atomic mass of carbon. Perhaps she has one too many carbon atoms in her expected molecule.

15 Gases, Liquids, and Solids

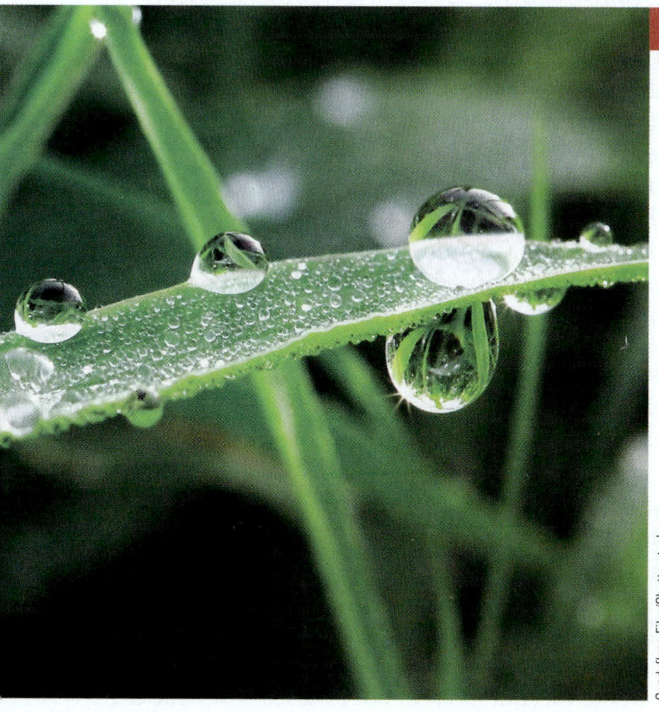

◀ Liquid drops adopt a spherical shape because of a macroscopic, measurable property known as surface tension. This property can be understood at the particulate level in terms of the strengths of the attractive forces among the particles that make up the liquid. Strong attractive forces lead to high surface tension.

CHAPTER CONTENTS

In Chapter 12, you studied chemical bonds, the attractive forces between atoms or ions that hold them together to form compounds. In this chapter, we focus on covalently bonded molecules and shift our attention to the attractive forces *between* the molecules. The strength of these attractive forces is largely responsible for the physical properties of compounds, including melting and boiling points. Thus, attractive forces between molecules determine whether a substance exists as a solid, liquid, or gas at a given temperature and pressure.

You first considered the particulate character of matter in the kinetic molecular theory in Section 2.3. The particles of a solid were described as holding in fixed position relative to one another. A degree of freedom is reached in the liquid state, in which particles move about among themselves but still remain together at the bottom of the container that holds them. As a gas, the particles gain complete independence from one another and fly about randomly to fill their containers.

In this chapter, we discuss the relationship among gases, liquids, and solids and the energy changes that accompany a change of state.

15.1 Does Liquid Water Exist Beyond Planet Earth?

Several sections of this chapter examine the *liquid* state of matter. As we will discuss in more detail in Section 15.7, liquid water is chemically unusual, with properties that don't fit the patterns followed by most other substances. But since liquid

Figure 15.1 The southern polar plain on Mars with the site of the liquid water lake enclosed in the highlight rectangle. The colors within the rectangle indicate the nature of the radar signal. The lake is located at the blue triangle at approximately the center of the rectangle. The white surface feature above (to the south of) the rectangle is Mars's south polar cap, which is made of both solid carbon dioxide and solid water.

Wikimedia Commons

water is so prevalent on Earth, we typically don't spend much time contemplating its uniqueness. One important characteristic of liquid water is that it is required for life, as we know it on Earth. Thus, as we investigate the universe to look for other forms of life, it is important to look for the presence of liquid water as well.

Our two nearest planetary neighbors, Venus and Mars, each likely had liquid water on their surfaces in the past. In 2016, a team of NASA scientists published the results of the first sophisticated, three-dimensional simulation of the history of Venus's atmosphere. They found that not only was it likely that Venus once had surface water but also that the Venusian ocean may have existed for two billion years. In 2018, a team of Italian scientists reported that radar measurements by the Mars Express spacecraft detected a 20-kilometer-wide lake of liquid water that exists under Mars's southern polar plane (**Figure 15.1**)! Could life exist in this Martian lake?

Other bodies in our solar system also appear to have liquid water. One example is Enceladus, Saturn's sixth-largest moon. In 2005, the unmanned robotic space probe *Cassini* flew over the south polar region of Enceladus and found geysers of what appeared to be steam and ice! *Cassini* was directed to fly through the plumes in 2008, discovering that they are composed of water, carbon dioxide, carbon monoxide, and carbon–hydrogen molecules. Given that all living organisms have molecules that include carbon atoms, the basic elements needed for life are present on Enceladus. **Figure 15.2** is a not-to-scale illustration of the model hypothesized to be the fundamental structure of Enceladus, with a rocky core, a liquid water ocean, and a solid water crust.

Is there evidence of liquid water existing beyond the solar system? A good answer to this question cannot be given at the current point in time. Certainly, *detection* of liquid water outside of our solar systems is slightly beyond our current capabilities. However, *models* of planets outside of the solar system indicate that planets with a radius about 2.5 times that of Earth likely contain water and 35% of the planets larger than Earth are likely to be water-rich. The James Webb Space Telescope (Section 8.1) should provide data that can refute or verify these models. In the words of Sara Seager, Professor of Planetary Science and Physics at the Massachusetts

Figure 15.2 A cutaway model of the fundamental structure of Enceladus, a moon of Saturn. The solid planetary core is illustrated in black and gray, its liquid ocean is show in blue, and its ice (solid water) crust is shown in white. Geysers emitting water and a mixture of carbon-atom-containing molecules are illustrated in the southern region of the moon.

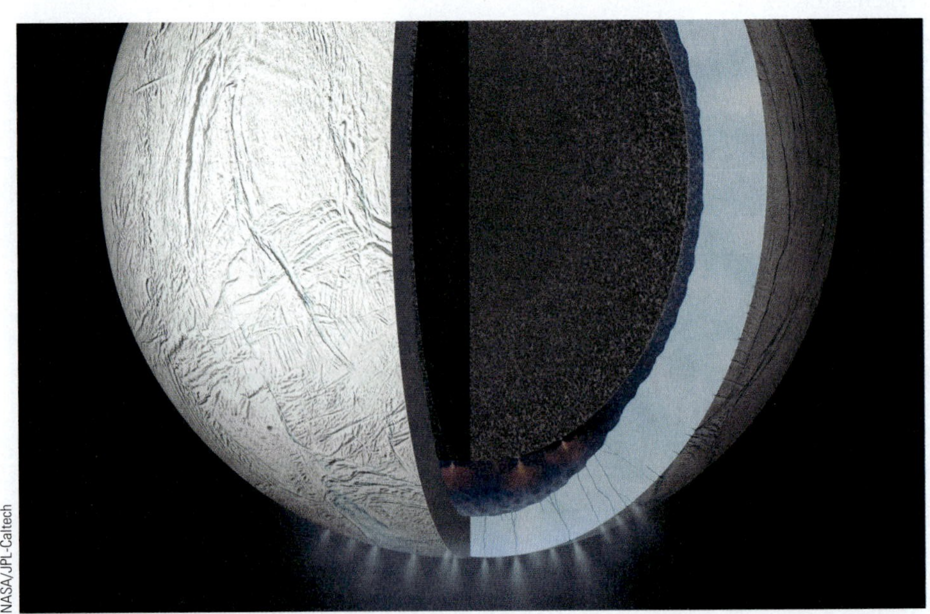

NASA/JPL-Caltech

15 Gases, Liquids, and Solids

Candyfloss Film/Shutterstock.com

◀ Liquid drops adopt a spherical shape because of a macroscopic, measurable property known as surface tension. This property can be understood at the particulate level in terms of the strengths of the attractive forces among the particles that make up the liquid. Strong attractive forces lead to high surface tension.

CHAPTER CONTENTS

In Chapter 12, you studied chemical bonds, the attractive forces between atoms or ions that hold them together to form compounds. In this chapter, we focus on covalently bonded molecules and shift our attention to the attractive forces *between* the molecules. The strength of these attractive forces is largely responsible for the physical properties of compounds, including melting and boiling points. Thus, attractive forces between molecules determine whether a substance exists as a solid, liquid, or gas at a given temperature and pressure.

You first considered the particulate character of matter in the kinetic molecular theory in Section 2.3. The particles of a solid were described as holding in fixed position relative to one another. A degree of freedom is reached in the liquid state, in which particles move about among themselves but still remain together at the bottom of the container that holds them. As a gas, the particles gain complete independence from one another and fly about randomly to fill their containers.

In this chapter, we discuss the relationship among gases, liquids, and solids and the energy changes that accompany a change of state.

15.1 Does Liquid Water Exist Beyond Planet Earth?

Several sections of this chapter examine the *liquid* state of matter. As we will discuss in more detail in Section 15.7, liquid water is chemically unusual, with properties that don't fit the patterns followed by most other substances. But since liquid

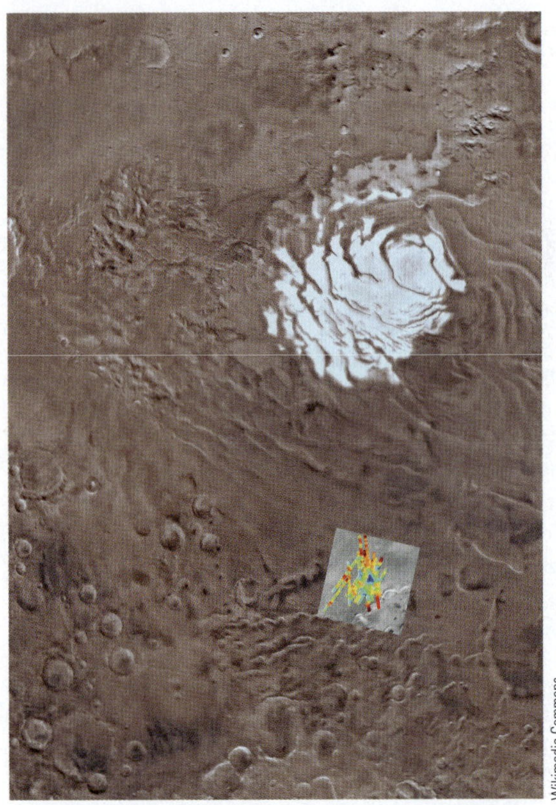

Figure 15.1 The southern polar plain on Mars with the site of the liquid water lake enclosed in the highlight rectangle. The colors within the rectangle indicate the nature of the radar signal. The lake is located at the blue triangle at approximately the center of the rectangle. The white surface feature above (to the south of) the rectangle is Mars's south polar cap, which is made of both solid carbon dioxide and solid water.

water is so prevalent on Earth, we typically don't spend much time contemplating its uniqueness. One important characteristic of liquid water is that it is required for life, as we know it on Earth. Thus, as we investigate the universe to look for other forms of life, it is important to look for the presence of liquid water as well.

Our two nearest planetary neighbors, Venus and Mars, each likely had liquid water on their surfaces in the past. In 2016, a team of NASA scientists published the results of the first sophisticated, three-dimensional simulation of the history of Venus's atmosphere. They found that not only was it likely that Venus once had surface water but also that the Venusian ocean may have existed for two billion years. In 2018, a team of Italian scientists reported that radar measurements by the Mars Express spacecraft detected a 20-kilometer-wide lake of liquid water that exists under Mars's southern polar plane (**Figure 15.1**)! Could life exist in this Martian lake?

Other bodies in our solar system also appear to have liquid water. One example is Enceladus, Saturn's sixth-largest moon. In 2005, the unmanned robotic space probe *Cassini* flew over the south polar region of Enceladus and found geysers of what appeared to be steam and ice! *Cassini* was directed to fly through the plumes in 2008, discovering that they are composed of water, carbon dioxide, carbon monoxide, and carbon–hydrogen molecules. Given that all living organisms have molecules that include carbon atoms, the basic elements needed for life are present on Enceladus. **Figure 15.2** is a not-to-scale illustration of the model hypothesized to be the fundamental structure of Enceladus, with a rocky core, a liquid water ocean, and a solid water crust.

Is there evidence of liquid water existing beyond the solar system? A good answer to this question cannot be given at the current point in time. Certainly, *detection* of liquid water outside of our solar systems is slightly beyond our current capabilities. However, *models* of planets outside of the solar system indicate that planets with a radius about 2.5 times that of Earth likely contain water and 35% of the planets larger than Earth are likely to be water-rich. The James Webb Space Telescope (Section 8.1) should provide data that can refute or verify these models. In the words of Sara Seager, Professor of Planetary Science and Physics at the Massachusetts

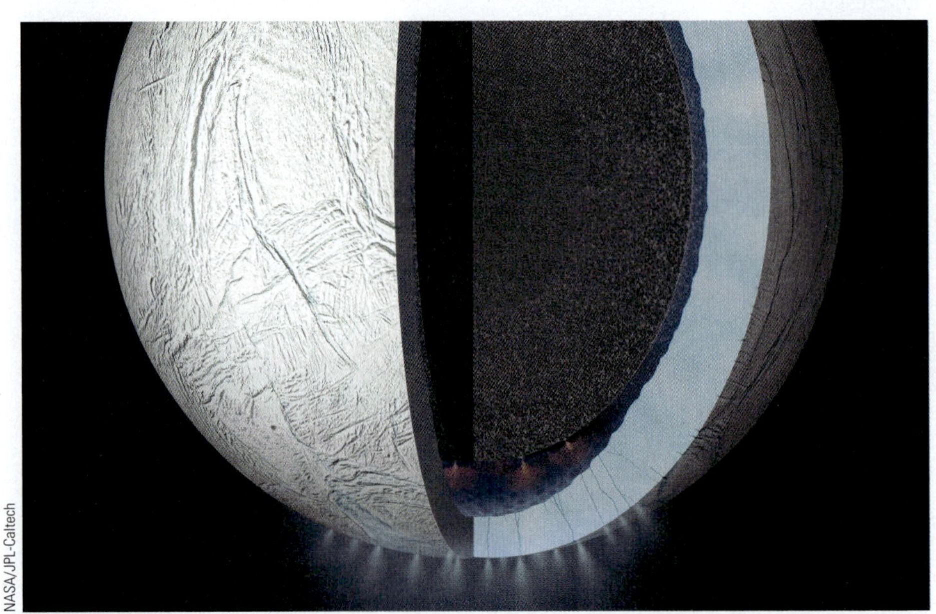

Figure 15.2 A cutaway model of the fundamental structure of Enceladus, a moon of Saturn. The solid planetary core is illustrated in black and gray, its liquid ocean is show in blue, and its ice (solid water) crust is shown in white. Geysers emitting water and a mixture of carbon-atom-containing molecules are illustrated in the southern region of the moon.

Institute of Technology (**Figure 15.3**), "It's amazing to think that the enigmatic intermediate-size exoplanets could be water worlds with vast amounts of water."

Sam Ogden/Science Source

Figure 15.3 Sara Seager, Professor of Planetary Science and Physics at the Massachusetts Institute of Technology. Her science research focuses on theory, computation, and data analysis of exoplanets, those that are outside of our solar system.

15.2 Dalton's Law of Partial Pressures

Goal 1 Given the partial pressure of each component in a mixture of gases, find the total pressure.

2 Given the total pressure of a gaseous mixture and the partial pressures of all components except one, or information from which those partial pressures can be obtained, find the partial pressure of the remaining component.

In Section 4.2, five macroscopic properties of gases are identified:

1. Gases may be compressed.
2. Gases may be expanded.
3. Gases have low densities.
4. Gases may be mixed in a fixed volume.
5. Gases exert constant pressure on the walls of their container uniformly in all directions.

These properties are explained by the kinetic molecular theory and the ideal gas model described in Chapters 4 and 14. A gas is made up of tiny molecules that are widely separated from one another so that they occupy the whole volume of the container that holds them. It is the vast, open space between molecules that makes it possible for gases to mix. If Gas A is added to Gas B in a rigid container (constant volume), the A molecules distribute themselves throughout the open space between the B molecules. The particle volume of the A and B molecules is negligible compared with the macroscopic volume occupied by the gas as a whole. One thing changes, though: pressure. It goes up.

Figure 15.4 illustrates an experiment. First, note that the volume of all three containers is the same. The gas pressure in the nitrogen vessel is 186 mm Hg, and the pressure in the oxygen vessel is 93 mm Hg, as shown. When the two samples are

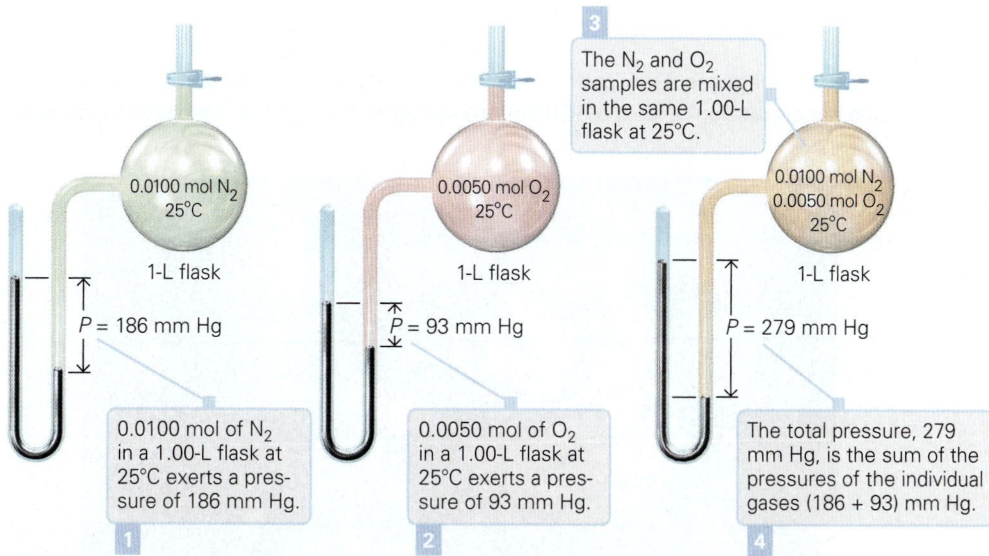

0.0100 mol N_2
25°C
1-L flask
P = 186 mm Hg
1
0.0100 mol of N_2 in a 1.00-L flask at 25°C exerts a pressure of 186 mm Hg.
0.0050 mol O_2
25°C
1-L flask
P = 93 mm Hg
2
0.0050 mol of O_2 in a 1.00-L flask at 25°C exerts a pressure of 93 mm Hg.
3
The N_2 and O_2 samples are mixed in the same 1.00-L flask at 25°C.
0.0100 mol N_2
0.0050 mol O_2
25°C
1-L flask
P = 279 mm Hg
4
The total pressure, 279 mm Hg, is the sum of the pressures of the individual gases (186 + 93) mm Hg.

Figure 15.4 Experimental apparatus to demonstrate Dalton's Law of Partial Pressures. Nitrogen gas exerts a pressure of 186 mm Hg. Oxygen gas exerts a pressure of 93 mm Hg. If the gases are combined in a container of the same volume at the same temperature, the total pressure is the sum of the individual pressures: 186 mm Hg + 93 mm Hg = 279 mm Hg.

mixed into a single vessel, the pressure exerted by the combined gases is the sum of the pressures of the individual gases, 186 mm Hg + 93 mm Hg = 279 mm Hg. In effect, each gas continues to exert the same pressure it exerted before the gases were mixed. After mixing, the individual gases are exerting their pressures in the same container. The total pressure is the sum of the two individual pressures.

John Dalton, of atomic theory fame, summed up these observations in what is now known as **Dalton's Law of Partial Pressures**: The total pressure exerted by a mixture of gases is the sum of the partial pressures of the gases in the mixture. The partial pressure of one gas in a mixture is the pressure that gas would exert if it alone occupied the same volume at the same temperature. Mathematically, this is

Dalton's atomic theory proposes that all matter is made up of tiny particles called *atoms*. His theory is described in Section 5.2.

$$P = p_1 + p_2 + p_3 + \ldots$$

where P is the total pressure and p_1, p_2, p_3, . . . are the partial pressures of gases 1, 2, 3, ... Notice that we use an uppercase P for total pressure and a lowercase p for partial pressure.

Active Example 15.1 Dalton's Law of Partial Pressures I

In a gas mixture the partial pressure of methane is 154 torr; of ethane, 178 torr; and of propane, 449 torr. Find the total pressure exerted by the mixture.

Think Before You Write The problem statement gives you the partial pressures of the components of a gas mixture, and it asks for total pressure. This indicates that you apply Dalton's Law of Partial Pressures.

Answers *Cover the left column with your cut-out shield. Reveal each answer only after you have written your own answer in the right column.*

$P = p_1 + p_2 + p_3 =$ 154 torr + 178 torr + 449 torr = 781 torr	Apply the relationship $P = p_1 + p_2 + p_3 + \ldots$. Set up and solve. 
You improved your understanding of Dalton's Law of Partial Pressures.	What did you learn by solving this Active Example?

Practice Exercise 15.1

A scientist sets up an artificial atmosphere with four gases. Their partial pressures are 0.78 bar nitrogen, 0.21 bar oxygen, 0.009 bar argon, and 0.025 bar water. (See the discussion of air in a SCUBA tank in Figure 15.5.) What is the total pressure?

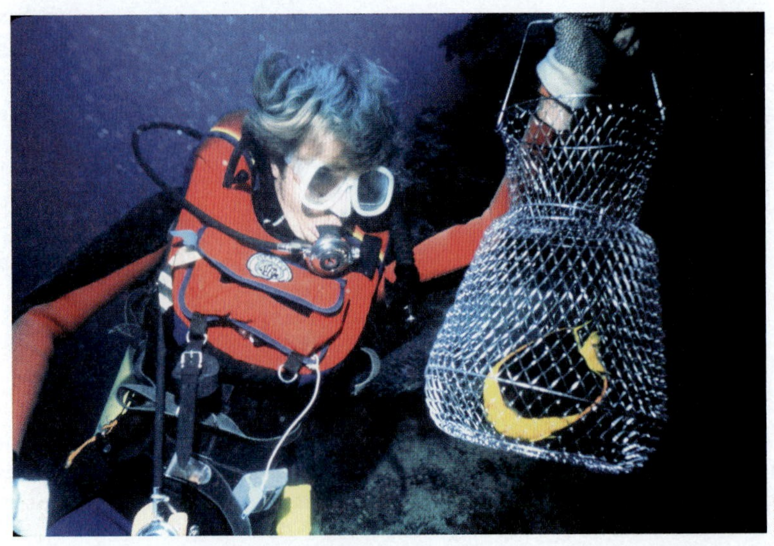

NOAA

Figure 15.5 SCUBA (self-contained underwater breathing apparatus) diving. The solubility of gases in the blood is proportional to their partial pressures. If air is compressed, the partial pressures of nitrogen and oxygen are increased, leading to problems such as nitrogen narcosis and oxygen toxicity. Helium is added to the breathing mixture for deep-sea diving.

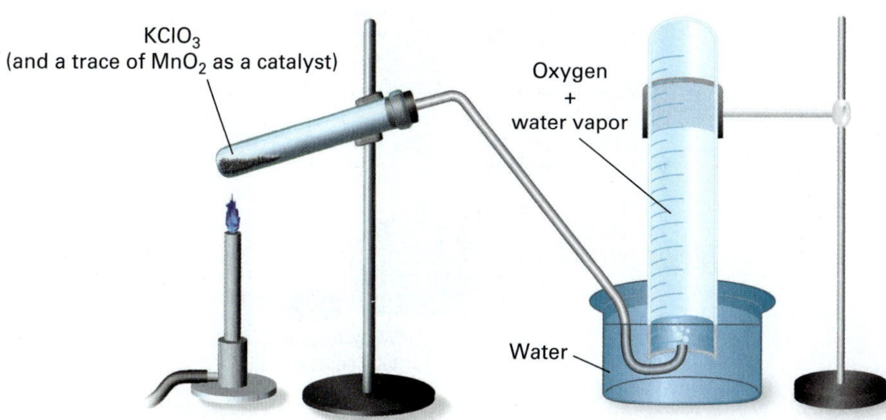

Figure 15.6 Laboratory preparation of oxygen. The heat applied to the test tube decomposes the solid $KClO_3$ into gaseous O_2 and solid KCl. The oxygen is directed into the bottom of an inverted bottle that is initially filled with water. As the oxygen accumulates in the bottle, it displaces the water and is saturated with water vapor until the bottle is filled with a mixture of oxygen and water vapor.

Gases generated in the laboratory may be collected by bubbling them through water, as shown in **Figure 15.6**. As the oxygen bubbles rise through the water, they become saturated with water vapor. The gas collected is therefore actually a mixture of oxygen and water vapor. The pressure exerted by the mixture is the sum of the partial pressure of the oxygen and the partial pressure of the water vapor (**Figure 15.7**). As you will see in Section 15.5, water vapor pressure depends only on temperature. Its values at different temperatures may be found by doing a search for "water vapor pressure table" on the Internet or by consulting a reference book.

Active Example 15.2 Dalton's Law of Partial Pressures II

Oxygen is generated for a laboratory experiment by bubbling the gas through water, as illustrated in Figures 15.6 and 15.7. The total pressure of the oxygen saturated with water vapor is 755 torr. The temperature of the gas mixture is 22°C, and water vapor pressure at that temperature is 19.8 torr. What is the partial pressure of the oxygen?

Think Before You Write You are given the total pressure of a mixture of gases, and you are also given the partial pressure of one of the components. You are asked to determine the partial pressure of the other component of the mixture. This should lead you to recognize that this is an application of Dalton's Law of Partial Pressures.

Answers *Cover the left column with your cut-out shield. Reveal each answer only after you have written your own answer in the right column.*

$P = p_{H_2O} + p_{O_2}$ Rearranging, $p_{O_2} = P - p_{H_2O} =$ 755 torr − 19.8 torr = 735 torr	Rewrite $P = p_1 + p_2 + p_3 + \ldots$ in terms of the total pressure of the mixture, the partial pressure of oxygen, and the partial pressure of water so that it specifically applies to this problem. Then rearrange the equation so that the unknown, p_{O_2}, is isolated on one side of the equals sign. Finish by substituting the measured quantities and solving.
You improved your understanding of Dalton's Law of Partial Pressures.	What did you learn by solving this Active Example? 

Practice Exercise 15.2

Calculate the partial pressure of oxygen (in atm) in a sample collected over water, given that the total pressure of the mixture is 0.98 atm and the partial pressure of water vapor is 16.5 mm Hg.

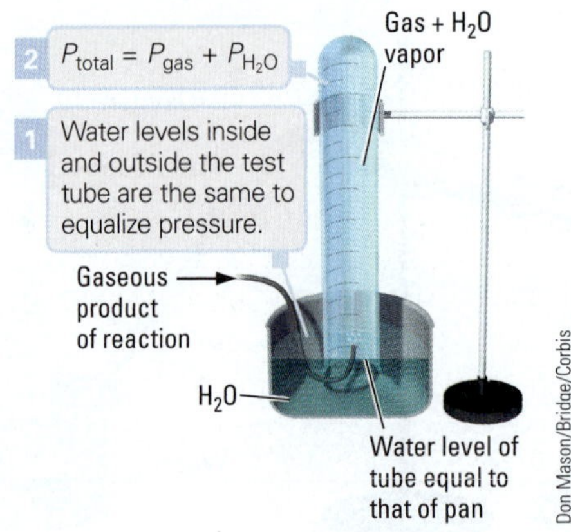

Figure 15.7 Applying Dalton's Law of Partial Pressures to a gas collected over water. When the water levels inside and outside of the tube are equalized, the pressure inside the tube is equal to atmospheric pressure, and thus P_{total} may be easily measured. The water vapor pressure depends on temperature, so its value can be looked up in a reference source. The pressure of the gaseous product is, therefore, the total pressure minus the water vapor pressure.

15.3 Properties of Liquids

Goal 3 Explain the differences between the physical behavior of liquids and gases in terms of the relative distances among particles and the effect of those distances on intermolecular attractions.

4 For two liquids, given comparative values of physical properties that depend on intermolecular attractions, predict the relative strengths of those attractions; or, given a comparison of the strengths of the intermolecular attractions, predict the relative values of physical properties that the attractions cause.

The properties of liquids are easy to observe and describe—more so than the properties of gases. To understand liquid properties, however, it is helpful to compare the structure of a liquid with the structure of a gas. In Chapter 4, you learned that gas particles are so far apart that attractive and repulsive forces between the particles are negligible. These forces are electrostatic in character. They are inversely related to the distance between the particles; the smaller the distance is, the stronger the forces are. In a liquid, particles are very close to one another. Consequently, the intermolecular attractions in a liquid are strong enough to affect its physical properties.

We can now compare the properties of liquids with five properties of gases that were listed in Section 4.2 (Figures 4.2 through 4.5):

1. *Gases may be compressed; liquids cannot.* Liquid particles are "touchingly close" to one another. There is no space between them, so they cannot be pushed closer, as in the compression of a gas.

2. *Gases may be expanded; liquids cannot.* The strong attractions between liquid particles hold them together; the volume of a liquid is constant, no matter the container size.

3. *Gases have low densities; liquids have relatively high densities.* Density is mass per unit volume—mass divided by volume. If the particles of a liquid are close together compared with the particles of a gas, a given number of liquid particles will occupy a much smaller volume than the same number of particles will occupy as a gas. A small denominator in the density ratio for a liquid means a higher value for the ratio.

4. *Gases may be mixed in a fixed volume; liquids cannot.* When one gas is added to another, the particles of the second gas occupy some of the space between the particles of the first gas. There is no space between particles of a liquid, so combining liquids must increase volume.

5. *Gases exert constant pressure on the walls of their container uniformly in all directions; the pressure in a liquid container increases with increasing depth.* For small samples of enclosed gases, gas pressure is independent of external factors such as gravitational forces. Liquid pressures depend on the depth of the liquid due to variation in weight at varying depth.

A liquid has several measurable properties with values that depend on intermolecular attractions, the tendency of the particles to stick together. In fact, if you think in terms of the "stick-togetherness" as equated with the strength of the intermolecular attractions, you can usually predict relative values of these properties for two liquids. The greater the stick-togetherness is, the stronger the intermolecular attractions are. We will now identify five of these properties.

Vapor Pressure The open space above any liquid contains some particles in the gaseous, or vapor, state. This is due to **evaporation** (**Figure 15.8**). The partial pressure exerted by these gaseous particles is called **vapor pressure**. If the gas space above the liquid is closed, the vapor pressure increases to a definite value called the **equilibrium vapor pressure**. In Section 15.5, you will study the mechanics by which that pressure is reached. Vapor pressure is inversely related to intermolecular attractions. If stick-togetherness is high between liquid particles, very few liquid particles "escape" into the gaseous state, so the vapor pressure is low.

Heat of Vaporization Energy must be transferred to a liquid to first overcome intermolecular attractions, then separate liquid particles from one another, and continue to keep them apart (**Figure 15.9**). The quantity of energy required to change one mole of a liquid to its vapor while at constant temperature and pressure is called **heat of vaporization**. The energy released in the opposite process, as vapor condenses to the liquid phase, is called **heat of condensation**. The greater the stick-togetherness is, the greater is the amount of energy transferred into or out of the liquid.

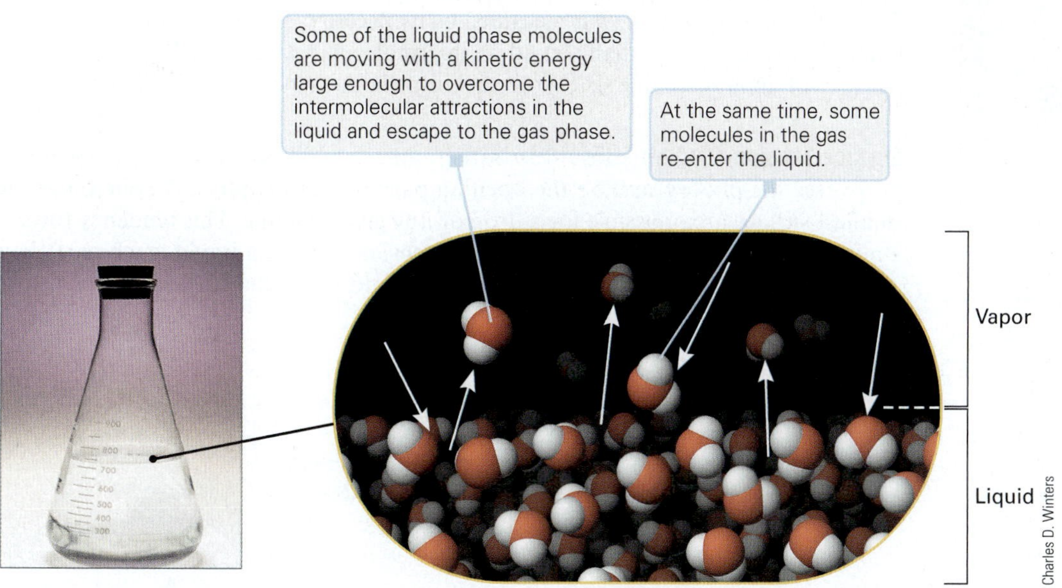

Some of the liquid phase molecules are moving with a kinetic energy large enough to overcome the intermolecular attractions in the liquid and escape to the gas phase.
At the same time, some molecules in the gas re-enter the liquid.
Vapor
Liquid
Charles D. Winters

Figure 15.8 Evaporation. Some molecules at the surface of the liquid phase have sufficient energy to escape and enter the vapor phase. Some molecules in the vapor phase will re-enter the liquid phase when they make contact with the surface. The partial pressure exerted by the molecules in the vapor phase is called *vapor pressure*.

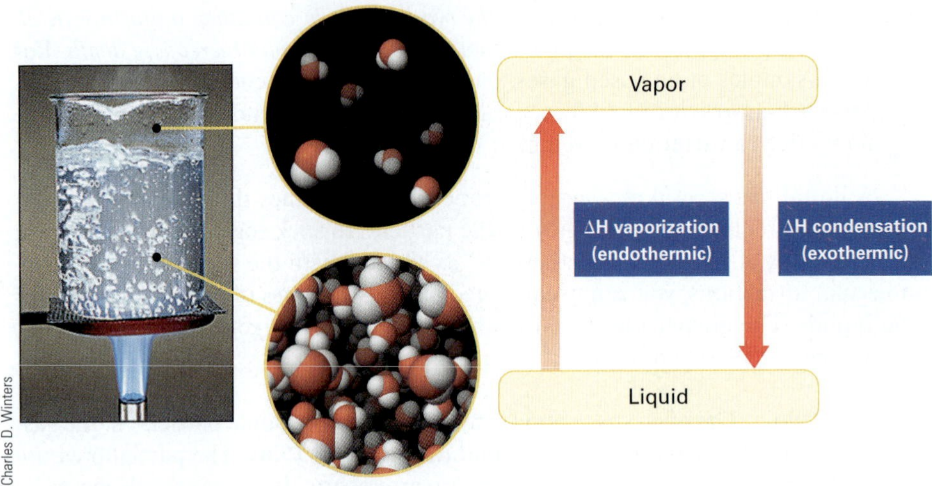

Figure 15.9 Heats of vaporization and condensation. Energy must be transferred to a system to overcome the attractive forces that are exerted among liquid molecules. When an equal quantity of vapor condenses to a liquid, an equal amount of energy is transferred out of the vapor.

Boiling Point Liquids can be changed to gases by boiling. The boiling process is discussed in more detail in Section 15.6. At the **boiling point**, the average kinetic energy of the liquid particles is high enough to overcome the forces of attraction that hold the particles in the liquid state. When stick-togetherness is high, it takes more motion (a higher temperature) to separate the particles within the liquid, where boiling occurs.

The trends in vapor pressure, molar heat of vaporization, and boiling point are shown for several substances in Table 15.1.

Viscosity Particles in a liquid are free to move about relative to one another; they "flow." Some liquids flow more easily than others. Water, for example, can be poured much more freely than syrup, and syrup pours more readily than honey (Figure 15.10). The ability of a liquid to flow is measured by its **viscosity**. Viscosity is an internal resistance to flow, and it is partially based on intermolecular attractions. When comparing particles of about the same size, more stick-togetherness means higher viscosity.

Figure 15.10 Viscosity. Honey pours relatively slowly, and thus it has a relatively high viscosity. This macroscopic characteristic indicates that there are relatively strong intermolecular attractions among the molecules that make up the mixture commonly known as honey.

Surface Tension When a liquid is broken into "small pieces," it forms spherical drops (see the photograph on the opening page of this chapter). A sphere has the smallest surface area possible for a drop of any given volume. This tendency toward a minimum surface is the result of **surface tension**. Within a liquid, each particle is attracted in all directions by the particles around it. At the surface, however, the

Table 15.1 Physical Properties of Liquids

Substance	Vapor Pressure at 20°C (torr)	Heat of Vaporization (kJ/mol)	Normal Boiling Point (°C)	Intermolecular Attractions
Mercury	0.0012	59	357	Strongest ↑
Water	17.5	41	100	
Benzene	75	31	80	
Ether	442	26	35	
Ethane	27,000	15	−89	Weakest

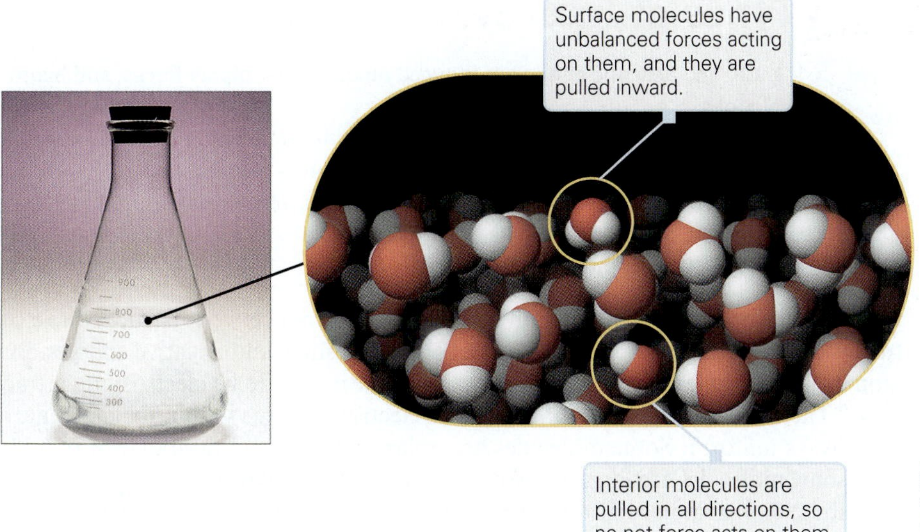

Surface molecules have unbalanced forces acting on them, and they are pulled inward.
Interior molecules are pulled in all directions, so no net force acts on them.
Charles D. Winters

Figure 15.11 Surface tension. Unbalanced attractive forces at the surface of a liquid pull surface molecules downward and sideways but not up. Molecules within the water are attracted in all directions. Surface tension is a result of the difference in these forces.

attraction is almost entirely downward, pulling the surface molecules into a sort of tight skin over a standing liquid. Similarly, in a spherical drop, the attraction at the surface is almost entirely inward. This is surface tension (**Figure 15.11**). Its effect in water may be seen when a needle floats if placed gently on a still surface, or when small bugs run across the surface of a quiet pond (**Figure 15.12**). High stick-togetherness at the surface means more resistance to anything that would break through or stretch that surface.

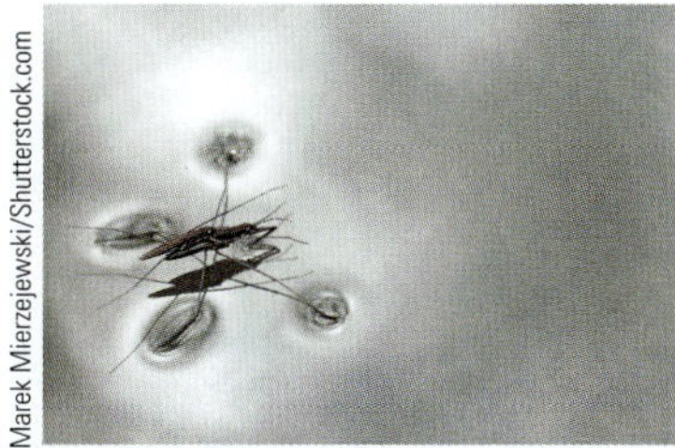

Figure 15.12 Insect walking on water. The surface tension of water creates a difficult-to-penetrate skin that will support small bugs or thin pieces of dense metals, such as a needle or razor blade. A bug literally runs on the water; it does not float in it.

a summary of... Properties of Liquids and Intermolecular Attractions

Vapor pressure: Liquids with relatively strong intermolecular attractions evaporate less readily, yielding lower vapor concentrations and therefore lower vapor pressures than liquids with weak intermolecular attractions.

Heat of vaporization: The heat of vaporization of a liquid with strong intermolecular attractions is higher than the heat of vaporization of a liquid with weak intermolecular attractions.

Boiling point: Liquids with strong intermolecular attractions require higher temperatures for boiling than liquids with weak intermolecular attractions.

Viscosity: Liquids with strong intermolecular attractions are generally more viscous than liquids with weak intermolecular attractions.

Surface tension: Liquids with strong intermolecular attractions have higher surface tension than liquids with weak intermolecular attractions.

✓ Target Check 15.1

a) What main difference between gases and liquids at the particulate level accounts for the large differences in their macroscopic properties?

b) Intermolecular attractions are stronger in A than in B, with all other potentially influencing factors being about equal. Which do you expect will have the higher surface tension, molar heat of vaporization, vapor pressure, boiling point, and viscosity?

c) X has a higher molar heat of vaporization than Y. Which do you expect will have a higher vapor pressure? Why?

15.4 Types of Intermolecular Forces

Goal 5 Identify and describe or explain induced dipole forces, dipole forces, and hydrogen bonds.

6 Given the structure of a molecule, or information from which it may be determined, identify the significant intermolecular forces present.

7 Given the molecular structures of two substances, or information from which they may be obtained, compare or predict relative values of physical properties that are related to them.

It was stated in the preceding section that attractive forces between particles are electrostatic in character; the attractions are between positive and negative charges. But atoms and molecules are electrically neutral. How can there be electrostatic attractions? The answer is that the *distribution* of electrical charge within the molecule is not always uniform. Some molecules are polar and some are nonpolar. ◀ In addition, some molecules are large and some are small. Molecular polarity and size both contribute to intermolecular attraction and therefore to physical properties.

▶ **P/Review** If you have drawn a Lewis diagram of a molecule (Section 13.2), you can predict that it is polar if either of these conditions exists: a central atom has a lone pair of electrons, or a central atom is bonded to atoms of different elements (see Section 13.6).

Three kinds of **intermolecular forces** can be traced to electrostatic attractions: dipole forces, induced dipole forces, and hydrogen bonds.

1. *Dipole forces.* A polar molecule is sometimes described as a **dipole**. The attraction between dipoles is between the positive pole of one molecule and the negative pole of another. Figure 15.13 shows the alignment of dipoles, one of several ways polar molecules attract one another.

   Table 15.2 compares the boiling points of four pairs of substances that have about the same molecular size, indicated approximately by their molar masses. In each pair, the boiling point of the substance with polar molecules is higher than the boiling point of the nonpolar substance. This is because polar molecules have stronger intermolecular attractions than nonpolar molecules.

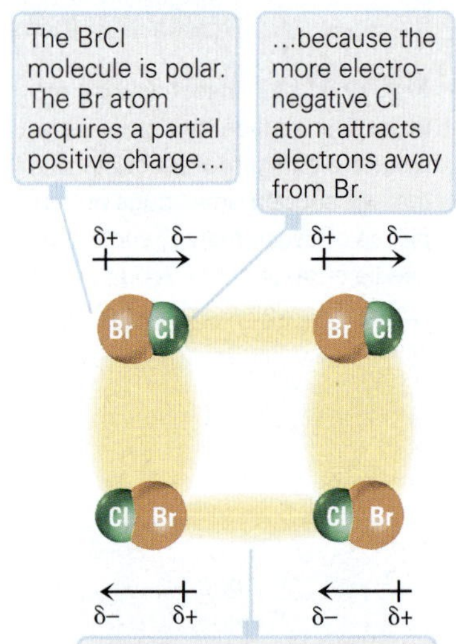
The BrCl molecule is polar. The Br atom acquires a partial positive charge...
...because the more electronegative Cl atom attracts electrons away from Br.
δ+ δ−
δ+ δ−
Br Cl
Br Cl
Cl Br
Cl Br
δ− δ+
δ− δ+
The partial negative region of one molecule is attracted to the partial positive region of the neighboring molecule. This results in dipole-dipole forces among the molecules.

Figure 15.13 Dipole forces. Molecules tend to arrange themselves by bringing oppositely charged regions close to one another and forcing similarly charged regions away from one another.

2. *Induced dipole forces.* Attractions between nonpolar molecules are called **induced dipole forces**. These are also called **dispersion forces**, **London forces**, or **London dispersion forces**. They are believed to be the result of shifting electron clouds within the molecules. If the electron movement in a molecule results in a temporary concentration of electrons at one side of the molecule, the molecule becomes a "temporary dipole." This is shown in the rightmost of the center pair of molecules in Figure 15.14. The electrons repel the electrons in the molecule next to it, pushing them to the far side of that molecule. The second molecule is thus "induced" to form a second temporary dipole (see the right pair in Figure 15.14). As long as these dipoles exist—a very small fraction of a second in each case—there is a weak attraction between them.

   The strength of induced dipole forces depends on the ease with which electron distributions can be induced to be distorted or "polarized." Large molecules, with many electrons and with electrons far removed from atomic nuclei, are more easily polarized than small molecules. Larger molecules also generally

Table 15.2 Boiling Points of Polar Versus Nonpolar Substances

Formula	Polar or Nonpolar	Molecular Mass (u)	Boiling Point (°C)	Formula	Polar or Nonpolar	Molecular Mass (u)	Boiling Point (°C)
N_2	Nonpolar	28	−196	GeH_4	Nonpolar	77	−88
CO	Polar	28	−192	AsH_3	Polar	78	−63
SiH_4	Nonpolar	32	−112	Br_2	Nonpolar	160	59
PH_3	Polar	34	−88	ICl	Polar	162	97

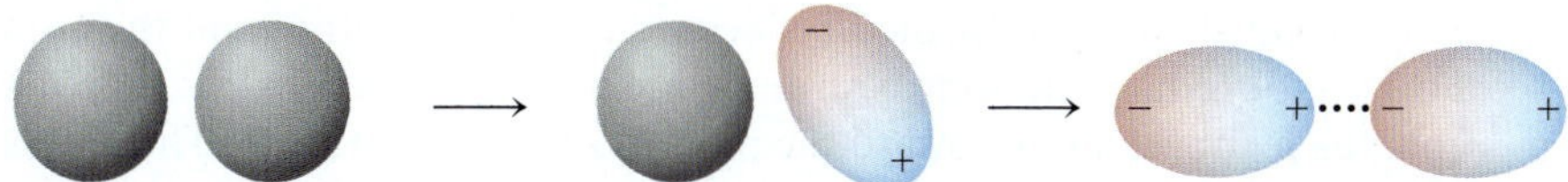

Two nonpolar atoms or molecules (depicted as having an electron cloud that has a time-averaged spherical shape).

Momentary attractions and repulsions between nuclei and electrons in neighboring molecules lead to induced dipoles.

Correlation of the electron motions between the two atoms or molecules (which are now dipoles) leads to a lower energy and stabilizes the system.

Figure 15.14 Induced dipole forces. Electron clouds in nonpolar molecules are constantly shifting. The temporary dipole in the right molecule in the center "induces" the molecule next to it to become another temporary dipole (right pair). The "instantaneous dipoles" are attracted to one another briefly. A small fraction of a second later, the clouds shift again and continue to interact with one another or with nearby molecules.

Charles D. Winters

Figure 15.15 Induced dipole forces and molecular size. Larger molecules, with a greater number of electrons, polarize more easily than smaller molecules. Bromine (*left*) exists as a liquid at room conditions, and iodine (*right*) is a solid. Both molecules are nonpolar and have induced dipole forces as the dominant intermolecular force. However, bromine molecules have 70 electrons, whereas iodine molecules have 106 electrons, so iodine molecules polarize more easily. Thus, the intermolecular attractive forces among iodine molecules are greater than among bromine molecules. This particulate-level difference is responsible for the macroscopic observations that iodine is a solid and bromine is a liquid at room temperature.

have greater mass. Consequently, intermolecular forces tend to increase with increasing molecular mass among otherwise similar substances (**Figure 15.15**). Notice in Table 15.2 the increase in boiling points for both polar and nonpolar molecules as molecular mass increases.

3. *Hydrogen bonds.* Some polar molecules have intermolecular attractions that are much stronger than ordinary dipole forces. These molecules always have a hydrogen atom bonded to an atom that is small and highly electronegative and that has at least one unshared pair of electrons. ▶ Nitrogen, oxygen, and fluorine are generally the only elements whose atoms satisfy these requirements. Study **Figure 15.16** to learn how to recognize hydrogen bonds.

◀ **P/Review** Electronegativity estimates the strength with which an atom attracts the pair of electrons that forms a bond between it and another atom. Covalent bonds are polar when there is an electronegativity difference between the bonded atoms. See Section 12.5.

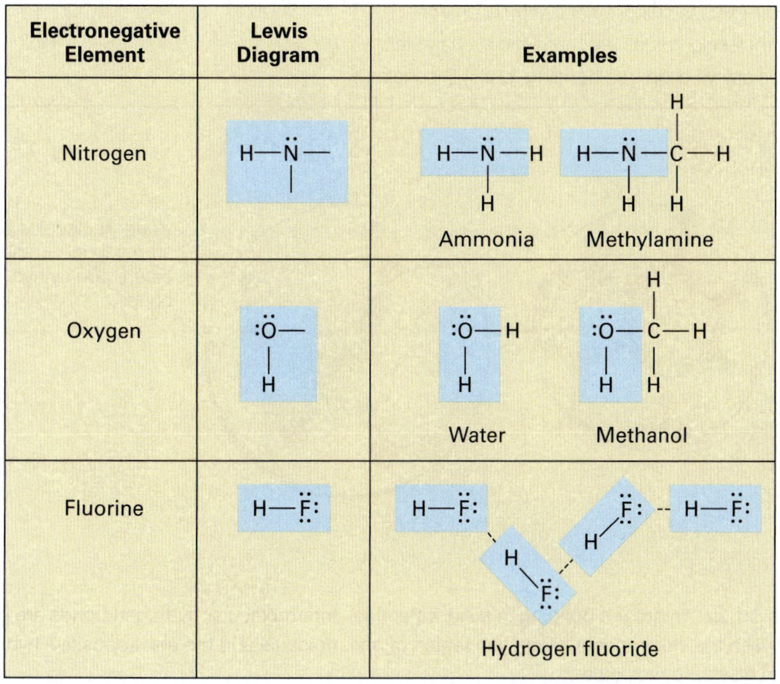

Electronegative Element	Lewis Diagram	Examples
Nitrogen	H—N̈— (with a bond below N)	H—N̈(H)—H Ammonia; H—N̈(H)—C(H)(H)—H Methylamine
Oxygen	:Ö—H (with a bond from O)	:Ö(H)—H Water; :Ö(H)—C(H)(H)—H Methanol
Fluorine	H—F̈:	H—F̈: ··· H—F̈: ··· H—F̈: ··· H—F̈: Hydrogen fluoride

Figure 15.16 Recognizing hydrogen bonding. Hydrogen bonds occur when a hydrogen atom is covalently bonded to a small atom that is highly electronegative and has one or more unshared electron pairs. Fluorine, oxygen, and nitrogen atoms fit this description. The hydrogen bond is between the atom of one of these elements in one molecule and the hydrogen atom of a nearby molecule. Hydrogen bonds between molecules are illustrated with dashed lines in the hydrogen fluoride example.

The covalent bond formed between the hydrogen atom and the atom of nitrogen, oxygen, or fluorine is strongly polar. The electron pair is shifted away from the hydrogen atom toward the more electronegative atom. This leaves the hydrogen nucleus—nothing more than a proton—as a small, highly concentrated region of positive charge at the edge of a molecule. The negative pole of another molecule, which is the region near an unshared electron pair on a nitrogen, oxygen, or fluorine atom, can get quite close to the hydrogen atom of the first molecule. This results in an extra strong attraction between the molecules. This kind of intermolecular attraction is a **hydrogen bond**.

Be sure to keep in mind that a hydrogen bond is an *intermolecular force*, an attraction between *different* molecules. It is not a covalent bond between atoms in the *same* molecule. The dotted lines in **Figure 15.17** represent hydrogen bonds between water molecules. Although a hydrogen bond is much stronger than an ordinary dipole–dipole force, it is roughly one-tenth as strong as a covalent bond between atoms of the same two elements.

Of the three kinds of intermolecular forces, hydrogen bonds are the strongest. When present between small molecules, hydrogen bonds are primarily responsible for the physical properties of a liquid. Dipole forces are the next strongest and induced dipole forces are the weakest of the three. Induced dipole forces are present between all molecules. In small molecules, induced dipole forces are important only when the other forces are absent. But between large molecules—molecules that contain many atoms or even few atoms that have many electrons—induced dipole forces are quite strong and often play the main role in determining physical properties.

Figure 15.18 summarizes the kinds of intermolecular forces and their effects on boiling points of similar compounds in three chemical families. We recommend that you study it carefully.

✓ Target Check 15.2

Identify the true statements and rewrite false statements to make them true.

a) Induced dipole forces are present only with nonpolar molecules.

b) All other things being equal, hydrogen bonds are stronger than dipole–dipole forces.

c) Polar molecules have a net electrical charge.

d) Intermolecular forces are magnetic in character.

e) H_2O displays hydrogen bonding, but H_2S does not.

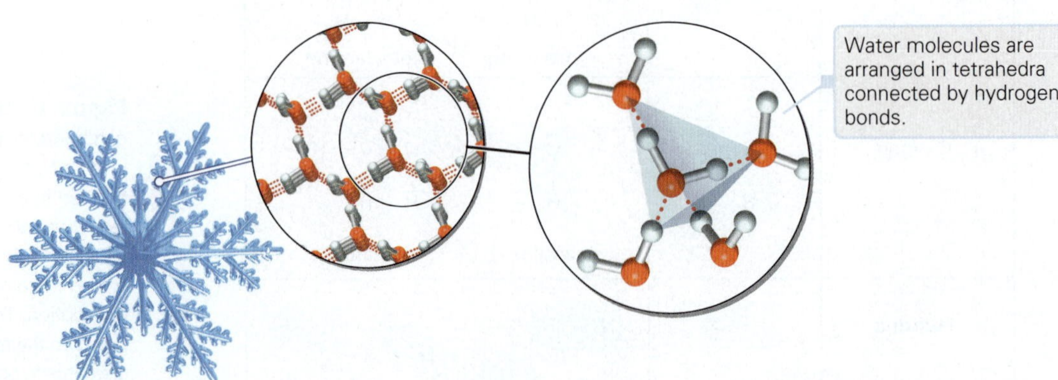

Water molecules are arranged in tetrahedra connected by hydrogen bonds.

Figure 15.17 Hydrogen bonding in solid water (ice). Intermolecular hydrogen bonds are present between the electronegative oxygen region of one molecule and the electropositive hydrogen region of a second molecule.

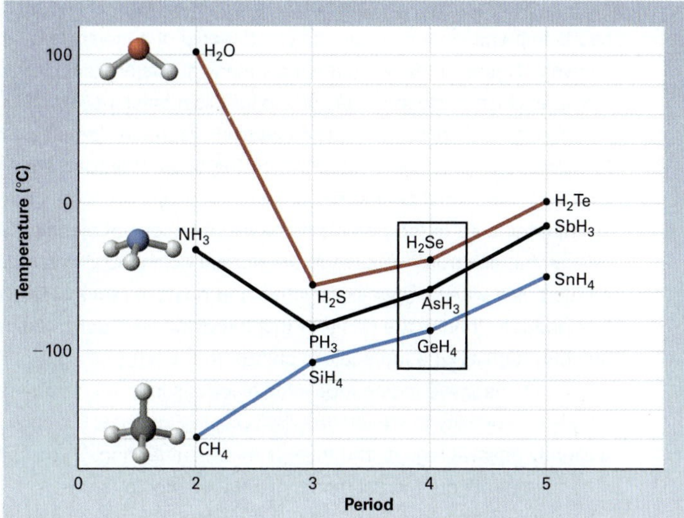

Figure 15.18 Intermolecular attractions illustrated by boiling points of hydrogen-containing compounds. The plot shows boiling point temperature (°C) versus period in the periodic table. Liquids with strong intermolecular attractions usually boil at higher temperatures than liquids with weak intermolecular attractions. These attractions are caused by induced dipole forces, dipole forces, and hydrogen bonding. Holding two of these variables essentially constant and changing the third, we can see how each variable affects the attractions by comparing the boiling points.

Induced dipole forces: The Group 4A/14 hydrogen-containing compounds (blue line) all have tetrahedral molecular geometries. They are nonpolar, and they have no hydrogen bonding. The only intermolecular forces are induced dipole forces. The molecules differ only in molecular size (mass), ranging from CH_4 (the smallest) to SnH_4 (the largest). The boiling points of the four compounds increase as their molecular sizes increase. Except for H_2O and NH_3, the same trend appears for Group 5A/15 (black line) and Group 6A/16 (red line) hydrogen-containing compounds. This suggests that, *other things being equal, intermolecular attractions increase as molecular size increases.*

Dipole forces: The molecules in the rectangle in the Period 4 hydrogen-containing compounds—H_2Se, AsH_3, and GeH_4—are about the same size (nearly equal molar mass), and none of them has hydrogen bonding. They differ most in polarity. GeH_4 has tetrahedral molecules; it is nonpolar. The trigonal pyramidal molecules of AsH_3 are polar but less so than bent H_2Se molecules. The least polar compound, GeH_4, has the lowest boiling point, and the most polar compound, H_2Se, has the highest boiling point. The same trend appears with the Period 3 and Period 5 hydrogen-containing compounds. This indicates that, *other things being equal, intermolecular attractions increase as molecular polarity increases.*

Hydrogen bonding: The high boiling points of H_2O and NH_3 violate the trends in which small molecules boil at lower temperatures than large molecules that are otherwise similar. H_2O and NH_3 are the only two substances shown that have hydrogen bonding. This indicates that, for small molecules in particular, *hydrogen bonding causes exceptionally strong intermolecular attractions.*

Target Check 15.3

Determine the molecular geometry and polarity of each of the following and, from that, identify the strongest intermolecular forces present.

a) CH_4 b) CO_2 c) OF_2 d) HOCl (the oxygen atom is central)

Target Check 15.4

Identify the molecule in each pair that you would expect to have the stronger intermolecular forces and state why.

a) CCl_4 or CBr_4 b) NH_3 or PH_3

Your Thinking

Thinking About

Mental Models

Particulate-level mental models of what *causes* each of the intermolecular forces are needed for a complete understanding of this concept. First, be sure that you grasp the distinction between *intra*molecular forces, which are chemical bonds between atoms within a single molecule, and *inter*molecular forces, the attractive forces that operate between whole molecules and other particles. Your "mental movie" needs to be based on an electron cloud model of a molecule or atom, such as the quantum model shown in Figure 11.18. No matter how many atoms are used to make up the particle, consider only the outer electron cloud of the particle for this mental model.

Consider Figure 15.14. The electron clouds in the left pair show the most common state for nonpolar molecules—one in which the outer electrons are evenly distributed throughout the molecule. Induced dipole forces are a result of a short-term, uneven shifting of this electron cloud so that one area of the particle has a slight excess of electrons and the opposite side is slightly deficient in electrons, relative to the normal balanced distribution. This makes the molecule just a little polar, with very weak negative and positive regions. The charged region of this molecule then induces (hence the name of this force) its neighbor to adjust its electron distribution. If the electron density of a molecule is shifted to the left (the right particle of the center pair in Figure 15.14), the negative region repels the electrons in the right side of its neighbor, inducing it to shift its electron density to the left (the right pair in Figure 15.14). Now, a slightly negative region is near a slightly positive region, and there is a weak attraction. If you can imagine this process, you have a good start at your mental model of induced dipole forces.

Now imagine a particle in which the uneven distribution of electrons is permanent. This is the case in polar molecules, as illustrated in Figure 15.13. The chlorine end of each molecule is the portion with a greater electron density and the accompanying partial negative charge. The region of a polar molecule with a partial negative charge is attracted to the positively charged end of a neighboring polar molecule.

The third type of mental model for intermolecular forces that you need to develop is to imagine how hydrogen bonds form. Before you think about the intermolecular forces, consider an individual molecule that has a hydrogen atom covalently bonded to oxygen, such as water. The bonding electrons in the O—H bond are displaced toward the oxygen atom, which is relatively small and highly electronegative. This leaves the hydrogen atom nucleus—a positively charged proton—at the end of the molecule with very little surrounding electron density. That positively charged region will be strongly attracted to the oxygen-atom region of a neighboring molecule, given that the oxygen atom not only has two unshared pairs of electrons but also has its bonding electrons shifted toward it and away from the hydrogen atom. The result is an intermolecular attractive force between the highly electronegative atom of one molecule and the highly electropositive hydrogen atom of a neighboring molecule.

The ability to imagine the particulate-level reasons for intermolecular attractive forces is a powerful tool in your "thinking as a chemist" knowledge stockpile. The more you understand this type of particulate-level behavior, the better you will understand the macroscopic properties that result from it.

15.5 Liquid–Vapor Equilibrium

Goal 8 Describe and explain the equilibrium between a liquid and its own vapor and the process by which it is reached.

In Section 4.5 we discussed how temperature is a measure of the average kinetic energy of the particles in a sample. The range of kinetic energies at a certain temperature is shown in **Figure 15.19**. Kinetic energy is plotted horizontally, and the fraction, or percentage, of the sample having a given kinetic energy is plotted vertically. The area beneath the curve represents all of the particles in the sample (100% of the sample). This type of graph is called a **kinetic energy distribution curve**.

To evaporate or vaporize, a molecule must be at the surface of a liquid. It also must have enough kinetic energy to overcome the attractions of other molecules that would hold it in the liquid state. If E in Figure 15.19 represents this minimum amount of kinetic energy—we will call it the *escape energy*—only those surface molecules having that energy or more can get away. The fraction, or percentage, of all surface molecules having that much energy is given by the area beneath the curve to the right of E.

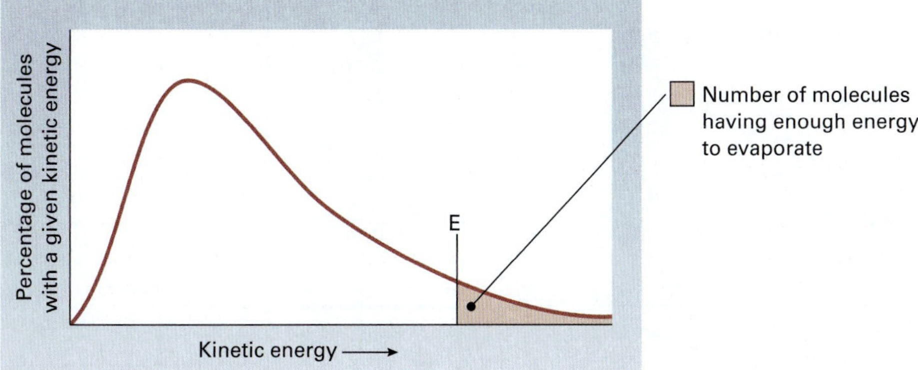

Figure 15.19 Kinetic energy distribution curve for a liquid at a given temperature. E is the escape energy, the minimum kinetic energy a molecule must have to break from the surface and evaporate into the gas phase. The total area between the curve and the horizontal axis represents the total number of molecules in the sample. The shaded area beneath the curve to the right of E represents the number of molecules having kinetic energy equal to or greater than E. At the temperature for which the graph is drawn, only a small fraction of all molecules has enough energy to evaporate.

The rate at which a particular liquid evaporates depends on two things, temperature and surface area. If we think in terms of a unit area and hold temperature constant, the vaporization rate is also constant. These conditions are assumed in the experiment described in **Figure 15.20**. Study that figure and its caption carefully before proceeding to the next paragraph, which is based on the illustration. The caption contains the information you need to satisfy Goal 8.

When a system such as that described in Figure 15.20 reaches the point at which the rates of change in opposite directions are equal, the system is said to be in **equilibrium**. ▶ It is important to note that once equilibrium is reached (Times 3 and 4 in Figure 15.20), the vapor concentration remains constant. ▶ On the *macroscopic* level, nothing is happening. On the *particulate* level, however, molecules are continually switching between the liquid and vapor states. Because of this constant activity, this kind of equilibrium is called a **dynamic equilibrium** (*dynamic* means "characterized by constant activity").

◀ **P/Review** The reversibility of chemical changes was first mentioned in Section 8.8. The dynamic equilibrium described here is for a physical change. Another physical equilibrium is described in Section 16.4. Chemical equilibria are those for which rates of reversible reactions in the forward and reverse directions are equal. Chemical equilibrium is examined in Section 18.6.

When the vapor concentration becomes constant at equilibrium, vapor pressure also becomes constant. The partial pressure exerted by a vapor in equilibrium with its liquid phase at a given temperature is the **equilibrium vapor pressure** of the substance at that temperature. Changes that occur in either direction, such as the change from a liquid to a vapor and the opposite change from a vapor to a liquid, are called **reversible changes**; if the change is chemical, it is a **reversible reaction**. Chemists write equations describing reversible changes with a double arrow, one pointing in each direction. For example, the reversible change between liquid benzene, $C_6H_6(\ell)$, and benzene vapor, $C_6H_6(g)$ (**Figure 15.21**), is represented by the following equation:

$$C_6H_6(\ell) \rightleftharpoons C_6H_6(g)$$

The term *humidity*, as it is commonly used in everyday language to describe weather, refers to what scientists call *relative humidity*, which is the ratio of the actual water vapor pressure to the equilibrium water vapor pressure at that temperature, $p_{H_2O}/p_{H_2O_{eq}}$. At high humidity, sweating is less effective at cooling our bodies because the evaporation rate and condensation rate of water in the air are nearly equal.

Thinking About Your Thinking

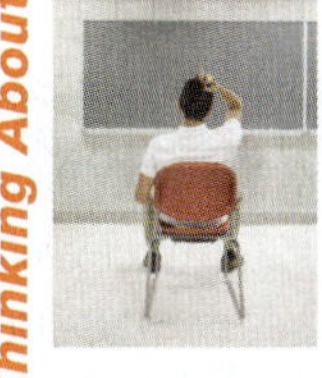

Mental Models

Developing a mental model of equilibrium processes is one of the most important thinking skills you will acquire in your chemistry coursework. Many chemical reactions that occur in living organisms and in the environment are reversible equilibrium processes.

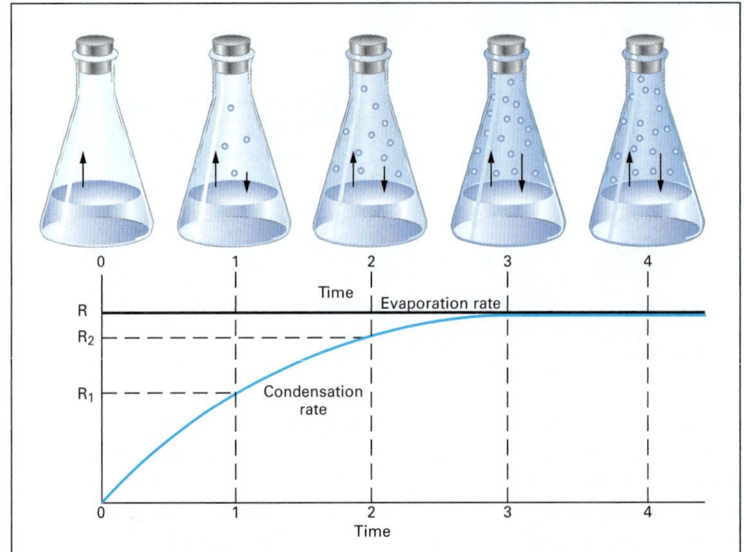

Figure 15.20 Liquid–vapor equilibrium. A liquid with weak intermolecular forces, such as benzene (Figure 15.21), is placed in an Erlenmeyer flask, which is then closed with a rubber stopper. At constant temperature and surface area, the liquid begins to evaporate at a constant rate. This rate is represented by the fixed length of the arrows pointing upward from the liquid in each view of the flask. It is also shown as the black horizontal line in the graph of evaporation rate versus time.

At first (Time 0), the movement of molecules is entirely in one direction, from the liquid to the vapor. As the concentration of molecules in the vapor builds up, an occasional molecule hits the surface and reenters the liquid. The change of state from gas to a liquid is called *condensation.* The rate of condensation per unit area at constant temperature (blue line in the graph) depends on the concentration of molecules in the vapor state. At Time 1, there will be a small number of molecules in the vapor state, so the condensation rate (R_1) will be more than zero but much less than the evaporation rate (R). This is shown by the condensation arrow being shorter than the evaporation arrow in the Time 1 flask.

As long as the rate of evaporation is greater than the rate of condensation, the vapor concentration will increase. Therefore, the rate of return from vapor to liquid increases with time (Time 2). Eventually, the rates of vaporization and condensation become equal (Time 3). The number of molecules moving from vapor to liquid in unit time just balances the number moving in the opposite direction. There is no further change in the vapor concentration, so the opposing rates remain equal (Time 4).

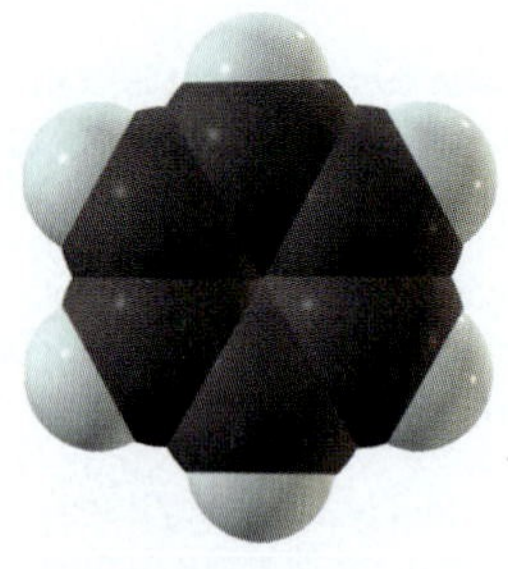

Figure 15.21 Benzene, C_6H_6, is an important substance in chemistry. Its three-dimensional structure is shown in the figure. Carbon atoms are black, and hydrogen atoms are white. It has weak intermolecular forces operating among its molecules, and therefore it develops a relatively high equilibrium vapor pressure. It is the starting point for producing many organic molecules, and it is a widely used solvent. It occurs naturally in petroleum. See Section 21.8.

Figure 15.20 is the starting point for your mental model. Turn that static series of "snapshots" into a dynamic "video clip" in your mind. Imagine the particles evaporating at a constant speed throughout the process. Regardless of the point in the process, the number of particles evaporating per unit of time remains the same. Now add the condensing particles to your mental video. The condensation speed depends on the number of particles in the space above the liquid. Initially, it is at zero, and it increases, rapidly at first, and then more slowly as equilibrium is approached, until the speed at which the particles return to the liquid is equal to the speed at which they leave the liquid. Equilibrium is established when the rate of evaporation equals the rate of condensation.

When you encounter equilibrium questions on examinations, use your mental model as you develop a written answer. Imagine the scenario posed in the question and give a written description of what you "see" happening in your mind.

The Effect of Temperature

Goal 9 Describe the relationship between vapor pressure and temperature for a liquid–vapor system in equilibrium; explain this relationship in terms of the kinetic molecular theory.

Many laboratory experiments show that vapor pressure increases as temperature rises. **Figure 15.22** shows how the vapor pressures of several substances change with temperature. Notice how a relatively small increase in temperature causes a large

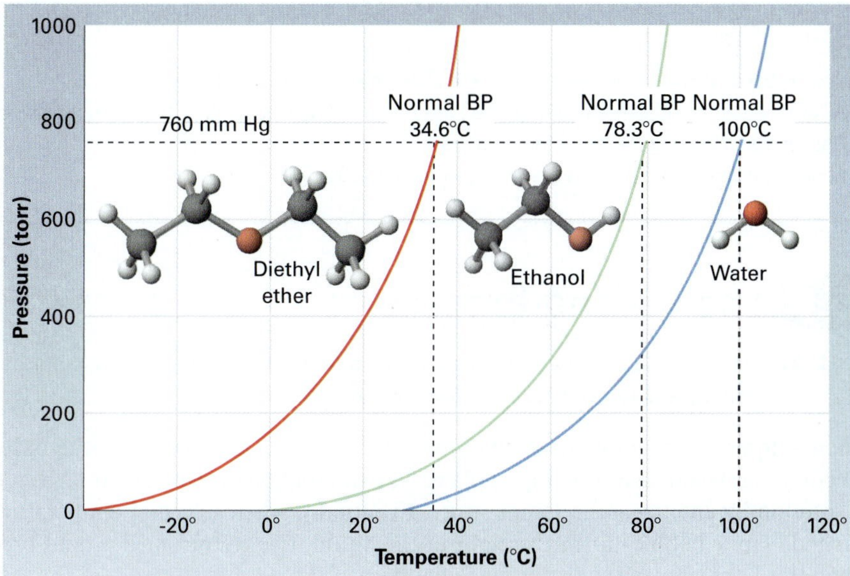

Figure 15.22 Vapor pressures of three liquids at different temperatures.

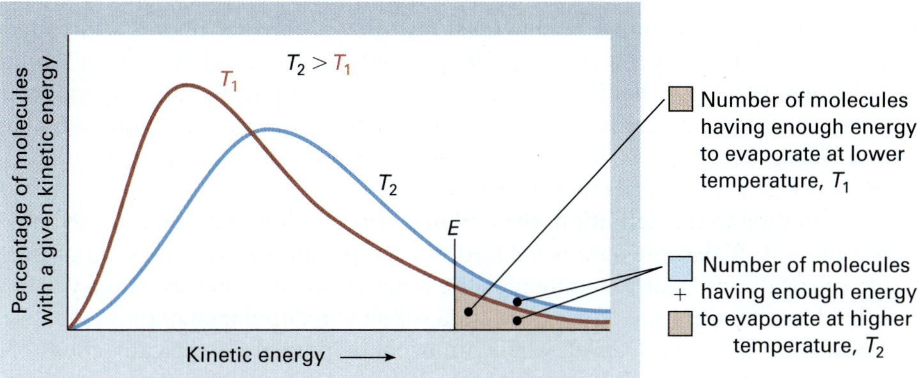

Figure 15.23 Kinetic energy distribution curves for the same liquid at two temperatures. Curve T_1 is for the lower temperature. As temperature rises from T_1 to T_2, the average kinetic energy increases. The curve shifts to the right and is spread out over a wider range. The area beneath the curve to the right of the escape energy, E, represents the fraction of the total number of molecules that have enough kinetic energy to evaporate. At T_1, only the fraction representing the area under the marked part of the red curve has the escape energy, but at higher temperature, T_2, that fraction is represented by the marked area under the blue curve.

increase in vapor pressure. The vapor pressure of water, for example, is 149 torr at 60°C (333 K) and 355 torr at 80°C (353 K). The vapor pressure more than doubles (increases by 138%), whereas absolute temperature increases by only about 6%. ▶

◀ **P/Review** The absolute temperature scale is based on the Celsius degree and setting the lowest temperature possible at zero kelvin: $T_K = T_{°C} + 273$. See Section 4.5.

Figure 15.23 explains the effect of temperature on vapor pressure. The kinetic energy distribution curve labeled T_1 gives the kinetic energy distribution at one temperature, and T_2 is the curve for a higher temperature. As in Figure 15.19, E is the escape energy, the minimum energy a molecule must have to evaporate into the gas phase. The area beneath the curve to the right of E is the fraction of the sample that has enough kinetic energy to evaporate. As illustrated, the area for T_2 is nearly twice the area for T_1. We conclude that it is an *increase in the number of molecules with enough energy to evaporate* that is responsible for higher vapor pressure at higher temperatures.

Target Check 15.5

Identify the true statements, and rewrite the false statements to make them true.

a) A liquid–vapor equilibrium is reached when the amount of liquid is equal to the amount of vapor.

b) Rate of evaporation depends on temperature.

c) Equilibrium vapor pressure is higher at higher temperatures.

15.6 The Boiling Process

Goal 10 Describe the process of boiling and the relationship among boiling point, vapor pressure, and surrounding pressure.

When a liquid is heated in an open container, bubbles form, usually at the base of the container where heat is being applied. The first bubbles are often air, driven out of solution by an increase in temperature. Eventually, when a certain temperature is reached, vapor bubbles form throughout the liquid, rise to the surface, and break. When this happens, we say the liquid is **boiling**.

In order for a stable bubble to form in a boiling liquid, the vapor pressure within the bubble must be high enough to push back the surrounding liquid and the atmosphere above the liquid (see Figure 15.24). The minimum temperature at which this can occur is called the *boiling point*: The boiling point is the temperature at which the vapor pressure of the liquid is equal to the pressure above its surface. Actually, the vapor pressure within a bubble must be a tiny bit greater than the surrounding pressure, which suggests that bubbles probably form in local "hot spots" within the boiling liquid. The boiling temperature at one atmosphere—the temperature at which the vapor pressure is equal to one atmosphere—is called the **normal boiling point**. Figure 15.22 shows that the normal boiling point of water is 100°C; of ethanol, 78.3°C, and of diethyl ether, 34.6°C.

According to the definition, the boiling point of a liquid depends on the pressure above it. If that pressure is reduced, the temperature at which the vapor pressure equals the lower surrounding pressure comes down also, and the liquid will boil at that lower temperature. This is why liquids boil at reduced temperatures at higher altitudes. In mile-high Denver, where atmospheric pressure is typically about 630 torr, water boils at 95°C. It is possible to boil water at room temperature by lowering the pressure in the space above it. When pressure is reduced to 20 torr, water boils at 22°C, which is "room temperature." A method for purifying a compound that might decompose or react with oxygen at its normal boiling point is to boil it at reduced temperature in a partial vacuum and then condense the vapor.

It is also possible to *raise* the boiling point of a liquid by *increasing* the pressure above it. The pressure cooker used in the kitchen takes advantage of this effect.

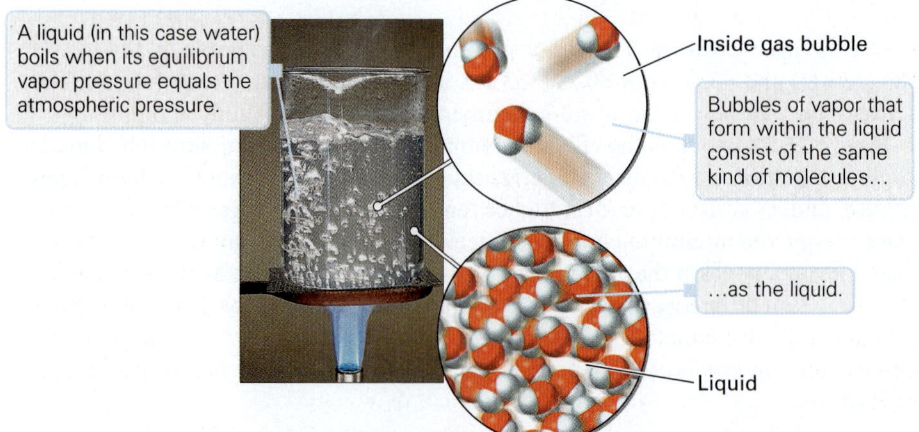
A liquid (in this case water) boils when its equilibrium vapor pressure equals the atmospheric pressure.
Inside gas bubble
Bubbles of vapor that form within the liquid consist of the same kind of molecules...
...as the liquid.
Liquid
Charles D. Winters

Figure 15.24 The boiling process. A liquid boils when its vapor pressure equals the pressure above it. The temperature at which this occurs is the boiling point.

By allowing the pressure to build up within the cooker, it is possible to reach temperatures as high as 110°C without boiling off the water. At this temperature, food cooks in about half the time required at 100°C.

Target Check 15.6

Are the following true or false?

a) Water can be made to boil at 15°C.

b) Bubbles can form anyplace in a boiling liquid.

15.7 Water—An "Unusual" Compound

Goal 11 Describe the typical relative density relationship between the solid and liquid phase of a substance, and explain why water is an exception to this trend.

Through much of this book, you have seen trends and regularities among physical and chemical properties. Many of these have been related to the periodic table. Predictions have been based on these trends. A prediction is not reliable, though, until it is confirmed in the laboratory. Sometimes a substance does not behave as it is expected to, and we have to look further, but most substances fit into regular patterns.

Water does not fit.

Water is so common, so much a part of our daily lives, that it is hard to think of it as unusual. But in terms of trends, unusual is exactly what water is. One example appears in Figure 15.18. Beginning at tellurium (Z = 52) in Group 6A/16, the boiling points of the hydrogen-containing compounds drop as the molecules become smaller, as expected: −4°C for H_2Te, −42°C for H_2Se, and −62°C for H_2S. If the trend continued, the boiling point of H_2O should be about −72°C. Instead, it is +100°C. And that is only one example of water's unusual behavior.

A particulate-level examination of the water molecule (**Figure 15.25**) gives us some clues to explain this unique behavior. Aside from fluorine, oxygen is the most electronegative non-noble gas element. Therefore, the electrons forming each bond between hydrogen and oxygen are drawn strongly toward the oxygen atom, resulting in two very polar bonds—more polar than the bonds in other hydrogen-containing compounds in the group. Furthermore, the 104.5° bond angle makes a strong dipole. Finally, add hydrogen bonding, which is probably the most important contributor to strong intermolecular attractions in water.

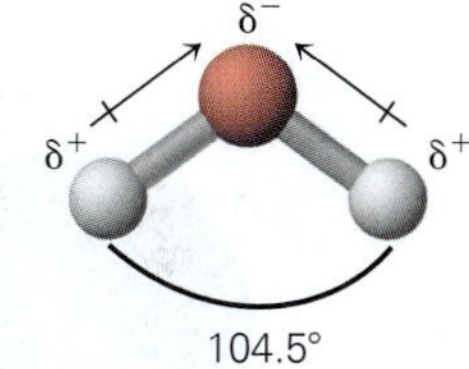

Figure 15.25 The water molecule. The geometry of the water molecule and the polarity of its bonds make water molecules highly polar. In addition, water displays strong hydrogen bonding. These three factors account for exceptionally strong intermolecular attractions that influence many properties of water.

Among molecules of comparable size, water has several other unusual properties. Exceptionally high surface tension and heat of vaporization are among them. Its vapor pressure is particularly low, even compared to larger molecules whose vapor pressures you would expect to be low. Check the compounds in Figure 15.22, for example. We don't usually think of water as viscous, but it is viscous when compared with substances with similar structures. Water dissolves a wider variety of gaseous, liquid, and solid substances than most solvents. This is explained in Chapter 16. Finally, the mere fact that water is a liquid at room conditions is unusual. It is one of a very small number of compounds without carbon that exist as liquids at normal temperatures and pressures.

Water's most visible unusual property is that its solid form—ice—floats on its liquid form. ▶ Almost all substances expand—become less dense—when heated, and they contract—become more dense—when cooled. Water becomes more dense as it is cooled—until it reaches 4°C (under normal atmospheric conditions). Below 4°C, it becomes less dense. When water freezes, there is about a 9% increase in volume as the molecules arrange themselves into an "open" crystal structure. Compare this with the closer packing the molecules have in the liquid state (see **Figure 15.26**). This expansion exerts enough force to break water pipes if the liquid is permitted to freeze in them.

> Water is about the only substance we normally encounter in the solid, liquid, and gas phases. An example of water in the gas phase is the water vapor in air, which is the humidity mentioned earlier.

Water is a most unusual compound. Much of life on Earth could not exist but for its molecular structure and the unique macroscopic properties that result.

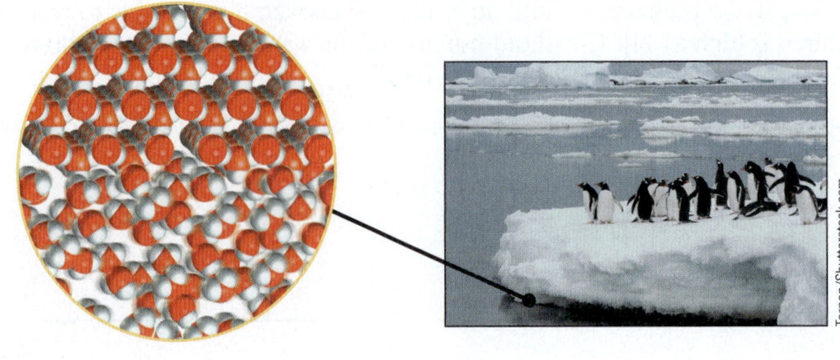

Figure 15.26 Fortunately for these penguins, water is one of the very few substances whose solid phase is *less* dense than the liquid. Water molecules in solid form (ice) are held in a crystal pattern that has voids between the molecules. When ice melts, the crystal collapses, the molecules are closer together, and the liquid is denser than the solid. This is why ice floats in water, a solid–liquid property shared by few other substances.

15.8 The Solid State

Goal 12 Distinguish between amorphous and crystalline solids.

A growing area of chemical research is the study of the **solid state** of matter. Many discoveries are being made and the traditional subdivisions of chemistry are being reorganized to accommodate the growth of interest in solids. Why is there such interest in solids? Materials such as high-temperature superconductors that conduct electricity with no resistance (**Figure 15.27**) and glass–ceramics that can be safely transferred directly from the stove to the refrigerator are examples of useful products that are driving research in the field known as **materials science**.

Figure 15.27 A magnet levitates above a super conductor. Materials science researchers are studying methods for preparing new substances with superconducting properties.

Solids can be classified based on their *macroscopic* properties. The differences on the macroscopic scale, however, result from differences at the *particulate* level. In particular, we can classify solids based on the forces holding the particles together. We will start by dividing solids into two categories based on the way the particles are arranged in the solid.

A solid with particles arranged in a geometric pattern that repeats itself over and over in three dimensions is a **crystalline solid** (**Figures 15.28[right]** and **15.29[a]**).

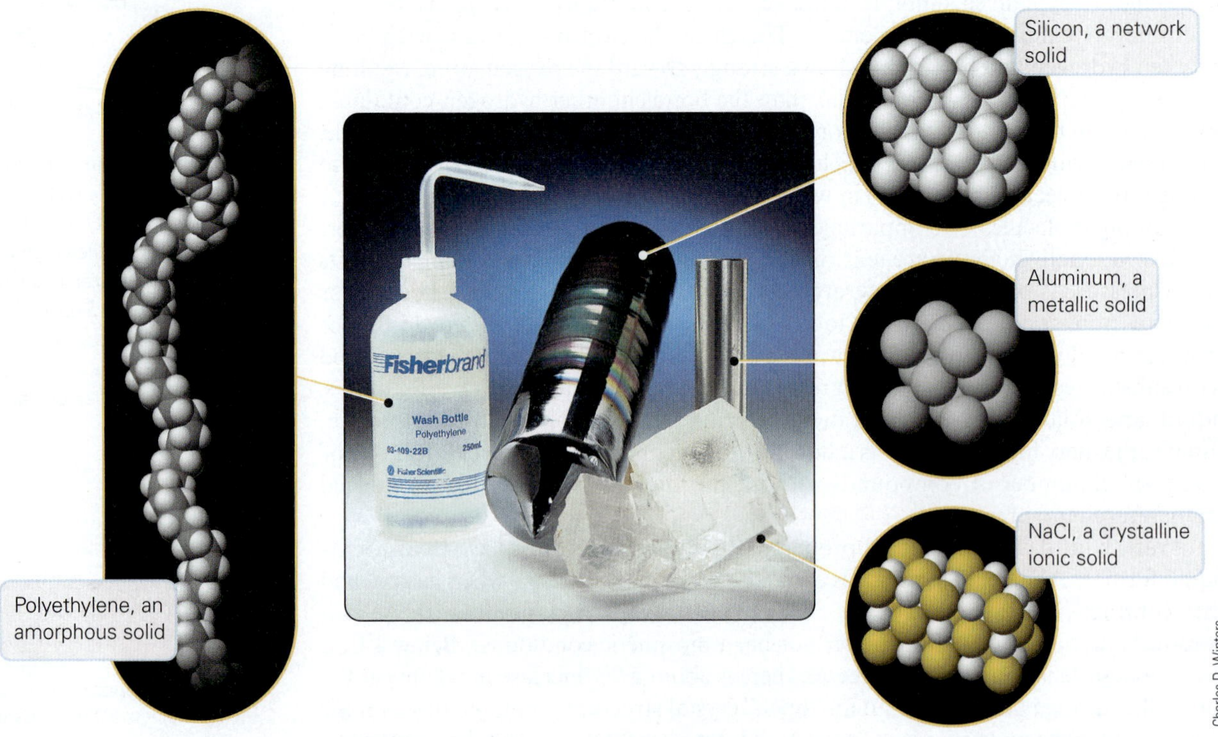

Figure 15.28 Amorphous (*left*) and crystalline (*right*) solids. An amorphous solid, such as the polyethylene bottle shown, has little long-range ordering at the particulate level. In contrast, each of the crystalline solids shown has an orderly, regularly repeating arrangement of the particles.

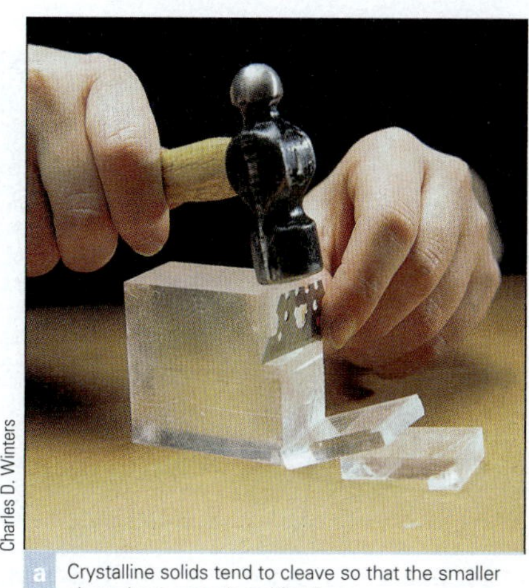
Charles D. Winters

a Crystalline solids tend to cleave so that the smaller pieces have smooth and flat faces.

Charles D. Winters

b Amorphous solids do not break along lines with predictable patterns.

Figure 15.29 Macroscopic properties of crystalline and amorphous solids. The presence or absence of a regular structure at the particulate level influences the macroscopic shape of the solid.

Each particle occupies a fixed position in the crystal. It can vibrate about that site but cannot move past its neighbors. The high degree of order often leads to large crystals that have a precise geometric shape. In ordinary table salt, we can distinguish small cubic crystals of sodium chloride. Large, beautifully formed crystals of minerals such as quartz (SiO_2) and fluorite (CaF_2) are found in nature.

In an **amorphous solid** such as glass, rubber, or plastic, there is no long-range ordering of the particles in the solid (**Figures 15.28[*left*]** and **15.29[b]**). Even though the arrangement around a particular point may appear to be in a regular pattern, the pattern does not repeat itself throughout the solid. From a structural standpoint, we may regard an amorphous solid as an intermediate between the solid and liquid states. In many amorphous solids, the particles have some freedom to move with respect to one another. The elasticity of rubber and the tendency of glass to flow when subjected to stress over a long period of time suggest that the particles in these materials are not rigidly fixed in position.

Crystalline solids have characteristic physical properties that identify them. Sodium chloride, for example, melts sharply at 801°C. This is in striking contrast to glass, an amorphous solid, which first softens and then slowly liquefies over a wide range of temperatures.

Target Check 15.7

Identify the main structural differences between crystalline solids and amorphous solids.

15.9 Types of Crystalline Solids

Goal 13 Distinguish among the following types of crystalline solids: ionic, molecular, covalent, and metallic.

Crystalline solids can be divided into the four classes named in Goal 13 on the basis of the types of forces that hold particles together in the crystal lattice.

Ionic Crystals Examples of ionic crystals are NaF, $CaCO_3$, AgCl, and NH_4Br. Oppositely charged ions are held together by strong electrostatic forces—ionic bonds. Ionic crystals are typically high melting and frequently water-soluble, and in the solid state, they have very low electrical conductivities. When ionic crystals are melted or dissolved in water, the ions become mobile, so the liquid or the solution conducts electricity readily. **Figures 15.30** and **15.31** illustrate the particulate-level structure of two ionic crystals, and **Figure 15.32** shows lead(II) sulfide at the macroscopic and particulate levels.

Figure 15.30 A particulate-level model of an ionic crystal. The gray spheres represent sodium ions, and the green spheres are chloride ions. The macroscopic physical properties of ionic crystals result from the strong electrostatic attractions among the ions.

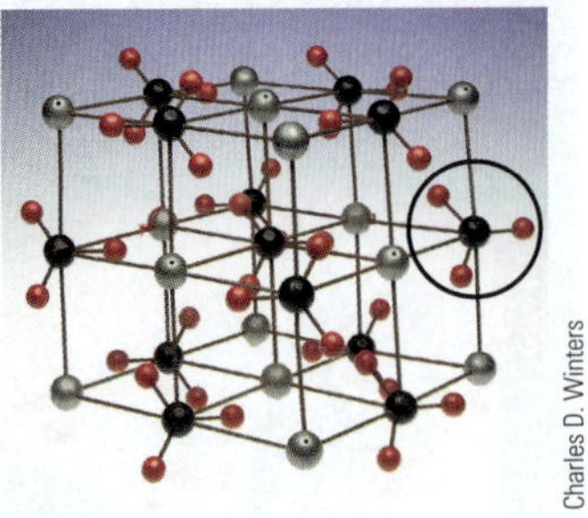

Figure 15.31 A particulate-level model of a calcium carbonate crystal. This is an example of an ionic crystal that has a polyatomic ion. The gray spheres represent calcium ions, Ca^{2+}. The black spheres with three red spheres attached are carbonate ions, CO_3^{2-}. The circle encloses one carbonate ion.

Figure 15.32 An ionic crystal. The chemical composition of the mineral galena is lead(II) sulfide.

Molecular Crystals Examples of molecular crystals include I_2 and ICl. Small, discrete molecules are held together by relatively weak intermolecular forces of the types discussed in Section 15.4. Molecular crystals are typically soft, low-melting, and generally (but not always) insoluble in water. They usually dissolve in nonpolar or slightly polar organic solvents such as carbon tetrachloride or chloroform ($CHCl_3$). Substances with a molecular crystal particulate-level structure, with rare exceptions, are nonconductors when pure, even in the liquid state.

Sulfur is another example of a molecular crystal. It has numerous forms at the molecular level, but the most common form of the solid is S_8 molecules, with eight atoms bonded in a ring (**Figure 15.33[a]**). When the solid is heated to a liquid under certain conditions, the molecules rearrange into long chains and the substance has the macroscopic characteristics of a plastic (**Figure 15.33[b]**).

Covalent Network Solids Examples of covalent network solids are diamond, C, and quartz, SiO_2. Atoms are covalently bonded to one another to form one large

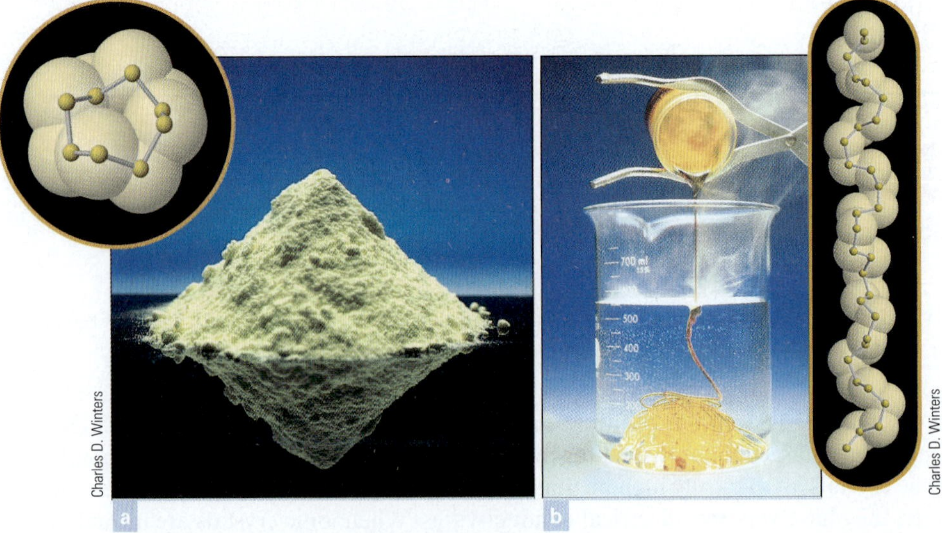

Figure 15.33 Two forms of sulfur. (a) One form of solid sulfur is an example of a molecular crystal in which distinct molecular units can be identified. (b) One liquid form consists of chains of sulfur atoms.

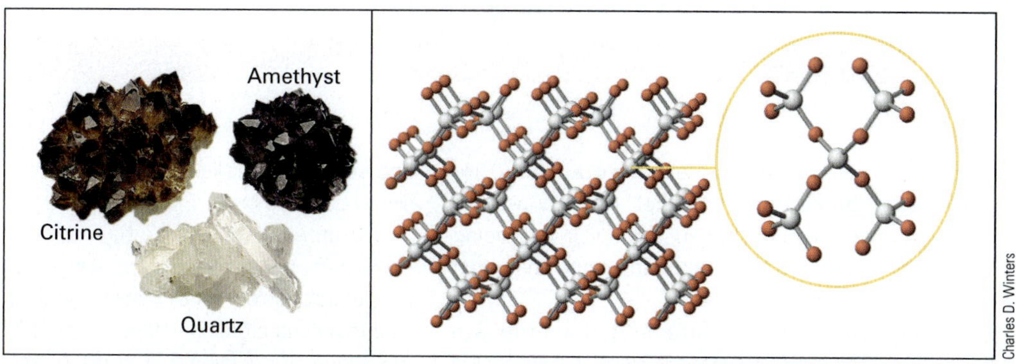

Figure 15.34 Forms of quartz crystals. Pure quartz, SiO_2, is colorless. Amethyst and citrine are mostly composed of silicon and oxygen, but small amounts of iron impurities make the crystals colored. Quartz is an example of a covalent network solid. Atoms are covalently bonded in a regularly repeating pattern, but there are no individual molecules of fixed size.

network of indefinite size. There are no small discrete molecules in network solids. In diamond, each carbon atom is covalently bonded to four other carbon atoms to give a structure that repeats throughout the entire crystal. The structure of silicon dioxide resembles that of diamond in that the atoms are held together by a continuous series of covalent bonds. Each silicon atom is bonded to four oxygen atoms, and each oxygen is bonded to two silicon atoms. **Figure 15.34** illustrates quartz crystals.

Covalent network solids, like ionic crystals, have high melting points. A very high temperature is needed to break the covalent bonds in crystals of quartz (about 1700°C) or diamond (about 3500°C). Covalent network solids are almost always insoluble in water or any common solvent. They are generally poor conductors of electricity in either the solid or liquid state.

Figure 15.35 Aluminum metal. This is an example of a metallic crystal.

Metallic Crystals Aluminum is an example of a metallic crystal (**Figure 15.35**). A simple model of bonding in a metal consists of a crystal of positive ions through which valence electrons move freely. This so-called electron-sea model of a metallic crystal is illustrated in Figure 12.19. ▶ Positively charged ions form the structural framework of the crystal; the valence electrons surrounding these ions are not tied down to any particular ion and therefore are not restricted to a particular location. It is because of these freely moving electrons that metals are excellent conductors of electricity.

◀ **P/Review** Metallic bonding results from the attractive forces between the positively charged metal ions in the crystal and the negatively charged electrons in the "electron sea." See Section 12.9, Figure 12.19.

Some metallic crystals can appear to be amorphous at the macroscopic level, but when observed at the microscopic level, they are found to be made of randomly oriented crystalline segments. We can see the tiny individual crystals with a microscope.

The general properties of the four kinds of crystalline solids are summarized in **Table 15.3.**

Table 15.3 General Properties of Crystals

Type	Examples	Properties
Ionic	KNO_3, NaCl, MgO	High-melting; generally water-soluble; brittle; conduct only when melted or dissolved in water
Molecular	$C_{10}H_8$, I_2	Low-melting; usually more soluble in organic solvents than in water; nonconductors in pure state
Network	SiO_2, C	Very high melting; insoluble in all common solvents; brittle; non- or semiconductors
Metallic	Cu, Fe	Wide range of melting points; insoluble in all common solvents; malleable, ductile; good electrical conductors

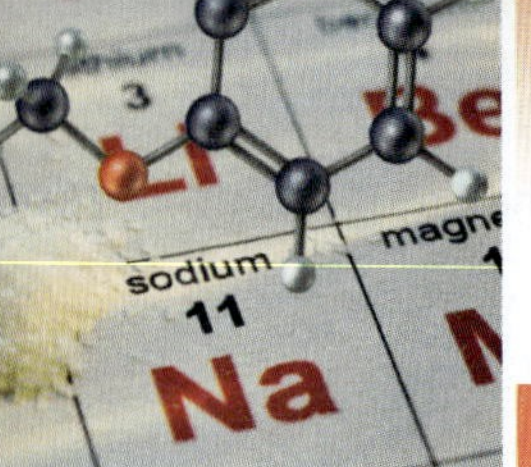

Everyday Chemistry 15.1

BUCKYBALLS

A long tradition in chemistry holds that the discoverer or inventor of new substances gives the new class of material its nonsystematic name. There are classes of organic compounds called *propellanes* (because their Lewis structures look like propellers), *basketanes* (like baskets), and *barrelanes* (like barrels).

In 1985, when chemists discovered (in soot, of all places) a stable C_{60} cluster whose Lewis structure resembled a geodesic dome, they gave the substance the name buckminsterfullerene after Buckminster Fuller, the inventor of the geodesic dome (**Figure 15.36**). Everybody now calls these clusters buckyballs because each cluster looks like a hollow soccer ball (**Figure 15.37**).

Chemists were amazed because C_{60} is a new allotrope, or form, of carbon. The graphite and diamond forms of carbon were discovered in prehistoric times, and a small amount of C_{60} was apparently being made every time a sooty fire burned on the planet. Nobody noticed buckyballs because the C_{60} was being oxidized (burned) by the same flame that created it.

Both graphite and diamond are high-melting network solids. A buckyball is a distinct cluster of carbon atoms, a molecule of carbon. The physical and chemical properties of buckyballs were completely undiscovered, and research laboratories around the world jumped into the fray.

Buckyballs have amazing physical properties. They are soft, like graphite. One research group is working to compress the hollow buckyball to about two-thirds of its original volume. Calculations predict that these squeezed buckyballs will be harder than diamond, the hardest known substance. An interesting experiment combining these ideas of softness and great hardness threw buckyballs against a steel surface at 17,000 miles per hour. Showing unprecedented resilience, the buckyballs simply bounced back.

The interior of a buckyball is large enough to hold an atom of any element in the periodic table. Researchers wasted no time in putting atoms of different metals in the center of buckyballs. The result was a new family of superconductors. Other teams are working on using buckyballs as the source of tiny ball bearings, lightweight batteries, and even superconducting wires that are just one cluster thick. A group at DuPont (home of Teflon) made buckyballs with carbon–fluorine bonds (like Teflon) and is experimenting with so-called *Teflonballs*.

Buckyballs are simply one member of the family of fullerenes, which ranges from about C_{20} up to at least C_{720}. Researchers have recently discovered "buckytubes," hollow needlelike tubes that nest within one another. Can we extend buckytubes to form buckyfibers? Are these the

Xuanlu Wang/Shutterstock.com

Figure 15.36 Geodesic dome at Science World in Vancouver, British Columbia, Canada. The philosopher, architect, and engineer R. Buckminster Fuller invented the geodesic dome. The architectural advantage of its linked hexagon and pentagon structure is that it can cover a large area without internal supports. The similar molecular structure of buckminsterfullerene inspired its discoverers to name the molecule in honor of Fuller.

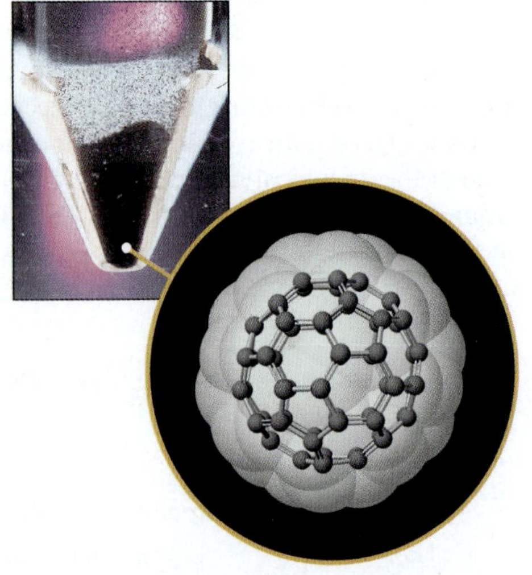

Charles D. Winters

Figure 15.37 Buckminsterfullerene, C_{60}. The photograph shows buckminsterfullerene, a black powder, in the tip of a pointed glass tube. The molecular structure of this form of the element carbon is similar to that of a geodesic dome and a soccer ball.

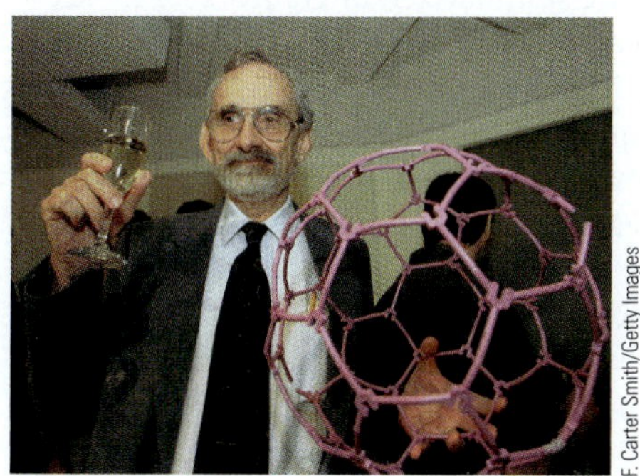

Figure 15.38 Robert F. Curl, Jr. (b. 1933).

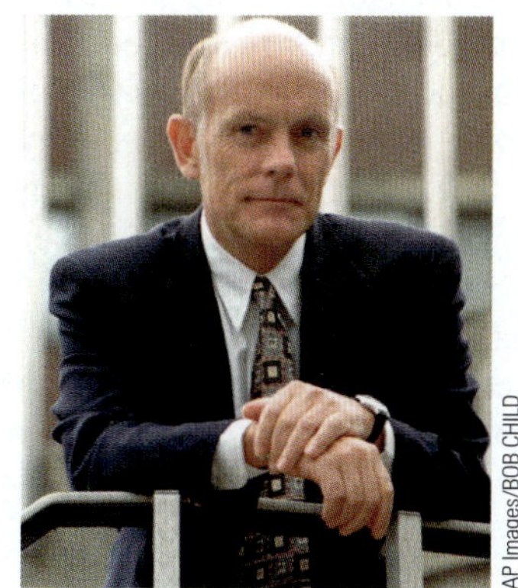

Figure 15.39 Richard E. Smalley (1943–2005).

Figure 15.40 Sir Harold W. Kroto (1939–2016).

first elementary building blocks of a new carbon-based technology? We're just beginning to find out.

Buckminsterfullerene is a superb modern example of the value of fundamental research to science in general and to applied areas in particular. Buckminsterfullerene was discovered by chemists doing experiments trying to determine the role of carbon in space and in the distant stars. The result led to dynamic new paths in chemistry, physics, and materials science here on Earth.

American chemists Robert F. Curl, Jr. (**Figure 15.38**) and Richard E. Smalley (**Figure 15.39**) and British chemist Sir Harold W. Kroto (**Figure 15.40**) were awarded the 1996 Nobel Prize in Chemistry for their discovery of the family of fullerenes.

Quick Quiz

1. **The discoverers of C_{60} gave it the nonsystematic name buckminsterfullerene, yet most chemists call the molecules buckyballs. Why?**
2. **What is an *allotrope*? What allotropes are discussed in this essay, and what element are they allotropes of?**

✓ Target Check 15.8

Identify the true statements, and rewrite the false statements to make them true.

a) A high-melting solid that conducts electricity is probably a metal.
b) Covalent network solids are usually good conductors of electricity.
c) A solid that melts at 152°C is probably an ionic crystal.
d) A soluble molecular crystal is a nonconductor of electricity but a good conductor when dissolved.

15.10 Energy and Change of State

Goal 14 Given two of the following, calculate the third: (a) mass of a pure substance changing between the liquid and vapor (gaseous) states; (b) heat of vaporization; (c) energy transferred.

15 Given two of the following, calculate the third: (a) mass of a pure substance changing between the solid and liquid states; (b) heat of fusion; (c) energy transferred.

Heat energy is transferred to or from a substance when it changes state. Energy must be transferred to a solid substance to cause it to **melt**, an endothermic change. Energy is transferred from a liquid when it **freezes**, an exothermic change. In the

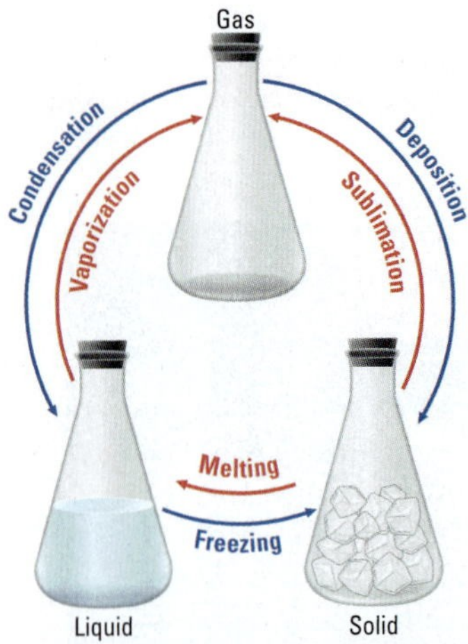

Figure 15.41 Changes of state. The changes labeled in blue are exothermic: Heat energy is transferred out of the substance as the process occurs. The changes labeled in red are endothermic: Heat energy is transferred into the substance as the process occurs.

change between a liquid and a gas, **vaporization** is endothermic and **condensation** is exothermic. Some substances can change directly between the solid and gas phases. When a gas changes to a solid, the process is exothermic and is called **deposition**. A solid changes to a gas in an endothermic process called **sublimation**. The names of the interconversions among states of matter are summarized in **Figure 15.41**.

In this section, you will learn how to calculate the quantity of energy transferred that accompanies a change of state. The SI unit of energy is the **joule (J)**. ◀ The joule is a small unit of energy. It is quite suitable for use with temperature changes (see the next section) and melting, but it is too small for the larger amounts of energy involved in boiling and chemical reactions. The kilojoule (kJ) is commonly used for these changes.

▶ **P/Review** Energy units are discussed more fully in Section 10.8. If a joule of energy is transferred to 1 g of liquid water, its temperature increases by about a quarter of a Celsius degree.

It has been found experimentally that the energy transfer required to vaporize a substance, q, is proportional to the amount of substance. Amount may be expressed as number of particles, grouped in moles, or as mass (m) in grams. The proportionality is changed into an equality by means of a proportionality constant, ΔH_{vap}, known as the **heat of vaporization**: ◀

▶ **P/Review** The symbol ΔH is used for change in enthalpy. Thermochemical equations, which include an enthalpy of reaction, $\Delta_r H$, term, were introduced in Section 10.9. The Greek Δ represents change. The sign of a Δ quantity is described more formally in Section 15.11.

$$q \propto m \xrightarrow{\text{the proportionality changes to an equality}} q = \Delta H_{vap} \times m$$

Solving for ΔH_{vap} gives the defining equation for heat of vaporization:

$$\Delta H_{vap} \equiv \frac{q}{m}$$

The units of heat of vaporization follow from the equation, energy units per mass units. If energy is expressed in kilojoules and mass is expressed in grams, the units of ΔH_{vap} are kJ/g. ◀ The heats of vaporization of several substances are given in the rightmost column of **Table 15.4**. ◀

In kJ/mol, ΔH_{vap} is the enthalpy change when 1 mole of substance is changed from a liquid to a vapor. This can be expressed in a thermochemical equation. For water, $H_2O(\ell) \rightarrow H_2O(g)$ $\Delta H_{vap} = 40.7$ kJ.

▶ **P/Review** In Section 15.3, we described heat of vaporization as the energy required to change 1 mole of a liquid to a gas. It was expressed in kJ/mol. In this section, we use the more convenient unit kJ/g to express heat of vaporization. Use the molar mass of a substance (g/mol) to convert between kJ/g and kJ/mol, if necessary.

When a vapor condenses to a liquid at the boiling point, the reverse energy change occurs. The energy change is then referred to as the **heat of condensation**. Values are the same as heats of vaporization, except that they are negative. This indicates that heat energy is transferred *from* the substance (an exothermic change) instead of being transferred *to* it. In solving a problem, use the negative of heat of vaporization if a gas is condensing.

Table 15.4 Heats of Fusion and Heats of Vaporization

Substance	Melting Point (°C)	Boiling Point (°C)	Heat of Fusion (J/g)	Heat of Vaporization (kJ/g)
Ag	962	2162	105	2.4
Al	660	2519	397	10.9
Au	1064	2856	63	1.7
Bi	271	1564	52	0.8
Cd	321	767	56	0.9
Cu	1084	2927	206	4.7
Fe	1538	2861	247	6.2
H_2O	0	100	333	2.26
Hg	39	357	11	0.3
Na	98	883	113	4.2
NaCl	801	1413	519	–
Ni	1455	2913	293	6.4
Pb	327	1749	23	0.9
Zn	420	907	112	1.8

Active Example 15.3 Calculating Heat of Vaporization

Calculate the heat of vaporization of a substance if 241 kJ is transferred to vaporize a 78.2-g sample.

Think Before You Write The problem statement gives you q and m, the two variables in the definition of heat of vaporization. Therefore, this is a straightforward algebra problem using $\Delta H_{vap} = q/m$.

Answers *Cover the left column with your cut-out shield. Reveal each answer only after you have written your own answer in the right column.*

Given: q = 241 kJ; m = 78.2 g **Wanted:** ΔH_{vap}, assume kJ/g $\Delta H_{vap} = \frac{q}{m} = \frac{241\ \text{kJ}}{78.2\ \text{g}} = 3.08\ \text{kJ/g}$ $\frac{24 \times 10^1}{8 \times 10^1} = 3$, OK.	Write the given quantities and wanted property and unit, and then substitute the given values into the equation and calculate the answer. Finish the solution with a **check** of your calculation.
You improved your understanding of heat of vaporization, and you improved your skill at solving problems that include heat of vaporization.	What did you learn by solving this Active Example?

Practice Exercise 15.3

A 7.3-g sample of zinc requires 13.1 kJ of energy to change from the liquid to the solid state. Determine the heat of vaporization of zinc.

The most common calculation is finding the energy transferred in changing the state of a given mass of material. In previous examples of defining equations and conversion factors, the property defined has been used as a conversion factor in the solution setup. This time, however, we use the equation form, $q = m \times \Delta H_{vap}$. This approach makes a neater package when change-of-state problems are combined with change-of-temperature problems, as they will be shortly.

Active Example 15.4 Using Heat of Vaporization to Determine Energy Transfer

How much energy must be transferred to vaporize 188 g of a liquid if its heat of vaporization is 1.13 kJ/g?

Think Before You Write *The heat of vaporization is the conversion factor used to convert between energy in kilojoules and mass in grams. You are given mass in grams, so it is a one-step conversion to determine the energy transferred in kilojoules.*

Answers *Cover the left column with your cut-out shield. Reveal each answer only after you have written your own answer in the right column.*

Given: 188 g; 1.13 kJ/g **Wanted:** Energy transferred (assume kJ) $q = m \times \Delta H_{vap} = 188\ \cancel{\text{g}} \times \frac{1.13\ \text{kJ}}{\cancel{\text{g}}} = 212\ \text{kJ}$ $188 \times 1.13 \approx 200 \times 1 = 200$, OK.	Solve the problem completely, from the **analyze** step to the **check** step.
You improved your understanding of heat of vaporization, and you improved your skill at solving problems that include heat of vaporization.	What did you learn by solving this Active Example?

Practice Exercise 15.4

What quantity of energy is transferred when 2.3 kg of ethanol condenses from the vapor phase to a liquid? For ethanol, $\Delta H_{vap} = 0.837$ kJ/g.

To melt a solid, energy must be transferred to the solid to overcome the forces that hold its particles together. Conceptually similar to heat of vaporization, **the heat of fusion, ΔH_{fus},** of a substance is the energy transfer required to melt 1 g of that substance. (*Fusion* is changing a solid to a liquid by the transfer of heat energy—that is, the melting process.) Heats of fusion are generally much smaller than heats of vaporization, so the usual units are joules per gram, J/g. Some typical heats of fusion are given in Table 15.4.

Just as condensation is the opposite of vaporization, freezing is the opposite of melting. The amount of energy transferred in freezing a sample is identical to the amount of energy transferred to melt that sample. Accordingly, **heat of solidification** is numerically equal to heat of fusion, but the sign is negative.

The heat transfer equation for the change between solid and liquid is like the equation for the liquid-to-gas change:

$$q = m \times \Delta H_{fus}$$

Calculation methods are just like those in vaporization problems.

Active Example 15.5 Using Heat of Fusion to Determine Energy Transfer

Calculate the quantity of heat transferred when 135 g of sodium freezes at its freezing point temperature. Express the answer in both joules and kilojoules.

Think Before You Write You are given mass, and you are asked to determine the energy transferred as sodium freezes. The heat of fusion of sodium is needed to solve this problem. It may be found in Table 15.4. That and $q = m \times \Delta H_{fus}$ are all you need.

Answers *Cover the left column with your cut-out shield. Reveal each answer only after you have written your own answer in the right column.*

Given: 135 g; ΔH_{fus} = 113 J/g (from Table 15.4) **Wanted:** q, heat transferred, in J and kJ $q = m \times (-\Delta H_{fus}) = 135\ \cancel{g} \times \frac{-113\ J}{\cancel{g}} = -1.53 \times 10^4\ J$ $135 \times -113 \approx 150 \times -100 = (-1.5 \times 10^2) \times 10^2 = -1.5 \times 10^4$, OK. $-1.53 \times 10^4\ \cancel{J} \times \frac{1\ kJ}{1000\ \cancel{J}} = -15.3\ kJ$ ◀ *The negative sign is applied to ΔH_{fus} because the metal is freezing, not melting.*	Solve the problem completely, from the **analyze** step to the **check** step.
You improved your understanding of heat of fusion, and you improved your skill at solving problems that include heat of fusion.	What did you learn by solving this Active Example?

Practice Exercise 15.5

How much transfer of energy is needed to melt 6.37 kg of aluminum at its melting point?

▶ **P/Review** You should be able to convert between values of three metric prefixes by memory: *kilo-*, *centi-*, and *milli-*. See Section 3.6. The correctness of a metric–metric conversion can be checked with the larger/smaller rule. A quantity may be expressed in a large number of small units (J) or a small number of large units (kJ).

15.11 Energy and Change of Temperature: Specific Heat

Goal 16 Given three of the following quantities, calculate the fourth: (a) energy transferred; (b) mass of a pure substance; (c) specific heat of that substance; (d) temperature change, or initial and final temperatures.

Let's look more closely at the algebraic sign of a change, a **delta quantity (Δ)**. A delta quantity is always calculated by subtracting the initial value from the final value. In this section, we will work with a change in temperature, ΔT. By definition:

$$\Delta T = T_{final} - T_{initial} = T_f - T_i$$

If the temperature rises from 20°C to 25°C, ΔT is positive:

$$\Delta T = T_f - T_i = 25°C - 20°C = 5°C$$

However, if the temperature falls from 25°C to 20°C, ΔT is negative:

$$\Delta T = T_f - T_i = 20°C - 25°C = -5°C$$

You have seen that heat energy must be transferred to a substance if it is to melt or boil. Energy is released in the opposite processes of freezing or condensing. Such changes of state are constant-temperature processes. The steam and the boiling water in a kettle are both at 100°C, and they remain there as long as the water boils. The water and the crushed ice in a glass are both at 0°C, and they remain there until all of the ice melts. However, what happens when energy is transferred but there is no change of state? The temperature changes.

Experiments indicate that the heat transferred, q, in heating or cooling a substance is proportional to both the mass of the sample, m, and its temperature change, ΔT. Combining these proportionalities yields ▶

▶ **P/Review** When a variable is proportional to two or more other variables, it is proportional to the product of those variables. The proportionality is changed into an equation with a proportionality constant. See Sections 3.11 and 4.7.

$$q \propto m \text{ and } q \propto \Delta T \xrightarrow{\text{combine the proportionalities}} q \propto m \times \Delta T$$

Introducing a proportionality constant, c,

$$q \propto m \times \Delta T \xrightarrow{\text{the proportionality changes to an equality}} q = m \times c \times \Delta T$$

The units of c may be found by solving the equation for that quantity:

$$q = m \times c \times \Delta T \xrightarrow{\text{divide both sides by } m \times \Delta T} c = \frac{q}{m \times \Delta T}$$

We therefore expect energy units divided by the product of mass units and temperature units. In this text, we will express values of c in J/g · °C. You may also encounter J/g · K in other sources.

Thinking About Your Thinking

Proportional Reasoning

Two pairs of proportionalities are expressed in the equation $q = m \times c \times \Delta T$: the heat transferred is proportional to the amount of substance *and* to the temperature change of that substance. Consider each of these separately. Think about the difference between having a drop of boiling water coming into contact with your skin versus an entire cup of boiling water. The temperature of the two samples of water is the same, but the mass is not. The amount of heat that transfers from the water to your body is greater when more water is spilled; heat transfer is proportional to mass.

Now hold constant the quantity variable and consider the effect of ΔT. A drop of 100°C water on your skin causes a more severe burn than a drop of 50°C water because the temperature change as your skin and the water exchange heat energy is greater for the 100°C drop. Heat transfer is proportional to temperature change.

Table 15.5 Selected Specific Heats

Substance	c (J/g · °C)
Elements	
Aluminum(s)	0.90
Cadmium(s)	0.23
Cadmium(ℓ)	0.27
Carbon	
Diamond(s)	0.51
Graphite(s)	0.71
Cobalt(s)	0.42
Copper(s)	0.38
Gold(s)	0.13
Gold(ℓ)	0.15
Iron(s)	0.45
Iron(ℓ)	0.45
Lead(s)	0.13
Magnesium(s)	1.02
Silicon(s)	0.71
Silver(s)	0.24
Silver(ℓ)	0.32
Sulfur(s)	0.74
Zinc(s)	0.39
Zinc(ℓ)	0.51
Compounds	
Acetone(ℓ)	2.17
Benzene(ℓ)	1.74
Carbon tetrachloride(ℓ)	0.85
Ethanol(ℓ)	2.44
Methanol(ℓ)	2.53
Water	
Solid (ice)	2.06
Liquid	4.18
Gas (steam)	2.00
Common Substances	
Concrete(s)	0.88
Glass(s)	0.84
Granite(s)	0.79
Wood(s)	1.76

The proportionality constant, c, is a property of a pure substance called its specific heat. **Specific heat** is the heat transfer required to change the temperature of 1 g of a substance by 1 degree Celsius. It measures the relative ease with which a substance may be heated or cooled. A substance with a low specific heat, such as aluminum, gains little energy in warming through a given temperature change compared with a substance with high specific heat, such as water (**Figure 15.42**). For example, when a slice of pizza wrapped in aluminum foil is removed from a hot oven, the foil cools much more rapidly than the pizza sauce (which is largely water). This is a result of the aluminum gaining less energy than the water when both were subjected to the same change in temperature. Specific heats of selected substances are given in **Table 15.5**.

A substance with high specific heat is best for retaining energy. At 4.18 J/g · °C, water has one of the highest specific heats of all substances. This is one reason that air temperatures near large bodies of water are usually higher in the winter and lower in the summer than nearby inland temperatures (**Figure 15.43**).

In specific heat problems, you always are given or have available three of the four factors in the equation $q = m \times c \times \Delta T$, and you solve algebraically for the fourth.

Charles D. Winters

Figure 15.42 Specific heats of water and three metals. The specific heat of liquid water, 4.18 J/g · °C, is much higher than that of the solid metals shown in this photograph: copper, 0.38 J/g · °C, aluminum, 0.90 J/g · °C, and iron, 0.45 J/g · °C. When the amount units are changed from grams to moles, you can see that the metals have very similar capacities to retain heat energy on a per mole basis: copper, 24 J/mol · °C; aluminum, 24 J/mol · °C; and iron, 25 J/mol · °C.

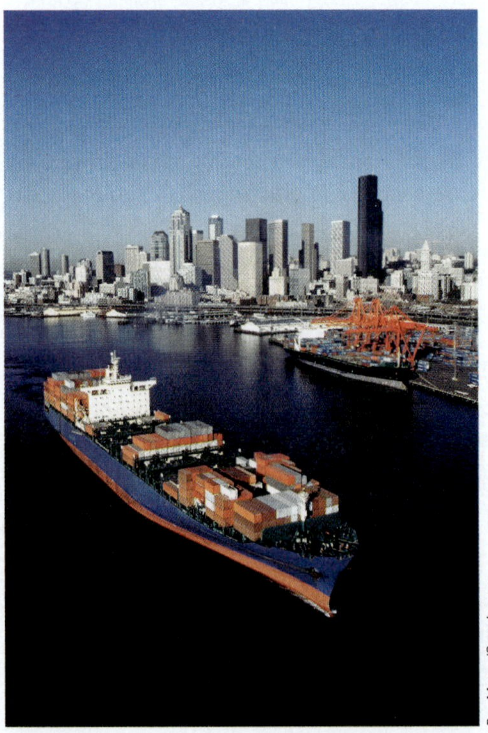

Don Mason/Getty Images

Figure 15.43 Effect of high specific heat of water. The ocean stores energy during the day and releases it at night, moderating the air temperature of coastal cities, such as Seattle, which is shown in this photograph.

Active Example 15.6 Using Specific Heat to Determine Energy Transfer

How much energy is transferred in raising the temperature of 475 g of water for a pot of tea from 14°C to 95°C? Answer in both joules and kilojoules.

Think Before You Write The value of q is wanted, and you are given m and data from which ΔT can be calculated. In problems like this, it is often assumed that you have access to a table of values of c. The equation $q = m \times c \times \Delta T$ is already solved for the wanted quantity, q, so the problem may be solved by direct substitution. For ΔT, substitute $T_f - T_i$.

Answers *Cover the left column with your cut-out shield. Reveal each answer only after you have written your own answer in the right column.*

Given: 475 g; T_i = 14°C; T_f = 95°C; c = 4.18 J/g · °C (from Table 15.5) **Wanted:** q in J and kJ $q = m \times c \times \Delta T = 475\,\text{g} \times \frac{4.18\ \text{J}}{\text{g} \cdot {}^\circ\text{C}} \times (95 - 14){}^\circ\text{C} = 1.6 \times 10^5\ \text{J}$ $475 \times 4.18 \times (95 - 14) \approx 500 \times 4 \times 80 = 2000 \times 80$ $= 160000 = 1.6 \times 10^5$, OK. $1.6 \times 10^5\ \text{J} \times \frac{1\ \text{kJ}}{1000\ \text{J}} = 1.6 \times 10^2\ \text{kJ}$ Larger value of smaller unit = Smaller value of larger unit, OK.	Solve the problem completely, from the **analyze** step to the **check** step. As always, include units in your calculation setup.
You improved your skill at solving problems with the $q = m \times c \times \Delta T$ relationship.	What did you learn by solving this Active Example?

Practice Exercise 15.6

Determine the amount of energy (in kJ) that must be transferred to increase the temperature of a 12.4 kg gold bar from 25°C to 59°C.

15.12 Change in Temperature Plus Change of State

Goal 17 Sketch, interpret, and identify regions in a graph of temperature versus energy for a pure substance over a temperature range from below the melting point to above the boiling point.

18 Given (a) the mass of a pure substance, (b) ΔH_{vap} and/or ΔH_{fus} of the substance, and (c) the average specific heat of the substance in the solid, liquid, and/or vapor state, calculate the total heat transferred in going from one state and temperature to another state and temperature.

If you were to take some ice (water in the solid state) from a freezer, place it in a flask, and then apply heat steadily, five things would happen:

1. The ice would warm to its melting point.
2. The ice would melt at constant temperature, the melting point.
3. The water would warm to its boiling point.
4. The water would boil at constant temperature, the boiling point.
5. The steam would become hotter.

The heat transfer for each of these steps can be calculated by the methods set forth in Sections 15.10 and 15.11 (see **Figure 15.44** to learn how heat-transfer data are collected in the laboratory). The specific heats of ice, water, and steam would be used for Steps 1, 3, and 5. The heat of fusion would be used for Step 2, and the heat

Figure 15.44 Coffee-cup calorimeter. Data for heat-transfer problems are obtained from calorimeters, highly insulated containers in which chemical or physical changes occur. This coffee-cup calorimeter is commonly used in introductory college chemistry laboratories.

of vaporization would be used for Step 4. The total heat transfer would be the sum of the five separate heat transfers.

Figure 15.45 illustrates this process by words, a graph, and the appropriate equation for each step. The shape of the temperature-versus-heat graph is typical for any pure substance. Refer to this illustration as you work through the next Active Example. It will help you to see clearly what is being done in each of the five steps. The general procedure for this kind of problem is:

how to... Calculate Total Heat Transfer for a Change in Temperature Plus a Change of State

Step 1: Sketch a graph having the shape shown in Figure 15.45. Mark the starting and ending points for the particular problem. Then mark the beginning and ending points of any change of state between the starting and ending points for the problem.

Step 2: Calculate the heat transfer, q, for each sloped and horizontal portion of the graph between the starting and ending points.

Step 3: Add the heat transfers calculated in Step 2. *Caution: Be sure the units are the same, either kilojoules or joules, for all numbers being added.*

TEMPERATURE-HEAT-ENERGY GRAPH FOR A PURE SUBSTANCE

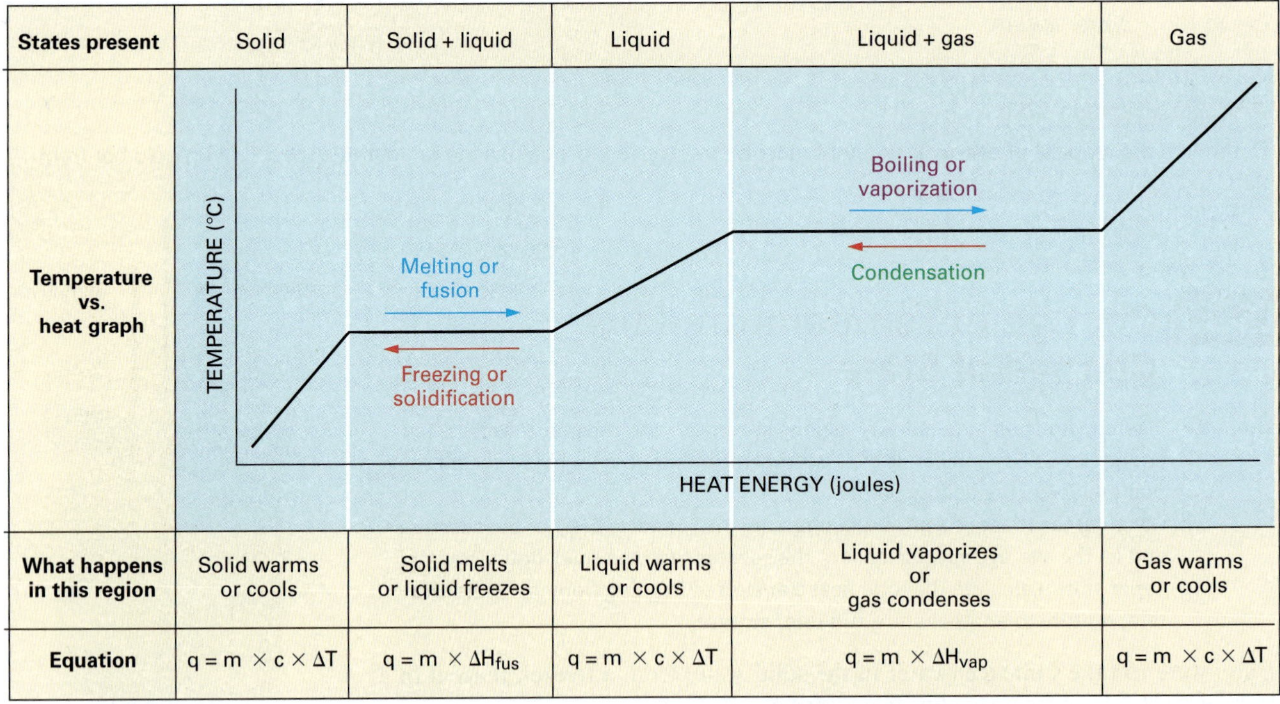

States present	Solid	Solid + liquid	Liquid	Liquid + gas	Gas
Temperature vs. heat graph					
What happens in this region	Solid warms or cools	Solid melts or liquid freezes	Liquid warms or cools	Liquid vaporizes or gas condenses	Gas warms or cools
Equation	$q = m \times c \times \Delta T$	$q = m \times \Delta H_{fus}$	$q = m \times c \times \Delta T$	$q = m \times \Delta H_{vap}$	$q = m \times c \times \Delta T$

Figure 15.45 Temperature–heat–energy graph for state and temperature changes. *Solid column:* When a solid below the freezing point is heated, temperature increases; when cooled, the temperature decreases. There is no change of state; the substance remains a solid. *Solid + liquid column:* At the melting point the solid melts as heat is transferred to the substance, or it freezes as heat is transferred out. Temperature remains constant during the change between solid and liquid. *Liquid column:* As heat is transferred to a liquid, its temperature increases; as heat is transferred out, the temperature goes down. There is no change of state; the substance remains a liquid. *Liquid + gas column:* At the boiling point the liquid boils as heat is transferred to a substance, or it condenses as heat is transferred out. Temperature remains constant during the change between liquid and gas. *Gas column:* As heat is transferred to a gas, its temperature increases; as heat is transferred out, the temperature goes down. There is no change of state; the substance remains a gas.

Active Example 15.7 Change in Temperature Plus Change of State I

Calculate the heat transferred when 45.0 g of steam, initially at 111°C, is changed to water at 25°C.

Think Before You Write Multiple-step change-in-temperature-plus-change-of-state problems are best planned by first drawing a sketch like Figure 15.45 with the temperatures from the problem specifically listed on the y-axis. You then use $q = m \times c \times \Delta T$ and $q = m \times \Delta H_{fus/vap}$ as appropriate for each step.

Answers *Cover the left column with your cut-out shield. Reveal each answer only after you have written your own answer in the right column.*

Begin by sketching the temperature–heat curve. Your sketch will be like Figure 15.45 but labeled specifically for the present Active Example. Use the letter A to label the initial temperature, 111°C, and the letters B and C to label the beginning and ending of the state changes between the initial temperature and the final temperature. Use the letter D to label the final temperature, 25°C. List each of the labeled temperatures of the y-axis of your graph.

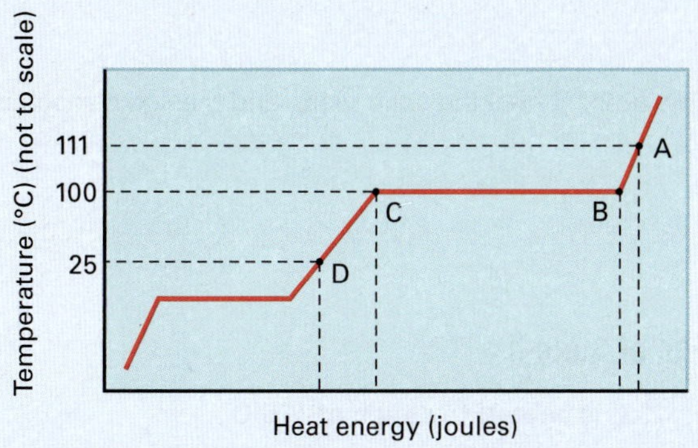

Point A is the starting point at 111°C. The horizontal line is drawn at 100°C, the boiling (and condensing) temperature of water. Points B and C are at the ends of the condensing line. Point D is liquid water at 25°C, the final temperature.

Your graph shows you that the heat transferred must be calculated for cooling steam from point A to point B, 111°C to 100°C ($q_{A\text{ to }B}$). Then it must be calculated for condensing the steam at 100°C ($q_{B\text{ to }C}$). Finally, q must be calculated for cooling water from 100°C to 25°C ($q_{C\text{ to }D}$). Calculate $q_{A\text{ to }B}$, cooling the 45.0 g of steam from 111°C to 100°C, using $q = m \times c \times \Delta T$. The value of c for steam is in Table 15.5. Express the answer in kilojoules.

$$q_{A\text{ to }B} = m \times c \times \Delta T = 45.0\,\cancel{g} \times \frac{2.00\text{ J}}{\cancel{g} \cdot \cancel{°C}} \times (100 - 111)\cancel{°C} \times \frac{1\text{ kJ}}{1000\text{ J}} = -0.99\text{ kJ}$$

$$45 \times 2.00 \times (100 - 111) \times \frac{1}{1000} \approx 50 \times 2 \times -10 \times 10^{-3} = -1\text{, OK.}$$

The negative value of the answer indicates that energy is being transferred out of the system; the change is exothermic. Your q value will always have the correct sign as long as you carefully apply the $T_{final} - T_{initial}$ rule for a Δ quantity.

The two significant figures in the answer come from $\Delta T = -11°C$.

Find $q_{B\text{ to }C}$, changing 45.0 g of steam at 100°C to liquid water at 100°C. Remember, energy is transferred *out* of the water in this process. You'll have to flip back to Table 15.4 to find the value of the heat of vaporization of water.

$$q_{B\text{ to }C} = m \times -\Delta H_{vap} = 45.0\,\cancel{g} \times \frac{-2.26\text{ kJ}}{\cancel{g}} = -102\text{ kJ}$$

$$45.0 \times -2.26 \approx 50 \times -2 = -100\text{, OK.}$$

The sign of the heat of vaporization is changed to negative to get the heat of condensation.

Cool the liquid water to 25°C, points C to D on your graph, using $q = m \times c \times \Delta T$ again, but with the value of the specific heat of liquid water.

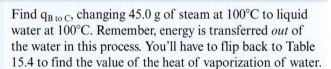

$$q_{C\text{ to }D} = m \times c \times \Delta T = 45.0\,\cancel{g} \times \frac{4.18\,\cancel{J}}{\cancel{g} \cdot \cancel{°C}} \times (25 - 100)\cancel{°C} \times \frac{1\text{ kJ}}{1000\,\cancel{J}} = -14\text{ kJ}$$

$$45.0 \times 4.18 \times (25 - 100) \times \frac{1}{1000} \approx 50 \times 4 \times -75 \times 10^{-3} = -15000 \times 10^{-3} = -15\text{, OK.}$$

$\Sigma q^* = q_{A\text{ to }B} + q_{B\text{ to }C} + q_{C\text{ to }D} = -0.99\text{ kJ} + (-102\text{ kJ}) + (-14\text{ kJ}) = -117\text{ kJ}$

$(-1) + (-102) + (-14) = -117$, OK.

You have calculated the energy for each of the three steps. Complete the problem by finding the sum of the individual steps.

You improved your skill at solving problems with a combination of the $q = m \times c \times \Delta T$ relationship and the $q = m \times \Delta H_{fus/vap}$ relationship.

What did you learn by solving this Active Example?

Practice Exercise 15.7

How much energy (in kJ) must be transferred to heat 75 g of solid zinc at 22°C until the solid melts and the liquid reaches a temperature of 555°C?

Active Example 15.8 Change in Temperature Plus Change of State II

Calculate the total heat transferred when 19.6 g of ice, initially at −12°C, is heated to steam at 115°C.

Think Before You Write You will plan the problem by drawing a sketch like Figure 15.45 with the temperatures from the problem specifically listed on the y-axis. You then use $q = m \times c \times \Delta T$ and $q = m \times \Delta H_{fus/vap}$ as appropriate for each step.

Answers *Cover the left column with your cut-out shield. Reveal each answer only after you have written your own answer in the right column.*

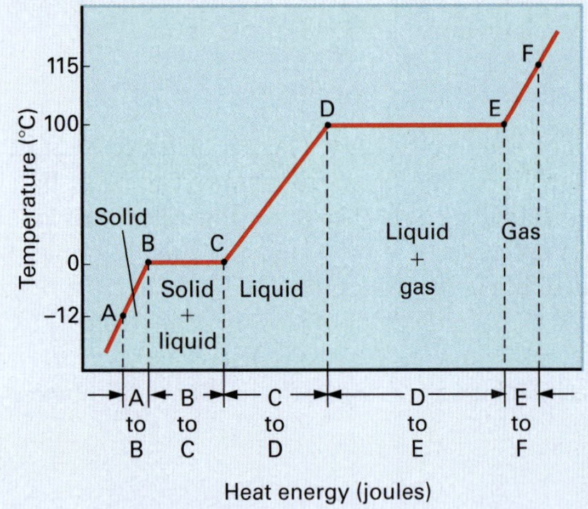

Start with a graph of temperature versus heat energy. Label all temperatures that will be included in calculations. Use A for the initial temperature of the ice, −12°C, B for the solid at 0°C, C for the liquid at 0°C, and so on. Indicate the state of the water (for example, solid, solid + liquid, liquid, and so on) on your graph.

*The Greek letter sigma, Σ, is used to indicate "the sum of" two or more quantities.

Active Example 15.7 Change in Temperature Plus Change of State I

Calculate the heat transferred when 45.0 g of steam, initially at 111°C, is changed to water at 25°C.

Think Before You Write Multiple-step change-in-temperature-plus-change-of-state problems are best planned by first drawing a sketch like Figure 15.45 with the temperatures from the problem specifically listed on the y-axis. You then use $q = m \times c \times \Delta T$ and $q = m \times \Delta H_{fus/vap}$ as appropriate for each step.

Answers *Cover the left column with your cut-out shield. Reveal each answer only after you have written your own answer in the right column.*

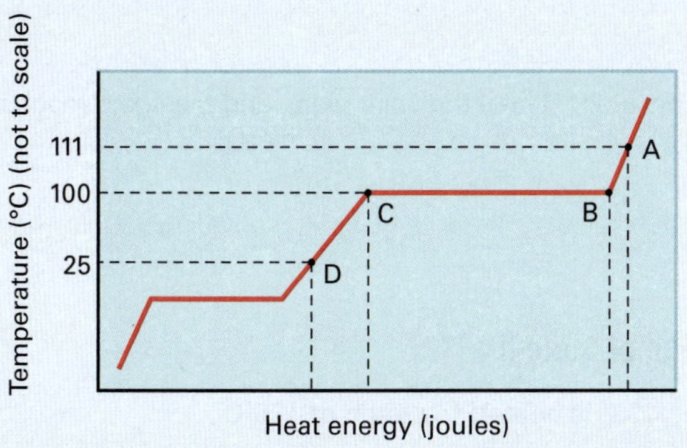

Point A is the starting point at 111°C. The horizontal line is drawn at 100°C, the boiling (and condensing) temperature of water. Points B and C are at the ends of the condensing line. Point D is liquid water at 25°C, the final temperature.

Begin by sketching the temperature–heat curve. Your sketch will be like Figure 15.45 but labeled specifically for the present Active Example. Use the letter A to label the initial temperature, 111°C, and the letters B and C to label the beginning and ending of the state changes between the initial temperature and the final temperature. Use the letter D to label the final temperature, 25°C. List each of the labeled temperatures of the y-axis of your graph.

$$q_{A\,to\,B} = m \times c \times \Delta T = 45.0\,\cancel{g} \times \frac{2.00\text{ J}}{\cancel{g} \cdot \cancel{°C}} \times (100 - 111)\cancel{°C}$$

$$\times \frac{1\text{ kJ}}{1000\text{ J}} = -0.99\text{ kJ}$$

$$45 \times 2.00 \times (100 - 111) \times \frac{1}{1000} \approx 50 \times 2 \times -10 \times 10^{-3}$$

$= -1$, OK.

The negative value of the answer indicates that energy is being transferred out of the system; the change is exothermic. Your q value will always have the correct sign as long as you carefully apply the $T_{final} - T_{initial}$ rule for a Δ quantity.

The two significant figures in the answer come from $\Delta T = -11°C$.

Your graph shows you that the heat transferred must be calculated for cooling steam from point A to point B, 111°C to 100°C ($q_{A\,to\,B}$). Then it must be calculated for condensing the steam at 100°C ($q_{B\,to\,C}$). Finally, q must be calculated for cooling water from 100°C to 25°C ($q_{C\,to\,D}$). Calculate $q_{A\,to\,B}$, cooling the 45.0 g of steam from 111°C to 100°C, using $q = m \times c \times \Delta T$. The value of c for steam is in Table 15.5. Express the answer in kilojoules.

$$q_{B\,to\,C} = m \times -\Delta H_{vap} = 45.0\,\cancel{g} \times \frac{-2.26\text{ kJ}}{\cancel{g}} = -102\text{ kJ}$$

$45.0 \times -2.26 \approx 50 \times -2 = -100$, OK.

The sign of the heat of vaporization is changed to negative to get the heat of condensation.

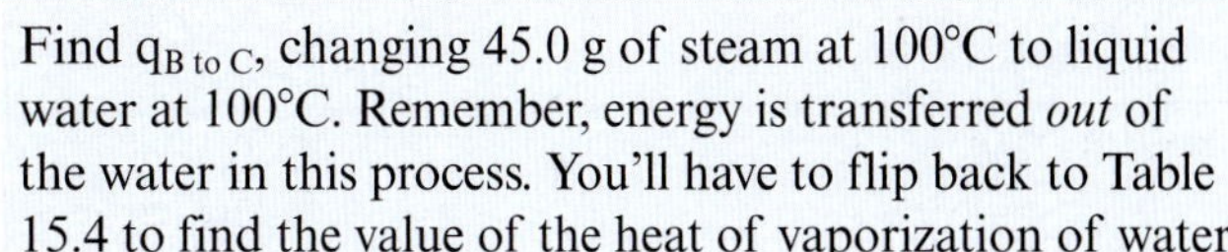

Find $q_{B\,to\,C}$, changing 45.0 g of steam at 100°C to liquid water at 100°C. Remember, energy is transferred *out* of the water in this process. You'll have to flip back to Table 15.4 to find the value of the heat of vaporization of water.

$$q_{C\,to\,D} = m \times c \times \Delta T = 45.0\,\cancel{g} \times \frac{4.18\,\cancel{J}}{\cancel{g} \cdot \cancel{°C}} \times (25 - 100)\cancel{°C} \times \frac{1\text{ kJ}}{1000\,\cancel{J}} = -14\text{ kJ}$$

$$45.0 \times 4.18 \times (25 - 100) \times \frac{1}{1000} \approx 50 \times 4 \times -75 \times 10^{-3}$$

$= -15000 \times 10^{-3} = -15$, OK.

Cool the liquid water to 25°C, points C to D on your graph, using $q = m \times c \times \Delta T$ again, but with the value of the specific heat of liquid water.

$\Sigma q^* = q_{A\ to\ B} + q_{B\ to\ C} + q_{C\ to\ D} = -0.99\ kJ + (-102\ kJ) + (-14\ kJ) = -117\ kJ$

$(-1) + (-102) + (-14) = -117$, OK.

You have calculated the energy for each of the three steps. Complete the problem by finding the sum of the individual steps.

You improved your skill at solving problems with a combination of the $q = m \times c \times \Delta T$ relationship and the $q = m \times \Delta H_{fus/vap}$ relationship.

What did you learn by solving this Active Example?

Practice Exercise 15.7

How much energy (in kJ) must be transferred to heat 75 g of solid zinc at 22°C until the solid melts and the liquid reaches a temperature of 555°C?

Active Example 15.8 Change in Temperature Plus Change of State II

Calculate the total heat transferred when 19.6 g of ice, initially at −12°C, is heated to steam at 115°C.

Think Before You Write You will plan the problem by drawing a sketch like Figure 15.45 with the temperatures from the problem specifically listed on the y-axis. You then use $q = m \times c \times \Delta T$ and $q = m \times \Delta H_{fus/vap}$ as appropriate for each step.

Answers *Cover the left column with your cut-out shield. Reveal each answer only after you have written your own answer in the right column.*

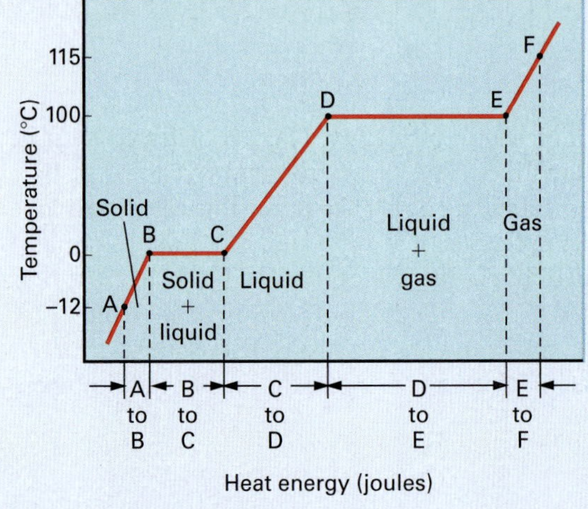

Start with a graph of temperature versus heat energy. Label all temperatures that will be included in calculations. Use A for the initial temperature of the ice, −12°C, B for the solid at 0°C, C for the liquid at 0°C, and so on. Indicate the state of the water (for example, solid, solid + liquid, liquid, and so on) on your graph.

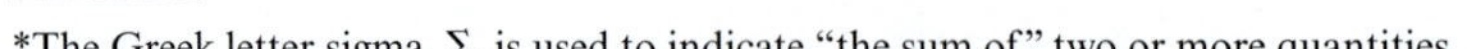

*The Greek letter sigma, Σ, is used to indicate "the sum of" two or more quantities.

$$q_{A\text{ to }B} = m \times c \times \Delta T = 19.6\ \cancel{g} \times \frac{2.06\ \cancel{J}}{\cancel{g} \cdot \cancel{°C}} \times [0 - (-12)]\cancel{°C} \times \frac{1\ kJ}{1000\ \cancel{J}} = 0.48\ kJ$$

$$19.6 \times 2.06 \times [0 - (-12)] \times \frac{1}{1000} \approx 20 \times 2 \times 12 \times 10^{-3} = 0.48,\ OK.$$

Determine $q_{A\text{ to }B}$, warming the ice from $-12°C$ to its melting point, using $q = m \times c \times \Delta T$ and the value of the specific heat of ice. Use kilojoules as the final energy unit for all steps in the problem.

$$q_{B\text{ to }C} = m \times \Delta H_{fus} = 19.6\ \cancel{g} \times \frac{333\ \cancel{J}}{\cancel{g}} \times \frac{1\ kJ}{1000\ \cancel{J}} = 6.53\ kJ$$

$$19.6 \times 333 \times \frac{1}{1000} \approx (2 \times 10^1) \times (3 \times 10^2) \times 10^{-3} = 6,\ OK.$$

Your graph indicates that the next calculation is $q_{B\text{ to }C}$, melting ice at 0°C, using $q = m \times \Delta H_{fus}$. Look up the heat of fusion of H_2O in Table 15.4 and complete this step. Remember that we are expressing all heat transfers in kilojoules.

$$q_{C\text{ to }D} = m \times c \times \Delta T = 19.6\ \cancel{g} \times \frac{4.18\ \cancel{J}}{\cancel{g} \cdot \cancel{°C}} \times [100 - 0]\cancel{°C} \times \frac{1\ kJ}{1000\ J} = 8.19\ kJ$$

$$19.6 \times 4.18 \times [100 - 0] \times \frac{1}{1000} \approx 20 \times 4 \times 10^2 \times 10^{-3} = 8,\ OK.$$

Continuing to follow the graph, the next region is $q_{C\text{ to }D}$, warming the liquid water from 0°C to 100°C, with $q = m \times c \times \Delta T$. Do the calculation.

$$q_{D\text{ to }E} = m \times \Delta H_{vap} = 19.6\ \cancel{g} \times \frac{2.26\ kJ}{\cancel{g}} = 44.3\ kJ$$

$$19.6 \times 2.26 \approx (2 \times 10^1) \times 2.25 = 4.5 \times 10^1 = 45,\ OK.$$

The next region on the graph is $q_{D\text{ to }E}$, vaporizing the liquid at 100°C, which requires the relationship $q = m \times \Delta H_{vap}$. Table 15.4 has values of heats of vaporization. Complete this step.

$$q_{E\text{ to }F} = m \times c \times \Delta T = 19.6\ \cancel{g} \times \frac{2.00\ \cancel{J}}{\cancel{g} \cdot \cancel{°C}} \times (115 - 100)\cancel{°C} \times \frac{1\ kJ}{1000\ \cancel{J}} = 0.59\ kJ$$

$$19.6 \times 2.00 \times (115 - 100) \times \frac{1}{1000} \approx 20 \times 2 \times 15 \times 10^{-3} = 600 \times 10^{-3} = 0.6,\ OK.$$

The only region remaining on the graph is $q_{E\text{ to }F}$, heating the steam from 100°C to 115°C, once again requiring $q = m \times c \times \Delta T$. Determine this heat energy transfer.

$\Sigma q = q_{A\ to\ B} + q_{B\ to\ C} + q_{C\ to\ D} + q_{D\ to\ E} + q_{E\ to\ F}$ $= 0.48\ kJ + 6.53\ kJ + 8.19\ kJ + 44.3\ kJ + 0.59\ kJ$ $= 60.1\ kJ$ $0.48 + 6.53 + 8.19 + 44.3 + 0.59 \approx 0.5 + 6.5 + 8.2 + 44.3 + 0.6 = 60.1$, OK.	In *Step 3* of our procedure, the individual heat transfers are added. All values are known in kilojoules, so use that unit and find the total heat energy transferred, as requested in the problem statement.
You improved your skill at solving problems with a combination of the $q = m \times c \times \Delta T$ relationship and the $q = m \times \Delta H_{fus/vap}$ relationship.	What did you learn by solving this Active Example?

Practice Exercise 15.8

Determine the energy transferred when a 50.0-g sample of steam at 133°C is cooled to ice at −22°C.

Candyfloss Film/Shutterstock.com

CHAPTER 15 IN REVIEW: GASES, LIQUIDS, AND SOLIDS

Goal 1 Given the partial pressure of each component in a mixture of gases, find the total pressure.

The **partial pressure** of a gas in a gaseous mixture is the pressure that gas alone would exert in the same volume at the same temperature.

Goal 2 Given the total pressure of a gaseous mixture and the partial pressures of all components except one, or information from which those partial pressures can be obtained, find the partial pressure of the remaining component.

The total pressure of a gas mixture is the sum of the partial pressures of all gases in that mixture, $P = p_1 + p_2 + p_3 + \ldots$. This is **Dalton's Law of Partial Pressures.**

Goal 3 Explain the differences between the physical behavior of liquids and gases in terms of the relative distances among particles and the effect of those distances on intermolecular attractions.

Important **properties of liquids** (and comparisons with gases) include the following:

1. Gases may be compressed; liquids cannot. Liquid particles are "touchingly close" to one another.
2. Gases expand to fill their containers; liquids do not. The strong attractions between liquid particles hold them together at the bottom of a container.
3. Gases have low densities; liquids have relatively high densities. If the particles of a liquid are close together compared with the particles of a gas, a given number of liquid particles will occupy a much smaller volume than the same number of particles will occupy as a gas.
4. Gases may be mixed in a fixed volume; liquids cannot. There is no space between particles of a liquid, so combining liquids must increase volume.
5. Gases exert constant pressure on the walls of their container uniformly in all directions; the pressure in a liquid container increases with increasing depth. Liquid pressures depend on the depth of the liquid due to variation in weight at varying depth.

Goal 4 For two liquids, given comparative values of physical properties that depend on intermolecular attractions, predict the relative strengths of those attractions; or, given a comparison of the strengths of the intermolecular attractions, predict the relative values of physical properties that the attractions cause.

Properties of liquids are related to intermolecular attractions:

Vapor pressure is the partial pressure of a vapor in equilibrium with its liquid state at a given temperature. Liquids with strong intermolecular attractions have lower vapor pressures than liquids with weak intermolecular attractions.

Heat of vaporization is the quantity of energy required to change one mole of a liquid to a gas while at constant temperature and pressure. Liquids with strong intermolecular attractions have higher heats of vaporization than liquids with weak intermolecular attractions.

Boiling point is the temperature at which vapor pressure becomes equal to the pressure above a liquid. Liquids with strong intermolecular attractions have higher boiling points than liquids with weak intermolecular attractions.

Viscosity is the resistance of a liquid to flow. Liquids with strong intermolecular attractions have higher viscosities than liquids with weak intermolecular attractions.

Surface tension is the force exerted on the molecules at the surface of a liquid by the molecules below the surface. Liquids with strong intermolecular attractions have greater surface tension than liquids with weak intermolecular attractions.

Goal 5 Identify and describe or explain induced dipole forces, dipole forces, and hydrogen bonds.

Induced dipole forces are comparatively weak intermolecular attractions between all molecules. They are the result of temporary dipoles caused by shifting electron density in molecules. Induced dipole forces vary directly with surface area and may be large if the molecules are large.

Dipole forces are electrostatic attractions between polar molecules (dipoles).

Exceptionally strong dipole-like forces called **hydrogen bonds** arise between molecules that have hydrogen atoms covalently bonded to a highly electronegative atom. This atom, usually nitrogen, oxygen, or fluorine, must have at least one unshared electron pair.

Goal 6 Given the structure of a molecule, or information from which it may be determined, identify the significant intermolecular forces present.

In general, a liquid with nonpolar molecules will have only **induced dipole forces** acting among the molecules. The larger the molecules, the greater the strength of the attractive forces. Liquids with polar molecules have **dipole forces** acting among the molecules. The more polar the molecules, the stronger the forces. Liquids with molecules with a hydrogen atom bonded to an atom that is small, highly electronegative, and that has at least one unshared pair of electrons have **hydrogen bonds** acting among the molecules.

Goal 7 Given the molecular structures of two substances, or information from which they may be obtained, compare or predict relative values of physical properties that are related to them.

All other things being equal, intermolecular attractive forces increase in the order: induced dipoles < dipole forces < hydrogen bonds. In large molecules, however, induced dipole forces can be the most important forces acting to determine values of physical properties.

Goal 8 Describe and explain the equilibrium between a liquid and its own vapor and the process by which it is reached.

Equilibrium is defined as the condition in which the rates of opposing changes are equal. In a liquid–vapor equilibrium, the rate of evaporation is equal to the rate of condensation. Such an equilibrium is achieved in a closed flask by starting with the movement of molecules in one direction, from liquid to vapor. The fraction of molecules with an energy greater than the escape energy, as illustrated on a **kinetic energy distribution curve**, will move into the vapor phase when they reach the liquid surface. The condensation rate is initially zero. As the number of molecules in the vapor state increases, the condensation rate increases. Simultaneously, the evaporation rate stays constant. As long as the rate of evaporation is greater than the rate of condensation, the vapor concentration will rise. Eventually, the evaporation and condensation rates become equal, and equilibrium is achieved.

Goal 9 Describe the relationship between vapor pressure and temperature for a liquid–vapor system in equilibrium; explain this relationship in terms of the kinetic molecular theory.

The partial pressure exerted by a vapor in equilibrium with a liquid is the **equilibrium vapor pressure** at the existing temperature. Equilibrium vapor pressure increases with increasing temperature because the shape of the kinetic energy distribution curve changes. At higher temperatures, a larger fraction of the liquid sample has enough energy to evaporate.

Goal 10 Describe the process of boiling and the relationship among boiling point, vapor pressure, and surrounding pressure.

The **boiling point** of a liquid is the temperature at which the vapor pressure of that liquid is equal to or slightly greater than the surrounding pressure. Normal boiling point is the boiling point at one atmosphere of pressure.

Goal 11 Describe the typical relative density relationship between the solid and liquid phase of a substance, and explain why water is an exception to this trend.

Water, a molecule necessary for life, breaks almost all the rules for predicting physical properties of liquids due to its extremely strong hydrogen bonding, which leads to exceptionally strong intermolecular attractions among its molecules. Some of the **unusual properties of water** include the following: an anomalously high boiling point, high surface tension, high heat of vaporization, low vapor pressure, high viscosity, an exceptional ability as a solvent, and its unusual state (liquid) at common temperatures and pressures. The fact that solid water (ice) floats on liquid water is also unusual; for most other substances, the solid phase is denser than the liquid phase.

Goal 12 Distinguish between amorphous and crystalline solids.	Solids can be classified based upon particle arrangement. A **crystalline solid** has its particles arranged in a repeating pattern. An **amorphous solid** has no long-range order among its particles.
Goal 13 Distinguish among the following types of crystalline solids: ionic, molecular, covalent, and metallic.	Crystalline solids can be classified based upon the forces that hold the particles together: **Ionic crystals** are composed of oppositely charged ions that are held together by strong ionic bonds. Ionic crystals typically have a high melting temperature, are frequently water soluble, and have very low electrical conductivities. **Molecular crystals** are made of small, discrete molecules held together by relatively weak intermolecular forces. Molecular crystals are typically soft, have a low melting temperature, and are generally insoluble in water. They are usually nonconductors. **Covalent network solids** are composed of atoms that are covalently bonded to each other to form a single, indefinite-sized network. Covalent network solids are almost always insoluble in any common solvent, are poor conductors of electricity, and have high melting points. **Metallic crystals** are made of a crystal lattice of positive ions through which valence electrons move freely. The freely moving electrons make metals excellent conductors of electricity. Metallic crystals are insoluble in common solvents, malleable, and ductile (can be pressed into thin sheets and drawn into wire), and they have a wide range of melting points.
Goal 14 Given two of the following, calculate the third: (a) mass of a pure substance changing between the liquid and vapor (gaseous) states; (b) heat of vaporization; (c) energy transferred.	In the change between a liquid and a gas, **vaporization** is endothermic and **condensation** is exothermic. The energy required to vaporize a substance, q, is proportional to the amount of substance: $\mathbf{q = \Delta H_{vap} \times m}$. The proportionality constant, ΔH_{vap}, is called the **heat of vaporization.** It is the energy transferred when one gram of a substance changes between the liquid and gaseous states.
Goal 15 Given two of the following, calculate the third: (a) mass of a pure substance changing between the solid and liquid states; (b) heat of fusion; (c) energy transferred.	Energy us transferred into a solid to **melt** it, an endothermic change. Energy is transferred out of a liquid when it **freezes**, an exothermic change. The energy transfer required to melt a substance, q, is proportional to the amount of substance: $\mathbf{q = \Delta H_{fus} \times m}$. The proportionality constant, ΔH_{fus}, is called the **heat of fusion.** It is the energy transferred when one gram of a substance changes between the liquid and solid states.
Goal 16 Given three of the following quantities, calculate the fourth: (a) energy transferred; (b) mass of a pure substance; (c) specific heat of the substance; (d) temperature change or initial and final temperatures.	The heat energy transferred, q, in heating or cooling a substance is proportional to both the mass of the sample, m, and its temperature change, ΔT: $\mathbf{q = m \times c \times \Delta T}$. The proportionality constant, c, is a property of a pure substance called its **specific heat.** It is the amount of energy needed to change the temperature of one gram of a substance by one degree Celsius.
Goal 17 Sketch, interpret, or identify regions in a graph of temperature versus energy for a pure substance over a temperature range from below the melting point to above the boiling point.	A graph of temperature versus heat energy for a pure substance has five sections: (1) a positively sloped section representing warming or cooling the solid, (2) a zero-slope section representing the solid melting or the liquid freezing, (3) a positively sloped section representing warming or cooling the liquid, (4) a zero-slope section representing the liquid vaporizing or the gas condensing, and (5) a positively sloped section representing warming or cooling the gas.
Goal 18 Given (a) the mass of a pure substance, (b) ΔH_{vap} and/or ΔH_{fus} of the substance, and (c) the average specific heat of the substance in the solid, liquid, and/or vapor state, calculate the total heat transferred in going from one state and temperature to another state and temperature.	To calculate the total heat transferred for a change in temperature plus a change in state: **1.** Sketch a graph of temperature versus heat energy for the substance. Mark the beginning and ending point for each phase of matter and each change of state. **2.** Calculate the heat transferred, q, for each sloped and horizontal portion of the graph between the starting and ending points. **3.** Add the heat transfers calculated in Step 2.

Key Terms

Most of the key terms and concepts and many others appear in the Glossary. Use your Glossary regularly.

amorphous solid p. 571
boiling p. 568
boiling point p. 558
condensation p. 576
covalent network solid p. 572
crystalline solid p. 570
Dalton's Law of Partial Pressures p. 554
delta quantity (Δ) p. 579
deposition p. 576
dipole p. 560
dispersion forces p. 560
dynamic equilibrium p. 565
equilibrium p. 565
equilibrium vapor pressure p. 557
evaporation p. 557

freeze p. 575
induced dipole forces p. 560
intermolecular forces p. 560
ionic crystals p. 571
heat of condensation p. 557
heat of fusion, ΔH_{fus} p. 578
heat of solidification p. 578
heat of vaporization, ΔH_{vap} p. 576
hydrogen bond p. 562
joule (J) p. 576
kinetic energy distribution curve p. 564
London dispersion forces p. 560
London forces p. 560
materials science p. 570
melt p. 575
metallic crystal p. 573
molecular crystal p. 572
normal boiling point p. 568
reversible changes p. 565
reversible reaction p. 565
solid state p. 570
specific heat p. 580
sublimation p. 576
surface tension p. 558
vapor pressure p. 557
vaporization p. 576
viscosity p. 558

Frequently Asked Questions

What is the best approach to take while reviewing the chapter?
In this chapter, we discuss the three common states of matter and the energy changes that occur when matter changes states. As you review this chapter, try to get the big picture. This is what will enable you to *explain* and *predict*, which are the key words in the early performance goals in this chapter.

I noticed that Chapter 18 is titled Chemical Equilibrium. *Does that have anything to do with the liquid–vapor equilibrium in Section 15.5?*
Yes, the concept of a dynamic equilibrium is introduced in the liquid–vapor system, and then it will be revisited again. Take the time to understand the idea of this equilibrium and you will find it much easier to understand the saturated solution equilibrium in Section 16.4. The ideas are almost the same. The concept comes up again in Chapter 18 on chemical equilibrium.

What is the best way to approach solving problems with a change in temperature plus a change of state?
Recognize that combination specific heat and change-of-state problems are a *group* of problems in which each problem is related to one short stretch on the horizontal axis of the temperature–heat curve (Figure 15.45). It helps a lot to sketch the graph, as you were guided to do in the last Active Examples of the chapter. Individually, the problems are relatively easy. Whatever is associated with a sloped line is to be solved by the specific heat equation, $q = m \times c \times \Delta T$. The calculation associated with a horizontal (change-of-state) line is $q = m \times \Delta H_{vap}$ or $q = m \times \Delta H_{fus}$. When you get all the q's calculated, add them for the final answer. When you reach this point, remember that joules cannot be added to kilojoules! Be sure all heat transfers are expressed in the same units before you add them, and be sure to apply the addition rule for significant figures.

Concept-Linking Exercises

Write a brief description of the relationships among each of the following groups of terms or phrases. Answers to the Concept-Linking Exercises are given at the end of the chapter.

1. Dipole forces, induced dipole forces, dispersion forces, London forces, London dispersion forces, hydrogen bonds, covalent bond, ionic bond
2. Stick-togetherness, vapor pressure, molar heat of vaporization, boiling point, viscosity, surface tension
3. Boiling point, normal boiling point, temperature, vapor pressure
4. Dynamic equilibrium, liquid–vapor equilibrium, evaporation rate, condensation rate, temperature, vapor pressure
5. Ionic crystal, molecular crystal, covalent network solid, metallic crystal
6. Amorphous solid, crystalline solid
7. Heat of vaporization, heat of condensation, heat of fusion, heat of solidification, specific heat

Small-Group Discussion Questions

Small-Group Discussion Questions are for group work, either in class or under the guidance of a leader during a discussion section.

1. Explain how the ideal gas model relates to Dalton's Law of Partial Pressures.
2. Give a particulate-level explanation for vapor pressure, heat of vaporization, boiling point, viscosity, and surface tension.
3. Draw Lewis diagrams and sketch models of electron clouds to illustrate each of the three types of intermolecular forces. Explain why each force occurs.
4. What are the differences among ionic bonds, metallic bonds, covalent bonds, and hydrogen bonds? Explain in as much detail as possible.
5. Sketch a plot of condensation rate versus time for a volatile liquid in a sealed container from the time the container is sealed until a dynamic liquid–vapor equilibrium is achieved. On the same axes, also plot evaporation rate versus time. Describe the factors that affect the condensation and evaporation rates.
6. Sketch a kinetic energy distribution curve for a liquid. Use the curve to explain why only some of the molecules in a

liquid sample have enough energy to evaporate. Then draw a second curve on the same axes for the same liquid at a higher temperature. Use the curves to explain why vapor pressure is higher at higher temperatures.

7. Sketch a pressure-versus-temperature curve to illustrate the nature of the change in vapor pressure with temperature for a typical liquid. If this liquid has induced dipole forces as the primary intermolecular forces, how would a curve for a liquid with dipole forces compare? Explain and draw this curve on your graph. Also explain the relative vapor pressure and draw a curve for a liquid with hydrogen bonding as the primary intermolecular force.

8. Define the term *boiling point*. Explain how it is possible to boil water at temperatures below and above 100°C. How is boiling point related to a vapor-pressure-versus-temperature curve for a liquid?

9. Distinguish between crystalline solids and amorphous solids at the macroscopic and particulate levels. Also distinguish among ionic crystals, molecular crystals, covalent network solids, and metallic crystals at both the macroscopic and particulate levels.

10. A student conducts a coffee-cup calorimetry experiment to determine the specific heat of an unknown metal. First, she measures 30.0 mL of room temperature water at 23.3°C into the calorimeter. She then adds 30.0 mL of water heated to 52.3°C to the calorimeter, and through graphical extrapolation, the instantaneous final temperature of the mixture is determined to be 36.2°C. She then measures 100.0 mL of water at 23.3°C to the dry calorimeter. She had a 72.45 g sample of the metal in a boiling water bath, which was at 98.7°C. She then transferred the metal to the water in the calorimeter, and the final temperature of the water and the metal was 25.9°C. Calculate the specific heat of the metal.

Questions, Exercises, and Problems

Interactive versions of these problems may be assigned by your instructor. Solutions for blue-numbered questions are at the end of the chapter. Questions other than those in the General Questions and More Challenging Problems sections are paired in consecutive odd–even number combinations; solutions for the odd numbered questions are also at the end of the chapter.

Section 15.2: Dalton's Law of Partial Pressures

1. A mixture of helium and argon occupies 1×10^2 L. The partial pressure of helium is 0.6 atm, and the partial pressure of argon is 0.4 atm. What are the partial volumes of the two gases? Explain your answer.

2. A mixture of argon and carbon dioxide gases is maintained in an 8.05 L flask at a temperature of 43°C. If the partial pressure of argon is 0.326 atm and the partial pressure of carbon dioxide is 0.234 atm, what is the total pressure in the flask?

3. Atmospheric pressure is the total pressure of the gaseous mixture called *air*. Atmospheric pressure is 749 torr on a day that the partial pressures of nitrogen, oxygen, and argon are 584 torr, 144 torr, and 19 torr, respectively. What is the partial pressure of all the other gases in the air on that day?

4. The stopcock connecting a 1.00 L bulb containing oxygen gas at a pressure of 540 torr and a 1.00 L bulb containing helium gas at a pressure of 776 torr is opened, and the gases are allowed to mix. Assuming that the temperature remains constant, what is the final pressure in the system?

5. A sample of "wet" hydrogen gas was collected over water (see Figure 15.6). The sample was adjusted so that it was at room temperature and pressure, 19°C and 733 mm Hg. Water vapor pressure at 19°C is 16.5 mm Hg. What is the partial pressure of the "dry" hydrogen gas?

6. Sodium metal reacts with water to produce hydrogen gas according to the following equation: $2\ Na(s) + 2\ H_2O(\ell) \rightarrow 2\ NaOH(aq) + H_2(g)$. In an experiment, the hydrogen gas is collected over water in a vessel where the total pressure is 745 torr and the temperature is 20°C, at which temperature the vapor pressure of water is 17.5 torr. Under these conditions, what is the partial pressure of hydrogen? If the wet hydrogen gas formed occupies a volume of 8.04 L, what amount in moles of hydrogen is formed?

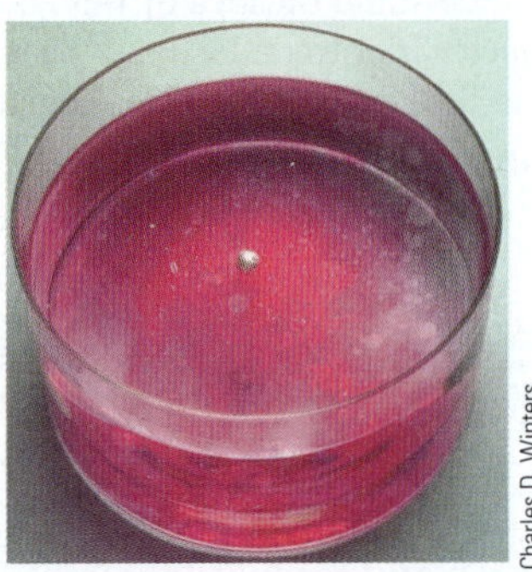

Sodium metal reacts with water, forming sodium ion, hydroxide ion, and hydrogen gas. A compound that changes from colorless to pink when it reacts with hydroxide ion was added to the water before the sodium.

Section 15.3: Properties of Liquids

7. Why will two gases mix with each other more quickly than two liquids?

8. Why is the liquid state of a substance denser than the gaseous state?

9. Why are intermolecular attractions stronger in the liquid state than in the gaseous state?

10. Explain why water is less compressible than air.

11. How do intermolecular attractive forces influence the boiling point of a pure substance?

12. How are intermolecular attractive forces and equilibrium vapor pressure related? Suggest a reason for this relationship.

13. Why does molar heat of vaporization depend on the strength of intermolecular forces?

14. What relationship exists between viscosity and intermolecular forces?

15. A tall, glass cylinder is filled to a depth of 1 m with water. Another tall, glass cylinder is filled to the same depth with syrup. Identical ball bearings are dropped into each tube at the same instant. In which tube will the ball bearings reach the bottom first? Explain your prediction in terms of viscosity and intermolecular attractive forces.

16. Which liquid is more viscous, water or motor oil? In which liquid do you suppose the intermolecular attractions are stronger? Explain.

17. If water spills on a laboratory desktop, it usually spreads over the surface, wetting any papers or books in its path. If mercury spills, it neither spreads nor makes paper wet, but forms little drops that are easy to combine into pools by pushing them together. Suggest an explanation for these facts in terms of the apparent surface tension and intermolecular attractive forces in mercury and water.

Charles Steele

Mercury drops. Notice that the larger the drop, the greater the degree to which it is flattened by the force of gravity.

18. A drop of honey and a drop of water of identical volumes are placed on a plate. The water drop forms a large, shallow pool, but the honey drop forms a circular blob with a much smaller diameter. Compare the surface tensions of the two liquids. Which liquid has stronger intermolecular attractive forces? Explain.

19. The level at which a duck floats on water is determined more by the thin oil film that covers its feathers than by a body density that is lower than the density of water. The water does not "mix" with the oil and therefore does not penetrate the feathers. If, however, a few drops of wetting agent are placed in the water near the duck, the poor duck will sink to its neck. State the effect of a wetting agent on surface tension and intermolecular attractions of water.

20. The cleansing ability of soap depends in large part on its ability to change the surface tension in water. How do you suppose soap affects the surface tension of water? Explain.

Questions 21 through 24: The table below gives the normal boiling and melting points for three nitrogen oxides.

	NO	N_2O	NO_2
Boiling point	−152°C	−88.5°C	+21.2°C
Melting point	−164°C	−90.8°C	−11.2°C

21. Which of the three oxides would you expect to have the highest molar heat of vaporization? Explain how you reached your conclusion.

22. Samples of the three substances are side-by-side at a temperature at which all are solids. The temperature is raised gradually. List the compounds in the order at which they will begin to melt.

23. Which of the three oxides would you expect to have a measurable vapor pressure at −90°C? Explain your answer.

24. Which of the three oxides would you expect to have the lowest viscosity at −90°C? Justify your conclusion.

Section 15.4: Types of Intermolecular Forces

25. Other things being equal, which produces stronger intermolecular attractions, induced dipole forces or dipole forces? What "other things being *un*equal" would reverse this order of attractions?

Eric Isselee/Shutterstock.com

In 2002, a team of scientists discovered that induced dipole forces among millions of tiny hairs on gecko feet and the surface are what allow these lizards to adhere to walls and ceilings.

26. Suggest a molecular structure in which hydrogen bonding may be present, but its contribution to intermolecular attractions is less than the contribution of induced dipole forces. Justify your suggestion.

27. What are the principal intermolecular forces in each of the following compounds: $NH(CH_3)_2$, CH_2F_2, C_3H_8?

28. Identify the principal intermolecular forces in each of the following compounds: NOCl, NH_2Cl, $SiCl_4$.

29. Compare dipole forces and hydrogen bonds. How are they different, and how are they similar?

30. Given an ionic compound and a polar molecular compound of about the same molar mass, which is likely to have the higher melting point? Why? In explaining your answer, identify the interparticle forces present and the roles they play.

Questions 31 through 34: On the basis of molecular size, molecular polarity, and hydrogen bonding, predict for each pair of compounds the one that has the higher boiling point. State the reason for your choice. Assume molecular size is related to molar mass.

31. CH_4 and NH_3

32. CH_4 and CCl_4

33. Ar and Ne

34. H_2S and PH_3

35. What feature of the hydrogen atom, when bonded to an appropriate second element, is largely responsible for the strength of hydrogen bonding between molecules?

36. Identify elements to which hydrogen atoms must be bonded if hydrogen bonding is to be a significant intermolecular attractive force. How are these elements different from other elements to which hydrogen might be bonded? Explain why the difference is important.

37. Of the three types of intermolecular forces, which one(s) (a) increase with molecular size; (b) account for the high melting point, boiling point, and other abnormal properties of water?

38. Of the three types of intermolecular forces, which one(s) operate(s) (a) in all molecular substances; (b) between all polar molecules?

39. Identify the intermolecular forces present in each of the following:

a) 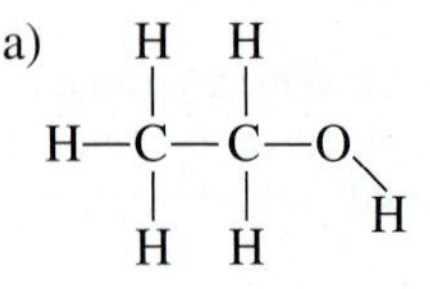

b) 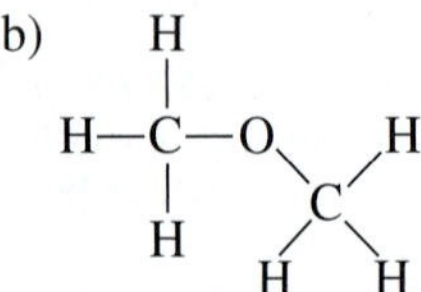

c) 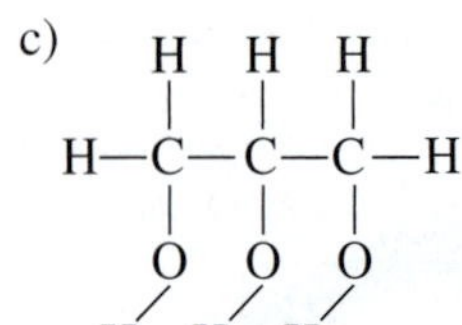

d) 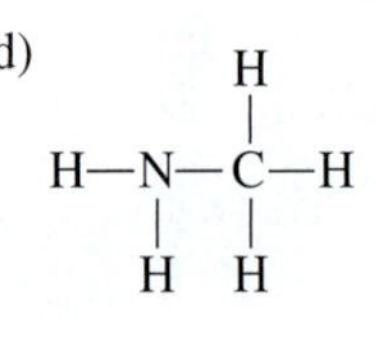

e)
```
    H  H
    |  |
H—C—C—H
    |  |
    H  H
```

40. Identify the intermolecular forces present in each of the following:

a) H—C≡N   b) O=C=O

c)
```
    P
  / | \
 H  H  H
```

d) 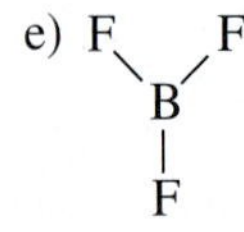

e)
```
F     F
  \  /
   B
   |
   F
```

41. Predict which compound, CO_2 or CS_2, has the higher melting and boiling points. Explain your prediction.

42. Select the best answer to complete the following statement. The boiling point of ICl (97°C) is higher than the boiling point of Br_2 (59°C) because:

a) the molecular mass of ICl is 162.4 u, whereas that of Br_2 is 159.8 u
b) there is hydrogen bonding in ICl, but not in Br_2
c) ICl is an ionic compound, whereas Br_2 is a molecular compound
d) ICl is polar, whereas Br_2 is nonpolar
e) induced dipole forces are much stronger for ICl than for Br_2

43. Predict which compound, CH_4 or CH_3F, has the higher vapor pressure as a liquid at a given temperature. Explain your prediction.

44. Predict which compound, SO_2 or CO_2, has the higher vapor pressure as a liquid at a given temperature. Explain your prediction.

Section 15.5: Liquid–Vapor Equilibrium

45. What is the meaning of *equilibrium*?

46. Why do we describe a liquid–vapor equilibrium as a *dynamic* equilibrium?

47. Explain why the rate of evaporation from a liquid depends on temperature.

48. Use the following vapor pressure data to answer the questions:

	Liquid	Vapor Pressure (torr)	Temperature (°C)
A	CH_3COOCH_3	400	40.0
B	C_7H_{16}	400	78.0

In which liquid are the intermolecular attractive forces the strongest? Explain. Will the vapor pressure of CH_3COOCH_3 at 78°C be higher or lower than 400 torr? Explain.

Section 15.6: The Boiling Process

49. The vapor pressure of a certain compound at 20°C is 906 torr. Is the substance a gas or a liquid at an external pressure of 760 torr? Explain.

50. Define *boiling point*. Draw a vapor-pressure-versus-temperature curve and locate the normal boiling point on it.

Charles D. Winters

A substance at its boiling point.

51. Liquid feed water is delivered to modern boilers at a temperature well above the normal boiling point of water. Explain how this is possible.

52. Normally, a gas may be condensed by cooling it. Suggest a second method and explain why it will work.

53. Explain why low-boiling liquids usually have low molar heats of vaporization.

54. Given the following vapor pressure data:

	Liquid	Vapor Pressure (torr)	Temperature (°C)
A	CS_2	400	28.0
B	C_8H_{18}	400	104.0

In which liquid are the intermolecular attractive forces the strongest? Explain. Which liquid would be expected to have the highest normal boiling point? Explain.

55. At 20°C, the vapor pressure of substance M is 520 torr; of substance N, 634 torr. Which substance will have the lower boiling point? The lower molar heat of vaporization?

56. The molar heat of vaporization of substance X is 34 kJ/mol; of substance Y, 27 kJ/mol. Which substance would be expected to have the higher normal boiling point? The higher vapor pressure at 25°C?

Section 15.8: The Solid State

57. Is ice a crystalline solid or an amorphous solid? On what properties do you base your conclusion?

58. Compare amorphous and crystalline solids in terms of structure. How do crystalline and amorphous solids differ in physical properties? Explain the difference.

Section 15.9: Types of Crystalline Solids

Questions 59 and 60: The physical properties of four solids are tabulated here. In each case, state whether the solid is most likely to be ionic, molecular, metallic, or a covalent network solid.

	Solid	Melting Point	Water Solubility	Conductivity (Pure)	Type of Solid
59.	A	2000°C	Insoluble	Nonconductor	______
	B	1050°C	Soluble	Nonconductor	______
60.	C	150°C	Insoluble	Nonconductor	______
	D	1450°C	Insoluble	Excellent	______

Section 15.10: Energy and Change of State

See Table 15.4 for heats of fusion and vaporization.

61. A student is to find the heat of vaporization of isopropyl alcohol (rubbing alcohol). She vaporizes 61.2 g of the liquid at its boiling point and measures the energy required as 44.8 kJ. What heat of vaporization does she report?

James McDowall/Shutterstock.com

A solution of isopropyl alcohol is commonly marketed as "rubbing alcohol" because it can be rubbed on skin as a cooling agent and antiseptic.

62. It is observed that 58.2 kJ of energy is transferred when a 26.4 g sample of an unknown liquid condenses. What is the heat of vaporization of the unknown liquid in kJ/g?

63. Calculate the energy transferred as 227 g of sodium vapor condenses.

64. The following information is given for benzene, C_6H_6, at 1 atm:

boiling point = 80°C $\quad \Delta H_{vap} = 0.393 \text{ kJ/g}$

melting point = 6°C $\quad \Delta H_{fus} = 127 \text{ J/g}$

What quantity of energy in kilojoules is needed to vaporize a 28.6-g sample of liquid benzene at its normal boiling point of 80°C?

65. 79.4 kJ was transferred due to the condensation of a sample of ethyl alcohol. If $\Delta H_{vap} = 0.880 \text{ kJ/g}$, what was the mass of the sample?

66. The following information is given for magnesium at 1 atm:

boiling point = 1090°C $\quad \Delta H_{vap} = 5.42 \text{ kJ/g}$

melting point = 649°C $\quad \Delta H_{fus} = 368 \text{ J/g}$

Heat is transferred to a sample of solid magnesium at its normal melting point of 649°C. What mass in grams of magnesium will melt if 10.5 kJ of energy is transferred?

67. Acetone, C_3H_6O, is a highly volatile solvent sometimes used as a cleansing agent prior to vaccination. It evaporates quickly from the skin, making the skin feel cold. What quantity of energy is transferred to 23.8 g of acetone as it evaporates if its molar heat of vaporization is 32.0 kJ/mol?

Gianni Diliberto/Getty Images

Acetone is sometimes used as the active ingredient in nail polish remover.

68. The following information is given for cadmium at 1 atm:

boiling point = 765°C $\quad \Delta H_{vap} = 0.890 \text{ kJ/g}$

melting point = 321°C $\quad \Delta H_{fus} = 54.4 \text{ J/g}$

What is the energy transferred, q, in kJ, for the process of freezing a 22.4-g sample of liquid cadmium at its normal melting point of 321°C?

69. Calculate the energy transferred when 3.30 kg of lead freezes.

70. How much energy must be transferred to melt 35.4 g of gold?

71. 36.9 g of an unknown metal transfers 2.51 kJ of energy in freezing. What is the heat of fusion of that metal?

72. An energy transfer of 7.08 kJ is required to melt 46.9 g of naphthalene, which can be used as mothballs. What is the heat of fusion of naphthalene?

73. A piece of zinc transfers 4.45 kJ while freezing. What is the mass of the sample?

74. Calculate the mass in grams of silver that can be changed from a solid to a liquid by the transfer of 11.3 kJ of energy to the solid at its melting point.

Section 15.11: Energy and Change of Temperature: Specific Heat

See Table 15.5 for specific heat values.

75. Samples of two different metals, A and B, have the same mass. The same amount of energy is transferred to each sample. The temperature of A increases by 11°C, and the sample of B increases by 13°C. Which metal has the higher specific heat? Explain your reasoning.

76. In the laboratory, a student finds that 27.6 J of energy must be transferred to increase the temperature of 10.7 g of gaseous xenon from 22.7°C to 39.9°C. What is the measured specific heat of xenon?

77. Find the quantity of energy transferred (in joules) as 467 g of zinc cools from 68°C to 31°C.

78. How much energy must be transferred to raise the temperature of 10.4 g of solid magnesium from 22.4°C to 39.5°C?

79. How much energy (kJ) is transferred when 2.30 kg of gold is cooled from 88°C to 22°C?

80. What is the quantity of energy transfer when the temperature of 10.1 g of solid silver is decreased from 38.1°C to 23.6°C?

81. The mass of a handful of copper coins is 144 g. The coins are at a temperature of 33°C. If 1.47 kJ of energy is transferred when they are tossed in a fountain and drop to the fountain's water temperature, what is that temperature?

Sascha Burkard/Shutterstock.com

A U.S. penny was made of pure copper until 1857. Today's penny is 2.5% copper and 97.5% zinc.

82. A sample of solid sulfur is heated with an electrical coil. If 105 J of energy is transferred to a 12.4-g sample initially at 24.4°C, what is the final temperature of the sulfur?

Section 15.12: Change in Temperature Plus Change of State

Questions 83 through 92: **Figure 15.46** *is a graph of temperature versus energy transferred to a sample of a pure substance. Assume that letters J through P on the horizontal and vertical axes represent measured quantities and that expressions such as R − S or X + Y + Z represent arithmetic operations to be performed with those quantities.*

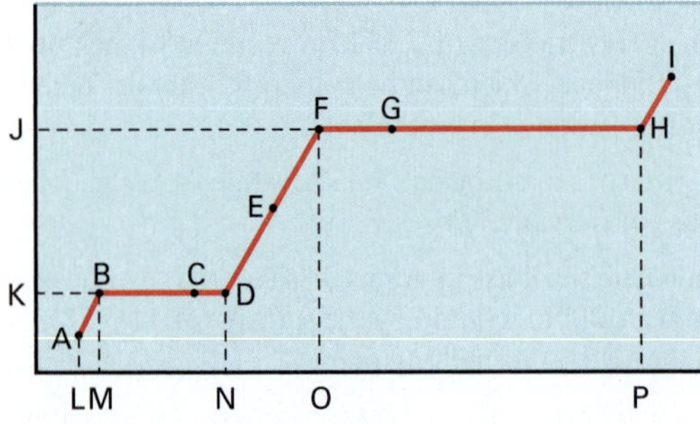

Figure 15.46 Graph of temperature (y-axis) versus heat energy transferred to the sample (x-axis) for a sample of a pure substance.

83. Identify by letter the boiling and freezing points in Figure 15.46.

84. What values are plotted, both vertically and horizontally?

85. Identify all points on the curve in Figure 15.46 where the substance is entirely gas.

86. Identify in Figure 15.46 all points on the curve where the substance is entirely liquid.

87. Identify in Figure 15.46 all points on the curve where the substance is partly solid and partly liquid.

88. Identify all points on the curve in Figure 15.46 where the substance is partly liquid and partly gaseous.

89. Describe the physical changes that occur as energy N − P is transferred out of the sample.

90. Describe what happens physically as the energy represented by N − M is transferred to the sample.

91. Using letters from the graph, show how you would calculate the energy transfer required to boil the liquid at its boiling point.

92. Using letters from the graph, write the expression for the energy transfer required to raise the temperature of the liquid from the freezing point to the boiling point.

93. A 127-g piece of ice is removed from a refrigerator at −11°C. It is placed in a bowl in which it melts and eventually warms to room temperature, 21°C. Calculate the amount of energy the sample has gained from the atmosphere.

94. The following information is given for *n*-pentane at 1 atm:

boiling point = 36.20°C $\quad \Delta H_{vap}$ (36.20°C) = 357.6 J/g

melting point = −129.7°C $\quad \Delta H_{fus}$ (−129.7°C) = 116.7 J/g

specific heat gas = 1.650 J/g · °C

specific heat liquid = 2.280 J/g · °C

A 26.10-g sample of liquid *n*-pentane is initially at −51.40°C. If the sample is heated at constant pressure (P = 1 atm), what quantity in kJ of energy must be transferred to raise the temperature of the sample to 65.30°C?

95. A home melting pot is used for a metal casting hobby. At the end of a work period, the pot contains 689 g of zinc at 552°C. How much energy will be transferred as the molten metal cools, solidifies, and cools further to room temperature, 21°C? Find the necessary data from the tables in the textbook.

Alexander Shingarev/Shutterstock.com

Electric melting pots for home use can heat metals past their melting points, and the liquid metals can be poured into molds and cooled back to the solid state.

96. The following information is given for chromium at 1 atm:

boiling point = 2672°C $\Delta H_{vap}(2672°C) = 5874\ J/g$

melting point = 1857°C $\Delta H_{fus}(1857°C) = 281.5\ J/g$

specific heat gas = $0.4600\ J/g \cdot °C$

specific heat liquid = $0.9370\ J/g \cdot °C$

A 36.90-g sample of solid chromium is initially at 1837°C. If the sample is heated at constant pressure (P = 1 atm), what quantity in kilojoules of heat energy must be transferred to raise the temperature of the sample to 2068°C?

97. A certain "white metal" alloy of lead, antimony, and bismuth melts at 264°C, and its heat of fusion is 29 J/g. Its average specific heat is $0.21\ J/g \cdot °C$ as a liquid and $0.27\ J/g \cdot °C$ as a solid. What quantity of energy transferred is required to heat 941 kg of the alloy in a melting pot from a starting temperature of 26°C to its operating temperature, 339°C?

98. The following information is given for bismuth at 1 atm:

boiling point = 1627°C $\Delta H_{vap}(1627°C) = 822.9\ J/g$

melting point = 271.0°C $\Delta H_{fus}(271.0°C) = 52.60\ J/g$

specific heat gas = $0.1260\ J/g \cdot °C$

specific heat liquid = $0.1510\ J/g \cdot °C$

A 22.80-g sample of liquid bismuth at 553.0°C is poured into a mold and allowed to cool to 28.0°C. What quantity of energy (in kilojoules) is transferred to the atmosphere in this process?

General Questions

99. Distinguish precisely and in scientific terms, the differences among items in each of the following groups.

a) Intermolecular forces, chemical bonds
b) Vapor pressure, equilibrium vapor pressure
c) Molar heat of vaporization, heat of vaporization
d) Dipole forces, induced dipole forces, dispersion forces, London forces, hydrogen bonds
e) Evaporation, vaporization, boiling, condensation
f) Fusion, solidification
g) Boiling point, normal boiling point
h) Amorphous solid, crystalline solid
i) Ionic, molecular, covalent network, metallic crystals
j) Heat of vaporization, heat of condensation
k) Heat of fusion, heat of solidification
l) Specific heat, heat of vaporization, heat of fusion

100. Classify each of the following statements as true or false.

a) Intermolecular attractions are stronger in liquids than in gases.
b) Substances with weak intermolecular attractions generally have low vapor pressures.
c) Liquids with high molar heats of vaporization usually are more viscous than liquids with low molar heats of vaporization.
d) A substance with a relatively high surface tension usually has a very low boiling point.
e) All other things being equal, hydrogen bonds are weaker than induced dipole or dipole forces.
f) Induced dipole forces become very strong between large molecules.
g) Other things being equal, nonpolar molecules have stronger intermolecular attractions than polar molecules.
h) The essential feature of a dynamic equilibrium is that the rates of opposing changes are equal.
i) Equilibrium vapor pressure depends on the concentration of a vapor above its own liquid.
j) The heat of vaporization is equal to the heat of fusion, but with opposite sign.
k) The boiling point of a liquid is a fixed property of the liquid.
l) If you break (shatter) an amorphous solid, it will break in straight lines, but if you break a crystalline solid, it will break in curved lines.
m) Ionic crystals are seldom soluble in water.
n) Molecular crystals are nearly always soluble in water.
o) The numerical value of heat of vaporization is always larger than the numerical value of heat of condensation.
p) The units of heat of fusion are $kJ/g \cdot °C$.
q) The temperature of water drops while it is freezing.
r) Specific heat is concerned with a change in temperature.

101. Identify the intermolecular attractions in CH_3OH and CH_3F. Which of the two substances do you expect will have the higher boiling point, and which will have the higher equilibrium vapor pressure? Justify your choices.

102. Under what circumstances might you find that a substance having only induced dipole forces is more viscous that a substance that exhibits hydrogen bonding?

More Challenging Problems

103. A liquid in a beaker is placed in an airtight cylinder. The liquid evaporates until equilibrium is established between the liquid and vapor states. The piston in the cylinder is suddenly adjusted to reduce the volume of the cylinder. If there is no change in temperature, what changes will occur, if any, in the rates of evaporation and condensation?

104. Three closed containers have identical volumes. A beaker containing a large quantity of ether, a highly volatile liquid, is placed in Container A. It evaporates until equilibrium is reached with a substantial amount of ether remaining. A beaker with a small amount of ether is placed in Container B. The ether all evaporates. A beaker with an intermediate amount of ether is placed in Container C. It evaporates until it reaches equilibrium with only a small amount of ether remaining. Compare the final ether vapor pressures in the three containers. Explain your answer.

105. The equilibrium vapor pressure of water at 24°C is 22.4 torr. A sealed flask contains air at 24°C and 757 torr and a glass vial filled with liquid water. The vial is broken, allowing some of the water to evaporate. What is the maximum pressure this system can reach?

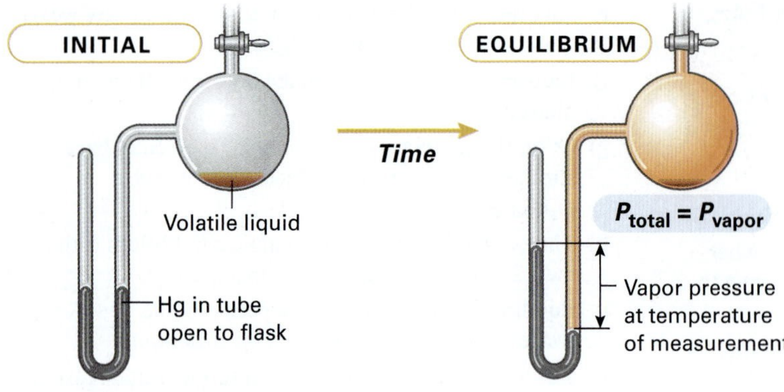

Figure 15.47 Measurement of vapor pressure. Initially, the flask contains the liquid whose vapor pressure is to be measured. The flask, tubes, and manometer above the mercury in the right leg are all at atmospheric pressure. The mercury in the left leg of the manometer is also at atmospheric pressure, so the mercury levels are the same in the two legs. To measure vapor pressure, evaporation occurs until equilibrium is reached. Vapor causes an increase in pressure that is measured directly by the difference in the mercury levels of the two legs of the manometer.

Questions 106 and 107 are based on the apparatus shown in **Figure 15.47**. *Study the caption that explains how vapor pressure is measured, and then answer the questions.*

106. Assume you are about to use the apparatus to determine the equilibrium vapor pressure of a volatile liquid. Would you expect the vapor pressure shown by the manometer to (a) increase uniformly until equilibrium is reached; (b) increase rapidly at first, and then slowly as equilibrium is reached; or (c) increase slowly at first and rapidly as equilibrium is approached? Justify your answer.

107. Suppose that all of the liquid initially introduced to the flask evaporated. Why and how could this occur? Explain in terms of evaporation and condensation rates. How would the vapor pressure shown by the manometer compare with the equilibrium vapor pressure at the existing temperature? Is there further action you can take to complete the vapor pressure measurement, or is it necessary to start over? Justify your answer.

108. An industrial process requires boiling a liquid whose boiling point is so high that maintenance costs on associated pumping equipment are prohibitive. Suggest a way this problem might be solved.

109. A calorimeter contains 72.0 g of water at 19.2°C. A 141-g piece of tin is heated to 89.0°C and dropped into the water. The entire system eventually reaches 25.5°C. Assuming all of the energy transferred to the water comes from the cooling of the tin—there is no energy loss to the calorimeter or the surroundings—calculate the specific heat of the tin.

110. A 54.1-g aluminum ice tray in a home refrigerator holds 408 g of water. Calculate the energy that must be transferred out of the tray and its contents to reduce the temperature from 17°C to 0°C, freeze the water, and drop the temperature of the tray and ice to −9°C. Assume the specific heat of aluminum remains constant over the temperature range involved.

111. The labels have come off the bottles of two white crystalline solids. You know one is sugar and the other is potassium sulfate. Suggest a safe test by which you could determine which is what.

112. The melting point of an amorphous solid is not always a definite value as it is for a pure substance. Suggest a reason for this.

113. Why does dew form overnight?

Dew on a spider web.

114. It is a hot summer day, and Chris wants a glass of lemonade. There is none in the refrigerator, so a new batch is prepared from freshly squeezed lemons. When finished, there is 175 g of lemonade at 23°C. That is not a very refreshing temperature, so it must be cooled with ice. But Chris doesn't like ice in lemonade! Therefore, just enough ice is used to cool the lemonade to 5°C. Of course, the ice will melt and reach the same temperature. If the ice starts at −8°C, and if the specific heat of lemonade is the same as that of water, what mass in grams of ice does Chris use? Assume there is no heat transfer to or from the surroundings. Answer in two significant figures.

Answers to Target Checks

1. (a) Gas particles are widely separated compared with liquid particles. (b) A will have the higher surface tension, molar heat of vaporization, boiling point, and viscosity, all of which increase in value with increasing strength of intermolecular forces. B will have the higher vapor pressure, a property that increases in value as the strength of intermolecular attractions decreases. (c) Y should have a higher vapor pressure. If X has a higher molar heat of vaporization than Y, it probably has stronger intermolecular attractions. That should cause X to have a lower vapor pressure.
2. b and e: true. a: Induced dipole forces are present between all molecules. c: Polar molecules have an unbalanced distribution of electrical charge, but the net charge is zero. d: Intermolecular forces are electrical in character.

3. (a) tetrahedral, nonpolar, induced dipole; (b) linear, nonpolar, induced dipole; (c) bent, polar, dipole; (d) bent, polar, hydrogen bonding. The intermolecular forces in (d) are the strongest.
4. (a) CBr_4 because a CBr_4 molecule is larger than a CCl_4 molecule and thus has stronger induced dipole forces. (b) NH_3, because it has hydrogen bonding and PH_3 does not.
5. b and c: true. a: A liquid–vapor equilibrium is reached when the rate of evaporation is equal to the rate of condensation.
6. Both true.
7. Structural particles in a crystalline solid are arranged in a regular geometric order. In an amorphous solid, the structural arrangement is irregular.
8. a: true. b: Covalent network solids are usually poor conductors of electricity. c: A solid that melts at 152°C is probably a molecular crystal. d: A soluble molecular crystal is a nonconductor of electricity both as a solid and when dissolved.

Solutions to Practice Exercises

1. 0.78 bar + 0.21 bar + 0.009 bar + 0.025 bar = 1.02 bar
2. $16.5 \text{ mm Hg} \times \frac{1 \text{ atm}}{760 \text{ mm Hg}} = 0.0217 \text{ atm};$

   $0.98 \text{ atm} - 0.0217 \text{ atm} = 0.96 \text{ atm}$
3. $\Delta H_{vap} = \frac{q}{m} = \frac{13.1 \text{ kJ}}{7.3 \text{ g}} = 1.8 \text{ kJ/g}$
4. $q = m \times -\Delta H_{vap} = 2.3 \text{ kg} \times \frac{-0.837 \text{ kJ}}{\text{g}} \times \frac{1000 \text{ g}}{\text{kg}}$

   $= -1.9 \times 10^3 \text{ kJ}$
5. $q = m \times \Delta H_{vap} = 6.37 \text{ kg} \times \frac{397 \text{ J}}{\text{g}} \times \frac{1000 \text{ g}}{\text{kg}} \times \frac{1 \text{ kJ}}{1000 \text{ J}}$

   $= 2.53 \times 10^3 \text{ kJ}$
6. $q = m \times c \times \Delta T = 12.4 \text{ kg} \times \frac{0.13 \text{ J}}{\text{g} \cdot {}^\circ\text{C}} \times (59 - 25)^\circ\text{C}$

   $\times \frac{1000 \text{ g}}{\text{kg}} \times \frac{1 \text{ kJ}}{1000 \text{ J}} = 55 \text{ kJ}$
7. $q = m \times c \times \Delta T = 75 \text{ g} \times \frac{0.39 \text{ J}}{\text{g} \cdot {}^\circ\text{C}} \times (420 - 22)^\circ\text{C}$

   $\times \frac{1 \text{ kJ}}{1000 \text{ J}} = 12 \text{ kJ}$

   $q = m \times \Delta H_{fus} = 75 \text{ g} \times \frac{112 \text{ J}}{\text{g}} \times \frac{1 \text{ kJ}}{1000 \text{ J}} = 8.4 \text{ kJ}$

   $q = m \times c \times \Delta T = 75 \text{ g} \times \frac{0.51 \text{ J}}{\text{g} \cdot {}^\circ\text{C}} \times (555 - 420)^\circ\text{C}$

   $\times \frac{1 \text{ kJ}}{1000 \text{ J}} = 5.2 \text{ kJ}$

   $12 \text{ kJ} + 8.4 \text{ kJ} + 5.2 \text{ kJ} = 26 \text{ kJ}$
8. $q = m \times c \times \Delta T = 50.0 \text{ g} \times \frac{2.00 \text{ J}}{\text{g} \cdot {}^\circ\text{C}} \times (100 - 133)^\circ\text{C}$

   $\times \frac{1 \text{ kJ}}{1000 \text{ J}} = -3.3 \text{ kJ}$

   $q = m \times -\Delta H_{vap} = 50.0 \text{ g} \times \frac{-2.26 \text{ kJ}}{\text{g}} = -113 \text{ kJ}$

   $q = m \times c \times \Delta T = 50.0 \text{ g} \times \frac{4.18 \text{ J}}{\text{g} \cdot {}^\circ\text{C}} \times (0 - 100)^\circ\text{C}$

   $\times \frac{1 \text{ kJ}}{1000 \text{ J}} = -20.9 \text{ kJ}$

   $q = m \times -\Delta H_{fus} = 50.0 \text{ g} \times \frac{-333 \text{ J}}{\text{g}} \times \frac{1 \text{ kJ}}{1000 \text{ J}} = -16.7 \text{ kJ}$

   $q = m \times c \times \Delta T = 50.0 \text{ g} \times \frac{2.06 \text{ J}}{\text{g} \cdot {}^\circ\text{C}} \times (-22 - 0)^\circ\text{C}$

   $\times \frac{1 \text{ kJ}}{1000 \text{ J}} = -2.3 \text{ kJ}$

   $(-3.3 \text{ kJ}) + (-113 \text{ kJ}) + (-20.9 \text{ kJ}) + (-16.7 \text{ kJ})$

   $+ (-2.3 \text{ kJ}) = -156 \text{ kJ}$

Answers to Concept-Linking Exercises

You may have found more relationships or relationships other than the ones given in these answers.

1. Dipole forces, induced dipole forces, and hydrogen bonds are all intermolecular attractions—forces that act *between* molecules. Induced dipole forces are also known as dispersion forces, London forces, and London dispersion forces. Covalent bonds are the forces that hold atoms together *within* a molecule. Ionic bonds are the forces that hold ions together *within* a crystal. Covalent and ionic bonds are much stronger than any of the intermolecular forces.
2. Many physical properties of liquids depend on intermolecular attractions, which can be thought of in terms of the stick-togetherness of particles. Strong intermolecular attractions lead to low vapor pressure, higher molar heat of vaporization, high boiling point, high viscosity, and high surface tension.
3. The boiling point of a liquid is the temperature at which its vapor pressure is equal to the pressure above its surface. The normal boiling point is the temperature at which the vapor pressure of the liquid is one atmosphere.
4. The situation that occurs when the condensation rate equals the evaporation rate in a liquid–vapor equilibrium is described as a condition of dynamic equilibrium. The equilibrium is described as dynamic because the particles are continually switching between the liquid and vapor states, although the concentration of particles in the vapor state remains constant. There is higher vapor pressure at higher temperatures because an increase in temperature leads to an increase in the number of particles with sufficient energy to evaporate.
5. Crystalline solids have an orderly arrangement of particles. Four kinds of crystalline solids are: (1) ionic crystals, ions held together by strong ionic bonds; (2) molecular crystals, molecules held together by weak intermolecular forces; (3) covalent network solids, in which atoms are held together by covalent bonds; and (4) metallic crystals, in which positive ions are held together in a sea of electrons. See Section 12.9, Figure 12.19.
6. Solids can be classified based on the arrangement of their particles. An amorphous solid is one in which there is no long-range ordering of the particles. In contrast, a crystalline solid has its particles arranged in a repeating, three-dimensional geometric pattern.
7. Heat of vaporization is the quantity of energy transfer required to vaporize (liquid → gas) 1 gram of a substance. Heat of condensation is the quantity of heat energy transferred when 1 gram of a substance condenses (gas → liquid). These terms are equal in magnitude but

opposite in sign for a given substance. Similarly, heat of fusion refers to the energy transfer required to melt (solid → liquid) 1 gram of a substance, and heat of solidification is the amount of energy transferred in freezing (liquid → solid) 1 gram of a substance. These terms are also equal in magnitude but opposite in sign. Specific heat is the amount of energy transfer required to change the temperature of 1 gram of a substance by 1 degree Celsius.

Solutions to Blue-Numbered Questions, Exercises, and Problems

1. Each gas fills the volume of the container, 1×10^2 L. A gas, either pure or a mixture, is made up of tiny particles that are widely separated from one another so that they occupy the whole volume of the container that holds them.
3. $p_{other} = P_{total} - p_{nitrogen} - p_{oxygen} - p_{argon} = 749$ torr − 584 torr − 144 torr − 19 torr = 2 torr
5. 733 mm Hg − 16.5 mm Hg = 717 mm Hg
7. Gas particles are very widely spaced; liquid particles are "touchingly close." When gases are mixed, the particles of one gas can occupy the empty spaces between the particles of the other. When liquids are mixed, particles must be "pushed aside" to make room for others.
9. Intermolecular attractions are stronger in liquids because of the lack of space between molecules.
11. The stronger the intermolecular attractions are, the greater the motion needed to separate the particles within the liquid and the higher the boiling point will be.
13. Energy transfer is required to overcome intermolecular attractions, separate liquid particles from one another, and keep them apart. As more energy transfer is required, the molar heat of vaporization will increase.
15. The ball bearing will reach the bottom of the water cylinder first. Molecules in syrup have stronger intermolecular attractions than those in water, so the syrup is more viscous. Syrup molecules are being pulled apart so that the ball bearing can pass through the liquid. This slows the rate of fall.
17. Intermolecular attractions are stronger in mercury than in water. Mercury therefore has a higher surface tension and clings to itself rather than spreading or penetrating paper.
19. The wetting agent reduces the surface tension of water, overcoming its intermolecular attractions and allowing it to penetrate the duck's feathers. The duck's buoyancy is reduced, and it sinks.
21. NO_2 has the highest boiling point, which suggests strong intermolecular attractions. It should also have the highest molar heat of vaporization.
23. Only N_2O is a liquid at −90°C, so it alone has a measurable equilibrium vapor pressure as that term is used in this chapter. NO is a gas, and its vapor pressure is its gas pressure. NO_2, a solid at −90°C, probably has a very small vapor pressure.
25. Other things being equal, dipole forces are stronger than induced dipole forces. Induced dipole forces are likely to be larger than dipole forces when the molecules are very large.
27. $NH(CH_3)_2$, hydrogen bonding; CH_2F_2, dipole; C_3H_8, induced dipole.
29. Dipole forces are attractive forces between polar molecules; hydrogen bonds are stronger dipole-like forces between polar molecules in which hydrogen is bonded to a highly electronegative element, usually nitrogen, oxygen, or fluorine.
31. NH_3 has the higher boiling point because it has relatively strong hydrogen bonds as the primary intermolecular force acting between its molecules, whereas CH_4 molecules have relatively weak induced dipole forces acting between them.
33. Argon has the higher boiling point because its atoms are larger than those of neon and thus the strength of the induced dipole forces in Ar are greater than in Ne.
35. The hydrogen atom consists of a single proton and a single electron. When the bonding electron pair is shifted away from the atom, the proton is responsible for the strength of the hydrogen bond. See the discussion on hydrogen bonds in Section 15.4.
37. (a) Induced dipoles; (b) Hydrogen bonding
39. Induced dipole forces act among all molecules. In (e), induced dipole forces are the major intermolecular force. In addition, dipole forces are present in (a) through (d) and the major intermolecular force in (b). Molecules (a), (c), and (d) have hydrogen bonding as the major intermolecular force.
41. CS_2 should have the higher melting and boiling points because its molecules are larger than otherwise similar CO_2 molecules, so the induced dipole forces are stronger in CS_2.
43. CH_4 should have the higher vapor pressure as a liquid at a given temperature because only weak induced dipoles are present. Relatively stronger dipole forces are present in CH_3F.
45. Rates of change in opposite directions are equal. See the discussion in Section 15.5.
47. At higher temperatures, a greater percentage of molecules in the liquid state have enough energy to vaporize into the gaseous state.
49. The vapor pressure of the compound exceeds the external pressure, so it is a gas. A compound boils, or changes from a liquid to a gas, when its vapor pressure equals the external pressure.
51. The water is delivered at very high pressure, pressure greater than the vapor pressure at the temperature of delivery.
53. Low-boiling liquids and a low heat of vaporization are both characteristic of relatively weak intermolecular attractions. A liquid with weak attractions would therefore exhibit both properties.
55. N should have both the lower boiling point and lower molar heat of vaporization.
57. Ice is a crystalline solid. Ice crystals have a definite geometric order, and ice melts at a definite, constant temperature.
59. A: Network solid. B: Ionic solid.

61. $\frac{44.8\text{ kJ}}{61.2\text{ g}} = 0.732\text{ kJ/g}$

63. $227\text{ g} \times \frac{4.2\text{ kJ}}{\text{g}} = 9.5 \times 10^2\text{ kJ}$

65. $79.4\text{ kJ} \times \frac{1\text{ g}}{0.880\text{ kJ}} = 90.2\text{ g}$

67. $23.8\text{ g } C_3H_6O \times \frac{1\text{ mol } C_3H_6O}{58.08\text{ g } C_3H_6O} \times \frac{32.0\text{ kJ}}{\text{mol } C_3H_6O} = 13.1\text{ kJ}$

69. $3.30\text{ kg} \times \frac{23\text{ kJ}}{\text{kg}} = 76\text{ kJ}$

71. $\frac{2.51\text{ kJ}}{36.9\text{ g}} \times \frac{1000\text{ J}}{\text{kJ}} = 68.0\text{ J/g}$

73. $4.45\text{ kJ} \times \frac{1000\text{ J}}{\text{kJ}} \times \frac{1\text{ g}}{112\text{ J}} = 39.7\text{ g}$

75. A. From $q = m \times c \times \Delta T$, when q and m are equal for two objects, c is inversely proportional to ΔT.

77. $q = 467\text{ g} \times \frac{0.39\text{ J}}{\text{g} \cdot °\text{C}} \times (31 - 68)°\text{C} = -6.7 \times 10^3\text{ J}$

79. $q = 2.30\text{ kg} \times \frac{1000\text{ g}}{\text{kg}} \times \frac{0.13\text{ J}}{\text{g} \cdot °\text{C}} \times (22 - 88)°\text{C} \times \frac{1\text{ kJ}}{1000\text{ J}}$
$= -2.0 \times 10^1\text{ kJ}$

81. $\Delta T = \frac{q}{m \times c} = 1.47\text{ kJ} \times \frac{1000\text{ J}}{\text{kJ}} \times \frac{1}{144\text{ g}} \times \frac{\text{g} \cdot °\text{C}}{0.38\text{ J}}$
$= 27°\text{C}; 33 - 27 = 6°\text{C}$

83. J (boiling point), K (freezing point)

85. H, I

87. C

89. Gas condenses at boiling point, J; liquid cools from boiling point, J, to freezing point K.

91. P – O

93. q (heat solid) $= 127\text{ g} \times \frac{2.06\text{ J}}{\text{g} \cdot °\text{C}} \times [0 - (-11)]°\text{C} \times \frac{1\text{ kJ}}{1000\text{ J}}$
$= 2.9\text{ kJ}$

q (melt solid) $= 127\text{ g} \times \frac{333\text{ J}}{\text{g}} \times \frac{1\text{ kJ}}{1000\text{ J}} = 42.3\text{ kJ}$

q (heat liquid) $= 127\text{ g} \times \frac{4.18\text{ J}}{\text{g} \cdot °\text{C}} \times (21 - 0)°\text{C} \times \frac{1\text{ kJ}}{1000\text{ J}}$
$= 11\text{ kJ}$

Total $q = 2.9\text{ kJ} + 42.3\text{ kJ} + 11\text{ kJ} = 56\text{ kJ}$

95. q (cool liquid) $= 689\text{ g} \times \frac{0.51\text{ J}}{\text{g} \cdot °\text{C}} \times (420 - 552)°\text{C}$
$\times \frac{1\text{ kJ}}{1000\text{ J}} = -46\text{ kJ}$

q (freeze liquid) $= 689\text{ g} \times \frac{-112\text{ J}}{\text{g}} \times \frac{1\text{ kJ}}{1000\text{ J}} = -77.2\text{ kJ}$

q (cool solid) $= 689\text{ g} \times \frac{0.39\text{ J}}{\text{g} \cdot °\text{C}} \times (21 - 420)°\text{C} \times \frac{1\text{ kJ}}{1000\text{ J}}$
$= -1.1 \times 10^2\text{ kJ}$

Total $q = -46\text{ kJ} + (-77.2\text{ kJ}) + (-1.1 \times 10^2\text{ kJ})$
$= -2.3 \times 10^2\text{ kJ}$

97. q (heat solid) $= 941\text{ kg} \times \frac{1000\text{ g}}{\text{kg}} \times \frac{0.27\text{ J}}{\text{g} \cdot °\text{C}} \times (264 - 26)°\text{C}$
$\times \frac{1\text{ kJ}}{1000\text{ J}} = 6.0 \times 10^4\text{ kJ}$

q (melt solid) $= 941\text{ kg} \times \frac{1000\text{ g}}{\text{kg}} \times \frac{29\text{ J}}{\text{g}} \times \frac{1\text{ kJ}}{1000\text{ J}}$
$= 2.7 \times 10^4\text{ kJ}$

q (heat liquid) $= 941\text{ kg} \times \frac{1000\text{ g}}{\text{kg}} \times \frac{0.21\text{ J}}{\text{g} \cdot °\text{C}} \times$
$(339 - 264)°\text{C} \times \frac{1\text{ kJ}}{1000\text{ J}} = 1.5 \times 10^4\text{ kJ}$

Total $q = 6.0 \times 10^4\text{ kJ} + 2.7 \times 10^4\text{ kJ} + 1.5 \times 10^4\text{ kJ}$
$= 10.2 \times 10^4\text{ kJ} = 1.02 \times 10^5\text{ kJ}$

100. True: a, c, f, h, i, r. False: b, d, e, g, j, k, l, m, n, o, p, q.

101. Both molecules have induced dipole and dipole forces. CH_3OH has hydrogen bonding and CH_3F does not. The molecules are about the same size. It is reasonable to predict stronger intermolecular forces in CH_3OH and therefore a higher boiling point, and CH_3F will have the higher equilibrium vapor pressure.

102. Large molecules having strong induced dipole forces may have stronger intermolecular attractive forces than small molecules with hydrogen bonding and therefore exhibit greater viscosity.

103. Reducing volume increases vapor concentration, which causes a temporary increase in the rate of condensation until equilibrium is once again established. Evaporation rate, which depends only on temperature, is not affected.

104. The final ether vapor pressure in Containers A and C is the greatest—the equilibrium vapor pressure. Container B has the lowest vapor pressure, having all evaporated before reaching equilibrium.

105. (757 + 22.4) torr = 779 torr

106. (b) Evaporation at constant rate begins immediately when liquid is introduced. At that time condensation rate is zero. Net rate of increase in vapor concentration is at a maximum, so rate of vapor pressure increase is maximum at start. Later condensation rate is more than zero but less than evaporation rate. Net rate of increase in vapor concentration is less than initially, so rate of vapor pressure increase is less than initially. At equilibrium, evaporation and condensation rates are equal. Vapor concentration and, therefore, vapor pressure remain constant.

107. All of the liquid evaporated before the vapor concentration was high enough to yield a condensation rate equal to the evaporation rate. At lower-than-equilibrium vapor concentration, the vapor pressure is lower than the equilibrium vapor pressure. More liquid must be introduced to the flask until some excess remains and the pressure stabilizes in order to measure equilibrium vapor pressure.

108. The liquid can be boiled by reducing the pressure and thus the boiling temperature.

110. $q_{Al} = 54.1\text{ g} \times \dfrac{0.90\text{ J}}{\text{g}\cdot{}^\circ\text{C}} \times (-9 - 17)^\circ\text{C} = -1.3 \times 10^3\text{ J}$

$= -1.3\text{ kJ}$

$q\text{ (cool water)} = 408\text{ g} \times \dfrac{4.18\text{ J}}{\text{g}\cdot{}^\circ\text{C}} \times (0 - 17)^\circ\text{C}$

$= -2.9 \times 10^4\text{ J} = -29\text{ kJ}$

$q\text{ (freeze water)} = 408\text{ g} \times \dfrac{-333\text{ J}}{\text{g}} = -1.36 \times 10^5\text{ J}$

$= -136\text{ kJ}$

$q\text{ (cool ice)} = 408\text{ g} \times \dfrac{2.06\text{ J}}{\text{g}\cdot{}^\circ\text{C}} \times (-9 - 0)^\circ\text{C} = -8 \times 10^3\text{ J}$

$= -8\text{ kJ}$

Total $q = -1.3\text{ kJ} + (-29\text{ kJ}) + (-136\text{ kJ}) + (-8\text{ kJ})$
$= -174\text{ kJ}$

111. Dissolve the compounds and check for electrical conductivity. The ionic potassium sulfate solute will conduct, whereas the molecular sugar solute will not.

112. Without a regular and uniform structure in an amorphous solid, some intermolecular forces are stronger than others. The weak forces are more easily overcome than the stronger forces, and thus a lower temperature is needed for melting in some parts of the solid than in other parts.

113. As temperature drops, the equilibrium vapor pressure drops below the atmospheric vapor pressure. The air becomes first saturated, and then supersaturated, and condensation (dew) begins to form.

114. Energy lost by lemonade = Energy gained by ice; Let M = Mass of ice

$175\text{ g} \times \dfrac{4.18\text{ J}}{\text{g}\cdot{}^\circ\text{C}} \times (23 - 5)^\circ\text{C} = \text{M g} \times \dfrac{2.06\text{ J}}{\text{g}\cdot{}^\circ\text{C}} \times 8^\circ\text{C}$

$+ \text{M g} \times \dfrac{333\text{ J}}{\text{g}} + \text{M g} \times \dfrac{4.18\text{ J}}{\text{g}\cdot{}^\circ\text{C}} \times 5^\circ\text{C}; \text{M} = 36\text{ g}$

16 Solutions

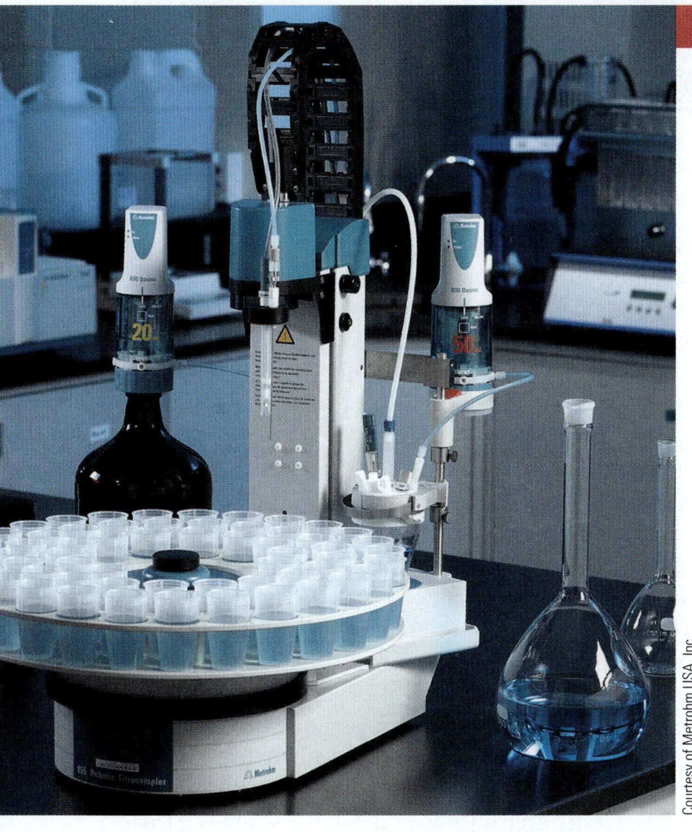

◀ Many chemical reactions commonly occur in a water solution, including those performed in research and instructional laboratories, industrial manufacturing plants, and even your home. The photograph shows an automated titrator. Titration is the very careful addition of one solution to another by a device that measures the volume of solution required to react with a measured amount of another dissolved substance. Titration is a tool used by many scientists, including those who work in environmental analysis, the food and pharmaceutical industries, the health sciences, and agriculture.

Courtesy of Metrohm USA, Inc.

CHAPTER CONTENTS

In this chapter, we examine mixtures made of substances dissolved in water. Such solutions are everywhere, both inside and outside the chemistry laboratory. All natural waters are solutions. What we call "freshwater" has a very low concentration of dissolved substances; the concentration is much higher with ocean water, or "saltwater." "Hard water" has calcium and magnesium salts dissolved in it. Rainwater is nearly pure, but even rainwater is a solution of atmospheric gases in very low concentrations. The small amount of oxygen dissolved in rivers, lakes, and oceans is mightily important to fish, which cannot survive without it.

16.1 Are There Earth-Like Oceans on Other Planets?

Seventy-one percent of Earth's surface is covered by the oceans. Being land animals, humans often don't fully realize that they spend the majority of their lives on less than 29% of Earth's surface, but indeed, Earth is more water-covered than it is not. All natural water contains dissolved substances, but the oceans contain a greater concentration of substances than freshwater. The primary substance dissolved in ocean water is sodium chloride, commonly known as table salt. If you've ever ingested a mouth full of seawater while in the ocean, you've tasted its saltiness.

NASA/JPL/DLR

Figure 16.1 The four largest moons of Jupiter in order of distance from the planet, (*left to right*) Io, Europa, Ganymede, and Callisto. Europa is about the same size as Earth's moon.

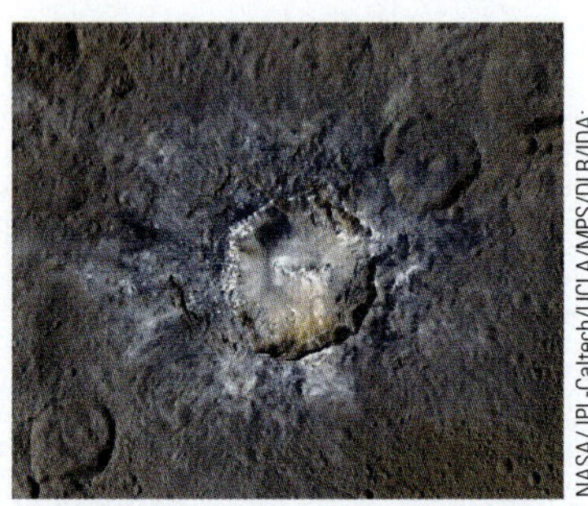

NASA/JPL-Caltech/UCLA/MPS/DLR/IDA;

Figure 16.2 The Haulani Crater on the dwarf planet Ceres. This photograph shows how a solution flowed from the crater's rim outward, leaving what used to be dissolved in the solution as a solid on the surface. A water solution was under the surface, and on impact, the water was changed to steam and evaporated into space.

In Section 15.1, you learned that oceans exist on other worlds. Are these oceans made up of seawater similar in composition to Earth's oceans? In 2019, scientists discovered that the subsurface ocean on Europa, the fourth largest moon of Jupiter (**Figure 16.1**), is likely a sodium chloride-containing ocean. Notice in the figure how the surface of Europa is unusually smooth, mostly free of the craters characterizing other moons. Beneath this smooth surface is a liquid ocean. Sodium chloride deposits are on the surface, and they are likely to have the subsurface ocean as their source.

Another potential ocean world is the dwarf planet Ceres (**Figure 16.2**), which is hypothesized to be about 25% water, some of which now may be in the liquid state or was liquid in the past. Ceres is the largest asteroid in the solar system, orbiting the sun in the asteroid belt. In 2017, a team of scientists analyzed data from the unmanned NASA spacecraft *Dawn,* collected as it orbited Ceres, finding a complex chemical environment on the surface. In addition to water, nitrogen-containing minerals, and carbonates, they discovered evidence of carbon–hydrogen molecules that formed on the planet itself!

The topic of this chapter is solutions, which are homogeneous mixtures. An ocean, whether on Earth or on another planet, is *not* a solution because it is not homogeneous, with variations in the concentrations of dissolved substances that depend on depth, evaporation rates, freshwater inflow, and other variables. However, any well-mixed individual sample of ocean water *is* a solution. On Earth, the oceans average 35 grams per liter of dissolved substances. In this chapter, you will learn the fundamentals of conceptualizing and problem solving about solutions, including knowledge of how they form, methods for expressing the concentration of dissolved substances, and how to solve stoichiometry problems that include one or more substances in the form of a water solution.

16.2 The Characteristics of a Solution

Goal 1 Define the term *solution*, and given a description of a substance, determine if it is a solution.

A **solution** is a homogeneous mixture. This implies uniform distribution of solution components so that a sample taken from any part of the solution will have the same composition as any other sample of the same solution. Two solutions made up of the same substances may, however, have different compositions. A solution of ammonia in water, for example, may contain 1% ammonia by mass, or 2%, 5%, 20.3%, on up to the 29% solution called *concentrated ammonia*. This leads to variable physical properties, which are determined by the composition of the mixture. ◀

▶ **P/Review** Pure substances have definite, unchanging physical and chemical properties. The properties of a mixture, however, depend on how much of each component is in the mixture. As the composition changes, so do the properties. This is shown in Section 2.5, Figure 2.14.

A solution may exist in any of the three states—gas, liquid, or solid. A sample of air is a gaseous solution made up of nitrogen, oxygen, argon, and other gases in small amounts. Like oxygen in water, the remaining dissolved carbon dioxide in a carbonated beverage that has gone flat is a familiar liquid solution of a gas.

Alcohol in water is an example of the solution of two liquids, and a sample of ocean water is a liquid solution of dissolved solids. ▶ Solid-state solutions are common in the form of metal alloys such as steel.

Technically, the oceans and the atmosphere are not solutions because they are not homogeneous. However, laboratory-sized samples of both satisfy the definition.

Particle size distinguishes solutions from other mixtures. Dispersed particles in solutions (which may be atoms, ions, or molecules) are very small—small enough so that a beam of light shining through the solution reveals no cloudiness. Particles in a solution do not settle on standing, and they are too small to see (**Figure 16.3**).

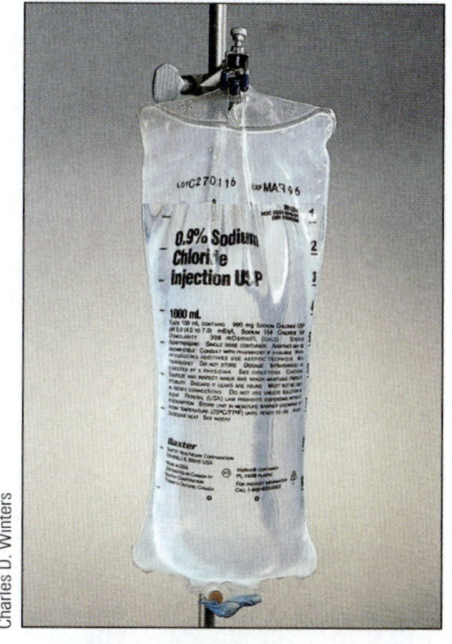

Charles D. Winters

Figure 16.3 Saline solution for injection. Dehydrated patients receive a sodium chloride (saline) solution that has a concentration of dissolved substances similar to the concentration in cells of the body. The dissolved sodium ions and chloride ions are too small to be seen.

 **Target Check 16.1**

Identify true statements, and rewrite the false statements to make them true.

a) At a given instant, a solution has a definite percentage composition.
b) The physical properties of a solution of A in B are variable.
c) A solution is always made up of two pure substances.
d) Different parts of a solution can be detected visually.

16.3 Solution Terminology

Goal 2 Distinguish among terms in the following groups: *solute* and *solvent*; *concentrated* and *dilute*; *solubility*, *saturated*, *unsaturated*, and *supersaturated*; *miscible* and *immiscible*.

In discussing solutions, we use a language of closely related and sometimes overlapping terms. We will now identify and define these terms.

Solute and Solvent

When solids or gases are dissolved in liquids, the solid or gas is said to be the **solute** and the liquid is the **solvent** (**Figure 16.4**). More generally, the solute is taken to be the substance present in a relatively small amount. The medium in which the solute is dissolved is the solvent. The distinction is not precise, however. Water is capable of dissolving more than its own mass of some solids, but water continues to be called the *solvent*. In alcohol–water solutions, either liquid may be the more abundant, and in a given context, either might be called the *solute* or *solvent*.

Charles D. Winters

Charles D. Winters

Figure 16.4 Solute and solvent. (a) When a solid is dissolved in water, the solid is the solute and water is the solvent. The solute in this photograph is solid copper(II) chloride. (b) The solution formed is a homogeneous mixture. Hydrated copper(II) ions and hydrated chloride ions are mixed among water molecules. At the macroscopic level, the solution has the same appearance and uniform composition throughout.

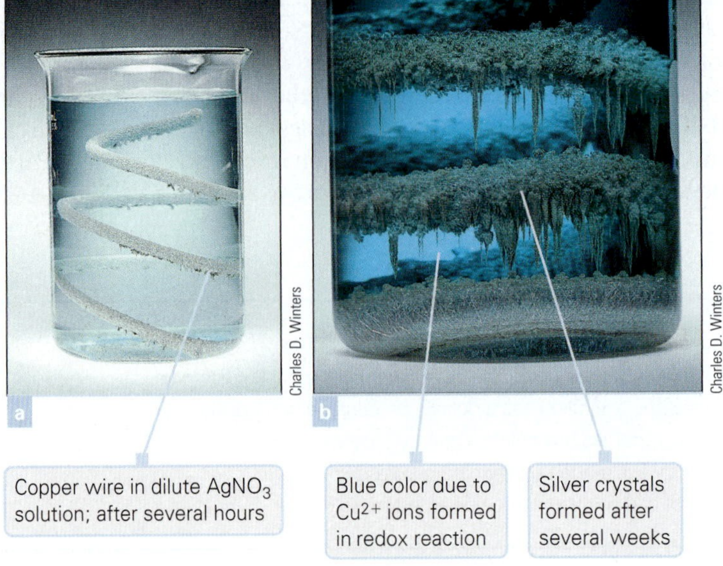

Figure 16.5 Concentrated and dilute. A copper wire reacts with an aqueous silver nitrate solution, forming silver metal and a copper(II) nitrate solution. The blue color of the solution is due to the hydrated copper(II) ions. (a) When the reaction has proceeded for a relatively short period of time, the color is light, indicating a dilute solution. (b) After a longer time span, the blue becomes much more intense, indicating a relatively large concentration of copper(II) ions. The solution is concentrated.

Concentrated and Dilute

A **concentrated** solution has a *relatively* large quantity of a specific solute per unit amount of solution, and a **dilute** solution has a *relatively* small quantity of the same solute per unit amount of solution (**Figure 16.5**). The terms compare concentrations of two solutions of the *same solute and solvent.* They carry no other quantitative meaning.

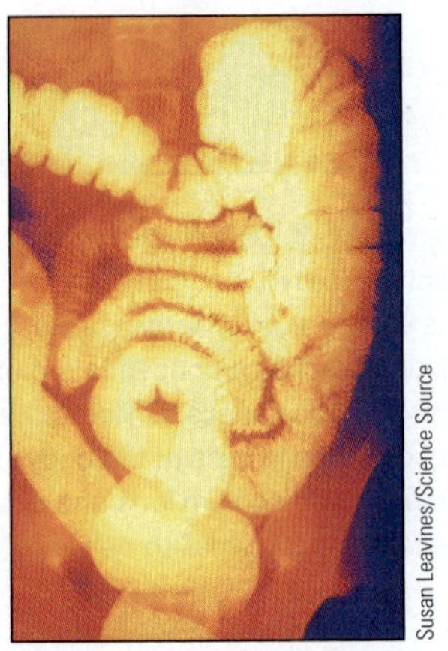

Figure 16.6 Barium sulfate suspension. Only 0.0002 g of barium sulfate dissolves in 100 mL of water at 20°C. Although soluble barium-ion-containing compounds are toxic to humans, barium sulfate is used for x-rays of the digestive tract because its solubility is very low and it improves the visibility of the tract on the image.

Solubility, Saturated, and Unsaturated

Solubility is a measure of how much solute will dissolve in a given amount of solvent at a given temperature (**Figure 16.6**). It is sometimes expressed by giving the mass in grams of solute that will dissolve in 100 g or 100 mL of solvent. A solution whose concentration is at the solubility limit for a given temperature is a **saturated solution**. If the concentration of a solute is less than the solubility limit, the solution is an **unsaturated solution**.

Supersaturated Solutions

Under carefully controlled conditions, a solution can be produced in which the concentration of the solute is greater than the normal solubility limit. Such a solution is said to be **supersaturated**. A supersaturated solution of sodium acetate, for example, may be prepared by dissolving 80 g of the salt in 100 g of water at about 50°C. If the solution is cooled to room temperature without stirring, shaking, or other disturbance, all 80 g of the solute will remain in solution, even though the solubility at 20°C is only 47 g per 100 g of water. The solution is unstable, however. A slight physical disturbance or the addition of a single crystal of solid sodium acetate can start crystallization. This proceeds quickly until the solution concentration reaches the saturated solution solubility limit (**Figure 16.7**).

Miscible and Immiscible

Miscible and *immiscible* are terms customarily limited to solutions of liquids in liquids. If two liquids dissolve in each other in all proportions, they are said to be **miscible** in each other (**Figure. 16.8[a]**). Alcohol and water, for example, are miscible liquids. Liquids that are insoluble in each other, such as oil and water, are **immiscible** (**Figure 16.8[b]**). Some liquid pairs will mix appreciably with each other but in limited proportions; they are said to be *partially miscible.*

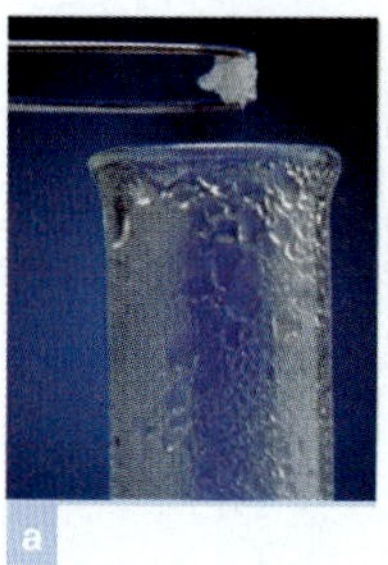

Figure 16.7 Supersaturated solution. A supersaturated solution of sodium acetate was prepared by dissolving solid sodium acetate in warm water. The solution was then allowed to cool without physical disturbance. At the macroscopic level, you cannot tell that the solution is supersaturated. (a) A small piece of solid sodium acetate is added to the solution. (b) The added solid sodium acetate initiates crystallization. (c and d) Crystallization continues as long as the solution exceeds its solubility limit at the present temperature of the system. (e) When the solution reaches its solubility limit, a large quantity of solid sodium acetate is in the flask. At the macroscopic level, it appears as if the crystal is static. However, as you will learn in the next section, the solid continues to dissolve but at a rate equal to the crystallization rate.

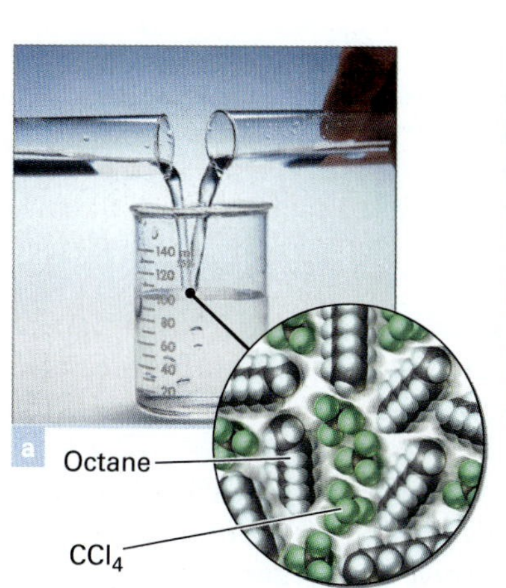

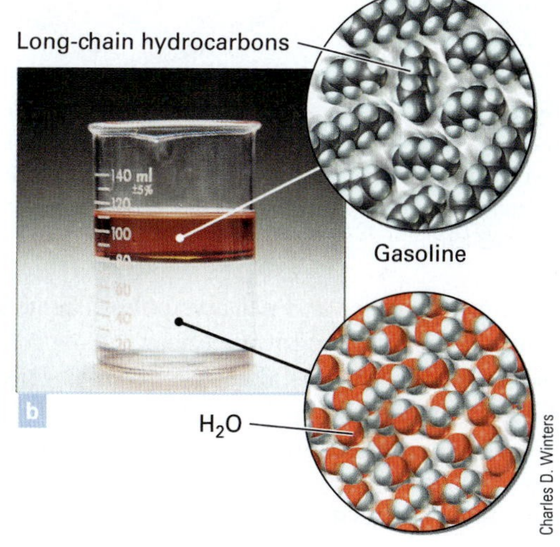

Figure 16.8 Miscible and immiscible. (a) Octane and carbon tetrachloride dissolve in each other, and thus they are miscible. (b) Gasoline (with some red dye added for visibility) and water are insoluble in each other. They are immiscible.

Target Check 16.2

A 0.100-g sample of A is dissolved in 1.00×10^3 mL of water, and 10.0 g of B is dissolved in 5.00×10^2 mL of water. Identify or explain why you cannot identify:

a) The solute and solvent in each solution
b) The solution that is more likely to be saturated
c) The "dilute" solution

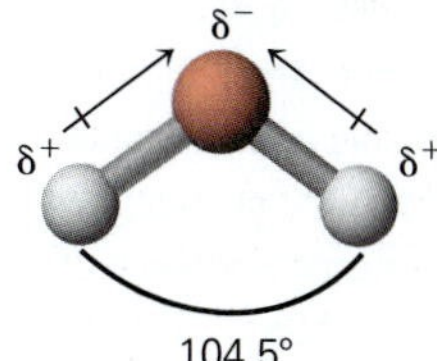

Figure 16.9 The polarity of the water molecule. The 104.5° H—O—H bond angle and the asymmetric distribution of the electrons bonding the hydrogen atoms (gray) to the oxygen atom (red) make water molecules highly polar. There is a partial negative charge near the oxygen atom, and there is a net partial positive charge centered between the hydrogen atoms. The lowercase Greek delta, δ, is used to indicate a partial charge.

◀ **P/Review** Determination of molecular polarity (Section 13.6) follows from an analysis of bond polarity (Section 12.5) and molecular geometry (Section 13.4).

16.4 The Formation of a Solution

Goal 3 Describe the formation of a saturated solution from the time excess solid solute is first placed into a liquid solvent.

4 Identify and explain the factors that determine the time required to dissolve a given amount of solute or to reach equilibrium.

Most solutions of interest to a chemist, biologist, or geologist are water solutions. The water molecule has unique particulate-level properties that make it a good macroscopic level solvent. Studies of the water molecule show that it is **polar** (Figure 16.9). ▶ A polar molecule is one with an asymmetrical (unbalanced) distribution of charge,

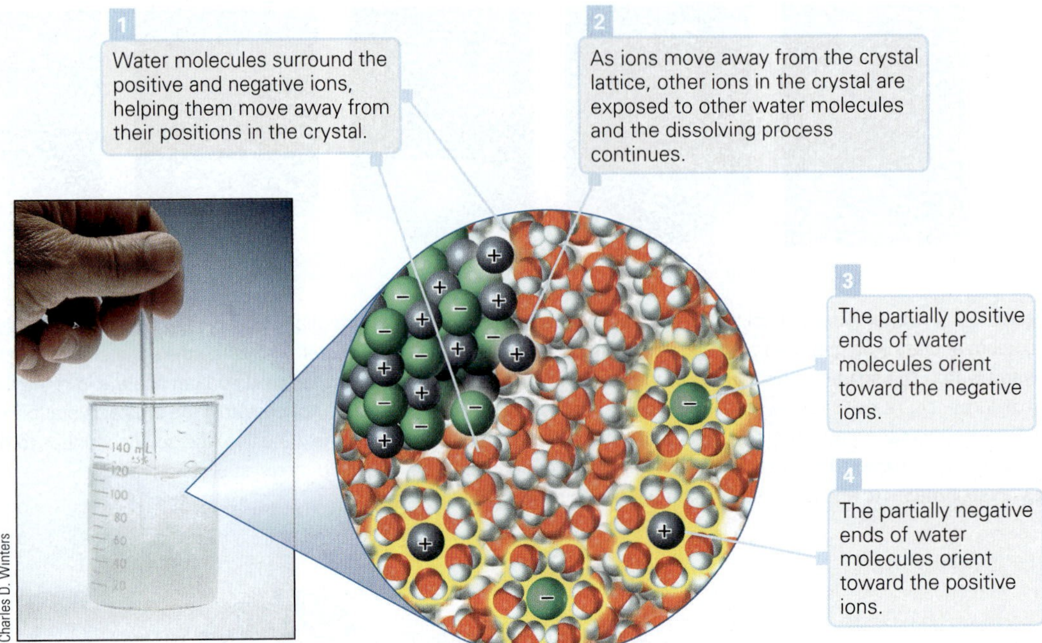

1
Water molecules surround the positive and negative ions, helping them move away from their positions in the crystal.
2
As ions move away from the crystal lattice, other ions in the crystal are exposed to other water molecules and the dissolving process continues.
3
The partially positive ends of water molecules orient toward the negative ions.
4
The partially negative ends of water molecules orient toward the positive ions.
Charles D. Winters

Figure 16.10 Dissolving an ionic solute in water. The negative ions of the solute are pulled from the crystal by relatively positive hydrogen regions in polar water molecules. Similarly, positive ions are attracted by the relatively negative oxygen region in the molecule. In solution, the ions are surrounded by those regions of the water molecules that have the opposite charge. These are hydrated ions.

resulting in positive and negative poles. The negatively charged electrons of the water molecule are net concentrated near the oxygen atom, which causes that area of the molecule to have a partial negative charge. The area midway between the two hydrogen atoms exhibits a partial positive charge.

When a soluble ionic crystal is placed in water, the negatively charged ions at the surface of the crystal are attracted by the positive region of the polar water molecules (Figure 16.10). A "tug of war" for the negative ions begins. Water molecules tend to pull them from the crystal, whereas neighboring positive ions tend to hold them in the crystal. In a similar way, positive ions at the surface of the crystal are attracted to the negative portion of the water molecules and are torn from the crystal. Once released, the ions are surrounded by the polar water molecules. Such ions are said to be **hydrated ions**.

The dissolving process is reversible. ◀ As the dissolved solute particles move randomly through the solution, they come into contact with the undissolved solute or with each other and crystallize—that is, return to the solid state. For NaCl, this process may be represented by the "reversible reaction" equation:

$$NaCl(s) \rightleftharpoons Na^+(aq) + Cl^-(aq)$$

▶ **P/Review** Reversible changes are those that can proceed in either direction, as suggested by the double arrow in the equation. In one direction, species on the left of the equation are reactants and species on the right are products; in the reverse direction, products are on the left and reactants are on the right. The change may be either chemical (Section 8.8) or physical (Section 15.5). Dissolving is a physical change. Though the formulas of the reactants and products look different, there has been no change at the particulate level. NaCl(s) is an assembly of Na^+ and Cl^- ions in a crystal, as shown in Section 12.3, Figure 12.7.

The rate per unit of surface area at which a solute dissolves depends primarily on temperature. The rate of crystallization per unit area depends primarily on the concentration of the solute at the crystal surface. When the rates of dissolving and crystallization become the same, the solution is saturated at its solubility limit at the existing temperature. A **dynamic equilibrium** is reached. ◀ The process is described in the caption of Figure 16.11.

▶ **P/Review** Equilibrium for crystallization/dissolving rates in a saturated solution is quite like the equilibrium for evaporation/condensation rates in a liquid–vapor equilibrium, which is discussed in Section 15.5. If Section 15.5 was not assigned to you earlier, you might find it helpful to study Figure 15.20 now.

The time required to dissolve a given amount of solute—or to reach equilibrium if excess solute is present—depends on several factors:

1. The dissolving process depends on surface area. A finely divided solid offers more surface area per unit of mass than a coarsely divided solid. Therefore, a finely divided solid dissolves more rapidly.

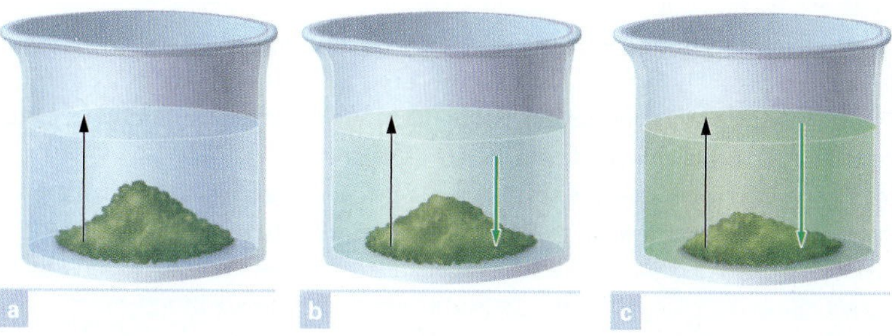

Figure 16.11 Development of equilibrium in forming a saturated solution. If temperature is held constant, the rate of dissolving per unit of solute surface area is constant. The constancy of the dissolving rate is indicated by the equal-length black arrows in the three views. The crystallization rate per unit of surface area increases as the solution concentration at the surface increases. In (a), when dissolving has just begun, solution concentration is zero, and the crystallization rate is zero. In (b), the solution concentration has risen to yield a crystallization rate above zero but less than the dissolving rate. The crystallization rate is shown by the green arrow. Since the dissolving rate is greater than the crystallization rate (black arrow is longer than the green arrow), the concentration of the solution is increasing. Concentration continues to increase until (c) when the crystallization rate has become equal to the dissolving rate. Equilibrium has been reached, and the solution is saturated.

2. In a still solution, concentration builds up at the solute surface, causing a higher crystallization rate than would be present if the solute was uniformly distributed. Stirring or agitating the solution prevents this buildup and maximizes the net dissolving rate. The net dissolving rate is the rate of dissolving minus the rate of crystallization. The "macro" dissolving rate—the effective rate we can see—is the combined result of two invisible particulate happenings.
3. At higher temperatures, particle movement is more rapid, thereby speeding up all physical processes.

Thinking About Your Thinking

Mental Models

You need to form two mental models as you study this section The first model is of the particulate-level process that occurs when a solid ionic solute is dissolved in water, as shown in Figure 16.10. First, imagine an ionic solid. Think about the positive ion–negative ion–positive ion–negative ion and so forth pattern in a simple ionic compound. Now imagine that this solid is in liquid water. A positive ion at the edge of the solid crystal will attract the negative portion of a polar water molecule, the end where the two lone pairs on the oxygen atom are exposed. If that ion breaks free from the crystal, the negative ends of a number of water molecules will quickly surround it. A similar process occurs for negative ions in the solid crystal and the positive ends of water molecules.

The second mental model that you need to develop is a "video in your mind" of the equilibrium process illustrated in Figure 16.11. Most importantly, you need to be able to picture in your mind each of the two processes, dissolving and crystallization, and the factors that affect each. The dissolving rate is essentially constant under the conditions mentioned in the caption of Figure 16.11. The crystallization rate increases with increasing solution concentration. Picture both of these processes happening simultaneously. Equilibrium occurs when the opposing rates are equal.

When you come to a homework or exam question that involves dissolving an ionic solute in water or the saturated solution equilibrium process, call your mental model into your working memory and then use your model to help answer the question. In this way, you are learning to think as a chemist.

✓ Target Check 16.3

Assume that the temperature remains constant while a solute dissolves until the solution becomes saturated. For a unit area,

a) Is the rate of dissolving when the solution is one-third saturated more than, equal to, or less than the rate of dissolving when the solution is two-thirds saturated?
b) Is the rate of crystallization when the solution is one-third saturated more than, equal to, or less than the rate of crystallization when the solution is two-thirds saturated?
c) Is the *net* rate of dissolving when the solution is one-third saturated more than, equal to, or less than the net rate of dissolving when the solution is two-thirds saturated?

16.5 Factors That Determine Solubility

Goal 5 Given the structural formulas of two molecular substances, or other information from which the strength of their intermolecular forces may be estimated, predict if they will dissolve appreciably in each other, and state the criteria on which your prediction is based.

6 Predict how and explain why the solubility of a gas in a liquid is affected by a change in the partial pressure of that gas over the liquid.

The extent to which a particular solute dissolves in a given solvent depends on three factors:

1. The strength of intermolecular forces within the solute, within the solvent, and between the solute and solvent
2. The partial pressure of a solute gas over a liquid solvent
3. The temperature

Intermolecular Forces

▶ **P/Review** Intermolecular forces fall into three categories: induced dipole forces, dipole forces, and hydrogen bonds. All other things being equal, these are listed in order of increasing strength. Molecular structure is primarily responsible for the types of intermolecular forces in any given substance. See Section 15.4 for further information on intermolecular forces.

Solubility, a macroscopic property, depends on intermolecular forces at the particulate level. ◀ Generally speaking, *if forces between molecules of A are about the same as the forces between molecules of B, A and B will probably dissolve in each other*. This is commonly paraphrased as *like dissolves like* (**Figure 16.12[a]**). On the other hand, if the intermolecular forces between molecules of A are quite different from the forces between molecules of B, it is unlikely that they will dissolve in each other (**Figure 16.12[b]**).

Consider, for example, hexane, C_6H_{14}, and decane, $C_{10}H_{22}$. Each substance has only induced dipole forces. The forces are roughly the same for the two substances, which are soluble in each other. Neither, however, is soluble in water or methanol, CH_3OH, two liquids that exhibit strong hydrogen bonding. But water and methanol are soluble in each other, again supporting the correlation between solubility and similar intermolecular forces.

Partial Pressure of Solute Gas over Liquid Solution

▶ **P/Review** The partial pressure of one gas in a mixture of gases is the pressure that one gas would exert if it alone occupies the same volume at the same temperature. See Section 15.2.

Changes in partial pressure of a solute gas over a liquid solution have a pronounced effect on the solubility of that gas. ◀ This is sometimes startlingly apparent when opening a bottle of a carbonated beverage (**Figure 16.13**). Such beverages are bottled under a carbon dioxide partial pressure that is slightly greater than one atmosphere, which increases the solubility of the gas. This is what is meant by "carbonated." As the pressure is released on opening, solubility decreases, which causes bubbles of carbon dioxide to escape from the solution.

In most dilute solutions, the solubility of a gaseous solute in a liquid is directly proportional to the pressure of the gas over the surface of the liquid (**Figure 16.14**). An equilibrium is reached that is similar to the liquid–vapor equilibrium described

Figure 16.12 Intermolecular forces and solubility. (a) Ethylene glycol, $HO-CH_2CH_2-OH$, is the primary ingredient in most automotive antifreeze solutions. At the particulate level, it has dipoles and exhibits hydrogen bonding, and thus it is soluble in water, which has similar intermolecular forces. (b) Motor oil is a mixture that primarily consists of nonpolar molecules made up of carbon atoms and hydrogen atoms. Induced dipole forces are the primary intermolecular forces, so motor oil is not soluble in water.

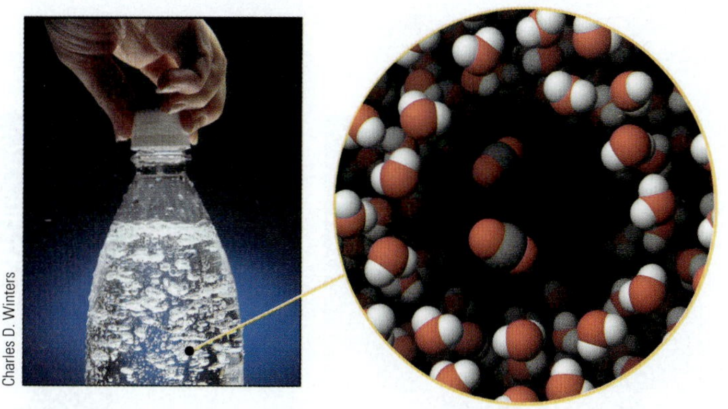

Figure 16.13 Carbonated beverages. Bottling under a high carbon dioxide partial pressure increases the solubility of the gas in the liquid solution. When you open the bottle, the partial pressure of carbon dioxide above the solution drops to become equal to atmospheric p_{CO_2}, dramatically decreasing the solubility of the gas, which escapes from the solution in the form of gas-phase bubbles that consist of carbon dioxide molecules (and a relatively small quantity of water molecules).

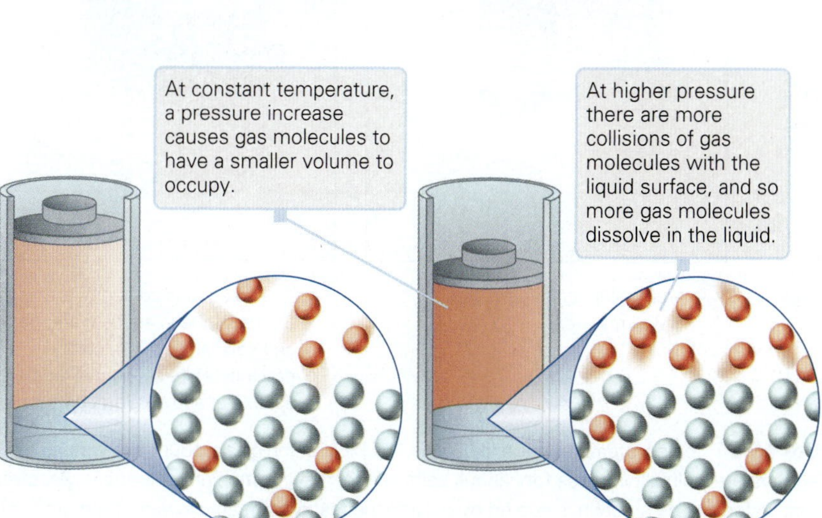

At constant temperature, a pressure increase causes gas molecules to have a smaller volume to occupy.
At higher pressure there are more collisions of gas molecules with the liquid surface, and so more gas molecules dissolve in the liquid.

Figure 16.14 Gas solubility and partial pressure. The solubility of a gas in a liquid is directly proportional to its partial pressure: solubility $\propto$ partial pressure.

in Section 15.5 and the solid-in-liquid equilibrium discussed in Section 16.4. Neither the partial pressure nor the total pressure caused by *other* gases affects the solubility of the solute gas. This is what would be expected for an ideal gas, in which all molecules are widely separated and completely independent.

Pressure has little or no effect on the solubility of solids or liquids in a liquid solvent. The gas pressure above a liquid surface has no effect on the equilibrium process that occurs as a solid solute dissolves in a liquid solvent.

Temperature

Temperature exerts a major influence on most chemical equilibria, including solution equilibria. Consequently, solubility depends on temperature. **Figure 16.15(a)** indicates that the solubility of most solids increases with rising temperature, but there are notable exceptions. The solubilities of gases in liquids, on the other hand, are generally lower at higher temperatures, as shown in **Figure 16.16**. The explanation of the relationship between temperature and solubility involves energy changes in the solution process, as well as other factors.

***a summary of...* Factors That Determine Solubility**

1. Substances with similar intermolecular forces will usually dissolve in one another.
2. The greater the partial pressure of a gas over a liquid solution, the more soluble the gas.
3. The solubility of most solids in liquids increases with increasing temperature, but there are exceptions. The solubility of most gases in liquids decreases with increasing temperature.

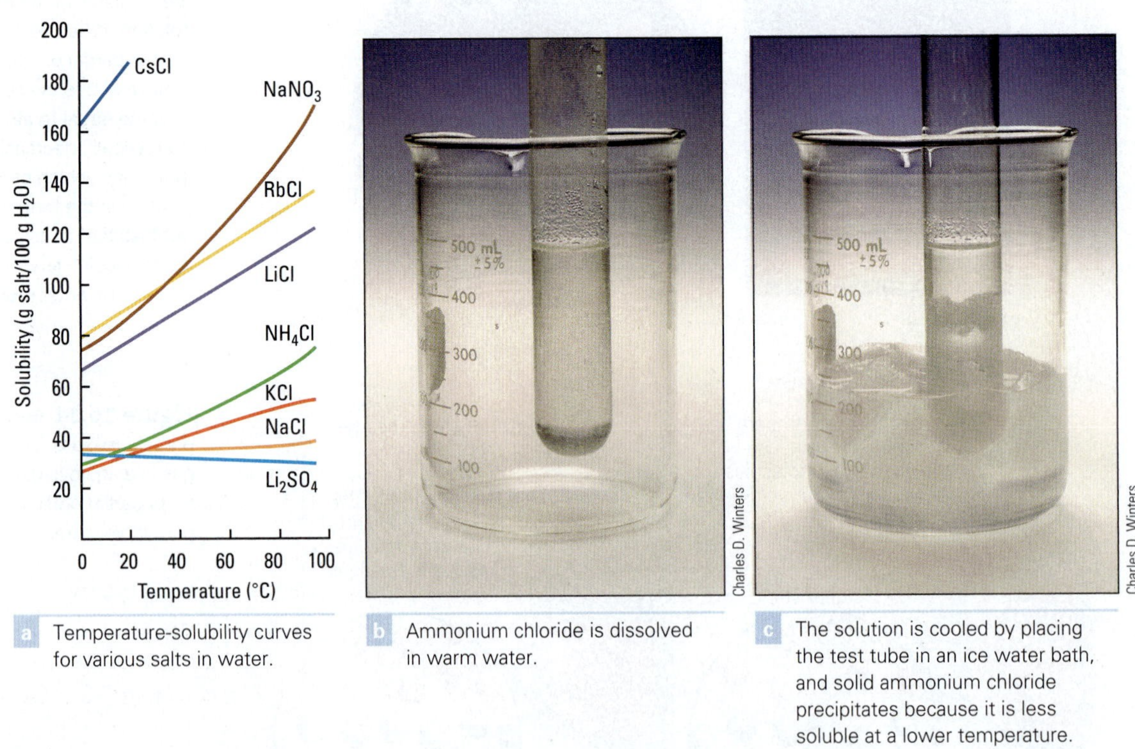

a Temperature-solubility curves for various salts in water.

b Ammonium chloride is dissolved in warm water.

c The solution is cooled by placing the test tube in an ice water bath, and solid ammonium chloride precipitates because it is less soluble at a lower temperature.

Figure 16.15 Temperature dependence of the solubility of ionic compounds in water. (a) Temperature–solubility curves for various salts in water. (b) Ammonium chloride is dissolved in warm water. (c) The solution is cooled by placing the test tube in an ice water bath, and solid ammonium chloride precipitates because it is less soluble at a lower temperature.

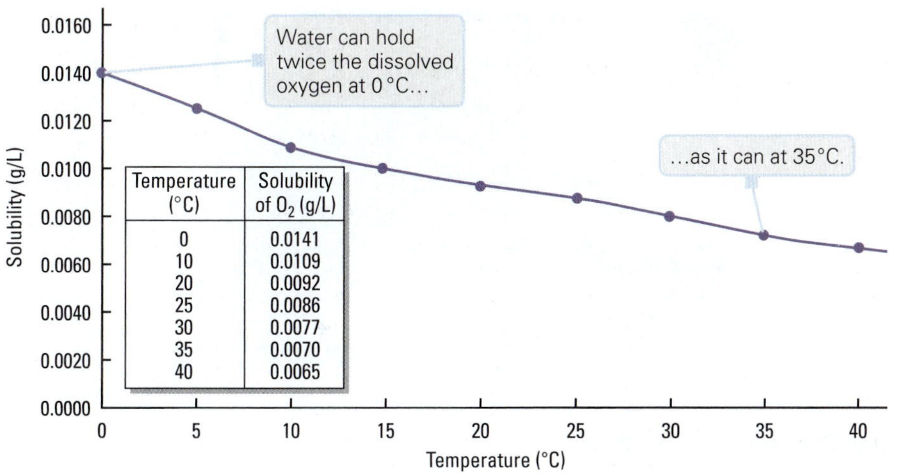

Temperature (°C)	Solubility of O_2 (g/L)
0	0.0141
10	0.0109
20	0.0092
25	0.0086
30	0.0077
35	0.0070
40	0.0065

Figure 16.16 Solubility of oxygen in water from 0°C to 40°C. Gases become less soluble in liquids with increasing temperature.

Target Check 16.4

If you are given the structural formulas of two substances, what would you look for to predict if one will dissolve in the other?

16.6 Solution Concentration: Percentage Concentration by Mass

Goal 7 Given the mass of solute and solvent or solution, calculate percentage concentration by mass.

8 Given mass of solution and percentage concentration by mass, calculate mass of solute and solvent.

The concentration of a solution tells us how much solute is present per given amount of solution or given amount of solvent. As a conversion factor, concentration has the form of a fraction. Amount of solute appears in the numerator and may be in grams, moles, or equivalents (eq), a unit we will describe in Section 16.9. Quantity of solvent or solution is in the denominator and may be in mass or volume units. In general, concentration is

$$\frac{\text{quantity of solute}}{\text{quantity of solution}} \quad or \quad \frac{\text{quantity of solute}}{\text{quantity of solvent}}$$

◀ **P/Review** Percentage concentration by mass corresponds with percentage composition by mass in Section 7.7, in which percentage calculations are discussed in general. When given mass quantities, you can find the percentage of any component of a solution with: % concentration by mass = (mass solute ÷ mass solution) × 100%. When the percentage of a component is known, it is a conversion factor between the mass of that component and the mass of the solution.

We begin with percentage concentration by mass. Percentage concentration is based on the ratio mass solute ÷ mass solution. We will refer to this as a **mass ratio**. The mass ratio is a decimal fraction that represents the mass in grams of solute in 1 g of solution. Therefore, 100 times the mass ratio is the mass in grams of solute in 100 g of solution—grams of solute per 100 g of solution—which is the definition of **percentage concentration by mass** (**Figure 16.17**): ▶

$$\%\ \text{concentration by mass} = \frac{\text{mass solute}}{\text{mass solution}} \times 100\%$$

$$= \frac{\text{mass solute}}{\text{mass solute} + \text{mass solvent}} \times 100\%$$

The concentrations of very dilute solutions are conveniently given in parts per million (ppm). A 0.0025% solution, which is 0.0025 parts per hundred, is equivalent to 25 ppm.

Be careful about the denominator. If a problem gives the mass of solute and mass of solvent, be sure to add them to get the mass of the solution.

Figure 16.17 Ingredients by weight label. The ingredients of liquid pesticides and pharmaceutical products are often listed as percentage by weight in the United States.

Active Example 16.1 Percentage Concentration by Mass

When 125 grams of a solution was evaporated to dryness, 42.3 grams of solute was recovered. What was the percentage concentration by mass of the solution?

Think Before You Write You are given both mass of solution and mass of solute, so this is a direct application of the definition of percentage concentration by mass.

Answers *Cover the left column with your cut-out shield. Reveal each answer only after you have written your own answer in the right column.*

Given: 125 g solution; 42.3 g solute **Wanted:** % concentration by mass $\%\text{ concentration by mass} = \frac{\text{mass solute}}{\text{mass solution}} \times 100\%$ $= \frac{42.3\text{ g}}{125\text{ g}} \times 100\% = 33.8\%$ $\frac{42.3}{125} \times 100 \approx \frac{40}{120} \times 100 = \frac{1}{3} \times 100 = 33$, OK.	Write the given quantity and wanted property, and then substitute the given values into the defining equation for percentage concentration by mass and calculate the answer. Finish the solution with a **check** of your calculation.
You improved your understanding of percentage concentration by mass, and you improved your skill at solving solution concentration problems.	What did you learn by solving this Active Example?

Practice Exercise 16.1

A solution is prepared by dissolving 2.00 kg of solute in enough solvent to make 50.0 kg of solution. What is the percentage concentration by mass of the solution?

Active Example 16.2 Preparation of a Solution of a Specified Percentage Concentration by Mass

You are to prepare 2.50×10^2 g of 7.00% by mass Na_2CO_3 solution. What mass in grams of sodium carbonate and what volume in milliliters of water do you use? (The density of water is 1.00 g/mL.)

Think Before You Write When a given quantity in a problem is a ratio, it almost always helps to write that quantity as a ratio. The 7.00% by mass Na_2CO_3 solution is $\frac{7.00\text{ g } Na_2CO_3}{100\text{ g solution}}$. This allows you to more clearly see the units of this given quantity.

Answers *Cover the left column with your cut-out shield. Reveal each answer only after you have written your own answer in the right column.*

Given: 2.50×10^2 g of 7.00% Na_2CO_3 solution
Wanted: g Na_2CO_3

g solution → g Na_2CO_3

$$\frac{7.00 \text{ g } Na_2CO_3}{100 \text{ g solution}}$$

Write the given quantity and the wanted units. Write the unit path that you will follow to determine the wanted quantity. Write the given ratio in the form of a fraction.

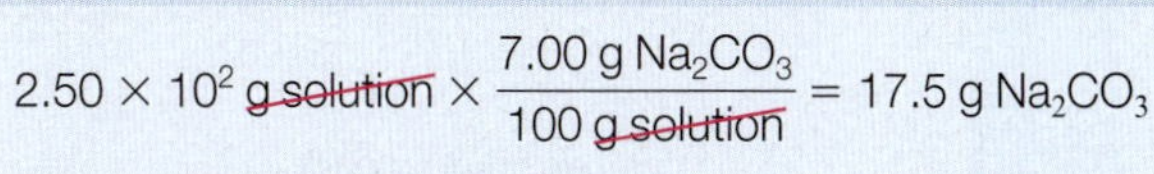

$$2.50 \times 10^2 \text{ g solution} \times \frac{7.00 \text{ g } Na_2CO_3}{100 \text{ g solution}} = 17.5 \text{ g } Na_2CO_3$$

$$250 \times \frac{7}{100} = (2.5 \times 8) - 2.5 = 20 - 2.5 = 17.5, \text{ OK.}$$

Use the percentage concentration by mass as a conversion factor to convert from g solution to g Na_2CO_3. Calculate just that quantity.

g H_2O = g solution − g solute =
2.50×10^2 g − 17.5 g Na_2CO_3 = 233 g H_2O
At 1.00 g/mL, 233 g H_2O = 233 mL H_2O

The mass of the solution is 2.50×10^2 g. The mass of the solute in the solution is 17.5 g. The rest is water. What is the mass of the water? What is the volume of that mass of water?

You improved your understanding of percentage concentration by mass, and you improved your skill at solving solution concentration problems.

What did you learn by solving this Active Example?

Practice Exercise 16.2

What mass of sodium chloride and what volume of water are needed to prepare 5.00×10^2 mL of a 0.89% by mass saline solution? (*Saline solution* is the everyday language phrase that refers to a sodium chloride solution.) The density of the solution is equal to the density of water to ± 0.01 g/mL.

16.7 Solution Concentration: Molarity

Goal 9 Given two of the following, calculate the third: amount in moles of solute (or data from which it may be found), volume of solution, molarity.

Concentrations of medicinal solutions are sometimes given in terms of weight/volume percent, the mass of solute per 100 mL of solution. The density of dilute solutions is very close to 1 g/mL, so the mass of 100 mL of solution is very close to 100 g. Thus, for solutions of 5% or less, mass percent and weight/volume percent are essentially equal.

In working with liquids, volume is easier to measure than mass. Therefore, a solution concentration based on volume is usually more convenient to use than one based on mass. **Molarity (M)** is the amount in moles of solute per volume in liters of solution. The defining equation is:

$$\text{M} \equiv \frac{\text{moles solute}}{\text{liter solution}} = \frac{\text{mol}}{\text{L}}$$

If a solution contains 0.755 mole of sulfuric acid per liter, we identify it as 0.755 M H_2SO_4. In words, it is "point seven-five-five molar sulfuric acid." In a calculation setup, we would write "0.755 mol H_2SO_4/L."

Notice that molarity is an equivalency that can be expressed as two conversion factors. The amount of solute in a sample of solution is proportional to the volume of the sample, and molarity is the proportionality constant. If both sides of the defining equation for molarity are multiplied by volume in L, the result is:

$$\text{Volume} \times \text{Molarity} = \text{V} \times \text{M} = \text{L} \times \frac{\text{mol}}{\text{L}} = \text{mol}$$

Everyday Chemistry 16.1

THE WORLD'S OCEANS: THE MOST ABUNDANT SOLUTION

Oceans cover 71% of Earth's surface and contain all but 2.5% of Earth's water (**Figure 16.18**). As land animals, we humans often fail to recognize the enormity of the oceans and the central role these aqueous solutions play in Earth's environment.

The composition of seawater varies depending on the location from which a sample is taken, but on average it is a 3.5% solution of a variety of dissolved solids. To be precise, an entire ocean does not meet the criterion of a homogeneous mixture and therefore is not a solution. Nonetheless, many environmental factors lead to a relatively thorough mixing effect in the world's oceans, and very large samples are, in fact, homogeneous. Scientists who have sampled seawater from across the world have verified this experimentally, finding that the samples are nearly identical when analyzed for the major solution components.

Let's consider the concentration of seawater more carefully. Percentage concentration by mass can be expressed as

$$\%\text{ concentration by mass} = \frac{\text{g solute}}{100\text{ g solution}}$$

Seawater is therefore 3.5 g of dissolved solids per 100 g of solution. At typical ocean temperatures, the density of seawater is 1.0 g/mL. We can therefore determine the mass of dissolved solids per liter:

$$\frac{3.5\text{ g solids}}{100\text{ g }\cancel{\text{seawater}}} \times \frac{1.0\text{ g }\cancel{\text{seawater}}}{1.0\text{ mL }\cancel{\text{seawater}}} \times \frac{1000\text{ mL }\cancel{\text{seawater}}}{1\text{ L seawater}} = \frac{35\text{ g solids}}{\text{L seawater}}$$

Imagine a 1-L soft drink bottle filled with seawater. There is 35 g (1.2 ounces) of dissolved solids in that bottle.

Chloride ions and sodium ions are by far the most abundant species dissolved in the oceans, making up about 85% of the dissolved solids. If you've ever gotten a mouthful of ocean water, you can recognize the salty taste characteristic of these two components of table salt. If we add magnesium ion, sulfate ion, calcium ion, and potassium ion to our list of dissolved solids in seawater, we account for 99% of the dissolved solids. Trace amounts of many other species are found in the oceans. Chemists have identified over 70 elements in seawater samples.

Sea salt is obtained by evaporating ocean water. One cubic foot of seawater yields more than 2 lb of salt. In comparison, the same-size sample from a typical freshwater lake yields less than an ounce of dissolved solids.

The high concentration of solute particles in the oceans can be traced to the global water cycle. Water from the atmosphere falls to Earth as rain or snow. This water dissolves some of the solids in Earth's surface, carries the dissolved solids into the ocean, and then leaves them there as the water evaporates back into the atmosphere (**Figure 16.19**).

Figure 16.18 Distribution of Earth's water. Almost all of it is found in the oceans. The total percentage of groundwater and freshwater in the liquid state is less than 1%.

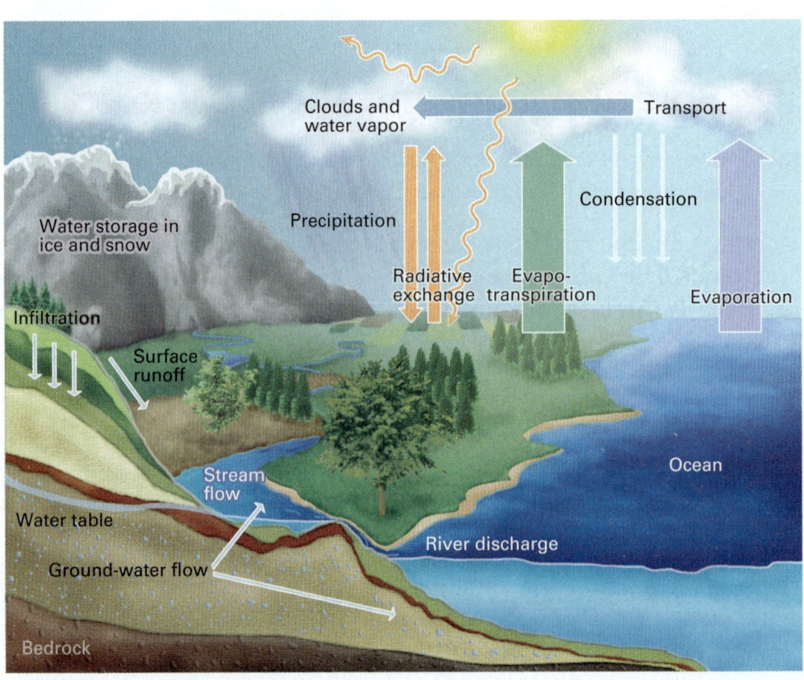

Figure 16.19 The global water cycle. The relatively high concentration of dissolved solids in the oceans results from this process.

Ocean water evaporates and is carried over landmasses by the winds. When the water is released from the atmosphere in the form of rain or snow, some finds its way into rivers, which mix the water with dissolved solids and eventually run to the seas. As the cycle continues, dissolved ions are continually deposited in the oceans.

All natural waters are solutions, and the size of the world's oceans makes them the most abundant aqueous solution on Earth (**Figure 16.20**). Seawater is among the most concentrated of the planet's natural waters, and its concentration continues to gradually increase over time. Chemists and other scientists study the oceans and their interaction with all aspects of the environment. Many concepts from chemistry are central to this research.

Quick Quiz

1. **What percentage of Earth's water is in the liquid state?**
2. **What volume in gallons of seawater contains 1 lb of dissolved solids?**

Juriah Mosin/Shutterstock.com

Figure 16.20 The ocean. The ocean is a liquid solution of many dissolved solids, covering more than 70% of the Earth's surface.

We will make good use of this equation in the solution dilution and solution stoichiometry sections later in this chapter.

All the calculation methods you have used before with conversion factors can be used with molarity. Notice that the units in the denominator are liters, but volume is often given in milliliters. If the volume is given in milliliters, you must convert it to liters as part of the solution setup.

Also notice that molarity includes one unit that is particulate (amount grouped in moles) and one unit that is macroscopic (volume, which can be measured). To be practical—to actually work with molarity in the laboratory—you must convert amount in moles to a macroscopic unit, typically mass in grams. The conversion factor, as you probably know from many uses by now, is molar mass in grams per mole.

Active Example 16.3 Calculation of Molarity

Calculate the molarity of a solution made by dissolving 15 grams of NaOH in water and diluting to 1.00×10^2 milliliters.

Think Before You Write The key to any "calculate the molarity" problem is to recognize that you need to determine both parts of the definition, the amount in moles of solute and the volume of solution in liters.

Answers *Cover the left column with your cut-out shield. Reveal each answer only after you have written your own answer in the right column.*

Given: 15 g NaOH **Wanted:** mol NaOH 1 mol NaOH = 40.00 g NaOH	Determine the amount in moles of solute first. **Analyze** that part of the problem by writing the given quantity and wanted unit, and **identify** the equivalency needed to solve the problem. 

$$15\ \cancel{\text{g NaOH}} \times \frac{1\ \text{mol NaOH}}{40.00\ \cancel{\text{g NaOH}}} = 0.38\ \text{mol NaOH}$$

$$\frac{15}{40} \approx \frac{15}{45} = 0.33\text{, OK.}$$

$$1.00 \times 10^2\ \cancel{\text{mL}} \times \frac{1\ \text{L}}{1000\ \cancel{\text{mL}}} = 0.100\ \text{L}$$

$$10^2 \div 10^3 = 10^{-1}\text{, OK.}$$

Construct the solution setup and calculate the amount in moles of NaOH. Also calculate the volume of solution in liters. **Check** both calculations.

$$\text{M} = \frac{\text{mol}}{\text{L}} = \frac{0.38\ \text{mol NaOH}}{0.100\ \text{L}} = 3.8\ \text{M NaOH}$$

$$0.38 \div 10^{-1} = 0.38 \times 10^1 = 3.8\text{, OK.}$$

You now have both amount in moles and volume in liters, the numerator and the denominator in the defining equation for molarity. Plug them into the definition and then calculate the answer. **Check** your calculation.

You improved your understanding of molarity, and you improved your skill at solving solution concentration problems.

What did you learn by solving this Active Example?

Practice Exercise 16.3

What is the molarity of a solution that was prepared by adding 10.4 grams of potassium chloride to enough water to make a total of 5.00×10^2 milliliters of solution?

Active Example 16.4 Calculation of Mass of Solute from Molarity and Solution Volume

What mass in grams of silver nitrate must be dissolved to prepare 5.00×10^2 mL of 0.150 M $AgNO_3$?

Think Before You Write When you have a ratio as a given quantity, write it as an equivalency: 0.150 mol $AgNO_3$ = 1 L solution. You can then clearly see the two conversion factors that result from the equivalency.

Answers *Cover the left column with your cut-out shield. Reveal each answer only after you have written your own answer in the right column.*

Given: 5.00×10^2 mL solution

0.150 mol $AgNO_3$ = 1 L solution

Wanted: g $AgNO_3$

mL soln → L soln → mol $AgNO_3$ → g $AgNO_3$

1 L soln = 1000 mL soln

0.150 mol $AgNO_3$ = 1 L solution

169.9 g $AgNO_3$ = 1 mol $AgNO_3$

Analyze the problem statement by listing the given and wanted, and then **identify** the equivalencies by first writing the unit path and then deriving the equivalency for each unit conversion. Don't change the equivalencies to conversion factors until the next step.

$$5.00 \times 10^2\ \cancel{\text{mL}} \times \frac{1\ \cancel{\text{L}}}{1000\ \cancel{\text{mL}}} \times \frac{0.150\ \cancel{\text{mol AgNO}_3}}{\cancel{\text{L}}} \times \frac{169.9\ \text{g AgNO}_3}{\cancel{\text{mol AgNO}_3}} = 12.7\ \text{g AgNO}_3$$

Construct the solution setup, changing the equivalencies directly to conversion factors in the setup, and **check** the value of the answer to be sure it makes sense.

Ocean water evaporates and is carried over landmasses by the winds. When the water is released from the atmosphere in the form of rain or snow, some finds its way into rivers, which mix the water with dissolved solids and eventually run to the seas. As the cycle continues, dissolved ions are continually deposited in the oceans.

All natural waters are solutions, and the size of the world's oceans makes them the most abundant aqueous solution on Earth (**Figure 16.20**). Seawater is among the most concentrated of the planet's natural waters, and its concentration continues to gradually increase over time. Chemists and other scientists study the oceans and their interaction with all aspects of the environment. Many concepts from chemistry are central to this research.

Quick Quiz

1. **What percentage of Earth's water is in the liquid state?**
2. **What volume in gallons of seawater contains 1 lb of dissolved solids?**

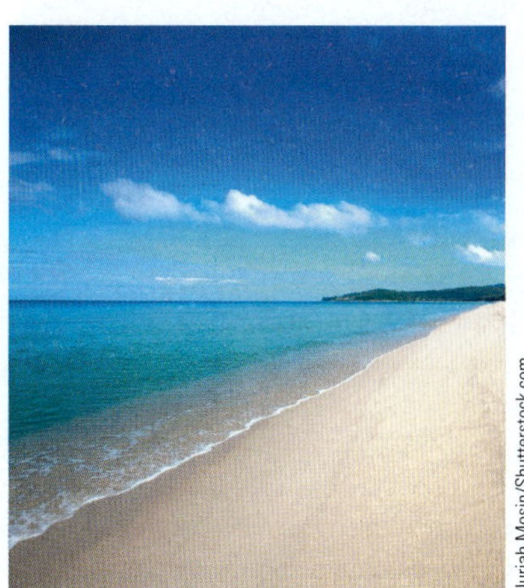

Figure 16.20 The ocean. The ocean is a liquid solution of many dissolved solids, covering more than 70% of the Earth's surface.

We will make good use of this equation in the solution dilution and solution stoichiometry sections later in this chapter.

All the calculation methods you have used before with conversion factors can be used with molarity. Notice that the units in the denominator are liters, but volume is often given in milliliters. If the volume is given in milliliters, you must convert it to liters as part of the solution setup.

Also notice that molarity includes one unit that is particulate (amount grouped in moles) and one unit that is macroscopic (volume, which can be measured). To be practical—to actually work with molarity in the laboratory—you must convert amount in moles to a macroscopic unit, typically mass in grams. The conversion factor, as you probably know from many uses by now, is molar mass in grams per mole.

Active Example 16.3 Calculation of Molarity

Calculate the molarity of a solution made by dissolving 15 grams of NaOH in water and diluting to 1.00×10^2 milliliters.

Think Before You Write The key to any "calculate the molarity" problem is to recognize that you need to determine both parts of the definition, the amount in moles of solute and the volume of solution in liters.

Answers *Cover the left column with your cut-out shield. Reveal each answer only after you have written your own answer in the right column.*

Given: 15 g NaOH **Wanted:** mol NaOH 1 mol NaOH = 40.00 g NaOH	Determine the amount in moles of solute first. **Analyze** that part of the problem by writing the given quantity and wanted unit, and **identify** the equivalency needed to solve the problem. 

$15\ \text{g NaOH} \times \frac{1\ \text{mol NaOH}}{40.00\ \text{g NaOH}} = 0.38\ \text{mol NaOH}$

$\frac{15}{40} \approx \frac{15}{45} = 0.33$, OK.

$1.00 \times 10^2\ \text{mL} \times \frac{1\ \text{L}}{1000\ \text{mL}} = 0.100\ \text{L}$

$10^2 \div 10^3 = 10^{-1}$, OK.

Construct the solution setup and calculate the amount in moles of NaOH. Also calculate the volume of solution in liters. **Check** both calculations.

$M = \frac{\text{mol}}{\text{L}} = \frac{0.38\ \text{mol NaOH}}{0.100\ \text{L}} = 3.8\ \text{M NaOH}$

$0.38 \div 10^{-1} = 0.38 \times 10^1 = 3.8$, OK.

You now have both amount in moles and volume in liters, the numerator and the denominator in the defining equation for molarity. Plug them into the definition and then calculate the answer. **Check** your calculation.

You improved your understanding of molarity, and you improved your skill at solving solution concentration problems.

What did you learn by solving this Active Example?

Practice Exercise 16.3

What is the molarity of a solution that was prepared by adding 10.4 grams of potassium chloride to enough water to make a total of 5.00×10^2 milliliters of solution?

Active Example 16.4 Calculation of Mass of Solute from Molarity and Solution Volume

What mass in grams of silver nitrate must be dissolved to prepare 5.00×10^2 mL of 0.150 M $AgNO_3$?

Think Before You Write When you have a ratio as a given quantity, write it as an equivalency: 0.150 mol $AgNO_3$ = 1 L solution. You can then clearly see the two conversion factors that result from the equivalency.

Answers *Cover the left column with your cut-out shield. Reveal each answer only after you have written your own answer in the right column.*

Given: 5.00×10^2 mL solution
0.150 mol $AgNO_3$ = 1 L solution
Wanted: g $AgNO_3$

mL soln → L soln → mol $AgNO_3$ → g $AgNO_3$

1 L soln = 1000 mL soln

0.150 mol $AgNO_3$ = 1 L solution

169.9 g $AgNO_3$ = 1 mol $AgNO_3$

Analyze the problem statement by listing the given and wanted, and then **identify** the equivalencies by first writing the unit path and then deriving the equivalency for each unit conversion. Don't change the equivalencies to conversion factors until the next step.

$$5.00 \times 10^2\ \text{mL} \times \frac{1\ \text{L}}{1000\ \text{mL}} \times \frac{0.150\ \text{mol AgNO}_3}{\text{L}} \times \frac{169.9\ \text{g AgNO}_3}{\text{mol AgNO}_3} = 12.7\ \text{g AgNO}_3$$

Construct the solution setup, changing the equivalencies directly to conversion factors in the setup, and **check** the value of the answer to be sure it makes sense.

$(5.00 \times 10^2) \times \frac{1}{1000} \times 0.150 \times 169.9 \approx (5 \times 10^2) \times 10^{-3}$

$\times (1.5 \times 10^{-1}) \times (2 \times 10^2) = (5 \times 1.5 \times 2) \times (10^2$

$\times 10^{-3} \times 10^{-1} \times 10^2) = 15 \times 10^0 = 15$, OK.

You improved your understanding of molarity, and you improved your skill at solving solution concentration problems.

What did you learn by solving this Active Example?

Practice Exercise 16.4

Determine the mass in grams of sodium sulfate dissolved in 2.00 L of a 0.12 M sodium sulfate solution.

A clearer understanding of molarity can be gained by mentally "preparing" a solution. Follow along visually in **Figure 16.21** as you study this paragraph. Consider the preparation of 250.0 mL of 0.0100 M potassium permanganate solution. First, you determine the mass of solute needed:

$$250.0\ \cancel{\text{mL}} \times \frac{1\ \cancel{\text{L}}}{1000\ \cancel{\text{mL}}} \times \frac{0.0100\ \cancel{\text{mol KMnO}_4}}{\cancel{\text{L}}} \times \frac{158.04\ \text{g KMnO}_4}{\cancel{\text{mol KMnO}_4}} = 0.395\ \text{g KMnO}_4$$

Now weigh out the 0.395 g $KMnO_4$ into a 250-mL volumetric flask (**Figure 16.21[1]**). Add *less than* 250 mL of water to the flask. After dissolving the solute (**Figure 16.21[2]**), add water to the 250-mL mark on the neck of the flask (**Figure 16.21[3]**). Finally, stopper and shake the flask to ensure thorough mixing (**Figure 16.21[4]**). Notice that molarity is based on the volume of the *solution*, not the volume of the *solvent*. This is why the solute is dissolved in less than 250 mL of water and then diluted to that volume.

Figure 16.22 contrasts a solution correctly made based on solution volume with one incorrectly made based on solvent volume. Remember, the denominator in the molarity fraction mol/L is liters of *solution*.

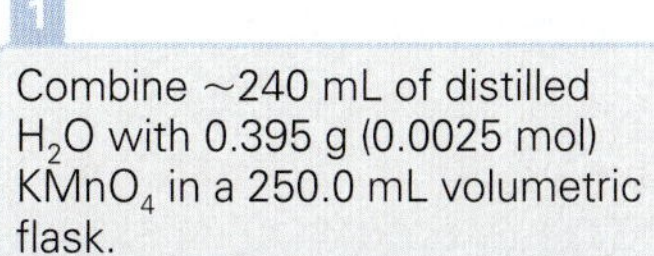

1
Combine ~240 mL of distilled H_2O with 0.395 g (0.0025 mol) $KMnO_4$ in a 250.0 mL volumetric flask.

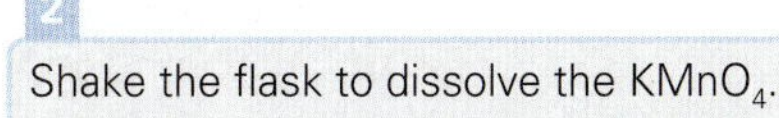

2
Shake the flask to dissolve the $KMnO_4$.

3
After the solid dissolves, add sufficient water to fill the flask to the mark etched in the neck, indicating a volume of 250.0 mL.

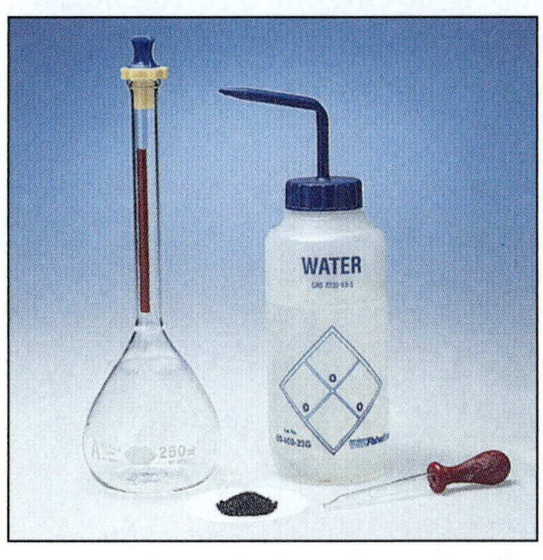

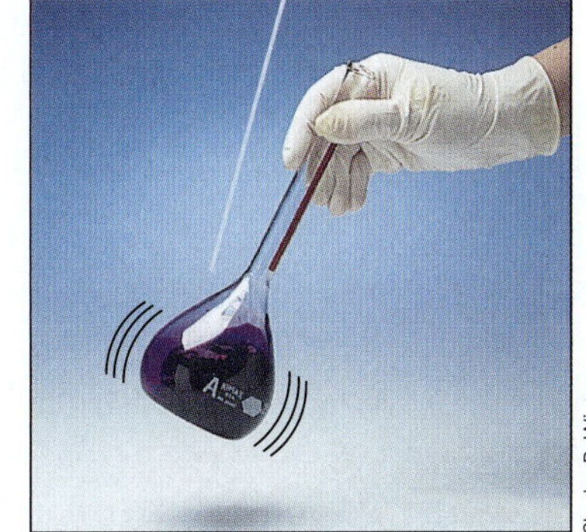

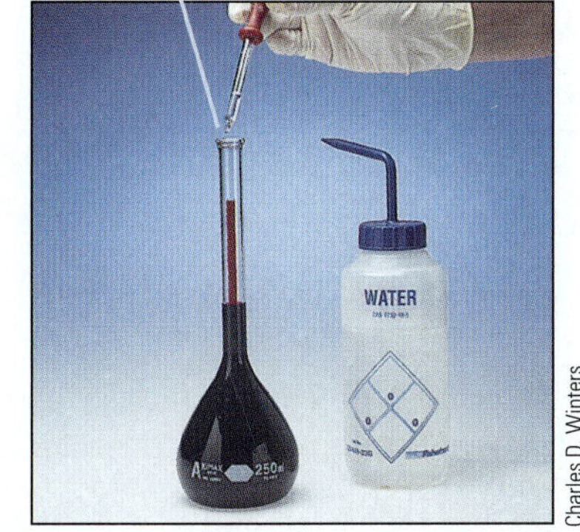

4
Shake the flask again to thoroughly mix its contents. The flask now contains 250.0 mL of 0.0100 M $KMnO_4$ solution.

Figure 16.21 Preparing a solution of specified molarity.

Figure 16.22 Solution versus solvent volume. Each solution contains 0.10 mole of potassium chromate. The solution on the left was made by adding the solute and then adding enough water to make a *solution* volume of 1000 mL. Note that the volume of this solution is exactly 1000 mL because it precisely reaches the etch mark on the volumetric flask. This is the appropriate procedure to make a solution of specified molarity. The solution on the right was made by adding 1000 mL of water—a *solvent* volume of 1000 mL—to the solute. Note that the *solution* volume in the flask on the right exceeds 1000 mL.

Charles D. Winters

Active Example 16.5 Calculation of Solution Volume from Molarity and Mass of Solute

Find the volume of a 1.40-M solution that contains 4.89 grams of ammonia. Answer in both liters and milliliters.

Think Before You Write Again, when you encounter molar concentration in a problem, think of it as an equivalency.

Answers *Cover the left column with your cut-out shield. Reveal each answer only after you have written your own answer in the right column.*

Given: 1.40 mol NH_3 = 1 L solution; 4.89 g NH_3 **Wanted:** solution volume in L and mL g NH_3 → mol NH_3 → L soln → mL soln 1 mol NH_3 = 17.03 g NH_3 1 L soln = 1.40 mol NH_3 1000 mL soln = 1 L soln	**Analyze** the problem statement by listing the given and wanted and then **identify** the equivalencies by first writing the unit path and then deriving the equivalency for each unit conversion. Don't change the equivalencies to conversion factors until the next step.
$4.89\ \cancel{\text{g NH}_3} \times \frac{1\ \cancel{\text{mol NH}_3}}{17.03\ \cancel{\text{g NH}_3}} \times \frac{1\text{ L}}{1.40\ \cancel{\text{mol NH}_3}} = 0.205\text{ L}$ $4.89 \times \frac{1}{17.03} \times \frac{1}{1.40} \approx 5 \times \frac{1}{15} \times \frac{1}{1.5} = \left(5 \times \frac{1}{15}\right) \times \frac{1}{1.5}$ $= \frac{1}{3} \times \frac{1}{1.5} = \frac{1}{4.5} \approx \frac{1}{5} = 0.2$, OK.	**Construct** the solution setup, changing the equivalencies directly to conversion factors in the setup, and **check** the value of the answer to be sure it makes sense.
You improved your understanding of molarity, and you improved your skill at solving solution concentration problems.	What did you learn by solving this Active Example?

Practice Exercise 16.5

A technician dissolves 5.29 g of ammonium sulfate in water to make a 0.10-M solution. What volume of solution did she make? Answer in milliliters.

16.8 Solution Concentration: Molality (Optional)

Goal 10 Given two of the following, calculate the third: moles of solute (or data from which it may be found), mass of solvent, molality.

Many physical properties are related to solution concentration expressed as **molality (m)**, the amount in moles of solute dissolved in 1 kg of solvent. ▶ The defining equation is:

$$m \equiv \frac{\text{mol solute}}{\text{kg solvent}}$$

If a solution contains 0.755 mole of acetic acid per kilogram of water, we identify it as 0.755 m CH_3COOH. In words, it is "point seven-five-five molal acetic acid." In a calculation setup, we would write "0.755 mol CH_3COOH/kg H_2O."

Like molarity, molality is defined as a ratio that leads to an equivalency and two conversion factors. Notice that the units in the denominator are kilograms, but the mass of solvent is usually given in grams. To be used in the defining equation, the solvent mass must be changed to kilograms.

Molality is not a convenient concentration unit to work with—not nearly as convenient as molarity. Measuring volume of solution, the liters in moles per liter, is easy, whereas neither moles of solute nor kilograms of solvent—the units in molality—can be measured directly in a solution that is already prepared. So why not use molarity all the time and forget about molality? Molarity is temperature dependent. As the volume of a solution changes on heating or cooling, its molarity changes—not a lot but enough so that properties that can be related to molality cannot be satisfactorily related to molarity.

Active Example 16.6 Calculation of Molality

Calculate the molality of a solution prepared by dissolving 15.0 grams of sugar, $C_{12}H_{22}O_{11}$ (Figure 16.23), in 3.50 × 10² milliliters of water. (The density of water is 1.00 g/mL.)

Think Before You Write The key to any "calculate the molality" problem is to recognize that you need to determine both parts of the definition, the amount in moles of solute and the mass of the solvent in kilograms.

Answers *Cover the left column with your cut-out shield. Reveal each answer only after you have written your own answer in the right column.*

Determine the amount in moles of solute first. **Analyze** that part of the problem by writing the given quantity and wanted unit, and **identify** the equivalency needed to solve the problem. **Construct** the solution setup, changing the equivalency directly to the conversion factor in the setup, and **check** the value of the answer to be sure it makes sense.

Given: 15.0 g $C_{12}H_{22}O_{11}$
Wanted: mol $C_{12}H_{22}O_{11}$

g $C_{12}H_{22}O_{11}$ → mol $C_{12}H_{22}O_{11}$

1 mol $C_{12}H_{22}O_{11}$ = 342.30 g $C_{12}H_{22}O_{11}$

$$15.0\ \text{g } C_{12}H_{22}O_{11} \times \frac{1\ \text{mol } C_{12}H_{22}O_{11}}{342.30\ \text{g } C_{12}H_{22}O_{11}} = 0.0438\ \text{mol } C_{12}H_{22}O_{11}$$

$$15.0 \times \frac{1}{342.30} \approx 15 \times \frac{1}{3 \times 10^2} = \frac{15}{3} \times 10^{-2} = 5 \times 10^{-2} = 0.05, \text{ OK.}$$

Complete all of the steps needed to complete the conversion of volume in milliliters of water to mass in kilograms of water.

Given: 3.50 × 10² mL H_2O
Wanted: kg H_2O

mL H_2O → g H_2O → kg H_2O

1.00 g H_2O = 1 mL H_2O

1 kg H_2O = 1000 g H_2O

$$3.50 \times 10^2\ \text{mL } H_2O \times \frac{1.00\ \text{g } H_2O}{\text{mL } H_2O} \times \frac{1\ \text{kg } H_2O}{1000\ \text{g } H_2O} = 0.350\ \text{kg } H_2O$$

$$3.50 \times 10^2 \times 10^{-3} = 3.50 \times 10^{-1} = 0.350, \text{ OK.}$$

Figure 16.23 Sugar production in the northern United States originates with growing sugar beets. The beets are sliced and placed in hot water, the resulting sugar juice is purified and concentrated, and the water is evaporated to yield table sugar, sucrose, $C_{12}H_{22}O_{11}$.

$$m \equiv \frac{\text{mol solute}}{\text{kg solvent}} = \frac{0.0438 \text{ mol } C_{12}H_{22}O_{11}}{0.350 \text{ kg } H_2O}$$

$$= 0.125 \text{ m } C_{12}H_{22}O_{11}$$

$$\frac{0.0438}{0.350} \approx \frac{4 \times 10^{-2}}{4 \times 10^{-1}} = 10^{-2} \times 10^{1} = 10^{-1} = 0.1, \text{ OK.}$$

You now have both amount in moles of solute and mass of solvent in kilograms, the numerator and the denominator in the defining equation for molality. Plug them into the definition and then calculate the answer. **Check** your calculation.

You improved your understanding of molality, and you improved your skill at solving solution concentration problems.

What did you learn by solving this Active Example?

Practice Exercise 16.6

A solution is prepared by dissolving 80.0 mg of potassium phosphate in 25 mL of water. What is the molality of the solution?

Active Example 16.7 Calculation of Mass of Solute from Molality and Mass of Solvent

What mass in grams of potassium chloride must be dissolved in 2.50×10^2 grams of water to make a 0.400 m solution?

Think Before You Write When you have a ratio as a given quantity, write it as an equivalency: 0.400 mol KCl = 1 kg H_2O. You can then clearly see the two conversion factors that result from the equivalency.

Answers *Cover the left column with your cut-out shield. Reveal each answer only after you have written your own answer in the right column.*

Given: 2.50×10^2 g H_2O; 0.400 mol KCl = 1 kg H_2O
Wanted: g KCl

g H_2O → kg H_2O → mol KCl → g KCl

1 kg H_2O = 1000 g H_2O

0.400 mol KCl = 1 kg H_2O

74.55 g KCl = 1 mol KCl

Analyze the problem statement by listing the given and wanted and then **identify** the equivalencies by first writing the unit path and then deriving the equivalency for each unit conversion. Don't change the equivalencies to conversion factors until the next step.

$$2.50 \times 10^2 \text{ g } \cancel{H_2O} \times \frac{1 \text{ kg } \cancel{H_2O}}{1000 \text{ g } \cancel{H_2O}} \times \frac{0.400 \text{ mol KCl}}{\text{kg } H_2O} \times \frac{74.55 \text{ g KCl}}{\text{mol KCl}} = 7.46 \text{ g KCl}$$

$$2.50 \times 10^2 \times \frac{1}{1000} \times 0.400 \times 74.55 \approx (2.5 \times 10^2) \times 10^{-3} \times (4 \times 10^{-1}) \times 75 = (2.5 \times 4 \times 75) \times (10^2 \times 10^{-3} \times 10^{-1}) = 750 \times 10^{-2} = 7.5, \text{ OK.}$$

Construct the solution setup, changing the equivalencies directly to conversion factors in the setup, and **check** the value of the answer to be sure it makes sense.

You improved your understanding of molality, and you improved your skill at solving solution concentration problems.

What did you learn by solving this Active Example?

Practice Exercise 16.7

What mass of calcium nitrate is needed to prepare a 0.75 m solution when mixed with 2.00 L of water?

16.9 Solution Concentration: Normality (Optional)

Goal 11 Given an equation for a neutralization reaction (or information from which it can be written), state the number of equivalents of acid or base per mole and calculate the equivalent mass of the acid or base.

12 Given two of the following, calculate the third: equivalents of acid or base (or data from which they may be found), volume of solution, normality.

A particularly convenient concentration in routine analytical work is **normality (N)**: the number of equivalents, eq, per liter of solution. ▶ (The equivalent will be defined shortly.) The defining equation is:

$$N \equiv \frac{\text{equivalents solute}}{\text{liter solution}} = \frac{\text{eq}}{\text{L}}$$

If a solution contains 0.755 equivalent of phosphoric acid per liter, we identify it as 0.755 N H_3PO_4. In words, it is "point seven-five-five normal phosphoric acid." In a calculation setup, we would write "0.755 eq H_3PO_4/L."

The defining equation for normality is very similar to the equation for molarity, M ≡ mol/L. Many of the calculations are similar, too. But to understand the difference, we must see what an equivalent is.

One **equivalent** of an acid is the quantity that yields 1 mole of hydrogen ions in a chemical reaction. One equivalent of a base is the quantity that reacts with 1 mole of hydrogen ions. Because hydrogen and hydroxide ions combine on a 1:1 ratio, 1 mole of hydroxide ions is 1 equivalent of base.

According to these statements, both 1 mole of HCl and 1 mole of NaOH are 1 equivalent. They yield, respectively, 1 mole of H^+ ions and 1 mole of OH^- ions. H_2SO_4, on the other hand, may have 2 equivalents per mole because it can release 2 moles of H^+ ions per mole of acid. Similarly, 1 mole of $Al(OH)_3$ may represent 3 equivalents because 3 moles of OH^- may react. The number of equivalents in a mole of an acid depends on a specific reaction, not just the number of moles of H's in a mole of the

It is common for most chemists to never use normality to express concentration. Since advocates of SI units have been pushing to make SI the only system of units used worldwide, they argue that the mole is a base unit under the SI system, whereas the equivalent is not even recognized. (Historically, the equivalent has an undisputed place in the evolution of chemical understanding.) In addition, the SI-only advocates argue that there is no problem that can be solved by normality that cannot be solved by molarity, so why have two systems? In the field—in the industrial laboratories in which so many practicing chemists work—normality is so convenient that there is unyielding resistance to change. Moreover, many academic chemists support that position. This is one of the major areas of disagreement among chemists.

compound. What counts is the number of H's *that react.* By controlling reaction conditions, phosphoric acid can have 1, 2, or, theoretically, 3 equivalents per mole:

$$NaOH(aq) + H_3PO_4(aq) \rightarrow NaH_2PO_4(aq) + H_2O(\ell) \qquad 1\ \text{eq acid/mol}$$

$$2\ NaOH(aq) + H_3PO_4(aq) \rightarrow Na_2HPO_4(aq) + 2\ H_2O(\ell) \qquad 2\ \text{eq acid/mol}$$

$$3\ NaOH(aq) + H_3PO_4(aq) \rightarrow Na_3PO_4(aq) + 3\ H_2O(\ell) \qquad 3\ \text{eq acid/mol}$$

There is 1 equivalent per mole of base in each of the preceding reactions. NaOH can have only 1 equivalent per mole because there is only 1 mole of OH^- in 1 mole of NaOH. It is noteworthy that *the number of equivalents of acid and base in each reaction are the same.* When NaH_2PO_4 is the product, 1 mol H_3PO_4 gives up only 1 eq H^+, and it reacts with 1 mol NaOH, which is 1 eq NaOH. When Na_2HPO_4 is the product, 1 mol H_3PO_4 yields 2 eq H^+, and it reacts with 2 mol NaOH, which is 2 eq NaOH. When Na_3PO_4 is the product, there are 3 eq H^+ and 3 mol NaOH, which is 3 eq NaOH.

Active Example 16.8 Normality I

State the number of equivalents of acid and base per mole in each of the following reactions: (a) $2\ HBr + Ba(OH)_2 \rightarrow BaBr_2 + 2\ H_2O$ and (b) $H_3C_6H_5O_7 + 2\ KOH \rightarrow K_2HC_6H_5O_7 + 2\ H_2O$

Think Before You Write You are interested only in the number of moles of H^+ or OH^- that *react*, not the number present, per mole of acid or base.

Answers *Cover the left column with your cut-out shield. Reveal each answer only after you have written your own answer in the right column.*

	Acid: eq/mol	Base: eq/mol
Reaction (a)	1	2
Reaction (b)	2	1

In (a), each mole of $Ba(OH)_2$ yields two OH^- ions, so there are 2 eq/mol. Each mole of HBr produces one H^+, so there is 1 eq/mol.

In (b), there could be 1, 2, or 3 equivalents of $H_3C_6H_5O_7$ per mole, depending on how many ionizable hydrogens are removed in the reaction. In this case, that number is two: $H_3C_6H_5O_7 \rightarrow 2\ H^+ + HC_6H_5O_7^{2-}$. The 2 H^+ ions removed combine with the 2 OH^- ions from 2 moles of KOH to form 2 H_2O molecules. In KOH, there is only one OH^- in a formula unit, so there can be only 1 eq/mol.

The formula of citric acid, $H_3C_6H_5O_7$, is written with its three ionizable hydrogens first. Insert the numbers of equivalents per mole for each reaction in the spaces here.

	Acid: eq/mol	Base: eq/mol
Reaction (a)		
Reaction (b)		

You improved your understanding of normality, and you improved your skill at solving solution concentration problems.

What did you learn by solving this Active Example?

Practice Exercise 16.8

Determine the number of equivalents of acid and base per mole in each of the following reactions:

(a) $Ca(OH)_2 + 2\ HI \rightarrow CaI_2 + 2\ H_2O$

(b) $NaOH + H_3C_9H_8NO_3 \rightarrow NaH_2C_9H_8NO_3 + H_2O$

The fact that the number of equivalents of *all* species in a chemical reaction is the same is *why* normality is a convenient concentration unit for analytical work. You will see how this becomes an advantage in Section 16.14.

Notice that, as with the three phosphoric acid neutralizations, *the number of equivalents of acid and base in each equation in Active Example 16.8 is the same.* In the first reaction, 2 mol HBr is 2 eq, and 1 mol $Ba(OH)_2$ is 2 eq. The "same number of equivalents of all reactants" idea extends to the product species, too. Once you find the number of equivalents of one species in a reaction, you have the number of equivalents of *all* species.

It is sometimes convenient to find the **equivalent mass (g/eq)** of a substance, the mass in grams per equivalent. Equivalent mass is similar to molar mass, g/mol. You can readily calculate equivalent mass by dividing molar mass by equivalents per mole:

$$\frac{\text{g/mol}}{\text{eq/mol}} = \frac{\text{g}}{\cancel{\text{mol}}} \times \frac{\cancel{\text{mol}}}{\text{eq}} = \text{g/eq}$$

Figure 16.24 Phosphoric acid is added to soft drinks to provide a tangy taste.

Blue Lemon Photo/Shutterstock.com

In ordinary acid–base reactions, there are 1, 2, or 3 equivalents per mole. It follows that the equivalent mass of an acid or a base is either (a) the same as, (b) one-half of, or (c) one-third of the molar mass. The molar mass of phosphoric acid is 97.99 g/mol. For the three reactions of phosphoric acid (**Figure 16.24**), the equivalent masses are:

$$NaH_2PO_4 \text{ produced: } \frac{97.99 \text{ g } H_3PO_4/\text{mol}}{1 \text{ eq } H_3PO_4/\text{mol}} = \frac{97.99 \text{ g } H_3PO_4}{1 \text{ eq } H_3PO_4}$$
$$= 97.99 \text{ g } H_3PO_4/\text{eq } H_3PO_4$$

$$Na_2HPO_4 \text{ produced: } \frac{97.99 \text{ g } H_3PO_4/\text{mol}}{2 \text{ eq } H_3PO_4/\text{mol}} = \frac{97.99 \text{ g } H_3PO_4}{2 \text{ eq } H_3PO_4}$$
$$= 49.00 \text{ g } H_3PO_4/\text{eq } H_3PO_4$$

$$Na_3PO_4 \text{ produced: } \frac{97.99 \text{ g } H_3PO_4/\text{mol}}{3 \text{ eq } H_3PO_4/\text{mol}} = \frac{97.99 \text{ g } H_3PO_4}{3 \text{ eq } H_3PO_4}$$
$$= 32.66 \text{ g } H_3PO_4/\text{eq } H_3PO_4$$

Active Example 16.9 Normality II

Calculate the equivalent masses of $Ba(OH)_2$, KOH, and $H_3C_6H_5O_7$ for the reactions in Active Example 16.8. Their molar masses are 171.3 g/mol $Ba(OH)_2$, 56.11 g/mol KOH, and 192.12 g/mol $H_3C_6H_5O_7$.

Think Before You Write Equivalent mass is the mass in grams per equivalent. You calculate equivalent mass by dividing molar mass by equivalents per mole.

Answers *Cover the left column with your cut-out shield. Reveal each answer only after you have written your own answer in the right column.*

$Ba(OH)_2$: $\frac{171.3 \text{ g } Ba(OH)_2}{2 \text{ eq } Ba(OH)_2} = 85.65 \text{ g } Ba(OH)_2/\text{eq } Ba(OH)_2$ KOH: $\frac{56.11 \text{ g KOH}}{1 \text{ eq KOH}} = 56.11 \text{ g KOH/eq KOH}$ $H_3C_6H_5O_7$: $\frac{192.12 \text{ g } H_3C_6H_5O_7}{2 \text{ eq } H_3C_6H_5O_7}$ $= 96.060 \text{ g } H_3C_6H_5O_7/\text{eq } H_3C_6H_5O_7$	Determine the three equivalent masses.
You improved your understanding of normality, and you improved your skill at solving solution concentration problems.	What did you learn by solving this Active Example?

Practice Exercise 16.9

Determine the equivalent masses of (a) HI and (b) $H_3C_9H_8NO_3$ for the reactions in Practice Exercise 16.8.

Just as molar mass makes it possible to convert in either direction between grams and moles, equivalent mass sets the path between grams and equivalents. In practice, it is often more convenient to use the fractional form for equivalent mass—the molar mass over the number of equivalents per mole. We use both setups in the next Active Example, but only the fractional setup thereafter. If your instructor emphasizes equivalent mass as a quantity, you should, of course, follow those instructions.

Figure 16.25 Barium hydroxide is used by analytical chemists in the determination of the concentration of acid solutions.

Active Example 16.10 Normality III

Calculate the number of equivalents in 68.5 g of $Ba(OH)_2$ (Figure 16.25).

Think Before You Write Equivalent mass is an equivalency that allows you to convert between the mass of a substance in grams and the number of equivalents.

Answers *Cover the left column with your cut-out shield. Reveal each answer only after you have written your own answer in the right column.*

Using equivalent mass, $68.5\ \text{g Ba(OH)}_2 \times \frac{1\ \text{eq Ba(OH)}_2}{85.65\ \text{g Ba(OH)}_2} = 0.800\ \text{eq Ba(OH)}_2$  Using the fractional setup, $68.5\ \text{g Ba(OH)}_2 \times \frac{2\ \text{eq Ba(OH)}_2}{171.3\ \text{g Ba(OH)}_2} = 0.800\ \text{eq Ba(OH)}_2$	The numbers you need are in Active Example 16.9. Complete the Active Example. 
You improved your understanding of normality, and you improved your skill at solving solution concentration problems.	What did you learn by solving this Active Example?

Practice Exercise 16.10

How many equivalents are in 205 g of HI (Practice Exercise 16.9)?

You are now ready to use the equivalent concept in normality problems.

Active Example 16.11 Normality IV

Calculate the normality of a solution that contains 2.50 grams of NaOH in 5.00×10^2 milliliters of solution.

Think Before You Write The key to any "calculate the normality" problem is to recognize that you need to determine both parts of the definition, the number of equivalents of solute and the volume of solution in liters.

Answers *Cover the left column with your cut-out shield. Reveal each answer only after you have written your own answer in the right column.*

Given: 2.50 g NaOH **Wanted:** eq NaOH 1 eq NaOH = 40.00 g NaOH *One mole of NaOH yields 1 mole of OH^- ions, so there is 1 equivalent of NaOH per mole.*	Take the first step by finding the number of equivalents of NaOH. **Analyze** that part of the problem by writing the given quantity and wanted unit, and **identify** the equivalency needed to solve the problem. 

$$2.50 \text{ g NaOH} \times \frac{1 \text{ eq NaOH}}{40.00 \text{ g NaOH}} = 0.0625 \text{ eq NaOH}$$

$$2.50 \times \frac{1}{40.00} = (2.5 \times 0.25 \times 10^{-1}) = 0.625 \times 10^{-1}$$
$$= 0.0625, \text{ OK.}$$

$$5.00 \times 10^2 \text{ mL} \times \frac{1 \text{ L}}{1000 \text{ mL}} = 0.500 \text{ L}$$

$$(5.00 \times 10^2) \times 10^{-3} = 5 \times 10^{-1} = 0.5, \text{ OK.}$$

Construct the solution setup, and calculate the number of equivalents of NaOH. Also calculate the volume of solution in liters. **Check** both calculations.

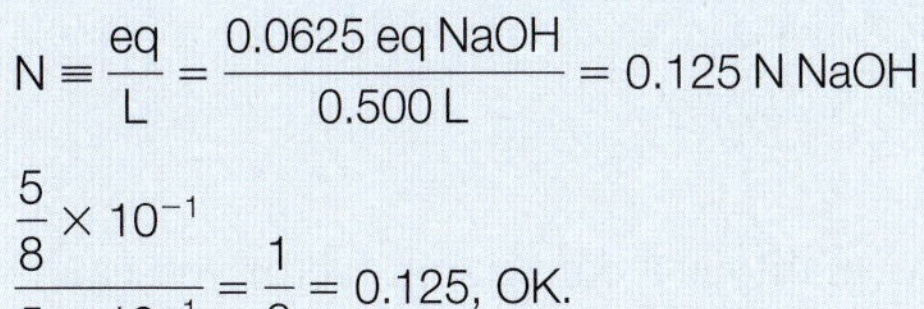

$$N \equiv \frac{\text{eq}}{\text{L}} = \frac{0.0625 \text{ eq NaOH}}{0.500 \text{ L}} = 0.125 \text{ N NaOH}$$

$$\frac{\frac{5}{8} \times 10^{-1}}{5 \times 10^{-1}} = \frac{1}{8} = 0.125, \text{ OK.}$$

You now have both equivalents and volume in liters, the numerator and the denominator in the defining equation for normality. Plug them into the definition and then calculate the answer. **Check** your calculation.

You improved your understanding of normality, and you improved your skill at solving solution concentration problems.

What did you learn by solving this Active Example?

Practice Exercise 16.11

Calculate the normality of a solution that contains 4.58 g of strontium hydroxide (strontium, Z = 38) in 2.50×10^2 mL of solution.

Just as molarity provides a way to convert in either direction between moles of solute and volume of solution, normality offers a unit path between equivalents of solute and volume of solution.

Active Example 16.12 Normality V

How many equivalents are in 18.6 milliliters of 0.856 N H_2SO_4?

Think Before You Write When you have a ratio as a given quantity, write it as an equivalency: 0.856 eq H_2SO_4 = 1 L solution. You can then clearly see the two conversion factors that result from the equivalency.

Answers *Cover the left column with your cut-out shield. Reveal each answer only after you have written your own answer in the right column.*

Given: 18.6 mL solution; 0.856 eq H_2SO_4 = 1 L solution
Wanted: eq H_2SO_4

mL → L → eq

1 L = 1000 mL

0.856 eq H_2SO_4 = 1 L

Analyze the problem statement by listing the given and wanted and then **identify** the equivalencies by first writing the unit path and then deriving the equivalency for each unit conversion. Don't change the equivalencies to conversion factors until the next step.

$$18.6\ \text{mL} \times \frac{1\ \text{L}}{1000\ \text{mL}} \times \frac{0.856\ \text{eq}\ H_2SO_4}{1\ \text{L}} = 0.0159\ \text{eq}\ H_2SO_4$$

$$18.6 \times \frac{1}{1000} \times 0.856 \approx (2 \times 10^1) \times 10^{-3} \times 0.8$$
$$= 1.6 \times 10^{-2} = 0.016,\ \text{OK.}$$

Construct the solution setup, changing the equivalencies directly to conversion factors in the setup, and **check** the value of the answer to be sure it makes sense.

You improved your understanding of normality, and you improved your skill at solving solution concentration problems.

What did you learn by solving this Active Example?

Practice Exercise 16.12

Determine the number of equivalents in 211 mL of 0.50 N hydrochloric acid.

Active Example 16.12 shows that normality is a conversion factor you can use to convert between volume and equivalents:

$$V \times N = \text{L} \times \frac{\text{eq}}{\text{L}} = \text{eq}$$

We will use this fact in Section 16.14.

Knowing how to prepare a solution of specified normality is another valuable skill.

Active Example 16.13 Normality VI

What mass in grams of phosphoric acid is needed to prepare 1.00×10^2 milliliters of 0.350 N H_3PO_4 for use in the reaction $H_3PO_4 + NaOH \rightarrow NaH_2PO_4 + H_2O$?

Think Before You Write Normality is an equivalency that allows you to convert between the given volume and number of equivalents. You need mass. Thus, the question is: How do I convert from equivalents to mass? The answer: equivalent mass.

Answers *Cover the left column with your cut-out shield. Reveal each answer only after you have written your own answer in the right column.*

Given: 1.00×10^2 mL; 0.350 eq H_3PO_4 = 1 L solution
Wanted: g H_3PO_4

$\text{mL} \rightarrow \text{L} \rightarrow \text{eq}\ H_3PO_4 \rightarrow \text{g}\ H_3PO_4$

1 L = 1000 mL

0.350 eq H_3PO_4 = 1 L

97.99 g H_3PO_4 = 1 eq H_3PO_4

Only 1 of the 3 available hydrogens in H_3PO_4 reacts, so there is only 1 eq/mol.

Analyze the problem statement by listing the given and wanted and then **identify** the equivalencies by first writing the unit path and then deriving the equivalency for each unit conversion. Don't change the equivalencies to conversion factors until the next step.

$$1.00 \times 10^2\ \text{mL} \times \frac{1\ \text{L}}{1000\ \text{mL}} \times \frac{0.350\ \text{eq}\ H_3PO_4}{1\ \text{L}} \times \frac{97.99\ \text{g}\ H_3PO_4}{1\ \text{eq}\ H_3PO_4} = 3.43\ \text{g}\ H_3PO_4$$

$$1.00 \times 10^2 \times \frac{1}{1000} \times 0.350 \times 97.99 \approx (1 \times 10^2) \times 10^{-3}$$
$$\times 0.35 \times (1 \times 10^2) = 0.35 \times (10^2 \times 10^{-3} \times 10^2)$$
$$= 0.35 \times 10^1 = 3.5,\ \text{OK.}$$

Construct the solution setup, changing the equivalencies directly to conversion factors in the setup, and **check** the value of the answer to be sure it makes sense.

You improved your understanding of normality, and you improved your skill at solving solution concentration problems.

What did you learn by solving this Active Example?

Practice Exercise 16.13

What mass in grams of phosphoric acid is needed to prepare 2.50×10^2 mL of 0.20 N H_3PO_4 for use in the reaction $H_3PO_4 + 2\ NaOH \rightarrow Na_2HPO_4 + 2\ H_2O$?

16.10 Solution Concentration: A Summary

The best guarantee of success in working with solution concentration is a clear understanding of the units in which it is expressed. You may take these units from concentration ratios:

$$\frac{\text{quantity of solute}}{\text{quantity of solution}} \quad or \quad \frac{\text{quantity of solute}}{\text{quantity of solvent}}$$

Table 16.1 makes these ratios specific for percentage concentration by mass, molarity, molality, and normality.

Thinking About Your Thinking

Proportional Reasoning

All of the solution concentration units introduced in this chapter are direct proportionalities. Percentage concentration by mass is a direct proportionality between mass of solute and mass of solution; molarity, between moles of solute and liters of solution; molality, between moles of solute and kilograms of solvent; and normality, between equivalents of solute and liters of solution. These proportional relationships allow you to think of solution concentration units as conversion factors between the two units in the fraction. Do you know mass of solution and need mass of solute? Use percentage concentration by mass. Do you know volume of solution and need amount in moles of solute? Use molarity. Thinking about solution concentration units in this way allows you to become more skilled at solving quantitative problems.

16.11 Dilution of Concentrated Solutions

Goal 13 Given any three of the following, calculate the fourth: (a) volume of concentrated solution, (b) molarity of concentrated solution, (c) volume of dilute solution, (d) molarity of dilute solution.

Some common acids and bases are available as concentrated solutions that are diluted to a lower concentration for use. To dilute a solution, you simply add

Table 16.1 Summary of Solution Concentrations

Name (Symbol)	Mathematical Form
Percentage by Mass (%)	$\frac{\text{g solute}}{\text{g solute} + \text{g solvent}} \times 100\%$
Molarity (M)	$\frac{\text{mol solute}}{\text{L solution}}$
Molality (m)	$\frac{\text{mol solute}}{\text{kg solvent}}$
Normality (N)	$\frac{\text{eq solute}}{\text{L solution}}$

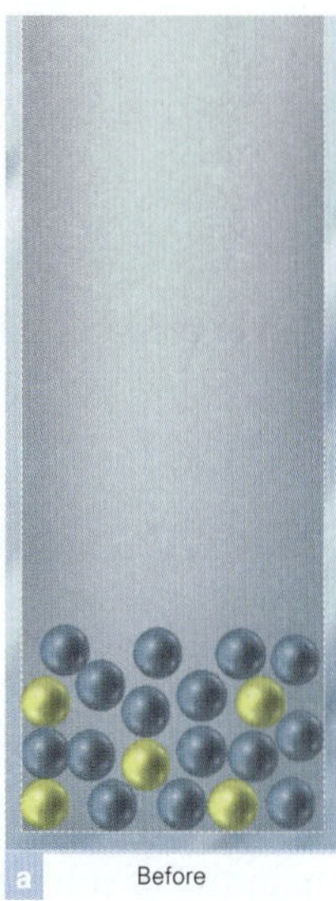

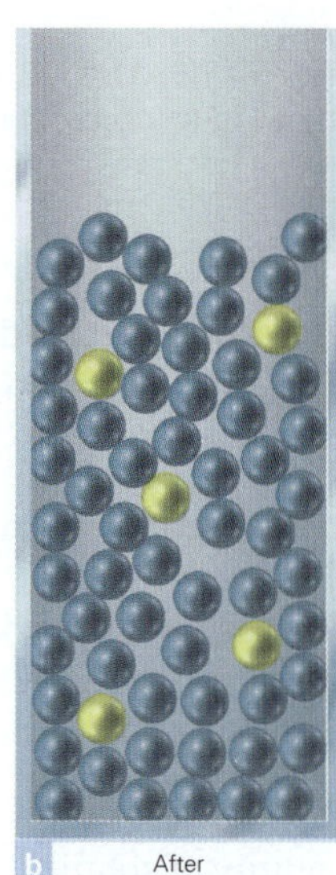

Figure 16.26 Dilution of a concentrated solution. Concentrated solutions are diluted by adding more *solvent* particles. The volume changes because of this addition. The number of *solute* particles, whether counted individually or grouped into moles, remains the same before (a) and after (b) the dilution. Thus, the number of moles of solute particles, n, is a constant: $V_c \times M_c = n = V_d \times M_d$.

more solvent. The amount in moles of solute remains the same, but it is distributed over a larger volume, as illustrated in **Figure 16.26**. The number of moles is V × M (volume × molarity). Using subscript *c* for the *c*oncentrated solution and subscript *d* for the *d*ilute solution, we obtain:

$$V_c \times M_c = n = V_d \times M_d$$

The $V_c \times M_c = V_d \times M_d$ relationship has two important applications. They are illustrated in the next two Active Examples.

Your Thinking

Mental Models

Figure 16.26 is drawn to help you form a particulate-level mental model of what happens when a solution is diluted by adding more solvent. Note, in particular, that the number of solute particles remains the same before and after dilution.

Active Example 16.14 Dilution of Concentrated Solutions I

What volume in milliliters of commercial hydrochloric acid, which is 11.6 molar, should you use to prepare 5.50 L of 0.500 molar hydrochloric acid?

Think Before You Write You are given the volume and concentration of the dilute solution (5.50 L of 0.500 M HCl) and the concentration of the concentrated solution (11.6 M). Thus, you have three of the four variables in $V_c \times M_c = V_d \times M_d$. You use algebra to determine the fourth.

Answers *Cover the left column with your cut-out shield. Reveal each answer only after you have written your own answer in the right column.*

The value of V_c is unknown. $V_c \times M_c = V_d \times M_d \xrightarrow{\text{divide both sides by } M_c} V_c = \frac{V_d \times M_d}{M_c}$	We've already **analyzed** the problem for you, and you have **identified** the algebraic relationship needed. What variable in $V_c \times M_c = V_d \times M_d$ is unknown? Solve the equation for that variable.

$$V_c = \frac{V_d \times M_d}{M_c} = \frac{5.50\,\cancel{L} \times 0.500\,\cancel{M}}{11.6\,\cancel{M}} \times \frac{1000\text{ mL}}{1\,\cancel{L}} = 237\text{ mL}$$

Solution of the equation yields the answer 0.237 L, but the question asks for milliliters. The final conversion changes the answer to the wanted units.

You can now **construct** the solution setup. Substitute and solve. Keep in mind that the question states, "How many milliliters...?"

$$\frac{5.50 \times 0.500 \times 1000}{11.6} \approx \frac{(55 \times 10^{-1}) \times 0.5 \times 10^3}{11}$$
$$= \frac{55}{11} \times 0.5 \times (10^{-1} \times 10^3) = 2.5 \times 10^2 = 250\text{, OK.}$$

Check to be sure that the value of the answer makes sense.

You improved your skill at solving problems involving dilution of concentrated solutions.

What did you learn by solving this Active Example?

Practice Exercise 16.14

Acetic acid is commonly sold commercially in a 17 M concentration. What volume (mL) of the concentrated acid is needed to make 0.800 L of 0.10 M acetic acid?

Figure 16.27 shows the procedure for preparation of a dilute solution.

Active Example 16.15 Dilution of Concentrated Solutions II

A student adds 50.0 mL of water to 25.0 mL of 0.881 M sodium hydroxide solution. What is the concentration of the diluted solution?

Think Before You Write You can recognize this as a solution dilution problem because you are adding a given amount of water to a given volume of a solution of specified concentration.

Answers *Cover the left column with your cut-out shield. Reveal each answer only after you have written your own answer in the right column.*

Given: $V_c = 25.0\text{ mL}$; $M_c = 0.881\text{ M}$;
$V_d = (50.0 + 25.0)\text{ mL} = 75.0\text{ mL}$
Wanted: M_d

$$V_c \times M_c = V_d \times M_d \xrightarrow{\text{divide both sides by } V_d} M_d = \frac{V_c \times M_c}{V_d}$$

The tricky part of this Active Example is the volume of the diluted solution. The problem states that 50.0 mL is added to 25.0 mL, giving the total volume of 75.0 mL.

Analyze the problem by writing the given and wanted, **identify** the algebraic relationship needed, and solve the equation for the wanted variable.

$$M_d = \frac{V_c \times M_c}{V_d} = \frac{25.0\,\cancel{mL} \times 0.881\text{ M}}{75.0\,\cancel{mL}} = 0.294\text{ M}$$

$$\frac{25.0 \times 0.881}{75.0} \approx \frac{25}{75} \times 0.9 = \frac{1}{3} \times 0.9 = 0.3\text{, OK.}$$

Construct the solution setup and **check** it.

You improved your skill at solving problems involving dilution of concentrated solutions.

What did you learn by solving this Active Example?

Practice Exercise 16.15

A 10.0-mL sample of 1.15 M chlorous acid is added to 1.00 L of water. Determine the new concentration of chlorous acid.

1
A 100.0-mL volumetric flask has been filled to the mark with a 0.100 M $K_2Cr_2O_7$ solution.
2
This is transferred to a 1.000-L volumetric flask.
3
All of the initial solution is rinsed out of the 100.0-mL flask and collected in the 1.000-L flask.
4
Distilled water is added to the 1.000-L flask until it is almost full, and then the flask is shaken to thoroughly mix the solution. Additional distilled water is added to bring the total volume to 1.000 L. The concentration of the now-diluted solution is 0.0100 M.
WATER
WATER
WATER
Charles D. Winters

Figure 16.27 Procedure for diluting a concentrated solution.

16.12 Solution Stoichiometry

Goal 14 Given the quantity of any species participating in a chemical reaction for which the equation can be written, find the quantity of any other species, either quantity being measured in (a) mass, (b) volume of solution at specified molarity, or (c) (if gases have been studied) volume of gas at a given temperature and pressure.

For any reaction whose equation can be written, the three steps for solving a stoichiometry problem are:

1. Convert the quantity of given species to amount of particles, grouped in moles.
2. Convert the amount in moles of given species to amount in moles of the wanted species.
3. Convert the amount in moles of wanted species to the quantity units required.

In Chapter 10, the quantities of given and wanted species in Steps 1 and 3 were measured in mass units, typically grams. Later in Chapter 10, we looked at how quantity of energy was related to mass of reactant or product. In Chapter 14, either or both quantities were measured in (a) mass or (b) volume of gas at a known temperature and pressure. In this section, either or both quantities may be measured in (a) mass, (b) volume of a gas at a specified temperature and pressure, or (c) volume of solution of known concentration.

It is critical for you to see these connections as a general relationship in which the quantities of given or wanted substances may be expressed in any combination of mass, energy, volume of solution of known concentration, or volume of gas at known temperature and pressure. This general stoichiometric pattern is illustrated in **Figure 16.28**.

General Stoichiometry Pattern

Measured macroscopic quantity *Reactant* ↔ Particulate quantity *Reactant* ↔ Particulate quantity *Product* ↔ Measured macroscopic quantity *Product*

Specific Applications

Energy & $\Delta_r H$ *Reactant*

Mass in grams & molar mass *Reactant*

PVT *or* gas volume & molar volume *Reactant*

Volume & solution molarity *Reactant*

Amount in moles *Reactant*

Mole ratio from balanced equation

Amount in moles *Product*

N_A

Quantity in number of particles *Reactant*

N_A

Quantity in number of particles *Product*

Energy & $\Delta_r H$ *Product*

Mass in grams & molar mass *Product*

PVT *or* gas volume & molar volume *Product*

Volume & solution molarity *Product*

Figure 16.28 Conversion between mass in grams and molar mass, energy and $\Delta_r H$, gas pressure, volume, and temperature or gas volume and molar volume, or volume and solution molarity, amount in moles, and quantity in number of particles for two species in a chemical change. In Chapter 10, we introduced the general stoichiometry pattern with mass, measured in grams. We then added thermochemical stoichiometry in Chapter 10. We then added gas volume at a known temperature and pressure as another measured macroscopic quantity in Chapter 14. This figure shows all stoichiometric relationships introduced in this textbook. In general, a measured macroscopic quantity is changed to the amount of particles in the sample, grouped in moles, and then amount in moles of one species in the chemical change is converted to amount in moles of another species using the mole ratio from the balanced equation. The resulting particulate quantity can be converted to a measured macroscopic quantity.

Active Example 16.16 Solution Stoichiometry I

Determine the mass in grams of lead(II) iodide that will precipitate (Figure 16.29) when excess potassium iodide solution is added to 50.0 mL of 0.811 M lead(II) nitrate.

Think Before You Write Consider the Specific Applications section of Figure 16.28. This problem starts you in the Volume and Solution Molarity *Reactant* box in the bottom left of the diagram. You want to get to the Mass in Grams and Molar Mass *Product* box on the right side. The figure shows that you need to first convert to amount in moles of reactant and then to amount in moles of product before arriving at your goal, mass (in grams) of product.

Answers *Cover the left column with your cut-out shield. Reveal each answer only after you have written your own answer in the right column.*

$Pb(NO_3)_2(aq) + 2\ KI(aq) \rightarrow PbI_2(s) + 2\ KNO_3(aq)$ **Given:** 50.0 mL solution; $0.811\ mol\ Pb(NO_3)_2 = 1\ L$ solution **Wanted:** g PbI_2 mL $Pb(NO_3)_2 \rightarrow$ L $Pb(NO_3)_2 \rightarrow$ mol $Pb(NO_3)_2 \rightarrow$ mol $PbI_2 \rightarrow$ g PbI_2	**Analyze** the problem. Write the reaction equation so that you can make the amount in moles of reactant to amount in moles of product conversion. Then list the given quantity, given equivalency, wanted units, and write a unit path. Then **identify** the equivalencies indicated by your unit

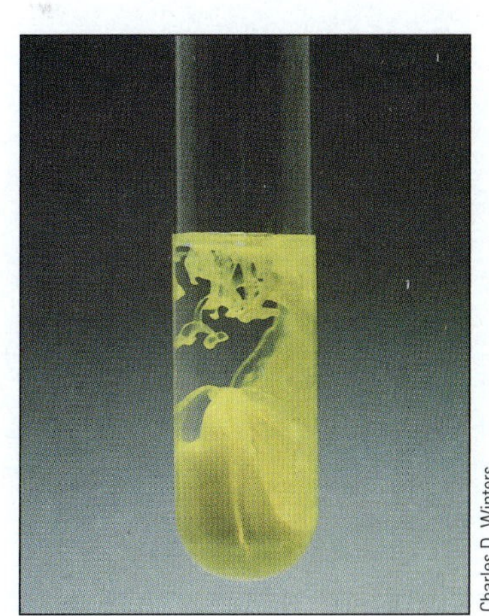
Charles D. Winters

Figure 16.29 Solid lead(II) iodide precipitates.

$1\ L\ Pb(NO_3)_2 = 1000\ mL\ Pb(NO_3)_2$

$0.811\ mol\ Pb(NO_3)_2 = 1\ L\ Pb(NO_3)_2$

$1\ mol\ PbI_2 = 1\ mol\ Pb(NO_3)_2$

$461.0\ g\ PbI_2 = 1\ mol\ PbI_2$

path. Once the volume of the solution is in liters, the defining equation for molarity (M ≡ mol/L) can be used as a conversion factor to find the number of moles. Save the setup until the next step.

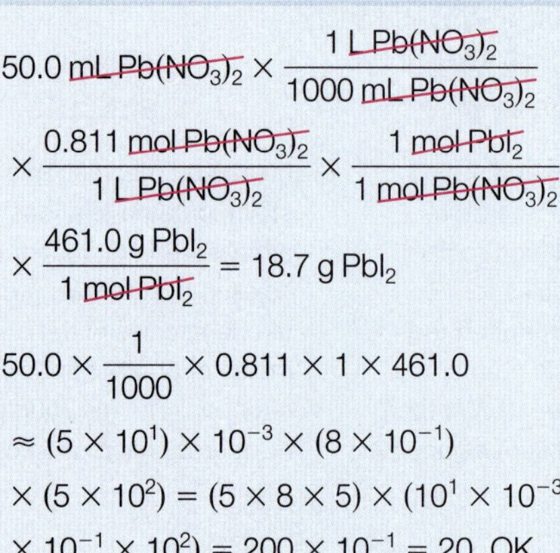

$$50.0\ mL\ Pb(NO_3)_2 \times \frac{1\ L\ Pb(NO_3)_2}{1000\ mL\ Pb(NO_3)_2} \times \frac{0.811\ mol\ Pb(NO_3)_2}{1\ L\ Pb(NO_3)_2} \times \frac{1\ mol\ PbI_2}{1\ mol\ Pb(NO_3)_2} \times \frac{461.0\ g\ PbI_2}{1\ mol\ PbI_2} = 18.7\ g\ PbI_2$$

$$50.0 \times \frac{1}{1000} \times 0.811 \times 1 \times 461.0$$

$$\approx (5 \times 10^1) \times 10^{-3} \times (8 \times 10^{-1}) \times (5 \times 10^2) = (5 \times 8 \times 5) \times (10^1 \times 10^{-3} \times 10^{-1} \times 10^2) = 200 \times 10^{-1} = 20, \text{OK.}$$

Construct the setup and calculate the final answer. **Check** the answer.

You improved your skill at solving stoichiometry problems in general, and solution stoichiometry problems in particular.

What did you learn by solving this Active Example?

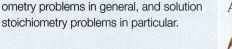

Practice Exercise 16.16

Excess cobalt(II) chloride solution is added to 25.0 mL of 0.34 M sodium hydroxide solution. Determine the mass of cobalt(II) hydroxide that precipitates.

Active Example 16.17 Solution Stoichiometry II

Calculate the volume in milliliters of 0.842 M sodium hydroxide required to precipitate as copper(II) hydroxide (Figure 16.30) all of the copper ions in 30.0 mL of 0.635 M copper(II) sulfate.

Think Before You Write You are given volume and molarity of one reactant, and you are asked for the volume of another reactant for which the molarity is given. Do *not* confuse this with a dilution problem that can be solved with the $V_c \times M_c = V_d \times M_d$ relationship! The key difference in this problem is that a chemical *reaction* occurs. This is a stoichiometry problem that requires you to follow the path shown in Figure 16.28.

Answers *Cover the left column with your cut-out shield. Reveal each answer only after you have written your own answer in the right column.*

$2\ NaOH(aq) + CuSO_4(aq) \rightarrow Cu(OH)_2(s) + Na_2SO_4(aq)$ **Given:** 30.0 mL solution; 0.635 mol $CuSO_4$ = 1000 mL solution **Wanted:** mL solution with a concentration ratio of 0.842 mol NaOH = 1000 mL mL $CuSO_4$ soln → mol $CuSO_4$ → mol NaOH → mL NaOH soln	**Analyze** the problem. Write the reaction equation so that you can make the amount in moles of reactant to amount in moles of product conversion. Then list the given quantity, given equivalency, wanted units, and write a unit path. Then **identify** the equivalencies indicated by your unit path. Think of molarity as mol/1000 mL. Volume of solution is often expressed in milliliters, so this shortcut helps to keep the setup concise.
$30.0\ \text{mL CuSO}_4 \times \frac{0.635\ \text{mol CuSO}_4}{1000\ \text{mL CuSO}_4} \times \frac{2\ \text{mol NaOH}}{1\ \text{mol CuSO}_4} \times \frac{1000\ \text{mL NaOH}}{0.842\ \text{mol NaOH}} = 45.2\ \text{mL NaOH}$ $30.0 \times \frac{0.635}{1000} \times 2 \times \frac{1000}{0.842} = 30 \times 0.635 \times 2 \times \frac{1}{0.842}$ $\approx (3 \times 10^1) \times (6 \times 10^{-1}) \times 2 \times \left(\frac{1}{8 \times 10^{-1}}\right)$ $= \left(3 \times 6 \times 2 \times \frac{1}{8}\right) \times (10^1 \times 10^{-1} \times 10^1)$ $= 4 \times 10^1 = 40$, OK.	**Construct** the setup and calculate the final answer. **Check** the answer.
You improved your skill at solving stoichiometry problems in general, and solution stoichiometry problems in particular.	What did you learn by solving this Active Example?

Practice Exercise 16.17

What volume of 0.105 M lithium nitrate must be added to 1.00×10^2 mL of 0.080 M sodium phosphate to completely precipitate out all of the phosphate ions as lithium phosphate?

Charles D. Winters

Figure 16.30 Copper(II) hydroxide is found in nature as part of the minerals malachite (green) and azurite (blue).

16.13 Titration Using Molarity

Goal 15 Given the volume of a solution that reacts with a known mass of a primary standard and the equation for the reaction, calculate the molarity of the solution.

16 Given the volumes of two solutions that react with each other in a titration, the molarity of one solution, and the equation for the reaction or information from which it can be written, calculate the molarity of the second solution.

One common laboratory operation in analytical chemistry is called **titration** (see Figure 16.31). Titration is the very careful addition of one solution to another by means of a device that can measure delivered volume precisely, such as a **buret**. The buret accurately measures the volume of solution required to react with a carefully measured amount of another dissolved substance. When that precise volume has been reached, an **indicator** dye changes color, and the flow from the buret is stopped. Phenolphthalein is a typical indicator for acid–base titrations. It is colorless in an acid solution and pink in a basic solution, as shown in Figure 16.31.

Titration can be used to **standardize** a solution, which means finding its concentration for use in later titrations. Sodium hydroxide cannot be weighed accurately because it absorbs moisture from the air and increases in weight during the weighing process. Therefore, it is not possible to prepare a sodium hydroxide solution by weighing whose molarity is known precisely. Instead, the solution is standardized against a weighed quantity of something that can be weighed accurately. Such a substance is called a **primary standard**. Oxalic acid dihydrate, $H_2C_2O_4 \cdot 2\ H_2O$, is an example. When used to standardize sodium hydroxide, the equation is $H_2C_2O_4(aq) + 2\ NaOH(aq) \rightarrow Na_2C_2O_4(aq) + 2\ H_2O(\ell)$. Notice that the water in the hydrate is not a part of the equation. When 1 mole of $H_2C_2O_4 \cdot 2\ H_2O$ dissolves, the hydrate water becomes a part of the solution and 1 mole of $H_2C_2O_4$ is available for reaction. The hydrate water must be taken into account in weighing the $H_2C_2O_4 \cdot 2\ H_2O$, however.

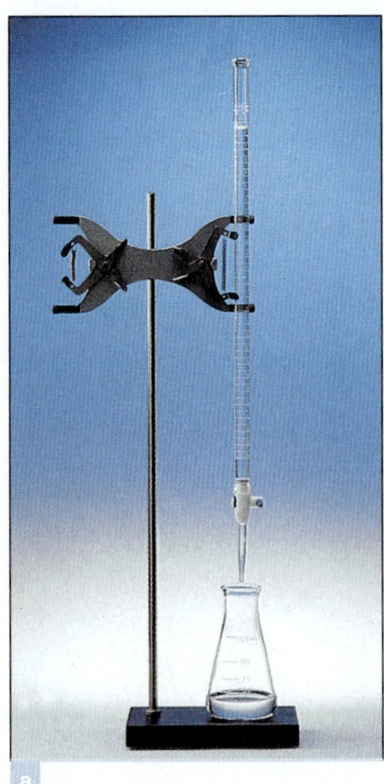

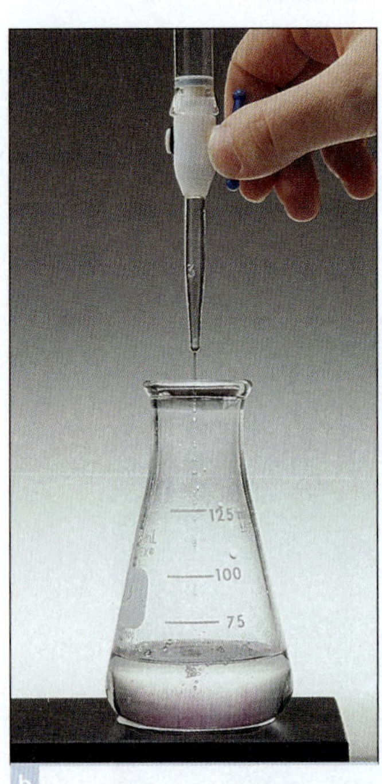

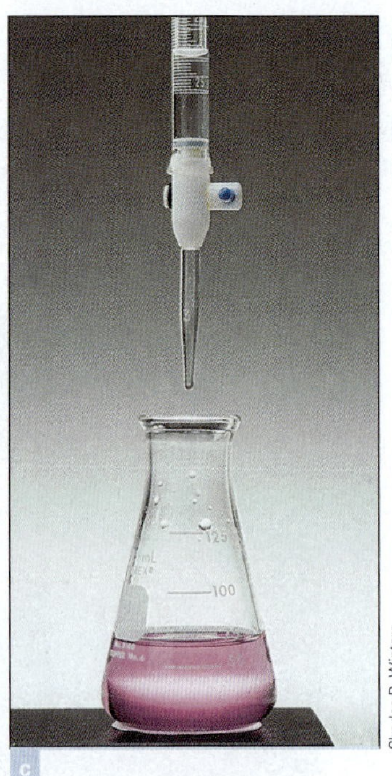

Figure 16.31 Titrating from a buret into a flask. (a) The flask contains an acidic solution of unknown concentration. A small quantity of dye is added to the acid. The dye is colorless in acidic solution. The buret is filled with a basic solution of known concentration. (b) By careful control of the valve, the chemist may deliver liquid from the buret to the flask in a steady stream, drop by drop, or in a single drop. (c) When the volume of base delivered by the buret reacts completely with the acid in the flask, the indicator dye changes color, signaling the chemist to stop the flow of base solution and read the delivered volume.

Active Example 16.18 Titration Using Molarity: Standardization of a Solution

A chemist dissolves 1.18 g $H_2C_2O_4 \cdot 2\,H_2O$ (126.07 g/mol) in water and titrates the solution with a solution of NaOH of unknown concentration. She determines that 28.3 mL NaOH(aq) is required to neutralize the acid. Calculate the molarity of the NaOH solution.

Think Before You Write A titration problem is a solution stoichiometry problem. The general plan for any titration problem is therefore outlined by Figure 16.28. You have the mass of the given species, which must be converted to amount in moles. The amount in moles of the given species is then converted to amount in moles of the wanted species.

Answers *Cover the left column with your cut-out shield. Reveal each answer only after you have written your own answer in the right column.*

$H_2C_2O_4(aq) + 2\,NaOH(aq) \rightarrow Na_2C_2O_4(aq) + 2\,H_2O(\ell)$

Given: 1.18 g $H_2C_2O_4 \cdot 2\,H_2O$ (126.07 g/mol)

Wanted: mol NaOH

g $H_2C_2O_4 \cdot 2\,H_2O \rightarrow$ mol $H_2C_2O_4 \rightarrow$ mol NaOH

1 mol $H_2C_2O_4$ = 126.07 g $H_2C_2O_4$

2 mol NaOH = 1 mol $H_2C_2O_4$

In this step, determine the amount of NaOH in moles. **Analyze** the problem. Write the reaction equation so that you can make the amount in moles of reactant to amount in moles of product conversion. Then list the given quantity, wanted units, and write a unit path. Then **identify** the equivalencies indicated by your unit path.

$$1.18\ \text{g}\ H_2C_2O_4 \cdot 2\,H_2O \times \frac{1\ \text{mol}\ H_2C_2O_4 \cdot 2\,H_2O}{126.07\ \text{g}\ H_2C_2O_4 \cdot 2\,H_2O} \times \frac{2\ \text{mol NaOH}}{1\ \text{mol}\ H_2C_2O_4} = 0.0187\ \text{mol NaOH}$$

$$1.18 \times \frac{1}{126.07} \times 2 \approx 1.25 \times \frac{1}{1.25 \times 10^2} \times 2 = 2 \times 10^{-2}$$

= 0.02, OK.

Construct the setup and calculate the final answer. **Check** the answer.

$$M \equiv \frac{\text{mol}}{\text{L}} = \frac{0.0187\ \text{mol NaOH}}{28.3\ \text{mL}} \times \frac{1000\ \text{mL}}{1\ \text{L}} = 0.661\ \text{M NaOH}$$

Molarity is moles per liter. Volume, given in milliliters, must be changed to liters.

You now have the numerator for the fraction in the equation that defines molarity, and you almost have the denominator. Complete the problem.

You improved your skill at solving stoichiometry problems in general and solution stoichiometry and titration problems in particular.

What did you learn by solving this Active Example?

Practice Exercise 16.18

A student dissolves 2.00 g of sodium carbonate in water and titrates the solution with a hydrochloric acid solution whose concentration is to be determined. The indicator changes color when 19.30 mL of acid have been titrated. What is the molar concentration of hydrochloric acid?

Once a solution is standardized, we may use it to find the concentration of another solution. This procedure is widely used in industrial laboratories.

Active Example 16.19 Titration Using Molarity: Using a Standardized Solution to Determine Molar Concentration

A 25.0-mL sample of an electroplating solution is analyzed for its sulfuric acid concentration. It takes 46.8 mL of the 0.661 M NaOH from Active Example 16.18 to neutralize the sample. Find the molarity of the acid.

Think Before You Write The problem statement gives you the volume and concentration of NaOH, so you can determine the amount in moles of NaOH. A balanced chemical equation gives you the information needed to find amount in moles of sulfuric acid. The definition of molarity completes the solution strategy.

Answers *Cover the left column with your cut-out shield. Reveal each answer only after you have written your own answer in the right column.*

$2\ NaOH(aq) + H_2SO_4(aq) \rightarrow Na_2SO_4(aq) + 2\ H_2O(\ell)$

Given: 46.8 mL solution; 0.661 mol NaOH = 1000 mL solution

Wanted: mol H_2SO_4

mL soln → mol NaOH → mol H_2SO_4

0.661 mol NaOH = 1000 mL solution

1 mol H_2SO_4 = 2 mol NaOH

Start by determining the amount of H_2SO_4 in moles. **Analyze** the problem. Write the reaction equation so that you can make the amount in moles of reactant to amount in moles of product conversion. Then list the given quantity, given equivalency, wanted units, and write a unit path. Then **identify** the equivalencies indicated by your unit path.

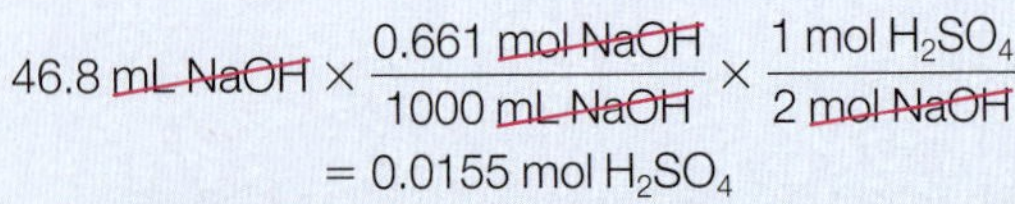

$$46.8\ \text{mL NaOH} \times \frac{0.661\ \text{mol NaOH}}{1000\ \text{mL NaOH}} \times \frac{1\ \text{mol } H_2SO_4}{2\ \text{mol NaOH}} = 0.0155\ \text{mol } H_2SO_4$$

$$46.8 \times \frac{0.661}{1000} \times \frac{1}{2} \approx (5 \times 10^1) \times (0.6 \times 10^{-3}) \times (5 \times 10^{-1})$$
$$= (5 \times 0.6 \times 5) \times (10^1 \times 10^{-3} \times 10^{-1}) = 15 \times 10^{-3}$$
$$= 0.015,\ \text{OK.}$$

Construct the setup and calculate the final answer. **Check** the answer.

$$M \equiv \frac{\text{mol}}{\text{L}} = \frac{0.0155\ \text{mol } H_2SO_4}{25.0\ \text{mL}} \times \frac{1000\ \text{mL}}{1\ \text{L}} = 0.620\ \text{M } H_2SO_4$$

There are three significant figures in both the numerator and the denominator of the equation, so there are three significant figures in the result. Even though your calculator does not display the zero in "0.620 M H_2SO_4," you must include it in the answer.

You now have the numerator for the fraction in the equation that defines molarity, and you almost have the denominator. Complete the problem.

You improved your skill at solving stoichiometry problems in general and solution stoichiometry and titration problems in particular.

What did you learn by solving this Active Example?

Practice Exercise 16.19

A 20.00-mL sample of a barium hydroxide solution is titrated to neutralization with 12.50 mL of the 1.95 M HCl solution from Practice Exercise 16.18. What is the molar concentration of the barium hydroxide solution?

Active Example 16.18 Titration Using Molarity: Standardization of a Solution

A chemist dissolves 1.18 g $H_2C_2O_4 \cdot 2\,H_2O$ (126.07 g/mol) in water and titrates the solution with a solution of NaOH of unknown concentration. She determines that 28.3 mL NaOH(aq) is required to neutralize the acid. Calculate the molarity of the NaOH solution.

Think Before You Write A titration problem is a solution stoichiometry problem. The general plan for any titration problem is therefore outlined by Figure 16.28. You have the mass of the given species, which must be converted to amount in moles. The amount in moles of the given species is then converted to amount in moles of the wanted species.

Answers *Cover the left column with your cut-out shield. Reveal each answer only after you have written your own answer in the right column.*

$H_2C_2O_4(aq) + 2\,NaOH(aq) \rightarrow Na_2C_2O_4(aq) + 2\,H_2O(\ell)$

Given: 1.18 g $H_2C_2O_4 \cdot 2\,H_2O$ (126.07 g/mol)

Wanted: mol NaOH

g $H_2C_2O_4 \cdot 2\,H_2O \rightarrow$ mol $H_2C_2O_4 \rightarrow$ mol NaOH

1 mol $H_2C_2O_4$ = 126.07 g $H_2C_2O_4$

2 mol NaOH = 1 mol $H_2C_2O_4$

In this step, determine the amount of NaOH in moles. **Analyze** the problem. Write the reaction equation so that you can make the amount in moles of reactant to amount in moles of product conversion. Then list the given quantity, wanted units, and write a unit path. Then **identify** the equivalencies indicated by your unit path.

$$1.18\ \text{g H}_2\text{C}_2\text{O}_4 \cdot 2\,\text{H}_2\text{O} \times \frac{1\ \text{mol H}_2\text{C}_2\text{O}_4 \cdot 2\,\text{H}_2\text{O}}{126.07\ \text{g H}_2\text{C}_2\text{O}_4 \cdot 2\,\text{H}_2\text{O}} \times \frac{2\ \text{mol NaOH}}{1\ \text{mol H}_2\text{C}_2\text{O}_4} = 0.0187\ \text{mol NaOH}$$

$$1.18 \times \frac{1}{126.07} \times 2 \approx 1.25 \times \frac{1}{1.25 \times 10^2} \times 2 = 2 \times 10^{-2}$$

= 0.02, OK.

Construct the setup and calculate the final answer. **Check** the answer.

$$M \equiv \frac{\text{mol}}{\text{L}} = \frac{0.0187\ \text{mol NaOH}}{28.3\ \text{mL}} \times \frac{1000\ \text{mL}}{1\ \text{L}} = 0.661\ \text{M NaOH}$$

Molarity is moles per liter. Volume, given in milliliters, must be changed to liters.

You now have the numerator for the fraction in the equation that defines molarity, and you almost have the denominator. Complete the problem.

You improved your skill at solving stoichiometry problems in general and solution stoichiometry and titration problems in particular.

What did you learn by solving this Active Example?

Practice Exercise 16.18

A student dissolves 2.00 g of sodium carbonate in water and titrates the solution with a hydrochloric acid solution whose concentration is to be determined. The indicator changes color when 19.30 mL of acid have been titrated. What is the molar concentration of hydrochloric acid?

Once a solution is standardized, we may use it to find the concentration of another solution. This procedure is widely used in industrial laboratories.

Active Example 16.19 Titration Using Molarity: Using a Standardized Solution to Determine Molar Concentration

A 25.0-mL sample of an electroplating solution is analyzed for its sulfuric acid concentration. It takes 46.8 mL of the 0.661 M NaOH from Active Example 16.18 to neutralize the sample. Find the molarity of the acid.

Think Before You Write The problem statement gives you the volume and concentration of NaOH, so you can determine the amount in moles of NaOH. A balanced chemical equation gives you the information needed to find amount in moles of sulfuric acid. The definition of molarity completes the solution strategy.

Answers *Cover the left column with your cut-out shield. Reveal each answer only after you have written your own answer in the right column.*

$2\ NaOH(aq) + H_2SO_4(aq) \rightarrow Na_2SO_4(aq) + 2\ H_2O(\ell)$

Given: 46.8 mL solution; 0.661 mol NaOH = 1000 mL solution

Wanted: mol H_2SO_4

mL soln $\rightarrow$ mol NaOH $\rightarrow$ mol H_2SO_4

0.661 mol NaOH = 1000 mL solution

1 mol H_2SO_4 = 2 mol NaOH

Start by determining the amount of H_2SO_4 in moles. **Analyze** the problem. Write the reaction equation so that you can make the amount in moles of reactant to amount in moles of product conversion. Then list the given quantity, given equivalency, wanted units, and write a unit path. Then **identify** the equivalencies indicated by your unit path.

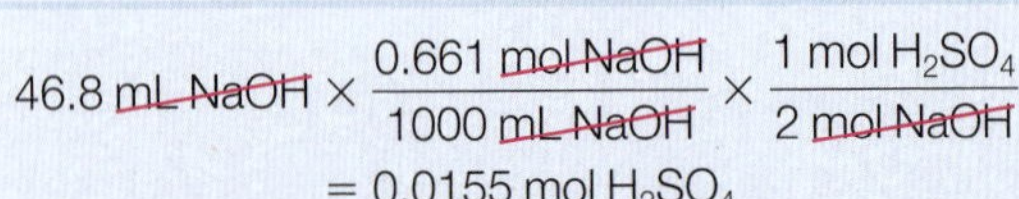

$$46.8\ \text{mL NaOH} \times \frac{0.661\ \text{mol NaOH}}{1000\ \text{mL NaOH}} \times \frac{1\ \text{mol } H_2SO_4}{2\ \text{mol NaOH}} = 0.0155\ \text{mol } H_2SO_4$$

$$46.8 \times \frac{0.661}{1000} \times \frac{1}{2} \approx (5 \times 10^1) \times (0.6 \times 10^{-3}) \times (5 \times 10^{-1})$$
$$= (5 \times 0.6 \times 5) \times (10^1 \times 10^{-3} \times 10^{-1}) = 15 \times 10^{-3}$$
$$= 0.015, \text{ OK.}$$

Construct the setup and calculate the final answer. **Check** the answer.

$$M \equiv \frac{\text{mol}}{\text{L}} = \frac{0.0155\ \text{mol } H_2SO_4}{25.0\ \text{mL}} \times \frac{1000\ \text{mL}}{1\ \text{L}} = 0.620\ M\ H_2SO_4$$

There are three significant figures in both the numerator and the denominator of the equation, so there are three significant figures in the result. Even though your calculator does not display the zero in "0.620 M H_2SO_4," you must include it in the answer.

You now have the numerator for the fraction in the equation that defines molarity, and you almost have the denominator. Complete the problem.

You improved your skill at solving stoichiometry problems in general and solution stoichiometry and titration problems in particular.

What did you learn by solving this Active Example?

Practice Exercise 16.19

A 20.00-mL sample of a barium hydroxide solution is titrated to neutralization with 12.50 mL of the 1.95 M HCl solution from Practice Exercise 16.18. What is the molar concentration of the barium hydroxide solution?

16.14 Titration Using Normality (Optional)

Goal 17 Given the volume of a solution that reacts with a known mass of a primary standard and the equation for the reaction, calculate the normality of the solution.

18 Given the volumes of two solutions that react with each other in a titration and the normality of one solution, calculate the normality of the second solution.

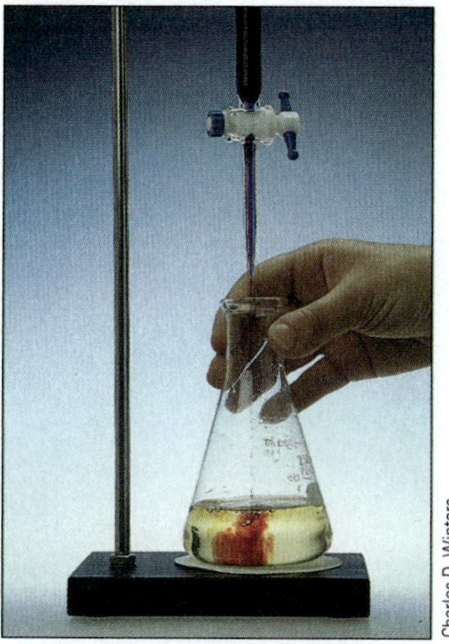

Figure 16.32 A purple solution containing permanganate ion, MnO_4^-, is titrated into an acidified solution that contains iron(II) ion. One product of the reaction is iron(III) ion, which has a pale yellow color in aqueous solution.

We noted earlier that normality is a convenient concentration unit in analytical work. This is because of the fact pointed out in Section 16.9: *The number of equivalents of all species in a reaction is the same.* Consequently, for an acid–base reaction,

$$\text{equivalents of acid} = \text{equivalents of base}$$

There are two ways to calculate the number of equivalents (eq) in a sample of a substance. If you know the mass of the substance and its equivalent mass, use equivalent mass as a conversion factor to get equivalents, as in Active Example 16.10. If the sample is a solution and you know its volume and normality, multiply one by the other: $V \times N = \text{eq}$.

We illustrate normality calculations by repeating Active Examples 16.18 and 16.19, which were solved with molarity. In the first Active Example in this section, a primary standard is used to standardize a solution. In the second Active Example, illustrating normality calculations, the standardized solution is used to find the concentration of another solution. Figure 16.32 shows a standardized solution of potassium permanganate (purple) being used in an analysis of a solution containing iron(II) ion.

Active Example 16.20 Titration Using Normality: Standardization of a Solution

A chemist dissolves 1.18 g $H_2C_2O_4 \cdot 2\,H_2O$ (126.07 g/mol) in water and titrates the solution with a solution of NaOH of unknown concentration. She determines that 28.3 mL NaOH(aq) is required to neutralize the acid. Calculate the normality of the NaOH solution for the complete neutralization: $H_2C_2O_4(aq) + 2\,NaOH \rightarrow Na_2C_2O_4(aq) + 2\,H_2O(\ell)$.

Think Before You Write Ordinarily, you include the molar mass of a substance as an equivalency when analyzing the problem. When working in normality, however, you use equivalent mass, g/eq, which can be expressed as a ratio of molar mass to equivalents per mole.

Answers *Cover the left column with your cut-out shield. Reveal each answer only after you have written your own answer in the right column.*

Given: 1.18 g $H_2C_2O_4 \cdot 2\,H_2O$ (126.07 g/mol)
Wanted: eq NaOH

$g\ H_2C_2O_4 \cdot 2\,H_2O \rightarrow eq\ H_2C_2O_4 \cdot 2\,H_2O$

$2\ eq\ H_2C_2O_4 = 126.07\ g\ H_2C_2O_4$

The equation shows that both ionizable hydrogens from $H_2C_2O_4$ are used, so there are 2 equivalents per mole of acid.

In this step, determine the equivalents of $H_2C_2O_4$. **Analyze** the problem. List the given quantity, wanted units, and write a unit path. Then **identify** the equivalency indicated by your unit path.

$1.18\ \text{g}\ H_2C_2O_4 \cdot 2H_2O \times \frac{2\ \text{eq NaOH or } H_2C_2O_4}{126.07\ \text{g}\ H_2C_2O_4 \cdot 2H_2O} = 0.0187\ \text{eq NaOH}$ $1.18 \times \frac{1}{126.07} \times 2 \approx 1.25 \times \frac{1}{1.25 \times 10^2} \times 2 = 2 \times 10^{-2} = 0.02$, OK. *Notice the "or" in the numerator. The number of equivalents is the same for all species in a reaction.*	To use the defining equation for normality, $N \equiv eq/L$, you must know, or be able to find, both the numerator and denominator quantities. The denominator results from a mL → L conversion from one of the givens. (We'll save this for the next step.) As for the numerator, you have already analyzed the setup needed to find equivalents of $H_2C_2O_4$. And what is that equal to? It equals the number of equivalents of NaOH. The reaction has the same number of equivalents of all species. If you can find one, you've found them all. Using only what you need from the foregoing information, **construct** the setup and calculate the number of equivalents of NaOH. **Check** the value of your answer.
$N \equiv \frac{eq}{L} = \frac{0.0187\ \text{eq NaOH}}{28.3\ \text{mL}} \times \frac{1000\ \text{mL}}{1\ \text{L}} = 0.661\ \text{N NaOH}$ *Normality is equivalents per liter. Volume, given in milliliters, must be changed to liters.*	You now have the numerator for the fraction in the equation that defines normality, and you almost have the denominator. Complete the problem.
You improved your skill at solving stoichiometry problems in general and solution stoichiometry and titration problems in particular.	What did you learn by solving this Active Example?

Practice Exercise 16.20

A student dissolves 2.00 g of sodium carbonate in water and titrates the solution with a hydrochloric acid solution whose concentration is to be determined. The indicator changes color when 19.30 mL of acid have been titrated. What is the normality of hydrochloric acid for the reaction $Na_2CO_3(aq) + 2\ HCl(aq) \rightarrow 2\ NaCl(aq) + H_2O(\ell) + CO_2(g)$?

Once you know the normality of one solution, you can use it to find the normality of another solution. The number of equivalents of a species in a reaction is the product of solution volume times normality. The number of equivalents of all species in a reaction is the same, so

$$V_1N_1 = eq = V_2N_2 \quad or \quad V_1N_1 = V_2N_2$$

in which subscripts 1 and 2 identify the reacting solutions. Solving for the second normality, we obtain

$V_1N_1 = V_2N_2$ is the key equation that make normality so useful in a laboratory that runs the same titrations again and again.

$$N_2 = \frac{V_1N_1}{V_2}$$

That's all it takes to calculate normality in this follow-up to Active Example 16.19.

Active Example 16.21 Titration Using Normality: Using a Standardized Solution to Determine Normality

A 25.0-mL sample of an electroplating solution is analyzed for its sulfuric acid concentration. It takes 46.8 mL of the 0.661 N NaOH from Active Example 16.20 to neutralize the sample. Find the normality of the acid.

Think Before You Write The problem statement gives you the volume and normality of one species in an acid–base reaction and the volume of another species for which you are asked to determine the normality. You need to solve $V_1N_1 = V_2N_2$ for the unknown quantity.

Answers *Cover the left column with your cut-out shield. Reveal each answer only after you have written your own answer in the right column.*

$$N_2 = \frac{V_1 N_1}{V_2} = \frac{46.8 \not{mL} \times 0.661\ N}{25.0 \not{mL}} = 1.24\ N\ H_2SO_4$$

$$\frac{46.8 \times 0.661}{25.0} \approx \frac{50 \times 0.6}{25} = 2 \times 0.6 = 1.2, \text{OK.}$$

Solve $V_1N_1 = V_2N_2$ for N_2, substitute, and determine the answer. **Check** your calculation.

You improved your skill at solving stoichiometry problems in general, and solution stoichiometry and titration problems in particular.

What did you learn by solving this Active Example?

Practice Exercise 16.21

A 20.00-mL sample of a barium hydroxide solution is titrated to neutralization with 12.50 mL of the 1.95 N HCl solution from Practice Exercise 16.20. What is the normality of the barium hydroxide solution?

The normality in Active Example 16.21, 1.24 N, is twice the molarity in Active Example 16.19, 0.620 M. This is always the case when there are 2 eq/mol. For an X molar solution,

$$\frac{X\ \text{mol}}{L} \times \frac{2\ \text{eq}}{\text{mol}} = 2X\ \text{eq/L}$$

When there is 1 equivalent per mole, the molarity and normality of the solution are the same:

$$\frac{X\ \text{mol}}{L} \times \frac{1\ \text{eq}}{\text{mol}} = X\ \text{eq/L}$$

16.15 Colligative Properties of Solutions (Optional)

Goal 19 Given (a) the molality of a solution, or data from which it may be found; (b) the normal freezing or boiling point of the solvent; and (c) the freezing or boiling point constant, find the freezing or boiling point of the solution.

20 Given the freezing point depression or boiling point elevation and the molality of a solution, or data from which they may be found, calculate the molal freezing point constant or molal boiling point constant.

21 Given (a) the mass of solute and solvent in a solution; (b) the freezing point depression or boiling point elevation, or data from which they may be found; and (c) the molal freezing/boiling point constant of the solvent, find the approximate molar mass of the solute.

A pure solvent has distinct physical properties, as does any pure substance. Introducing a solute into the solvent affects these properties. The properties of the solution depend on the relative amounts of solvent and solute. It has been found

Figure 16.33 Spreading salt on an icy street. One reason for putting salt on icy streets in winter is that some dissolves in whatever liquid is present. This lowers the freezing temperature and melts at least some of the ice or turns it to slush.

experimentally that, in *dilute* solutions of certain solutes, the *change* in some of these properties is proportional to the molal concentration of the solute particles. These are called **colligative properties**: solution properties that are determined only by the *number* of solute particles dissolved in a fixed quantity of solvent and not by the identity of the solute particles.

Freezing and boiling points of solutions are colligative properties. Perhaps the best known example of such a solution is the antifreeze used in automobile cooling systems. The solute that is dissolved in the radiator water reduces the freezing temperature well below the normal freezing point of pure water. It also raises the boiling point above the normal boiling point. Another example is found in cold climates where salt and other compounds are placed on sidewalks and roads to lower the freezing point by several degrees and help keep them safe to travel on (Figure 16.33).

The change in a freezing point is the **freezing point depression, ΔT_f** (Figure 16.34), and the change in a boiling point is the **boiling point elevation, ΔT_b**. The two proportionalities and their corresponding equations are as follows:

$$\Delta T_f \propto m \xrightarrow{\text{the proportionality changes to an equality}} \Delta T_f = K_f \times m$$

$$\Delta T_b \propto m \xrightarrow{\text{the proportionality changes to an equality}} \Delta T_b = K_b \times m$$

The proportionality constants, $\mathbf{K_f}$ and $\mathbf{K_b}$, are, respectively, the **molal freezing point depression constant** and the **molal boiling point elevation constant**. The freezing and boiling point constants are properties of the *solvent*, no matter what the solute may be. The freezing point constant for water is 1.86°C/m, and the boiling point constant is 0.51°C/m.

Chemists often take some liberties in units and algebraic signs when solving freezing and boiling point problems. Technically, °C · kg solvent/mole solute are the units for K_f or K_b. These units are usable, but they are awkward. The substitute, °C/m, is acceptable if we keep calculations based on the equations $\Delta T_f = K_f \times m$ and $\Delta T_b = K_b \times m$ separate from other calculations. We will follow that practice.

Again, technically, if the freezing point of the solvent is taken as the "initial" temperature and the always lower freezing point of the solution as the "final" temperature, ΔT_f and K_f must be negative quantities. In some texts, they are regarded as such. Most chemists, however, use the words depression and elevation to identify clearly the direction the temperature is changing and then treat both constants and the magnitude of temperature change as positive numbers. We follow this practice, too.

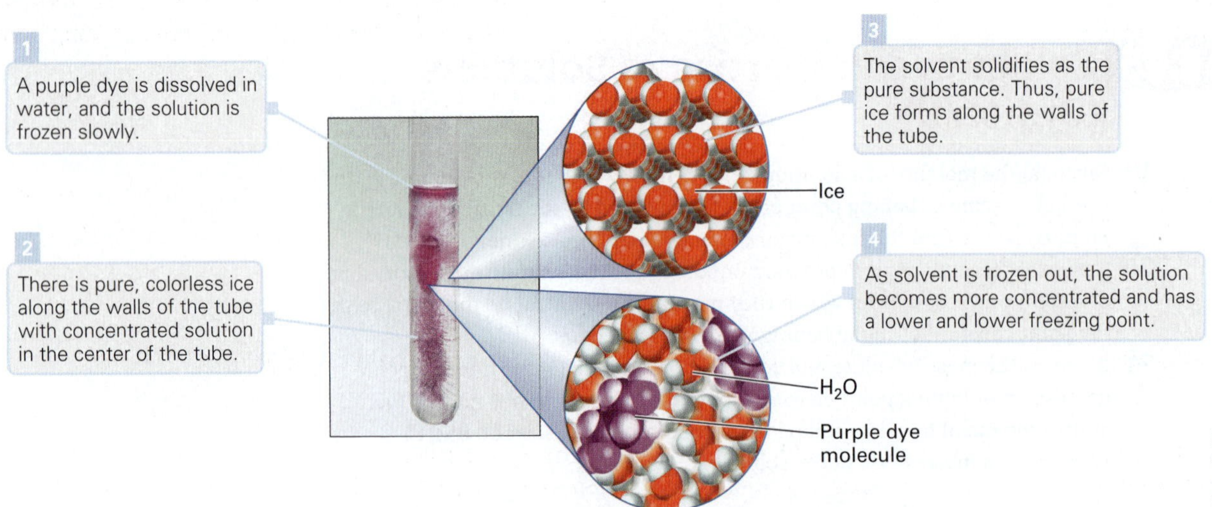

1
A purple dye is dissolved in water, and the solution is frozen slowly.
2
There is pure, colorless ice along the walls of the tube with concentrated solution in the center of the tube.
3
The solvent solidifies as the pure substance. Thus, pure ice forms along the walls of the tube.
Ice
4
As solvent is frozen out, the solution becomes more concentrated and has a lower and lower freezing point.
H2O
Purple dye molecule
Charles D. Winters

Figure 16.34 Freezing point depression. The change in freezing point of a solution is proportional to its molal concentration.

Active Example 16.22 Determination of Freezing Point from Molality Data

Determine the freezing point of a solution of 12.0 g urea, $CO(NH_2)_2$, in 2.50×10^2 g of water.

Think Before You Write You are given mass of a solute, urea, and mass of solvent, water. These are data from which molality can be calculated. You are asked to determine the freezing point. To use the equation $\Delta T_f = K_f \times m$, you need to change the given data to molality. The defining equation for molality is m ≡ mol solute/kg solvent (Section 16.8).

Answers *Cover the left column with your cut-out shield. Reveal each answer only after you have written your own answer in the right column.*

Given: 12.0 g $CO(NH_2)_2$
Wanted: mol $CO(NH_2)_2$

g $CO(NH_2)_2$ → mol $CO(NH_2)_2$

1 mol $CO(NH_2)_2$ = 60.06 g $CO(NH_2)_2$

Start by calculating the moles of solute, which is the numerator in the definition of molality. **Analyze** the problem by writing the given quantity and wanted unit, and write a unit path. Then **identify** the equivalency indicated by your unit path.

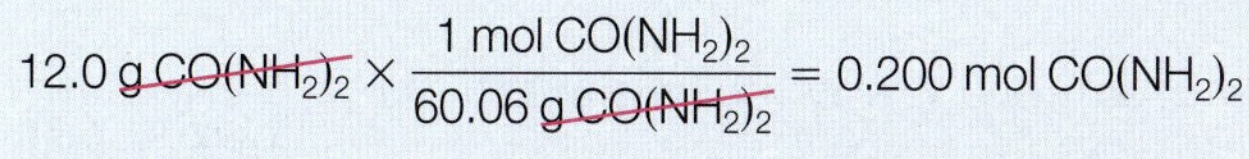

$$12.0\ \cancel{g\ CO(NH_2)_2} \times \frac{1\ mol\ CO(NH_2)_2}{60.06\ \cancel{g\ CO(NH_2)_2}} = 0.200\ mol\ CO(NH_2)_2$$

$$12.0 \times \frac{1}{60.06} \approx 12 \times \frac{1}{12 \times 5} = \frac{1}{5} = 0.2\text{, OK.}$$

Construct the setup and calculate the final answer. **Check** the answer.

$$m \equiv \frac{mol\ solute}{kg\ solvent} = \frac{0.200\ mol\ CO(NH_2)_2}{2.50 \times 10^2\ \cancel{g\ H_2O}} \times \frac{1000\ \cancel{g\ H_2O}}{1\ kg\ H_2O}$$
$$= 0.800\ m\ CO(NH_2)_2$$

Just as you changed milliliters of solution to liters to satisfy the defining equation for molarity, you must change grams of solvent into kilograms to satisfy the defining equation for molality.

Now calculate the molality of the solution by using its defining equation. Watch the units carefully.

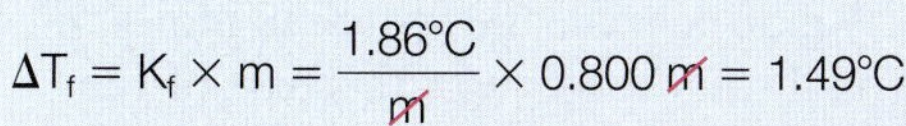

$$\Delta T_f = K_f \times m = \frac{1.86°C}{\cancel{m}} \times 0.800\ \cancel{m} = 1.49°C$$

Find the freezing point depression by substitution into the equation $\Delta T_f = K_f \times m$. K_f for water is 1.86°C/m.

0°C − 1.49°C = −1.49°C

The freezing point depression is 1.49°C. The normal freezing point of water is 0°C. State the freezing point of the solution.

You improved your understanding of colligative properties of solutions, and you improved your skill at solving freezing point depression and boiling point elevation problems.

What did you learn by solving this Active Example?

Practice Exercise 16.22

Calculate the boiling point of a solution made by dissolving 17.5 g of analine, $C_6H_5NH_2$, in 113 g of water.

Freezing point depression and boiling point elevation can be used to find the approximate molar mass of an unknown solute. The solution is prepared with measured masses of the solute and a solvent whose freezing or boiling point constant is known. The freezing point depression or boiling point elevation is found by experiment. The calculation procedure is as follows.

how to... Calculate the Molar Mass of a Solute from Freezing Point Depression or Boiling Point Elevation Data

Step 1: Calculate molality from $m = \Delta T_f/K_f$ or $m = \Delta T_b/K_b$. Express as mol solute/kg solvent.

Step 2: Using molality as a conversion factor between amount in moles of solute and mass in kilograms of solvent, find the amount in moles of solute.

Step 3: Use the defining equation for molar mass, $MM \equiv g/mol$, to calculate the molar mass of the solute.

Active Example 16.23 Determination of Molar Mass from Boiling Point Elevation

The molal boiling point elevation constant of benzene is 2.5°C/m. A solution of 15.2 g of unknown solute in 91.1 g benzene boils at a temperature 2.1°C higher than the boiling point of pure benzene. Estimate the molar mass of the solute.

Think Before You Write You are asked to find molar mass, which is a ratio. Thus, you need both the numerator—the mass of the solute (which is given)—and the denominator, the amount in moles of solute. To find amount in moles of solute, the boiling point elevation constant gets you from °C to molality, which is moles solute/kilogram solvent, and the mass in grams of solvent are given, so you can determine the amount in moles of solute.

Answers *Cover the left column with your cut-out shield. Reveal each answer only after you have written your own answer in the right column.*

$$\Delta T_b = K_b \times m \xrightarrow{\text{divide both sides by } K_b} m = \frac{\Delta T_b}{K_b} = \frac{2.1°C}{2.5°C/m}$$

$$= 2.1\cancel{°C} \times \frac{m}{2.5\cancel{°C}} = 0.84\ m = \frac{0.84\ \text{mol solute}}{\text{kg benzene}}$$

$$\frac{2.1}{2.5} \times \frac{4}{4} = \frac{8.4}{10} = 0.84, \text{OK.}$$

Calculate the molality of the solution (*Step 1*) from the equation $\Delta T_b = K_b \times m$.

Given: 91.9 g benzene
Wanted: mol solute

g benzene → kg benzene → mol solute

1 kg benzene = 1000 g benzene

0.84 mol solute = 1 kg benzene

$$91.1\ \cancel{\text{g benzene}} \times \frac{1\ \cancel{\text{kg benzene}}}{1000\ \cancel{\text{g benzene}}} \times \frac{0.84\ \text{mol solute}}{\cancel{\text{kg benzene}}} = 0.077\ \text{mol solute}$$

$$91.1 \times \frac{1}{1000} \times 0.84 \approx (9 \times 10^1) \times 10^{-3} \times (8 \times 10^{-1})$$
$$= 72 \times (10^1 \times 10^{-3} \times 10^{-1}) = 72 \times 10^{-2} = 0.072, \text{OK.}$$

Use the molality you calculated in *Step 1* to convert from the given 91.9 g of benzene to amount in moles of solute (*Step 2*). You'll want to **analyze** this part of the problem by writing the given quantity and wanted unit, write a unit path, **identify** the equivalencies, **construct** the setup, and perform a **check**.

$$MM \equiv \frac{g}{mol} = \frac{15.2\ g}{0.077\ mol} = 2.0 \times 10^2\ g/mol$$

$$\frac{15.2}{0.077} \approx 15 \times \frac{1}{7.5 \times 10^{-2}} = \frac{15}{7.5} \times 10^2 = 2 \times 10^2, \text{OK.}$$

In *Step 3*, you use the defining equation to find molar mass.

You improved your understanding of colligative properties of solutions, and you improved your skill at solving freezing point depression and boiling point elevation problems.

What did you learn by solving this Active Example?

Practice Exercise 16.23

The freezing point of cyclohexane is 6.50°C, and its molal freezing point depression constant is 20.2°C/m. When 2.00 g of a solute is added to 100.0 g of cyclohexane, the freezing point of the solution is 1.33°C. What is the molar mass of the solute?

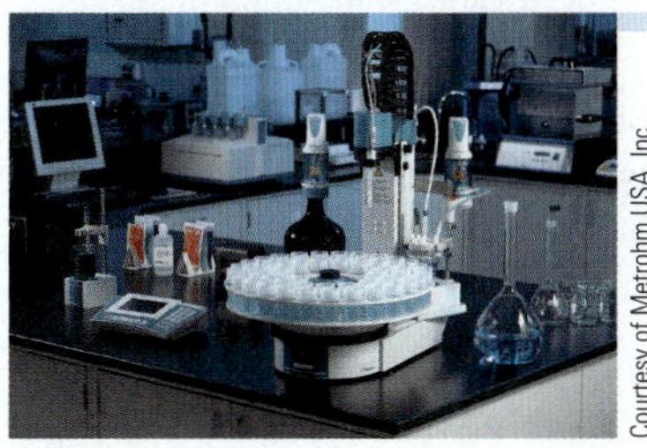

Courtesy of Metrohm USA, Inc.

CHAPTER 16 IN REVIEW: SOLUTIONS

Goal	Summary
Goal 1 Define the term *solution*, and given a description of a substance, determine if it is a solution.	A **solution** is a homogeneous mixture. Solutions have variable physical properties that depend on the composition of the mixture.
Goal 2 Distinguish among terms in the following groups: *solute* and *solvent*; *concentrated* and *dilute*; *solubility*, *saturated*, *unsaturated*, and *supersaturated*; *miscible* and *immiscible*.	The major component of a solution is called the **solvent;** the minor components are called the **solutes.** A **concentrated** solution has a relatively large quantity of solute per quantity of solvent; a **dilute** solution has a relatively small quantity of the same solute per quantity of solvent. A solution whose concentration is at the solubility limit for a given temperature is a **saturated** solution. If the concentration is less than the solubility limit, the solution is **unsaturated.** A **supersaturated** solution has a concentration greater than the normal solubility limit. Liquids are **miscible** if they dissolve in each other in all proportions. Liquids that are insoluble in each other are **immiscible**.
Goal 3 Describe the formation of a saturated solution from the time excess solid solute is first placed into a liquid solvent.	When a soluble ionic solid solute is placed in water, **polar** water molecules surround the ions, helping them move away from their positions in the crystal. In solution, the ions are **hydrated:** surrounded by water molecules. The **dissolving rate** is constant throughout the process. When dissolving has just begun, the **crystallization rate** is zero, and it increases until it equals the dissolving rate and a **dynamic equilibrium** is established.
Goal 4 Identify and explain the factors that determine the time required to dissolve a given amount of solute or to reach equilibrium.	A **finely divided solid** dissolves more rapidly because of greater surface area. **Stirring or agitating** a solution makes it dissolve more rapidly because of prevention of buildup of concentration at the solute surface. **Higher temperature** causes a solution to dissolve more rapidly because of faster particle movement.
Goal 5 Given the structural formulas of two molecular substances, or other information from which the strength of their intermolecular forces may be estimated, predict if they will dissolve appreciably in each other, and state the criteria on which your prediction is based.	Intermolecular forces fall into three categories: induced dipole forces, dipole forces, and hydrogen bonds. Substances with **similar intermolecular forces** will usually dissolve in one another.
Goal 6 Predict how and explain why the solubility of a gas in a liquid is affected by a change in the partial pressure of that gas over the liquid.	In most dilute solutions, **the solubility of a gas is directly proportional to the partial pressure of the gas over the surface of the liquid**.
Goal 7 Given the mass of solute and solvent or solution, calculate percentage concentration by mass.	$\text{\% concentration by mass} = \frac{\text{mass solute}}{\text{mass solution}} \times 100\% = \frac{\text{mass solute}}{\text{mass solute} + \text{mass solvent}} \times 100\%$
Goal 8 Given mass of solution and percentage concentration by mass, calculate mass of solute and solvent.	$\text{mass solution} \times \frac{\text{mass solute}}{\text{mass solution}} = \text{mass solute}$; mass solvent = mass solution − mass solute

Goal 9 Given two of the following, calculate the third: amount in moles of solute (or data from which it may be found), volume of solution, molarity.

Molarity is the amount in moles of solute in one liter of solution. Units of molarity are moles per liter, or mol/L. The symbol for molarity is M.

$$\text{M} \equiv \frac{\text{moles solute}}{\text{liter solution}} = \frac{\text{mol}}{\text{L}}$$

Goal 10 Given two of the following, calculate the third: moles of solute (or data from which it may be found), mass of solvent, molality.

Molality is moles of solute in one kilogram of solvent. The symbol for molality is m (note that molality is lower case m and molarity is upper case M).

$$\text{m} \equiv \frac{\text{mol solute}}{\text{kg solvent}}$$

Molality is used in situations where temperature independence is important. Neither the number of solute particles nor the mass of the solvent varies with temperature. In contrast, molarity is temperature dependent because the volume of a solution varies with temperature.

Goal 11 Given an equation for a neutralization reaction (or information from which it can be written), state the number of equivalents of acid or base per mole and calculate the equivalent mass of the acid or base.

One **equivalent** of an acid is defined as that amount of acid that yields one mole of hydrogen ion in a specific reaction. One equivalent of a base is that amount of base that reacts with one equivalent of an acid. The **equivalent mass** of a substance is the mass in grams of the substance per equivalent. To calculate equivalent mass, divide molar mass by equivalents per mole:

$$\frac{\text{g/mol}}{\text{eq/mol}} = \frac{\text{g}}{\text{mol}} \times \frac{\text{mol}}{\text{eq}} = \frac{\text{g}}{\text{eq}}, \text{ equivalent mass}$$

Goal 12 Given two of the following, calculate the third: equivalents of acid or base (or data from which they may be found), volume of solution, normality.

Normality is the number of equivalents of solute in one liter of solution. Units of normality are equivalents per liter, or eq/L. The symbol for normality is N.

$$\text{N} \equiv \frac{\text{equivalents solute}}{\text{liter solution}} = \frac{\text{eq}}{\text{L}}$$

Goal 13 Given any three of the following, calculate the fourth: (a) volume of concentrated solution, (b) molarity of concentrated solution, (c) volume of dilute solution, (d) molarity of dilute solution.

The key relationship for calculations involving dilution of concentrated solutions is

$\mathbf{M_c \times V_c = M_d \times V_d}$, in which M is molarity, V is volume, c is concentrated, and d is dilute.

Goal 14 Given the quantity of any species participating in a chemical reaction for which the equation can be written, find the quantity of any other species, either quantity being measured in (a) mass, (b) volume of solution at specified molarity, or (c) (if gases have been studied) volume of gas at given temperature and pressure.

For all **stoichiometry** problems, a macroscopic measurable quantity (mass, energy, volume of a gas at known pressure and temperature, or volume of a solution of known concentration) is changed to the amount in moles. The mole ratio from the balanced chemical equation is then used to change to the amount in moles of another species involved in the chemical change. Finally, the amount in moles is changed to a macroscopic measurable quantity. For solution stoichiometry, volume of solution and concentration, typically molarity, can be used to convert to amount in moles.

Goal 15 Given the volume of a solution that reacts with a known mass of a primary standard and the equation for the reaction, calculate the molarity of the solution.

Titration is the controlled and measured addition of one solution into another. Titration problems are solution stoichiometry problems. A **buret** is a device that measures delivered volumes precisely. An **indicator** is a substance that exhibits different color in solution at different solution acidities. A solution is **standardized** when its concentration is determined by reaction with a substance that can be weighed accurately, which is called a *primary standard*.

Goal 16 Given the volumes of two solutions that react with each other in a titration, the molarity of one solution, and the equation for the reaction or information from which it can be written, calculate the molarity of the second solution.

Once a solution is standardized, we may use it to find the concentration of another solution. Volume times molarity for the solution of known concentration yields amount in moles for that solute. A balanced chemical equation is used to convert to amount in moles of the other species. The definition of molarity, moles per liter, is used to calculate the molarity of the second solution.

Goal 17 Given the volume of a solution that reacts with a known mass of a primary standard and the equation for the reaction, calculate the normality of the solution.

In a reaction, **the numbers of equivalents of acid and base that react with each other are equal**. This idea of equal numbers of equivalents is the basis of the normality system. There are two ways to calculate the number of equivalents in a sample of a substance:

1. If you know the mass of the substance and its equivalent mass, use the equivalent mass as a conversion factor to get equivalents.
2. If the sample is a solution and you know its volume and normality, multiply one by the other: $V \times N = \text{eq}$.

Goal 18 Given the volumes of two solutions that react with each other in a titration and the normality of one solution, calculate the normality of the second solution.	For an acid–base titration, $\mathbf{V_{acid} \times N_{acid} = V_{base} \times N_{base}}$.
Goal 19 Given (a) the molality of a solution, or data from which it may be found; (b) the normal freezing or boiling point of the solvent; and (c) the freezing or boiling point constant, find the freezing or boiling point of the solution.	The properties of a solution that depend only on the number of solute particle present, without regard to their identity, are called *colligative properties.* **Freezing point depression and boiling point elevation** are colligative properties that are directly proportional to the molal concentration of any solute. These proportionality constants are called the molal freezing point constant and the molal boiling point constant, respectively. The values of these constants depend only on the chemical identity of the solvent in the solution. For freezing point depression, $\mathbf{\Delta T_f = K_f m}$. For boiling point elevation, $\mathbf{\Delta T_b = K_b m}$.
Goal 20 Given the freezing point depression or boiling point elevation and the molality of a solution, or data from which they may be found, calculate the molal freezing point constant or molal boiling point constant.	$\mathbf{K_f}$ and $\mathbf{K_b}$, are, respectively, the **molal freezing point depression constant and the molal boiling point elevation constant:** $\mathbf{K_f = \frac{\Delta T_f}{m}}$ **and** $\mathbf{K_b = \frac{\Delta T_b}{m}}$.
Goal 21 Given (a) the mass of solute and solvent in a solution; (b) the freezing point depression or boiling point elevation, or data from which they may be found; and (c) the molal freezing/boiling point constant of the solvent, find the approximate molar mass of the solute.	Freezing point depression and boiling point elevation measurements may be used to determine molar mass. To calculate the molar mass of a solute from freezing-point depression or boiling-point elevation data, 1. Calculate molality from $m = \Delta T_f/K_f$ or $m = \Delta T_b/K_b$. Express as mol solute/kg solvent. 2. Using molality as a conversion factor between moles of solute and kilograms of solvent, find the amount in moles of solute. 3. Use the defining equation for molar mass, MM = g/mol, to calculate the molar mass of the solute.

Key Terms

Most of the key terms and concepts and many others appear in the Glossary. Use your Glossary regularly.

buret p. 634
concentrated p. 604
dilute p. 604
dynamic equilibrium p. 606
hydrated ions p. 606
immiscible p. 604
indicator p. 634
mass ratio p. 611
miscible p. 604
molarity (M) p. 613
percentage concentration by mass p. 611
polar p. 605
primary standard p. 634
saturated solution p. 604
solubility p. 604
solute p. 603
solution p. 602
solvent p. 603
standardize p. 634
supersaturated p. 604
titration p. 634
unsaturated solution p. 604

Key Terms from Optional Sections:

boiling point elevation, ΔT_b (Sec. 16.15) p. 640
colligative properties (Sec. 16.15) p. 640
equivalent (Sec. 16.9) p. 621
equivalent mass (Sec. 16.9) p. 623
freezing point depression, ΔT_f (Sec. 16.15) p. 640
molal boiling point elevation constant, K_b (Sec. 16.15) p. 640
molal freezing point depression constant, K_f (Sec. 16.15) p. 640
molality (m) (Sec. 16.8) p. 619
normality (N) (Sec. 16.9) p. 621

Frequently Asked Questions

What is the best way to approach solution concentration problems?
To solve solution concentration problems easily, you must have a clear understanding of concentrations and the units in which they are expressed. Table 16.1 summarizes all the concentrations used in this chapter. Study carefully the concentrations that have been assigned to you. Then practice with the end-of-chapter questions until you have complete mastery of each performance goal.

How can I diagnose the difficulties I'm having with solution stoichiometry problems?
To understand solution stoichiometry, you must first understand both fundamental stoichiometry concepts and solution concentrations. If you have difficulty solving solution stoichiometry problems, ask yourself if you thoroughly understand: (a) writing chemical formulas from names, (b) calculating molar

masses from chemical formulas, (c) using molar mass to change from amount in moles to mass and from mass to amount in moles, (d) writing and balancing chemical equations, (e) using chemical equations to determine mole ratios, and (f) using molarity to convert from amount in moles to volume and from volume to amount in moles.

I'm using 22.7 L/mol to convert between amount in moles and volume in liters, but I'm getting the wrong answer. Why?
Once in a while a student is tempted to change between moles and liters by using 22.7 L/mol, or even worse, 22.7 mol/L. The number 22.7 is so convenient, and the units look like just what is needed. But they are not; 22.7 applies only to gases, not to solutions, and then only to gases at STP.

Concept-Linking Exercises

Write a brief description of the relationships among each of the following groups of terms or phrases. Answers to the Concept-Linking Exercises are given at the end of the chapter.

1. Solution, percentage composition, mixture, pure substance
2. Solubility, saturated, unsaturated, supersaturated
3. Solubility, intermolecular forces, partial pressure of a solute gas, temperature
4. Titration, indicator, standardization of a solution, primary standard, molarity

The following exercises are from optional sections.

5. Normality, equivalent, equivalent mass
6. Colligative properties, molality, freezing point depression, molal freezing point depression constant, boiling point elevation, molal boiling point elevation constant

Small-Group Discussion Questions

Small-Group Discussion Questions are for group work, either in class or under the guidance of a leader during a discussion section.

1. Both the ocean and the atmosphere can be classified as solutions in some situations and not as solutions in other situations. Explain.
2. Give specific examples of solutions in each of the following states: (a) gas in gas, (b) gas in liquid, (c) liquid in liquid, (d) solid in liquid, (e) solid in solid. For each example, identify the solute and solvent.
3. What are the similarities and differences between miscibility and solubility?
4. Do you expect table salt to dissolve in water and/or carbon tetrachloride (both are liquids at room temperature)? For each liquid, describe the dissolving process or explain why dissolving does not occur. Sketch the dissolving process at the particulate level.
5. Develop a list of 10 compounds, other than metals or metal alloys, which have intermolecular forces sufficiently different from that of water to make you confident that they will not dissolve in water. You may use your textbook to find compounds. Then develop a list of 10 compounds that you predict will dissolve in water. Give a brief justification for each of your choices.
6. If you were to be charged with manufacturing a brand of sugar that will be marketed as fast dissolving, what characteristics would you give it? In general, what characteristics speed the dissolving process? How do those characteristics affect the solubility of sugar?
7. In Section 16.6, we state that for solutions of 5% or less, mass percent and weight/volume percent are essentially equal. A student conducts an experiment to test this statement. He finds the following densities at 20°C: water, 0.9982 g/mL; 10.000 g sodium chloride in 100.00 g solution, 1.0726 g/mL; 10.000 g sugar (sucrose) in 100.00 g solution, 1.0400 g/mL. Do these data confirm or refute the statement?
8. Write a set of instructions sufficiently detailed so that a person who has no knowledge of chemistry can prepare 250.0 mL of a 0.10 M solution of sodium nitrate. Assume that you have standard chemistry laboratory equipment and glassware available.
9. (Optional Section 16.8) What are the similarities and differences between molarity and molality? What are the advantages and disadvantages of each unit?
10. (Optional Section 16.9) Why is equivalent concentration a convenient unit for analytical work? Explain how phosphoric acid can have between 1 and 3 equivalents per mole.
11. Explain how each solution concentration unit assigned in your course fits the general definition of a concentration ratio: quantity of solute per quantity of solution or quantity of solute per quantity of solvent. Also explain how each unit is a direct proportionality.
12. Write a set of instructions sufficiently detailed so that a person who has no knowledge of chemistry can prepare 100.0 mL of a 0.10 M solution of hydrochloric acid from a 12 M solution. Assume that you have standard chemistry laboratory equipment and glassware available.
13. Figure 16.28 summarizes stoichiometry, as it is presented in this introductory course. Four different measured macroscopic quantities can be changed to the same set of four quantities for any pair of substances involved in a chemical change. This yields 16 different combinations per pair of substances. Divide the 16 combinations among your group members, and find or write a problem for each of the combinations (for example, mass to gas volume and solution volume to quantity of energy).
14. Consider the titration of 0.100 M sodium hydroxide into 10.00 mL of 0.100 M sulfuric acid. Sketch a buret and a flask. Illustrate the titration process by illustrating the buret and flask (a) before the titration starts, (b) half way to neutralizing the acid, and (c) at the completion of the titration. Next, calculate the molar concentration

of the acid at points (b) and (c). Finally, construct particulate-level sketches of a tiny portion of the contents of the flask at each of the three points.

15. (Optional Section 16.14) For the same titration described in Question 14, calculate the normality of the acid at points (b) and (c), assuming that the acid is completely neutralized at the end of the titration.

16. (Optional Section 16.15) Define the term *colligative property*. Give two examples of solution properties that are colligative and two that are not colligative. Use conversion factors to determine the units of the molal boiling point elevation constant without using molality.

Questions, Exercises, and Problems

Interactive versions of these problems may be assigned by your instructor. Solutions for blue-numbered questions are at the end of the chapter. Questions other than those in the General Questions and More Challenging Problems sections are paired in consecutive odd-even number combinations; solutions for the odd numbered questions are also at the end of the chapter.

Section 16.2: The Characteristics of a Solution

1. Mixtures of gases are always true solutions. True or False? Explain why.

2. Every pure substance has a definite and fixed set of physical and chemical properties. A solution is prepared by dissolving one pure substance in another. Is it reasonable to expect that the solution will also have a definite and fixed set of properties that are different from the properties of either component? Explain your answer.

3. Can you see particles in a solution? If yes, give an example, and if no, say why.

4. What kinds of solute particles are present in a solution of an ionic compound? Of a molecular compound?

Section 16.3: Solution Terminology

5. Distinguish between the solute and solvent in each of the following solutions: (a) saltwater [NaCl(aq)]; (b) sterling silver (92.5% Ag, 7.5% Cu); (c) air (about 80% N_2, 20% O_2). On what do you base your distinctions?

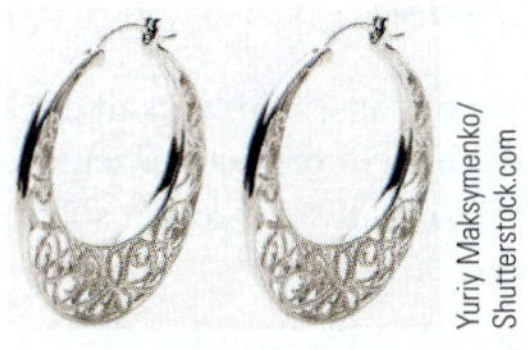

Yuriy Maksymenko/Shutterstock.com

Sterling silver is an alloy of 92.5% silver and 7.5% another metal, usually copper.

6. Explain why the distinction between solute and solvent is not clear for some solutions.

7. Would it be proper to say that a saturated solution is a concentrated solution? Or that a concentrated solution is a saturated solution? Point out the distinctions between these sometimes-confused terms.

8. Solution A contains 10 g of solute dissolved in 100 g of solvent, while solution B has only 5 g of a different solute per 100 g of solvent. Under what circumstances can solution A be classified as dilute and solution B as concentrated?

9. What happens if you add a very small amount of solid salt (NaCl) to each beaker described here? Include a statement comparing the *amount* of solid eventually found in the beaker with the amount you added: (a) a beaker containing *saturated* NaCl solution, (b) a beaker with *unsaturated* NaCl solution, (c) a beaker containing *supersaturated* NaCl solution.

Christine Myaskovsky/Cengage Learning

A saturated sodium chloride solution.

10. Suggest simple laboratory tests by which you could determine whether a solution is unsaturated, saturated, or supersaturated. Explain why your suggestions would distinguish among the different classifications.

11. In stating solubility, an important variable must be specified. What is that variable, and how does solubility of a solid solute usually depend on it?

12. Suggest units in which solubility might be expressed other than grams per 100 g of solvent.

13. When acetic acid, a clear, colorless liquid, and water are mixed, a clear, uniform, colorless liquid results. Is acetic acid soluble in water? Is acetic acid miscible in water? Explain your answers.

14. (a) The solubility of lead(II) chloride in water is 4.50 g per liter. If a lead(II) chloride solution had a concentration of 6.35 g per liter, would it be saturated, supersaturated, or unsaturated? (b) The solubility of calcium sulfate in water is 0.667 g per liter. If a calcium sulfate solution had a concentration of 3.47×10^{-3} g per liter, would it be relatively concentrated or dilute? (c) The liquid acetone and water dissolve in each other in all proportions. Are liquid acetone and water said to be miscible, immiscible, or partially miscible?

Section 16.4: The Formation of a Solution

15. How is it that both cations and anions, positively charged ions and negatively charged ions, can be hydrated by the same substance, water?

16. What does it mean to say that a solute particle is hydrated?

17. Explain why the dissolving process is reversible.

18. Describe the forces that promote the dissolving of a solid solute in a liquid solvent.

19. Describe the changes that occur between the time excess solute is placed into water and the time the solution becomes saturated.

20. Compare and contrast a dynamic equilibrium on both the particulate and the macroscopic levels. Pay particular attention to explaining why the equilibrium is dynamic.

21. At what time during the development of a saturated solution is the rate at which ions move from the aqueous phase to the solid phase [solute(aq) $\rightarrow$ solute(s)] greater than the rate from the solid phase to the aqueous phase [solute(s) $\rightarrow$ solute(aq)]?

22. Compare the rate at which ions pass from the solid phase to the aqueous phase with the rate at which they pass from the aqueous phase to the solid phase when the solution is unsaturated.

23. Why can't you prepare a supersaturated solution by adding more solute and stirring until it dissolved?

24. Why do people stir coffee after putting sugar into it? Would putting the coffee and sugar into a closed container and shaking be as effective as stirring?

Why is coffee stirred after adding sugar?

25. Identify three ways in which you can reduce the amount of time required to dissolve a given amount of solute in a fixed quantity of solvent.

26. Explain how each action in Question 25 speeds the dissolving process.

Section 16.5: Factors That Determine Solubility

Questions 27 through 30: Structural diagrams for several substances are given in **Table 16.2**.

Table 16.2 Lewis Diagrams for Questions 27 to 30

H—F: hydrogen fluoride	water
formic acid	tetrachloroethene
methylamine	tetrafluoromethane
hexane	ethanol
glycerine	benzene*
cyclohexane	dimethyl ether

*This diagram is not consistent with all of the properties of benzene, but it is adequate to predict benzene's ability to dissolve other substances or to be dissolved by other solvents.

27. Which of the following solutes do you expect to be more soluble in water than in cyclohexane: (a) formic acid, (b) benzene, (c) methylamine, (d) tetrafluoromethane? Explain your choice(s).

28. Which of the following solutes do you expect to be more soluble in water than in cyclohexane: (a) dimethyl ether, (b) hexane, (c) tetrachloroethene, (d) hydrogen fluoride? Explain your choice(s).

29. Which compound, glycerine or hexane, do you expect would be more miscible in water? Why?

30. Suppose you have a spot on some clothing and water will not take it out. If you have ethanol and cyclohexane available, which would you choose as the more promising solvent to try? Why?

31. On opening a bottle of carbonated beverage, many bubbles are released, and you hear the sound of escaping gas. This

suggests that the beverage is bottled under high pressure. Yet, for safety reasons, the pressure cannot be much more than one atmosphere. What gas do you suppose is in the small space between the beverage and the cap of a bottle of carbonated beverage before the cap is removed?

32. Determine whether each of the following statements is true or false:

a) The rate of crystallization when a solution is two-thirds saturated is higher than the rate of crystallization when a solution is one-third saturated.

b) The solubility of a solid in a liquid always increases with temperature.

c) The solubility of a gas in a liquid increases as the partial pressure of that gas over the surface of the liquid increases.

Section 16.6: Solution Concentration: Percentage Concentration by Mass

33. Calculate the percentage concentration by mass of a solution prepared by dissolving 2.32 g of calcium chloride in 81.0 g of water.

34. If 20.2 g of an aqueous solution of calcium nitrate contains 2.46 g of calcium nitrate, what is the percentage concentration by mass of calcium nitrate in the solution?

35. What mass in grams of ammonium nitrate must be weighed out to make 415 g of a 58.0% by mass solution? In what volume in milliliters of water should it be dissolved?

36. What mass in grams of copper(II) sulfate is there in 219 g of an aqueous solution that is 20.2% by mass copper(II) sulfate?

Section 16.7: Solution Concentration: Molarity

37. Potassium iodide can be the additive in "iodized" table salt. Calculate the molarity of a solution prepared by dissolving 2.41 g of potassium iodide in water and diluting to 50.0 mL.

Iodized salt is sodium chloride mixed with a small quantity of a soluble ionic compound that contains iodide or iodate ion.

38. A student weighs out a 4.80-g sample of aluminum bromide, transfers it to a 100-mL volumetric flask, adds enough water to dissolve it, and then adds water to the 100-mL mark. What is the molarity of aluminum bromide in the resulting solution?

39. A student dissolves 18.0 g of anhydrous nickel(II) chloride in water and dilutes it to 90.0 mL. He also dissolves 30.0 g of nickel(II) chloride hexahydrate in water and dilutes it to 90.0 mL. Identify the solution with the higher molar concentration and calculate its molarity.

40. The chemical name for the "hypo" used in photographic developing is sodium thiosulfate. It is sold as a pentahydrate, $Na_2S_2O_3 \cdot 5\,H_2O$. What is the molarity of a solution prepared by dissolving 1.2×10^2 g of this compound in water and diluting the solution to 1.25×10^3 mL?

41. Large quantities of silver nitrate are used in making photographic chemicals. Find the mass that must be used in preparing 2.50×10^2 mL of 0.058 M silver nitrate.

42. What mass in grams of zinc iodide must be dissolved to prepare 2.50×10^2 mL of a 0.150 M aqueous solution of the salt?

43. Potassium hydroxide is used in making liquid soap. What mass in grams would you use to prepare 2.50 L of 1.40 M potassium hydroxide?

44. You need to make an aqueous solution of 0.123 M manganese(II) acetate (acetate ion, $C_2H_3O_2^-$) for an experiment in lab, using a 250-mL volumetric flask. How much solid manganese(II) acetate should you add?

45. What volume of concentrated sulfuric acid, which is 18 molar, is required to obtain 5.19 mole of the acid?

46. What volume in milliliters of a 0.196 M aqueous solution of manganese(II) nitrate, $Mn(NO_3)_2$, must be measured to obtain 6.17 g of the salt?

47. If 0.132 M sodium chloride is to be the source of 8.33 g of dissolved solute, what volume of solution is needed?

48. Calculate the volume of concentrated ammonia solution, which is 15 molar, that contains 75.0 g of ammonia.

49. Calculate the amount in moles of silver nitrate in 55.7 mL of 0.204 M silver nitrate.

50. What amount in moles of solute are in 65.0 mL of 2.20 M sodium hydroxide?

51. Despite its intense purple color, potassium permanganate is used in bleaching operations. What amount in moles is in 25.0 mL of 0.0841 M $KMnO_4$?

Potassium permanganate has an intense purple color.

52. A student uses 29.3 mL of 0.482 M sulfuric acid to titrate a base of unknown concentration. What amount in moles of sulfuric acid reacts?

53. The density of 3.30 M potassium nitrate is 1.15 g/mL. What is its percentage concentration by mass?

54. An aqueous solution of 6.02 M hydrochloric acid has a density of 1.10 g/mL. What is the percent concentration by mass of hydrochloric acid in the solution?

Section 16.8: Solution Concentration: Molality (Optional)

55. Calculate the molal concentration of a solution of 44.9 g of naphthalene, $C_{10}H_8$, in 175 g of benzene, C_6H_6.

56. What is the molality of a solution prepared by dissolving 18.8 g of ammonium iodide in 4.50×10^2 mL of water? Assume that the density of water is 1.00 g/mL.

57. Diethylamine, $(CH_3CH_2)_2NH$, is highly soluble in ethanol, C_2H_5OH. Calculate the mass in grams of diethylamine that would be dissolved in 4.00×10^2 g of ethanol to produce 4.70 m $(CH_3CH_2)_2NH$.

58. What mass in grams of aluminum nitrate must be dissolved in 5.00×10^2 g of water to prepare a 0.121 m solution of the salt?

59. What volume in milliliters of water is needed to dissolve 97.7 mg sodium chloride in the preparation of a 2.80×10^{-3} m solution?

60. What mass in grams of water is required to dissolve 12.7 g of iron(III) chloride in order to produce a 0.172 m solution?

Penka Todorova Vitkova/Shutterstock.com

Wastewater treatment plant. Iron(III) chloride is used in wastewater treatment. Another use of iron(III) chloride is in the printed circuit board manufacturing process.

Section 16.9: Solution Concentration: Normality (Optional)

61. What is equivalent mass? Why can you state positively the equivalent mass of LiOH but not of H_2SO_4?

62. Explain why the number of equivalents in a mole of acid or base is not always the same.

63. State the number of equivalents in 1 mole of HNO_2; in 1 mole of H_2SeO_4 in the reaction $H_2SeO_4 \rightarrow H^+ + HSeO_4^-$. (Se is selenium, Z = 34.)

64. Give the number of equivalents of acid and base per mole in each of the following reactions:
a) $HNO_3 + KOH \rightarrow KNO_3 + H_2O$
b) $H_2S + 2\ NaOH \rightarrow Na_2S + 2\ H_2O$

65. State the maximum number of equivalents per mole of $Cu(OH)_2$; per mole of $Fe(OH)_3$.

66. What is the maximum number of equivalents in 1 mole of $Zn(OH)_2$ and RbOH? (Rb is rubidium, Z = 37.)

67. Calculate the equivalent masses of HNO_2 and H_2SeO_4 in Question 63.

68. Consider the reaction $HBr + KOH \rightarrow KBr + H_2O$. (a) What is the equivalent mass of KOH? (b) Calculate the number of equivalents in 29.5 g of KOH.

69. What are the equivalent masses of $Cu(OH)_2$ and $Fe(OH)_3$ in Question 65?

70. Consider the reaction $H_3PO_4 + 3\ KOH \rightarrow K_3PO_4 + 3\ H_2O$. (a) What is the equivalent mass of H_3PO_4? (b) Calculate the number of equivalents in 38.4 g of H_3PO_4.

71. What is the normality of the solution made when 2.25 g of potassium hydroxide is dissolved in water and diluted to 2.50×10^2 mL?

72. Calculate the normality of a solution that contains 3.46 g of H_3PO_4 in 909 mL of solution for the reaction $H_3PO_4 + 3\ KOH \rightarrow K_3PO_4 + 3\ H_2O$.

73. $NaHSO_4$ is used as an acid in the reaction $HSO_4^- \rightarrow H^+ + SO_4^{2-}$. What mass of $NaHSO_4$ must be dissolved in 7.50×10^2 mL of solution to produce 0.200 N $NaHSO_4$?

74. If 9.79 g of $NaHCO_3$ is dissolved in 5.00×10^2 mL of solution, what is the normality in the reaction $NaHCO_3 + HCl \rightarrow NaCl + H_2O + CO_2$?

75. A student dissolves 6.69 g $H_2C_2O_4$ in water, dilutes to 2.00×10^2 mL, and uses it in a reaction in which it ionizes as follows: $H_2C_2O_4 \rightarrow H^+ + HC_2O_4^-$. What is the normality of the solution?

76. It is desired to prepare 600.0 mL of 0.150 normal $Ba(OH)_2$ for use in the reaction $2\ HCl + Ba(OH)_2 \rightarrow BaCl_2 + 2\ H_2O$. What mass in grams of $Ba(OH)_2$ are needed?

77. What is the molarity of (a) 0.965 N sodium hydroxide, (b) 0.237 N H_3PO_4 in $H_3PO_4 + 2\ NaOH \rightarrow Na_2HPO_4 + 2\ H_2O$?

78. Consider the reaction $H_2CO_3 + 2\ NaOH \rightarrow Na_2CO_3 + 2\ H_2O$. (a) What is the normality of a 6.110×10^{-2} M H_2CO_3 solution? (b) What volume of this solution would contain 0.345 eq?

79. How many equivalents of solute are in 73.1 mL of 0.834 N NaOH?

80. How many equivalents are in 2.25 L of 0.871 N H_2SO_4?

81. What volume of 0.492 N $KMnO_4$ contains 0.788 eq?

82. Calculate the volume of 0.371 N HCl that contains 0.0385 eq.

Section 16.11: Dilution of Concentrated Solutions

83. What is the molarity of the acetic acid solution if 45.0 mL of 17 M $HC_2H_3O_2$ is diluted to 1.5 L?

84. A student dilutes 12.9 mL of a 10.9 M perchloric acid solution to a total volume of 2.50×10^2 mL. What is the concentration of the diluted solution?

85. How many milliliters of concentrated nitric acid, 16 M HNO_3, will you use to prepare 7.50×10^2 mL of 0.69 M HNO_3?

86. What volume in milliliters of 9.76 M hydrochloric acid solution should be used to prepare 2.00 L of 0.400 M HCl?

87. Calculate the volume of 18 M H_2SO_4 required to prepare 3.0 L of 2.9 N H_2SO_4 for the reactions in which the sulfuric acid is completely ionized.

88. A student adds 54.6 mL of water to 14.4 mL of a 0.791 M hydroiodic acid solution. What is the concentration of the diluted solution?

89. Calculate the normality of a solution prepared by diluting 15.0 mL of 15.0 mL of 15 M H_3PO_4 to 2.50×10^2 mL. The solution will be used in the reaction $H_3PO_4 + 2\ NaOH \rightarrow Na_2HPO_4 + 2\ H_2O$.

90. If 25.0 mL of 15 M HNO_3 is diluted to 4.00×10^2 mL, what is the normality of the diluted solution?

Section 16.12: Solution Stoichiometry

91. Calculate the mass in grams of magnesium hydroxide that will precipitate from 25.0 mL of 0.398 M magnesium chloride by the addition of excess sodium hydroxide solution.

92. What mass in grams of copper(II) hydroxide will precipitate when excess potassium hydroxide solution is added to 56.0 mL of 0.522 M copper(II) bromide solution?

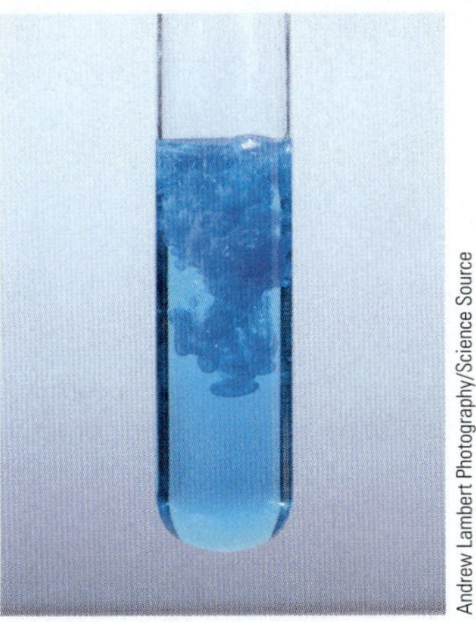

Andrew Lambert Photography/Science Source

Copper(II) hydroxide precipitates.

93. Calculate the mass of calcium phosphate that will precipitate when excess sodium phosphate solution is added to 100.0 mL of 0.130 M calcium nitrate.

94. What volume in milliliters of 0.464 M nitric acid solution is needed to neutralize 7.84 g of magnesium carbonate?

95. What volume in milliliters of 1.50 M sodium hydroxide solution must react with aluminum to yield 2.00 L of hydrogen, measured at 22°C and 789 torr, by the reaction $2\ Al + 6\ NaOH \rightarrow 2\ Na_3AlO_3 + 3\ H_2$? Assume complete conversion of reactants to products.

96. Calculate the volume of chlorine, measured at STP, that can be recovered from 50.0 mL of 1.20 M hydrochloric acid solution by the reaction $MnO_2 + 4\ HCl \rightarrow MnCl_2 + 2\ H_2O + Cl_2$, assuming complete conversion of reactants to products.

97. What volume of 0.842 M sodium hydroxide solution would react with 8.74 g of sulfamic acid, NH_2SO_3H, a solid acid with one replaceable hydrogen?

98. Calculate the volume in milliliters of 0.563 M barium hydroxide solution required to precipitate as calcium hydroxide all of the calcium ions in 111 mL of 0.658 M calcium iodide solution.

Section 16.13: Titration Using Molarity

99. The equation for a reaction by which a solution of sodium carbonate may be standardized is $2\ HC_7H_5O_2 + Na_2CO_3 \rightarrow 2\ NaC_7H_5O_2 + H_2O + CO_2$. A student determines that 5.038 g of $HC_7H_5O_2$ uses 51.89 mL of the sodium carbonate solution in the titration. Find the molarity of the sodium carbonate.

100. Potassium hydrogen phthalate is a solid, monoprotic acid frequently used in the laboratory as a primary standard. It has the formula $KHC_8H_4O_4$. This is often written in shorthand notation as KHP. If 25.0 mL of a potassium hydroxide solution is needed to neutralize 2.26 g of KHP, what is the molarity of the potassium hydroxide solution?

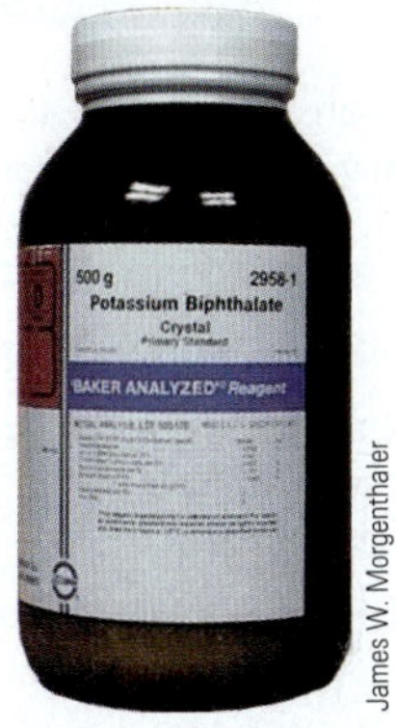

James W. Morgenthaler

Potassium hydrogen phthalate (sometimes called potassium biphthalate, as shown on this bottle) is an acid that is convenient to store and use because it is a solid.

101. A student is to titrate solid maleic acid, $H_2C_4H_2O_4$ (two replaceable hydrogens) with a KOH solution of unknown concentration. She dissolves 1.45 g of maleic acid in water and titrates 50.0 mL of the base to neutralize the acid. What is the molarity of the KOH solution?

102. Oxalic acid dihydrate is a solid, diprotic acid that can be used in the laboratory as a primary standard. Its formula is $H_2C_2O_4 \cdot 2\ H_2O$. A student dissolves 0.750 g of $H_2C_2O_4 \cdot 2\ H_2O$ in water and titrates the resulting solution with a solution of sodium hydroxide of unknown concentration. If 28.0 mL of the sodium hydroxide solution are required to neutralize the acid, what is the molarity of the sodium hydroxide solution?

103. A student finds that 37.80 mL of a 0.4052 M $NaHCO_3$ solution is required to titrate a 20.00-mL sample of sulfuric acid solution. What is the molarity of the acid? The reaction equation is $H_2SO_4 + 2\ NaHCO_3 \rightarrow Na_2SO_4 + 2\ H_2O + 2\ CO_2$.

104. The molarity of an aqueous solution of potassium hydroxide is determined by titration against a 0.138 M hydrobromic acid solution. If 27.4 mL of the base is required to neutralize 21.6 mL of hydrobromic acid, what is the molarity of the potassium hydroxide solution?

Section 16.14: Titration Using Normality (Optional)

105. What is the normality of the sodium carbonate solution in Question 99?

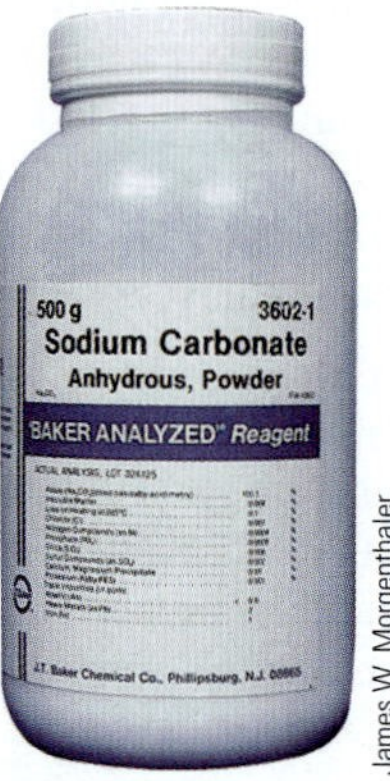

Sodium carbonate is often used as a primary standard in titrations.

106. What is the normality of the base in Question 100?

107. What is the normality of the base in Question 101?

108. Calculate the normality of the base in Question 102. Set up the problem completely from the data, not from the answer to Question 102.

109. Calculate the normality of a solution of sodium carbonate if a 25.0-mL sample requires 39.8 mL of 0.405 N sulfuric acid in a titration.

110. What is the normality of an acid if 12.8 mL is required to titrate 15.0 mL of 0.882 N sodium hydroxide?

111. A chemist finds that 42.2 mL of 0.402 N sodium hydroxide is required to titrate 50.0 mL of a solution of tartaric acid ($H_2C_4H_4O_6$) of unknown concentration. Find the normality of the acid.

112. The normality of an aqueous solution of perchloric acid is determined by titration with a 0.248 N potassium hydroxide solution. If 34.3 mL of potassium hydroxide are required to neutralize 29.3 mL of the acid, what is the normality of the perchloric acid solution?

113. When sodium hydroxide is titrated into phosphoric acid, 16.3 mL of 0.208 N sodium hydroxide is required for 20.0 mL of the phosphoric acid solution. Calculate the normality of the acid.

114. Oxalic acid dihydrate is a solid, diprotic acid that can be used in the laboratory as a primary standard. Its formula is $H_2C_2O_4 \cdot 2\,H_2O$. A student dissolves 0.523 g of $H_2C_2O_4 \cdot 2\,H_2O$ in water and titrates the resulting solution with a solution of potassium hydroxide of unknown concentration. If 32.8 mL of the potassium hydroxide solution are required to neutralize the acid, what is the normality of the potassium hydroxide solution?

115. A chemist dissolves 1.21 g of an organic compound that functions as a base in reaction with sulfuric acid in water and titrates it with 0.170 N sulfuric acid. What is the equivalent mass of the base if 30.7 mL of acid is required in the titration?

116. If 15.6 mL of 0.562 N sodium hydroxide is required to titrate a solution prepared by dissolving 0.631 g of an unknown acid, what is the equivalent mass of the acid?

Section 16.15: Colligative Properties of Solutions (Optional)

117. Is the partial pressure exerted by one component of a gaseous mixture at a given temperature and volume a colligative property? Justify your answer, pointing out in the process what classifies a property as "colligative."

118. The specific gravity of a solution of KCl is greater than 1.00. The specific gravity of a solution of NH_3 is less than 1.00. Is specific gravity a colligative property? Why or why not?

119. A student dissolves 27.2 g of aniline, $C_6H_5NH_2$, in 1.20×10^2 g of water. At what temperatures will the solution freeze and boil?

Aniline is used in the polyurethane manufacturing process. Materials made from polyurethane include foams (as pictured), spandex, hardwood floor coatings, dolly wheels, and many other end products.

120. The boiling point of benzene, C_6H_6, is 80.10°C at 1 atmosphere. K_b for benzene is 2.53°C/m. A nonvolatile molecular substance that dissolves in benzene is testosterone. If 10.14 g of testosterone, $C_{19}H_{28}O_2$ (288.4 g/mol), is dissolved in 231.0 g of benzene, what are the molality and the boiling point of the solution?

121. Calculate the freezing point of a solution of 2.12 g of naphthalene, $C_{10}H_8$, in 32.0 g of benzene, C_6H_6. Pure benzene freezes at 5.50°C, and its K_f = 5.10°C/m.

122. The freezing point of water is 0.00°C at 1 atmosphere. K_f for water is 1.86°C/m. A molecular substance that dissolves in water is antifreeze (ethylene glycol). If 11.35 g of antifreeze, CH_2OHCH_2OH (62.10 g/mol), is dissolved in 272.3 g of water, what are the molality and the boiling point of the solution?

123. What is the molality of a solution of an unknown solute in acetic acid if it freezes at 14.1°C? The normal freezing point of acetic acid is 16.6°C, and K_f = 3.90°C/m.

124. When 14.56 g of TNT, $C_7H_5N_3O_6$ (227.1 g/mol), is dissolved in 264.3 g of an organic solvent, the boiling point of the resulting solution is 0.200°C higher than that of the pure solvent. What are the molality of the solution and the value of K_b for the solvent?

125. A solution of 16.1 g of an unknown solute in 6.00×10^2 g of water boils at 100.28°C. Find the molar mass of the solute.

126. The boiling point of benzene, C_6H_6, is 80.10°C at 1 atmosphere. K_b for benzene is 2.53°C/m. In a laboratory experiment, students synthesized a new compound and found that when 11.5 g of the compound was dissolved in 246 g of benzene, the solution began to boil at 80.43°C. The compound was also found to be a nonvolatile molecular compound. What is the molecular mass that they determined for this compound?

127. When 12.4 g of an unknown solute is dissolved in 90.0 g of phenol, the freezing point depression is 9.6°C. Calculate the molar mass of the solute if K_f = 3.56°C/m for phenol.

128. The freezing point of water is 0.00°C at 1 atmosphere. K_f for water is 1.86°C/m. In a laboratory experiment, students synthesized a compound and found that when 11.2 g of the compound was dissolved in 2.80×10^2 g of water, the solution began to freeze at 21.12°C. The compound was also found to be a nonvolatile molecular compound. What is the molecular mass that they determined for this compound?

129. The normal freezing point of an unknown solvent is 28.7°C. A solution of 11.4 g of ethanol, C_2H_5OH, in 2.00×10^2 g of the solvent freezes at 22.5°C. What is the molal freezing point constant of the solvent?

130. When 19.77 g of glucose, $C_6H_{12}O_6$ (180.2 g/mol), is dissolved in 225.6 g of an organic solvent, the freezing point of the resulting solution is 1.06°C lower than that of the pure solvent. What is the molality of the solution? What is the value of K_f for the solvent?

General Questions

131. Distinguish precisely, and in scientific terms, the differences among items in each of the following pairs or groups.

a) Solute, solvent
b) Concentrated, dilute
c) Saturated, unsaturated, supersaturated
d) Soluble, miscible
e) (Optional) molality, molarity, normality
f) (Optional) molar mass, equivalent mass
g) (Optional) freezing point depression, boiling point elevation
h) (Optional) molal freezing point constant, molal boiling point constant

132. Determine whether each of the following statements is true or false:

a) The concentration is the same throughout a beaker of solution.
b) A saturated solution of solute A is always more concentrated than an unsaturated solution of solute B.
c) A solution can never have a concentration greater than its solubility at a given temperature.
d) A finely divided solute dissolves faster because more surface area is exposed to the solvent.
e) Stirring a solution increases the rate of crystallization.
f) Crystallization ceases when equilibrium is reached.
g) All solubilities increase at higher temperatures.
h) Increasing air pressure over water increases the solubility of nitrogen in the water.
i) An ionic solute is more likely to dissolve in a nonpolar solvent than in a polar solvent.
j) (Optional) The molarity of a solution changes slightly with temperature but the molality does not.
k) (Optional) If an acid and a base react on a 2:1 mole ratio, there are twice as many equivalents of acid as there are of base in the reaction.
l) The concentration of a primary standard is found by titration.
m) Colligative properties of a solution are independent of the kinds of solute particles, but they are dependent on particle concentration.

133. When you heat water on a stove, small bubbles appear long before the water begins to boil. What are they? Explain why they appear.

134. Antifreeze is put into the water in an automobile to prevent it from freezing in winter. What does the antifreeze do to the boiling point of the water, if anything?

Ethylene glycol, HO—CH_2—CH_2—OH, is the most widely used automotive antifreeze.

135. Does percentage concentration by mass of a solution depend on temperature?

More Challenging Problems

136. Suggest a way to separate two miscible liquids. Do you know of any widespread industrial process in which this is done?

137. In Chapter 2, we explain that physical properties must be employed to separate components of a mixture. Suggest a way to separate two immiscible liquids.

138. The text gives no explanation of forces that must be overcome during the dissolving process. Can you imagine what the forces might be?

139. Silver acetate has a solubility of 2.52 g/100 g water at 80°C and 1.02 g/100 g water at 20°C. How do you prepare a supersaturated solution of silver acetate? Why does crystallization not occur as soon as the ion concentration is greater than the concentration of a saturated solution?

140. Consider a sample of pure sugar that has been finely powdered. What advantage does it have over granular (crystalline) sugar in making bakery goods? Would there be any advantage or disadvantage in using it for sweetening coffee? Explain.

141. The density of 18.0% HCl is 1.09 g/mL. Calculate its molarity.

142. Methyl ethyl ketone, C_4H_8O is a solvent popularly known as MEK that is used to cement plastics. What mass in grams of MEK must be dissolved in 1.00×10^2 mL of benzene, density 0.879 g/mL, to yield a 0.254 molal solution?

143. Calculate the mass of $H_2C_2O_4 \cdot 2\,H_2O$ required to make 2.50×10^2 mL of 0.500 N $H_2C_2O_4$ that will be used in the reaction $H_2C_2O_4 + 2\,OH^- \rightarrow C_2O_4^{2-} + 2\,H_2O$.

144. Sodium carbonate decahydrate is used as a base in the reaction $CO_3^{2-} + 2\,H^+ \rightarrow CO_2 + H_2O$. Calculate the mass of the hydrate needed to prepare 1.00×10^2 mL of 0.500 N sodium carbonate.

145. The iron(III) ion content of a solution may be found by precipitating it as iron(III) hydroxide and then decomposing the hydroxide to iron(III) oxide by heat. What mass in grams of iron(III) oxide can be collected from 35.0 mL of 0.516 M iron(III) nitrate?

146. A student adds 25.0 mL of 0.350 M sodium hydroxide to 45.0 mL of 0.125 M copper(II) sulfate. What mass in grams of copper(II) hydroxide will precipitate?

147. A laboratory technician combines 25.0 mL of 0.269 M nickel(II) chloride with 30.0 mL 0.260 M potassium hydroxide. What mass in grams of nickel(II) hydroxide can precipitate?

148. An analytical procedure for finding the chloride ion concentration in a solution involves the precipitation of silver chloride: $Ag^+ + Cl^- \rightarrow AgCl$. What is the molarity of the chloride ion if 16.80 mL of 0.629 M silver nitrate (the source of silver ion) is needed to precipitate all of the chloride ion in a 25.00-mL sample of the unknown?

Charles Steele

Silver nitrate solution is added to an unknown chloride-ion solution, yielding a precipitate of silver chloride.

149. Calculate the hydroxide ion concentration in a 20.00-mL sample of an unknown if 14.75 mL 0.248 M sulfuric acid is used in a neutralization reaction.

150. A 694-mg sample of impure sodium carbonate was titrated with 41.24 mL of 0.244 M hydrochloric acid. Calculate the percentage of sodium carbonate in the sample.

151. A student received a 599-mg sample of a mixture of sodium hydrogen phosphate and sodium dihydrogen phosphate. She is to find the percentage of each compound in the sample. After dissolving the mixture, she titrated it with 19.58 mL 0.201 M sodium hydroxide. If the only reaction is $NaH_2PO_4 + NaOH \rightarrow Na_2HPO_4 + H_2O$, find the required percentages.

152. A chemist combines 60.0 mL of 0.322 M potassium iodide with 20.0 mL of 0.530 M lead(II) nitrate. (a) What mass in grams of lead(II) iodide will precipitate? (b) What is the final molarity of the potassium ion? (c) What is the final molarity of the lead(II) or iodide ion, whichever one is in excess?

153. A solution is defined as a homogeneous mixture. Is a small sample of air a solution? Is the atmosphere a solution?

154. If you know either the percentage concentration of a solution or its molarity, what additional information must you have before you can convert to the other concentration?

Formulas

34. $Ca(NO_3)_2$
36. $CuSO_4$
37. KI
38. $AlBr_3$
39. $NiCl_2 \cdot 6\,H_2O$
41. $AgNO_3$
42. ZnI_2
43. KOH
44. $Mn(C_2H_3O_2)_2$
45. H_2SO_4
47. NaCl
48. NH_3
49. $AgNO_3$
50. NaOH
52. H_2SO_4
53. KNO_3
54. HCl
56. NH_4I
58. $Al(NO_3)_3$
59. NaCl
60. $FeCl_3$
71. KOH

91. $Mg(OH)_2$, $MgCl_2$, NaOH

93. $Ca_3(PO_4)_2$, Na_3PO_4, $Ca(NO_3)_2$

97. NaOH

98. $Ba(OH)_2$, $Ca(OH)_2$, Ca^{2+}, CaI_2

100. KOH

102. NaOH

104. KOH, HBr

109. Na_2CO_3, H_2SO_4

110. NaOH

111. NaOH

112. $HClO_4$, KOH

113. NaOH, H_3PO_4

114. KOH

115. H_2SO_4

116. NaOH

144. $Na_2CO_3 \cdot 10\,H_2O$

145. $Fe(OH)_3$, Fe_2O_3, $Fe(NO_3)_3$

146. NaOH, $CuSO_4$, $Cu(OH)_2$

147. $NiCl_2$, KOH, $Ni(OH)_2$

148. $AgNO_3$

149. OH^-, H_2SO_4

150. Na_2CO_3, HCl

152. KI, $Pb(NO_3)_2$, PbI_2

Answers to Target Checks

1. True: (a) and (b). (c) A solution is always made up of two or more pure substances. (d) The different parts of a solution are too small to be detected visually.
2. (a) The solutes are A and B; water is the solvent. (b) Degree of saturation cannot be estimated without knowing the solubility of the compound. (c) The question has no meaning because the term *dilute* compares solutions of the same solute, not different solutes.
3. (a) Equal to. (b) Less than. (c) More than. The steadily increasing crystallization rate "subtracts from" the constant dissolving rate, reducing the net rate as time goes on. The net rate eventually reaches zero at equilibrium.
4. Main criterion: Do substances have similar intermolecular attractions? If yes, they are probably soluble. Look for similarities in polarity, hydrogen bonding capability, and size, each of which contributes to similar intermolecular forces.

Solutions to Practice Exercises

1. $$\frac{2.00\text{ kg}}{50.0\text{ kg}} \times 100\% = 4.00\%$$

2. $$5.00 \times 10^2\text{ mL solution} \times \frac{1.00\text{ g solution}}{1.00\text{ mL solution}} \times \frac{0.89\text{ g NaCl}}{100\text{ g solution}} = 4.5\text{ g NaCl}$$

$$5.00 \times 10^2\text{ g solution} - 4.5\text{ g NaCl} = 496\text{ g } H_2O = 496\text{ mL } H_2O$$

3. $$10.4\text{ g KCl} \times \frac{1\text{ mol KCl}}{74.55\text{ g KCl}} = 0.140\text{ mol KCl};$$

$$5.00 \times 10^2\text{ mL} \times \frac{1\text{ L}}{1000\text{ mL}} = 0.500\text{ L}$$

$$M = \frac{\text{mol}}{\text{L}} = \frac{0.140\text{ mol KCl}}{0.500\text{ L}} = 0.280\text{ M}$$

4. $$2.00\text{ L} \times \frac{0.12\text{ mol } Na_2SO_4}{\text{L}} \times \frac{142.04\text{ g } Na_2SO_4}{\text{mol } Na_2SO_4} = 34\text{ g } Na_2SO_4$$

5. $$5.29\text{ g } (NH_4)_2SO_4 \times \frac{1\text{ mol } (NH_4)_2SO_4}{132.14\text{ g } (NH_4)_2SO_4} \times \frac{1\text{ L}}{0.10\text{ mol } (NH_4)_2SO_4} \times \frac{1000\text{ mL}}{\text{L}} = 4.00 \times 10^2\text{ mL}$$

6. $$80.0\text{ mg } K_3PO_4 \times \frac{1\text{ g } K_3PO_4}{1000\text{ mg } K_3PO_4} \times \frac{1\text{ mol } K_3PO_4}{212.27\text{ g } K_3PO_4} = 3.77 \times 10^{-4}\text{ mol } K_3PO_4$$

$$25\text{ mL } H_2O \times \frac{1.00\text{ g } H_2O}{\text{mL } H_2O} \times \frac{1\text{ kg } H_2O}{1000\text{ g } H_2O} = 0.025\text{ kg } H_2O$$

$$m = \frac{\text{mol solute}}{\text{kg solvent}} = \frac{3.77 \times 10^{-4}\text{ mol } K_3PO_4}{0.025\text{ kg } H_2O} = 0.015\text{ m } K_3PO_4$$

7. $$2.00\text{ L } H_2O \times \frac{1000\text{ mL } H_2O}{\text{L } H_2O} \times \frac{1\text{ g } H_2O}{\text{mL } H_2O} \times \frac{1\text{ kg } H_2O}{1000\text{ g } H_2O} \times \frac{0.75\text{ mol } Ca(NO_3)_2}{\text{kg } H_2O} \times \frac{164.10\text{ g } Ca(NO_3)_2}{\text{mol } Ca(NO_3)_2} = 2.5 \times 10^2\text{ g } Ca(NO_3)_2$$

8. (a) 1 eq acid/mol and 2 eq base/mol; (b) 1 eq acid/mol and 1 eq base/mol

9. (a) $$\frac{127.9\text{ g HI}}{1\text{ eq HI}} = 127.9\text{ g HI/eq HI};$$

(b) $$\frac{181.19\text{ g } H_3C_9H_8NO_3}{1\text{ eq } H_3C_9H_8NO_3} = 181.19\text{ g } H_3C_9H_8NO_3/\text{eq } H_3C_9H_8NO_3$$

10. $$205\text{ g HI} \times \frac{1\text{ eq HI}}{127.9\text{ g HI}} = 1.60\text{ eq HI}$$

11. $$4.58\text{ g } Sr(OH)_2 \times \frac{2\text{ eq } Sr(OH)_2}{121.64\text{ g } Sr(OH)_2} = 0.0753\text{ eq } Sr(OH)_2$$

$$2.50 \times 10^2\text{ mL} \times \frac{1\text{ L}}{1000\text{ mL}} = 0.250\text{ L}$$

$$N \equiv \frac{\text{eq}}{\text{L}} = \frac{0.0753\text{ eq } Sr(OH)_2}{0.250\text{ L}} = 0.301\text{ N } Sr(OH)_2$$

12. $$211\text{ mL} \times \frac{1\text{ L}}{1000\text{ mL}} \times \frac{0.50\text{ eq HCl}}{\text{L}} = 0.11\text{ eq HCl}$$

13. $$2.50 \times 10^2\text{ mL} \times \frac{1\text{ L}}{1000\text{ mL}} \times \frac{0.20\text{ eq } H_3PO_4}{\text{L}} \times \frac{97.99\text{ g } H_3PO_4}{2\text{ eq } H_3PO_4} = 2.4\text{ g } H_3PO_4$$

14. $$V_c = \frac{V_d \times M_d}{M_c} = \frac{0.800\text{ L} \times 0.10\text{ M}}{17\text{ M}} \times \frac{1000\text{ mL}}{\text{L}} = 4.7\text{ mL}$$

15. $$M_d = \frac{V_c \times M_c}{V_d} = \frac{10.0\text{ mL} \times 1.15\text{ M}}{(10.0 + 1.00 \times 10^3)\text{ mL}} = 0.0114\text{ M}$$

16. $CoCl_2 + 2\ NaOH \rightarrow Co(OH)_2 + 2\ NaCl$

$$25.0\ mL\ NaOH \times \frac{1\ L\ NaOH}{1000\ mL\ NaOH} \times \frac{0.34\ mol\ NaOH}{L\ NaOH} \times$$

$$\frac{1\ mol\ CoCl_2}{2\ mol\ NaOH} \times \frac{129.83\ g\ CoCl_2}{mol\ CoCl_2} = 0.55\ g\ CoCl_2$$

17. $3\ LiNO_3(aq) + Na_3PO_4(aq) \rightarrow Li_3PO_4(s) + 3\ NaNO_3(aq)$

$$1.00 \times 10^2\ mL\ Na_3PO_4 \times \frac{0.080\ mol\ Na_3PO_4}{1000\ mL\ Na_3PO_4} \cdot$$

$$\frac{3\ mol\ LiNO_3}{1\ mol\ Na_3PO_4} \times \frac{1000\ mL\ LiNO_3}{0.105\ mol\ LiNO_3} = 2.3 \times 10^2\ mL\ LiNO_3$$

18. $Na_2CO_3(aq) + 2\ HCl(aq) \rightarrow 2\ NaCl(aq) + H_2CO_3(aq) \rightarrow 2\ NaCl(aq) + H_2O(\ell) + CO_2(g)$

$$2.00\ g\ Na_2CO_3 \times \frac{1\ mol\ Na_2CO_3}{105.99\ g\ Na_2CO_3} \times \frac{2\ mol\ HCl}{1\ mol\ Na_2CO_3}$$
$$= 0.0377\ mol\ HCl$$

$$M \equiv \frac{mol}{L} = \frac{0.0377\ mol\ HCl}{19.30\ mL} \times \frac{1000\ mL}{L} = 1.95\ M\ HCl$$

19. $Ba(OH)_2(aq) + 2\ HCl(aq) \rightarrow BaCl_2(aq) + 2\ H_2O(\ell)$

$$12.50\ mL\ HCl \times \frac{1.95\ mol\ HCl}{1000\ mL} \times \frac{1\ mol\ Ba(OH)_2}{2\ mol\ HCl}$$
$$= 0.0122\ mol\ Ba(OH)_2$$

$$M \equiv \frac{mol}{L} = \frac{0.0122\ mol\ Ba(OH)_2}{20.00\ mL} \times \frac{1000\ mL}{L}$$
$$= 0.610\ M\ Ba(OH)_2$$

20. $$2.00\ g\ Na_2CO_3 \times \frac{2\ eq\ HCl\ or\ Na_2CO_3}{105.99\ g\ Na_2CO_3} = 0.0377\ eq\ HCl$$

$$N \equiv \frac{eq}{L} = \frac{0.0377\ eq\ HCl}{19.30\ mL} \times \frac{1000\ mL}{L} = 1.95\ N\ HCl$$

21. $$N_2 = \frac{V_1N_1}{V_2} = \frac{12.50\ mL \times 1.95\ N}{20.0\ mL} = 1.22\ N\ Ba(OH)_2$$

22. $$17.5\ g\ C_6H_5NH_2 \times \frac{1\ mol\ C_6H_5NH_2}{93.13\ g\ C_6H_5NH_2} = 0.188\ mol\ C_6H_5NH_2$$

$$m \equiv \frac{mol\ solute}{kg\ solvent} = \frac{0.188\ mol\ C_6H_5NH_2}{113\ g\ H_2O} \times \frac{1000\ g\ H_2O}{kg\ H_2O}$$
$$= 1.66\ m\ CO(NH_2)_2$$

$$\Delta T_b = K_b \times m = \frac{0.51°C}{m} \times 1.66\ m = 0.85°C;$$

$$100.00°C + 0.85°C = 100.85°C$$

23. $$m = \frac{\Delta T_b}{K_b} = \frac{(6.50 - 1.33)°C}{20.2°C/m} = 0.256\ m$$

$$100.0\ g\ cyclohexane \times \frac{1\ kg\ cyclohexane}{1000\ g\ cyclohexane}$$

$$\times \frac{0.256\ mol\ solute}{kg\ cyclohexane} = 0.0256\ mol\ solute$$

$$\frac{2.00\ g}{0.0256\ mol} = 78.1\ g/mol$$

Answers to Concept-Linking Exercises

You may have found more relationships or relationships other than the ones given in these answers.

1. A solution is a homogeneous mixture, made up of two or more pure substances. Two solutions of the same substance may have different percentage compositions.
2. Solubility is a measure of how much solute will dissolve in a given amount of solvent at a given temperature. A solution holding that amount of solute is saturated; less than that amount, unsaturated; and more than that amount, supersaturated.
3. Solubility is a measure of how much solute will dissolve in a given amount of solvent at a given temperature. In general, substances with similar intermolecular forces will dissolve in one another. The greater the partial pressure of a solute gas over a liquid solvent is, the greater its solubility will be. The solubilities of most solids increase with increasing temperature. The solubilities of gases in liquids are generally lower at higher temperatures.
4. A primary standard may be weighed accurately, dissolved, and diluted to an accurately determined volume, yielding a solution whose concentration is known with a high degree of accuracy. It may be used to standardize—find the concentration of—a second solution by titration, using an indicator to signal when the two reactants are present in precisely the molar quantities by which they react. Concentration is expressed in molarity, moles of solute per liter of solution, mol/L.
5. Normality is the number of equivalents per liter of solution. For acids, 1 equivalent is the quantity that yields 1 mole of hydrogen ions in solution. For bases, 1 equivalent is the quantity that reacts with 1 mole of hydrogen ions. Equivalent mass is the number of grams per equivalent.
6. The depression of the freezing point and elevation of the boiling point of a solvent are proportional to the solute particle concentration of a solution expressed in molality, moles of solute particles per kilogram of solvent. The molal freezing point depression constant and molal boiling point elevation constant are proportionality constants for those proportionalities. These are colligative properties because they depend on solute particle concentration without regard to the identity of the solute particles.

Solutions to Blue-Numbered Questions, Exercises, and Problems

1. True. Provided that there is no chemical reaction, gases will always combine to form a homogeneous mixture, which is, by definition, a solution.

3. If particles were visible, there would be distinctly different phases and the mixture would not be homogeneous and therefore not a solution.

5. a) Salt is solute; water is solvent. When a solid is dissolved in a liquid, the solid is the solute and the liquid is the solvent. b) Copper is solute; silver is solvent. c) Oxygen is solute; nitrogen is solvent. For (b) and (c), the solute is the substance present in the smaller amount.

7. A saturated solution is at its solubility limit, so it is concentrated. A concentrated solution is not necessarily saturated. A concentrated solution contains a relatively large amount of solute; a saturated solution cannot hold more solute.

9. a) All salt added will be solid in the beaker. b) Some or all salt added will dissolve; little or no salt will be solid in the beaker. c) All added salt and some previously dissolved salt will be solid in the beaker.

11. Temperature. Usually, the higher the temperature, the higher the solubility.

13. Acetic acid is soluble in water because it is dispersed uniformly throughout the solution. It is also miscible, a term usually used to express the solubility of liquids in each other.

15. Cations are attracted to the negative portion of the water molecule; anions are attracted to the positive portion.

17. As dissolved solute particles move through the solution, they come into contact with each other and with undissolved solute and return to the solid state.

19. When dissolving begins, the crystallization rate is zero. Dissolving rate remains constant, and as the solution becomes more concentrated, crystallization rate increases. Concentration of the solute in solution continues to increase until the crystallization rate equals the dissolving rate.

21. Never.

23. A supersaturated solution is unstable, and a physical disturbance such as stirring will start crystallization.

25. Finely dividing a solid offers more surface area per unit of mass. Stirring or agitating the solution prevents concentration buildup at the solute surface, which minimizes crystallization rate. All physical processes speed up at higher temperatures because particle movement is more rapid.

27. (a) formic acid; (c) methylamine. Both compounds exhibit hydrogen bonding in water. Hexane and tetrafluoromethane are nonpolar and only exhibit induced dipole forces.

29. Glycerine exhibits hydrogen bonding, as does water, so they are miscible. Hexane is nonpolar.

31. Carbon dioxide, CO_2.

33. $\dfrac{2.32 \text{ g solute}}{(2.32 + 81.0) \text{ g solution}} \times 100\% = 2.78\%$

35. $415 \text{ g solution} \times \dfrac{58.0 \text{ g } NH_4NO_3}{100 \text{ g solution}} = 241 \text{ g } NH_4NO_3;$

$415 \text{ g solution} - 241 \text{ g } NH_4NO_3 = 174 \text{ g } H_2O = 174 \text{ mL } H_2O$

37. $2.41 \text{ g KI} \times \dfrac{1 \text{ mol KI}}{166.0 \text{ g KI}} = 0.0145 \text{ mol KI};$

$\dfrac{0.0145 \text{ mol KI}}{50.0 \text{ mL}} \times \dfrac{1000 \text{ mL}}{\text{L}} = 0.290 \text{ M KI}$

39. $18.0 \text{ g } NiCl_2 \times \dfrac{1 \text{ mol } NiCl_2}{129.59 \text{ g } NiCl_2} = 0.139 \text{ mol } NiCl_2;$

$30.0 \text{ g } NiCl_2 \cdot 6\,H_2O \times \dfrac{1 \text{ mol } NiCl_2 \cdot 6\,H_2O}{237.69 \text{ g } NiCl_2 \cdot 6\,H_2O}$
$= 0.126 \text{ mol } NiCl_2 \cdot 6\,H_2O$; The anhydrous compound has more moles and thus has the higher concentration:
$\dfrac{0.139 \text{ mol } NiCl_2}{90.0 \text{ mL}} \times \dfrac{1000 \text{ mL}}{\text{L}} = 1.54 \text{ M } NiCl_2$

41. $2.50 \times 10^2 \text{ mL} \times \dfrac{1 \text{ L}}{1000 \text{ mL}} \times \dfrac{0.058 \text{ mol } AgNO_3}{\text{L}} \times \dfrac{169.9 \text{ g } AgNO_3}{\text{mol } AgNO_3} = 2.5 \text{ g } AgNO_3$

43. $2.50 \text{ L} \times \dfrac{1.40 \text{ mol KOH}}{\text{L}} \times \dfrac{56.11 \text{ g KOH}}{\text{mol KOH}} = 196 \text{ g KOH}$

45. $5.19 \text{ mol } H_2SO_4 \times \dfrac{1 \text{ L}}{18 \text{ mol } H_2SO_4} = 0.29 \text{ L}$

47. $8.33 \text{ g NaCl} \times \dfrac{1 \text{ mol NaCl}}{58.44 \text{ g NaCl}} \times \dfrac{1 \text{ L}}{0.132 \text{ mol NaCl}} = 1.08 \text{ L}$

49. $55.7 \text{ mL} \times \dfrac{1 \text{ L}}{1000 \text{ mL}} \times \dfrac{0.204 \text{ mol } AgNO_3}{\text{L}} = 0.0114 \text{ mol } AgNO_3$

51. $25.0 \text{ mL} \times \dfrac{1 \text{ L}}{1000 \text{ mL}} \times \dfrac{0.0841 \text{ mol } KMnO_4}{\text{L}} = 0.00210 \text{ mol } KMnO_4$

53. $\dfrac{101.11 \text{ g } KNO_3}{\text{mol } KNO_3} \times \dfrac{3.30 \text{ mol } KNO_3}{\text{L}} \times \dfrac{1 \text{ L}}{1000 \text{ mL}} \times \dfrac{1 \text{ mL}}{1.15 \text{ g soln}} \times 100\% = 29.0\% \; KNO_3$

55. $44.9 \text{ g } C_{10}H_8 \times \dfrac{1 \text{ mol } C_{10}H_8}{128.16 \text{ g } C_{10}H_8} = 0.350 \text{ mol } C_{10}H_8;$

$\dfrac{0.350 \text{ mol } C_{10}H_8}{175 \text{ g}} \times \dfrac{1000 \text{ g}}{\text{kg}} = 2.00 \text{ m}$

57. $4.00 \times 10^2 \text{ g ethanol} \times \dfrac{1 \text{ kg ethanol}}{1000 \text{ g ethanol}} \times \dfrac{4.70 \text{ mol } (CH_3CH_2)_2NH}{\text{kg ethanol}} \times \dfrac{73.14 \text{ g } (CH_3CH_2)_2NH}{\text{mol } (CH_3CH_2)_2NH} = 138 \text{ g } (CH_3CH_2)_2NH$

59. $97.7 \text{ mg NaCl} \times \dfrac{1 \text{ g NaCl}}{1000 \text{ mg NaCl}} \times \dfrac{1 \text{ mol NaCl}}{58.44 \text{ mol NaCl}} \times \dfrac{1 \text{ kg } H_2O}{2.80 \times 10^{-3} \text{ mol NaCl}} \times \dfrac{1000 \text{ g } H_2O}{\text{kg } H_2O} \times \dfrac{1 \text{ mL } H_2O}{1 \text{ g } H_2O} = 597 \text{ mL } H_2O$

61. Equivalent mass is the mass of a substance that reacts with 1 mole of hydrogen or hydroxide ions. LiOH has 1 mole of OH^- ions, so equivalent mass = molar mass. H_2SO_4 can release 1 or 2 moles of H^+ ions, so equivalent mass = molar mass or 1/2 of molar mass.

63. 1 eq/mol HNO_2; 1 eq/mol H_2SeO_4

65. 2 eq/mol $Cu(OH)_2$; 3 eq/mol $Fe(OH)_3$

67. 47.02 g HNO_2/eq; 144.98 g H_2SeO_4/eq

69. 48.78 g $Cu(OH)_2$/eq; 35.62 g $Fe(OH)_3$/eq

71. $2.25 \text{ g KOH} \times \dfrac{1 \text{ eq KOH}}{56.11 \text{ g KOH}} = 0.0401 \text{ eq KOH};$

$\dfrac{0.0401 \text{ eq KOH}}{2.50 \times 10^2 \text{ mL}} \times \dfrac{1000 \text{ mL}}{\text{L}} = 0.160 \text{ N KOH}$

73. $7.50 \times 10^2\ mL \times \frac{1\ L}{1000\ mL} \times \frac{0.200\ eq\ NaHSO_4}{L}$

$\times \frac{120.06\ g\ NaHSO_4}{eq} = 18.0\ g\ NaHSO_4$

75. $6.69\ g\ H_2C_2O_4 \times \frac{1\ eq\ H_2C_2O_4}{90.04\ g\ H_2C_2O_4} = 0.0743\ eq\ H_2C_2O_4;$

$\frac{0.0743\ eq\ H_2C_2O_4}{2.00 \times 10^2\ mL} \times \frac{1000\ mL}{L} = 0.372\ N\ H_2C_2O_4$

77. (a) 0.965 M NaOH; (b) 0.119 M H_3PO_4

79. $73.1\ mL \times \frac{1\ L}{1000\ mL} \times \frac{0.834\ eq\ NaOH}{L} = 0.0610\ eq\ NaOH$

81. $0.788\ eq\ KMnO_4 \times \frac{1\ L}{0.492\ eq\ KMnO_4} = 1.60\ L$

83. $M_d = \frac{17\ M \times 45.0\ mL}{1.5\ L} \times \frac{1\ L}{1000\ mL} = 0.51\ M\ HC_2H_3O_2$

85. $V_c = \frac{0.69\ M \times 7.50 \times 10^2\ mL}{16\ M} = 32\ mL\ HNO_3$

87. $V_c = \frac{2.9\ eq/L \times 3.0\ L}{18\ mol/L} \times \frac{1\ mol}{2\ eq} \times \frac{1000\ mL}{L}$

$= 2.4 \times 10^2\ mL\ H_2SO_4$

89. $N_d = \frac{15\ mol/L \times 15.0\ mL}{2.50 \times 10^2\ mL} \times \frac{2\ eq}{mol} = 1.8\ NH_3PO_4$

91. $MgCl_2 + 2\ NaOH \rightarrow Mg(OH)_2 + 2\ NaCl$

$25.0\ mL \times \frac{1\ L}{1000\ mL} \times \frac{0.398\ mol\ MgCl_2}{L} \times$

$\frac{1\ mol\ Mg(OH)_2}{1\ mol\ MgCl_2} \times \frac{58.33\ g\ Mg(OH)_2}{mol\ Mg(OH)_2} = 0.580\ g\ Mg(OH)_2$

93. $2\ Na_3PO_4 + 3\ Ca(NO_3)_2 \rightarrow 6\ NaNO_3 + Ca_3(PO_4)_2$

$100.0\ mL \times \frac{1\ L}{1000\ mL} \times \frac{0.130\ mol\ Ca(NO_3)_2}{L} \times$

$\frac{1\ mol\ Ca_3(PO_4)_2}{3\ mol\ Ca(NO_3)_2} \times \frac{310.18\ g\ Ca_3(PO_4)_2}{mol\ Ca_3(PO_4)_2} = 1.34\ g\ Ca_3(PO_4)_2$

95. $2.00\ L \times \frac{273\ K}{295\ K} \times \frac{789\ torr}{1\ bar} \times \frac{1.013\ bar}{760\ torr} \times \frac{1\ mol\ H_2}{22.7\ L}$

$= 0.0857\ mol\ H_2$ *or*

$n = \frac{PV}{RT} = 789\ torr \times 2.00\ L \times \frac{mol \cdot K}{62.4\ L \cdot torr}$

$\times \frac{1}{295\ K} = 0.0857\ mol\ H_2$

$0.0857\ mol\ H_2 \times \frac{6\ mol\ NaOH}{3\ mol\ H_2} \times \frac{1\ L}{1.50\ mol\ NaOH}$

$\times \frac{1000\ mL}{L} = 114\ mL$

97. $NH_2SO_3H + NaOH \rightarrow NH_2SO_3Na + H_2O$

$8.74\ g\ NH_2SO_3H \times \frac{1\ mol\ NH_2SO_3H}{97.10\ g\ NH_2SO_3H} \times \frac{1\ mol\ NaOH}{1\ mol\ NH_2SO_3H}$

$\times \frac{1\ L}{0.842\ mol\ NaOH} \times \frac{1000\ mL}{L} = 107\ mL$

99. $5.038\ g\ HC_7H_5O_2 \times \frac{1\ mol\ HC_7H_5O_2}{122.12\ g\ HC_7H_5O_2} \times \frac{1\ mol\ Na_2CO_3}{2\ mol\ HC_7H_5O_2}$

$= 0.02063\ mol\ Na_2CO_3; \frac{0.02063\ mol\ Na_2CO_3}{51.89\ mL} \times \frac{1000\ mL}{L}$

$= 0.3976\ M\ Na_2CO_3$

101. $H_2C_4H_2O_4 + 2\ KOH \rightarrow K_2C_4H_2O_4 + 2\ H_2O$

$1.45\ g\ H_2C_4H_2O_4 \times \frac{1\ mol\ H_2C_4H_2O_4}{116.07\ g\ H_2C_4H_2O_4}$

$\times \frac{2\ mol\ KOH}{1\ mol\ H_2C_4H_2O_4} = 0.0250\ mol\ KOH$

$\frac{0.0250\ mol\ KOH}{50.0\ mL} \times \frac{1000\ mL}{L} = 0.500\ M\ KOH$

103. $37.80\ mL \times \frac{1\ L}{1000\ mL} \times \frac{0.4052\ mol\ NaHCO_3}{L}$

$\times \frac{1\ mol\ H_2SO_4}{2\ mol\ NaHCO_3} = 7.658 \times 10^{-3}\ mol\ H_2SO_4$

$\frac{7.658 \times 10^{-3}\ mol\ H_2SO_4}{20.00\ mL} \times \frac{1000\ mL}{L} = 0.3829\ M\ H_2SO_4$

105. At 2 eq/mol, 0.3976 Na_2CO_3 = 0.7952 N Na_2CO_3

107. At 1 eq/mol, 0.500 M KOH = 0.500 N KOH

109. $N_2 = \frac{0.405\ N \times 39.8\ mL}{25.0\ mL} = 0.645\ N\ Na_2CO_3$

111. $N_2 = \frac{0.402\ N \times 42.2\ mL}{50.0\ mL} = 0.339\ N\ H_2C_4H_4O_6$

113. $N_2 = \frac{0.208\ N \times 16.3\ mL}{20.0\ mL} = 0.170\ N\ H_3PO_4$ The normality of the acid depends on how many hydrogens react.

115. $\frac{1.21\ g}{30.7\ mL \times 0.170\ eq/L} \times \frac{1000\ mL}{L} = 232\ g/eq$

117. Partial pressure is a colligative property because it depends on the number of particles and is independent of their identity (for an ideal gas).

119. $27.2\ g\ C_6H_5NH_2 \times \frac{1\ mol\ C_6H_5NH_2}{93.13\ g\ C_6H_5NH_2}$

$= 0.292\ mol\ C_6H_5NH_2; \frac{0.292\ mol\ C_6H_5NH_2}{1.20 \times 10^2\ g\ H_2O}$

$\times \frac{1000\ g\ H_2O}{kg\ H_2O} = 2.43\ m$

$\Delta T_b = \frac{0.51°C}{m} \times 2.43\ m = 1.2°C; T_b = 101.2°C;$

$\Delta T_f = \frac{1.86°C}{m} \times 2.43\ m = 4.52°C; T_f = -4.52°C$

121. $2.12\ g\ C_{10}H_8 \times \frac{1\ mol\ C_{10}H_8}{128.16\ g\ C_{10}H_8} = 0.0165\ mol\ C_{10}H_8;$

$\frac{0.0165\ mol\ C_{10}H_8}{32.0\ g\ C_6H_6} \times \frac{1000\ g\ C_6H_6}{kg\ C_6H_6} = 0.516\ m$

$\Delta T_f = \frac{5.10°C}{m} \times 0.516\ m = 2.63°C;$

$T_f = 5.50°C - 2.63°C = 2.87°C$

123. $\Delta T_f = 16.6°C - 14.1°C = 2.5°C$;

$$m = 2.5°C \times \frac{1\ m}{3.90°C} = 0.64\ m$$

125. $m = (100.28 - 100.00)°C \times \frac{1\ m}{0.51°C} = 0.55\ m$;

$$6.00 \times 10^2\ g\ H_2O \times \frac{1\ kg\ H_2O}{1000\ g\ H_2O} \times \frac{0.55\ mol\ solute}{kg\ H_2O}$$

$$= 0.33\ mol\ solute; \frac{16.1\ g}{0.33\ mol} = 5.0 \times 10^1\ g/mol$$

127. $m = 9.6°C \times \frac{1\ m}{3.56°C} = 2.7\ m$; 90.0 g phenol

$$\times \frac{1\ kg\ phenol}{1000\ g\ phenol} \times \frac{2.7\ mol\ solute}{kg\ phenol} = 0.24\ mol\ solute;$$

$$\frac{12.4\ g}{0.24\ mol} = 52\ g/mol$$

129. $11.4\ g\ C_2H_5OH \times \frac{1\ mol\ C_2H_5OH}{46.07\ g\ C_2H_5OH} = 0.247\ mol\ C_2H_5OH$;

$$\frac{0.247\ mol\ C_2H_5OH}{2.00 \times 10^2\ g\ solvent} \times \frac{1000\ g\ solvent}{kg\ solvent} = 1.24\ m\ C_2H_5OH;$$

$$K_f = \frac{28.7°C - 22.5°C}{1.24\ m} = 5.0°C/m$$

132. True: a, d, h, j, m. False: b, c, e, f, g, i, k, l.

133. The bubbles are dissolved air (mostly nitrogen and oxygen) that become less soluble at higher temperatures.

134. It raises the boiling point.

135. No

136. Distillation is one method. It is used to separate petroleum into its components, which includes many products such as gases, gasoline, kerosene, fuel oil, lubricating oil, and asphalt.

138. Attractions between solute particles and attractions between solvent particles.

139. Dissolve more than 1.02 g of silver acetate at a temperature greater than 20°C; then cool the solution without disturbance. Crystallization does not occur at the solubility limit because solute particles are not properly organized for crystal formation.

140. Finely powdered pure sugar has more surface area than granular sugar and therefore dissolves more quickly. Both sweeten coffee equally.

141. $\frac{18.0\ g\ HCl}{100\ g\ soln} \times \frac{1\ mol\ HCl}{36.46\ g\ HCl} \times \frac{1.09\ g\ soln}{mL\ soln}$

$$\times \frac{1000\ mL\ soln}{L\ soln} = 5.38\ M\ HCl$$

142. $1.00 \times 10^2\ mL\ B \times \frac{0.879\ g\ B}{mL\ B} \times \frac{1\ kg\ B}{1000\ g\ B}$

$$\times \frac{0.254\ mol\ C_4H_8O}{kg\ B} \times \frac{72.10\ g\ C_4H_8O}{mol\ C_4H_8O} = 1.61\ g\ C_4H_8O$$

143. $2.50 \times 10^2\ mL \times \frac{1\ L}{1000\ mL} \times \frac{0.500\ eq\ H_2C_2O_4}{L}$

$$\times \frac{126.07\ g\ H_2C_2O_4 \cdot 2\ H_2O}{2\ eq\ H_2C_2O_4} = 7.88\ g\ H_2C_2O_4 \cdot 2\ H_2O$$

145. $2\ Fe(NO_3)_3 \rightarrow 2\ Fe(OH)_3 \rightarrow Fe_2O_3$; $35.0\ mL \times \frac{1\ L}{1000\ mL}$

$$\times \frac{0.516\ mol\ Fe(NO_3)_3}{L} \times \frac{1\ mol\ Fe_2O_3}{2\ mol\ Fe(NO_3)_3}$$

$$\times \frac{159.70\ g\ Fe_2O_3}{mol\ Fe_2O_3} = 1.44\ g\ Fe_2O_3$$

146. This limiting reactant problem is solved by the methods described in Sections 10.5 and 10.7.

$$2\ NaOH + CuSO_4 \rightarrow Cu(OH)_2 + Na_2SO_4$$

$$25.0\ mL \times \frac{1\ L}{1000\ mL} \times \frac{0.350\ mol\ NaOH}{L}$$

$$\times \frac{1\ mol\ Cu(OH)_2}{2\ mol\ NaOH} \times \frac{97.57\ g\ Cu(OH)_2}{mol\ Cu(OH)_2}$$

$$= 0.427\ g\ Cu(OH)_2$$

$$45.0\ mL \times \frac{1\ L}{1000\ mL} \times \frac{0.125\ mol\ CuSO_4}{L}$$

$$\times \frac{1\ mol\ Cu(OH)_2}{1\ mol\ CuSO_4} \times \frac{97.57\ g\ Cu(OH)_2}{mol\ Cu(OH)_2} = 0.549\ g\ Cu(OH)_2$$

0.427 g of $Cu(OH)_2$ will precipitate.

148. $16.80\ mL \times \frac{1\ L}{1000\ mL} \times \frac{0.629\ mol\ AgNO_3}{L}$

$$\times \frac{1\ mol\ Cl^-}{1\ mol\ AgNO_3} = 0.0106\ mol\ Cl^-$$

$$\frac{0.0106\ mol\ Cl^-}{25.00\ mL} \times \frac{1000\ mL}{L} = 0.424\ M\ Cl^-$$

150. $Na_2CO_3 + 2\ HCl \rightarrow 2\ NaCl + H_2O + CO_2$

$$41.24\ mL \times \frac{1\ L}{1000\ mL} \times \frac{0.244\ mol\ HCl}{L}$$

$$\times \frac{1\ mol\ Na_2CO_3}{2\ mol\ HCl} \times \frac{105.99\ g\ Na_2CO_3}{mol\ Na_2CO_3}$$

$$\times \frac{1000\ mg\ Na_2CO_3}{g\ Na_2CO_3} = 533\ mg\ Na_2CO_3$$

$$\frac{533\ mg\ Na_2CO_3}{694\ mg\ sample} \times 100\% = 76.8\%\ Na_2CO_3$$

151. $19.58\ mL \times \frac{0.201\ mmolNaOH}{mL\ NaOH} \times \frac{1\ mmol\ NaH_2PO_4}{1\ mmol\ NaOH}$

$$\times \frac{119.98\ mg\ NaH_2PO_4}{mmol\ NaH_2PO_4} = 472\ mg\ NaH_2PO_4$$

$$\frac{472\ mg}{599\ mg} \times 100\% = 78.8\%\ NaH_2PO_4; 100 - 78.8$$

$$= 21.2\%\ Na_2HPO_4$$

152. a) $2\ KI + Pb(NO_3)_2 \rightarrow PbI_2 + 2\ KNO_3$

$$60.0\ mL \times \frac{1\ L}{1000\ mL} \times \frac{0.322\ mol\ KI}{L} \times \frac{1\ mol\ PbI_2}{2\ mol\ KI} \times \frac{461.0\ g\ PbI_2}{mol\ PbI_2} = 4.45\ g\ PbI_2$$

$$20.0\ mL \times \frac{1\ L}{1000\ mL} \times \frac{0.530\ mol\ Pb(NO_3)_2}{L} \times \frac{1\ mol\ PbI_2}{1\ mol\ Pb(NO_3)_2} \times \frac{461.0\ g\ PbI_2}{mol\ PbI_2} = 4.89\ g\ PbI_2$$

4.45 g PbI_2 will precipitate

b)

	$2\ KI$	+	$Pb(NO_3)_2$	→	PbI_2	+	$2\ KNO_3$
Volume at start, mL	60.0		20.0				
Volume at start, L	0.0600		0.0200				
Molarity, mol/L	0.322		0.530				
Amount at start, mol	0.0193		0.0106				
Amount (mol) used (−), produced (+)	−0.0193		−0.00965		+0.00965		+0.0193
Amount at end, mol	0		0.0010		0.00965		0.0193

Total volume = 0.0600 L + 0.0200 L = 0.0800 L

$$\frac{0.0193\ mol\ KNO_3}{0.0800\ L} \times \frac{1\ mol\ K^+}{1\ mol\ KNO_3} = 0.241\ M\ K^+$$

c) $$\frac{0.0010\ mol\ Pb(NO_3)_2}{0.0800\ L} \times \frac{1\ mol\ Pb^{2+}}{1\ mol\ Pb(NO_3)_2} = 0.013\ M\ Pb^{2+}$$

153. A small sample of air is a homogeneous mixture and is therefore a solution. The atmosphere is a very tall sample that becomes less dense at higher elevations. The atmosphere is therefore not homogeneous; consequently, it is not a solution.

154. The density of a solution must be known in order to convert concentrations based on mass only (percentage, molality) to those based in volume (molarity, normality).

17 Acid–Base (Proton Transfer) Reactions

Charles D. Winters

◀ Did you know that your local grocery store is a major industrial outlet for acids and bases? It's a fact. They don't come in the kinds of bottles sold to college chemistry laboratories, but they are acids and bases nonetheless. If you doubt it, read some of the labels. The substances on the left in the photo are all acidic; those on the right are basic.

CHAPTER CONTENTS

Originally, the word *acid* described something that had a sour taste. The tastes of vinegar (acetic acid) and lemon juice (citric acid) are typical examples. Substances with such tastes have other common properties too. They impart certain colors to some substances, such as red cabbage juice, which is red in acid solutions; they react with carbonate ions and form carbon dioxide; they react with and neutralize a base; and they form hydrogen gas when they react with particular metals.

Traditionally, a base is something that tastes bitter. Bases impart a different color to some substances (red cabbage juice is yellow); they neutralize acids; and they form precipitates when added to solutions of many metal ions. Also, liquid solutions of bases feel slippery, or "soapy."

To understand these two distinct groups, we must ask: What particulate-level features in the chemical structure and composition of acids and bases are responsible for their characteristic macroscopic properties? Answering this question is the overarching goal of this chapter.

17.1 Is the Existence of Acid Molecules Exclusive to Earth?

This chapter is an introduction to fundamental concepts about a category of chemical reactions that have in common the transfer of a proton from one species to another. The species that is the proton source in such a reaction is called an *acid.* There are many subcategories of types of acids. One such subcategory is *amino acids*, which are so named because they have a group of atoms in the acid molecule called an *amine group* (—NH_2) and another group of atoms that includes the proton source, the acid group (—COOH). The hydrogen atom in the acid group releases just its proton when a proton-transfer reaction occurs. The electron of the hydrogen atom remains with the original molecule after this type of reaction occurs.

You are likely at least somewhat familiar with amino acids from a nutritional perspective. Perhaps you've heard of the essential amino acids, which are those that must be obtained from your diet because human bodies cannot transform other foods into these amino acids. The essential amino acids are available from a variety of food sources, including meat, fish, dairy, and vegetable sources such as legumes, whole grains, and nuts.

Functionally, amino acids are what DNA (deoxyribonucleic acid) directly codes for. DNA is the molecule that contains the genetic information for the functioning of all organisms. It is a polymer (*many parts*) molecule made up of nucleotide monomer (*one part*) units. There are four DNA nucleotides. The sequence of each set of three nucleotides codes for a specific amino acid. When a cell reads a DNA molecule, it is reading instructions for assembling a chain of amino acids. A long amino acid chain is a protein, and proteins serve many roles in an organism, including structural and functional roles. Thus, amino acids are an essential component of living organisms.

Is the existence of amino acids exclusive to Earth? In 1969, a meteorite that weighed over 200 pounds broke up into pieces as it fell to Earth over Australia. Fortunately, many fragments were preserved before significant contamination could take place (**Figure 17.1**). The initial analysis of non-mineral substances on the

Figure 17.1 A fragment of the Murchison meteorite. The name Murchison comes from the town in Australia over which it fell on September 28, 1969. The mineral composition of the meteorite indicates that it was formed relatively early in the history of the solar system, and since it contained a variety of carbon-containing molecules, such compounds were likely in existence well before life evolved on Earth.

meteorite indicated the presence of 6 amino acids that matched those known on Earth and *12 amino acids that do not occur on Earth!* Modern analysis has discovered 92 different amino acids that were in the meteorite, 73 of which do not occur naturally on Earth.

Many additional discoveries of amino acids originating beyond Earth have been made since 1969. For example, in 1994, an amino acid was detected in an interstellar gas and dust cloud, providing evidence that amino acid molecules can form on pieces of dust and eventually find their way to planets. In 2009, an amino acid was discovered for the first time on a comet by a team led by Jamie Cook of NASA (**Figure 17.2**). Since comets frequently landed on Earth early in its history, this could provide a clue about how the raw materials needed for life originated on Earth. Here in this chapter, you will begin to build your foundation of knowledge about acids in general. If your course covers Chapter 22 on Biochemistry, you will learn more about amino acids and how genetic information flows from DNA to amino acids and how these monomers are assembled to make protein polymers.

NASA/Goddard/Tim Childers

Figure 17.2 Dr. Jamie Elsila Cook is an astrochemist who works at NASA's Goddard Space Flight Center in Maryland. Her research is focused on detection and analysis of carbon-containing molecules originating beyond Earth.

17.2 The Arrhenius Theory of Acids and Bases (Optional)

Goal 1 Distinguish between an acid and a base according to the Arrhenius theory of acids and bases.

In 1884, Swedish scientist Svante Arrhenius (**Figure 17.3**) proposed that substances that are electrolytes—those that form electrically conducting solutions when dissolved in water—exist as charged particles in solution. These charged particles are called *ions*. Furthermore, Arrhenius hypothesized that an **acid** is a substance that increases the concentration of hydrogen ions in solution when dissolved in water, and a **base** is a substance that increases the concentration of hydroxide ions when dissolved in water.

As examples, when hydrogen chloride gas is bubbled through water, the concentration of hydrogen ion in the resulting solution increases. This can be explained if the hydrogen chloride molecules divide into hydrogen ions and chloride ions:

$$HCl(g) \xrightarrow{H_2O} H^+(aq) + Cl^-(aq)$$

When solid sodium hydroxide is dissolved in water, the concentration of hydroxide ion increases. Arrhenius reasoned that the ions in the crystal lattice of the solid must be released to move freely in water solution:

$$NaOH(s) \xrightarrow{H_2O} Na^+(aq) + OH^-(aq)$$

Edgar Fahs Smith Collection/Kislak Center for Special Collections, Rare Books and Manuscripts/University of Pennsylvania

Figure 17.3 Svante August Arrhenius (1859–1927). Arrhenius was awarded the 1903 Nobel Prize in Chemistry for proposing that ionic compounds can divide into oppositely charged particles in solution.

Thus, the *reason* why acid solutions have the macroscopic properties of sour taste, changing colors of some substances (**Figure 17.4[a]**), reaction with carbonate ion to form carbon dioxide gas (**Figure 17.4[b]**), reaction with some metals (**Figure 17.4[c]**), and so on, is because of reactions with the hydrogen ions in the acid solution. Correspondingly, the macroscopic properties of bases, such as bitter taste, production of a slippery feeling on skin, changing colors of some substances (Figure 17.4[a]), and so on, are explained by the chemical reactions of the hydroxide ion in the base solution.

Among the properties of some Arrhenius acids and bases is their ability to neutralize each other, forming a solution with the same hydrogen ion and hydroxide ion concentration as in pure water. This is explained at the particulate level by the net ionic equation for the reaction between a strong acid and a strong base:

$$H^+(aq) + OH^-(aq) \rightarrow H_2O(\ell)$$

A macroscopic property of acids is that they react with carbonate-ion-containing compounds to yield bubbles of carbon dioxide (Figure 17.4[b]). **Arrhenius acid–base theory** explains this by proposing that the carbon dioxide is the end result

a The juice of a red cabbage is normally blue-purple. On adding acid, the juice becomes more red. Adding base produces a yellow color.

b A piece of coral (mostly $CaCO_3$) dissolves in acid to give CO_2 gas.

c Zinc reacts with hydrochloric acid to produce zinc chloride and hydrogen gas.

Charles D. Winters

Figure 17.4 Some characteristic reactions of acids and bases. (a) Red cabbage juice changes color when added to acid or a base solution. (b) Substances containing carbonate ion, such as this piece of coral, react with acids to yield bubbles of carbon dioxide gas as one product of the reaction that occurs. (c) A number of metals react with acid solution to yield hydrogen gas as one product. In this photograph, zinc is reacting with hydrochloric acid.

The term *theory* is used in science to describe an explanation for a broad class of related phenomena. Thus, Arrhenius acid–base theory explains many specific acid–base reactions.

of the combination of hydrogen ions and carbonate ions. ◀ In Section 9.10, you learned that carbonic acid is one of the "unstable products" of an ion combination. It decomposes into carbon dioxide and water:

$$2\,H^+(aq) + CO_3^{2-}(aq) \rightarrow H_2CO_3(aq) \rightarrow CO_2(g) + H_2O(\ell)$$

Another macroscopic property of acids is that they react with some metals to form hydrogen gas (Figure 17.4[c]). This is explained by the reactivity of the hydrogen ion with metals above hydrogen in the activity series (Table 9.2):

$$Zn(s) + 2\,H^+(aq) \rightarrow H_2(g) + Zn^{2+}(aq)$$

A macroscopic property of solutions of bases is that they react with solutions of some metal ions to form solids. Arrhenius theory explains why. The net ionic equation for the precipitation of magnesium hydroxide is typical:

$$Mg^{2+}(aq) + 2\,OH^-(aq) \rightarrow Mg(OH)_2(s)$$

Target Check 17.1

According to the Arrhenius theory of acids and bases, how do you recognize an acid and a base?

17.3 The Brønsted–Lowry Theory of Acids and Bases

Goal 2 Given the equation for a Brønsted–Lowry acid–base reaction, or information from which it can be written, explain how or why it can be so classified.

3 Given the name or formula of a Brønsted–Lowry acid and the name or formula of a Brønsted–Lowry base, write the net ionic equation for the reaction between them.

An acid is a substance that increases the concentration of hydrogen ions in water solution. ▶ A base is a substance that increases the concentration of hydroxide ions when dissolved in water. When we examine the nature of acid–base *reactions*, we find two patterns. First, numerous reactions can be interpreted as the removal of a proton (hydrogen ion) from the acid molecule by a hydroxide ion. Second, substances other than hydroxide ions can remove protons from acid molecules.

◀ **P/Review** As the term *hydrogen ion* is used here, it is a proton. The most common hydrogen atom has one proton and one electron. In an acid–base reaction, the electron remains with the original reactant molecule; only the proton is removed. Its formula is H^+, which is a hydrogen ion. When in a water solution, this ion is hydrated—that is, chemically bonded to one or more water molecules. It is represented by the formula H_3O^+ and called a *hydronium ion*. See Section 6.9.

These observations were announced independently by Johannes N. Brønsted (**Figure 17.5**) and Thomas M. Lowry (**Figure 17.6**) in 1923. Their proposal is known by both their names. According to the **Brønsted–Lowry acid–base theory**, an acid–base reaction is a proton transfer reaction in which the proton is transferred from the acid to the base. An **acid** is a compound from which a proton can be removed, and a **base** is a compound that can remove a proton from an acid. According to this theory, anything that can take away a proton is a base. The hydroxide ion is only one example.

Lewis diagrams showing the formation of the hydronium ion when a hydrogen atom-containing molecule reacts with water (Section 6.9) illustrate a proton transfer reaction. Using nitric acid as an example, we have

Acid: H^+ source   Base: H^+ remover

$$HNO_3(aq) + H_2O(\ell) \longrightarrow H_3O^+(aq) + NO_3^-(aq)$$

Edgar Fahs Smith Collection/Kislak Center for Special Collections, Rare Books and Manuscripts/University of Pennsylvania

Figure 17.5 Johannes N. Brønsted (1879–1947) was primarily interested in studying thermochemistry, in addition to his more famous work with acids and bases.

The proton from the nitric acid molecule is transferred to one of the unshared electron pairs of the water molecule. The HNO_3 molecule is the acid, the proton source. The remover of the proton is the water molecule; water is a base in this reaction.

Another example of an acid–base reaction is the transfer of a proton from water to ammonia:

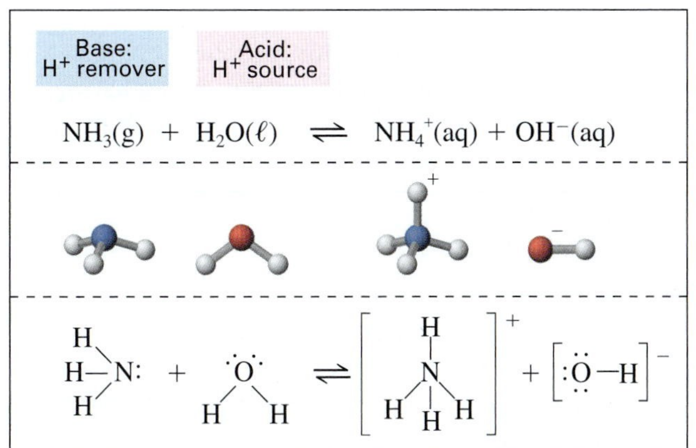

Edgar Fahs Smith Collection/Kislak Center for Special Collections, Rare Books and Manuscripts/University of Pennsylvania

Figure 17.6 Thomas M. Lowry (1874–1936) was the first professor of physical chemistry at Cambridge University.

This time water is an acid. It has a proton that can be taken by ammonia, the base. A substance that can behave as an acid in one case and a base in another, as water does in the two preceding reactions, is said to be **amphoteric**.

▶ **P/Review** A reversible reaction is one in which the products of the reaction can react with themselves to reproduce the original reactants. A double arrow in the equation for a reaction identifies it as reversible. Sometimes a longer arrow is used to point in the favored direction—to the major species in the reaction, those present in higher concentration—whereas the shorter arrow points to the minor species, those with lower concentrations.

In writing net ionic equations, only the major species are shown (see Section 9.4). Reversible reactions are introduced in Section 8.8, and they are used in discussing liquid–vapor equilibria in Section 15.5. Other important applications of reversible reactions appear in Chapters 18 and 19.

The double arrow in the equation for the ammonia–water reaction indicates that the reaction is reversible. ◀ It suggests correctly that the reaction reaches a chemical equilibrium. When a reversible reaction is read from left to right, the **forward reaction**, or the reaction in the **forward direction**, is described; from right to left, the change is the **reverse reaction**, or in the **reverse direction**. In 1 M $NH_3(aq)$, less than 1% of the ammonia is changed to ammonium ions. NH_3 is the major species in the solution, and NH_4^+ and OH^- are the minor species. The reverse reaction is therefore said to be favored.

HNO_3 is an acid, a proton source, in our first example reaction. NH_3 is a base, a proton remover, in the second reaction. Do you suppose that if we were to put HNO_3 and NH_3 together, the HNO_3 would release its proton to NH_3? The answer is yes. If NH_3 can take a proton from water, and if water can take a proton from HNO_3, then NH_3 can take a proton from HNO_3. Conventional and Lewis diagram equations describe the reaction:

Acid: H^+ source — Base: H^+ remover

$$HNO_3(aq) + NH_3(g) \longrightarrow NH_4^+(aq) + NO_3^-(aq)$$

Note that this is an acid–base reaction that does not directly involve either water or the hydroxide ion.

Reviewing the three acid–base reactions presented, we find that all fit into the general equation for a Brønsted–Lowry proton transfer reaction:

$$B + HA \rightarrow HB^+ + A^-$$

B — base: proton remover; HA — acid: proton source

In this equation, note that the charges are not "absolute" charges. Rather, they indicate that the acid species, in losing a proton, leaves a species having a charge one less than the acid and that the base, in gaining a proton, increases by one in charge.

Thinking About Your Thinking

Mental Models

Forming mental models of proton transfer reactions is a skill to develop as you study chemistry. This type of reaction literally occurs thousands of times per second in your body. Many common biological molecules are classified as acids and bases.

Lewis diagrams are good models to use when considering proton transfer reactions. Recall that a proton is symbolized as H^+, which is a hydrogen ion. The most common hydrogen atom isotope consists of a single proton (no neutron) and a single electron. If the electron is separated, what remains is nothing but a proton. This proton is what is transferred in acid–base reactions.

Take another look at the Lewis diagrams in the reactions illustrated in this section, this time looking for the common features of the bases and, in particular, the atom in the basic species that removes the proton. Do you see that the proton remover always has an unshared electron pair? Those negatively charged electrons are also part of a highly electronegative atom. Thus, the area in the species that acts as a base, or proton remover, has a greater negative charge than the remainder of the species. It attracts a positive charge.

Now revisit the Lewis diagrams of the acids in this section. Each has a hydrogen atom bonded to a highly electronegative atom. These bonding electrons are less attracted to the hydrogen atom than to the other atom in the bond. That "allows" the positively charged proton to "look for" a better source of opposite charge to be attracted to than the one it now has. When one is found, a proton transfer reaction occurs.

If you keep in mind that an acid is a proton source, a base is a proton remover, and an acid–base reaction is a proton transfer reaction, you will have an excellent mental model of proton transfer reactions.

Target Check 17.2

What is the difference between a Brønsted–Lowry acid and an Arrhenius acid? Between a Brønsted–Lowry base and an Arrhenius base? Are all Brønsted–Lowry bases also Arrhenius bases? Are all Arrhenius bases also Brønsted–Lowry bases? Explain.

Target Check 17.3

Compare and contrast proton transfer reactions with Brønsted–Lowry acid–base reactions.

17.4 The Lewis Theory of Acids and Bases (Optional)

Goal 4 Distinguish between a Lewis acid and a Lewis base. Given the structural formula of a molecule or ion, state if it can be a Lewis acid, a Lewis base, or both, and explain why.

5 Given the structural equation for a Lewis acid–base reaction, explain how or why it can be so classified.

Look again at the Lewis diagrams of the bases in Section 17.3. In each case, the structural feature of the base that permits it to remove the proton is an unshared pair of electrons. A hydrogen ion has no electron to contribute to the bond, but it is able to accept an electron pair to form the bond. According to the **Lewis acid–base theory**, a Lewis base is an electron pair donor and a Lewis acid is an electron pair acceptor.

A Lewis acid is not limited to a hydrogen ion. Another example of a Lewis acid is boron trifluoride, which behaves as a Lewis acid by accepting an unshared pair of electrons from ammonia:

$$BF_3 + :NH_3 \longrightarrow F_3B\text{—}NH_3$$

A + :B ⟶ A—B

Lewis acid (electron–pair acceptor) | Lewis base (electron–pair donor)

Charles D. Winters

Other examples of Lewis acid–base reactions appear in the formation of some complex ions and in organic reactions. You will study these in more-advanced courses.

Target Check 17.4

By the Brønsted–Lowry theory, water can be either an acid or a base. Can water be a Lewis acid? A Lewis base? Explain.

a summary of... **The Identifying Features of Acids and Bases**

The identifying features of acids and bases according to the three acid–base theories are summarized here:

Theory	Acid	Base
Arrhenius	Hydrogen ion concentration increase in water solution	Hydroxide ion concentration increase in water solution
Brønsted–Lowry	Proton source	Proton remover
Lewis	Electron pair acceptor	Electron pair donor

Our principal interest in acid–base chemistry is in aqueous solutions, in which the Brønsted–Lowry theory prevails. The balance of this chapter is limited to the proton transfer concept of acids and bases.

17.5 Conjugate Acid–Base Pairs

Goal 6 Define and identify conjugate acid–base pairs.

7 Given the formula of an acid or a base, write the formula of its conjugate base or acid.

In Section 17.3, we noted that the reaction between ammonia and water reaches equilibrium. The fact is that most acid–base reactions reach equilibrium. Accordingly, the general Brønsted–Lowry acid–base proton transfer reaction can be written with a double arrow to show that it is reversible. Look carefully at the reverse reaction, for which the labels for the reactants are printed in *italics*:

$$B + HA \rightleftharpoons HB^+ + A^-$$

B	HA	HB^+	A^-
base: proton remover	acid: proton source	*acid: proton source*	*base: proton remover*

Is not HB^+ releasing a proton to A^- in the reverse reaction? In other words, HB^+ is an acid in the reverse reaction, and A^- is a base. From this, we see that the products of any proton transfer acid–base reaction are *another* acid and base for the reverse reaction.

Combinations such as base B and acid HB^+ as well as acid HA and base A^- that result from an acid releasing a proton or a base removing one are called **conjugate acid–base pairs** (see Figure 17.7). In the preceding forward reaction, the acid, HA, releases a proton to the base, B, leaving the anion, A^-, and cation HB^+. HA and A^- are a conjugate acid–base pair. In the reverse reaction, acid HB^+ is the proton source for base A^-, restoring the original reactants, B and HA. This time, HB^+ and

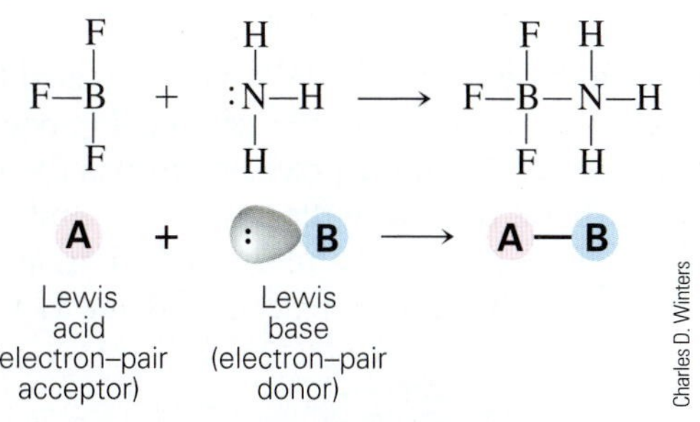

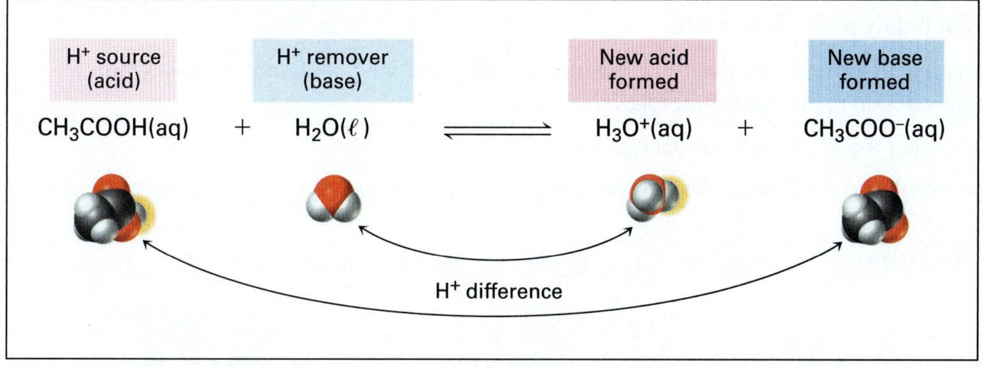

Figure 17.7 Conjugate acid–base pairs. A reversible proton transfer reaction—characterized by the exchange of a single proton, H^+—has two conjugate acid–base pairs.

B are a conjugate acid–base pair. Thus, we see that for every reversible proton transfer reaction, two conjugate acid–base pairs exist.

Let's apply these ideas to reaction of ammonia with water:

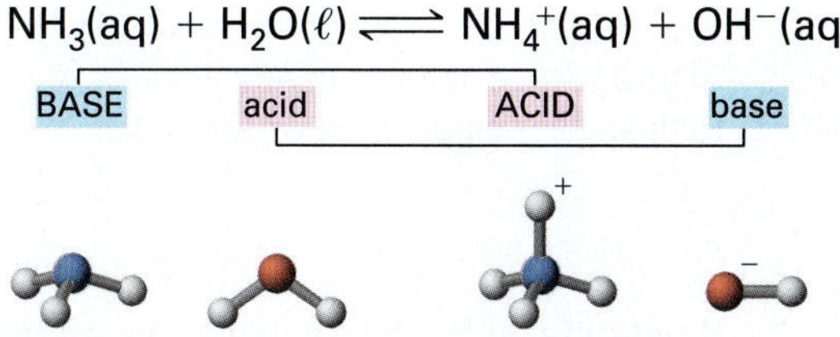

In the forward direction, H_2O is an acid and OH^- is its conjugate base. In the reverse direction, NH_4^+ is the acid and NH_3 is its conjugate base.

You can write the formula of the conjugate base of any acid simply by removing a proton. If HCO_3^- acts as an acid, its conjugate base is CO_3^{2-}. You can also write the formula of the conjugate acid of any base by adding a proton. If HCO_3^- is a base, its conjugate acid is H_2CO_3. Notice that HCO_3^- is amphoteric. Any amphoteric substance has both a conjugate base and a conjugate acid.

Active Example 17.1 Conjugate Acids and Bases

(a) Write the formula of the conjugate base of H_3PO_4. (b) Write the formula of the conjugate acid of $C_7H_5O_2^-$.

Think Before You Write Don't neglect to change the charge when adding or subtracting an H^+.

Answers *Cover the left column with your cut-out shield. Reveal each answer only after you have written your own answer in the right column.*

(a) $H_3PO_4 \rightarrow H^+ + H_2PO_4^-$, the conjugate base of H_3PO_4. (b) $C_7H_5O_2^- + H^+ \rightarrow HC_7H_5O_2$, the conjugate acid of $C_7H_5O_2^-$. *Remove a proton to get a conjugate base, and add one to get a conjugate acid.*	In (b), don't let $C_7H_5O_2^-$ confuse you just because it is unfamiliar. Just do what must be done to find the formula of a conjugate acid. Complete both parts.
You improved your understanding of conjugate acids and bases.	What did you learn by solving this Active Example?

Practice Exercise 17.1

What is the formula of the conjugate acid of the sulfate ion? What is the formula of the conjugate base of chlorous acid?

Active Example 17.2 Conjugate Acid–Base Pairs

Nitrous acid engages in a reaction with formate ion, CHO_2^-: $HNO_2(aq) + CHO_2^-(aq) \rightleftharpoons NO_2^-(aq) + HCHO_2(aq)$. Answer the questions about this reaction in the steps that follow.

Think Before You Write Notice that this is a proton transfer reaction.

Answers *Cover the left column with your cut-out shield. Reveal each answer only after you have written your own answer in the right column.*

HNO_2 is the acid; it releases a proton to CHO_2^-, the base, or proton remover.	For the forward reaction, identify the acid and the base.
$HCHO_2$ is the acid; it releases a proton to NO_2^-, the base, or proton remover.	Identify the acid and base for the reverse reaction.
NO_2^- is the conjugate *base* of HNO_2. *NO_2^- is the base that remains after the proton has been released by the acid.*	Identify the conjugate of HNO_2. Is it a conjugate acid or a conjugate base?
CHO_2^- and $HCHO_2$ make up the other conjugate acid–base pair. CHO_2^- is the base, and $HCHO_2$ is the conjugate acid—the species produced when the base takes a proton.	Identify the other conjugate acid–base pair and classify each species as the acid or the base.
You improved your understanding of conjugate acid–base pairs.	What did you learn by solving this Active Example?

Practice Exercise 17.2

Identify each conjugate acid–base pair in the reaction: $H_3PO_4(aq) + NH_3(aq) \rightleftharpoons H_2PO_4^-(aq) + NH_4^+(aq)$. Identify each acid and base.

17.6 Relative Strengths of Acids and Bases

Goal 8 Given a table of the relative strengths of acids and bases, arrange a group of acids or a group of bases in order of increasing or decreasing strength.

In Section 9.4, you learned to distinguish between the relatively few strong acids and the many weak acids. A **strong acid** is one that ionizes almost completely, whereas a **weak acid** ionizes but slightly. Hydrochloric acid is a strong acid; 0.01 M HCl is almost 100% ionized. Acetic acid is a weak acid; only 4% of the molecules ionize in 0.01 M $HC_2H_3O_2$.

Based on Brønsted–Lowry theory, an acid behaves as an acid by releasing protons. The more readily protons are surrendered, the stronger the acid. A base behaves as a base by removing protons. The greater the ability to remove protons is, the stronger the base will be.

Look at the ionization equations for hydrochloric and acetic acids, one written above the other, with the stronger acid first:

Stronger acid	$HCl \rightleftharpoons H^+ + Cl^-$	? base
Weaker acid	$HC_2H_3O_2 \rightleftharpoons H^+ + C_2H_3O_2^-$	? base

The conjugate bases of the acids are on the right-hand sides of the equations. What is the relative strength of the two bases? Which is stronger, Cl^- or $C_2H_3O_2^-$? If a chloride ion gained a proton, it would form HCl, a strong acid that would immediately lose that proton. The Cl^- ion has a weak attraction for protons. It is a **weak base**. If the acetate ion gained a proton, it would form $HC_2H_3O_2$, a weak acid that holds its proton relatively tightly. The $C_2H_3O_2^-$ ion has a stronger attraction for protons than the Cl^- ion. It is a stronger base. We can therefore complete the comparison between these acids and their conjugate bases:

Stronger acid	$HCl \rightleftharpoons H^+ + Cl^-$	Weaker base
Weaker acid	$HC_2H_3O_2 \rightleftharpoons H^+ + C_2H_3O_2^-$	Stronger base

Figure 17.8 lists many acids and bases in this way. Acid strength decreases from top to bottom, and the base strength increases. By referring to this table, we can compare the relative strengths of different acids and bases. Notice in Figure 17.8 that there are seven strong acids. Figure 17.8 also shows that there is only one **strong base**, hydroxide ion, OH^-.

Thinking About Your Thinking

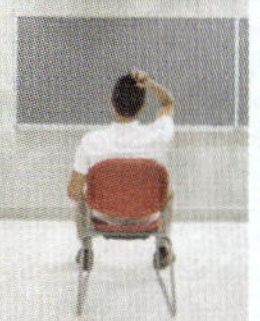

Classification

The classification thinking skill, as we discuss it in this book, is the understanding that different classification systems can be used for different purposes. In Chapter 9, we introduced the strong acid/strong base versus weak acid/weak base concept. This black-and-white view put all acids and bases into one or the other of these groups. In this section, we expand the acid and base strengths to include "shades of gray." We now recognize that acid and base strengths can be compared with each other. Instead of thinking with the absolute terms strong or weak, we think in the relative terms strong*er* or weak*er*.

Active Example 17.3 Relative Strengths of Acids

List the following acids in order of decreasing strength (strongest first): $HC_2O_4^-$, NH_4^+, H_3PO_4.

Think Before You Write This question assumes that you have access to a list of relative strengths of acids and bases, such as Figure 17.8. Be sure you understand how the table is organized.

Answers *Cover the left column with your cut-out shield. Reveal each answer only after you have written your own answer in the right column.*

$H_3PO_4 > HC_2O_4^- > NH_4^+$	Find the three acids among those shown in the table and list them from the strongest (first) to the weakest (last). 
You improved your understanding of relative strengths of acids and how to use a table of relative strengths of acids and bases.	What did you learn by solving this Active Example?

Practice Exercise 17.3

List the formulas of the following acids in order of increasing strength (weakest first): ammonium ion, nitrous acid, carbonic acid.

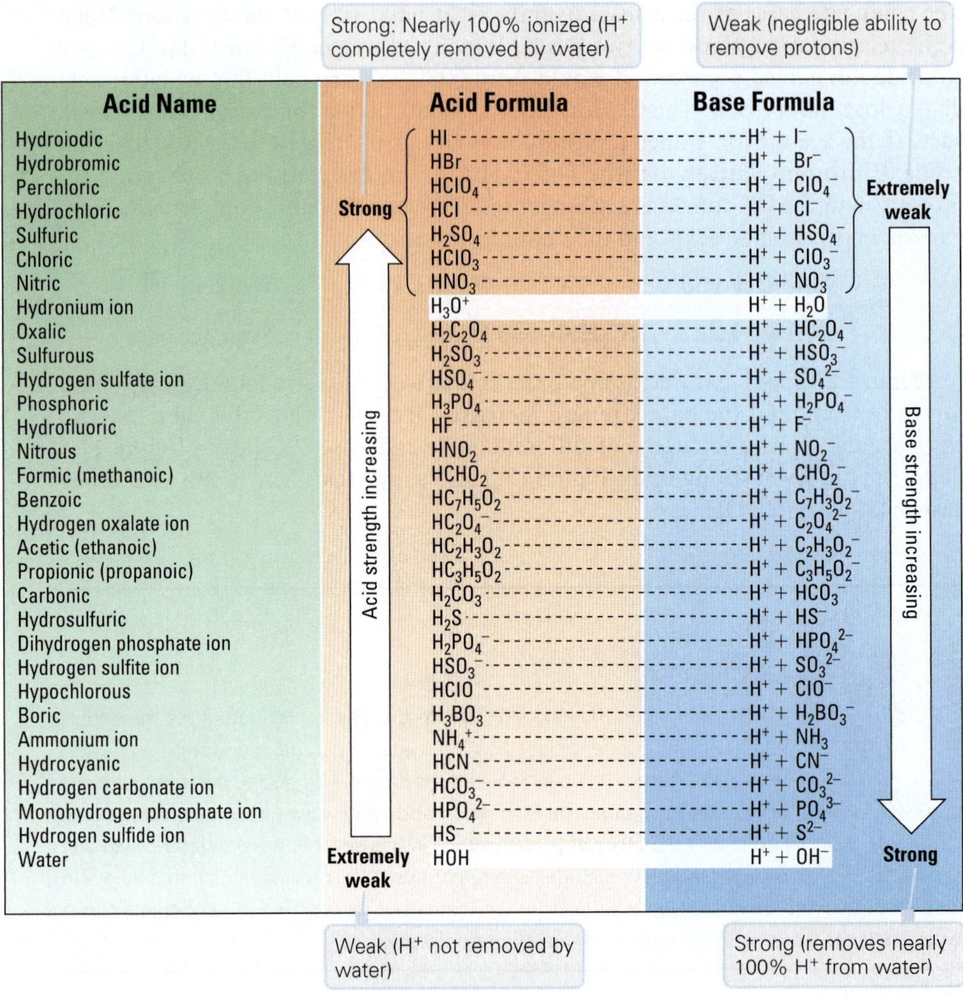

Acid Name	Acid Formula	Base Formula
Hydroiodic	HI	$H^+ + I^-$
Hydrobromic	HBr	$H^+ + Br^-$
Perchloric	$HClO_4$	$H^+ + ClO_4^-$
Hydrochloric	HCl	$H^+ + Cl^-$
Sulfuric	H_2SO_4	$H^+ + HSO_4^-$
Chloric	$HClO_3$	$H^+ + ClO_3^-$
Nitric	HNO_3	$H^+ + NO_3^-$
Hydronium ion	H_3O^+	$H^+ + H_2O$
Oxalic	$H_2C_2O_4$	$H^+ + HC_2O_4^-$
Sulfurous	H_2SO_3	$H^+ + HSO_3^-$
Hydrogen sulfate ion	HSO_4^-	$H^+ + SO_4^{2-}$
Phosphoric	H_3PO_4	$H^+ + H_2PO_4^-$
Hydrofluoric	HF	$H^+ + F^-$
Nitrous	HNO_2	$H^+ + NO_2^-$
Formic (methanoic)	$HCHO_2$	$H^+ + CHO_2^-$
Benzoic	$HC_7H_5O_2$	$H^+ + C_7H_5O_2^-$
Hydrogen oxalate ion	$HC_2O_4^-$	$H^+ + C_2O_4^{2-}$
Acetic (ethanoic)	$HC_2H_3O_2$	$H^+ + C_2H_3O_2^-$
Propionic (propanoic)	$HC_3H_5O_2$	$H^+ + C_3H_5O_2^-$
Carbonic	H_2CO_3	$H^+ + HCO_3^-$
Hydrosulfuric	H_2S	$H^+ + HS^-$
Dihydrogen phosphate ion	$H_2PO_4^-$	$H^+ + HPO_4^{2-}$
Hydrogen sulfite ion	HSO_3^-	$H^+ + SO_3^{2-}$
Hypochlorous	$HClO$	$H^+ + ClO^-$
Boric	H_3BO_3	$H^+ + H_2BO_3^-$
Ammonium ion	NH_4^+	$H^+ + NH_3$
Hydrocyanic	HCN	$H^+ + CN^-$
Hydrogen carbonate ion	HCO_3^-	$H^+ + CO_3^{2-}$
Monohydrogen phosphate ion	HPO_4^{2-}	$H^+ + PO_4^{3-}$
Hydrogen sulfide ion	HS^-	$H^+ + S^{2-}$
Water	HOH	$H^+ + OH^-$

Figure 17.8 Relative strengths of acids and bases.

Active Example 17.4 Relative Strengths of Bases

List the following bases in order of decreasing strength (strongest first): $HC_2O_4^-$, SO_3^{2-}, F^-.

Think Before You Write A list of relative strengths of acids and bases shows bases in order of strength, but be sure you know which end of the list is stronger and which is weaker.

Answers *Cover the left column with your cut-out shield. Reveal each answer only after you have written your own answer in the right column.*

$SO_3^{2-} > F^- > HC_2O_4^-$	Locate the ions in the Figure 17.8 table and list them in the requested order.
You improved your understanding of relative strengths of bases and how to use a table of relative strengths of acids and bases.	What did you learn by solving this Active Example?

Practice Exercise 17.4

List the formulas of the following bases in order of increasing strength (weakest first): sulfate ion, perchlorate ion, hydroxide ion.

The ion $HC_2O_4^-$ appears in both Active Examples 17.3 and 17.4, first as an acid and second as a base. It is the intermediate ion in the two-step ionization of oxalic acid, $H_2C_2O_4$. The $HC_2O_4^-$ ion is amphoteric.

17.7 Predicting Acid–Base Reactions

Goal 9 Given the formulas of a potential Brønsted–Lowry acid and a Brønsted–Lowry base, write the equation for the possible proton transfer reaction between them.

10 Given a table of relative strengths of acids and bases and information from which a proton transfer reaction equation between two species in the table may be written, write the equation and predict the direction in which the reaction will be favored.

A chemist likes to know if an acid–base reaction will occur when certain reactants are brought together. Obviously, there must be a potential proton source and proton remover—there can be no proton transfer reaction without both. From there, the decision is based on the relative strengths of the conjugate acid–base pairs. The stronger acid and base are the most reactive. They do what they must do to behave as an acid and a base. The weaker acid and base are more stable—less reactive. It follows that *the stronger acid will always surrender a proton to the stronger base, yielding the weaker acid and base as favored species at equilibrium.* **Figure 17.9** summarizes the proton transfer from the stronger acid to the stronger base, from the standpoint of positions in Figure 17.8.

Hydrogen sulfate ion, HSO_4^-, is a relatively strong acid that holds its proton weakly. Hydroxide ion, OH^-, is a strong base that attracts a proton strongly. If an HSO_4^- ion finds an OH^- ion, the OH^- will remove a proton from HSO_4^-:

$$HSO_4^-(aq) + OH^-(aq) \rightleftharpoons SO_4^{2-}(aq) + HOH(\ell)$$

Now identify the conjugate acid–base pairs. In the forward direction, the acid is HSO_4^-. Its conjugate base for the reverse reaction is SO_4^{2-}. Similarly, OH^- is the base in the forward reaction, and HOH is the conjugate acid for the reverse reaction. Let's show the acid–base roles for the different directions with the letters A for acid and B for base:

$$\underset{A}{HSO_4^-(aq)} + \underset{B}{OH^-(aq)} \rightleftharpoons \underset{B}{SO_4^{2-}(aq)} + \underset{A}{HOH(\ell)}$$

Now compare the two acids in strength. HSO_4^- is nearer the top of the list in Figure 17.8, a much stronger acid than water. Therefore, we label HSO_4^- with SA for stronger acid and water with WA for weaker acid. Similarly, compare the bases: OH^- is a stronger base (SB) than SO_4^{2-} (WB):

$$\underset{SA}{HSO_4^-(aq)} + \underset{SB}{OH^-(aq)} \rightleftharpoons \underset{WB}{SO_4^{2-}(aq)} + \underset{WA}{HOH(\ell)}$$

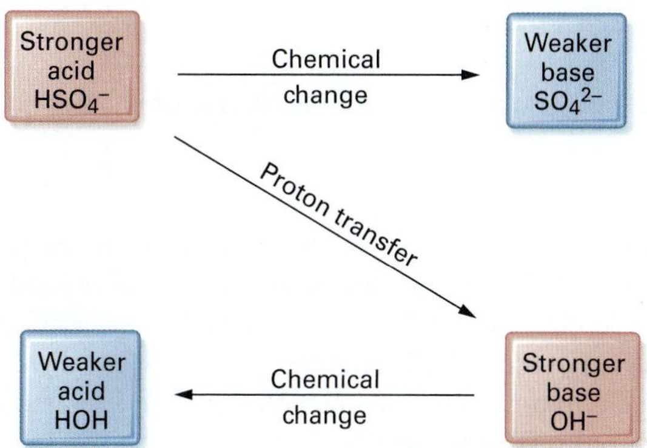

Figure 17.9 Predicting acid–base reactions from positions in Figure 17.8. The spontaneous chemical change always transfers a proton from the stronger acid to the stronger base, both shown in pink. The products of the reaction, the weaker acid and the weaker base, are shown in blue. The favored direction, forward or reverse, is the one that has the weaker acid and base as products.

As you see, the weaker combination is on the right-hand side in this equation. This indicates that the reaction is favored in the forward direction. The proton transfers spontaneously from the strong proton source to the strong proton remover. The products that are in greater abundance are the weaker conjugate base and conjugate acid.

The following procedure is recommended for predicting acid–base reactions.

how to... Predict the Favored Direction of an Acid–Base Reaction

Step 1: For a given pair of reactants, write the equation for the transfer of *one* proton from one species to the other. (Do not transfer two protons.)

Step 2: Label the acid and base on each side of the equation.

Step 3: Determine which side of the equation has *both* the weaker acid and the weaker base (they must both be on the same side). That side identifies the products in the favored direction.

Active Example 17.5 Predicting Acid–Base Reactions

Write the net ionic equation for the reaction between hydrofluoric acid and the sulfite ion, and predict which side will be favored at equilibrium.

Think Before You Write The reactants are HF, an acid, and SO_3^{2-}, a base. The reaction, therefore, will be a single-proton transfer reaction. The weaker (less reactive) species are favored at equilibrium.

Answers *Cover the left column with your cut-out shield. Reveal each answer only after you have written your own answer in the right column.*

Answer	Step
$HF(aq) + SO_3^{2-}(aq) \rightleftharpoons F^-(aq) + HSO_3^-(aq)$	The first step is to write the equation for the reaction. Complete this step. 
$HF(aq) + SO_3^{2-}(aq) \rightleftharpoons F^-(aq) + HSO_3^-(aq)$ A B B A *In each case, the acid is the species with a proton available to remove. A proton is taken by base SO_3^{2-} from acid HF in the forward reaction and by base F^- from acid HSO_3^- in the reverse reaction.*	Next, you need to identify the acid and base on each side of the equation. Do so with letters A and B, as in the preceding discussion. Add the labels to your equation in the box above. 
The forward reaction is favored at equilibrium. *HSO_3^- is a weaker acid than HF, and F^- is a weaker base than SO_3^{2-}. These species are the products in the favored direction.*	Finally, determine which reaction, forward or reverse, is favored at equilibrium. It is the side with the weaker acid and base. Refer to Figure 17.8. 
You improved your understanding of acid-base reactions and how to predict their favored direction.	What did you learn by solving this Active Example?

Practice Exercise 17.5

Write the net ionic equation for the reaction between nitrous acid and the chlorate ion, and predict which side will be favored at equilibrium.

Up to this point, most attention has been given to the direction in which an equilibrium is favored. This does not mean we can ignore the unfavored direction. Consider, for example, the reaction $NH_3(aq) + HOH(\ell) \rightleftharpoons NH_4^+(aq) + OH^-(aq)$. Although this reaction proceeds only slightly in the forward direction, many of the properties of household ammonia—its cleaning power, in particular—depend on the presence of OH^- ions (**Figure 17.10**).

The ion $HC_2O_4^-$ appears in both Active Examples 17.3 and 17.4, first as an acid and second as a base. It is the intermediate ion in the two-step ionization of oxalic acid, $H_2C_2O_4$. The $HC_2O_4^-$ ion is amphoteric.

17.7 Predicting Acid–Base Reactions

Goal 9 Given the formulas of a potential Brønsted–Lowry acid and a Brønsted–Lowry base, write the equation for the possible proton transfer reaction between them.

10 Given a table of relative strengths of acids and bases and information from which a proton transfer reaction equation between two species in the table may be written, write the equation and predict the direction in which the reaction will be favored.

A chemist likes to know if an acid–base reaction will occur when certain reactants are brought together. Obviously, there must be a potential proton source and proton remover—there can be no proton transfer reaction without both. From there, the decision is based on the relative strengths of the conjugate acid–base pairs. The stronger acid and base are the most reactive. They do what they must do to behave as an acid and a base. The weaker acid and base are more stable—less reactive. It follows that *the stronger acid will always surrender a proton to the stronger base, yielding the weaker acid and base as favored species at equilibrium.* **Figure 17.9** summarizes the proton transfer from the stronger acid to the stronger base, from the standpoint of positions in Figure 17.8.

Hydrogen sulfate ion, HSO_4^-, is a relatively strong acid that holds its proton weakly. Hydroxide ion, OH^-, is a strong base that attracts a proton strongly. If an HSO_4^- ion finds an OH^- ion, the OH^- will remove a proton from HSO_4^-:

$$HSO_4^-(aq) + OH^-(aq) \rightleftharpoons SO_4^{2-}(aq) + HOH(\ell)$$

Now identify the conjugate acid–base pairs. In the forward direction, the acid is HSO_4^-. Its conjugate base for the reverse reaction is SO_4^{2-}. Similarly, OH^- is the base in the forward reaction, and HOH is the conjugate acid for the reverse reaction. Let's show the acid–base roles for the different directions with the letters A for acid and B for base:

$$\underset{A}{HSO_4^-(aq)} + \underset{B}{OH^-(aq)} \rightleftharpoons \underset{B}{SO_4^{2-}(aq)} + \underset{A}{HOH(\ell)}$$

Now compare the two acids in strength. HSO_4^- is nearer the top of the list in Figure 17.8, a much stronger acid than water. Therefore, we label HSO_4^- with SA for stronger acid and water with WA for weaker acid. Similarly, compare the bases: OH^- is a stronger base (SB) than SO_4^{2-} (WB):

$$\underset{SA}{HSO_4^-(aq)} + \underset{SB}{OH^-(aq)} \rightleftharpoons \underset{WB}{SO_4^{2-}(aq)} + \underset{WA}{HOH(\ell)}$$

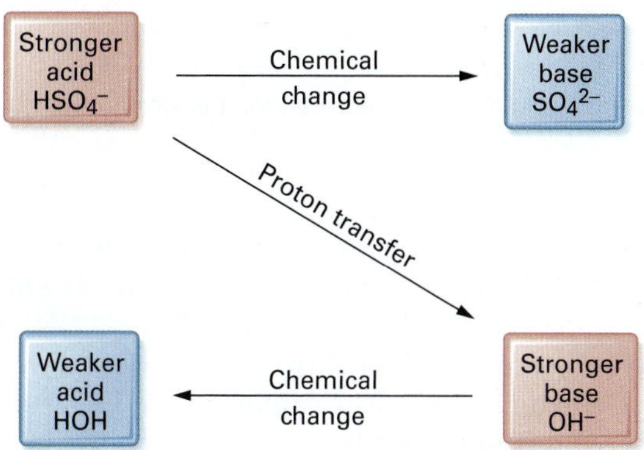

Figure 17.9 Predicting acid–base reactions from positions in Figure 17.8. The spontaneous chemical change always transfers a proton from the stronger acid to the stronger base, both shown in pink. The products of the reaction, the weaker acid and the weaker base, are shown in blue. The favored direction, forward or reverse, is the one that has the weaker acid and base as products.

As you see, the weaker combination is on the right-hand side in this equation. This indicates that the reaction is favored in the forward direction. The proton transfers spontaneously from the strong proton source to the strong proton remover. The products that are in greater abundance are the weaker conjugate base and conjugate acid.

The following procedure is recommended for predicting acid–base reactions.

how to... Predict the Favored Direction of an Acid–Base Reaction

Step 1: For a given pair of reactants, write the equation for the transfer of *one* proton from one species to the other. (Do not transfer two protons.)

Step 2: Label the acid and base on each side of the equation.

Step 3: Determine which side of the equation has *both* the weaker acid and the weaker base (they must both be on the same side). That side identifies the products in the favored direction.

Active Example 17.5 Predicting Acid–Base Reactions

Write the net ionic equation for the reaction between hydrofluoric acid and the sulfite ion, and predict which side will be favored at equilibrium.

Think Before You Write The reactants are HF, an acid, and SO_3^{2-}, a base. The reaction, therefore, will be a single-proton transfer reaction. The weaker (less reactive) species are favored at equilibrium.

Answers *Cover the left column with your cut-out shield. Reveal each answer only after you have written your own answer in the right column.*

$HF(aq) + SO_3^{2-}(aq) \rightleftharpoons F^-(aq) + HSO_3^-(aq)$	The first step is to write the equation for the reaction. Complete this step. 
$HF(aq) + SO_3^{2-}(aq) \rightleftharpoons F^-(aq) + HSO_3^-(aq)$ A B B A *In each case, the acid is the species with a proton available to remove. A proton is taken by base SO_3^{2-} from acid HF in the forward reaction and by base F^- from acid HSO_3^- in the reverse reaction.*	Next, you need to identify the acid and base on each side of the equation. Do so with letters A and B, as in the preceding discussion. Add the labels to your equation in the box above. 
The forward reaction is favored at equilibrium. *HSO_3^- is a weaker acid than HF, and F^- is a weaker base than SO_3^{2-}. These species are the products in the favored direction.*	Finally, determine which reaction, forward or reverse, is favored at equilibrium. It is the side with the weaker acid and base. Refer to Figure 17.8. 
You improved your understanding of acid–base reactions and how to predict their favored direction.	What did you learn by solving this Active Example?

Practice Exercise 17.5

Write the net ionic equation for the reaction between nitrous acid and the chlorate ion, and predict which side will be favored at equilibrium.

Up to this point, most attention has been given to the direction in which an equilibrium is favored. This does not mean we can ignore the unfavored direction. Consider, for example, the reaction $NH_3(aq) + HOH(\ell) \rightleftharpoons NH_4^+(aq) + OH^-(aq)$. Although this reaction proceeds only slightly in the forward direction, many of the properties of household ammonia—its cleaning power, in particular—depend on the presence of OH^- ions (**Figure 17.10**).

REDOX-BEFORE-ACID–BASE OPTION If you have already studied Chapter 19, you know about the electron-transfer character of oxidation–reduction reactions. There are several similarities between those reactions and the proton transfer reactions of this chapter. These are identified in the next section. If you have not yet studied Chapter 19, you may omit Section 17.8, which is repeated at the appropriate place in Chapter 19.

Charles D. Winters

Figure 17.10 Ammonia-based cleaning solution. Ammonia solution turns red litmus paper blue, indicating that ammonia is a base, removing protons from a small fraction of the water molecules to yield a low but detectable concentration of hydroxide ions.

17.8 Acid–Base Reactions and Redox Reactions Compared

Goal 11 (If Chapter 19 has been studied) Compare and contrast acid–base reactions with redox reactions.

We are now at the point where it will be useful to point out how acid–base reactions resemble redox reactions.

1. An acid–base reaction is a transfer of protons; a redox reaction is a transfer of electrons.
2. In both cases, the reactants are given special names to indicate their roles in the transfer process. An acid is a proton source; a base is a proton remover. A reducing agent is an electron source; an oxidizing agent is an electron remover.
3. Just as certain species can either provide or remove protons (e.g., HCO_3^- and H_2O) and thereby behave as an acid in one reaction and a base in another, certain species can either remove or provide electrons, acting as an oxidizing agent in one reaction and a reducing agent in another. An example is the Fe^{2+} ion, which can oxidize Zn atoms to Zn^{2+} in the reaction

$$Fe^{2+}(aq) + Zn(s) \rightarrow Zn^{2+}(aq) + Fe(s)$$

   Fe^{2+} can also reduce Cl_2 molecules to Cl^- ions in another reaction:

$$2\,Fe^{2+}(aq) + Cl_2(g) \rightarrow 2\,Cl^-(aq) + 2\,Fe^{3+}(aq)$$

4. Just as acids and bases may be classified as stronger or weaker depending on how readily they remove or provide protons, the relative strengths of oxidizing and reducing agents may be compared according to their tendencies to attract or release electrons.
5. Just as most acid–base reactions in solution reach a state of equilibrium, most aqueous redox reactions also reach equilibrium. Just as the favored side of an acid–base equilibrium equation can be predicted from acid–base strength, the favored side of a redox equilibrium equation also can be predicted from oxidizing agent–reducing agent strength.

17.9 The Water Equilibrium

Goal 12 Given the hydrogen ion or hydroxide ion concentration of water or a water solution, calculate the other value.

In the remaining sections of this chapter, you will be multiplying and dividing exponentials, taking the square root of an exponential, and working with logarithms. We will furnish brief comments on these operations as we come to them. For more detailed instructions, see Appendix I, Parts A and C.

One of the most critical equilibria in all of chemistry is represented by the last line in Figure 17.8, the ionization of water. Careful control of tiny traces of hydrogen ions and hydroxide ions marks the difference between success and failure in an untold number of industrial chemical processes. In biochemical systems, these concentrations are vital to survival.

Although pure water is generally regarded as a nonconductor, a sufficiently sensitive detector shows that even water contains a tiny concentration of ions. These ions come from the ionization of the water molecule:

$$2\,H_2O(\ell) \rightleftharpoons H_3O^+(aq) + OH^-(aq)$$

$$H{-}\ddot{O}{:}\;(H) + H{-}\ddot{O}{:}\;(H) \rightleftharpoons \left[H{-}\ddot{O}{-}H\;(H)\right]^+ + \left[{:}\ddot{O}{:}\;(H)\right]^-$$

Each chemical equilibrium has an **equilibrium constant (K)** that is calculated from the concentrations of one or more species in the equation. ◀ The equilibrium constant for water, or **water constant (K_w)**, at 25°C is:

▶ **P/Review** The equilibrium constant is defined in detail in Section 18.8. The value of an equilibrium constant is fixed at a specified temperature.

$$K_w = [H^+][OH^-] = 1.0 \times 10^{-14}$$

$K_w = 1.0 \times 10^{-14}$ to two significant figures only at 25°C. In this book, we assume that the temperature is at 25°C for all calculations that involve the water constant.

Enclosing a chemical symbol in square brackets is one way to represent the moles-per-liter (molar) concentration of that species. Thus, **$[H^+]$** and **$[OH^-]$** are the **molar concentration of the hydrogen ions** and **molar concentration of the hydroxide ions**, respectively.

At first, we will consider $K_w = 10^{-14}$ and work only with concentrations that can be expressed with whole number exponents. In the next section, after you have become familiar with the mathematical procedures, concentrations will be written in the usual scientific notation form, including coefficients.

The stoichiometry of the equilibrium constant expression for water indicates that the theoretical concentrations of hydrogen ions and hydroxide ions in pure water must be equal.* If $x = [H^+] = [OH^-]$, then substituting into the K_w expression gives:

$$x^2 = 10^{-14}$$

$$x = \sqrt{10^{-14}} = 10^{-7} \text{ moles/liter}$$

To take the square root of an exponential, divide the exponent by 2.

Thinking About Your Thinking

Equilibrium

The water equilibrium has the general mathematical form xy = constant. Compare $K_w = [H^+][OH^-]$ to the general form constant = xy. A characteristic of equilibrium relationships of the type constant = xy is that when x increases, y must decrease, and vice versa. The variables are inversely proportional.

Water or water solutions in which $[H^+] = [OH^-] = 10^{-7}$ M are **neutral solutions**—neither acidic nor basic. A solution in which $[H^+] > [OH^-]$ is an **acidic solution** (**Figure 17.11**); a solution in which $[OH^-] > [H^+]$ is a **basic solution** (**Figure 17.12**).

*Natural water is not pure. Dissolved minerals from the ground and gases from the atmosphere may cause variations as large as two orders of magnitude from the hydrogen ion concentration of pure water.

Marna G. Clarke/Cengage Learning; Antony McAulay/Shutterstock.com

Figure 17.11 Acid solutions found in the grocery store. Acidic solutions, such as cola, vinegar, and fruit juice, have $[H^+] > [OH^-]$.

Jeffrey B. Banke/Shutterstock.com

Figure 17.12 Basic solutions found in the grocery store. Basic solutions, such as most household cleaning solutions, have $[OH^-] > [H^+]$.

$K_w = [H^+] \times [OH^-]$ indicates an inverse relationship between $[H^+]$ and $[OH^-]$; if one concentration goes up, the other goes down. In fact, if we know either the hydrogen ion or hydroxide ion concentration, we can find the other by solving $K_w = [H^+] \times [OH^-]$ for the unknown.

Active Example 17.6 The Water Equilibrium

Determine the hydroxide ion concentration in a solution in which $[H^+] = 10^{-5}$ M. Is the solution acidic or basic?

Think Before You Write You are given $[H^+]$ and you want $[OH^-]$. Thus, to solve this problem, apply the relationship $K_w = [H^+] \times [OH^-] = 10^{-14}$.

Answers *Cover the left column with your cut-out shield. Reveal each answer only after you have written your own answer in the right column.*

$[H^+] \times [OH^-] = K_w \xrightarrow{\text{divide both sides by } [H^+]}$ $[OH^-] = \frac{K_w}{[H^+]} = \frac{10^{-14}}{10^{-5}} = 10^{-14-(-5)} = 10^{-9}\,M$ *To divide exponentials, subtract the denominator exponent from the numerator exponent. With negative exponents, the more negative the exponent, the smaller the value.*	Solve the equation for the wanted quantity, substitute, and solve. 
Since $[H^+] = 10^{-5}\,M > 10^{-9}\,M = [OH^-]$, the solution is acidic.	Is the solution acidic or basic? State the answer and explain.
You improved your understanding of the water equilibrium and calculations with $K_w = [H^+] \times [OH^-] = 10^{-14}$.	What did you learn by solving this Active Example? 

Practice Exercise 17.6

A solution has a hydroxide ion concentration of 10^{-8} M. Is the solution acidic or basic? Explain. What is the hydrogen ion concentration of the solution?

17.10 pH and pOH (Integer Values Only)

Goal 13 Given any one of the following, calculate the remaining three: hydrogen ion or hydroxide ion concentration expressed as 10 raised to an integral power or its decimal equivalent, pH, and pOH expressed as an integer.

A logarithm is a method for expressing an exponent. You know that $10^2 = 10 \times 10 = 100$. Contrast this with $(\log 100) = (\log 10^2) = 2$. In this example, 2 is the exponent. The logarithm of 10^2 is 2, the value of the exponent. An antilogarithm is the inverse of a logarithm. Thus, $(\text{antilog } 2) = 10^2$ because $(\log 10^2) = 2$. Logarithms and how to use a calculator to solve logarithm problems are discussed in Appendix I, Parts A and C.

Rather than express very small $[H^+]$ and $[OH^-]$ values in negative exponentials, chemists use **base-10 logarithms** in the form of "p" numbers. ◀ By this system, if Q is a number, then:

$$pQ = -\log Q$$

Applied to $[H^+]$ and $[OH^-]$, this becomes:

$$\mathbf{pH = -\log[H^+]} \quad \text{and} \quad \mathbf{pOH = -\log[OH^-]}$$

If the pH or pOH of a solution is known, the corresponding concentration is found by taking the **antilogarithm** of the negative of the "p" number:

$$[H^+] = \text{antilog}(-pH) = 10^{-pH}; [OH^-] = \text{antilog}(-pOH) = 10^{-pOH}$$

From these logarithmic relationships it follows that for a concentration expressed as an exponential, a "p" number is written simply by changing the sign of the exponent:

$$\text{if } [H^+] = 10^{-x}, \text{ then pH} = x; \text{ if } [OH^-] = 10^{-y}, \text{ then pOH} = y$$

Conversely, given the "p" number, the concentration is 10 raised to the opposite of that number:

$$\text{if pH} = a, \text{ then } [H^+] = 10^{-a}; \text{ if pOH} = b, \text{ then } [OH^-] = 10^{-b}$$

Active Example 17.7 pH and pOH

(a) The hydrogen ion concentration of a solution is 10^{-9} M. What is its pH?

(b) The pOH of a solution is 4. What is its hydroxide ion concentration?

Think Before You Write If $[H^+] = 10^{-x}$, then pH = x. If pOH = y, then $[OH^-] = 10^{-y}$.

Answers *Cover the left column with your cut-out shield. Reveal each answer only after you have written your own answer in the right column.*

(a) $[H^+] = 10^{-pH} = 10^{-9}$ M. pH = 9, the negative of the exponent of 10. (b) pOH = 4. $[OH^-] = 10^{-pOH} = 10^{-4}$ M. *The exponent of 10 is the negative of the pOH.*	Complete the problem.
You improved your understanding of the pH–$[H^+]$ and pOH–$[OH^-]$ relationships.	What did you learn by solving this Active Example?

Practice Exercise 17.7

What is the hydrogen ion concentration of a solution with a pH of 5? What is the pOH of a solution with a hydroxide ion concentration of 10^{-6} M?

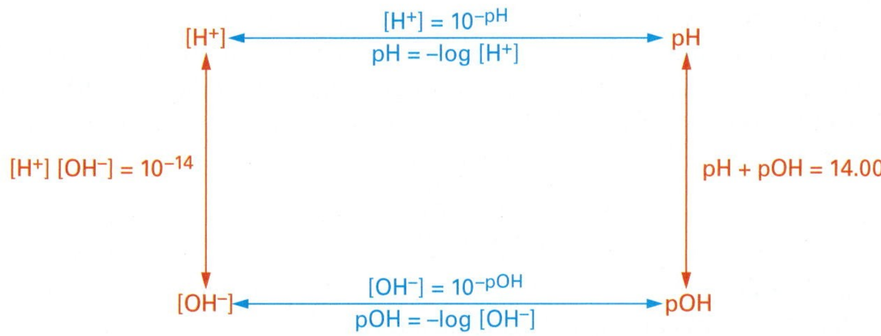

Figure 17.13 The "pH loop." Given the value for any corner of the pH loop, all other values may be calculated by progressing around the loop in either direction. Conversion equations are shown for each step.

The nature of the equilibrium constant and the logarithmic relationship between pH and pOH yield a simple equation that ties the two together. At 25°C,

$$\text{pH} + \text{pOH} = 14$$

If you know any one of the group consisting of pH, pOH, $[OH^-]$, or $[H^+]$, you can calculate the others. **Figure 17.13** is a "**pH loop**" that summarizes these calculations.

Active Example 17.8 The pH Loop I

What are the pH, pOH, $[OH^-]$, and $[H^+]$ of 0.01 M sodium hydroxide solution?

Think Before You Write To solve this problem, you need to work your way through the pH loop illustrated in Figure 17.13.

Answers *Cover the left column with your cut-out shield. Reveal each answer only after you have written your own answer in the right column.*

If 0.01 mol of NaOH is dissolved in 1 L of solution, $NaOH(s) \rightarrow Na^+(aq) + OH^-(aq)$, it forms 0.01 mol of $Na^+(aq)$ and 0.01 mol of $OH^-(aq)$, so the concentration of the hydroxide ion is 0.01 molar, 0.01 mol/L: $[OH^-] = 0.01\ M = 10^{-2}\ M$.	Start by stating the hydroxide ion concentration in a 0.01 M solution of sodium hydroxide. Explain your reasoning. 
$\text{pOH} = -\log[OH^-] = -\log 10^{-2} = 2$	Next, determine the pOH with the relationship $\text{pOH} = -\log[OH^-]$. 
$\text{pH} = 14 - \text{pOH} = 14 - 2 = 12$	Now that you have pOH, you can find pH with the relationship $\text{pH} + \text{pOH} = 14$. 
$[H^+] = 10^{-\text{pH}} = 10^{-12}\ M$	Finally, you can find the hydrogen ion concentration from $[H^+] = 10^{-\text{pH}}$. (We will discuss how to check your work following the example.) 
You improved your skill at performing pH loop calculations.	What did you learn by solving this Active Example?

Practice Exercise 17.8

What are the pH, pOH, $[OH^-]$, and $[H^+]$ of 0.001 M hydrochloric acid solution?

Two things from Active Example 17.8 are worthy of further discussion. First, if we extend the problem by one more step, we complete the full pH loop. We began with $[OH^-]$ and went counterclockwise through pOH, pH, and $[H^+]$. The $[H^+]$ of 10^{-12} M can be converted to $[OH^-]$:

$$[OH^-] = \frac{K_w}{[H^+]} = \frac{10^{-14}}{10^{-12}} = 10^{-2}\ M$$

This is the same as the starting $[OH^-]$. Completing the loop may, therefore, be used to check the correctness of other steps in the process.

The second observation from Active Example 17.8 is that the loop may be circled in either direction. Starting with $[OH^-] = 10^{-2}$ and moving clockwise, we obtain

$$[H^+] = \frac{K_w}{[OH^-]} = \frac{10^{-14}}{10^{-2}} = 10^{-12}\ M$$

It follows that pH = 12 and pOH = 14 − 12 = 2, the same results reached by circling the loop in the opposite direction.

You should now be able to make a complete trip around the pH loop on your own.

Active Example 17.9 The pH Loop II

The pH of a solution is 3. Calculate the pOH, $[H^+]$, and $[OH^-]$ in any order. Confirm your result by calculating the starting pH—by completing the loop.

Think Before You Write You may go either way around the loop, but complete it whichever way you choose, making sure you return to the starting point.

Answers *Cover the left column with your cut-out shield. Reveal each answer only after you have written your own answer in the right column.*

Counterclockwise From pH = 3, $[H^+] = 10^{-3}$ M $[OH^-] = \frac{10^{-14}}{10^{-3}} = 10^{-11}$ M From $[OH^-] = 10^{-11}$ M, pOH = 11 pH = 14 − 11 = 3 **Clockwise** pOH = 14 − 3 = 11 From pOH = 11, $[OH^-] = 10^{-11}$ M $[H^+] = \frac{10^{-14}}{10^{-11}} = 10^{-3}$ M From $[H^+] = 10^{-3}$ M, pH = 3	Complete the problem.
You improved your skill at performing pH loop calculations.	What did you learn by solving this Active Example?

Practice Exercise 17.9

Calculate the pH, $[H^+]$, and $[OH^-]$ of a solution with pOH = 6. Verify the correctness of your values by calculating the starting pOH.

▲ A sea slug excretes the strong acid sulfuric acid in self-defense.

▲ The tartness of lemons and oranges comes from the weak acid citric acid. The acid is found widely in nature and in many consumer products.

▲ Caffeine is a well-known stimulant and a weak base.

Figure 17.14 Acids and bases in biochemical systems.

Most of the solutions we work with in the laboratory and in biochemical systems have pH values between 0 and 14 (**Figure 17.14**). This corresponds to H^+ concentrations between 10^0 and 10^{-14} M, as shown in **Figure 17.15**.

Let's now work on developing a feeling for what pH means. pH is a measure of acidity. It is an inverse sort of measurement; the higher the acidity is, the lower the pH will be, and vice versa. Figure 17.15 brings out this relationship.

$[H^+]$	$[H^+]$	pH	Acidity or Bacisity*
1.0	10^0	0	Strongly acidic pH < 4
0.1	10^{-1}	1	
0.01	10^{-2}	2	
0.001	10^{-3}	3	
0.0001	10^{-4}	4	
0.0001	10^{-4}	4	Weakly acidic $4 \leq pH < 6$
0.00001	10^{-5}	5	
0.000001	10^{-6}	6	
0.000001	10^{-6}	6	Neutral (or near neutral) $6 \leq pH < 8$
0.0000001	10^{-7}	7	
0.00000001	10^{-8}	8	
0.00000001	10^{-8}	8	Weakly basic $8 \leq pH < 10$
0.000000001	10^{-9}	9	
0.0000000001	10^{-10}	10	
0.0000000001	10^{-10}	10	Strongly basic $10 \leq pH$
0.00000000001	10^{-11}	11	
0.000000000001	10^{-12}	12	
0.0000000000001	10^{-13}	13	
0.00000000000001	10^{-14}	14	

a

*Ranges of acidity and basicity are arbitrary.

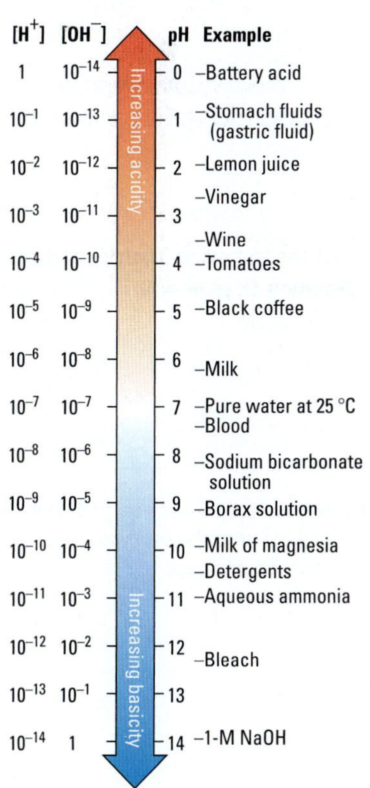

Figure 17.15 (a) A table of pH and hydrogen ion concentration, and (b) the pH values of common solutions.

On examining Figure 17.15, we see that each pH unit represents a hydrogen-ion-concentration factor of 10. Thus, a solution of pH = 2 is 10 times as acidic as a solution with pH = 3 and 100 times as acidic as a solution of pH = 4. In general, the relative acidity in terms of $[H^+]$ is 10^x, where x is the difference between the two pH measurements. From this, we conclude that a 0.1 M solution of a strong acid, with pH = 1, is 1 million times as acidic as a neutral solution, with pH = 7. (One million is based on the pH difference, 7 − 1 = 6. As an exponential, $10^6 = 1{,}000{,}000$.)

If you understand the idea behind pH, you will be able to make some comparisons.

Active Example 17.10 Ranking Solutions in Order of Acidity and the pH Loop

Arrange the following solutions in order of decreasing acidity (that is, highest $[H^+]$ first, lowest last): Solution A, pH = 8; Solution B, pOH = 4; Solution C, $[H^+] = 10^{-6}$ M; Solution D, $[OH^-] = 10^{-5}$ M.

Think Before You Write To list these solutions in any order, all values should be converted to the same basis: pH, pOH, $[H^+]$, or $[OH^-]$. The question asks for a list in terms of acidity, $[H^+]$; therefore, it is the most logical value to use for comparisons in this problem.

Answers *Cover the left column with your cut-out shield. Reveal each answer only after you have written your own answer in the right column.*

A: $[H^+] = 10^{-pH} = 10^{-8}$ M B: pH = 14 − pOH = 14 − 4 = 10; $[H^+] = 10^{-pH} = 10^{-10}$ M C: $[H^+] = 10^{-6}$ M D: $[H^+] = \frac{K_w}{[OH^-]} = \frac{10^{-14}}{10^{-5}} = 10^{-9}$ M	Go around the loop and find $[H^+]$ for each solution.
Most acidic Least acidic $10^{-6} > 10^{-8} > 10^{-9} > 10^{-10}$ C A D B	In arranging these $[H^+]$ values in decreasing order, remember that the exponents are negative.
You improved your understanding of acidity, and you improved your skill at performing pH loop calculations.	What did you learn by solving this Active Example?

Practice Exercise 17.10

List the following in order of increasing acidity: Solution A: $[H^+] = 10^{-2}$ M; Solution B: $[OH^-] = 10^{-10}$; Solution C: pH = 7; Solution D: pOH = 9.

Various methods are used to measure pH in the laboratory. Acid–base indicators (Figure 17.16) were first mentioned in the introduction to this chapter. Each indicator is effective over a specific pH range. Paper strips impregnated with an

Figure 17.16 A universal indicator is a solution that shows a wide range of colors as pH varies.

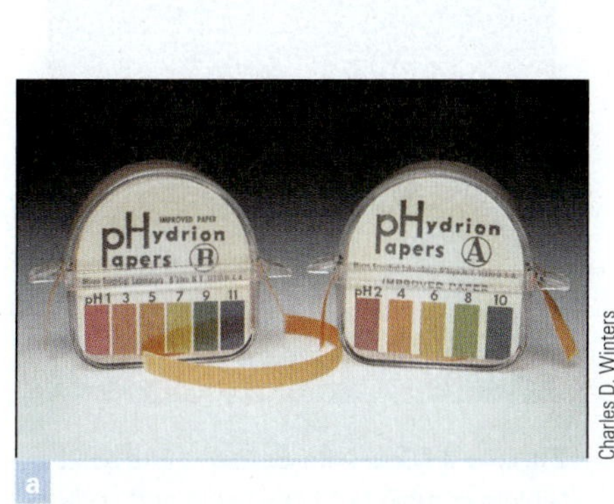

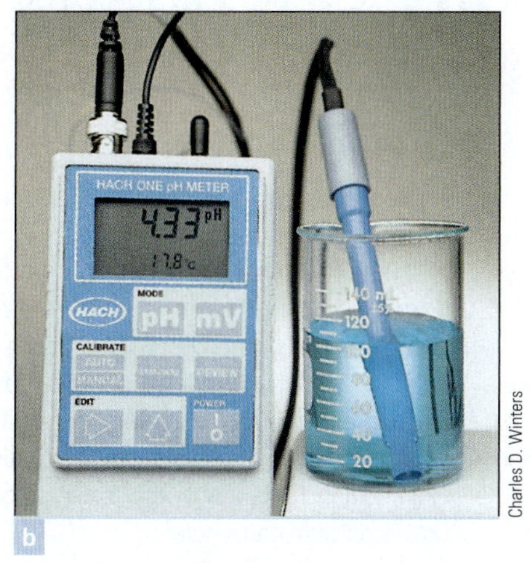

Figure 17.17 Measurement of pH. (a) The color imparted to papers impregnated with certain dyes can be used for approximate measurements of pH. (b) A pH meter is a voltmeter calibrated to measure pH.

Figure 17.18 Arnold O. Beckman (1900–2004) invented the pH meter in 1934. Arnold O. Beckman High School in Irvine, California is named in his honor.

indicator dye that changes color gradually over a limited pH range are used for rough measurements (Figure 17.17[a]). More accurate measurements are made with pH meters (Figures 17.17[b] and 17.18).

17.11 Non-Integer pH–[H$^+$] and pOH–[OH$^-$] Conversions (Optional)

Goal 14 Given any one of the following, calculate the remaining three: hydrogen ion concentration, hydroxide ion concentration, pH, and pOH.

Real-world solutions do not come neatly packaged in concentrations that can be expressed as whole-number powers of 10. [H$^+$] is apt to have a value such as 2.7×10^{-4} M; the pH of a solution is likely to be 11.10 (Figure 17.19). A scientist must be able to convert from each of these to the other.

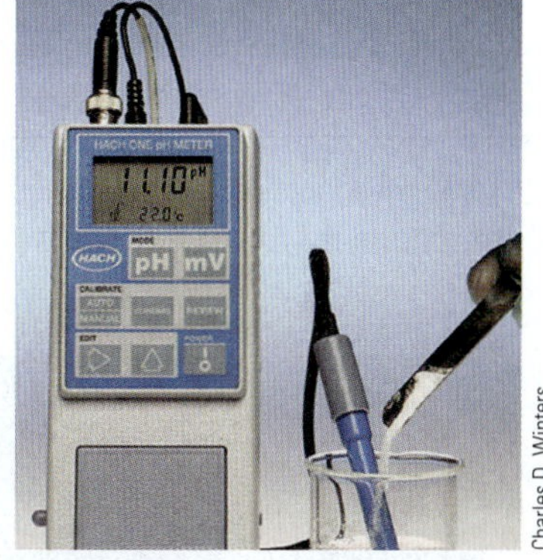

Figure 17.19 Creating a basic solution. When solid sodium carbonate is added to water, the pH of the resulting solution is greater than 7, indicating that the solution is basic.

A pH number is a logarithm. Table 17.1 shows the logarithms of 3.45 multiplied by five different powers of 10: 0, 1, 2, 8, and 12. One column shows the value of the logarithm to seven decimals. Another shows the logarithms rounded off to the correct number of significant figures. Notice three things:

1. The mantissa of the logarithm—the number to the right of the decimal in the Value column—is always the same, 0.5378191. This is the logarithm of 3.45, the coefficient for each entry in the Scientific Notation column in Table 17.1.
2. The characteristic of the logarithm—the number to the left of the decimal in the Value column—is the same as the exponent in the Scientific Notation column.

Table 17.1 Logarithms and Scientific Notation

Number		Logarithm	
Decimal Form	Scientific Notation	Value	Rounded Off
3.45	3.45×10^0	0.5378191	0.538
34.5	3.45×10^1	1.5378191	1.538
345	3.45×10^2	2.5378191	2.538
345,000,000	3.45×10^8	8.5378191	8.538
3,450,000,000,000	3.45×10^{12}	12.5378191	12.538

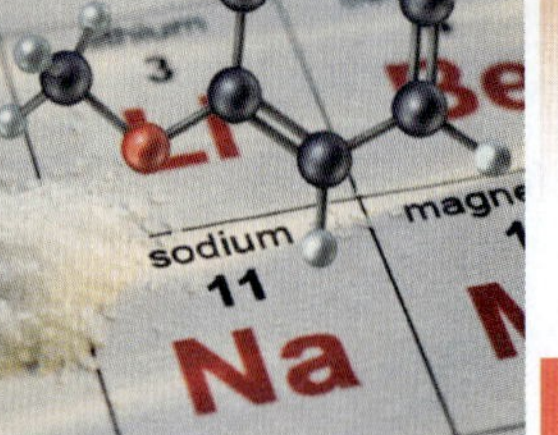

Everyday Chemistry 17.1

ACID–BASE REACTIONS

The introductory photograph for this chapter shows a few items commonly found in U.S. households. They are organized into two groups. The items on the left (cola, lemon, and cherries) all have acidic solutions. The items on the right (bleach, lye, and borax) all dissolve in water to form basic solutions. This type of categorization also applies to most other household foods and cleaners. Foods tend to be acids, and cleaning solutions tend to be bases. Let's consider some acid–base reactions that occur with common acidic or basic substances.

The pH of the fluid in your stomach is about 1, which is much more acidic than the foods that you eat (see Figure 17.15(b) for some food pH values). The acidity level of the contents of your stomach is primarily controlled by the natural formation of hydrochloric acid, which is believed to be in the stomach as an antibacterial agent and possibly as a pre-digestion agent for some foods. Under some conditions, such as when you ingest certain foods and beverages, the acidic solution in your stomach can pass into the esophagus, the tube that carries food into your stomach. When this happens, you experience a burning sensation behind your breastbone, commonly known as heartburn.

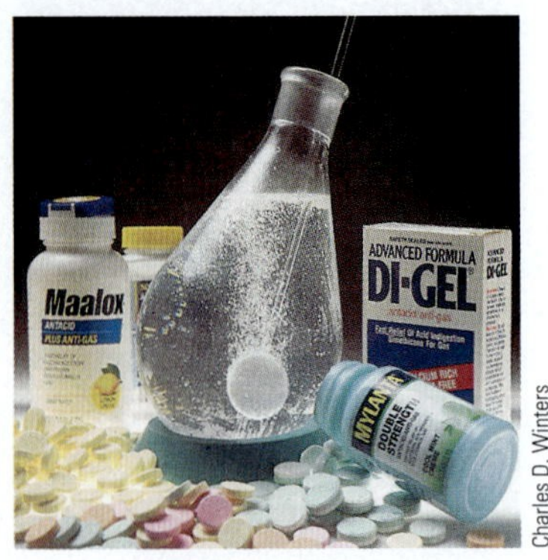

Figure 17.20 Common commercial antacids. Sales of antacids in the United States are estimated to exceed 1 billion dollars per year. About 25 million adults in the United States suffer from heartburn daily, and over 60 million are estimated to have at least one heartburn incident per month.

Antacids are used for heartburn relief (**Figure 17.20**). They are bases (a base is an anti-acid, or the chemical opposite of an acid), and they work via an acid–base reaction that neutralizes some of the acid in your esophagus and stomach. The active ingredient in Tums® and other similar products is calcium carbonate:

$$CaCO_3(s) + 2\,H^+(aq) \rightarrow Ca^{2+}(aq) + H_2O(\ell) + CO_2(g)$$

Alka-Seltzer® (**Figure 17.21**) contains sodium hydrogen carbonate:

$$NaHCO_3(s) + H^+(aq) \rightarrow Na^+(aq) + H_2O(\ell) + CO_2(g)$$

Rolaids® used to feature dihydroxyaluminum sodium carbonate as its active ingredient:

$$NaAl(OH)_2CO_3(s) + 4\,H^+(aq) \rightarrow Na^+(aq) + Al^{3+}(aq) + 3\,H_2O(\ell) + CO_2(g)$$

However, now they are usually made of calcium carbonate or a combination of calcium carbonate and magnesium hydroxide.

Note that no matter the antacid, an acid–base reaction occurs that yields the unstable "H_2CO_3," which decomposes to $H_2O(\ell)$ and $CO_2(g)$ (see Section 9.10). After ingesting these antacids, you normally experience a "burp," by which you expel the carbon dioxide gas.

Cleaning solutions are usually bases. This is because they are designed to decompose fatty molecules (which do not dissolve in water) into ions, which do dissolve in water. A typical fat consists of a long carbon–hydrogen chain with a $-COOH$ of atoms at the end. $CH_3(CH_2)_{12}COOH$ is an example. In this molecule, 14 carbon atoms are bonded to one another in a long chain. The first 13 are bonded to hydrogen atoms and their neighboring carbon atoms. At the end of the chain, the last carbon atom is double-bonded to an oxygen atom and single-bonded to an $-O-H$ group. This last hydrogen atom is acidic and takes part in acid–base reactions. Thus, the reaction between this fatty molecule and the hydroxide ion in a cleaning solution is

$$CH_3(CH_2)_{12}COOH(s) + OH^-(aq) \rightarrow CH_3(CH_2)_{12}COO^-(aq) + H_2O(\ell)$$

Not only is the product conjugate base of the reactant fatty acid molecule now soluble in water but it also actually aids in cleaning!

Proton transfer reactions are among the most common of all reaction types. Although the transfer of a proton is a very simple particulate-level process, its macroscopic applications are countless. You will learn about many additional acid–base reactions throughout your studies in biology and chemistry.

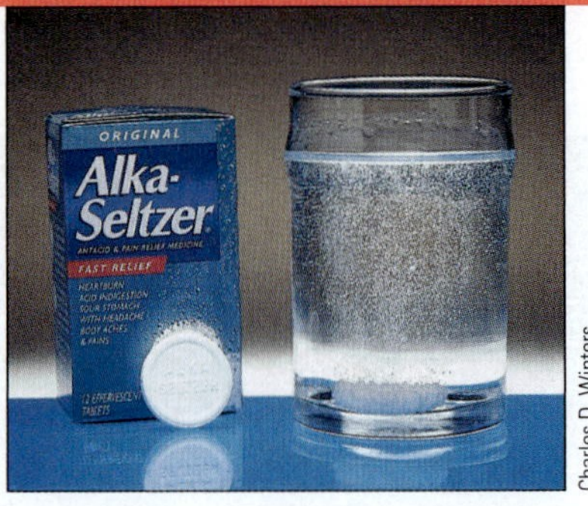

Figure 17.21 Alka-Seltzer®. The solid tablets contain both citric acid and sodium hydrogen carbonate, which cannot react until they are dissolved in solution. Once dissolved, an acid–base reaction takes place, forming bubbles of gaseous carbon dioxide.

Quick Quiz

1. **What acid is present in human stomachs? What is its pH? What purpose does it serve?**
2. **Many antacids can be used to relieve heartburn, but there is one common product formed when they neutralize some of the stomach acid. What is the product of the reaction, and what reactants react to form it?**

3. All numbers in the first two columns have three significant figures. This appears in both the decimal form of the number and in the coefficient when the number is written in exponential notation. Item 2 shows that the digits to the left of the decimal in a logarithm—the characteristic—are related only to the exponent of the number when it is written in scientific notation.

Item 2 shows that the digits to the left of the decimal in a logarithm—the characteristic—are related only to the exponent of the number when it is written in scientific notation. They have nothing to do with the coefficient of the number, which is where significant figures are expressed. Therefore, *in a logarithm, the digits to the left of the decimal are not counted as significant figures. Counting significant figures in a logarithm begins at the decimal point.*

The significant figures in a number written in scientific notation are the significant figures in the coefficient. These show up in the *mantissa* of the logarithm. To be correct in significant figures, the coefficient and the mantissa must have the same number of digits. The correctly rounded-off logarithms are in the right-hand column of Table 17.1. All numbers in that column are written in three significant figures. Only the digits after the decimal point are significant.

In working with pH, you will be finding logarithms of numbers smaller than 1. These logarithms are negative. The sign is changed to positive when the logarithm is written as a "p" value. Try one. Find log 3.45×10^{-6}. Enter log (3.45×10^{-6}) into your calculator. The display should read −5.462180905. If 3.45×10^{-6} represented [H^+], the pH is the opposite of −5.462180905, which is 5.462 when rounded off to three significant figures and with the sign changed.*

In the explanation section of the Active Example that follows, the calculator sequence is given in detail. Although calculator keys may be marked differently, the procedure is essentially the same for most common calculators.

Active Example 17.11 Non-Integer pH–[H^+] and pOH–[OH^-] Conversions I

Calculate the pH of an orange juice solution if [H^+] = 3.6×10^{-4} M.

Think Before You Write Recognize that this problem assumes that you know the relationship pH = −log [H^+].

Answers *Cover the left column with your cut-out shield. Reveal each answer only after you have written your own answer in the right column.*

pH = −log (3.6×10^{-4}) = 3.44 (see **Figure 17.22**).

Press	Display
log	log ( )
3.6	log (3.6)
EE	log (3.6 × 10)
+/−	log (3.6 × 10⁻)
4	log (3.6×10^{-4})
=	−3.443697499
+/−	3.443697499

Because of the given two-significant-figure hydrogen (hydronium) ion concentration, the answer should be rounded off to two significant figures, 3.44.

Enter "log (3.6×10^{-4})" into your calculator, hit the equals sign, and change the sign.

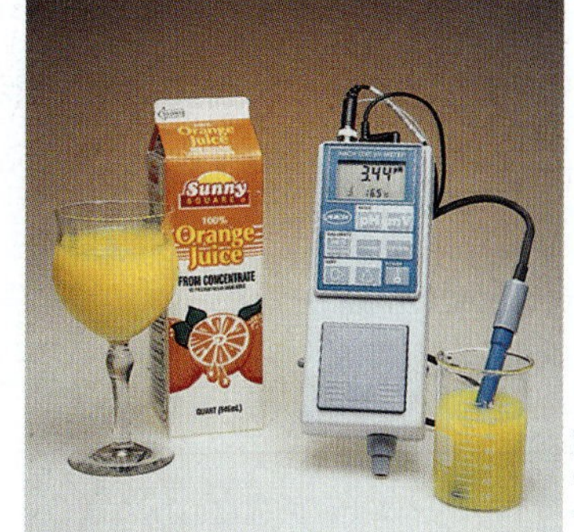

Charles D. Winters

Figure 17.22 Measuring the pH of orange juice. Real-world solutions almost always have non-integer pH values.

*The mantissa appears to be different here from what it was in Table 17.1, but really it is not when you trace its origin: log (3.45×10^{-6}) = log 3.45 + log 10^{-6} = 0.538 + (−6) = −5.462.

You improved your skill at performing non-integer pH loop calculations.	What did you learn by solving this Active Example?

Practice Exercise 17.11

What is the pH of a solution with a hydrogen ion concentration of 3×10^{-5} M?

Active Example 17.12 Non-Integer pH–[H^+] and pOH–[OH^-] Conversions II

Find the pOH of a solution if its hydroxide ion concentration is 7.9×10^{-5} M.

Think Before You Write The "p" value for any concentration is found by taking the log of the concentration and changing the sign: pOH = −log [OH^-].

Answers *Cover the left column with your cut-out shield. Reveal each answer only after you have written your own answer in the right column.*

pOH = −log (7.9 × 10^{-5}) = 4.10 (two significant figures) *The two significant figures in the given concentration leads to a two-significant-figure "p" value, which is two digits after the decimal point.*	Complete the problem. 
You improved your skill at performing non-integer pH loop calculations.	What did you learn by solving this Active Example? 

Practice Exercise 17.12

The hydroxide ion concentration of a solution is 6.86×10^{-7} M. What is its pOH?

Active Example 17.13 Non-Integer pH–[H^+] and pOH–[OH^-] Conversions III

The pOH of a solution is 6.24. Find [OH^-].

Think Before You Write Recognize that this problem assumes that you know the relationship [OH^-] = antilog (−pOH).

Answers *Cover the left column with your cut-out shield. Reveal each answer only after you have written your own answer in the right column.*

[OH^-] = antilog (−6.24) = $10^{-6.24}$ = 5.8 × 10^{-7} M *The pOH is given to two significant figures, so the hydroxide ion concentration is rounded off to two significant figures to yield the final answer, 5.8 × 10^{-7} M.*	Calculate the answer. 
You improved your skill at performing non-integer pH loop calculations.	What did you learn by solving this Active Example?

Practice Exercise 17.13

What is the hydroxide ion concentration of a solution with pOH = 11.5?

Active Example 17.14 Non-Integer pH–$[H^+]$ and pOH–$[OH^-]$ Conversions IV

Find the hydrogen ion concentration of a solution if its pH is 11.62.

Think Before You Write pH is given, and $[H^+]$ is wanted, so you apply the relationship $[H^+]$ = antilog (–pH).

Answers *Cover the left column with your cut-out shield. Reveal each answer only after you have written your own answer in the right column.*

$[H^+]$ = antilog (–11.62) = $10^{-11.62}$ = 2.4×10^{-12} M	Determine $[H^+]$.
You improved your skill at performing non-integer pH loop calculations.	What did you learn by solving this Active Example?

Practice Exercise 17.14

What is the hydrogen ion concentration of a solution with pH = 2.335?

Active Example 17.15 Non-Integer pH–$[H^+]$ and pOH–$[OH^-]$ Conversions V

$[OH^-]$ = 5.2×10^{-9} M for a certain solution. Calculate, in order, pOH, pH, and $[H^+]$, and then complete the pH loop by recalculating $[OH^-]$ from $[H^+]$.

Think Before You Write You've practiced each individual non-integer pH loop calculation in Active Examples 17.11 through 17.14. This is your opportunity to complete the full pH loop on your own.

Answers *Cover the left column with your cut-out shield. Reveal each answer only after you have written your own answer in the right column.*

pOH = –log (5.2×10^{-9}) = 8.28 pH = 14.00 – 8.28 = 5.72 $[H^+]$ = antilog (–5.72) = $10^{-5.72}$ = 1.9×10^{-6} M $[OH^-] = \frac{1.0 \times 10^{-14}}{1.9 \times 10^{-6}} = 5.3 \times 10^{-9}$ M *The variation between 5.2×10^{-9} M and 5.3×10^{-9} M comes from rounding off in expressing intermediate answers. If the calculator sequence is completed without rounding off, the loop returns to 5.2×10^{-9} M for $[OH^-]$.*	Take it all the way.
You improved your skill at performing non-integer pH loop calculations.	What did you learn by solving this Active Example?

Practice Exercise 17.15

What are the pOH, hydroxide ion concentration, and hydrogen ion concentration of a solution with pH = 8.8? Verify your answers by recalculating pH.

Charles D. Winters

CHAPTER 17 IN REVIEW: ACID–BASE (PROTON TRANSFER) REACTIONS

Goal 1 Distinguish between an acid and a base according to the Arrhenius theory of acids and bases.

An Arrhenius **acid** is a substance that increases the concentration of hydrogen ions in solution when dissolved in water, and an Arrhenius **base** is a substance that increases the concentration of hydroxide ions when dissolved in water.

Goal 2 Given the equation for a Brønsted–Lowry acid–base reaction, or information from which it can be written, explain how or why it can be so classified.

According to the **Brønsted–Lowry theory**, an acid–base reaction involves a transfer of a proton from one substance, the acid, to another, the base. An acid is a proton source; a base is a proton remover. When writing an equation, there must be a Brønsted–Lowry base whenever there is a Brønsted–Lowry acid.

Goal 3 Given the name or formula of a Brønsted–Lowry acid and the name or formula of a Brønsted–Lowry base, write the net ionic equation for the reaction between them.

Acid–base reactions are **reversible;** they reach a state of equilibrium according to the general equation $HA + B \rightleftharpoons A^- + HB^+$.

Goal 4 Distinguish between a Lewis acid and a Lewis base. Given the structural formula of a molecule or ion, state if it can be a Lewis acid, a Lewis base, or both, and explain why.

According to the **Lewis theory** of acids and bases, an acid is an electron-pair acceptor and a base is an electron-pair donor.

Goal 5 Given the structural equation for a Lewis acid–base reaction, explain how or why it can be so classified.

Structurally, the most common features of the **Lewis acids** introduced in this course is that they tend to be positive ions or species with a central atom with less than a full octet of valence electrons. **Lewis bases** tend to be negative ions or species with central atoms with one or more unshared electron pairs.

Goal 6 Define and identify conjugate acid–base pairs.

Two substances whose formulas differ only by a proton (a hydrogen ion) are a **conjugate acid–base pair**.

Goal 7 Given the formula of an acid or a base, write the formula of its conjugate base or acid.

If an acid has the general form HA, its conjugate base has the general form A^-. If a base has the general form B, its conjugate acid has the general form HB^+.

Goal 8 Given a table of the relative strengths of acids and bases, arrange a group of acids or a group of bases in order of increasing or decreasing strength.

The more readily protons are surrendered, the stronger the acid. The stronger the ability to remove protons is, the stronger the base will be. The **relative strengths** of Brønsted–Lowry acids and bases may be decided from their positions in a table of relative strengths of acids and bases.

Goal 9 Given the formulas of a potential Brønsted–Lowry acid and a Brønsted–Lowry base, write the equation for the possible proton transfer reaction between them.

Proton transfer reactions involve two conjugate acid–base pairs. Strong acids release protons readily; weak acids do not. Strong bases remove protons readily; weak bases do not. When a proton transfer reaction reaches equilibrium, the proton transfer yields the weaker conjugate acid and the weaker conjugate base.

Goal 10 Given a table of relative strengths of acids and bases and information from which a proton transfer reaction equation between two species in the table may be written, write the equation and predict the direction in which the reaction will be favored.

To **predict the favored direction of an acid–base reaction:**

1. For a given pair of reactants, write the equation for the transfer of *one* proton from one species to the other.
2. Label the acid and base on each side of the equation.
3. Determine which side of the equation has *both* the weaker acid and the weaker base. That side identifies the products in the favored direction.

Goal 11 (If Chapter 19 has been studied) Compare and contrast acid–base reactions with redox reactions.

Acid–base reactions resemble redox reactions in the following ways:

1. Transfer of a subatomic particle: An acid–base reaction is a transfer of protons; a redox reaction is a transfer of electrons.
2. Special names for the species involved in the transfer: An acid is a proton source; a base is a proton remover. A reducing agent is an electron source; an oxidizing agent is an electron remover.

	3. Species that can both remove and release subatomic particles, depending on relative strength. 4. Varying strengths of abilities to remove and release subatomic particles. 5. Most systems are equilibrium reactions; the favored direction of reaction can be predicted by comparison of relative strengths.
Goal 12 Given the hydrogen ion or hydroxide ion concentration of water or a water solution, calculate the other value.	Water itself is both a weak acid and a weak base. $\mathbf{K_w = [H^+][OH^-] = 1.0 \times 10^{-14}}$ at 25°C. A neutral water solution has pH = 7.00 at 25°C. An acidic water solution has pH less than 7.00 at 25°C; a basic water solution has pH greater than 7.00 when measured at 25°C.
Goal 13 Given any one of the following, calculate the remaining three: hydrogen ion or hydroxide ion concentration expressed as 10 raised to an integral power or its decimal equivalent, pH, and pOH expressed as an integer.	Because concentrations of $H^+(aq)$ or $OH^-(aq)$ are usually small but can vary over wide ranges, they are usually expressed in scientific notation and as logarithms. $\mathbf{pH = -log\,[H^+]}$ and $pOH = -\log [OH^-]$. From these equations, you can obtain $\mathbf{[H^+] = 10^{-pH}}$ and $[OH^-] = 10^{-pOH}$. In water solutions at 25°C, pH + pOH = 14.00.
Goal 14 Given any one of the following, calculate the remaining three: hydrogen ion concentration, hydroxide ion concentration, pH, and pOH.	The new skills needed to complete this optional section are not chemical but mathematical. The ideas concerning pH, pOH, $[H^+]$, and $[OH^-]$ are the same as in Section 17.10; only the numbers have been changed. Note that in a logarithm, the digits to the left of the decimal are not counted as significant figures. Counting significant figures in a logarithm begins at the decimal point.

Key Terms

Most of the key terms and concepts and many others appear in the Glossary. Use your Glossary regularly.

acid p. 663
acidic solution p. 676
amphoteric p. 665
antilogarithm p. 678
base p. 663
basic solution p. 676
Brønsted–Lowry acid–base theory p. 665
conjugate acid–base pairs p. 668
equilibrium constant (K) p. 676
forward reaction (forward direction of reaction) p. 666
logarithm (base-10) p. 678
molar concentration of the hydrogen ions or $[H^+]$ p. 676
molar concentration of the hydroxide ions or $[OH^-]$ p. 676
neutral solutions p. 676
pH = $-$log $[H^+]$ p. 678
pH loop p. 679
pOH = $-$log $[OH^-]$ p. 678
reverse reaction (reverse direction of reaction) p. 666
strong acid p. 670
strong base p. 671
water constant (K_w) p. 676
weak acid p. 670
weak base p. 671

Key Terms from Optional Sections:

Arrhenius acid–base theory (Sec. 17.2) p. 663
Lewis acid–base theory (Sec. 17.4) p. 667

Frequently Asked Questions

How do you write the formulas of conjugate acids and bases?
You can always write the conjugate base of an acid by removing one H from the acid formula and reducing the acid charge by one. Conversely, to write the formula of the conjugate acid of a base, add one H to the base formula and increase the charge by one.

How do I write the products in an equation for a Brønsted–Lowry acid–base reaction?
When writing the equation for a Brønsted–Lowry acid–base reaction, transfer only one proton to get the correct conjugate acid and base on the opposite side of the equation. Once the equation is written, each conjugate acid–base pair has the acid on one side and the base on the other side. The acid and base on the same side of the equation are *not* a conjugate pair.

Why does small *pH correspond with a* large *hydrogen ion concentration?*
A "p" number is the opposite (in sign) of the exponent of 10 when an ion concentration is given in exponential form. The exponents are always negative in common solutions, so "p" numbers are positive. That gives an inverse character to the concentration of an ion and its "p" number. For H^+, the larger the $[H^+]$, the more acidic the solution and the smaller the pH.

What potential mathematical pitfall should be avoided in this chapter?
Be careful of negative exponents. The more negative the exponent, the smaller the value. Even though 4 is larger than 3, -4 is smaller than -3. Therefore, 10^{-4} is smaller than 10^{-3}. Thinking of a number line may help.

Concept-Linking Exercises

Write a brief description of the relationships among each of the following groups of terms or phrases. Answers to the Concept-Linking Exercises are given at the end of the chapter.

Terms from optional Sections 17.2 and 17.4 are in italics. Exercise 11 covers terms related to logarithms, and it is also optional.

1. *Arrhenius acid*, Brønsted–Lowry acid, *Lewis acid*
2. *Arrhenius base*, Brønsted–Lowry base, *Lewis base*
3. Brønsted–Lowry acid–base theory, proton transfer reaction, proton source, proton remover, amphoteric
4. Reversible, forward direction, reverse direction, left, right, favored direction
5. *Lewis acid–base theory, electron pair donor, electron pair acceptor*
6. Conjugate acid, conjugate base, conjugate acid–base pair
7. Strong acid, weak acid
8. Water equilibrium, 10^{-14}, $[H^+]$, water constant, $[OH^-]$, K_w
9. pH, pOH, $[H^+]$, $[OH^-]$
10. Logarithm, antilogarithm
11. *Characteristic, mantissa, coefficient, exponential, logarithm*

Small-Group Discussion Questions

Small-Group Discussion Questions are for group work, either in class or under the guidance of a leader during a discussion section.

1. Although there are no "ammonium hydroxide" particles in a container of aqueous ammonia, bottles are often labeled as such. This traditional label likely originated from the Arrhenius acid–base model. Explain why Arrhenius theory requires hydroxide ions in solution, and discuss why it would lead to this name.
2. Hydrogen chloride is a stable, gaseous species at common conditions. When the gas is bubbled through water, a solution of hydrochloric acid is formed, in which the major species are hydrated hydrogen and chloride ions. How can it be that hydrogen chloride gas is "content" to exist (energetically stable) as a molecular compound until it comes into contact with water? Answer in terms of the Brønsted–Lowry model. Use the hydrogen chloride and water reaction as an illustration and explain the Brønsted–Lowry definitions of *acid* and *base*.
3. Classify each of the following categories as being either Lewis acids or Lewis bases. Give specific examples of at least three species from each category. Illustrate Lewis acid–base reactions from at least one species from each category. Can any category contain both Lewis acids and Lewis bases: a molecule with a double bond, positive ions, a molecule with an unoccupied orbital in the same principal energy level as its valence electrons, anions, a molecule with a polar double bond, a molecule with an unshared electron pair?
4. Write the formula of both the conjugate acid *and* the conjugate base of each of the following: water, ammonia, hydroxide ion, hydrogen sulfate ion.
5. Classify each of the acids in Figure 17.8 as strong or weak, according to the classification scheme introduced in Chapter 9. What acid demarcates the division between strong and weak? Rewrite the reactions in the vicinity of the division between strong and weak using a water molecule on the left side of the equation; for example, the reaction with HI would be rewritten as $HI + H_2O \rightleftharpoons H_3O^+ + I^-$. Explain why the acid you identified as separating the strong acids from the weak acids acts as the separation point.
6. Sketch particulate-level illustrations of a strong acid solution and a weak acid solution.
7. Sketch particulate-level illustrations of a strong base solution and a weak base solution.
8. At 30°C, the value of K_w, the water constant, is 1.5×10^{-14}. Determine the pH, pOH, $[H^+]$, and $[OH^-]$ for water at 30°C.
9. Rank the following solutions from lowest to highest pH: pure water, natural rainwater, seawater, vinegar, carbonated beverages, sodium hydroxide solution. Explain your reasoning for each solution.
10. What is the pH of a solution with a hydrogen ion concentration of 1.0 M?

Questions, Exercises, and Problems

Interactive versions of these problems may be assigned by your instructor. Solutions for blue-numbered questions are at the end of the chapter. Questions other than those in the General Questions and More Challenging Problems sections are paired in consecutive odd-even number combinations; solutions for the odd numbered questions are also at the end of the chapter.

Sections 17.2–17.4: The Arrhenius Theory of Acids and Bases (Optional); The Brønsted–Lowry Theory of Acids and Bases; The Lewis Theory of Acids and Bases (Optional)

1. Identify at least two of the classical properties of acids and two of bases. For one acid property and one base property, show how it is related to the ion associated with an acid or base.

2. In the following net ionic equations, identify each reactant as either a Brønsted–Lowry acid or a Brønsted–Lowry base. Also identify the proton source and the proton remover.

   a) $HCN(aq) + H_2O(\ell) \rightarrow CN^-(aq) + H_3O^+(aq)$

   b) $CH_3NH_2(aq) + H_2O(\ell) \rightarrow CH_3NH_3^+(aq) + OH^-(aq)$

3. Distinguish between an Arrhenius base and a Brønsted–Lowry base. Are the two concepts in agreement? Justify your answer.

4. a) Write a net ionic equation to show that hypochlorous acid behaves as a Brønsted–Lowry acid in water.
   b) The substance triethanolamine is a weak nitrogen-containing base, like ammonia. Write a net ionic equation to show that triethanolamine, $C_6H_{15}O_3N$, behaves as a Brønsted–Lowry base in water.

5. Explain or illustrate by an example what is meant by identifying a Lewis acid as an electron pair acceptor and a Lewis base as an electron pair donor.

6. Classify each of the following substances into the appropriate category: (1) Lewis acid; (2) Lewis base; (3) substance that can act as either a Lewis acid or Lewis base; (4) substance that is neither a Lewis acid nor a Lewis base. (a) CCl_4, (b) F^-, (c) Cu^{2+}, (d) Fe^{3+}, (e) O_2

7. When diethyl ether reacts with boron trifluoride, a covalent bond forms between the molecules. Describe the reaction from the standpoint of the Lewis acid–base theory, based on the following "structural" equation:

```
      ..                              ..
     :F:                             :F:
  ..  |      ..                   ..  |   ..
 :F — B  +  :O — C2H5  ⟶         :F — B — O — C2H5
  ..  |      |                    ..  |   |
     :F:    C2H5                     :F: C2H5
      ..                              ..
```

8. Aluminum chloride, $AlCl_3$, behaves more as a molecular compound than an ionic one. This is illustrated in its ability to form a fourth covalent bond with a chloride ion: $AlCl_3 + Cl^- \rightarrow AlCl_4^-$. From the Lewis diagram of the aluminum chloride molecule and the electron configuration of the chloride ion, show that this is an acid–base reaction in the Lewis sense, and identify the Lewis acid and the Lewis base.

Imageman/Shutterstock.com

Aluminum chloride is a white solid at room conditions, although it is typically manufactured mixed with some yellow iron(III) chloride. Its appearance is that of an ionic compound.

Section 17.5: Conjugate Acid–Base Pairs

9. Give the formula of the conjugate base of HF and $H_2PO_4^-$. Give the formula of the conjugate acid of NO_2^- and $H_2PO_4^-$.

10. Write the formulas of each of the following: (a) the conjugate base of CH_3COOH, (b) the conjugate acid of CN^-, (c) the conjugate base of H_2S, (d) the conjugate acid of HCO_3^-.

11. For the reaction $HSO_4^-(aq) + C_2O_4^{2-}(aq) \rightleftharpoons SO_4^{2-}(aq) + HC_2O_4^-(aq)$, identify the acid and the base on each side of the equation—that is, the acid and base for the forward reaction and the acid and base for the reverse reaction.

12. In the following net ionic equation, identify each species as either a Brønsted–Lowry acid or a Brønsted–Lowry base: $HF(aq) + CN^-(aq) \rightarrow F^-(aq) + HCN(aq)$. Identify the conjugate of each reactant and state whether it is a conjugate acid or a conjugate base.

13. Identify the conjugate acid–base pairs in Question 11.

14. In the following net ionic equation, identify each species as either a Brønsted–Lowry acid or a Brønsted–Lowry base: $CH_3COO^-(aq) + HS^-(aq) \rightarrow CH_3COOH(aq) + S^{2-}(aq)$. Identify the conjugate of each reactant and state whether it is a conjugate acid or a conjugate base.

15. For the reaction $HNO_2(aq) + C_3H_5O_2^-(aq) \rightleftharpoons NO_2^-(aq) + HC_3H_5O_2(aq)$, identify both conjugate acid–base pairs.

16. Identify both conjugate acid–base pairs in the reaction $HClO(aq) + CO_3^{2-}(aq) \rightleftharpoons ClO^-(aq) + HCO_3^-(aq)$.

17. Identify the conjugate acid–base pairs in $NH_4^+(aq) + HPO_4^{2-}(aq) \rightleftharpoons NH_3(aq) + H_2PO_4^-(aq)$.

18. Identify the conjugate acid–base pairs in $H_2PO_4^-(aq) + HCO_3^-(aq) \rightleftharpoons HPO_4^{2-}(aq) + H_2CO_3(aq)$.

Section 17.6: Relative Strengths of Acids and Bases

Refer to Figure 17.8 when answering questions in this section.

19. What is the difference between a strong base and a weak base, according to the Brønsted–Lowry concept? Give two examples of strong bases and two examples of weak bases.

20. What is the difference between a strong acid and a weak acid, according to the Brønsted–Lowry concept? Give two names and formulas of strong acids and two of weak acids.

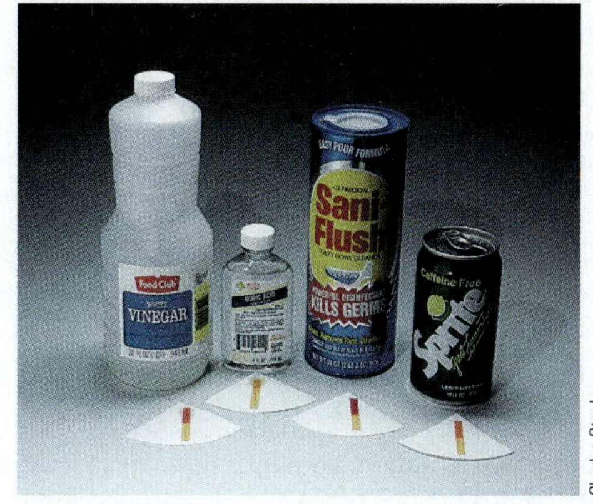

Charles Steele

Many household substances are weak acids, such as (*left to right*) vinegar, insecticides, toilet bowl cleaners, and soft drinks.

21. List the following acids in order of their increasing strength (weakest acid first): $HC_2O_4^-$, H_2SO_3, H_2O, $HClO$.

22. Select the specified acid from each of the following groups:
   a) Strongest of acetic acid, CH_3COOH; hydrogen carbonate ion, HCO_3^-; sulfurous acid, H_2SO_3
   b) Weakest of hydrofluoric acid, HF; phosphoric acid, H_3PO_4; sulfurous acid, H_2SO_3

23. List the following bases in order of their decreasing strength (strongest base first): CN^-, H_2O, HSO_3^-, ClO^-, Cl^-.

24. Select the specified *conjugate base* from each of the following groups of acids: (a) Weakest conjugate base of phosphoric acid, H_3PO_4; acetic acid, CH_3COOH; hydrogen carbonate ion, HCO_3^- ; (b) Strongest conjugate base of hydrosulfuric acid, H_2S; sulfurous acid, H_2SO_3; hydrocyanic acid, HCN.

Section 17.7: Predicting Acid–Base Reactions

For each acid and base given in this section, complete a proton transfer equation for the transfer of one proton. Using Figure 17.8, predict the direction in which the resulting equilibrium will be favored.

25. $HC_3H_5O_2(aq) + PO_4^{3-}(aq) \rightleftharpoons$

26. $HCN(aq) + CH_3COO^-(aq) \rightleftharpoons$

27. $HSO_4^-(aq) + CO_3^{2-}(aq) \rightleftharpoons$

28. $F^-(aq) + HCO_3^-(aq) \rightleftharpoons$

29. $H_2CO_3(aq) + NO_3^-(aq) \rightleftharpoons$

30. $H_3PO_4(aq) + CN^-(aq) \rightleftharpoons$

31. $NO_2^-(aq) + H_3O^+(aq) \rightleftharpoons$

32. $H_2BO_3^-(aq) + NH_4^+(aq) \rightleftharpoons$

33. $HSO_4^-(aq) + HC_2O_4^-(aq) \rightleftharpoons$

34. $HPO_4^{2-}(aq) + HC_2H_3O_2(aq) \rightleftharpoons$

Section 17.8: Acid–Base Reactions and Redox Reactions Compared

35. Explain how a strong acid is similar to a strong reducing agent. Explain how a strong base compares with a strong oxidizing agent.

36. Show how redox and acid–base reactions parallel each other—how they are similar and also what makes them different.

Section 17.9: The Water Equilibrium

37. Of what significance is the very small value of 10^{-14} for K_w, the ionization equilibrium constant for water?

38. How is it that we can classify water as a nonconductor of electricity and yet talk about the ionization of water? If it ionizes, why does it not conduct?

39. $[H^+] = 10^{-5}$ M and $[OH^-] = 10^{-9}$ M in a certain solution. Is the solution acidic, basic, or neutral? How do you know?

40. An aqueous solution has a hydrogen ion concentration of 10^{-4} M. (a) What is the hydroxide ion concentration in this solution? (b) Is this solution acidic, basic, or neutral?

41. What is $[OH^-]$ in 0.01 M HCl? (Hint: Begin by finding $[H^+]$ in 0.01 M HCl.)

42. An aqueous solution has a hydroxide ion concentration of 10^{-11} M. (a) What is the hydrogen ion concentration in this solution? (b) Is this solution acidic, basic, or neutral?

Section 17.10: pH and pOH (Integer Values Only)

43. In which classification—strongly acidic, weakly acidic, strongly basic, weakly basic, neutral, or close to neutral—does each of the following solutions belong: (a) pH = 7, (b) pH = 9, (c) pOH = 3?

44. Identify the approximate ranges of the pH scale that could be classified as strongly acidic, weakly acidic, strongly basic, weakly basic, neutral, or close to neutral.

Charles D. Winters

How would you classify the pH of the liquid within this fish (pH = 10.21)?

45. If the pH of a solution is 8.6, is the solution acidic or basic? How do you reach your conclusion? List in order the pH values of a solution that is neutral, one that is basic, and one that is acidic.

46. Select any integer from 1 to 14 and explain what is meant by saying that this number is the pH of a certain solution.

Questions 47 through 54: The pH, pOH, $[H^+]$, or $[OH^-]$ is given. Find each of the other values. Classify each solution as strongly acidic, weakly acidic, neutral (or close to neutral), weakly basic, or strongly basic.

47. pH = 5

48. pOH = 6

49. $[OH^-] = 10^{-1}$ M

50. $[H^+] = 0.1$ M

51. pOH = 4

52. $[OH^-] = 10^{-2}$ M

53. $[H^+] = 10^{-9}$ M

54. pH = 4

Section 17.11: Non-integer pH–$[H^+]$ and pOH–$[OH^-]$ Conversions (Optional)

Questions 55 through 62: The pH, pOH, $[H^+]$, or $[OH^-]$ of a solution is given. Find each of the other values.

55. $[OH^-] = 2.5 \times 10^{-10}$ M

56. pH = 6.62

57. The pH meter pictured here is measuring the pH of a 0.20 M ammonium nitrate solution. Determine the pOH, $[H^+]$, and $[OH^-]$ of the solution.

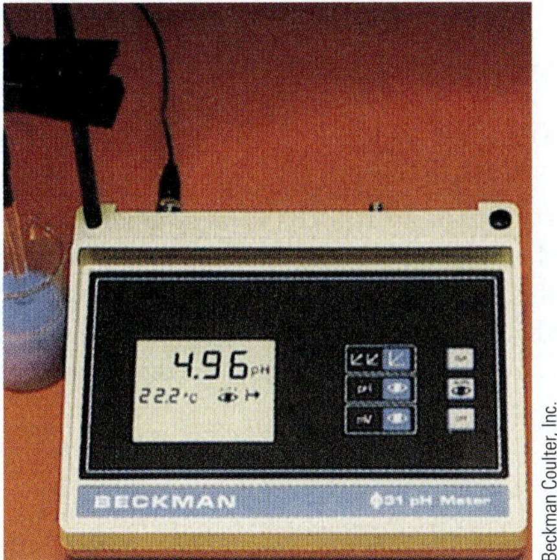

Beckman Coulter, Inc.

58. $[OH^-] = 1.1 \times 10^{-11}$ M

59. $[H^+] = 2.8 \times 10^{-1}$ M

60. pOH = 5.54

61. pOH = 7.40

62. $[H^+] = 7.2 \times 10^{-2}$ M

General Questions

63. Distinguish precisely, and in scientific terms, the differences among items in each of the following pairs.
   a) Acid and base—by Arrhenius theory
   b) Acid and base—by Brønsted–Lowry theory
   c) Acid and base—by Lewis theory
   d) Forward reaction, reverse reaction
   e) Acid and conjugate base, base and conjugate acid
   f) Strong acid, weak acid
   g) Strong base, weak base
   h) $[H^+]$, $[OH^-]$
   i) pH and pOH

64. Classify each of the following statements as true or false:
   a) All Brønsted–Lowry acids are Arrhenius acids.
   b) All Arrhenius bases are Brønsted–Lowry bases, but not all Brønsted–Lowry bases are Arrhenius bases.
   c) HCO_3^- is capable of being amphoteric.
   d) HS^- is the conjugate base of S^{2-}.
   e) If the species on the right side of an ionization equilibrium are present in greater abundance than those on the left, the equilibrium is favored in the forward direction.
   f) NH_4^+ cannot act as a Lewis base.
   g) Weak bases have a weak attraction for protons.
   h) The stronger acid and the stronger base are always on the same side of a proton transfer reaction equation.
   i) A proton transfer reaction is always favored in the direction that yields the stronger acid.
   j) A solution with pH = 9 is more acidic than one with pH = 4.
   k) A solution with pH = 3 is twice as acidic as one with pH = 6.
   l) A pOH of 4.65 expresses the hydroxide ion concentration of a solution in three significant figures.

65. Theoretically, can there be a Brønsted–Lowry acid–base reaction between OH^- and NH_3? If not, why not? If yes, write the equation.

66. Explain what *amphoteric* means. Give an example of an amphoteric substance, other than water, that does not contain carbon.

67. Very small concentrations of ions other than hydrogen and hydroxide are sometimes expressed with "p" numbers. Calculate pCl in a solution for which $[Cl^-] = 7.49 \times 10^{-8}$ M.

68. Can a substance that does not have hydrogen atoms be a Brønsted–Lowry acid? Explain.

69. Why do you suppose chemists prefer the pH scale to expressing hydrogen ion concentration directly? In other words, what is the advantage of saying pH = 4 rather than $[H^+] = 10^{-4}$ M?

70. What is the bromide ion concentration in a solution if a report gives the pBr as 7.2?

More Challenging Questions

71. Suggest a reason why the acid strength decreases with each step in the ionization of phosphoric acid:
$H_3PO_4 \rightarrow H^+ + H_2PO_4^-$
$H_2PO_4^- \rightarrow H^+ + HPO_4^{2-}$
$HPO_4^{2-} \rightarrow H^+ + PO_4^{3-}$

72. Theoretically, can there be a Brønsted–Lowry acid–base reaction between SO_4^{2-} and F^-? If not, why not? If yes, write the equation.

73. Sodium carbonate is among the most widely used industrial bases. How can it be a base when it does not contain a hydroxide ion? Write the equation for a reaction that demonstrates its character as a base.

74. Nonmetal oxides can react with water to form acids. For example, carbon dioxide reacts with water to form carbonic acid: $CO_2 + H_2O \rightarrow H_2CO_3$. What acid forms as the result of the reaction of sulfur trioxide and water? Write the equation for the reaction.

75. Metal oxides can react with water to form bases. Write an equation to show how calcium oxide can react with water to form a base. Name the base.

76. Nitrogen dioxide is emitted in automobile exhaust. Explain how this can contribute to acid rain, which is rain with a low pH.

Stefan Redel/Shutterstock.com

Automobile engines create nitrogen dioxide, a pollutant found in exhaust.

Answers to Target Checks

1. In water, an acid produces an increase in the concentration of H^+ ion and a base yields an increase in the concentration of OH^- ion.
2. An Arrhenius acid is a substance that causes an increase in the concentration of hydrogen ions in water solution. A Brønsted–Lowry acid is a compound from which a proton can be removed. The Arrhenius increase in hydrogen ion concentration occurs because water molecules remove protons from acid molecules. Both acid concepts consider what happens to a proton. With Arrhenius theory, the focus is on the increase in hydrogen ion concentration. With Brønsted–Lowry theory, the emphasis is on the reaction that causes the increase in hydrogen ion concentration.

   An Arrhenius base is a substance that causes an increase in the concentration of hydroxide ions in water solution. A Brønsted–Lowry base is a compound that removes a proton. When a Brønsted–Lowry base removes a proton from a water molecule, a hydroxide ion is formed; the concentration of the hydroxide ion increases. With Arrhenius theory, the focus is on the increase in hydroxide ion concentration. With Brønsted–Lowry theory, the emphasis is on the reaction or physical change that causes the increase in hydroxide ion concentration.

   All Arrhenius bases are Brønsted–Lowry bases since hydroxide ion concentration increases because the base species removes a proton from a water molecule.
3. The term *proton transfer reaction* describes what happens in a Brønsted–Lowry acid–base reaction.
4. Water can be a Lewis base because it has unshared electron pairs. It cannot be a Lewis acid because it has no vacant orbital to receive an electron pair from a Lewis base.

Solutions to Practice Exercises

1. HSO_4^-; ClO_2^-
2. H_3PO_4 (acid) and $H_2PO_4^-$ (base); NH_3 (base) and NH_4^+ (acid)
3. $NH_4^+ < H_2CO_3 < HNO_2$
4. $ClO_4^- < SO_4^{2-} < OH^-$
5. $HNO_2 + ClO_3^- \rightleftharpoons NO_2^- + HClO_3$, reverse
6. $[H^+] = \frac{K_w}{[OH^-]} = \frac{10^{-14}}{10^{-8}} = 10^{-6}$ M; acidic because $[H^+] > [OH^-]$
7. $[H^+] = 10^{-5}$ M; pOH = 6
8. pH = 3; pOH = 11; $[OH^-] = 10^{-11}$ M; $[H^+] = 10^{-3}$ M
9. pH = 8; $[H^+] = 10^{-8}$ M; $[OH^-] = 10^{-6}$ M; pOH = 6
10. Solution C ($[H^+] = 10^{-7}$ M) < Solution D ($[H^+] = 10^{-5}$ M) < Solution B ($[H^+] = 10^{-4}$ M) < Solution A ($[H^+] = 10^{-2}$ M)
11. $pH = -\log(3 \times 10^{-5}) = 4.5$
12. $pOH = -\log(6.86 \times 10^{-7}) = 6.164$
13. $[OH^-] = \text{antilog}(-11.5) = 10^{-11.5} = 3 \times 10^{-12}$ M
14. $[H^+] = \text{antilog}(-2.335) = 10^{-2.335} = 4.62 \times 10^{-3}$ M
15. $pOH = 14.00 - 8.8 = 5.2$; $[OH^-] = \text{antilog}(-5.2) = 10^{-5.2} = 6 \times 10^{-6}$ M; $[H^+] = \text{antilog}(-8.8) = 2 \times 10^{-9}$ M; $pH = -\log(2 \times 10^{-9}) = 8.7$

Answers to Concept-Linking Exercises

You may have found more relationships or relationships other than the ones given in these answers.

1. Arrhenius and Brønsted–Lowry acids are both associated with the hydrogen ion, or proton. An Arrhenius acid is a substance that when added to water, increases the hydrogen ion concentration. A Brønsted–Lowry acid is a substance that has a proton that can be removed by a base. A Lewis acid is any species that has a vacant valence orbital and can accept an electron pair, which includes the hydrogen ion.
2. The solution of an Arrhenius base contains hydroxide ions. A Brønsted–Lowry base removes protons in an acid–base reaction. A Lewis base has a pair of unshared electrons that can form a bond with a substance that has an empty valence orbital.
3. The Brønsted–Lowry acid–base theory identifies an acid–base reaction as a proton transfer reaction in which a proton is transferred from a proton source, the acid, to the proton remover, the base. A substance that can be a proton source or a proton remover is amphoteric.
4. A reversible reaction is one in which the products react and re-form the reactants, as the equation is written. Reading the equation from left to right is the forward direction; reading from right to left is the reverse direction. The direction in which the products are the species present in higher concentration is the favored direction.
5. According to the Lewis acid–base theory, a base is any species that has an unshared electron pair, an electron pair donor, that may form a bond with a species having a vacant valence orbital, which is an acid or an electron pair acceptor.
6. An acid is a substance that can release a proton in a reaction. The species that remains is the conjugate base of the original acid. Together they constitute a conjugate acid–base pair.
7. A strong acid releases protons easily in a proton transfer reaction and readily engages in an acid–base reaction. A weak acid holds its protons and tends not to engage in proton transfer reactions.
8. The water equilibrium is the reaction in which water dissociates into hydrogen ions and hydroxide ions: $H_2O(\ell) \rightleftharpoons H^+(aq) + OH^-(aq)$. The product of the ion concentrations in an aqueous solution equals 10^{-14} at 25°C and is called the *water constant*, K_w: $K_w = [H^+] \times [OH^-] = 10^{-14}$.
9. Hydrogen ion concentration, $[H^+]$, is often expressed in terms of pH, the negative of the logarithm of the concentration: $pH = -\log[H^+]$. Similarly, pOH is the negative of the logarithm of the hydroxide ion concentration: $pOH = -\log[OH^-]$.
10. The logarithm of a number, N, is the exponent, x, to which a base (10 in a decimal system) must be raised to be equal

to the original number: $N = 10^x$. An antilogarithm, M, is the number produced when the base is raised to the power of a given logarithm, y: $10^y = M$.

11. A number, N, written in exponential notation has the form $C \times 10^x$ in which C is the coefficient and 10^x is the exponential. If the logarithm of N is written in decimal form, the number to the left of the decimal is called the *characteristic*, and what follows the decimal is the mantissa. The characteristic of the logarithm expresses the exponent in the exponential of the number, and the mantissa is the logarithm of the coefficient.

Solutions to Blue-Numbered Questions, Exercises, and Problems

1. The classical properties of acids and bases are listed in the chapter introduction. As an example of how a property relates to the ion associated with it, an acid–base neutralization is $H^+ + OH^- \rightarrow H_2O$.

3. An Arrhenius base is a substance that, when dissolved in water, causes an increase in the concentration of hydroxide ions. A Brønsted–Lowry base is a substance that can remove a proton from an acid. When the acid is water, as it is in this chapter, Brønsted–Lowry bases will cause an increase in the concentration of hydroxide ions. The two concepts are in agreement, but the perspective on which they focus ($[OH^-]$ vs. proton transfer reaction) differs.

5. In the reaction shown here, $AlCl_3$, a Lewis acid, accepts an electron pair from Cl^-, a Lewis base, in a Lewis acid–Lewis base neutralization reaction.

7. BF_3 is a Lewis acid because the empty valence orbital of the boron atom accepts an electron pair from the oxygen atom in $C_2H_5OC_2H_5$, a Lewis base because it donates the electron pair.

9. F^-; HPO_4^{2-}; HNO_2; H_3PO_4

11. Acids: HSO_4^- (forward) and $HC_2O_4^-$ (reverse); bases: $C_2O_4^{2-}$ (forward) and SO_4^{2-} (reverse)

13. HSO_4^- and SO_4^{2-}; $HC_2O_4^-$ and $C_2O_4^{2-}$

15. HNO_2 and NO_2^-; $HC_3H_5O_2$ and $C_3H_5O_2^-$

17. NH_4^+ and NH_3; $H_2PO_4^-$ and HPO_4^{2-}

19. A strong base has a strong attraction for protons, whereas a weak base has little attraction for protons. Stronger bases are at the bottom of the right column in Figure 17.8, and weaker bases are at the top.

21. $H_2O < HClO < HC_2O_4^- < H_2SO_3$

23. $CN^- > ClO^- > HSO_3^- > H_2O > Cl^-$

25. $HC_3H_5O_2 + PO_4^{3-} \rightleftharpoons C_3H_5O_2^- + HPO_4^{2-}$ Forward

27. $HSO_4^- + CO_3^{2-} \rightleftharpoons SO_4^{2-} + HCO_3^-$ Forward

29. $H_2CO_3 + NO_3^- \rightleftharpoons HCO_3^- + HNO_3$ Reverse

31. $NO_2^- + H_3O^+ \rightleftharpoons HNO_2 + H_2O$ Forward

33. $HSO_4^- + HC_2O_4^- \rightleftharpoons H_2SO_4 + C_2O_4^{2-}$ Reverse
$HSO_4^- + HC_2O_4^- \rightleftharpoons SO_4^{2-} + H_2C_2O_4$ Reverse

35. A strong acid releases protons readily; a strong reducing agent releases electrons readily. A strong base attracts protons strongly; a strong oxidizing agent attracts electrons strongly.

37. The very small value for K_w indicates that water ionizes to a very small extent.

39. An acidic solution has a higher H^+ concentration than OH^- concentration. The solution is therefore acidic: $10^{-5} > 10^{-9}$.

41. 10^{-12} M

43. (a) neutral; (b) weakly basic; (c) strongly basic

45. Basic. Water solutions with pH = 7 ($[H^+] = [OH^-] = 10^{-7}$ M) are neutral, pH > 7 ($[H^+] < 10^{-7}$ M and $[OH^-] > 10^{-7}$ M) are basic, and those with pH < 7 ($[H^+] > 10^{-7}$ M and $[OH^-] < 10^{-7}$ M) are acidic.

	pH	pOH	$[H^+]$ (M)	$[OH^-]$ (M)	
47.	5	9	10^{-5}	10^{-9}	Weakly acidic
49.	13	1	10^{-13}	10^{-1}	Strongly basic
51.	10	4	10^{-10}	10^{-4}	Strongly basic
53.	9	5	10^{-9}	10^{-5}	Weakly basic
55.	4.40	9.60	4.0×10^{-5}	2.5×10^{-10}	
57.	4.96	9.04	1.1×10^{-5}	9.1×10^{-10}	
59.	0.55	13.45	2.8×10^{-1}	3.5×10^{-14}	
61.	6.60	7.40	2.5×10^{-7}	4.0×10^{-8}	

64. True: a, c, e, f, g, h. False: b, d, i, j, k, l

65. Yes, ammonia has a proton, so its proton theoretically can be removed by hydroxide ion: $OH^- + NH_3 \rightarrow H_2O + NH_2^-$. Additionally, hydroxide ion has a proton and ammonia has an unshared electron pair, so the opposite reaction is also theoretically possible: $OH^- + NH_3 \rightarrow O_2^- + NH_4^+$.

66. An amphoteric substance can be an acid by losing a proton or a base by gaining a proton. HX^- is a general formula of an amphoteric substance: $HX^- \rightarrow H^+ + X^{2-}$ (acid reaction); $HX^- + H^+ \rightarrow H_2X$ (base reaction). Examples of HX^- without carbon include HSO_4^- and $H_2PO_4^-$.

67. pCl = 7.126

68. No. A Brønsted–Lowry acid is a proton source. A proton is a hydrogen ion. If there are no hydrogen atoms in a substance, it cannot release a hydrogen ion to another species.

69. Chemists generally prefer the relative simplicity of the pH scale. Most people find that working with numbers such as pH = 4.0 is much more convenient than the equivalent hydrogen ion concentration 1×10^{-4} M or 0.0001 M.

70. $[Br^-]$ = antilog (−7.2) = 6×10^{-8} M

71. When a proton is removed from an H_3X species, a single positive charge is being pulled away from a particle with a single minus charge, H_2X^-. When a proton is removed from an H_2X^- species, a single positive charge is being pulled away from a particle with a double minus charge, HX^{2-}. The loss of the second proton is energetically more difficult, so H_2X^- is a weaker acid than H_3X.

72. There can be no proton transfer without a proton—an H^+ ion.

73. Carbonate ion is a proton remover: $H^+ + CO_3^{2-} \rightarrow HCO_3^-$

74. $SO_3 + H_2O \rightarrow H_2SO_4$, sulfuric acid

75. $CaO + H_2O \rightarrow Ca(OH)_2$, calcium hydroxide

76. The nitrogen dioxide can react with water to form nitrous and nitric acid: $2\ NO_2(g) + H_2O(\ell) \rightarrow HNO_2(aq) + HNO_3(aq)$. It is also possible for the nitrogen dioxide to combine with oxygen and water in the atmosphere to form nitric acid: $4\ NO_2(g) + 2\ H_2O(\ell) + O_2(g) \rightarrow 4\ HNO_3(aq)$.

18 Chemical Equilibrium

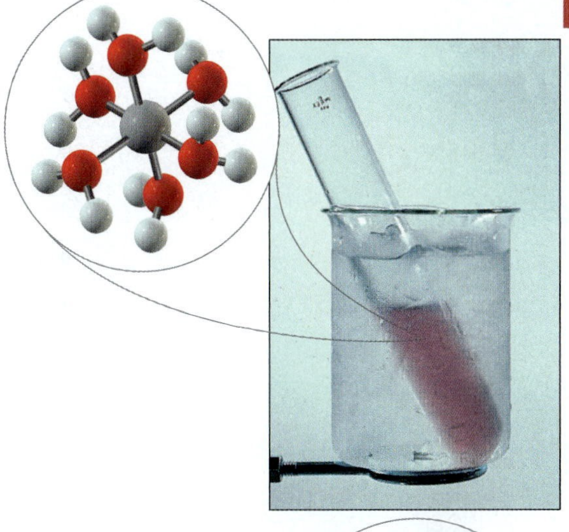

◀ One condition always found in a liquid–vapor, solute–solution, or chemical equilibrium is that the change is reversible. The solution in this test tube changes color depending on its temperature. When heated, the solution turns blue. When cooled, the solution turns pink. The solution will change color repeatedly when moved between the cold and warm water baths. The macroscopic-level color change occurs because the temperature change causes a chemical change. In this chapter, you will learn about the effect of temperature—and other conditions—on a chemical equilibrium.

Charles Steele

CHAPTER CONTENTS

If you have not yet studied Chapters 15 and 16, or if it has been some time since you studied those chapters, you should review Sections 15.5 and 16.4 before beginning this chapter. Pay particular attention to Figures 15.20 and 16.12 and their captions. In studying the liquid–vapor equilibrium in Chapter 15 and the solute–solution equilibrium in Chapter 16, you learned that equilibrium exists when forward and reverse rates are the same in a reversible physical change. Similarly, in this chapter you will see that chemical equilibrium exists when the forward and reverse rates are the same in a reversible chemical change.

18.1 What Patterns Characterize Reversible Chemical Equilibrium Reactions?

Consider the simple, hypothetical reversible reaction $A \rightleftharpoons B$. For every unit of time (e.g., second, decisecond, millisecond), 10% of A reacts to form B and 5% of B reacts to form A. **Table 18.1** tracks the reaction progress at each unit of time, starting with 100 molecules of A and 0 molecules of B. All numbers are rounded to whole numbers using the rounding guidelines from Section 3.8.

Table 18.1 Number of Molecules of A and B for A $\rightleftharpoons$ B with 10% of A Reacting Per Unit of Time and 5% of B Reacting Per Unit of Time

	Initial				Final	
Time	# A	A Reacting	# B	B Reacting	# A	# B
0	100		0			
1	100	−10 + 1	0 + 10	−1	91	9
2	91	−9 + 1	9 + 9	−1	83	17
3	83	−8 + 1	17 + 8	−1	76	24
4	76	−8 + 2	24 + 8	−2	70	30
5	70	−7 + 2	30 + 7	−2	65	35
6	65	−7 + 2	35 + 7	−2	60	40
7	60	−6 + 2	40 + 6	−2	56	44
8	56	−6 + 3	44 + 6	−3	53	47
9	53	−5 + 3	47 + 5	−3	51	49
10	51	−5 + 3	49 + 5	−3	49	51
11	49	−5 + 3	51 + 5	−3	47	53
12	47	−5 + 3	53 + 5	−3	45	55
13	45	−5 + 3	55 + 5	−3	43	57
14	43	−4 + 3	57 + 4	−3	42	58
15	42	−4 + 3	58 + 4	−3	41	59
16	41	−4 + 3	59 + 4	−3	40	60
17	40	−4 + 3	60 + 4	−3	39	61
18	39	−4 + 3	61 + 4	−3	38	62
19	38	−4 + 3	62+ 4	−3	37	63
20	37	−4 + 3	63 + 4	−3	36	64
21	36	−4 + 3	64 + 4	−3	35	65
22	35	−4 + 3	65 + 4	−3	34	66
23	34	−3 + 3	66 + 3	−3	34	66
24	34	−3 + 3	66 + 3	−3	34	66
25	34	−3 + 3	66 + 3	−3	34	66

We'll provide detail about each unit of time in Table 18.1 for the first couple of time intervals to help you follow the table. After the first time interval (*Time* = 1), 10% of A reacts to form B: $0.10 \times 100 = 10$. The number of B molecules increases from 0 to 10: $0 + 10 = 10$. Five percent of B reacts to form A: $0.05 \times 10 = 0.5 = 1$. The *net* amount of A that reacts is therefore 9 (*A Reacting* column): $-10 + 1 = -9$. The #*A Final* is therefore $100 - 10 + 1 = 91$. The # *B Final* is therefore $0 + 10 - 1 = 9$. After the first time interval, the reaction mixture consists of 91 A molecules and 9 B molecules.

We therefore start the second time interval with *Initial #A* = 91 and *Initial #B* = 9. We now again go through the reversible reaction cycle where 10% of A reacts to form B and 5% of B reacts to form A. Ten percent of the 91 A molecules change to B: $0.10 \times 91 = 9.1 = 9$. This results in $9 + 9 = 18$ B molecules. Five percent of these react back to A: $0.05 \times 18 = 0.9 = 1$. This leaves *Final #A* = $91 - 9 + 1 = 83$ and *Final #B* = $9 + 9 - 1 = 17$.

This logic repeats for each time interval. For each unit of time, 10% of A changes to B and 5% of B changes to A.

Now that you've seen the logic of this simulation, let's turn our focus to the results, as summarized in the *Final #A* and *Final #B* columns. What patterns are observed? Your experience with stoichiometry likely led you to notice that the sum of *Final #A* + *Final #B* = 100, without exception. This must be true given the 1:1 stoichiometry of $A \rightleftharpoons B$. Every time an A molecule is destroyed, one B molecule forms, and vice versa. Since the simulation started with 100 molecules, the molecule count will remain at 100 molecules, even as the mixture of A and B changes.

Another pattern is that the *difference* in *Final #A* and *Final #B* for each time interval decreases with time. For example, the difference in *Final #B* between Time 2 and Time 1 is $17 - 9 = 8$. Then the differences in the next six time intervals are 7–6–5–5–4–3 and so on until the difference goes to zero. This occurs because a fixed percentage of A reacts every time interval. As the number of A molecules decreases, the number of A molecules that react decreases, and the difference in number of A molecules decreases with each time interval. Relatedly, as B is being produced at a slower *rate*, where rate is the ratio of number of molecules reacting per unit of time, the rate of B reacting to form A must also slow.

Another pattern is that from Time 22 on, the number of A and B molecules remain unchanged. When this occurs, the system $A \rightleftharpoons B$ is said to be at *equilibrium*, the overarching topic of this chapter. The *equi-* part of the term *equilibrium* implies that something is *equal*, but with 34 A molecules and 66 B molecules, what is equal is *not* the number of A and B molecules. What *is* equal in this equilibrium? It is the *rate of reaction in opposite directions.* With each time interval, 3 A molecules change to B and 3 B molecules change to A. This is called a *dynamic*, or constantly changing, *equilibrium*. At the macroscopic level, the concentrations of A and B are static, but at the particulate level, A is continually being changed to B and B is being changed to A but at equal rates. We now investigate in further detail how and why a chemical equilibrium occurs.

18.2 The Character of an Equilibrium

Goal 1 Identify a chemical equilibrium by the conditions it satisfies.

The simulation in Section 18.1 demonstrated the character of an equilibrium. Additionally, a careful review of the liquid–vapor equilibrium in Section 15.5 and the solution equilibrium in Section 16.4 reveals four conditions that are true for every equilibrium:

1. *The change is reversible and can be represented by an equation with a double arrow* ($\rightleftharpoons$). In a reversible change, the substances on the left side of the equation produce the substances on the right, and those on the right change back into the substances on the left (**Figure 18.1**). These are the forward and reverse reactions, respectively.
2. *The equilibrium system is closed—closed in the sense that no substance can enter or leave the immediate vicinity of the equilibrium.* All substances on either side of an equilibrium equation must remain to form the substances on the other side in a **closed system**.
3. *The equilibrium is dynamic.* The reversible changes occur continuously at the particulate level in a state of **dynamic equilibrium**, even though there is no

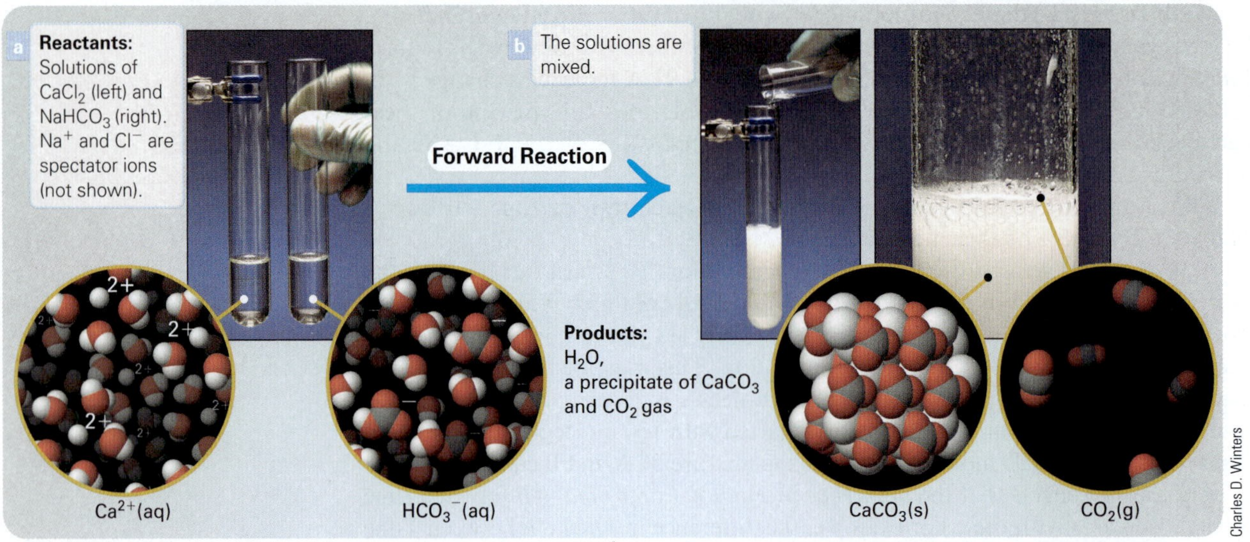

a **Reactants:** Solutions of $CaCl_2$ (left) and $NaHCO_3$ (right). Na^+ and Cl^- are spectator ions (not shown).

Forward Reaction

b The solutions are mixed.

Products: H_2O, a precipitate of $CaCO_3$ and CO_2 gas

$Ca^{2+}(aq)$ $HCO_3^-(aq)$ $CaCO_3(s)$ $CO_2(g)$

Charles D. Winters

Equilibrium Equation: $Ca^{2+}(aq) + 2\ HCO_3^-(aq) \rightleftharpoons CaCO_3(s) + CO_2(g) + H_2O(\ell)$

d The $CaCO_3$ dissolves when the solution has been saturated with CO_2.

c The reaction can be reversed by bubbling CO_2 gas into the $CaCO_3$ suspension.

Reverse Reaction

Elapsing time...

$Ca^{2+}(aq) + 2\ HCO_3^-(aq)$

$CaCO_3(s) + CO_2(g) + H_2O(\ell)$

Charles D. Winters

Figure 18.1 The reversibility of a chemical equilibrium. (a) Solutions that contain calcium ion and hydrogen carbonate ion are prepared. (b) The solutions are mixed, and the reaction proceeds in the forward direction. Solid calcium carbonate, carbon dioxide gas, and liquid water are formed. (c) Carbon dioxide gas is added. (d) The reaction proceeds in the reverse direction, visibly destroying the solid calcium carbonate, yielding calcium ions and hydrogen carbonate ions.

macroscopic appearance of change. By contrast, items in a static equilibrium are stationary, without motion, as an object hanging on a string.

4. *The things that are* equal *in an equilibrium are the forward rate of change (from left to right in the equation) and the reverse rate of change (from right to left).* Note in particular that the amounts of substances present in an equilibrium are not necessarily equal.

 **Target Check 18.1**

Can a solution in a beaker that is open to the atmosphere contain a chemical equilibrium? Explain.

18.3 The Collision Theory of Chemical Reactions

Goal 2 Distinguish between reaction-producing molecular collisions and molecular collisions that do not yield reactions.

If two molecules are to react chemically, it is reasonable to expect that they must come into contact with each other. What we see as a chemical reaction is the overall effect of a huge number of individual collisions between reacting particles. This view of chemical change is the **collision theory of chemical reactions**.

Figure 18.2 examines three kinds of molecular collisions for the imaginary reaction $A_2 + B_2 \rightarrow 2\,AB$. If the collision is to produce a reaction, the bond between A atoms in A_2 must be broken; similarly, the bonds in the B_2 molecules must be broken. It takes energy to break these bonds. This energy comes from the kinetic energy of the molecules just before they collide. In other words, there must be a violent, bond-breaking collision. This is most apt to occur if the molecules are moving at high speed. Figure 18.2(a) depicts a reaction-producing collision.

Thinking About Your Thinking

Mental Models

Figure 18.2 is a great starting point from which you can build a mental model of the collision theory of chemical reactions. Be sure to distinguish between successful and unsuccessful collisions. Figure 18.2(a) illustrates a successful collision. Try to picture this in motion in your mind. There are two essential requirements for a reaction-producing collision. First, the particles must have enough kinetic energy to overcome the electrical repulsions that occur as the

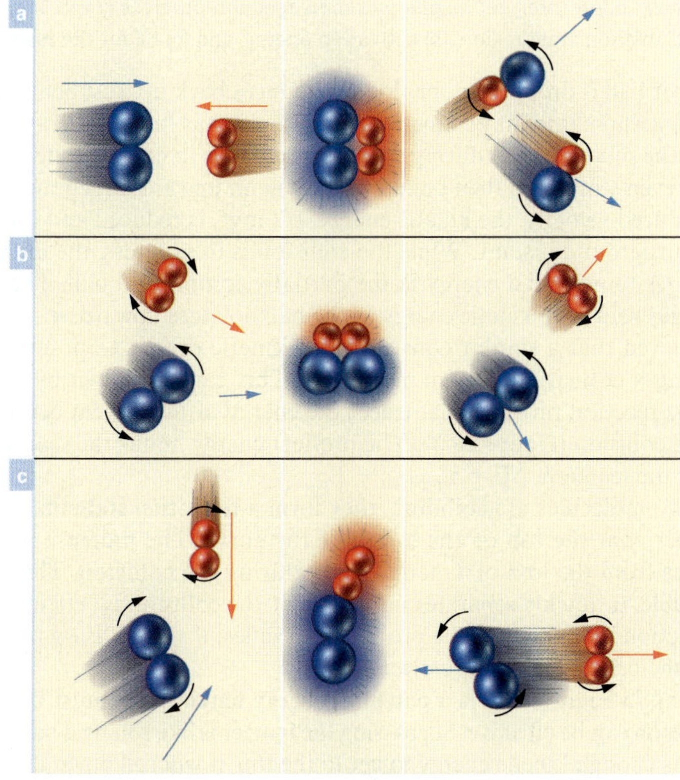

Figure 18.2 Molecular collisions and chemical reactions. (a) Reaction-producing collision between molecules A_2 and B_2, which have sufficient energy and proper orientation. (b) Collision has proper orientation but not enough kinetic energy. There is no reaction. (c) Glancing collision has enough kinetic energy but poor orientation. There is no reaction.

negatively charged electrons on the edge of one molecule repel the electrons of the other molecule. Second, the reactant particles must approach one another in an orientation that "lines up" the constituent atoms so that the product molecules will be formed.

In Figure 18.2(b), notice the middle frame in particular. Can you see how the orientation of the collision is such that the product molecules will form? If such a collision is not successful, it must be because the kinetic energy of the molecules is not great enough to overcome the electrostatic repulsions. Figure 18.2(c) shows another ineffective collision. It does not produce product molecules because the alignment of the colliding particles is such that the product cannot form. Be sure to imagine what these unsuccessful collisions would look like in three dimensions.

Not all collisions result in a reaction; in fact, most do not. If the colliding molecules do not have enough kinetic energy to break the bonds, the original molecules simply repel one another with the same identity they had before the collision (**Figure 18.2[b]**). Sometimes they may have enough kinetic energy but only "sideswipe" one another and move off unchanged (**Figure 18.2[c]**). Other sufficiently energetic collisions may have an orientation that pushes atoms in the original molecules closer together rather than pulling them apart.

To summarize: For an individual collision to result in a reaction, the particles must have (1) enough kinetic energy and (2) the proper orientation. The rate of a particular reaction depends on the frequency of effective collisions.

 **Target Check 18.2**

Draw three kinds of molecular collisions that will not result in a chemical change.

18.4 Energy Changes During a Molecular Collision

Goal 3 Sketch and/or interpret an energy versus reaction progress graph. Identify the (a) transition state region, (b) activation energy, and (c) ΔE for the reaction.

When a rubber ball is dropped to the floor, it bounces back up. Just before it hits the floor, it has a certain amount of kinetic energy. It also has kinetic energy as it leaves the floor on the rebound. But during the time the ball is in contact with the floor, it slows down, even stops, and then builds velocity in an upward direction. During the period of reduced velocity, the kinetic energy, $1/2\ mv^2$, is reduced and even reaches zero at the turnaround instant. While the collision is in progress, the initial kinetic energy changes to potential energy in the partially compressed ball. The potential energy changes back into kinetic energy as the ball bounces upward. ◀

▶ **P/Review** The change from kinetic energy to potential energy and then back to kinetic energy for a bouncing ball is an example of the Law of Conservation of Energy (Section 2.10).

It is believed that a similar conversion of kinetic energy to potential energy occurs during a collision between molecules. This can be shown by a graph of energy versus reaction progress that traces the energy of the system before, during, and after the collision (**Figure 18.3**). The product energy minus the reactant energy is the ΔE for the reaction, $\Delta E = E_{products} - E_{reactants}$.

When two molecules are colliding, they form a **transition state** that has a high potential energy at the top of the hump of the curve. The increase in potential energy comes from the loss of kinetic energy during the collision. The transition state is unstable. It quickly separates into parts. If the collision is "effective" in producing a reaction, the parts will be product molecules; if the collision is ineffective, they will be the original reactant molecules.

The hump in Figure 18.3 is a **potential energy barrier** that must be surpassed before a collision can be effective. Surpassing the barrier is like rolling a ball over a hill. If the ball has enough kinetic energy to get to the top, it will roll down the other side

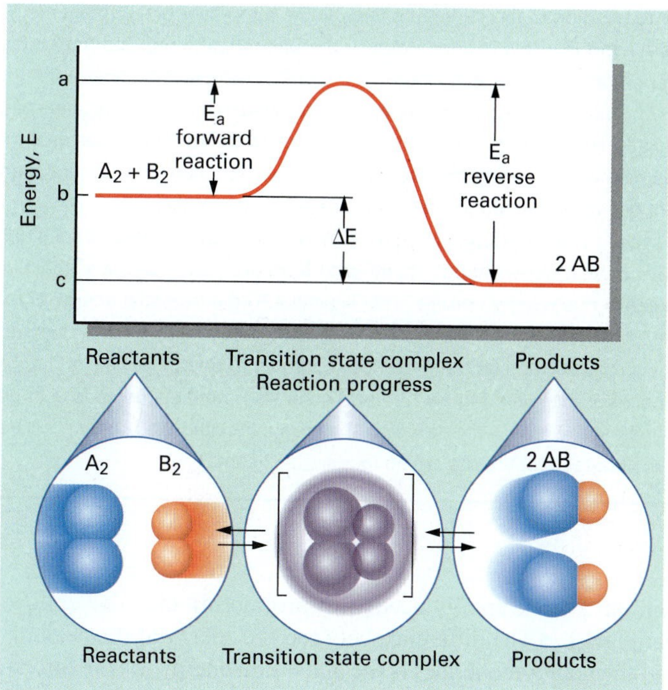

Figure 18.3 Energy versus reaction progress graph for the reaction $A_2 + B_2 \rightarrow 2\ AB$. E_a represents the activation energy for the forward and reverse reactions, as indicated. (Energy values a, b, and c are reference points in an end-of-chapter question and a *Thinking About Your Thinking* feature.)

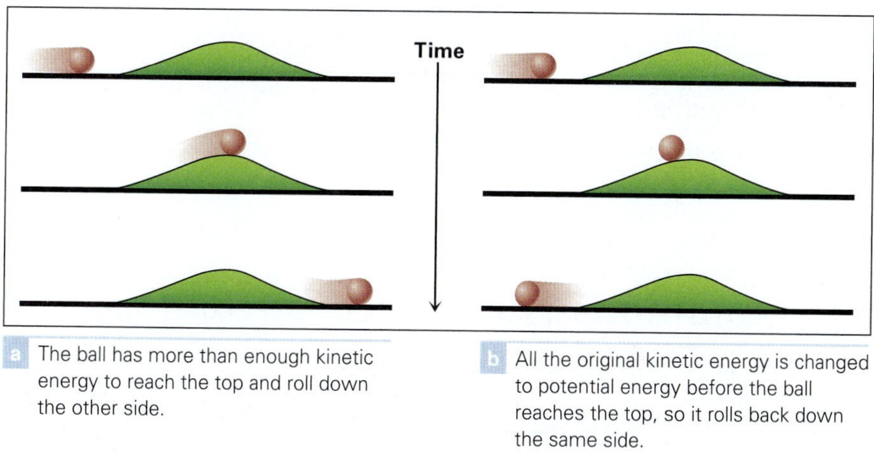

a The ball has more than enough kinetic energy to reach the top and roll down the other side.

b All the original kinetic energy is changed to potential energy before the ball reaches the top, so it rolls back down the same side.

Figure 18.4 Potential energy barrier. As the ball rolls toward the hill (potential energy barrier), its speed (kinetic energy) determines whether or not it will pass over the hill.

(**Figure 18.4[a]**). This corresponds to an effective collision in which the colliding molecules have enough kinetic energy to get over the potential energy barrier. However, if the ball does not have enough kinetic energy to reach the top of the hill, it rolls back to where it came from (**Figure 18.4[b]**). If the colliding particles do not have enough kinetic energy to meet the potential energy requirement, the collision is ineffective.

Thinking About Your Thinking

Mental Models

Figure 18.3 is designed to help you see the relationship between a graph, which is a mathematical construct, and what actually happens at the particulate level that results in the relationship expressed in the graph. This type of thinking frequently occurs in the sciences, and thus it is among the best ways of understanding scientific concepts. The key is to match your mental model of the actual process to the graph.

The horizontal portion of the curve at energy value b represents the total energy of the separate reactants A_2 and B_2. This is analogous to the ball rolling along the flat surface in Figure 18.4. The rising portion of the curve in Figure 18.3 represents increasing energy. As the two molecules approach one another, the potential energy of the system increases because of electrostatic repulsions. Similarly, in Figure 18.4, the potential energy of the ball increases as it moves up the hill. Imagine the molecules coming together and how that increases the potential energy as the negative edge of one molecule repels the negative edge of the other.

When the system energy reaches its maximum, energy a on the Figure 18.3 graph, the reacting particles have moved close enough together to form the transition state, the highest-energy species that exists in this reaction profile. This is similar to the potential energy of the ball at the top of the hill in Figure 18.4(a). As the product molecules move apart, the energy of the system decreases, shown on the graph as the drop from energy a to energy c. This is like the ball rolling down the other side of the hill in Figure 18.4(a). The ball's potential energy is changed to kinetic energy. As you look at the downward slope on the graph, imagine the product molecules moving apart until the electrostatic repulsions are no longer significant.

▶ **P/Review** Activation energy is similar to the escape energy in the evaporation of a liquid, as described in Section 15.5. Only molecules with more than a certain minimum kinetic energy are able to tear away from the bulk of the liquid and change to the vapor state.

The minimum kinetic energy needed to produce an effective collision is called **activation energy (E_a).** ◀ The difference between the energy at the peak of the Figure 18.3 curve and the reactant energies is the activation energy for the forward reaction. Similarly, the difference between the energy at the peak and the product energies is the activation energy for the reverse reaction.

 **Target Check 18.3**

In what sense is activation energy a "barrier"?

▶ **P/Review** The vapor pressure of a liquid rises as temperature increases. The boiling point of a liquid is the temperature at which its vapor pressure is equal to the pressure above the liquid. A higher pressure over the liquid requires a higher vapor pressure—and therefore a higher temperature—to boil the liquid. See Section 15.6.

▶ **P/Review** In principle, the plot in Figure 18.5 is the same curve as in Section 15.5, Figure 15.23, which we used to explain the effect of temperature on the vapor pressure of a liquid. The total area beneath the curve represents the entire sample, so it is the same for all temperatures. At higher temperatures, the curve flattens and shifts to the right. The *average* kinetic energy that corresponds with temperature is found along the x-axis.

18.5 Conditions That Affect the Rate of a Chemical Reaction

The Effect of Temperature on Reaction Rate

Goal 4 State and explain the relationship between reaction rate and temperature.

Chemical reactions are faster at higher temperatures. This can be seen in the kitchen in several ways. Food is refrigerated to slow down the chemical changes that occur in spoiling. A pressure cooker reduces the time needed to cook some items in boiling water because water boils at a higher temperature under increased pressure. ◀ The opposite effect is seen in open cooking at high altitudes, where reduced atmospheric pressure allows water to boil at lower temperatures. High-altitude cooking is slower—the result of reduced reaction rates.

Figure 18.5 explains the effect of temperature on reaction rates. ◀ The blue curve labeled 25°C gives the kinetic energy distribution among the particles in a sample at one temperature, and the red 75°C curve represents the distribution at a higher temperature. E_a is the activation energy, the minimum kinetic energy a particle must have to enter into a reaction-producing collision. It is the same at both temperatures. Only the fraction of the particles in the sample represented by the area beneath the curve to the right of E_a is able to react. Compare the areas to the right of E_a. As illustrated, the fraction that is able to react is about twice as much for 75°C as for 25°C. The rate of reaction is therefore higher at the higher temperature.

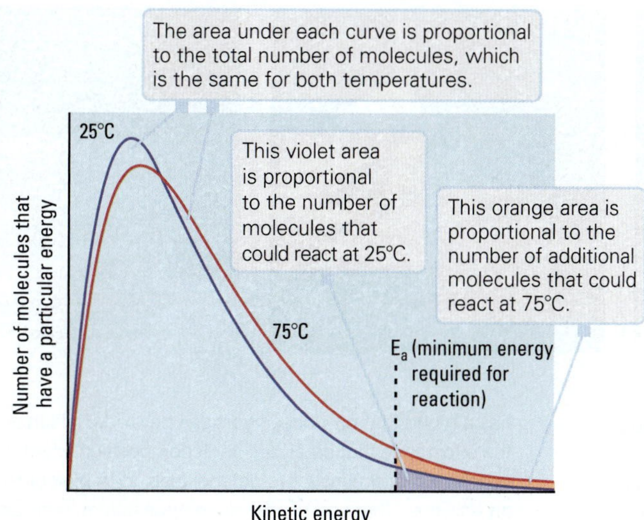

Figure 18.5 Kinetic energy distribution curves at two temperatures. E_a is the activation energy, the minimum kinetic energy required for a reaction-producing collision. Only the fraction of molecules represented by the total area beneath the curve to the right of the activation energy has enough kinetic energy to react. At 25°C, that area is shaded in violet, and at 75°C, that area is the sum of the orange and violet shading. The larger fraction of molecules that is able to react, represented by the summed areas, is responsible for the higher reaction rate at higher temperatures.

The Effect of a Catalyst on Reaction Rate

Goal 5 Using an energy versus reaction progress graph, explain how a catalyst affects reaction rate.

Driving from one city to another over rural roads and through towns takes a certain amount of time. If an interstate highway were built between the cities, you would have available an alternative route that would be much faster. This is what a **catalyst** does. It provides an alternative "route" for reactants to change to products. The activation energy with the catalyst is lower than the activation energy without the catalyst. The result is that a larger fraction of the molecules in a sample is able to enter into reaction-producing collisions, so the reaction rate increases. This is illustrated in **Figure 18.6**.

Catalysts exist in several different forms, and the precise function of many catalysts is a subject of active research. Some catalysts are mixed in the reacting chemicals,

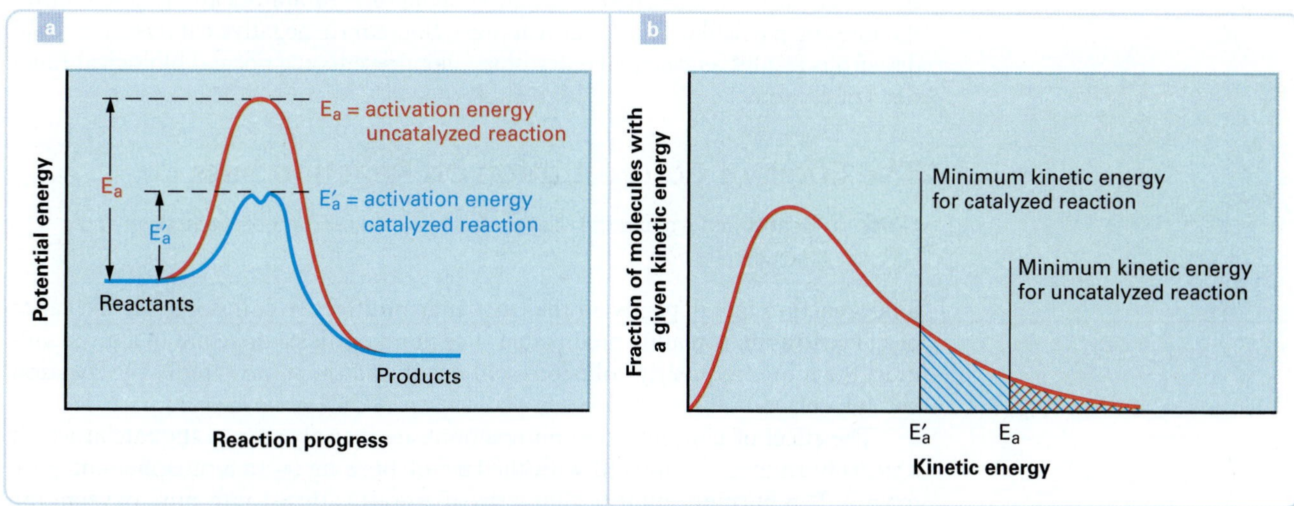

Figure 18.6 The effect of a catalyst on activation energy and reaction rate. (a) A catalyst provides a way for a reaction to occur with a lower activation energy (blue curve) than the same reaction without a catalyst (red curve). Molecules with a lower kinetic energy are therefore able to pass over the potential energy barrier. (b) The crosshatched area in each color represents the fraction of the total sample with enough energy to engage in reaction-producing collisions. The catalyzed area (blue) is much larger than the uncatalyzed area (red), so the catalyzed reaction rate is faster.

Charles D. Winters; Thomas Eisner/Daniel Aneshansley/Cornell University; Charles D. Winters

Figure 18.7 Catalytic decomposition of hydrogen peroxide, $2\ H_2O_2 \xrightarrow{MnO_2} 2\ H_2O + O_2$. (a) Solid MnO_2 catalyzes the decomposition of a 30% solution of H_2O_2. The chemical change is accompanied by the release of so much heat that some of the product H_2O occurs as steam. (b) The bombardier beetle shoots a hot liquid solution from a "gun" at the rear of its body. The beetle has a gland that produces hydrogen peroxide, which is mixed with the substance that catalyzes its decomposition when it becomes necessary for the insect to defend itself. (c) A biological catalyst—an enzyme—that speeds the decomposition of hydrogen peroxide solution occurs naturally in potatoes. You can see oxygen gas bubbling to the surface.

whereas others do no more than provide a surface on which the reaction may occur. In either case, the catalyst is not permanently affected by the reaction. Some catalysts undergo a chemical change, but during the course of the reaction they are regenerated in exactly the same amount that was present at the start.

A catalytic reaction that appears often in introductory chemistry laboratories is making oxygen by decomposing hydrogen peroxide, H_2O_2. Manganese(IV) oxide, MnO_2, is mixed in as a catalyst (**Figure 18.7[a]**). A well-known industrial process that utilizes a catalyst is the catalytic cracking of crude oil in which large hydrocarbon molecules are broken down into simpler and more useful products in the presence of a catalyst (**Figure 18.8**). Biological reactions are controlled by catalysts called *enzymes* (**Figure 18.7[c]**).

Some substances interfere with a normal reaction path from reactants to products, forcing the reaction to a higher activation energy route that is slower. Such substances are called negative catalysts or **inhibitors**. Inhibitors are used to control the rates of particular industrial reactions. Sometimes negative catalysts can have disastrous results, as when mercury poisoning prevents the normal biological functions of enzymes.

The Effect of Concentration on Reaction Rate

Goal 6 Identify and explain the relationship between reactant concentration and reaction rate.

Since reaction rate depends on the frequency of effective collisions, the influence of concentration is readily predictable. The more particles there are in a given volume, the more frequently collisions will occur and the more rapidly the reaction will take place.

The effect of concentration on reaction rate is easily seen in the rate at which objects burn in air compared with the rate of burning in an atmosphere of pure oxygen. If a burning splint (a thin strip of wood) is thrust into pure oxygen, the burning becomes brighter, more vigorous, and much faster. In fact, the typical laboratory test for oxygen is to ignite a splint, blow it out, and then, while there is still a faint glow, place it in oxygen. It immediately bursts back into full flame and burns vigorously (**Figure 18.9**). Charcoal, phosphorus, and other substances may be used in place of the splint.

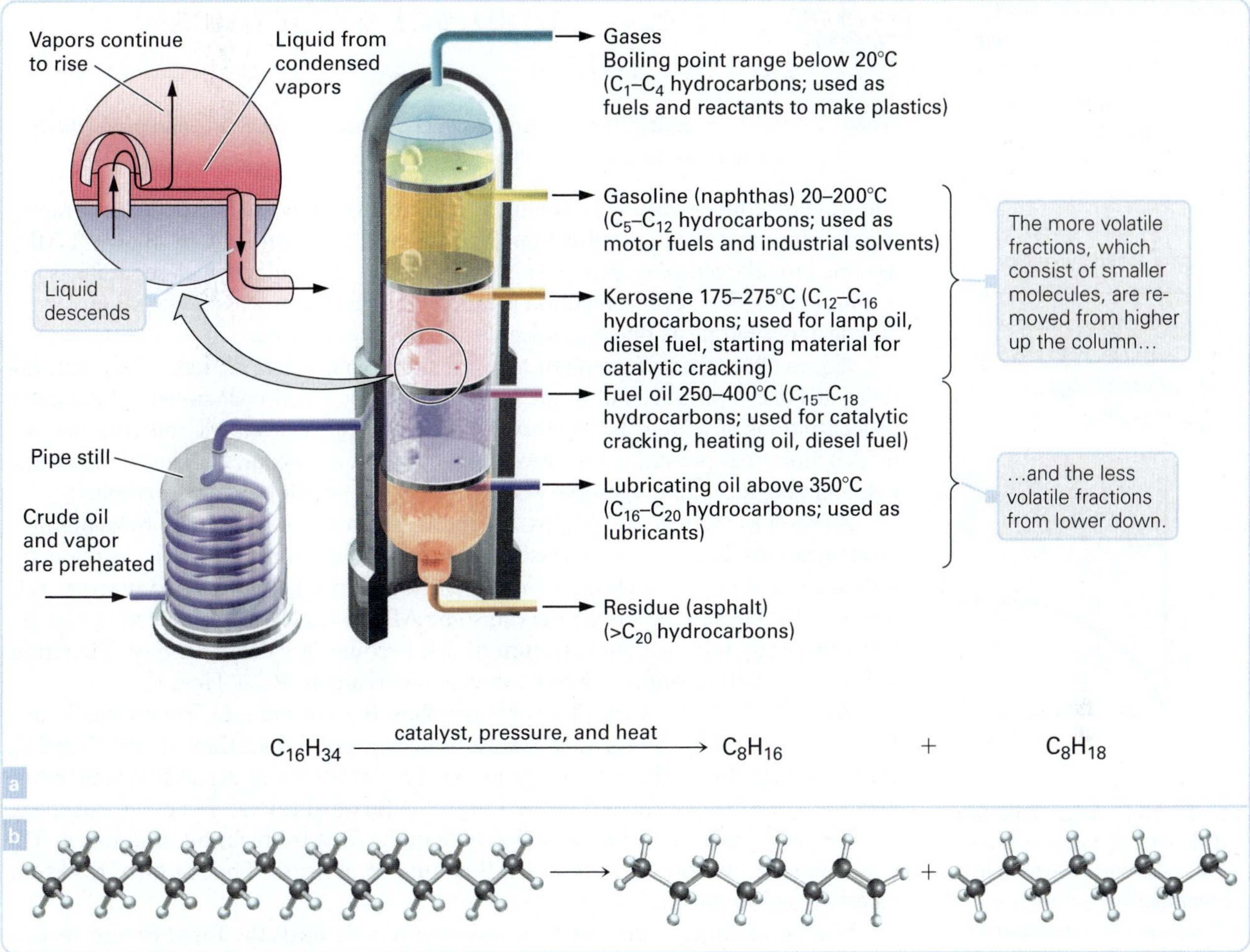

Vapors continue to rise

Liquid from condensed vapors

Liquid descends

Pipe still

Crude oil and vapor are preheated

Gases
Boiling point range below 20°C
(C_1–C_4 hydrocarbons; used as fuels and reactants to make plastics)

Gasoline (naphthas) 20–200°C
(C_5–C_{12} hydrocarbons; used as motor fuels and industrial solvents)

Kerosene 175–275°C (C_{12}–C_{16} hydrocarbons; used for lamp oil, diesel fuel, starting material for catalytic cracking)

Fuel oil 250–400°C (C_{15}–C_{18} hydrocarbons; used for catalytic cracking, heating oil, diesel fuel)

Lubricating oil above 350°C
(C_{16}–C_{20} hydrocarbons; used as lubricants)

Residue (asphalt)
($>C_{20}$ hydrocarbons)

The more volatile fractions, which consist of smaller molecules, are removed from higher up the column...

...and the less volatile fractions from lower down.

a

$$C_{16}H_{34} \xrightarrow{\text{catalyst, pressure, and heat}} C_8H_{16} + C_8H_{18}$$

b

Figure 18.8 Catalytic cracking. (a) Crude oil, a complex mixture, is distilled into simpler mixtures (fractions) based on their boiling points. In general, there is more demand for the gasoline fraction than for kerosene. (b) After distillation, the larger (12 to 16 carbons) molecules in the kerosene fraction can be decomposed into smaller (5 to 12 carbons) molecules that make up gasoline. This is accomplished by catalytic cracking, a process that uses a catalyst.

Summary of Effects on Reaction Rates

We have identified three factors that influence the rate of a chemical reaction, temperature, catalysis, and concentration. In relating these factors to equilibrium considerations, we will examine only concentration and temperature. These variables affect forward and reverse reaction rates differently. A catalyst, on the other hand, has the same effect on both forward and reverse rates. Therefore, a catalyst does not alter chemical equilibrium. A catalyst does cause a system to reach equilibrium more quickly.

Charles D. Winters

Figure 18.9 The glowing splint test for oxygen.

✓ Target Check 18.4

a) What happens to a reaction rate as temperature drops? Give two explanations for the change. State which one is more important and explain why.

b) How does a catalyst affect reaction rates?

c) Compare reaction rates when a given reactant concentration is high with the rate when the concentration is low. Explain the difference.

▶ **P/Review** The development of a liquid–vapor pressure equilibrium is described in Section 15.5. An equilibrium between excess solute and a saturated solution is examined in Section 16.4. Both equilibria involve physical changes in which one of the two opposing rates remains constant until the other catches up with it. Here we study how a chemical equilibrium develops. This time, both forward and reverse rates change as equilibrium is reached.

18.6 The Development of a Chemical Equilibrium ◀

Goal 7 Trace and explain the changes in concentrations of reactants and products that lead to a chemical equilibrium.

The role of concentration in chemical equilibrium may be illustrated by tracing the development of an equilibrium. The forward reaction in $A_2 + B_2 \rightleftharpoons 2\ AB$ is assumed to take place by the collision of A_2 and B_2 molecules, which separate as two AB molecules. The reverse reaction is exactly the reverse process: Two AB molecules collide and separate as one A_2 molecule and one B_2 molecule.

Figure 18.10 shows a graph of forward and reverse reaction rates versus time. Initially, at Time 0, pure A_2 and B_2 are introduced to the reaction chamber. At the initial concentrations of A_2 and B_2, the forward reaction begins at rate F_0. Initially, there are no AB molecules present, so the reverse reaction cannot occur. At Time 0 the reverse reaction rate, R_0, is zero. Observe where these points are plotted on the graph.

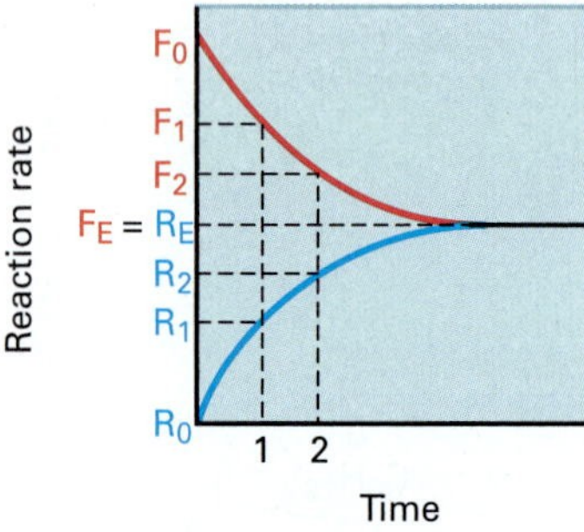

Figure 18.10 Changes in reaction rates during the development of a chemical equilibrium. The forward rate is represented by the upper curve in red, and the reverse rate is shown by the lower curve in blue.

As soon as the reaction begins, A_2 and B_2 are consumed, thereby reducing their concentrations in the reaction vessel. As these reactant concentrations decrease, the forward reaction rate declines. Consequently, at Time 1, the forward reaction rate drops to F_1. During the same interval, some AB molecules are produced by the forward reaction, and the concentration of AB becomes greater than zero. Therefore, the reverse reaction begins, with the reverse rate rising to R_1 at Time 1.

At Time 1, the forward rate is greater than the reverse rate. Therefore, A_2 and B_2 are consumed by the forward reaction more rapidly than they are produced by the reverse reaction. The net change in the concentrations of A_2 and B_2 is therefore decreasing, causing a further reduction in the forward rate at Time 2. Conversely, the reverse reaction uses AB more slowly than the forward reaction produces it. The net change in the concentration of AB is thus an increase. This, in turn, raises the reverse reaction rate at Time 2.

Similar changes occur over successive intervals until the forward and reverse rates eventually become equal. At this point a dynamic equilibrium is established. From this analysis we may state the following generalization:

For any reversible reaction in a closed system, whenever the opposing reactions are occurring at different rates, the faster reaction will gradually become slower, and the slower reaction will become faster. Finally, the reaction rates become equal, and equilibrium is established.

18.7 Le Chatelier's Principle

Oesper Collection in the History of Chemistry, University of Cincinnati

Figure 18.11 Henry Louis Le Chatelier (1850–1936). In addition to being a talented scientist, Le Chatelier was an educational reformer and an excellent teacher.

In the last section, you saw how concentrations and reaction rates interact with one another until the rates become equal and equilibrium is reached. In this section, we start with a system already at equilibrium and see what happens to it when the equilibrium is upset. To "upset" an equilibrium, you must somehow make the forward and reverse reaction rates unequal, at least temporarily. One way to do this is to change the concentration of at least one substance in the system. A gaseous equilibrium can often be upset simply by changing the volume of the container. If you change the temperature of a chemical equilibrium, you will also make the forward and reverse reaction rates unequal.

How an equilibrium responds to a disturbance can be predicted from the concentration and temperature effects already considered. The predictions may be summarized in **Le Chatelier's Principle**, developed by French scientist Henri Louis Le Chatelier (**Figure 18.11**), which says that if an equilibrium system is subjected to change, processes occur that tend to partially counteract the initial change, thereby bringing the system to a new position of equilibrium.

We will now see how Le Chatelier's Principle explains three different equilibrium changes.

The Concentration Effect

Goal 8 Given the equation for a chemical equilibrium, or information from which it can be written, predict the direction in which the equilibrium will shift because of a change in the concentration of one species.

The reaction of hydrogen and iodine to produce hydrogen iodide comes to equilibrium with hydrogen iodide as the favored species—that is, the species having the higher concentration. The equal forward and reverse reaction rates are shown by the equal-length arrows in the equation:

$$H_2(g) + I_2(g) \rightleftharpoons 2\ HI(g)$$

The sizes of the formulas represent the relative concentrations of the different species in the reaction. [HI] is greater than $[H_2]$ and $[I_2]$, which are equal. ▶

◀ **P/Review** Enclosing the formula of a substance in brackets represents its concentration in moles per liter. Brackets are used for hydrogen ion and hydroxide ion concentrations, $[H^+]$ and $[OH^-]$, in Section 17.7.

If more HI is forced into the system, [HI] is increased. This raises the rate of the reverse reaction, in which HI is a reactant. This is indicated by the longer arrow from right to left:

$$H_2(g) + I_2(g) \rightleftharpoons 2\ HI(g)$$

Because the rates are no longer equal, the equilibrium is destroyed.

Now the changes described in italics at the end of the last section begin. The unequal reaction rates cause the system to shift in the direction of the faster rate—to the left, or in the reverse direction. As a result, H_2 and I_2 are made by the reverse reaction faster than they are used by the forward reaction. Their concentrations increase, so the forward rate increases. Simultaneously, HI is used faster than it is produced, reducing both [HI] and the reverse reaction rate. Eventually, the rates become equal (note arrow lengths) at an intermediate value, and a new equilibrium is reached:

$$H_2(g) + I_2(g) \rightleftharpoons 2\ HI(g)$$

The sizes of the formulas indicate that all three concentrations are larger than they were originally, although [HI] has come down from the maximum it reached just after it was added.

The preceding example shows *why* an equilibrium shifts when it is disturbed in terms of concentrations and reaction rates. Le Chatelier's Principle makes it possible to predict the direction of a shift, forward or reverse, without such a detailed analysis. Just remember that the shift is always in the direction that tries to return the disturbed substance toward its original condition. **Figure 18.12** illustrates the Le Chatelier concentration effect.

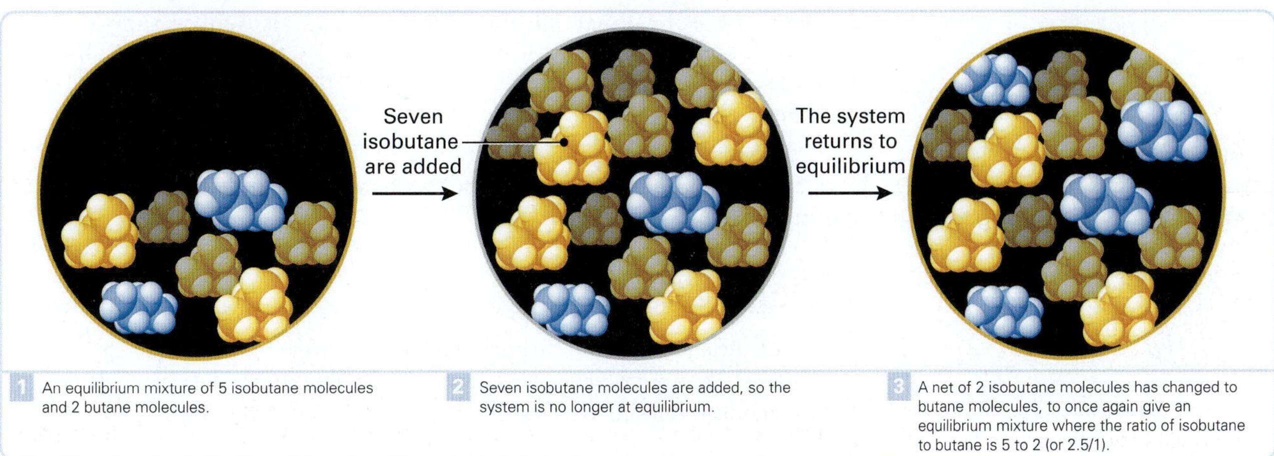

Seven
isobutane
are added
The system
returns to
equilibrium
1 An equilibrium mixture of 5 isobutane molecules and 2 butane molecules.
2 Seven isobutane molecules are added, so the system is no longer at equilibrium.
3 A net of 2 isobutane molecules has changed to butane molecules, to once again give an equilibrium mixture where the ratio of isobutane to butane is 5 to 2 (or 2.5/1).

Figure 18.12 The Le Chatelier concentration effect. The shift is in the direction that returns the disturbed concentration ratio to its equilibrium value (described in Section 18.8).

Active Example 18.1 Le Chatelier Concentration Effect I

The system $N_2(g) + 3\ H_2(g) \rightleftharpoons 2\ NH_3(g)$ is at equilibrium. Use Le Chatelier's Principle to predict the direction in which the equilibrium will shift if ammonia is withdrawn from the reaction chamber.

Think Before You Write The equilibrium disturbance is clearly stated: Ammonia is withdrawn. In which direction, forward or reverse, must the reaction shift to *counteract* the removal of ammonia—that is, to *produce* more ammonia to replace some of what has been taken away?

Answers *Cover the left column with your cut-out shield. Reveal each answer only after you have written your own answer in the right column.*

The shift will be in the *forward* direction. Ammonia is the product of the forward reaction and therefore will be partially restored to its original concentration by a shift in the forward direction.	Choose the direction of the shift and explain your reasoning. 
You improved your understanding of Le Chatelier's Principle.	What did you learn by solving this Active Example? 

Practice Exercise 18.1

Nitrogen and oxygen gases are in equilibrium with nitrogen monoxide gas: $N_2(g) + O_2(g) \rightleftharpoons 2\ NO(g)$. In which direction will the equilibrium shift if oxygen is removed from the reaction container?

Active Example 18.2 Le Chatelier Concentration Effect II

Predict the direction, forward or reverse, of a Le Chatelier shift in the equilibrium $CH_4(g) + 2\ H_2S(g) \rightleftharpoons 4\ H_2(g) + CS_2(g)$ caused by each of the following: (a) increase $[H_2S]$; (b) reduce $[CS_2]$; (c) increase $[H_2]$. Briefly justify each answer.

Think Before You Write The problem statement makes it clear that this is a Le Chatelier's Principle problem.

Answers *Cover the left column with your cut-out shield. Reveal each answer only after you have written your own answer in the right column.*

a) Forward, counteracting partially the increase in $[H_2S]$ by consuming some of it. b) Forward, restoring some of the CS_2 removed. c) Reverse, consuming some of the added H_2.	Choose the direction of each shift and explain your reasoning. 
You improved your understanding of Le Chatelier's Principle.	What did you learn by solving this Active Example? 

Practice Exercise 18.2

Consider the equilibrium system $N_2O_4(g) \rightleftharpoons 2\ NO_2(g)$. State the direction of the Le Chatelier shift for (a) adding N_2O_4; (b) removing N_2O_4; (c) adding NO_2; and (d) removing NO_2.

The Volume Effect

Goal 9 Given the equation for a chemical equilibrium involving one or more gases, or information from which it can be written, predict the direction in which the equilibrium will shift because of a change in the volume of the system.

A change in the volume of an equilibrium system that includes one or more gases changes the concentration of those gases. Usually, there is a Le Chatelier shift that partially offsets the initial change—there is one exception, as you will see shortly. Both the change and the adjustment involve the pressure caused by the entire system.

According to the kinetic molecular theory (Section 4.3), the pressure exerted by a gas is the combined result of billions upon billions of collisions of molecules hitting the walls of the container that holds the gas. If the frequency of these collisions increases, pressure increases; if frequency is reduced, pressure is reduced. One way to increase the frequency is to reduce the volume of the container. The molecules have shorter distances to travel before hitting the walls, so the frequency goes up. Conversely, if volume is increased, frequency and, therefore, pressure are reduced. The relationship is an inverse proportionality, expressed by Boyle's Law, which you studied in Section 4.6: $P \propto 1/V$.

There is another way to change the frequency of collisions between molecules and the walls of a container: change the number of gas molecules in the container. More particles mean more collisions in a given period of time, and therefore higher pressure; fewer particles yield lower pressure. This relationship is a direct proportionality: $P \propto n$, where n represents the amount of gas molecules in moles.

Whenever one quantity (P) is proportional to two other quantities (1/V and n), it is proportional to the product of those quantities:

$$P \propto \frac{1}{V} \times n \quad P \propto \frac{n}{V}$$

The ratio n/V is an expression of the concentration of the gas, commonly expressed in moles per liter. ▶

If the volume of a gas is reduced, the denominator in the n/V concentration ratio becomes smaller and the fraction becomes larger; thus, pressure increases. Le Chatelier's Principle calls for a shift that will partially counteract the change—to reduce pressure. What change can reduce the pressure of the system at the new volume? The numerator in the n/V concentration ratio must be reduced; there must be fewer gaseous molecules in the system. In general:

*If a gaseous equilibrium is compressed, the increased pressure will be partially relieved by a shift in the direction of fewer gaseous molecules; if the system is expanded, the reduced pressure will be partially restored by a shift in the direction of more gaseous molecules (***Figure 18.13***).*

The coefficients of gases in an equation are in the same proportion as the number of gaseous molecules. ▶ We use this fact in predicting Le Chatelier shifts caused by volume changes. The following Active Example shows how.

◀ **P/Review** If you have studied Chapter 14, you will recognize that the text discussion leads to the ideal gas equation (Section 14.4), PV = nRT. Solving for P at constant temperature gives

$$P = RT \times \frac{n}{V} = k \times \frac{n}{V}$$

where k is a constant. If k is a proportionality constant, it follows that

$$P \propto \frac{n}{V}$$

which expresses the concentration of the gas.

◀ **P/Review** The proportional relationship between equation coefficients and the number of gaseous molecules is used in stoichiometry problems when converting between volumes of different gases measured at the same temperature and pressure (Section 14.10).

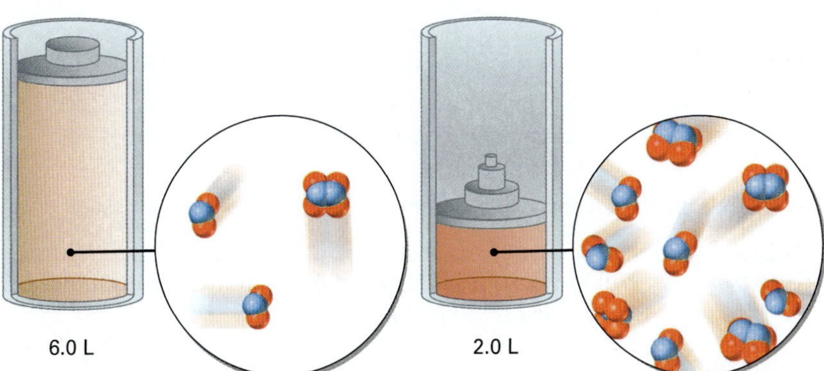

Figure 18.13 The Le Chatelier volume effect. The system $N_2O_4(g) \rightleftharpoons 2\ NO_2(g)$ is initially at equilibrium under conditions at which the ratio of N_2O_4 to NO_2 is 1:2 (*left*). The system is compressed to one-third of the original volume (*right*). Le Chatelier's Principle predicts that the system will respond by shifting in the direction of fewer gaseous molecules, which is in the reverse direction. This shift is verified by counting the molecules; the ratio of N_2O_4 to NO_2 is now 4:5. As NO_2 reacts to form N_2O_4, the total number of particles in the system is reduced, partially offsetting the increased pressure.

Active Example 18.3 Le Chatelier Volume Effect I

Predict the direction of the shift resulting from an expansion in the volume of the container holding the equilibrium system $2\ SO_2(g) + O_2(g) \rightleftharpoons 2\ SO_3(g)$.

Think Before You Write Even though the problem statement does not directly mention Le Chatelier's Principle, you should recognize that it describes an equilibrium and asks what happens when that equilibrium is disturbed.

Answers *Cover the left column with your cut-out shield. Reveal each answer only after you have written your own answer in the right column.*

Boyle's Law indicates that pressure is inversely proportional to volume, so the total pressure will be less at the larger volume.	Will total pressure increase or decrease as a result of expansion? Explain.
Since pressure is proportional to concentration, $P \propto n/V$, the number of molecules, n, must increase to raise the pressure.	The Le Chatelier shift must make up some of that lost pressure. How? By changing the number of gaseous molecules in the system. Will it take more molecules or fewer to raise pressure? Explain.
The reaction must shift in the *reverse* direction. *2 molecules → 3 molecules to increase the total number of molecules.*	Examine the reaction equation. Notice that a forward shift has *three* gas-phase reactant molecules, two SO_2 and one O_2, forming *two* gas-phase SO_3 product molecules. The reverse shift has *two* reactant molecules yielding *three* product molecules. In which direction will the reaction shift?
You improved your understanding of Le Chatelier's Principle.	What did you learn by solving this Active Example? 

Practice Exercise 18.3

In which direction will the reaction $CH_3OH(g) \rightleftharpoons CO(g) + 2\ H_2(g)$ shift when the container volume is decreased?

Active Example 18.4 Le Chatelier Volume Effect II

The volume occupied by the equilibrium $SiF_4(g) + 2\ H_2O(g) \rightleftharpoons SiO_2(s) + 4\ HF(g)$ is reduced. Predict the direction of the shift in the position of equilibrium.

Think Before You Write The problem statement describes an equilibrium system that is subjected to a disturbance. This requires the application of Le Chatelier's Principle.

Answers *Cover the left column with your cut-out shield. Reveal each answer only after you have written your own answer in the right column.*

Fewer. If volume is reduced, pressure increases. Increased pressure is counteracted by fewer molecules.	Will the shift be in the direction of more gaseous molecules or fewer? Explain. 
The shift is in the reverse direction. Gas-phase molecules change from 4 on the right to 3 on the left. *Note the molecule change is 4 to 3, not 5 to 3. Only gaseous molecules are involved in pressure adjustments, so the $SiO_2(s)$ doesn't count.*	Predict the direction of the shift and justify your prediction by stating the numerical change in molecules from the equation. 

You improved your understanding of Le Chatelier's Principle.	What did you learn by solving this Active Example? 

Practice Exercise 18.4

The system $2\ NH_3(g) + CO_2(g) \rightleftharpoons N_2CH_4O(s) + H_2O(g)$ is at equilibrium. The volume of the reaction container is doubled. Explain how the system will respond.

Active Example 18.5 Le Chatelier Volume Effect III

Returning to the familiar $H_2(g) + I_2(g) \rightleftharpoons 2\ HI(g)$, predict the direction of the shift that will occur because of a volume increase.

Think Before You Write An equilibrium system is subjected to a change. Le Chatelier's Principle applies.

Answers *Cover the left column with your cut-out shield. Reveal each answer only after you have written your own answer in the right column.*

There will be no shift, because each side of the equation has two gaseous molecules. *When the number of gaseous molecules is the same on both sides of the equilibrium equation, neither the number of molecules nor the pressure can be changed by a shift in equilibrium. Increasing or decreasing the volume has no effect on the equilibrium.*	Take it all the way. 
You improved your understanding of Le Chatelier's Principle.	What did you learn by solving this Active Example? 

Practice Exercise 18.5

The reaction $CuO(s) + H_2(g) \rightleftharpoons Cu(s) + H_2O(g)$ takes place in a closed container. The container volume is reduced to 1/10 of the original volume. How is the equilibrium affected? Explain.

The Temperature Effect

Goal 10 Given a thermochemical equation for a chemical equilibrium, or information from which it can be written, predict the direction in which the equilibrium will shift because of a change in temperature.

A change in temperature of a chemical system at equilibrium will change both forward and reverse reaction rates, but the rate changes are not equal. The equilibrium is, therefore, destroyed temporarily. The events that follow are again predictable by Le Chatelier's Principle.

A thermochemical equation is one that includes a change in energy. It can be written in two ways: with the enthalpy-of-reaction term, $\Delta_r H$, to the right of the conventional equation or with the energy term included as if it was a reactant or a product (Section 10.9). Including the energy term in the thermochemical equation, rather than showing $\Delta_r H$ separately, makes it easier to predict the Le Chatelier effect of a change in temperature. In the equation, we can think of energy as we would a substance being "added" or "removed."

Figure 18.14 gives a visible example of Le Chatelier's Principle as it relates to temperature.

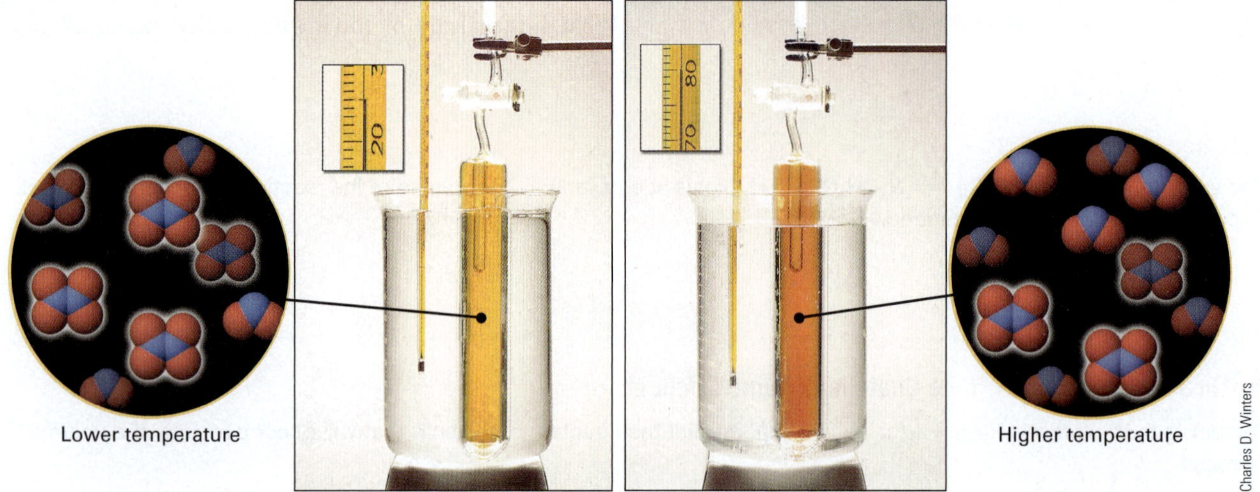

Figure 18.14 The gas-phase equilibrium for the reaction 2 $NO_2(g) \rightleftharpoons N_2O_4(g)$ + heat. The tubes contain the same total amount of gas. NO_2 is brown, whereas N_2O_4 is colorless. The left tube, at 25°C, contains very little brown gas, indicating that the equilibrium is shifted in the forward direction. The concentration of N_2O_4 molecules is relatively high. At higher temperature, 80°C (right tube), the gas is much darker because brown NO_2 is formed as the reaction shifts in the reverse direction. The N_2O_4 concentration has decreased.

Active Example 18.6 Le Chatelier Temperature Effect I

If the temperature of the equilibrium $PCl_5(g) \rightleftharpoons PCl_3(g) + Cl_2(g)$ + 92.5 kJ is increased, predict the direction of the Le Chatelier shift.

Think Before You Write In order to raise the temperature of something, we must heat it. We therefore interpret an increase in the temperature as the "addition of heat" and a lowering of temperature as the "removal of heat." In applying Le Chatelier's Principle to a thermochemical equation, we may regard the heat in much the same manner as we regard any chemical species in the equation.

Answers *Cover the left column with your cut-out shield. Reveal each answer only after you have written your own answer in the right column.*

The equilibrium must shift in the *reverse* direction to "use up" some of the added heat.	If heat is *added* to the equilibrium system shown, in which direction must it shift to *use up*, or *consume*, some of the heat that was added? (If chlorine were *added*, in which direction would the equilibrium shift to use up some of the chlorine?) Forward or reverse? Explain.
You improved your understanding of Le Chatelier's Principle.	What did you learn by solving this Active Example?

Practice Exercise 18.6

The system 2 $SO_2(g) + O_2(g) \rightleftharpoons$ 2 $SO_3(g)$ + 198 kJ comes to equilibrium, and then the temperature is decreased. How will the system respond?

Active Example 18.7 Le Chatelier Temperature Effect II

The thermal decomposition of limestone reaches the following equilibrium: $CaCO_3(s)$ + 176 kJ $\rightleftharpoons CaO(s) + CO_2(g)$. Predict the direction this equilibrium will shift if the temperature is reduced: forward or reverse. Justify your answer.

Think Before You Write The existing equilibrium is disrupted by the temperature change. Le Chatelier's Principle allows you to predict the direction of the shift.

Answers *Cover the left column with your cut-out shield. Reveal each answer only after you have written your own answer in the right column.*

Reverse. Reduction in temperature is interpreted as the removal of heat. The reaction will respond to replace some of the heat removed—as an exothermic reaction. Heat is produced as the reaction proceeds in the reverse direction.	No additional help is needed. Write your response here. 
You improved your understanding of Le Chatelier's Principle.	What did you learn by solving this Active Example?

Practice Exercise 18.7

Predict how the equilibrium will shift when the temperature is increased for the system $N_2(g) + O_2(g) + 181\ kJ \rightleftharpoons 2\ NO(g)$.

18.8 The Equilibrium Constant

Goal 11 Given any chemical equilibrium equation, or information from which it can be written, write the equilibrium constant expression.

If 1.000 mole of $H_2(g)$ and 1.000 mole of $I_2(g)$ are introduced into a 1.000-L reaction vessel, they will react to produce the following equilibrium:

$$H_2(g) + I_2(g) \rightleftharpoons 2\ HI(g)$$

At a temperature of 440°C, analysis at equilibrium will show hydrogen and iodine concentrations of 0.218 mol/L and a hydrogen iodide concentration of 1.564 mol/L.

Working in the opposite direction, if gaseous hydrogen iodide is introduced to a reaction chamber at 440°C at an initial concentration of 2.000 mol/L, it will decompose by the reverse reaction. The system will eventually come to equilibrium with $[H_2] = [I_2] = 0.218$ mol/L and $[HI] = 1.564$ mol/L, exactly the same equilibrium concentrations as in the first example. This illustrates the experimental fact that the position of an equilibrium when under the same conditions is the same, regardless of the direction from which it is approached.

As a third example, if the initial hydrogen iodide concentration is half as much, 1.000 mol/L, the equilibrium concentrations will be half of what they were when starting from double the hydrogen iodide concentration: $[H_2] = [I_2] = 0.109$ mol/L and $[HI] = 0.782$ mol/L.

If the experiment is repeated at the same temperature, starting this time with hydrogen at 1.000 mol/L and iodine at 0.500 mol/L, the equilibrium concentrations will be $[H_2] = 0.532$ mol/L, $[I_2] = 0.032$ mol/L, and $[HI] = 0.936$ mol/L.

The data of these four equilibria are summarized in **Table 18.2**. Analysis of these data leads to the observation that at equilibrium, the ratio $\frac{[HI]^2}{[H_2][I_2]}$ has a value of 51.5 for all four equilibria. For that matter, this ratio of equilibrium concentrations is always the same regardless of the initial concentrations of hydrogen, iodine, and hydrogen iodide, as long as the temperature is held at 440°C.

Data from countless other equilibrium systems show a similar regularity that defines the **equilibrium constant, K**: For any equilibrium at a given temperature, the ratio of the product of the concentrations of the species on the right side of the equilibrium equation, each raised to a power equal to its coefficient in the equation, to the corresponding product of the concentrations of the species on the left side of the equation, each raised to a power equal to its coefficient in the equation, is a constant.*

*The development of the equilibrium constant expression can be duplicated by a rigorous theoretical derivation. Theory and experiment support each other completely in this area.

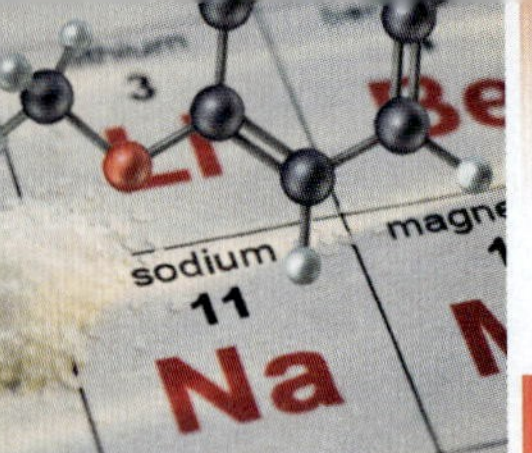

Everyday Chemistry 18.1

FERTILIZATION OF THE WORLD'S CROPS

Have you ever wondered how bread is made? Have you pondered how fruits and vegetables are grown? Have you ever thought about where taco shells come from? The ultimate source of all of these foods is farms. Each time you eat a plant product, unless it came from someone's backyard, there's a farmer responsible for planting and growing the crop that ultimately became your food.

The molecules that make up plants are composed mostly of carbon, hydrogen, and oxygen atoms. Plants obtain these atoms from chemical reactions in which the reactants are atmospheric carbon dioxide and water from the air or ground. A smaller, but still essential, component of the molecules is nitrogen. Chemistry plays a central role in providing a usable source of nitrogen for agriculture.

Since the atmosphere is about 78% nitrogen, it is logical to turn to the air for this element. Alas, the process is not that simple. Elemental nitrogen, N_2, has a very strong and stable triple bond between the two atoms. This makes it very unreactive. The challenge for chemists is to break the bond and put the nitrogen atoms into other substances that more easily release them to form molecules in plants.

Farmers are well aware of natural sources of nitrogen fertilizers. Animal waste, composted household garbage and plant clippings, and blood meal (dried blood from slaughtered animals) have long been known as effective fertilizers. However, beginning in about 1900, the world's demand for fertilizer began to exceed the supply. Local natural sources, plus a large guano deposit in Chile, were no longer adequate to supply nitrogen-based fertilizer to meet the growing worldwide demand.

Fritz Haber (**Figure 18.15**), a German-born chemist, solved the problem. The Haber synthesis, as it is now called, is based on the equilibrium:

$$N_2 + 3\,H_2 \rightleftharpoons 2\,NH_3 + 92\text{ kJ}$$

Nitrogen was obtained by distilling the air. Hydrogen was made by blowing steam over a bed of hot coke (mostly made of carbon), which produced hydrogen and carbon monoxide.

Haber's unique insight was to solve the problem of how to get the nitrogen-plus-hydrogen reaction to work. The key was finding the proper combination of temperature and pressure, as well as the right catalyst, to make the synthesis succeed. Haber had learned a great deal about energy relationships in chemical change, the combustion of hydrocarbons, oxidation–reduction reactions, and chemical equilibrium before he began his quest to perform the ammonia synthesis. He then applied his copious knowledge and searched for the proper conditions through extensive experimentation, finally finding that an iron catalyst, a temperature of about 500°C, and a pressure of about 200 atm produced adequate yields under safe conditions (**Figure 18.16**).

Science and Society/Superstock

Figure 18.15 Awarding the Nobel Prize to Fritz Haber (1868–1934) was controversial because of his work in developing and deploying poison gases in World War I.

Table 18.2 Equilibria of $H_2(g) + I_2(g) \rightleftharpoons 2\,HI(g)$ at 440°C

Experiment Number	Initial $[H_2]$	Initial $[I_2]$	Initial $[HI]$	Equilibrium $[H_2]$	Equilibrium $[I_2]$	Equilibrium $[HI]$	$\frac{[HI]^2}{[H_2][I_2]}$
1	1.000	1.000	0	0.218	0.218	1.564	51.5
2	0	0	2.000	0.218	0.218	1.564	51.5
3	0	0	1.000	0.109	0.109	0.782	51.5
4	1.000	0.500	0	0.532	0.032	0.936	51.5

Thus, for $H_2(g) + I_2(g) \rightleftharpoons 2\,HI(g)$ at 440°C,

$$K = \frac{[HI]^2}{[H_2][I_2]} = 51.5$$

Figure 18.17 illustrates the equilibrium constant concept as applied to the hydrogen–iodine–hydrogen iodide system.

Haber was awarded the 1918 Nobel Prize in Chemistry for his work.

A liquid ammonia solution can be sprayed on soil directly as a relatively inexpensive fertilizer. The ammonia can also be further changed into the solids ammonium sulfate, ammonium phosphate, or ammonium nitrate, which are other common fertilizers. Modern society would not be the same without Fritz Haber's contribution of a way to fertilize the world's crops.

Quick Quiz

1. **What reaction is the Haber synthesis based upon, and what is the most important application of the synthesis?**
2. **Write a balanced equation for the reaction that is used to make the hydrogen that is then used as a reactant in the Haber ammonia synthesis.**

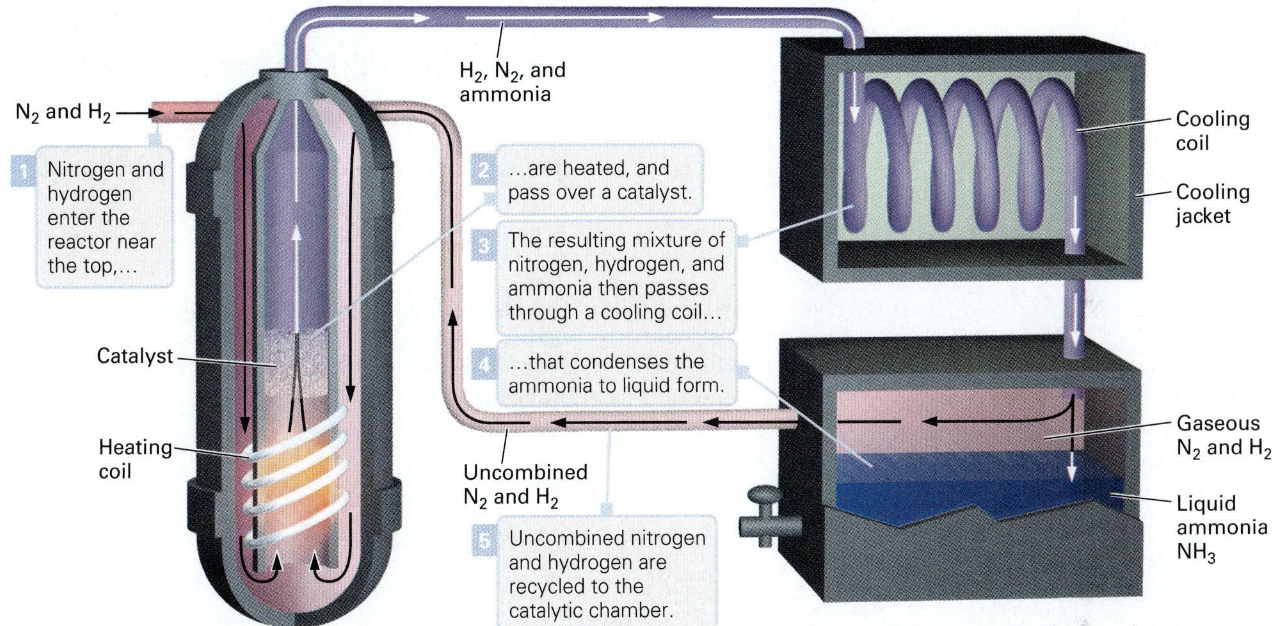

N_2 and H_2
1 Nitrogen and hydrogen enter the reactor near the top,...
Catalyst
Heating coil
H_2, N_2, and ammonia
2 ...are heated, and pass over a catalyst.
3 The resulting mixture of nitrogen, hydrogen, and ammonia then passes through a cooling coil...
4 ...that condenses the ammonia to liquid form.
Uncombined N_2 and H_2
5 Uncombined nitrogen and hydrogen are recycled to the catalytic chamber.
Cooling coil
Cooling jacket
Gaseous N_2 and H_2
Liquid ammonia NH_3

Figure 18.16 Schematic diagram of the process for the industrial production of ammonia from nitrogen and hydrogen.

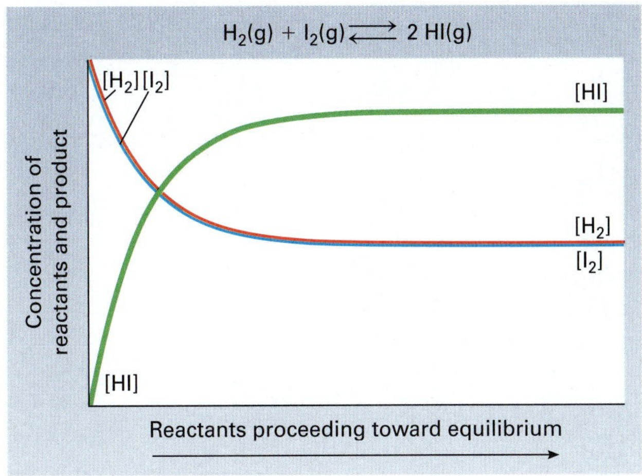

$H_2(g) + I_2(g) \rightleftarrows 2\,HI(g)$
$[H_2]\,[I_2]$
[HI]
$[H_2]$
$[I_2]$
[HI]
Concentration of reactants and product
Reactants proceeding toward equilibrium

Figure 18.17 The equilibrium constant. The ratio $[HI]^2/[H_2][I_2]$ is the same at equilibrium at any given temperature regardless of the initial concentrations of any species in the system.

The definition of K sets the procedure by which any equilibrium constant expression may be written. For the general equilibrium, $a\,A + b\,B \rightleftharpoons c\,C + d\,D$, where A, B, C, and D are chemical formulas and a, b, c, and d are their coefficients in the equilibrium equation:

1. Write in the *numerator* the concentration of each species on the *right-hand* side of the equation. $K = \frac{[C][D]}{}$
2. For each species on the right-hand side of the equation, use its coefficient in the equation as an exponent. $K = \frac{[C]^c[D]^d}{}$
3. Write in the *denominator* the concentration of each species on the *left-hand* side of the equation. $K = \frac{[C]^c[D]^d}{[A][B]}$
4. For each species on the left-hand side of the equation, use its coefficient in the equation as an exponent. $K = \frac{[C]^c[D]^d}{[A]^a[B]^b}$

If the equation were written in the reverse, $c\,C + d\,D \rightleftharpoons a\,A + b\,B$, the equilibrium constant would be the reciprocal of the preceding constant, $K = \frac{[A]^a[B]^b}{[C]^c[D]^d}$. Every equilibrium constant expression must be associated with a specific equilibrium equation.

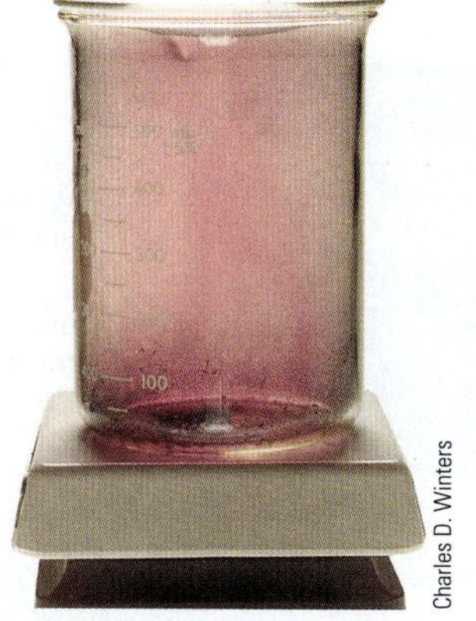

Figure 18.18 Iodine gas has a violet-pink color.

Active Example 18.8 The Equilibrium Constant Expression I

Write equilibrium constant expressions for each of the following equilibria:

a) $2\,HI(g) \rightleftharpoons H_2(g) + I_2(g)$ (**Figure 18.18**)

b) $HI(g) \rightleftharpoons \frac{1}{2}\,H_2(g) + \frac{1}{2}\,I_2(g)$

c) $2\,Cl_2(g) + 2\,H_2O(g) \rightleftharpoons 4\,HCl(g) + O_2(g)$

Think Before You Write The definition of K sets the procedure by which the equilibrium constant expression is written.

Answers *Cover the left column with your cut-out shield. Reveal each answer only after you have written your own answer in the right column.*

a) $K = \frac{[H_2][I_2]}{[HI]^2}$

Notice that this is the reciprocal of the equilibrium constant developed in the text, when the equation was written $H_2(g) + I_2(g) \rightleftharpoons 2\,HI(g)$. Its numerical value at 440°C is 1/51.5, or 0.0194.

b) $K = \frac{[H_2]^{1/2}[I_2]^{1/2}}{[HI]}$

The value of this equilibrium constant is not the same as in (a), 0.0194. It is, in fact, the square root of 0.0194, or 0.139. This emphasizes why we must associate any equilibrium constant expression with a specific chemical equation.

c) $K = \frac{[HCl]^4\,[O_2]}{[Cl_2]^2\,[H_2O]^2}$

The procedure for writing the equilibrium constant expression is the same no matter how complex the equation may be.

Once you are confident that you understand the definition of K, write all three K expressions.

You improved your skill at writing equilibrium constant expressions.	What did you learn by solving this Active Example?

Practice Exercise 18.8

Write the equilibrium constant expression for $4\ NH_3(g) + 7\ O_2(g) \rightleftharpoons 4\ NO_2(g) + 6\ H_2O(g)$.

So far, all equilibrium constant expressions have been for equilibria in which all substances are gases. An equilibrium may also have solids, liquids, or dissolved substances as part of its equation. Solute concentrations are variable, and they appear in equilibrium constant expressions just like the concentrations of gases. If a liquid solvent or a solid is part of an equilibrium, however, its concentration is essentially constant. Its concentration is therefore omitted in the equilibrium constant expression. (We can, if you wish, say that its constant value is "included" in the value of K.) Remember:

When writing an equilibrium constant expression, use only the concentrations of gases, (g), or dissolved substances, (aq). Do not include solids, (s), or liquids, (ℓ).

In Section 17.9, this rule was applied to the ionization of water:

$$H_2O(\ell) \rightleftharpoons H^+(aq) + OH^-(aq) \qquad K_w = [H^+][OH^-]$$

K_w has no denominator because the species on the left-hand side of the equilibrium equation is a liquid. This example also shows the common practice of using a subscript to identify a constant for a particular kind of equilibrium. We will describe other subscripts shortly.

Active Example 18.9 The Equilibrium Constant Expression II

Write the equilibrium constant expression for each of the following:

a) $CaCO_3(s) \rightleftharpoons CaO(s) + CO_2(g)$

b) $Li_2CO_3(s) \rightleftharpoons 2\ Li^+(aq) + CO_3^{2-}(aq)$

c) $4\ H_2O(g) + 3\ Fe(s) \rightleftharpoons 4\ H_2(g) + Fe_3O_4(s)$

d) $HF(aq) \rightleftharpoons H^+(aq) + F^-(aq)$

e) $NH_3(aq) + H_2O(\ell) \rightleftharpoons NH_4^+(aq) + OH^-(aq)$

Think Before You Write The equilibrium constant expression does not include solids or liquids.

Answers *Cover the left column with your cut-out shield. Reveal each answer only after you have written your own answer in the right column.*

a) $K = [CO_2]$

b) $K = [Li^+]^2\,[CO_3^{2-}]$

c) $K = \dfrac{[H_2]^4}{[H_2O]^4}$

d) $K = \dfrac{[H^+]\,[F^-]}{[HF]}$

e) $K = \dfrac{[NH_4^+\,][OH^-]}{[NH_3]}$

Write the five expressions.

You improved your skill at writing equilibrium constant expressions.

What did you learn by solving this Active Example?

Practice Exercise 18.9

Write the equilibrium constant expressions for the following:

a) $2\ H_2O(\ell) \rightleftharpoons 2\ H_2(g) + O_2(g)$

b) $CoCl_2(s) + 6\ H_2O(g) \rightleftharpoons CoCl_2 \cdot 6\ H_2O(s)$

18.9 The Significance of the Value of K

Goal 12 Given an equilibrium equation and the value of the equilibrium constant, identify the direction in which the equilibrium is favored.

By definition, an equilibrium constant is a ratio—a fraction. The numerical value of an equilibrium constant may be very large, very small, or anyplace in between. Even though there is no defined intermediate range, equilibria with constants between 0.01 and 100 (10^{-2} to 10^{2}) will have appreciable quantities of all species present at equilibrium.

To see what is meant by "very large" or "very small" K values, consider an equilibrium similar to the hydrogen iodide system studied in Section 18.8. If we substitute chlorine for iodine, the equilibrium equation is $H_2(g) + Cl_2(g)$ (**Figure 18.19**) $\rightleftharpoons 2\ HCl(g)$. At 25°C,

$$K = \frac{[HCl]^2}{[H_2][Cl_2]} = 2.4 \times 10^{33}$$

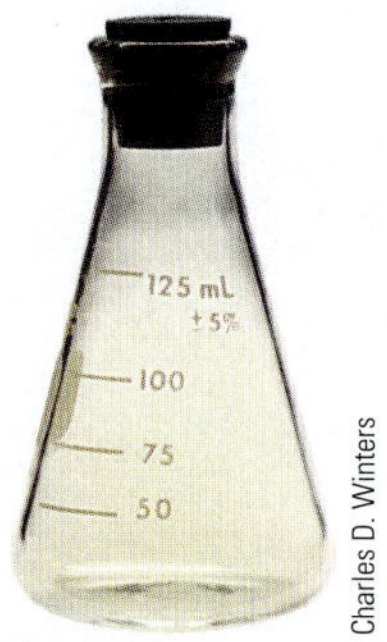

Figure 18.19 Chlorine gas has a greenish-yellow color.

This is a very large number—10 billion times larger than the number of particles in a mole! The only way an equilibrium constant ratio can become so huge is for the concentration of one or more reacting species to be very close to zero. If the denominator of a ratio is nearly zero, the value of the ratio will be very large. A near-zero denominator and large K mean the equilibrium is favored overwhelmingly in the forward direction.

By contrast, if the equilibrium constant is very small, it means the concentration of one or more of the species on the right-hand side of the equation is nearly zero. This puts a near-zero number in the numerator of K, and the equilibrium is strongly favored in the reverse direction.

Thinking About Your Thinking

Equilibrium

The concept discussed in this section requires the equilibrium thinking skill. The value of the equilibrium constant, K, sets the *ratio* of product to reactant concentrations for an equilibrium system. As the value of the numerator increases, the value of the denominator must also increase to keep their *ratio* constant, and vice versa, while at a fixed temperature.

a summary of... **The Significance of the Value of K**

If an equilibrium constant is very large (>100), the forward reaction is favored; if the constant is very small (<0.01), the reverse reaction is favored. If the constant is neither large nor small, appreciable quantities of all species are present at equilibrium.

Target Check 18.5

For the following reactions, determine whether the forward reaction is favored, the reverse reaction is favored, or if appreciable quantities of all species are present at equilibrium. Justify each answer.

a) $HBr(aq) \rightleftharpoons H^{+}(aq) + Br^{-}(aq)$ $K = 1 \times 10^{9}$

b) $Sn^{2+}(aq) + 4\ Cl^{-}(aq) \rightleftharpoons SnCl_4^{2-}(aq)$ $K = 3.0 \times 10^{1}$

18.10 Equilibrium Calculations: An Introduction (Optional)

Equilibrium calculations cover a wide range of problem types. A thorough understanding of these calculations is essential to understanding many chemical phenomena in the laboratory, industry, and living organisms. We will sample only a few in the remaining sections of this chapter.

You should write two things as a first step in solving any equilibrium problem. First is the equilibrium equation. Second is the equilibrium constant expression.

The numbers used in the equilibrium constant expression are concentrations in moles per liter. These concentrations are sometimes given, and sometimes you must figure them out from the available information. The figuring-out process uses the mole relationships expressed in the coefficients in an equation. For example, suppose you were to prepare two 500-mL (0.5-L) solutions. In the first, you dissolve 0.1 mole of NaCl; in the second, you dissolve 0.1 mole of $CaCl_2$. The molarities of both solutions would be:

$$M \equiv \frac{\text{mol}}{\text{L}} = \frac{0.1\ \text{mol}}{0.5\ \text{L}} = 0.2\ \text{mol/L}$$

But the concentrations used in equilibrium problems are usually ion concentrations. What are the Na^{+}, Ca^{2+}, and Cl^{-} concentrations in the two solutions? To answer these questions, we must examine the dissolving equations:

$$NaCl(s) \rightarrow Na^{+}(aq) + Cl^{-}(aq) \quad \text{and} \quad CaCl_2(s) \rightarrow Ca^{2+}(aq) + 2\ Cl^{-}(aq)$$

The NaCl equation shows that one mole of NaCl yields one mole of Na^{+} ion and one mole of Cl^{-} ion. Therefore, the ion concentrations are the same as the solute concentration:

$$\frac{0.2\ \cancel{\text{mol NaCl}}}{\text{L}} \times \frac{1\ \text{mol Na}^{+}}{1\ \cancel{\text{mol NaCl}}} = 0.2\ \text{mol Na}^{+}/\text{L}$$

and

$$\frac{0.2\ \cancel{\text{mol NaCl}}}{\text{L}} \times \frac{1\ \text{mol Cl}^{-}}{1\ \cancel{\text{mol NaCl}}} = 0.2\ \text{mol Cl}^{-}/\text{L}$$

The equation for dissolving $CaCl_2$ shows that one mole of $CaCl_2$ yields one mole of Ca^{2+} and *two* moles of Cl^{-}. Therefore, the chloride ion concentration is *twice* as large as the calcium ion concentration:

$$\frac{0.2\ \cancel{\text{mol CaCl}_2}}{\text{L}} \times \frac{1\ \text{mol Ca}^{2+}}{1\ \cancel{\text{mol CaCl}_2}} = 0.2\ \text{mol Ca}^{2+}/\text{L}$$

and

$$\frac{0.2\ \cancel{\text{mol CaCl}_2}}{\text{L}} \times \frac{2\ \text{mol Cl}^{-}}{1\ \cancel{\text{mol CaCl}_2}} = 0.4\ \text{mol Cl}^{-}/\text{L}$$

In the Active Examples that follow, we will use "twice as large" and similar reasoning without showing calculation setups.

Let's also look at the units of equilibrium constants. For reasons beyond the scope of this course, a rigorous treatment of equilibrium constant calculations results in an equilibrium constant with no units. Therefore, we omit units on K values.

Now you are ready to solve some equilibrium constant problems.

18.11 Equilibrium Calculations: Solubility Equilibria (Optional)

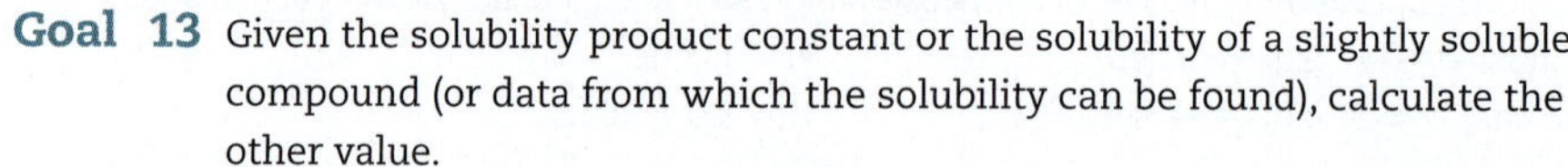

Goal 13 Given the solubility product constant or the solubility of a slightly soluble compound (or data from which the solubility can be found), calculate the other value.

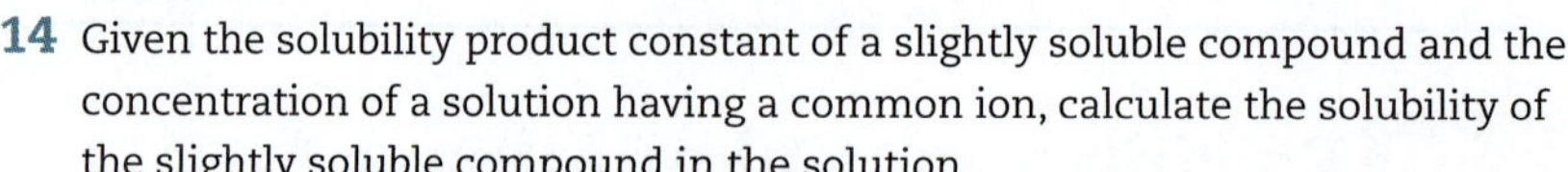

14 Given the solubility product constant of a slightly soluble compound and the concentration of a solution having a common ion, calculate the solubility of the slightly soluble compound in the solution.

PhotoSmart/Shutterstock.com

Figure 18.20 The stalagmites and stalactites found in many caves form as dissolved minerals precipitate from solution. These minerals are low-solubility compounds.

In Section 9.8, you used solubility Table 9.3 and the solubility guidelines in Figure 9.18 to predict whether or not a precipitate would form when ionic solutions are combined. The footnote to Table 9.3 said that for purposes of writing net ionic equations, a compound that did not dissolve to a concentration of 0.1 mole per liter of water would be considered slightly soluble or nearly insoluble. It is appropriate, then, to refer to these collectively as *low-solubility compounds* (**Figure 18.20**).

The equilibrium equation for dissolving a low-solubility compound is very similar to the equation for the ionization of water. (See the discussion in Section 17.9.) For silver chloride, for example, the equilibrium and K equations are:

$$AgCl(s) \rightleftharpoons Ag^{+}(aq) + Cl^{-}(aq) \qquad K_{sp} = [Ag^{+}][Cl^{-}]$$

The equilibrium constant for a low-solubility compound is the **solubility product constant, K_{sp}**. A K_{sp} expression has no denominator because the only species on the left side of the equation is a solid.

The rules for writing an equilibrium constant expression from Section 18.8 indicate that solids and liquids are not included.

Active Example 18.10 Calculation of K_{sp} from Molar Solubility Data

The chloride ion concentration of a saturated solution of silver chloride is 1.3×10^{-5} M. Calculate K_{sp} for silver chloride.

Think Before You Write As discussed in Section 18.10, the first step in solving any equilibrium problem is to write the equilibrium equation and then the associated equilibrium constant expression. You are being asked to calculate a K_{sp} value. This means that you need to determine the values of the concentrations in the K_{sp} expression.

Answers *Cover the left column with your cut-out shield. Reveal each answer only after you have written your own answer in the right column.*

$AgCl(s) \rightleftharpoons Ag^{+}(aq) + Cl^{-}(aq)$ $K_{sp} = [Ag^{+}][Cl^{-}]$ **Given:** 1.3×10^{-5} M Cl^{-} **Wanted:** K_{sp} of AgCl	**Analyze** the problem by writing the equation and identifying the K_{sp} expression, and write the given quantity and wanted value.
 $[Ag^{+}] = 1.3 \times 10^{-5}$ M *The equilibrium equation shows that equal amounts in moles of silver and chloride ions are released when silver chloride dissolves. Their concentrations are therefore equal.*	One of the two concentrations you need to be able to calculate K_{sp} is given. But you need $[Ag^{+}]$, too. What is it? (*Hint:* Think about the NaCl discussion at the beginning of Section 18.10.)
$K_{sp} = [Ag^{+}][Cl^{-}] = (1.3 \times 10^{-5})(1.3 \times 10^{-5}) = 1.7 \times 10^{-10}$	You have the equation, and you have the numbers to use. **Construct** the setup and answer.

$(1.3 \times 10^{-5})(1.3 \times 10^{-5}) = (1.3 \times 1.3) \times (10^{-5} \times 10^{-5})$

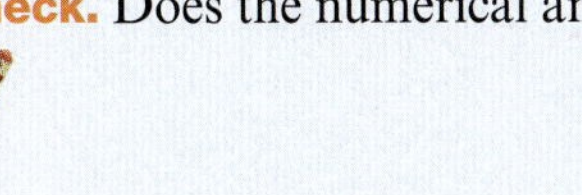

$\approx (1 \times 2) \times 10^{-10}$

$= 2 \times 10^{-10}$, OK.

The calculated answer appears to be correct.

Check. Does the numerical answer make sense?

You improved your skill at solving solubility equilibrium problems.

What did you learn by solving this Active Example?

Practice Exercise 18.10

The concentration of copper(II) ion in a saturated solution of copper(II) iodate is 3.3×10^{-3} M. Determine the K_{sp} for copper(II) iodate.

Active Example 18.11 Calculation of K_{sp} from Mass Per Volume Solubility Data

The solubility of magnesium fluoride (Figure 18.21) is 73 mg/L. What is its K_{sp}?

Think Before You Write The key to finding a K_{sp} value is to write the K_{sp} expression. You then need to determine the molar concentrations in the expression.

Answers *Cover the left column with your cut-out shield. Reveal each answer only after you have written your own answer in the right column.*

Given: 73 mg MgF_2
Wanted: mol MgF_2

mg MgF_2 → g MgF_2 → mol MgF_2

1 g MgF_2 = 1000 mg MgF_2

1 mol MgF_2 = 62.31 g MgF_2

This time the solution concentration is given in mg/L. It must be converted to molarity, mol/L. **Analyze** this part of the problem and **identify** the equivalencies needed to convert milligrams of magnesium fluoride to moles.

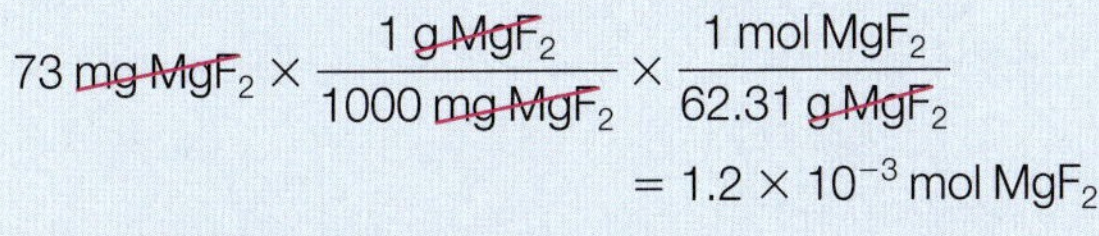

$$73 \text{ mg MgF}_2 \times \frac{1 \text{ g MgF}_2}{1000 \text{ mg MgF}_2} \times \frac{1 \text{ mol MgF}_2}{62.31 \text{ g MgF}_2} = 1.2 \times 10^{-3} \text{ mol MgF}_2$$

$M = 1.2 \times 10^{-3}$ mol/L MgF_2

Construct the setup to calculate the number of moles of magnesium fluoride in the solution, and then write the mol/L concentration.

$MgF_2(s) \rightleftharpoons Mg^{2+}(aq) + 2\,F^-(aq)$

$K_{sp} = [Mg^{2+}][F^-]^2$

$[Mg^{2+}] = 1.2 \times 10^{-3}$ M

$[F^-] = 2 \times [Mg^{2+}] = 2 \times (1.2 \times 10^{-3}\text{ M}) = 2.4 \times 10^{-3}$ M

According to the equilibrium equation, one mole of Mg^{2+} is produced for every mole of MgF_2 that dissolves. The molarity of Mg^{2+} is therefore equal to the molarity of the solution. The equation also shows that twice as many fluoride ions are produced as magnesium ions. It follows that the molarity of F^- is twice the molarity of Mg^{2+}.

Now you know the molarity of the solution in moles of MgF_2 per liter. But you need the ion concentrations. Think back to the examination of the calcium chloride solution at the beginning of Section 18.10. How were those ion concentrations determined? What are both the magnesium ion and fluoride ion concentrations in 1.2×10^{-3} M MgF_2? Before answering, write the equilibrium equation for MgF_2 and the K_{sp} expression equation. They will help you see where you are going.

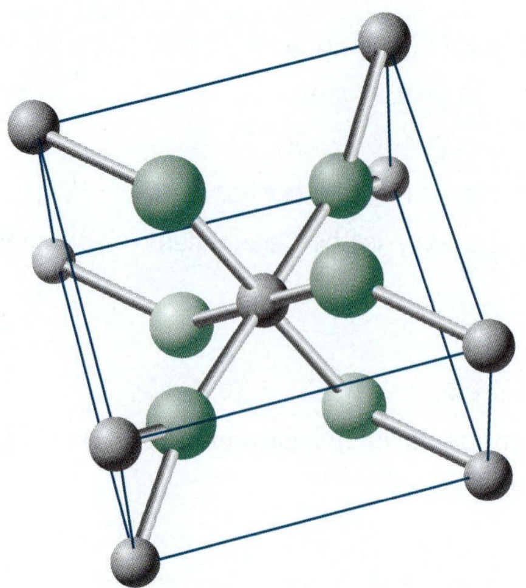

Figure 18.21 The three-dimensional particulate-level structure of magnesium fluoride. The gray spheres represent magnesium ions, and the green spheres represent fluoride ions.

$K_{sp} = [Mg^{2+}][F^-]^2 = (1.2 \times 10^{-3})(2.4 \times 10^{-3})^2 = 6.9 \times 10^{-9}$	You now have both ion concentrations. Substitute into K_{sp} equation and calculate the answer.
$(1.2 \times 10^{-3})(2.4 \times 10^{-3})^2 = (1.2 \times 10^{-3})(2.4 \times 10^{-3})$ $(2.4 \times 10^{-3}) = (1.2 \times 2.4 \times 2.4) \times (10^{-3} \times 10^{-3} \times 10^{-3})$ $\approx 1 \times 2 \times 3 \times 10^{-9} = 6 \times 10^{-9}$, OK.	**Check.** Estimate the numerical answer without a calculator.
You improved your skill at solving solubility equilibrium problems.	What did you learn by solving this Active Example?

Practice Exercise 18.11

A chemist measures the concentration of a saturated copper(I) bromide solution as 0.029 g/L at 25°C. Determine the K_{sp} value of copper(I) bromide.

Solubility product constants have already been determined for most common low-solubility ionic compounds. Their values may be found on the Internet and in reference books. They are used in several kinds of problems, one of which is the reverse of the last two Active Examples.

Active Example 18.12 Calculation of Solubility from K_{sp} I

Calculate the solubility of zinc carbonate in (a) mol/L and (b) g/100 mL. $K_{sp} = 1.4 \times 10^{-11}$ for zinc carbonate.

Think Before You Write You are given the value of K_{sp} for a low-solubility solid, and you are asked to find the solubility in two different concentration ratios. This Active Example will lead you through the approach to solving K_{sp}-to-solubility problems.

Answers *Cover the left column with your cut-out shield. Reveal each answer only after you have written your own answer in the right column.*

$ZnCO_3(s) \rightleftharpoons Zn^{2+}(aq) + CO_3^{2-}(aq)$ $K_{sp} = [Zn^{2+}][CO_3^{2-}] = 1.4 \times 10^{-11}$	Begin with the equilibrium equation and K_{sp} expression.

$[Zn^{2+}] = [CO_3^{2-}]$ *The equation shows that equal numbers of Zn^{2+} and CO_3^{2-} ions are formed when $ZnCO_3$ dissolves. Therefore, their concentrations must be equal.*	If $ZnCO_3$ is the only source of both Zn^{2+} and CO_3^{2-} ions, what can you conclude about their relative concentrations?
$K_{sp} = [Zn^{2+}][CO_3^{2-}] = s \times s = s^2 = 1.4 \times 10^{-11}$ $s = \sqrt{1.4 \times 10^{-11}} = 3.7 \times 10^{-6}\ M = [Zn^{2+}] = [CO_3^{2-}]$ *On most calculators, you find the square root of a number by pressing the $\sqrt{\ }$ key and entering the number and hitting the equals sign.*	Algebra is applied to find the solubility of $ZnCO_3$. Let the letter s represent $[Zn^{2+}]$ at equilibrium: $s = [Zn^{2+}]$. Because $[Zn^{2+}] = [CO_3^{2-}]$, s is also equal to $[CO_3^{2-}]$. Substitute s for the two ion concentrations in the K_{sp} equation and calculate its value.
Given: 100 mL **Wanted:** g $ZnCO_3$ mL → L → mol $ZnCO_3$ → g $ZnCO_3$ 1 L = 1000 mL 125.39 g $ZnCO_3$ = 1 mol $ZnCO_3$ $100\ \cancel{mL} \times \frac{1\ \cancel{L}}{1000\ \cancel{mL}} \times \frac{3.7 \times 10^{-6}\ \cancel{mol\ ZnCO_3}}{\cancel{L}} \times \frac{125.39\ g\ ZnCO_3}{1\ \cancel{mol\ ZnCO_3}} = 4.6 \times 10^{-5}\ g\ ZnCO_3$ *The solubility of zinc carbonate is 3.7×10^{-6} mol/L or 4.6×10^{-5} g/100 mL.*	You now know the solubility of $ZnCO_3$ in moles per liter, which answers part (a). To find it in grams per 100 mL, as requested in part (b), you must ask yourself how many grams of $ZnCO_3$ are in 100 mL of 3.7×10^{-6} M $ZnCO_3$? **Analyze** the problem statement, write the unit path and **identify** the equivalencies, and **construct** the solution.
$100 \times \frac{1}{1000} \times 3.7 \times 10^{-6} \times 125.39 = 10^2 \times 10^{-3} \times 3.7 \times 10^{-6} \times 1.2539 \times 10^2 = (3.7 \times 1.2539) \times (10^2 \times 10^{-3} \times 10^{-6} \times 10^2) \approx (4 \times 1.25) \times 10^{-5} = 5 \times 10^{-5}$, OK.	**Check** the numerical answer for the g/100 mL concentration without a calculator.
You improved your skill at solving solubility equilibrium problems.	What did you learn by solving this Active Example?

Practice Exercise 18.12

Determine the g/100 mL solubility of barium sulfate, which has a solubility product constant of 1.5×10^{-9}.

Suppose a soluble carbonate, such as Na_2CO_3, is added to the saturated solution of zinc carbonate in Active Example 18.12. What happens to the solubility of $ZnCO_3$? What does Le Chatelier's Principle predict? $[CO_3^{2-}]$ would no longer be equal to $[Zn^{2+}]$ because the CO_3^{2-} ion would be coming from two sources. According to Le Chatelier's Principle, the equilibrium should shift in the direction that would use up some of the added CO_3^{2-}. That is the reverse direction. Less $ZnCO_3$ would dissolve; its solubility would be reduced. Let's see if that prediction is confirmed by calculation.

Notice that we did not include in $[CO_3^{2-}]$ any carbonate ion that came from the small amount of $ZnCO_3$ that dissolved. In water, that concentration was only 0.0000037 mol/L, and Le Chatelier's Principle predicts an even smaller amount in the Na_2CO_3 solution. By rules of significant figures, carbonate from $ZnCO_3$ is negligible when added to carbonate from Na_2CO_3: 0.010 + less than 0.0000037 = 0.010.

Active Example 18.13 Calculation of Solubility from K_{sp} II: Presence of a Common Ion

Calculate the solubility of $ZnCO_3$ ($K_{sp} = 1.4 \times 10^{-11}$) in 0.010 M Na_2CO_3. Answer in moles per liter.

Think Before You Write You are given the value of K_{sp} for a low-solubility solid, and you are asked to find its solubility. However, the solution in which the solid is dissolving contains carbonate ion, one of the products of dissolving the low-solubility solid.

Answers *Cover the left column with your cut-out shield. Reveal each answer only after you have written your own answer in the right column.*

Assuming that the Na_2CO_3 is completely dissolved, what is $[CO_3^{2-}]$ in 0.010 M Na_2CO_3? (*Hint:* Look back on the Ca^{2+} ion discussion at the beginning of Section 18.10.)

$[CO_3^{2-}] = 0.010$ M in 0.010 M Na_2CO_3

In $Na_2CO_3(s) \rightleftharpoons 2\ Na^+(aq) + CO_3^{2-}(aq)$, the amount in moles of carbonate ion in solution is the same as the amount in moles of sodium carbonate dissolved.

Let's look again at the equilibrium and solubility product equations from Active Example 18.12:

$$ZnCO_3(s) \rightleftharpoons Zn^{2+}(aq) + CO_3^{2-}(aq)$$

$$K_{sp} = [Zn^{2+}][CO_3^{2-}] = 1.4 \times 10^{-11}$$

Zn^{2+} ion comes only from $ZnCO_3$. $[Zn^{2+}]$ is, therefore, the same as the solubility of zinc carbonate. You already have values for K_{sp} and $[CO_3^{2-}]$. The zinc ion concentration is the only unknown in the solubility product equation. Complete all of the remaining steps except for the check.

Given: $K_{sp} = [Zn^{2+}][CO_3^{2-}] = 1.4 \times 10^{-11}$
$[CO_3^{2-}] = 0.010$ M
Wanted: $[Zn^{2+}]$

$$[Zn^{2+}] = \frac{K_{sp}}{[CO_3^{2-}]} = \frac{1.4 \times 10^{-11}}{0.010} = 1.4 \times 10^{-9}\ \text{M}$$

Solving the solubility product equation for $[Zn^{2+}]$ and substituting known values of K_{sp} and $[CO_3^{2-}]$ yields a zinc carbonate solubility that is, indeed, smaller than its solubility in water: 3.7×10^{-6} M, from Active Example 18.12. Le Chatelier's Principle is confirmed.

Make sure the powers of 10 **check.**

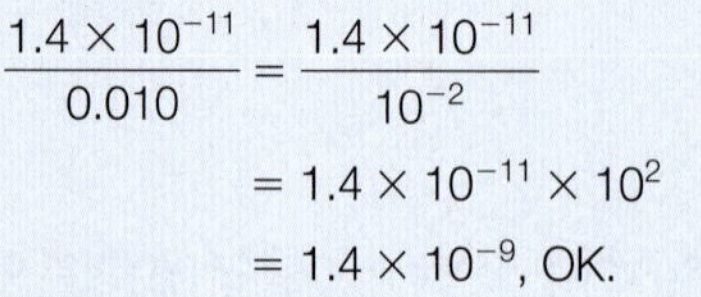

$$\frac{1.4 \times 10^{-11}}{0.010} = \frac{1.4 \times 10^{-11}}{10^{-2}} = 1.4 \times 10^{-11} \times 10^{2} = 1.4 \times 10^{-9}\text{, OK.}$$

What did you learn by solving this Active Example?

You improved your skill at solving solubility equilibrium problems.

Practice Exercise 18.13

What is the molar solubility of calcium fluoride in 0.020 M sodium fluoride solution? K_{sp} for calcium fluoride is 4.0×10^{-11}.

Reducing the concentration of an ion or the solubility of a compound by adding an ion that is already present is an example of the common ion effect, as shown in Figure 18.22.

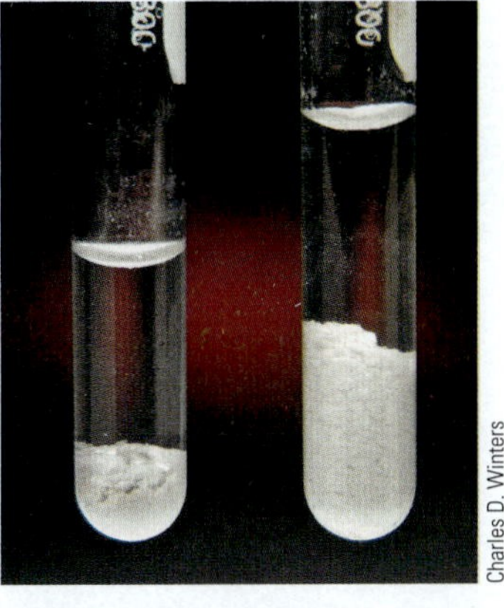

Figure 18.22 The common ion effect. The tube on the left contains a saturated solution of silver acetate with excess solid silver acetate at the bottom. A few drops of a solution of silver nitrate are added, providing a source of additional silver ion, and the result is shown in the tube on the right. The solubility of the solid is reduced by the addition of a common ion.

18.12 Equilibrium Calculations: Ionization Equilibria (Optional)

Note: Calculations in this section include converting between non-integer pH values and $[H^+]$ as described in optional Section 17.11.

Goal 15 Given the formula of a weak acid, HA, write the equilibrium equation for its ionization and the expression for its acid constant, K_a.

16 Given any two of the following three values for a weak acid, HA, calculate the third: (a) the initial concentration of the acid; (b) the pH of the solution, the percentage dissociation of the acid, or $[H^+]$ or $[A^-]$ at equilibrium; (c) K_a for the acid.

17 For a weak acid, HA, given K_a, $[H^+]$, and $[A^-]$, or information from which they may be obtained, calculate the pH of the buffer produced.

18 Given K_a for a weak acid, HA, determine the ratio between [HA] and $[A^-]$ that will produce a buffer of specified pH.

In Section 9.4, you learned that weak acids ionize only slightly when dissolved in water. If HA is the formula of a weak acid, its ionization equation and equilibrium constant expression are:

$$HA(aq) \rightleftharpoons H^+(aq) + A^-(aq) \qquad K_a = \frac{[H^+][A^-]}{[HA]}$$

The equilibrium constant is the **acid constant, K_a**. The undissociated molecule is the major species in the solution, and the H^+ ion and the conjugate base of the acid, A^-, are the minor species. ▶

◀ **P/Review** The major species in a solution are those that are present in higher concentrations, and the minor species have relatively low concentrations (Section 9.4). The conjugate base of an acid is the species that remains after the acid has lost its proton, H^+ (Section 17.5).

The ionization of a weak acid is usually so small that it is negligible compared with the initial concentration of the acid. For example, if a 0.12 M acid is 3.0% ionized, the amount ionized is $0.030 \times 0.12 = 0.0036$ mol/L. When this is subtracted from the initial concentration and rounded off according to the rules of significant figures, the result is $0.12 - 0.0036 = 0.1164 = 0.12$. (This is similar to the negligible addition noted in the margin at the end of Active Example 18.13.)

In more advanced courses, you will learn how to determine whether or not the ionization has a negligible effect on the initial concentration. In this book, we assume that all ionization concentrations are negligible *when subtracted from the initial concentration*. The ion concentrations by themselves, however, are not negligible. If the ionization of HA is the only source of H^+ and A^- in the equilibrium $HA(aq) \rightleftharpoons H^+(aq) + A^-(aq)$, then $[H^+] = [A^-]$. (This is the same as $[Ag^+]$ being equal to $[Cl^-]$ in Active Example 18.10.) This makes it possible to calculate the percentage ionization and K_a from the pH of a weak acid whose molarity has been determined by titration. ▶

$[H^+] = 10^{-pH}$. Converting between non-integer pH values and $[H^+]$ is discussed in Section 17.11. Briefly, one calculator procedure for changing pH to $[H^+]$ is as follows:

1. Press 10^x or INV log.
2. Change the sign to minus.
3. Enter the pH.
4. Push =.

If this procedure does not work with your calculator, do an Internet search for the instructions for your model of calculator or seek human help.

Active Example 18.14 Calculation of Percentage Ionization and K_a from Molar Concentration and pH

A 0.13 M solution of an unknown acid has a pH of 3.12. Calculate its percentage ionization and K_a.

Think Before You Write To find K_a, the acid constant expression, $K_a = \frac{[H^+][A^-]}{[HA]}$, will be the key relationship. Percentage ionization will be found by applying the percentage concept, (parts of A ÷ total parts) × 100%. ▶ In this case, it will be (acid molecules ionized ÷ total acid molecules) × 100%.

In Section 7.7, the general formula for calculating the percentage of any part A, in the total of all parts is introduced:

$$\%\ A = \frac{\text{parts of A}}{\text{total parts}} \times 100\%$$

Answers *Cover the left column with your cut-out shield. Reveal each answer only after you have written your own answer in the right column.*

Given: 0.13 M; pH = 3.12 **Wanted:** % ionization; K_a	List the two given quantities and the two quantities that are wanted. This will make clear what is known and what is wanted.
$[H^+] = 10^{-pH} = 10^{-3.12} = 7.6 \times 10^{-4}$ M	You will need $[H^+]$ to find the value of K_a, so determine the value of $[H^+]$ from the given pH.
$[A^-] = [H^+] = 7.6 \times 10^{-4}$ M *Because the ionization of HA is the only appreciable source for both H^+ and A^-, the ionization equation tells us that their concentrations are equal.*	Another species in the solution has the same concentration as $H^+(aq)$. Identify the species. Recall that the dissociation of a weak acid has the general pattern $HA(aq) \rightleftharpoons H^+(aq) + A^-(aq)$.
$K_a = \frac{[H^+][A^-]}{[HA]} = \frac{(7.6 \times 10^{-4})(7.6 \times 10^{-4})}{0.13}$ $= 4.4 \times 10^{-6}$	You now have all the values needed to calculate K_a. Complete the calculation.
$\frac{7.6 \times 10^{-4}}{0.13} \times 100\% = 0.58\%$ ionization	Now you can calculate the percentage ionization. Of the 0.13 mole of HA initially in 1 L of solution, only 7.6×10^{-4} mole ionized. What percentage of 0.13 is 7.6×10^{-4}?
0.13 − 0.00076 0.13 The assumption is valid according to the addition/subtraction rule for significant figures.	We assumed that the acid concentration, 0.13 M, remained the same after the weak acid underwent ionization. Check that assumption. Is 0.13 M − 7.6×10^{-4} M = 0.13 M?
You improved your skill at solving ionization equilibrium problems.	What did you learn by solving this Active Example?

Practice Exercise 18.14

A student measures a pH of 2.99 for a 0.12 M solution of an unknown weak acid. What is the K_a and percentage ionization of the acid?

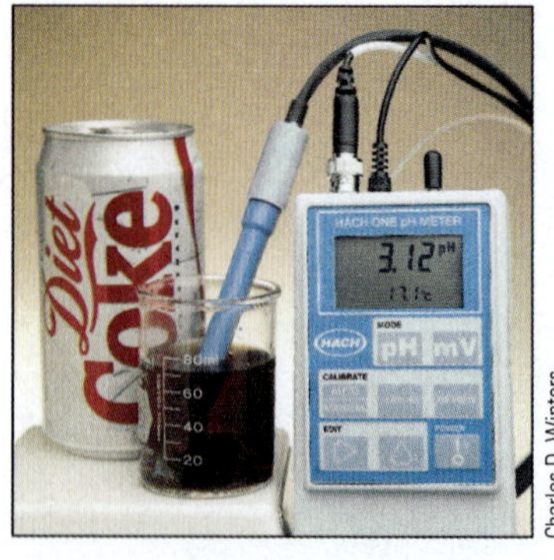

Figure 18.23 Measuring the pH of cola. Note that the solution is quite acidic (pH = 3.12). Nonetheless, the weak acid in the solution is only slightly ionized.

Just as the K_{sp} values of low-solubility solids are listed on the Internet and in reference books (Section 18.11), so are the K_a values of most weak acids. They can be used to determine $[H^+]$ and the pH of solutions of those acids (**Figure 18.23**). If the ionization of the acid is the only appreciable source of H^+ and A^- ions, we can start with the acid constant equation, multiply both sides by [HA], and substitute $[H^+]$ for its equal $[A^-]$:

$$K_a = \frac{[H^+][A^-]}{[HA]} \xrightarrow{\text{Multiply both sides by [HA]}}$$

$$[HA]\,K_a = [H^+]\,[A^-] \xrightarrow{[H^+] = [A^-]\text{, so substitute } [H^+] \text{ for } [A^-]}$$

$$[HA]\,K_a = [H^+]\,[H^+] \xrightarrow{\text{Take the square root of each side}} [H^+] = \sqrt{[HA]\,K_a}$$

Active Example 18.15 Calculation of pH from Acid Concentration and K_a

What is the pH of a 0.20 molar solution of the acid in Active Example 18.14, for which $K_a = 4.4 \times 10^{-6}$?

Think Before You Write You are given an acid concentration and its K_a value. You are asked to find pH. The manipulation of the K_a expression, as shown immediately preceding this Active Example, yields the relationship needed to solve this problem.

Answers *Cover the left column with your cut-out shield. Reveal each answer only after you have written your own answer in the right column.*

Answers	
Given: $K_a = 4.4 \times 10^{-6}$; [HA] = 0.20 M **Wanted:** pH $[H^+] = \sqrt{[HA]\,K_a} = \sqrt{(0.20)(4.4 \times 10^{-6})} = 9.4 \times 10^{-4}$ M $pH = -\log[H^+] = -\log(9.4 \times 10^{-4}) = 3.03$	$[H^+]$ may be found by direct substitution into $[H^+] = \sqrt{[HA]\,K_a}$. Additionally, you will use $pH = -\log[H^+]$. List the given and wanted, and then solve the problem.
0.20 − 0.00094 0.20 The assumption is valid.	We assume that [HA] remains at 0.20 M at equilibrium. Check to see if the assumption is correct.
You improved your skill at solving ionization equilibrium problems.	What did you learn by solving this Active Example?

Practice Exercise 18.15

What is the pH of a 0.10 M solution of acetic acid ($K_a = 1.8 \times 10^{-5}$)?

Just as we can force solubility equilibria in the reverse direction by adding a common ion, so can we force a weak acid equilibrium by adding a soluble ionic compound of the acid anion. If we add A^- to the equilibrium in $HA(aq) \rightleftharpoons H^+(aq) + A^-(aq)$, $[A^-]$ increases and the equilibrium is shifted to the left, reducing $[H^+]$. To find the pH of such a solution, we solve $K_a = \frac{[H^+][A^-]}{[HA]}$ for $[H^+]$:

$$[H^+] = K_a \times \frac{[HA]}{[A^-]}$$

Neither [HA] nor $[A^-]$ is changed significantly by the ionization of HA, which is even smaller than its ionization in water.

Active Example 18.16 Calculation of pH from Acid Concentration and K_a in the Presence of a Common Ion

Find the pH of a 0.20 M solution of the acid in Active Examples 18.14 and 18.15 ($K_a = 4.4 \times 10^{-6}$) if the solution is also 0.15 M in A^-.

Think Before You Write You are given three of the four variables in $K_a = \frac{[H^+][A^-]}{[HA]}$. You are asked to determine the remaining variable, $[H^+]$, and then change it to pH.

Answers *Cover the left column with your cut-out shield. Reveal each answer only after you have written your own answer in the right column.*

Given: $K_a = 4.4 \times 10^{-6}$; [HA] = 0.20 M; $[A^-]$ = 0.15 M
Wanted: pH

$$[H^+] = K_a \times \frac{[HA]}{[A^-]} = 4.4 \times 10^{-6} \times \frac{0.20}{0.15} = 5.9 \times 10^{-6}\text{ M}$$

$$pH = -\log[H^+] = -\log(5.9 \times 10^{-6}) = 5.23$$

Direct substitution into $[H^+] = K_a \times \frac{[HA]}{[A^-]}$ gives $[H^+]$, and pH follows. Complete the problem.

$$\begin{array}{r} 0.20 \\ -\ 0.00|00059 \\ \hline 0.20 \end{array}$$

Yes, the acid concentration is the same.

Is the initial acid concentration the same as the equilibrium concentration?

You improved your skill at solving ionization equilibrium problems.

What did you learn by solving this Active Example?

Practice Exercise 18.16

What is the pH of a 0.10 M solution of acetic acid ($K_a = 1.8 \times 10^{-5}$) that is also 0.10 M in acetate ion?

The solution in Active Example 18.16 is a **buffer solution**, or, more simply, a **buffer**. A buffer is a solution that resists changes in pH because it contains relatively high concentrations of both a weak acid and a weak base, as shown in **Figure 18.24.** The acid is able to consume any OH^- that may be added, and the base can react with H^+, both without significant change in either [HA] or $[A^-]$.

For example, if 0.001 mole of HCl was dissolved in 1 L of water, the pH would be 3. If the same amount of HCl was added to a liter of the buffer in Active Example 18.16, it would react with 0.001 mole of the A^- present. The new concentration of A^- would be 0.15 − 0.001 = 0.15 M, unchanged according to the rules of significant figures. As additional 0.001 mole of HA would be formed in the reaction, but that added to 0.20 is still 0.20 M. In other words, the $[HA]/[A^-]$ ratio would be unchanged, so the $[H^+]$ and pH would also be unchanged.

This all suggests that a buffer can be tailor-made for any pH simply by adjusting the $[HA]/[A^-]$ ratio to the proper value. Solving $K_a = \frac{[H^+][A^-]}{[HA]}$ for that ratio gives

$$K_a = \frac{[H^+]\,[A^-]}{[HA]} \xrightarrow{\text{Divide both sides by }[H^+]} \frac{K_a}{[H^+]} = \frac{[A^-]}{[HA]} \xrightarrow{\text{Take the inverse of both sides}} \frac{[H^+]}{K_a} = \frac{[HA]}{[A^-]}$$

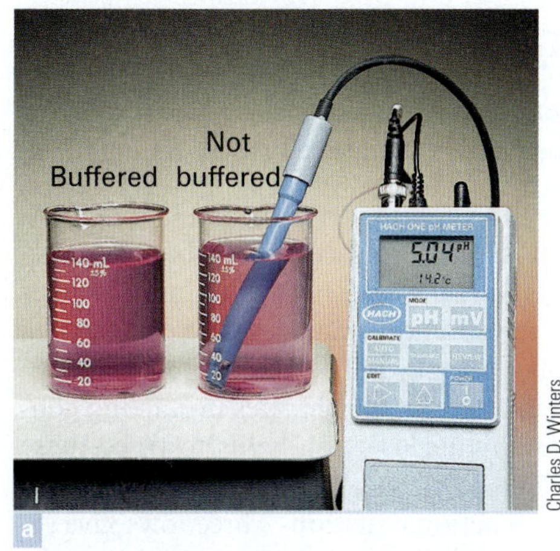

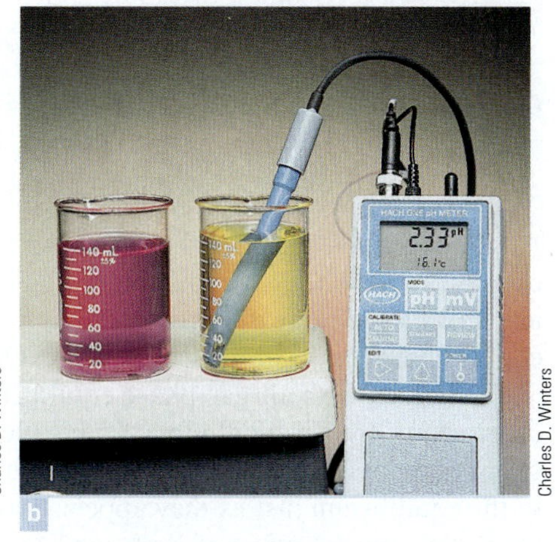

Figure 18.24 A buffer solution. (a) Two solutions of approximately the same pH are prepared. An indicator is added to both, showing that the pH of the solutions is about the same because the colors are the same. A pH meter measures the pH of the solution that is not buffered. (b) A few milliliters of 0.10 M hydrochloric acid solution are added to both solutions. The color of the indicator in the buffered solution does not change, indicating that the pH remains approximately the same. The solution that is not buffered drops significantly in pH, as shown by the indicator color change and the reading on the pH meter.

Active Example 18.17 Calculation of the Acid/Base Ratio for a Buffer of a Given K_a and Specified pH

What $[HA]/[A^-]$ ratio is necessary to produce a buffer with a pH of 5.00 if $K_a = 4.4 \times 10^{-6}$?

Think Before You Write $K_a = \frac{[H^+][A^-]}{[HA]}$. Since you are being asked for $[HA]/[A^-]$, there are only three variables in the equation. K_a is given and pH can be changed to $[H^+]$, so this is an algebra problem of the type: Given two of the three variables in an equation, find the third.

Answers *Cover the left column with your cut-out shield. Reveal each answer only after you have written your own answer in the right column.*

Given: pH = 5.00; $K_a = 4.4 \times 10^{-6}$ **Wanted:** $[HA]/[A^-]$ $\frac{[H^+]}{K_a} = \frac{[HA]}{[A^-]} = \frac{10^{-5.00}}{4.4 \times 10^{-6}} = 2.3$	Convert the pH to $[H^+]$, then plug that value and K_a into $\frac{[H^+]}{K_a} = \frac{[HA]}{[A^-]}$ to get the answer. You don't even need a calculator for the pH $\rightarrow$ $[H^+]$ conversion. 
If pH is 5.00, $[H^+] = 10^{-pH} = 10^{-5.00}$ M. This could be written simply as 10^{-5}, but the two zeros are held to show that all numbers are to two significant figures.	Explain why we did not calculate pH first and then plug it into the equation. Why did we write $10^{-5.00}$ instead of 10^{-5}? 
You improved your skill at solving ionization equilibrium problems.	What did you learn by solving this Active Example? 

Practice Exercise 18.17

What $[HA]/[A^-]$ ratio will produce an acetic acid–acetate ion buffer with a pH of 4.50? $K_a = 1.8 \times 10^{-5}$ for acetic acid.

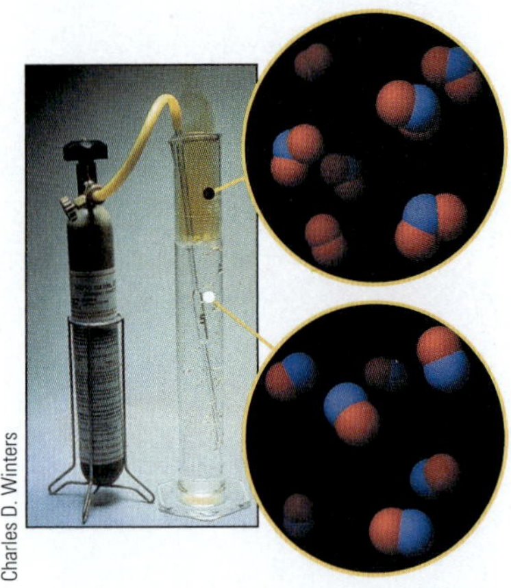
Charles D. Winters

Figure 18.25 NO reacts with O_2 to form NO_2. Gaseous NO from the cylinder is bubbled through water so that you can see that it is colorless. When the NO bubbles meet the colorless O_2 in the air above the surface of the water, NO and O_2 react to form dark reddish-brown NO_2.

18.13 Equilibrium Calculations: Gaseous Equilibria (Optional)

Goal 19 Given equilibrium concentrations of species in a gas phase equilibrium, or information from which they can be found, and the equation for the equilibrium, calculate the value of the equilibrium constant.

All the equilibria considered thus far in Sections 18.11 and 18.12 have been in aqueous solution. When an equilibrium involves only gases, the calculation principles and stoichiometric reasoning are the same as in solution equilibria. For example, an equilibrium system can be set up for $2\ NO(g) + O_2(g) \rightleftharpoons 2\ NO_2(g)$ (**Figure 18.25**).

However, with all-gas-phase equilibrium systems, the changes in starting concentrations are not negligible. In problem solving, it often helps to trace these changes by assembling them into a table. The columns are headed by the species in the equilibrium just as they appear in the reaction equation. Three rows give (1) the initial concentration of each substance, (2) the change in concentrations as the system reaches equilibrium, and (3) the equilibrium concentrations. You will learn to use this type of table in the next two Active Examples.

Active Example 18.18 Calculation of the Equilibrium Constant for a Gaseous Equilibrium I

A student places 0.052 mole of NO and 0.054 mole of O_2 in a 1.00-L vessel at a certain temperature. They react until equilibrium is reached according to the equation $2\ NO(g) + O_2(g) \rightleftharpoons 2\ NO_2(g)$. At equilibrium, $[NO_2] = 0.028$ M. Calculate K.

Think Before You Write You are given information from which the mol/L starting concentrations of the reactants can be determined, and you are given the equilibrium concentration of one species. You are asked to determine the value of the equilibrium constant, which will require determination of the equilibrium concentrations of all species.

Answers *Cover the left column with your cut-out shield. Reveal each answer only after you have written your own answer in the right column.*

	$2\ NO(g)$	+	$O_2(g)$	$\rightleftharpoons$	$2\ NO_2(g)$
mol/L at start	0.052		0.054		0.000
mol/L change, + or −					
mol/L at equilibrium					0.028

From the equilibrium equation, $2\ NO(g) + O_2(g) \rightleftharpoons 2\ NO_2(g)$, we have set up a table. Begin by inserting all given data. Assume that the initial concentration of NO_2 is zero, as implied but not directly specified by the problem statement. You'll be able to make four entries into the table.

	$2\ NO(g)$	+	$O_2(g)$	$\rightleftharpoons$	$2\ NO_2(g)$
mol/L at start					
mol/L change, + or −					
mol/L at equilibrium					

	$2\ NO(g)$	+	$O_2(g)$	$\rightleftharpoons$	$2\ NO_2(g)$
mol/L at start	0.052		0.054		0.000
mol/L change, + or −					+0.028
mol/L at equilibrium					0.028

To get any entry in the change line (middle row), find a species whose initial and final concentrations are known. In this case, $[NO_2]$ starts at 0.000 M and reaches 0.028 M. Its change is therefore +0.028 M. Insert that value into the table above.

	$2\,NO(g)$	+	$O_2(g)$	$\rightleftharpoons$	$2\,NO_2(g)$
mol/L at start	0.052		0.054		0.000
mol/L change, + or −	−0.028		−0.014		+0.028
mol/L at equilibrium					0.028

You can use stoichiometry to find the changes in the other two species. The coefficients in the equation show that the reaction that produces 0.028 mol/L NO_2 must use 0.028 mol/L NO because both coefficients are the same. Half as much O_2 must react because its coefficient is half the others, which is 0.014 mol/L O_2. Add these values to the table above.

	$2\,NO(g)$	+	$O_2(g)$	$\rightleftharpoons$	$2\,NO_2(g)$
mol/L at start	0.052		0.054		0.000
mol/L change, + or −	−0.028		−0.014		+0.028
mol/L at equilibrium	0.024		0.040		0.028

The final concentrations of the reactants are found by subtracting the amounts used from the starting concentrations (or adding them algebraically, as they appear in the table). Complete the table above.

$$2\,NO(g) + O_2(g) \rightleftharpoons 2\,NO_2(g)$$

$$K = \frac{[NO_2]^2}{[NO]^2[O_2]} = \frac{(0.028)^2}{(0.024)^2(0.040)} = 34$$

The equilibrium constant, K, may now be calculated by substituting the equilibrium concentrations from the third line of the table into the equilibrium constant expression.

$$\frac{1}{0.040} = \frac{1}{4 \times 10^{-2}} = \frac{1}{4} \times \frac{1}{10^{-2}} = 0.25 \times 10^2 = 25$$

The value of K appears to be correct.

The $(0.028)^2$ in the numerator essentially cancels the $(0.024)^2$ in the denominator. Thus, the value of K is a little larger than 1/0.040. What is the value of 1/0.040?

You improved your skill at solving gaseous equilibrium problems.

What did you learn by solving this Active Example?

Practice Exercise 18.18

A scientific researcher places 6.0 moles of sulfur dioxide and 4.0 moles of oxygen into a 1.0-L reaction vessel. At equilibrium, the vessel contains 4.0 moles of sulfur trioxide. Calculate K for $2\,SO_2(g) + O_2(g) \rightleftharpoons 2\,SO_3(g)$.

Active Example 18.19 Calculation of the Equilibrium Constant for a Gaseous Equilibrium II

A researcher injects 2.00 moles of hydrogen iodide gas into a 1.00-L reaction vessel. It decomposes according to the equation $2\,HI(g) \rightleftharpoons H_2(g) + I_2(g)$. At equilibrium, 1.60 mole of hydrogen iodide remains. Calculate the value of the equilibrium constant at the temperature of this experiment.

Think Before You Write You are given information from which the starting and equilibrium concentrations of the reactant can be determined. You are asked to determine the value of the equilibrium constant, which will require determination of the equilibrium concentrations of all species.

Answers *Cover the left column with your cut-out shield. Reveal each answer only after you have written your own answer in the right column.*

Complete as much of the following table as possible, based on the given data.

	$2\ HI(g)$	$\rightleftharpoons$	$H_2(g)$	+	$I_2(g)$
mol/L at start					
mol/L change, + or −					
mol/L at equilibrium					

	$2\ HI(g)$	$\rightleftharpoons$	$H_2(g)$	+	$I_2(g)$
mol/L at start	2.00		0.00		0.00
mol/L change, + or −					
mol/L at equilibrium	1.60				

The initial concentration minus the change must be equal to the equilibrium concentration. This fact will allow you to determine the mol/L change for HI(g). You can then use the reaction stoichiometry to complete the mol/L change line for hydrogen and iodine. Insert those three values into the table above.

	$2\ HI(g)$	$\rightleftharpoons$	$H_2(g)$	+	$I_2(g)$
mol/L at start	2.00		0.00		0.00
mol/L change, + or −	−0.40		+0.20		+0.20
mol/L at equilibrium	1.60				

The +0.20 mol/L change for $H_2(g)$ and $I_2(g)$ comes from the coefficients in the balanced chemical equation:

$$0.40\ \cancel{mol\ HI} \times \frac{1\ mol\ H_2}{2\ \cancel{mol\ HI}} = 0.20\ mol\ H_2$$

$$0.40\ \cancel{mol\ HI} \times \frac{1\ mol\ I_2}{2\ \cancel{mol\ HI}} = 0.20\ mol\ I_2$$

Complete the problem by determining the mol/L at equilibrium for the product species. Once you've entered the equilibrium concentrations into the table, you can write the K expression, substitute the equilibrium concentration values, and solve for K.

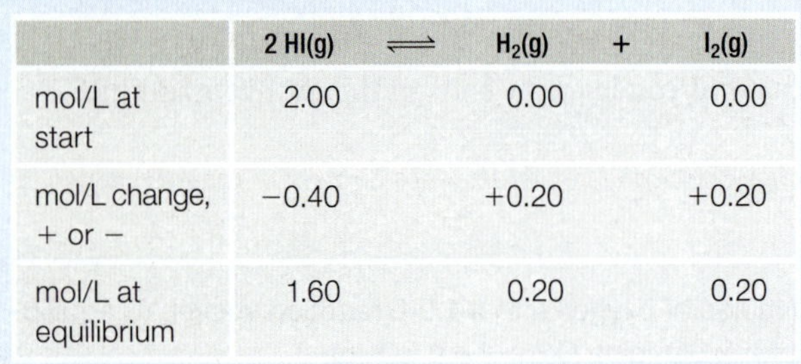

	$2\ HI(g)$	$\rightleftharpoons$	$H_2(g)$	+	$I_2(g)$
mol/L at start	2.00		0.00		0.00
mol/L change, + or −	−0.40		+0.20		+0.20
mol/L at equilibrium	1.60		0.20		0.20

$$2\ HI(g) \rightleftharpoons H_2(g) + I_2(g)$$

$$K = \frac{[H_2]\ [I_2]}{[HI]^2} = \frac{(0.20)\ (0.20)}{(1.60)^2} = 0.016$$

Check the value of K.

$$\frac{(0.20)\ (0.20)}{(1.60)^2} = \frac{(2 \times 10^{-1})\ (2 \times 10^{-1})}{1.6 \times 1.6} \approx \frac{2 \times 2 \times 10^{-2}}{1 \times 2}$$

$$= 2 \times 10^{-2} = 0.02\text{, OK.}$$

What did you learn by solving this Active Example?

You improved your skill at solving gaseous equilibrium problems.

Practice Exercise 18.19

An experiment is designed where 9.00 moles of carbon monoxide and 1.50 moles of chlorine are placed in a 3.00-L reaction chamber. When equilibrium is reached according to $CO(g) + Cl_2(g) \rightleftharpoons COCl_2(g)$, there are 6.30 moles of chlorine in the chamber. Calculate K.

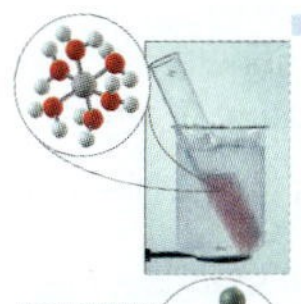

CHAPTER 18 IN REVIEW: CHEMICAL EQUILIBRIUM

Charles Steele

Goal 1 Identify a chemical equilibrium by the conditions it satisfies.

Four conditions characterize every **equilibrium**:

1. The change is reversible and can be represented by an equation with a double arrow.
2. The equilibrium system is "closed"—closed in the sense that no substance can enter or leave the immediate vicinity of the equilibrium.
3. The equilibrium is **dynamic.**
4. The things that are equal in an equilibrium are the forward rate of change (from left to right in the equation) and the reverse rate of change (from right to left).

Goal 2 Distinguish between reaction-producing molecular collisions and molecular collisions that do not yield reactions.

According to the **collision theory of chemical reactions**, all chemical reactions start with molecular collisions, but not all molecular collisions give a chemical reaction. The particles must have (1) enough kinetic energy and (2) the proper orientation.

Goal 3 Sketch and/or interpret an energy versus reaction progress graph. Identify the (a) transition state region, (b) activation energy, and (c) ΔE for the reaction.

An **energy versus reaction progress graph** shows potential energies of reactants, the transition state, and products in a reaction, and the activation energy as the reaction proceeds in either direction, as well as ΔE for the reaction. The graph illustrates the **activation energy, E_a**, for the forward and reverse reactions: the minimum kinetic energy needed to produce an effective collision.

Goal 4 State and explain the relationship between reaction rate and temperature.

Reaction rates are higher at higher temperatures because a larger fraction of the sample has enough kinetic energy to participate in reaction-producing collisions. This may be illustrated with a **kinetic energy distribution curve.** The energy of collision must be enough to overcome the mutual repulsion of the valence electrons of the reacting particles.

Goal 5 Using an energy versus reaction progress graph, explain how a catalyst affects reaction rate.

A **catalyst** increases reaction rate by providing an alternative reaction path with a lower activation energy. A larger fraction of the molecules in a sample is able to enter into reaction-producing collisions, so the reaction rate increases.

Goal 6 Identify and explain the relationship between reactant concentration and reaction rate.

Collision rates are higher at higher concentrations, so reaction rates are higher at higher reactant concentrations.

Goal 7 Trace and explain the changes in concentrations of reactants and products that lead to a chemical equilibrium.

For any reversible reaction in a closed system, whenever the opposing reactions are occurring at different rates, the faster reaction will gradually become slower, and the slower reaction will become faster. Finally, the reaction rates become equal, and **equilibrium** is established.

Goal 8 Given the equation for a chemical equilibrium, or information from which it can be written, predict the direction in which the equilibrium will shift because of a change in the concentration of one species.

Le Chatelier's Principle states that if an equilibrium system is subjected to change, processes occur that tend to partially counteract the initial change, thereby bringing the system to a new position of equilibrium. When the concentration of a species in an equilibrium reaction is changed, a shift will occur in the direction that tries to return the substance disturbed to its original condition.

Goal 9 Given the equation for a chemical equilibrium involving one or more gases, or information from which it can be written, predict the direction in which the equilibrium will shift because of a change in the volume of the system.

If a gaseous equilibrium is compressed, the increased pressure will be partially relieved by a shift in the direction of fewer gaseous molecules; if the system is expanded, the reduced pressure will be partially restored by a shift in the direction of more gaseous molecules.

Goal 10 Given a thermochemical equation for a chemical equilibrium, or information from which it can be written, predict the direction in which the equilibrium will shift because of a change in temperature.

A **thermochemical equation** is one that includes a change in energy. Including the energy term in a thermochemical equation, rather than showing $\Delta_r H$ separately, makes it easier to predict the Le Chatelier effect of a change in temperature. In the equation, think of energy as you would a substance being added or removed. An increase in temperature is interpreted as the "addition of heat," and a lowering of temperature is the "removal of heat."

Goal 11 Given any chemical equilibrium equation, or information from which it can be written, write the equilibrium constant expression.

An equilibrium system can be described by an **equilibrium constant, K**. The constant K is a concentration ratio; the form of the ratio depends on how the equilibrium equation is written. For the general reaction a A + b B $\rightleftharpoons$ c C + d D, $K = \frac{[C]^c[D]^d}{[A]^a[B]^b}$. When writing an equilibrium constant expression, use only the concentrations of gases, (g), or dissolved substances, (aq). Do not include solids, (s), or liquids, (ℓ).

Goal 12 Given an equilibrium equation and the value of the equilibrium constant, identify the direction in which the equilibrium is favored.

If an equilibrium constant is very large (>100), the forward reaction is favored; if the constant is very small (<0.01), the reverse reaction is favored. If the constant is neither large nor small, appreciable quantities of all species are present at equilibrium.

Goal 13 Given the solubility product constant or the solubility of a slightly soluble compound (or data from which the solubility can be found), calculate the other value.

No ionic compound is completely insoluble. The equilibrium constant for a low-solubility compound is the **solubility product constant, K_{sp}**. The solubility product constant for any ionic compound has the form

$$A_xB_y(s) \rightleftharpoons x\,A^{y+}(aq) + y\,B^{x-}(aq) \qquad K_{sp} = [A^{y+}]^x[B^{x-}]^y$$

The first two steps in solving any solubility product constant problem are the same:

1. Write the equilibrium equation for the reaction.
2. Write the solubility product constant expression.

To find K_{sp} from solubility:

1. Determine the molar concentrations of the ions in solution.
2. Substitute into the K_{sp} expression and solve.

To find solubility from K_{sp}:

1. Assign a variable to represent one of the ionic species in the equilibrium.
2. Determine the concentration of the other ionic species in terms of the same variable.
3. Substitute the concentrations from Steps 1 and 2 into the K_{sp} expression, equate to the K_{sp} value, and solve.

Goal 14 Given the solubility product constant of a slightly soluble compound and the concentration of a solution having a common ion, calculate the solubility of the slightly soluble compound in the solution.

The solubility of a low-solubility substance is reduced when a common ion—one already present in the solution—is introduced from another source. This is called the **common ion effect.**

Goal 15 Given the formula of a weak acid, HA, write the equilibrium equation for its ionization and the expression for its acid constant, K_a.

If HA is the formula of a weak acid, its ionization equation and equilibrium constant expression are

$$HA(aq) \rightleftharpoons H^+(aq) + A^-(aq) \qquad K_a = \frac{[H^+][A^-]}{[HA]}$$

The equilibrium constant is the **acid constant, K_a**. The undissociated molecule is the major species in the solution, and the H^+ ion and the conjugate base of the acid, A^-, are the minor species. Weak acids ionize only slightly when dissolved in water. The ionization of a weak acid is usually so small that it is negligible compared with the initial concentration of the acid.

Goal 16 Given any two of the following three values for a weak acid, HA, calculate the third: (a) the initial concentration of the acid; (b) the pH of the solution, the percentage dissociation of the acid, or $[H^+]$ or $[A^-]$ at equilibrium; (c) K_a for the acid.

Percentage ionization, as with all other percentage concepts, is the ratio of the part to the whole, multiplied by 100:

$$\%\ \text{ionization} = \frac{\text{amount of solute ionized}}{\text{total solute present}} \times 100\%.$$

To determine K_a from pH or percent ionization of an acid with a given concentration, determine the values of $[H^+]$ and $[A^-]$ with $[H^+] = 10^{-pH}$, substitute the values into the K_a expression, and solve. To determine percent ionization and pH from known molarity and K_a, if the ionization of the acid is the only source of H^+ and A^- ions, start with the acid constant equation, multiply both sides by [HA], and substitute $[H^+]$ for its equal $[A^-]$: $K_a[HA] = [H^+][A^-] = [H^+]^2$; $[H^+] = \sqrt{K_a[HA]}$.

Goal 17 For a weak acid, HA, given K_a, $[H^+]$, and $[A^-]$, or information from which they may be obtained, calculate the pH of the buffer produced.

A **buffer** is a solution that resists changes in pH because it contains relatively high concentrations of both a weak acid and a weak base. The acid is able to consume any OH^- that may be added, and the base can react with H^+, both without significant change in either [HA] or $[A^-]$. To find the pH of a buffer solution, solve the acid constant expression for $[H^+]$: $[H^+] = K_a \times \frac{[HA]}{[A^-]}$.

Goal 18 Given K_a for a weak acid, HA, determine the ratio between [HA] and $[A^-]$ that will produce a buffer of specified pH.

A buffer can be tailor-made for any pH simply by adjusting the $[HA]/[A^-]$ ratio to the proper value. Solving the acid constant expression for that ratio gives $\frac{[HA]}{[A^-]} = \frac{[H^+]}{K_a}$.

Goal 19 Given equilibrium concentrations of species in a gas phase equilibrium, or information from which they can be found, and the equation for the equilibrium, calculate the value of the equilibrium constant.

When an equilibrium involves only gases, the changes in the starting concentrations are not negligible. It often helps to trace these changes by assembling them into a table. The columns are headed by the species in the equilibrium just as they appear in the reaction equation. The three lines give the initial concentration of each substance, the change in concentration as the system reaches equilibrium, and the equilibrium concentration:

	a A	+	b B	$\rightleftharpoons$	c C	+	d D
mol/L at start							
mol/L change, + or −							
mol/L at equilibrium							

Key Terms

Most of the key terms and concepts and many others appear in the Glossary. Use your Glossary regularly.

activation energy (E_a) p. 704
catalyst p. 705
closed system p. 699
collision theory of chemical reactions p. 701
dynamic equilibrium p. 699
equilibrium constant (K) p. 715
inhibitor p. 706
Le Chatelier's Principle p. 708
potential energy barrier p. 702
transition state p. 702

Key Terms from Optional Sections:

acid constant (K_a) (Sec. 18.12) p. 727
buffer (buffer solution) (Sec. 18.12) p. 730
solubility product constant (K_{sp}) (Sec. 18.11) p. 722

Frequently Asked Questions

What is equal in an equilibrium?
We are used to thinking about "equal" in terms of amounts of physical things. Therefore, it is understandable that students sometimes resist the idea of *equilibrium* being applied to a system in which amounts may differ by many orders of magnitude. In A $\rightleftharpoons$ B, there may be millions of times more A than B in the system, but if the *rate* at which A becomes B is the same as the rate at which B becomes A, there is an equilibrium between them.

What overarching advice do you have regarding equilibrium problems?
Whenever you are trying to analyze or understand an equilibrium, write the equilibrium equation. It gives you a visual image of what's happening. If you are solving a quantitative equilibrium problem, also write the equilibrium constant expression. With those equations in black and white before you, it's much easier to reason through a Le Chatelier shift or to know where to put the numbers you must use to calculate an answer. In all equilibrium constant problems, find the values of the concentrations first and then substitute as the K expression requires. Keep the steps separate.

What is temperature, how does it affect an equilibrium, and why does a small change in temperature have so drastic an effect on reaction rate?
An understanding of Figure 18.5 will help you answer these questions. Realize that temperature is a measure of kinetic energy and that kinetic energy is measured along the horizontal axis of the graph, not the vertical axis. Therefore, the curve that stretches out farther to the right represents the higher energy and the higher temperature. This flattens the curve somewhat, making it look "lower," but that "lower" is not lower temperature. The main reason higher temperatures speed reaction rates is the resultant increase in the portion of the sample with enough kinetic energy to react, represented by the shaded portion of the graph under the curve and far to the right.

If you heat an equilibrium that is endothermic as the equation is written, the forward reaction rate increases. What happens to the reverse rate? Does it increase, decrease, or remain the same?
The reverse reaction rate also increases. All reaction rates are higher at higher temperatures. But the increases in the forward and reverse rates are not the same. That's what destroys the equilibrium—temporarily.

(Optional Section 18.11) In K_{sp} problems, it seems wrong to double the concentration of an ion and then also square it. How can that be correct?
Sometimes students hesitate to find one ion concentration by doubling another and then squaring the higher concentration when calculating K. You did this with the fluoride ion

concentration from MgF_2 in Active Example 18.11: $[F^-]$ is twice $[Mg^{2+}]$, and $[F^-]$ must be squared in calculating K_{sp}, which is $[Mg^{2+}][F^-]^2$. Finding $[F^-]$ is an independent step; so is squaring it when substituting into the K_{sp} expression.

It is no coincidence that $[F^-]$ is twice $[Mg^{2+}]$, but that is true if both ions come only from the solute. If they came from different sources, they would not have that 2:1 relationship, but you would still square $[F^-]$ in using K_{sp}.

Concept-Linking Exercises

Write a brief description of the relationships among each of the following groups of terms or phrases. Answers to the Concept-Linking Exercises are given at the end of the chapter.

1. Reversible reaction, equilibrium, forward rate, reverse rate
2. Collision theory of chemical reactions, energy versus reaction progress graph, transition state, activation energy
3. Temperature, catalyst, concentration, reaction rate
4. Le Chatelier's Principle, temperature, gas volume, concentration
5. Equilibrium constant, K, K_{sp}, K_a

Small-Group Discussion Questions

Small-Group Discussion Questions are for group work, either in class or under the guidance of a leader during a discussion section.

1. You place two equally filled glasses of water on a table in your kitchen. You leave one open and cover the other with tightly sealed plastic wrap. The next day, the open glass contains less water than the covered glass. Do you have a liquid–vapor equilibrium in either system? Explain, citing the four conditions that are true for every equilibrium.
2. A quantitative model for the collision theory of chemical reactions can be constructed. The rate constant for a reaction is a proportionality constant that influences the speed of the reaction. It can be broken down into three factors. Each of these three factors, in turn, affects the speed of a reaction. In symbols, $k = p \times Z \times f$, where k is the rate constant. The factor p accounts for the orientation of the colliding molecules. The Z factor accounts for the number of collisions per unit time when comparing equal concentrations of reactants. The f factor accounts for the fractions of collisions with sufficient energy for a reaction-producing collision. For each of the three factors, draw a particulate-level depiction of a favorable situation versus an unfavorable situation for a high-speed reaction process. Explain each pair of sketches in words.
3. Consider a simple equilibrium system $A \rightleftharpoons B$. The reaction proceeds in the forward direction at the rate of 10% of the number of A molecules per minute—that is, 10% of the total number of molecules of A change to B per minute. The reverse reaction also proceeds at the same rate, 10% of the B molecules react to for A molecules per minute. Thus, if we start with 100 molecules of A and no B, after 1 minute, 10 molecules of A change to B, but 10% of the 10 molecules of B change back to A, which is 1 molecule. Therefore, after 1 minute, the system will consist of 91 A molecules and 9 B molecules. Continue this analysis for this system, recording the number of A and B molecules after each minute until the system comes to equilibrium. What happens to the system at equilibrium? What if you start with 100 B molecules and no A molecules? How does this equilibrium system compare with the one where you started with 100 molecules of A?
4. Consider the chemical system illustrated in Figure 18.17. (a) If the initial concentrations of hydrogen and iodine are 1.0 M, estimate the value of the equilibrium constant for the reaction as written. (b) Is the forward rate of reaction or the reverse rate of reaction greater at each of the following points: (i) before the two concentration lines cross, (ii) at the time the two lines cross, and (iii) after the two lines cross?
5. Derive a mathematical statement that expresses the relationship among the values of the equilibrium constants for the following reactions: (a) $3\ H_2(g) + N_2(g) \rightleftharpoons 2\ NH_3(g)$; (b) $2\ NH_3(g) \rightleftharpoons 3\ H_2(g) + N_2(g)$; and (c) $3/2\ H_2(g) + 1/2\ N_2(g) \rightleftharpoons NH_3(g)$.
6. For systems not at equilibrium, it can be useful to calculate a concentration ratio known as the reaction quotient, Q. The reaction quotient has the same form as the equilibrium constant. For the general reaction, $a\,A + b\,B \rightleftharpoons c\,C + d\,D$, $Q = \frac{[C]^c\ [D]^d}{[A]^a\ [B]^b}$. Q = K only at equilibrium. If Q is larger than K, in which direction will the reaction shift to reach equilibrium? What if Q is smaller than K? Explain both answers.
7. If you breathe into a bag, you will feel light-headed and possibly pass out if you continue long enough. Explain this phenomenon, given the fact that carbonic acid in your blood is part of an equilibrium system that involves exhaled carbon dioxide: $H_2CO_3 \rightleftharpoons H_2O + CO_2$.
8. Consider the reaction $C(s) + H_2O(g) + \text{heat} \rightleftharpoons CO(g) + H_2(g)$. State how each of the following will affect the equilibrium concentrations of each reactant and product: (a) remove steam; (b) add hydrogen; (c) increase the temperature; (d) expand the volume of the container; (e) add carbon; (f) introduce a catalyst; (g) add helium to the container (helium does not react with any species in the equilibrium mixture).
9. Compare and contrast the macroscopic characteristics of an equilibrium system with the particulate-level characteristics of the system. Choose at least one physical equilibrium and one chemical equilibrium system to illustrate your comparison.

Questions, Exercises, and Problems

Interactive versions of these problems may be assigned by your instructor. Solutions for blue-numbered questions are at the end of the chapter. Questions other than those in the General Questions and More Challenging Problems sections are paired in consecutive odd-even number combinations; solutions for the odd numbered questions are also at the end of the chapter.

Section 18.2: The Character of an Equilibrium

1. What does it mean when an equilibrium is described as *dynamic*? Compare an equilibrium that is dynamic with one that is static.
2. What things are equal in an equilibrium? Give an example.
3. Undissolved table salt is in contact with a saturated salt solution in (a) a sealed container and (b) an open beaker. Which system, if either, can reach equilibrium? Explain your answer.
4. What is meant by saying that an equilibrium is confined to a "closed system"?
5. A garden in a park has a fountain that discharges water into a pond. The pond overflows into a stream that cascades to the bottom of a small pool. The water is then pumped up into the fountain. Is this system a dynamic equilibrium? Explain.
6. A river flows into a lake formed by a dam. Water flows through the dam's spillways as the river continues downstream. The water level of the lake is constant. Is this system a dynamic equilibrium? Explain.

Andrew Zarivny/Shutterstock.com

Is the lake formed by the construction of a dam an equilibrium system?

Section 18.3: The Collision Theory of Chemical Reactions

7. Explain why a molecular collision can be sufficiently energetic to cause a reaction, yet no reaction occurs as a result of that collision.
8. According to the collision theory of chemical reactions, what two conditions must be satisfied if a molecular collision is to result in a reaction?

Section 18.4: Energy Changes During a Molecular Collision

Assume heat to be the only form of reaction energy in the following questions. This makes ΔE equal to the $\Delta_r H$ discussed in Section 10.9.

9. Sketch an energy versus reaction progress graph for an endothermic reaction. Include a, b, and c points on the vertical axis and use them for the algebraic expressions of ΔE and the activation energy.
10. In the reaction for which Figure 18.3 is the energy versus reaction progress graph, is ΔE for the reaction positive or negative? Is the reaction exothermic or endothermic? Use the letters a, b, and c on the vertical axis of Figure 18.3 to state algebraically the ΔE and activation energy of the reaction.
11. Assuming the reaction described by Figure 18.3 to be reversible, compare the signs of the activation energies for the forward and reverse reactions. Which is positive, which is negative, or are they the same, and if so, are they positive or negative?
12. Assume that the reaction described by Figure 18.3 is reversible. Compare the magnitude of the activation energies for the forward and reverse reactions described by Figure 18.3. Which is greater, or are they equal?
13. What is a "transition state"? Why is it that we cannot list the physical properties of the species represented as a transition state?
14. Explain the significance of activation energy. For two reactions that are identical in all respects except activation energy, identify the reaction that would have the higher rate and explain why.

Section 18.5: Conditions That Affect the Rate of a Chemical Reaction

15. State the effect of a temperature increase and a temperature decrease on the rate of a chemical reaction. Explain each effect.
16. "At a given temperature, only a small fraction of the molecules in a sample has sufficient kinetic energy to engage in a chemical reaction." What is the meaning of that statement?
17. Suppose that two substances are brought together under conditions that cause them to react and reach equilibrium. Suppose that in another vessel the same substances and a catalyst are brought together, and again equilibrium is reached. How are the processes alike, and how are they different?

18. What is a catalyst? Explain how a catalyst affects reaction rates.

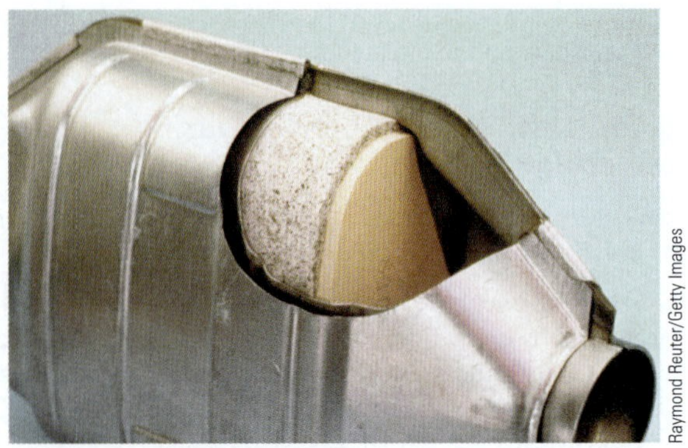
Raymond Reuter/Getty Images

A cutaway view of an automotive catalytic converter.

19. For the hypothetical reaction $A + B \rightarrow C$, what will happen to the reaction rate if the concentration of A is increased *and* the concentration of B is decreased? Explain.

20. For the hypothetical reaction $A + B \rightarrow C$, what will happen to the rate of reaction if the concentration of A is increased without changing the concentration of B? What will happen if the concentration of B is decreased without changing the concentration of A? Explain why in both cases.

Section 18.6: The Development of a Chemical Equilibrium

If nitrogen and hydrogen are brought together at the proper temperature and pressure, they will react until they reach equilibrium: $N_2(g) + 3\ H_2(g) \rightleftharpoons 2\ NH_3(g)$. Answer the following questions with regard to the establishment of that equilibrium.

21. When will the reverse reaction rate be at a maximum: at the start of the reaction, after equilibrium has been reached, or at some point in between?

22. When will the forward reaction rate be at a maximum: at the start of the reaction, after equilibrium has been reached, or at some point in between?

23. On a single set of coordinate axes, sketch graphs of the forward reaction rate versus time and the reverse reaction rate versus time from the moment the reactants are mixed to a point beyond the establishment of equilibrium.

24. What happens to the concentrations of each of the three species between the start of the reaction and the time equilibrium is reached?

Section 18.7: Le Chatelier's Principle

25. If the system $2\ SO_2(g) + O_2(g) \rightleftharpoons 2\ SO_3(g)$ is at equilibrium and the concentration of oxygen is reduced, predict the direction in which the equilibrium will shift. Justify or explain your prediction.

26. Consider the following system at equilibrium at 500 K: $PCl_3(g) + Cl_2(g) \rightleftharpoons PCl_5(g)$. When some chlorine is removed from the equilibrium system at constant temperature, in what direction must the reaction shift? What will happen to the concentration of PCl_3?

27. If additional oxygen is pumped into the equilibrium system $4\ NH_3(g) + 5\ O_2(g) \rightleftharpoons 4\ NO(g) + 6\ H_2O(g)$, in which direction will the reaction shift? Justify your answer.

28. Consider the following system at equilibrium at 723 K: $N_2(g) + 3\ H_2(g) \rightleftharpoons 2\ NH_3(g)$. When some ammonia is removed from the equilibrium system at constant temperature, in what direction must the reaction shift? What will happen to the concentration of hydrogen?

29. Predict the direction of the shift for the equilibrium $Cu(NH_3)_4^{2+}(aq) \rightleftharpoons Cu^{2+}(aq) + 4\ NH_3(aq)$ if the concentration of ammonia were reduced. Explain your prediction.

Charles D. Winters

Ammonia solution (colorless) was added to a solution of hydrated copper(II) ions (light blue, *bottom*), yielding $Cu(NH_3)_4^{2+}(aq)$ (dark blue, *top*).

30. Consider the following system at equilibrium at 600 K: $COCl_2(g) \rightleftharpoons CO(g) + Cl_2(g)$. When some $COCl_2(g)$ is added to the equilibrium system at constant temperature, in what direction must the reaction shift? What will happen to the concentration of CO?

31. A container holding the equilibrium $4\ H_2(g) + CS_2(g) \rightleftharpoons CH_4(g) + 2\ H_2S(g)$ is enlarged. Predict the direction of the Le Chatelier shift. Explain.

32. Consider the following system at equilibrium at 298 K: $2\ NOBr(g) \rightleftharpoons 2\ NO(g) + Br_2(g)$. If the volume of the equilibrium system is suddenly decreased at constant temperature, in what direction must the reaction shift? What will happen to the amount of bromine?

33. In what direction will $CO(g) + H_2O(g) \rightleftharpoons CO_2(g) + H_2(g)$ shift as a result of a reduction in volume? Explain.

34. Consider the following system at equilibrium at 1.15×10^3 K: $2\ SO_2(g) + O_2(g) \rightleftharpoons 2\ SO_3(g)$. If the volume of the equilibrium system is suddenly increased at constant temperature, in what direction must the reaction shift? What will happen to the amount of oxygen?

35. Which direction of the equilibrium $2\ NO_2(g) \rightleftharpoons N_2O_4(g) + 59.0$ kJ will be favored if the system is cooled? Explain.

36. Consider the following system at equilibrium at 500 K: $PCl_5(g) + 87.9\ kJ \rightleftharpoons PCl_3(g) + Cl_2(g)$. If the temperature of the equilibrium system is suddenly increased, in what direction must the reaction shift? What will happen to the concentration of chlorine?

37. If your purpose were to increase the yield of SO_3 in the equilibrium $SO_2(g) + NO_2(g) \rightleftharpoons SO_3(g) + NO(g) + 41.8$ kJ, would you use the highest or lowest operating temperature possible? Explain.

38. Consider the following system at equilibrium at 698 K: $H_2(g) + I_2(g) \rightleftharpoons 2\ HI(g)\ \Delta_rH = -10.4$ kJ. If the

temperature of the equilibrium system is suddenly decreased, in what direction must the reaction shift? What will happen to the concentration of iodine?

39. The solubility of calcium hydroxide is low; it reaches about 2.4×10^{-2} M at saturation. In acid solutions, with many H^+ ions present, calcium hydroxide is quite soluble. Explain this fact in terms of Le Chatelier's Principle. (*Hint:* Recall what you know of reactions in which molecular products are formed.)

40. Consider the following system at equilibrium at 350 K: $CH_4(g) + CCl_4(g) \rightleftharpoons 2\ CH_2Cl_2(g)$ $\Delta_rH = 18.8$ kJ. The production of $CH_2Cl_2(g)$ is favored by which of the following: (a) increasing the temperature; (b) increasing the pressure (by changing the volume); (c) increasing the volume; (d) adding CH_2Cl_2; (e) removing CCl_4?

Section 18.8: The Equilibrium Constant

For each equilibrium equation shown, write the equilibrium constant expression.

41. $CO(g) + H_2O(g) \rightleftharpoons CO_2(g) + H_2(g)$
42. $NH_4HS(s) \rightleftharpoons NH_3(g) + H_2S(g)$
43. $C(s) + H_2O(g) \rightleftharpoons CO(g) + H_2(g)$
44. $2\ SO_2(g) + O_2(g) \rightleftharpoons 2\ SO_3(g)$
45. $Zn_3(PO_4)_2(s) \rightleftharpoons 3\ Zn^{2+}(aq) + 2\ PO_4^{3-}(aq)$
46. $HF(aq) + H_2O(\ell) \rightleftharpoons H_3O^+(aq) + F^-(aq)$
47. $HNO_2(aq) + H_2O(\ell) \rightleftharpoons H_3O^+(aq) + NO_2^-(aq)$
48. $(CH_3)_3N(aq) + H_2O(\ell) \rightleftharpoons (CH_3)_3NH^+(aq) + OH^-(aq)$
49. $Cu(NH_3)_4^{2+}(aq) \rightleftharpoons Cu^{2+}(aq) + 4\ NH_3(aq)$
50. $Mg(OH)_2(s) \rightleftharpoons Mg^{2+}(aq) + 2\ OH^-(aq)$
51. The equilibrium between nitrogen monoxide, oxygen, and nitrogen dioxide may be expressed in the equation $2\ NO(g) + O_2(g) \rightleftharpoons 2\ NO_2(g)$. Write the equilibrium constant expression for this equation. Then express the same equilibrium in at least two other ways and write the equilibrium constant expression for each. Are the constants numerically equal? Cite some evidence to support your answer.

Charles D. Winters

Nitrogen monoxide and oxygen, both colorless gases, react to form reddish-brown nitrogen dioxide.

52. "The equilibrium constant expression for a given reaction depends on how the equilibrium equation is written." Explain the meaning of that statement. You may, if you wish, use the equilibrium equation $N_2(g) + 3\ H_2(g) \rightleftharpoons 2\ NH_3(g)$ to illustrate your explanation.

Section 18.9: The Significance of the Value of K

53. If sodium cyanide solution is added to silver nitrate solution, the following equilibrium will be reached: $Ag^+(aq) + 2\ CN^-(aq) \rightleftharpoons Ag(CN)_2^-(aq)$. For this equilibrium $K = 5.6 \times 10^{18}$. In which direction is the equilibrium favored? Justify your answer.

54. For the following equilibrium system, $K = 4.86 \times 10^{-5}$ at 298 K: $HClO(aq) + F^-(aq) \rightleftharpoons ClO^-(aq) + HF(aq)$. Assume that you start with equal concentrations of HClO and F^- and that no ClO^- or HF is initially present. Which of the following best describes the equilibrium system: (a) the forward reaction is favored at equilibrium; (b) the reverse reaction is favored at equilibrium; (c) appreciable quantities of all species are present at equilibrium?

55. A certain equilibrium has a very small equilibrium constant. In which direction, forward or reverse, is the equilibrium favored? Explain.

56. Acetic acid, $HC_2H_3O_2$, is a soluble weak acid. When placed in water, it partially ionizes and reaches equilibrium. Write the equilibrium equation for the ionization. Will the equilibrium constant be large or small? Justify your answer.

Questions 57 and 58: In Chapter 9, we discussed how to identify major and minor species and how to write net ionic equations. These skills are based on the solubility of ionic compounds, the strengths of acids, and the stability of certain ion combinations. Use these ideas to predict the favored direction of each equilibrium given. In each case, state whether you expect the equilibrium concentration to be large or small.

57. a) $H_2SO_3(aq) \rightleftharpoons H_2O(\ell) + SO_2(aq)$
b) $H^+(aq) + C_2H_3O_2^-(aq) \rightleftharpoons HC_2H_3O_2(aq)$

58. a) $HCl(aq) \rightleftharpoons H^+(aq) + Cl^-(aq)$
b) $BaSO_4(s) \rightleftharpoons Ba^{2+}(aq) + SO_4^{2-}(aq)$

Section 18.11: Equilibrium Calculations: Solubility Equilibria (Optional)

59. $Co(OH)_2$ dissolves in water to the extent of 3.7×10^{-6} mol/L. Find its K_{sp}.

60. A student measures the molar solubility of magnesium fluoride in a water solution to be 1.19×10^{-3} M. Based on her data, what is the solubility product constant for this compound?

61. If 2.50×10^2 mL of water will dissolve only 8.7 mg of silver carbonate, what is the K_{sp} of Ag_2CO_3?

62. The solubility of lead(II) sulfate in a water solution is measured as 3.95×10^{-3} g/100 mL. Based on these data, what is the solubility product constant for this compound?

63. Find the moles per liter and grams per 100 mL solubility of silver iodate, $AgIO_3$, if its $K_{sp} = 2.0 \times 10^{-8}$.

64. $K_{sp} = 8.7 \times 10^{-9}$ for calcium carbonate. Calculate its solubility in (a) moles per liter and (b) grams per 100 mL.

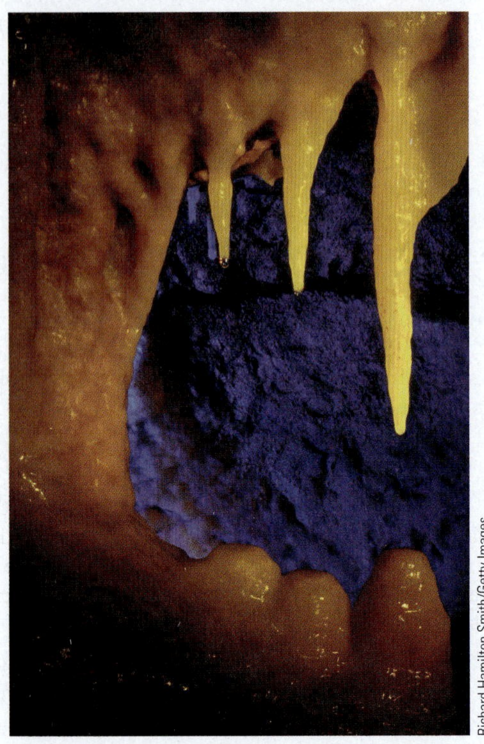
Richard Hamilton Smith/Getty Images

Stalactites and stalagmites in limestone caves are primarily composed of calcium carbonate.

65. Find the solubility (mol/L) of $Mn(OH)_2$ if its $K_{sp} = 1.0 \times 10^{-13}$.

66. K_{sp} for manganese(II) hydroxide is 4.6×10^{-14}. What is the molar solubility of manganese(II) hydroxide in a water solution?

67. How many grams of calcium oxalate will dissolve in 2.5×10^2 mL of 0.22 M $Na_2C_2O_4$ if $K_{sp} = 2.4 \times 10^{-9}$ for CaC_2O_4?

68. K_{sp} for silver hydroxide is 2.0×10^{-8}. Calculate the molar solubility of silver hydroxide in 0.103 M NaOH.

Section 18.12: Equilibrium Calculations: Ionization Equilibria (Optional)

69. The pH of 0.22 M $HC_4H_5O_3$ (acetoacetic acid) is 2.12. Find its K_a and percent ionization.

70. A student measured the pH of a 0.48 M aqueous solution of nitrous acid as 1.85. From these data, calculate the percentage ionization and the K_a value for this acid.

71. Find the pH of 0.35 M $HC_2H_3O_2$, acetic acid ($K_a = 1.8 \times 10^{-5}$).

72. What is the pH of a 0.501 M aqueous solution of hypochlorous acid, for which $K_a = 3.5 \times 10^{-8}$?

73. A student dissolves 24.0 g of sodium acetate, $NaC_2H_3O_2$, in 5.00×10^2 mL of 0.12 M $HC_2H_3O_2$ ($K_a = 1.8 \times 10^{-5}$). Calculate the pH of the solution.

74. What is the pH of a solution that contains 0.18 M sodium acetate (acetate ion, $C_2H_3O_2^-$) and 0.21 M acetic acid? K_a for acetic acid is 1.8×10^{-5}.

75. Find the ratio of $[HC_2H_3O_2]/[C_2H_3O_2^-]$ that will yield a buffer in which pH = 4.25 ($K_a = 1.8 \times 10^{-5}$).

76. What concentration ratio of hydrocyanic acid to cyanide ion, $[HCN]/[CN^-]$, will produce a buffer solution with a pH of 9.71? K_a for hydrocyanic acid is 4.0×10^{-10}.

Section 18.13: Equilibrium Calculations: Gaseous Equilibria (Optional)

77. A scientist introduces 0.351 mol of CO and 1.340 mol of Cl_2 into a reaction chamber having a volume of 3.00 L. When equilibrium is reached according to the equation $CO(g) + Cl_2(g) \rightleftharpoons COCl_2(g)$, there is 1.050 mol of Cl_2 in the chamber. Calculate K.

78. Consider reaction of phosphorus trichloride and chlorine at 530 K: $PCl_3(g) + Cl_2(g) \rightleftharpoons PCl_5(g)$. When 9.0×10^{-2} mole of $PCl_3(g)$ and 0.11 mole of $Cl_2(g)$ are added to a 1.00 L container, the equilibrium concentration of $PCl_5(g)$ is 5.3×10^{-2} M. Use these data to calculate the equilibrium constant for this reaction.

General Questions

79. Distinguish precisely, and in scientific terms, the differences among items in each of the following pairs.
   a) Reaction, reversible reaction
   b) Open system, closed system
   c) Dynamic equilibrium, static equilibrium
   d) Transition state, activation energy
   e) Catalyzed reaction, uncatalyzed reaction
   f) Catalyst, inhibitor
   g) Buffered solution, unbuffered solution

80. Classify each of the following statements as true or false.
   a) Some equilibria depend on a steady supply of a reactant in order to maintain the equilibrium.
   b) Both forward and reverse reactions continue after equilibrium is reached.
   c) Every time reactant molecules collide with proper orientation, there is a reaction.
   d) Potential energy during a collision is greater than potential energy before or after the collision.
   e) The properties of a transition state are between those of the reactants and products.
   f) Activation energy is positive for both the forward and reverse reactions.
   g) Kinetic energy is changed to potential energy during a collision.
   h) An increase in temperature speeds the forward reaction and slows the reverse reaction.
   i) A catalyst changes the steps by which a reaction is completed.
   j) An increase in the concentration of a substance on the right-hand side of an equation speeds the reverse reaction rate.
   k) An increase in the concentration of a substance in an equilibrium increases the reaction rate in which the substance is a product.

l) Reducing the volume of a gaseous equilibrium shifts the equilibrium in the direction of fewer gaseous molecules.

m) Raising temperature results in a shift in the forward direction of an endothermic equilibrium.

n) The value of an equilibrium constant depends on temperature.

o) A large K indicates that an equilibrium is favored in the reverse direction.

81. At Time 1, two molecules are about to collide. At Time 2, they are in the process of colliding, and their form is that of the transition state. Compare the sum of their kinetic energies at Time 1 with the kinetic energy of the transition state at Time 2. Explain your conclusions.

82. List three things you might do to increase the rate of the reverse reaction for which Figure 18.3 is the energy versus reaction progress graph.

More Challenging Questions

83. The Haber process for making ammonia by direct combination of the elements is described by the equation $N_2(g) + 3\ H_2(g) \rightleftharpoons 2\ NH_3(g) + 92\ kJ$. If a manufacturer wants to make the greatest amount of ammonia in the least time, removing product as the reaction proceeds, is the manufacturer more likely to conduct the reaction at (a) high pressure or low pressure, (b) high temperature or low temperature? Explain your choice in each case.

Ammonia is the second most produced fertilizer chemical in the United States. It is manufactured in plants like this.

84. Under proper conditions, the reaction in Question 83 will reach equilibrium. Is the manufacturer likely to conduct the reaction under those conditions, that is, at equilibrium? Explain.

85. The reaction in Question 83 has a yield of about 98% at 200°C and 1000 atm. Commercially, the reaction is performed at about 500°C and 350 atm, conditions at which the yield is only about 30%. Suggest why operation at the lower yield is economically more favorable.

86. The solubility of calcium hydroxide is low enough to be listed as "insoluble" in solubility tables, but it is much more soluble than most of the other ionic compounds that are similarly classified. Its K_{sp} is 5.5×10^{-6}

a) Write the equation for the equilibrium to which the K_{sp} is related.

b) If you had such an equilibrium, name at least two substances or general classes of substances that might be added to (1) reduce the solubility of $Ca(OH)_2$ and (2) increase its solubility. Justify your choices.

c) Without adding calcium or hydroxide ion, name a substance or class of substances that would, if added, (1) increase $[OH^-]$ and (2) reduce $[OH^-]$. Justify your choices.

87. The table here lists several "disturbances" that may or may not produce a Le Chatelier shift in the equilibrium $4\ NH_3(g) + 7\ O_2(g) \rightleftharpoons 6\ H_2O(g) + 4\ NO_2(g) + \text{energy}$. If a shift will result, place *F* in the Shift column if the shift is in the forward direction or place *R* if it is in the reverse direction. If the disturbance is an immediate change in the concentration of any species in the equilibrium, place in the concentration column of that substance *I* if the change is an increase or *D* if it is a decrease. Then determine what will happen to the concentrations of the other species because of the shift and insert *I* or *D* for increase or decrease. If there is no Le Chatelier shift, write *None* in the Shift column and leave the other columns blank.

Disturbance	Shift	$[NH_3]$	$[O_2]$	$[H_2O]$	$[NO_2]$
Add NO_2					
Reduce temperature					
Add N_2					
Remove NH_3					
Add a catalyst					

88. Some systems at equilibrium are exothermic, and some are endothermic. Is this statement always true, sometimes true, or never true? Explain your answer.

89. The equilibrium constant, K, can have many values for any individual equilibrium system. Why or how?

90. An all-gas-phase equilibrium can be reached between sulfur dioxide and oxygen on one side of the equation and sulfur trioxide on the other. Write two equilibrium constant expressions for this equilibrium, one of which has a value greater than 1 and one with a value less than 1. It is not necessary to state which is which.

91. "Hard" water has a high concentration of calcium and magnesium ions. Focusing on the calcium ion, a common home water-softening process is based on a reversible chemical change that can be expressed by $Na_2Ze(s) + Ca^{2+}(aq) \rightleftharpoons CaZe(s) + 2\ Na^+(aq)$. Ze^{2-} is the *zeolite ion*, a complex arrangement of silicate and aluminate groups. Na_2Ze represents a solid resin that is like an ionic compound composed of sodium ions and zeolite ions; CaZe is the corresponding calcium compound. When, during the day, is this system most likely to reach equilibrium? Why doesn't it reach equilibrium and stay there? Periodically it is necessary to "recharge" the water softener by running salt water, NaCl(aq), through it. Why is this necessary?

What concept discussed in this chapter does the recharging process illustrate?

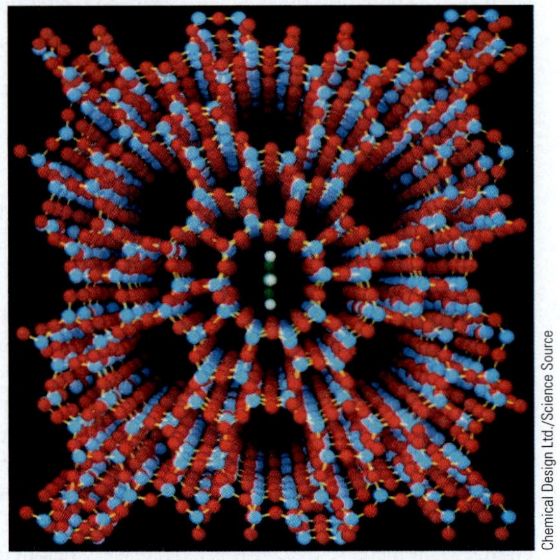

Zeolite ions are used in home water softeners because they have a porous but rigid structure that is weakly attracted to cations, readily allowing an exchange among different ions.

Answers to Target Checks

1. An equilibrium between a solution and excess undissolved solute can be contained in an open beaker if there is no significant evaporation of solvent (see Section 16.4).
2. Your drawings may have shown "glancing" collisions, collisions that have insufficient kinetic energy and/or collisions with improper orientations, all of which do not produce a chemical change.
3. A barrier is something that prevents or limits some event. Activation energy limits a reaction to the fraction of the intermolecular collisions that have enough kinetic energy and proper orientation.
4. (a) As temperature drops, the fraction of collisions with enough kinetic energy to meet the activation energy requirement drops significantly, which reduces reaction rate. Collision frequency also drops, which reduces reaction rate slightly. (b) A catalyst increases reaction rate by providing a reaction path with a lower activation energy. (c) At higher concentrations, collisions are more frequent, so reaction rate increases.
5. (a) The forward reaction is favored. (b) Appreciable quantities of all species are present at equilibrium.

Solutions to Practice Exercises

1. The shift will be in the reverse direction.
2. (a) forward, (b) reverse, (c) reverse, (d) forward
3. The reaction will shift in the reverse direction.
4. The equilibrium will shift in the reverse direction to create more gaseous molecules.
5. The equilibrium is not affected. One mole of gas is on each side of the equation, so a volume change does not affect the system.
6. The equilibrium will shift in the forward direction.
7. The equilibrium will shift in the forward direction.
8. $K = \dfrac{[NO_2]^4[H_2O]^6}{[NH_3]^4[O_2]^7}$
9. a) $K = [H_2]^2[O_2]$

   b) $K = \dfrac{1}{[H_2O]^6}$
10. $Cu(IO_3)_2(s) \rightleftharpoons Cu^{2+}(aq) + 2\,IO_3^-(aq)$ $K_{sp} = [Cu^{2+}][IO_3^-]^2 = (3.3 \times 10^{-3}) \times [2 \times (3.3 \times 10^{-3})]^2 = 1.4 \times 10^{-7}$
11. $\dfrac{0.029\text{ g CuBr}}{\text{L}} \times \dfrac{1\text{ mol CuBr}}{143.45\text{ g CuBr}} = 2.0 \times 10^{-4}\text{ mol/L CuBr}$

    $CuBr(s) \rightleftharpoons Cu^+(aq) + Br^-(aq)$ $K_{sp} = [Cu^+][Br^-] = (2.0 \times 10^{-4})(2.0 \times 10^{-4}) = 4.0 \times 10^{-8}$
12. $BaSO_4(s) \rightleftharpoons Ba^{2+}(aq) + SO_4^{2-}(aq)$

    $K_{sp} = [Ba^{2+}][SO_4^{2-}] = s \times s = s^2 = 1.5 \times 10^{-9}$

    $s = \sqrt{1.5 \times 10^{-9}} = 3.9 \times 10^{-5}$

    $100\text{ mL} \times \dfrac{1\text{ L}}{1000\text{ mL}} \times \dfrac{3.9 \times 10^{-5}\text{ mol BaSO}_4}{\text{L}} \times \dfrac{233.4\text{ g BaSO}_4}{\text{mol BaSO}_4} = 9.1 \times 10^{-4}\text{ g BaSO}_4$

    $9.1 \times 10^{-4}\text{ g/100 mL BaSO}_4$
13. $CaF_2(s) \rightleftharpoons Ca^{2+}(aq) + 2\,F^-(aq)$ $K_{sp} = [Ca^{2+}][F^-]^2 = 4.0 \times 10^{-11}$

    $[Ca^{2+}] = \dfrac{K_{sp}}{[F^-]^2} = \dfrac{4.0 \times 10^{-11}}{(0.020)^2} = 1.0 \times 10^{-7}\text{ M}$
14. $[H^+] = [A^-] = 10^{-pH} = 10^{-2.99} = 1.0 \times 10^{-3}\text{ M}$

    $K_a = \dfrac{[H^+][A^-]}{[HA]} = \dfrac{(1.0 \times 10^{-3})(1.0 \times 10^{-3})}{0.12} = 8.3 \times 10^{-6}$

    $\dfrac{1.0 \times 10^{-3}}{0.12} \times 100\% = 0.83\%$ ionized
15. $[H^+] = \sqrt{[HA]K_a} = \sqrt{(0.10)(1.8 \times 10^{-5}} = 1.3 \times 10^{-3}\text{ M}$

    $pH = -\log[H^+] = -\log(1.3 \times 10^{-3}) = 2.89$
16. $[H^+] = K_a \times \dfrac{[HA]}{[A^-]} = 1.8 \times 10^{-5} \times \dfrac{0.10}{0.10} = 1.8 \times 10^{-5}$

    $pH = -\log[H^+] = -\log(1.8 \times 10^{-5}) = 4.74$
17. $\dfrac{[H^+]}{K_a} = \dfrac{[HA]}{[A^-]} = \dfrac{10^{-4.50}}{1.8 \times 10^{-5}} = 1.8$
18.

	$2\,SO_2$	+	O_2	$\rightleftharpoons$	$2SO_3$
mol/L at start	6.0		4.0		0.0
mol/L change, + or −	−4.0		−2.0		+4.0
mol/L at equilibrium	2.0		2.0		4.0

$K = \dfrac{[SO_3]^2}{[SO_2]^2[O_2]} = \dfrac{(4.0)^2}{(2.0)^2(2.0)} = 2.0$

19.

	CO	+	Cl_2	⇌	$COCl_2$
mol/L at start	3.00		5.00		0.00
mol/L change, + or −	−2.90		−2.90		+2.90
mol/L at equilibrium	0.10		2.10		2.90

$$K = \frac{[COCl_2]}{[CO]\,[Cl_2]} = \frac{2.90}{(0.10)\,(2.10)} = 14$$

Answers to Concept-Linking Exercises

You may have found more relationships or relationships other than the ones given in these answers.

1. A reversible reaction reaches equilibrium if the forward reaction rate is equal to the reverse reaction rate.
2. The collision theory of chemical reactions explains that reactant particles must collide in order to react. An energy versus reaction progress graph traces the energy of the system before, during, and after collision. The transition state is the temporary high-energy combination of reacting particles during a collision. Activation energy is the increase in energy as reactant particles change to the transition state.
3. Temperature, a catalyst, and reactant concentrations are three variables that determine the rate of a chemical reaction.
4. Le Chatelier's Principle states that if an equilibrium system is disturbed in a way that makes the rates of forward and reverse reactions unequal, the equilibrium adjusts itself to counteract the disturbance partially until a new equilibrium is established. Temperature and concentration are two such disturbances. A change in gas volume also disturbs an equilibrium if the coefficients on opposite sides of the equation are unequal for species in the gas state.
5. An equilibrium constant (symbol K) is a ratio of the product of the concentrations of each species on the right-hand side of an equilibrium equation, each raised to a power equal to its coefficient in the equilibrium equation, divided by the comparable product of the concentrations of the species on the left-hand side of the equilibrium equation, each raised to a power equal to its coefficient in the equilibrium equation. K_{sp} identifies an equilibrium constant for a solubility equilibrium; *sp* means "solubility product." K_a identifies an acid equilibrium; *a* is for "acid."

Solutions to Blue-Numbered Questions, Exercises, and Problems

1. In a dynamic equilibrium, opposing changes continue to occur at equal rates. An equilibrium in which nothing is changing—a book resting on a table, for example—is called a static equilibrium.

3. Both systems can reach equilibrium. At equilibrium, the salt dissolves at the same rate at which it crystallizes. Whether the container is open or closed is of negligible importance.

5. The system is not an equilibrium because energy must be supplied constantly to keep it in operation. Also, the water is circulating, not moving reversibly in two opposing directions.

7. There will be no reaction if the orientation of the colliding particles is unfavorable.

9. $\Delta E = b - c$; $E_a = a - c$

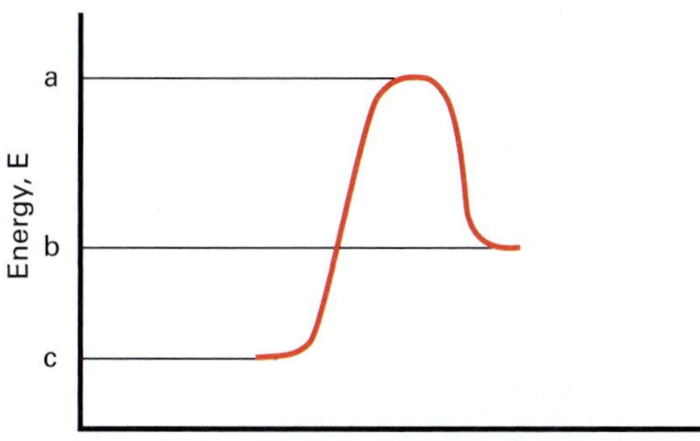

11. E_a (forward) $= a - b$; E_a (reverse) $= a - c$. Both activation energies are positive. Point a is the highest on the curve; $a > b$ and $a > c$.

13. A transition state is an unstable intermediate species formed during a collision of two reacting particles. The properties of a transition state cannot be described because the complex decomposes almost as soon as it forms.

15. At a higher temperature, a larger fraction of the molecules has enough kinetic energy to engage in a reaction-producing collision, so reaction rates are higher. Also, collisions are more frequent. At low temperature, a smaller fraction of the collisions produce reactions, and there are fewer collisions, so the reaction rate is slower.

17. The equilibria are identical. Equilibrium will be reached more quickly in the system with the catalyst.

19. An increase in the concentration of A will increase the reaction rate, whereas a decrease in the concentration of B will decrease the reaction rate. The net effect depends on the size of the two changes.

21. The reverse reaction rate reaches its maximum at equilibrium.

23. See Figure 18.10.

25. If O_2 concentration is decreased, equilibrium will shift in the reverse direction, the direction in which more O_2 will be produced.

27. If O_2 concentration is increased, equilibrium will shift in the forward direction, the direction in which more O_2 will be consumed.

29. Removal of NH_3 will shift the equilibrium forward, the direction in which some additional NH_3 will be produced.

31. Increasing volume decreases the pressure. The equilibrium shifts to the left (5 moles of gases on the left vs. 3 moles of gases on the right) to cause a pressure increase.

33. It will not shift because there will be no change in the total number of molecules of gases.

35. If heat is removed, the equilibrium will shift in the direction that produces heat, the forward direction.

37. Cool the system. Removing heat causes the reaction to shift in the direction that produces heat, the forward direction. That increases the SO_3 yield.

39. $Ca(OH)_2(s) \rightleftharpoons Ca^{2+}(aq) + 2\,OH^-(aq)$ is the equilibrium equation. H^+ ions from the acid combine with OH^- ions to form water molecules. This reduces the OH^- ion concentration and causes a forward shift in the equilibrium. The process continues until all the $Ca(OH)_2$ is dissolved.

41. $K = \dfrac{[CO_2][H_2]}{[CO][H_2O]}$

43. $K = \dfrac{[CO][H_2]}{[H_2O]}$

45. $K = [Zn^{2+}]^3[PO_4^{\,3-}]^2$

47. $K = \dfrac{[H_3O^+][NO_2^{\,-}]}{[HNO_2]}$

49. $K = \dfrac{[Cu^{2+}][NH_3]^4}{[Cu(NH_3)_4^{\,2+}]}$

51. The equilibrium constant expression for the given equation is $K = \dfrac{[NO_2]^2}{[NO]^2[O_2]}$. If the equation is written in reverse, the equilibrium constant expression is inverted. If different sets of coefficients are used, both the expression and its numerical value change. For example: $NO + \frac{1}{2}O_2 \rightleftharpoons NO_2$, $K_1 = \dfrac{[NO_2]}{[NO][O_2]^{1/2}}$; $4\,NO + 2\,O_2 \rightleftharpoons 4\,NO_2$, $K_2 = \dfrac{[NO_2]^4}{[NO]^4[O_2]^2}$. The equilibrium constants are not equal: $K_2 = K^2 = K_1^4$.

53. The equilibrium will be favored in the forward direction. If K is very large, at least one factor in the denominator must be very small, indicating that at least one reactant has been almost completely consumed.

55. The equilibrium will be favored in the reverse direction. A very small equilibrium constant results when the concentration of one species on the right-hand side of the equation is very small.

57. (a) The equilibrium is favored in the forward direction. H_2SO_3 is one of the acids that is unstable, decomposing to H_2O and SO_2, as indicated. K is large for this equilibrium. (b) The equilibrium is favored in the forward direction. $HC_2H_3O_2$ is a weak acid. Nearly all of the ions will combine to form the molecule.

59. $[Co^{2+}] = 3.7 \times 10^{-6}$; $[OH^-] = 2 \times (3.7 \times 10^{-6}) = 7.4 \times 10^{-6}$; $K_{sp} = [Co^{2+}][OH^-]^2 = (3.7 \times 10^{-6})(7.4 \times 10^{-6})^2 = 2.0 \times 10^{-16}$

61. $\dfrac{8.7\ \text{mg}\ Ag_2CO_3}{250\ \text{mL}} \times \dfrac{1\ \text{g}\ Ag_2CO_3}{1000\ \text{mg}\ Ag_2CO_3} \times \dfrac{1\ \text{mol}\ Ag_2CO_3}{275.8\ \text{g}\ Ag_2CO_3} \times \dfrac{1000\ \text{mL}}{\text{L}} = 1.3 \times 10^{-4}\ \text{M}\ Ag_2CO_3$

$[Ag_2CO_3] = [CO_3^{\,2-}] = 1.3 \times 10^{-4}$ M; $[Ag^+] = 2 \times (1.3 \times 10^{-4}\ \text{M}) = 2.6 \times 10^{-4}$ M; $K_{sp} = [Ag^+]^2[CO_3^{\,2-}] = (2.6 \times 10^{-4})^2(1.3 \times 10^{-4}) = 8.8 \times 10^{-12}$

63. $K_{sp} = [Ag^+][IO_3^-]$; $[Ag^+] = [IO_3^-] = s$; $s^2 = 2.0 \times 10^{-8}$; $s = 1.4 \times 10^{-4}$ M

$\dfrac{1.4 \times 10^{-4}\ \text{mol}\ AgIO_3}{\text{L}} \times \dfrac{282.8\ \text{g}\ AgIO_3}{\text{mol}\ AgIO_3} \times 0.100\ \text{L} = 4.0 \times 10^{-3}$ g/100 mL (0.100 L = 100 mL)

65. $K_{sp} = [Mn^{2+}][OH^-]^2$; $[Mn^{2+}] = s$; $[OH^-] = 2s$; $(s)(2s)^2 = 4s^3 = 1.0 \times 10^{-13}$; $s = 2.9 \times 10^{-5}$ M

67. $K_{sp} = [Ca^{2+}][C_2O_4^{\,2-}] = [Ca^{2+}](0.22) = 2.4 \times 10^{-9}$; $[Ca^{2+}] = 1.1 \times 10^{-8}$ M

$2.50 \times 10^2\ \text{mL} \times \dfrac{1.1 \times 10^{-8}\ \text{mol}\ CaC_2O_4}{1000\ \text{mL}} \times \dfrac{128.10\ \text{g}\ CaC_2O_4}{\text{mol}\ CaC_2O_4} = 3.5 \times 10^{-7}\ \text{g}\ CaC_2O_4$

69. $[H^+] = 10^{-2.12} = 7.6 \times 10^{-3}$ M; $K_a = \dfrac{[H^+][C_4H_5O_3^{\,-}]}{[HC_4H_5O_3]} = \dfrac{(7.6 \times 10^{-3})^2}{0.22} = 2.6 \times 10^{-4}$; $\dfrac{7.6 \times 10^{-3}}{0.22} \times 100\% =$ 3.5% ionized

71. $K_a = \dfrac{[H^+][C_2H_3O_2^{\,-}]}{[HC_2H_3O_2]} = \dfrac{[H^+]^2}{[HC_2H_3O_2]}$; $[H^+] = \sqrt{(1.8 \times 10^{-5})(0.35)} = 2.5 \times 10^{-3}$ M; $pH = -\log(2.5 \times 10^{-3}) = 2.60$

73. $24.0\ \text{g}\ NaC_2H_3O_2 \times \dfrac{1\ \text{mol}\ NaC_2H_3O_2}{82.03\ \text{g}\ NaC_2H_3O_2} = 0.293\ \text{mol}\ NaC_2H_3O_2$; $5.00 \times 10^2\ \text{mL} \times \dfrac{1\ \text{L}}{1000\ \text{mL}} = 0.500$ L

$\dfrac{0.293\ \text{mol}\ NaC_2H_3O_2}{0.500\ \text{L}} = 0.586\ \text{M}\ NaC_2H_3O_2 = 0.586\ \text{M}\ C_2H_3O_2^{\,-}$; $[H^+] = 1.8 \times 10^{-5} \times \dfrac{0.12}{0.586} = 3.7 \times 10^{-6}$; $pH = -\log(3.7 \times 10^{-6}) = 5.43$

75. $\dfrac{[HC_2H_3O_2]}{[C_2H_3O_2^{\,-}]} = \dfrac{10^{-4.25}}{1.8 \times 10^{-5}} = 3.1$

77. [CO] at start $= \dfrac{0.351\ \text{mol}}{3.00\ \text{L}} = 0.117$ M; $[Cl_2]$ at start $= \dfrac{1.340\ \text{mol}}{3.00\ \text{L}} = 0.447$ M; $[Cl_2]$ at end $= \dfrac{1.050\ \text{mol}}{3.00\ \text{L}} = 0.350$ M

	CO(g)	+	Cl_2(g)	$\rightleftharpoons$	$COCl_2$ (g)
Initial	0.117		0.447		0
Reacting	−0.097		−0.097		+0.097
Equilibrium	0.020		0.350		0.097

$K = \dfrac{[COCl_2]}{[CO][Cl_2]} = \dfrac{0.097}{(0.020)(0.350)} = 14$

80. True: b, d, f, g, i, j, l, m, n. False: a, c, e, h, k, o.

81. Kinetic energies are greater at Time 1 because at Time 2 some of that energy has been converted to potential energy in the transition state.

83. a) High pressure to force reaction to the smaller number of gaseous product molecules. b) High temperature, at which all reaction rates are faster.

84. A manufacturer cannot use an equilibrium, which is a closed system from which no product can be removed.

85. The higher temperature is used to speed the reaction rate to an acceptable level. Lower pressure is dictated by limits of mechanical design and safety.

86. (a) $Ca(OH)_2(s) \rightleftharpoons Ca^{2+}(aq) + 2\,OH^-(aq)$. (b) (1) Adding a strong base or soluble calcium compound would increase $[OH^-]$ and $[Ca^{2+}]$, respectively, causing a shift in the reverse direction and reducing the solubility of $Ca(OH)_2$. (2) Adding an acid to reduce $[OH^-]$ by forming water, adding a cation that will reduce $[OH^-]$ by precipitation, or adding an anion whose calcium salt is less soluble than $Ca(OH)_2$ would cause a shift in the forward direction, increasing the solubility of $Ca(OH)_2$. (c) (1) Any anion that reacts with calcium ion to form an ionic compound that is less soluble than $Ca(OH)_2$ will cause a forward shift, increasing $[OH^-]$. (2) An acid that will form water with OH^- or a cation that will precipitate OH^- will reduce $[OH^-]$.

87. Add NO_2: R–I–I–D–I. Reduce temperature: F–D–D–I–I. Add N_2: None. Remove NH_3: R–D–I–D–D. Add a catalyst: None.

88. The "truth" of the statement depends on how you interpret it. A system *at equilibrium* is neither endothermic nor exothermic. The system is closed; heat energy neither enters nor leaves. The thermochemical *equation* for the equilibrium is endothermic in one direction and exothermic in the other direction. Both directions describe the same equilibrium.

89. Temperature affects reaction rates in forward and reverse directions differently. Therefore, the value of an equilibrium constant depends on temperature. If you change the temperature, the value of K changes.

90. $2\,SO_2(g) + O_2(g) \rightleftharpoons 2\,SO_3(g)$ and $2\,SO_3(g) \rightleftharpoons 2\,SO_2(g) + O_2(g)$ both express the equilibrium described. The K expression for one equation is the reciprocal of the other (write the equilibrium constant expression for each equation to see this, if necessary). If any fraction is greater than 1, its reciprocal is less than 1. Put another way, if the numerator is greater than the denominator, the fraction is greater than 1. In the reciprocal, the numerator is less than the denominator, so the fraction is less than 1.

91. The system can reach equilibrium only when it is closed—that is, when no water is running in the house; it cannot be reached in an open system as hard water enters the softener and soft water leaves. $[Ca^{2+}]$ is relatively high in the hard water that enters the softener. By Le Chatelier's Principle, the reaction is favored in the forward direction in which Ca^{2+} ions in the water are replaced by Na^+ ions. This also means the Na^+ ions in the resin are replaced by Ca^{2+} ions from the water. Eventually, the Na^+ ions are used up and must be replenished. This is done by running water with a high sodium ion concentration through the softener. This forces the reaction in the reverse direction, again according to Le Chatelier's Principle. Na^+ ions replace the Ca^{2+} ions on the resin, and the Ca^{2+} ions are flushed down the drain.

19 Oxidation–Reduction (Electron Transfer) Reactions

Charles D. Winters

◀ All common batteries convert chemical energy into electrical energy by means of an oxidation–reduction reaction. A battery can be made by placing a strip of copper and a strip of zinc into a lemon! This lemon battery produces enough current to power a small clock. In this chapter, you will learn how a chemical reaction can generate an electric current.

CHAPTER CONTENTS

In this chapter, you will study oxidation–reduction reactions. You will see that redox reactions, as they are also known, are electron transfer reactions. As you study this chapter, look for parallels to the proton transfer reactions you studied in Chapter 17. Just as a proton is transferred from one species to another in an acid–base reaction, one or more electrons are transferred from one species to another in an oxidation–reduction reaction. You will find that looking for parallels between proton transfer reactions and electron transfer reactions will not only help you learn about redox reactions, but it will also help you better understand acid–base reactions.

19.1 How do You Power a Vehicle on the Surface of the Moon?

NASA

Figure 19.1 The Apollo 15 Lunar Roving Vehicle. The vehicle is about 10 feet long and 6 feet wide with a weight of 460 pounds. It was able to carry up to 1600 pounds. The total cost for the four vehicles that were assembled was $38 million, or $13 million per LRV that was transported to the Moon.

On the final three manned missions to the moon in 1971 and 1972, the flights of Apollo 15, 16, and 17 transported vehicles specifically designed to be driven on the surface of the moon. These vehicles were called *Lunar Roving Vehicles* (LRVs) (**Figure 19.1**), and they allowed astronauts to travel at speeds up to 10 mph over distances up to five miles from their base, the Lunar Module spacecraft. The vehicles needed the ability to carry two men, equipment, and samples while traveling up and down hills on the lunar surface in an atmosphere-free environment while tolerating a wide temperature range. What energy source could power such a vehicle?

The LRVs were powered by batteries. The batteries not only powered the motors at each wheel but they also powered the communications relay and the television camera. Two batteries made of silver oxide, zinc, and potassium hydroxide solution were on the vehicle. The net overall reaction that occurred in the batteries was $Zn(s) + Ag_2O(s) \rightarrow ZnO(s) + 2\ Ag(s)$. The source of energy from this reaction is not heat, however. The battery is engineered so that electrons flow through an external circuit as the reaction proceeds. Where do these electrons come from? Notice that the reactant zinc consists of individual, neutral atoms. The product zinc oxide is a combination of zinc ions and oxide ions. Zn atoms change to Zn^{2+} ions as the reaction proceeds. Thus, the zinc metal is the electron source. Where do the electrons go? Notice that the reactant silver oxide contains silver ions. The product is neutral silver atoms. Thus, the silver ions are what remove electrons from the zinc atoms. This type of reaction is an electron transfer reaction, the topic of this chapter. A battery changes chemical energy to electrical energy.

Harrison "Jack" Schmitt (**Figure 19.2**), the only scientist who flew on an Apollo mission (Ph.D. in geology), reflected on the importance of the LRVs by saying "...the Lunar Rover proved to be the reliable, safe and flexible lunar exploration vehicle we expected it to be. Without it, the major scientific discoveries of Apollo 15, 16, and 17 would not have been possible, and our current understanding of lunar evolution would not have been possible."

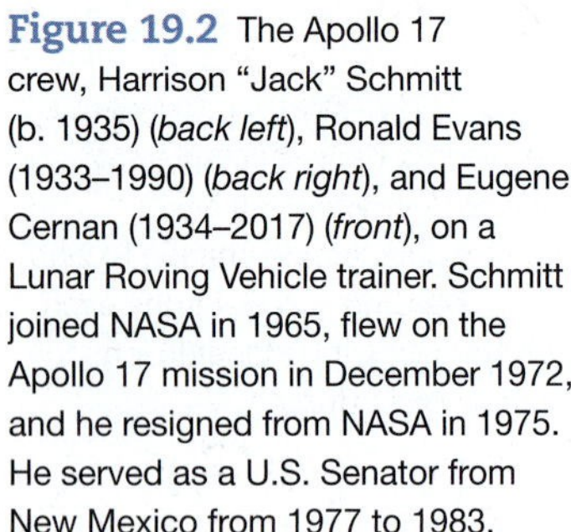

NASA

Figure 19.2 The Apollo 17 crew, Harrison "Jack" Schmitt (b. 1935) (*back left*), Ronald Evans (1933–1990) (*back right*), and Eugene Cernan (1934–2017) (*front*), on a Lunar Roving Vehicle trainer. Schmitt joined NASA in 1965, flew on the Apollo 17 mission in December 1972, and he resigned from NASA in 1975. He served as a U.S. Senator from New Mexico from 1977 to 1983.

19.2 Electron Transfer Reactions

Goal 1 Describe and explain oxidation and reduction in terms of electron transfer.

2 Given an oxidation half-reaction equation and a reduction half-reaction equation, combine them to form a net ionic equation for an oxidation–reduction reaction.

In the system in **Figure 19.3**, a strip of zinc is immersed in a solution containing zinc ions, and a strip of copper is placed in a solution that includes copper(II) ions. The solutions are connected by a **salt bridge** containing a solution of an ionic compound whose ions are not involved in the net chemical change. The two metal strips, which are called **electrodes** because they are the substances that electrons travel to and from, are connected by a wire. A light bulb in the external circuit is lit because of a flow of electrons from the zinc electrode to the copper electrode, and its brightness depends on the force that moves the electrons through the circuit. The system is known as an **electrochemical cell.**

Where do the electrons entering the light bulb come from, and where do they go when they leave the bulb? Four measurable observations answer those questions. After the cell has operated for a period of time, (1) the mass of the zinc electrode decreases, (2) the Zn^{2+} concentration increases, (3) the mass of the copper electrode increases, and (4) the Cu^{2+} concentration decreases. The first two observations indicate that neutral zinc atoms lose two electrons to become zinc ions. Stated another way, zinc atoms are being divided into zinc ions and two electrons:

$$Zn(s) \rightarrow Zn^{2+}(aq) + 2\ e^-$$

The electrons flow through the wire and the light bulb to the copper electrode, where they join a copper(II) ion to become a copper atom:

$$Cu^{2+}(aq) + 2\ e^- \rightarrow Cu(s)$$

The chemical change that occurs at the zinc electrode is oxidation: **oxidation** is defined as the loss of electrons. The reaction is described as a **half-reaction** because it cannot occur by itself. There must be a second half-reaction. The electrons lost by

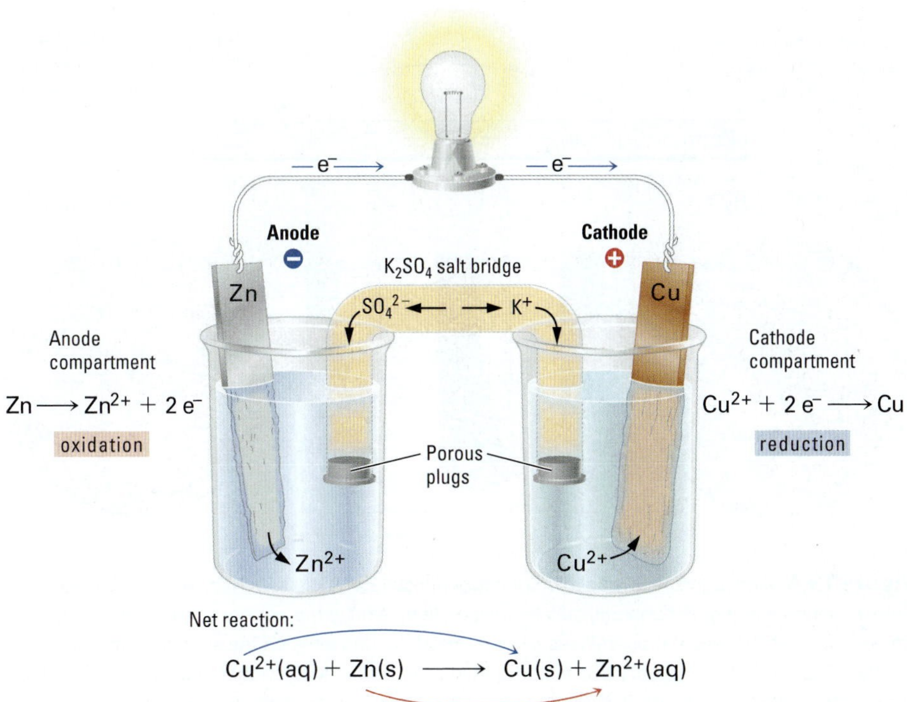

Figure 19.3 An electrochemical cell. In this system, a chemical reaction generates an electric current. Zinc atoms change to zinc ions, releasing two electrons each, which react with copper(II) ions to form copper atoms.

the substance oxidized must have someplace to go. In this case, they go to the copper(II) ion, which is reduced: **reduction** is a gain of electrons.

$Zn(s) \rightarrow Zn^{2+}(aq) + 2\,e^-$ and $Cu^{2+}(aq) + 2\,e^- \rightarrow Cu(s)$ are examples of **half-reaction equations**. If the half-reaction equations are added algebraically, the result is the net ionic equation for the oxidation–reaction (redox) reaction:

$$\begin{array}{r} Zn(s) \rightarrow Zn^{2+}(aq) + 2e^- \\ Cu^{2+}(aq) + 2e^- \rightarrow Cu(s) \\ \hline Zn(s) + Cu^{2+}(aq) \rightarrow Zn^{2+}(aq) + Cu(s) \end{array}$$

This chemical change is an **electron transfer reaction**. Electrons have been transferred from zinc atoms to copper(II) ions. Notice that although no electrons appear in the final equation, the electron transfer character of the reaction is quite clear in the half-reactions. Notice also that the number of electrons lost by one species is exactly equal to the number of electrons gained by the other.

If there is no need for the electrical energy that can be derived from this cell, the same reaction can be performed by simply dipping a strip of zinc into a solution containing copper(II) ions (**Figure 19.4**). A coating of copper atoms quickly forms on the surface of the zinc. If the copper atoms are washed off the zinc and the zinc is weighed, its mass will be less than it was at the beginning. The concentration of copper(II) ions in the solution goes down, and zinc ions appear. The half-reaction and net ionic equations are exactly as they are for the system connected to a light bulb.

This same reaction was used in Active Example 9.6: "A reaction occurs when a piece of zinc is dipped into a solution of copper(II) nitrate. Write the conventional, total ionic, and net ionic equations." The conventional equation is a single replacement equation:

$$Zn(s) + Cu(NO_3)_2(aq) \rightarrow Cu(s) + Zn(NO_3)_2(aq)$$

$Zn(s) + Cu^{2+}(aq) \rightarrow Cu(s) + Zn^{2+}(aq)$ is the net ionic equation produced in Active Example 9.6.

All of the single replacement redox reactions encountered in Chapters 8 and 9 can be analyzed in terms of half-reactions. For example,

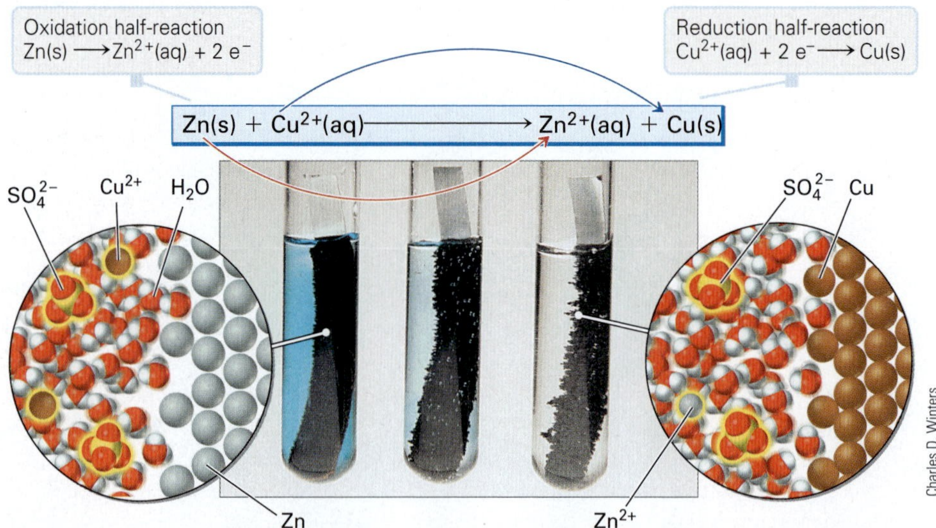

Figure 19.4 Zinc reacts with a solution of copper(II) sulfate. A bluish-silvery-white solid zinc strip is dipped into a blue copper(II) sulfate solution (*left*). Notice the copper coating that forms immediately. As the reaction progresses, copper metal accumulates on the surface of the zinc strip (*center*). The hydrated copper(II) ions impart a blue color to the solution that fades as they react to form copper atoms (*left to right*). Hydrated zinc and sulfate ions are colorless in solution.

1. The evolution of hydrogen gas on adding iron to hydrochloric acid (Section 9.6) (**Figure 19.5**):

$$\begin{array}{ll} \text{Reduction:} & 2\,H^+(aq) + \cancel{2e^-} \rightarrow H_2(g) \\ \text{Oxidation:} & Fe(s) \rightarrow Fe^{2+}(aq) + \cancel{2e^-} \\ \hline \text{Redox:} & 2\,H^+(aq) + Fe(s) \rightarrow H_2(g) + Fe^{2+}(aq) \end{array}$$

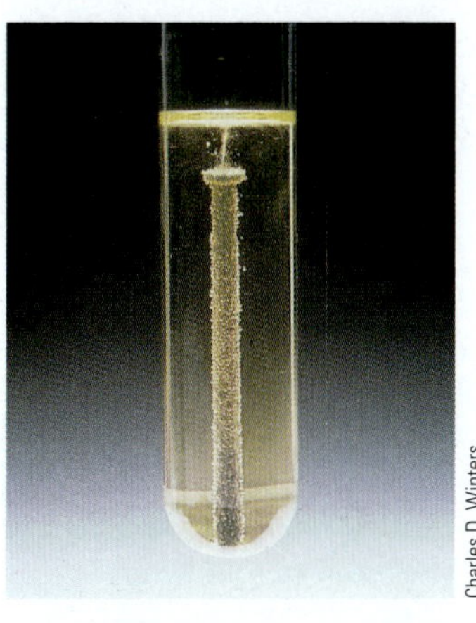

Figure 19.5 Iron metal is added to a hydrochloric acid solution. Hydrogen gas is bubbling from the solution.

2. The preparation of bromine by bubbling chlorine gas through a solution of sodium bromide (**Figure 19.6**):

$$\begin{array}{ll} \text{Reduction:} & Cl_2(g) + \cancel{2e^-} \rightarrow 2\,Cl^-(aq) \\ \text{Oxidation:} & 2\,Br^-(aq) \rightarrow Br_2(\ell) + \cancel{2e^-} \\ \hline \text{Redox:} & Cl_2(g) + 2\,Br^-(aq) \rightarrow 2\,Cl^-(aq) + Br_2(\ell) \end{array}$$

3. The reduction of silver ion (Section 8.9, Active Example 8.9 and Section 9.6, Active Example 9.8) by placing copper into a silver nitrate solution (**Figure 19.7**):

$$\begin{array}{ll} \text{Reduction:} & 2\,Ag^+(aq) + \cancel{2e^-} \rightarrow 2\,Ag(s) \\ \text{Oxidation:} & Cu(s) \rightarrow Cu^{2+}(aq) + \cancel{2e^-} \\ \hline \text{Redox:} & 2\,Ag^+(aq) + Cu(s) \rightarrow 2\,Ag(s) + Cu^{2+}(aq) \end{array}$$

The development of the equation $2\,Ag^+(aq) + Cu(s) \rightarrow 2\,Ag(s) + Cu^{2+}(aq)$ (the redox equation from Item 3 immediately preceding this paragraph) needs special comment. The reduction equation with the lowest whole-number coefficients for silver ion is $Ag^+(aq) + e^- \rightarrow Ag(s)$. Because two moles of electrons are lost in the oxidation reaction, *two moles of electrons must be gained in the reduction reaction.* As has already been noted, the number of electrons lost by one species must equal the number gained by the other species. It is, therefore, necessary to multiply the usual Ag^+ reduction equation by 2 to bring about this equality in electrons gained and lost. The two electrons then cancel when the half-reaction equations are added.

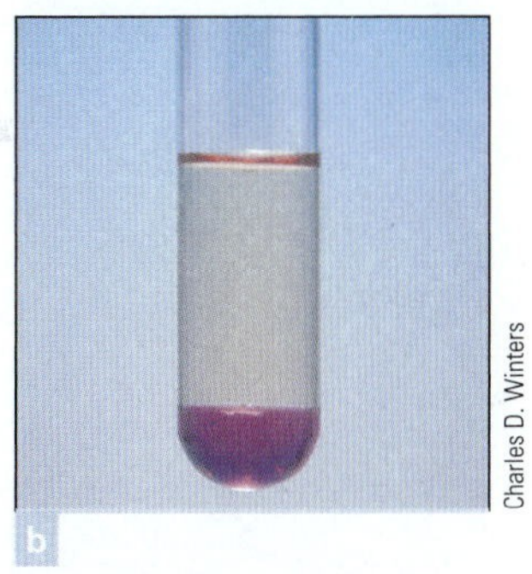

Figure 19.6 Preparation of bromine. (a) Chlorine gas is bubbled through a solution of sodium bromide. (b) The liquid bromine product is extracted by adding carbon tetrachloride. The top water layer and the bottom carbon tetrachloride layer both contain dissolved bromine, but it is more concentrated in the bottom.

Thinking About Your Thinking

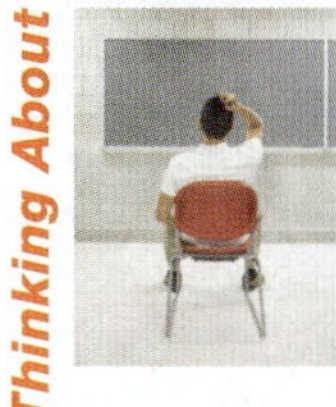

Mental Models

As with the proton transfer reactions of Chapter 17, working on the formation of a mental model of electron transfer reactions is a critical step toward thinking as a chemist. Both types of transfer reactions abound in biological systems. Students who study biochemistry are often amazed when they learn that a large fraction of the chemical changes that occur in living cells can be classified as either a proton transfer or an electron transfer reaction.

We can use Figures 19.3 and 19.4 as a starting point to help you form your mental movie about how electron transfer reactions occur at the particulate level. Look back and refamiliarize yourself with these two figures with a focus on what happens at the macroscopic level in Figure 19.3 and a focus on the particulate level changes in Figure 19.4. Now recall the electron-sea model of metallic bonding (Section 12.9), by which a metal is pictured as a crystal structure of metal ions immersed in a sea of electrons. Imagine the metallic zinc electrode in this way: zinc ions in a sea of electrons.

If one of those Zn^{2+} ions moves from the electrode into solution, its electrons are left behind. The charge balance in the metal is upset. There is more negative charge than positive. Negatively charged electrons repel other negatively charged electrons, creating a "flow" of electrons from the zinc electrode, through the connecting wires and light bulb, to the metallic copper electrode. These give the copper electrode a small negative charge that attracts it to the copper(II) ions

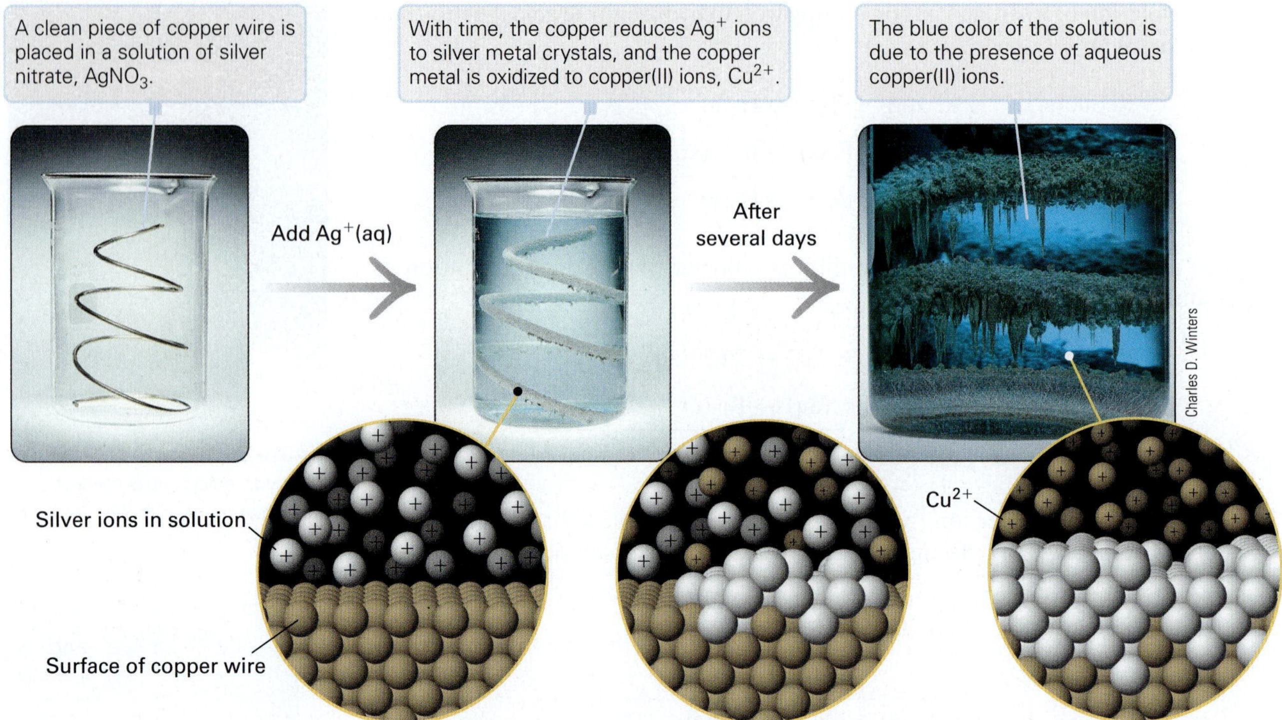
A clean piece of copper wire is placed in a solution of silver nitrate, $AgNO_3$.
With time, the copper reduces Ag^+ ions to silver metal crystals, and the copper metal is oxidized to copper(II) ions, Cu^{2+}.
The blue color of the solution is due to the presence of aqueous copper(II) ions.
Add Ag^+(aq)
After several days
Charles D. Winters
Silver ions in solution
Surface of copper wire
Cu^{2+}

Figure 19.7 The reduction of silver ion to solid silver and the oxidation of solid copper to copper(II) ion.

in the solution. Each copper(II) ion is joined by two electrons to become a copper atom that is deposited on the electrode.

Since a positively charged copper(II) ion is removed from the solution, the cell compartment has a change imbalance, with an excess of negative charge. This can be corrected by the flow of negatively charged SO_4^{2-} ions out of the compartment and into the salt bridge or with a positively charged ion (or ions) from the salt bridge moving into the compartment. In either case, the electrical charge in the salt bridge becomes out of balance, with a negative charge.

The compartment with zinc ions and sulfate ions compensates for the negative charge in the salt bridge. Either negative ions from the salt bridge enter this compartment or positive ions from the compartment enter the salt bridge. In both cases, the charge balance in the zinc sulfate solution is upset, with too many negative charges. This induces the formation of a zinc ion from the zinc electrode, which takes us back to the beginning of the cycle.

Run the cycle through your mind a few times until it becomes very clear. In particular, be sure that your mental model includes a flow of electrons (actually, more of a "jiggle" from neighbor to neighbor) through the metal electrodes and the connecting wires and the light bulb. This charge is compensated for by a flow of ions (again, it's actually a "jiggling" of neighbors) through the aqueous solutions and into and out of the salt bridge.

Active Example 19.1 Addition of Half-Reaction Equations I

Combine the following half-reactions to produce a balanced redox reaction equation: $Co^{2+}(aq) + 2\,e^- \rightarrow Co(s)$ and $Sn(s) \rightarrow Sn^{2+}(aq) + 2\,e^-$. Indicate which half-reaction is an oxidation reaction and which is a reduction.

Think Before You Write The key to adding half-reaction equations with the goal of producing the net ionic equation for a redox reaction is to get equal numbers of electrons on opposite sides of the equation.

Answers *Cover the left column with your cut-out shield. Reveal each answer only after you have written your own answer in the right column.*

Reduction: $Co^{2+}(aq) + \cancel{2 e^-} \rightarrow Co(s)$ Oxidation: $Sn(s) \rightarrow Sn^{2+}(aq) + \cancel{2 e^-}$ Redox: $Co^{2+}(aq) + Sn(s) \rightarrow Co(s) + Sn^{2+}(aq)$	Write each half-reaction equation and literally show the cancellation of electrons that leads to your balanced redox reaction equation. Oxidation is the loss of electrons; reduction is the gain of electrons. 
You improved your understanding of electron transfer reactions, and you improved your skill at adding half-reaction equations to produce a balanced redox equation.	What did you learn by solving this Active Example? 

Practice Exercise 19.1

Combine the following half-reactions to produce a balanced redox equation: $Cu(s) \rightarrow Cu^+(aq) + e^-$ and $Al^{3+}(aq) + 3\ e^- \rightarrow Al(s)$. Identify the oxidation half-reaction and the reduction half-reaction.

Active Example 19.2 Addition of Half-Reaction Equations II

Combine the following half-reactions to produce a balanced redox reaction equation: $Fe^{2+}(aq) \rightarrow Fe^{3+}(aq) + e^-$ and $Al^{3+}(aq) + 3\ e^- \rightarrow Al(s)$. Identify the oxidation half-reaction and the reduction half-reaction.

Think Before You Write The number of electrons on the left of the reaction arrow must be equal to the number of electrons on the right.

Answers *Cover the left column with your cut-out shield. Reveal each answer only after you have written your own answer in the right column.*

Oxidation: $3\ Fe^{2+}(aq) \rightarrow 3\ Fe^{3+}(aq) + \cancel{3 e^-}$ Reduction: $Al^{3+}(aq) + \cancel{3 e^-} \rightarrow Al(s)$ Redox: $Al^{3+}(aq) + 3\ Fe^{2+}(aq) \rightarrow Al(s) + 3\ Fe^{3+}(aq)$ *It is necessary to multiply the oxidation half-reaction equation by 3 in order to balance the electrons gained and lost.*	Do what must be done to balance the electrons. 
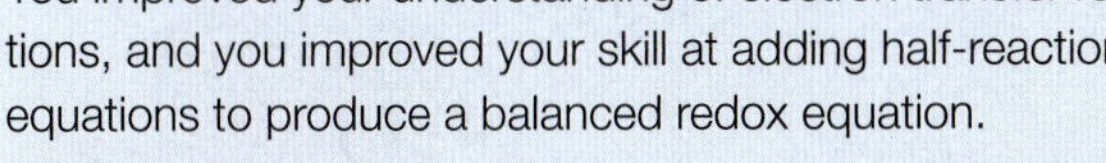 You improved your understanding of electron transfer reactions, and you improved your skill at adding half-reaction equations to produce a balanced redox equation.	What did you learn by solving this Active Example?

Practice Exercise 19.2

Combine the following half-reactions to produce a balanced redox reaction equation: $Sn^{2+}(aq) \rightarrow Sn^{4+}(aq) + 2\ e^-$ and $AgCl(s) + e^- \rightarrow Ag(s) + Cl^-(aq)$. Indicate which half-reaction is an oxidation reaction and which is a reduction.

Another reaction involving iron and aluminum introduces an additional technique.

Active Example 19.3 Addition of Half-Reaction Equations III

Arrange and modify the following half-reactions as necessary so that they add up to produce a balanced redox equation: $Fe^{2+}(aq) + 2\ e^- \rightarrow Fe(s)$ and $Al(s) \rightarrow Al^{3+}(aq) + 3\ e^-$. Identify the oxidation half-reaction and the reduction half-reaction.

Think Before You Write The electrons used in one half-reaction must equal the electrons produced in the other.

Answers *Cover the left column with your cut-out shield. Reveal each answer only after you have written your own answer in the right column.*

Reduction: $3\ Fe^{2+}(aq) + 6\ e^- \rightarrow 3\ Fe(s)$ Oxidation: $2\ Al(s) \rightarrow 2\ Al^{3+}(aq) + 6\ e^-$ Redox: $3\ Fe^{2+}(aq) + 2\ Al(s) \rightarrow 2\ Al^{3+}(aq) + 3\ Fe(s)$ *In this Active Example, electrons are transferred two at a time in the iron half-reaction and three at a time in the aluminum half-reaction. The simplest way to equate these is to take the iron half-reaction three times and the aluminum half-reaction twice. This gives six electrons for both half-reactions. The number of electrons lost in the oxidation half-reaction must always equal the number of electrons gained in the reduction half-reaction.*	Two electrons are transferred for each atom of iron and three per atom of aluminum. In what ratio must the atoms be used to equate the electrons gained and lost? Multiply and then add the half-reaction equations accordingly. 
You improved your understanding of electron transfer reactions, and you improved your skill at adding half-reaction equations to produce a balanced redox equation.	What did you learn by solving this Active Example?

Practice Exercise 19.3

Arrange and modify the following half-reactions to produce a balanced redox equation: $Mg(s) \rightarrow Mg^{2+}(aq) + 2\ e^-$ and $Cr^{3+}(aq) + 3\ e^- \rightarrow Cr(s)$. Identify the oxidation half-reaction and the reduction half-reaction.

19.3 Voltaic and Electrolytic Cells

Goal 3 Distinguish among electrolytic cells, voltaic cells, and galvanic cells.

4 Describe and identify the parts of an electrolytic or voltaic (galvanic) cell and explain how it operates.

Figure 19.8 illustrates the same copper–zinc system first shown in Figure 19.3, but the light bulb of Figure 19.3 has been replaced with a voltmeter, a device that

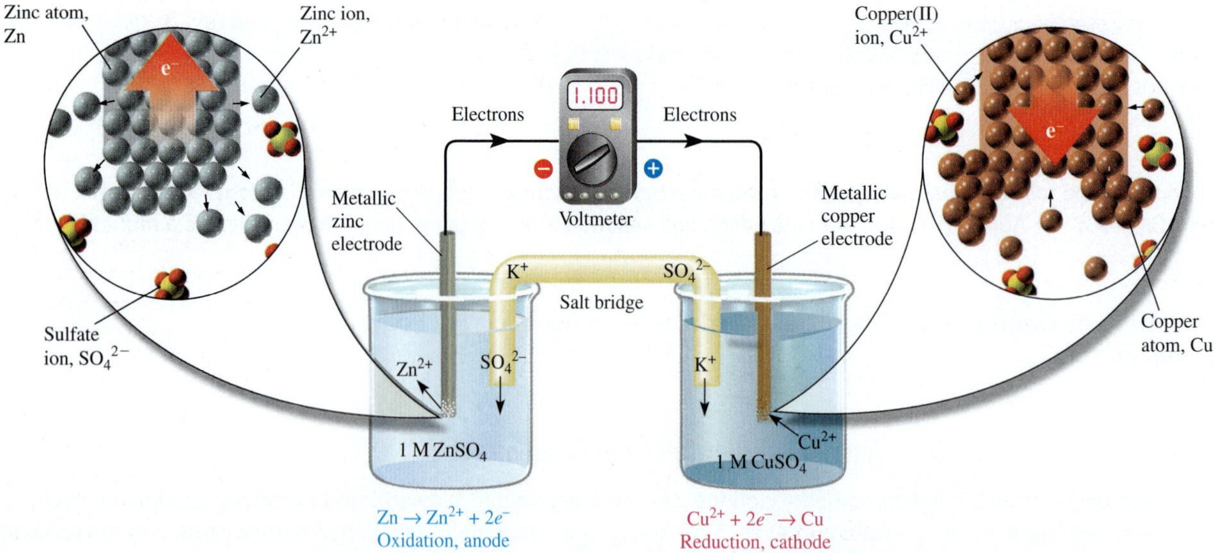

Figure 19.8 A voltaic cell, also called a galvanic cell, consists of two different electrodes immersed in two different ionic solutions, the electrolytes. It produces a spontaneous flow of electricity through both the electrolyte and the external circuit. Chemical changes occur at the electrodes of both half-cells. The electrode at which oxidation occurs is the anode, and the electrode at which reduction occurs is the cathode.

measures the force pushing electrons through the wire. This system is an example of a **voltaic cell**, which is also sometimes called a **galvanic cell**. A voltaic cell is one in which a spontaneous chemical reaction occurs, producing electricity and supplying it to an external circuit.

All voltaic cells are characterized by an electron transfer reaction that occurs on its own without an external energy source. The **anode** is the electrode at which oxidation occurs. Since an oxidation reaction yields electrons, the anode has a negative charge. Reduction occurs at the **cathode**, which has a positive charge. Electrons flow through the external circuit from anode to cathode. The two **half-cells** are connected by a salt bridge that maintains electrical neutrality in the solutions by allowing the passage of ions while preventing the mixing of the solutions.

The familiar alkaline battery used for flashlights, toys, and other electrical devices is a voltaic cell. When size is critical, as in calculators and watches, a silver oxide cell may be used. All of these cells "run down" and must be replaced when the chemical reactions in them reach equilibrium. The "ni–cad" (nickel–cadmium) voltaic cell runs down too, but unlike the others, it can be recharged (**Figure 19.9**). Lithium batteries, although relatively expensive, are now frequently used in portable consumer electronic devices because of their lighter weight (**Figure 19.10**). Lithium has the lowest density of any metal. Another familiar rechargeable battery is the lead storage battery used in automobiles.

Photo Intrigue/Shutterstock.com

Figure 19.9 Nickel–cadmium (ni–cad) batteries. The fundamental components of these batteries are nickel oxide, nickel hydroxide, nickel metal, cadmium hydroxide, cadmium metal, and potassium hydroxide.

FabrikaSimf /Shutterstock.com

Figure 19.10 Lithium batteries. These batteries are typically made from lithium metal and manganese(IV) oxide.

Another type of electrochemical cell is essentially the opposite of a voltaic cell; instead of supplying electricity, it requires an external electrical source. An **electrolytic cell** is made up of a container holding an ionic liquid or an ionic solution called an **electrolyte** and two electrodes (**Figure 19.11**). When the electrodes are connected to an outside source of electricity, they become charged, one positively and one negatively. Ions in the electrolyte move to the opposite charged electrode, where chemical reactions occur. The movement of ions is an electric current. The whole process is called **electrolysis.** ▶

◀ **P/Review** The operation of an electrolytic cell is first described in Section 9.2. An electrolyte must contain charged ions for the cell to conduct a current when it is connected to an outside source of electricity.

With the exception of barium, electrolytic cells are used to industrially produce all of the Group 1A/1 and 2A/2 metals. Sodium chloride is electrolyzed commercially in an apparatus called the *Downs cell* to produce sodium and chlorine. **Figure 19.12** shows the small-scale electrolysis of sodium chloride. Chlorine also comes from the electrolysis of sodium chloride *solutions*, after which the used electrolyte is evaporated to recover sodium hydroxide. Many common objects are made of metals that are electroplated with copper, nickel, chromium, zinc, tin, silver, gold, and other elements (**Figure 19.13**). The electrodeposits not only add beauty to the final product but they also protect the underlying base metal from corrosion.

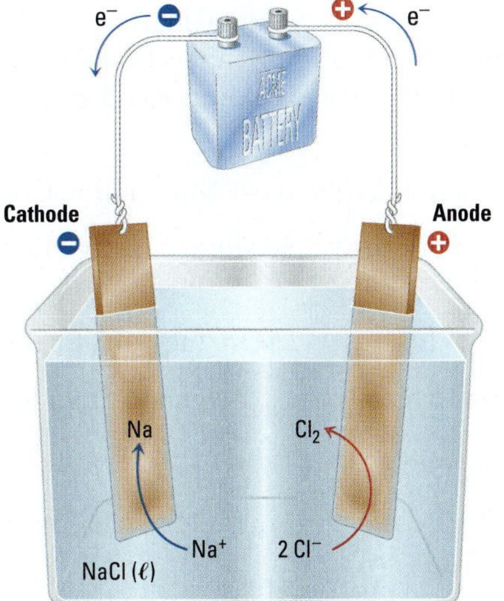

Figure 19.11 An electrolytic cell. This cell contains sodium chloride heated past its melting point, forming liquid (or molten) sodium chloride. An external energy source, such as a battery, causes a flow of electric current through the cell and an external circuit. The cells consist of two identical electrodes immersed in liquid. The sodium ions in the liquid are attracted to the cathode, where they react with electrons, forming sodium metal. The chloride ions are attracted to the anode, where they are oxidized to chlorine gas.

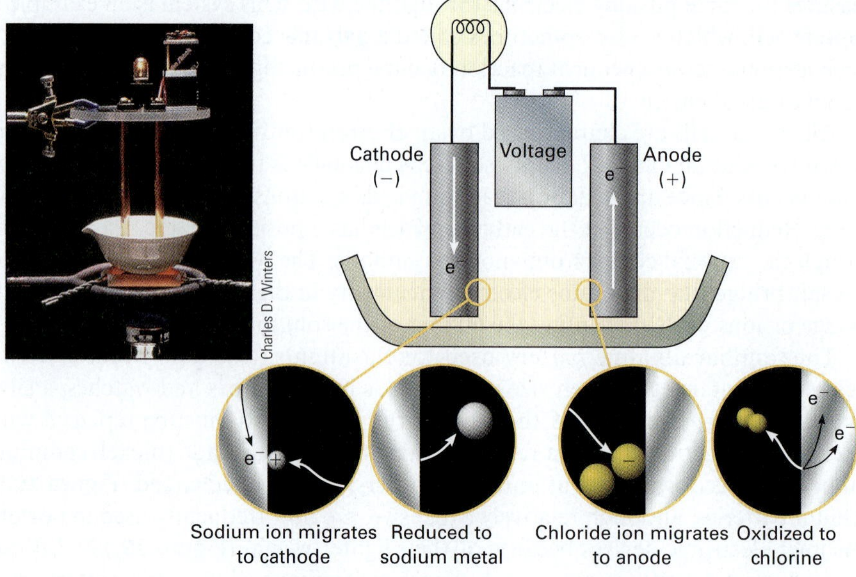

Figure 19.12 Electrolysis of sodium chloride. Solid sodium chloride is heated past its melting point. A battery causes the flow of electric current, pumping electrons to the cathode where sodium ions are reduced to sodium metal. Electrons are pumped from the anode where chlorine ions are oxidized to chlorine gas.

Figure 19.13 Electroplating. An electrolytic cell is used to add a coating of a relatively expensive metal to a less expensive base metal. The base metal for the Oscar statues shown here is brittanium, an alloy (solid solution) of tin, copper, and antimony. They are then sequentially electroplated with copper, nickel, silver, and, finally, 24-karat gold.

Target Check 19.1

Are the batteries used to power television remote controls best classified as electrolytic, voltaic, or galvanic cells? Explain.

19.4 Oxidation Numbers and Redox Reactions

Goal 5 Given the formula of an element, molecule, or ion, or information from which the formula can be determined, assign an oxidation number to each element in the formula.

6 Describe and explain oxidation and reduction in terms of change in oxidation numbers.

The redox reactions that we have discussed up to this point have been relatively simple ones involving only two reactants. With equations such as $Zn(s) + Cu^{2+}(aq) \rightarrow Cu(s) + Zn^{2+}(aq)$, we can see at a glance which species has gained and which has lost electrons. Some oxidation–reduction reactions are not so readily analyzed. Consider, for example, a reaction that is sometimes used in the general chemistry laboratory to prepare chlorine gas (**Figure 19.14**) from hydrochloric acid:

$$MnO_2(s) + 4\,H^+(aq) + 2\,Cl^-(aq) \rightarrow Mn^{2+}(aq) + Cl_2(g) + 2\,H_2O(\ell)$$

Also consider the reaction in a lead–acid storage battery that powers the starter to start an automobile:

$$Pb(s) + PbO_2(s) + 4\,H^+(aq) + 2\,SO_4^{2-}(aq) \rightarrow 2\,PbSO_4(s) + 2\,H_2O(\ell)$$

It is by no means obvious which species are gaining and losing electrons in these reactions.

"Electron bookkeeping" in more complex redox reactions is accomplished by using an **oxidation number**, which is a number assigned to each element in a species that is used to keep track of electrons. By following a set of rules, we may assign oxidation numbers to each element in a molecule or ion. The rules are as follows:

Figure 19.14 Laboratory preparation of chlorine. There are multiple methods to prepare small quantities of chlorine. In this reaction, NaCl, $K_2Cr_2O_7$, and H_2SO_4 react, forming Cl_2 as one product.

a summary of... **Oxidation Number Assignment Rules**

1. **The oxidation number of each atom in any elemental substance is zero (0).**
   Examples: Ox. No. of each hydrogen atom in $H_2 = 0$; Ox. No. of He = 0; Ox No. of each phosphorus atom in $P_4 = 0$
2. **The oxidation number of a monatomic ion is the same as the charge on the ion.**
   Examples: Ox. No. of $Na^+ = +1$; Ox. No. of $Mg^{2+} = +2$; Ox. No. of $N^{3-} = -3$
3. **The oxidation number of combined oxygen is −2.**
   Examples: Ox. No. of oxygen atom in $Na_2O = -2$; Ox. No. of oxygen atom in MgO = −2; Ox. No. of oxygen atom in $CH_3OH = -2$
   Exceptions: Oxygen atom in peroxides = −1 (e.g., H_2O_2); Oxygen atom in superoxides = −½ (e.g., KO_2); Oxygen atom in $OF_2 = +2$
4. **The oxidation number of combined hydrogen is +1.**
   Examples: Ox. No. of each hydrogen atom in $H_2SO_4 = +1$; Ox. No. of each hydrogen atom in $CH_4 = +1$; Ox. No. of hydrogen atom in HF = +1
   Exception: Ox. No. of a hydride ion, H^-, is −1 (e.g., NaH)
5. **In any molecular or ionic species, the sum of the oxidation numbers of all atoms in a molecule or polyatomic ion is equal to the charge on the species.**
   Examples are illustrated in the Active Examples that follow.

Active Example 19.4 Oxidation Number Assignment I

What are the oxidation numbers of the elements in CO_2?

Think Before You Write The problem asks you to assign oxidation numbers. The Oxidation Number Assignment Rules need to be applied.

Answers *Cover the left column with your cut-out shield. Reveal each answer only after you have written your own answer in the right column.*

Oxygen, −2	Oxidation Number Assignment Rules 1, 2, and 4 do not apply. Rule 3 gives one of the two oxidation numbers required. What is it? 
+4 *Note that Rule 5 requires that the sum of the oxidation numbers of atoms in the formula unit be equal to the charge on the unit, which is 0 for CO_2. There are two oxygen atoms, each at −2. The total contribution of oxygen is 2 × (−2), or −4. The sum of −4 plus the oxidation number of carbon is equal to 0. Carbon must therefore be +4.*	Carbon can exist in several different oxidation states. You can decide which one by applying Rule 5. What is that oxidation number? 
You improved your skill at assigning oxidation numbers. *This skill will be used later to identify oxidation and reduction in redox reactions.*	What did you learn by solving this Active Example? 

Practice Exercise 19.4

Assign the oxidation number to each element in ammonia.

There is a systematic way to reach the same conclusion that you might find helpful in more complicated examples. Applied to CO_2:

Write the formula with space in between the symbols of the elements or ions. Place the oxidation number of each element or ion beneath its symbol. Use n for the unknown oxidation number.	C O_2 n −2
Multiply each oxidation number by the number of atoms of that element in the formula unit.	C O_2 n 2(−2)
Add the oxidation numbers, set them equal to the charge on the species, and solve for the unknown oxidation number. In this case, n = +4.	C O_2 n + 2(−2) = 0 n = 4

Active Example 19.5 Oxidation Number Assignment II

Find the oxidation number of (a) S in SO_4^{2-} and (b) Cr in $HCr_2O_7^-$.

Think Before You Write The problem statement clearly is asking for application of the Oxidation Number Assignment Rules.

Answers *Cover the left column with your cut-out shield. Reveal each answer only after you have written your own answer in the right column.*

(a) S O_4 n + 4(−2) = −2 n = 6 (b) H Cr_2 O_7 1 + 2(n) + 7(−2) = −1 n = 6	Take it all the way to the final answers.
You improved your skill at assigning oxidation numbers.	What did you learn by solving this Active Example?

Practice Exercise 19.5

Assign oxidation numbers to elements in each ion: (a) carbonate ion, (b) sulfite ion.

Let's glance back at some of the equations in Section 19.3 to see if there is a regularity between oxidation and reduction and the change in oxidation number. **Table 19.1** summarizes the changes in three oxidation and three reduction half reactions that you've seen previously. Notice that for every oxidation half-reaction, the oxidation number increases. Conversely, for every reduction half-reaction, the oxidation number goes down.

We can now state a broader definition of oxidation and reduction. *Oxidation is an increase in oxidation number; reduction is a decrease in oxidation number.* These definitions are more useful in identifying the elements oxidized and reduced

Table 19.1 Summary of Selected Oxidation Reduction Reactions

Oxidation Half-Reaction	Oxidation Number Change	Reduction Half-Reaction	Oxidation Number Change
$Fe \rightarrow Fe^{2+} + 2\,e^-$	$0 \rightarrow +2$; increase	$2\,H^+ + 2\,e^- \rightarrow H_2$	$+1 \rightarrow 0$; reduction
$2\,Br^- \rightarrow Br_2 + 2\,e^-$	$-1 \rightarrow 0$; increase	$Cl_2 + 2\,e^- \rightarrow 2\,Cl^-$	$0 \rightarrow -1$; reduction
$Fe^{2+} \rightarrow Fe^{3+} + e^-$	$+2 \rightarrow +3$; increase	$Al^{3+} + 3\,e^- \rightarrow Al$	$+3 \rightarrow 0$; reduction

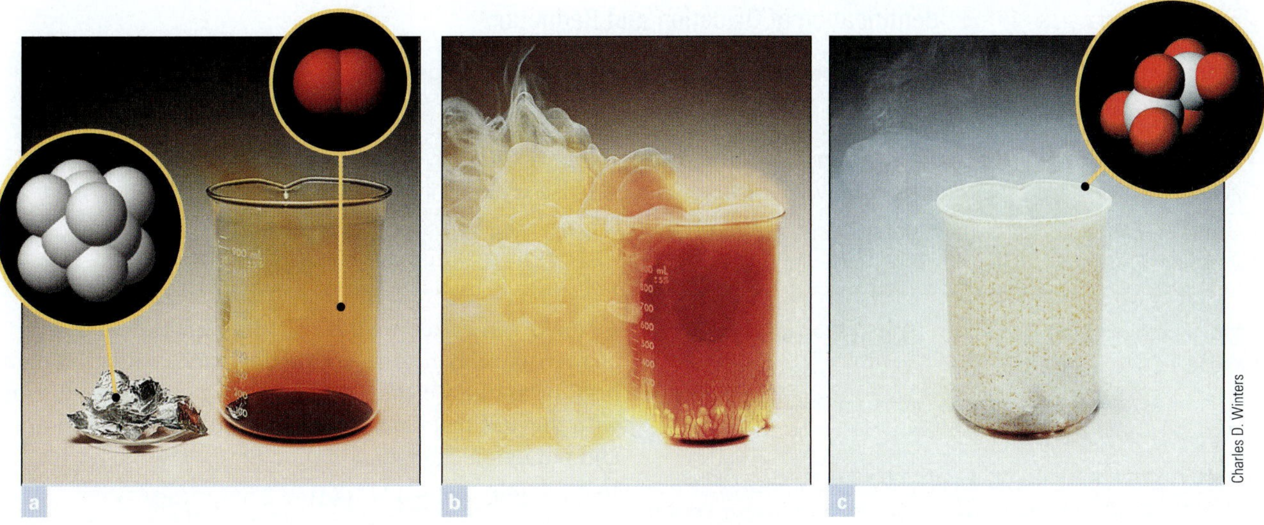

Figure 19.15 Oxidation numbers of reacting elements must change. (a) The oxidation numbers of aluminum and bromine are zero. (b) In a vigorous oxidation–reduction reaction, $2\ Al(s) + 3\ Br_2(\ell) \rightarrow 2\ Al_2Br_6(s)$, aluminum is oxidized and bromine is reduced. (c) The product of the reaction is solid Al_2Br_6 in which the oxidation number of aluminum is +3 and of bromine, −1.

when the electron transfer is not apparent. All you must do is find the elements that change their oxidation numbers and determine the direction of each change. One element must increase, and the other must decrease. (This corresponds with one species losing electrons while another gains.)

There are some techniques that enable you to spot quickly an element that changes oxidation number or to dismiss quickly some elements that do not change. These are as follows:

1. An element that is in its elemental state must change. As an element on one side of the equation, its oxidation number is 0; as anything other than an element on the other side, its oxidation number is *not* 0 (**Figure 19.15**).
2. In other than elemental form, hydrogen is +1 and oxygen is −2. Unless they are elements on one side, they do not change. (In more advanced courses, you will have to be alert to the hydride, peroxide, and superoxide exceptions noted in the oxidation number rules.)
3. A Group 1A/1 or 2A/2 element has only one oxidation state other than 0. If it does not appear as an element, it does not change. This observation is helpful when you must find the element oxidized or reduced in a conventional equation.

Additional exceptions to the oxidation number rules may also be introduced in more advanced courses. Follow the advice of your instructor about which exceptions, if any, you are to learn in the course in which this textbook is being used.

We will now use these ideas and your ability to assign oxidation numbers to find the elements oxidized and reduced in the equation of a reaction used to prepare small quantities of chlorine gas in the laboratory:

$$MnO_2(s) + 4\ H^+(aq) + 2\ Cl^-(aq) \rightarrow Mn^{2+}(aq) + Cl_2(g) + 2\ H_2O(\ell)$$

Chlorine is an element on the right, so it must be something else on the left. It is the chloride ion, Cl^-. The oxidation number change is −1 to 0, an *increase*, so chloride is *oxidized.*

Neither hydrogen nor oxygen appears as elements, so we conclude that they do not change oxidation state. That leaves manganese. Its oxidation state is +4 in MnO_2 on the left and +2 as Mn^{2+} on the right. This is a *decrease*, from +4 to +2, so manganese is *reduced.*

Active Example 19.6 Identification of Oxidation and Reduction

Determine the element oxidized and the element reduced in a lead–acid storage battery: $Pb(s) + PbO_2(s) + 4\ H^+(aq) + 2\ SO_4^{2-}(aq) \rightarrow 2\ PbSO_4(s) + 2\ H_2O(\ell)$.

Think Before You Write You can assign oxidation numbers to identify elements oxidized and reduced in a redox reaction. Oxidation is an increase in oxidation number, and reduction is a decrease in oxidation number.

Answers *Cover the left column with your cut-out shield. Reveal each answer only after you have written your own answer in the right column.*

Lead is both oxidized (0 in Pb to +2 in $PbSO_4$) and reduced (+4 in PbO_2 to +2 in $PbSO_4$). *The oxidation of Pb(s) can be spotted quickly because it is an element on the left.* *You might have thought sulfur to be the element reduced, but its oxidation state is +6 in the sulfate ion whether the ion is by itself on the left or part of a solid ionic compound on the right.*	To make this easier, we'll rewrite the equation in the answer space. Beneath the equation, write oxidation numbers for as many elements as necessary until you come up with the pair that changes. Then identify the oxidation and reduction changes. $Pb(s) + PbO_2(s) + 4\ H^+(aq) + 2\ SO_4^{2-}(aq) \rightarrow$ $2\ PbSO_4(s) + 2\ H_2O(\ell)$
You improved your understanding of electron transfer reactions, and you improved your skill at identifying oxidation and reduction in a redox reaction.	What did you learn by solving this Active Example?

Practice Exercise 19.6

Identify the element oxidized and the element reduced in the corrosion of iron: $2\ Fe(s) + O_2(g) + 2\ H_2O(\ell) \rightarrow 2\ Fe^{2+}(aq) + 4\ OH^-(aq)$.

Even though the oxidation number concept is very useful for keeping track of what the electrons are doing in a redox reaction, we should emphasize that it has been invented to meet a need. It has no experimental basis. Unlike the charge of a monatomic ion, the oxidation number of an atom in a molecule or polyatomic ion cannot be measured in the laboratory. It is all very good to talk about "+4 manganese" in MnO_2 or "+6 sulfur" in the SO_4^{2-} ion but take care not to fall into the trap of thinking that the elements in these species actually carry positive charges equal to their oxidation numbers.

a summary of... Definitions of Oxidation and Reduction

	Oxidation	Reduction
Change in electrons	Loss of electrons	Gain of electrons
Change in oxidation number	Increase	Decrease (reduction)

Target Check 19.2

What is the Oxidation Number Assignment Rule for each of the following: (a) elemental substance, (b) oxygen in a compound (some exceptions), (c) molecule, (d) monatomic ion, (e) hydrogen in a compound (one exception), (f) polyatomic ion?

19.5 Oxidizing Agents and Reducing Agents

Goal 7 Given a redox equation, or information from which the equation may be written, identify the oxidizing agent, the reducing agent, the element oxidized, and the element reduced.

The two essential reactants in a redox reaction are given special names to indicate the roles they play. The species that removes electrons—that is, the species that is itself reduced—is referred to as an **oxidizing agent**. The species from which the electrons are removed so reduction of another element can occur is called a **reducing agent**. Thus, the reducing agent is itself oxidized. For example, in the equation

Sometimes oxidizing agents are simply referred to as *oxidizers*, and reducing agents may also be called *reducers*.

$$Fe_2O_3(s) + 2\,Al(s) \rightarrow 2\,Fe(\ell) + Al_2O_3(s) \text{ (Figure 19.16)}$$

Fe^{3+} (in Fe_2O_3) takes electrons from Al—it *oxidizes* Al to Al^{3+}—and therefore Fe_2O_3 is the oxidizing agent. Note that an oxidizing or a reducing agent is the compound itself even if only one of its ions undergoes oxidation or reduction. Conversely, Al is the source of the electrons taken by Fe^{3+}—it causes Fe^{3+} to be reduced to Fe—and is therefore the reducing agent.

Another example is

$$MnO_2(s) + 4\,H^+(aq) + 2\,Cl^-(aq) \rightarrow Mn^{2+}(aq) + Cl_2(g) + 2\,H_2O(\ell)$$

Cl^- is the reducing agent, reducing manganese from +4 to +2. The oxidizing agent is MnO_2—the whole compound, not just the Mn; it oxidizes chlorine from −1 to 0.

The following Active Example summarizes the redox concepts introduced thus far in this chapter.

Figure 19.16 The reaction between iron(III) oxide and aluminum releases so much energy that *liquid* iron is formed as one of the products. Iron(III) oxide is the oxidizing agent, and aluminum is the reducing agent.

Active Example 19.7 Identification of Oxidation, Reduction, Oxidizing Agent, and Reducing Agent

Consider the redox reaction $5\,NO_3^-(aq) + 3\,As(s) + 2\,H_2O(\ell) \rightarrow 5\,NO(g) + 3\,AsO_4^{3-}(aq) + 4\,H^+(aq)$. (a) Determine the oxidation number of each element in each species. (b) Identify the element oxidized, the element reduced, the oxidizing agent, and the reducing agent.

Think Before You Write Part (a) asks you to apply the Oxidation Number Assignment Rules, and part (b) asks you to evaluate the oxidation number changes.

Answers *Cover the left column with your cut-out shield. Reveal each answer only after you have written your own answer in the right column.*

(a) NO_3^-: N = +5, O = −2; As: 0; H_2O: H = +1, O = −2; NO: N = +2, O = −2; AsO_4^{3-}: As = +5, O = −2; H^+: +1

(b) The element As is oxidized, increasing in oxidation number from 0 to +5. N is reduced, decreasing in oxidation number from +5 to +2. NO_3^- is the oxidizing agent, removing electrons from As. As is the reducing agent, furnishing electrons to NO_3^-.

Remember that elements are oxidized and reduced, but the entire species is the oxidizing or reducing agent. Answer both parts.

You improved your understanding of electron transfer reactions, and you improved your skill at identifying oxidation, reduction, the oxidizing agent, and the reducing agent in a redox reaction.

What did you learn by solving this Active Example?

Practice Exercise 19.7

Consider the reaction of copper and nitric acid: $3\ Cu(s) + 8\ H^+(aq) + 2\ NO_3^-(aq) \rightarrow 3\ Cu^{2+}(aq) + 2\ NO(g) + 4\ H_2O(\ell)$. (a) Determine the oxidation number of each element in each species. (b) Identify the element oxidized, the element reduced, the oxidizing agent, and the reducing agent.

19.6 Strengths of Oxidizing Agents and Reducing Agents

Goal 8 Distinguish between strong and weak oxidizing agents.

9 Given a table of the relative strengths of oxidizing and reducing agents, arrange a group of oxidizing agents or a group of reducing agents in order of increasing or decreasing strength.

An oxidizing agent earns its title by its ability to take electrons from another substance. A **strong oxidizing agent** has a strong attraction for electrons. Conversely, a **weak oxidizing agent** attracts electrons only slightly. ◀ The strength of a reducing agent is measured by its ability to give up electrons. A **strong reducing agent** releases electrons readily, whereas a **weak reducing agent** holds on to its electrons.

One way to measure the relative strengths of oxidizing agents is to use them as electrodes in a galvanic cell and observe the voltage produced, as shown in Figure 19.8. In more advanced courses, Table 19.2 has an additional column in which voltmeter readings are recorded.

Table 19.2 lists oxidizing agents in order of decreasing strength on the left side of the equation and lists reducing agents in order of increasing strength on the right side. The strongest oxidizing agent shown is fluorine, F_2, located at the top of the left column (**Figure 19.17**). Chlorine, Cl_2, listed below gold(III) ion, is used as a disinfectant in water supplies because of its ability to oxidize harmful organic matter (**Figure 19.18**). Notice that all equations in Table 19.2 are written as reduction half-reactions.

ACID–BASE-BEFORE-REDOX OPTION If you have already studied Chapter 17, you might find it interesting to compare Table 19.2 with Section 17.6, Figure 17.8. Figure 17.8 lists acids in order of decreasing strength on the left and bases in order of increasing strength on the right. In Figure 17.8, the substances are listed according to their tendencies to release and take protons; in Table 19.2 the substances are listed according to their tendencies to take or give electrons. Just as Figure 17.8 enables us to write acid–base reaction equations and predict the direction that will be favored at equilibrium, Table 19.2 enables us to do the same for redox reactions. These and other similarities between redox (electron transfer) and acid–base (proton transfer) reactions are summarized in Section 19.8.

Target Check 19.3

Why does a strong oxidizing agent become a weak reducing agent when it gains an electron?

19.7 Predicting Redox Reactions

Goal 10 Given a table of the relative strengths of oxidizing agents, and information from which an electron transfer reaction equation between two species in the table may be written, write the equation and predict the direction in which the reaction will be favored.

In Section 9.6, you wrote net ionic equations for possible single-replacement redox reactions. You used the activity series given in Table 9.2—reprinted in the margin

Table 19.2 Relative Strengths of Oxidizing and Reducing Agents

	Oxidizing Agent		Reducing Agent	
Stronger Oxidizing Agents	$F_2(g) + 2\,e^-$	$\rightleftharpoons$	$2\,F^-(aq)$	Weaker Reducing Agents
	$Au^+(aq) + e^-$	$\rightleftharpoons$	$Au(s)$	
	$Au^{3+}(aq) + 3\,e^-$	$\rightleftharpoons$	$Au(s)$	
	$Cl_2(g) + 2\,e^-$	$\rightleftharpoons$	$2\,Cl^-(aq)$	
	$O_2(g) + 4\,H^+(aq) + 4\,e^-$	$\rightleftharpoons$	$2\,H_2O(\ell)$	
	$Br_2(\ell) + 2\,e^-$	$\rightleftharpoons$	$2\,Br^-(aq)$	
	$NO_3^-(aq) + 4\,H^+ + 3\,e^-$	$\rightleftharpoons$	$NO(g) + 2\,H_2O(\ell)$	
	$Ag^+(aq) + e^-$	$\rightleftharpoons$	$Ag(s)$	
	$Fe^{3+}(aq) + e^-$	$\rightleftharpoons$	$Fe^{2+}(aq)$	
	$I_2(s) + 2\,e^-$	$\rightleftharpoons$	$2\,I^-(aq)$	
	$Cu^+(aq) + e^-$	$\rightleftharpoons$	$Cu(s)$	
	$Cu^{2+}(aq) + 2\,e^-$	$\rightleftharpoons$	$Cu(s)$	
	$Cu^{2+}(aq) + e^-$	$\rightleftharpoons$	$Cu^+(aq)$	
	$2\,H^+(aq) + 2\,e^-$	$\rightleftharpoons$	$H_2(g)$	
	$Pb^{2+}(aq) + 2\,e^-$	$\rightleftharpoons$	$Pb(s)$	
	$Sn^{2+}(aq) + 2\,e^-$	$\rightleftharpoons$	$Sn(s)$	
	$Ni^{2+}(aq) + 2\,e^-$	$\rightleftharpoons$	$Ni(s)$	
	$Co^{2+}(aq) + 2\,e^-$	$\rightleftharpoons$	$Co(s)$	
	$Cd^{2+}(aq) + 2\,e^-$	$\rightleftharpoons$	$Cd(s)$	
	$Fe^{2+}(aq) + 2\,e^-$	$\rightleftharpoons$	$Fe(s)$	
	$Zn^{2+}(aq) + 2\,e^-$	$\rightleftharpoons$	$Zn(s)$	
	$Al^{3+}(aq) + 3\,e^-$	$\rightleftharpoons$	$Al(s)$	
	$Mg^{2+}(aq) + 2\,e^-$	$\rightleftharpoons$	$Mg(s)$	
	$Na^+(aq) + e^-$	$\rightleftharpoons$	$Na(s)$	
	$Ca^{2+}(aq) + 2\,e^-$	$\rightleftharpoons$	$Ca(s)$	
	$Sr^{2+}(aq) + 2\,e^-$	$\rightleftharpoons$	$Sr(s)$	
	$Ba^{2+}(aq) + 2\,e^-$	$\rightleftharpoons$	$Ba(s)$	
	$Rb^+(aq) + e^-$	$\rightleftharpoons$	$Rb(s)$	
	$K^+(aq) + e^-$	$\rightleftharpoons$	$K(s)$	
Weaker Oxidizing Agents	$Li^+(aq) + e^-$	$\rightleftharpoons$	$Li(s)$	Stronger Reducing Agents

Table 9.2 Activity Series

Li, K, Ba, Sr, Ca, Na	Will replace H_2 from liquid water, steam, or acid
Mg, Al, Mn, Zn, Cr	Will replace H_2 from steam or acid
Fe, Ni, Sn, Pb	Will replace H_2 from acid
H_2	
Sb, Cu, Hg, Ag, Pd, Pt, Au	Will not replace H_2 from liquid water, steam, or acid

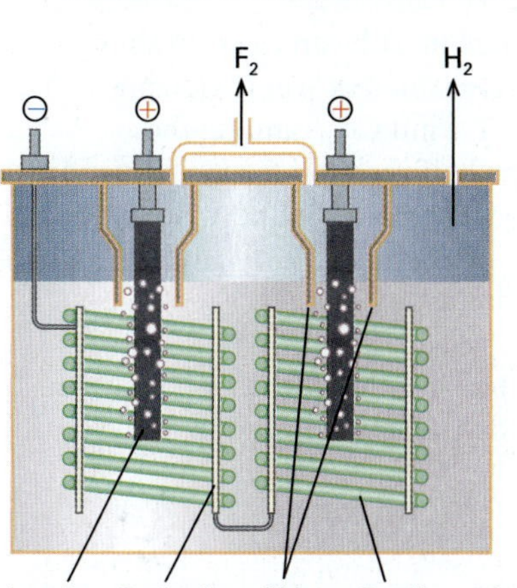

Figure 19.17 Industrial production of fluorine. Gaseous fluorine and hydrogen are the products of the electrolysis of a molten mixture of potassium fluoride and hydrogen fluoride. Great caution must be used in manufacturing, storing, and using fluorine due to its strength as an oxidizing agent.

Figure 19.18 The reaction of solid iron with chlorine gas, forming iron(III) chloride. In this redox reaction, electrons are transferred from iron to chlorine. Chlorine is a stronger oxidizing agent than the iron(III) ion, and iron is a stronger reducing agent than chloride ion.

next to Table 19.2—to predict whether or not the reaction would occur. In Table 9.2, the elements are listed in order of decreasing reactivity; the most active element is at the top. Any element in the activity series will react with and replace the dissolved ion of any element beneath it in the series.

Most metals in Table 9.2 are included in the list of reducing agents in the right-hand column of Table 19.2. Moreover, their order, in terms of reactivity, is exactly the same in the two tables. (The order appears to be inverted because the reactivity *increases* as you go down the column in Table 19.2, and it is just the opposite of Table 9.2. The most active metal in Table 19.2 is at the bottom.) Now you can understand that the arrangement of the activity series is determined by the energetic favorability with which atoms of the element release electrons and function as reducing agents.

Thinking About *Your Thinking*

Classification

Different classification systems can be used for different purposes. As you categorize your knowledge about chemistry, you need to reorganize your thinking when new information leads to a more efficient system. You will also find that there are strengths and weaknesses of various classification systems, and some may work better in one context, whereas others work better in another context.

Section 19.7 provides new information that will cause you to reorganize how you think about the activity series introduced in Section 9.6. You can now see that the activity series is just a subset of elements selected from Table 19.2. The activity series is a list of reducing agents, listed from strongest to weakest. The information found in the activity series is all taken from the table of relative strengths.

The next Active Example shows how to use the information in Table 19.2 to write redox reaction equations.

Active Example 19.8 Writing a Redox Equation Using a Relative Strengths Table

Write the net ionic equation for the redox reaction between the cobalt(II) ion, Co^{2+}, and metallic silver, Ag.

Think Before You Write The problem statement does not give you half-reaction equations. Thus, you should use Table 19.2 (or the equivalent from another resource) as a source of half-reactions.

Answers *Cover the left column with your cut-out shield. Reveal each answer only after you have written your own answer in the right column.*

Reduction: $Co^{2+}(aq) + 2\,e^- \rightleftharpoons Co(s)$	In any redox reaction, there must be an electron source (reducing agent) and an electron remover (oxidizing agent). Look at Table 19.2. Find Co^{2+} among the oxidizing agents. Take the half-reaction for the reduction of cobalt(II) ion directly from the table and write it down. 
Oxidation: $Ag(s) \rightleftharpoons Ag^+(aq) + e^-$	To obtain the *oxidation* half-reaction for silver, you need to *reverse* the reduction half-reaction found in Table 19.2. Write it down.

Reduction:	$Co^{2+}(aq) + 2e^- \rightleftharpoons Co(s)$	Multiply the oxidation equation by 2 to equalize electrons gained and lost and add it to the reduction equation to complete the net ionic equation.
2 × Oxidation:	$2\ Ag(s) \rightleftharpoons 2\ Ag^+(aq) + 2e^-$	
Redox:	$Co^{2+}(aq) + 2\ Ag(s) \rightleftharpoons Co(s) + 2\ Ag^+(aq)$	

You improved your understanding of electron transfer reactions, and you improved your skill at writing redox equations.	What did you learn by solving this Active Example?

Practice Exercise 19.8

Write the net ionic equation for the redox reaction between aqueous tin(II) ion and solid sodium.

The double arrows in Table 19.2 and Active Example 19.8 indicate that the reactions are reversible, as are most redox reactions that occur in aqueous solution. When a reversible reaction equation is read from left to right, the **forward reaction**, or the reaction in the forward direction, is described; from right to left, the change is the **reverse reaction**, or in the reverse direction.

In an aqueous solution redox reaction, a strong oxidizing agent takes electrons from a strong reducing agent to produce weaker oxidizing and reducing agents. The reversible reactions proceed in both directions until equilibrium is reached. At that time, the concentrations of the weaker oxidizing and reducing agents are greater than the concentrations of the stronger oxidizer and reducer. The reaction is said to be **favored** in the direction pointing to the weaker oxidizing and reducing agents.

Table 19.2 enables us to predict the favored direction of a redox reaction. This is illustrated in **Figure 19.19**. ▶ Applied to the reaction $2\ H^+(aq) + Fe(s) \rightleftharpoons H_2(g) + Fe^{2+}(aq)$, the positions in the table establish H^+ and Fe as the stronger oxidizing agent and reducing agent, respectively. The reaction is favored in the forward direction, yielding the weaker oxidizing agent and reducing agent, Fe^{2+} and H_2, respectively.

◀ **P/Review** Figure 19.19 is strikingly similar to Section 17.7, Figure 17.9, which describes the transfer of protons in an acid–base reaction.

Active Example 19.9 Predicting the Favored Direction of a Redox Reaction

In which direction, forward or reverse, will the redox reaction in Active Example 19.8 be favored? $Co^{2+}(aq) + 2\ Ag(s) \rightleftharpoons Co(s) + 2\ Ag^+(aq)$. Explain.

Think Before You Write Use Table 19.2 to identify the stronger and weaker oxidizing and reducing agents. Electrons will transfer from the stronger reducing agent to the stronger oxidizing agent, forming the weaker species.

Answers *Cover the left column with your cut-out shield. Reveal each answer only after you have written your own answer in the right column.*

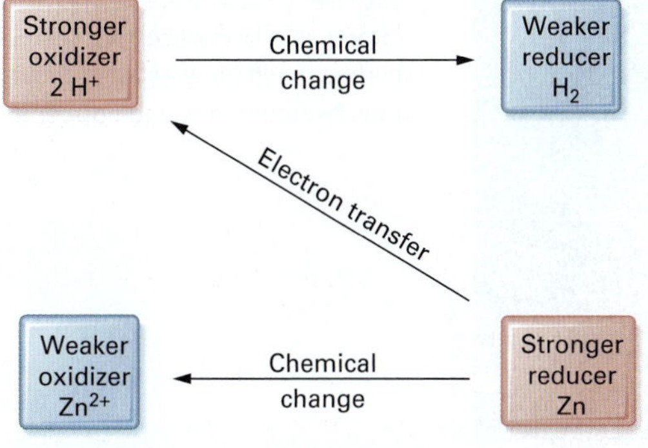

Figure 19.19 Predicting redox reactions from positions in Table 19.2. The spontaneous chemical change always transfers one or more electrons from the stronger reducing agent to the stronger oxidizing agent, both shown in pink. The products of the reaction, the weaker oxidizing and reducing agents, are shown in blue. The favored direction—forward or reverse—has the weaker oxidizing and reducing agents as products.

The reverse direction will be favored. Ag^+ is a stronger oxidizing agent than Co^{2+} and is therefore able to take electrons from cobalt atoms. Also, cobalt atoms are a stronger reducing agent than silver atoms and therefore readily release electrons to Ag^+. The favored direction is toward the weaker reducing and oxidizing agents, Ag and Co^{2+}.

State the correct direction and explain.

You improved your understanding of electron transfer reactions, and you improved your skill at predicting the favored direction of a redox reaction.

What did you learn by solving this Active Example?

Practice Exercise 19.9

In which direction will the redox reaction in Practice Exercise 19.8 be favored: $Sn^{2+}(aq) + 2\,Na(s) \rightleftharpoons Sn(s) + 2\,Na^+(aq)$?

Active Example 19.10 Predicting Redox Reactions I

Write the redox reaction equation for the reaction of metallic copper and a strong acid, H^+, forming copper(II) ions and hydrogen gas, and indicate the direction that is favored.

Think Before You Write A Relative Strengths Table, such as Table 19.2, is needed to write the equation and predict the favored direction.

Answers *Cover the left column with your cut-out shield. Reveal each answer only after you have written your own answer in the right column.*

Reduction: $2\,H^+(aq) + \cancel{2e^-} \rightleftharpoons H_2(g)$

Oxidation: $Cu(s) \rightleftharpoons Cu^{2+}(aq) + \cancel{2e^-}$

Redox: $2\,H^+(aq) + Cu(s) \rightleftharpoons H_2(g) + Cu^{2+}(aq)$

The reverse reaction is favored (**Figure 19.20**).

You will have to locate both half-reaction equations, reverse the appropriate equation to match the given reactants and products, and compare relative strengths. Write the half-reactions and sum them to produce the redox reaction equation. Indicate the favored direction of the reaction.

You improved your understanding of electron transfer reactions, and you improved your skill at writing and predicting the favored direction of a redox reaction.

What did you learn by solving this Active Example?

Practice Exercise 19.10

Will a reaction occur if a strip of aluminum is placed in a solution that contains nickel(II) ions? If yes, write the net ionic equation for the reaction. If no, explain.

Figure 19.20 Metallic copper does not react with a strong acid to release hydrogen because copper and hydrogen ion are weaker reducing and oxidizing agents than hydrogen gas and copper ion, respectively.

One of the properties of acids listed in Section 17.2 is their ability to release hydrogen gas on reaction with certain metals. Judging from Active Example 19.10, copper is not among those metals. The metals that do release hydrogen are the reducers below hydrogen in the right column of Table 19.2. But there is more to the reactions between metals and acids that meets the eye.

Active Example 19.11 Predicting Redox Reactions II

Write the equation between lead and nitric acid and predict which direction is favored.

Think Before You Write Lead is in Table 19.2, but you will search in vain for HNO_3. The major species present in nitric acid solution, H^+ and NO_3^-, are found in one equation, though (recall that nitric acid is a strong acid). We will comment on the imbalance between hydrogen ions and nitrate ions shortly.

Answers *Cover the left column with your cut-out shield. Reveal each answer only after you have written your own answer in the right column.*

2 × Reduction:

$$2\,NO_3^-(aq) + 8\,H^+(aq) + \cancel{6\,e^-} \rightleftharpoons 2\,NO(g) + 4\,H_2O(\ell)$$

3 × Oxidation:

$$3\,Pb(s) \rightleftharpoons 3\,Pb^{2+}(aq) + \cancel{6\,e^-}$$

Redox:

$$2\,NO_3^-(aq) + 8\,H^+(aq) + 3\,Pb(s) \rightleftharpoons 2\,NO(g) + 4\,H_2O(\ell) + 3\,Pb^{2+}(aq)$$

The forward reaction is favored.

Don't worry about those missing nitrate ions, the six unaccounted for from the eight moles of HNO_3 that furnished the eight H^+. They are present as spectators.

This reaction summarizes our equation-writing methods to this point. Take it all the way.

You improved your understanding of electron transfer reactions, and you improved your skill at writing and predicting the favored direction of a redox reaction.

What did you learn by solving this Active Example?

Practice Exercise 19.11

A silver wire is placed in a hydrochloric acid solution. Write the net ionic equation and predict the favored direction.

ACID–BASE-BEFORE-REDOX OPTION If you have already studied Chapter 17, you probably have noticed the several similarities between electron transfer reactions (redox) and proton transfer reactions (acid–base) that were mentioned earlier. These are summarized in Section 19.7, which is repeated at the appropriate place in Chapter 17.

19.8 Redox Reactions and Acid–Base Reactions Compared

Goal 11 Compare and contrast redox reactions with acid–base reactions.

We are now at the point where it will be useful to point out how acid–base reactions resemble redox reactions:

1. An acid–base reaction is a transfer of protons; a redox reaction is a transfer of electrons.

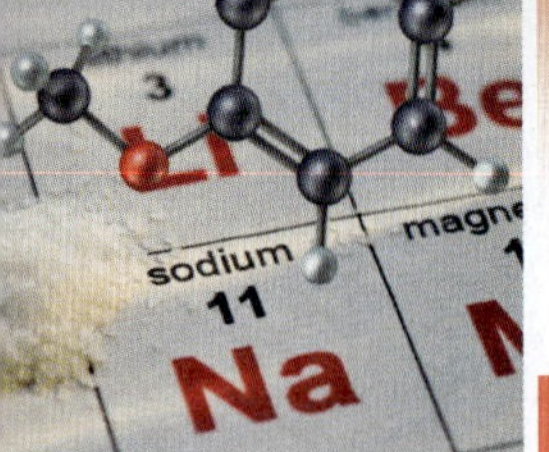

Everyday Chemistry 19.1

BATTERIES

Batteries are among the most common everyday applications of chemistry. Smart phones, watches and clocks, television remotes, toys, calculators, computers, smoke detectors, and automobiles are among the many common products that rely on battery power. Most citizens of industrialized nations can recall times when they became very aware of the role of batteries in their lives—those times when they encountered a dead battery!

Batteries are voltaic (or galvanic) cells. (Technically, the term *battery* only applies to a set of voltaic cells, but, in common usage, it describes both an individual cell and a set.) The same basic principles govern the operation of most common batteries. The reaction that occurs within a battery is an electron transfer, or redox, reaction. As with all redox reactions, an oxidizing agent and a reducing agent are required as well as electrodes and an electrolyte. The relative strengths of the oxidizing agent and reducing agent determine the characteristics of the battery (**Figure 19.21**).

Alkaline batteries (**Figure 19.22**) are the power source in television remote controls, many toys, and digital cameras. These are often classified by their electrical potential and shape and size, such as the round AA or AAA 1.5-volt batteries or the rectangular 9-volt batteries. The anode of an alkaline battery is a gel that contains zinc in contact with a potassium hydroxide solution. The presence of the hydroxide ion is what gives these batteries their name. The cathode is a mixture of manganese(IV) oxide and graphite. The reactions are:

Anode:

$$Zn(s) + \cancel{2\,OH^-(aq)} \rightleftharpoons ZnO(s) + \cancel{H_2O(\ell)} + \cancel{2\,e^-}$$

Cathode:

$$2\,MnO_2(s) + \cancel{H_2O(\ell)} + \cancel{2\,e^-} \rightleftharpoons Mn_2O_3(s) + \cancel{2\,OH^-(aq)}$$

Redox:

$$Zn(s) + 2\,MnO_2(s) \rightleftharpoons ZnO(s) + Mn_2O_3(s)$$

Charles D. Winters

Figure 19.21 There are many sizes and shapes of batteries. The chemical reactions used to provide electrical energy vary, too. Nonetheless, all batteries are based on electron transfer reactions.

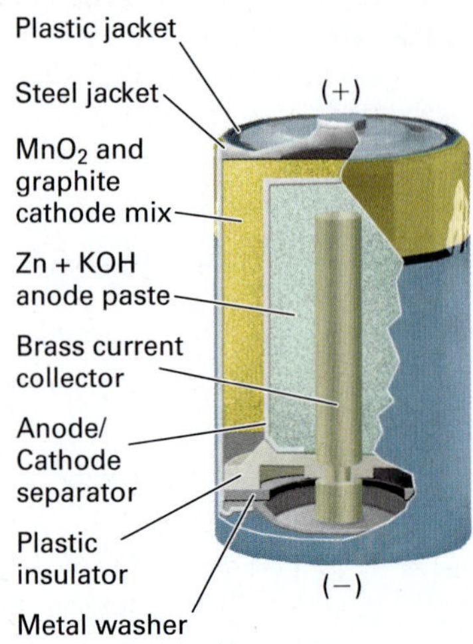

Figure 19.22 A cutaway view of an alkaline battery. It is estimated that over 10 billion alkaline batteries are produced each year.

Lead storage batteries (**Figure 19.23**) are most frequently used to start cars. The most common configuration within such a battery is a series of six cells, each with a potential of 2 volts, for a total battery potential of 12 volts. The name *lead battery* is derived from both the anode, which is a collection of lead plates, and the cathode, lead(IV) oxide plates. The relatively high density of lead, 11.3 g/cm^3, makes lead storage batteries heavy. The electrolyte in these batteries is a sulfuric acid solution. The oxidation–reduction reactions are:

Anode:

$$Pb(s) + SO_4^{2-}(aq) \rightleftharpoons PbSO_4(s) + \cancel{2\,e^-}$$

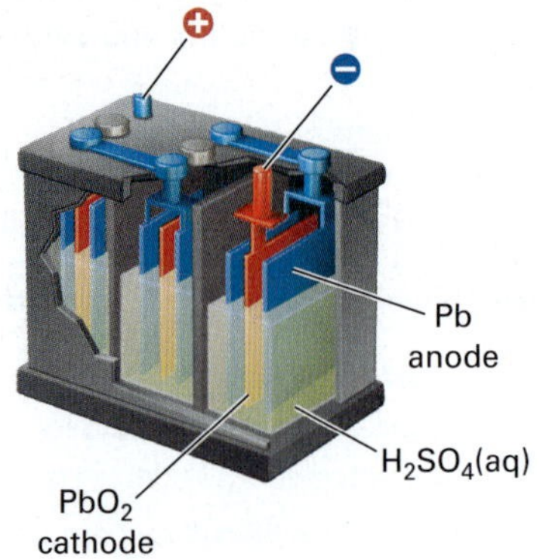

Figure 19.23 A cutaway view of a lead storage battery. Nearly every automobile in the world uses a lead battery to power its starter.

Cathode:

$$PbO_2(s) + 4\,H^+(aq) + SO_4^{2-}(aq) + \cancel{2\,e^-} \rightleftharpoons PbSO_4(s) + 2\,H_2O(\ell)$$

Redox:

$$Pb(s) + PbO_2(s) + 4\,H^+(aq) + 2\,SO_4^{2-}(aq) \rightleftharpoons 2\,PbSO_4(s) + 2\,H_2O(\ell)$$

One of the most important features of a lead–acid storage battery is its ability to be recharged. A running engine powers an alternator that is designed to pump electrons through the battery, forcing the chemical reaction to run backward, regenerating the reactants:

$$2\,PbSO_4(s) + 2\,H_2O(\ell) \rightleftharpoons Pb(s) + PbO_2(s) + 4\,H^+(aq) + 2\,SO_4^{2-}(aq)$$

From this point forward in your life, whenever you encounter a battery, we hope you will "see" the voltaic cell that lies beneath the metal or plastic casing. The electric power provided to your cell phone, television remote, car starter, or any other battery-powered device comes from electron transfer reactions.

Quick Quiz

1. **How many batteries are within a few steps of you right now? How many times each day do you rely on a battery as a source of electrical energy?**
2. **How is chemical energy changed into electrical energy in a battery?**

2. In both cases, the reactants are given special names to indicate their roles in the transfer process. An acid is a proton source; a base is a proton remover. A reducing agent is an electron source; an oxidizing agent is an electron remover.
3. Just as certain species can either provide or remove protons (e.g., HCO_3^- and H_2O) and thereby behave as an acid in one reaction and a base in another, certain species can either remove or provide electrons, acting as an oxidizing agent in one reaction and a reducing agent in another. An example is the Fe^{2+} ion, which can oxidize Zn atoms to Zn^{2+} in the reaction

$$Fe^{2+}(aq) + Zn(s) \rightarrow Zn^{2+}(aq) + Fe(s)$$

Fe^{2+} can also reduce Cl_2 molecules to Cl^- ions in another reaction:

$$2\,Fe^{2+}(aq) + Cl_2(g) \rightarrow 2\,Cl^-(aq) + 2\,Fe^{3+}(aq)$$

4. Just as acids and bases may be classified as stronger or weaker depending on how readily they remove or provide protons, the relative strengths of oxidizing and reducing agents may be compared according to their tendencies to attract or release electrons.
5. Just as most acid–base reactions in solution reach a state of equilibrium, most aqueous redox reactions also reach equilibrium. Just as the favored side of an acid–base equilibrium can be predicted from acid–base strengths, the favored side of a redox equilibrium also can be predicted from oxidizing agent–reducing agent strengths.

19.9 Writing Redox Equations (Optional)

Goal 12 Given the before and after formulas of species containing elements that are oxidized and reduced in an acidic solution, write the oxidation and reduction half-reaction equations and the net ionic equation for the reaction.

Thus far, we have considered only redox reactions for which the oxidation and reduction half-reactions are known. We are not always this fortunate. Sometimes we know only the species that contain the elements actually oxidized or reduced. We call the half-reaction equation that contains only these oxidized and reduced species a **skeleton equation.** Considering nitric acid, for example, suppose we know only that the product of the reduction of nitric acid is NO(g). How do we get from this information to the reduction half-reaction in Table 19.2?

The steps for writing a half-reaction equation in an acidic solution are listed in the following *how to* ..... Each step is illustrated for the NO_3^--to-NO change in Active Example 19.11.

how to... **Write Redox Equations for Half-Reactions in Acidic Solutions**

1. After identifying the element oxidized or reduced, write a skeleton half-reaction equation with the element in its original form (element, monatomic ion, or part of a polyatomic ion or compound) on the left and in its final form on the right.	$NO_3^-(aq) \rightleftharpoons NO(g)$
2. Balance the element oxidized or reduced.	Nitrogen is already balanced, one nitrogen atom on each side.
3. Balance elements other than hydrogen or oxygen, if any.	There are none.

4. Balance oxygen by adding water molecules when necessary.	There are three oxygens on the left and one on the right. Two water molecules are needed on the right: $NO_3^-(aq) \rightleftharpoons NO(g) + 2\ H_2O(\ell)$
5. Balance hydrogen by adding H^+ when necessary.	There are four hydrogens on the right and none on the left. Four hydrogen ions are needed on the left: $4\ H^+(aq) + NO_3^-(aq) \rightleftharpoons NO(g) + 2\ H_2O(\ell)$
6. Balance charge by adding electrons to the more positive side.	Total charge on the left is (4+) + (1−) = 3+; on the right, it is zero. Three electrons are needed on the left: $3\ e^- + 4\ H^+(aq) + NO_3^-(aq) \rightleftharpoons NO(g) + 2\ H_2O(\ell)$
7. Recheck the equation to be sure it is balanced in both atoms and charge.	Atoms: 4 H, 1 N, and 3 O on each side. Charge: (3−) + (4+) + (1−) = 0 balances 0 on the right.

Lloyd Sutton/Alamy Stock Photo

Figure 19.24 Large deposits of sulfur have been discovered in the United States, Mexico, and Poland. After the element is mined, it is shaped into large blocks for shipment.

Notice that these instructions are for redox half-reactions in *acidic* solutions. The procedure is somewhat similar with basic solutions, but we will omit that procedure in this introductory text.

When you have developed both of the half-reaction equations for a redox reaction, proceed as in the earlier Active Examples.

Active Example 19.12 Writing Redox Equations I

Write the net ionic equation for the redox reaction between iodide ion and sulfate ion in an acidic solution. The products are iodine and sulfur (Figure 19.24). The skeleton equation is $I^-(aq) + SO_4^{2-}(aq) \rightleftharpoons I_2(s) + S(s)$.

Think Before You Write We will guide you through the procedure for writing redox equations in acidic solutions in this Active Example.

Answers *Cover the left column with your cut-out shield. Reveal each answer only after you have written your own answer in the right column.*

Sulfur is reduced from an oxidation number of +6 to 0. Iodine is oxidized from −1 to 0.	First, assign oxidation numbers to each element, and then identify the element reduced and the element oxidized. 
$2\ I^-(aq) \rightleftharpoons I_2(s) + 2\ e^-$ *(This one happens to be in Table 19.2.)*	Balance atoms first, then charges, for the oxidation half reaction, $I^-(aq) \rightleftharpoons I_2(s)$. 
$SO_4^{2-}(aq) \rightleftharpoons S(s) + 4\ H_2O(\ell)$ *Four oxygen atoms in a sulfate ion require four water molecules.*	Now start working on the reduction half-reaction, $SO_4^{2-}(aq) \rightleftharpoons S(s)$. Sulfur is already in balance. The only other element is oxygen. According to Step 4, you balance oxygen by adding the necessary water molecules. Complete that step. 
$8\ H^+(aq) + SO_4^{2-}(aq) \rightleftharpoons S(s) + 4\ H_2O(\ell)$ *Eight hydrogen atoms on the right require adding 8 H^+ to the left.*	Next comes the hydrogen balancing of the sulfate ion–sulfur half-reaction, using H^+ ions. 

$$6\,e^- + 8\,H^+(aq) + SO_4^{2-}(aq) \rightleftharpoons S(s) + 4\,H_2O(\ell)$$

On the left, there are eight positive charges from hydrogen ion and two negative charges from sulfate ion, a net of 6+. On the right, the net charge is zero. Charge is balanced by adding six electrons to the left (positive) side.

To finish the sulfate ion–sulfur half-reaction, add to the positive side the electrons that will bring the charges into balance.

Reduction:

$$\cancel{6e^-} + 8\,H^+(aq) + SO_4^{2-}(aq) \rightleftharpoons S(s) + 4\,H_2O(\ell)$$

3 × Oxidation: $$6\,I^-(aq) \rightleftharpoons 3\,I_2(s) + \cancel{6e^-}$$

Redox:

$$8\,H^+(aq) + SO_4^{2-}(aq) + 6\,I^-(aq) \rightleftharpoons S(s) + 4\,H_2O(\ell) + 3\,I_2(s)$$

Now that you have the two half-reaction equations, finish writing the net ionic equation as you did before by equalizing the number of electrons and adding the half-reactions.

The atoms balance: six I, four O, one S, and eight H atoms on each side.

The charge balances: [(8+) + (2−) + (6−)] = 0 on the left. The charge is also 0 on the right.

Check to make sure that the equation is indeed balanced in both atoms and charge.

You improved your understanding of electron transfer reactions, and you improved your skill at writing redox reactions.

What did you learn by solving this Active Example?

Practice Exercise 19.12

Aqueous chromate ion, $CrO_4^{2-}(aq)$, and hydrogen sulfide gas react to produce aqueous chromium(III) ion and solid sulfur in an acidic solution. Write the net ionic equation for the reaction.

Active Example 19.13 Writing Redox Equations II

The permanganate ion, MnO_4^-, is a strong oxidizing agent that oxidizes chloride ion to chlorine in an acidic solution (Figure 19.25). Manganese ends up as a monatomic manganese(II) ion. Write the net ionic equation for the redox reaction.

Think Before You Write This is a challenging Active Example but watch how it falls into place when you follow the *how to...* procedure that has been outlined.

Answers *Cover the left column with your cut-out shield. Reveal each answer only after you have written your own answer in the right column.*

$$MnO_4^-(aq) + Cl^-(aq) \rightleftharpoons Cl_2(g) + Mn^{2+}(aq)$$

To be sure you have correctly interpreted the question, begin by writing an unbalanced skeleton equation. Put the identified reactants on the left side and the identified products on the right side.

$$2\,Cl^-(aq) \rightleftharpoons Cl_2(g) + 2\,e^-$$

With a switch from iodine to chlorine, this is the same oxidation half-reaction as in the last Active Example.

The oxidation half-reaction is easiest. Write and balance it next.

Figure 19.25 A purple solution containing permanganate ion is poured into a colorless acidic solution that contains chloride ion.

$MnO_4^-(aq) \rightleftharpoons Mn^{2+}(aq)$	Now write the formulas of the starting and ending species for the reduction half-reaction on opposite sides of a double arrow. 
$8\ H^+(aq) + MnO_4^-(aq) + 5\ e^- \rightleftharpoons Mn^{2+}(aq) + 4\ H_2O(\ell)$ *Here's how we did it:* *Oxygen:* $MnO_4^-(aq) \rightleftharpoons Mn^{2+}(aq) + 4\ H_2O(\ell)$ *(four waters for four oxygen atoms)* *Hydrogen:* $8\ H^+(aq) + MnO_4^-(aq) \rightleftharpoons Mn^{2+}(aq) + 4\ H_2O(\ell)$ *(eight* H^+ *for four waters)* *Charge: (8+) + (1−) = 7+ on the left; 2+ on the right. Charge is balanced by adding five negatives on the left, or five electrons, arriving at the final half-reaction equation.*	Can you take the skeleton reduction equation to a complete half-reaction equation? First balance oxygen, then balance hydrogen, and then balance charge. 
2 × Reduction: $16\ H^+(aq) + 2\ MnO_4^-(aq) + \cancel{10\ e^-} \rightleftharpoons 2\ Mn^{2+}(aq) + 8\ H_2O(\ell)$ 5 × Oxidation: $10\ Cl^-(aq) \rightleftharpoons 5\ Cl_2(g) + \cancel{10\ e^-}$ Redox: $16\ H^+(aq) + 2\ MnO_4^-(aq) + 10\ Cl^-(aq) \rightleftharpoons 2\ Mn^{2+}(aq) + 8\ H_2O(\ell) + 5\ Cl_2(g)$	Rewrite and combine the half-reaction equations to produce the net ionic equation. 
The atoms balance: 16 H, 2 Mn, 8 O, and 10 Cl. The charge balances: (16+) + (2−) + (10−) = 4+ on the left. 2(2+) = 4+ on the right. *Can you imagine a trial-and-error approach to an equation such as this?*	Check to make sure the equation is balanced in both atoms and charge. 
You improved your understanding of electron transfer reactions, and you improved your skill at writing redox reactions.	What did you learn by solving this Active Example? 

Practice Exercise 19.13

Gaseous sulfur dioxide is bubbled through an acidic solution that contains permanganate ion, MnO_4^-. Aqueous sulfate ion and manganese(II) ion form. Develop the half-reaction equations and combine them to produce the net ionic equation.

Charles D. Winters

CHAPTER 19 IN REVIEW: OXIDATION–REDUCTION (ELECTRON TRANSFER) REACTIONS

Goal 1 Describe and explain oxidation and reduction in terms of electron transfer.	**Oxidation** is a loss of electrons and an increase in oxidation number. **Reduction** is a gain of electrons and a reduction in oxidation number.
Goal 2 Given an oxidation half-reaction equation and a reduction half-reaction equation, combine them to form a net ionic equation for an oxidation–reduction reaction.	Oxidation–reduction (redox) reactions can be divided into an oxidation **half-reaction** equation and a reduction half-reaction equation. Addition of these equations gives a balanced equation for a redox reaction.
Goal 3 Distinguish among electrolytic cells, voltaic cells, and galvanic cells.	A **voltaic cell** is a cell in which an electrical potential is developed by a spontaneous chemical change. It is also called a **galvanic cell.** An **electrolytic cell** is a cell in which electrolysis occurs as a result of an externally applied electrical potential.
Goal 4 Describe and identify the parts of an electrolytic or voltaic (galvanic) cell and explain how it operates.	An electrolytic cell is made up of a container holding an ionic solution called an **electrolyte** and two **electrodes.** Ions in the electrolyte move to oppositely charged electrodes, where chemical reactions occur. The **anode** is the electrode at which oxidation occurs; reduction occurs at the **cathode.** In a voltaic cell, chemical changes occur at the electrodes, causing electricity to flow in an outside circuit. The solutions in each half-cell are connected by a **salt bridge.**
Goal 5 Given the formula of an element, molecule, or ion, or information from which the formula can be determined, assign an oxidation number to each element in the formula.	**Oxidation numbers** are assigned as follows: an element is 0; a monatomic ion is the charge on the ion; combined oxygen is -2 (peroxides, -1; superoxides, $-1/2$; OF_2, $+2$); combined hydrogen is -1 (H^-, -1); the sum of oxidation numbers of all atoms in a polyatomic species is equal to its charge.
Goal 6 Describe and explain oxidation and reduction in terms of change in oxidation numbers.	Oxidation is an increase in oxidation number. Reduction is a decrease in oxidation number.
Goal 7 Given a redox equation, or information from which the equation may be written, identify the oxidizing agent, the reducing agent, the element oxidized, and the element reduced.	The species that removes electrons in a redox reaction—that is, the species that is itself **reduced**—is referred to as an **oxidizing agent**. The species from which the electrons are removed so that reduction of another element can occur is called a **reducing agent**. Thus, the reducing agent is itself **oxidized.**
Goal 8 Distinguish between strong and weak oxidizing agents.	A **strong oxidizing agent** has a strong attraction for electrons. A **weak oxidizing agent** attracts electrons only slightly. A **strong reducing agent** releases electrons readily. A **weak reducing agent** holds on to its electrons.
Goal 9 Given a table of the relative strengths of oxidizing and reducing agents, arrange a group of oxidizing agents or a group of reducing agents in order of increasing or decreasing strength.	Table 19.2 lists oxidizing agents in order of decreasing strength on the left side of the half-reaction equation and lists reducing agents in order of increasing strength on the right side.
Goal 10 Given a table of the relative strengths of oxidizing agents, and information from which an electron transfer reaction equation between two species in the table may be written, write the equation and predict the direction in which the reaction will be favored.	When a reversible reaction equation is read from left to right, the **forward reaction**, or the reaction in the **forward direction**, is described; from right to left, the change is the **reverse reaction**, or in the **reverse direction**. A strong oxidizing agent takes electrons from a strong reducing agent to produce weaker oxidizing and reducing agents. The reaction is said to be **favored** in the direction pointing to the weaker oxidizing and reducing agents. Use Table 19.2 to identify the stronger and weaker oxidizing and reducing agents.

Goal 11 Compare and contrast redox reactions with acid–base reactions.

Acid–base reactions resemble redox reactions in the following ways:

1. Transfer of a subatomic particle: An acid–base reaction is a transfer of protons; a redox reaction is a transfer of electrons.
2. Special names for the species involved in the transfer: An acid is a proton source; a base is a proton remover. A reducing agent is an electron source; an oxidizing agent is an electron remover.
3. Species that can both remove and release subatomic particles, depending on relative strength.
4. Varying strengths of abilities to remove and release subatomic particles.
5. Most systems are equilibrium reactions; the favored direction of reaction can be predicted by comparison of relative strengths.

Goal 12 Given the before and after formulas of species containing elements that are oxidized and reduced in an acidic solution, write the oxidation and reduction half-reaction equations and the net ionic equation for the reaction.

To **write a redox equation for a half-reaction in acidic solution:**

1. After identifying the element oxidized or reduced, write a skeleton half-reaction equation with the element in its original form on the left and in its final form on the right.
2. Balance the element oxidized or reduced.
3. Balance elements other than hydrogen or oxygen, if any.
4. Balance oxygen by adding water molecules when necessary.
5. Balance hydrogen by adding H^+ when necessary.
6. Balance charge by adding electrons to the more positive side.
7. Recheck the equation to be sure it is balanced in both atoms and charge.

Key Terms

Most of the key terms and concepts and many others appear in the Glossary. Use your Glossary regularly.

anode p. 757
cathode p. 757
electrochemical cell p. 751
electrodes p. 751
electrolysis p. 757
electrolyte p. 757
electrolytic cell p. 757
electron transfer reaction p. 752
favored (direction of an equilibrium reaction) p. 767
forward reaction (forward direction of reaction) p. 767
galvanic cell p. 757
half-cells p. 757
half-reaction p. 751
half-reaction equation p. 752
oxidation p. 751
oxidation number p. 759
oxidizing agent p. 763
reducing agent p. 763
reduction p. 752
reverse reaction (reverse direction of reaction) p. 767
salt bridge p. 751
skeleton equation p. 771
strong oxidizing agent p. 764
strong reducing agent p. 764
voltaic cell p. 757
weak oxidizing agent p. 764
weak reducing agent p. 764

Frequently Asked Questions

How can I avoid confusing the definitions of oxidation and reduction?
Oxidation and reduction are so closely related and similarly defined that they are easy to confuse. *Oxidation* is not a common word outside its chemical sense but *reduction* is—and we can take advantage of it. If something is reduced, it gets smaller. If an element is reduced, its oxidation number becomes smaller. Oxidation is the opposite. Watch out for negative numbers that get smaller: "getting smaller" means becoming more negative, as from −1 to −3. "Becoming more negative" helps in the gain-or-loss-of-electron definition, too. "Becoming more negative" means getting more negative charge, or gaining negatively charged electrons. Oxidation is the opposite—that is, losing electrons.

When referring to oxidizing and reducing agents, is it the element or the entire species that acts as the agent?
Students sometimes summarize the relationship between species oxidized/reduced with oxidizing/reducing agents by saying, "Whatever is oxidized is the reducing agent, and whatever is reduced is the oxidizing agent." *Caution:* This is true for a *monatomic species only.* If the element being oxidized/reduced is part of a polyatomic species, the entire compound or ion is the reducing/oxidizing agent.

(Optional Section 19.8) I've found a different method for balancing complicated redox reactions, and I get the same answers as those in the book. Is it OK to use this method?
There are several ways to balance complicated redox equations. We have shown you only one, and that is only for acidic solutions. If your instructor prefers another method, by all means, use it. Whatever method you use, it takes practice to prefect it. You may question this while learning, but many students report that once they get the hang of it, balancing redox equations is fun!

Concept-Linking Exercises

Write a brief description of the relationships among each of the following groups of terms or phrases. Answers to the Concept-Linking Exercises are given at the end of the chapter.

1. Electrolytic cell, electrolyte, voltaic cell, electrolysis, galvanic cell, electrode
2. Oxidation, reduction, oxidation half-reaction, reduction half-reaction
3. Oxidation, reduction, oxidation number
4. Oxidizing agent, reducing agent, oxidizer, reducer
5. Strong oxidizing agent, weak oxidizing agent, strong reducing agent, weak reducing agent

Small-Group Discussion Questions

Small-Group Discussion Questions are for group work, either in class or under the guidance of a leader during a discussion section.

1. The oxidation–reduction process can be described as burning and "unburning." Determine which is which, and explain why these terms can be applied.
2. A voltaic cell is constructed with metal M in a solution containing M^{2+} ions and metal Me in a solution containing Me^{2+} ions. The metals are attached via a conducting wire, and the solutions are connected by a salt bridge. As the cell runs, the color of the solution containing M^{2+} gets lighter, and the color of the solution containing Me^{2+} gets darker. (a) Sketch this cell. (b) Write an equation for each half-reaction. (c) On your sketch, indicate the direction in which electrons travel through the wire. (d) On your sketch, indicate the direction in which anions travel through the salt bridge and solutions. (e) Label the anode and cathode on your sketch, and indicate whether each is positive or negative. (f) What is oxidized and what is reduced? (g) What are the oxidizing and reducing agents? (h) Describe what happens to the mass of each metal as the cell runs.
3. Figure 19.11 is a macroscopic-level illustration of the electrolysis of liquid sodium chloride. Sketch a particulate level illustration of the chemical changes that occur as this cell runs.
4. Vitamins such as vitamins C and E are often described as good for you because they are antioxidants. Which of the terms that were introduced in this chapter (oxidized, reduced, oxidizing agent, reducing agent) describe the functions of antioxidants in the body? Explain.
5. Experiments are conducted with four metals, A, B, C, and D, and solutions of their ions. When solid D is added to solutions of the ions of the other metals, metallic A, B, and C are formed. When metal A is added to a solution containing ions of C, ions of A form along with metallic C. When hydrochloric acid is dropped on each metal, B and D react to form hydrogen gas. A and C do not react with hydrochloric acid. List the metals A, B, C, and D in order of their strength as reducing agents, and explain your reasoning.
6. Steel objects can be coated with zinc in a process called *galvanization*. In terms of the relative strengths of oxidizing and reducing agents, why is this done?
7. A fuel cell is an electrochemical cell that has a continual supply of reactants. An example is a hydrogen–oxygen fuel cell, where the half-reaction $H_2(g) \rightleftharpoons 2\,H^+(aq) + 2\,e^-$ occurs at the anode and the half-reaction $O_2(g) + 4\,H^+(aq) + 4\,e^- \rightleftharpoons 2\,H_2O(\ell)$ occurs at the cathode. (a) What are the fuels in a hydrogen–oxygen fuel cell? (b) Write the net ionic equation for the overall reaction that occurs in a fuel cell? (c) How does a fuel cell resemble a battery? How is it different? (d) What are the oxidation and reduction half-reactions in a hydrogen–oxygen fuel cell? (e) What advantage do the products of a hydrogen–oxygen fuel cell have over the products emitted by a conventional gasoline-burning engine? (f) What are the disadvantages of a hydrogen–oxygen fuel cell as compared with a gasoline engine?
8. Plants produce simple sugar by using energy from the sun in a process known as photosynthesis, which can be summarized as $6\,CO_2 + 6\,H_2O + \text{energy} \rightarrow C_6H_{12}O_6 + 6\,O_2$. Animals metabolize simple sugar as a source of energy: $C_6H_{12}O_6 + 6\,O_2 \rightarrow 6\,CO_2 + 6\,H_2O + \text{energy}$. Explain the photosynthesis and metabolism of simple sugar in terms of oxidation and reduction. How would life on Earth be altered if photosynthesis did not occur?
9. In Figure 19.11, which illustrates the electrolysis of liquid sodium chloride, we show a battery as a power source. It would have been more intuitive to show the apparatus plugged into a standard electrical outlet, but such a power source does not work for electrolysis because it supplies alternating current. Why can't alternating current be used for electrolytic cells? What do you suppose is used in industry as a power source for large-scale electroplating?

Questions, Exercises, and Problems

Interactive versions of these problems may be assigned by your instructor. Solutions for blue-numbered questions are at the end of the chapter. Questions other than those in the General Questions and More Challenging Problems sections are paired in consecutive odd-even number combinations; solutions for the odd numbered questions are also at the end of the chapter.

Section 19.2: Electron Transfer Reactions

1. Using any example of a redox reaction, explain why such reactions are described as electron transfer reactions.
2. Identify the species oxidized and the species reduced in the following electron transfer reaction: $Mg(s) + Co^{2+}(aq) \rightleftharpoons$

$Mg^{2+}(aq) + Co(s)$. As the reaction proceeds, electrons are transferred from ______ to ______.

3. Classify each of the half-reaction equations given as an oxidation or reduction half-reaction equation:
   a) $Zn \rightleftharpoons Zn^{2+} + 2\,e^-$
   b) $2\,H^+ + 2\,e^- \rightleftharpoons H_2$
   c) $Fe^{2+} \rightleftharpoons Fe^{3+} + e^-$
   d) $NO + 2\,H_2O \rightleftharpoons NO_3^- + 4\,H^+ + 3\,e^-$
4. Classify each of the half-reaction equations given as an oxidation or reduction half-reaction equation:
   a) $2\,Cl^- \rightleftharpoons Cl_2 + 2\,e^-$
   b) $Na \rightleftharpoons Na^+ + e^-$
   c) $Sn^{2+} \rightleftharpoons Sn^{4+} + 2\,e^-$
   d) $O_2 + 4\,H^+ + 4\,e^- \rightleftharpoons 2\,H_2O$
5. Classify each of the half-reaction equations given as an oxidation or a reduction half-reaction equation:
   a) Dissolving ozone in water: $O_3 + H_2O + 2\,e^- \rightleftharpoons O_2 + 2\,OH^-$
   b) Tarnishing of silver: $2\,Ag + S^{2-} \rightleftharpoons Ag_2S + 2\,e^-$
6. Classify each of the half-reaction equations given as an oxidation or a reduction half-reaction equation:
   a) Dissolving gold ($Z = 79$): $Au + 4\,Cl^- \rightleftharpoons AuCl_4^- + 3\,e^-$
   b) One side of an automobile battery: $PbO_2 + SO_4^{2-} + 4\,H^+ + 2\,e^- \rightleftharpoons PbSO_4 + 2\,H_2O$

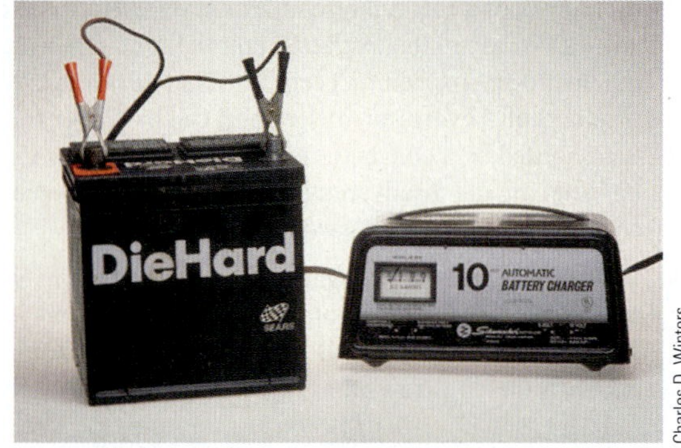

Charles D. Winters

When an automobile battery is charged, the chemical change runs in the opposite direction as when the battery is discharging.

7. Combine the following half-reaction equations to produce a balanced redox equation: $Cr \rightleftharpoons Cr^{3+} + 3\,e^-$ and $Cl_2 + 2\,e^- \rightleftharpoons 2\,Cl^-$
8. For the electron transfer reaction $2\,Ag(s) + Br_2(\ell) \rightleftharpoons 2\,Ag^+(aq) + 2\,Br^-(aq)$, write the oxidation half-reaction and the reduction half-reaction.
9. The half-reactions that take place at the electrodes of an alkaline cell, widely used in television remote controls and other devices, are: $NiOOH + H_2O + e^- \rightleftharpoons Ni(OH)_2 + OH^-$ and $Cd + 2\,OH^- \rightleftharpoons Cd(OH)_2 + 2\,e^-$. Which equation is for the oxidation half-reaction? Write the overall equation for the cell.

Charles D. Winters

Alkaline batteries are manufactured in a variety of sizes.

10. Identify each of the following half-reactions as either an oxidation half-reaction or a reduction half-reaction: $Fe^{2+}(aq) \rightleftharpoons Fe^{3+}(aq) + e^-$; $Cl_2(g) + 2\,e^- \rightleftharpoons 2\,Cl^-(aq)$. Write a balanced equation for the overall redox reaction.

Section 19.3: Voltaic and Electrolytic Cells

11. List as many things in your home as you can that are operated by voltaic cells.
12. How does electrolysis differ from the passage of electric current through a wire?
13. Can a galvanic cell operate an electrolytic cell? Explain.
14. Can an electrolytic cell operate a voltaic cell? Explain.

Section 19.4: Oxidation Numbers and Redox Reactions

15. Give the oxidation number of the element whose symbol is underlined:
    a) $\underline{Al}^{3+}$, $\underline{S}^{2-}$, $\underline{S}O_3^{2-}$, $Na_2\underline{S}O_4$
    b) $\underline{C}O_2$, $\underline{Cr}_2O_7^{2-}$, $\underline{Cl}^-$
16. Give the oxidation number of the element whose symbol is underlined:
    a) $\underline{N}_2O_3$, $\underline{N}O_3^-$, $\underline{Cr}O_4^{2-}$, $NaH_2\underline{P}O_4$
    b) $H_3\underline{As}O_4$, $\underline{N}H_2OH$, $\underline{Ni}$

Questions 17–22: (1) identify the element experiencing oxidation or reduction, (2) state "oxidized" or "reduced," and (3) show the change in oxidation number. Example: $2\,Cl^- \rightleftharpoons Cl_2 + 2\,e^-$. Chlorine oxidized from −1 to 0.

17. a) $Br_2 + 2\,e^- \rightleftharpoons 2\,Br^-$
    b) $Pb^{2+} + 2\,H_2O \rightleftharpoons PbO_2 + 4\,H^+ + 2\,e^-$
18. a) $Cu^{2+} + 2\,e^- \rightleftharpoons Cu$
    b) $Co^{3+} + e^- \rightleftharpoons Co^{2+}$
19. a) $8\,H^+ + IO_4^- + 8\,e^- \rightleftharpoons I^- + 4\,H_2O$
    b) $4\,H^+ + O_2 + 4\,e^- \rightleftharpoons 2\,H_2O$
20. a) $H_2O + SO_3^{2-} \rightleftharpoons SO_4^{2-} + 2\,H^+ + 2\,e^-$
    b) $PH_3 \rightleftharpoons P + 3\,H^+ + 3\,e^-$

21. a) $NO_2 + H_2O \rightleftharpoons NO_3^- + 2\,H^+ + e^-$
b) $2\,Cr^{3+} + 7\,H_2O \rightleftharpoons Cr_2O_7^{2-} + 14\,H^+ + 6\,e^-$

22. a) $2\,HF \rightleftharpoons F_2 + 2\,H^+ + 2\,e^-$
b) $MnO_4^{2-} + 2\,H_2O + 2\,e^- \rightleftharpoons MnO_2 + 4\,OH^-$

Section 19.5: Oxidizing Agents and Reducing Agents

23. Identify the oxidizing and reducing agents in $Br_2 + 2\,I^- \rightleftharpoons 2\,Br^- + I_2$.

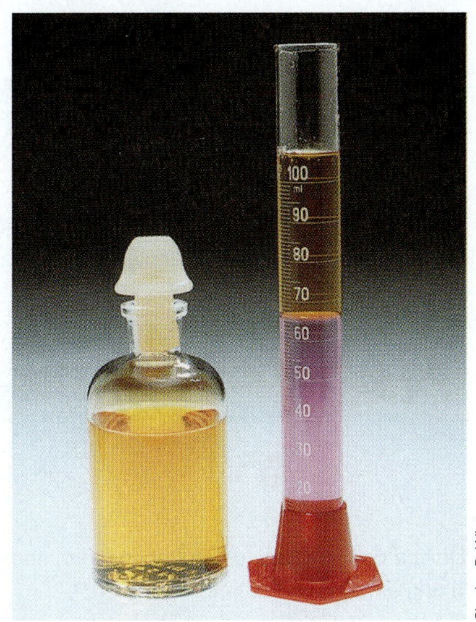

When an aqueous bromine solution (*orange*) is added to a solution containing iodide ion, bromide ion and iodine (*purple*) are formed. Two different solvents are used to separate the products.

24. For the redox reaction $SO_4^{2-} + Ni^{2+} \rightleftharpoons SO_2 + NiO_2$, assign oxidation numbers and use them to identify the element oxidized, the element reduced, the oxidizing agent, and the reducing agent.

25. What is the oxidizing agent in the equation for the lead storage battery: $Pb + PbO_2 + 4\,H^+ + 2\,SO_4^{2-} \rightleftharpoons 2\,PbSO_4 + 2\,H_2O$? What does it oxidize? Name the reducing agent and the species it reduces.

26. For the redox reaction $3\,Mn^{2+} + 2\,HNO_3 + 2\,H_2O \rightleftharpoons 3\,MnO_2 + 2\,NO + 6\,H^+$, assign oxidation numbers and use them to identify the element oxidized, the element reduced, the oxidizing agent, and the reducing agent.

Section 19.6: Strengths of Oxidizing Agents and Reducing Agents

27. Which is the stronger reducer, Zn or Fe^{2+}? On what basis do you make your decision? What is the significance of one reducer being stronger than another?

28. Consider the following half-reactions: $F_2(g) + 2\,e^- \rightleftharpoons 2\,F^-(aq)$ and $Cu^{2+}(aq) + e^- \rightleftharpoons Cu^+(aq)$ and $Fe^{2+}(aq) + 2\,e^- \rightleftharpoons Fe(s)$. Identify the strongest and weakest oxidizing agents and the strongest and weakest reducing agents.

29. Arrange the following oxidizers in order of increasing strength—that is, the weakest oxidizing agent first: Na^+, Br_2, Fe^{2+}, Cu^{2+}.

30. Consider the following half-reactions: $Cd^{2+}(aq) + 2\,e^- \rightleftharpoons Cd(s)$; $Pb^{2+}(aq) + 2\,e^- \rightleftharpoons Pb(s)$; $Cl_2(g) + 2\,e^- \rightleftharpoons 2\,Cl^-(aq)$. Identify the strongest and weakest oxidizing agents and the strongest and weakest reducing agents.

Section 19.7: Predicting Redox Reactions

In this section, write the redox equation for the redox reactants given, using Table 19.2 as a source of the required half-reactions. Then predict the direction in which the reaction will be favored at equilibrium.

31. $Br_2 + I^- \rightleftharpoons$

32. $Ni^{2+} + Cd \rightleftharpoons$

33. $H^+ + Br^- \rightleftharpoons$

34. $Cl_2 + Fe \rightleftharpoons$

35. $NO + H_2O + Fe^{2+} \rightleftharpoons$

36. $H^+ + Mg \rightleftharpoons$

Section 19.8: Redox Reactions and Acid–Base Reactions Compared

37. Explain how a strong acid is similar to a strong reducing agent. Explain how a strong base compares with a strong oxidizing agent.

38. Show how redox and acid–base reactions parallel each other—how they are similar, and also what makes them different.

Section 19.9: Writing Redox Equations (Optional)

In this section, each skeleton equation identifies an oxidizer and a reducer, as well as the oxidized and reduced products of the redox reaction. Write separate oxidation and reduction half-reaction equations, assuming that the reaction takes place in an acidic solution, and add them to produce a balanced redox equation.

39. $S_2O_3^{2-} + Cl_2 \rightleftharpoons SO_4^{2-} + Cl^-$

40. $Zn + Sb_2O_5 \rightleftharpoons Zn^{2+} + SbO^+$

41. $Sn + NO_3^- \rightleftharpoons H_2SnO_3 + NO_2$

42. $Ni^{2+} + Sn^{2+} \rightleftharpoons NiO_2 + Sn$

43. $C_2O_4^{2-} + MnO_4^- \rightleftharpoons CO_2 + Mn^{2+}$

44. $Cl^- + NO_3^- \rightleftharpoons ClO_3^- + NO$

45. $Cr_2O_7^{2-} + NH_4^+ \rightleftharpoons Cr_2O_3 + N_2$

46. $SO_4^{2-} + Hg \rightleftharpoons SO_2 + Hg^{2+}$

47. $As_2O_3 + NO_3^- \rightleftharpoons AsO_4^{3-} + NO$

48. $Pb^{2+} + Zn^{2+} \rightleftharpoons Zn + PbO_2$

General Questions

49. Distinguish precisely, and in scientific terms, the differences among items in each of the following pairs.
a) Oxidation, reduction (in terms of electrons)
b) Half-reaction equation, net ionic equation
c) Oxidation, reduction (in terms of oxidation numbers)
d) Oxidizing agent (oxidizer), reducing agent (reducer)
e) Electron transfer reaction, proton transfer reaction
f) Strong oxidizing agent, weak oxidizing agent
g) Strong reducing agent, weak reducing agent
h) Atom balance, charge balance (in equations)

50. Classify each of the following statements as true or false:
    a) Oxidation and reduction occur at the electrodes in a voltaic cell.
    b) The sum of the oxidation numbers in a molecular compound is zero, but in an ionic compound that sum may or may not be zero.
    c) The oxidation number of oxygen is the same in all of the following: O^{2-}, $HClO_3$, $S_2O_3^{2-}$, NO_2.
    d) The oxidation number of alkali metals is always -1.
    e) A substance that gains electrons is oxidized.
    f) A strong reducing agent has a strong attraction for electrons.
    g) The favored side of a redox equilibrium equation is the side with the weaker oxidizer and reducer.

51. One of the properties of acids listed in Chapter 17 is "the ability to react with certain metals and release hydrogen." Why is this property limited to *certain* metals? Identify two metals that do not release hydrogen from an acid and two that do.

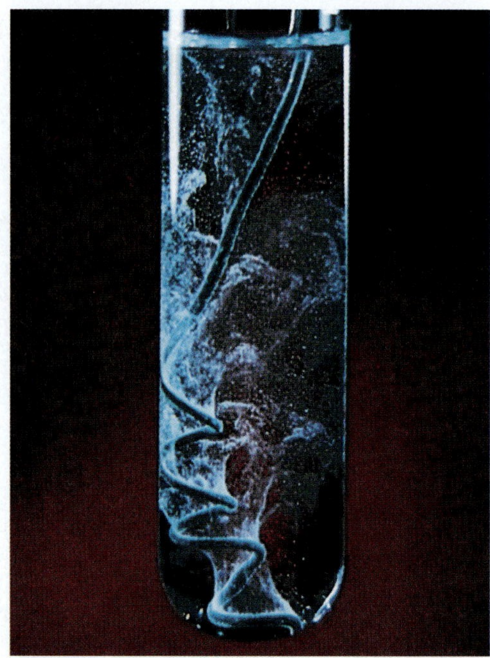

Aluminum metal reacts with hydrochloric acid, releasing bubbles of hydrogen gas.

52. The questions based on Section 19.9 identify reactants that engage in a redox reaction and their oxidized and reduced products. You were to write the redox equation. Nowhere among the reactants and products do you find a water molecule or hydrogen ion. Yet, when writing the equation, you added these species. Why is this permissible?

53. There is a fundamental difference between the electrolytic cell in Figure 19.11 and the voltaic cell in Figure 19.8. What is that difference?

54. It is sometimes said that in a redox reaction, the oxidizing agent is reduced and the reducing agent is oxidized. Is this statement (a) always correct, (b) never correct, or (c) sometimes correct? If you select (b) or (c), give an example in which the statement is incorrect.

More Challenging Questions

55. As an example of an electrolytic cell, the text states: "Sodium chloride is electrolyzed commercially in an apparatus called the *Downs cell* to produce sodium and chlorine." This is a high-temperature operation; the electrolyte is liquid (molten) NaCl. Write the half-reaction equations for the changes taking place at each electrode. Is the electrode at which sodium is produced the anode or the cathode?

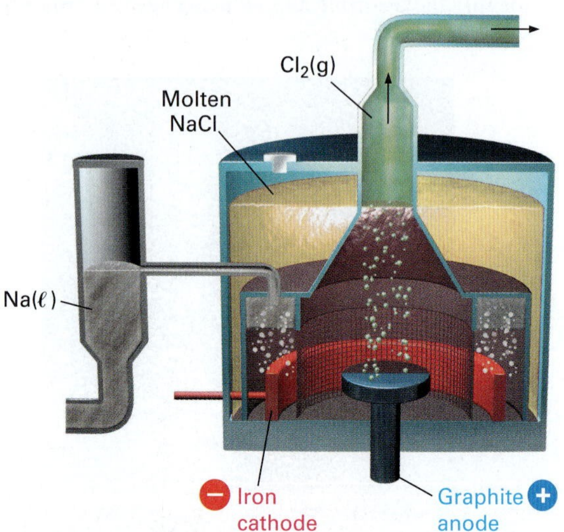

The Downs cell electrolyzes molten (melted) sodium chloride, producing sodium and chlorine.

56. Examine **Figure 19.26.** Assume that the diagram represents one of the earliest known electroplating systems in which both electrodes are copper and the electrolyte is an acidic solution of copper(II) sulfate. The net operation of this system effectively transfers copper atoms from one electrode to the other.

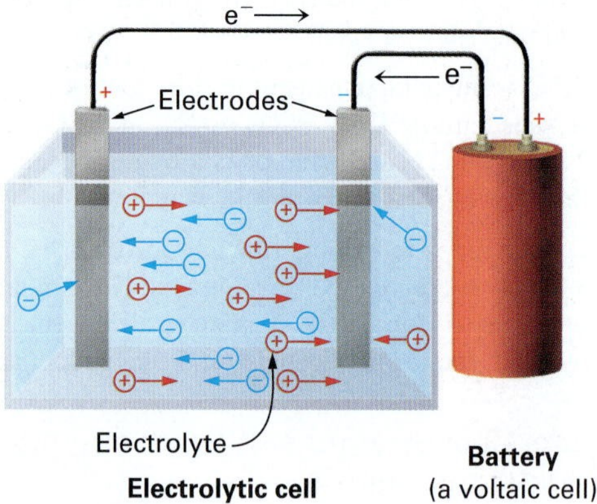

Figure 19.26 An electroplating system with copper electrodes and a solution of copper(II) sulfate as the electrolyte.

a) Write the oxidation and reduction half-reaction equations for the reactions that occur at the copper electrodes in Figure 19.26. Number your equations (1) and (2). (*Hint:* Read the introduction to this question very carefully. It contains what you need to write these equations.)

b) State which half-reaction equation, (1) or (2), represents the change at the electrode labeled "+" in Figure 19.26 and which is the change at the "−" electrode.

c) Which electrode, "+" or "−," is the anode and which is the cathode in Figure 19.26?

d) What ion or ions are carrying the charge through the electrolyte? From which electrode and to which electrode are they moving?

e) Ordinarily, this system operates at close to 100% efficiency, which means that 100 g of copper dissolved at one electrode yields 100 g of copper deposited at the other electrode. Sometimes, however, bubbles appear at one electrode, and the masses of copper dissolved and deposited are not quite equal. Considering everything in the solution, can you (1) identify the bubbles, (2) identify the electrode at which they appear, (3) write a half-reaction equation for what is occurring at that electrode, and (4) summarize in one sentence your answers to (1), (2), and (3)?

57. Marine equipment made of iron or copper alloys and through which seawater passes sometimes has inexpensive and replaceable zinc plugs sticking into the water stream. Can you suggest a reason for this? Explain your answer.

58. Have you ever put a piece of metal in your mouth, perhaps a paper clip of the end of a metal pencil, touched it to a metal filling in your teeth, and "tasted" the electricity produced? Whether you have done this or not, it does happen. Can you explain this phenomenon?

Answers to Target Checks

1. All common batteries, including those used to power television remote controls, are voltaic (galvanic) cells because they cause an electric current to flow in an external circuit rather than simply carrying current from an external source, which is the role of an electrolytic cell.

2. (a) 0, (b) −2, (c) the sum of the oxidation numbers of atoms in a molecule equals zero, (d) equal to ion charge, (e) +1, (f) ion charge equal to sum of oxidation number of elements in the ion.

3. A strong oxidizing agent is strong because it has a strong attraction for electrons. It is able to take them from many other species, thereby oxidizing them. Once it has gained an electron, it does not give it up readily, which it must do to function as a reducing agent. It therefore reduces few other species; consequently, it is a weak reducing agent.

Answers to Practice Exercises

1. Oxidation: $Cu(s) \rightleftharpoons Cu^{+}(aq) + e^{-}$

   Reduction: $Ag^{+}(aq) + e^{-} \rightleftharpoons Ag(s)$

   Redox: $Cu(s) + Ag^{+}(aq) \rightleftharpoons Cu^{+}(aq) + Ag(s)$

2. Oxidation: $Sn^{2+}(aq) \rightleftharpoons Sn^{4+}(aq) + 2\,e^{-}$

   Reduction: $2\,AgCl(s) + 2\,e^{-} \rightleftharpoons 2\,Ag(s) + 2\,Cl^{-}(aq)$

   Redox: $Sn^{2+}(aq) + 2\,AgCl(s) \rightleftharpoons Sn^{4+}(aq) + 2\,Ag(s) + 2\,Cl^{-}(aq)$

3. Reduction: $Cr^{3+}(aq) + 3\,e^{-} \rightleftharpoons Cr(s)$

   Oxidation: $Mg(s) \rightleftharpoons Mg^{2+}(aq) + 2\,e^{-}$

   Redox: $2\,Cr^{3+}(aq) + 3\,Mg(s) \rightleftharpoons 2\,Cr(s) + 3\,Mg^{2+}(aq)$

4. N: −3, H: +1

5. (a) CO_3^{2-}: C = +4; O = −2; (b) SO_3^{2-}: S = +4, O = −2

6. Iron is oxidized from 0 in Fe(s) to +2 in $Fe^{2+}(aq)$. Oxygen is reduced from 0 in $O_2(g)$ to −2 in $OH^{-}(aq)$.

7. (a) Cu: 0; H^{+}: +1; NO_3^{-}: N = +5, O = −2; Cu^{2+}: +2; NO: N = +2, O = −2; H_2O: H = +1, O = −2

   (b) Cu is oxidized, increasing in oxidation number from 0 to +2. N is reduced, decreasing in oxidation number from +5 to +2. NO_3^{-} is the oxidizing agent, removing electrons from Cu. Cu is the reducing agent, furnishing electrons to NO_3^{-}.

8. Reduction: $Sn^{2+}(aq) + 2\,e^{-} \rightleftharpoons Sn(s)$

   2 × Oxidation: $2\,Na(s) \rightleftharpoons 2\,Na^{+}(aq) + 2\,e^{-}$

   Redox: $Sn^{2+}(aq) + 2\,Na(s) \rightleftharpoons Sn(s) + 2\,Na^{+}(aq)$

9. Forward. $Sn^{2+}(aq)$ is a stronger oxidizing agent than $Na^{+}(aq)$. Na(s) is a stronger reducing agent than Sn(s).

10. $2\,Al(s) + 3\,Ni^{2+}(aq) \rightleftharpoons 2\,Al^{3+}(aq) + 3\,Ni(s)$

11. $2\,Ag(s) + 2\,H^{+}(aq) \rightleftharpoons 2\,Ag^{+}(aq) + H_2(g)$; favored in the reverse direction

12. $2\,CrO_4^{2-}(aq) + 3\,H_2S(g) + 10\,H^{+}(aq) \rightleftharpoons 2\,Cr^{3+}(aq) + 3\,S(s) + 8\,H_2O(\ell)$

13. $5\,SO_2(g) + 2\,MnO_4^{-}(aq) + 2\,H_2O(\ell) \rightleftharpoons 5\,SO_4^{2-}(aq) + 2\,Mn^{2+}(aq) + 4\,H^{+}(aq)$

Answers to Concept-Linking Exercises

You may have found more relationships or relationships other than the ones given in these answers.

1. An electric current passing through a solution—the electrolyte—is electrolysis. Current enters and leaves the electrolyte through electrodes immersed in the solution. If the current flows spontaneously, the system is called a voltaic or galvanic cell. If current must come from an outside source, the system is an electrolytic cell.

2. Oxidation is a loss of electrons. An oxidation half-reaction equation describes oxidation and has electrons as a product of the reaction. Reduction is a gain of electrons. A reduction half-reaction equation describes reduction and has electrons as one of the reactants.

3. An oxidation number is a number assigned to an element by rules to account for electrons transferred in redox reactions. An increase in oxidation number identifies oxidation of the element, and a decrease identifies reduction.

4. An oxidizing agent, also known as an oxidizer, oxidizes another species by taking electrons from it. A reducing agent, or reducer, reduces another species by giving electrons to it.

5. A strong oxidizing agent has a strong attraction for electrons and is able to draw electrons from many species; a weak oxidizing agent attracts electrons weakly and takes them from few other species. A strong reducing agent has a weak hold on electrons and gives them up easily to many species; a weak reducing agent holds its electrons tightly and gives them up to few other species.

Solutions to Blue-Numbered Questions, Exercises, and Problems

1. See the discussion in Section 19.2.
3. Oxidation: a, c, d. Reduction: b.
5. (a) Reduction; (b) oxidation
7. $$2\,Cr \rightleftharpoons 2\,Cr^{3+} + 6\,e^-$$
$$3\,Cl_2 + 6\,e^- \rightleftharpoons 6\,Cl^-$$
$$3\,Cl_2 + 2\,Cr \rightleftharpoons 2\,Cr^{3+} + 6\,Cl^-$$
9. The second equation is the oxidation half-reaction equation. Overall: $2\,NiOOH + 2\,H_2O + Cd \rightleftharpoons 2\,Ni(OH)_2 + Cd(OH)_2$
11. Examples include any item that runs on batteries, such as watches and calculators.
13. Yes. A galvanic cell causes current to flow through an external circuit by electrochemical action. An electrolytic cell is a cell through which a current driven by an external source passes.
15. (a) +3, −2, +4, +6; (b) +4, + 6, −1
17. (a) Bromine reduced from 0 to −1; (b) Lead oxidized from +2 to +4
19. (a) Iodine reduced from +7 to −1; (b) Oxygen reduced from 0 to −2
21. (a) Nitrogen oxidized from +4 to +5; (b) Chromium oxidized from +3 to +6
23. Bromine is the oxidizing agent, and the iodide ion is the reducing agent.
25. PbO_2 is the oxidizing agent, oxidizing Pb. Pb is the reducing agent, reducing the lead in PbO_2.
27. From Table 19.2, Zn is a stronger reducer than Fe^{2+}. A strong reducer releases electrons to an oxidizer more readily than a weak reducer releases them.
29. $Na^+ < Fe^{2+} < Cu^{2+} < Br_2$
31. $Br_2 + 2\,I^- \rightleftharpoons 2\,Br^- + I_2$; forward reaction favored
33. $2\,H^+ + 2\,Br^- \rightleftharpoons H_2 + Br_2$; reverse reaction favored
35. $2\,NO + 4\,H_2O + 3\,Fe^{2+} \rightleftharpoons 2\,NO_3^- + 8\,H^+ + 3\,Fe$; reverse reaction favored
37. A strong acid releases protons readily; a strong reducer releases electrons readily. A strong base attracts protons strongly; a strong oxidizer attracts electrons strongly.
39. $$S_2O_3^{2-} + 5\,H_2O \rightleftharpoons 2\,SO_4^{2-} + 10\,H^+ + 8\,e^-$$
$$4\,Cl_2 + 8\,e^- \rightleftharpoons 8\,Cl^-$$
$$S_2O_3^{2-} + 5\,H_2O + 4\,Cl_2 \rightleftharpoons 2\,SO_4^{2-} + 10\,H^+ + 8\,Cl^-$$
41. $$4\,NO_3^- + 8\,H^+ + 4\,e^- \rightleftharpoons 4\,NO_2 + 4\,H_2O$$
$$Sn + 3\,H_2O \rightleftharpoons H_2SnO_3 + 4\,H^+ + 4\,e^-$$
$$4\,NO_3^- + 4\,H^+ + Sn \rightleftharpoons 4\,NO_2 + H_2O + H_2SnO_3$$
43. $$2\,MnO_4^- + 16\,H^+ + 10\,e^- \rightleftharpoons 2\,Mn^{2+} + 8\,H_2O$$
$$5\,C_2O_4^{2-} \rightleftharpoons 10\,CO_2 + 10\,e^-$$
$$2\,MnO_4^- + 16\,H^+ + 5\,C_2O_4^{2-} \rightleftharpoons 2\,Mn^{2+} + 8\,H_2O + 10\,CO_2$$
45. $$Cr_2O_7^{2-} + 8\,H^+ + 6\,e^- \rightleftharpoons Cr_2O_3 + 4\,H_2O$$
$$2\,NH_4^+ \rightleftharpoons N_2 + 8\,H^+ + 6\,e^-$$
$$Cr_2O_7^{2-} + 2\,NH_4^+ \rightleftharpoons Cr_2O_3 + 4\,H_2O + N_2$$
47. $$4\,NO_3^- + 16\,H^+ + 12\,e^- \rightleftharpoons 4\,NO + 8\,H_2O$$
$$3\,As_2O_3 + 15\,H_2O \rightleftharpoons 6\,AsO_4^{3-} + 30\,H^+ + 12\,e^-$$
$$3\,As_2O_3 + 7\,H_2O + 4\,NO_3^- \rightleftharpoons 6\,AsO_4^{3-} + 14\,H^+ + 4\,NO$$
50. True: a, c, g. False: b, d, e, f.
51. This "property of an acid" is more correctly described as the property of an acid (hydrogen ion) acting as an oxidizing agent. The H^+ ion reacts only with metals whose ions are weaker oxidizing agents, located below hydrogen in Table 19.2.
52. All reactants were in acidic solutions. Water is available in large amounts in any aqueous solution, as is H^+ in an acidic solution.
53. In the electrolytic cell, the force that moves the charges through the circuit is *outside* the cell. In the voltaic cell, the cell itself is the *source* of the force that moves the charged particles.
54. The statement is (c) sometimes correct. In a simple element → monatomic ion redox reaction, the statement is correct. The *element* oxidized or reduced can always be identified by a change in oxidation number. The oxidizing or reducing *agent*, however, is a *species* that contains the element being oxidized or reduced, and it may be an element, a monatomic ion, a polyatomic ion (such as MnO_4^-), or a compound.
55. Oxidation occurs at the anode: $2\,Cl^- \rightleftharpoons Cl_2 + 2\,e^-$. Reduction occurs at the cathode: $Na^+ + e^- \rightleftharpoons Na$.
56. Your (1) and (2) numbers may be the reverse of ours. (a) (1) $Cu \rightleftharpoons Cu^{2+} + 2\,e^-$. (2) $Cu^{2+} + 2\,e^- \rightleftharpoons Cu$. (b) (1) occurs at the "+" electrode and (2) at the "−" electrode. (c) "+" is the anode, where oxidation occurs. "−" is the cathode, where reduction occurs. (d) Charge is carried through the electrolyte by Cu^{2+} ions moving from anode to cathode. (H^+ and SO_4^{2-} ions also move but without an identifiable "flow" of charge that is responsible for electrolysis in the solution.) (e) The bubbles are hydrogen. They come from the only ion that can be "deposited" as a gas, H^+. This occurs at the cathode, the same electrode at which copper is deposited. Any electrons that are used to reduce H^+ ions instead of copper ions cause the mass of copper deposited to be less than the mass of copper dissolved. (What happens to the concentration of Cu^{2+} ion in the solution?* The answer to this question appears at the bottom of the page.) In one sentence: Hydrogen gas, instead of copper, is reduced (deposited) at the cathode according to the half-reaction $2\,H^+(aq) + 2\,e^- \rightleftharpoons H_2(g)$.
57. Zinc is used to prevent harmful galvanic action (corrosion) in the equipment. The conditions for galvanic action are present: two metals in contact with an electrolyte (seawater) and in metal-to-metal contact with each other. If one of the metals is going to corrode, it will be the strongest reducing agent of the group. Zinc is a stronger reducing agent than iron or copper. When it corrodes, it is simply and cheaply replaced.
58. The moisture on your tongue becomes an electrolyte for the passage of electric current from one metal to the other. The tingle you "taste" is caused by that current.

*The concentration of copper increases over time.

20 Nuclear Chemistry

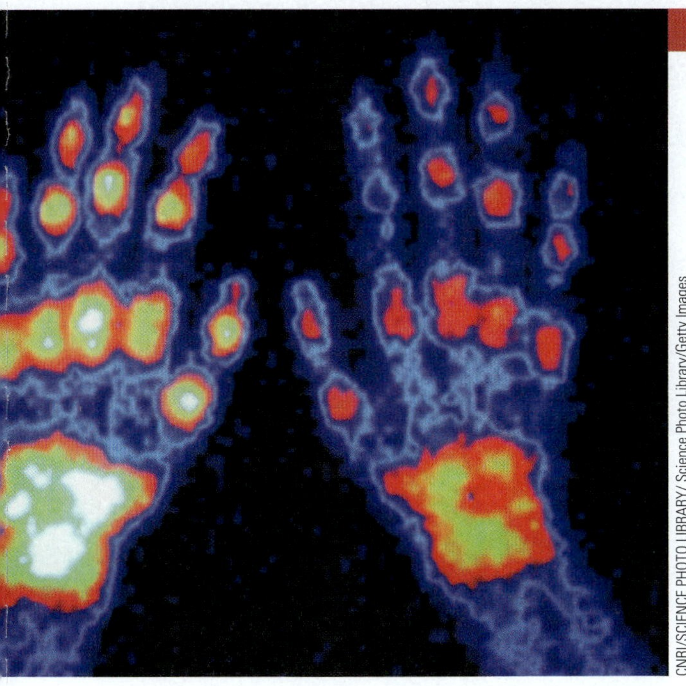

CNRI/SCIENCE PHOTO LIBRARY/ Science Photo Library/Getty Images

◀ Nuclear chemistry has many medicinal applications. This is a gamma camera scan of the hands of a person with extensive arthritis. A gamma camera is a device that detects gamma rays, which is a high-energy form of electromagnetic radiation that originates in an atomic nucleus. A radioactive substance was injected into the hands before the scan. This substance concentrates in inflamed tissue. The brighter areas indicate greater radiation emission. Notice how the right hand is more severely affected than the left.

CHAPTER CONTENTS

The study of the reactions of atomic nuclei by chemists is known as *nuclear chemistry*. In this chapter, you will learn about the nucleus and nuclear reactions. In practice, scientists in many fields study the nucleus. Much of the research in 21st century physics has been directed toward understanding the forces that hold the nucleus together. Geologists use nuclear reactions to understand changes that occur on Earth. Astronomers study the nuclear reactions that occur in stars. The interaction between radiation emitted from nuclear reactions and living systems is considered to be part of biology and medicine. You will see many examples in this chapter demonstrating that chemistry is truly the central science (see Figure 1.7).

20.1 How Did Marie Curie Find Happiness in Difficult Working Conditions?

> *...the Director permitted us to use an abandoned shed.... Its glass roof did not afford complete shelter against rain; the heat was suffocating in summer, and the bitter cold of winter was only a little lessened by the iron stove.... Yet it was in this miserable old shed that we passed the best and happiest years of our life, devoting our entire days to our work.*
>
> —*Marie Curie (1867–1934)*

How could the happiest years of one's life be spent working in a miserable old shed (**Figure 20.1**)? Understanding the answer to this question yields insight into the mind of the first person in the world to understand the cause of *radioactivity* (which

Figure 20.1 The "miserable old shed" at Sorbonne University in Paris, France, in which radium was first isolated.

will be explained in more detail in Section 20.3), a major topic in this chapter. Based on how uranium compound samples could electrify air in measurable proportion to the radioactivity of the element, Marie Curie formulated the hypothesis that the energy needed to induce the air to ionize came from *the uranium atoms themselves.* This was a bold hypothesis at this point in history because most scientists still hypothesized that the atom was indivisible. ◀

▶ **P/Review** A postulate of Dalton's atomic theory was that atoms are indivisible (Section 5.2). The word *atom* comes from the Greek *atomos*, which means indivisible.

Marie and her husband and fellow scientist, Pierre (**Figure 20.2**), began systematic testing of ores to see if any besides those containing uranium were radioactive, finding that what was then called *pitchblende ore* is more radioactive than pure uranium. Pitchblende is a mixture of many compounds, so they used chemical changes to separate the components of the mixture, keeping the most radioactive fractions and repeatedly separating them to try to isolate the source of radioactivity. Eventually, they were able to isolate *almost* pure bismuth ($Z = 83$) that was strongly radioactive. Since completely pure bismuth exhibited no radioactivity, they knew that the "impurity" in their bismuth sample was an as-yet-undiscovered radioactive element. They named it polonium in honor of Marie Curie's native country of Poland. About five months after discovering polonium, they isolated a radioactive fraction from pitchblende ore that was mostly nonradioactive barium, but that exhibited high radioactivity. They named the new element in this fraction radium, Latin for *ray.*

To unequivocally prove the existence of polonium and radium, the Curies needed to isolate the elements. This is where the miserable old shed and the happiest years of their lives came into play. The shed was their workroom for isolating radium from the pitchblende ore. The process required boiling, mixing, and stirring the ore in acid solution in huge barrels using long iron rods. From each ton of ore (2000 pounds), about 1/2000 pound of radium was eventually produced. Although the work was dangerous, physically demanding, and uncomfortable, Marie Curie was so excited about the potential for her results to be groundbreaking that she felt that the joy of potential for scientific discovery outweighed the discomfort of the treacherous and exhausting work.

Marie Curie was awarded the 1903 Nobel Prize in Physics for the discovery of radioactivity, sharing the prize with her husband, Pierre, and Henri Becquerel (whose role is described in Section 20.2). Marie was also awarded the 1911 Nobel Prize in Chemistry for her discovery of polonium and radium. By 1920, Marie began to suffer with medical issues. That year, she wrote to her sister, "Perhaps radium has

Figure 20.2 Pierre (*left*) and Marie (*right*) Curie with the Curie electrometer used to measure radioactivity. The electrometer was invented by Pierre and his brother, Jacques, about 15 years before it became the key piece of technology used in the discovery of radioactivity.

something to do with these troubles, but it cannot be affirmed with certainty." She passed away in 1934 at the age of 66 from aplastic anemia, a rare disease associated with the lack of production of a sufficient number of blood cells. One cause of this disease is now known to be exposure to radioactive materials.

20.2 The Dawn of Nuclear Chemistry

Goal 1 Define and describe radioactivity.

Serendipity. This pleasant-sounding word refers to finding valuable things you are not looking for—an accidental discovery, in other words. Serendipity has been a part of many scientific discoveries, but what happened to Henri Becquerel (**Figure 20.3**) in 1896 stands above them all. What he stumbled across affects your life and the life of every living organism on this planet. Becquerel discovered radioactivity.

Becquerel became interested in the penetrating power of x-rays soon after they were discovered—also accidentally—by Wilhelm Roentgen in 1895. Becquerel was also interested in phosphorescence, the phenomenon by which substances called *phosphors* glow after exposure to light. He wondered if phosphorescent light could penetrate black paper as x-rays can. His plan was to put a uranium-containing phosphor on top of unexposed film wrapped in black paper and place them in sunlight. The paper would prevent the film from being exposed by the sunlight so that if it was exposed at all, it would have to be from phosphorescent rays passing through the paper.

Alas, the day Becquerel chose for his experiment was cloudy. After waiting in vain for the sun to come out, he put his assembled material into a drawer to await the next sunny day. After several overcast days, Becquerel decided to develop the film to see if the initial cloudy day experiment had caused even a trace of phosphorescent light to penetrate the paper. To his amazement, the film was highly exposed! The only explanation for this was that some sort of rays were leaving the uranium compound continuously, passing through the paper, and exposing the film. Sunlight and phosphorescence had nothing to do with the result.

Figure 20.3 Antoine Henri Becquerel (1852–1908) shared the 1903 Nobel Prize in Physics with Marie Sklodowska Curie and Pierre Curie for his discovery of spontaneous radioactivity.

▶ **P/Review** Rutherford used alpha particles in his investigation of the atom (Section 5.4). Alpha particles are one of the three types of radioactive emissions discussed in the next section.

This is how Becquerel discovered **radioactivity**, the spontaneous emission of particles and/or electromagnetic radiation resulting from the decay, or breaking up, of an atomic nucleus. Of course, Becquerel did not know about the nucleus in 1896. In fact, Ernest Rutherford used one of those particles coming from a radioactive source in his 1911 experiment that led to the discovery of the nucleus (see Figure 5.9). ◀

20.3 Radioactivity

Goal 2 Name, identify from a description, and describe three types of radioactive emissions.

When the uranium ore from Becquerel's experiments—or any other radioactive substance—is placed within a lead block with a small tunnel, as illustrated in **Figure 20.4**, the emitted radiation is directed into a narrow beam because it does not pass through lead. If that beam is aimed into an electric field, it divides into three beams, indicating that radioactive emissions are made up of three different products.

One product, called an **alpha particle**, or an **α-particle**,* is attracted to the negatively charged plate, indicating that it has a positive charge. The α-particle has little penetrating power; it can be stopped by the outer layer of skin or a sheet of paper (**Figure 20.5**). Alpha particles are now known to be nuclei of helium atoms, having the nuclear symbol $^{4}_{2}\text{He}$. ◀ Since alpha particles are the *nuclei* of helium atoms only, two neutrons and two protons bound into a single particle, without the two electrons of a neutral helium atom, they have a 2+ charge. The emission of an alpha particle is an **alpha decay reaction**, or simply an **alpha decay**.

▶ **P/Review** An isotope is an atom with a specific number of neutrons and protons. If Sy is the symbol of an element, the nuclear symbol of an isotope of the element is:

$$^{\text{mass number}}_{\text{atomic number}}\text{Sy}$$

The name of an isotope is the name of the element followed by the mass number. For example, $^{238}_{92}\text{U}$ is uranium-238. (See Section 5.5.)

A different kind of radioactive emission also is a beam of particles, but these particles are negatively charged and therefore attracted to the positively charged plate (see Figure 20.4). Called **beta particles**, or **β-particles**, they have been identified as electrons. The nuclear symbol for a beta particle is $^{0}_{-1}\text{e}$, indicating zero mass number and a 1− charge. β-particles have considerably more penetrating power than α-particles, but they can be stopped by a sheet of lead or aluminum about 5 mm thick (see Figure 20.5). The emission of a beta particle is a **beta decay reaction**, or **beta decay**.

▶ **P/Review** The electromagnetic spectrum, which was described in Section 11.2, Figure 11.5, includes gamma rays, x-rays, ultraviolet and infrared rays, visible light, microwaves, and radio and television waves.

Another kind of radiation is the **gamma ray**, or **γ-ray**. Gamma rays are not particles but electromagnetic radiation, similar to x-rays but with shorter wavelengths and higher energies. ◀ Because of their high energy and lack of mass, gamma rays have high penetrating power. They can be stopped only by thick layers of lead or heavy concrete walls, as shown in Figure 20.5. Because they do not have an electric charge, gamma rays are not deflected by an electric field (see Figure 20.4).

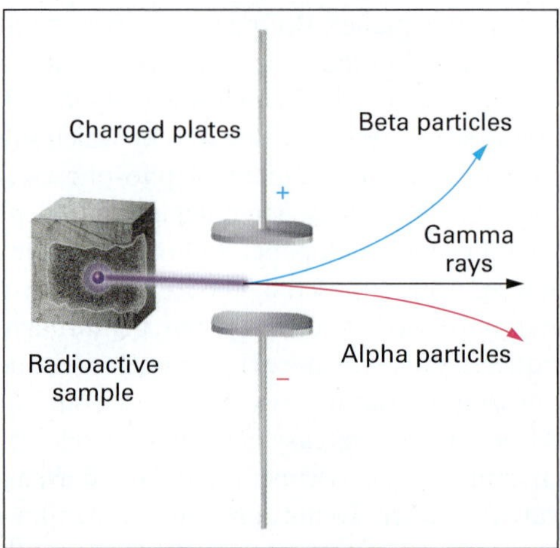

Figure 20.4 Alpha (α), beta (β), and gamma (γ) radioactive emissions. Alpha and beta particles are deflected by an electric field but gamma rays are not. The alpha particles are deflected toward the negative plate, which means that they have a positive charge. The beta particles are deflected toward the positive plate to a greater degree than the alpha particles, indicating that they have less mass than alpha particles and a negative charge. Since gamma emissions are not affected by the electric field, they are electrically neutral.

*α, β, and γ are the Greek lowercase letters alpha, beta, and gamma.

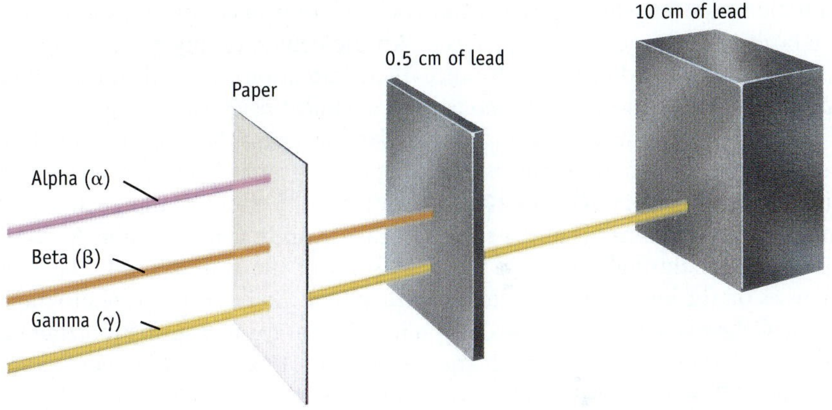

Figure 20.5 Penetrating ability of radioactive emissions. Relatively large and highly charged alpha particles can be stopped by a sheet of paper. Beta particles penetrate paper, but they are stopped by moderately thick sheets of metals such as lead (0.5 cm = 0.2 in.). Gamma radiation has no mass or charge, and it penetrates paper and sheets of metals, but it can be stopped by thick blocks of metals (10 cm = 4 in.).

Thinking About Your Thinking

Mental Models

Forming mental models of the three common products of spontaneous radioactive decay will help you think about and understand radioactivity. An alpha particle is a helium nucleus, two protons and two neutrons (and no electrons). Therefore, it has a 2+ charge and is attracted to negatively charged objects. It is relatively bulky and highly charged for a product of radioactive decay, so it has the least penetrating power. A beta particle is an electron, with a 1− charge. It is attracted to positively charged objects, and its penetrating power is greater than that of an alpha particle because it has less of a charge and it is much smaller. Finally, gamma radiation is a packet of electromagnetic radiation, similar to visible light but with much higher energy. It has no charge, so it is not influenced by an electrically charged object. It has very high penetrating power because it is a form of energy; it is not a particle. If you can now picture each type of radioactive emission in your mind, you have a good start at understanding radioactivity. Use your mental models as you answer questions and solve problems at the end of the chapter.

Radioactive substances can be harmful, but some have become valuable tools in industry, research, and medicine. In the following sections, we will discuss briefly some of these applications.

✓ Target Check 20.1

a) Write the nuclear symbol and electrical charge of an alpha particle and a beta particle.
b) List alpha, beta, and gamma rays in order of decreasing penetrating power.

20.4 The Detection and Measurement of Radioactivity

Goal 3 Identify the function of a Geiger counter and describe how it operates.

When alpha, beta, or gamma radiation collides with an atom or molecule, some of the radiation energy is transferred to the target particle. The collision changes the electron arrangement in the target. If a relatively small amount of energy is transferred, an electron in the target may be excited to a higher energy level. As the electron drops back to its original lower energy level, it releases electromagnetic energy.

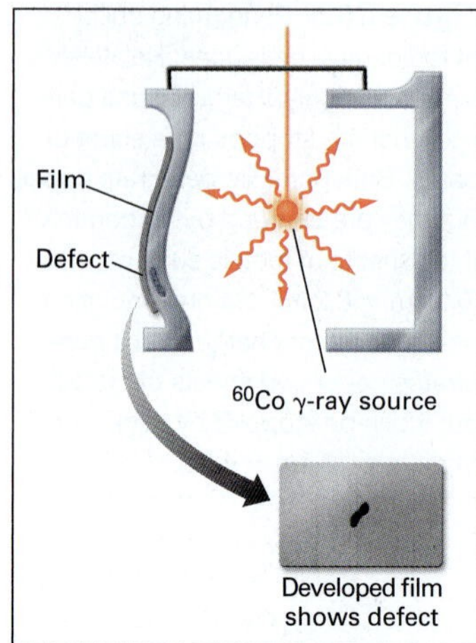

Figure 20.6 A practical use of photographic film to detect radiation. A gamma ray source is placed on one side of a cast metal part and film is placed on the other. If the part has a defect not visible to the eye, the gamma ray will reveal the defect by exposing the film more strongly at that location.

If radiation transfers enough energy to knock an electron completely out of the target, a positively charged ion is produced. Air molecules, or any gaseous molecules, can be ionized by a radioactive substance. If the radiation strikes chemically bonded atoms, it may break those bonds and cause a chemical reaction.

There are several ways to detect radioactivity. Perhaps the most obvious, but not necessarily the most convenient, is exposing photographic film, the very property that led to its discovery (**Figure 20.6**). Another is the cloud chamber, an enclosed container that contains air and a supersaturated vapor, usually water. As ionizing radiation passes through the cloud chamber, some of the air is ionized. Water vapor condenses on the ions, leaving a cloudlike track that can be seen and photographed.

The **Geiger counter** is the best-known instrument for measuring ionizing radiation. It consists of a tube filled with argon gas, called a *Geiger-Müller tube*, as shown in **Figure 20.7**. The gas is ionized by radiation passing through a thin glass window, permitting an electrical discharge between two electrodes. The current may be measured quantitatively on a meter. Some Geiger counters emit a click when radiation is detected. Geiger counters are used to measure alpha and beta radiation.

Geiger counters do not measure gamma rays effectively because there is insufficient particle density in a gas to guarantee interaction with a neutral gamma ray. A **scintillation counter** (**Figure 20.8**) uses a transparent solid, and a solid has a higher particle density than a gas. Higher density makes interaction with a gamma ray more likely. Particles in the solid absorb energy and then release some of it in flashes of light, which can be measured. This light emission, which may continue for some time after exposure to gamma rays stops, is **phosphorescence**. It is what Becquerel was looking for when he discovered radioactivity.

Two instruments used in medicine for detecting radioactivity are the gamma camera and the scanner (**Figure 20.9**). The gamma camera is placed over a target area and takes a snapshot of it. The scanner moves while taking many pictures, each picture showing a "slice" of the area under study. These pictures may be combined to give a three-dimensional view of the interior of an organ. The photograph on the first page of this chapter shows a gamma camera scan, which is also known as a *scintigram*. These processes are simple, usually cause no discomfort to the patient, and are often used instead of exploratory surgery.

Target Check 20.2

How do Geiger counters and scintillation counters "measure" radioactivity? Precisely, what do they count? Do they count "radiations" or do they count something that is proportional to radiation? Suggest units for radioactivity as it would be measured by either of these counters.

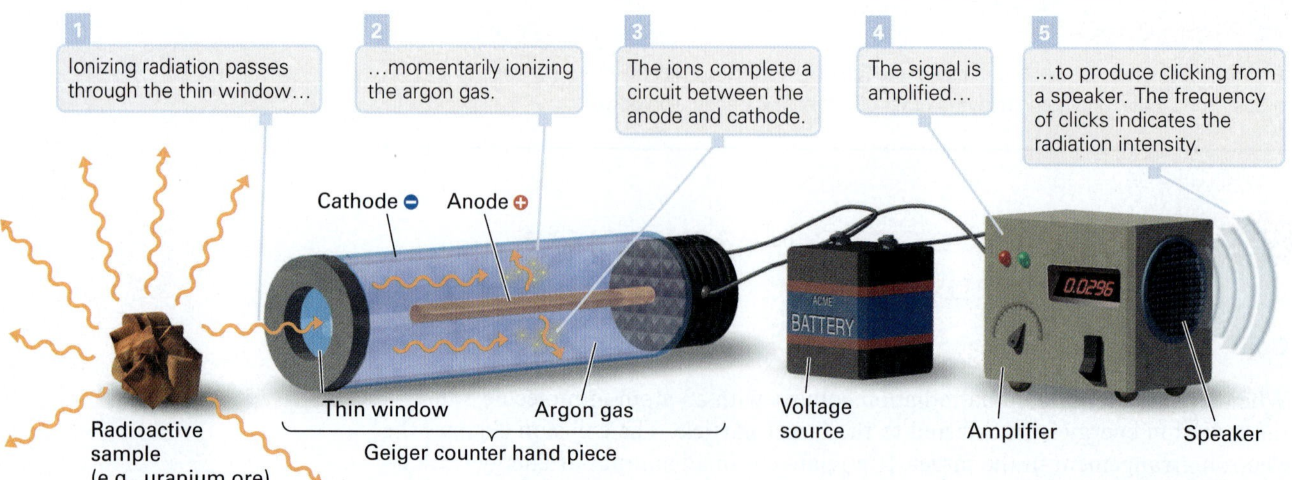

Figure 20.7 How a Geiger counter works.

20.5 The Effects of Radiation on Living Systems

Goal 4 Explain how exposure to radiation may harm or help living systems.

Shortly after radioactivity was discovered, some people hoped that radiation had certain curative powers. Radium compounds were made, and radium solutions were bottled and sold for drinking and bathing. This was before people knew the harmful effects of radiation exposure—and some early users of these "cures" paid a dear price for acquiring that knowledge. Today's medical practitioners are much wiser. They have devised sophisticated ways to use radioisotopes to examine patients, diagnose their illnesses, and treat their disorders. Industrial safety experts have also found methods to detect if workers have been exposed to unhealthy levels of radiation exposure (Figure 20.10).

Radioisotopes are atoms with an unstable nucleus. They are constantly changing, either into isotopes of the same element or into isotopes of different elements. When that happens, energy and a subatomic particle are given off. The new isotope may also be radioactive, leading to further decay, or it may be a stable isotope of some element.

The harmful effects of radiation on living systems come from its ability to break chemical bonds and thereby destroy healthy tissue. Obviously, the wise course is to avoid exposure to destructive radiation. Destructive radiation also has its good side: It destroys *unhealthy* tissue, too. Selective radiation of unhealthy tissue can eliminate it or reduce it. This is the key to radiation therapy (Figure 20.11). The trick is to do it without damaging surrounding healthy tissue as well or at least to be sure that the benefits of destroying unhealthy tissue outweigh the risks of destroying the good tissue. Modern practitioners have learned to do this fairly well, but there is still much room for improvement in targeting only the cancerous cells while avoiding the destruction of healthy cells.

The unit most commonly used to express radiation exposure is the roentgen equivalent man, usually abbreviated as **rem**.* More than 600 rem in one dose are fatal. The radiation therapy used in fighting cancers gives a *total* dose *over time* in the 4000- to 7000-rem range. This radiation indeed damages healthy tissue surrounding a tumor, but healthy cells are more capable of repairing themselves than are malignant cells.

Most of the radiation you encounter is measured not in rem but in *milli*rem (mrem). For example, a chest x-ray is about 10 mrem, a full body x-ray series is

Courtesy of Beckman Coulter, Inc

Figure 20.8 Scintillation counter. This instrument is especially useful for detecting gamma radiation because it uses a high-density transparent solid, in contrast with the low-density gas in a Geiger counter. Radiation interacts with the solid, resulting in flashes of light, which are counted by a detector.

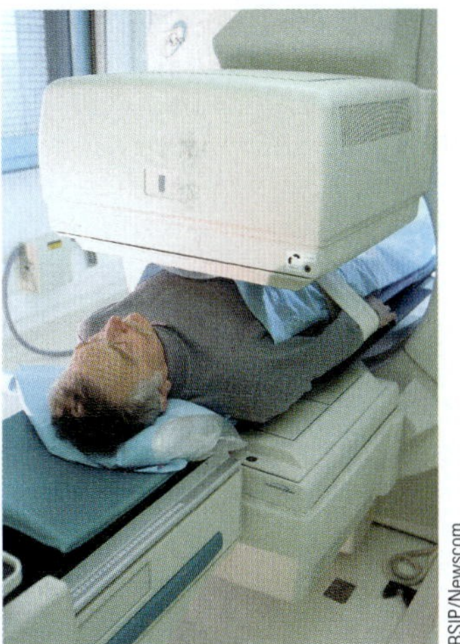

BSIP/Newscom

Figure 20.9 Gamma camera scanner. Two gamma cameras scan the patient, one from above and one from below. In this model, the table moves past the cameras to scan the whole body.

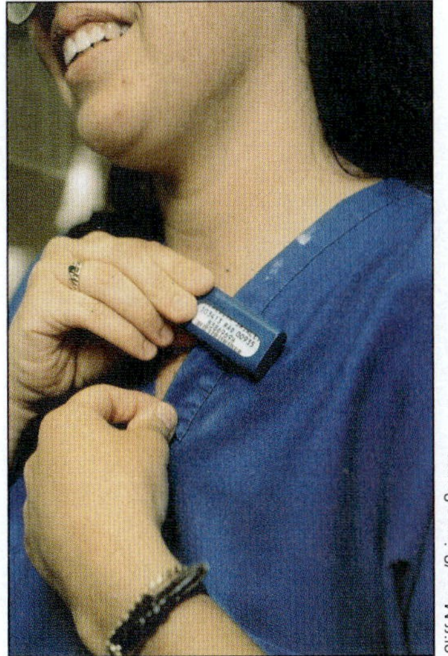

Cliff Moore/Science Source

Figure 20.10 Film badge. Workers in industries in which exposure to radiation is known to be above background levels are required to monitor their cumulative exposure dosage.

*The roentgen, R, is the amount of ionizing radiation that generates 2.09×10^9 ion pairs in 1 cm^3 of dry air. In SI units, 1 R = 0.00993 J/kg of air.

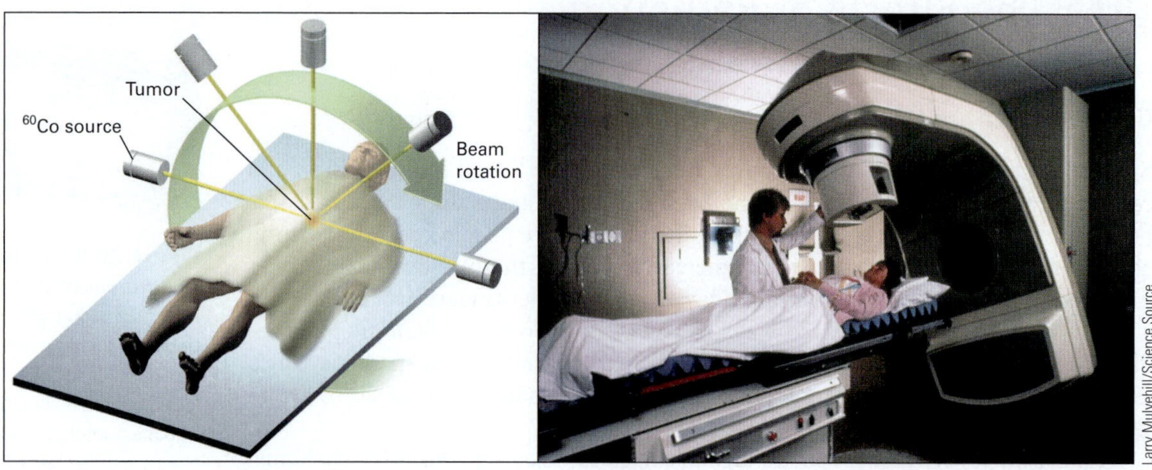

Figure 20.11 Radiation therapy. A cobalt-60 radiation source emits gamma radiation, which is directed at a cancerous tumor. The beam is rotated to minimize the destruction of healthy tissue and concentrate the radiation at the tumor.

about 1000 mrem, and a dental x-ray is about 1.5 mrem. The federal standard for occupational exposure is 5000 mrem/year.

Sources of radiation are everywhere. Terrestrial radiation is emitted from rocks, soil, and water, and even some of the atoms inside of us are a source of internal radiation. Although the atmosphere acts as a filter between us and the rest of the universe, we are constantly bombarded by cosmic radiation from sources outside Earth's atmosphere. The average U.S. exposure to natural sources of radiation is estimated at 320 mrem/year. **Figure 20.12** shows where this radiation comes from. Although it is not shown in the figure, exposure to the largest source of "natural" radiation (by far!) is voluntary for about 17% of Americans: tobacco smoke. At 1300 mrem per year, this is more than three times the dose for nonsmokers. Unfortunately for smokers, the alpha emitter $^{210}_{84}Po$ is found in the tobacco plant, and when inhaled, alpha particles can be very harmful.

Figure 20.12 Typical radiation experienced by people in the United States each year.

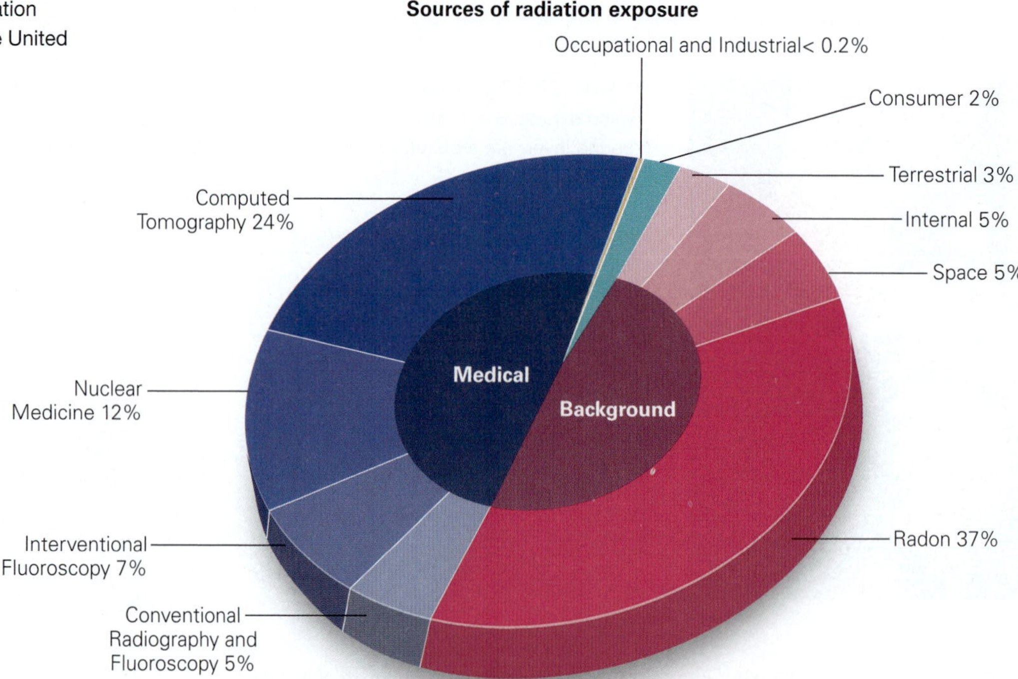

If you chose to live in a radiation-proof home to escape radiation originating from outer space, you would still have to contend with the radiation that is naturally in your body. If you weigh 150 lb, you have about 225 g of potassium ions in your body. Potassium ions participate in nerve conduction and the contraction of muscles, including the heart. Without potassium ions, you don't live. The natural abundance of ${}^{40}_{19}K$ is 0.0118% of all potassium ions. A 150-lb person, therefore, has about 4.1×10^{20} radioactive ${}^{40}_{19}K$ ions in his or her body. (You might like to confirm this statement by calculation. We'll show the calculation setup after the Target Check answers at the end of the chapter.)

Target Check 20.3

What is the easiest way you can protect yourself from the harmful effects of "natural" sources of radiation?

20.6 Half-Life

Goal **5** Describe and illustrate what is meant by the half-life of a radioactive substance.

6 Given the starting quantity of a radioactive substance, Figure 20.13, and two of the following, calculate the third: half-life, elapsed time, quantity of isotope remaining.

The rate at which a radioactive substance decays is measured by its **half-life**, the time it takes for one half of the radioactive atoms in a sample to decay. Each radioisotope has its own unique half-life, commonly written as $\mathbf{t_{1/2}}$. The units of $t_{1/2}$ are time units per half-life, or time/half-life. Time may be expressed in seconds, minutes, hours, days, or years.

Figure 20.13 shows a graph of the fraction of an original sample that remains (vertical axes) after a number of half-lives (horizontal axis). The vertical axis at the left has a conventional scale, giving values in decimal fractions. The scale values on the axis on the right are the fractions that remain after each half-life period: 1/2 after the first half-life; 1/2 of 1/2, or 1/4 after the second; 1/2 of 1/4, or 1/8 after the third; and so forth. Thus, the fraction of a sample that is still present after n half-lives is $(1/2)^n$. If S is the starting quantity and R is the amount that remains after n half-lives, then:

$$R = S \times (1/2)^n$$

Active Example 20.1 Half-Life I

The half-life of ${}^{210}_{83}Bi$ is 5.0 days. If you begin with 16 g of ${}^{210}_{83}Bi$, what mass will you have left 25 days (5 half-lives) later?

Think Before You Write You are given the half-life of an isotope and a starting quantity. You are asked for the amount that remains. This should prompt you to think of the relationship that states that the amount that remains (R) after n half-lives is $R = S \times (1/2)^n$, where S is the starting quantity.

Answers *Cover the left column with your cut-out shield. Reveal each answer only after you have written your own answer in the right column.*

Given: $S = 16\ g\ {}^{210}_{83}Bi$; n = 5 half-lives **Wanted:** g ${}^{210}_{83}Bi$ remaining (R) $R = S \times (1/2)^n = 16\ g\ {}^{210}_{83}Bi \times (1/2)^5 = 0.50\ g\ {}^{210}_{83}Bi$	With many types of calculators, to multiply a number by a number raised to a power, enter the number (16), press the multiply symbol (×), enter the number to be raised to a power (0.5 for 1/2), press x^y, enter the power (5), and then press =. Solve the problem.

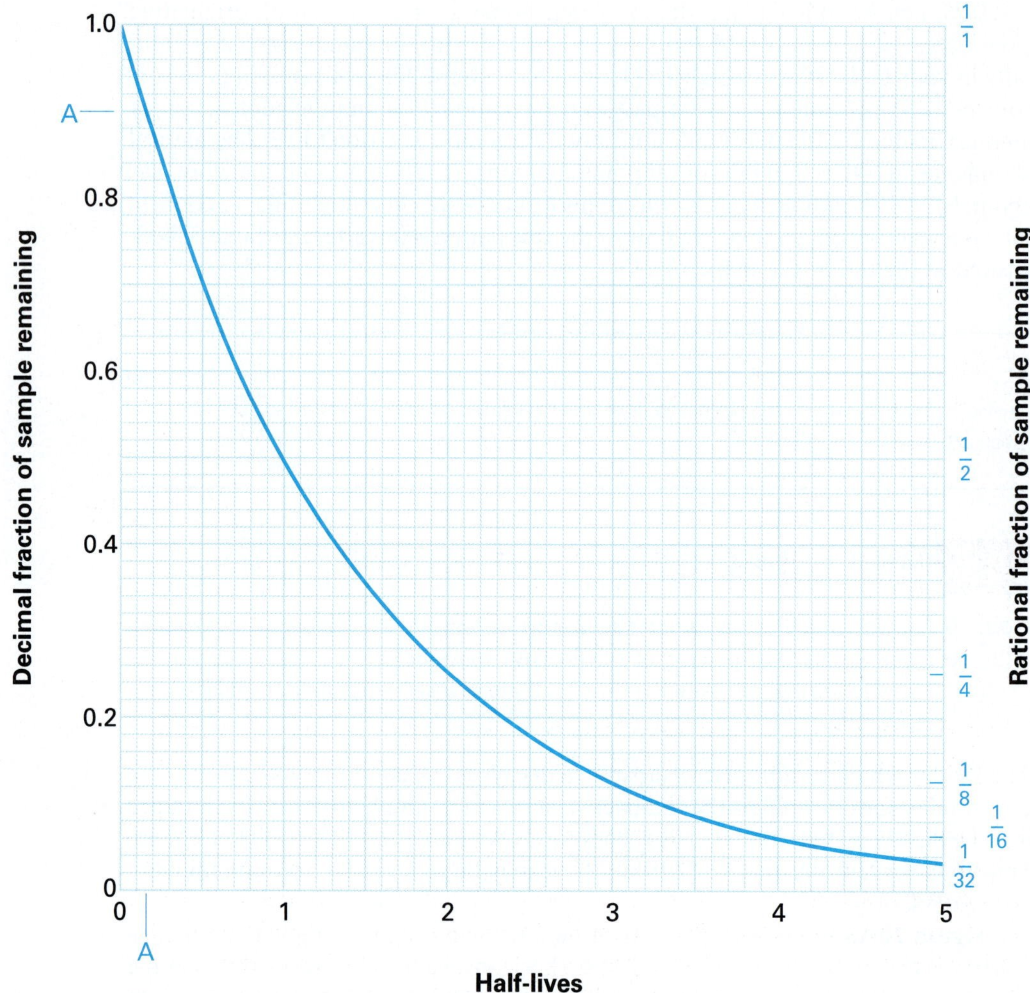

Figure 20.13 Half-life decay curve for a radioactive substance.

You improved your understanding of the half-life concept, and you improved your skill at solving half-life problems.	What did you learn by solving this Active Example?

Practice Exercise 20.1

The half-life of bismuth-205 is 15 days. A 32-g sample of $^{205}_{83}Bi$ is stored for 45 days. What mass of bismuth-205 remains?

Active Example 20.2 Half-Life II

The half-life of $^{45}_{19}K$ is 18 minutes. If you have a sample containing 2.1×10^3 micrograms (μg) of this isotope at noon, how many micrograms will remain at 3 o'clock in the afternoon?

Think Before You Write This is a two-step problem. The mass in micrograms of $^{45}_{19}K$ that remains is calculated by $R = S \times (1/2)^n$. To use that equation, however, you need the number of half-lives, n. Because n is proportional to time, that number can be calculated with the appropriate conversion factors.

Answers *Cover the left column with your cut-out shield. Reveal each answer only after you have written your own answer in the right column.*

Given: 3 hours (noon to 3 PM)
Wanted: half-lives (n)

h → min → half-lives

60 min = 1 h

1 half-life = 18 min

$$3\ \cancel{h} \times \frac{60\ \cancel{min}}{1\ \cancel{h}} \times \frac{1\ \text{half-life}}{18\ \cancel{min}} = 10\ \text{half-lives} = n$$

Plan and solve the problem to the point at which you calculate the number of half-lives. The conversion factor given in the problem statement is 1 half-life = 18 minutes.

Given: S = 2.1×10^3 μg; n = 10 half-lives
Wanted: μg $^{45}_{19}$K remaining (R)

$R = S \times (1/2)^n = 2.1 \times 10^3\ \mu g\ ^{45}_{19}K \times (1/2)^{10} = 2.1\ \mu g\ ^{45}_{19}K$

Now the problem is just like Active Example 20.1. Calculate the answer.

You improved your understanding of the half-life concept, and you improved your skill at solving half-life problems.

What did you learn by solving this Active Example?

Practice Exercise 20.2

The half-life of $^{38}_{19}$K is 7.6 minutes. The mass of potassium-38 in a sample is measured as 9.5×10^3 μg at 9:06 AM. What mass of potassium-38 will exist at 11:00 AM?

To find the half-life of a radioactive isotope, you must determine starting and remaining quantities over a measured period of time. One way to interpret these data is to express the numbers as the fraction of the sample remaining, R/S. This is the vertical axis in Figure 20.13. Start with the fraction on the vertical axis, project horizontally to the curve, and then project vertically to the number of half-lives on the horizontal axis. Divide time by half-lives to get the half-life of the substance: time/half-lives = $t_{1/2}$.

You will use this procedure to solve the next Active Example.

Active Example 20.3 Half-Life III

The mass of radioactive $^{125}_{51}$Sb in a sample is found to be 8.623 g. The sample is set aside for 157 days, which is 0.430 year. At that time, the sample contains 7.762 g of $^{125}_{51}$Sb. Find the half-life of $^{125}_{51}$Sb in years.

Think Before You Write You are given the initial and final mass of a radioactive substance, as well as the period of time needed for the decay. You are asked for the half-life. You will find R/S, use Figure 20.13 to identify the equivalent number of half-lives, and then determine time units per half-life.

Answers *Cover the left column with your cut-out shield. Reveal each answer only after you have written your own answer in the right column.*

Given: S = 8.623 g; R = 7.762 g
Wanted: Fraction remaining, R/S

R/S = 7.762 g ÷ 8.623 g = 0.9002

First, find the fraction of the isotope remaining after 0.430 year. The sample began with 8.623 g but only 7.762 g remain. What fraction of 8.623 g is 7.762 g? Calculate that value.

0.16 half-life has passed.

"A" on the vertical axis of the graph corresponds to 0.90, the fraction of radioisotope remaining. Moving horizontally to the curve and projecting down to the horizontal axis—again "A"—gives 0.16 half-life.

Use Figure 20.13 to determine how many half-lives have passed when 0.9002 of the original amount—90% of the starting mass—remains.

0.430 year ÷ 0.16 half-life = 2.7 years/half-life	You now have the number of years, 0.430 years, and the number of half-lives, 0.16 half-life. Calculate the years per half-life.
You improved your understanding of the half-life concept, and you improved your skill at solving half-life problems.	What did you learn by solving this Active Example?

Practice Exercise 20.3

A sample containing silicon-31 contains 251 mg of the isotope. The sample contains 113 mg of the isotope 3.0 hours after the initial mass measurement. What is the half-life of silicon-31?

You may wonder how you can solve Active Example 20.3 without Figure 20.13. Specifically, this means finding the number of half-lives, n, algebraically. The procedure is as follows:

The logarithm power rule is $\log x^y = y \log x$.

Solve $R = S \times (1/2)^n$ for $(0.5)^n$: $(0.5)^n = R/S$

Take the logarithm of both sides: $\log[(0.5)^n] = n \log 0.5 = \log(R/S)$

Solve for n: $n = \dfrac{\log(R/S)}{\log 0.5}$

Substitute values and calculate: $n = \dfrac{\log(7.762/8.623)}{\log 0.5}$

$$= \frac{\log 0.9002}{\log 0.5} = 0.1517$$

Then, 0.430 year/0.1517 half-life = 2.83 years/half-life. The small difference between the final values comes from the two-significant-figure reading of 0.16 half-life from Figure 20.13.

Bettmann/Getty Images

Figure 20.14 Willard Libby (1908–1980) was awarded the 1960 Nobel Prize in Chemistry for leading the team of scientists that developed radiocarbon dating.

In the mid- and late-1940s, a team of scientists led by the American chemist Willard Libby (**Figure 20.14**) used the half-life rate of decay of radioactive substances to form the basis of **radiocarbon dating**, by which scientists estimate the age of fossils. Carbon is found in all living organisms. Most carbon atoms are carbon-12; however, a small portion of the carbon in atmospheric carbon dioxide is carbon-14, a radioactive isotope with a half-life of 5.73×10^3 years. When a plant or animal is alive, it takes in this isotope from its environment, whereas the same isotope in the organism is disappearing by nuclear disintegration. A "steady-state" situation exists while the organism lives, maintaining a constant ratio—the ratio found in the atmosphere—of carbon-14 to carbon-12. When the organism dies, the disintegration of carbon-14 continues, but its intake stops. This leads to a gradual reduction in the ratio of $^{14}_{6}C$ to $^{12}_{6}C$. By measuring the ratio and the amount of $^{14}_{6}C$ now present in a sample, we can calculate the $^{14}_{6}C$ present when the organism died. The half-life decay curve in Figure 20.13 is then used to calculate the age of the sample.

Active Example 20.4 Half-Life IV

Ötzi the Iceman (Figure 20.15) is a Neolithic hunter whose frozen remains were found in the Similaun Glacier on the Austrian/Italian border in the Tyrolean Alps. Analysis of the Iceman shows that for every 15.3 units of $^{14}_{6}C$ present at the time of his death, 8.1 remain today. How old is the Iceman?

Think Before You Write This problem assumes that you know or can look up the half-life of carbon-14, 5.73×10^3 years. If you know the number of half-lives that have passed for a substance, you can use the half-life of the substance to convert into time units. The problem statement gives you values to determine R/S, and Figure 20.13 allows you to convert between R/S and number of half-lives.

Answers *Cover the left column with your cut-out shield. Reveal each answer only after you have written your own answer in the right column.*

Given: S = 15.3 units; R = 8.1 units **Wanted:** n (number of half-lives) R/S = 8.1 units ÷ 15.3 units = 0.53 Matching number of half-lives from Figure 20.13: 0.93 half-life	The quantities 15.3 units and 8.1 units identify starting and remaining amounts of radioactive isotope in the sample. Use these quantities and Figure 20.13 to determine the number of half-lives that have elapsed since the death of the Iceman.
Given: 0.93 half-life **Wanted:** years half-life → years 5.73×10^3 years = 1 half-life $0.93\ \text{half-life} \times \dfrac{5.73 \times 10^3\ \text{years}}{1\ \text{half-life}} = 5.3 \times 10^3\ \text{years}$	The half-life of carbon-14 is 5.73×10^3 years. How many years are in 0.93 half-life? 
You improved your understanding of the half-life concept, and you improved your skill at solving half-life problems.	What did you learn by solving this Active Example?

Practice Exercise 20.4

What percentage of carbon-14 remains in a sample that is measured as 1.7×10^4 years old?

MediaWorldImages/Alamy Stock Photo

Figure 20.15 Ötzi the Iceman. He was found in a glacier in the Tyrolean Alps near the Austrian/Italian border in 1991. He is believed to have lived in about 3300 BCE. He was so well preserved that scientists were able to examine the contents of his digestive system, finding that he was an omnivore.

Carbon dating has produced evidence of modern humans' presence on Earth as long ago as 200,000 years, although modern behavior and culture evolved much more recently. Similar dating techniques are also applied to mineral deposits. Analyses of geological deposits have identified rocks with an estimated age of 3.5 billion to 4.3 billion years, the latter figure being close to scientists' estimate of the age of Earth, 4.5 billion years. The oldest moon rocks also indicate an age of about 4.5 billion years.

20.7 Natural Radioactive Decay Series—Nuclear Equations

Goal 7 Describe a natural radioactive decay series.

8 Given the identity of a radioactive isotope and the particle it emits, write a nuclear equation for the emission.

▶ **P/Review** The number of protons in an atom determines what element it is. All atoms of a specific element have the same number of protons, which is the same as the atomic number (Section 5.5).

When a radioisotope emits an alpha or beta particle, there is a **transmutation** of an element—that is, a change from one element to another. This means that the remaining isotope has a different atomic number—a different number of protons—from the original isotope. ◀ A nuclear change—actually a nuclear reaction—has occurred. The original substance (isotope) has been destroyed and a new substance (isotope) has been formed.

The emission of a gamma ray does not change the elemental identity of the nucleus, even though energy is released. In that sense, a gamma emission, by itself, is not a nuclear change. Therefore, for the remainder of this chapter, we will consider only alpha and beta emissions in nuclear reactions.

Just as chemists write chemical equations to describe chemical changes, they write nuclear equations to describe nuclear changes. A nuclear equation shows the reactant isotopes or particles on the left of an arrow and the product isotopes or particles on the right. The first step in the natural radioactive series first observed by Becquerel is an alpha decay reaction. In it, a ${}^{238}_{92}\text{U}$ nucleus disintegrates, or decays, into a ${}^{4}_{2}\text{He}$ nucleus (alpha particle) and a ${}^{234}_{90}\text{Th}$ nucleus. The nuclear equation is:

$$ {}^{238}_{92}\text{U} \rightarrow {}^{4}_{2}\text{He} + {}^{234}_{90}\text{Th} $$

Notice that this equation is balanced in both mass number and atomic number. The total number of neutrons and protons is 238, the mass number of the uranium isotope. The total mass number of the two products is 234 + 4, again 238. This accounts for all the protons and neutrons in the reactant, ${}^{238}_{92}\text{U}$. In terms of protons only, the 92 in a uranium nucleus are accounted for by 90 in the thorium nucleus plus 2 in the helium nucleus. *A nuclear equation is balanced if the sums of the mass numbers on the two sides of the equation are equal and if the sums of the atomic numbers are equal.*

The ${}^{234}_{90}\text{Th}$ nucleus resulting from the disintegration of uranium-238 is also radioactive. In a beta decay reaction, it emits a beta particle, ${}^{0}_{-1}\text{e}$, and produces an isotope of protactinium, ${}^{234}_{91}\text{Pa}$:

$$ {}^{234}_{90}\text{Th} \rightarrow {}^{0}_{-1}\text{e} + {}^{234}_{91}\text{Pa} $$

In a beta particle emission, the mass numbers of the reactant and product isotopes are the same, whereas the atomic number increases by 1. Although the actual process is more complex, it appears as if a neutron divides into a proton and an electron, and the electron is ejected.

The two disintegrations described are only the first 2 of 14 steps that begin with ${}^{238}_{92}\text{U}$. There are eight α-particle emissions and six β-particle emissions, leading ultimately to a stable isotope of lead, ${}^{206}_{82}\text{Pb}$. This entire **natural radioactive decay series** is described in **Figure 20.16**. There are other natural disintegration series. One begins with ${}^{234}_{90}\text{Th}$ and ends with ${}^{208}_{82}\text{Pb}$, and another begins with ${}^{235}_{92}\text{U}$ and ends with ${}^{207}_{82}\text{Pb}$.

Active Example 20.5 Nuclear Equations I

Write the nuclear equation for the changes that occur in the uranium-238 disintegration series when ${}^{226}_{88}\text{Ra}$ ejects an α-particle. Ra is the symbol for radium, one of the elements discovered by Marie and Pierre Curie in their study of radioactivity.

Think Before You Write In a nuclear equation, one product will be the particle ejected. The mass number of the other product will be such that, when added to the mass number of the ejected particle, the total will be the mass number of the original isotope.

Answers *Cover the left column with your cut-out shield. Reveal each answer only after you have written your own answer in the right column.*

222 *The reactant isotope has a mass number of 226. It emits a particle having a mass number of 4. This leaves 226 − 4 = 222 as the mass number of the remaining particle.*	What is the mass number of the second product of the emission of an alpha particle from a ${}^{226}_{88}\text{Ra}$ nucleus?

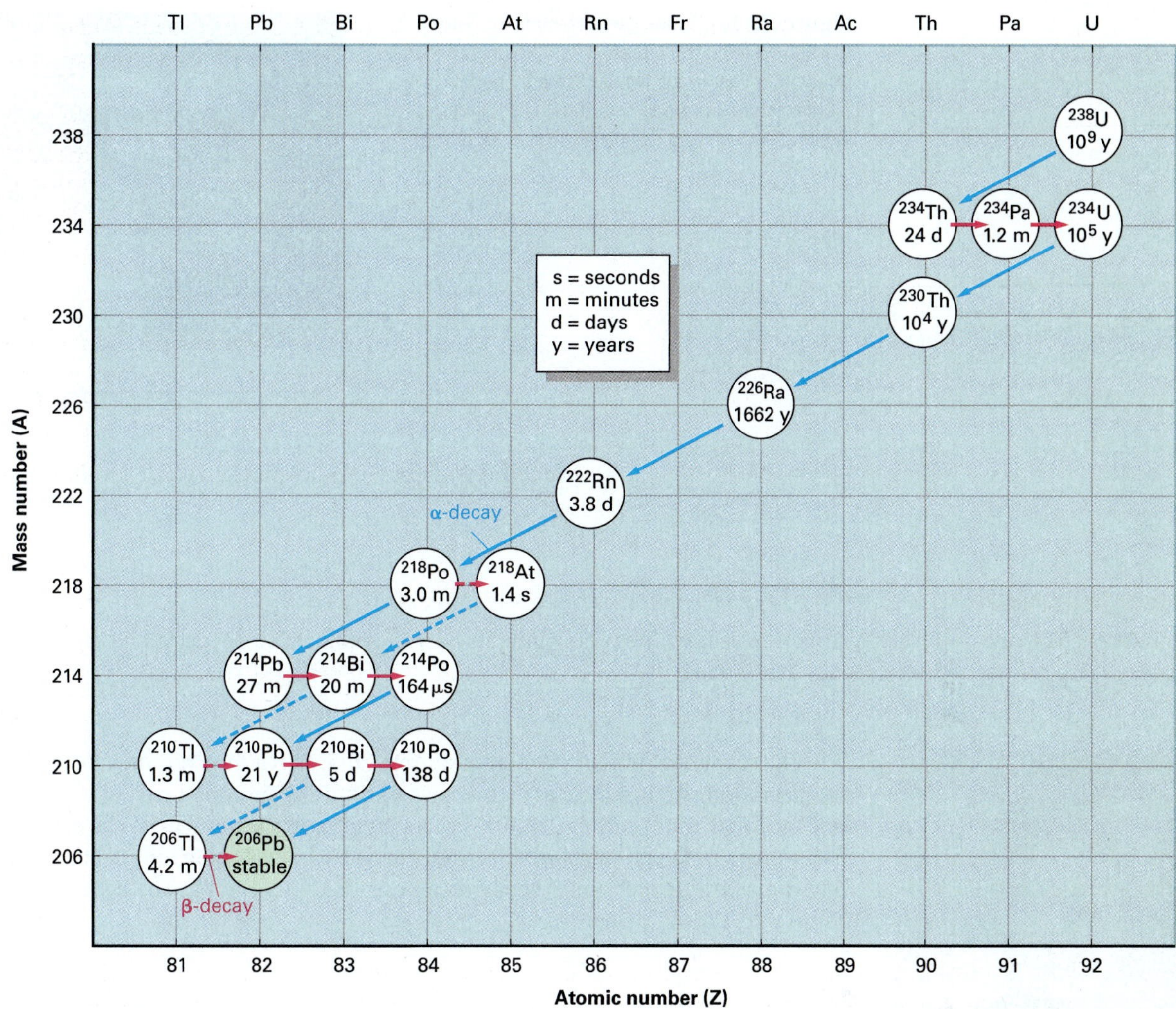

Figure 20.16 Radioactive decay series. This series begins with $^{238}_{92}U$ and, after eight alpha emissions and six beta emissions, produces $^{206}_{82}Pb$ as a stable end product. Branching occurs at some points in the chain; the more prevalent decay is indicated with a solid line, and the less prevalent decay is indicated with a dashed line.

86 *If two protons are emitted from a nucleus having 88 protons, 86 will remain.*	Determine the atomic number of the second product particle. The atomic number of the starting isotope is 88. It emitted a particle having two protons. How many protons are left in the nucleus of the other product?
$^{226}_{88}Ra \rightarrow ^{4}_{2}He + ^{222}_{86}Rn$	You now know the mass number and the atomic number of the second product of an alpha-particle emission from $^{226}_{88}Ra$. Using a periodic table, you can find the elemental symbol of this product and assemble all three symbols into the required nuclear equation.

Charles D. Winters

Figure 20.17 Radon gas detection kit. Radon is believed to be the cause of 21,000 lung cancer deaths per year in the United States. The Environmental Protection Agency and the U.S. Surgeon General recommend testing all homes for radon.

You improved your skill at writing nuclear equations.	What did you learn by solving this Active Example?

Practice Exercise 20.5

Radon-222, $^{222}_{86}Rn$, decays by alpha particle emission. The product of this nuclear change also decays by alpha particle emission. Write the nuclear equation for each decay.

The second product in Active Example 20.5, $^{222}_{86}Rn$, is a radioactive isotope of the gas radon. Radon-222 further decays by emitting an alpha particle. You should be concerned about radon-222 that enters building basements from the soil outside. Exposure to radon gas may increase your chances of developing lung cancer. The U.S. Environmental Protection Agency recommends that all houses be tested for radon (**Figure 20.17**).

Time Life Pictures/Getty Images

Figure 20.18 Marie Curie (1867–1934) who was awarded the 1903 Nobel Prize in Physics (with husband, Pierre, and Henri Becquerel) for the discovery of radioactivity and the 1911 Nobel Prize in Chemistry for the discovery of radium and polonium, the isolation of pure radium, and the study of radium compounds. She was the mother of Irene and Eve (Irene Curie-Joliot shared with her husband, Frederic, the 1935 Nobel Prize in Chemistry). Marie Curie was the first woman to teach at the Sorbonne (1906) and the first woman to be appointed as a full professor at the Sorbonne (1908).

Active Example 20.6 Nuclear Equations II

Write the nuclear equation for the emission of a β-particle from $^{210}_{83}Bi$.

Think Before You Write The method is the same for any nuclear equation. Remember that the beta particle, $^{0}_{-1}e$, has zero mass number and an effective atomic number of −1. Both mass number and atomic number must be conserved in the equation.

Answers *Cover the left column with your cut-out shield. Reveal each answer only after you have written your own answer in the right column.*

$^{210}_{83}Bi \rightarrow {}^{0}_{-1}e + {}^{210}_{84}Po$ *In the emission of a β-particle, the mass number of the radioactive isotope and the product isotope are the same. The product isotope has an atomic number greater by one than the radioactive isotope, an increase of one proton. Po is the symbol that corresponds to atomic number 84. The element is polonium, the other element discovered by the Curies in their investigation of radioactivity. The name of the element was selected to honor Madame Curie's native Poland (**Figure 20.18**).*	Write the equation. 

You improved your skill at writing nuclear equations.	What did you learn by solving this Active Example?

Practice Exercise 20.6

Cesium-137, $^{137}_{55}Cs$, did not exist in Earth's atmosphere until humans started testing nuclear weapons. It decays by beta emission with a half-life of 30 years. Write the nuclear equation for the decay of cesium-137.

20.8 Nuclear Reactions and Ordinary Chemical Reactions Compared

Goal 9 List or identify four ways in which nuclear reactions differ from ordinary chemical reactions.

Now that you have seen the nature of a nuclear change and the type of equation by which it is described, we will compare nuclear reactions with the chemical reactions you have studied. There are four important areas of contrast:

1. In ordinary chemical reactions, the chemical properties of an element depend only on the electrons outside the nucleus, and the chemical properties are essentially the same for all isotopes of the element. The nuclear properties of the various isotopes of an element are quite different, however. In the radioactive decay series beginning with uranium-238, $^{234}_{90}Th$ emits a β-particle, whereas a bit farther down the line $^{230}_{90}Th$ ejects an α-particle. Both $^{214}_{82}Pb$ and $^{210}_{82}Pb$ are β-particle emitters toward the end of the series, whereas the final product, $^{206}_{82}Pb$, has a stable nucleus, emitting neither alpha nor beta particles nor gamma rays.
2. Radioactivity is independent of the state of chemical combination of the radioactive isotope. The decomposition of $^{210}_{83}Bi$ occurs for atoms of that particular isotope whether they are in pure elemental bismuth, combined in bismuth(III) chloride, $BiCl_3$, bismuth(III) sulfate, $Bi_2(SO_4)_3$, or any other bismuth compound, or if they happen to be present in the low melting alloy used in sprinkler systems for fire protection in large buildings.
3. Nuclear reactions result in the formation of different elements because of changes in the number of protons in the nucleus of an atom. In ordinary chemical reactions, the atoms keep their elemental identities while changing from one compound as a reactant to another as a product.
4. Both nuclear and ordinary chemical changes involve energy, but the amount of energy for a given amount of reactant in a nuclear change is enormous—greater by several orders of magnitude, or multiples of 10—compared with the energies of ordinary chemical reactions.

20.9 Nuclear Bombardment and Induced Radioactivity

Goal 10 Define and identify nuclear bombardment reactions.
11 Distinguish natural radioactivity from induced radioactivity produced by bombardment reactions.
12 Define and identify transuranium elements.

Natural radioactive decay is an example of how to achieve the alchemist's get-rich-quick dream of converting one element to another. But the natural process for uranium does not yield the gold coveted by the alchemist; rather, it produces the element lead, with which the dreamer wanted to begin the transmutation. The question remained after radioactivity was discovered: Can we *initiate* the transmutation of one ordinarily stable element into another?

In 1919, Ernest Rutherford (see Section 5.4) answered *yes* to that question. He found that he could bombard the nuclei of nitrogen atoms with a beam of alpha particles from a radioactive source, producing atoms of oxygen-17 and hydrogen atoms:

$$^{14}_{7}\text{N} + ^{4}_{2}\text{He} \rightarrow ^{17}_{8}\text{O} + ^{1}_{1}\text{H}$$

The oxygen-17 isotope produced is stable; the experiment did not yield any radioactive isotopes. Similar experiments were conducted with other elements, using high-speed alpha particles as atomic "bullets." Scientists found that most of the elements up to potassium can be changed to other elements by **nuclear bombardment**. None of the isotopes produced were radioactive.

Hulton Archive/Getty Images

Figure 20.19 Marie, Irène, and Pierre Curie (*left to right*). Irène Joliot-Curie shared the 1935 Nobel Prize in Chemistry with her husband, Frédéric Joliot, for their synthesis of new radioactive elements.

One experiment during this period was first thought to yield a nuclear particle that had emitted some sort of high-energy radiation, perhaps a gamma ray. In 1932, James Chadwick correctly interpreted the experiment, and in doing so, he became the first person to identify the neutron. The reaction comes from bombarding a beryllium atom with a high-energy α-particle:

$$^{9}_{4}\text{Be} + ^{4}_{2}\text{He} \rightarrow ^{12}_{6}\text{C} + ^{1}_{0}\text{n}$$

$^{1}_{0}\text{n}$ is the nuclear symbol for the neutron, with zero charge and a mass number of 1.

Two years later, in 1934, Irène Curie (**Figure 20.19**) (a daughter of Marie and Pierre Curie) and her husband, Frédéric Joliot, used high-energy α-particles to produce the first synthetic radioisotope. Their target was boron-10; the product was a radioactive nitrogen nucleus:

$$^{10}_{5}\text{B} + ^{4}_{2}\text{He} \rightarrow ^{13}_{7}\text{N} + ^{1}_{0}\text{n}$$

Because this radioisotope is not found in nature, its decay is an example of **induced** or **artificial radioactivity**. When $^{13}_{7}\text{N}$ decays, it emits a particle having the mass of an electron and a charge equal to that of an electron, except that it is positive. This "positive electron" is called a **positron**, and it is represented by the symbol $^{0}_{1}\text{e}$. The decay equation is:

$$^{13}_{7}\text{N} \rightarrow ^{12}_{6}\text{C} + ^{0}_{1}\text{e}$$

Today, thousands of radioisotopes have been produced in laboratories all over the world. Many of these isotopes have been made in different kinds of **particle accelerators**, which use electric fields to increase the kinetic energy of the charged particles that bombard nuclei (**Figure 20.20**). Particle accelerators are manufactured in two basic designs: linear and circular. Among the earliest and best-known

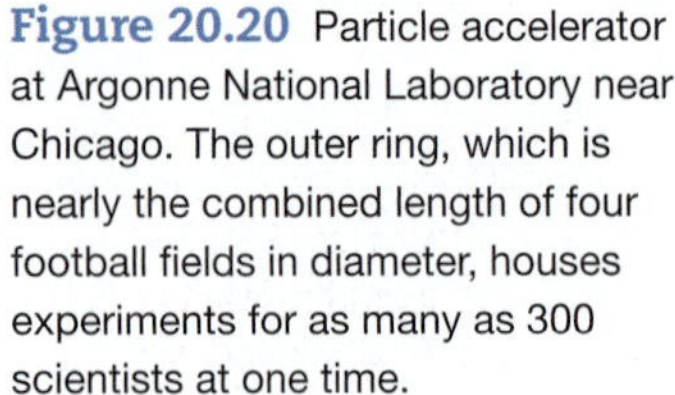

Figure 20.20 Particle accelerator at Argonne National Laboratory near Chicago. The outer ring, which is nearly the combined length of four football fields in diameter, houses experiments for as many as 300 scientists at one time.

Argonne National Laboratory

accelerators is the cyclotron, so named because of its circular shape. It was invented by Ernest Lawrence at the University of California, Berkeley, who won the 1939 Nobel Prize in Physics for his efforts.

One of the areas of research with bombardment reactions has been the production of elements that do not exist in nature. Except in trace quantities, no natural elements having atomic numbers greater than 92 have ever been discovered. In 1940, it was found that uranium-238 is capable of capturing a neutron:

$$^{238}_{92}U + ^{1}_{0}n \rightarrow ^{239}_{92}U$$

The newly formed isotope is unstable, progressing through two successive β-particle emissions, yielding isotopes of the elements having atomic numbers 93 and 94:

$$^{239}_{92}U \rightarrow ^{0}_{-1}e + ^{239}_{93}Np \text{ (neptunium)}$$

$$^{239}_{93}Np \rightarrow ^{0}_{-1}e + ^{239}_{94}Pu \text{ (plutonium)}$$

Neptunium, plutonium, and all the other synthetic elements having atomic numbers greater than 92 are called the **transuranium elements**. All transuranium isotopes are radioactive, and some with very short half-lives have only been briefly isolated in extremely small amounts. Some of the bombardments yielding transuranium products use relatively high mass isotopes as bullets. For example, einsteinium-247 is produced by bombarding uranium-238 with ordinary nitrogen nuclei:

$$^{238}_{92}U + ^{14}_{7}N \rightarrow ^{247}_{99}Es + 5\,^{1}_{0}n$$

Target Check 20.4

a) What is produced in a nuclear bombardment reaction?

b) What property must a particle have if it is to be used in a particle accelerator?

20.10 Uses of Radioisotopes

Goal 13 Identify and describe uses for synthetic radioisotopes.

Today, there are possibly thousands of uses for synthetic radioisotopes. The best known of these are in medicine. People are not usually aware of others, although at times they may be close at hand. For example, do you have a smoke detector in your home? Battery-powered smoke detectors use a chip of americium-241, $^{241}_{95}Am$. The americium ionizes the air in the detector, which causes a small current to flow through the air. When smoke enters, it reduces the current flow and sets off the alarm, which is powered by a battery (**Figure 20.21**). With a half-life of 458 years, the americium doesn't need changing every year as the battery does.

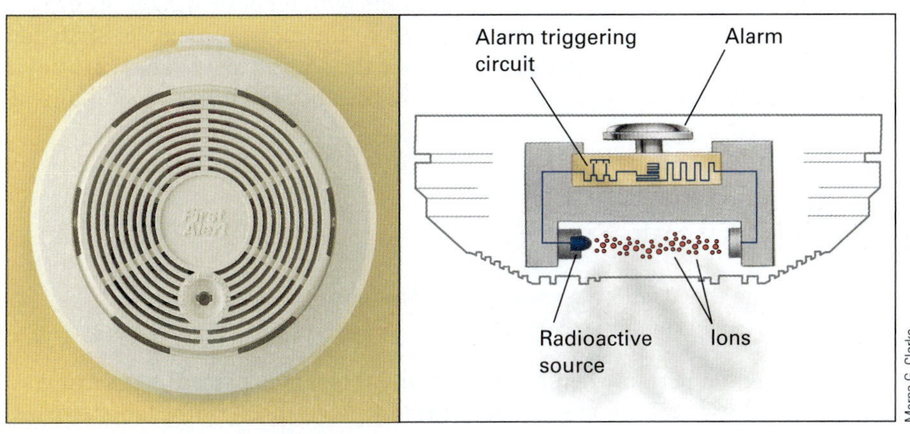

Figure 20.21 Smoke detector. The radioactive source in a smoke detector is americium-241 oxide, embedded in a gold foil matrix. Americium is an alpha and gamma emitter. The radiation ionizes the air in the ionization chamber, producing an electric current. When smoke particles interact with the ions, the current is reduced, and the alarm is triggered.

Figure 20.22 Using emissions from artificial radioisotopes for food preservation. The roll on the right was irradiated to preserve it. Both rolls were then left for two weeks, and the roll on the left spoiled.

Artificial radioisotopes are used in food preservation (**Figure 20.22**). Worldwide, more than 40 classes of foods are irradiated with gamma rays from cobalt-60, ${}^{60}_{27}Co$, or cesium-137, ${}^{137}_{55}Cs$. This process retards the growth of bacteria, molds, and yeasts in foods, just as heat pasteurization extends the shelf life of milk. Higher doses of gamma radiation sterilize foodstuffs, killing insects as well.

Industrial applications of radioisotopes include studies of piston wear and corrosion resistance. Petroleum companies use radioisotopes to monitor the progress of some oils through pipelines. The thickness of thin sheets of metal, plastic, and paper is subject to continuous production control through the use of a Geiger counter to measure the amount of radiation that passes through the sheet; the thinner the sheet is, the more radiation that the counter will detect. Quality control laboratories can detect small traces of radioactive elements in a metal part.

Scientific research is another major application of radioisotopes. Chemists use "tagged" atoms as *radioactive tracers* to study the mechanism, or series of individual steps, in complicated reactions. For example, by using water containing radioactive oxygen, scientists have determined that the oxygen in the glucose, $C_6H_{12}O_6$, formed in photosynthesis via the reaction

$$6\ CO_2(g) + 6\ H_2O(\ell) \rightarrow C_6H_{12}O_6(s) + 6\ O_2(g)$$

comes entirely from the carbon dioxide and that all oxygen from water is released as oxygen gas.

Archaeologists use neutron bombardment to produce radioactive isotopes in an artifact, which makes it possible to analyze the item without destroying it. Biologists employ radioactive tracers in the water absorbed by the roots of plants to study the rate at which the water is distributed throughout the plant system. These are but a few of the many ingenious applications that have been devised for this useful tool of science.

20.11 Nuclear Fission

Goal 14 Define and identify a nuclear fission reaction.
15 Define and identify a chain reaction.

In 1938, during the period when Nazi Germany was moving steadily toward war, dramatic and far-reaching events were taking place in German laboratories.

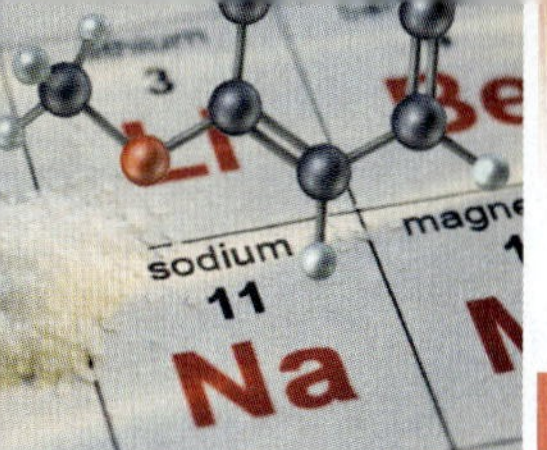

Everyday Chemistry 20.1

MEDICINE AND RADIOISOTOPES

Hospitals and larger medical clinics typically have a Department of Nuclear Medicine. This department is responsible for the production, use, and disposal of radioactive materials used at the medical facility (**Figure 20.23**). Medical uses of radioisotopes fall into two broad categories, diagnostic and therapeutic. A large hospital could use dozens of different radioisotopes in hundreds of diagnostic and therapeutic procedures.

Nearly all diagnostic radioisotopes emit gamma rays, which are easy to detect (**Figure 20.24**). A gamma ray is like a nuclear needle; it makes a quick, exceedingly narrow passage through the body, limiting damage to a small number of cells. This is one criterion such radioisotopes must meet. These radioisotopes must also have a short half-life to limit the time of the patient's exposure to radiation. The mechanism by which the body eliminates the radioisotope also must be known. Finally, the chemical behavior of the radioisotope must not interfere with normal body functions.

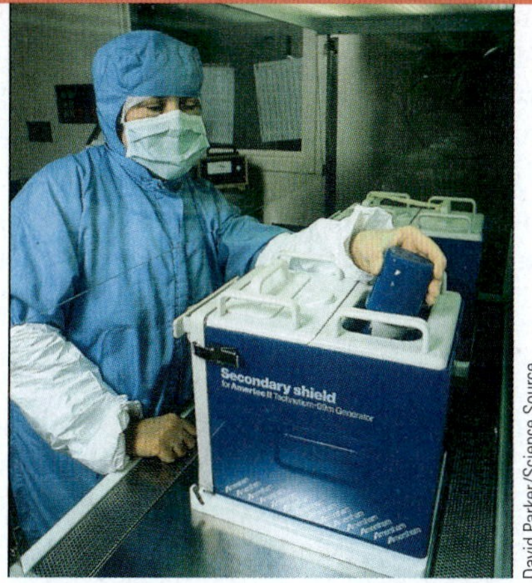

David Parker/Science Source

Figure 20.23 Radioisotope generator. The most widely used radioisotope generator converts molybdenum-99 to technetium-99m, in which the *m* indicates that the isotope is *m*etastable, which means that the nucleus is in a temporary high-energy state. The excess energy is given off in the form of gamma radiation, which is used for diagnostic procedures.

The goals of therapeutic uses of radioisotopes differ from the goals of diagnostic uses. Therapeutic radioisotopes are used to destroy abnormal—usually cancerous—cells as selectively as possible. Cell poisons such as radiation destroy abnormal cells more rapidly than normal cells because abnormal cells divide more quickly than normal cells. Therapeutic radioisotopes are usually alpha or beta emitters. These decay particles cause heavy damage confined to a small area, owing to their relatively low penetrating power. In the body, an alpha or beta emitter is a nuclear bull in a cellular china shop.

Quick Quiz

1. What four criteria must be met by a diagnostic radioisotope for it to be used medically?
2. Why are diagnostic radioisotopes usually gamma emitters, whereas therapeutic radioisotopes are usually beta emitters?

Diagnostic Radioisotopes

	Isotope	Half-Life	Emitted Particles	Uses
${}^{51}_{24}Cr$	Chromium-51	28 days	Gamma	Spleen imaging
${}^{59}_{26}Fe$	Iron-59	45 days	Beta, gamma	Bone marrow function
${}^{99m}_{43}Tc$	Technetium-99m	6 hours	Gamma	Bone, brain, liver, spleen imaging
${}^{131}_{53}I$	Iodine-131	8 days	Beta, gamma	Thyroid functioning
${}^{201}_{81}Tl$	Thallium-201	13 days	Beta, gamma	Heart imaging

Therapeutic Radioisotopes

	Isotope	Half-Life	Emitted Particles	Uses
${}^{32}_{15}P$	Phosphorus-32	14 days	Beta	Treatment of some leukemias, widespread carcinomas
${}^{60}_{27}Co$	Cobalt-60	5.3 years	Beta, gamma	External radiation source for cancer treatment
${}^{90}_{39}Y$	Yttrium-90	64 hours	Beta, gamma	Implanted in tumors
${}^{131}_{53}I$	Iodine-131	8 days	Beta, gamma	Treatment of thyroid cancer
${}^{226}_{88}Ra$	Radium-226	1620 years	Beta, gamma	Implanted in tumors

Figure 20.24 Thyroid gland imaging. (a) A healthy thyroid gland, located in the lower neck, is shaped like a bow tie. This image was produced by injecting the patient with a solution containing technetium-99m and scanning the neck with a gamma camera. (b) Hyperthyroidism, an overactive thyroid gland, is a common disease that has many causes. It can be treated with drugs, radioactive iodine treatment, or surgery.

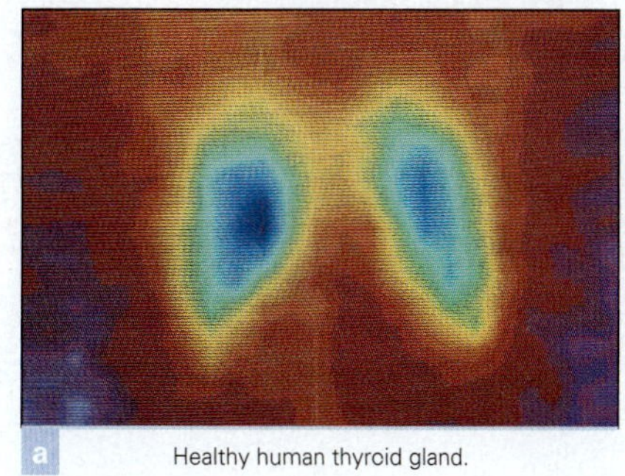

a Healthy human thyroid gland.

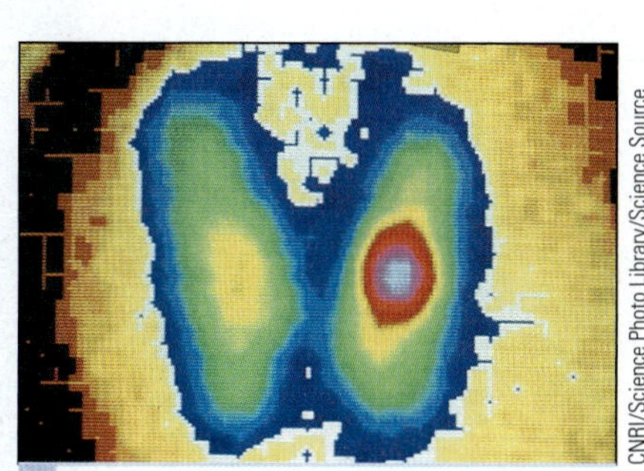

b Thyroid gland showing effect of hyperthyroidism.

CNRI/Science Photo Library/Science Source

Figure 20.25 Lise Meitner (1878–1968), one of a team of scientists who discovered nuclear fission. Meitner's work was recognized by naming element 109 meitnerium, symbol Mt, in her honor.

A team made up of Otto Hahn, Fritz Strassman, and Lise Meitner (**Figure 20.25**) was working with neutron bombardment of uranium. They were finding surprises among the products of the reaction. Namely, the products contained atoms of barium, krypton, and other elements far removed in both atomic mass and atomic number from the uranium atoms and neutrons used to produce them. The only possible explanation was difficult to believe: The uranium nucleus must be splitting into two nuclei of smaller mass. This kind of reaction is called **nuclear fission**.

In the fission of uranium-235, there are many products; it is not possible to write a single equation to show what happens. A representative equation is:

$$^{235}_{92}U + ^{1}_{0}n \rightarrow ^{94}_{38}Sr + ^{139}_{54}Xe + 3\,^{1}_{0}n$$

Note that it takes a neutron to initiate the reaction. Note also that the reaction produces *three* neutrons. If one or two of these collide with other fissionable uranium nuclei, there is the possibility of another fission or two. And the neutrons from those reactions can trigger others, repeatedly, as long as the supply of nuclei lasts. This is what is meant by a **chain reaction** (**Figure 20.26**) in which a nuclear product of the reaction becomes a nuclear reactant in the next step, thereby continuing the process.

The number of neutrons produced in the fission of $^{235}_{92}U$ varies with each reaction. Some reactions yield two neutrons per uranium atom; others, like that in Figure 20.26, yield three, and still others produce four or more. The average is about 2.5. If the quantity of uranium, or any other fissionable isotope, is large enough that most of the neutrons produced are captured within the sample, rather than escaping to the surroundings, the chain reaction will continue. The minimum quantity required for this purpose is called the **critical mass**.

Figure 20.26 Chain reaction started by the capture of a neutron. Each pair of isotopes shown is only one of the many different pairs that can be produced.

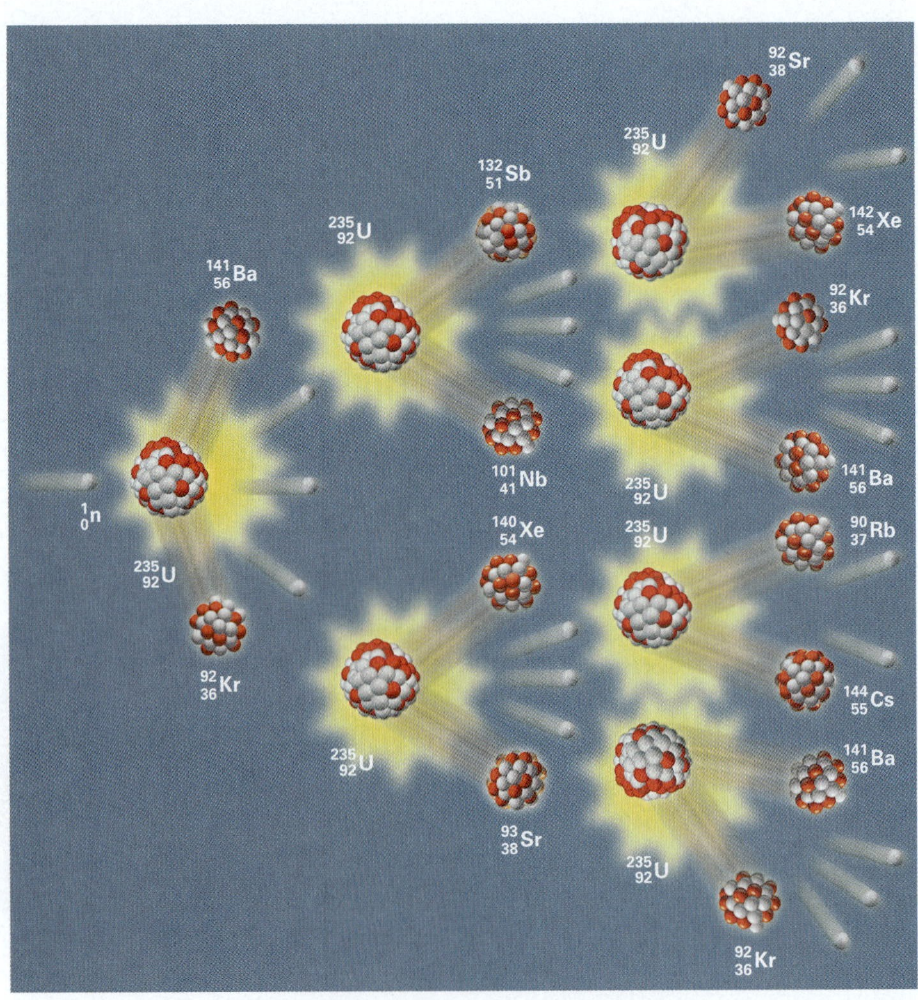

Everyday Chemistry 20.1

MEDICINE AND RADIOISOTOPES

Hospitals and larger medical clinics typically have a Department of Nuclear Medicine. This department is responsible for the production, use, and disposal of radioactive materials used at the medical facility (**Figure 20.23**). Medical uses of radioisotopes fall into two broad categories, diagnostic and therapeutic. A large hospital could use dozens of different radioisotopes in hundreds of diagnostic and therapeutic procedures.

Nearly all diagnostic radioisotopes emit gamma rays, which are easy to detect (**Figure 20.24**). A gamma ray is like a nuclear needle; it makes a quick, exceedingly narrow passage through the body, limiting damage to a small number of cells. This is one criterion such radioisotopes must meet. These radioisotopes must also have a short half-life to limit the time of the patient's exposure to radiation. The mechanism by which the body eliminates the radioisotope also must be known. Finally, the chemical behavior of the radioisotope must not interfere with normal body functions.

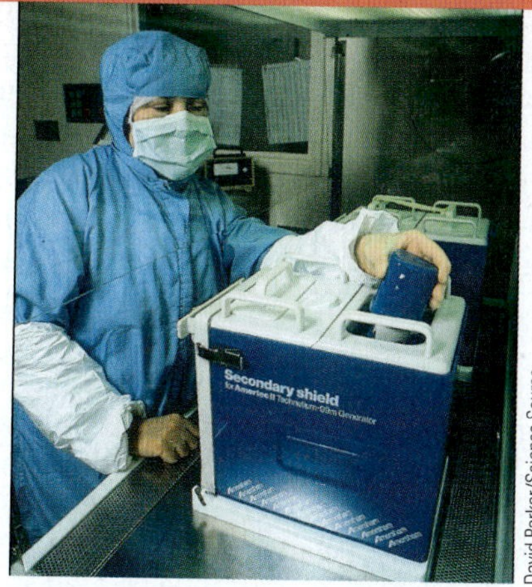

David Parker/Science Source

Figure 20.23 Radioisotope generator. The most widely used radioisotope generator converts molybdenum-99 to technetium-99m, in which the *m* indicates that the isotope is *m*etastable, which means that the nucleus is in a temporary high-energy state. The excess energy is given off in the form of gamma radiation, which is used for diagnostic procedures.

The goals of therapeutic uses of radioisotopes differ from the goals of diagnostic uses. Therapeutic radioisotopes are used to destroy abnormal—usually cancerous—cells as selectively as possible. Cell poisons such as radiation destroy abnormal cells more rapidly than normal cells because abnormal cells divide more quickly than normal cells. Therapeutic radioisotopes are usually alpha or beta emitters. These decay particles cause heavy damage confined to a small area, owing to their relatively low penetrating power. In the body, an alpha or beta emitter is a nuclear bull in a cellular china shop.

Diagnostic Radioisotopes

	Isotope	Half-Life	Emitted Particles	Uses
${}^{51}_{24}Cr$	Chromium-51	28 days	Gamma	Spleen imaging
${}^{59}_{26}Fe$	Iron-59	45 days	Beta, gamma	Bone marrow function
${}^{99m}_{43}Tc$	Technetium-99m	6 hours	Gamma	Bone, brain, liver, spleen imaging
${}^{131}_{53}I$	Iodine-131	8 days	Beta, gamma	Thyroid functioning
${}^{201}_{81}Tl$	Thallium-201	13 days	Beta, gamma	Heart imaging

Therapeutic Radioisotopes

	Isotope	Half-Life	Emitted Particles	Uses
${}^{32}_{15}P$	Phosphorus-32	14 days	Beta	Treatment of some leukemias, widespread carcinomas
${}^{60}_{27}Co$	Cobalt-60	5.3 years	Beta, gamma	External radiation source for cancer treatment
${}^{90}_{39}Y$	Yttrium-90	64 hours	Beta, gamma	Implanted in tumors
${}^{131}_{53}I$	Iodine-131	8 days	Beta, gamma	Treatment of thyroid cancer
${}^{226}_{88}Ra$	Radium-226	1620 years	Beta, gamma	Implanted in tumors

Quick Quiz

1. What four criteria must be met by a diagnostic radioisotope for it to be used medically?
2. Why are diagnostic radioisotopes usually gamma emitters, whereas therapeutic radioisotopes are usually beta emitters?

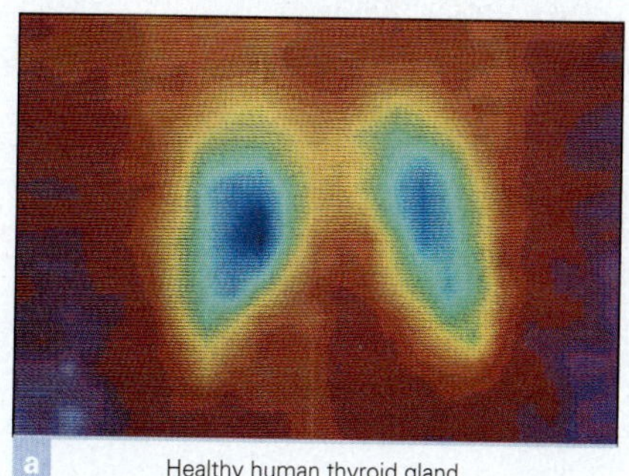

a Healthy human thyroid gland.

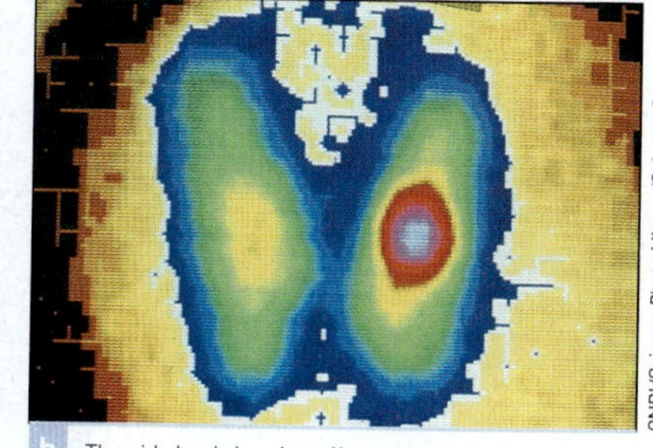

b Thyroid gland showing effect of hyperthyroidism.

CNRI/Science Photo Library/Science Source

Figure 20.24 Thyroid gland imaging. (a) A healthy thyroid gland, located in the lower neck, is shaped like a bow tie. This image was produced by injecting the patient with a solution containing technetium-99m and scanning the neck with a gamma camera. (b) Hyperthyroidism, an overactive thyroid gland, is a common disease that has many causes. It can be treated with drugs, radioactive iodine treatment, or surgery.

Figure 20.25 Lise Meitner (1878–1968), one of a team of scientists who discovered nuclear fission. Meitner's work was recognized by naming element 109 meitnerium, symbol Mt, in her honor.

A team made up of Otto Hahn, Fritz Strassman, and Lise Meitner (**Figure 20.25**) was working with neutron bombardment of uranium. They were finding surprises among the products of the reaction. Namely, the products contained atoms of barium, krypton, and other elements far removed in both atomic mass and atomic number from the uranium atoms and neutrons used to produce them. The only possible explanation was difficult to believe: The uranium nucleus must be splitting into two nuclei of smaller mass. This kind of reaction is called **nuclear fission**.

In the fission of uranium-235, there are many products; it is not possible to write a single equation to show what happens. A representative equation is:

$$^{235}_{92}U + ^{1}_{0}n \rightarrow ^{94}_{38}Sr + ^{139}_{54}Xe + 3\,^{1}_{0}n$$

Note that it takes a neutron to initiate the reaction. Note also that the reaction produces *three* neutrons. If one or two of these collide with other fissionable uranium nuclei, there is the possibility of another fission or two. And the neutrons from those reactions can trigger others, repeatedly, as long as the supply of nuclei lasts. This is what is meant by a **chain reaction** (**Figure 20.26**) in which a nuclear product of the reaction becomes a nuclear reactant in the next step, thereby continuing the process.

The number of neutrons produced in the fission of $^{235}_{92}U$ varies with each reaction. Some reactions yield two neutrons per uranium atom; others, like that in Figure 20.26, yield three, and still others produce four or more. The average is about 2.5. If the quantity of uranium, or any other fissionable isotope, is large enough that most of the neutrons produced are captured within the sample, rather than escaping to the surroundings, the chain reaction will continue. The minimum quantity required for this purpose is called the **critical mass**.

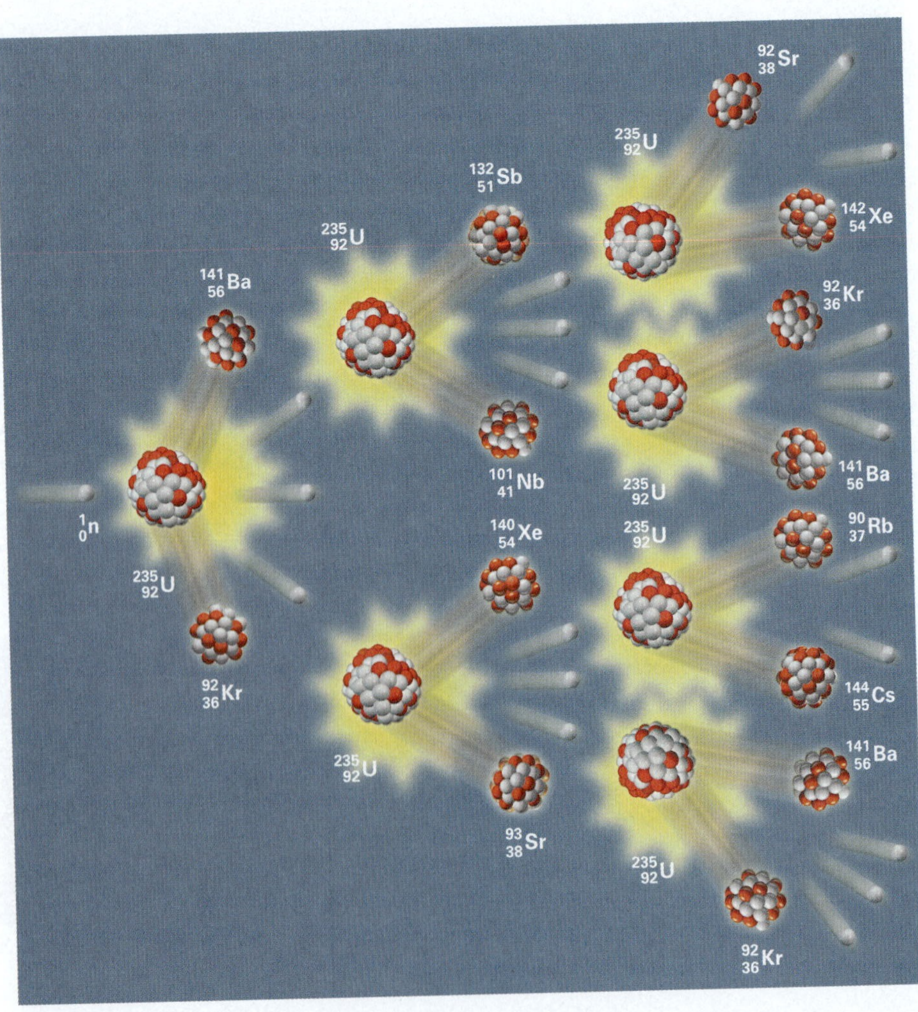

Figure 20.26 Chain reaction started by the capture of a neutron. Each pair of isotopes shown is only one of the many different pairs that can be produced.

Thinking About Your Thinking

Mental Models

Figure 20.26 is the starting point from which you can form a mental model of a nuclear chain reaction. On the left, you see the neutron, $^{1}_{0}n$, that is used to initiate the reaction. When it reacts with the uranium-235 nucleus, notice how two smaller nuclei (Ba and Kr) are produced and notice in particular the three neutrons that are also a product of the nuclear reaction. Now visually follow one neutron, the one moving down and to the right. It collides with another uranium-235 nucleus, producing two additional smaller nuclei and three additional neutrons. One neutron from an outside source gave rise to three neutrons, and these caused the formation of six neutrons (count the neutrons in the middle of the illustration), and these will in turn yield about 10 more neutrons, and so on. Now try to picture this process in your mind, as it would happen in three dimensions. As a test of your model, try to answer this question: Would a chain reaction be more likely to sustain itself in a sample in the shape of a sphere or in a sample of equal mass shaped like a sheet of paper? If you can mentally visualize the answer to this question, congratulations! You are well on your way to understanding nuclear fission.

Uranium-235 is capable of sustaining a chain reaction, but it makes up only 0.7% of all naturally occurring uranium. Therefore, it is not a very satisfactory source of nuclear fuel (Figure 20.27). An alternative is the plutonium isotope, $^{239}_{94}Pu$, produced from $^{238}_{92}U$, the most abundant uranium isotope. $^{239}_{94}Pu$ has a long half-life (2.4×10^4 years) and is fissionable. It has been used in the production of atomic bombs and is also used in some nuclear power plants to generate electrical energy. It is made in a **breeder reactor**, the name given to a device whose purpose is to produce fissionable fuel from nonfissionable isotopes.

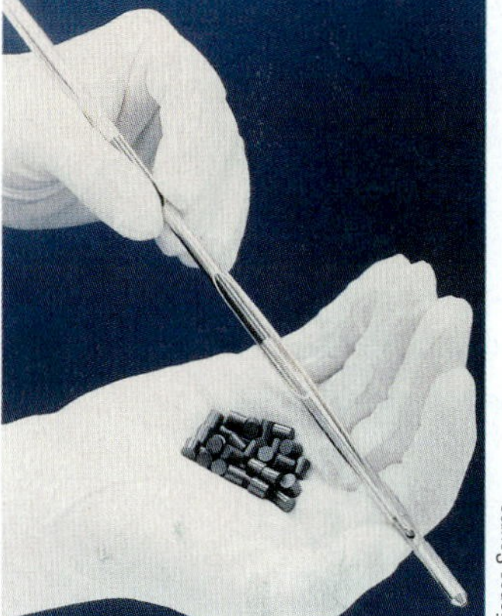

Science Source

Figure 20.27 Nuclear fuel. After mining, uranium is usually enriched to 3.5% $^{235}_{92}U$ in the form of uranium oxide, UO_2, powder. The powder is pressed into pellets, which are then inserted into tubes called fuel rods.

Target Check 20.5

How do the atomic numbers and mass numbers of the products of a fission reaction compare with those of the reactants?

20.12 Electrical Energy from Nuclear Fission

Goal 16 Describe how a nuclear power plant differs from a fossil-fueled power plant.

Except that which is produced by hydroelectric plants located on major rivers, most electrical energy comes from generators driven by steam. Traditionally, the steam comes from boilers fueled by oil, gas, or coal. The carbon dioxide produced from burning these fossil fuels and the uncertainties surrounding the availability and cost of petroleum bring attention to nuclear fission as an alternative energy source.

A photograph of the exterior of a **nuclear power plant** is shown in Figure 20.28, and a diagram of the operational portion of the interior of a nuclear power plant is shown in Figure 20.29. The turbine, generator, and condenser are similar to those found in any fuel-burning power plant. The nuclear fission reactor has three main components: the fuel elements, control rods, and moderator. The fuel elements are simply long trays that hold fissionable material in the reactor. As the fission reaction proceeds, fast-moving neutrons are released. These neutrons are slowed down by a moderator, which is water in the reactor illustrated. When the slower neutrons collide with more fissionable material, the reaction continues. The reaction rate is governed by cadmium or

Dobresum/Shutterstock.com

Figure 20.28 Nuclear power plant. The cooling towers are the most visibly prominent feature in nuclear power plants.

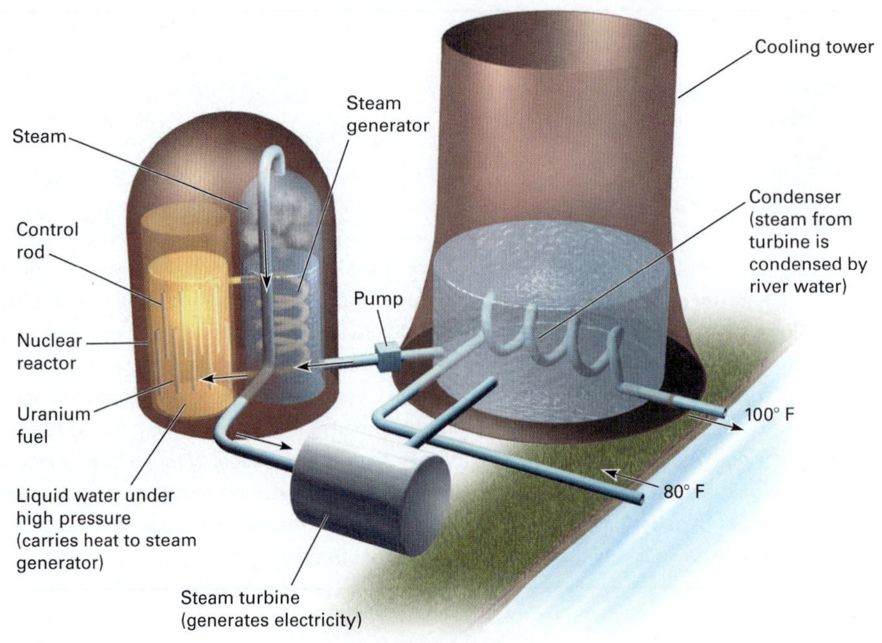

Figure 20.29 Schematic diagram of a nuclear power plant. Nuclear fission occurs in the reactor. Fission energy is used to heat water under pressure, which changes turbine water to steam in the steam generator. High-pressure steam drives the turbine, which in turn runs the electric generator that produces electric power. Spent steam from the turbine is changed to liquid water in the condenser and recycled back to the steam generator. Cooling water for the condenser comes from a cooling tower to which it is recycled. Make-up cooling water (to replace water lost to evaporation), or sometimes the cooling water itself, is drawn from a river, lake, or ocean.

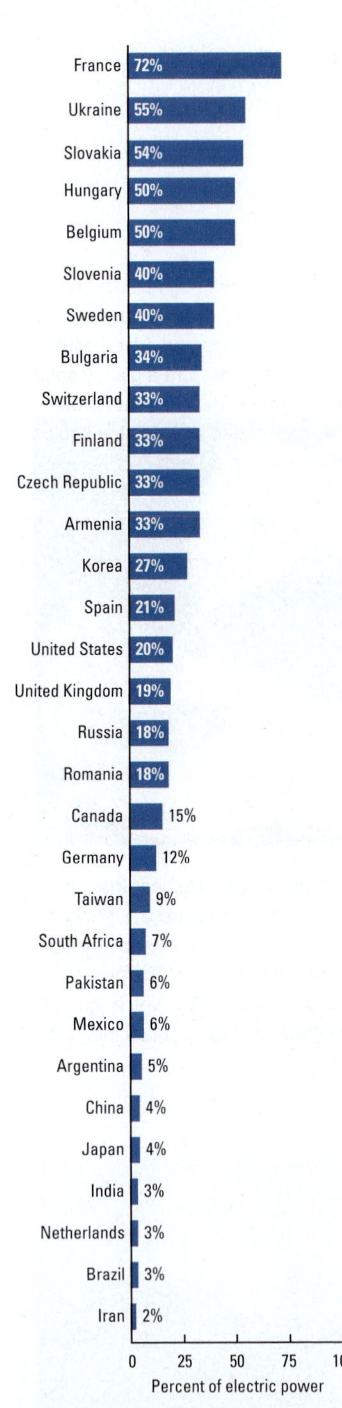

Figure 20.30 Percentage of electricity from nuclear power by country. The United States relies on nonnuclear sources of electrical energy much more than many of the other nations of the world.

boron control rods, which absorb excess neutrons. At times of peak power demand, the control rods are largely withdrawn from the reactor, permitting as many neutrons as necessary to find fissionable nuclei. When demand drops, the control rods are pushed in, absorbing neutrons and limiting the reaction.

The building and continued use of nuclear power plants face some opposition in the United States. The threat of an accident that might release large amounts of radiation over a densely populated area is the major concern. This fear became an actuality in Chernobyl, Republic of Ukraine, in 1986, when two water-cooling systems failed. The chain of events that followed, including fire and a nonnuclear explosion, led to the release of radioactive gases that spread over much of Europe and into Asia. Cooling system problems were also behind the Three Mile Island accident in Pennsylvania in 1979. In this incident, all radioactive substances were safely held within the reactor containment building, a safety feature in the design of U.S. power plants that is generally absent in the plants in the former Union of Soviet Socialist Republics (USSR).

Another concern is the potential for a natural disaster to damage a nuclear power plant and cause the release of radiation. The magnitude 9.0 earthquake centered near the east coast of Japan on March 11, 2011, and the associated tsunami triggered a nuclear emergency that reminded the world of the potential danger that may occur.

Whether getting energy from nuclear power plants is good or bad, ending the practice in the United States will not eliminate the danger. As Chernobyl demonstrated, the risk of accident is global and cannot be eliminated by a single nation. Recent statistics show that nuclear power plants produce about 20% of the electricity in the United States but over 50% of the electricity in France, Ukraine, and Slovakia (**Figure 20.30**). Even if Americans chose to reduce their energy demands by 20%, other nations have chosen nuclear power as their way to become energy-independent of imported oil. Nuclear power, for better or worse, is here to stay.

Even if a nuclear accident never occurs, there is still the problem of how and where to dispose of the dangerous radioactive wastes from nuclear reactors.

One method is to collect them in large containers that may be buried in caves deep below the Earth's surface. People who live and work near such disposal sites or along the routes by which the waste is transported are seldom enthusiastic about this solution. A preferred long-term storage form for radioactive waste involves vitrification, a method of trapping radioactive waste within a form of glass (**Figure 20.31**). However, this is more expensive than simply storing the unprocessed waste in containers.

Finally, there is fear that some irresponsible government may use nuclear fuel to manufacture nuclear weapons, spreading the threat of atomic warfare. More frightening is the possibility that some terrorist group might steal the materials needed to build a bomb. Although these threats cannot be removed from the world today, perhaps they would be lessened if the large-scale production of nuclear fuel for electric power was eliminated.

On the other side of all these concerns is, of course, the question, "If we do not build and operate nuclear power plants, how else will we meet the energy needs of the coming decades?" Perhaps the next section offers one answer, if it can be reached at all.

Science Photo Library/Science Source

Figure 20.31 Vitrification is one technique for storing nuclear waste. The liquid waste is heated to a high temperature and mixed with substances that form a glassy solid when cooled. This prevents problems associated with storing liquids in tanks that may leak.

20.13 Nuclear Fusion

Goal 17 Define and identify a nuclear fusion reaction.

There is nothing new about nuclear energy. Humans did not invent it. In fact, without knowing it, humanity has been dependent on its benefits since before the beginning of recorded time. In its common form, though, we do not call it nuclear energy. We call it solar energy—the energy that comes from the sun (**Figure 20.32**).

The energy that Earth derives from the sun comes from a type of nuclear reaction called **nuclear fusion**, in which two small nuclei combine to form a larger nucleus. The smaller nuclei are "fused" together, you might say. The typical fusion reaction believed to be responsible for the energy radiated by the sun is represented by the equation:

$$^{2}_{1}H + ^{3}_{1}H \rightarrow ^{4}_{2}He + ^{1}_{0}n$$

Fusion processes are, in general, more energetic than fission reactions. The fusion of one gram of hydrogen in the preceding reaction yields about four times as much energy as the fission of an equal mass of uranium-235. So far, people have been able to produce only one kind of large-scale fusion reaction, the explosion of a hydrogen bomb.

Research efforts are being made to develop a nuclear fusion reactor as a source of useful energy. It has several advantages over fission. It yields more energy per given quantity of fuel. The isotopes required for fusion are far more abundant than those needed for fission. Best of all, fusion yields no radioactive waste, removing both the need for extensive disposal systems and the danger of an accidental release of radiation into the atmosphere.

The main obstacle to be overcome before energy can be obtained from fusion is the extremely high temperature needed to start and sustain the reaction. The trigger for a hydrogen bomb is the heat generated by an atomic bomb. Furthermore, no substance known can hold fusion reactants at the needed temperature. Experiments on magnetic containment have been conducted, in which the fuel is suspended in a magnetic field. Energy is then added by pulsing laser beams into the magnetic bottle. Temperatures as high as 40 million degrees Celsius have been reported, but they could not be sustained.

Once the technological obstacles to using energy from fusion are overcome, time remains a serious problem. Even with continued research, it will be well into this century before a significant proportion of our energy needs could possibly be met by fusion.

NASA

Figure 20.32 The sun. The fundamental energy source of the sun is a series of reactions that can be summarized as the conversion of hydrogen into helium.

✓ Target Check 20.6

How do the products of a fusion reaction compare with the reactants?

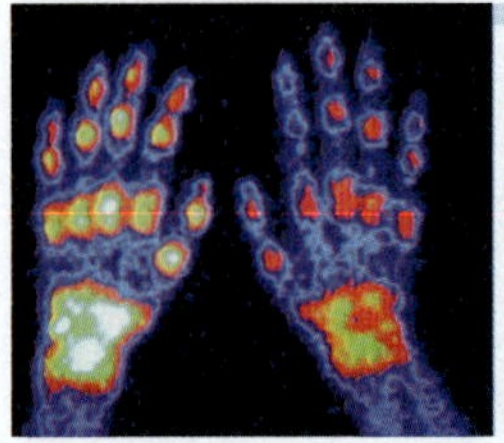

CNRI/SCIENCE PHOTO LIBRARY/ Science Photo Library/Getty Images

CHAPTER 20 IN REVIEW: NUCLEAR CHEMISTRY

Goal 1 Define and describe radioactivity.	**Radioactivity** is the spontaneous emission of particles or electromagnetic radiation resulting from the decay, or breaking up, of an atomic nucleus.
Goal 2 Name, identify from a description, and describe three types of radioactive emissions.	Three types of natural radioactivity, **alpha (α), beta (β)**, and **gamma (γ) rays**, can be formed when a nucleus decays. Alpha particles are much more massive than beta particles, and gamma rays have no mass. Alpha particles have a positive charge twice as large as the magnitude of the negative charge on beta particles; gamma rays have no charge. Penetrating power increases in the sequence alpha $<$ beta $<$ gamma.
Goal 3 Identify the function of a Geiger counter and describe how it operates.	The hand piece of a **Geiger counter** has a thin window through which ionizing radiation passes, entering a tube filled with argon gas. The gas is momentarily ionized, and the ions complete a circuit between two electrodes. The signal is amplified to produce clicking from a speaker and register on a meter. The frequency of the clicks indicates the radiation intensity.
Goal 4 Explain how exposure to radiation may harm or help living systems.	**Radioisotopes** are atoms with an unstable nucleus. The harmful effects of radiation on living systems come from its ability to break chemical bonds and thereby destroy healthy tissue. The key to radiation therapy is selective radiation of unhealthy tissue, which can eliminate it or reduce it.
Goal 5 Describe and illustrate what is meant by the half-life of a radioactive substance.	The rate at which a radioactive substance decays is measured by its **half-life**, the time it takes for one half of the radioactive atoms in a sample to decay.
Goal 6 Given the starting quantity of a radioactive substance, Figure 20.13, and two of the following, calculate the third: half-life, elapsed time, quantity of isotope remaining.	If S is the starting quantity of a radioactive substance and R is the amount that remains after n half-lives, then $\mathbf{R = S \times (1/2)^n}$. Figure 20.13 is a half-life decay curve for a radioactive substance, a plot of fraction of substance remaining versus half-lives elapsed.
Goal 7 Describe a natural radioactive decay series.	Products of radioactive decay may be other nuclei that undergo further decay, forming a **natural radioactive decay series**. When a radioisotope emits an alpha or beta particle, there is a **transmutation** of an element—that is, a change from one element to another. A decay series ends when a stable nucleus is formed.
Goal 8 Given the identity of a radioactive isotope and the particle it emits, write a nuclear equation for the emission.	An equation for radioactive decay is balanced for nuclear charge (number of protons) and nuclear mass (number of protons and neutrons). Emitted particles may be alpha particles, ${}^{4}_{2}He$, or beta particles, ${}^{0}_{-1}e$.
Goal 9 List or identify four ways in which nuclear reactions differ from ordinary chemical reactions.	There are four primary ways in which nuclear reactions differ from ordinary chemical reactions: **1.** Chemical properties are the same for all isotopes of an element. Nuclear properties of the isotopes of an element are quite different. **2.** Radioactivity is independent of the state of chemical combination of the radioactive isotope. **3.** Nuclear reactions can result in the formation of different elements. In chemical reactions, atoms keep their elemental identities. **4.** The amount of energy per quantity of reactant for nuclear changes is much larger than for chemical changes.
Goal 10 Define and identify nuclear bombardment reactions.	A **nuclear bombardment** reaction occurs when one nuclide is bombarded, or continuously exposed to a stream of particles, by another.
Goal 11 Distinguish natural radioactivity from induced radioactivity produced by bombardment reactions.	**Induced or artificial radioactivity** is generated when a stable nuclide is made radioactive by combination with another nuclide.
Goal 12 Define and identify transuranium elements.	The elements having atomic numbers greater than 92 are called the **transuranium elements.** All of these elements are radioactive.
Goal 13 Identify and describe uses for synthetic radioisotopes.	**Synthetic radionuclides** have many uses, including medical applications; ionizing air in a smoke detector; killing bacteria, molds, and yeasts to preserve food; and industrial and scientific applications.

Goal 14 Define and identify a nuclear fission reaction.	A heavier nucleus that splits into lighter nuclei undergoes nuclear **fission**.
Goal 15 Define and identify a chain reaction.	A **chain reaction** occurs when a product of one reaction is a reactant in the next step of the reaction pathway, thereby continuing the process. The minimum quantity of fissionable isotope required for this purpose is called the **critical mass**.
Goal 16 Describe how a nuclear power plant differs from a fossil-fueled power plant.	The turbine, generator, and condenser in a **nuclear power plant** are similar to those found in any fuel-burning power plant. The primary difference is the source of energy used to turn the turbines: nuclear fuel versus fossil fuel (or water from a dam flowing). The nuclear fission reactor has three main components: the fuel elements, control rods, and moderator.
Goal 17 Define and identify a nuclear fusion reaction.	Two light nuclei that are joined to form a heavier nucleus undergo nuclear **fusion**.

Key Terms

Most of the key terms and concepts and many others appear in the Glossary. Use your Glossary regularly.

alpha decay reaction, alpha decay p. 786
alpha particle, α-particle p. 786
artificial radioactivity p. 800
beta decay reaction, beta decay p. 786
beta particle, β-particle p. 786
breeder reactor p. 805
chain reaction p. 804
critical mass p. 804
gamma ray, γ-ray p. 786
Geiger counter p. 788
half-life, $t_{1/2}$ p. 791
induced radioactivity p. 800
natural radioactive decay series p. 796
nuclear bombardment p. 800
nuclear fission p. 804
nuclear fusion p. 807
nuclear power plant p. 805
particle accelerator p. 800
phosphorescence p. 788
positron p. 800
radioactivity p. 786
radiocarbon dating p. 794
radioisotopes p. 789
rem p. 789
scintillation counter p. 788
transmutation p. 796
transuranium elements p. 801

Frequently Asked Questions

What is the best overarching study approach for this chapter?
Much of the information in this chapter is conceptual in nature. It is important to carefully learn the meaning of terms and phrases. Once you have mastered these, your study strategy should focus on learning the relationships among the concepts.

What is the best overarching study approach for problems involving the half-life of radioisotopes?
Many seemingly different problem types arise from $R = S \times (1/2)^n$, and the half-life decay curve (Figure 20.13). Practicing many end-of-chapter problems will help you learn the variations of the problems based on this equation and graph.

How does writing a nuclear equation compare with writing a conventional chemical equation?
It is important to understand that writing a nuclear equation is different from writing a conventional chemical equation. The key to writing a nuclear equation is balancing both the sums of the mass numbers and the sums of the atomic numbers. In order to do this, you must memorize the mass number and atomic number of the alpha and beta particles.

How do I avoid confusing the similar terms fission *and* fusion*?*
The terms *fission* and *fusion* can be easy to reverse in one's mind. *Fission* was originally assigned to the process because it comes from the Latin word meaning "to split." *Fusion* describes "fusing together" of nuclei.

Concept-Linking Exercises

Write a brief description of the relationships among each of the following groups of terms or phrases. Answers to the Concept-Linking Exercises are given at the end of the chapter.

1. Alpha particles, beta particles, gamma rays, charge, penetrating power
2. Cloud chamber, Geiger counter, scintillation counter
3. Radiocarbon dating, half-life, carbon-14, carbon-12
4. Natural radioactive decay series, uranium, lead, alpha emission, beta emission
5. Nuclear bombardment, particle accelerator, transuranium elements
6. Nuclear fission, nuclear fusion, nuclear power plant, solar energy
7. Chain reaction, critical mass, breeder reactor

Small-Group Discussion Questions

Small-Group Discussion Questions are for group work, whether in class or under the guidance of a leader during a discussion section.

1. If an isotope has a neutron-to-proton ratio that is high relative to the ratio typically found in stable isotopes, is it most likely to decay by alpha, beta, or gamma emission? How will an isotope with a low neutron-to-proton ratio be likely to decay?

2. What are the limitations on carbon dating? For example, can it be used to date such artifacts as wooden caskets, ancient coins, animal bones, rocks, or scrolls? Can carbon dating be used for the bones of an animal that died 10 years ago? Can it be used for objects that are millions of years old?

3. Draw particulate-level sketches of the following processes: (a) the radioactive decay of tritium (hydrogen-3) via beta emission; (b) the nuclear bombardment that led to the discovery of the neutron, in which alpha particles bombarded beryllium-9, yielding carbon-12 and a neutron; and (c) the representative process for nuclear fusion, the reaction of hydrogen-2 and hydrogen-3 to yield helium-4 and a neutron.

4. The natural radioactive decay series that begins with uranium-235 and ends with lead-207 undergoes 11 reactions: alpha decay (hereafter α), beta decay (hereafter β), α, β, α, α, α, α, β, α, and β. Write the equation for each step.

5. Two types of radioactive decay not mentioned in the chapter are positron emission and electron capture. A positron has a charge opposite of an electron but the same mass. It has the symbol ${}^{0}_{+1}e$. Electron capture is the capture of an electron by a nucleus. It is believed that the captured electron usually comes from the $n = 1$ or $n = 2$ levels. When electron capture occurs, the electron is a reactant. Write balanced nuclear equations for the following processes: (a) Polonium-207 undergoes positron decay. (b) Carbon-11 decays by positron emission. (c) Oxygen-15 undergoes positron emission. (d) Potassium-38 emits a positron as it decomposes. (e) Rubidium-81 captures an orbital electron. (f) Beryllium-7 captures an extranuclear electron. (g) Gallium-67 decays by electron capture. (h) Iron-55 captures an orbital electron.

6. The famous Einstein equation $E = \Delta m \times c^2$ can be used to calculate the quantity of energy needed to separate a nucleus into its constituent parts. Consider the decomposition of 1 mole of hydrogen-2 (commonly called deuterium) nuclei: ${}^{2}_{1}H$ (2.01410 g/mol) $\rightarrow$ ${}^{1}_{1}H$ (1.007825 g/mol) + ${}^{1}_{0}n$ (1.008665 g/mol). (a) Calculate the mass defect, Δm, in g/mol. (b) Calculate the quantity of energy needed to separate the proton and neutron of a deuterium nucleus. (*Hint:* $1\ J = 1\ kg \cdot m^2/s^2$) (c) What mass of hydrogen would need to be burned according to $2\ H_2(g) + O_2(g) \rightarrow 2\ H_2O(g) + 484\ kJ$ in order to provide enough energy to decompose 1 mole of deuterium into subatomic particles? (d) What volume of hydrogen gas at room temperature is required to decompose 1 mole of deuterium into subatomic particles? (e) Is deuterium stable under ordinary conditions? Explain.

7. (a) Pure uranium metal is 0.7% uranium-235, with the rest being mostly uranium-238. Uranium-238 does not ordinarily undergo fission. Use this information to explain why natural uranium deposits do not undergo chain reactions. (b) The key to building a uranium bomb is to separate uranium-235 from the other naturally occurring isotopes. In the World War II era, that was accomplished by reacting uranium with fluorine to make gaseous uranium hexafluoride. The compound was then allowed to diffuse through a small opening into an empty chamber. Which uranium compound would move fastest in the gaseous state? Which side of a chamber, the one in which the gas was originally placed or the one on the other side of the hole, would become enriched in uranium-235? Explain.

8. Fission of 1 mole of uranium-235 releases 2×10^{10} kJ. Burning 1 ton of coal releases 2×10^{7} kJ. What weight of coal needs to be burned to equal the energy output of the fission of 1 lb of uranium?

9. List advantages and disadvantages of both fossil-fueled and nuclear-powered electrical power plants. Which is the better choice?

10. In Chapter 2, we described the Law of Conservation of Mass. Here, in this chapter, we illustrate nuclear changes in which the identity of an element is changed and matter is converted into energy. Why do scientists believe that the Law of Conservation of Mass is true?

Questions, Exercises, and Problems

Interactive versions of these problems may be assigned by your instructor. Solutions for blue-numbered questions are at the end of the chapter. Questions other than those in the General Questions and More Challenging Problems sections are paired in consecutive odd-even number combinations; solutions for the odd-numbered questions are at the end of the chapter.

Sections 20.2 and 20.3: The Dawn of Nuclear Chemistry and Radioactivity

1. What is an isotope?

2. (a) What is the nuclear symbol for a beta particle? (b) What is the name for the Greek letter γ? (c) Of the

radioactive emissions alpha, beta, and gamma, which is the most penetrating and which is the least penetrating?

3. *Decay* is a term used to describe what happens to a radioactive nucleus. What does *decay* mean in this sense?

4. Which of the following characterizes an alpha ray (choose all that apply): (a) It is composed of electrons; (b) it is composed of helium nuclei; (c) it is electromagnetic radiation; (d) it carries a positive charge; (e) it is a product of natural radioactive decay?

5. Compare the three forms of radioactive emissions in terms of mass number, electrical charge, and penetrating power.

6. Which of the following characterizes a beta ray (choose all that apply): (a) It is electromagnetic radiation; (b) it is composed of electrons; (c) it is attracted to the negatively charged plate in an electric field; (d) it is a product of natural radioactive decay; (e) it carries a negative charge?

Section 20.4: The Detection and Measurement of Radioactivity

7. What happens, or might happen, when an emission from a radioactive substance collides with an atom or molecule? Is this harmful? Explain.

8. Radiation is sometimes described as "ionizing radiation." What does this mean? Is all radiation ionizing? Justify your answer.

9. What is a Geiger counter? How does it work?

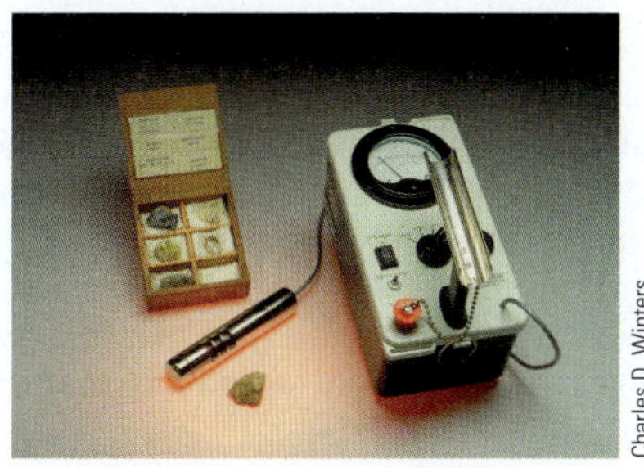
Charles D. Winters

A Geiger counter is used to detect ionizing radiation.

10. Identify some properties of radioactive emissions that are used in detecting and measuring them.

11. How do Geiger and scintillation counters differ in how they tell an observer that an object is radioactive? Can either or both be used to measure radiation as well as detect it? If so, precisely what is measured?

12. What is a scintillation counter? How does it work?

13. How do gamma cameras and scanners record the presence of radiation?

14. Distinguish between a gamma camera and a scanner. What is their principal advantage in medical applications compared with nonradiological procedures for similar purposes?

Section 20.5: The Effects of Radiation on Living Systems

15. Identify the greatest source of background radiation for U.S. citizens. What are the second and third greatest sources?

16. Is background radiation dangerous? Should we be concerned about it? If so, what can you do about it?

Charles D. Winters

The United States Environmental Protection Agency estimates that radon exposure causes over 21,000 cancer deaths per year. All homes should be tested for radon.

Section 20.6: Half-Life

17. The radioactivity of a sample has dropped to 1/4 of its original intensity. How many half-lives have passed?

18. What is meant by the half-life of a radioactive substance?

19. What fraction of a radioisotope remains after the passage of seven half-lives?

20. How many half-lives have passed when a radioactive substance has lost 15/16 of its radioactivity?

21. Calculate the mass of radioisotope in a sample that will be left after 33 minutes if the sample originally has 12.9 g of that radioisotope. The half-life of the radioisotope is 11.0 minutes.

22. Chromium-51 is a radioisotope that is used to assess the lifetime of red blood cells. The half-life of chromium-51 is 27.7 days. If you begin with 86.9 mg of chromium-51, how many milligrams will you have left after 55.4 days have passed?

23. One of the more hazardous radioactive isotopes in the fallout of atomic bombs is strontium-90, for which the half-life is 28 years. If 654 g $^{90}_{38}Sr$ fall on a family farm on the day a child is born in 2019, how many grams will still be on the land when the farmer's granddaughter is born in 2075? How about when the granddaughter marries on the same farm in 2095?

24. $^{214}_{83}Bi$ is a radioactive isotope of the element bismuth with a half-life of 19.7 minutes. What percentage of a stored sample of this isotope would be lost due to radioactive decay in a 551-second period?

25. The half-life of $^{208}_{81}Tl$ is 3.1 minutes. A 84.6-g sample is studied in the laboratory.
    a) What mass in grams of the isotope will remain after 12 minutes?

b) In how many minutes will the mass of ${}^{208}_{81}Tl$ be 3.48 g?

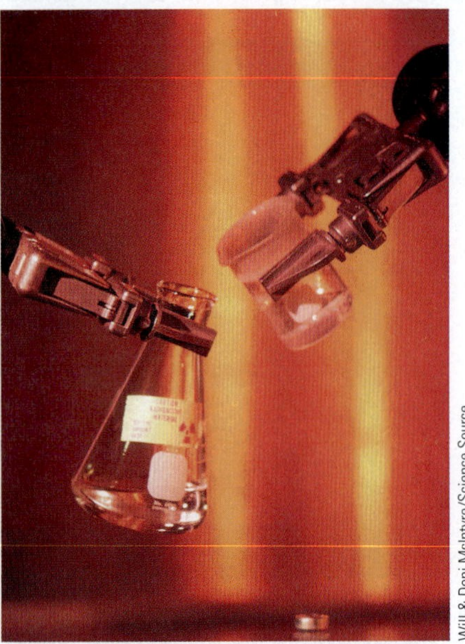
Will & Deni McIntyre/Science Source

Remotely operated robotic arms are used to maneuver dangerously radioactive samples.

26. If the mass of a sample of radioactive cesium-137 decays from 71.9 mg to 20.8 mg in 54.0 years, what is the half-life of cesium-137?

27. Uranium-235, the uranium isotope used in making the first atomic bomb, is the starting point of a natural radioactivity series. The next isotope in the series is thorium-231. At the beginning of a test period, a sample contained 9.53 g of the thorium isotope. After 83.2 hours only 1.05 g of the original isotope remained. What is the half-life of thorium-231?

28. The half-life of the radioactive isotope thorium-234 is 24.1 days. How long will it take for the mass of a sample of thorium-234 to decay from 84.5 mg to 20.9 mg?

29. When excavating for the foundation of a new building, a contractor uncovered human skeletons in what turned out to be a burial ground from an ancient civilization. They were taken to a nearby university and submitted to radiocarbon dating analysis. It was found that the bones emit radiation at a rate of 55% of the rate of a living organism. How many years ago did the specimen die? (Use 5.73×10^3 years as the half-life of carbon-14.)

30. The radioactive isotope carbon-14 is used for radiocarbon dating. The half-life of carbon-14 is 5.73×10^3 years. A wooden artifact in a museum has a ${}^{14}C$ to ${}^{12}C$ ratio that is 0.775 times that found in living organisms. Estimate the age of the artifact.

Section 20.7: Natural Radioactive Decay Series—Nuclear Equations

31. What happens to the nucleus of an atom that experiences an alpha decay reaction? Compare the final isotope with the original isotope. Does the element undergo transmutation?

32. What happens to the nucleus of an atom that experiences a beta decay reaction? Compare the final isotope with the original isotope. Does the element undergo transmutation?

33. Write nuclear equations for the beta emissions of ${}^{228}_{89}Ac$ and ${}^{212}_{83}Bi$.

34. You may use the Table of Elements, as necessary, to answer these questions. (a) When the isotope bismuth-214 undergoes alpha decay, what are the name and symbol of the product isotope? (b) When the isotope iron-59 undergoes beta decay, what are the name and symbol of the product isotope?

35. Write nuclear equations for the alpha decay of ${}^{216}_{84}Po$ and ${}^{234}_{92}U$.

36. You may use the Table of Elements, as necessary, to answer these questions. Write balanced nuclear equations for the following: (a) The isotope radon-222 undergoes alpha emission. (b) The isotope bismuth-214 undergoes beta emission.

Section 20.8: Nuclear Reactions and Ordinary Chemical Reactions Compared

37. Why is it possible to speak of the "chemical properties of lead" but not the "nuclear chemical properties of lead"?

38. How do the chemical properties of carbon-12 compare with the chemical properties of carbon-14? If there is a difference, explain why.

39. The radioactivity of a sample of dirt containing uranium compounds records 5000 counts per minute when measured with a Geiger counter. The sample is treated physically to isolate the uranium compound, which is then decomposed chemically into pure uranium. If you disregard any loss of radioactivity because of decay during the purification process, will the pure uranium still radiate at 5000 counts per minute, or will it be more or less than 5000? Explain your answer.

40. A fundamental idea of Dalton's atomic theory is that atoms of an element can be neither created nor destroyed. We now know that this is not always true. Specifically, it is not true for uranium and lead atoms as they appear in nature. Are the numbers of these atoms increasing or decreasing? Explain.

Section 20.9: Nuclear Bombardment and Induced Radioactivity

41. Distinguish between nuclear reactions that begin spontaneously and those that begin with nuclear bombardment. What is nuclear bombardment?

42. What distinguishes induced radioactivity from natural radioactivity?

43. Which of the following particles can be accelerated in particle accelerators and which cannot: electrons, protons, neutrons, positrons, alpha particles? Which property of the particle(s) governed your choice?

44. What does a particle accelerator do, and how does it do it?

A cyclotron is a particle accelerator in which particles travel in a nearly circular path. The cyclotron shown is emitting a proton beam.

45. Compare the atomic numbers of all elements that are naturally radioactive with the atomic numbers of elements that exhibit artificial radioactivity.

46. What are the transuranium elements? What property is associated with all transuranium elements? Do you know of any practical application of transuranium elements, or are they mostly laboratory curiosities, useful primarily in research?

47. Are elements in the lanthanide series of elements in the periodic table transuranium elements? What about elements in the actinide series?

48. Is the element with atomic number 118 a transuranium element? Do you expect it to be radioactive? What sort of chemical properties do you expect it to have?

49. Complete each nuclear bombardment equation by supplying the nuclear symbol for the missing species.

a) $^{44}_{20}\text{Ca} + ^{1}_{1}\text{H} \rightarrow ? + ^{1}_{0}\text{n}$

b) $^{252}_{98}\text{Cf} + ^{10}_{5}\text{B} \rightarrow 5\,^{1}_{0}\text{n} + ?$

c) $^{106}_{46}\text{Pd} + ^{4}_{2}\text{He} \rightarrow ^{109}_{47}\text{Ag} + ?$

50. Complete each nuclear bombardment equation by supplying the nuclear symbol for the missing species.

a) $? + ^{4}_{2}\text{He} \rightarrow ^{12}_{6}\text{C} + ^{1}_{0}\text{n}$

b) $^{114}_{48}\text{Cd} + ^{2}_{1}\text{H} \rightarrow ? + ^{1}_{1}\text{H}$

c) $^{235}_{92}\text{U} + ^{1}_{0}\text{n} \rightarrow ? + ^{92}_{38}\text{Sr} + 2\,^{1}_{0}\text{n}$

Sections 20.11–20.13: Nuclear Fission; Electrical Energy from Nuclear Fission; Nuclear Fusion

51. How are fission reactions like fusion reactions, and how are they different?

52. Can radioactive decay be classified as nuclear fission? Why or why not?

53. What is a chain reaction? What essential feature must be present in a nuclear reaction before it can become a chain reaction?

54. $^{235}_{92}\text{U} + ^{1}_{0}\text{n} \rightarrow ^{92}_{38}\text{Sr} + ^{139}_{54}\text{Xe} + 3\,^{1}_{0}\text{n}$ is referred to as a "representative equation," indicating that it is only one of several possible equations for the fission of $^{235}_{92}\text{U}$ when it is bombarded by a single neutron. Another fission of the same reactants yields the isotopes $^{144}_{55}\text{Cs}$ and $^{90}_{37}\text{Rb}$. Write the equation for that reaction.

55. Can a fusion reaction be a chain reaction? Why or why not?

56. Starting a chain reaction is one thing; keeping it going is another. What is required if a chain reaction is to continue? By what term is this requirement identified?

57. What advantages do fusion reactions have over fission reactions as a source of nuclear power? If fusion reactions are more desirable than fission reactions, why don't we use them instead of fission reactions?

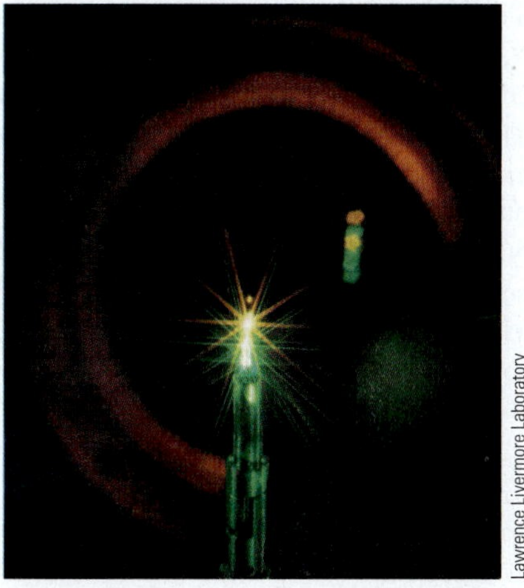

Small-scale fusion reactions are possible with today's technology. The photograph shows the reaction of hydrogen-2 and hydrogen-3 in a 1-mm capsule.

58. List some of the advantages and disadvantages of nuclear power plants compared with other sources of electrical energy. In your opinion, do the advantages outweigh the disadvantages?

General Questions

59. Distinguish precisely, and in scientific terms, the differences among items in each of the following pairs or groups.

a) Alpha, beta, and gamma radiation

b) x-rays, γ-rays

c) α-particle, β-particle

d) Natural and induced radioactivity

e) Chemical reaction, nuclear reaction

f) Isotope, radioisotope

g) Element, transuranium element

h) Nuclear fission, nuclear fusion

i) Atomic bomb, hydrogen bomb

60. Classify each of the following statements as true or false.

a) A radioactive atom decays in the same way whether or not the atom is chemically bonded in a compound.

b) The chemical properties of a radioactive atom of an element are different from the chemical properties of a nonradioactive atom of the same element.

c) α-particles have more penetrating power than β-particles.

d) α- and β-emissions are particles, but a γ-ray is an "energy ray."

e) Radioactivity is a nuclear change that has no effect on the electrons in nearby atoms.

f) The number of protons in a nucleus changes when it emits a beta particle.

g) The mass number of a nucleus changes in an alpha emission but not in a beta emission.

h) Synthetic radioisotopes have no application in everyday life or industry; they are used only for scientific research purposes.

i) The first transmutations were achieved by the alchemists.

j) Radioisotopes can be made by bombarding a nonradioactive isotope with atomic nuclei or subatomic particles.

k) The atomic numbers of products of a fission reaction are smaller than the atomic number of the original nucleus.

l) Nuclear power plants are a safe source of electrical energy.

m) The main obstacle to developing nuclear fusion as a source of electrical energy is a shortage of nuclei to serve as "fuel."

61. A major form of fuel for nuclear reactors used to produce electrical energy is a fissionable isotope of plutonium. Plutonium is a transuranium element. Why is this element used instead of a fissionable isotope that occurs in nature?

62. Why is half-life used for measuring rate of decay rather than the time required for the complete decay of a radioactive isotope?

63. A ton of high-grade coal has an energy output of about 2.5×10^7 kJ. The energy released in the fission of 1 mole of $^{235}_{92}U$ is about 2.0×10^{10} kJ. How may tons of coal could be replaced by 1 pound of uranium-235, assuming the materials and the technology were available?

Ivars Linards Zolnerovics/Shutterstock.com

The energy released by burning huge quantities of coal is equal to the energy released by the fission of a relatively tiny quantity of uranium.

More Challenging Questions

64. Loss of mass is not a satisfactory way to express rate of decay, but we used it in Active Examples 20.1 through 20.3 because it is the easiest to visualize. However, we chose the words for these Active Examples very precisely; carefully read, they are correct. Now, why is loss of mass usually unsatisfactory as a measure of rate of decay? Suggest a better way, and explain its advantage.

65. Suppose you have a radioisotope, A, that goes through a two-step decay sequence, first to B and then to C, which is stable. Suppose also that the half-life from A to B is six days, and the half-life from B to C is one day. Predict by listing in declining order, greatest to smallest, the amounts of A, B, and C that will be present (a) at the end of 6 days and (b) at the end of 12 days. Explain your prediction.

66. A fragment of cloth is believed to have been worn by a person at about the beginning of the Common Era. Analysis shows that radiation from the fragment is 22.7 units, whereas radiation from a living specimen is 29.0 units when measured with the same instrument. Is it possible that the fragment might have come from the period predicted? Justify your answer.

67. A sample of pure calcium chloride is prepared in a laboratory. A small but measurable amount of the calcium in the compound is made up of calcium-47 atoms, which are beta emitters with a half-life of 4.35 days. The compound is securely stored for a week in an inert atmosphere. When it is used at the end of that period, it is no longer pure. Why? With what element would you expect it to be contaminated?

68. Two uranium sulfides have the formulas US and US_2. A laboratory worker prepares 50.0 g of each compound. If the uranium in each compound has all the isotopes of uranium in their normal distribution in nature, which compound, if either, will exhibit the greater amount of radioactivity? If 0.5 mole of each compound is prepared, which compound (if either) will be more radioactive? Explain both answers.

69. Compare the representative equation for a fission reaction, $^{235}_{92}U + ^{1}_{0}n \rightarrow ^{92}_{38}Sr + ^{139}_{54}Xe + 3\ ^{1}_{0}n$, with the equation you wrote to answer Question 54. Which of the two reactions is more likely to contribute to a chain reaction? Explain.

Answers to Target Checks

1. a) α: $^{4}_{2}He$, 2+; β: $^{0}_{-1}e$, 1−; b) Gamma > Beta > Alpha

2. Geiger and scintillation counters "count" individual radioactive emissions by measuring electric current (Geiger) or light intensity (scintillation) produced by the radiation. These are both proportional to the intensity of radiation and are interpreted in that way. This intensity can be expressed as number of counts per unit of time.

3. Test your house for radon and don't smoke.

4. (a) Nuclear bombardment reactions produce isotopes that do not exist in nature. (b) To be accelerated, a particle must have an electrical charge.

5. Isotopes produced in a fission reaction have smaller atomic numbers and smaller mass numbers than the starting isotope.

6. An isotope produced by a fusion reaction has a larger atomic number and mass number than the starting isotopes.

Number of Radioactive ${}^{40}_{19}K$ Atoms in a 150-Pound Person

$$225\text{ g K} \times \frac{6.02 \times 10^{23}\text{ K atoms}}{39.10\text{ g K}} \times \frac{0.0118\ {}^{40}_{19}\text{K atoms}}{100\text{ K atoms}}$$

$$= 4.09 \times 10^{20}\ {}^{40}_{19}\text{K atoms}$$

Answers to Practice Exercises

1. 45 days ÷ 15 days/half-life = 3 half-lives; $R = S \times (1/2)^n = 32\text{ g } {}^{205}_{83}\text{Bi} \times (1/2)^3 = 4.0\text{ g } {}^{205}_{83}\text{Bi}$
2. $54\text{ min} + 60\text{ min} = 114\text{ min}; 114\text{ min} \times \frac{1\text{ half-life}}{7.6\text{ min}} =$ 15 half-lives
   $R = S \times (1/2)^n = 9.5 \times 10^3\ \mu\text{g } {}^{38}_{19}\text{K} \times (1/2)^{15} = 0.29\ \mu\text{g } {}^{38}_{19}\text{K}$
3. R/S = 113 mg ÷ 251 mg = 0.450; From Figure 20.13, this is 1.2 half-lives
   3.0 h ÷ 1.2 half-lives = 2.5 h/half-life
4. $1.7 \times 10^4\text{ years} \times \frac{1\text{ half-life}}{5.73 \times 10^3\text{ years}} = 3.0\text{ half-lives}$
   3.0 half-lives corresponds to $(0.5)^3 = 0.13$ remaining, or 13%
5. ${}^{222}_{86}\text{Rn} \rightarrow {}^{4}_{2}\text{He} + {}^{218}_{84}\text{Po};\ {}^{218}_{84}\text{Po} \rightarrow {}^{4}_{2}\text{He} + {}^{214}_{82}\text{Pb}$
6. ${}^{137}_{55}\text{Cs} \rightarrow {}^{0}_{-1}\text{e} + {}^{137}_{56}\text{Ba}$

Answers to Concept-Linking Exercises

You may have found more relationships or relationships other than the ones given in these answers.

1. Alpha and beta particles and gamma rays are products of radioactive decay. Alpha particles have a positive charge, beta particles have a negative charge, and gamma radiation is uncharged. The penetrating power of these emissions increases in the order: alpha < beta < gamma.
2. A cloud chamber, Geiger counter, and scintillation counter are all instruments used to detect radioactivity. A cloud chamber is a container filled with air and a supersaturated vapor. When radiation passes through the container, some air molecules ionize and vapor condenses on them, leaving a visible vapor trail. Radiation detection in a Geiger counter occurs in a tube filled with a gas. When radiation passes through the tube, gas molecules ionize and produce a current that can be quantitatively measured by a detector. A scintillation counter uses a transparent solid to detect radiation. When radiation passes through the solid, the particles absorb energy and then release it in the form of flashes of light, which can be quantitatively counted by a detector.
3. Radiocarbon dating is used to determine the age of fossils. It is based on measuring the ratio of carbon-14 to carbon-12 in the fossil. When an organism is alive, the ${}^{14}C$:${}^{12}C$ ratio in the organism is the same as in the environment. Carbon-14 is radioactive and slowly decays, with a half-life of 5.73×10^3 years. It is continually replenished in a living organism, however, so the ${}^{14}C$:${}^{12}C$ ratio remains constant as long as the organism is alive. Carbon-12 is a stable isotope, so it does not decay. When the organism dies, radioactive ${}^{14}C$ continues to decay, but it is no longer replenished. The age of a fossil can be determined by comparing the ${}^{14}C$:${}^{12}C$ ratio in the fossil to that in the environment.
4. A natural radioactive decay series describes the fate of natural unstable isotopes, from the starting radioactive isotope to the final stable isotope. Uranium-238 is an example of a natural radioactive isotope. It decays to lead-206 through a series of eight alpha emissions and six beta emissions.
5. Nuclear bombardment is the process by which two particles are made to collide with sufficient kinetic energy to form new particles. A particle accelerator is an instrument used to carry out bombardment reactions. The transuranium elements—those with higher atomic numbers than that of uranium, Z = 92—can be produced via nuclear bombardment in a particle accelerator.
6. Nuclear fission is a nuclear reaction in which a large nucleus splits into two smaller nuclei. A nuclear power plant uses the energy released in a fission reaction to generate electricity. Nuclear fusion is a nuclear reaction in which two small nuclei combine to form a larger nucleus. Solar energy results from fusion reactions.
7. A chain reaction has one of its own reactants as a product, allowing the original reaction to continue. The minimum quantity of matter necessary for a chain reaction to continue is its critical mass. A breeder reactor is a nuclear reactor in which fissionable fuel is produced from nonfissionable isotopes.

Solutions to Blue-Numbered Questions, Exercises, and Problems

1. An isotope is a specific kind of atom that has a specific nuclear composition.
3. It refers to the spontaneous decomposition of the nucleus.
5. Alpha particles are helium nuclei with a mass number of 4 and a 2+ charge. Beta particles are electrons with a mass number of 0 and a 1− charge. Gamma rays are high-energy electromagnetic radiation having no mass and no charge. Penetration power increases in the order alpha < beta < gamma.
7. The collision of a radioactive emission with an atom or molecule may rearrange the electrons in the target, possibly ionizing it and causing a potentially harmful chemical change.
9. A Geiger counter is a device for detecting and measuring radiation. Figure 20.7 describes how it works.
11. A Geiger counter "clicks" when a radioactive emission is detected; a scintillation counter counts pulses of light generated by radiation. Both devices can measure radiation as well as detect it. The Geiger counter actually measures electric current; the scintillation counter measures light pulses. Both devices express their measurements as counts per unit time.
13. A gamma camera is immobile while it takes a picture, creating essentially a two-dimensional image of an object. The scanner moves as it takes many pictures. Computer enhancement and combination of many pictures allow three-dimensional-like images to be constructed.
15. Figure 20.12 shows that medical procedures cumulatively count for about half of a person's exposure. Radon is second, and space and internal sources tie for third.

17. Two half-lives have passed. $1/4 = (1/2)^2$, where the exponent is the number of half-lives.

19. $(1/2)^7 = 1/128$

21. $R = 12.9 \text{ g} \times (1/2)^3 = 1.6 \text{ g}$

23. From 2019 to 2075 is 56 years, or two half-lives. Hence, $R = 654 \text{ g} \times (1/2)^2 = 1.6 \times 10^2$ g. From 2019 to 2095 is 76 years, or $76/28 = 2.7$ half-lives. Hence, $R = 654 \text{ g} \times (1/2)^{2.7} = 1.0 \times 10^2$ g.

25. (a) 12 min ÷ (3.1 min/half-life) = 3.9 half-lives. From the graph, R/S = 0.067. 0.067 × 84.6 g = 5.7 g remain. By the equation, $R = 84.6 \times (1/2)^{3.9} = 5.7$ g remains. (b) R/S = 3.48 g ÷ 84.6 g = 0.0411. From the graph, this is 4.6 half-lives. 4.6 half-lives × 3.1 min/half-life = 14 min.

27. R/S = 1.05 g ÷ 9.53 g = 0.110. From the graph, this is 3.25 half-lives. 83.2 h ÷ 3.25 half-lives = 25.6 h/half-life = $t_{1/2}$.

29. R/S = 0.55. From the graph, this is 0.86 half-life. 0.86 half-life × 5.73×10^3 years/half-life = 4.9×10^3 years.

31. The original nucleus disintegrates into a ${}^{4}_{2}\text{He}$ nucleus and another nucleus. The final isotope has two fewer protons and two fewer neutrons than the original. This is a transmutation, a change from one element to another, because the number of protons changes.

33. ${}^{228}_{89}\text{Ac} \rightarrow {}^{0}_{-1}\text{e} + {}^{228}_{90}\text{Th}$; ${}^{212}_{83}\text{Bi} \rightarrow {}^{0}_{-1}\text{e} + {}^{212}_{84}\text{Po}$

35. ${}^{216}_{84}\text{Po} \rightarrow {}^{4}_{2}\text{He} + {}^{212}_{82}\text{Pb}$; ${}^{234}_{92}\text{U} \rightarrow {}^{4}_{2}\text{He} + {}^{230}_{90}\text{Th}$

37. "Nuclear chemical properties of lead" is meaningless for two reasons. First, the chemical properties of all isotopes of lead are the same. Second, the nuclear properties of lead isotopes are specific for each individual isotope.

39. The count will remain at 5000/minute. The radioactivity of an element is independent of the form of the element, whether it is a pure element or in a compound.

41. Radioactivity is spontaneous, whereas nuclear bombardment reactions are produced by projecting a nuclear particle into another nuclear particle.

43. Electrons, protons, positrons, and alpha particles can be accelerated in particle accelerators. The particle must have an electrical charge to be accelerated.

45. Natural versus artificial radioactivity is not a function of atomic number. Isotopes of many elements are naturally radioactive, and artificially radioactive isotopes of many elements can be created.

47. All elements in the lanthanide series have atomic numbers less than 92, so none is a transuranium element. The elements in the actinide series with atomic numbers greater than 92 are transuranium elements.

49. (a) ${}^{44}_{21}\text{Sc}$; (b) ${}^{257}_{103}\text{Lr}$; (c) ${}^{1}_{1}\text{H}$

51. Both fission and fusion reactions result in a release of energy. In fission reactions, a larger nucleus splits into smaller nuclei. In a fusion reaction, two small nuclei combine to form a larger nucleus.

53. A chain reaction is a reaction that has as a product one of its own reactants; that product becomes a reactant, thereby allowing the original reaction to continue. For a chain reaction to continue, there must be enough fissionable material to react with the neutrons given off.

55. No: The products of a fusion reaction are not one of the reactants.

57. Nuclear fusion is more promising than fission as an energy source because it produces more energy per given amount of fuel. Fusion fuel is more abundant, and fusion reactions generate no hazardous radioactive waste. Fusion's major drawback is the extremely high temperature needed to initiate the process.

60. True: a, d, f, g, j, k. False: b, c, e, h, i, m. The answer to l is left to you.

61. Natural fissionable isotopes are rare. Plutonium-234 is produced from the most abundant uranium isotope, uranium-238.

62. Presumably, it takes an infinite time for all of a sample of radioactive matter to decay.

63. $1 \text{ lb} \times \dfrac{454 \text{ g}}{1 \text{ lb}} \times \dfrac{1 \text{ mol U}}{235 \text{ g U}} \times \dfrac{2.0 \times 10^{10} \text{ kJ}}{235 \text{ g U}} \times \dfrac{1 \text{ ton coal}}{2.5 \times 10^7 \text{ kJ}} = 1.5 \times 10^3 \text{ tons coal}$

64. Loss of mass of a specific isotope is difficult to measure because its decay product is another isotope that typically is mixed with the reactant. Rate of decay is best measured by Geiger counters or other devices described in Section 20.4 because their measurements are independent of the mass of the sample or the compound in which the radioisotope is found.

65. (a) A > C > B. At the end of one half-life, half of the original A would remain, leaving the other half to be divided between B and C. Half of B disintegrates in one day, so more than half of what was produced in days 1 through 5 has passed along to C. (b) C > A > B. At the end of two half-lives, A is down to 1/4 of the starting amount. Most of the 3/4 that disintegrated has passed through B to C.

66. R/S = 22.7 units ÷ 29.0 units = 0.783. From Figure 20.13, this corresponds to 0.35 half-life. The half-life of carbon-14 is 5.73×10^3 years. 0.35 half-lives × 5.73×10^3 years/half-life = 2.0×10^3 years. Radiocarbon dating indicates that the cloth is about 2000 years old, which places it at the beginning of the Common Era.

67. Emission of a beta particle would change a calcium atom into a scandium atom: ${}^{47}_{20}\text{Ca} \rightarrow {}^{0}_{-1}\text{e} + {}^{47}_{21}\text{Sc}$.

68. US is $\dfrac{238}{(238 + 32)} \times 100\% = 88\%$ uranium by mass. US_2 is 79% uranium by mass. Thus, for samples of equal mass, there are more uranium atoms in US than in US_2. Only the radioactive element, uranium, contributes to radioactivity, so US will exhibit the greater amount of radioactivity. The radioactivity of 0.5 mole of US will be the same as that of 0.5 mole of US_2 because both samples contain the same number of uranium atoms.

69. The representative equation for a fission reaction, ${}^{235}_{92}\text{U} + {}^{1}_{0}\text{n} \rightarrow {}^{94}_{38}\text{Sr} + {}^{139}_{54}\text{Xe} + 3\,{}^{1}_{0}\text{n}$, is more apt to be a chain reaction. Three neutrons per uranium atom are produced in the process in that reaction; two neutrons per uranium atom are produced in the process in Question 54. The greater the number of neutrons, the more likely it is that there will be a chain reaction.

21 Organic Chemistry

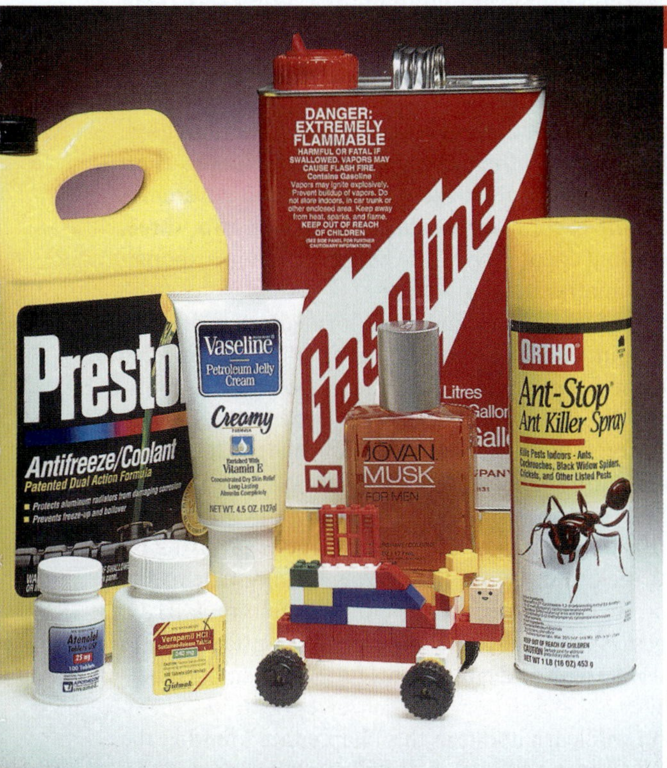

Charles D. Winters

◀ The starting material for everything you see in this photograph other than the metal cans is natural gas, petroleum, or coal. Many common products that we use every day in modern society come from these substances. How can so many diverse substances be made from such a small collection of naturally occurring starting materials? How does the transformation take place? We will investigate the answers to these questions in this chapter.

CHAPTER CONTENTS

This chapter is a brief survey of organic chemistry, the study of the properties and reactions of carbon compounds. Some instructors might seek mastery of specific chemical concepts in some or all areas of this chapter, and other instructors may believe that most topics are presented too briefly to set for them the kind of performance goals used in other chapters. Thus, we offer two alternatives. Goals are introduced throughout the chapter in the usual manner. We also offer the following chapter-wide performance goals:

Goal **A** Distinguish between organic and inorganic chemistry.

B Define the term *hydrocarbon*.

C Distinguish between saturated and unsaturated hydrocarbons.

D Write, recognize, or otherwise identify (a) the structural unit, or functional group, (b) the general formula, and (c) the molecular or structural formulas and/or names of specific examples of the following classes of organic compounds: alkanes, alkenes, alkynes, cycloalkanes, aromatic hydrocarbons, alcohols, ethers, aldehydes, ketones, carboxylic acids, esters, amines, and amides.

E Define and give examples of isomerism.

F Define and give examples of monomers and polymers.

21.1 Are There Organic Molecules in Space?

The subject of this chapter is *organic chemistry.* The essence of organic chemistry is that it is the study of molecules that include carbon atoms. If an entire named branch of chemistry is devoted to the study of molecules that include the atoms of just one element, that one element must have chemical properties that are distinct from all other elements. Indeed, carbon atoms are quite unique. What sets them apart from atoms of other elements is their ability to bond within molecules. Carbon atoms are "just right" to form covalent bonds that are strong enough to be stable yet weak enough to undergo chemical change under a variety of relatively mild conditions. Carbon is a nonmetal, so it forms covalent compounds rather than ionic compounds, yet it is more metallic—more to the left on the periodic table—than most of the other nonmetals. Thus, carbon almost exclusively is found covalently bonded to other atoms, and often, one carbon atom will bond to one, two, three, and even four other atoms. The bonding capability of the carbon atom is responsible for major aspects of the nature of matter in the universe.

It is estimated that the first organic molecules may have formed when the universe was less than 20 million years old. (Recall from Section 5.1 that the universe is presently 14 *b*illion years old.) The first organic molecule detected by humans beyond Earth was the simple species CH. This molecule is highly reactive and would immediately react on Earth to form something else, but in the low matter density of the interstellar medium, it can remain unreacted for long time periods.

The location in the universe where the widest variety of organic molecules have been detected other than on Earth is the giant molecular cloud Sagittarius B2 (**Figure 21.1**), which is located near the center of our Milky Way Galaxy. One of the most interesting molecules that have been detected in this cloud is ethanol, C_2H_6O, the intoxicant in alcoholic beverages! Another organic molecule in Sagittarius B2 is ethyl formate, $C_3H_6O_2$, which smells like rum and contributes to the taste of raspberries. Raspberry rum exists in space!

Many of the molecules that you will learn about in this chapter are found in the interstellar medium. In Sections 21.4 through 21.10, you will study hydrocarbons, molecules that are made up of only carbon and hydrogen atoms. Some hydrocarbons that have been detected in space include methane, CH_4, ethylene, C_2H_4, and benzene, C_6H_6. In Sections 21.11 through 21.15, you will study organic compounds containing oxygen and nitrogen. Some molecules within this category that have been found in the interstellar medium include methanol, CH_3OH, acetaldehyde, CH_3CHO, and urea, $(NH_2)_2CO$. Even high-atom-count molecules such as buckminsterfullerene, C_{60} (discussed in Everyday Chemistry 15.1), have been detected in space (**Figure 21.2**)!

Figure 21.1 The center of the Milky Way Galaxy as seen by the Spitzer Space Telescope. Sgr is an abbreviation for Sagittarius.

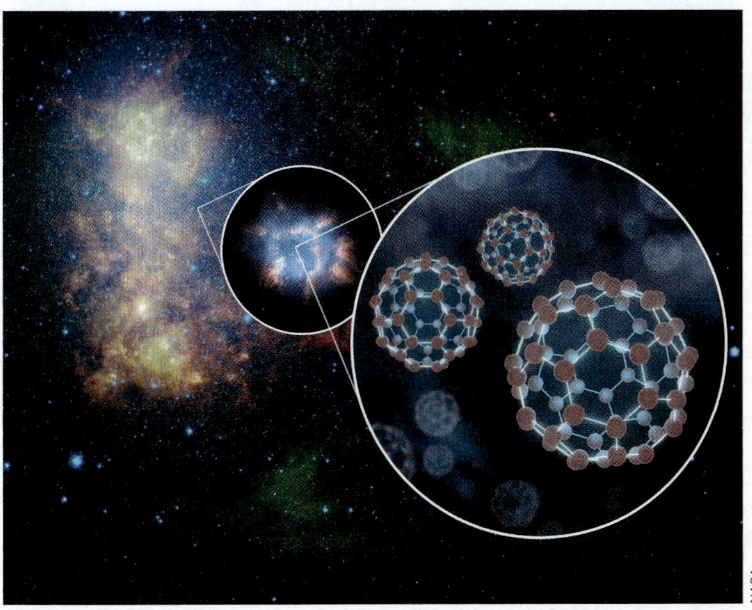

Figure 21.2 Buckminsterfullerene, C_{60}, in the Small Magellanic Cloud. The background is an infrared photograph of the Small Magellanic Cloud, a dwarf galaxy near our Milky Way Galaxy. The first zoom-in illustration shows a planetary nebula, a shell of gas and dust that was generated as a result of the death of a star. The second closeup shows C_{60} molecules in the planetary nebula.

The variety of carbon-atom-containing molecules in the interstellar medium is testimony to the importance of the bonding capability of carbon atoms. Covalently bonded carbon in molecules is found throughout the universe. In this chapter, you will begin to build your foundation of knowledge of organic chemistry—the chemistry of carbon-atom-containing molecules.

If your instructor does not express a preference between the two types of performance goals, we suggest you use the specific numbered in-chapter performance goals to guide your study.

21.2 The Nature of Organic Chemistry

Goal 1 Distinguish between organic and inorganic compounds.

In the early development of chemistry, the logical starting point was a study of substances that occur in nature. As in the organization of any body of knowledge, substances were grouped by certain common macroscopic characteristics. One system assigned substances to groups labeled as animal, vegetable, and mineral. Minerals and the compounds that may be derived from them originally made up the area of study commonly known as *inorganic chemistry*. Animal and vegetable substances were considered part of organic chemistry, which was originally defined as the chemistry of living organisms, including those compounds directly derived from living organisms by natural processes of decay.

In 1828, Friedrich Wöhler (**Figure 21.3**) heated ammonium cyanide, which everyone agreed is not organic, and obtained urea, which is found in urine, which everyone agreed *is* organic. The definitions of inorganic and organic chemistry had to be changed. The definitions changed from the macroscopic level—living versus nonliving—to the particulate level. Today, **organic chemistry** is known as the chemistry of carbon compounds. Inorganic chemistry is the study of all the other elements and their compounds. These definitions are loose, however, as there is considerable overlap between the two subdivisions of chemistry. In fact, carbon-atom-containing ions and molecules with no carbon–hydrogen bonds are usually classified as inorganic. Carbonates, cyanides, and oxides of carbon are examples.

Figure 21.3 German scientist Friedrich Wöhler (1800–1882) can be considered as the founding father of the science of organic chemistry. He showed that a "life force" was not required for producing compounds found in living organisms. They can be synthesized in the laboratory.

Organic chemistry is not "different" from inorganic chemistry. All of the chemical principles that apply to inorganic compounds, such as bonding, reaction rates, and equilibrium, apply equally to organic compounds. What truly makes organic chemistry unique is the ability of carbon atoms to bond to one another strongly to form long chain and ring systems. This results in the theoretical possibility of an almost unlimited number of organic compounds. This theoretical possibility is borne out in practice, too. In 1965, the Chemical Abstracts Service* began assigning a registry number to each new substance reported. The total number of classified chemical compounds has now passed 150 million, and the majority of these substances contain carbon. More than 15,000 new substances are registered each day.

Target Check 21.1

a) Which of the following compounds are classified as organic, and which are classified as inorganic: CH_4, NaCl, CH_3NH_2, CO?

b) The term *organic* has several meanings in everyday language, all of which are different from its meaning in chemistry. State an everyday definition of the term *organic*, and give an example of how it is used. Compare this with the chemical definition of the term.

21.3 The Molecular Structure of Compounds

Goal 2 Given Lewis diagrams, ball-and-stick models, or space-filling models of two or more organic molecules with the same molecular formula, distinguish between isomers and different orientations of the same molecule.

Table 21.1 summarizes the covalent bonding properties of carbon, hydrogen, oxygen, nitrogen, and the halogens, the elements most frequently found in organic compounds. ◀ All the molecular geometries are indicated—the three-dimensional shapes and, where constant, the actual bond angles. Of particular significance is the number of covalent bonds that atoms of the different elements usually form. This is determined by the electron configuration of the atom. The bonding relationships of these elements are basic to your understanding of the structure of organic compounds.

▶ **P/Review** Table 21.1 is derived largely from Section 13.4, Table 13.2. The earlier table includes illustrations that show how molecular structure is related to electron pair geometry and number of atoms bonded to the central atom.

Table 21.1 Bonding in Organic Compounds

Element	Number of Bonds	Molecular Bond Geometry			
		Single Bond	Double Bond	Double Bond	Triple Bond
Carbon	4	C (four single bonds) Tetrahedral: 109.5° angles	C= (two single bonds) Trigonal Planar: 120° angles	=C= Linear: 180° angle	—C≡ Linear: 180° angle
Hydrogen	1	H—			
Halogens	1	:Ẍ̤—			
Oxygen	2	:Ö: (two single bonds) Bent structure	:Ö=		
Nitrogen	3	N̈ (three single bonds) Trigonal pyramidal structure	N̈ (single and double bond) Angular structure		:N≡

*Chemical Abstracts Service compiles a database of information on chemical compounds and articles that have appeared in original research journals. It is a division of the American Chemical Society.

When carbon forms four single bonds, they are arranged tetrahedrally around the carbon atom; the molecular geometry is tetrahedral (**Figure 21.4**). Recall from Chapter 13 that it is not possible to represent this three-dimensional shape accurately in a two-dimensional sketch. Thus, the four bonds radiating from each carbon atom in

```
    x   z   y
    |   |   |
x—C—C—C—y
    |   |   |
    x   z   y
```

all form tetrahedral bond angles (109.5°). Furthermore, all three x positions are geometrically equal, the three y positions are equal, and the two z positions are equal. Also, note that if you flip the molecule left-to-right, the x positions are equivalent to the y positions.

Let's examine this relationship between two-dimensional sketches and three-dimensional molecules a bit more carefully. Compare the following two structural diagrams for C_3H_6BrCl and the ball-and-stick diagrams beneath them:

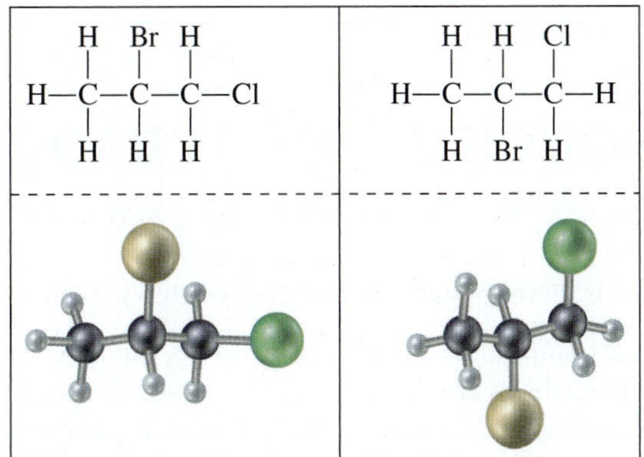

Are these the same molecule or isomers? One type of isomer is composed of compounds with the same molecular formula but different atom connectivities. ▶ The two compounds have the same molecular formulas, but do they have different atom connectivities? One way to answer this question is to consider whether a ball-and-stick model of the molecule would have to be taken apart and reassembled to form the other compound. Careful examination of the model diagrams shows that they have the same structure; the atoms are bonded in exactly the same way. The chlorine atom is on an end carbon, and the bromine atom is on the middle carbon atom. The diagrams are different views of the same molecule.

The molecule

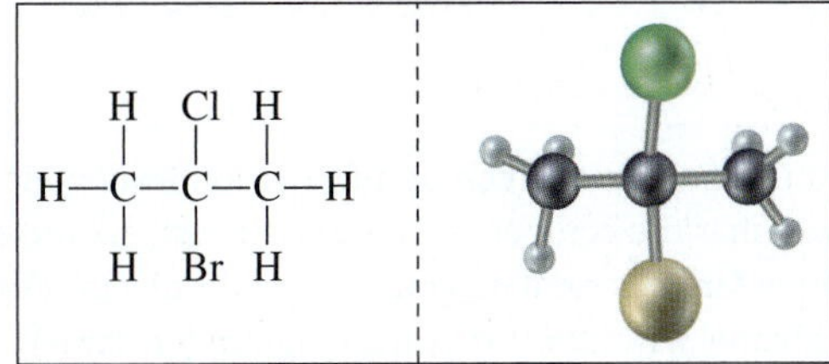

is an isomer of the molecule in the preceding box. Its molecular formula is still C_3H_6BrCl, but this time the chlorine and bromine atoms are both bonded to the central carbon—it differs in atom connectivity.

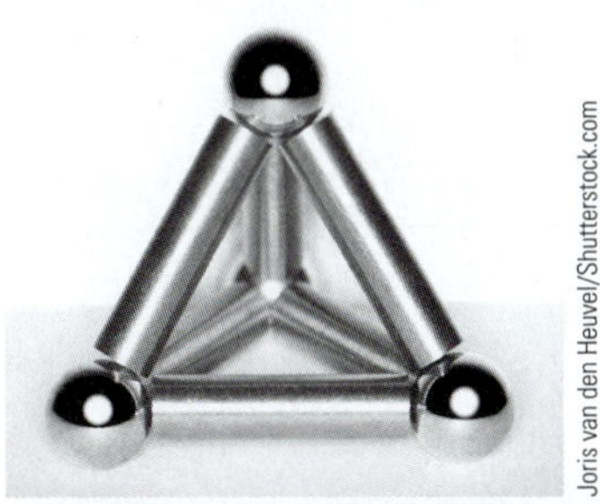

Joris van den Heuvel/Shutterstock.com

Charles D. Winters

Figure 21.4 Tetrahedral models. The metal figure is a tetrahedron. Its four faces are identical equilateral triangles. The model of methane, CH_4, has a tetrahedral structure. The carbon atom is in the middle of the tetrahedron, and a hydrogen atom is found at each of the four vertices.

◀ **P/Review** Isomers are distinctly different compounds, each with its own set of physical and chemical properties. See Section 13.2.

Thinking About Your Thinking

Mental Models

There are many structural diagrams and drawings of molecular models throughout this chapter. The mental translation between a structural diagram and the three-dimensional molecule it represents is essential to understanding organic chemistry. As you study this chapter, try to imagine a model of each molecule for which a structural diagram is drawn. If you have a model kit, build some of the molecules and rotate the model in your hands, looking at it from many angles. Practice rotating mental models of molecules in your mind. The more clearly you can see organic molecules in your mind, the better you will understand this area of chemistry.

Target Check 21.2

Thus far in this section, we have shown two isomers of the molecule C_3H_6BrCl. There are three more isomers. Draw their structural diagrams.

21.4 THROUGH 21.10: HYDROCARBONS

Goal 3 Distinguish between saturated and unsaturated hydrocarbons (or other compounds).

4 Distinguish between alkanes, alkenes, and alkynes (Sections 21.4 and 21.5).

The simplest organic compounds are the **hydrocarbons**, binary compounds of carbon and hydrogen. Hydrocarbons are divided into two major classes, the aliphatic hydrocarbons and the aromatic hydrocarbons. We will examine three groups of aliphatic hydrocarbons: alkanes, alkenes, and alkynes. The features that distinguish aliphatic and aromatic hydrocarbons will be identified when we study the latter group. As you will see shortly, these classifications are all based on molecular structure.

Table 21.1 shows that a carbon atom is able to form four bonds to a maximum of four other atoms. A hydrocarbon in which each carbon atom is bonded to the maximum of four other atoms is **saturated** in the sense that there is no room for the carbon atom to form a bond to another atom. Table 21.1 also shows that if a carbon atom is double- or triple-bonded, it is bonded to three or two other atoms. This is fewer than the maximum number of atoms with which the carbon atom is capable of bonding. Such a hydrocarbon is **unsaturated**.

21.4 Saturated Hydrocarbons: The Alkanes and Cycloalkanes

Goal 5 Given a formula of a hydrocarbon or information from which it can be written, determine whether the compound can be a normal alkane or a cycloalkane.

6 Given the name (or structural diagram) of a normal, branched, or halogen-substituted alkane, write the structural diagram (or name).

7 Given the name (or structural diagram) of a cycloalkane, write the structural diagram (or name).

In an **alkane**, each carbon atom forms single bonds to four other atoms, and there are no multiple bonds. Most of the carbon atoms in an alkane are arranged in a continuous chain in which all bond angles are 109.5°, the tetrahedral angle. If *all* carbon atoms in the molecule are in a continuous chain, the compound is a **normal alkane**.

In other alkanes, some carbon atoms appear as "branches" off the main chain. They are isomers of the normal alkane having the same number of carbon atoms.

The first 10 alkanes are shown in **Table 21.2**. Careful examination of the formulas shows that they have the form C_nH_{2n+2}, where n is the number of carbon atoms. Note that the difference from one alkane to the next is one carbon and two hydrogen atoms, a $-CH_2-$ structural unit. A series of compounds in which each member differs from the members before it by a $-CH_2-$ unit is called a **homologous series.**

There are three ways to write the formula of an alkane. Using octane, the alkane with eight carbons, as an example, there is the molecular formula, C_8H_{18}. A molecular formula gives no information as to how the atoms are arranged. A Lewis diagram that omits unshared electron pairs, commonly referred to as a **structural formula** or **structural diagram**, shows that arrangement. A compromise between them is the **line formula**, which includes each CH_2 unit in the alkane chain. The line formula for octane is $CH_3CH_2CH_2CH_2CH_2CH_2CH_2CH_3$. A long line formula is usually shortened by grouping the CH_2 units: $CH_3(CH_2)_6CH_3$. This is called a **condensed formula**.

Table 21.2 also serves to introduce organic nomenclature. Note that each alkane is named by combining a prefix and a suffix. The prefix indicates the number of carbon atoms in the compound, as shown in the third and fourth columns. The suffix identifying an alk*ane* is *-ane.* Thus, the name of methane comes from combining the prefix *meth-,* indicating one carbon, with the suffix *-ane*, indicating an alkane. The same prefixes are used to name the other organic compounds and groups as well.

Although there is only one possible structure for methane, ethane, and propane, as shown in **Figure 21.5**, there are two possible isomers of butane, shown in **Figure 21.6(a)**. Pentane has three isomers (**Figure 21.6[b]**). There are 5 isomeric hexanes, 9 heptanes, and 75 possible decanes. It is possible to draw over 300,000 isomeric structures for $C_{20}H_{42}$ and more than 100 million for $C_{30}H_{62}$. It is not surprising

Table 21.2 The Alkane Series

Molecular Formula	Name	Number of Carbon Atoms	Prefix	Physical State at 25°C
CH_4	Methane	1	Meth-	Gas
C_2H_6	Ethane	2	Eth-	Gas
C_3H_8	Propane	3	Prop-	Gas
C_4H_{10}	Butane	4	But-	Gas
C_5H_{12}	Pentane	5	Pent-	Liquid
C_6H_{14}	Hexane	6	Hex-	Liquid
C_7H_{16}	Heptane	7	Hept-	Liquid
C_8H_{18}	Octane	8	Oct-	Liquid
C_9H_{20}	Nonane	9	Non-	Liquid
$C_{10}H_{22}$	Decane	10	Dec-	Liquid

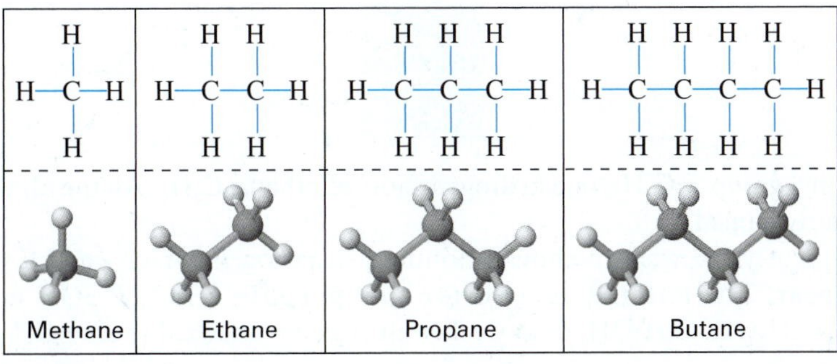

Figure 21.5 Lewis diagrams and ball-and-stick models of the four simplest alkanes. There is a tetrahedral orientation of all four bonds around each carbon atom.

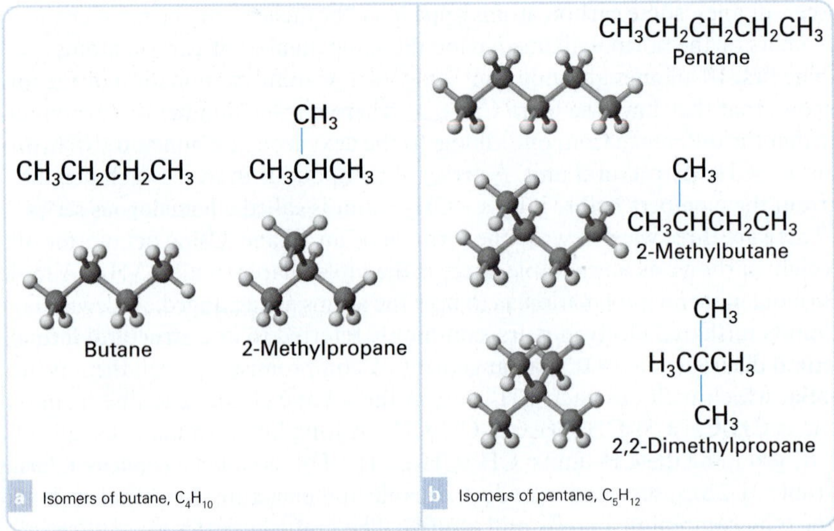

Figure 21.6 Lewis diagrams and ball-and-stick models of isomers of butane and pentane. (a) Each isomer of butane is composed of 4 carbon atoms and 10 hydrogen atoms. (b) Each isomer of pentane contains 5 carbon atoms and 12 hydrogen atoms.

that many of these compounds have not yet been prepared and identified! This does give you some idea, though, why there are so many organic compounds.

Alkyl Groups

An **alkyl group** is an alkane from which one hydrogen atom has been removed. If, on paper, we remove a hydrogen atom from methane, CH_4, we get $-CH_3$, where the dash indicates a bond that the alkyl is able to form with another atom. This $-CH_3$ group, appearing in the structural formula of a compound (attached to a carbon atom other than a terminal carbon), is called a **methyl group**. This term is made up of the prefix *meth-* for one carbon (Table 21.2) and the suffix *-yl*, which is applied to alk*yl* groups. If we compare two compounds,

```
    H   H   H   H   H                 H   H       H       H   H
    |   |   |   |   |                 |   |       |       |   |
H—C—C—C—C—C—H     and     H—C—C———C———C—C—H
    |   |   |   |   |                 |   |       |       |   |
    H   H   H   H   H                 H   H       |       H   H
                                                  |
                                                H—C—H
                                                  |
                                                  H
```

we see that the shaded H in the first compound has been replaced by a $-CH_3$ group, or methyl group, in the second. If the replacement group has two carbon atoms,

```
    H   H       H       H   H
    |   |       |       |   |
H—C—C———C———C—C—H
    |   |       |       |   |
    H   H       |       H   H
                |
              H—C—H
                |
              H—C—H
                |
                H
```

it is an ethyl group, $-C_2H_5$, one hydrogen short of ethane, C_2H_6. All the alkyl groups are similarly named.

Frequently, we wish to show a bonding situation in which *any* alkyl group may appear. The letter R is used for this purpose. Thus, R$-$OH could be CH_3OH, C_2H_5OH, C_3H_7OH, or any other alkyl group attached to an $-$OH group.

Naming the Alkanes by the IUPAC System

We are now ready to describe the International Union of Pure and Applied Chemistry (IUPAC) system of naming isomers of the alkanes, as well as other compounds we will encounter shortly.

how to... Name Alkanes

Step 1: Identify the longest continuous chain. We call this the **parent chain.** For example, in the compound having the skeleton structure,

```
C—C—C—C—C
      |
      C
      |
      C
```

the longest chain is six carbons long, a hexane, not five as you might first expect. This is readily apparent if we number the carbon atoms in the original representation of the structure and in an equivalent layout:

```
                                   C
                                   |
6   5   4   3             6   5   4   3   2   1
C—C—C—C—C                 C—C—C—C—C—C
            |
            C2
            |
            C1
```

Step 2: Identify the number of the carbon atom to which the alkyl group (or other species) is bonded to the parent chain. In the example compound, this is the *third* carbon, as shown. Counting always begins from the end of the chain that places the branch on the *lowest*-number carbon atom possible.

Step 3: Identify the branched group (or other species). In this example, the branch is a methyl group, $-CH_3$.

```
              H
              |
C—C—C—C—C—H
          |   |
          C   H
          |
          C
```

These three items of information are combined to produce the name of the compound, *3-methylhexane*. The 3 comes from the third carbon (*Step 2*); methyl comes from the branch group (*Step 3*); and hexane is the parent alkane (*Step 1*).

Sometimes a branch appears more than once in a single compound. The situation is governed by the following rule:

Step 4: If the same alkyl group, or other species, appears more than once, indicate the number of appearances by di-, tri-, tetra-, and so forth, and show the location of each branch by number. For example,

```
    C   C
    |   |
C—C—C—C—C
```

is 2,3-dimethylpentane. To write the structural formula for 2,2,5-trimethylhexane, we would establish a six-carbon skeleton and attach methyl groups as required, two to the second carbon and one to the fifth:

```
  C           C
  |           |
C—C—C—C—C—C
  |
  C
```

Twice earlier we referred to "other species"—species other than an alkyl group that might be attached to a hydrocarbon. The most common species other than an alkyl is a halogen atom. When attached to a hydrocarbon, halogen (Group 7A/17) atoms are named fluoro- (—F), chloro- (—Cl), bromo- (—Br), and iodo- (—I). If two chlorine atoms take the places of the methyl groups in 2,3-dimethylpentane earlier described, the compound would be 2,3-dichloropentane. This leads to the next nomenclature rule.

Step 5: If two or more different alkyl groups, or other species, are attached to the parent chain, they are named in alphabetical order. By this rule, the compound,

```
    Cl  Br
    |   |
C—C—C—C—C
```

is 3-bromo-2-chloropentane. The skeleton diagram for 2,2-dibromo-4-chloroheptane is

```
  Br    Cl
  |     |
C—C—C—C—C—C—C
  |
  Br
```

Charles D. Winters

Figure 21.7 A disposable butane lighter. Butane boils at $-20.5°C$ at atmospheric pressure; it is pressurized in lighters so that most of the fuel exists in the liquid state.

▶ **P/Review** In Sections 15.3 and 15.4, we discussed the relationships between intermolecular forces and physical properties. Substances with strong intermolecular forces tend to have lower vapor pressures and higher boiling points, heats of vaporization, viscosity, and surface tension. The forces, in increasing strength, were classified as (a) induced dipole forces, (b) dipole forces, and (c) hydrogen bonds. Induced dipole forces increase with molecular size, and with large molecules they may exert more influence on physical properties than dipole forces and hydrogen bonds.

▶ **P/Review** Section 18.5, Figure 18.8 illustrates the distillation of crude oil into fractions based on boiling point ranges. The gasoline fraction includes C_5–C_{12} hydrocarbons.

Physical Properties of the Alkanes

All alkane molecules are nonpolar. In Chapter 15, we stated that intermolecular forces between nonpolar molecules increase with increasing molecular size. ◀ As a result, larger molecules have higher melting and boiling points. Among normal alkanes—those with continuous chains—compounds having fewer than five carbon atoms have the weakest intermolecular attractions. They have low boiling points and are gases at room temperature. All are used as fuels; methane is the main constituent of natural gas, and butane is often used in disposable lighters (**Figure 21.7**).

Intermolecular forces are stronger between larger alkanes from C_5H_{12} to $C_{17}H_{36}$. These higher boiling compounds are liquids at room temperature (**Figure 21.8**). Several of the lower molar mass liquid alkanes containing 5 to 12 carbon atoms are present in gasoline. ◀ Diesel fuel and lubricating oils are made up largely of higher molar mass liquid alkanes. Alkanes with molar masses greater than 300 g/mol are normally solids at room temperature (**Figure 21.9**).

Cycloalkanes

If a hydrogen atom is removed from each end carbon of a normal alkane, the two end carbons can bond to each other to form a ring of carbon atoms. This ring compound, in which all carbon atoms are saturated, is a **cycloalkane**. Cycloalkanes can form with a minimum of three carbon atoms; however, the resulting 60° bond angles are severely strained from the normal tetrahedral angle, so cyclopropane is unstable. Cyclobutane, the four-carbon ring system, is also unstable (**Figure 21.10**). The more common cycloalkanes are those whose bond angles are close to or equal to the tetrahedral angle, such as cyclopentane, the five-membered ring system, and cyclohexane, the six-membered ring system.

Three structural diagrams and a ball-and-stick diagram of cyclohexane are as follows:

```
   H   H                   CH2
 H  \ /  H              /       \
  \  C  /            H2C         CH2
H—C     C—H            |         |
H—C     C—H          H2C         CH2
  /  C  \               \       /
 H  / \  H                 CH2
   H   H
```

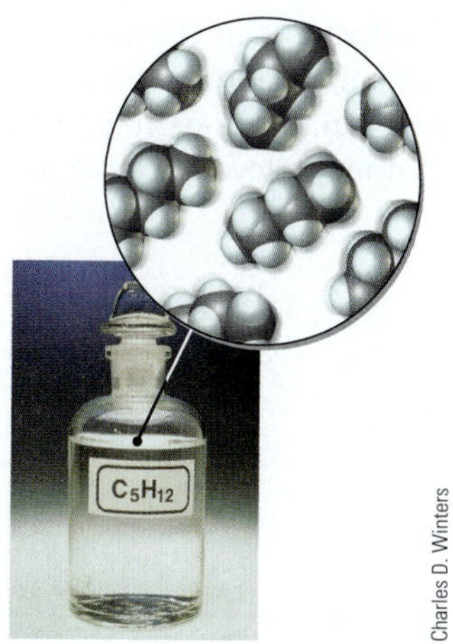

Charles D. Winters

Figure 21.8 Pentane, C_5H_{12}, boils at 36°C, so it is a liquid at room temperature and atmospheric pressure.

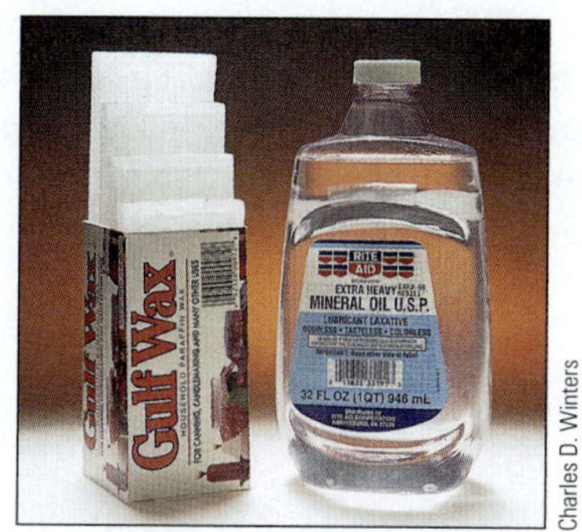

Charles D. Winters

Figure 21.9 Solid paraffin wax and liquid mineral oil. Paraffin wax is a mixture of straight-chain alkane molecules with 20 or more carbon atoms. Mineral oil is also a mixture, mainly of alkane molecules with 17 to 50 carbon atoms, but it includes other hydrocarbons as well.

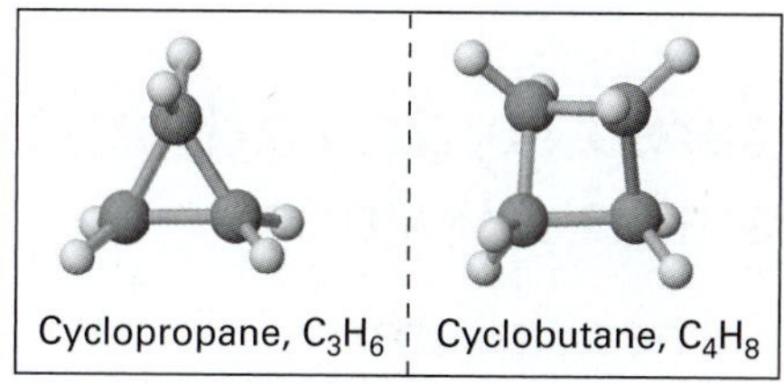

Figure 21.10 Unstable cycloalkanes. The normal tetrahedral bond angle for a carbon–carbon bond is 109.5°. In cyclopropane, the C—C bond angles are 60°. Orbital overlap is poor, so the bonds are relatively weak and the molecule is very reactive. Cyclobutane has C—C bond angles of 88° because the carbons are not in the same plane (they would be 90° if the carbon atoms were coplanar). Again, the large deviation from the normal tetrahedral bond angle causes the molecule to be relatively unstable.

The left diagram is the most complete. The middle diagram illustrates the common practice of using $-CH_2-$ when a carbon bonded to two hydrogen atoms is also bonded to two other atoms, as in a saturated chain. The diagram at the right is a skeleton diagram in which the vertex of each angle shows the relative location of a carbon atom. It is understood in such diagrams that each carbon atom forms as many additional bonds to hydrogen atoms as are necessary to bring its total number of bonds to four. All structural diagrams are equivalent to the ball-and-stick diagram on the right.

To name a cycloalkane, apply the prefix *cyclo-* to the name of the open chain alkane with the same number of carbon atoms. The preceding diagrams have six carbon atoms in the ring, hence the name cyclohexane. If one or more alkyl groups or other species, such as a halogen, takes the place of one of the invisible hydrogen atoms in the ring, alkane nomenclature rules for branched species apply. Thus,

CH_2 CH_3
H_2C CH
H_2C CH_2
CH_2

is methylcyclohexane. Note that the methyl carbon in the skeleton diagram is not shown. Like the vertex in the polygon, a bond with no symbol at its end is understood to be a carbon atom that forms additional bonds up to a total of four.

When two or more substitutions appear in a cycloalkane, their locations are determined by number. The ring carbon to which one substituent is bonded is number 1. Other numbers are assigned to give the lowest numbers possible to the other locations. Thus, the following compound is 1,3-dimethylcyclohexane, not 1,5-dimethylcyclohexane:

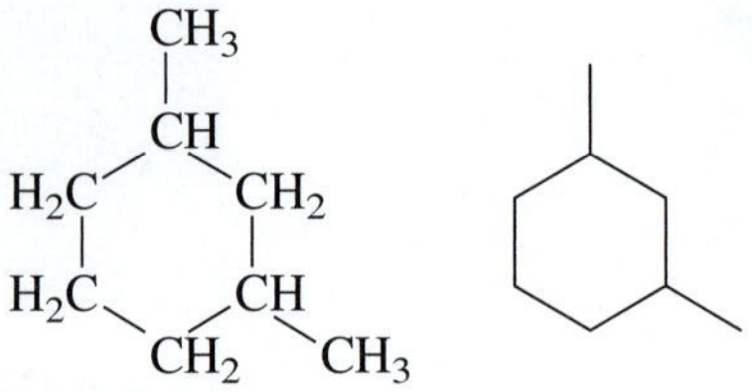

```
           CH3
           |
           CH
      H2C/    \CH2
       |        |
      H2C\    /CH
           CH2    \CH3
```

Target Check 21.3

a) Identify the alkanes among C_7H_{16}, C_5H_{10}, $C_{11}H_{22}$, C_9H_{20}.

b) Write the formula of the alkyl group derived from pentane.

c) Write a structural diagram of 3,3-difluoro-4-iododecane.

d) Write a structural diagram of 1,1-diethylcyclohexane.

21.5 Unsaturated Hydrocarbons: The Alkenes and Alkynes

Structure and Nomenclature

Goal 8 Given a formula of a hydrocarbon or information from which it can be written, determine whether the compound can be an alkene or an alkyne.

9 Given the name (or structural diagram) of an alkene or an alkyne, write the structural diagram (or name).

In a saturated hydrocarbon, each carbon atom is bonded to four other atoms. Hydrocarbons in which two or more carbon atoms are (1) connected by a double or triple bond and (2) bonded to fewer than four other atoms are unsaturated.

If one hydrogen atom, complete with its electron, is removed from each of two adjacent carbon atoms in an alkane (diagram [a] in the middle of this paragraph), each carbon is left with a single unpaired electron (diagram [b]). These electrons may then form a second bond between the two carbon atoms (diagram [c]):

Words or chemical symbols are sometimes placed above or above and below the arrow of an equation to indicate a substance whose presence or removal is necessary for a reaction to proceed or to identify a reaction condition.

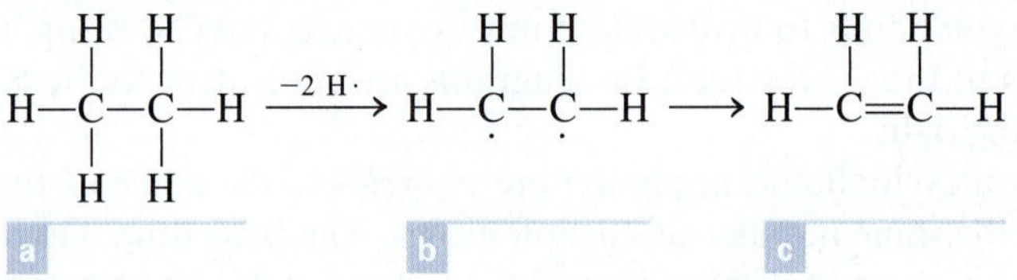

```
   H  H                     H  H              H  H
   |  |                     |  |              |  |
H—C—C—H   —2 H·→   H—C—C—H   →   H—C=C—H
   |  |                     ·  ·
   H  H
   a                        b                 c
```

Each carbon atom is now bonded to three other atoms. An aliphatic hydrocarbon that contains at least one carbon–carbon double bond is called an **alkene**. Figure 21.11(a) illustrates the simplest alkene, which has the common name ethylene (*eth-* = 2 carbon atoms + *-ylene* = alkene).

Removal of another hydrogen atom from each of the double-bonded carbon atoms in an alkene yields a triple bond:

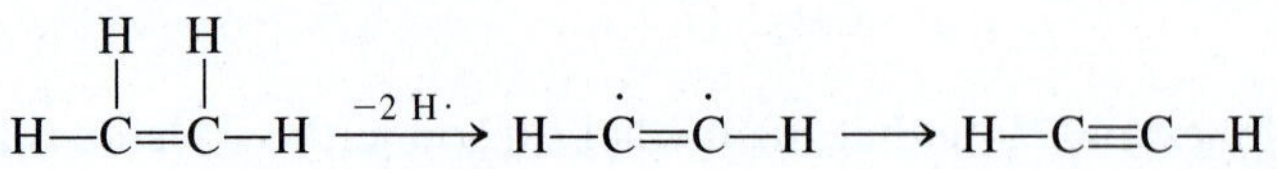

```
   H  H                     ·  ·
   |  |                     
H—C=C—H   —2 H·→   H—C=C—H   →   H—C≡C—H
```

Each carbon atom is now bonded to two other atoms. An aliphatic hydrocarbon in which two carbon atoms are triple-bonded to each other is called an **alkyne**.

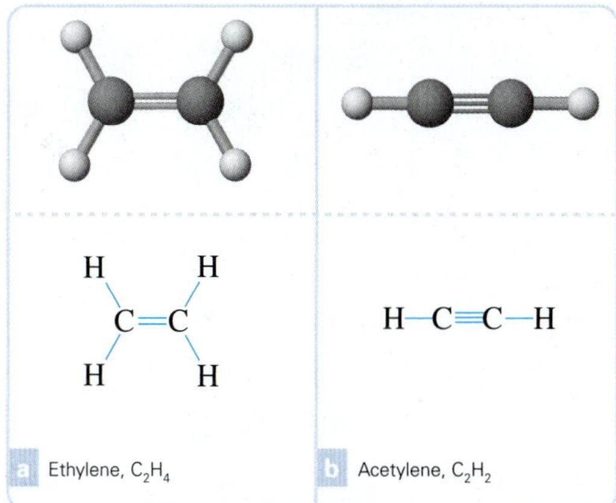

Figure 21.11 Ball-and-stick models of the first members of (a) the alkene and (b) alkyne hydrocarbon series: ethylene, C_2H_4, and acetylene, C_2H_2.

Acetylene, the common name for the most common alkyne, is illustrated in **Figure 21.11(b)**.

Both the alkenes and the alkynes make up new homologous series. Just as with the alkanes, each series may be extended by adding $-CH_2-$ units. Longer chains may have more than one multiple bond, but we will not consider such compounds in this text (**Figure 21.12**). The general formula for an alkene is C_nH_{2n}, and for an alkyne, C_nH_{2n-2}.

Table 21.3 gives the names and formulas of some of the simpler unsaturated hydrocarbons. The IUPAC nomenclature system for the alkenes matches that of the alkanes. The suffix designating the alk*ene* hydrocarbon series is *-ene*, just as *-ane* identifies an alk*ane.* For example, pentene is C_5H_{10}, hexene is C_6H_{12}, and octene is C_8H_{16}. The common names for the alkenes are produced similarly, except that the suffix is *-ylene.* These names for the smaller alkenes are often used instead of the IUPAC names: C_2H_4 is often called *ethylene*, C_3H_6 is propylene, and C_4H_8 is butylene.

Acetylene, C_2H_2, is the first member of the alkyne series (**Figure 21.13**). Despite its *-ene* ending, acetylene is an alkyne, not an alkene. The IUPAC system is used in common practice for all alkynes except acetylene. By the IUPAC system, the ending *-yne* is used to indicate the presence of a triple bond. So, for the alkynes with two, three, and four carbon atoms, the IUPAC names are ethyne, propyne, and butyne, respectively.

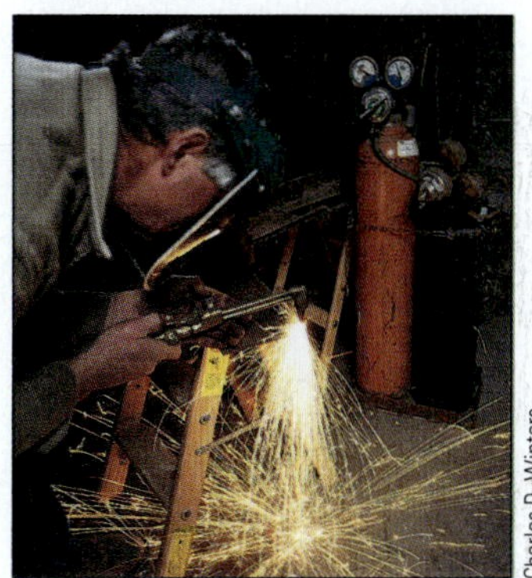

Figure 21.13 Oxyacetylene torch. Acetylene is the fuel in a welder's torch. It reacts with the oxygen in the air in a highly exothermic reaction, producing a 3000°C flame, which is hot enough to cut through and weld steel.

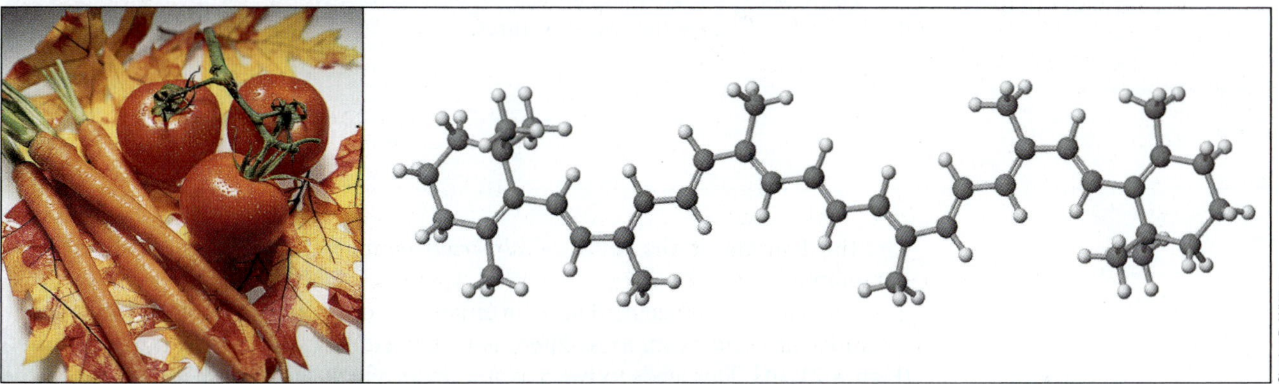

Figure 21.12 Carotene. An example of a molecule with multiple double bonds is carotene, which has 11 carbon–carbon double bonds. The carotene molecule plays an indirect role in photosynthesis by transferring the energy it receives from sunlight to the chlorophyll molecule. Carotenes primarily exist in two forms, alpha-carotene and beta-carotene. Beta-carotene is found in yellow, orange, and green fruits, vegetables, and plants, such as carrots, tomatoes, and leaves.

Table 21.3 Unsaturated Hydrocarbons

Hydrocarbon Series	n	Formulas		Names	
		Molecular	Structural	IUPAC	Common
Alkenes, C_nH_{2n}	2	C_2H_4	H H \ / C=C / \ H H	Ethene	Ethylene
	3	C_3H_6	H H \ | C=C—C—H / | | H H H	Propene	Propylene
	4	C_4H_8	H H H \ | | C=C—C—C—H / | | | H H H H	1-Butene	Butylene
Alkynes, C_nH_{2n-2}	2	C_2H_2	H—C≡C—H	Ethyne	Acetylene
	3	C_3H_4	H | H—C≡C—C—H | H	Propyne	—

Isomerism among the Unsaturated Hydrocarbons

Goal 10 Identify and distinguish between *cis* and *trans* geometric isomers.

There are two possible isomers of butyne. They differ by the position of the multiple bond. Melting and boiling points are given below the structures to show that these are indeed different molecules.

	—C≡C—C—C— (carbons 1 2 3 4)	—C—C≡C—C— (carbons 1 2 3 4)
	1-butyne	2-butyne
melting point	−122.5°C	−32.3°C
boiling point	+8.1°C	+27°C

There are three distinct isomers of butene. Depending on the location of the double bond, we have the two structures:

—C=C—C—C— (carbons 1 2 3 4)
1-butene

—C—C=C—C— (carbons 1 2 3 4)
2-butene

The third isomer is the result of *cis–trans* isomerism, as discussed in the next paragraph.

The part of a molecule that is on either side of a single bond may rotate freely around that bond as an axis. There is restricted rotation around a double bond (**Figure 21.14**). This leads to two possible arrangements around a double bond. The two methyl groups can be on the *same* side of the double bond, as in *cis*-2-butene, or on opposite sides, as in *trans*-2-butene. *Cis*- and *trans*- are prefixes meaning, respectively, "on this side" and "across." Remember *trans* by associating it with a word such as transcontinental, meaning across a continent. Another name for *cis–trans* isomers is **geometric isomers**.

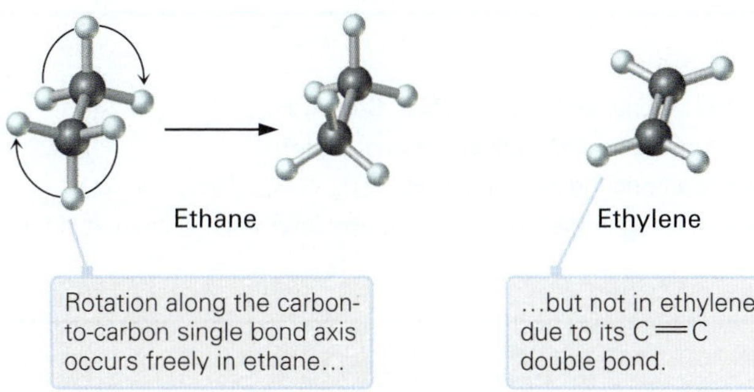

Figure 21.14 Restriction of rotation about a double bond. In general, the groups of atoms on either side of a single bond have the ability to rotate freely about the bond, but the presence of a double bond essentially locks the atoms in place, making geometric isomerism possible.

With their melting and boiling points to show they are different substances, the three butene isomers therefore are:

	C1=C2–C3–C4	C1 and C4 on same side of C2=C3	C1 and C4 on opposite sides of C2=C3
	1-butene	*cis*-2-butene	*trans*-2-butene
melting point	−185.4°C	−138.9°C	−105.6°C
boiling point	−6.3°C	+3.7°C	+0.9°C

Figure 21.15 shows another example of *cis–trans* isomers and the differences in their physical properties.

The compound 1-butene shows that the carbon chain is numbered through the multiple bond to give the multiple bond the lowest possible number. The compound

$$-\overset{5}{C}-\overset{4}{C}-\overset{3}{C}=\overset{2}{C}-\overset{1}{C}-$$

is 2-pentene because the double bond is attached to the *second* carbon, counting from the right.

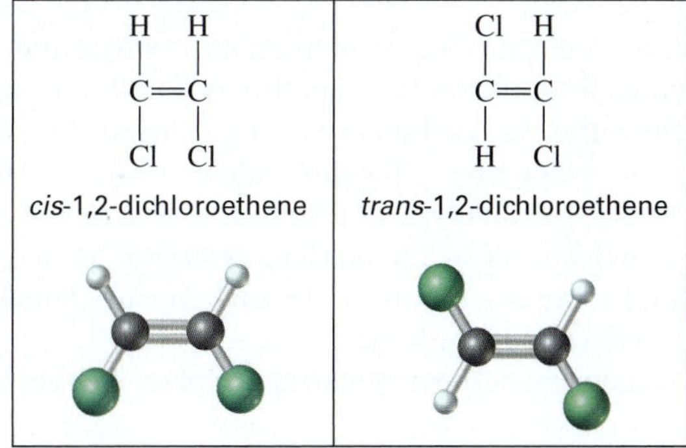

Figure 21.15 Physical properties of *cis*- and *trans*-1,2-dichloroethene. The boiling point of the *cis* isomer is 60°C, and its melting point is –80°C. The boiling and melting points of the *trans* isomer are 48°C and –50°C, respectively. If rotation could occur around the double bond, the boiling and melting point of the isomers would be the same. Since the physical properties of the two isomers are different, there must be two different molecules. Rotation about the double bond must be restricted to account for the two isomers because atom connectivity is otherwise the same.

Target Check 21.4

a) Identify the alkenes among the following: C_4H_6, C_2H_6, C_7H_{12}, C_8H_{16}.
b) Write a structural formula for *trans*-difluoroethene, $C_2H_2F_2$.
c) Identify the alkynes among the following: C_4H_6, C_2H_6, C_7H_{12}, C_8H_{16}.
d) How many isomeric straight-chain pentynes are possible? Write a structural formula for 2-pentyne.

21.6 Aromatic Hydrocarbons

Goal 11 Distinguish between aliphatic and aromatic hydrocarbons.
12 Given the name (or structural diagram) of an alkyl- or halogen-substituted benzene compound, write the structural diagram (or name).

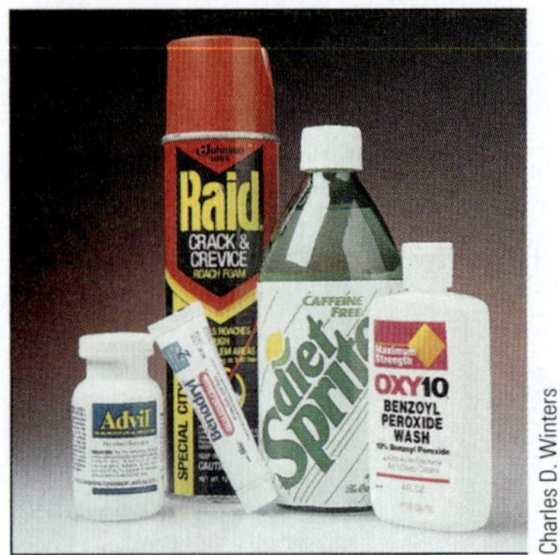

Figure 21.16 Consumer products containing compounds that contain a benzene ring. The ibuprofen in Advil, the propoxur in Raid, the diphenhydramine hydrochloride in Benadryl, the sodium benzoate in Sprite, and the benzoyl peroxide in Oxy-10 all have at least one benzene ring in their molecular structures.

Initially, aliphatic compounds were associated with oils and fats, which contain large carbon chains. By contrast, the term *aromatic* was associated with a series of compounds which, for some, were found in pleasant-smelling substances such as cloves, vanilla, wintergreen, and cinnamon. Ultimately, chemists found that the key structure in **aromatic hydrocarbons** is the benzene ring (Figure 21.16). Now, any hydrocarbon that does not contain a benzene ring—a hydrocarbon that is not an aromatic hydrocarbon—is an **aliphatic hydrocarbon**.

The simplest aromatic hydrocarbon is benzene, C_6H_6. Chemists struggled for decades to find a structural diagram that is consistent with its physical and chemical properties, but without success. Common forms and its space-filling model are:

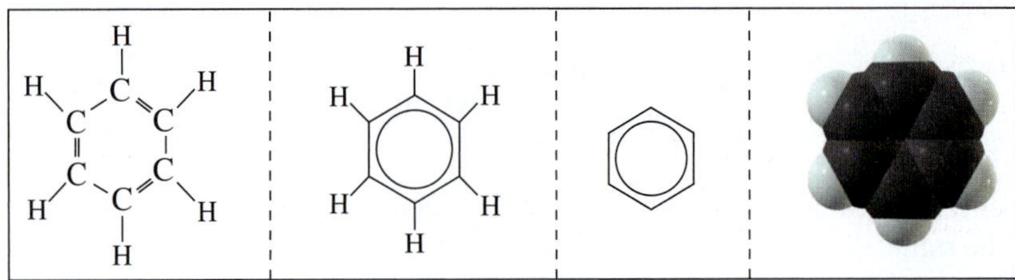

The structure at the left satisfies the octet rule and predicts a planar molecule with 120° bond angles, which benzene has, as shown in the model. However, the bonds in benzene are identical rather than alternating single and double bonds. The diagrams in the center show the three "second-bond" bonding electrons of the double bond as a circle within the carbon atom ring. This means that those electrons are delocalized, that is, belonging to the molecule as a whole rather than to specific carbon–carbon bonds. It is this type of diagram that is most commonly used. As with cycloalkanes, each corner of the benzene molecule has a carbon atom, but it forms only one bond to an atom outside the carbon ring. Unless indicated otherwise, this atom is assumed to be hydrogen.

An alkyl group, halogen, or other species may replace hydrogen on a benzene ring:

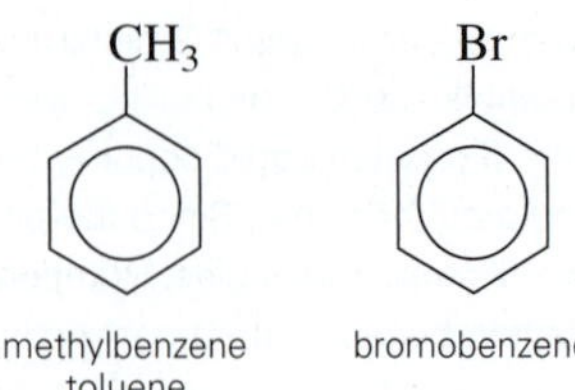

If two bromines substitute for hydrogens on the same ring, we must consider three possible isomers:

1,2-dibromobenzene
o-dibromobenzene

1,3-dibromobenzene
m-dibromobenzene

1,4-dibromobenzene
p-dibromobenzene

Two names are given for each isomer. The number system is the same as the system used to identify positions on a cyclohexane ring. It is more formal and serves any number of substituents. The other system uses the names *ortho*-dibromobenzene, *meta*-dibromobenzene, and *para*-dibromobenzene. *Ortho*-, *meta*-, and *para*- are prefixes commonly used when two hydrogens have been replaced from the benzene ring. Relative to position X, the other positions are shown here:

The physical properties of benzene and its derivatives are quite similar to those of other hydrocarbons: These compounds are nonpolar, insoluble in polar solvents such as water but generally soluble in nonpolar solvents. ▶ In fact, derivatives of benzene are widely used as the solvent for many nonpolar organic compounds. Like other hydrocarbons of comparable molar mass, benzene is a liquid at room temperature.

▶ **P/Review** Substances with intermolecular attractions that are roughly equal are most apt to be soluble in one another (Section 16.5). Similar molecular polarity contributes to similar intermolecular attractions.

Target Check 21.5

a) Write the structural formula of 1,3,5-trifluorobenzene.

b) Mothballs are made of *p*-dichlorobenzene (the Department of Health and Human Services has determined that *p*-dichlorobenzene may reasonably be anticipated to be a carcinogen). Write the structural formula for this substance.

21.7 Summary of the Categories of Hydrocarbons

The five categories of hydrocarbons we have considered are summarized in **Table 21.4**.

Table 21.4 Summary of Hydrocarbons

Type	Name	Formula	Saturation	Representative Structure
Aliphatic open-chain	Alkane	C_nH_{2n+2}	Saturated	—C—
	Alkene	C_nH_{2n}	Unsaturated	C=C
	Alkyne	C_nH_{2n-2}	Unsaturated	—C≡C—
Aliphatic cyclic	Cycloalkane	C_nH_{2n}	Saturated	
Aromatic	—	—	Unsaturated	

Tim Wright/Getty Images

Figure 21.17 Coal. Although coal is an important starting material from which many products of the chemical industry are made, its primary use is as a fuel in electrical power plants.

21.8 Sources and Preparation of Hydrocarbons

Goal 13 State the three types of fossil fuels from which most hydrocarbons are derived.

Almost all hydrocarbons are derived from fossil fuels: coal, natural gas, and petroleum. These substances are natural products that have resulted from the decay of plants and animals that lived millions of years ago. We are familiar with natural gas as the most common fuel for home heating and cooking with gas stoves. This fuel is primarily methane, CH_4, and a smaller but significant portion of ethane, C_2H_6, plus small amounts of other substances. Coal used to be a major source of benzene; benzene is actually a by-product of the preparation of coke from coal (**Figure 21.17**). Other aromatic hydrocarbons are also recovered from the same process. Today, coal is primarily used as a fuel, and most benzene comes from petroleum.

Petroleum is by far the largest source of the vast number of products broadly known as *petrochemicals*. Raw petroleum is a mixture of hydrocarbons that contain up to 40 carbon atoms per molecule. These large molecules are not useful in their natural form, so they are broken into smaller molecules in petroleum refineries (**Figure 21.18**; also see Figure 18.8). *Catalytic cracking* essentially "cracks" the long carbon chains into shorter molecules of 5 to 10 carbon atoms. ◀ *Fractional distillation* separates hydrocarbons into "fractions" that boil at different temperatures. Alkanes of up to 4 or 5 carbon atoms per molecule may be obtained in pure form by this method. The boiling points of larger alkanes are too close for their complete separation, so chemical methods must be used to obtain pure products.

▶ **P/Review** A *catalytic* process is one that occurs in the presence of a catalyst. In Section 18.5, a catalyst is defined as a substance that speeds the rate of reaction without being permanently affected.

Alkanes are prepared by several industrial and laboratory methods. One of the more important is the catalytic **hydrogenation** of an alkene. Hydrogenation is the reaction of a substance with hydrogen. The general reaction of the hydrogenation of an alkene is:

$$C_nH_{2n} + H_2 \xrightarrow{\text{catalyst}} C_nH_{2n+2}$$

Tungphoto/Shutterstock.com

Figure 21.18 Petroleum refinery towers. Raw petroleum, or crude oil, is the starting point from which a huge number of products are made. The first step in the refining process is to distill the oil mixture into other mixtures based on boiling point ranges. These fractions are then further refined and processed.

Unsaturated hydrocarbons are often prepared commercially from compounds derived originally from alkanes. Alkenes, for example, are produced from the *dehydration* of alcohols (Section 21.11) or the *dehydrohalogenation* of an alkyl halide. These two impressive terms describe very similar processes that are quite simple, at least in principle. Dehydration is the removal of a water molecule; dehydrohalogenation is the removal of a hydrogen atom and a halogen atom. For example, a water molecule may be separated from propyl alcohol, C_3H_7OH, to make propylene:

$$\mathrm{CH_3{-}CH_2{-}CH_2OH} \xrightarrow{H_2SO_4} \mathrm{CH_3{-}CH{=}CH_2} + \mathrm{HOH}$$

propyl alcohol → propylene + water

An alkyl halide is an alkane in which a halogen atom has been substituted for a hydrogen atom; viewed in another way, an alkyl halide is an alkyl group bonded to a halogen. The molecule is attacked with base in the presence of an alcohol to produce an alkene:

$$\mathrm{CH_3{-}CH_2{-}CH_2X} + \mathrm{KOH} \xrightarrow{\text{alcohol}} \mathrm{CH_3{-}CH{=}CH_2} + \mathrm{KX} + \mathrm{HOH}$$

propyl halide + base → propylene + ionic compound + water

Acetylene is the only alkyne produced commercially in large quantities. It is manufactured in a two-step process in which calcium oxide reacts with coke (carbon) at high temperatures to produce calcium carbide and carbon monoxide:

$$CaO(s) + 3\,C(s) \rightarrow CaC_2(s) + CO(g)$$

Calcium carbide then reacts with water to produce acetylene:

$$CaC_2(s) + 2\,H_2O(\ell) \rightarrow C_2H_2(g) + Ca(OH)_2(s)$$

21.9 Chemical Reactions of Hydrocarbons

Goal 14 Given the reactants in an addition or substitution reaction between (a) an alkane, alkene, alkyne, or benzene and (b) a hydrogen or halogen molecule, predict the products of the reaction.

The combustibility—ability to burn in air—of the hydrocarbons is probably the chemical reaction most important to modern society (**Figure 21.19**). As components of liquid and gaseous fuels, hydrocarbons are among the most heavily processed and distributed chemical products in the world. When burned in an excess of air, the end products are water and carbon dioxide. For example,

$$CH_4(g) + 2\,O_2(g) \rightarrow 2\,H_2O(\ell) + CO_2(g)$$

The reactivity of saturated hydrocarbons is significantly different from those of the unsaturated hydrocarbons. By opening a multiple bond in an alkene or alkyne, the compound is capable of reacting by addition, simply by adding atoms of some elements to the molecule. This is called an **addition reaction**. By contrast, an alkane molecule is literally saturated; there is no more room for an atom to join the molecule without first removing a hydrogen atom. A reaction in which a hydrogen atom in an alkane is replaced by an atom of another element is called a **substitution reaction**.

Both alkanes and alkenes undergo **halogenation reactions**—reaction with a halogen. These reactions serve to show the difference between addition and substitution reactions.

Figure 21.19 Combustion of natural gas. Hydrocarbons, such as natural gas, burn in air, releasing a relatively large amount of energy for ordinary chemical change. This energy is then used to generate electricity, operate engines, heat factories, businesses, and homes, and, in general, to provide energy for most of the conveniences of modern society.

Addition Reaction

$$\begin{array}{c} \text{H}\quad\text{H}\quad\text{H} \\ |\quad\;\;|\quad\;\;| \\ \text{H—C—C=C—H} \\ | \qquad\qquad\; \\ \text{H} \qquad\qquad\; \end{array} + \text{Cl}_2 \xrightarrow{\text{CH}_2\text{Cl}_2} \begin{array}{c} \text{H}\quad\text{H}\quad\text{H} \\ |\quad\;\;|\quad\;\;| \\ \text{H—C—C—C—H} \\ |\quad\;\;|\quad\;\;| \\ \text{H}\quad\text{Cl}\quad\text{Cl} \end{array}$$

propylene — chlorine — 1,2-dichloropropane

Substitution Reaction

$$\begin{array}{c} \text{H}\quad\text{H}\quad\text{H} \\ |\quad\;\;|\quad\;\;| \\ \text{H—C—C—C—H} \\ |\quad\;\;|\quad\;\;| \\ \text{H}\quad\text{H}\quad\text{H} \end{array} + \text{Cl}_2 \xrightarrow{\text{heat or light}} \begin{array}{c} \text{H}\quad\text{H}\quad\text{H} \\ |\quad\;\;|\quad\;\;| \\ \text{H—C—C—C—H} \\ |\quad\;\;|\quad\;\;| \\ \text{H}\quad\text{H}\quad\text{Cl} \end{array} + \text{HCl}$$

propane — chlorine — 1-chloropropane — hydrogen chloride

The substituted chlorine atom may appear on either an end carbon atom or the middle carbon; the actual product is usually a mixture of 1-chloropropane and 2-chloropropane.

Normally, addition reactions are more readily accomplished than substitution reactions. This is hinted at in the reaction conditions specified earlier. The addition of a halogen to an alkene will occur easily at room temperature, whereas the substitution of a halogen for a hydrogen in an alkane requires either high temperature or ultraviolet light. This shows that unsaturated hydrocarbons are more reactive than saturated hydrocarbons.

Sheila Fitzgerald/Shutterstock.com

Laurin Rinder/Shutterstock.com

Figure 21.20 Hydrogenation of an alkene produces an alkane. Vegetable oils have double bonds that are converted to saturated solid cooking fats by hydrogenation.

Hydrogenation is also an addition reaction. We have already indicated that the hydrogenation of an alkene may be used to produce an alkane (**Figure 21.20**). Hydrogenation of an alkyne is a stepwise process, which may often be controlled to give the intermediate alkene as a product:

$$\underset{\text{alkyne}}{R{-}C{\equiv}C{-}R'} \xrightarrow[\text{catalyst}]{H_2} \underset{\text{alkene}}{R{-}\overset{H}{\overset{|}{C}}{=}\overset{H}{\overset{|}{C}}{-}R'} \xrightarrow[\text{catalyst}]{H_2} \underset{\text{alkane}}{R{-}\underset{H}{\underset{|}{\overset{H}{\overset{|}{C}}}}{-}\underset{H}{\underset{|}{\overset{H}{\overset{|}{C}}}}{-}R'}$$

R′ in the diagram may be the same alkyl as R, or it may be a different alkyl.

Perhaps the most significant—and surprising—chemical property of benzene is that, despite its high degree of unsaturation, *it does not normally engage in addition reactions.* The classic 19th century chemical test for double bonds was a reaction with bromine to give addition products (**Figure 21.21**). Benzene does not give this reaction. The most important reaction of benzene itself is the substitution reaction in which one hydrogen is displaced from the benzene ring. Several substances maybe used for substitution, including the halogens:

$$\underset{\text{benzene}}{C_6H_6} + Cl_2 \xrightarrow{FeCl_3} \underset{\text{chlorobenzene}}{C_6H_5Cl} + HCl$$

Substitutions with nitric and sulfuric acids yield, respectively, nitrobenzene and benzenesulfonic acid. Second substitutions on the same ring are possible, although more difficult to bring about. Substitution reactions may also be performed on benzene derivatives, such as toluene, yielding isomers of nitrotoluene, for example. A triple nitro substitution produces 2,4,6-trinitrotoluene, better known simply as TNT.

Target Check 21.6

Complete the following equations:

a) $$H{-}\underset{H}{\underset{|}{\overset{H}{\overset{|}{C}}}}{-}\underset{H}{\underset{|}{\overset{H}{\overset{|}{C}}}}{-}\overset{H}{\overset{|}{C}}{=}\overset{H}{\overset{|}{C}}{-}H + Br_2 \xrightarrow{CH_2Cl_2}$$

b) $$H{-}\underset{H}{\underset{|}{\overset{H}{\overset{|}{C}}}}{-}\underset{H}{\underset{|}{\overset{H}{\overset{|}{C}}}}{-}\underset{H}{\underset{|}{\overset{H}{\overset{|}{C}}}}{-}\underset{H}{\underset{|}{\overset{H}{\overset{|}{C}}}}{-}\underset{H}{\underset{|}{\overset{H}{\overset{|}{C}}}}{-}\underset{H}{\underset{|}{\overset{H}{\overset{|}{C}}}}{-}H + Cl_2 \xrightarrow{\text{light}}$$

c) $$C_6H_6 + Cl_2 \xrightarrow{FeCl_3} \qquad + HCl$$

Charles D. Winters

A few minutes →

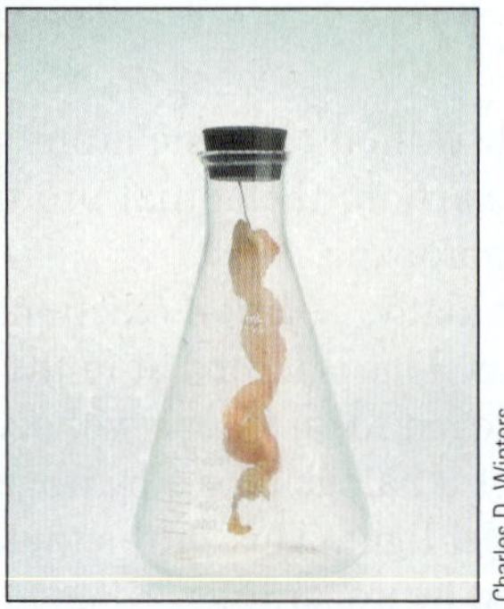
Charles D. Winters

Figure 21.21 Addition reaction of bromine with alkenes. Chemists used to test for the presence of double bonds by reaction with bromine. Since bromine easily undergoes an addition reaction with alkenes, the disappearance of the red-brown color of bromine indicates that it has reacted. These photographs show that the bromine color disappears in the presence of bacon. We conclude that this is evidence of alkenes in bacon.

21.10 Uses of Hydrocarbons

It is impossible to overstate the importance of hydrocarbons in modern society (**Figure 21.22**). Alkanes move nations, both literally and figuratively—literally in transportation and figuratively in world politics. The oil crises of the 1970s, Operation Desert Storm in 1991, and the Iraq War of 2003–2011 are events in which petroleum had a major impact on all industrialized nations. We rely heavily on petroleum products for the energy to heat homes, generate electricity, move people and goods, and manufacture almost everything we use, including many things derived directly from alkanes.

How close to you right now are hydrocarbons or products derived from hydrocarbons? Well, unless you're sitting naked as you study this book, you're probably wearing some. If everything you have on is not made of wool, cotton, silk, rubber, hemp, bone, or leather, you are almost certainly wearing something synthetic that started as a hydrocarbon. And even if you are wearing clothes derived entirely from natural products, the processes by which the natural fibers were converted into wearable clothing used hydrocarbons in some form.

Look around you. How many things do you see that are made of or derived from hydrocarbons? To help you, look at **Figure 21.23**. The chart starts with

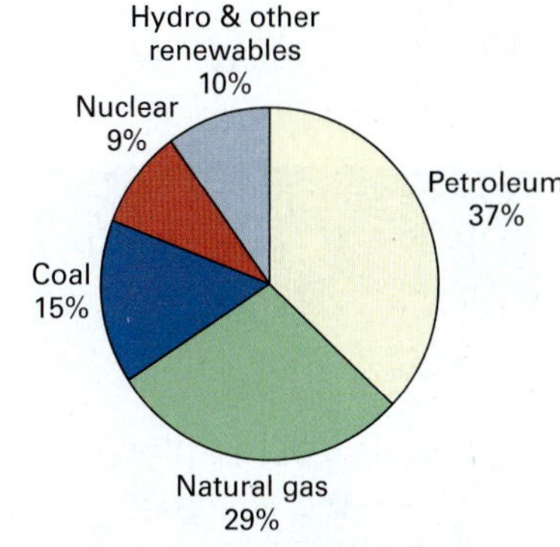

2016 energy consumption = 97.4 quadrillion Btu (97.4×10^{15} Btu)

Figure 21.22 U.S. energy sources in 2016. Petroleum, natural gas, and coal account for 81% of the total energy consumption.

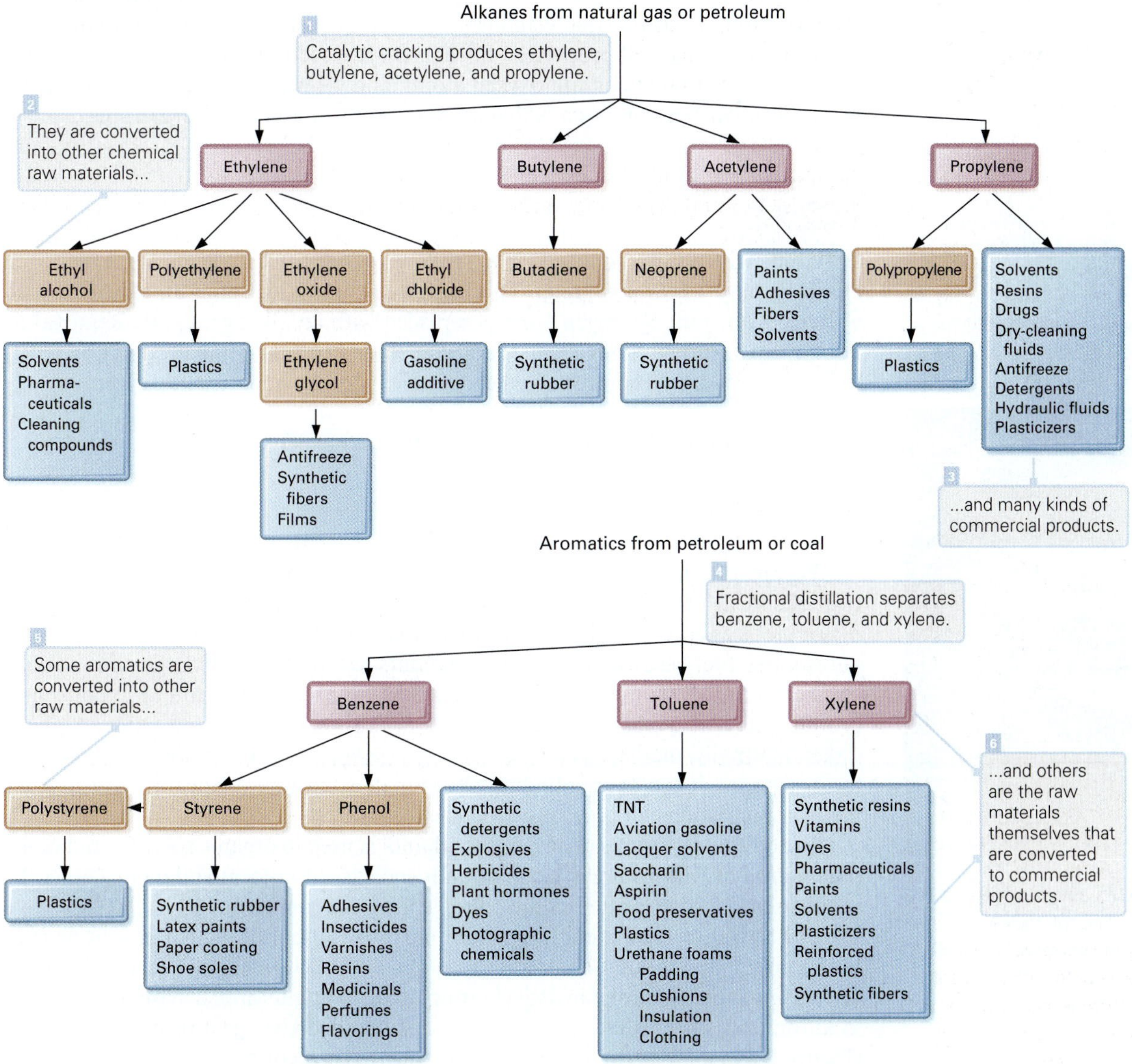

Alkanes from natural gas or petroleum

1 Catalytic cracking produces ethylene, butylene, acetylene, and propylene.

2 They are converted into other chemical raw materials...

Ethylene → Ethyl alcohol → Solvents, Pharmaceuticals, Cleaning compounds; Polyethylene → Plastics; Ethylene oxide → Ethylene glycol → Antifreeze, Synthetic fibers, Films; Ethyl chloride → Gasoline additive

Butylene → Butadiene → Synthetic rubber

Acetylene → Neoprene → Synthetic rubber; Paints, Adhesives, Fibers, Solvents

Propylene → Polypropylene → Plastics; Solvents, Resins, Drugs, Dry-cleaning fluids, Antifreeze, Detergents, Hydraulic fluids, Plasticizers

3 ...and many kinds of commercial products.

Aromatics from petroleum or coal

4 Fractional distillation separates benzene, toluene, and xylene.

5 Some aromatics are converted into other raw materials...

Benzene → Styrene → Polystyrene → Plastics; Styrene → Synthetic rubber, Latex paints, Paper coating, Shoe soles; Phenol → Adhesives, Insecticides, Varnishes, Resins, Medicinals, Perfumes, Flavorings; Synthetic detergents, Explosives, Herbicides, Plant hormones, Dyes, Photographic chemicals

Toluene → TNT, Aviation gasoline, Lacquer solvents, Saccharin, Aspirin, Food preservatives, Plastics, Urethane foams (Padding, Cushions, Insulation, Clothing)

Xylene → Synthetic resins, Vitamins, Dyes, Pharmaceuticals, Paints, Solvents, Plasticizers, Reinforced plastics, Synthetic fibers

6 ...and others are the raw materials themselves that are converted to commercial products.

Figure 21.23 Some of the organic chemicals obtained from fossil fuels and their uses as raw materials.

alkanes, takes you through other hydrocarbons, and ends at the useful products in the bottom box in each column. Can you not see, touch, or smell at least one of those end products at this very instant?

Where on Earth can you go to be totally separate from hydrocarbons? Almost nowhere. How about deep in the dense woods? No, the wonderful smell of a pine forest is from naturally occurring hydrocarbons called (for good reasons) *pinenes.* On an all-wooden boat on an ocean, out of sight of land? Perhaps, if the boat isn't painted with an oil-based paint commonly used on marine vessels. In the middle of a desert, on top of Mount Everest, or at the North or South Pole? Perhaps, but wherever it is, you'd better be dressed in clothing that was manufactured in the 18th century!

Charles D. Winters

Figure 21.24 Some common alcohols. Carburetor cleaner contains methanol, or methyl alcohol, CH_3OH. The alcohol in alcoholic beverages is ethanol, or ethyl alcohol, C_2H_5OH. Rubbing alcohol is 2-propanol, or isopropyl alcohol, C_3H_7OH.

21.11 THROUGH 21.15: ORGANIC COMPOUNDS CONTAINING OXYGEN AND NITROGEN

After carbon and hydrogen, the third element most commonly found in organic compounds is oxygen. Capable of forming two bonds (See Table 21.1), an oxygen atom serves as a connecting link between two other atoms or, double-bonded, usually to carbon, as a terminal atom in a functional group. A **functional group** is an atom or group of atoms that establishes the identity of a class of compounds and determines its chemical properties. In the next five sections, we will look at several functional groups that contain oxygen atoms, and we will look briefly at two classes of nitrogen-bearing compounds that are very important in living organisms.

21.11 Alcohols and Ethers

Although the unshared electron pairs are not shown on the oxygen atom in these structural diagrams, they play a key role in governing the physical and chemical properties of alcohols and ethers.

Beginning with a water molecule in the following diagrams, if we remove a hydrogen atom, we have a **hydroxyl group**. This functional group identifies an **alcohol**, which is formed when the hydrogen atom is replaced with an alkyl group, R. Removal of both hydrogen atoms from a water molecule leaves the functional group of an **ether**. The ether molecule is completed when two alkyl groups bond to the oxygen.

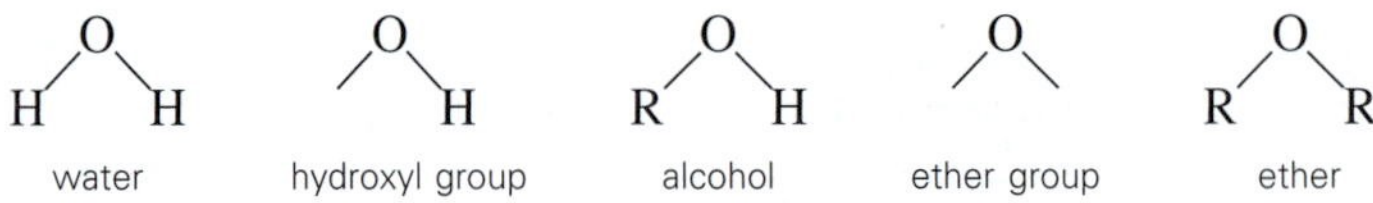

Figure 21.24 shows three common alcohols. Methanol, CH_3OH, is an industrial chemical with production measured in the billions of pounds annually. Its major use is as a building block for other molecules. It is also used as an alternative automotive fuel because its combustion emits less pollution than gasoline, and it can be made from sources other than petroleum. Methanol is found in consumer products such as windshield washer fluids, windshield deicers, and paint strippers. Taken internally, methanol can be a deadly poison, and even in doses as small as a few teaspoons, it can cause blindness. Fortunately, prompt medical treatment after methanol ingestion can prevent its tragic effects.

In addition to its uses in beverages, ethanol is used in organic solvents and in the preparation of various organic compounds such as chloroform and diethyl ether. Its production is also measured in the billions of pounds annually.

Other widely used alcohols include isopropyl alcohol, which is sold as rubbing alcohol, and *n*-butanol, used in lacquers in the automobile industry. Alcohols containing more than one hydroxyl group are also common. Permanent antifreeze in automobiles is ethylene glycol, which has two hydroxyl groups in the molecule (**Figure 21.25**). Glycerin, or glycerol, a trihydroxyl alcohol, has many uses in the manufacturing of drugs, cosmetics, explosives, and other chemicals (**Figure 21.26**).

Charles D. Winters

Figure 21.25 Engine antifreeze and coolant. The substance marketed as "permanent antifreeze" (because it can be used in both winter and summer) is a solution with ethylene glycol, $HO{-}CH_2CH_2{-}OH$, as its primary component. A molecule with two hydroxyl groups is called a *diol*.

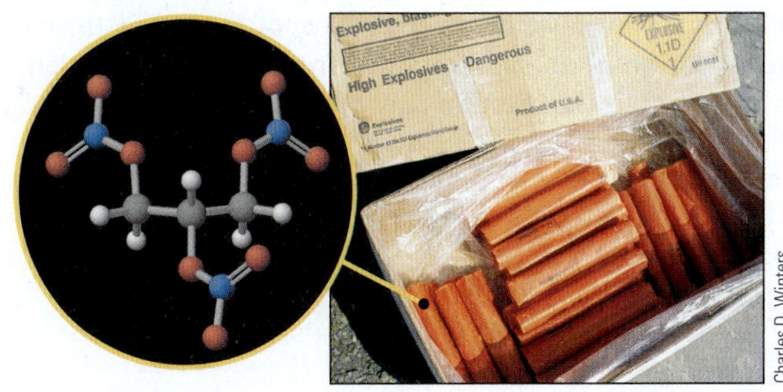

Charles D. Winters

Figure 21.26 Nitroglycerin. Glycerin, also called *glycerol* or *1,2,3-propanetriol*, reacts with nitric acid to yield nitroglycerin, an explosive. When nitroglycerin is mixed with an absorbent material, forming dynamite, it is less sensitive to shocks than liquid nitroglycerin alone. Nitroglycerin is also used medicinally to dilate the blood vessels of the heart and relieve chest pain.

The word *ether* generally makes people think of the anesthetic that is so identified. This compound is diethyl ether, or simply ethyl ether; its line formula is $C_2H_5{-}O{-}C_2H_5$. Recently, its isomer, methyl propyl ether (Neothyl), has been gaining popularity as an anesthetic. It has fewer objectionable aftereffects than diethyl ether. Diethyl ether is used as a solvent for dissolving fats from foods and animal tissue in the laboratory. Because of its great combustibility, diethyl ether is also used as a cold weather starting fluid for automobiles.

Structures and Names of Alcohols and Ethers

Goal 15 Identify the structural formulas of the functional groups that distinguish alcohols and ethers.

16 Given the name (or structural diagram) of an alcohol or ether, write the structural diagram (or name).

Figures 21.27 and **21.28** are models of an alcohol and an ether, respectively. See if you can locate the functional group in each model. Alcohols are sometimes described by their common name, which originate in the name of the alkyl group to which the hydroxyl group is bonded. This system names the alkyl group, followed by the word *alcohol*. Thus, CH_3OH is *methyl alcohol*, and C_2H_5OH is *ethyl alcohol*. Under IUPAC nomenclature rules for alcohols, the *e* at the end of the corresponding alkane is replaced with the suffix *-ol* and the result is the name of the alcohol. Thus, methyl alcohol becomes *methanol*, and ethyl alcohol becomes *ethanol*.

Propyl alcohol has two isomers:

```
     H   H   H              H      H      H
     |   |   |              |      |      |
 H—C—C—C—H          H—C——C——C—H
     |   |   |              |      |      |
     H   H   OH             H      OH     H
  n-propyl alcohol        isopropyl alcohol
    1-propanol               2-propanol
```

These isomers are distinguished by stating the number of the carbon atom to which the hydroxyl group is bonded. Accordingly, *n*-propyl alcohol becomes *1-propanol* and isopropyl alcohol is formally designated *2-propanol*.

All ethers are called *ether* and are identified specifically by first naming the two alkyl groups that are bonded to the functional group, followed by the word *ether*. If the groups are identical, the prefix *di-* may be used, as in diethyl ether (Figure 21.28).

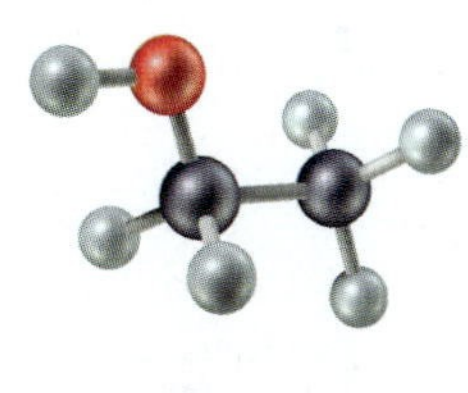

Figure 21.27 Models of ethanol (also called *ethyl alcohol*), C_2H_5OH.

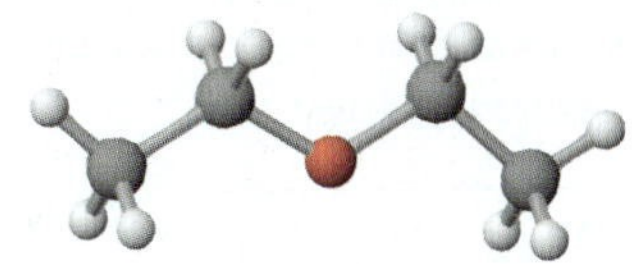

Figure 21.28 Model of diethyl ether, $C_2H_5OC_2H_5$.

Sources and Preparation of Alcohols and Ethers

Goal 17 Given the reactants (or products) of a dehydration reaction between two alcohols, predict the products (or reactants) of the reaction.

Charles D. Winters

Figure 21.29 Wine production. Yeast metabolizes sugar to obtain energy, producing ethanol as a product of the chemical change. Exposure to oxygen must be avoided during fermentation to prevent oxidation of the ethanol to acetic acid (Section 21.13). Home fermenting and brewing carboys are fitted with an airlock to allow carbon dioxide to escape while preventing oxygen from entering.

Hydration of Alkenes The major industrial source of several of the most important alcohols is the hydration of alkenes obtained from cracking petroleum. Beginning with ethylene, for example, the reaction may be summarized as follows:

$$\begin{array}{c} \mathrm{H}\quad\mathrm{H} \\ |\quad\;\; | \\ \mathrm{H{-}C{=}C{-}H} \\ \text{ethylene} \end{array} + \mathrm{HOH} \longrightarrow \begin{array}{c} \mathrm{H}\quad\mathrm{H} \\ |\quad\;\; | \\ \mathrm{H{-}C{-}C{-}H} \\ |\quad\;\; | \\ \mathrm{H}\quad\mathrm{OH} \\ \text{ethanol} \end{array}$$

Fermentation of Carbohydrates Making ethanol by fermenting sugars in the presence of yeast is probably the oldest synthetic chemical process known (**Figure 21.29**):

$$\underset{\text{glucose (sugar)}}{C_6H_{12}O_6} \xrightarrow{\text{Yeast}} 2\ CO_2 + 2\ \underset{\text{ethanol}}{C_2H_5OH}$$

A solution that is 95% ethyl alcohol (190 proof) may be obtained from the final mixture by fractional distillation. The mixture also yields two other products of commercial value today: 1-butanol and acetone (Section 21.12).

Under properly controlled conditions, ethers can be prepared by dehydrating alcohols. At 140°C, and with constant alcohol addition to replace the ether as it distills from the mixture, diethyl ether is formed from two molecules of ethanol:

$$\begin{array}{c} \mathrm{H}\quad\mathrm{H} \\ |\quad\;\; | \\ \mathrm{H{-}C{-}C{-}O{-}H} \\ |\quad\;\; | \\ \mathrm{H}\quad\mathrm{H} \\ \text{ethanol} \end{array} + \begin{array}{c} \mathrm{H}\quad\mathrm{H} \\ |\quad\;\; | \\ \mathrm{H{-}O{-}C{-}C{-}H} \\ |\quad\;\; | \\ \mathrm{H}\quad\mathrm{H} \\ \text{ethanol} \end{array} \longrightarrow \begin{array}{c} \mathrm{H}\quad\mathrm{H}\qquad\mathrm{H}\quad\mathrm{H} \\ |\quad\;\; |\qquad\; |\quad\;\; | \\ \mathrm{H{-}C{-}C{-}O{-}C{-}C{-}H} \\ |\quad\;\; |\qquad\; |\quad\;\; | \\ \mathrm{H}\quad\mathrm{H}\qquad\mathrm{H}\quad\mathrm{H} \\ \text{diethyl ether} \end{array} + \mathrm{HOH}$$

Physical and Chemical Properties of Alcohols and Ethers

Goal 18 Given the molecular structures of a set of alcohols or ethers, or information from which they may be obtained, predict relative values of boiling points or solubility in water.

The structural similarity between water and alcohols suggests similar intermolecular forces and therefore similar physical properties. One- to three-carbon-atom alcohols are liquids with boiling points ranging from 65°C to 97°C. These are comparable to the boiling point of water but well above the boiling points of alkanes of about the same molar mass. This is largely because of hydrogen bonding. ◀ Hydrogen bonding also accounts for the complete miscibility (solubility) between one- to three-carbon-atom alcohols and water. Solubility drops off sharply as the alkyl chain lengthens and the molecule assumes more of the character of the parent alkane. As usual, boiling points rise with increasing molecular size.

▶ **P/Review** Hydrogen bonding is the attraction between molecules in which a hydrogen atom is bonded to a nitrogen, oxygen, or fluorine atom that has at least one unshared pair of electrons (Section 15.4).

Ether molecules are less polar than alcohol molecules, and there is no opportunity for hydrogen bonding between them. Therefore, intermolecular attractions are lower, as are their respective boiling points. Ethers with two and three carbons are gases at room conditions. Diethyl ether, with four carbons, is a volatile liquid that boils at 35°C. The solubility of ether molecules in water is about the same as the solubility of alcohols that are isomers, probably due to the hydrogen bonding between the ether molecules and water molecules.

The chemical properties of alcohols are essentially the chemical properties of the functional group, —OH. In some reactions, the C—OH bond is broken, separating the entire hydroxyl group. In other reactions, the O—H bond in the hydroxyl group is broken.

Ethers are relatively unreactive compounds, quite resistant to attack by active metals, strong bases, and oxidizing agents. However, they are highly flammable and must be handled cautiously in the laboratory.

Target Check 21.7

a) Write the name and structural formula of the functional group that identifies an alcohol.
b) Write the structural formula of the functional group that identifies an ether.
c) Draw line formulas for all compounds with the formula C_3H_8O. Identify these as alcohols or ethers.

21.12 Aldehydes and Ketones

Goal 19 Identify the structural formulas of the functional groups that distinguish aldehydes and ketones.

20 Given the name (or structural diagram) of an aldehyde or ketone, write the structural diagram (or name).

21 Write structural diagrams to show how a specified aldehyde or ketone is prepared from an alcohol.

Aldehydes and **ketones** are characterized by the **carbonyl group,**

If at least one hydrogen atom is bonded to the carbonyl carbon, the compound is an aldehyde, RCHO; if two alkyl groups are attached, the compound is a ketone, R—CO—R′:

The simplest carbonyl compound is formaldehyde, H—CO—H, which has two hydrogen atoms bonded to the carbonyl carbon. If a methyl group replaces one of the hydrogens of formaldehyde, the result is acetaldehyde, CH_3—CO—H. Replacement of both formaldehyde hydrogens with methyl groups yields acetone, CH_3—CO—CH_3:

Figure 21.30 shows models of formaldehyde and acetone.

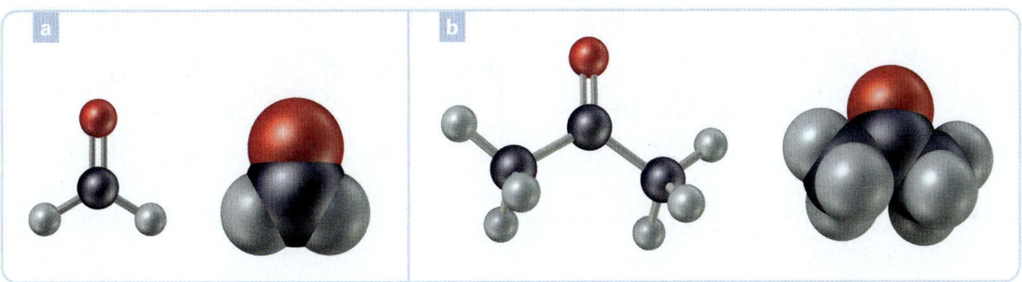

Figure 21.30 Models of (a) formaldehyde, H—CO—H, the simplest aldehyde, and (b) acetone, CH_3—CO—CH_3, the simplest ketone.

Many aldehydes are best known by their common names. The IUPAC nomenclature system for aldehydes employs the name of the parent hydrocarbon, substituting the suffix *-al* for the final *e* to identify the compound as an aldehyde. Thus, the IUPAC name for formaldehyde is methanal, for acetaldehyde, ethanal, and so forth.

Ketones are named by one of two systems. The first duplicates the method of naming ethers: identify each alkyl group attached to the carbonyl group, followed by the class name, ketone. Accordingly, methyl ethyl ketone has the following structure:

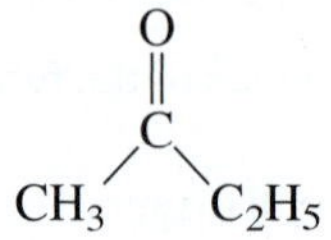

$$CH_3-\overset{O}{\overset{\|}{C}}-C_2H_5$$

Under the IUPAC system, the number of carbons in the longest chain carrying the carbonyl carbon establishes the hydrocarbon base, which is followed by *-one* to identify the ketone as the class of compound. Methyl ethyl ketone, having four carbons, is called *butanone*. Two isomers of pentanone would be 2-pentanone and 3-pentanone, the number being used to designate the carbonyl carbon:

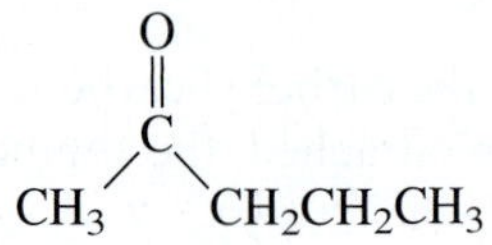

$$CH_3-\overset{O}{\overset{\|}{C}}-CH_2CH_2CH_3 \qquad CH_3CH_2-\overset{O}{\overset{\|}{C}}-CH_2CH_3$$

2-pentanone 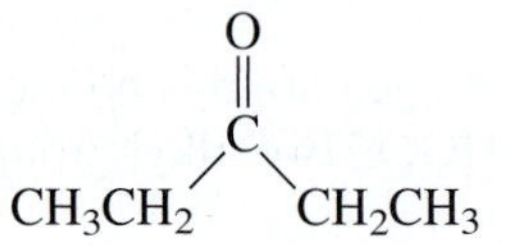 3-pentanone

Aldehydes and ketones may be prepared by oxidation of alcohols. If the product is to be a ketone, the alcohol must be a *secondary* alcohol, in which the hydroxyl group is bonded to a carbon that is attached to two other carbon atoms:

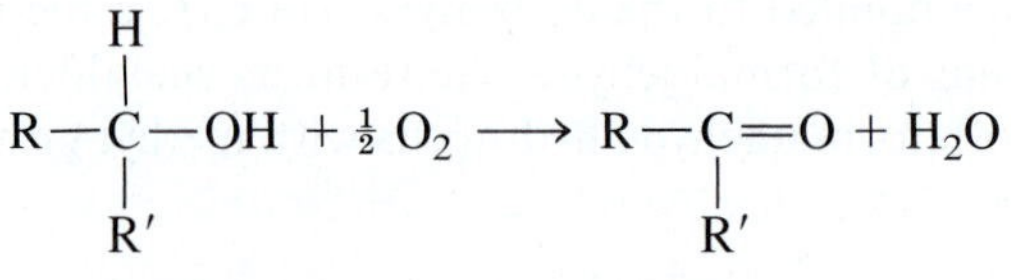

$$R-\underset{R'}{\underset{|}{\overset{H}{\overset{|}{C}}}}-OH + \tfrac{1}{2}\,O_2 \longrightarrow R-\underset{R'}{\underset{|}{C}}=O + H_2O$$

secondary alcohol ketone

Care must be taken not to overoxidize aldehyde preparations, since aldehydes are easily oxidized to carboxylic acids (see Section 21.13).

Aldehydes and ketones may also be produced by the hydration of alkynes. If the triple bond is on the last carbon in a carbon atom chain, an aldehyde is produced; if the bond is between internal carbons, the result is a ketone. A typical reaction is the commercial preparation of acetaldehyde:

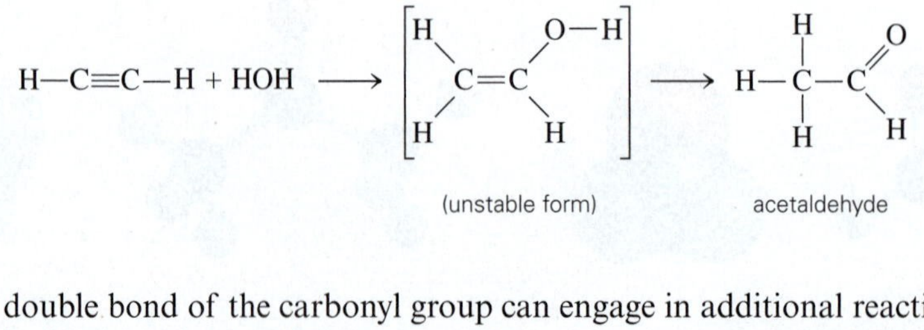

$$H-C\equiv C-H + HOH \longrightarrow \left[\begin{matrix} H & & O-H \\ & C=C & \\ H & & H \end{matrix}\right] \longrightarrow H-\underset{H}{\underset{|}{\overset{H}{\overset{|}{C}}}}-C\begin{matrix} {}^{\diagup\!\!\diagup O} \\ {}_{\diagdown H} \end{matrix}$$

(unstable form) acetaldehyde

The double bond of the carbonyl group can engage in additional reactions, just like the double bond in the alkenes. One such reaction is catalytic hydrogenation of ketones to secondary alcohols, in which the hydroxyl group is bonded to a carbon atom *within* the chain:

$$\text{R}-\overset{\overset{\text{R}'}{|}}{\text{C}}=\text{O} + \text{H}_2 \xrightarrow{\text{catalyst}} \text{R}-\underset{\underset{\text{H}}{|}}{\overset{\overset{\text{R}'}{|}}{\text{C}}}-\text{O}-\text{H}$$

ketone — secondary alcohol

Oxidation reactions occur quite readily with aldehydes but are resisted by ketones. When an aldehyde is oxidized, the product is a carboxylic acid:

$$\text{R}-\overset{\overset{\text{H}}{|}}{\text{C}}=\text{O} + \tfrac{1}{2}\,\text{O}_2 \longrightarrow \text{R}-\text{C}\begin{matrix}\diagup \text{OH} \\ \diagdown\!\!\diagdown \text{O}\end{matrix}$$

aldehyde — carboxylic acid

Formaldehyde is probably the best-known carbonyl compound. Large quantities are made into substances such as the binder in particle board. Since formaldehyde has been cited as a carcinogen, its once-common use in preserving biological specimens has vanished. Acetaldehyde (ethanal) is used in manufacturing organic compounds such as acetic acid and ethyl acetate. Other aldehydes you have probably encountered are benzaldehyde (almond flavor), cinnamaldehyde (cinnamon flavor), and vanillin (vanilla flavor) (**Figure 21.31**).

Acetone is the most commercially important ketone, with billions of pounds produced yearly in the United States. It is a solvent used in manufacturing other organic chemicals, drugs, and explosives. Acetone is also found in paint remover and nail polish remover. Methyl ethyl ketone (MEK) is used in the petroleum industry, as a lacquer solvent, and in nail polish remover. Most organic compounds whose names end in *-one* are ketones. You may be familiar with the anti-inflammatory cortisone, or the sex hormones progesterone and testosterone; all are ketones (**Figure 21.32**).

✓ Target Check 21.8

a) Write the name and structural formula of the functional group that identifies an aldehyde or ketone.

b) Write structural diagrams that show the difference between an aldehyde and a ketone.

c) Write structural diagrams for all possible aldehyde or ketone isomers of $C_5H_{10}O$; identify the aldehydes and the ketones.

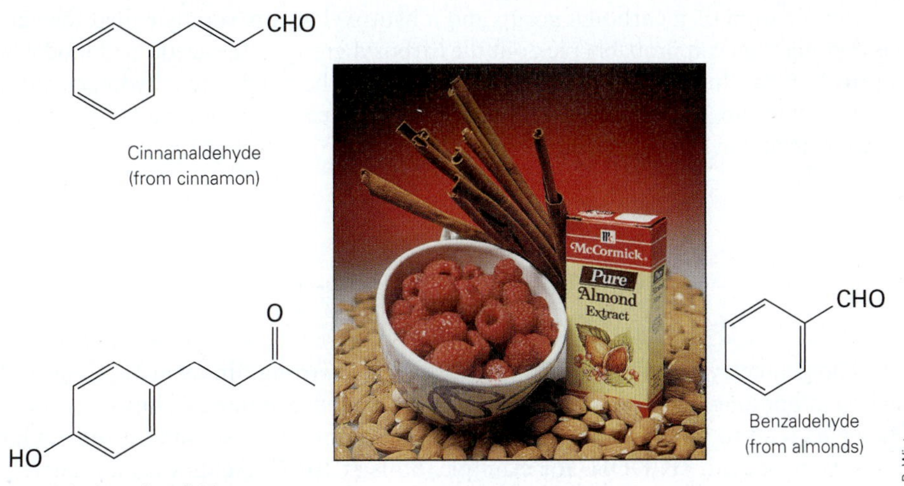

Figure 21.31 Odors due to aldehydes and ketones. Raspberries contain the ketone 4-(4-hydroxyphenyl)-2-butanone, which is called *raspberry ketone*. The aroma of cinnamon is due to cinnamaldehyde, and almond scent is due to benzaldehyde, and as indicated by their name, both are aldehydes.

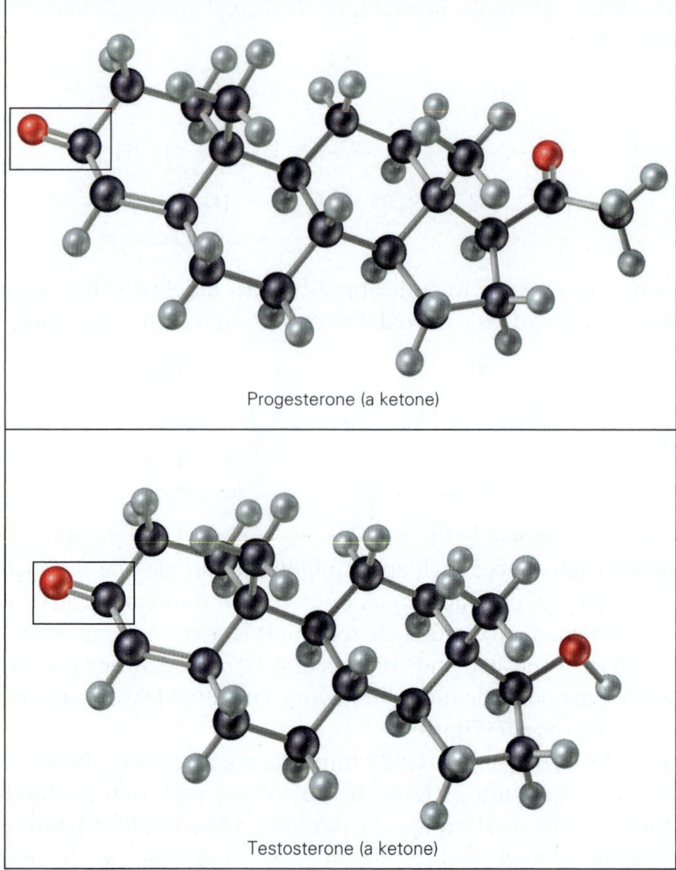

Figure 21.32 The steroids progesterone and testosterone. Steroid molecules have four rings: three cyclohexane rings and one cyclopentane ring (Section 22.5). Progesterone (*top*) and testosterone (*bottom*) also have ketone functional groups (*highlighted*). Both molecules are human sex hormones.

21.13 Carboxylic Acids and Esters

Goal 22 Identify the structural formulas of the functional groups that distinguish carboxylic acids and esters.

23 Given the name (or structural diagram) of a carboxylic acid or ester, write the structural diagram (or name).

24 Given the reactants (or products) of an esterification reaction, predict the products (or reactants) of the reaction.

As you have seen, oxidation of an aldehyde produces a **carboxylic acid**, the general formula of which is frequently shown as RCOOH. The functional group, −COOH, is a combination of a carbonyl group and a hydroxyl group, rightly called the **carboxyl group**. You can probably pick out the carboxyl group in the acetic acid model in **Figure 21.33(a)**. In an **ester**, the carboxyl carbon may be bonded to a hydrogen atom or an alkyl group, and the carboxyl hydrogen is replaced by another alkyl group, as shown here and in **Figure 21.33(b)**:

$$\underset{\text{carboxylic acid}}{R-\overset{\overset{\displaystyle O}{\|}}{C}-O-H} \qquad \underset{\text{ester}}{R-\overset{\overset{\displaystyle O}{\|}}{C}-O-R'}$$

The geometry of the carboxyl group results in strong dipole attractions and hydrogen bonding between molecules. As a consequence, boiling points of carboxylic acids tend to be high compared with other compounds of similar molecular mass. Formic acid, HCOOH, for example, boils at 101°C. Acids with a relatively small number of carbons are completely miscible in water but solubility drops off as the aliphatic chain lengthens and the molecule behaves more like a hydrocarbon.

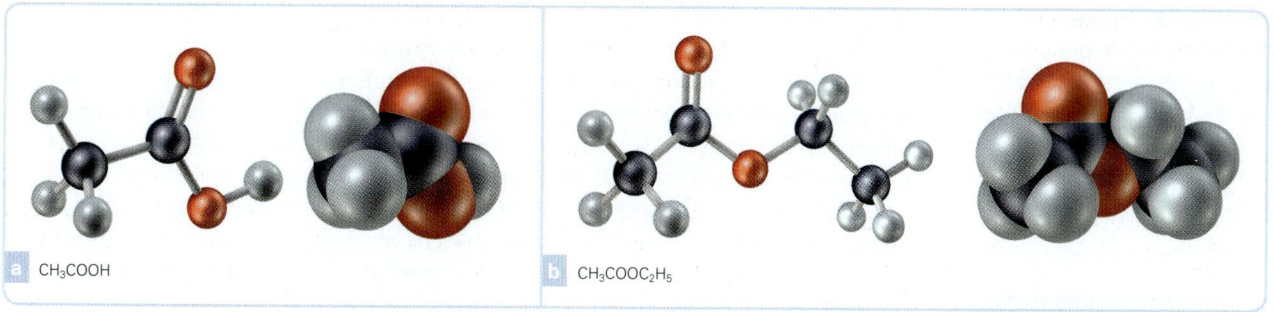

Figure 21.33 Models of (a) acetic acid, CH_3COOH, and (b) ethyl acetate, $CH_3COOC_2H_5$.

Acetic acid (**Figure 21.34**) is produced by the stepwise oxidation of ethanol, first to acetaldehyde and then to acetic acid ▶:

Potassium permanganate, $KMnO_4$, is a strong oxidizing agent.

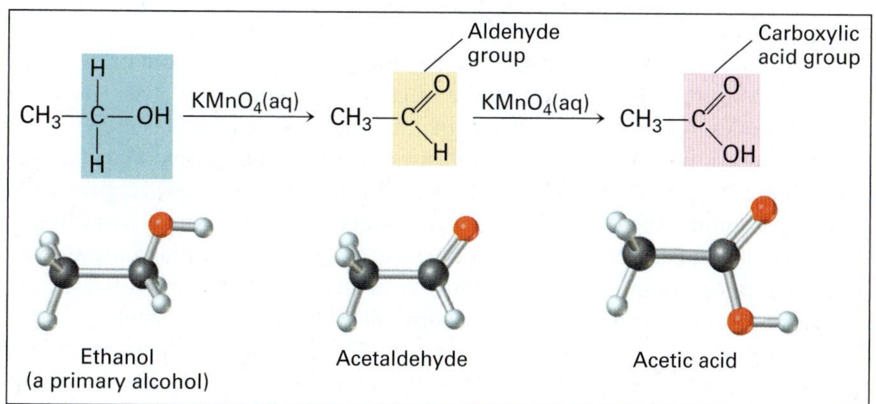

Carboxylic acids are weak acids that release a proton from the carboxyl group on ionization.* Acetic acid, for example, ionizes in water as follows:

$$CH_3COOH(aq) \rightleftharpoons CH_3COO^-(aq) + H^+(aq)$$

The ionization takes place, but slightly: Only about 1% of the acetic acid molecules ionize. The solution consists primarily of molecular CH_3COOH. The percentage ionization is not a factor in acetic acid participating in typical acid reactions such as neutralization:

$$CH_3COOH(aq) + OH^-(aq) \rightarrow H_2O(\ell) + CH_3COO^-(aq)$$

and the release of hydrogen on reaction with a metal:

$$2\,CH_3COOH(aq) + Ca(s) \rightarrow 2\,CH_3COO^-(aq) + Ca^{2+}(aq) + H_2(g)$$

Metal acetate salts may be obtained by evaporating the resulting solutions to dryness.

The reaction between an acid and an alcohol is called **esterification**. The products of the reaction are an ester and water. A typical esterification reaction is the following:

$$\underset{\substack{\text{acetic acid}\\ \textit{carboxylic acid}}}{H{-}\overset{H}{\underset{H}{C}}{-}\overset{O}{\overset{\|}{C}}{-}O{-}H} + \underset{\substack{\text{methanol}\\ \textit{alcohol}}}{H{-}O{-}\overset{H}{\underset{H}{C}}{-}H} \rightleftharpoons \underset{\substack{\text{methyl acetate}\\ \textit{ester}}}{H{-}\overset{H}{\underset{H}{C}}{-}\overset{O}{\overset{\|}{C}}{-}O{-}\overset{H}{\underset{H}{C}}{-}H} + HOH$$

Charles D. Winters

Figure 21.34 Acetic acid is responsible for the odor and taste of vinegar. Distilled white vinegar (as opposed to cider or malt vinegar, for example) is made from grain alcohol and distilled to remove all compounds but acetic acid and water. The label indicates that this brand has "5% Acidity," indicating that it is a 5% solution of acetic acid.

*In more advanced study of organic reactions, the term *acid* is also used in reference to Lewis acids (Section 17.4). This is why the adjective *carboxylic* is used to identify an organic acid containing the carboxyl group.

▶ **P/Review** A Brønsted–Lowry acid–base reaction involves the removal of a proton from the acid by the base. An ammonia molecule, with an unshared electron pair on the nitrogen, is the proton remover—the base—in this reaction. See Section 17.3.

Note how the water molecule is formed: *The acid contributes the entire hydroxyl group, whereas the alcohol furnishes only the hydrogen.*

The names of esters are derived from the parent alcohol and acid. The first term is the alkyl group associated with the alcohol; the second term is the name of the anion derived from the acid. In the preceding example, methanol (methyl alcohol) yields *methyl* as the first term, and acetic acid yields *acetate* as the second term.

Carboxylic acids engage in typical proton transfer acid–base-type reactions with ammonia to produce ionic compounds. ◀ The ammonium compound so produced may then be heated, which causes it to lose a water molecule. The resulting product is called an *amide.* Compared with the original acid, an amide substitutes an $-NH_2$ group for the $-OH$ group of the acid (Section 21.14):

$$\underset{\text{acid}}{R-\overset{\overset{\displaystyle O}{\|}}{C}-O-H} + NH_3 \longrightarrow \underset{\text{ionic compound}}{\left[R-\overset{\overset{\displaystyle O}{\|}}{C}-O\right]^- NH_4^+} \xrightarrow{\text{heat}} \underset{\text{amide}}{R-\overset{\overset{\displaystyle O}{\|}}{C}-NH_2} + H_2O$$

Figure 21.35 Sourdough bread. The distinctive tangy taste of sourdough bread results from the work of the yeast and lactobacteria that also produce the carbon dioxide that makes the bread rise. Acetic acid and lactic acid, both carboxylic acids, are produced during the fermentation, giving sourdough bread its sour taste.

Formic acid and acetic acid are the two most important carboxylic acids. Formic acid is a source of irritation in the bites of ants and other insects or in the scratch of nettles. A liquid with a sharp, irritating odor, formic acid is used in manufacturing esters, ionic compounds, and plastics. Acetic acid is present in a concentration of about 5% in vinegar and is responsible for its odor and taste (**Figure 21.35**). Acetic acid is among the least expensive organic acids and therefore is a raw material in many commercial processes that require a carboxylic acid. Sodium acetate is one of several common compounds of carboxylic acids. It is used to control the acidity of chemical processes and in preparing soaps and pharmaceutical agents.

Ethyl acetate and butyl acetate are two of the relatively few esters produced in large quantity. Both are used as solvents, particularly in manufacturing lacquers. Other esters are used in the plastics industry and some find application in medicinal fields. Esters are responsible for the odors of most fruits and flowers, leading to their use in the food and perfume industries (**Figure 21.36**).

Figure 21.36 Odors due to esters. The odors of most fruits and many perfumes are due to molecules that contain the ester functional group.

✓ Target Check 21.9

a) Write the name and structural formula of the functional group that defines a carboxylic acid.
b) Describe in words the reactants and products of an esterification reaction.
c) Write structural diagrams for all possible carboxylic acid or ester isomers of $C_4H_8O_2$; identify the carboxylic acid and the esters.

21.14 Amines and Amides

Goal 25 Identify the structural formulas of the functional groups that distinguish amines and amides.

26 Given the name (or structural diagram) of an amine or amide, write the structural diagram (or name).

27 Given the structural diagram of an amine, or information from which it can be written, classify the amine as primary, secondary, or tertiary.

28 Given the reactants of a reaction between a carboxylic acid and an amine, predict the products of the reaction.

Amines are organic derivatives of ammonia, NH_3. An amine is formed by replacing one, two, or all three hydrogens in an ammonia molecule with an alkyl group. The number of hydrogens replaced determines whether an amine is primary (one hydrogen replaced), secondary (two), or tertiary (three) (**Figure 21.37**). IUPAC names are rarely used for amines; they are named in practice by identifying in alphabetical order the alkyl groups that are bonded to the nitrogen atom, using appropriate prefixes if two or three identical groups are present, followed by the suffix *-amine.* Here are some examples:

$$\mathrm{H{-}\ddot{N}{-}H}\ \text{(with H on N)} \qquad \mathrm{CH_3{-}\ddot{N}{-}H}\ \text{(with H on N)} \qquad \mathrm{CH_3{-}\ddot{N}{-}C_2H_5}\ \text{(with H on N)} \qquad \mathrm{CH_3{-}\ddot{N}{-}C_2H_5}\ \text{(with CH}_3\text{ on N)}$$

ammonia | methylamine, primary amine | ethylmethylamine, secondary amine | ethyldimethylamine, tertiary amine

Models of ammonia and the different methylamines are shown in **Figure 21.38**.

Dimethylamine and trimethylamine are used in making anion-exchange resins. Many chemical products such as dyes, drugs, herbicides, fungicides, soaps, insecticides, and photographic developers are made from amines. The aromatic amine aniline (phenylamine) is used in dye making.

An **amide** is a derivative of a carboxylic acid in which the hydroxyl part of the carboxyl group is replaced by an $-NH_2$, $-NHR$, or $-NR_2$ group. For example,

$$\mathrm{CH_3{-}C({=}O){-}OH} \quad \text{becomes} \quad \mathrm{CH_3{-}C({=}O){-}NH_2}$$

acetic acid | acetamide

by substitution of the $-NH_2$ for the $-OH$, as shown. An amide is commonly named by replacing the *-ic* or *-oic acid* name of the acid with *amide.*

The amide structure appears in a biochemical system, proteins, as a connecting link between amino acids. The linkage is commonly called a **peptide linkage**. This linkage has the form:

$$\mathrm{R{-}C({=}O){-}N(H){-}C}$$

peptide linkage

An **amino acid** is an acid in which an amine group is substituted for a hydrogen atom in the molecule. The amino acids in protein structure have the following general formula:

$$\mathrm{R{-}C(H)(NH_2){-}C({=}O){-}OH}$$

in which the amine and carboxyl groups are bonded to the same carbon atom. The peptide linkage is formed when the carboxyl group of one amino acid and the amine group of another combine by removing a water molecule:

$$\underset{\text{Amino acid 1}}{\mathrm{H_2N{-}C(R)(H){-}C({=}O){-}OH}} + \underset{\text{Amino acid 2}}{\mathrm{H{-}N(H){-}C(R)(H){-}C({=}O){-}OH}} \longrightarrow \underset{\text{Dipeptide}}{\mathrm{H_2N{-}C(R)(H){-}C({=}O){-}N(H){-}C(R)(H){-}C({=}O){-}OH}} + \mathrm{H_2O}$$

Peptide linkages

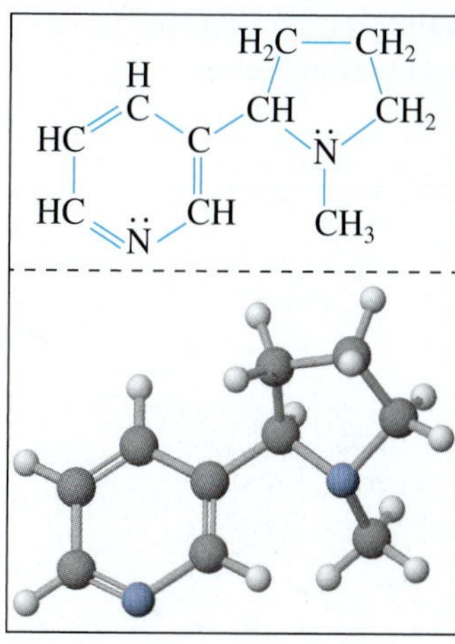

Figure 21.37 Nicotine, the stimulant in tobacco, is a tertiary amine.

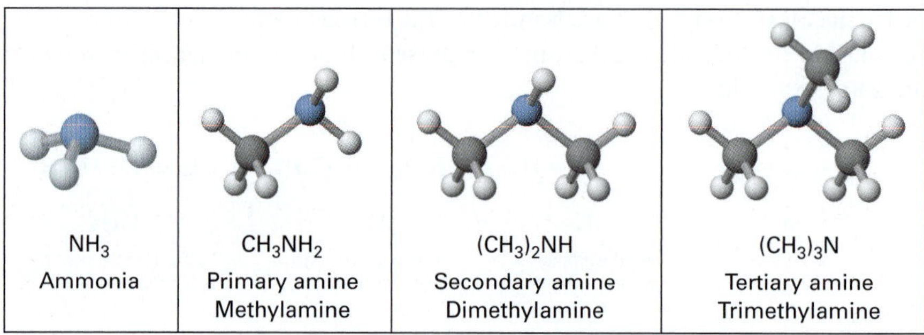

Figure 21.38 Models of ammonia and the methylamines.

Every living thing has proteins, which are chains of amino acids held together by amide linkages (peptide linkages). There are 20 different amino acids commonly found in proteins. Typical proteins are huge molecules with molar masses ranging from about 34,500 to 50,000,000 g/mol. These proteins perform many functions in a living system. Proteins are discussed in greater detail in Chapter 22.

Target Check 21.10

a) Write the structural formula of the functional group that defines an amine.
b) Write the structural formula of the functional group that defines an amide.

21.15 Summary of the Organic Compounds of Carbon, Hydrogen, Oxygen, and Nitrogen

The eight types of organic compounds of carbon, hydrogen, oxygen, and nitrogen you have studied are summarized in **Table 21.5**.

Table 21.5 Summary of Organic Compounds

Compound Class	General Formula	Functional Group	Names*
Alcohol	R—OH	—OH	Alkyl group + *alcohol*: methyl alcohol Alkane prefix + *-ol*: methanol
Ether	R—O—R′	O (bonded to two groups)	Name both alkyl groups + *ether*: ethyl methyl ether Alkyl group + *-oxy-* + alkane: methoxyethane
Aldehyde	R—CHO	C(=O)—H	Common prefix + *-aldehyde*: formaldehyde Alkane prefix + *-al*: methanal
Ketone	R—CO—R′	C=O (bonded to two groups)	Name both alkyl groups + *ketone*: methyl ethyl ketone; methyl *n*-propyl ketone (Number carbonyl carbon) + alkane prefix + *-one*: butanone; 2-pentanone
Carboxylic Acid	R—COOH	C(=O)—OH	Common name + *acid*: formic acid Alkane prefix + *-oic* + *acid*: methanoic acid
Ester	R—CO—OR′	C(=O)—OR′	Alcohol alkyl group + acid anion: methyl acetate Alcohol alkyl group + acid alkane prefix + *-oate*: methyl ethanoate
Amine	RNH_2 R_2NH R_3N	—N— (with a third bond)	Name alkyl group(s) + *amine*: methylamine Amino− + alkane: aminomethane
Amide	R—$CONH_2$	C(=O)—NH_2	Common acid prefix + *-amide*: formamide Alkane prefix + *-amide*: methanamide

*Common name followed by IUPAC name, with examples of each.

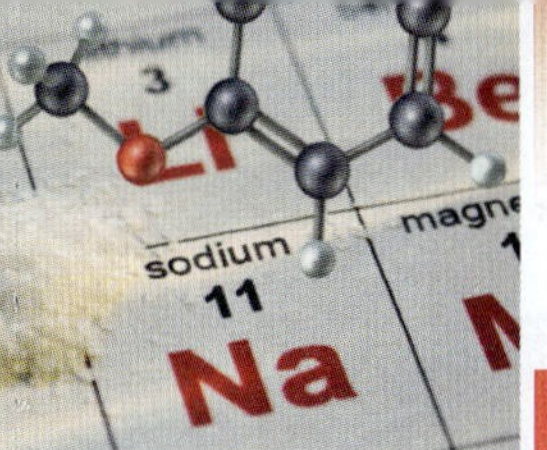

Everyday Chemistry 21.1

"IN WHICH THE SHAPE'S THE THING . . ."

You've probably seen the terms *saturated, monounsaturated,* and *polyunsaturated* used on food labels to describe oils and fats in what we eat. Saturated fats (**Figure 21.39**) are cylindrical and tend to pack into masses that are solids at human body temperature. Unsaturated fats and oils have a *cis* configuration around at least one double bond. These structures have a bend in them (**Figure 21.40**) and do not pack as easily as straight structures. As a result, the unsaturated fats and oils are liquids at human body temperature.

Hydrogenation of vegetable oils (seen on food labels as "partially hydrogenated vegetable oils") changes some *cis* fatty acids into saturated fats and also isomerizes some *cis* double bonds into *trans* double bonds. These *trans* double bonds give a cylindrical molecule (**Figure 21.41**), like saturated fats. Unfortunately, cylindrical fats, either saturated or *trans* unsaturated, are the fats that lead to blocked arteries. The bend in the molecules makes all the difference in the blood.

Quick Quiz

1. What particulate-level feature of the molecules that make up certain types of fats causes the molecules to tend to accumulate in arteries?
2. What type of bond—single, double, or triple—is responsible for putting a bend in fats and oils? What configuration about that bond distinguishes whether or not there is a bend?

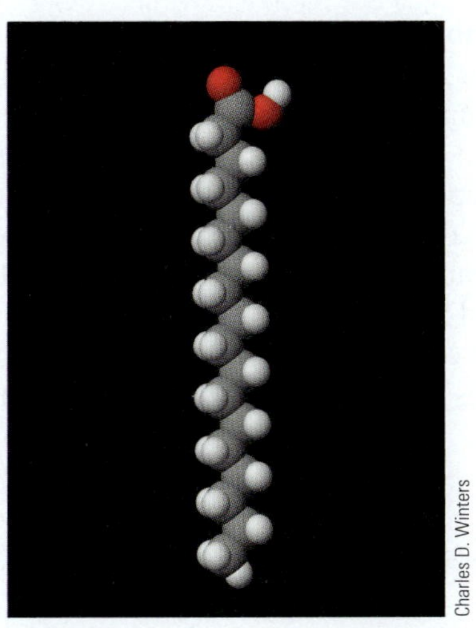

Charles D. Winters

Figure 21.39 Space-filling model of stearic acid, $CH_3(CH_2)_{16}COOH$. Notice that the molecule is cylindrical and straight.

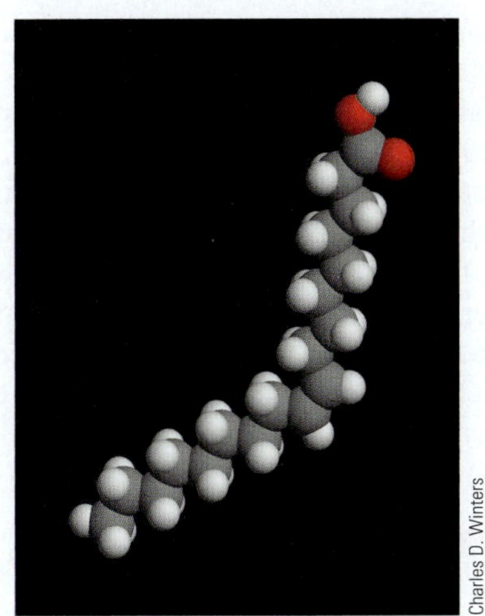

Charles D. Winters

Figure 21.40 Space-filling model of *cis*-oleic acid, $CH_3(CH_2)_7CH{=}CH(CH_2)_7COOH$. The molecule bends at the *cis* double bond.

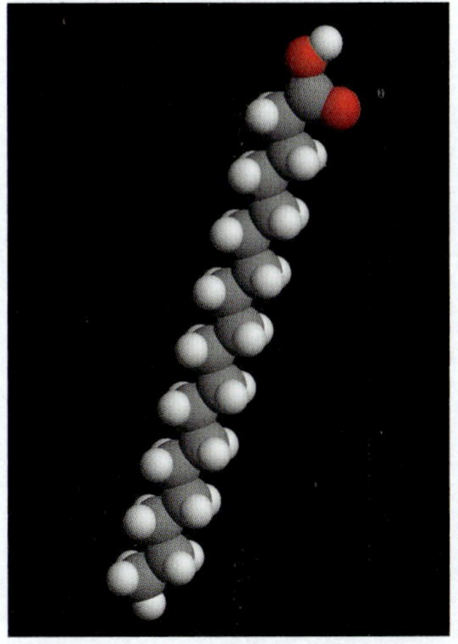

Charles D. Winters

Figure 21.41 Space-filling model of *trans*-oleic acid (elaidic acid). The molecule is straight, similar to a saturated fatty acid.

Thinking About Your Thinking

Classification

Table 21.5 summarizes a classification scheme for organic compounds based on functional groups, an atom or group of atoms in an organic molecule that is primarily responsible for its chemical properties. Thinking in this way—classifying molecules by functional group—is central to the organization of organic chemistry. Literally millions of different organic molecules are known to exist, and millions of reactions can be used to synthesize or transform these molecules. A body of knowledge so vast must have a classification scheme to give it organization. Organic functional groups are used by all chemists, everywhere in the world, to organize their knowledge.

Charles D. Winters

Charles D. Winters

Charles D. Winters

Figure 21.42 Plastics. Chemists use the word *polymers* to describe what in everyday language we call *plastics*.

21.16 THROUGH 21.17: POLYMERS

The word *plastics* is commonly used to describe a large number of familiar substances, all of which were first developed in the 20th century (**Figure 21.42**). These materials are made up of huge molecules having molecular masses that sometimes run into the millions.

One way to make a large molecule is to connect many small chemical units, somewhat like links being joined to form a chain. Each link is called a **monomer**, which, from the Greek, means "having one part." The resulting chain is called a **polymer**, which means "having many parts." The process by which polymers are formed is called **polymerization**. Polymers are classified according to the method by which they are formed, and these two classes guide the organization of the final two sections of this chapter.

21.16 Chain-Growth Polymers

Goal 29 Given the structural diagram or name of an ethylene-like monomer, predict the structural diagram of the product chain-growth polymer; given the structural diagram of a chain-growth polymer, predict the structural diagram and/or the name of the ethylene-like monomer from which it can be formed.

Alkene monomers form **chain-growth polymers**, which means that the monomers add to a growing polymer chain in sequential addition reactions in a chain-reaction process. The only product is the polymer. The polymerization of an alkene molecule begins when one bond of the double bond between carbon atoms is broken. This generates a new monomer with a broken bond. Each successive carbon atom then has an unshared electron with which to form a bond with a neighboring molecule. The chain continues to build indefinitely in the following way:

$$\begin{array}{c} \mathrm{H}\quad\mathrm{H} \\ |\quad\;\;| \\ \mathrm{C{=}C} \\ |\quad\;\;| \\ \mathrm{H}\quad\mathrm{R} \end{array} + \begin{array}{c} \mathrm{H}\quad\mathrm{H} \\ |\quad\;\;| \\ \mathrm{C{=}C} \\ |\quad\;\;| \\ \mathrm{H}\quad\mathrm{R} \end{array} + \begin{array}{c} \mathrm{H}\quad\mathrm{H} \\ |\quad\;\;| \\ \mathrm{C{=}C} \\ |\quad\;\;| \\ \mathrm{H}\quad\mathrm{R} \end{array} \longrightarrow \begin{array}{c} \mathrm{H}\quad\mathrm{H} \\ |\quad\;\;| \\ \cdot\mathrm{C{-}C}\cdot \\ |\quad\;\;| \\ \mathrm{H}\quad\mathrm{R} \end{array} + \begin{array}{c} \mathrm{H}\quad\mathrm{H} \\ |\quad\;\;| \\ \cdot\mathrm{C{-}C}\cdot \\ |\quad\;\;| \\ \mathrm{H}\quad\mathrm{R} \end{array} + \begin{array}{c} \mathrm{H}\quad\mathrm{H} \\ |\quad\;\;| \\ \cdot\mathrm{C{-}C}\cdot \\ |\quad\;\;| \\ \mathrm{H}\quad\mathrm{R} \end{array} \longrightarrow$$

$$\begin{array}{c} \mathrm{H}\quad\mathrm{H}\quad\mathrm{H}\quad\mathrm{H}\quad\mathrm{H}\quad\mathrm{H} \\ |\quad\;\;|\quad\;\;|\quad\;\;|\quad\;\;|\quad\;\;| \\ \mathrm{{-}C{-}C{-}C{-}C{-}C{-}C{-}} \\ |\quad\;\;|\quad\;\;|\quad\;\;|\quad\;\;|\quad\;\;| \\ \mathrm{H}\quad\mathrm{R}\quad\mathrm{H}\quad\mathrm{R}\quad\mathrm{H}\quad\mathrm{R} \end{array}$$

If R in the monomer is hydrogen, the monomer is ethylene, $CH_2{=}CH_2$, and the polymer is polyethylene. Ethylene is at the heart of the petrochemical industry. Its annual production in the United States is measured in billions of pounds. From that mass, billions of pounds of polyethylene are produced, equating to dozens of pounds per person each year (**Figure 21.43**). Polyethylene with a density in the 0.91−0.94 g/cm^3 range is called *low-density polyethylene* (coded for recycling as LDPE) (**Figure 21.44[a]**). It is a supple material that folds and bends easily because the intermolecular (induced dipole) forces between the branched carbon chains are weak. These properties make it ideal for sandwich bags and trash bags.

Induced dipole forces increase as branching decreases and molecular size increases. Polymers with long, straight chains, called *high-density polyethylene* (coded for recycling as HDPE), produce stronger interchain attractions and a corresponding increase in mechanical strength (**Figure 21.44[b]**). Plastic milk bottles are made of polyethylene with a density that exceeds 0.94 g/cm^3.

When covalent bonds form *between* carbon chains, a cross-linked polymer results and the physical properties change sharply (**Figure 21.44[c]**). Cross-linked polyethylene is used for plastic screw caps on soda bottles. This plastic is rigid enough to mold as a solid and has enough mechanical strength to hold the screw thread needed to tighten the cap on the bottle.

If R in a $CH_2{=}CHR$ monomer is a methyl group, $-CH_3$, the monomer is propylene, $CH_2{=}CHCH_3$, and the polymer is polypropylene (coded for recycling as PP). In comparison with $CH_2{=}CH_2$, the slightly larger R group in polypropylene increases attractive forces between chains. Polypropylene is used to make rigid packaging such as crates, textiles such as carpets, technical parts such as car bumpers, films such as food packaging, consumer products such as furniture, and many other common final products.

Halogen-substituted alkenes can also polymerize. These molecules are polar, so the resulting polymer has dipole attractions in addition to induced dipole forces. A familiar example of such a polymer is polyvinyl chloride (coded for recycling as PVC), which is made from the monomer $CH_2{=}CHCl$. Polyvinyl chloride is the third most produced plastic, after polyethylene and polypropylene. It is used to manufacture products such as water distribution and sewer pipes, coatings for wires and cables, clothing, blood- and urine-collection containers, and many other consumer products. **Table 21.6** lists more ethylene-based monomers that give addition polymers.

The attractive forces between polymer chains that are not cross-linked are weaker than covalent bonds. As a result, these polymers can easily be recycled by being melted to break *only* the interchain attractive forces, and then the liquid is molded into new shapes. These moldable polymers are called **thermoplastics** and account for over 85% of all plastics sold. To be recycled and reused efficiently, the different thermoplastics must be separated by type, as indicated by the recycling codes in **Table 21.7**. When processed correctly, the resulting recycled plastic costs about half the price of new plastic. Thermoplastics do not degrade in landfills, but they shouldn't be put there in the first place. Polymers are too valuable to bury.

James Holmes/Zedcor/Science Source

Figure 21.43 Polyethylene film. Sheets are manufactured by squeezing the liquid through a thin opening and inflating it with air.

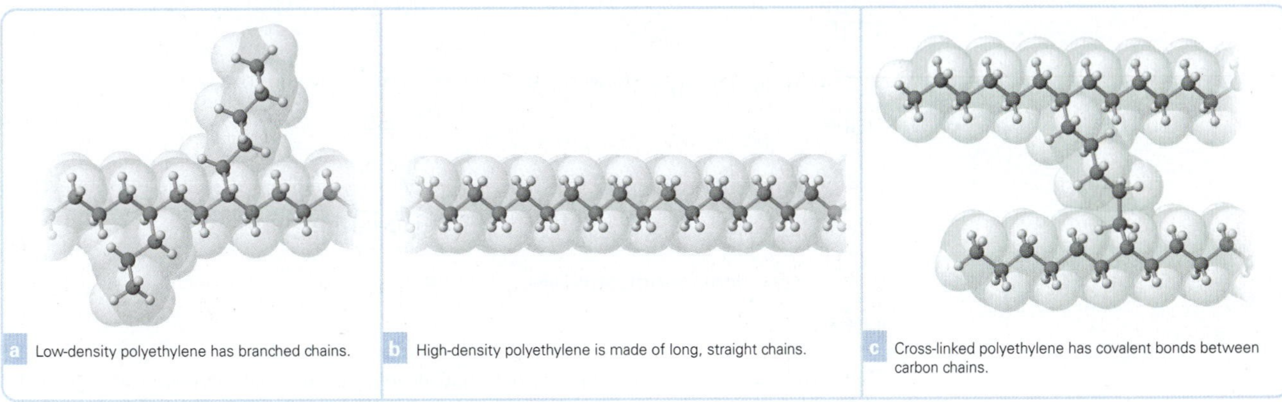
a Low-density polyethylene has branched chains.
b High-density polyethylene is made of long, straight chains.
c Cross-linked polyethylene has covalent bonds between carbon chains.

Figure 21.44 Molecular structure of forms of polyethylene.

Table 21.6 Common Polymers from Ethylene-like Monomers

Monomer Formula	Polymer Formula	Polymer Names	Uses
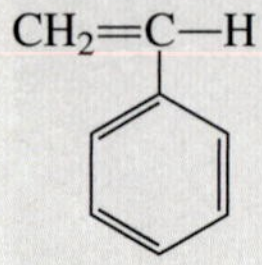 $CH_2{=}C(C_6H_5){-}H$ Styrene	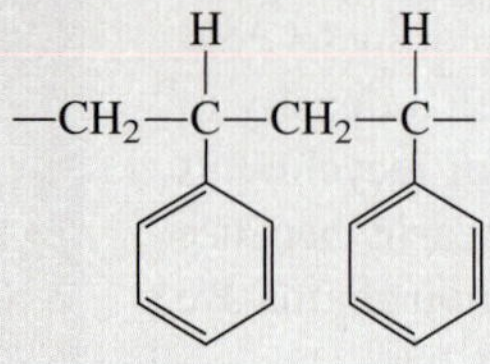 $-CH_2-CH(C_6H_5)-CH_2-CH(C_6H_5)-$	Polystyrene Styrofoam	Insulation, packaging
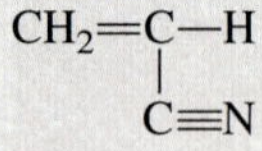 $CH_2{=}C(C{\equiv}N){-}H$ Acrylonitrile	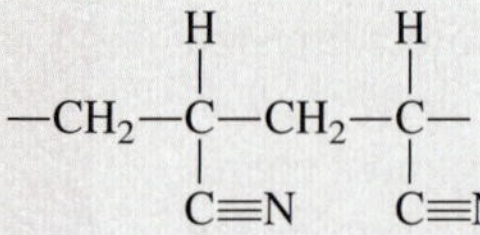 $-CH_2-CH(C{\equiv}N)-CH_2-CH(C{\equiv}N)-$	Polyacrylonitrile Orlon, Acrilan	Fabrics, rugs
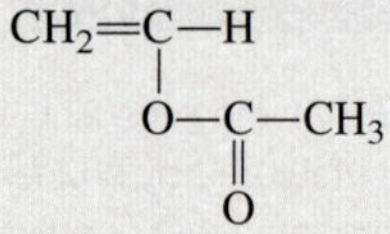 $CH_2{=}C(O{-}C({=}O){-}CH_3){-}H$ Vinyl acetate	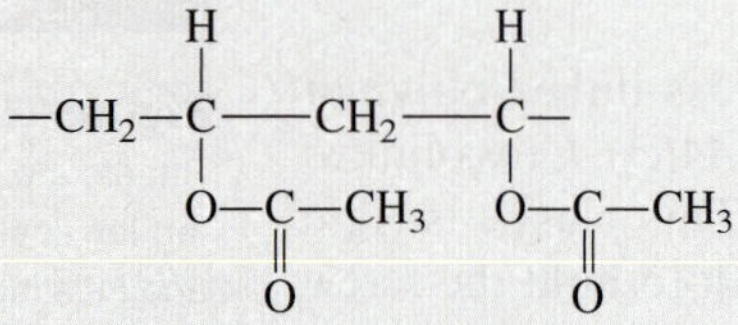 $-CH_2-CH(O{-}C({=}O){-}CH_3)-CH_2-CH(O{-}C({=}O){-}CH_3)-$	Polyvinylacetate PVA	Chewing gum, paint, glues, safety glass
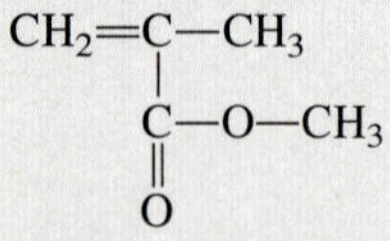 $CH_2{=}C(C({=}O){-}O{-}CH_3){-}CH_3$ Methyl methacrylate	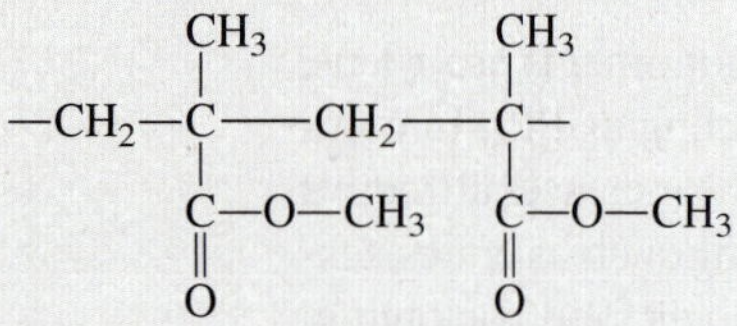 $-CH_2-C(CH_3)(C({=}O){-}O{-}CH_3)-CH_2-C(CH_3)(C({=}O){-}O{-}CH_3)-$	Lucite Plexiglass	Contact lenses, molded transparent objects
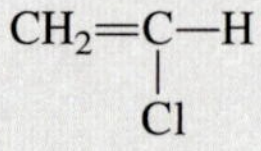 $CH_2{=}C(Cl){-}H$ Vinyl chloride	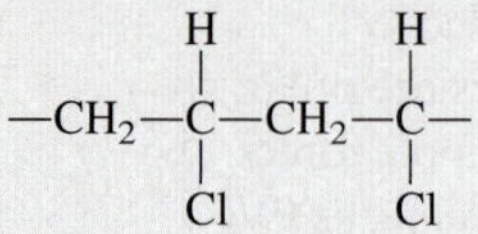 $-CH_2-CH(Cl)-CH_2-CH(Cl)-$	Polyvinyl chloride PVC, Tedlar (from polyvinyl fluoride)	Floor tile, pipe
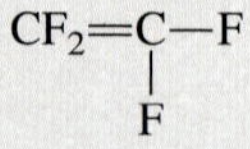 $CF_2{=}C(F){-}F$ Tetrafluoroethylene	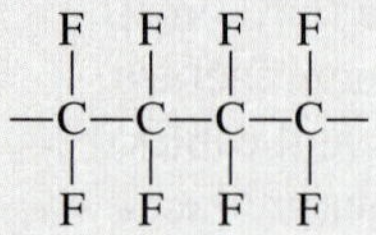 $-CF_2-CF_2-CF_2-CF_2-$	Teflon	Coatings, gaskets, bearings
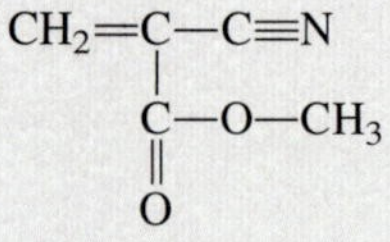 $CH_2{=}C(C({=}O){-}O{-}CH_3){-}C{\equiv}N$ Methyl α-cyanoacrylate	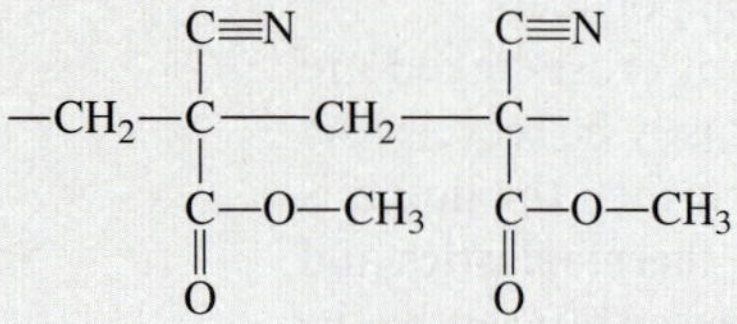 $-CH_2-C(C{\equiv}N)(C({=}O){-}O{-}CH_3)-CH_2-C(C{\equiv}N)(C({=}O){-}O{-}CH_3)-$	Superglue Crazy Glue	Adhesives, battlefield "stitches"

Table 21.7 Plastic Container Recycling Codes

Code	Plastic	Code	Plastic
1 PETE	Polyethylene terephthalate (PET)*	5 PP	Polypropylene
2 HDPE	High-density polyethylene	6 PS	Polystyrene
3 V	Poly(vinyl chloride) (PVC)*	7 OTHER	All other resins and layered multimaterial
4 LDPE	Low-density polyethylene		

*Bottle codes are different from standard industrial identification to avoid confusion with registered trademarks.

✓ Target Check 21.11

A certain monomer can be written as:

$$C_6H_5\text{—}CH{=}CH_2$$

$CH_2{=}CH$

Draw a section of the expected chain-growth polymer. Show at least three monomer units. This polymer is used as insulation and in packaging.

21.17 Step-Growth Polymers

Goal 30 Given the structural diagrams of a dicarboxylic acid and a dialcohol, predict the structural diagram of the product step-growth polymer; given the structural diagram of a step-growth polymer, predict the structural diagrams of the dicarboxylic acid and dialcohol from which it can be formed.

In Section 21.13, you learned that carboxylic acids react with alcohols to form esters. You also learned that carboxylic acids react with ammonia and amines to form amides. In each reaction, a molecule of water is split out in a **condensation reaction**. This is one mechanism by which **step-growth polymers** form.

In general, step-growth polymers are produced by the stepwise reaction of monomers with the growing chain. Each bond is formed independently of the others. Polymer chemists use condensation reactions to form polyesters and polyamides. However, to form the polymer chain by repeated condensation reactions, you must use a *di*carboxylic acid (two carboxyl groups) and a *di*alcohol. An example is the reaction of terephthalic acid and ethylene glycol:

$$HO\text{—}\overset{O}{\overset{\|}{C}}\text{—}C_6H_4\text{—}\overset{O}{\overset{\|}{C}}\text{—}OH + H\text{—}OCH_2CH_2OH \longrightarrow$$

terephthalic acid + ethylene glycol ⟶

$$\text{—}\overset{O}{\overset{\|}{C}}\text{—}C_6H_4\text{—}\overset{O}{\overset{\|}{C}}\text{—}OCH_2CH_2\text{—}O\text{—} + HOH$$

PET water

The linear polyester formed here is polyethylene terephthalate (PET) (coded for recycling as PETE). It has a density of 1.4 g/cm^3. Because the ester group is polar, there are dipole–dipole attractions between polymer chains, which are fairly strong. As a result, the most common use of PET polymers is as synthetic fibers (**Figure 21.45**). The second most common use of PET polymers is to manufacture plastic bottles, such as those commonly used for soft drinks. PET polymers are also manufactured in the form of threads and yarns, and when used in clothing, they are called *polyester*.

The ethylene glycol dialcohol used in the preceding reaction reacts to form linear chains. If you use an alcohol with three hydroxyl groups, such as glycerol, HO–CH$_2$–CH(OH)–CH$_2$–OH, the hydroxyl group of the middle carbon can form

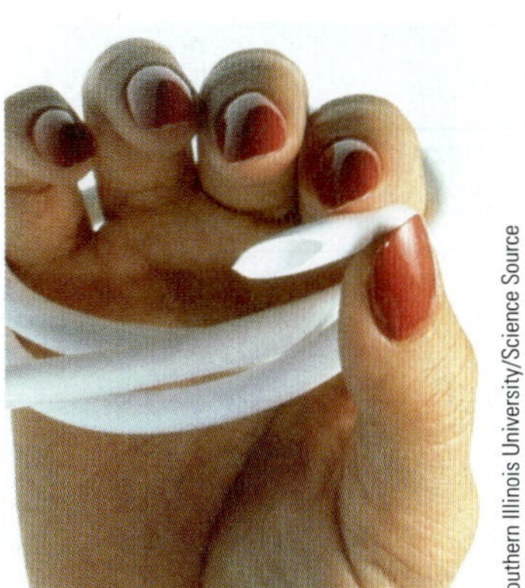

Figure 21.45 Dacron is a polyethylene terephthalate (PET) fiber. It is used in medicine to replace tissues because it is well tolerated by the body. The photograph shows a synthetic artery made of dacron.

ester bonds at an angle to the main chain, creating a cross-linked polyester. Formation of cross-links is the most effective way of making a polymer with mechanical strength:

$$\mathrm{HOCH_2CH(OH)CH_2OH} + \mathrm{HO{-}C(=O){-}C_6H_4{-}C(=O){-}OCH_2CH(OH)CH_2OH} \longrightarrow$$

$$\mathrm{{-}OCH_2CH(O{-})CH_2O{-}C(=O){-}C_6H_4{-}C(=O){-}OCH_2CH(O{-})CH_2O{-}} + H_2O$$

cross-linked polyester

Cross-linked polyesters are called *polyester resins*. The resin produced in the reaction here is called *glyptal*. It is used in the manufacturing of paints that coat and seal surfaces.

The best-known synthetic polymers are the nylons, a family of polyamide polymers. The first nylon was the brainchild of Wallace Carothers (1896–1937) who was hired away from Harvard University in 1928 by the DuPont Company. Carothers was asked to develop a substitute for silk. Silk was known to be a protein, so Carothers's group studied ways of making amide bonds. In 1935, they prepared a product that they named nylon 66 (**Figure 21.46**). The reactants were a dicarboxylic acid, adipic acid, and a diamine, hexamethylenediamine. Each monomer has six carbons, leading to the name nylon 66:

$$\underset{\text{adipic acid}}{\mathrm{HO{-}C(=O)CH_2CH_2CH_2CH_2C(=O){-}OH}} + \underset{\text{hexamethylenediamine}}{\mathrm{H{-}N(H)CH_2CH_2CH_2CH_2CH_2CH_2N(H){-}H}} \longrightarrow$$

$$\underset{\text{nylon 66}}{\mathrm{{-}C(=O)CH_2CH_2CH_2CH_2C(=O){-}N(H)CH_2CH_2CH_2CH_2CH_2CH_2N(H){-}}} + \underset{\text{water}}{\mathrm{HOH}}$$

The hydrogen bonding that can occur between amide linkages produces strong interchain attractive forces, which yield nylon fibers that have good tensile strength and therefore can be used for clothing (**Figure 21.47**). The first widespread use of nylon was in the production of toothbrush bristles and women's stockings. World War II started soon after nylon stockings became popular, so the use of nylon shifted to manufacturing of parachutes and ropes. Today, nylon continues to be used in clothing, in part, because it drapes well and can be easily dyed. Many types of cookware are made from nylon such as spatulas and tongs. Nylon is also widely used in the automotive industry and the electronic industry. Current U.S. production of nylon is measured in billions of pounds annually.

Target Check 21.12

A portion of the condensation polymer Kodel is given here. Give the structure of the starting materials from which Kodel can be made.

$$-\overset{\displaystyle O}{\overset{\|}{C}}-C_6H_4-\overset{\displaystyle O}{\overset{\|}{C}}-O-CH_2-C_6H_{10}-CH_2-O-\overset{\displaystyle O}{\overset{\|}{C}}-C_6H_4-\overset{\displaystyle O}{\overset{\|}{C}}-O-CH_2-C_6H_{10}-CH_2-O-$$

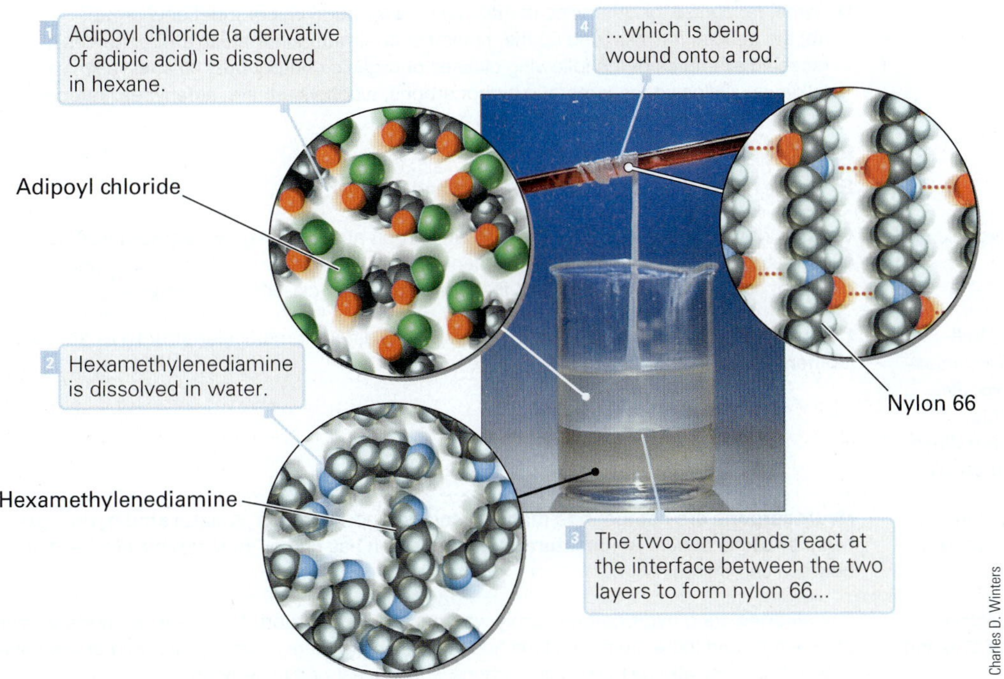

Figure 21.46 Preparation of nylon 66.

Hydrogen bonds to adjacent nylon 66 molecules.

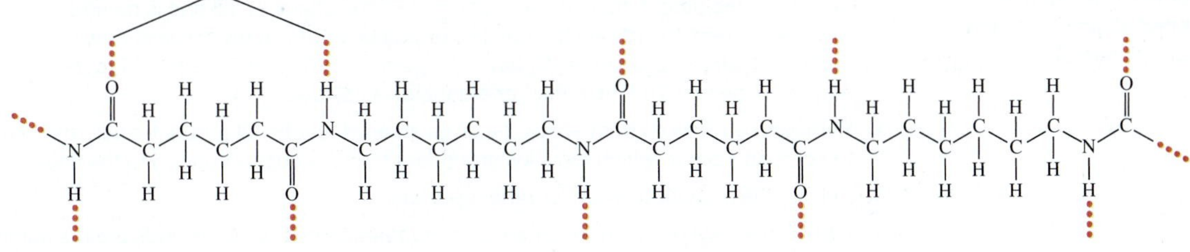

Figure 21.47 Hydrogen bonding in nylon 66. This property is responsible for the ability of nylon fibers to stretch.

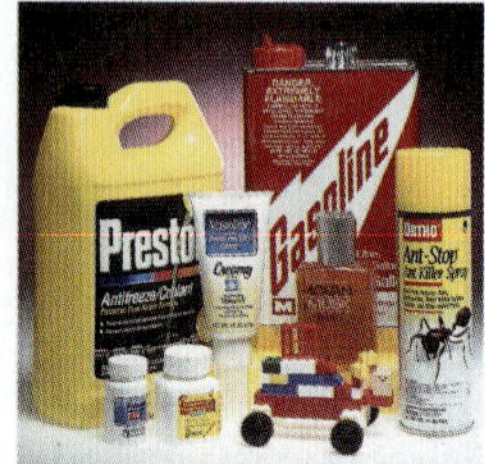
Charles D. Winters

CHAPTER 21 IN REVIEW: ORGANIC CHEMISTRY

The **overarching goals** for this chapter are the following:

A. Distinguish between organic and inorganic chemistry.

B. Define the term *hydrocarbon.*

C. Distinguish between saturated and unsaturated hydrocarbons.

D. Write, recognize, or otherwise identify (a) the structural unit or functional group, (b) the general formula, and (c) the molecular or structural formulas and/or names of specific examples of the following classes of organic compounds: alkanes, alkenes, alkynes, cycloalkanes, aromatic hydrocarbons, alcohols, ethers, aldehydes, ketones, carboxylic acids, esters, amines, and amides.

E. Define and give examples of isomerism.

F. Define and give examples of monomers and polymers.

Goal 1 Distinguish between organic and inorganic compounds.

Organic compounds are those that contain carbon atoms. **Inorganic compounds** have no carbon atoms; they are made up of elements other than carbon. Carbonates, cyanides, and oxides of carbon contain carbon, but they are traditionally classified as inorganic.

Goal 2 Given Lewis diagrams, ball-and-stick models, or space-filling models of two or more organic molecules with the same molecular formula, distinguish between isomers and different orientations of the same molecule.

Compounds with the same molecular formula but different molecular structures are called **isomers.**

Goal 3 Distinguish between saturated and unsaturated hydrocarbons (or other compounds).

Hydrocarbons are compounds made of carbon and hydrogen. A **saturated** hydrocarbon has only single bonds. An **unsaturated** hydrocarbon has one or more double or triple bonds between carbon atoms.

Goal 4 Distinguish between alkanes, alkenes, and alkynes (Sections 21.4 and 21.5).

The **alkanes** are a hydrocarbon family where each carbon atom forms single bonds to four other atoms and there are no multiple bonds. The **alkenes** are hydrocarbons with at least one double bond; the **alkynes** are hydrocarbons with at least one triple bond.

Goal 5 Given a formula of a hydrocarbon or information from which it can be written, determine whether the compound can be a normal alkane or a cycloalkane.

A hydrocarbon with all carbon atoms in the molecule in a continuous chain is a **normal alkane.** The normal and branched alkanes have the general formula C_nH_{2n+2}, where n is the number of carbon atoms in the compound. **Cycloalkanes** have all carbon–carbon single bonds, with at least some of the carbon atoms forming a ring.

Goal 6 Given the name (or structural diagram) of a normal, branched, or halogen-substituted alkane, write the structural diagram (or name).

To **name an alkane:**

1. Identify as the parent chain the longest continuous chain. An alkane is named by using a prefix to indicate the number of carbon atoms in the longest chain: *meth-* = 1, *eth-* = 2, *prop-* = 3, *but-* = 4, *pent-* = 5, *hex-* = 6, *hept-* = 7, *oct-* = 8, *non-* = 9, *dec-* = 10. The suffix *-ane* denotes an alkane.
2. Removing an H atom from an alkane gives an **alkyl** functional group. Identify by number the carbon atom to which the alkyl group (or other species) is bonded to the chain.
3. Identify the branched group (or other species).
4. If the same alkyl group, or other species, appears more than once, indicate the number of appearances by di-, tri-, tetra-, and so on, and show the location of each branch by number.
5. If two or more different alkyl groups, or other species, are attached to the parent chain, they are named in alphabetical order.

Goal 7 Given the name (or structural diagram) of a cycloalkane, write the structural diagram (or name).

Cycloalkanes are named according to the number of carbon atoms in the ring with the prefix *cyclo-*.

Goal 8 Given a formula of a hydrocarbon or information from which it can be written, determine whether the compound can be an alkene or an alkyne.

The **alkenes** are hydrocarbons with at least one double bond; the **alkynes** are hydrocarbons with at least one triple bond.

Goal 9 Given the name (or structural diagram) of an alkene or an alkyne, write the structural diagram (or name).

The IUPAC nomenclature system for the alkenes and alkynes matches that of the alkanes. The suffix *-ene* denotes an alkene. The suffix *-yne* denotes an alkyne. In naming unsaturated compounds, the position of the double or triple bond is specified by number.

Goal 10 Identify and distinguish between *cis* and *trans* geometric isomers.

Double bonds can give **geometric isomers,** also called *cis–trans isomers*. Two alkyl groups can be on the same side (*cis*) or on opposite sides (*trans*) of the double bond.

Goal 11 Distinguish between aliphatic and aromatic hydrocarbons.

Any hydrocarbon that does not contain a benzene ring is an **aliphatic** hydrocarbon. A hydrocarbon with one or more benzene rings, which may be substituted, is an **aromatic** hydrocarbon.

Goal 12 Given the name (or structural diagram) of an alkyl- or halogen-substituted benzene compound, write the structural diagram (or name).

The carbon on a benzene ring to which a functional group is bonded is identified by number. *Ortho-, meta-,* and *para-* prefixes are also used for disubstituted rings.

Goal 13 State the three types of fossil fuels from which most hydrocarbons are derived.

Almost all hydrocarbons are derived from coal, natural gas, or petroleum.

Goal 14 Given the reactants in an addition or substitution reaction between (a) an alkane, alkene, alkyne, or benzene and (b) a hydrogen or halogen molecule, predict the products of the reaction.

An alkene or alkyne is capable of reacting via an **addition reaction**, where atoms of an element (or compound) are added to the unsaturated hydrocarbon. A reaction in which a hydrogen atom in an alkane is replaced by an atom of another element is a **substitution reaction**.

Goal 15 Identify the structural formulas of the functional groups that distinguish alcohols and ethers.

The general formula of an **alcohol** is R—OH. The general formula for an **ether** is R—O—R'.

```
   O          O
  / \        / \
 R   H      R   R'
 alcohol    ether
```

Goal 16 Given the name (or structural diagram) of an alcohol or ether, write the structural diagram (or name).

The suffix *-ol* denotes an alcohol. The word *ether* preceded by the names of the attached alkyl groups is the name of the ether.

Goal 17 Given the reactants (or products) of a dehydration reaction between two alcohols, predict the products (or reactants) of the reaction.

Ethers can be prepared by dehydrating alcohols. The —H from the hydroxyl group of one molecule reacts with the —OH hydroxyl group of another molecule, producing an R—O—R' ether and a water molecule.

Goal 18 Given the molecular structures of alcohols or ethers, or information from which they may be obtained, predict relative values of boiling points or solubility in water.

In general, the boiling points of alcohols increase as the number of carbon atoms increases. The boiling point of an alcohol is always much higher than that of the alkane with the same number of carbon atoms. The 1-to-3-carbon alcohols are completely soluble in water and the solubility drops off as the alkyl chain lengthens. The boiling points of ethers are lower than the corresponding alcohols. The solubility of ethers in water is about the same as the solubility of the isomeric alcohols.

Goal 19 Identify the structural formulas of the functional groups that distinguish aldehydes and ketones.

Aldehydes and **ketones** have a carbon atom double bonded to an oxygen atom. In an aldehyde, the carbon is at the end of the chain; in a ketone, the carbon is inside the chain.

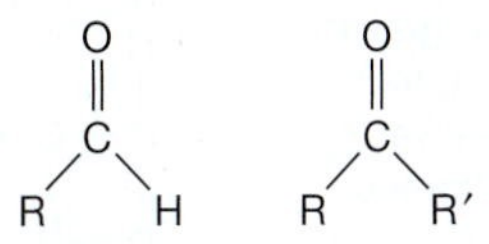

```
    O           O
    ||          ||
    C           C
   / \         / \
  R   H       R   R'
 aldehyde    ketone
```

Goal 20 Given the name (or structural diagram) of an aldehyde or ketone, write the structural diagram (or name).

The IUPAC system uses the suffix *-al* for aldehydes and the suffix *-one* for ketones. Ketones may also be named like ethers but using the word ketone.

Goal 21 Write structural diagrams to show how a specified aldehyde or ketone is prepared from an alcohol.

Aldehydes and ketones are prepared by oxidation of alcohols or hydrations of alkynes. Aldehydes are themselves easily oxidized; ketones resist oxidation. Aldehydes and ketones may be reduced to alcohols.

Goal 22 Identify the structural formulas of the functional groups that distinguish carboxylic acids and esters.

The general formula of a **carboxylic acid** is RCOOH. The functional group, —COOH, is a combination of a carbonyl group and a hydroxyl group called a **carbonyl group.** In an **ester,** the carboxyl hydrogen is replaced by another alkyl group, RCOOR′.

$$\mathrm{R{-}C({=}O){-}O{-}H} \qquad \mathrm{R{-}C({=}O){-}O{-}R'}$$

carboxylic acid ester

Goal 23 Given the name (or structural diagram) of a carboxylic acid or ester, write the structural diagram (or name).

IUPAC uses the suffix *-oic* and the word *acid* to denote carboxylic acids. Esters have two word names. The first word is the alkyl group from the alcohol and the second is the anion derived from the acid. (Remember *-ic* → *-ate* in anions of acids.)

Goal 24 Given the reactants (or products) of an esterification reaction, predict the products (or reactants) of the reaction.

The reaction between an acid and an alcohol is called **esterification.** The products of the reaction are an ester and water. The acid contributes the entire hydroxyl group, while the alcohol furnishes only the hydrogen.

Goal 25 Identify the structural formulas of the functional groups that distinguish amines and amides.

Amines are organic derivatives of ammonia, NH_3. An amine is formed by replacing one, two, or three hydrogens in an ammonia molecule with an alkyl group. An **amide** is a derivative of a carboxylic acid in which the hydroxyl part of the carboxyl group is replaced by an NH_2 group.

Goal 26 Given the name (or structural diagram) of an amine or amide, write the structural diagram (or name).

The IUPAC system names amines like ethers but using the word *amine*. Name an amide by replacing the *-oic acid* suffix with the word *amide*.

Goal 27 Given the structural diagram of an amine, or information from which it can be written, classify the amine as primary, secondary, or tertiary.

The number of ammonia hydrogens replaced determines whether an amine is primary (one hydrogen replaced), secondary (two), or tertiary (three).

Goal 28 Given the reactants of a reaction between a carboxylic acid and an amine, predict the products of the reaction.

A carboxylic acid and an amine react by removing a water molecule:

$$\mathrm{H_2N{-}CH(R){-}C({=}O){-}OH + H{-}NH{-}CH(R){-}C({=}O){-}OH \longrightarrow H_2N{-}CH(R){-}C({=}O){-}NH{-}CH(R){-}C({=}O){-}OH + H_2O}$$

Amino acid 1 Amino acid 2 Dipeptide

Peptide linkages

Goal 29 Given the structural diagram or name of an ethylene-like monomer, predict the structural diagram of the product chain-growth polymer; given the structural diagram of a chain-growth polymer, predict the structural diagram and/or the name of the ethylene-like monomer from which it can be formed.

Small molecules called **monomers** join together to form **polymers. Chain-growth polymers** are formed by repeated addition reactions of an alkene monomer to give an alkane-like polymer chain. For example, polyethylene is formed from ethylene monomers:

$$\mathrm{CH_2{=}CHR + CH_2{=}CHR + CH_2{=}CHR \longrightarrow \cdot CH_2{-}CHR\cdot + \cdot CH_2{-}CHR\cdot + \cdot CH_2{-}CHR\cdot \longrightarrow {-}CH_2{-}CHR{-}CH_2{-}CHR{-}CH_2{-}CHR{-}}$$

Goal 8 Given a formula of a hydrocarbon or information from which it can be written, determine whether the compound can be an alkene or an alkyne.	The **alkenes** are hydrocarbons with at least one double bond; the **alkynes** are hydrocarbons with at least one triple bond.
Goal 9 Given the name (or structural diagram) of an alkene or an alkyne, write the structural diagram (or name).	The IUPAC nomenclature system for the alkenes and alkynes matches that of the alkanes. The suffix *-ene* denotes an alkene. The suffix *-yne* denotes an alkyne. In naming unsaturated compounds, the position of the double or triple bond is specified by number.
Goal 10 Identify and distinguish between *cis* and *trans* geometric isomers.	Double bonds can give **geometric isomers,** also called *cis–trans isomers*. Two alkyl groups can be on the same side (*cis*) or on opposite sides (*trans*) of the double bond.
Goal 11 Distinguish between aliphatic and aromatic hydrocarbons.	Any hydrocarbon that does not contain a benzene ring is an **aliphatic** hydrocarbon. A hydrocarbon with one or more benzene rings, which may be substituted, is an **aromatic** hydrocarbon.
Goal 12 Given the name (or structural diagram) of an alkyl- or halogen-substituted benzene compound, write the structural diagram (or name).	The carbon on a benzene ring to which a functional group is bonded is identified by number. *Ortho-, meta-,* and *para-* prefixes are also used for disubstituted rings.
Goal 13 State the three types of fossil fuels from which most hydrocarbons are derived.	Almost all hydrocarbons are derived from coal, natural gas, or petroleum.
Goal 14 Given the reactants in an addition or substitution reaction between (a) an alkane, alkene, alkyne, or benzene and (b) a hydrogen or halogen molecule, predict the products of the reaction.	An alkene or alkyne is capable of reacting via an **addition reaction**, where atoms of an element (or compound) are added to the unsaturated hydrocarbon. A reaction in which a hydrogen atom in an alkane is replaced by an atom of another element is a **substitution reaction**.
Goal 15 Identify the structural formulas of the functional groups that distinguish alcohols and ethers.	The general formula of an **alcohol** is R—OH. The general formula for an **ether** is R—O—R'. R–O–H (alcohol) R–O–R′ (ether)
Goal 16 Given the name (or structural diagram) of an alcohol or ether, write the structural diagram (or name).	The suffix *-ol* denotes an alcohol. The word *ether* preceded by the names of the attached alkyl groups is the name of the ether.
Goal 17 Given the reactants (or products) of a dehydration reaction between two alcohols, predict the products (or reactants) of the reaction.	Ethers can be prepared by dehydrating alcohols. The —H from the hydroxyl group of one molecule reacts with the —OH hydroxyl group of another molecule, producing an R—O—R' ether and a water molecule.
Goal 18 Given the molecular structures of alcohols or ethers, or information from which they may be obtained, predict relative values of boiling points or solubility in water.	In general, the boiling points of alcohols increase as the number of carbon atoms increases. The boiling point of an alcohol is always much higher than that of the alkane with the same number of carbon atoms. The 1-to-3-carbon alcohols are completely soluble in water and the solubility drops off as the alkyl chain lengthens. The boiling points of ethers are lower than the corresponding alcohols. The solubility of ethers in water is about the same as the solubility of the isomeric alcohols.
Goal 19 Identify the structural formulas of the functional groups that distinguish aldehydes and ketones.	**Aldehydes** and **ketones** have a carbon atom double bonded to an oxygen atom. In an aldehyde, the carbon is at the end of the chain; in a ketone, the carbon is inside the chain. R–C(=O)–H (aldehyde) R–C(=O)–R′ (ketone)
Goal 20 Given the name (or structural diagram) of an aldehyde or ketone, write the structural diagram (or name).	The IUPAC system uses the suffix *-al* for aldehydes and the suffix *-one* for ketones. Ketones may also be named like ethers but using the word ketone.

Goal 21 Write structural diagrams to show how a specified aldehyde or ketone is prepared from an alcohol.

Aldehydes and ketones are prepared by oxidation of alcohols or hydrations of alkynes. Aldehydes are themselves easily oxidized; ketones resist oxidation. Aldehydes and ketones may be reduced to alcohols.

Goal 22 Identify the structural formulas of the functional groups that distinguish carboxylic acids and esters.

The general formula of a **carboxylic acid** is RCOOH. The functional group, —COOH, is a combination of a carbonyl group and a hydroxyl group called a **carbonyl group.** In an **ester,** the carboxyl hydrogen is replaced by another alkyl group, RCOOR′.

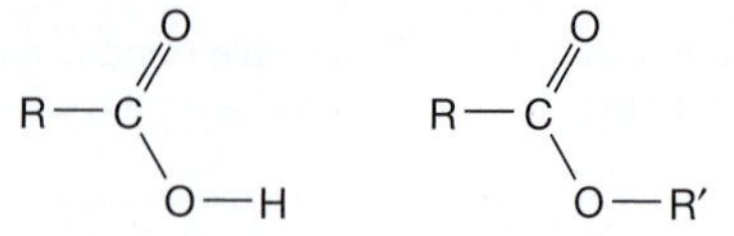

```
        O              O
       //             //
  R—C            R—C
       \              \
        O—H            O—R'
  carboxylic acid      ester
```

Goal 23 Given the name (or structural diagram) of a carboxylic acid or ester, write the structural diagram (or name).

IUPAC uses the suffix *-oic* and the word *acid* to denote carboxylic acids. Esters have two word names. The first word is the alkyl group from the alcohol and the second is the anion derived from the acid. (Remember *-ic* → *-ate* in anions of acids.)

Goal 24 Given the reactants (or products) of an esterification reaction, predict the products (or reactants) of the reaction.

The reaction between an acid and an alcohol is called **esterification.** The products of the reaction are an ester and water. The acid contributes the entire hydroxyl group, while the alcohol furnishes only the hydrogen.

Goal 25 Identify the structural formulas of the functional groups that distinguish amines and amides.

Amines are organic derivatives of ammonia, NH_3. An amine is formed by replacing one, two, or three hydrogens in an ammonia molecule with an alkyl group. An **amide** is a derivative of a carboxylic acid in which the hydroxyl part of the carboxyl group is replaced by an NH_2 group.

Goal 26 Given the name (or structural diagram) of an amine or amide, write the structural diagram (or name).

The IUPAC system names amines like ethers but using the word *amine*. Name an amide by replacing the *-oic acid* suffix with the word *amide*.

Goal 27 Given the structural diagram of an amine, or information from which it can be written, classify the amine as primary, secondary, or tertiary.

The number of ammonia hydrogens replaced determines whether an amine is primary (one hydrogen replaced), secondary (two), or tertiary (three).

Goal 28 Given the reactants of a reaction between a carboxylic acid and an amine, predict the products of the reaction.

A carboxylic acid and an amine react by removing a water molecule:

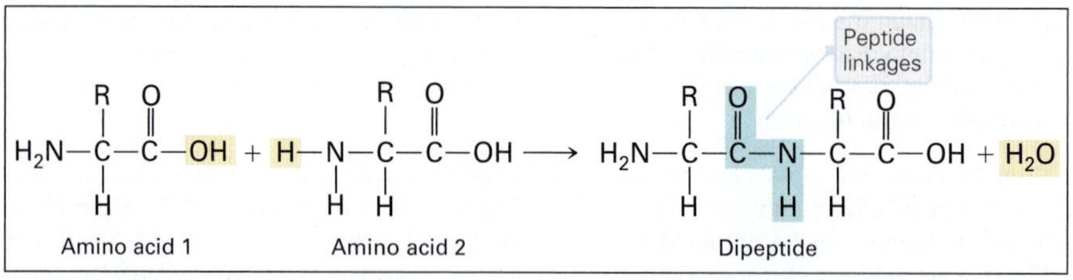

```
                                                          Peptide linkages
     R  O              R  O                      R  O     R  O
     |  ||             |  ||                     |  ||    |  ||
H2N—C—C—OH + H—N—C—C—OH ⟶ H2N—C—C—N—C—C—OH + H2O
     |            |  |                           |     |  |
     H            H  H                           H     H  H
  Amino acid 1      Amino acid 2                   Dipeptide
```

Goal 29 Given the structural diagram or name of an ethylene-like monomer, predict the structural diagram of the product chain-growth polymer; given the structural diagram of a chain-growth polymer, predict the structural diagram and/or the name of the ethylene-like monomer from which it can be formed.

Small molecules called **monomers** join together to form **polymers. Chain-growth polymers** are formed by repeated addition reactions of an alkene monomer to give an alkane-like polymer chain. For example, polyethylene is formed from ethylene monomers:

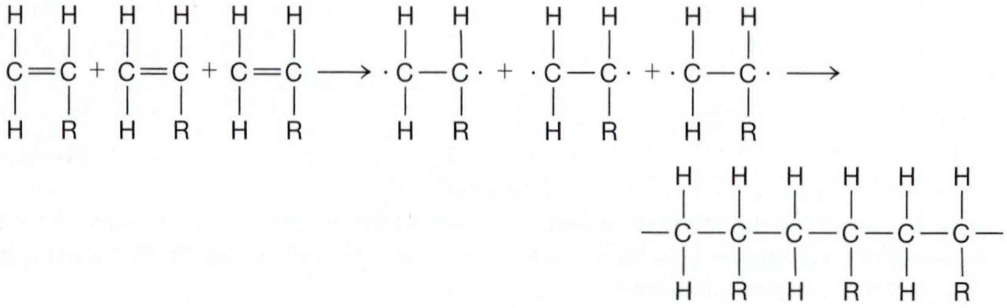

```
H  H    H  H    H  H        H  H       H  H       H  H
|  |    |  |    |  |        |  |       |  |       |  |
C=C + C=C + C=C ⟶ ·C—C· + ·C—C· + ·C—C· ⟶
|  |    |  |    |  |        |  |       |  |       |  |
H  R    H  R    H  R        H  R       H  R       H  R

                                  H  H  H  H  H  H
                                  |  |  |  |  |  |
                                —C—C—C—C—C—C—
                                  |  |  |  |  |  |
                                  H  R  H  R  H  R
```

Goal 30 Given the structural diagrams of a dicarboxylic acid and a dialcohol, predict the structural diagram of the product step-growth polymer; given the structural diagram of a step-growth polymer, predict the structural diagrams of the dicarboxylic acid and dialcohol from which it can be formed.

One type of **step-growth polymer** is formed by repeated condensation reactions to give a polymer chain with repeated ester or amide functional groups. For example, nylon 66 is formed from a 6-carbon dicarboxylic acid and a 6-carbon diamine:

$$\underset{\text{adipic acid}}{HO-\overset{O}{\overset{\|}{C}}CH_2CH_2CH_2CH_2\overset{O}{\overset{\|}{C}}-OH} + \underset{\text{hexamethylenediamine}}{H-\overset{H}{\overset{|}{N}}CH_2CH_2CH_2CH_2CH_2CH_2\overset{H}{\overset{|}{N}}-H} \longrightarrow$$

$$\underset{\text{nylon 66}}{-\overset{O}{\overset{\|}{C}}CH_2CH_2CH_2CH_2\overset{O}{\overset{\|}{C}}-\overset{H}{\overset{|}{N}}CH_2CH_2CH_2CH_2CH_2CH_2\overset{H}{\overset{|}{N}}-} + \underset{\text{water}}{HOH}$$

Key Terms

Most of the key terms and concepts and many others appear in the Glossary. Use your Glossary regularly.

addition reaction p. 835
alcohol p. 838
aldehydes p. 841
aliphatic hydrocarbon p. 832
alkane p. 822
alkene p. 828
alkyne p. 828
alkyl group p. 824
amide p. 847
amines p. 846
amino acid p. 847
aromatic hydrocarbons p. 832
carbonyl group p. 841
carboxyl group p. 844
carboxylic acid p. 844
chain-growth polymers p. 850
condensation reaction p. 853
condensed formula p. 823
cycloalkane p. 826
ester p. 844
ether p. 838
esterification p. 845
functional group p. 838
geometric isomers p. 830
halogenation reactions p. 835
homologous series p. 823
hydrocarbons p. 822
hydrogenation p. 834
hydroxyl group p. 838
ketones p. 841
line formula p. 823
methyl group p. 824
monomer p. 850
normal alkane p. 822
organic chemistry p. 819
parent chain p. 825
peptide linkage p. 847
polymer p. 850
polymerization p. 850
saturated p. 822
step-growth polymers p. 853
structural diagram p. 823
structural formula p. 823
substitution reaction p. 835
thermoplastics p. 851
unsaturated p. 822

Frequently Asked Questions

I've noticed that this chapter is different from most of the other chapters in the book. Should I change my study approach for this chapter?
The study of organic chemistry is quite different from the study of most other topics in this book. For example, you may have already noticed that there are no calculations in this chapter. You will have to make some modifications to your "normal" study techniques to accommodate the differences in this chapter.

How should I review the chapter after I've studied each section?
The most effective way to review material often differs from the most effective way to first learn material. This is especially true with organic chemistry. The best way to first learn the subject matter is in little groups, as we have presented it. However, the best way to review the material is to look at the big picture. Our summary of the hydrocarbons (Table 21.4) and summary of the organic functional groups (Table 21.5) will be particularly useful in your chapter review.

Do you have any additional study hints for the organic chemistry chapter?
Another useful general approach to studying organic chemistry can be summarized as "memorize, then apply." The general formulas of the functional groups must be memorized, and they must be memorized before you can predict the products of organic reactions. As an example, even though we present both the alcohol functional group and the reactions of alcohols in the alcohols and ethers section, you should review by learning all functional groups and then learning all reactions for which you are responsible. Many students find it useful to place information to be memorized on flash cards. For example, to memorize functional groups, one flash card would have ester on one side and R—CO—OR′ on the other side.

Concept-Linking Exercises

Write a brief description of the relationships among each of the following groups of terms or phrases. Answers to the Concept-Linking Exercises are given at the end of the chapter.

1. Hydrocarbons, saturated, unsaturated
2. Alkane, normal alkane, cycloalkane, homologous series
3. Structural formula, structural diagram, line formula, condensed formula, Lewis diagram
4. Alkene, alkyne, *cis, trans,* geometric isomers
5. Aliphatic compounds, aromatic compounds, benzene ring, delocalized electrons
6. Catalytic cracking, fractional distillation, catalytic hydrogenation
7. Addition reaction, substitution reaction, halogenation reaction, hydrogenation reaction
8. Functional group, hydroxyl group, carbonyl group, carboxyl group
9. Amide, peptide linkage, amino acid
10. Monomer, polymer, chain-growth polymer, step-growth polymer

Small-Group Discussion Questions

Small-Group Discussion Questions are for group work, whether in class or under the guidance of a leader during a discussion section.

1. Why do the vast majority of known chemical compounds contain carbon?
2. Draw structural diagrams of the five isomers of C_6H_{14}.
3. There are eight isomers of five-carbon alkyl groups. Draw the structural diagram and state the IUPAC name of each.
4. Which of the following have *cis–trans* isomers? Draw structural diagrams of each, including both *cis–trans* isomers when applicable: (a) $CH_3CH_2CH_2OH$, (b) C_2H_2, (c) $CH_3CH{=}CH_2$, (d) $BrCH{=}CHBr$, (e) $BrCH{=}CHCl$, (f) $CH_3CH_2CH{=}CHCH_3$, (g) $(CH_3)_2C{=}C(CH_3)CH_2CH_3$, (h) $(CH_3)_2C{=}CHCH_3$.
5. Name each of the following molecules. In parts (e) and (f), use the fact that the following molecule is named toluene:

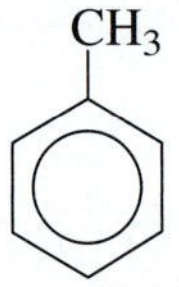

CH_3

a) Cl, Cl

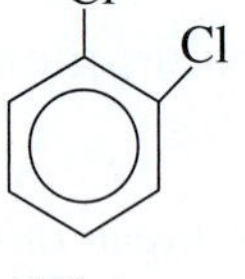

b) CH_3, CH_3

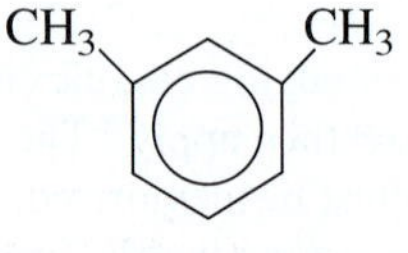

c) Br, CH_3, CH_3

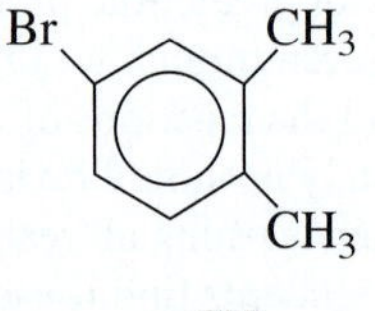

d) CH_3, CH_3, H_3C, CH_3

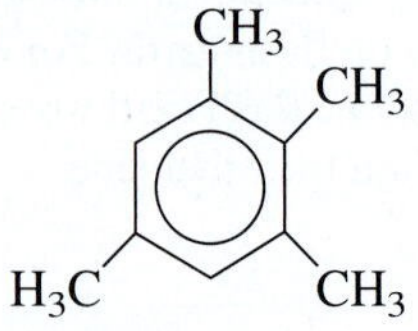

e) CH_3, Br

f) Cl, CH_3, Cl

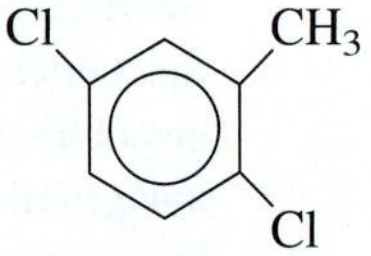

6. Define each of the following terms, and then construct an organizing scheme that allows you to classify molecules into the appropriate categories: hydrocarbon, aliphatic, aromatic, open chain, cyclic, saturated, unsaturated.
7. Predict the products of the following reactions. The structure 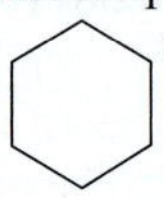 represents C_6H_{12}, with six carbons bonded to one another. Each carbon has single bonds to two hydrogen atoms.

a) H Br + KOH $\xrightarrow{\text{alcohol}}$ H H

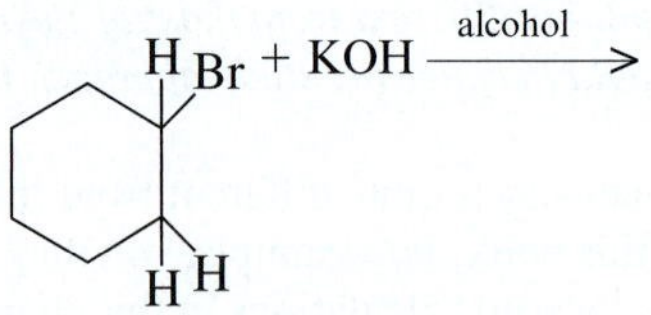

b) CH_3 OH $\xrightarrow{H_2SO_4}$

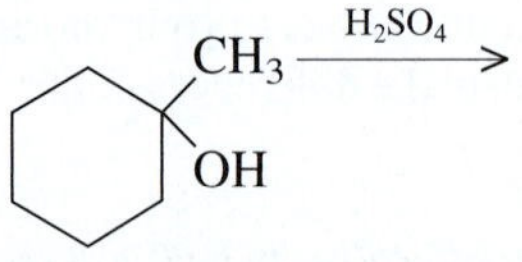

8. Predict the products of the following reactions.

a) H H C=C H H + Cl_2 $\xrightarrow{CH_2Cl_2}$

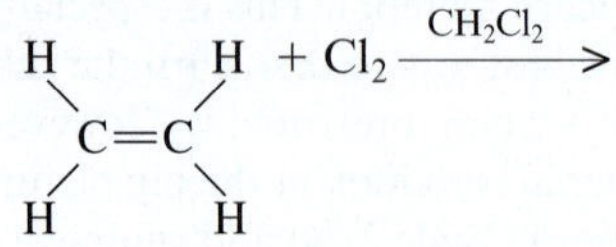

b) $CH_3CH_2CH_2CH{=}CH_2 + HCl \rightarrow$ (1-chloropentane is *not* formed)

c) H_3C / H_3C $\rangle C{=}CH_2$ + HCl → (1-chloro-2-methylpropane is *not* formed)

d) (1-methylcyclohexene: ring with CH_3 and H on the double bond) + HBr → (1-bromo-2-methylcyclohexane is *not* formed)

e) $CH_3CH_2CH{=}CHCH_3$ + HBr → (a mixture of two products is formed)

f) From the reactions in parts (b) through (e), can you describe the rule that governs the addition of HX to an alkene?

9. Construct a list of things that you can see at this moment that are made of or derived from hydrocarbons.

10. Provide the name for each of the following.

a) CH_3OH

b) CH_3CH_2OH

c) $H_3C{-}C(OH)(CH_3){-}CH_2CH_2CH_3$

d) $H_3C{-}C(CH_3)_2{-}OH$

e) $CH_3OCH_2CH_3$

f) $CH_3CH_2OCH_2CH_3$

g) (cyclopentyl)$-OCH_2CH_2CH_3$

11. Draw the structural diagrams of: (a) formaldehyde; (b) ethanal; (c) propanal; (d) 2-ethyl-4-methylpentanal; (e) acetone; (f) 3-hexanone; (g) 4-hexen-2-one; (h) 2,4-hexanedione.

12. Draw the structural diagrams of the carboxylic acid and alcohol from which each ester is formed.

a) $HC({=}O){-}O{-}CH_2CH(CH_3)_2$

b) $CH_3CH_2CH_2CH_2{-}C({=}O){-}O{-}CH(CH_3)_2$

c) (cyclohexyl)$-O{-}C({=}O){-}CH_2CH_2CH(CH_3)_2$

d) (2-hydroxyphenyl, with OH)$-C({=}O){-}OCH_3$

e) $CH_3CH_2CH_2{-}C({=}O){-}OCH_2CH_3$

13. Draw a structural diagram of each of the following:

a) A secondary amine with three carbon atoms

b) A tertiary amine with four carbon atoms

c) An amide with two carbon atoms

d) An amino acid with two carbon atoms

14. For each compound class that follows: (a) write a general formula; (b) draw the structure of its functional group; (c) draw the structural diagrams of two molecules in the class; and (d) name each of the molecules drawn in part (c): alcohol, ether, aldehyde, ketone, acid, ester, amine, amide.

15. Draw a section of the chain-growth polymer formed from each monomer listed and include at least four monomer units: (a) ethylene, $CH_2{=}CH_2$; (b) propylene, $CH_2{=}CHCH_3$; (c) vinyl chloride, $CH_2{=}CHCl$; (d) 1,1-dichloroethylene, $CH_2{=}CCl_2$; (e) tetrafluoroethylene, $CF_2{=}CF_2$; (f) acrylonitrile, $CH_2{=}CHCN$; (g) styrene, $CH_2{=}CHC_6H_5$; (h) ethyl acrylate, $CH_2{=}CHCOOCH_2CH_3$.

16. Draw structural diagrams of the monomers from which each type of the following step-growth polymers are formed:

a) Polyamides, $-C({=}O){-}R{-}C({=}O){-}\ddot{N}(H){-}R'{-}\ddot{N}(H){-}$

b) Polyesters, $-C({=}O){-}R{-}C({=}O){-}O{-}R'{-}O{-}$

c) Polycarbonates, $-O{-}C({=}O){-}O{-}R{-}$

d) Polyurethanes, $-C({=}O){-}\ddot{N}(H){-}R{-}\ddot{N}(H){-}C({=}O){-}O{-}R'{-}O{-}$

Questions, Exercises, and Problems

Interactive versions of these problems may be assigned by your instructor. Solutions for blue-numbered questions are at the end of the chapter. Questions other than those in the General Questions and More Challenging Problems sections are paired in consecutive odd-even number combinations; solutions for the odd-numbered questions are at the end of the chapter.

Section 21.3: The Molecular Structure of Compounds

1. Is the cyanide ion, CN^-, or the carbonate ion an *organic compound*? What about the acetate ion, CH_3COO^-?
2. Compare the original and modern definitions of *organic chemistry*. Why was the definition changed? Which definition includes the other?
3. What is the bond angle around a carbon atom with four single bonds? What word describes this geometry?
4. What is the bond angle around a carbon atom with a double bond and two single bonds? What word describes this molecular geometry?

Section 21.4: Saturated Hydrocarbons: The Alkanes and Cycloalkanes

5. What is a *hydrocarbon*? Which, among the following, are hydrocarbons: CH_3OH; C_3H_4; C_8H_{10}; $CH_3CH_2CH_3$?

Charles D. Winters

A number of common products are made of hydrocarbons.

6. What is the everyday (non-chemistry) definition of *saturated*? Use structural diagrams to show how the terms *saturated* and *unsaturated* are logically applied to hydrocarbons.
7. Write the molecular formulas of the alkanes having 11 and 21 carbon atoms. How did you arrive at these formulas?
8. Explain why an alkane is an example of a homologous series.
9. What are isomers?
10. Draw structural formulas for all the hydrocarbons with the formula C_4H_{10}.

One isomer of C_4H_{10} has its carbon atoms arranged in a continuous chain.

11. Write the molecular formula, line formula, condensed formula, and structural formula of the normal alkane with seven carbon atoms.
12. Write the molecular formula, the line formula, and the condensed formula for the normal alkane having 12 carbon atoms.
13. Write the molecular formula that represents the alkyl groups having two and four carbon atoms.
14. What is the name of the CH_3CH_2- group?
15. (a) Write the molecular formula of butane. (b) What is the name of $C_{10}H_{22}$?
16. Give the molecular formula of each of the following compounds: propane, cyclobutane, methane.
17. What is the IUPAC name of the molecule whose carbon skeleton is shown here?

```
    C—C
    |
C—C—C—C—C—C
        |
        C
```

18. What is the molecular formula for 2,2,3-trimethylbutane?
19. Draw the carbon skeleton of 2,3-dimethylpentane.
20. Give the IUPAC name of the molecule whose carbon skeleton is shown here.

```
  C   C
  |   |
C—C—C—C—C
  |
  C
```

21. Both 1,1,1- and 1,1,2-trichloroethane are used industrially as fat and grease solvents. Draw the structural diagrams of these isomers.
22. Draw structural formulas for all of the compounds that would be named as a bromobutane.
23. Is the general formula of a cycloalkane the same as the general formula of an alkane,C_nH_{2n+2}? Draw any structural diagrams to illustrate your answer.
24. How does a cycloalkane differ from a normal alkane?
25. Draw the skeleton diagram of cyclopentane.
26. Name the cycloalkane whose structural diagram resembles a square.
27. Draw a structural diagram of 1-chloro-2-iodocyclopentane.
28. Draw a structural diagram of 1-bromo-3-methylcyclohexane.

Questions 29–32: Write the names of the compounds whose carbon skeletons are given.

29.

```
        C
        |
C—C—C—C—C
    |
    C
    |
    C—C
```

30.

```
C
|
C—C—C—C—C
    |
    C—C
      |
    C—C—C
```

31.

$$\begin{array}{c} \quad\quad\quad\quad\ \ Cl \\ \quad\quad\quad\quad\ \ | \\ Br-C-C-C-C \\ \quad\quad\quad\quad\ \ | \\ \quad\quad\quad\quad\ \ Cl \end{array}$$

32.

$$\begin{array}{c} \quad\ \ Cl \quad\quad\quad F \\ \quad\ \ | \quad\quad\quad\ | \\ C-C-C-C-C-C \\ \quad\ \ | \quad\quad\quad\ | \\ \quad\ \ Cl \quad\quad\quad F \end{array}$$

33. Draw the skeleton diagram of 1-chloro-2-iodocyclopentane.

34. Draw the skeleton diagram of 1-bromo-3-methylcyclohexane.

35. Name the molecule whose carbon skeleton is drawn here.

Cl, CH_2CH_3 (cyclohexane ring)

36. Give the IUPAC name of the molecule whose carbon skeleton is shown here.

$CH_2CH_2CH_3$, F (cyclopentane ring)

Section 21.5: Unsaturated Hydrocarbons: The Alkenes and Alkynes

37. What is the difference in bonding and in the general molecular formula between an alkene and an alkane with the same number of carbon atoms?

38. Classify each of the following as an alkane, an alkene, or an alkyne and as saturated or unsaturated: $CHCCH_3$, CH_2CHCH_3, $CH_3(CH_2)_3CH_3$.

39. Draw the structural formula of trichloroethene, a metal degreaser that was once used to decaffeinate coffee. Why isn't the IUPAC name for this substance 1,1,2-trichloroethene?

Buena Vista Pictures/courtesy Everett Collection

The 1998 film *A Civil Action* tells the story of how trichloroethene contaminated the aquifer in Woburn, Massachusetts in the 1980s. The actor John Travolta played the role of the plaintiff's attorney.

40. Give the molecular formula for each of the following compounds: 2-butyne, propyne, 2-butene.

41. Draw the structural formula and explain in words the differences between *cis*-3-heptene and *trans*-3-heptene.

42. Are *cis–trans* isomers possible for $CH_3CH_2CH{=}CHCH_2CH_3$? If they are, write the IUPAC names of the isomers. If *cis–trans* isomers are not possible, write the IUPAC name of the compound.

43. The sex pheromone of the common housefly is *cis*-9-tricosene, where tricosene is the IUPAC name of a 23-carbon alkene. Draw the condensed formula of this molecule, marketed under the name Muscalure.

dcb/Shutterstock.com

Pheromones are substances secreted by one member of a species that affect the behavior of another. Female houseflies emit a sex pheromone that attracts male flies.

44. A molecule marketed as Disparlure, the sex pheromone of the gypsy moth, is produced from the molecule shown here. Give the IUPAC name of this molecule. (An 18-carbon alkene is an octadecene.)

CH_3, H, C, H_3C, $(CH_2)_4$, CH_3, $(CH_2)_9$, C=C, H, H

45. Give the IUPAC name of the following molecule:

$$\begin{array}{c} CH_3 \\ | \\ CH_3-C-C{\equiv}C-CH_2-CH_2-CH_3 \\ | \\ CH_3 \end{array}$$

46. Give the IUPAC name of the following molecule:

$$\begin{array}{l} H-C{\equiv}C-CH-CH_2-CH_3 \\ \quad\quad\quad\quad\ \ | \\ \quad\quad\quad\quad CH_2-CH_2-CH_3 \end{array}$$

Section 21.6: Aromatic Hydrocarbons

47. Dimethylbenzenes have the common name xylene. Draw all possible xylene isomers and give their IUPAC names.

48. Draw the structural formulas for all of the compounds that would be named as a bromochlorobenzene.

49. Name the molecule given here.

Cl, Br (benzene ring)

50. Name the molecule given here.

F, Br (substituents on a benzene ring)

Section 21.9: Chemical Reactions of Hydrocarbons

51. Draw skeletal formulas for all the possible dichloro substitution products of propane and give their IUPAC names.

52. Draw skeletal formulas for all the possible dichloro substitution products of butane and give their IUPAC names.

53. Draw skeletal formulas for all the possible dichloro addition products of (normal) butene.

54. Draw skeletal formulas for all the possible dibromo addition products of the pentenes.

55. Write an equation for the hydrogenation of 2-butene. Does the *cis* or *trans* geometry of the butene starting material make a difference in the products obtained?

56. Would you get a different product from the hydrogenation of 1-butene than from the hydrogenation of *trans*-2-butene? Explain.

Section 21.11: Alcohols and Ethers

57. Write the Lewis structures for all possible isomers with the formula $C_4H_{10}O$. Identify them as alcohols or ethers.

Diethyl ether is one of the isomers with the formula $C_4H_{10}O$.

58. Show how alcohols and ethers are structurally related to water.

59. Explain why ethers with formula $C_4H_{10}O$ have boiling points between 32°C and 39°C, whereas alcohols with the same formula have boiling points between 82°C and 118°C.

60. Explain why the ether with formula C_2H_6O is very slightly soluble in water, whereas the alcohol with the same formula is infinitely soluble in water.

61. Write the structural formula for 2-hexanol.

62. Give the IUPAC name of

$$\begin{array}{ccccccccccc} & & H & & H & & H & & OH & & H \\ & & | & & | & & | & & | & & | \\ H & - & C & - & C & - & C & - & C & - & C & - & H \\ & & | & & | & & | & & | & & | \\ & & H & & H & & H & & H & & H \end{array}$$

63. Write the structural formula for butyl ethyl ether.

64. Write the structural formula for ethyl propyl ether.

65. Write a structural equation showing how dipropyl ether might be prepared from an alcohol.

66. How could diisopropyl ether be prepared from an alcohol? The isopropyl group is

$$\begin{array}{l} \quad\quad\ \ CH_3 \\ \quad\quad / \\ -CH \\ \quad\quad \backslash \\ \quad\quad\ \ CH_3 \end{array}$$

Section 21.12: Aldehydes and Ketones

67. Write structural formulas for propanal and propanone.

68. What is the molecular formula for dimethyl ketone?

69. Use structural formulas to prepare acetone by oxidation of an alcohol.

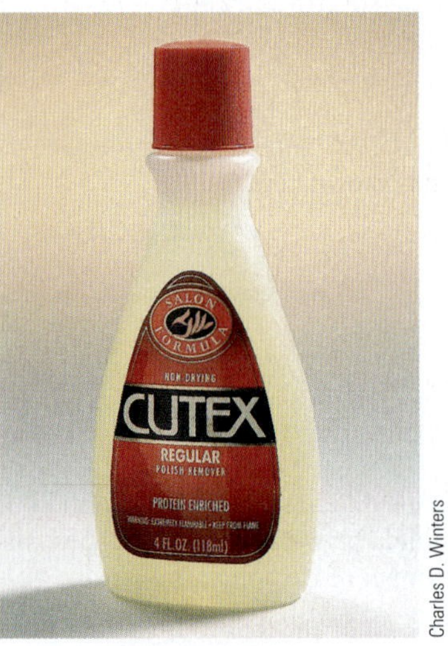

Charles D. Winters

Acetone is the primary component in some nail polish remover solutions.

70. Use structural formulas to prepare butanal by oxidation of an alcohol.

Section 21.13: Carboxylic Acids and Esters

71. Write the structural formula for hexanoic acid, a wretched-smelling substance found in goat sweat.

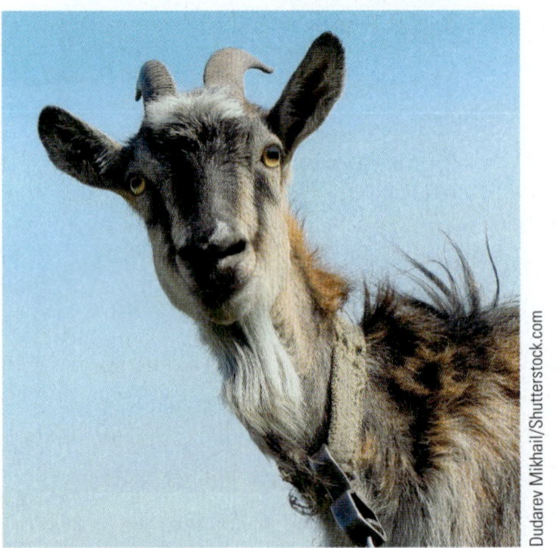

Dudarev Mikhail/Shutterstock.com

A trivial name for hexanoic acid is caproic acid, which is derived from the Latin word *caper*, meaning "goat."

72. What is the molecular formula for formic acid?

73. Write the equation for the reaction between propanoic acid and ethanol, and name the ester formed in this reaction.

74. Write the equation for the reaction between acetic acid and 1-propanol. Name the ester formed in this reaction. Do the same for the reaction between acetic acid and 2-propanol.

Section 21.14: Amines and Amides

75. Give Lewis diagrams for all amines with the formula C_2H_7N and name them.

76. Give structural formulas for all amines with the formula C_3H_9N and name them.

77. Classify the amines from Question 75 as primary, secondary, or tertiary.

78. Classify the amines from Question 76 as primary, secondary, or tertiary.

79. Write the equation for the reaction between propanoic acid and ammonia. Name the product and give its functional group.

80. Write the equation for the reaction between propanoic acid and diethylamine.

Section 21.16: Chain-Growth Polymers

81. Draw three repeating units of the chain-growth polymer made from the following monomer:

$$\mathrm{HHC{=}CHBr}\quad (\text{H, H on top; } \mathrm{C{=}C};\ \text{H, Br below})$$

82. Draw three repeating units of the chain-growth polymer made from the following monomer:

$$\mathrm{HHC{=}CFF}\quad (\text{H, F on top; } \mathrm{C{=}C};\ \text{H, F below})$$

83. The chain-growth polymer shown here is used for ropes, fabrics, and indoor-outdoor carpeting. Give the structure of the monomer from which this polymer was made.

$$\left[\mathrm{CH_2{-}CH(CH_3){-}CH_2{-}CH(CH_3){-}CH_2{-}CH(CH_3)}\right]_n$$

84. The chain-growth polymer shown here is used to thicken motor oil. Give the structure of the monomer from which this polymer was made.

$$\left[\mathrm{CH_2{-}C(CH_3)_2{-}CH_2{-}C(CH_3)_2{-}CH_2{-}C(CH_3)_2}\right]_n$$

85. Draw three repeating units of the chain-growth polymer made from chlorotrifluoroethene.

86. Plastic laboratory ware is usually a chain-growth polymer made from 4-methyl-1-pentene. Draw three repeating units of this polymer.

Section 21.17: Step-Growth Polymers

87. Lexan is a polycarbonate ester step-growth condensation polymer that is transparent and nearly unbreakable. It is used in "bulletproof" windows (a 1-inch-thick Lexan plate will stop a .38-caliber bullet fired from 12 feet), football and motorcycle helmets, and the visors in astronauts' helmets. It is made from the two monomers shown here. Draw two repeating units of this polymer.

$$\mathrm{HO{-}C({=}O){-}OH}\qquad \mathrm{HO{-}C_6H_4{-}C(CH_3)_2{-}C_6H_4{-}OH}$$

88. Draw two repeating units of the polyester formed from the two monomers given here.

$$\mathrm{HO{-}C({=}O){-}CH_2{-}C({=}O){-}OH}\qquad \mathrm{HO{-}CH_2{-}CH_2{-}CH_2{-}OH}$$

89. Kevlar is an aramid, a long-chain synthetic polyamide in which at least 85% of the amide linkages are attached directly to two aromatic rings. Because of its great mechanical strength, Kevlar is used in "bulletproof" clothing and in radial tires. The two monomers used to produce Kevlar are shown here. Draw two repeating units of the Kevlar polymer.

$$\mathrm{HO{-}C({=}O){-}C_6H_4{-}C({=}O){-}OH}\qquad \mathrm{H_2N{-}C_6H_4{-}NH_2}$$

90. Nomex is a type of nylon called an *aramid*. It has great heat resistance and is used in flame-resistant clothing worn by firefighters and race car drivers. The two monomers for Nomex are below. Draw two repeating units of the Nomex polymer.

$$\mathrm{HO{-}C({=}O){-}C_6H_4{-}C({=}O){-}OH}\ (meta)\qquad \mathrm{H_2N{-}C_6H_4{-}NH_2}\ (meta)$$

Johnny Habell/Shutterstock.com

Firefighters wear gear made with Nomex fibers because of their flame-resistant properties.

91. Give the monomers from which the following step-growth condensation polymer can be made.

$$\left[-\overset{\overset{\large O}{\|}}{C}-C_6H_4-\overset{\overset{\large O}{\|}}{C}-O-CH_2-\underset{\underset{\large CH_3}{|}}{\overset{\overset{\large H}{|}}{C}}-O- \right]_n$$

92. Give the monomers from which the following nylon polymer can be made.

$$\left[-\overset{\overset{\large O}{\|}}{C}-CH_2-\overset{\overset{\large O}{\|}}{C}-\overset{\overset{\large H}{|}}{N}-CH_2-CH_2-\overset{\overset{\large H}{|}}{N}- \right]_n$$

93. A leading nylon used in Europe is nylon 6, shown here. Nylon 6 is made by polymerization of a *single, difunctional* reactant. Draw the structural diagram of this reactant.

$$\left[-\overset{\overset{\large O}{\|}}{C}-CH_2-CH_2-CH_2-CH_2-CH_2-\overset{\overset{\large H}{|}}{N}- \right]_n$$

94. Stanyl is nylon 46, widely used in Europe. Draw the structural diagrams of the Stanyl monomers.

$$\left[-\overset{\overset{\large O}{\|}}{C}-CH_2-CH_2-CH_2-CH_2-\overset{\overset{\large O}{\|}}{C}-\overset{\overset{\large H}{|}}{N}-CH_2-CH_2-CH_2-CH_2-\overset{\overset{\large H}{|}}{N}- \right]_n$$

General Questions

95. Distinguish precisely, and in scientific terms, the differences among items in each of the following pairs or groups.
   a) Organic chemistry, inorganic chemistry
   b) Saturated and unsaturated hydrocarbons
   c) Alkanes, alkenes, alkynes
   d) Normal alkane, branched alkane
   e) *Cis* and *trans* isomers
   f) Structural formula, condensed (line) formula, molecular formula
   g) Addition reaction, substitution reaction
   h) Alkane, alkyl group
   i) Monomer, polymer
   j) Aliphatic hydrocarbon, aromatic hydrocarbon
   k) Ortho-, meta-, para-
   l) Alcohol, aldehyde, carboxylic acid
   m) Hydroxyl group, carbonyl group, carboxyl group
   n) Primary, secondary, tertiary alcohols
   o) Alcohol, ether
   p) Aldehyde, ketone
   q) Carboxylic acid, ester
   r) Carboxylic acid, amide
   s) Amine, amide
   t) Primary, secondary, tertiary amine
   u) Chain-growth polymer, step-growth polymer

96. Classify each of the following statements as true or false.
   a) To be classified as organic, a compound must be or have been part of a living organism.
   b) Carbon atoms normally form four bonds in organic compounds.
   c) Only an unsaturated hydrocarbon can engage in an addition reaction.
   d) Members of a homologous series differ by a distinct structural unit.
   e) Alkanes, alkenes, and alkynes are unsaturated hydrocarbons.
   f) Alkyl groups are a class of organic compounds.
   g) Isomers have the same molecular formulas but different structural formulas.
   h) *Cis–trans* isomerism appears among alkenes but not alkynes.
   i) All aliphatic hydrocarbons are unsaturated.
   j) Aromatic hydrocarbons have a ring structure.
   k) An alcohol has one alkyl group bonded to an oxygen atom, and an ether has two.
   l) Carbonyl groups are found in alcohols and aldehydes.
   m) An ester is an aromatic hydrocarbon, made by the reaction of an alcohol with a carboxylic acid.
   n) An amine has one, two, or three alkyl groups substituted for hydrogens in an ammonia molecule.
   o) An amide has the structure of a carboxylic acid, except that $-NH_2$ replaces $-OH$ in the carboxyl group.
   p) A peptide linkage arises when a water molecule forms from a hydrogen from the $-NH_2$ group of one amino acid molecule and an $-OH$ from another amino acid molecule.
   q) Forming cross-links in a polymer makes the polymer more likely to possess low mechanical strength.
   r) Chain-growth polymers are made by a reaction that gives off water as a second product.
   s) Nylon is an example of a step-growth condensation polymer.

97. What is the difference in bonding and in general molecular formula between an alkene and a cycloalkane with the same number of carbon atoms?

98. Draw all isomers of C_4H_8.

99. Why is the delocalized structure (I) for benzene more appropriate than the cyclohexatriene structure (II)?

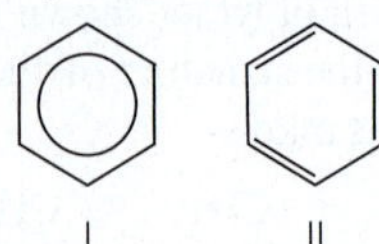

100. Show that the following statement is true: "Every alcohol with two or more carbons is an isomer of at least one ether."

101. Polymers are typically described as having a range of molecular masses, such as "polyvinyl alcohol, molecular mass 31,000–50,000 u." Why do polymers not have unique molecular masses, even though the monomer starting materials do have unique molecular masses?

102. A sample of polystyrene has an average molecular mass of 1.8×10^6 u. If a single styrene molecule has the formula C_8H_8, about how many styrene molecules are in a chain of this polystyrene?

focal point/Shutterstock.com

Foamed polystyrene containers are used for packaging hot food and for packing materials.

More Challenging Problems

103. Explain why aldehydes are unstable with regard to oxidation, whereas ketones are stable. (*Hint:* Look at the number of hydrogen atoms bonded to the carbon that is bonded to the oxygen.)

104. Chemists often use different isotopes such as oxygen-18 to chart the path of organic reactions. If acetic acid reacts with methanol that contains only oxygen-18, show where the oxygen-18 atom exists in the ester product.

105. Can trimethylamine react with a carboxylic acid to form an amide? Explain why or why not using chemical equations.

106. Write three repeating units of the addition polymer that can be made from acetylene. This material, called *polyacetylene*, conducts electricity because of the alternating single-bond/double-bond pattern in the main chain.

107. Amides are often made by reaction of a carboxylic acid and an amine. This reaction is both acid-catalyzed and reversible. Use these facts to explain why nylon stockings have a very short wear life in cities where acid rain is common.

Answers to Target Checks

1. (a) CH_4 and CH_3NH_2 are organic; NaCl and CO are inorganic. CO is inorganic because organic compounds have carbon–hydrogen bonds. (b) Some everyday definitions of the term *organic* include (1) food grown with "natural" fertilizers, (2) derived from living organisms, (3) organized, and (4) essential. Note that none of these everyday definitions is based on consideration of the particulate-level composition of matter.

2.

```
  Br H  Cl          H  Cl Br          H  H  Cl
  |  |  |           |  |  |           |  |  |
H—C—C—C—H         H—C—C—C—H         H—C—C—C—H
  |  |  |           |  |  |           |  |  |
  H  H  H           H  H  H           H  H  Br
```

3. (a) C_7H_{16} and C_9H_{20} are normal alkanes. C_5H_{10} and $C_{11}H_{22}$ could be cycloalkanes. (b) $-C_5H_{11}$.

(c)

```
  H  H  F  I  H  H  H  H  H  H
  |  |  |  |  |  |  |  |  |  |
H—C—C—C—C—C—C—C—C—C—C—H
  |  |  |  |  |  |  |  |  |  |
  H  H  F  H  H  H  H  H  H  H
```

(d) CH_3CH_2 and CH_2CH_3 on the same carbon of a cyclohexane ring

4. (a) C_8H_{16} is the only molecule that can be a straight-chain alkene. C_4H_6 and C_7H_{12} could be cycloalkenes.

(b)

```
F       H
 \     /
  C=C
 /     \
H       F
```

(c) C_4H_6 and C_7H_{12} can be alkynes. (d) The isomeric pentynes are 1-pentyne and 2-pentyne. The structural formula for 2-pentyne is

```
  H          H  H
  |          |  |
H—C—C≡C—C—C—H
  |          |  |
  H          H  H
```

5.

1,3,5-trifluorobenzene p-dichlorobenzene

6. (a)

```
  H  H  H  H
  |  |  |  |
H—C—C—C—C—H
  |  |  |  |
  H  H  Br Br
```

(b)

```
  H  H  H  H  H  H
  |  |  |  |  |  |
H—C—C—C—C—C—C—H + HCl
  |  |  |  |  |  |
  H  H  H  H  H  Cl
```

You may have correctly substituted the chlorine atom for any of the hydrogen atoms on the reactant molecule or indicated that the product is a mixture of the possible substitution products.

(c) Cl (chlorobenzene)

7. (a) Hydroxyl group, $-OH$

(b) $-O-$, in which both bonds from oxygen are to carbon atoms.

(c) Alcohols: $CH_3CH_2CH_2-OH$; $CH_3CH(OH)CH_3$

Ether: $CH_3-O-CH_2CH_3$

8. (a) Carbonyl group: $\geq C{=}O$

(b) Aldehyde: $R(H)C{=}O$

Ketone: $R(R')C{=}O$

(c)

$CH_3CH_2CH_2CH_2-C(=O)-H$ aldehyde $CH_3CH_2-CH(CH_3)-C(=O)-H$ aldehyde

$CH_3CH_2CH_2-C(=O)-CH_3$ ketone $CH_3-CH(CH_3)-CH_2-C(=O)-H$ aldehyde

$CH_3CH_2-C(=O)-CH_2-CH_3$ ketone $CH_3-C(CH_3)_2-C(=O)-H$ aldehyde

9. (a) Carboxyl group: $-C(=O)-O-H$

(b) acid + alcohol → ester + water

(c)

$CH_3CH_2CH_2-C(=O)-OH$ acid $CH_3CH_2-C(=O)-O-CH_3$ ester $H-C(=O)-O-CH_2CH_2CH_3$ ester

$CH_3-CH(CH_3)-C(=O)-OH$ acid $CH_3-C(=O)-O-CH_2CH_3$ ester $H-C(=O)-O-CH(CH_3)-CH_3$ ester

10. (a) $R_1-N(R_3)-R_2$

One or two of the Rs may be H.

(b) $-C(=O)-NH_2$

11. $-CH_2-CH(C_6H_5)-CH_2-CH(C_6H_5)-CH_2-CH(C_6H_5)-$

12. $HO-C(=O)-C_6H_4-C(=O)-OH + HO-CH_2-C_6H_{10}-CH_2-OH$

Answers to Concept-Linking Exercises

You may have found more relationships or relationships other than the ones given in these answers.

1. Hydrocarbons are binary compounds of carbon and hydrogen. A hydrocarbon is saturated when each carbon is single-bonded to four other atoms. An unsaturated hydrocarbon has two or more carbon atoms double- or triple-bonded to other carbon atoms.

2. Each carbon atom in an alkane forms a single bond to four other atoms. There are no multiple bonds. A normal alkane is one in which all carbon atoms are in a continuous chain; there are no branches off the main chain. A homologous series is a series of compounds in which each member differs from the one before it by a $-CH_2-$ unit. The alkane series: CH_4, CH_3CH_3, $CH_3CH_2CH_3$, and so on, is an example of a homologous series. A cycloalkane is a molecule composed of a ring of carbon atoms in which all carbon atoms are saturated.

3. A Lewis diagram shows how atoms are arranged and bonded in a molecule or ion. It is known as a structural formula or a structural diagram when unshared electron pairs are omitted. A line formula is a "shorthand" Lewis diagram in which bonds between a chain carbon and atoms or other branches from the chain are not shown precisely. When the CH_2 units in a line formula are grouped together, the result is a condensed formula.

4. An alkene is an aliphatic hydrocarbon in which two carbon atoms are double-bonded to each other. Alkene isomers that exist because of the double-bond geometry are called *cis–trans isomers* or *geometric isomers*. If identical groups of atoms are on the same side of the double bond, they are *cis* to one another. Identical groups on opposite sides of the double bond are *trans.* An alkyne is an aliphatic hydrocarbon in which two carbon atoms are triple-bonded to each other.

5. Aromatic compounds are those that contain the benzene ring. These compounds have delocalized bonding electrons—that is, bonding electrons that belong to the molecule as a whole rather than to a specific bond in the benzene ring. Non-aromatic hydrocarbons are aliphatic hydrocarbons.

6. Catalytic cracking and fractional distillation are processes used in petroleum refining. Catalytic cracking is the process in which long-chain hydrocarbons are split into shorter molecules. Hydrocarbons are separated into fractions that boil at different temperatures in the process of fractional distillation. Alkanes are prepared by catalytic hydrogenation of an alkene, in which the alkene reacts with hydrogen in the presence of a catalyst to form the corresponding alkane.

7. An addition reaction adds atoms across a double or triple bond. Alkanes undergo substitution reactions by which a hydrogen atom is replaced by another atom. Halogenation reactions are addition or substitution reactions with a halogen (Group 7A/17 element). Hydrogenation is an addition reaction with hydrogen.

8. A functional group is an atom or group of atoms that establishes the identity of a class of compounds and determines its chemical properties. The hydroxyl group, $-OH$, identifies alcohols. The carbonyl group, $\overset{O}{\overset{\|}{C}}$ (with bonds to two groups), characterizes both aldehydes ($R-CHO$) and ketones ($R-C(=O)-R'$). The carboxyl group, $-C(=O)-O-H$, distinguishes carboxylic acids.

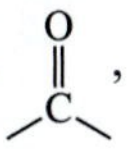

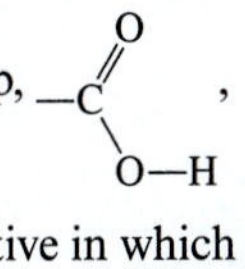

9. An amide is a carboxylic acid derivative in which the hydroxyl part of the carboxyl group is replaced with an NH_2, NHR, or NR_2 group. An amino acid is an acid in which an amine group is substituted for a hydrogen atom in the molecule. The amide functional group appears in proteins as a peptide linkage between amino acids.

10. A polymer is a multiple part chemical compound formed by bonding two or more one-part chemical species known as monomers. A chain-growth polymer results from combining alkene monomers in such a way that one bond in the double bond is broken, leaving each monomer with an unshared electron that is used to form a bond with the neighboring molecule. A step-growth polymer results from the reaction of dicarboxylic acid molecules with dialcohol molecules. The $-OH$ group from the acid combines with the $-H$ of the hydroxyl group from the alcohol to form a water molecule, and the remainder of the molecules bond to form the polymer.

Solutions to Blue-Numbered Questions, Exercises, and Problems

1. The cyanide ion, CN^-, and the carbonate ion, CO_3^{2-}, are not organic because they do not contain carbon–hydrogen bonds. The acetate ion, $C_2H_3O_2^-$, is organic.

3. 109.5°; tetrahedral

5. A hydrocarbon is a compound made up of carbon and hydrogen atoms. C_3H_4, C_8H_{10}, and $CH_3CH_2CH_3$ are hydrocarbons.

7. $C_{11}H_{24}$; $C_{21}H_{44}$. Alkanes have the general formula C_nH_{2n+2}.

9. Isomers are compounds having the same molecular formula but different structural formulas.

11. C_7H_{16} is the molecular formula; $CH_3CH_2CH_2CH_2CH_2CH_2CH_3$ is the line formula; $CH_3(CH_2)_5CH_3$ is the condensed formula. The structural formula is:

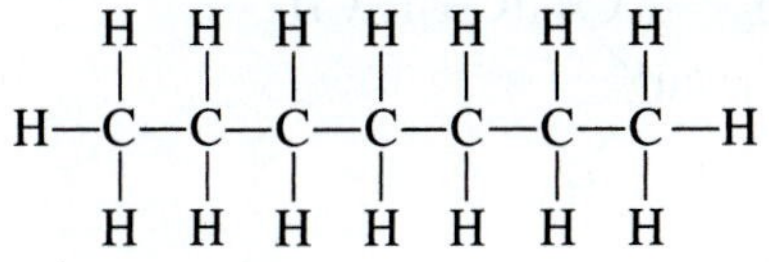

```
    H   H   H   H   H   H   H
    |   |   |   |   |   |   |
H—C—C—C—C—C—C—C—H
    |   |   |   |   |   |   |
    H   H   H   H   H   H   H
```

13. C_2H_5- C_4H_9-

15. (a) C_4H_{10}; (b) decane

17. 2-methyl-4-ethylhexane

19. 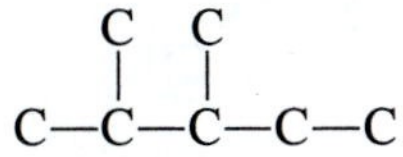
```
    C   C
    |   |
C—C—C—C—C
```

21. 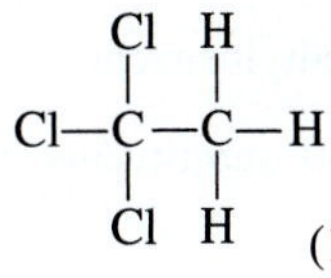
```
    Cl  H               H   Cl
    |   |               |   |
Cl—C—C—H          Cl—C—C—H
    |   |               |   |
    Cl  H   (1,1,1)     Cl  H   (1,1,2)
```

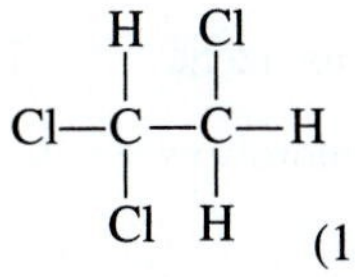

23. The general formula of a cycloalkane is C_nH_{2n}. Cyclobutane, C_4H_8, is an example:
```
   H   H
   |   |
H—C—C—H
   |   |
H—C—C—H
   |   |
   H   H
```

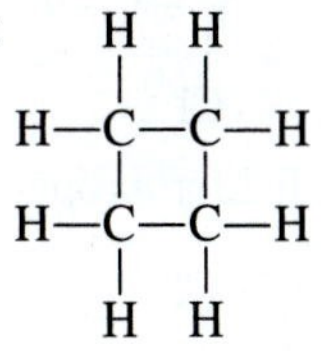

25. 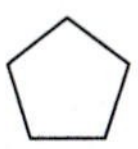 (cyclopentane line structure: pentagon)

27. 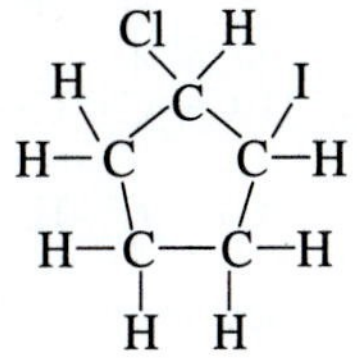
```
       Cl  H
   H    \ /    I
    \    C    /
H—C         C—H
     |       |
  H—C——C—H
     |    |
     H    H
```

29. 4-ethylheptane

31. 1-bromo-3,3-dichlorobutane

33. Cl I

34. 1-chloro-2-ethylcyclohexane

35. An alkene has one or more double bonds; an alkane has only single bonds. The general formula for an alkane is C_nH_{2n+2}, for an alkene, C_nH_{2n}.

39. Cl Cl C=C Cl H

This molecule can have only two additional atoms attached to each of its two carbons, so the first two chlorines go on one carbon atom. The third chlorine must go on the other carbon atom, thus, 1,1,2 is the only possible arrangement. It does not need to be specified.

41. H H C=C CH_3CH_2 $CH_2CH_2CH_3$ H $CH_2CH_2CH_3$ C=C CH_3CH_2 H

Think of a line drawn between the two lines that depict a double bond. In *cis*-3-heptene, the hydrogen atoms attached to the double-bonded carbons are on the same side of that line. In *trans*-3-heptene, the hydrogen atoms attached to the double-bonded carbons are on opposite sides of that line.

43. $CH_3(CH_2)_6CH_2$ $CH_2(CH_2)_{11}CH_3$ C=C H H

45. 2,2-dimethyl-3-heptyne

47. CH_3 CH_3 CH_3 CH_3 CH_3 CH_3

1,2-dimethylbenzene 1,3-dimethylbenzene 1,4-dimethylbenzene

49. 1-bromo-4-chlorobenzene. Because both substituents are halogen atoms, the lower number is given to the first halogen in the alphabet.

51. Cl C—C—C Cl Cl Cl C—C—C Cl C—C—C Cl Cl Cl C—C—C

1,1-dichloropropane 1,2-dichloropropane 2,2-dichloropropane 1,3-dichloropropane

53. Cl Cl C—C—C—C Cl Cl C—C—C—C

55.

$$H_3C—\underset{H}{C}=\underset{H}{C}—CH_3 + H_2 \xrightarrow{\text{catalyst}} H_3C—\underset{H}{\overset{H}{C}}—\underset{H}{\overset{H}{C}}—CH_3$$

Because there is only one straight-chain butane molecule, it makes no difference if the starting material is *cis*- or *trans*-2-butene.

57.

```
   H  H  H  H               H  H  H  H
   |  |  |  |               |  |  |  |
H—C—C—C—C—OH          H—C—C—C—C—H both alcohols
   |  |  |  |               |  |  |  |
   H  H  H  H               H  H  OH H

   H CH3 H                  H CH3 H
   |  |  |                  |  |  |
H—C—C—C—OH            H—C—C—C—H both alcohols
   |  |  |                  |  |  |
   H  H  H                  H  OH H

   H  H     H  H            H  H  H     H
   |  |     |  |            |  |  |     |
H—C—C—O—C—C—H         H—C—C—C—O—C—H both ethers
   |  |     |  |            |  |  |     |
   H  H     H  H            H  H  H     H
```

59. Although ethers have two carbon–oxygen bonds that are polar, the dipoles of these two bonds almost cancel each other by geometry. The forces of attraction in an ether are therefore weak dipole–dipole. An alcohol has an −OH group. The proton in the −OH group on one alcohol molecule can hydrogen bond with the lone pair electrons of the oxygen atom in the −OH group of another alcohol molecule. The higher the forces of attraction, the higher the boiling point will be.

61.

```
   H  OH H  H  H  H
   |  |  |  |  |  |
H—C—C—C—C—C—C—H
   |  |  |  |  |  |
   H  H  H  H  H  H
```

63.

```
   H  H  H  H     H  H
   |  |  |  |     |  |
H—C—C—C—C—O—C—C—H
   |  |  |  |     |  |
   H  H  H  H     H  H
```

65.

```
   H  H  H                H  H  H               H  H  H     H  H  H
   |  |  |                |  |  |               |  |  |     |  |  |
H—C—C—C—OH + H—O—C—C—C—H  ⟶  H—C—C—C—O—C—C—C—H + HOH
   |  |  |                |  |  |               |  |  |     |  |  |
   H  H  H                H  H  H               H  H  H     H  H  H
```

67.

```
   H  H    O              H  O  H
   |  |   //              |  ‖  |
H—C—C—C               H—C—C—C—H
   |  |    \              |     |
   H  H     H             H     H
```

69.

```
   H  OH H                      H  O  H
   |  |  |                      |  ‖  |
H—C—C—C—H + ½O2  ⟶  H—C—C—C—H + H2O
   |  |  |                      |     |
   H  H  H                      H     H
```

71.

```
   H  H  H  H  H  O
   |  |  |  |  |  ‖
H—C—C—C—C—C—C
   |  |  |  |  |   \
   H  H  H  H  H    OH
```

73.

```
   H  H  O              H  H             H  H  O     H  H
   |  |  ‖              |  |             |  |  ‖     |  |
H—C—C—C—OH + H—O—C—C—H  ⟶  H—C—C—C—O—C—C—H + HOH
   |  |                 |  |             |  |        |  |
   H  H                 H  H             H  H        H  H
```

The organic reaction product is ethyl propanoate, an ester.

75.

```
    H       H                    H   H
    |   ..  |                 .. |   |
H—C—N—C—H             H—N—C—C—H
    |   |   |                 |  |   |
    H   H   H                 H  H   H

 dimethylamine              ethylamine
```

77. Dimethylamine is a secondary amine; ethylamine is a primary amine.

79.

```
    H  H  O                           H  H  O
    |  |  ||          ..              |  |  ||  ..
H—C—C—C—OH + H—N—H ⟶ H—C—C—C—N—H + HOH
    |  |              |               |  |      |
    H  H              H               H  H      H
```

The organic reaction product is propanamide, an amide.

81.

```
 [H  H   H  H   H  H ]
 [|  |   |  |   |  | ]
—[C—C—C—C—C—C]—
 [|  |   |  |   |  | ]
 [H  Br  H  Br  H  Br]
```

83.

```
H  H
|  |
C=C
|  |
H  CH3
```

85.

```
 [Cl F  Cl F  Cl F]
 [|  |  |  |  |  |]
—[C—C—C—C—C—C]—
 [|  |  |  |  |  |]
 [F  F  F  F  F  F]
```

87.

```
 [O             CH3              O              CH3             ]
 [||            |                ||             |               ]
—[C—O—(C6H4)—C—(C6H4)—O—C—O—(C6H4)—C—(C6H4)—O]—
 [              |                               |               ]
 [              CH3                             CH3             ]
```

89.

```
 [O           O                         O           O                       ]
 [||          ||                        ||          ||                      ]
—[C—(C6H4)—C—NH—(C6H4)—NH—C—(C6H4)—C—NH—(C6H4)—NH]—
```

91.

```
    O          O                         H
    ||         ||                        |
HO—C—(C6H4)—C—OH          HO—CH2—C—OH
                                         |
                                         CH3
```

93.

```
    O
    ||
HO—C—CH2—CH2—CH2—CH2—CH2—NH2
```

96. True: b, c, d, g, h, j, k, n, o, p, s. False: a, e, f, i, l, m, q, r.

97. An alkene and a cycloalkane with the same number of carbon atoms have the same molecular formula, C_nH_{2n}. The alkene has a double bond between two carbon atoms, and there is no closed loop of carbon atoms. The cycloalkane has only single bonds, and the carbon atoms are assembled in a closed ring.

99. Experimentally, there is only one type of carbon–carbon bond in benzene, not two. Benzene also does *not* undergo the addition reactions typical of alkenes. The delocalized structure reminds us that benzene is different from both alkanes and alkenes.

101. The number of monomers in different polymers varies, so they have no definite molecular mass.

102. $\dfrac{1.8 \times 10^6 \text{ u}}{\text{polymer}} \times \dfrac{1 \text{ molecule}}{104 \text{ u}} = 1.7 \times 10^4 \text{ molecules/polymer}$

103. For an aldehyde to be oxidized to a carboxylic acid, an oxygen atom must be inserted between the carbonyl carbon and the hydrogen bonded to it. If there is no hydrogen atom bonded to the carbonyl atom, as in a ketone, there can be no oxidation.

104. The esterification equation is:

$$CH_3{-}\overset{\overset{\large O}{\|}}{C}{-}OH + H\overset{*}{O}{-}CH_3 \longrightarrow CH_3{-}\overset{\overset{\large O}{\|}}{C}{-}\overset{*}{O}{-}CH_3 + HOH$$

The asterisk on the oxygen atom in methanol identifies the oxygen-18 atom. It is the presence of the radioactive oxygen in the ester product, rather than in the water product, that shows that the water molecule is made up from a hydroxyl from the acid and only a hydrogen atom from the alcohol.

105. Look at the equation describing formation of a peptide linkage:

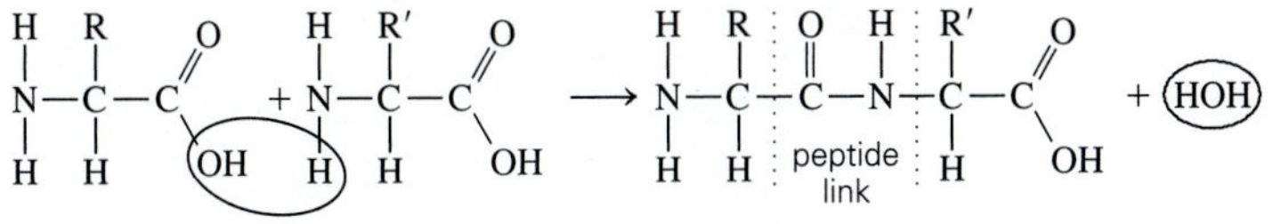

$$\text{H}_2\text{N}{-}\text{CHR}{-}\text{C(=O)}{-}\boxed{\text{OH}} + \boxed{\text{H}}{-}\text{NH}{-}\text{CHR}'{-}\text{C(=O)}{-}\text{OH} \longrightarrow \text{H}_2\text{N}{-}\text{CHR}{-}\underbrace{\text{C(=O)}{-}\text{NH}}_{\text{peptide link}}{-}\text{CHR}'{-}\text{C(=O)}{-}\text{OH} + \text{(HOH)}$$

This condensation reaction requires loss of a water molecule, one hydrogen of which must come from the amine. A tertiary amine has no hydrogens to lose. Trimethylamine (or any tertiary amine) *cannot* form an amide.

106. $\left[\!-CH{=}CH{-}CH{=}CH{-}CH{=}CH-\!\right]_n$

107. Because the acid-catalyzed condensation reaction in which amides are made is reversible, the amides can be decomposed in the presence of an acid catalyst and water. Acid rain gives precisely that combination.

22 Biochemistry

Kenneth Eward/BioGrafx/Science Source

◀ DNA, or deoxyribonucleic acid, is the figurative bridge between biology and chemistry. Scientists who study biological molecules, such as DNA, proteins, and carbohydrates, are known as biochemists. Chemically, DNA molecules consist of two strands that are polymers of nucleic acids (acids prevalently found in the nuclei of cells), held together in a vine-like helix by hydrogen bonds. Biologically, DNA molecules are the storehouse of instructions for the development of all cellular forms of life. Remarkably, the structure of DNA was first elucidated only about seven decades ago, yet the study of biological molecules has now grown to be the most active area of chemical research.

CHAPTER CONTENTS

This chapter is a brief survey of biochemistry, the chemistry of life. As in Chapter 21, some instructors may prefer the following chapter-wide performance goals rather than the more specific numbered goals within the chapter:

Goal A Identify, describe the distinguishing features of, and give an example of a molecule in each of the four major classes of biological molecules: proteins, carbohydrates, lipids, and nucleic acids.

B Identify the monomers and describe how the polymeric molecules are assembled in the following: proteins, carbohydrates, and nucleic acids.

C Describe how an enzyme functions as a biological catalyst.

D Describe the process by which protein molecules are constructed from the information encoded in DNA molecules.

If you have studied Chapter 21, especially Sections 21.13 through 21.17, you are ready to begin studying this chapter. If you haven't studied Chapter 21 yet, learn the content of at least those last five sections. Return here with four ideas firmly understood: (1) the general formula of a carboxylic acid is R—COOH and the functional group, —COOH, is a combination of a carbonyl group and a hydroxyl group, rightly called the *carboxyl group* ▶; (2) an amine is formed by replacing one, two, or all three hydrogens in an ammonia molecule with an alkyl group; (3) large molecules can be made by bonding many smaller molecules, somewhat like links can be joined together to form a chain; and (4) the physical and chemical properties

The symbol R stands for an atom or a group of atoms that is attached to the rest of the molecule.

of large molecules can be researched and predicted, just as the properties of small molecules are researched and predicted.

Biochemistry is the application of the scientific method to the study of molecules found in living organisms. Although there are no clear dividing lines separating biology and biochemistry or biochemistry and chemistry, biochemists usually focus their research on matter between the cellular and molecular levels (**Figure 22.1**). The science of living organisms at a level larger than the single cell is generally the realm of traditional biology, and molecular-level research on nonbiological molecules and some submolecular research is traditional chemistry. The study of the tiniest bits of matter, those that are smaller than subatomic particles, is classified as modern physics.

In this chapter, we organize the study of biochemistry via introduction of the major classes of biological macromolecules. A **macromolecule** is a molecule with a relatively high molecular mass (*macro-* means large). Macromolecules typically are made from molecules of relatively low molecular mass that are bonded to one another by a common chemical reaction repeated over and over again, forming the macromolecule.

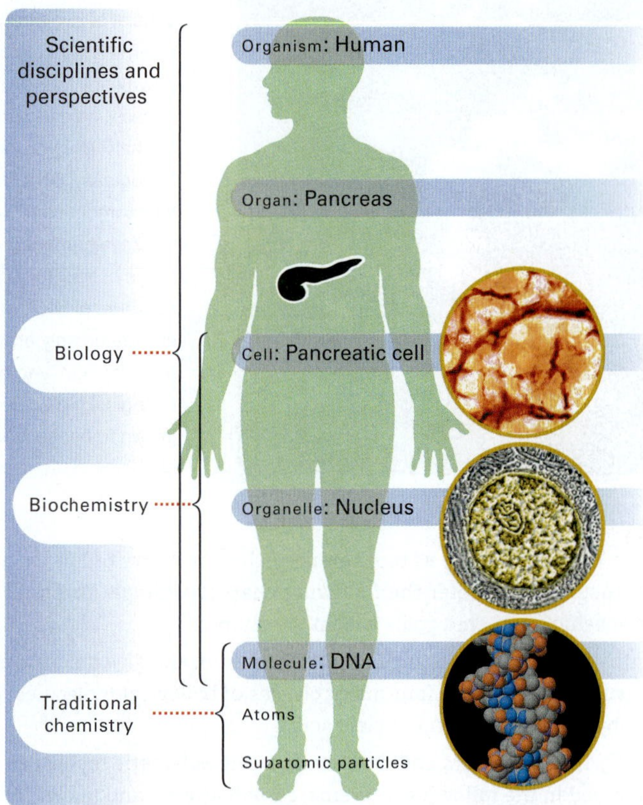

Figure 22.1 Scientific disciplines: biology, biochemistry, and traditional chemistry. Biochemists are scientists who seek answers to questions about the chemistry of molecules found in living organisms. Most biochemists use chemical techniques to study biologically relevant systems ranging from as large as a single cell to as small as a macromolecule.

Thinking About Your Thinking

Classification

Recall the classification system used by organic chemists that is described in Chapter 21. Organic molecules are organized according to their functional groups. Biochemists also use functional groups as a classification system, but because biochemical molecules are usually much larger than organic molecules, biochemists also employ a broader classification scheme for the macromolecules they study. The four major classes of biological molecules are (1) proteins (Sections 22.2 and 22.3); (2) carbohydrates (Section 22.4); (3) lipids (Section 22.5); and (4) nucleic acids (Section 22.6). Keep this broad classification system in mind as you study this chapter, and use it to organize your understanding of biochemistry.

22.1 Is There Life on Other Planets?

Atoms of all but five of the elements are made in stars. Hydrogen and helium were made in the Big Bang. Lithium, beryllium, and boron were formed from cosmic ray interaction with the lighter elements. Mid-sized elements such as carbon were formed by nuclear fusion in the interior of stars. Additional elements were synthesized when supernovae happened; when hypergiant stars die, they scatter heavy elements across the galaxy (**Figure 22.2**).

The most common elements in the Milky Way Galaxy—elements made by the Big Bang and in stars—are hydrogen, helium, oxygen, and carbon. Helium is nonreactive; it will not form chemical bonds. The most abundant elements in the human body are hydrogen, oxygen, and carbon. This match is no coincidence. We are made of elements that the universe is made of. Given that we are made of the most common ingredients that exist and given that carbon is almost always found bonded to other atoms in molecules, sharing electrons, it is logical that the molecules of life everywhere in the universe are based on carbon. Given the vastness of the universe and the commonality of ingredients across the universe, odds are that life has evolved in many places in addition to Earth. In this chapter, you will study the molecules upon which life is based, seeing that hydrogen, oxygen, and carbon are often the only elements present in large biomolecules.

Source: ESA/Hubble & NASA

Figure 22.2 Supernova remnant SNR B0519–69.0. This photograph, taken by the Hubble Telescope, shows the remnants that remain after a white dwarf star 150,000 light years away went supernova about 600 years ago, scattering elements throughout the region.

22.2 Amino Acids and Proteins

Goal 1 Given a table of Lewis diagrams of amino acids and their corresponding three-letter and one-letter abbreviations, draw the Lewis diagram of a polypeptide from its abbreviation.

2 Given a Lewis diagram of a polypeptide or information from which it may be written, identify the N-terminus amino acid and the C-terminus amino acid.

3 Explain the meaning of the terms *primary*, *secondary*, *tertiary*, and *quaternary structure* as they apply to proteins.

4 Describe how hydrogen bonding results in (a) α-helix and (b) β-pleated sheet secondary protein structures.

The word *protein* originally comes from the Greek word *prōtos,* meaning "first." Prōtos evolved to the Greek word *prōteios,* meaning "primary," and then to the German word protein. Thus, a protein is a biological substance of first or primary importance. Why are proteins so important within the world of biological molecules?

It is estimated that a single mammalian cell contains 10,000 different protein molecule types and more than a *billion* total protein molecules! Why are there so many protein molecules in a cell? They are the primary agents that perform the duties encoded by the instructions in our genes. Proteins are involved in almost every biochemical process.

Proteins are polymers composed of one or more chains of amino acids, the monomers. The primary difference between different protein molecules is their sequence of amino acids. An **amino acid** is a molecule that contains both an amine group and a carboxylic acid group. ◀ In the amino acid monomers in proteins, the amine group and carboxyl group are attached to the same carbon atom:

▶ **P/Review** $-NH_2$ is the amine group in an amino acid and $-COOH$ is the carboxyl group. These structures are introduced in Sections 21.14 and 21.13, respectively.

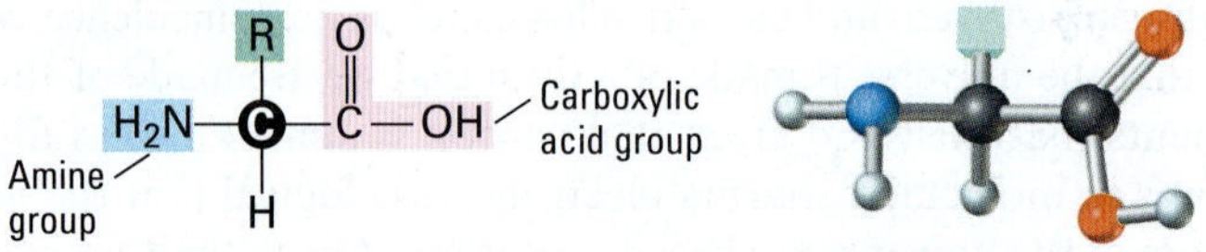

R O
$H_2N-C-C-OH$
H
Amine group
Carboxylic acid group

Amino acids differ by the identity of the R group.

For most organisms, the DNA of their species codes for 20 amino acids. These are given in **Table 22.1**. Note that they may be classified based on the polarity of the R group. Also note in Table 22.1 that each amino acid may be identified in three ways: by its complete name, by a three-letter abbreviation, and by a single-letter abbreviation.

Primary Protein Structure

Biochemists typically consider the structure of a protein molecule at four levels. The most fundamental level is the sequence of amino acids in the chain, which is called the **primary structure** of the protein.

Two amino acids can react with each other in two possible ways, carboxyl group of amino acid 1 with amine group of amino acid 2, or carboxyl group of amino acid 2 with amine group of amino acid 1. For example, the reaction of glycine (R group = H) and alanine (R group = CH_3) can give either glycylalanine (abbreviated as Gly-Ala or G-A):

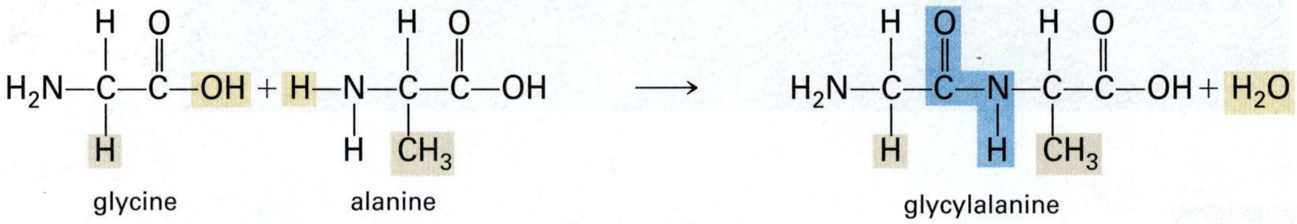

$$H_2N-CH(H)-C(=O)-OH + H-NH-CH(CH_3)-C(=O)-OH \longrightarrow H_2N-CH(H)-C(=O)-NH-CH(CH_3)-C(=O)-OH + H_2O$$

glycine alanine glycylalanine

or alanylglycine (abbreviated as Ala-Gly or A-G):

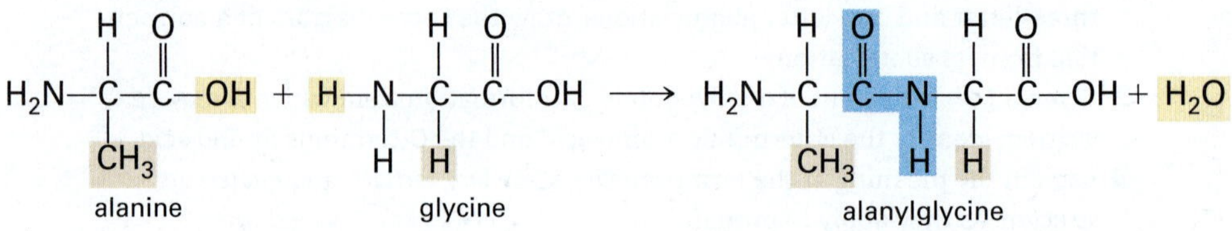

$$H_2N-CH(CH_3)-C(=O)-OH + H-NH-CH(H)-C(=O)-OH \longrightarrow H_2N-CH(CH_3)-C(=O)-NH-CH(H)-C(=O)-OH + H_2O$$

alanine glycine alanylglycine

The bond between the amino acids is called a **peptide linkage**. A **peptide** is a short chain of amino acid monomers bonded together by peptide linkages. A peptide is formed when the hydroxyl part of the carboxyl group of one amino acid molecule reacts with the hydrogen atom of the amine group of another amino acid molecule to form a molecule of water. ◀ The two amino acid molecules are then

▶ **P/Review** The peptide linkage was introduced in Section 21.14. It is formed by the reaction of the $-COOH$ group of one amino acid and the $-NH_2$ group of another.

Table 22.1 The 20 Amino Acids Commonly Found in Proteins*

Amino Acid	Abbreviation 3-letter	Abbreviation 1-letter	Structure	Amino Acid	Abbreviation 3-letter	Abbreviation 1-letter	Structure
Nonpolar R Groups							
Glycine	Gly	G	$H-CH-COOH$ NH_2	†Isoleucine	Ile	I	$CH_3-CH_2-CH-CH-COOH$ $CH_3 \quad NH_2$
Alanine	Ala	A	$CH_3-CH-COOH$ NH_2	Proline	Pro	P	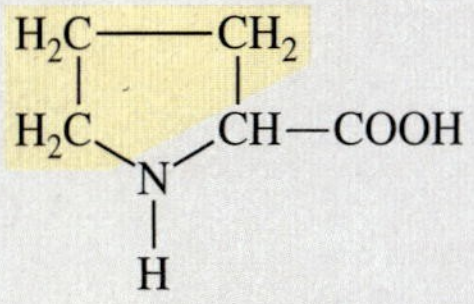 H_2C-CH_2 $H_2C \quad CH-COOH$ N H
†Valine	Val	V	$CH_3-CH-CH-COOH$ $CH_3 \quad NH_2$	†Phenylalanine	Phe	F	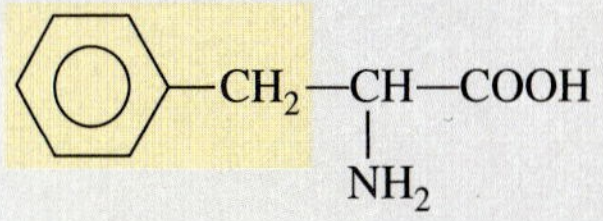 $C_6H_5-CH_2-CH-COOH$ NH_2
†Leucine	Leu	L	$CH_3-CH-CH_2-CH-COOH$ $CH_3 \quad NH_2$	†Methionine	Met	M	$CH_3-S-CH_2CH_2-CH-COOH$ NH_2
				†Tryptophan	Trp	W	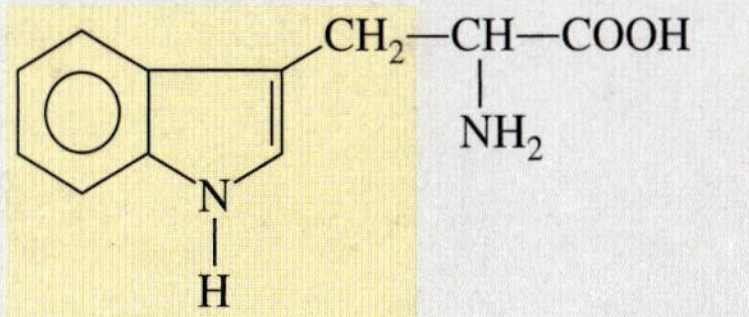 (indole ring)$-CH_2-CH-COOH$ NH_2 N H
Polar but Neutral R Groups							
Serine	Ser	S	$HO-CH_2-CH-COOH$ NH_2	Asparagine	Asn	N	$H_2N-C(=O)-CH_2-CH-COOH$ NH_2
†Threonine	Thr	T	$CH_3-CH-CH-COOH$ $OH \quad NH_2$	Glutamine	Gln	Q	$H_2N-C(=O)-CH_2CH_2-CH-COOH$ NH_2
Cysteine	Cys	C	$HS-CH_2-CH-COOH$ NH_2	Tyrosine	Tyr	Y	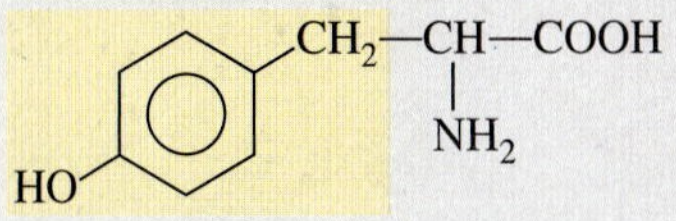 $HO-C_6H_4-CH_2-CH-COOH$ NH_2
Acidic R Groups				***Basic R Groups***			
Glutamic acid	Glu	E	$HO-C(=O)-CH_2CH_2-CH-COOH$ NH_2	†Lysine	Lys	K	$H_2N-CH_2CH_2CH_2CH_2-CH-COOH$ NH_2
Aspartic acid	Asp	D	$HO-C(=O)-CH_2-CH-COOH$ NH_2	‡Arginine	Arg	R	$H_2N-C(=NH)-NH-CH_2CH_2CH_2-CH-COOH$ NH_2
				‡Histidine	His	H	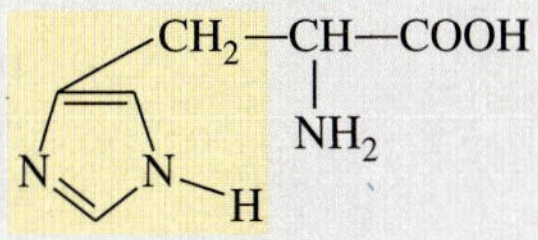 (imidazole ring: N, $N-H$)$-CH_2-CH-COOH$ NH_2

*The R group in each amino acid is highlighted.

†Essential amino acids that must be part of the human diet. The other amino acids can be synthesized by the body.

‡Growing children also require arginine and histidine in their diet.

bonded through the nitrogen atom. The shortest peptide is a **dipeptide**, a two-amino-acid peptide, as illustrated here:

$$H_2N{-}C(R)(H){-}C(=O){-}OH + H{-}N(H){-}C(R)(H){-}C(=O){-}OH \longrightarrow H_2N{-}C(R)(H){-}C(=O){-}N(H){-}C(R)(H){-}C(=O){-}OH + H_2O$$

amino acid 1 — amino acid 2 — dipeptide

Peptide linkage

If a third amino acid, valine (R group = $CH(CH_3)_2$) is added to the end of the dipeptides mentioned earlier, four **tripeptides** are possible. One is glycylalanylvaline, Gly-Ala-Val, or G-A-V:

$$NH_2{-}CH(H){-}C(=O){-}NH{-}CH(CH_3){-}C(=O){-}NH{-}CH(C(H)(CH_3)_2){-}C(=O){-}OH$$

Another is valylglycylalanine, Val-Gly-Ala, or V-G-A:

$$NH_2{-}CH(C(H)(CH_3)_2){-}C(=O){-}NH{-}CH(H){-}C(=O){-}NH{-}CH(CH_3){-}C(=O){-}OH$$

Another is valylalanylglycine, Val-Ala-Gly, or V-A-G:

$$NH_2{-}CH(C(H)(CH_3)_2){-}C(=O){-}NH{-}CH(CH_3){-}C(=O){-}NH{-}CH(H){-}C(=O){-}OH$$

The fourth possible tripeptide is alanylglycylvaline, Ala-Gly-Val, or A-G-V:

$$NH_2{-}CH(CH_3){-}C(=O){-}NH{-}CH(H){-}C(=O){-}NH{-}CH(C(H)(CH_3)_2){-}C(=O){-}OH$$

Building on these examples, it is logical to conclude that the number of possible proteins, made from only 20 amino acids, is theoretically infinite. A continuous unbranched chain of bonded amino acids is called a **polypeptide**. A molecule is called a **protein** when it is made up of one or more polypeptides more than approximately 50 amino acids long (we use the term *approximately* because the dividing line between polypeptides and proteins is not sharp).

The end of a polypeptide or protein with the free amine group is called the **N-terminus**. When a protein is synthesized in a cell, amino acids are added to the carbonyl end of the growing chain. When the chain is completed, the end of the molecule with the free carboxyl group is called the **C-terminus**. To match how polypeptide chains are synthesized within cells, the convention is to write a peptide sequence with the N-terminus on the left and C-terminus on the right, as we have done with the examples mentioned earlier.

Target Check 22.1

List the tripeptides formed when alanine, valine, and leucine are mixed and peptide bonds are induced to form in all possible combinations. Use the one-letter abbreviations for the amino acids given in Table 22.1.

Secondary Protein Structure

Review the Lewis structures of the tripeptides mentioned earlier, observing in particular the pattern in the atoms in the chain with the peptide linkages. This is the long chain that extends horizontally left to right. We will call this the polypeptide **backbone chain**.

The backbone chain atoms in a polypeptide follow the pattern: nitrogen–single bond–carbon–single bond–carbon–single bond, repeating the −N−C−C− pattern for each additional amino acid monomer. Since all atoms in this chain of atoms are single bonded, there is freedom of rotation about the single bonds. Also keep in mind that, in spite of the Lewis structures mentioned earlier showing the atoms in a straight line, the electron-pair geometry about each bond is tetrahedral, so the backbone chain is arranged in a zigzag pattern.

The atomic-level composition of a polypeptide therefore allows the atoms in the backbone chain to become close enough to one another to interact via attractive intermolecular forces. Which of the intermolecular forces is going to be the most significant in determining the three-dimensional shape of a peptide or protein? The strongest force: hydrogen bonding. The electropositive hydrogen atoms attached to nitrogen atoms in the backbone chain are attracted to the electronegative oxygen atoms in the chain or a neighboring chain. The local structure of the polypeptide chain that results from these hydrogen bonding interactions is called the **secondary structure** of the polypeptide.

There are two primary patterns of secondary structure in polypeptides. Both structures reflect a maximum amount of hydrogen bonding—that is, all of the hydrogen bond donors and acceptors in the polypeptide backbone chain are hydrogen-bonded. The term **conformation** is used to describe each of the structural arrangements possible for a given set of atoms. The most stable conformations possible result from maximizing the hydrogen bonding in polypeptides.

Figure 22.3 illustrates the two primary patterns of secondary structure at the particulate level. **Figure 22.3[a]** illustrates an **alpha-helix**, or **α-helix**. Look at the illustration, paying attention in particular to the hydrogen bonds. Note how each backbone chain −N−H group is hydrogen-bonded to a backbone chain −C=O group. To achieve this maximum amount of hydrogen bonding, the backbone chain must twist into a coiled conformation called a *helix*. A **helix** is a curved object in three-dimensional space. You've probably seen a coiled spring—it is an example of a helix.

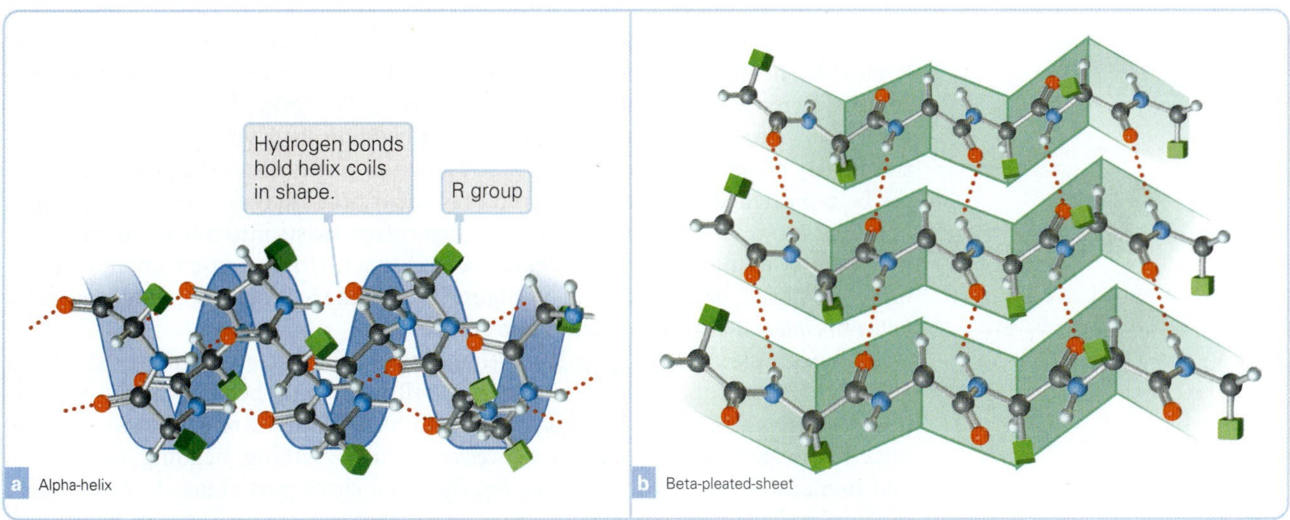

Figure 22.3 Alpha-helix and beta-pleated sheet secondary structures. R groups are represented in these ball-and-stick models with green blocks. The α-helix structure is a continuous series of loops. The β-pleated sheet structure is like a curtain but with sharp angles, like pleats, instead of rounded curves. In both the α-helix and β-pleated sheet structures, the helices and sheets are stabilized by hydrogen bonding between N−H and O=C groups.

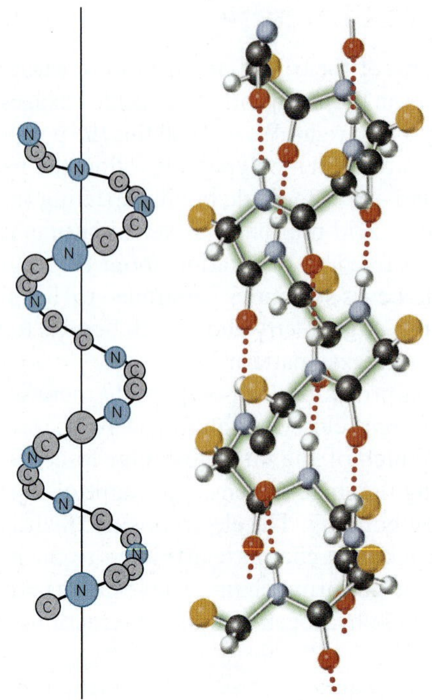

Figure 22.4 Alpha-helix protein structure. The backbone chain atoms are in a right-handed spiral conformation. This arrangement results from the pattern of hydrogen bonding between the −N−H groups and the −C=O groups.

Figure 22.4 is a particulate-level illustration of a polypeptide made from 11 amino acids. The illustration on the right in Figure 22.4 of an α-helix structure is similar to Figure 22.3[a], except that it is rotated by 90°. Again, note how each backbone chain −N−H group is hydrogen-bonded (red dots) to a backbone chain −C=O group. If all of the atoms other than those in the backbone chain are removed from the illustration, the illustration on the left in Figure 22.4 results. This makes it easier to see the −N−C−C− pattern where the chain coils as if it were climbing a right-handed screw thread.

An example of a polypeptide with an alpha-helix secondary structure is alpha-keratin, which is what hair, finger- and toenails, and the outermost skin layer is primarily composed of. The coiled alpha-helical structure at the particulate level leads to stiff, strong, difficult-to-permeate body parts at the macroscopic level.

The other primary pattern of secondary structure, the **beta-pleated sheet**, or **β-pleated sheet**, is illustrated in **Figure 22.3[b]**. First note that it has a significantly difference appearance than the alpha helix. The term *pleated sheet* is used to describe its structure because it appears like a bedding sheet with folds that occur in a regularly repeating pattern, otherwise known as pleats. Both the alpha helix and the beta-pleated sheet secondary structures occur because they maximize hydrogen bonding. However, in the alpha helix, a *single chain* twists into a helix to maximize hydrogen bonding, whereas in a beta-pleated sheet, either a single chain can fold back upon itself, or, as illustrated in Figure 22.3[b], the hydrogen bonding may occur *between chains* that run parallel to one another.

An example of a polypeptide with a beta-pleated sheet secondary structure is fibroin. You probably have not heard of fibroin, but when combined with the protein sericin, the product is silk, which you likely have heard of. Silkworms produce silk for their cocoons. Humans weave silk into clothing, bedding, upholstery, and numerous other products. The beta-pleated sheet particulate-level structure of fibroin yields macroscopically strong fibers that are not very elastic. Woven silk reflects light in such a way that the fabric has a characteristic sheen.

A third type of secondary structure exists, but it is not a pattern. It is called a **random coil** because there is no repeating pattern of organization for the polypeptide chain. Instead, the lack of interaction of relatively strong intermolecular forces yields a random arrangement for the coil. Most commonly, random coils are

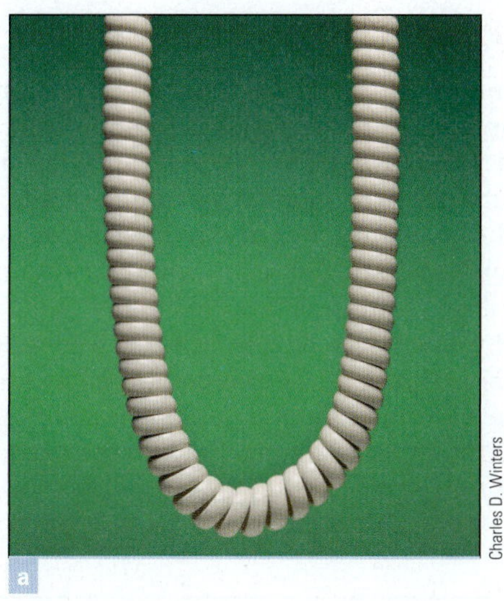

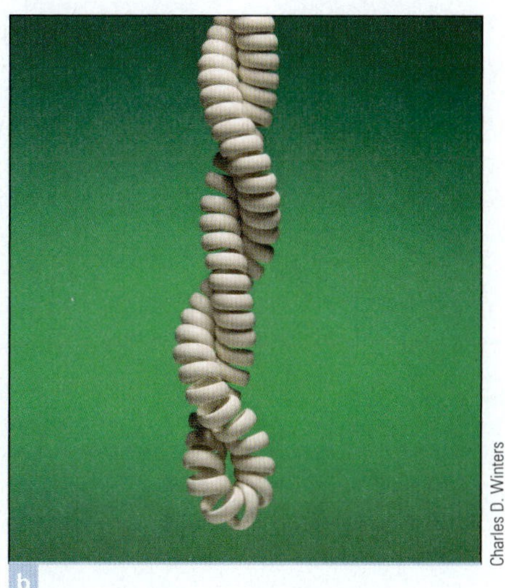

Figure 22.5 Macroscopic analogy for protein structure. (a) If the cord was made by fitting together small pieces, the sequence of those pieces would be the primary structure. The coiling of the cord is the secondary structure. The U-shaped loop is the beginning of a tertiary structure. (b) The twists and turns of the cord make up a more complicated tertiary structure.

in a loosely balled formation. Many polypeptide chains consist of all three types of secondary structure in the same chain, mostly in the form of a random coil, with smaller sections of alpha helix and beta-pleated sheet structure.

Tertiary Protein Structure

Figure 22.5 illustrates a macroscopic analogy for the **tertiary structure** of a protein, its three-dimensional shape. If the cord was assembled by fitting together small cord pieces, the sequence of the pieces would represent the primary structure. **Figure 22.5[a]** shows that the cord is twisted into a helix, which is its secondary structure. The cord also curls back on itself, beginning a tertiary structure. **Figure 22.5[b]** shows that the curly cord has itself been twisted, yielding its tertiary structure.

Figure 22.6 illustrates the *causes* of the tertiary structure of proteins: interactions between the R groups in the protein chain. Observe the S—S bond highlighted in yellow. This **disulfide bond** is one example of how the tertiary structure of a protein may be stabilized. The amino acid cysteine contains a sulfur–hydrogen bond in its R group. Two cysteines in a protein can form a disulfide bond, which stabilizes the protein structure much like a cross-link causes the strengthening of the structure of a polymer. ▶ Many other noncovalent intermolecular forces contribute to the stability of the tertiary structure of a protein, as shown in Figure 22.6.

The perm hairstyle, or more formally, permanent wave, is accomplished by breaking and reforming the disulfide linkages in the protein in the hair. Straight hair is first wrapped on a roller; the disulfide bonds are then broken by applying the perm solution and then allowed to reform when a neutralizer solution is applied. The new disulfide bonds give curl to the hair.

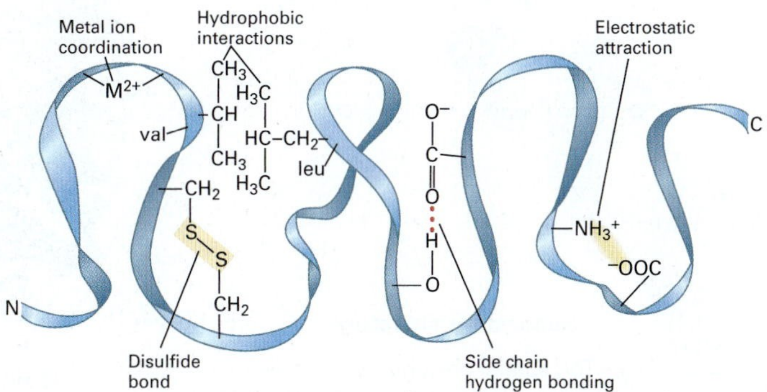

Figure 22.6 Tertiary structure of proteins. Covalent bonds between sulfur atoms are called *disulfide bonds* (*bottom left*). Other weaker noncovalent interactions also serve to stabilize proteins (*left to right*): Metal ions, such as Mg^{2+} and Zn^{2+} (symbolized in general as M^{2+}), can interact with negatively charged ionized forms of amino acid functional groups. Hydrophobic (*fear of water*) interactions are due to the energetic favorability of nonpolar groups to cluster together in the absence of water. Hydrogen bonding can occur between side chain R groups. At common biological pH, many functional groups exist as positively charged and negatively charged ions, which are then subject to electrostatic attractions.

Quaternary Protein Structure

Some proteins are composed of more than one polypeptide chain. The term **quaternary structure** refers to the number of chains and how these chains are arranged in relation to one another. The atoms in each individual chain in a protein are held together by covalent bonds, but the chains themselves are attracted to one another only by intermolecular forces. Returning to the telephone cord analogy, you can think of proteins with quaternary structure as composed of more than one telephone cord. The first protein for which the complete primary, secondary, tertiary, and quaternary structure was known is hemoglobin (**Figure 22.7**). It is composed of four polypeptide chains. Hemoglobin is the molecule found in vertebrate red blood cells that serves to transport oxygen throughout the body. Each human red blood cell contains 300,000,000 hemoglobin molecules!

Target Check 22.2

a) What is the *secondary structure* of a polypeptide chain? What is its cause?

b) What is the *tertiary structure* of a protein? What is its cause?

c) What is the *quaternary structure* of a protein? What is its cause?

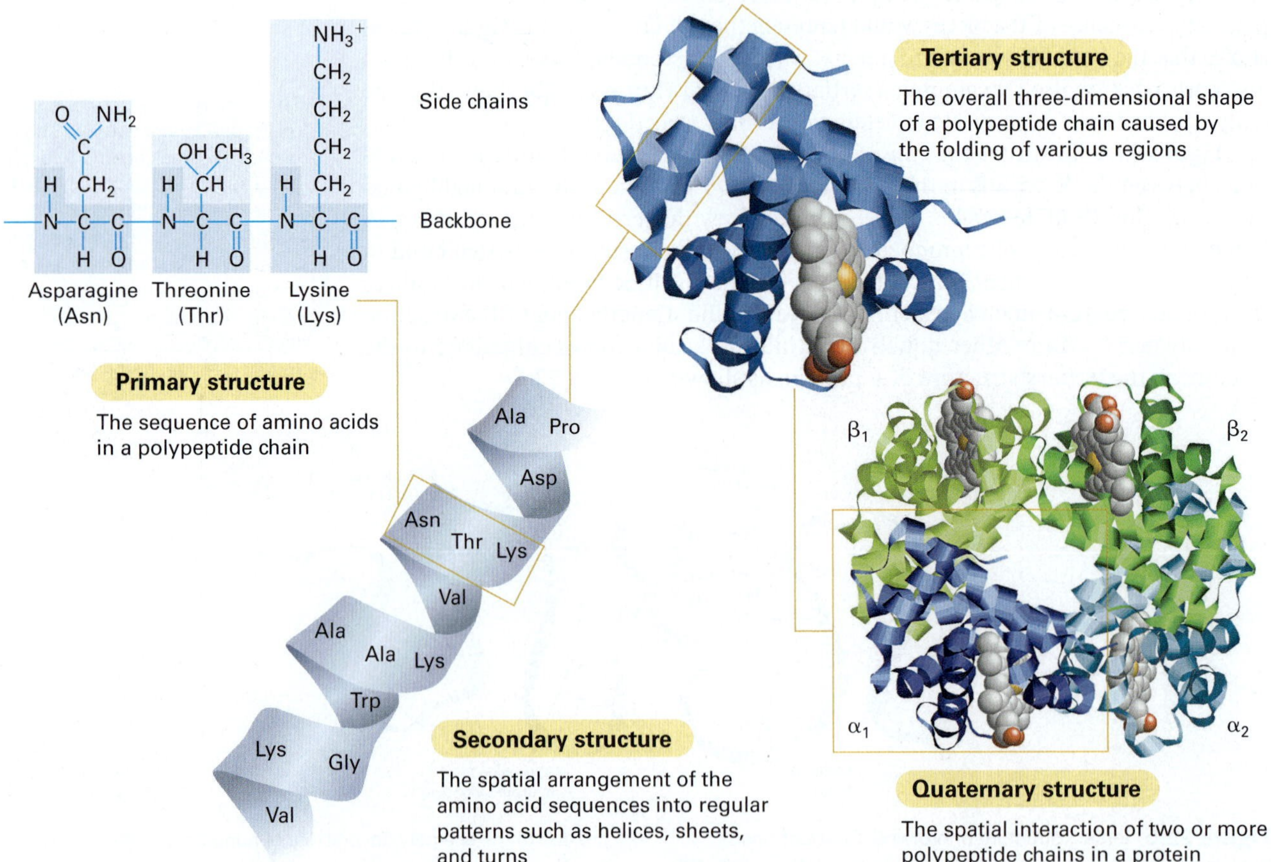

Figure 22.7 Primary, secondary, tertiary, and quaternary structure of hemoglobin. Four polypeptide chains are shown in the quaternary structure illustration, two in blue and two in green, labeled α_1, α_2, β_1, and β_2. The three-dimensional arrangement of those four polypeptide chains is the quaternary structure of the protein.

22.3 Enzymes

Goal 5 Define the term *enzyme*.

6 Define the following terms as they apply to enzymes: *substrate*, *active site*, *inhibitor*.

7 Use the induced fit model to explain enzyme activity.

Recall from Section 9.7 that complete combustion of a carbon–hydrogen–oxygen compound yields carbon dioxide and water. For example, the combustion reaction of glucose, $C_6H_{12}O_6$, is as follows:

$$C_6H_{12}O_6(s) + 6\,O_2(g) \rightarrow 6\,CO_2(g) + 6\,H_2O(\ell)$$

This equation also describes the overall process living organisms use to produce the energy needed to live. Sustaining life requires a slow combustion. Sugars are eaten (glucose is the simplest sugar) and oxygen is breathed in; carbon dioxide and water are exhaled.

How do living organisms perform and control the chemical reactions of a combustion process at body temperature? Each chemical reaction in a living system is controlled by a catalyst called an **enzyme**, which is a molecule that increases the rate of a chemical reaction without itself undergoing a permanent change. Life in its present form is not possible without enzymes.

Most enzymes are proteins. Enzymes are relatively large molecules in comparison with the size of the molecules they catalyze (**Figure 22.8[1]**). The molecule that the enzyme acts upon is called a **substrate** (the green molecule in Figure 22.8[1]). The increase in reaction rate for an enzyme-catalyzed reaction in comparison with the uncatalyzed reaction is very large: It is not unusual for a catalyzed reaction to be a billion times faster or even more. Most enzymes are highly specific, with a shape that allows them to catalyze only one reaction type. Accordingly, over 2000 different enzymes are currently known to exist, responsible for catalyzing more than 5000 reaction types.

Catalysis is initiated when a substrate binds to the surface of an enzyme. This takes place at a specific location on the larger enzyme molecule called its **active site** (Figure 22.8[1]). The geometry of the active site is critical: It must approximately fit the portion of the substrate that binds to the enzyme. The atoms in the active

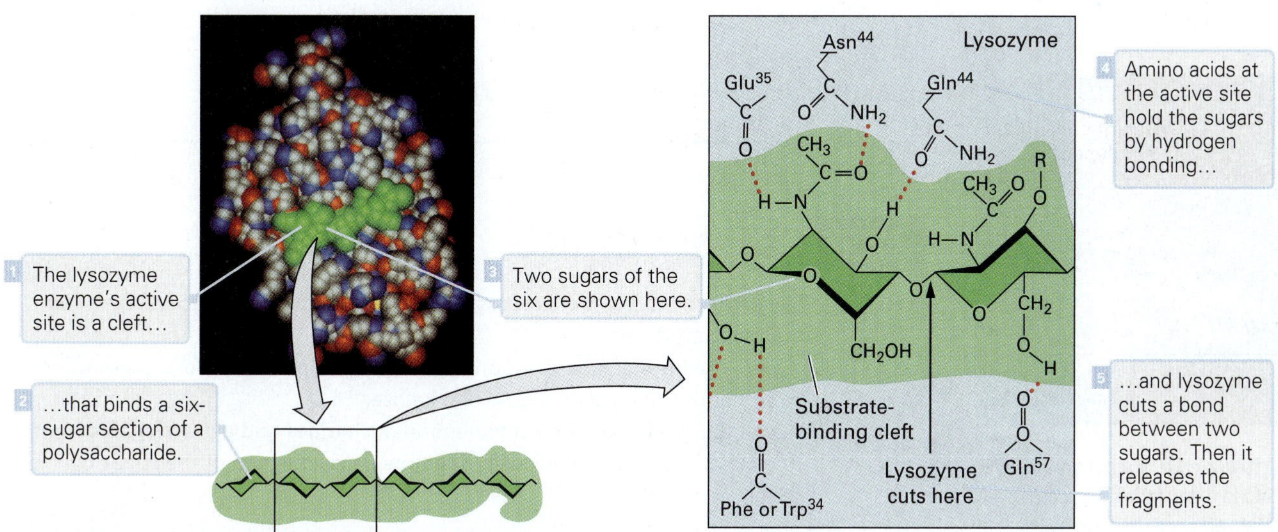

Figure 22.8 An enzyme and its active site. The large enzyme molecule illustrated is lysozyme, nature's bacteria killer. It catalyzes the splitting of polysaccharide molecules (large carbohydrate polymers that will be discussed in Section 22.4) found in the cell walls of bacteria. When the cell walls are sufficiently disintegrated, the bacteria burst. The lysozyme molecule has an indentation that serves as its active site, shown here with a portion of a green polysaccharide molecule within it.

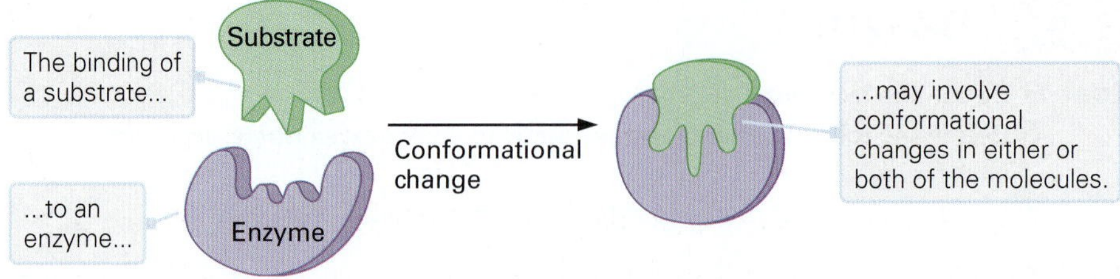
The binding of a substrate...
Substrate
...to an enzyme...
Enzyme
Conformational change
...may involve conformational changes in either or both of the molecules.

Figure 22.9 Induced fit model of enzyme activity. The close but imperfect match between the shape of the substrate and the active site of the enzyme dynamically changes to become an exact match as the two molecules bind. This is called a *conformational change*, which is a change in the three-dimensional arrangement of the atoms in a molecule.

site are also a critical aspect of the enzyme because the nature of the interactions between the enzyme and substrate must allow binding. **Figure 22.8[4]** shows hydrogen bonding as the mechanism of binding for the reaction illustrated.

Experimental measurements show that the shape and size of the active site of an enzyme changes as the substrate begins to bind. This is one piece of the puzzle that has led to the development of a theory known as the **induced fit model**: The shape of the substrate is a close but not exact match to the shape of the active site of the enzyme (**Figure 22.9**). As the substrate binds to the enzyme, both molecules change shape. The molecules responsible for binding are in continual motion. The distortion of the shape of the substrate and the molecular motion likely leads to the breaking of a bond in the substrate, yielding products. After the products are released, the enzyme returns to its original shape and is ready to catalyze another reaction.

The discussion of the induced fit model thus far has been focused on assuming that the enzyme converts the substrate to products, which then move out of the active site. Some substrates may bind to the active site and remain bound. Such a molecule is called an **inhibitor**. It renders an enzyme inactive. An example of a medical application of use of an inhibitor is in the treatment of acquired immunodeficiency syndrome (AIDS). An enzyme called *HIV* (human immunodeficiency virus) *protease* is necessary for the production of new virus particles. Scientists have now developed pharmaceutical drugs that deliver HIV protease inhibitors that cause the reduction of viral production.

Target Check 22.3

a) What is an *enzyme*?

b) What is a *substrate*?

c) What is "induced" in the *induced fit model* of enzyme activity?

Charles D. Winters

Figure 22.10 Carbohydrates are found in many foods such as breads, pastas, and rice. In general, it is believed that the carbohydrates in refined foods such as white bread are a less healthy source of energy-providing carbohydrates than those found in unprocessed foods such as whole grains.

22.4 Carbohydrates

Goal

8 Distinguish among monosaccharides, disaccharides, oligosaccharides, and polysaccharides.

9 Given a Lewis diagram of a monosaccharide in open chain form, determine whether the sugar is an aldose or a ketose.

10 Given the Lewis diagrams of two monosaccharides and a description of the bond linking the molecules, draw the Lewis diagram of the resulting disaccharide.

Carbohydrates are molecules that were originally thought to be "hydrates of of carbon," $C_x(H_2O)_y$. They occur in a wide variety of plant-based foods (**Figure 22.10**). We now know that carbohydrates are not hydrates of carbon and they do not even necessarily have to fit the general $C_x(H_2O)_y$ formula, although many do. **Carbohydrate** is now the name for a class of molecules, which are aldehydes or ketones with two or more —OH groups. We will organize this section in order of increasing

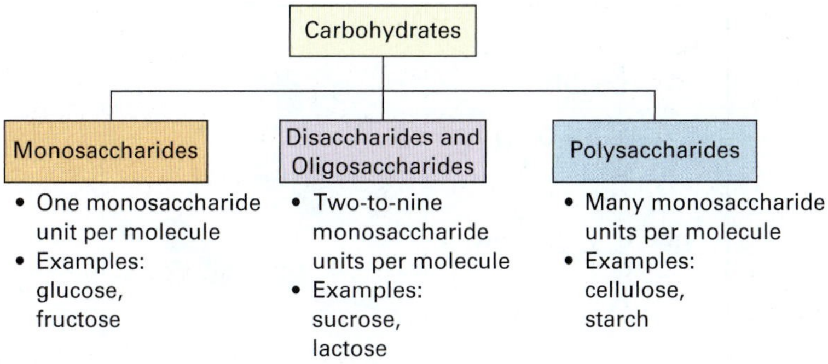

Carbohydrates

Monosaccharides	Disaccharides and Oligosaccharides	Polysaccharides
• One monosaccharide unit per molecule • Examples: glucose, fructose	• Two-to-nine monosaccharide units per molecule • Examples: sucrose, lactose	• Many monosaccharide units per molecule • Examples: cellulose, starch

Figure 22.11 Carbohydrate classification.

carbohydrate size, beginning with the simplest group, monosaccharides, which are carbohydrates that cannot be changed into simpler compounds by reaction with water. Next, we will discuss disaccharides, two-monosaccharide-unit molecules, and oligosaccharides, three-to-approximately-nine-monosaccharide-unit molecules. The section concludes with a discussion of polysaccharides, which have at least 10 monosaccharide units and generally have many, many more than 10 (**Figure 22.11**).

Monosaccharides

A **monosaccharide** is a carbohydrate that cannot be changed into simpler compounds by reaction with water. Monosaccharides serve as the most fundamental carbohydrate molecule; other categories of carbohydrates are synthesized by bonding monosaccharides to one another. Monosaccharides have the formula $(CH_2O)_n$, where $n \geq 3$.

In monosaccharides, every carbon atom but one is bonded to a hydroxyl group. The remaining carbon atom is double-bonded to an oxygen atom, forming a carbonyl group. If the carbonyl group is at the end of the carbon chain, the monosaccharide is an **aldose**, derived from the aldehyde functional group. If the carbonyl group is within the chain, the monosaccharide is a **ketose**, derived from the ketone functional group. ▶ The general ending *-ose* denotes the name of a carbohydrate.

▶ **P/Review** Aldehydes and ketones are characterized by the carbonyl group. An aldehyde has a hydrogen atom bonded to the carbon atom of a carbonyl group, and a ketone has two alkyl groups attached to the carbonyl carbon. See Section 21.12.

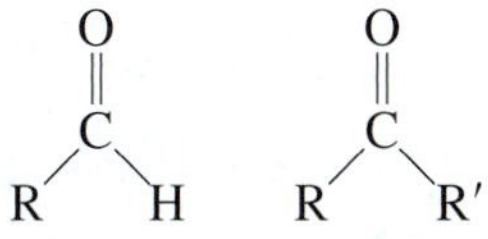

$$\underset{\text{aldehyde}}{R-\overset{\overset{\displaystyle O}{\|}}{C}-H} \qquad \underset{\text{ketone}}{R-\overset{\overset{\displaystyle O}{\|}}{C}-R'}$$

Monosaccharides are all colorless crystalline solids at room temperature. As you would expect from a molecule with so many polar —OH groups, hydrogen bonding makes the monosaccharides water-soluble. Solutions of monosaccharides generally have a sweet taste, with differing levels of sweetness. Accordingly, carbohydrates are often referred to as **sugars**. Monosaccharides may be referred to as **simple sugars**.

The most important energy source in living organisms is **glucose**, $C_6H_{12}O_6$ (**Figure 22.12**). Generally, glucose reacts with the oxygen in inhaled air, ultimately yielding carbon dioxide and water, transferring energy to specialized energy-carrier molecules that participate in reactions that require energy. The glucose concentration in human blood is typically about one gram per liter.

Monosaccharides exist as an equilibrium mixture of open chain and cyclic forms in water solutions. ▶ However, an equilibrium mixture is more than 99% cyclic and less than 1% open chain. We will show illustrations of both the open chain molecule and its more common cyclic form. In the diagram of the cyclic form,

▶ **P/Review** In Section 18.2, we stated that for reversible chemical changes in a closed system, a dynamic equilibrium occurs where the forward rate of change is equal to the reverse rate of change. The amounts of substances in an equilibrium are not necessarily equal.

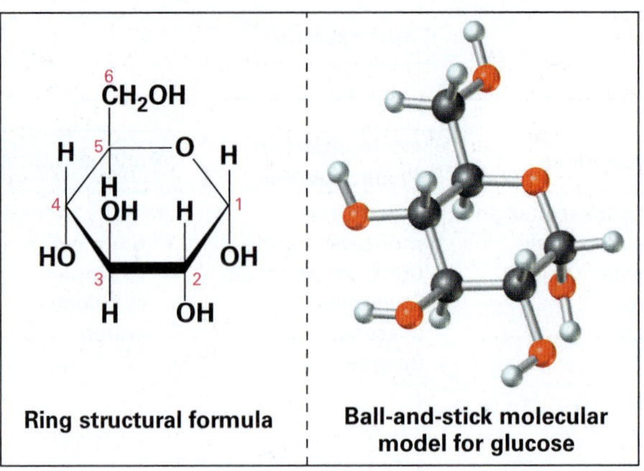

Figure 22.12 A ring structural formula and a ball-and-stick molecular model of the glucose molecule. Compare the diagram with the model to see how the diagram represents the actual molecular structure shown in the model.

there is an (unwritten) carbon atom at each point where straight lines meet. The two forms of glucose are the following:

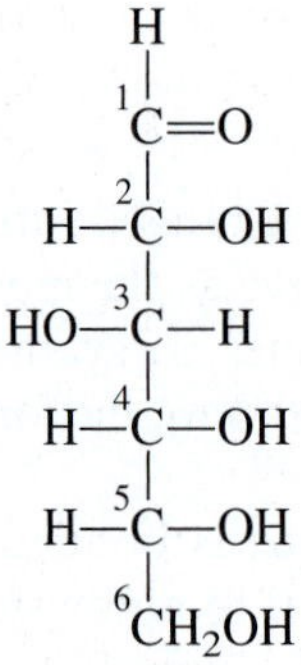

glucose 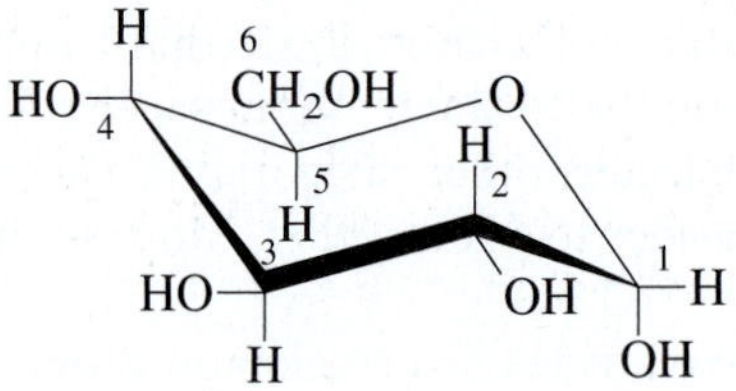 α-glucose

Glucose may exist in two different configurations that vary at carbon-1, designated alpha (α) and beta (β). Consider the structures of α-glucose and β-glucose:

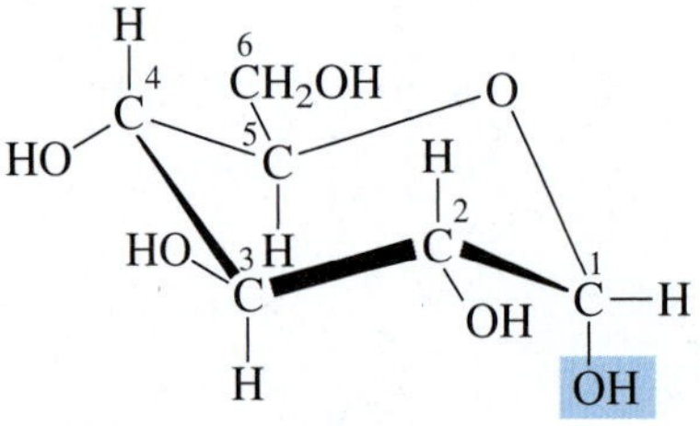

α-glucose Axial β-glucose Equatorial

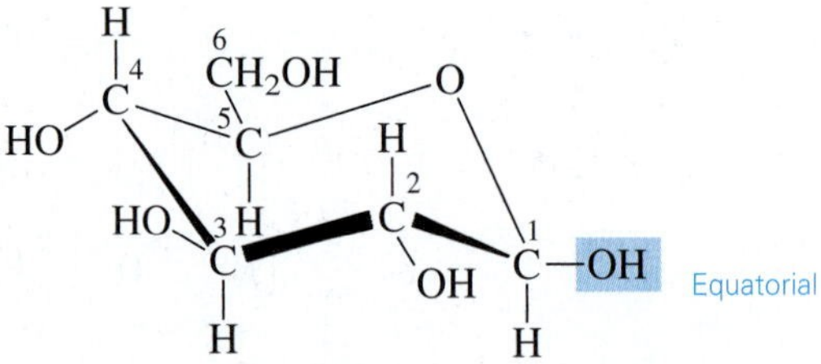

First, look at the number of oxygen atoms bonded to carbon-1. There are two. Now look at the number of oxygen atoms bonded to the other carbon atoms. They are bonded to one oxygen atom. Carbon-1 is the only carbon atom in glucose that is bonded to two oxygen atoms. Now look at the hydroxyl group bonded to carbon-1. The bond connecting the −OH to carbon-1 in α-glucose is nearly vertical. In β-glucose, the bond connecting the −OH to carbon-1 is nearly horizontal. This difference is important when considering the bonding of glucose units.

A widely consumed monosaccharide that occurs naturally in honey, fruits, vegetables, and grains (along with glucose) is **fructose**, $C_6H_{12}O_6$. Note that it has the same chemical formula as glucose, but a different arrangement of atoms, as illustrated in the forthcoming diagram. Fructose is relatively inexpensive to isolate and purify, and it is sweeter than glucose or table sugar, and therefore it is often used in commercial production of foods and beverages. High-fructose corn syrup (HFCS) is a well-known sweetener used in soft drinks, breakfast cereals, and other processed foods as a less expensive alternative to table sugar. The manufacturing process for HFCS converts some of the natural glucose and table sugar in corn to fructose. There are serious concerns about the effect on human health of HFCS—and all other added sugars—as widely used ingredients in foods and beverages.

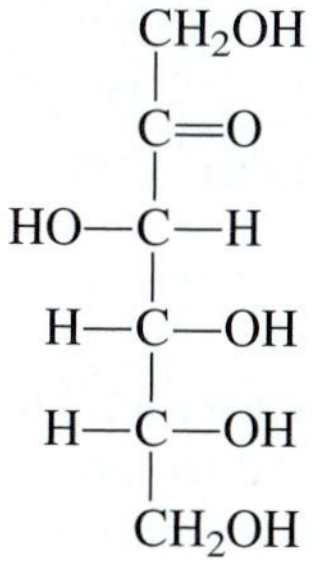

CH_2OH, C=O, HO—C—H, H—C—OH, H—C—OH, CH_2OH

fructose

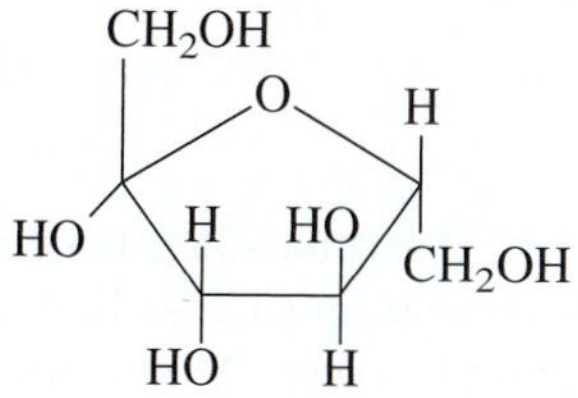

CH_2OH, O, H, HO, H, HO, CH_2OH, HO, H

fructose

Disaccharides and Oligosaccharides

A **disaccharide** is a two-monosaccharide-unit molecule, and an **oligosaccharide** is a three-to-approximately-nine-monosaccharide-unit molecule. To form a disaccharide, two monosaccharides are bonded together in a condensation reaction (Section 21.17) that results in the formation of covalent bonds to an oxygen atom, linking the monosaccharides, also yielding a water molecule, in the general form X−OH + OH−Y → X− O −Y + HOH. The oxygen atom linking the two monomers is shaded in blue here, and we will continue to use blue shading for the remainder of this section to help you identify atoms that link monomers.

Sucrose is the most recognizable carbohydrate to most people because it is table sugar. Although chemists use the word *sugar* to refer generally to all carbohydrates, it specifically refers to sucrose in everyday language. Glucose and fructose are the monosaccharides that are bonded to form sucrose:

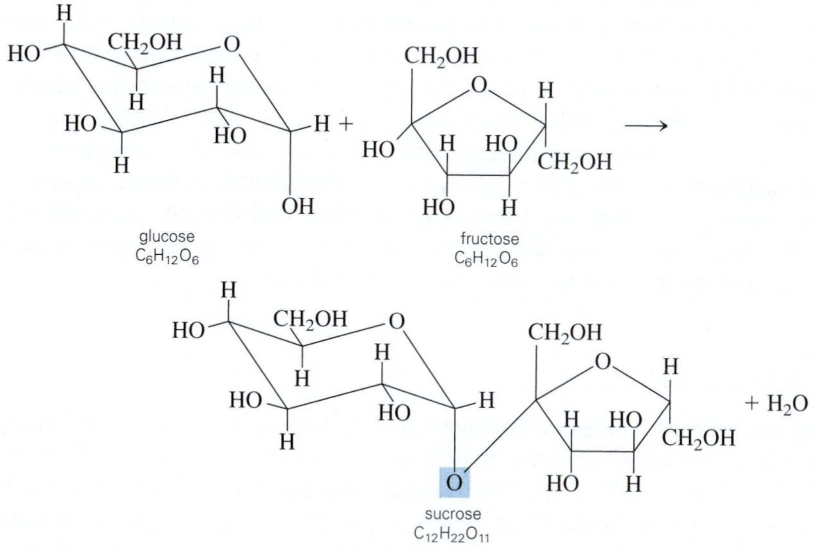

glucose $C_6H_{12}O_6$ + fructose $C_6H_{12}O_6$ →

sucrose $C_{12}H_{22}O_{11}$ + H_2O

Note that the formula of sucrose, $C_{12}H_{22}O_{11}$, is one water molecule short of the sum of the formulas of glucose and fructose, $C_6H_{12}O_6 + C_6H_{12}O_6$. Sucrose is refined from plant sources, both sugar beets and sugar cane.

Lactose makes up the majority of the carbohydrate in mammal milk. Human milk is about 7% by mass lactose and cow's milk is about 5% by mass lactose. The monosaccharides that bond to form lactose are galactose and glucose.

Galactose is an isomer of glucose, so both monosaccharides have the formula $C_6H_{12}O_6$, and after the disaccharide is formed and the water molecule is split out, lactose has the formula $C_{12}H_{22}O_{11}$:

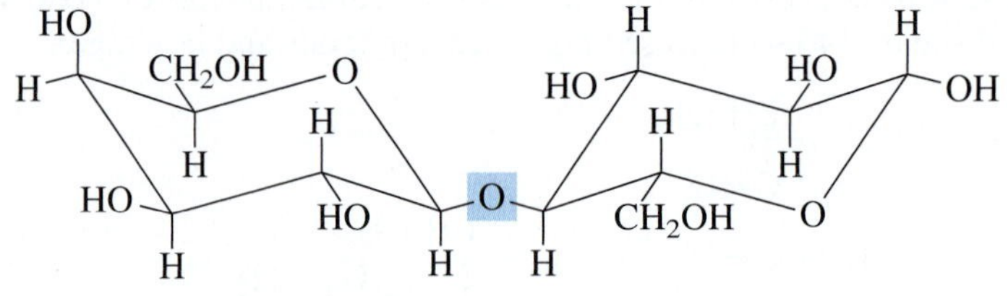

lactose

A third important disaccharide is **maltose**, which is found in malt, a substance that is made by allowing grain such as barley or wheat to sprout, or germinate, due to exposure to water and oxygen. Malt is most famously known as one of the ingredients of beer. Maltose is formed as a product of the condensation reaction of two glucose molecules, $C_6H_{12}O_6 + C_6H_{12}O_6 \rightarrow C_{12}H_{22}O_{11} + HOH$:

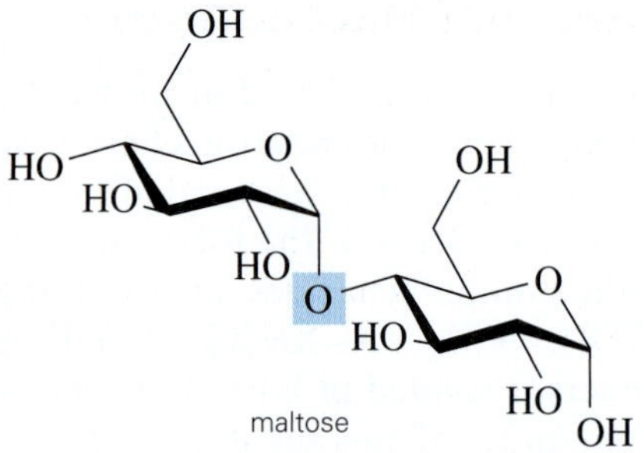

maltose

Ingested disaccharides are digested to monosaccharides before being absorbed in the small intestine. The digestion process involves an enzyme that specifically matches the disaccharide. For sucrose, the process is $C_6H_{11}O_5{-}O{-}C_6H_{11}O_5 + H_2O \xrightarrow{\text{Sucrase}} C_6H_{12}O_6 + C_6H_{12}O_6$. This is a hydrolysis reaction, which is the opposite of a condensation reaction. The enzyme that speeds up the digestion of sucrose is named sucrase. Similarly, the enzymes that catalyze the reactions of lactose and maltose are named lactase and maltase, respectively. You may have heard of lactase (**Figure 22.13**). It is available commercially as a food supplement for adults who have lost the ability to produce the enzyme.

You probably associate monosaccharides and disaccharides with the basic taste called **sweetness**. Sweetness in foods has an important role in human evolutionary history, where associating sweet taste with desirable food likely helped early humans to prefer energy-rich foods. The relative sweetness of the mono- and disaccharides introduced thus far in this section are given in **Table 22.2**.

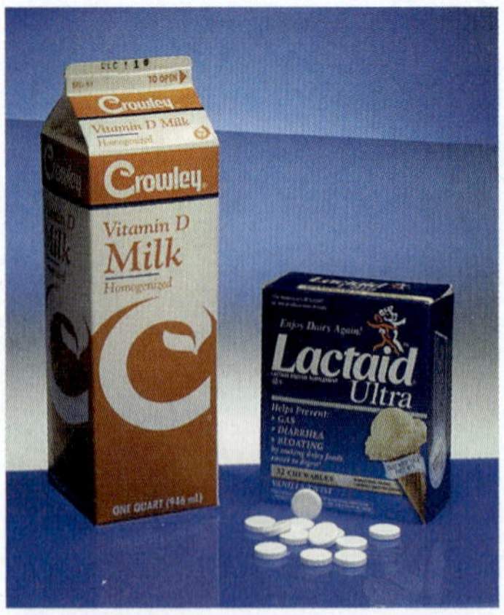

Charles D. Winters

Figure 22.13 Milk and Lactaid. It is estimated that about 65% of the world's adult population is lactose intolerant to some degree. It is most prevalent in people of East Asian descent. The active ingredient in the Lactaid dietary supplement in the photograph is the lactase enzyme.

Polysaccharides

A **polysaccharide** is a molecule that consists of a large number of bonded monosaccharide units, ranging between a minimum of about 10 monosaccharide units to a maximum of about 3000 units. The units are linked end-to-end to form long chains, and they can also form branches. Some polysaccharides are made up of only one type of monomer, and others consist of more than one type of monomer. Accordingly, polysaccharide formulas are variable, but many generally have a formula that has the form $C_x(H_2O)_y$.

Cellulose is an example of a **structural polysaccharide**, which means that its primary biological function is to provide a relatively rigid supporting framework for a plant or animal. Cellulose is the most abundant carbon-containing molecule on Earth. Cotton is about 90% cellulose. Paper is made by separating the cellulose in

Table 22.2 Relative Sweetness of Selected Monosaccharides and Disaccharides* (Relative to Sucrose = 100)

Carbohydrate	Relative Sweetness
Fructose	120–170
Sucrose	100
Glucose	74
Maltose	33
Lactose	16

*Sweetness is temperature dependent. The higher the temperature, the lower the relative sweetness.

wood from the other components, bleaching it (except for cardboard and paperboard), and drying it. A cellulose molecule is made up of about 300 to 10,000 glucose monomers bonded in a straight chain. The —OH groups on each glucose monomer can hydrogen bond to the oxygen atoms on a neighboring chain to form a rigid structure. Hydrogen bonding also occurs within chains. Note how the glucose monomers align in a straight chain with —OH groups extending outward to maximize hydrogen bonding:

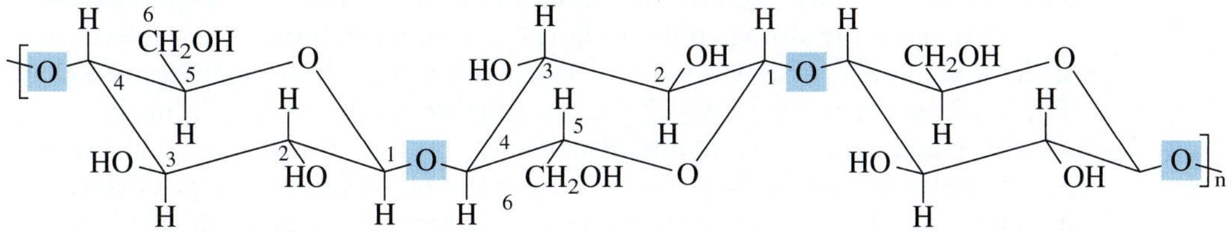

cellulose

Another category of polysaccharides are those whose biological purpose is food storage. They are called **storage polysaccharides**, and they serve as the storehouse of chemical energy. **Starch** is the storage polysaccharide for plants. It is made up of two polysaccharides: in general, 15% to 30% amylose and 85% to 70% amylopectin.

Amylose is a polysaccharide made of glucose monomers, typically 300 to 3000 units in a continuous chain. The chain typically does not stay straight, however. It has a secondary structure similar to that of a protein α-helix where it twists into a helical coil. Most commonly, there are six glucose monomers per turn. The chain has the structure:

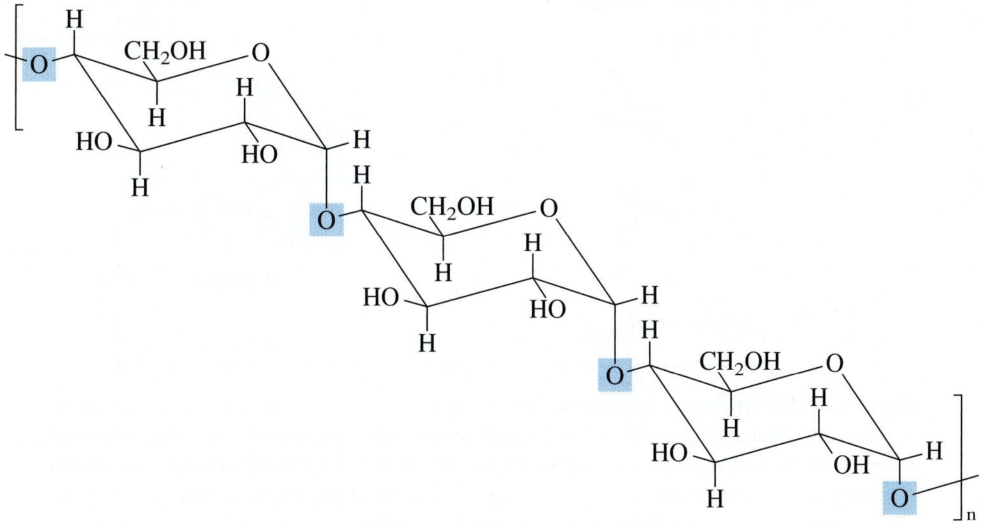

amylose

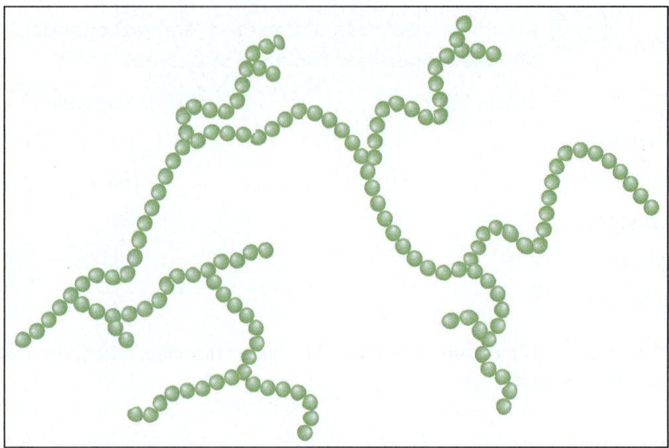

Figure 22.14 A molecular model of the plant starch amylopectin. Each sphere represents a glucose monomer. The monomers are linked in linear chains by one type of link and branches occur every 24-to-30 glucose units by a link of a different nature.

Compare the amylose structure with the cellulose structure. The monomer units are the same, but the nature of the links differ.

The other form of plant starch, **amylopectin**, is also formed from glucose monomers and also has the same links as amylose. However, amylopectin is a much larger molecule, typically with 2000 to 200,000 glucose units. Furthermore, amylopectin has numerous branches off the main chain. These branches typically occur every 24-to-30 glucose units, and branches can form off of the branches (**Figure 22.14**).

Glycogen is the storage polysaccharide for animals, fungi, and bacteria. It is structurally similar to amylopectin except that it is more branched (**Figure 22.15**). The branches occur every 8-to-12 glucose units, and a glycogen molecule has up to 120,000 glucose units. In animals, glycogen is stored in the liver and skeletal muscles. The glycogen in the liver is used to maintain blood glucose levels. Glycogen in muscles is used to satisfy the energy needs of the specific muscle in which it is stored.

Target Check 22.4

a) How many monomer units are in a molecule of each of the following: monosaccharide, disaccharide, oligosaccharide, and polysaccharide?

b) Each of the following carbohydrates are central to human biology. Explain the primary function of each: glucose, sucrose, and glycogen.

c) What is the commonality among cellulose, starch, and glycogen?

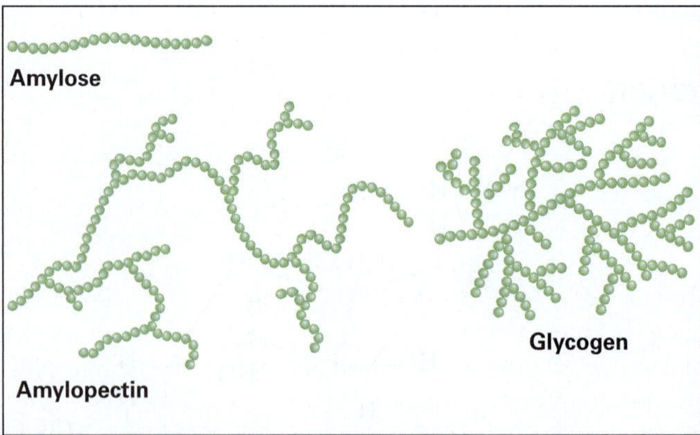

Figure 22.15 Amylose, amylopectin, and glycogen. Amylose is made up of glucose monomers in a straight chain arrangement, and hydrogen bonding within the chain causes the molecule to exist as a helix. Amylopectin forms when glucose monomers are linked in straight chains and branches occur every 24-to-30 units. Glycogen is made up of glucose monomers arranged as straight chains with branches that occur every 8-to-12 units.

22.5 Lipids

Goal 11 State the defining characteristics of four types of lipids: fats and oils, phospholipids, waxes, and steroids.

12 Identify the structural property that distinguishes fats from oils.

13 Describe the structure of a phospholipid in terms of the polarity of its head and tails.

14 Describe the general structure of most waxes.

15 Identify the structural feature common to all steroid molecules.

Lipids are molecules found in living organisms and are insoluble in water but soluble in nonpolar solvents. They have many biological functions including energy storage, serving as a structural component of cell membranes, and as a signaling molecule to allow cell-to-cell communication. We will present four categories of lipid structure: (1) fats and oils, (2) phospholipids, (3) waxes, and (4) steroids.

Fats and Oils

Most fats and oils are **triglycerides**. The term *triglyceride* is based on the roots *tri-*, indicating three fatty acids, and *-glyceride*, indicating the molecule **glycerol**:

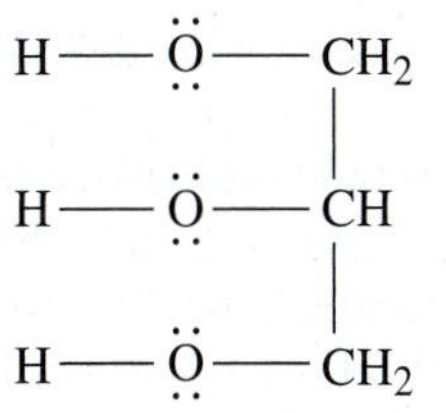

$$
\begin{array}{l}
H—\ddot{\underset{..}{O}}—CH_2 \\
\qquad\quad\;\; | \\
H—\ddot{\underset{..}{O}}—CH \\
\qquad\quad\;\; | \\
H—\ddot{\underset{..}{O}}—CH_2
\end{array}
$$

A **fatty acid** is an aliphatic (Section 21.6) carboxylic acid (Section 21.13) that may be either saturated—no C=C double bonds—with the general formula $CH_3(CH_2)_nCOOH$ or unsaturated with one or more C=C double bonds. Each –OH group on glycerol reacts with the –COOH group of a fatty acid: R–OH + HOOC–R′ → R–O–O–R′ + HOH. Thus, the resulting product is a triester, for example:

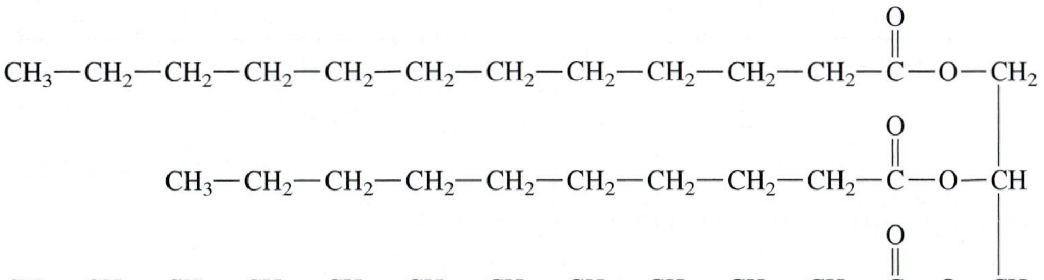

$$
\begin{array}{r}
CH_3—CH_2—CH_2—CH_2—CH_2—CH_2—CH_2—CH_2—CH_2—CH_2—CH_2—\overset{\overset{\displaystyle O}{\|}}{C}—O—CH_2 \\
CH_3—CH_2—CH_2—CH_2—CH_2—CH_2—CH_2—CH_2—CH_2—\overset{\overset{\displaystyle O}{\|}}{C}—O—CH \;\; \\
CH_3—CH_2—CH_2—CH_2—CH_2—CH_2—CH_2—CH_2—CH_2—CH_2—CH_2—\overset{\overset{\displaystyle O}{\|}}{C}—O—CH_2
\end{array}
$$

Classification of triglycerides as fats or oils is not rigidly defined. In general, a **fat** is a mixture of triglycerides with the fatty acid portions of the molecules mostly saturated and made up of relatively longer chains. An **oil** is a mixture of triglycerides with the fatty acid portions of the molecules mostly unsaturated and longer chain or saturated and short chain. When the fatty acid portion of a triglyceride is made up of saturated long chains, the molecule tends to exhibit stronger intermolecular forces, and the melting temperature is higher. Accordingly, fats are generally solids at room temperature. Shorter or unsaturated

Charles D. Winters

Figure 22.16 Fats and oils. Both are structurally triglycerides. In cooking, classification as a fat or oil usually is based on the state of the substance at room temperature. The liquids are therefore cooking oils, and the solid butter is a fat.

Table 22.3 Approximate Fatty Acid Composition of Common Fats and Oils

Fat or Oil	% Saturated (no double bonds)	% Monosaturated (1 double bond)	% Polyunsaturated (> 1 double bond)
Human	35	55	10
Butter	66	31	4
Margarine, soft	18	37	45
Margarine, stick	21	46	33
Coconut	92	6	2
Palm kernel	81	18	1
Palm	47	43	10
Peanut	18	48	34
Olive	14	77	9
Corn	13	25	62
Canola	6	58	36

fatty acid chains in a triglyceride lead to weaker intermolecular forces and lower melting temperatures. Thus, oils are usually liquids at room temperature (**Figure 22.16**).

High levels of triglycerides in the blood are correlated to narrowing of the arteries due to a buildup of plaque. This can lead to heart attacks, strokes, and kidney disease. **Table 22.3** lists the percentage of saturated fatty acids, monounsaturated fatty acids, and polyunsaturated fatty acids in some common dietary fats and oils. This table compares human body fat with butter and margarines and then with several vegetable oils.

Phospholipids

Phospholipids are a class of lipids that have a glycerol "backbone," two fatty acid residues, and a phosphate group:

polar head

$$
\begin{array}{rl}
 & CH_2-O-\overset{\overset{\displaystyle O}{\|}}{P}-O^- \\
 & \ \,|\qquad\qquad\ \ |\\
 & \ \,|\qquad\qquad\ \ O-CH_2-CH_2-\overset{+}{N}(CH_3)_3 \\
CH_3-(CH_2)_4-CH{=}CH-CH_2-CH{=}CH-(CH_2)_7-\overset{\overset{\displaystyle O}{\|}}{C}-O- & CH \\
 & \ \,| \\
CH_3-(CH_2)_{16}-\overset{\overset{\displaystyle O}{\|}}{C}-O- & CH_2
\end{array}
$$

two nonpolar tails

A phospholipid has a polar head and two long, nonpolar tails. Phospholipids are found in all animal and vegetable cells as part of cell membranes (**Figure 22.17**). The phospholipid shown here is commonly called a **lecithin**, which is a substance that will interact with both water and fat. They are used widely in the cosmetic and food industries as an *emulsifier*, a substance that holds two immiscible liquids together as a suspension. Lecithins are found in many food products, such as chocolate, ice cream, and margarine.

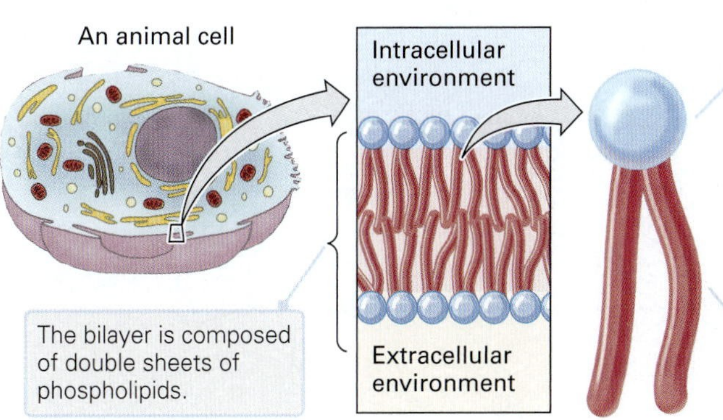
An animal cell
Intracellular environment
Extracellular environment
The bilayer is composed of double sheets of phospholipids.
The polar head of the phospholipid is attracted to water molecules. Fatty acid chains are in the boundary between outside and inside the cell.
The nonpolar fatty acid chains are attracted to each other and are oriented away from water molecules that are inside or outside the cell.

Figure 22.17 Phospholipids in cell membranes. The polar head is represented by a blue sphere, and the nonpolar tails are shown in red.

Waxes

The general structure for many **waxes** is that of an ester:

$$R{-}\overset{\overset{\displaystyle O}{\|}}{C}{-}O{-}R'$$

where R is an alkyl group attached to the carboxylic acid carbon and R′ is an alkyl group attached to the alcohol oxygen. The R groups are generally 10 or more carbon atoms long. Their relatively high molecular masses lead to waxes occurring in the solid state at room temperature. The long carbon chains also make them insoluble in water. Biologically, waxes most commonly serve to retain water, as on plant leaves, or to repel water, as on bird feathers.

Charles D. Winters

Figure 22.18 Beeswax is secreted by worker honeybees and used to build honeycomb cells.

Beeswax is produced by worker honeybees and used to build honeycomb cells (**Figure 22.18**). It is a mixture whose major components are monoesters, diesters (the molecule has two ester functional groups), hydrocarbons, and fatty acids. In the esters, the carboxylic acid R group tends to be about 15 carbons long, and the alcohol R′ group is typically about 30 carbons long. Beeswax is a solid no matter the weather conditions, with a melting point range of 62°C to 65°C.

Lanolin is secreted by glands in the skin of wool-bearing animals, such as sheep. Lanolin is a mixture that primarily consists of long-chain esters. The number of carbons in the chains varies from about 10 to 40. Lanolin has a melting point range of about 36°C to 44°C, depending on its purity. Lanolin is very effective at smoothing and moisturizing human skin, and thus it is valued in the cosmetic and health-care product industries.

Carnauba wax is secreted by the leaves of the Brazilian carnauba palm tree. It is a mixture of monoesters, diesters, carboxylic acids, and long-carbon-chain alcohols. The carbon chains on the esters vary from about 23 to 34 carbons in length. It has a melting point range of 82°C to 86°C, which is the highest of the common natural waxes. The carboxylic acids in the wax have hydroxyl groups that tie together adjacent carbon chains by hydrogen bonding. This makes carnauba wax relatively hard in comparison to other waxes. It is used as a paper coating to give a glossy look to paperback book covers, magazines, and some forms of print advertising. Carnauba wax used to be considered as among the best waxes to protect and shine car paint, but it is losing popularity as the quality of easier-to-apply synthetic polymer polishes improves.

Steroids

A **steroid** molecule has a core structural feature of 17 carbon atoms arranged in four fused rings: three six-carbon cyclohexane rings and one five-carbon cyclopentane ring. Different steroids have different atoms attached to the core. The major categories of steroids are illustrated in **Figure 22.19**.

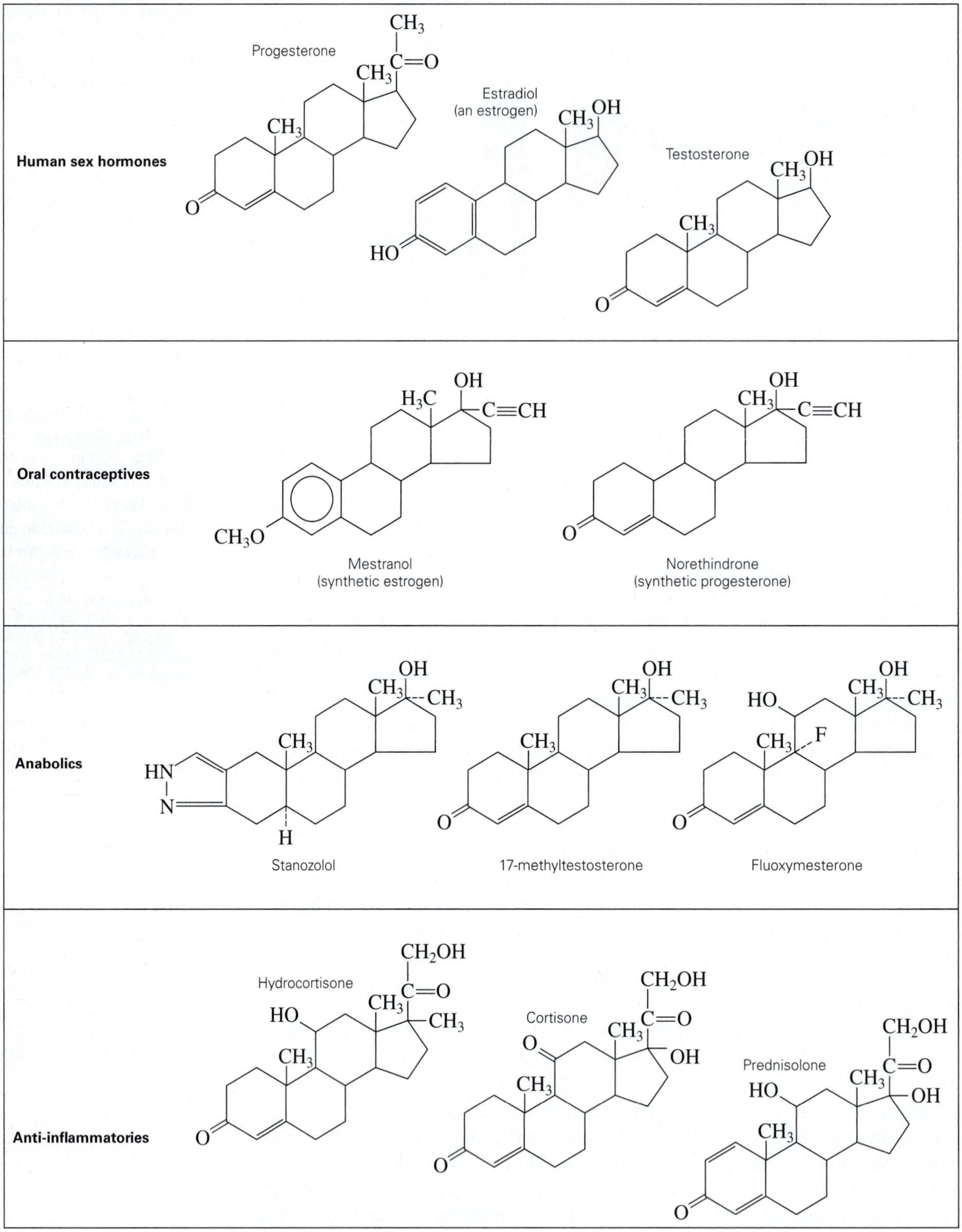

Human sex hormones

Progesterone

Estradiol (an estrogen)

Testosterone

Oral contraceptives

Mestranol (synthetic estrogen)

Norethindrone (synthetic progesterone)

Anabolics

Stanozolol

17-methyltestosterone

Fluoxymesterone

Anti-inflammatories

Hydrocortisone

Cortisone

Prednisolone

Figure 22.19 Steroids. In mammals, male and female sex hormones are steroids. Oral contraceptives are synthetic steroids. Anabolic steroids increase muscle mass. Adrenocorticoid hormones such as cortisone have amazing anti-inflammatory properties. Hundreds of steroid-based drugs are available by prescription in the United States.

The most abundant steroid in the human body is **cholesterol**:

HO

cholesterol

Cholesterol has two essential biological functions in humans. First, it is present in cell membranes where it regulates the fluidity of the membranes. Second, it is the starting-point reactant for the body's synthesis of other steroids. The current version of the U.S. Department of Health and Human Services and Department of Agriculture's *Dietary Guidelines for Americans* recommends minimization of dietary cholesterol because the body can synthesize all of the cholesterol it needs. Common dietary sources of cholesterol include eggs, dairy products, and meats. Plant foods contain negligible dietary cholesterol (**Figure 22.20**). Eating patterns that include high intake of dietary cholesterol appear to lead to increased risk of cardiovascular disease, but scientific study of nutrition is complex and continued research is needed.

Figure 22.20 Peanut butter nutrition facts. Peanuts are harvested from plants, and plants change the cholesterol they synthesize almost immediately to other molecules, resulting in a negligible quantity of cholesterol being within them at any given moment. Note that the Nutrition Facts label lists 0 mg of cholesterol. Current U. S. Food and Drug Administration law allows foods with less than 2 mg of cholesterol per serving to be labeled as "cholesterol free."

Target Check 22.5

a) Some margarine brands offer three different types of margarine: (1) a thick liquid in a squeeze bottle; (2) a soft solid in a tub; and (3) a harder solid sold in sticks. Describe the relative amounts of saturated and unsaturated fatty acids you would expect to find in these three different types of margarine.

b) Cocoa butter, obtained from chocolate, has, for a vegetable product, a relatively high melting point of 35°C. Does cocoa butter contain many or few saturated fatty acids? Explain.

c) What structural feature is characteristic of a steroid?

22.6 Nucleic Acids

Goal 16 Describe the biological roles of DNA and RNA.

17 Describe the three components of a nucleotide.

18 Draw Lewis diagrams of adenine, cytosine, guanine, thymine, and uracil.

19 Determine whether any two DNA or RNA nitrogen bases are complementary.

20 Describe the process by which a protein molecule is formed.

Nucleic acids were so named historically because they were discovered initially within the nuclei of cells and because they had the same phosphate groups as in phosphoric acid. There are two types of nucleic acids in all living systems. Today, the term **nucleic acid** refers to these two kinds of molecules. **Deoxyribonucleic acid (DNA)** stores genetic information and transmits that information to the next generation during cell division. **Ribonucleic acid (RNA)** assists in the process by serving as a messenger and as a switching engine to transfer the correct amino acid during protein synthesis. DNA molecules have the highest molar masses of any molecules in a living system, up to *several billion* g/mol. DNA is perhaps the most intensively studied molecule type since its structure was discovered about 70 years ago by James Watson and Francis Crick (**Figure 22.21**). RNA molecules have molar masses of about 30,000 g/mol.

Figure 22.21 James Watson (b. 1928) (*left*) and Francis Crick (1916–2004) (*right*). Watson and Crick shared the 1962 Nobel Prize with Maurice Wilkins for their discovery of the structure of DNA.

Nucleic acid monomers are called **nucleotides**. Each nucleotide has three parts: (1) a nitrogen-containing cyclic molecule called a *base* (the five bases are shown in **Figure 22.22**); (2) a sugar, either ribose in RNA or deoxyribose in DNA (**Figure 22.23**); and (3) one or more phosphate groups, attached to the hydroxyl groups of the sugar.*

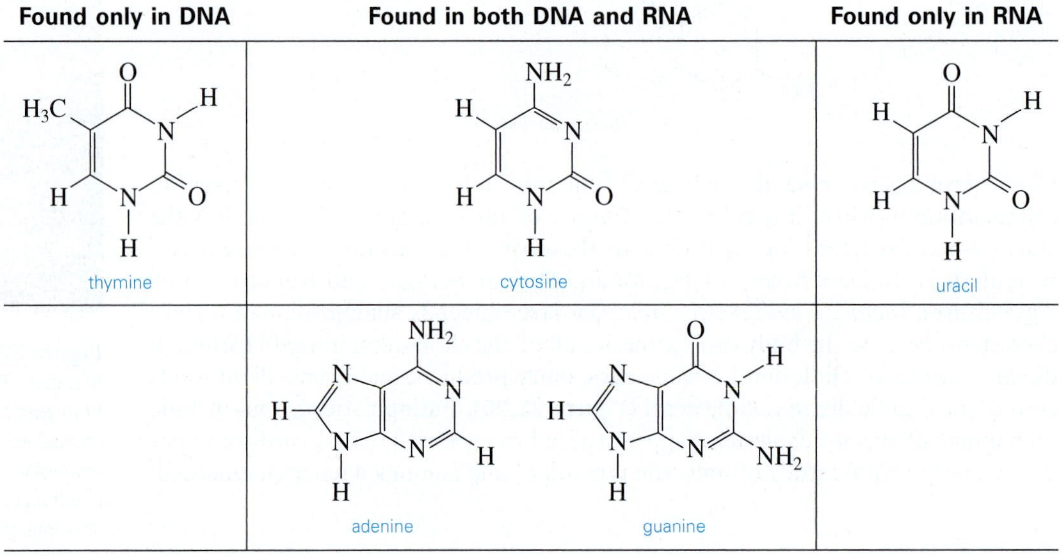

Found only in DNA
Found in both DNA and RNA
Found only in RNA
thymine
cytosine
uracil
adenine
guanine

Figure 22.22 The nitrogen-containing bases in DNA and RNA.

ribose
α-D-ribose
a
2-deoxyribose
β-D-2-deoxyribose
b

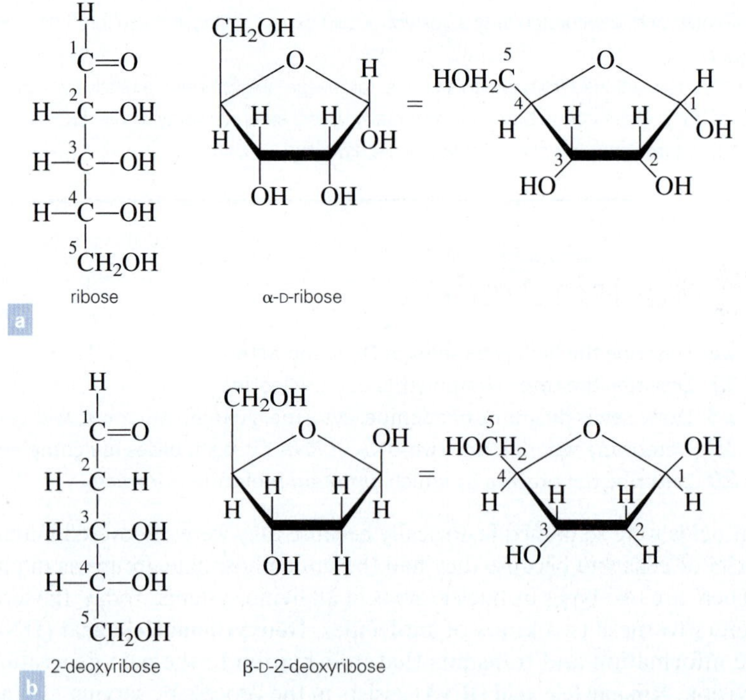

Figure 22.23 Ribose and deoxyribose. (a) Ribose is a five-carbon sugar with the formula $C_5H_{10}O_5$ and is a component of RNA. (b) Deoxyribose may be derived from ribose by the loss of an oxygen atom. Accordingly, its formula is $C_5H_{10}O_4$. It is a component of DNA.

---

*If no phosphate groups are attached to the sugar, the two-component base–sugar combination is called a *nucleoside*.

A DNA or RNA strand has a backbone that consists of alternating sugars and phosphate groups, as shown in **Figure 22.24**. The primary structure of DNA or RNA is the order of nucleotides joined by covalent bonds between carbon-5 in the sugar of one nucleotide and the phosphate group on carbon-3 of the sugar of another nucleotide. (See Figure 22.24 to identify the 5′ and the 3′ positions.) The secondary structure of DNA is determined by hydrogen bonds between base pairs on *different* DNA molecules.

Maximum hydrogen bonding occurs between thymine (T) on one DNA strand and adenine (A) on another and between cytosine (C) on one DNA strand and guanine (G) on another (**Figure 22.25**). This gives the famous double helix proposed

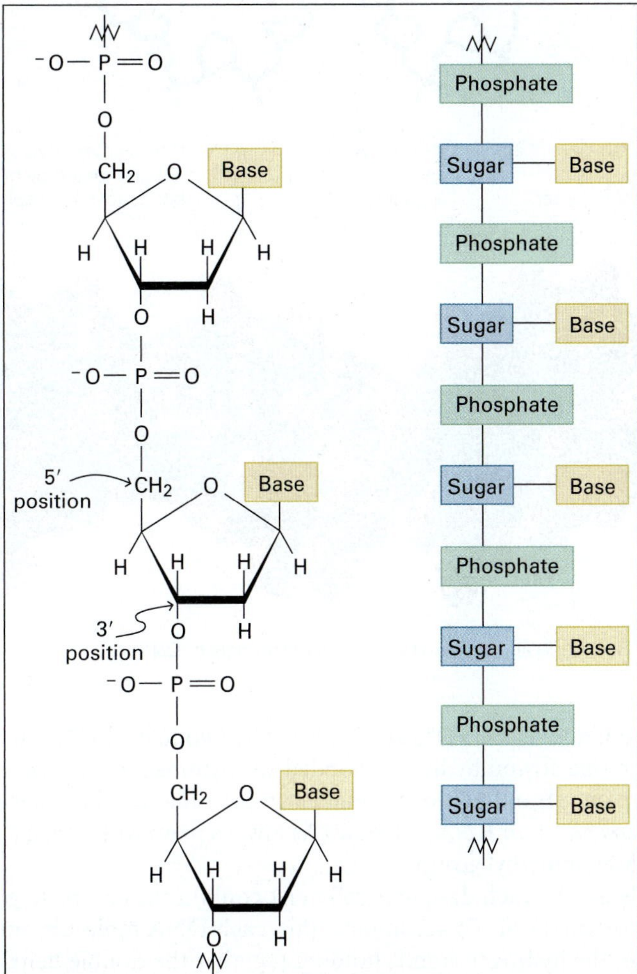

Figure 22.24 The left drawing shows the sugar–phosphate chain in DNA. The right drawing shows how DNA is a nucleotide polymer. Each nucleotide has three parts: phosphate–sugar–base.

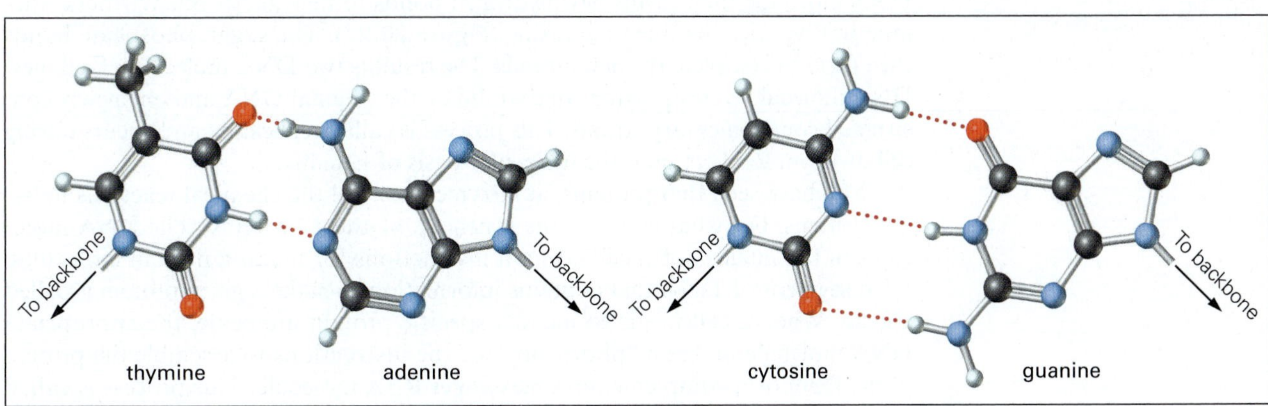

Figure 22.25 Hydrogen bonding between DNA base pairs. The DNA double helix is held together by thymine-adenine (T-A) and cytosine-guanine (C-G) hydrogen bonds.

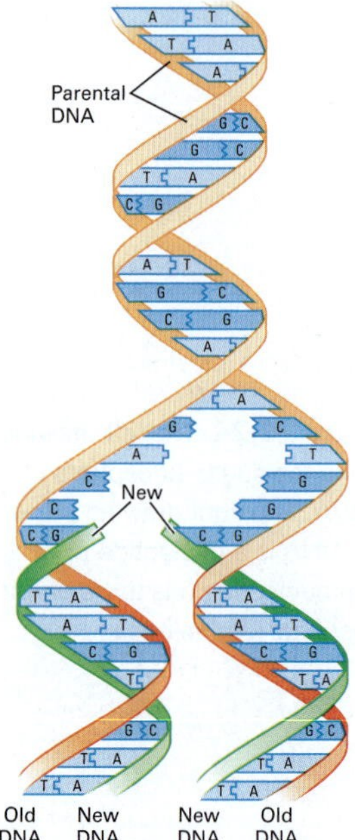
Parental DNA
New
Old DNA
New DNA
New DNA
Old DNA

Figure 22.27 DNA replication. As the DNA double helix (*orange*) unwinds, each strand serves as a template on which complementary subunits (*green*) are assembled in the cell. A, Adenine; T, thymine; G, guanine; C, cytosine.

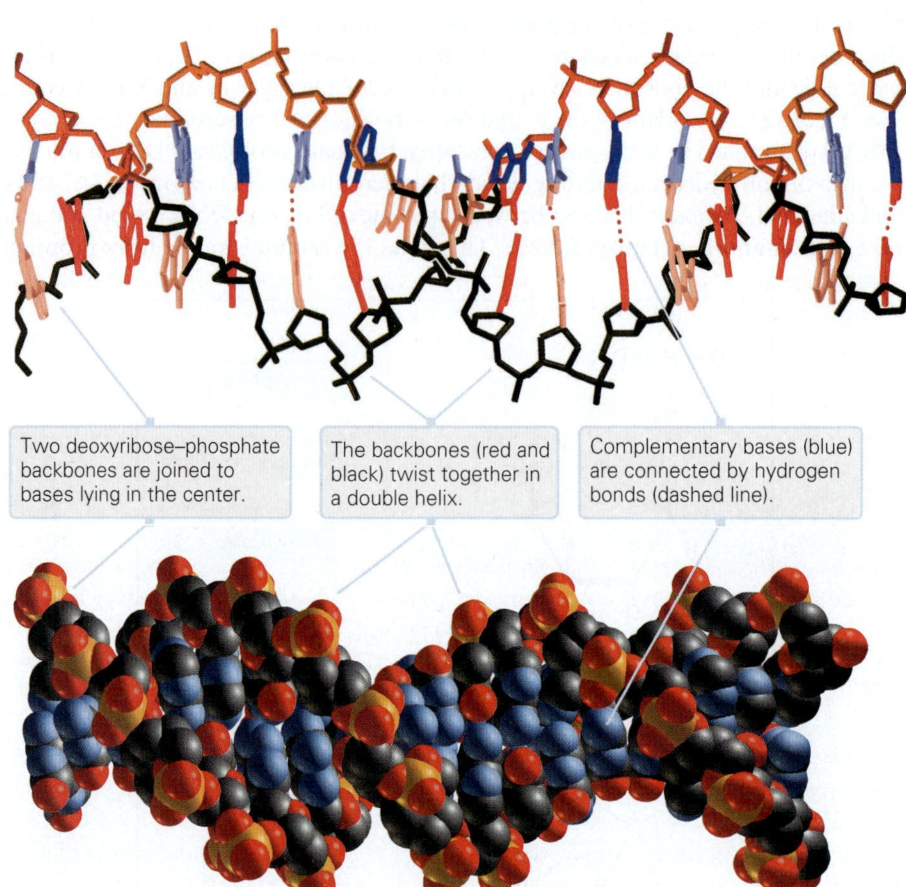
Two deoxyribose–phosphate backbones are joined to bases lying in the center.
The backbones (red and black) twist together in a double helix.
Complementary bases (blue) are connected by hydrogen bonds (dashed line).

Figure 22.26 Backbone and space-filling molecular models of DNA.

by Watson and Crick in 1953 (**Figure 22.26**). The two helical DNA strands—with all adenines on one strand hydrogen-bonded to thymines to the other strand and all guanines on one strand hydrogen-bonded to cytosines on the other strand—are termed *complementary.* In RNA, adenine (A) always pairs with uracil (U). Uracil is a thymine without a methyl group.

When cells divide, each daughter cell must contain the complete genetic information of the original cell. To accomplish this, each DNA molecule must duplicate itself. To do so, the hydrogen bonds holding together the double helix are broken, yielding two separate complementary single strands. ◀ Each strand of the original DNA molecule then forms new hydrogen bonds to new nucleotide partners, adenine to thymine, guanine to cytosine (**Figure 22.27**). The sugar–phosphate bonds then form to complete the new strands. The result is two DNA molecules: Each new DNA molecule is formed from one strand of the original DNA and one newly constructed complementary strand. This process is called **replication** and occurs during cell division. Replication is the molecular basis of heredity.

Hydrogen bonds are weaker than covalent bonds, and thus DNA can be unwound without damage.

You have seen that proteins, as enzymes, control the chemical reactions in living systems, but what controls the synthesis of these proteins? The DNA molecules in the nucleus of a cell contain instructions for making protein molecules. Each segment of DNA that contains information to make a given protein is called a **gene**. When a cell needs to make a specific protein molecule, the appropriate DNA molecule makes a "photocopy" of the instructions to assemble the protein in the form of a complementary **messenger RNA** molecule. This process is called **transcription**.

The messenger RNA leaves the cell nucleus and travels to a **ribosome**, which is a structure outside the nucleus where protein synthesis occurs. Ribosomes bind to the messenger RNA chain. Amino acids are carried to the ribosome by small molecules of **transfer RNA**. Each transfer RNA molecule carries with it a specific amino acid, and to maximize interchain hydrogen bonding, the transfer RNA bonds to a complementary site on the messenger RNA. As molecules of transfer RNA line up along the messenger RNA, adjacent amino acids carried by the transfer RNA form amide bonds with each other. As these amide bonds form, each amino acid is separated from its transfer RNA, and each transfer RNA is then separated from the larger messenger RNA. When the final transfer RNA is separated from the messenger RNA, a molecule of protein remains. This process is called **translation**.

Thinking About Your Thinking

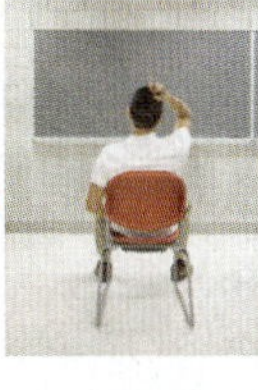

Mental Models

Figures 22.22 through 22.27 are designed to help you form a mental model of the DNA molecule. Each figure provides a different level of detail at which you need to understand the structure and function of DNA. The figures begin with "close-ups" (Figures 22.22 and 22.23) and then "zoom out" until you see the DNA molecule from a distance in Figure 22.26.

Figures 22.22 and 22.23 have you consider the bases and sugars in DNA at the most detailed level, showing the atoms and the bonds that connect them. Figure 22.24 then shows the sugar–phosphate backbone to which the DNA nitrogen bases are bonded. The backbone is shown as atoms and bonds in the left part of Figure 22.24, and then the fundamental elements of DNA structure are shown as a schematic drawing on the right. Compare the two representations carefully. Figure 22.25 emphasizes the geometry that allows hydrogen bonding between complementary base pairs.

In the top part of Figure 22.26, the sugar–phosphate backbone is in red, and the bases are shown in blue. The bottom part shows you a space-filling model. Here, your perspective is of a segment of the whole molecule.

Figure 22.27 is a schematic drawing similar to Figure 22.26 (*top*), but its purpose is to show you how the structure of DNA leads to the mechanism for its duplication.

Using different models to represent the same concept at different levels of detail is a common way of thinking in science. These models of DNA are an application of this way of thinking.

The diagram here is the central dogma of molecular biology, as expressed by Francis Crick in 1958. It still works, and it is a good summary for this section.

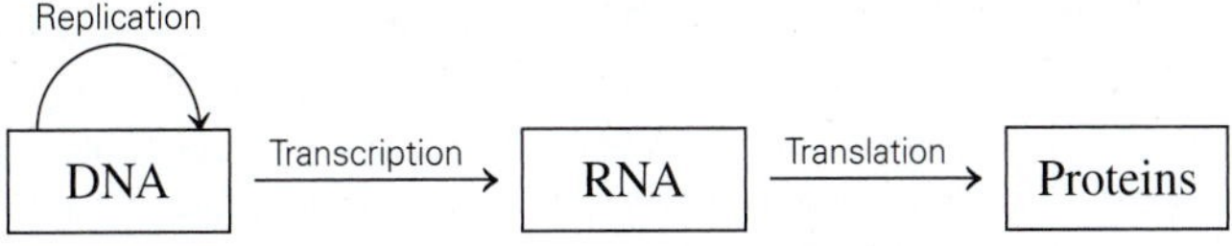

 **Target Check 22.6**

a) Describe the three components of a nucleotide.

b) How does the structure of an RNA nucleotide differ from that of a DNA nucleotide?

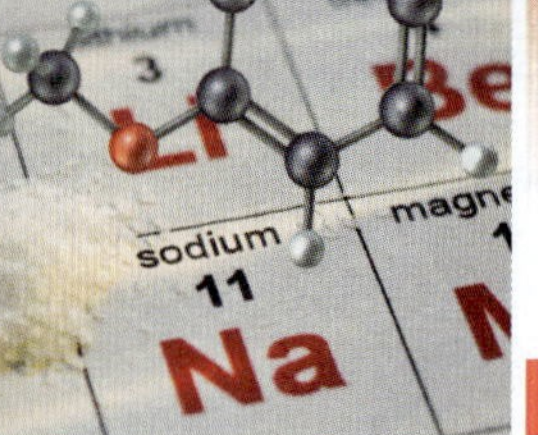

Everyday Chemistry 22.1

DESIGNER GENES

Since the earliest days of human civilization, it has been well known that children tend to resemble their parents. How are these traits passed from generation to generation? Scientist and friar Gregor Mendel (1822–1884) investigated this question with his now famous pea plant experiments. These experiments provided the beginnings of the field of study known as classical genetics. Mendel's "atom of inheritance," now called a *gene*, was the hypothetical construct he used to explain how physical characteristics were passed through families. Mendel learned that two genes were associated with each trait, one inherited from each parent. Mendel also found that not all genes are created equally. The trait associated with a dominant gene was expressed when paired with a recessive gene.

We now understand genes on the particulate level. A gene is a sequence of nucleotides along a DNA strand (Section 22.6) (**Figure 22.28**). The information in genes is coded by the four-letter DNA alphabet. A group of three letters specifies a certain amino acid. For example, the DNA sequence T-A-C codes for the amino acid tyrosine. Amino acids are assembled to form peptides and eventually proteins (Section 22.2). Each gene codes for one specific protein.

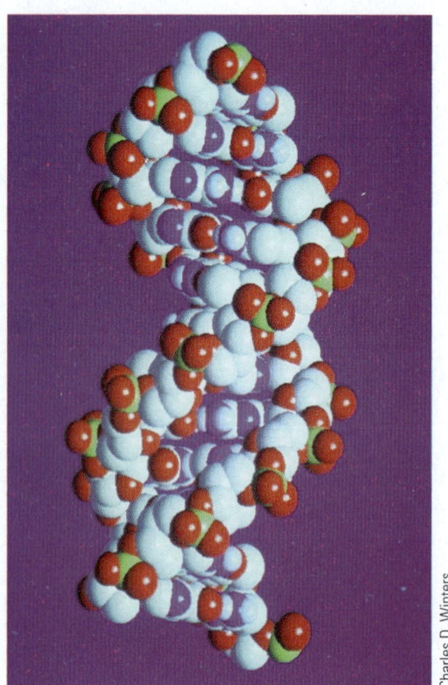

Charles D. Winters

Figure 22.28 A space-filling molecular model of a segment of DNA.

The genetic code is universal among all life on the earth. All living organisms, including the bacteria that live in your intestines, the plants that produce the air you breathe, your pet cat, and you, share the same genetic code.

The universal genetic code serves as a set of instructions that any cell can read. Take a gene coding for a certain protein out of a human cell and place it in a bacterial cell, and the bacterium faithfully produces that protein. This is the basis of the process known as genetic engineering. Of course, it is not a simple process to move genes from one cell to another, but our knowledge of how to accomplish this continues to grow; it is an active area of research around the world. Genetically engineered products are continually becoming more common. The insulin used by people with diabetes is produced by genetically engineered bacteria. You can probably find genetically engineered products at your local supermarket. Strawberries have been designed to be more resistant to frost, and tomatoes have been engineered to resist rot.

Human society has been practicing a form of genetic engineering for all of recorded history, but it has been called *selective breeding*. Cows are bred to produce more milk, dogs are bred to have aesthetically pleasing features, and crops are bred to have better yields. The difference between then and now is that today we are learning to control genes at the particulate level.

Manipulating bacterial cells is not a particularly controversial subject. Changing the genes in human cells is another matter (**Figure 22.29**). In 2003, scientists completed the first major stage of a $3 billion project with the goal of producing a complete map of all human genes. This investigation is known as the Human Genome Project. A genome is a description of all the genes in an organism. The genome sequence has been completed. Humans have approximately 22,000 protein-coding genes.

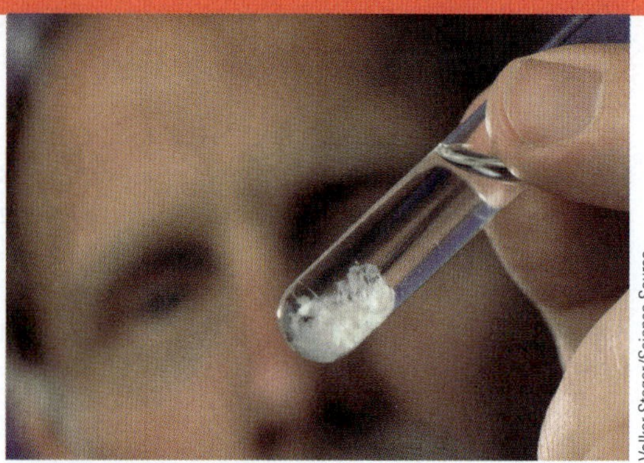

Volker Steger/Science Source

Figure 22.29 A scientist precipitates solid DNA from solution.

It will be a great day when the gene that codes for a genetic disease such as sickle cell anemia can be identified and altered to prevent the disease. But if we have the knowledge about how to change this gene, we will also have the ability to manipulate other genes. Along with an understanding of the human genome come serious ethical issues. If a genetic predisposition for depression is identified in a fetus, what steps, if any, should parents take? Should they be allowed to have their babies be genetically engineered?

Questions such as these about genetic engineering will be issues for society in your lifetime. We encourage you to learn all you can about the interplay of science, society, ethics, and morals, because decisions around such issues will have a profound effect on the future of the world.

Quick Quiz

1. **What is a gene? What role do genes serve for an organism?**
2. **What is genetic engineering?**

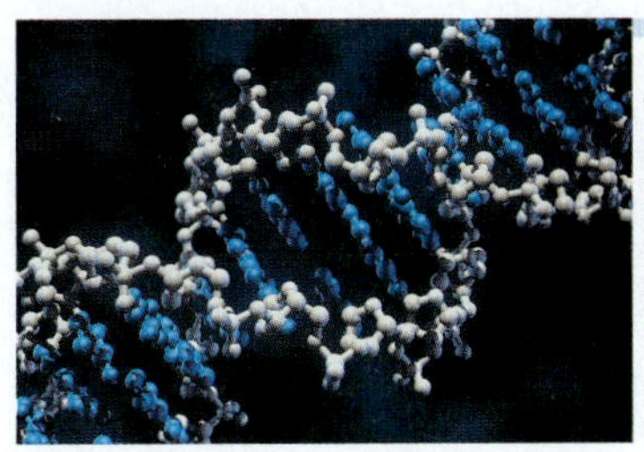
Kenneth Eward/BioGrafx/Science Source

CHAPTER 22 IN REVIEW: BIOCHEMISTRY

The **overarching goals** for this chapter are the following:

A. Identify, describe the distinguishing features of, and give an example of a molecule in each of the four major classes of biological molecules: proteins, carbohydrates, lipids, and nucleic acids.

B. Identify the monomers and describe how the polymeric molecules are assembled in the following: proteins, carbohydrates, and nucleic acids.

C. Describe how an enzyme functions as a biological catalyst.

D. Describe the process by which protein molecules are constructed from the information encoded in DNA molecules.

Goal 1 Given a table of Lewis diagrams of amino acids and their corresponding three-letter and one-letter abbreviations, draw the Lewis diagram of a polypeptide from its abbreviation.

The bond between amino acids is called a **peptide linkage**, which is formed when the hydroxyl part of the carboxyl group of an amino acid molecule reacts with a hydrogen of the $-NH_2$ group of another amino acid molecule to form a molecule of water.

Goal 2 Given a Lewis diagram of a polypeptide or information from which it may be written, identify the N-terminus amino acid and the C-terminus amino acid.

An **amino acid** is a molecule that contains both an amine group and a carboxylic acid group. The amino acid with the free carboxyl group at the end of a polypeptide chain is called the **C-terminus**; the amino acid with the free amine group is called the **N-terminus**.

Goal 3 Explain the meaning of the terms *primary*, *secondary*, *tertiary*, and *quaternary structure* as they apply to proteins.

The **primary structure** of a protein is its sequence of amino acids. The **secondary structure** of a protein is the local spatial layout of the amino acid backbones. The **tertiary structure** of a protein describes the spatial arrangement of the entire polypeptide chain. The **quaternary structure** of a protein describes how its polypeptide chains are arranged to form the protein molecule.

Goal 4 Describe how hydrogen bonding results in (a) α-helix and (b) β-pleated sheet secondary protein structures.

Secondary structures allow for maximum hydrogen bonding and the greatest stability. The **α-helix** structure is a continuous series of loops. The **β-pleated sheet** structure is like a curtain, but with sharp angles, like pleats.

Goal 5 Define the term *enzyme.*

An **enzyme** is a molecule that increases the rate of a chemical reaction without itself undergoing a permanent change.

Goal 6 Define the following terms as they apply to enzymes: *substrate*, *active site*, *inhibitor*.

A **substrate** is a molecule upon which an enzyme acts; it binds at the enzyme's **active site**. Enzyme **inhibitors** compete with the substrate for the active site.

Goal 7 Use the induced fit model to explain enzyme activity.

In the **induced fit model**, the shape of the substrate is a close, but not exact, match to the shape of the active site of the enzyme. As the substrate binds to the enzyme, either or both molecules change shape slightly.

Goal 8 Distinguish among monosaccharides, disaccharides, oligosaccharides, and polysaccharides.

A **monosaccharide** is a carbohydrate that cannot be changed into simpler compounds by reaction with water. A **disaccharide** is a two-monosaccharide-unit molecule, and an **oligosaccharide** is a three-to-approximately-nine-monosaccharide-unit molecule. A **polysaccharide** is a molecule that consists of a large number of bonded monosaccharide units, ranging between a minimum of about 10 monosaccharide units to a maximum of about 3000 units.

Goal 9 Given a Lewis diagram of a monosaccharide in open chain form, determine whether the sugar is an aldose or a ketose.

In monosaccharides, every carbon atom but one is bonded to a hydroxyl group. The remaining carbon atom is double-bonded to an oxygen atom, forming a carbonyl group. If the carbonyl group is at the end of the carbon chain, the monosaccharide is an **aldose**, derived from the aldehyde functional group. If the carbonyl group is within the chain, the monosaccharide is a **ketose**, derived from the ketone functional group.

Goal 10 Given the Lewis diagrams of two monosaccharides and a description of the bond linking the molecules, draw the Lewis diagram of the resulting disaccharide.

Two monosaccharides are combined by a reaction between two $-OH$ groups, resulting in a disaccharide with an $-O-$ bond between the monomers and a water molecule.

Goal 11 State the defining characteristics of four type of lipids: fats and oils, phospholipids, waxes, and steroids.	Most fats and oils are **triglycerides**. **Phospholipids** are a class of lipids that have an alcohol "backbone," two fatty acid residues, and a phosphate group. The general structure for many **waxes** is that of an ester with R groups that are generally 10 or more carbon atoms long. A **steroid** molecule has a core structural feature of 17 carbon atoms arranged in four fused rings: three six-carbon cyclohexane rings and one five-carbon cyclopentane ring.
Goal 12 Identify the structural property that distinguishes fats from oils.	In general, a **fat** is a mixture of triglycerides with the fatty acid portions of the molecules mostly saturated and made up of relatively longer chains. An **oil** is a mixture of triglycerides with the fatty acid portions of the molecules mostly unsaturated and longer chain or saturated and short chain.
Goal 13 Describe the structure of a phospholipid in terms of the polarity of its head and tails.	A phospholipid has a polar head and two long, nonpolar tails.
Goal 14 Describe the general structure of most waxes.	The general structure for many **waxes** is that of an ester with R groups that are generally 10 or more carbon atoms long.
Goal 15 Identify the structural feature common to all steroid molecules.	A **steroid** molecule has a core structural feature of 17 carbon atoms arranged in four fused rings: three six-carbon cyclohexane rings and one five-carbon cyclopentane ring.
Goal 16 Describe the biological roles of DNA and RNA.	**Deoxyribonucleic acid (DNA)** stores genetic information and transmits that information to the next generation during cell division. **Ribonucleic acid (RNA)** assists in the process by serving as a messenger and as a switching engine to transfer the correct amino acid during protein synthesis.
Goal 17 Describe the three components of a nucleotide.	Each nucleotide has three parts: (1) a nitrogen-containing cyclic molecule called a *base*; (2) a sugar, either ribose in RNA or deoxyribose in DNA; and (3) one or more phosphate groups, attached to the hydroxyl groups of the sugar.
Goal 18 Draw Lewis diagrams of adenine, cytosine, guanine, thymine, and uracil.	There are five nucleic acid nitrogen bases: thymine (T), cytosine (C), adenine (A), guanine (G), and uracil (U). They are illustrated in Figure 22.22.
Goal 19 Determine whether any two DNA or RNA nitrogen bases are complementary.	In DNA, adenine (A) and thymine (T) form a complementary pair; cytosine (C) and guanine (G) form the other complementary pair. In RNA, adenine (A) always pairs with uracil (U); uracil is a thymine without a methyl group.
Goal 20 Describe the process by which a protein molecule is formed.	In **translation**, messenger RNA is decoded and individual amino acids are brought by different transfer RNA molecules to be assembled into proteins.

Key Terms

Most of the key terms and concepts and many others appear in the Glossary. Use your Glossary regularly.

primary structure (protein) p. 878
protein p. 880
quaternary structure (protein) p. 884
random coil p. 882
replication p. 900
ribonucleic acid (RNA) p. 897
ribosome p. 901
secondary structure (protein) p. 881
simple sugars p. 887
starch p. 891
steroid p. 895
storage polysaccharide p. 891
structural polysaccharide p. 890
substrate p. 885
sucrose p. 889
sugars p. 887
sweetness p. 890
tertiary structure (protein) p. 883
transcription p. 900
transfer RNA p. 901
translation p. 901
triglyceride p. 893
tripeptide p. 880
waxes p. 895

Frequently Asked Questions

What are the commonalities among the classes of biological molecules?
This chapter is organized around the four major classes of biological molecules. Recognize what proteins, carbohydrates, and nucleic acids have in common: They are assembled from simple monomer units. Proteins are assembled from amino acids, carbohydrates are assembled from monosaccharides, and nucleic acids are assembled from nucleotides. *Lipid* is a catchall classification that includes fats, oils, phospholipids, waxes, steroids, and some other molecules. Organize your study into these four categories.

What are the key things to learn about proteins?
Amino acids are the monomers of proteins. Learn how amino acids combine to make peptide chains. Learn the differences among primary, secondary, tertiary, and quaternary protein structures. Enzymes are proteins that catalyze biochemical reactions. The induced fit model will help you understand how an enzyme works.

What are the key things to learn about carbohydrates?
Monosaccharides are carbohydrates, as are the complex carbohydrates that can be built from them. Take some time to understand the three-dimensional Lewis diagrams used to represent carbohydrates. Learn how monosaccharides react to form a disaccharide.

What are the key things to learn about nucleic acids?
Nucleotides are the monomers of nucleic acids. C and G (cytosine and guanine) are similar looking letters; they form a complementary base pair. A and T (adenine and thymine) are letters near the beginning and end of the alphabet; they form a complementary base pair. Uracil (U), near the end of the alphabet, substitutes for thymine, also near the end of the alphabet, in RNA. The summary at the end of Section 22.6 is a good place to start your study of how the information encoded in DNA is used to assemble proteins.

Concept-Linking Exercises

Write a brief description of the relationships among each of the following groups of terms or phrases. Answers to the Concept-Linking Exercises are given at the end of the chapter.

1. Primary, secondary, tertiary, quaternary protein structure
2. α-helix, β-pleated sheet, hydrogen bonding
3. Enzyme, enzyme substrate, active site, induced fit model
4. Monosaccharide, disaccharide, oligosaccharide, polysaccharide, carbohydrate
5. Fat, oil, phospholipid, glycerol, triglyceride
6. DNA, RNA, proteins, translation, transcription, replication

Small-Group Discussion Questions

Small-Group Discussion Questions are for group work, whether in class or under the guidance of a leader during a discussion section.

1. Does the amino acid glycine have isomers? If so, draw the structure of each. Does alanine have isomers? If so, draw the structure of each isomer. Draw the structure of all isomers (if applicable) of glycine and alanine using wedge-and-dash diagrams. Without drawing the isomers, determine whether or not the other 18 amino acids commonly found in proteins have isomers. Use structural diagrams to illustrate the reaction of alanine and glycine in all possible combinations to form a dipeptide.
2. What is an enzyme and how does it work? How many examples can you identify of products in your home that contain enzymes?
3. Expand the carbohydrate classification scheme illustrated in Figure 22.11 to include as many examples as possible in each category. Draw the structures and/or describe the characteristics of as many of the examples as you can.
4. Define each of the following and draw structural diagrams of two examples from each category: fat, oil, phospholipid, wax, steroid.

5. In their classic 1953 article published in *Nature* magazine that first described the structure of DNA, Watson and Crick stated, "It has not escaped our notice that the specific pairing we have postulated immediately suggests a possible copying mechanism for the genetic material." What pairing, what copying mechanism, and what genetic material were they writing of? Explain in as much detail as you can.

Questions, Exercises, and Problems

Interactive versions of these problems may be assigned by your instructor. Solutions for blue-numbered questions are at the end of the chapter. Questions other than those in the General Questions and More Challenging Problems sections are paired in consecutive odd-even number combinations; solutions for the odd-numbered questions are at the end of the chapter.

Section 22.2: Amino Acids and Proteins

1. Draw the Lewis diagram that illustrates the general structure of an amino acid.

2. How are amino acids subdivided by the nature of their R groups?

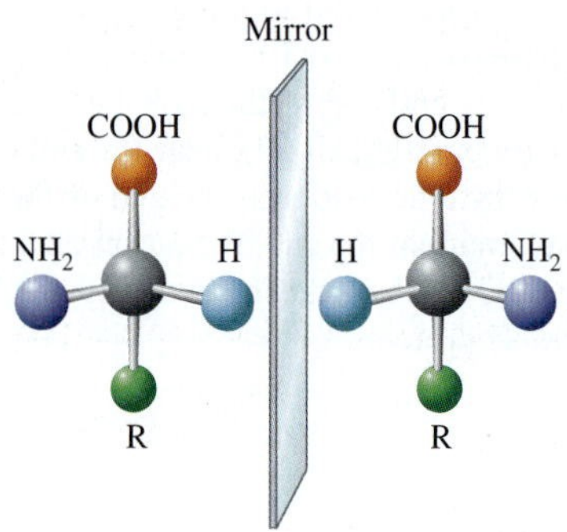

Nineteen of the 20 common amino acids occur in two forms: one the mirror image of the other.

Questions 3–8: You may use Table 22.1 if you wish.

3. Which of the 20 amino acids commonly found in proteins have an aromatic R group?

4. How does the R group in proline differ from the other R groups of amino acids commonly found in proteins?

5. Give in words the name of the tripeptides abbreviated as V-T-I and I-V-T. Give the name of the C-terminus amino acid in each tripeptide.

6. Give in words the name of the tetrapeptides abbreviated as S-F-G-Y and Y-F-G-S. Give the name of the N-terminus amino acid in each tetrapeptide.

7. Write the Lewis diagram of the tripeptide abbreviated as C-G-F.

8. Write the Lewis diagram of the tripeptide abbreviated as A-C-F.

9. How does tertiary protein structure differ from quaternary protein structure?

10. How does secondary protein structure differ from tertiary protein structure?

11. Describe in words the hydrogen bonding that occurs in a protein having an α-helix secondary structure.

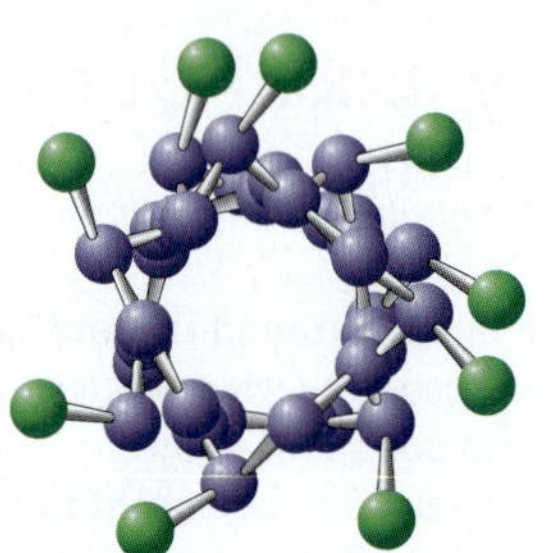

A top view of an α-helix, looking down the long axis, shows how the side chains (*green*) point away from the axis.

12. Describe in words the hydrogen bonding that occurs in a protein having a β-pleated sheet secondary structure between adjacent polypeptide chains.

Section 22.3: Enzymes

13. To what class of biological macromolecules do enzymes belong?

Thomas M Perkins/Shutterstock.com

Enzymes in baker's yeast catalyze the reaction of sugars in the bread dough to form the carbon dioxide gas that causes bread to rise.

14. Chemists have isolated, purified, studied, and reported on thousands of different enzymes. Why does a living system need so many enzymes?

15. What is an enzyme substrate?

16. What is enzyme specificity?

17. How does an enzyme affect the activation energy (Section 18.4) of a reaction?

18. How does an enzyme differ from an inorganic laboratory catalyst such as MnO_2 in the following reaction?

$$2\ KClO_3(s) \xrightarrow{MnO_2,\ heat} 2\ KCl(s) + 3\ O_2(g)$$

19. Many enzymes work best at temperatures near normal human body temperature, 37°C. What happens to the rates of enzyme-catalyzed reactions when you run a fever?

20. Many enzymes undergo *feedback inhibition*, in which a high concentration of enzyme reaction product slows down or stops enzyme activity. Compare the action of a furnace thermostat to enzyme feedback inhibition.

Section 22.4: Carbohydrates

21. Give an example of an aldose sugar.

22. Give an example of a ketose sugar.

23. Examine the Lewis structures of α-glucose and β-glucose. What is the difference between these two glucose isomers?

24. What are the four general classes of saccharides? How do they differ?

25. To which saccharide class do the following belong: sucrose, glycogen, and fructose?

26. To which saccharide class do the following belong: cellulose, ribose, and lactose?

27. Name the simple sugars in lactose.

28. Name the simple sugars in sucrose.

29. Benedict's test is a classic test for some sugars. In this test, an aldehyde is oxidized to a carboxylic acid, and a color change occurs. Which mono- and disaccharide that you have studied would give a positive Benedict's test?

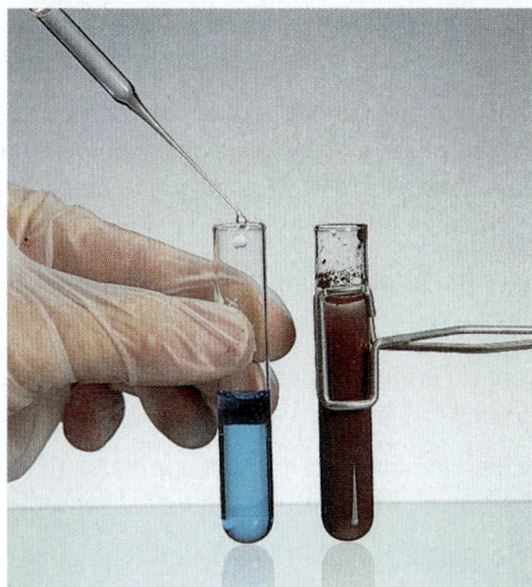

In a Benedict's test, blue Benedict's solution turns red in the presence of the aldehyde groups of sugar molecules.

30. A Tollens' test oxidizes aldehydes to carboxylic acids but does not react with ketones. Which monosaccharides and disaccharides discussed in this chapter would give a negative Tollens' test?

31. What are the monosaccharides that react to form the disaccharide maltose?

32. Cellobiose is a plant disaccharide made from two glucose molecules. What is its chemical formula?

33. What are the similarities and differences in the function of amylopectin and glycogen? What is the difference in their molecular structure?

34. What is the function of the polysaccharide glycogen? Where is it stored in animal bodies?

Section 22.5: Lipids

35. What physical property do the four classes of lipids share?

36. What are the four classes of lipids?

37. Are oils usually obtained from animal sources or plant sources?

38. What physical property differentiates fats and oils?

39. Use the letters A, B, and C to stand for three different fatty acids. Draw out all possible triglycerides you can make from these acids and one glycerol molecule.

40. Explain why the human body contains only fatty acids with an even number of carbon atoms. (*Hint:* Fatty acids are built from acetic acid molecules.)

41. Use the letters A and B to stand for two different fatty acids in a phospholipid. Draw out all possible phospholipids you can make from these two fatty acids.

42. R in beeswax (the group connected to the carbon) may be $C_{25}H_{51}-$, $C_{26}H_{53}-$, or $C_{27}H_{55}-$. R′ in beeswax (the group connected to the oxygen) may be $C_{30}H_{61}-$, $C_{31}H_{63}-$, or $C_{32}H_{65}-$. Draw the condensed structure of the nine possible molecules that are classified as beeswax.

Section 22.6: Nucleic Acids

43. Give a brief written explanation of the function of DNA and of RNA.

44. Name the two types of nucleic acid polymers present in cells.

45. Draw the Lewis diagram for uracil.

46. Draw the Lewis diagram for thymine.

47. Draw the Lewis diagrams for adenine and thymine.

48. Draw the Lewis diagrams for guanine and cytosine.

49. Draw the Lewis diagram for the sugar ribose. How does this sugar differ from the sugar deoxyribose?

50. Draw the Lewis diagram for the sugar deoxyribose.

51. Draw the nucleoside adenosine, which is an adenine–ribose combination.

52. Draw the nucleoside deoxyadenosine, which is an adenine–deoxyribose combination.

53. Draw the Lewis diagram for the DNA fragment that is *complementary* to the guanine-thymine-adenine DNA fragment.

54. Single-letter codes are often used to stand for the base in a nucleic acid. Draw the Lewis diagram for a DNA fragment having the bases guanine-thymine-adenine, abbreviated as G-T-A.

55. Although RNA is single-stranded, the strand sometimes folds back on itself to give a complementary portion.

What would be the complementary portion of the RNA fragment having the bases uracil-cytosine-guanine?

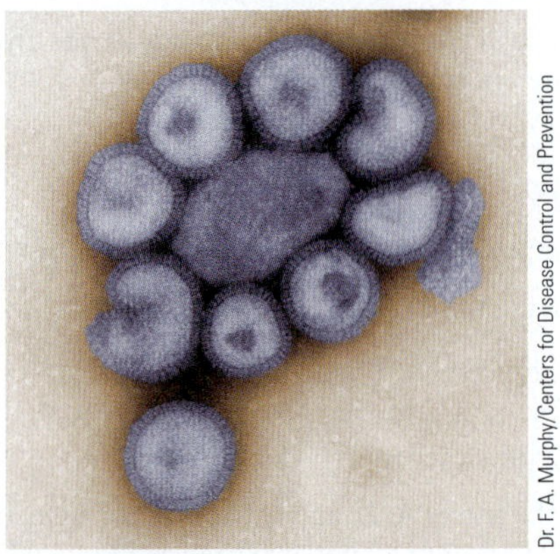

A virus is a particle that contains DNA or RNA that is surrounded by a coat of protein that can replicate only in a host cell. The swine influenza virus shown in this electron microscope image is an RNA virus.

56. Draw the Lewis diagram for an RNA fragment having the bases uracil-cytosine-guanine, abbreviated as U-C-G.

57. Describe in words the role in protein synthesis of transfer RNA.

58. Describe in words the role in protein synthesis of messenger RNA.

General Questions

59. Distinguish precisely, and in scientific terms, the differences between items in each of the following pairs.

a) Secondary protein structure, tertiary protein structure
b) α-helix, β-pleated sheet
c) Reversible enzyme inhibitors, irreversible enzyme inhibitors
d) α-glucose, β-glucose
e) Fats, oils
f) Saturated fatty acids, unsaturated fatty acids
g) Triglycerides, phospholipids
h) Triglycerides, waxes
i) Waxes, steroids
j) Nucleoside, nucleotide
k) Thymine, uracil
l) α-helix, double helix

60. Classify each of the following statements as true or false:

a) The most common primary protein structures are the α-helix and the β-pleated sheet.
b) Sulfur–sulfur bonds called disulfide linkages are important in the tertiary protein structure of many proteins.
c) An enzyme substrate is the product of an enzyme-catalyzed reaction.
d) The catalytic properties of an enzyme exist at its active site.
e) Carbohydrates are hydrates of carbon—that is, compounds that consist of carbon and water molecules.
f) Glucose is also called blood sugar.
g) Sucrose is a glucose–fructose combination.
h) Fats and oils are distinguished by their state of matter at room temperature.
i) Steroids have three cyclopentane rings and one cyclohexane ring fused.
j) Nucleic acid monomers are called nucleotides.
k) DNA is a nucleotide (phosphate–sugar–base) polymer.

61. Biochemists use 120 g/mol as the "average molar mass" of a single amino acid in a protein. Hemoglobin is a small protein that transports oxygen from the lungs to the capillaries. The molar mass of hemoglobin is about 64,500 g/mol. Approximately how many amino acids make up hemoglobin?

62. The tobacco mosaic virus is a small virus that has been crystallized in pure form. One complete virus has a molar mass of about 40 million g/mol, of which 38 million grams is protein. Use the data in Question 61 to determine the approximate number of amino acids in the proteins of a tobacco mosaic virus.

As its name implies, the tobacco mosaic virus infects the tobacco plant, causing a visible mosaic pattern on its leaves. It was the first virus for which the structure was known in detail.

63. Which of the following biological molecules are polymers: cellulose, proteins, DNA, starch, and RNA?

64. What are the monomer units for the polymers listed in Question 63?

65. With what element does protein supply us that carbohydrates or fats or oils do not?

66. What element is found in DNA and RNA but not in proteins?

More Challenging Problems

67. The structure of nylon 66 is given in Section 21.17. Use this structure to explain why nylon fabric is used in waterproof windbreakers, whereas cotton, a cellulose material, is used when moisture must be absorbed.

68. Rayon is an ester made by treating cotton with acetic acid. Use the cellulose structure given in Section 22.4 to draw a Lewis structure for a short rayon "molecule." (*Hint:* An alcohol–acid esterification reaction occurs randomly.)

69. The base content of a sample of pure mammalian DNA was analyzed to be 21% guanine. In this DNA sample, what is the percentage of cytosine? What is the percentage of adenine? What is the percentage of thymine? Explain.

70. RNA, unlike DNA, is a single-stranded macromolecule. Would you expect the percentage guanine in RNA to equal the percentage cytosine? Why or why not?

Answers to Target Checks

1. A-V-L, A-L-V, V-A-L, V-L-A, L-V-A, L-A-V.
2. (a) The local structure of the polypeptide chain that results from hydrogen bonding interactions. (b) The three-dimensional shape of a protein that results from interactions between the R groups in the protein chain. (c) The number of chains in a protein and how these chains are arranged in relation to one another, resulting from attractive intermolecular forces.
3. (a) An enzyme is a molecule that increases a reaction rate without itself undergoing permanent change. (b) A substrate is the molecule that is acted upon by an enzyme; the reactant in an enzyme-catalyzed reaction. (c) The substrate and enzyme are induced, or influenced, to change shape as the substrate begins to bind.
4. (a) Monosaccharide = 1; disaccharide = 2; oligosaccharide = 3 to 9; polysaccharide = 10 and up. (b) Glucose is an energy molecule found in blood. Sucrose is table sugar and an energy source in food. Glycogen is the energy storage polysaccharide. (c) Cellulose, starch, and glycogen share a common monomer, glucose.
5. (a) The squeezable liquid margarine is the lowest of the three in saturated (higher melting point) fatty acids and highest in unsaturated (lower melting point) fatty acids. The soft solid has more saturated fatty acids than the liquid and fewer unsaturated fatty acids than the liquid. The stick margarine is highest in saturated fatty acids and lowest in unsaturated fatty acids. It most resembles butter, also sold in sticks. (b) Judging by its relatively high melting point, cocoa butter contains many saturated fatty acids. (c) Steroids are characterized by four fused rings, as shown here: 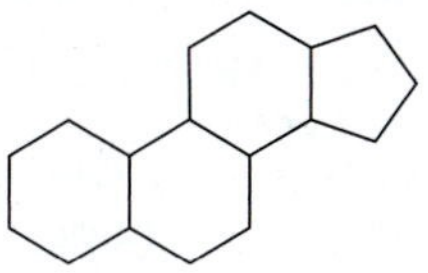
6. (a) A nucleotide is composed of: (1) a nitrogen-containing base, (2) a sugar, and (3) phosphate groups. (b) RNA uses the sugar ribose and the bases adenine, cytosine, guanine, and uracil. DNA uses the sugar deoxyribose and the bases adenine, cytosine, guanine, and thymine.

Answers to Concept-Linking Exercises

You may have found more relationships or relationships other than the ones given in these answers.

1. Primary protein structure is the linear amino acid residue sequence. Secondary structure is the regular local conformation maintained by hydrogen bonding. Tertiary structure describes the three-dimensional arrangement of the fully folded polypeptide chain. Quaternary structure describes how two or more polypeptide chains combine to make a protein.
2. The two most common secondary protein structures are the α-helix and the β-pleated sheet. Both structures reflect a maximum amount of hydrogen bonding. The α-helix secondary structure is formed by hydrogen bonds within a chain; the β-pleated sheet secondary structure is formed by hydrogen bonds between adjacent chains in different molecules or when a single chain folds back on itself.
3. Enzymes are biochemical catalysts; they speed up the rate of a biochemical reaction. The enzyme substrate is the reactant in the enzyme-catalyzed reaction. The active site is that part of the enzyme to which the substrate binds during the reaction. The induced fit model is used to explain enzyme activity. The shapes of the substrate and enzyme are modified as the molecules bind and induce a fit.
4. Monosaccharides, disaccharides, oligosaccharides, and polysaccharides are carbohydrates: aldehydes or ketones with two or more —OH groups. A monosaccharide is the simplest carbohydrate, one that cannot be broken down to simpler carbohydrates. A disaccharide is a chemical combination of two monosaccharides; an oligosaccharide is a three-to-approximately-nine-monosaccharide-unit molecule; a polysaccharide is a combination of many monosaccharides.
5. Fats and oils have a glycerol "backbone" residue bonded to two fatty acid residues, forming a type of molecule known as a triglyceride. If a macroscopic sample of a triglyceride is solid at room temperature, it is classified as a fat; if a sample is liquid at room temperature, it is classified as an oil. A phospholipid is similar to a fat or oil, but its components are an alcohol backbone, fatty acid residues, and a phosphate group.
6. Replication is the process by which two new DNA molecules form by building a new complementary strand on each strand of the original molecule. Transcription is the process by which a segment of a DNA molecule is copied in the form of a complementary messenger RNA molecule. The messenger RNA molecule then bonds to complementary transfer RNA molecules, each with a specific amino acid. This process is called *translation*. The amino acids then bond to form a protein.

Answers to Blue-Numbered Questions, Exercises, and Problems

1.
```
      R  :O:
  ..  |   ||  ..
H—N—C—C—O—H
   |   |      ..
   H   H
```

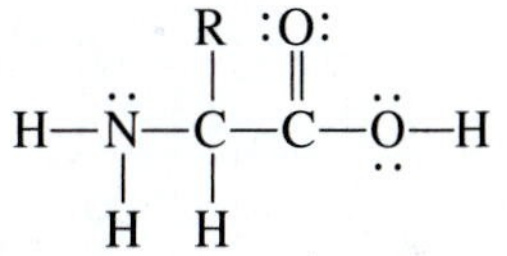

3. Phenylalanine, tyrosine, and tryptophan

5. V-T-I is valylthreonylisoleucine. I-V-T is isoleucylvalylthreonine. The C terminal acid in V-T-I is isoleucine, in I-V-T, threonine.

7.
```
                                  C6H5 (benzene ring)
                                   |
     SH                            |
     |                             |
     CH2  O       H   O            CH2  O
     |    ||      |   ||           |    ||
NH2—CH—C—NH—CH—C—NH—CH—C—OH
```

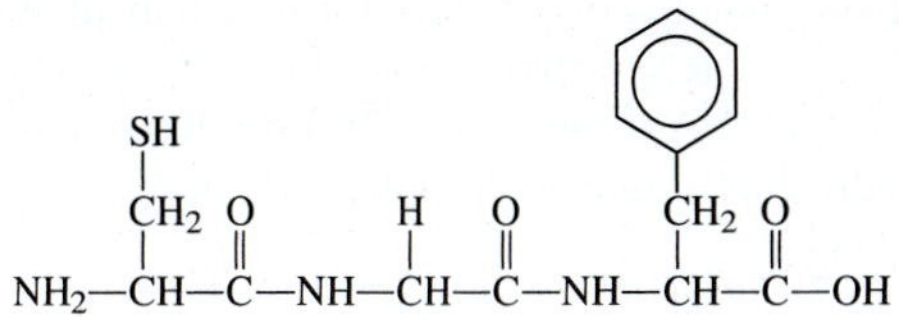

9. Tertiary protein structure describes the overall three-dimensional shape of a polypeptide chain caused by the folding of various regions. Quaternary protein structure describes how multiple polypeptide chains are arranged in relation to one another.

11. The α-helix secondary structure involves hydrogen bonding between the hydrogen attached to the peptide link nitrogen and a peptide link oxygen of an amino acid farther down the *same* protein chain.

13. Enzymes are usually proteins.

15. An enzyme substrate is a reactant that the enzyme helps change to product in the enzyme-catalyzed reaction.

17. They lower the activation energy of a reaction.

19. When you run a fever, enzyme-catalyzed reactions run faster than at normal body temperature. This may help the body fight off illness more quickly.

21. Examples of aldose sugars are glucose, ribose, and deoxyribose. Aldose usually refers only to monosaccharides.

23. The two glucose isomers differ only at carbon-1. On each structure, find carbon-1, which is the only carbon bonded to two oxygen atoms. In α-glucose, the OH attached to carbon-1 is vertical, either pointed down or up. In β-glucose, the OH attached to carbon-1 is horizontal, or nearly so.

25. Sucrose is a disaccharide, glycogen is a polysaccharide, and fructose is a monosaccharide.

27. Galactose and glucose.

29. Glucose, ribose, deoxyribose, and lactose would give a positive Benedict's test because all these sugars can have an aldehyde group in the open-chain form.

31. Maltose is formed as a product of the condensation reaction of two glucose molecules.

33. Amylopectin is one of the forms of plant starch. It is a storage polysaccharide for plants. Glycogen is the storage polysaccharide for animals, fungi, and bacteria. Glycogen is structurally similar to amylopectin except that it is more branched.

35. Lipids are molecules found in living organisms and are insoluble in water but soluble in nonpolar solvents.

37. Plant sources.

39.

A—O—CH_2	A—O—CH_2	B—O—CH_2	B—O—CH_2
B—O—CH	C—O—CH	A—O—CH	C—O—CH
C—O—CH_2	B—O—CH_2	C—O—CH_2	A—O—CH_2

41.

CH_2—O—P(=O)(—O^-)—O—X^+	CH_2—O—P(=O)(—O^-)—O—X^+
A—O—CH	B—O—CH
B—O—CH_2	A—O—CH_2

43. DNA is the storehouse of genetic information in all life forms. Messenger RNA carries instructions for protein synthesis from DNA to ribosomes. Transfer RNA delivers specific individual amino acids to the ribosome.

45. See Figure 22.22.

47. See Figure 22.22.

49. See the Lewis diagrams in Section 22.6. Ribose has a hydroxyl group at carbon 2; deoxyribose does not.

51.

53. C–A–T:

55. A–G–C.

57. Transfer RNA picks up an amino acid molecule and carries it to a protein being synthesized by a ribosome.

60. True: b, d, f, g, h, j, k. False: a, c, e, i.

61. $\frac{64{,}500 \text{ g}}{\text{mole hemoglobin}} \times \frac{1 \text{ mole amino acid}}{120 \text{ g}} = 538 \frac{\text{amino acids}}{\text{hemoglobin}}$

63. All are polymers.

64. Glucose is the monomer in cellulose; amino acids, in proteins; nucleotides (adenine, cytosine, guanine, thymine), in DNA; glucose, in starch; and nucleotides (adenine, cytosine, guanine, uracil), in RNA.

65. Nitrogen is found in all proteins; if you picked sulfur (from cysteine or methionine), that is also true.

66. Phosphorus

67. The coil of the α-helix and the strong hydrogen bonding within the protein chains keep water from soaking into the nylon; there is no further hydrogen bonding to be made by the water to the nylon. In cotton, however, there is little hydrogen bonding between the cellulose chains, so water molecules can form hydrogen bonds with (and therefore soak into) the hydroxyl groups on the sugars that make up cellulose.

68.

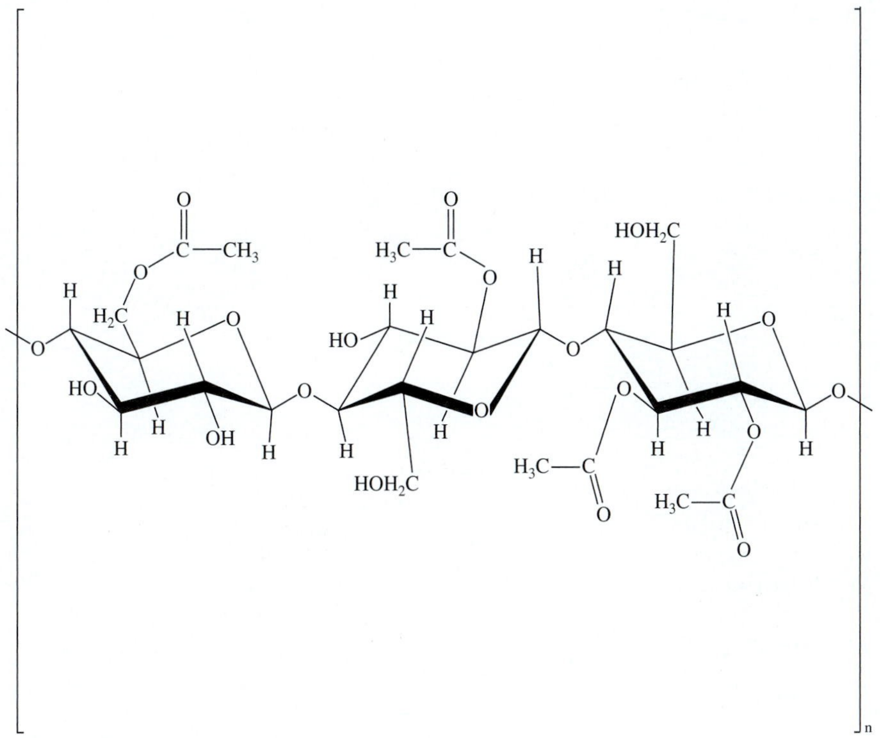

69. Because guanine and cytosine are complementary base pairs in DNA, there must also be 21% guanine. If guanine + cytosine = 42%, then adenine and thymine must equal 58%. Adenine is then 29%, as is thymine.

70. Because RNA is single-stranded, there is no complementary RNA strand. As a result, the percentage of guanine has no relationship to the percentage of cytosine.

Appendix I

CHEMICAL CALCULATIONS

A beginning student in chemistry is assumed to have developed arithmetic and algebraic calculation skills in earlier mathematics classes. Often chemistry is the first occasion for the algebra skills to be put to the test of practical application. Experience shows that many students who may have learned these skills at one time, but have not used them regularly, can profit from a review of basic concepts. Others can benefit from a handy reference to the calculation techniques used in chemistry. This section of the Appendix is intended to meet these needs.

Part A The Scientific Calculator

Every chemistry student uses a scientific calculator to perform chemical calculations. A suitable calculator can (1) add, subtract, multiply, and divide; (2) perform these operations in scientific notation; (3) work with logarithms; and (4) raise any base to a power. Calculators that can perform these operations usually have other capabilities, too, such as squares and square roots, trigonometric functions, shortcuts for pi and percentage, enclosures, statistical features, and different levels of storage and recall.

Your instructor likely will have guidelines about the type of calculator that is permitted for use on quizzes and exams in your course. We strongly suggest that you use the calculator that is permitted in your chemistry course, if it differs from the calculator you use in your math course, for all end-of-chapter questions, exercises, and problems, even if you prefer to use your math calculator. Using your chemistry calculator for every calculation in your chemistry course will help you maximize your efficiency in using the calculator.

If you haven't yet purchased a calculator for your chemistry course, we have another suggestion: purchase a solar-powered calculator. This will prevent having to deal with a dead battery. Furthermore, it will keep dead batteries out of the landfill.

Different calculator brands vary in details, particularly when some keys are used for more than one function. Some calculators offer an option on the number of digits to be displayed after the decimal point. With or without such an option, the number varies on different calculators. Accordingly, answers in this book may differ slightly from yours. In general, it is most convenient to do an Internet search for specific directions on these or other variations that may appear.

One precautionary note: *A calculator should never be used as a substitute for thinking.* If a problem is simple and can be solved mentally, do it "in your head." You will make fewer mistakes. If you use your calculator, *think* your way through each problem and estimate the answer mentally. Suggestions on approximating answers are given throughout the book, beginning in Chapter 3. If the calculator answer appears reasonable, round it off properly and write it down. Then run through the calculation again to be sure you haven't made a keyboard error. Your calculator is an obedient and faithful servant that will do exactly what you tell it to do, but it is not responsible for the mistakes you make in your instructions.

The suggestion was made that you round off your answer properly before recording it. The reason for this is that calculator answers to many problems are limited in length only by the display. For example, $273 \div 45.6 = 5.986842105$ on one calculator. Some calculators can show more numbers and therefore do. Usually, only the first three or four digits have meaning, and the others should be discarded. Procedures for deciding how many digits to write are given in Section 3.8 on significant figures.

Part B Arithmetic and Algebra

We present here a brief review of arithmetic and algebra to the point that it is used or assumed in this text. Formal mathematical statement and development are avoided. The only purpose of this section is to refresh your memory in those areas where it may be needed.

1. ADDITION. a + b. Example:

$$2 + 3 = 5$$

The result of an addition is a **sum**.

2. SUBTRACTION. a − b. Example:

$$5 - 3 = 2$$

Subtraction may be thought of as the addition of a negative number. In that sense, $a - b = a + (-b)$. Example:

$$5 - 3 = 5 + (-3) = 2$$

The result of a subtraction is a **difference**.

3. MULTIPLICATION. $a \times b = ab = a \cdot b = a(b) = (a)(b) = b \times a = ba = b \cdot a = b(a)$. The foregoing all mean that **factor** a is to be multiplied by factor b. Reversing the sequence of the factors, $a \times b = b \times a$, indicates that factors may be taken in any order when two or more are multiplied together. Examples:

$$2 \times 3 = 2 \cdot 3 = 2(3) = (2)(3) = 3 \times 2 = 3 \cdot 2 = 3(2) = 6$$

The result of a multiplication is a **product**.

Grouping of Factors (a)(b)(c) = (ab)(c) = (a)(bc). Factors may be grouped in any way in multiplication. Example:

$$(2)(3)(4) = (2 \times 3)(4) = (2)(3 \times 4) = 24$$

Multiplication by 1 $n \times 1 = n$. If any number is multiplied by 1, the product is the original number. Examples:

$$6 \times 1 = 6 \quad 3.25 \times 1 = 3.25$$

Multiplication of Fractions

$$\frac{a}{b} \times \frac{c}{d} \times \frac{e}{f} = \frac{ace}{bdf}$$

If two or more fractions are to be multiplied, the product is equal to the product of the numerators divided by the product of the denominators. Example:

$$4 \times \frac{9}{2} \times \frac{1}{6} = \frac{4}{1} \times \frac{9}{2} \times \frac{1}{6} = \frac{4 \times 9 \times 1}{1 \times 2 \times 6} = \frac{36}{12} = 3$$

4. DIVISION. $a \div b = a/b = \frac{a}{b}$. The foregoing all mean that a is to be divided by b. Example:

$$12 \div 4 = 12/4 = \frac{12}{4} = 3$$

The result of a division is a **quotient**.

Special Case If any number is divided by the same or an equal number, the quotient is equal to 1. Examples:

$$\frac{4}{4} = 1 \quad \frac{8-3}{4+1} = \frac{5}{5} = 1 \quad \frac{n}{n} = 1$$

Division by 1 $\frac{n}{1} = n$. If any number is divided by 1, the quotient is the original number. Examples:

$$\frac{6}{1} = 6 \quad \frac{3.25}{1} = 3.25$$

From this it follows that any number may be expressed as a fraction having 1 as the denominator. Examples:

$$4 = \frac{4}{1} \quad 9.12 = \frac{9.12}{1} \quad m = \frac{m}{1}$$

5. RECIPROCALS. If n is any number, the reciprocal of n is $\frac{1}{n}$; if $\frac{a}{b}$ is any fraction, the reciprocal of $\frac{a}{b}$ is $\frac{b}{a}$. The first part of the foregoing sentence is actually a special case of the second part: If n is any number, it is equal to $\frac{n}{1}$. Its reciprocal is therefore $\frac{1}{n}$.

Inverse A reciprocal is sometimes referred to as the inverse (more specifically, the multiplicative inverse) of a number. This is because the product of any number multiplied by its reciprocal equals 1. Examples:

$$2 \times \frac{1}{2} = \frac{2}{2} = 1; \quad n \times \frac{1}{n} = \frac{n}{n} = 1$$

$$\frac{4}{3} \times \frac{3}{4} = \frac{12}{12} = 1: \quad \frac{m}{n} \times \frac{n}{m} = \frac{mn}{mn} = 1$$

Division may be regarded as multiplication by a reciprocal:

$$a \div b = \frac{a}{b} = a \times \frac{1}{b}$$

$$\text{Example: } 6 \div 2 = \frac{6}{2} = 6 \times \frac{1}{2} = 3$$

$$a \div \frac{b}{c} = \frac{a}{b/c} = a \times \frac{c}{b}$$

$$\text{Example: } 6 \div \frac{2}{3} = \frac{6}{2/3} = 6 \times \frac{3}{2} = 9$$

6. SUBSTITUTION. If $d = b + c$, then $a(b + c) = ad$. Any number or expression may be substituted for its equal in any other expression. Example:

$$7 = 3 + 4. \text{ Therefore, } 2(3 + 4) = 2 \times 7.$$

7. "CANCELLATION." $\frac{ab}{ca} = \frac{ab}{ac} = \frac{b}{c}$. The process commonly called *cancellation* is actually a combination of grouping of factors (see 3), substitution (see 6) of 1 for a number divided by itself (see 4), and multiplication by 1 (see 3). Note the steps in the following examples:

$$\frac{xy}{yz} = \frac{yx}{yz} = \left(\frac{y}{y}\right) \times \left(\frac{x}{z}\right) = 1 \times \frac{x}{z} = \frac{x}{z}$$

$$\frac{24}{18} = \frac{6 \times 4}{6 \times 3} = \left(\frac{6}{6}\right) \times \left(\frac{4}{3}\right) = 1 \times \frac{4}{3} = \frac{4}{3}$$

Note that only *factors,* or *multipliers,* can be canceled. There is no cancellation in $\frac{a+b}{a+c}$.

8. ASSOCIATIVE PROPERTIES. An arithmetic operation is associative if the numbers can be grouped in any way.

Addition $a + (b - c) = (a + b) - c$. Addition, including subtraction, is associative. Example:

$$3 + 4 - 5 = (3 + 4) - 5 = 3 + (4 - 5) = 2$$

Multiplication $(a \times b)/c = a \times (b/c)$. Multiplication, including division (multiplication by an inverse), is associative. (See "Grouping of Factors" under Multiplication.) Example:

$$4 \times 6/2 = (4 \times 6)/2 = 4 \times (6/2) = 12$$

9. EXPONENTIALS. An exponential has the form B^p, where B is the **base** and p is the **power** or **exponent**. An exponential indicates the number of times the base is used as a factor in multiplication. For example, 10^3 means 10 is to be used as a factor 3 times:

$$10^3 = 10 \times 10 \times 10 = 1000.$$

Negative Exponents A negative exponent tells the number of times a base is used as a divisor. For example, 10^{-3} means 10 is used as a divisor 3 times:

$$10^{-3} = \frac{1}{10} \times \frac{1}{10} \times \frac{1}{10} = \frac{1}{10^3} = 0.001.$$

Sign of the Exponent An exponential may be moved in either direction between the numerator and denominator by changing the sign of the exponent:

$$\frac{1}{2^3} = 2^{-3} \quad 3^4 = \frac{1}{3^{-4}}$$

Multiplication of Exponentials Having the Same Base $a^m \times a^n = a^{(m+n)}$. To multiply exponentials, add the exponents. Example:

$$10^3 \times 10^4 = 10^7$$

Division of Exponentials Having the Same Base $a^m \div a^n = \frac{a^m}{a^n} = a^{m-n}$. To divide exponentials, subtract the denominator exponent from the numerator exponent. Example:

$$10^7 \div 10^4 = \frac{10^7}{10^4} = (10^7)(10^{-4}) = 10^{(7-4)} = 10^3$$

Zero Power $a^0 = 1$. Any base raised to the zero power equals 1. The fraction $\frac{a^m}{a^m} = 1$ because the numerator is the same as the denominator. By division of exponentials, $\frac{a^m}{a^m} = a^{(m-m)} = a^0$.

Raising a Product to a Power $(ab)^n = a^n \times b^n$. When the product of two or more factors is raised to some power, each factor is raised to that power. Example:

$$(2 \times 5y)^3 = 2^3 \times 5^3 \times y^3 = 8 \times 125 \times y^3 = 1000y^3$$

Raising a Fraction to a Power $\left(\frac{a}{b}\right)^n = \frac{a^n}{b^n}$. When a fraction is raised to some power, the numerator and denominator are both raised to that power. Example:

$$\left(\frac{2x}{5}\right)^3 = \frac{2^3 x^3}{5^3} = \frac{8x^3}{125} = 0.064x^3$$

Square Root of Exponentials $\sqrt{a^{2n}} = a^n$. To find the square root of an exponential, divide the exponent by 2. Example:

$$\sqrt{10^6} = 10^3.$$

If the exponent is odd, see the discussion in Square Root of a Product.

Square Root of a Product $\sqrt{ab} = \sqrt{a} \times \sqrt{b}$. The square root of the product of two numbers equals the product of the square roots of the numbers. Example:

$$\sqrt{9 \times 10^{-6}} = \sqrt{9} \times \sqrt{10^{-6}} = 3 \times 10^{-3}$$

Using this principle, by adjusting a decimal point, you may take the square root of an exponential having an odd exponent. Example:

$$\sqrt{10^5} = \sqrt{10 \times 10^4} = \sqrt{10} \times \sqrt{10^4} = 3.16 \times 10^2$$

The same technique may be used in taking the square root of a number expressed in exponential notation. Example:

$$\sqrt{1.8 \times 10^{-5}} = \sqrt{18 \times 10^{-6}}$$
$$= \sqrt{18} \times \sqrt{10^{-6}} = 4.2 \times 10^{-3}$$

10. SOLVING AN EQUATION FOR AN UNKNOWN QUANTITY. Most problems in this book can be solved by the use of conversion factors. There are times, however, when algebra must be used, particularly in relation to the gas laws. Solving an equation for an unknown involves rearranging the equation so that the unknown is the only item on one side and only known quantities are on the other. "Rearranging" an equation may be done in several ways, but the important thing is that *whatever is done to one side of the equation must also be done to the other.* The resulting relationship remains an equality, a true equation. Among the operations that may be performed on both sides of an equation are addition, subtraction, multiplication, division, and raising to a power, which includes taking a square root.

In the following examples a, b, and c represent known quantities, and x is the unknown. The object in each case it to solve the equation for x. The steps of the algebraic solution are shown, as well as the operation performed on both sides of the equation. Each example is accompanied by a practice problem that is solved by the same method. You should be able to solve the problem, even if you have not yet reached that point in the book where such a problem is likely to appear. Answers to these practice problems are at the end of this Appendix.

Example 1. $x + a = b$. Solve for x.

Subtract a: $x + a - a = b - a$

Simplify: $x = b - a$

Practice 1.

If $P = p_{O_2} + p_{H_2O}$, find p_{O_2} when P = 748 torr and p_{H_2O} = 24 torr.

Example 2(a). $ax = b$. Solve for x.

Divide by a, which is called the **coefficient** of x: $\frac{ax}{a} = \frac{b}{a}$

Simplify: $x = \frac{b}{a}$

Practice 2.

At a certain temperature, PV = k. If P = 1.23 atm and k = 1.62 L · atm, find V.

Note: Example 2(a) is an example of dividing both sides of the equation by the coefficient of x, which is the same as multiplying both sides by the inverse of the coefficient. This is best seen in a more complex example:

Example 2(b). $\frac{ax}{b} = \frac{c}{d}$ Solve for x.

Isolate the coefficient of x: $\frac{a}{b}x = \frac{c}{d}$

Multiply by the inverse of the coefficient of x: $\frac{b}{a} \times \frac{a}{b}x = \frac{c}{d} \times \frac{b}{a}$

Simplify: $x = \frac{cb}{da}$

Example 3. $\frac{x}{a} = b$. Solve for x.

Multiply by a: $\frac{ax}{a} = ba$

Simplify: $x = ba$

Practice 3.

In a fixed volume, $\frac{P}{T} = k$. Find P if $k = \frac{2.4 \text{ torr}}{K}$ and $T = 300$ K.

Example 4. $\frac{b}{ax} = \frac{d}{c}$. Solve for x.

Invert both sides of the equation: $\frac{ax}{b} = \frac{c}{d}$

Multiply by the inverse of the coefficient of x: $\frac{b}{a} \times \frac{ax}{b} = \frac{b}{a} \times \frac{c}{d}$

Simplify: $x = \frac{bc}{ad}$

Practice 4.

For gases at constant volume, $\frac{P_1}{T_1} = \frac{P_2}{T_2}$. If $P_1 = 0.80$ atm at $T_1 = 320$ K, at what value of T_2 will $P_2 = 1.00$ atm?

Example 5. $\frac{a}{b + x} = c$. Solve for x.

Multiply by (b + x): $(b + x)\frac{a}{b + x} = c\,(b + x)$

Simplify: $a = c\,(b + x)$

Divide by c: $\frac{a}{c} = \frac{c\,(b + x)}{c}$

Simplify: $\frac{a}{c} = b + x$

Subtract b: $\frac{a}{c} - b = x$

Practice 5.

In what mass of water in grams must you dissolve 20.0 grams of salt to make a 25.0% solution? The formula is

$$\frac{\text{g salt}}{\text{g salt} + \text{g water}} \times 100 = \%.$$

Part C Logarithms

Most of what you need to know about logarithms is explained at the points in the text where they are used. Comments here are limited to basic information needed to support the text explanations.

The common (base-10) **logarithm** of a number is the power, or exponent, to which 10 must be raised to be equal to the number. Expressed mathematically,

$$\text{If } N = 10^x, \text{ then } \log N = \log 10^x = x$$

The number 100 may be written as the base, 10, raised to the second power: $100 = 10^2$. According to the preceding equation, 2 is the logarithm of 10^2, or 100. Similarly, if $1000 = 10^3$, $\log 1000 = \log 10^3 = 3$. And if $0.0001 = 10^{-4}$, $\log 0.0001 = \log 10^{-4} = -4$.

Just as the powers to which 10 can be raised may be either positive or negative, so may logarithms be positive or negative. The changeover occurs at the value 1, which is 10^0. The logarithm of 1 is therefore 0. It follows that the logarithms of numbers greater than 1 are positive, and logarithms of numbers less than 1 are negative.

The powers to which 10 may be raised are not limited to integers. For example, 10 can be raised to the 2.45 power: $10^{2.45}$. The logarithm of $10^{2.45}$ is 2.45. Such a logarithm is made up of two parts. The digit or digits to the left of the decimal are the **characteristic**. The characteristic reflects the size of the number; it is related to the exponent of 10 when the number is expressed in exponential notation. The characteristic is 2 in 2.45. The digits to the right of the decimal make up the **mantissa**, which is the logarithm of the coefficient of the number when written in exponential notation. In 2.45 the mantissa is 0.45.

The number that corresponds to a given logarithm is its **antilogarithm**. In $N = 10^x$, the antilogarithm of x is 10^x, or N. The antilogarithm of 2 is 10^2, or 100. The antilogarithm of 2.45 is $10^{2.45}$. The value of the antilogarithm of 2.45 can be found on a calculator, as described in Part A of this Appendix: antilog $2.45 = 10^{2.45} = 2.8 \times 10^2$. In this exponential form of the antilogarithm of 2.45, the characteristic, 2, is the exponent of 10, and the mantissa, 0.45, is the logarithm of the coefficient, 2.8. In terms of significant figures, the mantissa matches the coefficient.

Because logarithms are exponents, they are governed by the rules of exponents given in Section 9 of Part B of this Appendix, Exponentials. For example, the product of two exponentials to the same base is the base raised to a power equal to the sum of the exponents: $a^m \times a^n = a^{m+n}$. The exponents are added. Accordingly, exponents to the base 10 (logarithms) are added to get the logarithm of the product of two numbers: $10^m \times 10^n = 10^{m+n}$. Thus,

$$\log ab = \log a + \log b$$

In a similar fashion, the logarithm of a quotient is the logarithm of the dividend minus the logarithm of the divisor (or the logarithm of the numerator minus the logarithm of the denominator if the expression is written as a fraction):

$$\log \frac{a}{b} = \log a - \log b$$

Log $10^x = x$ is the basis for converting between pH and hydrogen ion concentration between any "p" number and its corresponding value in exponential notation. This is the only application of logarithms in this text. In more advanced chemistry courses, you will encounter applications of $\log (a \div b) = \log a - \log b$ and others that are beyond the scope of this discussion.

Ten is not the only base for logarithms. Many natural phenomena, both chemical and otherwise, involve logarithms to the **base *e***, which is 2.718.... Logarithms to base *e* are known as **natural logarithms**. Their value is 2.303 times greater than a base 10 logarithm. Physical chemistry relationships that appear in base *e* may be converted to base 10 by the 2.303 factor, although modern calculators make it just as easy to work in base *e* as in base 10. The "p" concept, however, uses base 10 by definition.

Part D Estimating Calculation Results

If you estimate answers before accepting the number displayed on a calculator, a large percentage of potential calculation errors will never occur. Challenge every answer. Be sure it is reasonable before you write it down.

There is no single "right way" to estimate an answer. As your mathematical skills grow, you will develop techniques that are best for you. You will also find that one method works best on one kind of problem and another method on another problem. The ideas that follow should help you get started in this practice.

In general, estimating a calculated results involves rounding off the given numbers and mentally calculating the approximate answer. For example, if you multiply 325 by 8.36 on your calculator, you will get 2717. To see if this answer is reasonable, you might round off 325 to 300, and round 8.36 up to 9. The problem then becomes 300×9, which you can calculate mentally to $(3 \times 9) \times 100$, which is 27×100, or 2700. This is reasonably close to your calculator answer of 2717. Even so, you should key the numbers into your calculator again, time permitting, to be sure you haven't made a small error in pressing the correct buttons.

If your calculator answer for the preceding problem had been 1254.5, or 38.875598, or 27,170, your estimated 2700 would signal that an error was made. These three numbers represent common calculation errors. The first comes from a mistake that may arise anytime the number is transferred from one position (your paper) to another (your calculator). It is called *transposition* and appears when two numerals are changed in position. In this case, the error is calculating 325×3.86 (instead of 8.36) = 1254.3.

The answer 38.875598 comes from pressing the wrong function key: $325 \div 8.36 = 38.875598$. This answer is so unreasonable—$325 \times 8.36 =$ about 39—that no mental arithmetic should be necessary to tell you it is wrong. But calculators speak to some students with a mystic authority they would never dare to challenge. On one occasion, neither student nor instructor could figure out why or how, on an exam, a student used a calculator to divide 428 by 0.01, and then wrote down 7290 for the answer, when all he had to do was move the decimal two places!

The answer 27,170 ($= 325 \times 83.6$) results from a decimal error, as might arise from transposing the decimal point in a number or putting an incorrect number of zeros in numbers like 0.00123 or 123,000. Decimal errors are also apt to appear through an incorrect use of scientific notation, either with or without a calculator.

Scientific notation is a valuable aid in estimating results. For example, in calculating $41{,}300 \times 0.00524$, you can regard both numbers as falling between 1 and 10 for a quick calculation of the coefficient: $4 \times 5 = 20$. Then, thinking of the exponents, changing 41,300 to 4 moves the decimal four places left, so the exponential is 10^4. Changing 0.0524 to 5 has the decimal moving two places right, so the exponential is 10^{-2}. Adding exponents gives $4 + (-2) = 2$. The estimated answer is 20×10^2, or 20×100, which is 2000. On the calculator it comes out to 2164.12.

Another "trick" you can use is to move the decimals in such a way that the moves cancel and at the same time the problem is simplified. In $41{,}300 \times 0.0524$, the decimal in the first factor can be moved two places left (divide by 100), and in the second factor two places right (multiply by 100). Dividing and multiplying by 100 is the same as multiplying by 100/100, which is equal to 1. The problem simplifies to 413×5.24, which is easily estimated as $400 \times 5 = 2000$.

You can use the same technique to simplify fractions, too. $371{,}000 \div 6240$ can be simplified to $371 \div 6.24$ by moving the decimal three places left in both numerator and denominator, which is the same as multiplying by 1 in the form $0.001 \div 0.001$. An estimated $360 \div 6$ gives 60 as the approximate answer. The calculator answer is 59.455128. Similarly, $0.000406 \div 0.000839$ becomes about $4 \div 8$, or 0.5; by calculator, the answer is 0.4839042.

Answers to Practice Problems

1. $p_{O_2} = P - p_{H_2O} = 748 - 24 = 724 \text{ torr}$

2. $V = \dfrac{k}{P} = \dfrac{1.62 \text{ L} \cdot \text{atm}}{1.23 \text{ atm}} = 1.32 \text{ L}$

3. $P = kT = \dfrac{2.4 \text{ torr}}{\text{K}} \times 300 \text{ K} = 720 \text{ torr}$

4. $T_2 = \dfrac{T_1 P_2}{P_1} = \dfrac{320 \text{ K} \times 1.00 \text{ atm}}{0.80 \text{ atm}} = 400 \text{ K}$

5. $\dfrac{\text{g salt}}{\text{g salt} + \text{g water}} \times 100 = \%$

$$\text{g salt} \times 100 = \% \,(\text{g salt} + \text{g water})$$

$$\frac{\text{g salt} \times 100}{\%} = \text{g salt} + \text{g water}$$

$$\text{g water} = \frac{\text{g salt} \times 100}{\%} - \text{g salt}$$

$$= \frac{20.0 \times 100}{25.0} - 20.0 = 60.0$$

Appendix II

THE SI SYSTEM OF UNITS

The International System of Units, known as *Système International d'Unités* in French and therefore abbreviated as SI, is a system that has been developed and refined over many decades with the goal of standardization of units of measurement worldwide. For example, it is inconvenient when length in horse racing is expressed in furlongs, American football length is expressed in yards, American surveyors measure length in chains, scientists who study crystallography express length in ångströms, sailors use nautical miles for distance, airline pilots express altitude in feet, and so on. The SI convention is to always use just one unit—the meter—as the unit of length, no matter what country you live in and no matter what profession you practice.

The SI system uses metric units. The metric system itself has a choice of tens of units for any physical quantity. Returning to the length example, in the metric system, length can be expressed in centimeters, millimeters, kilometers, megameters, and so on. Thus, the SI system standardizes the metric system as well.

The conceptual aim of the SI system is to select the smallest set of units possible to serve as base units, which are those that can serve as the basis from which all other units can be derived. For example, a length unit and a time unit must be selected because length and time are different types of quantities. But there is no volume unit in the set of base units because volume is length times width times height, and thus volume can be expressed as the product of the length base unit for each of the three dimensions. It turns out that seven base units are necessary, and we will define each of these later.

The foundation of the SI system is a set of constants that have their basis in naturally occurring, measurable quantities. These quantities have been measured, and the most reliable measurements known have been changed to defined constants. In turn, the seven base units are defined in terms of the constants.

The Constants

Seven constants are used to define the base units. Their values are set by definition, but the definitions are not arbitrary; they are based on natural phenomena and measurement of their values. We describe these constants in alphabetical order. We first name the constant, then state the definition, and then we briefly explain the natural basis that was used as the source of the definition.

Avogadro Constant (N_A): $6.02214076 \times 10^{23}$/mole

The Avogadro constant is based upon the number of carbon-12 atoms in 12 grams of carbon-12.

Boltzmann Constant (k): 1.380649×10^{-23} joules/kelvin

The Boltzmann constant is the gas constant R divided by the Avogadro constant. Thus, it is the gas constant per individual molecule rather than per mole of molecules. The average kinetic energy of an ideal gas is proportional to its absolute temperature; in symbols, this may be written as $KE_{ave} \propto T$. The Boltzmann constant is the proportionality constant that changes this proportionality to an equality: $KE_{ave} = \mathbf{k} \times T$.

Elementary Charge (e): $1.602176634 \times 10^{-19}$ coulombs

Elementary charge is the amount of charge in one electron.

Hyperfine Transition Frequency of Cesium-133 ($\Delta\nu_{Cs}$): 9,192,631,770 hertz

The element cesium (Z = 55) is a Group 1A/1 metal. Only one isotope of cesium is stable, cesium-133. Because it is a Group 1A/1 element, cesium has just one valence electron. The spin of this electron can be aligned with the spin of the nucleus, or it can have opposite spin. There is a very tiny energy difference between the two spin states. The frequency of electromagnetic radiation that causes the electron to transition between these states is the hyperfine transition frequency.

Luminous Efficacy of Monochromatic Radiation of Frequency 540×10^{12} hertz (K_{cd}): 683 lumens per watt

Efficacy is the ability to produce an effect. Luminous is how well the eye perceives light. Thus, luminous efficacy is a measure of how well a light source produces visible light per unit of power. Monochromatic radiation refers to electromagnetic radiation of one color. The frequency 540×10^{12} hertz is visible green light about where the human eye has a maximum response to light.

Planck Constant (h): 6.62607015 × 10^{-34} joule seconds

The energy of a photon of electromagnetic radiation is directly proportional to its frequency: $E \propto \nu$. The Planck constant is the proportionality constant that changes this proportionality to an equality: $E = h \times \nu$.

Speed of Light in a Vacuum (c): 299,792,458 meters per second

The speed at which photons of electromagnetic radiation travel in empty space.

The constants used to define the SI base units are summarized in **Table AP.1**.

Table AP.1 The Defining Constants of the International System of Units

Defining Constant	Symbol	Quantity
Avogadro constant	N_A	$6.02214076 \times 10^{23}$/mol
Boltzmann constant	k	1.380649×10^{-23} J/K
Elementary charge	e	$1.602176634 \times 10^{-19}$ C
Hyperfine transition frequency of cesium-133	$\Delta\nu_{Cs}$	9,192,631,770 Hz
Luminous efficacy of monochromatic radiation of frequency 540×10^{12} hertz	K_{cd}	683 lm/W
Planck constant	h	$6.62607015 \times 10^{-34}$ J · s
Speed of light in a vacuum	c	299,792,458 m/s

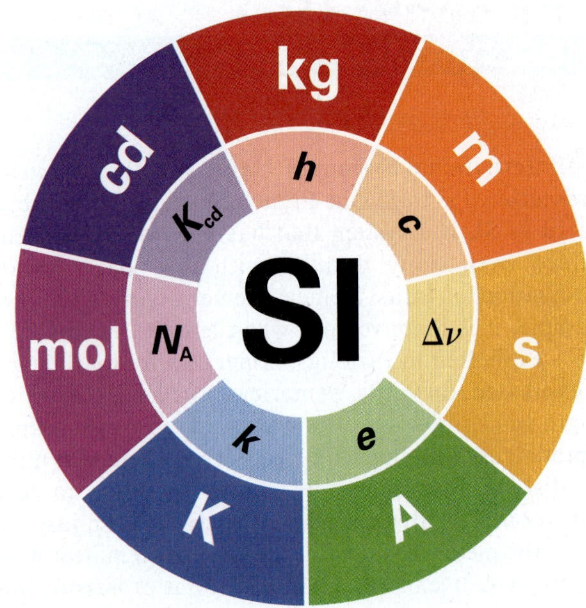

Figure AP.1 The seven SI base units and the seven constants that define them. The base units form the outer ring, and the constants, the inner ring.

SI Base Units

Figure AP.1 illustrates the relationship between the seven constants and the seven SI base units. The constants are used to define the base units. As we did with the constants, we describe these base units in alphabetical order. We first state the physical quantity, then the base unit used to express that quantity, and then we state the constants that define the base unit.

Electric Current: Ampere (A)

The ampere is defined based on the constant value of elementary charge (e), $1.602176634 \times 10^{-19}$ coulombs (C). 1 C = 1 A × s. The definition of a second (s) is given later.

Luminous Intensity: Candela (cd)

The candela is defined based on the constant value of luminous efficacy of monochromatic radiation of frequency 540×10^{12} hertz (K_{cd}), 683 lumens (lm) per watt. 1 lm = 1 cd · sr. A steradian (sr) is a dimensionless unit that expresses solid angle. A watt is a joule per second. A joule, in turn, is 1 kg · m^2/s^2. The kilogram, meter, and second are all base units that are defined later.

Thermodynamic Temperature: Kelvin (K)

The kelvin is defined based on the constant value of the Boltzmann Constant (k), 1.380649×10^{-23} joules per kelvin (J/K). 1 J = 1 kg · m^2/s^2. The kilogram, meter, and second are all base units that are defined later.

Mass: Kilogram (kg)

The kilogram is defined based on the constant value of the Planck constant (h), $6.62607015 \times 10^{-34}$ joule seconds (J · s). 1 J = 1 kg · m^2/s^2. The meter and second are base units that are defined later.

Length: Meter (m)

The meter is defined based on the constant value of the speed of light in a vacuum (c), 299,792,458 meters per second (m/s). The second is a base unit that is defined later.

Amount of Substance: Mole (mol)

The mole is defined based on the constant value of the Avogadro constant (N_A), $6.02214076 \times 10^{23}$/mole.

Time: Second (s)

The second is defined based on the constant value of the hyperfine transition frequency of cesium-133 ($\Delta\nu_{Cs}$), 9,192,631,770 hertz. A hertz is an inverse second, 1/s.

The SI base units are summarized in **Table AP.2**.

Table AP.2 SI Base Units

Physical Quantity	Base Unit	Symbol
Electric current	ampere	A
Luminous intensity	candela	cd
Thermodynamic temperature	kelvin	K
Mass	kilogram	kg
Length	meter	m
Amount of substance	mole	mol
Time	second	s

Derived Units

In the International System of Units all physical quantities are expressed in the base units listed in Table AP.2 or in combinations of those units. The combinations are called derived units. For example, the density of a substance is found by dividing the mass of a sample in kilograms—the base unit of mass—by its volume in cubic meters. The base unit of length is the meter, and volume is length times width times height, and therefore cubic meters is the volume unit. The resulting SI units of density are kilograms per cubic meter, kg/m^3. Selected derived units used in chemistry are given in **Table AP.3**.

Table AP.3 Selected SI Derived Units

Physical Quantity	Name of Unit	Symbol	Definition
Area	square meter	m^2	
Concentration	mole per cubic meter	mol/m^3	
Density	kilogram per cubic meter	kg/m^3	
Energy	joule	J	$kg \cdot m^2/s^2$
Frequency	hertz	Hz	1/s
Force	newton	N	$kg \cdot m/s^2$
Pressure	pascal	Pa	$kg/m \cdot s^2$
Specific heat	joule per kilogram kelvin	$J/kg \cdot K$	$m^2/s^2 \cdot K$
Temperature	degree Celsius	°C	$T_K - 273.15$
Velocity	meter per second	m/s	
Volume	cubic meter	m^3	

Glossary

$\propto$ is (directly) proportional to.
$\equiv$ is exactly equal to; is defined as.
Σ the sum of all values of.
22.7 L/mol the molar volume of an ideal gas at IUPAC standard temperature and pressure, 0°C and 1 bar.
6.02×10^{23} the number of units in one mole.
absolute pressure the sum of gauge pressure and atmospheric pressure.
absolute temperature scale *see Kelvin temperature scale.*
absolute zero the lowest temperature. At this temperature, 0 K, equal to −273.15°C or −459.67°F, molecular motion is at a minimum.
acid (equilibrium) constant (K_a) *see equilibrium constant.*
acid a substance that yields hydrogen (hydronium) ions in aqueous solution (Arrhenius definition); a substance from which a proton can be removed via a chemical reaction (Brønsted–Lowry definition); a substance that forms covalent bonds by accepting a pair of electrons (Lewis definition).
acid anion a negatively charged ion that has a proton that can be removed by reaction with a base.
acidic solution an aqueous solution in which the hydrogen-ion concentration is greater than the hydroxide-ion concentration; a solution in which the pH is less than 7.
actinoids elements 90 (Th) through 103 (Lr).
activation energy (E_a) the energy barrier that must be overcome to start a chemical reaction.
active site the location on an enzyme that binds the substrate.
activity series a list of metal elements and hydrogen in order of reactivity in a single-replacement oxidation–reduction reaction.
actual yield the measured yield of a product of a chemical change, determined by experiment or experience.
addition polymer a polymer formed by monomers binding to each other without forming any other product.
addition reaction the reaction of an organic compound with another compound to form a single product.
adenosine triphosphate (ATP) the molecule involved in transferring chemical energy within cells.
alcohol an organic compound consisting of an alkyl group and at least one hydroxyl group, having the general formula ROH.
aldehyde a compound consisting of a carbonyl group bonded to a hydrogen on one side and a hydrogen, alkyl, or aryl group on the other, having the general formula RCHO.

The abbreviation q.v. stands for the Latin quod vide, literally meaning "which see." It tells you that a term in a definition is also another entry in the glossary.

algebra a branch of mathematics in which symbols are used to represent quantities in equations.
aldose any of a class of monosaccharides containing an aldehyde group.
aliphatic hydrocarbon an alkane, alkene, or alkyne.
alkali metal a metal from Group 1A/1 of the periodic table.
alkaline basic; having a pH greater than 7.
alkaline earth metal a metal from Group 2A/2 of the periodic table.
alkane a saturated hydrocarbon containing only single bonds, in which each carbon atom is bonded to four other atoms.
alkene an unsaturated hydrocarbon containing a double bond, in which each carbon atom that is double-bonded is bonded to a maximum of three atoms.
alkyl group an alkane hydrocarbon group lacking one hydrogen atom, having the general formula C_nH_{2n+1}, and frequently symbolized by the letter R.
alkyne an unsaturated hydrocarbon containing a triple bond, in which each carbon atom that is triple-bonded is bonded to a total of two atoms.
alloy a solid mixture of two or more elements that has macroscopic metallic properties.
alpha (α) particle the nucleus of a helium atom, often emitted in nuclear disintegration.
alpha decay reaction a nuclear transformation in which an alpha particle is one product of the change.
alpha (α) helix a coiled secondary protein structure that is stabilized by hydrogen bonds.
amide a derivative of a carboxylic acid in which the hydroxyl group is replaced by an $-NH_2$ group, having the general formula $RCONH_2$.
amine an ammonia derivative in which one or more hydrogens are replaced by an alkyl group.
amino acid a carboxylic acid containing both an amine group and a variable alkyl group R. Amino acids of the general form $RCH(NH_2)COOH$ can form amide bonds with each other to form peptides and proteins.
Amonton's Law the pressure exerted by a fixed amount of gas at constant volume is directly proportional to Kelvin temperature, $P \propto T$.
amorphous without definite structure or shape.
amphoteric, amphiprotic pertaining to a substance that can act as an acid or a base.
amylopectin a highly branched, high molar mass polysaccharide that is one of two main forms of plant starch.
amylose a mostly unbranched, soluble polysaccharide that is one of two main forms of plant starch.

analyze (a problem statement) the process of determining a given quantity and its property and a wanted unit and its property.
angstrom (ångström) a length unit equal to 10^{-10} m.
angular (or bent) (molecular geometry) arrangement of three electron pairs and two bonded atoms surrounding a central atom in a molecule or ion to form a 120° angle between the two bonded atoms and the central atom.
anhydride (anhydrous) a substance that is without water or from which water has been removed.
anhydrous compound a substance that is without water or from which water has been removed.
anion a negatively charged ion.
anode the electrode at which oxidation occurs in an electrochemical cell.
antilogarithm the number whose logarithm is a given number.
aqueous pertaining to water.
aqueous solution a solution in which the solvent is water.
argon core the electron configuration of argon: $1s^22s^22p^63s^23p^6$; symbolized as [Ar] in an electron configuration of a species with more electrons than argon.
aromatic hydrocarbon a hydrocarbon containing a benzene ring.
Arrhenius acid–base theory *see acid and base.*
artificial radioactivity *see induced radioactivity.*
atmosphere (pressure unit) a unit of pressure based on atmospheric pressure at sea level and capable of supporting a mercury column 760 mm high.
atom the smallest particle of an element that can combine with atoms of other elements to form chemical compounds.
atomic mass the average mass of the atoms of an element compared with an atom of carbon-12 at exactly 12 atomic mass units. Also called *atomic weight.*
atomic mass unit (u) a unit of mass that is exactly 1/12 of the mass of an atom of carbon-12.
atomic number (Z) the number of protons in an atom of an element.
atomic weight *see atomic mass.*
average kinetic energy for an ideal gas with all particles at equal mass, the energy due to particle motion, which is directly proportional to absolute temperature.
Avogadro's Law the volume of a gas at constant temperature and pressure is proportional to the number of particles.
Avogadro constant; Avogadro's number $6.02214076 \times 10^{23}$/mole; historically based on the number of carbon atoms in exactly 12 grams of carbon-12; the number of units in 1 mole.
balanced equation an equation describing a chemical reaction that has the same number of atoms of each element and the same total charge for both reactants and products.
ball-and-stick model a three-dimensional representation of a molecule that uses balls to represent atoms and sticks to represent electron pairs.
bar (pressure unit) 100 kilopascals.
barometer a laboratory device for measuring atmospheric pressure.
base a substance that yields hydroxide ions in aqueous solution (Arrhenius definition); a substance that removes protons in chemical reaction (Brønsted–Lowry definition); a substance that forms covalent bonds by donating a pair of electrons (Lewis definition); mathematically, any positive real number not equal to one that has the form b in the function kb^e.
base units seven units used to express the seven base quantities that make up the fundamental foundation of the International System of Units.
basic solution an aqueous solution in which the hydroxide-ion concentration is greater than the hydrogen-ion concentration; a solution in which the pH is greater than 7.
beeswax a wax that is a mixture of several substances including esters, acids, and hydrocarbons with fatty acid esters of straight chain alcohols as the major components.
bent (or angular) (molecular geometry) arrangement of four electron pairs and two bonded atoms surrounding a central atom in a molecule or ion to form a 109.5° angle between the two bonded atoms and the central atom.
benzene ring a planar six-carbon structural unit connected by delocalized electrons that is found in aromatic hydrocarbons, including the benzene molecule.
beta (β) particle a high-energy electron, often emitted in nuclear disintegration.
beta (β)-pleated sheet a pleated-sheet secondary protein structure stabilized by hydrogen bonds.
beta decay reaction a nuclear transformation in which a beta particle is one product of the change.
binary molecular compound a molecular compound consisting of two elements.
biochemistry the study of life on a molecular level.
boiling the process by which a liquid undergoes a constant temperature transition to a gas because of the transfer of heat energy to the substance.
boiling point the temperature at which vapor pressure becomes equal to the pressure above a liquid; the temperature at which vapor bubbles form spontaneously any place within a liquid.
boiling-point elevation the difference between the boiling point of a solution and the boiling point of the pure solvent.
bombardment (nuclear) the striking of a target nucleus by an atomic particle, causing a nuclear change.
bond angle the angle formed by the bonds between two atoms that are bonded to a common central atom.
bond *see chemical bond.*
bonding electrons the electrons transferred or shared in forming chemical bonds; valence electrons.
Boyle's Law the pressure of a fixed quantity of a gas at constant temperature is inversely proportional to volume, $P \propto (1/V)$.

breeder reactor a nuclear reactor designed to create new fissionable nuclear fuel from nonfissionable isotopes.
Brønsted–Lowry acid–base theory *see acid and base.*
buffer a solution that resists a change in pH.
buret a glass tube of uniform width calibrated to accurately measure volume of liquid delivered through an adjustable-flow stopcock at the bottom of the tube.
calorie a unit of heat energy equal to 4.184 joules.
Calorie (food, kcal) a unit of heat energy equal to 4184 joules.
calorimeter a laboratory device for measuring heat energy transfer.
carbohydrates a class of organic compounds consisting mainly of polyhydroxy aldehydes or ketones. Carbohydrates form the supporting tissue of plants and serve as food for animals, including people.
carbonyl group an organic functional group, C=O, characteristic of aldehydes and ketones.
carboxyl group an organic functional group, $-COOH$, characteristic of carboxylic acids.
carboxylic acid an organic acid containing the carboxyl group, having the general formula RCOOH.
carnauba wax a wax that is a mixture of several substances including esters of fatty acids, fatty alcohols, acids, and hydrocarbons; its hardness is due to carboxylic acid components with hydroxyl groups that tie together adjacent chains by hydrogen bonding.
catalyst a substance that increases the rate of a chemical reaction by lowering activation energy. The catalyst is either a non-participant in the reaction, or it is regenerated. *See also inhibitor.*
cathode the negative electrode in a cathode ray tube; the electrode at which reduction occurs in an electrochemical cell.
cation a positively charged ion.
cell, electrochemical a device in which either an electrical potential is developed by a spontaneous chemical change or electrolysis occurs as a result of an externally applied electrical potential.
cell, electrolytic a cell in which electrolysis occurs as a result of an externally applied electrical potential.
cell, galvanic *see cell, voltaic.*
cell, voltaic a cell in which an electrical potential is developed by a spontaneous chemical change. Also called a galvanic cell.
cellulose a polysaccharide that serves as the major structural component of plants.
Celsius scale a system of temperature measurement based on assignment of 0°C to the freezing point of water and 100°C to the boiling point of water, with 100 equally divided degrees between the reference points (historical); a system of measurement based on assignment of −273.15°C to absolute zero and 273.16 kelvin = 0.01°C (modern).
central atom an atom in a molecule or ion that is bonded to two or more other atoms.
chain-growth polymer a polymer formed as monomers add to a growing polymer chain in sequential addition reactions in a chain-reaction process.
chain reaction a reaction that has as a product one of its own reactants; that product becomes a reactant, thereby allowing the original reaction to continue.
change of color a difference in the composition of a substance such that a different range of wavelengths of the electromagnetic spectrum are reflected from it; one form of evidence that a chemical reaction has occurred.
charge cloud *see electron cloud.*
charge density the amount of electric charge per unit volume.
Charles's Law the volume of a fixed quantity of a gas at constant pressure is directly proportional to absolute temperature, $V \propto T$.
check (a solution) the process of determining if the value of a wanted quantity is reasonable and reflecting on what was learned by solving a problem.
chemical bond a general term that sometimes includes all of the electrostatic attractions among atoms, molecules, and ions, but more often refers to covalent and ionic bonds. *See covalent bond, ionic bond.*
chemical change a change in which one or more substances disappear and one or more new substances form.
chemical equation a symbolic representation of chemical change, with the formulas of the beginning substances to the left of an arrow that points to the formulas of the substances formed.
chemical family (families) groups of elements that have similar chemical properties because of similar valence electron configuration, appearing in the same column in the periodic table.
chemical formula *see formula, chemical.*
chemical nomenclature a system of names used in chemistry.
chemical properties the types of chemical change a substance is able to experience.
chemical reaction *see chemical change.*
chemical symbol *see symbol (chemical).*
classification (thinking skill) recognition of the possible ways that a collection of objects can be arranged according to similarities or differences in shared characteristics and/or qualities.
closed system a sample of matter that is isolated so that it cannot exchange matter with the rest of the universe.
cloud chamber a device in which condensation tracks form behind radioactive emissions as they travel through a supersaturated vapor.
coefficient in a chemical equation, numbers used to equalize the number of atoms before and after the chemical change; mathematically, the quantity c when a number is written in exponential notation in the form $c \times 10^e$.
colligative properties physical properties of mixtures that depend on the concentration of particles, irrespective of their identity.

collision theory of chemical reactions a particulate-level model of chemical reactivity that features sufficient kinetic energy and proper orientation as necessary conditions for a reaction-producing collision.
colloid a nonsettling dispersion of aggregated ions or molecules intermediate in size between the particles in a true solution and those in a suspension.
combination (synthesis) reaction a reaction in which two or more substances combine to form a single product.
Combined Gas Law the volume of a fixed quantity of a gas is proportional to temperature and inversely proportional to pressure, $V \propto (T/P)$.
combustion the process of burning.
common ion effect the percentage ionization of a weak acid or low-solubility ionic compound is decreased when an ion in common with the acid or ionic compound is introduced to the solution.
compound a pure substance that can be broken down into two or more other pure substances by a chemical change.
concentrated adjective for a solution with a relatively large amount of solute per given quantity of solvent or solution.
condensation the act of condensing.
condensation polymer a polymer formed through condensation reactions (q.v.).
condensation reaction a chemical change in which two molecules or functional groups react to form a larger molecule and a small molecule, such as water.
condense to change from a vapor to a liquid or solid.
condensed formula a symbolic representation of an organic compound that expands the fundamental formula to suggest the arrangement of atoms while still fitting on a standard line of text.
conductor a substance that readily conveys electricity.
conjugate acid–base pair a Brønsted–Lowry acid and the base derived from it when it loses a proton, or a Brønsted–Lowry base and the acid developed from it when it removes a proton.
construct (a solution) the process of setting up a series of multiplied conversion factors or solving an algebraic equation and substituting quantities for variables and solving for the answer.
continuous spectrum the band of colors that results from electromagnetic radiation emissions over a range of wavelengths.
conventional equation a chemical equation written with conventional formulas for soluble aqueous ionic compounds and strong acids.
conversion factor the relationship between different units of measurement that express the same quantity, written in the form of a fraction.
coordinate covalent bond a bond in which both bonding electrons are furnished by only one of the bonded atoms.
copolymer a polymer formed from two or more different monomers.
coulomb a unit of electrical charge; ampere times second.
covalent bond the chemical bond between two atoms that share a pair of electrons.
covalent network solid a crystalline solid made of atoms that are connected by a network of covalent bonds, essentially forming one large molecule.
critical mass the minimum quantity of fissionable material needed to sustain a nuclear chain reaction.
crystal lattice *see crystalline solid.*
crystalline solid a solid in which the ions and/or molecules are arranged in a definite geometric pattern.
C-terminal the end of a polypeptide or protein chain terminated by an amino acid with a free carboxyl group.
cubic centimeter (cm^3) a unit of volume equal to the volume of a cube with a 1 cm length, width, and height.
cycloalkane an alkane (q.v.) that has one or more rings of carbon atoms.
Dalton's atomic theory a model of matter based on the hypothesis that each element is made of indivisible particles called atoms.
Dalton's Law of Partial Pressures the total pressure exerted by a mixture of gases is the sum of the partial pressures (q.v.) of the gases in the mixture.
decompose to change chemically into simpler substances.
decomposition reaction a reaction in which a single compound breaks down into simpler substances.
defining equation an equation used to establish the definition of a unit or a property of a substance.
delocalized electrons electrons in a molecule that are not restricted to remaining near a single atom or between two atoms in a covalent bond.
delta quantity (Δ) final value minus initial value.
density the ratio of mass of a substance to its volume.
deoxyribonucleic acid (DNA) a large nucleotide (q.v.) polymer found in the cell nucleus. DNA contains genetic information and controls protein synthesis.
deposition the process by which a gas undergoes transition directly to a solid without entering the liquid phase because of the transfer of heat energy out of the substance.
derived unit a unit of measurement derived from the seven base units in the International System of Units.
diatomic having two atoms.
dilute adjective for a solution with a relatively small amount of solute per given quantity of solvent or solution; verb meaning to reduce the concentration of a solution by adding solvent.
dipole a polar molecule.
dipole forces a type of intermolecular attractive force of intermediate relative strength that occurs between the positive pole of one polar molecule and the negative pole of another.
diprotic acid an acid capable of yielding two protons per molecule in complete ionization.

directly proportional two quantities that have a constant ratio; expressed as $y \propto x$ and $y = kx$, where k is a nonzero constant.
disaccharide a sugar composed of two monosaccharides (q.v.).
discrete discontinuous; individually distinct.
dispersion forces weak electrical attractions between molecules, temporarily produced by the shifting of internal electrons.
dissolve to pass into solution.
distillation the process of separating components of a mixture by boiling off and condensing the more volatile component.
distilled water water that has been purified by distillation (q.v.).
disulfide linkages sulfur-sulfur covalent bonds that strengthen the tertiary structure of many proteins.
double bond a covalent chemical bond formed by the sharing of two pairs of electrons between two bonded atoms.
double-replacement equation (double-replacement reaction) a chemical equation with the form $AX + BY \rightarrow AY + BX$, where the reactants are two compounds, and the ions appear to exchange partners.
dynamic equilibrium a state in which opposing changes occur at equal rates, resulting in zero net change over a period of time.
electric charge a property of matter that causes a force of attraction or repulsion; occurs as two types, positive and negative; carried by some subatomic particles, occurring in discrete units.
electrochemical cell *see cell, electrochemical.*
electrodes conductors by which electric charges enter or leave electrolytes.
electrolysis the passage of electric charge through an electrolyte.
electrolyte a substance that, when dissolved, yields a solution that conducts electricity; a solution or other medium that conducts electricity by ionic movement.
electrolytic cell *see cell, electrolytic.*
electromagnetic radiation energy in the form of electric and magnetic waves, including gamma rays; x-rays; ultraviolet, visible, and infrared light; microwaves; and radio waves.
electromagnetic spectrum the range of frequencies and corresponding wavelengths of electromagnetic waves.
electron subatomic particle carrying a unit negative charge and having a mass of 9.1×10^{-28} gram, or 1/1837 of the mass of a hydrogen nucleus, found outside the nucleus of the atom.
electron cloud the region of space around or between atomic nuclei that is occupied by electrons.
electron configuration the orbital arrangement of electrons in ions or atoms.
electron dot symbols *see Lewis diagram.*
electron-pair angle the angle formed by any two electron pairs in a molecule or ion and the central atom between them.
electron-pair geometry the arrangement of electron pairs around a central atom in a molecule or ion.
electron orbit the circular or elliptical path supposedly followed by an electron around an atomic nucleus, according to the Bohr theory of the atom.
electron orbital a mathematically described region within an atom in which there is a high probability that an electron will be found.
electron-pair geometry a description of the distribution of bonding and unshared electron pairs around a bonded atom.
electron-pair repulsion the principle that electron-pair geometry is the result of repulsion between electron pairs around a bonded atom, causing them to be as far apart as possible.
electron-sea model a particulate-level model of a metallic crystal that features a definite crystal pattern of positive ions with valence electrons moving relatively freely among the ions.
electron transfer reaction a chemical change in which one or more electrons are transferred from one substance, the reducing agent, to another, the oxidizing agent.
electronegativity a scale of the relative ability of an atom of one element to attract the electron pair that forms a single covalent bond with an atom of another element.
electrostatic force the force of attraction or repulsion between electrically charged objects.
element a pure substance that cannot be decomposed into other pure substances by ordinary chemical means.
elemental symbol *see symbol (chemical).*
empirical formula a formula that represents the lowest integral ratio of atoms of the elements in a compound.
endothermic reaction a change in which energy is transferred from the surroundings to the system, resulting in a positive $\Delta_r H$, an increase in enthalpy of the system.
energy the ability to do work or transfer heat.
enthalpy the heat content of a chemical system.
enthalpy of reaction, $\Delta_r H$ *see heat of reaction.*
enzyme a molecule that catalyzes chemical reactions.
equation a mathematical statement indicating that quantities or values are the same; a symbolic representation of a chemical change that illustrates how atoms are conserved while (a) reacting substance(s) (is) are transformed into (a) product(s).
equilibrium *see dynamic equilibrium.*
equilibrium (thinking skill) reasoning in the form $ab = cd$ or $ab = k$.
equilibrium constant (K) with reference to an equilibrium equation, the ratio in which the numerator is the product of concentrations of the species on the right-hand side of the equation, each raised to a power corresponding to its coefficient in the equation, and the denominator is the corresponding product of the species on the left side of the equation; symbol: K, K_c, or K_{eq}.
equilibrium vapor pressure *see vapor pressure.*

equivalency an expression stating that two quantities with different units represent the same property.
equivalent the quantity of an acid (or base) that yields or reacts with one mole of H^+ (or OH^-) in a chemical reaction; the quantity of a substance that gains or loses one mole of electrons in a redox reaction.
equivalent mass mass in grams per equivalent (q.v.).
ester an organic compound formed by the reaction between a carboxylic acid and an alcohol, having the general formula R—CO—OR′.
esterification the reaction between a carboxylic acid and an alcohol, yielding an ester and water.
ether an organic compound in which two alkyl groups are bonded to the same oxygen, having the general formula R—O—R′.
evaporation the process by which a substance transitions from the liquid phase to the gas phase at a temperature less than the boiling point.
exact numbers values that have no associated uncertainty because they are counted or established by definition.
excess reactant the reactant(s) in a chemical reaction that remain when the reaction is complete.
excited state the state of an atom in which one or more electrons have absorbed energy—becoming "excited"—to raise them to energy levels above ground state.
exothermic reaction a reaction that results in the transfer of energy to its surroundings.
exponent a number, as e in 10^e, denoting the power to which another number (the base) is to be raised.
exponential a number, called the base, raised to some power, called the exponent.
exponential notation *see exponential notation, standard and scientific notation.*
exponential notation, standard a method of writing numbers in the form: $a.bcd \times 10^c$, where a.bcd is a number equal to or greater than 1 and less than 10.
Fahrenheit scale a system of temperature measurement based on assignment of 32°F to the freezing point of water and 212°F to the boiling point of water, with 180 equally divided degrees between the reference points (historical); a system of measurement based on assignment of −273.15°C to absolute zero, 273.16 kelvin = 0.01°C, and $T_{°F} - 32 = 1.8T_{°C}$ (modern).
family *see chemical family.*
fat an ester formed from glycerol and three fatty acids. Fats are solids at room temperature. Also called triacylglycerols or triglycerides.
fatty acid a long chain carboxylic acid, typically having between 10 and 24 carbon atoms.
favored direction (of an equilibrium reaction) the side (reactants or products) of a chemical equilibrium with the weaker acid and base or weaker oxidizing and reducing agents.
field *see force field.*
filtration the process of passing a liquid (or gas) through a filter to separate components of a mixture based on relative particle sizes.
first ionization energy the energy required to remove an electron from a neutral atom.
fission a nuclear reaction in which a large nucleus splits into two smaller nuclei.
force field a region of space in which a force is effectively operative.
formal model (thinking skill) a mental representation of a natural system with parts that are abstract entities that have to be imagined and hold the same relationship to one another as the real thing.
formation of a gas a chemical change that yields a gaseous product; one form of evidence that a chemical reaction has occurred.
formation of a solid product a chemical change that yields a solid product; one form of evidence that a chemical reaction has occurred.
formula, chemical a combination of chemical symbols and subscript numbers that represents the elements in a pure substance and the ratio in which the atoms of the different elements appear.
formula mass (weight) the mass in u of one formula unit of a substance; the molar mass of a formula unit of a substance.
formula unit a real (molecular) or hypothetical (ionic) unit particle represented by a chemical formula.
forward (direction of) reaction the chemical change that occurs as read from left to right in a reversible reaction equation.
fractional distillation the separation of a mixture into fractions whose components boil over a given temperature range.
freezing the process by which a liquid undergoes a constant temperature transition to a solid because of the transfer of heat energy out of the substance.
freezing point depression the difference between the freezing point of a solution and the freezing point of the pure solvent.
fructose a ketone (q.v.) monosaccharide (q.v.) with the formula $C_6H_{12}O_6$ that is the sugar found in fruits and honey and widely used in industry as a sweetener in the form of high-fructose corn syrup.
functional group an atom or a group of atoms that establishes the identity of a class of compounds and determines its chemical properties.
fusion the process of melting; also, a nuclear reaction in which two small nuclei combine to form a larger nucleus.
galvanic cell *see cell, voltaic.*
gamma (γ) ray a high-energy electromagnetic emission in radioactive disintegration.
gas the state of matter characterized by particles that are independent and have a macroscopic-level variable shape and volume.
gas constant (R) the proportionality constant in the ideal gas equation; 0.0821 L · atm/mol · K = 62.4 L · torr/mol · K; the Avogadro constant times the Boltzmann constant.
gauge pressure the pressure above atmospheric pressure.

Gay-Lussac's Law the pressure exerted by a fixed quantity of gas at constant volume is directly proportional to absolute temperature, $P \propto T$.
Geiger counter (Geiger-Müller counter) an electrical device for detecting and measuring the intensity of radioactive emission.
gene typically, a sequence of nucleotides in a segment of DNA that contains the information needed to produce a protein.
geometric isomers two compounds having the same molecular formulas but different geometric configurations around a structurally rigid bond.
given quantity in a quantitative problem solution setup, the quantity given in the problem statement that will need to be converted to an equivalent amount of another unit.
glucose an aldehyde (q.v.) monosaccharide (q.v.) with the formula $C_6H_{12}O_6$ that is the sugar used as an energy source by most living organisms.
glycogen a polysaccharide (q.v.) that is the quick-acting carbohydrate reserve in mammals; animal starch.
gram (g) 1/1000 the mass of a kilogram (q.v.).
ground state the state of an atom in which all electrons occupy the lowest possible energy levels.
group (periodic table) the elements making up a vertical column in the periodic table.
half-cell an electrode and an electrolyte that comprise half of an electrochemical cell.
half-life ($t_{1/2}$) the time required for the disintegration of one-half of the radioactive atoms in a sample.
half-reaction the oxidation or reduction half of an oxidation–reduction reaction.
half-reaction equation a symbolic representation of either the oxidation or reduction half of an oxidation–reduction reaction.
halide ion F^-, Cl^-, Br^-, I^-, or At^-.
halogen family the name of the chemical family consisting of fluorine, chlorine, bromine, iodine, and astatine; any member of the halogen family.
halogenation reaction the reaction of an organic compound with a halogen.
heat the form in which energy is transferred from a substance at higher temperature to one at lower temperature.
heat of fusion (ΔH_{fus}) (solidification) the heat transferred when one gram of a substance changes between a solid and a liquid at constant pressure and temperature. *See also molar heat of fusion (solidification).*
heat of reaction, $\Delta_r H$ the change of enthalpy (q.v.) in a chemical reaction.
heat of vaporization (ΔH_{vap}) (condensation) the heat transferred when one gram of a substance changes between a liquid and a vapor at constant pressure and temperature. *See also molar heat of vaporization (condensation).*
heterogeneous mixture having a nonuniform composition, usually with visibly different parts or phases.
homogeneous mixture having a uniform appearance and uniform properties throughout.
homologous series a series of compounds in which each member differs from the one next to it by the same structural unit.
hydrate a crystalline solid that contains water of hydration.
hydrated hydrogen ion *see hydronium ion.*
hydrated ion an ion in solution surrounded by water molecules.
hydrocarbon an organic compound consisting of only carbon and hydrogen.
hydrogenation addition of hydrogen to a double or triple bond to produce a saturated product.
hydrogen bond an intermolecular force (attraction) between a hydrogen atom in one molecule and a highly electronegative atom (fluorine, oxygen, or nitrogen) of another polar molecule; the polar molecule may be of the same substance containing the hydrogen, or of a different substance.
hydronium ion a hydrated hydrogen ion, H_3O^+.
hydroxyl group an organic functional group, $-OH$, characteristic of alcohols.
hypothesis a proposed explanation for a pattern observed in the natural world.
ideal gas a hypothetical gas that behaves according to the ideal gas model (q.v.) over all ranges of temperature and pressure.
ideal gas equation the equation $PV = nRT$ that relates quantitatively the pressure, volume, quantity, and temperature of an ideal gas.
ideal gas law *see ideal gas equation.*
ideal gas model a representation of a gas as identical, volumeless particles that move in straight lines, undergo collisions with no total loss of energy, and do not attract or repel other particles.
ideal yield the amount of product formed from the complete conversion of the given amount of reactant to product.
identify (equivalencies or an equation) the process of extracting equivalencies from a problem statement or recalling, looking up, or deriving equivalencies or an algebraic equation needed to solve a quantitative problem.
immiscible insoluble (usually used only in reference to liquids).
inches of mercury (in. Hg) (pressure unit) the pressure exerted by a 1-inch circular column of mercury 1 inch in height at 32°F and 9.80865 m/s^2 gravitational acceleration.
indicator a substance that changes from one color (or colorless) to another, used to signal the end of a titration.
induced dipole force a type of intermolecular attractive force of varying relative strength, depending on molecular size, that occurs between nonpolar molecules because of shifting electron clouds within the molecules.

induced fit model a model used to explain enzyme function that features a close match between the shape of the substrate and the active site of the enzyme that becomes slightly modified to an exact fit as the two molecules bind.
induced radioactivity decay of a radionuclide that was formed from a previously stable nucleus that artificially was made radioactive.
inhibitor a substance added to a chemical reaction to retard its rate; sometimes called a negative catalyst; a molecule that decreases enzyme activity.
inorganic chemistry the chemistry of all chemical compounds except the hydrocarbons and most of their derivatives.
International System of Units (abbreviated SI from the French *Le Système International d'Unités*) a subset of metric units that is used to express physical quantities in terms of seven base units and in combinations of those units.
intermolecular force attractive force that operates between molecules in a substance or mixture.
inversely proportional two quantities that have a constant product; expressed as $y \propto 1/x$ and $xy = k$, where k is a nonzero constant.
ion an atom or group of covalently bonded atoms that is electrically charged because of an excess or deficiency of electrons relative to the number of protons.
ion-combination reaction when two solutions are combined, the formation of a precipitate or molecular compound by a cation from one solution and an anion from the second solution.
ionic bond the chemical bond arising from the electrostatic attraction forces between oppositely charged ions in an ionic compound.
ionic compound a compound composed of ions that are attracted to one another by ionic bonds (q.v.).
ionic crystal a crystalline solid (q.v.) composed of oppositely charged ions held together by strong electrostatic forces.
ionic equation a chemical equation in which dissociated compounds are shown in ionic form.
ionizable hydrogen a hydrogen atom in a molecule or ion that can be removed by reaction with a base.
ionization the formation of an ion from a molecule or atom.
ionization energy the energy required to remove an electron from an atom or ion.
isoelectronic having the same electron configuration.
isomers two or more compounds having the same molecular formulas but different structural formulas and different physical and chemical properties.
isotopes two or more atoms of the same element that have different atomic masses because of different numbers of neutrons.
IUPAC International Union of Pure and Applied Chemistry.
joule (J) the SI energy unit, defined as a force of one newton applied over a distance of one meter; $1\ J = 1\ kg \cdot m^2/s^2$; 1 joule = 0.239 calorie.
K the symbol for the kelvin, the absolute temperature unit; the symbol for an equilibrium constant. K_a is the constant for the ionization of a weak acid; K_{sp} is the constant for the equilibrium between a slightly soluble ionic compound and a saturated solution of its ions; K_w is the constant for the ionization of water.
kelvin the SI unit of temperature. The kelvin is defined based on the constant value of the Boltzmann constant (k), 1.380649×10^{-23} joules/kelvin (J/K). $1\ J = 1\ kg \cdot m^2/s^2$.
Kelvin temperature scale an absolute temperature scale with 0 K at absolute zero, or −273.15°C, and the magnitude of the kelvin unit as 1/273.16 of the difference between absolute zero and the triple point of water, 273.16 K.
ketone a compound consisting of a carbonyl group bonded on each side to an alkyl group, having the general formula R—CO—R′.
ketose any of a class of monosaccharides containing a ketone group.
kilogram (kg) the SI unit of mass. The kilogram is defined based on the constant value of the Planck constant (h), $6.62607015 \times 10^{-34}$ joule seconds (J · s). $1\ J = 1\ kg \cdot m^2/s^2$.
kilocalorie (kcal) 1000 calories (q.v.); a food Calorie (q.v.).
kilojoule (kJ) 1000 joules (q.v.).
kilopascal (kPa) 1000 pascals (q.v.).
kinetic energy energy of motion; translational kinetic energy is equal to ½ × mass × (velocity)^2.
kinetic energy distribution curve a plot of percentage of molecules with a given kinetic energy versus kinetic energy that illustrates the nature of the range of kinetic energies of the molecules in a sample.
kinetic molecular theory the general theory (q.v.) that all matter consists of particles in constant motion, with different degrees of freedom distinguishing among solids, liquids, and gases.
kinetic theory of gases the portion of the kinetic molecular theory that describes gases and from which the model of an ideal gas is developed.
lactose a disaccharide made from galactose and glucose with the formula $C_{12}H_{22}O_{11}$ that is also called milk sugar because it sweetens the milk of mammals.
lanolin a wax that is a complex mixture of esters of 33 high molar mass alcohols and 36 fatty acids; also called *wool fat*.
lanthanoids elements 58 (Ce) through 71 (Lu).
law a statement that describes a pattern of regularity in nature.
Law of Combining Volumes when gases at the same temperature and pressure react, the reactant and product volumes are in a ratio of small whole numbers.
Law of Conservation of Energy the quantity of energy within an isolated system does not change.

Law of Conservation of Mass the total mass of the reactants in a chemical change is equal to the total mass of the products.

Law of Conservation of Mass and Energy the total quantity of mass and energy in the universe is fixed and does not change.

Law of Constant Composition *see Law of Definite Composition.*

Law of Definite Composition any compound is always made up of elements in the same proportion by mass.

Law of Multiple Proportions when two elements combine to form more than one compound, the different weights of one element that combine with the same weight of the other element are in a simple ratio of whole numbers.

Le Chatelier's Principle if an equilibrium system is subjected to a change, processes occur that tend to counteract partially the initial change, thereby bringing the system to a new position of equilibrium.

lecithin any of a group of phospholipids that have an alcohol backbone, two fatty acid residues, a phosphate group, and a quaternary saturated amine in the head group.

Lewis acid–base theory *see acid and base.*

Lewis diagram, formula, structure, or symbol a diagram representing the valence electrons and covalent bonds in an atomic or molecular species.

light the portion of the electromagnetic spectrum (q.v.) visible to the human eye.

limiting reactant the reactant first totally consumed in a reaction, thereby determining the maximum yield possible.

linear arrangement of two electron pairs surrounding a central atom in a molecule or ion to form a 180° angle between the pairs and the central atom (electron pair geometry); arrangement of two electron pairs and two bonded atoms surrounding a central atom in a molecule or ion to form a 180° angle between the bonded atoms and the central atom (molecular geometry).

line formula a condensed formula representing an organic compound that includes each CH_2 unit in an alkane chain.

line spectrum the spectral lines that appear when light emitted from a sample is analyzed in a spectroscope.

lipid any of a group of compounds that are found in living organisms, insoluble in water, and soluble in nonpolar solvents.

liquid the state of matter characterized by particles in contact with one another that move freely among themselves and have macroscopic-level variable shape and constant volume.

liter (L) 0.001 m^3 (exactly).

local conformation the arrangement in space of atoms in a molecule, with consideration restricted to a segment of a larger molecule.

localized electrons electrons in a molecule that are restricted to remaining near a single atom or between two atoms in a covalent bond.

logarithm the power to which a base must be raised to produce a given number.

London (dispersion) forces *see induced dipole forces.*

lone pairs a pair of valence electrons in a molecule that are not used for bonding.

macromolecular crystal a crystal made up of a large but indefinite number of atoms covalently bonded to each other to form a huge molecule. Also called a *network solid.*

macromolecule a polymeric molecule with molar mass $\geq$ 5000 g/mol.

macroscopic consideration of matter on a scale observable by the human eye; it is measurable with conventional apparatus.

main group elements elements from one of the A Groups (IUPAC Groups 1–2 and 13–18) of the periodic table.

major species in an acid solution, the species present in greatest abundance.

manometer a laboratory device for measuring gas pressure.

mass a property reflecting the quantity of matter in a sample.

mass-energy equivalence mass and energy are two forms of the same quantity, related by: Energy = mass × (speed of light)^2.

mass-to-mass stoichiometry path mass of given → amount in moles of given → amount in moles of wanted → mass of wanted.

mass number (A) the total number of protons plus neutrons in the nucleus of an atom.

mass ratio a quotient of masses.

mass spectrometer a laboratory device in which a flow of gaseous ions may be analyzed in regard to their charge and/or mass.

materials science the field of scientific study that investigates the particulate-level structure of materials and their macroscopic properties.

matter that which has mass.

melt the process by which a solid undergoes a constant temperature transition to a liquid because of the transfer of heat energy to the substance.

mental model (thinking skill) a mental representation of a natural system with parts that hold the same relationship to one another as the real thing, including abstract entities that have to be imagined.

messenger RNA an RNA molecule transcribed from a DNA template that carries genetic information to the site of protein synthesis.

metal a substance that possesses metallic properties, such as luster, ductility, malleability, and good conductivity of heat and electricity; an element that loses electrons to form monatomic cations.

metallic bond forces of attraction between delocalized electrons in a metallic crystal and the metal ions.
metallic character exhibition of the physical and chemical properties of metals.
metallic crystal a crystalline solid made of an orderly, repeating pattern of positive ions through which delocalized valence electrons move relatively freely.
metalloid an element that has both metallic and nonmetallic properties.
meter (m) the SI unit of length. The meter is defined based on the constant value of the speed of light in a vacuum (c), 299,792,458 meters per second.
methyl group the one-carbon alkyl structural group $-CH_3$.
metric system a system of measurement used by most of the world that is based on a small number of basic units and a standard set of prefixes that represent multiples of 10.
microscopic consideration of matter on a scale not observable by the human eye but observable with a classic microscope.
milliliter (mL) 0.001 liter.
millimeter of mercury (mm Hg) (pressure unit) the pressure exerted at the base of a column of mercury 1 mm high.
minor species in an acid solution, the species present in lesser abundance.
miscible soluble (usually used only in reference to liquids).
mixture a sample of matter containing two or more pure substances.
model a representation of something else.
molal boiling point elevation constant the ratio of boiling point elevation to molality for a dilute ideal solution.
molal freezing point depression constant the ratio of freezing point depression to molality for a dilute ideal solution.
molality solution concentration expressed in moles of solute per kilogram of solvent.
molar concentration of hydrogen ions ($[H^+]$) the ratio of amount of hydrogen ion in a solution, expressed in moles, to the volume of the solution, expressed in liters.
molar concentration of hydroxide ions ($[OH^-]$) the ratio of amount of hydroxide ion in a solution, expressed in moles, to the volume of the solution, expressed in liters.
molar heat of fusion (solidification) the heat energy transferred when one mole of a substance changes between a solid and a liquid at constant temperature and pressure.
molar heat of vaporization (condensation) the heat energy transferred when one mole of a substance changes between a liquid and a vapor at constant temperature and pressure.
molarity (M) solution concentration expressed in moles of solute per liter of solution.
molar mass (weight) the mass of one mole of any substance.
molar volume the volume occupied by one mole, usually of a gas.
molar volume method (to solve a gas stoichiometry problem) first use the ideal gas equation to determine the molar volume at the given temperature and pressure, and then use the molar volume to calculate the wanted quantity.
mole (mol) the SI unit of amount of substance. The mole is defined based on the constant value of the Avogadro constant (N_A), $6.02214076 \times 10^{23}$/mole.
molecular compound a compound whose fundamental particles are molecules rather than ions.
molecular crystal a molecular solid in which the molecules are arranged according to a definite geometric pattern.
molecular formula a description of the composition of a molecule that lists each element in the molecule by chemical symbol and the number of atoms of each element, if more than one, with a subscript after the symbol.
molecular geometry a description of the shape of a molecule.
molecular mass (weight) the number that expresses the average mass of the molecules of an element or compound compared to the mass of an atom of carbon-12 at a value of exactly 12 u; the average mass of the molecules of a compound expressed in atomic mass units.
molecular product a product of a chemical change that exists in molecular form.
molecule the smallest unit particle of a pure substance that can exist independently and possess the identity of the substance.
monatomic having only one atom.
monomer the individual chemical structural unit from which a polymer may be developed.
monoprotic acid an acid capable of yielding one proton per molecule in complete ionization.
monosaccharide the simplest sugars, which cannot be converted to smaller carbohydrates; one sugar unit per molecule.
multiple bond a covalent chemical bond formed by the sharing of two or more pairs of electrons between two bonded atoms.
natural radioactive decay series the sequence of nuclear reactions that occur as a naturally occurring radioactive substance decays to a stable isotope.
negative catalyst *see inhibitor.*
neon core the electron configuration of neon: $1s^22s^22p^6$; symbolized as [Ne] in an electron configuration of a species with more electrons than neon.
net ionic equation an ionic equation from which all spectators have been removed.
network solid a crystal made up of a large, indefinite number of atoms covalently bonded to each other to form a huge molecule. Also called a *macromolecular crystal.*
neutralization reaction the reaction between an acid and a base to form an ionic compound and water; any reaction between an acid and a base.

neutral molecule a molecule with no net charge because the number of protons equals the number of electrons.
neutral solution a solution with pH = 7.
neutron an electrically neutral subatomic particle having a mass of 1.7×10^{-24} gram, which is almost the same as the mass of a hydrogen atom, approximately equal to the mass of a proton, or 1 atomic mass unit, found in the nucleus of the atom.
newton the SI unit of force, equal to $kg \cdot m^2/s^2$.
noble gases the name of the chemical family of relatively unreactive elemental gases appearing in Group 8A/18 of the periodic table.
nomenclature a system of names used in a particular science.
nonconductor a substance that does not readily convey electricity.
nonelectrolyte a substance that, when dissolved, yields a solution that is a nonconductor of electricity; a solution or other fluid that does not conduct electricity by ionic movement.
nonmetal a substance that possesses nonmetallic properties, such as being dull and brittle, and a poor conductivity of heat and electricity; an element that forms covalent bonds or gains electrons to form monatomic anions.
nonpolar pertaining to a bond or molecule having a symmetrical distribution of electric charge.
nonpolar covalent bond a bond in which bonding electrons are shared equally.
normal alkane a straight chain alkane (q.v.).
normal boiling point the temperature at which a substance boils in an open vessel at one atmosphere pressure.
normality solution concentration in equivalents (q.v.) per liter.
N-terminal the end of a polypeptide or protein chain terminated by an amino acid with a free amine group.
nuclear bombardment direction of high-energy particles such as protons, neutrons, and alpha particles at a larger nucleus.
nuclear charge the electrical charge due to the protons in the nucleus of an atom.
nuclear fission a nuclear reaction of a large nucleus, splitting into two nuclei of smaller mass, accompanied by the release of a large quantity of energy in comparison with the energy transferred in chemical change.
nuclear fusion a nuclear reaction of two small nuclei, combining to form a nucleus of larger mass, accompanied by the release of a relatively large quantity of energy in comparison with the energy transferred in nuclear fission.
nuclear model of the atom a model of the atom that features a relatively tiny, dense nucleus that contains most of the mass of the atom and all of its positive charge.
nuclear power plant an electrical generating facility that utilizes a nuclear reactor as a source of heat to change water into steam, which turns turbines that produce electricity.
nuclear symbol a symbol for an isotope of an element in the form ${}^{\text{mass number}}_{\text{atomic number}}\text{Sy}$, where Sy is the chemical symbol of the element.
nucleotide a compound consisting of a nitrogen-containing base, a sugar, and one or more phosphate groups. Nucleotides are the monomers for the polymers DNA and RNA. DNA contains the sugar deoxyribose; RNA contains the sugar ribose.
nucleus the extremely dense central portion of the atom that contains the neutrons and protons that constitute nearly all the mass of the atom and all of the positive charge.
nuclide an atomic nucleus, typically identified by its atomic number and mass number.
octet of electrons the valence electron configuration ns^2np^6; eight electrons surrounding an atom in a Lewis diagram.
octet rule the general guideline that atoms tend to form stable bonds by sharing or transferring electrons until the atom is surrounded by a total of eight electrons.
oil an ester formed from glycerol and three fatty acids. Oils are liquids at room temperature. Also called *triacylglycerols* or *triglycerides.*
orbit *see electron orbit.*
orbital *see electron orbital.*
organic chemistry the chemistry of carbon compounds.
overlap (orbital) the merging of two atomic orbitals and their associated electrons to create a covalent bond.
oxidation chemical reaction with oxygen; a chemical change in which the oxidation number (state) of an element is increased; also, the loss of electrons in a redox reaction.
oxidation number a number assigned to each element in a compound, ion, or elemental species by a set of rules. Its two main functions are to organize and simplify the study of oxidation–reduction reactions and to serve as a base for one branch of chemical nomenclature.
oxidation–reduction reaction a chemical change in which electrons are transferred from one species to another.
oxidation state *see oxidation number.*
oxidizer, oxidizing agent the substance that takes electrons from another species in a chemical change, thereby oxidizing it.
oxoacid an acid that contains oxygen.
oxoanion an anion that contains oxygen.
oxyacid an acid that contains oxygen.
oxyanion an anion that contains oxygen.
partial pressure the pressure one component of a mixture of gases would exert if it alone occupied the same volume as the mixture at the same temperature.
particle accelerator a device that uses electrical fields to increase the kinetic energy of charged particles that bombard nuclei.
particulate consideration of matter on a scale too small to be observable with the human eye or a conventional microscope; it must be modeled.
pascal (pressure unit) one newton per square meter.

Pauli exclusion principle the principle that says, in effect, that no more than two electrons can occupy the same orbital.
peptide an amino acid polymer typically containing 50 or fewer amino acids. Also called *polypeptide.*
peptide linkage a covalent bond formed by the reaction of the carboxyl group of one amino acid and the amine group of another, yielding a peptide and water.
percent the amount of one part of a mixture per 100 total parts in the mixture.
percentage composition by mass the percentage by mass of each element in a compound.
percentage concentration by mass grams of solute per 100 grams of solution.
percentage yield the actual yield of a chemical reaction expressed as a percentage of theoretical yield.
periods horizontal rows of the periodic table.
periodic table of the elements a table of chemical elements arranged in order of increasing atomic number with elements with similar chemical properties arranged in vertical columns.
pH a way of expressing hydrogen-ion concentration; the negative of the logarithm of the hydrogen-ion concentration.
pH loop a series of equations that describe the relationship among pH, pOH, $[OH^-]$, and $[H^+]$.
phase a visibly distinct part of a heterogeneous sample of matter.
phospholipid a class of lipids that have an alcohol backbone, two fatty acid residues, and a phosphate group, with a polar head and a long, nonpolar tail.
phosphorescence emission of light by a substance after absorbing energy from a source of electromagnetic radiation; emission may continue for some time after exposure to the energy source stops.
photon a massless quantum particle that carries the electromagnetic force.
physical change a change in the physical form of a substance without changing its chemical identity.
physical properties properties of a substance that can be observed and measured without changing the substance chemically.
planetary model of the atom a model of the atom that features electrons orbiting the nucleus as planets orbit the sun.
plum pudding model the model of the atom proposed after the discovery of the electron and before the discovery of the nucleus that is analogous to a plum pudding, with negatively charged electron particles (plums) embedded in a positively charged spherical "gel" (pudding).
pOH a way of expressing hydroxide-ion concentration; the negative logarithm of the hydroxide-ion concentration.
polar pertaining to a bond or molecule having an asymmetrical distribution of electric charge.
polar covalent bond a bond in which bonding electrons are shared but shared unequally.
polar molecule a molecule in which there is an asymmetrical distribution of charge, resulting in positive and negative poles.
polyatomic pertaining to a species consisting of more than one atom; usually said of polyatomic ions.
polyatomic ions ions that consist of more than one atom.
polycrystalline a solid made of small crystals arranged in a random or a directed order.
polymer a chemical compound formed by bonding two or more monomers. Polymers with molar mass $\geq$ 5000 g/mol are also called *macromolecules.*
polymerization the reaction in which monomers combine to form polymers.
polypeptide an amino acid polymer typically containing fewer than about 50 amino acid residues.
polyprotic acid an acid capable of yielding more than one proton per molecule on complete ionization.
polysaccharide complex carbohydrates made up of many monosaccharides; many sugar units per molecule.
positron a subatomic particle with a charge of +1 and the same mass as an electron.
potential energy energy possessed by a body by virtue of its position in an attractive and/or repulsive force field.
potential energy barrier *see activation energy.*
pounds per square inch (psi) a force of 1 pound applied to an area of 1 square inch.
precipitate a solid that forms when two solutions are mixed.
precipitation the process of formation of a solid product as a result of a chemical change.
pressure force per unit area.
primary standard a soluble solid of reasonable cost used in a titration that is very stable and pure, preferably with a high molar mass, that can be weighed accurately.
primary structure the amino acid sequence in a protein.
principal energy level(s) the main energy levels within the electron arrangement in an atom. They are quantized by a set of integers beginning at $n = 1$ for the lowest level, $n = 2$ for the next, and so forth; also called the principal quantum number.
principal quantum number (n) *see principal energy levels.*
probability (thinking skill) comprehension of the probabilistic nature of appropriate natural relationships and sampling procedures.
product a substance formed as a result of a chemical change.
proportionality constant the nonzero constant k in the equation that expresses the relationship between two variables x and y, $y = kx$.
proportional reasoning (thinking skill) recognizing relationships of the type $y \propto x$, $y \propto 1/x$, $y = mx$, and $y = m(1/x)$, and comparison of proportions.
protein an amino acid polymer typically containing more than 50 amino acids.

proton a subatomic particle carrying a unit positive charge and having a mass of 1.7×10^{-24} gram, which is almost the same as the mass of a hydrogen atom, almost the same as the mass of a neutron, or 1 atomic mass unit, found in the nucleus of the atom.
proton transfer reaction a chemical change in which a proton is transferred from one substance, the acid, to another, the base.
pure substance a sample consisting of only one kind of matter, either compound or element.
quantity a property of an object or a sample that can be expressed as an amount, or the number of objects in a sample; quantity = value × unit.
quantization of energy the existence of certain discrete electron energy levels within an atom such that electrons may have any one of these energies but no energy between two such levels.
quantize *see quantization of energy.*
quantized energy levels the energies of electrons in atoms are restricted to specific values; they may never be between two of those values.
quantum jump (leap) movement of an electron within an atom from one energy level to another without measurable passage through energy levels in between.
quantum mechanical model of the atom an atomic concept that recognizes four quantum numbers by which the characteristics of electrons and orbitals may be described.
quantum mechanics a branch of science that describes the quantum behavior of photons and subatomic particles.
quaternary structure the arrangement of polypeptide chains in a protein composed of more than one chain.
R a symbol used to designate any alkyl group; the gas constant, having a value of 0.0821 L · atm/mol · K.
radioactive decay series a description of a series of nuclear reactions, starting from a radioactive isotope and ending with a stable isotope.
radioactivity the spontaneous emission of rays and/or particles from an atomic nucleus.
radiocarbon dating a method of determining the age of the remains of plants and animals that is based on comparing the carbon-14-to-carbon-12 ratio in the atmosphere to the same ratio in the object to be dated.
radioisotope a radioactive isotope (q.v.).
radionuclide a radioactive atomic nucleus.
reactants substances that will be destroyed in a chemical change.
reactivity the relative capacity of a species to undergo chemical change.
redox a term coined from *red*uction–*ox*idation to refer to oxidation–reduction reactions.
reducer, reducing agent the substance that loses electrons to another species, thereby reducing it.
reduction a chemical change in which the oxidation number (state) of an element is reduced; also, the gain of electrons in a redox reaction.
rem roentgen equivalent in man; a unit of measure of radiation dose that accounts for both the absorbed dose and a quality factor that depends on the type of radiation.
replication (DNA) the biochemical process by which two new DNA molecules are each formed from one strand of original DNA and one newly constructed complementary strand.
representative elements *see main group elements.*
resonance hybrid a species that has an actual structure that is the average of two or more Lewis structures.
resonance structures two or more equivalent Lewis structures for a molecule or ion that are created by changing only the positions of the electrons, in which the actual species is an average of the resonance structures.
reverse (direction of) reaction the chemical change that occurs as read from right to left in a reversible reaction equation.
reversible reaction a chemical reaction in which the products may react to re-form the original reactants.
ribonucleic acid (RNA) a nucleotide (q.v.) polymer found in the nucleus and other parts of the cell.
ribosome structure found in the cell outside the nucleus where protein synthesis occurs based on the coding of the messenger RNA.
round off to express as a number with fewer digits.
Rutherford scattering experiments experiments conducted by Ernest Rutherford and collaborators that provided evidence for the nuclear model of the atom.
rule of octet *see octet rule.*
salt the product of a neutralization reaction other than water; an ionic compound that does not contain the hydrogen ion, H^+, the oxide ion, O^{2-}, or the hydroxide ion, OH^-.
salt bridge an apparatus that connects the two half cells in a voltaic cell so that the cell maintains electrical neutrality without bulk mixing of solutions.
saturated hydrocarbon a hydrocarbon that contains only single bonds, in which each carbon atom is bonded to four other atoms.
saturated solution a solution of such concentration that it is or would be in a state of equilibrium with excess solute present.
scientific method the typically cyclical procedure used to create knowledge about the natural world, characterized by observation and measurement, proposing a hypothesis, testing and revising the hypothesis, and combining a group of related hypotheses into a theory.
scientific model *see model.*
scientific notation a method of writing numbers in the form: a.bcd × 10^e.
scintillation counter an instrument used to measure radioactivity that is based on a transparent solid that phosphoresces after absorbing energy from radiation.
second ionization energy the energy required to remove an electron from an ion with a 1+ charge.

secondary structure the regular local conformational structures α-helicies and β-pleated sheets in a protein that reflect a maximum amount of hydrogen bonding.

semimetal *see metalloid.*

seven common strong acids nitric acid, sulfuric acid, hydrochloric acid, hydrobromic acid, hydroiodic acid, chloric acid, and perchloric acid.

significant figure rule for addition and subtraction round off an answer to the first column (hundredths, tenths, etc.) that has an uncertain digit.

significant figure rule for multiplication and division round off the answer to the same number of significant figures as the smallest number of significant figures in any measured quantity.

significant figures the digits in a measurement that are known to be accurate plus one uncertain digit.

silk fibroin a type of protein consisting of layers of antiparallel β-pleated sheets.

single bond a covalent chemical bond formed by the sharing of one pair of electrons between two bonded atoms.

single-replacement equation (single-replacement reaction) a chemical equation with the form $A + BX \rightarrow AX + B$, in which the reactants are an element and a compound, and the element appears to replace one of the ions in the compound.

SI units units of measurement associated with the International System of Units.

skeleton equation an oxidation–reduction half-reaction equation that contains only the oxidized and reduced species.

solid the state of matter characterized by particles vibrating in fixed positions and macroscopic-level constant shape and constant volume.

solubility the quantity of solute that will dissolve in a given quantity of solvent or in a given quantity of solution, at a specified temperature, to establish an equilibrium between the solution and excess solute; frequently expressed in grams of solute per 100 grams of solvent.

solubility product constant (K_{sp}) *under K, see K_{sp}.*

soluble a property of a substance that refers to its ability to dissolve in a specified solvent.

solute the substance dissolved in the solvent, sometimes not clearly distinguishable from the solvent (q.v.), but usually the lesser of the two.

solution a homogeneous mixture of two or more substances of molecular or ionic particle size, the concentration of which may be varied, usually within certain limits.

solution inventory a precise identification of the chemical species present in a solution, in contrast with the solute from which they may have come; that is, sodium ions and chloride ions, rather than sodium chloride.

solvent the medium in which the solute is dissolved; *see solute.*

space-filling model a three-dimensional representation of a molecule representing its outer boundaries using spheres of different color to represent different atoms.

***s, p, d,* and *f* sublevels** for atoms other than hydrogen, the principal energy levels are split into sublevels designated *s, p, d,* and *f,* in order of increasing energy for a particular principal energy level.

species a generic term for any chemical particle, such as atom, ion, or molecule.

specific gravity the ratio of the density of a substance to the density of some standard, usually water at 4°C.

specific heat the quantity of heat energy that must be transferred to raise the temperature of one gram of a substance one degree Celsius.

spectator ions ions present in a solution where a reaction occurs but do not themselves react.

spectators *see spectator ions.*

spectrometer a laboratory instrument used to analyze spectra.

spectrum (plural: spectra) the result of dispersing a beam of light into its component colors; also the result of dispersing a beam of gaseous ions into its component particles, distinguished by mass and electric charge.

speed of light (c) the speed of all electromagnetic radiation in open space: 299,792,458 m/s.

spontaneous pertaining to a change that appears to take place by itself, without outside influence.

stable pertaining to that which does not change spontaneously.

standard atmosphere of pressure *see atmosphere.*

standard exponential notation *see exponential notation, standard.*

standard temperature and pressure (STP) arbitrarily defined conditions of temperature (0°C) and pressure (1 bar) (IUPAC definition) at which gas volumes and quantities may be measured and/or compared.

standardize determination of the concentration of a solution to be used in a titration by titrating it against a primary standard.

starting formula in a formal procedure for balancing a chemical equation, the formula with the largest number of atoms.

states of matter the four principal conditions in which matter typically exists: gas, liquid, solid, and plasma.

state symbols symbols used in chemical equations to indicate the physical state of the species: (s) for solid, (ℓ) for liquid, (g) for gas, and (aq) for aqueous.

static electricity an accumulation of stationary electric charge.

step-growth polymers a polymer produced by the stepwise reaction of monomers with the growing chain, where each bond is formed independently of the others.

steroid any of a group of lipids that have 17 carbon atoms arranged in three cyclohexane rings and one cyclopentane ring fused together within the larger molecule.

stoichiometric amount the amount in moles of a substance.

stoichiometry the quantitative relationships among the substances involved in a chemical reaction, established by the equation for the reaction.

STP abbreviation for standard temperature and pressure. *See standard temperature and pressure.*
strong acid an acid that ionizes almost completely in aqueous solution; an acid that transfers its protons readily.
strong base a highly soluble compound that produces $OH^-(aq)$ when dissolved in H_2O; a base that has a strong attraction for protons.
strong electrolyte a substance that, when dissolved, yields a solution that is a good conductor of electricity because of nearly complete ionization or dissociation.
strong oxidizer (oxidizing agent) an oxidizer that has a strong attraction for electrons.
strong reducer (reducing agent) a reducer that releases electrons readily.
structural formula (diagram) a Lewis diagram (q.v.) without unshared electron pairs.
subatomic particle a particle smaller than an atom; in chemistry, electron, neutron, and proton.
sublevels the energy levels into which the principal energy levels are divided according to the quantum mechanical model of the atom; usually specified *s*, *p*, *d*, and *f*.
sublimation the process by which a solid undergoes transition directly to a gas without entering the liquid phase because of the transfer of heat energy to the substance.
substitution reaction a reaction in which a hydrogen atom in an alkane is replaced by an atom of another element.
substrate the reactant molecule that an enzyme helps convert to products.
sucrose a disaccharide made from glucose and fructose with the formula $C_{12}H_{22}O_{11}$ that is commonly used as table sugar.
supersaturated pertaining to a state of solution concentration that is greater than the equilibrium concentration (solubility) at a given temperature and/or pressure.
surface tension a property of the surface of a liquid that minimizes the area of the surface and causes the surface to act like an elastic membrane.
suspension a mixture that gradually separates by settling.
symbol (chemical) a one- or two-letter abbreviation used to represent in writing the name of a chemical element.
synthesis reaction a reaction in which two or more substances combine to form a single product.
terminal atom (in a Lewis diagram) an atom that is bonded to only one other atom.
tertiary structure the three-dimensional structure of a protein.
tetrahedral related to a tetrahedron; usually used in reference to the orientation of four covalent bonds radiating from a central atom toward the vertices of a tetrahedron, or to the 109.5° angle formed by any two corners of the tetrahedron and the central atom as its vertex; arrangement of four electron pairs surrounding a central atom in a molecule or ion to form a 109.5° angle between any two pairs and the central atom (electron-pair geometry); arrangement of four electron pairs and four bonded atoms surrounding a central atom in a molecule or ion to form a 109.5° angle between any two bonded atoms and the central atom (molecular geometry).
tetrahedron a regular four-sided solid having congruent equilateral triangles as its four faces.
theory a related set of hypotheses that explain a broad group of related phenomena.
thermal pertaining to heat.
thermochemical equation a chemical equation that includes an energy term, or for which $\Delta_r H$ is indicated.
thermochemical stoichiometry stoichiometry that includes the energy involved in a chemical reaction, with proportionalities defined by the thermochemical equation.
thermoplastic a polymer that can be easily melted and molded into new shapes because it has weak attractive forces between non-cross-linked polymer chains.
thermoset a polymer with a relatively rigid structure because of cross-links between carbon chains.
third ionization energy the energy required to remove an electron from an ion with a 2+ charge.
titration the controlled and measured addition of one solution into another.
torr a unit of pressure essentially equal to the pressure unit millimeter of mercury (mm Hg).
total ionic equation a chemical equation written with ion formulas for soluble aqueous ionic compounds and strong acids.
total ionization the removal of all ionizable hydrogen from an acid molecule or ion.
transcription synthesis of RNA where the genetic information encoded in DNA is copied to a complementary messenger RNA molecule.
transfer RNA a small RNA molecule that carries with it a specific amino acid that is transferred to a polypeptide chain as the transfer RNA molecule bonds to a complementary site on messenger RNA in a ribosome.
transition elements element from one of the B groups (IUPAC Groups 3–12) of the periodic table.
transition metals *see transition elements.*
transition state an intermediate molecular species presumed to be formed during the interaction (collision) of reacting molecules in a chemical change.
translation the process that occurs in a ribosome where the genetic information in messenger RNA is decoded to produce a protein molecule.
transmutation conversion of an atom from one element to another by means of a nuclear change.
transuranium elements elements whose atomic numbers are greater than 92.
trigonal (triangular) planar arrangement of three electron pairs surrounding a central atom in a molecule or ion to form a 120° angle between any two pairs and the central atom (electron-pair geometry); arrangement of three electron pairs and three bonded atoms surrounding a central atom in a molecule or ion to form a 120° angle between any two bonded atoms and the central atom (molecular geometry).

trigonal (triangular) pyramidal arrangement of four electron pairs and three bonded atoms surrounding a central atom in a molecule or ion to form a 109.5° angle between any two bonded atoms and the central atom (molecular geometry).
triple bond a covalent chemical bond formed by the sharing of three pairs of electrons between two bonded atoms.
triprotic acid an acid capable of yielding three protons in complete ionization.
uncertain digit the digit in a measured quantity that cannot be accurately measured and is estimated; the last digit written when expressing a measured quantity.
uncertainty (in measurement) that which is not accurately measurable.
unit a standardized amount of a quantity used to express magnitudes of that quantity; quantity = value × unit.
unit path in a quantitative problem solution setup, the sequence of units to be followed to convert between the given quantity and the wanted quantity.
universal gas constant (R) *see gas constant (R).*
unsaturated hydrocarbon a hydrocarbon that contains one or more multiple bonds.
unsaturated solution a solution with a concentration that is less than the solubility limit.
valence electrons the highest-energy *s* and *p* electrons in an atom, which determine the bonding characteristics of an atom of an element.
valence shell electron-pair repulsion theory (VSEPR) a model used to predict and explain molecular geometry that is based on mutual repulsion among regions of electron density (electron pairs or multiple bonds) in the valence shell of a central atom in a molecule or ion and the resulting tendency of the regions to minimize potential energy by distributing themselves as far away from each other as possible.
value an amount; quantity = value × unit.
van der Waals forces a general term for all kinds of weak intermolecular attractions.
vapor a gas.
vaporize, vaporization changing from a solid or liquid to a gas.
vapor pressure the pressure or partial pressure exerted by a vapor that is in contact with its liquid phase. The term often refers to the pressure or partial pressure of a vapor that is in equilibrium with its liquid state at a given temperature.
viscosity the resistance of a liquid to flow.
volatile that which vaporizes easily.
voltaic cell *see cell, voltaic.*
wanted quantity in a quantitative problem solution setup, the quantity asked for in the problem statement that will result from an equivalent amount of another unit.
water (equilibrium) constant (K_w) $K_w = [H^+][OH^-] = 1.0 \times 10^{-14}$ at 25°C.
water of crystallization, water of hydration water molecules that are included as structural parts of crystals formed from aqueous solutions.
wave equation an equation that describes the propagation of waves.
wave-particle duality all matter and energy possess both wavelike and particle-like properties.
wax various substances that are mixtures that contain esters of fatty acids and are solid at room temperature, malleable, and insoluble in water.
weak acid an acid that ionizes only slightly in aqueous solution; an acid that does not release its protons readily.
weak base a base that dissociates only slightly in aqueous solution; a base that has a weak attraction for protons.
weak electrolyte a substance that, when dissolved, yields a solution that is a poor conductor of electricity because of limited ionization or dissociation.
weak oxidizer (oxidizing agent) an oxidizer that has a weak attraction for electrons.
weak reducer (reducing agent) a reducer that does not release electrons readily.
wedge-and-dash diagram a structural diagram (q.v.) that includes wedge-shaped lines to indicate atoms in front of the plane of the page and dashed lines to indicate atoms behind the plane of the page so that the three-dimensional structure of the species is illustrated.
weight a measure of the force of gravitational attraction.
yield the amount of product from a chemical reaction.
Z atomic number.

Index

Page numbers in *italics* indicate figures and illustrations; page numbers followed by "t" indicate tables.

D

E

F

M

N

M

N

O

P